Some Fundamental Constants[a]

Quantity	Symbol	Value[b]
Atomic mass unit	u	$1.660\,540\,2(10) \times 10^{-27}$ kg $931.434\,32(28)$ MeV/c^2
Avogadro's number	N_A	$6.002\,136\,7(36) \times 10^{23}$ (mol)$^{-1}$
Bohr magneton	$\mu_B = \dfrac{e\hbar}{2m_e}$	$9.274\,015\,4(31) \times 10^{-24}$ J/T
Bohr radius	$a_0 = \dfrac{\hbar^2}{m_e e^2 k}$	$0.529\,177\,249(24) \times 10^{-10}$ m
Boltzmann's constant	$k_b = R/N_A$	$1.380\,658(12) \times 10^{-23}$ J/K
Compton wavelength	$\lambda_c = \dfrac{h}{m_e c}$	$2.426\,310\,58(22) \times 10^{-12}$ m
Deuteron mass	m_d	$3.343\,586\,0(20) \times 10^{-27}$ kg $2.013\,553\,214(24)$ u
Electron mass	m_e	$9.109\,389\,7(54) \times 10^{-31}$ kg $5.485\,799\,03(13) \times 10^{-4}$ u $0.510\,999\,06(15)$ MeV/c^2
Electron-volt	eV	$1.602\,177\,33(49) \times 10^{-19}$ J
Elementary charge	e	$1.602\,177\,33(49) \times 10^{-19}$ C
Gas constant	R	$8.314\,510(70)$ J/K·mol
Gravitational constant	G	$6.672\,59(85) \times 10^{-11}$ N·m²/kg²
Hydrogen ground state	$E_0 = \dfrac{m_e e^4 k_e^2}{2\hbar^2} = \dfrac{e^2 k_e}{2a_0}$	$13.605\,698(40)$ eV
Josephson frequency-voltage ratio	$2e/h$	$4.835\,976\,7(14) \times 10^{14}$ Hz/V
Magnetic flux quantum	$\Phi_0 = \dfrac{h}{2e}$	$2.067\,834\,61(61) \times 10^{-15}$ Wb
Neutron mass	m_n	$1.674\,928\,6(10) \times 10^{-27}$ kg $1.008\,664\,904(14)$ u $939.565\,63(28)$ MeV/c^2
Nuclear magneton	$\mu_n = \dfrac{e\hbar}{2m_p}$	$5.050\,786\,6(17) \times 10^{-27}$ J/T
Permeability of free space	μ_0	$4\pi \times 10^{-7}$ N/A² (exact)
Permittivity of free space	$\epsilon_0 = 1/\mu_0 c^2$	$8.854\,187\,817 \times 10^{-12}$ C²/N·m² (exact)
Planck's constant	h $\hbar = h/2\pi$	$6.626\,075(40) \times 10^{-34}$ J·s $1.054\,572\,66(63) \times 10^{-34}$ J·s
Proton mass	m_p	$1.672\,623(10) \times 10^{-27}$ kg $1.007\,276\,470(12)$ u $938.272\,3(28)$ MeV/c^2
Quantized Hall resistance	h/e^2	$25812.805\,6(12)$ Ω
Rydberg constant	R_H	$1.097\,373\,153\,4(13) \times 10^7$ m^{-1}
Speed of light in vacuum	c	$2.997\,924\,58 \times 10^8$ m/s (exact)

[a] These constants are the values recommended in 1986 by CODATA, based on a least-squares adjustment of data from different measurements. For a more complete list, see Cohen, E. Richard, and Barry N. Taylor, *Rev. Mod. Phys.* **59**:1121, 1987.

[b] The numbers in parentheses for the values below represent the uncertainties in the last two digits.

PRINCIPLES OF

Physics

R A Y M O N D A. S E R W A Y

James Madison University

S A U N D E R S G O L D E N S U N B U R S T S E R I E S

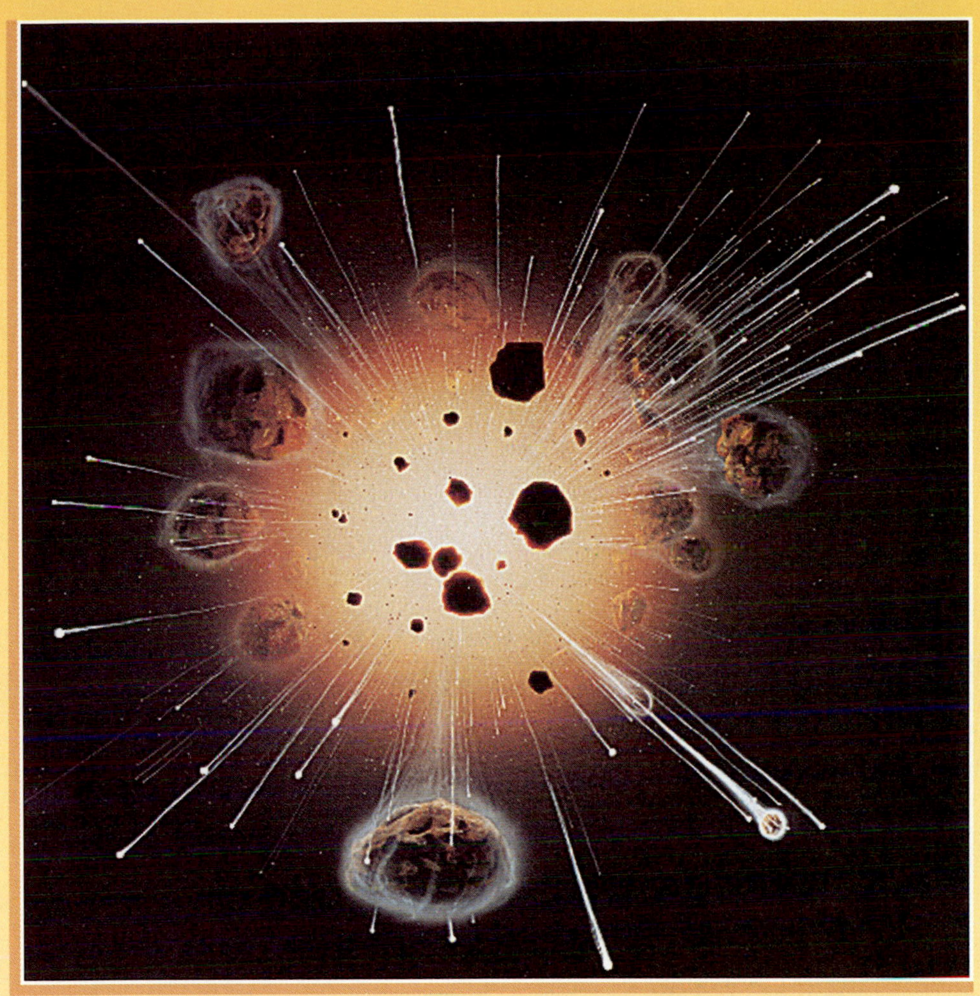

PRINCIPLES OF Physics

SAUNDERS COLLEGE PUBLISHING
Harcourt Brace College Publishers

Fort Worth Philadelphia San Diego New York Orlando Austin San Antonio Toronto Montreal London Sydney Tokyo

Text Typeface: New Baskerville
Compositor: Progressive Typographers, Inc.
Publisher: John Vondeling
Developmental Editor: Jennifer Bortel
Managing Editor: Carol Field
Project Editor: Sally Kusch
Copy Editor: Mary Patton
Manager of Art and Design: Carol Bleistine
Art Director: Carol Bleistine
Art and Design Coordinator: Sue Kinney
Text Designer: Tracy Baldwin
Cover Designer: Lawrence R. Didona
Text Artwork: Rolin Graphics
Layout Artwork: Dorothy Chattin
Director of EDP: Tim Frelick
Production Manager: Charlene Squibb
Marketing Manager: Marjorie Waldron

Cover Credit: West Coast of Florida. © Bruce Byers 1992/FPG International

Printed in the United States of America

PRINCIPLES OF PHYSICS

ISBN 0-03-097715-0

Library of Congress Catalog Card Number: 93-087099

4567 032 987654321

Preface

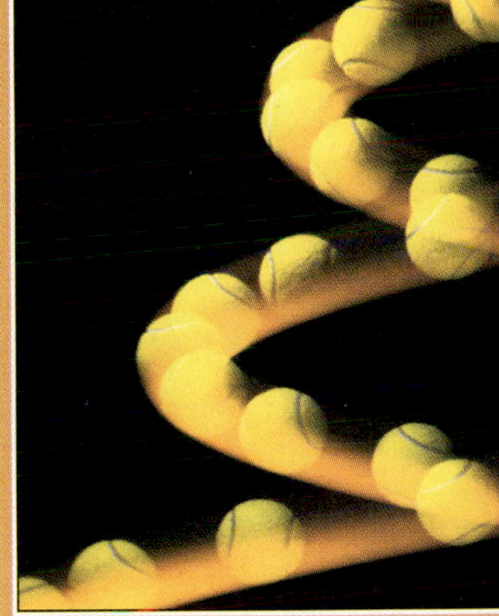

Principles of Physics is designed for a one-year introductory calculus-based physics course for science and engineering students. This project was conceived because of the well-known problem that we continue to wrestle with in teaching the introductory calculus-based physics course. The course content (hence the size of textbooks) continues to grow, while the number of contact hours with students has either dropped or remained unchanged. Furthermore, traditional one-year courses cover little if any 20th century physics.

In preparing this book, I was influenced by recent attempts to reform this course, especially the efforts of the Introductory University Physics Project (IUPP) sponsored by the American Association of Physics Teachers and the American Institute of Physics. The primary goals and guidelines of this project are:

- reduce course content, following the "less may be more" theme
- incorporate contemporary physics naturally into the course
- organize the course in the context of one or more "story lines"
- treat all student constituents equitably

Recognizing a need for a textbook that could meet these guidelines several years ago, I studied the various proposed IUPP models and the many reports from IUPP committees. Eventually, I became actively involved in the review and planning of one specific model initially developed at the U. S. Air Force Academy, entitled "A Particles Approach to Introductory Physics." I spend part of the summer of 1990 at the academy working with Colonel James Head and Lt. Col. Rolf Enger, the primary authors of the IUPP model, and with other members of that department. This most useful collaboration was the starting point of this project.

In my opinion, the IUPP model developed at the U. S. Air Force Academy and modified by a team of people over the last few years, was an excellent choice for designing a reformed curriculum for several reasons:

- It is an evolutionary approach (rather than a revolutionary approach), which should meet the current demands of the physics community;
- it deletes many topics in classical physics (such as fluid mechanics, statics, and optical instruments), and places much less emphasis on rigid body motion, circuits, and thermodynamics;
- it introduces some topics in 20th century physics, such as special relativity, fission, fusion, energy and momentum quantization, and the Bohr model of the hydrogen atom, early in the textbook;
- it makes a deliberate attempt to show the unity of physics. For example, when fundamental interactions and the field concept are introduced in Chapter 6, gravitational, electric, and magnetic fields are discussed in the same context.

Textbook Features

The textbook includes many features which are intended to enhance its usefulness to both the student and instructor:

Style

I have attempted to write the book in a style that is clear, logical, and succinct. At the same time, my relaxed writing style is meant to make the reading more enjoyable and less intimidating.

Previews

Most chapters begin with a chapter preview, which includes a brief discussion of chapter objectives and content. This feature enables the student to better understand how the topics covered in that chapter fit into the overall structure and objectives of the course.

Mathematical Level

Calculus is introduced gradually, where it is needed, keeping in mind that many students often take a course in calculus concurrently. Most steps are shown when basic equations are developed, and reference is made to mathematical appendices provided at the end of the text. Vector products are covered where they are needed in physical applications.

Worked Examples

Every chapter includes several worked examples of varying difficulty which are intended to reinforce conceptual understanding and to serve as models for solving end-of-chapter problems.

Worked Example Exercises

Many examples are followed immediately by exercises with answers that represent extensions of the worked examples. These exercises are intended to encourage students to practice problem solving for immediate reinforcement, and to test the student's understanding of concepts and problem-solving skills. Students who work through these exercises on a regular basis should find the end-of-chapter problems less intimidating.

Important Statements and Equations

Most important statements and definitions are set in blue print for added emphasis and ease of review. Important equations are highlighted with a tan screen for review or reference.

Illustrations

The text material, worked examples, and end-of-chapter questions and exercises are accompanied by numerous figures, photographs, and tables. Full color is used to add clarity to the figures and to make the visual presentations as realistic as possible. Three-dimensional effects are produced with the use of color air-brushed areas, where appropriate. Vectors are color-coded, and curves in *xy*-plots are drawn in color. Color photographs have been carefully selected, and their accompanying captions have been written to serve as an added instructional tool.

Margin Notes

Comments and references to the descriptions of specific equations are presented in the margins, where they help locate and review important statements, concepts, and equations.

Units

The international system of units (SI) is used throughout the text. The British engineering system of units is used only to a limited extent in chapters on mechanics, heat, and thermodynamics.

Problem-Solving Strategies and Hints

I have included general strategies and hints for solving the types of problems featured in both the worked examples and in the end-of-chapter problems. This feature will help students identify important steps and understand the logic for solving specific classes of problems.

Summaries

Each chapter contains a summary that reviews the important concepts and equations discussed in that chapter.

Questions and Conceptual Exercises

A list of questions and conceptual exercises is provided at the end of each chapter. The questions, which require verbal responses, provide the student with a means of self-testing the concepts presented in the chapter. Others could serve as a basis for initiating classroom discussions. Conceptual exercises, identified by blue numbers, are intended to test the students' ability to perform order-of-magnitude calculations and to apply concepts they have learned to real-life situations without performing detailed calculations.

Problems

End-of-chapter Problems An extensive set of problems is included at the end of each chapter. Answers to odd-numbered problems are given at the end of the book; these pages have colored edges for ease of location. For the convenience of both the student and the instructor, about two thirds of the problems are keyed to specific sections of the chapter. The remaining problems, labeled "Additional Problems," are not keyed to specific sections. Problems are marked to indicate one of three levels of difficulty. Straightforward problems are numbered in black, intermediate problems are numbered in blue, and the most challenging problems are numbered in magenta.

To accommodate the various tools that instructors are currently using in the classroom, we have included some problems that use laboratory data and others that can be explored and solved using computers, and such tools as spreadsheets and similar software.

Interactive Physical Problems This textbook brings more than 100 worked examples and end-of-chapter problems to life with real-time simulations and animations using the highly acclaimed program Interactive Physics II, developed by Knowledge Revolution. These examples and problems are marked with the icon . This powerful and intuitive simulation program allows students and instructors to quickly build and explore physics experiments on the computer. These simulations can also be used by the instructor in the classroom or laboratory to help students understand physics concepts by developing visualization skills. Experiments are easily created by first drawing objects on the screen, defining parameters such as mass, friction, elasticity, and spring constants. The simulation is started by simply clicking the RUN button. The simulation engine calculates the motion of the defined system and displays it in smooth animation. The results can be measured and analyzed in graphical, digital, tabular, and bar graph formats. The acquired data can also be exported to a spreadsheet of your choice for other types of analyses.

Instructors will be provided with a set of three disks containing Interactive Physics templates designed for use with this text. The first two disks, containing about 50 simulations, were prepared by the staff at Knowledge Revolution, while the third disk, containing about 50 simulations, was prepared by the author. A Macintosh version and a Windows version of the program are available.

Spreadsheet Problems and Examples Selected problems and examples throughout the text have been keyed with a box to indicate the availability of spreadsheet templates to accompany these problems on a separate disk. This is an option (see Student Ancillaries); the problems can also be solved analytically without using the spreadsheet tool. The spreadsheets will be helpful for working some of the more difficult problems. Written instructions for use of the spreadsheets, together with the spreadsheet templates on disk, will be included in Supplement II. The templates can be used with a variety of programs such as Lotus 1-2-3.

Calculator/Computer Problems Numerical problems that can best be solved with the use of programmable calculators or computers are given in a selected number of chapters. These will be useful in those courses where the instructor wishes to put programming skills to practice.

Laboratory Problems Selected problems throughout the text, denoted by a triangle, use experimental data to challenge students' problem-solving skills and to compare textbook models of physical phenomena with real life data.

Appendices and Endpapers Several appendices are provided at the end of the text. Most of the appendix material represents a review of mathematical techniques used in the text, including scientific notation, algebra, geometry, trigonometry, differential calculus, and integral calculus. References to these appendices are made throughout the text.

Ancillaries

Student Ancillaries

New Options to Help Students Learn Physics Saunders College Publishing is offering some new and unique items to enhance the classroom experience. These ancillaries will allow instructors to customize the textbook to their students' needs. Any of the these ancillaries may be shrinkwrapped with the text at a reduced price:

Life Science Applications for Physics by Jerry S. Faughn, Eastern Kentucky University, and Raymond A. Serway Provides examples and readings from the biological sciences as they relate to physics. Topics include "Friction in Human Joints," "Physics of the Human Circulatory System," and "Ultrasound and its Applications."

Spreadsheet Modeling for Physics by David A. Stetser, Miami University, and Colonel James Head, United States Air Force Academy This workbook and accompanying disk contain problems and examples keyed to both *Principles of Physics* and *Physics for Scientists and Engineers.* The workbook describes how to use spreadsheets for solving physics problems and includes an introduction to modeling particle dynamics using a personal computer.

Advanced Problems Workbook by Harry H. Bingham, University of California, Berkeley Offers a collection of challenging problems and examples for more advanced students. Solutions to these problems appear in a separate instructor's manual.

Pocket Guide to Principles of Physics by V. Gordon Lind, Utah State University This 5 × 7 handbook provides formulas and helpful hints at a glance and gives students a convenient reference booklet—an invaluable companion when it comes to reviewing concepts and solving problems in the text.

Student Solutions Manual by Steve Van Wyk, Chapman College, Ralph McGrew, Broome Community College, and Raymond A. Serway Offers detailed solutions to approximately one third of the end-of-chapter problems.

Study Guide by John R. Gordon, James Madison University, and Raymond A. Serway Contains chapter objectives; a skills section that reviews mathematical techniques and suggested approaches to problem-solving methodology; and notes from selected chapter sections, including a glossary of important terms, theorems, and concepts.

Physics Problem Set for Interactive Physics and Discovery Exercises for Interactive Physics by Raymond A. Serway and Knowledge Revolution Two disks in Macintosh or IBM format with 50 to 80 problems and examples taken from the text can be packaged with the *Discovery Exercises* workbook by Jon Staib, James Madison University. The workbook features directed exercises that make use of the examples and problems on the disks. To obtain a student version of the Interactive II program, write to Knowledge Revolution, 15 Brush Place, San Francisco, CA 94103-3903.

So You Want to Take Physics: A Preparatory Course with Calculus by Rodney Cole, University of California, Davis Can be used as a primary text in a separate course preceding the physics course, or as a supplement to the calculus-based physics course. The text offers explanations, worked examples, and a review of techniques to students who need additional preparation. The friendly, straightforward style encourages students to experiment with mathematics as it is used in physics. An instructor's manual is available with this title.

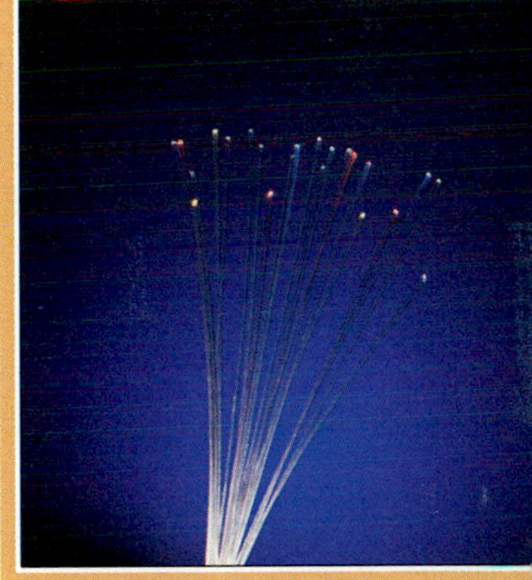

Mathematical Methods for Introductory Physics with Calculus by Ronald C. Davidson, Princeton University This brief book is designed for students who find themselves unable to keep pace in their physics class because of a lack of familiarity with the necessary mathematical tools. *Mathematical Methods* provides a brief overview of all the various types of mathematical topics that may be needed in an introductory-level physics course through the use of many worked examples and exercises.

Instructor Ancillaries

Instructor's Manual with Complete Solutions by Steve Van Wyk, Ralph McGrew, and Raymond A. Serway This manual consists of complete solutions to all of the problems in the text, and answers to even-numbered problems. All solutions have been carefully reviewed for accuracy.

Printed Test Bank by Myron Schneiderwent, University of Wisconsin—Superior Offers over 1500 multiple choice *and* conceptually-oriented critical thinking questions.

Computerized Test Bank Available for the IBM PC, Macintosh, and Apple II computers, the computerized version of the printed test bank also contains over 1500 test questions. The test bank enables the instructor to create many unique tests and permits the editing of questions as well as addition of new questions. The software program solves all problems and prints each answer on a separate grading key.

Overhead Transparency Acetates 150 transparencies with approximately 300 full-color figures from the text. An improved production process yields crisper figures and more legible labels for easier viewing in even large lecture halls.

Transparency Masters with Selected Solutions by Raymond A. Serway Provides solutions selected from the student solutions manual; suitable for creating transparency acetates for classroom use.

Physics Laboratory Manual by David Loyd Supplements the learning of basic physical principles while introducing laboratory procedures and equipment. Each chapter of the laboratory manual includes a pre-laboratory assignment, objectives, an equipment list, the theory behind the experiment, experimental procedures, graphs, and questions. A laboratory report is provided for each experiment so the student can record data, calculations, and experimental results.

Instructor's Manual to accompany Physics Laboratory Manual Each chapter contains a discussion of the experiment, teaching hints, answers to selected questions, and a post-laboratory quiz with short answer and essay questions. We have also included a list of the suppliers of scientific equipment and a summary of the equipment needed for all the laboratory experiments in the manual.

Saunders Physics Videodisc, Version 2 This exclusive set of two videodiscs contains 70 physics demonstrations and over 500 still images from *Principles of Physics, Physics for Scientists and Engineers,* and a variety of other sources.

LectureActive presentation software to accompany the Saunders Physics Videodisc allows the instructor to combine video, animation, and graphics of his or her choosing from the videodisc with lecture notecards to create a complete lecture. The instructor has the ability to locate images on the videodisc that pertain to a particular topic, modify both audio and video images (including video captioning that permits character labeling directly on the video screen), attach a video segment or still image to a notecard on the computer screen, and add lecture notes from other sources, so previous work is not lost to the user. With *LectureActive*, the lecture can be read directly from the computer screen, while the computer runs images from the videodisc that the instructor has chosen, or the instructor can print the lecture notes that have accompanying barcodes to call up the selected video. Available in Macintosh or IBM.

Physics Demonstrations Videotape by J. C. Sprott, University of Wisconsin, Madison
This video features two hours of demonstrations divided into 12 primary topics. Each topic contains between four and nine demonstrations for a total of 70 physics demonstrations.

Teaching Options

Although many topics found in traditional textbooks have been omitted from this textbook, instructors may find that the current text still contains more material than can be covered in a two-semester sequence. For this reason, I would like to offer the following suggestions. If you wish to place more emphasis on contemporary topics in physics, you should consider omitting Chapters 13, 14, 15, 24, 25 and 26. On the other hand, if you wish to follow a more traditional approach which places more emphasis on classical physics, you could omit Chapters 10, 12, 29, 30, 31, and 32. Either approach can be used without any loss in continuity. Other teaching options would fall somewhere in between these two extremes by deleting appropriate optional sections labeled with asterisks (*).

Reviewers

As any author well knows, a project of this magnitude would be an impossible task without the assistance of many talented and dedicated individuals whose comments, criticisms, and suggestions are most useful in preparing a finished product. I am very grateful to the following reviewers of the manuscript for their valuable feedback: Edward Adelson, Ohio State University; Subash Antani, Edgewood College; Harry Bingham, University of California, Berkeley; Laurie M. Brown, Northwestern University; Anthony Buffa, California Polytechnic State University, San Luis Obispo; Gordon Emslie, University of Alabama at Huntsville; Donald Erbschloe, United States Air Force Academy, Gerald Hart, Moorhead State University; Joey Huston, Michigan State University; David Judd, Broward Community College; V. Gordon Lind, Utah State University; Barrett Lowe; David Markowitz, University of Connecticut; Roy Middleton, University of Pennsylvania; Clement J. Moses, Utica College of Syracuse University; Anthony Novaco, Lafayette College; Desmond Penny, Southern Utah University; Prabha Ramakrishnan, North Carolina State University; Rogers Redding, University of North Texas; J. Clinton Sprott, University of Wisconsin at Madison; B. Cecil Thompson, University of Texas at Arlington.

Acknowledgments

I am indebted to Colonel James Head and Lt. Col. Rolf Enger of the U. S. Air Force Academy for inspiring me to begin work on this project while I was at the Academy during the summer of 1990, and for their warm hospitality during my visit. They and their colleagues designed the original IUPP model upon which this textbook is based. Special thanks go to Steve Van Wyk and Ralph McGrew for assembling the problem sets, writing new problems, and checking solutions to all problems, and for their contributions in preparing the Instructor's Manual and Student Solutions Manual. I am especially grateful to my colleague John R. Gordon, for his diligence in preparing the Student Study Guide, and to Linda Miller for typing the Student Study Guide. I appreciate the efforts of Jon Staib for preparing the workbook that accompanies the Interactive Physics Problem Sets, and Ray Smith for reviewing and improving the Interactive Physics II simulations. Sarah Evertson was very helpful in locating many of the photographs used in this text. During the development of this text, I benefited from many useful discussions with my colleagues and other physics instructors, including Don Chodrow, Jerry Faughn, John R. Gordon, Dorn Peterson, Joseph Rudmin, Gerald Taylor, Robert Bauman, and Clem Moses. I owe a debt of gratitude to Irene Nunes for clarifying much of the material in this manuscript and for eliminating many inconsistencies, redundancies, and other nightmares that authors must face. Special thanks and recognition go to the professional staff at Saunders College Publishing for their careful attention to detail and for their dedication to producing a first-quality textbook, especially Jennifer Bortel, Sally Kusch, Lloyd Black, Carol Bleistine, Tim Frelick and Marjorie Waldron. I sincerely appreciate the efforts of my publisher and good friend John Vondeling, whose enthusiasm, wisdom, and incredible sense of timing never ceases to amaze me.

Finally, I thank my entire family for their endless love, understanding, and continued encouragement, especially my beautiful and devoted wife Elizabeth, who is always at my side to share my joys and sorrows.

Ray Serway
Harrisonburg, VA

December 1993

I feel it is appropriate to offer some words of advice which should be of benefit to you, the student. Before doing so, I will assume that you have read the preface, which describes the various features of the text that will help you through the course.

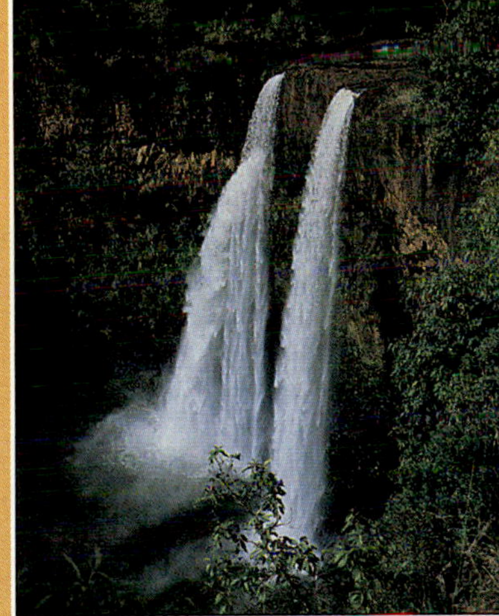

To the Student

How to Study

Very often instructors are asked "How should I study physics and prepare for examinations?" There is no simple answer to this question, but I would like to offer some suggestions based on my own experiences in learning and teaching over the years.

First and foremost, maintain a positive attitude towards the subject matter, keeping in mind that physics is the most fundamental of all natural sciences. Other science courses that follow will use the same physical principles, so it is important that you understand and be able to apply the various concepts and theories discussed in the text.

Concepts and Principles

It is essential that you understand the basic concepts and principles *before* attempting to solve assigned problems. This is best accomplished through a careful reading of the textbook before attending your lecture on that material. In the process, it is useful to jot down certain points which are not clear to you. Take careful notes in class, and then ask questions pertaining to those ideas that require clarification. Keep in mind that few people are able to absorb the full meaning of scientific material after one reading. Several readings of the text and notes may be necessary. Your lectures and laboratory work should supplement the text and clarify some of the more difficult material. You should reduce memorization of material to a minimum. Memorizing passages from a text, equations, and derivations does not necessarily mean you understand the material. Your understanding of the material will be enhanced through a combination of efficient study habits, discussions with other students and instructors, and your ability to solve the problems in the text. Ask questions whenever you feel it is necessary.

Study Schedule

It is important to set up a regular study schedule, preferably on a daily basis. Make sure to read the syllabus for the course and adhere to the schedule set by your instructor. The lectures will be much more meaningful if you read the corresponding textual material before attending the lecture. As a general rule, you should devote about two hours of study time for every hour in class. If you are having trouble with the course, seek the advice of the instructor or students who have taken the course. You may find it necessary to seek further instruction from experienced students. Very often, instructors will offer review sessions in addition to regular class periods. It is important that you avoid the practice of delaying study until a day or two before an exam. More often than not, this will lead to disastrous results. Rather than an all-night study session, it is better to briefly review the basic concepts and equations, followed by a good night's rest. If you feel in need of additional help in understanding the concepts, preparing for exams, or in problem-solving, we suggest that you acquire a copy of the Student Study Guide which accompanies the text and should be available at your college bookstore.

Use the Features

You should make full use of the various features of the text discussed in the preface. For example, marginal notes are useful for locating and describing important equations and concepts, while important statements and definitions are highlighted in color. Many useful tables are contained in appendices, but most are incorporated in the text where they are used most often. Appendix B is a convenient review of mathematical techniques. Answers to odd-numbered problems are provided in the study guide. Exercises (with answers), which follow some worked examples, represent extensions of those examples, and in most cases you are expected to perform a simple calculation. Their purpose is to test your problem-solving skills as you read through the text. Problem-Solving Strategies and Hints are included in selected chapters throughout the text to give you additional information to help you solve problems. An overview of the entire text is given in the table of contents, while the index will enable you to locate specific material quickly. Footnotes are sometimes used to supplement the discussion or to cite other references on the subject. Many chapters include problems that require the use of programmable calculators, computers, or simulation software (Interactive Physics II). Problems and examples marked with boxes can be solved either analytically or with the use of spreadsheets available from your instructor. These are intended for those courses that place some emphasis on numerical methods. You may want to develop appropriate programs for some of these problems even if they are not assigned by your instructor.

After reading a chapter, you should be able to define any new quantities introduced in that chapter, and discuss the principles and assumptions that were used to arrive at certain key relations. The chapter summaries and the review sections of the study guide should help you in this regard. In some cases, it will be necessary to refer to the index of the text to locate certain topics. You should be able to correctly associate with each physical quantity a symbol used to represent that quantity and the unit in which the quantity is specified. Furthermore, you should be able to express each important relation in a concise and accurate prose statement.

The Importance of Problem Solving

R.P. Feynman, Nobel laureate in physics, once said. "You do not know anything until you have practiced." In keeping with this statement, I strongly advise you to develop the skills necessary to solve a wide range of problems. Your ability to solve problems will be one of the main tests of your knowledge of physics; therefore, you should try to solve as many problems as possible. It is essential that you understand basic concepts and principles before attempting to solve problems.

It is good practice to try to find alternate solutions to the same problem. For example, problems in mechanics can be solved using Newton's laws, but very often an alternative method using energy considerations is more direct. You should not deceive yourself into thinking you understand the problem after seeing its solution in class. You must be able to solve the problem and similar problems on your own.

The method of solving problems should be carefully planned. A systematic plan is especially important when a problem involves several concepts. First, read the problem several times until you are confident you understand what is being asked. Look for any key words that will help you interpret the problem, and perhaps allow you to make certain assumptions. Your ability to interpret the question properly is an integral part of problem solving. You should acquire the habit of writing down the information given in a problem, and decide what quantities need to be found. You might want to construct a table listing quantities given, and quantities to be found. This procedure is sometimes used in the worked examples of the text. After you have decided on the method you feel is appropriate for the situation, proceed with your solution. General problem-solving strategies of this type are included in the text and are highlighted by a light blue screen.

I often find that students fail to recognize the limitations of certain formulas or physical laws in a particular situation. It is very important that you understand and remember the assumptions which underlie a particular theory or formalism. For example, certain equations in kinematics apply only to a particle moving with constant acceleration. These equations are not valid for situations in which the acceleration is not constant, such as the motion of an object connected to a spring, or the motion of an object through a fluid.

General Problem-Solving Strategy

Most courses in general physics require the student to learn the skills of problem solving, and examinations are largely composed of problems that test such skills. This brief section describes some useful ideas which will enable you to increase your accuracy in solving problems, enhance your understanding of physical concepts, eliminate initial panic or lack of direction in approaching a problem, and organize your work. One way to help accomplish these goals is to adopt a problem-solving strategy. Many chapters in this text will include a section labeled "Problem-Solving Strategies and Hints" which should help you through the "rough spots."

In developing problem-solving strategies, five basic steps are commonly used.

1. Draw a suitable diagram with appropriate labels and coordinate axes if needed.
2. As you examine what is being asked in the problem, identify the basic physical principle (or principles) that are involved, listing the knowns and unknowns.
3. Select a basic relationship or derive an equation that can be used to find the unknown, and solve the equation for the unknown symbolically.
4. Substitute the given values along with the appropriate units into the equation.
5. Obtain a numerical value for the unknown. The problem is verified and receives a check mark if the following questions can be properly answered:

Do the units match? Is the answer reasonable? Is the plus or minus sign proper or meaningful?

One of the purposes of this strategy is to promote accuracy. Properly drawn diagrams can eliminate many sign errors. Diagrams also help to isolate the physical principles of the problem. Symbolic solutions and carefully labeled knowns and unknowns will help eliminate other careless errors. The use of symbolic solutions should help you think in terms of the physics of the problem. A check of units at the end of the problem can indicate a possible algebraic error. The physical layout and organization of your problem will make the final product more understandable and easier to follow. Once you have developed an organized system for examining problems and extracting relevant information, you will become a more confident problem solver.

Example

A person driving in a car at a speed of 20 m/s applies the brakes and stops in a distance of 100 m. What was the acceleration of the car?

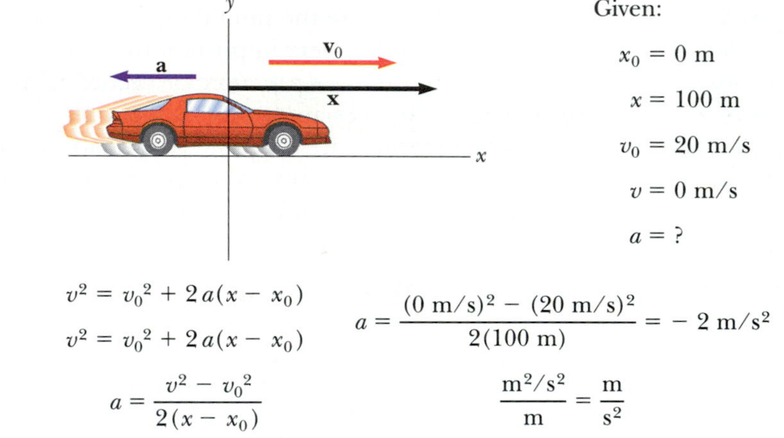

$$v^2 = v_0{}^2 + 2a(x - x_0)$$

$$v^2 = v_0{}^2 + 2a(x - x_0)$$

$$a = \frac{v^2 - v_0{}^2}{2(x - x_0)}$$

$$a = \frac{(0\ \text{m/s})^2 - (20\ \text{m/s})^2}{2(100\ \text{m})} = -2\ \text{m/s}^2$$

$$\frac{\text{m}^2/\text{s}^2}{\text{m}} = \frac{\text{m}}{\text{s}^2}$$

Experiments

Physics is a science based upon experimental observations. In view of this fact, I recommend that you try to supplement the text through various type of "hands-on" experiments, either at home or in the laboratory. These can be used to test ideas and models discussed in class or in the text. For example, the common "Slinky" toy is excellent for studying traveling waves; a ball swinging on the end of a long string can be used to investigate pendulum motion; various masses attached to the end of a vertical spring or rubber band can be used to determine their elastic nature; an old pair of Polaroid sunglasses and some discarded lenses and magnifying glass are the components of various experiments in optics; you can get an approximate measure of the acceleration of graivty by dropping a ball from a known height by simply measuring the time of its fall with a stopwatch. The list is endless. When physical models are not available, be imaginative and try to develop models of your own.

An Invitation to Physics

It is my sincere hope that you too will find physics an exciting and enjoyable experience, and that you will profit from this experience, regardless of your chosen profession. Welcome to the exciting world of physics.

The scientist does not study nature because it is useful; he studies it because he delights in it, and he delights in it because it is beautiful. If nature were not beautiful, it would not be worth knowing, and if nature were not worth knowing, life would not be worth living.

HENRI POINCARÉ

Contents Overview

Contents

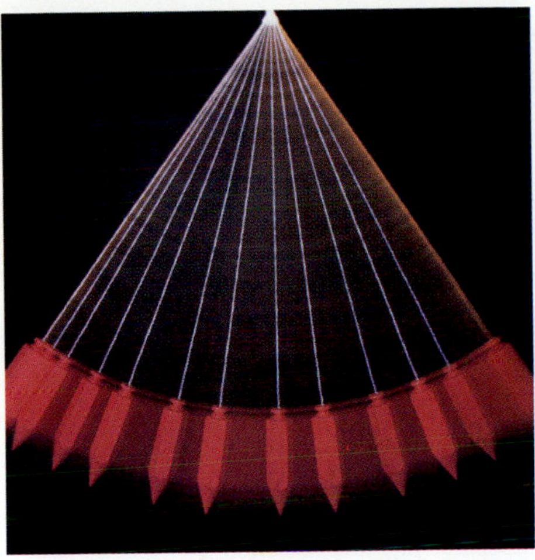

11 Rotational Motion 294

12 Orbital Motions and the Hydrogen Atom 328

13 Temperature and the Kinetic Theory of Gases 355

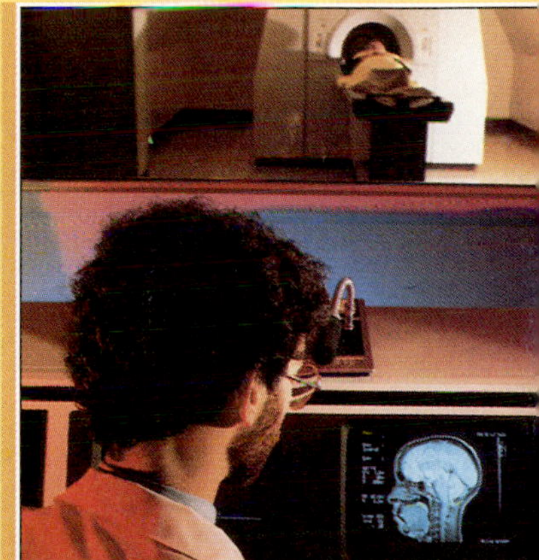

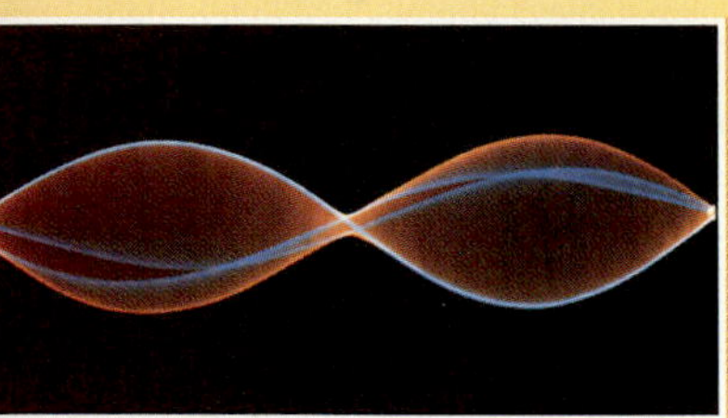

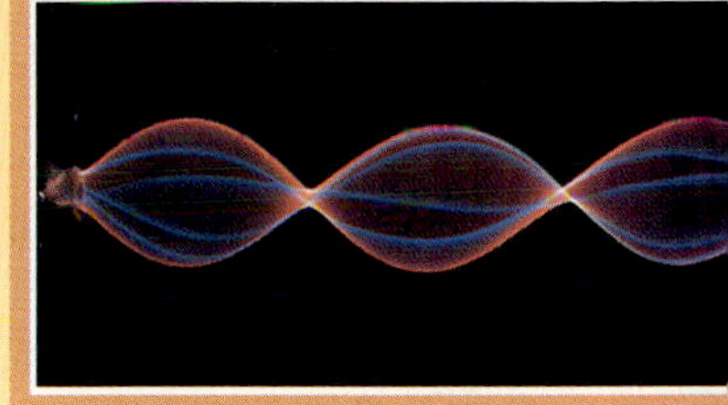

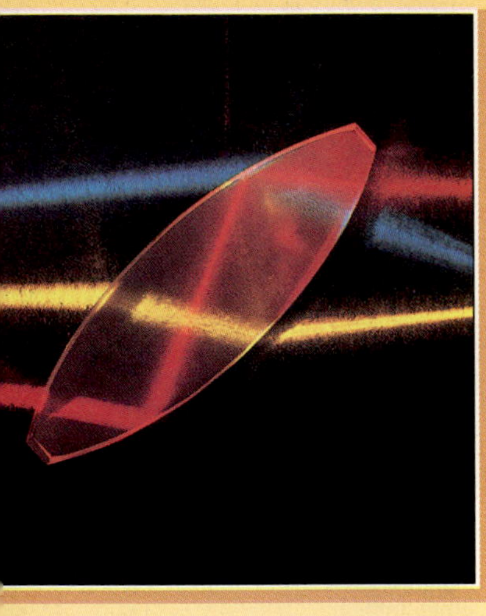

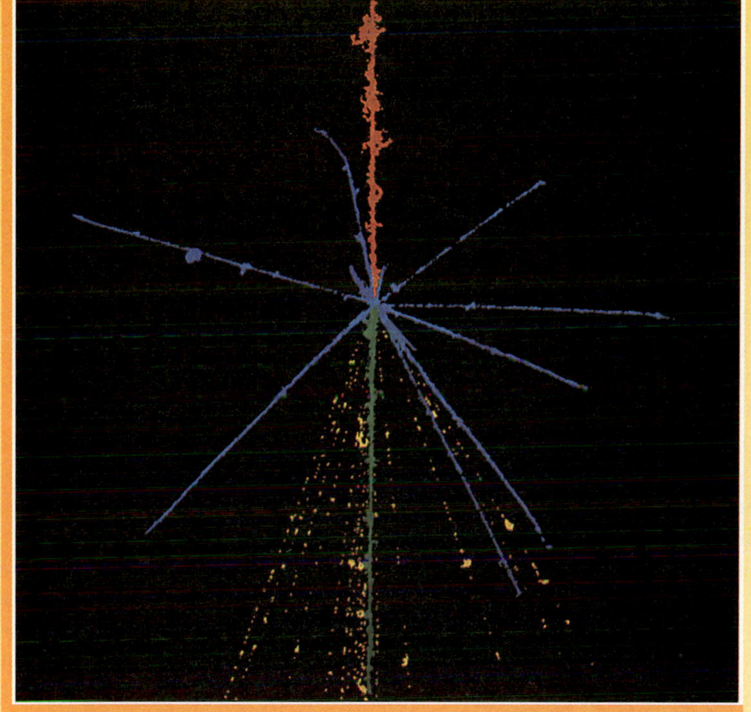

The power of a cosmic ray is shown in this photograph taken in 1950 by Cecil Powell, the English physicist who pioneered the use of photographic emulsions to record the tracks of electrically charged particles. *(C. Powell, P. Fowler, and D. Perkins/Science Photo Library)*

An Invitation to Physics

Physics, the most fundamental physical science, is concerned with the basic principles of the Universe. It is the foundation upon which the other physical sciences—astronomy, chemistry, and geology—are based. The beauty of physics lies in the simplicity of its fundamental theories and in the manner in which just a small number of basic concepts, equations, and assumptions can alter and expand our view of the world around us.

The myriad physical phenomena in our world are parts of one or more of the following five areas of physics:

1. Classical mechanics, which is concerned with the motion of objects moving at speeds that are low compared to the speed of light
2. Relativity, which is a theory describing particles moving at any speed, even those whose speeds approach the speed of light
3. Thermodynamics, which deals with heat, temperature, and the behavior of large numbers of particles
4. Electromagnetism, which involves the theory of electricity, magnetism, and electromagnetic fields
5. Quantum mechanics, a theory dealing with the behavior of submicroscopic particles as well as the macroscopic world

When a discrepancy between theory and experiment arises in any of these areas, new theories and experiments must be formulated to remove the discrepancy. Many times a theory is satisfactory under limited conditions; a more general theory might be satisfactory without such limitations. A classic example is Newton's laws of motion, which accurately describe the motions of bodies at low speeds but do not apply to objects moving at speeds comparable to the speed of light. The special theory of relativity developed by Albert Einstein (1879–1955) successfully predicts the motions of objects at speeds approaching the speed of light and hence is a more general theory of motion.

Classical physics, developed prior to 1900, includes the theories, concepts, laws, and experiments in classical mechanics, thermodynamics, and electromagnetism.

Galileo Galilei (1564–1642) made significant contributions to classical mechanics through his work on the laws of motion with constant acceleration. In the same era, Johannes Kepler (1571–1630) used astronomical observations to develop empirical laws for the motions of planetary bodies.

The most important contributions to classical mechanics, however, were provided by Isaac Newton (1642–1727), who developed classical mechanics as a systematic theory and was one of the originators of the calculus as a mathematical tool. Although major developments in classical physics continued in the 18th century, thermodynamics and electricity and magnetism were not developed until the latter part of the 19th century, principally because the apparatus for controlled experiments was either too crude or unavailable until then. Although many electric and magnetic phenomena had been studied earlier, it was the work of James Clerk Maxwell (1831–1879) that provided a unified theory of electromagnetism. In this text we shall treat the various disciplines of classical physics in separate sections; however, we will see that the disciplines of mechanics and electromagnetism are basic to all the branches of classical and modern physics.

A new era in physics, usually referred to as *modern physics,* began near the end of the 19th century. Modern physics developed mainly because of the discovery that many physical phenomena could not be explained by classical physics. The two most important developments in this modern era were the theories of relativity and quantum mechanics. Einstein's theory of relativity completely revolutionized the traditional concepts of space, time, and energy. Among other things, Einstein's theory corrected Newton's laws of motion in describing the motion of objects moving at speeds comparable to the speed of light. The theory of relativity also assumes that the speed of light is the upper limit of the speed of an object or signal and shows that mass and energy are related. Quantum mechanics was formulated by a number of distinguished scientists to provide descriptions of physical phenomena at the atomic level.

Scientists continually work at improving our understanding of fundamental laws, and new discoveries are made every day. In many research areas there is a great deal of overlap among physics, chemistry, and biology. The many technological advances in recent times are the result of the efforts of many scientists, engineers, and technicians. Some of the most notable recent developments are (1) unmanned space missions and manned moon landings, (2) microcircuitry and high-speed computers, and (3) sophisticated imaging techniques used in scientific research and medicine. The impacts of such developments and discoveries on our society have indeed been great, and it is very likely that future discoveries and developments will be exciting, challenging, and of great benefit to humanity.

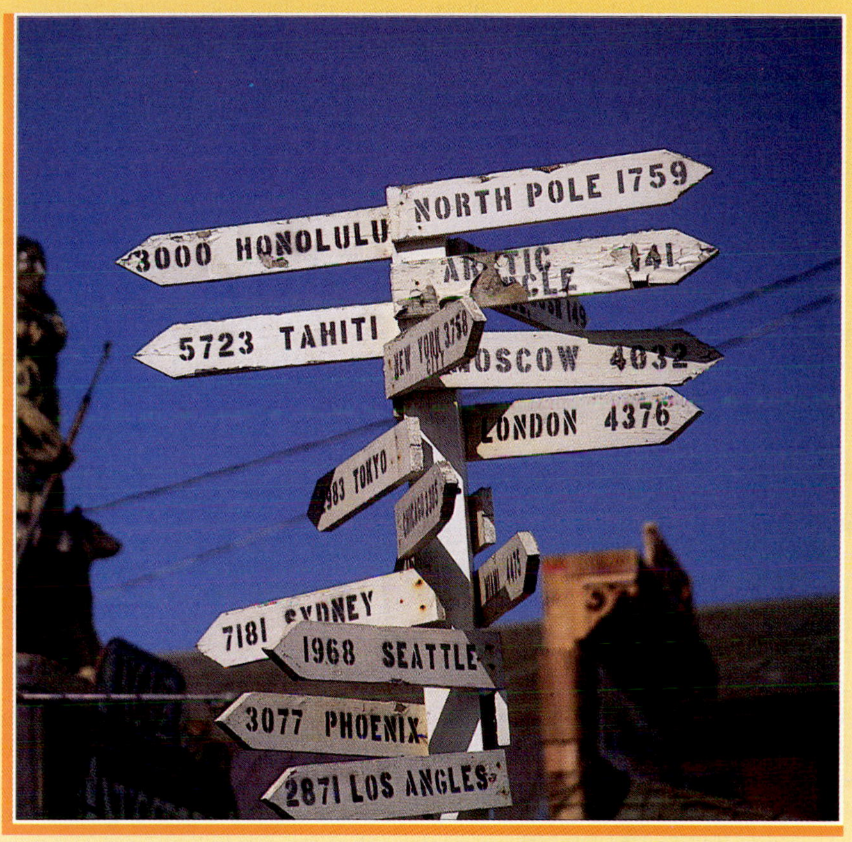

This interesting set of directional arrows was photographed in Nome, Alaska. A particular location relative to Nome is specified by the direction of an arrow and the distance to that location. This is one example of a vector quantity. We shall examine this and other vector quantities in this chapter. *(David Bartruff, FPG International)*

Introduction and Vectors

CHAPTER 1

The goal of physics is to provide a quantitative understanding of certain basic phenomena that occur in our Universe. Physics is a science based on experimental observations and mathematical analyses. The main objective behind such experiments and analyses is to develop theories that explain the phenomenon being studied and to relate those theories to other established theories. Fortunately, it is possible to explain the behavior of various physical systems using relatively few fundamental laws. Analytical procedures require the expression of those laws in the language of mathematics, the tool that provides a bridge between theory and experiment. In this chapter we shall discuss a few mathematical concepts and techniques that will be used throughout the text.

Since future chapters will be concerned with the laws of physics, it is necessary to provide clear definitions of the basic quantities involved in these laws. For example, such physical quantities as force, velocity, volume, and acceleration can be described in terms of more fundamental quantities. In the next several chapters we shall encounter three of the latter: **length** (L), **time** (T), and **mass** (M). Later in the text we shall have to add units for two other standard quantities to our list, one for temperature (the kelvin), the other for electric current (the ampere). In our study of mechanics, however, we shall be concerned only with the units of mass, length, and time.

Some physical quantities require the measurement of more than one attribute. For example, physical quantities that have both numerical and directional properties—such as force, velocity, and acceleration—are represented by vectors. A large part of this chapter is concerned with vector algebra and general properties of vectors. The addition and subtraction of vectors will be discussed, together with some common applications to physical situations. Since vectors will be used throughout this text, it is imperative that you understand both their graphical and their algebraic properties.

1.1 Standards of Length, Mass, and Time

If we are to report the results of a measurement of a certain quantity to someone who wishes to reproduce this measurement, a unit for the quantity must be defined. For example, it would be meaningless for a visitor from another planet to talk to us about a length of 8 "glitches" if we did not know the meaning of the unit "glitch." On the other hand, if someone familiar with our system of measurement and weights reports that a wall is 2.0 meters high and our unit of length is defined to be 1.0 meter, we then know that the height of the wall is twice our fundamental unit of length. Likewise, if we are told that a person has a mass of 75 kilograms and our unit of mass is defined as 1.0 kilogram, then that person has a mass 75 times larger than our fundamental unit of mass. In 1960 an international committee agreed on a system of definitions and standards to describe fundamental physical quantities. It is called the **SI system** (Système International) of units. Its units of length, mass, and time are the meter, kilogram, and second, respectively.

Length

In A.D. 1120 the king of England decreed that the standard of length in his country would be the yard, and that the yard would be precisely equal to the distance from the tip of his nose to the end of his outstretched arm. Similarly, the original standard for the foot adopted by the French was the length of the royal foot of King Louis XIV. This standard prevailed until 1799, when the legal standard of length in France became the meter, defined as one ten-millionth of the distance from the equator to the North Pole.

Many other systems have been developed in addition to those just discussed, but the advantages of the French system have caused it to prevail in most countries and in scientific circles everywhere. As recently as 1960, the length of the meter was still defined as the distance between two lines on a specific bar of platinum-iridium alloy stored under controlled conditions. This standard was abandoned for several reasons, a principal one being that the limited accuracy with which the separation between the lines can be determined does not meet the current requirements of

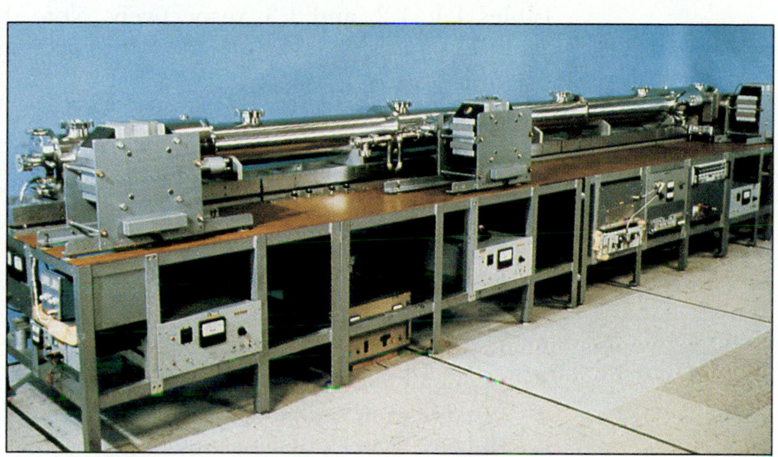

(*Left*) The National Standard Kilogram No. 20, an accurate copy of the International Standard Kilogram kept at Sèvres, France, is housed under a double bell jar in a vault at the National Bureau of Standards. (*Right*) The primary frequency standard (an atomic clock) at the National Bureau of Standards. This device keeps time with an accuracy of about 3 millionths of a second per year. *(Photos courtesy of National Bureau of Standards, U.S. Department of Commerce)*

science and technology. Until recently, the meter was defined as 1 650 763.73 wavelengths of orange-red light emitted from a krypton-86 lamp. However, in October 1983 the **meter** was redefined to be *the distance traveled by light in a vacuum during a time of 1/299 792 458 second.* In effect, this latest definition establishes that the speed of light in a vacuum is 299 792 458 meters per second.

Mass

The SI unit of mass, the **kilogram,** is defined as *the mass of a specific platinum-iridium alloy cylinder kept at the International Bureau of Weights and Measures at Sèvres, France.* At this point, we should add a word of caution. Most beginning students of physics tend to confuse the physical quantities called weight and mass. For the present we shall not discuss the distinction between them; they will be clearly defined in later chapters. For now you should note that they are distinctly different quantities.

A triple beam balance is used to make accurate mass measurements. This particular balanced system contains 215 g of potassium ferrocyanide. *(Michael Dalton, Fundamental Photographs)*

Time

Before 1960 the standard of time was defined in terms of the average length of a solar day in the year 1900. (A solar day is the time interval between successive appearances of the Sun at the highest point it reaches in the sky each day.) The basic unit of time, the second, was defined to be $(1/60)(1/60)(1/24) = 1/86\,400$ of the average solar day. In 1967 the second was redefined to take advantage of the great precision obtainable with a device known as an atomic clock, which uses the characteristic frequency of the cesium-133 atom as the "reference clock." The **second** is now defined as *9 192 631 770 times the period of one oscillation of the cesium atom.*

Approximate Values for Length, Mass, and Time

Approximate values of various lengths, masses, and time intervals are presented in Tables 1.1, 1.2, and 1.3, respectively. Note the wide range of values for these quantities.[1] You should study the tables and get a feel for what is meant by a mass

[1] If you are unfamiliar with the use of powers of ten (scientific notation), you should review Appendix B.1.

Table 1.1
Approximate Values of Some Measured Lengths

	Length (m)
Distance from Earth to most remote quasar known	1.4×10^{26}
Distance from Earth to most remote normal galaxies known	4×10^{25}
Distance from Earth to nearest large galaxy (M 31 in Andromeda)	2×10^{22}
Distance from Sun to nearest star (Proxima Centauri)	4×10^{16}
One lightyear	9.46×10^{15}
Mean orbit radius of Earth	1.5×10^{11}
Mean distance from Earth to Moon	3.8×10^{8}
Distance from equator to North Pole	1×10^{7}
Mean radius of Earth	6.4×10^{6}
Typical altitude of orbiting Earth satellite	2×10^{5}
Length of a football field	9.1×10^{1}
Length of a housefly	5×10^{-3}
Size of smallest dust particles	1×10^{-4}
Size of cells of most living organisms	1×10^{-5}
Diameter of a hydrogen atom	1×10^{-10}
Diameter of a uranium nucleus	1.4×10^{-14}
Diameter of a proton	1×10^{-15}

Table 1.2
Masses of Various Bodies (Approximate Values)

	Mass (kg)
Universe	10^{52}
Milky Way galaxy	10^{42}
Sun	2×10^{30}
Earth	6×10^{24}
Moon	7×10^{22}
Shark	3×10^{2}
Human	7×10^{1}
Frog	1×10^{-1}
Mosquito	1×10^{-5}
Bacterium	1×10^{-15}
Hydrogen atom	1.67×10^{-27}
Electron	9.11×10^{-31}

Table 1.3
Approximate Values of Some Time Intervals

	Interval (s)
Age of the Universe	5×10^{17}
Age of the Earth	1.3×10^{17}
Time since the fall of the Roman empire	5×10^{12}
Average age of a college student	6.3×10^{8}
One year	3.2×10^{7}
One day (time for one revolution of Earth about its axis)	8.6×10^{4}
Time between normal heartbeats	8×10^{-1}
Period[a] of audible sound waves	1×10^{-3}
Period of typical radio waves	1×10^{-6}
Period of vibration of an atom in a solid	1×10^{-13}
Period of visible light waves	2×10^{-15}
Duration of a nuclear collision	1×10^{-22}
Time for light to cross a proton	3.3×10^{-24}

[a] A period is defined as the time interval required for one complete vibration.

of 100 kilograms, for example, or by a time interval of 3.2×10^7 seconds. Systems of units commonly used are the SI system, in which the units of length, mass, and time are the meter (m), kilogram (kg), and second (s), respectively; the cgs or Gaussian system, in which the units of length, mass, and time are the centimeter (cm), gram (g), and second, respectively; and the British engineering system (sometimes called the conventional system), in which the units of length, mass, and time are the foot (ft), slug, and second, respectively. Throughout most of this text we shall use SI units since they are almost universally accepted in science and industry. We will make limited use of conventional units in the study of classical mechanics.

Some of the most frequently used prefixes for the powers of ten and their abbreviations are listed in Table 1.4. For example, 10^{-3} m is equivalent to 1 millimeter (mm), and 10^3 m is 1 kilometer (km). Likewise, 1 kg is 10^3 g, and 1 megavolt (MV) is 10^6 volts.

Table 1.4
Some Prefixes for Powers of Ten

Power	Prefix	Abbreviation
10^{-18}	atto	a
10^{-15}	femto	f
10^{-12}	pico	p
10^{-9}	nano	n
10^{-6}	micro	μ
10^{-3}	milli	m
10^{-2}	centi	c
10^{-1}	deci	d
10^{3}	kilo	k
10^{6}	mega	M
10^{9}	giga	G
10^{12}	tera	T
10^{15}	peta	P
10^{18}	exa	E

1.2 Density and Atomic Mass

A fundamental property of any substance is its **density** ρ (Greek letter rho), defined as *mass per unit volume* (a table of the letters in the Greek alphabet is provided at the back of the book):

$$\rho \equiv \frac{m}{V} \qquad [1.1]$$

For example, aluminum has a density of 2.70×10^3 kg/m^3, and lead has a density of 11.3×10^3 kg/m^3. A list of densities for various substances is given in Table 1.5.

The difference in density between aluminum and lead is due in part to their different *atomic masses;* the atomic mass of lead is 207, and that of aluminum is 27. However, the ratio of atomic masses, $207/27 = 7.67$, does not correspond to the ratio of densities, $11.3/2.70 = 4.19$. The discrepancy is due to the difference in atomic spacings and atomic arrangements in the crystal structures of the two types of elements.

All ordinary matter consists of atoms, and each atom is made up of electrons and a nucleus. Practically all of the mass of an atom is contained in the nucleus, which consists of protons and neutrons. Thus, we can understand why the atomic masses of the elements differ. The mass of a nucleus is measured relative to the mass of an atom of carbon-12 (this isotope of carbon has six protons and six neutrons).

The mass of ^{12}C is defined to be exactly 12 atomic mass units (u), where $1\text{ u} = 1.6605402 \times 10^{-27}$ kg. In these units, the proton and neutron have masses of about 1 u. More precisely,

$$\text{Mass of proton} = 1.0073\text{ u}$$

$$\text{Mass of neutron} = 1.0087\text{ u}$$

The mass of the nucleus of ^{27}Al is approximately 27 u. In fact, a more precise measurement shows that the nuclear mass is always slightly *less* than the combined mass of the protons and neutrons making up the nucleus. The processes of nuclear fission and nuclear fusion are based on this mass difference.

Table 1.5
Densities of Various Substances

Substance	Density ρ (kg/m^3)
Platinum	21.45×10^3
Gold	19.3×10^3
Uranium	18.7×10^3
Lead	11.3×10^3
Copper	8.93×10^3
Iron	7.86×10^3
Aluminum	2.70×10^3
Magnesium	1.75×10^3
Water	1.00×10^3
Air	0.0013×10^3

One **mole** (mol) of any element (or compound) consists of Avogadro's number, N_A, of molecules of the substance. Avogadro's number, $N_A = 6.02 \times 10^{23}$, was defined so that one mole of carbon-12 atoms would have a mass of exactly 12 g. A mole of one element differs in mass from a mole of another. For example, one mole of aluminum has a mass of 27 g, and one mole of lead has a mass of 207 g. But one mole of aluminum contains the same number of atoms as one mole of lead, since there are 6.02×10^{23} atoms in one mole of *any* element. The mass per atom, which is *not* equal to the atomic mass, is then given by

Atomic mass

$$m = \frac{\text{atomic mass}}{N_A} \qquad [1.2]$$

For example, the mass of an aluminum atom is

$$m = \frac{27 \text{ g/mol}}{6.02 \times 10^{23} \text{ atoms/mol}} = 4.5 \times 10^{-23} \text{ g/atom}$$

Note that 1 u is equal to N_A^{-1} g.

Example 1.1 How Many Atoms in the Cube?

A solid cube of aluminum (density 2.7 g/cm³) has a volume of 0.20 cm³. How many aluminum atoms are contained in the cube?

Solution Since density equals mass per unit volume, the mass of the cube is

$$\rho V = (2.7 \text{ g/cm}^3)(0.20 \text{ cm}^3) = 0.54 \text{ g}$$

To find the number of atoms, N, we can set up a proportion using the fact that one mole of aluminum (27 g) contains 6.02×10^{23} atoms:

$$\frac{6.02 \times 10^{23} \text{ atoms}}{27 \text{ g}} = \frac{N}{0.54 \text{ g}}$$

$$N = \frac{(0.54 \text{ g})(6.02 \times 10^{23} \text{ atoms})}{27 \text{ g}} = 1.2 \times 10^{22} \text{ atoms}$$

1.3 Dimensional Analysis

The word *dimension* has a special meaning in physics. It usually denotes the physical nature of a quantity. Whether a distance is measured in units of feet or meters or furlongs, it is a distance. We say its dimension is *length.*

The symbols that will be used to specify length, mass, and time are L, M, and T, respectively. We will often use brackets [ ] to denote the dimensions of a physical quantity. For example, in this notation the dimensions of velocity, v, are written $[v] = \text{L/T}$, and the dimensions of area, A, are $[A] = \text{L}^2$. The dimensions of area, volume, velocity, and acceleration are listed in Table 1.6, along with their units in the three common systems. The dimensions of other quantities, such as force and energy, will be described as they are introduced in the text.

In many situations, you may be faced with having to derive or check a specific formula. Although you may have forgotten the details of the derivation, there is a useful and powerful procedure called *dimensional analysis* that can be used as a

Table 1.6
Dimensions of Area, Volume, Velocity, and Acceleration

System	Area (L^2)	Volume (L^3)	Velocity (L/T)	Acceleration (L/T^2)
SI	m^2	m^3	m/s	m/s^2
cgs	cm^2	cm^3	cm/s	cm/s^2
British engineering	ft^2	ft^3	ft/s	ft/s^2

consistency check, to assist in the derivation, or to check your final expression. This procedure should always be used and should help minimize the rote memorization of equations. Dimensional analysis makes use of the fact that *dimensions can be treated as algebraic quantities.* That is, quantities can be added or subtracted only if they have the same dimensions. Furthermore, the terms on both sides of an equation must have the same dimensions. By following these simple rules, you can use dimensional analysis to help determine whether or not an expression has the correct form, because the relationship can be correct only if the dimensions on the two sides of the equation are the same.

To illustrate this procedure, suppose you wish to derive a formula for the distance x traveled by a car in a time t if the car starts from rest and moves with constant acceleration a. In Chapter 2 we shall find that the correct expression for this special case is $x = \frac{1}{2}at^2$. Let us check the validity of this expression from a dimensional analysis approach.

The quantity x on the left side has the dimension of length. In order for the equation to be dimensionally correct, the quantity on the right side must also have the dimension of length. We can perform a dimensional check by substituting the basic dimensions for acceleration, L/T^2, and time, T, into the equation. That is, the dimensional form of the equation $x = \frac{1}{2}at^2$ can be written as

$$L = \frac{L}{\cancel{T^2}} \cdot \cancel{T^2} = L$$

The units of time cancel as shown, leaving the unit of length.

Example 1.2 Analysis of an Equation

Show that the expression $v = v_0 + at$ is dimensionally correct, where v and v_0 represent velocities, a is acceleration, and t is a time interval.

Solution Since

$$[v] = [v_0] = \frac{L}{T}$$

and the dimensions of acceleration are L/T^2, the dimensions of at are

$$[at] = \frac{L}{T^{\cancel{2}}} \cdot \cancel{T} = \frac{L}{T}$$

and the expression is dimensionally correct. On the other hand, if the expression were given as $v = v_0 + at^2$, it would be dimensionally *incorrect.* Try it and see!

1.4 Conversion of Units

Sometimes it is necessary to convert units from one system to another. Conversion factors between SI and conventional units of length are as follows:

$$1 \text{ mile} = 1609 \text{ m} = 1.609 \text{ km} \qquad 1 \text{ ft} = 0.3048 \text{ m} = 30.48 \text{ cm}$$

$$1 \text{ m} = 39.37 \text{ in.} = 3.281 \text{ ft} \qquad 1 \text{ in.} = 0.0254 \text{ m} = 2.54 \text{ cm}$$

A more complete list of conversion factors can be found in Appendix A.

Units can be treated as algebraic quantities that can cancel each other. For example, suppose we wish to convert 15.0 in. to centimeters. Since 1 in. = 2.54 cm (exactly), we find that

$$15.0 \text{ in.} = (15.0 \cancel{\text{in.}})\left(2.54 \frac{\text{cm}}{\cancel{\text{in.}}}\right) = 38.1 \text{ cm}$$

Example 1.3 The Density of a Cube

The mass of a solid cube is 856 g, and each edge has a length of 5.35 cm. Determine the density ρ of the cube in SI units.

Solution Since 1 g = 10^{-3} kg and 1 cm = 10^{-2} m, the mass, m, and volume, V, in SI units are given by

$$m = 856 \cancel{\text{g}} \times 10^{-3} \text{ kg}/\cancel{\text{g}} = 0.856 \text{ kg}$$

$$V = L^3 = (5.35 \cancel{\text{cm}} \times 10^{-2} \text{ m}/\cancel{\text{cm}})^3$$

$$= (5.35)^3 \times 10^{-6} \text{ m}^3 = 1.53 \times 10^{-4} \text{ m}^3$$

Therefore

$$\rho = \frac{m}{V} = \frac{0.856 \text{ kg}}{1.53 \times 10^{-4} \text{ m}^3} = 5.60 \times 10^3 \text{ kg/m}^3$$

(*Left*) Conversion of miles to kilometers. (*Right*) This modern car speedometer gives speed readings in both miles per hour and kilometers per hour. You should confirm the conversion between the two units for a few readings on the dial. *(Paul Silverman, Fundamental Photographs)*

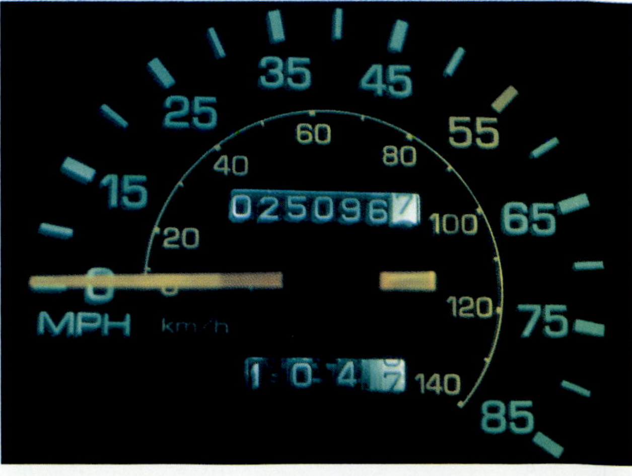

1.5 Order-of-Magnitude Calculations

It is often useful to compute an approximate answer to a given physical problem even where little information is available. This answer can then be used to determine whether or not a more precise calculation is necessary. Such an approximation is usually based on certain assumptions, which must be modified if greater precision is needed. Thus, we will sometimes refer to an *order of magnitude* of a certain quantity as the power of ten of the number that describes that quantity. Usually, when an order-of-magnitude calculation is made, the results are reliable to within a factor of 10. If a quantity increases in value by three orders of magnitude, this means that its value increases by a factor of $10^3 = 1000$.

The spirit of attempting order-of-magnitude calculations, sometimes referred to as "guesstimates" or "ball-park figures," is captured by the following quotation: "Make an estimate before every calculation, try a simple physical argument . . . before every derivation, guess the answer to every puzzle. Courage: no one else needs to know what the guess is."[2]

Example 1.4 The Number of Atoms in a Solid

Estimate the number of atoms in 1 cm^3 of a solid.

Solution From Table 1.1 we note that the diameter of an atom is about 10^{-10} m. Thus, if in our model we assume that the atoms in the solid are solid spheres of this diameter, then the volume of each sphere is about 10^{-30} m^3 (more precisely, volume $= 4\pi r^3/3 = \pi d^3/6$, where $r = d/2$). Therefore, since 1 $cm^3 = 10^{-6}$ m^3, the number of atoms in the solid is on the order of $10^{-6}/10^{-30} = 10^{24}$ atoms.

A more precise calculation would require knowledge of the density of the solid and the mass of each atom. However, our estimate agrees with the more precise calculation to within a factor of 10.

Example 1.5 How Much Gas Do We Use?

Estimate the number of gallons of gasoline used by all U.S. cars each year.

Solution Since there are about 200 million people in the United States, an estimate of the number of cars in the country is 40 million (assuming one car and five people per family). We must also estimate that the average distance traveled per year is 10 000 miles. If we assume gasoline consumption of 0.05 gal/mi, each car uses about 500 gal/year. Multiplying this by the total number of cars in the United States gives an estimated total consumption of 2×10^{10} gal, which corresponds to a yearly consumer expenditure of over $20 billion! This is probably a low estimate since we haven't accounted for commercial consumption and for such factors as two-car families.

[2] E. Taylor and J. A. Wheeler, *Spacetime Physics,* San Francisco, W. H. Freeman, 1966, p. 60.

1.6 Significant Figures

When one performs measurements on certain quantities, the accuracy of the measured values can vary; that is, the true values are known only to be within the limits of the experimental uncertainty. The value of the uncertainty can depend on various factors such as the quality of the apparatus, the skill of the experimenter, and the number of measurements performed.

Suppose that in a laboratory experiment we are asked to measure the area of a rectangular plate, using a meter stick as a measuring instrument. Let us assume that the accuracy to which we can measure a particular dimension of the plate is ± 0.1 cm. If the length of the plate is measured to be 16.3 cm, we can claim only that its length lies somewhere between 16.2 cm and 16.4 cm. In this case, we say that the measured value has three significant figures. Likewise, if the plate's width is measured to be 4.5 cm, the actual value lies between 4.4 cm and 4.6 cm. This measured value has only two significant figures. Note that the significant figures include the first estimated digit. Thus, we could write the measured values as 16.3 ± 0.1 cm and 4.5 ± 0.1 cm.

Suppose now that we would like to find the area of the plate by multiplying the two measured values together. If we were to claim that the area is $(16.3 \text{ cm})(4.5 \text{ cm}) = 73.35 \text{ cm}^2$, our answer would be unjustifiable since it contains four significant figures, which is greater than the number of significant figures in either of the measured lengths. A good "rule of thumb" to use as a guide in determining the number of significant figures that can be claimed is as follows: **When multiplying several quantities, the number of significant figures in the final answer is the same as the number of significant figures in the *least* accurate of the quantities being multiplied, where "least accurate" means "having the lowest number of significant figures." The same rule applies to division.**

Applying this rule to the multiplication example above, we see that the answer for the area can have only two significant figures since the dimension 4.5 cm has only two significant figures. Thus, we can claim only that the area is 73 cm^2, realizing that the value can range between $(16.2 \text{ cm})(4.4 \text{ cm}) = 71 \text{ cm}^2$ and $(16.4 \text{ cm})(4.6 \text{ cm}) = 75 \text{ cm}^2$.

The presence of zeros in an answer may be misinterpreted. For example, suppose the mass of an object is measured to be 1500 g. This value is ambiguous because it is not known whether the last two zeros are being used to locate the decimal point or whether they represent significant figures in the measurement. In order to remove this ambiguity, it is common to use scientific notation to indicate the number of significant figures. In this case, we express the mass as 1.5×10^3 g if there are two significant figures in the measured value and 1.50×10^3 g if there are three significant figures. Likewise, a number such as 0.00015 should be expressed in scientific notation as 1.5×10^{-4} if it has two significant figures or as 1.50×10^{-4} if it has three significant figures. The three zeros between the decimal point and the digit 1 in the number 0.00015 are not counted as significant figures because they are present only to locate the decimal point. In general, a **significant figure** is a reliably known digit, other than a zero used to locate the decimal point.

For addition and subtraction, the number of decimal places must be considered. **When numbers are added (or subtracted), the number of decimal places in the result should equal the smallest number of decimal places of any term in the** sum. For example, if we wish to compute 123 + 5.35, the answer is 128 and not 128.35. As another example, if we compute the sum 1.0001 + 0.0003 = 1.0004, the result has the correct number of decimal places; consequently it has five significant figures even though one of the terms in the sum, 0.0003, has only one significant figure. Likewise, if we perform the subtraction 1.002 − 0.998 = 0.004, the result has three decimal places in accordance with the rule, but only one significant figure.

Throughout this book, we shall generally assume that the given data are precise enough to yield an answer having three significant figures. Thus, if we state that a jogger runs a distance of 5 m, it is to be understood that the distance covered is 5.00 m. Likewise, if the speed of a car is given as 23 m/s, its value is understood to be 23.0 m/s.

Example 1.6 Installing a Carpet

A carpet is to be installed in a room whose length is measured to be 12.71 m (four significant figures) and whose width is measured to be 3.46 m (three significant figures). Find the area of the room.

Solution If you multiply 12.71 m by 3.46 m on your calculator, you will get an answer of 43.9766 m^2. How many of these numbers should you claim? Our rule of thumb for multiplication tells us that you can claim only the number of significant figures in the least accurate of the quantities being measured. In this example, we have only three significant figures in our least accurate measurement, so we should express our final answer as 44.0 m^2. Note that in the answer given, we used a general rule for rounding off numbers, which states that the last digit retained is to be increased by 1 if the first digit dropped was equal to 5 or greater.

1.7 Coordinate Systems and Frames of Reference

Many aspects of physics deal in some way or another with locations in space. For example, the mathematical description of the motion of an object requires a method for describing the position of the object. Thus, it is fitting that we first discuss how to describe the position of a point in space. It is done by means of coordinates. A point on a line can be located with one coordinate; a point in a plane is located with two coordinates; and three coordinates are required to locate a point in space.

A coordinate system used to specify locations in space consists of

1. A fixed reference point *O*, called the origin
2. A set of specified axes or directions with an appropriate scale and labels on the axes
3. Instructions that tell us how to label a point in space relative to the origin and axes

One convenient coordinate system that we will use frequently is the *cartesian coordinate system,* sometimes called the *rectangular coordinate system.* Such a system in

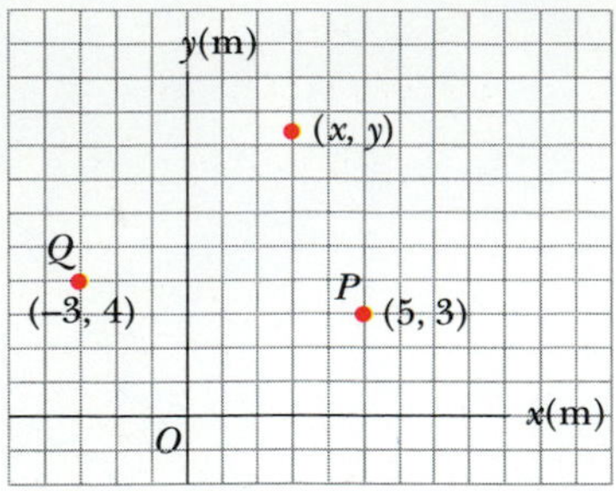

Figure 1.1
Designation of points in a cartesian coordinate system. Every point is labeled with coordinates (x, y).

two dimensions is illustrated in Figure 1.1. An arbitrary point in this system is labeled with the coordinates (x, y). Positive x is taken to the right of the origin, and positive y is upward from the origin. Negative x is to the left of the origin, and negative y is downward from the origin. For example, the point P, which has coordinates (5, 3), may be reached by going first 5 meters to the right of the origin and then 3 meters above the origin. Similarly, the point Q has coordinates (−3, 4), which correspond to going 3 meters to the left of the origin and 4 meters above the origin.

Sometimes it is more convenient to represent a point in a plane by its *plane polar coordinates* (r, θ), as in Figure 1.2a. In this coordinate system, r is the length of the line from the origin to the point, and θ is the angle between that line and a fixed axis, usually the positive x axis, with θ measured counterclockwise. From the right triangle in Figure 1.2b, we find $\sin\theta = y/r$ and $\cos\theta = x/r$. (A review of trigonometric functions is given in Appendix B.4.) Therefore, starting with plane polar coordinates, one can obtain the cartesian coordinates through the equations

$$x = r\cos\theta \qquad [1.3]$$

$$y = r\sin\theta \qquad [1.4]$$

Furthermore, it follows that

$$\tan\theta = \frac{y}{x} \qquad [1.5]$$

and

$$r = \sqrt{x^2 + y^2} \qquad [1.6]$$

You should note that these expressions relating the coordinates (x, y) to the coordinates (r, θ) apply only when θ is defined as in Figure 1.2a, where positive θ is an angle measured *counterclockwise* from the positive x axis. Other choices are made in navigation and astronomy. If the reference axis for the polar angle θ is chosen to be other than the positive x axis, or the sense of increasing θ is chosen differently, then the corresponding expressions relating the two sets of coordinates will change.

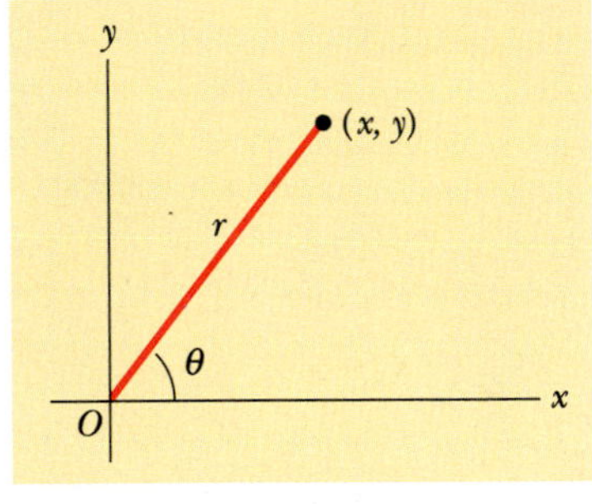

(a)

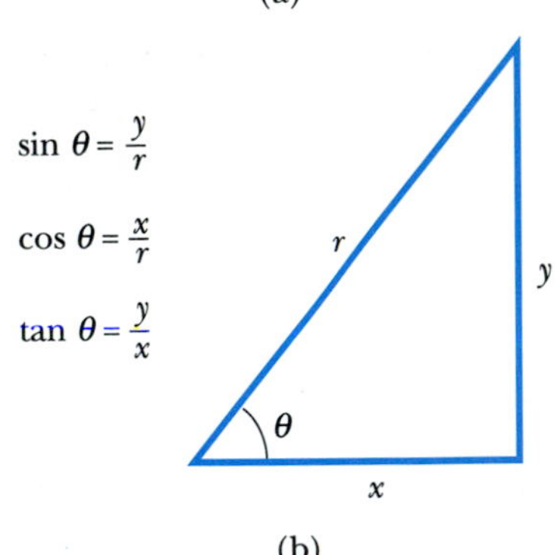

(b)

Figure 1.2
(a) The plane polar coordinates of a point are represented by the distance r and the angle θ.
(b) The right triangle used to relate (x, y) to (r, θ).

1.8 Problem-Solving Strategy

Most courses in general physics require the student to learn the skills of problem solving, and examinations usually include problems that test such skills. This brief section describes some useful ideas that will enable you to increase your accuracy in solving problems, enhance your understanding of physical concepts, eliminate initial panic or lack of direction in approaching a problem, and organize your work. One way to help accomplish these goals is to adopt a problem-solving strategy. Many chapters in this text will include a section labeled "Problem-Solving Strategies" that should help you through the rough spots.

Six basic steps are commonly used to develop a problem-solving strategy:

1. Read the problem carefully at least twice. Be sure you understand the nature of the problem before proceeding further.

2. Draw a suitable diagram with appropriate labels and coordinate axes if needed.
3. As you examine what is being asked in the problem, identify the basic physical principle or principles that are involved, listing the knowns and unknowns.
4. Select a basic relationship or derive an equation that can be used to find the unknown, and symbolically solve the equation for the unknown.
5. Substitute the given values, along with the appropriate units, into the equation.
6. Obtain a numerical value for the unknown. The problem is verified if the following questions can be properly answered: Do the units match? Is the answer reasonable? Is the plus or minus sign proper or meaningful?

A MENU FOR PROBLEM SOLVING

Read Problem → Draw Diagram → Identify Data → Choose Equation(s) → Solve Equation(s) → Evaluate and Check Answer

One of the purposes of this strategy is to promote accuracy. Properly drawn diagrams can eliminate many sign errors. Diagrams also help to isolate the physical principles of the problem. Symbolic solutions and carefully labeled knowns and unknowns will help eliminate other careless errors. The use of symbolic solutions should help you think in terms of the physics of the problem. A check of units at the end of the problem can indicate a possible algebraic error. The physical layout and organization of your problem will make the final product more understandable and easier to follow. Once you have developed an organized system for examining problems and extracting relevant information, you will become a more confident problem solver.

1.9 Vectors and Scalars

Each of the physical quantities that we shall encounter in this text can be placed in one of two categories: it is either a scalar or a vector. A scalar is a quantity that is completely specified by a number with appropriate units. That is,

> A **scalar** has only magnitude and no direction. On the other hand, a **vector** is a physical quantity that must be specified by both magnitude and direction.

Jennifer pointing in the right direction. *(Photo by Raymond A. Serway)*

The number of apples in a basket is an example of a scalar quantity. If you are told there are 38 apples in the basket, this completes the required information; no specification of direction is required. Other examples of scalars are temperature,

The number of apples in the basket is one example of a scalar quantity. Can you think of other examples? *(Superstock)*

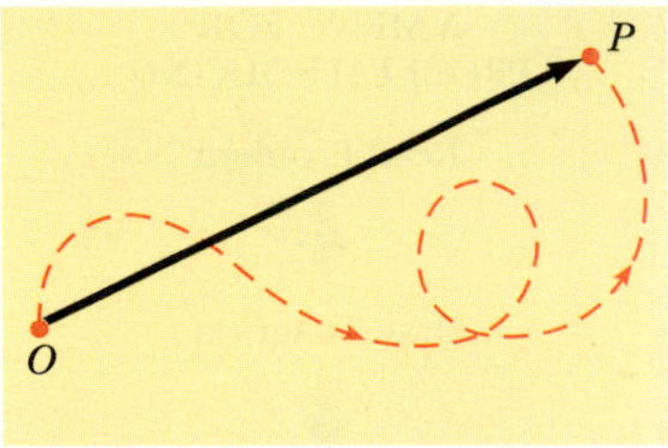

Figure 1.3
As a particle moves from O to P along the broken line, its displacement vector is the arrow drawn from O to P.

volume, mass, and time intervals. The rules of ordinary arithmetic are used to manipulate scalar quantities.

Force is an example of a vector quantity. To completely describe the force on an object, we must specify both the direction of the applied force, the magnitude of the force, and the location of the force. When the motion (velocity) of an object is described, we must specify both the speed and the direction of its motion.

Another simple example of a vector quantity is the **displacement** of a particle, defined as its *change in position.* Suppose the particle moves from some point O to a point P along a straight path, as in Figure 1.3. We represent this displacement by drawing an arrow from O to P, where the tip of the arrow represents the direction of the displacement and the length of the arrow represents the magnitude of the displacement. If the particle travels along some other path from O to P, such as the broken line in Figure 1.3, its displacement is still OP. The vector displacement along any indirect path from O to P is defined as being equivalent to the displacement represented by the direct path from O to P. The magnitude of the displacement is the shortest distance between the end points. Thus, *the displacement of a particle is completely known if its initial and final coordinates are known.* The path need not be specified. In other words, *the displacement is independent of the path,* if the end points of the path are fixed.

It is important to note that the *distance* traveled by a particle is distinctly different from its displacement. The distance traveled (a scalar quantity) is the length of the path, which in general can be much greater than the magnitude of the displacement (see Fig. 1.3).

If the particle moves along the x axis from position x_i to position x_f, as in Figure 1.4, its displacement is given by $x_f - x_i$. (The indices i and f refer to the initial and final values.) We use the Greek letter delta (Δ) to denote the *change* in a quantity. Therefore, we define the change in the position of the particle (the displacement) as

Definition of displacement along a line

$$\Delta x \equiv x_f - x_i \qquad [1.7]$$

From this definition we see that Δx is positive if x_f is greater than x_i and negative if x_f is less than x_i. For example, if a particle changes its position from $x_i = -3$ units to $x_f = 5$ units, its displacement is 8 units.

Many physical quantities in addition to displacement are vectors. They include velocity, acceleration, force, and momentum, all of which will be defined in later chapters. In this text we will use boldface letters, such as **A**, to represent arbitrary vectors. Another common method of notating vectors with which you should be familiar is to use an arrow over the letter: $\vec{A}$.

The magnitude of the vector **A** is written A or, alternatively, $|\mathbf{A}|$. The magnitude of a vector has physical units, such as meters for displacement or meters per second for velocity, as discussed earlier. Vectors combine according to special rules, which will be discussed in Sections 1.10 and 1.11.

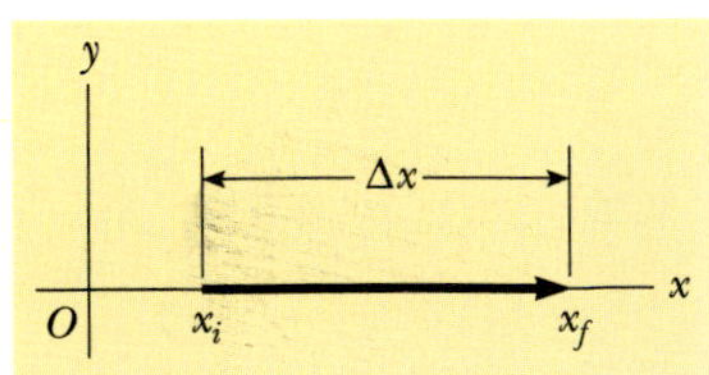

Figure 1.4
A particle moving along the x axis from x_i to x_f undergoes a displacement $\Delta x = x_f - x_i$.

1.10 Some Properties of Vectors

Equality of Two Vectors. Two vectors **A** and **B** are defined to be equal if they have the same magnitude and the same direction. That is, $\mathbf{A} = \mathbf{B}$ only if $A = B$ *and* **A** and **B** act in parallel. For example, all the vectors in Figure 1.5 are equal even

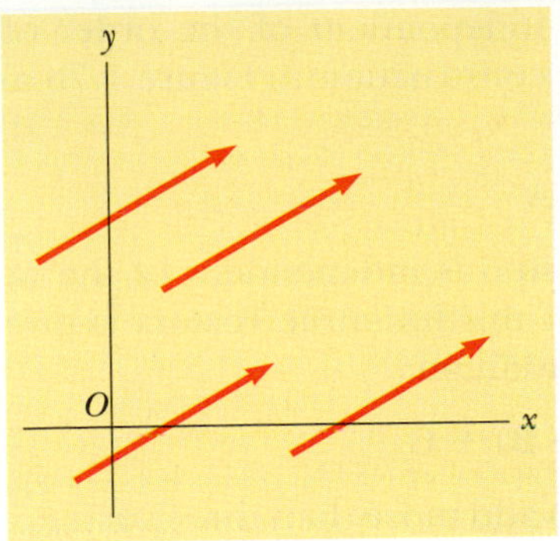

Figure 1.5
Four representations of the same vector.

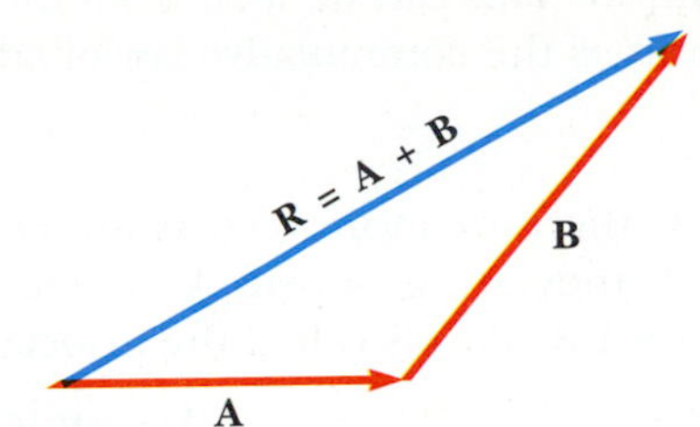

Figure 1.6
When vector **A** is added to vector **B**, the resultant **R** is the blue vector that runs from the tail of **A** to the tip of **B**.

though they have different starting points. This property allows us to translate a vector parallel to itself in a diagram without affecting the vector. In fact, for most purposes, any vector can be moved parallel to itself without affecting the vector.

Addition. When two or more vectors are added together, they must *all* have the same units. For example, it would be meaningless to add a velocity vector to a displacement vector since they are different physical quantities. Scalars obey the same rule. For example, it would be meaningless to add time intervals and temperatures.

The rules for vector sums are conveniently described by geometric methods. To add vector **B** to vector **A**, first draw vector **A**, with its magnitude represented by a convenient scale, on graph paper and then draw vector **B** to the same scale with its tail starting from the tip of **A**, as in Figure 1.6. The *resultant vector* **R** = **A** + **B** is the blue vector drawn from the tail of **A** to the tip of **B**. This is known as the *triangle method of addition.* An alternative graphical procedure for adding two vectors, known as the *parallelogram rule of addition,* is shown in Figure 1.7a. In this construction, the tails of the two vectors **A** and **B** are together, and the resultant vector **R** is the diagonal of a parallelogram formed with **A** and **B** as its sides.

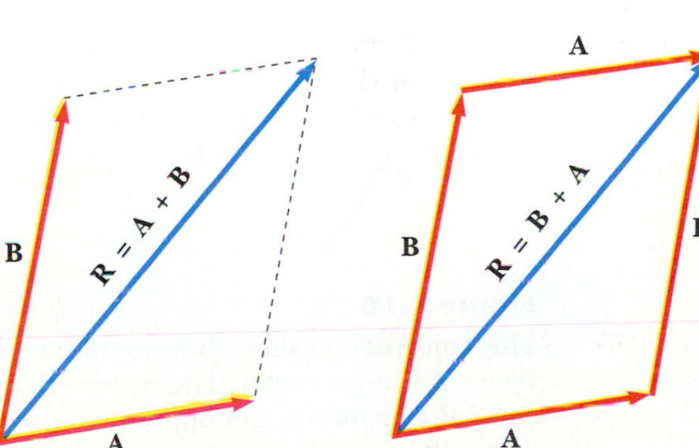

Figure 1.7
(a) In this construction, the resultant **R** is the diagonal of a parallelogram with sides **A** and **B**. (b) This construction shows that **A** + **B** = **B** + **A**.

When two vectors are added, the sum is independent of the order of the addition. This can be seen from the geometric construction in Figure 1.7b and is known as the **commutative law of addition:**

Commutative law

$$\mathbf{A} + \mathbf{B} = \mathbf{B} + \mathbf{A} \tag{1.8}$$

If three or more vectors are added, their sum is independent of the way in which they are grouped. A geometric proof of this for three vectors is given in Figure 1.8. This is called the **associative law of addition:**

Associative law

$$\mathbf{A} + (\mathbf{B} + \mathbf{C}) = (\mathbf{A} + \mathbf{B}) + \mathbf{C} \tag{1.9}$$

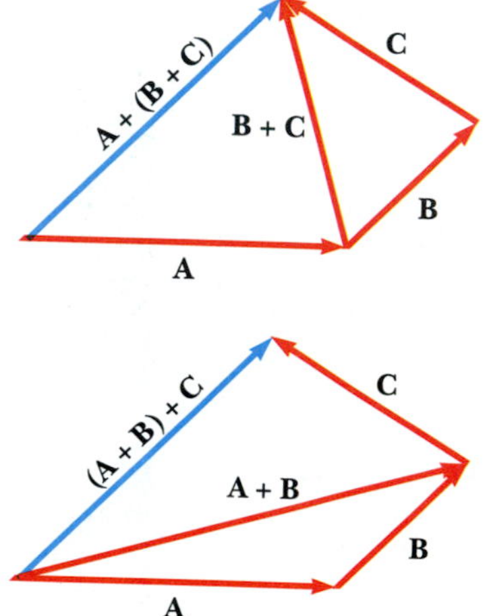

Figure 1.8
Geometric constructions for verifying the associative law of addition.

Geometric constructions can also be used to add more than three vectors. This is shown in Figure 1.9 for the case of four vectors. The resultant vector sum $\mathbf{R} = \mathbf{A} + \mathbf{B} + \mathbf{C} + \mathbf{D}$ is *the vector that completes the polygon.* In other words, **R** is *the vector drawn from the tail of the first vector to the tip of the last vector.* Again, the order of the summation is unimportant.

Thus we conclude that *a vector is a quantity that has both magnitude and direction and also obeys the laws of vector addition* as described in Figures 1.6 to 1.9.

Negative of a Vector. The negative of the vector **A** is defined as the vector that, when added to **A**, gives zero for the vector sum. That is, $\mathbf{A} + (-\mathbf{A}) = 0$. The vectors **A** and $-\mathbf{A}$ have the same magnitude but opposite directions.

Subtraction of Vectors. The operation of vector subtraction makes use of the definition of the negative of a vector. We define the operation $\mathbf{A} - \mathbf{B}$ as vector $-\mathbf{B}$ added to vector **A**:

$$\mathbf{A} - \mathbf{B} = \mathbf{A} + (-\mathbf{B}) \tag{1.10}$$

The geometric construction for subtracting two vectors is shown in Figure 1.10.

Multiplication of a Vector by a Scalar. If a vector **A** is multiplied by a positive scalar quantity m, the product $m\mathbf{A}$ is a vector that has the same direction as **A** and magnitude mA. If m is a negative scalar quantity, the vector $m\mathbf{A}$ is directed opposite **A**. For example, the vector $5\mathbf{A}$ is five times as great in magnitude as **A** and has the

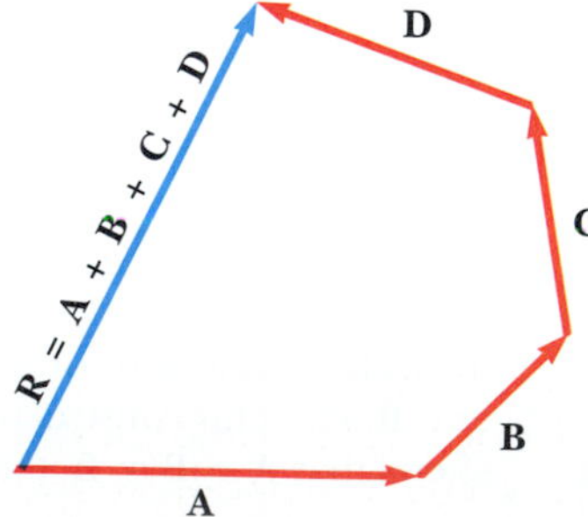

Figure 1.9
Geometric construction for summing four vectors. The resultant vector **R** in blue completes the polygon.

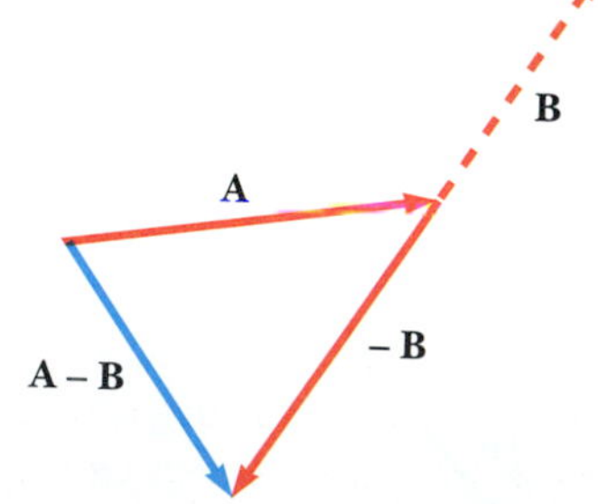

Figure 1.10
This construction shows how to subtract vector **B** from vector **A**. The vector $-\mathbf{B}$ is equal in magnitude and opposite to the vector **B**.

same direction as A. On the other hand, the vector $-\frac{1}{3}\mathbf{A}$ has one-third the magnitude of A and points in the direction opposite A (because of the negative sign).

Multiplication of Two Vectors. Two vectors A and B can be multiplied to produce either a scalar or a vector quantity. The *scalar product* (or dot product) $\mathbf{A}\cdot\mathbf{B}$ is a scalar quantity equal to $AB \cos \theta$, where θ is the angle between A and B. The vector product (or cross product) $\mathbf{A} \times \mathbf{B}$ is a vector quantity whose magnitude is equal to $AB \sin \theta$. We shall discuss these products more fully in Chapters 7 and 11, where they are first used.

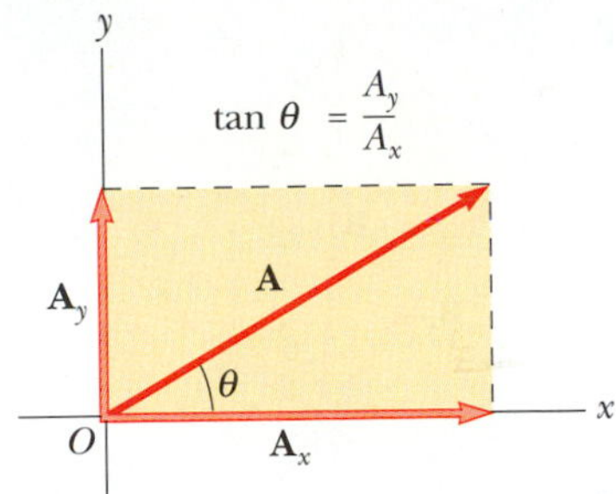

Figure 1.11
Any vector A lying in the xy plane can be represented by its component vectors $\mathbf{A}_x$ and $\mathbf{A}_y$, where $\mathbf{A} = \mathbf{A}_x + \mathbf{A}_y$.

1.11 Components of a Vector and Unit Vectors

The geometric method of adding vectors is not the recommended procedure in situations where great precision is required or in three-dimensional problems. In this section we describe a method of adding vectors that makes use of the *projections* of a vector along the axes of a rectangular coordinate system. These projections are called the **component vectors.** Any vector can be completely described by its component vectors.

Consider a vector A lying in the xy plane and making an arbitrary angle θ with the positive x axis, as in Figure 1.11. The vector A can be expressed as the sum of two other vectors $\mathbf{A}_x$ and $\mathbf{A}_y$, called the **component vectors** of A. The component vector $\mathbf{A}_x$ represents the projection of A along the x axis, while $\mathbf{A}_y$ represents the projection of A along the y axis. From Figure 1.11 we see that $\mathbf{A} = \mathbf{A}_x + \mathbf{A}_y$. We will often refer to the components of a vector A, written as A_x and A_y. The components of a vector can be positive or negative. The component A_x is positive if $\mathbf{A}_x$ points along the positive x axis and is negative if $\mathbf{A}_x$ points along the negative x axis. The same is true for the component A_y.

A_x negative, A_y positive	A_x positive, A_y positive
A_x negative, A_y negative	A_x positive, A_y negative

Figure 1.12
The signs of the components of a vector A depend on the quadrant in which the vector is located.

From Figure 1.11 and the definition of the sine and cosine of an angle, we see that $\cos \theta = A_x/A$ and $\sin \theta = A_y/A$. Hence, the components of A are given by

$$A_x = A \cos \theta$$
$$A_y = A \sin \theta \quad [1.11]$$

Components of the vector A

These components form two sides of a right triangle, the hypotenuse of which has a magnitude A. Thus, it follows that the magnitude of A and its direction are related to its components through the expressions

$$A = \sqrt{A_x^2 + A_y^2} \quad [1.12]$$

Magnitude of A

$$\tan \theta = \frac{A_y}{A_x} \quad [1.13]$$

Direction of A

To solve for θ, we can write $\theta = \tan^{-1}(A_y/A_x)$, which is read "$\theta$ equals the angle the tangent of which is the ratio A_y/A_x." *Note that the signs of the components A_x and A_y depend on the angle θ.* For example, if $\theta = 120°$, A_x is negative and A_y is positive. On the other hand, if $\theta = 225°$, both A_x and A_y are negative. Figure 1.12 summarizes the signs of the components when A lies in the various quadrants.

If you choose reference axes or an angle other than those shown in Figure 1.11, the components of the vector must be modified accordingly. In many applications it is more convenient to express the components of a vector in a coordinate

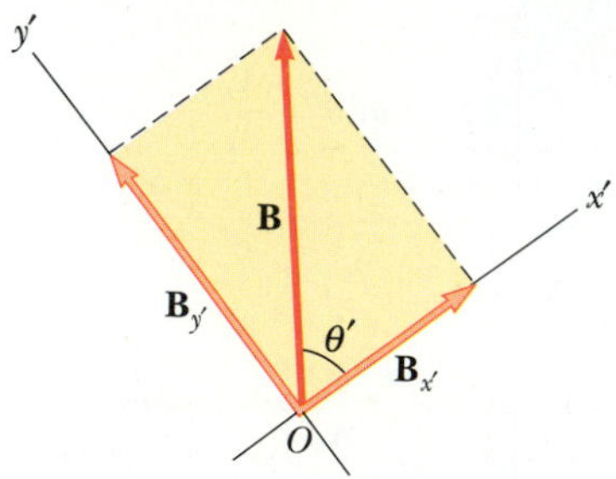

Figure 1.13
The component vectors of **B** in a coordinate system that is tilted.

system having axes that are not horizontal and vertical but are still perpendicular to each other. Suppose a vector **B** makes an angle θ' with the x' axis defined in Figure 1.13. The components of **B** along these axes are given by $B_{x'} = B\cos\theta'$ and $B_{y'} = B\sin\theta'$, as in Equation 1.9. The magnitude and direction of **B** are obtained from expressions equivalent to Equations 1.10 and 1.11. Thus, we can express the components of a vector in *any* coordinate system that is convenient for a particular situation.

The components of a vector differ when viewed from different coordinate systems. Furthermore, the components of a vector can change with respect to a fixed coordinate system if the vector changes in magnitude, orientation, or both.

Vector quantities are often expressed in terms of unit vectors. A **unit vector** is *a dimensionless vector one unit in length* used to specify a given *direction.* Unit vectors have no other physical significance. They are used simply as a convenience in describing a direction in space. We will use the symbols **i**, **j**, and **k** to represent unit vectors pointing in the x, y, and z directions, respectively. Thus, the unit vectors **i**, **j**, and **k** form a set of mutually perpendicular vectors as shown in Figure 1.14a, where the magnitude of the unit vectors equals unity; that is, $|\mathbf{i}| = |\mathbf{j}| = |\mathbf{k}| = 1$.

Consider a vector **A** lying in the xy plane, as in Figure 1.14b. The product of the component A_x and the unit vector **i** is the vector $A_x\mathbf{i}$ parallel to the x axis with magnitude A_x. Likewise, $A_y\mathbf{j}$ is a vector of magnitude A_y parallel to the y axis. Thus, the unit-vector notation for the vector **A** is written

$$\mathbf{A} = A_x\mathbf{i} + A_y\mathbf{j} \qquad [1.14]$$

The vectors $A_x\mathbf{i}$ and $A_y\mathbf{j}$ are the vector components of **A**. These should not be confused with A_x and A_y, which we shall always refer to as the components of **A**.

Now suppose we wish to add vector **B** to vector **A**, where **B** has components B_x and B_y. The procedure for performing this sum is to simply add the x and y components separately. The resultant vector $\mathbf{R} = \mathbf{A} + \mathbf{B}$ is therefore

$$\mathbf{R} = (A_x + B_x)\mathbf{i} + (A_y + B_y)\mathbf{j} \qquad [1.15]$$

Thus, the components of the resultant vector are given by

$$R_x = A_x + B_x$$
$$R_y = A_y + B_y \qquad [1.16]$$

The magnitude of **R** and the angle it makes with the x axis can then be obtained from its components using the relationships

$$R = \sqrt{R_x^2 + R_y^2} = \sqrt{(A_x + B_x)^2 + (A_y + B_y)^2} \qquad [1.17]$$

$$\tan\theta = \frac{R_y}{R_x} = \frac{A_y + B_y}{A_x + B_x} \qquad [1.18]$$

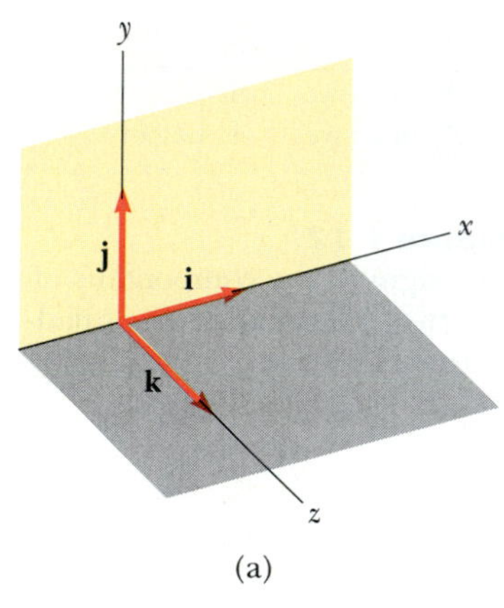

(a)

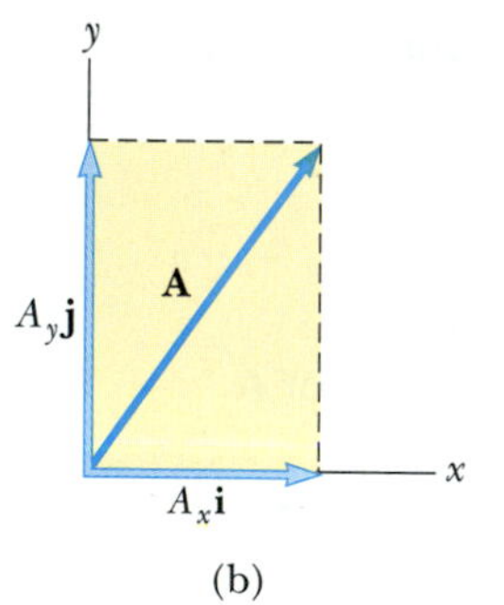

(b)

Figure 1.14
(a) The unit vectors, **i**, **j**, and **k** are directed along the x, y, and z axes, respectively. (b) A vector **A** lying in the xy plane has component vectors $A_x\mathbf{i}$ and $A_y\mathbf{j}$ where A_x and A_y are the components of **A**.

The procedure just described for adding two vectors **A** and **B** using the component method can be checked using a geometric construction, as in Figure 1.15. Again you must take note of the *signs* of the components when using either the algebraic or the geometric method.

The extension of these methods to three-dimensional vectors is straightforward. If **A** and **B** both have x, y, and z components, we express them in the form

$$\mathbf{A} = A_x\mathbf{i} + A_y\mathbf{j} + A_z\mathbf{k}$$
$$\mathbf{B} = B_x\mathbf{i} + B_y\mathbf{j} + B_z\mathbf{k}$$

The sum of A and B is

$$\mathbf{R} = \mathbf{A} + \mathbf{B} = (A_x + B_x)\mathbf{i} + (A_y + B_y)\mathbf{j} + (A_z + B_z)\mathbf{k} \qquad [1.19]$$

Thus, the resultant vector also has a z component, given by $R_z = A_z + B_z$. The same procedure can be used to sum up three or more vectors.

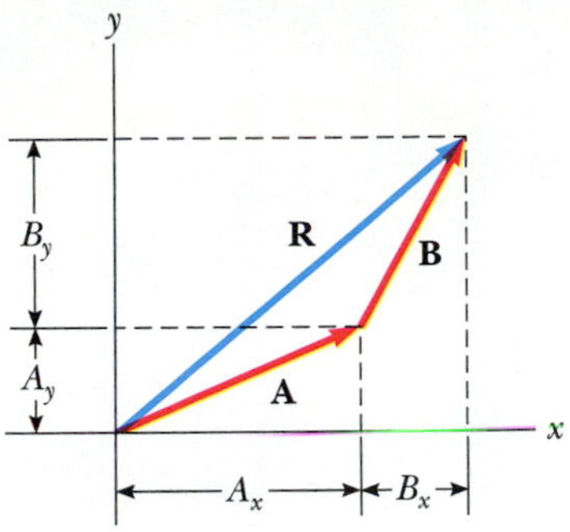

Figure 1.15 A geometric construction showing the relation between the components of the resultant **R** of two vectors and the individual component vectors.

Problem-Solving Strategy: Adding Vectors, Using Components

When two or more vectors are to be added, the following steps are recommended:

1. Select a coordinate system.
2. Draw a sketch of the vectors to be added (or subtracted), with a label on each vector.
3. Find the x and y components of all vectors.
4. Find the resultant components (the algebraic sum of the components) in both the x and y directions.
5. Use the Pythagorean theorem to find the magnitude of the resultant vector.
6. Use a suitable trigonometric function to find the angle the resultant vector makes with the x axis.

Example 1.7 The Sum of Two Vectors

Find the sum of two vectors **A** and **B** lying in the xy plane and given by

$$\mathbf{A} = 2\mathbf{i} + 2\mathbf{j} \quad \text{and} \quad \mathbf{B} = 2\mathbf{i} - 4\mathbf{j}$$

Solution Note that $A_x = 2$, $A_y = 2$, $B_x = 2$, and $B_y = -4$. Therefore, the resultant vector **R** is given by

$$\mathbf{R} = \mathbf{A} + \mathbf{B} = (2 + 2)\mathbf{i} + (2 - 4)\mathbf{j} = 4\mathbf{i} - 2\mathbf{j}$$

or

$$R_x = 4, \; R_y = -2$$

The magnitude of **R** is given by

$$R = \sqrt{R_x^2 + R_y^2} = \sqrt{(4)^2 + (-2)^2} = \sqrt{20} = 4.47$$

Many examples in this text will be followed by an exercise. The purpose of such exercises is to test your understanding of the example by asking you to do a calculation or answer some other question related to the example. The answer to an exercise will be provided at its end when appropriate. Here is your first exercise, related to Example 1.7.

Exercise Find the angle θ that the resultant vector **R** makes with the positive x axis.

Answer 333°.

Example 1.8 The Resultant Displacement

A particle undergoes three consecutive displacements, given by $\mathbf{d}_1 = (\mathbf{i} + 3\mathbf{j} - \mathbf{k})$ cm, $\mathbf{d}_2 = (2\mathbf{i} - \mathbf{j} - 3\mathbf{k})$ cm, and $\mathbf{d}_3 = (-\mathbf{i} + \mathbf{j})$ cm. Find the resultant displacement of the particle.

Solution

$$\mathbf{R} = \mathbf{d}_1 + \mathbf{d}_2 + \mathbf{d}_3$$
$$= (1 + 2 - 1)\mathbf{i} + (3 - 1 + 1)\mathbf{j} + (-1 - 3 + 0)\mathbf{k}$$
$$= (2\mathbf{i} + 3\mathbf{j} - 4\mathbf{k}) \text{ cm}$$

That is, the resultant displacement vector has components $R_x = 2$ cm, $R_y = 3$ cm, and $R_z = -4$ cm. Its magnitude is

$$R = \sqrt{R_x^2 + R_y^2 + R_z^2} = \sqrt{(2 \text{ cm})^2 + (3 \text{ cm})^2 + (-4 \text{ cm})^2} = 5.39 \text{ cm}$$

Example 1.9 Taking a Hike

A hiker begins a trip by first walking 25 km due southeast from her base camp. On the second day she walks 40 km in a direction 60° north of east, at which point she discovers a forest ranger's tower.

(a) Determine the components of the hiker's displacements in the first and second days.

Solution If we denote the displacement vectors on the first and second days by **A** and **B**, respectively, and use the camp as the origin of coordinates, we get the vectors shown in Figure 1.16. Displacement **A** has a magnitude of 25.0 km and is 45° southeast. Its components are

$$A_x = A\cos(-45°) = (25 \text{ km})(0.707) = 17.7 \text{ km}$$

$$A_y = A\sin(-45°) = -(25 \text{ km})(0.707) = -17.7 \text{ km}$$

The negative value of A_y indicates that the y coordinate decreased in this displacement. The signs of A_x and A_y are also evident from Figure 1.16.

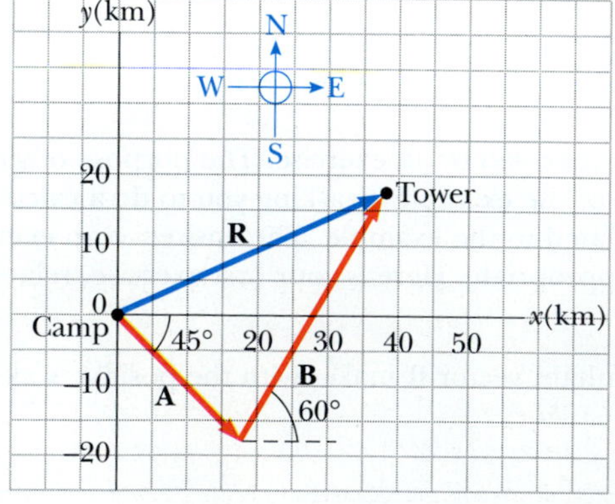

Figure 1.16 (Example 1.9) The total displacement of the hiker is the vector $\mathbf{R} = \mathbf{A} + \mathbf{B}$.

The second displacement, **B**, has a magnitude of 40.0 km and is 60° north of east. Its components are

$$B_x = B\cos 60° = (40\text{ km})(0.50) = \boxed{20.0\text{ km}}$$

$$B_y = B\sin 60° = (40\text{ km})(0.866) = \boxed{34.6\text{ km}}$$

(b) Determine the components of the hiker's total displacement for the trip.

Solution The resultant displacement for the trip, **R** = **A** + **B**, has components given by

$$R_x = A_x + B_x = 17.7\text{ km} + 20.0\text{ km} = \boxed{37.7\text{ km}}$$

$$R_y = A_y + B_y = -17.7\text{ km} + 34.6\text{ km} = \boxed{16.9\text{ km}}$$

In unit-vector form, we can write the total displacement **R** = (37.7**i** + 16.9**j**) km.

Exercise Determine the magnitude and direction of the total displacement.

Answer 41.3 km, 24.1° north of east from the base camp.

Summary

Mechanical quantities can be expressed in terms of three fundamental quantities, *length, mass,* and *time,* which in the SI system have the units *meters* (m), *kilograms* (kg), and *seconds* (s), respectively. It is often useful to use the *method of dimensional analysis* to check equations and to assist in deriving expressions.

The **density** of a substance is defined as its *mass per unit volume.* Different substances have different densities mainly because of differences in their atomic masses and atomic arrangements.

The number of molecules in one mole of any element or compound, called **Avogadro's number** (N_A), is 6.02×10^{23}.

Vectors are quantities that have both magnitude and direction and obey the vector law of addition. **Scalars** are quantities that have only magnitude.

Two vectors **A** and **B** can be added using either the triangle method or the parallelogram rule. In the triangle method (Fig. 1.6), the vector **C** = **A** + **B** runs from the tail of **A** to the tip of **B**. In the parallelogram method (Fig. 1.7a), **C** is the diagonal of a parallelogram having **A** and **B** as its sides.

The x component of the vector **A**, A_x, is equal to its projection along the x axis of a coordinate system as in Figure 1.11, where $A_x = A\cos\theta$ and θ is the angle **A** makes with the x axis. Likewise, the y component of **A**, A_y, is its projection along the y axis, where $A_y = A\sin\theta$. The resultant of two or more vectors can be found by resolving all vectors into their x and y components, adding their resultant x and y components, and then using the Pythagorean theorem to find the magnitude of the resultant vector. The angle that the resultant vector makes with the x axis can be found by the use of a suitable trigonometric function.

If a vector **A** has an x component equal to A_x and a y component equal to A_y, the vector can be expressed in unit-vector form as $\mathbf{A} = A_x\mathbf{i} + A_y\mathbf{j}$. In this notation, **i** is a unit vector in the positive x direction and **j** is a unit vector in the positive y direction. Since **i** and **j** are unit vectors, $|\mathbf{i}| = |\mathbf{j}| = 1$.

Questions and Conceptual Exercises*

1. What types of natural phenomena could serve as alternative time standards?
2. The height of a horse is sometimes given in units of "hands." Why is this a poor standard of length?
3. Express the following quantities using the prefixes given in Table 1.4: (a) 3×10^{-4} m, (b) 5×10^{-5} s, (c) 72×10^{2} g.
4. Does a dimensional analysis give any information on constants of proportionality that may appear in an algebraic expression? Explain.
5. Suppose that two quantities A and B have different dimensions. Determine which of the following arithmetic operations *could* be physically meaningful: (a) $A + B$, (b) A/B, (c) $B - A$, (d) AB.
6. What accuracy is implied in an order-of-magnitude calculation?
7. Apply an order-of-magnitude calculation to an everyday situation you might encounter. For example, how far do you walk or drive each day?
8. Estimate your age in seconds.
9. Estimate the masses of various objects around you in grams or in kilograms. If a scale is available, check your estimates.
10. Is it possible to use length, density, and time as three fundamental units rather than length, mass, and time? If so, what could be used as a standard of density?
11. A book is moved once around the perimeter of a tabletop with dimensions 1 m × 2 m. If the book ends up at its initial position, what is its displacement? What is the distance traveled?
12. If **B** is added to **A**, under what condition does the resultant vector have a magnitude equal to $A + B$? Under what conditions is the resultant vector equal to zero?
13. Can the magnitude of a particle's displacement be greater than the distance traveled? Explain.
14. The magnitudes of two vectors **A** and **B** are $A = 5$ units and $B = 2$ units. Find the largest and smallest values possible for the resultant vector $\mathbf{R} = \mathbf{A} + \mathbf{B}$.
15. A vector **A** lies in the xy plane. For what orientations of **A** will both of its rectangular components be negative? For what orientations will its components have opposite signs?
16. Can a vector have a component equal to zero and still have a nonzero magnitude? Explain.
17. If one of the components of a vector is not zero, can its magnitude be zero? Explain.
18. If the component of vector **A** along the direction of vector **B** is zero, what can you conclude about the two vectors?
19. If $\mathbf{A} = \mathbf{B}$, what can you conclude about the components of **A** and **B**?
20. Can the magnitude of a vector have a negative value? Explain.
21. If $\mathbf{A} + \mathbf{B} = 0$, what can you say about the components of the two vectors?
22. Which of the following are vectors and which are not: force, temperature, the volume of water in a can, the ratings of a TV show, the height of a building, the velocity of a sports car, the age of the Universe?
23. Under what circumstances would a nonzero vector lying in the xy plane have components that were equal in magnitude?
24. Is it possible to add a vector quantity to a scalar quantity? Explain.
25. Two vectors have unequal magnitudes. Can their sum be zero? Explain.
26. (a) What is the resultant displacement of a walk of 80 m followed by a walk of 125 m when both displacements are in the eastward direction? (b) What is the resultant displacement in a situation in which the 125-m walk is in the direction opposite the 80-m walk?
27. While traveling along a straight interstate highway you notice that the mile marker reads 260. You travel until you reach the 150-mile marker and then retrace your path to the 175-mile marker. What is the magnitude of your resultant displacement from the 260-mile marker?
28. A submarine dives at an angle of 30° with the horizontal and follows a straight-line path for a total distance of 50 m. How far is the submarine below the surface of the water?
29. A roller coaster travels 135 ft at an angle of 40° above the horizontal. How far does it move horizontally and vertically?

*Conceptual Exercises are numbered in blue.

Problems

Section 1.2 Density and Atomic Mass

1. Calculate the density of a solid cube that measures 5 cm on each side and has a mass of 350 g.
2. The standard kilogram is a platinum-iridium cylinder 39 mm in height and 39 mm in diameter. What is the density of the material?
3. The planet Jupiter has an average radius 10.95 times the average radius of the Earth and a mass 317.4 times that of the Earth. Calculate the ratio of Jupiter's mass density to the mass density of the Earth.
4. Calculate the mass of an atom of (a) helium, (b) iron, (c) lead. Give your answers in atomic mass units and in grams. The atomic weights of the atoms given are 4, 56, and 207, respectively.
5. Assume that it takes 7 minutes to fill a 30-gal gasoline tank. (a) Calculate the rate at which the tank is filled, in gallons per second. (b) Calculate the rate at which the tank is filled, in cubic meters per second. (c) Determine the time, in hours, required to fill a one-cubic-meter volume at the same rate. (1 U.S. gal = 231 in.^{3})
6. A flat, circular plate of copper has a radius of 0.243 m and a mass of 62 kg. What is the thickness of the plate?

Section 1.3 Dimensional Analysis

7. Show that the expression $x = vt + \frac{1}{2}at^2$ is dimensionally correct, where x is a coordinate and has units of length, v is velocity, a is acceleration, and t is time.
8. Show that the equation $v^2 = v_0{}^2 + 2ax$ is dimensionally consistent, where v and v_0 represent velocities, a is acceleration, and x is a distance.
9. The period T of a simple pendulum is measured in time units and is given by

$$T = 2\pi\sqrt{\frac{\ell}{g}}$$

where ℓ is the length of the pendulum and g is the acceleration due to gravity in units of length, divided by the square of time. Show that this equation is dimensionally consistent.

10. The consumption of natural gas by a company satisfies the empirical equation $V = 1.5t + 0.008t^2$, where V is the volume in millions of cubic feet and t the time in months. Express this equation in units of cubic feet and seconds. Put the proper units on the coefficients. Assume a month is 30 days.
11. Newton's law of universal gravitation is given by

$$F = G\frac{Mm}{r^2}$$

Here F is the force of gravity, M and m are masses, and r is a length. Force has the units kg·m/s^2. What are the SI units of the proportionality constant G?

Section 1.4 Conversion of Units

12. A rectangular building lot is 100.0 ft by 150.0 ft. Determine the area of this lot in square meters.
13. A section of land has an area of 1 square mile and contains 640 acres. Determine the number of square meters in 1 acre.
14. A solid piece of lead has a mass of 23.94 g and a volume of 2.10 cm^3. From these data, calculate the density of lead in SI units (kg/m^3).
15. The mass of the Sun is about 1.99×10^{30} kg, and the mass of a hydrogen atom, of which the Sun is mostly composed, is 1.67×10^{-27} kg. How many atoms are there in the Sun?
16. Using the fact that the speed of light in free space is about 3.00×10^8 m/s, determine how many miles a pulse from a laser beam will travel in one hour.
17. (a) Find a conversion factor to convert from miles per hour to kilometers per hour. (b) Until recently, federal law mandated that highway speeds would be 55 mi/h. Use the conversion factor of part (a) to find the speed in kilometers per hour. (c) The maximum highway speed has been raised to 65 mi/h in some places. In kilometers per hour, how much increase is this over the 55-mi/h limit?
18. (a) How many seconds are there in a year? (b) If one micrometeorite (a sphere with a diameter of 10^{-6} m) struck each square meter of the Moon each second, how many years would it take to cover the Moon to a depth of 1 m? (*Hint:* Consider a cubic box on the Moon 1 m on a side, and find how long it would take to fill the box.)
19. One gallon of paint (volume = 3.78×10^{-3} m^3) covers an area of 25 m^2. What is the thickness of the paint on the wall?
20. The base of a pyramid covers an area of 13 acres (1 acre = 43 560 ft^2) and has a height of 481 ft. The volume of a pyramid is given by the expression $V = (1/3)Bh$, where B is the area of the base and h is the

Figure 1.17 (Problems 20 and 21) *(© Will and Deni McIntyre/Photo Researcher, Inc.)*

height. Find the volume of this pyramid in cubic meters.

21. The pyramid described in Problem 20 contains approximately 2 million stone blocks that average 2.50 tons each. Find the weight of this pyramid in pounds.
22. The nearest star is about 4×10^{13} km away. If our Sun (diameter = 1.4×10^9 m) were represented by a cherry pit 7 mm in diameter, determine the distance to the next cherry pit.
23. You can obtain a rough estimate of the size of a molecule by the following simple experiment. Let a droplet of oil spread out on a smooth water surface. The resulting "oil slick" will be approximately one molecule thick. If an oil droplet of mass 9.0×10^{-7} kg and density 918 kg/m^3 spreads out into a circle of radius 41.8 cm on the water surface, calculate the diameter of an oil molecule.
24. From the fact that the average density of the Earth is 5.5 g/cm^3 and its mean radius is 6.37×10^6 m, compute the mass of the Earth.
25. One cubic meter (1.0 m^3) of aluminum has a mass of 2.70×10^3 kg, and 1.0 m^3 of iron has a mass of 7.86×10^3 kg. Find the radius of a solid aluminum sphere that will balance a solid iron sphere of radius 2.0 cm on an equal-arm balance.

Section 1.5 Order-of-Magnitude Calculations

26. Assuming 60 heartbeats a minute, estimate the total number of times the heart of a human beats in a lifetime of 70 years.
27. Estimate the amount of motor oil used by all cars in the United States each year and its cost to the consumers.
28. Soft drinks are commonly sold in aluminum containers. Estimate the number of such containers thrown away and/or recycled each year by U.S. consumers. Approximately how many tons of aluminum does this represent?
29. Army engineers in 1946 determined the distance from the Earth to the Moon by using radar. If the interval from the time at which a signal was sent out from their radar to the time at which it was received back was 2.56 s, what is the distance from the Earth to the Moon? (The speed of radar waves is 3×10^8 m/s).
30. The United States consumes petroleum at a rate of about 6×10^9 barrels per year. Assuming a barrel has a length of 1 m, compare the length of 6 billion barrels, laid end to end, with the coast-to-coast distance of the United States (about 4000 km).
31. A high fountain of water is situated at the center of a circular pool as in Figure 1.18. A student walks around the pool and estimates its circumference to be 150 m. Next, the student stands at the edge of the pool and uses a protractor to gauge the angle of elevation of the top of the fountain to be 55°. How high is the fountain?

Figure 1.18 (Problem 31)

32. Estimate the number of piano tuners living in New York City. This problem was posed by the physicist Enrico Fermi, who was well known for making order-of-magnitude calculations.

Section 1.6 Significant Figures

33. Calculate (a) the circumference of a circle of radius 3.5 cm and (b) the area of a circle of radius 4.65 cm.
34. Carry out the following arithmetic operations: (a) the sum of the numbers 756, 37.2, 0.83, and 2.5: (b) the product 3.2×3.563; (c) the product $5.6 \times \pi$.
35. How many significant figures are there in (a) 78.9 ± 0.2, (b) 3.788×10^9, (c) 2.46×10^{-6}, (d) 0.0053?
36. The *radius* of a solid sphere is measured to be (6.50 ± 0.20) cm, and its mass is measured to be (1.85 ± 0.02) kg. Determine the density of the sphere, in kilograms per cubic meter, and the uncertainty in the density.

Section 1.7 Coordinate Systems and Frames of Reference

37. Two points in the xy plane have cartesian coordinates (2.0 −4.0) and (−3.0, 3.0), where the units are meters. Determine (a) the distance between these points and (b) their polar coordinates.
38. A point in the xy plane has cartesian coordinates (−3.0, 5.0) m. What are the polar coordinates of this point?
39. The polar coordinates of a point are $r = 5.50$ m and $\theta = 240°$. What are the cartesian coordinates of this point?
40. Two points in a plane have polar coordinates (2.50 m, 30°) and (3.80 m, 120°). Determine (a) the cartesian coordinates of these points and (b) the distance between them.

Section 1.9 Vectors and Scalars
Section 1.10 Some Properties of Vectors

41. A surveyor estimates the distance across a river by the following method. Beginning directly across from a tree on the opposite bank, the surveyor walks 100 m along the riverbank, then sights across to the tree. The angle from his baseline (the line he followed when he walked) to the tree is 35°. How wide is the river?
42. A person walks along a circular path of radius 5 m, around one half of the circle. (a) Find the magnitude of the displacement vector. (b) How far did the person walk? (c) What is the magnitude of the displacement if the circle is completed?
43. A force vector $\mathbf{F}_1$ of magnitude 6 N acts at the origin in a direction 30° above the positive x axis. A second force vector $\mathbf{F}_2$ of magnitude 5 N acts at the origin in the direction of the positive y axis. Graphically find the magnitude and direction of the resultant force vector $\mathbf{F}_1 + \mathbf{F}_2$.
44. Each of the displacement vectors **A** and **B** shown in Figure 1.19 has a magnitude of 3 m. Graphically find (a) **A** + **B**, (b) **A** − **B**, (c) **B** − **A**, (d) **A** − 2**B**.

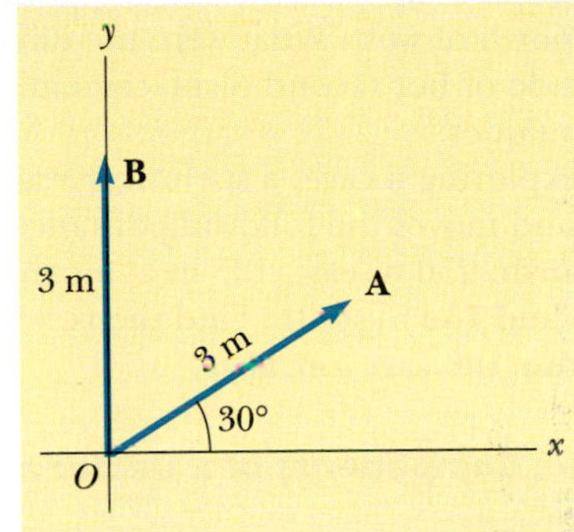

Figure 1.19 (Problem 44)

45. A roller coaster moves 200 ft horizontally, then rises 135 ft at an angle of 30° above the horizontal. It then travels 135 ft at an angle of 40° downward. At the end of this movement, what is its displacement from its starting point? Use graphical techniques.
46. The driver of a car, obviously lost, drives 3 km north, 2 km northeast (45° east of north), 4 km west, and then 3 km southeast (45° east of south). Where does he end up relative to his starting point? Work out your answer graphically. Check by using components. (The car is not near the North Pole or the South Pole.)
47. Find the horizontal and vertical components of the 100-m displacement of a superhero who flies from the top of a tall building along the path shown in Figure 1.20.

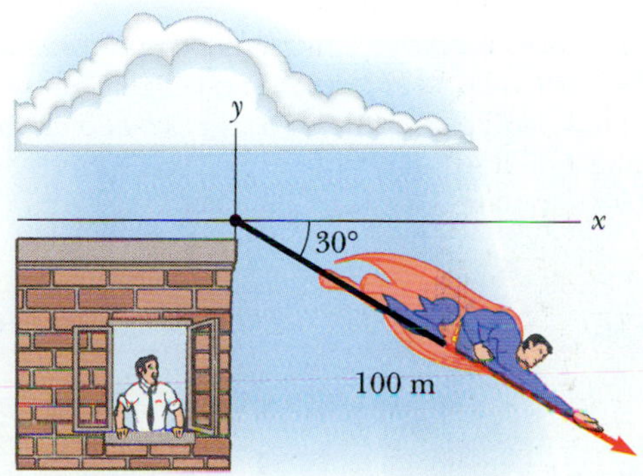

Figure 1.20 (Problem 47)

48. Walking along streets laid out in a square grid pattern, a child walks 2 blocks west, 3 blocks north, then 2 blocks west. (a) Find the total distance traveled. (b) Find the net displacement (magnitude and direction) from the starting point.
49. A jogger runs 100 m due west, then changes direction for the second leg of the run. At the end of the run, she is 175 m away from the starting point at an angle of 15° north of west. What were the direction and magnitude of her second displacement? Use graphical techniques.
50. While exploring a cave, a spelunker starts at the entrance and moves the following distances. She goes 75 m north, 250 m east, 125 m at an angle 30° north of east, and 150 m south. Find the resultant displacement from the cave entrance.

Section 1.11 Components of a Vector and Unit Vectors

51. A vector has an x component of -25 units and a y component of 40 units. Find the magnitude and direction of this vector.
52. Vector **B** has x, y, and z components of 4, 6, and 3, respectively. Calculate the magnitude of **B** and the angles that **B** makes with the coordinate axes.
53. Three vectors are given by $\mathbf{A} = \mathbf{i} + 3\mathbf{j}$; $\mathbf{B} = 2\mathbf{i} - \mathbf{j}$, and $\mathbf{C} = 3\mathbf{i} + 5\mathbf{j}$. Find (a) the sum of the three vectors and (b) the magnitude and direction of the resultant vector.
54. At the 1993 Superbowl, quarterback Troy Aikman of the Dallas Cowboys took the ball from the line of scrimmage, ran backward for 10 yards, then ran sideways parallel to the line of scrimmage for 15 yards. At this point he threw a 50-yard forward pass straight downfield perpendicular to the line of scrimmage. What was the magnitude of the football's resultant displacement?
55. A jet airliner moving initially at 300 mph due east enters a region where the wind is blowing at 100 mph in a direction 30° north of east. What are the new velocity and direction of the aircraft?
56. A novice golfer on the green takes three strokes to sink the ball. The successive displacements are 4 m due north, 2 m northeast, and 1 m 30° west of south. Starting at the same initial point, an expert golfer could make the hole in what single vector displacement?
57. A particle undergoes two displacements. The first has a magnitude of 150 cm and makes an angle of 120° with the positive x axis. The *resultant* displacement has a magnitude of 140 cm and is directed at an angle of 35° to the positive x axis. Find the magnitude and direction of the second displacement.
58. An airplane starting from airport A flies 300 km east, then 350 km 30° west of north, and then 150 km north to arrive finally at airport B. There is no wind on this day. (a) In what direction should the pilot fly to travel directly from A to B? (b) How far will the pilot travel in this direct flight?
59. Three vectors are oriented as shown in Figure 1.22, where $|\mathbf{A}| = 20$, $|\mathbf{B}| = 40$, and $|\mathbf{C}| = 30$ units. Find (a) the x and y components of the resultant vector and (b) the magnitude and direction of the resultant vector.

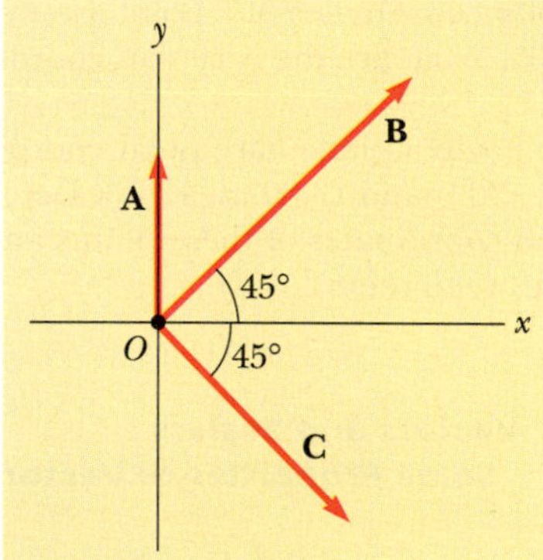

Figure 1.22 (Problem 59)

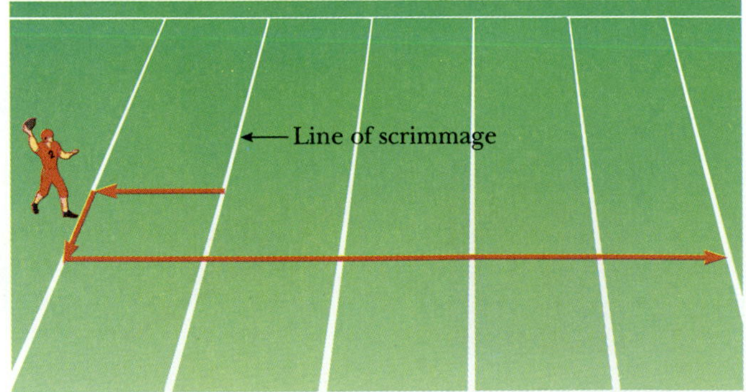

Figure 1.21 (Problem 54). Photo: Troy Aikman in action at the 1993 Superbowl. *(Rick Rickman, Duomo).*

Additional Problems

60. The eye of a hurricane passes over Grand Bahama Island. It is moving in a direction 60.0° north of west with a speed of 41 km/h. Three hours later the course of the hurricane suddenly shifts due north, and its speed slows to 25 km/h. How far is the hurricane from Grand Bahama 4.5 hours after it passes over the island?

61. A useful fact is that there are about $\pi \times 10^7$ s in one year. Use a calculator to find the percentage error in this approximation.
Note: percent error
$$= \frac{|\text{assumed value} - \text{true value}|}{\text{true value}} \times 100$$

62. Assume that there are 50 million passenger cars in the United States and that the average fuel consumption is 20 mi/gal of gasoline. If the average distance traveled by each car is 10 000 miles/year, how much gasoline would be saved per year if average fuel consumption could be increased to 25 mi/gal?

63. The basic function of the carburetor of an automobile is to atomize the gasoline and mix it with air to promote rapid combustion. As an example, assume that 30 cm^3 of gasoline is atomized into N spherical droplets, each with a radius of 2.0×10^{-5} m. What is the total surface area of these N spherical droplets?

△64. The data in the following table represent measurements of the masses and dimensions of solid cylinders of aluminum, copper, brass, tin, and iron. Use these data to calculate the densities of the substances. Compare your results for aluminum, copper, and iron with those given in Table 1.5.

Substance	Mass (g)	Diameter (cm)	Length (cm)
Aluminum	51.5	2.52	3.75
Copper	56.3	1.23	5.06
Brass	94.4	1.54	5.69
Tin	69.1	1.75	3.74
Iron	216.1	1.89	9.77

65. An air-traffic controller notices two aircraft on his radar screen. The first is at altitude 800 m, horizontal distance 19.2 km, and 25° south of west. The second aircraft is at altitude 1100 m, horizontal distance 17.6 km, and 20° south of west. What is the distance between the two aircraft? (Place the x axis west, the y axis south, and the z axis vertical.)

66. A vector is given by $\mathbf{R} = 2\mathbf{i} + \mathbf{j} + 3\mathbf{k}$. Find (a) the magnitudes of the x, y, and z components, (b) the magnitude of $\mathbf{R}$, and (c) the angles between $\mathbf{R}$ and the x, y, and z axes.

67. A person going for a walk follows the path shown in Figure 1.23. The total trip consists of four straight-line paths. At the end of the walk, what is the person's resultant displacement, measured from the starting point?

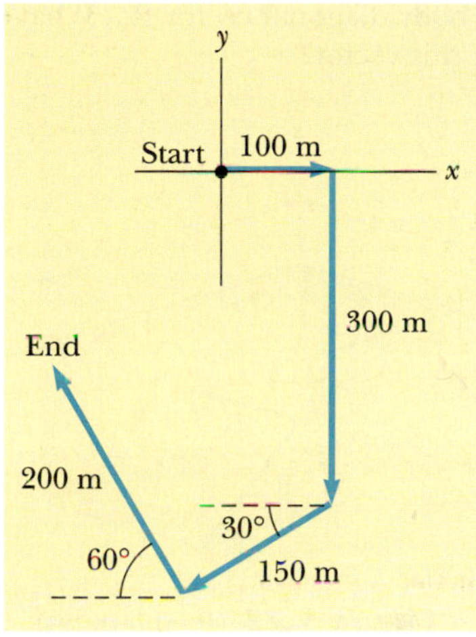

Figure 1.23 (Problem 67)

68. Two people pull on a stubborn mule, as shown by the helicopter view in Figure 1.24. Find (a) the single force which is equivalent to the two forces shown,

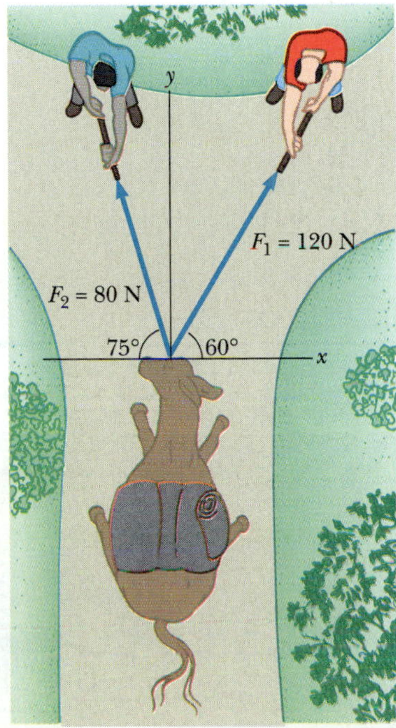

Figure 1.24 (Problem 68)

and (b) the force that a third person would have to exert on the mule to make the net force equal to zero.

69. A rectangular parallelepiped has dimensions a, b, and c, as in Figure 1.25. (a) Obtain a vector expression for the face diagonal vector $\mathbf{R}_1$. What is the magnitude of this vector? (b) Obtain a vector expression for the body diagonal vector $\mathbf{R}_2$. What is the magnitude of this vector?

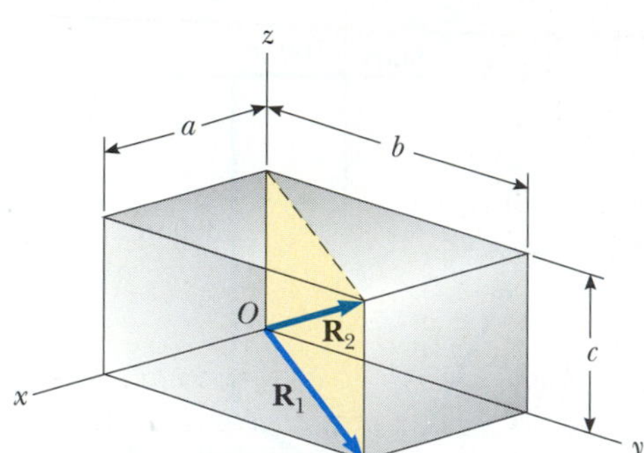

Figure 1.25 (Problem 69)

70. The distance from the Sun to the nearest star is on the order of 10^{17} m. The Milky Way galaxy is roughly a disk of radius 10^{21} m and thickness 10^{19} m. Estimate the number of stars in the Milky Way. Assume the distance between the Sun and the nearest neighbor is typical.

By Parker and Hart

By permission of John Hart and Field Enterprises, Inc.

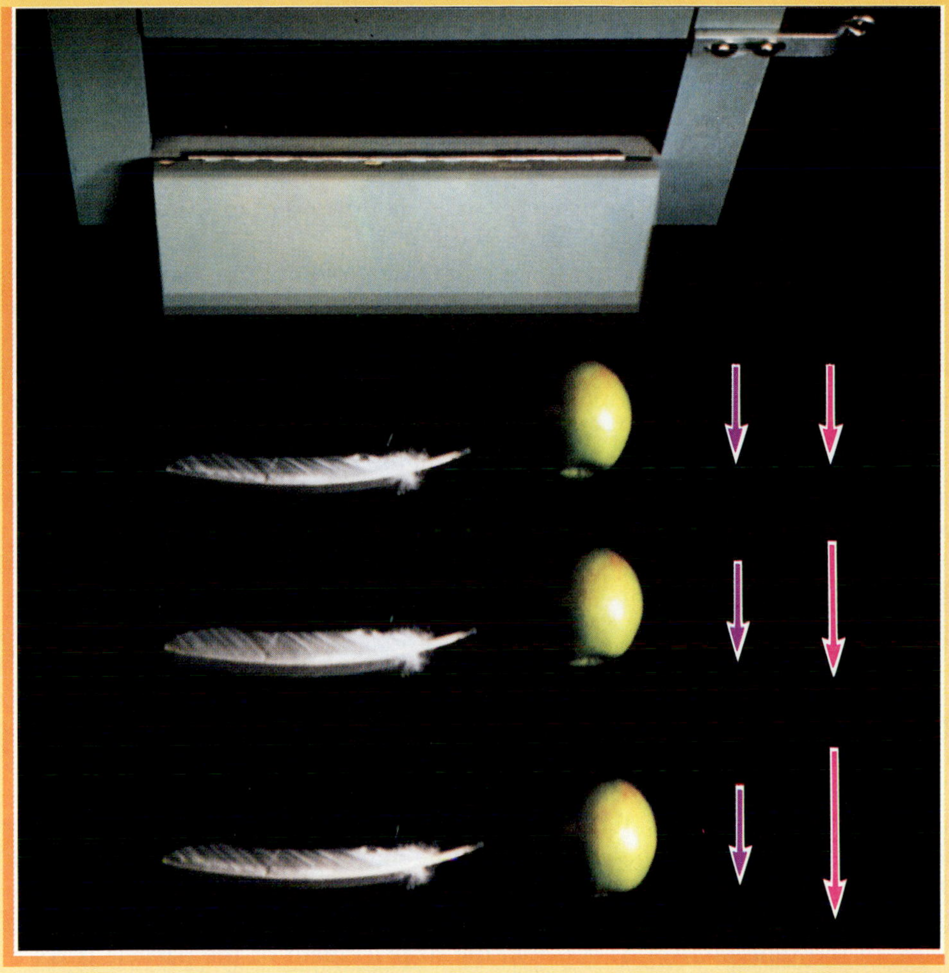

An apple and a feather, released from rest in a 4-ft vacuum chamber, fall at the same rate, regardless of their masses. Neglecting air resistance, all objects fall to the Earth with the same acceleration of magnitude 9.8 m/s^2, as indicated by the violet arrows in this multiflash photograph. The velocity of the two objects increases linearly with time, as indicated by the series of red arrows.

Motion in One Dimension

CHAPTER 2

Dynamics is the study of the motions of objects and the relationships of those motions to such concepts as force and mass. Before beginning our study of dynamics, however, it is convenient to describe motion using the concepts of space and time, without regard to the causes of the motion; this is a portion of mechanics called *kinematics.* In this chapter we shall consider motion along a straight line, that is, one-dimensional motion. Starting with the concept of displacement, discussed in Chapter 1, we shall define velocity and acceleration. Using these concepts, we shall proceed to study the motion of objects undergoing constant acceleration. In Chapter 3 we shall extend our discussion to two-dimensional motion.

From everyday experience we recognize that motion represents continuous change in the position of an object. The movement of an object through space may be accompanied by the rotation or vibration of the object. Such motions can be quite complex. However, it is sometimes possible to simplify matters by temporarily neglecting the internal motions of the moving object. In many situations, an object can be treated as a *particle* if the only motion being considered is translation through space.

Although an idealized particle is a mathematical point with no size, we can sometimes perform useful calculations by representing macroscopic objects as particles. For example, if we wish to describe the motion of the Earth around the Sun, we can approximate the Earth by treating it as a particle and attain reasonable accuracy in a prediction of the Earth's orbit. This approximation is justified because the radius of the Earth's orbit is large compared with the dimensions of the Earth and Sun. On the other hand, we could not use a particle description to explain the internal structure of the Earth or such phenomena as tides, earthquakes, and volcanic activity. On a much smaller scale, it is possible to explain the pressure exerted by a gas on the walls of a container by treating the gas molecules as particles. However, the particle description of the gas molecules is generally inadequate for understanding those properties of the gas that depend on the internal motions (vibrations) and rotations of the gas molecules.

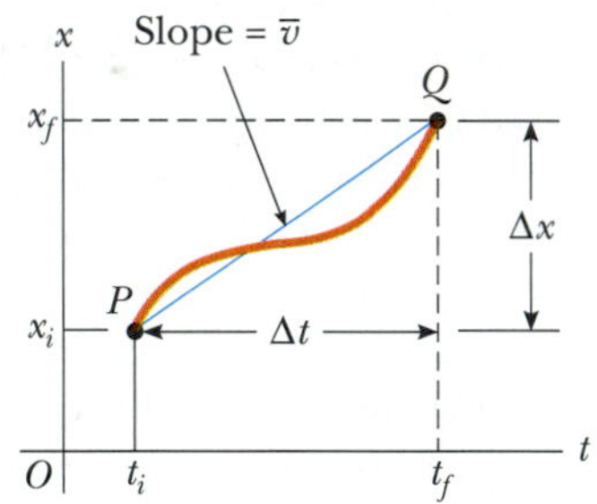

Figure 2.1
Position-time graph for a particle moving along the x axis. The average velocity $\bar{v}$ in the interval $\Delta t = t_f - t_i$ is obtained from the slope of the straight line connecting the points P and Q.

2.1 Average Velocity

The motion of a particle is completely known if the position of the particle in space is known at all times. Consider a particle moving along the x axis from point P to point Q. Let its position at point P be x_i at some time t_i, and let its position at point Q be x_f at time t_f. At times other than t_i and t_f, the position of the particle between these two points may vary, as in Figure 2.1. Such a plot is often called a *position-time graph*. In the time interval $\Delta t = t_f - t_i$, the displacement of the particle is $\Delta \mathbf{x} = x_f - x_i$. (Recall that displacement is defined as the change in the position of the particle, which is equal to its final position value minus its initial position value.) Using unit vector notation, the **displacement vector** can be expressed as $\Delta \mathbf{x} \equiv (x_f - x_i)\mathbf{i}$.

Average velocity

The x component of the **average velocity** of the particle, $\bar{\mathbf{v}}$, is defined as the ratio of its displacement vector, $\Delta\mathbf{x}$, and the time interval, Δt:

$$\bar{\mathbf{v}} \equiv \frac{\Delta \mathbf{x}}{\Delta t} = \frac{(x_f - x_i)\mathbf{i}}{t_f - t_i} \qquad [2.1]$$

From this definition we see that the average velocity has the dimensions of length divided by time (L/T)—m/s in SI units and ft/s in conventional units. The average velocity is *independent* of the path taken between the points P and Q. This is true because the average velocity is proportional to the displacement, $\Delta\mathbf{x}$, which in turn depends only on the initial and final coordinates of the particle. It therefore follows that if a particle starts at some point and returns to the same point via any path, its average velocity for this trip is zero, since its displacement along such a path is zero. Displacement should not be confused with the distance traveled, since the distance traveled for any motion is clearly nonzero. Thus, average velocity gives us no details of the motion between points P and Q. (How we evaluate the velocity at some instant in time is discussed in the next section.) Finally, note that the average velocity in one dimension can be positive or negative, depending on the sign of the displacement vector. (The time interval, Δt, is always positive.) If the x coordinate of the particle increases in time (that is, if $x_f > x_i$), then $\Delta \mathbf{x}$ is positive and $\bar{\mathbf{v}}$ is positive. This corresponds to an average velocity in the positive x direction. On the other hand, if the coordinate decreases in time ($x_f < x_i$), $\Delta \mathbf{x}$ is negative; hence, $\bar{\mathbf{v}}$ is negative. This corresponds to an average velocity in the negative x direction.

The average velocity can also be interpreted geometrically. A straight line drawn between the points P and Q in Figure 2.1 forms the hypotenuse of a right triangle of height Δx and base Δt. The slope of this line is the ratio $\Delta x/\Delta t$. Therefore, we see that the *average* velocity of the particle during the time interval t_i to t_f is equal to the "slope" of the straight line joining the initial and final points on the space-time graph. (The word *slope* will often be used in reference to the graphs of physical data. Regardless of what data are plotted, the word *slope* will represent the ratio of the change in the quantity represented on the vertical axis to the change in the quantity represented on the horizontal axis.)

Example 2.1 Calculate the Average Velocity

A particle moving along the x axis is located at $x_i = 12$ m at $t_i = 1$ s and at $x_f = 4$ m at $t_f = 3$ s. Find its displacement and average velocity during this time interval.

Solution The displacement is

$$\Delta \mathbf{x} = (x_f - x_i)\mathbf{i} = (4\text{ m} - 12\text{ m})\mathbf{i} = -8\mathbf{i}\text{ m}$$

The average velocity is

$$\bar{\mathbf{v}} = \frac{\Delta \mathbf{x}}{\Delta t} = \frac{(x_f - x_i)\mathbf{i}}{t_f - t_i} = \frac{(4\text{ m} - 12\text{ m})\mathbf{i}}{3\text{ s} - 1\text{ s}} = -4\mathbf{i}\text{ m/s}$$

Since the displacement is negative for this time interval, we conclude that the particle has moved to the left, toward decreasing values of x.

2.2 Instantaneous Velocity

We would like to be able to define the velocity of a particle at a particular instant of time, rather than just during a finite interval of time. The velocity of a particle at any instant of time, or at some point on a space-time graph, is called the **instantaneous velocity.** This concept is especially important when the average velocity is *not constant* through different time intervals.

Consider the motion of a particle between the two points P and Q on the space-time graph shown in Figure 2.2. As the point Q is brought closer and closer to the point P, the time intervals ($\Delta t_1, \Delta t_2, \Delta t_3, \ldots$) get progressively smaller. The average velocity for each time interval is given by the slope of the appropriate dotted line in Figure 2.2. As the point Q approaches P, the time interval approaches zero, but at the same time the slope of the dotted line approaches that of

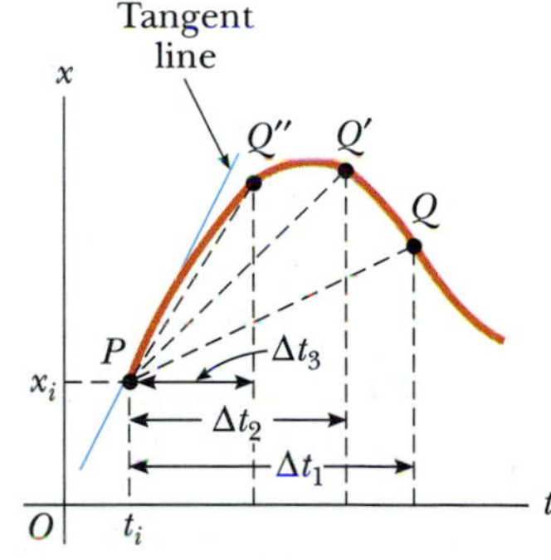

Figure 2.2 Position-time graph for a particle moving along the x axis. As the time intervals, starting at t_i, get smaller and smaller, the average velocity for one interval approaches the slope of the line tangent at P. The instantaneous velocity at P is obtained from the slope of the blue tangent line at the time t_i.

the blue line tangent to the curve at the point P. The slope of the line tangent to the curve at P gives the *instantaneous velocity* at the time t_i. In other words,

the instantaneous velocity, v, equals the limiting value of the ratio $\Delta \mathbf{x}/\Delta t$ as Δt approaches zero:[1]

$$\mathbf{v} \equiv \lim_{\Delta t \to 0} \frac{\Delta \mathbf{x}}{\Delta t} \quad [2.2]$$

In the calculus notation, this limit is called the *derivative* of x with respect to t, written $d\mathbf{x}/dt$:

The instantaneous vector is the derivative of the displacement vector with respect to time.

$$\mathbf{v} \equiv \lim_{\Delta t \to 0} \frac{\Delta \mathbf{x}}{\Delta t} = \frac{d\mathbf{x}}{dt} \quad [2.3]$$

The instantaneous velocity can be positive, negative, or zero. When the slope of the space-time graph is positive, such as at the point P in Figure 2.3, v is positive. At point R, v is negative since the slope is negative. Finally, the instantaneous velocity is zero at the peak Q (the turning point), where the slope is zero. *From here on, we shall usually use the word* velocity *to designate instantaneous velocity.*

The **instantaneous speed** of a particle is defined as the magnitude of the instantaneous velocity vector. Hence, by definition, *speed* can never be negative.

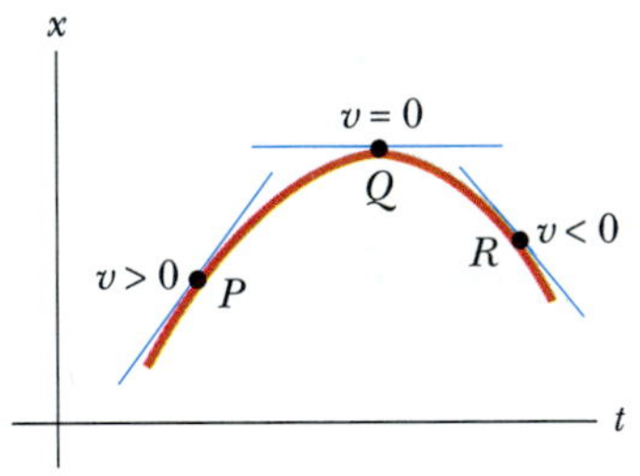

x
v = 0
Q
R
v < 0
v > 0
P
t

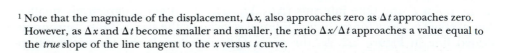

Figure 2.3
In the position-time graph shown here, the velocity is positive at P, where the slope of the tangent line is positive; the velocity is zero at Q, where the slope of the tangent line is zero; and the velocity is negative at R, where the slope of the tangent line is negative.

Example 2.2 Average and Instantaneous Velocity

A particle moves along the x axis. Its x coordinate varies with time according to the expression $x = -4t + 2t^2$, where x is in meters and t is in seconds. The position-time graph for this motion is shown in Figure 2.4. Note that the particle moves in the negative x direction for the first second of motion, stops instantaneously at $t = 1$ s, and then heads back in the positive x direction for $t > 1$ s.

(a) Determine the displacement of the particle in the time intervals $t = 0$ to $t = 1$ s and $t = 1$ s to $t = 3$ s.

Solution In the first time interval we set $t_i = 0$ and $t_f = 1$ s. Since $x = -4t + 2t^2$, we get for the first displacement

$$\begin{aligned}\Delta \mathbf{x}_{01} &= (x_f - x_i)\mathbf{i} \\ &= [-4(1) + 2(1)^2]\mathbf{i} - [-4(0) + 2(0)^2]\mathbf{i} \\ &= -2\mathbf{i} \text{ m}\end{aligned}$$

Likewise, in the second time interval we can set $t_i = 1$ s and $t_f = 3$ s. Therefore, the displacement in this interval is

[1] Note that the magnitude of the displacement, Δx, also approaches zero as Δt approaches zero. However, as Δx and Δt become smaller and smaller, the ratio $\Delta x/\Delta t$ approaches a value equal to the *true* slope of the line tangent to the x versus t curve.

$$\begin{aligned}\Delta \mathbf{x}_{13} &= (x_f - x_i)\mathbf{i} \\ &= [-4(3) + 2(3)^2]\mathbf{i} - [-4(1) + 2(1)^2]\mathbf{i} \\ &= \boxed{8\mathbf{i}\ \text{m}}\end{aligned}$$

Note that these displacements can also be read directly from the position-time graph (Fig. 2.4).

(b) Calculate the average velocity in the time intervals $t = 0$ to $t = 1$ s and $t = 1$ s to $t = 3$ s.

Solution In the first time interval, $\Delta t = t_f - t_i = 1$ s. Therefore, using Equation 2.1 and the results from (a) gives

$$\bar{\mathbf{v}}_{01} = \frac{\Delta \mathbf{x}_{01}}{\Delta t} = \frac{-2\mathbf{i}\ \text{mi}}{1\ \text{s}} = \boxed{-2\mathbf{i}\ \text{m/s}}$$

Likewise, in the second time interval, $\Delta t = 2$ s; therefore,

$$\bar{\mathbf{v}}_{13} = \frac{\Delta \mathbf{x}_{13}}{\Delta t} = \frac{8\mathbf{i}\ \text{mi}}{2\ \text{s}} = \boxed{4\mathbf{i}\ \text{m/s}}$$

These values (the coefficients of **i**) agree with the slopes of the lines joining these points in Figure 2.3.

(c) Find the instantaneous velocity of the particle at $t = 2.5$ s.

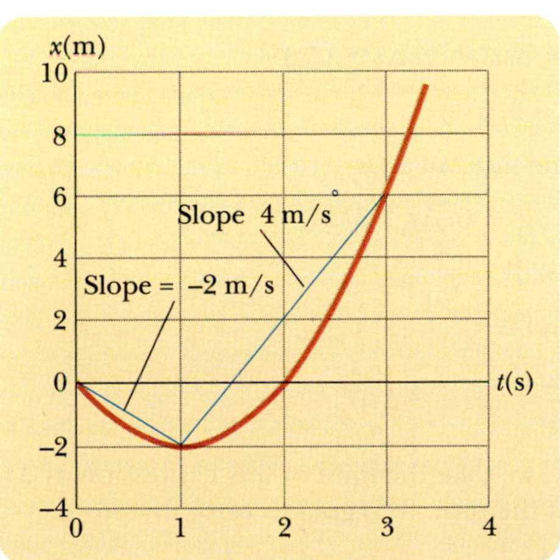

Figure 2.4 (Example 2.2) Position-time graph for a particle having an x coordinate that varies in time according to $x = -4t + 2t^2$.

Solution By measuring the slope of the position-time graph at $t = 2.5$ s, we find that $\mathbf{v} = 6\mathbf{i}$ m/s. (You should show that the velocity is $-4\mathbf{i}$ m/s at $t = 0$ and zero at $t = 1$ s.) Do you see any symmetry in the motion? For example, does the speed ever repeat itself?

Example 2.3 The Limiting Process

The position of a particle moving along the x axis varies in time according to the expression $x = 3t^2$, where x is in meters, 3 is in meters per second per second, and t is in seconds. Find the velocity in terms of t at any time.

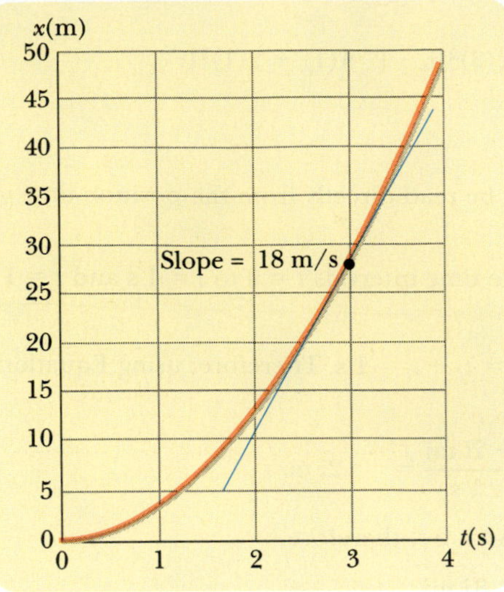

Figure 2.5 (Example 2.3) Position-time graph for a particle having an x coordinate that varies in time according to $x = 3t^2$. Note that the instantaneous velocity at $t = 3$ s is obtained from the slope of the blue line tangent to the curve at this point.

Solution The position-time graph for this motion is shown in Figure 2.5. We can compute the velocity at any time t by using the definition of the instantaneous velocity. If the initial coordinate of the particle at time t is $x_i = 3t^2$, then the coordinate at a later time, $t + \Delta t$, is

$$x_f = 3(t + \Delta t)^2 = 3[t^2 + 2t\,\Delta t + (\Delta t)^2]$$
$$= 3t^2 + 6t\,\Delta t + 3(\Delta t)^2$$

Therefore, the displacement in the time interval Δt is

$$\Delta\mathbf{x} = (x_f - x_i)\mathbf{i} = [3t^2 + 6t\,\Delta t + 3(\Delta t)^2 - 3t^2]\mathbf{i}$$
$$= (6t\,\Delta t + 3(\Delta t)^2)\mathbf{i}$$

The average velocity in this time interval is

$$\bar{\mathbf{v}} = \frac{\Delta\mathbf{x}}{\Delta t} = (6t + 3\,\Delta t)\mathbf{i}$$

To find the instantaneous velocity, we take the limit of this expression as Δt approaches zero. In doing so, we see that the term $3\,\Delta t$ goes to zero; therefore,

$$\mathbf{v} = \lim_{\Delta t \to 0} \frac{\Delta\mathbf{x}}{\Delta t} = 6t\mathbf{i} \text{ m/s}$$

Notice that this expression gives us the velocity at *any* general time t. It tells us that $\mathbf{v}$ is increasing linearly in time. It is then a straightforward matter to find the velocity at some specific time from the expression $\mathbf{v} = 6t\mathbf{i}$. For example, at $t = 3$ s, the velocity is $\mathbf{v} = 6(3)\mathbf{i} = 18\,\mathbf{i}$ m/s. Again, this can be checked from the slope of the graph (the blue line) at $t = 3$ s.

The limiting process can also be examined numerically. For example, we can compute the displacement and average velocity for various time intervals beginning at $t = 3$ s, using the expressions for $\Delta\mathbf{x}$ and $\bar{\mathbf{v}}$. The results of such calculations are given in Table 2.1. Notice that as the time intervals get smaller and smaller, the average velocity more nearly approaches the value of the instantaneous velocity at $t = 3$ s, namely, 18 m/s.

Table 2.1
Displacement and Average Velocity for Various Time Intervals for the Function $x = 3t^2$ (the intervals begin at $t = 3$ s)

Δt (s)	Δx (m)	$\Delta x/\Delta t$ (m/s)
1.00	21	21
0.50	9.75	19.5
0.25	4.69	18.8
0.10	1.83	18.3
0.05	0.9075	18.15
0.01	0.1803	18.03
0.001	0.018003	18.003

2.3 Acceleration

When the velocity of a particle changes with time, the particle is said to be *accelerating*. For example, the speed of a car increases when you "step on the gas." The car slows down when you apply the brakes and changes direction when you turn the wheel. However, we need a more precise definition of acceleration than this.

Suppose a particle moving along the x axis has a velocity v_i at time t_i and a velocity v_f at time t_f.

The **average acceleration** of a particle in the time interval $\Delta t = t_f - t_i$ is defined as the ratio $\Delta \mathbf{v}/\Delta t$, where $\Delta \mathbf{v} = \mathbf{v}_f - \mathbf{v}_i$ is the *change* in velocity in this time interval:

$$\bar{a} \equiv \frac{v_f - v_i}{t_f - t_i} = \frac{\Delta v}{\Delta t} \qquad [2.4]$$

Average acceleration

Acceleration is a vector quantity having dimensions of length divided by (time)^2, or L/T^2. Some of the common units of acceleration are meters per second per second (m/s^2) and feet per second per second (ft/s^2).

In some situations, the value of the average acceleration may be different for different time intervals. It is therefore useful to define the **instantaneous acceleration** as the limit of the average acceleration as Δt approaches zero. This concept is analogous to the definition of instantaneous velocity discussed in the previous section. If we imagine that the point Q is brought closer and closer to the point P in Figure 2.6 and take the limit of the ratio $\Delta v/\Delta t$ as Δt approaches zero, we get the *instantaneous acceleration:*

$$a \equiv \lim_{\Delta t \to 0} \frac{\Delta v}{\Delta t} = \frac{dv}{dt} \qquad [2.5]$$

Instantaneous acceleration

That is, the instantaneous acceleration equals the derivative of the velocity with respect to time, which by definition is the slope of the velocity-time graph. One can interpret the derivative of the velocity with respect to time as the *time rate of change*

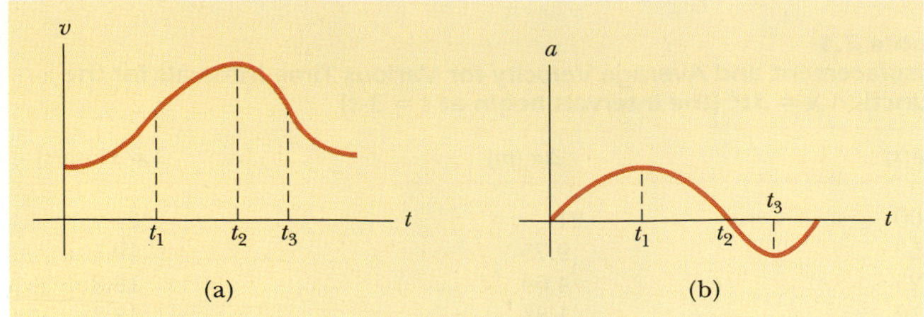

Figure 2.6
The instantaneous acceleration can be obtained from the velocity-time graph (a). At each instant, the acceleration in the a versus t graph (b) equals the slope of the line tangent to the v versus t curve.

of velocity. Again you should note that if **a** is positive, the acceleration is in the positive x direction, whereas negative **a** implies acceleration in the negative x direction. *From now on we shall use the term* acceleration *to mean instantaneous acceleration.*

Since $\mathbf{v} = d\mathbf{x}/dt$, the acceleration can also be written

$$\mathbf{a} = \frac{d\mathbf{v}}{dt} = \frac{d}{dt}\left(\frac{d\mathbf{x}}{dt}\right) = \frac{d^2\mathbf{x}}{dt^2} \qquad \textbf{[2.6]}$$

That is, the acceleration equals the *second derivative* of the displacement with respect to time.

Figure 2.6 shows how the acceleration-time curve can be derived from the velocity-time curve. In these sketches, the acceleration at any time is simply the slope of the velocity-time graph at that time. Positive values of the acceleration correspond to those points where the velocity is increasing in the positive x direction. The acceleration reaches a maximum at time t_1, when the slope of the velocity-time graph is a maximum. The acceleration then goes to zero at time t_2, when the velocity is a maximum (that is, when the velocity is momentarily not changing and the slope of the v versus t graph is zero). Finally, the acceleration is negative when the velocity in the positive x direction is decreasing in time.

Galileo performing demonstrations of balls rolling down a grooved inclined plane. As a ball rolled down the incline, Galileo carefully measured its position at the end of equal time intervals and showed that the distance traveled was proportional to the square of the elapsed time. This painting by Giuseppe Bezzouli is located in the Zoological Museum in Florence, Italy. *(Art Resource)*

As an example of the computation of acceleration, consider the car pictured in Figure 2.7. In this case, the velocity of the car has changed from an initial value of $30\mathbf{i}$ m/s to a final value of $15\mathbf{i}$ m/s in a time interval of 2 s. The average acceleration during this time interval is

$$\bar{\mathbf{a}} = \frac{15\mathbf{i}\ \text{m/s} - 30\mathbf{i}\ \text{m/s}}{2.0\ \text{s}} = -7.5\mathbf{i}\ \text{m/s}^2$$

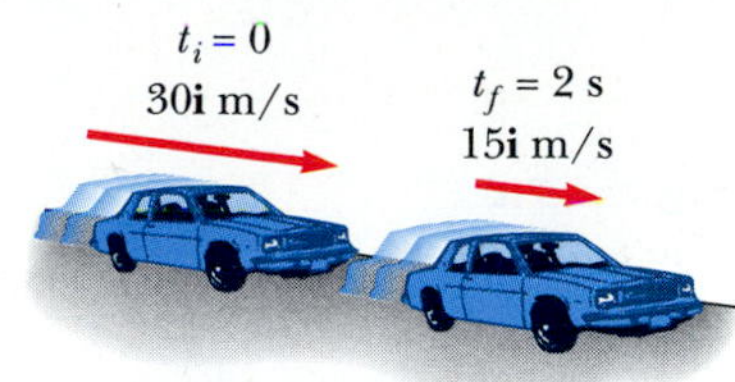

Figure 2.7 The velocity of the car decreases from $30\mathbf{i}$ m/s to $15\mathbf{i}$ m/s in a time interval of 2 s.

The minus sign in this example indicates that the acceleration vector is in the negative x direction (to the left). For the case of motion in a straight line, the direction of the velocity of an object and the direction of its acceleration are related as follows. *When the object's velocity and acceleration are in the same direction, the object is speeding up in that direction.* On the other hand, *when the object's velocity and acceleration are in opposite directions, the speed of the object decreases in time.*

Example 2.4 Average and Instantaneous Acceleration

The velocity of a particle moving along the x axis varies in time according to the expression $\mathbf{v} = (40 - 5t^2)\mathbf{i}$ m/s, where t is in seconds.

(a) Find the average acceleration in the time interval $t = 0$ to $t = 2$ s.

Solution The velocity-time graph for this function is given in Figure 2.8. The velocities at $t_i = 0$ and $t_f = 2$ s are found by substituting these values of t into the expression given for the velocity:

$$\mathbf{v}_i = (40 - 5t_i^2)\mathbf{i}\ \text{m/s} = [40 - 5(0)^2]\mathbf{i}\ \text{m/s} = 40\mathbf{i}\ \text{m/s}$$

$$\mathbf{v}_f = (40 - 5t_f^2)\mathbf{i}\ \text{m/s} = [40 - 5(2)^2]\mathbf{i}\ \text{m/s} = 20\mathbf{i}\ \text{m/s}$$

Therefore, the average acceleration in the specified time interval, $\Delta t = t_f - t_i = 2$ s, is given by

$$\bar{\mathbf{a}} = \frac{\mathbf{v}_f - \mathbf{v}_i}{t_f - t_i} = \frac{(20 - 40)\mathbf{i}\ \text{m/s}}{(2 - 0)\text{s}} = -10\mathbf{i}\ \text{m/s}^2$$

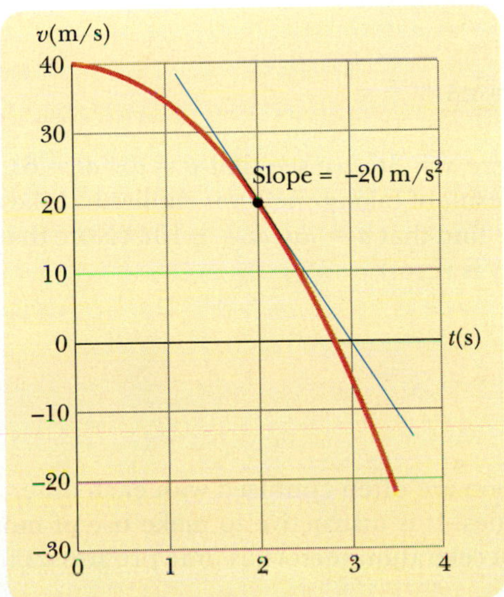

Figure 2.8 (Example 2.4) The velocity-time graph for a particle moving along the x axis according to the relation $v = (40 - 5t^2)$ m/s. Note that the acceleration at $t = 2$ s is obtained from the slope of the blue tangent line at that time.

The negative sign is consistent with the fact that the slope of the line joining the initial and final points on the velocity-time graph is negative.

(b) Determine the acceleration at $t = 2$ s.

Solution The velocity at time t is given by $\mathbf{v}_i = (40 - 5t^2)\mathbf{i}$ m/s, and the velocity at time $t + \Delta t$ is given by

$$\mathbf{v}_f = 40\mathbf{i} - 5(t + \Delta t)^2\mathbf{i} = [40 - 5t^2 - 10t\,\Delta t - 5(\Delta t)^2]\mathbf{i}$$

Therefore, the change in velocity over the time interval Δt is

$$\Delta\mathbf{v} = \mathbf{v}_f - \mathbf{v}_i = [-10t\,\Delta t - 5(\Delta t)^2]\mathbf{i} \text{ m/s}$$

Dividing this expression by Δt and taking the limit of the result as Δt approaches zero, we get the acceleration at *any* time t:

$$\mathbf{a} = \lim_{\Delta t \to 0} \frac{\Delta \mathbf{v}}{\Delta t} = \lim_{\Delta t \to 0} (-10t - 5\,\Delta t)\mathbf{i} = -10t\mathbf{i} \text{ m/s}^2$$

Therefore, at $t = 2$ s we find that

$$\mathbf{a} = (-10)(2)\mathbf{i} \text{ m/s}^2 = -20\mathbf{i} \text{ m/s}^2$$

This result can also be obtained by measuring the slope of the velocity-time graph at $t = 2$ s. Note that the acceleration is not constant in this example. Situations involving constant acceleration will be treated in Section 2.5.

So far we have evaluated the derivatives of a function by starting with the definition of the function and then taking the limit of a specific ratio. Those of you familiar with the calculus should recognize that there are specific rules for taking the derivatives of functions. These rules, which are listed in Appendix B.6, enable us to evaluate derivatives quickly.

Suppose x is proportional to some power of t, such as

$$x = At^n$$

where A and n are constants. (This is a very common functional form.) The derivative of x with respect to t is given by

$$\frac{dx}{dt} = nAt^{n-1}$$

Applying this rule to Example 2.3, where $x = 3t^2$, we see that $v = dx/dt = 6t$, and $a = dv/dt = 6$ in agreement with our result of taking the limit explicitly. Likewise, in Example 2.4, where $v = 40 - 5t^2$, we find that $a = dv/dt = -10t$. (Note that the rate of change of any constant quantity is zero.)

2.4 Motion Diagrams

The concepts of velocity and acceleration are often confused with each other, but in fact they are quite different quantities. It is instructive to make use of motion diagrams to describe the velocity and acceleration vectors as time progresses while

Figure 2.9 (a) Motion diagram for an object whose constant acceleration is in the direction of its velocity. The velocity vector at each instant is indicated by a red arrow, and the constant acceleration vector by a violet arrow. (b) Motion diagram for an object whose constant acceleration is in the direction *opposite* the velocity at each instant.

an object is in motion. In order not to confuse these two vector quantities in Figure 2.9, we use red for velocity vectors and violet for acceleration vectors. The vectors are sketched at several instants during the motion of the object, and the time intervals between adjacent positions are assumed to be equal.

Figure 2.9a represents a car moving to the right with a constant positive acceleration. In this case, note that the velocity vector increases in time. Because the car speeds up as it travels to the right, its displacement between adjacent positions increases as time progresses. If the car moves initially to the right with a constant negative acceleration (that is, a constant deceleration), as in Figure 2.9b, the velocity vector decreases in time and eventually reaches zero. (This type of motion is exhibited by a car that skids to a stop after applying its brakes.) In this case, the car slows as it moves to the right, so its displacement between adjacent positions decreases as time progresses. From this diagram we see that the acceleration and velocity vectors are *not* in the same direction.

You should be able to construct a motion diagram for a particle that moves initially to the left with a constant positive or negative acceleration. You should also construct appropriate motion diagrams after completing the mathematical solutions to kinematic problems, to see if your answers are consistent with the diagrams.

2.5 One-Dimensional Motion with Constant Acceleration

If the acceleration of a particle varies in time, the motion can be complex and difficult to analyze. A very common and simple type of one-dimensional motion occurs when the acceleration is constant, or uniform. In this case the average acceleration equals the instantaneous acceleration. Consequently, the velocity increases or decreases at the same rate throughout the motion.

If we replace $\bar{a}$ with a in Equation 2.4, we find that

$$a = \frac{v_f - v_i}{t_f - t_i}$$

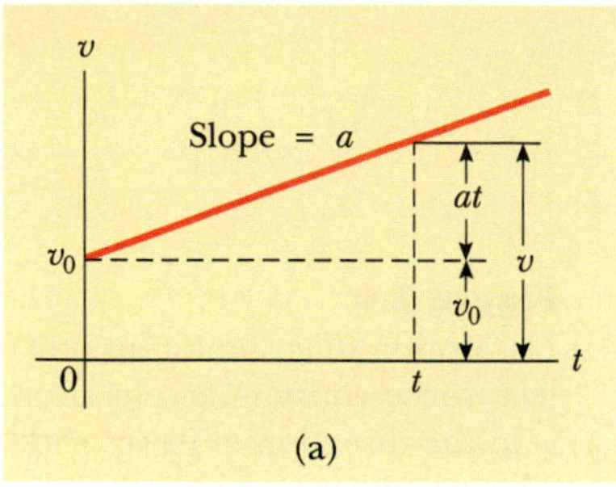

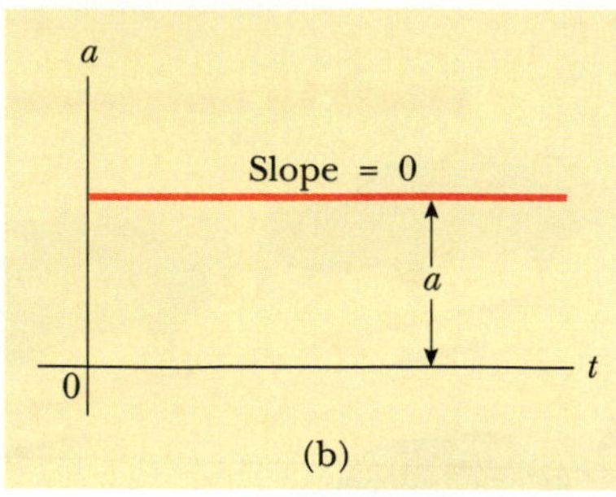

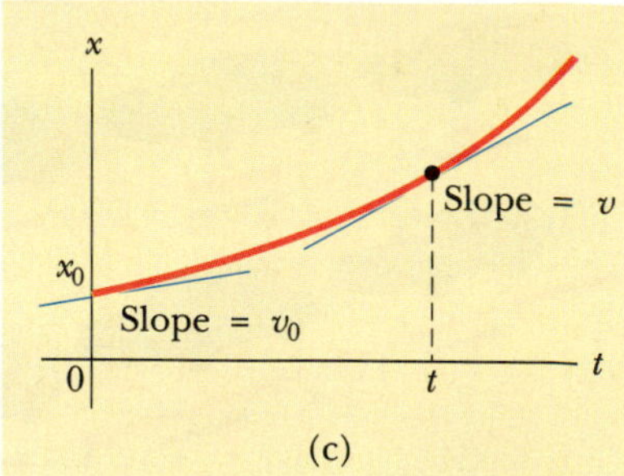

Figure 2.10
A particle moving along the x axis with constant acceleration a; (a) the velocity-time graph, (b) the acceleration-time graph, and (c) the space-time graph.

Since we are dealing with motion in one dimension, we shall drop the boldface notation for vectors and use a minus sign when the vector is in the negative x direction. For convenience, let $t_i = 0$ and t_f be any arbitrary time t. Also, let $v_i = v_0$ (the initial velocity at $t = 0$) and $v_f = v$ (the velocity at any arbitrary time t). With this notation, we can express the acceleration as

$$a = \frac{v - v_0}{t}$$

or

Velocity as a function of time

$$v = v_0 + at \qquad \text{(for constant } a\text{)} \qquad [2.7]$$

This expression enables us to predict the velocity at *any* time t if the initial velocity, acceleration, and elapsed time are known. A graph of velocity versus time for this motion is shown in Figure 2.10a. The graph is a straight line, the slope of which is the acceleration, a, consistent with the fact that $a = dv/dt$ is a constant. From this graph and from Equation 2.7, we see that the velocity at any time t is the sum of the initial velocity, v_0, and the change in velocity, at. The graph of acceleration versus time (Fig. 2.10b) is a straight line with a slope of zero, since the acceleration is constant. Note that if the acceleration were negative (the particle slowing down), the slope of Figure 2.10a would be negative.

Because the velocity varies linearly in time according to Equation 2.7, we can express the average velocity in any time interval as the arithmetic mean of the initial velocity, v_0, and the final velocity, v:

$$\bar{v} = \frac{v_0 + v}{2} \qquad \text{(for constant } a\text{)} \qquad [2.8]$$

Note that this expression is useful only when the acceleration is constant, that is, when the velocity varies linearly with time.

We can now use Equations 2.1 and 2.8 to obtain the displacement as a function of time. Again we choose $t_i = 0$, at which time the initial position is $x_i = x_0$. This gives

$$\Delta x = \bar{v}\,\Delta t = \left(\frac{v_0 + v}{2}\right) t$$

or

$$x - x_0 = \tfrac{1}{2}(v + v_0)t \qquad \text{(for constant } a\text{)} \qquad [2.9]$$

We can obtain another useful expression for the displacement by substituting Equation 2.7 into Equation 2.9:

$$x - x_0 = \tfrac{1}{2}(v_0 + v_0 + at)t$$

$$x - x_0 = v_0 t + \tfrac{1}{2}at^2 \quad \text{(for constant } a\text{)} \qquad [2.10]$$

Displacement as a function of time

The validity of this expression can be checked by differentiating it with respect to time, to give

$$v = \frac{dx}{dt} = \frac{d}{dt}(x_0 + v_0 t + \tfrac{1}{2}at^2) = v_0 + at$$

Finally, we can obtain an expression that does not contain the time by substituting the value of t from Equation 2.7 into Equation 2.9. This gives

$$x - x_0 = \tfrac{1}{2}(v_0 + v)\left(\frac{v - v_0}{a}\right) = \frac{v^2 - v_0^2}{2a}$$

$$v^2 = v_0^2 + 2a(x - x_0) \quad \text{(for constant } a\text{)} \qquad [2.11]$$

Velocity as a function of displacement

A position-time graph for motion under constant acceleration, assuming positive a, is shown in Figure 2.10c. Note that the curve representing Equation 2.10 is a parabola. The slope of the tangent to this curve at $t = 0$ equals the initial velocity, v_0, and the slope of the tangent line at any time t equals the velocity at that time.

If motion occurs in which the acceleration is *zero*, then we see that

$$\left.\begin{aligned} v &= v_0 \\ x - x_0 &= vt \end{aligned}\right\} \text{when } a = 0$$

That is, when the acceleration is zero, the velocity remains constant and the displacement changes linearly with time.

Equations 2.7 through 2.11 are five *kinematic expressions that may be used to solve any problem in one-dimensional motion with constant acceleration.* Keep in mind that these relationships were derived from the definitions of velocity and acceleration, together with some simple algebraic manipulations and the requirement that the acceleration be constant. It is often convenient to choose the initial position of the particle as the origin of the motion, so that $x_0 = 0$ at $t = 0$. In such a case, the displacement is simply x.

The four kinematic equations that are used most often are listed in Table 2.2 for convenience. The choice of which kinematic equation or equations you should use in a given situation depends on what is known beforehand. Sometimes it is necessary to use two of these equations to solve for two unknowns, such as the

Table 2.2
Kinematic Equations for Motion in a Straight Line Under Constant Acceleration

Equation	Information Given by Equation
$v = v_0 + at$	Velocity as a function of time
$x - x_0 = \frac{1}{2}(v + v_0)t$	Displacement as a function of velocity and time
$x - x_0 = v_0 t + \frac{1}{2}at^2$	Displacement as a function of time
$v^2 = v_0^2 + 2a(x - x_0)$	Velocity as a function of displacement

Note: Motion is along the x axis. At $t = 0$, the position of the particle is x_0 and its velocity is v_0.

displacement and velocity at some instant. For example, suppose the initial velocity, v_0, and acceleration, a, are given. You can then find (1) the velocity after a time t has elapsed, using $v = v_0 + at$, and (2) the displacement after a time t has elapsed, using $x - x_0 = v_0 t + \frac{1}{2}at^2$. You should recognize that the quantities that vary during the motion are velocity, displacement, and time.

You will get a great deal of practice in the use of these equations by solving exercises and problems. Many times you will discover that there is more than one method for obtaining a solution.

Problem-Solving Strategy: Accelerated Motion

The following procedure is recommended for solving problems that involve accelerated motion:

1. Make sure all the units in the problem are consistent. That is, if distances are measured in meters, be sure that velocities have units of meters per second and accelerations have units of meters per second per second.
2. Choose a coordinate system.
3. Make a list of all the quantities given in the problem and a separate list of those to be determined.
4. Think about what is going on physically in the problem, and then select from the list of kinematic equations the one or ones that will enable you to determine the unknowns.
5. Construct an appropriate motion diagram, and check to see if your answers are consistent with the diagram.

Example 2.5 The Indianapolis 500

A racing car starting from rest in the pits accelerates at a rate of 5 m/s². What is the velocity of the car after it has traveled 100 ft?

Solution Refer to the preceding Problem-Solving Strategy to see how its steps are applied to this example. First, you must ensure that the units are consistent. The units in the statement of this problem are *not* consistent. If we choose to leave the distance at 100 ft, we must change the length dimension of the units of acceleration from meters to feet. As an alternative, we can leave the units of acceleration as meters per second per second and convert the distance traveled to meters. Let's do the latter. The table of conversion factors in Appendix A of this text gives 1 ft = 0.305 m; thus, 100 ft = 30.5 m.

Second, you must choose a coordinate system. A convenient one is shown in Figure 2.11. The origin of the coordinate system is the initial location of the car, and the positive direction is to the right. Using this convention, we require that velocities, accelerations, and displacements to the right be positive, and vice versa.

Next, you will find it convenient to make a list of the quantities given in the problem and a separate list of those to be determined:

Given	To Be Determined
$v_0 = 0$	v
$a = +5\text{ m/s}^2$	
$x = +30.5\text{ m}$	

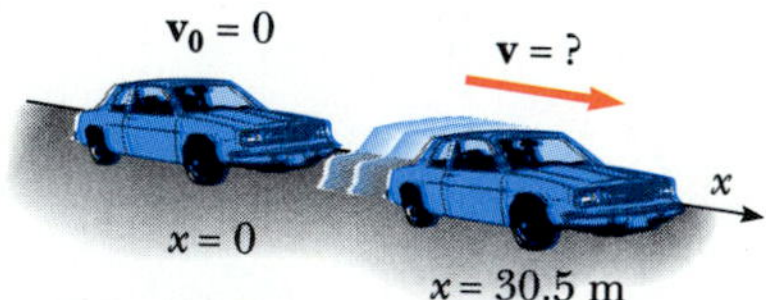

Figure 2.11 (Example 2.5)

The final step is to select from the kinematic equations (Table 2.2) those that will allow you to determine the unknowns. In our present case, the equation

$$v^2 = v_0^2 + 2ax$$

is the best choice since it will give us a value for v directly:

$$v^2 = (0)^2 + 2(5\ \text{m/s}^2)(30.5\ \text{m}) = 305\ \text{m}^2/\text{s}^2$$

from which

$$v = \sqrt{305\ \text{m}^2/\text{s}^2} = \pm 17.5\ \text{m/s}$$

Since the car is moving to the right, we choose +17.5 m/s as the correct solution for v.

Exercise Solve this same problem using $x = v_0 t + \frac{1}{2}at^2$ to find t and then using the expression $v = v_0 + at$ to find v.

Example 2.6 Accelerating an Electron

An electron in the cathode ray tube of a TV set enters a region where it accelerates uniformly from a speed of 3×10^4 m/s to a speed of 5×10^6 m/s in a distance of 2 cm.

(a) How long is the electron in this region where it accelerates?

Solution Taking the direction of motion to be along the x axis, we can use Equation 2.9 to find t, since the displacement and velocities are known:

$$x - x_0 = \tfrac{1}{2}(v_0 + v)t$$

$$t = \frac{2(x - x_0)}{v_0 + v} = \frac{2(2 \times 10^{-2}\ \cancel{\text{m}})}{(3 \times 10^4 + 5 \times 10^6)\ \cancel{\text{m}}/\text{s}}$$

$$= 7.95 \times 10^{-9}\ \text{s}$$

(b) What is the acceleration of the electron in this region?

Solution To find the acceleration, we can use $v = v_0 + at$ and the results from (a):

$$a = \frac{v - v_0}{t} = \frac{(5 \times 10^6 - 3 \times 10^4)\ \text{m/s}}{7.95 \times 10^{-9}\ \text{s}}$$

$$= 6.25 \times 10^{14}\ \text{m/s}^2$$

We also could have used Equation 2.11 to obtain the acceleration, since the velocities and displacement are known. Try it! Although a is very large in this example, the acceleration occurs over a very short time interval and is a typical value for such charged particles in acceleration.

Example 2.7 A "Catch-Up" Problem

A car traveling at a constant speed of 30 m/s (≈67 mi/h) passes a trooper hidden behind a billboard. One second after the speeding car passes the billboard, the trooper sets off in chase with a constant acceleration of 3.0 m/s². How long does it take the trooper to overtake the speeding car?

Solution To solve this problem algebraically, let us write an expression for the position of each vehicle as a function of time. It is convenient to choose the origin at the position of the billboard and take $t = 0$ as the time the trooper begins to move. At that instant the speeding car has already traveled a distance of 30 m, since it travels at a constant speed of 30 m/s. Thus, the initial position of the speeding car is given by $x_0 = 30$ m. Since the car moves with constant speed, its acceleration is zero, and applying Equation 2.10 gives

$$x_C = 30 \text{ m} + (30 \text{ m/s})t$$

Note that at $t = 0$, this expression does give the car's correct initial position, $x_C = x_0 = 30$ m. Likewise, for the trooper who starts from the origin at $t = 0$, we have $x_0 = 0$, $v_0 = 0$, and $a = 3.0 \text{ m/s}^2$. Hence, the position of the trooper versus time is given by

$$x_T = \tfrac{1}{2}at^2 = \tfrac{1}{2}(3.0 \text{ m/s}^2)t^2$$

The trooper overtakes the car at the instant that $x_T = x_C$, or

$$\tfrac{1}{2}(3.0 \text{ m/s}^2)t^2 = 30 \text{ m} + (30 \text{ m/s})t$$

This gives the quadratic equation

$$1.5t^2 - 30t - 30 = 0$$

whose positive solution is $t = 21$ s. Note that in this time interval, the trooper travels a distance of about 660 m.

Exercise This problem can also be easily solved graphically. On the *same* graph, plot the position versus time for each vehicle, and from the intersection of the two curves determine the time at which the trooper overtakes the speeding car.

2.6 Freely Falling Bodies

It is well known that all objects, when dropped, fall toward the Earth with nearly constant acceleration. There is a legend that Galileo Galilei first discovered this fact by observing that two different weights dropped simultaneously from the Leaning Tower of Pisa hit the ground at approximately the same time. Although there is some doubt that this particular experiment was carried out, it is well established that Galileo did perform many systematic experiments on objects moving on inclined planes. Through careful measurements of distances and time intervals, he was able to show that the displacement of an object starting from rest is proportional to the square of the time the object is in motion. This observation is consistent with one of the kinematic equations we derived for motion under constant acceleration (Eq. 2.10). Galileo's achievements in the science of mechanics paved the way for Newton in his development of the laws of motion.

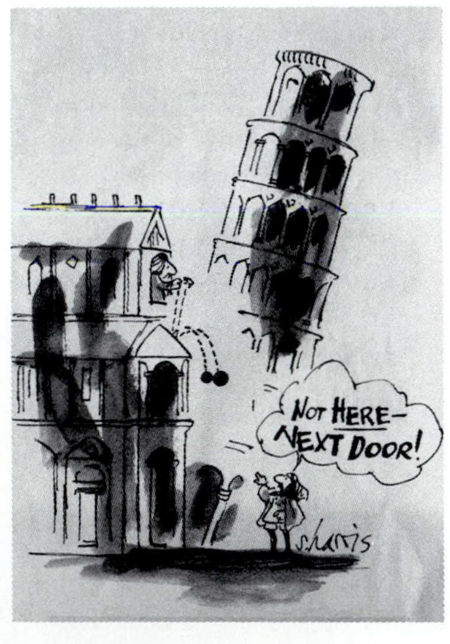

You might want to try the following experiment. Drop a coin and a crumpled-up piece of paper simultaneously from the same height. In the absence of air resistance, the two experience the same motion and hit the floor at the same time. In a real (nonideal) experiment, air resistance cannot be neglected. In the idealized case, however, where air resistance *is* neglected, such motion is referred to as *free fall.* If this same experiment could be conducted in a good vacuum, where air friction is truly negligible, the paper and coin would fall with the same acceleration, regardless of the shape or weight of the paper. This point is illustrated very convincingly on page 31, in the photograph of the apple and feather falling in a vacuum. On August 2, 1971, such an experiment was conducted on the Moon by astronaut David Scott. He simultaneously released a geologist's hammer and a falcon's feather, and in unison they fell to the lunar surface. This demonstration surely would have pleased Galileo!

We shall denote the free-fall acceleration with the symbol **g**. The magnitude of **g** decreases with increasing altitude. Furthermore, there are slight variations in the magnitude of **g** with latitude. However, at the surface of the Earth the magnitude of **g** is approximately 9.80 m/s^2, or 980 cm/s^2, or 32 ft/s^2. Unless stated otherwise, we shall use the value 9.80 m/s^2 when doing calculations. Furthermore, we shall assume that the vector **g** is directed downward toward the center of the Earth.

Free-fall acceleration $g = 9.80$ m/s^2

When we use the expression *freely falling body,* we do not necessarily mean an object dropped from rest.

> A freely falling body is an object moving freely under the influence of gravity only, regardless of its initial motion. Objects thrown upward or downward and those released from rest are all falling freely once they are released!

It is important to emphasize that any freely falling object experiences an acceleration directed downward, as shown in the multiflash photograph of a falling billiard ball (Fig. 2.12). This is true regardless of the initial motion of the object. An object thrown upward (or downward) experiences the same acceleration as an object released from rest.

> Once they are in free fall, all objects have a downward acceleration equal to the free-fall acceleration.

If we neglect air resistance and assume that the gravitational acceleration does not vary with altitude, then the motion of a freely falling body is equivalent to

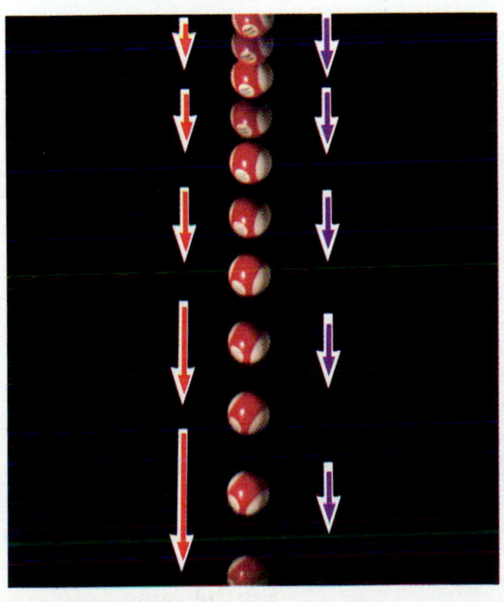

(a)

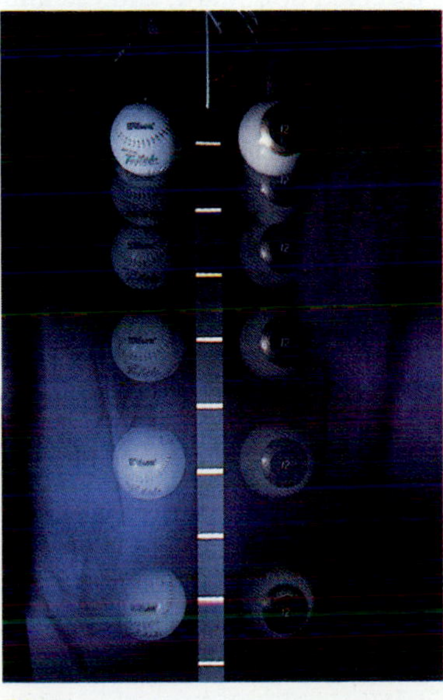

(b)

Figure 2.12 *(Left)* A multiflash photograph of a falling billiard ball. As the ball falls, the spacing between successive images increases, indicating that the ball accelerates downward. The motion diagram shows that the ball's velocity (red arrows) increases with time while its acceleration (violet arrows) remains constant. *(Right)* A multiflash photograph of two freely falling balls released simultaneously. The larger ball on the left is a shot put (mass 5.4 kg), and the ball on the right is a baseball (mass 0.23 kg). The distance between markers is 10 cm. Note that although the objects have different masses, they fall at the same rate. (Over short distances such as this, air resistance is negligible.) (a, © Richard Megna 1990, Fundamental Photographs. b, Courtesy of Henry Leap and Jim Lehman) motion in one dimension under constant acceleration. Therefore, the equations developed in Section 2.5 for objects moving with constant acceleration can be applied. The only modification that we need to make in these equations for freely falling bodies is to note that the motion is in the vertical direction (the y direction) rather than the horizontal direction (the x direction), and that the acceleration is downward and has a magnitude of 9.80 m/s². Thus, for a freely falling body we always take $a = -g = -9.80 \text{ m/s}^2$, where the minus sign means that the acceleration of the body is downward.

Example 2.8 Look Out Below!

A golf ball is released from rest at the top of a very tall building. Neglecting air resistance, calculate the position and velocity of the ball after 1, 2, and 3 s.

Solution We choose our coordinates so that the starting point of the ball is at the origin ($y_0 = 0$ at $t = 0$) and remember that we have defined y to be positive upward. Since $v_0 = 0$, and $a = -g = -9.80 \text{ m/s}^2$, Equations 2.7 and 2.10 become

$$v = at = (-9.80 \text{ m/s}^2)t$$

$$y = \tfrac{1}{2}at^2 = \tfrac{1}{2}(-9.80 \text{ m/s}^2)t^2$$

where t is in seconds, v is in meters per second, and y is in meters. These expressions give the velocity and displacement at any time t after the ball is released. Therefore, at $t = 1$ s,

$$v = (-9.80 \text{ m/s}^2)(1 \text{ s}) = \boxed{-9.80 \text{ m/s}}$$

$$y = \tfrac{1}{2}(-9.80 \text{ m/s}^2)(1 \text{ s})^2 = \boxed{-4.90 \text{ m}}$$

Likewise, at $t = 2$ s, we find that $v = -19.6$ m/s and $y = -19.6$ m. Finally, at $t = 3$ s, $v = -29.4$ m/s and $y = -44.1$ m. The minus signs for v indicate that the velocity vector is directed downward, and the minus signs for y indicate displacement in the negative y direction.

Exercise Calculate the position and velocity of the ball after 4 s.

Answer −78.4 m, −38.4 m/s.

Example 2.9 Not a Bad Throw for a Rookie!

A stone is thrown from the top of a building with an initial velocity of 20 m/s straight upward. The building is 50 m high, and the stone just misses the edge of the roof on its way down, as in Figure 2.13. Determine (a) the time needed for the stone to reach its maximum height, (b) the maximum height, (c) the time needed for the stone to return to the level of the thrower, (d) the velocity of the stone at this instant, and (e) the velocity and position of the stone at $t = 5$ s.

Solution

(a) To find the time necessary to reach the maximum height, use Equation 2.7, $v = v_0 + at$, noting that $v = 0$ at maximum height:

$$20 \text{ m/s} + (-9.80 \text{ m/s}^2)t_1 = 0$$

$$t_1 = \frac{20 \text{ m/s}}{9.80 \text{ m/s}^2} = \boxed{2.04 \text{ s}}$$

(b) This value of time can be substituted into Equation 2.10, $y = v_0 t + \frac{1}{2}at^2$, to give the maximum height measured from the position of the thrower:

$$y_{max} = (20 \text{ m/s})(2.04 \text{ s}) + \tfrac{1}{2}(-9.80 \text{ m/s}^2)(2.04 \text{ s})^2 = \boxed{20.4 \text{ m}}$$

(c) When the stone is back at the height of the thrower, the y coordinate is zero. From the expression $y = v_0 t + \frac{1}{2}at^2$ (Eq. 2.10), with $y = 0$, we obtain the expression

$$20t - 4.9t^2 = 0$$

This is a quadratic equation and has two solutions for t. The equation can be factored to give

$$t(20 - 4.9t) = 0$$

One solution is $t = 0$, corresponding to the time the stone starts its motion. The other solution—the one we are after—is $t = 4.08$ s.

(d) The value for t found in (c) can be inserted into $v = v_0 + at$ (Eq. 2.7) to give

$$v = 20 \text{ m/s} + (-9.80 \text{ m/s}^2)(4.08 \text{ s}) = \boxed{-20.0 \text{ m/s}}$$

Note that the velocity of the stone when it arrives back at its original height is equal in magnitude to its initial velocity but opposite in direction. This indicates that the motion is symmetric.

(e) From $v = v_0 + at$ (Eq. 2.7), the velocity after 5 s is

$$v = 20 \text{ m/s} + (-9.80 \text{ m/s}^2)(5 \text{ s}) = \boxed{-29.0 \text{ m/s}}$$

We can use $y = v_0 t + \frac{1}{2}at^2$ (Eq. 2.10) to find the position of the particle at $t = 5$ s:

$$y = (20 \text{ m/s})(5 \text{ s}) + \tfrac{1}{2}(-9.80 \text{ m/s}^2)(5 \text{ s})^2 = \boxed{-22.5 \text{ m}}$$

Exercise Find the velocity of the stone just before it hits the ground.

Answer -37 m/s.

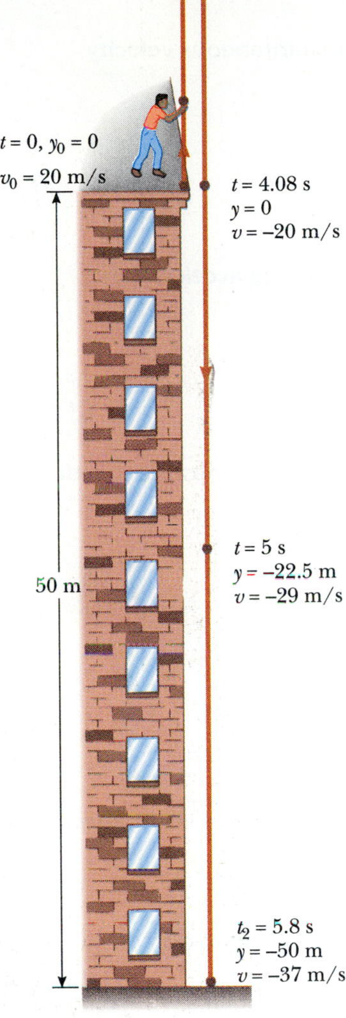

Figure 2.13 (Example 2.9) Position and velocity versus time for a freely falling particle initially thrown upward with a velocity $v_0 = 20$ m/s. (Not drawn to scale.)

Summary

The **average velocity** of a particle during some time interval is equal to the ratio of the displacement vector, $\Delta \mathbf{x}$, and the time interval, Δt:

$$\overline{\mathbf{v}} \equiv \frac{\Delta \mathbf{x}}{\Delta t} \qquad [2.1]$$

Average velocity

The **instantaneous velocity** of a particle is defined as the limit of the ratio $\Delta \mathbf{x}/\Delta t$ as Δt approaches zero.

Instantaneous velocity

$$v \equiv \lim_{\Delta t \to 0} \frac{\Delta x}{\Delta t} = \frac{dx}{dt} \qquad [2.3]$$

The **speed** of a particle is defined as the magnitude of the instantaneous velocity vector.

The **average acceleration** of a particle during some time interval is defined as the ratio of the change in its velocity, Δv, and the time interval, Δt:

Average acceleration

$$\bar{a} \equiv \frac{\Delta v}{\Delta t} \qquad [2.4]$$

The **instantaneous acceleration** is equal to the limit of the ratio $\Delta v/\Delta t$ as $\Delta t \to 0$. By definition, this equals the derivative of v with respect to t, or the time rate of change of the velocity:

Instantaneous acceleration

$$a \equiv \lim_{\Delta t \to 0} \frac{\Delta v}{\Delta t} = \frac{dv}{dt} \qquad [2.5]$$

The slope of the tangent to the x versus t curve at any instant gives the instantaneous velocity of the particle.

The slope of the tangent to the v versus t curve gives the instantaneous acceleration of the particle.

The **equations of kinematics** for a particle moving along the x axis with uniform acceleration a (constant in magnitude and direction) are

Equations of kinematics

$$v = v_0 + at \qquad [2.7]$$

$$x - x_0 = \tfrac{1}{2}(v_0 + v)t \qquad [2.9]$$

$$x - x_0 = v_0 t + \tfrac{1}{2}at^2 \qquad [2.10]$$

$$v^2 = v_0^2 + 2a(x - x_0) \qquad [2.11]$$

(constant a only)

Freely falling body

A body falling freely experiences an acceleration directed toward the center of the Earth. If air friction is neglected, and if the altitude of the motion is small compared with the Earth's radius, then one can assume that the free-fall acceleration, g, is constant over the range of motion, where g is equal to 9.80 m/s^2, or 32 ft/s^2. Assuming y to be positive upward, the acceleration is given by $-g$, and the equations of kinematics for a body in free fall are the same as those already given, with the substitutions $x \to y$ and $a \to -g$.

Questions and Conceptual Exercises

1. Average velocity and instantaneous velocity are generally different quantities. Can they ever be equal for a specific type of motion? Explain.
2. If the average velocity is nonzero for some time interval, does this mean that the instantaneous velocity is never zero during this interval? Explain.
3. If the average velocity equals zero for some time interval Δt and if $v(t)$ is a continuous function, show that the instantaneous velocity must go to zero some time in this interval. (A sketch of x versus t might be useful in your proof.)
4. Is it possible to have a situation in which the velocity and acceleration have opposite signs? If so, sketch a velocity-time graph to prove your point.
5. If the velocity of a particle is nonzero, can its acceleration ever be zero? Explain.
6. If the velocity of a particle is zero, can its acceleration be nonzero? Explain.

7. Can the equations of kinematics (Eqs. 2.7 through 2.11) be used in a situation where the acceleration varies in time? Can they be used when the acceleration is zero?
8. A ball is thrown vertically upward. What are its velocity and acceleration when it reaches its maximum altitude? What is its acceleration just before it strikes the ground?
9. A stone is thrown upward from the top of a building. Does the stone's displacement depend on the location of the origin of the coordinate system? Does the stone's velocity depend on the origin? (Assume that the coordinate system is stationary with respect to the building.) Explain.
10. A child throws a marble in the air with an initial speed v_0. Another child drops a ball at the same instant. Compare the accelerations of the two objects while they are in flight.
11. A student at the top of a building of height h throws one ball upward with an initial speed v_0 and then throws a second ball downward with the same initial speed. How do the final velocities of the balls compare when they reach the ground?
12. Can the instantaneous velocity of an object ever be greater in magnitude than the average velocity? Can it ever be less?
13. If a car is traveling eastward, can its acceleration be westward? Explain.
14. If the average velocity of an object is zero in some time interval, what can you say about the displacement of the object for that interval?
15. Two cars are moving in the same direction in parallel lanes along a highway. At some instant, the velocity of car A exceeds the velocity of car B. Does this mean that the acceleration of A is greater than that of B? Explain.
16. A ball is thrown upward. While the ball is in the air, (a) does its acceleration increase, decrease, or remain constant? (b) Describe what happens to its velocity.
17. Car A traveling south from New York to Miami has a speed of 25 m/s. Car B traveling west from New York to Chicago also has a speed of 25 m/s. Are their velocities equal? Explain.
18. The motion of the Earth's crustal plates is described by a model referred to as *Plate tectonic motion.* Measurements indicate that coastal portions of southern California have northward plate tectonic motions of 2.5 cm per year. Estimate the time it would take for this motion to carry southern California to Alaska.
19. Galileo experimented with balls rolling down inclined planes in order to reduce the acceleration along the plane and thus reduce the rate of descent of the balls. Suppose the angle that the incline makes with the horizontal is θ. How would you expect the acceleration along the plane to decrease as θ decreases? What specific trigonometric dependence on θ would you expect for the acceleration?
20. A ball rolls in a straight line along the horizontal direction. Using motion diagrams (or multiflash photographs) as in Figure 2.10, describe the velocity and acceleration of the ball for each of the following situations: (a) The ball moves to the right at a constant speed. (b) The ball moves from right to left and continually slows down. (c) The ball moves from right to left and continually speeds up. (d) The ball moves to the right, first speeding up at a constant rate, and then slowing down at a constant rate.
21. A rapidly growing plant doubles in height each week. At the end of the 25th day, the plant reaches the height of the building. At what time was the plant one fourth the height of the building?
22. A pebble is dropped into a water well, and the splash is heard 16 s later, as illustrated in the cartoon strip on p. 57. What is the *approximate* distance from the rim of the well to the water's surface?

Problems

Section 2.1 Average Velocity

1. The position of a pinewood derby car was observed at various times; the results are summarized in the following table. Find the average velocity of the car for (a) the first second, (b) the last 3 seconds, and (c) the entire period of observation.

x (m)	0	2.3	9.2	20.7	36.8	57.5
t (s)	0	1.0	2.0	3.0	4.0	5.0

2. A motorist drives north for 35 minutes at 85 km/h and then stops for 15 minutes. He then continues north, traveling 130 km in 2 h. (a) What is his total displacement? (b) What is his average velocity?
3. The displacement versus time for a certain particle moving along the x axis is shown in Figure 2.14. Find the average velocity in the time intervals (a) 0 to 2 s, (b) 0 to 4 s, (c) 2 s to 4 s, (d) 4 s to 7 s, (e) 0 to 8 s.
4. A jogger runs in a straight line, with an average speed of 5 m/s for 4 min, and then with an average speed of 4 m/s for 3 min. (a) What is her total displace-

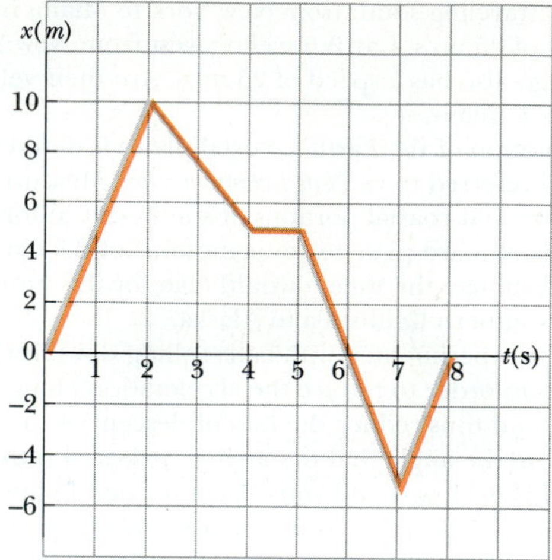

Figure 2.14 (Problem 3)

ment? (b) What is her average speed during this time?

5. An athlete swims the length of a 50-m pool in 20 s and makes the return trip to the starting position in 22 s. Determine his average velocity in (a) the first half of the swim, (b) the second half of the swim, and (c) the round trip.

Section 2.2 Instantaneous Velocity

6. A position-time graph for a particle moving along the x axis is shown in Figure 2.15. (a) Find the average velocity in the time interval $t = 1.5$ s to $t = 4$ s. (b) Determine the instantaneous velocity at $t = 2$ s by measuring the slope of the tangent line shown in the graph. (c) At what value of t is the velocity zero?

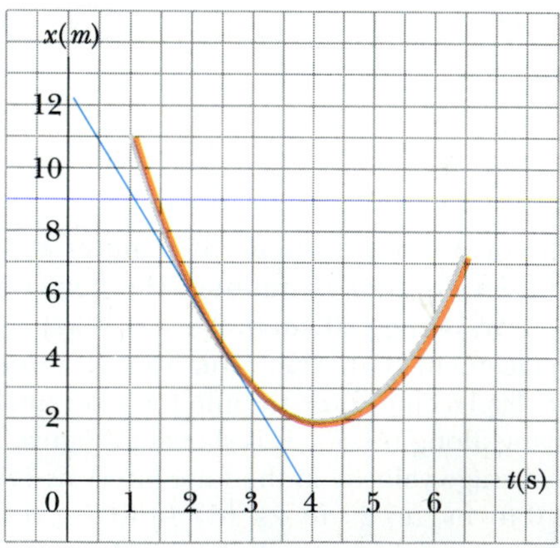

Figure 2.15 (Problem 6)

7. At $t = 1$ s, a particle moving with constant velocity is located at $x = -3$ m, and at $t = 6$ s the particle is located at $x = 5$ m. (a) From this information, plot the position as a function of time. (b) Determine the velocity of the particle from the slope of this graph.

8. (a) Use the data in Problem 1 to construct a smooth graph of position versus time. (b) By constructing tangents to the $x(t)$ curve, find the instantaneous velocity of the car at several instants. (c) Plot the instantaneous velocity versus time and, from this, determine the average acceleration of the car. (d) What was the initial velocity of the car?

9. Find the instantaneous velocity of the particle described in Figure 2.14 at the following times: (a) $t = 1$ s, (b) $t = 3$ s, (c) $t = 4.5$ s, and (d) $t = 7.5$ s.

10. The position-time graph for a particle moving along the z axis is as shown in Figure 2.16. Determine whether the velocity is positive, negative, or zero at times (a) t_1, (b) t_2, (c) t_3, (d) t_4.

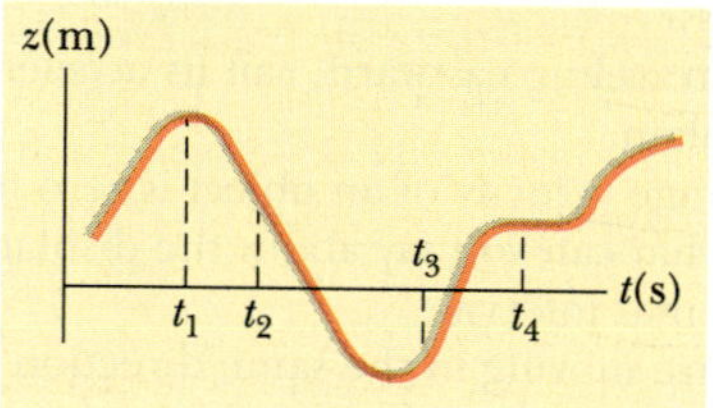

Figure 2.16 (Problem 10)

Section 2.3 Acceleration

11. A particle is moving with a velocity $\mathbf{v}_0 = 60\mathbf{i}$ m/s at $t = 0$. Between $t = 0$ and $t = 15$ s the velocity decreases uniformly to zero. What was the acceleration during this 15-s interval? What is the significance of the sign of your answer?

12. A 50-g superball traveling at 25 m/s bounces off a brick wall and rebounds at 22 m/s. A high-speed camera records this event. If the ball is in contact with the wall for 3.5 ms, what is the magnitude of the average acceleration of the ball during this time interval? (Note: 1 ms = 10^{-3} s.)

13. A velocity-time graph for an object moving along the x axis is shown in Figure 2.17. (a) Plot a graph of the acceleration versus time. (b) Determine the average acceleration of the object in the time intervals $t = 5$ s to $t = 15$ s and $t = 0$ to $t = 20$ s.

14. The velocity of a particle as a function of time is

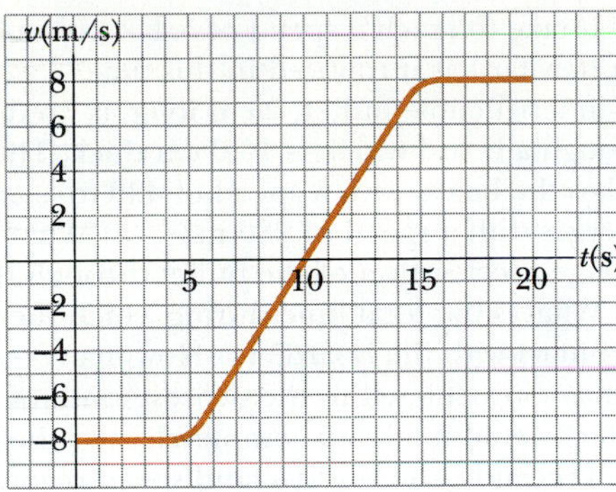

Figure 2.17 (Problem 13)

shown in Figure 2.18. At $t = 0$, the particle is at $x = 0$. (a) Sketch the acceleration as a function of time. (b) Determine the average acceleration of the particle in the time interval $t = 2$ s to $t = 8$ s. (c) Determine the instantaneous acceleration of the particle at $t = 4$ s.

15. A particle moves along the x axis according to the equation $x = 2 + 3t - t^2$, where x is in meters and t is in seconds. At $t = 3$ s, find (a) the position of the particle, (b) its velocity, and (c) its acceleration.

16. When struck by a club, a golf ball initially at rest acquires a speed of 31 m/s. If the ball is in contact with the club for 1.17 ms, what is the magnitude of the average acceleration of the ball?

17. Figure 2.19 shows a graph of v versus t for the motion of a motorcyclist as he starts from rest and moves along the road in a straight line. (a) Find the average acceleration for the time interval $t_0 = 0$ to $t_1 = 6$ s. (b) Estimate the time at which the acceleration has its greatest positive value and the value of the acceleration at that instant. (c) When is the acceleration zero? (d) Estimate the maximum negative value of the acceleration and the time at which it occurs.

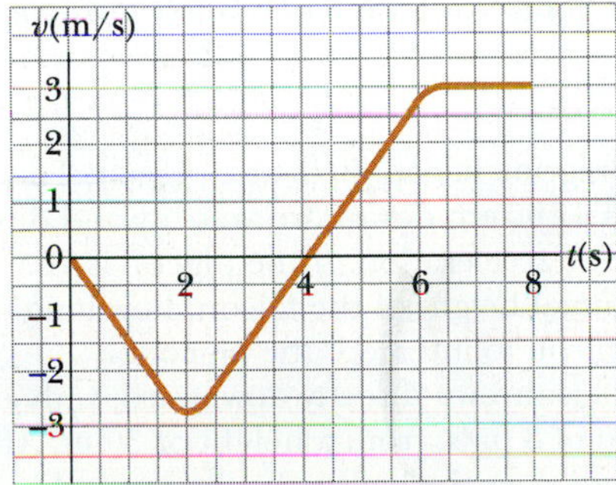

Figure 2.18 (Problem 14)

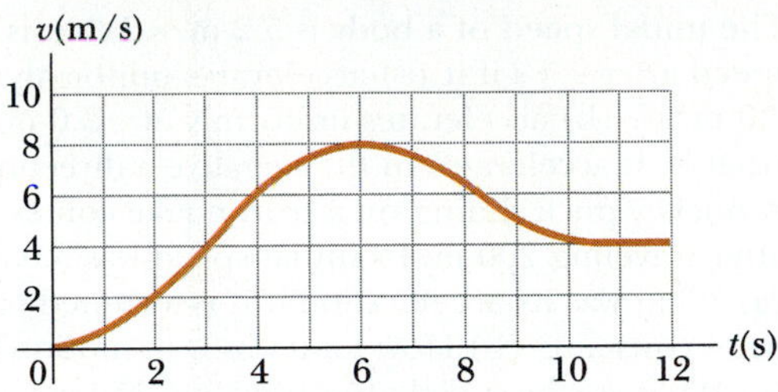

Figure 2.19 (Problem 17)

Section 2.4 Motion Diagrams
Section 2.5 One-Dimensional Motion with Constant Acceleration

18. A particle travels in the positive x direction for 10 s at a constant speed of 50 m/s. It then accelerates uniformly to a speed of 80 m/s in the next 5 s. Find (a) the average acceleration of the particle in the first 10 s, (b) its average acceleration in the interval $t = 10$ s to $t = 15$ s, (c) the total displacement of the particle between $t = 0$ and $t = 15$ s, and (d) its average speed in the interval $t = 10$ s to $t = 15$ s.

19. A body moving with uniform acceleration has a velocity of 12 cm/s in the positive x direction when its x coordinate is 3 cm. If its x coordinate 2 s later is −5 cm, what is the magnitude of its acceleration?

20. Figure 2.20 represents part of the performance data of a car owned by a proud physics student. (a) Calculate from the graph the total distance traveled. (b) What distance does the car travel between the times $t = 10$ s and $t = 40$ s? (c) Draw a graph of its acceleration versus time between $t = 0$ and $t = 50$ s. (d) Write an equation for x as a function of time for each phase of the motion represented by (i) *Oa*, (ii) *ab*, (iii) *bc*. (e) What is the average velocity of the car between $t = 0$ and $t = 50$ s?

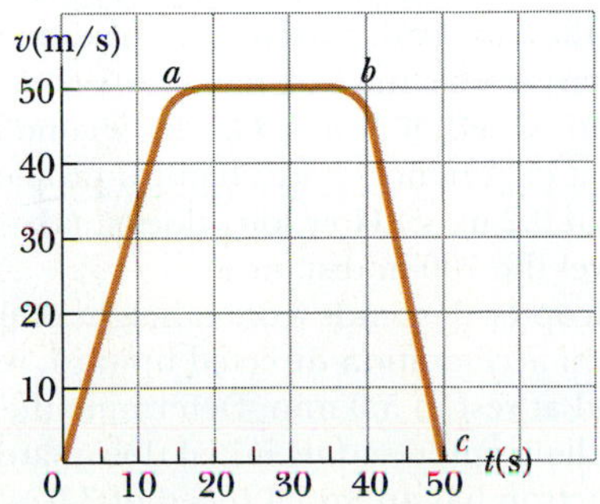

Figure 2.20 (Problem 20)

21. The initial speed of a body is 5.2 m/s. What is its speed after 2.5 s if it (a) accelerates uniformly at 3.0 m/s^2? (b) accelerates uniformly at -3.0 m/s^2 (that is, it accelerates in the negative x direction)?
22. A hockey puck sliding on a frozen lake comes to rest after traveling 200 m. Its initial speed was 3.0 m/s. (a) What was its acceleration if it is assumed to have been constant? (b) How long was it in motion? (c) What was its speed after traveling 150 m?
23. A jet plane lands with a speed of 100 m/s and can accelerate at a maximum rate of -5.0 m/s^2 as it comes to rest. (a) From the instant the plane touches the runway, what is the minimum time needed before it can come to rest? (b) Can this plane land on a small tropical island airport where the runway is 0.80 km long?
24. A car and train move together along parallel paths at 25 m/s. The car then undergoes a uniform acceleration of -2.5 m/s^2 because of a red light and comes to rest. It remains at rest for 45 s, then accelerates back to a speed of 25 m/s at a rate of 2.5 m/s^2. How far behind the train is the car when it reaches the speed of 25 m/s, assuming that the speed of the train has remained 25 m/s?
25. A drag racer starts her car from rest and accelerates at 10 m/s^2 for the entire distance of 400 m ($\frac{1}{4}$ mile). (a) How long did it take the race car to travel this distance? (b) What is the speed of the race car at the end of the run?
26. A locomotive slows from a speed of 26 m/s to zero in 18 s. What distance does it travel?
27. An advertisement claims that a certain brand of disc brake can stop a car traveling at a speed of 88 km/h in 5.0 s. Determine the acceleration and compare this to the acceleration due to gravity.
28. Two express trains begin moving 5 minutes apart. Starting from rest, each is capable of a maximum speed of 160 km/h after uniformly accelerating over a distance of 2.0 km. (a) What is the acceleration of each train? (b) How far ahead is the first train when the second one starts? (c) How far apart are they when they are both traveling at maximum speed?
29. A go-cart travels the first half of a 100-m track with a constant speed of 5 m/s. On the second half of the track, it experiences a mechanical problem and slows down at 0.2 m/s^2. How long does it take the go-cart to travel the 100-m distance?
30. A helicopter descends from a height of 600 m with uniform acceleration directed upward, reaching the ground at rest in 5.0 min. Determine the acceleration of the helicopter and its initial downward velocity.
31. An electron has an initial speed of 3.0×10^5 m/s. If it undergoes an acceleration of 8.0×10^{14} m/s^2 in the direction of the electron's motion, (a) how long will it take to reach a speed of 5.4×10^5 m/s, and (b) how far has it traveled in this time?
32. An indestructible bullet, 2 cm long, is fired straight through a board which is 10.0 cm thick. The bullet strikes the board with a speed of 420 m/s and emerges with a speed of 280 m/s. (a) What is the average acceleration of the bullet through the board? (b) What is the total time during which the bullet is in contact with the board? (c) How many thicknesses of board (calculated to the nearest tenth of a centimeter) would it take to completely stop the bullet?
33. Until recently, the world's land speed record was held by Colonel John P. Stapp, USAF. On March 19, 1954, he rode a rocket-propelled sled that moved down the track at 632 mi/h. He and the sled were safely brought to rest in 1.4 s. Determine (a) the negative acceleration he experienced and (b) the distance he traveled during this negative acceleration.

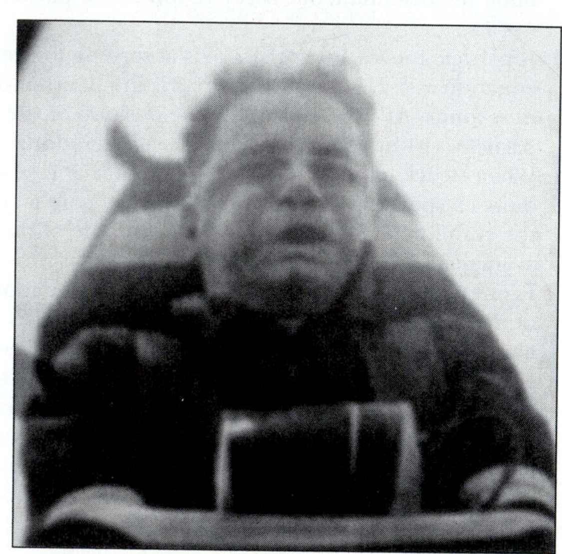

Figure 2.21 (Problem 33).
Col. John Stapp on rocket sled. *(Photri, Inc.)*

34. A peregrine falcon dives at a pigeon. The falcon starts with zero downward velocity and dives with the free-fall acceleration. If the pigeon is 76.0 m below the initial height of the falcon, how long does it take the falcon to intercept the pigeon?
35. A ball is thrown directly downward, with an initial speed of 8 m/s, from a height of 30 m. After what interval does the ball strike the ground?
36. A student throws a set of keys vertically upward to her sorority sister, who is in a window 4.0 m above. The

keys are caught 1.5 s later by the sister's outstretched hand. (a) With what initial velocity were the keys thrown? (b) What was the velocity of the keys just before they were caught?

37. A hot air balloon is traveling vertically upward at a constant speed of 5.0 m/s. When it is 21.0 m above the ground, a package is released from the balloon. (a) For how long after being released is the package in the air? (b) What is the velocity of the package just before impact with the ground? (c) Repeat (a) and (b) for the case of the balloon descending at 5.0 m/s.

(Problem 37). Hot air balloon over Carefree, Arizona. *(Russell Schlepman)*

38. A ball is thrown vertically upward from the ground with an initial speed of 15 m/s. (a) How long does it take the ball to reach its maximum altitude? (b) What is its maximum altitude? (c) Determine the velocity and acceleration of the ball at $t = 2$ s.

39. A ball thrown vertically upward is caught by the thrower after 20 s. Find (a) the initial velocity of the ball and (b) the maximum height it reaches.

40. A baseball is hit so that it travels straight upward after being struck by the bat. A fan observes that it takes 3 s for the ball to reach its maximum height. Find (a) its initial velocity and (b) the height reached by the ball. Ignore the effects of air resistance.

41. An astronaut standing on the Moon drops a hammer, letting it fall 1 m to the surface. The lunar gravity produces a constant acceleration of 1.62 m/s^2. Upon returning to Earth, the astronaut again drops the hammer, letting it fall to the ground from a height of 1 m with an acceleration of 9.80 m/s^2. Compare the times of fall in the two situations.

42. A stone falls from rest from the top of a high cliff. A second stone is thrown downward from the same height 2.0 s later, with an initial speed of 30 m/s. If the two stones hit the ground below simultaneously, how high is the cliff?

43. A daring cowboy sitting on a tree limb wishes to drop vertically onto a horse galloping under the tree. The speed of the horse is 10 m/s, and the distance from the limb to the saddle is 3 m. (a) What must be the horizontal distance between the saddle and limb when the cowboy makes his move? (b) How long is the cowboy in the air?

Additional Problems

44. A motorist is traveling at 18.0 m/s when he sees a deer in the road 38.0 m ahead. (a) If the maximum negative acceleration of the vehicle is -4.5 m/s^2, what is the maximum reaction time, Δt, of the motorist that will allow him to avoid hitting the deer? (b) If his reaction time is 0.30 s, how fast will he be traveling when he hits the deer?

45. The position of a softball tossed vertically upward is described by the equation $y = 7t - 4.9t^2$, where y is in meters and t in seconds. Find (a) the initial speed v_0 at $t_0 = 0$, (b) the velocity at $t = 1.26$ s, and (c) the acceleration of the ball.

46. An inquisitive physics student and mountain climber climbs a 50-m cliff that overhangs a calm pool of water. He throws two stones vertically downward, 1 s apart, and observes that they cause a single splash. The first stone has an initial speed of 2 m/s. (a) How long after release of the first stone do the two stones hit the water? (b) What initial velocity must the second stone have if they are to hit simultaneously? (c) What is the speed of each stone at the instant the two hit the water?

47. A "superball" is dropped from a height of 2 m above the ground. On the first bounce the ball reaches a height of 1.85 m, where it is caught. Find the velocity of the ball (a) just as it makes contact with the ground and (b) just as it leaves the ground on the bounce. (c) Neglecting the time the ball spends in contact with the ground, find the total time required

for the ball to go from the dropping point to the point where it is caught.

48. A Cessna 150 aircraft has a lift-off speed of about 125 km/h. (a) What minimum constant acceleration does this require if the aircraft is to be airborne after a take-off run of 250 m? (b) What is the corresponding take-off time? (c) If the aircraft continues to accelerate at this rate, what speed will it reach 25 s after it begins to roll?

49. One runner covered the 100-m dash in 10.3 s. Another runner came in second at a time of 10.8 s. Assuming that the runners traveled at their average speeds for the entire distance, determine the separation between the two runners when the winner crossed the finish line.

50. A falling object requires 1.50 s to travel the last 30 m before hitting the ground. From what height above the ground did it fall?

51. A young woman named Kathy Kool buys a sports car that can accelerate at the rate of 16 ft/s^2. She decides to test the car by dragging with another speedster, Stan Speedy. Both start from rest, but experienced Stan leaves the starting line 1 s before Kathy. If Stan moves with a constant acceleration of 12 ft/s^2 and Kathy maintains an acceleration of 16 ft/s^2, find (a) the time it takes Kathy to overtake Stan, (b) the distance she travels before she catches him, and (c) the speeds of both cars at the instant she overtakes him.

52. A hockey player takes a slap shot at a puck at rest on the ice. The puck glides over the ice for 10 ft without friction, at which point it runs over rough ice. The puck then accelerates at a uniform rate of -20 ft/s^2. If the velocity of the puck is 40 ft/s after the puck has traveled 100 ft from the point of impact, (a) what is the average acceleration imparted to the puck as it is struck by the hockey stick? (Assume that the duration of contact is 0.01 s.) (b) How far, in all, does the puck travel before coming to rest? (c) What is the total time the puck is in motion, neglecting contact time?

53. A speeder passes a parked police car at 105 km/h. The police car starts from rest with a uniform acceleration of 2.44 m/s^2. How far does the speeder get before being overtaken by the police car?

54. In 1987 Art Boileau won the Los Angeles Marathon, 26 miles and 385 yards, in 2 h, 13 min, 9 s. (a) Find his average speed in meters per second and in miles per hour. (b) At the 21-mile marker, Boileau had a 2.50-min lead on the second-place runner, who later crossed the finish line 30 s after Boileau. Assume that Boileau maintained his constant average speed and that the two runners were running at the same speed when Boileau passed the 21-mile marker. Find the average acceleration (in meters per second squared) that the second-place contestant had during the remaining race, after Boileau passed the 21-mile marker.

55. A rock is dropped from rest into a well. (a) If the sound of the splash is heard 2.40 s later, how far below the top of the well is the surface of the water? The speed of sound in air (for the existing temperature) is 336 m/s. (b) If the travel time for the sound is neglected, what percentage error is introduced when the depth of the well is calculated?

56. A rocket is fired vertically upward with an initial velocity of 80 m/s. It accelerates upward at 4 m/s^2 until it reaches an altitude of 1000 m. At that point, its engines fail and the rocket goes into free flight, with an acceleration of -9.80 m/s^2. (a) How long is the rocket in motion? (b) What is its maximum altitude? (c) What is its velocity just before it collides with the Earth? (*Hint:* Consider the motion while the engine is operating separate from the free-fall motion.)

57. In a 100-m race, Maggie and Judy cross the finish line in a dead heat, both taking 10.2 s. Accelerating uniformly, Maggie took 2.0 s and Judy 3.0 s to attain maximum speed, which they maintained for the rest of the race. (a) What was the acceleration of each sprinter? (b) What were their respective maximum speeds? (c) Which sprinter was ahead at the 6-s mark, and by how much?

58. In the first hour of travel, a train moves with a speed v, in the next half hour it has a speed $3v$, in the next 90 min it travels with a speed $v/2$, and in the final 2 h it travels with a speed $v/3$. (a) Plot the speed-time graph for this trip. (b) How far does the train travel in this trip? (c) What is the average speed of the train over the entire trip?

59. The engineer on a commuter train can minimize the time t between two stations by accelerating (a_1 = 0.1 m/s^2) for a time t_1, then causing a negative acceleration ($a_2 = -0.5$ m/s^2) by using his brakes for a time t_2. Since the stations are only 1 km apart, the train never reaches its maximum velocity. Find the minimum time of travel t, and the time t_1.

60. In order to protect his food from hungry bears, a boy scout raises his food pack, of mass m, with a rope that is thrown over a tree limb at height h above his hands. He walks away from the vertical rope with constant velocity v_0, holding the free end of the rope in his hands (Fig. 2.22). (a) Show that the velocity v of the food pack is $x(x^2 + h^2)^{-1/2}v_0$ where x is the distance he has walked away from the vertical rope. (b) Show that the acceleration a of the food pack is $h^2(x^2 + h^2)^{-3/2}v_0{}^2$. (c) What values do the acceleration and velocity v have shortly after the boy scout

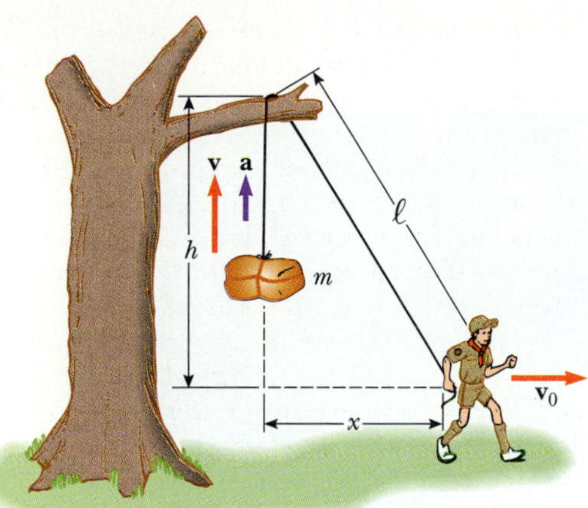

Figure 2.22 (Problem 60)

leaves the vertical rope? (d) What values do the velocity and acceleration approach as the distance x continues to increase?

61. Two objects, A and B, are connected by a rigid rod that has a length L. The objects slide along perpendicular guide rails, as shown in Figure 2.23. If A slides to the left with a constant speed v, find the velocity of B when $\alpha = 60°$.

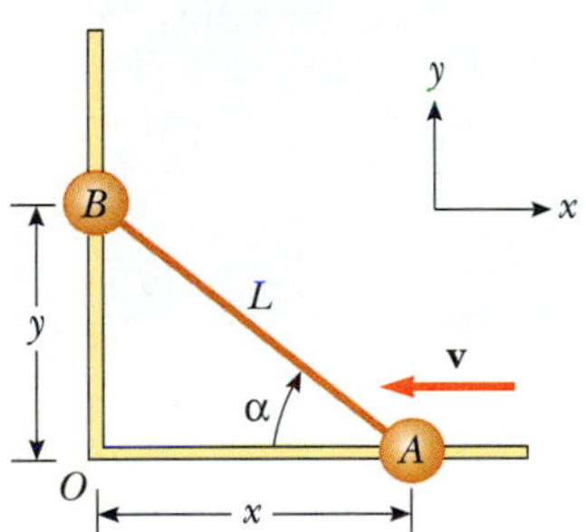

Figure 2.23 (Problem 61)

Calculator/Computer Problems

62. In Problem 60, let the height h equal 6 m and the velocity v_0 equal 2 m/s. Assume that the food pack starts from rest on a ledge over a cliff 6 m below the boy scout's hands. (a) Tabulate and plot the velocity-time graph. (b) Tabulate and plot the acceleration-time graph. (Let the range of time be from 0 s to 6 s and the time intervals be 0.5 s.)

63. A particle undergoes a varying acceleration. The velocity is measured at 0.5-s intervals and is tabulated as follows. (a) Determine the average acceleration in each interval. (b) Use a numerical integration procedure to determine the position of the particle at the end of each time interval. Assume the initial position of the particle is zero.

t (s)	0	0.5	1.0	1.5	2.0	2.5	3.0	3.5	4.0	4.5	5.0
v (m/s)	0	1	3	4.5	7.0	9.5	10.5	12	14	15	17.5

64. The acceleration of a particle moving along the x axis varies with position according to the expression

$$a = a_0 e^{-bx}$$

where $a_0 = 3\ \text{m/s}^2$ and $b = 1\ \text{m}^{-1}$. If the particle starts at rest from the origin, use a numerical integration method to find the position of the particle at $t = 2.37$ s. The accuracy of your calculation should be at least 1%.

65. A particle moving along the x axis undergoes an acceleration given by $a = \sqrt{3 + t^3}\ \text{m/s}^2$. Use a numerical integration method to find the position and velocity of the particle at $t = 5.7$ s, to within 1% accuracy.

See question 22

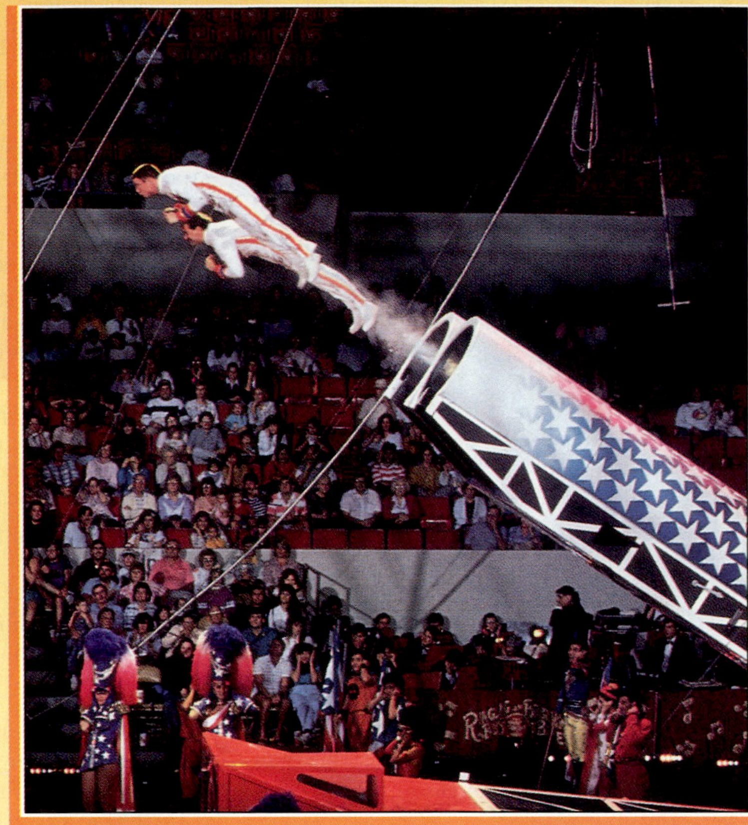

The circus stuntmen being shot out of cannons are human projectiles. Neglecting air resistance, they move in parabolic paths until they land in a strategically placed net. What initial condition(s) will determine where the catching net should be placed? *(Ringling Brothers Circus)*

Motion in Two Dimensions

CHAPTER 3

In this chapter we deal with the kinematics of a particle moving in a plane, or two-dimensional motion. Some common examples of motion in a plane are the motions of projectiles and satellites and the motion of charged particles in uniform electric fields. We begin by showing that velocity and acceleration are vector quantities. As in the case of one-dimensional motion, we shall derive the kinematic equations for two-dimensional motion from the fundamental definitions of displacement, velocity, and acceleration. As special cases of motion in two dimensions, we shall treat motion in a plane with constant acceleration and uniform circular motion.

3.1 The Displacement, Velocity, and Acceleration Vectors

In Chapter 2 we found that a particle's motion along a straight line is completely known if its position is known as a function of time. Now let us extend this idea to the motion of a particle in the *xy* plane. We begin by describing the position of a particle with a *position vector* **r**, drawn from the origin of some reference frame to the particle located in the *xy* plane, as in Figure 3.1. At time t_i the particle is at the point *P*, and at some later time t_f the particle is at *Q*. As the particle moves from *P* to *Q* in the time interval $\Delta t = t_f - t_i$, the position vector changes from $\mathbf{r}_i$ to $\mathbf{r}_f$. Because $\mathbf{r}_f = \mathbf{r}_i + \Delta\mathbf{r}$, the **displacement vector** for the particle is given by

Definition of the displacement vector

$$\Delta\mathbf{r} \equiv \mathbf{r}_f - \mathbf{r}_i \qquad [3.1]$$

The direction of $\Delta\mathbf{r}$ is indicated in Figure 3.1. Note that the displacement vector equals the difference between the final position vector and the initial position vector. As we see from Figure 3.1, the magnitude of the displacement vector is *less* than the distance traveled whenever the particle's path from *P* to *Q* is curved. When the path is a straight line, however, the magnitude of the displacement is equal to the distance traveled.

We now define the average velocity of the particle during the time interval Δt as the ratio of the displacement to the time interval for this displacement:

Average velocity

$$\bar{\mathbf{v}} \equiv \frac{\Delta\mathbf{r}}{\Delta t} \qquad [3.2]$$

Since the displacement is a vector and the time interval is a scalar, we conclude that the average velocity is a *vector* quantity directed along $\Delta\mathbf{r}$. Note that the average velocity between points *P* and *Q* is *independent of the path* between the two points. This is because the average velocity is proportional to the displacement, which in turn depends only on the initial and final position vectors. As in the case of one-dimensional motion, we conclude that if a particle starts its motion at some point and returns to this point via any path, its average velocity for this trip is zero since its displacement is zero.

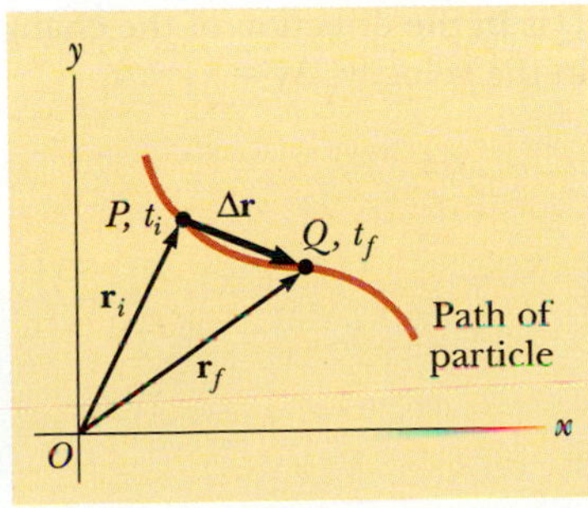

Figure 3.1
A particle moving in the *xy* plane is located with the position vector **r**, drawn from the origin to the particle. The displacement of the particle as it moves from *P* to *Q* in the time interval $\Delta t = t_f - t_i$ is equal to the vector $\Delta\mathbf{r} = \mathbf{r}_f - \mathbf{r}_i$.

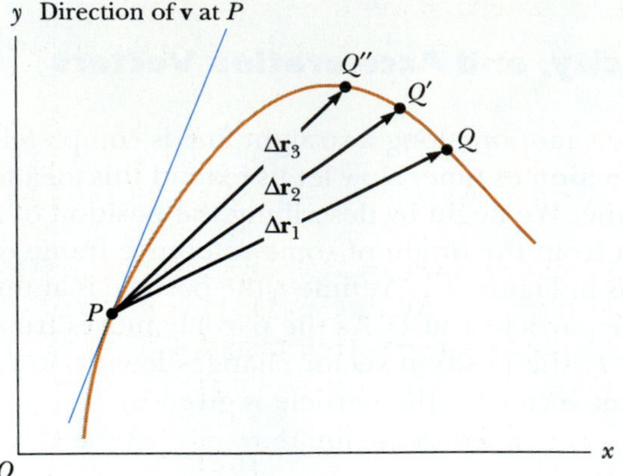

Figure 3.2
As a particle moves between two points, its average velocity is in the direction of the displacement vector $\Delta\mathbf{r}$. As the point Q moves closer to P, the direction of $\Delta\mathbf{r}$ approaches that of the line tangent to the curve at P. By definition, the instantaneous velocity at P is in the direction of this tangent line.

Consider again the motion of a particle between two points in the *xy* plane, as in Figure 3.2. As the time intervals become smaller and smaller, the displacements, $\Delta\mathbf{r}_1$, $\Delta\mathbf{r}_2$, $\Delta\mathbf{r}_3$, . . . , get progressively smaller and the direction of the displacement approaches that of the line tangent to the path at the point P.

The **instantaneous velocity, v,** is defined as the limit of the average velocity, $\Delta\mathbf{r}/\Delta t$, as Δt approaches zero:

Instantaneous velocity

$$\mathbf{v} \equiv \lim_{\Delta t \to 0} \frac{\Delta\mathbf{r}}{\Delta t} = \frac{d\mathbf{r}}{dt} \qquad [3.3]$$

That is, the instantaneous velocity equals the derivative of the position vector with respect to time. The direction of the velocity vector is along a line that is tangent to the path of the particle and in the direction of motion. This is illustrated in Figure 3.3 for two points along the path. The magnitude of the instantaneous velocity vector is called the *speed.* Note that Equation 3.3 is a logical application of differentiation as developed in the study of calculus.

As the particle moves from P to Q along some curved path, its instantaneous velocity vector changes from $\mathbf{v}_i$ at time t_i to $\mathbf{v}_f$ at time t_f (Figure 3.3).

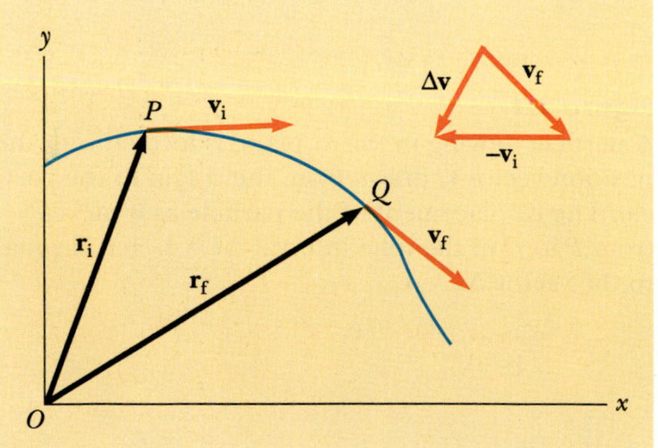

Figure 3.3
The average acceleration vector, $\bar{\mathbf{a}}$, for a particle moving from P to Q is in the direction of the change in the velocity, $\Delta\mathbf{v} = \mathbf{v}_f - \mathbf{v}_i$.

The **average acceleration** of the particle as it moves from P to Q is defined as the ratio of the change in the instantaneous velocity vector, $\Delta\mathbf{v}$, to the elapsed time, Δt:

$$\overline{\mathbf{a}} \equiv \frac{\mathbf{v}_\mathrm{f} - \mathbf{v}_\mathrm{i}}{t_f - t_i} = \frac{\Delta\mathbf{v}}{\Delta t} \qquad [3.4]$$

Average acceleration

Since the average acceleration is the ratio of a vector, $\Delta\mathbf{v}$, and a scalar, Δt, we conclude that $\overline{\mathbf{a}}$ is a vector quantity directed along $\Delta\mathbf{v}$. As is indicated in Figure 3.3, the direction of $\Delta\mathbf{v}$ is found by adding the vector $-\mathbf{v}_\mathrm{i}$ (the negative of $\mathbf{v}_\mathrm{i}$) to the vector $\mathbf{v}_\mathrm{f}$, since by definition $\Delta\mathbf{v} = \mathbf{v}_\mathrm{f} - \mathbf{v}_\mathrm{i}$.

The **instantaneous acceleration, a,** is defined as the limiting value of the ratio $\Delta\mathbf{v}/\Delta t$ as Δt approaches zero:

$$\mathbf{a} \equiv \lim_{\Delta t \to 0} \frac{\Delta\mathbf{v}}{\Delta t} = \frac{d\mathbf{v}}{dt} \qquad [3.5]$$

Instantaneous acceleration

In other words, the instantaneous acceleration equals the first derivative of the velocity vector with respect to time.

It is important to recognize that a particle can accelerate in several ways. First, the magnitude of the velocity vector (the speed) may change with time. Second, a particle accelerates when the direction of the velocity vector changes with time (a curved path) even though the speed may be constant. Finally, acceleration may be due to a change in both the magnitude and the direction of the velocity vector.

3.2 Motion in Two Dimensions with Constant Acceleration

Consider the two-dimensional motion of a particle with constant acceleration (that is, we assume that the magnitude and direction of the acceleration remain unchanged during the motion). A particle in motion can be described by its position vector, $\mathbf{r}$. The position vector for a particle moving in the xy plane can be written

$$\mathbf{r} = x\mathbf{i} + y\mathbf{j} \qquad [3.6]$$

where x, y, and $\mathbf{r}$ change with time as the particle moves, as in Figure 3.3. If the position vector is known, the velocity of the particle can be obtained from Equations 3.3 and 3.6, which give

$$\mathbf{v} = \frac{d\mathbf{r}}{dt} = \frac{dx}{dt}\mathbf{i} + \frac{dy}{dt}\mathbf{j}$$

$$\mathbf{v} = v_x\mathbf{i} + v_y\mathbf{j} \qquad [3.7]$$

Because $\mathbf{a}$ is a constant, its components, a_x and a_y, are also constants. Therefore, we can apply the equations of kinematics to both the x and y components of the velocity vector. Using Equation 2.7 ($v = v_0 + at$), we can substitute $v_x = v_{x0} + a_xt$ and $v_y = v_{y0} + a_yt$ into Equation 3.7 to obtain

$$\mathbf{v} = (v_{x0} + a_x t)\mathbf{i} + (v_{y0} + a_y t)\mathbf{j}$$
$$= (v_{x0}\mathbf{i} + v_{y0}\mathbf{j}) + (a_x\mathbf{i} + a_y\mathbf{j})t$$

Velocity vector as a function of time

$$\mathbf{v} = \mathbf{v}_0 + \mathbf{a}t \quad [3.8]$$

This result states that the velocity of a particle at some time t equals the vector sum of its initial velocity, $\mathbf{v}_0$, and the additional velocity, $\mathbf{a}t$, acquired in the time t as a result of its constant acceleration.

Similarly, from Equation 2.10 we know that the x and y coordinates of a particle moving with constant acceleration are

$$x = x_0 + v_{x0}t + \tfrac{1}{2}a_x t^2 \qquad \text{and} \qquad y = y_0 + v_{y0}t + \tfrac{1}{2}a_y t^2$$

Substituting these expressions into Equation 3.6 gives

$$\mathbf{r} = (x_0 + v_{x0}t + \tfrac{1}{2}a_x t^2)\mathbf{i} + (y_0 + v_{y0}t + \tfrac{1}{2}a_y t^2)\mathbf{j}$$
$$= (x_0\mathbf{i} + y_0\mathbf{j}) + (v_{x0}\mathbf{i} + v_{y0}\mathbf{j})t + \tfrac{1}{2}(a_x\mathbf{i} + a_y\mathbf{j})t^2$$

Position vector as a function of time

$$\mathbf{r} = \mathbf{r}_0 + \mathbf{v}_0 t + \tfrac{1}{2}\mathbf{a}t^2 \quad [3.9]$$

This equation says that the displacement vector $\mathbf{r} - \mathbf{r}_0$ is the vector sum of a displacement $\mathbf{v}_0 t$, arising from the initial velocity of the particle, and a displacement $\frac{1}{2}\mathbf{a}t^2$, resulting from the uniform acceleration of the particle.

Example 3.1 Motion in a Plane

A particle moves in the xy plane with an x component of constant acceleration only, given by $a_x = 4\ \text{m/s}^2$. The particle starts from the origin at $t = 0$, with an initial velocity having an x component of 20 m/s and a y component of −15 m/s.

(a) Determine the components of velocity as a function of time and the total velocity vector at any time.

Solution Since $v_{x0} = 20$ m/s and $a_x = 4\ \text{m/s}^2$, the equations of kinematics give

$$v_x = v_{x0} + a_x t = (20 + 4t)\ \text{m/s}$$

Also, since $v_{y0} = -15$ m/s and $a_y = 0$,

$$v_y = v_{y0} = -15\ \text{m/s}$$

Therefore, using the preceding results and noting that the velocity vector $\mathbf{v}$ has two components, we get

$$\mathbf{v} = v_x\mathbf{i} + v_y\mathbf{j} = [(20 + 4t)\mathbf{i} - 15\mathbf{j}]\ \text{m/s}$$

We could also obtain this result using Equation 3.8 directly, noting that $\mathbf{a} = 4\mathbf{i}\ \text{m/s}^2$ and $\mathbf{v}_0 = (20\mathbf{i} - 15\mathbf{j})$ m/s. Try it!

(b) Calculate the velocity and speed of the particle at $t = 5$ s.

Solution With $t = 5$ s, the result from (a) gives

$$\mathbf{v} = \{[20 + 4(5)]\mathbf{i} - 15\mathbf{j}\}\ \text{m/s} = (40\mathbf{i} - 15\mathbf{j})\ \text{m/s}$$

That is, at $t = 5$ s, $v_x = 40$ m/s and $v_y = -15$ m/s. The speed is defined as the magnitude of **v**, or

$$|\mathbf{v}| = v = \sqrt{v_x^2 + v_y^2} = \sqrt{(40)^2 + (-15)^2}\ \text{m/s} = 42.7\ \text{m/s}$$

(*Note:* v is larger than v_0. Why?)

The angle θ that **v** makes with the x axis can be calculated using the fact that $\tan\theta = v_y/v_x$, or

$$\theta = \tan^{-1}\left(\frac{v_y}{v_x}\right) = \tan^{-1}\left(\frac{-15}{40}\right) = -20.6^\circ$$

(c) Determine the x and y coordinates at any time t and the displacement vector at this time.

Solution Since at $t = 0$, $x_0 = y_0 = 0$, the expressions for the x and y coordinates, the equations of kinematics give

$$x = v_{x0}t + \tfrac{1}{2}a_x t^2 = (20t + 2t^2)\ \text{m}$$

$$y = v_{y0}t = (-15t)\ \text{m}$$

Therefore, the displacement vector at any time t is given by

$$\mathbf{r} = x\mathbf{i} + y\mathbf{j} = [(20t + 2t^2)\mathbf{i} - 15t\mathbf{j}]\ \text{m}$$

Thus, for example, at $t = 5$ s, $x = 150$ m and $y = -75$ m, or $\mathbf{r} = (150\mathbf{i} - 75\mathbf{j})$ m. It follows that the distance of the particle from the origin to this point is the magnitude of the displacement, or

$$|\mathbf{r}| = r = \sqrt{(150)^2 + (-75)^2}\ \text{m} = 168\ \text{m}$$

Note that this is *not* the distance the particle travels in this time. Can you determine this distance from the available data?

Exercise Obtain **r** by applying Equation 3.9 directly, with $\mathbf{v}_0 = (20\mathbf{i} - 15\mathbf{j})$ m/s and $\mathbf{a} = 4\mathbf{i}$ m/s^2.

3.3 Projectile Motion

Assumptions of projectile motion

Anyone who has observed a baseball (or, for that matter, any object thrown into the air) in motion has observed projectile motion. For an arbitrary direction of the initial velocity, the ball moves in a curved path. This very common form of motion is surprisingly simple to analyze if the following two assumptions are made: (1) the free-fall acceleration, **g**, is constant over the range of motion and is directed downward,[1] and (2) the effect of air resistance is negligible.[2] With these assumptions,

[1] This approximation is reasonable as long as the range of motion is small compared with the radius of the Earth (6.4×10^6 m). In effect, this approximation is equivalent to assuming that the Earth is flat within the range of motion considered.

[2] This approximation is generally *not* justified, especially at high velocities. In addition, the spin of a projectile, such as a baseball, can give rise to some very interesting effects associated with aerodynamic forces (for example, a curve thrown by a pitcher).

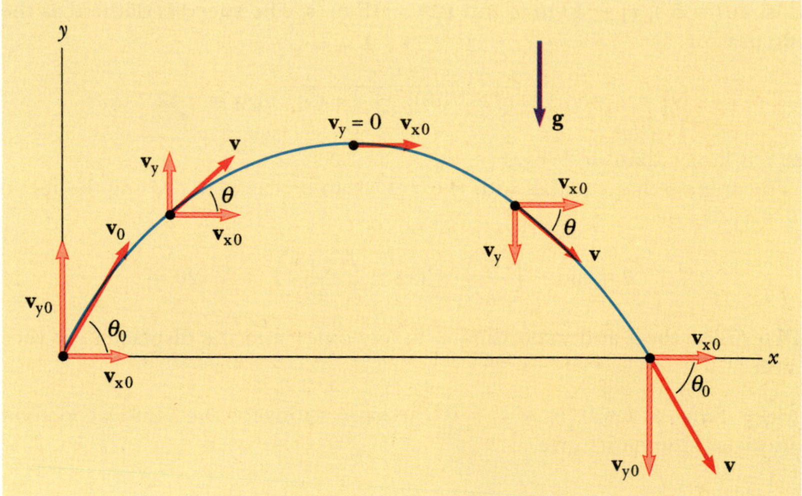

Figure 3.4
The parabolic trajectory of a projectile that leaves the origin with a velocity $\mathbf{v}_0$. Note that the velocity vector, $\mathbf{v}$, changes with time. However, the x component of the velocity vector, $\mathbf{v}_{x0}$, remains constant in time. Also, $\mathbf{v}_y = 0$ at the peak.

we shall find that the path of a projectile, which we call its *trajectory,* is *always* a parabola. *We shall use these assumptions throughout this chapter.*

If we choose our reference frame so that the y direction is vertical and positive upward, then $a_y = -g$ (as in one-dimensional free fall) and $a_x = 0$ (since air friction is neglected). Furthermore, let us assume that at $t = 0$ the projectile leaves the origin ($x_0 = y_0 = 0$) with a velocity $\mathbf{v}_0$, as in Figure 3.4. If the vector $\mathbf{v}_0$ makes an angle θ_0 with the horizontal, where θ_0 is called the projection angle, then from the definitions of the cosine and sine functions and Figure 3.4 we have

$$\cos\theta_0 = v_{x0}/v_0 \qquad \text{and} \qquad \sin\theta_0 = v_{y0}/v_0$$

Therefore, the initial x and y components of velocity are given by

$$v_{x0} = v_0\cos\theta_0 \qquad \text{and} \qquad v_{y0} = v_0\sin\theta_0$$

Substituting these expressions into Equations 3.8 and 3.9 with $a_x = 0$ and $a_y = -g$ gives the velocity components and coordinates for the projectile at any time t:

Horizontal velocity component

$$v_x = v_{x0} = v_0\cos\theta_0 = \text{constant} \qquad [3.10]$$

Vertical velocity component

$$v_y = v_{y0} - gt = v_0\sin\theta_0 - gt \qquad [3.11]$$

Horizontal position component

$$x = v_{x0}t = (v_0\cos\theta_0)t \qquad [3.12]$$

Vertical position component

$$y = v_{y0}t - \tfrac{1}{2}gt^2 = (v_0\sin\theta_0)t - \tfrac{1}{2}gt^2 \qquad [3.13]$$

From Equation 3.10 we see that v_x remains constant in time and is equal to the initial x component of velocity, since there is no horizontal component of acceleration. Also, for the y motion we note that v_y and y are identical to the expressions for the freely falling body discussed in Chapter 2. In fact, *all* of the equations of kinematics developed in Chapter 2 are applicable to projectile motion.

If we solve for t in Equation 3.12 and substitute that expression for t into Equation 3.13, we find that

$$y = (\tan\theta_0)x - \left(\frac{g}{2v_0^2\cos^2\theta_0}\right)x^2 \qquad [3.14]$$

which is valid for the angles in the range $0 < \theta_0 < \pi/2$. This equation is of the form $y = ax - bx^2$, which describes a parabola that passes through the origin. Thus, we have proved that the trajectory of a projectile is a parabola. Note that the trajectory is *completely* specified if v_0 and θ_0 are known.

One can obtain the speed, v, as a function of time for the projectile by noting that Equations 3.10 and 3.11 give the x and y components of velocity at any instant. Therefore, by definition, since v is equal to the magnitude of **v**,

$$v = \sqrt{v_x^2 + v_y^2} \qquad [3.15]$$

Speed

Also, since the velocity vector is tangent to the path at any instant, as shown in Figure 3.4, the angle θ that **v** makes with the horizontal can be obtained from v_x and v_y through the expression

$$\tan\theta = \frac{v_y}{v_x} \qquad [3.16]$$

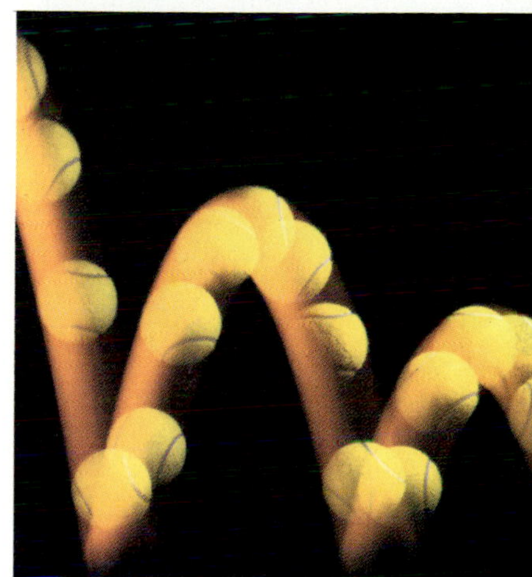

Multiflash photograph of a tennis ball undergoing several bounces off a hard surface. Note the parabolic path of the ball following each bounce. *(© Richard Megna 1992, Fundamental Photographs)*

The vector expression for the position vector as a function of time for the projectile follows directly from Equation 3.9, with $\mathbf{a} = \mathbf{g}$:

$$\mathbf{r} = \mathbf{v}_0 t + \tfrac{1}{2}\mathbf{g}t^2$$

This expression is equivalent to Equations 3.12 and 3.13 and is plotted in Figure 3.5. Note that it is consistent with Equation 3.13, since the expression for **r** is a vector equation and $\mathbf{a} = \mathbf{g} = -g\mathbf{j}$ when the upward direction is taken to be positive. It is interesting to note that the motion can be considered the superposition of the term $\mathbf{v}_0 t$, which would be the displacement if no acceleration were present, and the term $\frac{1}{2}\mathbf{g}t^2$, which arises from the free-fall acceleration. In other words, if there were no acceleration, the particle would continue to move along a straight path in the direction of $\mathbf{v}_0$. Therefore, the vertical distance—$\frac{1}{2}gt^2$—through which the particle "falls," measured from the straight line, is that of a freely falling body. *We conclude that projectile motion is the superposition of two motions: (1) the motion of a freely falling body in the vertical direction with constant acceleration and (2) uniform motion in the horizontal direction with constant velocity.*

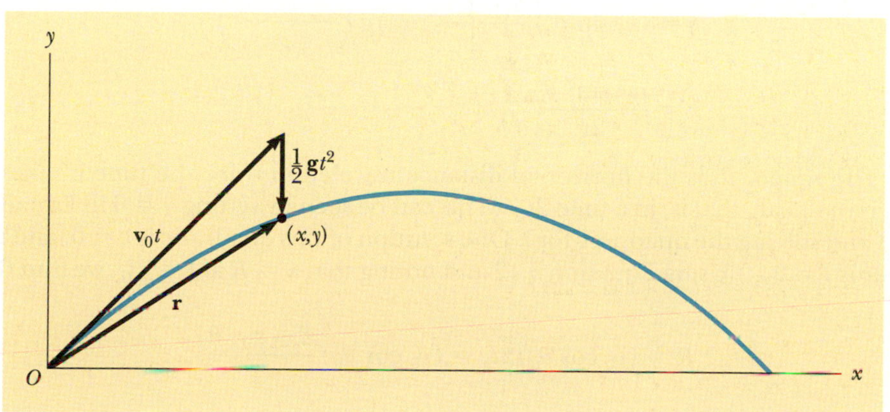

Figure 3.5
The displacement vector, **r**, of a projectile having an initial velocity at the origin of $\mathbf{v}_0$. The vector $\mathbf{v}_0 t$ would be the displacement of the projectile if it experienced no acceleration, and the vector $\frac{1}{2}gt^2$ is its vertical displacement in the time t.

A welder can cut holes through a heavy metal construction beam with a hot torch. Note that the sparks generated in this process follow parabolic paths, as one would expect, since they are projectiles moving with free-fall acceleration. *(The Telegraph Colour Library, FPG International)*

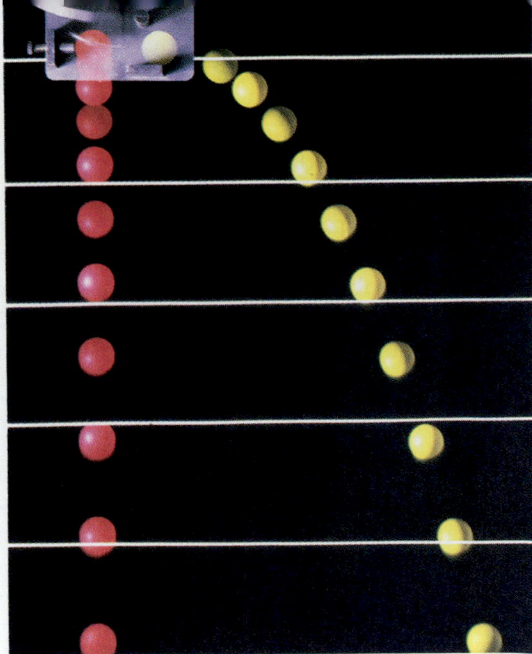

This multiflash photograph of two balls released simultaneously illustrates both free fall (yellow ball) and projectile motion (red ball). Can you explain why both balls reach the floor simultaneously? *(© Richard Megna 1990, Fundamental Photographs)*

Special Case: Horizontal Range and Maximum Height of a Projectile. Let us assume that a projectile is fired from the origin at $t = 0$ with a positive v_y component, as in Figure 3.6. Two special points are interesting to analyze: the peak whose cartesian coordinates are $(R/2, h)$ and the point with coordinates $(R, 0)$. The distance R is called the *horizontal range* of the projectile, and h is its *maximum height.* Let us find h and R in terms of v_0, θ_0, and g.

We can determine h by noting that at the peak, $v_y = 0$. Therefore, Equation 3.11 can be used to determine the time, t_1, it takes to reach the peak:

$$t_1 = \frac{v_0 \sin \theta_0}{g}$$

Substituting this expression for t_1 into Equation 3.13 gives h in terms of v_0 and θ_0:

$$h = (v_0 \sin \theta_0) \frac{v_0 \sin \theta_0}{g} - \tfrac{1}{2} g \left(\frac{v_0 \sin \theta_0}{g} \right)^2$$

Maximum height of projectile

$$h = \frac{v_0{}^2 \sin^2 \theta_0}{2g} \qquad [3.17]$$

The range, R, is the horizontal distance traveled in twice the time it takes to reach the peak, that is, in a time $2t_1$. (This can be seen by setting $y = 0$ in Equation 3.13 and solving the quadratic for t. One solution of this quadratic is $t = 0$, and the other is $t = 2t_1$.) Using Equation 3.12 and noting that $x = R$ at $t = 2t_1$, we find that

$$R = (v_0 \cos \theta_0) 2t_1 = (v_0 \cos \theta_0) \frac{2v_0 \sin \theta_0}{g}$$

$$R = \frac{2v_0^2 \sin \theta_0 \cos \theta_0}{g}$$

Since $\sin 2\theta = 2 \sin \theta \cos \theta$, R can be written in the more compact form

$$R = \frac{v_0^2 \sin 2\theta_0}{g} \qquad [3.18]$$

Range of projectile

Keep in mind that Equations 3.17 and 3.18 are useful for calculating h and R only if v_0 and θ_0 are known and only for a symmetric path, as shown in Figure 3.6 (which means that only $\mathbf{v}_0$ has to be specified). The general expressions given by Equations 3.10 through 3.13 are the *most important* results, since they give the coordinates and velocity components of the projectile at *any* time t.

The maximum value of R from Equation 3.18 is $R_{max} = v_0^2/g$. This result follows from the fact that the maximum value of $\sin 2\theta_0$ is unity, which occurs when $2\theta_0 = 90°$. Therefore, we see that R is a maximum when $\theta_0 = 45°$, as would be expected if air friction were neglected.

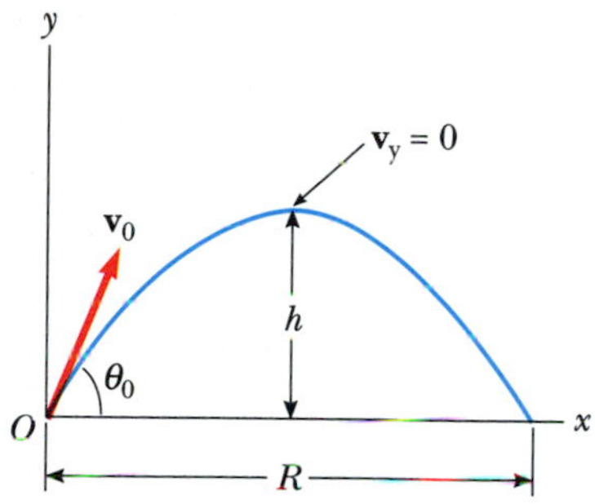

Figure 3.6
A projectile fired from the origin at $t = 0$ with an initial velocity $\mathbf{v}_0$. The maximum height of the projectile is h, and its horizontal range is R.

Before we look at some numerical examples dealing with projectile motion, let us pause to summarize what we have learned so far about this kind of motion:

1. Provided air resistance is negligible, the horizontal component of velocity, v_x, remains constant since there is no horizontal component of acceleration.
2. The vertical component of acceleration is equal to the free-fall acceleration, $\mathbf{g}$.
3. The vertical component of velocity, v_y, and the displacement in the y direction are identical to those of a freely falling body.
4. Projectile motion can be described as a superposition of the two motions in the x and y directions.

Problem-Solving Strategy: Projectile Motion

We suggest that you use the following approach to solving projectile motion problems:

1. **Select a coordinate system.**
2. **Resolve the initial velocity vector into x and y components.**
3. **Treat the horizontal motion and the vertical motion independently.**
4. **Follow the techniques for solving problems with constant velocity to analyze the horizontal motion of the projectile.**
5. **Follow the techniques for solving problems with constant acceleration to analyze the vertical motion of the projectile.**

Example 3.2 The Long Jump

A long-jumper leaves the ground at an angle of 20° to the horizontal and at a speed of 11 m/s.

(a) How far does he jump? (Assume that the motion of the long-jumper is equivalent to that of a particle.)

Solution His horizontal motion is described by using Equation 3.12:

$$x = (v_0 \cos \theta_0)t = (11 \text{ m/s})(\cos 20°)t$$

The value of x can be found if t, the total time of the jump, is known. We can find t using the expression $v_y = v \sin \theta_0 - gt$ by noting that at the top of the jump the vertical component of velocity goes to zero:

$$v_y = v_0 \sin \theta_0 - gt$$

$$0 = (11 \text{ m/s}) \sin 20° - (9.80 \text{ m/s}^2)t_1$$

$$t_1 = 0.384 \text{ s}$$

Note that t_1 is the time interval required to reach the *top* of the jump. Because of the symmetry of the vertical motion, an identical time interval passes before the jumper returns to the ground. Therefore, the *total time* in the air is $t = 2t_1 = 0.768$ s. Substituting this value for t into the expression for x gives

$$x = (11 \text{ m/s})(\cos 20°)(0.768 \text{ s}) = 7.94 \text{ m}$$

(b) What is the maximum height reached?

Solution The maximum height reached is found using Equation 3.13, with $t = t_1 = 0.384$ s:

$$y_{max} = (v_0 \sin \theta_0)t_1 - \tfrac{1}{2}gt_1^2$$

$$y_{max} = (11 \text{ m/s})(\sin 20°)(0.384 \text{ s}) - \tfrac{1}{2}(9.80 \text{ m/s}^2)(0.384 \text{ s})^2 = 0.722 \text{ m}$$

The assumption that the motion of the long-jumper is that of a projectile is an oversimplification of the situation. Nevertheless, the values obtained are reasonable.

Exercise Use Equations 3.17 and 3.18 to find the maximum height and horizontal range of the long jumper.

Example 3.3 Just a "Stone's Throw" From Here

A stone is thrown from the top of a building upward, at an angle of 30° to the horizontal and with an initial speed of 20 m/s, as in Figure 3.7. The height of the building is 45 m.

(a) How long is the stone "in flight"?

Solution The initial x and y components of the velocity are

$$v_{x0} = v_0 \cos \theta_0 = (20 \text{ m/s})(\cos 30°) = 17.3 \text{ m/s}$$

$$v_{y0} = v_0 \sin \theta_0 = (20 \text{ m/s})(\sin 30°) = 10 \text{ m/s}$$

To find t, we can use $y = v_{y0}t - \frac{1}{2}gt^2$ (Eq. 3.13) with $y = -45$ m and $v_{y0} = 10$ m/s (we have chosen the top of the building as the origin, as shown as in Figure 3.7):

$$-45 \text{ m} = (10 \text{ m/s})t - \tfrac{1}{2}(9.80 \text{ m/s}^2)t^2$$

Solving the quadratic equation for t gives, for the positive root, $t = 4.22$ s. Does the negative root have any physical meaning? (Can you think of another way of finding t from the information given?)

(b) What is the speed of the stone just before it strikes the ground?

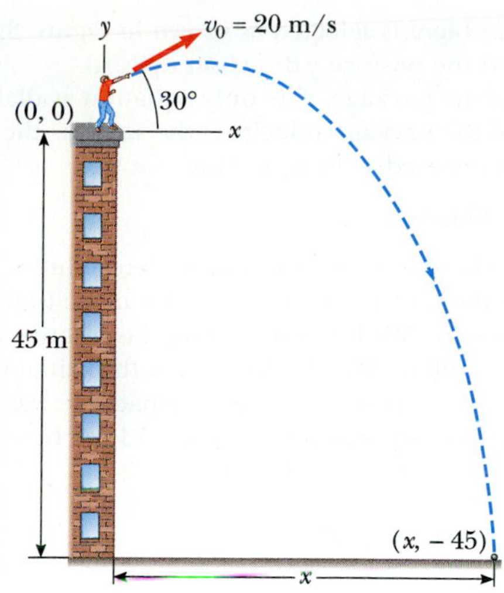

Figure 3.7 (Example 3.3)

Solution The y component of the velocity just before the stone strikes the ground can be obtained using the equation $v_y = v_{y0} - gt$ (Eq. 3.11) with $t = 4.22$ s:

$$v_y = 10 \text{ m/s} - (9.80 \text{ m/s}^2)(4.22 \text{ s}) = -31.4 \text{ m/s}$$

Since $v_x = v_{x0} = 17.3$ m/s, the required speed is given by

$$v = \sqrt{v_x^2 + v_y^2} = \sqrt{(17.3)^2 + (-31.4)^2} \text{ m/s} = \boxed{35.9 \text{ m/s}}$$

Exercise Where does the stone strike the ground?

Answer 73 m from the base of the building.

Example 3.4 The Stranded Explorers

An Alaskan rescue plane drops a package of emergency rations to a stranded explorer, as shown in Figure 3.8. If the plane is traveling horizontally 100 m above the ground, at 40 m/s, where does the package strike the ground relative to the point at which it was released?

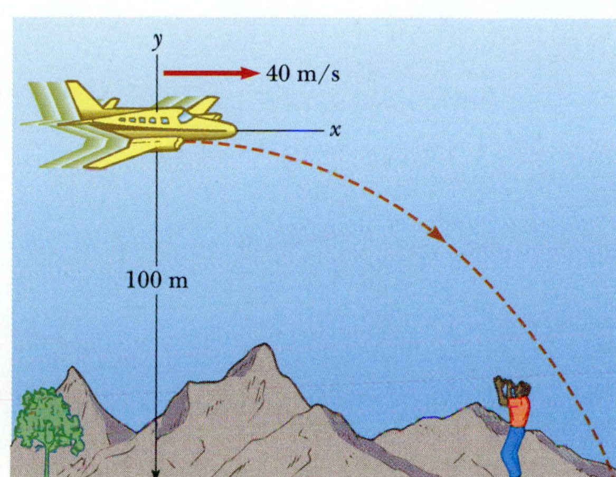

Figure 3.8 (Example 3.4) To an observer on the ground, a package released from the rescue plane travels along the path shown.

Solution The coordinate system for the problem is selected as shown in Figure 3.8, with the positive x direction to the right and the positive y direction upward.

Consider first the horizontal motion of the package. The only equation available to us is $x = v_{x0}t$. The initial x component of the package velocity is the same as the velocity of the plane when the package was released—40 m/s. Thus, we have

$$xv_{x0}t = (40 \text{ m/s})t$$

If we know t, the length of time the package is in the air, we can determine x, the distance traveled by the package along the horizontal. To find t, we move to the equations for the vertical motion of the package. We know that at the instant the package hits the ground its y coordinate is -100 m. We also know that the initial velocity of the package in the vertical direction, v_{y0}, is zero because the package was released with only a horizontal component of velocity. From Equation 3.13, we have

$$y = -\tfrac{1}{2}gt^2$$

$$-100 \text{ m} = -\tfrac{1}{2}(9.80 \text{ m/s}^2)t^2$$

$$t^2 = 20.4 \text{ s}^2$$

$$t = 4.51 \text{ s}$$

The value for the time of flight substituted into the equation for the x coordinate gives

$$x = (40 \text{ m/s})(4.51 \text{ s}) = 180 \text{ m}$$

Exercise What are the horizontal and vertical components of the velocity of the package just before it hits the ground?

Answer $v_x = 40$ m/s; $v_y = -44.1$ m/s.

3.4 Uniform Circular Motion

Figure 3.9a shows an object moving in a circular path with *constant linear speed v*. It is often surprising to students to find that *even though the object moves at a constant speed, it still has an acceleration.* To see why, consider the defining equation for average acceleration, $\bar{\mathbf{a}} = \Delta\mathbf{v}/\Delta t$.

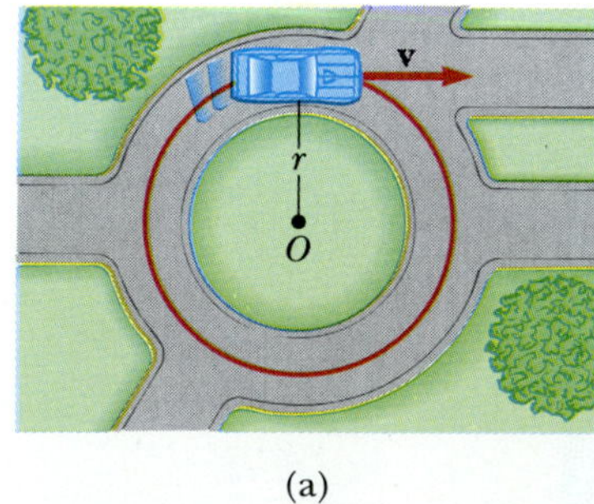

(a)

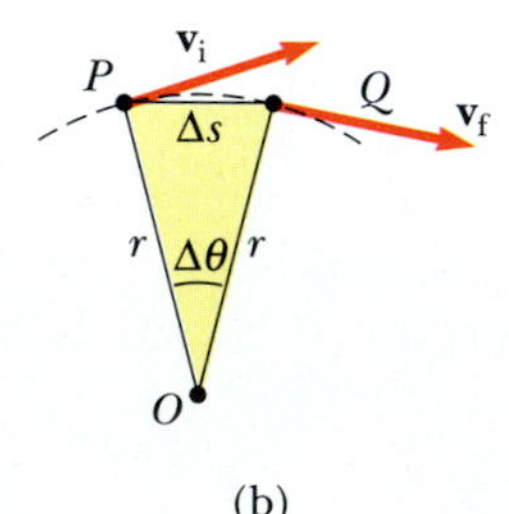

(b)

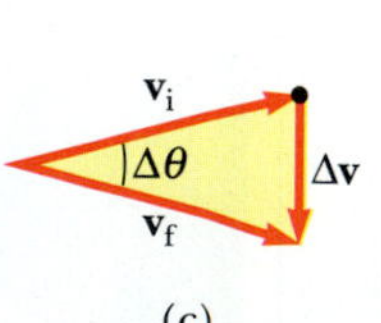

(c)

Figure 3.9 (a) Circular motion of an object moving with a constant speed. (b) As the particle moves from P to Q, the direction of its velocity vector changes from $\mathbf{v}_i$ to $\mathbf{v}_f$. (c) The construction for determining the direction of the change in velocity, $\Delta\mathbf{v}$, which is toward the center of the circle.

Note that the acceleration depends on *the change in the velocity vector.* Because velocity is a vector, there are two ways in which an acceleration can be produced: by a change in the *magnitude* of the velocity and by a change in the *direction* of the velocity. It is the latter situation that occurs for an object moving in a circular path with constant speed. The velocity vector is always tangent to the path of the particle. We shall show that the acceleration vector in this case is perpendicular to the path and always points toward the center of the circle. An acceleration of this nature is called a **centripetal** (center-seeking) **acceleration,** and its magnitude is given by

$$a_r = \frac{v^2}{r} \qquad [3.19]$$

Magnitude of centripetal acceleration

To derive Equation 3.19, consider Figure 3.9b. Here an object is seen first at point P with velocity $\mathbf{v}_i$ at time t_i, then at point Q with velocity $\mathbf{v}_f$ at a later time t_f. Let us also assume here that $\mathbf{v}_i$ and $\mathbf{v}_f$ differ only in direction; their magnitudes are the same (that is, $v_i = v_f = v$). In order to calculate the acceleration, let us begin with the defining equation for average acceleration (Eq. 3.4):

$$\bar{\mathbf{a}} = \frac{\mathbf{v}_f - \mathbf{v}_i}{t_f - t_i} = \frac{\Delta \mathbf{v}}{\Delta t}$$

This equation indicates that we must vectorially subtract $\mathbf{v}_i$ from $\mathbf{v}_f$, where $\Delta\mathbf{v} = \mathbf{v}_f - \mathbf{v}_i$ is the change in the velocity. Since $\mathbf{v}_i + \Delta\mathbf{v} = \mathbf{v}_f$, the vector $\Delta\mathbf{v}$ can be found graphically as shown by the vector triangle in Figure 3.9c. Note that when Δt is very small, Δs and $\Delta\theta$ are also very small. In this case, $\mathbf{v}_f$ will be almost parallel to $\mathbf{v}_i$, and the vector $\Delta\mathbf{v}$ will be approximately perpendicular to them, pointing toward the center of the circle.

Now consider the triangle in Figure 3.9b, which has sides Δs and r. This triangle and the one with sides Δv and v in Figure 3.9c are similar. (Two triangles are *similar* if the angle between any two sides is common to both triangles and if the ratio of lengths of these sides is the same.) This enables us to write a relationship for the lengths of the sides:

$$\frac{\Delta v}{v} = \frac{\Delta s}{r}$$

This equation can be solved for Δv, and the expression so obtained can be substituted into $\bar{a} = \Delta v/\Delta t$ to give $\bar{a}\Delta t = v\,\Delta s/r$, or

$$\bar{a} = \frac{v}{r}\frac{\Delta s}{\Delta t}$$

Now imagine that points P and Q in Figure 3.9b become extremely close together. In this case $\Delta\mathbf{v}$ points toward the center of the circular path, and because the acceleration is in the direction of $\Delta\mathbf{v}$, it too is toward the center. Furthermore, as the two points P and Q approach each other, Δt approaches zero, and the ratio $\Delta s/\Delta t$ approaches the speed v. Hence, in the limit $\Delta t \rightarrow 0$ the acceleration is

$$a_r = \frac{v^2}{r}$$

Thus, we conclude that in uniform circular motion the acceleration is directed inward toward the center of the circle and has magnitude v^2/r. You should show that the dimensions of a_r are L/T^2—as required because this is a true acceleration.

In many situations it is convenient to describe the motion of a particle moving with constant speed in a circle of radius r in terms of the **period,** T, which is defined as the time required for one complete revolution. In the time T the particle moves a distance of $2\pi r$, which is equal to the circumference of the particle's circular path. Therefore, since its speed is equal to the circumference of the circular path divided by the period, or $v = 2\pi r/T$, it follows that

Period

$$T \equiv \frac{2\pi r}{v} \qquad [3.20]$$

3.5 Tangential and Radial Acceleration in Curvilinear Motion

Let us consider the motion of a particle along a curved path where the velocity changes in both direction and magnitude, as described in Figure 3.10. In this situation, the velocity of the particle is always tangent to the path; however, the acceleration vector **a** is now at some angle to the path. As the particle moves along the curved path in Figure 3.10, we see that the direction of the total acceleration vector, **a**, changes from point to point just as it does in circular motion. This vector can be resolved into two component vectors: a radial component vector, $\mathbf{a}_r$, and a tangential component vector, $\mathbf{a}_t$. That is, the *total* acceleration vector, **a**, can be written as the vector sum of these component vectors:

Total acceleration

$$\mathbf{a} = \mathbf{a}_r + \mathbf{a}_t \qquad [3.21]$$

The tangential acceleration arises from the change in the speed of the particle, and its absolute magnitude is given by

Tangential acceleration

$$a_t = \frac{d|\mathbf{v}|}{dt} \qquad [3.22]$$

The radial acceleration is due to the time rate of change in direction of the velocity vector and has an absolute magnitude given by

Centripetal acceleration

$$a_r = \frac{v^2}{r} \qquad [3.23]$$

where r is the radius of curvature of the path at the point in question. Since $\mathbf{a}_r$ and $\mathbf{a}_t$ are perpendicular component vectors of **a**, it follows that $a = \sqrt{a_r^2 + a_t^2}$. As in the case of uniform circular motion, $\mathbf{a}_r$ always points toward the center of curvature, as shown in Figure 3.10. Also, at a given speed, a_r is large when the radius of curvature is small (as at points P and Q in Fig. 3.10) and small when r is large (such as at point R). The direction of $\mathbf{a}_t$ is either the same as that of **v** (if v is increasing) or opposite that of **v** (if v is decreasing).

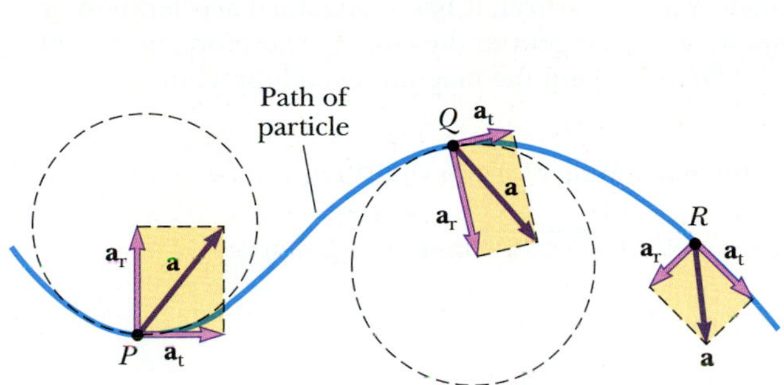

Figure 3.10
The motion of a particle along an arbitrary curved path lying in the *xy* plane. If the velocity vector *v* (always tangent to the path) changes in direction and magnitude, the component vectors of the acceleration of the particle are a tangential vector, $\mathbf{a}_t$, and a radial vector, $\mathbf{a}_r$.

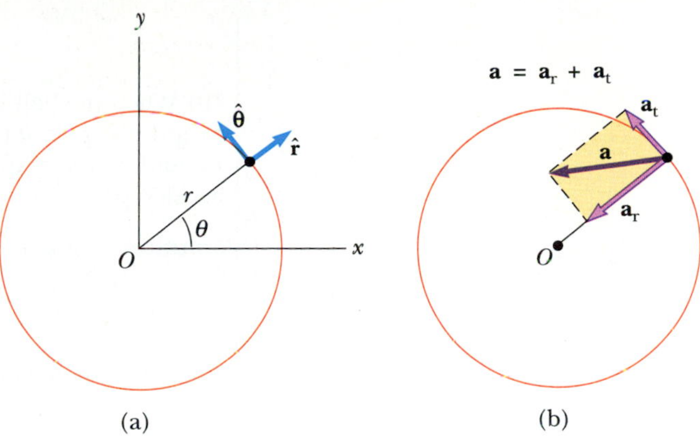

Figure 3.11
(a) Description of the unit vectors $\hat{\mathbf{r}}$ and $\hat{\boldsymbol{\theta}}$. (b) The total acceleration *a* of a particle rotating in a circle consists of a radial component vector, $\mathbf{a}_r$, directed toward the center of rotation, and a tangential component vector, $\mathbf{a}_t$. The component vector $\mathbf{a}_t$ is zero if the speed is constant.

Note that in the case of uniform circular motion, where v is constant, $a_t = 0$ and the acceleration is always radial, as we described in Section 3.4. Furthermore, if the direction of v doesn't change, then there is no radial acceleration and the motion is one-dimensional ($a_r = 0$, $a_t \neq 0$).

It is convenient to write the acceleration of a particle moving in a circular path in terms of unit vectors. We can do this by defining the unit vectors $\hat{\mathbf{r}}$ and $\hat{\boldsymbol{\theta}}$, where $\hat{\mathbf{r}}$ is *a unit vector directed radially outward along the radius vector,* from the center of curvature, and $\hat{\boldsymbol{\theta}}$ is a *unit vector tangent to the circular path,* as in Figure 3.11a. The vectors $\hat{\mathbf{r}}$ and $\hat{\boldsymbol{\theta}}$ are always perpendicular to each other. The direction of $\hat{\boldsymbol{\theta}}$ is that of increasing θ, where θ is measured counterclockwise from the positive *x* axis. Note that both $\hat{\mathbf{r}}$ and $\hat{\boldsymbol{\theta}}$ "move along with the particle" and so vary in time relative to a stationary observer. Using this notation, we can express the total acceleration as

$$\mathbf{a} = \mathbf{a}_t + \mathbf{a}_r = \frac{d|\mathbf{v}|}{dt}\,\hat{\boldsymbol{\theta}} - \frac{v^2}{r}\,\hat{\mathbf{r}} \qquad [3.24]$$

These vectors are described in Figure 3.11b. The negative sign for $\mathbf{a}_r$ indicates that it is always directed radially inward, *opposite* the unit vector $\hat{\mathbf{r}}$.

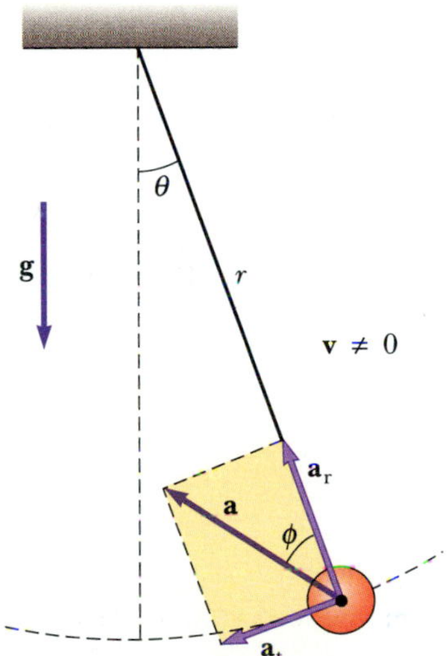

Figure 3.12
(Example 3.6) Circular motion of a ball tied on a string of length *r*. The ball swings in a vertical plane, and its acceleration, a, has a radial component vector, $\mathbf{a}_r$, and a tangential component vector, $\mathbf{a}_t$.

Example 3.5 The Swinging Ball

A ball tied to the end of a string 0.50 m in length swings in a vertical circle under the influence of gravity, as in Figure 3.12. When the string makes an angle of $\theta = 20°$ with the vertical, the ball has a speed of 1.5 m/s.

(a) Find the radial component of acceleration at this instant.

Solution Since $v = 1.5$ m/s and $r = 0.5$ m, we find that

$$a_r = \frac{v^2}{r} = \frac{(1.5\ \text{m/s})^2}{0.50\ \text{m}} = \boxed{4.5\ \text{m/s}^2}$$

(b) When the ball is at an angle θ to the vertical, it has a tangential acceleration of magnitude $g \sin \theta$ (the component of g tangent to the circle). Therefore, at $\theta = 20°$ we find that $a_t = g \sin 20° = 3.36\ \text{m/s}^2$. Find the magnitude and direction of the *total* acceleration at $\theta = 20°$.

Solution Since $\mathbf{a} = \mathbf{a}_r + \mathbf{a}_t$, the magnitude of a at $\theta = 20°$ is given by

$$a = \sqrt{a_r^2 + a_t^2} = \sqrt{(4.5)^2 + (3.36)^2}\ \text{m/s}^2 = \boxed{5.6\ \text{m/s}^2}$$

If ϕ is the angle between $\mathbf{a}$ and the string, then

$$\phi = \tan^{-1}\frac{a_t}{a_r} = \tan^{-1}\left(\frac{3.36\ \text{m/s}^2}{4.5\ \text{m/s}^2}\right) = \boxed{37°}$$

Note that all of the vectors—$\mathbf{a}$, $\mathbf{a}_t$, and $\mathbf{a}_r$—change in direction *and* magnitude as the ball swings through the circle. When the ball is at its lowest elevation ($\theta = 0$), $a_t = 0$ since there is no tangential component of $\mathbf{g}$ at this angle, and a_r is a *maximum* since v is a maximum. When the ball is at its highest position ($\theta = 180°$), a_t is again zero but a_r is a minimum, since v is a minimum. Finally, in the two horizontal positions ($\theta = 90°$ and $270°$), $|\mathbf{a}_t| = g$ and a_r is somewhere between its minimum and maximum values.

3.6 Relative Velocity and Relative Acceleration

Observers in different frames of reference may make different measurements of a moving particle's displacement, velocity, and acceleration. That is, measurements taken by two observers moving with respect to each other generally do not agree.

For example, if two cars are moving in the same direction with speeds of 50 mi/h and 60 mi/h, a passenger in the slower car will claim that the speed of the faster car relative to that of the slower car is 10 mi/h. Of course, a stationary observer will measure the speed of the faster car to be 60 mi/h. This simple example demonstrates that velocity measurements differ in different frames of reference.

Another simple example is a package dropped from an airplane that is flying with constant velocity parallel to the Earth. An observer in the moving airplane would describe the motion of the package as a straight line toward the Earth. On the other hand, an observer on the ground would view the trajectory of the package as a parabola. Relative to the ground, the package has a vertical component of velocity (resulting from its free-fall acceleration and equal to the velocity measured by the observer in the airplane) *and* a horizontal component of velocity (given to it by the airplane's motion). If the airplane continues to move horizontally with the same velocity, the package will hit the ground directly beneath the airplane (assuming friction is neglected)!

To conceptualize a more general situation, consider a particle located at the point P in Figure 3.13. Imagine that the motion of this particle is being described by two observers, one in reference frame S, fixed with respect to the Earth, and another in reference frame S', moving to the right relative to S with a constant

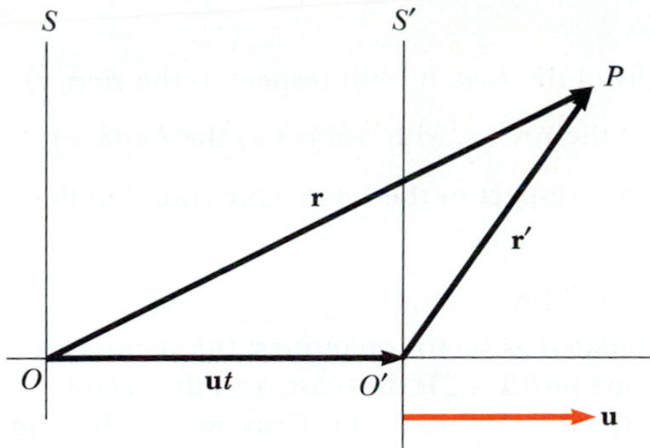

Figure 3.13
A particle located at the point P is described by two observers, one in the fixed frame of reference, S, and the other in the frame S', which moves to the right with a constant velocity $\mathbf{u}$. The vector $\mathbf{r}$ is the particle's position vector relative to S, and $\mathbf{r}'$ is the position vector relative to S'.

velocity $\mathbf{v}$. (Relative to an observer in S', S moves to the left with a velocity $-\mathbf{v}$.) The location of an observer in his own frame of reference is irrelevant in this discussion, but to be definite we can place the observer at the origin.

We label the position of the particle with respect to the S frame with the position vector $\mathbf{r}$ and label its position relative to the S' frame with the vector $\mathbf{r}'$, at some time t. If the origins of the two reference frames coincide at $t = 0$, then the vectors $\mathbf{r}$ and $\mathbf{r}'$ are related to each other through the expression $\mathbf{r} = \mathbf{r}' + \mathbf{v}t$, or

Galilean coordinate transformation

$$\mathbf{r}' = \mathbf{r} - \mathbf{v}t \qquad [3.25]$$

That is, in a time t the S' frame is displaced to the right by an amount $\mathbf{v}t$.

If we differentiate Equation 3.25 with respect to time and note that $\mathbf{v}$ is constant, we get

Galilean velocity transformation

$$\frac{d\mathbf{r}'}{dt} = \frac{d\mathbf{r}}{dt} - \mathbf{v}$$

$$\mathbf{u}' = \mathbf{u} - \mathbf{v} \qquad [3.26]$$

where $\mathbf{u}'$ is the velocity of the particle observed in the S' frame and $\mathbf{u}$ is the velocity observed in the S frame. Equations 3.25 and 3.26 are known as *Galilean transformation equations.*

Although observers in two different reference frames will measure different velocities for the particles, they will measure the *same acceleration* when $\mathbf{v}$ is constant. This can be seen by taking the time derivative of Equation 3.26, which gives

$$\frac{d\mathbf{u}'}{dt} = \frac{d\mathbf{u}}{dt} - \frac{d\mathbf{v}}{dt}$$

But $d\mathbf{v}/dt = 0$, since $\mathbf{v}$ is constant. Therefore, we conclude that $\mathbf{a}' = \mathbf{a}$ since $\mathbf{a}' = d\mathbf{u}'/dt$ and $\mathbf{a} = d\mathbf{u}/dt$. That is, *the acceleration of the particle measured by an observer in the Earth's frame of reference will be the same as that measured by any other observer moving with* constant velocity *with respect to the first observer.*

Example 3.6 Crossing a River

A boat heading due north crosses a wide river with a speed of 10 km/h relative to the water. The river has a uniform velocity of 5 km/h due east. Determine the velocity of the boat with respect to an observer on the riverbank.

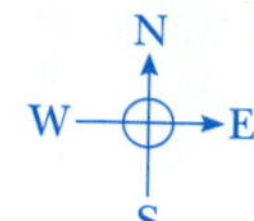

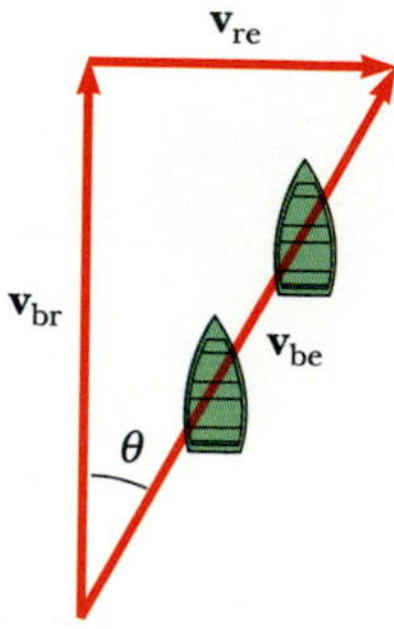

Figure 3.14
(Example 3.7)

Solution We have

$\mathbf{v}_{br} = 10\mathbf{j}$ km/h due north (The velocity of the *boat,* b, with respect to the *river,* r)

$\mathbf{v}_{re} = 5\mathbf{i}$ km/h due east (The velocity of the *river,* r, with respect to the *Earth,* e)

and we want $\mathbf{v}_{be}$, the velocity of the *boat* with respect to the *Earth.* Our equation becomes

$$\mathbf{v}_{be} = \mathbf{v}_{br} + \mathbf{v}_{re}$$

The terms in the equation must be manipulated as vector quantities; the vectors are shown in Figure 3.14. The quantity $\mathbf{v}_{br}$ is due north, $\mathbf{v}_{re}$ is due east, and the vector sum of the two, $\mathbf{v}_{be}$, is at an angle θ, as defined in Figure 3.14a. Thus, we see that the speed of the boat with respect to the Earth can be found, from the Pythagorean theorem, as

$$v_{be} = \sqrt{(v_{br})^2 + (v_{re})^2} = \sqrt{(10\text{ km/h})^2 + (5\text{ km/h})^2} = 11.2\text{ km/h}$$

and the direction of $\mathbf{v}_{be}$ is

$$\theta = \tan^{-1}\left(\frac{v_{re}}{v_{br}}\right) = \tan^{-1}\frac{5}{10} = 26.6°$$

Therefore, the boat will be traveling at a speed of 11.2 km/h in the direction 63.4° north of east with respect to the Earth.

Exercise If the width of the river is 3 km, find the time it takes the boat to cross the river.

Answer 30 min.

Example 3.7 Which Way Should the Boat Head?

If the boat of the preceding example travels with the same speed of 10 km/h relative to the water and is to travel due north, as in Figure 3.15, what should be its heading?

Solution We know

$\mathbf{v}_{br}$ = the velocity of the *boat,* b, with respect to the *river,* r

$\mathbf{v}_{re}$ = the velocity of the *river,* r, with respect to the *Earth,* e

and we need $\mathbf{v}_{be}$, the velocity of the *boat,* b, with respect to the *Earth,* e. We have

$$\mathbf{v}_{be} = \mathbf{v}_{br} + \mathbf{v}_{re}$$

The relationship between these three quantities, shown in Figure 3.15, agrees with our intuitive guess that the boat must head upstream in order to be pushed directly northward across the water. The speed v_{be} can be found, from the Pythagorean theorem, as

$$v_{be} = \sqrt{(v_{br})^2 - (v_{re})^2} = \sqrt{(10\text{ km/h})^2 - (5\text{ km/h})^2} = 8.66\text{ km/h}$$

and the direction of $\mathbf{v}_{be}$ is

$$\theta = \tan^{-1}\frac{v_{re}}{v_{be}} = \tan^{-1}\left(\frac{5}{8.66}\right) = 30°$$

where θ is west of north.

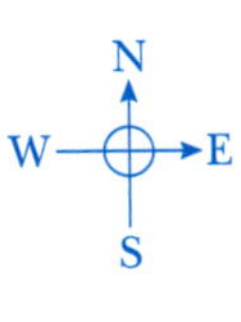

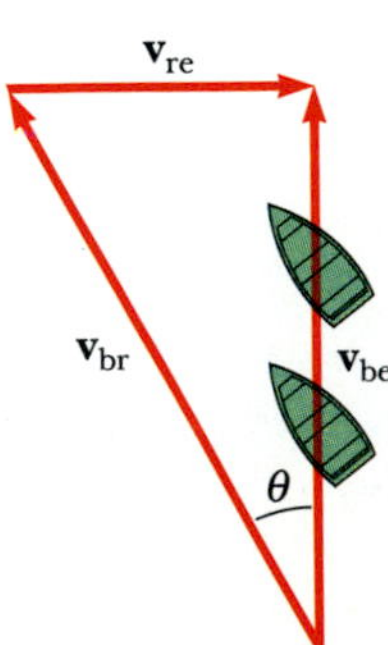

Figure 3.15
(Example 3.8)

Exercise If the width of the river is 3 km, how long does it take the boat to cross the river?

Answer 20.8 min.

Summary

If a particle moves with *constant* acceleration **a** and has a velocity $\mathbf{v}_0$ and position vector $\mathbf{r}_0$ at $t = 0$, its velocity and position vector at some later time t are given by

Velocity vector as a function of time

$$\mathbf{v} = \mathbf{v}_0 + \mathbf{a}t \qquad [3.8]$$

Position vector as a function of time

$$\mathbf{r} = \mathbf{r}_0 + \mathbf{v}_0 t + \tfrac{1}{2}\mathbf{a}t^2 \qquad [3.9]$$

For two-dimensional motion in the xy plane under constant acceleration, these vector expressions are equivalent to two component expressions, one for the motion along x with an acceleration $\mathbf{a}_x$ and one for the motion along y with an acceleration $\mathbf{a}_y$.

Projectile motion is two-dimensional motion under constant acceleration, where $a_x = 0$ and $a_y = -g$. In this case, if $x_0 = y_0 = 0$, the components of Equations 3.8 and 3.9 reduce to

Projectile motion equations

$$v_x = v_{x0} = \text{constant} \qquad [3.10]$$

$$v_y = v_{y0} - gt \qquad [3.11]$$

$$x = v_{x0}t \qquad [3.12]$$

$$y = v_{y0}t - \tfrac{1}{2}gt^2 \qquad [3.13]$$

where $v_{x0} = v_0 \cos \theta_0$, $v_{y0} = v_0 \sin \theta_0$ is the initial speed of the projectile, and θ_0 is the angle $\mathbf{v}_0$ makes with the positive x axis. Note that these expressions give the velocity components (and hence the velocity vector) and the coordinates (and hence the position vector) at *any* time t that the projectile is in motion.

It is useful to think of projectile motion as the superposition of two motions: (1) uniform motion in the x direction, where v_x remains constant, and (2) motion in the vertical direction, subject to a constant downward free-fall acceleration of magnitude $g = 9.80\ \text{m/s}^2$.

A particle moving in a circle of radius r with constant speed v undergoes a centripetal (or radial) acceleration, $\mathbf{a}_r$, because the direction of **v** changes in time. The magnitude of $\mathbf{a}_r$ is given by

Magnitude of centripetal acceleration

$$a_r = \frac{v^2}{r} \qquad [3.19]$$

and its direction is always toward the center of the circle.

If a particle moves along a curved path in such a way that the magnitude and direction of **v** change in time, the particle has an acceleration vector that can be described by two component vectors: (1) a radial component vector, $\mathbf{a}_r$, arising from the change in direction of **v**, and (2) a tangential component vector, $\mathbf{a}_t$, arising from the change in magnitude of **v**. The magnitude of $\mathbf{a}_r$ is v^2/r, and the magnitude of $\mathbf{a}_t$ is $d|\mathbf{v}|dt$.

The velocity of a particle, **u**, measured in a fixed frame of reference, S, is related to the velocity of the same particle, $\mathbf{u}'$, measured in a moving frame of reference, S', by

Galilean velocity transformation

$$\mathbf{u}' = \mathbf{u} - \mathbf{v} \qquad [3.26]$$

where **v** is the velocity of S' relative to S.

Questions and Conceptual Exercises

1. If the average velocity of a particle is zero in some time interval, what can you say about the displacement of the particle for that interval?
2. If you know the position vectors of a particle at two points along its path and also know the time it took to get from one point to the other, can you determine the particle's instantaneous velocity? its average velocity? Explain.
3. Describe a situation in which the velocity of a particle is perpendicular to the position vector.
4. Can a particle accelerate if its speed is constant? Can it accelerate if its velocity is constant? Explain.
5. Explain whether or not the following particles have an acceleration: (a) a particle moving in a straight line with constant speed and (b) a particle moving around a curve with constant speed.
6. Correct the following statement: "The racing car rounds the turn at a constant velocity of 90 miles per hour."
7. Determine which of the following moving objects would exhibit an approximate parabolic trajectory: (a) a ball thrown in an arbitrary direction, (b) a jet airplane, (c) a rocket leaving the launching pad, (d) a rocket a few minutes after launch with failed engines, (e) a tossed stone moving to the bottom of a pond.
8. A student argues that as a satellite orbits the Earth in a circular path, it moves with a constant velocity and therefore has no acceleration. The professor claims that the student is wrong since the satellite must have a centripetal acceleration as it moves in its circular orbit. What is wrong with the student's argument?
9. What is the fundamental difference between the unit vectors $\hat{\mathbf{r}}$ and $\hat{\boldsymbol{\theta}}$, defined in Figure 3.12, and the unit vectors **i** and **j**?
10. At the end of a pendulum's arc, its velocity is zero. Is its acceleration also zero at that point?
11. If a rock is dropped from the top of a sailboat's mast, will it hit the deck at the same point whether the boat is at rest or in motion at constant velocity?
12. A stone is thrown upward from the top of a building. Does the stone's displacement depend on the location of the origin of the coordinate system? Does the stone's velocity depend on the location of the origin?
13. Is it possible for a vehicle to travel around a curve without accelerating? Explain.
14. An object moves in a circular path with constant speed v. (a) Is the velocity of the object constant? (b) Is its acceleration constant? Explain.
15. As a projectile moves in its parabolic path, is there any point along its path where the velocity and acceleration are (a) perpendicular to each other? (b) parallel to each other?
16. A projectile is fired at some angle to the horizontal with some initial speed v_0, and air resistance is neglected. Is the projectile a freely falling body? What is its acceleration in the vertical direction? What is its acceleration in the horizontal direction?
17. State which of the following quantities, if any, remain constant as a projectile moves through its parabolic trajectory: (a) speed, (b) acceleration, (c) horizontal component of velocity, (d) vertical component of velocity.
18. The maximum range of a projectile occurs when it is launched at an angle of 45° with the horizontal, if air resistance is neglected. If air resistance is not neglected, will the optimum angle be greater or less than 45°? Explain.
19. A ball is tossed upward in the air by a passenger on a train that is moving with a constant velocity. Describe the path of the ball as seen by the passenger. Describe its path as seen by a stationary observer outside the train. How would these observations change if the train were accelerating along the track?
20. A person drops a spoon on a train that is moving with constant velocity. What is the acceleration of the spoon relative to (a) the train? (b) the Earth?

21. Describe how a driver can steer a car traveling at constant speed so that (a) the acceleration is zero or (b) the magnitude of the acceleration remains constant.
22. An ice skater is executing a "figure eight," consisting of two equal, tangent circular paths. Throughout the first loop she increases her speed uniformly, and during the second loop she moves at a constant speed. Make a sketch of her acceleration vector at several points along the path of motion.
23. Construct motion diagrams showing the velocity and acceleration of a projectile at several points along its path if (a) the projectile is fired horizontally, and (a) the projectile is fired at an angle θ with the horizontal.
24. A baseball is thrown such that its initial x and y components of velocity are known. Neglecting air resistance, describe how you would calculate, at the instant the ball reaches the top of its trajectory, (a) its coordinates, (b) its velocity, and (c) its acceleration. How would these results change if air resistance were taken into account?
25. A projectile is fired at an angle of 30° with the horizontal with some initial speed. If a second projectile is fired with the same initial speed, what other projectile angle would give the same horizontal range? Neglect air resistance.
26. A projectile is fired on the Earth with some initial velocity. Another projectile is fired on the Moon with the *same* initial velocity. Neglecting air resistance, which projectile has the greater range? Which reaches the greater altitude? (Note that the free-fall acceleration on the Moon is about 1.6 m/s^2.)
27. A coin on a table is given an initial horizontal velocity such that it ultimately leaves the end of the table and hits the floor. At the instant the coin leaves the end of the table, a ball is released from the same height and falls to the floor. Explain why the two objects hit the floor simultaneously, even though the coin has an initial velocity.

Problems

Section 3.1 The Displacement, Velocity, and Acceleration Vectors

1. Suppose that the trajectory of a particle is given by $\mathbf{r}(t) = x(t)\mathbf{i} + y(t)\mathbf{j}$ with $x(t) = at^2 + bt$ and $y(t) = ct + d$, where a, b, c, and d are constants that have appropriate dimensions. What displacement does the particle undergo between $t = 1$ s and $t = 3$ s?
2. Suppose that the position vector function for a particle is given as $\mathbf{r}(t) = x(t)\mathbf{i} + y(t)\mathbf{j}$, with $x(t) = at + b$ and $y(t) = ct^2 + d$, where $a = 1$ m/s, $b = 1$ m, $c = 1/8$ m/s^2, and $d = 1$ m. (a) Calculate the average velocity during the time interval from $t = 2$ s to $t = 4$ s. (b) Determine the velocity and the speed at $t = 2$ s.
3. A motorist drives south at 20 m/s for 3 min, then turns west and travels at 25 m/s for 2 min, and finally travels northwest at 30 m/s for 1 min. For this 6-min trip, find (a) the net vector displacement of the motorist, (b) the motorist's average speed, and (c) the average velocity of the motorist.
4. A golf ball is hit off a tee at the edge of a cliff. Its x and y coordinates versus time are given by the following expressions:

$$x = (18 \text{ m/s})\,t \quad \text{and} \quad y = (4 \text{ m/s})\,t - (4.9 \text{ m/s}^2)\,t^2$$

(a) Write a vector expression for the position $\mathbf{r}$ versus time t, using the unit vectors $\mathbf{i}$ and $\mathbf{j}$. By taking derivatives, repeat for (b) the velocity vector $\mathbf{v}$ versus time and (c) the acceleration vector $\mathbf{a}$ versus time. (d) Find the x and y coordinates of the golf ball at $t = 3$ s. Using the unit vectors $\mathbf{i}$ and $\mathbf{j}$, write expressions for (e) the velocity $\mathbf{v}$ and (f) the acceleration $\mathbf{a}$ at the instant $t = 3$ s.

Section 3.2 Motion in Two Dimensions with Constant Acceleration

5. At $t = 0$, a particle moving in the xy plane with constant acceleration has a velocity of $\mathbf{v}_0 = (3\mathbf{i} - 2\mathbf{j})$ m/s at the origin. At $t = 3$ s, the particle's velocity is $\mathbf{v} = (9\mathbf{i} + 7\mathbf{j})$ m/s. Find (a) the acceleration of the particle and (b) its coordinates at any time t.
6. A particle starts from rest at $t = 0$ at the origin and moves in the xy plane with a constant acceleration of $\mathbf{a} = (2\mathbf{i} + 4\mathbf{j})$ m/s^2. After a time t has elapsed, determine (a) the x and y components of velocity, (b) the coordinates of the particle, and (c) the speed of the particle.
7. A fish swimming in a horizontal plane has velocity $\mathbf{v}_0 = (4\mathbf{i} + \mathbf{j})$ m/s at a point in the ocean whose displacement from a certain rock is $\mathbf{r}_0 = (10\mathbf{i} - 4\mathbf{j})$ m. After the fish swims with constant acceleration for 20.0 s, its velocity is $\mathbf{v} = (20\mathbf{i} - 5\mathbf{j})$ m/s. (a) What are

the components of the acceleration? (b) What is the direction of the acceleration with respect to unit vector **i**? (c) Where is the fish at $t = 25$ s, and in what direction is it moving?

8. The vector position of a particle varies in time according to the expression $\mathbf{r} = (3\mathbf{i} - 6t^2\mathbf{j})$ m. (a) Find expressions for the velocity and acceleration as functions of time. (b) Determine the particle's position and velocity at $t = 1$ s.

9. A particle initially located at the origin has an acceleration of $\mathbf{a} = 3\mathbf{j}$ m/s² and an initial velocity of $\mathbf{v}_0 = 5\mathbf{i}$ m/s. Find (a) the vector position and velocity at any time t and (b) the coordinates and speed of the particle at $t = 2$ s.

Section 3.3 Projectile Motion

(Neglect air resistance in all problems.)

10. A student stands at the edge of a cliff and throws a stone horizontally over the edge with a speed of 18 m/s. The cliff is 50 m above a flat horizontal beach, as shown in Figure 3.16. How long after being released does the stone strike the beach below the cliff? With what speed and angle of impact does it land?

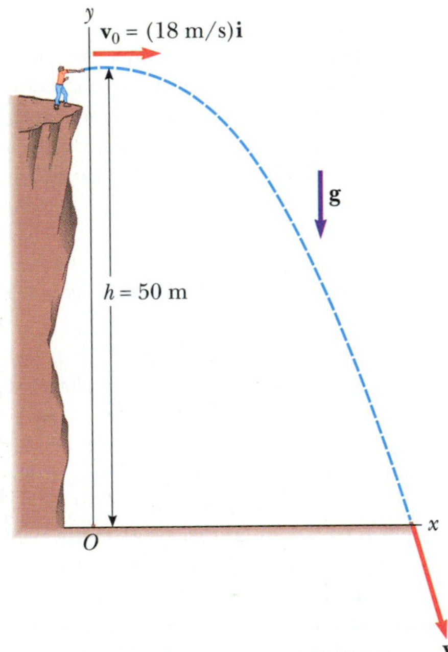

Figure 3.16 (Problem 10)

11. In a local bar, a customer slides an empty beer mug on the counter for a refill. The bartender is momentarily distracted and does not see the mug, which slides off the counter and strikes the floor 1.4 m from the base of the counter. If the height of the counter is 0.86 m, (a) with what velocity did the mug leave the counter, and (b) what was the direction of the mug's velocity just before it hit the floor?

12. A student decides to measure the muzzle velocity of the pellets from his BB gun. He points the gun horizontally. On a vertical wall a distance x away from the gun, a target is placed. The shots hit the target a vertical distance y below the gun. (a) Show that the position of the pellet when traveling through the air is given by $y = Ax^2$, where A is a constant. (b) Express the constant A in terms of the initial velocity and the free-fall acceleration, (c) If $x = 3.0$ m and $y = 0.21$ m, what is the initial speed of the BB?

13. Superman is flying at treetop level near Paris when he sees the Eiffel Tower elevator start to fall (the cable snaps). His x-ray vision tells him Lois Lane is inside. If Superman is 1 km away from the tower, and the elevator falls from a height of 240 m, how long does he have to save Lois, and what must his average velocity be?

14. A golfer wants to drive a golf ball a distance of 310 yards (283 m). If the 4-wood launches the ball at 15° above the horizontal, what must be the initial speed of the ball to achieve the required distance? (Ignore air friction and use $g = 9.80$ m/s².)

15. A ball is thrown horizontally from the top of a building 35 m high. The ball strikes the ground at a point 80 m from the base of the building. Find (a) the time the ball is in flight, (b) its initial velocity, and (c) the x and y components of velocity just before the ball strikes the ground.

16. During the 1968 Olympics in Mexico City, Bob Beamon executed a record long jump. The horizontal

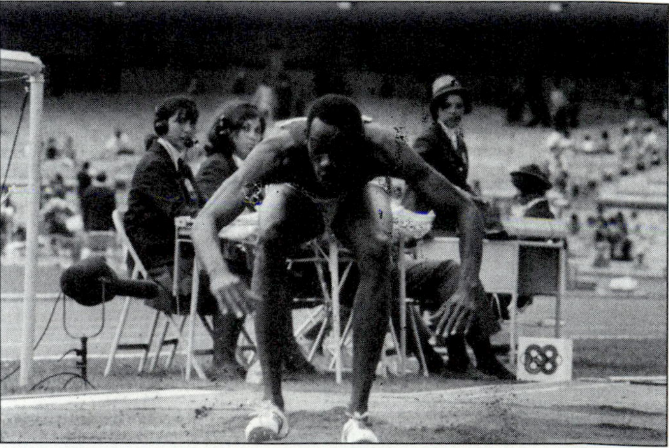

Figure 3.17 (Problem 16). Bob Beamon landing after his record jump. *(AP Wirephoto)*

distance he achieved was 8.90 m. His center of gravity started at an elevation of 1.0 m, reached a maximum height of 1.90 m, and finished at 0.15 m. From these data determine (a) his time of flight, (b) his horizontal and vertical velocity components at the takeoff time, and (c) his takeoff angle.

17. A place kicker must kick a football from a point 36 m (about 40 yards) from the goal, and the ball must clear the crossbar, which is 3.05 m high. When kicked, the ball leaves the ground with a speed of 20.0 m/s at an angle of 53° to the horizontal. (a) By how much does the ball clear or fall short of clearing the crossbar? (b) Does the ball approach the crossbar while still rising or while falling?

18. A fireman, 50 m away from a burning building, directs a stream of water from a fire hose at an angle of 30° above the horizontal. If the speed of the stream is 40 m/s, at what height will the stream of water strike the building?

19. It has been said that in his youth George Washington threw a silver dollar across a river. Assuming that the river was 75 m wide, (a) what *minimum initial* speed was necessary to get the coin across the river, and (b) how long was the coin in flight?

20. A tennis player standing 12.6 m from the net hits the ball at 3° above the horizontal. To clear the net, the ball must rise at least 0.33 m. If the ball just clears the net at the apex of its trajectory, how fast was the ball moving when it left the racket?

21. An artillery shell is fired with an initial velocity of 300 m/s at 55° above the horizontal. It explodes on a mountainside 42 s after firing. What are the x and y coordinates of the shell where it explodes, relative to its firing point?

22. The fastest recorded pitch in major-league baseball, thrown by Nolan Ryan in 1974, was clocked at 100.8 mi/h. If a pitch were thrown horizontally with this velocity, how far would the ball fall vertically by the time it reached home plate, 60 ft away?

23. An astronaut on a strange planet finds that she can jump a *maximum* horizontal distance of 15 m if her initial speed is 3 m/s. What is the free-fall acceleration on the planet?

24. A ball is tossed from an upper-story window of a building. The ball is given an initial velocity of 8 m/s at an angle of 20° below the horizontal. It strikes the ground 3 s later. (a) How far horizontally from the base of the building does the ball strike the ground? (b) Find the height from which the ball was thrown. (c) How long does it take the ball to reach a point 10 m below the level of launching? Ignore air friction.

Section 3.4 Uniform Circular Motion

25. Find the acceleration of a particle moving with a constant speed of 8 m/s in a circle 2 m in radius.

26. Young David who slew Goliath experimented with slings before tackling the giant. He found that he could revolve a sling of length 0.6 m at the rate of 8 rev/s. If he increased the length to 0.9 m, he could revolve the sling only 6 times per second. (a) Which rate of rotation gives the greater linear speed? (b) What is the centripetal acceleration at 8 rev/s? (c) What is the centripetal acceleration at 6 rev/s?

27. An athlete rotates a 1-kg discus along a circular path of radius 1.06 m. The maximum speed of the discus is 20 m/s. Determine the magnitude of the maximum radial acceleration of the discus.

(Problem 22). Nolan Ryan throwing his fast ball. *(Michael Layton, Duomo)*

Figure 3.18 (Problem 27). Discus thrower Pam Dukes. *(David Madison, Duomo)*

28. The orbit of the Moon about the Earth is approximately circular, with a mean radius of 3.84×10^8 m. It takes 27.3 days for the Moon to complete one revolution about the Earth. Find (a) the mean orbital speed of the moon and (b) its centripetal acceleration.
29. In the spin cycle of a washing machine, the tub, of radius 0.30 m, rotates uniformly at the rate of 630 rpm. What is the maximum linear speed with which water leaves the machine?
30. A tire 0.5 m in radius rotates at a constant rate of 200 rev/min. Find the speed and magnitude and acceleration of a small stone lodged in the tread of the tire (on its outer edge).

Section 3.5 Tangential and Radial Acceleration in Curvilinear Motion

31. Figure 3.19 represents the total acceleration of a particle moving clockwise in a circle of radius 2.5 m at a given instant of time. At this instant, find (a) the centripetal acceleration, (b) the speed of the particle, and (c) its tangential acceleration.

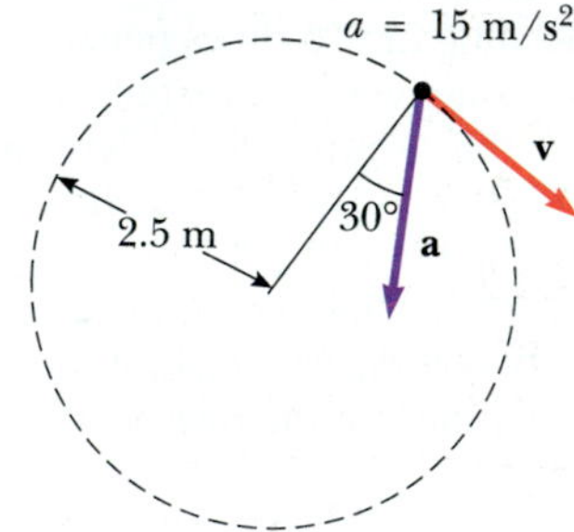

Figure 3.19 (Problem 31)

32. An automobile whose speed is increasing at a rate of 0.6 m/s² travels along a circular road of radius 20 m. When the instantaneous speed of the automobile is 4 m/s, find (a) the tangential acceleration component, (b) the centripetal acceleration component, and (c) the magnitude and direction of the total acceleration.
33. A train slows down as it rounds a sharp horizontal turn, slowing from 90 km/h to 50 km/h in the 15 s that it takes to round the bend. The radius of the curve is 150 m. Compute the acceleration at the moment the train speed reaches 50 km/h, assuming it continues to slow down at this time.
34. At some instant of time, a particle moving counter clockwise in a circle of radius 2 m has a speed of 8 m/s, and its total acceleration is directed as shown in Figure 3.20. At this instant, determine (a) the centripetal acceleration of the particle, (b) the tangential acceleration, and (c) magnitude of the total acceleration.

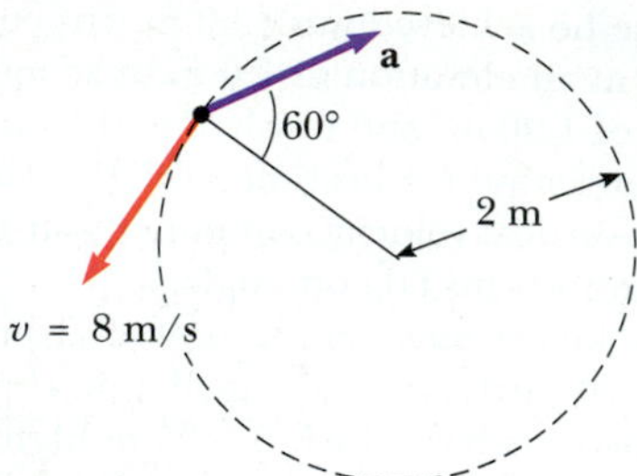

Figure 3.20 (Problem 34)

35. A student swings a ball attached to the end of a 0.6-m-long string in a vertical circle. The ball moves 4.3 m/s at its highest point and 6.5 m/s at its lowest point. Find the acceleration of the ball at (a) its highest point and (b) its lowest point.

Section 3.6 Relative Velocity and Relative Acceleration

36. The current in a river has a steady speed of 0.5 m/s. A student swims upstream a distance of 1 km and then swims back to her starting point. If the student can swim at a speed of 1.2 m/s in still water, how long does the round trip in the river take? Compare this with the time required for a trip of equal distance in still water.
37. Two canoeists in identical canoes exert the same effort paddling in a river. One paddles directly upstream (and moves upstream), whereas the other paddles directly downstream. An observer on the riverbank reckons their speeds to be 1.2 m/s and 2.9 m/s. How fast is the river flowing?
38. A boat crosses a river with a width w = 160 m, in which the current flows with a uniform speed of 1.5 m/s. The steersman maintains a bearing (i.e., the direction in which his boat points) perpendicular to the river and a throttle setting to give a constant speed of 2 m/s with respect to the water. (a) What is the velocity of the boat relative to a stationary shore observer? (b) How far downstream from the initial position is the boat when it reaches the opposite shore?
39. The pilot of an airplane notes that the compass indicates a heading due west. The speed of the airplane relative to the air is 150 km/h. If there is a wind of 30 km/h toward the north, find the velocity of the airplane relative to the ground.
40. A car travels due east with a speed of 50 km/h. Rain is falling vertically with respect to the Earth. The traces of the rain on the side windows of the car

make an angle of 60° with the vertical. Find the velocity of the rain with respect to (a) the car and (b) the Earth.

41. A child in danger of drowning in a river is being carried downstream by a current that flows uniformly with a speed of 2.5 km/h. The child is 0.6 km from shore and 0.8 km upstream of a boat landing when a rescue boat sets out from the landing. (a) If the boat proceeds at its maximum speed of 20 km/h with respect to the water, what heading relative to the shore should the boatman take? (b) What angle does the boat velocity **v** make with the shore? (c) How long will it take the boat to reach the child?

42. A science student is riding on a flatcar of a train that is traveling along a straight horizontal track at a constant speed of 10 m/s. The student throws a ball into the air along a path that he judges to make an initial angle of 60° with the horizontal and to be in line with the track. The student's professor, who is standing on the ground nearby, observes the ball to rise vertically. How high does the ball rise?

Additional Problems

43. At $t = 0$ a particle leaves the origin with a velocity of 6 m/s in the positive y direction. Its acceleration is given by $\mathbf{a} = (2\mathbf{i} - 3\mathbf{j})$ m/s². When the particle reaches its *maximum* y coordinate, its y component of velocity is zero. At this instant, find (a) the velocity of the particle and (b) its x and y coordinates.

44. A car is parked on a steep incline overlooking the ocean, where the incline makes an angle of 37° with the horizontal. The negligent driver leaves the car in neutral, and the parking brakes are defective. The car rolls from rest down the incline with a constant acceleration of 4 m/s² and travels 50 m to the edge of the cliff. The cliff is 30 m above the ocean. Find (a) the speed of the car when it reaches the cliff and the time it takes to get there, (b) the velocity of the car when it lands in the ocean, (c) the total time the car is in motion, and (d) the position of the car when it lands in the ocean, relative to the base of the cliff.

45. A batter hits a pitched baseball 1 m above the ground, imparting to it a speed of 40 m/s. The resulting line drive is caught on the fly by the left fielder 60 m from home plate, with his glove 1 m above the ground. If the shortstop, 45 m from home plate and in line with the drive, were to jump straight up to make the catch instead of allowing the left fielder to make the play, how high above the ground would his glove have to be?

46. The initial speed of a cannonball is 200 m/s. If it is fired at a target that is at a horizontal distance of 2 km from the cannon, find (a) the two projected angles that will result in a hit and (b) the total time of flight for each of the two trajectories found in (a).

47. A particle has velocity components

$$v_x = +4 \text{ m/s} \qquad v_y = -(6 \text{ m/s}^2)t + 4 \text{ m/s}$$

Calculate the speed of the particle and the direction $\theta = \tan^{-1}(v_y/v_x)$ of the velocity vector at $t = 2$ s.

48. The x and y coordinates of a particle are given by

$$x = 2 \text{ m} + (3 \text{ m/s})t \qquad y = x - (5 \text{ m/s}^2)t^2$$

How far from the origin is the particle at (a) $t = 0$; (b) $t = 2$ s?

49. The astronaut orbiting the Earth in the photograph is preparing to dock with a spinning Westar VI satellite. The satellite is in a circular orbit 600 km above the Earth's surface, where the free-fall acceleration is 8.21 m/s². The radius of the Earth is 6400 km. Determine the speed of the satellite and the time required to complete one orbit around the Earth.

(Problem 49) *(Courtesy of NASA)*

50. A rocket is launched at an angle of 53° to the horizontal with an initial speed of 100 m/s. For 3 s it moves along its initial line of motion with an acceleration of 30 m/s². Then its engines fail and the rocket proceeds to move in free fall. Find (a) the maximum altitude reached by the rocket, (b) its total time of flight, and (c) its horizontal range.

51. A boat requires 2 min to cross a river that is 150 m wide. The boat's speed relative to the water is 3 m/s, and the river current flows at a speed of 2 m/s. At what upstream or downstream points could the boat reach the opposite shore in 2 min?

52. A home run is hit in such a way that the baseball just clears a wall 21 m high, located 130 m from home plate. The ball is hit at an angle of 35° to the horizontal, and air resistance is negligible. Find (a) the initial speed of the ball, (b) the time it takes the ball to reach the wall, and (c) the velocity components and the speed of the ball when it reaches the wall. (Assume the ball is hit at a height of 1 m above the ground.)

53. A daredevil is shot out of a cannon at 45° to the horizontal with an initial speed of 25 m/s. A net is a horizontal distance of 50 m from the cannon. At what height above the cannon should the net be placed in order to catch the daredevil?

54. The position of a particle as a function of time t is described by

$$\mathbf{r} = (bt)\mathbf{i} + (c - dt^2)\mathbf{j} \qquad b = 2 \text{ m/s}$$

$$c = 5 \text{ m} \qquad d = 1 \text{ m/s}^2$$

(a) Express y in terms of x, and sketch the trajectory of the particle. What is the shape of the trajectory? (b) Derive a vector relation for the velocity. (c) At what time ($t > 0$) is the velocity vector perpendicular to the position vector?

55. A bomber is flown horizontally, with a ground speed of 275 m/s, at an altitude of 3000 m over level terrain. Neglect the effects of air resistance. (a) How far from the point that is vertically under the point of release will a bomb hit the ground? (b) If the plane maintains its original course and speed, where will it be when the bomb hits the ground? (c) For the preceding conditions, at what angle from the vertical at the point of release must the telescopic bomb sight be set so that the bomb will hit the target seen in the sight at the time of release?

56. A diver launches herself from a diving board 3 m above the water with a speed of 2 m/s, at an angle of 60° with the horizontal. Determine the amount of time she is in the air.

57. A rifle is aimed horizontally toward the center of a target 100 m away, but the bullet strikes 10 cm below the center. Calculate the speed of the bullet just as it emerges from the rifle.

58. A football is thrown toward a receiver with an initial speed of 20 m/s, at an angle of 30° above the horizontal. At that instant, the receiver is 20 m from the quarterback. In what direction and with what constant speed should the receiver run in order to catch the football at the level at which it was thrown?

59. A student who can swim at a speed of 1.5 m/s in still water wishes to cross a river that has a current of velocity 1.2 m/s toward the south. The width of the river is 50 m. (a) If the student starts from the west bank of the river, in what direction should she head in order to swim directly across the river? How long will this trip take? (b) If she heads due east, how long will it take to cross the river? (*Note*: The student travels farther than 50 m in this case.)

60. A sailboat sails for 1 hour at 4 km/h on a steady compass heading of 40° east of north. The sailboat is simultaneously carried along by a current. At the end of the hour the boat is 6.12 km from its starting point. The line from the starting point to the boat lies 60° east of north. Find the components of the velocity of the water.

61. A sailor aims his rowboat toward an island situated 2 km east and 3 km north of his starting position. After an hour of rowing he sees the island due west. He then aims the boat in the direction opposite his first direction, rows for another hour, and ends up 4 km east of his starting position. He correctly deduces that the current is from west to east. (a) What is the speed of the current? (b) Show that the boat's velocity relative to the shore for the first hour can be expressed as $\mathbf{u} = (4 \text{ km/h})\mathbf{i} + (3 \text{ km/h})\mathbf{j}$, where $\mathbf{i}$ is directed east and $\mathbf{j}$ is directed north.

62. After delivering his toys in the usual manner, Santa decides to have some fun and slide down an icy roof, as in Figure 3.21. He starts from rest at the top of the roof, which is 8 m in length, and accelerates at the rate of 5 m/s². The edge of the roof is 6 m above a soft snowbank, on which Santa lands. Find (a) Santa's velocity components when he reaches the snowbank, (b) the total time he is in motion, and (c) the distance d between the house and the point where he lands in the snow.

Figure 3.21 (Problem 62)

63. A skier leaves the ramp of a ski jump with a velocity of 10 m/s, 15° above the horizontal, as in Figure 3.22. The slope is inclined at 50°, and air resistance is negligible. Find (a) the distance from the ramp to where the jumper lands and (b) the velocity compo-

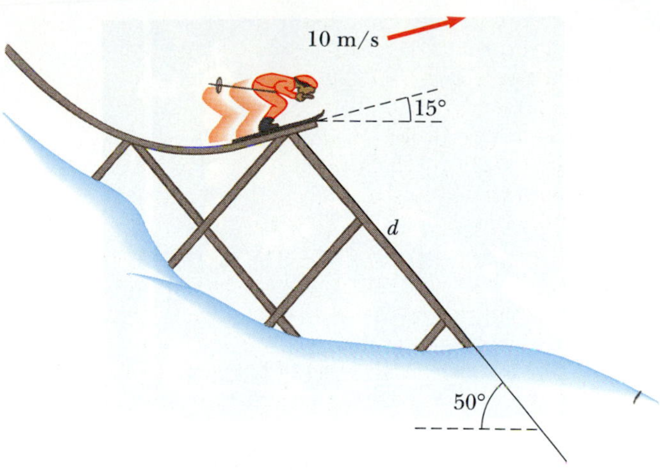

Figure 3.22 (Problem 63)

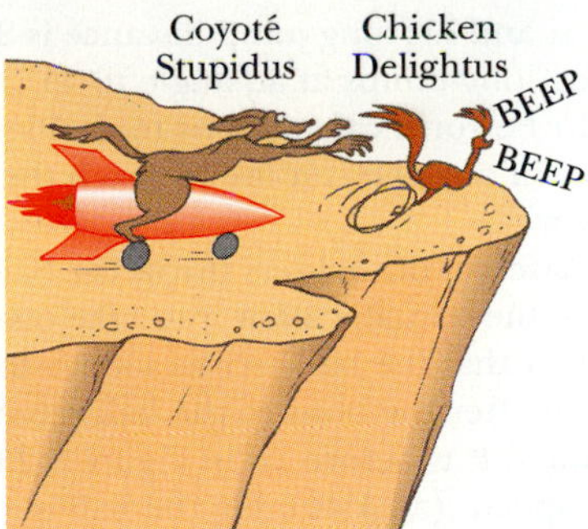

Figure 3.24 (Problem 66)

nents just before the landing. (How do you think the results might be affected if air resistance were included? Note that jumpers lean forward in the shape of an airfoil, with their hands at their sides, to increase their distance. Why does this work?)

64. An enemy ship is on the east side of a mountain island, as shown in Figure 3.23. The enemy ship can maneuver to within 2500 m of the 1800-m-high mountain peak and can shoot projectiles with an initial speed of 250 m/s. If the western shoreline is horizontally 300 m from the peak, what are the distances from the western shore at which a ship can be safe from the bombardment of the enemy ship?

65. A hawk is flying horizontally at 10.0 m/s in a straight line, 200 m above the ground. A mouse it has been carrying is released from its grasp. The hawk continues on its path at the same speed for 2 seconds before attempting to retrieve its prey. To accomplish the retrieval, it dives in a straight line at constant speed and recaptures the mouse 3.0 m above the ground. (a) Assuming no air resistance, find the diving speed of the hawk. (b) What angle did the hawk make with the horizontal during its descent? (c) For how long did the mouse "enjoy" free fall?

66. The determined coyote is out once more to try to capture the elusive roadrunner. The coyote wears a pair of Acme jet-powered roller skates, which provide a constant horizontal acceleration of 15 m/s^2 (Fig. 3.24). The coyote starts off at rest 70 m from the edge of a cliff at the instant the roadrunner zips past him in the direction of the cliff. (a) If the roadrunner moves with constant speed, determine the minimum speed he must have in order to reach the cliff before the coyote. (b) If the cliff is 100 m above the base of a canyon, determine where the coyote lands in the canyon (assume his skates are still in operation when he is in "flight"). (c) Determine the coyote's velocity components just before he lands in the canyon. (As usual, the roadrunner saves himself by making a sudden turn at the cliff.)

67. An Olympic decathlon star, who happens to be a bright physics student, is trapped on the roof of a burning building with a pencil, paper, and pocket calculator. He has 15 min to decide whether to jump to the next building, either by running at top speed horizontally off the edge or by using the long-jump technique. The next building is horizontally 30 ft away and vertically 10 ft below. His 100-m dash time

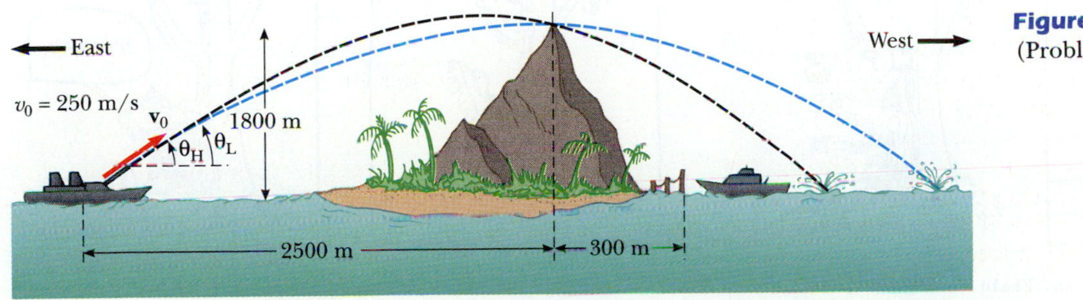

Figure 3.23 (Problem 64)

is 10.3 s, and his long-jump distance is 25.5 ft. (Assume he long-jumps at an angle to 45° above the horizontal.) Perform calculations to decide which method (if any) he can use to reach the other building safely.

68. When baseball outfielders throw the ball in from the outfield, they usually allow it to take one bounce on the theory that the ball arrives sooner this way. Suppose that after the bounce the ball rebounds at the same angle θ as before, as in Figure 3.25, but loses half its speed. (a) Assuming the ball is always thrown with the same initial speed, at what angle θ should the ball be thrown in order to go the same distance D with one bounce (blue path) as one thrown upward at 45° with no bounce (green path)? (b) Determine the ratio of the times for the one-bounce and no-bounce throws.

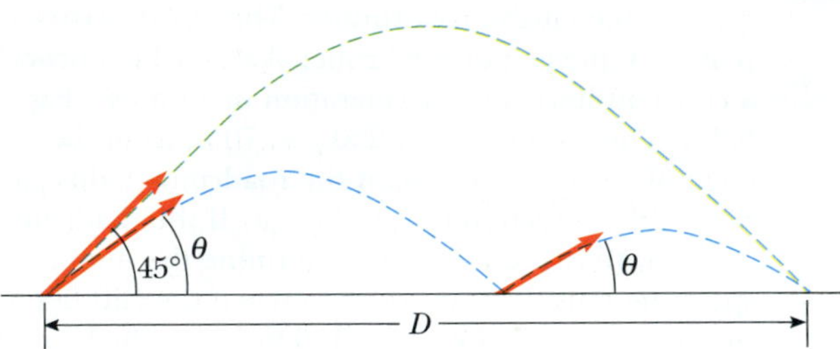

Figure 3.25 (Problem 68)

69. In a very popular lecture demonstration, a projectile is fired, leaving the gun at the same time the target is dropped from rest, as in Figure 3.26. Show that if the gun is initially aimed at the target, the projectile will hit the target.

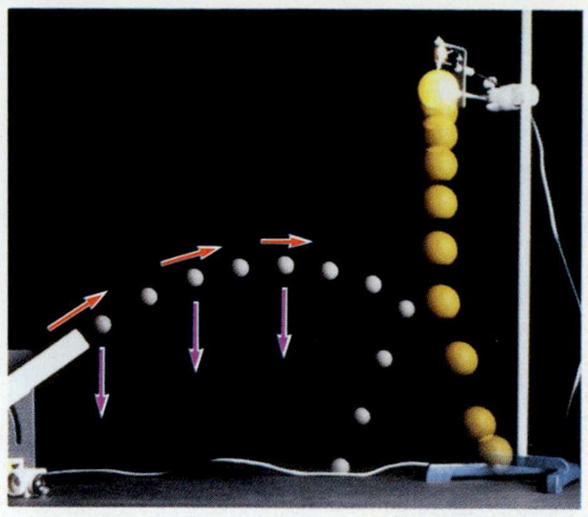

Figure 3.26 (Problem 69)

Calculator/Computer Problems

70. A projectile is fired from the origin with an initial speed of v_0 at an angle of θ_0 to the horizontal. Write programs that will enable you to tabulate the projectile's x and y coordinates, displacement, x and y components of velocity, and speed as functions of time. Tabulate these values for inputs of $v_0 = 50$ m/s and $\theta_0 = 60°$ at time intervals of 0.2 s, until a total time of 4.4 s is reached.

71. A ball bearing is dropped from the point $x = 4$ m, $y = 2$ m. At the same moment, a second bearing is launched from $x = 0$, $y = 0$ at an angle of 20° above the positive x axis at a speed of 6 m/s. Determine (a) the minimum distance between the bearings and (b) the time at which this minimum occurs. *Suggestion:* If you cannot solve this problem analytically, you may wish to write and run a short computer program that locates the minimum of the *square* of the distance between the bearings.

The Wizard of Id by Parker and Hart

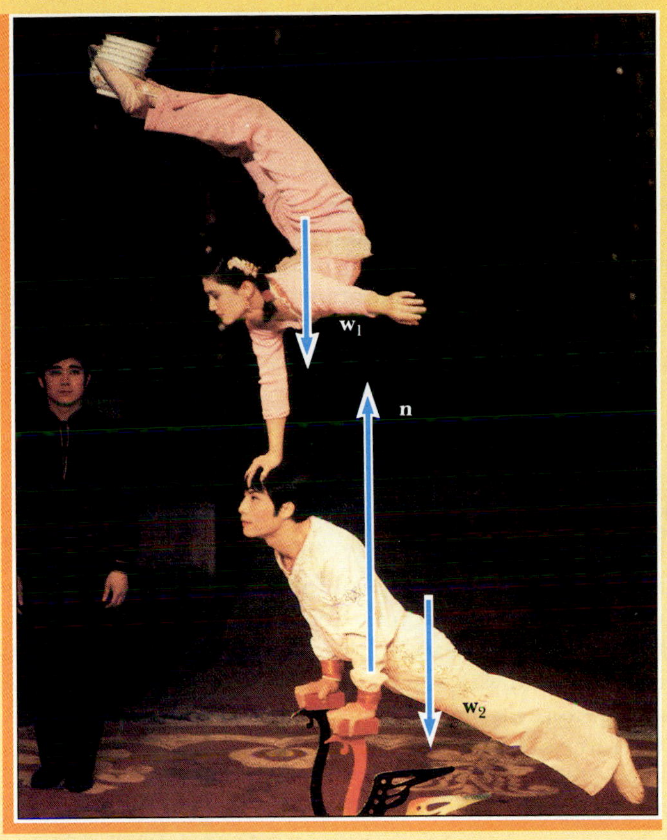

The Chinese acrobats in this difficult formation represent a balanced mechanical system. The external forces acting on the system, as shown by the blue vectors, are the weights of the acrobats, $\mathbf{w}_1$ and $\mathbf{w}_2$, and the upward force of the support $\mathbf{n}$ on the lower acrobat. The vector sum of these external forces must be zero in such a balanced system. Later we shall learn that a quantity called the net external torque acting on such a balanced system must also be zero. *(J. P. Lafont, Sygma)*

The Laws of Motion

CHAPTER 4

In the preceding two chapters on kinematics, we described the motions of particles based on the definitions of displacement, velocity, and acceleration. In addition, we would like to be able to answer specific questions related to the causes of motion, such as "What mechanism causes motion?" and "Why do some objects accelerate at higher rates than others?" In this chapter we shall describe the change in motion of particles using the concepts of force and mass. We shall then discuss the three fundamental laws of motion, which are based on experimental observations and were formulated about three centuries ago by Sir Isaac Newton.

We shall also discuss *force laws,* which describe how to calculate the force on an object if its environment is known. We shall see that, although the force laws are simple in form, they successfully explain a wide variety of phenomena and experimental observations. They, together with the laws of motion, are the foundations of classical mechanics.

4.1 Introduction to Classical Mechanics

The purpose of classical mechanics is to provide a connection between the motion of a body and the forces acting on it. Keep in mind that classical mechanics deals with objects that are large compared with the dimensions of atoms ($\approx 10^{-10}$ m) and that move at speeds much lower than the speed of light (3×10^8 m/s).

It is possible to describe the acceleration of an object in terms of the resultant force acting on the object (the interaction of the object with its environment) and the object's mass—a measure of its tendency to resist an acceleration when a force acts on it.

4.2 The Concept of Force

Everyone has a basic understanding of the concept of force as a result of everyday experiences. When you push or pull an object, you exert a force on it. You exert a force when you throw or kick a ball. In these examples, the word *force* is associated with the result of muscular activity and with some change in the state of motion of an object. Forces do not always cause an object to move, however. For example, as you sit reading this book, the force of gravity acts on your body, and yet you remain stationary. You can push on a block of stone and yet fail to move it.

What force (if any) causes a distant star to drift freely through space? Newton answered such questions by stating that the change in velocity of an object is caused by unbalanced forces. Therefore, if an object moves with uniform motion (constant velocity), no force is required to maintain the motion. Since only a force can cause a change in velocity, we can think of force as that which causes a body to accelerate.

A body accelerates due to an external force.

Now consider a situation in which several forces act simultaneously on an object. In this case, the object accelerates only if the *net force* acting on it is not equal to zero. We shall often refer to the net force as the *resultant force* or as the *unbalanced force. If the net force is zero, the acceleration is zero and the velocity of the object remains constant.* That is, if the net force acting on the object is zero, either the object will be at rest or it will move with constant velocity. *When a body has constant velocity, or is at rest, it is said to be in equilibrium.*

Definition of equilibrium

This chapter is concerned with the relation between the force on an object and the acceleration of that object. If you pull on a spring, as in Figure 4.1a, the spring stretches. If the spring is calibrated, the distance it stretches can be used to measure the strength of the force. If a child pulls hard enough on a cart to overcome friction, as in Figure 4.1b, the cart moves. When a football is kicked, as in Figure 4.1c, it is both deformed and set in motion. These are all examples of a class of forces called *contact forces.* A contact force is the result of physical contact between two objects.

Another class of forces does not involve physical contact between two objects. Early scientists, including Newton, were uneasy with the concept of forces that act

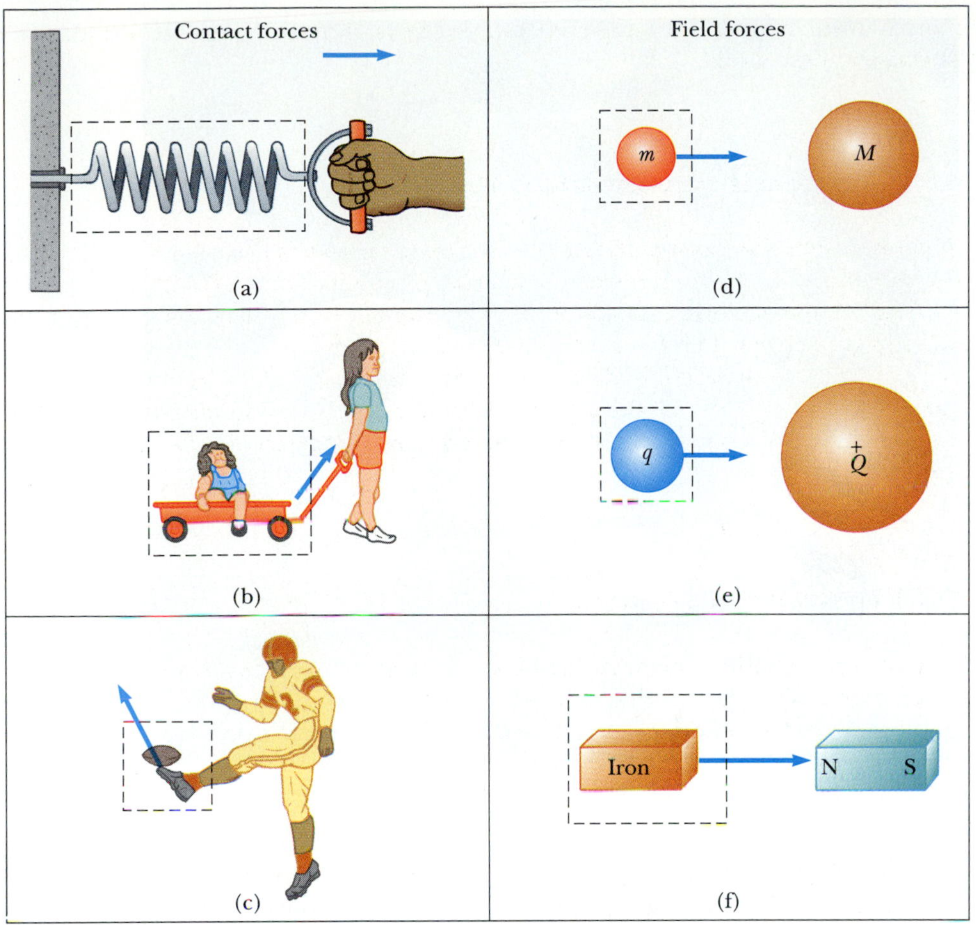

Figure 4.1 Some examples of forces applied to various objects. In each case a force is exerted on the particle or object within the boxed area. The environment external to the boxed area provides this force.

between two disconnected bodies. To overcome this conceptual problem, Michael Faraday (1791–1867) introduced the concept of a *field,* and the corresponding forces are called *field forces.* According to this approach, when mass m_1 is placed at some point P near a second mass, m_2, one can say that m_1 interacts with m_2 by virtue of the gravitational field that exists at P. The field at P is created by the mass m_2. Thus, the force of gravitational attraction between two objects, illustrated in Figure 4.1d, is an example of a field force. This force keeps objects bound to the Earth and gives rise to what we commonly call the *weight* of the object. The planets of our Solar System are under the action of gravitational forces. Another common example of a field force is the electric force that one electric charge exerts on another, as in Figure 4.1e. The two charges might be an electron and a proton forming a hydrogen atom. A third example of a field force is the force exerted by a bar magnet on a piece of iron, as shown in Figure 4.1f. The forces between subatomic particles are also field forces but are usually very short-range. They are the dominating interactions for particle separations on the order of 10^{-15} m.

The distinction between contact forces and field forces is not as sharp as it might seem. At the atomic level, the so-called contact forces are actually due to

A football is set in motion as a result of the contact force **F** on it due to the kicker's foot. The ball is distorted in the short time it is in contact with the foot. *(Ralph Cowan, Tony Stone Worldwide)*

Fundamental forces in nature

repulsive forces between electric charges, which themselves are field forces. In developing models for macroscopic phenomena, it is convenient to use both classifications of forces. However, the only known *fundamental* forces in nature are (1) gravitational attractions between objects because of their masses, (2) electromagnetic forces between electric charges at rest or in motion, (3) strong nuclear forces between subatomic particles, and (4) weak nuclear forces (the so-called weak interaction), which arise in certain radioactive decay processes. These are all field forces. In classical physics we shall be concerned only with gravitational and electromagnetic forces.

It is sometimes convenient to use the deformation of a spring to measure force. Springs that elongate in proportion to an applied force are said to obey *Hooke's law.* Such springs can be constructed and calibrated to measure unknown forces. Suppose a force is applied vertically to a spring that obeys Hooke's law and has a fixed upper end, as in Figure 4.2a. We can calibrate the spring by defining the unit force, $\mathbf{F}_1$, as the force that produces an elongation of 1 cm. If a force $\mathbf{F}_2$, applied horizontally as in Figure 4.2b, produces an elongation of 2 cm, the magnitude of $\mathbf{F}_2$ is 2 units. If the two forces $\mathbf{F}_1$ and $\mathbf{F}_2$ are applied simultaneously, as in Figure 4.2c, the elongation of the spring is $\sqrt{5} = 2.24$ cm. The single force **F** that would produce this same elongation is the vector sum of $\mathbf{F}_1$ and $\mathbf{F}_2$, as described in Figure 4.2c. That is, $|\mathbf{F}| = \sqrt{F_1^2 + F_2^2} = \sqrt{5}$ units, and its direction is $\theta = \arctan(-0.5) = -26.6°$. *Because forces are vectors, you must use the rules of vector addition to get the resultant force on a body.*

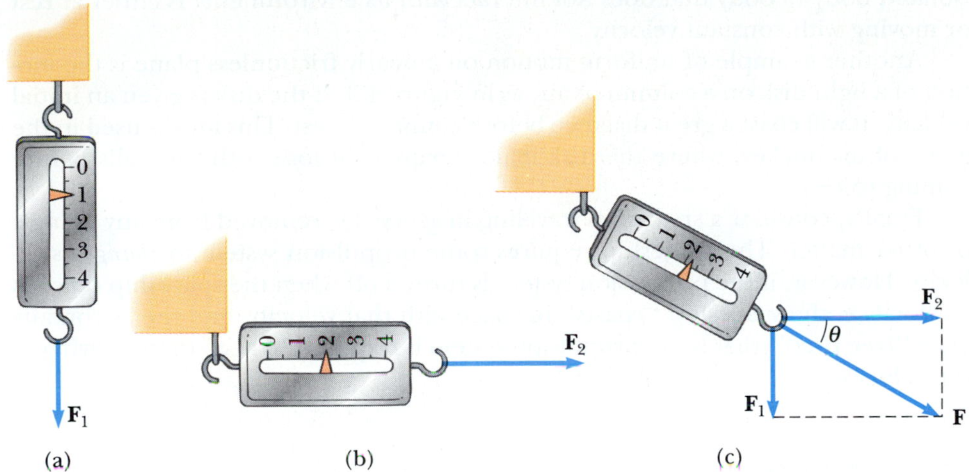

Figure 4.2 The vector nature of a force is tested with a spring scale. (a) The downward vertical force $\mathbf{F}_1$ elongates the spring 1 unit. (b) The horizontal force $\mathbf{F}_2$ elongates the spring 2 units. (c) The combination of $\mathbf{F}_1$ and $\mathbf{F}_2$ elongates the spring $\sqrt{1^2 + 2^2} = \sqrt{5}$ units.

4.3 Newton's First Law and Inertial Frames

Before we state Newton's first law, consider the following simple experiment. Suppose a book is lying on a table. Obviously, the book will remain at rest in the absence of any influences. Now imagine that you push the book with a horizontal force great enough to overcome the usual force of friction between the book and table. The book can then be set in motion with constant velocity if the applied force is equal in magnitude to, and in the direction opposite, the friction force. If the applied force exceeds the force of friction, the book accelerates. If you stop applying the force, the book stops sliding after moving a short distance because the force of friction retards its motion (causes a negative acceleration). Now imagine that the book is pushed across a smooth, highly waxed floor. The book will again come to rest once the force is no longer applied, but not as quickly as before. If you can imagine the motion of the book on a horizontal, frictionless surface, the book, once set in motion, will slide until it hits the wall.

Before about 1600, scientists felt that the natural state of matter was the state of rest. Galileo was the first to take quite a different approach. He devised thought experiments, such as an object moving on a frictionless surface, and concluded that it is not the nature of an object to stop once set in motion: rather, it is an object's nature to resist deceleration and acceleration. In his words, "Any velocity, once imparted to a moving body, will be rigidly maintained as long as the external causes of retardation are removed."

This new approach to motion was later formalized by Newton in a statement that has come to be known as **Newton's first law of motion:**

A statement of Newton's first law

> **An object at rest will remain at rest, and an object in motion will continue in motion with a constant velocity (that is, with constant speed in a straight line), unless it experiences a net external force.**

In simpler terms, we can say that *when the net force on a body is zero, its acceleration is zero.* That is, when $\Sigma \mathbf{F} = 0$, then $\mathbf{a} = 0$. From the first law, we conclude that an

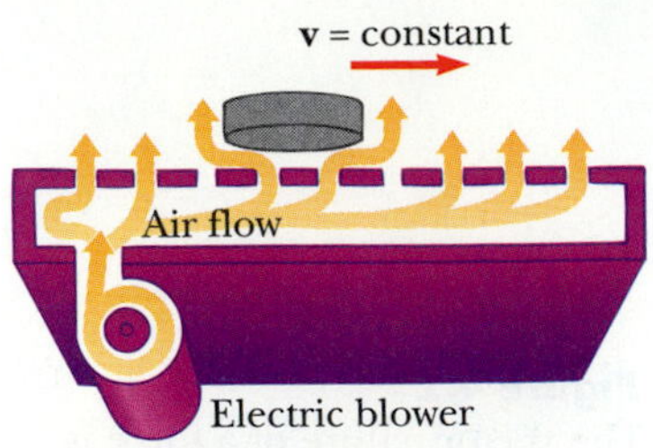

Figure 4.3
A disk moving on a column of air is an example of uniform motion, that is, motion in which the acceleration is zero.

isolated body (a body that does not interact with its environment) is either at rest or moving with constant velocity.

Another example of uniform motion on a nearly frictionless plane is the motion of a light disk on a column of air, as in Figure 4.3. If the disk is given an initial velocity, it will coast a great distance before coming to rest. This idea is used in the game of air hockey, where the disk makes many collisions with the walls before coming to rest.

Finally, consider a spaceship traveling in space, far removed from any planets or other matter. The spaceship requires some propulsion system to *change* its velocity. However, if the propulsion system is turned off when the spaceship reaches a velocity **v**, the spaceship "coasts" in space with that velocity, and the astronauts get a "free ride" (that is, no propulsion system is required to keep them moving at the velocity **v**).

Inertial Frames

Newton's first law, sometimes called the *law of inertia,* defines a special set of reference frames called *inertial frames.*

Inertial frame

An **inertial frame of reference** is one in which Newton's first law is valid.

Any reference frame that moves with constant velocity with respect to an inertial frame is itself an inertial frame. A reference frame that moves with constant velocity relative to the distant stars is the best approximation of an inertial frame. The Earth is not an inertial frame because of its orbital motion about the Sun and rotational motion about its own axis. As the Earth travels in its nearly circular orbit about the Sun, it experiences a centripetal acceleration of about 4.4×10^{-3} m/s^2 toward the Sun. In addition, since the Earth rotates about its own axis once every 24 h, a point on the equator experiences an additional centripetal acceleration of 3.37×10^{-2} m/s^2 toward the center of the Earth. However, these accelerations are small compared with g and so can often be neglected. In most situations *we shall assume that a frame on or near the Earth's surface is an inertial frame.*

Thus, if an object is moving with constant velocity, an observer in one inertial frame (say, one at rest with respect to the object) will claim that the acceleration and the resultant force on the object are zero. An observer in *any other* inertial frame will also find that $\mathbf{a} = 0$ and $\mathbf{F} = 0$ for the object. According to the first law, a body at rest and one moving with constant velocity are equivalent. Unless stated otherwise, we shall usually write the laws of motion with respect to an observer "at rest" in an inertial frame.

4.4 Inertial Mass

Inertia

If an attempt is made to change the velocity of an object, the object will resist the change. **Inertia** is a property of matter; in practical terms it is a measure of the tendency of an object to remain at rest or in uniform motion. For instance, consider two large, solid cylinders of equal size, one balsa wood and the other steel. If you were to push the cylinders along a horizontal, rough surface, it would certainly take more effort to get the steel cylinder rolling. Likewise, once they were in motion, it would require more effort to bring the steel cylinder to rest. Therefore, we say that the steel cylinder has more inertia than the balsa wood cylinder.

Mass is used to measure inertia, and the SI unit of mass is the kilogram. The greater the mass of a body, the less it will accelerate (change its state of motion) under the action of an applied force.

It is important to remember that mass should not be confused with weight. *Mass and weight are two different quantities.* The weight of a body is equal to the force of gravity acting on the body, and so the weight varies with location. For example, a person who weighs 180 lb on Earth weighs only about 30 lb on the Moon. On the other hand, the mass of a body is the same everywhere. An object having a mass of 2 kg on Earth also has a mass of 2 kg on the Moon.

Mass and weight are different quantities

A quantitative measurement of mass can be made by comparing the accelerations that a given force produces on different bodies. Suppose a force acting on a body of mass m_1 produces an acceleration $\mathbf{a}_1$, and the *same force* acting on a body of mass m_2 produces an acceleration $\mathbf{a}_2$. The ratio of the two masses is defined as the *inverse* ratio of the magnitudes of the accelerations produced by the same force:

$$\frac{m_1}{m_2} \equiv \frac{a_2}{a_1} \qquad [4.1]$$

If one mass is standard and known—say, 1 kg—the mass of an unknown object can be obtained from acceleration measurements. For example, if the standard 1-kg mass undergoes an acceleration of 3 m/s^2 under the influence of some force, a 2-kg mass will undergo an acceleration of 1.5 m/s^2 under the action of the same force.

Mass is an inherent property of a body, independent of the body's surroundings and of the method used to measure it. It is an experimental fact that *mass is a scalar quantity.* Finally, *mass is a quantity that obeys the rules of ordinary arithmetic.* That is, several masses can be combined in simple numerical fashion. For example, if you combine a 3-kg mass with a 5-kg mass, their total mass is 8 kg. This can be verified experimentally by comparing the acceleration of each object produced by a known force with the acceleration of the combined system using the same force.

Isaac Newton (1642–1727), an English physicist and mathematician, was one of the most brilliant scientists in history. Before the age of 30, he formulated the basic concepts and laws of mechanics, discovered the law of universal gravitation, and invented the mathematical methods of calculus. As a consequence of his theories, Newton was able to explain the motions of the planets, the ebb and flow of the tides, and many special features of the motions of the Moon and Earth. He also interpreted many fundamental observations concerning the nature of light. His contributions to physical theories dominated scientific thought for two centuries and remain important today. *(Giraudon/Art Resource)*

4.5 Newton's Second Law

Newton's first law explains what happens to an object when the resultant of all external forces on it is zero. In such an instance, the object either remains at rest or moves in a straight line with constant speed. Newton's second law answers the question of what happens to an object that has a nonzero resultant force acting on it.

Imagine that you are pushing a block of ice across a smooth horizontal surface, so that frictional forces can be neglected. When you exert some horizontal force $\mathbf{F}$, the block moves with some acceleration $\mathbf{a}$. If you apply a force twice as large, the acceleration doubles. If you increase the applied force to $3\mathbf{F}$, the original acceleration is tripled, and so on. From such observations, we conclude that *the acceleration of an object is directly proportional to the resultant force acting on it.*

As stated in the preceding section, the acceleration of an object also depends on its mass. This can be understood by considering the following set of experi-

Table 4.1
Units of Force, Mass, and Acceleration[a]

System of Units	Mass	Acceleration	Force
SI	kg	m/s^2	$N = kg \cdot m/s^2$
cgs	g	cm/s^2	$dyne = g \cdot cm/s^2$
British engineering	slug	ft/s^2	$lb = slug \cdot ft/s^2$

[a] $1\ N = 10^5\ dyne = 0.225\ lb$.

ments. If you apply a force **F** to a block of ice on a frictionless surface, the block will undergo some acceleration **a**. If the mass of the block is doubled, the same applied force will produce an acceleration **a**/2. If the mass is tripled, the same applied force will produce an acceleration **a**/3, and so on. We conclude that *the acceleration of an object is inversely proportional to its mass.*

These observations are summarized in **Newton's second law:**

$\mathbf{a} \propto \frac{\Sigma \mathbf{F}}{m}$

The acceleration of an object is directly proportional to the resultant force acting on it and inversely proportional to its mass.

Thus, we can relate mass and force through the following mathematical statement of Newton's second law:[1]

Newton's second law

$$\sum \mathbf{F} = m\mathbf{a} \tag{4.2}$$

You should note that Equation 4.2 is a *vector* expression and hence is equivalent to the following three component equations:

Newton's second law—component form

$$\sum F_x = ma_x \qquad \sum F_y = ma_y \qquad \sum F_z = ma_z \tag{4.3}$$

Units of Force and Mass

The SI unit of force is the newton, which is defined as the force that, when acting on a 1-kg mass, produces an acceleration of 1 m/s².

From this definition and Newton's second law, we see that the newton can be expressed in terms of the fundamental units of mass, length, and time:

Definition of a newton

$$1\ N \equiv 1\ kg \cdot m/s^2 \tag{4.4}$$

The units of force, mass, and acceleration are summarized in Table 4.1. Most of the calculations we shall make in our study of mechanics will be in SI units. Conversion factors between the three systems are given in Appendix A.

[1] Equation 4.2 is valid only when the speed of the object is much less than the speed of light. We will treat the relativistic situation in Chapter 10.

Example 4.1 An Accelerating Hockey Puck

A hockey puck with a mass of 0.3 kg slides on the horizontal frictionless surface of an ice rink. Two forces act on the puck as shown in Figure 4.4. The force $\mathbf{F}_1$ has a magnitude of 5 N, and $\mathbf{F}_2$ has a magnitude of 8 N. Determine the acceleration of the puck.

Solution The resultant force in the x direction is

$$\sum F_x = F_{1x} + F_{2x} = F_1 \cos 20° + F_2 \cos 60°$$
$$= (5\text{ N})(0.940) + (8\text{ N})(0.500) = 8.70\text{ N}$$

The resultant force in the y direction is

$$\sum F_y = F_{1y} + F_{2y} = -F_1 \sin 20° + F_2 \sin 60°$$
$$= -(5\text{ N})(0.342) + (8\text{ N})(0.866) = 5.22\text{ N}$$

Now we can use Newton's second law in component form to find the x and y components of acceleration:

$$a_x = \frac{\Sigma F_x}{m} = \frac{8.70\text{ N}}{0.3\text{ kg}} = 29.0\text{ m/s}^2$$

$$a_y = \frac{\Sigma F_y}{m} = \frac{5.22\text{ N}}{0.3\text{ kg}} = 17.4\text{ m/s}^2$$

The acceleration has a magnitude of

$$a = \sqrt{(29.0)^2 + (17.4)^2}\text{ m/s}^2 = 33.8\text{ m/s}^2$$

and its direction is

$$\theta = \tan^{-1}(a_y/a_x) = \tan^{-1}(17.4/29.0) = 31.0°$$

relative to the positive x axis.

Exercise Determine the components of a third force that, when applied to the puck, will cause it to be in equilibrium.

Answer $F_x = -8.70\text{ N}$, $F_y = -5.22\text{ N}$.

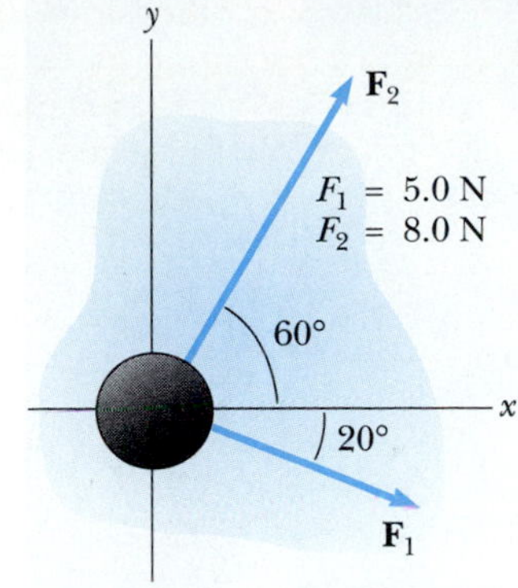

Figure 4.4 (Example 5.1) An object moving on a frictionless surface accelerates in the direction of the *resultant* force, $\mathbf{F}_1 + \mathbf{F}_2$.

4.6 Weight

We are well aware of the fact that all objects are attracted to the Earth. The force exerted by the Earth on an object is called the **weight** of the object, w. This force is directed toward the center of the Earth.[2] More precisely, the weight of an object is the resultant gravitational force on the object and is actually due to all other bodies in the Universe!

[2] This statement is a simplification in that it ignores the fact that the mass distribution of the Earth is not perfectly spherical.

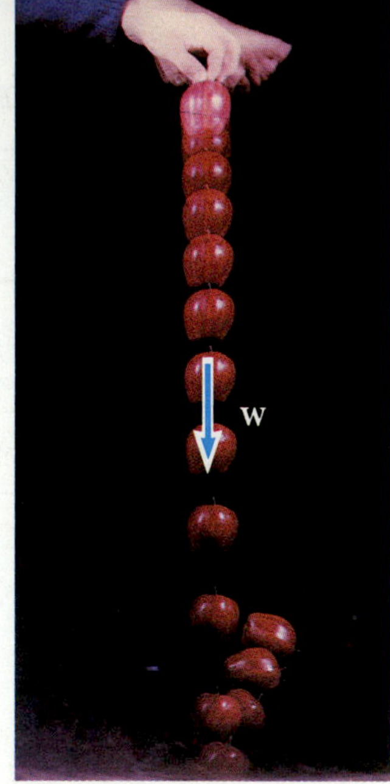

Figure 4.5
The only force acting on an object in free fall is the weight, w. Newton's second law applied in the vertical direction gives $w = mg$.

We have seen that a freely falling body experiences an acceleration **g** directed toward the center of the Earth. Applying Newton's second law, $\mathbf{F} = m\mathbf{a}$, to the freely falling body (Fig. 4.5), with $\mathbf{a} = \mathbf{g}$ and $\mathbf{F} = \mathbf{w}$, gives

$$\mathbf{w} = m\mathbf{g} \quad [4.5]$$

Since weight depends on g, it varies with geographic location. Bodies weigh less at higher altitudes than at sea level because g decreases with increasing distance from the center of the Earth. Hence, weight, unlike mass, is not an inherent property of a body. For example, if a body has a mass of 70 kg, then the magnitude of its weight in a location where $g = 9.80 \text{ m/s}^2$ is $mg = 686$ N (about 154 lb). At the top of a mountain where $g = 9.76 \text{ m/s}^2$, the body's weight would be 683 N. This corresponds to a decrease of about 0.4 lb. Thus, if you want to lose weight without going on a diet, climb a mountain or weigh yourself at 30 000 ft during a flight on a jet airplane.

Since $\mathbf{w} = m\mathbf{g}$, we can compare the masses of two bodies by measuring their weights with a spring scale or a chemical balance. That is, at a given location the ratio of the weights of two bodies equals the ratio of their masses.

4.7 Newton's Third Law

Newton's **third law** states:

A statement of Newton's third law

If two bodies interact, the force exerted on body 1 by body 2 is equal in magnitude but opposite in direction to the force exerted on body 2 by body 1.

That is,

$$\mathbf{F}_{12} = -\mathbf{F}_{21} \quad [4.6]$$

This law, which is illustrated in Figure 4.6, is equivalent to stating that *forces always occur in pairs,* or that *a single isolated force cannot exist.* The force that body 1 exerts on body 2 is sometimes called the *action force,* and the force of body 2 on body 1 is called the *reaction force.* In reality, either force can be labeled the action or reaction force. *The action force is equal in magnitude to the reaction force and opposite in direction. In all cases, the action and reaction forces act on different objects.* For example, the force acting on a freely falling projectile is its weight, $\mathbf{w} = m\mathbf{g}$. This equals the force of the Earth on the projectile. The reaction to this force is the force of the projectile on the Earth, $\mathbf{w}' = -\mathbf{w}$. The reaction force, $\mathbf{w}'$, must accelerate the Earth toward the projectile just as the action force, **w**, accelerates the projectile toward the Earth. However, since the Earth has such a large mass, its acceleration due to this reaction force is negligibly small.

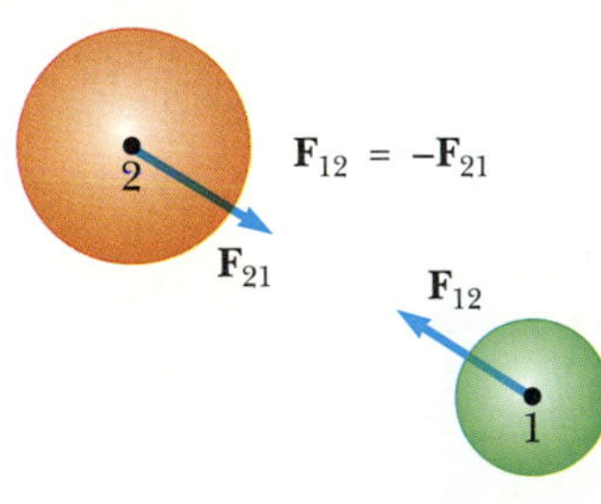

Figure 4.6
Newton's third law. The force of body 1 on body 2 is equal to and opposite the force of body 2 on body 1.

One example of Newton's third law in action is shown in Figure 4.7. You directly experience the law if you slam your fist against a wall or kick a football with your bare foot. You should be able to identify the action and reaction forces in these cases.

The weight of a body, **w**, is equal to the force the Earth exerts on the body. If the body is a block at rest on a table, as in Figure 4.8a, the reaction force to **w** is the

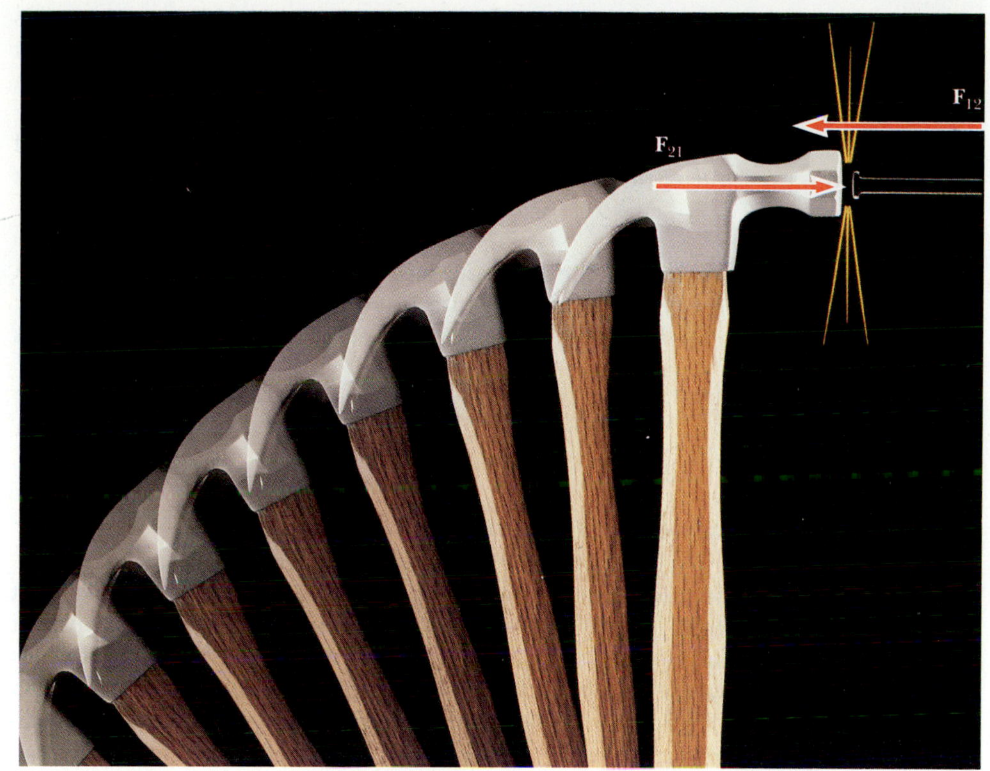

Figure 4.7
This is an example of Newton's third law in action. The force of the hammer on the nail ($\mathbf{F}_{21}$) is equal in magnitude and opposite in direction to the force of the nail on the hammer ($\mathbf{F}_{12}$). *(John Gillmoure, The Stock Market)*

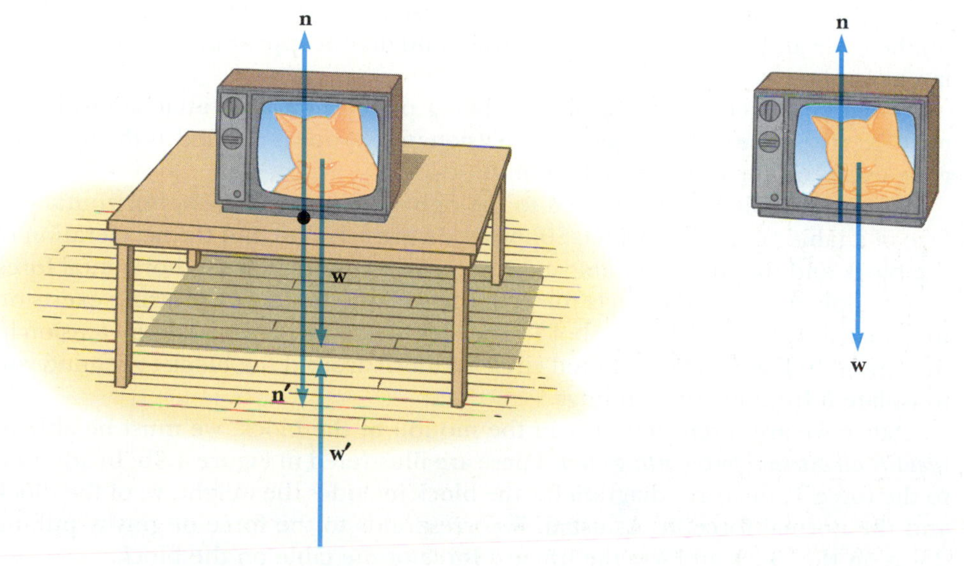

Figure 4.8
When a TV set is sitting on a table, the forces acting on the set are the normal force, **n**, and the force of gravity, **w**, as illustrated in (b). The reaction to **n** is the force of the TV set on the table, **n**′. The reaction to **w** is the force of the TV set on the Earth, **w**′.

force the block exerts on the Earth, $\mathbf{w}'$. The block does not accelerate, since it is held up by the table. The table, therefore, exerts on the block an upward action force, $\mathbf{n}$, called the **normal force.**[3] This is the force that prevents the block from falling through the table; it can have any value needed, up to the point of breaking the table. The normal force balances the weight and provides equilibrium. The reaction to $\mathbf{n}$ is the force of the block on the table, $\mathbf{n}'$. Therefore, we conclude that

$$\mathbf{w} = -\mathbf{w}' \quad \text{and} \quad \mathbf{n} = -\mathbf{n}'$$

Note that the forces acting on the block are $\mathbf{w}$ and $\mathbf{n}$, as in Figure 4.8b. When treating the motion of a body, we shall be interested only in such external forces. From the first law, we see that since the block is in equilibrium ($\mathbf{a} = 0$), it follows that $w = n = mg$.

4.8 Some Applications of Newton's Laws

In this section we present some simple applications of Newton's laws to bodies that are either in equilibrium ($\mathbf{a} = 0$) or moving linearly under the action of constant external forces. For our model, we shall assume that the bodies behave as particles so that we need not worry about rotational motions. In this section we shall also neglect the effects of friction for those problems involving motion. This is equivalent to stating that the surfaces are *frictionless*. Finally, we shall usually neglect the masses of any ropes involved in the problems. In this approximation, the magnitude of the force exerted at any point along the rope is the same at all points along the rope.

When we apply Newton's laws to a body, we shall be interested only in those external forces that act *on the body*. For example, in Figure 4.8 the only external forces acting on the block are $\mathbf{n}$ and $\mathbf{w}$. The reactions to these forces, $\mathbf{n}'$ and $\mathbf{w}'$, act on the table and on the Earth, respectively, and do not appear in Newton's second law as applied to the block.

Tension

When an object such as a block is being pulled by a rope attached to it, the rope exerts a force on the object. The **tension** in the rope is defined as the force the rope exerts on the object attached to it.

Consider a block being pulled to the right on the frictionless, horizontal surface of a table, as in Figure 4.9a. Suppose you are asked to find the acceleration of the block and the force the table exerts on it. First, note that the horizontal force being applied to the block acts through the string. The force the string exerts on the block is denoted by the symbol $\mathbf{T}$. The magnitude of $\mathbf{T}$ is equal to the tension in the string. In Figure 4.9a, a dotted circle is drawn around the block to remind you to isolate it from its surroundings.

Since we are interested only in the motion of the block, we must be able to *identify all external forces acting on it*. These are illustrated in Figure 4.9b. In addition to the force $\mathbf{T}$, the force diagram for the block includes the weight, $\mathbf{w}$, of the block and the normal force, $\mathbf{n}$. As usual, $\mathbf{w}$ corresponds to the force of gravity pulling down on the block and $\mathbf{n}$ is the upward force of the table on the block.

[3] The word *normal* is used because the direction of $\mathbf{n}$ is always *perpendicular* to the surface.

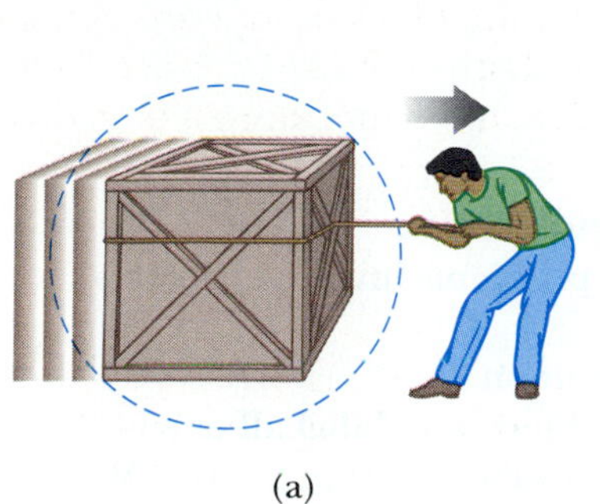
(a)

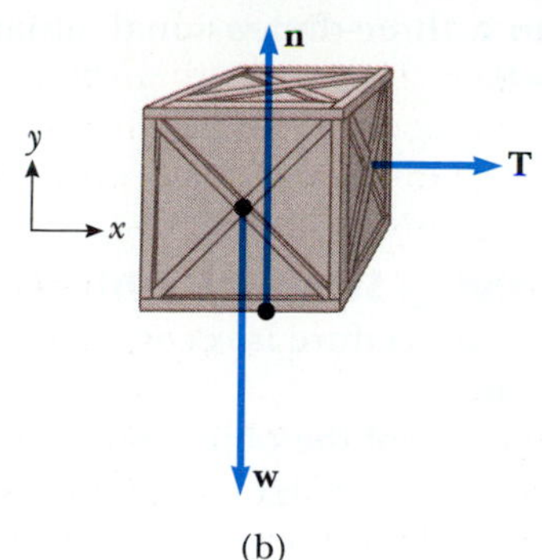

(b)

Figure 4.9
(a) A block being pulled to the right on a frictionless surface. (b) The free-body diagram that represents the external forces on the block.

The construction of such a diagram, called a **free-body diagram,** is an important step in applying Newton's laws. The *reactions* to the forces we have listed—namely, the force of the string on the hand, the force of the block on the Earth, and the force of the block on the table—are not included in the free-body diagram since they act on *other* bodies and not on the block.

Free-body diagrams are important when applying Newton's laws.

Now let us apply Newton's second law to the block. First we must choose an appropriate coordinate system. In this case it is convenient to use the coordinate system shown in Figure 4.9b, with the x axis horizontal and the y axis vertical. We can apply Newton's second law in the x direction, y direction, or both, depending on what we are asked to find in the problem. In addition, we may be able to use the equations of motion for constant acceleration that are found in Chapter 2. However, you should use these equations only when the acceleration is constant. For example, if the force T in Figure 4.9 is constant, then it follows that the acceleration in the x direction is also constant, because $a_x = T/m$. Hence, if we need to find the displacement or the velocity of the block at some instant of time, we can use the equations of motion with constant acceleration.

Objects in Equilibrium and Newton's First Law

Objects that are either at rest or moving with constant velocity are said to be in equilibrium, and Newton's first law is a statement of one condition that must be true for equilibrium conditions to prevail. In equation form, this condition of equilibrium can be expressed as

$$\sum \mathbf{F} = 0 \quad [4.7]$$

First condition of equilibrium

This statement signifies that the *vector* sum of all the forces (the net force) acting on an object in equilibrium is zero.

Usually, the problems we encounter in our study of equilibrium will be more easily solved if we work with Equation 4.7 in terms of the components of the external forces acting on an object. By this we mean that, in a two-dimensional problem, the sum of all the external forces in the x and y directions must separately equal zero; that is,

$$\sum F_x = 0 \qquad \sum F_y = 0 \quad [4.8]$$

This set of equations is often referred to as the **first condition for equilibrium.** We shall not consider three-dimensional problems in this text, but the extension of

Equations 4.8 to a three-dimensional situation can be made by adding a third equation, $\Sigma F_z = 0$.

Problem-Solving Strategy: Objects in Equilibrium

The following procedure is recommended for problems involving objects in equilibrium:

1. Make a sketch of the object under consideration.
2. Draw a free-body diagram for the *isolated* object, and label all external forces acting on the object. Assume a direction for each force. If you select a direction that leads to a negative sign in your solution for a force, do not be alarmed; this merely means that the direction of the force is the opposite of what you assumed.
3. Resolve all forces into x and y components, choosing a convenient coordinate system.
4. Use the equations $\Sigma F_x = 0$ and $\Sigma F_y = 0$. Remember to keep track of the signs of the various force components.
5. Application of Step 4 leads to a set of equations with several unknowns. Solve the simultaneous equations for the unknowns in terms of the known quantities.

Example 4.2 A Traffic Light at Rest

A traffic light weighing 100 N hangs from a cable tied to two other cables fastened to a support, as in Figure 4.10a. The upper cables make angles of 37° and 53° with the horizontal. Find the tension in the three cables.

Solution First we construct a free-body diagram for the traffic light, as in Figure 4.10b. The tension in the vertical cable, T_3, supports the light, and so we see that $T_3 = w = 100$ N. Now we construct a free-body diagram for the knot that holds the three cables together, as in Figure 4.10c. This is a convenient point to choose because all forces in question act at this point. We choose the coordinate axes as shown in Figure 4.10c, and resolve the forces into their x and y components:

Force	x component	y component
$\mathbf{T}_1$	$-T_1 \cos 37°$	$T_2 \sin 37°$
$\mathbf{T}_2$	$T_2 \cos 53°$	$T_2 \sin 53°$
$\mathbf{T}_3$	0	-100 N

The first condition for equilibrium gives us the equations

$$(1) \qquad \sum F_x = T_2 \cos 53° - T_1 \cos 37° = 0$$

$$(2) \qquad \sum F_y = T_1 \sin 37° + T_2 \sin 53° - 100\text{ N} = 0$$

From (1) we see that the horizontal components of $\mathbf{T}_1$ and $\mathbf{T}_2$ must be equal in magnitude, and from (2) we see that the sum of the vertical components of $\mathbf{T}_1$ and $\mathbf{T}_2$ must balance the weight of the light. We can solve (1) for T_2 in terms of T_1 to give

$$T_2 = T_1 \left(\frac{\cos 37°}{\cos 53°}\right) = 1.33 T_1$$

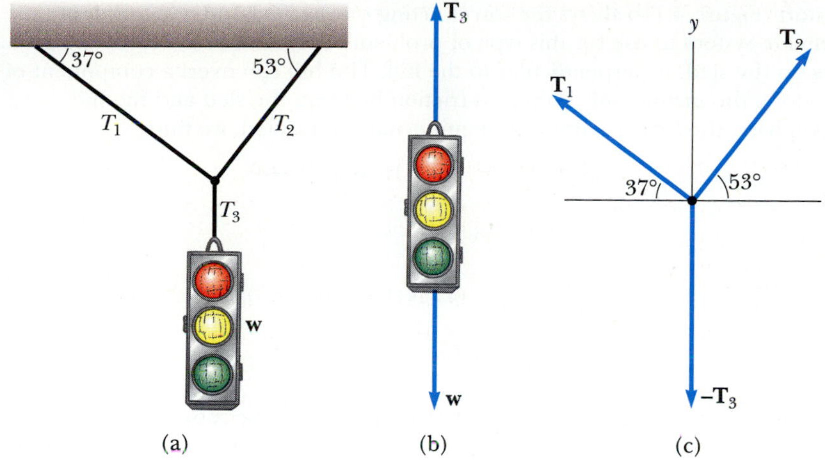

Figure 4.10 (Example 4.2) (a) A traffic light suspended by cables. (b) Free-body diagram for the traffic light. (c) Free-body diagram for the knot in the cable.

This value for T_2 can be substituted into (2) to give

$$T_1 \sin 37° + (1.33T_1)(\sin 53°) - 100 \text{ N} = 0$$

$$T_1 = 60.0 \text{ N}$$

$$T_2 = 1.33T_1 = 79.8 \text{ N}$$

Exercise In what situation will $T_1 = T_2$?

Answer When the supporting cables make equal angles with the horizontal support.

Example 4.3 A Sled on a Slick Hill

A child holds a sled at rest on a frictionless snow-covered hill, as shown in Figure 4.11a. If the sled weighs 77 N, find the force the child must exert on the rope and the force the hill exerts on the sled.

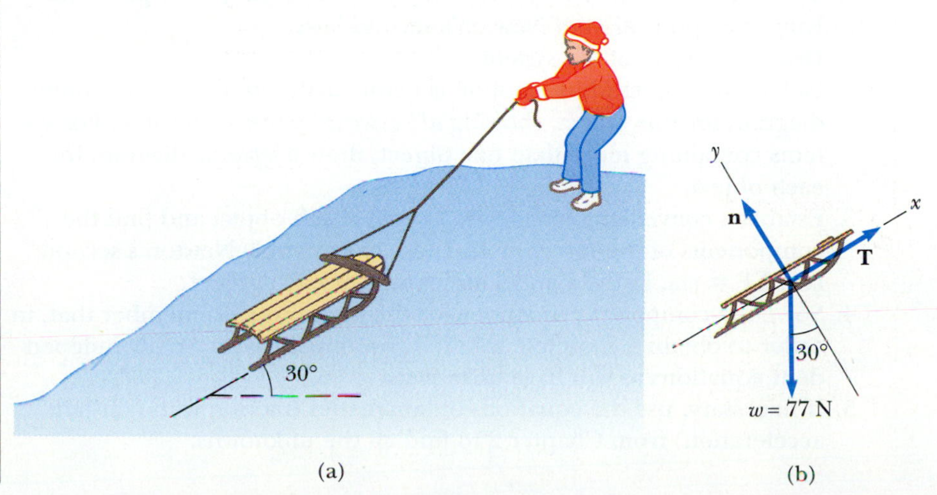

Figure 4.11 (Example 4.3) (a) A child holding a sled on a frictionless hill. (b) Free-body diagram for the sled.

Solution Figure 4.11b shows the forces acting on the sled and a convenient coordinate system to use for this type of problem. Note that **n**, the force the ground exerts on the sled, is perpendicular to the hill. The hill can exert a component of force along the incline only if there is friction between the sled and the hill.

Applying the first condition for equilibrium to the sled, we find

$$\sum F_x = T - (77\text{ N})(\sin 30°) = 0$$

$$T = 38.5\text{ N}$$

$$\sum F_y = n - (77\text{ N})(\cos 30°) = 0$$

$$n = 66.7\text{ N}$$

Note that n is *less* than the weight of the sled in this case, because the sled is on an incline and **n** is equal to and opposite the component of weight perpendicular to the incline.

Exercise What happens to the normal force as the angle of incline increases?

Answer It decreases.

Exercise Under what conditions would the normal force equal the weight of the sled?

Answer If the sled were on a horizontal surface and the applied force were either zero or along the horizontal.

Accelerating Objects and Newton's Second Law

In a situation in which a net force is acting on an object, the object is accelerated, and we must use Newton's second law in order to determine the features of the motion. The representative problems and suggestions that follow should help you to solve problems of this kind.

Problem-Solving Strategy: Newton's Second Law

The following procedure is recommended when dealing with problems involving the application of Newton's second law:

1. **Draw a diagram of the system.**
2. **Isolate the object whose motion is being analyzed. Draw a free-body diagram for this object, showing *all external forces acting on it.* For systems containing more than one object, draw a *separate* diagram for each object.**
3. **Establish convenient coordinate axes for each object and find the components of the forces along those axes. Apply Newton's second law, $\Sigma \mathbf{F} = m\mathbf{a}$, in the x and y directions for each object.**
4. **Solve the component equations for the unknowns. Remember that, in order to obtain a complete solution, you must have as many independent equations as you have unknowns.**
5. **If necessary, use the equations of kinematics (motion with constant acceleration) from Chapter 2 to find all the unknowns.**

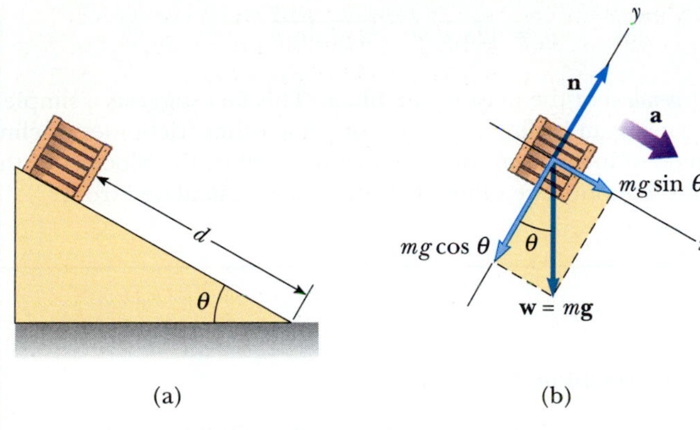

Figure 4.12 (Example 4.4) (a) A block sliding down a frictionless incline. (b) Free-body diagram for the block. Note that the magnitude of its acceleration along the incline is $g \sin\theta$.

Example 4.4 A Block on a Smooth Incline

A block of mass m is placed on a frictionless, inclined plane of angle θ, as in Figure 4.12a.

(a) Determine the acceleration of the block after it is released.

Solution The free-body diagram for the block is shown in Figure 4.12b. The only forces on the block are the normal force **n**, acting perpendicular to the plane, and the weight **w**, acting vertically downward. *It is convenient to choose the coordinate axes with x along the incline and y perpendicular to it.* Then we replace the weight vector with a component of magnitude $mg \sin\theta$ along the positive x axis and a component of magnitude $mg\cos\theta$ in the negative y direction. Applying Newton's second law in component form, while noting that $a_y = 0$, gives

$$(1) \qquad \sum F_x = mg\sin\theta = ma_x$$

$$(2) \qquad \sum F_y = n - mg\cos\theta = 0$$

From (1) we see that the acceleration along the incline is provided by the component of weight down the incline:

$$(3) \qquad a_x = \boxed{g\sin\theta}$$

From (2) we conclude that the component of weight perpendicular to the incline is *balanced* by the normal force; that is, $n = mg\cos\theta$. Note that the acceleration given by (3) is *independent* of the mass of the block—it depends only on the angle of inclination and on g.

Special Cases When $\theta = 90°$, $a = g$ and $n = 0$. This case corresponds to the block in free fall. When $\theta = 0$, $a_x = 0$ and $n = mg$ (its maximum value).

(b) Suppose the block is released from rest at the top, and the distance from the front edge of the block to the bottom is d. How long does it take the front edge of the block to reach the bottom, and what is its speed just as it gets there?

Solution Since $a_x =$ constant, we can apply the equation $x - x_0 = v_{x0}t + \frac{1}{2}a_xt^2$ (Eq. 2.10) to the block. Since the displacement $x - x_0 = d$ and $v_{x0} = 0$, we get $d = \frac{1}{2}a_xt^2$, or

$$(4) \qquad t = \sqrt{\frac{2d}{a_x}} = \boxed{\sqrt{\frac{2d}{g\sin\theta}}}$$

Also, since $v_x^2 = v_{x0}^2 + 2a_x(x - x_0)$ (Eq. 2.11) and $v_{x0} = 0$, we find that $v_x^2 = 2a_xd$, or

$$(5) \qquad v_x = \sqrt{2a_x d} = \sqrt{2gd \sin \theta}$$

Again, t and v_x are *independent* of the mass of the block. This fact suggests a simple method of measuring g using an inclined air track or some other frictionless incline. Simply measure the angle of inclination, the distance traveled by the block, and the time it takes to reach the bottom. The value of g can then be calculated from (4) and (5).

Example 4.5 Atwood's Machine

When two unequal masses are hung vertically over a light, frictionless pulley as in Figure 4.13a, the arrangement is called *Atwood's machine.* The device is sometimes used in the laboratory to measure the free-fall acceleration. Calculate the acceleration of the two masses and the tension in the string in terms of g.

Solution The free-body diagrams for the two masses are shown in Figure 4.13b, where we assume that $m_2 > m_1$. When Newton's second law is applied to m_1, with a upward for this mass, we find

$$(1) \qquad \sum F_y = T - m_1 g = m_1 a$$

Similarly, for m_2 we find

$$(2) \qquad \sum F_y = T - m_2 g = -m_2 a$$

The negative sign on the right-hand side of (2) indicates that m_2 accelerates downward, in the negative y direction.

When (2) is subtracted from (1), T drops out and we get

$$-m_1 g + m_2 g = m_1 a + m_2 a$$

$$(3) \qquad a = \left(\frac{m_2 - m_1}{m_1 + m_2}\right) g$$

If (3) is substituted into (1), we get

$$(4) \qquad T = \left(\frac{2m_1 m_2}{m_1 + m_2}\right) g$$

Special Cases When $m_1 = m_2$, $a = 0$ and $T = m_1 g = m_2 g$, as we would expect for the balanced case. Also, if $m_2 \gg m_1$, $a \approx g$ (a freely falling body) and $T \approx 2m_1 g$.

Exercise Find the acceleration and tension of an Atwood's machine in which $m_1 = 2$ kg and $m_2 = 4$ kg.

Answer $a = 3.27 \text{ m/s}^2$, $T = 26.1$ N.

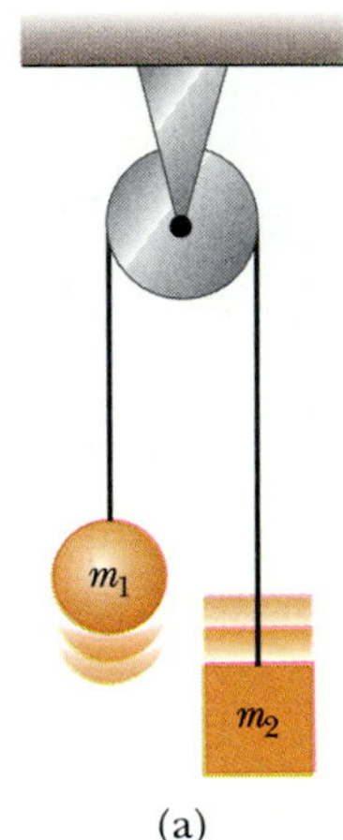

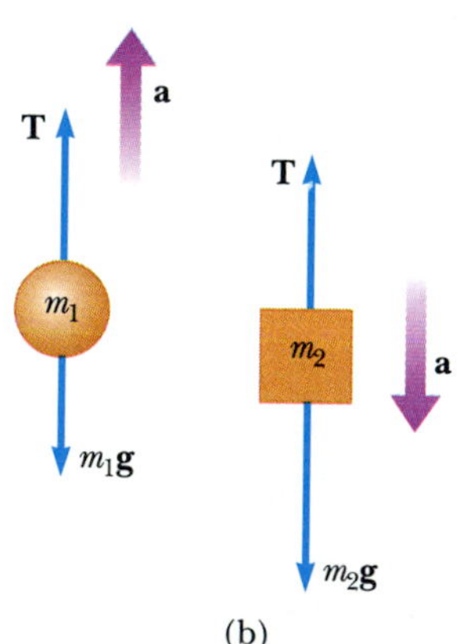

Figure 4.13 (Example 4.5) Atwood's machine. (a) Two masses connected by a light string over a frictionless pulley. (b) Free-body diagrams for m_1 and m_2.

Example 4.6 Two Connected Objects

Two unequal masses are attached by a light string that passes over a light, frictionless pulley as in Figure 4.14a. The block of mass m_2 lies on a frictionless incline of angle θ. Find the acceleration of the two masses and the tension in the string.

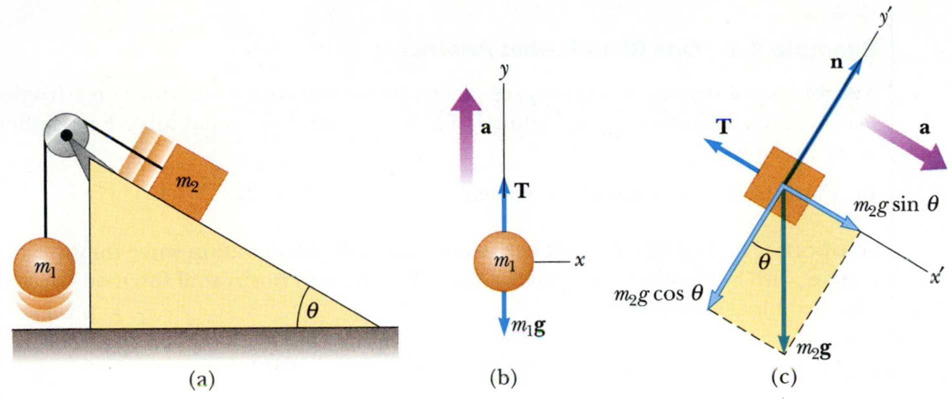

Figure 4.14 (Example 4.6) (a) Two masses connected by a light string over a frictionless pulley. (b) Free-body diagram for m_1. (c) Free-body diagram for m_2 (the incline is frictionless).

Solution Since the two masses are connected by a string (which we assume doesn't stretch), they have accelerations of the same magnitude. The free-body diagrams for the two masses are shown in Figure 4.14b and 4.14c. Application of Newton's second law in component form to m_1, assuming that **a** is upward for this mass, gives the equations of motion for m_1:

$$(1) \qquad \sum F_x = 0$$

$$(2) \qquad \sum F_y = T - m_1 g = m_1 a$$

Note that in order for a to be positive, it is necessary that $T > m_1 g$.

Now, for m_2 it is convenient to choose the positive x' axis along the incline as in Figure 4.14c. Application of Newton's second law in component form to m_2, assuming that **a** is down the incline for this block, gives the equations of motion for m_2:

$$(3) \qquad \sum F_{x'} = m_2 g \sin\theta - T = m_2 a$$

$$(4) \qquad \sum F_{y'} = n - m_2 g \cos\theta = 0$$

Expressions (1) and (4) provide no information regarding the acceleration. However, if we solve (2) and (3) simultaneously for the unknowns a and T, we get

$$(5) \qquad a = \frac{m_2 g \sin\theta - m_1 g}{m_1 + m_2}$$

When this is substituted into (2), we find

$$(6) \qquad T = \frac{m_1 m_2 g(1 + \sin\theta)}{m_1 + m_2}$$

Note that m_2 accelerates down the incline (the $+x'$ direction) if $m_2 \sin\theta$ exceeds m_1 (that is, if a is positive, as we assumed). If m_1 exceeds $m_2 \sin\theta$, the acceleration of m_2 is up the incline and downward for m_1. Also note that the result for the acceleration, (5), can be interpreted as the resultant unbalanced force on the system divided by the total mass of the system.

Exercise If $m_1 = 10$ kg, $m_2 = 5$ kg, and $\theta = 45°$, find the acceleration.

Answer $a = -4.22$ m/s^2, where the negative sign indicates that m_2 accelerates up the incline.

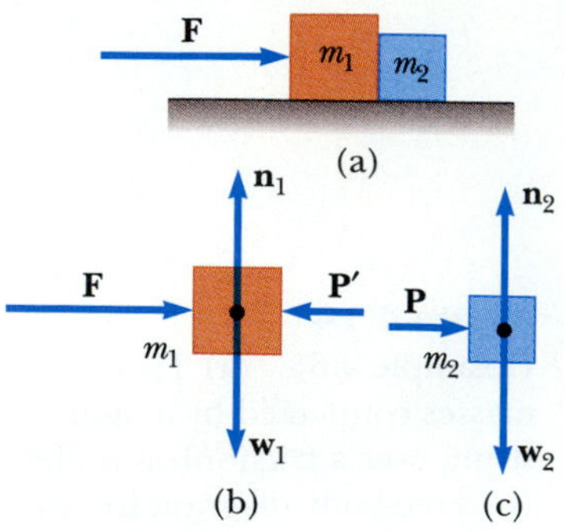

Figure 4.15
(Example 4.7)

Example 4.7 One Block Pushes Another

Two blocks of masses m_1 and m_2 are placed in contact with each other on a frictionless, horizontal surface, as in Figure 4.15a. A constant horizontal force **F** is applied to m_1 as shown.

(a) Find the acceleration of the system.

Solution The two blocks must experience the *same* acceleration since they are in contact with each other. Because the force **F** is the *only* horizontal force on the system (the two blocks), we have

$$\sum F_x(\text{system}) = F = (m_1 + m_2)a$$

$$(1) \qquad a = \frac{F}{m_1 + m_2}$$

(b) Determine the magnitude of the contact force between the two blocks.

Solution To solve this part of the problem, it is necessary to first construct a free-body diagram for each block shown in Figures 4.15b and 4.15c, where the contact force is denoted by **P**. From Figure 4.15c we see that the only horizontal force acting on m_2 is the contact force **P** (the force of m_1 acting on m_2), which is to the right. Applying Newton's second law to m_2 gives

$$(2) \qquad \sum F_x = P = m_2 a$$

Substituting the value of the acceleration a from (1) into (2) gives

$$(3) \qquad P = m_2 a = \left(\frac{m_2}{m_1 + m_2}\right)F$$

Note from this result that the contact force **P** is *less* than the applied force **F**. This inequality is consistent with the fact that the force required to accelerate m_2 alone must be *less* than the force required to produce the same acceleration for the system of two blocks.

It is instructive to check this expression for P by considering the forces acting on m_1, shown in Figure 4.15b. The horizontal forces acting on m_1 are the applied force **F** to the right and the contact force **P′** to the left (the force of m_2 on m_1). From Newton's third law, **P′** is the reaction to **P**, so $|\mathbf{P}'| = |\mathbf{P}|$. Applying Newton's second law to m_2 gives

$$(4) \qquad \sum F_x = F - P' = F - P = m_2 a$$

Substituting the value of a from (2) into (4) gives

$$P = F - m_2 a = F - \frac{m_1 F}{m_1 + m_2} = \left(\frac{m_2}{m_1 + m_2}\right)F$$

This agrees with (3), as it must.

Exercise If $m_1 = 4$ kg, $m_2 = 3$ kg, and $F = 9$ N, find the acceleration of the system and the magnitude of the contact force.

Answer $a = 1.29\ \text{m/s}^2$; $P = 3.86$ N.

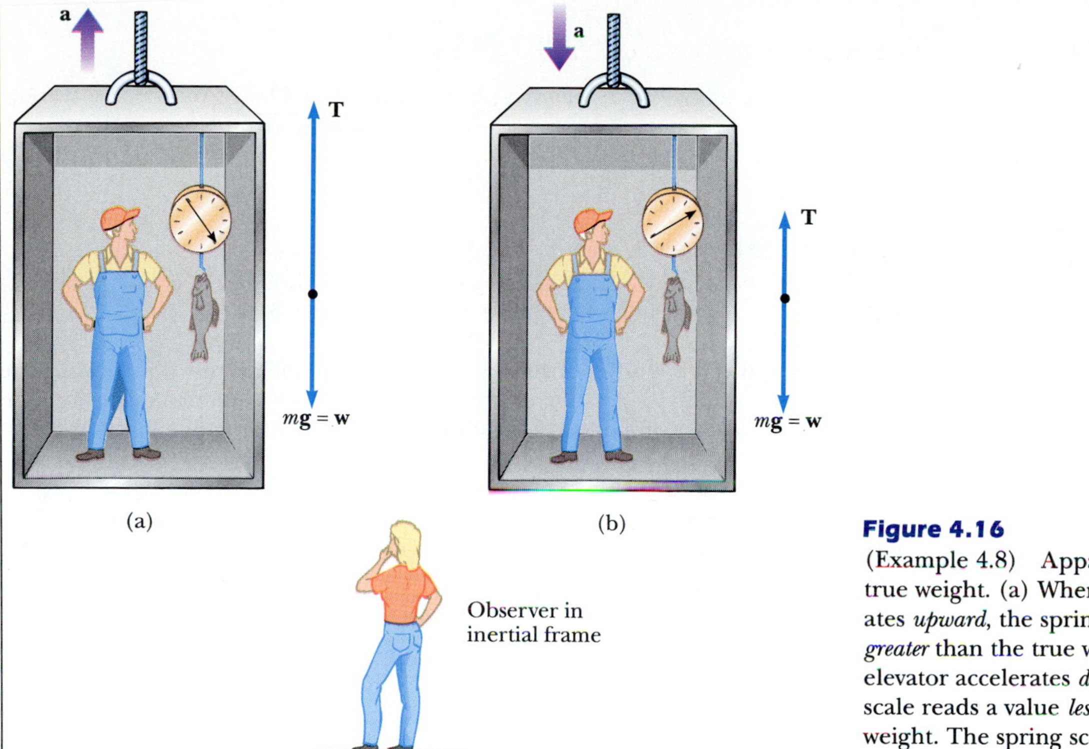

Figure 4.16
(Example 4.8) Apparent weight versus true weight. (a) When the elevator accelerates *upward,* the spring scale reads a value *greater* than the true weight. (b) When the elevator accelerates *downward,* the spring scale reads a value *less* than the true weight. The spring scale reads the *apparent* weight.

Example 4.8 Weighing a Fish in an Elevator

A person weighs a fish on a spring scale attached to the ceiling of an elevator, as shown in Figure 4.16. Show that if the elevator accelerates or decelerates, the spring scale reads a weight different from the true weight of the fish.

Solution The external forces acting on the fish are its true weight, **w**, and the upward constraint force, **T**, exerted on it by the scale. By Newton's third law, T is also the reading of the spring scale. If the elevator is at rest or moving at constant velocity, then the fish is not accelerating and $T = w = mg$ (where $g = 9.80 \text{ m/s}^2$). If the elevator moves upward with an acceleration **a** relative to an observer outside the elevator in an inertial frame (Fig. 4.16a), then the second law applied to the fish of mass m gives the total force **F** on the fish:

$$(1) \qquad \sum F = T - w = ma \qquad \text{(if } \mathbf{a} \text{ is upward)}$$

Likewise, if the elevator accelerates downward as in Figure 4.16b, Newton's second law applied to the fish becomes

$$(2) \qquad \sum F = T - w = -ma \qquad \text{(if } \mathbf{a} \text{ is downward)}$$

Thus, we conclude from (1) that the scale reading, T, is greater than the true weight, w, if **a** is upward. From (2) we see that T is less than w if **a** is downward.

For example, if the true weight of the fish is 40 N, and a is 2 m/s² upward, then the scale reading is

$$T = ma + mg = mg\left(\frac{a}{g} + 1\right)$$

$$= w\left(\frac{a}{g} + 1\right) = (40\text{ N})\left(\frac{2\text{ m/s}^2}{9.80\text{ m/s}^2} + 1\right)$$

$$= \boxed{48.2\text{ N}}$$

If a is 2 m/s² downward, then

$$T = -ma + mg = mg\left(1 - \frac{a}{g}\right) = \boxed{31.8\text{ N}}$$

Hence, if you buy a fish in an elevator, make sure the fish is weighed while the elevator is at rest or accelerating downward! Furthermore, note that one cannot determine the *direction* of motion of the elevator from the information given here.

Special Cases If the cable breaks, then the elevator falls freely and $a = -g$. Since $w = mg$, we see from (1) that the apparent weight, T, is zero; that is, the fish appears to be weightless. If the elevator accelerates *downward* with an acceleration *greater* than g, the fish (along with the person in the elevator) will eventually hit the ceiling, since the acceleration of the fish will still be that of a freely falling body relative to an outside observer.

More on Free-Body Diagrams

In order to successfully apply Newton's second law to a mechanical system, you must first be able to recognize all the forces acting on the system. That is, you must be able to construct the correct free-body diagram. The importance of this cannot be overemphasized. An incorrect free-body diagram will most likely result in the wrong solution to the problem. Figure 4.17 presents a number of mechanical systems with their corresponding free-body diagrams. You should examine these carefully and then proceed to construct free-body diagrams for other systems described in the problems. When a system contains more than one element, it is important that you construct a free-body diagram for *each* element.

As usual, **F** denotes some applied force, $\mathbf{w} = m\mathbf{g}$ is the weight, **n** denotes a normal force, and **T** is the force of tension.

Summary

Newton's first law

Newton's first law states that a body at rest will remain at rest, and a body in uniform motion in a straight line will maintain that motion, unless an external resultant force acts on the body.

Newton's second law

Newton's second law states that the acceleration of an object is directly proportional to the resultant force acting on the object and inversely proportional to the object's mass. If the mass of the body is constant, the net force equals the product of the mass and its acceleration, or $\Sigma\mathbf{F} = m\mathbf{a}$.

Inertial frame

Newton's first and second laws are valid in an inertial frame of reference. An **inertial frame** is one in which Newton's first law is valid.

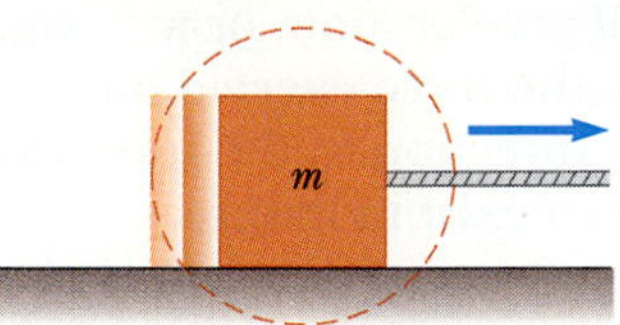

A block pulled to the right on a smooth, horizontal surface

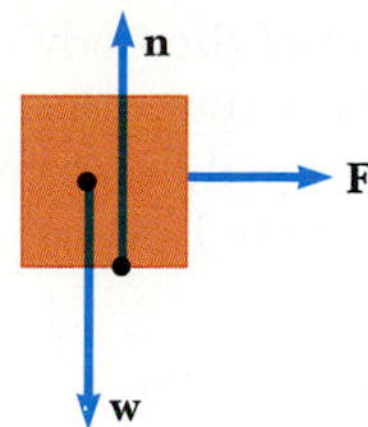

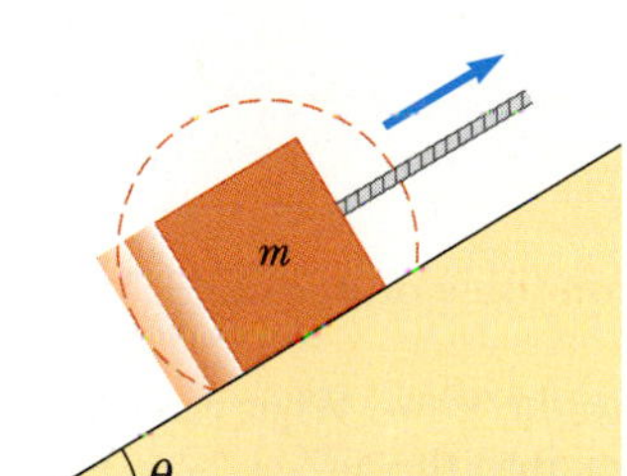

A block being pulled up a smooth incline

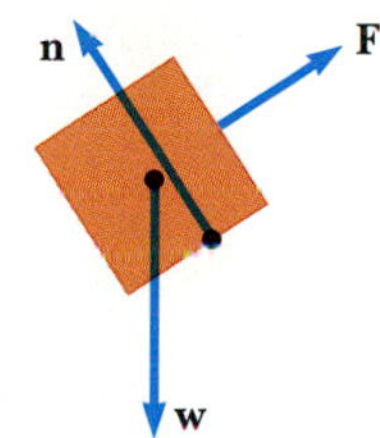

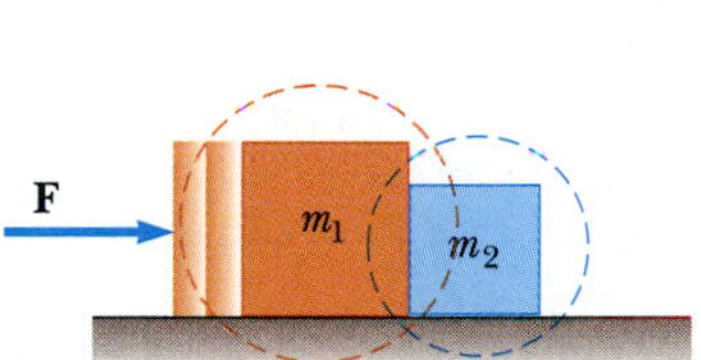

Two blocks in contact, pushed to the right on a smooth surface

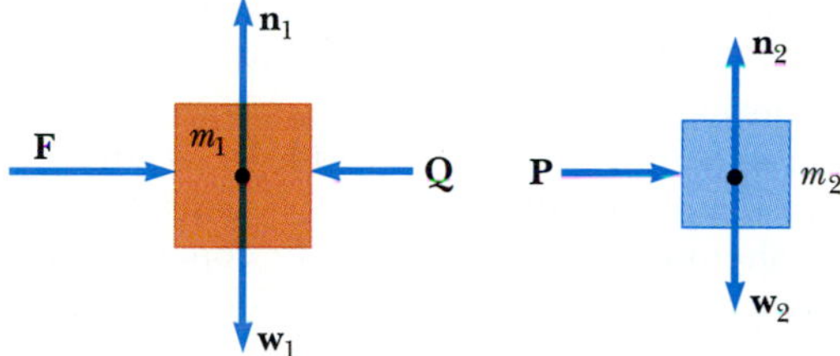

Note: **P** = −**Q** since they are an action-reaction pair

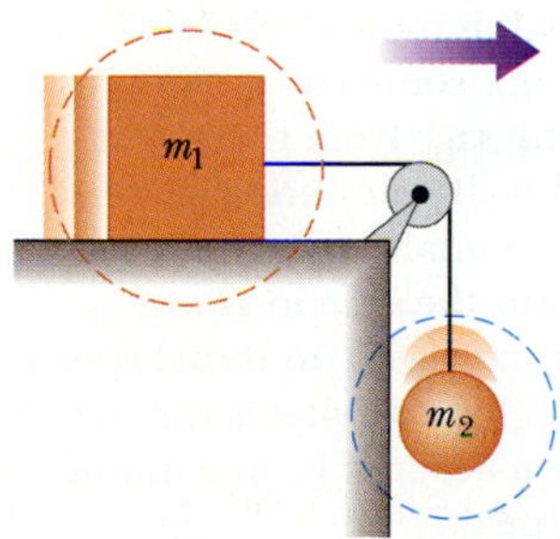

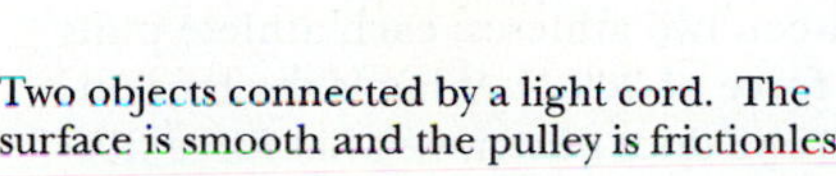

Two objects connected by a light cord. The surface is smooth and the pulley is frictionless

Figure 4.17
Various mechanical configurations (*left*) and the corresponding free-body diagrams (*right*).

Mass is a scalar quantity.

Weight The **weight** of a body is the force exerted on the body by the Earth. It is equal to the product of the body's mass and the free-fall acceleration, or $w = mg$.

Newton's third law **Newton's third law** states that if two bodies interact, the force exerted on body 1 by body 2 is equal in magnitude but opposite in direction to the force exerted on body 2 by body 1. Thus, an isolated force cannot exist in nature.

Questions and Conceptual Problems

1. If an object is at rest, can we conclude that no external forces are acting on it?
2. If gold were sold by weight, would you rather buy it in Denver or in Death Valley? If it were sold by mass, at which of the two locations would you prefer to buy it? Why?
3. A passenger sitting in the rear of a bus claims that he was injured when the driver slammed on the brakes, causing a suitcase to come flying toward the passenger from the front of the bus. If you were the judge in this case, what disposition would you make? Why?
4. A space explorer is in a spaceship moving through space, far from any planet or star. She notices a large rock, taken as a specimen from an alien planet, floating around the cabin of the spaceship. Should she push it gently toward a storage compartment or kick it toward the compartment? Why?
5. How much does an astronaut weigh out in space, far from any planet?
6. The observer in the elevator of Example 4.8 would claim that the "weight" of the fish is T, the scale reading. This is obviously wrong. Why does this observation differ from that of a person outside the elevator, at rest with respect to the elevator?
7. Identify the action-reaction pairs in the following situations: a man takes a step; a snowball hits a girl in the back; a baseball player catches a ball; a gust of wind strikes a window.
8. While a football is in flight, what forces act on it? What are the action-reaction pairs while the football is being kicked and while it is in flight?
9. A ball is held in a person's hand. (a) Identify all the external forces acting on the ball and the reaction to each. (b) If the ball is dropped, what force is exerted on it while it is falling? Identify the reaction force in this case. (Neglect air resistance.)
10. A child pulls a wagon with some force, causing it to accelerate. Newton's third law says that the wagon exerts an equal and opposite reaction force on the child. How can the wagon accelerate?
11. A rubber ball is dropped onto the floor. What force causes the ball to bounce back into the air?
12. What is wrong with the statement "Since the car is at rest, there are no forces acting on it"? How would you correct this sentence?
13. If you have ever taken a ride in an elevator of a high-rise building, you may have experienced the nauseating sensation of "heaviness" and "lightness" depending on the direction of a. Explain these sensations. Are we truly weightless in free fall?
14. In an attempt to define Newton's third law, a student states that the action and reaction forces are equal and opposite each other. If this is the case, how can there ever be a net force on an object?
15. The force of gravity is twice as great on a 20-N rock as on a 10-N rock. Why doesn't the 20-N rock have a greater free-fall acceleration?
16. Is it possible to have motion in the absence of a force? Explain.
17. Is there any relation between the net force acting on an object and the direction in which it moves? Explain.
18. The mayor of a city decides to fire some city employees because they will not remove the sags from the cables that support the city traffic lights. If you were a lawyer, what defense would you give on behalf of the employees? Who do you think would win the case in court?
19. A 0.15-kg baseball is thrown upward with an initial speed of 20 m/s. If air resistance is neglected, what is the net force on the ball (a) when it reaches half its maximum height? (b) when it reaches its peak?
20. In a tug-of-war between two athletes, each athlete pulls on the rope with a force of 200 N. What is the tension in the rope? If the rope does not move what force does each athlete exert against the ground?
21. If a car is traveling eastward with a constant speed of 20 m/s, what is the resultant force acting on it?

22. Suppose the head of a hammer is loose, and you wish to tighten it. In terms of inertia, explain how you can accomplish this by banging the bottom of the handle (rather than the hammerhead) against a hard surface.
23. Suppose a truck loaded with sand accelerates at 0.5 m/s^2 on a highway. If the driving force on the truck remains constant, what happens to the truck's acceleration if its trailer leaks sand at a constant rate through a hole in its bottom?
24. As a rocket is fired from a launching pad, its speed and acceleration increase with time as its engines continue to operate. Explain why this occurs even though the thrust of the engines remains constant.
25. If a small sports car collides head-on with a massive truck, which vehicle experiences the greater impact force? Which vehicle experiences the greater acceleration? Explain.
26. Draw a free-body diagram for each of the following objects: (a) a projectile in motion in the presence of air resistance, (b) a rocket leaving the launch pad with its engines operating, (c) an athlete running along a horizontal track.

Problems

Sections 4.1 Through 4.7

1. A force, **F**, applied to an object of mass m_1 produces an acceleration of 3 m/s^2. The same force applied to a second object of mass m_2 produces an acceleration of 1 m/s^2. (a) What is the value of the ratio m_1/m_2? (b) If m_1 and m_2 are combined, find their acceleration under the action of the force **F**.
2. A 6-kg object undergoes an acceleration of 2 m/s^2. (a) What is the magnitude of the resultant force acting on the object? (b) If this same force is applied to a 4-kg object, what acceleration does it produce?
3. A force of 10 N acts on a body of mass 2 kg. What are (a) the body's acceleration, (b) its weight in newtons, and (c) its acceleration if the force is doubled?
4. A 3-kg particle starts from rest and moves a distance of 4 m in 2 s, under the action of a single, constant force. Find the magnitude of the force.
5. A 5.0-g bullet leaves the muzzle of a rifle with a speed of 320 m/s. What average force is exerted on the bullet while it is traveling down the 0.82-m-long barrel of the rifle? Assume the bullet's acceleration is constant.
6. A pitcher releases a baseball of weight 1.4 N with a speed of 32 m/s by uniformly accelerating his arm for 0.09 s. If the ball starts from rest, (a) through what distance does the ball accelerate before its release? (b) What average force is exerted on the ball to produce this acceleration?
7. A 3-kg mass undergoes an acceleration given by **a** = (2**i** + 5**j**) m/s^2. Find the resultant force, **F**, and its magnitude.
8. Verify the following conversions: (a) 1 N = 10^5 dynes, (b) 1 N = 0.225 lb.
9. A woman weighs 120 lb. Determine (a) her weight in newtons and (b) her mass in kilograms.
10. (a) A car with a mass of 850 kg is moving to the right with a constant speed of 1.44 m/s. What is the force on the car? (b) What is the force on the car if it is moving to the left?
11. What is the mass of an astronaut whose weight on the Moon is 115 N? The acceleration due to gravity on the Moon is 1.63 m/s^2.
12. If a man weighs 900 N on Earth, what would he weigh on Jupiter, where the acceleration due to gravity is 25.9 m/s^2?
13. The engine on a 0.2-kg model airplane exerts a forward force on the plane of 10 N. If the plane accelerates at 2 m/s^2, what is the magnitude of the resistive force due to the wind acting on the airplane?
14. An 1800-kg car is traveling in a straight line with a speed of 25 m/s. What is the magnitude of the constant horizontal force that is needed to bring the car to rest in a distance of 80 m?
15. A boat moves through the water with two forces acting on it. One is a 2000-N forward push by the motor; the other is an 1800-N resistive force due to the water. (a) What is the acceleration of the 1000-kg boat? (b) If it starts from rest, how far will it move in 10 s? (c) What will be its speed at the end of this time?
16. Forces of 10 N north, 20 N east, and 15 N south are simultaneously applied to a 4-kg mass. Obtain the object's acceleration.
17. A heavy freight train has a mass of 15 000 metric tons. If the locomotives can exert pull of 750 000 N, how long does it take to increase the speed from 0 to 80 km/h?

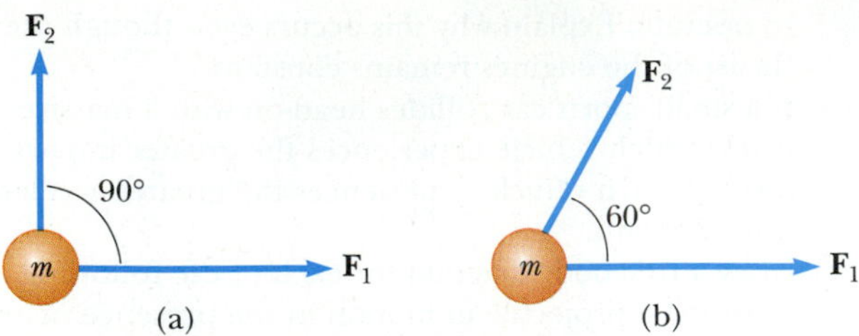

Figure 4.18 (Problem 18)

18. Two forces, $\mathbf{F}_1$ and $\mathbf{F}_2$, act on a 5-kg mass. If $F_1 =$ 20 N and $F_2 = 15$ N, find the accelerations in (a) and (b) of Figure 4.18.
19. A constant force changes the speed of an 80-kg sprinter from 3 m/s to 4 m/s in 0.5 s. (a) Calculate the magnitude of the acceleration of the sprinter. (b) Obtain the magnitude of the force. (c) Determine the magnitude of the acceleration of a 50-kg sprinter experiencing the same force. (Assume linear motion.)
20. A 4-kg object has a velocity of $3\mathbf{i}$ m/s at one instant. Eight seconds later, its velocity is $(8\mathbf{i} + 10\mathbf{j})$ m/s. Assuming the object was subject to a constant net force, find (a) the components of the force and (b) its magnitude.
21. A barefoot field-goal kicker imparts a speed of 30 m/s to a football initially at rest. If the football has a mass of 0.5 kg and the time of contact with the football is 0.025 s, what is the force exerted on the foot?
22. A 2-ton truck provides an acceleration of 3 ft/s^2 to a 5-ton trailer. If the truck exerts the same force on the road while pulling a 15-ton trailer, what acceleration results?
23. An electron of mass 9.1×10^{-31} kg has an initial speed of 3.0×10^5 m/s. It travels in a straight line, and its speed increases to 7.0×10^5 m/s in a distance of 5.0 cm. Assuming its acceleration is constant, (a) determine the force on the electron and (b) compare this force with the weight of the electron, which we neglected.
24. A 15-lb block rests on the floor. (a) What force does the floor exert on the block? (b) If a rope is tied to the block and run vertically over a pulley, and the other end is attached to a free-hanging 10-lb weight, what is the force of the floor on the 15-lb block? (c) If we replace the 10-lb weight in part (b) with a 20-lb weight, what is the force of the floor on the 15-lb block?

Section 4.8 Some Applications of Newton's Laws

25. It is amateur night at the tightrope walkers' convention, and a 600-N performer finds himself in the awkward position shown in Figure 4.19. If the angle between the rope and the horizontal is 8°, find the tension in the rope on either side of the performer.

Figure 4.19 (Problem 25)

26. A 150-N bird feeder is supported by three cables as shown in Figure 4.20. Find the tension in each cable.

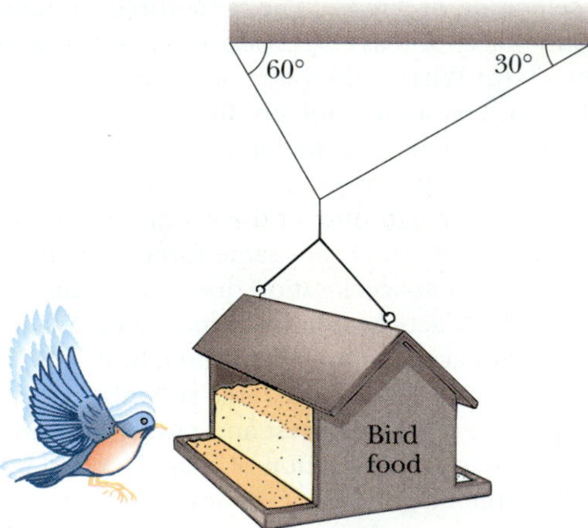

Figure 4.20 (Problem 26)

27. A 1-kg mass is observed to accelerate at 10 m/s^2 in a direction 30° north of east (Fig. 4.21). One of the two forces acting on the mass has a magnitude of 5 N and is directed north. Determine the magnitude and direction of the second force acting on the mass.

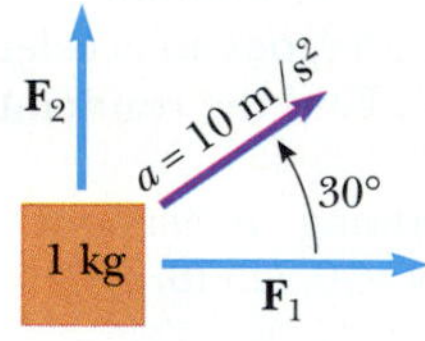

Figure 4.21 (Problem 27)

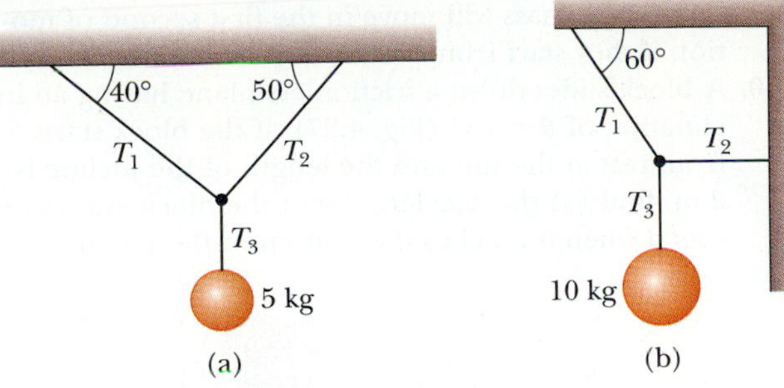

Figure 4.22 (Problem 28)

28. Find the tension in each cord of the systems described in Figure 4.22. (Neglect the masses of the cords.)
29. The distance between two telephone poles is 50 m. When a 1-kg bird lands on the telephone wire midway between the poles, the wire sags 0.2 m. How much tension does the bird produce in the wire? Ignore the weight of the wire.
30. The systems shown in Figure 4.23 are in equilibrium. If the spring scales are calibrated in newtons, what do they read? (Neglect the masses of the pulleys and strings, and assume the incline is frictionless.)

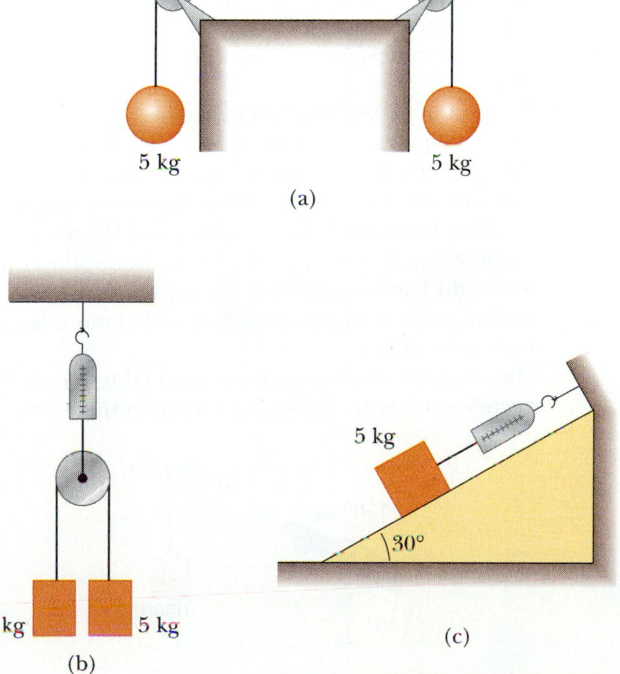

Figure 4.23 (Problem 30)

31. A 0.15-kg baseball moving at 20 m/s strikes the glove of a catcher. The glove recoils a distance of 8 cm. What average force does the glove exert on the ball? What average force does the ball exert on the glove?
32. A train has a mass of 5.22×10^6 kg and is moving with a speed of 90.0 km/h. The engineer applies the brakes, which results in a net backward force of 1.87×10^6 N on the train. The brakes are held on for 30.0 s. (a) What is the new speed of the train? (b) How far does the train travel during this period?
33. The parachute on a 900-kg race car opens at the end of a quarter-mile run, when the car is traveling at 35 m/s. What total retarding force must be supplied by the parachute to stop the car in a distance of 1000 m?
34. A 5-kg bucket of water is raised from a well by a rope attached to the bucket. If the upward acceleration of the bucket is 3 m/s², find the force exerted on the bucket by the rope.
35. Two people pull as hard as they can on ropes attached to a boat that has a mass of 200 kg. If they pull in the same direction, the boat has an acceleration of 1.52 m/s² to the right. If they pull in opposite directions, the boat has an acceleration of 0.518 m/s² to the left. What is the force exerted by each person on the boat? (Disregard any other forces on the boat.)
36. A 5-kg mass placed on a frictionless, horizontal table is connected to a cable that passes over a pulley and then is fastened to a hanging 10-kg mass, as in Figure 4.24. Find the acceleration of the two objects and the tension in the string.

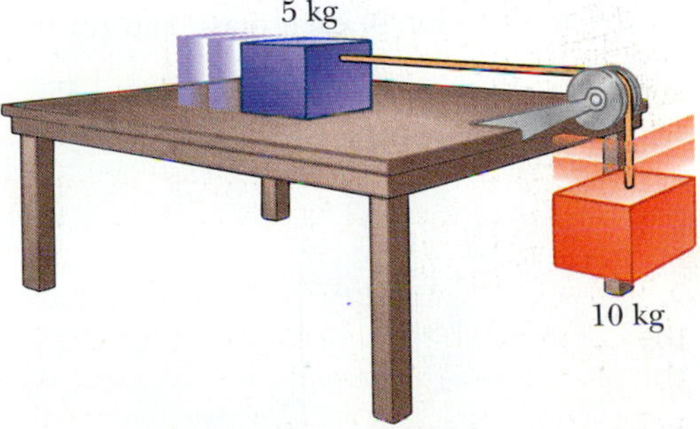

Figure 4.24 (Problem 36)

37. A 2000-kg sailboat experiences an eastward force of 3000 N from the ocean tide and a wind force against its sails of magnitude 6000 N, directed toward the northwest (45° north of west). What are the magnitude and direction of the resultant acceleration?

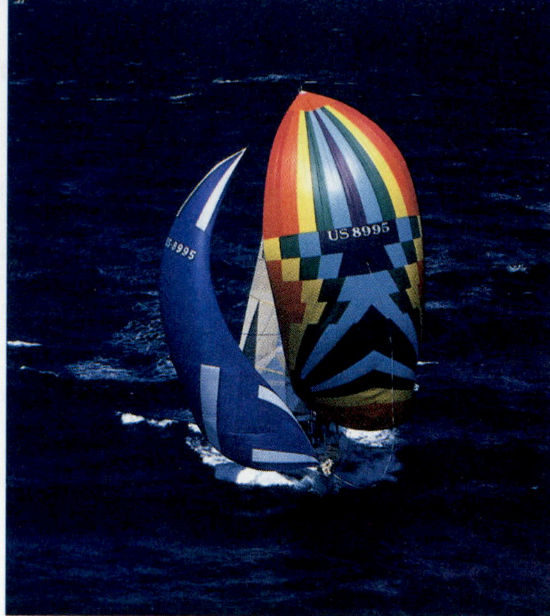

(© Doug Peebles/Index Stock International)

38. Two masses, m_1 and m_2, situated on a frictionless, horizontal surface are connected by a light string. A force, F, is exerted on one of the masses to the right (Fig. 4.25). Determine the acceleration of the system and the tension, T, in the string.

Figure 4.25 (Problem 38)

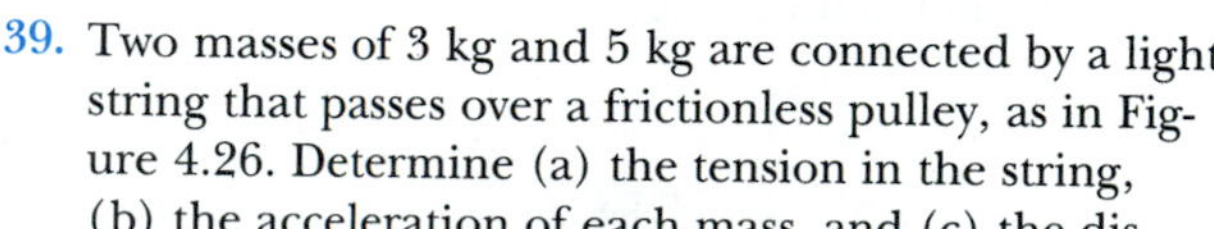

39. Two masses of 3 kg and 5 kg are connected by a light string that passes over a frictionless pulley, as in Figure 4.26. Determine (a) the tension in the string, (b) the acceleration of each mass, and (c) the distance each mass will move in the first second of motion if they start from rest.

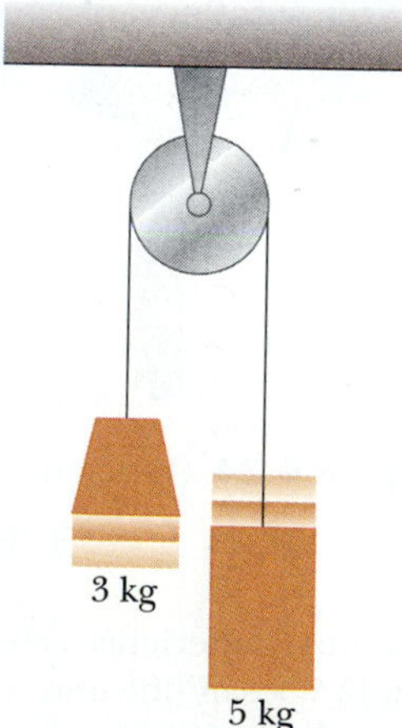

Figure 4.26 (Problem 39)

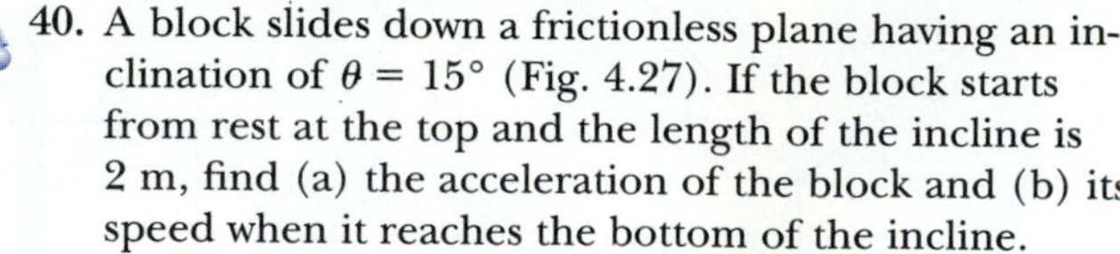

40. A block slides down a frictionless plane having an inclination of $\theta = 15°$ (Fig. 4.27). If the block starts from rest at the top and the length of the incline is 2 m, find (a) the acceleration of the block and (b) its speed when it reaches the bottom of the incline.

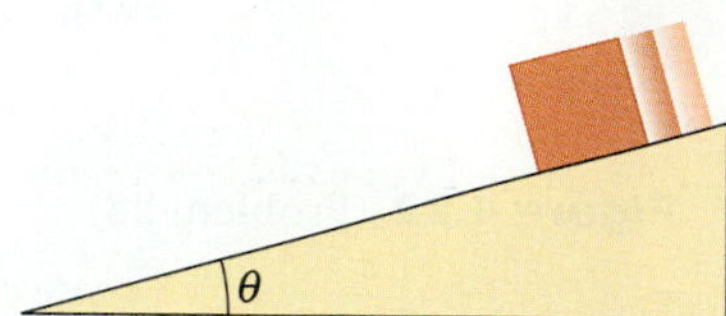

Figure 4.27 (Problems 40 and 41)

41. A block is given an initial velocity of 5 m/s up a frictionless 20° incline (Fig. 4.27). How far up the incline does the block slide before coming to rest?

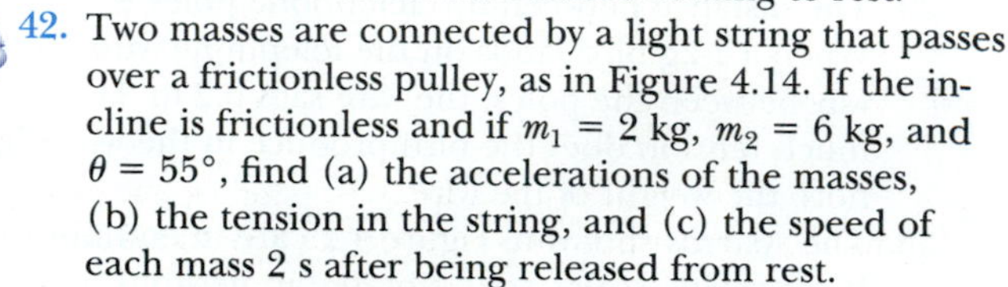

42. Two masses are connected by a light string that passes over a frictionless pulley, as in Figure 4.14. If the incline is frictionless and if $m_1 = 2$ kg, $m_2 = 6$ kg, and $\theta = 55°$, find (a) the accelerations of the masses, (b) the tension in the string, and (c) the speed of each mass 2 s after being released from rest.

43. A 3.0-kg mass is moving in a plane, with its x and y coordinates given by $x = 5t^2 - 1$ and $y = 3t^3 + 2$ (x and y are in meters and t is in seconds). Find the magnitude of the net force acting on this mass at $t = 2.0$ s.

44. A net horizontal force, $F = A + Bt^3$, acts on a 2-kg object, where $A = 5.0$ N and $B = 2.0$ N/s^3. What is the horizontal speed of this object 4 s after it starts from rest?

45. A car of mass 1500 kg is being pulled up a loading ramp inclined at 30° with the horizontal, as in Figure 4.28. The car is attached to a cable, which passes over a frictionless pulley to a 10 000 N counterweight. Find (a) the tension in the cable, and (b) the acceleration of the system. (c) What mass should the counterweight have in order for the car to move down the incline at an acceleration of 2 m/s^2? (Ignore any effects of friction.)

46. Mass m_1 on a frictionless horizontal table is connected to mass m_2 through a very light pulley, P_1,

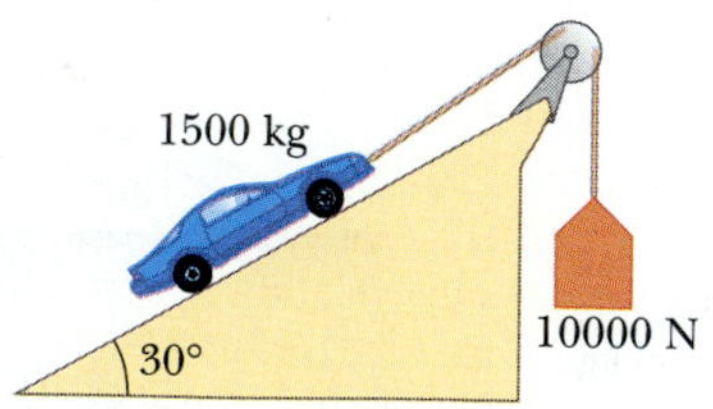

Figure 4.28 (Problem 45)

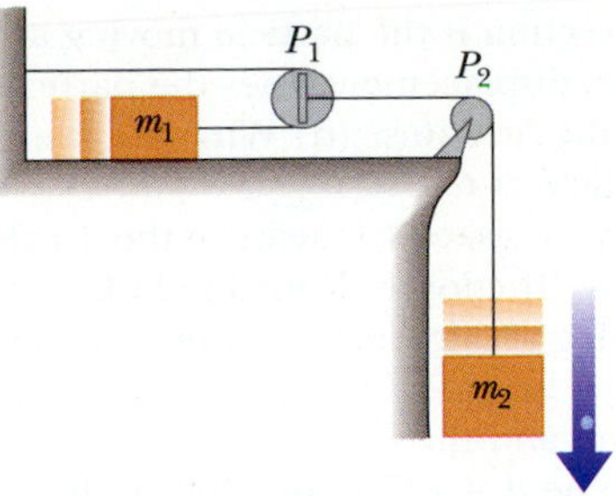

Figure 4.29 (Problem 46)

and a light fixed pulley, P_2, as shown in Figure 4.29. (a) If a_1 and a_2 are the accelerations of m_1 and m_2, respectively, what is the relation between these accelerations? Express (b) the tensions in the strings and (c) the accelerations a_1 and a_2 in terms of the masses m_1, and m_2, and g.

Additional Problems

47. (a) What is the resultant force exerted by the two cables supporting the traffic light in Figure 4.30? (b) What is the weight of the light?

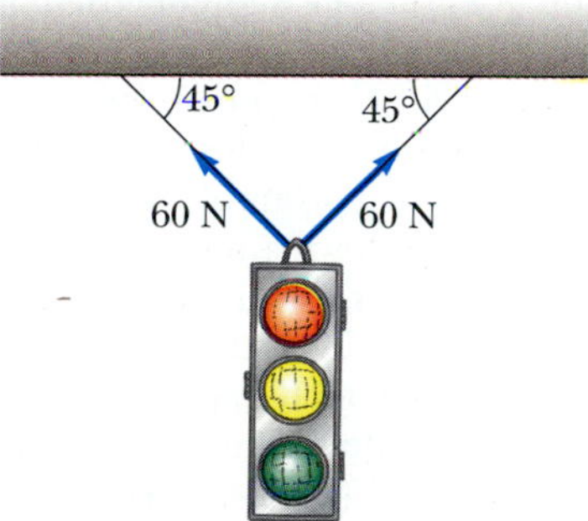

Figure 4.30 (Problem 47)

48. Some baseball pitchers are capable of throwing a fast ball at 100 mi/h. The pitcher achieves this speed by moving his arm through a distance of about 1.5 m. What average force must he exert on the 0.15-kg ball during this time?
49. A car is at rest at the top of a driveway that has a slope of 20°. If the brake of the car is released with the car in neutral gear, find (a) the acceleration of the car down the drive and (b) the time it takes for the car to reach the street, 10 m away.
50. A 2000-kg car is slowed down uniformly from a speed of 20 m/s to 5 m/s in a time of 4 s. What force did the road on the car during this time, and how far did the car travel during the deceleration?

Figure 4.31 (Problem 52)

51. An elevator accelerates upward at 1.5 m/s². If the elevator has a mass of 200 kg, find the tension in the supporting cable.
52. Big Al remembered from high school physics that pulleys can be used as aids in lifting heavy objects. Al designed the frictionless pulley system in Figure 4.31 to lift a safe to a second-floor office. The safe weighs 400 lb, and Al can pull with a force of 240 lb. (a) Will Big Al be able to raise the safe? (b) What is the maximum weight Big Al can lift using his pulley system? (*Note:* The large pulley is fastened by a yoke to the rope Big Al is pulling.)
53. The largest-caliber antiaircraft gun operated by the Luftwaffe during World War II was the 12.8-cm Flak 40. This weapon fired a 25.8-kg shell with a muzzle speed of 880 m/s. What propulsive force was necessary to attain the muzzle speed within the 6.0-m barrel? (Assume constant acceleration and neglect the Earth's gravitational effect.)
54. Three blocks are in contact with each other on a frictionless, horizontal surface, as in Figure 4.32. A horizontal force **F** is applied to m_1. If $m_1 = 2$ kg, $m_2 = 3$ kg, $m_3 = 4$ kg, and $F = 18$ N, find (a) the accel-

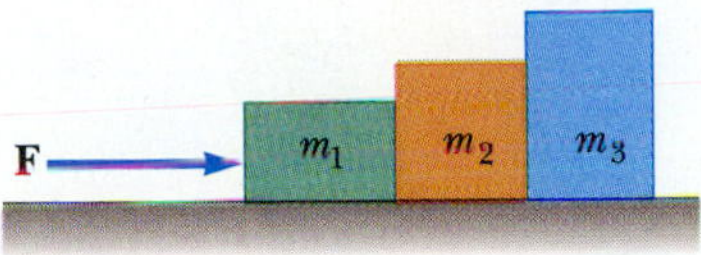

Figure 4.32 (Problem 54)

eration of the blocks, (b) the *resultant* force on each block, and (c) the magnitudes of the contact forces between the blocks.

55. A high diver of mass 70 kg jumps off a board 10 m above the water. If his downward motion is stopped 2 s after he enters the water, what average upward force did the water exert on the diver?

56. An inventive child named Pat wants to reach an apple in a tree without climbing the tree. Sitting in a chair connected to a rope that passes over a frictionless pulley (Fig. 4.33), Pat pulls on the loose end of the rope with such a force that the spring scale reads 250 N. Pat's true weight is 320 N, and the chair weighs 160 N. (a) Draw free-body diagrams for Pat and the chair considered as separate systems, and another diagram for Pat and the chair considered as one system. (b) Show that the acceleration of the system is *upward* and find its magnitude. (c) Find the force Pat exerts on the chair.

Figure 4.33 (Problem 56)

57. Two forces, $\mathbf{F}_1 = (-6\mathbf{i} - 4\mathbf{j})$ N and $\mathbf{F}_2 = (-3\mathbf{i} + 7\mathbf{j})$ N, act on a particle of mass 2 kg that is initially at rest at coordinates (−2 m, +4 m). (a) What are the components of the particle's velocity at $t = 10$ s? (b) In what direction is the particle moving at $t = 10$ s? (c) What displacement does the particle undergo during the first 10 s? (d) What are the coordinates of the particle at $t = 10$ s?

△58. A student is asked to measure the acceleration of a cart on a "frictionless" inclined plane as in Figure 4.12, using an air track, a stopwatch, and a meter stick. The height of the incline is measured to be 1.774 cm, and the total length of the incline is measured to be $d = 127.1$ cm. Hence, the angle of inclination θ is determined from the relation $\sin\theta = 1.774/127.1$. The cart is released from rest at the top of the incline, and its displacement along the incline, x, is measured versus time, where $x = 0$ refers to the initial position of the cart. For x values of 10.0 cm, 20.0 cm, 35.0 cm, 50.0 cm, 75.0 cm, and 100 cm, the measured times to undergo these displacements (averaged over five runs) are 1.02 s, 1.53 s, 2.01 s, 2.64 s, 3.30 s, and 3.75 s, respectively. Construct a graph of x versus t^2, and perform a linear least-squares fit to the data. Determine the acceleration of the cart from the slope of this graph, and compare it with the value you would get using $a' = g\sin\theta$, where $g = 9.80\ \text{m/s}^2$.

59. One of the great dangers to mountain climbers is the *avalanche,* in which a mass of snow and ice breaks loose and goes on an essentially frictionless "ride" down the mountain on a cushion of compressed air. If you were on a 30° mountain slope and an avalanche started 400 m up the slope, how much time would you have to get out of the way?

60. A van accelerates down a hill (Fig. 4.34), going from rest to 30 m/s in 6 s. During the acceleration, a toy ($m = 0.1$ kg) hangs by a string from the van's ceiling. The acceleration is such that the string remains perpendicular to the ceiling. Determine (a) the angle θ and (b) the tension in the string.

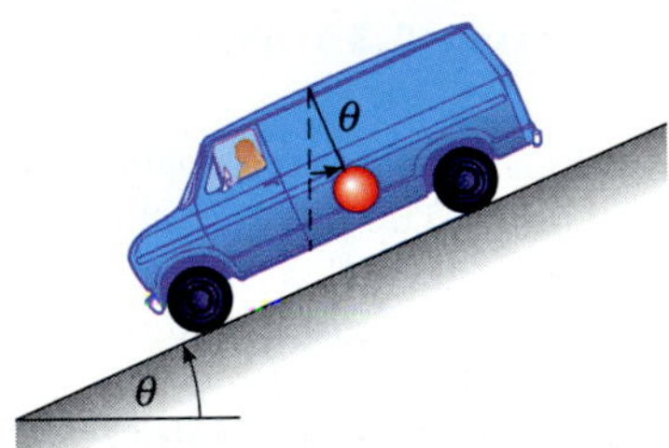

Figure 4.34 (Problem 60)

This calf-roping scene, taken in Steamboat, Colorado, is a standard rodeo event. The external forces acting on the horse are the force of friction between the horse and ground, the force of gravity, the tension force of the rope attached to the calf, the force of the cowboy on the horse, and the upward force of the ground. Can you identify the forces acting on the calf? *(FourbyFive, Inc.)*

More Applications of Newton's Laws

CHAPTER 5

In Chapter 4 we introduced Newton's laws of motion and applied them to some linear motion situations in which we were able to neglect frictional forces. In this chapter we shall expand our investigation to systems moving in the presence of friction forces. These systems include objects moving on rough surfaces and objects moving through viscous media such as liquids and air. In one section we discuss how numerical methods can be used to solve such "real-world" problems as motion in which the resistive force is velocity-dependent. We also apply Newton's laws to the dynamics of circular motion. Finally, we conclude this chapter with a brief discussion of an object moving in a frame that is accelerating with respect to the observer—in other words, a noninertial reference frame.

5.1 Forces of Friction

When a body is moving on a rough surface or through a viscous medium such as air or water, resistance to the motion occurs because of the interaction between the body and its surroundings. We call such resistance a **force of friction.** Forces of friction are very important in our everyday lives. They allow us to walk and run and are necessary for the motion of wheeled vehicles.

Consider a block on a table, such as that in Figure 5.1a. If we apply an external horizontal force **F**, acting to the right, to the block, the block will remain stationary if **F** is not too large. The force that keeps the block from moving acts to the left and is called the force of static friction, $\mathbf{f}_s$. As long as the block is not moving, $\mathbf{f}_s = \mathbf{F}$. Thus, if **F** is increased, $\mathbf{f}_s$ also increases. Likewise, if **F** decreases, $\mathbf{f}_s$ also decreases. Experiments show that the frictional force arises from the nature of the two surfaces: because of their roughness, contact is made only at a few points, as shown in the "magnified" view of the surface in Figure 5.1a. Actually, the frictional force is much more complicated than presented here, since it ultimately involves forces between atoms or molecules where the surfaces are in contact.

If we increase the magnitude of **F** enough, as in Figure 5.1b, the block eventually begins to move. When the block is on the verge of moving, f_s is a maximum.

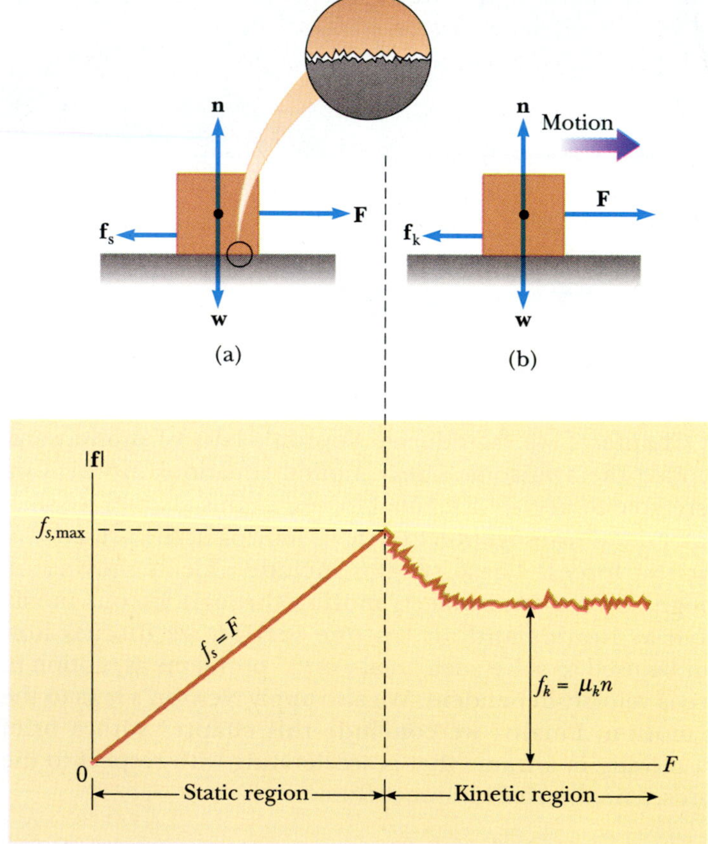

Figure 5.1 The force of friction, **f**, between a block and a rough surface is opposite the applied force, **F**. (a) The force of static friction equals the applied force. (b) When the applied force exceeds the force of kinetic friction, the block accelerates to the right. (c) A graph of the magnitude of the frictional force versus the applied force. Note that $f_{s,\max} > f_k$.

When F exceeds $f_{s,\max}$, the block moves, accelerating to the right. When the block is in motion, the frictional force becomes less than $f_{s,\max}$ (Fig. 5.1c). We call the frictional force for an object in motion the **force of kinetic friction, $\mathbf{f}_k$**. The unbalanced force in the x direction, $\mathbf{F} - \mathbf{f}_k$, produces an acceleration to the right. If $F = f_k$, the block moves to the right with constant speed. If the applied force is removed, then the frictional force acting to the left decelerates the block and eventually brings it to rest.

The force of kinetic friction is less than $f_{s,\max}$ for the following reason. When the object is stationary, the contact points between the object and the surface are said to be cold-welded. While the object is in motion, these small welds can no longer form, and the frictional force decreases.

Experimentally, one finds that both $f_{s,\max}$ and f_k are proportional to the normal force acting on the block and depend on the nature of the two surfaces that are in contact. The experimental observations can be summarized as follows:

1. The force of static friction between any two surfaces in contact is opposite the applied force and can have the values

$$f_s \leq \mu_s n \qquad [5.1]$$

where the dimensionless constant μ_s is called the **coefficient of static friction** and n is the normal force. The equality in Equation 5.1 holds when the block is on the verge of slipping, that is, when $f_s = f_{s,\max} \equiv \mu_s n$. The inequality holds when the applied force is less than this value. In general, when an object is at rest relative to a surface, the frictional force always acts in such a way as to maintain a velocity of zero relative to the surface.

2. The force of kinetic friction is opposite the direction of motion and is given by

$$f_k = \mu_k n \qquad [5.2]$$

where μ_k is the **coefficient of kinetic friction.**

3. The values of μ_k and μ_s depend on the nature of the surfaces, but μ_k is generally less than μ_s. Typical values of μ range from around 0.01 to 1.5. Table 5.1 lists some reported values.

Table 5.1
Coefficients of Friction[a]

	μ_s	μ_k
Steel on steel	0.74	0.57
Aluminum on steel	0.61	0.47
Copper on steel	0.53	0.36
Rubber on concrete	1.0	0.8
Wood on wood	0.25–0.5	0.2
Glass on glass	0.94	0.4
Waxed wood on wet snow	0.14	0.1
Waxed wood on dry snow	—	0.04
Metal on metal (lubricated)	0.15	0.06
Ice on ice	0.1	0.03
Teflon on Teflon	0.04	0.04
Synovial joints in humans	0.01	0.003

[a] All values are approximate.

Example 5.1 The Sliding Hockey Puck

A hockey puck is given an initial speed of 20 m/s on a frozen pond, as in Figure 5.2. The puck remains on the ice and slides 120 m before coming to rest. Determine the coefficient of kinetic friction between the puck and the ice.

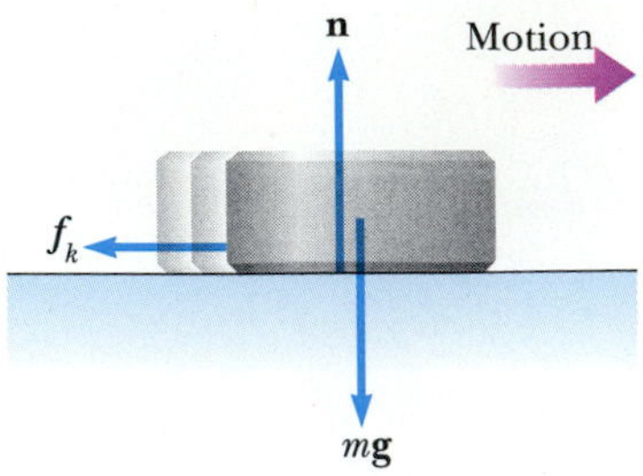

Solution The acceleration of the puck can be found from $v^2 = v_0^2 + 2ax$, with the final speed v equal to zero, the initial speed $v_0 = 20$ m/s, and the distance traveled $x = 120$ m.

$$v^2 = v_0^2 + 2ax$$
$$0 = (20 \text{ m/s})^2 + 2a(120 \text{ m})$$
$$a = -1.67 \text{ m/s}^2$$

The negative sign means that the acceleration is to the left, *opposite* the direction of the velocity.

The magnitude of the force of kinetic friction can be determined using $f_k = \mu_k n$, where n is found from $\Sigma F_y = 0$ as follows:

$$\sum F_y = n - w = 0$$
$$n = w = mg$$

Thus,

$$f_k = \mu_k n = \mu_k mg$$

Now we apply Newton's second law along the horizontal direction. We shall take the positive direction toward the right.

$$\sum F_x = -f_k = ma$$
$$-\mu_k \cancel{m}g = \cancel{m}(-1.67 \text{ m/s}^2)$$
$$\mu_k = \frac{1.67 \text{ m/s}^2}{9.80 \text{ m/s}^2} = 0.170$$

Figure 5.2 (Example 5.1) *After* the puck is given an initial velocity, the external forces acting on it are the weight, mg, the normal force, **n**, and the force of kinetic friction, $\mathbf{f}_k$.

Example 5.2 Connected Objects

A ball and a cube are connected by a light string that passes over a frictionless pulley, as in Figure 5.3a. The coefficient of sliding friction between the cube and the surface is 0.30. Find the acceleration of the two objects and the tension in the string.

Solution First, let us isolate each object in Figure 5.3a and determine the external forces on each. Newton's second law applied to the cube in component form, with the positive x direction to the right, gives

$$\sum F_x = T - f_k = (4 \text{ kg})a$$
$$\sum F_y = n - (4 \text{ kg})g = 0$$

Since $f_k = \mu_k n$ and $n = mg = (4 \text{ kg})(9.80 \text{ m/s}^2) = 39.2$ N, we have $f_k = \mu_k n = (0.30)(39.2 \text{ N}) = 11.8$ N. Therefore.

$$(1) \qquad T = f_k + (4 \text{ kg})a = 11.8 \text{ N} + (4 \text{ kg})a$$

Now we apply Newton's second law to the ball moving in the vertical direction, where the downward direction is selected as positive:

$$\sum F_y = (7 \text{ kg})g - T = (7 \text{ kg})a$$

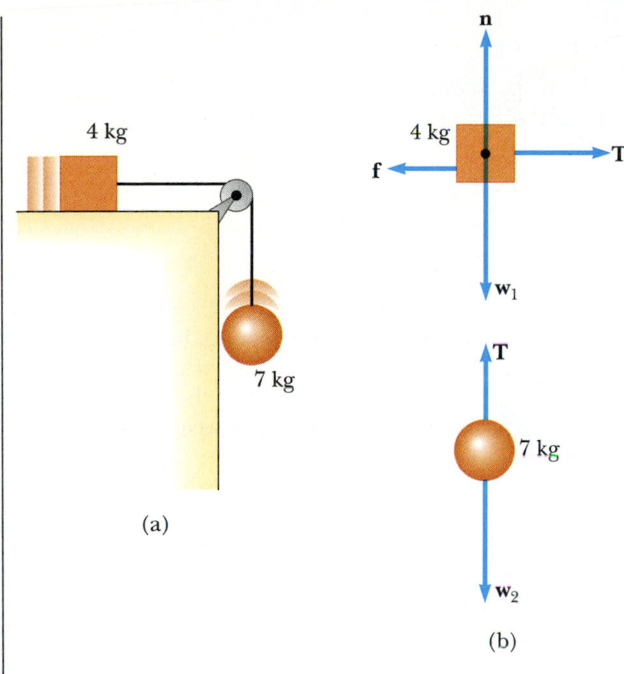

Figure 5.3 (Example 5.2) (a) Two objects connected by a light string that passes over a frictionless pulley. (b) Free-body diagrams for the objects.

or

$$(2) \qquad T = 68.6\text{ N} - (7\text{ kg})a$$

Subtracting (1) from (2) eliminates T:

$$56.8\text{ N} - (11\text{ kg})a = 0$$

$$a = \boxed{5.16\text{ m/s}^2}$$

When this value for the acceleration is substituted into (1), we get

$$T = \boxed{32.5\text{ N}}$$

5.2 Newton's Second Law Applied to Uniform Circular Motion

As these cyclists in the Tour de France negotiate a curve on a flat racing track, the centripetal force is provided by the force of static friction between the tires and the track surface. *(Michel Gouverneur, Photo News, Gamma Sport)*

Solving problems involving friction is just one of many applications of Newton's second law. Let us now apply Newton's second law to another common situation: uniform circular motion. In Chapter 3, Section 3.4, we found that a particle moving with uniform speed v in a circular path of radius r experiences an acceleration of magnitude

$$a_r = \frac{v^2}{r}$$

Because the velocity vector, **v**, changes direction continuously during the motion, the acceleration vector, $\mathbf{a}_r$, is directed toward the center of the circle and is called centripetal acceleration. Furthermore, $\mathbf{a}_r$ is always perpendicular to **v**.

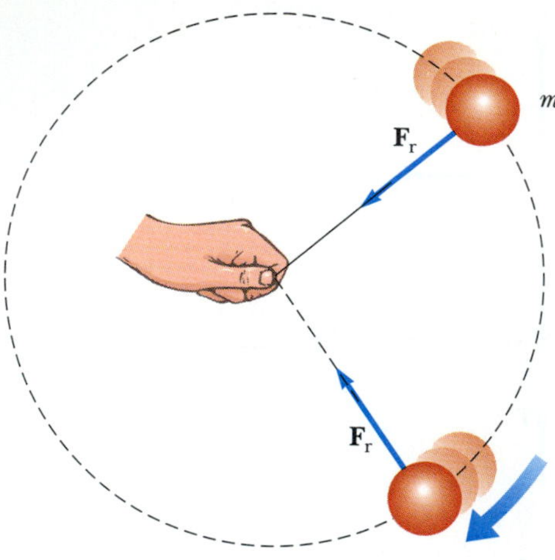

Figure 5.4
An overhead view of a ball moving in a horizontal circular path. A force $\mathbf{f}_r$ directed toward the center of the circle keeps the ball moving in the circle with a constant speed.

Consider a ball of mass m that is tied to a string of length r and is being whirled in a horizontal circular path on a tabletop, as in Figure 5.4. Let us assume that the ball moves with constant speed. The inertia of the ball would tend to maintain the motion in a straight-line path; however, the string prevents this motion by exerting a force on the ball that makes it follow a circular path. This force is directed along the length of the string toward the center of the circle, as shown in Figure 5.4, and is an example of a class of forces called **centripetal forces.** If we apply Newton's second law along the radial direction, we find that the required centripetal force is

Centripetal force

$$F_r = ma_r = m\frac{v^2}{r} \qquad [5.3]$$

This skateboarder is executing some daring maneuvers as he moves in a spiral path inside a cylindrical pipe. The centripetal force acting on the skateboarder is the normal force acting toward the center of the pipe. *(Eric Sander, Gamma-Liaison)*

Like the centripetal acceleration, the centripetal force acts toward the center of the circular path followed by the ball. Because they act toward the center of rotation, centripetal forces cause a change in the direction of the velocity. Centripetal forces are no different from any other forces we have encountered. The term *centripetal* is used simply to indicate that *the force is directed toward the center of a circle.* In the case of a ball rotating at the end of a string, the tension force provides the centripetal force. For a satellite in a circular orbit around the Earth, the force of gravity is the centripetal force. The centripetal force acting on a car rounding a curve on a flat road is the force of friction between the tires and the pavement, and so forth.

Regardless of the example used, if the centripetal force acting on an object vanishes, the object no longer moves in its circular path; instead it moves along a straight-line path tangent to the circle. This idea is illustrated in Figure 5.5 for the case of the ball whirling in a circle at the end of a string. If the string breaks at some instant, the ball moves along a straight-line path that is tangent to the circle at the point where the string broke.

In general, a body can move in a circular path under the influence of friction, the gravitational force, or a combination of forces. Let us consider some examples of uniform circular motion. In each case, be sure to recognize the external force or forces that cause the body to move in its circular path.

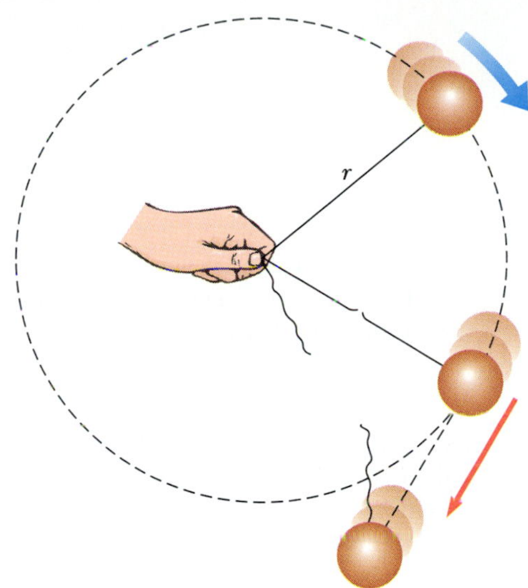

Figure 5.5
When the string breaks, the ball moves in the direction tangent to the circular path.

Example 5.3 How Fast Can It Spin?

A ball of mass 0.5 kg is attached to the end of a cord whose length is 1.5 m. The ball is whirled in a horizontal circle as in Figure 5.5. If the cord can withstand a maximum tension of 50 N, what is the maximum speed the ball can have before the cord breaks?

Solution Because the centripetal force in this case is equal to the tension T in the cord, Equation 5.3 gives

$$T = m\frac{v^2}{r}$$

Solving for v, we have

$$v = \sqrt{\frac{Tr}{m}}$$

The maximum speed that the ball can have corresponds to the maximum value of the tension. Hence, we find

$$v_{max} = \sqrt{\frac{T_{max}r}{m}} = \sqrt{\frac{(50 \text{ N})(1.5 \text{ m})}{0.5 \text{ kg}}} = 12.2 \text{ m/s}$$

Exercise Calculate the tension in the cord if the speed of the ball is 5 m/s.

Answer 8.33 N.

A 10-hour exposure showing circular star trails around the south celestial pole. The circular trails are due to the rotation of the Earth about its axis (the camera is in motion) while the stars remain fixed. *(© Anglo-Australian Telescope Board 1980)*

Example 5.4 The Conical Pendulum

A small body of mass m is suspended from a string of length L. The body revolves with constant speed v in a horizontal circle of radius r, as in Figure 5.6. Since the string sweeps out the surface of a cone, the system is known as a *conical pendulum.* Find the speed of the body and the period of revolution, T_P.

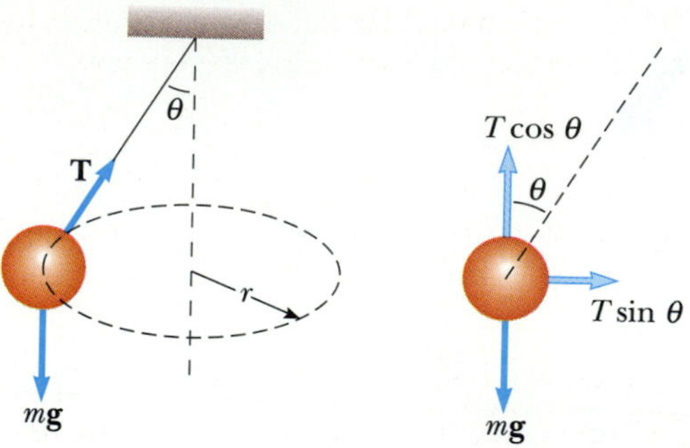

Figure 5.6 (Example 5.4) The conical pendulum and its free-body diagram.

Solution The free-body diagram for the mass m is shown in Figure 5.6, where the tension force, T, has been resolved into a vertical component, $T\cos\theta$, and a component $T\sin\theta$ acting toward the center of rotation. Since the body does not accelerate in the vertical direction, the vertical component of the tension force must balance the weight. Therefore,

$$(1) \qquad T\cos\theta = mg$$

Since the centripetal force in this example is provided by the horizontal component, $T\sin\theta$, from Newton's second law we get

$$(2) \qquad T\sin\theta = ma_r = \frac{mv^2}{r}$$

By dividing (2) by (1), we eliminate T and find that

$$\tan\theta = \frac{v^2}{rg}$$

But from the geometry we note that $r = L\sin\theta$; therefore,

$$v = \sqrt{rg\tan\theta} = \sqrt{Lg\sin\theta\tan\theta}$$

The period of revolution, T_P (not to be confused with the tension T), is given by

$$(3) \qquad T_P = \frac{2\pi r}{v} = \frac{2\pi r}{\sqrt{rg\tan\theta}} = 2\pi\sqrt{\frac{L\cos\theta}{g}}$$

The intermediate algebraic steps used in obtaining (3) are left to the reader. Note that T_P is independent of m! If we take $L = 1.00$ m and $\theta = 20°$, we find, using (3), that

$$T_P = 2\pi\sqrt{\frac{(1.00\text{ m})(\cos 20°)}{9.80\text{ m/s}^2}} = 1.95\text{ s}$$

Is it physically possible to have a conical pendulum with $\theta = 90°$?

Example 5.5 What Is the Maximum Speed of the Car?

A 1500-kg car moving on a flat road negotiates a curve with a 35-m radius, as in Figure 5.7. If the coefficient of static friction between the tires and the pavement is 0.50, find the maximum speed at which the car can make the turn successfully.

Solution In this case, the centripetal force which enables the car to remain in its circular path is the force of static friction. Hence, from Equation 5.3 we have

$$(1) \qquad f_s = m\frac{v^2}{r}$$

The maximum speed at which the car can round the curve corresponds to the speed at which it is on the verge of skidding outward. At this point the friction force has its maximum value, given by

$$f_{s,\max} = \mu n$$

Because the magnitude of the normal force equals the weight in this case, we find

$$f_{s,\max} = \mu mg = (0.5)(1500\text{ kg})(9.80\text{ m/s}^2) = 7350\text{ N}$$

Substituting this value into (1), we find that the maximum speed is

$$v_{\max} = \sqrt{\frac{f_{s,\max}r}{m}} = \sqrt{\frac{(7350\text{ N})(35\text{ m})}{1500\text{ kg}}} = 13.1\text{ m/s}$$

Exercise On a wet day, the car described in this example begins to skid on the curve when its speed reaches 8 m/s. What is the coefficient of static friction in this case?

Answer 0.187.

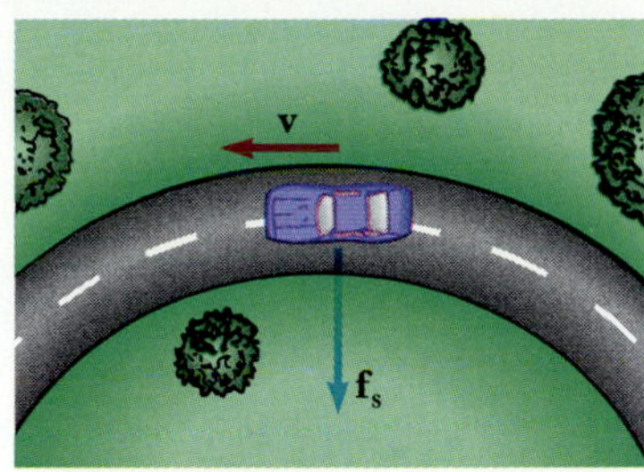

Figure 5.7 (Example 5.5) The force of static friction directed toward the center of the circular arc keeps the car moving in a circle.

Example 5.6 Let's Go Loop-the-loop

A pilot of mass m in a jet aircraft executes a "loop-the-loop" maneuver like that illustrated in Figure 5.8a. In this flying pattern, the aircraft moves in a vertical circle of radius 2.70 km at a *constant speed* of 225 m/s. Determine the force of the seat on the pilot at (a) the bottom of the loop and (b) the top of the loop. Express the answers in terms of the weight of the pilot, mg.

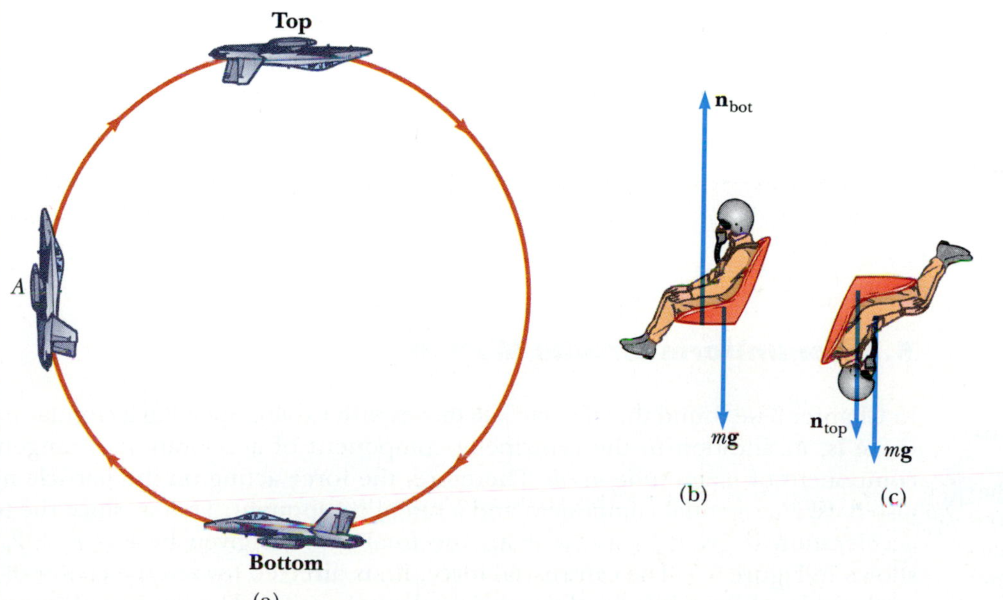

Figure 5.8 (Example 5.6)

Solution

(a) The free-body diagram for the pilot at the bottom of the loop is shown in Figure 5.8b. The only forces acting on the pilot are the downward force of gravity, mg, and the upward force, $\mathbf{n}_{\text{bot}}$, exerted by the seat. Since the net force upward which provides the centripetal acceleration is $n_{\text{bot}} - mg$, Newton's second law for the radial direction gives

$$n_{\text{bot}} - mg = m\frac{v^2}{r}$$

$$n_{\text{bot}} = mg + m\frac{v^2}{r} = mg\left[1 + \frac{v^2}{rg}\right]$$

Substituting the values given for the speed and radius, $v = 225$ m/s and $r = 2.70 \times 10^3$ m, gives

$$n_{\text{bot}} = mg\left[1 + \frac{(225 \text{ m/s})^2}{(2.70 \times 10^3 \text{ m})(9.80 \text{ m/s}^2)}\right] = 2.91\, mg$$

Hence, the force of the seat on the pilot is *greater* than the true weight, mg, by a factor of 2.91.

(b) The free-body diagram for the pilot at the top of the loop is shown in Figure 5.8c. At this point, both the weight and the force of the seat on the pilot, $\mathbf{n}_{\text{top}}$, act *downward,* so the net force downward which provides the centripetal acceleration has a magnitude $n_{\text{top}} + mg$. Applying Newton's second law gives

$$n_{\text{top}} + mg = m\frac{v^2}{r}$$

$$n_{\text{top}} = m\frac{v^2}{r} - mg = mg\left[\frac{v^2}{rg} - 1\right]$$

$$n_{\text{top}} = mg\left[\frac{(225 \text{ m/s})^2}{(2.70 \times 10^3 \text{ m})(9.80 \text{ m/s}^2)} - 1\right] = 0.91\, mg$$

In this case, the force of the seat on the pilot is a factor of 0.91 times the true weight. Hence, the pilot will feel lighter at the top of the loop.

Exercise Calculate the centripetal force on the pilot if the aircraft is at point A in Figure 5.8a, midway up the loop.

Answer $n_A = 1.91\, mg$ directed to the right.

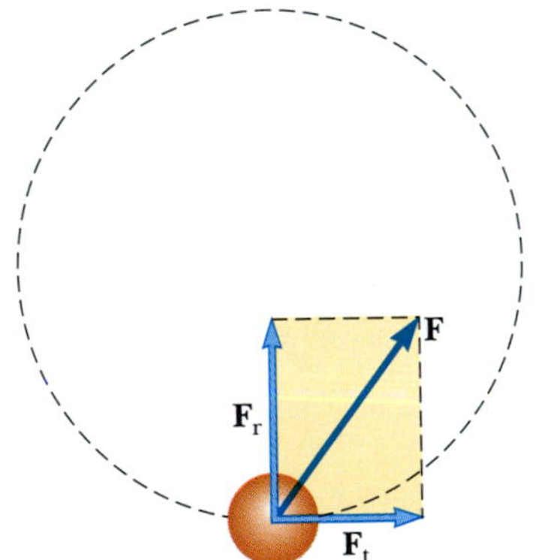

Figure 5.9 When the force acting on a particle has a tangential component, F_t, its speed changes. The total force on the particle in this case is the vector sum of the tangential force and the centripetal force; that is, $\mathbf{F} = \mathbf{F}_t + \mathbf{F}_r$.

5.3 Nonuniform Circular Motion

In Chapter 3 we found that if a particle moves with varying speed in a circular path, there is, in addition to the centripetal component of acceleration, a tangential component of magnitude dv/dt. Therefore, the force acting on the particle must also have a tangential component and a radial component. That is, since the total acceleration is given by $\mathbf{a} = \mathbf{a}_r + \mathbf{a}_t$, the total force is given by $\mathbf{F} = \mathbf{F}_r + \mathbf{F}_t$, as shown in Figure 5.9. The centripetal force, $\mathbf{F}_r$, is directed toward the center of the circle and is responsible for the centripetal acceleration. The tangential force, $\mathbf{F}_t$, is responsible for the tangential acceleration, which causes the speed of the particle to change with time. The following example demonstrates this type of motion.

Example 5.7 Follow the Rotating Ball

A small sphere of mass m is attached to the end of a cord of length R, which rotates in a *vertical* circle about a fixed point O, as in Figure 5.10a. Determine the tension in the cord at any instant that the speed of the sphere is v when the cord makes an angle θ with the vertical.

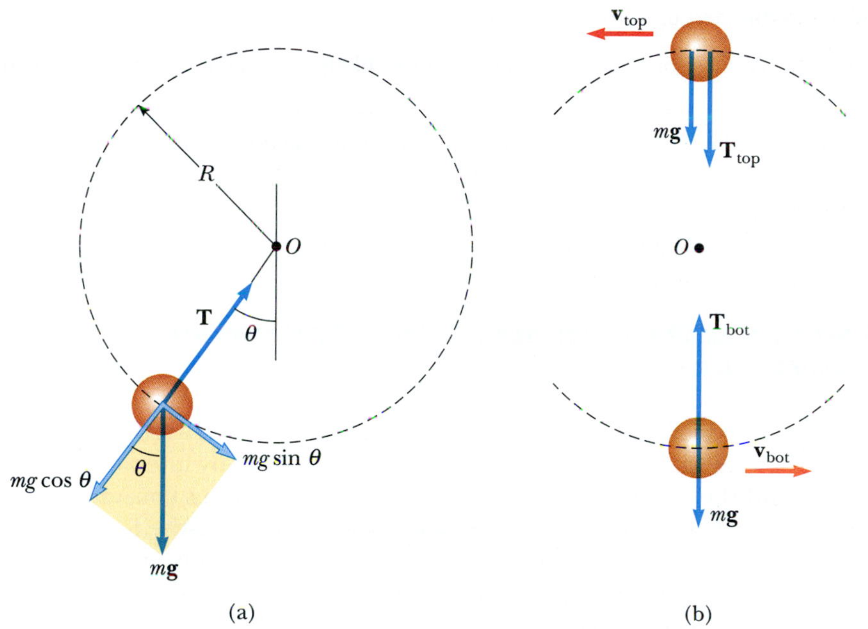

Figure 5.10 (Example 5.7) (a) Forces acting on a sphere of mass m connected to a cord of length R and rotating in a vertical circle centered at O. (b) Forces acting on the sphere when it is at the top and bottom of the circle. Note that the tension at the bottom is a maximum and the tension at the top is a minimum.

Solution First we note that the speed is *not* uniform, since there is a tangential component of acceleration arising from the weight of the sphere. From the free-body diagram in Figure 5.10a, we see that the only forces acting on the sphere are the weight, $m\mathbf{g}$, and the constraint force (equal to the tension force), $\mathbf{T}$. Now we resolve $m\mathbf{g}$ into a tangential component, $mg\sin\theta$, and a radial component, $mg\cos\theta$. Applying Newton's second law to the forces in the tangential direction gives

$$\sum F_t = mg\sin\theta = ma_t$$

$$(1) \qquad a_t = g\sin\theta$$

This component causes v to change in time, since $a_t = dv/dt$. Applying Newton's second law to the forces in the radial direction and noting that both $\mathbf{T}$ and $\mathbf{a}_r$ are directed toward O, we get

$$\sum F_r = T - mg\cos\theta = \frac{mv^2}{R}$$

$$(2) \qquad T = m\left(\frac{v^2}{R} + g\cos\theta\right)$$

Limiting Cases At the *top* of the path, where $\theta = 180°$, we see from (2) that since $\cos 180° = -1$,

$$T_{\text{top}} = m\left(\frac{v_{\text{top}}^2}{R} - g\right)$$

This is the *minimum* value of T. Note that at this point $a_t = 0$, and therefore the acceleration is radial and directed downward, as in Figure 5.10b.

At the *bottom* of the path, where $\theta = 0$, again from (2) we see that since $\cos 0 = 1$,

$$T_{\text{bot}} = m\left(\frac{v_{\text{bot}}^2}{R} + g\right)$$

This is the *maximum* value of T. Again, at this point $a_t = 0$, and the acceleration is radial and directed upward.

Exercise At what orientation is the cord most likely to break if the average speed increases?

Answer At the bottom of the path, where T has its maximum value.

5.4 Motion in the Presence of Velocity-Dependent Resistive Forces

In the preceding section we described the interaction between a moving object and the surface along which it moves. We completely ignored any interaction between the object and the *medium* through which it moves. Now let us consider the effect of a medium such as a liquid or gas. The medium exerts a **resistive** force, **R**, on the object moving through it. The magnitude of this force depends on such factors as the speed of the object, and the direction of **R** is always opposite the direction of motion of the object relative to the medium. Generally, the magnitude of the resistive force increases with increasing speed. Some examples are the air resistance associated with moving vehicles (sometimes called drag) and the viscous forces that act on objects moving through a liquid.

The resistive force can have a complicated speed dependence. In the following discussions, we shall consider two situations. First, we shall assume that the resistive force is proportional to the speed; this is the case for objects that fall through a liquid with low speed and for very small objects, such as dust particles, that move through air. Second, we shall treat situations in which the resistive force is proportional to the square of the speed of the object; large objects moving through air in free fall experience such a force.

Resistive Force Proportional to Velocity

If we assume that the resistive force acting on an object that is moving through a viscous medium is proportional to the object's velocity, then the resistive force can be expressed as

$$\mathbf{R} = -b\mathbf{v} \qquad [5.4]$$

where v is the velocity of the object and b is a constant that depends on the properties of the medium and on the shape and dimensions of the object. If the object is a sphere of radius r, then b is proportional to r.

Consider a sphere of mass m released from rest in a liquid, as in Figure 5.11a. Assuming the only forces acting on the sphere are the resistive force, $-b\mathbf{v}$, and the weight, $m\mathbf{g}$, let us describe the sphere's motion.[1]

Applying Newton's second law to the vertical motion, choosing the downward direction to be positive, and noting that $\Sigma F_y = mg - bv$, we get

$$mg - bv = ma = m\frac{dv}{dt}$$

where the acceleration is downward. Simplifying the above expression gives

$$\frac{dv}{dt} = g - \frac{b}{m}v \qquad [5.5]$$

Equation 5.5 is called a *differential equation*, and the methods of solving such an equation may not be familiar to you as yet. However, note that initially, when $v = 0$, the resistive force is zero and the acceleration, dv/dt, is simply g. As t increases, the resistive force increases and the acceleration *decreases.* Eventually, the acceleration becomes zero when the resistive force *equals* the weight. At this point, the sphere reaches its *terminal speed*, v_t, and from then on it continues to move with zero acceleration. The terminal speed can be obtained from Equation 5.5 by setting $a = dv/dt = 0$. This gives

$$mg - bv_t = 0 \qquad \text{or} \qquad v_t = \frac{mg}{b}$$

The expression for v that satisfies Equation 5.5 with $v = 0$ at $t = 0$ is

$$v = \frac{mg}{b}(1 - e^{-bt/m}) = v_t(1 - e^{-t/\tau}) \qquad [5.6]$$

This function is plotted in Figure 5.11b. The time $\tau = m/b$ is the time it takes the object to reach 63% of its terminal speed. We can check that Equation 5.6 is a solution to Equation 5.5 by direct differentiation:

$$\frac{dv}{dt} = \frac{d}{dt}\left(\frac{mg}{b} - \frac{mg}{b}e^{-bt/m}\right) = -\frac{mg}{b}\frac{d}{dt}e^{-bt/m} = ge^{-bt/m}$$

Substituting this expression and Equation 5.6 into Equation 5.5 shows that our solution satisfies the differential equation.

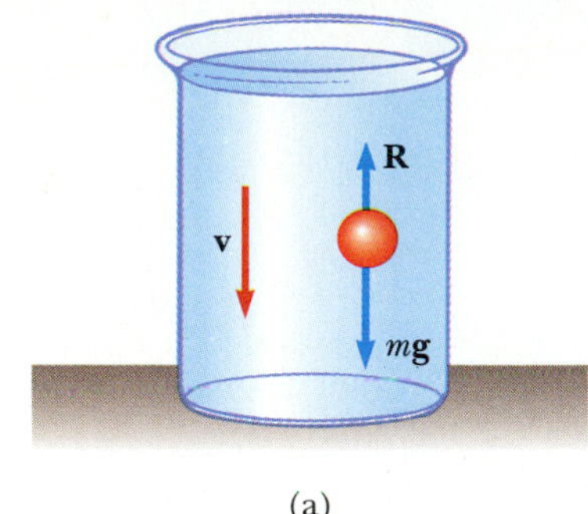

(a)

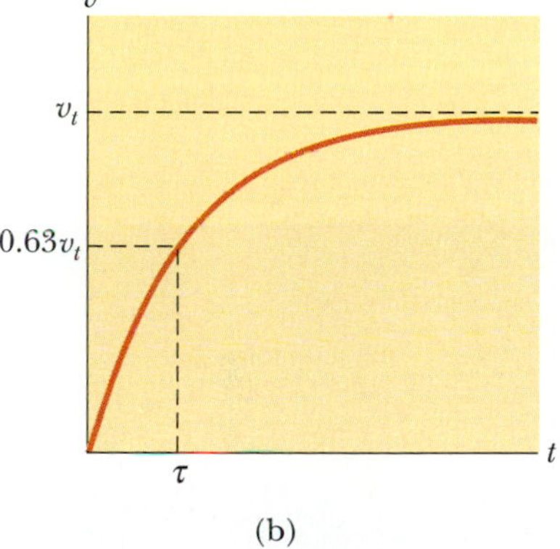

(b)

Figure 5.11
(a) A small sphere falling through a viscous fluid. (b) The velocity-time graph for an object falling through a viscous medium. The object reaches a maximum, or terminal, speed, v_t, and τ is the time it takes to reach $0.63v_t$.

Example 5.8 A Sphere Falling in Oil

A small sphere with a mass of 2 g is released from rest in a large vessel filled with oil. The sphere reaches a terminal speed of 5 cm/s. Determine the time τ and the time it takes the sphere to reach 90% of its terminal speed.

Solution Since the terminal speed is given by $v_t = mg/b$, the coefficient b is

[1] There is also a *buoyant* force, which is constant and equal to the weight of the displaced fluid. This force will change the effective weight of the sphere only by a constant factor, so we can ignore it here.

$$b = \frac{mg}{v_t} = \frac{(2\text{ g})(980\text{ cm/s}^2)}{5\text{ cm/s}} = 392\text{ g/s}$$

Therefore, the time τ is given by

$$\tau = \frac{m}{b} = \frac{2\text{ g}}{392\text{ g/s}} = 5 \times 10^{-3}\text{ s}$$

The speed of the sphere as a function of time is given by Equation 5.6, $v = v_t(1 - e^{-t/\tau})$. To find the time t it takes the sphere to reach a speed of $0.90\ v_t$, we set $v = 0.90\ v_t$ into the expression and solve for t:

$$0.90v_t = v_t(1 - e^{-t/\tau})$$

$$1 - e^{-t/\tau} = 0.90$$

$$e^{-t/\tau} = 0.10$$

$$-\frac{t}{\tau} = -2.30$$

$$t = 2.30\tau = 2.30(5.10 \times 10^{-3}\text{ s})$$

$$= 11.7 \times 10^{-3}\text{ s} = 11.7\text{ ms}$$

Exercise Calculate the speed the sphere would have at $t = 11.7$ ms if air resistance were not present, and compare this value with the true speed at that instant.

Answer 11.5 cm/s, compared with the true speed of 4.50 m/s.

(Top) Aerodynamic car. Streamlined bodies are used for sports cars and other vehicles to reduce air drag and increase fuel efficiency. (Bottom) By spreading their arms and legs out from their bodies while keeping the planes of their bodies parallel to the ground, sky divers experience maximum air drag, resulting in a terminal speed of about 60 m/s. *(Top, © 1992 Dick Kelley; Bottom, Tom Sanders, The Stock Market)*

Air Drag

For large objects moving at high speeds through air, such as airplanes, sky divers, and baseballs, the drag force is approximately proportional to the *square* of the speed. In these situations, the magnitude of the drag force can be expressed as

$$R = \tfrac{1}{2}D\rho A v^2 \qquad \textbf{[5.7]}$$

where ρ is the density of air, A is the cross-sectional area of the moving object measured in a plane perpendicular to its motion, and D is a dimensionless empirical quantity called the *drag coefficient.* The drag coefficient has a value of about 0.5 for spherical objects but can be as high as 2 for irregularly shaped objects.

Consider an airplane in flight that experiences such a drag force. Equation 5.7 shows that the drag force is proportional to the density of air and hence decreases with decreasing air density. Since air density decreases with increasing altitude, the drag force on an airplane flying at a given speed must also decrease with increasing altitude. Furthermore, if the plane's speed is doubled, the drag force increases by a factor of 4. In order to maintain this increased speed, the propulsive force also increases by a factor of 4.

Now let us analyze the motion of a mass in free fall subject to an upward air drag force given by $R = \frac{1}{2}D\rho Av^2$. Suppose a mass m is released from rest from the position $y = 0$, as in Figure 5.12. The mass experiences two external forces: the weight, mg, downward and the drag force, $\mathbf{R}$, upward. (There is also an upward buoyant force, which we will neglect.) Hence, the magnitude of the net force is given by

$$F_{\text{net}} = mg - \tfrac{1}{2}D\rho A v^2 \qquad \textbf{[5.8]}$$

Table 5.2
Terminal Speeds for Objects Falling Through Air

Object	Mass (kg)	Area (m^2)	v_t (m/s)[a]
Sky diver	75	0.7	60
Baseball (radius 3.66 cm)	0.145	4.2×10^{-3}	33
Golf ball (radius 2.1 cm)	0.046	1.4×10^{-3}	32
Hailstone (radius 0.5 cm)	4.8×10^{-4}	7.9×10^{-5}	14
Raindrop (radius 0.2 cm)	3.4×10^{-5}	1.3×10^{-5}	9

[a] The drag coefficient, D, is assumed to be 0.5 in each case.

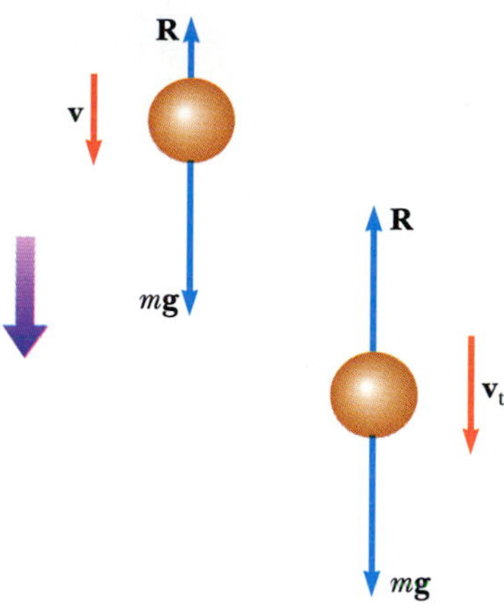

Figure 5.12
An object falling through air experiences a drag force, **R**, and the force of gravity, $m\mathbf{g}$. The object reaches terminal speed (on the right) when the net force is zero, that is, when $\mathbf{R} = m\mathbf{g}$. Before this occurs, the acceleration varies with speed according to Equation 5.9.

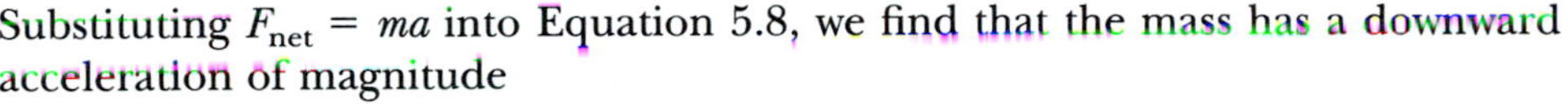

Substituting $F_{\text{net}} = ma$ into Equation 5.8, we find that the mass has a downward acceleration of magnitude

$$a = g - \left(\frac{D\rho A}{2m}\right)v^2 \qquad [5.9]$$

Again, we can calculate the terminal speed, v_t, using the fact that when the weight is balanced by the drag force, the net force is zero and therefore the acceleration is zero. Setting $a = 0$ in Equation 5.9 gives

$$g - \left(\frac{D\rho A}{2m}\right)v_t^2 = 0$$

Terminal speed

$$v_t = \sqrt{\frac{2mg}{D\rho A}} \qquad [5.10]$$

Using this expression, we can determine how the terminal speed depends on the dimensions of the object. Suppose the object is a sphere of radius r. In this case, $A \propto r^2$ and $m \propto r^3$ (since the mass is proportional to the volume). Therefore, $v_t \propto \sqrt{r}$. That is, as r increases, the terminal speed increases with the square root of the radius.

Table 5.2 lists the terminal speeds for several objects falling through air.

5.5 Numerical Modeling in Particle Dynamics[2]

As we have seen in this and the preceding chapter, the study of the dynamics of a particle focuses on describing the position, velocity, and acceleration as functions of time. Cause-and-effect relationships exist between these quantities: a velocity causes position to change, an acceleration causes velocity to change, and an acceleration is the direct result of applied forces. Therefore, the study of motion usually begins with an evaluation of the net force on the particle.

In this section we confine our discussion to one-dimensional motion, so boldface notation will not be used for the vector quantities. If a particle of mass m moves under the influence of a net force F, Newton's second law tells us that the

[2] The author is most grateful to Colonel James Head of the U.S. Air Force Academy for preparing this section. A more extensive treatment of this topic by Colonel Head is included in a supplement entitled *Numerical Methods in Physics.*

acceleration of the particle is given by $a = F/m$. In general, we can then obtain the solution to a dynamics problem by using the following procedure:

1. Sum all the forces on the particle to get the net force F.
2. Use this force to determine the acceleration, using $a = F/m$.
3. Use this acceleration to determine the velocity from $dv/dt = a$.
4. Use this velocity to determine the position from $dx/dt = v$.

Example 5.9 An Object Falling in a Vacuum

We can illustrate the procedure just described by considering a particle falling in vacuum under the influence of the force of gravity, as in Figure 5.13.

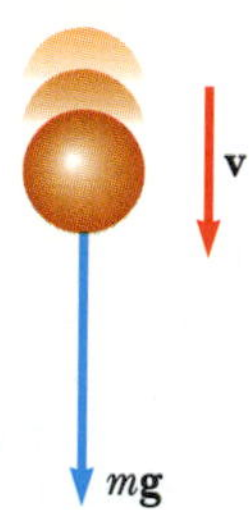

Figure 5.13
(Example 5.9)

Solution Applying Newton's second law, we set the sum of the external forces equal to the mass of the particle times its acceleration (taking upward to be the positive direction):

$$F = ma = -mg$$

Thus, $a = -g$, a constant. Since $a = dv/dt$, the resulting differential equation for velocity is $dv/dt = -g$, which may be integrated to give

$$v(t) = v_0 - gt$$

Then, since $v = dx/dt$, the position of the particle is obtained from another integration, which yields the well-known result

$$x(t) = x_0 + x_0/v_0 t - \tfrac{1}{2}gt^2$$

In this expression, x_0 and v_0 represent the position and velocity of the particle at $t = 0$.

The procedure just described is straightforward for some physical situations, such as the one described in Example 5.9. In the "real-world," however, complications often arise which make analytical solutions for many practical situations difficult and perhaps beyond the mathematical abilities of introductory physics students. For example, the force may depend on the position of the particle, as in cases where variation in gravity with height must be taken into account. The force may also vary with velocity, as we have seen in cases of resistive forces caused by motion through a viscous medium. The force may depend on both position and velocity, as in the case of an object falling through air where the drag force depends on velocity and on height (air density). In rocket motion the mass changes with time, so even if the force is constant, the acceleration is not.

Another complication arises because the equations relating acceleration, velocity, position, and time are differential, not algebraic, equations. Differential equations are usually solved using integral calculus and other special techniques which introductory students may not have mastered.

So how does one proceed to solve real-world problems without advanced mathematics? The answer is to solve such problems on personal computers, using elementary numerical methods. The simplest of these is the Euler method, named after the Swiss mathematician Leonhard Euler (1707–1783).

The Euler Method

In the **Euler method** of solving differential equations, derivatives are approximated by finite differences. Considering a small increment of time, Δt, the relationship between acceleration and velocity may be approximated as

$$a(t) = \frac{\Delta v}{\Delta t} = \frac{v(t + \Delta t) - v(t)}{\Delta t}$$

Then the velocity of the particle at the end of the period Δt is approximately equal to the velocity at the beginning of the period, plus the acceleration during the interval multiplied by Δt:

$$v(t + \Delta t) = v(t) + a(t)\Delta t \qquad [5.11]$$

Since the acceleration is a function of time, this estimate of $v(t + \Delta t)$ will be accurate only if the time interval Δt is short enough that the change in acceleration during it is very small (as will be discussed later).

The position can be found in the same manner:

$$v(t) = \frac{x(t + \Delta t) - x(t)}{\Delta t}$$

so

$$x(t + \Delta t) = x(t) + v(t)\Delta t \qquad [5.12]$$

It may be tempting to add the term $\frac{1}{2}a(\Delta t)^2$ to this result to make it look like the familiar kinematics equation, but this term is not included in the Euler method of integration because Δt is assumed to be so small that Δt^2 is nearly zero.

If the acceleration at any instant t is known, the particle's velocity and position at $(t + \Delta t)$ can be calculated from Equations 5.11 and 5.12. The calculation can then proceed in a series of finite steps to determine the velocity and position at any later time. The acceleration is determined by the net force acting on the object,

$$a(x, v, t) = \frac{F(x, v, t)}{m} \qquad [5.13]$$

which may depend explicitly on the position, velocity, or time.

It is convenient to set up the numerical solution to this kind of problem by numbering the steps and entering the calculations in a table. Table 5.3 illustrates how to do this in an orderly way.

Table 5.3
The Euler Method for Solving Dynamics Problems

Step	Time	Position	Velocity	Acceleration
0	t_0	x_0	v_0	$a_0 = F(x_0, v_0, t_0)/m$
1	$t_1 = t_0 + \Delta t$	$x_1 = x_0 + v_0\,\Delta t$	$v_1 = v_0 + a_0\,\Delta t$	$a_1 = F(x_1, v_1, t_1)/m$
2	$t_2 = t_1 + \Delta t$	$x_2 = x_1 + v_1\,\Delta t$	$v_2 = v_1 + a_1\,\Delta t$	$a_2 = F(x_2, v_2, t_2)/m$
3	$t_3 = t_2 + \Delta t$	$x_3 = x_2 + v_2\,\Delta t$	$v_3 = v_2 + a_2\,\Delta t$	$a_3 = F(x_3, v_3, t_3)/m$
	⋮	⋮	⋮	⋮
n	t_n	x_n	v_n	a_n

The equations provided in the table can be entered into a spreadsheet and the calculations performed row by row to determine the velocity, position, and acceleration as functions of time. The calculations can also be done by a program written in BASIC, PASCAL, or FORTRAN, or with commercially available mathematics packages for personal computers. Many small time increments can be taken, and accurate results can usually be obtained with the help of a computer. Graphs of velocity versus time or position versus time can be displayed to help you visualize the motion.

The Euler method has the advantage that the dynamics is not obscured—the fundamental relationships of acceleration to force, velocity to acceleration, and position to velocity are clearly evident. Indeed, these relationships form the heart of the calculations. There is no need to use advanced mathematics, and the basic physics governs the dynamics.

The Euler method is completely reliable for infinitesimally small time increments, but for practical reasons a finite increment size must be chosen. In order for the finite difference approximation of Equation 5.11 to be valid, the time increment must be small enough that the acceleration does not change appreciably during it. We can determine an appropriate size for the time increment by thinking about the particular problem that is being investigated. The criterion for the size of the time increment may need to be changed during the course of the motion. In practice, however, we usually choose a time increment appropriate to the initial conditions of the problem and use the same value throughout the calculations.

The size of the time increment influences the accuracy of the result, but unfortunately it is not easy to determine the accuracy of a solution by the Euler method without a knowledge of the correct analytical solution. One method of determining the accuracy of the numerical solution is to repeat the calculations with a smaller time increment and compare results. If the two calculations agree to a certain number of significant figures, one can assume that the results are correct to that precision. The student is encouraged to read the supplement entitled *Numerical Methods in Physics,* which provides several examples of real-world physics problems whose solutions were obtained using the Euler method.

*5.6 Motion in Accelerated Frames

When Newton's laws of motion were introduced in Chapter 4, we emphasized that they are valid when observations are made in an *inertial* frame of reference. Usually motions are analyzed using inertial reference frames, but there are cases in which an accelerating frame is more convenient. In this section we analyze how an observer in a noninertial frame (one that is accelerating) would attempt to apply Newton's second law.

If a particle moves with an acceleration **a** relative to an observer in an inertial frame, then the inertial observer may use Newton's second law and correctly claim that $\Sigma \mathbf{F} = m\mathbf{a}$. An observer in an accelerated frame who is trying to apply Newton's second law to the motion of the particle must introduce *fictitious* forces to make the law work. They *appear* to be real in the accelerating frame. However, we emphasize that these fictitious forces *do not* exist when the motion is observed

Fictitious forces

in an inertial frame. The fictitious forces are used only in an accelerating frame but *do not* represent real forces on the body. (By "real" forces, we mean the interaction of the body with its environment.) If the fictitious forces are properly defined in the accelerating frame, then the description of motion in this frame will be equivalent to the description by an inertial observer who considers only real forces.

In order to better understand the motion of a rotating system, consider a car traveling along a highway at a high speed and approaching a curved exit ramp, as in Figure 5.14. As the car takes the sharp left turn onto the ramp, a person sitting in the passenger seat slides to the right and hits the door. At that point, the force of the door keeps him from being ejected from the car. What causes the passenger to move toward the door? A popular but *improper* explanation is that some mysterious force pushes him outward. (It is often called the "centrifugal" force, but we shall not use that term since it often creates confusion.) The passenger invents this fictitious force in order to explain what is going on in his accelerated frame of reference.

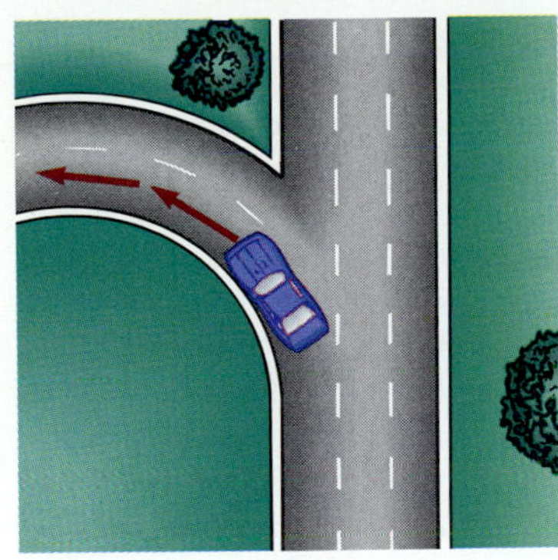

Figure 5.14 A car approaching a curved exit ramp.

The phenomenon is correctly explained as follows. Before the car enters the ramp, both it and the passenger are moving in a straight-line path. As the car enters the ramp and travels a curved path, the passenger tends to move along the original straight-line path. This is in accordance with Newton's first law: the natural tendency of a body is to continue moving in a straight line. If a sufficiently large centripetal force (toward the center of curvature) acts on the passenger, he moves in a curved path along with the car. The origin of this centripetal force is the force of friction between the passenger and the car seat. If this frictional force is not large enough, however, the passenger slides to the right (that is, away from the center of curvature) as the car turns under him. Eventually the passenger encounters the door, which provides a large enough centripetal force to enable him to follow the same curved path as the car. In summary, he slides toward the door not because of some mysterious outward force but because *there is no centripetal force large enough to enable him to travel along the circular path followed by the car.*

Example 5.10 Fictitious Force in a Rotating System

An observer in a rotating system is one example of a noninertial observer. Suppose a block of mass m lying on a horizontal, frictionless turntable is connected to a string as in Figure 5.15. According to an inertial observer, if the block rotates uniformly, it undergoes a centripetal acceleration v^2/r, where v is its tangential speed. The inertial observer concludes that this centripetal acceleration is provided by the tension in the string, T, and writes Newton's second law, $T = mv^2/r$.

According to a noninertial observer attached to the turntable, the block is at rest. Therefore, in applying Newton's second law, this observer introduces a fictitious *outward* force called the *centrifugal force,* of magnitude mv^2/r. According to the noninertial observer, this "centrifugal" force balances the force of tension, and therefore $T - mv^2/r = 0$.

Be careful when using fictitious forces to describe physical phenomena. Remember that fictitious forces, such as centrifugal force, are used *only* in noninertial, or accelerated, frames of reference. When solving problems in dynamics, it is generally best to use an inertial frame.

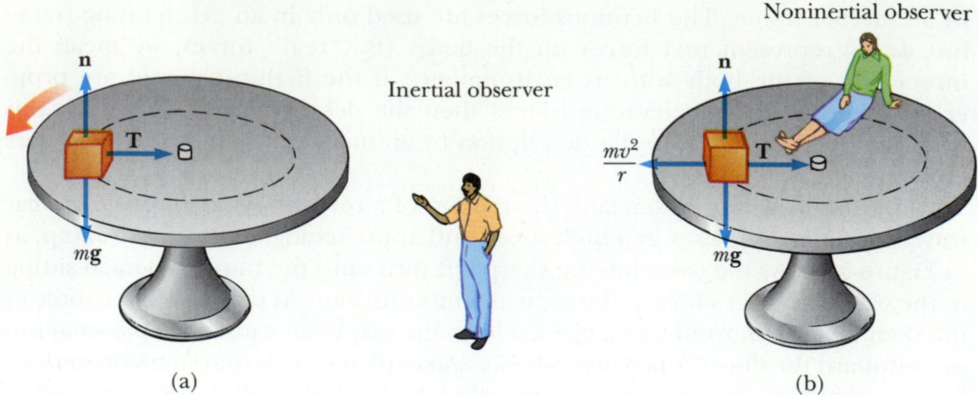

Figure 5.15 (Example 5.10) A block of mass m connected to a string tied to the center of a rotating turntable. (a) The inertial observer claims that the centripetal force is provided by the force of tension, **T**. (b) The noninertial observer claims that the block is not accelerating, and therefore he introduces a fictitious centrifugal force, mv^2/r, which acts outward and balances the tension.

Summary

Forces of friction

The **maximum force of static friction,** $\mathbf{f}_{s,\max}$, between a body and a surface is proportional to the normal force acting on the body. This maximum force occurs when the body is on the verge of slipping. In general, $f_s \le \mu_s n$, where μ_s is the *coefficient of static friction* and **n** is the normal force. When a body slides over a rough surface, the *force of kinetic friction,* $\mathbf{f}_k$, is opposite the motion and is also proportional to the normal force. The magnitude of this force is given by $f_k = \mu_k n$, where μ_k is the *coefficient of kinetic friction.* Usually, $\mu_k < \mu_s$.

Newton's second law, applied to a particle moving in **uniform circular motion,** states that the net force in the radial direction must equal the product of the mass and the centripetal acceleration:

Uniform circular motion

$$F_r = ma_r = \frac{mv^2}{r} \qquad [5.3]$$

A particle moving in nonuniform circular motion has both a centripetal acceleration and a tangential component of acceleration. In the case of a particle rotating in a vertical circle, the force of gravity provides the tangential acceleration and part or all of the centripetal acceleration.

Velocity-dependent resistive forces

A body moving through a liquid or gas experiences a **resistive force** that is velocity-dependent. This resistive force, which opposes the motion, generally increases with velocity. The force depends on the shape of the body and on the properties of the medium through which the body is moving. In the limiting case for a falling body, when the resistive force balances the weight ($a = 0$), the body reaches its **terminal speed.**

Fictitious forces

An observer in a noninertial (accelerated) frame of reference must introduce **fictitious forces** when applying Newton's second law in that frame. If these fictitious forces are properly defined, the description of motion in the noninertial frame will be equivalent to that made by an observer in an inertial frame. However, the observers in the two different frames will not agree on the causes of the motion.

Questions and Conceptual Problems

1. Although the frictional force between two surfaces may decrease as the surfaces are smoothed, the force will again increase if the surfaces are made extremely smooth and flat. How do you explain this?
2. Why is it that the frictional force involved in the rolling of one body over another is less than that involved in a sliding motion?
3. A massive metal object on a rough metal surface may undergo contact welding to that surface. Discuss how this affects the frictional forces between the object and the surface.
4. Suppose you are driving a car along a highway at a high speed. Why should you avoid slamming on your brakes if you want to stop in the shortest possible distance?
5. The driver of a speeding empty truck slams on the brakes and skids to a stop through a distance *d*. (a) If the truck's mass were doubled by a heavy load, what would be the truck's "skidding distance"? (b) If the initial speed of the truck were halved, what would be the truck's "skidding distance"?
6. If you push on a heavy box which is at rest, it requires some force **F** to start its motion. However, once the box is sliding, it requires a *smaller* force to maintain its motion. Why is this so?
7. What causes a rotary lawn sprinkler to turn?
8. Because the Earth rotates about its axis and about the Sun, it is a noninertial frame of reference. Assuming the Earth is a uniform sphere, why would the *apparent weight* of an object be greater at the poles than at the equator?
9. Explain why the Earth is not spherical in shape and bulges at the equator.
10. How would you explain the force that pushes a rider toward the side of a car as the car rounds a corner?
11. When an airplane does an inside loop-the-loop in a vertical plane, at what point does the pilot appear to be heaviest? What is the constraint force acting on the pilot?
12. A sky diver in free fall reaches terminal speed. After the parachute opens, what parameters change to decrease this terminal speed?
13. Why is it that an astronaut in a space capsule orbiting the Earth experiences a feeling of weightlessness?
14. A pail of water can be whirled in a vertical path so that no water is spilled. Why does the water stay in, even when the pail is above your head?
15. Imagine that you attach a heavy object to one end of a spring and then whirl the spring and object in a horizontal circle (by holding the free end of the spring). Does the spring stretch? If so, why? Discuss this in terms of centripetal force.
16. It has been suggested that rotating cylinders about 16 km in length and 8 km in diameter be placed in space and used as colonies. The purpose of the rotation would be to simulate gravity for the inhabitants. Explain how this would work.
17. Why do pilots tend to black out when pulling out of steep dives?
18. Cite an example of a situation in which an automobile driver can have a centripetal acceleration but no tangential acceleration.
19. Is it possible for a car to move in a circular path in such a way that it has a tangential acceleration but no centripetal acceleration?
20. Analyze the motion of a rock dropped into water in terms of its speed and acceleration as it falls. Assume that a resistive force acts on the rock and increases as the speed increases.
21. Consider a sky diver falling through air *before* reaching terminal speed. As the speed of the sky diver increases, what happens to her acceleration?
22. Centrifuges are often used in dairies to separate the cream from the milk. Which remains on the inside?
23. We often think of the brakes and the gas pedal as the devices that accelerate a car. Could a steering wheel also fall into this category? Explain.
24. Suppose that a baseball and a softball are dropped from an airplane. Which has the higher terminal speed? Which experiences the greater acceleration before reaching terminal speed—say, one second after they are released?
25. Recently, experimental trains have been designed to travel around the curves of conventional tracks at high speeds (about 150 mi/h), thereby reducing travel time by about 35%. The greater speed is made possible by a control mechanism that tilts the train as it negotiates a curve. Discuss why tilting enables a train to travel faster around curves.

(Question 25) A tilting high-speed train in Sweden. *(Gamma)*

26. Consider a small raindrop and a large raindrop falling through the atmosphere. Compare their terminal speeds. What are their accelerations when they reach terminal speed?
27. A sky diver is being pulled by the Earth with a force of 850 N. If the force of air resistance has a magnitude of 320 N, what additional force is needed for the sky diver to maintain a constant speed?
28. A roller coaster travels through a vertical loop-the-loop in a circular path. Draw vectors showing the directions of the instantaneous velocity and the net force on the roller coaster at (a) the lowest point and (b) the highest point of the loop.
29. An object executes circular motion with a constant speed whenever a net force of constant magnitude acts perpendicular to the velocity. What happens to the speed if the force is not perpendicular to the velocity?
30. On long journeys, jet aircraft usually fly at high altitudes of about 30 000 ft. What is the main advantage of flying at these altitudes from an economic viewpoint?

Problems

Section 5.1 Forces of Friction

1. A 25-kg block is initially at rest on a horizontal surface. A horizontal force of 75 N is required to set the block in motion. After it is in motion, a horizontal force of 60 N is required to keep it moving with constant speed. From this information, find the coefficients of static and kinetic friction.
2. Assume that the coefficient of friction between the wheels of a race car and the track is 1. If the car starts from rest and accelerates at the maximum possible constant rate for 400 m ($\frac{1}{4}$ mile), what is its speed at the end of the race?
3. A racing car accelerates uniformly from 0 to 80 mi/h in 8 s. The external force that accelerates the car is the frictional force between the tires and the road. If the tires do not spin, determine the *minimum* coefficient of friction between the tires and the road.

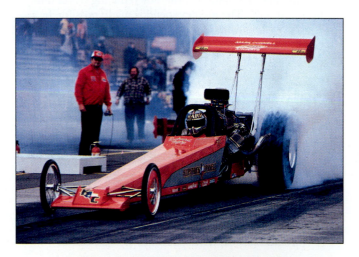

(Problem 3) *(Patrick Behar, Photo Researchers, Inc.)*

4. A car is traveling at 50 mi/h on a horizontal highway. (a) If the coefficient of friction between the road and the tires on an icy day is 0.1, what is the *minimum* distance in which the car will stop? (b) What is the stopping distance when the surface is dry and $\mu = 0.6$?
5. A block moves up a 45° incline with constant speed, under the action of a force of 15 N applied *parallel* to the incline. If the coefficient of kinetic friction is 0.3, determine (a) the weight of the block and (b) the minimum force required to allow the block to move *down* the incline at constant speed.
6. A 9-kg hanging weight is connected by a string over a pulley to a 5-kg block that is sliding on a flat table (Fig. 5.16). If the coefficient of sliding friction is 0.2, find the tension in the string.

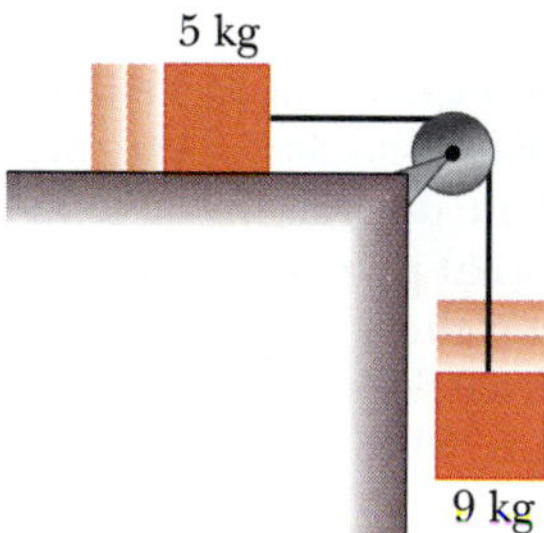

Figure 5.16 (Problem 6)

7. A child stands on the surface of a frozen pond, 12 m from the shore. If the coefficient of static friction between her boots and the ice is 0.05, determine the minimum time required for the child to walk to the shore without slipping.
8. A boy drags his 60-N sled at constant speed up a 15° hill. He does so by pulling with a 25-N force on a

rope attached to the sled. (a) If the rope is inclined at 35° to the horizontal, what is the coefficient of kinetic friction between the sled and the snow? (b) At the top of the hill, he jumps on the sled and slides down the hill. What is his acceleration down the slope?

9. A 3-kg block starts from rest at the top of a 30° incline and slides a distance of 2 m down the incline in 1.5 s. Find (a) the acceleration of the block, (b) the coefficient of kinetic friction between the block and the plane, (c) the frictional force acting on the block, and (d) the speed of the block after it has slid 2 m.

10. In order to determine the coefficients of friction between rubber and various surfaces, a student uses a rubber eraser and an incline. In one experiment the eraser slips down the incline when the angle of inclination is 36° and then moves down the incline with constant speed when the angle is reduced to 30°. From these data, determine the coefficients of static and kinetic friction for this experiment.

11. A block is placed on a plane that is inclined at 60° with respect to the horizontal. If the block slides down the plane with an acceleration of $g/2$, determine the coefficient of kinetic friction between the block and the plane.

12. Determine the stopping distance for a skier with a speed of 20 m/s (Fig. 5.17). Assume $\mu_k = 0.18$ and $\theta = 5°$.

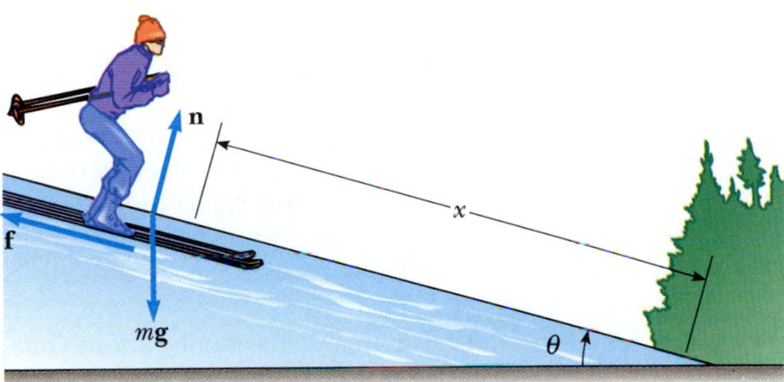

Figure 5.17 (Problem 12)

13. A woman at an airport is towing her 20-kg suitcase at constant speed by pulling on a strap at an angle of θ above the horizontal (Fig. 5.18). She pulls on the strap with a 35-N force, and the friction force on the suitcase is 20 N. (a) What angle does the strap make with the horizontal? (b) What normal force does the ground exert on the suitcase?

14. The parachute on a race car of weight 8820 N opens at the end of a quarter-mile run when the car is traveling at 55 m/s. What is the total retarding force required to stop the car in a distance of 1000 m?

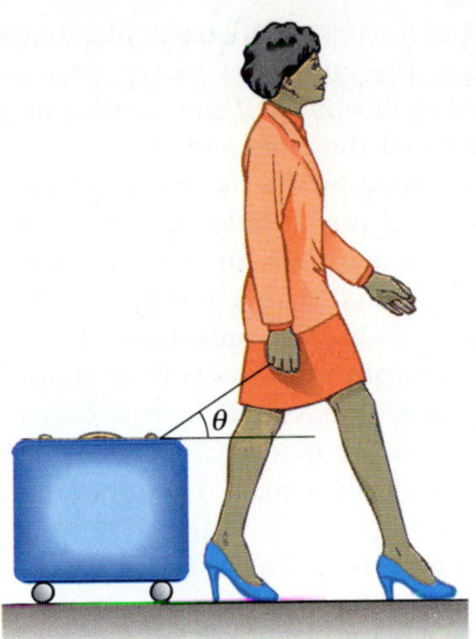

Figure 5.18 (Problem 13)

15. A rifle bullet with a mass of 12 g, traveling with a speed of 400 m/s, strikes a large wooden block and penetrates it to a depth of 15 cm. Determine the magnitude of the frictional force (assumed constant) that acts on the bullet.

Section 5.2 Newton's Second Law Applied to Uniform Circular Motion

16. A radio-controlled toy car completes one lap around a circular track (a distance of 200 m) in 25 s. (a) What is the average speed? (b) If the mass of the car is 1.5 kg, what is the magnitude of the centripetal force that keeps it in a circle?

17. What centripetal force is required to keep a 1.5-kg mass moving in a circle of radius 0.4 m at a speed of 4 m/s?

18. In a hydrogen atom, the electron in orbit about the proton feels an attraction of about 8.20×10^{-8} N. If the radius of the orbit is 5.3×10^{-11} m, what is the frequency, in revolutions per second? (See inside the front cover of this book for additional data.)

19. A 3-kg mass attached to a light string rotates in circular motion on a horizontal, frictionless table. The radius of the circle is 0.8 m, and the string can support a mass of 25 kg before breaking. What range of speeds can the mass have without breaking the string?

20. An automobile moves at constant speed over the crest of a hill. The arc made by the driver is part of a vertical circle of radius 18 m. At the top of the hill, the driver notices that she barely remains in contact with the seat. Find the speed of the vehicle.

21. A crate of eggs is positioned in the middle of the

flatbed of a truck as the truck negotiates a curve in the road. The curve may be regarded as an arc of a circle of radius 35 m. If the coefficient of static friction between the crate and the flatbed of the truck is 0.6, what must be the maximum speed of the truck if the crate is not to slide during the maneuver?

22. A 1500-kg car rounds an unbanked curve with a radius of 52 m at a speed of 12 m/s. What minimum coefficient of friction must exist between the road and tires to prevent the car from slipping?

23. An air puck of mass 0.25 kg is tied to a string and allowed to revolve in a circle of radius 1.0 m on a horizontal, frictionless table. The other end of the string passes through a hole in the center of the table, and a 1.0-kg mass is tied to it. The suspended mass remains in equilibrium while the puck on the tabletop revolves. (a) What is the tension in the sting? (b) What is the centripetal force acting on the puck? (c) What is the speed of the puck?

24. The speed of the tip of the minute hand of a town clock is 1.75×10^{-3} m/s. (a) What is the speed of the tip of the second hand, of the same length? (b) What is the centripetal acceleration of the tip of the second hand?

25. A coin is placed 30 cm from the center of a rotating, horizontal turntable. The coin is observed to slip when its speed is 50 cm/s. (a) What provides the centripetal force when the coin is stationary relative to the turntable? (b) What is the coefficient of static friction between the coin and the turntable?

Section 5.3 Nonuniform Circular Motion

26. A car traveling on a straight road at 9.0 m/s goes over a hump in the road. The hump may be regarded as an arc of a circle of radius 11.0 m. (a) What is the apparent weight of a 600-N woman in the car as she rides over the hump? (b) What must be the speed of the car over the hump if she is to experience weightlessness? (The apparent weight must be zero.)

27. A pail half-filled with water is rotated in a vertical circle of radius 1 m. What is the minimum speed of the pail at the top of the circle if no water is to spill out?

28. A hawk flies in a horizontal arc of radius 12 m at a constant speed of 4 m/s. (a) Find the centripetal acceleration of the hawk. (b) The hawk continues to fly along the same horizontal arc but increases its speed at the rate of 1.2 m/s^2. Find the hawk's acceleration (magnitude and direction) under these conditions.

29. A 40-kg child sits in a conventional swing of length 3 m, supported by two chains. If the tension in each chain is 350 N when the swing is at its lowest point, find (a) the child's speed at the lowest point and (b) the force of the seat on the child at the lowest point. (Neglect the mass of the seat.)

30. A 0.40-kg object is swung in a circular path in a vertical plane, on a string 0.5 m long. If a constant speed of 4.0 m/s is maintained, what is the tension in the string when the object is at the top of the circle?

31. A Ferris wheel with radius 20 m makes 1 revolution every 9.0 seconds. What force does a 55-kg passenger exert on the seat when she is at the top of the Ferris wheel?

32. A roller-coaster vehicle has a mass of 500 kg when fully loaded with passengers (Fig. 5.19). (a) If the vehicle has a speed of 20 m/s at point A, what is the force of the track on the vehicle at this point? (b) What is the maximum speed the vehicle can have at B and still remain on the track?

33. Tarzan ($m = 85$ kg) tries to cross a river by swinging from a vine. The vine is 10 m long, and his speed at the bottom of the swing (as he just clears the water) is 8 m/s. Tarzan doesn't know that the vine has a breaking strength of 1000 N. Does he make it safely across the river?

34. The Six Flags Great America amusement park in Gur-

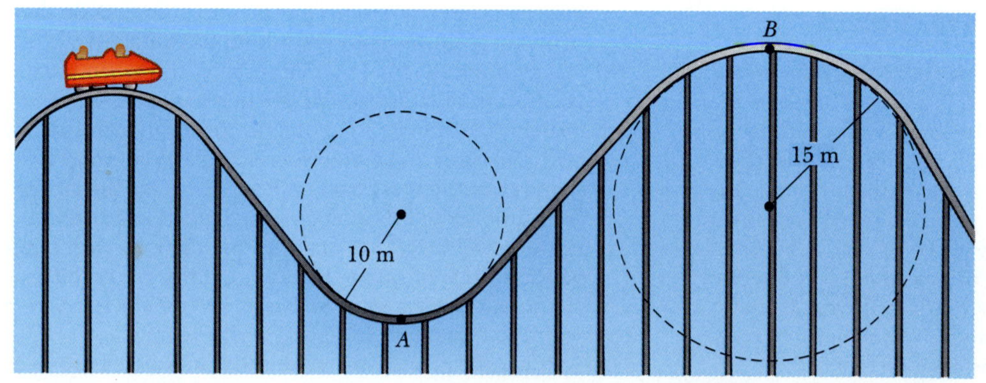

Figure 5.19
(Problem 32)

Figure 5.20
(Problem 34) *(Frank Cezus, FPG International)*

(Problem 35) *(Guy Sauvage, Photo Researchers, Inc.)*

nee, Illinois, has a roller coaster that incorporates some of the latest design technology and some basic physics. The vertical loop, instead of being circular, is shaped like a teardrop (Fig. 5.20). The cars ride on the inside of the loop at the top, and the speeds are high enough to ensure that the cars remain on the track. The biggest loop is 40 m high (about 130 ft), with a maximum speed of 31 m/s (nearly 70 mph) at the bottom.[3] Suppose the speed at the top is 13.0 m/s and the corresponding centripetal acceleration is $2g$. (a) What is the radius of the teardrop of the arc at the top? (b) If the total mass of the roller coaster at the top of the loop is M, what force does the rail exert on it at the top? (c) Suppose, instead, that the roller coaster has a circular loop of radius 20 m. If the cars have the same speed, 13 m/s at the top, what is the centripetal acceleration at the top? (d) Comment on the normal force at the top in this situation.

*Section 5.4 Motion in the Presence of Velocity-Dependent Resistive Forces

35. A sky diver of mass 80 kg jumps from a slow-moving aircraft and reaches a terminal speed of 50 m/s. (a) What is the acceleration of the sky diver when her speed is 30 m/s? What is the drag force on the diver when her speed is (b) 50 m/s? (c) 30 m/s?

36. A small piece of Styrofoam packing material is dropped from a height of 2.0 m above the ground. Until the terminal speed is reached, the acceleration of the piece of Styrofoam is given by $a = -g + bv$, where g is the free-fall acceleration, v the speed, and b a constant. When the piece has fallen 0.5 m, its terminal speed is reached, and it takes an extra 5 s to reach the ground. (a) What is the numerical value of the constant b? (b) What is the acceleration at $t = 0$? (c) What is the acceleration when the speed is 0.15 m/s?
37. The driver of a motorboat cuts the engine when the speed is 10 m/s, and the boat coasts to rest. The equation governing the motion of the motorboat during this period is $v = v_0 e^{-ct}$, where v is the speed at time t, v_0 is the initial speed, and c is a constant. At $t = 20$ s the speed is 5 m/s. (a) Find the constant c. (b) What is the speed at $t = 40$ s? (c) Differentiate the preceding expression for $v(t)$ and thus show that the acceleration of the boat is proportional to the speed at any time.
38. (a) Estimate the terminal speed of a wooden sphere (density 0.83 g/cm^3) moving in air if its radius is 8.0 cm. (b) From what height would a freely falling object reach this speed in the absence of air resistance?
39. A small, spherical bead of mass 3 g is released from rest at $t = 0$ in a bottle of liquid shampoo. The terminal speed, v_t, is observed to be 2 cm/s. Find (a) the value of the constant b in Equation 5.6, (b) the time τ it takes to reach $0.63v_t$, and (c) the value of the retarding force when the bead reaches terminal speed.

[3] From the *New York Times*, Aug. 2, 1988.

Section 5.5 Numerical Modeling in Particle Dynamics

40. A 3-g leaf is dropped from a height of 2.0 m above the ground. Assume the net downward force on the leaf is $F = mg - bv$, where the drag factor is $b = 0.03$ kg/s. (a) Calculate the terminal speed of the leaf. (b) Use Euler's method of numerical analysis to find the speed and position of the leaf, as functions of time, from the instant it is released until 99% of terminal speed is reached. (*Hint:* Try $\Delta t = 0.005$ s.)
41. A hailstone of mass 4.8×10^{-4} kg falls through the air and experiences a net force given by

$$F = -mg + Dv^2$$

where $D = 2.5 \times 10^{-5}$ kg/m. (a) Calculate the terminal speed of the hailstone. (b) Use Euler's method of numerical analysis to find the speed and position of the hailstone at 0.2-s intervals, taking the initial speed to be zero. Continue the calculation until the hailstone reaches 99% of terminal speed.
42. Wind-tunnel measurements show that a baseball can be suspended nearly motionless in an upward-directed, 95-mi/h vertical airstream. Thus, the terminal speed of a baseball is 95 mi/h (42.5 m/s). A baseball has a mass of 5 ounces (142 g). (a) If a baseball experiences a drag force $R = Dv^2$, what is the value of the coefficient D? (b) What is the magnitude of the drag force when the baseball is traveling at 80 mi/h (36 m/s)? (c) Use a numerical method to determine the motion of a baseball thrown vertically upward at an initial speed of 80 mi/h. How high does the ball go? How long is it in the air? What speed does it have when it hits the ground? (d) If the terminal speed is 95 mi/h, should we believe reports of a pitcher who is said to throw a 100-mi/h fastball?
43. A 50-kg parachutist jumps from an airplane and falls to Earth with a drag force proportional to the square of the speed, $R = Dv^2$. Take $D = 0.2$ kg/m (with the parachute closed) and $D = 20$ kg/m (with the chute open). (a) Determine the terminal speed of the parachutist in both configurations, before and after the chute is opened. (b) Set up a numerical analysis of the motion and compute the speed and position as functions of time, assuming the jumper begins the descent at 1000 m above the ground and is in free fall for 10 s before opening the parachute. (*Hint:* When the parachute opens, a sudden large acceleration takes place; a smaller time step may be necessary in this region.)
44. Consider a 10-kg projectile launched with an initial speed of 100 m/s, at an angle of 35° elevation. The resistive force is $\mathbf{R} = -b\mathbf{v}$, where $b = 10$ kg/s. (a) Use a numerical method to determine the horizontal and vertical positions of the projectile as functions of time. (b) What is the range of this projectile? (c) Determine the elevation angle that gives the maximum range for the projectile. (*Hint:* Adjust the elevation angle by trial and error to find the greatest range.)
45. A baseball is subject to a drag force of $|\mathbf{R}| = Dv^2$ opposite to its velocity (see Problem 42 for data). It is batted with an initial speed of 110 mi/h (49 m/s). (a) At what angle should a batter hit the ball to achieve the greatest distance? (b) How long is the ball in the air? (c) How far can a good center fielder run while this ball is in the air?

*Section 5.6 Motion in Accelerated Frames

46. A ball is suspended from the ceiling of a moving car by a string 25 cm in length. An observer in the car notes that the ball deflects 6 cm from the vertical toward the rear of the car. What is the acceleration of the car?
47. A 3-kg mass hangs at one end of a rope that is attached to a support on a railroad car. When the car accelerates to the right, the cord makes an angle of 4° with the vertical, as shown in Figure 5.21. Find the acceleration of the car.

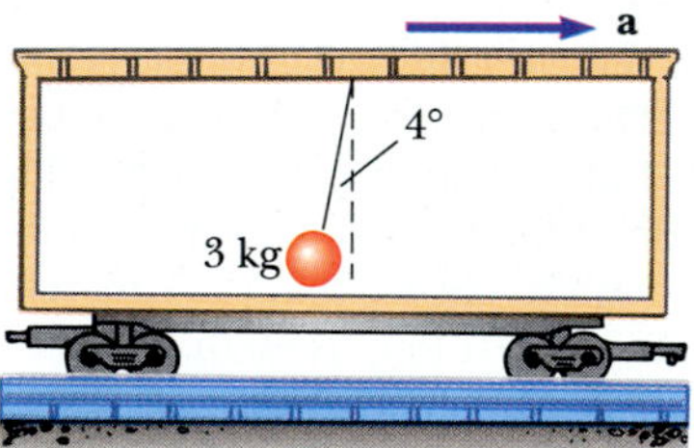

Figure 5.21 (Problem 47)

48. A 5-kg mass attached to a spring scale rests on a frictionless, horizontal surface as in Figure 5.22. The spring scale, attached to the front end of a boxcar,

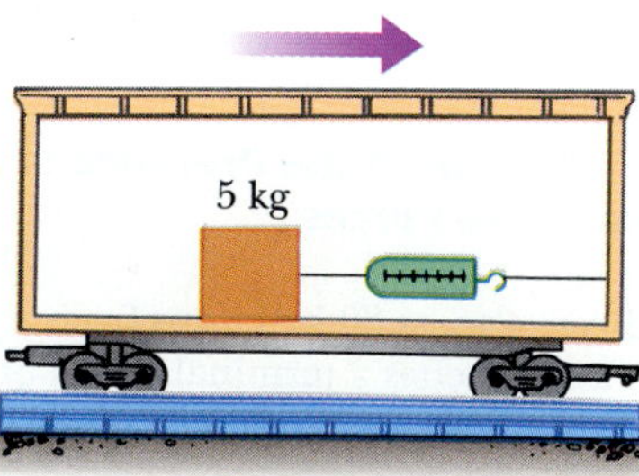

Figure 5.22 (Problem 48)

reads 18 N when the car is in motion. (a) If the spring scale reads zero when the car is at rest, determine the acceleration of the car. (b) What will the spring scale read if the car moves with constant velocity? (c) Describe the forces on the mass as observed by someone in the car and by someone at rest outside the car.

49. A merry-go-round turns completely each 12 s. If a 45-kg child sits on the horizontal floor of the merry-go-round 3 m from the center, find (a) the child's acceleration and (b) the horizontal force of friction that acts on the child. (c) What minimum coefficient of static friction would be necessary to keep the child from slipping?

50. A plumb bob does not hang exactly along a line directed to the center of the Earth's rotation. At 35° north latitude, how much does the plumb bob deviate from a radial line because of the rotation?

Additional Problems

51. A spinning ball of radius 5.0 cm slows uniformly in 0.3 s from 30 rev/min to rest. Compute the radial, tangential, and net accelerations of a point on the equator of the ball at the beginning of this time period.

52. In the Bohr model of the hydrogen atom, the speed of the electron is approximately 2.2×10^6 m/s. Find (a) the centripetal force acting on the electron as it revolves in a circular orbit of radius 0.53×10^{-10} m, (b) the centripetal acceleration of the electron, and (c) the number of revolutions per second made by the electron.

53. A 0.40-kg pendulum bob passes through the lowest part of its path with a speed of 8.2 m/s. What is the tension in the pendulum cable at this point if the pendulum is 80 cm long?

54. A conical pendulum is a bob moving in a horizontal circle at the end of a long wire (Fig. 5.23). The angle between the wire and the vertical does not change. Consider a conical pendulum, with an 80-kg bob, on a 10-m wire that makes an angle of 5° with the vertical. Determine (a) the tension in the wire and its horizontal and vertical components, and (b) the radial acceleration of the bob.

55. A small turtle, appropriately named Dizzy, is placed on a horizontal, rotating turntable 20 cm from its center. Dizzy's mass is 50 g, and the coefficient of static friction between his feet and the turntable is 0.3. Find (a) the *maximum* number of revolutions per second the turntable can have if Dizzy is to remain stationary relative to the turntable and (b) Dizzy's speed and radial acceleration when he is on the verge of slipping.

56. Because of the Earth's rotation about its axis, a point on the equator experiences a centripetal acceleration of $0.034\ m/s^2$, while a point at a pole experiences no centripetal acceleration. (a) Show that at the equator the gravitational force on an object (the true weight) must *exceed* the object's apparent weight. (b) What are the apparent weights at the equator and at the poles of a person having a mass of 75 kg? (Assume the Earth is a uniform sphere and take the free-fall acceleration to be $g = 9.800\ m/s^2$.)

57. An engineer wishes to design a curved exit ramp for a toll road in such a way that a car will not have to rely on friction to round the curve without skidding. Suppose that a typical car rounds the curve with a speed of 30 mi/h (13.4 m/s) and that the radius of the curve is 50 m. At what angle should the curve be banked? (See Fig. 5.24.)

58. A car rounds a banked curve as in Figure 5.24. The radius of curvature of the road is R, the banking angle is θ, and the coefficient of static friction is μ.

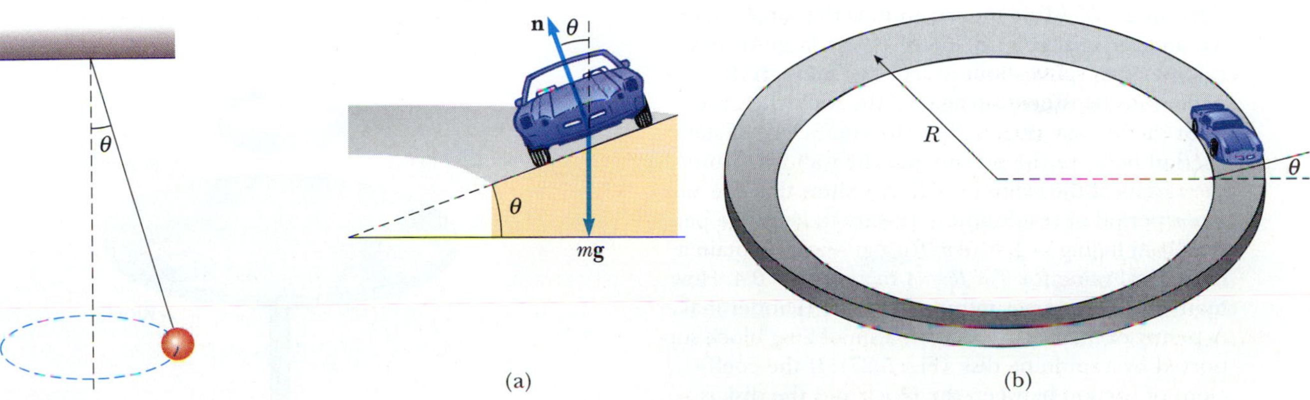

Figure 5.23 (Problem 54)

Figure 5.24 (Problems 57 and 58)

(a) Determine the *range* of speeds the car can have without slipping up or down the road. (b) Find the minimum value for μ so that the minimum speed is zero. (c) What range of speeds is possible if $R = 100$ m, $\theta = 10°$, and $\mu = 0.1$ (slippery conditions)?

59. A model airplane of mass 0.75 kg flies in a horizontal circle at the end of a 60-m control wire, with a speed of 35 m/s. Compute the tension in the wire if it makes a constant angle of 20° with the horizontal. The airplane is acted upon by the tension in the control line, its weight, and the aerodynamic lift, which acts at 20° inward from the vertical, as shown in Figure 5.25.

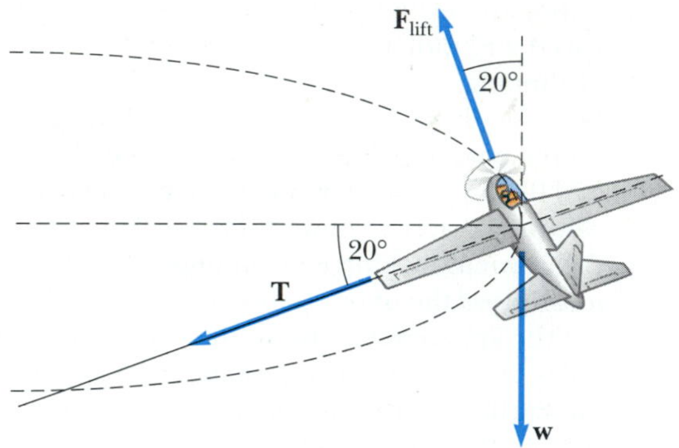

Figure 5.25 (Problem 59)

60. A student builds and calibrates an accelerometer, which she uses to determine the speed of her car around a certain highway curve. The accelerometer is a simple pendulum with a protractor, which she attaches to the roof of her car. Her friend observes that the pendulum hangs at an angle of 15° from the vertical when the car has a speed of 23.0 m/s. (a) What is the centripetal acceleration of the car as it rounds the curve? (b) What is the radius of the curve? (c) What is the speed of the car if the pendulum deflection is 9.0° while the car rounds the same curve?

61. An amusement park ride consists of a large vertical cylinder that spins about its axis fast enough that any person inside is held up against the wall when the floor drops away (Fig. 5.26). The coefficient of static friction between the person and the wall is μ_s, and the radius of the cylinder is R. (a) Show that the *maximum* period of revolution necessary to keep the person from falling is $T = (4\pi^2 R\mu_s/g)^{1/2}$. (b) Obtain a numerical value for T if $R = 4$ m and $\mu_s = 0.4$. How many revolutions per minute does the cylinder make?

62. A penny of mass 3.1 g rests on a small 20-g block supported by a spinning disk (Fig. 5.27). If the coefficient of friction between the block and the disk is 0.75 (static) and 0.64 (kinetic), and that between the penny and the block is 0.52 (static) and 0.45 (kinetic), what is the maximum angular speed, in revolutions per minute, the disk can have without the block or penny sliding on the disk?

Figure 5.26 (Problem 61)

63. A crate of weight w is pushed by a force F on a horizontal floor. (a) If the coefficient of static friction is μ_s and F is directed at angle ϕ *below* the horizontal, show that the minimum value of F that will move the crate is given by

$$F = \frac{\mu_s w \sec \phi}{1 - \mu_s \tan \phi}$$

(b) Find the minimum value of F that can produce motion when $\mu_s = 0.4$, $W = 100$ N, and $\phi = 0°, 15°, 30°, 45°$, and $60°$.

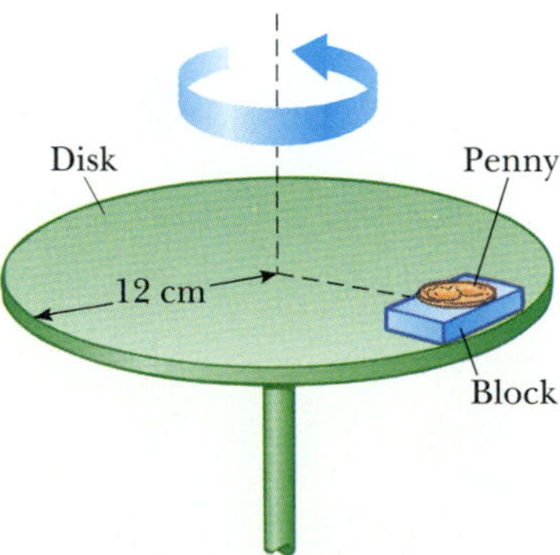

Figure 5.27 (Problem 62)

64. The expression $F = arv + br^2v^2$ gives the magnitude of the resistive force (in newtons) on a sphere of radius r (in meters) exerted by a stream of air with speed v (in meters per second), where a and b are constants with appropriate SI units. Their numerical values are $a = 3.1 \times 10^{-4}$ and $b = 0.87$. Using this formula, find the terminal speed for water droplets falling under their own weight in air, taking the following values for the drop radii: (a) 10 μm, (b) 100 μm, (c) 1 mm. Note that for (a) and (c) you can obtain accurate answers without solving a quadratic equation, by considering which of the two contributions to the air resistance is dominant and ignoring the lesser contribution.

65. An amusement park ride consists of a rotating circular platform, 8 m in diameter, from which "bucket seats" are suspended at the ends of 2.5-m chains (Fig. 5.28). When the system rotates, each chain holding a seat makes an angle of $\theta = 28°$ with the vertical. (a) What is the speed of the seat? (b) If a child of mass 40 kg sits in the 10-kg seat, what is the tension in the chain?

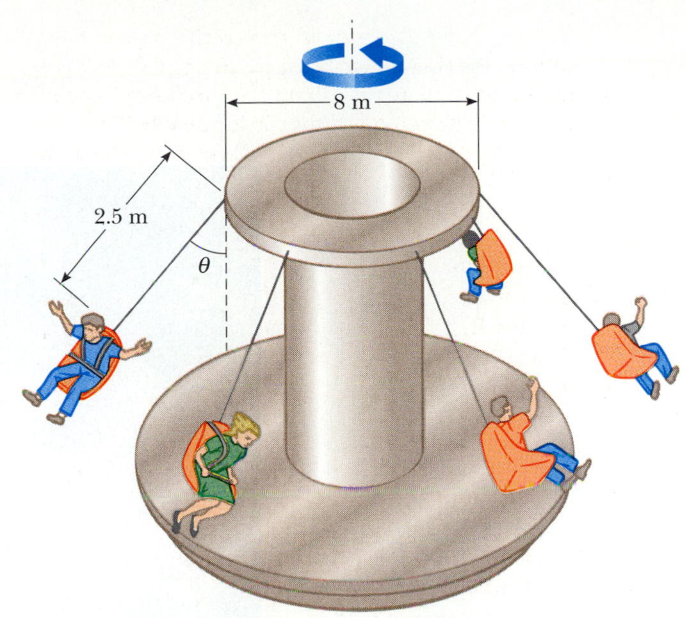

Figure 5.28 (Problem 65)

The aurora borealis (the Northern Lights) is rarely as colorful as seen in this dramatic photograph. The aurora is caused by the interaction between cosmic rays—energetic charged particles coming from the sun—and gas molecules in the Earth's upper atmosphere. Those cosmic rays trapped by the Earth's magnetic field over the North and South Poles collide with atmospheric molecules, causing them to emit visible light. *(Peka Parviainen, Science Photo Library.)*

Forces and Fields in Nature

CHAPTER 6

We begin this chapter with a discussion of the known fundamental forces in nature, some of their important properties, and their relative strengths. We then introduce the field concept, which is widely used in describing gravitational, electromagnetic, and nuclear interactions. Finally, we shall use the field concept to describe the motions of charged particles in uniform electric fields and in uniform magnetic fields.

6.1 Forces in Nature

In preceding chapters we introduced some forces that people experience every day, such as the force of gravity, which acts on all objects at or near the Earth's surface, and the normal force that acts on one object in contact with another. We also described the force of friction experienced by an object as it slides over a surface, and the drag force that acts on an object moving through a viscous medium. Other forces to be encountered later in our study of physics include the restoring force in a deformed spring, the electrostatic force between two charged objects, and the magnetic force between a magnet and a piece of iron.

You may think that the list of fundamental forces gets longer and longer, so that by the end of this course you will be totally overwhelmed. Fortunately, that is not the case. A force is considered fundamental if it is the result of a basic interaction between particles. In the year 1967, four fundamental forces were known; in order of decreasing strength, they are the **strong** force, the **electromagnetic** force, the **weak nuclear** force, and the **gravitational** force. We shall briefly describe them, then discuss the progress that has been made toward simplification since 1967.

The Gravitational Force

The gravitational force is the mutual force of attraction between any two masses in the Universe. We first encountered it when we addressed the concept of weight in Chapter 4, Section 4.6. It is interesting and rather curious that, although the gravitational force can be very strong between macroscopic objects, it is the weakest of all the fundamental forces. For example, the gravitational force between the electron and proton in the hydrogen atom is only about 10^{-47} N, whereas the electrostatic force between these same two particles is about 10^{-7} N.

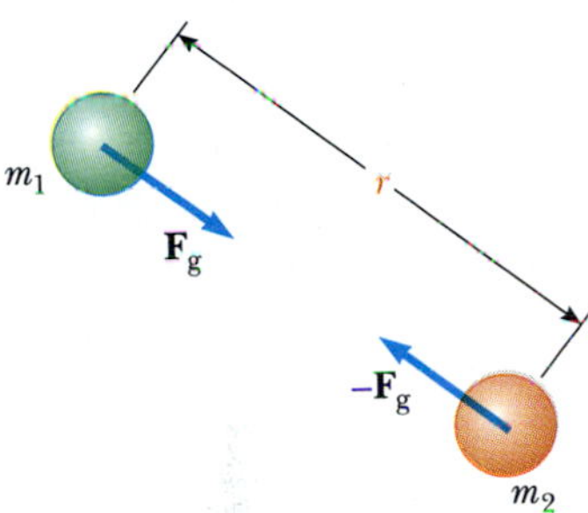

Figure 6.1
Two particles of masses m_1 and m_2 attract each other with a force of magnitude $Gm_1 m_2/r^2$.

Newton's law of gravitation states that *every particle in the Universe attracts every other particle with a force that is directly proportional to the product of the two masses and inversely proportional to the square of the distance between them.* If the particles have masses m_1 and m_2 and are separated by a distance r, as in Figure 6.1, the magnitude of the gravitational force is

Newton's universal law of gravitation

$$F_g = G\frac{m_1 m_2}{r^2} \qquad [6.1]$$

where G is the *universal gravitational constant,* with the value $G = 6.67 \times 10^{-11}$ $\text{N}\cdot\text{m}^2/\text{kg}^2$.

Example 6.1 Satellite Motion

Consider a satellite of mass m that is moving in a circular orbit about the Earth at a constant speed v and at an altitude h above the Earth's surface, as in Figure 6.2.

(a) Determine the speed of the satellite in terms of G, h, R_e (the radius of the Earth), and M_e (the mass of the Earth).

Solution Because the only external force on the satellite is the force of gravity, which acts toward the center of the Earth, we have

$$F_g = G\frac{M_e m}{r^2}$$

From Newton's second law and the fact that $r = R_e + h$, we get

$$G\frac{M_e m}{r^2} = m\frac{v^2}{r}$$

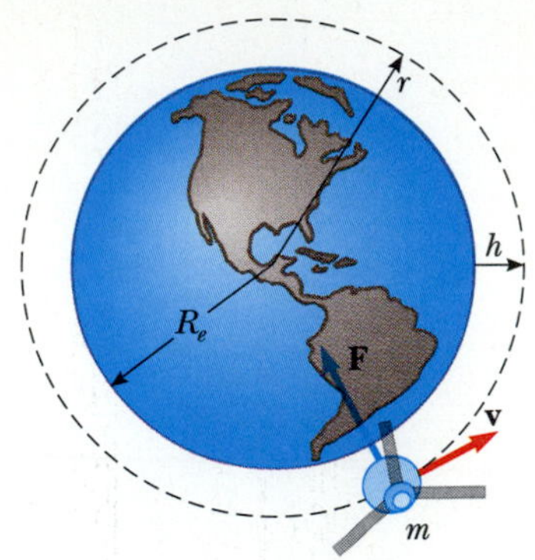

Figure 6.2
(Example 6.1) A satellite of mass m moving in a circular orbit of radius r around the Earth with constant speed v. The centripetal force is provided by the gravitational force between the satellite and the Earth.

or

$$v = \sqrt{\frac{GM_e}{r}} = \sqrt{\frac{GM_e}{R_e + h}} \quad [6.2]$$

(b) Determine the satellite's period of revolution, T (the time for one revolution about the Earth).

Solution Since the satellite travels a distance of $2\pi r$ (the circumference of the circle) in a time T, we find, using Equation 6.2, that

$$T = \frac{2\pi r}{v} = \frac{2\pi r}{\sqrt{GM_e/r}} = \left(\frac{2\pi}{\sqrt{GM_e}}\right) r^{3/2} \quad [6.3]$$

The planets move around the Sun in approximately circular orbits whose radii can be calculated from Equation 6.3, with M_e replaced by the mass of the Sun. The fact that the square of the period is proportional to the cube of the radius of the orbit was first recognized as an empirical relation based on planetary data. This topic is discussed in more detail in Chapter 12.

Exercise A satellite is in a circular orbit at an altitude of 1000 km. The radius of the Earth is 6.37×10^6 m. Find the speed of the satellite and the period of its orbit.

Answer 7.35×10^3 m/s; 6.31×10^3 s = 105 min.

The Electromagnetic Force

The electromagnetic force is the force that binds atoms and molecules in compounds to form ordinary matter. It is much stronger than the gravitational force, as we shall see shortly. The force that causes a rubbed comb to attract bits of paper and the force that a magnet exerts on an iron nail are electromagnetic forces. Essentially all forces at work in our macroscopic world (apart from the gravitational force) are manifestations of the electromagnetic force. For example, friction forces, contact forces, tension forces, and forces in elongated springs are consequences of electromagnetic forces between charged particles in proximity.

The electromagnetic force involves two types of particles: those with positive charge and those with negative charge. Unlike the gravitational force, which is *always* an attractive interaction, the electromagnetic force can be *either* attractive or repulsive, depending on the charges on the particles. In general, the electromagnetic force acts between two charged particles that are in relative motion.

When the charged particles are at rest relative to each other, we refer to the electromagnetic force between them as the *electrostatic* force. **Coulomb's law** expresses the magnitude of the electrostatic force F_e between two charged particles separated by a distance r:

Coulomb's law

$$F_e = k_e \frac{q_1 q_2}{r^2} \quad [6.4]$$

where q_1 and q_2 are the charges on the two particles, measured in coulombs (C), and k_e is the *Coulomb constant,* with the value $k_e = 8.99 \times 10^9\ \text{N}\cdot\text{m}^2/\text{C}^2$. Note that the electrostatic force has the same mathematical form as Newton's universal law of gravity (Eq. 6.1); however, charge replaces mass, and the constants are different. The electrostatic force is *attractive* if the two charges have opposite signs, and *repulsive* if the two charges have the same sign, as indicated in Figure 6.3. In other words, *opposite charges attract each other, and like charges repel each other.*

The smallest amount of *isolated* charge found in nature (so far) is the charge on an electron or proton. This fundamental unit of charge has the magnitude $e = 1.60 \times 10^{-19}$ C. Theories developed in the 1970s and 1980s propose that protons and neutrons are made up of smaller particles called **quarks,** which have charges of either $2e/3$ or $-e/3$ (discussed further in Chapter 32). Although experimental evidence has been found for such particles inside nuclear matter, *free* quarks have never been detected.

The most important properties of electric charges can be summarized as follows:

Properties of electric charges

1. Charges are scalar quantities; hence, their values are additive.
2. The two kinds of charges that exist in nature are labeled *positive* and *negative.* Charges of like sign repel each other, and charges of opposite sign attract each other.
3. The net charge on any object has discrete values; in other words, charge is said to be *quantized.* Any isolated elementary charged particle has a charge of $\pm e$. For example, electrons have a charge $-e$ and protons have a charge $+e$. (Neutrons have no charge.) Since e is the fundamental unit of charge, the net charge of an object is Ne, where N is an integer. For macroscopic objects, N is very large and the quantization of charge is not noticed.
4. Electric charge is always conserved. This means that in any kind of process—such as a collision event, a chemical reaction, or nuclear decay—the total charge of an isolated system remains constant.

Example 6.2 Which Force Is Strongest?

Two protons in the nucleus of an atom are separated by 10^{-15} m. Compare the electrostatic force between the two protons with the gravitational force between them.

Solution Since the charge on a proton is equal to $+1.6 \times 10^{-19}$ C, we calculate the electrostatic repulsion between the two protons, using Equation 6.4, to be

$$F_e = (8.99 \times 10^9 \text{ N}\cdot\text{m}^2/\text{C}^2)\frac{(1.6 \times 10^{-19}\text{ C})^2}{(1 \times 10^{-15}\text{ m})^2} = 230 \text{ N}$$

We calculate the gravitational force between the two protons, using Equation 6.1 and the fact that the proton mass is 1.67×10^{-27} kg:

$$F_g = (6.67 \times 10^{-11} \text{ N}\cdot\text{m}^2/\text{kg}^2)\frac{(1.67 \times 10^{-27}\text{ kg})^2}{(1 \times 10^{-15}\text{ m})^2} = 1.86 \times 10^{-34} \text{ N}$$

The corresponding ratio of the electrostatic force to the gravitational force is about 10^{36}. These results verify that the attractive gravitational force between the two protons is negligibly small relative to the repulsive electrostatic force between them.

The nucleus of an atom contains only protons and neutrons, and the force between the protons is always repulsive. What, then, is the source of the binding force that holds the nuclear constituents together? Read on for the answer.

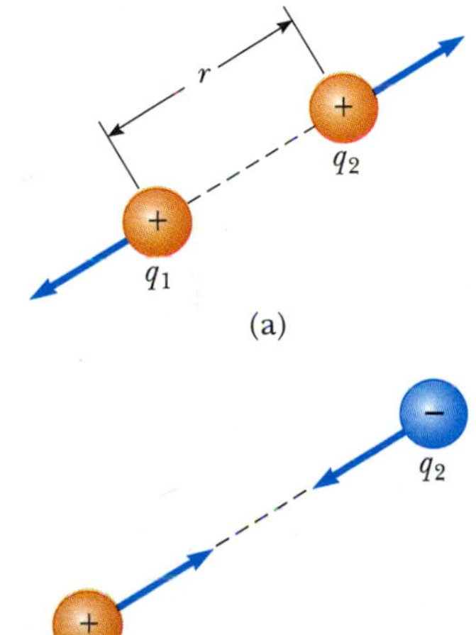

Figure 6.3
Two point charges separated by a distance r exert an electrostatic force on each other that is given by Coulomb's law. (a) When the charges are of the same sign, the force is repulsive. (b) When the charges are of opposite signs, the force is attractive.

The Strong Force

An atom, as we currently understand it, consists of an extremely dense positively charged nucleus surrounded by a cloud of negatively charged electrons, with the electrons attracted to the nucleus by the Coulomb force. Since all nuclei except those of hydrogen are combinations of positively charged protons and neutral neutrons (collectively called nucleons), why doesn't the repulsive electrostatic force between the protons cause nuclei to break apart? Clearly, there must be an attractive force that overcomes the strong electrostatic repulsive force and is responsible for the stability of nuclei. This great force that binds the nucleons to form a nucleus is called the **strong** force. It is the strongest known fundamental force. It should be noted that electrons and certain other particles are immune to the strong force. Unlike the gravitational and electromagnetic forces, which depend on distance in an inverse-square fashion, the strong force is extremely short-range; its strength decreases very rapidly outside the nucleus and is negligible for separations greater than about 10^{-14} m. For separations of about 10^{-15} m (a typical nuclear dimension), the strong force is about two orders of magnitude stronger than the electromagnetic force.

The Weak Force

The **weak** force is a short-range force that tends to produce instability in certain nuclei. It was first observed in naturally occurring radioactive substances and was later found to play a key role in most radioactive decay reactions. The weak force is about 10^{25} times *stronger* than the gravitational force and about 10^{12} times *weaker* than the electromagnetic force.

The Current View of Fundamental Forces

For years physicists have searched for a simplification scheme that would reduce the number of fundamental forces in nature. In 1967 physicists predicted that the electromagnetic force and the weak force, originally thought to be independent of each other and both fundamental, are in fact manifestations of one force, now called the **electroweak** force. The prediction was confirmed experimentally in 1984.

Scientists believe that the fundamental forces of nature are closely related to the origin of the Universe. The Big Bang theory states that the Universe began with a cataclysmic explosion 15 to 20 billion years ago. According to this theory, the first moments after the Big Bang saw such extremes of energy that all the fundamental forces were unified into one force. Physicists are continuing their search for connections among the known fundamental forces—connections that could eventually prove that the forces are all merely different forms of a single *superforce.* The recent success linking the electromagnetic and weak nuclear forces has spurred greater efforts at a unification scheme (yet to be proved) called the **Grand Unified Theory** (GUT). This fascinating subject continues to be at the forefront of physics.

6.2 The Gravitational Field

When Newton first published his theory of gravitation, his contemporaries found it difficult to accept the concept of a field force that could act at a distance. They asked how it was possible for two masses to interact even though they were not in contact with each other. Although Newton himself could not answer this question, his theory was considered a success because it satisfactorily explained the motions of the planets.

An alternative approach to describing the gravitational interaction is to introduce the concept of a **gravitational field, g,** at every point in space. When a particle of mass m is placed at a point where the field is $\mathbf{g}$, the particle experiences a force $\mathbf{F}_g = m\mathbf{g}$. In other words, the field $\mathbf{g}$ exerts a force on the particle. Hence, the gravitational field is defined by

Gravitational field

$$\mathbf{g} \equiv \frac{\mathbf{F}_g}{m} \qquad [6.5]$$

That is, the gravitational field at a point in space equals the gravitational force that a test mass experiences at that point divided by the mass. Consequently, if $\mathbf{g}$ is known at some point in space, a test particle of mass m experiences a gravitational force $m\mathbf{g}$ when placed at that point.

As an example, consider an object of mass m near the Earth's surface. The gravitational force on the object is directed toward the center of the Earth and has a magnitude mg. Thus we see that the gravitational field experienced by the object at some point has a magnitude equal to the acceleration of gravity at that point. Since the gravitational force on the object has a magnitude GM_em/r^2 (where M_e is the mass of the Earth), the field $\mathbf{g}$ at a distance r from the center of the Earth is given by

Gravitational field of the Earth

$$\mathbf{g} = \frac{\mathbf{F}_g}{m} = -\frac{GM_e}{r^2}\hat{\mathbf{r}} \qquad [6.6]$$

where $\hat{\mathbf{r}}$ is a unit vector pointing radially outward from the Earth, and the minus sign indicates that the field points toward the center of the Earth, as shown in Figure 6.4a. Note that the field vectors at different points surrounding the spherical mass vary in both direction and magnitude. In a small region near the Earth's surface, g is approximately constant and the downward field is uniform, as indicated in Figure 6.4b. This expression is valid at all points *outside* the Earth's surface, assuming that the Earth is spherical and that rotation can be neglected. At the Earth's surface, where $r = R_e$, $\mathbf{g}$ has a magnitude of 9.80 m/s^2.

The field concept is used in many other areas of physics. In fact, it was introduced by Michael Faraday (1791–1867) in his study of electromagnetism. In spite

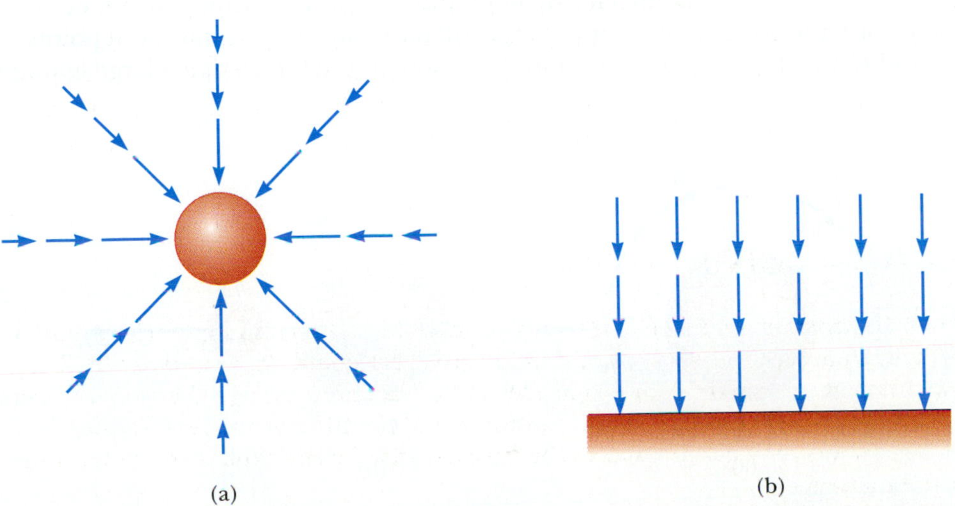

Figure 6.4
(a) The gravitational field vectors in the vicinity of a uniform spherical mass such as the Earth vary in both direction and magnitude. (b) The gravitational field vectors in a small region near the Earth's surface are uniform; that is, they match in both magnitude and direction.

of its abstract nature, the field concept is particularly useful for describing electric and magnetic interactions.

Gravitational, electric, and magnetic fields are *vector fields,* since a vector is associated with each point in space. Similarly, a *scalar field* is one in which a scalar quantity is used to describe each point in space. For example, the variation in temperature over a given region can be described by a scalar temperature field.

6.3 The Electric Field

The electric field at a point in space is defined in terms of the electric force acting on a test charge q_0 placed at that point. To be precise,

> **the electric field vector E at a point in space is defined as the electric force $\mathbf{F}_e$ acting on a positive test charge placed at that point, divided by the magnitude of the test charge, q_0:**

Definition of electric field

$$\mathbf{E} \equiv \frac{\mathbf{F}_e}{q_0} \qquad [6.7]$$

Thus, we can say that *an electric field exists at a point if a test charge at rest placed at that point experiences an electric force.* Note that **E** is the field *external* to the test charge —it is not the field produced by the test charge. The vector **E** has the SI units newtons per coulomb (N/C). The direction of **E** is in the direction of $\mathbf{F}_e$, since we have assumed that $\mathbf{F}_e$ acts on a positive test charge.

Once the electric field at some point is known, the electric force on *any* charged particle placed at that point can be calculated from Equation 6.7. Furthermore, the electric field is said to exist at some point (even empty space) regardless of whether a test charge is located at that point.

When Equation 6.7 is applied, we must assume that the test charge q_0 is small enough that it does not disturb the charge distribution which is responsible for the electric field. For instance, if a vanishingly small test charge q_0 is placed near a uniformly charged metallic sphere, as in Figure 6.5a, the charge on the metallic sphere, which produces the electric field, will remain uniformly distributed. Furthermore, the force $\mathbf{F}_e$ on the test charge will have the same magnitude at points *A*, *B*, and *C*, which are equidistant from the sphere. If the test charge is large enough

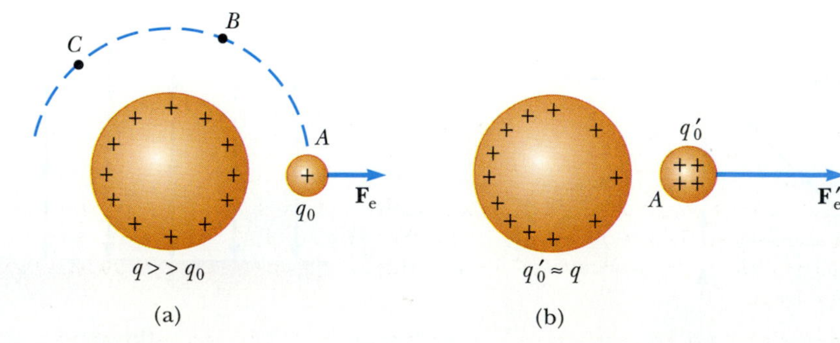

Figure 6.5
(a) When a small test charge, q_0, is placed near a conducting sphere of charge q (where $q \gg q_0$), the charge on the conducting sphere remains uniform. (b) If the test charge q_0' is on the order of the charge on the sphere, the charge on the sphere is nonuniform.

$(q_0' \gg q_0)$, as in Figure 6.5b, the charge on the metallic sphere will be redistributed and the ratio of the force to the test charge at point A will change: $(F_e'/q_0' \neq F_e/q_0)$. Furthermore, the distribution of charge on the sphere will change as the test charge is moved from point A to point B or C.

Consider a point charge q located a distance r from a positive test charge, q_0. According to Coulomb's law, the force on the test charge is

$$\mathbf{F} = k_e \frac{q q_0}{r^2} \hat{\mathbf{r}}$$

where $\hat{\mathbf{r}}$ is a unit vector that is directed away from q toward q_0 (Fig. 6.6). Since the electric field at the position of the test charge is defined by $\mathbf{E} = \mathbf{F}_e/q_0$, we find that the electric field *at the position of q_0 due to the charge q* is given by

$$\mathbf{E} = k_e \frac{q}{r^2} \hat{\mathbf{r}} \qquad [6.8]$$

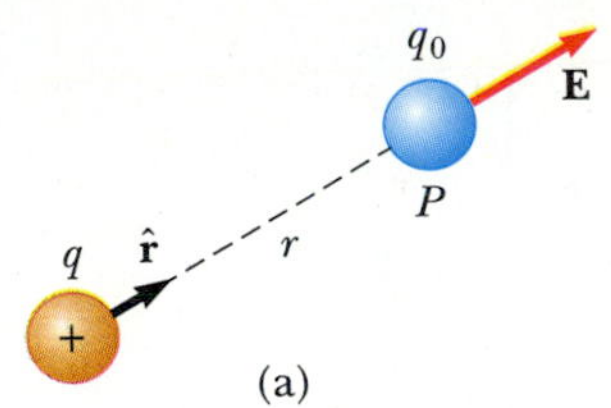

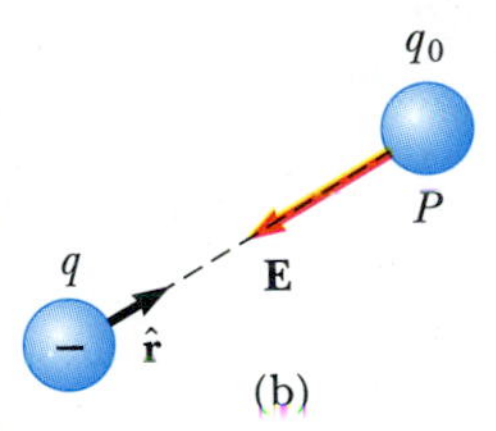

Figure 6.6
A test charge q_0 at the point P is at a distance r from the point charge q. (a) If q is positive, the electric field at P points radially *outward* from q. (b) If q is negative, the electric field at P points radially *inward* toward q.

If q is *positive,* as in Figure 6.6a, the electric field is directed radially *outward* from q. If q is *negative,* as in Figure 6.6b, the field is directed *toward* q.

In order to calculate the electric field due to a group of point charges, we first calculate the electric field vectors at the point P individually, using Equation 6.8, and then add them *vectorially*. In other words, the total electric field due to a group of charges equals the vector sum of the electric fields of all the charges. Thus, the electric field due to a group of charges (excluding the test charge q_0) can be expressed as

$$\mathbf{E} = k_e \sum_i \frac{q_i}{r_i^2} \hat{\mathbf{r}}_i \qquad [6.9]$$

where r_i is the distance from the ith charge, q_i, to the point P (the location of the test charge) and $\hat{\mathbf{r}}_i$ is a unit vector directed from q_i toward P.

Example 6.3 Electric Force on a Proton

Find the electric force on a proton placed in an electric field of 2×10^4 N/C, directed along the positive x axis. Compare the electric force on the proton to its weight.

Solution Since the charge on a proton is

$$+e = +1.60 \times 10^{-19}\ \text{C},$$

the electric force on it is

$$\mathbf{F}_e = e\mathbf{E} = (1.60 \times 10^{-19}\ \text{C})(2 \times 10^4 \mathbf{i}\ \text{N/C})$$
$$= 3.20 \times 10^{-15} \mathbf{i}\ \text{N}$$

where $\mathbf{i}$ is a unit vector in the positive x direction. The weight of the proton is calculated to be $mg = (1.67 \times 10^{-27}\ \text{kg})(9.8\ \text{m/s}^2) = 1.6 \times 10^{-26}$ N. Hence, we see that the magnitude of the gravitational force in this case is negligible compared with the electric force.

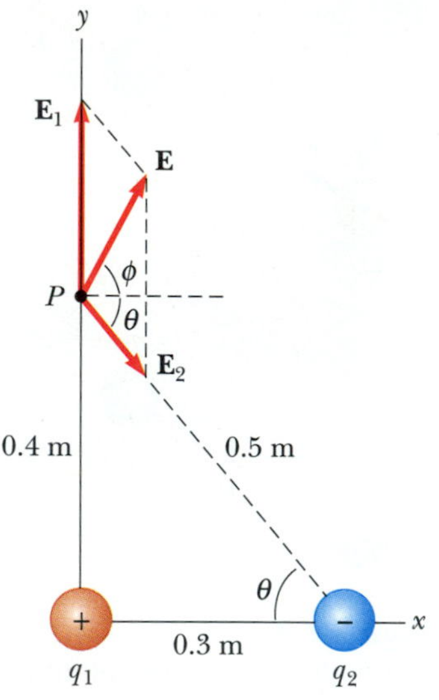

Figure 6.7
(Example 6.4) The total electric field E at P equals the vector sum $\mathbf{E}_1 + \mathbf{E}_2$, where $\mathbf{E}_1$ is the field due to the positive charge q_1 and $\mathbf{E}_2$ is the field due to the negative charge q_2.

Example 6.4 Electric Field Due to Two Charges

The charge $q_1 = 7\ \mu\text{C}$ is located at the origin, and a second charge, $q_2 = -5\ \mu\text{C}$ is located on the x axis 0.3 m from the origin (Figure 6.7). Find the electric field at the point P with coordinates (0, 0.4) m.

Solution First, let us find the magnitudes of the electric fields due to each charge. The fields $\mathbf{E}_1$ due to the 7-μC charge and $\mathbf{E}_2$ due to the -5-μC charge at P are shown in Figure 6.7. Their magnitudes are given by

$$E_1 = k_e \frac{|q_1|}{{r_1}^2} = \left(8.99 \times 10^9\ \frac{\text{N}\cdot\text{m}^2}{\text{C}^2}\right) \frac{(7 \times 10^{-6}\ \text{C})}{(0.4\ \text{m})^2}$$
$$= 3.93 \times 10^5\ \text{N/C}$$
$$E_2 = k_e \frac{|q_2|}{{r_2}^2} = \left(8.99 \times 10^9\ \frac{\text{N}\cdot\text{m}^2}{\text{C}^2}\right) \frac{(5 \times 10^{-6}\ \text{C})}{(0.5\ \text{m})^2}$$
$$= 1.80 \times 10^5\ \text{N/C}$$

The vector $\mathbf{E}_1$ has only a y component. The vector $\mathbf{E}_2$ has an x component, given by $E_2 \cos\theta = \frac{3}{5}E_2$, and a negative y component, given by $-E_2 \sin\theta = -\frac{4}{5}E_2$. Hence, we can express the vectors as

$$\mathbf{E}_1 = 3.93 \times 10^5\mathbf{j}\ \text{N/C}$$
$$\mathbf{E}_2 = (1.08 \times 10^5\mathbf{i} - 1.44 \times 10^5\mathbf{j})\ \text{N/C}$$

The resultant field $\mathbf{E}$ at P is the vector sum of $\mathbf{E}_1$ and $\mathbf{E}_2$:

$$\mathbf{E} = \mathbf{E}_1 + \mathbf{E}_2 = (1.08 \times 10^5\mathbf{i} + 2.49 \times 10^5\mathbf{j})\ \text{N/C}$$

From this result, we find that $\mathbf{E}$ has a magnitude of 2.71×10^5 N/C and makes an angle ϕ of 66° with the positive x axis.

Exercise Find the electric force on a test charge of 2×10^{-8} C placed at P.

Answer 5.42×10^{-3} N in the same direction as $\mathbf{E}$.

6.4 Motion of Charged Particles in a Uniform Electric Field

The motion of a charged particle in a uniform electric field is equivalent to that of a projectile in a uniform gravitational field. When a particle of charge q is placed in an electric field $\mathbf{E}$, the electric force on the charge is $q\mathbf{E}$. If this is the only force exerted on the charge, then Newton's second law applied to the charge gives

$$\mathbf{F} = q\mathbf{E} = m\mathbf{a}$$

where m is the mass of the charge and we assume that the speed of the charge is small compared with the speed of light. The acceleration of the particle is therefore

$$\mathbf{a} = \frac{q\mathbf{E}}{m} \qquad [6.10]$$

If $\mathbf{E}$ is uniform (that is, constant in magnitude and direction), the acceleration is a constant of the motion. If the charge is positive, the acceleration is in the direction

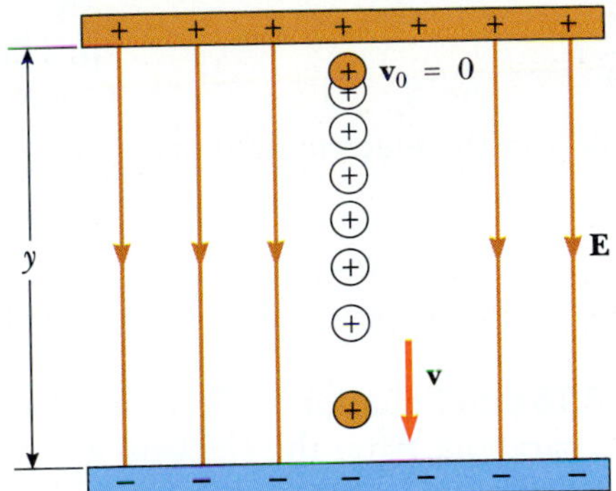

Figure 6.8
A positive point charge q in a uniform electric field **E** undergoes constant acceleration in the direction of the field. This is analogous to a mass falling in a uniform gravitational field.

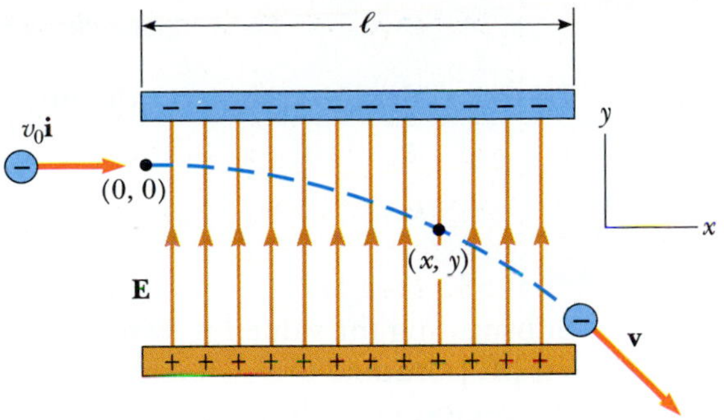

Figure 6.9
An electron is projected horizontally into a uniform electric field produced by two charged plates. The electron undergoes a downward acceleration (opposite **E**), and its motion its parabolic.

of the electric field. If the charge is negative, the acceleration is in the direction *opposite* the electric field.

Consider the special case of a positive point charge released from rest in a uniform electric field, directed along the negative y axis. Such a field can be produced by two parallel plates carrying equal and opposite charge, as in Figure 6.8. Let us describe the motion of the charge. In this case, the acceleration of the charge is also along the negative y axis, is constant, and has a magnitude given by $a = qE/m$. Since the acceleration is constant, we can apply the equations of kinematics in one dimension to this situation (Chapter 2). Taking the initial position of the charge to be at the origin and its initial velocity to be zero, we find that the coordinate and velocity of the charge after a time t has elapsed are

$$y = \tfrac{1}{2}at^2 = \frac{qE}{2m}t^2 \qquad [6.11]$$

$$v = at = \frac{qE}{m}t \qquad [6.12]$$

Note that the motion of a charged particle in a uniform electric field is analogous to that of a mass falling from rest in the presence of a uniform gravitational field.

Now consider the case of an electron of charge $-e$ projected horizontally, with an initial velocity $v_0\mathbf{i}$, into a uniform electric field directed upward, as in Figure 6.9. The charged plates that produce such a field could be the deflection plates of a cathode ray tube. Since the electric field **E** is in the positive y direction and the electron carries a negative charge, its acceleration is in the negative y direction. That is,

$$\mathbf{a} = -\frac{eE}{m}\mathbf{j} \qquad [6.13]$$

Because the acceleration is constant, we can apply the equations of kinematics in two dimensions (from Chapter 3) with $v_{x0} = v_0$ and $v_{y0} = 0$. The components of velocity of the electron after it has been in the electric field a time t are

$$v_x = v_0 = \text{constant} \qquad [6.14]$$

$$v_y = at = -\frac{eE}{m}t \quad [6.15]$$

Likewise, the electron's coordinates after a time t in the electric field are

$$x = v_0 t \quad [6.16]$$

$$y = \tfrac{1}{2}at^2 = -\tfrac{1}{2}\frac{eE}{m}t^2 \quad [6.17]$$

Substituting the value $t = x/v_0$ from Equation 6.16 into Equation 6.17, we see that y is proportional to x^2. Hence, the trajectory is a parabola. After the electron leaves the region of uniform electric field, it continues to move in a straight line with a speed $v > v_0$.

Note that we have neglected the gravitational force on the electron. This is a good approximation when dealing with atomic particles. For an electric field of magnitude 10^4 N/C, the ratio of the electric force, eE, to the gravitational force, mg, for the electron is $\approx 10^{14}$. The corresponding ratio for a proton is $\approx 10^{11}$.

Example 6.5 An Accelerated Electron

An electron enters a uniform electric field as in Figure 6.9, with $v_0 = 3 \times 10^6$ m/s and $E = 200$ N/C. The width of the plates is $\ell = 0.1$ m.

(a) Find the acceleration of the electron while in the electric field.

Solution Since the charge on the electron has a magnitude of 1.60×10^{-9} C and $m = 9.11 \times 10^{-31}$ kg, Equation 6.13 gives

$$\mathbf{a} = -\frac{eE}{m}\mathbf{j} = -\frac{(1.6 \times 10^{-19}\text{ C})(200\text{ N/C})}{9.11 \times 10^{-31}\text{ kg}}\mathbf{j}$$

$$= -3.51 \times 10^{13}\mathbf{j}\text{ m/s}^2$$

(b) Find the time it takes the electron to travel through the region of the electric field.

Solution The horizontal distance traveled by the electron while in the electric field is $\ell = 0.1$ m. Using $x = v_0 t$ with $x = \ell$, we find that the time spent in the electric field is

$$t = \frac{\ell}{v_0} = \frac{0.1\text{ m}}{3 \times 10^6\text{ m/s}} = 3.33 \times 10^{-8}\text{ s}$$

(c) What is the vertical displacement y of the electron while it is in the electric field?

Solution Using $y = \frac{1}{2}at^2$ and the results from (a) and (b), we find that

$$y = \tfrac{1}{2}at^2 = -\tfrac{1}{2}(3.51 \times 10^{13}\text{ m/s}^2)(3.33 \times 10^{-8}\text{ s})^2$$

$$= -0.0195\text{ m} = -1.95\text{ cm}$$

If the separation between the plates is smaller than this, the electron will strike the positive plate.

Exercise Find the speed of the electron as it emerges from the electric field.

Answer 3.22×10^6 m/s.

6.5 The Magnetic Field

Anyone who has used a magnet or a compass is familiar with magnetic interactions. In order to determine whether a magnetic field exists at some point P in space, we need a suitable probe. As with electric fields, let us select as our probe a charged particle. As we learned in Section 6.3, if a force (other than gravity) is exerted on the test charge while it is at rest at P, we conclude that an electrostatic field exists at that point. Now let us assume that the particle is moving with a velocity **v** as it passes P. *If the moving test charge experiences a force at P that it did not experience while at rest at that same point, we conclude that a magnetic field exists at P.* It is important to recognize that the magnetic field interacts with the test charge only if there is relative motion between the charge and the field. Later in the text (Chapter 19), we shall find that moving charges are the source of magnetic fields.

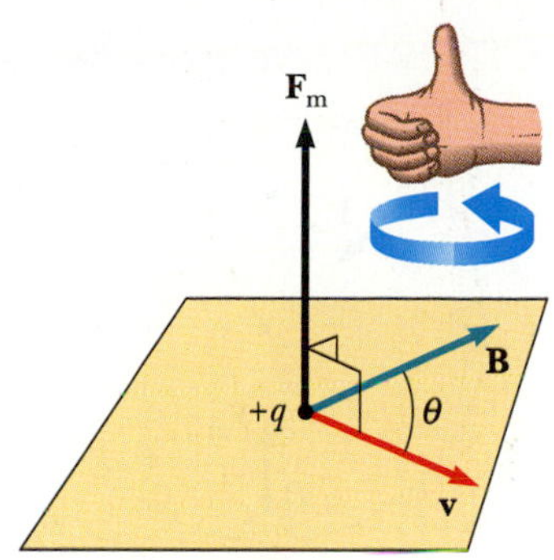

Figure 6.10
The magnetic force acting on a moving charged particle is directed perpendicularly to the plane formed by the vectors **v** and **B**. Thus, $\mathbf{F}_m$ is perpendicular to both **v** and **B**.

Let us denote the magnetic field with **B** and the magnetic force on the moving test charge with $\mathbf{F}_m$. The direction of the magnetic force is perpendicular to both **v** and the direction of the magnetic field, as in Figure 6.10. If the moving charge were negative, the magnetic force would be opposite that shown in Figure 6.10. The magnetic force can be expressed in the form

The magnetic force

$$\mathbf{F}_m = q\mathbf{v} \times \mathbf{B} \quad [6.18]$$

where $\mathbf{v} \times \mathbf{B}$ is called a vector product (or cross product). A convenient rule to use to determine the direction of $\mathbf{v} \times \mathbf{B}$ is the right-hand rule illustrated in Figure 6.10. The four fingers of the right hand are pointed along **v** and then "wrapped" into **B** through the angle θ. The direction of the erect right thumb is the direction of the vector $\mathbf{v} \times \mathbf{B}$. This rule holds for the cross product of any two vectors. We shall discuss the nature of the vector product in more detail in Chapter 11. The magnitude of the magnetic force is

$$F_m = qvB \sin \theta \quad [6.19]$$

where θ is the angle between **v** and **B**. From this expression, we see that the magnetic force is zero, when **v** is parallel to **B** ($\theta = 0$ or $180°$). Furthermore, the magnetic force has its maximum value when **v** is perpendicular to **B** ($\theta = 90°$). That is,

$$(F_m)_{\max} = qvB \quad [6.20]$$

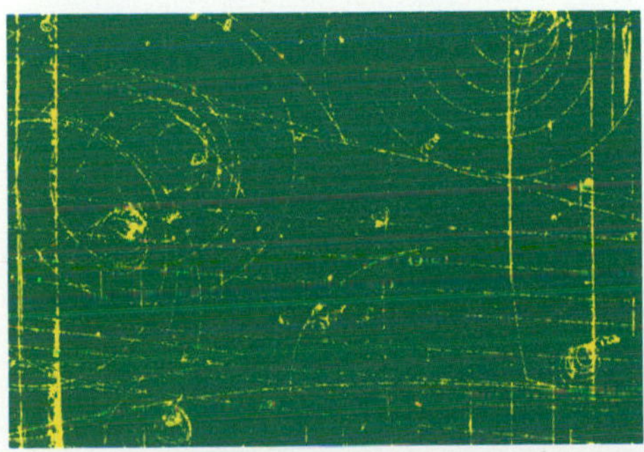

A color-enhanced photograph left by particles moving in a magnetic field. The curved tracks are due to charged particles that are deflected from their initial direction of motion by the magnetic force. *(Photo Researchers, Inc.)*

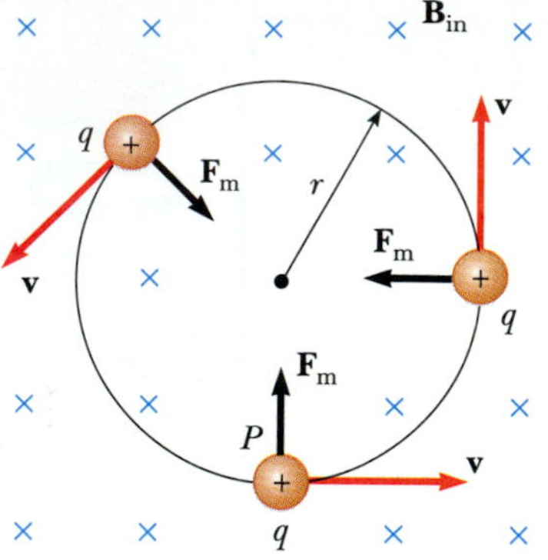

Figure 6.11
When the velocity of a charged particle is perpendicular to a uniform magnetic field, the particle moves in a circular path whose plane is perpendicular to **B**, which is directed into the page (the blue crosses represent the tail of the vector). The magnetic force, $\mathbf{F}_m$, on the charge is always directed toward the center of the circle, and its magnitude is qvB.

Equation 6.18 can be regarded as an operational definition of the magnetic field at a point in space. Note that the magnetic force is a sideways force perpendicular to the magnetic field, whereas the electric force is always in the direction of the electric field. Furthermore, the magnetic force acts on a charged particle only when the particle is in motion, whereas the electric force acts on a charged particle independent of its velocity.

The SI unit of the magnetic field is called the **tesla** (T), where 1 T = N·s/C·m.

6.6 Motion of Charged Particles in Uniform and Nonuniform Magnetic Fields

Consider a positively charged particle moving in a uniform magnetic field, with the initial velocity vector of the particle perpendicular to the field. Let us assume that the magnetic field is directed *into* the page, as indicated by the crosses in Figure 6.11, which represent the *tail* of the vector **B**. Later we shall use dots to represent the *tip* of a vector directed *out of* the page. You can see that *the charged particle moves in a circle whose plane is perpendicular to the magnetic field.* That is because the magnetic force is at right angles to **v** and **B** and has a constant magnitude equal to qvB. As the force $\mathbf{F}_m$ deflects the particle, the directions of **v** and $\mathbf{F}_m$ change continuously, as indicated in Figure 6.11. Therefore, the magnetic force changes only the direction of **v**, leaving the speed constant. The sense of rotation is counterclockwise for a positively charged particle, as shown in the figure. If the charge were negative, the sense of the rotation would be clockwise. Since the magnetic force is in the radial direction and its magnitude is qvB, Newton's second law applied to this circular motion gives

$$qvB = \frac{mv^2}{r}$$

$$r = \frac{mv}{qB} \qquad [6.21]$$

The period of the motion (the time required for one revolution) is equal to the circumference of the circle divided by the speed of the particle:

$$T = \frac{2\pi r}{v} = \frac{2\pi m}{qB} \qquad [6.22]$$

This result shows that the period does not depend on the speed of the particle or on the radius of its orbit.

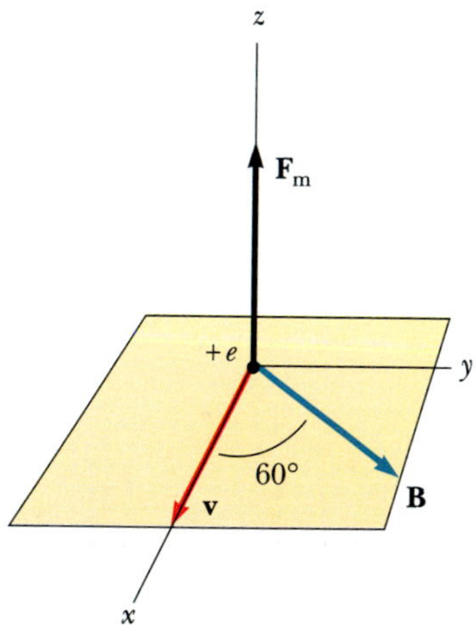

Figure 6.12
(Example 6.6) The magnetic force $\mathbf{F}_m$ on a proton is in the positive z direction when **v** and **B** lie in the xy plane.

Example 6.6 A Proton Moving in a Magnetic Field

A proton moves with a speed of 8×10^6 m/s along the x axis. It enters a region where there is a magnetic field of magnitude 2.5 T, directed at an angle of 60° to the x axis and lying in the xy plane (Fig. 6.12). Calculate the initial magnetic force and the acceleration of the proton.

Solution From Equation 6.19, we get

$$F_m = qvB \sin\theta$$

$$= (1.6 \times 10^{-19}\ \text{C})(8 \times 10^6\ \text{m/s})(2.5\ \text{T})(\sin 60°) = 2.77 \times 10^{-12}\ \text{N}$$

Since $\mathbf{v} \times \mathbf{B}$ is in the positive z direction and since the charge is positive, the force $\mathbf{F}$ is in the positive z direction.

Since the mass of the proton is 1.67×10^{-27} kg, its initial acceleration is

$$a = \frac{F_m}{m} = \frac{2.77 \times 10^{-12}\ \text{N}}{1.67 \times 10^{-27}\ \text{kg}} = 1.66 \times 10^{15}\ \text{m/s}^2$$

in the positive z direction.

Exercise Verify that the units of $\mathbf{F}_m$ in the preceding calculation for the magnetic force reduce to newtons.

When a charged particle moves in a nonuniform magnetic field, the motion is complex. For example, in a magnetic field that is strong at the ends and weak in the middle, as in Figure 6.13, the particle can oscillate back and forth between the end points. Such a field can be produced by two loops of wire carrying an electric current I, as in Figure 6.13 (current is the rate at which charge flows through a wire). In this case, a particle starting at one end spirals along the field lines until it reaches the other end, where it reverses its path and spirals back. Such a configuration is known as a *magnetic bottle* because charged particles can be trapped in it. Magnetic bottles have been used to confine very hot gases ($T > 10^6$ K) consisting of electrons and positive ions, known as **plasmas.** This type of plasma-confinement scheme could play a crucial role in achieving a controlled nuclear fusion process, which could supply humans with an almost endless source of energy. Unfortunately, the magnetic bottle has its problems. If a large number of particles is trapped, the confining force is small near the axis, and the particles "leak" from the system along the axis.

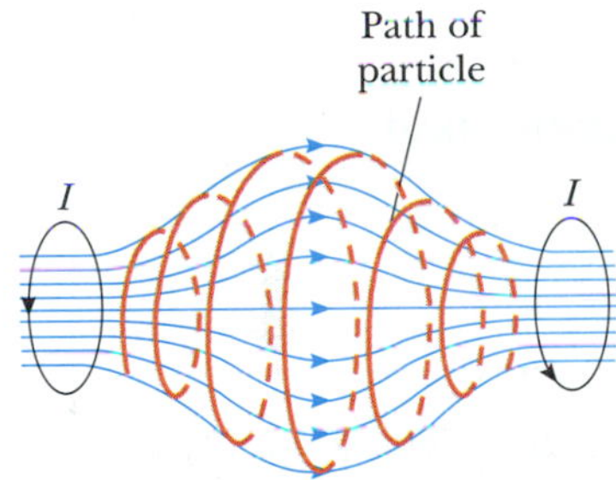

Figure 6.13
A charged particle moving in a nonuniform magnetic field represented by the blue lines (a magnetic bottle) spirals about the field (red path) and oscillates between the end points.

Summary

The fundamental forces in nature are the gravitational force, the electromagnetic force, the weak force, and the strong force. The gravitational force between any two particles of masses m_1 and m_2 is attractive and has a magnitude given by Newton's universal law of gravity:

$$F_g = G\frac{m_1 m_2}{r^2} \qquad [6.1]$$

The universal law of gravity

where G is the *universal gravitational constant;* $G = 6.67 \times 10^{-11}\ \text{N}\cdot\text{m}^2/\text{kg}^2$.

The electrostatic force between any two charges q_1 and q_2, at rest relative to each other and separated by a distance r, has a magnitude of

$$F_e = k_e\frac{q_1 q_2}{r^2} \qquad [6.4]$$

The electrostatic force between two charges

where q_1 and q_2 are measured in coulombs (C), and k_e is a constant called the *Coulomb constant;* $k_e = 8.99 \times 10^9 \text{ N}\cdot\text{m}^2/\text{C}^2$. When the charges have the same sign, the electrostatic force is repulsive. When the charges have opposite signs, the electrostatic force is attractive.

The gravitational field **g** at any point in space is defined as the gravitational force experienced by a test mass at that point, divided by the mass:

The gravitational field

$$\mathbf{g} \equiv \frac{\mathbf{F}_g}{m} \qquad [6.5]$$

The gravitational field an object experiences above the Earth's surface has a magnitude equal to GM_e/r^2, where r is the object's distance from the center of the Earth, M_e is the mass of the Earth, and the direction of **g** is toward the center of the Earth.

The electric field vector **E** at any point in space is defined as the electric force experienced by a positive test charge q_0 at that point, divided by the magnitude of the test charge:

Electric field

$$\mathbf{E} \equiv \frac{\mathbf{F}_e}{q_0} \qquad [6.7]$$

The vector **E** has the SI units newtons per coulomb (N/C).

The electric field due to some point charge q at a point a distance r from the point charge is

Electric field due to a point charge

$$\mathbf{E} = k_e \frac{q}{r^2}\hat{\mathbf{r}} \qquad [6.8]$$

where the field is directed away from the charge if q is positive and toward the charge if q is negative.

A charged particle moving in an electric field **E** experiences an electric force equal to $q\mathbf{E}$, and its acceleration is, from Newton's second law, equal to $q\mathbf{E}/m$. If the field is constant in magnitude and direction, the acceleration of the charge is a constant of the motion, and one can apply the equations of kinematics to describe the charge's motion.

The **magnetic force** on a charge q, moving with a velocity **v** in a magnetic field **B**, is

Magnetic force on a charged particle moving in a magnetic field

$$\mathbf{F}_m = q\mathbf{v} \times \mathbf{B} \qquad [6.18]$$

The magnetic force is in a direction perpendicular both to the velocity of the particle and to the field. The *magnitude* of the magnetic force is

$$F_m = qvB \sin\theta \qquad [6.19]$$

where θ is the angle between **v** and **B.**

The SI unit of **B** is the **tesla** (T).

If a charged particle moves in a uniform magnetic field so that the initial velocity of the particle is *perpendicular* to the field, the movement of the particle will be in a circle whose plane is *perpendicular* to the magnetic field. The radius r of the circular path is

$$r = \frac{mv}{qB} \qquad [6.21]$$

where m is the mass of the particle and q is its charge.

Questions and Conceptual Exercises

1. Estimate the gravitational force between you and a person 2 m away from you.
2. If the gravitational force on objects is directly proportional to their masses, why don't large masses fall with greater acceleration than small ones?
3. If someone told you that astronauts are weightless in orbit because they are beyond the pull of gravity, would you accept this statement? Explain.
4. In reality, the Earth rotates about its axis with a period of 24 h. Does this rotation affect the value of g? Explain.
5. What are the similarities and differences between Newton's universal law of gravitation and Coulomb's law?
6. A daring young scientist proposes a theory that people and objects are bound to the Earth by electric forces rather than by gravity. How could you prove this theory wrong?
7. Is it possible for an electric field to exist in empty space? Explain.
8. A "free" electron and "free" proton are placed in the same electric field. Compare the electric forces on them. Compare their accelerations.
9. A negative charge is placed in a region where the electric field is directed vertically upward. What is the direction of the electric force that is acting on this charge?
10. How would you experimentally distinguish between a gravitational field and an electric field?
11. An inflated toy balloon becomes slightly larger as it acquires an electric charge. Why?
12. At a given instant, a proton moves in the positive x direction in a region where there is a magnetic field in the negative z direction. What is the direction of the magnetic force? Will the proton continue to move in the positive x direction? Explain.
13. Suppose an electron is chasing a proton up this page when suddenly a magnetic field is turned on, perpendicular to the page. Describe the subsequent motions of the particles.
14. Why does the picture on a TV screen become distorted when a magnet is brought near the screen?
15. How can the motion of a moving charged particle be used to distinguish between a magnetic field and an electric field? Give a specific example to support your argument.
16. A proton, moving horizontally, enters a region where there is a uniform magnetic field perpendicular to the proton's velocity, as shown in Figure 6.14. Describe the proton's subsequent motion. How would an electron behave under the same circumstances?

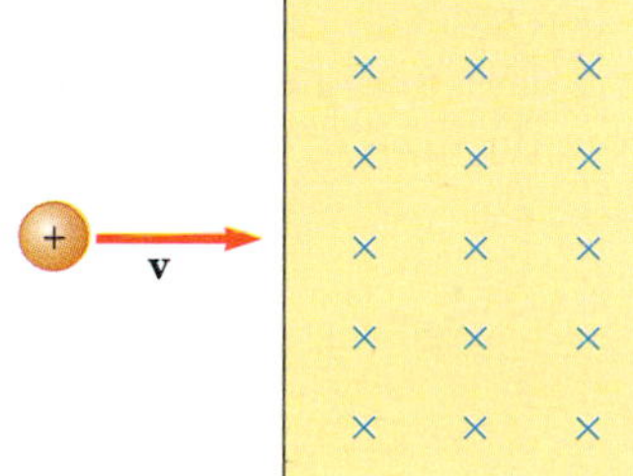

Figure 6.14 (Question 16)

17. Richard Feynmann once said that if two persons stood at arm's length from each other and each person had 1% more electrons than protons, the repulsive force between the two people would be enough to lift a "weight" equal to that of the entire Earth. Carry out an order-of-magnitude calculation to substantiate this assertion.

Problems

Section 6.1 Forces in Nature

1. Two identical, isolated particles, each of mass 2 kg, are separated by a distance of 30 cm. What is the magnitude of the gravitational force of one particle on the other?
2. Two stars of masses M and $4M$ are separated by a distance d. Determine the location of a point measured from M at which the net force on a third mass would be zero.
3. A satellite of mass 300 kg is in a circular orbit about the Earth at an altitude equal to the Earth's mean radius (see Example 6.1). Find (a) the satellite's orbital speed, (b) the period of its revolution, and (c) the gravitational force acting on it.
4. While two astronauts were on the surface of the

Moon, a third astronaut orbited the Moon. Assume the orbit to be circular and 100 km above the surface of the Moon. If the mass and radius of the Moon are 7.4×10^{22} kg and 1.7×10^{6} m, respectively, determine (a) the orbiting astronaut's acceleration, (b) the astronaut's orbital speed, and (c) the period (time for one revolution) of the orbit.

5. An astronaut weighs 140 N on the Moon's surface. When he is in a circular orbit about the Moon at an altitude above the Moon's surface equal to the Moon's radius, what gravitational force does the Moon exert on him?
6. Two small objects attract each other with a gravitational force of 1.0×10^{-8} N when they are separated by 20 cm. If the total mass of the two objects is 5.0 kg, what is the mass of each?
7. Plaskett's binary system consists of two stars that revolve in a circular orbit about a center of gravity midway between them. This means that the masses of the two stars are equal (Fig. 6.15). If the orbital velocity of each star is 220 km/s and the orbital period of each is 14.4 days, find the mass M of each star. (For comparison, the mass of our Sun is 1.99×10^{30} kg.)

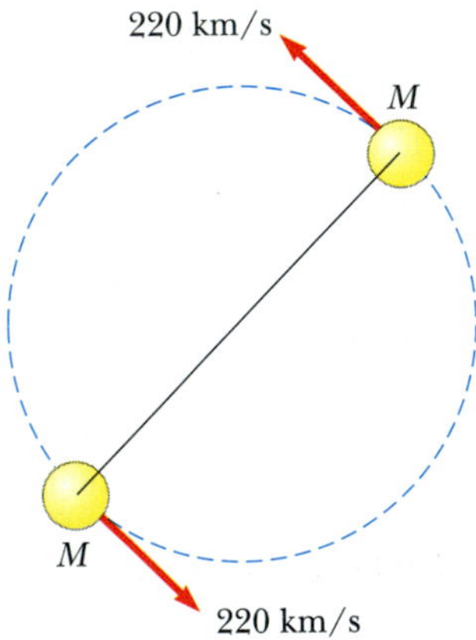

Figure 6.15 (Problem 7)

8. The force of gravity on the surface of the Moon is about one-sixth that on the surface of the Earth. If the radius of the Moon is about one-quarter that of the Earth, find the ratio of the average mass density of the Moon to the average mass density of the Earth.
9. Suppose that 1 g of hydrogen is separated into 6×10^{23} electrons and 6×10^{23} protons. Suppose also that the protons are placed at the Earth's north pole and the electrons are placed at the south pole. What is the resulting compressional force on the Earth?
10. Two protons in a molecule (each having a charge of $+1.6 \times 10^{-19}$ C) are separated by a distance of 3.8×10^{-10} m. Find the electrostatic force exerted by one proton on the other.
11. A 6.7-μC charge is located 5.0 m from a -8.4-μC charge. Find the electrostatic force exerted by one charge on the other.
12. Four point charges are situated at the corners of a square with sides of length a, as in Figure 6.16. Find the resultant force on the positive charge q.

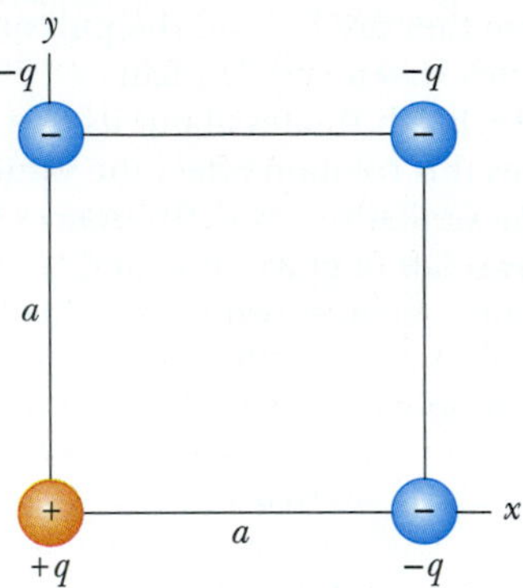

Figure 6.16 (Problem 12)

Section 6.2 The Gravitational Field

13. When a falling meteor is at a distance above the Earth's surface of 3 times the Earth's radius, what is its acceleration due to the Earth's gravity?
14. If the mass of Mars is 0.108 times that of Earth and its radius is 0.6 that of Earth, estimate the magnitude of the gravitational acceleration at the surface of Mars.
15. A space station, in the form of a large wheel 120 m in diameter, rotates to provide an "artificial gravity" of 3 m/s² for persons situated at the outer rim. Find the rotational frequency of the wheel (in revolutions per minute) that will produce this effect.
16. Isaac Newton was able to estimate a value for G, the universal gravitational constant, from the following data: the radius of the Earth is about 6400 km, the average density of rocks is about 5.5 g/cm³, and $g =$ 9.8 m/s² near the surface of the Earth. What value did Newton obtain for G?
17. At the moment of a total eclipse, the Moon lies along a line from the Earth to the Sun. If your normal weight is 600 N, how much is your weight decreased by the combined pull of the Sun and Moon?

$$M_{\text{Sun}} = 2.0 \times 10^{30} \text{ kg}, \; r_{\text{S-E}} = 1.5 \times 10^{8} \text{ km}$$
$$M_{\text{Moon}} = 7.4 \times 10^{22} \text{ kg}, \; r_{\text{M-E}} = 3.8 \times 10^{5} \text{ km}$$

18. Calculate the fractional difference $\Delta g/g$ in the acceleration due to gravity at points on the Earth's surface

nearest to and farthest from the Moon, taking into account the gravitational effect of the Moon. (This difference is responsible for the occurrence of the *lunar tides* on the Earth.)

Section 6.3 The Electric Field

19. In a thundercloud there may be electric charges of +40 C near the top of the cloud and −40 C near the bottom of the cloud. These charges are separated by about 2 km. What is the electric force between them?
20. An airplane is flying through a thundercloud at a height of 2000 m. (This is a very dangerous thing to do because of updrafts, turbulence, and the possibility of electric discharge.) If there are charge concentrations of +40 C at height 3000 m within the cloud and −40 C at height 1000 m, what is the magnitude of the electric field E at the aircraft?
21. An object having a net charge of 24 μC is placed in a uniform electric field of 610 N/C, directed vertically upward. What is the mass of this object if it "floats" in this electric field?
22. An alpha particle (with a charge of $+2e$) is sent at high speed toward a gold nucleus (charge $+79e$). What is the electric force acting on the alpha particle when it is at a distance of 2×10^{-14} m from the gold nucleus?
23. The nucleus of a hydrogen atom, a proton, sets up an electric field. The average distance between the proton and the electron of a hydrogen atom is about 5.1×10^{-11} m. What is the magnitude of the electric field at this distance from the proton?
24. Imagine that the Moon is held in its orbit about the Earth by electric forces rather than by gravitation (a *hypothetical* situation). What is the necessary electric charge, $-Q$, that would be on the Earth and the charge $+Q$ that would be on the Moon to hold the Moon in a circular orbit whose period is 27.3 days? The Earth-Moon distance is 384 000 km, and the mass of the Moon is 7.35×10^{22} kg.

Section 6.4 Motion of Charged Particles in a Uniform Electric Field

25. The electron gun in a television tube is to accelerate electrons from rest to 3.0×10^{7} m/s within a distance of 2.0 cm. What electric field is required?
26. An electron and a proton are placed at rest in an external electric field of 520 N/C. Calculate the speed of each particle after 48 nanoseconds.
27. A proton accelerates from rest in a uniform electric field of 640 N/C. At some later time, its speed is 1.20×10^{6} m/s. (a) Find the acceleration of the proton. (b) How long does it take the proton to reach this speed? (c) How far has it moved in this time?
28. An electron with a speed of 3×10^{6} m/s moves into a uniform electric field of magnitude 1000 N/C. The field is parallel to the electron's velocity and acts to decelerate the electron. How far does the electron travel before it is brought to rest?
29. A proton moving at 3×10^{4} m/s is projected at an angle of 30° above a horizontal plane. If an electric field of 400 N/C is acting downward, how long does it take the proton to return to the horizontal plane? (*Hint:* Ignore gravity.)
30. A proton and an electron both start from rest and from the same point in a uniform electric field of magnitude 370 N/C. How far apart are they after 1 μs? (Ignore the attraction between the electron and the proton. If you like, you might imagine the experiment to be tried with the proton only, and then repeated with the electron only.)
31. A proton has an initial velocity of 4.50×10^{5} m/s in the horizontal direction. It enters a uniform electric field of 9.60×10^{3} N/C, directed vertically. Ignoring any gravitational effects, find (a) the time it takes the proton to travel 5.0 cm horizontally, (b) the vertical displacement of the proton after it has traveled 5.0 cm horizontally, and (c) the horizontal and vertical components of the proton's velocity after it has traveled 5.0 cm horizontally.
32. An electron, traveling with an initial velocity equal to $8.6 \times 10^{5}\mathbf{i}$ m/s, enters a region of a uniform electric field given by $\mathbf{E} = 4.1 \times 10^{3}\mathbf{i}$ N/C. (a) Find the acceleration of the electron. (b) Determine the time it takes for the electron to come to rest after it enters the field. (c) How far does the electron move in the electric field before coming to rest?

Section 6.5 The Magnetic Field

33. A proton is moving at right angles to a magnetic field of 2 T. What speed does the proton have if the magnetic force on it has a magnitude of 6×10^{-11} N?
34. A proton is moving at right angles to a magnetic field of 0.1 T with a speed of 2×10^{7} m/s. Find the magnitude of the acceleration of the proton.
35. An electron moving along the positive x axis perpendicular to a magnetic field experiences a magnetic deflection in the negative y direction. What is the direction of the magnetic field over this region?
36. What force of magnetic origin is experienced by a proton moving north to south, with a speed equal to 4.8×10^{6} m/s, at a location where the vertical component of the Earth's magnetic field is 75 μT directed downward? In what direction is the proton deflected?

37. A proton moving with a speed of 4×10^6 m/s through a magnetic field of 1.7 T experiences a magnetic force of magnitude 8.2×10^{-13} N. What is the angle between the proton's velocity and the field?
38. A proton moves perpendicularly to a uniform magnetic field **B**, with a speed of 10^7 m/s, and experiences an acceleration of 2×10^{13} m/s^2 in the $+x$ direction when its velocity is in the $+z$ direction. Determine the magnitude and direction of the field.
39. Indicate the initial direction of the deflection of charged particles as they enter the magnetic fields, as shown in Figure 6.17.

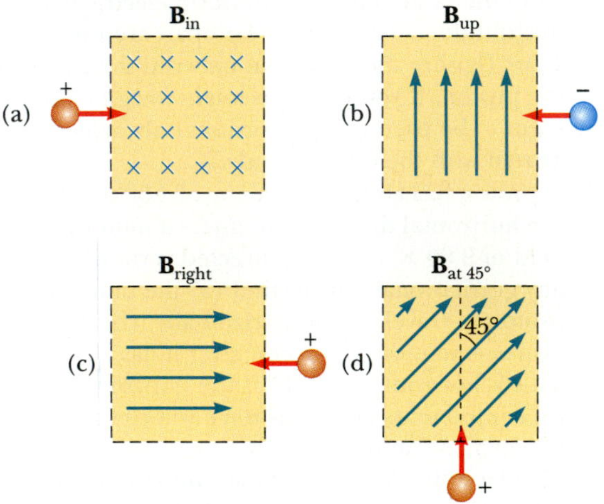

Figure 6.17 (Problem 39)

40. Sodium ions (Na^+), with a speed of 0.851 m/s, move through the bloodstream in the arm of a person standing near a large magnet. The magnetic field has a strength of 0.254 T and makes an angle of 51.0° with respect to the motion of the sodium ions. The arm contains 100 cm^3 of blood with 3.0×10^{20} Na^+ ions per cubic centimeter. If no other ions were present in the arm, what would be the magnetic force on the arm?

Section 6.6 Motion of Charged Particles in Uniform and Nonuniform Magnetic Fields

41. The magnetic field of the Earth at a certain location is directed vertically downward and has a magnitude of 0.5×10^{-4} T. A proton is moving horizontally westward in this field with a speed of 6.2×10^6 m/s. (a) What are the direction and magnitude of the magnetic force the field exerts on this charge? (b) What is the radius of the circular arc followed by this proton?
42. A singly charged positive ion has a mass of 3.2×10^{-26} kg. After being accelerated to a speed of 1.71×10^7 m/s, the ion enters a magnetic field of 0.92 T along a direction perpendicular to the direction of the field. Calculate the radius of the path of the ion in the field.
43. An electron moving at 2.65×10^7 m/s, perpendicularly to the Earth's magnetic field of 50 μT, has a circular trajectory. Determine (a) the radius of the trajectory and (b) the time required for the electron to complete one circle.
44. What magnetic field would be required to constrain an electron whose speed is 1.6×10^7 m/s to a circular path of radius 0.5 m?
45. A beam of protons (all with velocity v) emerges from a particle accelerator and is deflected in a circular arc with a radius of 0.45 m by a transverse uniform magnetic field of magnitude 0.80 T. (a) Determine the speed v of the protons in the beam. (b) What time is required for the deflection of a particular proton through an angle of 90°?
46. Electrons in a television tube are accelerated to a speed of 7.26×10^7 m/s, and the horizontal distance from the gun to a viewing screen is 35 cm. What is the deflection caused by the vertical component of the Earth's magnetic field (4×10^{-5} T), assuming that any change in the horizontal component of the beam velocity is negligible?
47. A cosmic-ray proton, traveling at half the speed of light (1.5×10^8 m/s), is headed directly toward the center of the Earth in the plane of the Earth's equator. Will it hit the Earth? As an estimate, assume that the Earth's magnetic field is 5×10^{-5} T and extends out one Earth diameter, or 1.3×10^7 m. Calculate the radius of curvature of the proton in this magnetic field.
48. A singly charged heavy ion is observed to complete five revolutions in a uniform magnetic field, of magnitude 5×10^{-2} T, in 1.50 ms. Calculate the (approximate) mass of the ion in kilograms.

Additional Problems

49. Determine the gravitational force that the Earth exerts on the Moon, which has a mass of 7.35×10^{22} kg and a radius of 1740 km. The Moon is 3.84×10^5 km from the Earth, center to center.
50. A satellite of mass 600 kg is in a circular orbit about the Earth at a height above the Earth equal to the Earth's mean radius. Find (a) the satellite's orbital speed, (b) the period of its revolution, and (c) the gravitational force acting on it.
51. Io, a small moon of the giant planet Jupiter, has an

orbital period of 1.77 days and an orbital radius of 4.22×10^5 km. From these data, and using the value of *G*, determine the mass of Jupiter.

52. A small 2-g plastic ball is suspended by a 20-cm-long string in a uniform electric field, as shown in Figure 6.18. If the ball is in equilibrium when the string makes a 15° angle with the vertical (as indicated), what is the net charge on the ball?

53. A proton moves in a circular orbit perpendicular to a uniform magnetic induction of 0.758 T. Find the time it takes the proton to make one pass around the orbit.

54. An electron travels a circular orbit of radius 1.77 m in a region of uniform magnetic induction of magnitude 0.664 T, perpendicular to the motion. If the magnetic field doubles, what is the new radius of the orbit?

55. At the equator, near the surface of the Earth, the magnetic field is approximately 50 μT northward, and the electric field is about 100 N/C downward. Find the gravitational, electric, and magnetic forces on an electron moving eastward in a straight line, at 6×10^6 m/s, in this environment.

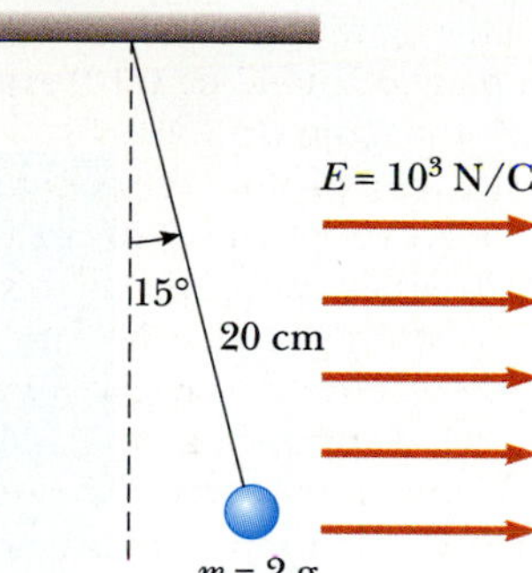

Figure 6.18 (Problem 52)

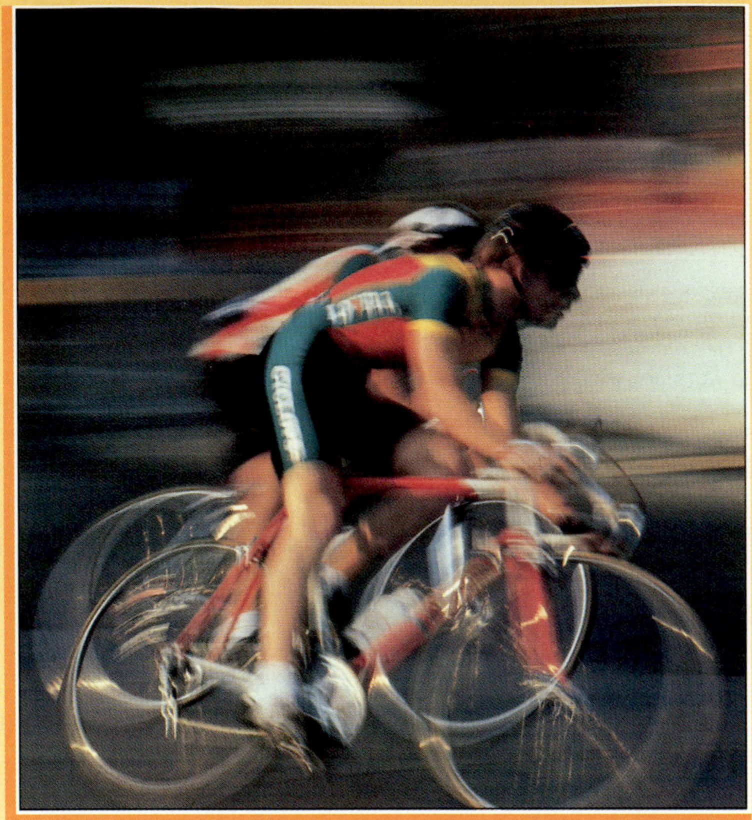

Cyclists are working hard and expending energy road-racing at top speed. *(© Index/Stock).*

Work and Energy

CHAPTER 7

Energy is present in the Universe in a variety of forms, including mechanical, electromagnetic, chemical, thermal, and nuclear. When energy is changed from one form to another, its total amount remains the same. That is, conservation of energy says that if an isolated system loses energy in some form, it will gain an equal amount of energy in other forms.

In this chapter we shall concern ourselves with the mechanical form of energy. We shall see that the concepts of work and energy can be applied to the dynamics of a mechanical system without resorting to Newton's laws. (However, it is important to note that the work-energy concepts are based upon Newton's laws and therefore do not involve any new physical principles.) This "energy approach" to describing motion is especially useful when the force acting on a particle is not constant; in such a case, the acceleration is not constant, and we cannot apply the kinematic equations we developed in Chapter 2. Particles in nature are often subject to forces that vary with the particles' positions. These forces include gravitational forces and the force exerted on a body attached to a spring. We shall describe techniques for treating such systems with the help of an extremely important development called the *work-energy theorem,* which is the central topic of this chapter.

We begin by defining work, a concept that provides a link between the concepts of force and energy.

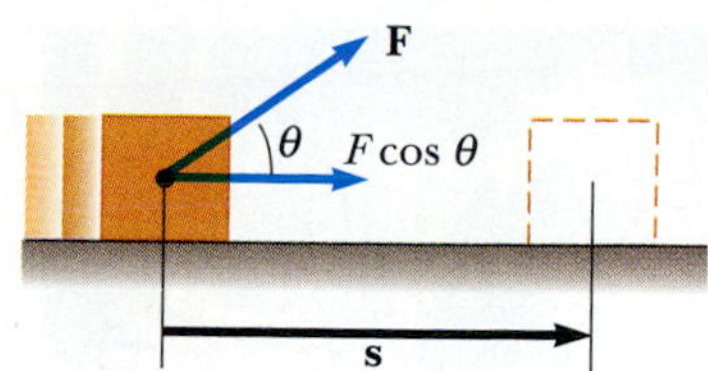

Figure 7.1
If a particle undergoes a displacement **s**, the work done by the force **F** is $(F\cos\theta)s$.

7.1 Work Done by a Constant Force

Almost all of the terms we have used thus far—*velocity, acceleration, force,* and so on—have conveyed the same meaning in physics as they do in everyday life. Now, however, we encounter a term whose meaning in physics is distinctly different from its everyday meaning. This new term is **work**. It can be defined with the help of Figure 7.1. Here we see a particle that undergoes a displacement **s** along a straight line while acted on by a constant force **F**, which makes an angle θ with **s**.

The work, W, done by an agent exerting a constant force is the product of the component of the force along the direction of displacement of the point of application of the force and the magnitude of the displacement:

$$W \equiv (F\cos\theta)s \qquad [7.1]$$

From this definition, we see that a force does no work on a particle if the particle does not move: if $s = 0$, Equation 7.1 gives $W = 0$. Also note from Equation 7.1 that the work done by a force is zero when the force is perpendicular to the displacement, as in Figure 7.2. That is, if $\theta = 90°$, then $W = 0$ because $\cos 90° = 0$. The work done by the force of gravity during the horizontal displacement is zero for the same reason. In general, the particle may be moving with a constant or varying velocity under the influence of several forces. In that case, since work is a scalar quantity, the total work done as the particle undergoes some displacement is the algebraic sum of the work done by each of the forces.

The sign of the work also depends on the direction of **F** relative to **s**. The work done by the applied force is positive when the vector associated with the component $F\cos\theta$ is in the *same direction* as the displacement. For example, when an object is lifted, the work done by the applied force is positive, since the lifting force is upward—that is, in the same direction as the displacement.

When the vector associated with the component $F\cos\theta$ is in the direction *opposite* the displacement, *W is negative.* In the case of the object being lifted, for instance, the work done by the gravitational force is negative. It is important to note that work is an energy transfer; if energy is transferred *to* the system (object), W is positive; if energy is transferred *from* the system, W is negative.

Figure 7.2
No work is done when a bucket of water is moved horizontally because the applied force **F** is perpendicular to the displacement.

If an applied force **F** acts along the direction of the displacement, then $\theta = 0$ and $\cos 0 = 1$. In this case, Equation 7.1 gives

$$W = Fs \qquad [7.2]$$

Work is a scalar quantity, and its units are force multiplied by length. Therefore, the SI unit of work is the **newton · meter** (N · m). Another name for the newton · meter is the **joule** (J). The unit of work in the cgs and British engineering

Athletes limbering up before rigorous exercise often push against a rigid object to condition their muscles. Although this athlete is pushing against the wall, no work is done on the wall since it does not move. *(Fourby-Five)*

Table 7.1
Units of Work in the Three Common Systems of Measurement

System	Unit of Work	Name of Combined Unit
SI	newton·meter (N·m)	joule (J)
cgs	dyne·centimeter (dyne·cm)	erg
British engineering (conventional)	foot·pound (ft·lb)	foot·pound (ft·lb)

systems is the **dyne·cm,** which is also called the **erg,** and ft·lb, respectively. These are summarized in Table 7.1. Note that $1\text{J} = 10^7$ ergs.

Example 7.1 Mr. Clean

A man cleaning his apartment pulls a vacuum cleaner with a force of magnitude $F = 50$ N at an angle of 30°, as shown in Figure 7.3. The vacuum cleaner is pulled a distance of 3 m. Calculate the work done by the 50-N force.

Solution We can use $W = (F\cos\theta)s$, with $F = 50$ N, $\theta = 30°$, and $s = 3$ m, to get

$$W_F = (50\text{ N})(\cos 30°)(3\text{ m}) = 130\text{ J}$$

Note that the normal force, **n**, the weight $m\mathbf{g}$, and the upward component of the applied force, 50 sin 30°, do *no* work because they are perpendicular to the displacement.

Exercise Find the work done by the man on the vacuum cleaner if he pulls it a distance of 3 m horizontally with a force $F = 50$ N.

Answer 150 J.

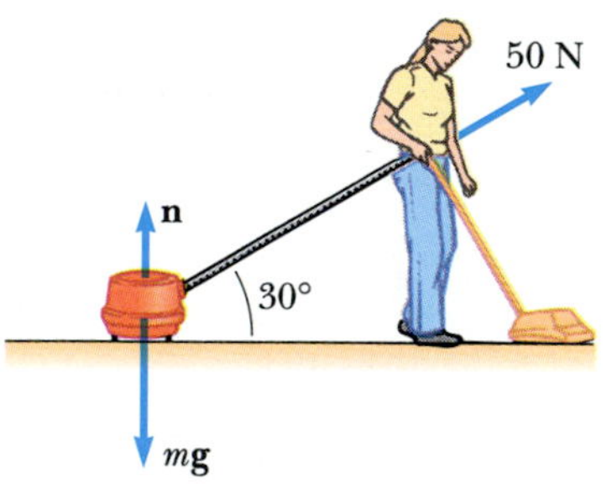

Figure 7.3
(Example 7.1) A vacuum cleaner being pulled at an angle of 30° with the horizontal.

7.2 The Scalar Product of Two Vectors

We have defined work as a *scalar* quantity given by the product of the magnitude of the displacement and the component of a force in the direction of the displacement. It is convenient to express Equation 7.1 in terms of a **scalar product** of the two vectors **F** and **s**. We write this scalar product **F·s**. Because of the dot symbol, the scalar product is often called the *dot product.* Thus, we can express Equation 7.1 as a scalar product:

Work expressed as a dot product

$$W = \mathbf{F}\cdot\mathbf{s} = Fs\cos\theta \qquad [7.3]$$

In other words, **F·s** (read "F dot s") is a shorthand notation for $Fs\cos\theta$.

In general, the scalar product of any two vectors A and B is defined as a scalar quantity equal to the product of the magnitudes of the two vectors and the cosine of the angle θ that is included between the directions of A and B:

$$\mathbf{A}\cdot\mathbf{B} \equiv AB\cos\theta \qquad [7.4]$$

Scalar product of any two vectors A and B

where θ is the smaller angle between **A** and **B**, as in Figure 7.5. Note that **A** and **B** need not have the same units.

In Figure 7.4 $B\cos\theta$ is the projection of **B** onto **A**. Therefore, the definition of $\mathbf{A}\cdot\mathbf{B}$ as given by Equation 7.4 can be considered as the product of the magnitude of **A** and the projection of **B** onto **A**.[1]

From Equation 7.4 we also see that the scalar product is *commutative*. That is,

The order of the dot product can be reversed

$$\mathbf{A}\cdot\mathbf{B} = \mathbf{B}\cdot\mathbf{A} \qquad [7.5]$$

Finally, the scalar product obeys the *distributive law of multiplication*, so that

$$\mathbf{A}\cdot(\mathbf{B}+\mathbf{C}) = \mathbf{A}\cdot\mathbf{B} + \mathbf{A}\cdot\mathbf{C} \qquad [7.6]$$

The dot product is simple to evaluate from Equation 7.4 when **A** is either perpendicular or parallel to **B**. If **A** is perpendicular to **B** ($\theta = 90°$), then $\mathbf{A}\cdot\mathbf{B} = 0$. (The equality $\mathbf{A}\cdot\mathbf{B} = 0$ also holds in the trivial case when either **A** or **B** is zero.) If **A** and **B** point in the same direction ($\theta = 0°$), then $\mathbf{A}\cdot\mathbf{B} = AB$. If **A** and **B** point in opposite directions ($\theta = 180°$), then $\mathbf{A}\cdot\mathbf{B} = -AB$. The scalar product is negative when $90° < \theta < 180°$.

The unit vectors **i**, **j**, and **k**, which were defined in Chapter 1, lie in the positive *x*, *y*, and *z* directions, respectively, of a right-handed coordinate system. Therefore, it follows from the definition of $\mathbf{A}\cdot\mathbf{B}$ that the scalar products of these unit vectors are given by

Dot products of unit vectors

$$\mathbf{i}\cdot\mathbf{i} = \mathbf{j}\cdot\mathbf{j} = \mathbf{k}\cdot\mathbf{k} = 1 \qquad [7.7]$$

$$\mathbf{i}\cdot\mathbf{j} = \mathbf{i}\cdot\mathbf{k} = \mathbf{j}\cdot\mathbf{k} = 0 \qquad [7.8]$$

Two vectors **A** and **B** can be expressed in component form as

$$\mathbf{A} = A_x\mathbf{i} + A_y\mathbf{j} + A_z\mathbf{k}$$

$$\mathbf{B} = B_x\mathbf{i} + B_y\mathbf{j} + B_z\mathbf{k}$$

Therefore, Equations 7.7 and 7.8 reduce the scalar product of **A** and **B** to

$$\mathbf{A}\cdot\mathbf{B} = A_xB_x + A_yB_y + A_zB_z \qquad [7.9]$$

In the special case where $\mathbf{A} = \mathbf{B}$, we see that

$$\mathbf{A}\cdot\mathbf{A} = A_x^2 + A_y^2 + A_z^2 = A^2$$

[1] This is equivalent to stating that $\mathbf{A}\cdot\mathbf{B}$ equals the product of the magnitude of **B** and the projection of **A** onto **B**.

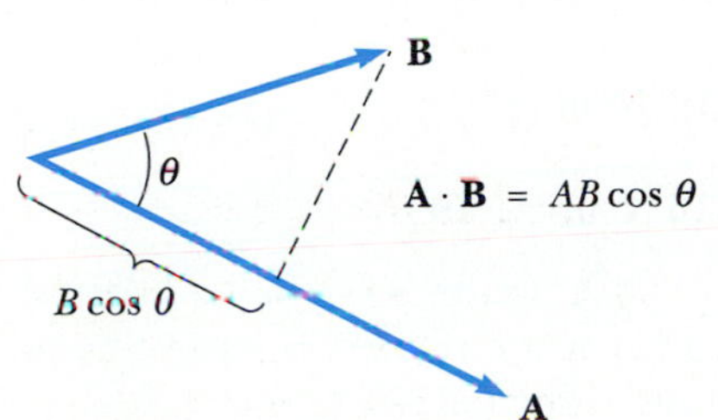

Figure 7.4
The scalar product $\mathbf{A}\cdot\mathbf{B}$ equals the magnitude of **A** multiplied by the projection of **B** onto **A**.

Example 7.2 The Scalar Product

The vectors **A** and **B** are given by **A** = 2**i** + 3**j** and **B** = −**i** + 2**j**.

(a) Determine the scalar product **A** · **B**.

$$\mathbf{A}\cdot\mathbf{B} = (2\mathbf{i} + 3\mathbf{j})\cdot(-\mathbf{i} + 2\mathbf{j})$$
$$= -2\mathbf{i}\cdot\mathbf{i} + 2\mathbf{i}\cdot 2\mathbf{j} - 3\mathbf{j}\cdot\mathbf{i} + 3\mathbf{j}\cdot 2\mathbf{j}$$
$$= -2 + 6 = \boxed{4}$$

where we have used the facts that **i** · **i** = **j** · **j** = 1 and **i** · **j** = **j** · **i** = 0. The same result is obtained using Equation 7.9 directly, where $A_x = 2$, $A_y = 3$, $B_x = -1$, and $B_y = 2$.

(b) Find the angle θ between **A** and **B**.

Solution The magnitudes of **A** and **B** are given by

$$A = \sqrt{A_x^2 + A_y^2} = \sqrt{(2)^2 + (3)^2} = \sqrt{13}$$
$$B = \sqrt{B_x^2 + B_y^2} = \sqrt{(-1)^2 + (2)^2} = \sqrt{5}$$

Using Equation 7.4 and the result from (a),

$$\cos\theta = \frac{\mathbf{A}\cdot\mathbf{B}}{AB} = \frac{4}{\sqrt{13}\,\sqrt{5}} = \frac{4}{\sqrt{65}}$$
$$\theta = \cos^{-1}\frac{4}{8.06} = 60.3°$$

Example 7.3 Work Done by a Constant Force

A particle moving in the xy plane undergoes a displacement **s** = (2**i** + 3**j**) m while a constant force, given by **F** = (5**i** + 2**j**) N, acts on the particle.

(a) Calculate the magnitudes of the displacement and the force.

Solution

$$s = \sqrt{x^2 + y^2} = \sqrt{(2)^2 + (3)^2} = \boxed{\sqrt{13}\text{ m}}$$

$$F = \sqrt{F_x^2 + F_y^2} = \sqrt{(5)^2 + (2)^2} = \boxed{\sqrt{29}\text{ N}}$$

(b) Calculate the work done by **F**.

Solution Substituting the expressions for **F** and **s** into Equation 7.3 and using Equations 7.7 and 7.8, we get

$$W = \mathbf{F}\cdot\mathbf{s} = (5\mathbf{i} + 2\mathbf{j})\cdot(2\mathbf{i} + 3\mathbf{j})\text{ N}\cdot\text{m}$$
$$= 5\mathbf{i}\cdot 2\mathbf{i} + 2\mathbf{j}\cdot 3\mathbf{j} = 16\text{ N}\cdot\text{m} = \boxed{16\text{ J}}$$

Exercise Calculate the angle between **F** and **s**.

Answer 34.5°.

7.3 Work Done by a Varying Force

Consider a particle being displaced along the x axis under the action of a varying force, as in Figure 7.5. The particle is displaced in the direction of increasing x from $x = x_i$ to $x = x_f$. In such a situation, we cannot use $W = (F \cos \theta)s$ to calculate the work done by the force, because this relationship applies only when **F** is constant in magnitude and direction. However, if we imagine that the particle undergoes a very small displacement, Δx, shown in Figure 7.5a, then the x component of the force, F_x, is approximately constant over this interval and we can express the work done by the force for this small displacement as

$$W_1 = F_x \, \Delta x \qquad [7.10]$$

This quantity is just the area of the shaded rectangle in Figure 7.6a. If we imagine that the F_x versus x curve is divided into a large number of such intervals, then the total work done for the displacement from x_i to x_f is approximately equal to the sum of a large number of such terms:

$$W \cong \sum_{x_i}^{x_f} F_x \, \Delta x$$

If the displacements Δx are allowed to approach zero, then the number of terms in the sum increases without limit, but the value of the sum approaches a definite value equal to the area under the curve bounded by F_x and the x axis. As you probably have learned in the calculus, this limit of the sum is called an **integral** and is represented by

$$\lim_{\Delta x \to 0} \sum_{x_i}^{x_f} F_x \, \Delta x = \int_{x_i}^{x_f} F_x \, dx$$

The limits on the integral, $x = x_i$ to $x = x_f$, define what is called a **definite integral.** (An *indefinite integral* is the limit of a sum over an unspecified interval. Appendix B.7 gives a brief description of integration.) This definite integral is numerically equal to the area under the F_x versus x curve between x_i and x_f. Therefore, we can express the work done by F_x for the displacement of the particle from x_i to x_f as

$$W = \int_{x_i}^{x_f} F_x \, dx \qquad [7.11]$$

This equation reduces to Equation 7.1 when $F_x = F \cos \theta$ is constant.

If more than one force acts on the particle, the total work done is just the work done by the resultant force. If we express the resultant force in the x direction as ΣF_x in the x direction, then the *net work* done as the particle moves from x_i to x_f is

$$W_{\text{net}} = \int_{x_i}^{x_f} \left(\sum F_x \right) dx \qquad [7.12]$$

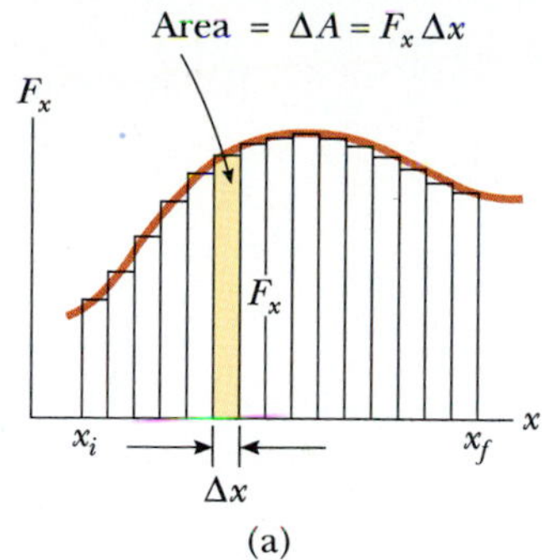

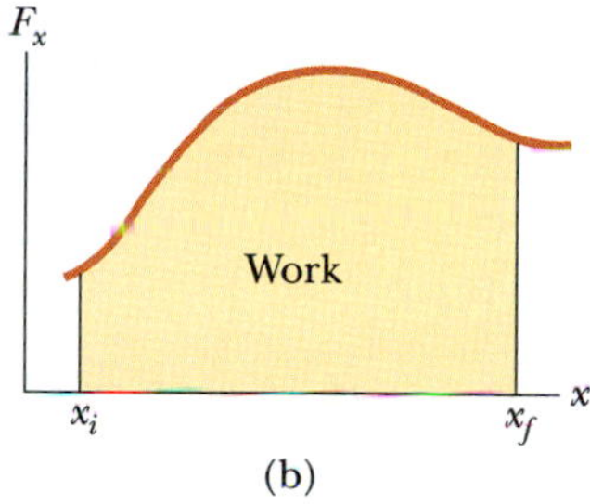

Figure 7.5
(a) The work done by the force F_x for the small displacement Δx is $F_x \, \Delta x$, which equals the area of the shaded rectangle. The total work done for the displacement from x_i to x_f is approximately equal to the sum of the areas of all the rectangles. (b) The work done by the variable force F_x as the particle moves from x_i to x_f is *exactly* equal to the area under this curve.

Example 7.4 Calculating Total Work Done From a Graph

A force acting on a particle varies with x as shown in Figure 7.6. Calculate the work done by the force as the particle moves from $x = 0$ to $x = 6$ m.

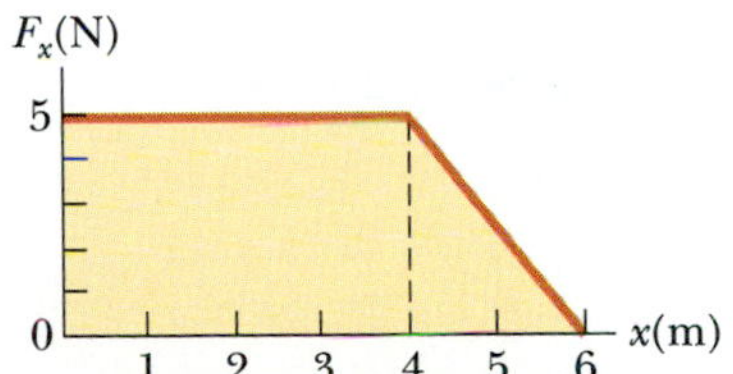

Figure 7.6
(Example 7.4) The force acting on a particle is constant for the first 4 m of motion and then decreases linearly with x from $x = 4$ m to $x = 6$ m. The net work done by this force is the area under this curve.

Solution The work done by the force is equal to the total area under the curve from $x = 0$ to $x = 6$ m. This area is equal to the area of the rectangular section from $x = 0$ to $x = 4$ m plus the area of the triangular section from $x = 4$ m to $x = 6$ m. The area of the rectangle is $(4)(5)$ N·m = 20 J, and the area of the triangle is equal to $\frac{1}{2}(2)(5)$ N·m = 5 J. Therefore, the total work done is 25 J.

Work Done by a Spring

A common physical system for which the force varies with position is shown in Figure 7.7. A mass on a horizontal, smooth surface is connected to a spring. If the spring is stretched or compressed a small distance from its unstretched, or equilibrium, configuration, the spring exerts a force on the mass given by

Spring force

$$F_s = -kx \quad [7.13]$$

where x is the displacement of the mass from the unstretched ($x = 0$) position and k is a positive constant called the *force constant* of the spring. As mentioned in Chapter 4, Section 4.2, this force law for springs is known as **Hooke's law.** Note that Hooke's law is valid only in the case of small displacements from equilibrium. The value of k is a measure of the stiffness of the spring. Stiff springs have large k values, and soft springs have small k values.

The negative sign in Equation 7.13 signifies that the force exerted by the spring is always directed *opposite* the displacement. For example, when $x > 0$, as in Figure 7.7a, the spring force is to the left, or negative. When $x < 0$, as in Figure 7.7c, the spring force is to the right, or positive. Of course, when $x = 0$, as in Figure 7.7b, the spring is unstretched and $F_s = 0$. Because the spring force always acts toward the equilibrium position, it is sometimes called a *restoring force.* Once the mass is displaced some distance x_m from equilibrium and then released, it moves from $-x_m$ through zero to $+x_m$. The details of the ensuing oscillating motion will be given in Chapter 21.

Suppose that the mass is pushed to the left a distance x_m from equilibrium, as in Figure 7.7c, and then released. Let us calculate the *work done by the spring force* as the mass moves from $x_i = -x_m$ to $x_f = 0$. Applying Equation 7.11 assuming the mass may be treated as a particle, we get

Work done by a spring

$$W_s = \int_{x_i}^{x_f} F_s\,dx = \int_{-x_m}^{0} (-kx)\,dx = \tfrac{1}{2}kx_m^2 \quad [7.14]$$

where we have used the indefinite integral $\int x\,dx = x^2/2$. That is, the work done by the spring force is positive, since the spring force is in the same direction as the displacement (both are to the right). However, if we consider the work done by the spring force as the mass moves from $x_i = 0$ to $x_f = x_m$, we find that $W_s = -\frac{1}{2}kx_m^2$, since for this part of the motion the displacement is to the right and the spring force is to the left. Therefore, the *net* work done by the spring force as the body moves from $x_i = -x_m$ to $x_f = x_m$ is *zero.*

If we plot F_s versus x, as in Figure 7.7d, we arrive at the same results. Note that the work calculated in Equation 7.14 is equal to the area of the shaded triangle in Figure 7.7d, with base x_m and height kx_m. This area is $\frac{1}{2}kx_m^2$.

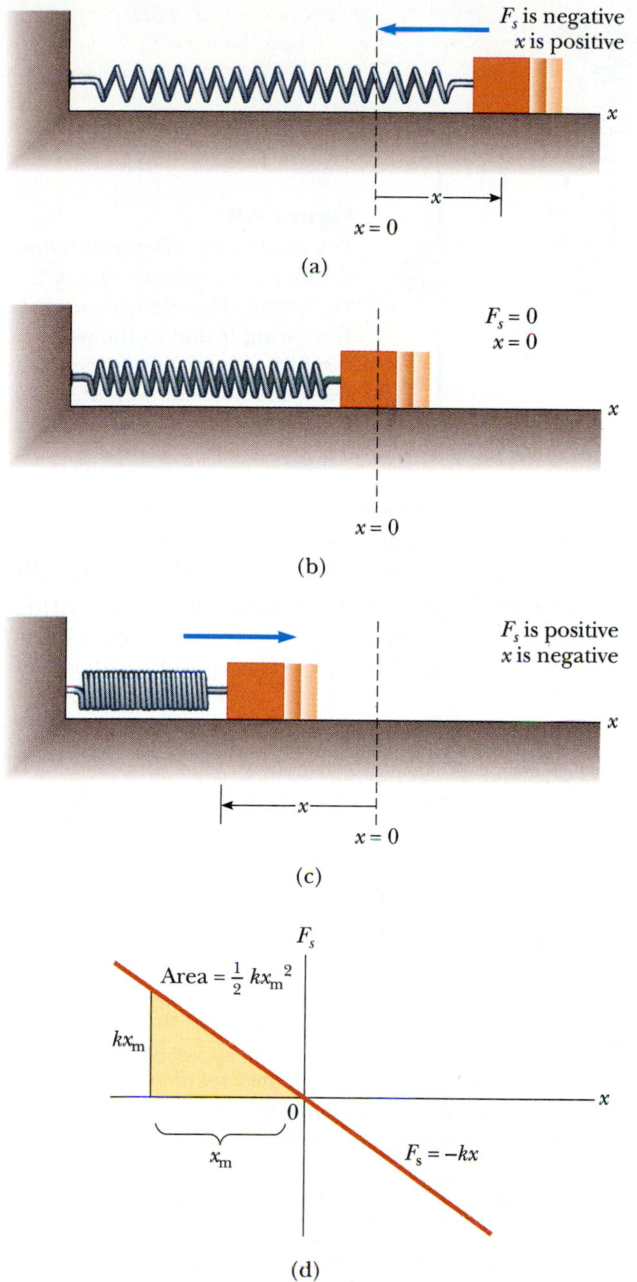

Figure 7.7
The force of a spring on a block varies with the block's displacement from the equilibrium position $x = 0$. (a) When x is positive (stretched spring), the spring force is to the left. (b) When x is zero, the spring force is zero (natural length of the spring). (c) When x is negative (compressed spring), the spring force is to the right. (d) Graph of $\mathbf{F}_s$ versus x for systems described above. The work done by the spring force as the block moves from $-x_m$ to 0 is the area of the shaded triangle, $\frac{1}{2}kx_m^2$.

If the mass undergoes an *arbitrary* displacement from $x = x_i$ to $x = x_f$, the work done by the spring force is

$$W_s = \int_{x_i}^{x_f} (-kx)\,dx = \tfrac{1}{2}kx_i^2 - \tfrac{1}{2}kx_f^2 \qquad [7.15]$$

From this equation we see that the work done is zero for any motion that ends where it began ($x_i = x_f$). We shall make use of this important result in Chapter 8, where we describe the motion of this system in more detail.

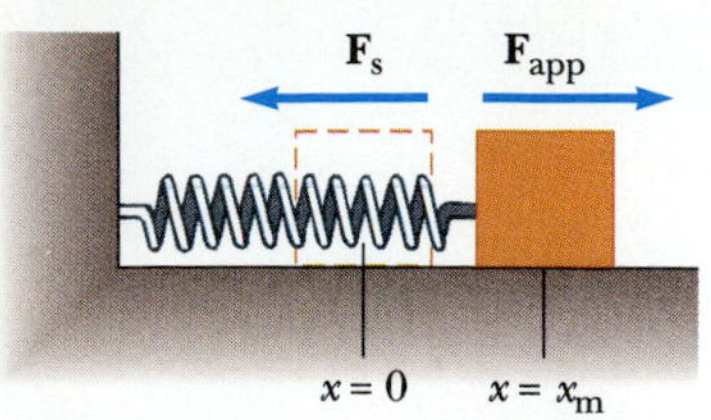

Figure 7.8
A block being pulled to the right on a frictionless surface by a force $\mathbf{F}_{app}$ from $x = 0$ to $x = x_m$. If the process is carried out very slowly, the applied force is equal and opposite to the spring force at all times.

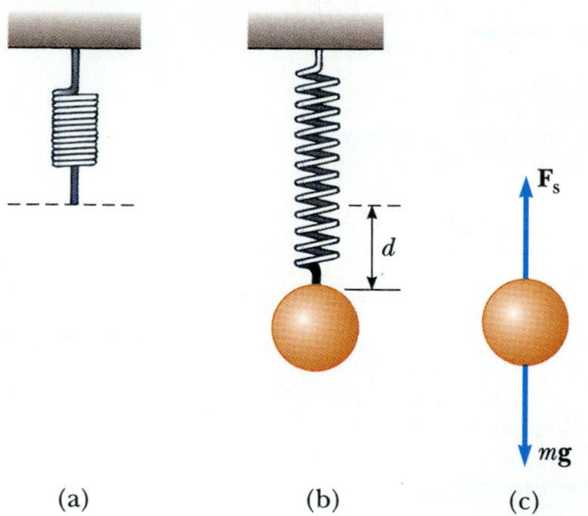

Figure 7.9
(Example 7.6) Determination of the force constant of a helical spring. The elongation d of the spring is due to the weight mg. Since the spring force upward balances the weight, it follows that $k = mg/d$.

Equations 7.14 and 7.15 describe the work done by the spring on the mass. Now let us consider the work done on the spring when it is stretched *very slowly* from $x_i = 0$ to $x_f = x_m$ by an *external agent,* as in Figure 7.8. This work can be easily calculated by noting that the *applied force,* $\mathbf{F}_{app}$, is equal to and opposite the spring force, $\mathbf{F}_s$, at any value of the displacement, so that $F_{app} = -(-kx) = kx$. Therefore, the work done on the spring by this applied force (the external agent) is

$$W_{F_{app}} = \int_0^{x_m} F_{app}\, dx = \int_0^{x_m} kx\, dx = \tfrac{1}{2}kx_m^{\ 2}$$

You should note that this work is equal to the negative of the work done by the spring force for this displacement.

Example 7.5 Measuring *k* for a Spring

A common technique used to measure the force constant of a spring is described in Figure 7.9. The spring is hung vertically, as shown in Figure 7.9a, and then a mass m is attached to the lower end of the spring, as in Figure 7.9b. The spring stretches a distance d from its equilibrium position under the action of the "load" mg. Since the spring force is upward, it must balance the weight mg downward when the system is at rest. In this case, we can apply Hooke's law to give $|\mathbf{F}_s| = kd = mg$, or

$$k = \frac{mg}{d}$$

For example, if a spring is stretched a distance of 2.0 cm by a mass of 0.55 kg, the force constant of the spring is

$$k = \frac{mg}{d} = \frac{(0.55\ \text{kg})(9.80\ \text{m/s}^2)}{2.0 \times 10^{-2}\ \text{m}} = 2.7 \times 10^2\ \text{N/m}$$

Example 7.6 Work Required to Stretch a Spring

One end of a horizontal Hooke's-law spring ($k = 80$ N/m) is held fixed while an external force is applied to the free end, stretching it from $x_0 = 0$ to $x_1 = 4$ cm.

(a) Find the work done by the external force.

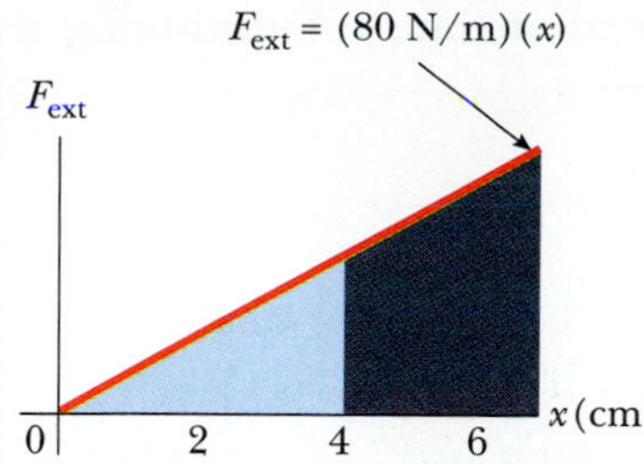

Figure 7.10
External force required to stretch a spring that obeys Hooke's law.

Solution We place the zero reference of the coordinate axis at the free end of the unstretched spring (Fig. 7.10). The external force is $F_{ext} = (80\text{ N/m})(x)$. The work done by F_{ext} is the area of the triangle from 0 to 4 cm:

$$W = \tfrac{1}{2}kx_1^2 = \tfrac{1}{2}\,(80\text{ N/m})\,(0.04\text{ m})^2 = \boxed{64\text{ mJ}}$$

(b) Find the additional work done in stretching the spring from $x_1 = 4$ cm to $x_2 = 7$ cm.

Solution The work done in stretching the spring the additional amount, from $x_1 = 0.04$ m to $x_2 = 0.07$ m, is the darker-shaded area between these limits. Geometrically, it is

$$W = \tfrac{1}{2}kx_2^2 - \tfrac{1}{2}kx_1^2 = \tfrac{1}{2}(80\text{ N/m})\,[(0.07\text{ m})^2 - (0.04\text{ m})^2] = \boxed{132\text{ mJ}}$$

Using calculus, we find that

$$W = \int_{x_1}^{x_2} F_{ext}\,dx = \int_{0.04\text{ m}}^{0.07\text{ m}} (80\text{ N/m})\,x\,dx = \tfrac{1}{2}(80\text{ N/m})\,(x^2)\Big|_{0.04\text{ m}}^{0.07\text{ m}}$$

$$W = \tfrac{1}{2}(80\text{ N/m})[(0.07\text{ m})^2 - (0.04\text{ m})^2] = \boxed{132\text{ mJ}}$$

7.4 Kinetic Energy and the Work-Energy Theorem

Solutions using Newton's second law can be difficult if the forces in the problem are complex. An alternative approach that enables us to understand and solve such motion problems is to relate the speed of the particle to its displacement under the influence of some net force. As we shall see in this section, if the work done by the net force on a particle can be calculated for a given displacement, the change in the particle's speed will be easy to evaluate.

Figure 7.11 shows a particle of mass m moving to the right under the action of a constant net force **F**. Because the force is constant, we know from Newton's second law that the particle will move with a constant acceleration **a**. If the particle is displaced a distance s, the work done by $\mathbf{F}_{net}$ is

$$W_{net} = F_{net}s = (ma)\,s \qquad [7.16]$$

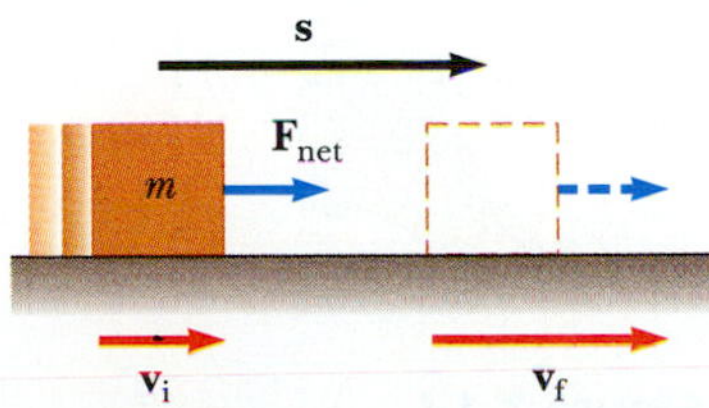

Figure 7.11
A particle undergoing a displacement and change in velocity under the action of a constant net force **F**.

In Chapter 2 (Eqs. 2.4 and 2.9) we found that the following relationships are valid when a particle undergoes constant acceleration:

$$s = \tfrac{1}{2}(v_i + v_f)\,t \qquad a_x = \frac{v_f - v_i}{t}$$

where v_i is the speed at $t = 0$ and v_f is the speed at time t. Substituting these expressions into Equation 7.16 gives

$$W_{\text{net}} = m\left(\frac{v_f - v_i}{t}\right)\frac{1}{2}(v_i + v_f)t$$

$$W_{\text{net}} = \frac{1}{2}mv_f^2 - \frac{1}{2}mv_i^2 \qquad [7.17]$$

The quantity $\frac{1}{2}mv^2$ represents the energy associated with the motion of a particle. It is so important that it has been given a special name—**kinetic energy.** The kinetic energy, K, of a particle of mass m moving with a speed v is defined as

Kinetic energy is energy associated with the motion of a particle

$$K \equiv \frac{1}{2}mv^2 \qquad [7.18]$$

Kinetic energy is a scalar quantity and has the same units as work. For example, a 1-kg mass moving with a speed of 4.0 m/s has a kinetic energy of 8.0 J. It is often convenient to write Equation 7.17 in the form

The work-energy theorem states that the work done on a particle equals the change in its kinetic energy

$$W_{\text{net}} = K_f - K_i = \Delta K \qquad [7.19]$$

That is, the work done by the constant net force $\mathbf{F}_{\text{net}}$ in displacing a particle equals the change in kinetic energy of the particle.

Equation 7.19 is an important result known as the **work-energy theorem.** For convenience, it was derived under the assumption that the net force acting on the particle was constant. A more general derivation would show that this equation is valid under all circumstances, including that of a variable force.

The work-energy theorem also says that the speed of the particle will increase if the net work done on it is positive, because the final kinetic energy will be greater than the initial kinetic energy. The speed will decrease if the net work is negative, because the final kinetic energy will be less than the initial kinetic energy. Notice that the speed and kinetic energy of a particle change only if work is done on the particle by some external force.

Consider the relationship between the work done on a particle and the change in its kinetic energy as expressed by Equation 7.19. Because of this connection, we can also think of kinetic energy as the work the particle can do in coming to rest. For example, suppose a hammer is on the verge of striking a nail, as in Figure 7.12. The moving hammer has kinetic energy and can do work on the nail. The work done on the nail appears as the product Fs, where F is the average force exerted on the nail by the hammer and s is the distance the nail is driven into the wall. However, the hammer and the nail are not particles, so part of the kinetic energy of the hammer goes into warming the hammer and nail, and all of the work done on the nail goes into warming the nail and wall and locally deforming the wall.

Figure 7.12
The moving hammer has kinetic energy and thus is able to do work on the nail, driving it into the wall.

Situations Involving Kinetic Friction

When dealing with a force acting on an extended object, one must be careful in calculating the work done by that force because the displacement of the object is generally not equal to the displacement of the point of application of that force. Suppose that an object of mass m sliding on a horizontal surface is acted on with a constant horizontal force $\mathbf{F}$ to the right and a kinetic frictional force $\mathbf{f}$ to the left, where $F > f$. In this case, the net force is to the right as in Figure 7.11, and the net work done on the object as it undergoes a displacement $\mathbf{s}$ to the right is

$$W_{\text{net}} = (\mathbf{F} - \mathbf{f}) \cdot \mathbf{s} = Fs - fs \qquad [7.20]$$

The quantity Fs is the work done on the object by the constant force **F**. The quantity $-fs$ is negative because the force of kinetic friction is opposite the displacement. However, it is incorrect to say that $-fs$ is the work done by the frictional force on the object. The work done by kinetic friction depends on both the displacement of the object and on the details of the motion between the initial and final positions. In fact, the work done by kinetic friction on an extended object cannot be explicitly evaluated because friction forces and their individual displacements are very complex.

Now suppose that a block moving on a horizontal surface, and given an initial horizontal velocity $\mathbf{v}_i$, slides a distance s before reaching a final velocity $\mathbf{v}_f$ as in Figure 7.13. The external force which causes the block to undergo an acceleration in the negative x direction is the force of kinetic friction, **f**, acting to the left, opposite the motion. The initial kinetic energy of the block is $\frac{1}{2}mv_i^2$ and its final kinetic energy is $\frac{1}{2}mv_f^2$. The change in kinetic energy of the block is equal to $-fs$. This can be shown by applying Newton's second law to the block. (Newton's second law gives the acceleration of the center of mass of any object regardless of how or where the forces act.) Since the net force on the block in the x direction is the friction force, Newton's second law gives $-f = ma$. Multiplying both sides of this expression by s, and using the expression $v_f^2 - v_i^2 = 2as$ for motion under constant acceleration gives $-fs = (ma)s = \frac{1}{2}mv_i^2 - \frac{1}{2}mv_f^2$ or

$$\Delta K = -fs \qquad [7.21]$$

This result says that the loss in kinetic energy of the block is equal to $-fs$, which corresponds to the energy dissipated by the force of kinetic friction. Part of this energy is transferred to internal energy of the block, and part is transferred from the block to the surface. In effect, the loss in kinetic energy of the block results in an increase in internal energy of both the block and surface in the form of thermal (heat) energy. For example, if the loss in kinetic energy of the block is 300 J, and 100 J appears as an increase in internal energy of the block, then the remaining 200 J must have been transferred from the block to the surface.

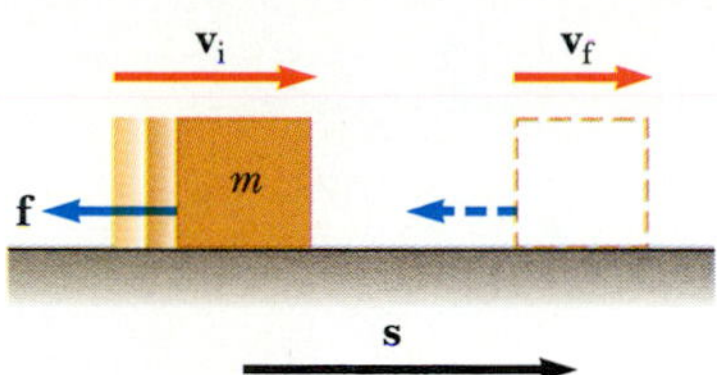

Figure 7.13
A block sliding to the right on a horizontal surface slows down in the presence of a force of kinetic friction acting to the left. The initial velocity of the block is $\mathbf{v}_i$, and its final velocity is $\mathbf{v}_f$. The normal force and force due to gravity are not included in the diagram since they do not influence the change in velocity of the block.

Example 7.7 A Block Pulled on a Smooth Surface

A 6-kg block initially at rest is pulled to the right along a frictionless horizontal surface by a constant, horizontal force of 12 N, as in Figure 7.14a. Find the speed of the block after it has moved a distance of 3 m.

Solution The weight is balanced by the normal force, and neither of these forces does work, since the displacement is horizontal. Because there is no friction, the resultant external force is the 12-N force. The work done by this force is

$$W = Fs = (12\text{ N})(3\text{ m}) = 36\text{ N}\cdot\text{m} = 36\text{ J}$$

Using the work-energy theorem and noting that the initial kinetic energy is zero, we get

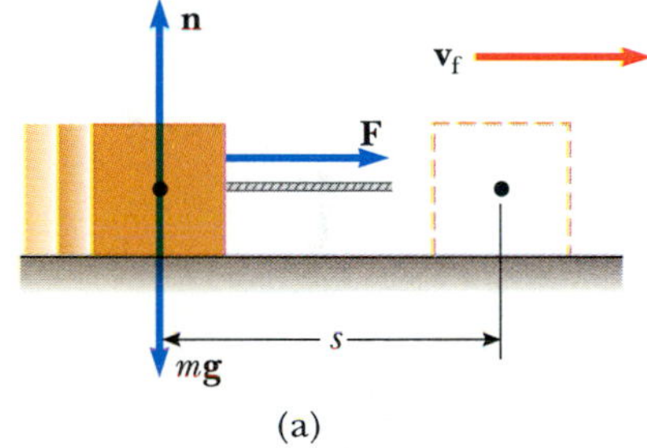

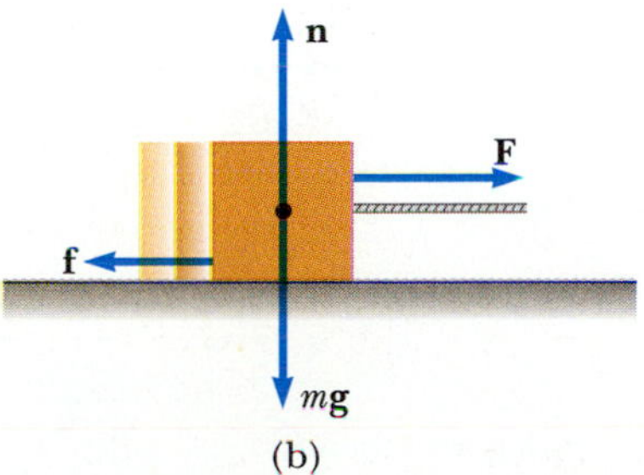

Figure 7.14
(a) Example 7.7. (b) Example 7.8.

[2] For more details on energy transfer situations involving forces of kinetic friction, see B. A. Sherwood and W. H. Bernard, *American Journal of Physics,* 52, 1001 (1984), and R. P. Bauman, *The Physics Teacher,* 30, 264 (1992).

$$W = K_f - K_i = \tfrac{1}{2}mv_f^2 - 0$$

$$v_f^2 = \frac{2W}{m} = \frac{2(36\text{ J})}{6\text{ kg}} = 12\text{ m}^2/\text{s}^2$$

$$v_f = \boxed{3.46\text{ m/s}}$$

Exercise Find the acceleration of the block, and determine its final speed, using the kinematic equation $v_f^2 = v_i^2 + 2as$.

Answer $a = 2\text{ m/s}^2$; $v_f = 3.46\text{ m/s}$.

Example 7.8 A Block Pulled on a Surface with Friction

Find the final speed of the block described in Example 7.7 if the coefficient of kinetic friction between the block and surface is 0.15.

Solution In this case, we must use the alternative equation for change of kinetic energy, ΔK. The net force applied to the block is the sum of the applied 12-N force and the frictional force, as in Figure 7.14b. Since the frictional force is in the direction opposite to the 12-N force, it must be subtracted. The magnitude of the frictional force is given by $f = \mu n = \mu mg$. Therefore the net force acting on the block is

$$F_{\text{net}} = 12\text{ N} - \mu mg = 12\text{ N} - (0.15)(6\text{ kg})(9.80\text{ m/s}^2)$$

$$= 12\text{ N} - 8.82\text{ N} = 3.18\text{ N}$$

Multiplying this constant force by the displacement gives

$$\Delta K = F_{\text{net}}s = (3.18\text{ N})(3\text{ m}) = 9.54\text{ J} = \tfrac{1}{2}mv_f^2$$

using the information that $v_i = 0$. Therefore

$$v_f^2 = \frac{2(9.54\text{ J})}{6\text{ kg}} = 3.18\text{ m}^2/\text{s}^2$$

$$v_f = \boxed{1.78\text{ m/s}}$$

Exercise Find the acceleration of the block from Newton's second law, and then use kinematics to determine the final speed of the block.

Answer $a = 0.530\text{ m/s}^2$; $v_f = 1.78\text{ m/s}$.

Example 7.9 A Mass-Spring System

A block of mass 1.6 kg is attached to a spring having a force constant of 10^3 N/m, as in Figure 7.8. The spring is compressed a distance of 2.0 cm, and the block is released from rest.

(a) Calculate the speed of the block as it passes through the equilibrium position $x = 0$, if the surface is frictionless.

Solution We use Equation 7.14 to find the work done by the spring with $x_m = -2.0\text{ cm} = -2 \times 10^{-2}\text{ m}$:

$$W_s = \tfrac{1}{2}kx_{m^2} = \tfrac{1}{2}(10^3\text{ N/m})(-2 \times 10^{-2}\text{ m})^2 = 0.20\text{ J}$$

Using the work-energy theorem with $v_i = 0$ gives

$$W_s = \tfrac{1}{2}mv_f^2 - \tfrac{1}{2}mv_i^2$$

$$0.20\text{ J} = \tfrac{1}{2}(1.6\text{ kg})v_f^2 - 0$$

$$v_f^2 = \frac{0.4\text{ J}}{1.6\text{ kg}} = 0.25\text{ m}^2/\text{s}^2$$

$$v_f = 0.50\text{ m/s}$$

(b) Calculate the speed of the block as it passes through the equilibrium position if a constant frictional force of 4.0 N retards its motion.

Solution The work done cannot be calculated in the presence of friction, but the alternative equation for change in kinetic energy is applicable and may be added to the kinetic energy found in the absence of friction. Considering only the frictional force, the kinetic energy lost due to friction is

$$-fs = -(4\text{ N})(2 \times 10^{-2}\text{ m}) = -0.08\text{ J}$$

The final kinetic energy, without this loss, was found in part (a) to be 0.20 J. Therefore the final kinetic energy in the presence of friction is

$$K_f = 0.20\text{ J} - 0.08\text{ J} = 0.12\text{ J} = \tfrac{1}{2}mv_f^2$$

$$\tfrac{1}{2}(1.6\text{ kg})v_f^2 = 0.12\text{ J}$$

$$v_f^2 = \frac{0.24\text{ J}}{1.6\text{ kg}} = 0.15\text{ m}^2/\text{s}^2$$

$$v_f = 0.39\text{ m/s}$$

Note that this value for v_f is *less* than that obtained in the frictionless case. Is this result sensible?

7.5 Power

From a practical viewpoint, for example, in a foot race, it is interesting to know not only the work done on an object, but also the rate at which the work is being done. The time rate of doing work is called **power**.

If an external force is applied to an object and if the work done by this force is W in the time interval Δt, then the **average power** during this interval is defined as the ratio of the work done to the time interval:

$$\overline{P} \equiv \frac{W}{\Delta t} \qquad [7.22]$$

The work done on the object contributes to increasing the energy of the object. A more general definition of power is the *time rate of energy transfer.* The **instantaneous power,** P, is the limiting value of the average power as Δt approaches zero:

$$P \equiv \lim_{\Delta t \to 0} \frac{W}{\Delta t} = \frac{dW}{dt} \qquad [7.23]$$

where we have represented the infinitesimal value of the work done by dW (even though it is not a change and therefore not a differential). We know from Equation 7.4 that $dW = \mathbf{F} \cdot d\mathbf{s}$. Therefore, the instantaneous power can be written

Instantaneous power

$$P = \frac{dW}{dt} = \mathbf{F} \cdot \frac{d\mathbf{s}}{dt} = \mathbf{F} \cdot \mathbf{v} \qquad [7.24]$$

where we have used the fact that $\mathbf{v} = d\mathbf{s}/dt$.

The SI unit of power is joules per second (J/s), also called a *watt* (W) (after James Watt):

The watt

$$1 \text{ W} = 1 \text{ J/s}$$

The symbol W for watt should not be confused with the symbol W for work.

The unit of power in the British engineering system is the horsepower (hp):

$$1 \text{ hp} = 550 \text{ ft} \cdot \text{lb/s} = 746 \text{ W}$$

A new unit of energy (or work) can now be defined in terms of the unit of power. One kilowatt-hour (kWh) is the energy converted or consumed in 1 h at the constant rate of 1 kW = 1000 J/s. Therefore,

$$1 \text{ kWh} = (10^3 \text{ W})(3600 \text{ s}) = (10^3 \text{ J/s})(3600 \text{ s}) = 3.6 \times 10^6 \text{ J}$$

It is important to realize that a kilowatt-hour is a unit of energy, not power. When you pay your electric bill, you are buying energy, and the amount of electricity used by an appliance is usually expressed in multiples of kilowatt-hours. For example, an electric bulb rated at 100 W would "consume" 3.6×10^5 J of energy in 1 h.

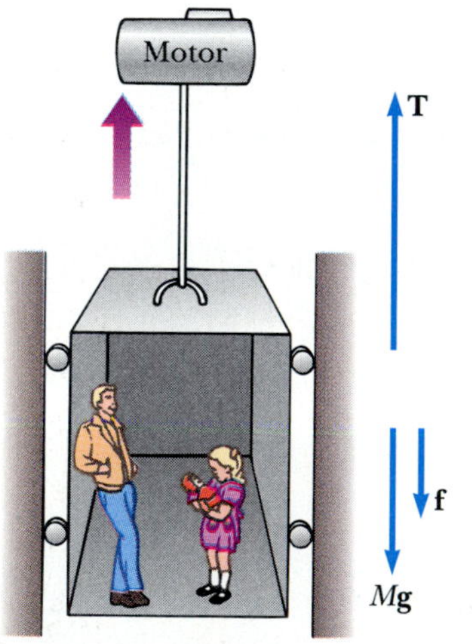

Figure 7.15
(Example 7.10) A motor provides a force T upward on the elevator. A frictional force f and the total weight Mg act downward.

Example 7.10 Power Delivered by an Elevator Motor

An elevator has a mass of 1000 kg and carries a maximum load of 800 kg. A constant frictional force $f = 4000$ N retards the elevator's motion upward, as in Figure 7.15.

(a) What must be the minimum power delivered by the motor to lift the elevator at a constant speed of 3 m/s?

Solution The motor must supply the force, T, that pulls the elevator upward. From Newton's second law and from the fact that $a = 0$ since v is constant, we get

$$T - f - Mg = 0$$

where M is the *total* mass (elevator plus load), equal to 1800 kg. Therefore,

$$\begin{aligned} T &= f + Mg \\ &= 4 \times 10^3 \text{ N} + (1.8 \times 10^3 \text{ kg})(9.80 \text{ m/s}^2) \\ &= 2.16 \times 10^4 \text{ N} \end{aligned}$$

Using Equation 7.24 and the fact that T is in the same direction as v,

$$P = \mathbf{T} \cdot \mathbf{v} = Tv$$
$$= (2.16 \times 10^4 \text{ N})(3 \text{ m/s}) = 6.48 \times 10^4 \text{ W}$$
$$= 64.8 \text{ kW} = \boxed{86.9 \text{ hp}}$$

(b) What power must the motor deliver at any instant if it is designed to provide an upward acceleration of 1.0 m/s^2?

Solution Applying Newton's second law to the elevator gives

$$T - f - Mg = Ma$$

$$T = M(a + g) + f$$
$$= (1.8 \times 10^3 \text{ kg})(1.0 + 9.80) \text{ m/s}^2 + 4 \times 10^3 \text{ N}$$
$$= 2.34 \times 10^4 \text{ N}$$

Therefore, using Equation 7.24, we get for the required power

$$P = Tv = \boxed{(2.34 \times 10^4 \, v) \text{ W}}$$

where v is the instantaneous speed of the elevator in meters per second. Hence, the power required increases with increasing speed.

Summary

The **work** done by a *constant* force **F** acting on a particle is defined as the product of the component of the force in the direction of the particle's displacement and the magnitude of the displacement. If the force makes an angle θ with the displacement **s**, the work done by **F** is

$$W \equiv (F \cos \theta) s \qquad [7.1]$$

Work done by a constant force

The **scalar, or dot, product** of any two vectors **A** and **B** is

$$\mathbf{A} \cdot \mathbf{B} \equiv AB \cos \theta \qquad [7.5]$$

Scalar product

where the result is a scalar quantity and θ is the angle included between the directions of the two vectors. The scalar product obeys the commutative and distributive laws.

The *work* done by a *varying* force acting on a particle moving along the x axis from x_i to x_f is

$$W \equiv \int_{x_i}^{x_f} F_x \, dx \qquad [7.11]$$

Work done by a varying force

where F_x is the component of force in the x direction. If several forces are acting on a particle, the net work done is the sum of the work done individually by all the forces.

The **kinetic energy** of a particle of mass m moving with a speed v is

$$K \equiv \tfrac{1}{2} m v^2 \qquad [7.18]$$

Kinetic energy

The **work-energy theorem** states that the net work done on a particle by external forces equals the change in kinetic energy of the particle:

Work-energy theorem

$$W_{\text{net}} = K_f - K_i = \tfrac{1}{2}mv_f^2 - \tfrac{1}{2}mv_i^2 \quad [7.19]$$

Average power is the time rate of doing work:

$$\overline{P} = \frac{W}{\Delta t} \quad [17.22]$$

If an agent applies a force **F** to a particle moving with a velocity **v**, the **instantaneous power** delivered by that agent is

Power

$$P \equiv \frac{dW}{dt} = \mathbf{F}\cdot\mathbf{v} \quad [7.24]$$

Questions and Conceptual Exercises

1. When a particle rotates in a circle, a *centripetal force* acts on it, directed toward the center of rotation. Why is it that this force does no work on the particle?
2. Explain why the work done by the force of sliding friction is negative when an object undergoes a displacement on a rough surface.
3. Is any direction associated with the dot product of two vectors?
4. If the dot product of two vectors is positive, does this imply that the vectors must have positive rectangular components?
5. As the load on a spring hung vertically is increased, one does not expect the F_s versus x curve to always remain linear, as in Figure 7.8d. Explain qualitatively what you expect for this curve as m is increased.
6. Can the kinetic energy of an object have a negative value?
7. If the speed of a particle is doubled, what happens to its kinetic energy?
8. What can be said about the speed of a particle if the net work done on the particle is zero?
9. An Earth satellite is in a circular orbit at an altitude of 500 km. Does the gravitational force acting on the satellite do any work? What does the work-energy theorem tell us about the speed of the satellite?
10. Can the average power ever equal the instantaneous power? Explain.
11. In Example 7.10, does the required power increase or decrease as the force of friction is reduced?
12. One bullet has twice the mass of a second bullet. If both are fired so that they have the same speed, which has more kinetic energy? What is the ratio of the kinetic energies of the two bullets?
13. When a punter kicks a football, is he doing any work on the ball while it is in contact with his toe? Is he doing any work on the ball after it loses contact with his toe? Are any forces doing work on the ball while it is in flight?
14. Cite two examples in which a force is exerted on an object without doing any work on the object.
15. Two sharpshooters fire .30-caliber rifles, using identical shells. The barrel of rifle A is 2 cm longer than that of rifle B. Which rifle has the higher muzzle speed? (*Hint:* The force of the expanding gases in the barrel accelerates the bullets.)
16. A team of furniture movers wishes to load a truck using a ramp from the ground to the rear of the truck. One of the movers claims that less work would be required to load the truck if the length of the ramp were increased, reducing the angle of the ramp with respect to the horizontal. Is his claim valid? Explain.
17. As a simple pendulum swings back and forth, the forces acting on the suspended mass are the force of gravity, the tension in the supporting cord, and air resistance. (a) Which of these forces, if any, do no work on the pendulum? (b) Which of these forces does negative work at all times during its motion? (c) Describe the work done by the force of gravity while the pendulum is swinging.
18. Estimate the work done by a major-league pitcher when he throws a baseball at 40 m/s (90 mi/h). The mass of a baseball is approximately 0.15 kg.
19. A catcher "gives" with the ball when she catches a 0.15-kg baseball moving at 40 m/s. If she moves her catcher's mitt through a distance of 2 cm, what is the average force acting on her hand?
20. Estimate the time it takes you to climb a flight of stairs. (For a ballpark estimate, assume you can travel upward a distance of 8 m in 6 s.) Then approximate the power required to perform this task. Express your answer in horsepower.

Problems

Section 7.1 Work Done by a Constant Force

1. If a man lifts a 20-kg bucket from a well and does 6 kJ of work, how deep is the well? Assume that the bucket comes to rest at the top.
2. A 65-kg woman climbs a flight of 20 stairs, each 23 cm high. How much work is done against the force of gravity in the process?
3. A tugboat exerts a constant force of 5000 N on a ship moving at constant speed through a harbor. How much work does the tugboat do on the ship in a distance of 3 km?

(Problem 3). *(© David Plowden/Photo Researchers, Inc.)*

4. Verify the following energy unit conversions: (a) 1 J = 10^7 ergs, (b) 1 J = 0.737 ft·lb.
5. A shopper in a supermarket pushes a cart with a force of 35 N directed at an angle of 25° downward from the horizontal. Find the work done by the shopper as she moves down a 50-m length of aisle.
6. Batman, whose mass is 80 kg, is holding onto the free end of a 12-m rope, the other end of which is fixed to a tree limb above. He puts the rope in motion as only Batman knows how, eventually getting it to swing enough that he can reach a ledge when the rope makes a 60° angle with the downward vertical. How much work is done against the force of gravity in this maneuver?

Section 7.2 The Scalar Product of Two Vectors

7. Two vectors are given by **A** = 4**i** + 3**j** and **B** = −**i** + 3**j**. Find (a) **A**·**B** and (b) the angle between **A** and **B**.
8. A vector is given by **A** = − 2**i** + 3**j**. Find (a) the magnitude of **A** and (b) the angle that **A** makes with the positive y axis. [In (b), use the definition of the scalar product.]
9. Vector **A** has a magnitude of 5 units, and **B** has a magnitude of 9 units. The two vectors make an angle of 50° with each other. Find **A**·**B**.
10. Given two arbitrary vectors **A** and **B**, show that **A**·**B** = $A_xB_x + A_yB_y + A_zB_z$. (*Hint:* Write **A** and **B** in unit vector form and use Eq. 7.8.)
11. For the three vectors **A** = 3**i** + **j** − **k**, **B** = − **i** + 2**j** + 5**k**, and **C** = 2**j** − 3**k**, find **C** · (**A** − **B**).
12. A force **F** = (6**i** − 2**j**) N acts on a particle that undergoes a displacement **s** = (3**i** + **j**) m. Find (a) the work done by the force on the particle and (b) the angle between **F** and **s**.
13. Vector **A** is 2 units long and points in the positive y direction. Vector **B** has a negative x component 5 units long, a positive y component 3 units long, and no z component. Find **A**·**B** and the angle between the vectors.
14. As a particle moves from the origin to (3**i** − 4**j**) m, it is acted upon by a force given by (4**i** − 5**j**) N. Calculate the work done by this force as the particle moves through the given displacement.
15. Find the angle between the two vectors given by **A** = − 5**i** − 3**j** + 2**k** and **B** = − 2**j** − 2**k**.
16. Using the definition of the scalar product, find the angles between the following pairs of vectors: (a) **A** = 3**i** − 2**j** and **B** = 4**i** − 4**j**, (b) **A** = − 2**i** + 4**j** and **B** = 3**i** − 4**j** + 2**k**, (c) **A** = **i** − 2**j** + 2**k** and **B** = 3**j** + 4**k**.

Section 7.3 Work Done by a Varying Force

17. A particle object moves from $x = 0$ to $x = 3$ m. If the resultant force acting on that object is in the x direction and varies as shown in Figure 7.16, determine the total work done on that object.

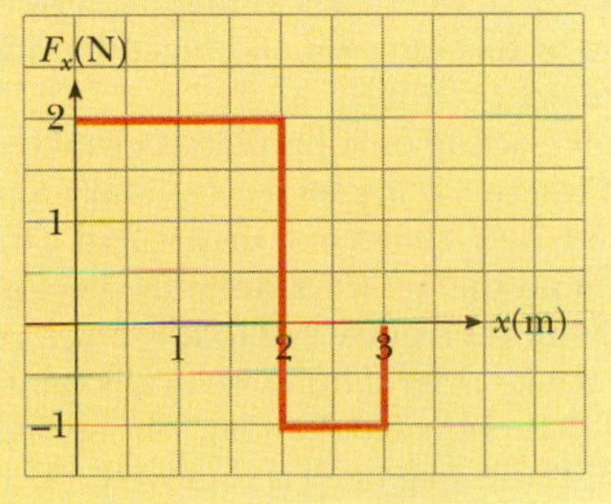

Figure 7.16 (Problem 17)

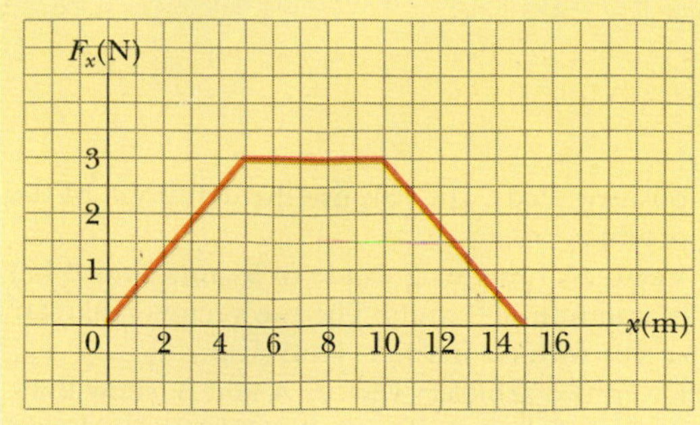

Figure 7.17 (Problem 18)

18. A body is subject to a force $\mathbf{F}_x$ that varies with position as in Figure 7.17. Find the work done by the force on the body as it moves (a) from $x = 0$ to $x = 5$ m, (b) from $x = 5$ m to $x = 10$ m, and (c) from $x = 10$ m to $x = 15$ m. (d) What is the total work done by the force over the distance $x = 0$ to $x = 15$ m?
19. The force acting on a particle varies as in Figure 7.18. Find the work done by the force as the particle moves (a) from $x = 0$ to $x = 8$ m, (b) from $x = 8$ m to $x = 10$ m, and (c) from $x = 0$ to $x = 10$ m.

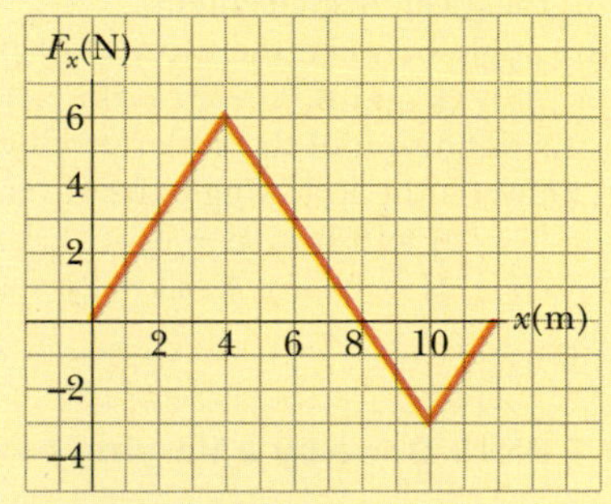

Figure 7.18 (Problem 19)

20. If an applied force varies with position according to $F_x = 3x^3 - 5$, where x is in meters, how much work is done by this force on an object that moves from $x = 4$ m to $x = 7$ m?
21. The force acting on a particle is given by $F_x = (8x - 16)$ N, where x is in meters. (a) Make a plot of this force versus x from $x = 0$ to $x = 3$ m. (b) From your graph, find the net work done by this force as the particle moves from $x = 0$ to $x = 3$ m.
22. When a 4-kg mass is hung vertically on a certain light spring that obeys Hooke's law, the spring stretches 2.5 cm. If the 4-kg mass is removed, (a) how far will the spring stretch if a 1.5-kg mass is hung on it, and (b) how much work must an external agent do to stretch the same spring 4.0 cm from its unstretched position?

(Problem 24). *(Gamma)*

23. A marine in the jungle finds himself in the middle of a swamp. It is estimated that the force F_x he must exert in the x direction in order to get out is $F_x = (1000 - 50x)$ N, where x is in meters. (a) Sketch the graph of F_x versus x. (b) What is the average force exerted in traveling from zero to x? How much work does he do in moving 20 m?
24. An archer pulls his bow string back 0.4 m by exerting a force that increases uniformly from zero to 230 N. (a) What is the equivalent spring constant of the bow? (b) How much work is done in pulling the bow?
25. The force required to stretch a Hooke's-law spring varies from zero to 50 N as one end of the spring is moved 12 cm from its unstressed position. (a) Find the force constant k of the spring. (b) Find the work done in stretching the spring.
26. If it takes 4 J of work to stretch a Hooke's-law spring 10 cm beyond its unstressed length, determine the additional work required to stretch it another 10 cm.

Section 7.4 Kinetic Energy and the Work-Energy Theorem

27. A 0.6-kg particle has a speed of 2 m/s at point **A** and kinetic energy of 7.5 J at point **B**. What are (a) its kinetic energy at point A, (b) its speed at point **B**?, and (c) the total work done on the particle as it moves from **A** to **B**?
28. A 0.3-kg ball has a speed of 15 m/s. (a) What is its kinetic energy? (b) If its speed is doubled, what is its kinetic energy?
29. Calculate the kinetic energy of a 1000-kg satellite orbiting the Earth at a speed of 7×10^3 m/s.
30. Two men push a 2500-kg car from rest to a speed v, doing 5000 J of work in the process. During this time,

(Problem 30). *(© Renee Lynn/Photo Researchers, Inc.)*

the car moves 25 m. Neglecting friction between the car and the road, and internal friction of the car, (a) what is the final speed, v, of the car? (b) What is the horizontal force exerted on the car?

31. A 3-kg mass has an initial velocity $\mathbf{v}_0 = (6\mathbf{i} - 2\mathbf{j})$ m/s. (a) What is its kinetic energy at this time? (b) Find the *change* in its kinetic energy if its velocity changes to $(8\mathbf{i} + 4\mathbf{j})$ m/s. (*Hint:* Remember that $v^2 = \mathbf{v} \cdot \mathbf{v}$.)
32. A 15-g bullet is accelerated in a rifle barrel 72 cm long to a speed of 780 m/s. Use the work-energy relation to find the average force that acts on the bullet while it is being accelerated.
33. A 100-kg sled is dragged by a team of dogs a distance of 2 km over a horizontal surface at a constant velocity. If the coefficient of friction between the sled and snow is 0.15, find (a) the work done by the team of dogs and (b) the energy lost due to friction.

(Problem 33). *(Gamma)*

34. A horizontal force of 150 N is used to push a 40-kg box a distance of 6 m on a horizontal surface. If the box moves at constant speed, find (a) the work done by the 150-N force, (b) the energy lost due to friction, and (c) the coefficient of kinetic friction between the box and surface.
35. A 15-kg block is dragged over a horizontal surface by a constant force of 70 N acting at an angle of 20° above the horizontal. The block is displaced 5 m, and the coefficient of kinetic friction between the block and surface is 0.3. Find the work done by (a) the 70-N force, (b) the normal force, and (c) the force of gravity. (d) Calculate the energy lost due to friction.
36. A cart loaded with bricks has a total mass of 18 kg and is pulled at constant speed by a rope. The rope is inclined at 20° above the horizontal, and the cart moves a distance of 20 m on a horizontal surface. The coefficient of kinetic friction between the cart and surface is 0.5. (a) What is the tension of the rope? (b) How much work is done on the cart by the rope? (c) What is the energy lost due to friction?
37. A bullet with a mass of 5 g and a speed of 600 m/s strikes a tree and penetrates it to a depth of 4 cm. (a) Use work and energy considerations to find the average frictional force that stops the bullet. (b) Assuming that the frictional force is constant, determine how much time elapsed between the moment the bullet entered the tree and the moment it stopped.
38. A block of mass 12 kg slides from rest down a frictionless 35° incline and is stopped by a strong spring with $k = 3.0 \times 10^4$ N/m. The block slides a total distance $d = 3.0$ m from the point of release to the point where it comes to rest against the spring. When the block comes to rest, how far has the spring been compressed?
39. A sled of mass m on a frozen pond is given a kick, which imparts to it an initial speed $v_0 = 2$ m/s. The coefficient of kinetic friction between the sled and the ice is $\mu_k = 0.1$. Find the distance the sled moves before coming to rest.
40. A crate of mass 10 kg is pulled up an incline with an initial speed of 1.5 m/s. The pulling force is 100 N parallel to the incline, which makes an angle of 20° with the horizontal. If the coefficient of kinetic friction is 0.4, and the crate is pulled a distance of 5 m, (a) how much work is done by gravity? (b) How much energy is lost because of friction? (c) How much work is done by the 100-N force? (d) What is the change in kinetic energy of the crate? (d) What is the speed of the crate after it has been pulled 5 m?

Section 7.5 Power

41. A 700-N marine in basic training climbs a 10-m vertical rope, at uniform speed, in 8 s. What is his effective power output?

42. Water flows over a section of Niagara Falls at a rate of 1.2×10^6 kg/s and falls 50 m. How many 60-W bulbs can be lit with this power?
43. A weight lifter lifts 250 kg through 2 m in 1.5 s. What is his effective power output?
44. A 1500-kg car accelerates uniformly from rest to a speed of 10 m/s in 3 s. Neglecting friction between car and highway and within the car, find (a) the work done on the car in this time, (b) the average power delivered by the engine in the first 3 s, and (c) the instantaneous power delivered by the engine at $t = 2$ s.
45. A certain automobile engine delivers a power of 30 hp (2.24×10^4 W) to its wheels when the car is moving at a constant speed of 27 m/s ($\approx$ 60 mi/h). What is the resistive force acting on the car at that speed?
46. A skier of mass 70 kg is pulled up a slope by a motor-driven cable. (a) How much work is required to pull him a distance of 60 m up a 30° slope (assumed frictionless) at a constant speed of 2 m/s? (b) What horsepower motor is required to perform this task?
47. A 65-kg athlete runs a distance of 600 m up a mountainside that is inclined at 20° to the horizontal. He performs this feat in 80 s. Assuming that air resistance is negligible, (a) how much work is done against gravity, and (b) what is his power output during the run?
48. A car of weight 2500 N operating at a rate of 130 kW develops a maximum speed of 31 m/s on a level, horizontal road. Assuming that the resistive force (due to friction and air resistance) remains constant, (a) what is the car's maximum speed on an incline of 1 in 20 (i.e., if $\sin \theta = 0.05$), and (b) what is its power output on a 1-in-10 incline if it is traveling at 10 m/s?

Additional Problems

49. A baseball outfielder throws a 0.15-kg baseball at a speed of 40 m/s and an initial angle of 30°. What is the kinetic energy of the baseball at the highest point of the trajectory?
50. While running, a person dissipates about 0.6 J of mechanical energy per step per kilogram of body mass. If a 60-kg runner dissipates a power of 70 W during a race, how fast is the person running? Assume a running step is 1.5 m long.
51. A raindrop ($m = 3.35 \times 10^{-5}$ kg) falls vertically at constant speed under the influence of the forces of gravity and drag (Fig. 7.19). In falling through 100 m, (a) what work is done by gravity? (b) how much kinetic energy is lost due to the drag force?
52. A 15-kg box is dragged at uniform speed up an incline that is 8.0 m long and makes a 15° angle with the horizontal. If the coefficient of friction between the box and the incline is 0.40, how much work is done by (a) the applied force, (b) the normal force, and (c) the gravitational force? (d) How much energy is lost due to friction?

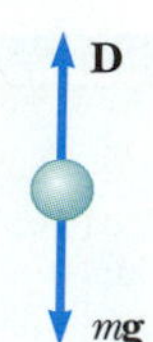

Figure 7.19 (Problem 51)

53. A bartender in a western saloon slides a bottle of rye on the horizontal counter to a cowboy at the other end of the bar, 7 m away. With what speed does he release the bottle if the coefficient of sliding friction is 0.1 and the bottle comes to rest in front of the cowboy?
54. The force $F_x(x)$, depicted graphically in Figure 7.20, acts on a particle as it moves along the x axis. Use graphical techniques to estimate the work done by this force as the particle on which it acts moves from $x = 0$ to $x = 8.0$ m.

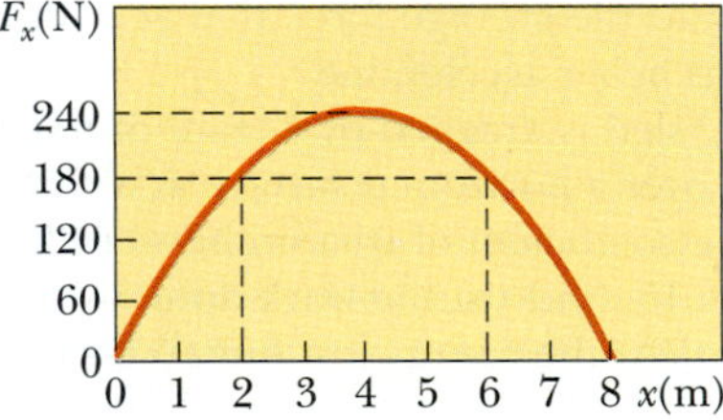

Figure 7.20 (Problem 54)

55. The direction of an arbitrary vector **A** can be completely specified with the angles α, β, and γ that the vector makes with the x, y, and z axes, respectively. If $\mathbf{A} = A_x\mathbf{i} + A_y\mathbf{j} + A_z\mathbf{k}$, (a) find expressions for $\cos \alpha$, $\cos \beta$, and $\cos \gamma$ (these are known as *direction cosines*), and (b) show that these angles satisfy the relation $\cos^2 \alpha + \cos^2 \beta + \cos^2 \gamma = 1$. (*Hint:* Take the scalar product of **A** with **i**, **j**, and **k** separately.)
56. A 4-kg particle moves along the x axis. Its position varies with time according to $x = t + 2t^3$, where x is in meters and t is in seconds. Find (a) the kinetic energy at any time t, (b) the acceleration of the particle and the force acting on it at time t, (c) the power being delivered to the particle at time t, and (d) the work done on the particle in the interval $t = 0$ to $t = 2$ s. (*Note:* $P = dW/dt$.)
57. A pile driver of mass 2100 kg is used to drive a steel I-beam into the ground. The mass falls freely from

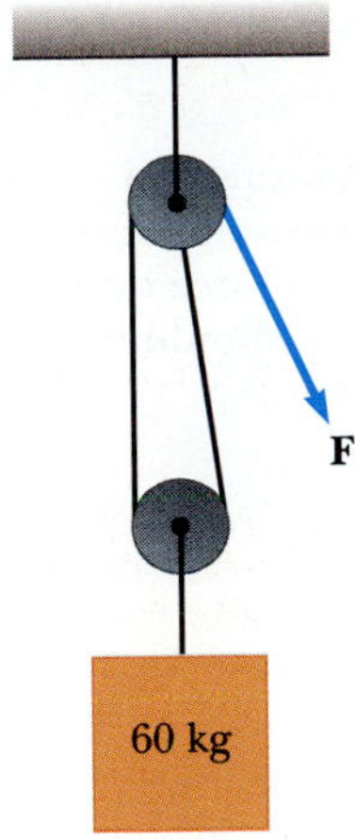

Figure 7.21 (Problem 61)

rest a distance of 5 m before contacting the beam, and it drives the beam 12 cm into the ground before coming to rest. Using the work-energy relation, calculate the average force that the beam exerts on the mass while the mass is brought to rest.

58. A rope tow that is pulling skiers up a 600-m-long 30° slope moves at 3 m/s and carries a maximum of 120 passengers at any one time. The average mass of each passenger is 80 kg. Neglecting friction, determine the power rating a motor must have in order to operate the tow under maximum load conditions.

59. A 200-g block is pressed against a spring of force constant 1400 N/m until the block compresses the spring 10 cm. The spring rests at the bottom of a ramp inclined at 60° to the horizontal. Determine how far up the incline the block moves before momentarily coming to rest, (a) if there is no friction between the block and the ramp and (b) if the coefficient of kinetic friction is 0.4.

60. A 0.4-kg particle slides on a horizontal, circular track 1.5 m in radius. It is given an initial speed of 8 m/s.

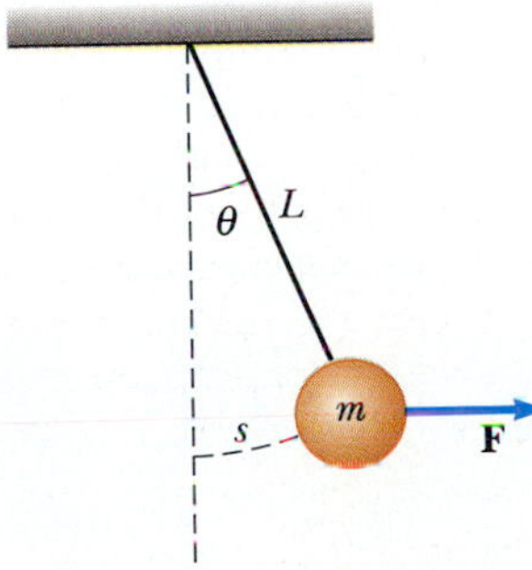

Figure 7.22 (Problem 62)

Figure 7.23 (Problem 63)

After one revolution, its speed drops to 6 m/s because of friction. (a) Find the kinetic energy lost because of friction in one revolution. (b) Calculate the coefficient of kinetic friction. (c) What is the total number of revolutions the particle makes before coming to rest?

61. A 60-kg load is raised by a two-pulley arrangement like that in Figure 7.21. How much work is done by the force **F** to raise the load 3 m if there is a frictional force of 20 N in each pulley? (The pulleys do not rotate, but the rope slides across each surface.)

62. A small sphere of mass m hangs from a string of length L, as in Figure 7.22. A variable horizontal force **F** is applied to the mass in such a way that it moves slowly from the vertical position until the string makes an angle θ with the vertical. Assuming the sphere is always in equilibrium, (a) show that $F = mg \tan \theta$, and (b) make use of the expression $W = \int \mathbf{F} \cdot d\mathbf{s}$ to show that the work done by the force **F** is equal to $mgL(1 - \cos \theta)$. (*Hint:* Note that $s = L\theta$, and so $ds = L\, d\theta$.)

63. The ball launcher in a pinball machine has a spring with a force constant of 1.2 N/cm (Fig. 7.23). The surface on which the ball moves is inclined 10° with respect to the horizontal. If the spring is initially compressed 5 cm, find the launching speed of a 100-g ball when the plunger is released. Friction and the mass of the plunger are negligible.

64. Suppose a car is modeled as a cylinder moving with a speed v, as in Figure 7.24. In a time Δt, a column of air of mass Δm must be moved a distance $v\,\Delta t$ and hence must be given a kinetic energy $\frac{1}{2}(\Delta m)v^2$. Using this model, show that the power loss due to air resistance is $\frac{1}{2}\rho A v^3$ and the drag force is $\frac{1}{2}\rho A v^2$, where ρ is the density of air.

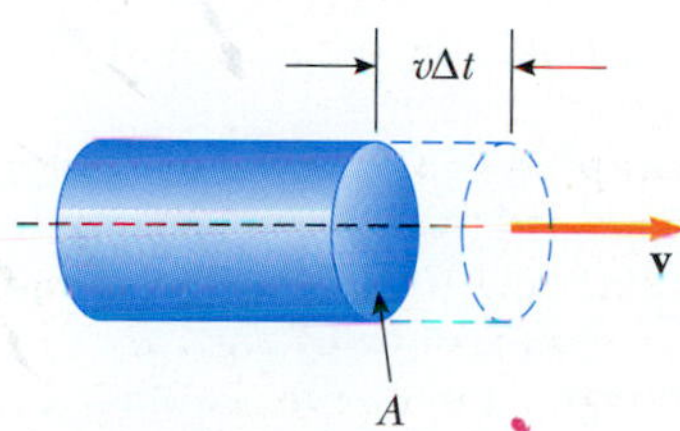

Figure 7.24 (Problem 64)

Calculator/Computer Problems

65. A 5-kg particle starts at the origin and moves along the x axis. The net force acting on the particle is measured at intervals of 1 m to be 27.0. 28.3, 36.9, 34.0, 34.5, 34.5, 46.9, 48.2, 50.0, 63.5, 13.6, 12.2, 32.7, 46.6, 27.9 (in newtons). Determine the total work done on the particle over this interval.
66. A 0.178-kg particle moves along the x axis from $x = 12.8$ m to $x = 23.7$ m, under the influence of a force given by

$$F = \frac{375}{x^3 + 3.75x}$$

where F is in newtons and x is in meters. Use a method of numerical integration to estimate the total work done by this force during this displacement. Your calculations should have an accuracy within 2% or less.

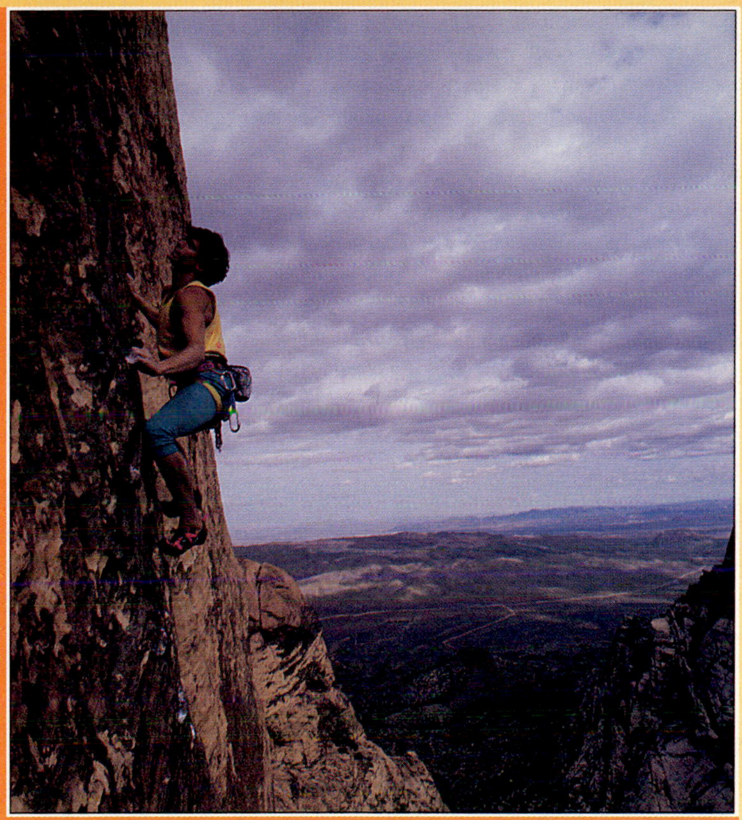

The skillful rock climber in this photograph moves cautiously up the mountain. As the climber ascends the mountain, his internal (biochemical) energy is transformed into gravitational potential energy. Can you identify other forms of energy transformations if he descends the mountain with the help of supporting ropes? *(Greg Epperson)*

Potential Energy and Conservation of Energy

CHAPTER 8

In Chapter 7 we introduced the concept of kinetic energy, which is associated with the motion of an object. We found that a particle's kinetic energy can change only if work is done on the particle. In this chapter we introduce another form of mechanical energy, called *potential energy,* that is associated with the position or configuration of an object. The potential energy of a system can be thought of as stored energy that can either be converted to kinetic energy or do work.

The concept of potential energy can be used only with a special class of forces called *conservative forces*. When only internal conservative forces, such as gravitational or spring forces, act within a system, the kinetic energy gained (or lost) by the system as its members change their relative positions is compensated by equal loss (or gain) in potential energy. This is an application known as the *law of conservation of energy*.

8.1 Potential Energy

In Chapter 7 we saw that an object with kinetic energy can do work on another object, as illustrated by the moving hammer driving a nail into the wall. Now we shall see that an object can also do work because of the energy resulting from its *position* in space.

As an object falls in a gravitational field, the field exerts a force on it in the direction of its motion, doing work on it, and thereby increasing its kinetic energy. Consider a brick dropped from rest directly above a nail in a board that is lying horizontally on the ground. When the brick is released, it falls toward the ground, gaining speed and therefore gaining kinetic energy. Due to its position in space, the brick has potential energy (it has the *potential* to do work), which is converted into kinetic energy as it falls. When the brick reaches the ground, it does work on the nail, driving it into the board. The energy that an object has due to its position in space is called **gravitational potential energy.** It is energy held by the gravitational field and transferred to the object as it falls.

Let us now derive an expression for the gravitational potential energy of an object at a given location in space. To do this, consider a block of mass m at an initial height y_i above the ground, as in Figure 8.1. With air resistance neglected, as the block falls, the only force that does work on it is the gravitational force, $m\mathbf{g}$. The work done by the gravitational force as the block undergoes a downward displacement $\mathbf{s}$ is given by the product of the downward force mg times the displacement, or

$$W_g = (m\mathbf{g}) \cdot \mathbf{s} = (-mg\mathbf{j}) \cdot (y_f - y_i)\mathbf{j} = mgy_i - mgy_f$$

We now define the quantity mgy to be the gravitational potential energy, U_g:

$$U_g \equiv mgy \qquad [8.1]$$

Thus, the gravitational potential energy associated with an object at any point in space is the product of the object's weight and its vertical coordinate. The origin of the coordinate system could be located at the surface of the Earth or at any other convenient point.

If we substitute U for the mgy terms in the expression for W_g, we have

$$W_g = U_i - U_f \qquad [8.2]$$

From this result, we see that the work done on any object by the force of gravity—that is, the energy transferred to the object from the gravitational field—is equal to the initial value of the potential energy minus the final value of potential energy. For convenience, the potential energy is often "assigned to" the object, which includes the gravitational field as part of the system. However, when this choice is made, one must be careful *not* to include the work done by the field on the object; the field no longer exerts an external force.

The units of gravitational potential energy are the same as those of work. That is, potential energy may be expressed in joules, ergs, or foot-pounds. Potential energy, like work and kinetic energy, is a scalar quantity.

Note that the gravitational potential energy associated with an object depends only on the vertical height of the object above the surface of the Earth. From this

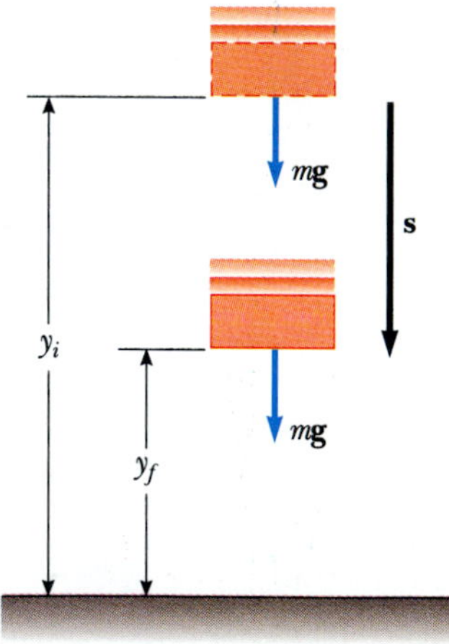

Figure 8.1 The work done by the gravitational force as the block falls from y_i to y_f is equal to $mgy_i - mgy_f$.

result, we see that the same amount of work is done on an object if it falls vertically to the Earth as if it starts at the same point and slides down a frictionless incline to the Earth. Also note that Equation 8.1 is valid only for objects near the surface of the Earth, where g is approximately constant.

In working problems involving gravitational potential energy, it is always necessary to choose a location at which to set the gravitational potential energy equal to zero. The choice of zero level is completely arbitrary, because the important quantity is the *difference* in potential energy, and this difference is independent of the choice of zero level.

It is often convenient to choose the surface of the Earth as the reference position for zero potential energy, but again this is not essential. Often, the statement of the problem suggests a convenient level to use. The following example illustrates this important point.

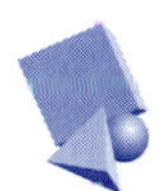

Example 8.1 Wax Your Skis

A novice skier of mass 60 kg is at the top of a slope like that in Figure 8.2. At the initial point, A, the skier is 10 m vertically above point B.

(a) Setting the zero level for gravitational potential energy at B, first find the gravitational potential energy of the skier at A and B and then find the difference in potential energy between these two points.

Solution The gravitational potential energy at B is zero by choice. Hence, the potential energy at the initial point is

$$U_i = mgy_i = (60\text{ kg})(9.80\text{ m/s}^2)(10\text{ m}) = \boxed{5880\text{ J}}$$

Since $U_f = 0$, the difference in potential energy is

$$U_i - U_f = 5880\text{ J} - 0\text{ J} = \boxed{5880\text{ J}}$$

(b) Repeat this problem with the zero level at point A.

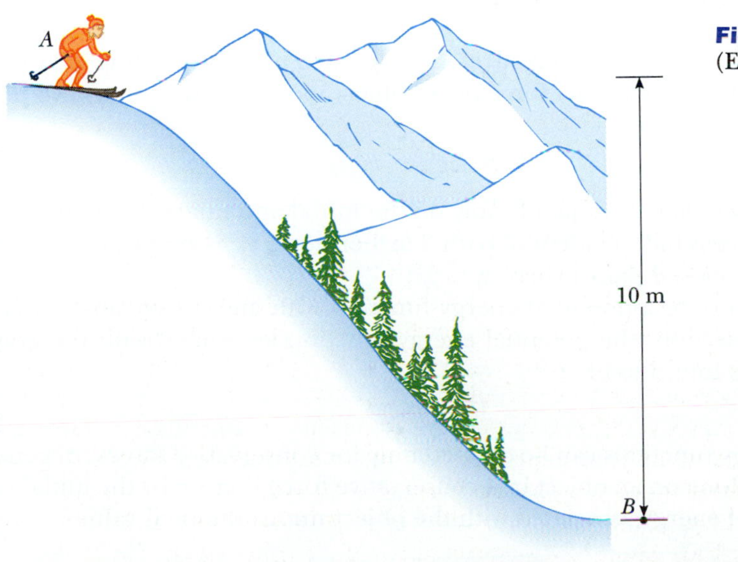

Figure 8.2 (Example 8.1)

Solution In this case, the initial potential energy is zero because of the choice of the reference level. The final potential energy is

$$U_f = mgy_f = (60\ \text{kg})(9.80\ \text{m/s}^2)(-10\ \text{m}) = \boxed{-5880\ \text{J}}$$

Note that the distance y_f is -10 m because the final point is 10 m *below* the zero reference level. The difference in potential energy is

$$U_i - U_f = 0\ \text{J} - (-5880\ \text{J}) = \boxed{5880\ \text{J}}$$

These calculations show that the potential energy of the skier at the top of the slope is greater than the potential energy at the bottom by 5880 J, *regardless of the zero level selected.*

Exercise If the zero level is selected to be midway down the slope, at a height of 5 m, find the initial potential energy, the final potential energy, and the difference in potential energy between points A and B.

Answer 2940 J, -2940 J, 5880 J.

8.2 Conservative and Nonconservative Forces

Forces found in nature can be divided into two categories: conservative and nonconservative. We shall describe the properties of conservative and nonconservative forces separately in this section.

Conservative Forces

A force is conservative if the work it does on an object moving between any two points is independent of the path taken by the object. The work done by a conservative force depends only on the initial and final coordinates of the object. A conservative force can also be defined in a second way. *A force is conservative if the work it does on an object moving through any closed path is zero.*

The force of gravity is conservative. As we learned in the preceding section, the work done by the gravitational force on an object moving between any two points near the Earth's surface is

$$W_g = mgy_i - mgy_f$$

From this, we see that W_g depends only on the initial and final coordinates of the object and hence is independent of path. Furthermore, W_g is zero when the object moves over any closed path (where $y_i = y_f$).

We can associate a potential energy function with any conservative force. In the preceding section, the potential energy function associated with the gravitational force was found to be

$$U_g = mgy$$

Potential energy functions can be defined only for conservative forces. In general, the work, W_c, done on an object by a conservative force is given by the initial value of the potential energy associated with the object minus the final value:

$$W_c = U_i - U_f \quad [8.3]$$

The gravitational potential energy is the energy stored in the gravitational field when the object is lifted against the field.

Another example of a conservative force is the force of a spring on an object attached to the spring, where the spring force is given by $F_s = -kx$. As we learned in Chapter 7 (Eq. 7.15), the work done by the spring force is

$$W_s = \tfrac{1}{2}kx_i^2 - \tfrac{1}{2}kx_f^2$$

where the initial and final coordinates of the object are measured from its equilibrium position, $x = 0$. Again we see that W_s depends only on the initial and final coordinates of the object and is zero for any closed path. Hence, the spring force is conservative. The **elastic potential energy** function associated with the spring force is defined by

$$U_s = \tfrac{1}{2}kx^2 \quad [8.4]$$

The elastic potential energy can be thought of as the energy stored in the deformed spring (one that is either compressed or stretched from its equilibrium position). To visualize this, consider Figure 8.3, which shows an undeformed spring on a frictionless, horizontal surface. When the block is pushed against the spring (Fig. 8.3b), compressing the spring a distance x, the elastic potential energy stored in the spring is $kx^2/2$. When the block is released, the spring snaps back to its original length and the stored elastic potential energy is transformed into kinetic energy of the block (Fig. 8.3c). The elastic potential energy stored in the spring is zero whenever the spring is undeformed ($x = 0$). Energy is stored in the

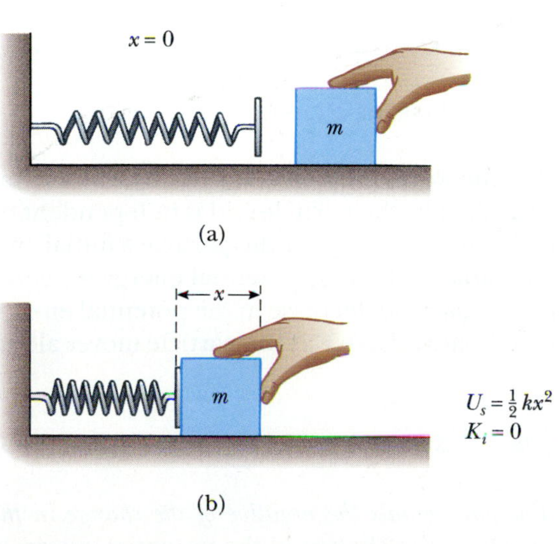

Figure 8.3
A block of mass m on a horizontal, frictionless surface is pushed against a spring and then released. If x is the compression in the spring as in (b), the elastic potential energy stored in the spring is $\frac{1}{2}kx^2$. This energy is transferred to the block in the form of kinetic energy as in (c).

spring only when the spring is either stretched or compressed. Furthermore, the elastic potential energy is a maximum when the spring has reached its maximum compression or extension (that is, when $|x|$ is a maximum). Finally, since the elastic potential energy is proportional to x^2, we see that U_s is always positive in a deformed spring. If the spring and object are taken together as the system, then no work is done as the spring changes length because the forces are internal.

Nonconservative Forces

Definition of a nonconservative force

A force is called *nonconservative* if it leads to dissipation of mechanical energy. For example, if you move an object on a horizontal surface, returning to the same location and same state of motion, but you found that it was necessary to do a net amount of work on the object, then something must have dissipated that energy transferred to the object. That dissipative force is recognized as friction between the surface and the object. Friction is a dissipative, or "nonconservative," force. By contrast, if the object is lifted, work is required, but that energy is recovered when the object is lowered. The gravitational force is a nondissipative or "conservative" force.

Suppose you displace a book between two points on a table. If the book is displaced in a straight line along the blue path between points A and B in Figure 8.4, the loss in mechanical energy due to friction is simply $-fd$, where d is the distance between the two points. However, if the book is moved along any other path between the two points, the loss in mechanical energy due to friction is greater (in absolute magnitude) than $-fd$. For example, the loss in mechanical energy due to friction along the red semicircular path in Figure 8.4 is equal to $-f(\pi d/2)$, where d is the diameter of the circle.

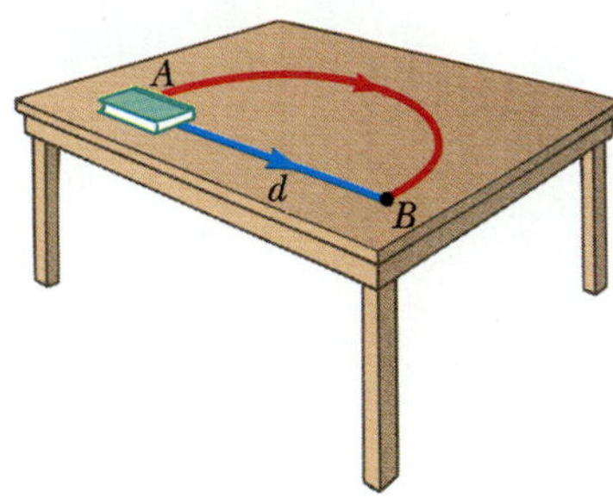

Figure 8.4
The loss in mechanical energy due to the force of friction depends on the path taken as the book is moved from A to B, and hence friction is a nonconservative force. The loss in mechanical energy is greater along the red path compared to the blue path.

8.3 Conservative Forces and Potential Energy

In the preceding section we found that the work done on a particle by a conservative force does not depend on the path taken by the particle and is independent of the particle's velocity. The work done is a function only of the particle's initial and final coordinates. Consequently, we are able to define a potential energy function U such that the work done on a particle equals the decrease in the potential energy of the particle. The work done by a conservative force **F** as the particle moves along the x axis can be expressed as[1]

$$W_c = \int_{x_i}^{x_f} F_x\,dx = -\Delta U = U_i - U_f \qquad [8.5]$$

That is, *the work done by a conservative force equals the negative of the change in the potential energy associated with that force,* where the change in the potential energy is defined as $\Delta U = U_f - U_i$. For example, the work, W_c, done by the gravitational field on an object, as the object is lowered in the field, is $U_i - U_f$, which is positive, showing that energy has been transferred from the gravitational field to the object. This energy may appear as kinetic energy of the falling object or may be transferred to something else. We can also express Equation 8.5 as

[1] For a general displacement, the work done in two or three dimensions also equals $U_i - U_f$, where $U = U(x, y, z)$. We write this formally as $W = \int_i^f \mathbf{F}\cdot d\mathbf{s} = U_i - U_f$.

$$\Delta U = U_f - U_i = -\int_{x_i}^{x_f} F_x \, dx \qquad [8.6]$$

Change in potential energy

where F_x is the component of **F** in the direction of the displacement: **F** is the force exerted by the field on the object. Therefore ΔU is negative when F_x and dx are in the same direction, as when an object is lowered in a gravitational field or a spring pushes an object toward equilibrium.

It is often convenient to establish some particular location, x_i, to be a reference point and measure all potential energy differences with respect to that point. We can then define the potential energy function as

$$U_f = -\int_{x_i}^{x_f} F_x \, dx + U_i \qquad [8.7]$$

Furthermore, as we discussed earlier, the value of U_i is often taken to be zero at some arbitrary reference point. It really doesn't matter what value we assign to U_i, since any value only shifts U_f by a constant, and it is only the *change* in potential energy that is physically meaningful. If the conservative force is known as a function of position, we can use Equation 8.7 to calculate the change in potential energy of a body as it moves from x_i to x_f. It is interesting to note that in one dimension a force is *always* conservative if it is a function of position only. This is generally not the case for motion involving two- or three-dimensional displacements, however.

The amount of mechanical energy dissipated by a nonconservative force depends on the path as an object moves from one position to another, and can also depend on the object's speed or on other quantities. Therefore, the work done is not simply a function of the initial and final coordinates of the particle. We conclude that there is no potential energy *function* associated with a nonconservative force.

8.4 Conservation of Energy

An object held at some height h above the floor has no kinetic energy, but, as we learned earlier, there is an associated gravitational potential energy equal to mgh, relative to the floor if the gravitational field is included as part of the system. If the object is dropped, it falls to the floor; as it falls, its speed and thus its kinetic energy increase while the potential energy decreases. If factors such as air resistance are ignored, whatever potential energy the object loses as it moves downward appears as kinetic energy. In other words, the sum of the kinetic and potential energies, called the *mechanical energy* E, remains constant in time. This is an example of the **law of conservation of energy**. For the case of a freely falling object, the law tells us that any increase (or decrease) in potential energy is accompanied by an equal decrease (or increase) in kinetic energy.

Since the total mechanical energy E is defined as the sum of the kinetic and potential energies, we can write

$$E \equiv K + U \qquad [8.8]$$

Total mechanical energy

Therefore, we can apply conservation of energy in the form $E_i = E_f$, or

$$K_i + U_i = K_f + U_f \qquad [8.9]$$

Conservation of energy

A formal statement of conservation of energy

More formally, *conservation of energy requires that the total mechanical energy of a system remains constant in any isolated system of objects which interact only through conservative forces.* It is important to note that Equation 8.9 is valid *provided* no energy is added to or removed from the system. Furthermore, there must be no nonconservative forces within the system.

Since mechanical energy E remains constant with time, $dE/dt = 0$. Taking the derivative of Equation 8.8 with respect to time,

$$\frac{dE}{dt} = 0 = \frac{dK}{dt} + \frac{dU}{dt} \qquad [8.10]$$

Since $K = \frac{1}{2}mv^2$, then

$$\frac{dK}{dt} = \frac{d}{dt}\left(\tfrac{1}{2}mv^2\right) = mv\frac{dv}{dt} = mva = F_x v$$

Applying the chain rule (see Appendix B.6) to dU/dt, we have

$$\frac{dU}{dt} = \frac{dU}{dx}\frac{dx}{dt} = \left(\frac{dU}{dx}\right)v$$

Substituting these expressions for dK/dt and dU/dt into Equation 8.10 gives

$$F_x v + \left(\frac{dU}{dx}\right)v = 0$$

or

Relation between a conservative force and potential energy

$$F_x = -\frac{dU}{dx} \qquad [8.11]$$

That is, *the conservative internal force acting on a part of the system equals the negative derivative of the potential energy associated with that system.*

We can easily check this relationship for the two instances already discussed. In the case of the deformed spring, $U_s = \frac{1}{2}kx^2$, and therefore

$$F_s = -\frac{dU_s}{dx} = -\frac{d}{dx}\left(\tfrac{1}{2}kx^2\right) = -kx$$

which corresponds to the restoring force exerted by the spring. In the case of an object located a distance y above some reference point, the gravitational potential energy function is given by $U_g = mgy$, and it follows from Equation 8.11 that $F_g = -mg$.

We now see that U is an important function; from it can be derived the conservative force acting in any system. Furthermore, Equation 8.11 should clarify the fact that adding a constant to the potential energy is unimportant, since the location of the reference point is arbitrary.

Equation 8.11 can also be written in the form $dU = -Fdx$, which, when integrated between the initial and final position values, gives

$$U_f - U_i = -\int_{x_i}^{x_f} F\,dx \qquad [8.12]$$

This result, which is identical to Equation 8.6, tells us that if the conservative force F acting on an object within a system is known as a function of x, we can calculate the *difference* in the potential energy associated with the object between the initial and final positions.

If more than one conservative force acts on the object, then a potential energy function is associated with *each* force. In such a case, we can apply the law of conservation of energy for the system as

Conservation of mechanical energy

$$K_i + \sum U_i = K_f + \sum U_f \quad [8.13]$$

where the number of terms in the sums equals the number of conservative forces present. For example, if a mass connected to a spring oscillates vertically, two conservative forces act on it: the spring force and the force of gravity. (We will discuss this situation later, in a worked example.)

If the force of gravity is the *only* force acting on a body, then the total mechanical energy of the body is constant. Therefore, the law of conservation of energy for a freely falling body can be written

Conservation of mechanical energy for a freely falling body

$$\tfrac{1}{2}mv_i^2 + mgy_i = \tfrac{1}{2}mv_f^2 + mgy_f \quad [8.14]$$

Example 8.2 Ball in Free Fall

A ball of mass m is dropped from a height h above the ground, as in Figure 8.5.

(a) Neglecting air resistance, determine the speed of the ball when it is at a height y above the ground.

Solution Since the ball is in free fall, the only force acting on it is the gravitational force. We can therefore use the law of conservation of energy. When the ball is released from rest at a height h above the ground, its kinetic energy is $K_i = 0$ and its potential energy is $U_i = mgh$, where the y coordinate is measured from ground level. When the ball is at a distance y above the ground, its kinetic energy is $K_f = \frac{1}{2}mv_f^2$ and its potential energy relative to the ground is $U_f = mgy$. Since mechanical energy is constant, we get

$$K_i + U_i = K_f + U_f$$
$$0 + mgh = \tfrac{1}{2}mv_f^2 + mgy$$
$$v_f^2 = 2g(h - y)$$
$$v_f = \sqrt{2g(h - y)}$$

(b) Determine the speed of the ball at y if it is given an initial speed v_i at the initial altitude h.

Solution In this case, the initial energy includes kinetic energy equal to $\frac{1}{2}mv_i^2$, and Equation 8.14 gives

$$\tfrac{1}{2}mv_i^2 + mgh = \tfrac{1}{2}mv_f^2 + mgy$$
$$v_f^2 = v_i^2 + 2g(h - y)$$
$$v_f = \sqrt{v_i^2 + 2g(h - y)}$$

This result is consistent with the expression $v_y^2 = v_{y0}^2 - 2g(y - y_0)$, from kinematics, where $y_0 = h$. Furthermore, this result is valid even if the initial velocity is at an angle to the horizontal (the projectile situation).

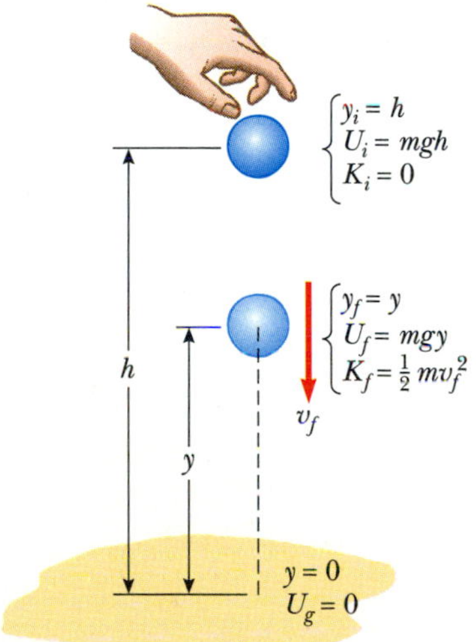

Figure 8.5 (Example 8.2) A ball is dropped from a height h above the floor. Initially, its total energy is potential energy, equal to mgh relative to the floor. At the elevation y, its total energy is the sum of kinetic and potential energies.

8.5 Changes in Mechanical Energy When Nonconservative Forces are Present

In real physical systems, nonconservative forces, such as friction, are usually present. Such forces remove mechanical energy from the system; therefore the total mechanical energy is not constant in these systems. In general, we cannot calculate the work done by nonconservative forces, but we can find the change in kinetic energy of the system, acted upon by a net force, using the net-force equation:

$$\int \mathbf{F}_{\text{net}} \cdot d\mathbf{x} = \Delta K \qquad [8.15]$$

Since the change in kinetic energy can be the result of many types of forces, it is convenient to separate ΔK into three parts:

1. The change in kinetic energy due to internal conservative forces, $\Delta K_{\text{int-c}}$
2. The change in kinetic energy due to internal nonconservative forces, $\Delta K_{\text{int-nc}}$
3. The change in kinetic energy due to external forces (conservative or nonconservative), ΔK_{ext}.

The first of these is $\Delta K_{\text{int-c}} = W_c = -\Delta U$. This is simply an exchange of potential energy into kinetic energy within the system. The second term, $\Delta K_{\text{int-nc}}$, may be positive or negative; if it represents internal friction, it will be negative; however, it can be positive (as in the case of a muscle operating on biochemical energy). The last term is work done if the external forces are conservative, but more generally is ΔK_{ext}, the change in kinetic energy of the system due to external forces. Thus, we have

$$\Delta K = \Delta K_{\text{int-c}} + \Delta K_{\text{int-nc}} + \Delta K_{\text{ext}}$$

or

$$\Delta K + \Delta U = \Delta K_{\text{int-nc}} + \Delta K_{\text{ext}} \qquad [8.16]$$

From this we see that if there are no internal nonconservative forces and no external forces acting on the system, then the right side of Equation 8.15 is zero, and the sum $K + U$ is constant, which is consistent with Equation 8.9. If internal nonconservative forces are present, they may increase or decrease the kinetic energy. Similarly, the external forces may either increase or decrease the kinetic energy of the system.

Problem-Solving Strategies

As we have seen, many problems in physics can be solved using the principle of conservation of energy. The following procedure should be used when you apply this principle:

1. **Define your system, which may consist of more than one object and may or may not include fields, springs, or other sources of potential energy.**
2. **Select a reference position for the zero point of potential energy (both gravitational and spring), and use this throughout your analysis.**

If there is more than one conservative force, write an expression for the potential energy associated with each force.

3. **Determine whether any nonconservative forces are present. Remember that if friction or air resistance is present, mechanical energy *is not constant.***
4. **If mechanical energy is *constant*, you can write the total initial energy, E_i, at some point as the sum of the kinetic and potential energy at that point. Then write an expression for the total final energy, $E_f = K_f + U_f$, at the final point that is of interest. Since mechanical energy is *constant*, you can equate the two total energies and solve for the quantity that is unknown.**
5. **If external forces or frictional forces are present (and thus mechanical energy is not *constant*), first write expressions for the total initial and total final energies. In this case, however, the total final energy differs from the total initial energy, the difference being the amount of energy dissipated by nonconservative forces. That is, you should apply Equation 8.15.**

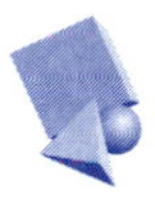

Example 8.3 Crate Sliding Down a Ramp

A 3-kg crate slides down a ramp at a loading dock. The ramp is 1 m in length, and inclined at an angle of 30°, as shown in Figure 8.6. The crate starts from rest at the top, experiences a constant frictional force of magnitude 5 N, and continues to move a short distance on the flat floor. Use energy methods to determine the speed of the crate just before it reaches the bottom of the ramp.

Solution Because $v_i = 0$, the initial kinetic energy is zero. If the y coordinate is measured from the bottom of the ramp, then $y_i = 0.50$ m. Therefore, the total mechanical energy of the crate at the top is all potential energy, given by

$$U_i = mgy_i = (3 \text{ kg})(9.80 \text{ m/s}^2)(0.50 \text{ m}) = 14.7 \text{ J}$$

When the crate reaches the bottom, its potential energy is zero because its elevation is $y_f = 0$. Therefore, the total mechanical energy at the bottom is all kinetic energy:

$$K_f = \tfrac{1}{2}mv_f^2$$

However, we cannot say that $U_i = K_f$ in this case, because there is an external nonconservative force that removes mechanical energy from the crate: the force of friction. In this case, $\Delta K_{\text{ext}} = -fs$, where s is the displacement along the ramp. (Re-

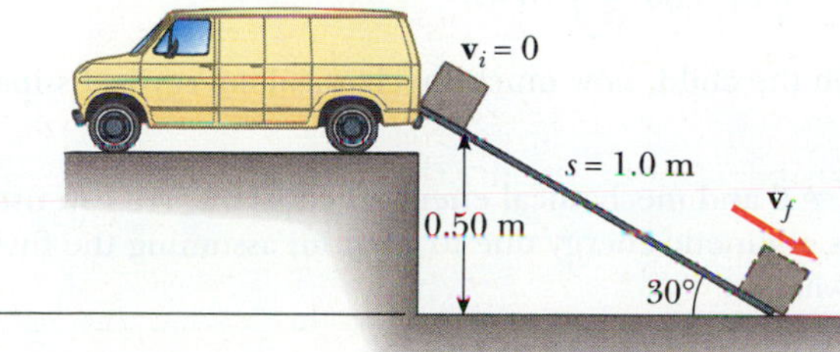

Figure 8.6
(Example 8.3) A crate slides down an incline under the influence of gravity. Its potential energy decreases while its kinetic energy increases.

call that the forces normal to the ramp do no work on the crate because they are perpendicular to the displacement.) With $f = 5$ N and $s = 1$ m, we have

$$\Delta K_{\text{ext}} = -fs = (-5\text{ N})(1\text{ m}) = -5\text{ J}$$

This says that some mechanical energy is lost because of the presence of the retarding frictional force. Applying Equation 8.15 gives

$$-fs = \tfrac{1}{2}mv_f^2 - mgy_i$$

$$\tfrac{1}{2}mv_f^2 = 14.7\text{ J} - 5\text{ J} = 9.7\text{ J}$$

$$v_f^2 = \frac{19.4\text{ J}}{3\text{ kg}} = 6.47\text{ m}^2/\text{s}^2$$

$$v_f = 2.54\text{ m/s}$$

Exercise Use Newton's second law to find the acceleration of the crate along the ramp, and the equations of kinematics to determine the final speed of the crate.

Answer 3.23 m/s²; 2.54 m/s.

Exercise If the ramp is assumed to be frictionless, find the final speed of the crate and its acceleration along the ramp.

Answer 3.13 m/s; 4.90 m/s².

Example 8.4 Motion on a Curved Track

A child of mass m takes a ride on an irregularly curved slide of height 6 m, as in Figure 8.7. The child starts from rest at the top.

(a) Determine the speed of the child at the bottom, assuming no friction is present.

Solution First, note that the normal force, n, does no work on the child since this force is always perpendicular to each element of the displacement. Furthermore, since there is no friction, mechanical energy is constant; $K + U =$ constant. If we measure the y coordinate from the bottom of the slide, then $y_i = h$, $y_f = 0$, and we get

$$K_i + U_i = K_f + U_f$$

$$0 + mgh = \tfrac{1}{2}mv_f^2 + 0$$

$$v_f = \sqrt{2gh}$$

Note that the result is the same as it would be if the child fell vertically through a distance h! For example, if $h = 6$ m, then

$$v_f = \sqrt{2gh} = \sqrt{2\left(9.80\ \frac{\text{m}}{\text{s}^2}\right)(6\text{ m})} = 10.8\text{ m/s}$$

(b) If a frictional force acts on the child, how much mechanical energy is dissipated by this force?

Solution In this case, $\Delta K_{\text{ext}} \neq 0$ and mechanical energy is constant. We can use Equation 8.16 to find the loss of kinetic energy due to friction, assuming the final velocity at the bottom is known:

$$\Delta K_{\text{ext}} = E_f - E_i = \tfrac{1}{2}mv_f^2 - mgh$$

For example, if $v_f = 8.0$ m/s, $m = 20$ kg, and $h = 6$ m, we find that

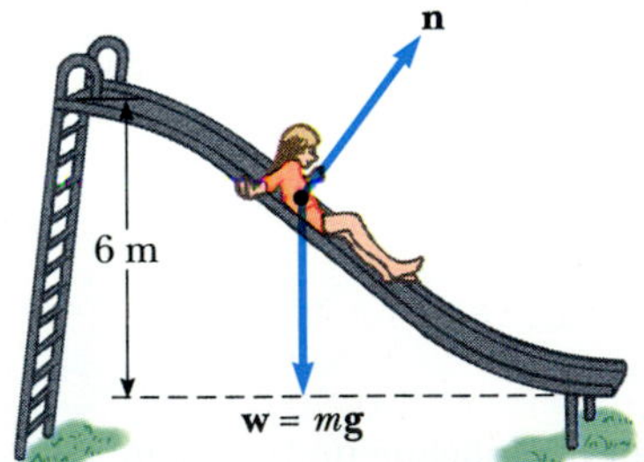

Figure 8.7 (Example 8.4) If the slide is frictionless, the speed of the child at the bottom depends only on the height of the slide and is independent of the shape of the slide.

$$\Delta K_{ext} = \tfrac{1}{2}(20\ \text{kg})(8.0\ \text{m/s})^2 - (20\ \text{kg})(9.80\ \text{m/s}^2)(6\ \text{m})$$

$$= -536\ \text{J}$$

Again, ΔK_{ext} is negative since friction is removing kinetic energy from the system. Note, however, that because the slide is curved, the normal force changes in magnitude and direction during the motion. Therefore, the frictional force, which is proportional to n, also changes during the motion. Do you think it would be possible to determine μ from these data?

Example 8.5 Let's Go Skiing

A skier starts from rest at the top of a frictionless incline of height 20 m, as in Figure 8.8. At the bottom of the incline, the skier encounters a horizontal surface where the coefficient of kinetic friction between the skis and the snow is 0.21. How far does the skier travel on the horizontal surface before coming to rest?

Solution First, let us calculate the speed of the skier at the bottom of the incline. Since the incline is frictionless, we can apply conservation of energy to find

$$v = \sqrt{2gh} = \sqrt{2(9.80\ \text{m/s}^2)(20\ \text{m})} = 19.8\ \text{m/s}$$

Now we apply the net force equation as the skier moves along the rough horizontal surface. The change in kinetic energy along the horizontal is $\Delta K_{ext} = -fs$, where s is the horizontal displacement. Therefore,

$$\Delta K_{ext} = -fs = K_f - K_i$$

To find the distance the skier travels before coming to rest, we take $K_f = 0$. Since $v_i = 19.8$ m/s, and the frictional force is given by $f = \mu n = \mu mg$, we get

$$-\mu mgs = -\tfrac{1}{2}mv_i^2$$

or

$$s = \frac{v_i^2}{2\mu g} = \frac{(19.8\ \text{m/s})^2}{2(0.21)(9.80\ \text{m/s}^2)} = 95.2\ \text{m}$$

Exercise Find the horizontal distance the skier travels before coming to rest if the incline has a coefficient of kinetic friction equal to 0.21.

Answer 40.3 m.

Figure 8.8 Example 8.5

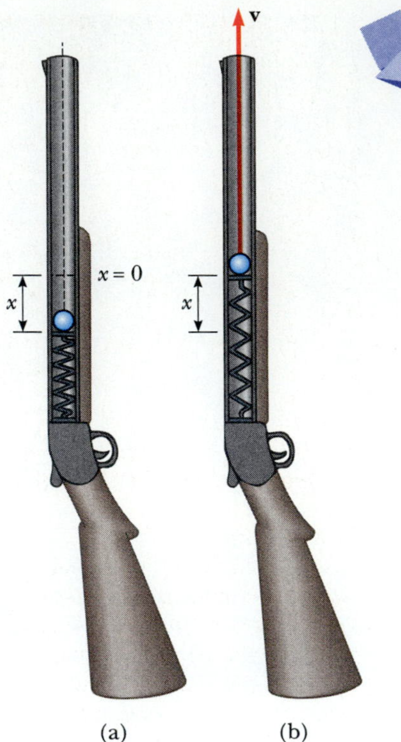

Figure 8.9 Example 8.6

Example 8.6 The Spring-Loaded Popgun

The launching mechanism of a toy gun consists of a spring of unknown spring constant (Fig. 8.9a). When the spring is compressed a distance of 0.12 m, the gun can launch a 20-g projectile to a maximum height of 20 m when fired vertically from rest. Neglect all resistive forces.

(a) Determine the value of the spring constant.

Solution Since the projectile starts at rest, the initial kinetic energy in the system is zero. If the reference of gravitational potential energy is set at the lowest position of the projectile, then the projectile's initial gravitational potential energy is also zero. Hence, the total initial energy of the system is the elastic potential energy stored in the spring, which is $kx^2/2$, where $x = 0.12$ m. Since the projectile rises to a maximum height of $h = 20$ m, its final gravitational potential energy is equal to mgh, its final kinetic energy is zero, and the final elastic potential energy is zero. Since there are no external or nonconservative forces, conservation of energy can be applied to give

$$\tfrac{1}{2}kx^2 = mgh$$

$$\tfrac{1}{2}k(0.12 \text{ m})^2 = (0.02 \text{ kg})(9.80 \text{ m/s}^2)(20 \text{ m})$$

$$k = \boxed{544 \text{ N/m}}$$

(b) Find the speed of the projectile as it moves through the equilibrium position of the spring (where $x = 0$), as shown in Figure 8.9b.

Solution Using the same reference level for the gravitational potential energy as in part (a), we see that the initial energy of the system is still the elastic potential energy $kx^2/2$. The final energy of the system when the projectile moves through the unstretched position of the spring consists of the kinetic energy of the projectile, $mv^2/2$, and the gravitational potential energy of the projectile, mgx. Hence, conservation of energy in this case gives

$$\tfrac{1}{2}kx^2 = \tfrac{1}{2}mv^2 + mgx$$

Solving for v gives

$$v = \sqrt{\frac{kx^2}{m} - 2gx}$$

$$= \sqrt{\frac{(544 \text{ N/m})(0.12 \text{ m})^2}{(0.02 \text{ kg})} - 2(9.80 \text{ m/s}^2)(0.12 \text{ m})}$$

$$= \boxed{19.7 \text{ m/s}}$$

Exercise What is the speed of the projectile when it is at a height of 10 m?

Answer 14.0 m/s.

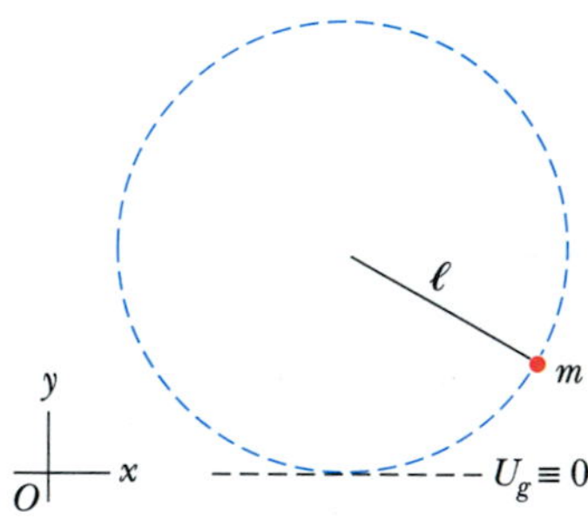

Figure 8.10 Example 8.7

Example 8.7 Ball on a String

A particle of mass m is attached to a string of length ℓ and set in motion in a vertical circular path, as in Figure 8.10. The only forces acting on the ball are the force of gravity and the tension force in the string. The speed v_t at the top of the circle is sufficient to keep the string under tension at all times during its motion. Find the speed of the ball at the bottom of the circle.

Solution Since the ball always moves at right angles to the direction of the tension force T (a force of constraint), the tension force does no work. The only force that does work on the particle is the conservative force of gravity, so we may apply the conservation-of-energy principle. Setting $U_g \equiv 0$ at the bottom, we have

$$E_{\text{top}} = E_{\text{bottom}}$$

$$(U_g)_t + K_t = (U_g)_b + K_b$$

$$mg(2\ell) + \tfrac{1}{2}mv_t^2 = 0 + \tfrac{1}{2}mv_b^2$$

Dividing by m and solving for v_b, we obtain

$$v_b = \pm\sqrt{4g\ell + v_t^2}$$

Twin Falls on the Island of Kauai, Hawaii. The potential energy of the water at the top of the falls is converted into kinetic energy at the bottom. In many locations, this energy is used to produce electrical energy. *(Bruce Byers, FPG International)*

8.6 Conservation of Energy in General

We have seen that the total mechanical energy of a system is constant when only conservative internal forces act within the system. Furthermore, we were able to associate a potential energy function with each conservative force. On the other hand, mechanical energy is lost when nonconservative forces, such as friction, are present.

We can generalize the energy conservation principle to include all forces acting on the system, both conservative and nonconservative. In the study of thermodynamics we shall find that mechanical energy can be transformed into internal energy of the system. For example, when a block slides over a rough surface, the mechanical energy lost is transformed into internal energy temporarily stored in the block and the surface, as evidenced by a measurable increase in the block's temperature. We shall see that, on a submicroscopic scale, this internal energy is associated with the vibration of atoms about their equilibrium positions. Such internal atomic motion has kinetic and potential energy, and so one can say that frictional forces arise fundamentally from conservative atomic forces. Therefore, if we include this increase in the internal energy of the system in our energy expression, the total energy is conserved.

This is just one example of how you can analyze an isolated system and always find that its total energy does not change, as long as you account for all forms of energy. That is, *energy can never be created or destroyed. Energy may be transformed from one form to another, but the total energy of an isolated system is always constant.* From a universal point of view, we can say that the *total energy of the universe is constant:* if one part of the universe gains energy in some form, another part must lose an equal amount of energy. No violation of this principle has been found.

Total energy is always conserved

Other examples of energy transformations include the energy carried by sound waves resulting from the collision of two objects, the energy radiated by an accelerating charge in the form of electromagnetic waves (a radio antenna), and the elaborate sequence of energy conversions in a thermonuclear reaction.

In subsequent chapters we shall see that the energy concept, and especially transformations of energy, unite the branches of physics. The subjects of mechan-

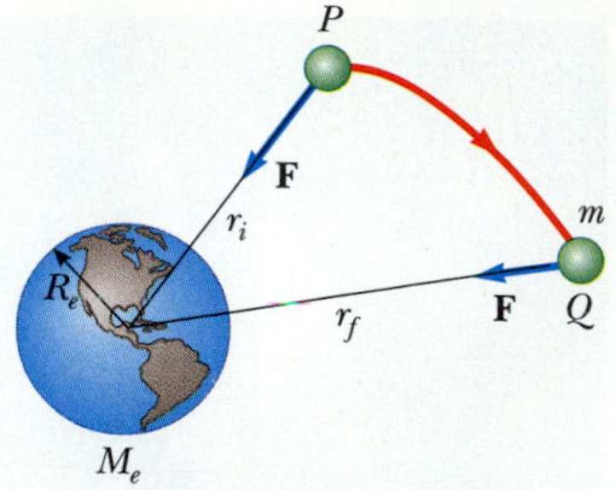

Figure 8.11 As a particle of mass m moves from P to Q above the Earth's surface, the potential energy changes according to Equation 8.18.

ics, thermodynamics, and electromagnetism cannot really be separated. In practical terms, all mechanical and electronic devices rely on energy transformations.

8.7 Gravitational Potential Energy Revisited

Earlier in this chapter we introduced the concept of gravitational potential energy, that is, the energy associated with the position of a particle. We emphasized the fact that the gravitational potential energy function, $U = mgy$, is valid only when a particle is near the Earth's surface. Since the gravitational force between two particles varies as $1/r^2$, it follows that the correct potential energy function of the system depends on the amount of separation between the particles.

Consider a particle of mass m moving between two points P and Q above the Earth's surface, as in Figure 8.11. From Chapter 6, Equation 6.1, the gravitational force acting on m is

$$\mathbf{F}_g = -\frac{GM_e m}{r^2}\hat{\mathbf{r}} \qquad [8.17]$$

where $\hat{\mathbf{r}}$ is a unit vector directed from the Earth to the particle, and the negative sign indicates that the force is attractive. This expression shows that the gravitational force is a *central force,* that is, one that depends only on the polar coordinate r. Furthermore, the gravitational force is conservative. Since the change in potential energy associated with a given displacement of the particle is defined as the negative of the work done by the conservative gravitational force during that displacement, Equation 8.12 gives

$$U_f - U_i = -\int_{r_i}^{r_f} F(r)\, dr = GM_e m \int_{r_i}^{r_f} \frac{dr}{r^2} = GM_e m \left[-\frac{1}{r}\right]_{r_i}^{r_f}$$

or

$$U_f - U_i = -GM_e m\left(\frac{1}{r_f} - \frac{1}{r_i}\right) \qquad [8.18]$$

As always, the choice of a reference point for the potential energy is completely arbitrary. It is customary to locate the reference point where the force is zero. Taking $U_i = 0$ at $r_i = \infty$, we obtain the important result

Gravitational potential energy $r > R_e$

$$U(r) = -\frac{GM_e m}{r} \qquad [8.19]$$

This equation applies to the Earth-particle system separated by a distance r, provided that $r > R_e$. The result is not valid for particles moving inside the Earth, where $r < R_e$. Because of our choice of U_i, the function $U(r)$ is always negative (Fig. 8.12).

Although Equation 8.19 was derived for the particle-Earth system, it can be applied to *any* two particles. That is, the gravitational potential energy associated with *any pair* of particles of masses m_1 and m_2 separated by a distance r is given by

$$U_g = -\frac{Gm_1 m_2}{r} \qquad [8.20]$$

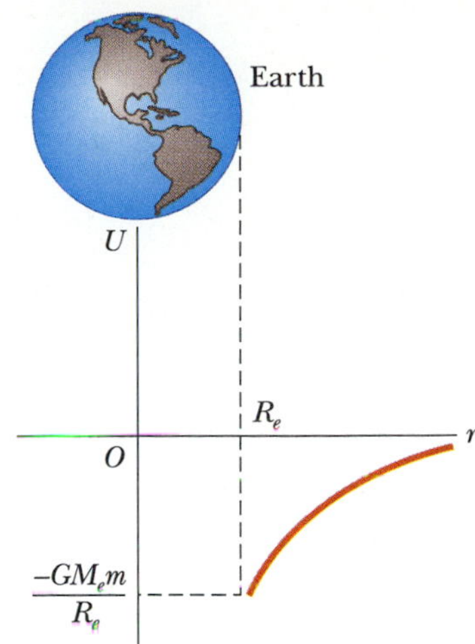

Figure 8.12
Graph of the gravitational potential energy, U_g, versus r for a particle above the Earth's surface. The potential energy goes to zero as r approaches ∞.

This expression also applies to larger objects *if they are spherically symmetric,* as first shown by Newton using integral calculus. Equation 8.20 shows that the gravitational potential energy for any pair of particles varies as $1/r$ (whereas the force between them varies as $1/r^2$). Furthermore, the potential energy is *negative,* since the force is attractive and we have taken the potential energy as zero when the particle separation is infinity. Because the force between the particles is attractive, we know that an external agent must do positive work to increase the separation between the two particles. The work done by the external agent produces an increase in the potential energy as the two particles are separated. That is, U_g becomes less negative as r increases. (Note that part of the work done can also produce a change in kinetic energy of the system. That is, if the work done in separating the particles exceeds the increase in potential energy, the excess energy is accounted for by the increase in kinetic energy of the system.) When the two particles are separated by a distance r, an external agent would have to supply an energy *at least* equal to $+Gm_1 m_2/r$ in order to separate the particles by an infinite distance.

It is convenient to think of the absolute value of the potential energy as the *binding energy* of the system. If the external agent supplies an energy *greater than* the binding energy, $Gm_1 m_2/r$, the additional energy of the system is in the form of kinetic energy when the particles are at an infinite separation.

We can extend this concept to three or more particles. In this case, the total potential energy of the system is the sum over all *pairs* of particles.[2] Each pair contributes a term of the form given by Equation 8.20. For example, if the system contains three particles, as in Figure 8.13, we find that

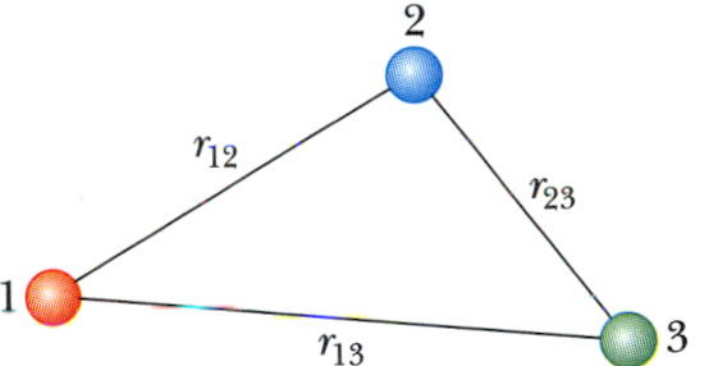

Figure 8.13
Diagram of three interacting particles.

$$U_{\text{total}} = U_{12} + U_{13} + U_{23} = -G\left(\frac{m_1 m_2}{r_{12}} + \frac{m_1 m_3}{r_{13}} + \frac{m_2 m_3}{r_{23}}\right) \qquad [8.21]$$

The absolute value of U_{total} represents the work needed to separate the particles by an infinite distance. If the system consists of four particles, there are six terms in the sum, corresponding to the six distinct pairs of interaction forces.

Example 8.8 The Change in Potential Energy

A particle of mass m is displaced through a small vertical distance Δy near the Earth's surface. Let us show that the general expression for the change in gravitational potential energy given by Equation 8.18 reduces to the familiar relationship $\Delta U_g = mg\,\Delta y$.

Solution We can express Equation 8.18 in the form

$$\Delta U_g = -GM_e m\left(\frac{1}{r_f} - \frac{1}{r_i}\right) = GM_e m\left(\frac{r_f - r_i}{r_i r_f}\right)$$

[2] The fact that one can add potential energy terms for all pairs of particles stems from the experimental fact that gravitational forces obey the superposition principle. That is, if $\Sigma \mathbf{F} = \mathbf{F}_{12} + \mathbf{F}_{13} + \mathbf{F}_{23} + \cdots$, then there exists a potential energy term for each interaction $\mathbf{F}_{ij}$.

If both the initial and final positions of the particle are close to the Earth's surface, then $r_f - r_i = \Delta y$ and $r_i r_f \approx R_e^2$. (Recall that r is measured from the center of the Earth.) Therefore, the *change* in potential energy becomes

$$\Delta U_g \approx \frac{GM_e m}{R_e^2} \Delta y = mg \, \Delta y$$

where we have used the fact that $g = GM_e/R_e^2$. Keep in mind that the reference point is arbitrary, since it is the *change* in potential energy that is meaningful.

8.8 Electric Potential Energy

In this brief section we calculate the potential energy of a charged particle in the presence of a second charged particle and then extend the result to the general case of a system of charged particles.

Let us begin by recalling that the magnitude of the electric (Coulomb) force between two charges q_1 and q_2 separated by a distance r (Fig. 8.14) is given by Equation 6.4 in Chapter 6:

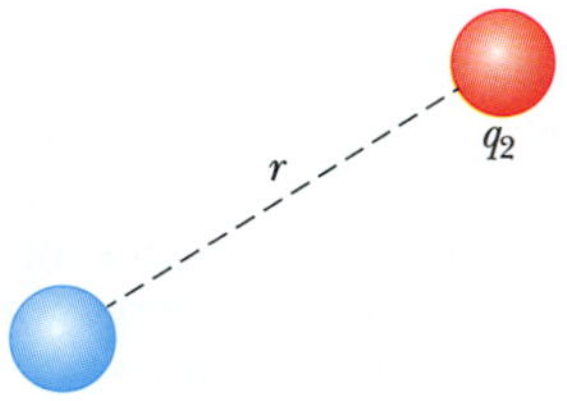

$$F_e = k_e \frac{q_1 q_2}{r^2}$$

r q_2 q_1

Figure 8.14
If two charges are separated by a distance r, the electric potential energy of the pair of charges is given by $k_e q_1 q_2 / r$.

where k_e is the Coulomb constant. Note that, since this expression has the same mathematical form as the gravitational force between two masses, the electric force is a conservative central force just like the gravitational force. Thus, the result for the change in potential energy associated with a given displacement of one of the charges has the same form as Equation 8.18, with the constant $-GM_e m$ replaced by the Coulomb constant $k_e q_1 q_2$:

$$U_f - U_i = k_e \frac{q_1 q_2}{r_f} - k_e \frac{q_1 q_2}{r_i} \qquad [8.22]$$

Again, if we choose infinity as our potential energy reference point ($r_i = \infty$), then $U_i = 0$. Thus, the electric potential energy associated with two charged particles separated by a distance r is

Electric potential energy for two charges

$$U_e = k_e \frac{q_1 q_2}{r} \qquad [8.23]$$

Note that the electric potential energy is positive if the charges have the *same* sign. This is consistent with the fact that like charges repel. It takes energy supplied by some external agent to bring two like charges from an infinite separation to some finite separation r. If two like charges initially separated by some distance r are released from rest, the charges repel each other, and their potential energy is transformed into kinetic energy as they move farther apart.

Now consider a system consisting of more than two charged particles. In this case, the total electric potential energy can be found by calculating U_e for *every pair* of charges and summing the results algebraically. In effect, potential energy (calculated from Eq. 8.23) is associated with *each pair* of charges. For example, the three charges in Figure 8.15 have the potential energy

$$U_e = k_e \frac{q_1 q_2}{r_{12}} + k_e \frac{q_1 q_3}{r_{13}} + k_e \frac{q_2 q_3}{r_{23}} \quad [8.24]$$

Thus, to find the total potential energy of a configuration of multiple charges, we algebraically sum the potential energy terms of *all possible combinations of two charges.*

Physically we can interpret this result as follows: Imagine that q_1 is fixed at the position shown in Chapter 6, Figure 6.28, with q_2 *and* q_3 at infinity. The work required to bring q_2 from infinity to its position near q_1 is $k_e q_1 q_2/r$, which is the first term in Equation 8.24. The last two terms in Equation 8.24 represent the work required to bring q_3 from infinity to its position near q_1 and q_2. In other words, U_e represents the total energy that would be required to establish the configuration shown in Figure 8.15, if all the charges were initially located at infinity. (You should show that the result is independent of the order in which the charges are assembled.)

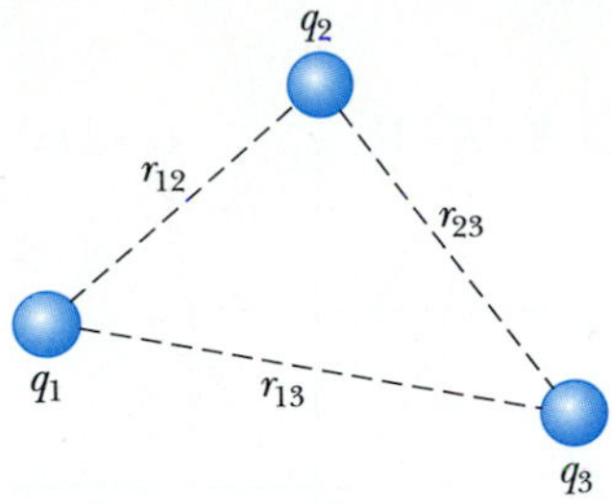

Figure 8.15 Three charges are fixed at the positions shown. The electric potential energy of this system of charges is given by Equation 8.24.

Example 8.9 Potential Energy of a System of Three Charges

A 5-μC charge is located at the origin, a second charge of $-2\ \mu$C is located on the x axis at the position (3, 0) m, and a third charge of 4 μC is located at the point P, whose coordinates are (0, 4) m, as shown in Figure 8.16. Taking $U_\infty = 0$, find the total electric potential energy of the system.

Solution We consider every possible combination of two charges and obtain

$$U_e = k_e \frac{q_1 q_2}{r_{12}} + k_e \frac{q_1 q_3}{r_{13}} + k_e \frac{q_2 q_3}{r_{23}}$$

Substituting,

$$U_e = 8.99 \times 10^9 \frac{\text{N}\cdot\text{m}^2}{\text{C}^2}\left(\frac{(5\times10^{-6}\text{ C})(-2\times10^{-6}\text{ C})}{3\text{ m}} + \frac{(5\times10^{-6}\text{ C})(4\times10^{-6}\text{ C})}{4\text{ m}} + \frac{(-2\times10^{-6}\text{ C})(4\times10^{-6}\text{ C})}{5\text{ m}}\right)$$

The result is

$$U_e = 6.05 \times 10^{-4}\text{ J}$$

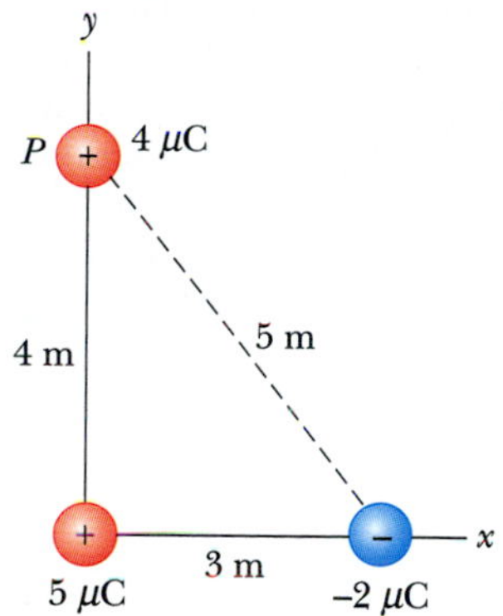

Figure 8.16 (Example 8.9)

Example 8.10 The Distance of Closest Approach

A proton moving with a speed of 1.5×10^7 m/s is headed straight toward a gold nucleus which is at rest in a target. Find the closest distance d between the proton and the gold nucleus.

Solution As the proton proceeds toward the gold nucleus on a head-on collision course, it will eventually stop momentarily at the distance of closest approach d, because of the Coulomb repulsive force, and reverse its path. Since the energy of the

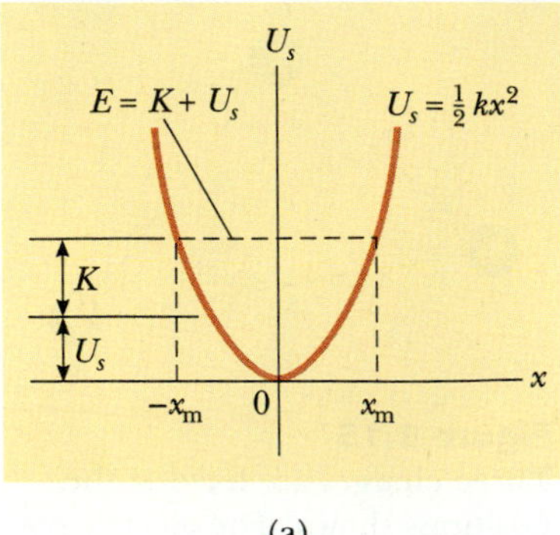

(a)

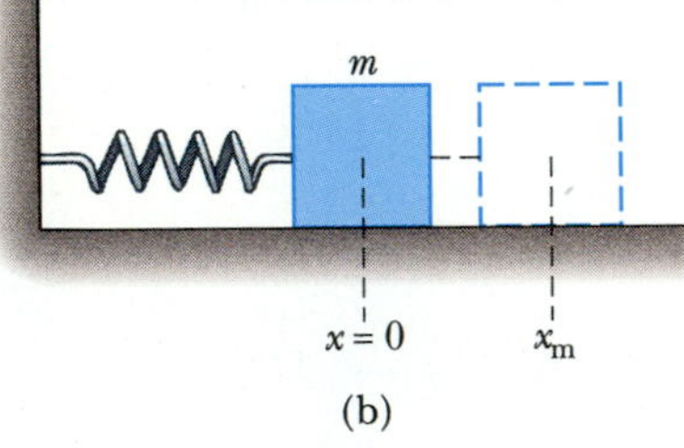

(b)

Figure 8.17
(a) The potential energy as a function of x for the mass-spring system shown in (b). The mass oscillates between the turning points, which have the coordinates $x = \pm x_m$. Note that the restoring force of the spring always acts toward $x = 0$, the position of stable equilibrium.

system (proton + gold nucleus) must be conserved, we can equate the initial kinetic energy of the proton (when it is very far from the gold nucleus) to the electric potential energy of the system at the point of closest approach (when the system has zero kinetic energy). That is,

$$\tfrac{1}{2}mv^2 = k_e\frac{q_1q_2}{r} = k_e\frac{(e)(Ze)}{d}$$

where Z is the charge number. Solving for d, and noting that $Z = 79$ for gold, gives

$$d = \frac{2k_eZe^2}{mv^2}$$

$$d = \frac{2(9\times10^9\ \text{N}\cdot\text{m}^2/\text{C}^2)\ (79)\ (1.6\times10^{-19}\ \text{C})^2}{(1.67\times10^{-27}\ \text{kg})(1.5\times10^7\ \text{m/s})^2} = 9.69\times10^{-12}\ \text{m}$$

This simple idea was used by Rutherford to estimate the radius of the gold nucleus, using alpha particles (the nuclei of helium atoms) rather than protons.

*8.9 Energy Diagrams and Stability of Equilibrium

The motion of a system can often be understood qualitatively through an analysis of the system's potential energy curve. Consider the potential energy function for the mass-spring system, given by $U_s = \frac{1}{2}kx^2$. This function is plotted versus x in Figure 8.17a. The spring force is related to U through Equation 8.11:

$$F_s = -\frac{dU_s}{dx} = -kx$$

That is, the force is equal to the negative of the *slope* of the U versus x curve. When the mass is placed at rest at the equilibrium position ($x = 0$), where $F = 0$, it will remain there unless some external force acts on it. If the spring is stretched from equilibrium, x is positive and the slope dU/dx is positive; therefore, F_s is negative and the mass accelerates back toward $x = 0$. If the spring is compressed, x is negative and the slope is negative; therefore, F_s is positive and again the mass accelerates toward $x = 0$.

Stable equilibrium

From this analysis, we conclude that the $x = 0$ position is one of **stable equilibrium.** That is, any movement away from this position results in a force that is directed back toward $x = 0$. In general, *positions of stable equilibrium correspond to those points for which $U(x)$ has a minimum value.*

From Figure 8.17 we see that if the mass is given an initial displacement x_m and released from rest, its total energy initially is the potential energy stored in the spring, given by $\frac{1}{2}kx_m^2$. As motion commences, the system acquires kinetic energy and loses an equal amount of potential energy. Since the total energy must remain constant, the mass oscillates between the two points $x = \pm x_m$, called the *turning points.* In fact, because no energy loss (no friction) takes place, the mass oscillates between $-x_m$ and $+x_m$ forever. (We shall discuss these oscillations further in Chapter 21). From an energy viewpoint, the energy of the system cannot exceed $\frac{1}{2}kx_m^2$; therefore, the mass must stop at these points and, because of the spring force, accelerate toward $x = 0$.

Another simple mechanical system that has a position of stable equilibrium is a ball rolling about in the bottom of a spherical bowl. If the ball is displaced from its lowest position, it always tends to return to that position when released.

Now consider an example in which the U versus x curve is as shown in Figure 8.18. In this case, $F_x = 0$ at $x = 0$, and so the particle is in equilibrium at this point. However, this is a position of **unstable equilibrium** for the following reason. Suppose the particle is displaced to the *right* ($x > 0$). Because the slope is negative for $x > 0$, $F_x = -dU/dx$ is positive and the particle accelerates away from $x = 0$. Now suppose the particle is displaced to the left ($x < 0$). In this case the force is *negative,* since the slope is positive for $x < 0$, and the particle again accelerates away from the equilibrium position. The $x = 0$ position in this situation is called a position of *unstable equilibrium* because, for any displacement from this point, the force pushes the particle farther away from equilibrium. In fact, the force pushes the particle toward a position of lower potential energy. A ball placed on the top of an inverted spherical bowl is obviously in a position of unstable equilibrium. If the ball is displaced slightly from the top and released, it will surely roll off the bowl. In general, *positions of unstable equilibrium correspond to those points for which $U(x)$ has a maximum value.*[3]

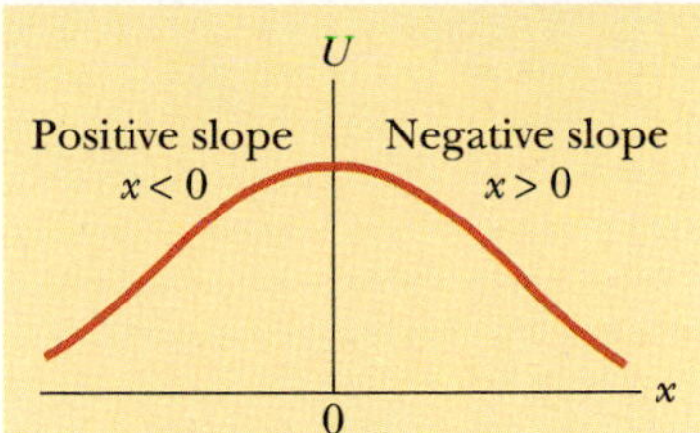

Figure 8.18 A plot of U versus x for a system that has a position of unstable equilibrium, located at $x = 0$. In this case, the force on the system for finite displacements is directed away from $x = 0$.

Neutral equilibrium

Finally, a situation may arise where U is constant over some region, and hence $F = 0$. This is called a position of **neutral equilibrium.** Small displacements from this position produce neither restoring nor disrupting forces. A ball lying on a flat horizontal surface is an example of an object in neutral equilibrium.

Example 8.11 Vibrations of a Diatomic Molecule

The potential energy function associated with the force between two atoms in a diatomic molecule may be approximated as follows:

$$U(x) = \frac{c}{x^9} - \frac{d}{x}$$

where c and d are positive constants (with units) and x is the distance between the atoms. A qualitative sketch of the $U(x)$ versus x curve is shown in Figure 8.19a. According to classical physics, at temperatures above absolute zero, the molecule undergoes vibrations along a line joining the two atoms. For a given total energy E, this motion changes from a minimum atomic separation a to a maximum atomic separation b, as shown in Figure 8.19b. As the temperature is lowered, the amplitudes of vi-

[3] You can mathematically test whether an extreme of U is stable or unstable by examining the sign of d^2U/dx^2.

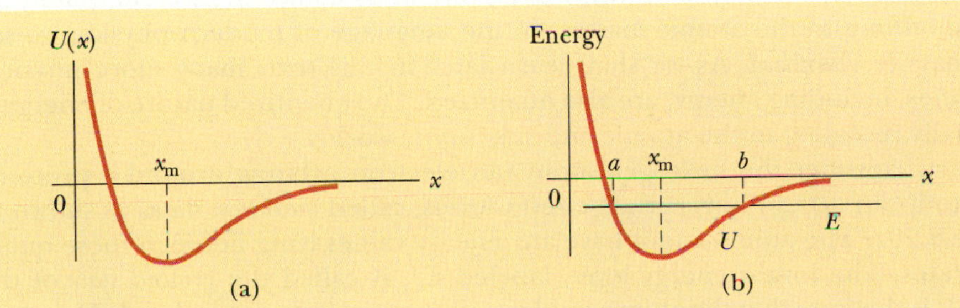

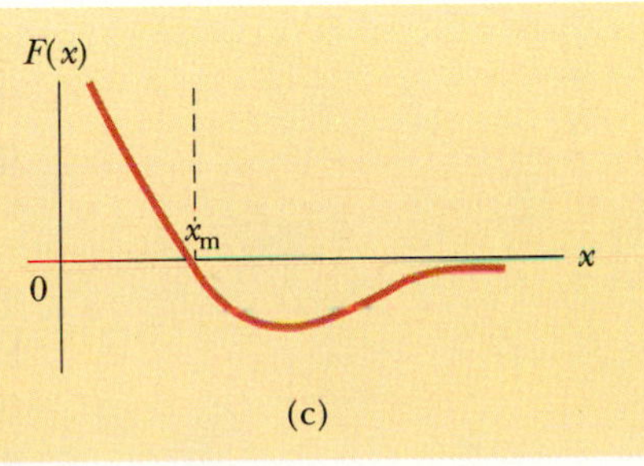

Figure 8.19 (Example 8.10)

bration become smaller and the atomic separation approaches the distance x_m, at which the potential energy U is a minimum.

(a) Find the force $F(x)$ between the atoms and sketch a curve of $F(x)$ versus x.

Solution Since the force is conservative, we have

$$F(x) = -\frac{dU}{dx} = -\frac{d}{dx}\left[\frac{c}{x^9} - \frac{d}{x}\right]$$

Taking the derivative gives

$$F(x) = \boxed{\frac{9c}{x^{10}} - \frac{d}{x^2}}$$

This force is sketched in Figure 8.19c. In drawing graphs such as this, it is helpful to remember that the force is proportional to the *negative slope* of the $U(x)$ versus x curve.

(b) Find the distance x_m.

Solution We find the distance x_m by noting that the slope of the potential energy curve is zero at this point. So we set the derivative of U equal to zero and solve for x_m.

$$\frac{dU}{dx} = 0$$

$$\left(-\frac{9c}{{x_m}^{10}} + \frac{d}{{x_m}^2}\right) = 0$$

Solving for x_m, we obtain

$${x_m}^8 = \frac{9c}{d}$$

$$x_m = \boxed{\left(\frac{9c}{d}\right)^{1/8}}$$

8.10 Quantization of Energy

As you may have learned in a chemistry course, all ordinary matter consists of atoms, and each atom consists of a collection of electrons and a nucleus. Thus, on the fine scale of the atomic world, we see that mass comes in discrete quantities corresponding to the atomic masses. In the language of modern physics, we say that mass is *quantized.* As we shall learn later in this text, many more physical quantities, including energy, are also quantized. The quantized nature of energy is especially revealing in the atomic and subatomic worlds.

First, consider the hydrogen atom (an electron orbiting around a proton). The atom can occupy only certain energy levels, called *quantum states,* as shown in Figure 8.20a. The atom cannot have any energy values lying between these quantum states. The lowest energy level, labeled E_0, is called the *ground state* of the atom. It is the state that the atom would usually occupy if it were isolated. The atom

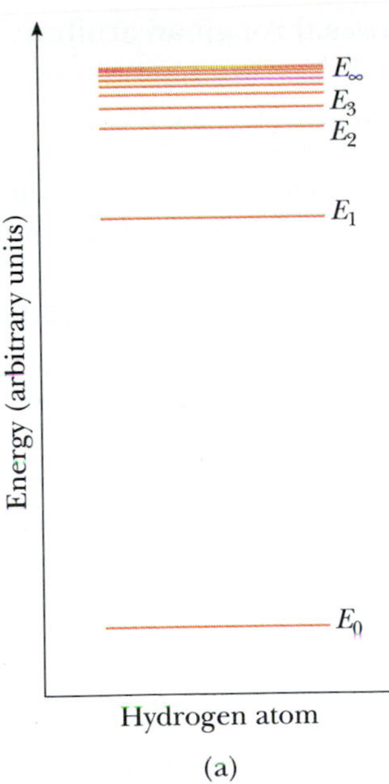

Figure 8.20
(a) Quantum states of the hydrogen atom. The lowest state, E_0, is the ground state. (b) The energy levels of an earth satellite are also quantized, but are so close together that they cannot be distinguished from each other.

could move to higher energy states by absorbing energy from some external source or by colliding with other atoms. The highest energy level shown in Figure 8.20a, E_∞, is the energy of the atom when the electron is completely removed from the proton; it is called the *ionization energy*. Note that the energy levels get closer together at the high end of the scale.

Next, consider a satellite in orbit about the Earth. If you were asked to describe the possible energies that the satellite could have, it would be reasonable (but incorrect) to say that the satellite could have any arbitrary energy you chose. Just like the energy of the hydrogen atom, *the energy of the satellite is quantized.* If you were to construct an energy level diagram for the satellite showing its allowed energies, the levels would be so close to one another, as in Figure 8.20b, it would be impossible to tell that they were not continuous. In other words, we have no way of experiencing quantization of energy in the macroscopic world; hence, we can ignore it when we describe everyday experiences.

Summary

The **gravitational potential energy** of a particle of mass m that is elevated a distance y near the Earth's surface is

$$U_g \equiv mgy \qquad [8.1]$$

Gravitational potential energy

The **elastic potential energy** stored in a spring of force constant k is

$$U_s \equiv \tfrac{1}{2}kx^2 \qquad [8.4]$$

Potential energy stored in a spring

A force is **conservative** if the work it does on a particle is independent of the path the particle takes between two given points. Alternatively, a force is conservative if the work it does is zero when the particle moves through an arbitrary closed path and returns to its initial position. A force that does not meet these criteria is said to be **nonconservative.**

A **potential energy** function U can be associated only with a conservative force. If a conservative force **F** acts on a particle that moves along the x axis from x_i to x_f, *the change in the potential energy equals the negative of the work done by that force:*

Change in potential energy

$$U_f - U_i = -\int_{x_i}^{x_f} F_x\, dx \qquad [8.6]$$

The **total mechanical energy of a system** is defined as the sum of the kinetic energy and potential energy:

Total mechanical energy

$$E \equiv K + U \qquad [8.8]$$

The **law of conservation of energy** states that if no external forces do work on the system, and there are no nonconservative forces, the total mechanical energy is constant:

Conservation of energy

$$K_i + U_i = K_f + U_f \qquad [8.9]$$

The change in total mechanical energy of a system equals the change in kinetic energy due to internal conservative forces, $\Delta K_{\text{int-nc}}$, plus the change in kinetic energy due to all external forces, ΔK_{ext}:

$$\Delta K + \Delta U = \Delta K_{\text{int-nc}} + \Delta K_{\text{ext}} \qquad [8.16]$$

The gravitational force is conservative, and therefore a potential energy function can be defined. The **gravitational potential energy** associated with two particles separated by a distance r is

Gravitational potential energy for a pair of particles

$$U_g = -\frac{Gm_1 m_2}{r} \qquad [8.20]$$

where U_g is taken to be zero at $r = \infty$. The total gravitational potential energy for a system of particles is the sum of energies for all pairs of particles, with each pair represented by a term of the form given by Equation 8.20.

The **electric potential energy of a pair of charges** separated by a distance r is

Electric potential energy of two charges

$$U_e = k_e \frac{q_1 q_2}{r} \qquad [8.23]$$

where k_e is the Coulomb constant. This represents the work required to bring the charges from an infinite separation to the separation r. The potential energy of a distribution of charges is obtained by summing terms such as Equation 8.23 over all *pairs* of particles.

Questions and Conceptual Exercises

1. Advertisements for the "Superball" once stated that the ball would rebound to a height greater than the height from which it was dropped. Is this possible?

2. A ball is dropped by a person at the top of a building, while another person at the bottom observes its motion. Will these two people always agree on the value of the

ball's potential energy? on the change in potential energy of the ball? on the kinetic energy of the ball?

3. Discuss the production and dissipation of mechanical energy in (a) lifting a weight, (b) holding the weight up, and (c) lowering the weight slowly. Include the muscles in your discussion.
4. Discuss the energy transformations that occur during a pole vault event. Ignore rotational motion.
5. Discuss the roles of kinetic and potential energy in the following sports: (a) baseball, (b) football, (c) tennis, (d) basketball, (e) track.
6. Many mountain roads are built so that they spiral around the mountain rather than go straight up the slope. Discuss this design from the viewpoint of energy and power.
7. A bowling ball is suspended from the ceiling of a lecture hall by a strong cord. The bowling ball is drawn away from its equilibrium position, then released from rest at the tip of the demonstrator's nose; the demonstrator remains stationary. Explain why she is not struck by the ball on its return swing. Would she be safe if the ball were given a slight push from the position of release?
8. Can the gravitational potential energy of an object ever have a negative value? Explain.
9. A pile driver is a device used to drive objects into the Earth by repeatedly dropping a heavy weight on them. By how much does the energy of a pile driver increase when the weight it drops is doubled? (Assume the weight is dropped from the same height each time.)
10. Our body muscles exert forces when we lift, push, run, jump, and so on. Are these forces conservative?
11. When nonconservative forces act on a system, does the total mechanical energy remain constant?
12. A block is connected to a spring that is suspended from the ceiling. If the block is set in motion and air resistance is neglected, describe the energy transformations that occur within the system consisting of the block and spring.
13. A ball rolls on a horizontal surface. Is the ball in stable, unstable, or neutral equilibrium?
14. What would the curve of U versus x look like if a particle were in a region of neutral equilibrium?
15. Discuss the energy transformations that occur during the operation of an automobile.
16. A ball is thrown straight up into the air. At what position is its kinetic energy a maximum? At what position is its gravitational potential energy a maximum?
17. Three identical balls are thrown from the top of a building, all with the same initial speed. One ball is thrown horizontally, the second at some angle above the horizontal, and the third at some angle below the horizontal. Neglecting air resistance, compare the speeds of the balls as they reach the ground.
18. Is any portion of a running high-jumper's kinetic energy converted into potential energy during the jump?
19. In the pole vault or high jump, why does the athlete attempt to keep her center of mass as low as possible near the top of the jump?
20. Give a physical explanation of the fact that the potential energy of a pair of like charges is positive, whereas the potential energy of a pair of unlike charges is negative.
21. An Olympic high jumper whose height is 2 m makes a record leap of 2.3 m over a horizontal bar. Estimate the speed with which he must leave the ground to perform this feat. (*Hint:* Estimate the position of his center of mass before he jumps, assuming he is in a horizontal position when he reaches the peak of his jump.)
22. Estimate the kinetic energy of an 80 000-kg airliner flying at a speed of 600 km/h.

Problems

Section 8.1 Potential Energy
Section 8.2 Conservative and Nonconservative Forces

1. What is the gravitational potential energy, relative to the ground, of a 0.15-kg baseball at the top of a 100-m-tall building?
2. A 2-kg ball is attached to the bottom end of a 1-m-long string hanging from the ceiling of a room. The height of the room is 3 m. What is the gravitational potential energy of the ball relative to (a) the ceiling, (b) the floor, and (c) a point at the same elevation as the ball?
3. A 1000-kg roller coaster is initially at the top of a rise, at point A. It then moves 135 ft, at an angle of 40° below the horizontal, to a lower point, B. (a) Choose point B to be the zero level for gravitational potential energy, and find the potential energy of the roller coaster at points A and B, and the difference in potential energy between these points. (b) Repeat part (a), setting the zero reference level at point A.
4. A 40-N child is in a swing that is attached to ropes

2 m long. Find the gravitational potential energy of the child relative to the child's lowest position when (a) the ropes are horizontal, (b) the ropes make a 30° angle with the vertical, and (c) the child is at the bottom of the circular arc.

5. A 3-kg particle moves from the origin to the position having coordinates $x = 5$ m and $y = 5$ m, under the influence of gravity acting in the negative y direction (Fig. 8.21). Using Equation 8.1, calculate the work done by gravity when the particle goes from O to C along the following paths: (a) OAC, (b) OBC, (c) OC. Your results should all be identical. Why?

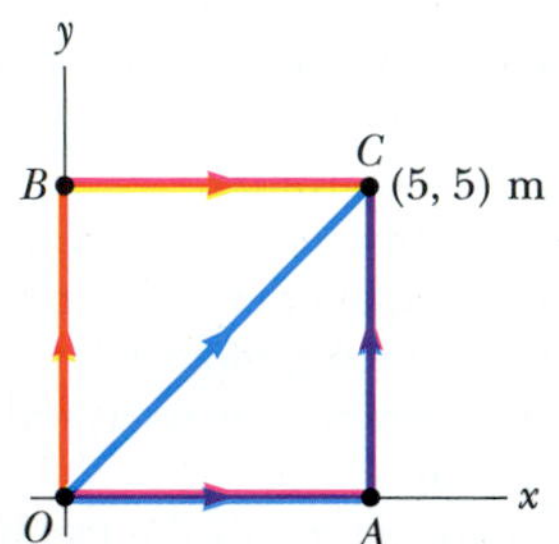

Figure 8.21 (Problem 5)

Section 8.3 Conservative Forces and Potential Energy
Section 8.4 Conservation of Energy

6. A 4-kg particle moves along the x axis under the influence of a single conservative force. If the work done on the particle is 80 J as the particle moves from $x = 2$ m to $x = 5$ m, find (a) the change in the particle's kinetic energy, (b) the change in potential energy, and (c) its speed at $x = 5$ m if it starts from rest at $x = 2$ m.

7. A single conservative force, $F_x = (2x + 4)$ N, acts on a 5-kg particle, where x is in meters. As the particle moves along the x axis from $x = 1$ m to $x = 5$ m, calculate (a) the work done by this force, (b) the change in the potential energy, and (c) the particle's kinetic energy at $x = 5$ m if its speed at $x = 1$ m is 3 m/s.

8. A single constant force $\mathbf{F} = (3\mathbf{i} + 5\mathbf{j})$ N acts on a 4-kg particle. (a) Calculate the work done by this force if the particle moves from the origin to the point with vector position $\mathbf{r} = (2\mathbf{i} - 3\mathbf{j})$ m. Does this result depend on the path? Explain. (b) What is the speed of the particle at $\mathbf{r}$ if its speed at the origin is 4 m/s? (c) What is the change in the potential energy?

9. Use conservation of energy to determine the final speed of a 5.0-kg mass attached to a light cord that passes over a massless, frictionless pulley and is attached to another mass of 3.5 kg, when the 5.0-kg mass has fallen (starting from rest) a distance of 2.5 m. (See Fig. 8.22.)

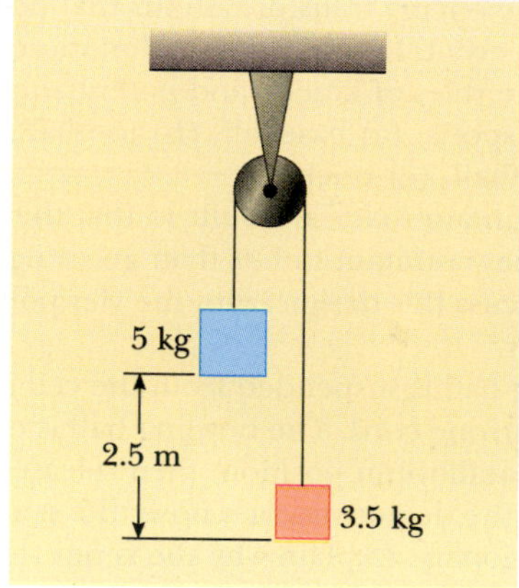

Figure 8.22 (Problem 9)

10. A bead slides without friction around a loop-the-loop (Fig. 8.23). If the bead is released from a height of $h = 3.5R$, what is its speed at point A? How large is the normal force on it if its mass is 5.0 g?

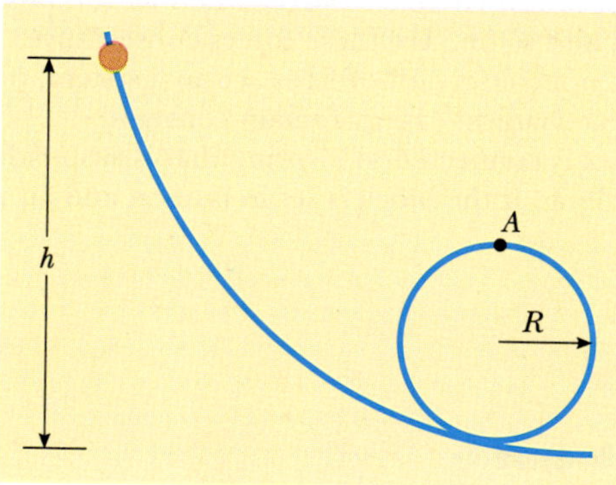

Figure 8.23 (Problem 10)

11. A particle of mass 0.5 kg is shot from P, as shown in Figure 8.24, with an initial velocity $\mathbf{v}_0$, having a horizontal component of 30 m/s. The particle rises to a maximum height of 20 m above P. Using conservation of energy, determine: (a) the vertical component of $\mathbf{v}_0$, (b) the work done by the gravitational force on the particle during its motion from P to B, and (c) the horizontal and vertical components of the velocity vector when the particle reaches B.

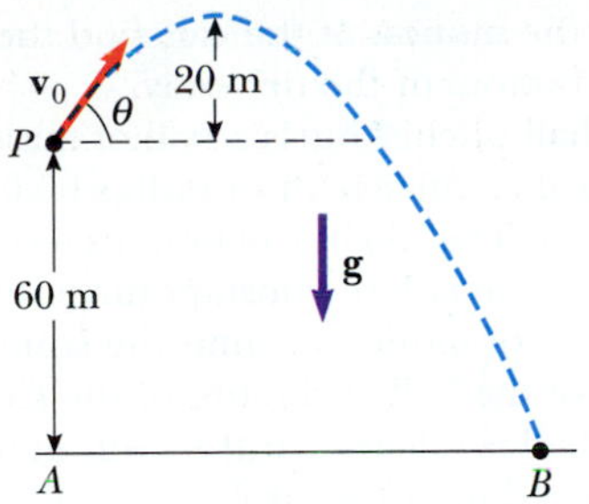

Figure 8.24 (Problem 11)

12. At time t_i, the kinetic energy of a particle is 30 J and the potential energy associated with the particle is 10 J. At some later time t_f, its kinetic energy is 18 J. (a) If only conservative forces act on the particle, what is the potential energy at time t_f? What is its total energy? (b) If the potential energy at time t_f is 5 J, are there any nonconservative forces or external forces acting on the particle? Explain.
13. A rocket is launched, at an angle of 53° to the horizontal, from altitude h with speed v_0. (a) Use energy methods to find the speed of the rocket when its altitude is $h/2$. (b) Find the x and y components of velocity when the rocket's altitude is $h/2$, using the fact that $v_x = v_{x0}$ is constant (since $a_x = 0$) and the result of part (a).
14. A simple 2.0-m-long pendulum is released from rest when the support string is at an angle of 25° from the vertical. What is the speed of the suspended mass at the bottom of the swing?
15. Dave Johnson, the bronze medalist at the 1992 Olympic decathlon in Barcelona, leaves the ground at the high jump with a vertical velocity of 6 m/s. Assuming his height is 2.3 m, how high can he jump?

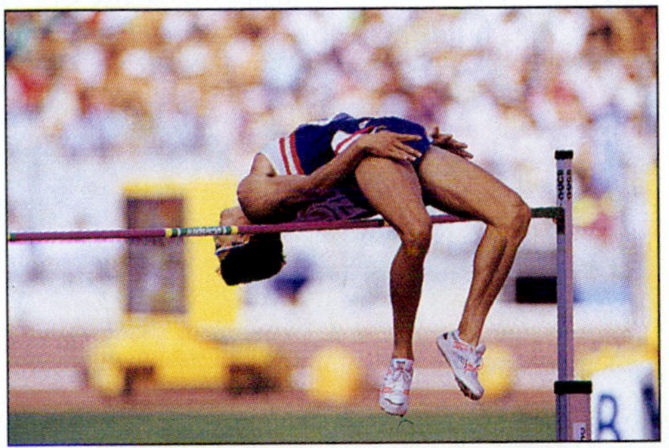

(Problem 15) *(FPG International)*

16. A 0.4-kg ball is thrown into the air and reaches a maximum altitude of 20 m. Taking its initial position as the point of zero potential energy and using energy methods, find (a) its initial speed, (b) its total mechanical energy, and (c) the ratio of its kinetic energy to its potential energy when its altitude is 10 m.

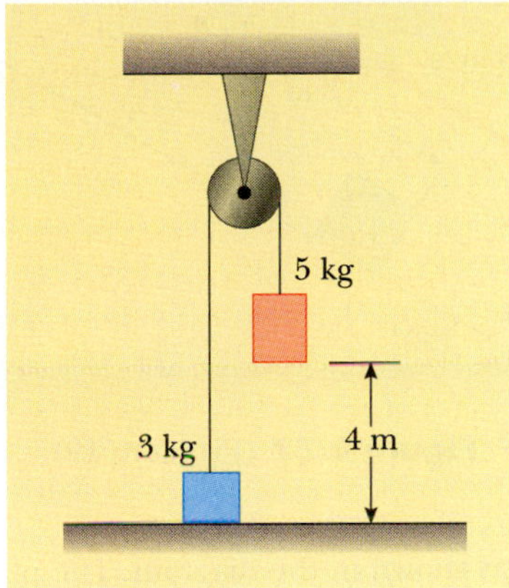

Figure 8.25 (Problem 17)

17. Two masses are connected by a light string that passes over a light, frictionless pulley as in Figure 8.25. The 5-kg mass is released from rest. Using the law of conservation of energy, (a) determine the velocity of the 3-kg mass just as the 5-kg mass hits the ground, and (b) find the maximum height to which the 3-kg mass will rise.
18. A child slides down the frictionless slide shown in Figure 8.26. In terms of R and H, at what height h will he lose contact with the section of radius R?

Figure 8.26 (Problem 18)

Section 8.5 Changes in Mechanical Energy When Nonconservative Forces are Present

19. A 5-kg block is set in motion up an inclined plane, as in Figure 8.27, with an initial speed of 8 m/s. The block comes to rest after traveling 3 m along the

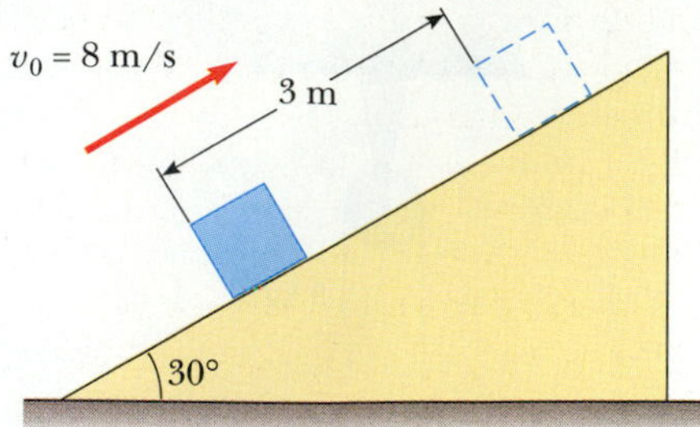

Figure 8.27 (Problem 19)

plane, as shown in the diagram. The plane is inclined at an angle of 30° to the horizontal. (a) Determine the change in kinetic energy. (b) Determine the change in potential energy. (c) Determine the frictional force on the block (assumed to be constant). (d) What is the coefficient of kinetic friction?

20. A block with a mass of 3 kg starts at a height of $h =$ 60 cm on a plane with an inclination angle of 30°, as shown in Figure 8.28. Upon reaching the bottom of the ramp, the block slides along a horizontal surface. If the coefficient of friction on both surfaces is $\mu_k =$ 0.20, how far does the block slide on the horizontal surface before coming to rest? (*Hint:* Divide the path into two straight-line parts.)

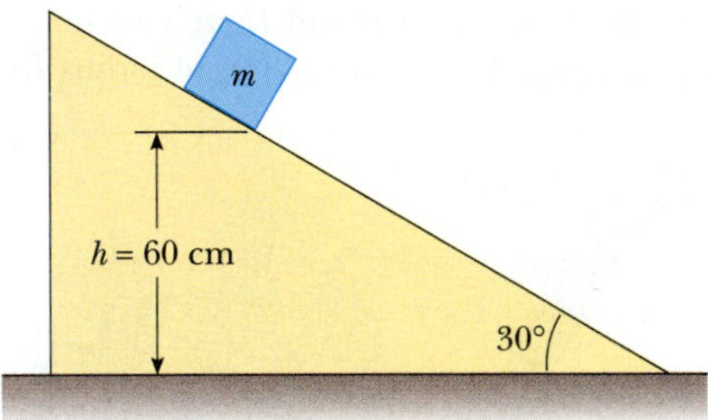

Figure 8.28 (Problem 20)

21. A parachutist of mass 50 kg jumps out of an airplane at a height of 1000 m. The parachute deploys, and she lands on the ground with a speed of 5 m/s. How much energy was lost to air friction during this jump?
22. A child starts from rest at the top of a slide of height 4 m. (a) What is her speed at the bottom if the incline is frictionless? (b) If she reaches the bottom with a speed of 6 m/s, what percentage of her total energy at the top of the slide was lost as a result of friction?
23. A 2000-kg car starts from rest at the top of a 5-m-long driveway that is sloped at an angle of 20° with the horizontal. If an average friction force of 4000 N impedes the motion of the car, find the speed of the car at the bottom of the driveway.
24. A softball pitcher hurls a ball of mass 0.25 kg around a vertical circular path of radius 60.0 cm before releasing it. The pitcher maintains a component of force on the ball of constant magnitude 30.0 N in the direction of motion around the complete path. The speed of the ball at the top of the circle is 15.0 m/s. If the ball is released at the bottom of the circle, what is its speed upon release?
25. A 70-kg diver steps off a 10-m tower and drops straight down into the water. If he comes to rest 5 m beneath the surface of the water, determine the average resistance force exerted on the diver by the water.
26. A force F_x, shown as a function of distance in Figure 8.29, acts on a 5-kg mass. If the particle starts from rest at $x = 0$ m, determine the speed of the particle at $x = 2$, 4, and 6 m.

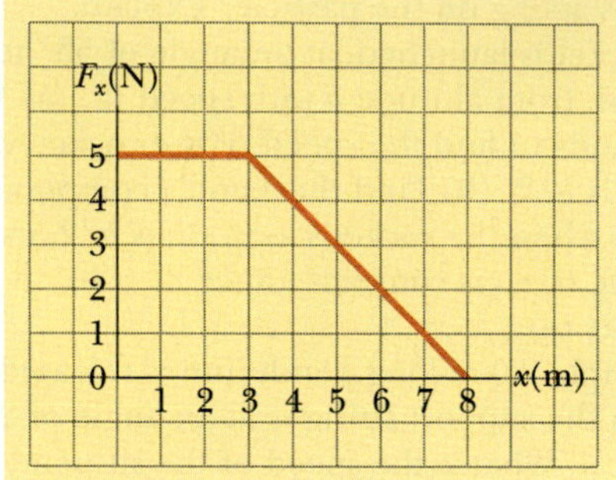

Figure 8.29 (Problem 26)

27. The coefficient of friction between the 3.0-kg object and the surface in Figure 8.30 is 0.40. The masses start from rest. What is the speed of the 5.0-kg mass when it has fallen a vertical distance of 1.5 m?

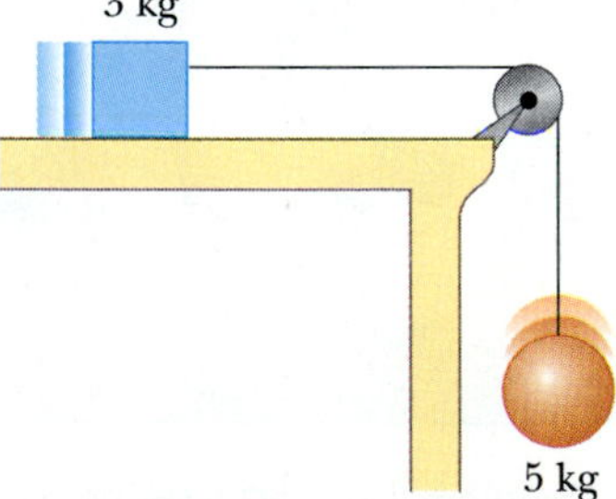

Figure 8.30 (Problem 27)

28. A toy gun uses a spring to project a 5.3-g soft rubber sphere. The spring constant is 8.0 N/m, the barrel of

the gun is 15 cm long, and there is a constant frictional force of 0.032 N between the barrel and the projectile. With what speed is the projectile launched from the barrel of the gun if the spring was compressed 5.0 cm?

29. A skier starts from rest at the top of a hill that is inclined at an angle of 10.5° to the horizontal. The hill is 200 m long, and the coefficient of friction between the snow and the skis is 0.075. At the bottom of the hill the snow is level and the coefficient of friction is unchanged. How far does the skier move along the horizontal portion of the snow before coming to rest?

Section 8.7 Gravitational Potential Energy Revisited

Assume $U = 0$ at $r = \infty$.

30. A satellite of the Earth has a mass of 100 kg and is at an altitude of 2×10^6 m. (a) What is the potential energy of the satellite-Earth system? (b) What is the magnitude of the force on the satellite?
31. A system consists of three particles, each of mass 5 g, located at the corners of an equilateral triangle with sides of 30 cm. (a) Calculate the potential energy of the system. (b) If the particles are released simultaneously, where will they collide?
32. How much energy is required to move a 1000-kg mass from the Earth's surface to an altitude twice the Earth's radius?
33. After our Sun exhausts its nuclear fuel, its ultimate fate is possibly to collapse to a *white dwarf* state, in which it has approximately the mass of the Sun but the radius of the Earth. Calculate (a) the average density of the white dwarf, (b) the acceleration due to gravity at its surface, and (c) the gravitational potential energy of a 1-kg object at its surface. (Take $U_g = 0$ at infinity.)
34. At the Earth's surface a projectile is launched straight up at a speed of 10 km/s. To what height will it rise? Ignore air resistance.

Section 8.8 Electric Potential Energy

Note: Assume that $U = 0$ at $r = \infty$ unless the statement of the problem requires otherwise.

35. What is the potential energy of two 8-μC charges located 2 m apart?
36. Two point charges, $Q_1 = +5$ nC and $Q_2 = -3$ nC, are separated by 35 cm. (a) What is the potential energy of the pair? What is the significance of the algebraic sign of your answer? (b) What is the electric field at a point midway between the charges?
37. The charge $q_1 = -9\ \mu$C is located at the origin, and a second charge, $q_2 = -1\ \mu$C, is located on the x axis at $x = 0.7$ m. Calculate the electric potential energy of this pair of charges.
38. The Bohr model of the hydrogen atom states that the electron can exist only in certain allowed orbits. The radius of each Bohr orbit is given by the expression $r = n^2(0.0529\text{ nm})$ where $n = 1, 2, 3, \ldots$. Calculate the electric potential energy of a hydrogen atom when the electron (a) is in the first allowed orbit, $n = 1$; (b) is in the second allowed orbit, $n = 2$; and (c) has escaped from the atom, $r = \infty$.
39. Calculate the energy required to assemble the array of charges shown in Figure 8.31, where $a = 0.20$ m, $b = 0.40$ m, and $q = 6\ \mu$C.

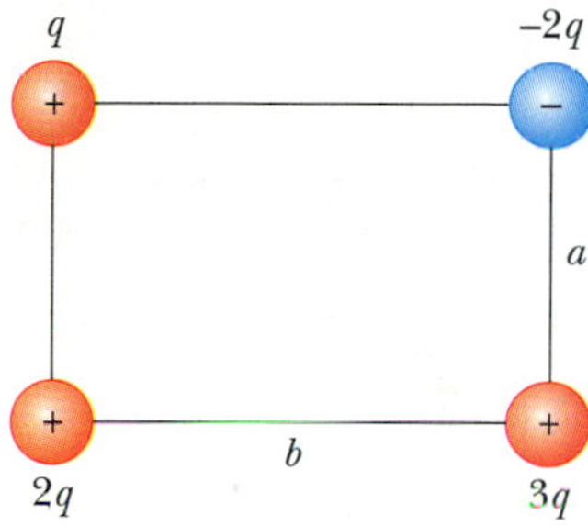

Figure 8.31 (Problem 39)

40. Show that the amount of work required to assemble four identical point charges of magnitude Q at the corners of a square with sides of length s is given by $5.41\ kQ^2/s$.
41. Four equal point charges of charge $q = +5\ \mu$C are located at the corners of a 30-cm by 40-cm rectangle. Calculate the electric potential energy stored in this charge configuration.
42. Four charges are located at the corners of a rectangle, as in Figure 8.32. How much energy would be expended in removing the two 4-μC charges to infinity?

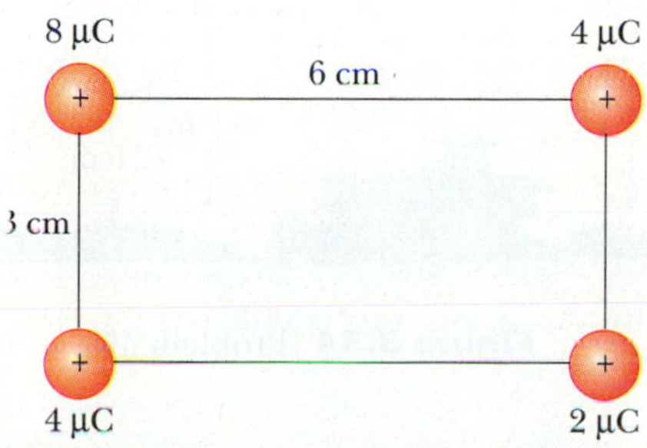

Figure 8.32 (Problem 42)

*Section 8.9 Energy Diagrams and Stability of Equilibrium

43. Consider the potential energy curve, $U(x)$ versus x, shown in Figure 8.33. (a) Determine whether the force F_x is positive, negative, or zero at the points indicated. (b) Indicate points of stable, unstable, or neutral equilibrium. (c) With reference to the potential energy curve in Figure 8.33, make a rough sketch of the curve F_x versus x from $x = 0$ to $x = 8$ m.

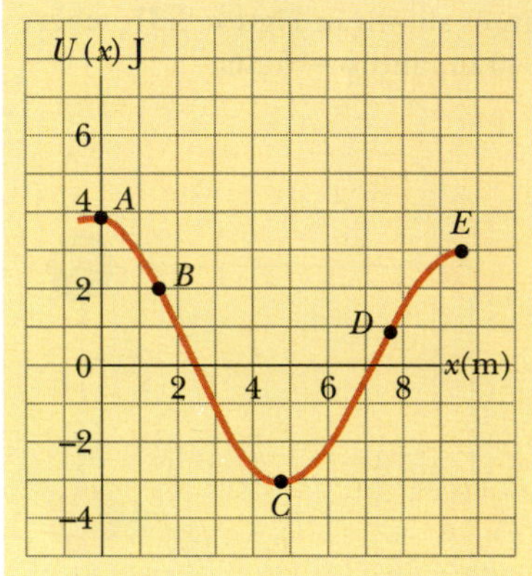

Figure 8.33 (Problem 43)

44. A right circular cone can be balanced on a horizontal surface in three different ways. Sketch these three equilibrium configurations, and identify them as positions of stable, unstable, or neutral equilibrium.
45. A particle of mass $m = 5$ kg is released from point A on the frictionless track shown in Figure 8.34. Determine (a) the speed of mass m at points B and C and (b) the net work done by the force of gravity in moving the particle from A to C.

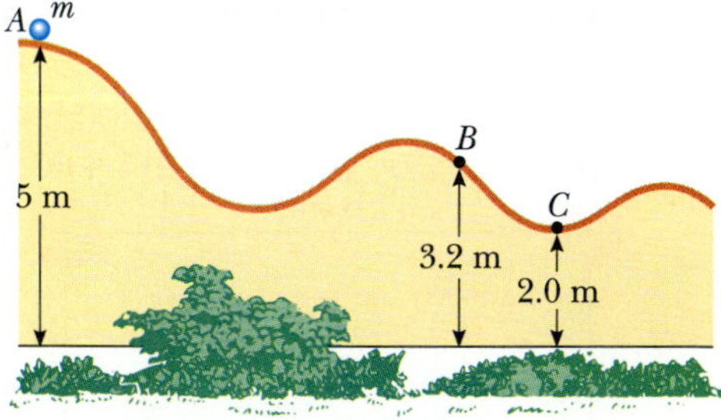

Figure 8.34 (Problem 45)

46. A hollow pipe has one or two extra weights attached to its inner surface, as shown in Figure 8.35. Explain why one is in unstable equilibrium, one is in neutral equilibrium, and one is in stable equilibrium. (In each diagram, O is the center of curvature and c.m. is the center of mass.)

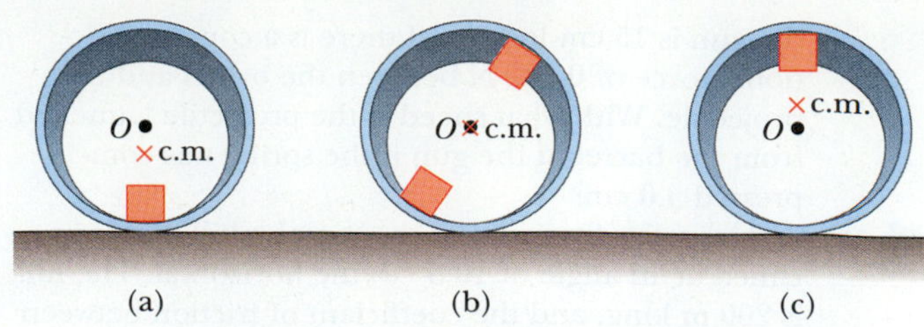

Figure 8.35 (Problem 46)

47. The potential energy of a two-particle system separated by a distance r is given by $U(r) = A/r$, where A is a constant. Find the radial force $\mathbf{F}_r$.
48. The potential energy function for a system is given by $U = ax^2 - bx$, where a and b are constants. (a) Find the force F_x associated with this potential energy function. (b) At what value of x is the force zero?

Additional Problems

49. A 200-g-particle is released from rest at point A along the diameter on the inside of a smooth hemispherical bowl of radius $R = 30$ cm (Fig. 8.36). Calculate (a) its gravitational potential energy at point A relative to point B, (b) its kinetic energy at point B, (c) its speed at point B, and (d) its kinetic energy and potential energy at point C.

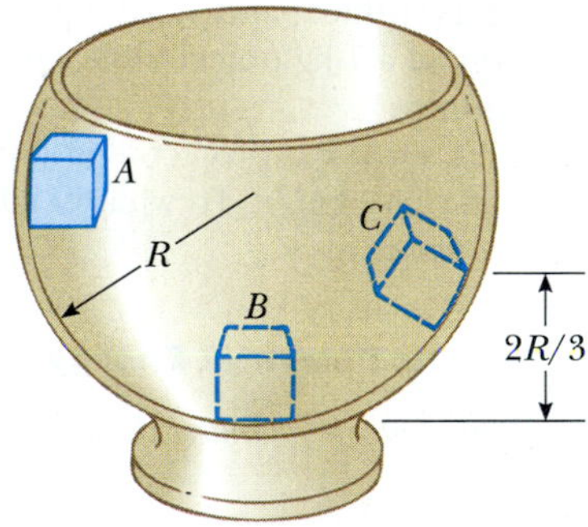

Figure 8.36 (Problems 49 and 50)

50. The particle described in Problem 49 (Fig. 8.36) is released from point A at rest, and the surface of the bowl is rough. The speed of the particle at point B is 1.5 m/s. (a) What is its kinetic energy at B? (b) How much energy is lost as a result of friction as the parti-

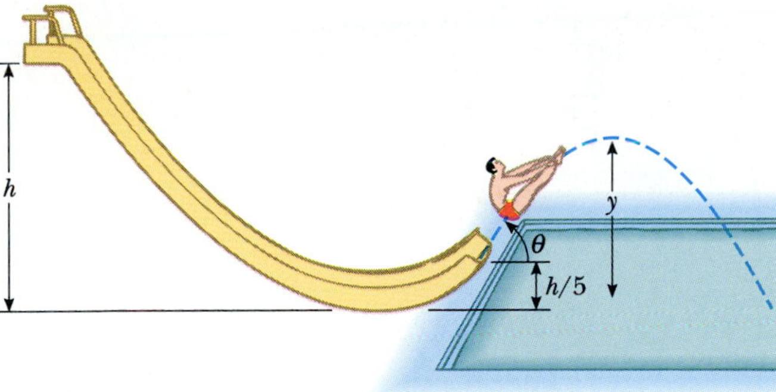

Figure 8.37 (Problem 52)

Figure 8.38 (Problem 54)

cle goes from A to B? (c) Is it possible to determine μ from these results in any simple manner? Explain.

51. A surprising demonstration involves dropping an egg from a third-floor window so that it lands on a 5-cm-thick foam-rubber pad without breaking. If a 56-g egg falls freely for 12 m, and the foam behaves like a spring of force constant 5760 N/m, by how much is the pad compressed?

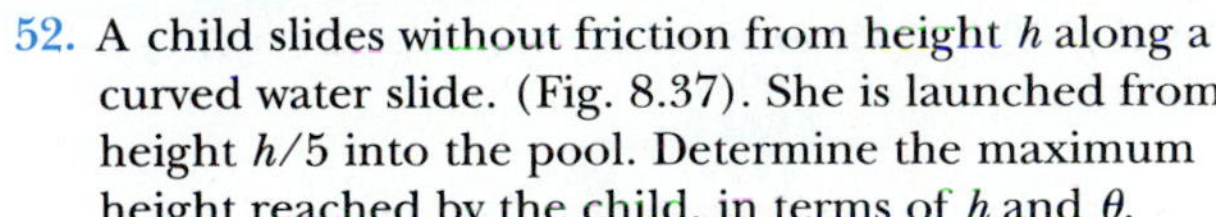

52. A child slides without friction from height h along a curved water slide. (Fig. 8.37). She is launched from height $h/5$ into the pool. Determine the maximum height reached by the child, in terms of h and θ.

53. The masses of the javelin, the discus, and the shot are 0.8 kg, 2.0 kg, and 7.2 kg, respectively, and record throws in the track events that use these objects are about 89 m, 69 m, and 21 m, respectively. Neglecting air resistance, (a) calculate the minimum initial kinetic energies that would produce these throws, and (b) estimate the average force exerted on each object during the throw, assuming the force acts over a distance of 2 m. (c) Do your results suggest that air resistance is an important factor?

54. A child's pogo stick (Fig. 8.38) stores energy in a spring ($k = 2.5 \times 10^4$ N/m). At position A ($x_1 = -0.1$ m) the spring compression is a maximum, and the child is momentarily at rest. At position B ($x = 0$) the spring is relaxed, and the child is moving upward. At position C the child is again momentarily at rest at the top of the jump. Assume that the combined mass of the child and the pogo stick is 25 kg. (a) Calculate the total energy of the system if both potential energies are zero at $x = 0$. (b) Determine x_2. (c) Calculate the speed of the child at $x = 0$. (d) Determine the value of x for which the kinetic energy of the system is a maximum. (e) Obtain the child's maximum upward velocity.

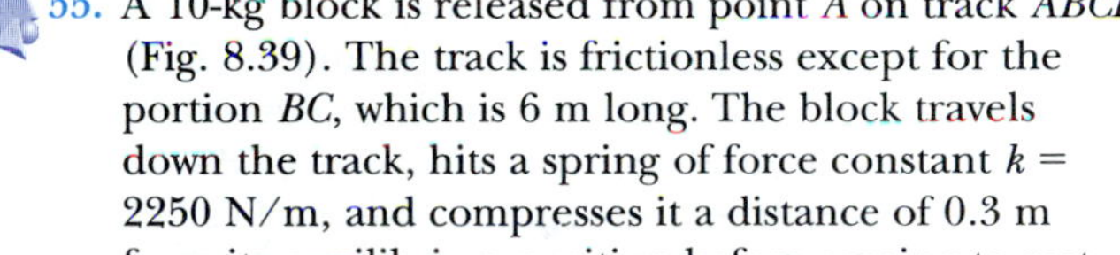

55. A 10-kg block is released from point A on track $ABCD$ (Fig. 8.39). The track is frictionless except for the portion BC, which is 6 m long. The block travels down the track, hits a spring of force constant k = 2250 N/m, and compresses it a distance of 0.3 m from its equilibrium position before coming to rest momentarily. Determine the coefficient of kinetic friction between the track portion BC and the block.

56. A 10-g mass is attached to the end of an unstressed, light, vertical spring ($k = 49$ N/m) and then dropped. (a) What is the maximum speed of the falling mass? (b) How far does the mass drop before coming to rest momentarily?

57. A 2-kg block situated on a rough incline is connected to a spring, of negligible mass, having a spring constant of 100 N/m (Fig. 8.40). The block is released

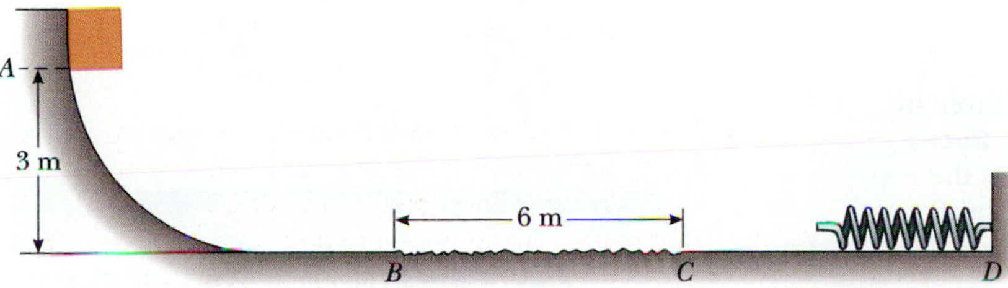

Figure 8.39 (Problem 55)

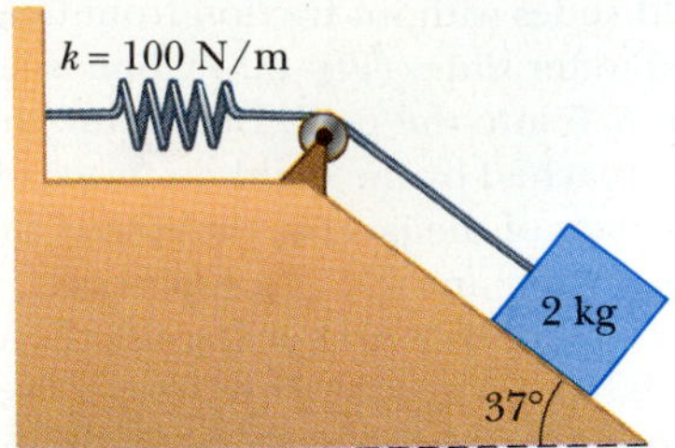

Figure 8.40 (Problems 57 and 58)

from rest when the spring is unstretched and the pulley is frictionless. The block moves 20 cm down the incline before coming to rest. Find the coefficient of kinetic friction between the block and the incline.

58. Suppose the incline is frictionless for the system described in Problem 57 (Fig. 8.40). The block is released from rest with the spring initially unstretched. (a) How far does it move down the incline before coming to rest? (b) What is the acceleration of the block when it reaches its lowest point? Is the acceleration constant? (c) Describe the energy transformations that occur during the descent of the block.

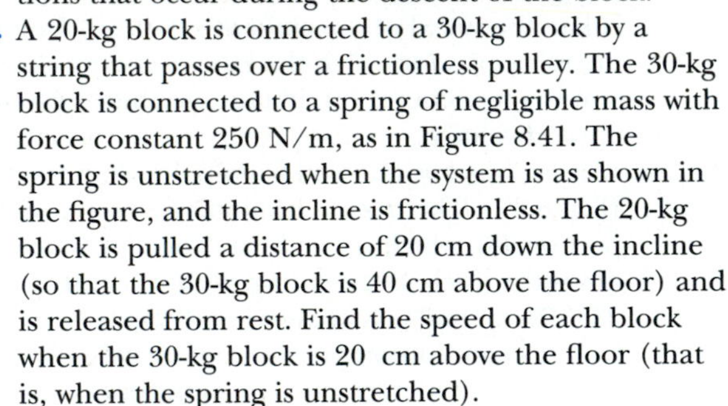

59. A 20-kg block is connected to a 30-kg block by a string that passes over a frictionless pulley. The 30-kg block is connected to a spring of negligible mass with force constant 250 N/m, as in Figure 8.41. The spring is unstretched when the system is as shown in the figure, and the incline is frictionless. The 20-kg block is pulled a distance of 20 cm down the incline (so that the 30-kg block is 40 cm above the floor) and is released from rest. Find the speed of each block when the 30-kg block is 20 cm above the floor (that is, when the spring is unstretched).

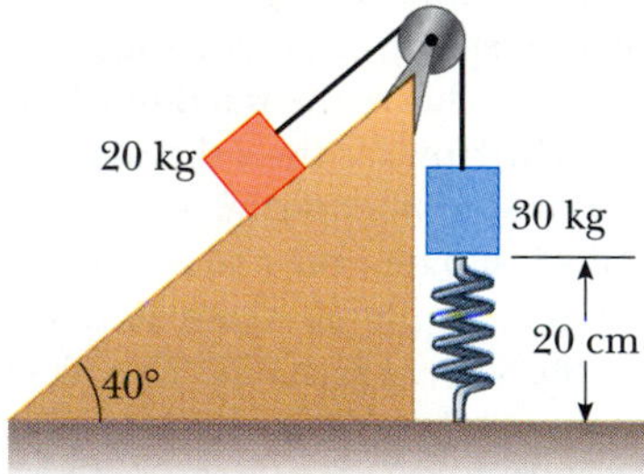

Figure 8.41 (Problem 59)

60. A potential energy function for a system is given by $U(x) = -x^3 + 2x^2 + 3x$. (a) Determine the force F_x as a function of x. (b) For what values of x is the force equal to zero? (c) Plot $U(x)$ versus x and F_x versus x, and indicate points of stable and unstable equilibrium.

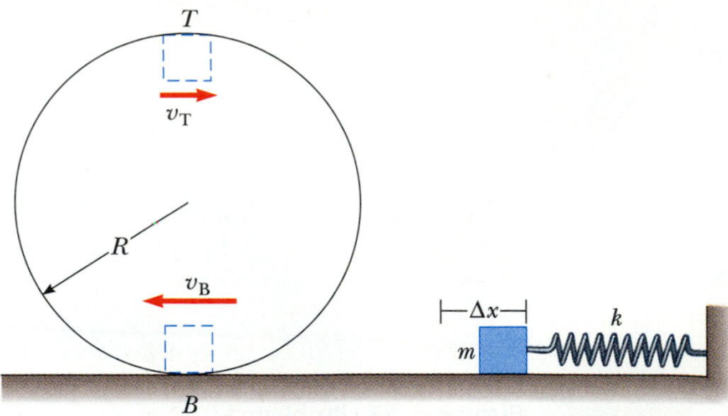

Figure 8.42 (Problem 61)

61. A block of mass 0.5 kg is pushed against a horizontal spring of negligible mass, compressing the spring a distance of Δx (Fig. 8.42). The spring constant is 450 N/m. When released, the block travels along a frictionless horizontal surface to point *B*, the bottom of a vertical circular track of radius $R = 1$ m, and continues to move up the track. The circular track is not frictionless. The speed of the block at the bottom of the vertical circular track is $v_B = 12$ m/s, and the block experiences an average frictional force of 7.0 N while sliding up the curved track. (a) What was the initial compression of the spring? (b) What is the speed of the block at the top of the circular track? (c) Does the block reach the top of the track, or does it fall off before reaching the top?

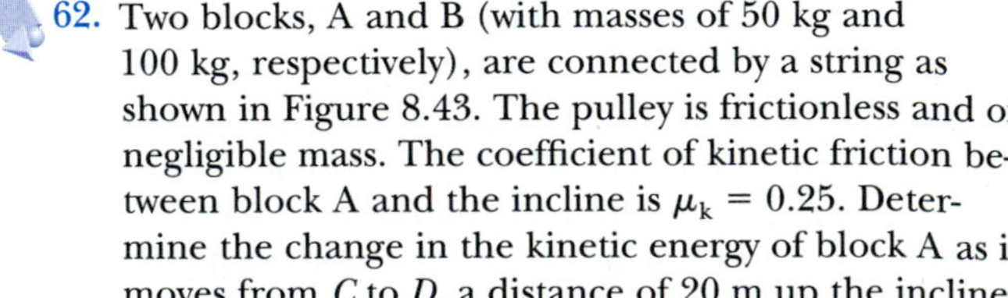

62. Two blocks, A and B (with masses of 50 kg and 100 kg, respectively), are connected by a string as shown in Figure 8.43. The pulley is frictionless and of negligible mass. The coefficient of kinetic friction between block A and the incline is $\mu_k = 0.25$. Determine the change in the kinetic energy of block A as it moves from *C* to *D*, a distance of 20 m up the incline.

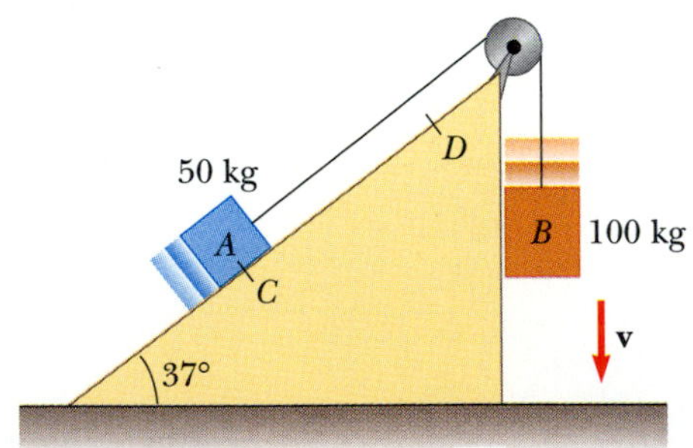

Figure 8.43 (Problem 63)

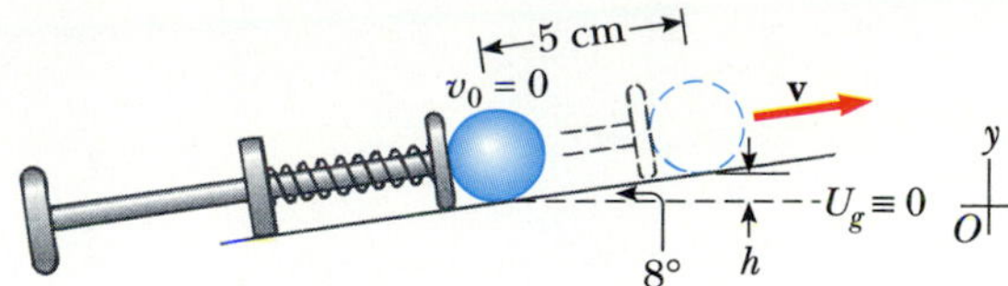

Figure 8.44 (Problem 63)

63. A pinball machine launches a 100-g ball with a spring-driven plunger (Fig. 8.44). The game board is inclined at 8° above the horizontal. Find the force constant k of the spring that will give the ball a speed of 80 cm/s when the plunger is released from rest, with the spring compressed 5 cm from its relaxed position. Assume that the plunger's mass and frictional effects are negligible.

64. A 1.0-kg mass slides to the right on a surface with coefficient of friction $\mu = 0.25$ (Fig. 8.45). It has a speed of $v_i = 3$ m/s when contact is made with a spring with spring constant $k = 50$ N/m. The mass comes to rest after the spring has been compressed a distance d. The mass is then forced toward the left by the spring, and it continues to move in that direction beyond the spring's unstretched position. Finally the mass comes to rest a distance D to the left of the unstretched spring. Find the following: (a) the compression distance d, (b) the speed v at the unstretched position, and (c) the distance D to the left of the unstretched spring where the mass comes to rest.

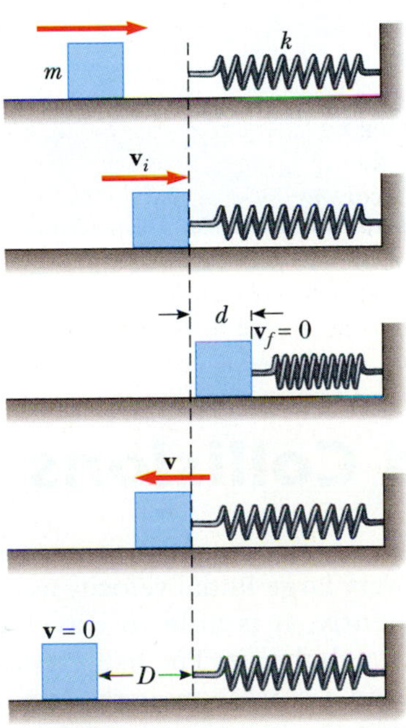

Figure 8.45 (Problem 64)

65. In the dangerous "sport" of bungee-jumping, a daring student jumps from a balloon with a specially designed elastic cord attached to his ankles, as shown in the photograph. The unstretched length of the cord is 25 m, the student weighs 700 N, and the balloon is 36 m above the surface of a river below. Calculate the required force constant of the cord if the student is to stop safely 4 m above the river.

Figure 8.46 (Problem 65). *(Gamma)*

66. A small object with charge −4 nC and mass 0.3 g moves freely after being released from rest 80 cm from a *fixed* spherical object with charge +5 μC and radius 6 cm. Find the speed of the 0.3-g object just before it collides with the sphere.

What would William Tell think? An apple being pierced by a 30-caliber bullet traveling at a supersonic speed of 900 m/s. This collision was photographed with a microflash stroboscope using an exposure time of 0.33 μs. The velocity of the bullet decreases from $\mathbf{v}_{1i}$ to $\mathbf{v}_{1f}$ because of the collision, so it loses kinetic energy. This loss in energy appears in the apple, which disintegrates completely after the collision. Note that the points of both entry and exit of the bullet are visually explosive. *(Shooting the Apple, 1964. © Harold Edgerton, Courtesy of Palm Press)*

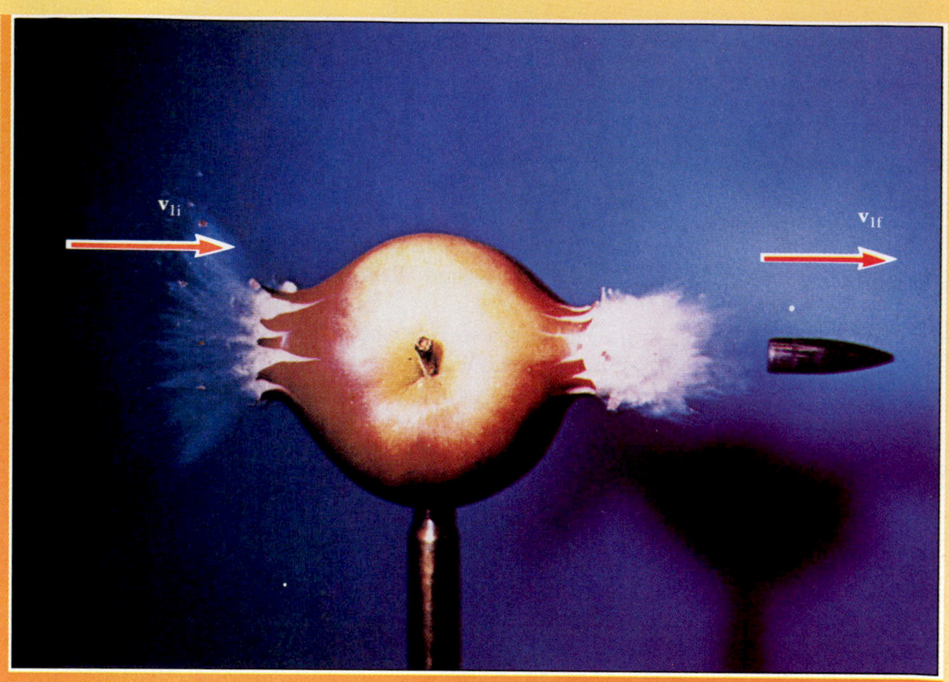

Momentum and Collisions

CHAPTER 9

Consider what happens when a golf ball is struck by a club. The ball is given a very large initial velocity as a result of the collision; consequently, it is able to travel more than a hundred meters through the air. The ball experiences a large change in velocity and a correspondingly large acceleration. Furthermore, because the ball experiences this acceleration over a very short time interval, the average force on it during the collision is very large. By Newton's third law, the club experiences a reaction force that is equal to and opposite the force on the ball. This reaction force produces a change in the velocity of the club. Since the club is much more massive than the ball, however, the change in velocity of the club is much less than the change in velocity of the ball.

One of the main objectives of this chapter is to enable you to understand and analyze such events. As a first step, we shall introduce the concept of *momentum,* a term that is often used in descriptions of objects in motion. For example, a very massive football player is often said to have a great deal of momentum as he runs down the field. A much less massive player, such as a halfback, can have equal or greater momentum if his speed is greater than that of a more massive player. This follows from the fact that momentum is defined as the product of mass and velocity.

The concept of momentum leads us to a second conservation law, that of conservation of momentum. This law is especially useful for treating problems that involve collisions between objects.

9.1 Linear Momentum and Its Conservation

The **linear momentum** of a particle of mass m moving with a velocity $\mathbf{v}$ is defined to be the product of the mass and velocity:[1]

$$\mathbf{p} \equiv m\mathbf{v} \qquad [9.1]$$

Definition of linear momentum of a particle

Since momentum equals the product of a scalar, m, and a vector, $\mathbf{v}$, it is a vector quantity. Its direction is along $\mathbf{v}$ and it has dimensions of ML/T. In the SI system, momentum has the units kg · m/s.

If a particle is moving in an arbitrary direction in three-dimensional space, $\mathbf{p}$ has three components, and Equation 9.1 is equivalent to the component equations given by

$$p_x = mv_x \qquad p_y = mv_y \qquad p_z = mv_z \qquad [9.2]$$

As you can see from its definition, the concept of momentum provides a quantitative distinction between heavy and light particles moving at the same velocity. For example, the momentum of a bowling ball moving at a speed of 10 m/s is much greater than that of a tennis ball moving at the same speed. Newton called the product $m\mathbf{v}$ *quantity of motion,* perhaps a more graphic description than *momentum,* which comes from the Latin word for movement.

We can relate the linear momentum of a particle to the resultant force acting on the particle by using Newton's second law of motion. In Chapter 4 we learned that Newton's second law can be written as $\sum \mathbf{F} = m\mathbf{a}$. However, this form applies only when the mass of the system remains constant. In situations where the mass is changing with time, one must use an alternative statement of Newton's second law: *The time rate of change of momentum of a particle is equal to the resultant force on the particle.*

$$\sum \mathbf{F} = \frac{d\mathbf{p}}{dt} \qquad [9.3]$$

Newton's second law for a particle

From Equation 9.3 we see that if the resultant force is zero, the time derivative of the momentum is zero, and therefore the momentum of any object must be constant. In other words, the linear momentum of an object is *constant* when $\sum \mathbf{F} = 0$.

[1] This expression is nonrelativistic and is valid only when $v \ll c$, where c is the speed of light.

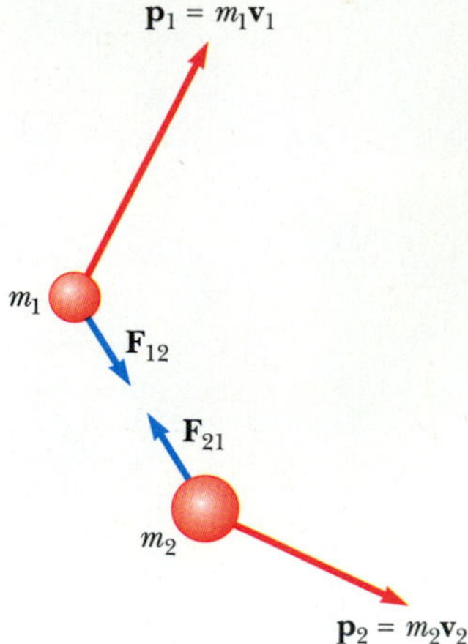

Figure 9.1
At some instant, the momentum of m_1 is $\mathbf{p}_1 = m_1\mathbf{v}_1$ and the momentum of m_2 is $\mathbf{p}_2 = m_2\mathbf{v}_2$. Note that $\mathbf{F}_{12} = -\mathbf{F}_{21}$.

Of course, if the particle is isolated (that is, if it does not interact with its environment), then by necessity $\Sigma\mathbf{F} = 0$ and $\mathbf{p}$ remains unchanged.

In a certain sense, conservation of linear momentum is just another way of stating Newton's first law. If an object is in motion, its linear momentum and, consequently, its velocity do not change unless an external force acts on the system. A good example is a rocket, its engines disengaged, coasting through space far from any planets.

Conservation of Linear Momentum for a Two-Particle System

To see how to apply the principle of *conservation of linear momentum,* consider a system of two particles that can interact with each other but are isolated from their surroundings (Fig. 9.1). That is, the particles may exert a force on each other, but no *external* forces are present. It is important to note the impact of Newton's third law on this analysis. Recall from Chapter 4 that Newton's third law states that the forces on these two particles are always equal in magnitude and opposite in direction; that is, *forces always occur in pairs.* Thus, if an *internal* force (say a gravitational force) acts on particle 1, then there must be a second *internal* force, equal in magnitude but opposite in direction, which acts on particle 2.

Suppose that at some instant, the momentum of particle 1 is $\mathbf{p}_1$ and the momentum of particle 2 is $\mathbf{p}_2$. Applying Newton's second law to each particle, we can write

$$\mathbf{F}_{12} = \frac{d\mathbf{p}_1}{dt} \quad \text{and} \quad \mathbf{F}_{21} = \frac{d\mathbf{p}_2}{dt}$$

where $\mathbf{F}_{12}$ is the force on particle 1 due to particle 2 and $\mathbf{F}_{21}$ is the force on particle 2 due to particle 1. (These forces could be gravitational forces, or they could have some other origin. The source of the forces isn't important for the present discussion.) Newton's third law tells us that $\mathbf{F}_{12}$ and $\mathbf{F}_{21}$ are equal in magnitude and opposite in direction. That is, they form an action-reaction pair, and $\mathbf{F}_{12} = -\mathbf{F}_{21}$. We can also express this condition as

$$\mathbf{F}_{12} + \mathbf{F}_{21} = 0$$

or as

$$\frac{d\mathbf{p}_1}{dt} + \frac{d\mathbf{p}_2}{dt} = \frac{d}{dt}(\mathbf{p}_1 + \mathbf{p}_2) = 0$$

Since the time derivative of the total momentum, $\mathbf{P} = \mathbf{p}_1 + \mathbf{p}_2$, is *zero,* we conclude that the *total* momentum, $\mathbf{P}$, must remain constant; that is,

$$\mathbf{P} = \mathbf{p}_1 + \mathbf{p}_2 = \text{constant} \qquad [9.4]$$

or, equivalently,

$$\mathbf{p}_{1i} + \mathbf{p}_{2i} = \mathbf{p}_{1f} + \mathbf{p}_{2f} \qquad [9.5]$$

where $\mathbf{p}_{1i}$ and $\mathbf{p}_{2i}$ are initial values and $\mathbf{p}_{1f}$ and $\mathbf{p}_{2f}$ are final values of the momentum during a time period, dt, over which the reaction pair interacts. Equation 9.5 in component form says that the total momenta in the x, y, and z directions are all *independently conserved;* that is,

$$P_{ix} = P_{fx} \qquad P_{iy} = P_{fy} \qquad P_{iz} = P_{fz} \qquad [9.6]$$

This result is known as the **conservation of linear momentum.** It is considered to be one of the most important laws of mechanics. We can state it as follows:

Conservation of momentum

Whenever two isolated, uncharged particles interact with each other, their total momentum remains constant.

That is, *the total momentum of an isolated system at all times equals its initial total momentum.*

We can also describe conservation of momentum in another way. Since we require that the system be isolated, no external forces are present, and the total momentum of the system remains constant. Therefore, momentum conservation is an alternative and more general statement of Newton's third law.

Notice that we have made no statement concerning the nature of the forces acting on the system. The only requirement was that the forces must be *internal* to the system. Thus, momentum is constant for a two-particle system *regardless* of the nature of the internal forces. One can use a similar and equivalent argument to show that the law of conservation of momentum also applies to a system of many particles.

Example 9.1 The Recoiling Cannon

A 3000-kg cannon rests on a frozen pond as in Figure 9.2. The cannon is loaded with a 30-kg cannonball and is fired horizontally. If the cannon recoils with a velocity of 1.8i m/s, what is the velocity of the cannonball just after it leaves the cannon?

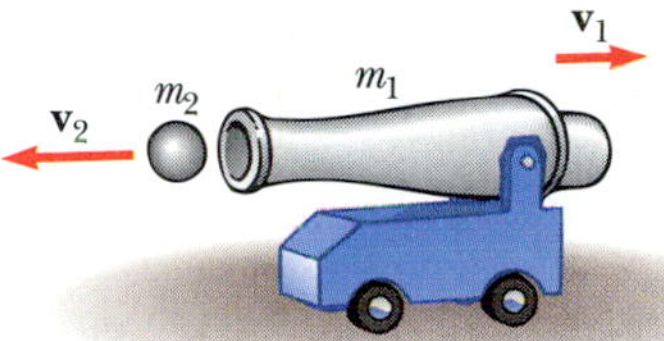

Figure 9.2 (Example 9.1) When the cannonball is fired to the left, the cannon recoils to the right.

Solution We take the system to consist of the cannonball and the cannon. Because of the force of gravity and the normal force, the system is not really isolated. However, both forces are directed perpendicularly to the motion of the system. Therefore, momentum is constant in the x direction since there are no external forces in this direction (assuming the surface is frictionless).

Because the cannon and cannonball are at rest before firing, the total momentum of the system is zero ($m_1\mathbf{v}_{1i} + m_2\mathbf{v}_{2i} = 0$). Therefore, the total momentum after firing must also be zero; that is,

$$m_1\mathbf{v}_{1f} + m_2\mathbf{v}_{2f} = 0$$

With $m_1 = 3000$ kg, $\mathbf{v}_{1f} = 1.8\mathbf{i}$ m/s, and $m_2 = 30$ kg, solving for $\mathbf{v}_{2f}$—the velocity of the cannonball—gives

$$\mathbf{v}_{2f} = -\frac{m_1}{m_2}\mathbf{v}_{1f} = -\left(\frac{3000 \text{ kg}}{30 \text{ kg}}\right)(1.8\mathbf{i} \text{ m/s}) = \boxed{-180\mathbf{i} \text{ m/s}}$$

The negative sign for $\mathbf{v}_{2f}$ indicates that the ball moves to the left after firing, in the direction opposite the movement of the cannon.

In the words of Newton's third law, for every force (to the left) on the ball, there is an equal but opposite force (to the right) on the cannon. Since the cannon is much more massive than the ball, the acceleration and consequent speed of the cannon are much smaller than the acceleration and speed of the ball.

Example 9.2 Decay of the Kaon at Rest

A meson is a nuclear particle that is more massive than an electron but less massive than a proton or neutron. One type of meson, called the neutral kaon (K^0), decays into a pair of charged pions (π^+ and π^-) that are oppositely charged but equal in

Figure 9.3 (Example 9.2) A kaon at rest decays spontaneously into a pair of oppositely charged pions. The pions move apart with momenta of equal magnitudes but opposite directions.

mass, as in Figure 9.3. A pion is a particle associated with the strong nuclear force that binds protons and neutrons together in the nucleus. Assuming the kaon is initially at rest, let us prove that after the decay, the two pions must have momenta that are equal in magnitude but opposite in direction.

Solution The decay of the kaon, represented in Figure 9.3, can be written

$$K^0 \rightarrow \pi^+ + \pi^-$$

If we let $\mathbf{p}^+$ be the momentum of the positive pion and $\mathbf{p}^-$ be the momentum of the negative pion after the decay, then the final linear momentum of the system can be written

$$\mathbf{P}_f = \mathbf{p}^+ + \mathbf{p}^-$$

Since the kaon is at rest before the decay, we know that $\mathbf{P}_i = 0$. Furthermore, since momentum is conserved, $\mathbf{P}_i = \mathbf{P}_f = 0$, so that

$$\mathbf{p}^+ + \mathbf{p}^- = 0$$

or

$$\mathbf{p}^+ = -\mathbf{p}^-$$

Thus, we conclude that the two linear momentum vectors of the pions are equal in magnitude and opposite in direction.

9.2 Impulse and Momentum

As we have seen, the momentum of a particle changes if a net force acts on the particle. Let us assume that a single force $\mathbf{F}$ acts on a particle and that this force may vary with time. According to Newton's second law, $\mathbf{F} = d\mathbf{p}/dt$, or

$$d\mathbf{p} = \mathbf{F}\,dt \qquad [9.7]$$

We can integrate this expression to find the change in the momentum of a particle. If the momentum of the particle changes from $\mathbf{p}_i$ at time t_i to $\mathbf{p}_f$ at time t_f, then integrating Equation 9.7 gives

Impulse of a force

$$\Delta\mathbf{p} = \mathbf{p}_f - \mathbf{p}_i = \int_{t_i}^{t_f} \mathbf{F}\,dt \qquad [9.8]$$

The quantity on the right side of Equation 9.8 is called the *impulse* of the force F for the time interval $\Delta t = t_f - t_i$. Impulse is a vector defined by

$$\mathbf{I} \equiv \int_{t_i}^{t_f} \mathbf{F}\,dt = \Delta\mathbf{p} \qquad [9.9]$$

Impulse-momentum theorem

That is, the **impulse** of the force **F** equals the change in the momentum of any object. This statement, known as the **impulse-momentum theorem,** is equivalent to Newton's second law. From this definition we see that impulse is a vector quantity having a magnitude equal to the area under the force-time curve, as described in Figure 9.4. In this figure it is assumed that the force varies in time in the general manner shown and is nonzero in the time interval $\Delta t = t_f - t_i$. The direction of the impulse vector is the same as the direction of the change in momentum. Impulse has the dimensions of momentum, ML/T. Note that impulse is *not* a property of the particle itself; rather, it is a measure of the degree to which an external force changes the momentum of the particle. Therefore, when we say that an impulse is given to a particle, it is implied that momentum is transferred from an external agent to that particle.

Since the force can generally vary in time as in Figure 9.4a, it is convenient to define a time-averaged force $\overline{\mathbf{F}}$, given by

$$\overline{\mathbf{F}} \equiv \frac{1}{\Delta t}\int_{ti}^{t_f} \mathbf{F}\,dt \qquad [9.10]$$

where $\Delta t = t_f - t_i$. Therefore, we can express Equation 9.9 as

$$\mathbf{I} = \Delta\mathbf{p} = \overline{\mathbf{F}}\,\Delta t \qquad [9.11]$$

This average force, described in Figure 9.4b, can be thought of as the constant force that would give the same impulse to the particle in the time interval Δt as the actual time-varying force gives over this same interval.

In principle, if **F** is known as a function of time, the impulse can be calculated from Equation 9.9. The calculation becomes especially simple if the force acting on the particle is constant. In this case, $\overline{\mathbf{F}} = \mathbf{F}$ and Equation 9.11 becomes

$$\mathbf{I} = \Delta\mathbf{p} = \mathbf{F}\,\Delta t \qquad [9.12]$$

In many physical situations, we shall use what is called the **impulse approximation:** *we assume that one of the forces exerted on a particle acts for a short time but is much greater than any other force present.* This approximation is especially useful in treating collisions, where the duration of the force is very short. When this approximation is made, we refer to the force as an *impulse force.* For example, when a baseball is struck with a bat, the duration of the collision is about 0.01 s, and the average force the bat exerts on the ball in this interval is typically several thousand newtons. This is much greater than the force of gravity, so the impulse approximation is justified. It is important to remember that $\mathbf{p}_i$ and $\mathbf{p}_f$ represent the momenta *immediately* before and after the collision, respectively. Therefore, in the impulse approximation very little motion of the particle takes place during the collision.

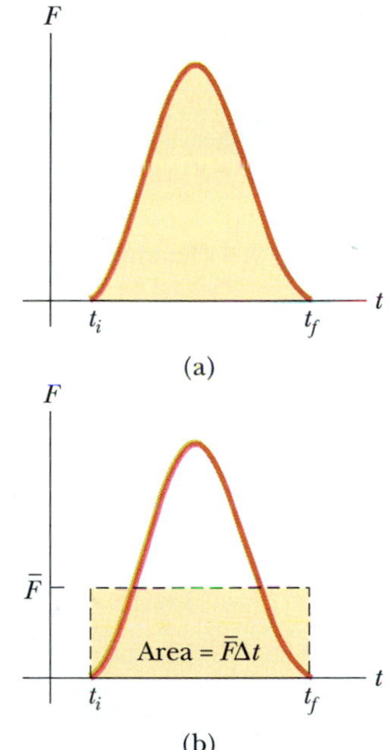

Figure 9.4
(a) A force acting on a particle may vary in time. The impulse is the area under the force-versus-time curve. (b) The average force (horizontal line) would give the same impulse to the particle in the time Δt as the real time-varying force described in (a).

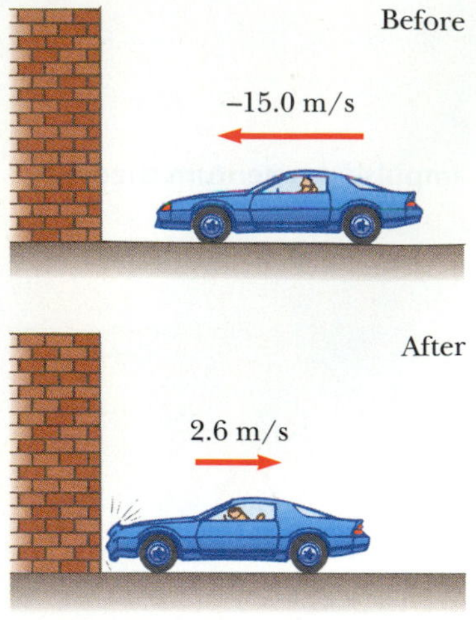

Figure 9.5 (Example 9.2)

Example 9.3 How Good Are Bumpers?

In a particular crash test, an automobile of mass 1500 kg collides with a wall as in Figure 9.5. The velocities of the automobile are $\mathbf{v}_i = -15.0\mathbf{i}$ m/s and $\mathbf{v}_f = 2.6\mathbf{i}$ m/s. If the collision lasts for 0.150 s, find the impulse due to the collision and the average force exerted on the automobile.

Solution The initial and final momenta of the automobile are given by

$$\mathbf{p}_i = m\mathbf{v}_i$$

$$= (1500 \text{ kg})(-15.0\mathbf{i} \text{ m/s}) = -2.25 \times 10^4\mathbf{i} \text{ kg}\cdot\text{m/s}$$

$$\mathbf{p}_f = m\mathbf{v}_f = (1500 \text{ kg})(2.6\mathbf{i} \text{ m/s}) = 0.39 \times 10^4\mathbf{i} \text{ kg}\cdot\text{m/s}$$

Hence, the impulse, which equals the change in momentum, is

$$\mathbf{I} = \Delta\mathbf{p} = \mathbf{p}_f - \mathbf{p}_i$$

$$= 0.39 \times 10^4\mathbf{i} \text{ kg}\cdot\text{m/s} - (-2.25 \times 10^4\mathbf{i} \text{ kg}\cdot\text{m/s})$$

$$\mathbf{I} = 2.64 \times 10^4\mathbf{i} \text{ kg}\cdot\text{m/s}$$

The average force exerted on the automobile is

$$\overline{\mathbf{F}} = \frac{\Delta\mathbf{p}}{\Delta t} = \frac{2.64 \times 10^4\mathbf{i} \text{ kg}\cdot\text{m/s}}{0.150 \text{ s}} = 1.76 \times 10^5\mathbf{i} \text{ N}$$

The photograph of a car crash test (an inelastic collision) illustrates that much of the car's initial kinetic energy is transformed into the energy it took to damage the vehicle. Why do safety belts and air bags help prevent serious injury in such collisions? *(Courtesy of General Motors)*

9.3 Collisions

In this section we shall use the law of conservation of momentum to describe what happens when two objects collide with each other. *The force due to the collision is assumed to be much larger than any external forces present.*

A collision may be the result of physical contact between two objects, as described in Figure 9.6a. This is a common observation when two macroscopic objects, such as two billiard balls or a baseball and a bat, collide. The notion of what we mean by *collision* must be generalized since "contact" on a submicroscopic scale is ill-defined and hence meaningless. More accurately, forces between two bodies arise from the electrostatic interaction of the electrons in the surface atoms of the bodies.

To understand this distinction between macroscopic and microscopic collisions, consider the collision of a proton with an alpha particle (the nucleus of the helium atom), such as occurs in Figure 9.6b. Because the two particles are positively charged, they repel each other.

When two particles of masses m_1 and m_2 collide, the collision forces may vary in time in a complicated way (Fig. 9.7). If $\mathbf{F}_{12}$ is the force on m_1 due to m_2, then the change in momentum of m_1 due to the collision is given by Equation 9.8:

$$\Delta\mathbf{p}_1 = \int_{t_i}^{t_f} \mathbf{F}_{12}\, dt$$

Likewise, if $\mathbf{F}_{21}$ is the force on m_2 due to m_1, the change in momentum of m_2 is

$$\Delta \mathbf{p}_2 = \int_{t_i}^{t_f} \mathbf{F}_{21}\, dt$$

However, Newton's third law states that the force on m_1 due to m_2 is equal to and opposite the force on m_2 due to m_1; that is, $\mathbf{F}_{12} = -\mathbf{F}_{21}$. (This is described graphically in Fig. 9.7.) Hence, we conclude that

$$\Delta \mathbf{p}_1 = -\Delta \mathbf{p}_2$$

$$\Delta \mathbf{p}_1 + \Delta \mathbf{p}_2 = 0$$

Since the total momentum of the system is $\mathbf{P} = \mathbf{p}_1 + \mathbf{p}_2$, we conclude that the *change* in the momentum of the system due to the collision is zero; that is,

$$\mathbf{P} = \mathbf{p}_1 + \mathbf{p}_2 = \text{constant}$$

This is precisely what we expect if no external forces are acting on the system (Section 9.2). However, the result is also valid if we consider the motion just before and just after the collision. Since the forces due to the collision are internal to the system, they do not affect the total momentum of the system. Therefore, for any type of collision, the total momentum of the system just before the collision equals the total momentum of the system just after the collision.

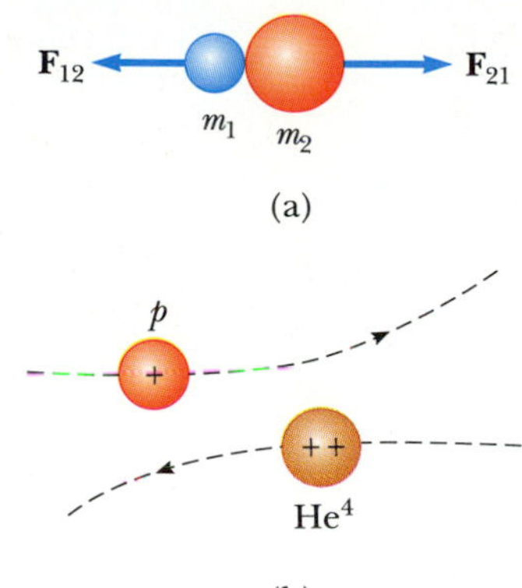

Figure 9.6
(a) The collision between two objects as the result of direct contact. (b) The collision between two charged particles.

Example 9.4 The Cadillac Versus the "Beetle"

A large luxury car with a mass of 1800 kg sitting at a traffic light is struck from the rear by a compact car with a mass of 900 kg. The two cars become entangled as a result of the collision. If the compact car was moving at a velocity of (20 m/s)**i** before the collision, what is the velocity of the entangled mass after the collision?

Solution The momentum before the collision is that of the compact car alone, because the large car was initially at rest. Thus, for the momentum before the collision we have

$$\mathbf{p}_i = m_1 \mathbf{v}_i = (900\text{ kg})(20\text{ m/s})\mathbf{i} = (1.80 \times 10^4\text{ kg}\cdot\text{m/s})\mathbf{i}$$

After the collision, the mass that moves is the sum of the masses of the large car and the compact car. The momentum of the combination is

$$\mathbf{p}_f = (m_1 + m_2)\mathbf{v}_f = (2700\text{ kg})(\mathbf{v}_f)$$

Equating the momentum before to the momentum after and solving for $\mathbf{v}_f$, the velocity of the wreckage, we have

$$\mathbf{v}_f = \frac{\mathbf{p}_i}{m_1 + m_2} = \frac{(1.80 \times 10^4\text{ kg}\cdot\text{m/s})\mathbf{i}}{2700\text{ kg}} = (6.67\text{ m/s})\mathbf{i}$$

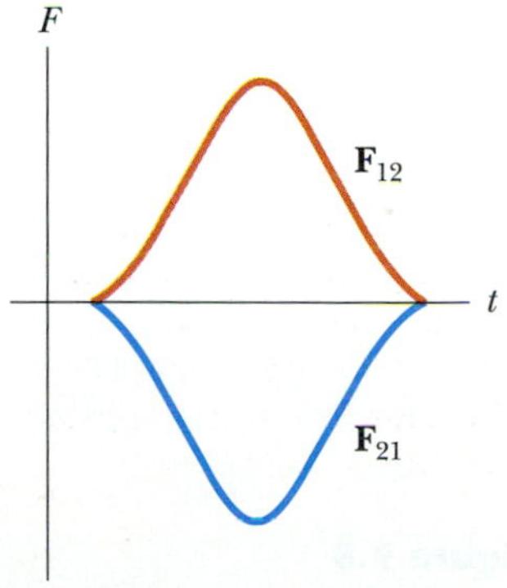

Figure 9.7
The force as a function of time for the two colliding particles described in Figure 9.6a. Note that $\mathbf{F}_{12} = -\mathbf{F}_{21}$.

Example 9.5 An Explosive Situation

A particle initially at rest explodes into two fragments whose masses are m_1 and m_2. How are the velocities of these two fragments related to each other?

Solution As with a collision, momentum does not change when an object explodes, because all explosive forces are internal forces. Thus,

$$\mathbf{p}_i = \mathbf{p}_f$$

Because the system's momentum is initially zero (the particle was at rest), the momentum remains zero after the explosion. Thus, this zero momentum must be distributed between the two fragments so that they carry away the same amounts of momentum in opposite directions from the explosion site. We have

$$0 = m_1\mathbf{v}_{1f} + m_2\mathbf{v}_{2f}$$

or

$$\mathbf{v}_{1f} = -\left(\frac{m_2}{m_1}\right)\mathbf{v}_{2f}$$

The minus sign tells us that the fragments move in opposite directions.

9.4 Elastic and Inelastic Collisions in One Dimension

As we have seen, momentum is conserved in any type of collision. Kinetic energy, however, is generally *not* constant in a collision because some is converted to thermal energy, internal elastic potential energy when the bodies deform, and rotational energy.

Inelastic collision

We define an **inelastic collision** as one in which *total kinetic energy is not constant (even though momentum is constant).* The collision of a rubber ball with a hard surface is inelastic because some of the ball's kinetic energy is lost when the ball deforms while in contact with the surface. When two objects collide and stick together after a collision, some kinetic energy is lost, and the collision is called **perfectly inelastic.** For example, if two vehicles collide and become entangled, as in Example 9.4, they move with some common velocity after the perfectly inelastic collision. If a meteorite collides with the Earth, it becomes buried, and the collision is perfectly inelastic.

Elastic collision

An **elastic collision** is defined as one in which *total kinetic energy is constant (as well as momentum).* Billiard-ball collisions and the collisions of air molecules with the walls of a container at ordinary temperatures are highly elastic. Real collisions in the macroscopic world, such as those between billiard balls, are only approximately elastic because some deformation and loss of kinetic energy takes place. Truly elastic collisions do occur, however, between atomic and subatomic particles. Elastic and perfectly inelastic collisions are *limiting* cases; most collisions fall into a category between them.

In the remainder of this section we treat collisions in one dimension and consider the two extreme cases—perfectly inelastic and elastic collisions. The important distinction between these two types of collisions is that *momentum is conserved in all cases, but kinetic energy is constant only in elastic collisions.*

Figure 9.8 A perfectly inelastic head-on collision between two particles.

Perfectly Inelastic Collisions

Consider two objects of masses m_1 and m_2, moving with initial velocities $\mathbf{v}_{1i}$ and $\mathbf{v}_{2i}$ along a straight line, as in Figure 9.8. If the two objects collide head on, stick together, and move with some common velocity $\mathbf{v}_f$ after the collision, the collision is perfectly inelastic. Since the total momentum before the collision equals the total momentum of the composite system after the collision, we have

$$m_1\mathbf{v}_{1i} + m_2\mathbf{v}_{2i} = (m_1 + m_2)\mathbf{v}_f \qquad [9.13]$$

$$\mathbf{v}_f = \frac{m_1\mathbf{v}_{1i} + m_2\mathbf{v}_{2i}}{m_1 + m_2} \qquad [9.14]$$

Elastic Collisions

Now consider two particles that undergo an elastic head-on collision (Fig. 9.9). In this case, both momentum and kinetic energy are constant; we can therefore write these conditions

$$m_1v_{1i} + m_2v_{2i} = m_1v_{1f} + m_2v_{2f} \qquad [9.15]$$

$$\tfrac{1}{2}m_1v_{1i}^2 + \tfrac{1}{2}m_2v_{2i}^2 = \tfrac{1}{2}m_1v_{1f}^2 + \tfrac{1}{2}m_2v_{2f}^2 \qquad [9.16]$$

where v is positive if a particle moves to the right and negative if it moves to the left.

In a typical problem involving elastic collisions, there are two unknown quantities, and Equations 9.15 and 9.16 can be solved simultaneously to find them. An alternative approach, employing a little mathematical manipulation of Equation 9.16, often simplifies the process. To see this, let's cancel the factor of $\frac{1}{2}$ in Equation 9.16 and rewrite it as

$$m_1(v_{1i}^2 - v_{1f}^2) = m_2(v_{2f}^2 - v_{2i}^2)$$

Here we have moved the terms containing m_1 to one side of the equation and those containing m_2 to the other. Next, let us factor both sides:

$$m_1(v_{1i} - v_{1f})(v_{1i} + v_{1f}) = m_2(v_{2f} - v_{2i})(v_{2f} + v_{2i}) \qquad [9.17]$$

We now separate the terms containing m_1 and m_2 for the conservation-of-momentum condition (Eq. 9.15) to get

$$m_1(v_{1i} - v_{1f}) = m_2(v_{2f} - v_{2i}) \qquad [9.18]$$

To obtain our final result, we divide Equation 9.17 by Equation 9.18 and get

$$v_{1i} + v_{1f} = v_{2f} + v_{2i}$$

or

$$v_{1i} - v_{2i} = -(v_{1f} - v_{2f}) \qquad [9.19]$$

This equation, in combination with the condition for conservation of momentum, can be used to solve problems dealing with perfectly elastic collisions. According to Equation 9.19, the relative velocity of the two objects before the collision, $v_{1i} - v_{2i}$, equals the negative of the relative velocity of the two objects after the collision, $-(v_{1f} - v_{2f})$.

Suppose that the masses and the initial velocities of both particles are known. Equations 9.15 and 9.16 can be solved for the final velocities in terms of the initial

Before collision

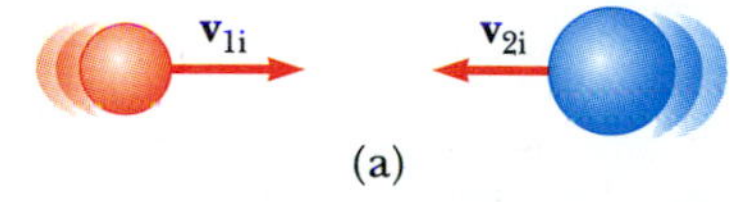

(a)

After collision

$\mathbf{v}_{1f}$ $\mathbf{v}_{2f}$

(b)

Figure 9.9 An elastic head-on collision between two particles.

velocities, since there are two equations and two unknowns. Solving for v_{1f} and v_{2f} gives

Elastic collision: relations between final and initial velocities

$$v_{1f} = \left(\frac{m_1 - m_2}{m_1 + m_2}\right) v_{1i} + \left(\frac{2m_2}{m_1 + m_2}\right) v_{2i} \qquad [9.20]$$

$$v_{2f} = \left(\frac{2m_1}{m_1 + m_2}\right) v_{1i} + \left(\frac{m_2 - m_1}{m_1 + m_2}\right) v_{2i} \qquad [9.21]$$

It is important to remember that the appropriate signs for v_{1i} and v_{2i} must be included in Equations 9.20 and 9.21, since velocities are vectors. For example, if m_2 is moving to the left initially, as in Figure 9.9, then v_{2i} is negative.

Let us consider some special cases. If $m_1 = m_2$, then we see that $v_{1f} = v_{2i}$ and $v_{2f} = v_{1i}$. That is, the particles exchange velocities if they have equal masses. This is what one observes in billiard-ball collisions.

If m_2 is initially at rest, $v_{2i} = 0$, and Equations 9.20 and 9.21 become

$$v_{1f} = \left(\frac{m_1 - m_2}{m_1 + m_2}\right) v_{1i} \qquad [9.22]$$

$$v_{2f} = \left(\frac{2m_1}{m_1 + m_2}\right) v_{1i} \qquad [9.23]$$

If m_1 is very large compared with m_2, we see from Equations 9.22 and 9.23 that $v_{1f} \approx v_{1i}$ and $v_{2f} \approx 2v_{1i}$. That is, when a very heavy particle collides head on with a very light one that is initially at rest, the heavy particle continues its motion unaltered after the collision, while the light particle rebounds with a velocity equal to about twice the initial velocity of the heavy particle. An example of this is the collision of a moving heavy atom, such as uranium, with a light atom, such as hydrogen.

If m_2 is much larger than m_1, and m_2 is initially at rest, then we find from Equations 9.22 and 9.23 that $v_{1f} \approx -v_{1i}$ and $v_{2f} \approx 0$. That is, when a very light particle collides head on with a very heavy particle that is initially at rest, the velocity of the light particle is reversed while the heavy particle remains approximately at rest. For example, imagine what happens when a marble hits a stationary bowling ball.

Example 9.6 Slowing Down Neutrons by Collisions

In a nuclear reactor, neutrons are produced when the $^{235}_{92}U$ nucleus undergoes fission. These neutrons are moving at high speeds, typically 10^7 m/s, and must be slowed down to about 10^3 m/s. Once the neutrons have slowed down, they have a high probability of producing another fission event and hence a sustained chain reaction. The high-speed neutrons can be slowed down by passing through a solid or liquid material called a *moderator.* The slowing-down process involves elastic collisions.

Let us show that a neutron can lose most of its kinetic energy if it collides elastically with a moderator containing light nuclei, such as deuterium and carbon. Hence, the moderator material is usually heavy water (D_2O) or graphite (which contains carbon nuclei).

Solution Let us assume that the moderator nucleus of mass m_2 is initially at rest and that the neutron of mass m_1 and initial velocity v_{1i} collides head on with it. Since momentum and energy are conserved, Equations 9.22 and 9.23 apply. The initial kinetic energy of the neutron is

$$K_{1i} = \frac{1}{2}m_1 v_{1i}^2$$

After the collision, the neutron has a kinetic energy given by $\frac{1}{2}m_1 v_{1f}^2$, where v_{1f} is given by Equation 9.22. We can express this energy as

$$K_{1f} = \frac{1}{2}m_1 v_{1f}^2 = \frac{m_1}{2}\left(\frac{m_1 - m_2}{m_1 + m_2}\right)^2 v_{1i}^2$$

Therefore, the *fraction* of the total kinetic energy possessed by the neutron *after* the collision is given by

$$(1) \qquad f_1 = \frac{K_{1f}}{K_{1i}} = \left(\frac{m_1 - m_2}{m_1 + m_2}\right)^2$$

From this result, we see that the final kinetic energy of the neutron is small when m_2 is close to m_1, and zero when $m_1 = m_2$.

We can use Equation 9.23 to calculate the kinetic energy of the moderator nucleus after the collision:

$$K_{2f} = \frac{1}{2}m_2 v_{2f}^2 = \frac{2m_1^2 m_2}{(m_1 + m_2)^2} v_{1i}^2$$

Hence, the fraction of the total kinetic energy transferred to the moderator nucleus is given by

$$(2) \qquad f_2 = \frac{K_{2f}}{K_{1i}} = \frac{4m_1 m_2}{(m_1 + m_2)^2}$$

Since the total energy is conserved, (2) can also be obtained from (1) with the condition that $f_1 + f_2 = 1$, so that $f_2 = 1 - f_1$.

Suppose that heavy water is used for the moderator. Collisions of the neutrons with deuterium nuclei in D_2O ($m_2 = 2m_1$) predict that $f_1 = 1/9$ and $f_2 = 8/9$. That is, 89% of the neutron's kinetic energy is transferred to the deuterium nucleus. In practice, the moderator efficiency is reduced because head-on collisions are very unlikely. How would the result differ if graphite were used as the moderator?

Example 9.7 A Two-Body Collision with Spring

A block of mass $m_1 = 1.60$ kg, moving to the right with a speed of 4.00 m/s on a frictionless, horizontal track, collides with a spring attached to a second block, of mass $m_2 = 2.10$ kg, that is moving to the left with a speed of 2.50 m/s (Fig. 9.10a). The

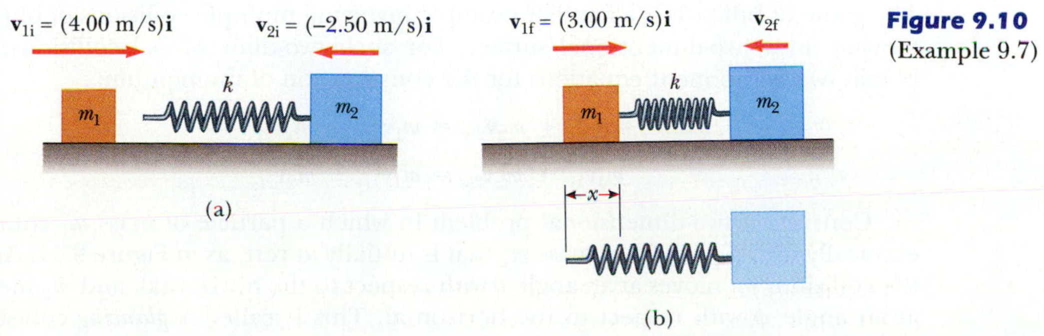

Figure 9.10 (Example 9.7)

spring constant is 600 N/m. For the instant when m_1 is moving to the right with a speed of 3.00 m/s, determine (a) the velocity of m_2 and (b) the distance x that the spring is compressed.

Solution

(a) First, note that the initial velocity of m_2 is $(-2.50\text{ m/s})\mathbf{i}$ because its direction is to the left. Since the total momentum of the system is constant, we have

$$m_1\mathbf{v}_{1i} + m_2\mathbf{v}_{2i} = m_1\mathbf{v}_{1f} + m_2\mathbf{v}_{2f}$$

$$(1.60\text{ kg})(4.00\text{ m/s})\mathbf{i} + (2.10\text{ kg})(-2.50\text{ m/s})\mathbf{i}$$
$$= (1.60\text{ kg})(3.00\text{ m/s})\mathbf{i} + (2.10\text{ kg})\mathbf{v}_{2f}$$

$$\mathbf{v}_{2f} = (-1.74\text{ m/s})\mathbf{i}$$

This result shows that m_2 is still moving leftward at that instant.

(b) To determine the compression in the spring, x, shown in Figure 9.10b, we can make use of conservation of energy since no friction forces are acting on the system. Thus, we have

$$\tfrac{1}{2}m_1v_{1i}^2 + \tfrac{1}{2}m_2v_{2i}^2 = \tfrac{1}{2}m_1v_{1f}^2 + \tfrac{1}{2}m_2v_{2f}^2 + \tfrac{1}{2}kx^2$$

Substitution of the given values and the result of part (a) into this expression gives

$$x = 0.173\text{ m}$$

Exercise Find the velocity of m_1 and the compression in the spring at the instant that m_2 is at rest.

Answer $(0.719\text{ m/s})\mathbf{i}$; 0.251 m.

9.5 Two-Dimensional Collisions

In Section 9.1 we showed that the total momentum of a system of particles is constant when the system is isolated (that is, when no external forces act on the system). For a general collision of two particles in three-dimensional space, the conservation-of-momentum law implies that the total momentum in each direction is constant. However, an important subset of collisions takes place in a plane. The game of billiards is a familiar example involving multiple collisions of objects moving on a two-dimensional surface. For such two-dimensional collisions, we obtain two component equations for the conservation of momentum:

$$m_1v_{1ix} + m_2v_{2ix} = m_1v_{1fx} + m_2v_{2fx}$$
$$m_1v_{1iy} + m_2v_{2iy} = m_1v_{1fy} + m_2v_{2fy}$$

Consider a two-dimensional problem in which a particle of mass m_1 collides elastically with a particle of mass m_2 that is initially at rest, as in Figure 9.11. After the collision, m_1 moves at an angle θ with respect to the horizontal, and m_2 moves at an angle ϕ with respect to the horizontal. This is called a *glancing* collision.

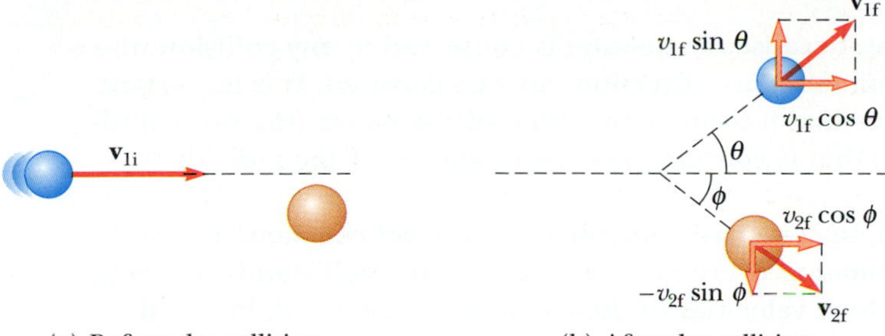

(a) Before the collision (b) After the collision

Figure 9.11
An elastic glancing collision between two particles.

Applying the law of conservation of momentum in component form, and noting that the total y component of momentum is zero, we get

$$x \text{ component:} \quad m_1 v_{1i} + 0 = m_1 v_{1f} \cos\theta + m_2 v_{2f} \cos\phi \qquad [9.24]$$

$$y \text{ component:} \quad 0 + 0 = m_1 v_{1f} \sin\theta - m_2 v_{2f} \sin\theta \qquad [9.25]$$

We now have two independent equations. So long as no more than two of the preceding quantities are unknown, we can solve the problem completely.

Because the collision is elastic, we can write a third equation for kinetic energy, in the form

Conservation of energy

$$\tfrac{1}{2} m_1 v_{1i}^2 = \tfrac{1}{2} m_1 v_{1f}^2 + \tfrac{1}{2} m_2 v_{2f}^2 \qquad [9.26]$$

If we know the initial velocity, v_{1i}, and the masses, we are left with four unknowns. Since we have only three equations, one of the four remaining quantities (v_{1f}, v_{2f}, θ, ϕ) must be given to determine the motion after the collision from conservation principles alone.

If the collision is inelastic, kinetic energy is *not* constant, and Equation 9.26 does *not* apply.

Problem-Solving Strategy: Collisions

The following procedure is recommended when dealing with problems involving collisions between two objects.

1. **Set up a coordinate system and define your velocities with respect to that system. It is convenient to have the x axis coincide with one of the initial velocities.**
2. **In your sketch of the coordinate system, draw all velocity vectors with labels and include all the given information.**
3. **Write expressions for the x and y components of the momentum of each object before and after the collision. Remember to include the appropriate signs for the components of the velocity vectors. For example, if an object is moving in the negative x direction, its x component of velocity must be taken to be negative. It is essential that you pay careful attention to signs.**
4. **Now write expressions for the *total* momenta in the x direction *before* and *after* the collision, and equate the two. Repeat this procedure for the total momenta in the y direction. These steps follow from**

the fact that, because momentum is conserved in any collision, the total momentum in any direction must be constant. It is important to emphasize that it is the momentum of the *system* (the two colliding objects) that is constant, not the momenta of the individual objects.

5. **If the collision is inelastic, kinetic energy is *not* constant, and additional information is probably required. If the collision is totally inelastic, the final velocities of the two objects are equal. Proceed to solve the momentum equations for the unknown quantities.**
6. **If the collision is elastic, kinetic energy is also constant, and you can equate the total kinetic energy before the collision to the total kinetic energy after the collision. This provides an additional relationship between the velocities.**

Example 9.8 Proton–Proton Collision

A proton collides in a perfectly elastic fashion with another proton that is initially at rest. The incoming proton has an initial speed of 3.5×10^5 m/s and makes a glancing collision with the second proton, as in Figure 9.11. (At close separations, the protons exert a repulsive electrostatic force on each other.) After the collision, one proton moves off at an angle of 37° to the original direction of motion, and the second deflects at an angle of ϕ to the same axis. Find the final speeds of the two protons and the angle ϕ.

Solution Both momentum and kinetic energy are constant in this glancing elastic collision. Since $m_1 = m_2$, $\theta = 37°$, and we are given $v_{1i} = 3.5 \times 10^5$ m/s, Equations 9.24, 9.25, and 9.26 become

$$v_{1f} \cos 37° + v_{2f} \cos \phi = 3.5 \times 10^5$$

$$v_{1f} \sin 37° - v_{2f} \sin \phi = 0$$

$$v_{1f}^2 + v_{2f}^2 = (3.5 \times 10^5)^2$$

Simultaneous solution of these three equations with three unknowns gives

$$v_{1f} = \boxed{2.80 \times 10^5 \text{ m/s}} \qquad v_{2f} = \boxed{2.11 \times 10^5 \text{ m/s}}$$

$$\phi = \boxed{53.0°}$$

It is interesting to note that $\theta + \phi = 90°$. This result is *not* accidental. *Whenever two equal masses collide elastically in a glancing collision and one of them is initially at rest, their final velocities are* always *at right angles to each other.*

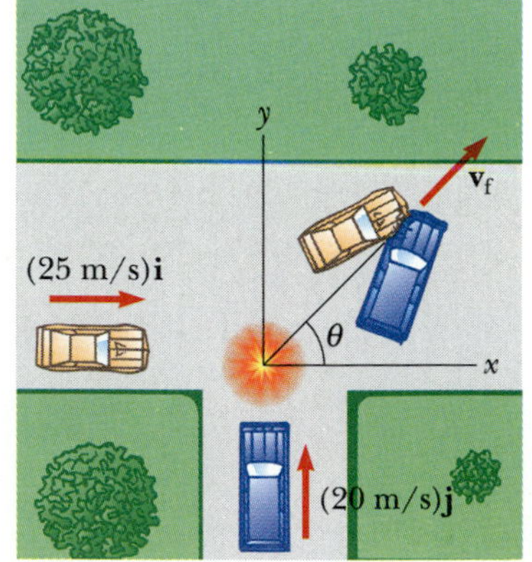

Figure 9.12 (Example 9.9) Top view of a car colliding with a van.

Example 9.9 Collision at an Intersection

A 1500-kg car, traveling east with a speed of 25 m/s, collides at an intersection with a 2500-kg van that is traveling north at a speed of 20 m/s, as shown in Figure 9.12. Find the direction and magnitude of the velocity of the wreckage after the collision, assuming that the vehicles undergo a perfectly inelastic collision (that is, they stick together).

Solution Let us choose east to be along the positive x direction and north to be along the positive y direction, as in Figure 9.12. Before the collision, the only object having momentum in the x direction is the car. Thus, the total initial momentum of the system (car plus van) in the x direction is

$$\sum \mathbf{p}_{xi} = (1500 \text{ kg})(25 \text{ m/s})\mathbf{i} = (37\ 500 \text{ kg}\cdot\text{m/s})\mathbf{i}$$

Now let us assume that the wreckage moves at an angle θ and speed v_f after the collision, as in Figure 9.12. The total momentum in the x direction after the collision is

$$\sum \mathbf{p}_{xf} = (4000 \text{ kg})(v_f \cos\theta)\mathbf{i}$$

Because momentum is constant in the x direction, we can equate these two equations to get

$$(1) \qquad 37\ 500 \text{ kg}\cdot\text{m/s} = (4000 \text{ kg})(v_f \cos\theta)$$

Similarly, the total initial momentum of the system in the y direction is that of the van, which has the value $(2500 \text{ kg})(20 \text{ m/s})\mathbf{j}$. Applying conservation of momentum to the y direction, we have

$$\sum \mathbf{p}_{yi} = \sum \mathbf{p}_{yf}$$

$$(2500 \text{ kg})(20 \text{ m/s})\mathbf{j} = (4000 \text{ kg})(v_f \sin\theta)\mathbf{j}$$

$$(2) \qquad 50\ 000 \text{ kg}\cdot\text{m/s} = (4000 \text{ kg})(v_f \sin\theta)$$

If we divide (2) by (1), we get

$$\tan\theta = \frac{50\ 000}{37\ 500} = 1.33$$

$$\theta = \boxed{53.1°}$$

When this angle is substituted into (2)—or, alternatively, into (1)—the value of v_f is

$$v_f = \frac{50\ 000 \text{ kg}\cdot\text{m/s}}{(4000 \text{ kg})(\sin 53°)} = \boxed{15.6 \text{ m/s}}$$

9.6 The Center of Mass

In this section we describe the overall motion of a mechanical system in terms of a very special point called the *center of mass* of the system. The mechanical system can be either a system of particles or any object. We shall see that the system moves as if all its mass were concentrated at the center of mass. Furthermore, if the resultant external force on the system is **F** and the total mass of the system is M, the center of mass moves with an acceleration given by $\mathbf{a} = \mathbf{F}/M$. That is, the system moves as if the resultant external force were applied to a single particle of mass M situated at the center of mass. This result was implicitly assumed in earlier chapters, where nearly all examples referred to the motions of extended objects.

Consider a system consisting of a pair of particles connected by a light, rigid rod (Fig. 9.13). The center of mass is somewhere on the rod and is closer to the larger mass. If a single force is applied at some point on the rod that is closer to the

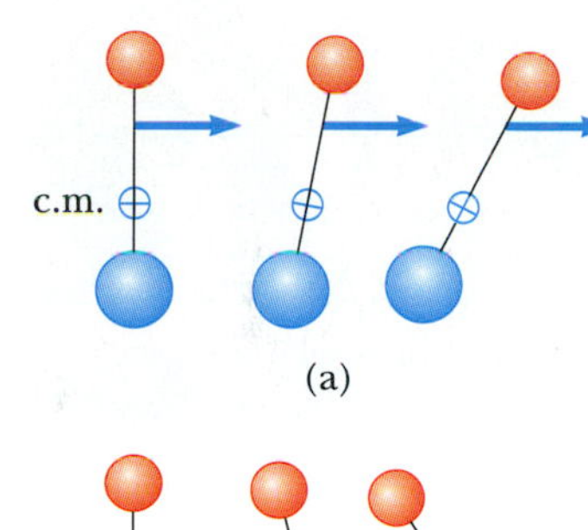

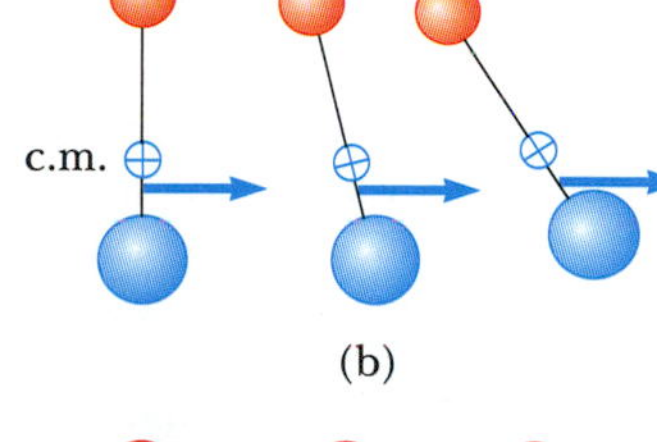

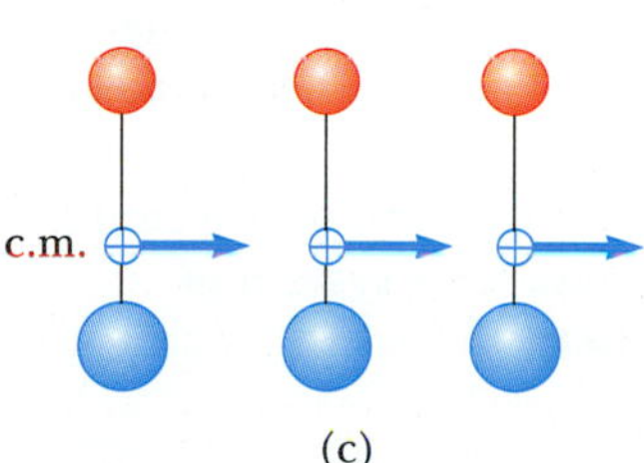

Figure 9.13 Two unequal masses are connected by a light, rigid rod. (a) The system rotates clockwise when a force is applied above the center of mass. (b) The system rotates counterclockwise when a force is applied below the center of mass. (c) The system moves in the direction of the applied force without rotating when the force is applied at the center of mass.

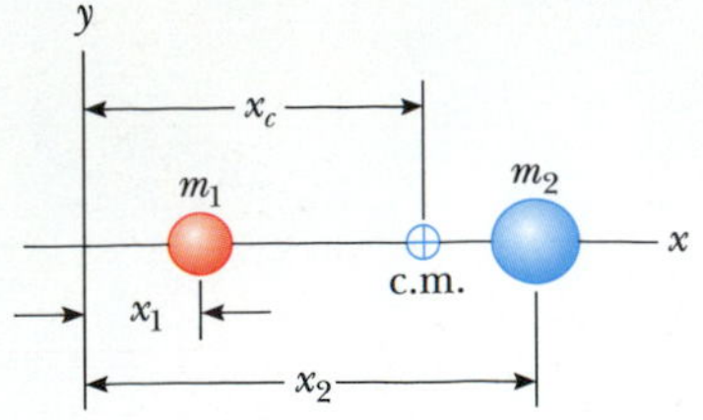

Figure 9.14
The center of mass of two particles on the x axis is located at x_c, a point between the particles, that is closer to the larger mass.

When this gymnast is in the balanced position on the beam (first frame), her center of mass must be directly above the midpoint of her feet. Describe the motion of her center of mass as she undergoes the exercise shown in the last two frames. *(© Jerry Wachter, Photo Researchers, Inc.)*

smaller mass than to the larger one, the system will rotate clockwise (Fig. 9.13a). If the force is applied at a point on the rod closer to the larger mass, the system will rotate counterclockwise (Fig. 9.13b). If the force is applied at the center of mass, the system will move in the direction of **F** without rotating (Fig. 9.13c). Thus, the center of mass can be easily located.

The position of the center of mass of a system can be described as the *average position* of the system's mass. For example, the center of mass of the pair of particles described in Figure 9.14 lies on the x axis, somewhere between the particles. The x coordinate of the center of mass in this case is defined to be

$$x_c \equiv \frac{m_1 x_1 + m_2 x_2}{m_1 + m_2} \qquad [9.27]$$

For example, if $x_1 = 0$, $x_2 = d$, and $m_2 = 2m_1$, we find that $x_c = \frac{2}{3}d$. That is, the center of mass lies closer to the more massive particle. If the two masses are equal, the center of mass lies midway between the particles.

We can extend the center-of-mass concept to a system of many particles in three dimensions. The x coordinate of the center of mass of n particles is defined to be

$$x_c \equiv \frac{m_1 x_1 + m_2 x_2 + m_3 x_3 + \ldots + m_n x_n}{m_1 + m_2 + m_3 + \ldots + m_n} = \frac{\Sigma m_i x_i}{\Sigma m_i} \qquad [9.28]$$

where x_i is the x coordinate of the ith particle and Σm_i is the *total mass* of the system. For convenience, we shall express the total mass as $M = \Sigma m_i$, where the sum runs over all n particles. The y and z coordinates of the center of mass are similarly defined by the equations

$$y_c \equiv \frac{\Sigma m_i y_i}{M} \quad \text{and} \quad z_c \equiv \frac{\Sigma m_i z_i}{M} \qquad [9.29]$$

The center of mass can also be located by its position vector, $\mathbf{r}_c$. The rectangular coordinates of this vector are x_c, y_c, and z_c, defined in Equations 9.28 and 9.29. Therefore,

Vector position of the center of mass for a system of particles

$$\mathbf{r} = x_c\mathbf{i} + y_c\mathbf{j} + z_c\mathbf{k} = \frac{\Sigma m_i x_i \mathbf{i} + \Sigma m_i y_i \mathbf{j} + \Sigma m_i z_i \mathbf{k}}{M}$$

$$\mathbf{r}_c \equiv \frac{\Sigma m_i \mathbf{r}_i}{M} \qquad [9.30]$$

where $\mathbf{r}_i$ is the position vector of the ith particle, defined by

$$\mathbf{r}_i \equiv x_i\mathbf{i} + y_i\mathbf{j} + z_i\mathbf{k}$$

The center of mass of a homogeneous, symmetric body must lie on an axis of symmetry. For example, the center of mass of a homogeneous rod must lie midway between the ends of the rod. The center of mass of a homogeneous sphere or a homogeneous cube must lie at the geometric center of the object. One can determine the center of mass of an irregularly shaped object, such as a wrench, experimentally by suspending the wrench from two different points (Fig. 9.15). The wrench is first hung from point A, and a vertical line AB is drawn (which can be established with a plumb bob) when the wrench is in equilibrium. The wrench is then hung from point C, and a second vertical line, CD, is drawn. The center of mass coincides with the intersection of these two lines. In fact, if the wrench is hung freely from any point, the vertical line through that point must pass through the center of mass.

Since a rigid body is a continuous distribution of mass, each portion is acted upon by the force of gravity. The net effect of all of these forces is equivalent to the effect of a single force, $M\mathbf{g}$, acting through a special point called the **center of gravity.** If $\mathbf{g}$ is constant over the mass distribution, then the center of gravity coincides with the center of mass. If a rigid body is pivoted at its center of gravity, it will be balanced in any orientation.

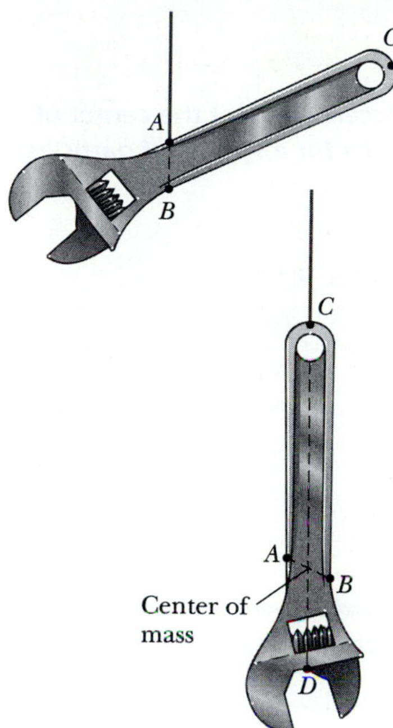

Figure 9.15 An experimental technique for determining the center of mass of a wrench. The wrench is hung freely from two different pivots, *A* and *C*. The intersection of the two vertical lines *AB* and *CD* locates the center of mass.

Example 9.10 Where Is the Center of Mass?

Three particles are located in a coordinate system, as shown in Figure 9.16. Find the location of the center of mass.

Solution The y coordinate of the center of mass is zero because all particles are on the x axis. To find the x coordinate of the center of mass, we use Equation 9.28:

$$x_c = \frac{\sum m_i x_i}{\sum m_i}$$

For the numerator, we find

$$\begin{aligned}\sum m_i x_i &= m_1 x_1 + m_2 x_2 + m_3 x_3 \\ &= (5\text{ kg})(-0.5\text{ m}) + (2\text{ kg})(0\text{ m}) + (4\text{ kg})(1\text{ m}) \\ &= 1.5\text{ kg}\cdot\text{m}\end{aligned}$$

The denominator is $\sum m_i = 11$ kg; therefore,

$$x_c = \frac{1.5\text{ kg}\cdot\text{m}}{11\text{ kg}} = 0.136\text{ m}$$

Exercise If a fourth particle of mass 2 kg is placed at the position $x = 0$ and $y = 0.25$ m, find the x and y coordinates of the center of mass for this system of four particles.

Answer $x_c = 0.115$ m; $y_c = 0.038$ m.

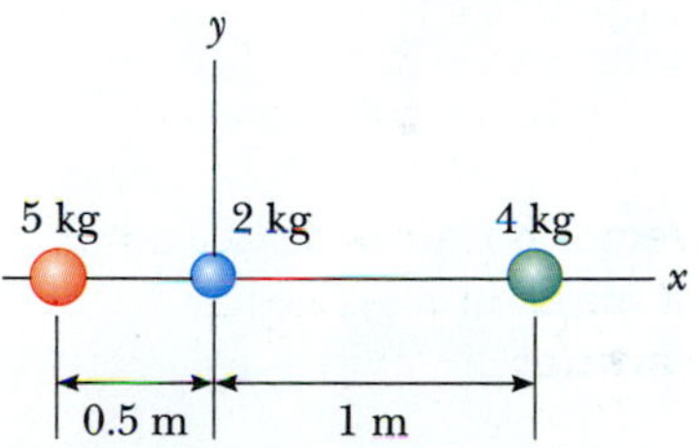

Figure 9.16 (Example 9.10) Locating the center of mass for a system of three particles.

9.7 Motion of a System of Particles

We can begin to understand the physical significance and utility of the center-of-mass concept by taking the time derivative of the position vector of the center of mass $\mathbf{r}_c$, given by Equation 9.30. Assuming that M remains constant—that is, no particles enter or leave the system—we get the following expression for the **velocity of the center of mass:**

Velocity of the center of mass

$$\mathbf{v}_c = \frac{d\mathbf{r}_c}{dt} = \frac{1}{M}\sum m_i \frac{d\mathbf{r}_i}{dt} = \frac{\sum m_i \mathbf{v}_i}{M} \qquad [9.31]$$

where $\mathbf{v}_i$ is the velocity of the ith particle. Rearranging Equation 9.31 gives

Total momentum of a system of particles

$$M\mathbf{v}_c = \sum m_i \mathbf{v}_i = \sum \mathbf{p}_i = \mathbf{P} \qquad [9.32]$$

This result tells us that *the total momentum of the system equals the total mass multiplied by the velocity of the center of mass*—in other words, the total momentum of a single particle of mass M moving with a velocity $\mathbf{v}_c$.

If we now differentiate Equation 9.32 with respect to time, we get the **acceleration of the center of mass:**

Acceleration of the center of mass for a system of particles

$$\mathbf{a}_c = \frac{d\mathbf{v}_c}{dt} = \frac{1}{M}\sum m_i \frac{d\mathbf{v}_i}{dt} = \frac{1}{M}\sum m_i \mathbf{a}_i \qquad [9.33]$$

Rearranging this expression and using Newton's second law, we get

$$M\mathbf{a}_c = \sum m_i \mathbf{a}_i = \sum \mathbf{F}_i \qquad [9.34]$$

where $\mathbf{F}_i$ is the force on particle i.

The forces on any particle in the system may include both external and internal forces. However, by Newton's third law, the force of particle 1 on particle 2, for example, is equal to and opposite the force of particle 2 on particle 1. When we sum over all internal forces in Equation 9.34, they cancel in pairs and the net force on the system is due *only* to external forces. Thus, we can write Equation 9.34 in the form

Newton's second law for a system of particles

$$\sum \mathbf{F}_{\text{ext}} = M\mathbf{a}_c = \frac{d\mathbf{P}}{dt} \qquad [9.35]$$

That is, the resultant external force on the system of particles equals the total mass of the system multiplied by the acceleration of the center of mass. If we compare this to Newton's second law for a single particle, we see that the center of mass moves like an imaginary particle of mass M under the influence of the resultant external force on the system. In the absence of external forces, the center of mass moves with uniform velocity, as in the case of the rotating wrench in Figure 9.17. If the resultant force acts along a line through the center of mass of an extended body such as the wrench, the body is accelerated without rotation, and its kinetic energy is associated entirely with its linear motion. If the resultant force does not act through the center of mass, the body will be rotationally accelerated, and the body will acquire a rotational kinetic energy in addition to the kinetic energy of its linear motion. The linear acceleration of the center of mass is the same in either case, as given by Equation 9.35.

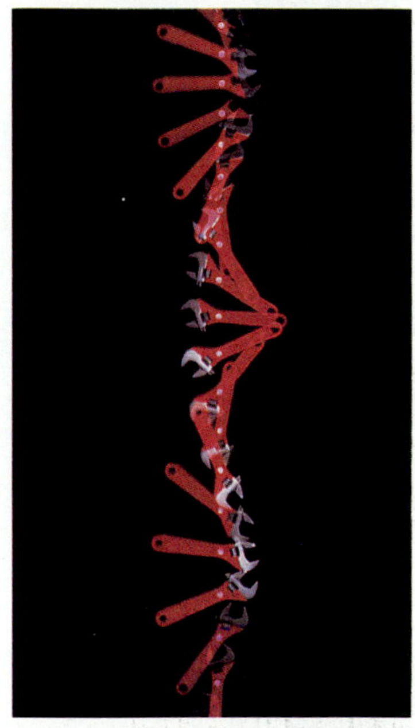

Figure 9.17 Multiflash photograph of a wrench moving on a horizontal surface. The center of mass of the wrench moves in a straight line as the wrench rotates about this point as marked with a white dot. *(Richard Megna, Fundamental Photographs)*

Finally, we see that if the resultant external force is zero, then from Equation 9.35 it follows that

$$\frac{d\mathbf{P}}{dt} = M\mathbf{a}_c = 0$$

so that

$$\mathbf{P} = M\mathbf{v}_c = \text{constant} \qquad \left(\text{when } \sum \mathbf{F}_{\text{ext}} = 0\right) \qquad [9.36]$$

That is, the total linear momentum of a system of particles is constant if no external forces are acting on the system. It follows that, for an *isolated* system of particles, both the total momentum and the velocity of the center of mass are constant in time. The law of conservation of momentum that was derived in Section 9.1 for a two-particle system is thus generalized to a many-particle system.

Suppose an isolated system consisting of two or more members is at rest. The center of mass of such a system will remain at rest unless acted upon by an external force. For example, consider a system, initially at rest, that is made up of a swimmer and a raft. When the swimmer dives off the raft, the center of mass of the system remains at rest (if we neglect the friction between raft and water). Furthermore, the momentum of the diver is equal in magnitude to the momentum of the raft, but opposite in direction.

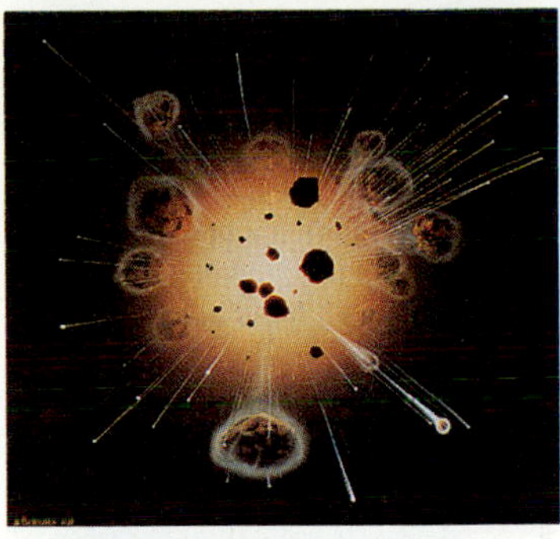

When an object explodes into many fragments, the center of mass of the fragments follows the same parabolic path that the object would have taken had there been no explosion. If the object is at rest before the explosion, the center of mass of the fragments after the explosion must also be at rest. *(Sally Bensusen, Science Photo Library).*

*9.8 Rocket Propulsion

When ordinary vehicles, such as automobiles and locomotives, are propelled, the driving force for the motion is that of friction. In the case of the automobile, the driving force is the force of the road on the car. A locomotive "pushes" against the tracks; hence, the driving force is the reaction force of the tracks on the locomotive. However, a rocket moving in space has no road or tracks to "push" against. Therefore, the source of the propulsion of a rocket must be different. *The operation of a rocket depends upon the law of conservation of momentum as applied to a system of particles, where the system is the rocket plus its ejected fuel.*

The propulsion of a rocket (Fig. 9.18) can be understood by first considering the mechnical system, consisting of a machine gun mounted on a cart on wheels. As the machine gun is fired, each bullet receives a momentum $m\mathbf{v}$ in some direction, where $\mathbf{v}$ is measured with respect to a stationary Earth frame. For each bullet that is fired, the gun and cart must receive a compensating momentum in the opposite direction. That is, the reaction force of the bullet on the gun accelerates the cart and gun. If n bullets are fired each second, then the average force on the gun is equal to $\mathbf{F}_{av} = nm\mathbf{v}$.

Figure 9.18
Lift-off of the space shuttle Columbia. Massive amounts of thrust are generated by the shuttle's liquid-fueled engines, aided by the two solid-fuel boosters. Many physical principles in the areas of mechanics, thermodynamics, and electricity and magnetism are involved in the lift-off. *(Courtesy of NASA).*

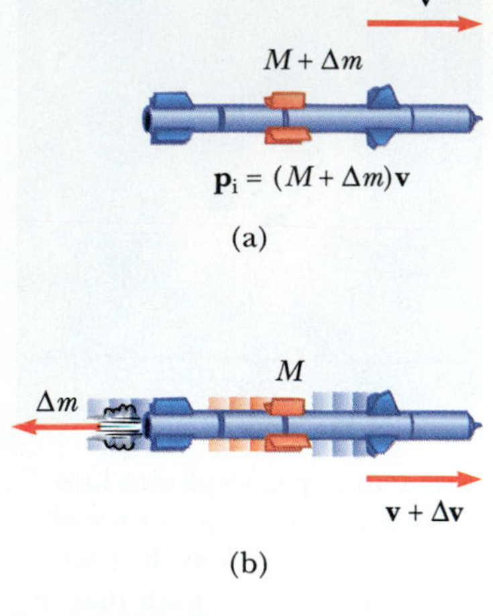

Figure 9.19
Rocket propulsion. (a) The initial mass of the rocket is $M + \Delta m$ at a time t, and its speed is v. (b) At a time $t + \Delta t$, the rocket's mass has reduced to M, and an amount of fuel Δm has been ejected. The rocket's speed increases by an amount Δv.

In a similar manner, as a rocket moves in free space (a vacuum), *its momentum changes when some of its mass is released in the form of ejected gases. Since the ejected gases acquire some momentum, the rocket receives a compensating momentum in the opposite direction.* Therefore, *the rocket is accelerated as a result of the "push," or thrust, from the exhaust gases.* In free space, the center of mass of the system moves uniformly, independent of the propulsion process. It is interesting to note that the rocket and machine gun represent cases of the *inverse* of an inelastic collision; that is, momentum is conserved, but the kinetic energy of the system is *increased* (at the expense of internal energy).

Suppose that at some time t, the momentum of the rocket plus the fuel is $(M + \Delta m)v$ (Fig. 9.19a). At some short time—Δt—later, the rocket ejects some fuel of mass Δm and the rocket's speed therefore increases to $v + \Delta v$ (Fig. 9.19b). If the fuel is ejected with a velocity v_e *relative to the rocket,* then the velocity of the fuel relative to a stationary frame of reference is $v - v_e$. Thus, if we equate the total initial momentum of the system with the total final momentum, we get

$$(M + \Delta m)v = M(v + \Delta v) + \Delta m(v - v_e)$$

Simplification of this expression gives

$$M\Delta v = v_e \Delta m$$

We can also arrive at this result by considering the system in the center-of-mass frame of reference—that is, a frame whose velocity equals the center-of-mass velocity. In this frame, the total momentum is zero; therefore, if the rocket gains a momentum $M\Delta v$ by ejecting some fuel, the exhaust gases obtain a momentum $v_e \Delta m$ in the *opposite* direction, and so $M\Delta v - v_e \Delta m = 0$. If we now take the limit as Δt goes to zero, the $\Delta v \rightarrow dv$ and $\Delta m \rightarrow dm$. Furthermore, the increase in the exhaust mass, dm, corresponds to an equal decrease in the rocket mass, so that $dm = -dM$. Note that dM is given a negative sign because it represents a decrease in mass. Using this fact, we get

$$M\,dv = -v_e\,dM \qquad [9.37]$$

Integrating this equation, and taking the initial mass of the rocket plus fuel to be M_i and the final mass of the rocket plus its remaining fuel to be M_f, we get

$$\int_{v_i}^{v_f} dv = -v_e \int_{M_i}^{M_f} \frac{dM}{M}$$

Basic expression for rocket propulsion

$$v_f - v_i = v_e \ln\left(\frac{M_i}{M_f}\right) \qquad [9.38]$$

This is the basic expression of rocket propulsion. First, it tells us that the increase in velocity is proportional to the exhaust velocity, v_e. Therefore, the exhaust velocity should be very high. Second, the increase in velocity is proportional to the logarithm of the ratio M_i/M_f. Therefore, this ratio should be as large as possible, which means that the rocket should carry as much fuel as possible.

The *thrust* on the rocket is the force exerted on it by the ejected exhaust gases. We can obtain an expression for the thrust from Equation 9.37:

$$\text{Thrust} = M\frac{dv}{dt} = \left| v_e \frac{dM}{dt} \right| \qquad [9.39]$$

Here we see that the thrust increases as the exhaust velocity increases and as the rate of change of mass (burn rate) increases.

Example 9.11 A Rocket in Space

A rocket moving in free space has a speed of 3×10^3 m/s. Its engines are turned on, and fuel is ejected in a direction opposite the rocket's motion at a speed of 5×10^3 m/s relative to the rocket.

(a) What is the speed of the rocket once its mass is reduced to one-half its mass before ignition?

Solution Applying Equation 9.38, we get

$$v_f = v_i + v_e \ln\left(\frac{M_i}{M_f}\right)$$

$$= 3 \times 10^3 + 5 \times 10^3 \ln\left(\frac{M_i}{0.5M_i}\right)$$

$$= 6.47 \times 10^3 \text{ m/s}$$

(b) What is the thrust on the rocket if it burns fuel at the rate of 50 kg/s?

Solution

$$\text{Thrust} = \left| v_e \frac{dM}{dt} \right| = (5 \times 10^3 \text{ m/s})(50 \text{ kg/s})$$

$$= 2.50 \times 10^5 \text{ N}$$

Summary

The **linear momentum** of any object of mass m moving with a velocity **v** is defined to be

$$\mathbf{p} \equiv m\mathbf{v} \qquad [9.1]$$

Conservation of momentum applied to two interacting objects states that if the two objects form an isolated system, their total momentum is constant regardless of the nature of the force between them. Therefore, the total momentum of the system at all times equals its initial total momentum:

$$\mathbf{p}_{1i} + \mathbf{p}_{2i} = \mathbf{p}_{1f} + \mathbf{p}_{2f} \qquad [9.5]$$

Conservation of momentum

The **impulse** of a force **F** on any object is equal to the change in the momentum of the object and is given by

$$\mathbf{I} = \int_{t_i}^{t_f} \mathbf{F}\, dt = \Delta\mathbf{p} \qquad [9.9]$$

Impulse-momentum theorem

This is called the **impulse-momentum theorem.**

Impulsive forces are forces that are very strong compared with other forces on the system. They usually act for a very short time, as in the case of collisions.

When two bodies collide, the total momentum of the system before the collision always equals the total momentum after the collision, regardless of the nature of the collision. An **inelastic collision** is a collision for which the kinetic energy is not constant, but momentum is. A perfectly inelastic collision is one in which the colliding bodies stick together after the collision. An **elastic collision** is one in which both momentum and kinetic energy are constant.

Elastic and inelastic collision

In a two- or three-dimensional collision, the components of momentum in each direction are conserved independently.

The **vector position of the center of mass of a system of particles** is defined as

Center of mass for a system of particles

$$\mathbf{r}_c \equiv \frac{\Sigma m_i \mathbf{r}_i}{M} \quad [9.30]$$

where $M = \Sigma m_i$ is the total mass of the system and $\mathbf{r}_i$ is the vector position of the ith particle.

The **velocity of the center of mass for a system of particles** is given by

Velocity of the center of mass

$$\mathbf{v}_c = \frac{\Sigma m_i v_i}{M} \quad [9.31]$$

The total momentum of a system of particles equals the total mass multiplied by the velocity of the center of mass, that is, $\mathbf{P} = M\mathbf{v}_c$.

Newton's second law applied to a system of particles is given by

Newton's second law for a system of particles

$$\sum \mathbf{F}_{\text{ext}} = M\mathbf{a}_c = \frac{d\mathbf{P}}{dt} \quad [9.35]$$

where $\mathbf{a}_c$ is the acceleration of the center of mass and the sum is over all external forces. Therefore, the center of mass moves like an imaginary particle of mass M under the influence of the resultant external force on the system. It follows from Equation 9.35 that the total momentum of the system is constant if no external forces are acting on the system.

Questions and Conceptual Exercises

1. If the velocity of a particle is doubled, by what factor is its momentum changed? What happens to its kinetic energy?
2. If two particles have equal kinetic energies, are their momenta necessarily equal? Explain.
3. Does a large force always produce a larger impulse on a body than a smaller force does? Explain.
4. If two objects collide and one is initially at rest, is it possible for both to be at rest after the collision? Is it possible for one to be at rest after the collision? Explain.
5. Explain how momentum is conserved when a ball bounces from a floor.
6. Is it possible to have a collision in which all of the kinetic energy is lost? If so, cite an example.
7. When a ball rolls down an incline, its momentum increases. Does this imply that momentum is not conserved? Explain.
8. Consider a perfectly inelastic collision between a car and a large truck. Which vehicle loses more kinetic energy as a result of the collision?
9. Can the center of mass of a body lie outside the body? If so, give examples.
10. A boy stands at one end of a floating raft that is stationary relative to the shore. He then walks to the opposite

end of the raft, away from the shore. What happens to the center of mass of the system (boy + raft)? Does the raft move? Explain.

11. A meter stick is balanced in a horizontal position, resting on the extended index fingers of the right and left hands. If the two fingers are brought together, the stick remains balanced and the two fingers always meet at the 50-cm mark regardless of their original positions (try it!). Carefully explain this observation.
12. A researcher tranquilizes a polar bear on a frictionless glacier. How might the researcher, knowing her own weight, be able to *estimate* the weight of the polar bear using a measuring tape and a rope?
13. A sharpshooter fires a rifle while standing with the butt of the gun against his shoulder. If the forward momentum of a bullet is the same as the backward momentum of the gun, why isn't it as dangerous to be hit by the gun as by the bullet?
14. Early in this century, Robert Goddard proposed sending a rocket to the Moon. Critics took the position that in a vacuum such as exists between the Earth and the Moon, the gases emitted by the rocket would have nothing to push against to propel the rocket. According to *Scientific American* (January 1975), Goddard placed a gun in a vacuum and fired a blank cartridge from it. (A blank cartridge fires only the wadding and hot gases of the burning gunpowder.) What happened when the gun was fired?
15. A pole-vaulter falls from a height of 5 m onto a foam-rubber pad. Estimate his velocity just before he reaches the pad. Could you calculate the force exerted on him due to the collision? Explain.
16. As a ball falls toward the Earth, its momentum increases. How would you reconcile this fact with the law of conservation of momentum?
17. Does the center of mass of a rocket in free space accelerate? Explain. Can the speed of a rocket exceed the exhaust speed of the fuel? Explain.
18. A skater is standing still on a frictionless ice rink. Her friend throws a Frisbee straight at her. In which of the following cases is the greatest momentum transferred to the skater? She (a) catches the Frisbee and holds it; (b) catches it momentarily but drops it vertically down; (c) catches the Frisbee momentarily and then throws it back to her friend.
19. As viewed by us on the Earth, the Moon revolves around the Earth. Is the Moon's linear momentum constant? Is its kinetic energy constant? Assume, for simplicity, that the Moon's orbit is perfectly circular.
20. A large sheet is held at its edges by two students in such a way that it forms a nearly vertical "net" to catch an object. A third student, who happens to be the star pitcher on the baseball team, throws a raw egg into the sheet (after which the egg is caught). Explain why the egg does not break in the process, regardless of the initial speed of the egg. (If you try this one, make sure the pitcher hits the target near its center, and do not allow the egg to fall on the floor after being caught.)
21. If a raw egg is dropped, it falls apart upon impact with the floor. However, if you drop a raw egg onto a thick foam rubber cushion from a height of about 1 m, the egg rebounds without breaking. Why is this possible? (In this demonstration, be sure to catch the egg after the first bounce.)
22. Locate the center of gravity for each of the following uniform objects: (a) a sphere, (b) a cube, (c) a cylinder, (d) a doughnut.
23. A magician places some dishes and silverware on a table covered by a tablecloth. The magician proceeds to rapidly remove the tablecloth, apparently without disturbing the items on top. Explain how this trick is possible on the basis of what you have learned in this chapter.
24. "During a high jump the athlete's center of gravity passes under the bar." Can this statement be true? Discuss.
25. An airbag is inflated when a collision occurs, which protects the passenger (the dummy, in this case) from serious injury. Why does the airbag soften the blow? Discuss the physics involved in this dramatic photograph.

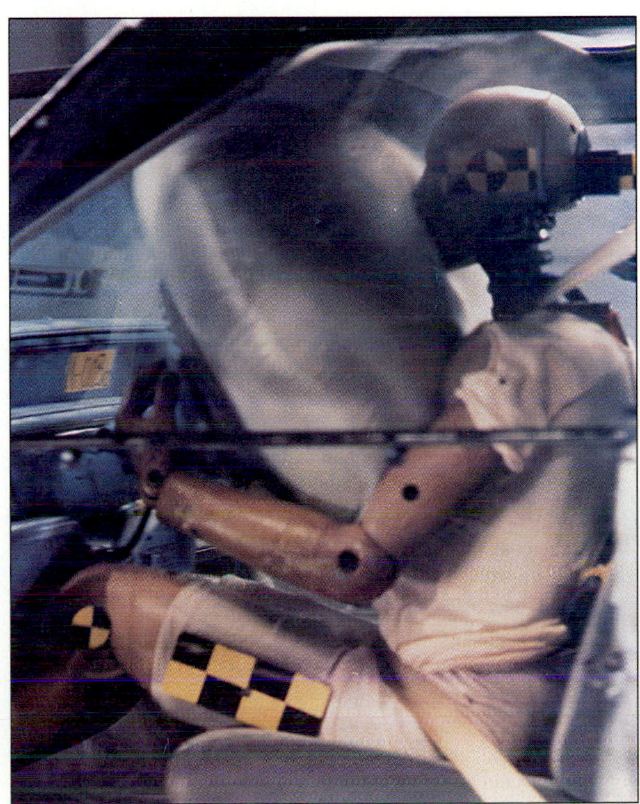

(Question 25) *(Courtesy of General Motors)*

Problems

Section 9.1 Linear Momentum and Its Conservation

1. The momentum of a 1250-kg car is equal to the momentum of a 5000-kg truck traveling at a speed of 10 m/s. What is the speed of the car?
2. A 3-kg particle has a velocity of $(3\mathbf{i} - 4\mathbf{j})$ m/s. Find its x and y components of momentum and the magnitude of the particle's total momentum.
3. Calculate the magnitude of the linear momentum for the following cases: (a) a proton moving at a speed of 5×10^6 m/s, (b) a 15-g bullet moving with a speed of 300 m/s; (c) a 75-kg sprinter running with a speed of 10 m/s; (d) the Earth (mass $= 5.98 \times 10^{24}$ kg) moving with an orbital speed equal to 2.98×10^4 m/s.
4. A 0.10-kg ball is thrown straight up into the air with an initial speed of 15 m/s. Find the momentum of the ball (a) at its maximum height and (b) halfway up to its maximum height.
5. A 1500-kg car moving with a speed of 15 m/s collides with a utility pole and is brought to rest in 0.3 s. Find the average force exerted on the car during the collision.
6. A 0.5-kg football is thrown with a speed of 15 m/s. A stationary receiver catches the ball and brings it to rest in 0.02 s. What is the average force exerted on the receiver?
7. A 40-kg child standing on a frozen pond throws a 0.5-kg stone to the east with a speed of 5 m/s. Neglecting friction between child and ice, find the recoil velocity of the child.
8. A rifle with a weight of 30 N fires a 5-g bullet with a speed of 300 m/s. (a) Find the recoil speed of the rifle. (b) A 700-N man holds the rifle firmly against his shoulder. Find the recoil speed of man and rifle.
9. A pitcher claims he can throw a baseball with as much momentum as a 3.0-g bullet moving with a speed of 1500 m/s. A baseball has a mass of 0.145 kg. What must be its speed if the pitcher's claim is valid?
10. Two blocks of masses M and $3M$ are placed on a horizontal, frictionless surface. A light spring is attached to one of them, and the blocks are pushed together with the spring between them (Fig. 9.20). A string holding them together is burned, after which the block of mass $3M$ moves to the right with a speed of 2 m/s. What is the speed of the block of mass M? (Assume the blocks are initially at rest.)
11. A 60-kg boy and a 40-kg girl, both wearing skates, face each other at rest. The boy pushes the girl, sending her eastward with a speed of 4 m/s. Describe the subsequent motion of the boy. (Neglect friction.)

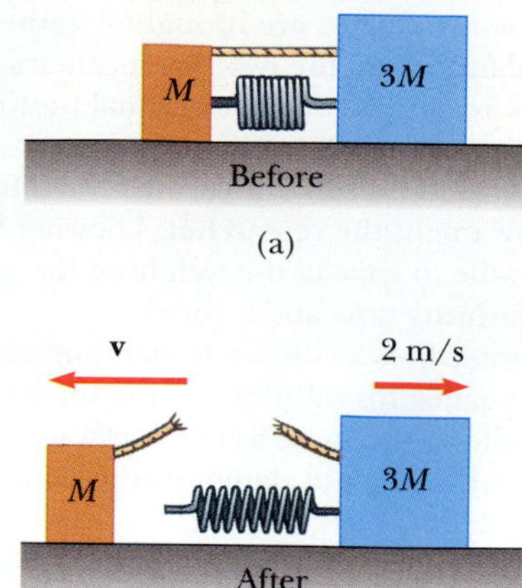

Figure 9.20 (Problem 10)

12. Identical air cars ($m = 200$ g) are equipped with identical linear springs ($k = 3000$ N/m). They move toward each other on a horizontal air track with speeds of 3.0 m/s and collide, compressing the springs (Fig. 9.21). Find the maximum compression of a spring.

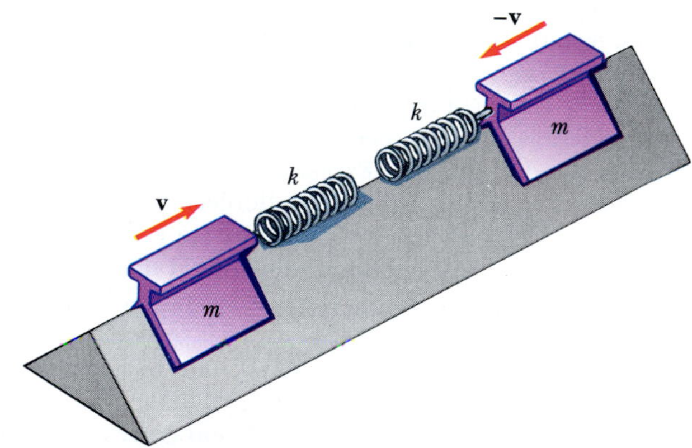

Figure 9.21 (Problem 12)

Section 9.2 Impulse and Momentum

13. The force F_x acting on a 2-kg particle varies in time as shown in Figure 9.22. Find (a) the impulse of the force, (b) the final velocity of the particle if it is ini-

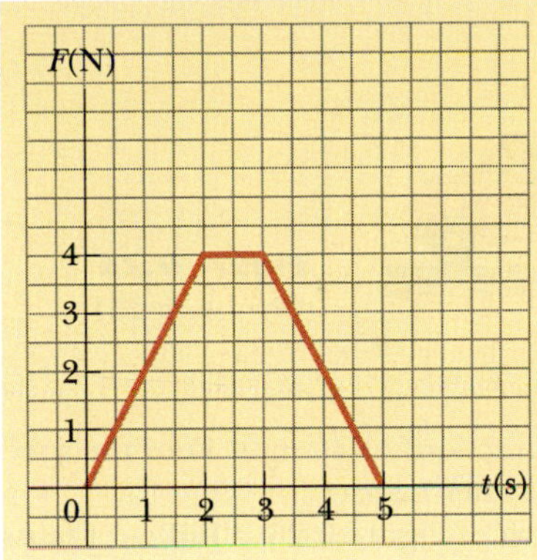

Figure 9.22 (Problems 13 and 14)

tially at rest, and (c) the final velocity of the particle if it is initially moving along the x axis with a velocity of -2 m/s.

14. Find the average force exerted on the particle described in Figure 9.22 for the time interval $t_i = 0$ to $t_f = 5$ s.
15. An estimated force-time curve for a baseball struck by a bat is shown in Figure 9.23. From this curve, determine (a) the impulse delivered to the ball, (b) the average force exerted on the ball, and (c) the peak force exerted on the ball.

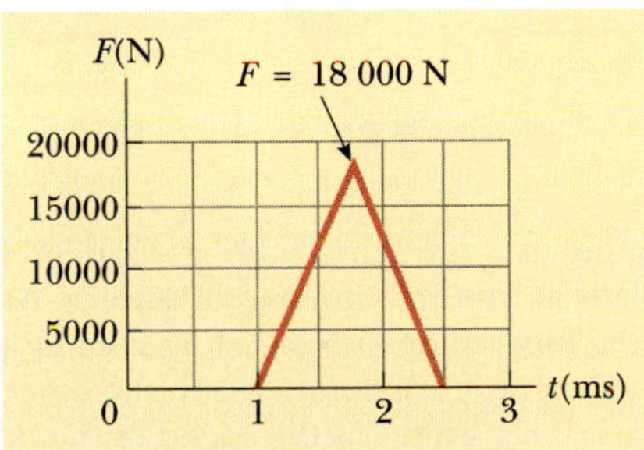

Figure 9.23 (Problem 15)

16. A car is stopped for a traffic signal. When the light turns green, the car accelerates, increasing its speed from zero to 5.20 m/s over a time interval of 0.832 s. What linear impulse and average force does a 70-kg passenger in the car experience?
17. A 0.15-kg baseball is thrown with a speed of 40 m/s. It is hit straight back at the pitcher with a speed of 50 m/s. (a) What is the impulse delivered to the baseball? (b) Find the average force exerted by the bat on the ball if the two are in contact for 2×10^{-3} s. Compare this with the weight of the ball and determine whether or not the impulse approximation is valid in this situation.
18. A tennis player receives a shot with the ball (0.06 kg) traveling horizontally at 50 m/s, and returns the shot with the ball traveling horizontally at 40 m/s in the opposite direction. What is the impulse delivered to the ball by the racket?
19. A 3.0-kg steel ball strikes a massive wall with a speed of 10 m/s, at an angle of 60° with the surface. It bounces off with the same speed and angle (Fig. 9.24). If the ball is in contact with the wall for 0.20 s, what is the average force exerted on the ball by the wall?

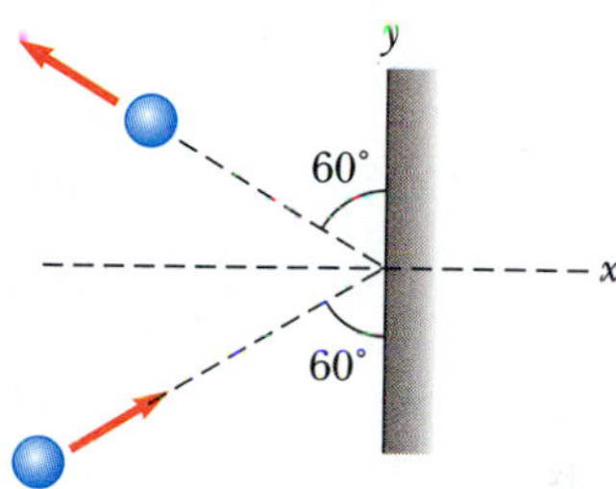

Figure 9.24 (Problem 19)

20. A child bounces a 50-g Superball on the sidewalk. The velocity of the Superball just before striking the sidewalk is -21 m/s (downward), and its velocity just after leaving the sidewalk is 19 m/s (upward). If the contact time with the sidewalk is (1/800) s, what is the average force exerted on the Superball by the sidewalk?

Section 9.3 Collisions
Section 9.4 Elastic and Inelastic Collisions in One Dimension

21. A 2.5-kg mass, moving initially with a speed of 10 m/s, makes a perfectly inelastic head-on collision with a 5-kg mass that is initially at rest. (a) Find the final velocity of the composite particle. (b) How much kinetic energy is lost in the collision?
22. A 2000-kg meteorite has a speed of 120 m/s just before colliding head-on with the Earth. Determine the recoil speed of the Earth (mass = 5.98×10^{24} kg).
23. A 10-g bullet is stopped in a 5-kg block of wood. The speed of the bullet-plus-wood combination immediately after the collision is 0.60 m/s. What was the original speed of thc bullet?

Figure 9.25 (Problem 25)

24. A 90-kg halfback running north with a speed of 10 m/s is tackled by a 120-kg opponent running south with a speed of 4 m/s. Assuming the collision to be perfectly inelastic and head-on, (a) calculate the velocity of the players just after the tackle and (b) determine the energy lost as a result of the collision. Can you account for the missing energy?
25. A 1200-kg car, traveling initially with a speed of 25 m/s in an easterly direction, crashes into the rear end of a 9000-kg truck that is moving in the same direction at 20 m/s (Fig. 9.25). The velocity of the car immediately after the collision is 18 m/s to the east. (a) What is the velocity of the truck immediately after the collision? (b) How much mechanical energy is lost in the collision? How do you account for this loss in energy?
26. A railroad car of mass 2.5×10^4 kg, moving with a speed of 4 m/s, collides and couples with three other coupled railroad cars, each of the same mass as the single car and moving in the same direction with an initial speed of 2 m/s. (a) What is the speed of the four entangled cars after the collision? (b) How much kinetic energy is lost in the collision?
27. A neutron in a reactor makes an elastic head-on collision with the nucleus of a carbon atom that is initially at rest. (a) What fraction of the neutron's kinetic energy is transferred to the carbon nucleus? (b) If the initial kinetic energy of the neutron is 1 MeV = 1.6×10^{-13} J, find its final kinetic energy and the kinetic energy of the carbon nucleus after the collision. (The mass of the carbon nucleus is about 12 times the mass of the neutron.)
28. A 75-kg ice skater, moving at 10 m/s, crashes into a stationary skater of equal mass. After the collision, the two skaters move as a unit at 5 m/s. The average force that a skater can experience without breaking a bone is 4500 N. If the impact time is 0.1 s, does a bone break?
29. Two billiard balls have velocities of 2.0 m/s and −0.5 m/s before they meet in an elastic head-on collision. What are their final velocities?
30. High-speed stroboscopic photographs show that the head of a golf club of mass 200 g is traveling at 55 m/s just before it strikes a 46-g golf ball at rest on a tee. After the collision, the club head travels (in the same direction) at 40 m/s. Find the speed of the golf ball just after impact.
31. As shown in Figure 9.26, a bullet of mass m and speed v passes completely through a pendulum bob of mass M. The bullet emerges with a speed of $v/2$. The pendulum bob is suspended by a stiff rod of length ℓ and negligible mass. What is the minimum value of v such that the pendulum bob will barely swing through a complete vertical circle?

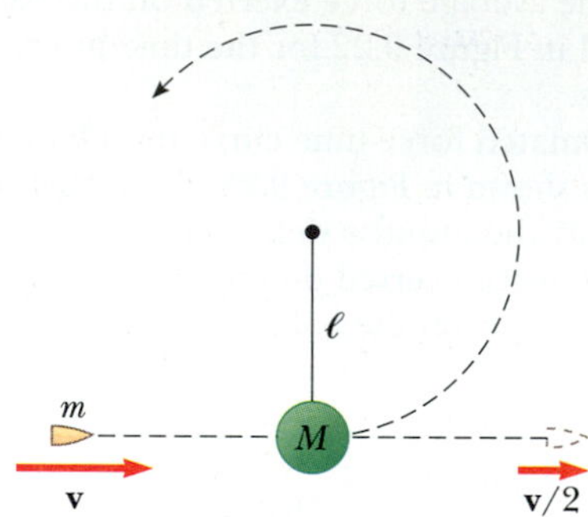

Figure 9.26 (Problem 31)

32. A 12-g bullet is fired into a 100-g wooden block that is initially at rest on a horizontal surface. After impact, the block slides 7.5 m before coming to rest. If the coefficient of friction between the block and the surface is 0.65, what was the speed of the bullet immediately before impact?
33. Consider a frictionless track ABC like that in Figure 9.27. A block of mass $m_1 = 5$ kg is released from A. It makes a head-on elastic collision at B with a block

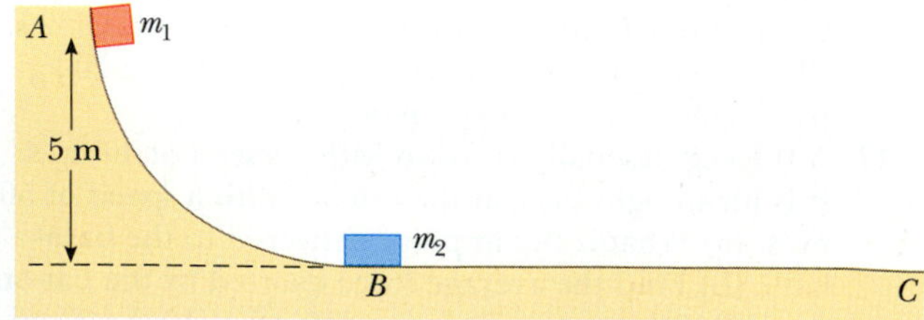

Figure 9.27 (Problem 33)

of mass $m_2 = 10$ kg that is initially at rest. Calculate the maximum height to which m_1 rises after the collision.

34. A 730-N man stands in the middle of a frozen pond of radius 5 m. He is unable to get to the south shore because of a lack of friction between his shoes and the ice. To overcome his difficulty, he throws his 1.2-kg physics textbook horizontally, at a speed of 5 m/s, toward the north shore. How long does it take him to reach the south shore?

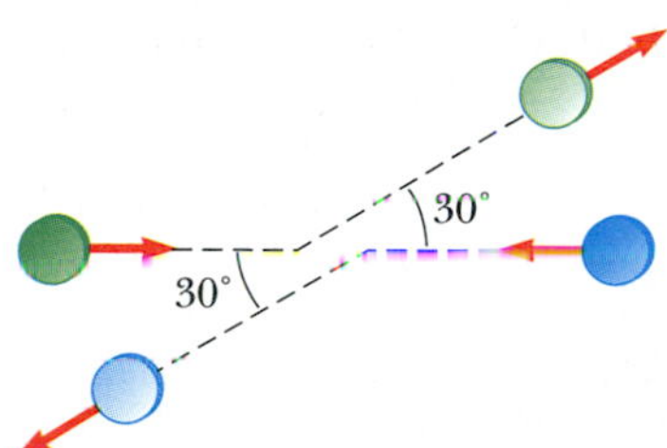

Figure 9.28 (Problem 34)

35. A 7-kg bowling ball collides head on with a 2-kg bowling pin. The pin flies forward with a speed of 3 m/s. If the ball continues forward with a speed of 1.8 m/s, what was the initial speed of the ball? Ignore rotation of the ball.

36. A 90-kg fullback, moving east with a speed of 5 m/s, is tackled by a 95-kg opponent who is running north with a speed of 3 m/s. If the collision is perfectly inelastic, calculate (a) the velocity of the players just after the tackle and (b) the energy lost as a result of the collision. Can you account for the missing energy?

37. An eastern kingbird and a bee are approaching each other at right angles. The bird has a mass of 0.125 kg and a speed of 0.6 m/s, and the bee has a mass of 5.0×10^{-3} kg and a speed of 15.0 m/s. If the bird catches the bee, what is the new speed of the bird?

38. The mass of the blue puck in Figure 9.28 is 20% greater than the mass of the green one. Before colliding, the pucks approach each other with equal and opposite momentum, and the green puck has an initial speed of 10.0 m/s. Find the speeds of the pucks after the collision if half the kinetic energy is lost during the collision.

39. A 3-kg mass with an initial velocity of 5**i** m/s collides with and sticks to a 2-kg mass with an initial velocity of -3**j** m/s. Find the final velocity of the composite mass.

40. A proton, moving with a velocity of v_0**i**, collides elastically with another proton that is initially at rest. If the two protons have the same speed after the collision, find (a) the speed of each proton after the collision in terms of v_0 and (b) the direction of the velocity vectors after the collision.

41. A billiard ball moving at 5 m/s strikes a stationary ball of the same mass. After the collision, the first ball moves at 4.33 m/s, at an angle of 30° with respect to its original line of motion. Assuming an elastic collision (and ignoring friction and rotational motion), find the magnitude and direction of the struck ball's velocity.

42. A 0.3-kg puck, initially at rest on a horizontal, frictionless surface, is struck by a 0.2-kg puck, which is initially moving along the x axis with a velocity of 2 m/s. After the collision, the 0.2-kg puck has a speed of 1 m/s at an angle of $\theta = 53°$ to the positive x axis (see Fig. 9.13). (a) Determine the velocity of the 0.3-kg puck after the collision. (b) Find the fraction of kinetic energy lost in the collision.

43. Two shuffleboard disks of equal mass, one orange and the other yellow, are involved in a perfectly elastic glancing collision. The yellow disk is initially at rest and is struck by the orange disk, which is moving with a speed of 5 m/s. After the collision, the orange disk moves in a direction that makes an angle of 37° with its initial direction of motion, and the velocity of the yellow disk is perpendicular to that of the orange disk (after the collision). Determine the final speed of each disk.

44. A 10-kg mass that is initially at rest explodes into three pieces. A 4.5-kg piece goes north at 20 m/s, and a 2-kg piece moves eastward at 60 m/s. (a) Determine the magnitude and direction of the velocity of the third piece. (b) Find the energy of the explosion.

45. An unstable nucleus of mass 17×10^{-27} kg, initially at rest, disintegrates into three particles. One of the particles, of mass 5.0×10^{-27} kg, moves along the y axis with a velocity of 6×10^{6} m/s. Another particle, of mass 8.4×10^{-27} kg, moves along the x axis with a velocity of 4×10^{6} m/s. Find (a) the velocity of the third particle and (b) the total energy given off in the process.

46. At an intersection, a 1500-kg car, traveling east with a speed of 20 m/s, collides with a 2500-kg van traveling south at a speed of 15 m/s. The vehicles undergo a perfectly inelastic collision, and the wreckage slides 6 m before coming to rest. Find the magnitude and direction of the constant force that decelerated them.

Section 9.6 The Center of Mass

47. Four objects are situated along the y axis as follows: a 2-kg object is at +3 m, a 3-kg object is at +2.5 m, a 2.5-kg object is at the origin, and a 4-kg object is at -0.5 m. Where is the center of mass of these objects?

48. A uniform T square used by a draftsman has dimen-

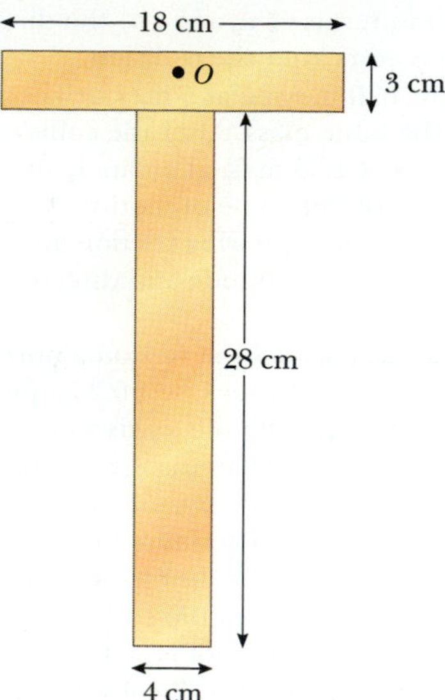

Figure 9.29 (Problem 48)

sions as shown in Figure 9.29. Locate the center of mass with respect to point *O*. (*Hint:* Note that the mass of each rectangular part is proportional to its area.)

49. A uniform carpenter's square has the shape of an L, as in Figure 9.30. Locate the center of mass relative to an origin at the lower left corner. (See the hint for Problem 48.)

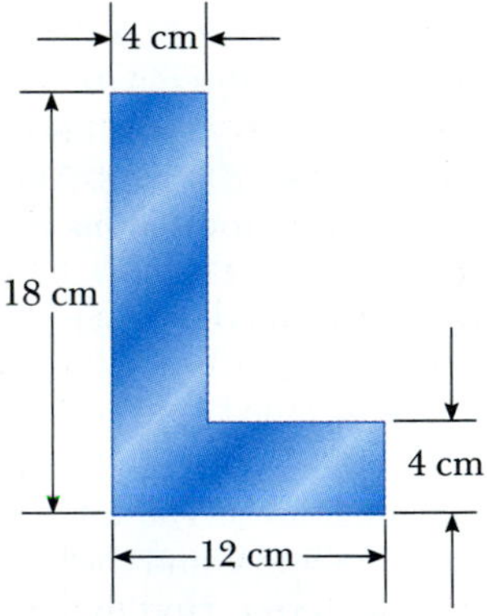

Figure 9.30 (Problem 49)

50. The mass of the Earth is 5.98×10^{24} kg, and the mass of the Moon is 7.36×10^{22} kg. The distance of separation, measured from their centers, is 3.84×10^8 m. Locate the center of mass of the Earth-Moon system as measured from the center of the Earth.

51. The mass of the Sun is 329 390 Earth masses, and the mean distance from the center of the Sun to the center of the Earth is 1.496×10^8 km. Treating the Earth and Sun as particles, with each mass concentrated at its respective geometric center, how far from the center of the Sun is the center of mass of the Earth-Sun system? Compare this distance with the mean radius of the Sun (6.960×10^5 km).

52. The separation between the hydrogen and chlorine atoms of the HC1 molecule is about 1.30×10^{-10} m. Locate the center of mass of the molecule as measured from the hydrogen atom. (Chlorine is 35 times more massive than hydrogen.)

53. A uniform piece of sheet steel is shaped as shown in Figure 9.31. Compute the *x* and *y* coordinates of the center of mass of the piece.

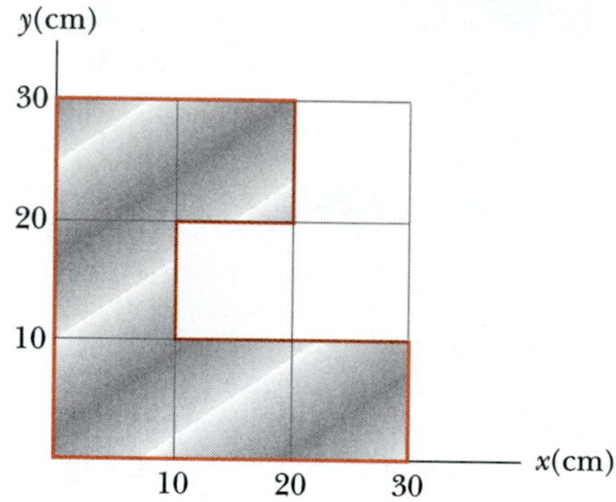

Figure 9.31 (Problem 53)

Section 9.7 Motion of a System of Particles

54. A 5-kg particle moves along the *x* axis with a velocity of 3 m/s. A 2-kg particle moves along the *x* axis with a velocity of −2.5 m/s. Find (a) the velocity of the center of mass and (b) the total momentum of the system.

55. A 2-kg particle has a velocity of (2**i** − 3**j**) m/s, and a 3-kg particle has a velocity of (**i** + 6**j**) m/s. Find (a) the velocity of the center of mass and (b) the total momentum of the system.

56. (a) A 3.0-g particle is moving toward a 7.0-g particle with a speed of 3.0 m/s. With what speed does each approach the center of mass? (b) What is the momentum of each particle, relative to the center of mass?

57. Romeo entertains Juliet by playing his guitar at the rear of their boat, which floats in still water. After the serenade, Juliet carefully moves to the rear of the boat (away from shore) to plant a kiss on Romeo's cheek. If the 80-kg boat is facing shore and the 55-kg Juliet moves 2.7 m toward the 77-kg Romeo, how far does the boat move toward shore?

*Section 9.8 Rocket Propulsion

58. A rocket engine consumes 80 kg of fuel per second. If the exhaust speed is 2.5×10^3 m/s, calculate the thrust on the rocket.
59. The first stage of a Saturn V space vehicle consumes fuel at the rate of 1.5×10^4 kg/s, with an exhaust speed of 2.6×10^3. (These are approximate figures.) (a) Calculate the thrust produced by these engines. (b) Find the vehicle's initial acceleration on the launch pad if its initial mass is 3×10^6 kg. [You must include the force of gravity to solve (b).]
60. A large rocket with an exhaust speed of $v_e = 3000$ m/s develops a thrust of 24 million newtons. (a) How much mass is being blasted out of the rocket exhaust per second? (b) What is the maximum speed the rocket can attain if it starts from rest in a force-free environment with $v_e = 3$ km/s, and if 90% of its initial mass is fuel?
61. Fuel aboard a rocket has a density of 1.4×10^3 kg/m^3 and is ejected with a speed of 3.0×10^3 m/s. If the engine is to provide a thrust of 2.5×10^6 N, what volume of fuel must be burned per second?
62. A rocket for use in deep space is to have the capability to boost a payload (plus the rocket frame and engine) of 3.0 metric tons to a speed of 10 000 m/s with an engine and fuel designed to produce an exhaust speed of 2000 m/s. (a) What amount of fuel-and-oxidizer mixture is required? (b) If a different fuel and engine design could give an exhaust speed of 5000 m/s, what amount of fuel-and-oxidizer mixture would be required for the same task? Comment.

Additional Problems

63. A golf ball (m = 46 g) is struck a blow that makes an angle of 45° with the horizontal. The drive lands 200 m away on a flat fairway. If the golf club and ball are in contact for 7 ms, what is the average force of impact? (Neglect air resistance effects.)
64. A 30-06 caliber hunting rifle fires a bullet of mass 0.012 kg, with a muzzle velocity of 600 m/s to the right. The rifle has a mass of 4.0 kg. (a) What is the recoil velocity of the rifle as the bullet leaves the rifle? (b) If the rifle is stopped by the hunter's shoulder in a distance of 2.5 cm, what is the average force exerted on the shoulder by the rifle? (c) If the hunter's shoulder is partially restricted from recoiling, would the force exerted on the shoulder be the same as in part (b)? Explain.

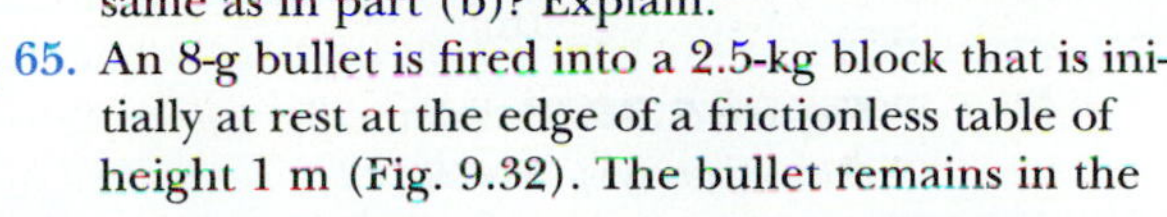

65. An 8-g bullet is fired into a 2.5-kg block that is initially at rest at the edge of a frictionless table of height 1 m (Fig. 9.32). The bullet remains in the block, and after impact the block lands 2 m from the bottom of the table. Determine the initial speed of the bullet.

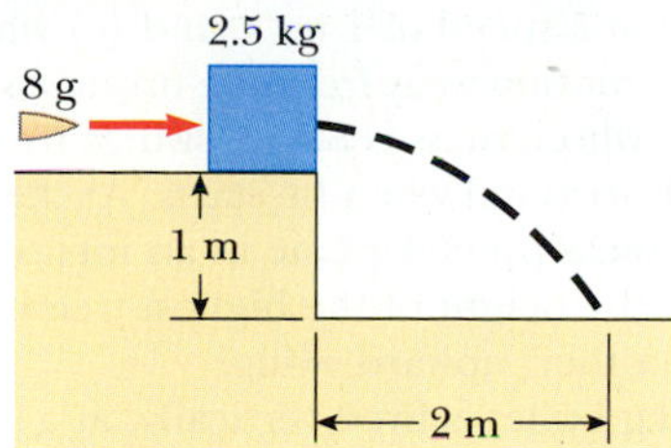

Figure 9.32 (Problem 65)

66. A uranium-238 nucleus (mass = 238 units) that is initially at rest decays, transforming into an alpha particle (mass = 4 units) and a residual thorium nucleus (mass = 234 units). If the alpha particle has a speed of 1.5×10^7 m/s, determine the recoil speed of the thorium nucleus.
67. A spacecraft is stationary in deep space when its rocket engine is ignited for a 100-s "burn." Hot gases are ejected at a constant rate of 150 kg/s, with a speed of 3000 m/s relative to the spacecraft. The initial mass of the spacecraft (plus fuel and oxidizer) is 25 000 kg. Determine the thrust of the rocket engine and the initial acceleration in units of g. What is the final speed of the spacecraft?
68. A 75-kg fire fighter slides down a pole while a constant frictional force of 300 N retards his motion. To cushion the fall, a horizontal 20-kg platform at the bottom of the pole is supported by a spring. The fire fighter starts from rest 4 m above the platform, and the spring constant is 4000 N/m. Find (a) the fire fighter's speed just before he collides with the platform and (b) the maximum distance the spring is compressed. (Assume the frictional force acts during the entire motion.)
69. A 70-kg astronaut is working on the engines of her ship, which is drifting through space with a constant velocity. Wishing to get a better view of the Universe, she pushes against the ship and soon finds herself 30 m behind it. Without a thruster, the only way for the astronaut to return to the ship is to throw her 0.5-kg wrench directly away from it. If she throws the wrench with a speed of 20 m/s, how long does it take her to reach the ship?
70. A pool ball rolling across a table at 1.5 m/s makes a head-on collision with an identical ball. Assuming the collision is elastic, find the speed of each ball after the collision (a) when the second ball is initially at rest, (b) when the second ball is in motion toward

the first at a speed of 1 m/s, and (c) when the second ball is in motion away from the first at a speed of 1 m/s.

71. Tarzan, whose mass is 80 kg, swings from a 3-m vine that is horizontal when he starts. At the bottom of his arc, he picks up 60-kg Jane in an inelastic collision. What is the height of the highest tree limb they can reach on their upward swing?

72. A 0.03-kg bullet is fired vertically, at a speed of 200 m/s, into a 0.15-kg baseball that is initially at rest. How high does the combination rise after the collision, assuming the bullet is embedded in the ball?

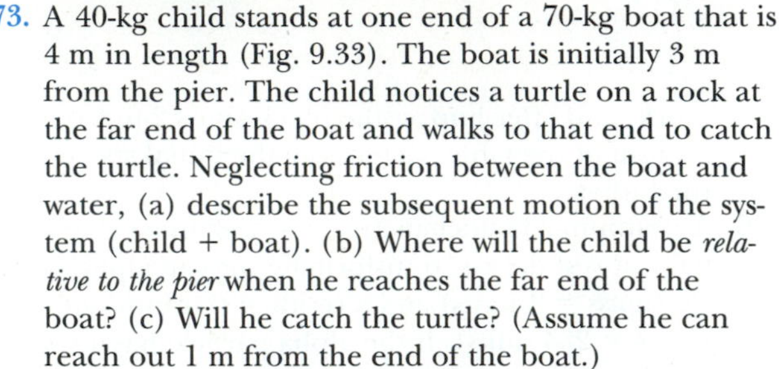

73. A 40-kg child stands at one end of a 70-kg boat that is 4 m in length (Fig. 9.33). The boat is initially 3 m from the pier. The child notices a turtle on a rock at the far end of the boat and walks to that end to catch the turtle. Neglecting friction between the boat and water, (a) describe the subsequent motion of the system (child + boat). (b) Where will the child be *relative to the pier* when he reaches the far end of the boat? (c) Will he catch the turtle? (Assume he can reach out 1 m from the end of the boat.)

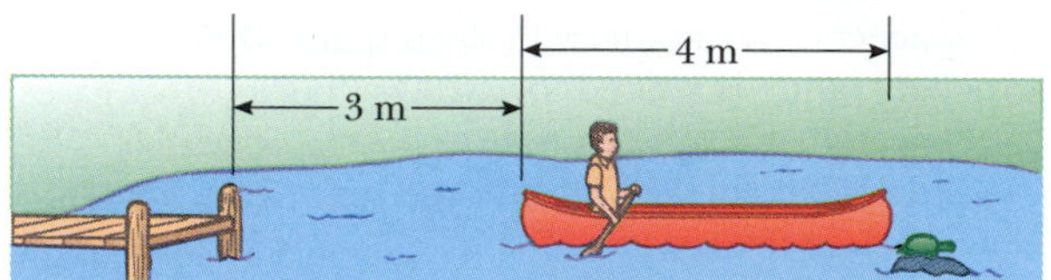

Figure 9.33 (Problem 73)

74. A right-triangular object of mass M has the dimensions shown in Figure 9.34. Locate the coordinates of the center of mass, assuming the object has a uniform mass per unit area.

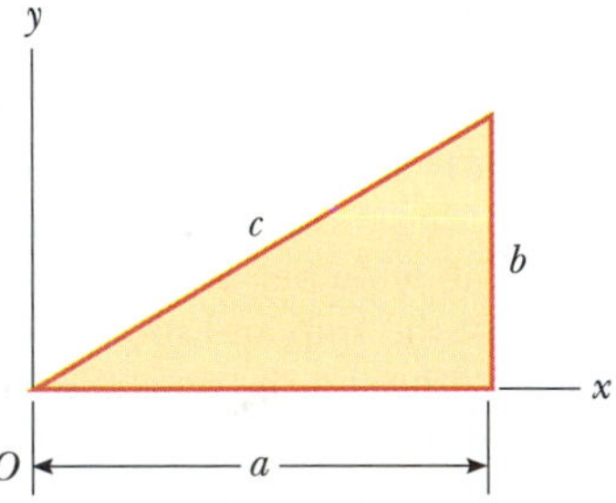

Figure 9.34 (Problem 74)

75. A 5-g bullet, moving with an initial speed of 400 m/s, is fired into and passes through a 1-kg block, as in Figure 9.35. The block, initially at rest on a frictionless, horizontal surface, is connected to a spring of force constant 900 N/m. If the block moves 5 cm to the right after impact, find (a) the speed at which the bullet emerges from the block and (b) the mechanical energy lost in the collision.

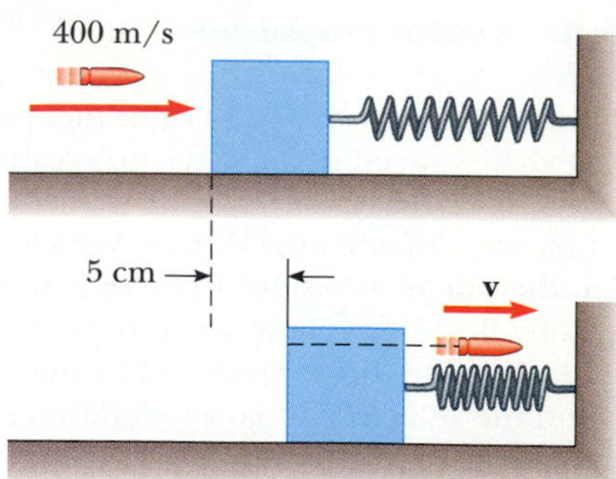

Figure 9.35 (Problem 75)

△76. A student performs a ballistic pendulum experiment, using an apparatus similar to that shown in the multiflash photo (Fig. 9.36a). She obtains the following data (averaged over a series of five measurements): $h = 8.68$ cm, $m_1 = 68.8$ g, $m_2 = 263$ g. (a) Determine

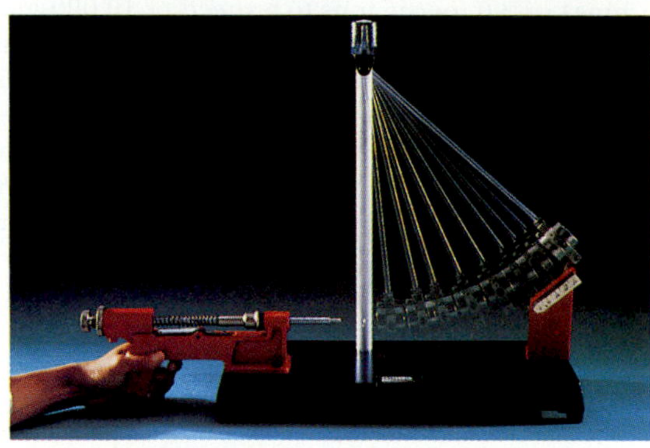

(a)

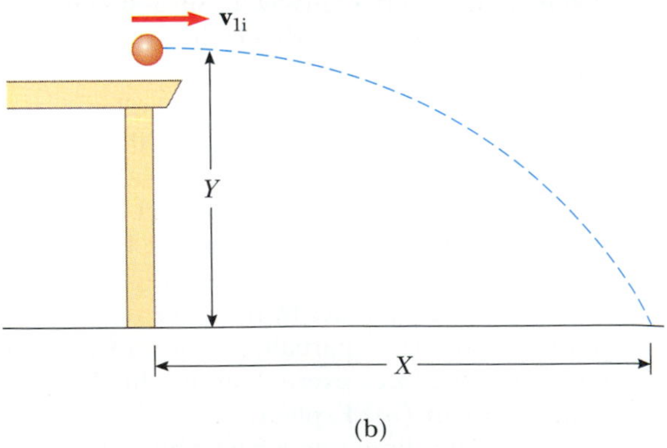

(b)

Figure 9.36 (Problem 76).
(Photo courtesy of CENCO)

the initial speed v_{1i}, of the projectile. (b) The second part of the student's experiment is to obtain v_{1i} by firing the same projectile horizontally (with the pendulum removed from the path of the projectile) and measuring its horizontal displacement, X, and vertical displacement, Y, before it strikes the floor (Fig. 9.36b). Show that the initial speed of the projectile is related to X and Y through the relation

$$v_{1i} = \frac{X}{\sqrt{2Y/g}}$$

What numerical value does the student obtain for v_{1i} based on her measured values of $X = 257$ cm and $Y = 85.3$ cm? What factors might account for the difference in this value compared to that obtained in part (a)?

Calculator/Computer Problem

77. Consider a head-on elastic collision between a moving particle of mass m_1 and an initially stationary particle of mass m_2 (see Example 9.6). (a) Plot f_2, the fraction of energy transferred to m_2, as a function of the ratio m_2/m_1, and show that f_2 reaches a maximum when $m_2/m_1 = 1$. (b) Perform an analytical calculation that verifies that f_2 is a maximum when $m_1 = m_2$.

Albert Einstein (1879–1955), one of the greatest physicists of all times, is best known for developing the theory of relativity. He is shown here in a playful mood riding a bicycle. The photograph was taken in 1933 in Santa Barbara, California. *(From the California Institute of Technology archives)*

Relativity and Its Applications

CHAPTER 10

Most of our everyday experiences and observations have to do with objects that move at speeds much lower than the speed of light. Newtonian mechanics and early ideas about space and time were formulated to describe the motions of such objects and, as we saw in earlier chapters, are very successful in describing a wide range of phenomena. Although Newtonian mechanics works very well at low speeds, it fails when applied to particles whose speeds approach that of light. Experimentally, the predictions of Newtonian theory at high speeds can be tested by accelerating an electron through a large electric potential difference. For example, it is possible to accelerate an electron to a speed of $0.99\,c$ (where c is the speed of light) by using a potential difference of several million volts. According to Newtonian mechanics, if the potential difference (as well as the corresponding energy) is increased by a factor of 4, then the speed of the electron should be doubled to $1.98c$. However, experiments show that the speed of the electron—as well as the speeds of all other particles in the Universe—always remains *less* than the speed of light, regardless of the size of the accelerating voltage. Because it places no upper limit on the speed that a particle can attain, Newtonian mechanics is contrary to modern experimental results and is clearly a limited theory.

In 1905, at the age of 26, Albert Einstein published his special theory of relativity. Regarding the theory, Einstein wrote:

> The relativity theory arose from necessity, from serious and deep contradictions in the old theory from which there seemed no escape. The strength of the new theory lies in the consistency and simplicity with which it solves all these difficulties, using only a few very convincing assumptions. . . .[1]

Although Einstein made many other important contributions to science, his theory of relativity alone represents one of the greatest intellectual achievements of the 20th century. With this theory, one can correctly predict experimental observations over the range of speeds from $v = 0$ to speeds approaching the speed of light. Newtonian mechanics, which was accepted for over 200 years, is in fact a special case of Einstein's theory. This chapter gives an introduction to the special theory of relativity, with emphasis on some of its consequences.

Special relativity covers such phenomena as the slowing down of clocks and the contraction of lengths in moving reference frames as measured by a stationary observer. We shall also discuss the relativistic forms of momentum and energy as well as some consequences of the famous mass-energy formula, $E = mc^2$. You may want to consult any of a number of excellent books on relativity for more details on the subject.[2]

In addition to its well known and essential role in contemporary theoretical physics, the relativity theory also has many practical applications, including the design of accelerators and other devices that utilize high-speed particles. We shall have occasion to use relativity in some subsequent chapters of this text, most often presenting only the outcome of relativistic effects.

10.1 The Principle of Relativity

In order to describe a physical event, it is necessary to establish a frame of reference, such as one that is fixed in the laboratory. You should recall from Chapter 4 that Newton's laws are valid in *all* inertial frames of reference. Since an inertial frame of reference is defined as one in which Newton's first law is valid, one can say that *an inertial system is a system in which a free body exhibits no acceleration.* Furthermore, any system moving with constant velocity with respect to an inertial system is also an inertial system. This means that the results of an experiment performed in

Inertial frame of reference

[1] A. Einstein and L. Infeld, *The Evolution of Physics,* New York, Simon and Schuster, 1961.

[2] The following books are recommended for more details on the special theory of relativity at the introductory level: E. F. Taylor and J. A. Wheeler, *Spacetime Physics,* San Francisco, W. H. Freeman, 1963; R. Resnick, *Introduction to Special Relativity,* New York, Wiley, 1968; and A. P. French, *Special Relativity,* New York, Norton, 1968. Other suggested readings are reprints on "Special Relativity Theory," published by the American Institute of Physics.

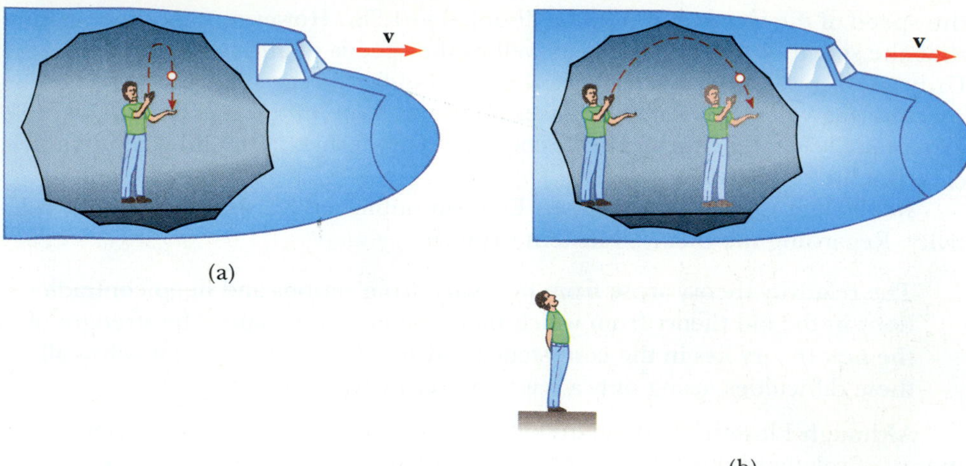

Figure 10.1
(a) The passenger sees the ball move in a vertical path when thrown upwards. (b) The Earth observer views the path of the ball to be a parabola.

a vehicle moving with uniform velocity will be identical to the results of the same experiment performed in the stationary laboratory.

According to the principle of Newtonian relativity, the laws of mechanics are the same in all inertial frames of reference.

Let us consider a common observation that illustrates the equivalence of the laws of mechanics in different inertial frames. Consider an airplane moving in flight with a constant velocity, as in Figure 10.1a. If a passenger in the airplane throws a ball straight up in the air, the passenger observes that the ball moves in a vertical path. The motion of the ball appears to be precisely the same as if the ball were thrown by a person at rest on Earth. The law of gravity and the equations of motion under constant acceleration are obeyed whether the airplane is at rest or in uniform motion. Now consider the same experiment viewed by an observer at rest on the Earth. This stationary observer sees the path of the ball as a parabola, as in Figure 10.1b. Furthermore, according to this observer, the ball has a velocity to the right equal to the velocity of the plane. Although the two observers disagree on certain aspects of the experiment, they agree that the motion of the ball obeys the law of gravity and the laws of motion. Thus, we draw the following important conclusion: there is no preferred frame of reference for describing the laws of mechanics.

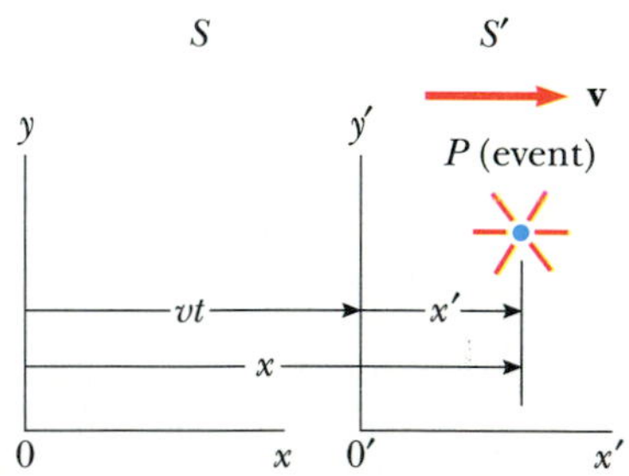

Figure 10.2
An event occurs at a point *P*. The event is observed by two observers in inertial frames *S* and *S*′, where *S*′ moves with a velocity **v** relative to *S*.

Suppose that some physical phenomenon, which we call an *event,* occurs in an inertial system—that is, a system that is not accelerating. The event's location and time of occurrence can be specified by the coordinates (x, y, z, t). We would like to be able to transform the space and time coordinates of the event from the first inertial system to another system that is moving with uniform relative velocity. This is accomplished by using what is called *Galilean transformation.*

Consider two inertial systems, S and S' (Fig. 10.2). The system S' moves with a constant velocity **v** along the xx' axis, where **v** is measured relative to S. We assume that an event occurs at the point P and that the origins of S and S' coincide at $t = 0$. The event might be a heartbeat or the flash of a flashbulb. An observer in system S would describe the event with space-time coordinates (x, y, z, t), whereas an observer in system S' would use (x', y', z', t'), to describe the same event. As we can see from Figure 10.2, these coordinates are related by the equations

$$\begin{aligned} x' &= x - vt \\ y' &= y \\ z' &= z \\ t' &= t \end{aligned} \qquad [10.1]$$

Galilean transformation of coordinates

These equations constitute what is known as a **Galilean transformation of coordinates.** Note that the fourth coordinate, time, is *assumed* to be the same in both inertial systems. That is, within the framework of classical mechanics, clocks are universal, so the time of an event clocked by an observer in S is the same as the time clocked for the same event by an observer in S'. Consequently, the time interval between two successive events should also be the same for both observers. Although this assumption may seem obvious, it turns out to be *incorrect* in situations in which v is comparable to the speed of light. This point represents one of the most profound differences between Newtonian concepts and the ideas contained in Einstein's theory of relativity.

Now suppose two events are separated by a distance Δx and a time interval Δt as measured by an observer in S. It follows from Equation 10.1 that the corresponding displacement $\Delta x'$ measured by an observer in S' is given by $\Delta x' = \Delta x - v\,\Delta t$, where Δx is the displacement measured by the observer in S. Since $\Delta t = \Delta t'$, we find that

$$\frac{\Delta x'}{\Delta t} = \frac{\Delta x}{\Delta t} - v$$

If the two events correspond to the passage of a moving object past two "milestones" in Frame S, then $\Delta x/\Delta t$ is the average velocity in S and $\Delta x'/\Delta t$ is the average velocity in S'. As Δt approaches zero, $\Delta x'/\Delta t$ becomes

$$u'_x = u_x - v \qquad [10.2]$$

Galilean addition law for velocities

where u_x and u'_x are the instantaneous velocities of the object relative to S and S', respectively.

This result, which is called the **Galilean addition law for velocities** (or Galilean velocity transformation), is used in everyday observations and is consistent with our intuitive notions of time and space. As we shall soon see, however, it leads to serious contradictions when applied to light waves.

10.2 The Michelson-Morley Experiment

Many experiments similar to the one described in the preceding section show us that the laws of mechanics are the same in all inertial frames of reference. When similar inquiries are made into the laws of other branches of physics, the results are contradictory. In particular, the laws of electricity and magnetism are found to depend on the frame of reference used. It might be argued that it is these laws that are wrong, but that is difficult to accept because the laws are in total agreement with all known experimental results. The Michelson-Morley experiment was an attempt to explain this dilemma.

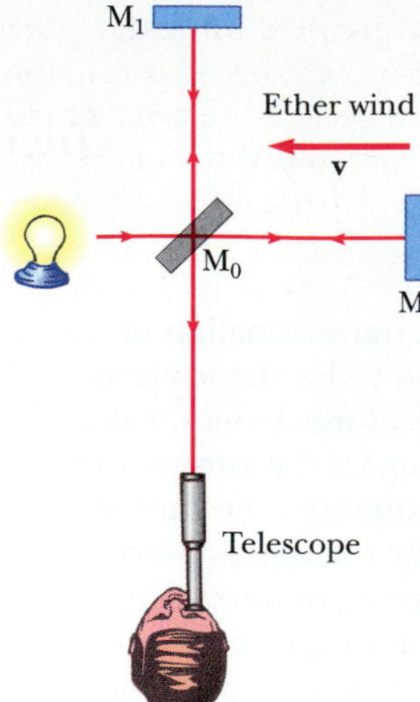

Figure 10.3
According to the ether wind theory, the speed of light should be $c - v$ as the beam approaches mirror M_2 and $c + v$ after reflection.

Albert A. Michelson (1852–1931) A German-American physicist, Michelson invented the interferometer and spent much of his life making accurate measurements of the speed of light. In 1907 he was the first American to be awarded the Nobel prize for his work in optics. His most famous experiment, conducted with Edward Morley in 1887 (which gave negative results), implied that it was impossible to measure the absolute velocity of the Earth with respect to a medium called the *ether*. Subsequent work by Einstein in his special theory of relativity wiped out the ether concept by assuming that the speed of light has the same value in all inertial reference frames. *(AIP Emilio Segrè Visual Archives, Michelson Collection)*

The experiment stemmed from a misconception early physicists had concerning the manner in which light propagates. The properties of mechanical waves, such as water and sound waves, were well known, and all of these waves require a medium to support the disturbances. In the 19th century, physicists thought that electromagnetic waves also required a medium through which to propagate. They proposed that such a medium existed, and they named it the **luminiferous ether.** This ether was assumed to be present everywhere, even in a vacuum, and to have the unusual property of being a massless but rigid medium (a strange concept indeed!). Additionally, it was found that the troublesome laws of electricity and magnetism took on their simplest forms in a frame of reference *at rest* with respect to the luminiferous ether; this frame was called the **absolute frame.** In any reference frame moving with respect to it, the laws of electricity and magnetism would remain valid but would have to be modified.

As a result of the importance attached to this absolute frame, experimental proof of its existence became of considerable interest in physics. However, all attempts to detect the presence of the ether (and hence the absolute frame) proved futile! The most famous experiment designed to show the presence of the ether was performed in 1887 by A. A. Michelson (1852–1931) and E. W. Morley (1838–1923).[3] The objective was to determine the velocity of the Earth with respect to the ether, and the experimental tool used was a device called the interferometer, shown in Figure 10.3.

Suppose one arm of an interferometer (M_0M_2 in Fig. 10.3) were aligned along the direction of the motion of the Earth through space. The Earth moving through the ether would be equivalent to the ether flowing past the Earth in the opposite direction. This "ether wind" blowing in the direction opposite the Earth's motion should cause the speed of light, as measured in the Earth's frame of reference, to be $c - v$ as the light approaches mirror M_2 in Figure 10.3, and $c + v$ after reflection, where c is the speed of light in the ether frame and v is the speed of the Earth through space and hence the speed of the ether wind. The incident and reflected beams of light would recombine, and an interference pattern would form.

[3] A. A. Michelson and E. W. Morley, *Am. J. Sci.* 134:333, 1887.

During the experiment, the interference pattern was observed while the interferometer was rotated through an angle of 90°. The idea was that this rotation would change the speed of the ether wind along the direction of the arms of the interferometer, and consequently the interference pattern would shift slightly but measurably. Measurements failed to show any change in the pattern! The Michelson-Morley experiment was repeated by other researchers under varying conditions and at different locations, but the result was always the same: *no interference-pattern shift of the magnitude required was ever observed.*

The negative result of the Michelson-Morley experiment not only contradicted the ether hypothesis; it also meant that it was impossible to measure the absolute (orbital) velocity of the Earth with respect to the ether frame. From a theoretical viewpoint, this meant that it was impossible to find the absolute frame. However, as we shall see in the next section, Einstein offered a postulate that places a different interpretation on the negative result. In later years, when more was known about the nature of light, the idea of an ether that permeated all space was relegated to the ash heap of worn-out concepts. Light is now understood to be *an electromagnetic wave that requires no medium for its propagation.* As a result, an ether in which light could travel became an unnecessary construct.

Modern versions of the Michelson-Morley experiment have compared the frequencies of resonant laser cavities of identical length, oriented at right angles to each other. Most recently, Doppler shift experiments using gamma rays emitted by a radioactive sample of ^{57}Fe have placed an upper limit of about 5 cm/s on ether wind velocity. These results have shown quite conclusively that the motion of the Earth has no effect on the speed of light!

10.3 Einstein's Postulates

In the preceding section we noted the serious contradiction between the invariance of the speed of light and the Galilean addition law for velocities. In 1905 Einstein proposed a theory that resolved this contradiction. It was to completely alter our notions of space and time.[4] His special theory of relativity deals with situations involving inertial reference frames and is based on two postulates. First, Einstein postulated that the laws of physics are the same in every inertial frame of reference. In his own words, "The same laws of electrodynamics and optics will be valid for all frames of reference for which the equations of mechanics hold good." This is, in effect, a generalization of Newton's principle of relativity, which applies only to the laws of mechanics.

The first postulate of relativity

Now consider Einstein's second postulate, which states that

> the speed of light has the same value for all observers, independent of their motion or of the motion of the light source.

The second postulate of relativity

Here we are faced with a fundamental problem. We can demonstrate the nature of the problem by considering a light pulse sent out by an observer in a boxcar, moving with a speed v (Fig. 10.4). The light pulse has a speed c relative to

[4] A. Einstein, "On the Electrodynamics of Moving Bodies," *Ann. Physik* 17:891 (1905). For English translations of this article and other publications by Einstein, see H. Lorentz, A. Einstein, H. Minkowski, and H. Weyl, *The Principle of Relativity,* Dover, 1958.

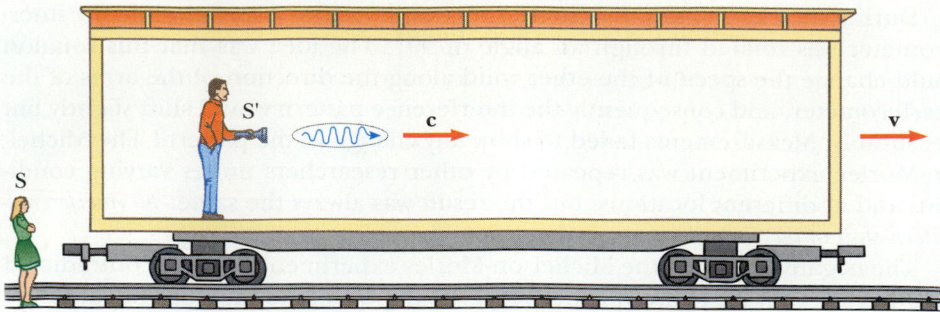

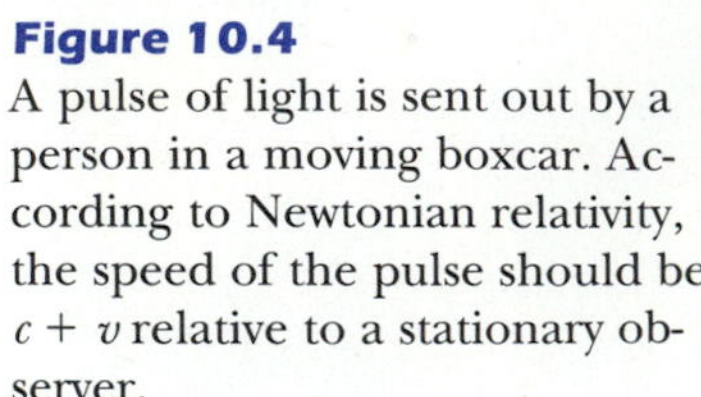
Figure 10.4
A pulse of light is sent out by a person in a moving boxcar. According to Newtonian relativity, the speed of the pulse should be $c + v$ relative to a stationary observer.

observer S' in the boxcar. According to the ideas of Newtonian relativity, the speed of the pulse relative to stationary observer S outside the boxcar should be $c + v$. This is in obvious contradiction to Einstein's second postulate, which states that the speed of the light pulse is the same for all observers. According to Einstein's theory, the stationary and moving observers should both measure the same speed for the light pulse. This conclusion seems strange because it contradicts our intuition, or what we often call common sense. However, common-sense ideas are based on everyday experiences, which do not involve speed-of-light measurements.

Although the Michelson-Morley experiment was performed before Einstein published his work on relativity, it is not clear whether Einstein was aware of the details of the experiment. Nonetheless, the second postulate explains the null result of the experiment because, in effect, the second postulate means that the premises of the experiment were incorrect. For example, in explaining the expected results, we stated that when light traveled against the ether wind its speed was $c - v$. However, if the state of motion of the observer or of the source has no influence on the value found for the speed of light, the value will always be measured to be c. Likewise, after reflection from the mirror, the light makes the return

Albert Einstein, one of the greatest physicists of all times, was born in Ulm, Germany. In 1905, at the age of 26, he published four scientific papers that revolutionized physics. One of these papers, for which he was awarded the 1921 Nobel Prize in physics, dealt with the photoelectric effect. Another was concerned with Brownian motion, the irregular motion of small particles suspended in a liquid. The remaining two papers were concerned with what is now considered his most important contribution of all, the special theory of relativity. In 1915, Einstein published his work on the general theory of relativity, which relates gravity to the structure of space and time. The most dramatic prediction of this theory is the degree to which light would be deflected by a gravitational field. Measurements made by astronomers on bright stars in the vicinity of the eclipsed Sun in 1919 confirmed Einstein's prediction, and Einstein suddenly became a world celebrity.

trip with a speed of c, not $c + v$. Thus, if we accept Einstein's second postulate, the motion of the Earth should not influence the interference pattern observed in the Michelson-Morley experiment, and a null result should be expected.

If we accept Einstein's theory of relativity, we must conclude that relative motion is unimportant in measuring the speed of light. At the same time, we must alter our common-sense notions of space and time and be prepared for some bizarre consequences.

The speed of light is the speed limit of the Universe.

10.4 Consequences of Special Relativity

Before we discuss the consequences of special relativity, we must first understand how an observer in an inertial reference frame describes an event. As mentioned earlier, an event is an occurrence described by three space coordinates and one time coordinate. In general, different observers in different inertial frames would describe the same event with different space-time coordinates.

The reference frame used to describe an event consists of a coordinate grid and a set of clocks situated at the grid intersections, as shown in Figure 10.5 in two dimensions. It is necessary that the clocks be synchronized. This can be accomplished in many ways with the help of light signals. For example, suppose an observer at the origin with a master clock sends out a pulse of light at $t = 0$. The light pulse takes a time r/c to reach a second clock, situated a distance r from the origin. Hence, the second clock will be synchronized with the clock at the origin if the second clock reads a time r/c at the instant the pulse reaches it. This procedure of synchronization assumes that the speed of light has the same value in all directions and in all inertial frames. Furthermore, the procedure concerns an event recorded by an observer in a specific inertial reference frame. An observer in some other inertial frame would assign different space-time coordinates to events, using another coordinate grid with another array of clocks.

Almost everyone who has dabbled even superficially with science is aware of some of the startling predictions that arise from Einstein's approach to relative motion. As we examine some of the consequences of relativity in the following four subsections, we shall find that they conflict with our basic notions of space and time.

Figure 10.5 In relativity, we use a reference frame consisting of a coordinate grid and a set of synchronized clocks.

Simultaneity and the Relativity of Time

A basic premise of Newtonian mechanics is that a universal time scale exists that is the same for all observers. In fact, Newton wrote that "absolute, true, and mathematical time, of itself, and from its own nature, flows equably without relation to anything external." Thus, Newton and his followers took simultaneity for granted. In his special theory of relativity, Einstein abandoned this assumption. According to him, *a time interval measurement depends on the reference frame in which the measurement is made.*

Einstein devised the following thought experiment to illustrate this point. A boxcar moves with uniform velocity, and two lightning bolts strike its ends as in Figure 10.6a, leaving marks on the boxcar and on the ground. The marks on the boxcar are labeled A' and B', and those on the ground are labeled A and B. An observer O' moving with the boxcar is midway between A' and B', and a ground observer O is midway between A and B. The events recorded by the observers are the light signals from the lightning bolts.

Let us assume that the two light signals reach O at the same time, as indicated in Figure 10.6b. This observer realizes that the light signals have traveled at the same speed over equal distances. Thus, O rightly concludes that the events at A and B occurred simultaneously. Now consider the same events as viewed by O'. By the time the light has reached observer O, O' has moved as indicated in Figure 10.6b. Thus, the light signal from B' has already swept past O', while the light from A' has not yet reached O'. According to Einstein, O' must find that light travels at the same speed measured by O. Therefore, O' concludes that the lightning struck the front of the boxcar before it struck the back. This thought experiment clearly demonstrates that the two events, which appear to O to be simultaneous, do not appear to O' to be simultaneous. In other words,

Simultaneity is not absolute

> two events that are simultaneous in one reference frame are in general not simultaneous in a second frame moving with respect to the first. That is, simultaneity is not an absolute concept but one that depends upon the state of motion of the observer.

At this point, you might wonder which observer is right concerning the two events. The answer is that *both are correct,* because the principle of relativity states that *there is no preferred inertial frame of reference.* Although the two observers reach different conclusions, each is correct in his or her own reference frame because the concept of simultaneity is not absolute. This, in fact, is the central point of relativity: any uniformly moving frame of reference can be used to describe events

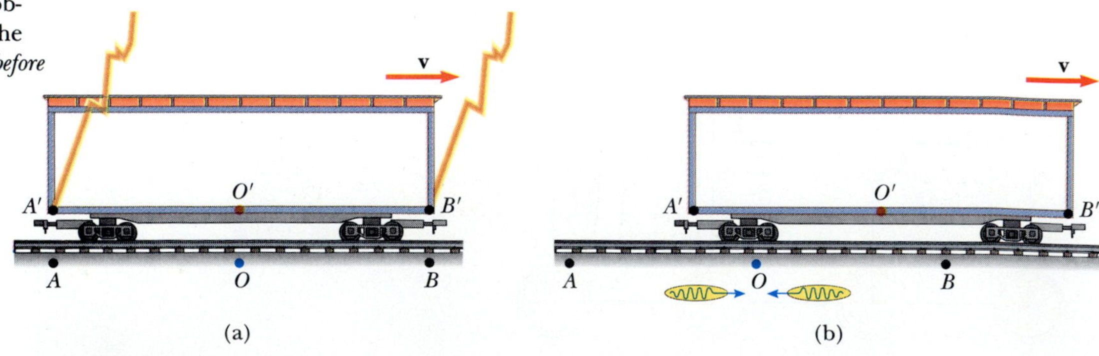

Figure 10.6 Two lightning bolts strike the end of a moving boxcar. (a) The events appear to be simultaneous to the stationary observer O, who is midway between A and B. (b) The events do not appear to be simultaneous to the observer O', who claims that the front of the train is struck *before* the rear.

and do physics. There is nothing wrong with the clocks and meter sticks used to perform measurements. It is simply that time intervals and length measurements depend on the state of motion of the observer. Observers in different inertial frames of reference will always measure different time intervals with their clocks and different distances with their meter sticks. They will, however, agree on the laws of physics in their respective frames, since these laws must be the same for all observers in uniform motion. It is the alteration of time and space that allows the laws of physics to be the same for all observers in uniform motion.

Time Dilation

The fact that observers in different inertial frames always measure different time intervals between a pair of events can be illustrated in another way by a vehicle moving to the right with a speed v, as in Figure 10.7a. A mirror is fixed to the ceiling of the vehicle, and the observer Liz at O', at rest in this system, holds a laser a distance d below the mirror. At some instant the laser emits a pulse of light directed toward the mirror (event 1), and at some later time, after reflecting from the mirror, the pulse arrives back at the laser (event 2). Liz carries a clock, C', which she uses to measure the time interval $\Delta t'$ between these two events. Because the light pulse has the speed c, the time it takes to travel from Liz to the mirror and back can be found from the definition of speed:

$$\Delta t' = \frac{\text{distance traveled}}{\text{speed of light}} = \frac{2d}{c} \qquad [10.3]$$

This time interval $\Delta t'$—measured by Liz, who, remember, is at rest in the moving vehicle—requires only a *single* clock, C', in this reference frame.

Now consider the same pair of events as viewed by an observer, Mark, standing at O in a stationary frame, as in Figure 10.7b. According to Mark, the mirror and laser are moving to the right with a speed of v. By the time the light pulse reaches the mirror, the mirror has moved a distance $v\,\Delta t/2$, where Δt is the time it takes the light to travel from O' to the mirror and back to O' as measured by Mark. In other words, Mark concludes that, because of the motion of the vehicle, if the light is to hit the mirror, the light must leave the laser at an angle with respect to the vertical direction. Comparing Figures 10.7a and 10.7b, we see that the light must travel farther for Mark than it does for Liz.

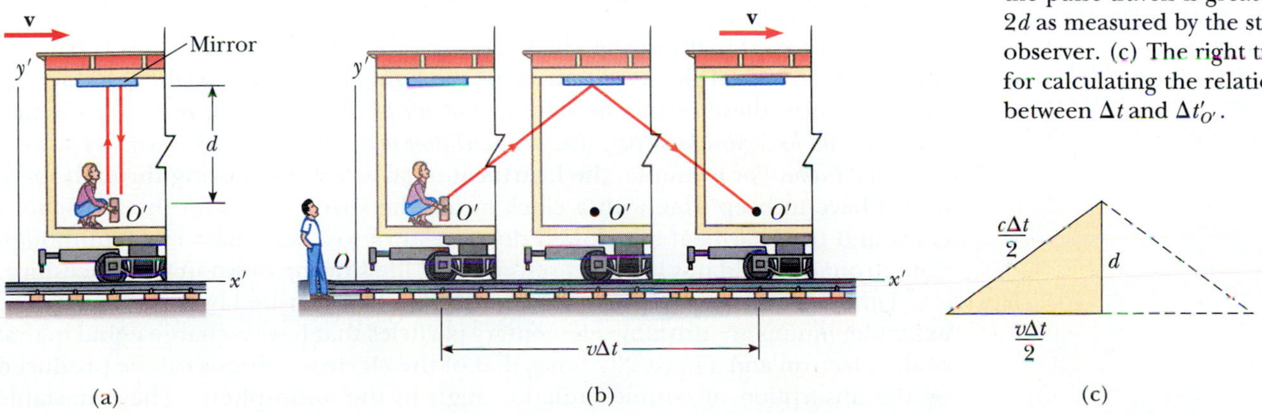

Figure 10.7
(a) A mirror is fixed to a moving vehicle, and a light pulse leaves O', a point that is at rest in the vehicle. (b) Relative to a stationary observer on Earth, the mirror and O' move with a speed v. Note that the distance the pulse travels is greater than $2d$ as measured by the stationary observer. (c) The right triangle for calculating the relationship between Δt and $\Delta t'_{O'}$.

According to the second postulate of special relativity, both observers must measure c for the speed of light. Since the light travels farther for Mark, it follows that the time interval Δt that he measures in the stationary frame is *longer* than the time interval $\Delta t'$ that Liz measures in the moving frame. To obtain a relationship between these two time intervals, it is convenient to use the right triangle shown in Figure 10.7c. The Pythagorean theorem gives

$$\left(\frac{c\,\Delta t}{2}\right)^2 = \left(\frac{v\,\Delta t}{2}\right)^2 + d^2$$

Solving for Δt gives

$$\Delta t = \frac{2d}{\sqrt{c^2 - v^2}} = \frac{2d}{c\sqrt{1 - \frac{v^2}{c^2}}} \qquad [10.4]$$

Because $\Delta t' = 2d/c$, we can express Equation 10.4 as

Time dilation

$$\Delta t = \frac{\Delta t'}{\sqrt{1 - \frac{v^2}{c^2}}} = \gamma\,\Delta t' \qquad [10.5]$$

where $\gamma = (1 - v^2/c^2)^{-1/2}$.

The two events observed by Mark occur at *different* positions. In order to measure the time interval Δt, he must therefore use *two synchronized clocks* situated at *different places* in his reference frame. Thus, the two situations are not symmetrical.

From Equation 10.5, we see that the time interval Δt measured by Mark in the stationary frame is *longer* than the time interval $\Delta t'$ measured by Liz in the moving frame (because γ is always greater than unity). This result leads us to the following conclusion:

According to a stationary observer, a moving clock runs more slowly than an identical stationary clock. This effect is known as time dilation.

The time interval $\Delta t'$ in Equation 10.5 is called the **proper time.** In general, proper time is defined as *the time interval between two events as measured by an observer who sees the events occur at the same place.* In our case, Liz measures the proper time. That is, *proper time is always the time measured with a single clock at rest in the frame in which the event occurs.*

We have seen that moving clocks run slow by a factor of γ^{-1}. This is true for ordinary mechanical clocks as well as for the light clock just described. In fact, we can generalize these results by stating that *all physical processes, including chemical reactions and biological processes, slow down relative to a stationary clock when they occur in a moving frame.* For example, the heartbeat of an astronaut moving through space would have to keep time with a clock inside the spaceship. Both the astronaut's clock and the heartbeat are slowed down relative to a stationary clock (although the astronaut would not have any sensation of life slowing down in the spaceship).

Time dilation is a real phenomenon that has been verified by experiments. For example, muons are unstable elementary particles that have a charge equal to that of the electron and a mass 207 times that of the electron. Muons can be produced by the absorption of cosmic radiation high in the atmosphere. These unstable particles have a lifetime of only 2.2 μs when measured in a reference frame that is at rest with respect to them. If we take 2.2 μs as the average lifetime of a muon and assume that the muon's speed is close to the speed of light, we find that these particles can travel a distance of only about 600 m before they decay into something else (Fig. 10.8a). Hence, they should not reach the Earth from the upper atmosphere, where they are produced. However, experiments show that a large number of muons *do* reach the Earth. The phenomenon of time dilation explains this effect. Relative to an observer on Earth, the muons have a lifetime equal to $\gamma\tau$, where $\tau = 2.2$ μs is the lifetime in a frame of reference traveling with the muons. For example, for $v = 0.99c$, $\gamma \approx 7.1$ and $\gamma\tau \approx 16$ μs. Hence, the average distance traveled, as measured by an observer on Earth, is $\gamma v\tau \approx 4800$ m (Fig. 10.8b).

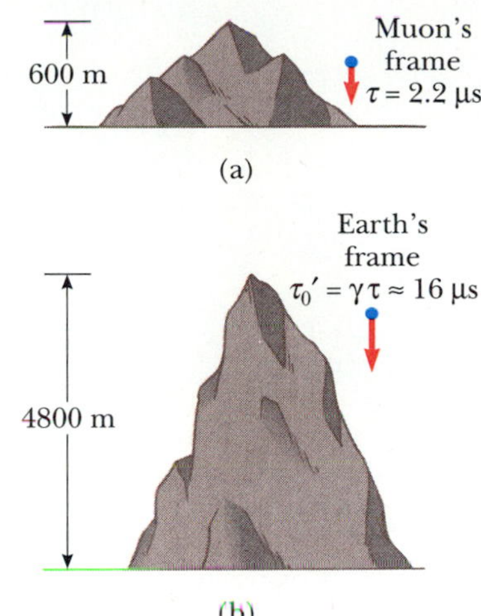

Figure 10.8
(a) The muons travel only about 600 m as measured in their reference frame, where their lifetime is about 2.2 μs. Because of time dilation, the muons' lifetime is longer as measured by an observer on Earth. (b) Muons traveling with a speed $0.99c$ travel 4800 m as measured by an observer on Earth.

In 1976, experiments with muons were conducted at the laboratory of the European Council for Nuclear Research (CERN) in Geneva. Muons were injected into a large storage ring and reached speeds of about $0.9994c$. Electrons produced by the decaying muons were detected by counters around the ring, enabling scientists to measure the decay rate and hence the lifetime of the muons. The lifetime of the moving muons was measured to be about 30 times as long as that of the stationary muons (Fig. 10.9), in agreement with the prediction of relativity to within two parts in a thousand.

The results of an experiment reported by Hafele and Keating provided direct evidence for the phenomenon of time dilation.[5] The experiment used stable cesium atomic clocks. Time intervals measured with four such clocks in jet flight were compared with time intervals measured by reference atomic clocks on Earth. (Because of the Earth's rotation about its axis, a ground-based clock is not in a true inertial frame.) In order to compare these results with the theory, many factors had to be considered, including periods of acceleration and deceleration relative to the Earth, variations in direction of travel, and the weaker gravitational field experienced by the flying clocks compared to the Earth-based clock. Their results were in good agreement with the predictions of the special theory of relativity. In their paper, Hafele and Keating report the following: "Relative to the atomic time scale of the U.S. Naval Observatory, the flying clocks lost 59 ± 10 ns during the eastward trip and gained 273 ± 7 ns during the westward trip. . . . These results

[5] J. C. Hafele and R. E. Keating, "Around the World Atomic Clocks: Relativistic Time Gains Observed," *Science,* July 14, 1972, p. 168.

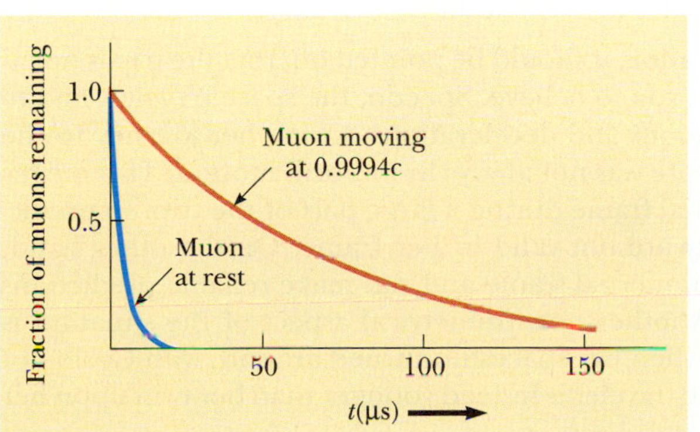

Figure 10.9
Decay curves for muons traveling at a speed of $0.9994c$ and for muons at rest.

provide an unambiguous empirical resolution of the famous clock paradox with macroscopic clocks.''

Example 10.1 **What Is the Period of the Pendulum?**

The period of a pendulum, defined as the time required for one complete oscillation, is measured to be 3.0 s in the inertial frame of the pendulum. What is the period when measured by an observer moving at a speed of $0.95c$ with respect to the pendulum?

Solution In this case, the proper time is 3 s. We can use Equation 10.5 to calculate the period measured by the moving observer.

$$T = \gamma T' = \frac{1}{\sqrt{1 - \dfrac{(0.95c)^2}{c^2}}} T' = (3.2)(3.0 \text{ s}) = \boxed{9.6 \text{ s}}$$

That is, the observer moving at a speed of $0.95c$ observes the pendulum as oscillating more slowly than does a stationary observer.

The Twin Paradox

An interesting consequence of time dilation is a phenomenon called the twin paradox. Consider an experiment involving 20-year-old twins, Speedo and Goslo. Speedo, the more venturesome twin, sets out on a journey toward a star 30 lightyears from Earth. Her spaceship can accelerate to a speed close to the speed of light. After reaching the star, Speedo becomes very homesick and immediately returns to Earth at the same high speed. Upon her return, she is shocked to find that many things have changed. Old cities have expanded and new cities have appeared. Lifestyles, people's appearances, and transportation systems have changed dramatically. Speedo's twin brother, Goslo, has aged to about 80 years old and is now wiser, feeble, and somewhat hard of hearing. Speedo, on the other hand, has aged only about ten years. This is because her bodily processes slowed down during her travels in space.

It is quite natural to raise the question ''Which twin actually traveled at a speed close to the speed of light?'' Herein lies the paradox: From Goslo's frame of reference, he was at rest while his sister Speedo traveled at a high speed. On the other hand, according to space traveler Speedo, it was she who was at rest while her brother zoomed away and then returned. This leads to contradictory viewpoints as to which twin aged.

To help you resolve this paradox, it should be pointed out that the trip is not as symmetrical as we may have led you to believe. Speedo, the space traveler, had to experience a series of accelerations and decelerations during her journey to the star and back home and therefore was not always in uniform motion. This means that she has been in a noninertial frame during a large part of the trip, so predictions based on special relativity are not valid in her frame. On the other hand, Goslo, on Earth, has been in an inertial frame and can make reliable predictions based on the special theory. Another nonsymmetrical aspect of the situation is Speedo's experience of forces when her spaceship turned around, with Goslo not subject to such forces. The space traveler is indeed younger than her twin upon her return to Earth.

Length Contraction

We have seen that measured time intervals are not absolute; that is, the time interval between two events depends on the frame of reference in which the interval is measured. Likewise, the measured distance between two points depends on the frame of reference. The **proper length** of an object is defined as *the length of the object measured in the reference frame in which the object is at rest.* The length of an object measured in a reference frame in which the object is moving is always less than the proper length. This effect is known as **length contraction.**

To understand length contraction quantitatively, let us consider a spaceship traveling at a speed of v from one star to another, and two observers, one in the ship and the other on Earth. The observer at rest on Earth (and also assumed to be at rest with respect to the two stars) measures the distance between the stars to be L_p, the proper length. According to this observer, the time it takes the spaceship to complete the voyage is $\Delta t = L_p/v$. What does the observer in the spaceship measure for the distance between the stars? Because of time dilation, the space traveler measures a smaller time of travel: $\Delta t' = \Delta t/\gamma$. The space traveler claims to be at rest and sees the destination star as moving toward the spaceship with speed v. Since the space traveler reaches the star in the time $\Delta t'$, he or she concludes that the distance, L, between the stars is shorter than L_p. This distance measured by the space traveler is

$$L = v\,\Delta t' = v\frac{\Delta t}{\gamma}$$

Since $L_p = v\,\Delta t$, we see that $L = L_p/\gamma$, or

$$L = L_p\left(1 - \frac{v^2}{c^2}\right)^{1/2} \qquad [10.6]$$

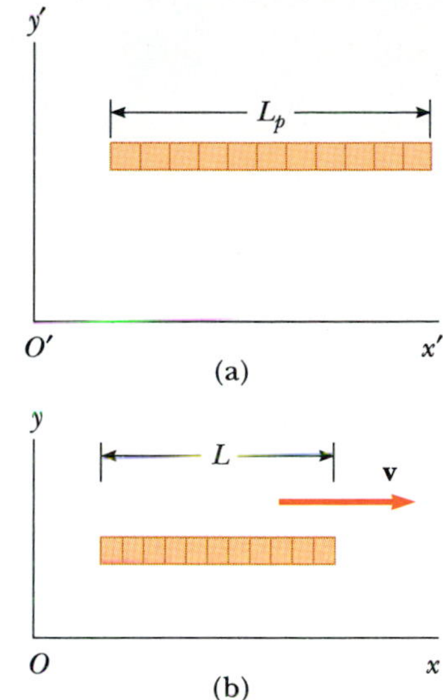

Figure 10.10 (a) A meter stick as viewed by an observer at rest with respect to it. (b) The same stick as seen by an observer moving with a speed v to the meter stick. The moving stick is always measured to be *shorter* than in its own rest frame by a factor $\sqrt{1 - v^2/c^2}$.

According to this result,

Length contraction

> if an observer at rest with respect to an object measures the object's length to be L_p (the proper length), an observer moving at a relative speed v with respect to the object will find it to be shorter than its proper length by the factor $(1 - v^2/c^2)^{1/2}$.

You should note that *the length contraction takes place only along the direction of motion.* For example, suppose a meter stick moves past a stationary Earth observer with a speed of v. The length of the stick, as measured by an observer in the frame attached to it (in other words, the stick is at rest relative to this observer), is the proper length L_p, as illustrated in Figure 10.10a. The length of the stick, L, as measured by the Earth observer in the stationary frame, is shorter than L_p by the factor $(1 - v^2/c^2)^{1/2}$.

Note that length contraction is a symmetrical effect: if the stick were at rest on Earth, an observer in the moving frame would measure its length to be shorter by the same factor $(1 - v^2/c^2)^{1/2}$.

It is important to emphasize that proper length and proper time are measured in *different* reference frames. As an example of this point, let us return to high-speed decaying muons (Fig. 10.8b). An observer in the muon's reference frame would measure the proper lifetime, while an Earth-based observer would measure the proper height of the mountain in Figure 10.10b. In the muon's reference frame there is no time dilation, but the distance of travel is observed to be shorter when measured in this frame. In the Earth observer's reference frame there is time dilation, but the distance of travel is measured to be the actual height of the mountain. Thus, when calculations on the muon are performed in both frames, one sees the effect of "offsetting penalties," and the outcome of the experiment is the same!

Example 10.2 The Contraction of a Spaceship

A spaceship is measured to be 100 m long by observer O', who is at rest relative to the ship. If this spaceship now flies by another observer, O, with a speed of $0.99c$, what length does this observer find?

Solution From Equation 10.6, the length measured by observer O is

$$L = L_p\sqrt{1 - \frac{v^2}{c^2}} = (100 \text{ m})\sqrt{1 - \frac{(0.99c)^2}{c^2}} = 14 \text{ m}$$

Exercise If the ship moves past observer O with a speed of $0.01c$, what length does the observer measure?

Answer 99.99 m.

Example 10.3 How High Is the Spaceship?

An observer on Earth sees a spaceship at an altitude of 435 m moving downward toward the Earth with a speed of $0.970c$. What is the altitude of the spaceship as measured by an observer in the ship at this time?

Solution The moving observer in the ship finds the altitude to be

$$L = L_p\sqrt{1 - \frac{v^2}{c^2}} = (435 \text{ m})\sqrt{1 - \frac{(0.970c)^2}{c^2}}$$

$$= 106 \text{ m}$$

Example 10.4 A Triangular Spaceship

A spaceship in the form of a triangle flies by an observer at a speed of $0.950c$. When the ship is at rest (Fig. 10.11a), the lengths x and y are found to be 50.0 m and 25.0 m, respectively. What is the shape of the ship as seen by an observer at rest, when the ship is moving in the direction shown in Figure 10.11b?

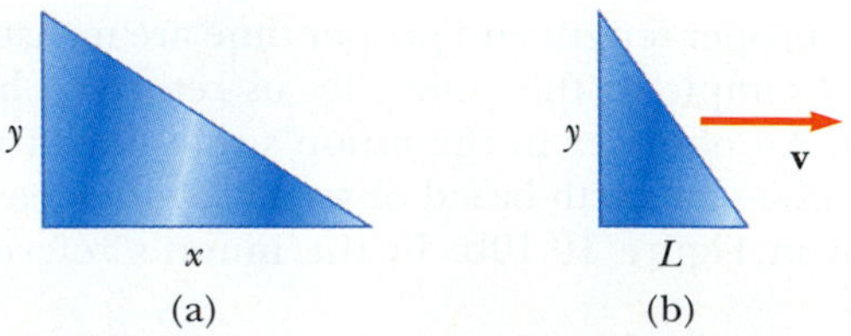

y
x
(a)

y
v
L
(b)

Figure 10.11 (Example 10.4) (a) When the spaceship is at rest, its shape is as shown. (b) To a stationary observer, the spaceship appears to look like this when it moves to the right with a speed v. Note that only its x dimension is contracted in this case.

Solution The observer sees the horizontal length of the ship to be contracted to

$$L = L_p\sqrt{1 - \frac{v^2}{c^2}} = (50.0 \text{ m})\sqrt{1 - \frac{(0.950c)^2}{c^2}}$$

$$= \boxed{15.6 \text{ m}}$$

The 25-m vertical height is unchanged because it is perpendicular to the direction of relative motion between the stationary observer and the spaceship. Figure 10.11b represents the shape of the spaceship as seen by the observer.

10.5 The Lorentz Transformation

Suppose an event that occurs at some point P is reported by two observers, one at rest in a frame that we can call S and another in a frame moving to the right with a speed v, as in Figure 10.12 which we can call S'. The observer in S reports the event with space-time coordinates (x, y, z, t), while the observer in S' reports the same event using the coordinates (x', y', z', t'). We would like to find a relation between these coordinates that is valid for all speeds. In Section 10.1 we found that the Galilean transformation of coordinates (Eq. 10.1) does not agree with experiment at speeds comparable to the speed of light.

The correct equations, which are valid for speeds ranging from $v = 0$ to $v = c$ and enable us to transform from S to S', are given by the **Lorentz transformation equations:**

$$\begin{aligned} x' &= \gamma(x - vt) \\ y' &= y \\ z' &= z \\ t' &= \gamma\left(t - \frac{v}{c^2}x\right) \end{aligned} \qquad [10.7]$$

Lorentz transformation for $S \rightarrow S'$

where γ has the same value as in Equation 10.5. The Lorentz transformation equations were originally derived in 1890 by H. A. Lorentz (1853–1928). However, it was Einstein who recognized their physical significance and took the bold step of interpreting them within the framework of the theory of relativity.

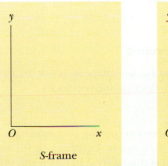

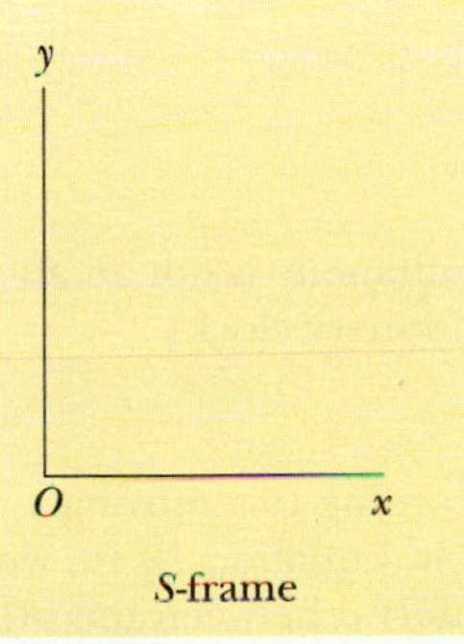

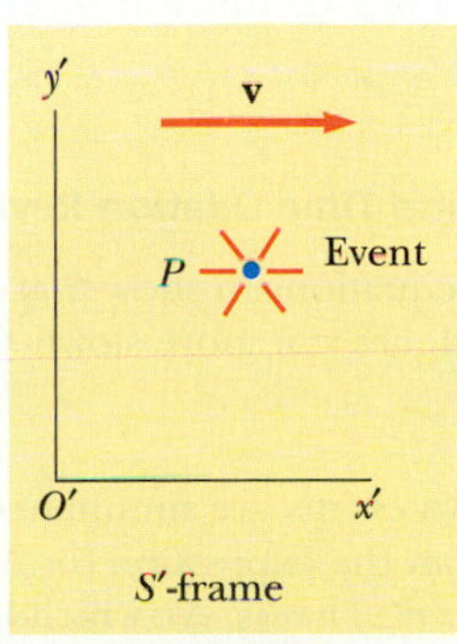

Figure 10.12 Representation of an event that occurs at some point P as observed by an observer at rest in the S frame and another observer in the S' frame, which is moving to the right with a speed v.

We see that the value for t' assigned to an event by the observer in S' depends both on the time t and on the coordinate x as measured by the observer in S. This is consistent with the notion that an event is characterized by four space-time coordinates (x, y, z, t). In other words, in relativity, space and time are not separate concepts, but rather are closely interwoven with each other. This is unlike the case of the Galilean transformation, in which $t = t'$.

If we wish to transform coordinates in the S' frame to coordinates in the S frame, we simply replace v with $-v$ and interchange the primed and unprimed coordinates in Equation 10.7. The resulting transformation is

Inverse Lorentz transformation for $S' \rightarrow S$

$$\begin{aligned} x &= \gamma(x' + vt') \\ y &= y' \\ z &= z' \\ t &= \gamma\left(t' + \frac{v}{c^2}x'\right) \end{aligned} \qquad [10.8]$$

When $v \ll c$, the Lorentz transformation should reduce to the Galilean transformation. To check this, note that as $v \rightarrow 0$, $v/c^2 \ll 1$ and $v^2/c^2 \ll 1$, so that Equation 10.7 reduces in this limit to the Galilean coordinate transformation equations, given by Equation 10.1:

$$x' = x - vt \qquad y' = y \qquad z' = z \qquad t' = t$$

In many situations, scientists want to know the *difference* in coordinates between two events or the time *interval* between two events as seen by two observers, one in S and one in S'. This can be accomplished by writing the Lorentz equations in a form suitable for describing pairs of events. Using Equations 10.7 and 10.8, we can express the differences between the four variables x, x', t, and t' in the form

$$\begin{aligned} \Delta x' &= \gamma(\Delta x - v\,\Delta t) \\ \Delta t' &= \gamma\left(\Delta t - \frac{v}{c^2}\Delta x\right) \end{aligned} \qquad S \rightarrow S' \qquad [10.9]$$

$$\begin{aligned} \Delta x &= \gamma(\Delta x' + v\Delta t') \\ \Delta t &= \gamma\left(\Delta t' + \frac{v}{c^2}\Delta x'\right) \end{aligned} \qquad S' \rightarrow S \qquad [10.10]$$

where $\Delta x' = x_2' - x_1'$ and $\Delta t' = t_2' - t_1$ are the differences measured by the observer in S', and $\Delta x = x_2 - x_1$ and $\Delta t = t_2 - t_1$ are the differences measured by the observer in S. (We have not included the expressions for relating the y and z coordinates, since they are unaffected by motion along the x direction.)

Example 10.5 Simultaneity and Time Dilation Revisited

Use the Lorentz transformation equations to show that (a) simultaneity is not an absolute concept and (b) moving clocks run more slowly than stationary clocks.

Solution

(a) **Simultaneity.** Suppose that two events are simultaneous according to a moving observer, O', so that $\Delta t' = 0$. From the expression for Δt given in Equation 10.10, we see that in this case, $\Delta t = \gamma v\,\Delta x'/c^2$. That is, Δt, which is the interval between the two events as measured by the stationary observer O, is nonzero, so the events do not appear to O to be simultaneous.

(b) **Time Dilation.** Suppose that observer O' finds that two events occur at the same place ($\Delta x' = 0$) but at different times ($\Delta t' \neq 0$). In this situation, the expression for Δt given in Equation 10.10 becomes $\Delta t = \gamma \Delta t'$. This is the equation for time dilation found earlier (Eq. 10.5), where $\Delta t' = \Delta t$ is the proper time measured by the single clock in O'.

Exercise Use the Lorentz transformation equations, in the form of Equations 10.9 and 10.10, to confirm length contraction. That is, $L = L_p/\gamma$.

The Lorentz Velocity Transformation

Let us now derive the Lorentz velocity transformation, which is the relativistic counterpart of the Galilean velocity transformation (Eq. 10.2). Suppose that an unaccelerated object is observed in the (moving) S' frame at x'_1 at time t'_1 and at x'_2 at time t'_2. Its speed, u'_x, measured in S' is

$$u'_x = \frac{x'_2 - x'_1}{t'_2 - t'_1} = \frac{dx'}{dt'} \qquad [10.11]$$

Using Equation 10.7, we have

$$dx' = \gamma(dx - v\,dt)$$

$$dt' = \gamma\left(dt - \frac{v}{c^2}\,dx\right)$$

Substituting these into Equation 10.11 gives

$$u'_x = \frac{dx'}{dt'} = \frac{dx - v\,dt}{dt - \dfrac{v}{c^2}\,dx} = \frac{\dfrac{dx}{dt} - v}{1 - \dfrac{v}{c^2}\dfrac{dx}{dt}}$$

But dx/dt is just the velocity u_x of the object measured in the stationary frame S, and so this expression becomes

$$u'_x = \frac{u_x - v}{1 - \dfrac{u_x v}{c^2}} \qquad [10.12]$$

Lorentz velocity transformation for $S \rightarrow S'$

Similarly, if the object has velocity components along y and z, the components in S' are given by

$$u'_y = \frac{u_y}{\gamma\left(1 - \dfrac{u_x v}{c^2}\right)} \quad \text{and} \quad u'_z = \frac{u_z}{\gamma\left(1 - \dfrac{u_x v}{c^2}\right)} \qquad [10.13]$$

When u_x and v are both much smaller than c (the nonrelativistic case), the denominator of Equation 10.12 approaches unity, and $u'_x \approx u_x - v$. This corresponds to the Galilean velocity transformations. In the other extreme, when $u_x = c$, Equation 10.12 becomes

$$u'_x = \frac{c - v}{1 - \frac{cv}{c^2}} = \frac{c\left(1 - \frac{v}{c}\right)}{1 - \frac{v}{c}} = c$$

From this result, we see that an object moving with a speed of c relative to an observer in one inertial reference frame also has the speed c relative to an observer in *any other* inertial reference frame—*independent* of the relative motions of S and S'. Note that this conclusion is consistent with Einstein's second postulate, namely, that the speed of light must be c with respect to all inertial frames of reference. Furthermore, the speed of an object can never exceed c. That is, the speed of light is the "ultimate" speed. We shall return to this point later, when we consider the energy of a particle.

To obtain u_x in terms of u'_x, we replace v with $-v$ in Equation 10.12 and interchange the roles of u_x and u'_x. This gives

Inverse Lorentz velocity transformation for $S' \rightarrow S$

$$u_x = \frac{u'_x + v}{1 + \frac{u'_x v}{c^2}} \quad [10.14]$$

Example 10.6 Relative Velocity of Spaceships

Two spaceships, A and B, are moving in *opposite* directions, as in Figure 10.13. An observer on the Earth measures the speed of A to be $0.75c$ and the speed of B to be $0.85c$. Find the velocity of B with respect to A.

Solution This problem can be solved by taking the S' frame as being attached to spacecraft A. As usual, call the (stationary) Earth frame S. Spacecraft B can be considered as moving to the left with a velocity of $u_x = -0.85c$ relative to the Earth observer. Hence, the velocity of B with respect to A can be obtained using Equation 10.12:

$$u'_x = \frac{u_x - v}{1 - \frac{u_x v}{c^2}} = \frac{-0.85c - 0.75c}{1 - \frac{(-0.85c)(0.75c)}{c^2}}$$

$$= \boxed{-0.98c}$$

The negative sign for u'_x indicates that spaceship B is moving in the negative x direction as observed by A. Note that the result is less than c. That is, a body whose speed is less than c in one frame of reference must have a speed less than c in *any other*

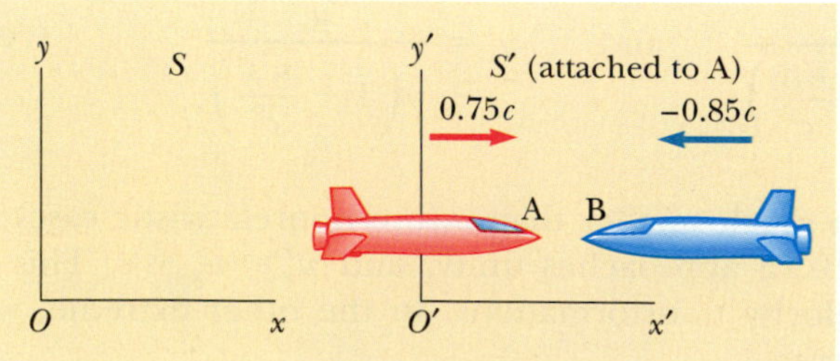

Figure 10.13 (Example 10.6) Two spaceships A and B move in *opposite* directions. The velocity of B relative to A is *less* than c and is obtained by using the relativistic velocity transformation.

frame. If the Galilean velocity transformation were used in this example, we would find that $u'_x = u_x - v = -0.85c - 0.75c = -1.6c$, which is a physical impossibility.

Example 10.7 A Speeding Motorcycle

Imagine a motorcycle rider moving with a speed of $0.8c$ past a stationary observer, as shown in Figure 10.14. If the rider tosses a ball in the forward direction with a speed of $0.7c$ relative to himself, what is the speed of the ball as seen by the observer?

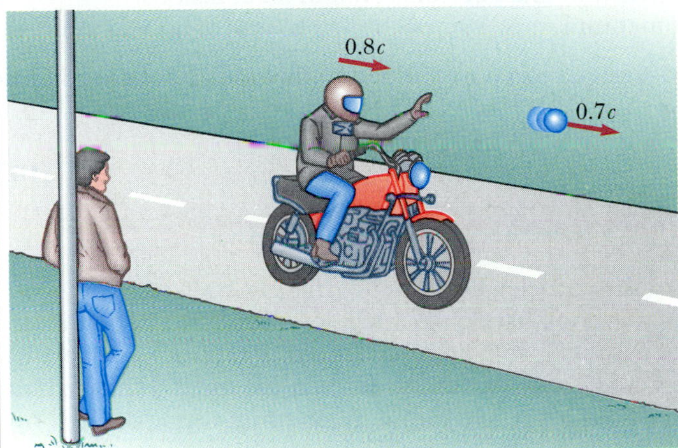

Figure 10.14 (Example 10.7) A motorcyclist traveling at speed $0.8c$ passes a stationary observer and throws a ball in the direction of motion with a speed of $0.7c$ relative to himself.

Solution In this situation, the speed of the motorcycle with respect to the stationary observer is $v = 0.8c$. The speed of the ball in the frame of reference of the motorcyclist is $0.7c$. Therefore, the speed, u_x, of the ball relative to the observer is

$$u_x = \frac{u'_x + v}{1 + \dfrac{u'_x v}{c^2}} = \frac{0.7c + 0.8c}{1 + \dfrac{(0.7c)(0.8c)}{c^2}} = 0.962c$$

Exercise Suppose that the motorcyclist, moving with a speed $0.8c$, turns on a light signal that moves away from him with a speed of c in the direction of the moving motorcycle. What would the observer measure for the speed of the light signal?

Answer c.

10.6 Relativistic Momentum

We have seen that the principle of relativity is satisfied if the Galilean transformation is replaced by the more general Lorentz transformation. Therefore, in order to properly describe the motions of material particles within the framework of special relativity, we must generalize Newton's laws and the definitions of momentum and energy. These generalized definitions reduce to the classical (nonrelativistic) definitions for $v \ll c$.

First, recall that the law of conservation of momentum requires that when two bodies collide, the total momentum remains constant, assuming the bodies are isolated (that is, they interact only with each other). Suppose the collision is described in a reference frame, S, in which the momentum is conserved. If the velocities in a second reference frame, S', are calculated using the Lorentz transformation and the classical definition of momentum, $p = m\mathbf{u}$, one finds that momentum is *not* conserved in the second reference frame.[6] However, because the laws of physics are the same in all inertial frames, the momentum must be conserved in all systems. In view of this condition, and assuming the Lorentz transformation is correct, we must modify the definition of relativistic momentum, **p**, to satisfy the following conditions:

1. The relativistic momentum must be conserved in all collisions.
2. The relativistic momentum must approach the classical value $m\mathbf{u}$ as $\mathbf{u} \to 0$.

The correct relativistic equation for momentum that satisfies these conditions is

Definition of relativistic momentum

$$\mathbf{p} \equiv \frac{m\mathbf{u}}{\sqrt{1 - \frac{u^2}{c^2}}} = \gamma m\mathbf{u} \qquad [10.15]$$

where **u** is the velocity of the particle. The proof of this generalized expression for **p** is beyond the scope of this book. When u is much less than c, the denominator of Equation 10.15 approaches unity, so **p** approaches $m\mathbf{u}$. Therefore, the relativistic equation for **p** reduces to the classical expression when u is small compared with c.

The relativistic force **F** on a particle whose momentum is **p** is defined by the expression

$$\mathbf{F} \equiv \frac{d\mathbf{p}}{dt} \qquad [10.16]$$

where **p** is given by Equation 10.15. This expression is identical to the classical statement of Newton's second law, which says that force equals the time rate of change of momentum.

Example 10.8 Momentum of an Electron

An electron, which has a mass of 9.11×10^{-31} kg, moves with a speed of $0.750c$. Find its relativistic momentum and compare this with the momentum calculated from the classical expression.

Solution Using Equation 10.15 with $u = 0.75c$, we have

$$p = \frac{mu}{\sqrt{1 - \frac{u^2}{c^2}}}$$

$$= \frac{(9.11 \times 10^{-31}\ \text{kg})(0.750 \times 3 \times 10^8\ \text{m/s})}{\sqrt{1 - \frac{(0.750c)^2}{c^2}}}$$

$$= 3.10 \times 10^{-22}\ \text{kg}\cdot\text{m/s}$$

[6] For particle velocity we use the symbol **u** rather than **v**, which is used for the relative velocity of two reference frames.

The incorrect classical expression would give

$$\text{Momentum} = mu = 2.05 \times 10^{-22}\ \text{kg} \cdot \text{m/s}$$

The correct relativistic result is 50% greater than the classical result!

It is left to an end-of-chapter problem (Problem 54) to show that the acceleration a of a particle decreases under the action of a constant force, in which case $a \propto (1 - u^2/c^2)^{3/2}$. Furthermore, as the velocity of a particle approaches c, the acceleration caused by any finite force approaches zero. Hence, it is impossible to accelerate a particle from rest to a speed equal to or greater than c.

10.7 Relativistic Energy

We have seen that the definition of momentum and the laws of motion require generalization to make them compatible with the principle of relativity. This implies that the definition of kinetic energy must also be modified.

In order to derive the relativistic form of the work-energy theorem (Eq. 7.19 in Chapter 7), let us start with the definition of the work done by a force, **F**, acting on a particle and make use of the definition of relativistic force (Eq. 10.16).

$$W = \int_{x_1}^{x_2} F\, dx = \int_{x_1}^{x_2} \frac{dp}{dt}\, dx$$

where we have assumed that the force and the motion of the particle are along the x axis. In order to perform this integration, we make repeated use of the chain rule for derivatives:

$$\left(\frac{dp}{dt}\right) dx = \left(\frac{dp}{du}\frac{du}{dt}\right) dx = \frac{dp}{du}\left(\frac{du}{dx}\frac{dx}{dt}\right) dx$$

$$= \frac{dp}{du}\, u \frac{du}{dx}\, dx = \frac{dp}{du}\, u\, du$$

Since p depends on u, according to Equation 10.15, we have

$$\frac{dp}{du} = \frac{d}{du}\frac{mu}{\sqrt{1 - \frac{u^2}{c^2}}} = \frac{m}{\left(1 - \frac{u^2}{c^2}\right)^{3/2}}$$

Using these results, we can express the work as

$$W = \int_0^u \frac{dp}{du}\, u\, du = \int_0^u \frac{mu}{\left(1 - \frac{u^2}{c^2}\right)^{3/2}}\, du$$

where we have assumed that the particle is accelerated from rest to some final velocity u. Evaluating the integral, we find that

$$W = \frac{mc^2}{\sqrt{1 - \frac{u^2}{c^2}}} - mc^2 \qquad [10.17]$$

View of the Large Electron-Positron (LEP) collider at CERN, the European center for particle physics near Geneva. LEP accelerates electrons and positrons to an energy of 50 GeV. It is housed in a circular tunnel 100 m underground and 27 km in circumference. The particles travel in opposite directions inside a narrow aluminum pipe which passes through large electromagnets. Test instruments sit on top of the magnets in this picture taken shortly before LEP started operations in July 1989. The long magnets are dipoles that bend the particles around the ring; the square blue magnets are quadrupoles that focus them in tight bunches. The particles collide at four points around the ring. *(David Parker, Science Photo Library)*

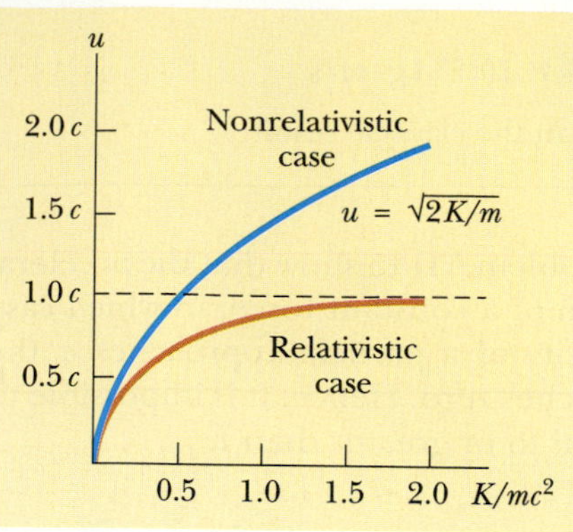

Figure 10.15
A graph comparing relativistic and nonrelativistic kinetic energy. The energies are plotted versus speed. In the relativistic case, u is always *less* than c.

Recall from Chapter 7, Equation 7.19, that the work-energy theorem states that the work done by a force acting on a particle equals the change in kinetic energy of the particle. Since the initial kinetic energy here is zero, we conclude that the work W is equivalent to the relativistic kinetic energy K; that is,

Relativistic kinetic energy

$$K = \frac{mc^2}{\sqrt{1 - \frac{u^2}{c^2}}} - mc^2 \qquad [10.18]$$

This equation has been confirmed by experiments using high-energy particle accelerators. At low speeds, where $u/c \ll 1$, Equation 10.18 should reduce to the classical expression $K = \frac{1}{2}mu^2$. We can check this by using the binomial expansion $(1 - x^2)^{-1/2} \approx 1 + \frac{1}{2}x^2 + \cdots$ for $x \ll 1$, where the higher-order powers of x are neglected in the expansion. In our case, $x = u/c$, so that

$$\frac{1}{\sqrt{1 - \frac{u^2}{c^2}}} = \left(1 - \frac{u^2}{c^2}\right)^{-1/2} \approx 1 + \frac{1}{2}\frac{u^2}{c^2} + \cdots$$

Substituting this into Equation 10.18 gives

$$K \approx mc^2\left(1 + \frac{1}{2}\frac{u^2}{c^2} + \cdots\right) - mc^2 = \tfrac{1}{2}mu^2$$

which agrees with the classical result. Figure 10.15 is a graph comparing the relativistic and nonrelativistic expressions. In the relativistic case the particle speed never exceeds c, regardless of the kinetic energy. The two curves are in good agreement when $u \ll c$.

It is useful to write the relativistic kinetic energy for a particle in the form

$$K = \gamma mc^2 - mc^2 \qquad [10.19]$$

where γ, as usual, equals $(1 - u^2/c^2)^{-1/2}$. The constant term mc^2, which is independent of the speed, is called the **rest energy,** E_0, of the particle. The term γmc^2, which depends on the particle speed, is therefore the sum of the kinetic and rest energies. We define γmc^2 to be the **total energy** E; that is,

Definition of total energy

$$E = \gamma mc^2 = K + mc^2 \qquad [10.20]$$

or

$$E = \frac{mc^2}{\sqrt{1 - \frac{u^2}{c^2}}} \quad [10.21]$$

Energy-mass relation

This, of course, is Einstein's famous mass-energy equation. The relation $E = \gamma mc^2 = \gamma E_0$ shows that *mass is a property of energy*. Furthermore, the result shows that a small mass corresponds to an enormous amount of energy. This concept is fundamental to much of the field of nuclear physics.

In many situations, the momentum or energy of a particle is known, rather than its speed. It is therefore useful to have an expression relating the total energy E to the relativistic momentum p. This is accomplished by using the expressions $E = \gamma mc^2$ and $p = \gamma mu$. By squaring these equations and subtracting, we can eliminate u (Problem 28). The result, after some algebra, is

$$E^2 = p^2c^2 + (mc^2)^2 \quad [10.22]$$

Energy-momentum relation

When the particle is at rest, $p = 0$ and $E = E_0 = mc^2$. That is, the total energy equals the rest energy. As we shall discuss later, it is well established that particles exist that have zero mass, such as photons. If we set $m = 0$ in Equation 10.22, we see that

$$E = pc \quad [10.23]$$

Energy-momentum relation for photons and neutrinos

This equation is an *exact* expression relating energy and momentum for photons and neutrinos, which always travel at the speed of light.

Finally, note that since the mass m of a particle is independent of the particle's motion, m must have the same value in all reference frames. On the other hand, the total energy and momentum of a particle depend on the reference frame in which they are measured, since they both depend on velocity. Because m is a constant, then, according to Equation 10.22, the quantity $E^2 - p^2c^2$ must have the same value in all reference frames. That is, $E^2 - p^2c^2$ is said to be *invariant* under a Lorentz transformation.

It is convenient to express the energy of electrons and other subatomic particles in electron volts (eV). An electron volt is the energy gained by an electron (or proton) when it is accelerated through a voltage of 1 volt = 1 joule per coulomb (a volt is the SI unit of electric potential). Since the charge on an electron is $|e| = 1.6 \times 10^{-19}$ C, the conversion factor from joules to electron volts is

$$1 \text{ eV} = 1.60 \times 10^{-19} \text{ J}$$

For example, the mass of an electron is 9.11×10^{-31} kg. Hence, its rest energy is

$$mc^2 = (9.11 \times 10^{-31} \text{ kg})(3.00 \times 10^8 \text{ m/s})^2 = 8.20 \times 10^{-14} \text{ J}$$

Converting this to electron volts, we have

$$mc^2 = (8.20 \times 10^{-14} \text{ J})(1 \text{ eV}/1.60 \times 10^{-19} \text{ J}) = 0.511 \text{ MeV}$$

where 1 MeV = 10^6 eV.

"Imagination is more important than knowledge."
Albert Einstein.

Example 10.9 The Energy of a Speedy Electron

An electron moves with a speed of $u = 0.850c$. Find its total energy and kinetic energy in electron volts.

Solution Using the fact that the rest energy of the electron is 0.511 MeV, together with Equation 10.21, we get

$$E = \frac{mc^2}{\sqrt{1 - \frac{u^2}{c^2}}} = \frac{0.511 \text{ MeV}}{\sqrt{1 - \frac{(0.85c)^2}{c^2}}}$$

$$= 1.90(0.511 \text{ MeV}) = \boxed{0.970 \text{ MeV}}$$

The kinetic energy is obtained by subtracting the rest energy from the total energy:

$$K = E - mc^2 = 0.970 \text{ MeV} - 0.511 \text{ MeV}$$

$$= \boxed{0.459 \text{ MeV}}$$

Example 10.10 The Energy of a Speedy Proton

The total energy of a proton is three times its rest energy. (a) Find the proton's rest energy in electron volts.

Solution

$$\text{Rest energy} = mc^2 = (1.67 \times 10^{-27} \text{ kg})(3 \times 10^8 \text{ m/s})^2$$

$$= (1.50 \times 10^{-10} \text{ J})(1 \text{ eV}/1.60 \times 10^{-19} \text{ J})$$

$$= \boxed{939 \text{ MeV}}$$

(b) What is the speed of the proton?

Solution Since the total energy E is three times the rest energy, Equation 10.21 gives

$$E = 3mc^2 = \frac{mc^2}{\sqrt{1 - \frac{u^2}{c^2}}}$$

$$3 = \frac{1}{\sqrt{1 - \frac{u^2}{c^2}}}$$

Solving for u gives

$$\left(1 - \frac{u^2}{c^2}\right) = \frac{1}{9} \quad \text{or} \quad \frac{u^2}{c^2} = \frac{8}{9}$$

$$u = \frac{\sqrt{8}}{3}c = \boxed{2.83 \times 10^8 \text{ m/s}}$$

(c) Determine the kinetic energy of the proton in electron volts.

Solution

$$K = E - mc^2 = 3mc^2 - mc^2 = 2mc^2$$

Since $mc^2 = 939$ MeV,

$$K = \boxed{1878 \text{ MeV}}$$

(d) What is the proton's momentum?

Solution We can use Equation 10.22 to calculate the momentum with $E = 3mc^2$:

$$E^2 = p^2c^2 + (mc^2)^2 = (3mc^2)^2$$

$$p^2c^2 = 9(mc^2)^2 - (mc^2)^2 = 8(mc^2)^2$$

$$p = \sqrt{8}\,\frac{mc^2}{c} = \sqrt{8}\,\frac{(939\text{ MeV})}{c} = 2656\,\frac{\text{MeV}}{c}$$

The unit of momentum is written MeV/c for convenience.

10.8 Applications of Special Relativity: Fission and Fusion

Einstein's special theory of relativity has been confirmed by a number of experiments. One important investigation, concerned with muon decay and time dilation in the muon's reference frame, was discussed in Section 10.4. This section describes other applications and further evidence of the theory.

One of the first predictions of special relativity that was experimentally confirmed is the variation of momentum with velocity. As early as 1909 experiments were performed on electrons, which can easily be accelerated to speeds close to c in the presence of a strong electric field. As you will recall from Chapter 6, Section 6.6, when an energetic electron enters a magnetic field **B** with its velocity vector perpendicular to **B**, a centripetal magnetic force of magnitude evB (Eq. 6.20) is exerted on the electron, causing it to move in a circle of radius r. In this situation, the relativistic momentum of the electron is given by $p = eB/r$. From the relativistic equivalent of Newton's second law, $\mathbf{F} = d\mathbf{p}/dt$, the variation in momentum with kinetic energy can be checked experimentally. The results of such experiments (on electrons and on other charged particles), performed with the use of high-energy accelerators, support the relativistic expressions.

Nuclear Fission

Nuclear fission occurs when a heavy nucleus, such as ^{235}U, splits—or fissions—into two smaller nuclei. In such a reaction, *the total mass and energy of the products is less than the original mass.* **Fission**

Nuclear fission was first observed in 1938 by Otto Hahn and Fritz Strassman, following some basic studies by Fermi. After bombarding uranium with neutrons, Hahn and Strassman discovered among the reaction products two medium-mass elements, barium and lanthanum. Shortly thereafter, Lisa Meitner and Otto Frisch explained what had happened. The uranium nucleus had split into two nearly equal fragments after absorbing a neutron. Such an occurrence was of considerable interest to physicists attempting to understand the nucleus, but it was to have even more far-reaching consequences. Measurements showed that about 208 MeV of energy are released in each fission event, and this fact was to affect the course of human history.

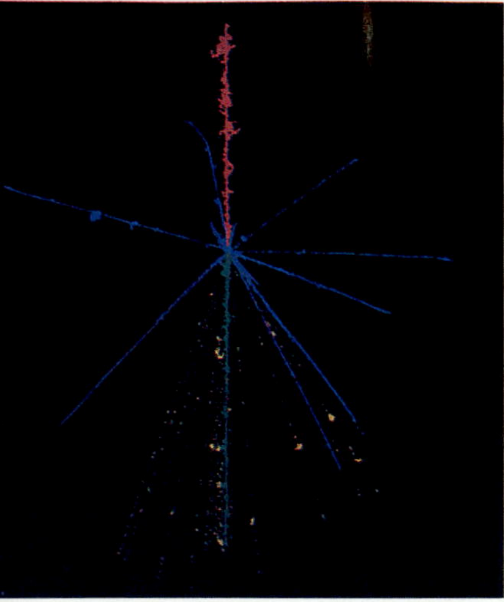

The power of a cosmic ray is shown in this false-color emulsion photo of a cosmic ray sulphur nucleus (red) colliding with a nucleus in the emulsion. The collision produces a spray of other particles: a fluorine nucleus (green), other nuclear fragments (blue), and 16 pions (yellow). The length of the sulphur track is 0.11 mm. The curlicues adorning the track of the sulphur nucleus are electrons which it has knocked out of atoms in passing. The photograph was taken in 1950 by Cecil Powell, the English physicist who pioneered the use of photographic emulsions to record the tracks of electrically charged particles. *(C. Powell, P. Fowler, and D. Perkins/Science Photo Library)*

The release of enormous quantities of energy in nuclear fission processes provides a demonstration of the relationship of mass to energy. All reactions that release energy, including the atomic and hydrogen bombs as well as chemical reactions, yield products with less total mass than the original materials. This is required by the fundamental equation, first offered by Einstein, $E = \gamma mc^2$, which applies, as written here, only for particles.

For example, in a nuclear reactor, the uranium nucleus undergoes fission, a reaction that results in several lighter fission products and a release of energy to the surroundings. In the case of ^{235}U (the parent nucleus), which undergoes spontaneous fission, the products are two lighter nuclei and two neutrons. The total mass of the products, after they have come to rest, is *less* than that of the parent nucleus by some amount Δm. The corresponding energy loss of the system, Δmc^2, associated with this mass difference—sometimes called the **disintegration energy** or the "Q" of the reaction—is exactly equal to the total kinetic energy acquired and then given up by the products. In peacetime applications, this energy is then used to produce heat and steam for the generation of electrical power.

It is important to recognize that such processes do *not* violate either the law of conservation of energy or the law of conservation of mass. It simply requires that mass γm for a particle be reinterpreted to be equal to E/c^2, and hence energy-dependent. It is then apparent that conservation of mass is a consequence of conservation of energy. Whatever energy (or mass) is lost by the reacting system is given up to the surroundings, as could be demonstrated with sufficiently careful measurements.

It is conventional to represent an element by $^{A}_{Z}X$, where X represents the chemical symbol for the element, A is the mass number, which is the number of nucleons (neutrons plus protons) in the nucleus, and Z is the atomic number, which is the number of protons in the nucleus. For example, $^{56}_{26}Fe$ has a mass number of 56 and an atomic number of 26; it therefore contains 26 protons and 30 neutrons. (Note that the number of neutrons in the nucleus is equal to A − Z.)

The fission of $^{235}_{92}U$ by neutron bombardment can be represented by

$$^{1}_{0}n + ^{235}_{92}U \longrightarrow ^{236}_{92}U^* \longrightarrow X + Y + \text{neutrons} \qquad \textbf{[10.24]}$$

where $^{236}_{92}U^*$ is an intermediate state that lasts for only about 10^{-12} s before splitting into the fission products X and Y, usually called **fission fragments.** Many combinations of X and Y exist that satisfy the requirements of conservation of mass-energy and charge. In uranium fission, about 90 different fragments can be formed. The fission process also results in the production and release of several neutrons per fission event, typically two or three (the average is 2.47). The fission fragments and the released neutrons have a great deal of kinetic energy after the fission event. The neutrons, when appreciably slowed by a **moderator,** can initiate additional fission events.

The breakup of the uranium nucleus can be compared to what happens to a drop of water when excess energy is added to it. All the atoms in the drop have some initial energy, but not enough to break up the drop. However, if enough energy is added to set the drop to vibrating, it will undergo elongation and compression until the amplitude of vibration becomes great enough to cause the drop to break. In the uranium nucleus, a similar process occurs (Fig. 10.16). The sequence of events is as follows:

1. The ^{235}U nucleus captures a thermal (slow-moving) neutron.

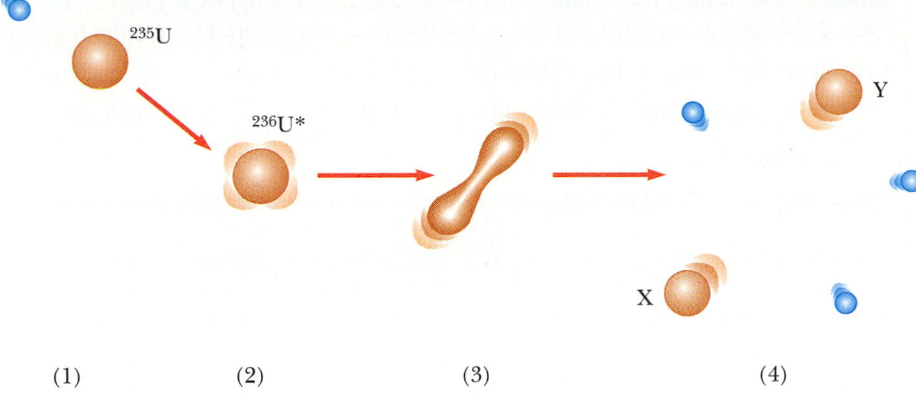

Figure 10.16 The stages in a nuclear fission event as described by the liquid-drop model of the nucleus.

2. This capture results in the formation of ^{236}U*, and the excess energy of this nucleus causes it to undergo violent oscillations.
3. The ^{236}U* nucleus becomes highly distorted, and the force of repulsion between protons in the two halves of the dumbbell shape tends to increase the distortion.
4. The nucleus splits into two fragments, emitting several neutrons in the process.

Let us estimate the disintegration energy, Q, released in a typical fission process. The binding energy per nucleon is about 7.6 MeV for heavy nuclei (those having a mass number approximately equal to 240) and about 8.5 MeV for nuclei of intermediate mass. This means that the nucleons in the fission fragments are more tightly bound to each other and have less mass than the nucleons in the original heavy nucleus. This decrease in mass per nucleon appears as released energy when fission occurs. The amount of energy released is (8.5 − 7.6) MeV per nucleon. Assuming a total of 240 nucleons, we find that the energy released per fission event is

$$Q = (240 \text{ nucleons})\left(8.5\,\frac{\text{MeV}}{\text{nucleon}} - 7.6\,\frac{\text{MeV}}{\text{nucleon}}\right) = 220 \text{ MeV}$$

This is indeed a very large amount of energy relative to the amount of energy released in chemical processes. For example, the energy released in the combustion of one molecule of octane—a petroleum compound used in gasoline engines—is about one-millionth the energy released in a single fission event!

Example 10.11 A Typical Fission Event

Determine the disintegration energy, Q, associated with the spontaneous fission of ^{236}U into the fragments ^{90}Rb and ^{143}Cs.

Solution The reaction equation must be written in order to determine the number of neutrons involved in the fission:

$$^{236}_{92}\text{U} \longrightarrow {}^{90}_{37}\text{Rb} + {}^{143}_{55}\text{Cs} + 3\,^{1}_{0}\text{n}$$

We deduce the fact that three neutrons are products of the reaction by balancing the mass numbers and atomic numbers on both sides of the reaction equation. From Appendix A.3 we obtain the atomic masses ^{236}U, 236.045 562 u; ^{143}Cs, 142.927 220 u; ^{90}Rb, 89.914 811 u; and n, 1.008 665 u. For this fission reaction, $Q = (\Delta m)c^2$ is given by

$$Q = (M_U - M_{Rb} - M_{Cs} - 3m_n)c^2$$

$$Q = [236.045\ 562\ \text{u} - 89.914\ 811\ \text{u} - 236.045\ 562\ \text{u} - 3(1.008\ 665\ \text{u})]c^2$$
$$= (0.1775\ \text{u})c^2$$

Converting to the conventional units of million electron volts (MeV), we have

$$Q = (0.1775\ \text{u})c^2\left[\frac{931.5\ \text{MeV}/c^2}{1\ \text{u}}\right] = \boxed{166\ \text{MeV}}$$

Example 10.12 The Energy Released in the Fission of ^{235}U

Calculate the total energy released if 1 kg of ^{235}U undergoes fission, taking the disintegration energy per event to be $Q = 208$ MeV (a more accurate value than the estimate given before).

Solution We need to know the number of nuclei in 1 kg of uranium. Since $A = 235$, we can use Avogadro's number to determine the number of nulei:

$$N = \left(\frac{6.02 \times 10^{23}\ \text{nuclei/mol}}{235\ \text{g/mol}}\right)(10^3\ \text{g})$$
$$= 2.56 \times 10^{24}\ \text{nuclei}$$

Hence, the total disintegration energy is

$$E = NQ = (2.56 \times 10^{24}\ \text{nuclei})\left(208\ \frac{\text{MeV}}{\text{nucleus}}\right)$$
$$= \boxed{5.32 \times 10^{26}\ \text{MeV}}$$

Since 1 MeV is equivalent to 4.45×10^{-20} kWh, $E = 2.37 \times 10^7$ kWh. This is enough energy to keep a 100-W lightbulb burning for about 30 000 years. Thus, 1 kg of ^{235}U is a relatively large amount of fissionable material.

Example 10.13 Fuel Consumed by a Nuclear Power Plant

A typical nuclear fission power plant produces about 1000 MW of electrical power. Assume that the plant has an overall efficiency of 40% and that each fission event produces 200 MeV of energy. Calculate the mass of ^{235}U consumed each day.

Solution If the electrical power output of 1000 MW is 40% of the power derived from fission reactions, the power output of the fission process is

$$\frac{1000\ \text{MW}}{0.40} = 2500\ \text{MW} = \left(2.5 \times 10^9\ \frac{\text{J}}{\text{s}}\right)\left(\frac{86\ 400\ \text{s}}{\text{d}}\right) = 2.16 \times 10^{14}\ \frac{\text{J}}{\text{d}}$$

The number of fissions per day is

$$\left(2.16 \times 10^{14}\ \frac{\text{J}}{\text{d}}\right)\left(\frac{1\ \text{fission}}{200 \times 10^6\ \text{eV}}\right)\left(\frac{1\ \text{eV}}{1.602 \times 10^{-19}\ \text{J}}\right) = 6.74 \times 10^{24}\ \text{d}^{-1}$$

This also is the number of ^{235}U nuclei used, so the mass of ^{235}U used per day is

$$\left(6.74 \times 10^{24} \frac{\text{nuclei}}{\text{d}}\right)\left(\frac{235 \text{ g/mol}}{6.02 \times 10^{23} \text{ nuclei/mol}}\right) = 2631 \text{ g/d} = \boxed{2.63 \text{ kg/d}}$$

In contrast, a coal-burning steam plant producing the same electrical power uses more than 6×10^6 kg/d of coal.

It is important to note that the isotope ^{235}U constitutes only 0.72% of the element natural uranium, whereas the isotope ^{238}U constitutes 99.27%. Since a higher concentration of ^{235}U is more practical as a reactor fuel, it is common to "enrich" uranium fuel to a ^{235}U concentration as high as 90%.

Nuclear Fusion

In order to understand the process of fusion, it is instructive to recall the concept of binding energy that was discussed in Chapter 8, Section 8.7, as it relates to two particles bound together by the attractive force of gravity. If the two particles are separated by some distance r, the binding energy is equal to the minimum energy required to separate the particles by an infinite distance. Likewise, the binding energy of any nucleus is the energy that is required to separate the nucleus into its individual nucleons. (Here the nucleons are bound together by the strong nuclear force.)

As we have learned, the total mass of a nucleus is always less than the sum of the masses of its individual nucleons. According to the Einstein mass-energy relationship, the binding energy of the nucleus is $(\Delta m)c^2$, where Δm is the mass difference. *The total energy of the bound system (the nucleus) is less than the total energy of the separated nucleons.* In most chemical processes, two atoms combine with the release of energy. Similarly, if two nuclei can be combined to form a larger nucleus, energy is released in the process. This is exactly what happens in the process of **nuclear fusion.** In this situation, two very light nuclei such as the hydrogen isotopes deuterium (^2H) and tritium (^3H) fuse together to form a heavier nucleus. In another example, a neutron and a proton can combine to form a deuteron, emitting a 2.2-MeV gamma ray in the process (Fig. 10.17).

All stars generate their energy through fusion processes. About 90% of the stars, including our own Sun, fuse hydrogen; some older stars fuse helium or other heavier elements. Stars are born in regions of space containing vast clouds of dust and gas. Recent mathematical models of these clouds indicate that star formation is triggered by shock waves passing through a cloud. The waves are similar to sonic

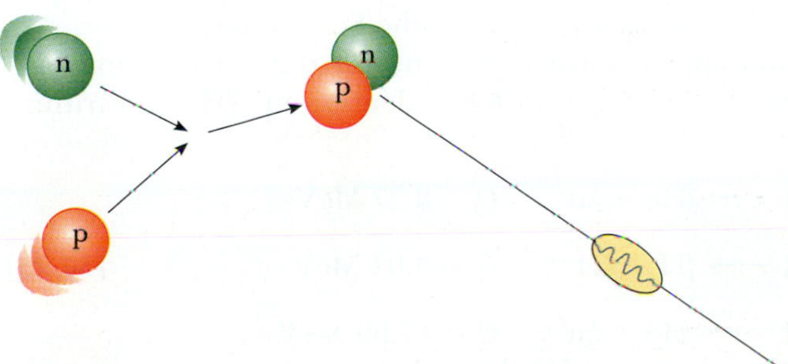

Figure 10.17 A 2.2 million eV gamma ray is emitted when a neutron and a proton combine to form a deuteron, indicating that the deuteron has a lower mass than the sum of the individual proton and neutron masses.

booms and are produced by events such as the explosion of a nearby star. The shock waves compress certain regions of the cloud, causing those regions to collapse under their own gravity. As the gas falls inward toward the center of a collapsing region, the atoms gain speed, causing the temperature of the gas to rise. Two conditions must be met before fusion reactions in the star can sustain the star's energy needs: (1) the temperature must be high enough (about 10^7 K for hydrogen) to allow the kinetic energy of the positively charged hydrogen nuclei to overcome their mutual Coulomb repulsion when they collide, and (2) the density of nuclei must be high enough to ensure a high probability of collision.

When fusion reactions occur in a star, the energy liberated eventually becomes sufficient to prevent further collapse of the star under its own gravity. The star then lives out the remainder of its life under a balance between the inward force of gravity that tends to cause it to collapse and the outward force due to temperature effects and radiation pressure. The proton–proton cycle is a series of three nuclear reactions believed to be the stages in the liberation of energy in our Sun and other stars rich in hydrogen. An overall view of the cycle is that four protons combine to form an alpha particle (the nucleus of the helium atom) and two positrons. (Positrons are particles with charge $+e$ and mass equal to that of the electron.) The total energy released is 27.7 MeV, about 6.9 MeV/nucleon, compared with an energy release in fission reactions of roughly 1 MeV/nucleon.

The energy-liberating fusion reactions we have briefly described are called **thermonuclear fusion reactions.** The hydrogen (fusion) bomb, first exploded in 1952, is an example of an uncontrolled thermonuclear fusion reaction.

The enormous amount of energy released in fusion reactions suggests the possibility of harnessing this energy for useful purposes here on Earth. A great deal of effort is currently under way to develop a sustained and controllable thermonuclear reactor—a fusion power reactor. Controlled fusion is often called the ultimate energy source because of the availability of its source of fuel: water. For example, if the isotope of hydrogen known as deuterium were used as the fuel, 0.6 g of it could be extracted from 1 gal of water at a cost of about 4 cents. Such rates would make the fuel costs of even an inefficient reactor almost insignificant. An additional advantage of fusion reactors is that comparatively few radioactive by-products are formed. In fact, the end product of the fusion of hydrogen nuclei is safe, nonradioactive helium. Unfortunately, a thermonuclear reactor that can deliver a net power output over a reasonable time interval is not yet a reality, and many difficulties must be overcome before a successful device is constructed.

As mentioned earlier, the Sun's energy is based, in part, upon a set of reactions in which ordinary hydrogen is converted to helium. Unfortunately, the proton–proton interaction is not suitable for use in a fusion reactor because it requires very high pressures and densities. The process works in the Sun only because of the extremely high density of protons in the Sun's interior.

The fusion reactions that appear most promising for the construction of a fusion power reactor involve the hydrogen isotopes deuterium (${}^2_1\mathrm{H}$) and tritium (${}^3_1\mathrm{H}$):

$$\begin{aligned} {}^2_1\mathrm{H} + {}^2_1\mathrm{H} &\longrightarrow {}^3_2\mathrm{He} + {}^1_0\mathrm{n} \qquad Q = 3.27\ \mathrm{MeV} \\ {}^2_1\mathrm{H} + {}^2_1\mathrm{H} &\longrightarrow {}^3_1\mathrm{H} + {}^1_1\mathrm{H} \qquad Q = 4.03\ \mathrm{MeV} \\ {}^2_1\mathrm{H} + {}^3_1\mathrm{H} &\longrightarrow {}^4_2\mathrm{He} + {}^1_0\mathrm{n} \qquad Q = 17.59\ \mathrm{MeV} \end{aligned} \tag{10.25}$$

where Q is the amount of energy released per reaction. Deuterium is available in almost unlimited quantities from our lakes and oceans and is very inexpensive to extract. Tritium, however, is radioactive and decays to ^3He. For this reason, tritium does not occur naturally to any great extent and must be artificially produced.

One possible scheme for extracting energy from a fusion reactor is to surround the reactor core with a lithium blanket. Energetic particles, such as neutrons, would be absorbed by the molten lithium, and the energy would then be transferred to a heat exchanger to generate steam. One advantage of this scheme is the formation of tritium from neutron-induced reactions in the lithium blanket; the tritium can then be recirculated into the reactor for fuel.

One of the major obstacles to obtaining energy from nuclear fusion is the fact that the Coulomb repulsive force between two charged nuclei must be overcome before the nuclei can fuse. The fundamental problem, then, is to give the two nuclei enough kinetic energy to overcome this repulsive force. This can be accomplished by heating the fuel to an extremely high temperature (about 10^8 K, an order of magnitude greater than the interior temperature of the Sun). As you might expect, such high temperatures are not easy to obtain in the laboratory or a power plant. At these high temperatures, the atoms are ionized, and the system consists of a collection of electrons and nuclei commonly referred to as a plasma.

In addition to high temperature requirements, two other critical factors determine whether or not a thermonuclear reactor will be successful: **plasma ion density,** n, and **plasma confinement time,** τ, the time interval during which the interacting ions are maintained at a temperature equal to or greater than the temperature required for the reaction. The density and the confinement time must both be great enough to ensure that more fusion energy will be released than is required to heat the plasma.

A rule called **Lawson's criterion** states that a net power output in a fusion reactor is possible under the following conditions:

$$\begin{aligned} n\tau &\geq 10^{14}\ \mathrm{s/cm^3} \qquad \text{Deuterium-tritium interaction} \\ n\tau &\geq 10^{16}\ \mathrm{s/cm^3} \qquad \text{Deuterium-deuterium interaction} \end{aligned} \tag{10.26}$$

In the last decade, dramatic progress has been made toward the development of a fusion reactor that confines the plasma in a magnetic field. (Chapter 6, Section 6.6, contains a discussion of the concept of magnetic field confinement.) The tokamak fusion test reactor (TFTR) at Princeton University has attained central ion temperatures of approximately 4×10^8 K, representing a fivefold increase since 1981. During this same period, plasma confinement times have increased from 0.02 s to about 1.4 s. Values obtained for $n\tau$ for the D-T reaction are well above 10^{13} s/cm^3, close to the value required by Lawson's criterion. In 1991, reaction rates of 6×10^{17} D-T fusions/s were reached in the JET tokamak fusion reactor at Abington, England, and 1×10^{17} D-D fusions/s were reported in the TFTR tokamak device at Princeton. An international collaborative team representing four major fusion programs is currently designing and building a fusion reactor called the International Thermonuclear Experimental Reactor (ITER). This facility is meant to address the remaining technological and scientific issues that will establish the feasibility of fusion power.

Photograph of the Princeton TFTR tokamak device prior to the installation of the external structure components. *(Courtesy of the Plasma Physics Laboratory at Princeton)*

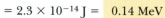

Example 10.14 The Fusion of Two Deuterons

The separation between two deuterons must be as little as about 10^{-14} m in order for the attractive nuclear force to overcome the repulsive Coulomb force. Calculate the electric potential energy due to the repulsive force.

Solution The electric potential energy associated with two charges separated by a distance r is given by Equation 8.23 in Chapter 8:

$$U_e = k_e \frac{q_1 q_2}{r}$$

where k_e is the Coulomb constant. For the case of two deuterons, $q_1 = q_2 = +e$, so that

$$U_e = k_e \frac{e^2}{r} = \left(8.99 \times 10^9 \frac{\text{N}\cdot\text{m}^2}{\text{C}^2}\right) \frac{(1.6 \times 10^{-19}\ \text{C})^2}{10^{-14}\ \text{m}}$$

$$= 2.3 \times 10^{-14}\ \text{J} = \boxed{0.14\ \text{MeV}}$$

*10.9 General Relativity[7]

Up to this point, we have sidestepped a curious puzzle. Mass has two seemingly different properties: a *gravitational attraction* for other masses and an *inertial* property that resists acceleration. To designate these two attributes, we shall use the subscripts g and i and write

Gravitational property	$W = m_g g$
Inertial property	$F = m_i a$

The numerical value for the gravitational constant G was chosen to make the magnitudes of m_g and m_i numerically equal. Regardless of how G is chosen, however, the strict *proportionality* of m_g and m_i has been established experimentally to an extremely high degree: a few parts in 10^{12}. Thus, it appears that gravitational mass and inertial mass may indeed be exactly proportional.

But why? They seem to involve two entirely different attributes: a force of mutual gravitational attraction between masses and the resistance of a single mass to being accelerated. This question, which puzzled Newton and many other physicists over the years, was answered when Einstein published his theory of gravitation, known as *general relativity,* in 1916. It is a mathematically complex theory; we shall be able to merely hint at its elegance and insight.

In Einstein's view, the remarkable coincidence that m_i and m_g seemed to be exactly proportional was evidence for a very intimate and basic connection between the two concepts. He pointed out that no *mechanical* experiment (such as dropping a mass) could distinguish between the two situations sketched in Figures 10.18a and 10.18b. In each case, a mass released by the observer's hand would undergo a downward acceleration of g relative to the floor.

[7] This section was taken from A. Hudson and R. Nelson, *University Physics,* 2d ed., Philadelphia, Saunders College Publishing, 1990, with permission of the publisher.

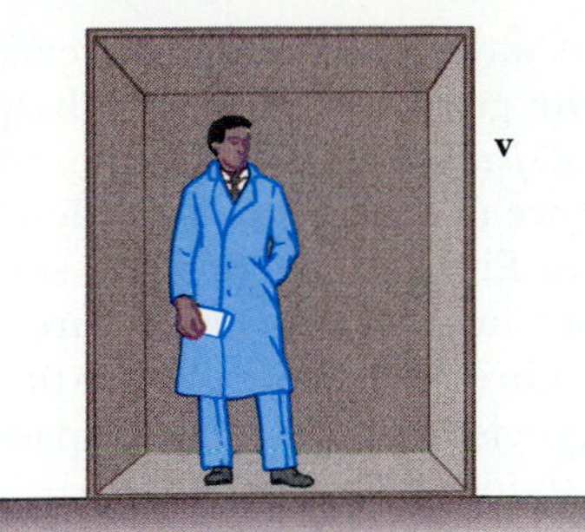

(a) The observer is at rest in a uniform gravitational field where the acceleration due to gravity is **g**.

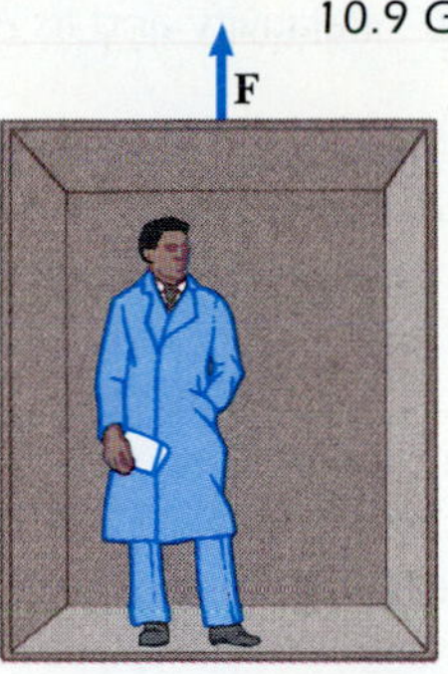

(b) The observer is in a region where gravity is negligible, but the frame of reference of the observer is accelerated through space (by the external force **F**) with an acceleration equal to **g**.

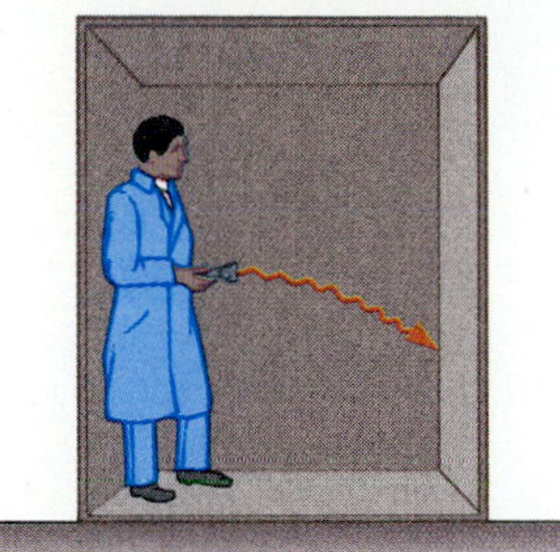

(c) If (a) and (b) are truly equivalent, as Einstein proposed, then a ray of light would be bent in a gravitational field. Such an effect has been experimentally verified by light and radio signals that pass close to the strong gravitaional field of the sun.

Figure 10.18
According to Einstein, the (a) and (b) frames of reference are equivalent in every way. No experiment of any sort could distinguish any difference.

Einstein carried this idea further to propose, as one of two fundamental postulates in his general theory of relativity, that *no* experiment, mechanical *or otherwise,* could distinguish between the two cases. This extension to include all phenomena (not just mechanical ones) has interesting consequences. For example, suppose that a pulse of light were sent horizontally across the box in Figure 10.18b. The pulse of light would have a trajectory that bent downward toward the floor as the box accelerated upward to meet it. Therefore, proposed Einstein, in case (a) a beam of light should be bent downward by the gravitational field. (No such bending is predicted in Newton's theory of gravitation.)

The two postulates of Einstein's **general relativity** are as follows:

Postulates of general relativity

1. All the laws of nature may be stated so that they have the same form for observers in any space-time frame of reference, whether accelerated or not.
2. In the neighborhood of any given point, a gravitational field is equivalent in every respect to an accelerated frame of reference in the absence of gravitational effects. (This is the *principle of equivalence.*)

The second postulate implies that gravitational mass and inertial mass are completely *equivalent,* not just proportional. What were thought to be two different types of mass are actually, in a basic sense, identical.

One interesting effect predicted by general relativity is that time scales are altered by gravity. A clock in the presence of gravity runs more slowly than one situated where gravity is negligible. Consequently, spectral lines emitted by atoms in the presence of a strong gravitational field are *red-shifted* to lower frequencies when compared with the same spectral emissions in a weak field. This gravitational red shift has been detected in spectral lines emitted by atoms in massive stars. It has also been verified on the Earth by comparisons of the frequency of gamma rays (a type of electromagnetic radiation) emitted from nuclei that are separated vertically by about 20 m.

The second postulate suggests that a gravitational field may be "transformed away" at any point if we choose an appropriately accelerated frame of reference—

a freely falling one. Einstein developed an ingenious way of describing the exact amount of acceleration that is necessary to make the gravitational field "disappear." He specified a certain quantity, the *curvature of space-time,* that describes the gravitational effect at every point. In fact, the curvature of space-time completely replaces Newton's gravitational theory. According to Einstein, there is no such thing as a gravitational force. Rather, the presence of a mass causes a curvature of space-time in the vicinity of the mass, and this curvature dictates the space-time path that all freely moving objects follow. As one physicist says: "Mass tells space-time how to curve; curved spacetime tells mass how to move."

If the concentration of mass becomes very great, as is believed to occur when a large star exhausts its nuclear fuel and collapses to a very small volume, a **black hole** may form. Here the curvature is so extreme that, within a certain distance from the center, all matter and light become trapped.

Summary

The two basic postulates of the **special theory of relativity** are as follows:

1. The laws of physics are the same in every inertial frame of reference.
2. The speed of light has the same value for all observers, independent of their motion and of the motion of the light source.

Three consequences of the special theory of relativity are as follows:

1. Events that are simultaneous for one observer are not simultaneous for another observer who is in motion relative to the first.
2. Clocks in motion relative to an observer appear to be slowed down by a factor γ. This is known as **time dilation.**
3. Lengths of objects in motion appear to be contracted in the direction of motion.

These three statements can be summarized by saying that duration, length, and simultaneity are not absolute concepts in relativity.

In order to satisfy these postulates, the Galilean transformations must be replaced by the **Lorentz transformations,** given by

Lorentz transformation for $S \rightarrow S'$

$$\begin{aligned} x' &= \gamma(x - vt) \\ y' &= y \\ z' &= z \\ t' &= \gamma\left(t - \frac{v}{c^2}x\right) \end{aligned} \qquad [10.7]$$

where $\gamma = (1 - v^2/c^2)^{-1/2}$.

In these equations, it is assumed that the primed system moves with a speed of v along the xx' axes.

The relativistic form of the **velocity transformation** is

$$u'_x = \frac{u_x - v}{1 - \dfrac{u_x v}{c^2}} \qquad [10.12]$$

Lorentz velocity transformation for $S \rightarrow S'$

where u_x is the speed of an object as measured in the S frame and u'_x is its speed measured in the S' frame.

The relativistic expression for the **momentum** of a particle moving with a velocity **u** is

$$\mathbf{p} \equiv \frac{m\mathbf{u}}{\sqrt{1 - \dfrac{u^2}{c^2}}} = \gamma m\mathbf{u} \qquad [10.15]$$

Momentum

The relativistic expression for the **kinetic energy** of a particle is

$$K = \gamma mc^2 - mc^2 \qquad [10.19]$$

Kinetic energy

where mc^2 is called the **rest energy** of the particle.

The total energy E of a particle is related to the mass through Einstein's famous **energy-mass relation:**

$$E = \gamma mc^2 = \frac{mc^2}{\sqrt{1 - \dfrac{u^2}{c^2}}} \qquad [10.21]$$

Energy-mass relation

The relativistic momentum is related to the total energy through the equation

$$E^2 = p^2c^2 + (mc^2)^2 \qquad [10.22]$$

Energy-momentum relation

Fission

Nuclear fission occurs when a heavy nucleus, such as ^{235}U, splits into two smaller fragments. Thermal neutrons can create fission in ^{235}U by the following process:

$${}^1_0\text{n} + {}^{235}_{92}\text{U} \longrightarrow {}^{236}_{92}\text{U}^* \longrightarrow \text{X} + \text{Y} + \text{neutrons} \qquad [10.24]$$

Fission of ^{235}U

where X and Y are the fission fragments and ^{236}U* is a compound nucleus in an excited state. On the average, 2.47 neutrons are released per fission event. The fragments and neutrons have a great deal of kinetic energy following the fission event. The energy released per fission event is about 208 MeV.

Fusion

In **nuclear fusion,** two light nuclei combine to form a heavier nucleus and release energy in the process. The major obstacle to obtaining useful energy from fusion is the large Coulomb repulsive force between the charged nuclei at close separations. Sufficient energy must be supplied to the particles to overcome this Coulomb barrier and thereby enable the nuclear attractive force to take over. The temperature required to produce fusion is of the order of 10^8 K. At such high temperatures, all matter is in the form of a **plasma,** which consists of positive ions and free electrons.

Confinement time

Lawson's criterion

Two critical parameters involved in fusion reactor design are **plasma ion density,** n, and **plasma confinement time,** τ. The confinement time is the time interval during which the interacting particles must be maintained at a temperature equal or greater than the critical ignition temperature. **Lawson's criterion** states that for the D-T reaction, $n\tau \geq 10^{14}$ s/cm^3, and for D-D, $n\tau \geq 10^{16}$ s/cm^3.

Questions and Conceptual Exercises

1. What two speed measurements will two observers in relative motion *always* agree upon?
2. An astronaut moves away from the Earth at a speed close to the speed of light. If an observer on Earth could make measurements of the astronaut's size and pulse rate, what changes (if any) would he or she measure? Would the astronaut measure any changes?
3. Two identically constructed clocks are synchronized. One is put in orbit around the Earth while the other remains on Earth. Which clock runs more slowly? When the moving clock returns to Earth, will the two clocks still be synchronized?
4. High-energy particles sometimes travel through a material medium faster than the speed of light in that medium. Does this violate the principles of relativity?
5. A spaceship in the shape of a sphere moves past an observer on Earth with a speed of $0.5c$. What shape does the observer see as the spaceship moves past?
6. Explain why it is necessary, when defining length, to specify that the positions of the ends of a rod are to be determined simultaneously.
7. When we say that a moving clock runs more slowly than a stationary one, does this imply that there is something physically unusual about the moving clock?
8. When we speak of time dilation, do we mean that time passes more slowly in moving systems or that it simply appears to do so?
9. List some ways our day-to-day lives would change if the speed of light were only 50 m/s.
10. Give a physical argument showing that it is impossible to accelerate an object of mass m to the speed of light, even with a continuous force acting on it.
11. It is said that Einstein, in his teenage years, asked the question "What would I see in a mirror if I carried it in my hands and ran at the speed of light?" How would you answer this question?
12. Suppose astronauts were paid according to the time they spent traveling in space. After a long voyage at a speed near that of light, some astronauts return home and open their pay envelopes. What is their reaction?
13. Some distant galaxies, observed as quasars, are receding from us at half the speed of light (or greater). What is the speed of the light we receive from these quasars?
14. How is it possible that photons of light, with zero mass, have momentum?
15. Imagine an astronaut on a trip to Sirius, which lies 8 lightyears from the Earth. Upon arrival at Sirius, the astronaut finds that the trip lasted 6 years. If the trip was made at a constant speed of $0.8c$, how can the 8-lightyear distance be reconciled with the 6-year duration?
16. Relativistic quantities must make a smooth transition to their Newtonian counterparts as the speed of a system becomes small with respect to the speed of light. Explain.
17. If the speed of a particle is doubled, what effect does it have on its momentum? (Assume that its initial speed is less than $c/2$.)

Problems

Section 10.1 The Principle of Relativity

1. A ball is thrown at 20 m/s inside a boxcar that is moving along the tracks at 40 m/s. What is the speed of the ball with respect to the ground if the ball is thrown (a) forward? (b) backward? (c) out the side door?
2. A 2000-kg car moving with a speed of 20 m/s collides with and sticks to a 1500-kg car that is at rest at a stop sign. Show that momentum is conserved in a reference frame that is moving with a speed of 10 m/s in the direction of the moving car.
3. In a laboratory frame of reference, an observer notes that Newton's second law is valid. Show that it is also valid for an observer moving at a constant speed relative to the laboratory frame.
4. Show that Newton's second law is not valid in a reference frame that is moving past the laboratory frame of Problem 3 with a constant acceleration.
5. A billiard ball of mass 0.3 kg moves with a speed of 5 m/s and collides elastically with a ball of mass 0.2 kg that is moving in the opposite direction with a speed of 3 m/s. Show that momentum is conserved in a frame of reference that is moving with a speed of 2 m/s in the direction of the second ball.

Section 10.4 Consequences of Special Relativity

6. How fast must a meter stick be moving if its length is observed to shrink to 0.5 m?
7. With what speed does a clock have to be moving in order to run at a rate that is one-half the rate of a clock at rest?
8. An astronaut at rest on Earth has a heart rate of 70 beats/min. When the astronaut is traveling in a spaceship at $0.9c$, what is his rate as measured (a) by an observer also in the ship and (b) by an observer at rest on the Earth?
9. An atomic clock moves at 1000 km/h for one hour as measured by an identical clock on Earth. How many nanoseconds slow will the moving clock be at the end of the one-hour interval?
10. Muons move in circular orbits at a speed of $0.9994c$ in a storage ring of radius 500 m. If a muon at rest decays into other particles after $T = 2.2\ \mu s$, how many trips around the storage ring do we expect the muons to make before they decay?
11. A spacecraft moves at a speed of $0.9c$. If its length is L_0 when measured from inside the spacecraft, what is its length, measured by a ground observer?
12. The cosmic rays of highest energy are protons, with kinetic energy of 10^{13} MeV. (a) How long would it take a proton of this energy to travel across the Milky Way galaxy, of diameter 10^5 lightyears, as measured in the proton's frame? (b) From the point of view of the proton, how many kilometers across is our galaxy?
13. A muon formed high in the Earth's atmosphere travels at speed $v = 0.99c$ for a distance of 4.6 km before it decays into an electron, a neutrino, and an antineutrino ($\mu^- \rightarrow e^- + \nu + \bar{\nu}$). (a) How long does the muon live, as measured in its reference frame? (b) How far does the muon travel, as measured in its frame?
14. If astronauts could travel at $v = 0.95c$, we on Earth would say it takes $(4.2/0.95) = 4.4$ years to reach Alpha Centauri, 4.2 lightyears away. The astronauts disagree. (a) How much time passes on the astronauts' clocks during a trip from Earth to Alpha Centauri? (b) What is the distance to Alpha Centauri, as measured by the astronauts?
17. Two jets of material fly away from the center of a radio galaxy in opposite directions. Both jets move at a speed of $0.75c$ relative to the galaxy. Determine the speed of one jet relative to the other.
18. A Klingon spaceship moves away from the Earth at a speed of $0.8c$ (Fig. 10.19). The starship *Enterprise* pursues at a speed of $0.9c$ relative to the Earth. Observers on Earth see the *Enterprise* overtaking the Klingon ship at a relative speed of $0.1c$. With what speed is the *Enterprise* overtaking the Klingon ship, as seen by the crew of the *Enterprise?*

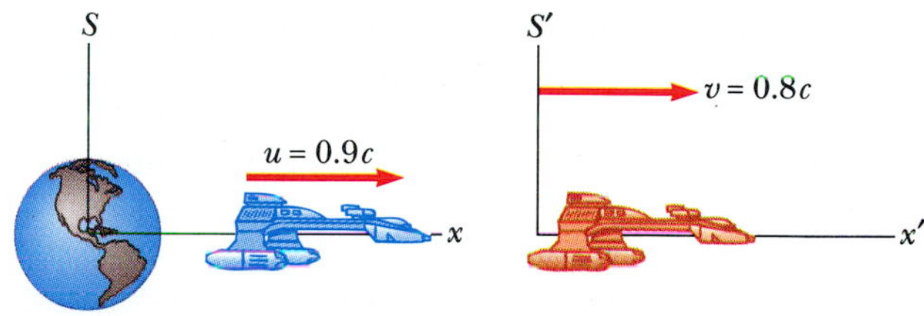

Figure 10.19
(Problem 18) The Earth is frame S. The Klingon ship is frame S'. The Enterprise is the object whose motion is followed from S and S'.

19. A friend travels by you at a high speed in a spaceship. He tells you that his ship is 20 m long and that the identical ship you are sitting in is 19 m long. According to your observations, (a) how long is your ship, (b) how long is your friend's ship, and (c) what is the speed of your friend's ship?
20. A moving rod is 2.0 m long, and its length is oriented at an angle of 30° with respect to the direction of motion. The rod has a speed of $0.995c$. (a) What is the proper length of the rod? (b) What is the orientation angle in the proper frame?
21. Observer A measures the lengths of two rods, one stationary, the other moving with a speed of $0.955c$. She finds that the rods have the same length. A second observer, B, travels along with the moving rod. What is the ratio of the length of A's rod to the length of B's rod, according to observer B?

Section 10.5 The Lorentz Transformation

15. For what value of v does $\gamma = 1.01$?
16. A certain quasar recedes from the Earth with a speed of $v = 0.87c$. A jet of material is ejected from the quasar toward the Earth with a speed of $0.55c$ relative to the quasar. Find the speed of the ejected material relative to the Earth.

Section 10.6 Relativistic Momentum

22. Calculate the momentum of an electron moving with a speed of (a) $0.01c$, (b) $0.5c$, (c) $0.9c$.
23. An electron has a momentum that is 90% greater than its classical momentum. (a) Find the speed of the electron. (b) How would your result change if the particle were a proton?

24. An electron has a speed of $v = 0.9c$. At what speed will a proton have a momentum equal to that of the electron?
25. An unstable particle at rest breaks into two fragments of *unequal* mass. The rest mass of the lighter fragment is 2.5×10^{-28} kg, and that of the heavier fragment is 1.67×10^{-27} kg. If the lighter fragment has a speed of $0.893c$ after the breakup, what is the speed of the heavier fragment?
26. A golf ball travels with a speed of 90 m/s. By what fraction does its relativistic momentum, p, differ from its classical value, mv? That is, find the ratio $(p - mv)/mv$.
27. The *nonrelativistic* expression for the momentum of a particle, $p = mv$, can be used if $v \ll c$. For what speed does the use of this formula give an error in the momentum of (a) 1 percent and (b) 10 percent?

Section 10.7 Relativistic Energy

28. Show that the energy-momentum relationship given by $E^2 = p^2c^2 + (mc^2)^2$ follows from the expressions $E = \gamma mc^2$ and $p = \gamma mu$.
29. A proton moves with the speed of $0.95c$. Calculate its (a) rest energy, (b) total energy, and (c) kinetic energy.
30. Find the speed at which the relativistic kinetic energy is two times the nonrelativistic value.
31. Find the speed of a particle whose total energy is twice its rest energy.
32. Determine the energy required to accelerate an electron from (a) $0.50c$ to $0.9c$ and (b) $0.90c$ to $0.99c$.
33. A spaceship of mass 10^6 kg is to be accelerated to $0.6c$. (a) How much energy does this require? (b) How many kilograms of matter will it take to provide this much energy?
34. The Sun radiates approximately 4×10^{26} J of energy into space each second. (a) How much mass is released as radiation each second? (b) If the mass of the Sun is 2×10^{30} kg, how long can the Sun survive if the energy release continues at the present rate?
35. When 1.0 g of hydrogen combines with 8.0 g of oxygen, 9.0 g of water is formed. During this chemical reaction, 2.86×10^5 J of energy is released. How much mass do the constituents of this reaction lose? Is the loss of mass likely to be detectable?
36. An unstable particle with a mass of 3.34×10^{-27} kg is initially at rest. The particle decays into two fragments that fly off with velocities of $0.987c$ and $-0.868c$. Find the rest masses of the fragments. (*Hint:* Conserve both mass-energy and momentum.)

Section 10.8 Applications of Special Relativity: Fission and Fusion

37. The annual energy requirement of the United States is of the order of 10^{20} J. (a) How many atoms of ^{235}U (with an energy release of 208 MeV per fission event) must be fissioned every second to meet this requirement? (b) How many kilograms of ^{235}U would be required each year?
38. How many grams of deuterium (atomic mass = 2.0141 u) must be fused to form helium (atomic mass = 4.0026 u) each second to produce 3 GJ of energy?
39. Suppose enriched uranium containing 3.4% of the fissionable isotope $^{235}_{92}$U is used as fuel for a ship. The water exerts an average frictional drag of 1.0×10^5 N on the ship. How far can the ship travel per kilogram of fuel? Assume that the energy released per fission event is 208 MeV and that the ship's engine has an efficiency of 20%.
40. The oceans have a volume of 317 million cubic miles and contain 1.32×10^{21} kg of water. Of all the hydrogen nuclei in this water, 0.0156% are deuterium. (a) If all of these deuterium nuclei were fused to helium via the first reaction in Equation 10.25, determine the total amount of energy that could be released. (b) Current world energy consumption is about 7×10^{12} W. If consumption were 100 times greater, how many years would the energy supply calculated in (a) last?

Additional Problems

41. An astronaut is traveling in a space vehicle that has a speed of $0.50c$ relative to the Earth. The astronaut measures his pulse rate at 75 per minute. Signals generated by the astronaut's pulse are radioed to Earth when the vehicle is moving perpendicularly to a line that connects the vehicle with an Earth observer. What pulse rate does the Earth observer measure? What would be the pulse rate if the speed of the space vehicle were increased to $0.99c$?
42. An astronaut wishes to visit the Andromeda galaxy (2 million lightyears away) in a one-way trip that will take 30 years in the spaceship's frame of reference. Assuming that his speed is constant, how fast must he travel relative to the Earth?
43. The net nuclear reaction inside the Sun is $4p \rightarrow He^4 + \Delta E$. If the rest mass of each proton is 938.2 MeV and the rest mass of the He^4 nucleus is 3727 MeV, calculate the percentage of the starting mass that is given off.

44. An electron has a speed of $0.75c$. Find the speed of a proton that has (a) the same kinetic energy as the electron; (b) the same momentum as the electron.

45. The average lifetime of a pi meson in its own frame of reference is 2.6×10^{-8} s. If the meson moves with a speed of $0.95c$, what are (a) its mean lifetime as measured by an observer on Earth and (b) the average distance it travels before decaying, as measured by an observer on Earth?

46. If you travel on a jet plane from New York to Los Angeles (4000 km air distance) at an average speed of 1000 km/h, how much younger are you on arrival than you would have been had you remained in New York during the time it took the plane to make the journey? (*Hint:* Note that Δt, the time that would have been spent in New York, is extremely close to $\Delta t'$, the time spent on the plane.)

47. A supertrain (rest length = 100 m) travels at a speed of $0.95c$ as it passes through a tunnel (rest length = 50 m). As seen by a trackside observer, is the train ever completely within the tunnel? If so, by how much?

48. An electron has a kinetic energy five times greater than its rest energy. Find (a) its total energy and (b) its speed.

49. A radioactive nucleus moves with a speed of v relative to a laboratory observer. The nucleus emits an electron in the positive x direction with a speed of $0.7c$ relative to the decaying nucleus and with a speed of $0.85c$ in the $+x$ direction relative to the laboratory observer. What is the value of v?

50. Energy reaches the upper atmosphere of the Earth from the Sun at the rate of 1.79×10^{17} W. If all of this energy were absorbed by the Earth and not re-emitted, how much would the mass of the Earth increase in one year?

51. The muon is an unstable particle that spontaneously decays into an electron and two neutrinos. If the number of muons at $t = 0$ is N_0, the number at time t is given by $N = N_0 e^{-t/\tau}$, where τ is the mean lifetime, equal to 2.2 μs. Suppose the muons move at a speed of $0.95c$ and there are 5×10^4 muons at $t = 0$. (a) What is the observed lifetime of the muons? (b) How many muons remain after 3 km of travel?

52. Suppose that noted astronomers conclude that our Sun is about to undergo a supernova explosion. In an effort to escape, we depart in a spaceship at $v = 0.8c$ and head toward the star Tau Ceti, 12 lightyears away. When we reach the midpoint (in space) of our journey, we see the supernova explosion of our Sun and, unfortunately, at the same instant we see the explosion of Tau Ceti. (a) In the spaceship's frame of reference, should we conclude that the two explosions occurred simultaneously? If not, which occurred first? (b) In a frame of reference in which the sun and Tau Ceti are at rest, did they explode simultaneously? If not, which occurred first?

53. Imagine that the entire Sun collapses to a sphere of radius R_g such that the work required to remove a small mass m from the surface would be equal to its rest energy mc^2. This radius is called the *gravitational radius* for the Sun. Find R_g. (It is believed that the ultimate fate of many stars is to collapse to their gravitational radii or smaller.)

54. A charged particle moves along a straight line in a uniform electric field E with a speed of v. If the motion and the electric field are both in the x direction, (a) show that the acceleration of the charge q in the x direction is given by

$$a = \frac{dv}{dt} = \frac{qE}{m}\left(1 - \frac{v^2}{c^2}\right)^{3/2}$$

(b) Discuss the significance of the dependence of the acceleration on the speed. (c) If the particle starts from rest at $x = 0$ at $t = 0$, how would you proceed to find the speed of the particle and its position after a time t has elapsed?

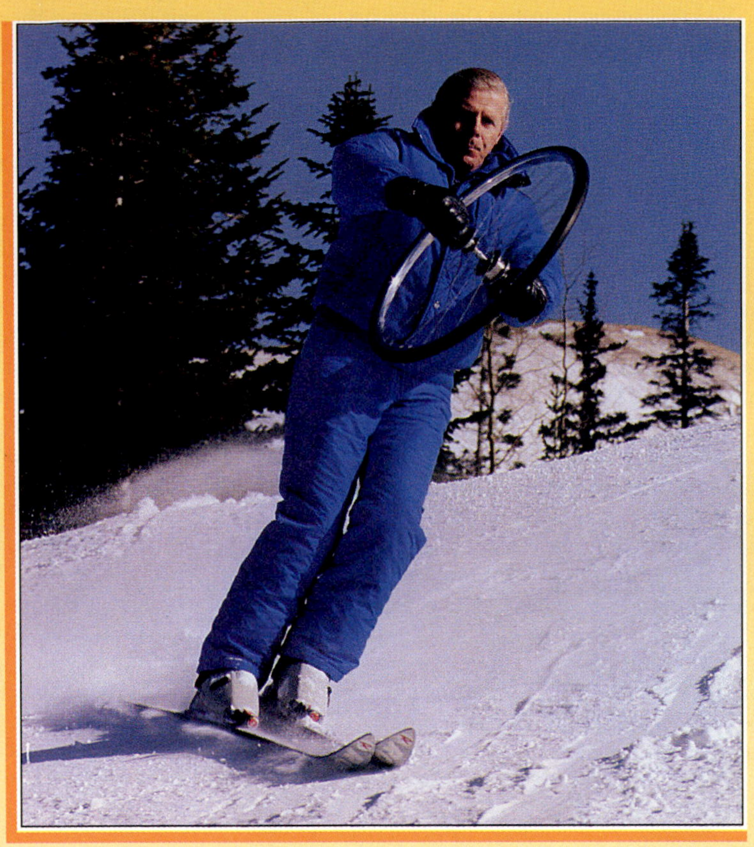

Derek Swinson, professor of physics at the University of New Mexico, demonstrating the "gyro-ski" technique. The skier initiates a turn by lifting the axle of the rotating bicycle wheel. The direction of the turn depends on whether the left or right hand is used to lift the axle from the horizontal. The system (the skier plus the rotating bicycle wheel) is trying to conserve angular momentum. *(Photograph courtesy of Derek Swinson)*

Rotational Motion

CHAPTER 11

In this chapter we investigate the dynamics of particles moving along a circular path about a fixed axis. We shall encounter such terms as angular displacement, angular velocity, angular acceleration, torque, and angular momentum and show how these quantities are useful in describing rotational motion. Next we shall define a vector product, which is a convenient mathematical tool for expressing such physical quantities as torque and angular momentum.

One of the central points of this chapter is to develop the concept of the angular momentum of a system of particles. By analogy with the conservation of linear momentum (Chapter 9), we shall find that the angular momentum of any isolated system is always constant. The results we derive here will enable us to understand the rotational motions of a diverse range of objects in our environment, from an electron orbiting a nucleus to clusters of galaxies orbiting a common center. Although the focus of this chapter is the circular motion of particles, the extension of this treatment to other objects is reasonably straightforward and is treated in an optional section.

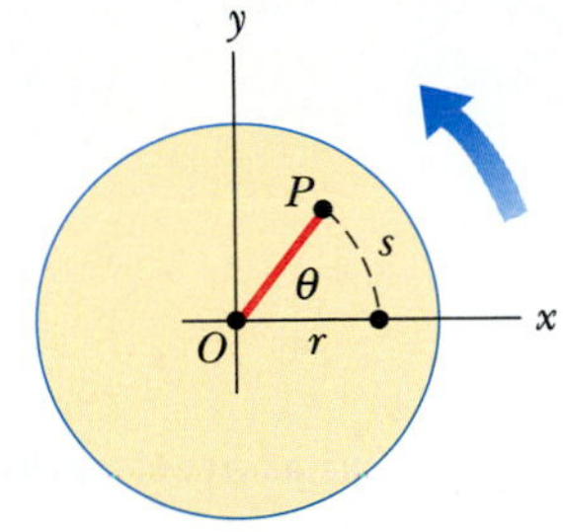

Figure 11.1 Rotation of a disk about a fixed axis through O perpendicular to the plane of the figure (the z axis). Note that a particle at P moves in a circular path of radius r centered at O.

11.1 Angular Velocity and Angular Acceleration

We began our study of linear motion by defining the terms "displacement," "velocity," and "linear acceleration." We will take the same basic approach now as we turn to the study of rotational motion. Let us begin by considering a circular disk rotating about a fixed axis that is perpendicular to the disk and goes through the point O (Fig. 11.1). A point P on the disk is at a fixed distance r from the origin and rotates about O in a circle of radius r. In fact, *every point on the disk undergoes circular motion about O.*

It is convenient to represent the position of the point P with its polar coordinates: (r, θ). In this representation, the only coordinate that changes in time is the angle θ; r remains constant. As a point on the disk moves along the circle of radius r from the positive x axis $(\theta = 0)$ to the point P, it moves through an arc of length s, which is related to the angular position θ through the relation $s = r\theta$, or

$$\theta = \frac{s}{r} \quad [11.1]$$

It is important to note the units of θ as expressed by Equation 11.1. The angle θ is the ratio of an arc length and the radius of the circle, and hence is a pure number. However, we commonly refer to the unit of θ as a **radian** (rad). One **radian** is the angle subtended by an arc length equal to the radius of the arc. Since the circumference of a circle is $2\pi r$, it follows that 360° corresponds to an angle of $2\pi r/r$ rad or 2π rad (one revolution). Hence, 1 rad $= 360°/2\pi \approx 57.3°$. To convert an angle in degrees to an angle in radians, we can use the fact that 2π radians $= 360°$; hence,

$$\theta\ (\text{rad}) = \frac{\pi}{180°}\,\theta\ (\text{deg})$$

For example, 60° equals $\pi/3$ rad, and 45° equals $\pi/4$ rad.

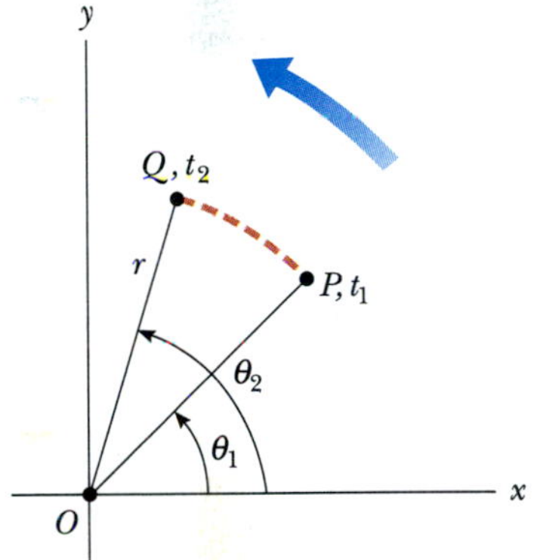

Figure 11.2 A particle on a rotating disk moves from P to Q along the arc of a circle. In the time interval $\Delta t = t_2 - t_1$, the radius vector r sweeps out an angle given by $\Delta\theta = \theta_2 - \theta_1$.

In Figure 11.2, as the particle travels from P to Q in a time Δt, the radius vector sweeps out an angle of $\Delta\theta = \theta_2 - \theta_1$, which equals the **angular displacement** during the time interval Δt. We define the **average angular speed** $\overline{\omega}$ (omega) as the ratio of this angular displacement to the time interval Δt:

$$\overline{\omega} \equiv \frac{\theta_2 - \theta_1}{t_2 - t_1} = \frac{\Delta\theta}{\Delta t} \quad [11.2]$$

By analogy with linear speed, the **instantaneous angular speed,** ω, is defined as the limit of the ratio in Equation 11.2 as Δt approaches zero:

Instantaneous angular speed

$$\omega \equiv \lim_{\Delta t \to 0} \frac{\Delta\theta}{\Delta t} = \frac{d\theta}{dt} \quad [11.3]$$

Angular speed has units of rad/s (or s^{-1} since radians are not dimensional). Let us adopt the convention that the fixed axis of rotation for the disk is the z axis, as in Figure 11.1. We shall take ω to be positive when θ is increasing (counterclockwise motion) and negative when θ is decreasing (clockwise motion).

If the instantaneous angular speed of a particle changes from ω_1 to ω_2 in the time interval Δt, the particle has an angular acceleration. The **average angular acceleration,** $\overline{\alpha}$ (alpha), of a rotating particle is defined as the ratio of the change in the angular speed to the time interval, Δt:

Average angular acceleration

$$\overline{\alpha} \equiv \frac{\omega_2 - \omega_1}{t_2 - t_1} = \frac{\Delta\omega}{\Delta t} \qquad [11.4]$$

By analogy with linear acceleration, the **instantaneous angular acceleration** is defined as the limit of the ratio $\Delta\omega/\Delta t$ as Δt approaches zero:

Instantaneous angular acceleration

$$\alpha \equiv \lim_{\Delta t \to 0} \frac{\Delta\omega}{\Delta t} = \frac{d\omega}{dt} \qquad [11.5]$$

Angular acceleration has units of rad/s^2 or s^{-2}. Note that α is positive when ω is increasing in time, and negative when ω is decreasing in time.

Let us now generalize our argument from the circular disk to any rigid body. A rigid body is any object whose elements remain fixed with respect to one another. **For rotation about a fixed axis, every particle on a rigid body, such as a circular disk, has the same angular velocity and the same angular acceleration.** That is, the quantities $\boldsymbol{\omega}$ and $\boldsymbol{\alpha}$ characterize the rotational motion of the *entire* rigid body. Using these quantities, we can greatly simplify the analysis of rigid-body rotation.

The angular displacement ($\boldsymbol{\theta}$), angular velocity ($\boldsymbol{\omega}$), and angular acceleration ($\boldsymbol{\alpha}$) are analogous to linear displacement (**x**), linear velocity (**v**), and linear acceleration (**a**), respectively, for the corresponding motion discussed in Chapter 2. The variables θ, ω, and α differ dimensionally from the variables x, v, and a, only by a length factor.

We have indicated how the signs for $\boldsymbol{\omega}$ and $\boldsymbol{\alpha}$ are determined, but we have not specified any direction in space associated with these vector quantities.[1] For rotation about a fixed axis, the only direction in space that uniquely specifies the rotational motion is the direction along the axis. However, we must also decide on the sense of these quantities—that is, whether they point into or out of the plane of Figure 11.1.

The direction of $\boldsymbol{\omega}$ is along the axis of rotation, which is the z axis in Figure 11.1. By convention, we take the direction of $\boldsymbol{\omega}$ to be *out of* the plane of the diagram when the rotation is counterclockwise and *into* the plane of the diagram when the rotation is clockwise. To further illustrate this convention, it is convenient to use the *right-hand rule* illustrated by Figure 11.3a. The four fingers of the right hand are wrapped in the direction of the rotation. The extended right thumb points in the direction of $\boldsymbol{\omega}$. Figure 11.3b illustrates that $\boldsymbol{\omega}$ is also in the direction of advance of a similarly rotating right-handed screw.

The sense of $\boldsymbol{\alpha}$ follows from its definition as $d\boldsymbol{\omega}/dt$. It is the same as $\boldsymbol{\omega}$ if the angular speed (the magnitude of $\boldsymbol{\omega}$) is increasing in time, and antiparallel to $\boldsymbol{\omega}$ if the angular speed is decreasing in time.

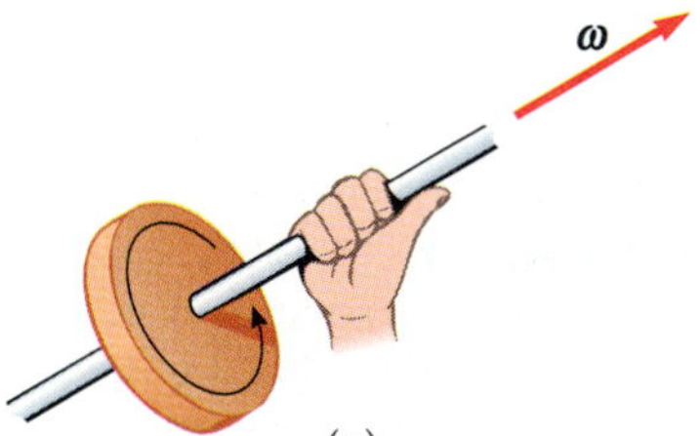

ω

(a)

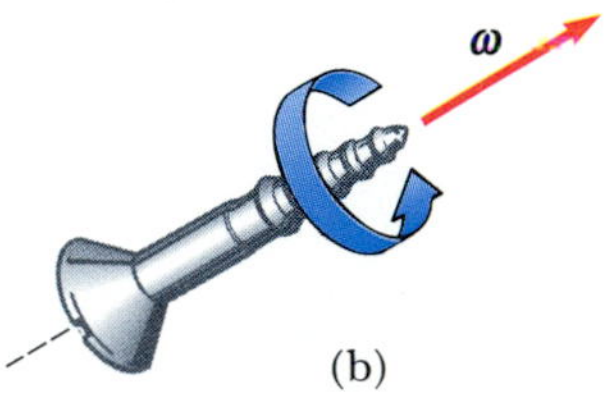

ω

(b)

Figure 11.3
(a) The right-hand rule for determining the direction of the angular velocity. (b) The direction of $\boldsymbol{\omega}$ is in the direction of advance of a right-handed screw.

[1] Although we do not verify it here, the instantaneous angular velocity and instantaneous angular acceleration are vector quantities, but the corresponding average values are not. This is because angular displacement is not a vector quantity for finite rotations.

Table 11.1
Comparison of Kinematic Equations for Rotational and Linear Motion Under Constant Acceleration

Rotational Motion About Fixed Axis with α = Constant. Variables: θ and ω	Linear Motion with a = Constant. Variables: x and v
$\omega = \omega_0 + \alpha t$	$v = v_0 + at$
$\theta = \theta_0 + \omega_0 t + \frac{1}{2}\alpha t^2$	$x = x_0 + v_0 t + \frac{1}{2}at^2$
$\theta = \theta_0 + \frac{1}{2}(\omega_0 + \omega)t$	$x = x_0 + \frac{1}{2}(v_0 + v)t$
$\omega^2 = \omega_0^2 + 2\alpha(\theta - \theta_0)$	$v^2 = v_0^2 + 2a(x - x_0)$

11.2 Rotational Kinematics

In our study of linear motion, we found that the simplest accelerated motion to analyze is motion under constant linear acceleration (Chapter 2). Likewise, for rotational motion about a fixed axis, the simplest accelerated motion to analyze is motion under constant angular acceleration. Therefore, we shall next develop kinematic relations for rotational motion under constant angular acceleration.

If we write Equation 11.5 in the form $d\omega = \alpha\, dt$ and let $\omega = \omega_0$ at $t_0 = 0$, we can integrate this expression directly:

$$\omega = \omega_0 + \alpha t \qquad (\alpha = \text{constant}) \qquad [11.6]$$

Likewise, substituting Equation 11.6 into Equation 11.3 and integrating once more (with $\theta = \theta_0$ at $t_0 = 0$), we get

$$\theta = \theta_0 + \omega_0 t + \tfrac{1}{2}\alpha t^2 \qquad [11.7]$$

If we eliminate t from Equations 11.6 and 11.7, we get

$$\omega^2 = \omega_0^2 + 2\alpha(\theta - \theta_0) \qquad [11.8]$$

Rotational kinematic equations (α = constant)

If we eliminate α, we obtain

$$\theta = \theta_0 + \tfrac{1}{2}(\omega_0 + \omega)t \qquad [11.9]$$

Notice that these kinematic expressions for rotational motion under constant angular acceleration are of the *same form* as those for linear motion under constant linear acceleration, with the substitutions $x \rightarrow \theta$, $v \rightarrow \omega$, and $a \rightarrow \alpha$ (Table 11.1). Furthermore, the expressions are valid for both rigid-body rotation about a *fixed* axis and particle motion about a *fixed* axis.

Example 11.1 A Rotating Wheel

A wheel rotates with a constant angular acceleration of 3.5 rad/s^2.

(a) If the angular speed of the wheel is 2.0 rad/s at $t_0 = 0$, through what angle does the wheel rotate in 2 s?

Solution

$$\theta - \theta_0 = \omega_0 t + \tfrac{1}{2}\alpha t^2$$

$$= \left(2.0\,\frac{\text{rad}}{\text{s}}\right)(2\text{ s}) + \tfrac{1}{2}\left(3.5\,\frac{\text{rad}}{\text{s}^2}\right)(2\text{ s})^2$$

$$= \boxed{11\text{ rad} = 630° = 1.75\text{ rev}}$$

(b) What is the angular speed at $t = 2$ s?

Solution

$$\omega = \omega_0 + \alpha t = 2.0\text{ rad/s} + \left(3.5\,\frac{\text{rad}}{\text{s}^2}\right)(2\text{ s})$$

$$= \boxed{9.0\text{ rad/s}}$$

We could also obtain this result using Equation 11.8 and the results of (a). Try it!

Exercise Find the angle through which the wheel rotates between $t = 2$ s and $t = 3$ s.

Answer 10.8 rad.

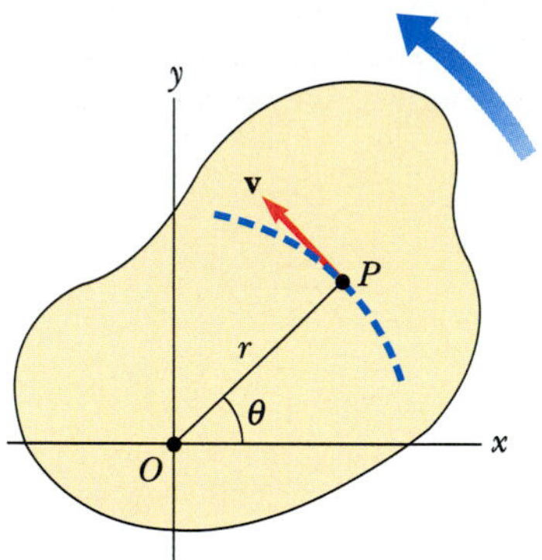

Figure 11.4
As a particle rotates about the fixed axis through O, the particle has a linear velocity **v** which is always tangent to the circular path of radius r.

11.3 Relations Between Angular and Linear Quantities

In this section we shall derive some useful relations between the angular speed and angular acceleration of a rotating particle and its linear speed and linear acceleration. Keep in mind that, when a rigid body rotates about a fixed axis, *every* particle of the body moves in a circle whose center is the axis of rotation.

Consider a particle rotating in a circle of radius r about the z axis, as in Figure 11.4. Since the particle moves along a circular path, its linear velocity vector **v** is always tangent to the path; hence, we often call this quantity *tangential velocity*. The magnitude of the tangential velocity of the particle is, by definition, ds/dt, where s is the distance traveled by the particle along the circular path. Recalling from Equation 11.1 that $s = r\theta$, and noting that r is a constant, we get

$$v = \frac{ds}{dt} = r\frac{d\theta}{dt}$$

Relationship between linear and angular speed

$$v = r\omega \qquad [11.10]$$

That is, the magnitude of the tangential velocity of the particle equals the distance of the particle from the axis of rotation multiplied by the particle's angular speed.

We can relate the angular acceleration of the particle to its tangential acceleration, a_t—which is the component of its acceleration tangent to the path of motion—by taking the time derivative of v:

$$a_t = \frac{dv}{dt} = r\frac{d\omega}{dt}$$

Relationship between tangential and angular acceleration

$$a_t = r\alpha \qquad [11.11]$$

That is, the tangential component of the linear acceleration of a particle undergoing circular motion equals the distance of the particle from the axis of rotation multiplied by the angular acceleration.

In Chapter 3 we found that a particle rotating in a circular path undergoes a centripetal, or radial, acceleration of magnitude v^2/r directed toward the center of rotation (Fig. 11.5). Since $v = r\omega$, we can express the centripetal acceleration of the particle as

$$a_r = \frac{v^2}{r} = r\omega^2 \quad [11.12]$$

The *total linear acceleration* of the particle is $\mathbf{a} = \mathbf{a}_t + \mathbf{a}_r$. Therefore, the magnitude of the total linear acceleration of the particle is

$$a = \sqrt{a_t{}^2 + a_r{}^2} = \sqrt{r^2\alpha^2 + r^2\omega^4} = r\sqrt{\alpha^2 + \omega^4} \quad [11.13]$$

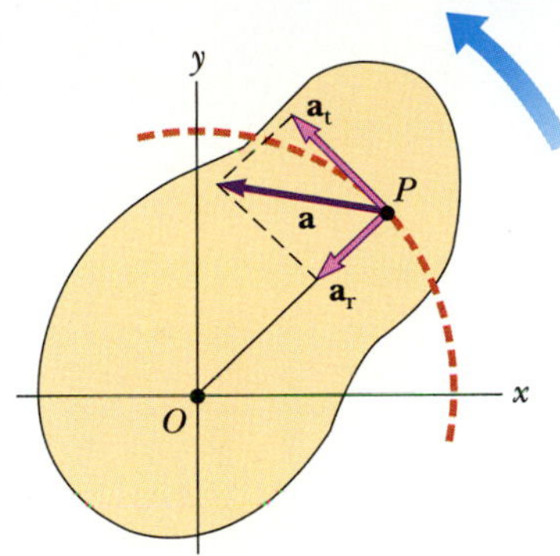

Figure 11.5 As a particle rotates about a fixed axis through O, the particle experiences a tangential component of acceleration, $\mathbf{a}_t$, and a centripetal component of acceleration, $\mathbf{a}_r$. The total linear acceleration of the particle is $\mathbf{a} = \mathbf{a}_t + \mathbf{a}_r$.

11.4 Rotational Kinetic Energy

Let us consider a rigid body as a collection of particles, and let us assume that the body rotates about the fixed z axis with an angular speed of ω (Fig. 11.6). Each particle of the body has some kinetic energy, determined by its mass and speed. If the mass of the ith particle is m_i and its speed is v_i, the kinetic energy of this particle is

$$K_i = \tfrac{1}{2}m_i v_i^2$$

To proceed further, we must recall that, although every particle in the rigid body has the same angular speed ω, the individual linear speeds depend on the distance r_i from the axis of rotation, according to the expression $v_i = r_i\omega$ (Eq. 11.10). The *total* kinetic energy of the rotating rigid body is the sum of the kinetic energies of the individual particles:

$$K = \sum K_i = \sum \tfrac{1}{2} m_i v_i^2 = \tfrac{1}{2}\sum m_i r_i^2 \omega^2$$

$$K = \tfrac{1}{2}\left(\sum m_i r_i^2\right)\omega^2$$

where we have factored ω^2 from the sum because it is common to every particle. The quantity in parentheses is called the **moment of inertia,** I:

$$I = \sum m_i r_i^2 \quad [11.14]$$

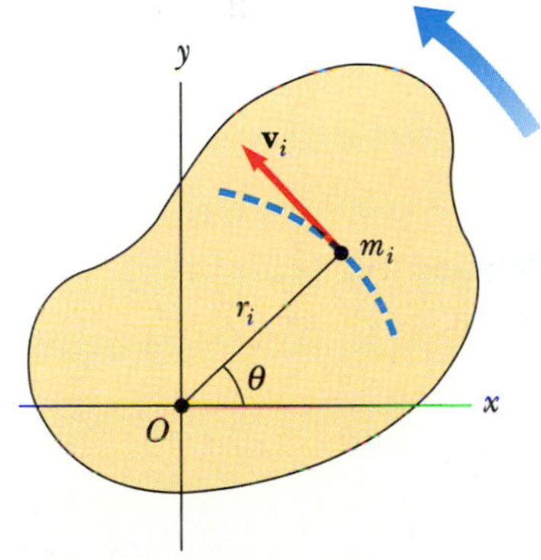

Figure 11.6 A particle rotating about the z axis with angular velocity ω. The kinetic energy of the particle of mass m_i is $\frac{1}{2}m_i v_i^2$. The total kinetic energy of the body is $\frac{1}{2}I\omega^2$.

Therefore, we can express the kinetic energy of the rotating rigid body (Eq. 11.13) as

$$K = \tfrac{1}{2}I\omega^2 \quad [11.15]$$

From the definition of moment of inertia, we see that it has dimensions of ML^2 (kg · m^2 in SI units). It plays the role of mass in *all* rotational equations. Although we shall commonly refer to the quantity $\frac{1}{2}I\omega^2$ as the **rotational kinetic energy,** it is not a new form of energy. It is ordinary kinetic energy, since it was derived from a sum over individual kinetic energies of the particles contained in the rigid body. However, the form of the kinetic energy given by Equation 11.15 is a convenient one for dealing with rotational motion, provided we know how to calculate I. It is important to recognize the analogy between kinetic energy associated with linear motion, $\frac{1}{2}mv^2$, and rotational kinetic energy, $\frac{1}{2}I\omega^2$. The quantities I and ω in rotational motion are analogous to m and v in linear motion, respectively.

Example 11.2 The Oxygen Molecule

Consider the diatomic molecule oxygen, O_2, which is rotating in the xy plane about the z axis passing through its center, perpendicular to its length. At room temperature, the "average" separation between the two oxygen atoms is 1.21×10^{-10} m (the atoms are treated as point masses).

(a) Calculate the moment of inertia of the molecule about the z axis.

Solution The mass of an oxygen atom is 2.66×10^{-26} kg, and the distance of each atom from the z axis is $d/2$. Therefore, the moment of inertia about the z axis is

$$I = \sum m_i r_i^2 = m\left(\frac{d}{2}\right)^2 + m\left(\frac{d}{2}\right)^2 = \frac{md^2}{2}$$

$$= \left(\frac{2.66 \times 10^{-26}}{2}\ \text{kg}\right)(1.21 \times 10^{-10}\ \text{m})^2$$

$$= \boxed{1.95 \times 10^{-46}\ \text{kg}\cdot\text{m}^2}$$

(b) If the molecule's angular speed about the z axis is 2.0×10^{12} rad/s, what is its rotational kinetic energy?

$$K = \tfrac{1}{2}I\omega^2 = \tfrac{1}{2}(1.95 \times 10^{-46}\ \text{kg}\cdot\text{m}^2)\left(2.0 \times 10^{12}\ \frac{\text{rad}}{\text{s}}\right)^2$$

$$= 3.89 \times 10^{-22}\ \text{J}$$

This is about one order of magnitude smaller than the average kinetic energy associated with the linear motion of the center of mass of the molecule at room temperature, which is about 6.2×10^{-21} J.

Example 11.3 Four Rotating Particles

Four particles are fixed at the corners of a frame of negligible mass lying in the xy plane (Fig. 11.7).

(a) If the rotation of the system occurs about the y axis with an angular speed of ω, find the moment of inertia about the y axis and the rotational kinetic energy about this axis.

Solution First, note that, because the particles of mass m lie on the y axis, they do not contribute to I_y (that is, $r_i = 0$ for these particles about this axis). Applying Equation 11.14, we get

$$I_y = \sum m_i r_i^2 = Ma^2 + Ma^2 = 2Ma^2$$

Therefore, the rotational kinetic energy about the y axis is

$$K = \tfrac{1}{2}I_y\omega^2 = \tfrac{1}{2}(2Ma^2)\omega^2 = \boxed{Ma^2\omega^2}$$

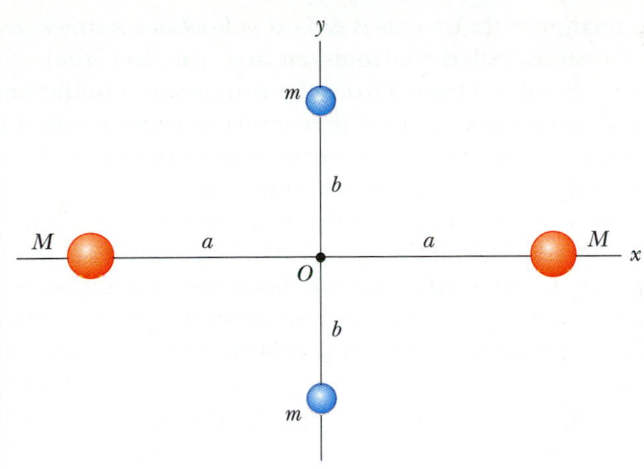

Figure 11.7
(Example 11.3) All particles are at a fixed separation as shown. The moment of inertia depends on the axis about which it is evaluated.

The fact that the masses m do not enter into this result makes sense, because these particles have no motion about the chosen axis of rotation and hence have no kinetic energy.

(b) Now suppose the system rotates in the xy plane about an axis through O (the z axis). Calculate the moment of inertia about the z axis and the rotational kinetic energy about this axis.

Solution Since r_i in Equation 11.14 is the *perpendicular* distance to the axis of rotation, we get

$$I_z = \sum m_i r_i^2 = Ma^2 + Ma^2 + mb^2 + mb^2$$

$$= 2Ma^2 + 2mb^2$$

$$K = \tfrac{1}{2} I_z \omega^2 = \tfrac{1}{2}(2Ma^2 + 2mb^2)\omega^2 = (Ma^2 + mb^2)\omega^2$$

Comparing the results for (a) and (b), we conclude that both the moment of inertia and the rotational kinetic energy of the system depend on the axis of rotation. In (b), we would expect the result to include all masses and distances, since all particles are in motion for rotation in the xy plane. Furthermore, the fact that the kinetic energy in (a) is smaller than that in (b) indicates that it would take less effort (work) to set the system into rotation about the y axis than about the z axis.

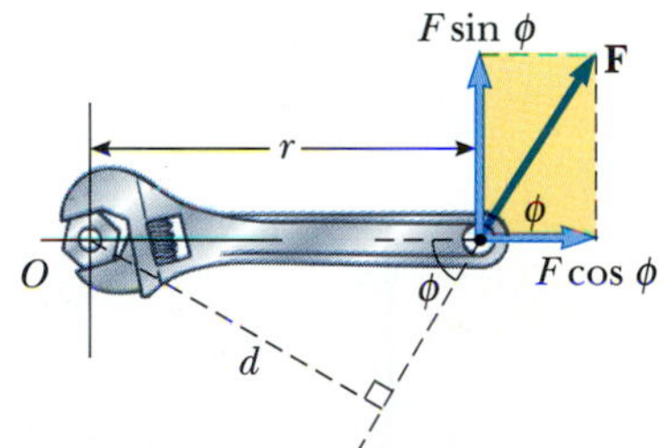

Figure 11.8
The force **F** has a greater rotating tendency about O as F increases and as the moment arm, d, increases. It is the component $F \sin \phi$ that tends to rotate the system about O.

11.5 Torque and the Vector Product

When a force is exerted on a rigid body pivoted about some axis, the body tends to rotate about that axis. The tendency of a force to rotate a body about some axis is measured by a vector quantity called **torque** ($\boldsymbol{\tau}$). Consider the wrench pivoted about the axis through O in Figure 11.8. The applied force **F** generally can act at an angle of ϕ to the horizontal. We define the magnitude of the torque, τ (Greek letter tau), resulting from the force **F** with the expression

$$\tau \equiv rF \sin \phi = Fd \qquad [11.16]$$

Definition of torque

Moment arm

It is very important to recognize that *torque is defined only when a reference axis is specified.* The quantity $d = r \sin \phi$, called the **moment arm** (or *lever arm*) of the force **F**, represents the perpendicular distance from the rotation axis to the line of action of **F**. Note that the only component of **F** that tends to cause a rotation is $F \sin \phi$, the component perpendicular to r. The horizontal component, $F \cos \phi$, passes through O and has no tendency to produce a rotation.

If two or more forces are acting on a rigid body, as in Figure 11.9, then each has a tendency to produce a rotation about the pivot at O. For example, $\mathbf{F}_2$ tends to rotate the body clockwise, and $\mathbf{F}_1$ tends to rotate the body counterclockwise. We shall use the convention that the sign of the torque resulting from a force is positive if its turning tendency is counterclockwise and negative if its turning tendency is clockwise. For example, in Figure 11.8, the torque resulting from $\mathbf{F}_1$, which has a moment arm of d_1, is *positive* and equal to $+F_1 d_1$; the torque from $\mathbf{F}_2$ is *negative* and equal to $-F_2 d_2$. Hence, the *net* torque acting on the rigid body about O is

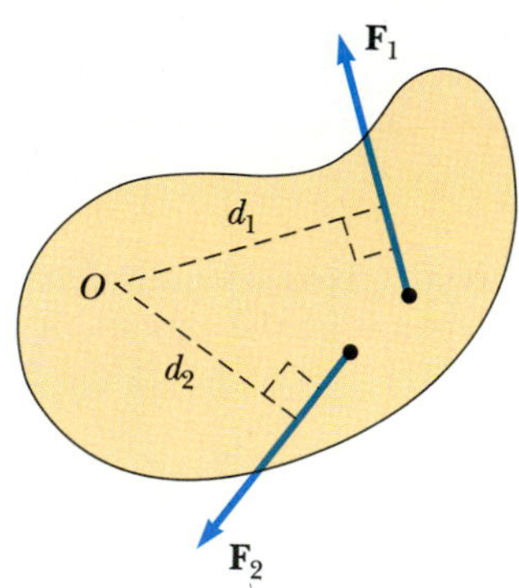

Figure 11.9
The force $\mathbf{F}_1$ tends to rotate the body counterclockwise about O, and $\mathbf{F}_2$ tends to rotate the body clockwise.

$$\tau_{\text{net}} = \tau_1 + \tau_2 = F_1 d_1 - F_2 d_2$$

From the definition of torque, we see that the rotating tendency increases as F increases and as d increases. For example, it is easier to close a door if we push at the doorknob rather than at a point close to the hinge. *Torque should not be confused with force.* Torque has units of force times length—N·m in SI units.

Now consider a force **F** acting on a particle located at the vector position **r** (Fig. 11.10). *The origin O is assumed to be in an inertial frame, so Newton's second law is valid.* The *magnitude* of the torque due to this force relative to the origin is, by definition, equal to $rF \sin \phi$, where ϕ is the angle between **r** and **F**. The axis about which **F** would tend to produce rotation is perpendicular to the plane formed by **r** and **F**. If the force lies in the *xy* plane, as in Figure 11.10, then the torque is represented by a vector parallel to the *z* axis. The force in Figure 11.9 creates a torque that tends to rotate the body counterclockwise looking down the *z* axis, and so the sense of $\boldsymbol{\tau}$ is toward increasing *z*, and $\boldsymbol{\tau}$ is in the positive *z* direction. If we reversed the direction of **F** in Figure 11.10, $\boldsymbol{\tau}$ would then be in the negative *z* direction. The torque involves two vectors, **r** and **F**, and is in fact defined to be equal to the *vector product*, or *cross product*, of **r** and **F**:

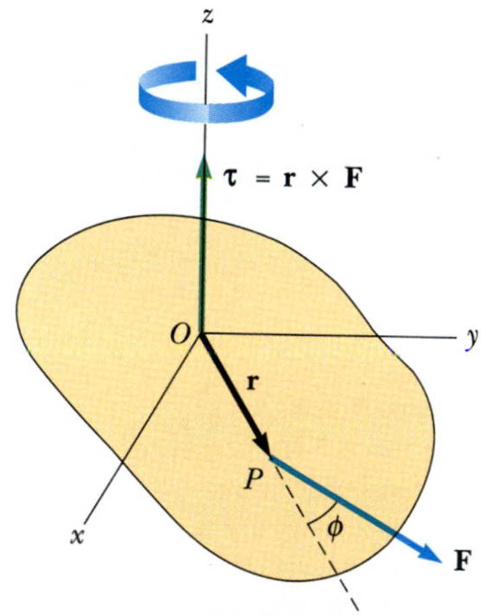

Figure 11.10
The torque vector $\boldsymbol{\tau}$ lies in a direction perpendicular to the plane formed by the position vector **r** and the applied force **F**.

$$\boldsymbol{\tau} \equiv \mathbf{r} \times \mathbf{F} \qquad [11.17]$$

We now give a formal definition of the vector product—a quantity we met briefly in Chapter 6. Given any two vectors **A** and **B**, the vector product $\mathbf{A} \times \mathbf{B}$ is defined as a third vector, **C**, the *magnitude* of which is $AB \sin \theta$, where θ is the angle included between **A** and **B**:

$$\mathbf{C} = \mathbf{A} \times \mathbf{B} \qquad [11.18]$$

$$C \equiv |\mathbf{C}| = |AB \sin \theta| \qquad [11.19]$$

Note that the quantity $AB \sin \theta$ is equal to the area of the parallelogram formed by **A** and **B**, as shown in Figure 11.11. The *direction* of $\mathbf{A} \times \mathbf{B}$ is perpendicular to the plane formed by **A** and **B**, and its sense is determined by the advance of a right-handed screw when the screw is turned from **A** to **B** through the angle θ. A more convenient rule to use for the direction of $\mathbf{A} \times \mathbf{B}$ is the right-hand rule illustrated in Figure 11.11. The four fingers of the right hand are pointed along **A** and then "wrapped" into **B** through the angle θ. The direction of the erect right thumb is

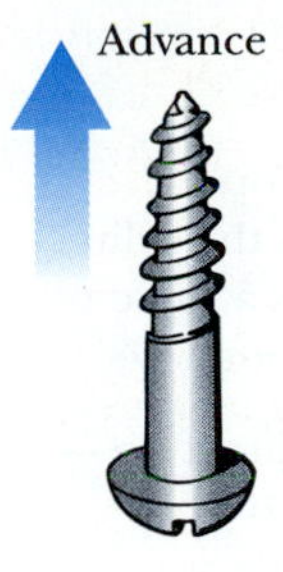

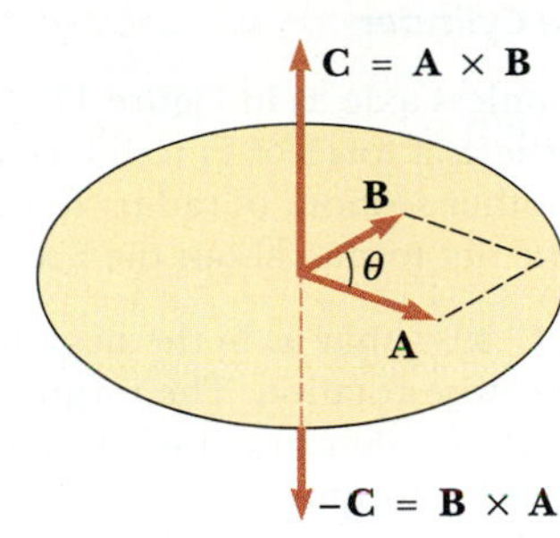

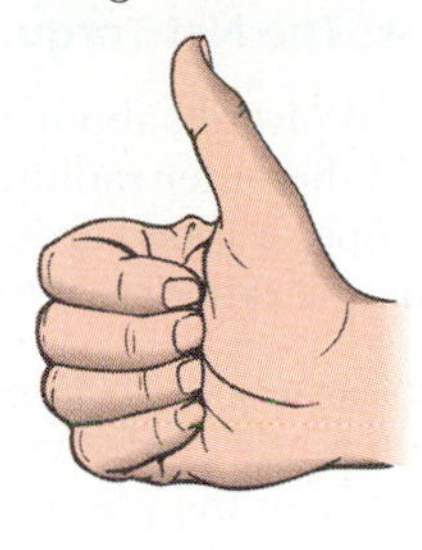

Figure 11.11 The vector product $\mathbf{A} \times \mathbf{B}$ is a third vector $\mathbf{C}$ having a magnitude $AB \sin \theta$ equal to the area of the parallelogram shown. The direction of $\mathbf{C}$ is perpendicular to the plane formed by A and B, and the sense of $\mathbf{C}$ is determined by the right-hand rule.

the direction of $\mathbf{A} \times \mathbf{B}$. Because of the notation, $\mathbf{A} \times \mathbf{B}$ is often read "A cross B"; hence the term "cross product."

Some properties of the vector product follow from its definition:

1. Unlike the case of the scalar product, the order in which the two vectors are multiplied in a cross product is important; that is

$$\mathbf{A} \times \mathbf{B} = -(\mathbf{B} \times \mathbf{A}) \qquad [11.20]$$

Properties of the vector product

   Therefore, if you change the order of the cross product, you must change the sign. One could easily verify this relation with the right-hand rule (Fig. 11.11).
2. If $\mathbf{A}$ is parallel to $\mathbf{B}$ ($\theta = 0°$ or $180°$), then $\mathbf{A} \times \mathbf{B} = 0$; therefore, it follows that $\mathbf{A} \times \mathbf{A} = 0$.
3. If $\mathbf{A}$ is perpendicular to $\mathbf{B}$, then $|\mathbf{A} \times \mathbf{B}| = AB$.

It is left to an end-of-chapter exercise to show, from Equations 11.18 and 11.19 and the definition of unit vectors, that the cross products of the rectangular unit vectors $\mathbf{i}$, $\mathbf{j}$, and $\mathbf{k}$ obey the following expressions:

Cross products of unit vectors

$$\begin{aligned} \mathbf{i} \times \mathbf{i} &= \mathbf{j} \times \mathbf{j} = \mathbf{k} \times \mathbf{k} = 0 \\ \mathbf{i} \times \mathbf{j} &= -\mathbf{j} \times \mathbf{i} = \mathbf{k} \\ \mathbf{j} \times \mathbf{k} &= -\mathbf{k} \times \mathbf{j} = \mathbf{i} \\ \mathbf{k} \times \mathbf{i} &= -\mathbf{i} \times \mathbf{k} = \mathbf{j} \end{aligned} \qquad [11.21]$$

Signs are interchangeable. For example, $\mathbf{i} \times (-\mathbf{j}) = -\mathbf{i} \times \mathbf{j} = -\mathbf{k}$.

The torque exerted on the nail by the hammer increases as the length of the handle (lever arm) increases. *(© Richard Megna, 1991, Fundamental Photographs)*

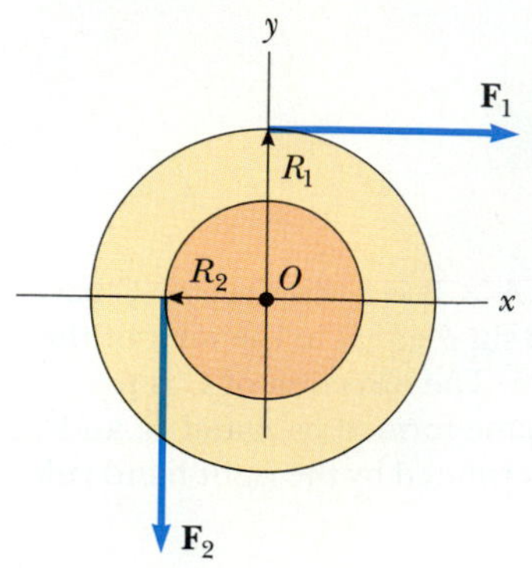
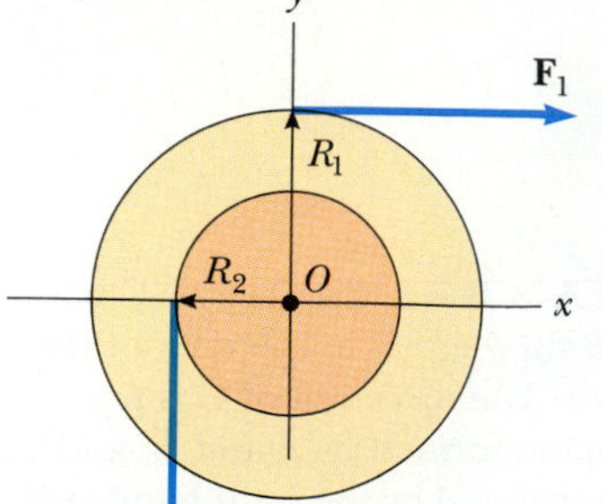

Figure 11.12
(Example 11.4) A solid cylinder pivoted about the z axis through O. The moment arm of $\mathbf{F}_1$ is R_1, and the moment arm of $\mathbf{F}_2$ is R_2.

Example 11.4 The Net Torque on a Cylinder

A solid cylinder is pivoted about a frictionless axle as in Figure 11.12. A rope wrapped around the outer radius, R_1, exerts a force of $\mathbf{F}_1$ to the right on the cylinder. A second rope wrapped around another section, of radius R_2, exerts a force of $\mathbf{F}_2$ downward on the cylinder. What is the net torque about the z axis through O?

Solution The torque due to $\mathbf{F}_1$ is $R_1F_1(-\mathbf{k})$—that is, in the negative z direction—indicating that it tends to produce a clockwise rotation. The torque due to $\mathbf{F}_2$ is $R_2F_2(+\mathbf{k})$—that is, in the positive z direction—because it tends to produce a counterclockwise rotation. Therefore, the net torque is

$$\boldsymbol{\tau}_{\text{net}} = \boldsymbol{\tau}_1 + \boldsymbol{\tau}_2 = R_1F_1(-\mathbf{k}) + R_2F_2(+\mathbf{k}) = (R_2F_2 - R_1F_1)\mathbf{k}$$

Exercise Suppose $F_1 = 5$ N, $R_1 = 1.0$ m, $F_2 = 6$ N, and $R_2 = 0.5$ m. What is the net torque, and which way will the cylinder rotate?

Answer $(-2\mathbf{k})$ N·m; clockwise.

Example 11.5 The Cross Product

Two vectors lying in the xy plane are given by the equations $\mathbf{A} = 2\mathbf{i} + 3\mathbf{j}$ and $\mathbf{B} = -\mathbf{i} + 2\mathbf{j}$. Find $\mathbf{A} \times \mathbf{B}$, and verify that $\mathbf{A} \times \mathbf{B} = -\mathbf{B} \times \mathbf{A}$.

Solution Use of Equation 11.21 for the cross product of unit vectors gives

$$\mathbf{A} \times \mathbf{B} = (2\mathbf{i} + 3\mathbf{j}) \times (-\mathbf{i} + 2\mathbf{j})$$

$$= 2\mathbf{i} \times 2\mathbf{j} + 3\mathbf{j} \times (-\mathbf{i}) = 4\mathbf{k} + 3\mathbf{k} = 7\mathbf{k}$$

(We have omitted the terms with $\mathbf{i} \times \mathbf{i}$ and $\mathbf{j} \times \mathbf{j}$, since they are zero.)

$$\mathbf{B} \times \mathbf{A} = (-\mathbf{i} + 2\mathbf{j}) \times (2\mathbf{i} + 3\mathbf{j})$$

$$= -\mathbf{i} \times 3\mathbf{j} + 2\mathbf{j} \times 2\mathbf{i} = -3\mathbf{k} - 4\mathbf{k} = -7\mathbf{k}$$

Therefore, $\mathbf{A} \times \mathbf{B} = -\mathbf{B} \times \mathbf{A}$.

Exercise Use the results of this example and Equation 11.19 to find the angle between $\mathbf{A}$ and $\mathbf{B}$.

Answer 60.3°.

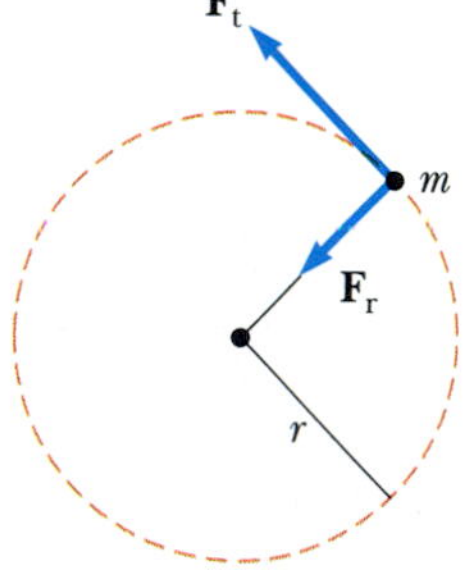

Figure 11.13
A particle of mass m rotating in a circle under the influence of a tangential force $\mathbf{F}_t$. A centripetal force $\mathbf{F}_r$ must also be present to maintain the circular motion.

11.6 Relation Between Torque and Angular Acceleration

In this section we shall show that the angular acceleration of a particle rotating about a fixed axis is proportional to the net torque acting about that axis. The ideas embodied in this situation can easily be extended to the case of a rigid body rotating about a fixed axis.

Consider a particle of mass m rotating in a circle of radius r under the influence of both a tangential force, $\mathbf{F}_t$, and a centripetal force, $\mathbf{F}_r$, as in Figure 11.13.

(The centripetal force, whose nature is not yet specified, *must* be present to keep the particle moving in its circular path.) According to Newton's second law, $\mathbf{F}_t = m\mathbf{a}_t$ and $\mathbf{F}_r = m\mathbf{a}_r$, where $\mathbf{a}_t$ is the tangential acceleration of the particle and $\mathbf{a}_r$ is its centripetal acceleration. The torque due to the centripetal force about the origin is zero since $\mathbf{F}_r$ is antiparallel to r; hence, $\mathbf{r} \times \mathbf{F}_r = 0$. The torque about the origin due to the tangential force is given by $\mathbf{r} \times \mathbf{F}_t$, but since r is perpendicular to $\mathbf{F}_t$, the magnitude of the torque is simply $F_t r$. Thus, the magnitude of the net torque on the particle is

$$\tau = F_t r = (ma_t)r$$

Since the tangential acceleration is related to the angular acceleration through Equation 11.10, $a_t = r\alpha$, the torque can be expressed as

$$\tau = (mr\alpha)r = (mr^2)\alpha$$

Recall from Equation 11.14 that the quantity mr^2 is the moment of inertia of the rotating mass about the z axis passing through the origin, so that

Relationship between net torque and angular acceleration

$$\sum \tau = I\alpha \qquad [11.22]$$

That is, *the net torque acting on the particle is proportional to its angular acceleration,* and the proportionality constant is the moment of inertia. It is important to note that $\Sigma\tau = I\alpha$ is the rotational analogue of Newton's second law of motion, $\Sigma F = ma$.

11.7 Angular Momentum

Again let us consider a particle of mass m, situated at the vector position r and moving with a momentum p, as shown in Figure 11.14. The **instantaneous angular momentum**, L, of the particle relative to the origin O is defined by the cross product of its instantaneous vector position and the instantaneous linear momentum, p:

$$\mathbf{L} = \mathbf{r} \times \mathbf{p} \qquad [11.23]$$

The SI units of angular momentum are $\text{kg} \cdot \text{m}^2\text{s}^{-1}$. It is important to note that both the magnitude and the direction of L depend on the choice of origin. The direction of L is perpendicular to the plane formed by r and p, and the sense of L is governed by the right-hand rule. For example, in Figure 11.14, r and p are assumed to be in the xy plane, and L points in the z direction. Since $\mathbf{p} = m\mathbf{v}$, the magnitude of L is

$$L = mvr \sin\phi \qquad [11.24]$$

where ϕ is the angle between r and p. It follows that L is zero when r is parallel to p ($\phi = 0°$ or $180°$). In other words, when the particle moves along a line that passes through the origin, it has zero angular momentum with respect to the origin. This is equivalent to stating that it has no tendency to rotate about the origin. On the other hand, if r is perpendicular to p ($\phi = 90°$), L is a maximum and equal to mrv. In this case, the particle has maximum tendency to rotate about the origin. In fact, at that instant the particle moves exactly as though it were on the rim of a wheel rotating about the origin in a plane defined by r and p. A particle has nonzero

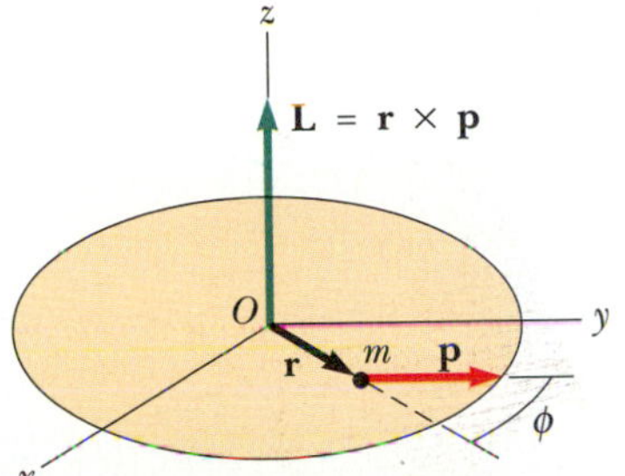

Figure 11.14
The angular momentum L of a particle of mass m and momentum p located at the position r is a vector given by $\mathbf{L} = \mathbf{r} \times \mathbf{p}$. The value of L depends on the choice of origin and is a vector perpendicular to both r and p.

This remarkable photograph of Hurricane Gladys about 150 mi southwest of Tampa, Florida, was taken from the Apollo 7 spacecraft at an altitude of 97 mi. The spiral-shaped, nonrigid mass of air undergoes rotation and has angular momentum. *(Courtesy of NASA)*

angular momentum about some point if the position vector of the particle measured from that point rotates about the point.

For linear motion, we found that the resultant force on a particle equaled the time rate of change of the particle's linear momentum (Eq. 9.7). We shall now show that Newton's second law implies that the resultant torque acting on a particle equals the time rate of change of the particle's angular momentum. Let us start by writing the torque on the particle in the form

$$\boldsymbol{\tau} = \mathbf{r} \times \mathbf{F} = \mathbf{r} \times \frac{d\mathbf{p}}{dt} \qquad [11.25]$$

where we have used the fact that $\mathbf{F} = d\mathbf{p}/dt$. Now let us differentiate Equation 11.23 with respect to time, using the chain rule.

$$\frac{d\mathbf{L}}{dt} = \frac{d}{dt}(\mathbf{r} \times \mathbf{p}) = \mathbf{r} \times \frac{d\mathbf{p}}{dt} + \frac{d\mathbf{r}}{dt} \times \mathbf{p}$$

It is important to adhere to the order of terms since $\mathbf{A} \times \mathbf{B} = -\mathbf{B} \times \mathbf{A}$.

The last term on the right in the preceding equation is zero because $\mathbf{v} = d\mathbf{r}/dt$ is parallel to $\mathbf{p}$. Therefore,

$$\frac{d\mathbf{L}}{dt} = \mathbf{r} \times \frac{d\mathbf{p}}{dt} \qquad [11.26]$$

Comparing Equations 11.25 and 11.26, we see that

Torque equals time rate of change of angular momentum

$$\boldsymbol{\tau} = \frac{d\mathbf{L}}{dt} \qquad [11.27]$$

This result, $\boldsymbol{\tau} = d\mathbf{L}/dt$, is the rotational analog of Newton's second law, $\mathbf{F} = d\mathbf{p}/dt$. Equation 11.27 says that the **torque** acting on a particle is equal to the time rate of change of the particle's angular momentum. It is important to note that Equation 11.27 is valid only if the origins of $\boldsymbol{\tau}$ and $\mathbf{L}$ are the *same.* Equation 11.27 is also valid when several forces are acting on the particle, in which case $\boldsymbol{\tau}$ is the *net* torque on the particle. *Furthermore, the expression is valid for any origin fixed in an inertial frame.* Of course, the same origin must be used in calculating all torques as well as the angular momentum.

A System of Particles

The total angular momentum, $\mathbf{L}$, of a system of particles about some point is defined as the vector sum of the angular momenta of the individual particles:

$$\mathbf{L} = \mathbf{L}_1 + \mathbf{L}_2 + \ . \ . \ . \ + \mathbf{L}_n = \sum \mathbf{L}_i$$

where the vector sum is over all of the n particles in the system.

Since the individual momenta of the particles may change in time, the total angular momentum may also vary in time. In fact, from Equations 11.25 and 11.26, we find that the time rate of change of the total angular momentum equals the vector sum of *all* torques, including those associated with internal forces between particles and those associated with external forces. However, the net torque associated with internal forces is zero. To understand this, recall that Newton's third law tells us that the internal forces occur in equal and opposite pairs that lie along the line of separation of each pair of particles. Therefore, the torque due to each action-reaction force pair is zero. By summation, we see that *the net internal torque vanishes.* Finally, we conclude that the total angular momentum can vary with time *only* if there is a net *external* torque on the system, so that we have

$$\sum \boldsymbol{\tau}_{\text{ext}} = \sum \frac{d\mathbf{L}_i}{dt} = \frac{d}{dt}\sum \mathbf{L}_i = \frac{d\mathbf{L}}{dt} \qquad [11.28]$$

That is, the time rate of change of the total angular momentum of the system about some origin in an inertial frame equals the net external torque acting on the system about that origin. Note that Equation 11.28 is the rotational analog of $\mathbf{F}_{\text{ext}} = d\mathbf{p}/dt$ (Eq. 9.35) for a system of particles.

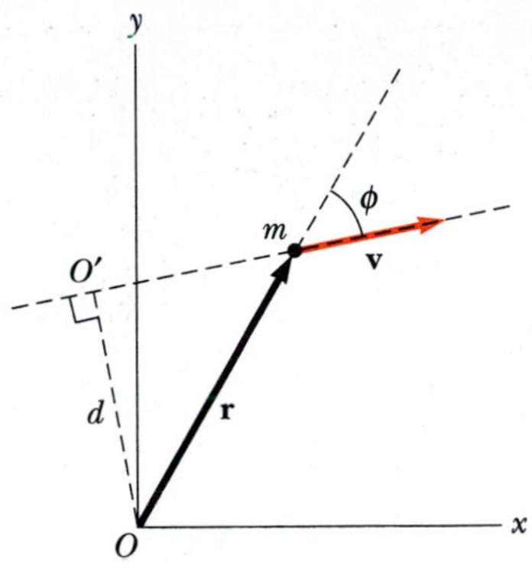

Figure 11.15 (Example 11.6) A particle moving in a straight line with a velocity **v** has an angular momentum relative to O equal in magnitude to mvd, where $d = r \sin \phi$ is the distance of closest approach to the origin. The vector $\mathbf{L} = \mathbf{r} \times \mathbf{p}$ points *into* the diagram in this case.

Example 11.6 Linear Motion

A particle of mass m moves in the xy plane with a velocity of **v** along a straight line (Fig. 11.15). What are the magnitude and direction of its angular momentum with respect to the origin O?

Solution From the definition of angular momentum, $\mathbf{L} = \mathbf{r} \times \mathbf{p} = rmv \sin \phi\,(-\mathbf{k})$. Therefore, the magnitude of **L** is

$$L = mvr \sin \phi = \boxed{mvd}$$

where $d = r \sin \phi$ is the distance of closest approach of the particle to the origin. The direction of **L** from the right-hand rule is *into* the diagram, and we can write the vector expression $\mathbf{L} = -(mvd)\mathbf{k}$.

Exercise What is the angular momentum of the particle with respect to the origin O' shown in Figure 11.15?

Answer Zero, because **r** is parallel to **p** in this case.

Example 11.7 Circular Motion

A particle moves in the xy plane in a circular path of radius r, as in Figure 11.16.

(a) Find the magnitude and direction of its angular momentum relative to O when its velocity is **v**.

Solution Since **r** is perpendicular to **v**, $\phi = 90°$ and the magnitude of **L** is simply

$$L = mvr \sin 90° = \boxed{mvr} \qquad \text{(for } \mathbf{r} \text{ perpendicular to } \mathbf{v}\text{)}$$

The direction of **L** is perpendicular to the plane of the circle, and the sense of **L** depends on the direction of **v**. If the sense of the rotation is counterclockwise, as in Figure 11.15, then by the right-hand rule, the direction of $\mathbf{L} = \mathbf{r} \times \mathbf{p}$ is *out of* the paper. Hence, we can write the vector expression $\mathbf{L} = (mvr)\mathbf{k}$. (If the particle were to move clockwise, **L** would point into the paper.)

(b) Find an alternative expression for L in terms of the angular velocity, ω.

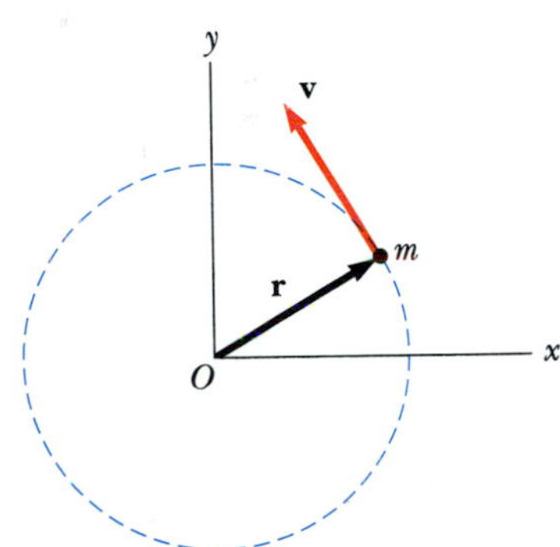

Figure 11.16 (Example 11.7) A particle moving in a circle of radius r has an angular momentum equal in magnitude to mvr relative to the center. The vector $\mathbf{L} = \mathbf{r} \times \mathbf{p}$ (not shown) points *out* of the diagram.

Solution Since, from Equation 11.9, $v = r\omega$ for a particle rotating in a circle, we can express L as

$$L = mvr = mr^2\omega = I\omega$$

where I is the moment of inertia of the particle about the z axis through O. Furthermore, in this case the angular momentum is in the *same* direction as the angular velocity vector, $\boldsymbol{\omega}$ (see Section 11.1), and so we can write $\mathbf{L} = I\boldsymbol{\omega} = I\omega\mathbf{k}$.

Exercise A car of mass 1500 kg moves on a circular race track of radius 50 m, with a speed of 40 m/s. What is the magnitude of its angular momentum relative to the center of the race track?

Answer $3.00 \times 10^6\ \text{kg}\cdot\text{m}^2/\text{s}$.

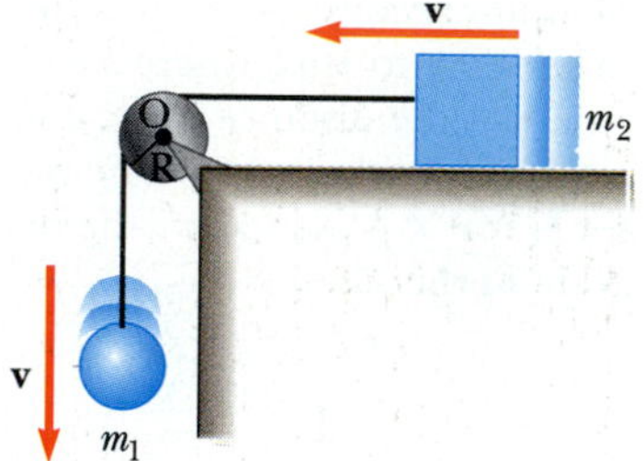

Figure 11.17 (Example 11.8)

Example 11.8 Two Connected Masses

This example illustrates that the net torque acting on a system equals the time rate of change of angular momentum of the system about some origin. Two masses, m_1 and m_2, are connected by a light cord that passes over a pulley of radius R whose mass is negligible (Fig. 11.17). The mass m_2 slides on a frictionless, horizontal surface. Let us determine the linear acceleration of the two masses, using the concepts of angular momentum and torque.

Solution First, let us calculate the angular momentum of the system, which consists of the two masses. (Note that if the pulley's mass were not neglected, the pulley would also possess angular momentum and become part of the system.) We shall calculate the angular momentum about an axis along the axle of the pulley through O. At the instant the two masses have a speed v, the magnitude of the angular momentum of m_1 is $m_1 vR$ and the magnitude of the angular momentum of m_2 is $m_2 vR$. Therefore, the magnitude of the total angular momentum of the system is

$$L = m_1 vR + m_2 vR = (m_1 + m_2)vR$$

Now let us evaluate the total external torque on the system about the axle. Because the force of the axle on the pulley has zero moment arm, this force does not contribute to the torque. Furthermore, the normal force acting on m_2 is balanced by the weight $m_2\mathbf{g}$, and hence these forces do not contribute to the torque. The external force $m_1\mathbf{g}$ produces a torque about the axle equal in magnitude to $m_1 gR$, where R is the moment arm of the force about the axle. This is the magnitude of the *total* external torque about O; that is, $\tau_{\text{ext}} = m_1 gR$. This result, together with the preceding result and the relation

$$\tau_{\text{ext}} = \frac{dL}{dt}$$

gives

$$m_1 gR = \frac{d}{dt}\Big[(m_1 + m_2)Rv\Big] = (m_1 + m_2)R\frac{dv}{dt}$$

Because $dv/dt = a$, we can solve this expression for a to get

$$a = \frac{m_1 g}{m_1 + m_2}$$

We did not include the force of tension in evaluating the net torque about the axle because this force is *internal* to the system under consideration. Only the *external* torques contribute to the change in angular momentum of the system.

11.8 Conservation of Angular Momentum

In Chapter 9 we found that the total linear momentum of a system of particles remains constant when the net external force acting on the system is zero. In rotational motion, we have an analogous conservation law that states that *the total angular momentum of a system remains constant if the net external torque acting on the system is zero.*

Since the net torque acting on the system equals the time rate of change of the system's angular momentum, we see that if

$$\sum \boldsymbol{\tau}_{\text{ext}} = \frac{d\mathbf{L}}{dt} = 0 \qquad [11.29]$$

then

$$\mathbf{L} = \text{constant} \qquad [11.30]$$

For a system of particles, we can write this conservation law as $\sum \mathbf{L}_i = \text{constant}$, where $\mathbf{L}_i$ is the angular momentum of the ith particle in the system, given by $\mathbf{L}_i = \mathbf{r}_i \times \mathbf{p}_i$. Hence, we can express the conservation of angular momentum for a system of particles as

$$\sum \mathbf{L}_i = \sum \mathbf{r}_i \times \mathbf{p}_i = \text{constant} \qquad [11.31]$$

From Equation 11.31 we derive a third conservation law to add to our list of conserved quantities. We can now state that *the total energy, linear momentum, and angular momentum of an isolated system all remain constant.*

If the system of "particles" under consideration is a rigid body rotating about a fixed axis, then the angular momentum of the body about that axis has a magnitude given by $L = I\omega$, where I is the moment of inertia about the axis. In this case, if the net external torque on the body is zero, we can express the conservation of angular momentum as $I\omega = \text{constant}$.

Although we do not prove it here, an important theorem concerns the angular momentum of a system of particles relative to the center of mass of the system. It can be stated as follows: **The net torque acting on a system of particles about the center of mass equals the time rate of change of angular momentum, regardless of the motion of the center of mass.** This theorem applies even if the center of mass is accelerating, provided that both τ and $\mathbf{L}$ are evaluated relative to the center of mass.

Many examples can be used to demonstrate conservation of angular momentum; some of them should be familiar to you. You may have observed a figure skater spinning. The angular speed of the skater increases as she pulls her hands and feet close to the trunk of her body, as in Figure 11.18. Neglecting friction between skater and ice, we see that there are no external torques on the skater.

Figure 11.18 Photograph of Kristi Yamaguchi, winner of a gold medal in the 1992 Olympics. Because angular momentum is conserved, her angular speed increases as she draws her hands and feet close to her body. *(PIC: Rick Stewart)*

The moment of inertia of her body decreases as her hands and feet are brought in. The resulting change in angular speed is accounted for as follows. Since angular momentum must be conserved, the product $I\omega$ remains constant, and a decrease in I causes a corresponding increase in ω.

An interesting astrophysical example of conservation of angular momentum occurs when, at the end of its lifetime, a massive star uses up all its fuel and collapses under the influence of gravitational forces, causing a gigantic outburst of energy called a supernova explosion. The best-studied example of a remnant of a supernova explosion is the Crab Nebula, a chaotic, expanding mass of gas (Fig. 11.19). Part of the star's mass is released into space, where it eventually condenses into new stars and planets. Most of what is left behind makes up a **neutron star,** an extremely dense sphere of matter with a diameter of about 10 km in comparison with the 10^6-km diameter of the original star. As the rotational inertia of the system decreases, the star's rotational speed increases. More than 300 rapidly rotating neutron stars have been identified, with periods of rotation ranging from 1.6 ms to 4 s. The neutron star is a most dramatic system—an object with a mass greater than the Sun, rotating about its axis a few times each second!

Figure 11.19
The Crab Nebula, in the constellation Taurus. This nebula is the remnant of a supernova explosion, which was seen on Earth in the year A.D. 1054. It is located some 6300 lightyears away and is approximately 6 lightyears in diameter, still expanding outward. *(National Optical Astronomy Observations)*

Example 11.9 A Rotating Ball on a Horizontal, Frictionless Surface

A ball of mass m on a horizontal, frictionless table is connected to a string that passes through a small hole in the table. The ball is set into circular motion of radius R, at which time its speed is v_0 (Fig. 11.20). If the string is pulled from the bottom so that the radius of the circular path is decreased to r, what is the final speed v of the ball?

Solution Let us take the torque about the center of rotation, O. Note that the force of gravity acting on the ball is balanced by the upward normal force, and so these forces cancel. The force, **F**, of the string on the ball (the centripetal force) acts toward the center of rotation, while the vector position **r** is directed away from O. Thus, we see that $\boldsymbol{\tau} = \mathbf{r} \times \mathbf{F} = 0$. Again, since $\boldsymbol{\tau} = d\mathbf{L}/dt = 0$, **L** is a constant of the motion. That is, $mv_0R = mvr$, or

$$v = \frac{v_0 R}{r}$$

From this result, we see that as r decreases, the speed v increases.

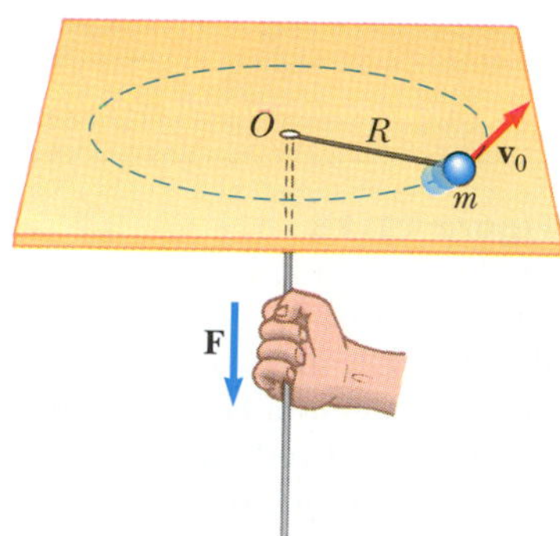

Figure 11.20
(Example 11.9)

*11.9 Quantization of Angular Momentum

We have seen that the concept of angular momentum is very useful for describing the motions of macroscopic systems. The concept is also valid and equally useful on a submicroscopic scale, and it has been used extensively in the development of modern theories of atomic, molecular, and nuclear physics. In these developments, it was found that the angular momentum of a system is a *fundamental* quantity. The word "fundamental" in this context implies that angular momentum is an inherent property of atoms, molecules, and their constituents.

In order to explain the results of a variety of experiments on atomic and molecular systems, we must assign discrete values to—in modern terminology, quantize—the angular momentum. These discrete values are multiples of a fundamental unit of angular momentum that equals $h/2\pi$, where h is called **Planck's constant.**

$$\text{Fundamental unit of angular momentum} = \frac{h}{2\pi} = 1.054 \times 10^{-34}\ \text{kg}\cdot\text{m}^2/\text{s}$$

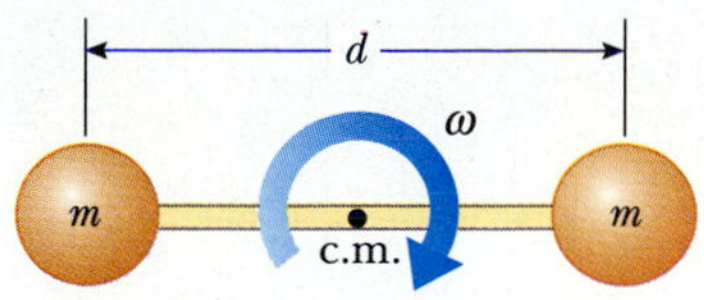

Figure 11.21
The rigid-rotor model of a diatomic molecule. The rotation occurs about the center of mass in the plane of the diagram.

Example 11.10 Estimating the Rotational Frequency of the Oxygen Molecule

Estimate the rotational frequency of the oxygen molecule, using the facts that the mass of an oxygen atom is 2.66×10^{-26} kg and that the "average" separation between the two oxygen atoms is 1.21×10^{-10} m.

Solution Consider the O_2 molecule as a rigid rotor—that is, two oxygen atoms separated by a fixed distance d and rotating about their center of mass (Fig. 11.21). To find the rotational frequency, first we must evaluate the angular momentum of the molecule about an axis through its center of mass, perpendicular to the plane of rotation.

$$L = L_1 + L_2 = mvr + mvr = 2mvr$$

In this example, $r = d/2 = 0.605 \times 10^{-10}$ m. Also, it is useful to use the relation $v = r\omega$ (Eq. 11.4), giving

$$L = 2mvr = 2mr^2\omega = 2(2.22 \times 10^{-26}\ \text{kg})(0.605 \times 10^{-10}\ \text{m})^2\omega$$
$$= (1.95 \times 10^{-46}\ \text{kg}\cdot\text{m}^2)\,\omega$$

We can now estimate the lowest rotational frequency of the molecule by equating the rotational angular momentum with the fundamental unit of angular momentum, $h/2\pi$.

$$L = (1.95 \times 10^{-46}\ \text{kg}\cdot\text{m}^2)\,\omega = \frac{h}{2\pi} = 1.054 \times 10^{-34}\ \text{kg}\cdot\text{m}^2/\text{s}$$

Therefore,

$$\omega = \frac{1.054 \times 10^{-34}\ \text{kg}\cdot\text{m}^2/\text{s}}{1.95 \times 10^{-46}\ \text{kg}\cdot\text{m}^2} = 5.41 \times 10^{11}\ \text{rad/s}$$

This result is in good agreement with measured rotational frequencies. The rotational frequencies are much lower than the vibrational frequencies of the molecule, which are typically of the order of 10^{13} rad/s.

Classical concepts and mechanical models can be useful in describing some features of atomic and molecular systems. However, as shown by the preceding example, a wide variety of submicroscopic phenomena can be explained only if one assumes discrete values of the angular momentum associated with a particular type of motion.

The Danish physicist Niels Bohr (1885–1962), in his theory of the hydrogen atom, suggested the radical idea of angular momentum quantization. Strictly classical models were unsuccessful in describing many properties of the hydrogen atom, such as the fact that the atom emits radiation at discrete frequencies. Bohr postulated that the electron could occupy only those orbits about the proton for which the orbital angular momentum was equal to $nh/2\pi$, where n is an integer, or quantum number. That is, Bohr made the bold postulate that the orbital angular momentum is quantized. From this simple model, one can estimate the rotational frequencies of the electron in the allowed orbits.

Although Bohr's theory provided some insight concerning the behavior of matter at the atomic level, it is basically incorrect. Subsequent developments in quantum mechanics from 1924 to 1930 provided models and interpretations that are still accepted.

Later developments in atomic physics indicated that the electron possesses another kind of angular momentum, called *spin,* which is another inherent property of the electron. The spin angular momentum is also restricted to discrete values. Later we shall return to this important property and discuss its great impact on modern physical science.

*11.10 Rotation of Rigid Bodies

In this section we shall investigate the rotational motion of rigid bodies, using as a basis much of what we have learned concerning the circular motion of particles. First we shall discuss the dynamics of a rigid body rotating about a fixed axis. Next we shall show how to determine the kinetic energy of a rotating rigid body and how its change is related to the work done by external forces. Finally we shall use the principle of conservation of energy to describe an object that rolls on a surface without slipping.

Rotational Dynamics

In Section 11.6 we learned that the magnitude of the net torque acting on a particle is given by Equation 11.22:

$$\tau_{\text{net}} = I\alpha$$

where I is the moment of inertia and α is the angular acceleration. Importantly, this result also applies to a rigid body of arbitrary shape rotating about a fixed axis, since the body can be regarded as an infinite number of mass elements of infinitesimal size, each rotating in a circle about the rotation axis. The important and strikingly simple result given by $\boldsymbol{\tau}_{\text{net}} = I\boldsymbol{\alpha}$ is in complete agreement with experimental observations. Its simplicity is a result of how the motion is described. **Although the points on a rigid body rotating about a fixed axis may not experience the same force, linear acceleration, or linear velocity, every point on the body has the same angular acceleration and angular velocity at any instant. Therefore, at any instant the rigid body as a whole is characterized by specific values for angular acceleration, net torque, and angular velocity.**

The specific form of I depends upon the axis of rotation and upon the size and shape of the body. The moments of inertia of rigid bodies can be evaluated by methods of integral calculus. Table 11.2 gives the moments of inertia for a number of bodies about specific axes. In all cases, note that I is proportional to the mass of the object and to the square of a geometric factor.

Table 11.2
Moments of Inertia of Homogeneous Rigid Bodies with Different Geometries

Hoop or cylindrical shell $I_c = MR^2$	Hollow cylinder $I_c = \frac{1}{2} M(R_1^2 + R_2^2)$
Solid cylinder or disk $I_c = \frac{1}{2} MR^2$	Rectangular plate $I_c = \frac{1}{12} M(a^2 + b^2)$
Long thin rod $I_c = \frac{1}{12} ML^2$	Long thin rod $I = \frac{1}{3} ML^2$
Solid sphere $I_c = \frac{2}{5} MR^2$	Thin spherical shell $I_c = \frac{2}{3} MR^2$

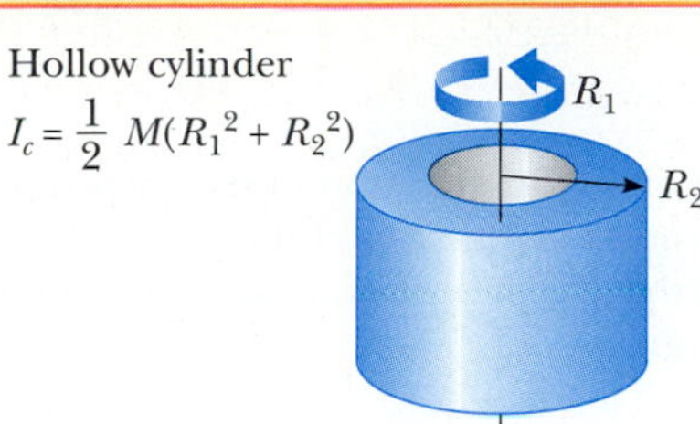

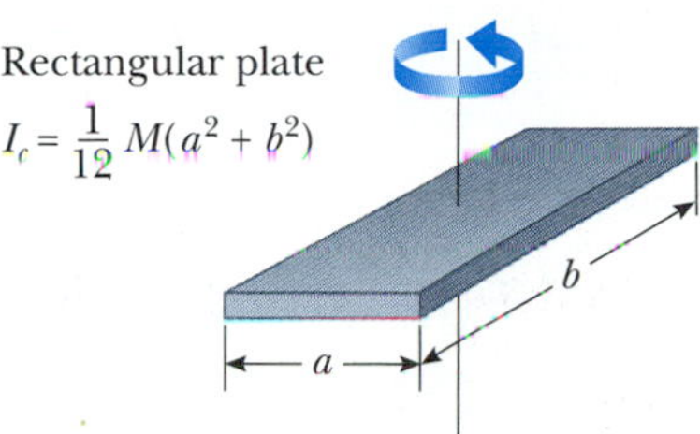

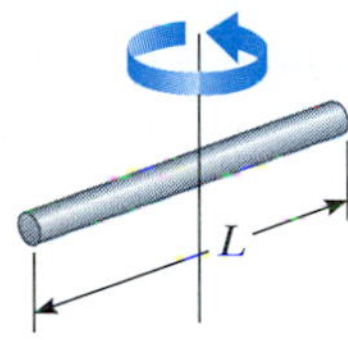

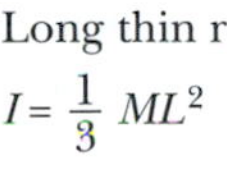

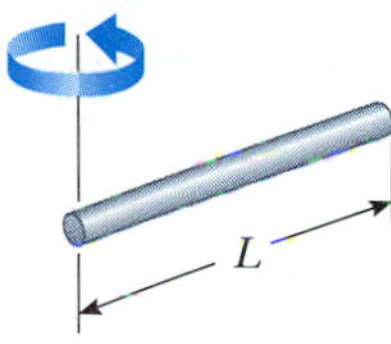

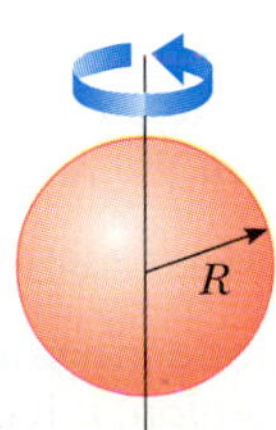

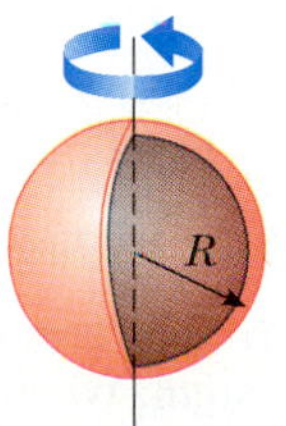

Example 11.11 Angular Acceleration of a Wheel

A wheel of radius R, mass M, and moment of inertia I is mounted on a frictionless, horizontal axle as in Figure 11.22. A light cord wrapped around the wheel supports a body of mass m. Calculate the linear acceleration of the suspended body, the angular acceleration of the wheel, and the tension in the cord.

Solution The torque acting on the wheel about its axis of rotation is $\tau = TR$. The weight of the wheel and the normal force of the axle on the wheel pass through the axis of rotation and produce no torque. Since $\tau = I\alpha$, we get

$$\tau = I\alpha = TR$$

$$\alpha = TR/I \qquad (1)$$

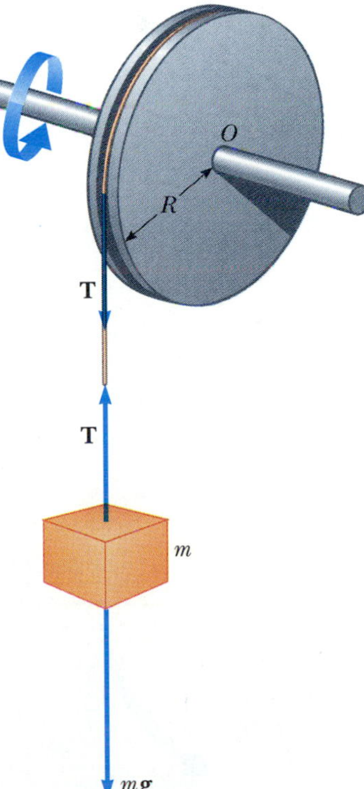

Figure 11.22 (Example 11.11) The cord attached to m is wrapped around the pulley, which produces a torque about the axle through O.

Now let us apply Newton's second law to the motion of the suspended mass m, making use of the free-body diagram (Fig. 11.21):

$$\sum F_y = T - mg = -ma$$

$$(2) \qquad a = \frac{mg - T}{m}$$

The linear acceleration of the suspended mass is equal to the tangential acceleration of a point on the rim of the wheel. Therefore, the angular acceleration of the wheel and this linear acceleration are related by $a = R\alpha$. This fact, together with (1) and (2), gives

$$a = R\alpha = \frac{TR^2}{I} = \frac{mg - T}{m}$$

$$T = \frac{mg}{1 + \dfrac{mR^2}{I}}$$

Likewise, solving for a and α gives

$$a = \frac{g}{1 + I/mR^2}$$

$$\alpha = \frac{a}{R} = \frac{g}{R + I/mR}$$

Exercise The wheel in Figure 11.22 is a solid disk of $M = 2.0$ kg, $R = 30$ cm, and $I = 0.09 \text{ kg}\cdot\text{m}^2$. The suspended object has a mass of $m = 0.5$ kg. Find the tension in the cord and the angular acceleration of the wheel.

Answer 3.27 N; 10.9 rad/s^2.

Example 11.12 A Rotating Rod Revisited

A uniform rod of length L and mass M is free to rotate on a frictionless pin through one end (Fig. 11.23). The rod is released from rest in the horizontal position.

(a) What is the angular velocity of the rod when it is at its lowest position?

Solution The answer can be obtained easily by considering the mechanical energy of the system. When the rod is in the horizontal position, it has no kinetic energy. Its potential energy relative to the lowest position of its center of mass (O') is $MgL/2$. When it reaches its lowest position, the energy is entirely kinetic energy, $\frac{1}{2}I\omega^2$, where I is the moment of inertia about the pivot. Since $I = \frac{1}{3}ML^2$ (Table 11.2) and since mechanical energy is constant, we have

$$\tfrac{1}{2}MgL = \tfrac{1}{2}I\omega^2 = \tfrac{1}{2}(\tfrac{1}{3}ML^2)\omega^2$$

$$\omega = \sqrt{\frac{3g}{L}}$$

For example, if the rod is a meter stick, we find that $\omega = 5.42$ rad/s.

(b) Determine the linear speed of the center of mass and the linear speed of the lowest point on the rod in the vertical position.

$$v_c = r\omega = \frac{L}{2}\omega = \tfrac{1}{2}\sqrt{3gL}$$

The lowest point on the rod has a linear speed equal to $2v_c = \sqrt{3gL}$.

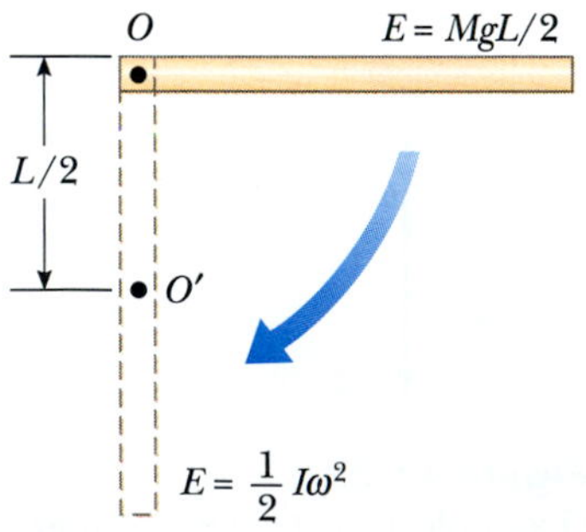

Figure 11.23 (Example 11.12) A uniform rigid rod pivoted at O rotates in a vertical plane under the action of gravity.

Work and Energy in Rotational Motion

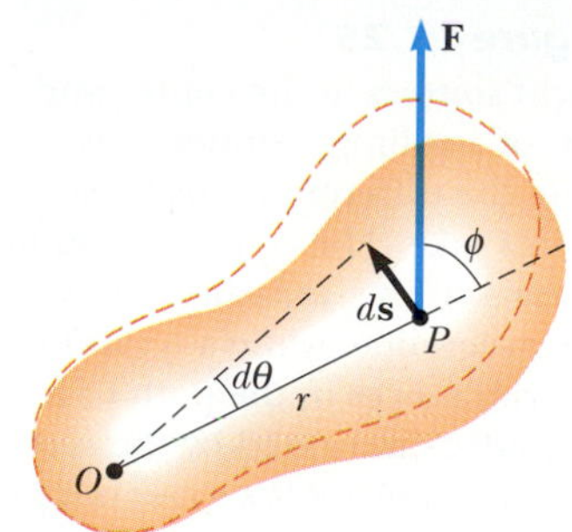

Figure 11.24
A rigid body rotates about an axis through O under the action of an external force **F** applied at P.

Consider a rigid body pivoted at the point O in Figure 11.24. Suppose a single external force, **F**, is applied at the point P and d**s** is the displacement of the point of application of the force. The work done by **F** as the body rotates through an infinitesimal distance $ds = r\,d\theta$ in a time dt is

$$dW = \mathbf{F} \cdot d\mathbf{s} = (F \sin \phi)\, r\, d\theta$$

where $F \sin \phi$ is the tangential component of **F**, or the component of the force along the displacement. Note from Figure 11.24 that *the radial component of* **F** *does no work because it is perpendicular to the displacement.*

Since the magnitude of the torque due to **F** about the origin was defined as $rF \sin \phi$, we can write the work done for the infinitesimal rotation in the form

$$dW = \tau\, d\theta \qquad [11.32]$$

The rate at which work is being done by **F** for rotation about the fixed axis is obtained by formally dividing the left and right sides of Equation 11.32 by dt:

$$\frac{dW}{dt} = \tau \frac{d\theta}{dt} \qquad [11.33]$$

But the quantity dW/dt is, by definition, the instantaneous power, P, delivered by the force. Furthermore, since $d\theta/dt = \omega$, Equation 11.33 reduces to

$$P = \frac{dW}{dt} = \tau\omega \qquad [11.34]$$

Power delivered to a rigid body

This expression is analogous to $P = Fv$ in the case of linear motion, and the expression $dW = \tau\, d\theta$ is analogous to $dW = F_x\, dx$.

In linear motion, we found the energy concept, and in particular the work-energy theorem, to be extremely useful in describing the motion of a system. The energy concept can be equally useful in simplifying the analysis of rotational motion. From what we learned of linear motion, we expect that for rotation of a symmetric object (such as a symmetric wheel) about a fixed axis, the work done by external forces will equal the change in the rotational kinetic energy. To show that this is in fact the case, let us begin with $\tau = I\alpha$. Using the chain rule from the calculus, we can express the torque as

$$\tau = I\alpha = I\frac{d\omega}{dt} = I\frac{d\omega}{d\theta}\frac{d\theta}{dt} = I\frac{d\omega}{d\theta}\,\omega$$

Rearranging this expression and noting that $\tau\, d\theta = dW$, we get

$$\tau\, d\theta = dW = I\omega\, d\omega$$

Integrating this expression, we get the total work done:

$$W = \int_{\theta_0}^{\theta} \tau\, d\theta = \int_{\omega_0}^{\omega} I\omega\, d\omega = \tfrac{1}{2}I\omega^2 - \tfrac{1}{2}I\omega_0^2 \qquad [11.35]$$

Work-energy theorem for rotational motion

Figure 11.25
Light sources at the center and rim of a rolling cylinder illustrate the different paths these points take. The center moves in a straight line, as indicated by the green line, while a point on the rim moves in the path of a cycloid, as indicated by the red curve. *(Courtesy of Henry Leap and Jim Lehman)*

Rolling Motion of a Rigid Body

Suppose a cylinder is rolling on a straight path as in Figure 11.25. The center of mass moves in a straight line, while a point on the rim moves in a more complex path corresponding to the path of a cycloid. Let us further assume that the cylinder of radius R is uniform and rolls on a surface with friction. As the cylinder rotates through an angle of θ, its center of mass moves a distance of $s = r\theta$. Therefore, the speed and acceleration of the center of mass for *pure rolling motion* are

$$v_c = \frac{ds}{dt} = R\frac{d\theta}{dt} = R\omega \qquad [11.36]$$

$$a_c = \frac{dv_c}{dt} = R\frac{d\omega}{dt} = R\alpha \qquad [11.37]$$

The linear velocities of various points on the rolling cylinder are illustrated in Figure 11.26. Note that the linear velocity of any point is in a direction perpendicular to the line from that point to the contact point. At any instant, the point P is at rest relative to the surface, since sliding does not occur.

A general point on the cylinder, such as Q, has both horizontal and vertical components of velocity. However, the points P and P' and the point at the center of mass are unique and of special interest. Relative to the surface on which the cylinder is moving, the center of mass moves with a speed of $v_c = R\omega$, whereas the contact point P has zero velocity. The point P' has a speed equal to $2v_c = 2R\omega$, since all points on the cylinder have the same angular speed.

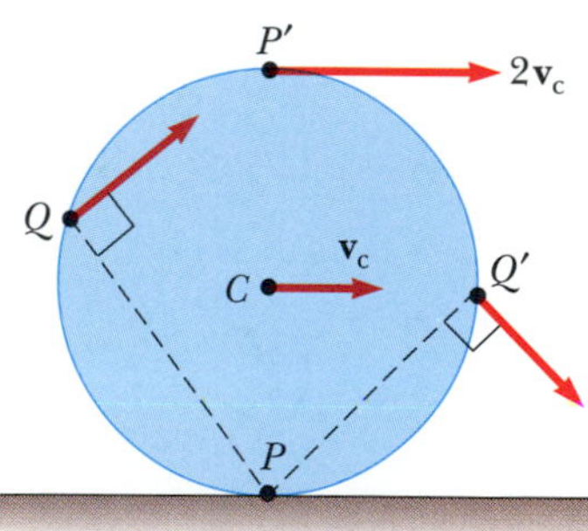

Figure 11.26
All points on a rolling body move in a direction perpendicular to an axis through the contact point P. The center of the body moves with a velocity $\mathbf{v}_c$, while the point P' moves with the velocity $2\mathbf{v}_c$.

We can express the total kinetic energy of the rolling cylinder as

$$K = \tfrac{1}{2} I_P \omega^2 \qquad [11.38]$$

where I_p is the moment of inertia about the axis through P.

A useful theorem called the **parallel axis theorem** enables us to calculate the moment of inertia, I_p, through any axis parallel to the axis through the center of mass of a body. This theorem states that

$$I_p = I_c + MR^2 \qquad [11.39]$$

where R is the distance from the center-of-mass axis to the parallel axis, and M is the total mass of the body. Substitution of this expression into Equation 11.38 gives

$$K = \tfrac{1}{2} I_c \omega^2 + \tfrac{1}{2} M R^2 \omega^2$$

$$K = \tfrac{1}{2} I_c \omega^2 + \tfrac{1}{2} M v_c^2 \qquad [11.40]$$

Total kinetic energy of a rolling body

where we have used the fact that $v_c = R\omega$.

We can think of Equation 11.40 as follows: The first term on the right, $\tfrac{1}{2} I_c \omega^2$, represents the rotational kinetic energy about the center of mass, and the term $\tfrac{1}{2} M v_c^2$ represents the kinetic energy the cylinder would have if it were just translating through space without rotating. Thus, we can say that

> the total kinetic energy of an object undergoing rolling motion is the sum of a rotational kinetic energy about the center of mass and the translational kinetic energy of the center of mass.

We can use energy methods to treat a class of problems concerning the rolling motion of a rigid body down a rough incline. We shall assume that the rigid body in Figure 11.27 does not slip and is released from rest at the top of the incline. Note that rolling motion is possible only if a frictional force is present between the object and the incline to produce a net torque about the center of mass. Despite the presence of friction, no loss of mechanical energy takes place because the contact point is at rest relative to the surface at any instant. On the other hand, if the rigid body were to slide, mechanical energy would be lost as motion progressed.

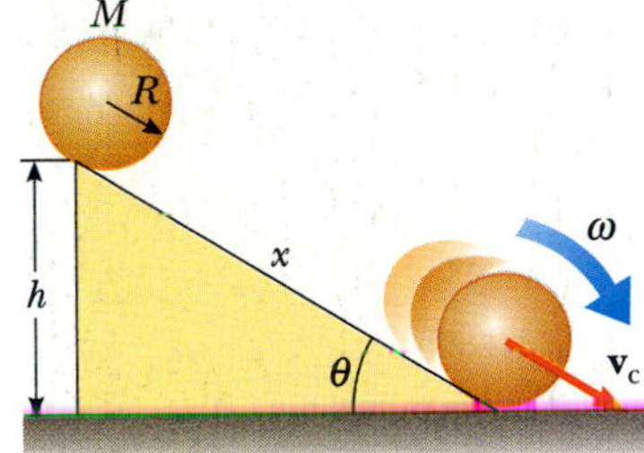

Figure 11.27 A round object rolling down an incline. Mechanical energy is constant if no slipping occurs and there is no rolling friction.

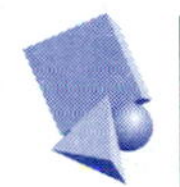

Example 11.13 A Race Between a Cylinder and a Sphere

A solid cylinder and a solid sphere, each of mass M and radius R, are released from rest at the top of an incline of height h and roll down the incline without sliding (Fig. 11.27). Determine the speeds of their centers of mass when they reach the bottom of the incline.

Solution Because the objects undergo pure rolling motion on the incline, we can apply the principle of conservation of energy by equating the initial potential energy, Mgh, to the final kinetic energy given by Equation 11.40:

$$Mgh = \tfrac{1}{2} I_c \omega^2 + \tfrac{1}{2} M v_c^2$$

Using the fact that $v_c = R\omega$ for pure rolling motion, we can write this as

$$Mgh = \tfrac{1}{2}\left(\frac{I_c}{R^2} + M\right) v_c^2$$

Solving for v_c gives

$$v_c = \left(\frac{2gh}{1 + I_c/MR^2}\right)^{1/2}$$

From Table 11.2, we find that for a solid cylinder $I_c = \tfrac{1}{2} MR^2$, and for a solid sphere $I_c = \tfrac{2}{5} MR^2$. Substitution of these values for I_c into the preceding expression gives

$$v_{\text{cylinder}} = \sqrt{\frac{4}{3} gh} \quad \text{and} \quad v_{\text{sphere}} = \sqrt{\frac{10}{7} gh}$$

The results show that the speed of the cylinder is less than that of the sphere, so the sphere wins the race. This is due to the fact that, compared to the sphere, the cylinder has a higher moment of inertia and therefore has more of its energy in the form of rotational kinetic energy and less in translational kinetic energy. The results are also quite interesting in that the speeds are independent of the masses and radii! The constant factors that appear in the expressions for the speed depend on the moment of inertia about the center of mass for the specific body.

Exercise If the object that is rolled down the rough incline is a hollow cylinder, what is the speed of its center of mass when it reaches the bottom?

Answer $\sqrt{gh}$.

Summary

The **instantaneous angular speed** of a particle rotating in a circle or of a rigid body rotating about a fixed axis is

Instantaneous angular velocity

$$\omega = \frac{d\theta}{dt} \quad [11.3]$$

where ω is in rad/s or in s^{-1}.

The **instantaneous angular acceleration** of a rotating body is

Instantaneous angular acceleration

$$\alpha = \frac{d\omega}{dt} \quad [11.5]$$

and has units of rad/s^2 or s^{-2}.

When a rigid body rotates about a fixed axis, every part of the body has the same angular velocity and the same angular acceleration. However, different parts of the body, in general, have different linear velocities and different linear accelerations.

If a particle (or body) undergoes rotational motion about a fixed axis under constant angular acceleration, α, one can apply equations of kinematics by analogy with kinematic equations for linear motion with constant linear acceleration:

Rotational kinematic equations

$$\omega = \omega_0 + \alpha t \quad [11.6]$$

$$\theta = \theta_0 + \omega_0 t + \tfrac{1}{2}\alpha t^2 \quad [11.7]$$

$$\omega^2 = \omega_0{}^2 + 2\alpha(\theta - \theta_0) \quad [11.8]$$

$$\theta = \theta_0 + \tfrac{1}{2}(\omega_0 + \omega)t \quad [11.9]$$

When a particle rotates about a fixed axis, the angular speed and angular acceleration are related to the linear speed and tangential linear acceleration through the relationships

Relationship between linear and angular speed

$$v = r\omega \quad [11.10]$$

Relationship between linear and angular acceleration

$$a_t = r\alpha \quad [11.11]$$

The **moment of inertia of a system of particles** is

Moment of inertia for a system of particles

$$I = \sum m_i r_i^2 \quad [11.14]$$

If a rigid body rotates about a fixed axis with angular speed ω, its **kinetic energy** can be written

$$K = \frac{1}{2} I\omega^2 \qquad [11.15]$$

Kinetic energy of a rotating rigid body

where I is the moment of inertia about the axis of rotation.

The **torque,** $\boldsymbol{\tau}$, due to a force **F** about an origin in an inertial frame is defined to be

$$\boldsymbol{\tau} \equiv \mathbf{r} \times \mathbf{F} \qquad [11.17]$$

Torque

Given two vectors **A** and **B**, their **cross product,** **A** × **B**, is a vector **C** h the magnitude

$$C \equiv |AB \sin\theta| \qquad [11.19]$$

Magnitude of the cross product

where θ is the angle between **A** and **B**. The direction of **C** is perpendicular to the plane formed by **A** and **B**, and its sense is determined by the right-hand rule. Some properties of the cross product include the facts that $\mathbf{A} \times \mathbf{B} = -\mathbf{B} \times \mathbf{A}$ and $\mathbf{A} \times \mathbf{A} = 0$.

The net torque acting on a particle is proportional to the angular acceleration of the particle, and the proportionality constant is the moment of inertia, I:

$$\sum \tau = I\alpha \qquad [11.22]$$

The **angular momentum,** **L**, of a particle of linear momentum $\mathbf{p} = m\mathbf{v}$ is

$$\mathbf{L} \equiv \mathbf{r} \times \mathbf{p} \qquad [11.23]$$

Angular momentum of a particle

where **r** is the vector position of the particle relative to an origin in an inertial frame. If ϕ is the angle between **r** and **p**, the magnitude of **L** is

$$L = mvr \sin\phi = I\omega \qquad [11.24]$$

The net **torque** acting on a particle is equal to the time rate of change of its angular momentum:

$$\sum \boldsymbol{\tau} = \frac{d\mathbf{L}}{dt} \qquad [11.27]$$

Relation between torque and angular momentum

The law of conservation of angular momentum states that the total angular momentum of a system remains constant if the net external torque acting on the system is zero:

$$\sum \boldsymbol{\tau}_{\text{ext}} = \frac{d\mathbf{L}}{dt} = 0 \qquad [11.29]$$

$$\mathbf{L} = \text{constant} \qquad [11.30]$$

Conservation of angular momentum

Angular momentum is an inherent property of atoms, molecules, and their constituents. Such systems have discrete (quantized) values of angular momenta that are integer multiples of $h/2\pi$, where h is Planck's constant.

The **total kinetic energy** of a rigid body, such as a cylinder, that is rolling on a rough surface without slipping equals the rotational kinetic energy about the body's center of mass, $\frac{1}{2} I_c\omega^2$, plus the translational kinetic energy of the center of mass, $\frac{1}{2} Mv_c^2$:

Total kinetic energy of a rolling body

$$K = \frac{1}{2}I_c\omega^2 + \frac{1}{2}Mv_c^2 \quad [11.40]$$

In this expression, v_c is the speed of the center of mass and $v_c = R\omega$ for pure rolling motion.

Questions and Conceptual Exercises

1. Are the kinematic expressions for θ, ω, and α valid when the angular displacement is measured in degrees instead of radians?
2. If a car's wheels are replaced with wheels of a larger diameter, will the reading of the speedometer change? Explain.
3. When a wheel of radius R rotates about a fixed axis, do all points on the wheel have the same angular velocity? Do they all have the same linear velocity? If the angular velocity is constant and equal to $\boldsymbol{\omega}_0$, describe the linear speeds and linear accelerations of the points at $r = 0$, $r = R/2$, and $r = R$.
4. What is the magnitude of the angular velocity, $\boldsymbol{\omega}$, of the second hand of a clock? What is the angular acceleration, α, of the second hand?
5. If you see an object rotating, is there necessarily a net torque acting on it?
6. Explain why changing the axis of rotation of an object changes the object's moment of inertia.
7. It is more difficult to do a sit-up with your hands behind your head than with your arms stretched out in front of you. Why?
8. In order for a helicopter to be stable as it flies, it must have two propellers. Why?
9. Suppose you remove two eggs from the refrigerator, one hard-boiled and the other uncooked. You wish to determine which is the hard-boiled egg without breaking the eggs. This can be done by spinning the two eggs on the floor and comparing the rotational motions. Which egg spins faster? Which rotates more uniformly? Explain.
10. Suppose the planets of the Solar System evolved from the condensation of a primordial gas. From the observation that all the planets revolve around the Sun in the same direction, what would you infer about the state of the gas prior to its condensation?
11. A student sits on a stool that is free to rotate about its vertical axis. The student and stool are set into rotation while the student, with outstretched arms, holds a pair of weights. If she suddenly drops the weights to the floor, what happens to her angular velocity? Explain.
12. A cat usually lands on its feet regardless of the position it is in when dropped. A slow-motion film of a cat falling shows that the upper half of its body twists in one direction while the lower half twists in the opposite direction. Why does this type of rotation occur?

(Question 12) A falling, twisting cat. *(Photo Researchers, Inc.)*

13. A ladder rests inclined against a wall. Would you feel safer climbing up the ladder if you were told that the floor is frictionless but the wall is rough, or that the wall is frictionless but the floor is rough? Justify your answer.
14. Stars originate as large bodies of slowly rotating gas. Because of gravity, these regions of gas slowly decrease in size. What happens to the angular speed of a star as it shrinks? Explain.
15. Often when a high diver wants to flip in midair, she draws her legs up against her chest. Why does this make her rotate faster? What should she do when she wants to come out of her flip?
16. As a tetherball winds around a pole, what happens to its angular speed? Explain.
17. Space colonies have been proposed that would consist of large cylinders placed in space. Engineers would simulate gravity in these cylinders by setting them into rotation about their long axes. Discuss the difficulties that would be encountered in attempting to set the cylinders into rotation.
18. If the net force acting on a system is zero, then is it necessarily true that the net torque on it is also zero?

19. Why does a tightrope walker carry a long pole to help stay in balance while walking a tightrope?
20. If global warming occurs over the next century, it is likely that the polar ice caps of the Earth will melt and the water will be distributed closer to the equator. How would this change the moment of inertia of the Earth? Would the length of the day (one revolution) increase or decrease?
21. Vector **A** is in the negative y direction, and vector **B** is in the negative x direction. What are the directions of (a) $\mathbf{A} \times \mathbf{B}$ and (b) $\mathbf{B} \times \mathbf{A}$?

Problems

Section 11.2 Rotational Kinematics

1. A wheel starts from rest and rotates, with constant angular acceleration, to an angular speed of 12 rad/s in a time of 3 s. Find (a) the angular acceleration of the wheel and (b) the angle, in radians, through which it rotates in this time.
2. The turntable of a record player rotates at the rate of $33\frac{1}{3}$ rev/min and takes 60 s to come to rest when switched off. Calculate (a) its angular acceleration and (b) the number of revolutions it makes before coming to rest.
3. What is the angular speed, in radians per second, of (a) the Earth in its orbit about the Sun? (b) the Moon in its orbit about the Earth?
4. A wheel rotates in such a way that its angular displacement in a time t is given by $\theta = at^2 + bt^3$, where a and b are constants. Determine equations for (a) the angular speed and (b) the angular acceleration, both as functions of time.
5. An electric motor, rotating a workshop grinding wheel at a rate of 100 rev/min, is switched off. Assume constant negative acceleration of magnitude 2 rad/s². (a) How long will it take for the grinding wheel to stop? (b) Through how many radians has the wheel turned during the time found in (a)?
6. The angular position of a point on a wheel can be described by $\theta = 5 + 10t + 2t^2$ rad. Determine the angular position, speed, and acceleration of the point at $t = 0$ and $t = 3$ s.
7. A car accelerates uniformly from rest and reaches a speed of 22 m/s in 9 s. The diameter of a tire is 58 cm. (a) Find the number of revolutions that a tire makes during this motion, assuming no slipping occurs. (b) What is the final rotational speed of a tire, in revolutions per second?
8. A dentist's drill starts from rest. After 3.2 s of constant angular acceleration, it turns at a rate of 2.51×10^4 rev/min. (a) Find the drill's angular acceleration. (b) Determine the angle (in radians) through which the drill rotates during this period.
9. The tub of a washer goes into its spin cycle, starting from rest and reaching an angular speed of 5.0 rev/s in 8.0 s. At this point the person doing the laundry opens the lid, and a safety switch turns off the washer. The tub slows to rest in 12.0 s. Through how many revolutions does the tub turn? Assume constant angular acceleration while the machine is starting and stopping.
10. A grinding wheel, initially at rest, is rotated for 8 s with constant angular acceleration $\alpha = 5$ rad/s². The wheel is then brought to rest, with uniform negative acceleration, in 10 revolutions. Determine the negative acceleration required and the time needed to bring the wheel to rest.
11. A centrifuge in a medical laboratory rotates at an angular speed of 3600 rev/min. When switched off, it rotates 50 times before coming to rest. Find the constant angular acceleration of the centrifuge.
12. An airliner arrives at the terminal, and the engines are shut off. The rotor of one of the engines has an initial clockwise angular speed of 2000 rad/s. The engine's rotation slows with an angular acceleration of magnitude 80.0 rad/s². (a) Determine the angular speed after 10.0 s. (b) How long does it take the rotor to come to rest?

Section 11.3 Relations Between Angular and Linear Quantities

13. A race car travels on a circular track of radius 250 m. If the car moves with a constant speed of 45 m/s, find (a) the angular speed of the car and (b) the magnitude and direction of the car's acceleration.
14. An automobile accelerates from zero to 30 m/s in 6 s. The wheels have diameters of 0.4 m. What is the angular acceleration of each wheel?
15. A discus thrower accelerates a discus from rest to a speed of 25 m/s by whirling it through 1.25 revolutions. Assume the discus moves on the arc of a circle 1 m in radius. (a) Calculate the final angular speed

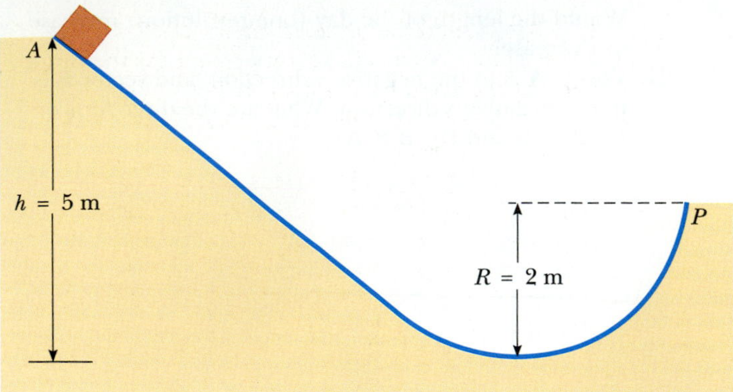

Figure 11.28 (Problem 18)

of the discus. (b) Determine the angular acceleration of the discus, assuming it to be constant. (c) Calculate the acceleration time. Neglect the spin of the discus about its center of mass.

16. A disk 8 cm in radius rotates at a constant rate of 1200 rev/min about its axis. Determine (a) the angular speed of the disk, (b) the linear speed at a point 3 cm from the disk's center, (c) the radial acceleration of a point on the rim, and (d) the distance a point on the rim moves in 2 s.

17. A car is traveling at 36 km/h on a straight road. The radius of the tires is 25 cm. Find the angular speed of one of the tires, with its axle taken as the axis of rotation.

18. A 6-kg block is released from A on a frictionless track, as shown in Figure 11.28. Determine the radial and tangential components of acceleration for the block at P.

Section 11.4 Rotational Kinetic Energy

19. The four particles in Figure 11.29 are connected by rigid rods of negligible mass. The origin is at the center of the rectangle. If the system rotates in the xy plane about the z axis with an angular speed of 6 rad/s, calculate (a) the moment of inertia of the system about the z axis and (b) the kinetic energy of the system.

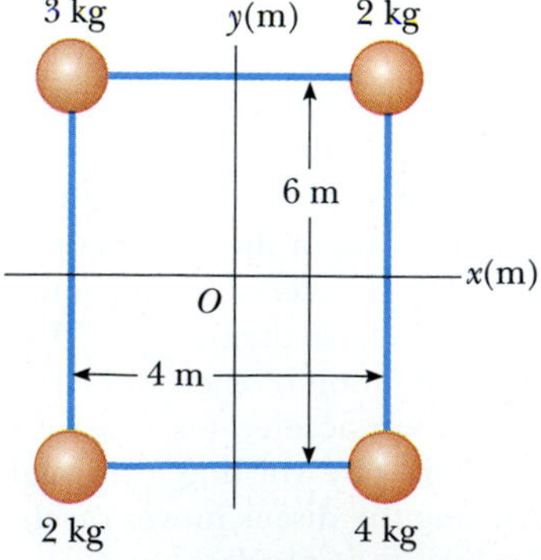

Figure 11.29 (Problem 19)

20. Three particles are connected by rigid rods, of negligible mass, lying along the y axis (Fig. 11.30). The system rotates about the x axis with an angular speed of 2 rad/s. (a) Find the moment of inertia about the x axis and the total kinetic energy evaluated from $\frac{1}{2}I\omega^2$. (b) Find the linear speed of each particle and the total kinetic energy evaluated from $\Sigma\frac{1}{2}m_iv_i^2$.

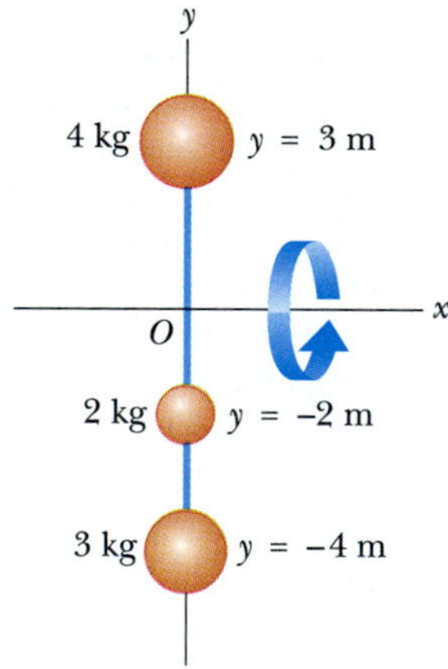

Figure 11.30 (Problem 20)

Section 11.5 Torque and the Vector Product

21. Find the net torque on the wheel in Figure 11.31, about the axle through O, if a = 10 cm and b = 25 cm.

22. Calculate the net torque (magnitude and direction) on the beam shown in Figure 11.32, about (a) an axis through O, perpendicular to the figure, and (b) an axis through C, perpendicular to the figure.

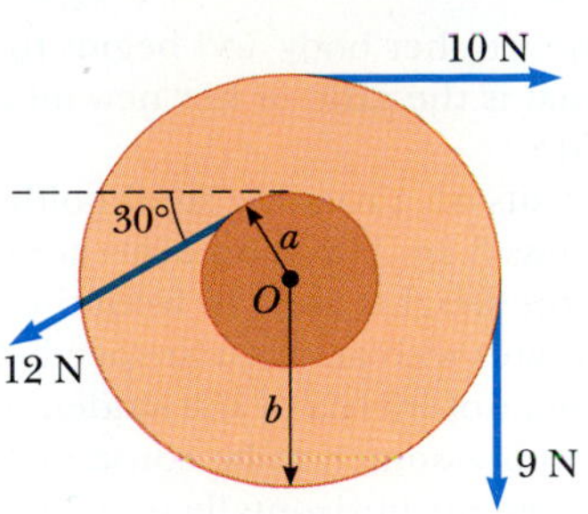

Figure 11.31 (Problem 21)

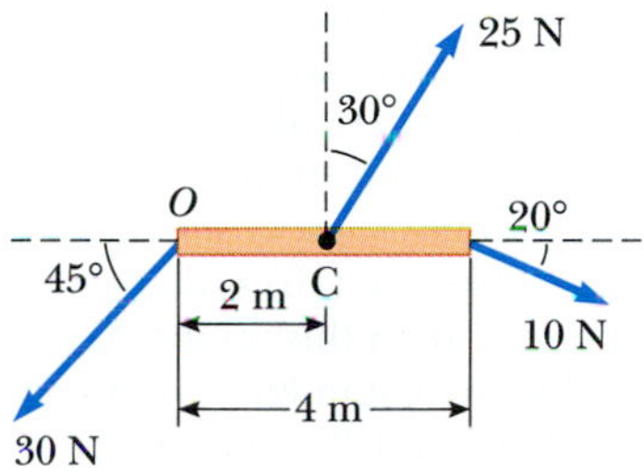

Figure 11.32 (Problem 22)

23. The fishing pole shown in Figure 11.33 is inclined to the horizontal at an angle of 20°. What is the torque exerted by the fish about an axis perpendicular to the page and passing through the hand of the person holding the pole?
24. A 48-kg diver stands at the end of a 3-m-long diving board. What torque does the weight of the diver produce about an axis that is perpendicular to the long dimension of the diving board and passing through its midpoint?
25. Two vectors are given by $\mathbf{A} = -3\mathbf{i} + 4\mathbf{j}$ and $\mathbf{B} = 2\mathbf{i} + 3\mathbf{j}$. Find (a) $\mathbf{A} \times \mathbf{B}$ and (b) the angle between **A** and **B**.
26. Given $\mathbf{M} = 6\mathbf{i} + 2\mathbf{j} - \mathbf{k}$ and $\mathbf{N} = 2\mathbf{i} - \mathbf{j} - 3\mathbf{k}$, calculate the vector product $\mathbf{M} \times \mathbf{N}$.
27. A particle is located at the vector position $\mathbf{r} = (\mathbf{i} + 3\mathbf{j})$ m, and the force acting on it is $\mathbf{F} = (3\mathbf{i} + 2\mathbf{j})$ N. What is the torque about (a) the origin and (b) the point having coordinates (0, 6) m?
28. If $|\mathbf{A} \times \mathbf{B}| = \mathbf{A} \cdot \mathbf{B}$, what is the angle between **A** and **B**?
29. A force of $\mathbf{F} = 2\mathbf{i} + 3\mathbf{j}$ (in newtons) is applied to an object that is pivoted about a fixed axis aligned along the z coordinate axis. If the force is applied at the point $\mathbf{r} = 4\mathbf{i} + 5\mathbf{j} + 0\mathbf{k}$ (in meters), find (a) the magnitude of the net torque about the z axis and (b) the direction of the torque vector, $\boldsymbol{\tau}$.

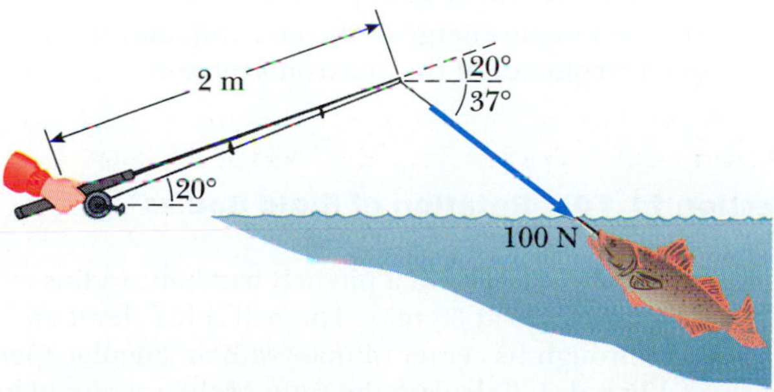

Figure 11.33 (Problem 23)

Section 11.6 Relation Between Torque and Angular Acceleration

30. The combination of an applied force and a frictional force produces a constant total torque of 36 N·m on a wheel rotating about a fixed axis. The applied force acts for 6 s, during which time the angular speed of the wheel increases from 0 to 10 rad/s. The applied force is then removed, and the wheel comes to rest in 60 s. Find (a) the moment of inertia of the wheel, (b) the magnitude of the frictional torque, and (c) the total number of revolutions of the wheel.
31. A model airplane whose mass is 0.75 kg is tethered by a wire so that it flies in a circle 30 m in radius. The airplane engine provides a thrust of 0.80 N perpendicular to the tethering wire. (a) Find the torque that the engine thrust produces about the center of the circle. (b) Find the angular acceleration of the airplane when it is in level flight. (c) Find the linear acceleration of the airplane tangent to its flight path. Neglect air drag.

Section 11.7 Angular Momentum

32. A 1.5-kg particle moves in the xy plane with a velocity of $\mathbf{v} = (4.2\mathbf{i} - 3.6\mathbf{j})$ m/s. Determine the particle's angular momentum when its position vector is $\mathbf{r} = (1.5\mathbf{i} + 2.2\mathbf{j})$ m.
33. The position vector of a 2-kg particle is given as a function of time by $\mathbf{r} = (6\mathbf{i} + 5t\mathbf{j})$ m. Determine the angular momentum of the particle as a function of time.
34. Two particles move in *opposite* directions along a straight line (Fig. 11.34). The particle of mass m moves to the right with a speed of v while the particle of mass $3m$ moves to the left with a speed of v. What is the *total* angular momentum of the system relative to (a) point A, (b) point O, and (c) point B?
35. A light, rigid rod 1 m in length rotates in the xy plane about a pivot through the rod's center. Two particles with masses of 4 kg and 3 kg are connected to its

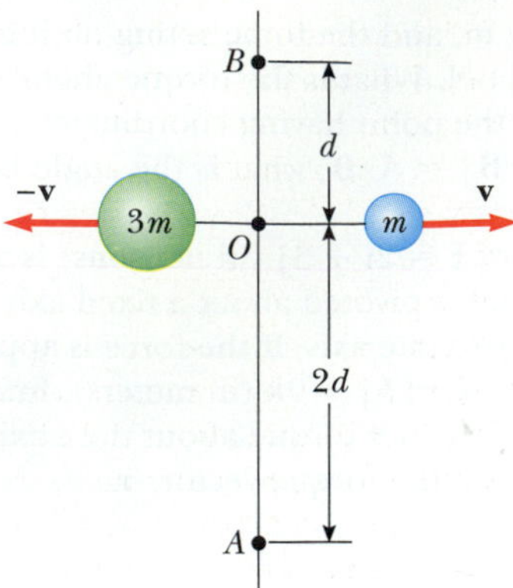

Figure 11.34 (Problem 34)

ends (Fig. 11.35). Determine the angular momentum of the system about the origin at the instant the speed of each particle is 5 m/s.

36. An airplane of mass 12 000 kg flies parallel to the ground at an altitude of 10 km with a constant speed of 175 m/s relative to the Earth. (a) What is the magnitude of the airplane's angular momentum relative to a ground observer who is directly below the airplane? (b) Does this value change as the airplane continues its motion along a straight line?

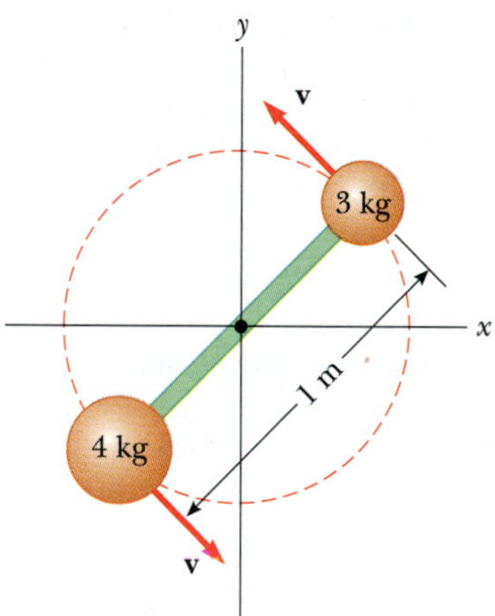

Figure 11.35 (Problem 35)

Section 11.8 Conservation of Angular Momentum
*Section 11.9 Quantization of Angular Momentum

37. A figure skater rotates about a vertical axis through her center with both arms and one leg extended. Her rotation rate is 0.5 rev/s. She then pulls her arms and leg in closer to her body and begins to spin at 1.7 rev/s. What is the ratio of her new moment of inertia to her old?

38. A student sits on a rotating stool holding two weights, each of mass 3 kg. When his arms are extended horizontally, the weights are 1 m from the axis of rotation and he rotates with an angular speed of 0.75 rad/s. The moment of inertia of the student plus stool is 3 $kg \cdot m^2$ and is assumed to be constant. The student pulls the weights horizontally to 0.3 m from the rotation axis. (a) Find the new angular speed of the student. (b) Find the kinetic energy of the student before and after the weights are pulled in.

39. A merry-go-round rotates at the rate of 0.2 rev/s with an 80-kg man standing 2 m from the axis of rotation. (a) What is the new angular velocity when the man walks to a point 1 m from the center? Assume the merry-go-round is a solid cylinder having a moment of inertia of 50 $kg \cdot m^2$. (b) Calculate the change in kinetic energy due to this moment. How do you account for this change in kinetic energy?

40. The ball in Figure 11.20 has a mass of 0.12 kg. The distance of the ball from the center of rotation is originally 40.0 cm, and the ball is moving with a speed of 80.0 cm/s. The string is pulled downward 15.0 cm through the hole in the frictionless table. Determine the work done on the ball. (*Hint:* Consider the change of kinetic energy.)

41. A woman whose mass is 60 kg stands at the rim of a horizontal turntable having a moment of inertia of 500 $kg \cdot m^2$ and a radius of 2 m. The system is initially at rest, and the turntable is free to rotate about a frictionless, vertical axle through its center. The woman then starts walking around the rim in a clockwise direction (looking downward), at a constant speed of 1.5 m/s relative to the Earth. (a) In what direction and with what angular speed does the turntable rotate? (b) How much work does the woman do to set the system into motion?

42. In the Bohr model of the hydrogen atom, the electron moves in a circular orbit of radius 0.529×10^{-10} m around the proton. Assuming the orbital angular momentum of the electron is equal to $h/2\pi$, calculate (a) the orbital speed of the electron, (b) the kinetic energy of the electron, and (c) the angular frequency of the electron's motion.

*Section 11.10 Rotation of Rigid Bodies

43. The center of mass of a pitched baseball (radius = 3.8 cm) moves at 38 m/s. The ball spins about an axis through its center of mass with an angular speed of 125 rad/s. Calculate the ratio of the rotational ki-

netic energy to the translational kinetic energy. Treat the ball as a uniform solid sphere.

44. An automobile tire, considered as a solid disk, has a radius of 25 cm and a mass of 6 kg. Find its rotational kinetic energy when it is rotating about an axis through its center at an angular speed of 2 rev/s.

45. A horizontal 800-N merry-go-round of radius 1.5 m is started from rest by a constant horizontal force of 50 N applied tangentially to the cylinder. Find the kinetic energy of the solid cylinder after 3 s.

46. A cylinder of mass 10 kg rolls without slipping. At the instant its center of mass has a speed of 10 m/s, determine (a) the translational kinetic energy of its center of mass, (b) the rotational kinetic energy about its center of mass; and (c) its total kinetic energy.

47. The top in Figure 11.36 has a moment of inertia of 4.0×10^{-4} kg·m² and is initially at rest. It is free to rotate about the stationary axis AA'. A string, wrapped around a peg along the axis of the top, is pulled in such a manner as to maintain a constant tension of 5.57 N. If the string does not slip while it is unwound from the peg, what is the angular speed of the top after 80 cm of string has been pulled off of the peg? (*Hint:* Consider the work done.)

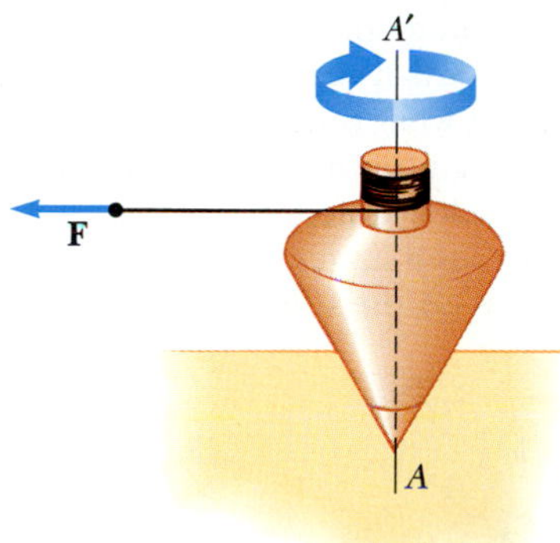

Figure 11.36 (Problem 47)

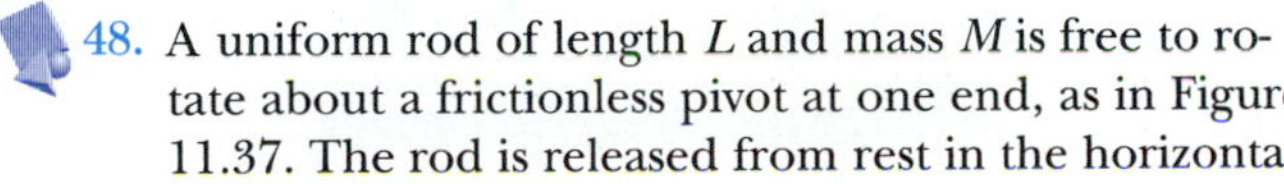

48. A uniform rod of length L and mass M is free to rotate about a frictionless pivot at one end, as in Figure 11.37. The rod is released from rest in the horizontal position. What are the *initial* angular acceleration of the rod and the *initial* linear acceleration of the right end of the rod?

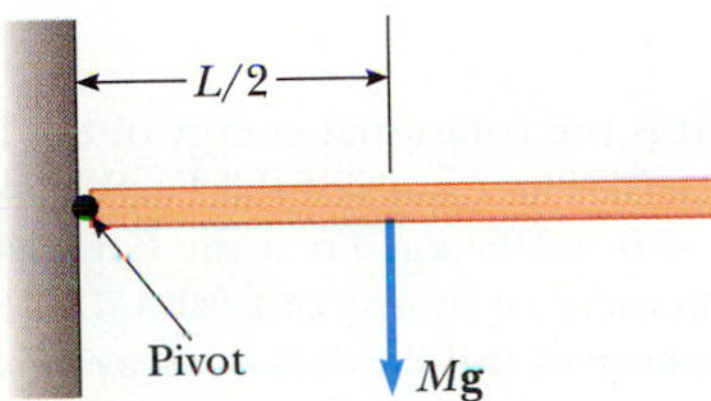

Figure 11.37 (Problem 48)

49. A uniform solid sphere rolls along a horizontal surface without slipping. What fraction of the sphere's total kinetic energy is in the form of rotational kinetic energy about its center of mass?

50. Two masses, m_1 and m_2, are connected to each other by a light cord that passes over two identical pulleys, each having a moment of inertia of I (Fig. 11.38). Find the acceleration of each mass and the tensions T_1, T_2, and T_3 in the cord. (Assume that no slipping occurs between the pulleys and cord.)

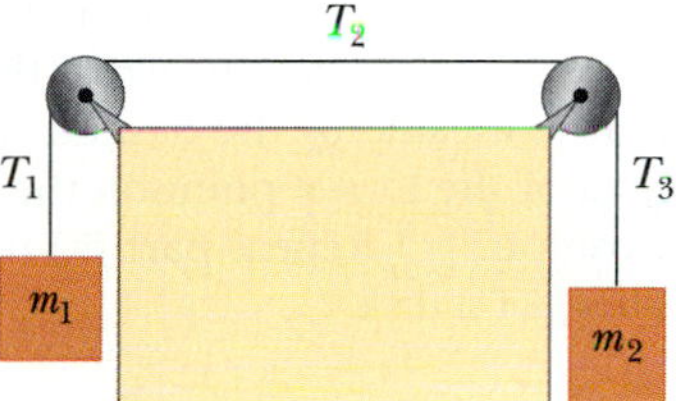

Figure 11.38 (Problem 50)

51. Consider two masses connected by a string that passes over a pulley having a moment of inertia of I about its axis of rotation, as in Figure 11.39. The string does not slip on the pulley, and the system is released from rest. Use the principle of conservation of energy to find the linear speeds of the masses after m_2 descends through a distance h, and the angular speed of the pulley at this time.

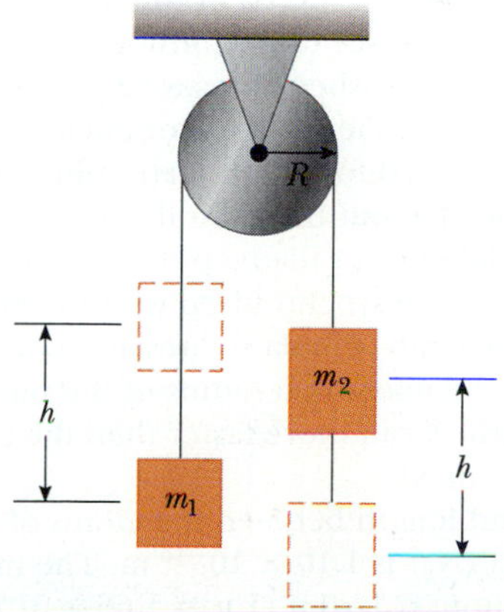

Figure 11.39 (Problem 51)

52. (a) A uniform solid disk of radius R and mass M is free to rotate on a frictionless pivot through a point

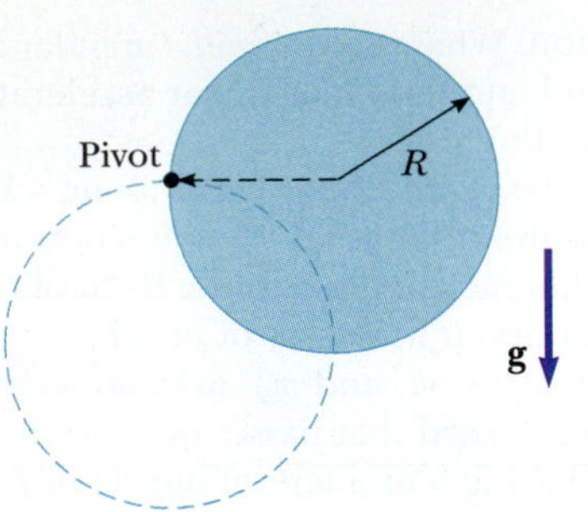

Figure 11.40 (Problem 52)

on its rim (Fig. 11.40). If the disk is released from rest in the position shown by the green circle, what is the speed of its center of mass when the disk reaches the position indicated by the dashed circle? (b) What is the speed of the lowest point on the disk in the dashed position? (c) Repeat part (a) using a uniform hoop instead of a disk.

Additional Problems

53. A grinding wheel is in the form of a uniform solid disk of radius 7 cm and mass 2 kg. It starts from rest and accelerates uniformly under the action of the constant torque of 0.6 N·m that the motor exerts on the wheel. (a) How long does the wheel take to reach its final operating speed of 1200 rpm? (b) Through how many revolutions does the grindstone turn while accelerating?
54. The net work done in accelerating a propeller from rest to an angular speed of 200 rad/s is 3000 J. What is the moment of inertia of the propeller?
55. A horizontal force of magnitude 6.5 N is exerted tangentially on a frisbee of mass 32 g and radius 14.3 cm. Assuming the frisbee is originally at rest and the force is exerted for 0.08 s, determine the frisbee's angular speed about the central axis when it is released.
56. A celestial object called a pulsar emits light in short bursts that are synchronized with its rotation. A pulsar in the Crab Nebula rotates at a rate of 30 rev/s. What is the maximum radius of the pulsar, if no part of its surface can move faster than the speed of light (3×10^8 m/s)?
57. The bond length between the atoms of a molecule of nitrogen (N_2) is 1.10×10^{-10} m. The mass of each nitrogen atom is 14.0 u (1 u = 1.66×10^{-27} kg). Find the moment of inertia about an axis passing through the molecule's center of mass, perpendicular to the line joining the two atoms.

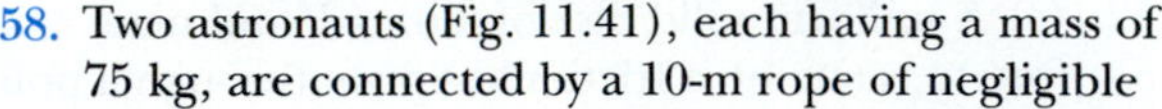

58. Two astronauts (Fig. 11.41), each having a mass of 75 kg, are connected by a 10-m rope of negligible mass. They are isolated in space, orbiting their center of mass at speeds of 5 m/s. (a) Calculate the magnitude of the angular momentum of the system by treating the astronauts as particles. (b) Calculate the kinetic energy of the system. By pulling in on the rope, the astronauts shorten the distance between them to 5 m. (c) What is the new angular momentum of the system? (d) What are the astronauts' new speeds? (e) What is the new kinetic energy of the system? (f) How much work is done by the astronauts in shortening the rope?

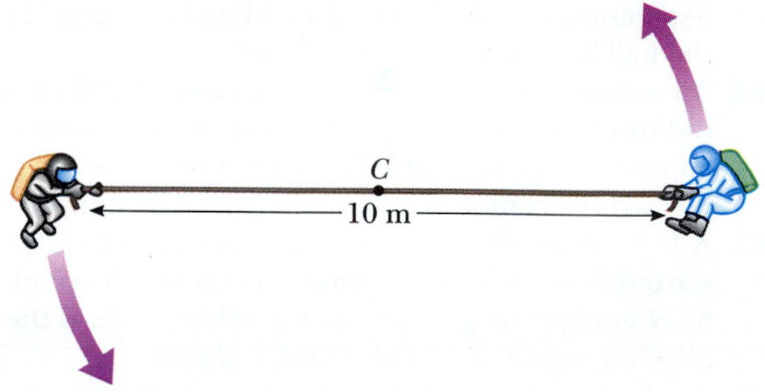

Figure 11.41 (Problem 58)

59. A small, solid sphere of mass m and radius r rolls, without slipping, along the track shown in Figure 11.42. If it starts from rest at the top of the track at a height h, (a) what is the minimum value of h (in terms of the radius of the loop R) such that the sphere completes the loop? You may assume that h and R are much larger than r. (b) What are the force components on the sphere at point P if $h = 3R$?

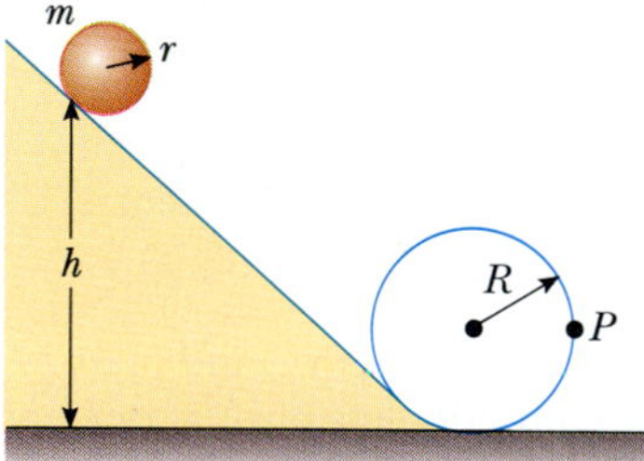

Figure 11.42 (Problem 59)

60. (a) What is the rotational energy of the Earth about its spin axis? The radius of the Earth is 6400 km, and its mass is 6×10^{24} kg. Treat the Earth as a sphere with a moment of inertia of $\frac{2}{5}MR^2$. (b) The rotational energy of the Earth is decreasing steadily because of tidal friction. Estimate the change in the rotational energy of the Earth in *one day*, given that the

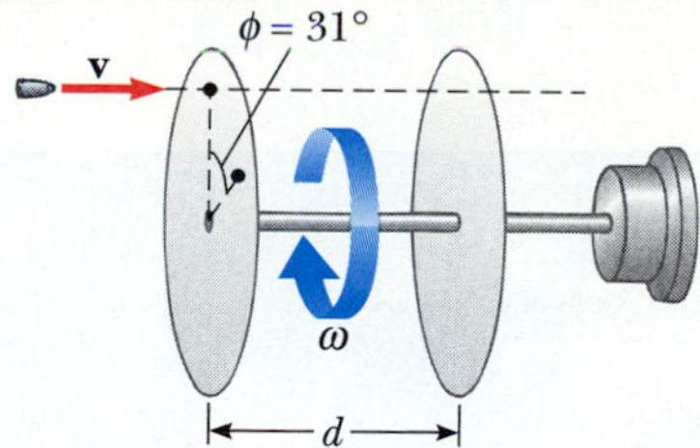

Figure 11.43 (Problem 61)

rotational period of the Earth decreases by about 10 μs each *year.*

61. We can determine the speed of a moving bullet by allowing the bullet to pass through two rotating paper disks that are mounted a distance d apart on the same axle (Fig. 11.43). From the angular displacement $\Delta\theta$ of the two bullet holes in the disks and the rotational speed ω of the disks, we can determine the speed of the bullet. Find the bullet speed for the following data: $d = 80$ cm, $\omega = 900$ rpm, and $\Delta\theta = 31°$.

62. The pulley shown in Figure 11.44 has radius R and moment of inertia I. One end of the mass m is connected to a spring of force constant k, and the other end is fastened to a cord wrapped around the pulley. The pulley axle and the incline are frictionless. If the pulley is wound counterclockwise so as to stretch the spring a distance d from its *unstretched* position, and then released from rest, find (a) the angular speed of the pulley when the spring is again unstretched and (b) a numerical value for the angular speed at this point if $I = 1$ kg·m², $R = 0.3$ m, $k = 50$ N/m, $m = 0.5$ kg, $d = 0.2$ m, and $\theta = 37°$.

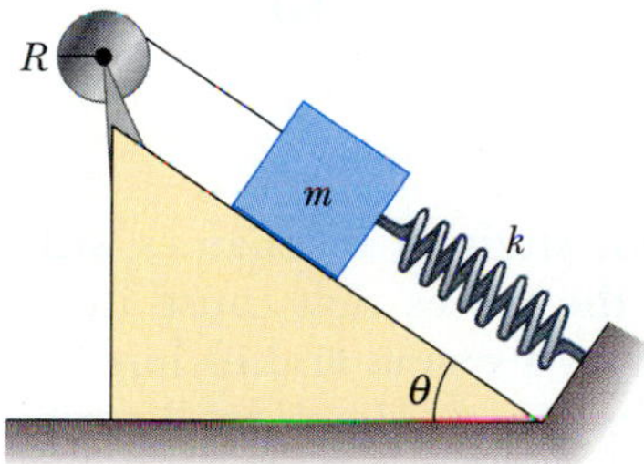

Figure 11.44 (Problem 62)

63. Two blocks, as shown in Figure 11.45, are connected by a string of negligible mass passing over a pulley of radius 0.25 m and moment of inertia I. The block on the incline is moving up with a constant acceleration of 2 m/s². (a) Determine T_1 and T_2, the tensions in the two parts of the string. (b) Find the moment of inertia of the pulley.

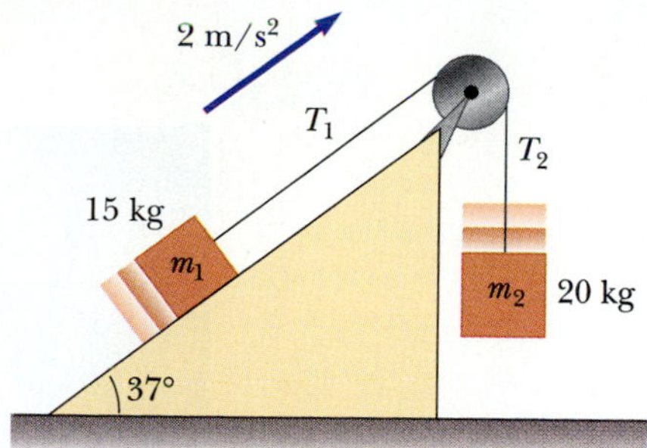

Figure 11.45 (Problem 63)

64. As a result of friction, the angular speed of a wheel changes with time according to

$$\frac{d\theta}{dt} = \omega_0 e^{-\sigma t}$$

where ω_0 and σ are constants. The angular speed changes from 3.5 rad/s at time $t = 0$ to 2.0 rad/s at time $t = 9.3$ s. Use this information to determine σ and ω_0. Then determine (a) the angular acceleration at $t = 3$ s, (b) the number of revolutions the wheel makes in the first 2.5 s, and (c) the number of revolutions the wheel makes before coming to rest.

65. An electric motor can accelerate a Ferris wheel of moment of inertia $I = 20\ 000$ kg·m² from rest to 10 rev/min in 12 s. When the motor is turned off, the Ferris wheel slows down from 10 to 8 rev/min in 10 s, due to frictional losses. Determine (a) the torque generated by the motor to bring the wheel to 10 rev/min and (b) the power needed to maintain the Ferris wheel's rotation speed of 10 rev/min.

66. A sphere rolls down an incline without sliding, as in Figure 11.26. (a) Use the results of Example 11.12 and kinematics to show that the acceleration of the sphere's center of mass is equal to $\frac{5}{7}\, g \sin\theta$. (b) Show that the same result can be obtained using dynamic methods. (*Hint:* Evaluate the torque about the sphere's center of mass, and note that $\tau = I\alpha$ still applies in the center-of-mass frame of reference.) (c) Explain why the acceleration is less than $g \sin\theta$.

This painting by the Englishman Joseph Wright (1734–1797) is entitled "A Philospher Lecturing on the Orrery." An orrery is a mechanical model of the solar system. Through his paintings, Wright explored the impact of new scientific and industrial developments on our view of the world. *(The Bridgeman Art Library)*

Orbital Motions and the Hydrogen Atom

CHAPTER 12

We began our study of mechanics with translational motion and the forces that cause it. If we know the forces acting on a system and its initial conditions, we can predict its future. However, describing motion in this manner is often tedious and time-consuming. Fortunately, we can often follow the simpler approach of using conservation principles. We easily solved many interesting problems involving motion by recognizing that certain fundamental quantities, such as energy and momentum, were conserved.

In this chapter we return to Newton's universal law of gravitation—one of the fundamental force laws in nature—and show how it, together with Newton's laws of motion, enables us to understand a variety of familiar orbital motions, such as the motions of planets and Earth satellites. Newton's theory of gravitation evolved out of his studies of the motions of the Moon and planets, for which he used the groundwork provided by Copernicus, Brahe, Kepler, and other astronomers. In examining this evolution, we shall once again make use of conservation of energy and angular momentum.

We conclude this chapter with a discussion of Niels Bohr's famous model of the hydrogen atom, which represents an interesting mixture of classical physics (Newton's laws of motion and the Coulomb interaction) and quantum physics (quantization of angular momentum). Although Bohr's theory contains ideas that are contrary to classical physics, his model successfully predicts the observed spectral lines of hydrogen.

12.1 Newton's Universal Law of Gravity Revisited

Prior to 1686, many data had been collected on the motions of the Moon and the planets, but a clear understanding of the forces that caused those motions was not yet attainable. In that year, Isaac Newton provided the key that unlocked the secrets of the heavens. He knew, from the first law, that a net force had to be acting on the Moon. If not, the Moon would move in a straight-line path rather than in its almost circular orbit. Newton reasoned that this force between Moon and Earth was an attractive force. He also concluded that there could be nothing special about the Earth-Moon system or the Sun and its planets that would cause gravitational forces to act on them alone. He wrote, "I deduced that the forces which keep the planets in their orbs must be reciprocally as the squares of their distances from the centers about which they revolve; and thereby compared the force requisite to keep the Moon in her orb with force of gravity at the surface of the Earth; and found them answer pretty nearly."

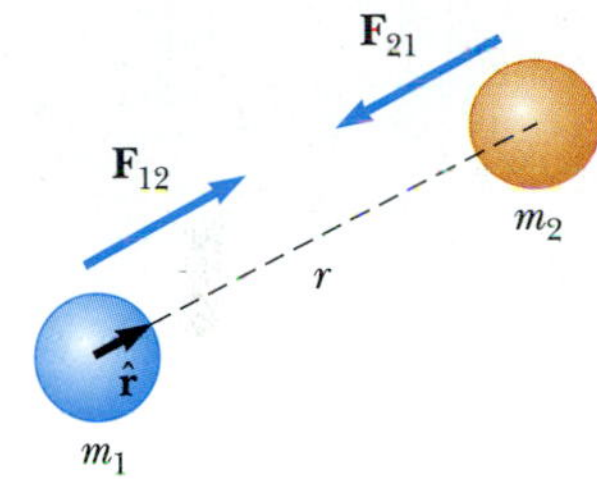

Figure 12.1 The gravitational force between two particles is attractive. The unit vector $\hat{\mathbf{r}}$ is directed from m_1 to m_2. Note that $\mathbf{F}_{12} = -\mathbf{F}_{21}$.

As you should recall from Chapter 6, every particle in the universe attracts every other particle with a force that is directly proportional to the product of their masses and inversely proportional to the square of the distance between them. If two particles have masses m_1 and m_2 and are separated by a distance r, the magnitude of the gravitational force between them is

Universal law of gravity

$$F_g = G\frac{m_1 m_2}{r^2} \qquad [12.1]$$

where G is the *gravitational constant,* whose value in SI units is

$$G = 6.672 \times 10^{-11}\,\frac{\mathrm{N\cdot m^2}}{\mathrm{kg^2}} \qquad [12.2]$$

The force law given by Equation 12.1 is often referred to as an **inverse-square law** because the magnitude of the force varies as the inverse square of the separation of the particles. We can express this force in vector form by defining a unit vector, $\hat{\mathbf{r}}$, directed from m_1 to m_2, as shown in Figure 12.1. The force on m_2 due to m_1 is

$$\mathbf{F}_{21} = -G\frac{m_1 m_2}{r^2}\,\hat{\mathbf{r}} \qquad [12.3]$$

The minus sign in Equation 12.3 indicates that m_2 is attracted to m_1, and so the force must be directed toward m_1. Likewise, by Newton's third law the force on m_1

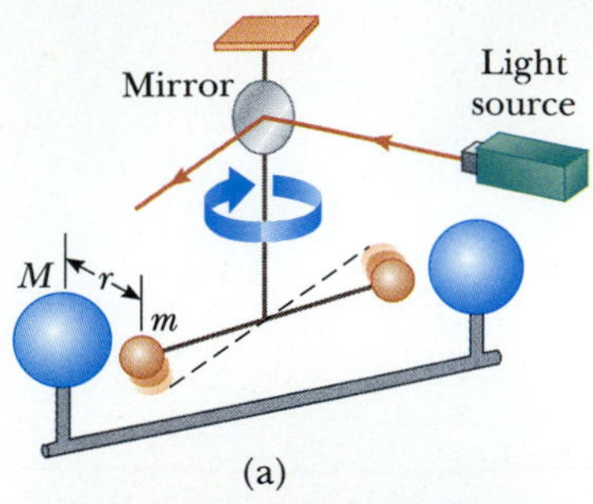

(a)

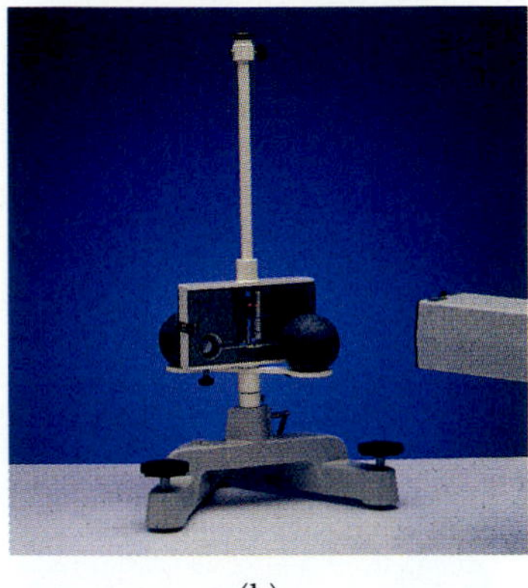
(b)

Figure 12.2
(a) Schematic diagram of the Cavendish apparatus for measuring G. The smaller spheres of mass m are attracted to the large spheres of mass M, and the rod rotates through a small angle. A light beam reflected from a mirror on the rotating apparatus measures the angle of rotation. (b) Photograph of a student Cavendish apparatus. *(Courtesy of PASCO Scientific)*

due to m_2, designated $\mathbf{F}_{12}$, is equal in magnitude to $\mathbf{F}_{21}$ and in the opposite direction. That is, these forces form an action-reaction pair, and $\mathbf{F}_{12} = -\mathbf{F}_{21}$.

The gravitational force exerted by a finite-size, spherically symmetric mass distribution on a particle outside the sphere is the same as if the entire mass of the sphere were concentrated at its center. For example, the force on a particle of mass m at the Earth's surface has the magnitude

$$F_g = G\frac{M_e m}{R_e^2}$$

where M_e is the Earth's mass and R_e is the Earth's radius. This force is directed toward the center of the Earth.

Measurement of the Gravitational Constant

The gravitational constant, G, was first measured in an important experiment by Sir Henry Cavendish in 1798. The apparatus he used consists of two small spheres, each of mass m, fixed to the ends of a light horizontal rod suspended by a thin wire, as in Figure 12.2a. Two large spheres, each of mass M, are then placed near the smaller spheres. The attractive force between the smaller and larger spheres causes the rod to rotate and twist the wire. If the system is oriented as shown in Figure 12.2a, the rod rotates clockwise when viewed from the top. The angle through which it rotates is measured by the deflection of a light beam that is reflected from a mirror attached to the wire. The experiment is carefully repeated with different masses at various separations. In addition to providing a value for G, the results show that the force is attractive, proportional to the product mM, and inversely proportional to the square of the distance r.

Example 12.1 Billiards, Anyone?

Three billiard balls, each of mass 0.3 kg, are placed on a table at the corners of a right triangle, as shown from overhead in Figure 12.3. Find the net gravitational force on ball m_1 due to the forces exerted by the other two balls.

Solution To find the net gravitational force on m_1, we first calculate the force exerted on m_1 by m_2. Then we find the force on m_1 due to m_3. Finally we add these two forces *vectorially* to obtain the net force on m_1.

The force exerted on m_1 by m_2, denoted by $\mathbf{F}_1$ in Figure 12.3, is upward. Its magnitude is calculated using Equation 12.1:

$$F_1 = G\frac{m_1 m_2}{r^2} = (6.67 \times 10^{-11}\ \mathrm{N\cdot m^2/kg^2})\ \frac{(0.3\ \mathrm{kg})(0.3\ \mathrm{kg})}{(0.4\ \mathrm{m})^2}$$

$$= 3.75 \times 10^{-11}\ \mathrm{N}$$

This result shows that the gravitational forces between the common objects that surround us have extremely small magnitudes.

Now let us calculate the gravitational force exerted on m_1 by m_3. This force, denoted by $\mathbf{F}_2$ in Figure 12.3, is directed to the right, and its magnitude is

$$F_2 = G\frac{m_1 m_3}{r^2} = (6.67 \times 10^{-11}\ \mathrm{N\cdot m^2/kg^2})\ \frac{(0.3\ \mathrm{kg})(0.3\ \mathrm{kg})}{(0.3\ \mathrm{m})^2}$$

$$= 6.67 \times 10^{-11}\ \mathrm{N}$$

The net gravitational force exerted on m_1 is found by adding $\mathbf{F}_1$ and $\mathbf{F}_2$ as *vectors*. The magnitude of this net force is

$$F = \sqrt{F_1^2 + F_2^2} = \sqrt{(3.75)^2 + (6.67)^2} \times 10^{-11}\ \text{N} = 7.65 \times 10^{-11}\ \text{N}$$

Exercise Find the direction of the resultant force on m_1.

Answer The vector **F** makes an angle of 29.3° with the line joining m_1 and m_3.

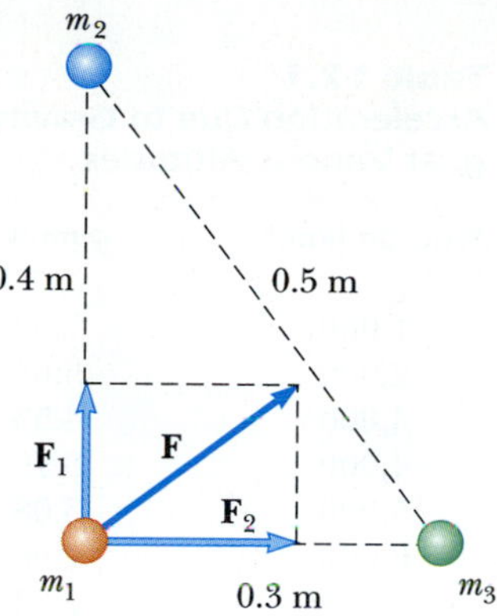

Figure 12.3 (Example 12.1)

Example 12.2 The Mass of the Earth

Use the gravitational force law to find an approximate value for the mass of the Earth.

Solution Figure 12.4, obviously not to scale, shows a baseball falling toward the Earth at a location where the acceleration due to gravity is g. We know from Chapter 6 that the gravitational force exerted on the baseball by the Earth is the same as the weight of the ball, $w = m_b g$. Since the force in the gravitational law is the weight of the ball, we find

$$m_b g = G\frac{M_e m_b}{R_e^2}$$

We can divide each side of this equation by m_b and solve for the mass of the Earth, M_e:

$$M_e = \frac{gR_e^2}{G}$$

The falling baseball is close enough to the Earth so that the distance of separation between the center of the ball and the center of the Earth can be taken as the radius of the Earth, 6.38×10^6 m. Thus, the mass of the Earth is

$$M_e = \frac{(9.80\ \text{m/s}^2)(6.38 \times 10^6\ \text{m})^2}{6.67 \times 10^{-11}\ \text{N}\cdot\text{m}^2/\text{kg}^2} = 5.98 \times 10^{24}\ \text{kg}$$

Example 12.3 Gravity and Altitude

Derive an expression that shows, for any point above the Earth's surface, how the acceleration due to gravity varies with distance from the center of the Earth.

Solution The falling baseball of Example 12.2 can be used here also. Now, however, assume that the ball is situated some arbitrary distance, $r > R_e$, from the Earth's center. The first equation in Example 12.2, with r replacing R_e and m_b removed from both sides, becomes

$$g = G\frac{M_e}{r^2}$$

This indicates that the acceleration due to gravity at a point above the Earth's surface decreases as the inverse square of the distance between the point and the center of the Earth. Our assumption of Chapter 2 that objects fall with a constant acceleration is obviously incorrect in light of this example. For short falls, however, the change in g is so small that neglecting the variation does not introduce a significant effect in our results.

Because the true weight of an object is mg, a change in the value of g produces a change in weight. For example, if you weigh 800 N at the surface of the Earth, you

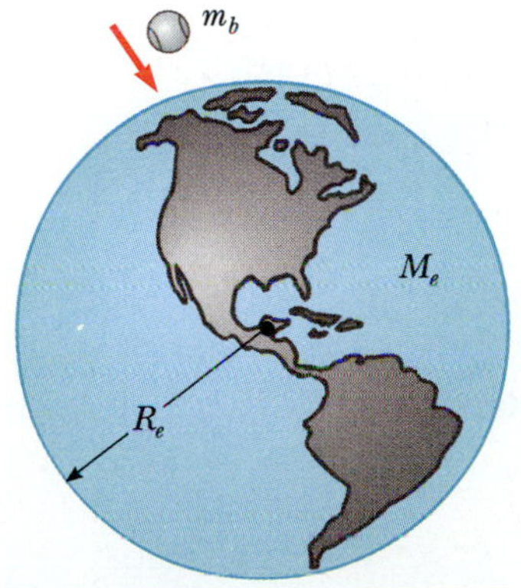

Figure 12.4 (Example 12.2) A baseball falling toward the Earth (not drawn to scale).

Table 12.1
Acceleration Due to Gravity, *g*, at Various Altitudes

Altitude (km)[a]	g (m/s^2)
1,000	7.33
2,000	5.68
3,000	4.53
4,000	3.70
5,000	3.08
6,000	2.60
7,000	2.23
8,000	1.93
9,000	1.69
10,000	1.49
50,000	0.13

[a] All values are distances above the Earth's surface.

will weigh only 200 N at a height above the Earth equal to the radius of the Earth. Also, if the distance of an object from the Earth becomes infinitely large, the true weight approaches zero. Values of g at various altitudes are given in Table 12.1.

Exercise If an object weighs 270 N at the Earth's surface, what will it weigh at an altitude equal to twice the radius of the Earth?

Answer 30 N.

Exercise Determine the magnitude of the acceleration due to gravity at an altitude of 500 km. By what percentage is the weight of a body reduced at this altitude?

Answer 8.43 m/s^2; 14%.

12.2 Kepler's Laws

The movements of the planets, stars, and other celestial bodies have been observed by people for thousands of years. Early in history, scientists regarded the Earth as the center of the Universe. This so-called geocentric model was elaborated and formalized by the Greek astronomer Claudius Ptolemy in the second century A.D. and was accepted for the next 1400 years. In 1543 the Polish astronomer Nicolaus Copernicus (1473–1543) suggested that the Earth and the other planets revolve in circular orbits about the Sun (the heliocentric hypothesis).

The Danish astronomer Tycho Brahe (1546–1601) made accurate astronomical measurements over a period of 20 years and provided the basis for the currently accepted model of the Solar System. It is interesting to note that these precise observations, made on the planets and 777 stars, were carried out with nothing more elaborate than a large sextant and compass; the telescope had not yet been invented.

The German astronomer Johannes Kepler, who was Brahe's assistant, acquired Brahe's astronomical data and spent about 16 years trying to deduce a mathematical model for the motions of the planets. After many laborious calculations, he found that Brahe's precise data on the revolution of Mars about the Sun provided the answer. Kepler's analysis first showed that the concept of circular orbits about the Sun had to be abandoned. He eventually discovered that the orbit of Mars could be accurately described by an ellipse with the Sun at one focus. He then generalized this analysis to include the motions of all planets. The complete analysis is summarized in three statements, known as **Kepler's laws:**

Kepler's laws

1. Every planet moves in an elliptical orbit with the Sun at one of the focal points.
2. The radius vector drawn from the Sun to any planet sweeps out equal areas in equal time intervals.
3. The square of the orbital period of any planet is proportional to the cube of the semimajor axis of the elliptical orbit.

Newton demonstrated that these laws were consequences of a simple force that exists between any two masses. Newton's universal law of gravity, together with his laws of motion, provides the basis for a full mathematical solution to the motion of planets and satellites. More important, Newton's universal law of gravity correctly describes the gravitational attractive force between *any* two masses.

12.3 The Universal Law of Gravity and the Motions of Planets

In formulating his universal law of gravity, Newton built on an observation which suggests that the gravitational force between two bodies is proportional to the inverse square of the separation. Let us compare the acceleration of the Moon in its orbit with the acceleration of an object falling near the Earth's surface, such as an apple (Fig. 12.5). Assume that the two accelerations have the same cause, namely, the gravitational attraction of the Earth. From the inverse-square law, Newton found that the acceleration of the Moon toward the Earth (centripetal acceleration) should be proportional to $1/r_m^2$, where r_m is the separation between centers of the Earth and Moon. Furthermore, the acceleration of the apple toward the Earth should be proportional to $1/R_e^2$, where R_e is the radius of the Earth. When the values $r_m = 3.84 \times 10^8$ m and $R_e = 6.37 \times 10^6$ m are used, the ratio of the Moon's acceleration, a_m, to the apple's acceleration, g, is predicted to be

$$\frac{a_m}{g} = \frac{(1/r_m)^2}{(1/R_e)^2} = \left(\frac{R_e}{r_m}\right)^2 = \left(\frac{6.37 \times 10^6 \text{ m}}{3.84 \times 10^8 \text{ m}}\right)^2 = 2.75 \times 10^{-4}$$

Therefore,

$$a_m = (2.75 \times 10^{-4})(9.80 \text{ m/s}^2) = 2.70 \times 10^{-3} \text{ m/s}^2$$

The centripetal acceleration of the Moon can also be calculated kinematically from a knowledge of the Moon's orbital period, T, where $T = 27.32$ days $= 2.36 \times 10^6$ s, and its mean distance from the Earth, r_m. In the time T, the Moon travels a distance $2\pi r_m$, which equals the circumference of its orbit. Therefore, its orbital speed is $2\pi r_m/T$, and its centripetal acceleration is

$$a_m = \frac{v^2}{r_m} = \frac{(2\pi r_m/T)^2}{r_m} = \frac{4\pi^2 r_m}{T^2} = \frac{4\pi^2(3.84 \times 10^8 \text{ m})}{(2.36 \times 10^6 \text{ s})^2} = 2.72 \times 10^{-3} \text{ m/s}^2$$

This agreement provides strong evidence that the inverse-square law of force is correct.

Johannes Kepler (1571–1630)
A German astronomer, Kepler is best known for developing the laws of planetary motion based on the careful observations of Tycho Brahe. Throughout his life, Kepler was side-tracked by mystic notions dating back to the ancient Greeks. For example, he believed in the "music of the spheres" notion proposed by Pythagoras in which each planet in its motion sounds out an exact musical note. After spending several years trying to work out a "regular-solid theory" of the planets, he concluded that the Copernican view of circular planetary orbits had to be abandoned. Instead, he found that the planetary orbits are ellipses with the sun always at one of the foci.

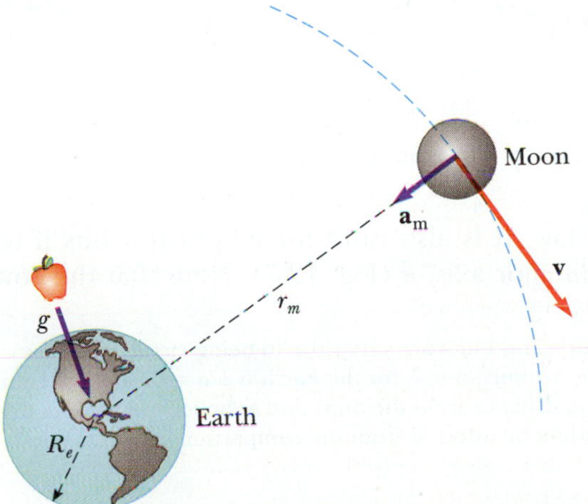

Figure 12.5
As the Moon revolves about the Earth, the Moon experiences a centripetal acceleration a_m directed toward the Earth. An object near the Earth's surface experiences an acceleration equal to g. (Dimensions are not to scale.)

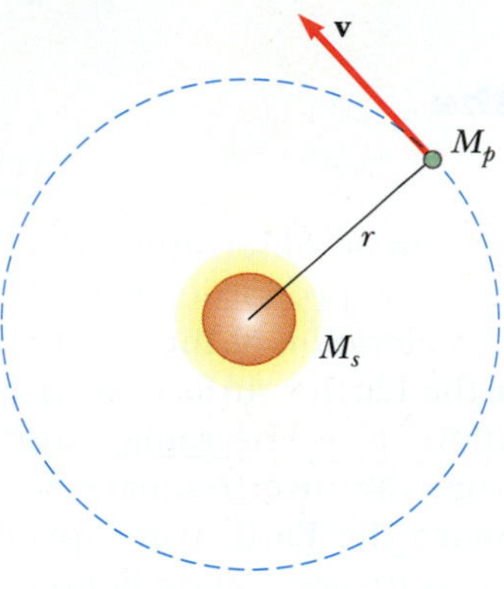

Figure 12.6
A planet of mass M_p moving in a circular orbit about the Sun. The orbits of all planets except Mars, Mercury, and Pluto are nearly circular.

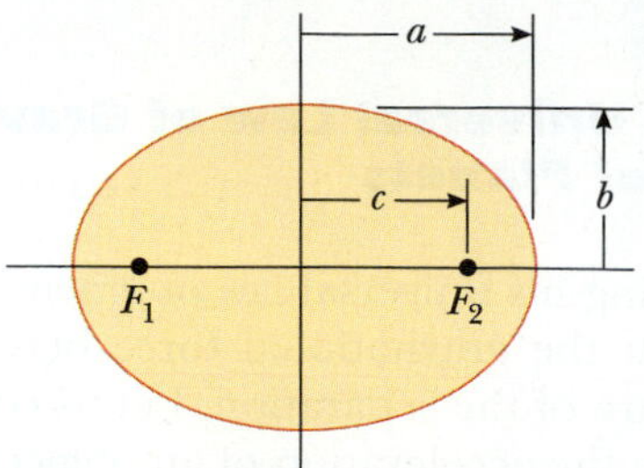

Figure 12.7
Plot of an ellipse. The semimajor axis has a length a, and the semiminor axis has a length b. The focal points are located at a distance c from the center, where $a^2 = b^2 + c^2$, and the eccentricity is defined as $e = c/a$.

Kepler's Third Law

Kepler's third law can be predicted from the inverse-square law for circular orbits.[1] Consider a planet of mass M_p which is assumed to be moving about the Sun (mass M_s) in a circular orbit, as in Figure 12.6. Since the gravitational force on the planet is equal to the centripetal force needed to keep it moving in a circle, we can use Equation 5.3 and write

$$\frac{GM_sM_p}{r^2} = \frac{M_pv^2}{r}$$

But the orbital speed of the planet is simply $2\pi r/T$, where T is the period; therefore, the preceding expression becomes

$$\frac{GM_s}{r^2} = \frac{(2\pi r/T)^2}{r}$$

Kepler's third law

$$T^2 = \left(\frac{4\pi^2}{GM_s}\right)r^3 = K_s r^3 \qquad [12.4]$$

where K_s is a constant given by

$$K_s = \frac{4\pi^2}{GM_s} = 2.97 \times 10^{-19}\ \mathrm{s^2/m^3}$$

Equation 12.4 is Kepler's third law. It is also valid for elliptical orbits if we replace r with the length of the semimajor axis, a (Fig. 12.7). Note that the con-

[1] The orbits of all planets except Mars, Mercury, and Pluto are very close to being circular. For example, the ratio of the semiminor axis to the semimajor axis for the Earth is $b/a = 0.99986$. Hence, the difference between perihelion (minimum distance from the Sun) and aphelion (maximum distance from the Sun) is much greater than might be inferred from the comparison of semimajor and semiminor axes.

Table 12.2
Useful Planetary Data[a]

Body	Mass (kg)	Mean Radius (m)	Period (s)	Distance from Sun (m)	$\frac{T^2}{r^3}\left(\frac{s^2}{m^3}\right)$
Mercury	3.18×10^{23}	2.43×10^{6}	7.60×10^{6}	5.79×10^{10}	2.97×10^{-19}
Venus	4.88×10^{24}	6.06×10^{6}	1.94×10^{7}	1.08×10^{11}	2.99×10^{-19}
Earth	5.98×10^{24}	6.37×10^{6}	3.156×10^{7}	1.496×10^{11}	2.97×10^{-19}
Mars	6.42×10^{23}	3.37×10^{6}	5.94×10^{7}	2.28×10^{11}	2.98×10^{-19}
Jupiter	1.90×10^{27}	6.99×10^{7}	3.74×10^{8}	7.78×10^{11}	2.97×10^{-19}
Saturn	5.68×10^{26}	5.85×10^{7}	9.35×10^{8}	1.43×10^{12}	2.99×10^{-19}
Uranus	8.68×10^{25}	2.33×10^{7}	2.64×10^{9}	2.87×10^{12}	2.95×10^{-19}
Neptune	1.03×10^{26}	2.21×10^{7}	5.22×10^{9}	4.50×10^{12}	2.99×10^{-19}
Pluto	$\approx 1.4 \times 10^{22}$	$\approx 1.5 \times 10^{6}$	7.82×10^{9}	5.91×10^{12}	2.96×10^{-19}
Moon	7.36×10^{22}	1.74×10^{6}	—	—	—
Sun	1.991×10^{30}	6.96×10^{8}	—	—	—

[a] For a more complete set of data, see, for example, the *Handbook of Chemistry and Physics,* Boca Raton, Florida, The Chemical Rubber Publishing Co.

stant of proportionality, K_s, is independent of the mass of the planet. Equation 12.4 is therefore valid for *any* planet. If we were to consider the orbit of a satellite about the Earth, such as the Moon, then the constant would have a different value, with the Sun's mass replaced by the Earth's mass. In this case, the proportionality constant would equal $4\pi^2/GM_e$.

Table 12.2 is a collection of useful planetary data. The last column verifies that T^2/r^3 is constant.

Example 12.4 An Earth Satellite

A satellite of mass m moves in a circular orbit about the Earth with a constant speed of v, at a height of $h = 1000$ km above the Earth's surface, as in Figure 12.8. (For clarity, this figure is not drawn to scale.) Find the orbital speed of the satellite. The Earth's radius is 6.38×10^6 m, and its mass is 5.98×10^{24} kg.

Solution The only external force on the satellite is the gravitational attraction exerted by the Earth. This force is directed toward the center of the satellite's circular path and is the centripetal force acting on the satellite. Since the force of gravity is GM_em/r^2, we find that

$$F_g = G\frac{M_em}{r^2} = m\frac{v^2}{r}$$

$$v^2 = \frac{GM_e}{r}$$

In this expression, the distance r is the Earth's radius plus the height of the satellite; that is, $r = R_e + h = 7.38 \times 10^6$ m, so that

$$v^2 = \frac{(6.67 \times 10^{-11}\ \mathrm{N\cdot m^2/kg^2})(5.98 \times 10^{24}\ \mathrm{kg})}{7.38 \times 10^6\ \mathrm{m}} = 5.40 \times 10^7\ \mathrm{m^2/s^2}$$

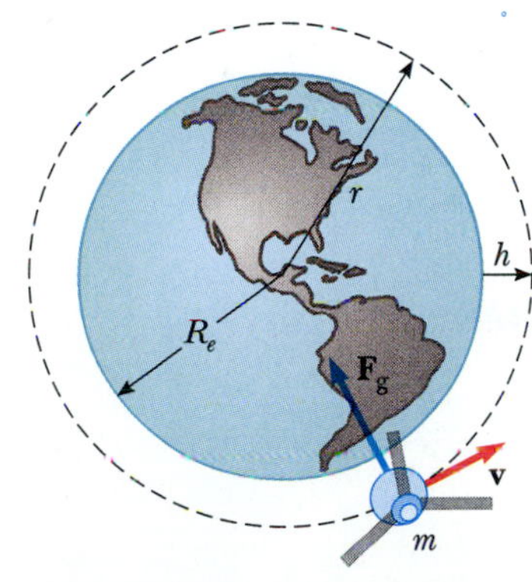

Figure 12.8 (Example 12.4) A satellite of mass m moving around the Earth in a circular orbit of radius r and with constant speed v. The centripetal force is provided by the gravitational force acting on the satellite (not drawn to scale).

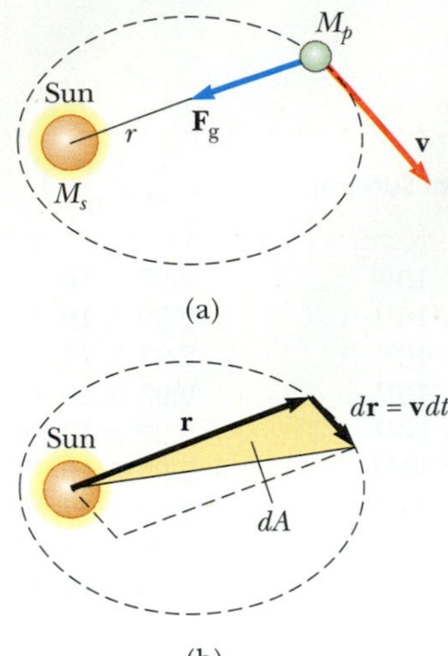

Figure 12.9 (a) The force acting on a planet acts toward the Sun, along the radius vector. (b) As a planet orbits the Sun, the area swept out by the radius vector in a time dt is equal to one half the area of the parallelogram formed by the vectors $\mathbf{r}$ and $d\mathbf{r} = \mathbf{v}\, dt$.

Therefore,

$$v = 7.35 \times 10^3 \text{ m/s} \approx 16\,400 \text{ mi/h}$$

Note that v *is independent of the mass of the satellite!*

Exercise Calculate the period of revolution, T, of the satellite.

Answer 105 min.

Kepler's Second Law and Conservation of Angular Momentum

Consider a planet of mass M_p moving about the Sun in an elliptical orbit (Fig. 12.9). The gravitational force acting on the planet is a central force, always along the radius vector, directed toward the Sun. The torque on the planet due to this central force is clearly zero, since $\mathbf{F}$ is parallel to $\mathbf{r}$. That is

$$\boldsymbol{\tau} = \mathbf{r} \times \mathbf{F} = \mathbf{r} \times F(r)\hat{\mathbf{r}} = 0$$

But recall that the torque equals the time rate of change of angular momentum; that is, $\boldsymbol{\tau} = d\mathbf{L}/dt$. Therefore,

because $\boldsymbol{\tau} = 0$, the angular momentum, L, of the planet is a constant of the motion:

$$\mathbf{L} = \mathbf{r} \times \mathbf{p} = m\mathbf{r} \times \mathbf{v} = \text{constant}$$

Since $\mathbf{L}$ is a constant of the motion, the planet's motion at any instant is restricted to the plane formed by $\mathbf{r}$ and $\mathbf{v}$.

We can relate this result to the following geometric consideration. In a time of dt, the radius vector $\mathbf{r}$ in Figure 12.9b sweeps out the area dA, which equals one-half the area $|\mathbf{r} \times d\mathbf{r}|$ of the parallelogram formed by the vectors $\mathbf{r}$ and $d\mathbf{r}$. Since the displacement of the planet in the time dt is given by $d\mathbf{r} = \mathbf{v}\, dt$, we get

$$dA = \tfrac{1}{2}|\mathbf{r} \times d\mathbf{r}| = \tfrac{1}{2}|\mathbf{r} \times \mathbf{v}\, dt| = \frac{L}{2m}\, dt$$

Kepler's second law

$$\frac{dA}{dt} = \frac{L}{2m} = \text{constant} \qquad [12.5]$$

where L and m are both constants of the motion. Thus, we conclude that

the radius vector from the Sun to any planet sweeps out equal areas in equal times.

It is important to recognize that this result, which is Kepler's second law, is a consequence of the fact that the force of gravity is a central force, which in turn implies that angular momentum of the planet is constant. Therefore, the law applies to *any* situation that involves a central force, whether inverse-square or not.

The inverse-square nature of the force of gravity is not revealed by Kepler's second law. Although we do not prove it here, Kepler's first law (as well as his third law) is a direct consequence of the fact that the gravitational force varies as $1/r^2$. That is, under an inverse-square force law, the orbits of the planets can be shown to be ellipses with the Sun at one focus.

Astronauts orbiting the Earth (1985) after recapturing the communications spacecraft Westar VI, which had been stranded since its initial deployment. Dale A. Gardner, left, holds a For Sale sign in light reference to the recaptured vehicle, while Joseph P. Allen IV, right, stands on an "arm" controlled by Dr. Anna L. Fisher inside Discovery's cabin. *(NASA photo)*

Example 12.5 **Motion in an Elliptical Orbit**

A planet of mass m moves in an elliptical orbit about the Sun (Fig. 12.10) with a perihelion of p and an aphelion of a. If the speed of the planet at p is v_p, what is its speed at a? Assume the distances r_a and r_p are known.

Solution The angular momentum of the planet relative to the Sun is $m\mathbf{r} \times \mathbf{v}$. At the points a and p, $\mathbf{v}$ is perpendicular to $\mathbf{r}$. Therefore, the magnitude of the angular momentum at these positions is $L_a = mv_ar_a$ and $L_p = mv_pr_p$. The direction of the angular momentum is out of the plane of the page. Since angular momentum is constant, we see that

$$mv_ar_a = mv_pr_p$$

$$v_a = \frac{r_p}{r_a} v_p$$

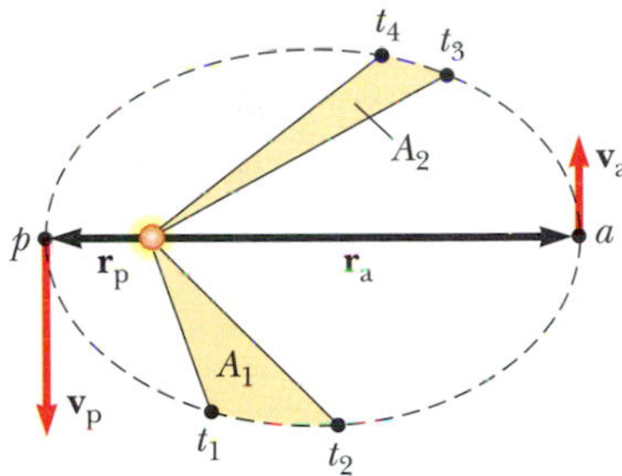

Figure 12.10 (Example 12.5) As a planet moves about the Sun in an elliptical orbit, its angular momentum is constant. Therefore, $mv_ar_a = mv_pr_p$, where the subscripts a and p represent aphelion and perihelion, respectively.

12.4 Energy Considerations in Planetary and Satellite Motion

Consider a body of mass m moving with a speed of v in the vicinity of a massive body of mass $M \gg m$. The system might be a planet moving around the Sun or a satellite in orbit around the Earth. If we assume that M is at rest in an inertial reference frame,[2] then the total energy, E, of the two-body system when the bodies are

[2] To see that this is reasonable, consider an object of mass m falling toward the Earth. Since the center of mass of the object-Earth system is stationary, it follows that $mv = M_ev_e$. Thus, the Earth acquires a kinetic energy equal to

$$\tfrac{1}{2}M_ev_e^2 = \tfrac{1}{2}\frac{m^2}{M_e}v^2 = \frac{m}{M_e}K$$

where K is the kinetic energy of the object. Since $M_e \gg m$, the kinetic energy of the Earth is negligible.

(Left) This image of Saturn, taken by the Voyager I spacecraft on October 18, 1980, was color-enhanced to increase the visibility of large, bright features in Saturn's North Temperate Belt. The distinct color difference between the North Equatorial Belt and other belts may be due to a thicker haze layer covering the northern belt. (Right) Jupiter and its four planet-size moons were photographed by Voyager I and assembled into this collage. The moons are not to scale, but are in their relative positions. Nine other smaller moons circle Jupiter. Not visible is Jupiter's faint ring of particles, seen for the first time by Voyager I. *(NASA photos)*

separated by a distance r is the sum of the kinetic energy of the mass m and the potential energy of the system:

$$E = K + U$$

Recall from Chapter 8, Equation 8.20, that the gravitational potential energy, U_g, associated with *any pair* of particles of masses m_1 and m_2 separated by a distance r is

$$U_g = -\frac{Gm_1 m_2}{r}$$

Therefore, in our case,

$$E = \tfrac{1}{2}mv^2 - \frac{GMm}{r} \qquad [12.6]$$

Furthermore, the total energy is constant if we assume that the system is isolated. Therefore, as the mass m moves from P to Q in Figure 12.11, the total energy remains constant and Equation 12.6 gives

$$E = \tfrac{1}{2}mv_i^2 - \frac{GMm}{r_i} = \tfrac{1}{2}mv_f^2 - \frac{GMm}{r_f} \qquad [12.7]$$

This result shows that E may be positive, negative, or zero, depending on the value of v. However, for a bound system, such as the Earth and Sun, E is necessarily *less than zero* if we use the arbitrary convention $U \to 0$ as $r \to \infty$. We can easily establish that $E < 0$ for the system consisting of a mass m moving in a circular orbit about a body of mass M. Newton's second law applied to the body of mass m gives

$$\frac{GMm}{r^2} = \frac{mv^2}{r}$$

Multiplying both sides by r and dividing by 2,

$$\tfrac{1}{2}mv^2 = \frac{GMm}{2r} \qquad [12.8]$$

Substituting this into Equation 12.6, we obtain

$$E = \frac{GMm}{2r} - \frac{GMm}{r}$$

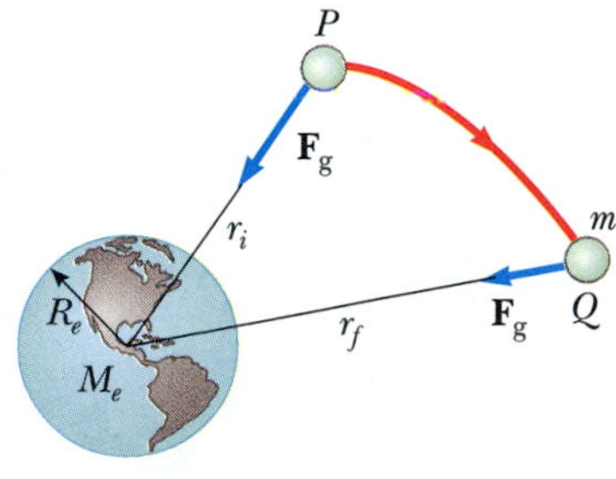

Figure 12.11
As a particle of mass m moves from P to Q above the Earth's surface, the total energy of the system remains constant.

$$E = -\frac{GMm}{2r} \qquad [12.9]$$

Total energy for circular orbits

This clearly shows that *the total energy must be negative in the case of circular orbits.* Note that *the kinetic energy is positive and equal to one-half the magnitude of the potential energy.* The absolute value of E is also equal to the binding energy of the system.

The total mechanical energy is also negative in the case of elliptical orbits. The expression for E for elliptical orbits is the same as Equation 12.9, with r replaced by the semimajor axis length, a. The total energy, the total angular momentum, and the total linear momentum of a planet-Sun system are constants of the motion.

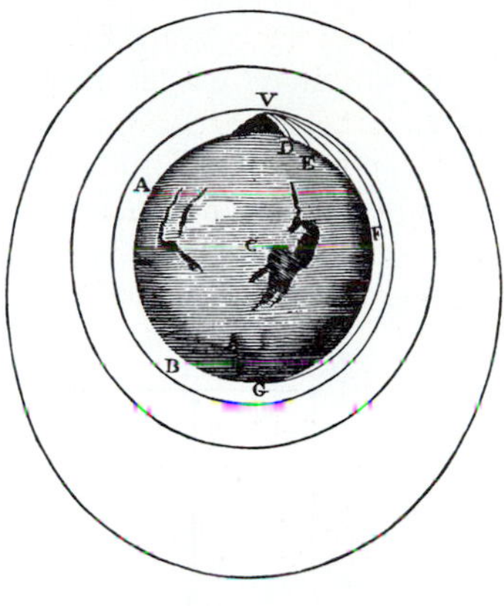

"... the greater the velocity ... with which (a stone) is projected, the farther it goes before it falls to the earth. We may therefore suppose the velocity to be so increased, that it would describe an arc of 1, 2, 5, 10, 100, 1000 miles before it arrived at the earth, till at last, exceeding the limits of the earth, it should pass into space without touching."—Newton, *System of the World.*

Example 12.6 Changing the Orbit of a Satellite

Calculate the work required to move an Earth satellite of mass m from a circular orbit of radius $2R_e$ to one of radius $3R_e$.

Solution Applying Equation 12.9, we get for the total initial and final energies

$$E_i = -\frac{GM_em}{4R_e} \qquad E_f = -\frac{GM_em}{6R_e}$$

Therefore, the work required to increase the energy of the system is equal to the final energy minus the initial energy of the system:

$$W = E_f - E_i = -\frac{GM_em}{6R_e} - \left(-\frac{GM_em}{4R_e}\right) = \frac{GM_em}{12R_e}$$

For example, if we take $m = 10^3$ kg, the work required is $W = 5.2 \times 10^9$ J, which is the energy equivalent of 39 gal of gasoline.

If we wish to determine how the energy is distributed after work is done on the system, we find from Equation 12.8 that the change in kinetic energy is $\Delta K = -GM_em/12R_e$ (it decreases), while the corresponding change in potential energy is $\Delta U = GM_em/6R_e$ (it increases). Thus, the work done on the system is given by $W = \Delta K + \Delta U = GM_em/12R_e$, as we calculated above. In other words, the work required is determined by the difference between $|\Delta U|$ and $|\Delta K|$; that difference is $\frac{1}{2}\Delta U$.

Example 12.7 A Satellite in an Elliptical Orbit

A satellite moves in an elliptical orbit about the Earth, as in Figure 12.12. The minimum and maximum distances from the surface of the Earth are 400 km and 3000 km. Find the speeds of the satellite at the apogee and perigee.

Solution Since the mass of the satellite is negligible compared with the Earth's mass, we take the center of mass of the Earth to be at rest. Gravity is a *central* force, and so the angular momentum of the satellite about the Earth's center of mass remains constant in time. With subscripts a and p for the apogee and perigee positions, conservation of angular momentum gives $L_p = L_a$, or

$$mv_pr_p = mv_ar_a$$

$$v_pr_p = v_ar_a \qquad [1]$$

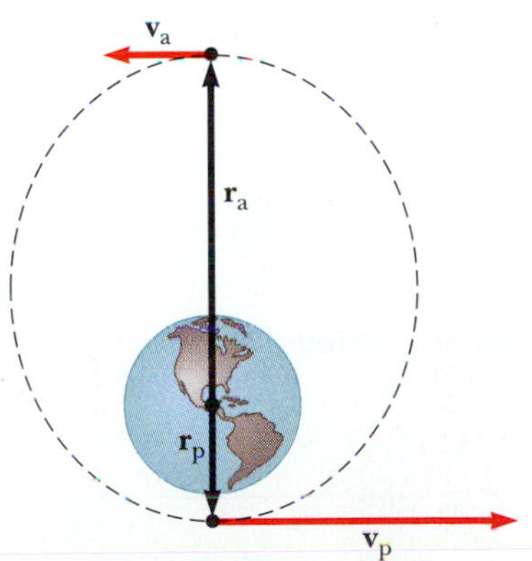

Figure 12.12 (Example 12.7) A satellite in an elliptical orbit about the Earth.

Applying conservation of energy, we obtain $E_p = E_a$, or

$$U_p + K_p = U_a + K_a$$

$$-G\frac{M_e m}{r_p} + \frac{1}{2}mv_p^2 = -G\frac{M_e m}{r_a} + \frac{1}{2}mv_a^2$$

$$2GM_e\left(\frac{1}{r_a} - \frac{1}{r_p}\right) = (v_a^2 - v_p^2) \quad [2]$$

Assuming the Earth's radius is 6.37×10^6 m and using the given data, we find that $r_a = 9.37 \times 10^6$ m and $r_p = 6.77 \times 10^6$ m. Since we know the numerical values of G, M_e, r_p, and r_a, we can use Equations (1) and (2) to determine the two unknowns, (v_p and v_a). Solving the equations simultaneously, we obtain

$$v_p = 8.25 \text{ km/s} \qquad v_a = 5.96 \text{ km/s}$$

Escape Speed

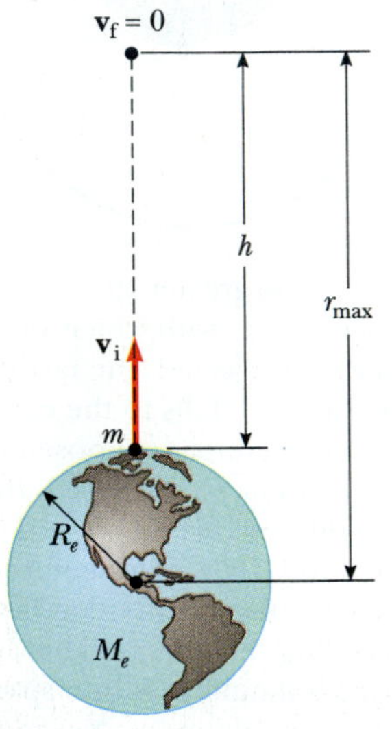

Figure 12.13
An object of mass m projected upward from the Earth's surface with an initial speed v_i reaches a maximum altitude h.

Suppose an object of mass m is projected vertically from the Earth's surface with an initial speed v_i, as in Figure 12.13. We can use energy considerations to find the minimum value of the initial speed such that the object will escape the Earth's gravitational field. Equation 12.6 gives the total energy of the object at any point when its speed and distance from the center of the Earth are known. At the surface of the Earth, where $v_i = v$, $r_i = R_e$. When the object reaches its maximum altitude, $v_f = 0$ and $r_f = r_{\text{max}}$. Because the total energy of the system is constant, substitution of these conditions into Equation 12.7 gives

$$\frac{1}{2}mv_i^2 - \frac{GM_e m}{R_e} = -\frac{GM_e m}{r_{\text{max}}}$$

Solving for v_i^2 gives

$$v_i^2 = 2GM_e\left(\frac{1}{R_e} - \frac{1}{r_{\text{max}}}\right) \quad [12.10]$$

Therefore, if the initial speed is known, this expression can be used to calculate the maximum altitude, h, since we know that $h = r_{\text{max}} - R_e$.

We are now in a position to calculate the minimum speed the object must have at the Earth's surface in order to escape from the influence of the Earth's gravitational field. This corresponds to the situation where the object can *just* reach infinity with a final speed of *zero*. Setting $r_{\text{max}} = \infty$ in Equation 12.10 and taking $v_i = v_{\text{esc}}$ (the escape speed), we get

Escape speed

$$v_{\text{esc}} = \sqrt{\frac{2GM_e}{R_e}} \quad [12.11]$$

Note that this expression for v_{esc} is independent of the mass of the object projected from the Earth. For example, a spacecraft has the same escape speed as a molecule. Furthermore, the result is independent of the *direction* of the velocity, provided the trajectory does not intersect the Earth. If the object is given an initial speed equal to v_{esc}, its *total* energy is equal to zero. This can be seen by noting that when $r = \infty$, the object's kinetic energy and its potential energy are both zero. If v_i is greater than v_{esc}, the *total* energy is greater than zero and the object has some residual kinetic energy at $r = \infty$.

Example 12.8 The Escape Speed of a Rocket

Calculate the escape speed from the Earth for a 5000-kg spacecraft, and determine the kinetic energy it must have at the Earth's surface in order to escape the Earth's gravitational field.

Solution Using Equation 12.11 with $M_e = 5.98 \times 10^{24}$ kg and $R_e = 6.37 \times 10^6$ m, we get

$$v_{\text{esc}} = \sqrt{\frac{2GM_e}{R_e}}$$

$$= \sqrt{\frac{2(6.67 \times 10^{-11}\ \text{N}\cdot\text{m}^2/\text{kg}^2)(5.98 \times 10^{24}\ \text{kg})}{6.37 \times 10^6\ \text{m}}}$$

$$= 1.12 \times 10^4\ \text{m/s}$$

This corresponds to about 25 000 mi/h.

The required kinetic energy of the spacecraft is

$$K = \tfrac{1}{2}mv_{\text{esc}}^2 = \tfrac{1}{2}(5 \times 10^3\ \text{kg})(1.12 \times 10^4\ \text{m/s})^2$$

$$= 3.14 \times 10^{11}\ \text{J}$$

Table 12.3 Escape Speeds for the Planets, the Moon, and the Sun

Planet	v_{esc} (km/s)
Mercury	4.3
Venus	10.3
Earth	11.2
Moon	2.3
Mars	5.0
Jupiter	60
Saturn	36
Uranus	22
Neptune	24
Pluto	1.1
Sun	618

Finally, note that Equations 12.10 and 12.11 can be applied to objects projected from *any* planet. That is, in general, the escape speed from any planet of mass M and radius R is given by

$$v_{\text{esc}} = \sqrt{\frac{2GM}{R}} \qquad [12.12]$$

Table 12.3 lists escape speeds for the planets, the Moon, and the Sun. Note that the values vary from 2.3 km/s for the Moon to about 618 km/s for the Sun. These results, together with some ideas from the kinetic theory of gases (Chapter 13), explain why some planets have atmospheres and others do not. As we shall see later, a gas molecule has an average kinetic energy that depends on its temperature. Lighter atoms, such as hydrogen and helium, have higher average speeds than do the heavier species. When the speed of the lighter atoms is not much less than the escape speed, a significant fraction of the molecules have a chance to escape from the planet. This mechanism also explains why the Earth does not retain hydrogen and helium molecules in its atmosphere but does retain much heavier molecules, such as oxygen and nitrogen. On the other hand, Jupiter has a very large escape speed (60 km/s), which enables it to retain hydrogen, the primary constituent of its atmosphere.

The bright star at the left is a supernova which exploded on February 23, 1987. The central portion of the photograph is the Large Magellanic Cloud, a galaxy which is 35 000 lightyears wide, and about 160 000 lightyears from the Earth. *(Royal Observatory, Edinburgh, Science Photo Library)*

Black Holes

In Chapter 11 we briefly described a rare event called a supernova—the catastrophic explosion of a very massive star. The material that remains in the central core of such an object continues to collapse, but the core's ultimate fate depends on its mass. If the core has a mass less than 1.4 times the mass of our Sun, it gradually cools down and ends its life as a black dwarf star. However, if the core's

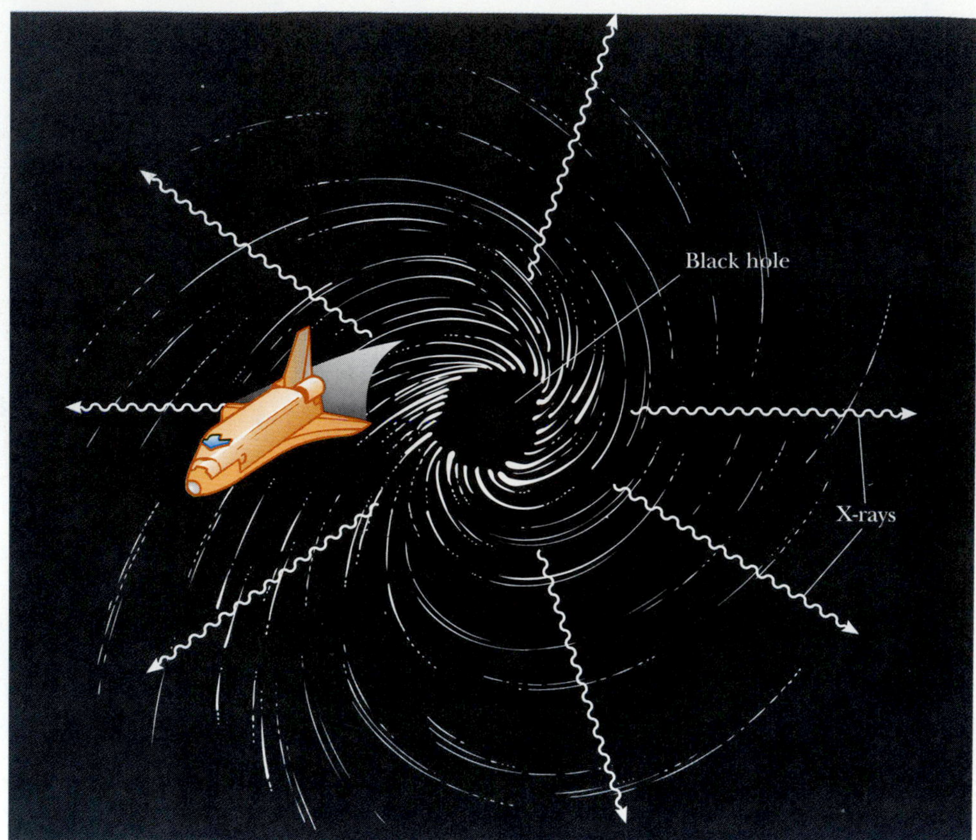

Figure 12.14
The gravitational field in the vicinity of a black hole is so strong that nothing can escape. Any object moving close to a black hole can emit x-rays.

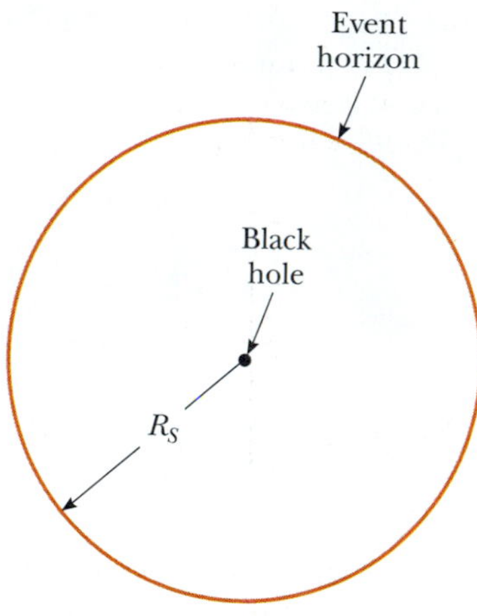

Figure 12.15
A black hole. The distance R_s equals the Schwarzschild radius. Any event occurring within the boundary of radius R_s, called the event horizon, is invisible to an outside observer.

mass is greater than this, the star may collapse further due to gravitational forces. What remains is a neutron star, a region compressed to a radius of about 10 km. (On Earth, a teaspoon of this material would weigh about 5 billion tons!)

An even more unusual star death may occur when the core has a mass greater than about three solar masses. No known forces in nature are strong enough to prevent the collapse of such a star. The collapse may continue until the star becomes a mere point in space, commonly referred to as a **black hole.** *In effect, black holes are remains of stars that have collapsed under their own weight.*

Objects such as spaceships can enter a black hole. Once in, they can never escape. As soon as an object enters a black hole, it is pulled deeper toward the center and is trapped forever (Fig. 12.14).

The escape speed from any spherical body can be calculated from Equation 12.12. If the escape speed exceeds the speed of light, c, radiation within the body (such as visible light) cannot escape, and the body appears to be black; hence the origin of the terminology "black hole." The critical radius, R_s, for which this occurs is called the **Schwarzschild radius** (Fig. 12.15). Taking $v_{esc} = c$ in Equation 12.12 and solving for R_s, we obtain $R_s = 2GM/c^2$. For example, the value for R_s for a black hole with a mass equal to that of the Sun is calculated to be 3.0 km (about 2 mi); a black hole whose mass equals that of the Earth has a radius of about 9 mm (about the size of a dime).

Although light from a black hole cannot escape, light from events taking place near the black hole should be visible. A companion star captured by the strong

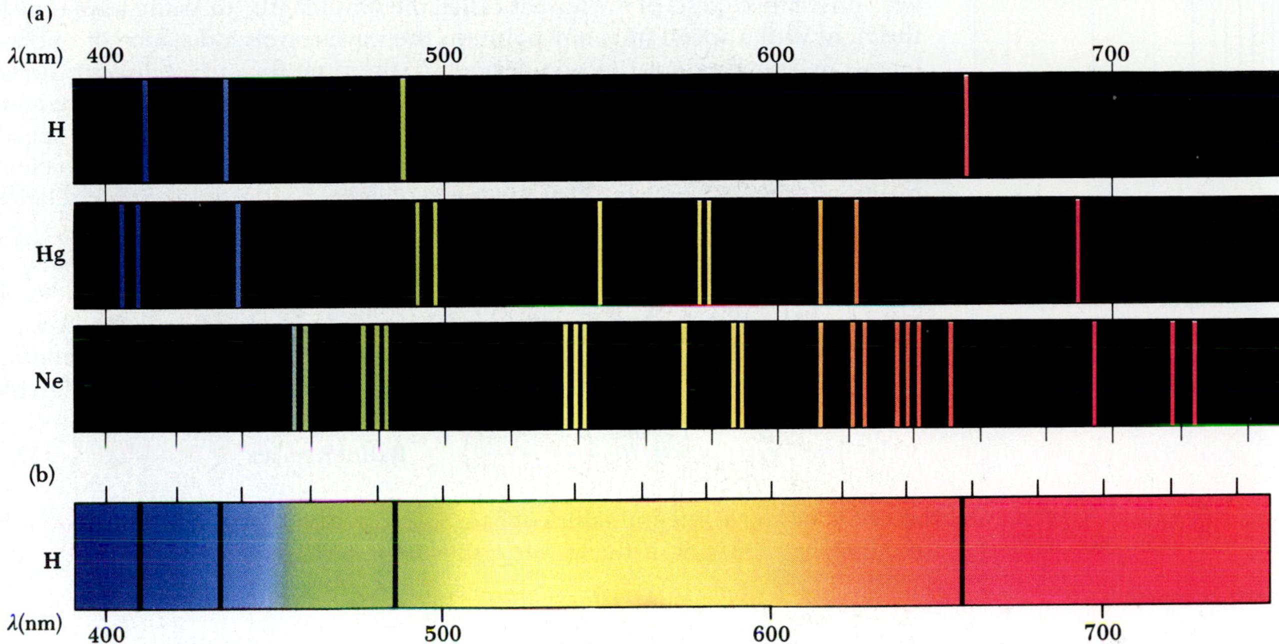

gravitational field of a black hole should emit x-rays (Fig. 12.14). Based on this reasoning, several candidates for black holes have been detected, the most famous being Cygnus X-1, the first x-ray source detected in the constellation Cygnus. There is also evidence that supermassive black holes exist at the centers of galaxies.

Figure 12.16
Visible spectra (a) Line spectra produced by emission in the visible range for the elements hydrogen, helium, and neon.
(b) The absorption spectrum for hydrogen. The dark absorption lines occur at the same wavelengths as the emission line: for hydrogen shown in (a). *(Whitten, K.W., K.D. Gailey, and R.E. Davis, General Chemistry, 4th ed., Saunders College Publishing 1992)*

12.5 Atomic Spectra and the Bohr Theory of Hydrogen

As you may have already learned in chemistry, the hydrogen atom is the simplest known atomic system, and an especially important one to understand. Much of what is learned about the hydrogen atom (which consists of one proton and one electron) can be extended to other single-electron ions such as He^+ and Li^{2+}. Furthermore, a thorough understanding of the physics underlying the hydrogen atom can then be used to describe more complex atoms and the periodic table of the elements.

Suppose an evacuated glass tube is filled with hydrogen (or some other gas). If a voltage applied between metal electrodes in the tube is great enough to produce an electric current in the gas, the tube emits light whose color is characteristic of the gas (this is how a neon sign works). When the emitted light is analyzed with a device called a spectroscope, a series of discrete lines is observed, each line corresponding to a different wavelength, or color, of light. Such a series of spectral lines is commonly referred to as an **emission spectrum.** The wavelengths contained in a given line spectrum are characteristic of the element emitting the light (Fig. 12.16). Because no two elements emit the same line spectrum, this phenomenon represents a marvelous and reliable technique for identifying elements in a substance.

As you shall learn in more detail in Chapter 22, a wave is any disturbance that moves from one location to another. A common form of periodic wave is the sinusoidal wave, whose shape is depicted in Figure 12.17. The distance between

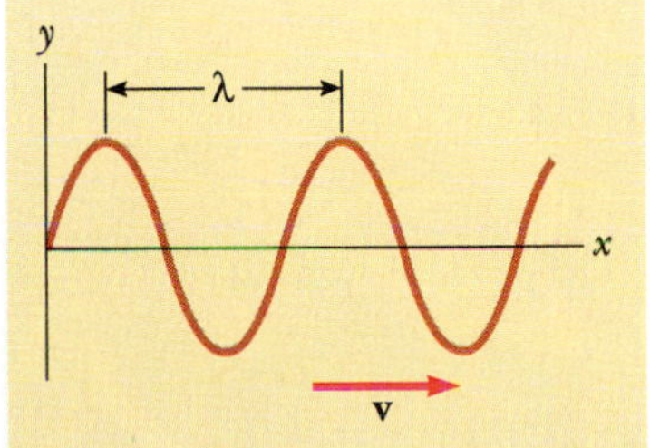

Figure 12.17
A sinusoidal wave traveling to the right. Any point on the wave moves a distance of one wavelength, λ, in a time equal to the period of the wave.

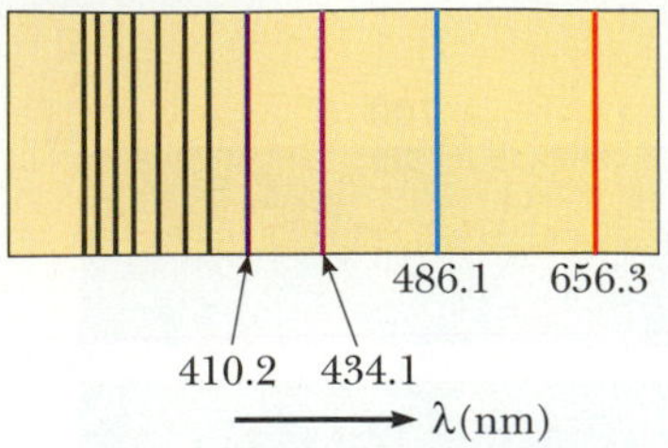

Figure 12.18
A series of spectral lines for atomic hydrogen. The prominent lines labeled are part of the Balmer series.

two consecutive crests of the wave is called the **wavelength,** λ. As the wave travels to the right with a speed of v, any point on the wave travels a distance of one wavelength in a time interval of one period, T (the time for one cycle), so the wave speed is given by $v = \lambda / T$. The inverse of the period, $1/T$, is called the **frequency,** f, of the wave; it represents the number of cycles per second. Thus, the speed of the wave is often written as $v = \lambda f$. In this section, since we shall deal with electromagnetic waves—which travel at the speed of light, c—the appropriate relation is

$$c = \lambda f \qquad [12.13]$$

The emission spectrum of hydrogen shown in Figure 12.18 includes four prominent lines that occur at wavelengths of 656.3 nm, 486.1 nm, 434.1 nm, and 41.0.2 nm. In 1885, Johann Balmer (1825–1898) found that the wavelengths of these and less prominent lines can be described by this simple empirical equation:

$$\frac{1}{\lambda} = R_H\left(\frac{1}{2^2} - \frac{1}{n^2}\right) \qquad \text{Balmer series} \qquad [12.14]$$

where n may have integral values of 3, 4, 5, . . . , and R_H is a constant, now called the **Rydberg constant.** If the wavelength is in meters, R_H has the value

Rydberg constant

$$R_H = 1.0973732 \times 10^7 \text{ m}^{-1}$$

The first line in the Balmer series, at 656.3 nm, corresponds to $n = 3$ in Equation 12.14; the line of 486.1 nm corresponds to $n = 4$; and so on.

In addition to emitting light at specific wavelengths, an element can also absorb light at specific wavelengths. The spectral lines corresponding to this process form what is known as an **absorption spectrum.** An absorption spectrum can be obtained by passing a continuous radiation spectrum (one containing all wavelengths) through a vapor of the element being analyzed. The absorption spectrum consists of a series of dark lines superimposed on the otherwise continuous spectrum. Each line in the absorption spectrum of a given element has been found experimentally to coincide with a line in the emission spectrum of the element. That is, if hydrogen is the absorbing vapor, dark lines appear at the visible wavelengths 656.3 nm, 486.1 nm, 434.1 nm, and 410.2 nm.

At the beginning of the 20th century, scientists were perplexed by the failure of classical physics to explain the characteristics of spectra. Why did atoms of a given element emit only certain lines? Furthermore, why did the atoms absorb only

Niels Bohr (1885–1962), a Danish physicist, proposed the first quantum model of the atom. He was an active participant in the early development of quantum mechanics and provided much of its philosophical framework. During the 1920s and 1930s, Bohr headed the Institute for Advanced Studies in Copenhagen. The institute was a magnet for many of the world's best physicists and provided a forum for the exchange of ideas. Bohr, a firm believer in doing physics on a "man-to-man" basis, was always the initiator of probing questions, reflections, and discussions with his guests.

When Bohr visited the United States in 1939 to attend a scientific conference, he brought news that the fission of uranium had been discovered by Hahn and Strassman in Berlin. The results, confirmed by other scientists shortly thereafter, were the foundations of the atomic bomb developed in the United States during World War II. After the war, Bohr committed himself to many human issues, including the development of peaceful uses of atomic energy.

Bohr was awarded the 1922 Nobel prize in physics for his investigation of the structure of atoms and of the radiation emanating from them.

those wavelengths which they emitted? In 1913 Bohr provided an explanation of atomic spectra that includes some features of the currently accepted theory. Using the simplest atom, hydrogen, Bohr described a model of what he thought must be the atom's structure. His model of the hydrogen atom contains some classical features as well as some revolutionary postulates that could not be justified within the framework of classical physics.[3] The basic assumptions of the Bohr theory as it applies to the hydrogen atom are as follows:

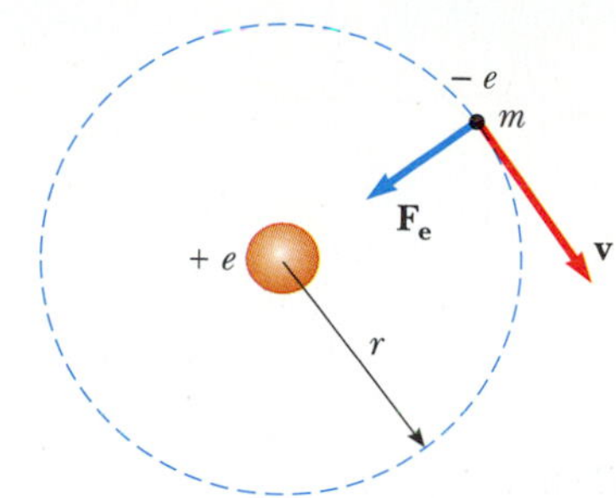

Figure 12.19 Diagram representing Bohr's model of the hydrogen atom, in which the orbiting electron is allowed to be only in specific orbits of discrete radii.

Assumptions of the Bohr theory

1. The electron moves in circular orbits about the proton under the influence of the Coulomb force of attraction, as in Figure 12.19. In this case, the Coulomb force is the centripetal force.
2. Only certain electron orbits are stable. These are orbits in which the hydrogen atom does not emit energy in the form of radiation. Hence, the total energy of the atom remains constant, and classical mechanics can be used to describe the electron's motion.
3. Radiation is emitted by the hydrogen atom when the electron "jumps" from a more energetic initial state to a lower state. The "jump" cannot be visualized or treated classically. In particular, the frequency, f, of the radiation emitted in the jump is related to the change in the atom's energy and is *independent of the frequency of the electron's orbital motion.* The frequency of the emitted radiation is

$$E_i - E_f = hf \qquad [12.15]$$

where E_i is the energy of the initial state, E_f is the energy of the final state, h is Planck's constant, and $E_i > E_f$.
4. The size of the allowed electron orbits is determined by a condition imposed on the electron's orbital angular momentum: the allowed orbits are those for which the electron's orbital angular momentum about the nucleus is an integral multiple of $\hbar = h/2\pi$.

$$mvr = n\hbar \qquad n = 1, 2, 3, \ldots \qquad [12.16]$$

Using these four assumptions, we can calculate the allowed energy levels and emission wavelengths of the hydrogen atom. Recall from Chapter 8, Equation 8.23, that the electrical potential energy of the system shown in Figure 12.19 is given by $U_e = -k_e e^2/r$, where k_e is the Coulomb constant, e is the charge on the electron, and r is the electron-proton separation. Thus, the total energy of the atom, which contains both kinetic and potential energy terms, is

$$E = K + U_e = \tfrac{1}{2}mv^2 - k_e\frac{e^2}{r} \qquad [12.17]$$

Applying Newton's second law to this system, we see that the Coulomb attractive force on the electron, $k_e e^2/r^2$ (Eq. 6.4), must equal the mass times the centripetal acceleration ($a = v^2/r$) of the electron:

$$\frac{k_e e^2}{r^2} = \frac{mv^2}{r}$$

[3] The Bohr model can be applied successfully to such hydrogen-like ions as singly ionized helium and doubly ionized lithium, but the theory does not properly describe the spectra of more complex atoms and ions.

From this expression, we find the kinetic energy to be

$$K = \tfrac{1}{2}mv^2 = \frac{k_e e^2}{2r} \qquad [12.18]$$

Substituting this value of K into Equation 12.17, we find that the total energy of the atom is

$$E = \frac{k_e e^2}{2r} \qquad [12.19]$$

Note that the total energy is negative, indicating a bound electron-proton system. This means that energy in the amount of $k_e e^2/2r$ must be added to the atom just to remove the electron and make the total energy zero. An expression for r, the radius of the allowed orbits, can be obtained by eliminating v by substitution between Equations 12.16 and 12.18:

Radii of Bohr orbits in hydrogen

$$r_n = \frac{n^2\hbar^2}{mk_e e^2} \qquad n = 1, 2, 3, \ldots \qquad [12.20]$$

This result shows that the radii have discrete values, or are *quantized.*

The orbit for which $n = 1$ has the smallest radius; it is called the **Bohr radius,** a_0, and has the value

$$a_0 = \frac{\hbar^2}{mk_e e^2} = 0.0529 \text{ nm} \qquad [12.21]$$

The first three Bohr orbits are shown to scale in Figure 12.20.

The quantization of the orbit radii immediately leads to energy quantization. This can be seen by substituting $r_n = n^2 a_0$ into Equation 12.19. The allowed energy levels are found to be

$$E_n = -\frac{k_e e^2}{2a_0}\left(\frac{1}{n^2}\right) \qquad n = 1, 2, 3, \ldots \qquad [12.22]$$

Insertion of numerical values into Equation 12.22 gives

$$E_n = -\frac{13.6}{n^2}\text{ eV} \qquad n = 1, 2, 3, \ldots \qquad [12.23]$$

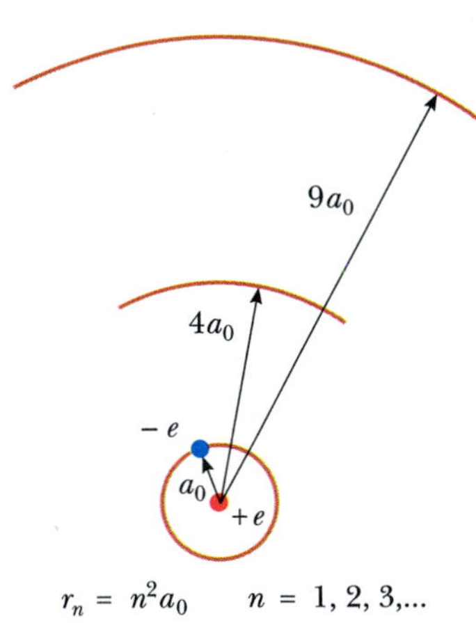

Figure 12.20
The first three Bohr orbits for hydrogen.

As we learned in Chapter 8, Section 8.10, the lowest stationary state is called the **ground state.** It has $n = 1$ and an energy of $E_1 = -13.6$ eV. The next state, the **first excited state,** has $n = 2$ and an energy of $E_2 = E_1/2^2 = -3.4$ eV. Figure 12.21 is an energy level diagram showing the energies of these discrete energy states and the corresponding quantum numbers. The uppermost level, corresponding to $n = \infty$ (or $r = \infty$) and $E = 0$, represents the state for which the electron is removed from the atom. Recall from Section 8.10 that the minimum energy required to ionize the atom (that is, to completely remove an electron in the ground state from the proton's influence) is called the **ionization energy.** As can be seen from Figure 12.21, the ionization energy for hydrogen, based on Bohr's calculation, is 13.6 eV. This constituted another major achievement for the Bohr theory, since the ionization energy for hydrogen had already been measured to be precisely 13.6 eV.

Figure 12.21 also shows other spectral series (the Lyman series and the Paschen series) that were found after Balmer's discovery. These spectra obey other empirical formulas, which are reconciled with the Bohr model.

Equation 12.22, together with Bohr's third postulate, can be used to calculate the frequency of the radiation that is emitted when the electron jumps from an outer orbit to an inner orbit:

$$f = \frac{E_i - E_f}{h} = \frac{k_e e^2}{2a_0 h}\left(\frac{1}{n_f^2} - \frac{1}{n_i^2}\right) \quad [12.24]$$

Since the quantity being measured is wavelength, it is convenient to convert frequency to wavelength, using $c = f\lambda$, to get

Emission wavelengths of hydrogen

$$\frac{1}{\lambda} = \frac{f}{c} = \frac{k_e e^2}{2a_0 hc}\left(\frac{1}{n_f^2} - \frac{1}{n_i^2}\right) \quad [12.25]$$

The remarkable fact is that the *theoretical* expression, Equation 12.25, is identical to a generalized form of the empirical relations discovered by Balmer and others (see Eq. 12.14),

$$\frac{1}{\lambda} = R_H\left(\frac{1}{n_f^2} - \frac{1}{n_i^2}\right) \quad [12.26]$$

provided that the combination of constants $k_e e^2/2\,a_0\,hc$ is equal to the experimentally determined Rydberg constant. After Bohr demonstrated the agreement of these two quantities to a precision of about 1%, it was soon recognized as the crowning achievement of his new theory of quantum mechanics. Furthermore, Bohr showed that all of the spectral series for hydrogen have a natural interpretation in his theory. Figure 12.21 shows these spectral series as transitions between energy levels.

Bohr immediately extended his model for hydrogen to other elements in which all but one electron had been removed. Ionized elements such as He^+, Li^{2+}, and Be^{3+} were suspected to exist in hot stellar atmospheres, where frequent atomic collisions occur with enough energy to completely remove one or more atomic electrons. Bohr showed that many mysterious lines observed in the Sun and several stars could not be due to hydrogen, but were correctly predicted by his theory if attributed to singly ionized helium.

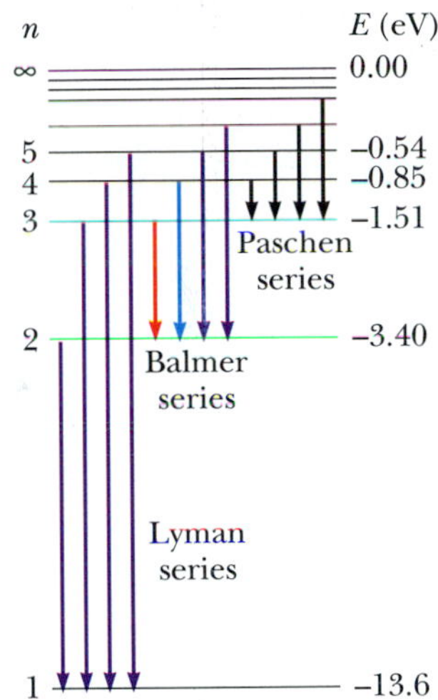

Figure 12.21
An energy level diagram for hydrogen. The discrete allowed energies are plotted on the vertical axis. Nothing is plotted on the horizontal axis, but the horizontal extent of the diagram is made large enough to show allowed transitions. Quantum numbers are given on the left.

Example 12.9 An Electronic Transition in Hydrogen

The electron in the hydrogen atom makes a transition from the $n = 2$ energy state to the ground state (corresponding to $n = 1$). Find the wavelength and frequency of the emitted radiation.

Solution We can use Equation 12.26 directly to obtain λ, with $n_i = 2$ and $n_f = 1$:

$$\frac{1}{\lambda} = R_H\left(\frac{1}{n_f^2} - \frac{1}{n_i^2}\right)$$

$$\frac{1}{\lambda} = R_H\left(\frac{1}{1^2} - \frac{1}{2^2}\right) = \frac{3R_H}{4}$$

$$\lambda = \frac{4}{3R_H} = \frac{4}{3(1.097 \times 10^{-7}\ \text{m}^{-1})}$$

$$= 1.215 \times 10^{-7}\ \text{m} = \boxed{121.5\ \text{nm}} \quad \text{(ultraviolet)}$$

Since $c = f\lambda$, the frequency of the radiation is

$$f = \frac{c}{\lambda} = \frac{3.00 \times 10^8 \text{ m/s}}{1.215 \times 10^{-7} \text{ m}} = 2.47 \times 10^{15} \text{ s}^{-1}$$

Exercise What is the wavelength of the radiation emitted by hydrogen when the electron makes a transition from the $n = 3$ state to the $n = 1$ state?

Answer $\frac{9}{8R_H} = 102.6$ nm.

Bohr's Correspondence Principle

In our study of relativity in Chapter 10, we found that newtonian mechanics cannot be used to describe phenomena that occur at speeds approaching the speed of light. Newtonian mechanics is a special case of relativistic mechanics and is usable only when v is much less than c. Similarly, *quantum mechanics is in agreement with classical physics where the energy differences between quantized levels vanish.* This principle, first set forth by Bohr, is called the **correspondence principle.**

For example, consider an electron orbiting the hydrogen atom with $n >$ 10 000. For such large values of n, the energy differences between adjacent levels approach zero and the levels are nearly continuous. Consequently, the classical model is reasonably accurate in describing the system for large values of n. According to the classical picture, the frequency of the light emitted by the atom is equal to the frequency of revolution of the electron in its orbit about the nucleus. Calculations show that for $n >$ 10 000, this frequency differs from that predicted by quantum mechanics by less than 0.015%.

Summary

Universal law of gravity

Newton's **universal law of gravity** states that the gravitational force of attraction between any two particles of masses m_1 and m_2 separated by a distance r has the magnitude

$$F_g = G\frac{m_1 m_2}{r^2} \qquad [12.1]$$

where G is the gravitational constant, 6.672×10^{-11} N·m²/kg².

Kepler's laws

Kepler's laws of planetary motion state the following:

1. Every planet moves in an elliptical orbit with the Sun at one of the focal points.
2. The radius vector drawn from the Sun to any planet sweeps out equal areas in equal time intervals.
3. The square of the orbital period of any planet is proportional to the cube of the semimajor axis for the elliptical orbit.

Kepler's second law is a consequence of the fact that the force of gravity is a *central force.* This implies that the angular momentum of the planet-Sun system is a constant of the motion.

Kepler's first and third laws are a consequence of the inverse-square nature of the universal law of gravity. Newton's second law, together with the force law given by Equation 12.1, verifies that the period, T, and radius, r, of the orbit of a planet about the Sun are related by

$$T^2 = \left(\frac{4\pi^2}{GM_s}\right) r^3 \qquad [12.4]$$

Kepler's third law

where M_s is the mass of the Sun. Most planets have nearly circular orbits about the Sun. For elliptical orbits, Equation 12.4 is valid if r is replaced by the semi-major axis, a.

If an isolated system consists of a particle of mass m moving with a speed of v in the vicinity of a massive body of mass M, the *total energy* of the system is constant and is

$$E = \tfrac{1}{2}mv^2 - \frac{GMm}{r} \qquad [12.6]$$

If m moves in a circular orbit of radius r about M, where $M \gg m$, *the total energy of the system is*

$$E = -\frac{GMm}{2r} \qquad [12.9]$$

Total energy for circular orbits

The total energy is negative for any bound system—that is, one in which the orbit is closed, such as a circular or an elliptical orbit.

The minimum speed an object must have to escape the gravitational field of a uniform sphere of mass M and radius R is

$$v_{\text{esc}} = \sqrt{\frac{2GM}{R}} \qquad [12.12]$$

Escape speed

The Bohr model of the atom is successful in describing the spectra of atomic hydrogen and hydrogen-like ions. One of the basic assumptions of the model is that the electron can exist only in discrete orbits such that the angular momentum mvr is an integral multiple of $h/2\pi = \hbar$. Assuming circular orbits and a simple coulombic attraction between the electron and proton, the energies of the quantum states for hydrogen are calculated to be

$$E_n = -\frac{k_e e^2}{2a_0}\left(\frac{1}{n^2}\right) \qquad [12.22]$$

Allowed energies of the hydrogen atom

where k_e is the Coulomb constant, e is electronic charge, n is an integer called a *quantum number,* and $a_0 = 0.0529$ nm is the **Bohr radius.**

If the electron in the hydrogen atom makes a transition from an orbit whose quantum number is n_i to one whose quantum number is n_f, where $n_f < n_i$, the frequency of the radiation emitted by the atom, is

$$f = \frac{k_e e^2}{2a_0 h}\left(\frac{1}{n_f^2} - \frac{1}{n_i^2}\right) \qquad [12.24]$$

Frequency of radiation emitted from hydrogen

Using $E = hf = hc/\lambda$, one can calculate the wavelengths of the radiation for various transitions in which there is a change in quantum number, $n_i \rightarrow n_f$. The calculated wavelengths are in excellent agreement with observed line spectra.

Questions and Conceptual Exercises

1. If the gravitational force on an object is directly proportional to its mass, why don't large masses fall with greater acceleration than small ones?
2. The mass of the Moon was known before any human could travel there and measure the acceleration of falling objects on its surface. How was this mass determined?
3. The gravitational force that the Sun exerts on the Moon is about twice as great as the gravitational force that the Earth exerts on the Moon. Why doesn't the Sun pull the Moon away from the Earth during a total eclipse of the Sun?
4. Does the escape speed of a rocket depend on its mass? Explain.
5. Compare the energies required for a 10^5-kg spacecraft and a 10^3-kg satellite to reach the Moon.
6. Explain why it takes more fuel for a spacecraft to travel from the Earth to the Moon than for the return trip. Estimate the difference.
7. Is the magnitude of the potential energy associated with the Earth-Moon system greater than, less than, or equal to the kinetic energy of the Moon relative to the Earth?
8. Explain carefully why no work is done on a planet as it moves in a circular orbit around the Sun, even though a gravitational force is acting on the planet. What is the *net* work done on a planet during each revolution as it moves around the Sun in an elliptical orbit?
9. With reference to Figure 12.10, consider the area swept out by the radius vector in the time intervals $t_2 - t_1$ and $t_4 - t_3$. Under what condition is A_1 equal to A_2?
10. If A_1 equals A_2 in Figure 12.10, is the average speed of the planet in the time interval $t_2 - t_1$ less than, equal to, or greater than its average speed in the time interval $t_4 - t_3$?
11. At what position in its elliptical orbit is the speed of a planet a maximum? At what position is the speed a minimum?
12. If you were given the mass and radius of planet X, how would you calculate the acceleration due to gravity on the surface of this planet?
13. If a hole could be dug to the center of the Earth, do you think that the force on a mass m would still obey Equation 12.1 there? What do you think the force on m would be at the center of the Earth?
14. Henry Cavendish, in his 1798 experiment, was said to have "weighed the Earth." Explain this statement.
15. The *Voyager* spacecraft was accelerated toward escape speed from the Sun by the gravitational force from the giant planet Jupiter. How is this possible? Wouldn't the additional speed gained on approach be lost as the spacecraft receded away?
16. The *Apollo 13* spaceship developed trouble in the oxygen system about halfway to the Moon. Why did the mission continue on around the Moon and then return home, rather than immediately turn back to Earth?
17. A possible antisatellite weapon has been discussed: another satellite would be launched into a contrary orbit at the same height, but in the opposite direction. If this new satellite, filled with copper wire, were exploded, would the copper chaff hitting the original satellite do any damage?
18. Consider a satellite in a circular orbit about the Earth. By reference to the appropriate relations, explain whether the following quantities remain constant or change during the motion: (a) kinetic energy, (b) gravitational potential energy, (c) linear momentum, (d) angular momentum. Now consider a satellite in an elliptical orbit. Which of the preceding quantities are constant and which are variable?
19. Show that a billion-ton black hole has a Schwarzschild radius comparable to the radius of the proton (1 ton $\approx 10^3$ kg, proton radius $\approx 10^{-15}$ m).
20. Discuss the similarities and differences between the classical description of planetary motion and the Bohr model of the hydrogen atom.
21. The Bohr theory of the hydrogen atom is based upon several assumptions. Discuss those assumptions and their significance. Do any of them contradict classical physics?
22. Suppose that the electron in the hydrogen atom obeyed classical mechanics rather than quantum mechanics. Why should such a "hypothetical" atom emit a continuous spectrum rather than the observed line spectrum?
23. Explain the significance behind the fact that the total energy of the atom in the Bohr model is negative.

Problems

Section 12.1 Newton's Universal Law of Gravity Revisited

1. The masses of the Earth and Sun are 5.98×10^{24} kg and 1.99×10^{30} kg, respectively. Assuming the orbit of the Earth is circular, with a period of 365 days, determine (a) the radius of the Earth's orbit about the Sun and (b) the force of the Earth on the Sun.
2. Which exerts a greater force of gravitational attraction on objects on the Earth: the Moon or the Sun? Calculate these forces on a 1-kg mass.
3. The gravitational field on the surface of the Moon is about one-sixth that on the surface of the Earth. If the radius of the Moon is about one-quarter that of the Earth, find the ratio of the average mass density of the Moon to the average mass density of the Earth.
4. In the Bohr model of the ground state of the hydrogen atom, the electron and proton are separated by 5×10^{-11} m. If the only force between them were gravitational, what would be the period of revolution of the electron orbiting around the proton? The masses of the electron and proton are $m_e = 9.11 \times 10^{-31}$ kg and $m_p = 1.67 \times 10^{-27}$ kg.
5. Two ocean liners, each with a mass of 40 000 metric tons, are moving on parallel courses, 100 m apart. What is the magnitude of the acceleration of one of the liners toward the other due to the mutual gravitational attraction? (Treat the liners as spheres.)

Section 12.2 Kepler's Laws
Section 12.3 The Universal Law of Gravity and the Motions of Planets

6. Given that the Moon's period about the Earth is 27.32 days and the Earth-Moon distance is 3.84×10^8 m, estimate the mass of the Earth. Assume the orbit is circular. Why do you suppose your estimate is high?
7. The *Explorer VIII* satellite, placed into orbit November 3, 1960, to investigate the ionosphere, had the following orbit parameters: perihelion altitude, 459 km; aphelion altitude, 2289 km (both distances above the Earth's surface); period, 112.7 min. Find the ratio v_p/v_a.
8. Io, a small moon of the giant planet Jupiter, has an orbital period of 1.77 days and an orbital radius of 4.22×10^5 km. From these data, determine the mass of Jupiter.
9. At its aphelion the planet Mercury is 6.99×10^{10} m from the Sun, and at its perihelion it is 4.60×10^{10} m from the Sun. If its orbital speed is 3.88×10^4 m/s at the aphelion, what is its orbital speed at the perihelion?
10. Geosynchronous satellites orbit the Earth at a distance of 42 000 km from the Earth's center. Their angular velocity at this height is the same as the rotation of the Earth, so they appear stationary at certain locations in the sky. What is the force acting on a 1000-kg satellite at this height?
11. A "synchronous" satellite, which always remains above the same point on a planet's equator, is put in orbit around Jupiter to study the famous red spot. Jupiter rotates once every 9.9 h. Use the data of Table 12.2 to find the altitude of such an orbiting satellite on Jupiter.
12. Halley's comet approaches the Sun to within 0.57 A.U. (1 A.U. = 150×10^6 km), and its orbital period is 75.6 years. How far from the Sun will Halley's comet travel before it starts its return journey (Fig. 12.22)?
13. A satellite is in a circular orbit just above the surface of the Moon. (The radius of the Moon is 1738 km.) (a) What is the acceleration of the satellite? (b) What

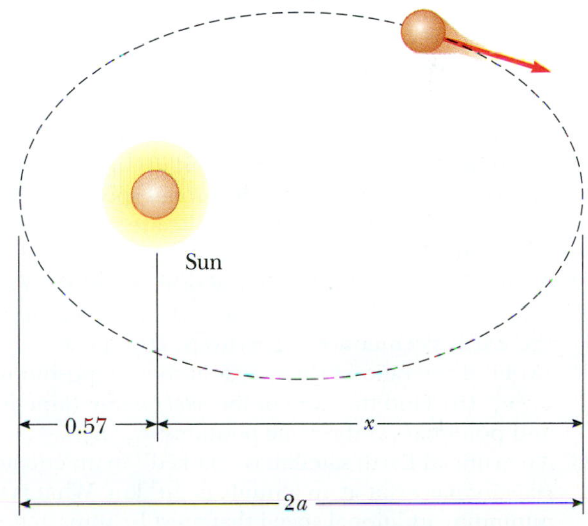

Figure 12.22 (Problem 12)

is the speed of the satellite? (c) What is the period of the satellite orbit?

14. A satellite of mass 200 kg is placed in Earth orbit at a height of 200 km above the surface. (a) Assuming a circular orbit, how long does the satellite take to complete one orbit? (b) What is the satellite's speed? (c) What is the minimum energy necessary to place this satellite in orbit (assuming no air friction)?
15. A satellite of the Earth has a mass of 100 kg and is at an altitude of 2×10^6 m. (a) What is the potential energy of the satellite at this location? (b) What is the magnitude of the gravitational force on the satellite?
16. A satellite orbits the Earth in an elliptical path in a time of 6×10^3 s. If the perihelion of the orbit is at a distance of 2×10^5 m from the Earth's *surface,* how far is the aphelion from the Earth? (*Hint:* First calculate the length of the semimajor axis. The radius of the Earth is 6.37×10^6 m.)

Section 12.4 Energy Considerations in Planetary and Satellite Motion

17. A spaceship is fired from the Earth's surface with an initial speed of 2.0×10^4 m/s. What will its speed be when it is very far from the Earth? (Neglect friction.)
18. The escape speed from the surface of the Earth is 11.2 km/s. Estimate the escape speed for a spacecraft from the surface of the Moon. The Moon has a mass $\frac{1}{81}$ that of Earth and a radius $\frac{1}{4}$ that of Earth.
19. (a) Calculate the minimum energy required to send a 3000-kg spacecraft from the Earth to a distant point in space where Earth's gravity is negligible. (b) If the journey is to take three weeks, what *average* power will the engines have to supply?
20. A 1000-kg satellite orbits the Earth at an altitude of 100 km. It is desired to increase the altitude of the orbit to 200 km. How much energy must be added to the system to effect this change in altitude?
21. A satellite moves in a circular orbit just above the surface of a planet. Show that the orbital speed, v, and the escape speed v_{esc} of the satellite are related by the expression $v_{esc} = \sqrt{2}v$.
22. A satellite moves in an elliptical orbit about the Earth such that, at perigee and apogee, the distances from the Earth's center are, respectively, D and $4D$. (a) Find the ratio of the speeds at the two positions: v_p/v_a. (b) Find the ratio of the *total* energy (kinetic and potential) at the same positions: E_p/E_a.
23. An artificial Earth satellite is "parked" in an equatorial circular orbit at an altitude of 10^3 km. What is the minimum additional speed that must be imparted to the satellite if it is to escape Earth's gravitational attraction? How does this compare with the minimum escape speed for leaving from the Earth's surface?
24. (a) What is the minimum speed necessary for a spacecraft to escape the Solar System, starting at the Earth's orbit? (b) *Voyager 1* achieved a maximum speed of 125 000 km/h on its way to photograph Jupiter. Beyond what distance from the Sun is this speed sufficient to escape the Solar System?
25. The planet Uranus has a mass about 14 times the Earth's mass, and its radius is equal to about 3.7 Earth radii. (a) By setting up ratios with the corresponding Earth values, find the acceleration due to gravity at the cloud tops of Uranus. (b) Ignoring the rotation of the planet, find the minimum escape speed from Uranus.
26. In a certain double-star system, the two stars rotate in circular orbits about their common center of mass. The stars are spherical; they have the same density, ρ; and their radii are R and $2R$. Their centers are a distance of $5R$ apart. Find the period, T, of the stars' orbital motion in terms of ρ, R, and G.

Section 12.5 Atomic Spectra and the Bohr Theory of Hydrogen

27. For a hydrogen atom in its ground state, use the Bohr model to compute (a) the orbital speed of the electron, (b) the kinetic energy of the electron, and (c) the electrical potential energy of the atom.
28. Radiation is emitted from a hydrogen atom, which undergoes a transition from the $n = 3$ state to the $n = 2$ state. Calculate (a) the energy, (b) the wavelength, and (c) the frequency of the emitted radiation.
29. How much energy is required to ionize hydrogen (a) when it is in the ground state? (b) when it is in the state for which $n = 3$?
30. Show that the speed of the electron in the nth Bohr orbit in hydrogen is given by

$$v_n = \frac{k_e e^2}{n\hbar}$$

31. A hydrogen atom is in its first excited state ($n = 2$). Using the Bohr theory of the atom, calculate (a) the radius of the orbit, (b) the linear momentum of the electron, (c) the angular momentum of the electron, (d) the kinetic energy, (e) the potential energy, and (f) the total energy.
32. (a) Calculate the angular momentum of the Moon due to its orbital motion about the Earth. In your calculation, use 3.84×10^8 m as the average Earth-Moon distance and 2.36×10^6 s as the period of the Moon in its orbit. (b) Determine the corresponding quantum number. (c) By what fraction would the Earth-Moon distance have to be increased to increase the quantum number by 1?

Additional Problems

33. Show that the escape speed from the surface of a planet of uniform density is directly proportional to the radius of the planet.
34. *Voyagers 1* and *2* surveyed the surface of Jupiter's moon Io and photographed active volcanoes spewing liquid sulfur to heights of 70 km above the surface of this moon. Estimate the speed with which the liquid sulfur left the volcano. Io's mass is 8.9×10^{22} kg, and its radius is 1820 km.
35. A cylindrical habitat in space, 6 km in diameter and 30 km long, has been proposed (by G. K. O'Neill, 1974). Such a habitat would have cities, land, and lakes on the inside surface and air and clouds in the center. This would all be held in place by rotation of the cylinder about its long axis. How fast would the cylinder have to rotate to imitate a 1-g gravity field at the walls of the cylinder?
36. While approaching a planet circling a distant star, a space traveler determines the planet's radius to be half that of the Earth. After landing on the surface, the traveler finds the acceleration due to gravity to be twice that on the Earth's surface. Find the mass of the planet, M_p, in terms of the mass of the Earth, M_e.
37. In introductory physics laboratories, a typical Cavendish balance for measuring the gravitational constant G uses lead spheres of masses 1.5 kg and 15 g whose centers are separated by about 4.5 cm. Calculate the gravitational force between these spheres, treating each sphere as a point mass at the center of the sphere.
38. As an astronaut, you observe a small planet to be spherical. After landing on the planet, you set off and walk straight ahead, to return to your spacecraft from the opposite side after completing a lap of 25 km. You then hold a hammer and feather at a height of 1.4 m, release them, and observe them to descend together to the surface in 29.2 s. Determine the mass of the planet.
39. For any planet, comet, or asteroid orbiting the Sun, Kepler's third law may be written $T^2 = kr^3$, where T is the orbital period and r is the semimajor axis of the orbit. (a) What is the value of k if T is measured in years and r is measured in A.U.s? One astronomical unit (A.U.) is the mean distance from the Earth to the Sun. (b) Use this new value of k to quickly find the orbital period of Jupiter if its mean radius from the Sun is 5.2 A.U.
40. A "treetop satellite" is a satellite that orbits just above the surface of a spherical object, assumed to offer no air resistance. (a) Find the period of a treetop satellite of Earth. (b) Prove that its speed is given by $v = \sqrt{4\pi G\rho/3}$, where ρ is the density of the planet.

41. Two hypothetical planets of masses m_1 and m_2 and radii r_1 and r_2, respectively, are at rest when they are an infinite distance apart. Because of their gravitational attraction, they head toward each other on a collision course. (a) When their center-to-center separation is d, find the speed of each planet and their *relative* velocity. (b) Find the kinetic energy of each planet *just* before they collide if $m_1 = 2 \times 10^{24}$ kg, $m_2 = 8 \times 10^{24}$ kg, $r_1 = 3 \times 10^6$ m, and $r_2 = 5 \times 10^6$ m. (*Hint:* Note that both energy and momentum are conserved.)

42. The maximum distance from the Earth to the Sun (at the aphelion) is 1.521×10^{11} m, and the distance of closest approach (at the perihelion) is equal to 1.471×10^{11} m. If the Earth's orbital speed at the perihelion is 3.027×10^4 m/s, determine (a) the Earth's orbital speed at the aphelion, (b) the kinetic and potential energy at the perihelion, and (c) the kinetic and potential energy at the aphelion. Is the total energy constant? (Neglect the effect of the Moon and other planets.)
43. After a supernova explosion, a star may undergo a gravitational collapse to reach an extremely dense state and become what is known as a neutron star, in which all the electrons and protons are squeezed together to form neutrons. A neutron star with a mass about equal to that of the Sun would have a radius of about 10 km. Find (a) the acceleration due to gravity at its surface, (b) the weight of a 70-kg man at its surface, and (c) the energy required to remove a neutron of mass 1.67×10^{-27} kg from its surface to infinity.
44. When the *Apollo 11* spacecraft orbited the Moon, its mass was 9.979×10^3 kg, its period was 119 min, and its mean distance from the Moon's center was 1.849×10^6 m. Assuming its orbit to be circular and the Moon to be a uniform sphere, find (a) the mass of the Moon, (b) the orbital speed of the spacecraft, and (c) the minimum energy required for the craft to leave the orbit and escape the Moon's gravity.
45. X-ray pulses from Cygnus X-1, a celestial x-ray source, have been recorded during high-altitude rocket flights. The signals can be interpreted as originating when a blob of ionized matter orbits a black hole with a period of 5 ms. If the blob were in a circular orbit about a black hole whose mass was 20 times the mass of the Sun, what would be the radius of the orbit?
46. Studies of the relationship of the Sun to its local galaxy, the Milky Way, have revealed that the Sun is near the outer edge of the galactic disc, about 30 000 light years from the center. Furthermore, it has been found that the Sun has an orbital speed of approximately 250 km/s around the galactic center. (a) What is the period of the Sun's galactic

motion? (b) What is the approximate mass of the Milky Way galaxy? Using the fact that the Sun is a typical star of mass 2×10^{30} kg, estimate the number of stars in our local galaxy.

47. Four possible transitions for a hydrogen atom are as follows:

(A) $n_i = 2$; $n_f = 5$ (B) $n_i = 5$; $n_f = 3$

(C) $n_i = 7$; $n_f = 4$ (D) $n_i = 4$; $n_f = 7$

(a) Which transition emits the shortest wavelength photon? (b) In which transition does the atom gain the most energy? (c) In which transition(s) does the atom lose energy?

48. (a) Determine the amount of work (in joules) that must be done on a 100-kg payload to elevate it to a height of 1000 km above the Earth's surface. (b) Determine the amount of additional work that is required to put the payload into circular orbit at this elevation.

49. Two stars of masses M and m, separated by a distance of d, revolve in circular orbits about their center of mass (Fig. 12.23). Show that each star has a period given by

$$T^2 = \frac{4\pi^2}{G(M + m)} d^3$$

(*Hint:* Apply Newton's second law to each star, and note that the center-of-mass condition requires that $Mr_2 = mr_1$, where $r_1 + r_2 = d$.)

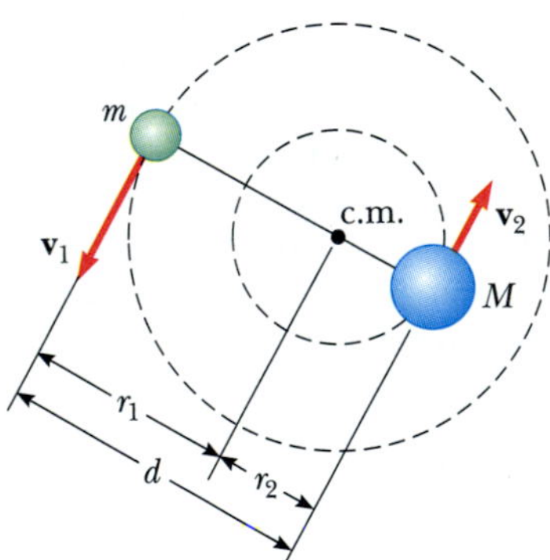

Figure 12.23 (Problem 49)

50. In 1978, astronomers at the U.S. Naval Observatory discovered that the planet Pluto has a moon, called Charon, that eclipses the planet every 6.4 days. If, from observation, the center-to-center separation between Pluto and Charon is 19 700 km, find the total mass $(M + m)$ of Pluto and its moon. (*Hint:* Use the result of Problem 49.)

51. In an effort to explain large meteor collisions with the Earth, scientists have postulated the existence of a solar companion star (which some call Nemesis) that is extremely dim and extremely far from the Sun. It moves through the axis of the cloud of comets, disturbing their orbits, every 3×10^7 years. If this hypothetical star has a mass 0.2 times the Sun's mass, determine the star's average distance from the Sun. $M_{Sun} = 2 \times 10^{30}$ kg.

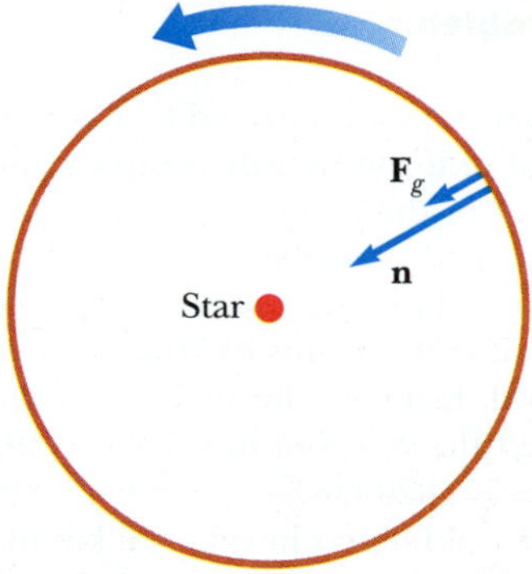

Figure 12.24 (Problem 52). *(Ringworld)*

52. In Larry Niven's science fiction novel *Ringworld,* a solid ring of material rotates about a star (Fig. 12.24). The rotational speed of the ring is 1.25×10^6 m/s, and its radius is 1.53×10^{11} m. The inhabitants of the ring world experience a normal contact force, N. Acting alone, this normal force would produce an inward acceleration of 9.90 m/s². Additionally, the star at the center of the ring exerts a gravitational force on the ring and its inhabitants. (a) Show that the total centripetal acceleration of the inhabitants is 10.2 m/s². (b) The difference between the total acceleration and the acceleration provided by the normal force is due to the gravitational attraction of the central star. Show that the mass of the star is approximately 10^{32} kg.

53. A rocket is given an initial speed vertically upward of $v_0 = 2\sqrt{Rg}$ at the surface of the Earth, which has radius R and surface free-fall acceleration g. The rocket motors are then cut off, and thereafter the rocket coasts under the action of gravitational forces only. (Ignore atmospheric friction and the Earth's rotation.) Derive an expression for the subsequent speed, v, as a function of the distance, r, from the center of the Earth in terms of g, R, and r.

54. Three stars, each of mass M, are situated so that each is at a vertex of an equilateral triangle with sides of length d. They are rotating in a plane about their common center of mass, keeping their relative distances constant because of their mutual gravitational attraction. Find the period T of this rotational motion in terms of M, G, and d.

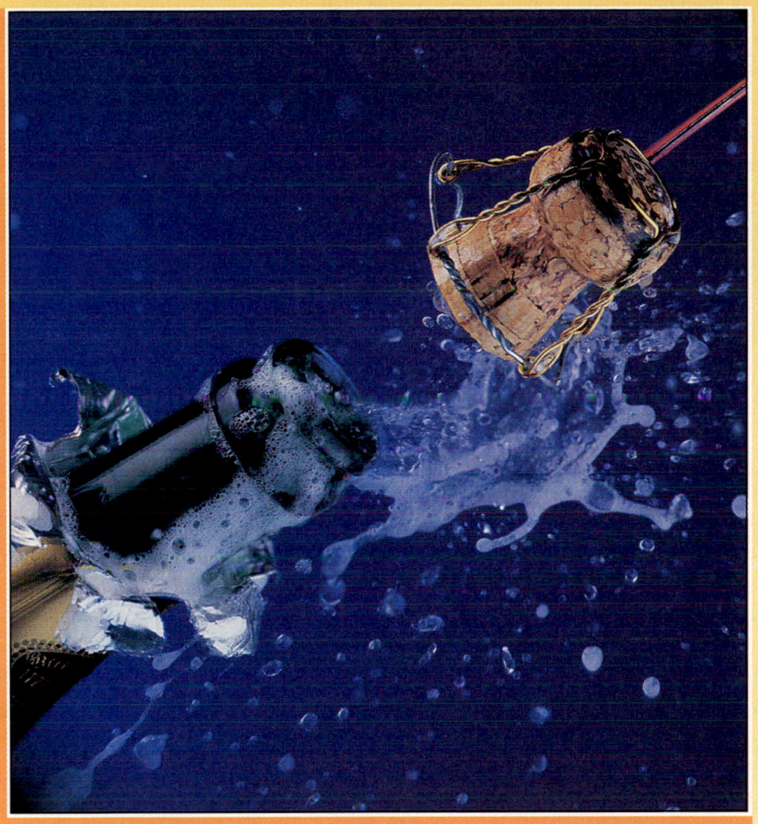

After the bottle is shaken, the cork is blown off. Contrary to common belief, shaking the champagne bottle does not increase the CO_2 pressure inside. Since the temperature of the bottle and its contents remain constant, the equilibrium pressure does not change, as can be shown by replacing the cork with a pressure gauge. Shaking the bottle displaces some CO_2 from the "head space" to bubbles within the liquid which become attached to the walls. If the bubbles remain attached to the walls, when the bottle is opened, the bubbles beneath the liquid level rapidly expand, expelling liquid in the process. So blame the mess on the bubbles. *(Steve Niedorf, The IMAGE Bank)*

Temperature and the Kinetic Theory of Gases

CHAPTER 13

Our study thus far has focused mainly on Newtonian mechanics, which explains a wide range of phenomena, such as the motion of baseballs, rockets, and planets. We now turn to the study of thermodynamics, which is concerned with the concepts of heat and temperature. As we shall see, thermodynamics is very successful in explaining the bulk properties of matter and the correlation between these properties and the mechanics of atoms and molecules.

Have you ever wondered how a refrigerator cools, or what types of transformations occur in an automobile engine or what happens to the kinectic energy of an object once the object comes to rest? The laws of thermodynamics and the concepts of heat and temperature enable us to answer such practical questions.

Many things can happen to an object when it is heated. Its size changes slightly, but it may also melt, boil, ignite, or even explode. The outcome depends upon the composition of the object, the degree to which it is heated, and its environment. In general, thermodynamics must concern itself with the physical and chemical transformations of matter in all of its forms: solid, liquid, gas, and plasma.

This chapter concludes with a study of ideal gases. We shall approach this study on two levels. The first will examine ideal gases on the macroscopic scale. Here we shall be concerned with the relationships among such quantities as pressure, volume, and temperature. On the second level, we shall examine gases on a microscopic scale, using a model that pictures the components of a gas as small particles. The latter approach, called the kinetic theory of gases, will help us to understand what happens on the atomic level to affect such macroscopic properties as pressure and temperature.

13.1 Pressure

Consider a fluid (a liquid or gas) at rest in a container. The fluid exerts a force perpendicular to the walls of the container at all points, and perpendicular to the surface of any object immersed in the fluid.

The force the fluid exerts on a surface is usually described by the pressure, which has units of force per unit area. The pressure at a specific point in a fluid can be measured with the device shown in Figure 13.1, which consists of an evacuated cylinder enclosing a light piston connected to a spring. As the device is submerged in a fluid, the fluid presses down on the top of the piston and compresses the spring until the inward force of the fluid is balanced by the outward spring force. The spring is calibrated in advance: a known force is applied to it, compressing it a given distance. If F is the magnitude of the normal force on the piston and A is the area of the piston, then the **pressure,** P, on the surface area is defined as the ratio of force to area:

Definition of pressure

$$P \equiv \frac{F}{A} \quad [13.1]$$

The pressure in a fluid is not the same at all points. To define the pressure at a specific point, consider a small area element, δA. If the force perpendicular to that element is F, then the pressure at that point is

$$P = \lim_{\delta A \to 0} \frac{F}{\delta A} = \frac{dF}{dA} \quad [13.2]$$

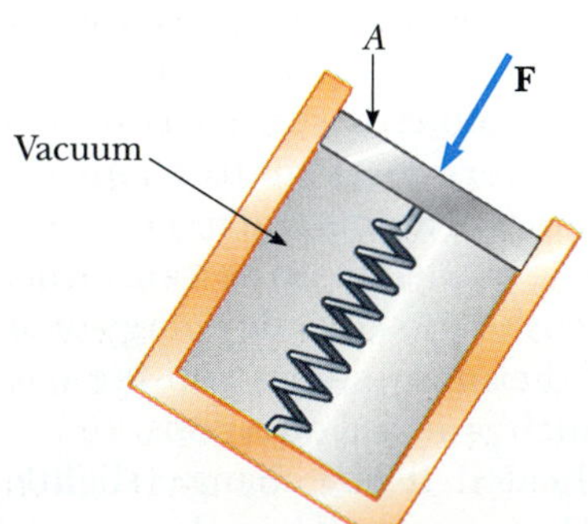

Figure 13.1
A simple device for measuring pressure.

Since pressure is force per unit area, it has units of N/m^2 in the SI system. Another name for the SI unit of pressure is **pascal** (Pa).

$$1 \text{ Pa} \equiv 1 \text{ N/m}^2 \qquad [13.3]$$

Pascal is the SI unit of pressure

One atmosphere (atm) of pressure, which is atmospheric pressure at sea level, is equal to 1.013×10^5 Pa.

13.2 Temperature and the Zeroth Law of Thermodynamics

We often associate the concept of **temperature** with how hot or cold an object feels when we touch it. Thus, our senses provide us with a qualitative indication of temperature. However, our senses are unreliable and often misleading. For example, if we remove a metal ice tray and a package of frozen vegetables from the freezer, the ice tray feels colder to the hand than the vegetables even though the two are at the same temperature. This is because metal is a better heat conductor than cardboard, and so the ice tray conducts heat from our hand more efficiently than does the cardboard package. What we need is a reliable and reproducible method for establishing the relative "hotness" or "coldness" of objects. Scientists have developed a variety of thermometers for making such quantitative measurements.

We are all familiar with the fact that two objects at different initial temperatures may eventually reach some intermediate temperature when placed in contact with each other. For example, if two soft drinks, one hot and the other cold, are placed in an insulated container, the two eventually reach an equilibrium temperature, with the cold one warming up and the hot one cooling off. Likewise, if a cup of hot coffee is cooled with an ice cube, the ice eventually melts and the coffee's temperature decreases.

In order to understand the concept of temperature, it is useful to first define two often-used phrases, *thermal contact* and *thermal equilibrium.* To grasp the meaning of thermal contact, imagine two objects placed in an insulated container so that they interact with each other but not with the rest of the world. If the objects are at different temperatures, energy is exchanged between them. The energy exchanged between objects because of a temperature difference is called **heat.** We shall examine the concept of heat in more detail in Chapter 14. For purposes of the current discussion, we shall assume that two objects are in **thermal contact** with each other if heat can be exchanged between them. **Thermal equilibrium** is the situation in which two objects in thermal contact with each other cease to have any exchange of heat.

Now consider two objects, A and B, that are not in thermal contact with each other, and a third object, C, that will be our thermometer. We wish to determine whether or not A and B would be in thermal equilibrium with each other, once placed in thermal contact. The thermometer (object C) is first placed in thermal contact with A until thermal equilibrium is reached. At that point, the thermometer's reading remains constant, and we record it. The thermometer is then placed in thermal contact with B, and its reading is recorded after thermal equilibrium is reached. If the two readings are the same, then A and B are in thermal equilibrium with each other.

We can summarize these results in a statement known as the **zeroth law of thermodynamics** (the law of equilibrium):

Zeroth law of thermodynamics

If bodies A and B are separately in thermal equilibrium with a third body C, then A and B will be in thermal equilibrium with each other if placed in thermal contact.

This statement, insignificant and obvious as it may seem, is easily proved experimentally and is very important because it can be used to define temperature. We can think of temperature as the property that determines whether or not an object is in thermal equilibrium with other objects. *Two objects in thermal equilibrium with each other are at the same temperature.*

13.3 Thermometers and Temperature Scales

Thermometers are devices used to define and measure the temperature of a system. All thermometers make use of a change in some physical property with temperature. Some of the physical properties used are (1) the change in volume of a liquid, (2) the change in length of a solid, (3) the change in pressure of a gas held at constant volume, (4) the change in volume of a gas held at constant pressure, (5) the change in electric resistance of a conductor, and (6) the change in color of a very hot object. For a given substance, a temperature scale can be established based on any one of these physical quantities.

The most common thermometer in everyday use consists of a mass of liquid —usually mercury or alcohol—that expands into a glass capillary tube when heated (Fig. 13.2). In this case the physical property is the change in volume of a liquid. One can define any temperature change to be proportional to the change in length of the liquid column. The thermometer can be calibrated by placing it in thermal contact with some natural systems that remain at constant temperature. One such system is a mixture of water and ice in thermal equilibrium at atmospheric pressure, which is defined to have a temperature of zero degrees Celsius, written 0°C; this temperature is called the ice point of water. Another commonly used system is a mixture of water and steam in thermal equilibrium at atmospheric pressure; its temperature is 100°C, the steam point of water. Once the liquid levels in the thermometer have been established at these two points, the column is divided into 100 equal segments, each denoting a change in temperature of one Celsius degree.

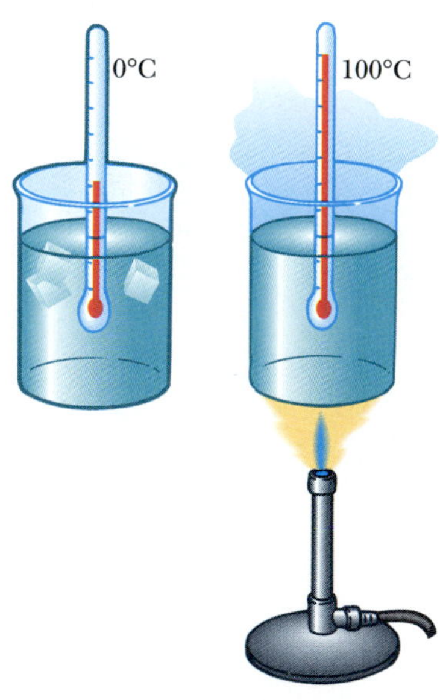

Figure 13.2 Schematic diagram of a mercury thermometer. As a result of thermal expansion, the level of the mercury rises as the mercury is heated from 0°C (the ice point) to 100°C (the steam point).

Thermometers calibrated in this way present problems when extremely accurate readings are needed. For instance, an alcohol thermometer calibrated at the ice and steam points of water might agree with a mercury thermometer only at the calibration points. Because mercury and alcohol have different thermal expansion properties, when one indicates a temperature of 50°C, say, the other may indicate a slightly different value. The discrepancies between different types of thermometers are especially large when the temperatures to be measured are far from the calibration points.

An additional practical problem of any thermometer is its limited temperature range. A mercury thermometer, for example, cannot be used below the freezing point of mercury, which is −39°C. To surmount these problems, we need a universal thermometer whose readings are independent of the substance used. The gas thermometer approaches this requirement.

The Constant-Volume Gas Thermometer and the Kelvin Scale

In a **gas thermometer**, the temperature readings are nearly independent of the substance used in the thermometer. One type of gas thermometer is the constant-volume unit shown in Figure 13.3. The physical property used in this device is the pressure variation with temperature of a fixed volume of gas. When the constant-volume gas thermometer was developed, it was calibrated using the ice and steam points of water, as follows. (A different calibration procedure, to be discussed shortly, is now used.) The gas flask was inserted into an ice bath, and mercury reservoir B was raised or lowered until the volume of the confined gas was at some value, indicated by the zero point on the scale. The height h, the difference between the levels in the reservoir and column A, indicated the pressure in the flask at 0°C. The flask was inserted into water at the steam point, and reservoir B was readjusted until the height in column A was again brought to zero on the scale, ensuring that the gas volume was the same as it had been in the ice bath (hence the designation "constant-volume"). This gave a value for the pressure at 100°C. These pressure and temperature values were then plotted on a graph, as in Figure 13.4. The line connecting the two points serves as a calibration curve for measuring unknown temperatures. If we wanted to measure the temperature of a substance, we would place the gas flask in thermal contact with the substance and adjust the column of mercury until the gas took on its specified volume. The height of the mercury column would tell us the pressure of the gas, and we could then find the temperature of the substance from the graph.

Now suppose that temperatures are measured with various gas thermometers containing different gases. Experiments show that the thermometer readings are nearly independent of the type of gas used, so long as the gas pressure is low and the temperature is well above the point at which the gas liquifies (Fig. 13.5). The agreement among thermometers using different gases improves as the pressure is reduced.

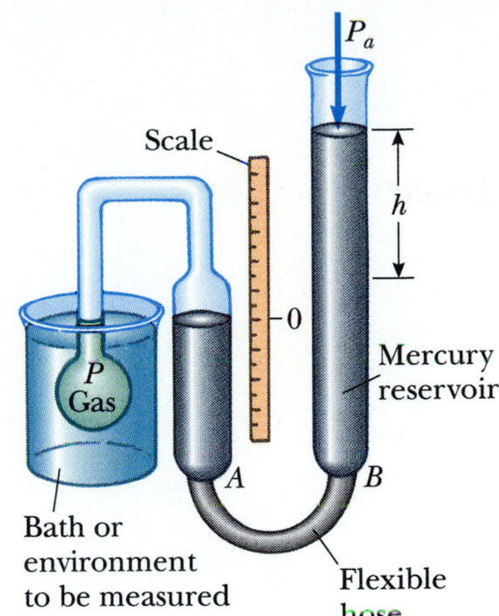

Figure 13.3 A constant-volume gas thermometer measures the pressure of the gas contained in the flask immersed in the bath. The volume of gas in the flask is kept constant by raising or lowering reservoir B such that the mercury level in column A remains constant.

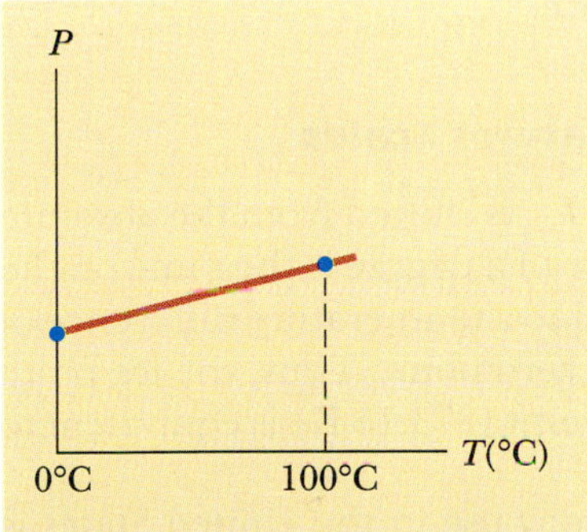

Figure 13.4 A typical graph of pressure versus temperature taken with a constant-volume gas thermometer. The dots represent known reference temperatures (the ice point and the steam point).

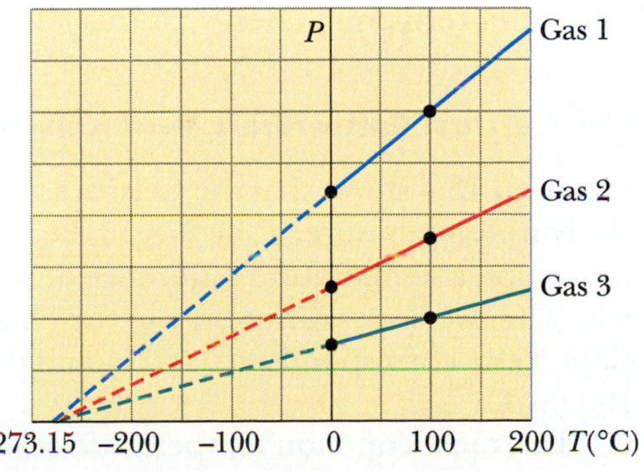

Figure 13.5 Pressure versus temperature for dilute gases. Note that, for all gases, the pressure extrapolates to zero at the unique temperature of −273.15°C.

If you extend the curves in Figure 13.5 back toward negative temperatures, you find, in every case, that the pressure is zero when the temperature is −273.15°C. This significant temperature is used as the basis for the Kelvin temperature scale, which sets −273.15°C as its zero point (0 K). The size of a degree on the Kelvin scale is identical to the size of a degree on the Celsius scale. Thus, the relationship that enables one to convert between these temperatures is

$$T_C = T - 273.15 \quad [13.4]$$

where T_C is the **Celsius temperature** and T is the **Kelvin temperature** (sometimes called the **absolute temperature**).

Early gas thermometers made use of ice and steam points according to the procedure just described. However, these points are experimentally difficult to duplicate. For this reason, a new procedure based on a single fixed point was adopted in 1954 by the International Committee on Weights and Measures. The **triple point of water,** which corresponds to *the single temperature and pressure at which water, water vapor, and ice can coexist in equilibrium,* was chosen as a convenient and reproducible reference temperature for the Kelvin scale. It occurs at a temperature of 0.01°C and a pressure of 4.58 mm of mercury. The temperature at the triple point of water on the Kelvin scale has been assigned a value of 273.16 kelvin (K). Thus, the SI unit of temperature, the **kelvin,** is *defined as 1/273.16 of the temperature of the triple point of water.*

The kelvin

Figure 13.6 shows the Kelvin temperature for various physical processes and structures. The temperature 0 K is often referred to as **absolute zero,** and as Figure 13.6 shows, this temperature has never been achieved, although laboratory experiments have come close.

What would happen to a substance if its temperature could reach 0 K? As Figure 13.5 indicates, the pressure it exerted on the walls of its container would be zero. In Section 13.6 we shall show that the pressure of a gas is proportional to the kinetic energy of the molecules of that gas. Thus, according to classical physics, the kinetic energy of the gas would go to zero, and there would be no motion at all of the individual components of the gas; hence, the molecules would settle out on the bottom of the container. Quantum theory, to be discussed in Chapter 29, modifies this statement to indicate that there would be some residual energy, called the zero-point energy, at this low temperature.

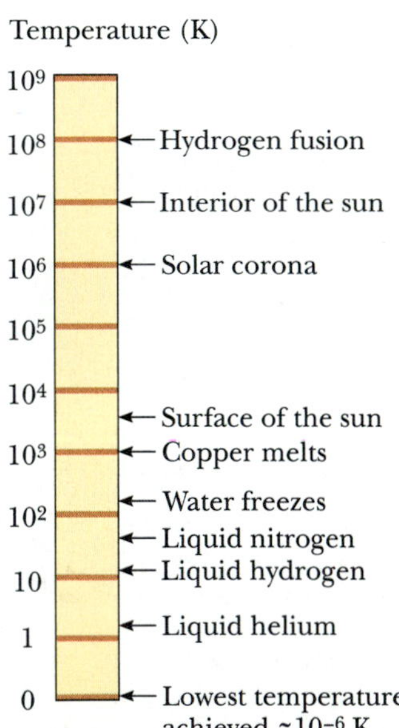

Figure 13.6
Absolute temperature at which various selected physical processes take place.

The Celsius, Fahrenheit, and Kelvin Temperatures Scales

Equation 13.4 shows that the Celsius temperature, T_C, is shifted from the absolute (Kelvin) temperature, T, by 273.15. Because the size of a degree is the same on the two scales, a temperature difference of 5°C, is equal to a temperature difference of 5 K. The two scales differ only in the choice of the zero point. Thus, the ice point (273.15 K) corresponds to 0.00°C, and the steam point (373.15 K) is equivalent to 100.00°C.

The most common temperature scale in everyday use in the United States is the **Fahrenheit scale.** This scale sets the temperature of the ice point at 32°F and the temperature of the steam point at 212°F. The relationship between the Celsius and Fahrenheit temperature scales is

$$T_F = \tfrac{9}{5}T_C + 32 \quad [13.5]$$

Equation 13.5 can easily be used to find a relationship between changes in temperature on the Celsius and Fahrenheit scales. It is left as a problem for you to show that if the Celsius temperature changes by ΔT_C, the Fahrenheit temperature changes by an amount ΔT_F, given by

$$\Delta T_F = \tfrac{9}{5}\,\Delta T_C \qquad [13.6]$$

Example 13.1 Converting Temperatures

On a day when the temperature reaches 50°F, what is the temperature in degrees Celsius and in kelvin?

Solution Let us solve Equation 13.5 for T_C and substitute $T_F = 50°\text{F}$:

$$T_C = \tfrac{5}{9}(T_F - 32) = \tfrac{5}{9}(50 - 32) = \boxed{10°\text{C}}$$

From Equation 13.4, we find that

$$T = T_C + 273.15 = \boxed{283.15\text{ K}}$$

Exercise On a hot summer day, the temperature is reported to be 30°C. What is the temperature in Fahrenheit degrees and in kelvin?

Answer 102°F; 303.15 K.

Example 13.2 Heating a Pan of Water

A pan of water is heated from 25°C to 80°C. What is the change in its temperature on the Kelvin scale and on the Fahrenheit scale?

Solution From Equation 13.4, we see that the change in temperature on the Celsius scale equals the change on the Kelvin scale. Therefore,

$$\Delta T = \Delta T_C = 80 - 25 = 55°\text{C} = \boxed{55\text{ K}}$$

From Equation 13.6, we find that the change in temperature on the Fahrenheit scale is greater than the change on the Celsius scale by the factor $\frac{9}{5}$:

$$\Delta T_F = \tfrac{9}{5}\,\Delta T_C = \tfrac{9}{5}(55) = \boxed{99°\text{F}}$$

13.4 Thermal Expansion of Solids and Liquids

Our discussion of the liquid thermometer made use of one of the best-known changes that occurs in a substance: as its temperature increases, its volume increases. (As we shall see shortly, in some materials the volume decreases when the temperature increases.) This phenomenon, known as **thermal expansion,** plays an important role in numerous applications. For example, thermal expansion joints must be included in buildings, concrete highways, and bridges to compensate for changes in dimensions with temperature variations.

Thermal expansion joints are used to separate sections of roadways on bridges. Without these joints, the surfaces would buckle due to thermal expansion on very hot days, or crack due to contraction on very cold days.

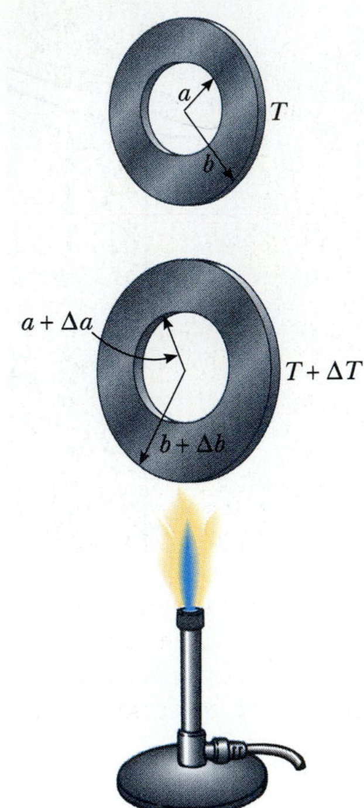

Figure 13.7
Thermal expansion of a homogeneous metal washer. As the washer is heated, all dimensions increase. (The expansion is exaggerated in this figure.)

The overall thermal expansion of an object is a consequence of the change in the average separation between its constituent atoms or molecules. To understand this, consider how the atoms in a solid substance behave. At ordinary temperatures, the atoms vibrate about their equilibrium positions with an amplitude of about 10^{-11} m, and the average spacing between the atoms is about 10^{-10} m. As the temperature of the solid increases, the atoms vibrate with larger amplitudes and the average separation between them increases. Consequently, the solid expands. If the thermal expansion of an object is sufficiently small compared with the object's initial dimensions, then the change in any dimension is, to a good approximation, dependent on the first power of the temperature change.

Suppose an object has an initial length of L_0 along some direction at some temperature. The length increases by ΔL for the change in temperature ΔT. Experiments show that, when ΔT is small enough, ΔL is proportional to ΔT and to L_0:

$$\Delta L = \alpha L_0 \, \Delta T \qquad [13.7]$$

or

$$L - L_0 = \alpha L_0 (T - T_0) \qquad [13.8]$$

where L is the final length, T is the final temperature, and the proportionality constant α is called the **average coefficient of linear expansion** for a given material and has units of $(°\text{C})^{-1}$.

It may be helpful to think of a thermal expansion as a magnification or a photographic enlargement. For example, as a metal washer is heated (Fig. 13.7), all dimensions, including the radius of the hole, increase according to Equation 13.7. Table 13.1 lists the average coefficient of linear expansion for various materials. Note that for these materials α is positive, indicating an increase in length with increasing temperature. This is not always the case. For example, some substances, such as calcite ($CaCO_3$), expand along one dimension (positive α) and contract along another (negative α) with increasing temperature.

Table 13.1
Average Coefficients of Linear Expansion for Some Materials Near Room Temperature

Material	Average Coefficient of Linear Expansion $[(°\text{C})^{-1}]$	Material	Average Coefficient of Volume Expansion $[(°\text{C})^{-1}]$
Aluminum	24×10^{-6}	Ethyl alcohol	1.12×10^{-4}
Brass and bronze	19×10^{-6}	Benzene	1.24×10^{-4}
Copper	17×10^{-6}	Acetone	1.5×10^{-4}
Glass (ordinary)	9×10^{-6}	Glycerin	4.85×10^{-4}
Glass (Pyrex)	3.2×10^{-6}	Mercury	1.82×10^{-4}
Lead	29×10^{-6}	Turpentine	9.0×10^{-4}
Steel	11×10^{-6}	Gasoline	9.6×10^{-4}
Invar (Ni-Fe alloy)	0.9×10^{-6}	Air	3.67×10^{-3}
Concrete	12×10^{-6}	Helium	3.665×10^{-3}

Because the linear dimensions of an object change with temperature, it follows that surface area and volume also change with temperature. Consider a square having an initial length of L_0 on a side and therefore an initial area of $A_0 = L_0^2$. As the temperature is increased, the length of each side increases to

$$L = L_0 + \alpha L_0 \,\Delta T$$

The new area, $A = L^2$, is

$$L^2 = (L_0 + \alpha L_0 \,\Delta T)(L_0 + \alpha L_0 \,\Delta T) = L_0^2 + 2\alpha L_0^2 \,\Delta T + \alpha^2 L_0^2 \,(\Delta T)^2$$

The last term in this expression contains the quantity $\alpha \,\Delta T$ raised to the second power. Because $\alpha \,\Delta T$ is much less than unity, squaring it makes it even smaller. Therefore, we can neglect this term to get a simpler expression:

$$A = L^2 = L_0^2 + 2\alpha L_0^2 \,\Delta T$$

$$A = A_0 + 2\alpha A_0 \,\Delta T$$

or

$$\Delta A = A - A_0 = \gamma A_0 \,\Delta T \qquad [13.9]$$

where $\gamma = 2\alpha$. The quantity γ is called the **average coefficient of area expansion.**

By a similar procedure, we can show that the *increase in volume* of an object accompanying a change in temperature is

$$\Delta V = \beta V_0 \,\Delta T \qquad [13.10]$$

where β, the **average coefficient of volume expansion,** is given by $\beta = 3\alpha$.

As Table 13.1 indicates, each substance has its own characteristic coefficients of expansion. For example, when the temperatures of a brass rod and a steel rod of equal length are raised by the same amount from some common initial value, the brass rod expands more than the steel rod because brass has a larger coefficient of expansion than steel. A simple device called a bimetallic strip that utilizes this principle is found in practical devices such as thermostats. The strip is made by securely bonding two different metals together. As the temperature of the strip increases, the two metals expand by different amounts, and the strip bends as in Figure 13.8.

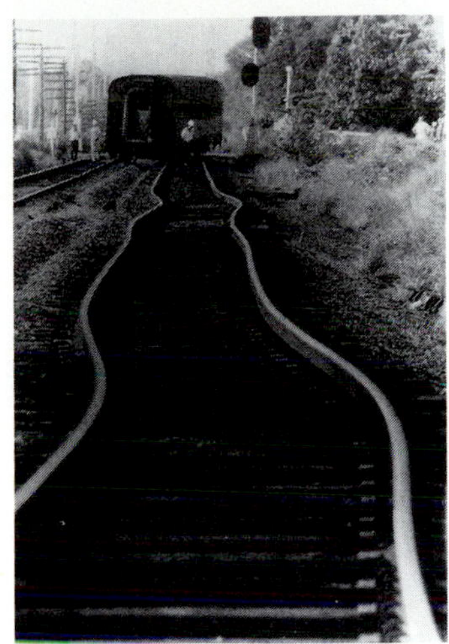

Thermal expansion: the extreme heat of a July day in Asbury Park, New Jersey, caused these railroad tracks to buckle. *(Wide World Photos)*

Example 13.3 Expansion of a Railroad Track

A steel railroad track has a length of 30 m when the temperature is 0°C. What is its length on a hot day, when the temperature is 40°C?

Solution If we use Table 13.1 and Equation 13.7, and note that the change in temperature is 40°C, we find that the *increase* in length is

$$\Delta L = \alpha L_0 \,\Delta T = [11 \times 10^{-6} (°\mathrm{C})^{-1}](30 \text{ m})(40°\mathrm{C}) = 0.013 \text{ m}$$

Therefore, the track's length at 40°C is 30.013 m.

Exercise What is the length of the same railroad track on a cold winter day, when the temperature is 0°F?

Answer 29.994 m.

Figure 13.8
The bimetallic strip bends as it is heated by a flame. It returns to its original shape when cooled to room temperature. Which way would it bend if it were cooled? *(Courtesy of CENCO)*

Example 13.4 Does the Hole Get Bigger or Smaller?

A hole of cross-sectional area 100 cm² is cut in a piece of steel at 20°C. What is the area of the hole if the steel is heated from 20°C to 100°C?

Solution A hole in a substance expands in exactly the same way as would a piece of the substance having the same shape as the hole. The change in the area of the hole can be found by using Equation 13.9.

$$\Delta A = \gamma A_0 \, \Delta T = [22 \times 10^{-6}(^\circ\text{C})^{-1}](100 \text{ cm}^2)(80^\circ\text{C}) = 0.18 \text{ cm}^2$$

Therefore, the area of the hole at 100°C is

$$A = A_0 + \Delta A = \boxed{100.18 \text{ cm}^2}$$

The Unusual Behavior of Water

Liquids generally increase in volume with increasing temperature and have volume expansion coefficients about ten times greater than those of solids. Water is an exception to this rule, as we can see from its density-versus-temperature curve in Figure 13.9. As the temperature increases from 0°C to 4°C, water contracts and thus its density increases. Above 4°C, water expands with increasing temperature. In other words, the density of water reaches a maximum value of 1000 kg/m³ at 4°C.

We can use this unusual thermal expansion behavior of water to explain why a pond freezes at the surface. When the atmospheric temperature drops from, say, 7°C to 6°C, the water at the surface of the pond also cools and consequently decreases in volume. This means that the surface water is denser than the water below it, which has not cooled and decreased in volume. As a result, the surface water sinks and warmer water from below is forced to the surface to be cooled. When the atmospheric temperature is between 4°C and 0°C, however, the surface

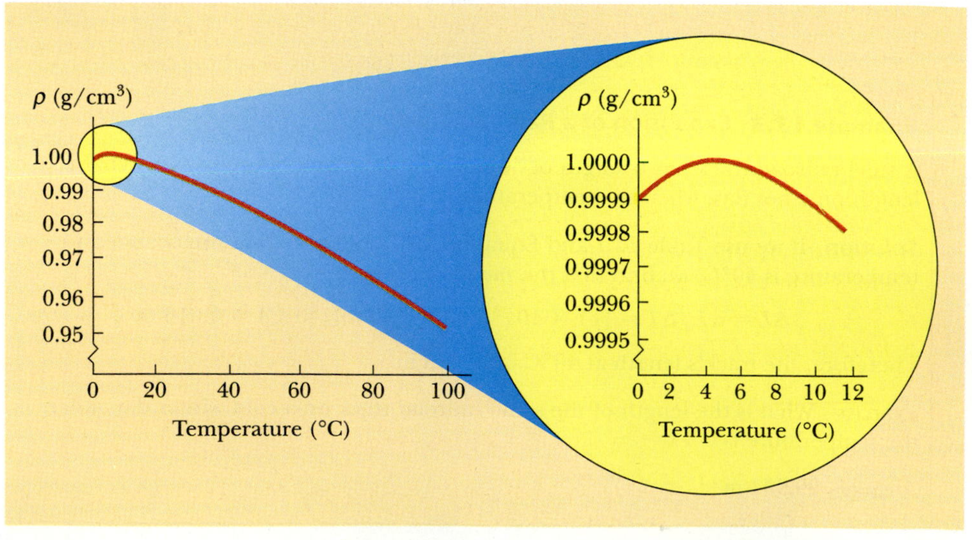

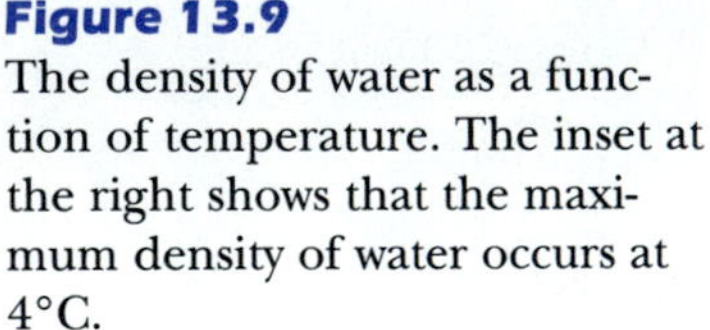
Figure 13.9 The density of water as a function of temperature. The inset at the right shows that the maximum density of water occurs at 4°C.

water expands as it cools, becoming less dense than the water below it. The mixing process stops, and eventually the surface water freezes. As the water freezes, the ice remains on the surface because ice is less dense than water. The ice continues to build up on the surface, while water near the bottom of the pool remains at 4°C. If this did not happen, fish and other forms of marine life would not survive.

13.5 Macroscopic Description of an Ideal Gas

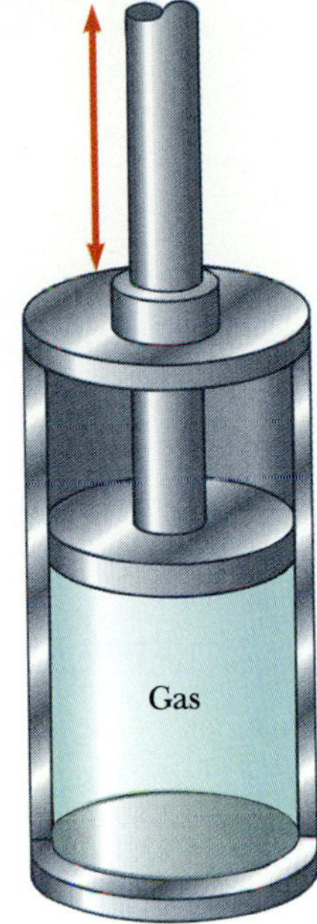

Figure 13.10
A gas confined to a cylinder whose volume can be varied with a movable piston.

In this section we shall be concerned with the properties of a gas of mass m confined to a container of volume V, pressure P, and temperature T. It is useful to know how these quantities are related. In general, the equation that interrelates these quantities, called the *equation of state,* is very complicated. However, if the gas is maintained at a very low pressure (or low density), the equation of state is found experimentally to be quite simple. Such a low-density gas is commonly referred to as an **ideal gas.** Most gases at room temperature and atmospheric pressure behave approximately as ideal gases.

It is convenient to express the amount of gas in a given volume in terms of the number of moles, n. As we learned in Section 1.2, one mole of any substance is that mass of the substance that contains Avogadro's number, $N_A = 6.022 \times 10^{23}$, of molecules. The number of moles, n, of a substance is related to its mass, m, through the expression

$$n = \frac{m}{M} \qquad [13.11]$$

where M is the molar mass of the substance, usually expressed in grams per mole. For example, the molar mass of molecular oxygen, O_2, is 32.0 g/mol. Therefore, the mass of one mole of oxygen is 32.0 g.

Now suppose an ideal gas is confined to a cylindrical container whose volume can be varied by means of a movable piston, as in Figure 13.10. We shall assume that the cylinder does not leak, and so the mass (or the number of moles) remains constant. For such a system, experiments provide the following information. First, when the gas is kept at a constant temperature, its pressure is inversely proportional to the volume (Boyle's law). Second, when the pressure of the gas is kept constant, the volume is directly proportional to the temperature (the law of Charles and Gay-Lussac). These observations can be summarized by the following **equation of state for an ideal gas:**

$$PV = nRT \qquad [13.12]$$

Ideal gas law

In this expression, called the **ideal gas law,** R is a constant for a specific gas that can be determined from experiments, and T is the absolute temperature in kelvin. Experiments on several gases show that as the pressure approaches zero, the quantity PV/nT approaches the same value of R for all gases. For this reason, R is called the **universal gas constant.** In SI units, where pressure is expressed in pascals and volume in cubic meters, the product PV has units of newton-meters, or joules, and R has the value

$$R = 8.31 \text{ J/mol} \cdot \text{K} \qquad [13.13]$$

The universal gas constant

If the pressure is expressed in atmospheres and the volume in liters ($1 \text{ L} = 10^3 \text{ cm}^3 = 10^{-3} \text{ m}^3$), then R has the value

$$R = 0.0821\ \text{L}\cdot\text{atm/mol}\cdot\text{K}$$

Using this value of R and Equation 13.12, one finds that the volume occupied by 1 mol of any gas at atmospheric pressure and 0°C (273 K) is 22.4 L.

The ideal gas law is often expressed in terms of the total number of molecules, N. Since the total number of molecules equals the product of the number of moles and Avogadro's number, N_A, we can write Equation 13.12 as

$$PV = nRT = \frac{N}{N_A}RT$$

$$PV = Nk_BT \qquad [13.14]$$

where k_B is called **Boltzmann's constant** and has the value

Boltzmann's constant

$$k_B = \frac{R}{N_A} = 1.38 \times 10^{-23}\ \text{J/K} \qquad [13.15]$$

Example 13.5 Squeezing a Tank of Gas

Pure helium gas is admitted into a leakproof cylinder containing a movable piston. The initial volume, pressure, and temperature of the gas are 15 liters, 2 atm, and 300 K. If the volume is decreased to 12 liters and the pressure increased to 3.5 atm, find the final temperature of the gas. Assume it behaves like an ideal gas.

Solution Because no gas escapes from the cylinder, the number of moles remains constant; therefore, use of $PV = nRT$ at the initial and final points gives

$$\frac{P_iV_i}{T_i} = \frac{P_fV_f}{T_f}$$

where i and f refer to the initial and final values. Solving for T_f, we get

$$T_f = \left(\frac{P_fV_f}{P_iV_i}\right)(T_i) = \frac{(3.5\ \text{atm})(12\ \text{liters})}{(2\ \text{atm})(15\ \text{liters})}(300\ \text{K}) = 420\ \text{K}$$

Example 13.6 Heating a Bottle of Air

A sealed glass bottle at 27°C contains air at atmospheric pressure and has a volume of 30 cm^3. It is then tossed into an open fire. When the temperature of the air in the bottle reaches 200°C, what is the pressure inside the bottle? Assume any volume changes of the bottle are small enough to be negligible.

Solution This example is approached in the same fashion as was Example 13.5. We start with the expression

$$\frac{P_iV_i}{T_i} = \frac{P_fV_f}{T_f}$$

Since the initial and final volumes of the gas are assumed equal, this expression reduces to

$$\frac{P_i}{T_i} = \frac{P_f}{T_f}$$

Before evaluating the final pressure, we must convert the given temperatures to kelvin: $T_i = 27°C = 300$ K and $T_f = 200°C = 473$ K. Thus,

$$P_f = \left(\frac{T_f}{T_i}\right)(P_i) = \left(\frac{473 \text{ K}}{300 \text{ K}}\right)(1 \text{ atm}) = \boxed{1.58 \text{ atm}}$$

Obviously, the higher the temperature, the higher the pressure exerted by the trapped air. Of course, if the pressure rises high enough, the bottle will shatter.

Exercise In this example we neglected the change in volume of the bottle. If the coefficient of volume expansion for glass is 27×10^{-6} $(°C)^{-1}$, find the magnitude of this volume change.

Answer 0.14 cm^3.

Example 13.7 The Volume of One Mole of Gas

Verify that 1 mol of oxygen occupies a volume of 22.4 liters at 1 atm and 0°C.

Solution Let us solve the ideal gas equation for V:

$$V = \frac{nRT}{P}$$

In our problem, the mass of the gas, m, is assumed to be 1 mol, M. Thus,

$$n = \frac{m}{M} = 1 \text{ mol}$$

Let us convert the temperature to the Kelvin scale and substitute into the ideal gas equation:

$$V = \frac{nRT}{P} = \frac{(1 \text{ mol})(0.0821 \text{ liter}\cdot\text{atm/mol}\cdot\text{K})(273 \text{ K})}{1 \text{ atm}} = \boxed{22.4 \text{ liters}}$$

This result has general validity: *one mole of any gas at standard temperature and pressure (STP) occupies a volume of 22.4 liters.*

Exercise Repeat the calculation in this example problem for hydrogen gas to show that it also occupies a volume of 22.4 liters at STP.

13.6 The Kinetic Theory of Gases

In the preceding section we discussed the properties of an ideal gas, using such quantities as pressure, volume, number of moles, and temperature. In this section we shall show that these macroscopic properties can be understood on the basis of what is happening on the atomic (microscopic) scale. In addition, we shall reexamine the ideal gas law in terms of the behavior of the individual molecules which make up the gas.

Because the molecular interactions in a gas are much weaker than those in solids and liquids, our present discussion will be restricted to the molecular behavior of gases.

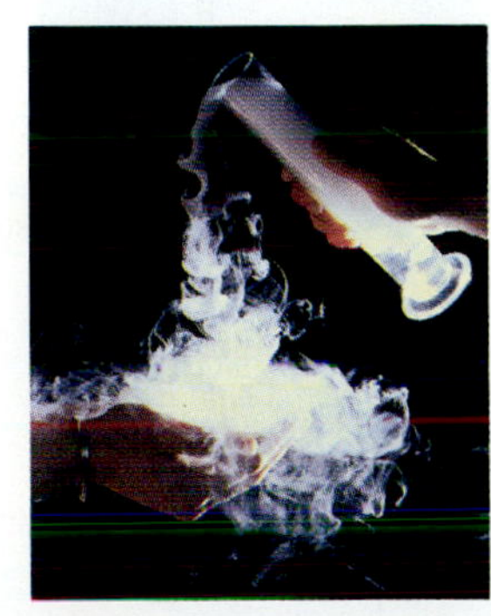

The glass vessel contains dry ice (solid carbon dioxide). The white cloud is carbon dioxide vapor, which is denser than air and hence falls from the vessel as shown. *(© R. Folwell/Science Photo Library)*

Molecular Model for the Pressure of an Ideal Gas

Assumptions of kinetic theory for an ideal gas

We shall first use the kinetic theory of gases to show that the pressure a gas exerts on the walls of its container is a consequence of the collisions of the gas molecules with the walls. We make the following assumptions:

1. *The number of molecules is large, and the average separation between them is large compared with their dimensions.* This means that the molecules occupy a negligible volume in the container.
2. *The molecules obey Newton's laws of motion, but as a whole they move randomly.* By "randomly" we mean that any molecule can move in any direction with any speed. We also assume that the distribution of speeds does not change in time, despite the collisions between molecules. That is, at any given moment, a certain percentage of molecules move at high speeds, a certain constant percentage move at low speeds, and so on.
3. *The molecules undergo elastic collisions with each other and with the walls of the container.* Thus, in the collisions *both kinetic energy and momentum are constant.*
4. *The forces between molecules are negligible except during a collision.* The forces between molecules are short-range, so the molecules interact with each other only during collisions.
5. *The gas under consideration is a pure substance;* that is, *all molecules are identical.*

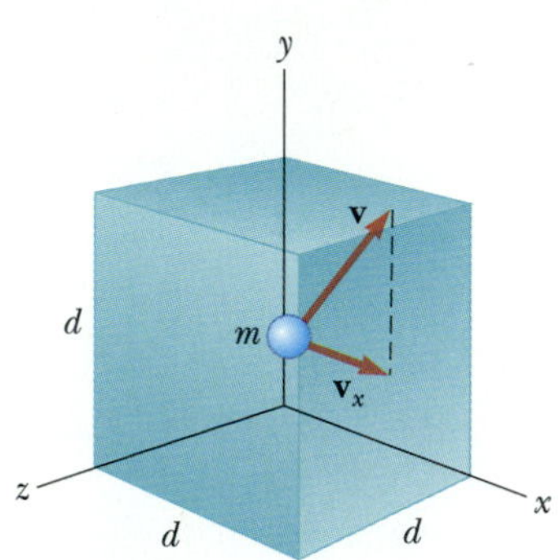

Figure 13.11 A cubical box with sides of length d containing an ideal gas. The molecule shown moves with velocity v.

Although we often picture an ideal gas as consisting of single atoms, molecular gases exhibit equally good approximations to ideal behavior at low pressures. Effects of molecular rotations or vibrations have no effect, on the average, on the motions considered here.

Now let us derive an expression for the pressure of N molecules of an ideal gas in a container of volume V. The container is a cube with edges of length d (Fig. 13.11). We shall focus our attention on one of these molecules, of mass m and assumed to be moving so that its component of velocity in the x direction is v_x (Fig. 13.12). As the molecule collides elastically with any wall, its velocity is reversed. Since the momentum, p, of the molecule is mv_x before the collision and $-mv_x$ after the collision, the *change in momentum of the molecule is*

$$\Delta p_x = mv_f - mv_i = -mv_x - mv_x = -2mv_x$$

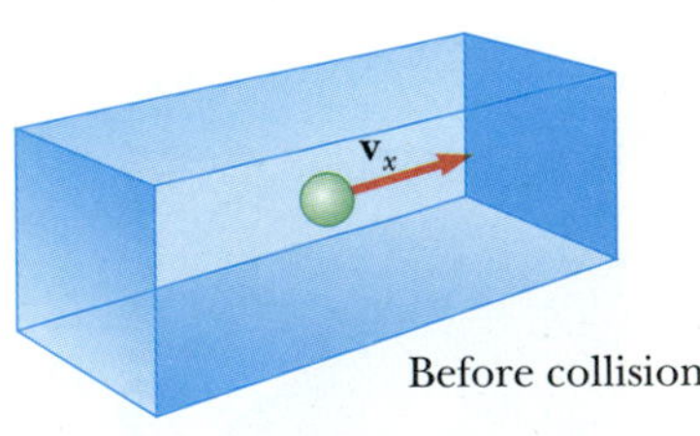

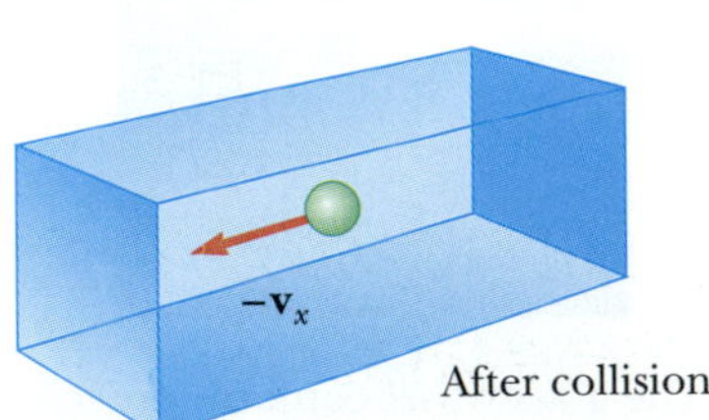

Figure 13.12 A molecule collides elastically with the wall of the container. Its x component of momentum is reversed, and momentum is imparted to the wall.

Because the momentum of the system consisting of the wall and the molecule must be constant, we see that, since the change in momentum of the molecule is $-2mv_x$, the change in momentum of the wall must be $2mv_x$. Applying the impulse-momentum theorem (Eq. 9.9) to the wall gives

$$F_1\,\Delta t = \Delta p = 2mv_x$$

In order for the molecule to make two collisions with the same wall, it must travel a distance of $2d$ in the x direction. Therefore, the time interval between two collisions with the same wall is

$$\Delta t = \frac{2d}{v_x}$$

The substitution of this result into the impulse-momentum equation enables us to express the force of a molecule on the wall:

$$F_1 = \frac{2mv_x}{\Delta t} = \frac{2mv_x^2}{2d} = \frac{mv_x^2}{d}$$

The total force exerted on the wall by all the molecules is found by adding the forces exerted by the individual molecules:

$$F = \frac{m}{d}\,(v_{x1}{}^2 + v_{x2}{}^2 + \cdots + v_{xN}{}^2)$$

In this equation, v_{x1} is the x component of velocity of molecule 1, v_{x2} is the x component of velocity of molecule 2, and so on. The summation terminates when we reach molecule N because there are N molecules in the container.

To proceed further, note that the average value of the square of the velocity in the x direction for the N molecules is

$$\overline{v_x^2} = \frac{v_{x1}{}^2 + v_{x2}{}^2 + \cdots + v_{xN}{}^2}{N}$$

Thus, the total force on the wall can be written

$$F = \frac{m}{d}\,N\overline{v_x^2}$$

Now let us focus on one molecule in the container and say that this molecule has velocity components v_x, v_y, and v_z. The Pythagorean theorem relates the square of the velocity to the square of these components:

$$v^2 = v_x^2 + v_y^2 + v_z^2$$

Hence, the average value of v^2 for all the molecules in the container is related to the average values of v_x^2, v_y^2, and v_z^2 according to the expression

$$\overline{v^2} = \overline{v_x^2} + \overline{v_y^2} + \overline{v_z^2}$$

However, the total is the same in any direction because the motion is completely random. Therefore,

$$\overline{v_x^2} = \overline{v_y^2} = \overline{v_z^2}$$

and we have

$$\overline{v^2} = 3\overline{v_x^2}$$

Thus, the total force on the wall is

$$F = \frac{N}{3}\left(\frac{m\overline{v^2}}{d}\right)$$

This expression allows us to find the pressure exerted on the wall:

$$P = \frac{F}{A} = \frac{F}{d^2} = \frac{1}{3}\left(\frac{N}{d^3}\,m\overline{v^2}\right) = \frac{1}{3}\left(\frac{N}{V}\right)(m\overline{v^2})$$

$$P = \frac{2}{3}\left(\frac{N}{V}\right)\left(\frac{1}{2}\,m\overline{v^2}\right) \qquad \textbf{[13.16]}$$

Pressure of an ideal gas

This result shows that *the pressure is proportional to the number of molecules per unit volume and to the average translational kinetic energy of the molecules,* $\frac{1}{2}m\overline{v^2}$. With this simplified model of an ideal gas, we have arrived at an important result that relates the large-scale quantity of pressure to an atomic quantity, the average value of the square of the molecular speed. Thus, we have a key link between the atomic world and the large-scale world.

You should note that Equation 13.16 verifies some features of pressure that are probably familiar to you. One way to increase the pressure inside a container is to increase the number of molecules per unit volume in the container. You do this when you add air to a tire. The pressure in the tire can also be increased by increasing the average translational kinetic energy of the molecules in the tire. As we shall see shortly, this can be accomplished by increasing the temperature of the gas inside the tire. That is why the pressure inside a tire increases as the tire heats up during long trips. The continuous flexing of the tires as they move along the road surface generates heat that is transferred to the air inside the tires, increasing the air's temperature, which in turn produces an increase in pressure.

Molecular Interpretation of Temperature

We can obtain some insight into the meaning of temperature by first writing Equation 13.16 in the form

$$PV = \tfrac{2}{3}N(\tfrac{1}{2}m\overline{v^2})$$

Let us now compare this with the equation of state for an ideal gas:

$$PV = Nk_BT$$

Recall that the equation of state is based on experimental facts concerning the macroscopic behavior of gases. Equating the right sides of these expressions, we find that

Temperature is proportional to average kinetic energy

$$T = \frac{2}{3k_B}(\tfrac{1}{2}m\overline{v^2}) \qquad [13.17]$$

That is, *temperature is a direct measure of average molecular kinetic energy.*

By rearranging Equation 13.17, we can relate the translational molecular kinetic energy to the temperature:

Average kinetic energy per molecule

$$\tfrac{1}{2}m\overline{v^2} = \tfrac{3}{2}k_BT \qquad [13.18]$$

That is, average translational kinetic energy per molecule is $\frac{3}{2}k_BT$. Since $\overline{v_x^2} = \frac{1}{3}\overline{v^2}$, it follows that

$$\tfrac{1}{2}m\overline{v_x^2} = \tfrac{1}{2}k_BT \qquad [13.19]$$

In a similar manner, for the y and z motions it follows that

$$\tfrac{1}{2}m\overline{v_y^2} = \tfrac{1}{2}k_BT \qquad \text{and} \qquad \tfrac{1}{2}m\overline{v_z^2} = \tfrac{1}{2}k_BT$$

Theorem of equipartition of energy

Thus, each translational degree of freedom contributes an equal amount of energy to the gas, namely, $\frac{1}{2}k_BT$. (In general, "degrees of freedom" refers to the number of independent means by which a molecule can possess energy.) A generalization of this result, known as the **theorem of equipartition of energy,** says that the energy of a system in thermal equilibrium is equally divided among all degrees of freedom.

The total translational kinetic energy of N molecules of gas is simply N times the average energy per molecule, which is given by Equation 13.18:

Total kinetic energy of *N* molecules

$$E = N(\tfrac{1}{2}m\overline{v^2}) = \tfrac{3}{2}Nk_BT = \tfrac{3}{2}nRT \qquad [13.20]$$

Table 13.2
Some Root-Mean-Square Speeds

Gas	Molar Mass, M (kg/mol)	v_{rms} at 20°C (m/s)
H_2	2.02×10^{-3}	1902
He	4.0×10^{-3}	1352
H_2O	18×10^{-3}	637
Ne	20.1×10^{-3}	603
N_2 or CO	28×10^{-3}	511
NO	30×10^{-3}	494
CO_2	44×10^{-3}	408
SO_2	48×10^{-3}	390

where we have used $k_B = R/N_A$ for Boltzmann's constant and $n = N/N_A$ for the number of moles of gas. This result, together with Equation 13.16, implies that the pressure exerted by an ideal gas depends only on the number of molecules per unit volume and the temperature.

The square root of $\overline{v^2}$ is called the *root-mean-square* (rms) *speed* of the molecules. From Equation 13.18 we get, for the rms speed,

$$v_{\text{rms}} = \sqrt{\overline{v^2}} = \sqrt{\frac{3k_BT}{m}} = \sqrt{\frac{3RT}{M}} \qquad [13.21]$$

Root-mean-square speed

where M is the molar mass in kg/mol. This expression shows that, at a given temperature, lighter molecules move faster, on the average, than heavier molecules. For example, hydrogen, with a molar mass of 2×10^{-3} kg/mol, moves four times as fast as oxygen, whose molar mass is 32×10^{-3} kg/mol. The rms speed is not the speed at which a gas molecule moves across a room, since such a molecule undergoes several billion collisions per second with other molecules under standard conditions.

Table 13.2 lists the rms speeds for various molecules at 20°C.

Example 13.8 **A Tank of Helium**

A tank of volume 0.3 m^3 contains 2 mol of helium gas at 20°C. Assume the helium behaves like an ideal gas.

(a) Find the total internal energy of the system.

Solution Using Equation 13.20 with $n = 2$ and $T = 293$ K, we get

$$E = \tfrac{3}{2}nRT = \tfrac{3}{2}(2 \text{ mol})(8.31 \text{ J/mol}\cdot\text{K})(293 \text{ K})$$

$$= 7.30 \times 10^3 \text{ J}$$

(b) What is the average kinetic energy per molecule?

Solution From Equation 13.18, we see that the average kinetic energy per molecule is

$$\tfrac{1}{2}m\overline{v^2} = \tfrac{3}{2}k_BT = \tfrac{3}{2}(1.38 \times 10^{-23}\text{ J/K})(293\text{ K})$$

$$= 6.07 \times 10^{-21}\text{ J}$$

Exercise Using the fact that the molar mass of helium is 4×10^{-3} kg/mol, determine the rms speed of the atoms at 20°C.

Answer 1.35×10^3 m/s.

Summary

The **zeroth law of thermodynamics** states that if two objects, A and B, are separately in thermal equilibrium with a third object, then A and B are in thermal equilibrium with each other.

The **pressure,** P, in a fluid is the force per unit area that the fluid exerts on any surface:

Average pressure

$$P \equiv \frac{F}{A} \qquad [13.1]$$

In the SI system, pressure has units of N/m², and 1 N/m² = 1 pascal (Pa).

The relationship between T_C, the *Celsius temperature,* and T, the *Kelvin (absolute) temperature,* is

$$T_C = T - 273.15 \qquad [13.4]$$

The relationship between the *Fahrenheit* and *Celsius* temperatures is

$$T_F = \tfrac{9}{5}T_C + 32 \qquad [13.5]$$

When a substance is heated, it generally expands. If an object has an initial length of L_0 at some temperature and undergoes a change in temperature of ΔT, its length changes by the amount ΔL, which is proportional to the object's initial length and the temperature change:

Equation of linear thermal expansion

$$\Delta L = \alpha L_0\,\Delta T \qquad [13.7]$$

The parameter α is called the **average coefficient of linear expansion.**

The change in area of a substance is given by

$$\Delta A = \gamma A_0\,\Delta T \qquad [13.9]$$

where γ is the **average coefficient of area expansion** and is equal to 2α.

The change in volume of most substances is proportional to the initial volume, V_0, and the temperature change, ΔT:

$$\Delta V = \beta V_0\,\Delta T \qquad [13.10]$$

where β is the **average coefficient of volume expansion** and is equal to 3α.

An **ideal gas** is one that obeys the equation

Equation of state for an ideal gas

$$PV = nRT \qquad [13.12]$$

where P is the pressure of the gas, V is its volume, n is the number of moles of gas, R is the universal gas constant (8.31 J/mol·K), and T is the absolute temperature in kelvin. A real gas at very low pressures behaves approximately as an ideal gas.

The **pressure** of N molecules of an ideal gas contained in a volume V is given by

$$P = \frac{2}{3}\left(\frac{N}{V}\right)\left(\frac{1}{2}m\overline{v^2}\right) \qquad [13.16]$$

Pressure and molecular kinetic energy

where $\frac{1}{2}m\overline{v^2}$ is the **average kinetic energy per molecule.**

The average kinetic energy of the molecules of a gas is directly proportional to the absolute temperature of the gas:

$$\frac{1}{2}m\overline{v^2} = \frac{3}{2}k_BT \qquad [13.18]$$

Average kinetic energy per molecule

where k_B is **Boltzmann's constant** (1.38×10^{-23} J/K).

The **root-mean-square** (rms) speed of the molecules of gas is

$$v_{\text{rms}} = \sqrt{\frac{3k_BT}{m}} = \sqrt{\frac{3RT}{M}} \qquad [13.21]$$

Root-mean-square speed

Questions and Conceptual Exercises

1. A woman wearing high-heeled shoes is invited into a home in which the kitchen has a newly installed vinyl floor covering. Why should the homeowner be concerned?
2. Is it possible for two objects to be in thermal equilibrium if they are not in contact with each other? Explain.
3. Why should the amalgam used in dental fillings have the same coefficient of expansion as a tooth? What would occur if the two were mismatched?
4. A piece of copper is dropped into a beaker of water. If the water's temperature rises, what happens to the temperature of the copper? When will the water and copper be in thermal equilibrium?
5. If a jar lid is screwed on too tightly to remove, it is sometimes possible to loosen the lid after holding it under hot water. Explain.
6. Estimate the force of the atmosphere on a person's chest in view of the fact that atmospheric pressure is about 10^5 Pa.
7. A microwave oven is used to heat food in a sealed pouch. Why should the pouch be pricked with a fork before it is heated?
8. What does the ideal gas law predict about the volume of a gas at absolute zero? Why is this prediction incorrect?
9. If a helium-filled balloon is placed in a freezer, will its volume increase, decrease, or remain the same?
10. What happens to a helium-filled rubber balloon released into the air? Does it expand or contract? Does it stop rising at some height?
11. Explain why a column of mercury in a thermometer first descends slightly and then rises when placed in hot water.
12. A steel wheel bearing has an inside diameter 0.1 mm smaller than the diameter of the axle. How can it be made to fit onto the axle without removing any material?
13. Markings to indicate length are placed on a steel tape in a room at a temperature of 22°C. Are measurements made with the tape on a day when the temperature is 27°C too long, too short, or accurate? Defend your answer.
14. Determine the number of grams in one mole of each of the following gases: (a) hydrogen, (b) helium, and (c) carbon monoxide.
15. Why is it necessary to use absolute temperature when using the ideal gas law?
16. An inflated rubber balloon filled with air is immersed in a flask of 77-K liquid nitrogen. Describe what happens to the balloon, assuming that it remains flexible while being cooled.
17. Two identical cylinders at the same temperature each contain the same kind of gas. If the volume of cylinder A is three times greater than the volume of cylinder B, what can you say about the relative pressures in the cylinders?

Figure 13.13 (Question 18). *(Courtesy of CENCO)*

18. When the metal ring and metal sphere in Figure 13.13 are both at room temperature, the sphere does not fit through the ring. After the ring is heated, the sphere can pass through the ring. Why does this occur?
19. After food is cooked in a pressure cooker, why is it very important to cool the container with cold water before attempting to remove the lid?
20. A gas consists of a mixture of He and N_2 molecules. Do the lighter He molecules travel faster than the N_2 molecules? Explain.
21. Ideal gas is contained in a vessel at a temperature of 300 K. The temperature is increased to 900 K. (a) By what factor does the rms speed of each molecule change? (b) By what factor does the pressure in the vessel change?

Problems

Section 13.1 Pressure

1. Use the facts that normal atmospheric pressure is 1.013×10^5 Pa and the Earth's radius is 6.37×10^6 m to estimate the mass of the Earth's atmosphere.
2. The window of a certain underwater sea vessel is designed to withstand a water pressure of 8 atmospheres. If the window is a circle of diameter 50 cm, what is the force of the water on the window?

Section 13.3 Thermometers and Temperature Scales

3. The pressure in a constant-volume gas thermometer is 0.700 atm at 100°C and 0.512 atm at 0°C. (a) What is the temperature when the pressure is 0.0400 atm? (b) What is the pressure at 450°C?
4. A constant-volume gas thermometer is calibrated in dry ice (−80°C) and in boiling ethyl alcohol (78°C). The two pressures are 0.900 atm and 1.635 atm. (a) What value of absolute zero does the calibration yield? (b) What pressures would be found at the freezing and boiling points of water?
5. Convert the following temperatures to Celsius and kelvin: (a) the normal human body temperature, 98.6°F; (b) the air temperature on a cold day, −5°F.
6. Show that the temperature −40° is unique in that it has the same numerical value on the Celsius and Fahrenheit scales.
7. Show that if the temperature on the Celsius scale changes by ΔT_C, the Fahrenheit temperature changes by ΔT_F, which equals $\frac{9}{5}\,\Delta T_C$.
8. The melting point of gold is 1064°C, and the boiling point is 2660°C. (a) Express these temperatures in kelvin. (b) Compute the difference of these temperatures in Celsius degrees and kelvin degrees, and compare the two numbers.
9. Liquid nitrogen has a boiling point of −195.81°C at atmospheric pressure. Express this temperature in (a) degrees Fahrenheit, and (b) kelvin.
10. The highest recorded temperature on Earth was 136°F, at Azizia, Libya, in 1922. The lowest recorded temperature was −127°F, at Vostok Station, Antarctica, in 1960. Express these temperature extremes in degrees Celsius.
11. The temperature of one northeastern state varies from 105°F in the summer to −25°F in winter. Express this range of temperature in degrees Celsius.
12. The boiling point of sulfur is 444.60°C. The melting point is 586.1°F below the boiling point. (a) Determine the melting point in degrees Celsius. (b) Find the melting and boiling points in degrees Fahrenheit.

Section 13.4 Thermal Expansion of Solids and Liquids (Use Table 13.1)

13. The New River Gorge bridge in West Virginia is a steel arch bridge 518 m in length. How much does its length change between temperature extremes of −20°C and 35°C?
14. A copper steam pipe is 2 m long and is installed in a basement when the temperature is 20°C. What is the length of the pipe when it carries steam at 120°C?
15. A copper telephone wire is strung, with little sag, be-

tween two poles that are 35 m apart. How much longer is the wire on a summer day, with $T_C = 35°C$, than on a winter day, with $T_C = -20°C$?

16. The volume coefficient of expansion for carbon tetrachloride is 5.81×10^{-4} $(°C)^{-1}$. If a 50-gal steel container is filled completely with carbon tetrachloride when the temperature is 10°C, how much will spill over when the temperature rises to 30°C?

17. A brass ring of diameter 10.00 cm at 20°C is heated and slipped over an aluminum rod of diameter 10.01 cm at 20°C. Assume the coefficients of linear expansion are constant. (a) To what temperature must this combination be cooled to separate it? Is that temperature attainable? (b) What if the aluminum rod were 10.02 cm in diameter?

18. The concrete sections of a certain superhighway are designed to be 25 m long. The sections are poured and cured at 10°C. What minimum spacing should the engineer leave *between the sections* to eliminate "buckling" if the concrete is to reach a temperature of 50°C?

19. A square hole (8.0 cm on a side) is cut in a sheet of copper. Calculate the change in the area of this hole if the temperature of the sheet is increased by 50 K.

20. A pair of eyeglass frames is made of epoxy plastic. At room temperature (assume 20.0°C) the frames have circular lens holes 2.2 cm in radius. To what temperature must the frames be heated in order to insert lenses 2.21 cm in radius? Assume that the coefficient of linear expansion for epoxy is 1.3×10^{-4} $(°C)^{-1}$.

21. An automobile fuel tank is filled to the brim with 45 liters (11.9 gal) of gasoline at 10°C. Immediately afterward the vehicle is parked in the sun, where the temperature is 35°C. How much gasoline overflows from the tanks as a result of the expansion? (Neglect the expansion of the tank.)

22. A volumetric Pyrex glass flask is calibrated at 20°C. It is filled to the 100-mL mark with 35°C acetone. (a) What is the volume of the acetone when it is cooled to 20°C? (b) How significant is the change in volume of the flask itself?

23. The active element of a certain laser is made of a glass rod 30 cm long by 1.5 cm in diameter. If the temperature of the rod increases by 65°C, find the increase in the rod's (a) length, (b) diameter, and (c) volume. ($\alpha = 9 \times 10^{-6} (°C)^{-1}$.)

Section 13.5 Macroscopic Description of an Ideal Gas

24. Gas is contained in an 8-liter vessel at a temperature of 20°C and a pressure of 9 atm. (a) Determine the number of moles of gas in the vessel. (b) How many molecules are there in the vessel?

25. A 50-g sample of dry ice (solid CO_2) is placed in a 4-liter container. The system is sealed and allowed to reach room temperature (20°C). By how much does the pressure increase inside the container when the dry ice turns to gas?

26. A helium-filled balloon has a volume of 1 m^3. As it rises in the Earth's atmosphere, its volume expands. What is its new volume (in cubic meters) if its original temperature and pressure were 20°C and 1 atm and its final temperature and pressure are −40°C and 0.1 atm?

27. An auditorium has the dimensions 10 m × 20 m × 30 m. How many molecules of air are needed to fill the auditorium at 20°C and 1 atm pressure?

28. A sample of ideal gas with a molar mass of 4 g/mol is in a thermodynamic state, with $P = 1.2$ atm, $V = 8.8$ L, and $T = 85°C$. Determine the mass of this sample of gas.

29. The mass of a hot air balloon and its cargo (not including the air inside) is 200 kg. The air outside is at a temperature of 10°C and a pressure of 1 atm = 1.013×10^5 Pa. The volume of the balloon is 400 m^3. To what temperature must the air in the balloon be heated before the balloon will lift off? (Air density at 10°C is 1.25 kg/m^3.)

30. A tank with a volume of 0.1 m^3 contains helium gas at a pressure of 150 atm. How many balloons can the tank blow up if each filled balloon is a sphere 30 cm in diameter, at an absolute pressure of 1.2 atm?

31. An automobile tire is inflated with air that is originally at 10°C and normal atmospheric pressure. During the process, the air is compressed to 28% of its original volume and the temperature is increased to 40°C. What is the tire pressure? After the car is driven at high speed, the tire air temperature rises to 85°C and the interior volume of the tire increases by 2%. What is the new tire pressure? Express each answer in pascals.

32. Nine grams of water are placed in a 2-L pressure cooker and heated to 500°C. What is the pressure inside the container? ($R = 0.082$ L-atm/mol·K.)

33. One mole of oxygen gas is at a pressure of 6 atm and a temperature of 27°C. (a) If the gas is heated at constant volume until the pressure triples, what is the final temperature? (b) If the gas is heated until both the pressure and volume are doubled, what is the final temperature?

34. A swimmer has 0.82 L of dry air in his lungs when he dives into a lake. Assuming the pressure of the dry air is 95% of the external pressure at all times, what is the volume of the dry air at a depth of 10.0 m? Assume that atmospheric pressure at the surface is equal to 1.013×10^5 Pa.

35. An air bubble has a volume of 1.5 cm^3 when released by a submarine 100 m below the surface of a lake.

What is the volume of the bubble when it reaches the surface? Assume that the temperature of the air in the bubble remains constant during ascent.

36. A weather balloon is designed to expand to a maximum radius of 20 m when in flight at its working altitude, where the air pressure is 0.03 atm and the temperature is 200 K. If the balloon is filled at atmospheric pressure and 300 K, what is its radius at lift-off?

37. (a) Estimate the total mass of air inside a typical-size house on a day when the temperature inside is 0°F. (Assume a molecular weight of 28.8 g/mol for air.) (b) How much mass must enter or leave the house if the temperature increases to 100°F?

Section 13.6 The Kinetic Theory of Gases

38. Use the definition of Avogadro's number to find the mass of a helium atom.

39. A sealed cubical container 20 cm on a side contains three times Avogadro's number of molecules at a temperature of 20°C. Find the force exerted by the gas on one of the walls of the container.

40. (a) Calculate the rms speed of an H_2 molecule when the temperature is 100°C. (b) Repeat the same calculation for an N_2 molecule.

41. A cylinder contains a mixture of helium and argon gas in equilibrium at a temperature of 150°C. (a) What is the average kinetic energy of each type of molecule? (b) What is the rms speed of each type of molecule?

42. What is the temperature at which the rms speed of nitrogen molecules equals the rms speed of helium at 20°C?

43. At what temperature would the rms speed of helium atoms (mass $= 6.66 \times 10^{-27}$ kg) equal (a) the escape speed from Earth, 1.12×10^4 m/s, and (b) the escape speed from the Moon, 2.37×10^3 m/s?

44. In a period of 1 s, 5×10^{23} nitrogen molecules strike a wall of area 8 cm^2. If the molecules move with a speed of 300 m/s and strike the wall head-on in a perfectly elastic collision, find the pressure exerted on the wall. (The mass of one N_2 molecule is 4.68×10^{-26} kg.)

45. The temperature of the Sun's interior is approximately 2×10^7 K. Find (a) the average translational kinetic energy of a proton in the Sun's interior and (b) the proton's root-mean-square speed.

Additional Problems

46. A gold ring has an inner diameter of 2.168 cm at a temperature of 15°C. Determine its diameter at 100°C. ($\alpha_{gold} = 1.42 \times 10^{-5}/(°C^{-1})$.)

47. Determine the change in length of a 20-m railroad track made of steel if the temperature is changed from −15°C to +35°C. ($\alpha_{steel} = 1.1 \times 10^{-5}/°C^{-1}$.)

48. At what Fahrenheit temperature are the Kelvin and Fahrenheit temperatures numerically equal?

49. The density of gasoline is 730 kg/m^3 at 0°C. Its volume expansion coefficient is $9.6 \times 10^{-4}/°C^{-1}$. If one gallon of gasoline occupies 0.0038 m^3, how many extra kilograms of gasoline would you get if you bought 10 gallons of gasoline at 0°C rather than at 20°C from a pump not temperature-compensated?

50. What is the average translational kinetic energy of a hydrogen molecule ($m = 3.35 \times 10^{-27}$ kg) in an ideal gas at 20°C? Of a xenon molecule ($m = 2.18 \times 10^{-25}$ kg)? What is the root mean square speed of each?

51. The rectangular plate shown in Figure 13.14 has an area, A, equal to ℓw. If the temperature increases by ΔT, show that the increase in area is given by $\Delta A = 2\alpha A\,\Delta T$, where α is the coefficient of linear expansion. What approximation does this expression assume? (*Hint:* Note that each dimension increases according to $\Delta\ell = \alpha\ell\,\Delta T$.)

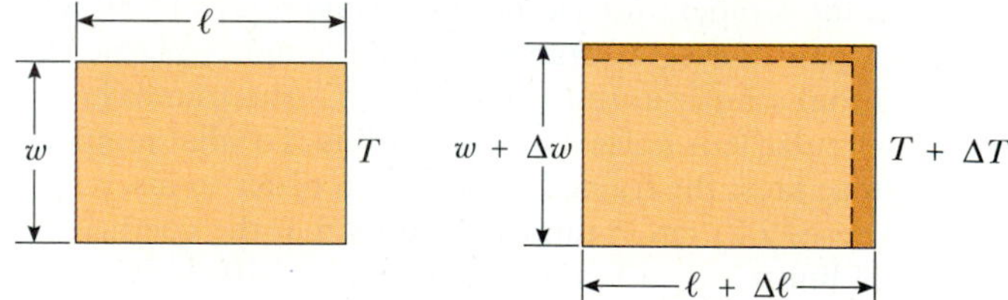

Figure 13.14 (Problem 51)

52. A mercury thermometer is constructed as in Figure 13.15. The capillary tube has a diameter of 0.004 cm, and the bulb has a diameter of 0.25 cm. Neglecting the expansion of the glass, and assuming the bulb is spherical, find the change in height of the mercury column for a temperature change of 30°C.

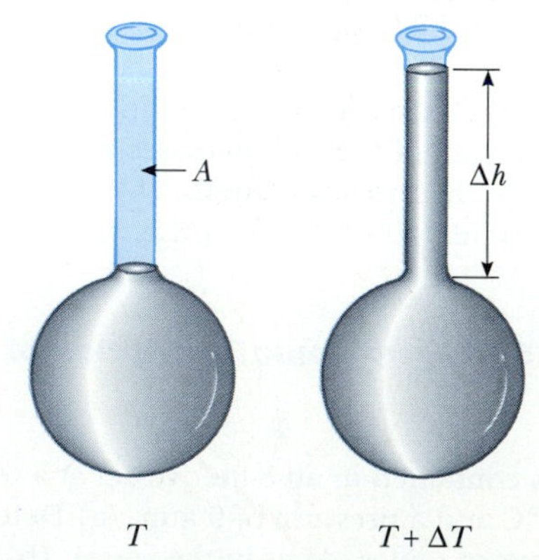

Figure 13.15 (Problems 52 and 54).

53. A fluid has a density ρ. (a) Show that the *fractional* change in density for the change in temperature ΔT is given by $\Delta\rho/\rho = -\beta\,\Delta T$. What does the negative sign signify? (b) Fresh water has a maximum density of 1.000 g/cm^3 at 4°C. At 10°C, its density is 0.9997 g/cm^3. What is β for water over this temperature interval?

54. A liquid with a coefficient of volume expansion β just fills a spherical shell of volume V at a temperature of T (Fig. 13.15). The shell is made of a material that has a coefficient of linear expansion of α. The liquid is free to expand into a capillary of cross-sectional area A at the top. (a) If the temperature increases by ΔT, show that the liquid rises in the capillary by the amount Δh given by $\Delta h = (V/A)(\beta - 3\alpha)\,\Delta T$. (b) For a typical system, such as a mercury thermometer, why is it a good approximation to neglect the expansion of the shell?

55. A steel ball bearing is 4.000 cm in diameter at 20°C. A bronze plate has a hole in it that is 3.994 cm in diameter at 20°C. What common temperature must they have in order that the ball just squeeze through the hole?

56. Show that the equation of state of an ideal gas can be written as $PM = \rho RT$, where M is its molecular weight.

57. A vertical cylinder of cross-sectional area A is fitted with a tight-fitting, frictionless piston of mass m (Fig. 13.16). (a) If there are n moles of an ideal gas in the cylinder at a temperature of T, determine the height, h, at which the piston will be in equilibrium under its own weight. (b) What is the value for h if n = 0.2 mol, T = 400 K, A = 0.008 m^2, and m = 20 kg?

58. A sphere 20 cm in diameter contains an ideal gas at 1 atm pressure and 20°C. As the sphere is heated to 100°C, gas is allowed to escape. The valve is closed and the sphere is placed in an ice-water bath at 0°C. (a) How many moles of gas escape from the sphere as it warms? (b) What is the pressure in the sphere when it is in the ice water?

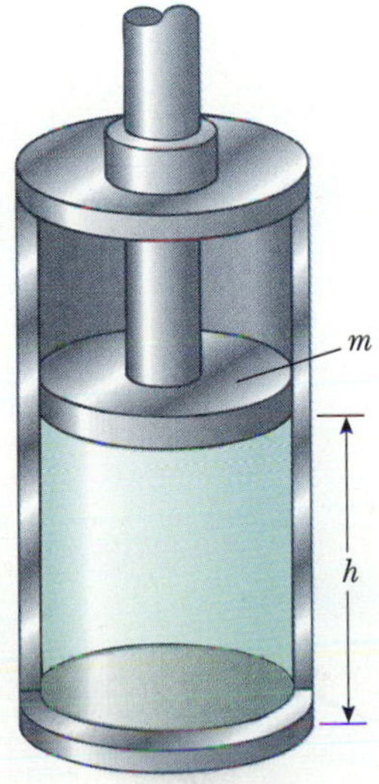

Figure 13.16 (Problem 57)

59. An expandable cylinder has its top connected to a spring of constant 2×10^3 N/m (see Fig. 13.17). The cylinder is filled with 5 L of gas with the spring relaxed at a pressure of 1 atm and a temperature of 20°C. (a) If the lid has a cross-sectional area of 0.01 m^2 and negligible mass, how high will the lid rise when the temperature is raised to 250°C? (b) What is the pressure of the gas at 250°C?

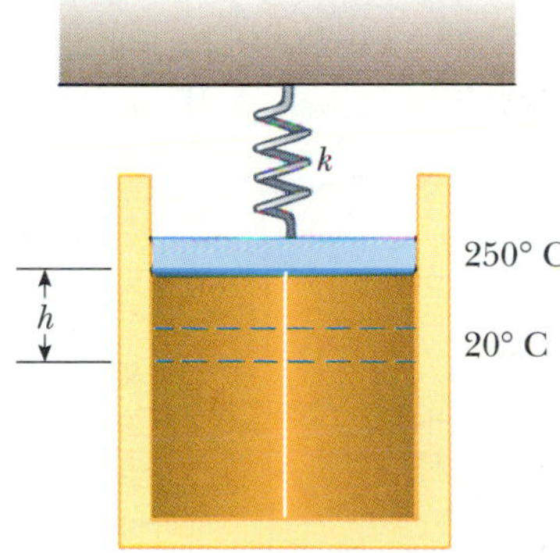

Figure 13.17 (Problem 59)

60. A bimetallic strip is made of two thin dissimilar metals bonded together. As they are heated, the metal with the larger coefficient of linear expansion expands more than the other, forcing the bar into an arc, with the outer metal (in blue) having the larger arc length (see Fig. 13.18). (a) Derive an expression for the subtended angle θ, as a function of the initial

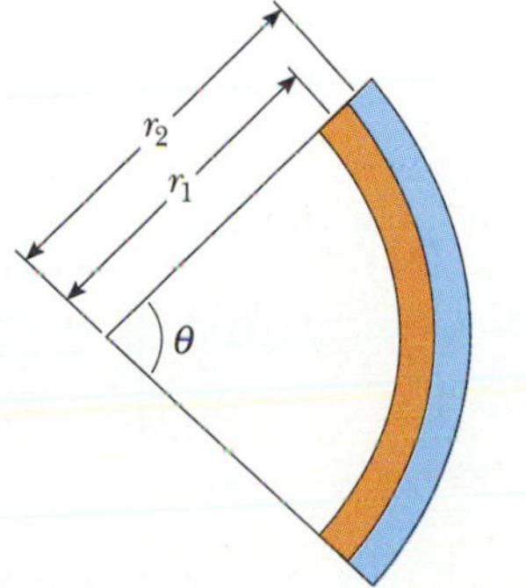

Figure 13.18 (Problem 60)

length of the rods, their coefficients of linear expansion, the change in temperature, and the separation of centers ($\Delta r = r_2 - r_1$). (b) Show that the subtended angle goes to zero when ΔT goes to zero or the two coefficients of expansion become equal. (c) What happens if the bar is cooled?

61. A hollow aluminum cylinder is to be fitted over a steel piston. At 20°C the inside diameter of the cylinder is 99% of the outside diameter of the piston. To what common temperature should the two pieces be heated in order that the cylinder just fit over the piston?

By permission of John Hart and Field Enterprises, Inc.

This German steel worker is pouring molten steel, whose temperature is about 1600°C, into a mold. The properties of various types of steel depend on their composition and the specific processes used in their fabrication. *(Gerd Ludwig, Woodfin Camp & Associates)*

Heat and the First Law of Thermodynamics

CHAPTER 14

Until about 1850, the fields of heat and mechanics were considered to be two distinct branches of science, and the law of conservation of energy seemed to describe only certain kinds of mechanical systems. Mid-19th-century experiments performed by the Englishman James Joule (1818–1889) and others showed that energy may be added to (or removed from) a system either as heat or as work done on (or by) the system. Now thermal energy is treated as a form of energy that can be transformed into mechanical energy. Once the concept of energy was broadened to include thermal energy, the law of conservation of energy emerged as a universal law of nature.

This chapter focuses on the concept of heat, the first law of thermodynamics, and some important applications. The first law of thermodynamics is merely the law of conservation of energy. It tells us only that an increase in one form of energy must be accompanied by a decrease in some other form of energy. The first law places no restrictions on the types of energy conversions that can occur. Furthermore, it makes no distinction between heat and work. According to the first law, a system's internal energy can be increased either by transfer of thermal energy to the system or by work done on the system. An important difference between thermal energy and mechanical energy is not evident from the first law: it is possible to convert work completely to thermal energy but impossible to convert thermal energy completely to mechanical energy in a process at constant temperature.

14.1 Heat Flow and Internal Energy

A major distinction must be made between internal energy, thermal energy, and heat. **Internal energy** is all of the energy belonging to a system while it is stationary (neither translating nor rotating), including nuclear energy, chemical energy, and strain energy (as for a compressed or stretched spring), as well as thermal energy. **Thermal energy** is that portion of the internal energy that changes when the temperature of the system changes. **Thermal energy transfer** is the transfer of thermal energy caused by a temperature difference between the system and its surroundings, *which may or may not change the amount of thermal energy in the system.*

James Prescott Joule (1818–1889) A British physicist, Joule received some formal education in mathematics, philosophy, and chemistry from John Dalton but was in large part self-educated. Joule's most active research period, from 1837 through 1847, led to the establishment of the principle of conservation of energy and the equivalence of heat and other forms of energy. His study of the quantitative relationship between electrical, mechanical and chemical effects of heat culminated in his announcement in 1843 of the amount of work required to produce a unit of heat, called the mechanical equivalent of heat. *(By kind permission of the President and Council of the Royal Society)*

In practice, the term "heat" is used to mean both thermal energy and thermal energy transfer. Hence, one must always examine the context of the term "heat" to determine its intended meaning.

In the previous chapter we showed that the thermal energy of a monatomic ideal gas is associated with the internal motion of its atoms. In this special case, the thermal energy is simply the kinetic energy on a microscopic scale; the higher the temperature of the gas, the greater the kinetic energy of the atoms and the greater the thermal energy of the gas. More generally, however, thermal energy includes other forms of molecular energy, such as rotational energy and vibrational kinetic and potential energy.

As an analogy, consider the distinction between work and energy that we discussed in Chapter 7. The work done on (or by) a system is a measure of the energy transferred between the system and its surroundings, whereas the mechanical energy of the system (kinetic and/or potential) is a consequence of its motion and coordinates. Thus, when a person does work on a system, energy is transferred from the person to the system. It makes no sense to talk about the work *of* a system—one should refer only to the *work done on or by a system* when some process has occurred in which energy has been transferred to or from the system. Likewise, it makes no sense to use the term *heat* unless energy has been transferred as a result of a temperature difference.

It is also important to recognize that energy can be transferred between two systems even when no thermal energy transfer occurs. For example, when a gas is compressed by a piston, the gas is warmed and its thermal energy increases, but there is no transfer of thermal energy; if the gas then expands rapidly, it cools and its thermal energy decreases, but there is no transfer of thermal energy to the surroundings. In each case, energy is transferred to or from the system as work, but appears within the system as an increase or decrease of thermal energy. The changes in internal energy in these examples are equal to the changes in thermal energy, and are measured by corresponding changes in temperature.

Units of Heat

Before it was understood that thermal energy is a form of energy, scientists defined heat in terms of the temperature changes it produced in a body. Hence, the unit of energy called the **calorie** (cal) was defined as *the amount of heat necessary to raise the temperature of 1 g of water by 1° C.*[1] (Note that the "Calorie," with a capital C, used to denote the chemical energy content of foods, is actually a kilocalorie.) Likewise, the unit of heat in the British system was the **British thermal unit** (Btu), defined as *the heat required to raise the temperature of 1 lb of water from 63°F to 64°F.*

Definition of the calorie

Since heat is now recognized as a form of energy, scientists are increasingly using the SI unit of energy, the *joule* (J), for heat. In this textbook, heat will usually be measured in joules. Each calorie can be translated, by experimental measurement, into joules. For example, the 15° calorie has been measured to be approximately 4.1850 J. We use the conversion factor

$$1 \text{ cal} = 4.186 \text{ J} \qquad [14.1]$$

Mechanical equivalent of heat

This relation is known, for purely historical reasons, as the **mechanical equivalent of heat.**

Example 14.1 Losing Weight the Hard Way

A student eats a dinner rated at 2000 (food) Calories. He wishes to do an equivalent amount of work in the gymnasium by lifting a 50-kg mass. How many times must he raise the mass to expend this much energy? Assume that he raises it a distance of 2 m each time and that no work is done when it is dropped to the floor.

Solution Since 1 (food) Calorie = 10^3 cal, the work required is 2×10^6 cal. Converting this to joules, we have, for the total work required,

$$W = (2 \times 10^6 \text{ cal})(4.186 \text{ J/cal}) = 8.37 \times 10^6 \text{ J}$$

The work done in lifting the mass a distance of h once is equal to mgh, and the work done in lifting it n times is $nmgh$. We equate this to the total work required:

$$W = nmgh = 8.37 \times 10^6 \text{ J}$$

Since $m = 50$ kg and $h = 2$ m, we get

$$n = \frac{8.37 \times 10^6 \text{ J}}{(50 \text{ kg})(9.80 \text{ m/s}^2)(2 \text{ m})} = 8.54 \times 10^3 \text{ times}$$

If the student is in good shape and lifts, say, once every 5 s, it will take him about 12 h to perform this feat. Clearly, it is much easier to lose weight by dieting.

14.2 Specific Heat

The quantity of heat energy required to raise the temperature of a given mass of a substance by some amount varies with the substance. For example, the heat required to raise the temperature of 1 kg of water by 1°C is about 4190 J, but the heat

[1] Careful measurements showed that energy depends somewhat on temperature, hence leading to a variety of calories, such as the 15° calorie, the 20° calorie, the mean calorie, etc.

required to raise the temperature of 1 kg of copper by 1°C is only 387 J. Every substance requires a unique amount of heat to change the temperature of 1 kg of it by 1°C, and this number is referred to as the **specific heat** of the substance. Table 14.1 lists specific heats for several substances.

Suppose that a quantity Q of thermal energy is transferred to m kg of a substance, thereby changing its temperature by ΔT. The **specific heat,** c, of the substance is defined as

Specific heat

$$c \equiv \frac{Q}{m\,\Delta T} \qquad [14.2]$$

From this definition, we can express the thermal energy transferred, Q, between a system of mass m and its surroundings for the temperature change ΔT as

$$Q = mc\,\Delta T \qquad [14.3]$$

For example, the energy required to raise the temperature of 0.5 kg of water by 3°C is (0.5 kg)(4186 J/kg · °C)(3°C) = 6280 J. Note that when the temperature increases, ΔT and Q are taken to be *positive,* corresponding to thermal energy flowing *into* the system. When the temperature decreases, ΔT and Q are *negative* and thermal energy flows *out of* the system.

When specific heats are measured, the values obtained are also found to depend on the conditions of the experiment. In general, measurements made at constant pressure are different from those made at constant volume. For solids and liquids, the difference between the two values is usually no more than a few percent and is often neglected. For gases, the difference between the two values is significant.

Note from Table 14.1 that water has the highest specific heat of any substance with which we are likely to come in contact routinely. The high specific heat of water is responsible for the moderate temperatures found in regions near large

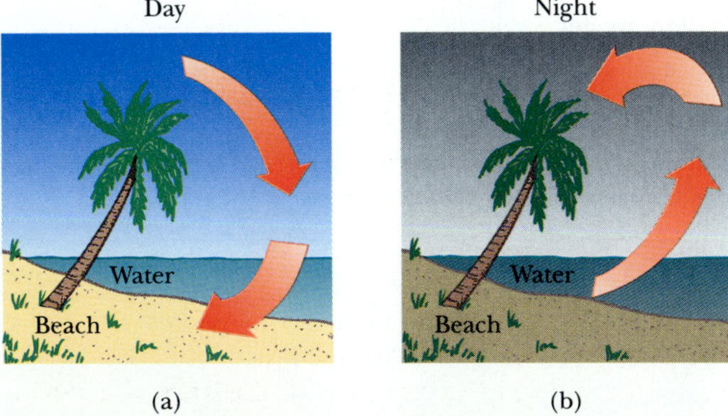

Figure 14.1
Circulation of air at the beach. (a) On a hot day, the air above the warm sand warms faster than the air above the cooler water. The cooler air over the water moving toward the beach displaces the rising warmer air. (b) At night, the sand cools more rapidly than the water and hence the air currents reverse their directions.

Table 14.1
Specific Heats of Some Substances at 25°C and Atmospheric Pressure

Substance	Specific Heat, *c* J/kg · °C	Specific Heat, *c* cal/g · °C
Elemental Solids		
Aluminum	900	0.215
Beryllium	1830	0.436
Cadmium	230	0.055
Copper	387	0.0924
Germanium	322	0.077
Gold	129	0.0308
Iron	448	0.107
Lead	128	0.0305
Silicon	703	0.1.68
Silver	234	0.056
Other Solids		
Brass	380	0.092
Wood	1700	0.41
Glass	837	0.200
Ice (−5°C)	2090	0.50
Marble	860	0.21
Liquids		
Alcohol (ethyl)	2400	0.58
Mercury	140	0.033
Water (15°C)	4186	1.00

bodies of water. As the temperature of a body of water decreases during the winter, the water gives off heat to the air, which carries the heat landward when prevailing winds are favorable. For example, the prevailing winds off the western coast of the United States are toward the land, and the heat liberated by the Pacific Ocean as it cools keeps coastal areas much warmer than they would be otherwise. This explains why the western coastal states generally have more favorable winter weather than the eastern coastal states, where the winds do not carry heat toward land.

The fact that the specific heat of water is higher than that of sand is responsible for the pattern of air flow at a beach. During the day, the Sun adds roughly equal amounts of energy to beach and water, but the lower specific heat of sand causes the beach to reach a higher temperature than the water. Consequently, the air above the land reaches a higher temperature than that above the water, and cooler air from above the water farther out is drawn in to displace this rising hot air, resulting in a breeze from ocean to land during the day. Because the hot air gradually cools as it rises, it subsequently sinks, setting up the circulating pattern shown in Figure 14.1a. During the night, the sand cools more quickly than the water, and the circulating pattern reverses itself because the hotter air is now over the water (Fig. 14.1b). (The offshore and onshore breezes are certainly well known to sailors.)

Conservation of Energy: Calorimetry

Situations in which mechanical energy is converted to thermal energy occur frequently. We shall look at some in the examples following this section and in the problems at the end of the chapter, but most of our attention here will be directed toward a particular kind of conservation-of-energy situation. In problems using the procedure we shall describe, called *calorimetry* problems, only the thermal energy transfer between the system and its surroundings is considered.

One technique for measuring the specific heat of a solid or liquid is simply to heat the substance to some known temperature, place it in a vessel containing water of known mass and temperature, and measure the temperature of the water after equilibrium is reached. Since a negligible amount of mechanical work is done in the process, the law of conservation of energy requires that the thermal energy that leaves the warmer substance (of unknown specific heat) equals the heat that enters the water.[2] Devices in which this thermal energy transfer occurs are called **calorimeters.**

For example, suppose that m_x is the mass of a substance whose specific heat we wish to determine, c_x its specific heat, and T_x its initial temperature. Let m_w, c_w, and T_w represent corresponding values for the water. If T is the final equilibrium temperature after everything is mixed, then from Equation 14.3 we find that the thermal energy gained by the water is $m_w c_w (T - T_w)$ and the thermal energy lost by the substance of unknown c is $-m_x c_x (T - T_x)$. Assuming that the combined system (water + unknown) does not lose or gain any thermal energy, it follows that the thermal energy gained by the water must equal the thermal energy lost by the unknown (conservation of energy):

$$m_w c_w (T - T_w) = -m_x c_x (T - T_x)$$

Solving for c_x gives

$$c_x = \frac{m_w c_w (T - T_w)}{m_x (T_x - T)} \qquad [14.4]$$

Example 14.2 Cooling a Hot Ingot

A 0.05-kg ingot of metal is heated to 200°C and then dropped into a calorimeter containing 0.4 kg of water that is initially at 20°C. If the final equilibrium temperature of the mixed system is 22.4°C, find the specific heat of the metal.

Solution Because the heat lost by the ingot equals the heat gained by the water, we can write

$$m_x c_x (T_i - T_f) = m_w c_w (T_f - T_i) \quad (0.05\text{ kg})(c_x)(200°\text{C} - 22.4°\text{C})$$
$$= (0.4\text{ kg})(4186\text{ J/kg}\cdot°\text{C})(22.4°\text{C} - 20°\text{C})$$

from which we find that

$$c_x = 453\text{ J/kg}\cdot°\text{C}$$

[2] For precise measurements, the container holding the water should be included in our calculations, since it also exchanges heat. This would require a knowledge of its mass and composition. However, if the mass of the water is large compared with that of the container, we can neglect the thermal energy gained by the container. Furthermore, precautions must be taken in such measurements to minimize thermal energy transfer between the system and the surroundings.

The ingot is most likely iron, as can be seen by comparing this result with the data in Table 14.1.

Exercise What is the total thermal energy transferred to the water as the ingot is cooled?

Answer 4020 J.

Example 14.3 Fun Time for a Cowhand

A cowhand fires a silver bullet of mass 2 g, with a muzzle speed of 200 m/s, into the pine wall of a saloon. Assume that all the thermal energy generated by the impact remains with the bullet. What is the temperature change of the bullet?

Solution The initial kinetic energy of the bullet is

$$\tfrac{1}{2}mv^2 = \tfrac{1}{2}(2 \times 10^{-3}\text{ kg})(200\text{ m/s})^2 = 40\text{ J}$$

Nothing in the environment is hotter than the bullet, so the bullet gains no thermal energy. Its temperature increases because the 40 J of kinetic energy becomes 40 J of extra internal energy. The temperature change will be the same as if 40 J of thermal energy were transferred from a stove to the bullet, and we consider this imaginary process to compute ΔT from

$$Q = mc\,\Delta T = 40\text{ J}$$

Since the specific heat of silver is 234 J/kg · °C (Table 14.1), we get

$$\Delta T = \frac{Q}{mc} = \frac{40\text{ J}}{(2 \times 10^{-3}\text{ kg})(234\text{ J/kg}\cdot{}^\circ\text{C})} = 85.5^\circ\text{C}$$

Exercise Suppose the cowhand runs out of silver bullets and fires a lead bullet of the same mass and velocity into the wall. What is the temperature change of this bullet?

Answer 156°C.

14.3 Latent Heat and Phase Changes

A substance usually undergoes a change in temperature when thermal energy is transferred between the substance and its surroundings. There are situations, however, in which the transfer of thermal energy does not result in a change in temperature. This is the case whenever the physical characteristics of the substance change from one form to another, commonly referred to as a **phase change.** Some common phase changes are solid to liquid (melting), liquid to gas (boiling), and a change in crystalline structure of a solid. All such phase changes involve a change in internal energy.

The thermal energy transfer required to change the phase of a given mass, m, of a pure substance is

$$Q = mL \qquad [14.5]$$

Latent heat

where L is called the **latent heat** ("hidden" heat) of the substance and depends on the nature of the phase change as well as on the properties of the substance. Heat

Table 14.2
Heats of Fusion and Vaporization

Substance	Melting Point (°C)	Heat of Fusion J/kg	Heat of Fusion (cal/g)	Boiling Point (°C)	Heat of Vaporization J/kg	Heat of Vaporization (cal/g)
Helium	−269.65	5.23×10^3	(1.25)	−268.93	2.09×10^4	(4.99)
Nitrogen	−209.97	2.55×10^4	(6.09)	−195.81	2.01×10^5	(48.0)
Oxygen	−218.79	1.38×10^4	(3.30)	−182.97	2.13×10^5	(50.9)
Ethyl alcohol	−114	1.04×10^5	(24.9)	78	8.54×10^5	(204)
Water	0.00	3.34×10^5	(79.7)	100.00	2.26×10^6	(540)
Sulfur	119	3.81×10^4	(9.10)	444.60	3.26×10^5	(77.9)
Lead	327.3	2.45×10^4	(5.85)	1750	8.70×10^5	(208)
Aluminum	660	3.97×10^5	(94.8)	2450	1.14×10^7	(2720)
Silver	960.80	8.82×10^4	(21.1)	2193	2.33×10^6	(558)
Gold	1063.00	6.44×10^4	(15.4)	2660	1.58×10^6	(377)
Copper	1083	1.34×10^5	(32.0)	1187	5.06×10^6	(1210)

of fusion, L_f, is the term used when the phase change is from solid to liquid, and **heat of vaporization,** L_v, is the term used when the phase change is from liquid to gas.[3] For example, the heat of fusion for water at atmospheric pressure is 3.33×10^5 J/kg, and the heat of vaporization of water is 2.26×10^6 J/kg. The latent heats of different substances vary considerably, as is seen in Table 14.2.

Consider, for example, the thermal energy required to convert a 1-g block of ice at −30°C to steam (water vapor) at 120°C. Figure 14.2 indicates the experimental results obtained when thermal energy is gradually added to the ice. Let us examine each portion of the curve separately.

This device, called Hero's engine, was invented around 150 B.C. by Hero in Alexandria. When water is boiled in the flask, which is suspended by a cord, steam exits through two tubes at the sides of the flask (in opposite directions), creating a torque that rotates the flask. *(Courtesy of CENCO)*

Part A. During this portion of the curve, we are changing the temperature of the ice from −30°C to 0°C. Since the specific heat of ice is 2090 J/kg · °C, we can calculate the amount of thermal energy added from Equation 14.3:

$$Q = m_i c_i \,\Delta T = (10^{-3}\text{ kg})(2090\text{ J/kg}\cdot{}^\circ\text{C})(30^\circ\text{C}) = 62.7\text{ J}$$

Part B. When the ice reaches 0°C, the ice/water mixture remains at this temperature—even though thermal energy is being added—until all the ice melts to become water at 0°C. The thermal energy required to melt 1 g of ice at 0°C is, from Equation 14.5,

$$Q = mL_f = (10^{-3}\text{ kg})(3.33 \times 10^5\text{ J/kg}) = 333\text{ J}$$

Part C. Between 0°C and 100°C, nothing surprising happens. No phase change occurs in this region. The thermal energy added to the water is being used to

[3] When a gas cools, it eventually returns to the liquid phase, or *condenses.* The thermal energy transfer during the process is called the *heat of condensation,* and it equals in magnitude the heat of vaporization. When a liquid cools, it eventually solidifies, and the *heat of solidification* equals the heat of fusion.

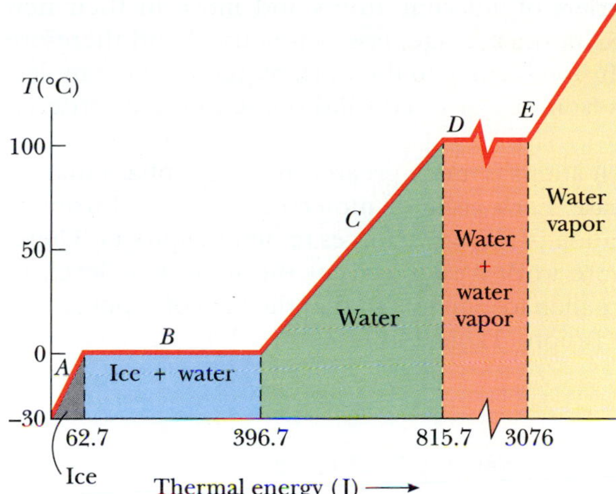

Figure 14.2 A plot of temperature versus thermal energy added when 1 g of ice initially at −30°C is converted to steam at 120°C.

increase its temperature. The amount of thermal energy necessary to increase the temperature from 0°C to 100°C is

$$Q = m_w c_w \Delta T = (10^{-3}\text{ kg})(4.19 \times 10^3\text{ J/kg}\cdot{}^\circ\text{C})(100^\circ\text{C}) = 4.19 \times 10^2\text{ J}$$

Part D. At 100°C, another phase change occurs as the water changes from water at 100°C to steam at 100°C. Just as in Part B, the water/steam mixture remains at 100°C—even though thermal energy is being added—until all the liquid has been converted to steam. The thermal energy required to convert 1 g of water to steam at 100°C is

$$Q = mL_v = (10^{-3}\text{ kg})(2.26 \times 10^6\text{ J/kg}) = 2.26 \times 10^3\text{ J}$$

Part E. On this portion of the curve, thermal energy is being added to the steam with no phase change occurring. The thermal energy that must be added to raise the temperature of the steam to 120°C is

$$Q = m_s c_s \Delta T = (10^{-3}\text{ kg})(2.01 \times 10^3\text{ J/kg}\cdot{}^\circ\text{C})(20^\circ\text{C}) = 40.2\text{ J}$$

The *total amount of thermal energy* that must be added to change one gram of ice at −30°C to steam at 120°C is therefore about 3.11×10^3 J. Conversely, to cool one gram of steam at 120°C down to the point at which we have ice at −30°C, we must remove 3.11×10^3 J of thermal energy.

Phase changes can be described in terms of a rearrangement of molecules when heat is added to or removed from a substance. Consider first the liquid-to-gas phase change. The molecules in a liquid are close together, and the forces between them are stronger than those between the molecules of a gas, which are far apart. Therefore, work must be done on the liquid, at the molecular level, against these attractive molecular forces in order to separate the molecules. The heat of vaporization is the amount of thermal energy that must be added to the liquid to accomplish this.

Similarly, at the melting point of a solid, we imagine that the amplitude of vibration of the atoms about their equilibrium position becomes large enough to allow the atoms to pass the barriers of adjacent atoms and move to their new positions. The new locations are, on the average, less symmetrical and therefore have higher energy. The heat of fusion is equal to the work required, at the molecular level, to transform the mass from the ordered solid phase to the disordered liquid phase.

The average distance between atoms is much greater in the gas phase than in either the liquid or the solid phase. Each atom or molecule is removed from its neighbors, without the compensation of attractive forces to new neighbors. Therefore, it is not surprising that more work is required, at the molecular level, to vaporize a given mass of substance than to melt it, and that the heat of vaporization is much greater than the heat of fusion (Table 14.2).

Problem-Solving Strategy: Calorimetry Problems

If you are having difficulty with calorimetry problems, consider the following factors.

1. **Be sure your units are consistent throughout. For instance, if you are using specific heats in cal/g · °C, be sure that masses are in grams and temperatures are Celsius throughout.**
2. **Losses and gains in thermal energy are found by using $Q = mc\,\Delta T$ only for those intervals in which no phase changes are occurring. The equations $Q = mL_f$ and $Q = mL_v$ are to be used only when phase changes *are* taking place.**
3. **Often sign errors occur in heat loss = heat gain equations. One way to check your equation is to examine the signs of all ΔTs that appear in it.**

Example 14.4 Boiling Liquid Helium

Liquid helium has a very low boiling point, 4.2 K, and a very low heat of vaporization, 2.09×10^4 J/kg (Table 14.2). A constant power of 10 W (1 W = 1 J/s) is transferred to a container of liquid helium from an immersed electric heater. At this rate, how long does it take to boil away 1 kg of liquid helium?

Solution Since $L_v = 2.09 \times 10^4$ J/kg, we must supply 2.09×10^4 J of energy to boil away 1 kg. The power supplied to the helium is 10 W = 10 J/s. That is, in 1 s, 10 J of energy is transferred to the helium. Therefore, the time it takes to transfer 2.09×10^4 J is

$$t = \frac{2.09 \times 10^4\ \text{J}}{10\ \text{J/s}} = 2.09 \times 10^3\ \text{s} \approx 35\ \text{min}$$

In contrast, 1 kg of liquid nitrogen ($L_v = 2.01 \times 10^5$ J/kg) would boil away in about 5.6 h with the same power input.

Exercise If 10 W of power is supplied to 1 kg of water at 100°C, how long will it take for the water to boil away completely?

Answer 62.8 h.

14.4 Work and Thermal Energy in Thermodynamic Processes

In the macroscopic approach to thermodynamics, we describe the *state* of a system with such variables as pressure, volume, temperature, and internal energy. The number of macroscopic variables needed to characterize a system depends on the system's nature. For a homogeneous system, such as a gas containing only one type of molecule, usually only two variables are needed. However, it is important to note that a *macroscopic state* of an isolated system can be specified only if the system is in thermal equilibrium internally. In the case of a gas in a container, internal thermal equilibrium requires that every part of the container be at the same pressure and temperature.

Consider gas contained in a cylinder fitted with a movable piston (Fig. 14.3). In equilibrium, the gas occupies a volume V and exerts a uniform pressure P on the cylinder walls and piston. If the piston has a cross-sectional area of A, the force exerted by the gas on the piston is $F = PA$. Now let us assume that the gas expands **quasi-statically,** that is, slowly enough to allow the system to remain essentially in thermodynamic equilibrium at all times. As the piston moves up a distance of dy, the work done by the gas on the piston is

$$dW = F\,dy = PA\,dy$$

Since $A\,dy$ is the increase in volume of the gas dV, we can express the work done as

$$dW = P\,dV \qquad [14.6]$$

Since the gas expands, dV is positive and the work done by the gas is positive, whereas if the gas is compressed, dV is negative, indicating that the work done by

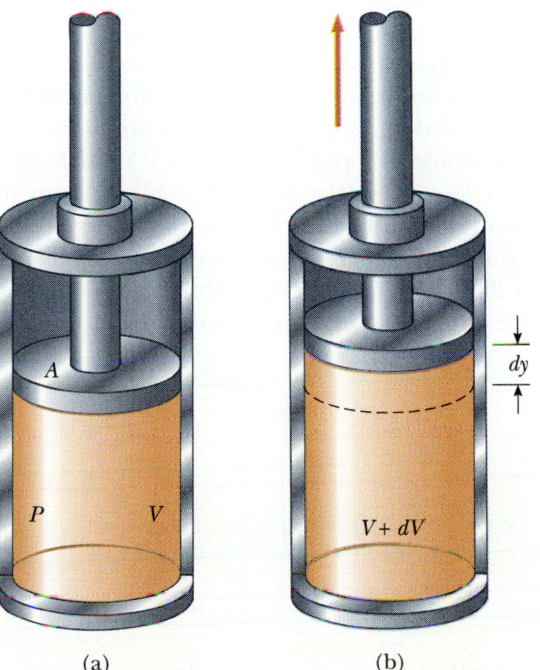

Figure 14.3
Gas contained in a cylinder at a pressure P does work on a moving piston as the system expands from a volume V to a volume $V + dV$.

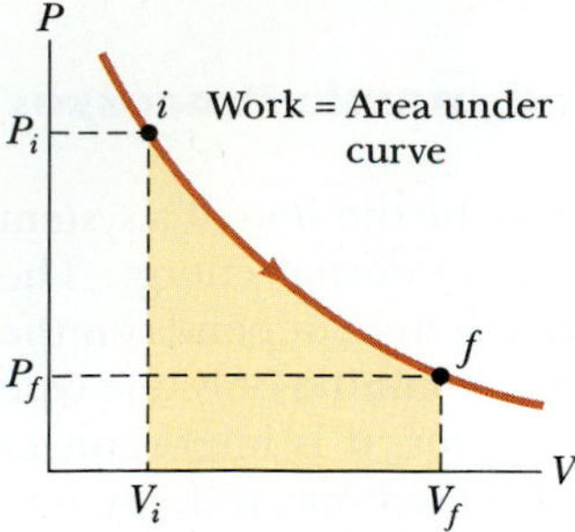

Figure 14.4
A gas expands quasi-statically (slowly) from state i to state f. The work done by the gas equals the shaded area under the PV curve.

the gas is negative. (In the latter case, negative work can be interpreted as being work done *on* the gas.) Clearly, the work done by the gas is zero when the volume remains constant. The total work done by the gas as its volume changes from V_i to V_f is given by the integrating Equation 14.6:

$$W = \int_{V_i}^{V_f} P\,dV \qquad [14.7]$$

To evaluate this integral, one must know how the pressure varies during the expansion process. (Note that a *process* is *not* specified merely by the initial and final states. Rather, a process is a *fully specified* change in state of a system.) In general, the pressure is not constant throughout the expansion process, but depends on the volume and temperature. If the pressure and volume are known at each step of the process, the states of the gas can be represented as a curve on a *PV* diagram, as in Figure 14.4. The work done in the expansion from the initial state to the final state is the area under the curve.

Work equals area under the curve in a *PV* diagram

As Figure 14.4 shows, the work done in the expansion from the initial state to the final state depends on the path taken between the two states. To illustrate this important point, consider several different paths connecting i and f (Fig. 14.5). In the process depicted in Figure 14.5a, the pressure of the gas is first reduced from P_i to P_f by cooling at constant volume, V_i, and the gas then expands from V_i to V_f at constant pressure, P_f. The work done along this path is $P_f(V_f - V_i)$. In Figure 14.5b, the gas first expands from V_i to V_f at constant pressure, P_i, and then its pressure is reduced to P_f at constant volume, V_f. The work done along this path is $P_i(V_f - V_i)$, which is greater than that for the process described in Figure 14.5a. Finally, for the process described in Figure 14.5c, where both P and V change continuously, the work done has some value intermediate between the values ob-

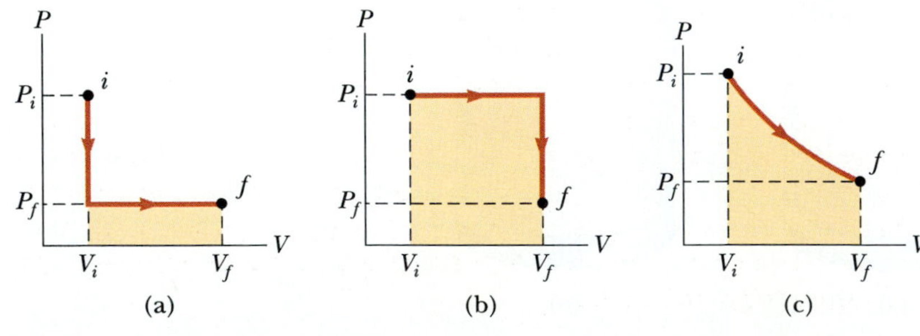

Figure 14.5
The work done by a gas as it is taken from an initial state to a final state depends on the path between these states.

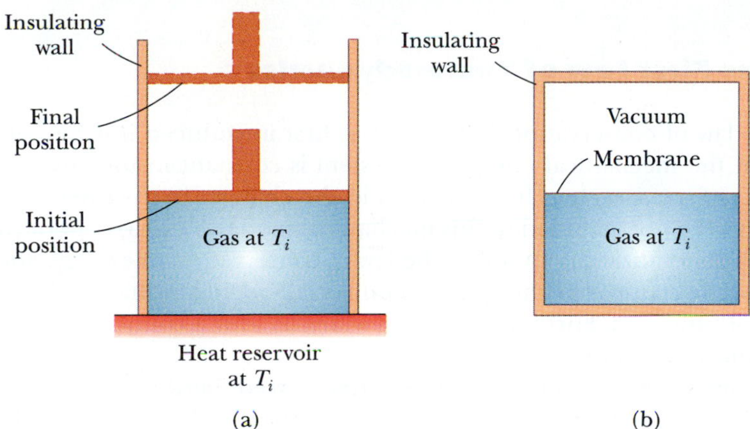

Figure 14.6
(a) A gas at temperature T_i expands slowly by absorbing thermal energy from a reservoir at the same temperature. (b) A gas expands rapidly into an evacuated region after a membrane is broken.

tained in the first two processes. To evaluate the work in this case, the shape of the PV curve must be known. Therefore, we see that the work done by a system depends on the process by which the system goes from the initial to the final state. In other words, the work done depends on the initial, final, and intermediate states of the system.

Work done depends on the path between the initial and final states

In a similar manner, the thermal energy transferred into or out of the system is also found to depend on the process. This can be demonstrated by the situations depicted in Figure 14.6. In each case, the gas has the same initial volume, temperature, and pressure and is assumed to be ideal. In Figure 14.6a, the gas is in thermal contact with a heat reservoir. If the pressure of the gas is infinitesimally greater than atmospheric pressure, the gas, because it absorbs thermal energy from the reservoir, expands and causes the piston to rise. During this expansion to some final volume, V_f, and final pressure P_f, sufficient thermal energy to maintain a constant temperature of T_i is transferred from the reservoir to the gas.

Now consider the thermally insulated system shown in Figure 14.6b. When the membrane is broken, the gas expands rapidly into the vacuum until it occupies a volume of V_f and is at a pressure of P_f. In this case, the gas does no work since there is no movable piston. Furthermore, no thermal energy is transferred through the thermally insulated wall, which we call an *adiabatic wall*, and so the temperature remains at T_i. This process is often referred to as **adiabatic free expansion** or simply *free expansion*. In general, an adiabatic process is one in which no heat is transferred between the system and its surroundings.

Free expansion of a gas

The initial and final states of the ideal gas in Figure 14.6a are identical to the initial and final states in Figure 14.6b, but the paths are different. In the first case, thermal energy is transferred slowly to the gas, and the gas does work on the piston. In the second case, no thermal energy is transferred and the work done is zero. Therefore, we conclude that *thermal energy transfer, like work, depends on the initial, final, and intermediate states of the system.* Furthermore, since thermal energy transfer and work depend on the path, neither quantity is determined by the end points of a thermodynamic process.

14.5 The First Law of Thermodynamics

When the law of conservation of energy was first introduced in Chapter 8, it was stated that the mechanical energy of a system is constant in the absence of nonconservative forces, such as friction. That is, the changes in the internal energy of the system were not included in this mechanical model. The first law of thermodynamics is a generalization known as the law of conservation of energy, and encompasses possible changes in energy. It is a universally valid law that can be applied to all kinds of processes. Furthermore, it provides us with a connection between the microscopic and macroscopic worlds.

Change in internal energy

We have seen that energy can be transferred between a system and its surroundings in two ways. One is work done by (or on) the system which requires that there be a macroscopic displacement of the point of application of a force (or pressure). The other mode is thermal energy transfer, which occurs through random molecular collisions. Each of these represents a change of energy of the system, and therefore usually results in measurable changes in the macroscopic variables of the system, such as pressure, temperature, and volume of a gas.

To put these ideas on a more quantitative basis, suppose a system undergoes a change from an initial state to a final state. During this change, positive Q is the thermal energy transferred *to* the system, and positive W is the work done *by* the system.[4] As an example, suppose the system is a gas whose pressure and volume change from P_i, V_i to P_f, V_f. If the quantity $Q - W$ is measured for various paths connecting the initial and final equilibrium states (that is, for various *processes*), one finds that it is the same for *all* paths connecting the initial and final states. We conclude that the quantity $Q - W$ is determined completely by the initial and final states of the system, and we call it the *change in the energy of the system.* Although Q and W both depend on the path, *$Q - W$ is independent of the path.* If we represent the energy function with the letter U, then the *change* in energy, $\Delta U = U_f - U_i$, can be expressed as

First-law equation

$$\Delta U = U_f - U_i = Q - W \quad [14.8]$$

where all quantities must have the same energy units. Equation 14.8 is known as the **first-law equation** and is a key equation to many applications.

When a system undergoes an infinitesimal change in state, where a small amount of thermal energy dQ is transferred and a small amount of work dW is done, the energy also changes by a small amount dU. Thus, for infinitesimal processes we can express the first-law equation as[5]

First-law equation for infinitesimal changes

$$dU = dQ - dW \quad [14.9]$$

[4] Note that this is a change of notation from that previously followed, where positive W represented work done *on* the system; negative W was work done *by* the system.

[5] It should be noted that dQ and dW are not true differential quantities, although dU is a true differential. In fact, dQ and dW are not differentials of any definable quantity. For further details on this point, see R.P. Bauman, *Modern Thermodynamics and Statistical Mechanics,* New York, Macmillan, 1992.

On the microscopic level, the internal energy of a system includes the kinetic and potential energies of the molecules making up the system. Part of the energy, U, is the internal energy of the system. In thermodynamics, we do not concern ourselves with the specific form of the internal energy. An analogy can be made between the internal energy of a system and the potential energy function associated with a body moving under the influence of gravity without friction. The potential energy function is independent of the path, and it is only the changes in it that are of concern. Likewise, the change in internal energy of a thermodynamic system is what matters, since only differences are defined. Because absolute values are not defined, any reference state can be chosen for the internal energy.

Let us look at some special cases in which the only changes in energy will be changes in internal energy. First consider an *isolated system,* that is, one that does not interact with its surroundings. In this case, no thermal energy transfer takes place and the work done is zero; hence, the internal energy remains constant. That is, since $Q = W = 0$, $\Delta U = 0$ and so $U_i = U_f$. We conclude that *the internal energy of an isolated system remains constant.*

For isolated systems, *U* remains constant

Next consider the case in which a system (one not isolated from its surroundings) is taken through a **cyclic process,** that is, one that originates and ends at the same state. In this case, the change in the internal energy must again be *zero* and therefore the thermal energy added to the system must equal the work done during the cycle. That is, in a cyclic process,

Cyclic process

$$\Delta U = 0 \qquad \text{and} \qquad Q = W$$

Note that *the net work done per cycle equals the area enclosed by the path representing the process on a PV diagram.* As we shall see in Chapter 15, cyclic processes are very important in describing the thermodynamics of *heat engines*—devices in which some part of the thermal energy input is extracted as mechanical work.

If a process occurs in which the work done is zero, then the change in internal energy equals the thermal energy entering or leaving the system. If thermal energy enters the system, Q is positive and the internal energy increases. For a gas, we can associate this increase in internal energy with an increase in the kinetic energy of the molecules. On the other hand, if a process occurs in which the thermal energy transferred is zero and work is done by the system, then the magnitude of the change in internal energy equals the negative of the work done by the system. That is, the internal energy of the system decreases. For example, if a gas is compressed with no thermal energy transferred (by a moving piston, say), the work done by the gas is negative and the internal energy again increases. This is because kinetic energy is transferred from the moving piston to the gas molecules.

No practical distinction exists between thermal energy transfer and work on a microscopic scale. Each can produce a change in the internal energy of a system. Although the macroscopic quantities Q and W are *not* properties of a system, they are related to changes of the internal energy of a stationary system through the first-law equation. Once a process, or path, is defined, Q and W can be either calculated or measured, and the change in internal energy can be found from the first-law equation. One of the important consequences of the first law is the existence of a quantity we called internal energy, the value of which is determined by the state of the system. The internal energy function is therefore called a *state function.*

14.6 Some Applications of the First Law of Thermodynamics

Adiabatic process

In order to apply the first law of thermodynamics to specific systems, it is useful to first define some common thermodynamic processes. As we learned in Section 14.4, an adiabatic process is defined as one during which no thermal energy enters or leaves the system; that is, $Q = 0$. An adiabatic process can be achieved either by thermally insulating the system from its surroundings (as in Fig. 14.6b) or by performing the process rapidly. Applying the first-law equation in this case, we see that

First-law equation applied to an adiabatic process

$$\Delta U = -W \quad [14.10]$$

From this result, we see that when a gas expands adiabatically, W is positive, so ΔU is negative and the temperature of the gas decreases. Conversely, the gas temperature is raised when the gas is compressed adiabatically.

Adiabatic processes are very important in engineering practice. Common applications include the expansion of hot gases in an internal combustion engine, the liquefaction of gases in a cooling system, and the compression stroke in a diesel engine.

The *free expansion* depicted in Figure 14.6b is an adiabatic process in which no work is done on or by the gas. Since $Q = 0$ and $W = 0$, we see from the first law that $\Delta U = 0$ for this process. That is, *the initial and final internal energies of a gas are equal in an adiabatic free expansion.* As we shall see in Chapter 15, the internal energy of an ideal gas depends only on its temperature. Thus, we would expect no change in temperature during an adiabatic free expansion. This is in accord with experiments performed at low pressures. Careful experiments with real gases at high pressures show a slight increase or decrease in temperature after the expansion.

Isobaric process (constant pressure)

A process that occurs at constant pressure is called an **isobaric process.** When such a process occurs, the thermal energy transfer and the work done may or may not be zero. For example, in an isobaric expansion, the work done is simply the pressure multiplied by the change in volume, or $P(V_f - V_i)$.

Isovolumetric process

A process that takes place at constant volume is called an **isovolumetric process.** In such a process, the work done is clearly zero. Hence, from the first-law equation we see that, in an isovolumetric process,

First-law equation applied to an isovolumetric process

$$\Delta U = Q \quad [14.11]$$

This equation tells us that, *if thermal energy is added to a system kept at constant volume, all of the thermal energy goes into increasing the internal energy of the system.* When a mixture of gasoline vapor and air explodes in the cylinder of an engine, the temperature and pressure rise suddenly because the cylinder volume doesn't change appreciably during the short duration of the explosion.

Isothermal process (constant temperature)

A process that occurs at constant temperature is called an **isothermal process.** The internal energy of an ideal gas is a function of temperature only. Hence, in an isothermal process of a stationary ideal gas, $\Delta U = 0$.

Isothermal Expansion of an Ideal Gas

Suppose an ideal gas is allowed to expand quasi-statically at constant temperature as described by the PV diagram in Figure 14.7. (Isothermal expansion can be achieved by placing the gas in good thermal contact with a heat reservoir at the same temperature, and carrying out the expansion very slowly, as in Figure 14.6a.) The curve is a hyperbola with the equation $PV =$ constant. Let us calculate the work done by the gas in the expansion from state i to state f.

This work is given by Equation 14.7. Since the gas is ideal and the process is quasi-static, we can apply $PV = nRT$ for each point on the path. Therefore, we have

$$W = \int_{V_i}^{V_f} P\,dV = \int_{V_i}^{V_f} \frac{nRT}{V}\,dV$$

Since T is constant in this case, it can be removed from the integral, along with n and R:

$$W = nRT \int_{V_i}^{V_f} \frac{dV}{V} = nRT \ln V \Big]_{V_i}^{V_f}$$

To evaluate the integral, we used $\int \frac{dx}{x} = \ln x$. Thus, we find

$$W = nRT \ln\left(\frac{V_f}{V_i}\right) \qquad [14.12]$$

Work done in an isothermal process

Numerically, this work equals the shaded area under the PV curve in Figure 14.7. Since the gas expands, $V_f > V_i$ and the work done by the gas is positive, as we would expect. If the gas is compressed, then $V_f < V_i$ and the work done by the gas is negative. In the next chapter we shall find that the internal energy of an ideal gas depends only on temperature. Hence, for an isothermal process $\Delta U = 0$, and from the first law we conclude that the thermal energy given up by the reservoir (and transferred to the gas) equals the work done by the gas, or $Q = W$.

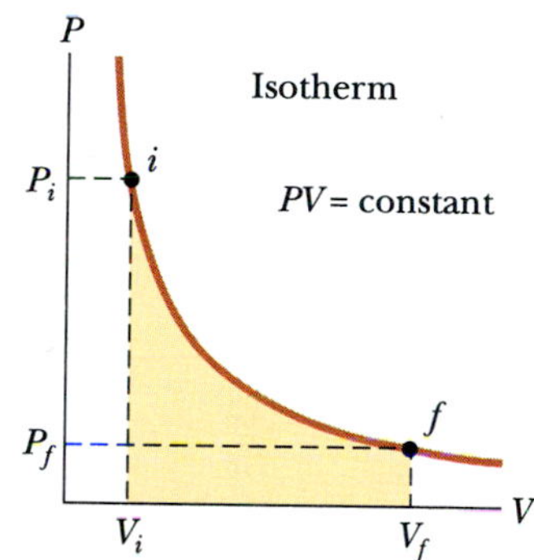

Figure 14.7
The PV diagram for an isothermal expansion of an ideal gas from an initial state to a final state. The curve is a hyperbola.

The Boiling Process

Suppose that a liquid of mass m vaporizes at constant pressure P. Its volume in the liquid state is V_ℓ, and its volume in the vapor state is V_v. Let us find the work done in the expansion and the change in internal energy of the system.

Since the expansion takes place at constant pressure, the work done by the system is

$$W = \int_{V_\ell}^{V_v} P\,dV = P\int_{V_\ell}^{V_v} dV = P(V_v - V_\ell)$$

The thermal energy that must be transferred to the liquid to vaporize all of it is equal to $Q = mL_v$, where L_v is the heat of vaporization of the liquid. Using the first-law equation and the preceding result, we get

$$\Delta U = Q - W = mL_v - P(V_v - V_\ell) \qquad [14.13]$$

Example 14.5 Boiling Water

One gram of water occupies a volume of 1 cm³ at atmospheric pressure. When this amount of water is boiled, it becomes 1671 cm³ of steam. Calculate the change in internal energy for this process.

Solution Since the heat of vaporization of water is 2.26×10^6 J/kg at atmospheric pressure (Table 14.2), the heat required to boil 1 g is

$$Q = mL_v = (1 \times 10^{-3}\ \text{kg})(2.26 \times 10^6\ \text{J/kg}) = 2260\ \text{J}$$

The work done by the system is positive and equal to

$$W = P(V_v - V_\ell)$$
$$= (1.013 \times 10^5\ \text{N/m}^2)[(1671 - 1) \times 10^{-6}\ \text{m}^3]$$
$$= 169\ \text{J}$$

Hence, the change in internal energy is

$$\Delta U = Q - W = 2260\ \text{J} - 169\ \text{J} = 2091\ \text{J}$$

The internal energy of the system *increases* since ΔU is positive. We see that most (93%) of the thermal energy that is transferred to the liquid goes into increasing the internal energy. Only a small fraction (7%) goes into work done on the surroundings.

Example 14.6 Thermal Energy Transferred to a Solid

The internal energy of a solid also increases when thermal energy is transferred to it from its surroundings.

A 1-kg bar of copper is heated at atmospheric pressure. Its temperature increases from 20°C to 50°C.

(a) Find the work done by the copper.

Solution The change in volume of the copper can be calculated using Equation 13.7 and the volume expansion coefficient for copper taken from Table 13.1 (remembering that $\beta = 3\alpha$):

$$\Delta V = \beta V \Delta T = [5.1 \times 10^{-5}\ (°\text{C})^{-1}](50°\text{C} - 20°\text{C})V = 1.5 \times 10^{-3}\ V$$

But the volume is equal to m/ρ, and the density of copper is $8.92 \times 10^3\ \text{kg/m}^3$. Hence,

$$\Delta V = (1.5 \times 10^{-3})\left(\frac{1\ \text{kg}}{8.92 \times 10^3\ \text{kg/m}^3}\right) = 1.7 \times 10^{-7}\ \text{m}^3$$

Since the expansion takes place at constant pressure, the work done is

$$W = P\Delta V = (1.013 \times 10^5\ \text{N/m}^2)(1.7 \times 10^{-7}\ \text{m}^3)$$
$$= 1.9 \times 10^{-2}\ \text{J}$$

(b) What quantity of thermal energy is transferred to the copper?

Solution Taking the specific heat of copper from Table 14.1 and using Equation 14.3, we find that the thermal energy transferred is

$$Q = mc\Delta T = (1\ \text{kg})(387\ \text{J/kg}\cdot°\text{C})(30°\text{C})$$
$$= 1.16 \times 10^4\ \text{J}$$

(c) What is the increase in internal energy of the copper?

Solution From the first law of thermodynamics, the increase in internal energy is

$$\Delta U = Q - W = 1.16 \times 10^4 \text{ J}$$

Note that almost *all* of the thermal energy transferred goes into increasing the internal energy. The fraction of thermal energy that is used to do work against the atmosphere is only about 10^{-6}! Hence, in the thermal expansion of a solid or a liquid, the small amount of work done is usually ignored.

Summary

Thermal energy transfer is a form of energy transfer that takes place as a consequence of a temperature difference. The **internal energy** of a substance is a function of the state of the substance and generally increases with increasing temperature.

The **calorie** is the amount of heat necessary to raise the temperature of 1 g of water by 1°C. The 15° calorie is related to the joule by experiment, yielding approximately 4.186 J/cal. This relationship was historically called the mechanical equivalent of heat.

The thermal energy required to change the temperature of a substance by ΔT is

Thermal energy required to raise the temperature of a substance

$$Q = mc\,\Delta T \qquad [14.3]$$

where m is the mass of the substance and c is its **specific heat.**

The thermal energy required to change the phase of a pure substance of mass m is

Latent heat

$$Q = mL \qquad [14.5]$$

The parameter L is called the **latent heat** of the substance and depends on the nature of the phase change and the properties of the substance.

A **quasi-static process** is one that proceeds slowly enough to allow the system to always be in a state of equilibrium.

The work done by a gas as its volume changes from some initial value, V_i, to some final value, V_f, is

Work done by a gas

$$W = \int_{V_i}^{V_f} P\,dV \qquad [14.7]$$

where P is the pressure, which may vary during the process. In order to evaluate W, the nature of the process must be specified—that is, P and V must be known during each step of the process. The work done depends on the initial, final, and intermediate states. In other words, W depends on the path taken between the initial and final states.

First law of thermodynamics

The **first law of thermodynamics** is the law of conservation of energy. It includes all forms of energy, including thermal energy.

When a stationary system undergoes a change from one state to another, the change in its internal energy, ΔU, is given by

First-law equation

$$\Delta U = Q - W \qquad [14.8]$$

where Q is the thermal energy transferred into (or out of) the system and W is the work done by (or on) the system. Although Q and W both depend on the path taken from the initial state to the final state, the quantity ΔU is path-independent.

An isolated system is one that does not interact with its surroundings. The internal energy of an isolated system remains constant.

In a **cyclic process** (one that originates and terminates at the same state), $\Delta U = 0$, and therefore $Q = W$.

An **adiabatic process** is one in which no thermal energy is transferred between the system and its surroundings ($Q = 0$). In this case, the first-law equation gives $\Delta U = -W$. In the **adiabatic free expansion** of a gas, $Q = 0$ and $W = 0$, and so $\Delta U = 0$.

An **isobaric process** is one that occurs at constant pressure. The work done in such a process is $P\Delta V$.

An **isovolumetric process** is one that occurs at constant volume. No work (of expansion) is done in such a process.

An **isothermal process** is one that occurs at constant temperature. The work done by an ideal gas during an isothermal process is

Work done in an isothermal process

$$W = nRT \ln\left(\frac{V_f}{V_i}\right) \qquad [14.12]$$

Questions and Conceptual Exercises

1. Clearly distinguish among temperature, thermal energy, thermal energy transfer, and internal energy.
2. When a sealed Thermos bottle full of hot coffee is shaken, what are the changes, if any, in (a) the temperature of the coffee and (b) its internal energy?
3. Use the first law of thermodynamics to explain why the total energy of an isolated system is always constant.
4. Ethyl alcohol has about one-half the specific heat of water. If equal masses of alcohol and water in separate beakers are supplied with the same amount of thermal energy, compare the temperature increases of the two liquids.
5. A small crucible is taken from a 200°C oven and immersed in a tub of water at room temperature (this process is often referred to as quenching). Estimate the final equilibrium temperature.
6. The U.S. penny is now made of copper-coated zinc. Can a calorimetric experiment be devised to test for the metal content in a collection of pennies? If so, describe the procedure you would use.
7. What is wrong with the statement "Given any two bodies, the one with the higher temperature contains more thermal energy"?
8. Pioneers stored fruits and vegetables in underground cellars. Discuss as fully as possible the reasons for this practice.
9. In winter, the pioneers mentioned in Question 8 stored an open barrel of water alongside their produce. Why?
10. What is the major problem that arises in measuring specific heats if a sample with a temperature above 100°C is placed in water?
11. The air temperature above coastal areas is profoundly influenced by the large specific heat of water. One reason is that the thermal energy released during the cooling of 1 cubic meter of water by 1°C raises the temperature of an enormously larger volume of air by 1°C. Estimate this volume of air. (The specific heat of air is approximately 1.0 kJ/kg·°C. Take the density of air to be 1.3 kg/m^3.)

Problems

Section 14.1 Heat Flow and Internal Energy

1. An 80-kg weight-watcher wishes to climb a mountain to work off the equivalent of a large piece of chocolate cake, rated at 700 (food) Calories. How high must the person climb?
2. Water at the top of Niagara Falls has a temperature of 10°C. If it falls a distance of 50 m and all of its potential energy goes into heating the water, calculate the temperature of the water at the bottom of the falls. (Neglect cooling of the water by evaporation.)
3. Joule's famous experiment for measuring the mechanical equivalent of heat is illustrated in Figure 14.8. A rotating paddle wheel is immersed in water in a thermally insulated container. As the weights fall a distance of h, the loss in potential energy of the weights is $2mgh$, and this energy is used to heat the water. If the two masses are 1.5 kg each, and the vessel contains 200 g of water, what is the increase in the temperature of the water after the masses fall a distance of 3 m?

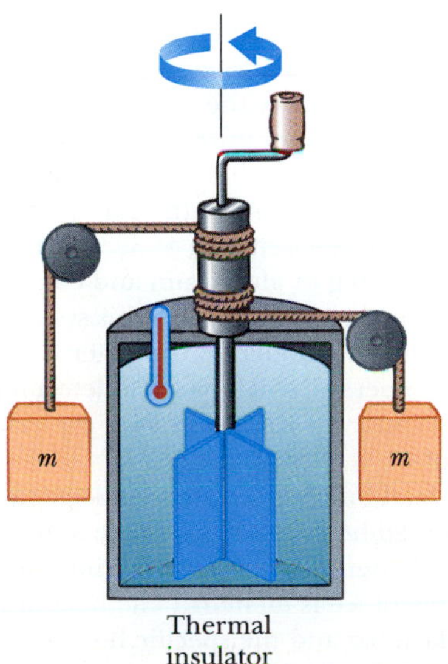

Figure 14.8
(Problem 3) The falling weights rotate the paddles, causing the temperature of the water to rise.

4. A 50-g sample of copper is at 25°C. If 1200 J of thermal energy is added to it, what is the final temperature of the copper?
5. The temperature of a silver bar rises by 10.0°C when it absorbs 1.23 kJ of thermal energy. The mass of the bar is 525 g. Determine the specific heat of silver.
6. How many joules of energy are required to raise the temperature of 100 g of gold from 20°C to 100°C?
7. A 1.5-kg iron horseshoe initially at 600°C is dropped into a bucket containing 20 kg of water at 25°C. What is the final temperature of the horseshoe? (Neglect the heat capacity of the container.)

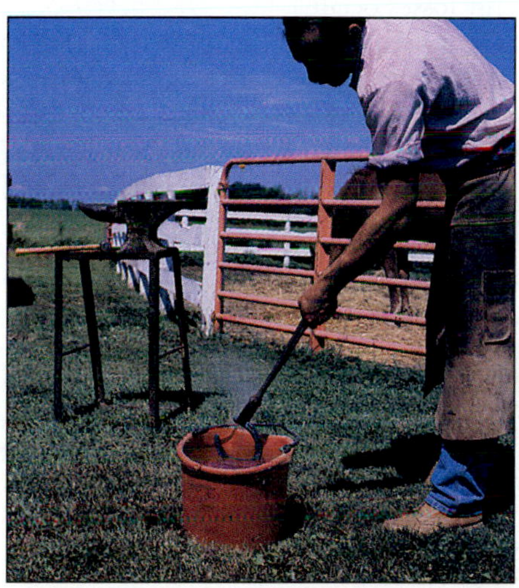

(Problem 7) A "red-hot" horseshoe being cooled in water.
(© Chris Sorensen 1993)

8. What is the final equilibrium temperature when 10 g of milk at 10°C is added to 160 g of coffee at 90°C? (Assume the heat capacities of the two liquids are the same as that of water, and neglect the heat capacity of the container.)
9. Copper pellets, each of mass 1 g, are heated to 100°C. How many pellets must be added to 500 g of water that is initially at 20°C to make the final equilibrium temperature 25°C? (Neglect the heat capacity of the container.)
10. A student inhales air at a temperature of 22°C and exhales air at 37°C. The average volume of air in one breath is 200 cm^3. Make a rough estimate of the amount of thermal energy absorbed in one day by the air breathed by the student. The density of air is approximately 1.25 kg/m^3, and the specific heat of air is 1000 $J/kg \cdot °C$.

11. A 1.5-kg copper block is given an initial speed of 3 m/s on a rough horizontal surface. Because of friction, the block comes to rest. (a) If 85% of its initial kinetic energy is converted to thermal energy within the block, calculate the increase in temperature of the block. (b) What happens to the remaining energy?
12. An aluminum cup of mass 200 g contains 800 g of water in thermal equilibrium at 80°C. The combination of cup and water is cooled uniformly so that the temperature decreases by 1.5 C° per minute. At what rate is thermal energy being removed? Express your answer in watts.
13. If 90 g of molten lead at 327.3°C is poured into a 300-g casting made of iron and initially at 20°C, what is the final temperature of the system? (Assume no energy losses occur.)
14. An aluminum calorimeter of mass 100 g contains 250 g of water. The calorimeter and water are in thermal equilibrium at 10°C. Two metallic blocks are placed in the water. One is a 50-g piece of copper at 80°C; the other has a mass of 70 g and is originally at a temperature of 100°C. The entire system stabilizes at a final temperature of 20°C. (a) Determine the specific heat of the unknown sample. (b) Determine the material of the unknown, using Table 14.1.
15. A 3-g copper penny at 25°C drops a distance of 50 m to the ground. (a) If 60% of the penny's initial potential energy goes into increasing its internal energy, determine its final temperature. (b) Does the result depend on the mass of the penny? Explain.
16. Lake Erie contains roughly 4×10^{11} m^3 of water. (a) How much thermal energy is required to raise the temperature of that volume of water from 11°C to 12°C? (b) Approximately how many years would it take to supply this amount of thermal energy by using the full output of a 1000-MW electric power plant?

Section 14.3 Latent Heat and Phase Changes

17. How much thermal energy is required to change a 40-g ice cube from ice at −10°C to steam at 110°C?
18. Determine the final state when 20 g of 0°C ice and 10 g of 100°C steam are mixed together in an insulated container.
19. Steam at 100°C is added to ice at 0°C. (a) Find the amount of ice melted and the final temperature when the mass of steam is 10 g and the mass of ice is 50 g. (b) Repeat when the mass of steam is 1 g and the mass of ice is 50 g.
20. A 40-g block of ice is cooled to −78°C. It is added to 560 g of water in an 80-g copper calorimeter at a temperature of 25°C. Determine the final temperature. (If all the ice does not melt, determine how much ice is left.) Remember that the ice must first warm to 0°C, melt, and then continue warming as water. The specific heat of ice is 0.500 cal/g · °C.
21. A 1-kg block of copper at 20°C is dropped into a large vessel of liquid nitrogen at 77 K. How many kilograms of nitrogen boil away by the time the copper reaches 77 K? (The specific heat of copper is 0.092 cal/g · °C. The heat of vaporization of nitrogen is 48 cal/g.)
22. How much thermal energy, in kilocalories, is required to vaporize a 1-g ice cube at 0°C? The heat of fusion of ice is 80 cal/g. The heat of vaporization of water is 540 cal/g.
23. One liter of water at 30°C is used to make iced tea. How much ice at 0°C must be added to the tea to lower its temperature to 10°C?
24. A 50-g copper calorimeter contains 250 g of water at 20°C. How much steam must be condensed into the water to make the final temperature of the system 50°C?
25. In an insulated vessel, 250 g of ice at 0°C is added to 600 g of water at 18°C. (a) What is the final temperature of the system? (b) How much ice remains?
26. A 50-g ice cube at −20°C is dropped into a container of water at 0°C. How much water freezes onto the ice?
27. An iron nail is driven into a block of ice by a single blow of a hammer. The hammerhead has a mass of 0.5 kg and an initial speed of 2.0 m/s. Nail and hammer are at rest after the blow. How much ice melts? Assume the temperature of the nail is 0.0°C before and after.
28. A beaker of water sits in the sun until it reaches an equilibrium temperature of 30°C. The beaker is made of 100 g of aluminum and contains 180 g of water. In an attempt to cool this system down, 100 g of ice at 0°C is added to the water. (a) Determine the final temperature. If $T = 0$°C, determine how much ice remains. (b) Repeat this for the case in which 50 g of ice is used.
29. A 3-g lead bullet is traveling at a speed of 240 m/s when it embeds in a block of ice at 0°C. If all the thermal energy generated goes into melting ice, what quantity of ice is melted? (The heat of fusion for ice is 80 kcal/kg and the specific heat of lead is 0.03 kcal/kg · °C.)

Section 14.4 Work and Thermal Energy in Thermodynamic Processes

30. Gas in a container is at a pressure of 1.5 atm and a volume of 4 m^3. What is the work done by the gas if (a) it expands at constant pressure to twice its initial

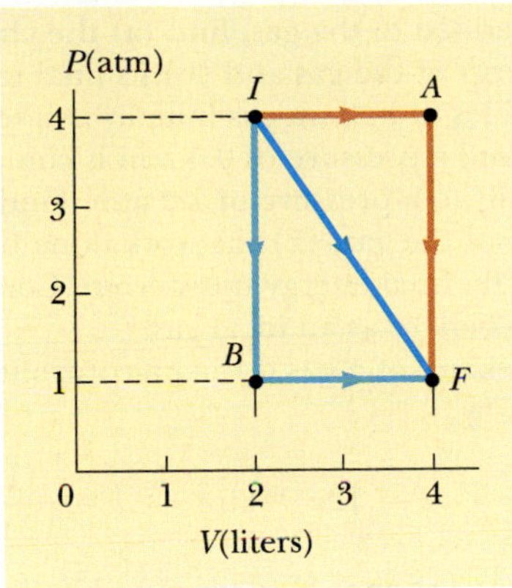

Figure 14.9 (Problem 31)

volume? (b) it is compressed at constant pressure to one quarter of its initial volume?

31. A gas expands from I to F along three possible paths, as indicated by Figure 14.9. Calculate the work, in joules, done by the gas along the paths IAF, IF and IBF.

32. (a) Determine the work done by a fluid that expands from i to f as indicated in Figure 14.10. (b) How much work is performed by the fluid if it is compressed from f to i along the same path?

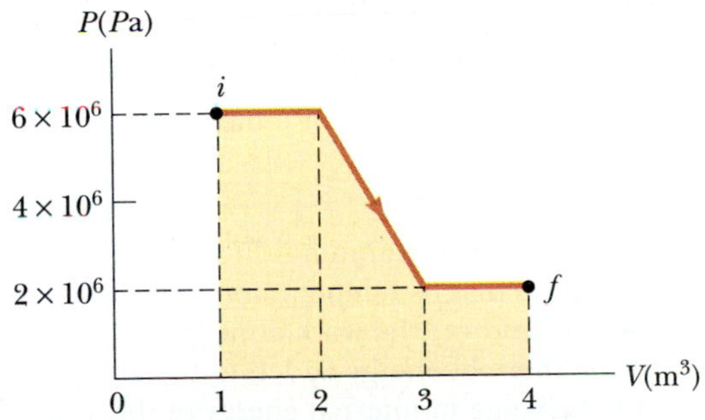

Figure 14.10 (Problem 32)

33. A sample of ideal gas is expanded to twice its original volume of 1 m^3 in a quasi-static process for which $P = \alpha V^2$, with $\alpha = 5.0$ atm/m^6, as shown in Figure 14.11. How much work was done by the expanding gas?

34. An ideal gas at STP (1 atm and 0°C) is taken through a process wherein the volume is expanded from 25 L to 80 L. During this process the pressure varies inversely as the volume squared, $P = 0.5aV^{-2}$. (a) Determine the constant a in standard SI units. (b) Find the final pressure and temperature. (c) Determine a general expression for the work done by the gas during this process. (d) Compute the actual work, in joules, done by the gas in this process.

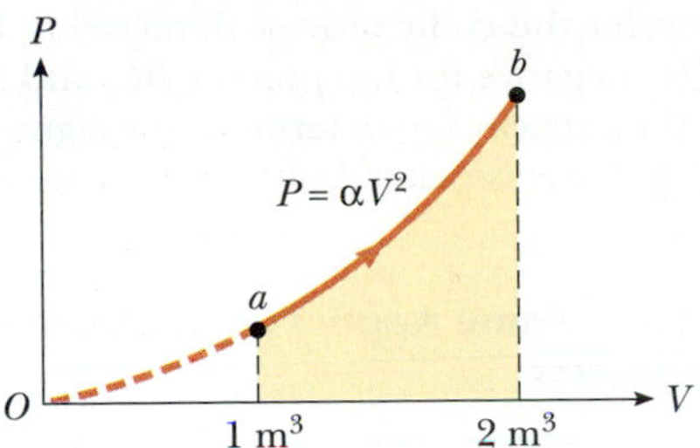

Figure 14.11 (Problem 33)

35. One mole of an ideal gas does 3000 J of work on its surroundings as it expands isothermally to a final pressure of 1 atm and volume of 25 L. Determine (a) the initial volume and (b) the temperature of the gas.

Section 14.5 The First Law of Thermodynamics

36. A gas is compressed at a constant pressure of 0.8 atm from a volume of 9 L to a volume of 2 L. In the process, 400 J of thermal energy flows out of the gas. (a) What is the work done by the gas? (b) What is the change in internal energy of the gas?

37. A thermodynamic system undergoes a process in which its internal energy decreases by 500 J. If at the same time, 220 J of work is done on the system, find the thermal energy transferred to or from the system.

38. A gas is taken through the cyclic process described in Figure 14.12. (a) Find the net thermal energy transferred to the system during one complete cycle. (b) If the cycle is reversed—that is, the process goes along $ACBA$—what is the net thermal energy transferred per cycle?

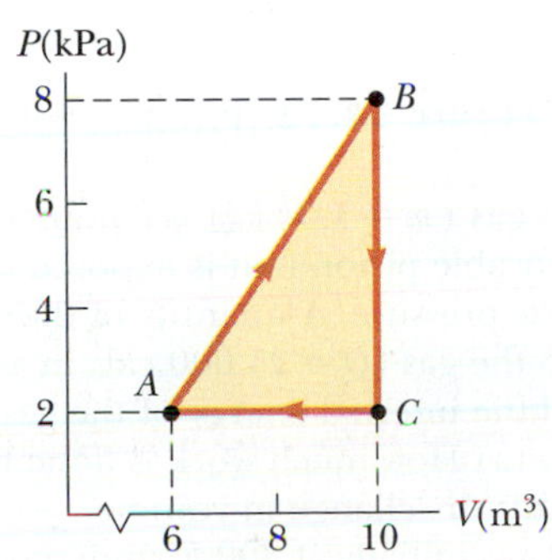

Figure 14.12 (Problems 38 and 39)

39. Consider the cyclic process depicted in Figure 14.12. If Q is negative for the process BC, and ΔU is negative for the process CA, determine the signs of Q, W, and ΔU that are associated with each process.

Section 14.6 Some Applications of the First Law of Thermodynamics

40. Five moles of an ideal gas expands isothermally at 127°C to four times its initial volume. Find (a) the work done by the gas and (b) the thermal energy flow into the system, both in joules.
41. How much work is done by the steam when 1 mol of water at 100°C boils and becomes 1 mol of steam at 100°C and 1 atm pressure? Determine the change in internal energy of the steam as it vaporizes. Consider the steam to be an ideal gas.
42. One mole of gas initially at a pressure of 2 atm and a volume of 0.3 L has an internal energy that may be set equal to 91 J. In its final state, the pressure is 1.5 atm, the volume is 0.8 L, and the internal energy equals 182 J. For the three paths IAF, IBF, and IF in Figure 14.13, calculate (a) the work done by the gas and (b) the net thermal energy transferred in the process.

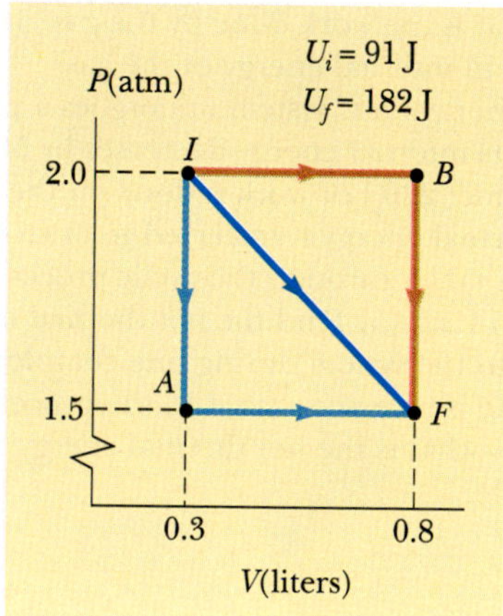

Figure 14.13 (Problem 42)

43. Nitrogen gas (m = 1.00 kg) is confined in a cylinder with a movable piston that is exposed to normal atmospheric pressure. A quantity of thermal energy is added to the gas (Q = 25 000 cal) in an isobaric process, and the internal energy of the gas increases by 8000 cal. (a) How much work is done by the gas? (b) What is the change in volume?
44. An ideal gas initially at 300 K undergoes an isobaric expansion at a pressure of 2.5 kPa. If the volume increases from 1 m^3 to 3 m^3 and 12 500 J of thermal energy is added to the gas, find (a) the change in internal energy of the gas and (b) its final temperature.
45. Two moles of helium gas initially at a temperature of 300 K and a pressure of 0.4 atm is compressed isothermally to a pressure of 1.2 atm. Find (a) the final volume of the gas, (b) the work done by the gas, and (c) the thermal energy transferred. Consider the helium to behave as an ideal gas.
46. The pressure of a gas during a controlled expansion is given by

$$P = 12e^{-bV} \text{ atm} \qquad b = \frac{1}{12 \text{ m}^3}$$

where the volume, V, is expressed in cubic meters (Fig. 14.14). Determine the work performed when the gas expands from V = 12 m^3 to V = 36 m^3.

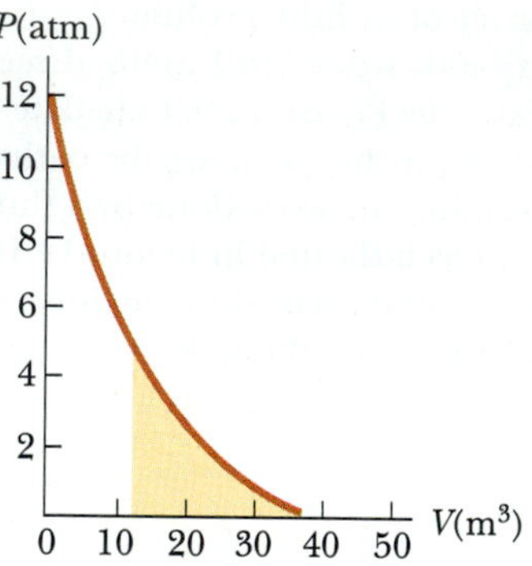

Figure 14.14 (Problem 46)

47. A 1-kg block of aluminum is heated at atmospheric pressure so that its temperature increases from 22°C to 40°C. Find (a) the work done by the aluminum, (b) the thermal energy added to the aluminum, and (c) the change in internal energy of the aluminum.
48. Helium with an initial volume of 1 liter (10^{-3} m^3) and an initial pressure of 10 atm expands to a final volume of 1 m^3. The relationship between pressure and volume during the expansion is PV = constant. Determine (a) the value of the constant, (b) the final pressure, and (c) the work done by the helium during the expansion.

Additional Problems

49. One hundred grams of liquid nitrogen at 77 K is stirred into a beaker containing 200 g of 5°C water. If the nitrogen leaves the solution as soon as it turns to gas, how much water freezes? (The heat of vaporiza-

tion of nitrogen is 6.09 cal/g, and the heat of fusion of water is 80 cal/g.)

50. How much water at 20°C is needed to melt 1 kg of solid mercury at −39°C? (The heat of fusion of mercury is 2.8 cal/g.)

51. A Styrofoam container used as a picnic cooler contains a block of ice at 0°C. If 225 g of ice melts in 1 hour, how much thermal energy per second is passing through the walls of the container? (The heat of fusion of ice is 3.33×10^5 J/kg.)

52. In braking an automobile, the friction between the brake drums and brake shoes converts the car's kinetic energy into thermal energy. If a 1500-kg automobile traveling at 30 m/s brakes to a halt, how much does the temperature rise in each of the four 8-kg brake drums? Neglect other energy losses, such as warming of tires. (The specific heat of each iron brake drum is 448 J/kg · °C.)

53. A water heater runs on solar power. If the solar collector has an area of 6 m^2, and the power per unit area delivered by sunlight is 1000 W/m^2, how long does it take to increase the temperature of 1 m^3 of water from 20°C to 60°C?

54. A 75-kg cross-country skier moves across snow so that the coefficient of friction between skis and snow is 0.2. Assume that all the snow beneath his skis is at 0°C and that all the thermal energy generated by friction is added to the snow, which sticks to his skis until melted. How far would he have to ski to melt 1 kg of snow?

(Problem 54) A cross-country skier. *(Nathan Bilow, Leo de Wys, Inc.)*

55. Around a crater formed by an iron meteorite, 75.0 kg of rock has melted under the impact of the meteorite. The rock has a specific heat of 0.8 kcal/kg · °C, a melting point of 500°C, and a latent heat of fusion of 48.0 kcal/kg. The original temperature of the ground was 0.0°C. If the meteorite hit the ground with a terminal speed of 600 m/s, what was its minimum mass? Assume no energy loss to either the surrounding unmelted rock or the atmosphere during the impact. Disregard the heat capacity of the meteorite.

56. A *flow calorimeter* is an apparatus used to measure the specific heat of a liquid. The technique is to measure the temperature difference between the input and output points of a flowing stream of the liquid while adding thermal energy at a known rate. In one particular experiment, a liquid of density 0.78 g/cm^3 flows through the calorimeter at the rate of 4 cm^3/s. At steady state, a temperature difference of 4.8°C is established between the input and output points when thermal energy is supplied at the rate of 30 J/s. What is the specific heat of the liquid?

57. One mole of an ideal gas is contained in a cylinder with a movable piston. The initial pressure, temperature, and volume are P_0, V_0, and T_0, respectively. Find the work done by the gas for the following processes and show each process on a *PV* diagram: (a) an isobaric compression in which the final volume is one-half the initial volume, (b) an isothermal compression in which the final pressure is four times the initial pressure, (c) an isovolumetric process in which the final pressure is triple the initial pressure.

58. A gas expands from a volume of 2 m^3 to a volume of 6 m^3 along two different paths, as shown in Figure 14.15. The thermal energy added to the gas along the path *IAF* is equal to 4×10^5 cal. Find (a) the work done by the gas along the path *IAF*, (b) the work done along the path *IF*, (c) the change in internal energy of the gas, and (d) the thermal energy transferred in the process along the path *IF*.

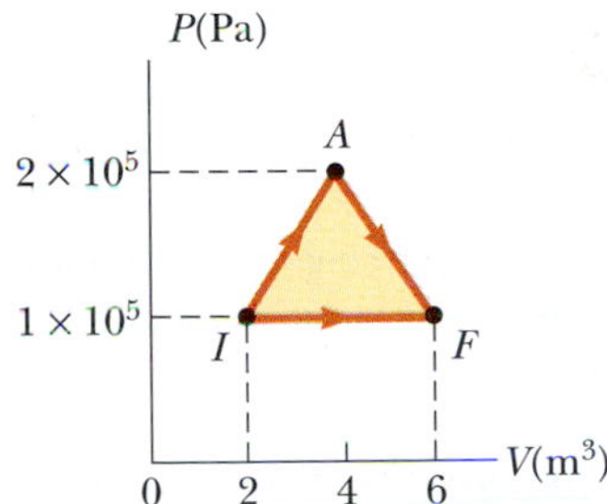

Figure 14.15 (Problem 58)

59. A 5-g lead bullet traveling at 300 m/s strikes a flat steel plate and stops. If the collision is inelastic, does the bullet melt? Lead has a melting point of 327°C, a specific heat of 0.128 J/g · °C, and a heat of fusion of 24.5 J/g.

60. An ideal gas initially at pressure P_0, volume V_0, and temperature T_0 is taken through a cycle as shown in Figure 14.16. (a) Find the net work done by the gas

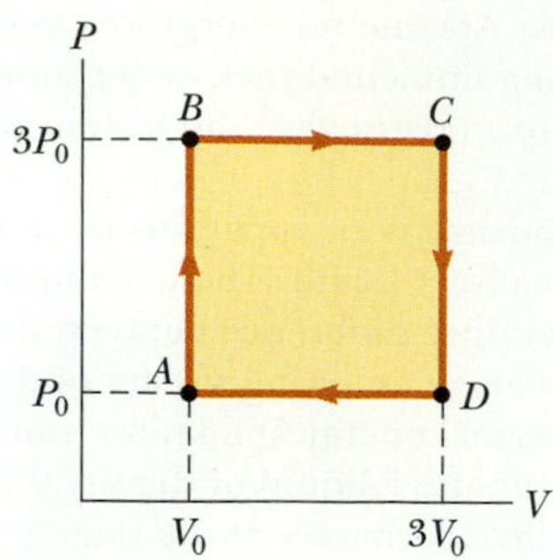

Figure 14.16 (Problem 60)

per cycle. (b) What is the net thermal energy added to the system per cycle? (c) Obtain a numerical value for the net work done per cycle for one mole of gas that is initially at 0°C.

61. Using the data in Example 14.4 and Table 14.2, calculate the change in internal energy when 2 cm^3 of liquid helium at 4.2 K is converted to helium gas at 273.15 K and atmospheric pressure. (Assume that the molar heat capacity of helium gas is 24.9 J/mol·K, and note that 1 cm^3 of liquid helium is equivalent to 3.1×10^{-2} moles.)

62. An iron plate is held against an iron wheel so that a sliding frictional force of 50 N acts between the two pieces of metal. The relative speed at which the two surfaces slide over each other is 40 m/s. (a) Calculate the rate at which mechanical energy is converted to thermal energy. (b) The plate and the wheel have masses of 5 kg each, and each receives 50% of the thermal energy. If the system is run as described for 10 s and each object is then allowed to reach a uniform internal temperature, what is the resultant temperature increase?

63. In a showdown on the streets of Laredo, the good guy drops a 5-g silver bullet, at a temperature of 20°C, into a 100-cm^3 cup of water at 90°C. Simultaneously, the bad guy drops a 5-g copper bullet, at the same initial temperature, into an identical cup of water. Which one ends the showdown with the coolest cup of water in the west? Neglect any thermal energy transfer into or out of the container.

64. A passive solar home uses a wall of concrete to collect solar energy. The wall has an absorbing area of 20.0 m^2 and a mass of 1.0×10^4 kg. During a "solar day" the wall is illuminated for 6.0 h, with an average power per unit area of 400 W/m^2. The wall stores 30% of the energy. (a) If concrete has a specific heat of 920 J/kg·°C, what is the heat capacity of the wall? (Heat capacity is defined as the mass of an object times its specific heat.) (b) What is the energy stored by the wall during a "solar day"? (c) If the initial temperature of the wall is 15°C, what is its temperature at the end of the "solar day"?

Figure 14.17 (Problem 67)

65. Calculate the temperature increase in 2 kg of water when it is heated with an 800-W immersion heater for a period of 5 min.

66. A 60.0-kg runner dissipates 300 W of power while running a marathon. Assuming that 10.0% of the runner's energy is dissipated in the muscle tissue and that the excess thermal energy is primarily removed from the body by sweating, determine the volume of bodily fluid (assume it is water) lost per hour. (At 37.0°C the heat of vaporization of water is equal to 575 kcal/kg.)

67. A "solar cooker" consists of a curved reflecting mirror that focuses sunlight onto the object to be heated (Fig. 14.17). The solar power per unit area reaching the Earth at some location is 600 W/m^2, and a small solar cooker has a diameter of 0.6 m. Assuming that 40% of the incident energy is converted into thermal energy, how long would it take to completely boil off 0.5 liters of water that is initially at 20°C? (Neglect the heat capacity of the container.)

△68. A student obtains the following data in a method-of-mixtures experiment designed to measure the specific heat of aluminum:

Initial temperature of water and calorimeter: 70°C
Mass of water: 0.400 kg
Mass of calorimeter: 0.040 kg
Specific heat of calorimeter: 0.63 kJ/kg·°C
Initial temperature of aluminum: 27°C
Mass of aluminum: 0.200 kg
Final temperature of mixture: 66.3°C

Use these data to determine the specific heat of aluminum. Your result should be within 15% of the value listed in Table 14.1.

69. An ideal gas is enclosed in a cylinder with a movable piston on top of it. The piston has a mass of 8000 g and an area of 5 cm^2 and is free to slide up and down, keeping the pressure of the gas constant. How much work is done as the temperature of 0.2 mol of the gas is raised from 20°C to 300°C?

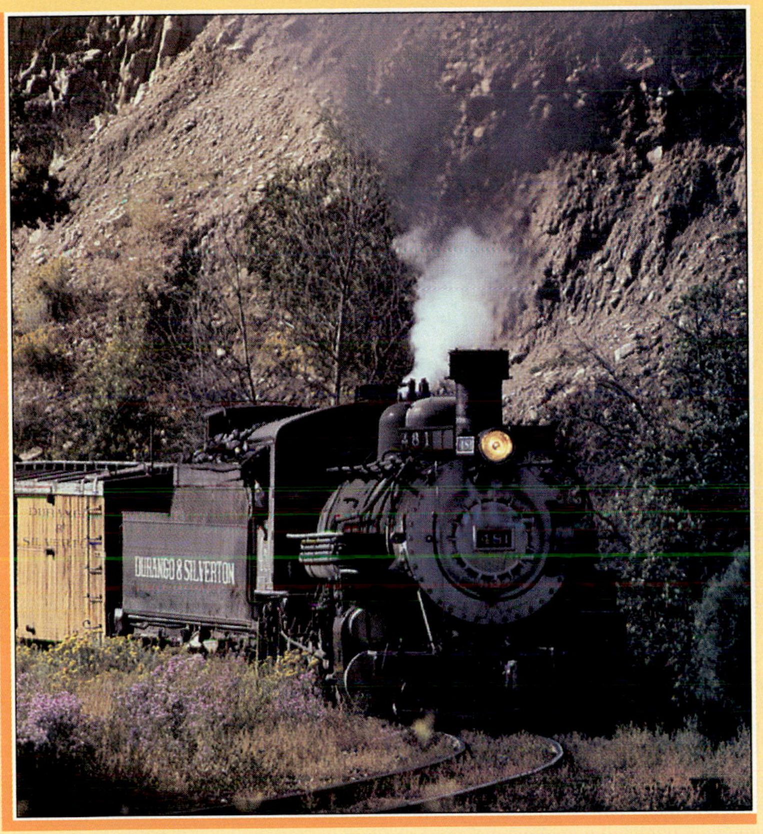

This steam-driven locomotive runs from Durango to Silverton, Colorado. Early steam-driven locomotives obtained their energy by burning wood or coal. The heat generated produces the steam, which powers the locomotive. Modern trains use electricity or diesel fuel to power their locomotives. All heat engines extract heat from a burning fuel and convert only a fraction of this energy to mechanical energy. *(© Lois Moulton, Tony Stone Worldwide, Ltd.)*

Heat Engines, Entropy and the Second Law of Thermodynamics

CHAPTER 15

The first law of thermodynamics, studied in the last chapter, is a statement of conservation of energy, generalized to include heat as a form of energy transfer. This law tells us only that an increase in one form of energy must be accompanied by a decrease in some other form of energy. It places no restrictions on the types of energy conversions that can occur. Furthermore, it makes no fundamental distinction between heat and work: according to the first law, the internal (thermal) energy of a body can be increased either by adding heat to it or by doing work on it. An important difference exists between heat and work, however, that is not evident from the first law. One manifestation of this difference is the fact that it is impossible to convert thermal energy into mechanical energy in an isothermal process.

Contrary to what the first law implies, only certain types of energy conversions can take place. The second law of thermodynamics establishes which natural processes can, and which cannot, occur. The following are examples of processes that are consistent with the first law of thermodynamics but proceed in an order governed by the second law.

1. When two objects at different temperatures are placed in thermal contact with each other, thermal energy always flows from the warmer to the cooler object, never from the cooler to the warmer.
2. A rubber ball dropped to the ground bounces several times and eventually comes to rest, but a ball lying on the ground never begins bouncing on its own.
3. An oscillating pendulum eventually comes to rest because of collisions with air molecules and friction at the point of suspension, and the initial mechanical energy is converted to thermal energy; the reverse conversion of energy never occurs.

These are all *irreversible processes,* that is, processes that occur naturally in only one direction. No irreversible process ever runs backward, because if it did, it would violate the second law of thermodynamics.[1]

From an engineering viewpoint, perhaps the most important application of the second law of thermodynamics is the limited efficiency of heat engines. The second law says that a machine capable of continuously converting thermal energy in a cyclic process *completely* to other forms of energy cannot be constructed.

15.1 Heat Engines and The Second Law of Thermodynamics

A **heat engine** is a device that converts thermal energy to other useful forms, such as electrical or mechanical energy. In a typical process for producing electricity in a power plant, for instance, coal or some other fuel is burned, and the thermal energy produced is used to convert water to steam. This steam is directed at the blades of a turbine, setting it into rotation. Finally, the mechanical energy associated with this rotation is used to drive an electric generator. Another heat engine, the internal combustion engine in your automobile, extracts thermal energy from a burning fuel and converts a fraction of this energy to mechanical energy.

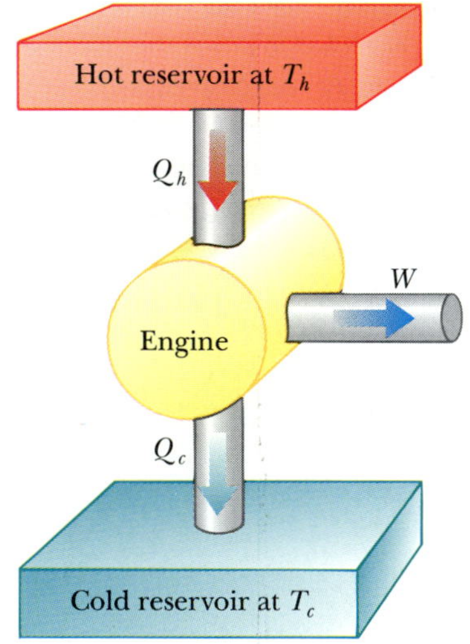

Figure 15.1 Schematic representation of a heat engine. The engine receives heat Q_h from the hot reservoir, expels heat Q_c to the cold reservoir, and does work W.

A heat engine carries some working substance through a cyclic process during which (1) thermal energy is absorbed from a source at a high temperature, (2) work is done by the engine, and (3) thermal energy is expelled by the engine to a source at a lower temperature. As an example, consider the operation of a steam engine in which the working substance is water. The water is carried through a cycle in which it first evaporates into steam in a boiler and then expands against a piston. After the steam is condensed with cooling water, it is returned to the boiler, and the process is repeated.

It is useful to represent a heat engine schematically as in Figure 15.1. The engine absorbs a quantity of heat, Q_h, from the hot reservoir, does work W, and then gives up heat Q_c to the cold reservoir. Because the working substance goes through a cycle, its initial and final internal energies are equal, so $\Delta U = 0$. Hence, from the first-law equation we see that *the net work, W, done by a heat engine equals the*

[1] To be more precise, we should say that the set of events in the time-reversed sense is highly improbable. From this viewpoint, events in one direction are vastly more probable than those in the opposite direction.

net thermal energy flowing into it. As we can see from Figure 15.1, $Q_{net} = Q_h - Q_c$; therefore,

$$W = Q_h - Q_c \qquad [15.1]$$

where Q_h and Q_c are taken to be positive quantities.

If the working substance is a gas, *the net work done for a cyclic process is the area enclosed by the curve representing the process on a PV diagram.* This is shown for an arbitrary cyclic process in Figure 15.2.

The **thermal efficiency,** e, of a heat engine is defined as the ratio of the net work done to the thermal energy absorbed at the higher temperature during one cycle:

Thermal efficiency

$$e \equiv \frac{W}{Q_h} = \frac{Q_h - Q_c}{Q_h} = 1 - \frac{Q_c}{Q_h} \qquad [15.2]$$

We can think of the efficiency as the ratio of what you get (mechanical energy) to what you give (thermal energy at the higher temperature). Equation 15.2 shows that a heat engine has 100% efficiency ($e = 1$) only if $Q_c = 0$—that is, if no thermal energy is expelled to the cold reservoir. In other words, a heat engine with perfect efficiency would have to convert all of the absorbed thermal energy to mechanical work. One of the consequences of the second law of thermodynamics is that this is impossible.

Second law of thermodynamics

The **second law of thermodynamics** can be stated as follows: *It is impossible to construct a heat engine that, operating in a cycle, produces no other effect than the absorption of thermal energy from a reservoir and the performance of an equal amount of work.*

This form of the second law is useful in understanding the operation of heat engines. With reference to Equation 15.2, the second law says that, during the operation of a heat engine, W can never be equal to Q_h or, alternatively, that some heat, Q_c, must be rejected to the environment. As a result, it is theoretically impossible to construct an engine that works with 100% efficiency.

Our assessment of the first two laws of thermodynamics can be summed up as follows: the first law says *we cannot get more energy out of a cyclic process than the amount of thermal energy we put in,* and the second law says *we cannot break even because we must put more thermal energy in, at the higher temperature, than the net amount of work output.*

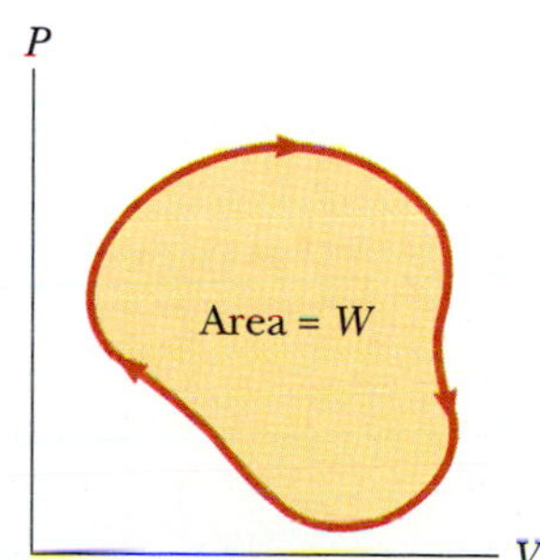

Figure 15.2 The *PV* diagram for an arbitrary cyclic process. The net work done equals the area enclosed by the curve.

Example 15.1 The Efficiency of an Engine

Find the efficiency of an engine that introduces 2000 J of heat during the combustion phase and loses 1500 J at exhaust.

Solution The efficiency of the engine is given by Equation 15.2 as

$$e = 1 - \frac{Q_c}{Q_h} = 1 - \frac{1500\text{ J}}{2000\text{ J}} = 0.25, \text{ or } 25\%$$

Exercise If an engine has an efficiency of 20% and loses 3000 J to the cooling water, how much work is done by the engine?

Answer 750 J.

15.2 Reversible and Irreversible Processes

In the next section we shall discuss a theoretical heat engine that is the most efficient engine possible. In order to understand its nature, we must first examine the meaning of reversible and irreversible processes. A **reversible** process is one that can be performed so that, at its conclusion, both the system and its surroundings have been returned to their exact initial conditions. A process that does not satisfy these requirements is **irreversible.**

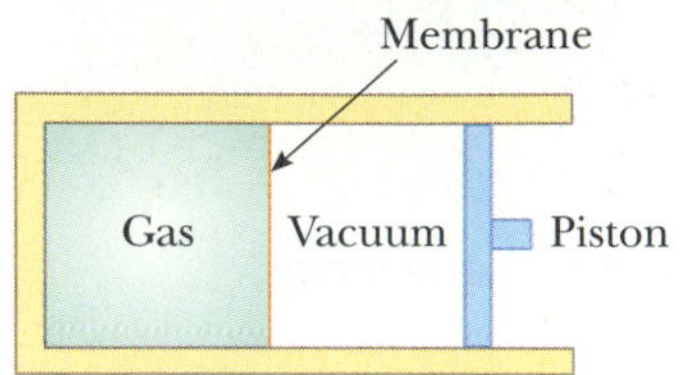

Figure 15.3
Free expansion of a gas.

All natural processes are known to be irreversible. From the endless number of examples that could be selected, let us examine the free expansion of a gas, already discussed in Section 14.4, and show that it cannot be reversible. The gas is contained in an insulated container, as shown in Figure 15.3, with a membrane separating the gas from a vacuum. If the membrane is punctured, the gas expands freely into the vacuum. Because the gas does not exert a force through a distance on the surroundings, it does no work as it expands. In addition, no thermal energy is transferred to or from the gas since the container is insulated from its surroundings. Thus, in this process, the system has changed but the surroundings have not. Now imagine that we try to reverse the process by first compressing the gas to its original volume. Let's say an engine is being used to force the piston inward. Note, however, that this action is changing both the system and surroundings. The surroundings are changing because work is being done by an outside agent on the system, and the system is changing because the compression is increasing the temperature of the gas. We can lower the temperature of the gas by allowing it to come into contact with an external heat reservoir. Although this second procedure returns the gas to its original state, the surroundings are again affected because thermal energy is added to the surroundings. If this thermal energy could somehow be used to drive the engine and compress the gas, the system and its surroundings could be returned to their initial states. However, our statement of the second law says that this extracted thermal energy cannot be completely converted into mechanical energy isothermally. We must conclude that a reversible process has not occurred.

Although real processes are always irreversible, some are *almost* reversible. If a real process occurs very slowly so that the system is virtually always in equilibrium, the process can be considered reversible. For example, imagine compressing a gas

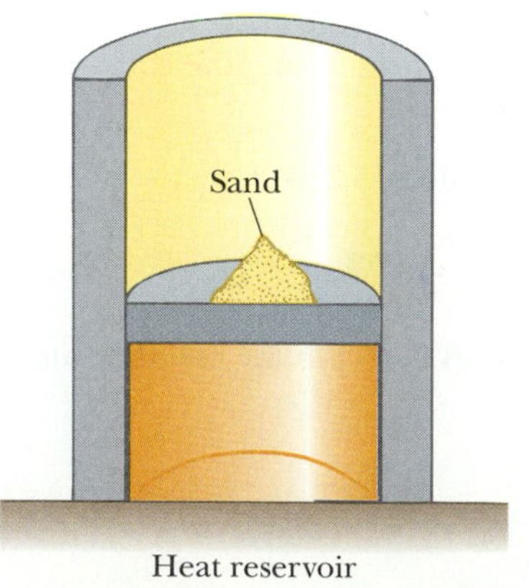

Figure 15.4
A gas in thermal contact with a heat reservoir is compressed slowly by dropping grains of sand onto the piston. The compression is isothermal and reversible.

very slowly by dropping some grains of sand onto a frictionless piston as in Figure 15.4. The compression process can be reversed by the placement of the gas in thermal contact with a heat reservoir. The pressure, volume, and temperature of the gas are well defined during this isothermal compression. Each added grain of sand represents a change to a new equilibrium state. The process can be reversed by the slow removal of grains of sand from the piston.

A general characteristic of a reversible process is that there can be no dissipative effects present, such as turbulence or friction, that convert mechanical energy into thermal energy. In reality, such effects are impossible to eliminate completely, and hence it is not surprising that real processes in nature are irreversible.

Sadi Carnot (1796–1832)
A French physicist, Carnot is considered to be the founder of the science of thermodynamics. Some of his notes found after his death indicate that he was the first to recognize the relationship between work and heat.

15.3 The Carnot Engine

In 1824 a French engineer named Sadi Carnot (1796–1832) described a theoretical engine, now called a *Carnot engine,* that is of great importance from both practical and theoretical viewpoints. He showed that a heat engine operating in an ideal, reversible cycle—called a Carnot cycle—between two heat reservoirs is the most efficient engine possible. Such an ideal engine establishes an upper limit on the efficiencies of all engines. *That is, the net work done by a working substance taken through the Carnot cycle is the greatest amount of work possible for a given amount of thermal energy supplied to the substance at the upper temperature.*

To describe the Carnot cycle, we shall assume that the substance working between temperatures T_c and T_h is an ideal gas contained in a cylinder with a movable piston at one end. The cylinder walls and the piston are thermally nonconducting. Four stages of the Carnot cycle are shown in Figure 15.5, and Figure 15.6 is the *PV* diagram for the cycle. The cycle consists of two adiabatic and two isothermal processes, all reversible.

1. The process $A \rightarrow B$ is an isothermal expansion at temperature T_h, in which the gas is placed in thermal contact with a heat reservoir at temperature T_h (Fig. 15.5a). During the process, the gas absorbs heat Q_h from the reservoir and does work W_{AB} in raising the piston.
2. In the process $B \rightarrow C$, the base of the cylinder is replaced by a thermally nonconducting wall and the gas expands adiabatically; that is, no thermal energy enters or leaves the system (Fig. 15.5b). During the process, the temperature falls from T_h to T_c and the gas does work W_{BC} in raising the piston.
3. In the process $C \rightarrow D$, the gas is placed in thermal contact with a heat reservoir at temperature T_c (Fig. 15.5c) and is compressed isothermally at temperature T_c. During this time, the gas expels heat Q_c to the reservoir, and the work done on the gas is W_{CD}.
4. In the final stage, $D \rightarrow A$, the base of the cylinder is again replaced by a thermally nonconducting wall (Fig. 15.5d) and the gas is compressed adiabatically. The temperature of the gas increases to T_h, and the work done on the gas is W_{DA}.

Carnot showed that the thermal efficiency of a Carnot engine is

$$e_c = \frac{T_h - T_c}{T_h} = 1 - \frac{T_c}{T_h} \qquad [15.3]$$

Carnot efficiency

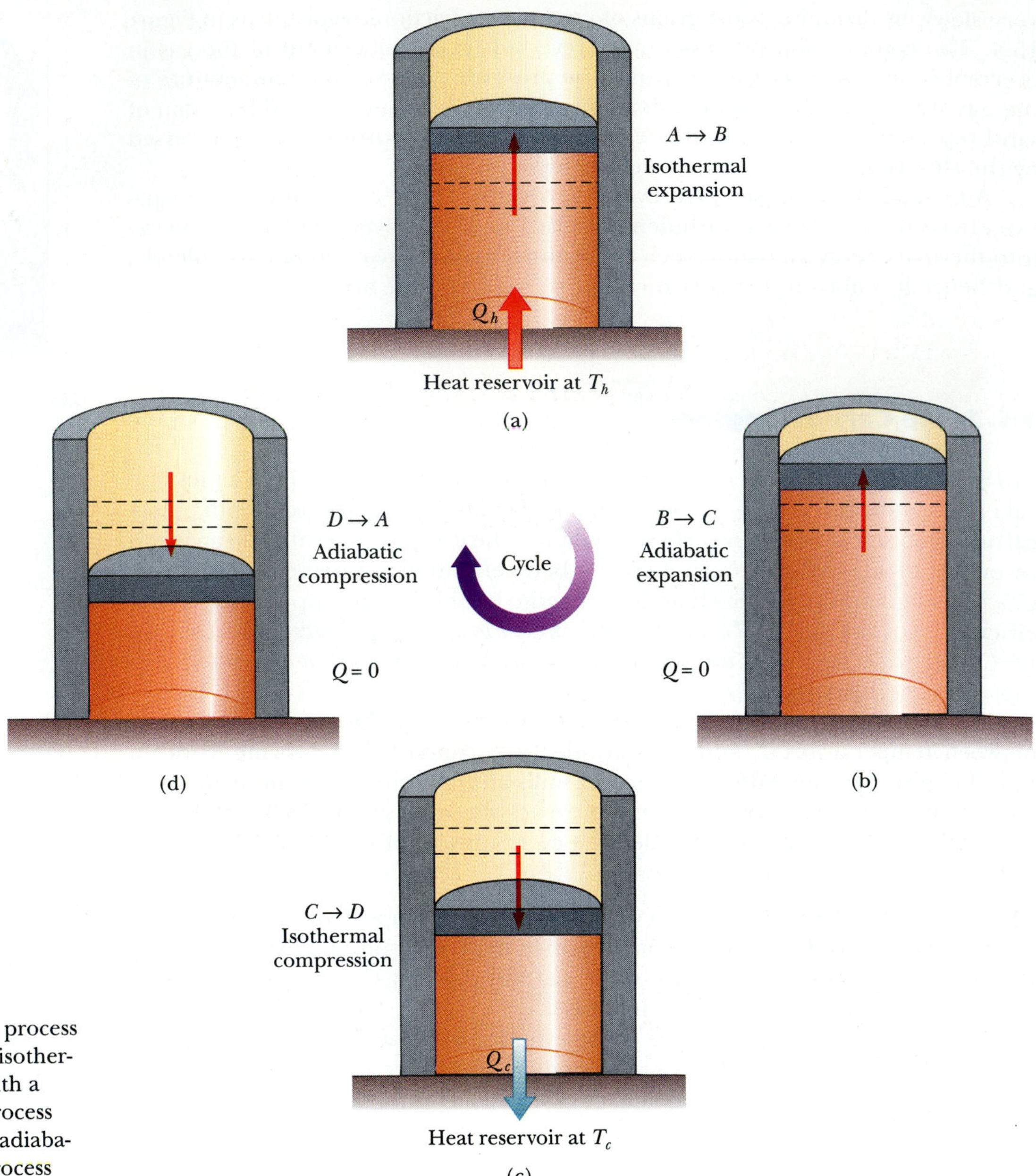

Figure 15.5
The Carnot cycle. (a) In process $A \rightarrow B$, the gas expands isothermally while in contact with a reservoir at T_h. (b) In process $B \rightarrow C$, the gas expands adiabatically ($Q = 0$). (c) In process $C \rightarrow D$, the gas is compressed isothermally while in contact with a reservoir at $T_c < T_h$. (d) In process $D \rightarrow A$, the gas is compressed adiabatically. The upward arrows on the piston in (a) and (b) indicate sand being removed during the expansions, and the downward arrows in (c) and (d) indicate the addition of sand during the compressions.

It is important to note that in the equation, T must be in kelvin. From this result, we see that all Carnot engines operating between the same two temperatures have the same efficiency. Furthermore, the efficiency of a Carnot engine operating between two temperatures is greater than the efficiency of any real engine operating between the same two temperatures.

Equation 15.3 can be applied to any working substance operating in a Carnot cycle between two heat reservoirs. According to this result, the efficiency is zero if $T_c = T_h$, as one would expect. The efficiency increases as T_c is lowered and as T_h is increased. However, the efficiency can be unity (100%) only if $T_c = 0$ K. Such reservoirs are not available, and so the maximum efficiency is always less than unity. In most practical cases, the cold reservoir is near room temperature, about 300 K. Therefore, one usually strives to increase the efficiency by raising the temperature of the hot reservoir. *All real engines are less efficient than the Carnot engine because they are all subject to practical difficulties, including friction, but especially the need to operate irreversibly to complete a cycle in a brief time period.*

Figure 15.6 The *PV* diagram for the Carnot cycle. The net work done, W, equals the net heat received in one cycle, $Q_h - Q_c$. Note that $\Delta U = 0$ for the cycle.

Example 15.2 The Steam Engine

A steam engine has a boiler that operates at 500 K. The heat changes water to steam, which drives the piston. The temperature of the exhaust is that of the outside air, about 300 K. What is the maximum thermal efficiency of this steam engine?

Solution From the expression for the efficiency of a Carnot engine, we find the maximum thermal efficiency for any engine operating between these temperatures:

$$e_c = 1 - \frac{T_c}{T_h} = 1 - \frac{300 \text{ K}}{500 \text{ K}} = \boxed{0.4, \text{ or } 40\%}$$

This is the highest theoretical efficiency of the engine. In practice, the efficiency is considerably lower. (Note that the process described is *not* cyclic.)

Exercise Determine the maximum work the engine can perform in each cycle of operation if it absorbs 200 J of thermal energy from the hot reservoir during each cycle.

Answer 80 J.

Example 15.3 The Carnot Efficiency

The highest theoretical efficiency of a gasoline engine, based on the Carnot cycle, is 30%. If this engine expels its gases into the atmosphere, which has a temperature of 300 K, what is the temperature in the cylinder immediately after combustion?

Solution The Carnot efficiency is used to find T_h:

$$e_c = 1 - \frac{T_c}{T_h}$$

$$T_h = \frac{T_c}{1 - e_c} = \frac{300 \text{ K}}{1 - 0.3} = \boxed{429 \text{ K}}$$

Actual gas engines operate on a cycle significantly different from the Carnot cycle and therefore have lower maximum possible efficiencies.

Exercise If the heat engine absorbs 837 J of heat from the hot reservoir during each cycle, how much work can it perform in each cycle?

Answer 251 J.

A model steam engine equipped with a built-in horizontal boiler. The water is heated electrically, which generates steam that is used to power the electric generator at the left. *(Courtesy of CENCO)*

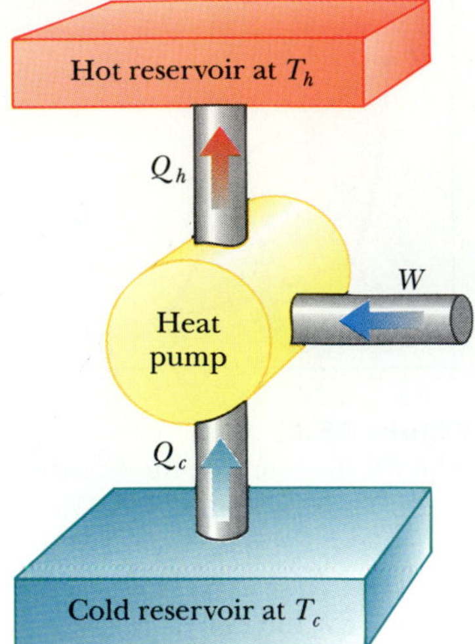

Figure 15.7
Schematic diagram of a heat pump, which absorbs heat Q_c from the cold reservoir and expels heat Q_h to the hot reservoir. The work done on the heat pump is W.

15.4 Heat Pumps and Refrigerators

A heat pump is a mechanical device that moves thermal energy from a region at lower temperature to a region at higher temperature. Heat pumps have long been popular for cooling and are now becoming increasingly popular for heating purposes as well. In the heating mode, a circulating coolant fluid absorbs heat from the outside and releases it to the interior of the structure. The fluid is usually in the form of a low-pressure vapor when in the coils of the exterior part of the unit, where it absorbs heat from either the air or the ground. This gas is then compressed into a hot, high-pressure vapor and enters the interior part of the unit where it condenses to a liquid and releases its stored thermal energy. An air conditioner is simply a heat pump installed backward, with "exterior" and "interior" interchanged.

Figure 15.7 is a schematic representation of a heat pump. The outside temperature is T_c, the inside temperature is T_h, and the heat absorbed by the circulating fluid is Q_c. The compressor does work W on the fluid, and the thermal energy transferred from the pump into the structure is Q_h.

The effectiveness of a heat pump, in its heating mode, is described in terms of a number called the **coefficient of performance,** COP. This is defined as the ratio of the heat transferred into the hot reservoir and the work required to transfer that heat:

$$\text{COP(heat pump)} \equiv \frac{\text{heat transferred}}{\text{work done by pump}} = \frac{Q_h}{W} \qquad [15.4]$$

If the outside temperature is 25°F or higher, the COP for a heat pump is about 4. That is, the heat transferred into the house is about four times greater than the work done by the compressor in the heat pump. However, as the outside temperature decreases, it becomes more difficult for the heat pump to extract sufficient heat from the air, and the COP drops. In fact, the COP can fall below unity for temperatures below the midteens.

A Carnot-cycle heat engine run in reverse constitutes a heat pump–in fact, the heat pump with the highest possible coefficient of performance for the temperatures between which it operates. The maximum coefficient of performance is

$$\text{COP}_c \text{ (heat pump)} = \frac{T_h}{T_h - T_c}$$

Although heat pumps are relatively new products in heating, the refrigerator has been a standard appliance in homes for years. The refrigerator works much like a heat pump, except that it cools its interior by pumping heat from the food storage compartments into the warmer air outside. During its operation, a refrigerator removes heat Q_c from the interior of the refrigerator, and in the process its motor does work W. The coefficient of performance of a refrigerator or of a heat pump used in its cooling cycle is

$$\text{COP(refrigerator)} = \frac{Q_c}{W} \qquad [15.5]$$

An efficient refrigerator is one that removes the greatest amount of thermal energy from the cold reservoir with the least amount of work. Thus, a good refrigerator should have a high coefficient of performance, typically 5 or 6.

The highest possible coefficient of performance is again that of a refrigerator whose working substance is carried through the Carnot heat-engine cycle in reverse:

$$\text{COP}_c\ (\text{refrigerator}) = \frac{T_c}{T_h - T_c}$$

As the difference between temperatures of the two reservoirs approaches zero, the theoretical coefficient of performance of a Carnot heat pump approaches infinity. In practice, the low temperature of the cooling coils and the high temperature at the compressor limit the COP to values below 10.

15.5 An Alternative Statement of the Second Law

The first law of thermodynamics says that energy is conserved in every process. There is more to the story than that, however, and the second law adds the finishing touches. To illustrate what we mean, suppose you wish to cool off a hot piece of pizza by placing it on a block of ice. You will certainly be successful, because in every like situation that has ever been encountered, heat transfer has always taken place from a hot object to a cooler one. If you wait a short while, some amount of heat—500 J, say—will be removed from the pizza and added to the ice. Energy is conserved, and the first law is satisfied. Yet, nothing in the first law says that this heat transfer could not proceed in the opposite direction. (Imagine your astonishment if someday you place a piece of hot pizza on ice and 500 J of heat moves from the ice to the pizza!) It is the second law that determines the directions of such natural phenomena.

An analogy can be made with the impossible sequence of events seen in a movie film running backward—such as a person diving out of a swimming pool, an apple rising from the ground and latching onto the branch of a tree, or a pot of hot water becoming colder as it rests over an open flame. Such events occurring backward in time are impossible because they violate the second law of thermodynamics. *Real processes have a preferred direction of time,* often called the *arrow of time.*

The second law of thermodynamics

The second law has been stated in many different ways, but all the statements can be shown to be equivalent. Which form you use depends on the application you have in mind. For example, if you were concerned about the heat transfer between pizza and ice, you might choose to concentrate on the second law in this form: *Heat will not flow spontaneously from a cold object to a hot object.* At first glance, this statement of the second law seems to be radically different from that in Section 15.1. The two are, in fact, equivalent in all respects. Although we shall not prove it here, it can be shown that if either statement is false, so is the other.

15.6 Entropy

The zeroth law of thermodynamics involves the concept of temperature, and the first law involves the concept of internal (thermal) energy. Temperature and internal energy are both state functions; that is, they can be used to describe the thermodynamic state of a system. Another state function, this one related to the second law of thermodynamics, is the **entropy function,** S. In this section we define entropy on a macroscopic scale as it was first expressed by the German physicist Rudolf Clausius (1822–1888) in 1865.

Consider any infinitesimal process for a system between two equilibrium states. If dQ_r is the amount of thermal energy that would be transferred *if the system had followed a reversible path,* then the change in entropy, *regardless of the actual path followed,* is equal to this amount of thermal energy transferred along the reversible path divided by the absolute temperature of the system:

Clausius' definition of change in entropy

$$dS = \frac{dQ_r}{T} \tag{15.6}$$

The subscript r on the term dQ_r is a reminder that the thermal energy transfer is to be measured along a reversible path, even though the system may actually have followed some irreversible path. When thermal energy is absorbed by the system, dQ_r is positive and hence the entropy increases. When heat is expelled by the system, dQ_r is negative and the entropy decreases. Note that Equation 15.6 defines not entropy, but rather the *change* in entropy. This is consistent with the fact that a change in state always accompanies heat transfer. Hence, the meaningful quantity in a description of a process is the *change* in entropy.

Entropy originally found its place in thermodynamics, but its importance grew tremendously as the field of physics called statistical mechanics developed, because this method of analysis provided an alternative way of interpreting entropy. In statistical mechanics, a substance's behavior is described in terms of the statistical behavior of its atoms and molecules. One of the main outcomes of this treatment is the principle that *isolated systems tend toward disorder, and entropy is a measure of that disorder.*

For example, consider the molecules of a gas in the air in your room. If all the gas molecules moved together like soldiers marching in step, they would be in a very ordered state. It is also an unlikely state. If you could see the molecules, you would see that they actually move haphazardly in all directions, bumping into one another, changing speed upon collision, some going fast, some slowly. This is a highly disordered state. It is also highly possible that the actual state of the system is such a highly disordered state.

The cause of this perpetual drive toward disorder is easily seen. For any given energy of the system, only certain states are possible, or accessible. Among those states, it is assumed that all are equally probable. However, when such possible states are examined, it is found that far more of them are disordered states than ordered states. Because each of the states is equally probable, it is highly probable that the actual state will be one of the highly disordered states, or more precisely, that the actual state will be one in which the system moves between states of equivalent amounts of disorder. In the following discussion, therefore, the term *state* will indicate a collection of states of equivalent amounts of disorder.

The second law yet again

All physical processes tend toward more probable states for the system and its surroundings. The more probable state is always one of higher disorder. Because entropy is a measure of disorder, an alternative way of saying this is that *the entropy of the Universe increases in all natural processes.* This statement is yet another way of stating the second law of thermodynamics.

To calculate the change in entropy for a finite process, we must recognize that T is generally not constant. If dQ_r is the thermal energy transferred reversibly when the system is at a temperature of T, then the change in entropy in an arbitrary reversible process between an initial state and a final state is

$$\Delta S = \int_i^f dS = \int_i^f \frac{dQ_r}{T} \quad \text{(reversible path)} \qquad [15.7]$$

Changes in entropy for a finite process

Although we do not prove it here, the change in entropy of a system that goes from one state to another has the same value for *all* paths connecting the two states. That is, the change in entropy of a system depends only on the properties of the initial and final equilibrium states.

In the case of a *reversible, adiabatic* process, no heat is transferred between the system and its surroundings, and therefore $\Delta S = 0$. Since no change in entropy occurs, such a process is often referred to as an **isentropic process.**

$\Delta S = 0$ for a reversible, adiabatic process

Consider the changes in entropy that occur in a Carnot heat engine operating between the temperatures T_c and T_h. In one cycle, the engine absorbs heat Q_h from the hot reservoir and rejects heat Q_c to the cold reservoir. Thus, the total change in entropy for one cycle is

$$\Delta S = \frac{Q_h}{T_h} - \frac{Q_c}{T_c}$$

where the negative sign in the second term represents the fact that heat Q_c is expelled by the system. For a Carnot cycle,

$$\frac{Q_c}{Q_h} = \frac{T_c}{T_h}$$

Using this result in the preceding expression for ΔS, we find that the total change in entropy for a Carnot engine operating in a cycle is *zero.* That is,

$$\Delta S = 0$$

Change in entropy for a Carnot cycle is zero

Now consider a system taken through an arbitrary reversible cycle. Since the entropy function is a state function and hence depends only on the properties of a given equilibrium state, we conclude that $\Delta S = 0$ for *any* cycle. In general, we can write this condition in the mathematical form

$$\oint \frac{dQ_r}{T} = 0 \qquad [15.8]$$

$\Delta S = 0$ for any cycle

where the symbol $\oint$ indicates that the integration is over a *closed* path.

Another important property of entropy is the fact that the entropy of the Universe is not changed by a reversible process. This can be understood by noting that two bodies, A and B, that interact with each other reversibly must always be in thermal equilibrium with each other. That is, their temperatures must always be equal. Therefore, when a small amount of heat, dQ, is transferred from A to B, the increase in entropy of B is dQ/T, while the corresponding change in entropy of A is $-dQ/T$. Thus, the total change in entropy of the system (A + B) is zero, and the entropy of the Universe is unaffected by the reversible process.

As a special case, the following example shows how to calculate the change in entropy of a substance that undergoes a phase change.

Example 15.4 **Change in Entropy—Melting Process**

A solid substance with a latent heat of fusion of L_f melts at the temperature T_m, the equilibrium melting point.

Rudolph Clausius (1822–1888). "I propose . . . to call S the entropy of a body, after the Greek word 'transformation.' I have designedly coined the word 'entropy' to be similar to energy, for these two quantities are so analogous in their physical significance, that an analogy of denominations seems to be helpful." *(AIP Niels Bohr Library, Lande Collection)*

Calculate the change in entropy that occurs when m grams of this substance is melted.

Solution Let us assume that the melting process occurs so slowly that it can be considered reversible. In that case the temperature can be regarded as constant and equal to T_m. Making use of Equation 15.7 and the fact that the heat of fusion $Q = mL_f$, we find that

$$\Delta S = \int \frac{dQ_r}{T} = \frac{1}{T_m}\int dQ = \frac{Q}{T_m} = \frac{mL_f}{T_m}$$

Note that we can remove T_m from the integral since the process is isothermal. Also, the quantity Q is the total heat required to melt the substance and is equal to mL_f (Section 14.3).

Exercise Calculate the change in entropy when 0.30 kg of lead melts at 327°C. Lead has a heat of fusion of 24.5 kJ/kg.

Answer $\Delta S = 12.3\ \text{J/K}$.

15.7 Entropy Changes in Irreversible Processes

By definition, calculation of the change in entropy requires information about a reversible path connecting the initial and final equilibrium states. In order to calculate changes in entropy for real (irreversible) processes, we must first recognize that the entropy function (like internal energy) depends only on the *state* of the system. That is, entropy is a state function. Hence, the change in entropy when a system moves between any two equilibrium states depends only on the initial and final states. Experimentally one finds that the entropy change is the same for all processes that can occur between a given set of initial and final states. It is possible to show that if this were not the case, the second law of thermodynamics would be violated.

In view of the fact that the entropy of a system depends only on the state of the system, we can now calculate entropy changes for irreversible processes between two equilibrium states. We can do this by devising a reversible process (or series of reversible processes) between the same two equilibrium states and computing $\int dQ_r/T$ for the reversible process. The entropy change for the irreversible process is the same as that for the reversible process between the same two equilibrium states. In irreversible processes, it is critically important to distinguish between Q, the actual thermal energy transfer in the process, and Q_r, the thermal energy that would have been transferred along a reversible path. It is only the second that gives the correct value for entropy change. For example, as we will see, if an ideal gas expands adiabatically into a vacuum, $Q = 0$, but $\Delta S \neq 0$ because $Q_r \neq 0$. The reversible path, between the same two states is the reversible, isothermal expansion that gives $\Delta S > 0$.

As we shall see in the following examples, change in entropy for the system plus its surroundings is always positive for an irreversible process. In general, the total entropy (and disorder) always increases in irreversible processes. From these considerations, the second law of thermodynamics can be stated as follows: *The total entropy of an isolated system that undergoes a change cannot decrease.* Furthermore, if the process is *irreversible,* the total entropy of an isolated system always *increases.* On the other hand, in a reversible process, the total entropy of an isolated system remains constant.

When dealing with interacting bodies that are not isolated from the environment, remember that the increase of entropy applies for the system *and* its surroundings. When two bodies interact in an irreversible process, the increase in entropy of one part of the system is greater than the decrease in entropy of the other part. Hence, we conclude that the change in entropy of the Universe must be greater than zero for an irreversible process and equal to zero for a reversible process. Ultimately, the entropy of the Universe should reach a maximum value. At this point the Universe will be in a state of uniform temperature and density. All physical, chemical, and biological processes will cease, since a state of perfect disorder implies that no energy is available for doing work. This gloomy state of affairs is sometimes referred to as the heat death of the Universe.

Heat Conduction

Consider the reversible transfer of thermal energy Q from a hot reservoir at temperature T_h to a cold reservoir at temperature T_c. Since the cold reservoir absorbs thermal energy Q, its entropy increases by Q/T_c. At the same time, the hot reservoir loses thermal energy Q and its entropy decreases by Q/T_h. The increase in entropy of the cold reservoir is greater than the decrease in entropy of the hot reservoir, since T_c is less than T_h. Therefore, the total change in entropy of the system (and of the Universe) is greater than zero:

$$\Delta S_U = \frac{Q}{T_c} - \frac{Q}{T_h} > 0$$

Example 15.5 Which Way Does the Heat Flow?

A large cold object is at 273 K, and a large hot object is at 373 K. Show that it is impossible for a small amount of heat energy, say 8 J, to be transferred from the cold object to the hot object without decreasing the entropy of the Universe.

Solution We assume that, during the heat transfer, the two objects do not undergo a temperature change. This is not a necessary assumption; it is used to avoid reliance on the techniques of integral calculus. The process as described is irreversible, so we must find an equivalent reversible process. It is sufficient to assume that the hot and cold objects are connected by a poor thermal conductor whose temperature spans the range from 273 K to 373 K, which transmits thermal energy but whose state does not change during the process. Then the thermal energy transfer to or from each object is reversible, and we may set $Q = Q_r$. The entropy change of the hot object is

$$\Delta S_h = \frac{Q}{T_h} = \frac{8\text{ J}}{373\text{ K}} = 0.0214\text{ J/K}$$

The cold object loses heat, and its entropy change is

$$\Delta S_c = \frac{Q}{T_c} = \frac{-8\text{ J}}{273\text{ K}} = -0.0293\text{ J/K}$$

The net entropy change of the Universe is

$$\Delta S_U = \Delta S_c + \Delta S_h = -0.0079\text{ J/K}$$

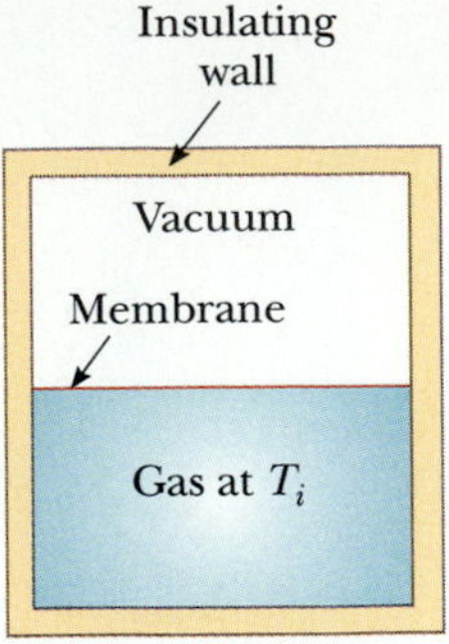

Figure 15.8
Free expansion of a gas. When the partition separating the gas from the evacuated region is ruptured, the gas expands freely and irreversibly so that it occupies a greater final volume. The container is thermally insulated from its surroundings, so $Q = 0$.

This is in violation of the concept that the entropy of the Universe always increases in natural processes. That is, *the spontaneous transfer of heat from a cold to a hot object cannot occur.*

Exercise In the preceding example, suppose that 8 J of heat is transferred from the hot to the cold object. What is the net entropy change of the Universe?

Answer $+0.0079$ J/K.

Free Expansion

An ideal gas in an insulated container initially occupies a volume of V_i (Fig. 15.8). A partition separating the gas from an evacuated region is suddenly broken so that the gas expands (irreversibly) to the volume V_f. Let us find the change in entropy of the gas and the Universe.

The process is clearly neither reversible nor quasi-static. The work done by the gas against the vacuum is zero, and since the walls are insulating, no heat is transferred during the expansion. That is, $W = 0$ and $Q = 0$. Using the first law, we see that the change in internal energy is zero; therefore, $U_i = U_f$. Since the gas is ideal, U depends on temperature only, so we conclude that $T_i = T_f$.

To apply Equation 15.7, we must find Q_r; that is, we must find an equivalent reversible path that shares the same initial and final states. A simple choice is an isothermal, reversible expansion in which the gas pushes slowly against a piston. Since T is constant in this process, Equation 15.7 gives

$$\Delta S = \int \frac{dQ_r}{T} = \frac{1}{T}\int_i^f dQ_r$$

But $\int dQ_r$ is simply the work done by the gas during the reversible isothermal expansion from V_i to V_f, which is given by Equation 14.12. Using this result, we find that

$$\Delta S = nR \ln \frac{V_f}{V_i} \qquad [15.9]$$

Since $V_f > V_i$, we conclude that ΔS is positive, and so both the entropy and the disorder of the gas (and the Universe) increase as a result of the irreversible, adiabatic expansion.

Example 15.6 Free Expansion of a Gas

Calculate the change in entropy of 2 mol of an ideal gas that undergoes a free expansion to three times its initial volume.

Solution Using Equation 15.9 with $n = 2$ and $V_f = 3V_i$, we find that

$$\begin{aligned}\Delta S &= nR \ln \frac{V_f}{V_i} \\ &= (2\ \text{mol})(8.31\ \text{J/mol}\cdot\text{K}) \ln 3 \\ &= 18.3\ \text{J/K}\end{aligned}$$

Irreversible Heat Transfer

A substance of mass m_1, specific heat c_1, and initial temperature T_1 is placed in thermal contact with a second substance of mass m_2, specific heat c_2, and initial temperature T_2, where $T_2 > T_1$. The two substances are contained in an insulated box so that no heat is lost to the surroundings. The system is allowed to reach thermal equilibrium. What is the total entropy change for the system?

First, let us calculate the final equilibrium temperature, T_f. Energy conservation requires that the heat lost by one substance equal the heat gained by the other. Since, by definition, $Q = mc\,\Delta T$ for each substance, we get $Q_1 = -Q_2$, or

$$m_1 c_1\,\Delta T = -m_2 c_2\,\Delta T$$

$$m_1 c_1 (T_f - T_1) = -m_2 c_2 (T_f - T_2)$$

Solving for T_f gives

$$T_f = \frac{m_1 c_1 T_1 + m_2 c_2 T_2}{m_1 c_1 + m_2 c_2} \qquad [15.10]$$

Note that $T_1 < T_f < T_2$, as expected.

The process is irreversible because the system goes through a series of nonequilibrium states. During such a transformation, the temperature at any time is not well defined. However, we can imagine that the hot body, initially at temperature T_i, is slowly cooled to temperature T_f by contact with a series of reservoirs differing infinitesimally in temperature, the first reservoir being at T_i and the last at T_f. Such a series of very small changes in temperature would approximate a reversible process. Applying Equation 15.7 and noting that $dQ = mc\,dT$ for an infinitesimal change, we get

$$\Delta S = \int_1 \frac{dQ_1}{T} + \int_2 \frac{dQ_2}{T} = m_1 c_1 \int_{T_1}^{T_f} \frac{dT}{T} + m_2 c_2 \int_{T_2}^{T_f} \frac{dT}{T}$$

where we have assumed that the specific heats remain constant. Integrating, we find

$$\Delta S = m_1 c_1 \ln \frac{T_f}{T_1} + m_2 c_2 \ln \frac{T_f}{T_2} \qquad [15.11]$$

Change in entropy for a heat transfer process

where T_f is given by Equation 15.10. If Equation 15.10 is substituted into Equation 15.11, it can be shown that one of the terms in Equation 15.11 will always be positive and the other negative. (You may want to verify this for yourself.) However, the positive term will always be larger than the negative term, resulting in a positive value for ΔS. Thus, we conclude that the entropy of the system and of the Universe increases in this irreversible process.

Finally, you should note that Equation 15.11 is valid only when no mixing occurs between the two substances in thermal contact. If the substances are liquids or gases, and mixing occurs, Equation 15.11 applies only if the two fluids are identical, as in the following example.

Example 15.7 Calculating ΔS for a Mixing Process

One kilogram of water at 0°C is mixed with an equal mass of water at 100°C. After equilibrium is reached, the mixture has a uniform temperature of 50°C. What is the change in entropy of the system?

Solution The change in entropy can be calculated from Equation 15.11, using the values $m_1 = m_2 = 1$ kg, $c_1 = c_2 = 4186$ J/kg·K, $T_1 = 0°C$ (= 273 K), $T_2 = 100°C$ (= 373 K), and $T_f = 50°C$ (= 323 K).

$$\Delta S = m_1 c_1 \ln \frac{T_f}{T_2} + m_2 c_2 \ln \frac{T_f}{T_2}$$

$$= (1\text{ kg})(4186\text{ J/kg}\cdot\text{K}) \ln \left(\frac{323\text{ K}}{273\text{ K}}\right) + (1\text{ kg})(4186\text{ J/kg}\cdot\text{K}) \ln \left(\frac{323\text{ K}}{373\text{ K}}\right)$$

$$= 704\text{ J/K} - 602\text{ J/K} = \boxed{102\text{ J/K}}$$

That is, as a result of this irreversible process, the increase in entropy of the cold water is greater than the decrease in entropy of the warm water. Consequently, the increase in entropy of the system is 102 J/K.

15.8 Entropy on a Microscopic Scale[2]

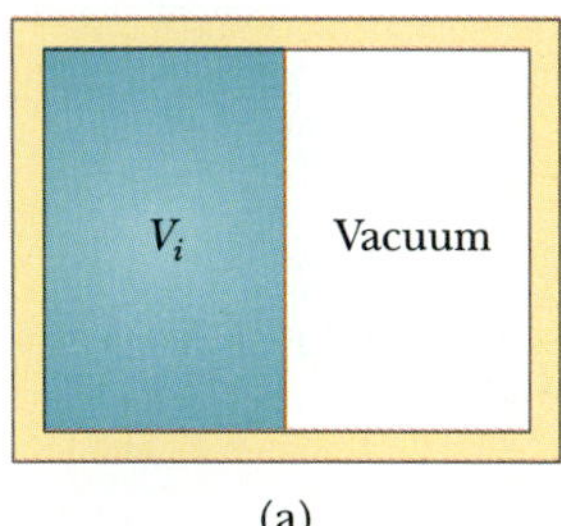

(a)

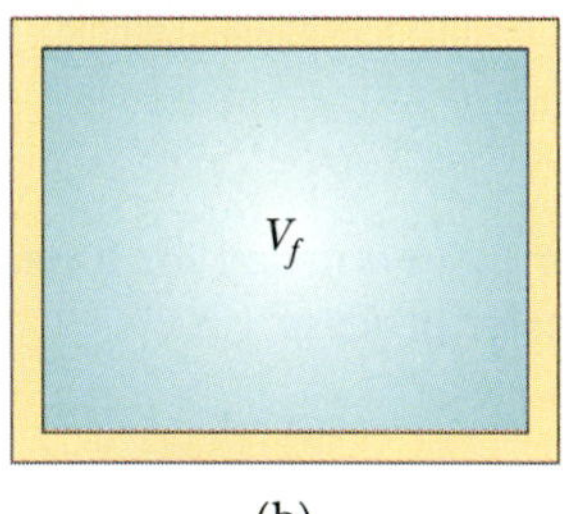

(b)

Figure 15.9
In a free expansion, the gas is allowed to expand into a larger volume that previously was a vacuum.

As we have seen, entropy can be approached via macroscopic concepts, using parameters such as pressure and temperature. Entropy can also be treated from a microscopic viewpoint through statistical analysis of molecular motions. We shall use a microscopic model to investigate the free expansion of an ideal gas discussed in the preceding section.

In the kinetic theory of gases, gas molecules are taken to be particles moving in a random fashion. Suppose all the gas is confined to the volume V_i (Fig. 15.9a). When the partition separating V_i from the larger container is removed, the molecules eventually are distributed in some fashion throughout the greater volume, V_f (Fig. 15.9b). The exact nature of the distribution is a matter of probability.

Such a probability can be determined by first finding the probabilities for the variety of molecular locations involved in the free expansion process. The instant after the partition is removed (and before the molecules have had a chance to rush into the other half of the container), all the molecules are in the initial volume. Now let us estimate the probability of the molecules' arriving at a particular configuration through natural random motions into some larger volume V. Assume each molecule occupies some microscopic volume, V_m. Then the total number of possible locations of that molecule, in a macroscopic initial volume, V_i, is the ratio, $W_i = V_i/V_m$, which is a huge number. We assume each of these locations is equally probable.

As more molecules are added to the system, the number of possible states multiply together. Neglecting the very small probability of two molecules trying to occupy the same location, each molecule may go into any of the (V_i/V_m) locations, so the number of possibilities is $W_i^N = (V_i/V_m)^N$. Although a large number of the possible states would be considered unusual, the total number of states is so enormous that these unusual states have a negligible probability of occurring. Hence, the number of equivalent states is proportional to $(V_i/V_m)^N$. Similarly, if the volume is increased to V_f, the number of equivalent states increases to $W_f^N = (V_f/V_m)^N$. Hence, the probabilities are $P_i = cW_i^N$ and $P_f = cW_f^N$, where the constant, c, has been left undetermined. The ratio of these probabilities is

[2] This section was adapted from A. Hudson and R. Nelson, *University Physics*, Philadelphia, Saunders College Publishing, 1990, used with permission of the publisher.

$$\frac{W_f}{W_i} = \frac{\left(\frac{V_f}{V_m}\right)^N}{\left(\frac{V_i}{V_m}\right)^N} = \left(\frac{V_f}{V_i}\right)^N$$

If we now take the natural logarithm of this equation and multiply by Boltzmann's constant, k_B, we find

$$k_B \ln\left(\frac{W_f}{W_i}\right) = nN_A k_B \ln\left(\frac{V_f}{V_i}\right)$$

where we write the number of molecules, N, as nN_A, the number of moles times Avogadro's number. From Section 13.7 we know that $N_A k_B$ is the universal gas constant, R, so this equation may be written as

$$k_B \ln W_f - k_B \ln W_i = nR \ln\left(\frac{V_f}{V_i}\right) \qquad [15.12]$$

From thermodynamic considerations (specifically, Eq. 15.9) we know that when n mol of a gas undergoes a free expansion from V_i to V_f, the change in entropy, ΔS, is

$$S_f - S_i = nR \ln\left(\frac{V_f}{V_i}\right) \qquad [15.13]$$

Note that the right-hand sides of Equations 15.12 and 15.13 are identical. We thus make the following connection between *entropy* and *probability:*

Entropy (microscopic definition)

$$S \equiv k_B \ln W \qquad [15.14]$$

Although our discussion used the specific example of the free expansion of an ideal gas, a more rigorous development of the statistical interpretation of entropy leads to the same conclusion. *Entropy is a measure of microscopic disorder.*

Imagine that a container of gas has molecules with speeds above the mean value on the left side, and molecules with speeds lower than the mean value on the right side (an *ordered* arrangement), as in Figure 15.10a. Compare with an even mixture of fast- and slow-moving molecules like that in Figure 15.10b (a *disordered* arrangement). You might expect the ordered arrangement to be very unlikely because random motions tend to mix the slow- and fast-moving molecules uniformly. Yet each of these arrangements, *individually,* is equally probable. However, there are far more disordered arrangements than ordered arrangements, so *collectively,* the ordered arrangements are much less probable.

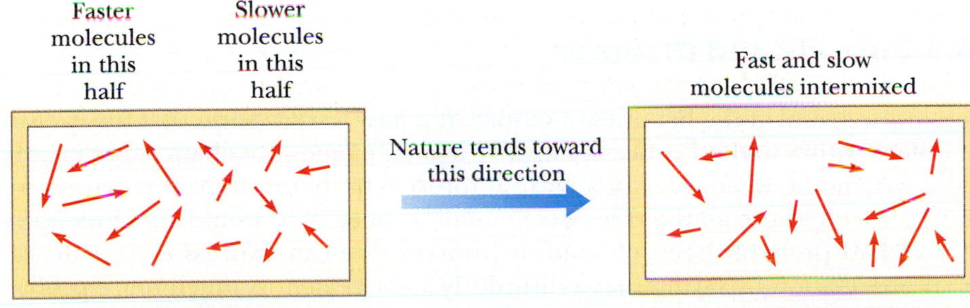

Figure 15.10
A box of gas in two equally probable states of molecular motions that may exist. (a) An *ordered* arrangement; one of a *few,* and therefore *collectively unlikely,* set. (b) A *disordered* arrangement; one of *many,* and therefore a *collectively likely* set.

Let us estimate this probability for 100 molecules. The chance of any one molecule's being in the left part of the container as a result of random motion is 1/2. If the molecules move independently, the probability of 50 faster molecules being found in the left part at any instant is $(1/2)^{50}$. Likewise, the probability of the remaining 50 slower molecules being found in the right part at any instant is $(1/2)^{50}$. Therefore, the probability of finding this fast-slow separation through random motion is the product $(1/2)^{50}\,(1/2)^{50} = (1/2)^{100}$, which corresponds to about 1 chance in 10^{30}. When this calculation is extrapolated from 100 molecules to a mole of gas (about 10^{23} molecules), the ordered arrangement is found to be *extremely* improbable!

Example 15.8 Free Expansion of an Ideal Gas—Revisited

Again consider the free expansion of an ideal gas. Let us verify that the macroscopic and microscopic approaches lead to the same conclusion. Suppose that one mole of gas undergoes a free expansion to four times its initial volume (V_i). The initial and final temperatures are, of course, the same. (a) Using a macroscopic approach, calculate the entropy change. (b) Find the probability, P_i, that all the molecules will, through random motions, be found simultaneously in the original volume, V_i. (c) Using the probability considerations of part (b), calculate the change in entropy, ΔS, for the free expansion and show that it agrees with part (a).

Solution

(a) From Equation 15.9 we obtain

$$\Delta S = nR\ln\left(\frac{V_f}{V_i}\right) = (1)R\ln\left(\frac{4V_i}{V_i}\right) = R\ln 4$$

(b) The number of states available to a single molecule in V_i is $W_i = (V_i/V_m)$. For one mole (N_A), the number of available states is

$$P_i = cW_i^{N_A} = c\left(\frac{V_i}{V_m}\right)^{N_A}$$

(c) The number of states for all N_A molecules in the volume $V_f = 4V_i$ is

$$P_f = cW_f^{N_A} = c\left(\frac{V_f}{V_m}\right)^{N_A} = c\left(\frac{4V_i}{V_m}\right)^{N_A}$$

From Equation 15.14 we obtain

$$\Delta S = k_B\ln W_f - k_B\ln W_i = k_B\ln\left(\frac{W_f}{W_i}\right) = k_B\ln(4^{N_A}) = N_A k_B\ln 4 = R\ln 4$$

This answer is the same as that in part (a), which dealt with macroscopic parameters.

*15.9 Entropy and Disorder

As you look around at the beauties of nature, it is easy to recognize that the events taking place within natural processes involve a large element of chance. For example, the spacing between trees in a natural forest is quite random. If you were to discover a forest where all the trees were equally spaced, you would conclude that the forest had probably been planted by humans. We can express the results of such an observation by saying that **a disorderly arrangement is much more probable that an orderly one if the laws of nature are allowed to act without interference.**

Table 15.1
Possible Results of Drawing Four Marbles from a Bag

End Result	Possible Draws	Total Number of Same Results
All R	RRRR	1
1G, 3R	RRRG, RRGR, RGRR, GRRR	4
2G, 2R	RRGG, RGRG, GRRG, RGGR, GRGR, GGRR	6
3G, 1R	GGGR, GGRG, GRGG, RGGG	4
All G	GGGG	1

We already found that entropy is a measure of the disorder in a system through the relation $S = k_B \ln W$, where W is the number of available states over which the molecules can be distributed and is therefore a measure of the disorder, or randomness, of the system. We can further explore the meaning of this relation by considering another specific example. Imagine that you have a bag of 100 marbles, of which 50 are red and 50 are green. You are allowed to draw four marbles from the bag according to the following rules. Draw one marble, record its color, return it to the bag, and draw another. Continue this process until four marbles have been drawn. Because each marble is returned to the bag before another is drawn, the probability of drawing a red marble is always the same as that of drawing a green one. All the possible outcomes are shown in Table 15.1. For example, the result RRGR means that you drew a red marble on the first draw, a red one on the second, a green one on the third, and a red one on the fourth. As this table indicates, there is only one way to draw four red marbles. However, there are four possible sequences that could produce one green and three red marbles, six sequences that could produce two green and two red, four sequences that could produce three green and one red, and one sequence that could produce all green. The most likely outcome—two red and two green marbles—corresponds to the most disordered state. This is considered disorder simply because we cannot distinguish between individual molecules of the same color, or between individual molecules. The probability that you would draw four red or four green marbles, these being the most ordered states, is much lower. From Equation 15.16 we see that the state with the greatest disorder has the highest entropy because it is most probable. On the other hand, the most ordered states (all red and all green) are the least likely to occur and are the states of lowest entropy. The outcome of the draw can range from a highly ordered state, which has the lowest entropy, to a highly disordered state, which has the highest entropy. Thus, entropy can be seen as an index of how far the system has progressed from an ordered to a disordered state.

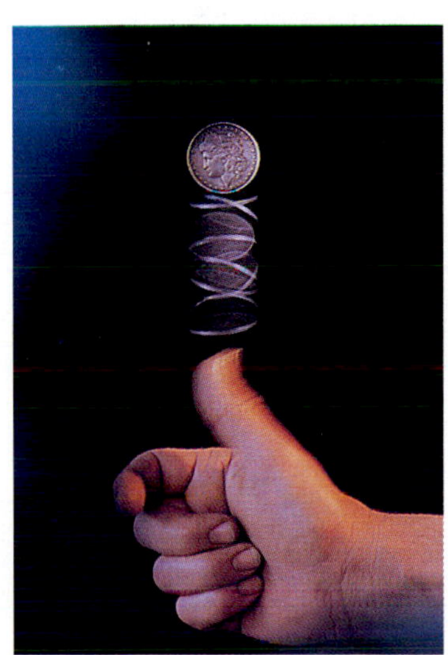

The flip of a coin is a simple example of a probabilistic event. *(© Fred Lyon, Photo Researchers, Inc.)*

Summary

Real processes proceed in directions governed by the second law of thermodynamics.

A **heat engine** is a device that converts thermal energy to other useful forms of energy. The net work done by a heat engine in carrying a substance through a cyclic process ($\Delta U = 0$) is

Work done by a heat engine

$$W = Q_h - Q_c \quad [15.1]$$

where Q_h is the heat absorbed from a hot reservoir and Q_c is the heat rejected to a cold reservoir.

The **thermal efficiency,** e, of a heat engine is defined as the ratio of the net work done to the thermal energy absorbed per cycle from the higher temperature reservoir:

Thermal efficiency

$$e \equiv \frac{W}{Q_h} = 1 - \frac{Q_c}{Q_h} \quad [15.2]$$

The **second law of thermodynamics** can be stated as follows: *It is impossible to construct a heat engine that, operating in a cycle, produces no effect other than the absorption of thermal energy from a reservoir and the performance of an equal amount of work.*

A **reversible** process is one that can be performed so that, at its conclusion, both the system and its surroundings have been returned to their exact initial conditions. A process that does not satisfy these requirements is **irreversible.**

The *efficiency of a heat engine* operating in the **Carnot cycle** is given by

Efficiency of a Carnot engine

$$e_c = 1 - \frac{T_c}{T_h} \quad [15.3]$$

where T_c is the absolute temperature of the cold reservoir and T_h is the absolute temperature of the hot reservoir.

No real heat engine operating between the temperatures T_c and T_h can be more efficient than an engine operating reversibly in a Carnot cycle between the same two temperatures.

The second law of thermodynamics states that when real (irreversible) processes occur, the degree of disorder in the system plus the surroundings increases. When a process occurs in an isolated system, ordered energy is converted into disordered energy. The measure of disorder in a system is called **entropy,** S.

The **change in entropy,** dS, of a system moving between two equilibrium states is

Change in entropy between two equilibrium states

$$dS = \frac{dQ_r}{T} \quad [15.6]$$

The change in entropy of a system moving between two equilibrium states is

$$\Delta S = \int_i^f \frac{dQ_r}{T} \quad [15.7]$$

The value of ΔS is the same for all reversible paths connecting the initial and final states.

The change in entropy for any reversible, cyclic process is zero, and when such a process occurs, the entropy of the Universe remains constant.

Entropy is a state function; that is, it depends on the state of the system. The change in entropy for a system undergoing a real (irreversible) process between two equilibrium states is the same as that for a reversible process between the same states.

In an irreversible process, the total entropy of an isolated system always increases. In general, the total entropy (and disorder) always increases in any irreversible process. Furthermore, the change in entropy of the Universe is greater than zero for an irreversible process.

From a microscopic viewpoint, **entropy,** S, is defined as

$$S \equiv k_B \ln W \qquad [15.14]$$

Entropy, S

where k_B is Boltzmann's constant and W is the number of (microscopic) states available to the system in its (macroscopic) state. Because of the statistical tendency of systems to proceed toward states of greater probability and greater disorder, all natural processes are irreversible and increase entropy. Thus, *entropy is a measure of microscopic disorder.*

Questions and Conceptual Exercises

1. What are some factors that affect the efficiency of automobile engines?
2. The first law says we cannot get more out of a process than we put in, but the second law says we cannot break even. Explain.
3. Is it possible to cool a room by leaving the door of a refrigerator open? What happens to the temperature of a room in which an air conditioner is left running on a table in the middle of the room?
4. A steam-driven turbine is one major component of an electric power plant. Why is it advantageous to increase the temperature of the steam as much as possible?
5. Is it possible to construct a heat engine that creates no thermal pollution?
6. Electrical energy can be converted to heat energy with an efficiency of 100%. Why is this number misleading with regard to heating a home? That is, what other factors must be considered in comparing the cost of electric heating with the cost of hot-air or hot-water heating?
7. Discuss three common examples of natural processes that involve an increase in entropy. Be sure to account for all parts of each system under consideration.
8. Discuss the change in entropy of a gas that expands between the same two states (a) at constant temperature; (b) adiabatically.
9. Suppose the waste heat at a power plant is exhausted to a pond of water. Could the efficiency of the plant be increased by refrigerating the water?
10. An engineer claims to have developed an engine that takes in 70 000 J of heat at 500 K and expels 20 000 J at 300 K, with 10 000 J of work being done. Would this be worth investing in? Why or why not?
11. A living system, such as a tree, combines unorganized molecules (CO_2, H_2O), using sunlight, to produce leaves and branches. Is this reduction of entropy in the tree a violation of the second law of thermodynamics?
12. A diesel engine operates with a compression ratio of about 15. A gasoline engine has a compression ratio of about 6. Which engine runs hotter? Which engine (at least theoretically) can operate more efficiently?
13. In Israel, solar ponds have been constructed in which the Sun's energy is concentrated near the bottom of a salty pond. With the proper layering of salt in the water, convection is prevented, and temperatures of 100°C may be reached. Can you make a guess as to the maximum efficiency with which useful energy can be extracted from the pond?
14. The vortex tube (Fig. 15.11) is a T-shaped device that takes in compressed air at 20 atm and 20°C and produces cold air at −20°C out one flared end and +60°C hot air out the other flared end. Does the operation of this device violate the second law of thermodynamics? (For further information, see R. Hilsch, *Rev. Sci. Instr.* 18:108, 1947; or Y. Soni and W. Thompson, *Trans. A.S.M.E.*, May 1975, p. 316.)

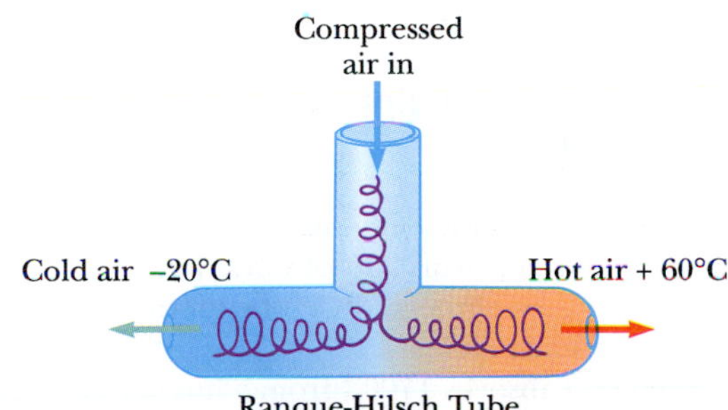

Figure 15.11 (Question 14)

15. All natural processes are irreversible. Give some examples of irreversible processes that occur in nature.
16. A thermodynamic process occurs in which the entropy of a system changes by −8.0 J/K. According to the second law of thermodynamics, what can you conclude about the entropy change of the environment?
17. Coins in the bottom of a box show 10 heads and 10 tails. The box is shaken, and there are 7 heads and 13 tails. Has the second law of thermodynamics been violated? Explain.
18. How could you increase the entropy of one mole of a

metal that is at room temperature? How could you decrease its entropy?

19. A heat pump is to be installed in a region where the average outdoor temperature in the winter months is −20°C. In view of this, why would it be advisable to place the outdoor compressor coils deep in the ground? Why are heat pumps not commonly used for heating in cold climates?
20. Suppose your roommate is "Mr. Clean" and tidies up your messy room after a big party. Since more order is being created by your roommate, does this represent a violation of the second law of thermodynamics?
21. If you shake a jar full of jelly beans of different sizes, the larger jelly beans tend to appear near the top, while the smaller ones tend to fall to the bottom. Why does this occur? Does the process violate the second law of thermodynamics?
22. The device in Figure 15.12, called a thermoelectric converter, is essentially a thermocouple driven by a temperature difference. In the left-hand photograph, both "legs" of the device are at the same temperature, and no electrical energy is produced. However, when one leg is at a higher temperature than the other, as in the right-hand photograph, electrical energy is produced as the device extracts energy from the hot reservoir and drives a small electric motor. (a) Why does the temperature differential produce electrical energy in this demonstration? (b) In what sense does this intriguing experiment demonstrate the second law of thermodynamics?

Figure 15.12 (Question 22) *(Courtesy of PASCO Scientific Co.)*

Problems

Section 15.1 Heat Engines and the Second Law of Thermodynamics

1. In each cycle, a heat engine absorbs 375 J of thermal energy and performs 25 J of work. Find (a) the efficiency of the engine and (b) the thermal energy expelled in each cycle.
2. An engine absorbs 1700 J from a hot reservoir and expels 1200 J to a cold reservoir in each cycle. (a) What is the efficiency of the engine? (b) How much work is done in each cycle? (c) What is the power output of the engine if each cycle lasts for 0.3 s?
3. A heat engine performs 200 J of work in each cycle and has an efficiency of 30%. For each cycle of operation, (a) how much thermal energy is absorbed and (b) how much thermal energy is expelled?
4. The heat absorbed by an engine is three times greater than the work it performs. (a) What is its thermal efficiency? (b) What fraction of the heat absorbed is expelled to the cold reservoir?
5. A particular engine has a power output of 5 kW and an efficiency of 25%. If the engine expels 8000 J of thermal energy in each cycle, find (a) the heat absorbed in each cycle and (b) the time for each cycle.

Section 15.3 The Carnot Engine

6. A heat engine operates between two reservoirs at temperatures of 20°C and 300°C. What is the maximum possible efficiency for this engine?
7. A Carnot engine has a power output of 150 kW. The engine operates between two reservoirs at 20°C and 500°C. (a) How much thermal energy is absorbed per hour? (b) How much thermal energy is lost per hour?
8. One of the most efficient engines ever built operates between 430°C and 1870°C. Its actual efficiency is 42%. (a) What is its maximum theoretical efficiency? (b) How much power does the engine deliver if it absorbs 1.4×10^5 J of thermal energy each second? (c) Compare this observed efficiency with the predicted efficiency for real engines.

$$e = \frac{\sqrt{T_h} - \sqrt{T_c}}{\sqrt{T_h}}$$

9. In a steam turbine, steam at 800°C enters and is exhausted at 120°C. What is the (maximum) efficiency of this turbine?
10. The efficiency of a 1000-MW nuclear power plant is 33%; that is, 2000 MW of heat is rejected to the environment for every 1000 MW of electrical power produced. If a river of flow rate 10^6 kg/s were used to transport the excess thermal energy away, what would be the average temperature increase of the river?
11. A steam engine is operated in a cold climate where the exhaust temperature is at 0°C. (a) Calculate the theoretical maximum efficiency of the engine using an intake steam temperature of 100°C. (b) If, instead, superheated steam at 200°C is used, find the engine's maximum possible efficiency.

(Problem 11) A steam-driven locomotive. *(Paul Chesley, Tony Stone Worldwide)*

12. A heat engine operating between 200°C and 80°C achieves 20% of the maximum possible efficiency. What energy input will enable the engine to perform 10^4 J of work?
13. An ideal gas is taken through a Carnot cycle. The isothermal expansion occurs at 250°C, and the isothermal compression takes place at 50°C. If the gas absorbs 1200 J of heat during the isothermal expansion, find (a) the heat expelled to the cold reservoir in each cycle and (b) the net work done by the gas in each cycle.
14. The exhaust temperature of a Carnot heat engine is 300°C. What is the intake temperature if the efficiency of the engine is 30%?
15. A heat engine operates in a Carnot cycle between 80°C and 350°C. It absorbs 21 000 J of thermal energy per cycle from the hot reservoir. The duration of each cycle is 1 s. (a) What is the maximum power output of this engine? (b) How much thermal energy does it expel in each cycle?
16. A power plant that uses the temperature gradient in the ocean is to operate between 20°C (surface-water temperature) and 5°C (temperature of water at a depth of about 1 km). (a) What is the maximum efficiency of such a system? (b) If the power output of the plant is 75 MW, how much thermal energy is absorbed per hour? (c) In view of your answer to (a), do you think such a system is worthwhile (considering that there is no charge for fuel)?
17. A heat pump, shown in Figure 15.13, is essentially a heat engine run backward. It extracts thermal energy from colder air outside and deposits it in a warmer room. Typically, the ratio of the actual thermal energy entering the room to the work done by the device's motor is 10% of the theoretical maximum. Determine the thermal energy entering the room per joule of work done by the motor if the inside temperature is 20.0°C and the outside temperature is −5.0°C.

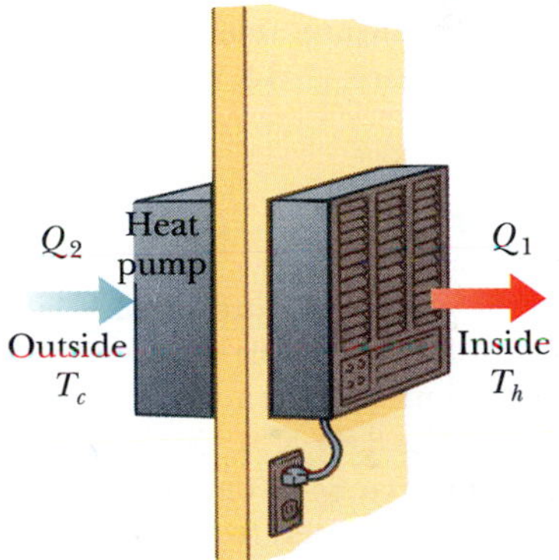

Figure 15.13 (Problem 17)

18. Suppose a heat engine is connected to two heat reservoirs, one a pool of molten aluminum (660°C) and the other a block of solid mercury (−38.9°C). The engine runs by freezing 1.0 g of aluminum and melting 15.0 g of mercury during each cycle. The heat of fusion of aluminum is 3.97×10^5 J/kg; the fusion of mercury is 1.18×10^4 J/kg. What is the efficiency of this engine?
19. An electric generating plant has a power output of 500 MW. The plant uses steam at 600°C and exhausts water at 40°C. The system operates with one-half the maximum (Carnot) efficiency. (a) At what rate is heat expelled to the environment? (b) If the waste heat goes into a river whose flow rate is 1.2×10^6 kg/s, what is the rise in temperature of the river?

Section 15.4 Heat Pumps and Refrigerators

20. An ideal refrigerator (or heat pump) is equivalent to a Carnot engine running in reverse. That is, heat Q_c is absorbed from a cold reservoir and heat Q_h is rejected to a hot reservoir. (a) Show that the work that must be supplied to run the refrigerator is given by

$$W = \frac{T_h - T_c}{T_c} Q_c$$

(b) Show that the coefficient of performance of the ideal refrigerator is given by

$$\text{COP} = \frac{T_c}{T_h - T_c}$$

21. What is the coefficient of performance of a refrigerator that operates with Carnot efficiency between temperatures −3°C and +27°C?
22. What is the maximum possible coefficient of performance of a heat pump that brings thermal energy from the outdoors at −3°C into a +22°C house?
23. How much work is required, using an ideal Carnot refrigerator, to remove 1 J of thermal energy from helium gas at 4 K and reject this thermal energy to a room-temperature (293-K) environment?

Section 15.6 Entropy

24. What is the change in entropy (ΔS), its normal melting point, when one mole of silver (108 g) is melted at its normal melting point (961°C)? The heat of fusion for silver is 8.84×10^4 J/kg.
25. Calculate the change in entropy of 250 g of water when it is slowly heated from 20°C to 80°C. (*Hint:* Note that $dQ = mc\ dT$.)
26. An ice tray contains 500 g of water at 0°C. Calculate the change in entropy of the water as it freezes completely and slowly at 0°C.
27. At a pressure of 1 atm, liquid helium boils at 4.2 K. The latent heat of vaporization is 20.5 kJ/kg. Determine the entropy change (per kilogram) resulting from vaporization.
28. An avalanche of ice and snow, of mass 1000 kg, slides downhill a vertical distance of 200 m. What is the total entropy change if the mountain air temperature is −3°C?

Section 15.7 Entropy Changes in Irreversible Processes

29. A 1500-kg car traveling at 20 m/s crashes into a concrete wall. If the air temperature is 20°C, what is the total entropy change?

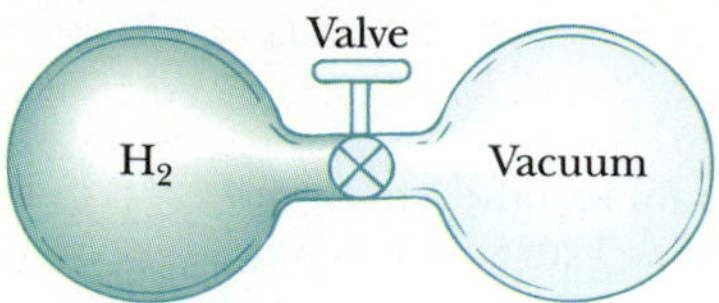

Figure 15.14 (Problem 31)

30. The surface of the Sun is approximately 5700 K, and the temperature of the Earth's surface is about 290 K. What entropy change occurs when 1000 J of heat energy is transferred from the Sun to the Earth?
31. One mole of H_2 gas is contained in the left side of the container (Fig. 15.14; equal volumes left and right). The right side is evacuated. When the valve is opened, hydrogen gas streams into the right side. What is the final entropy change? Does the temperature of the gas change?
32. A 2-liter container has a center partition that divides it into two equal parts, as shown in Figure 15.15. The left side contains H_2 gas, and the right side contains O_2 gas. Both gases are at room temperature and 1 atmosphere. The partition is removed and the gases are allowed to mix. What is the entropy increase?
33. A 2-kg block moving with an initial speed of 5 m/s slides on a table and is stopped by the force of friction. Assuming the table and air remain at a temperature of 20°C, calculate the entropy change of the Universe.
34. A 70-kg log falls from a height of 25 m into a lake. If the log, the lake, and the air are all at 300 K, find the change in entropy of the Universe for this process.
35. If 200 g of water at 20°C is mixed with 300 g of water at 75°C, find (a) the final equilibrium temperature of the mixture and (b) the change in entropy of the system.

Section 15.8 Entropy on a Microscopic Scale
Section 15.9 Entropy and Disorder

36. Prepare a table like Table 13.1 for the following occurrence. You toss four coins into the air simultaneously. Record the results of your tosses in terms of

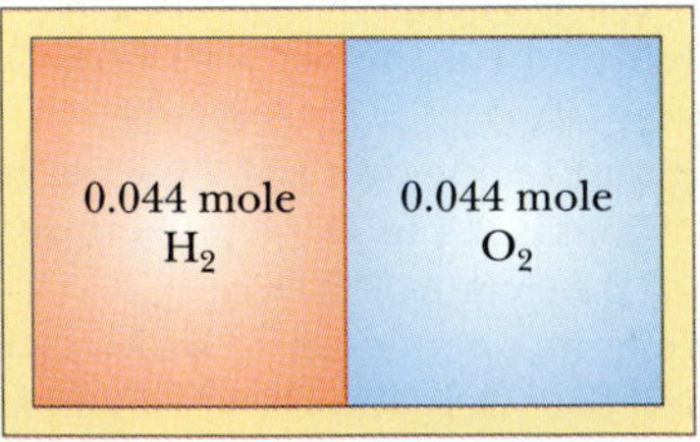

Figure 15.15 (Problem 32)

the numbers of heads and tails that result. For example, HHTH and HTHH are two possible ways in which three heads and one tail can be achieved. (a) On the basis of your table, what is the most probable result of a toss? In terms of entropy, (b) what is the most ordered state, and (c) what is the most disordered?

37. Repeat the procedure used to construct Table 13.1 (a) for the case in which you draw three marbles from your bag rather than four and (b) for the case in which you draw five rather than four.
38. If you toss two dice, what is the total number of ways in which you can obtain (a) a 12 and (b) a 7?
39. Consider the standard deck of 52 playing cards that have been thoroughly shuffled. (a) What is the probability of drawing the ace of spades in one draw? (b) What is the probability of drawing any ace? (c) What is the probability of drawing any spade?
40. What is the probability of drawing a heart *or* the queen of spades in one draw from a full deck? (These are penalty cards in the game of Hearts.)

Additional Problems

41. On a cold day, a heat pump is used to extract thermal energy from the outside air at −5°C and bring it into the +20°C house. What is the maximum possible coefficient of performance of the heat pump, and what minimum amount of work must be done to bring 1000 J of thermal energy into the house?
42. On a hot day, a Florida home is kept cool by an air conditioner. The outside temperature is 36°C, and the interior air is at 18°C. If 50 000 kJ/h of thermal energy is removed from the house, what is the minimum power that must be provided to the air conditioner?
43. Every second at Niagara Falls, some 5000 m^3 of water falls a distance of 50 m. What is the increase in entropy per second due to the falling water? (Assume a 20°C environment).

(Problem 43) A dramatic photograph of Niagara Falls. *(Jan Kopec, Tony Stone Worldwide)*

44. A heat engine operates between two reservoirs at $T_2 = 600$ K and $T_1 = 350$ K. It absorbs 1000 J of thermal energy from the higher-temperature reservoir and performs 250 J of work. Find (a) the entropy change of the Universe, ΔS_U, for this process and (b) the work, W, that could have been done by an ideal Carnot engine operating between these two reservoirs. (c) Show that the difference between the work done in parts (a) and (b) is $T_1 \Delta S_U$.
45. A house loses heat through the exterior walls and roof at a rate of 5000 J/s = 5 kW when the interior temperature is 22°C and the outside temperature is −5°C. Calculate the electric power required to maintain the interior temperature at 22°C for the following two cases. (a) The electric power is used in electric resistance heaters (which convert all of the electricity supplied into heat). (b) The electric power is used to drive an electric motor that operates the compressor of a heat pump (which has a coefficient of performance equal to 60% of the Carnot cycle value).
46. One mole of an ideal monatomic gas is taken through the cycle shown in Figure 15.16. The process AB is a reversible isothermal expansion. Calculate (a) the net work done by the gas, (b) the heat added to the gas, (c) the heat expelled by the gas, and (d) the efficiency of the cycle.

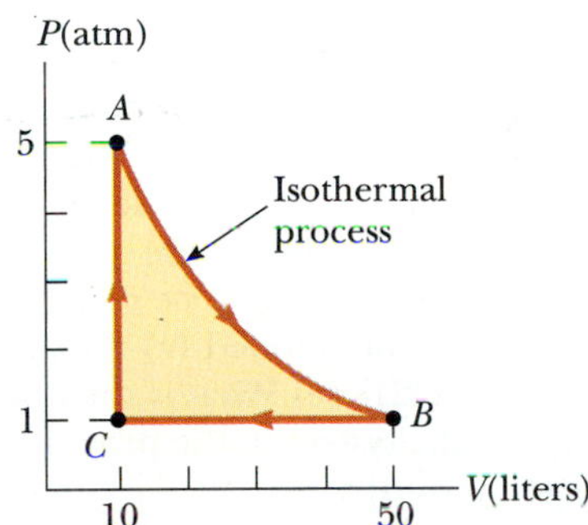

Figure 15.16 (Problem 46)

47. An athlete whose mass is 70 kg drinks 16 ounces (453.6 g) of refrigerated water. The water is at a temperature of 35°F. (a) Neglecting the temperature change of the body that results from the water intake (so that the body is regarded as a reservoir at 98.6°F), find the entropy increase of the entire system. (b) Assume that the entire body is cooled by the drink and that the average specific heat of a human is equal to the specific heat of liquid water. Neglect-

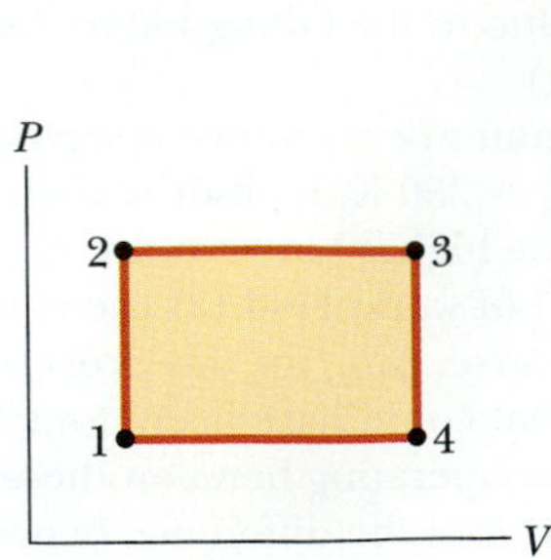

Figure 15.17 (Problem 50)

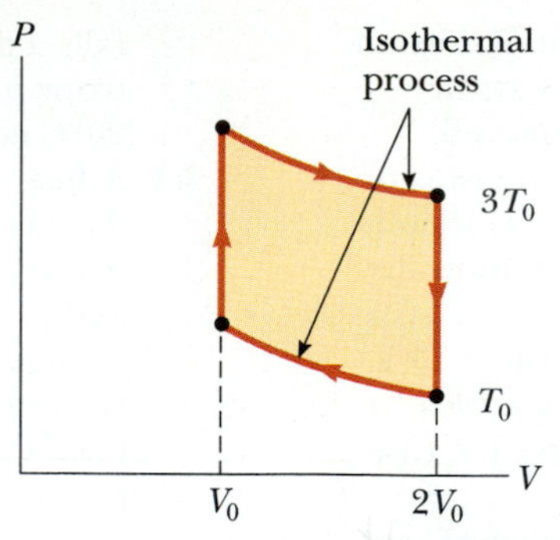

Figure 15.18 (Problem 51)

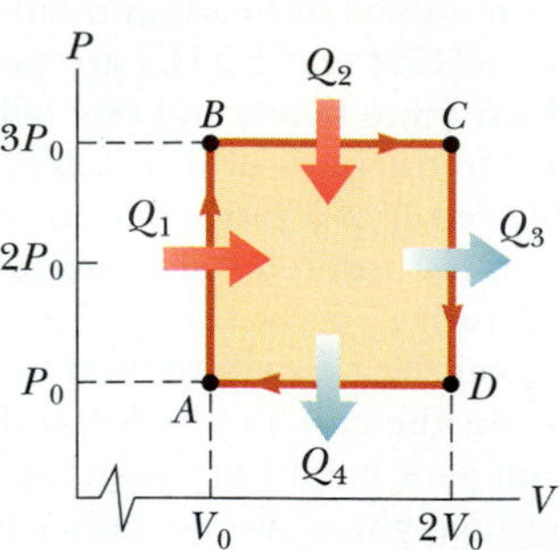

Figure 15.19 (Problem 53)

ing any other heat transfers and any metabolic heat release, find the athlete's temperature after she drinks the cold water, given an initial body temperature of 98.6°F. Under *these* assumptions, what is the entropy increase of the entire system? Compare this result with (a).

48. A refrigerator has a coefficient of performance of 3. The ice tray compartment is at −20°C, and the room temperature is 22°C. The refrigerator can convert 30 g of water at 22°C to 30 g of ice at −20°C each minute. What input power is required? Ignore any cooling by the refrigerator, and give your answer in watts.

49. What is the minimum amount of work that must be done to extract 400 J of heat from a massive object at 0°C while rejecting heat to a hot reservoir at a temperature of 20°C?

50. A gas follows the path 123 on the *PV* diagram in Figure 15.17 and 418 J of heat flows into the system. Also, 167 J of work is done. (a) What is the internal energy change of the system? (b) How much thermal energy flows into the system if the process follows the path 143? The work done by the gas along this path is 63 J. What net work would be done on or by the system if the system followed (c) the path 12341? (d) the path 14321? (e) What is the change in internal energy of the system in the processes described in parts (c) and (d)?

51. Figure 15.18 represents n mol of an ideal monatomic gas being taken through a reversible cycle that consists of two isothermal processes at temperatures $3T_0$ and T_0 and two constant-volume processes. For each cycle, determine, in terms of n, R, and T_0, (a) the net heat transferred to the gas and (b) the efficiency of an engine operating in this cycle.

52. An electric power plant has an overall efficiency of 20%. The plant is to deliver 150 MW of power to a city, and its turbines use coal as the fuel. The burning coal produces steam at 190°C, which drives the turbines. This steam is then condensed into water at 25°C by passing it through cooling coils in contact with river water. (a) How many metric tons of coal does the plant consume each day (1 metric ton = 10^3 kg)? (b) What is the total cost of the fuel per year if the delivered price is $8/metric ton?

53. One mole of a monatomic ideal gas is taken through the reversible cycle shown in Figure 15.19. At point A, the pressure, volume, and temperature are P_0, V_0, and T_0, respectively. In terms of R and T_0, find (a) the total heat entering the system per cycle, (b) the total thermal energy leaving the system per cycle, (c) the efficiency of an engine operating in this reversible cycle, and (d) the efficiency of an engine operating in a Carnot cycle between the same temperature extremes for this process.

54. The Stirling engine described in Figure 15.20a operates between the isotherms T_1 and T_2, where $T_2 > T_1$. Assuming that the operating gas is an ideal monatomic gas, calculate the efficiency of an engine whose constant-volume processes occur at the volumes V_1 and V_2.

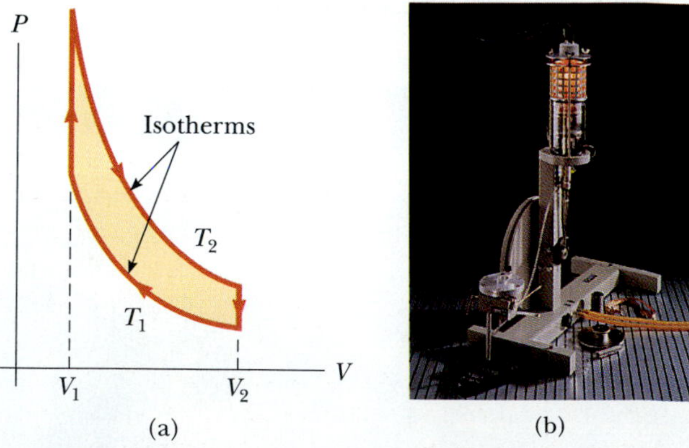

Figure 15.20 (Problem 54)
(a) An idealized cycle of a Stirling heat engine, patented in 1827. (b) Model of a Stirling engine. *(Courtesy of CENCO)*

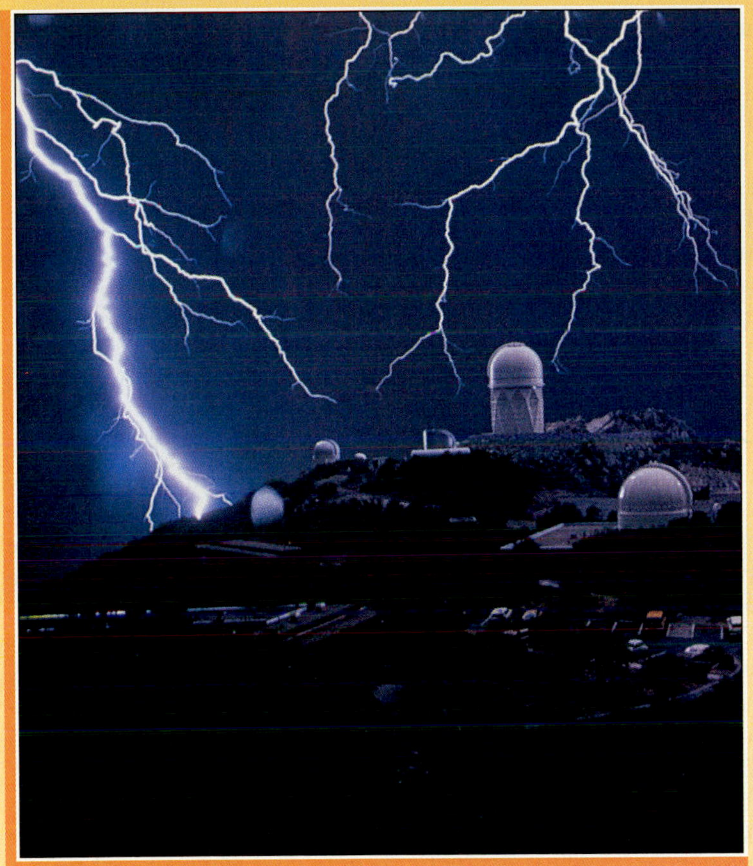

This dramatic one-minute exposure captures multiple lightning bolts illuminating Kitt Peak National Observatory in Arizona, illustrating electrical breakdown in the atmosphere. *(© Gary Ladd 1972)*

Electric Forces and Electric Fields

CHAPTER 16

In Chapter 6 we investigated the nature of the electromagnetic force between charged particles. This chapter begins with a review of some of the basic properties of the electrostatic force, as well as some properties of the electric field associated with stationary charged particles. Our study of electrostatics then continues with the concept of an electric field that is associated with a continuous charge distribution and the effect of this field on other charged particles.

Two methods for calculating the electric fields associated with continuous charge distributions are discussed. The first method uses Coulomb's law and integral calculus, and the second uses *Gauss's law.* An outgrowth of Coulomb's law, Gauss's law is much more convenient for calculating electric fields of highly symmetric charge distributions. Furthermore, Gauss's law serves as a guide for understanding more complicated problems.

16.1 Historical Overview

In this chapter we continue our study of electric and magnetic phenomena. We *strongly suggest* that you review Chapter 6, Sections 6.1 and 6.3, before continuing. The laws of electricity and magnetism play a central role in the operation of devices such as radios, televisions, electric motors, computers, high-energy accelerators, and a host of electronic devices used in medicine. More fundamental, however, is the fact that the interatomic and intermolecular forces that are responsible for the formation of solids and liquids are electric in origin. Furthermore, as we learned in Chapter 6, such forces as the pushes and pulls between objects and the elastic force in a spring arise from electric forces at the atomic level.

Chinese documents suggest that magnetism was recognized as early as about 2000 B.C. The ancient Greeks observed electric and magnetic phenomena possibly as early as 700 B.C. They found that a piece of amber, when rubbed, became electrified and attracted pieces of straw or feathers. The existence of magnetic forces was known from observations that pieces of a naturally occurring stone called *magnetite* (Fe_3O_4) were attracted to iron. (The word *electric* comes from the Greek word for amber, *elecktron.* The word *magnetic* comes from *Magnesia,* on the coast of Turkey where magnetite was found.)

In 1600, the Englishman William Gilbert discovered that electrification was not limited to amber but was a general phenomenon. Scientists went on to electrify a variety of objects, including chickens and people! Experiments by Charles Coulomb in 1785 confirmed the inverse-square force law for electricity (Eq. 6.8 in Chapter 6).

It was not until the early part of the 19th century that scientists established that electricity and magnetism are related phenomena. In 1820, Hans Oersted discovered that a compass needle, which is magnetic, is deflected when placed near a circuit carrying an electric current. In 1831, Michael Faraday in England and, almost simultaneously, Joseph Henry in the United States showed that, when a wire is moved near a magnet (or, equivalently, when a magnet is moved near a wire), an electric current is observed in the wire. In 1873, James Clerk Maxwell used these observations and other experimental facts as a basis for formulating the laws of electromagnetism as we know them today. Shortly thereafter (around 1888), Heinrich Hertz verified Maxwell's predictions by producing electromagnetic waves in the laboratory. This achievement was followed by such practical developments as radio and television.

Maxwell's contributions to the science of electromagnetism were especially significant because the laws he formulated are basic to *all* forms of electromagnetic phenomena. His work is comparable in importance to Newton's discovery of the laws of motion and the theory of gravitation.

16.2 Properties of Electric Charges

A number of simple experiments demonstrate the existence of electrostatic forces. For example, after running a comb through your hair, you will find that the comb attracts bits of paper. The attractive electrostatic force is often strong enough to suspend the bits. The same effect occurs with other rubbed materials, such as glass or rubber.

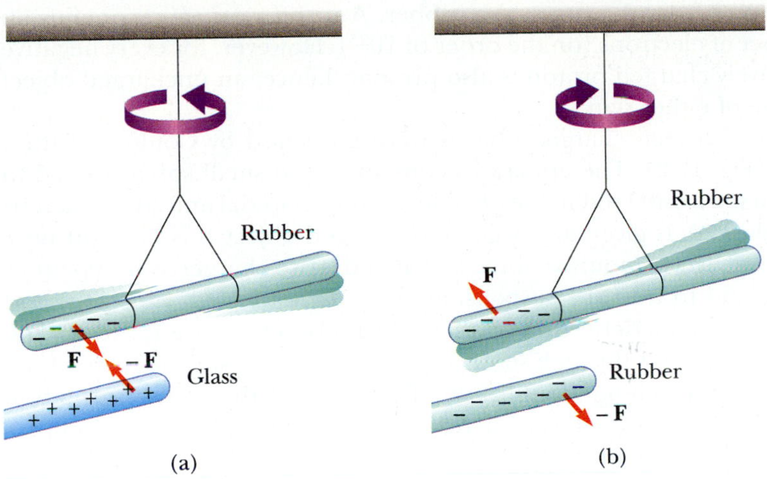

Figure 16.1
(a) A negatively charged rubber rod, suspended by a thread, is attracted to a positively charged glass rod. (b) A negatively charged rubber rod is repelled by another negatively charged rubber rod.

Another simple experiment is to rub an inflated balloon with wool (or across your hair). On a dry day, the rubbed balloon will stick to the wall of a room, often for hours. When materials behave this way, they are said to have become **electrically charged.** You can give your body an electric charge by walking across a wool rug or by sliding across a car seat. You can then feel, and remove, the charge on your body by lightly touching another person. Under the right conditions, a visible spark is seen when you touch, and a slight tingle is felt by both parties. (Such an experiment works best on a dry day because excessive moisture can provide a pathway for charge to leak off a charged object.)

Experiments also demonstrate that there are two kinds of electric charge, which were given the names **positive** and **negative** by Benjamin Franklin (1706–1790). Figure 16.1 illustrates the interactions of the two charges. A hard rubber (or plastic) rod that has been rubbed with fur (or an acrylic material) is suspended by a piece of string. When a glass rod that has been rubbed with silk is brought near the rubber rod, the rubber rod is attracted toward the glass rod (Fig. 16.1a). If two charged rubber rods (or two charged glass rods) are brought near each other, as in Figure 16.1b, the force between them is repulsive. This observation demonstrates that the rubber and glass have different kinds of charge. We use the convention suggested by Franklin, wherein the electric charge on the glass rod is called positive and that on the rubber rod is called negative. On the basis of observations such as these, we conclude that *like charges repel one another, and unlike charges attract one another.*

A natural tendency exists for charge to be transferred between unlike materials. Rubbing the two materials together only increases the area of contact and thus enhances the charge transfer process.

Another important characteristic of electric charge is that *it is always conserved.* That is, when two initially neutral objects are charged by being rubbed together, charge is not created in the process. The objects become charged because *negative charge is transferred* from one object to the other. One object gains some amount of negative charge while the other loses an equal amount of negative charge and hence is left with a positive charge. For example, when a glass rod is rubbed with silk, as in Figure 16.2, the silk obtains a negative charge that is equal in magnitude to the positive charge on the glass rod as negatively charged electrons are transferred from the glass to the silk. Likewise, when rubber is rubbed with fur, elec-

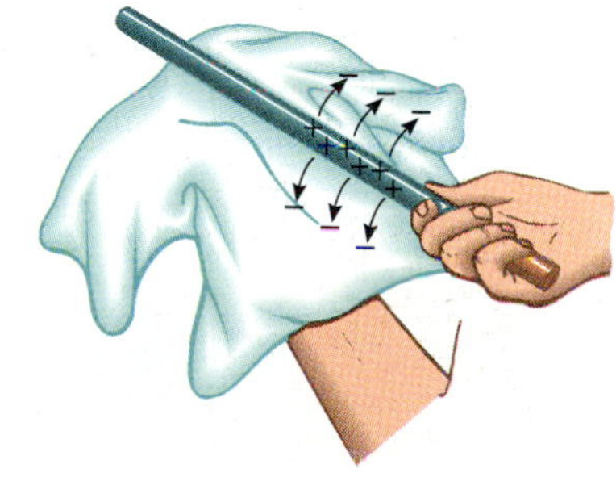

Figure 16.2
When a glass rod is rubbed with silk, electrons are transferred from the glass to the silk. Because of conservation of charge, each electron adds negative charge to the silk, and an equal positive charge is left behind on the rod. Also, because the charges are transferred in discrete bundles, the charges on the two objects are $\pm e$ or $\pm 2e$ or $\pm 3e$, and so on.

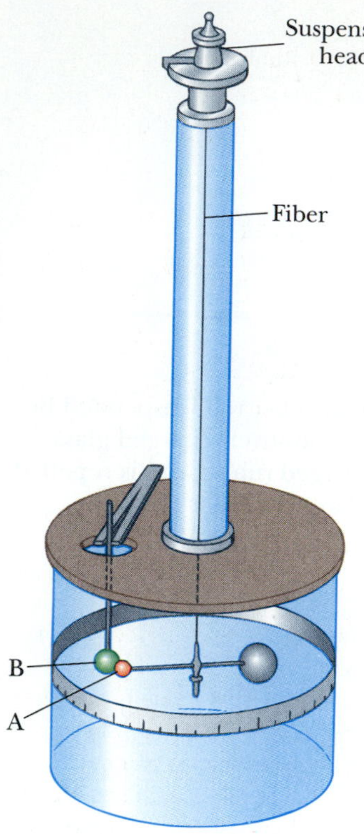

Figure 16.3
Coulomb's torsion balance was used to establish the inverse-square law for the electrostatic force between two charges.

trons are transferred from the fur to the rubber. An *uncharged object* contains an enormous number of electrons (on the order of 10^{23}). However, for every negative electron, a positively charged proton is also present; hence, an uncharged object has no net charge of either sign.

Electric forces between charged objects were measured by Coulomb with a torsion balance (Fig. 16.3). The apparatus consists of two small spheres fixed to the ends of a light horizontal rod made of an insulating material and suspended by a silk thread. Sphere A is given a charge, and charged object B is brought near sphere A. The attractive or repulsive force between the two charged objects causes the rod to rotate and to twist the suspension. The angle through which the rod rotates is measured via the deflection of a light beam by a mirror attached to the suspension. The rod rotates through some angle against the restoring force of the twisted thread before reaching equilibrium. The value of the angle of rotation increases as the charge on the objects increases. Thus, the angle of rotation provides a quantitative measure of the electric force of attraction or repulsion.

From our discussion thus far, and from what we learned in Chapter 6, we conclude that electric charge has the following important properties:

1. **Two kinds of charges exist in nature, with the property that unlike charges attract one another and like charges repel one another.**
2. **The force between charges varies as the inverse square of their separation.**
3. **Charge is conserved.**
4. **Charge is quantized.**

16.3 Insulators and Conductors

It is convenient to classify substances in terms of their ability to conduct electrical charge.

Conductors are materials in which electric charges move freely, and insulators are materials in which electric charges do not move freely.

Materials such as glass, rubber, and lucite are insulators. When such materials are charged by rubbing, only the rubbed area becomes charged, and there is no tendency for charge to move to other regions of the material. In contrast, materials such as copper, aluminum, and silver are good conductors. When such materials are charged in some small region, the charge readily distributes itself over the entire surface of the material. If you hold a copper rod in your hand and rub it with wool or fur, it will not attract a small piece of paper. This might suggest that a metal cannot be charged. However, if you hold the copper rod by an insulating handle and then rub, the rod will remain charged and attract the piece of paper. In the first case, the electric charges produced by rubbing readily move from the copper through your body and finally to Earth. In the second case, the insulating handle prevents the flow of charge to Earth.

Metals are good conductors

Semiconductors are a third class of materials, and their electrical properties are somewhere between those of insulators and those of conductors. Silicon and germanium are well-known examples of semiconductors that are widely used in the fabrication of a variety of electronic devices. The electrical properties of semiconductors can be changed over many orders of magnitude by adding controlled amounts of certain foreign atoms to the materials.

Charging by Induction

When a conductor is connected to Earth by means of a conducting wire or pipe, it is said to be **grounded.** For present purposes, the Earth can be considered an infinite reservoir for electrons; this means that it can accept or supply an unlimited number of electrons. With this in mind, we can understand how to charge a conductor by a process known as **induction.**

Consider a negatively charged rubber rod brought near a neutral (uncharged) conducting sphere that is insulated so that there is no conducting path to ground (Fig. 16.4). The repulsive force between the electrons in the rod and those in the sphere causes a redistribution of charge on the sphere so that some electrons move to the side of the sphere farthest away from the rod. The region of the sphere nearest the rod has an excess of positive charge because of the migration of electrons away from this location. If a grounded conducting wire is then connected to the sphere, as in Figure 16.4b, some of the electrons leave the sphere and travel to the Earth. If the wire to ground is then removed (Fig. 16.4c), the conducting sphere is left with an excess of induced positive charge. Finally, when the rubber rod is removed from the vicinity of the sphere (Fig. 16.4d), the induced positive charge remains on the ungrounded sphere. This excess positive charge becomes uniformly distributed over the surface of the ungrounded sphere because of the repulsive forces among the like charges and the high mobility of charge carriers in a metal.

In the process of inducing a charge on the sphere, the charged rubber rod loses none of its negative charge, since it never came in contact with the sphere. *Charging an object by induction requires no contact with the object inducing the charge.* This is in contrast to charging an object by rubbing, which does require contact between the two objects.

A process very similar to charging by induction in conductors also takes place in insulators. In most neutral atoms and molecules, the center of positive charge coincides with the center of negative charge. However, in the presence of a charged object, these centers may shift slightly, resulting in more positive charge on one side of the molecule than on the other. This effect is known as **polarization.** The realignment of charge within individual molecules produces an induced charge on the surface of the insulator, as shown in Figure 16.5a. With these concepts in mind, you should be able to explain why a comb that has been rubbed through hair attracts bits of neutral paper, or why a balloon that has been rubbed against your clothing can stick to a neutral wall.

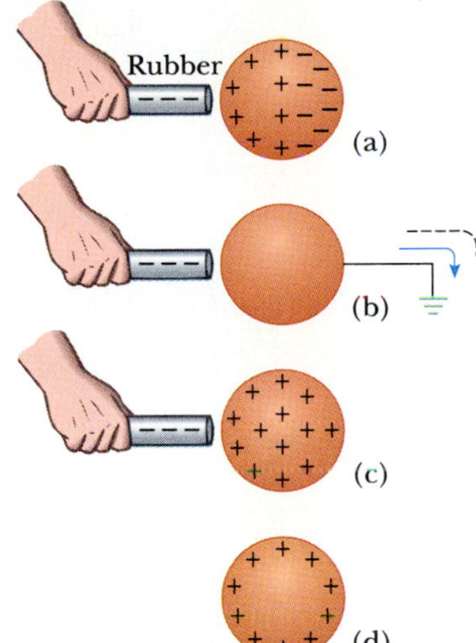

Figure 16.4
Charging a metal object by induction. (a) The charge on a neutral metal sphere is redistributed when a charged rubber rod is placed near the sphere. (b) The sphere is grounded, and some of the electrons leave the sphere through the ground wire. (c) The ground connection is removed, and the sphere is left with excess positive charge. (d) When the rubber rod is moved away, the sphere becomes uniformly charged.

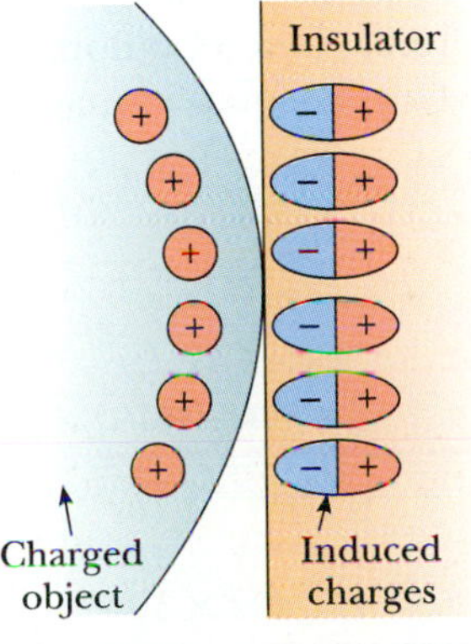

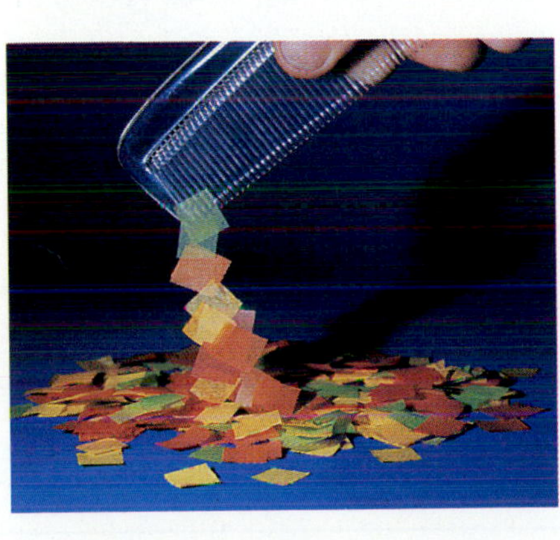

Figure 16.5
(a) A charged object induces charges on the surface of an insulator. (b) A charged comb attracts bits of paper because charges are displaced in the paper. Note that the paper is neutral but polarized.

16.4 Coulomb's Law Revisited

In Chapter 6 we introduced Coulomb's law, which describes the electrostatic force between two stationary charged particles. When the charges are of the same sign, the force on either charge is repulsive, and when the charges are of opposite sign, the force is attractive (Fig. 16.6). The magnitude of the electrostatic force between two charges separated by a distance of r is

Coulomb's law

$$F = k_e \frac{|q_1||q_2|}{r^2} \quad [16.1]$$

where k_e (= 8.99×10^9 N·m²C²) is the Coulomb constant. The constant k_e is also written

$$k_e = \frac{1}{4\pi\epsilon_0}$$

where the constant ϵ_0 is known as the *permittivity of free space* and has the value

Permittivity of free space

$$\epsilon_0 = 8.8542 \times 10^{-12}\ \mathrm{C^2/N \cdot m^2} \quad [16.2]$$

The charge of an electron or proton has the magnitude $|e| = 1.60 \times 10^{-19}$ C. Therefore, 1 C of charge is equal to the charge of 6.25×10^{18} electrons (that is, $1/e$). This can be compared with the number of free electrons in 1 cm³ of copper, which is on the order of 10^{23}. Note that 1 C is a substantial amount of charge. In typical electrostatic experiments, where a rubber or glass rod is charged by friction, a net charge on the order of 10^{-6} C (= 1 μC) is obtained. In other words, only a very small fraction of the total available charge is transferred between the rod and the rubbing material.

The charges and masses of the electron, proton, and neutron are given in Table 16.1.

When dealing with Coulomb's force law, you must remember that force is a *vector* quantity and must be treated accordingly. Furthermore, note that *Coulomb's law applies exactly only to point charges or particles.* The electrostatic force on q_2 due to q_1, written $\mathbf{F}_{21}$, can be expressed in vector form as

$$\mathbf{F}_{21} = k_e \frac{q_1 q_2}{r^2}\hat{\mathbf{r}} \quad [16.3]$$

where $\hat{\mathbf{r}}$ is a unit vector directed from q_1 to q_2 as in Figure 16.6a. Since Coulomb's law obeys Newton's third law, the electric force on q_2 due to q_1 is equal in magni-

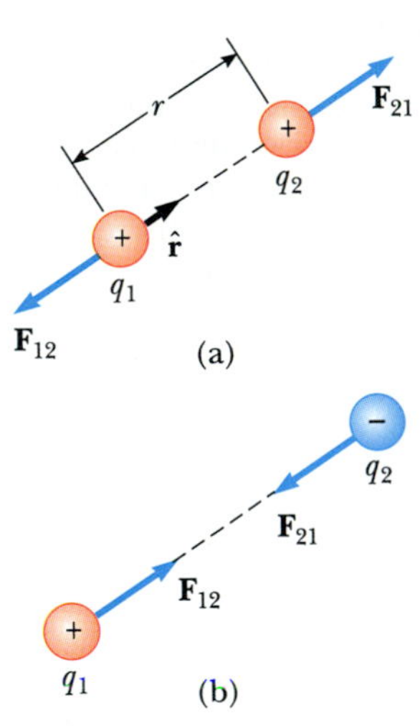

Figure 16.6
Two point charges separated by a distance of r exert an electrostatic force on each other, given by Coulomb's law. Note that the force on q_1 is equal to and opposite the force on q_2. (a) When the charges are of the same sign, the force is repulsive. (b) When the charges are of opposite signs, the force is attractive.

Table 16.1
Charge and Mass of the Electron, Proton, and Neutron

Particle	Charge (C)	Mass (kg)
Electron (e)	$-1.6021917 \times 10^{-19}$	9.1095×10^{-31}
Proton (p)	$+1.6021917 \times 10^{-19}$	1.67261×10^{-27}
Neutron (n)	0	1.67492×10^{-27}

tude to the force on q_2 due to q_1 and in the opposite direction; that is, $\mathbf{F}_{12} = -\mathbf{F}_{21}$. Finally, from Equation 16.5 we see that if q_1 and q_2 have the same sign, the product q_1q_2 is positive and the force is repulsive, as in Figure 16.6a. If q_1 and q_2 are of opposite sign, as in Figure 16.6b, the product q_1q_2 is negative and the force is attractive.

When more than two charges are present, the force between any pair is given by Equation 16.5. Therefore, the resultant force on any one charge equals the *vector* sum of the forces due to the various individual charges. This principle of *superposition* as applied to electrostatic forces is an experimentally observed fact. For example, if there are four charges, then the resultant force on particle 1 due to particles 2, 3, and 4 is given by

$$\mathbf{F}_1 = \mathbf{F}_{12} + \mathbf{F}_{13} + \mathbf{F}_{14}$$

Charles Coulomb (1736–1806), the great French physicist for whom the unit of electric charge called the *coulomb* was named.

Example 16.1 Where Is the Resultant Force Zero?

Three charges lie along the x axis as in Figure 16.7. The positive charge $q_1 = 15\ \mu\text{C}$ is at $x = 2$ m, and the positive charge $q_2 = 6\ \mu\text{C}$ is at the origin. Where must a *negative* charge, q_3, be placed on the x axis so that the resultant force on it is zero?

Solution Since q_3 is negative and both q_1 and q_2 are positive, the forces $\mathbf{F}_{31}$ and $\mathbf{F}_{32}$ are both attractive, as indicated in Figure 16.7. If we let x be the coordinate of q_3, then the forces $\mathbf{F}_{31}$ and $\mathbf{F}_{32}$ have magnitudes

$$F_{31} = k_e \frac{|q_3||q_1|}{(2-x)^2} \quad \text{and} \quad F_{32} = k_e \frac{|q_3||q_2|}{x^2}$$

If the resultant force on q_3 is zero, then $\mathbf{F}_{32}$ must be equal to and opposite $\mathbf{F}_{31}$, or

$$k_e \frac{|q_3||q_2|}{x^2} = k_e \frac{|q_3||q_1|}{(2-x)^2}$$

Since k_e and q_3 are common to both sides, they cancel, and we find that

$$(2-x)^2|q_2| = x^2|q_1|$$

$$(4 - 4x + x^2)(6 \times 10^{-6}\ \text{C}) = x^2(15 \times 10^{-6}\ \text{C})$$

Solving this quadratic equation for x, we find that $x = 0.775$ m. Why is the negative root not acceptable?

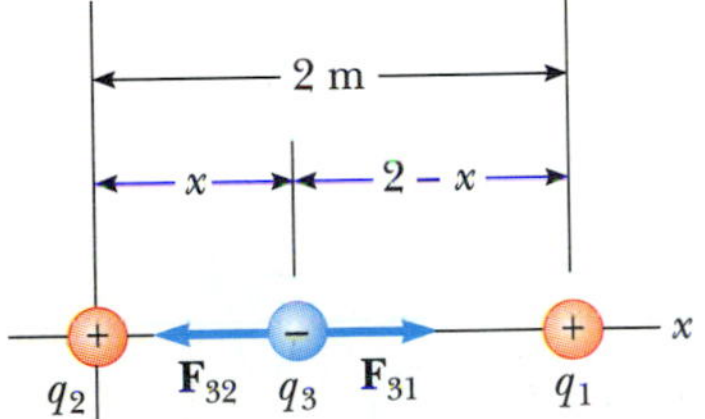

Figure 16.7 (Example 16.1) Three point charges are placed along the x axis. The charge q_3 is negative, whereas q_1 and q_2 are positive. If the net force on q_3 is zero, then the force on q_3 due to q_1 must be equal to and opposite the force on q_3 due to q_2.

Example 16.2 The Hydrogen Atom

The electron and proton of a hydrogen atom are separated by a distance of (on the average) approximately 5.3×10^{-11} m. Find the magnitudes of the electrostatic force and the gravitational force between them.

Solution From Coulomb's law, we find that the attractive electrical force has the magnitude

$$F_e = k_e \frac{|e|^2}{r^2} = 8.99 \times 10^9 \frac{\text{N}\cdot\text{m}^2}{\text{C}^2} \frac{(1.6 \times 10^{-19}\ \text{C})^2}{(5.3 \times 10^{-11}\ \text{m})^2}$$

$$= 8.2 \times 10^{-8}\ \text{N}$$

Using Newton's universal law of gravity (Eq. 11.1) and Table 16.1, we find that the gravitational force has the magnitude

$$F_g = G\frac{m_e m_p}{r^2}$$

$$= \left(6.67 \times 10^{-11} \frac{\text{N}\cdot\text{m}^2}{\text{kg}^2}\right) \times \frac{(9.11 \times 10^{-31}\text{ kg})(1.67 \times 10^{-27}\text{ kg})}{(5.3 \times 10^{-11}\text{ m})^2}$$

$$= 3.6 \times 10^{-47}\text{ N}$$

The ratio $F_e/F_g \approx 3 \times 10^{39}$. Thus, the gravitational force between charged atomic particles is negligible compared with the electrostatic force.

16.5 Electric Fields

The concept of an electric field associated with one or more point charges was introduced in Chapter 6. In this section we shall briefly review this concept as it applies to point charges, and then show that it naturally leads to a procedure for calculating the electric field associated with a continuous charge distribution.

Electric Field Due to Point Charges—Revisited

As you will recall, the electric field vector **E** at any location is defined as the electrostatic force **F** acting on a positive test charge placed at that location, divided by the magnitude of the test charge, q_0. That is, $\mathbf{E} \equiv \mathbf{F}/q_0$ (Eq. 6.7 in Chapter 6). If the test charge is located a distance of r from the point charge q, according to Coulomb's law the electrostatic force on the test charge has a magnitude of $k_e q q_0/r^2$. Hence, the electric field at the position of q_0 due to the charge q is

$$\mathbf{E} = k_e \frac{q}{r^2}\hat{\mathbf{r}} \qquad [16.4]$$

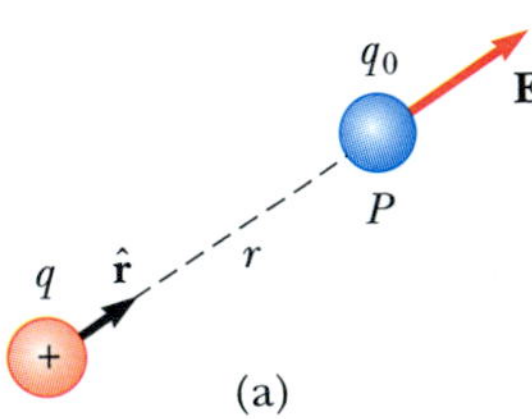

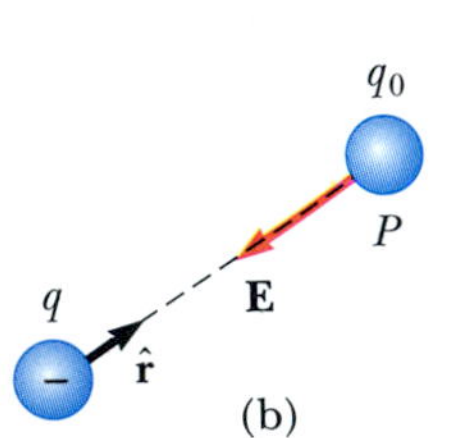

Figure 16.8
A test charge of q_0 at the point P is at a distance of r from the point charge q. (a) If q is positive, the electric field at P points radially *outward* from q. (b) If q is negative, the electric field at P points radially *inward* toward q.

where $\hat{\mathbf{r}}$ is a unit vector directed away from q toward q_0, as in Figure 16.8. Note that if q is positive, as in Figure 16.8a, the electric field is directed away from it. If q is negative, as in Figure 16.8b, the electric field is directed toward it. In order to calculate the electric field due to a group of point charges, one can apply the superposition principle, which says that the total electric field due to a group of charges equals the *vector sum* of the electric fields of all charges. That is,

$$\mathbf{E} = k_e \sum_i \frac{q_i}{r_i^2}\hat{\mathbf{r}}_i \qquad [16.5]$$

where r_i is the distance from the ith charge, q_i, to the point P (the location of the test charge) and $\hat{\mathbf{r}}_i$ is a unit vector directed from q_i toward P.

Example 16.3 The Electric Field of a Dipole

An **electric dipole** consists of a positive charge, q, and a negative charge, $-q$, separated by a distance of $2a$, as in Figure 16.9. Find the electric field, **E**, due to these charges along the y axis at the point P, which is a distance of y from the origin. Assume that $y \gg a$.

Solution At P, the fields $\mathbf{E}_1$ and $\mathbf{E}_2$ due to the two charges are equal in magnitude, since P is equidistant from the two equal and opposite charges. The total field $\mathbf{E} = \mathbf{E}_1 + \mathbf{E}_2$, where the magnitudes of $\mathbf{E}_1$ and $\mathbf{E}_2$ are given by

$$E_1 = E_2 = k_e \frac{q}{r^2} = k_e \frac{q}{y^2 + a^2}$$

The y components of $\mathbf{E}_1$ and $\mathbf{E}_2$ cancel each other. The x components are equal, since they are both along the x axis. Therefore, $\mathbf{E}$ lies along the x axis and has a magnitude equal to $2E_1 \cos\theta$. From Figure 16.9 we see that $\cos\theta = a/r = a/(y^2 + a^2)^{1/2}$. Therefore,

$$E = 2E_1 \cos\theta = 2k_e \frac{q}{(y^2 + a^2)} \frac{a}{(y^2 + a^2)^{1/2}}$$

$$= k_e \frac{2qa}{(y^2 + a^2)^{3/2}}$$

Using the approximation $y \gg a$, we can neglect a^2 in the denominator and write

$$E \approx k_e \frac{2qa}{y^3}$$

Thus, we see that along the y axis the field of a *dipole* at a distant point varies as $1/r^3$, whereas the more slowly varying field of a *point charge* goes as $1/r^2$. This is because at distant points, the fields of the two equal and opposite charges almost cancel each other. The $1/r^3$ variation in E for the dipole is also obtained for a distant point along the x axis (Problem 16) and for a general distant point. The dipole is a good model of many molecules, such as HCl.

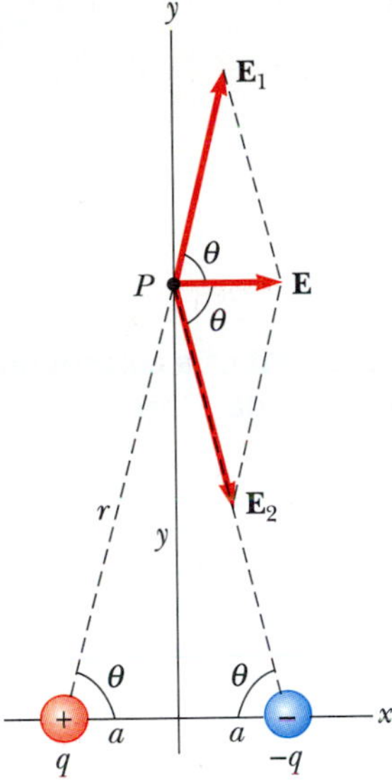

Figure 16.9
(Example 16.3) The total electric field $\mathbf{E}$ at P due to two equal and opposite charges (an electric dipole) equals the vector sum $\mathbf{E}_1 + \mathbf{E}_2$. The field $\mathbf{E}_1$ is due to the positive charge q, and $\mathbf{E}_2$ is the field due to the negative charge $-q$.

Electric Field Due to Continuous Charge Distributions

In most practical situations (such as an object charged by rubbing), the average separation between charges is small compared with their distances from the field point. In such cases, the system of charges can be considered *continuous.* That is, we imagine that the system of closely spaced charges is equivalent to a total charge that is continuously distributed through some volume or over some surface.

To evaluate the electric field of a continuous charge distribution, the following procedure is used. First, we divide the charge distribution into small elements, each of which contains a small charge, Δq, as in Figure 16.10. Next, we use Coulomb's law to calculate the electric field due to one of these elements at a point, P. Finally, we evaluate the total field at P due to the charge distribution by summing the contributions of all the charge elements (that is, by applying the superposition principle).

The electric field at P due to one element of charge Δq is given by

$$\Delta \mathbf{E} = k_e \frac{\Delta q}{r^2} \hat{\mathbf{r}}$$

where r is the distance from the element to point P and $\hat{\mathbf{r}}$ is a unit vector directed from the element toward P. The total electric field at P due to all elements in the charge distribution is approximately

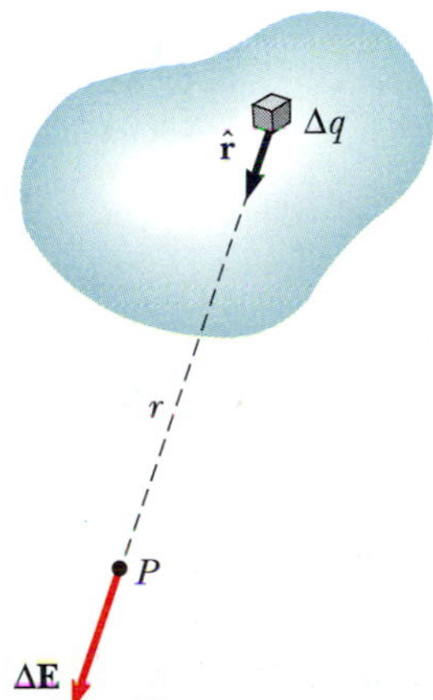

Figure 16.10
The electric field at P due to a continuous charge distribution is the vector sum of the fields due to all the elements, Δq, of the charge distribution.

$$\mathbf{E} \approx k_e \sum_i \frac{\Delta q_i}{r_i^2} \hat{\mathbf{r}}_i$$

where the index i refers to the ith element in the distribution. If the separation between elements in the charge distribution is small compared with the distance to P, the charge distribution can be approximated as continuous. Therefore, the total field at P in the limit $\Delta q_i \to 0$ becomes

Electric field of a continuous charge distribution

$$\mathbf{E} = k_e \lim_{\Delta q_i \to 0} \sum_i \frac{\Delta q_i}{r_i^2} \hat{\mathbf{r}}_i = k_e \int \frac{dq}{r^2} \hat{\mathbf{r}} \qquad [16.6]$$

where the integration is a *vector* operation and must be treated with caution. We shall illustrate this type of calculation with several examples in which we shall assume that the charge is *uniformly* distributed on a line or a surface or throughout some volume. When performing such calculations, it is convenient to use the concept of a charge density along with the following notations:

If a charge, Q, is uniformly distributed throughout a volume, V, the *charge per unit volume*, ρ, is defined by

Volume charge density

$$\rho \equiv \frac{Q}{V} \qquad [16.7]$$

where ρ has units of coulombs per cubic meter.

If a charge, Q, is uniformly distributed on a surface of area A, the *surface charge density*, σ, is defined by

Surface charge density

$$\sigma \equiv \frac{Q}{A} \qquad [16.8]$$

where σ has units of coulombs per square meter.

Finally, if a charge Q is uniformly distributed along a line of length ℓ, the *linear charge density*, λ, is defined by

Linear charge density

$$\lambda \equiv \frac{Q}{\ell} \qquad [16.9]$$

where λ has units of coulombs per meter.

If the charge is *nonuniformly* distributed over a volume, surface, or line, we must express the charge densities as

$$\rho = \frac{dQ}{dV} \qquad \sigma = \frac{dQ}{dA} \qquad \lambda = \frac{dQ}{d\ell}$$

where dQ is the amount of charge in a small volume, surface, or length element.

Example 16.4 The Electric Field Due to a Charged Rod

A rod of length ℓ has a uniform positive charge per unit length of λ and a total charge of Q. Calculate the electric field at a point, P, along the axis of the rod, a distance of d from one end (Fig. 16.11).

Solution For this calculation, the rod is taken to be along the x axis. Divide the rod into segments, each of length Δx and each carrying a charge of Δq. Because the charge distribution is uniform, the ratio $\Delta q/\Delta x$ is equal to the ratio Q/ℓ: $\Delta q/\Delta x = Q/\ell = \lambda$. Therefore, the charge Δq is $\Delta q = \lambda\, \Delta x$.

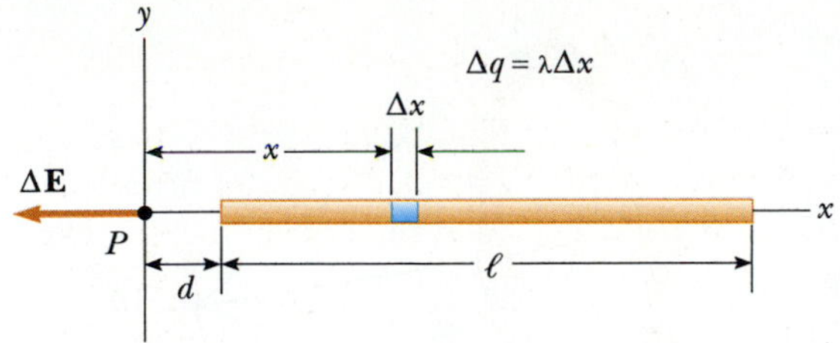

Figure 16.11 (Example 16.4) The electric field at P due to a uniformly charged rod lying along the x axis. The field at P due to the segment of charge Δq is $k_e\,\Delta q/x^2$. The total field at P is the vector sum over all segments of the rod.

The field $\Delta \mathbf{E}$ due to this segment at the point P is in the negative x direction, and its magnitude is given by

$$\Delta E = k_e \frac{\Delta q}{x^2} = k_e \frac{\lambda\,\Delta x}{x^2}$$

Note that each charge element produces a field in the negative x direction, and so the problem of summing the contributions of all the elements is particularly simple in this case. The total field at P due to all segments of the rod, which are at different distances from P, is given by Equation 16.6, which in this case becomes

$$E = \int_d^{\ell+d} k_e\lambda \frac{dx}{x^2}$$

where the limits on the integral extend from one end of the rod ($x = d$) to the other ($x = \ell + d$). Since k and λ are constants, they can be removed from the integral. Thus, we find that[1]

$$\begin{aligned} E &= k_e\lambda \int_d^{\ell+d} \frac{dx}{x^2} = k_e\lambda \left[-\frac{1}{x}\right]_d^{\ell+d} \\ &= k_e\lambda\left(\frac{1}{d} - \frac{1}{\ell + d}\right) \\ &= \frac{k_e Q}{d(\ell + d)} \end{aligned}$$

where we have used the fact that the total charge $Q = \lambda\ell$. From this result we see that, if P is *far* from the rod ($d \gg \ell$), then ℓ in the denominator can be neglected, and $E \approx k_e Q/d^2$. This is just the form you would expect for a point charge. Therefore, at large distances from the rod, the charge distribution appears to be a point charge of magnitude Q. The use of the limiting technique ($d \rightarrow \infty$) is often a good method for checking a theoretical formula.

Example 16.5 The Electric Field of a Uniform Ring of Charge

A thin ring of radius a has a uniform positive charge per unit length, with a total charge of Q. Calculate the electric field along the axis of the ring at a point, P, lying a distance x from the center of the ring (Fig. 16.12a).

[1] It is important that you understand the procedure used to carry out integrations such as this. First, choose a charge element whose parts are all equidistant from the point at which the field strength is being calculated. Next, express the charge element Δq in terms of the other variables within the integral (in this example there is one variable, x). In examples that have spherical or cylindrical symmetry, the variable will be a radial coordinate.

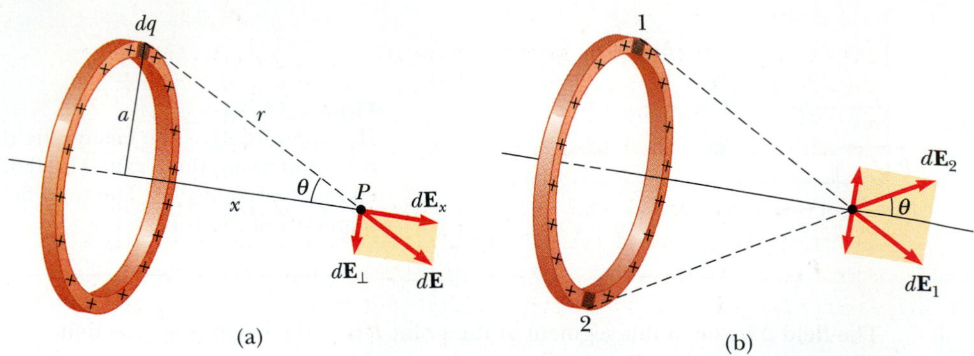

Figure 16.12
(Example 16.5) A uniformly charged ring of radius a. (a) The field at P on the x axis due to an element of charge dq. (b) The total electric field at P is along the x axis. Note that the perpendicular component of the electric field at P due to segment 1 is canceled by the perpendicular component due to segment 2.

Solution The magnitude of the electric field at P due to the element of charge dq is

$$dE = k_e \frac{dq}{r^2}$$

This field has an x component, $dE_x = dE \cos \theta$, along the axis of the ring and a component, $dE_\perp$, perpendicular to the axis. But as we see in Figure 16.12b, the resultant field at P must lie along the x axis, since the perpendicular components sum to zero. That is, the perpendicular component of any element is canceled by the perpendicular component of an element on the opposite side of the ring. Since $r = (x^2 + a^2)^{1/2}$ and $\cos \theta = x/r$, we find that

$$dE_x = dE \cos \theta = \left(k_e \frac{dq}{r^2} \right) \frac{x}{r} = \frac{k_e x}{(x^2 + a^2)^{3/2}}\, dq$$

In this case, all segments of the ring give the *same* contribution to the field at P because they are all equidistant from this point. We now use this result and Equation 16.6 to find the total field at P:

$$E_x = \int \frac{k_e x}{(x^2 + a^2)^{3/2}}\, dq = \frac{k_e x}{(x^2 + a^2)^{3/2}}\, Q$$

This result shows that the field is zero at $x = 0$. Does this surprise you?

Exercise Show that, at large distances from the ring ($x \gg a$), the electric field along the axis approaches that of a point charge of magnitude Q.

Problem-Solving Strategy and Hints

1. **Units: When performing calculations that involve the Coulomb constant, k_e ($= 1/4\pi\epsilon_0$), charges must be in coulombs and distances in meters. If they appear in other units, you must convert them.**
2. **Applying Coulomb's law to point charges: It is important to use the superposition principle properly when dealing with a collection of**

interacting charges. When several charges are present, the resultant force on any one is the *vector sum* of the forces due to the individual charges. Be very careful in the algebraic manipulation of vector quantities. It may be useful to review the material on vector addition in Chapter 1.

3. **Calculating the electric field of point charges: Remember that the superposition principle can be applied to electric fields, which are also vector quantities. To find the total electric field at a given point, first calculate the electric field at that point due to each individual charge. The resultant field at the point is the vector sum of the fields due to the individual charges.**
4. **Continuous charge distributions: When you are confronted with problems that involve a continuous distribution of charge, the vector sums for evaluating the total electric field at some point must be replaced by vector integrals. The charge distribution is divided into infinitesimal pieces, and the vector sum is carried out by integrating over the entire charge distribution. Examples 16.4 and 16.5 demonstrate such procedures.**
5. **Symmetry: Whenever dealing with either a distribution of point charges or a continuous charge distribution, take advantage of any symmetry in the system to simplify your calculations.**

16.6 Electric Field Lines

A convenient aid for visualizing electric field patterns is to draw lines pointing in the same direction as the electric field vector at any point. These lines, called **electric field lines,** are related to the electric field in any region of space in the following manner:

1. The electric field vector, **E**, is *tangent* to the electric field line at each point.
2. The number of lines per unit area through a surface that is perpendicular to the lines is proportional to the strength of the electric field in that region. Thus, **E** is large where the field lines are close together and small where they are far apart.

These properties are illustrated in Figure 16.13. The density of lines through surface A is greater than the density of lines through surface B. Therefore, the electric field is more intense on surface A than on surface B. Furthermore, the field drawn in Figure 16.13 is nonuniform since the lines at different locations point in different directions.

Some representative electric field lines for a single positive point charge are shown in Figure 16.14a. Note that in this two-dimensional drawing we show only the field lines that lie in the plane containing the point charge. The lines are actually directed radially outward in *all* directions from the charge, somewhat like the needles of a porcupine. Since a positive test charge placed in this field would be repelled by the charge q, the lines are directed radially away from q. Similarly, the electric field lines for a single negative point charge are directed toward the

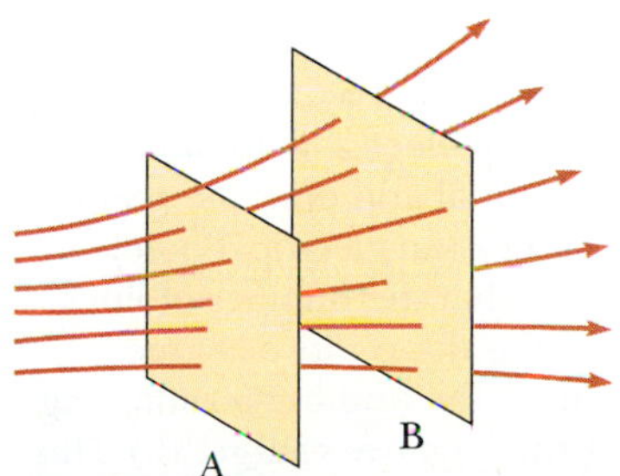

Figure 16.13
Electric field lines penetrating two surfaces. The magnitude of the field is greater on surface A than on surface B.

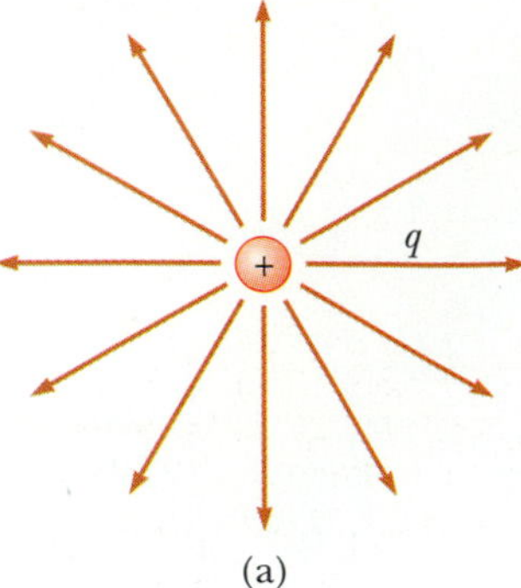

(a)

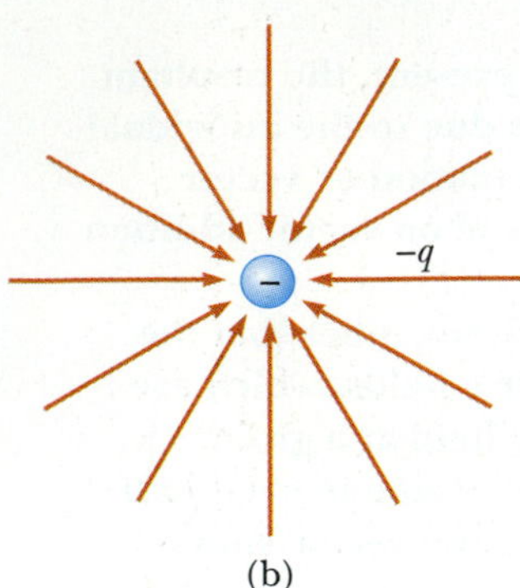

(b)

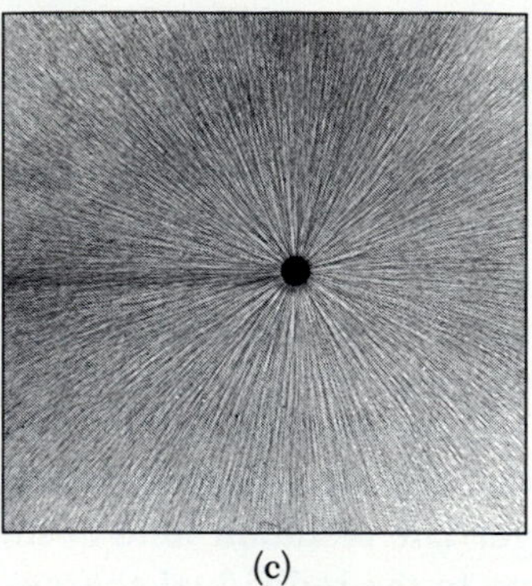

(c)

Figure 16.14 The electric field lines for a point charge. (a) For a positive point charge, the lines are radially outward. (b) For a negative point charge, the lines are radially inward. Note that the figures show only those field lines that lie in the plane containing the charge. (c) The dark areas are small pieces of thread suspended in oil, which align with the electric field produced by a small charged conductor at the center. *(Photo courtesy of Harold M. Waage, Princeton University)*

charge (Fig. 16.14b). In either case, the lines are radial and extend to infinity. Note that the lines are closer together as they get nearer the charge, indicating that the strength of the field is increasing.

The rules for drawing electric field lines for any charge distribution are as follows:

1. **The lines must begin on positive charges (or at infinity) and must terminate either on negative charges or, in the case of an excess of positive charge, at infinity.**
2. **The number of lines drawn leaving a positive charge or approaching a negative charge is proportional to the magnitude of the charge.**
3. **No two field lines can cross each other.**

Is this visualization of the electric field in terms of field lines consistent with Coulomb's law? To answer this question, consider an imaginary spherical surface of radius r, concentric with the charge. From symmetry, we see that the magnitude of the electric field is the same everywhere on the surface of the sphere. The number of lines, N, that emerge from the charge is equal to the number that penetrate the spherical surface. Hence, the number of lines per unit area on the sphere is $N/4\pi r^2$ (where the surface area of the sphere is $4\pi r^2$). Since E is proportional to the number of lines per unit area, we see that E varies as $1/r^2$. This is consistent with the result obtained from Coulomb's law, that is, $E = k_e q/r^2$.

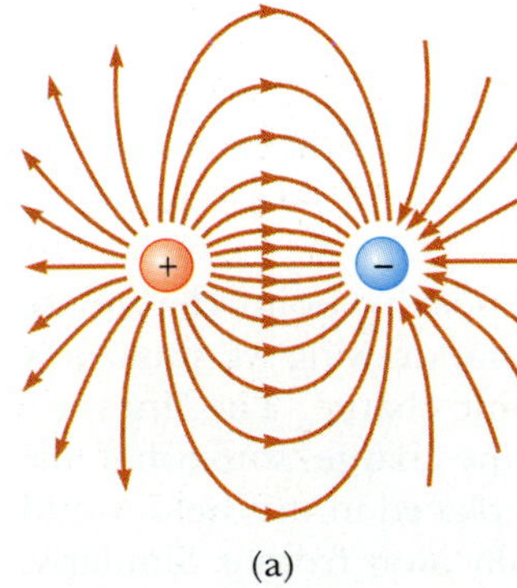

(a)

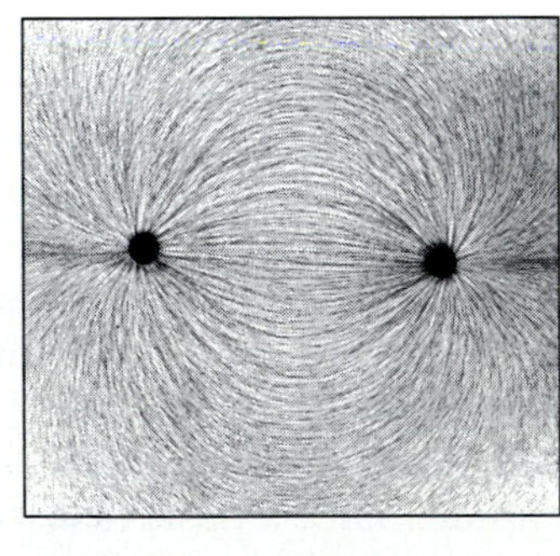

(b)

Figure 16.15 (a) The electric field lines for two equal and opposite point charges (an electric dipole). Note that the number of lines leaving the positive charge equals the number terminating at the negative charge. (b) This photograph was taken using small pieces of thread suspended in oil, which align with the electric field. *(Photo courtesy of Harold M. Waage, Princeton University)*

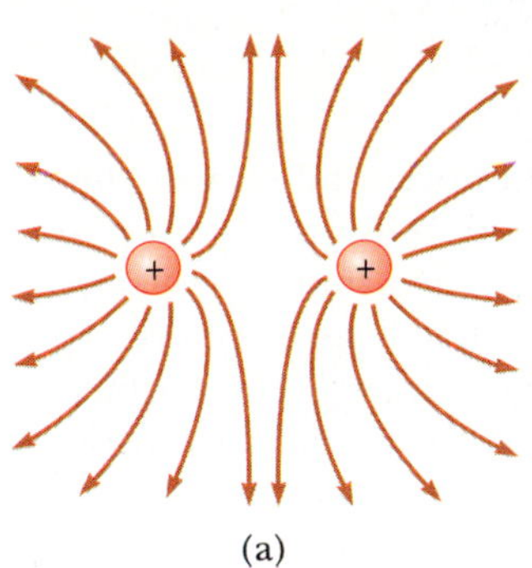

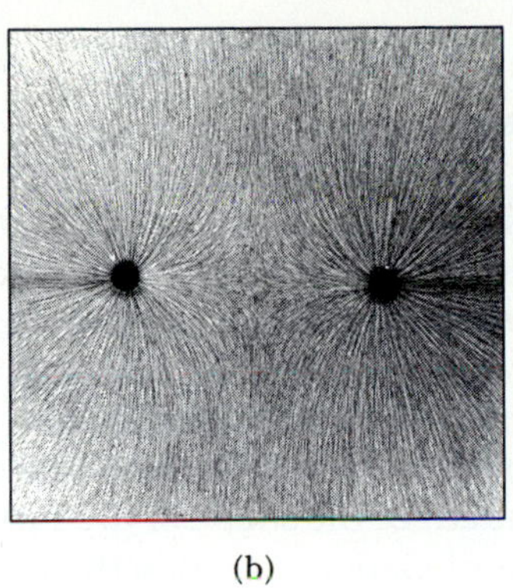

Figure 16.16
(a) The electric field lines for two positive point charges. (b) This photograph was taken using small pieces of thread suspended in oil, which align with the electric field. *(Photo courtesy of Harold M. Waage, Princeton University)*

It is important to note that electric field lines are not material objects. They are used only to provide a qualitative description of the electric field. One problem with this model is the fact that one always draws a finite number of lines from each charge, which makes it appear as if the field were quantized and acted only in a certain direction. The field, in fact, is continuous—existing at every point. Another problem with this model is the danger of getting the wrong impression from a two-dimensional drawing of field lines used to describe a three-dimensional situation.

Since charge is quantized, the number of lines leaving any material object must be $0, \pm C'e, \pm 2C'e, \ldots$, where C' is an arbitrary (but fixed) proportionality constant. Once C' is chosen, the number of lines is not arbitrary. For example, if object 1 has charge Q_1 and object 2 has charge Q_2, then the ratio of numbers of lines is $N_2/N_1 = Q_2/Q_1$.

The electric field lines for two point charges of equal magnitude but opposite signs (the electric dipole) are shown in Figure 16.15. In this case, the number of lines that begin at the positive charge must equal the number that terminate at the negative charge. At points very near the charges, the lines are nearly radial. The high density of lines between the charges indicates a region of strong electric field. The attractive nature of the force between the charges can also be seen from Figure 16.15.

Figure 16.16 shows the electric field lines in the vicinity of two equal positive point charges. Again, close to either charge the lines are nearly radial. The same number of lines emerges from each charge because the charges are equal in magnitude. At great distances from the charges, the field is approximately equal to that of a single point charge of magnitude $2q$. The bulging out of the electric field lines between the charges indicates the repulsive nature of the electric force between like charges.

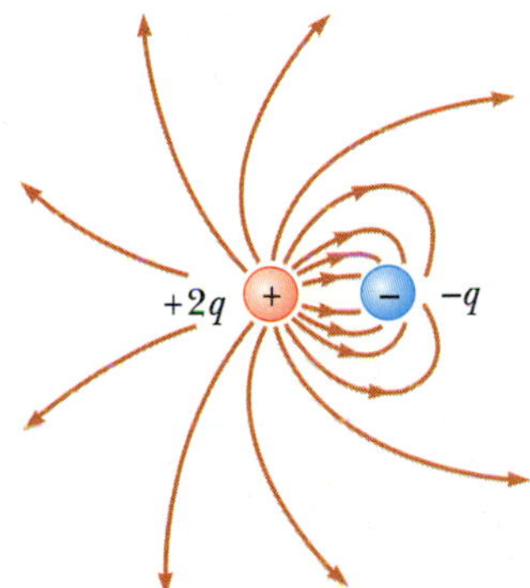

Figure 16.17
The electric field lines for a point charge of $+2q$ and a second point charge of $-q$. Note that two lines leave the charge $+2q$ for every line that terminates on $-q$.

Finally, in Figure 16.17 is a sketch of the electric field lines associated with the positive charge $+2q$ and the negative charge $-q$. In this case, the number of lines leaving the charge $+2q$ is twice the number terminating on the charge $-q$. Hence, only half of the lines that leave the positive charge end at the negative charge. The remaining half terminate on a negative charge we assume to be located at infinity. At great distances from the charges (great compared with the charge separation), the electric field lines are equivalent to those of a single charge, $+q$.

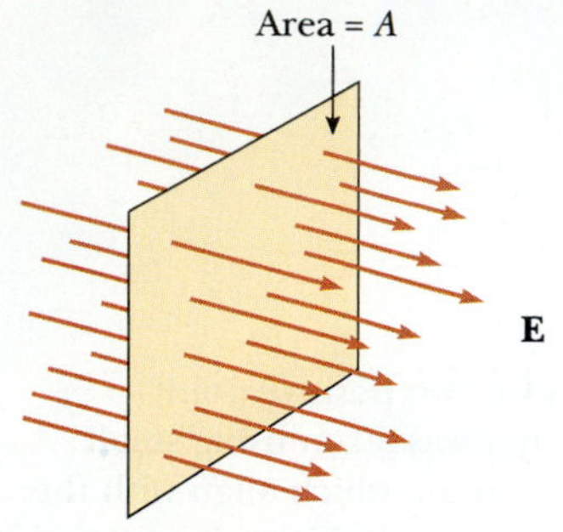

Figure 16.18
Field lines of a uniform electric field penetrating a plane of area A perpendicular to the field. The electric flux, Φ, through this area is equal to EA.

16.7 Electric Flux

Now that we have described the concept of electric field lines qualitatively, let us use a new concept, electric flux, to approach electric field lines on a quantitative basis. *Electric flux is a measure of the number of electric field lines penetrating some surface.* When the surface being penetrated is a closed surface (such as a sphere) that encloses some net charge, the number of lines that go through the surface is proportional to the net charge within the surface. The number of lines through the surface is independent of the shape of the surface enclosing the charge. This is essentially a statement of Gauss's law, described in the next section.

First consider an electric field that is uniform in both magnitude and direction, as in Figure 16.18. The field lines penetrate a plane rectangular surface of area A, which is perpendicular to the field. Recall that the number of lines per unit area is proportional to the magnitude of the electric field. Therefore, the number of lines penetrating the surface of area A is proportional to the product EA. The product of the electric field strength, E, and a surface area, A, perpendicular to the field is called the **electric flux,** Φ:

$$\Phi = EA \qquad [16.10]$$

From the SI units of E and A, we see that electric flux has the units $N \cdot m^2/C$.

If the surface under consideration is not perpendicular to the field, the number of lines (the flux) through it must be less than that given by Equation 16.10. This can be easily understood by considering Figure 16.19 where the normal to the surface of area A is at an angle of θ to the uniform electric field. Note that the number of lines that cross this area is equal to the number that cross the projected area A', which is perpendicular to the field. From Figure 16.19 we see that the two areas are related by $A' = A \cos \theta$. Since the flux through area A equals the flux through A', we conclude that the desired flux is

$$\Phi = EA \cos \theta \qquad [16.11]$$

From this result, we see that the flux through a surface of fixed area has the maximum value, EA, when the surface is perpendicular to the field (in other words, when the *normal* to the surface is parallel to the field, that is, $\theta = 0°$); the flux is zero when the surface is parallel to the field (when the normal to the surface is perpendicular to the field, that is, $\theta = 90°$).

In more general situations, the electric field may vary over the surface in question. Therefore, our definition of flux given by Equation 16.11 has meaning only over a small element of area. Consider a general surface divided up into a

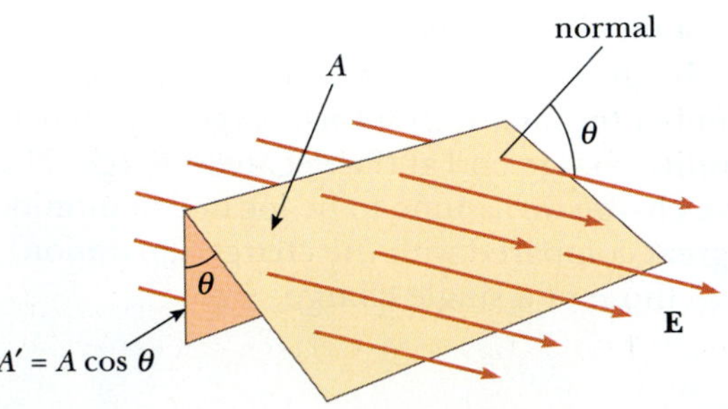

Figure 16.19
Field lines for a uniform electric field through an area, A, that is at an angle θ to the field. Since the number of lines that go through the shaded area, A', is the same as the number of lines that go through A, we conclude that the flux through A' is equal to the flux through A and is given by $\Phi = EA \cos \theta$.

large number of small elements, each of area ΔA. The variation in the electric field over the element can be neglected if the element is small enough. It is convenient to define a vector, $\Delta \mathbf{A}_i$, whose magnitude represents the area of the ith element and whose direction is *defined to be perpendicular* to the surface, as in Figure 16.20. The electric flux, $\Delta\Phi_i$, through this small element is

$$\Delta\Phi_i = E_i\,\Delta A_i \cos\theta = \mathbf{E}_i \cdot \Delta\mathbf{A}_i$$

where we have used the definition of the scalar product of two vectors ($\mathbf{A}\cdot\mathbf{B} = AB\cos\ \theta$). By summing the contributions of all elements, we obtain the total flux through the surface.[2] If we let the area of each element approach zero, then the number of elements approaches infinity and the sum is replaced by an integral. Therefore, the *general definition of electric flux* is

$$\Phi \equiv \lim_{\Delta A_i \to 0} \sum \mathbf{E}_i \cdot \Delta\mathbf{A}_i = \int_{\text{surface}} \mathbf{E}\cdot d\mathbf{A} \qquad \textbf{[16.12]}$$

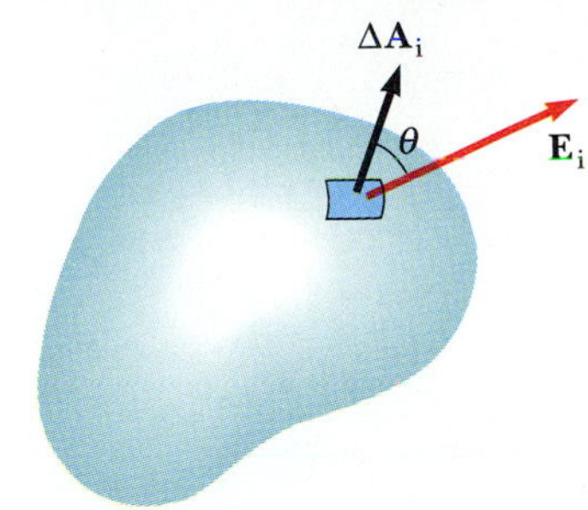

Figure 16.20
A small element of a surface of area $\Delta\mathbf{A}_i$. The electric field makes an angle θ with the normal to the surface (the direction of $\Delta\mathbf{A}_i$), and the flux through the element is equal to $E_i\ \Delta A_i$ cos Δ.

Equation 16.12 is a surface integral, which must be evaluated over the surface in question. In general, the value of Φ depends both on the field pattern and on the specified surface.

We shall usually be interested in evaluating the flux through a *closed surface.* (A closed surface is defined as one that divides space into an inside region and an outside region, so that movement cannot take place from one region to the other without penetrating the surface. The surface of a sphere, for example, is a closed surface.) Consider the closed surface in Figure 16.21. Note that the vectors $\Delta\mathbf{A}_i$ point in different directions for the various surface elements. At each point, these vectors are *perpendicular* to the surface and, by convention, always point *outward.* At the elements labeled ① and ②, E is outward and $\theta < 90°$; hence, the flux $\Delta\Phi = \mathbf{E}\cdot\Delta\mathbf{A}$ through these elements is positive. For elements such as ③, where the field lines are directed into the surface, $\theta > 90°$ and the flux becomes negative because of the cos θ factor. (Element ③ is on the side of the surface we cannot see; hence the dashed lines representing the vectors $\mathbf{E}_3$ and $\Delta\mathbf{A}_3$.) The total, or net, flux through the surface is proportional to the net number of lines penetrating the surface (where the net number means *the number leaving the volume surrounding the surface minus the number entering the surface*). If more lines are leaving the surface than entering, the net flux is positive. If more lines enter than leave the surface, the net flux is negative. (The net flux can be nonzero only when some net charge is contained within the closed surface.) Using the symbol $\oint$ to represent an *integral over a closed surface,* we can write the net flux, Φ_c, through a closed surface

$$\Phi_c = \oint \mathbf{E}\cdot d\mathbf{A} = \oint E_n\, dA \qquad \textbf{[16.13]}$$

where E_n represents the component of the electric field perpendicular to the surface and the subscript c denotes a closed surface.

Evaluating the net flux through a closed surface can be very cumbersome. However, if the field is perpendicular to the surface at each point and constant in magnitude, the calculation is straightforward. The following example illustrates this point.

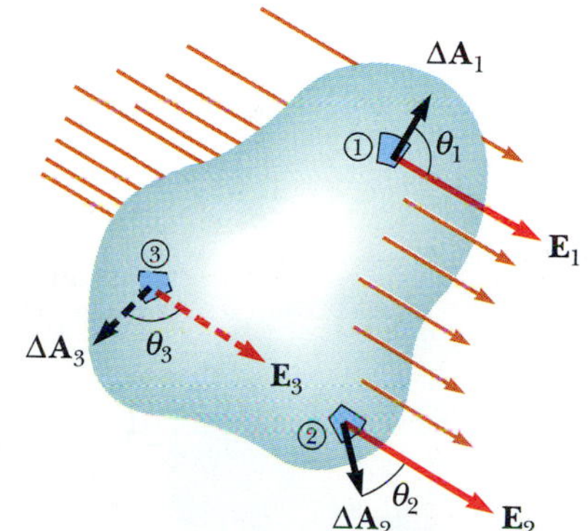

Figure 16.21
A closed surface in an electric field. The area vectors $\Delta\mathbf{A}_i$ are, by convention, normal to the surface and point outward. The flux through an area element can be positive (elements ① and ②) or negative (element ③). Element ③ is on the back side of the surface.

[2] It is important to note that drawings with field lines can have inaccuracies, since a small area (depending on its location) may happen to have too many or too few penetrating lines. At any rate, it is stressed that the basic definition of electric flux is $\int \mathbf{E}\cdot d\mathbf{A}$. The use of lines is only an aid for visualizing the concept.

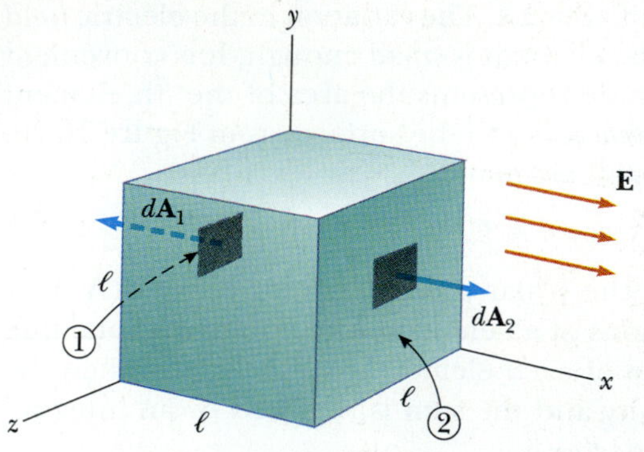

Figure 16.22
(Example 16.6) A hypothetical surface in the shape of a cube in a uniform electric field parallel to the x axis. The net flux through the surface is zero.

Example 16.6 Flux Through a Cube

Consider a uniform electric field, **E**, oriented in the x direction. Find the net electric flux through the surface of a cube with edges of length ℓ, oriented as in Figure 16.22.

Solution The net flux can be evaluated by summing up the fluxes through each face of the cube. First, note that the flux through *four* of the faces is zero, since **E** is perpendicular to $d\mathbf{A}$ on these faces. In particular, the orientation of $d\mathbf{A}$ is perpendicular to **E** for the two faces parallel to the xz plane in Figure 16.22. Therefore, $\theta = 90°$, so $\mathbf{E}\cdot d\mathbf{A} = E\,dA\cos 90° = 0$. The fluxes through the planes parallel to the yx plane are also zero for the same reason.

Now consider the faces parallel to the yz plane. The net flux through these faces is given by

$$\Phi_c = \int_1 \mathbf{E}\cdot d\mathbf{A} + \int_2 \mathbf{E}\cdot d\mathbf{A}$$

For the face labeled ①, **E** is constant and inward, and $d\mathbf{A}$ is outward ($\theta = 180°$), so we find that the flux through this face is

$$\int_1 \mathbf{E}\cdot d\mathbf{A} = \int_1 E\,dA\cos 180° = -E\int_1 dA$$
$$= -EA = -E\ell^2$$

since the area of each face is $A = \ell^2$.

For face ②, **E** is constant and outward and in the same direction as $d\mathbf{A}\,(\theta = 0°)$, so the flux through this face is

$$\int_2 \mathbf{E}\cdot d\mathbf{A} = \int_2 E\,dA\cos 0° = E\int_2 dA = +EA = E\ell^2$$

Hence, the net flux over all faces is zero, since

$$\Phi_c = -E\ell^2 + E\ell^2 = 0$$

16.8 Gauss's Law

In this section we describe a general relation between the net electric flux through a closed surface (often called a *gaussian surface*) and the charge *enclosed* by the surface. This relation, known as *Gauss's law,* is of fundamental importance in the study of electrostatic fields.

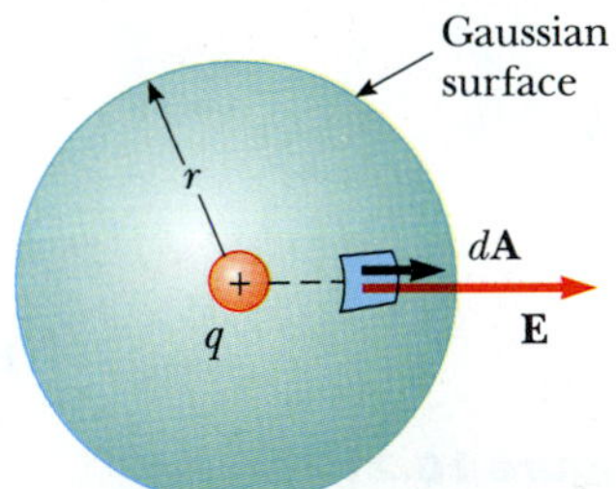

Figure 16.23
A spherical surface of radius r surrounding a point charge, q. When the charge is at the center of the sphere, the electric field is normal to the surface and constant in magnitude everywhere on the surface.

First, let us consider a positive point charge, q, located at the center of a sphere of radius r, as in Figure 16.23. From Equation 16.4, we know that the magnitude of the electric field everywhere on the surface of the sphere is $E = k_e q/r^2$. Furthermore, the field lines radiate outward, and hence are perpendicular (or normal) to the surface at each point. That is, at each point on the surface, **E** is parallel to the vector $\Delta\mathbf{A}_i$ representing the local element of area ΔA_i. Therefore,

$$\mathbf{E}\cdot\Delta\mathbf{A}_i = E_n\,\Delta A_i = E\,\Delta A_i$$

and from Equation 16.14 we find that the net flux through the gaussian surface is

$$\Phi_c = \oint E_n\,dA = \oint E\,dA = E\oint dA$$

since E is constant over the surface and is given by $E = k_e q/r^2$. Furthermore, for a spherical gaussian surface, $\oint dA = A = 4\pi r^2$ (the surface area of a sphere). Hence, the net flux through the gaussian surface is

$$\Phi_c = \frac{k_e q}{r^2}(4\pi r^2) = 4\pi k_e q$$

Recalling that $k_e = 1/4\pi\epsilon_0$, we can write this in the form

$$\Phi_c = \frac{q}{\epsilon_0} \qquad [16.14]$$

Note that this result, which is independent of r, says that the net flux through a spherical gaussian surface is proportional to the charge q *inside* the surface. The fact that the flux is independent of the radius is a consequence of the inverse-square dependence of the electric field given by Equation 16.4. That is, E varies as $1/r^2$, but the area of the sphere varies as r^2. Their combined effect produces a flux that is independent of r.

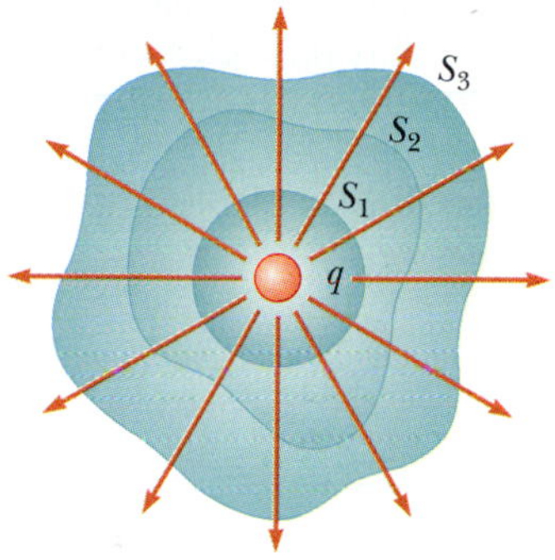

Figure 16.24
Closed surfaces of various shapes surrounding a charge, q. Note that the net electric flux through each surface is the same.

Now consider several closed surfaces surrounding a charge, q, as in Figure 16.24. Surface S_1 is spherical, whereas surfaces S_2 and S_3 are nonspherical. The flux that passes through surface S_1 has the value q/ϵ_0. As we discussed in the preceding section, the flux is proportional to the number of electric field lines passing through that surface. The construction in Figure 16.24 shows that the number of electric field lines through the spherical surface S_1 is equal to the number of electric field lines through the nonspherical surfaces S_2 and S_3. Therefore, it is reasonable to conclude that the net flux through any closed surface is independent of the shape of that surface. (One can prove that this is the case if $E \propto 1/r^2$.) In fact, *the net flux through any closed surface surrounding the point charge q is given by* q/ϵ_0.

Now consider a point charge located *outside* a closed surface of arbitrary shape, as in Figure 16.25. As you can see from this construction, electric field lines enter the surface, and field lines leave it. However, *the number of electric field lines entering the surface equals the number leaving the surface.* Therefore, we conclude that *the net electric flux through a closed surface that surrounds no charge is zero.* If we apply this result to Example 16.6, we can easily see that the net flux through the cube is zero, since it was assumed there was no charge inside the cube.

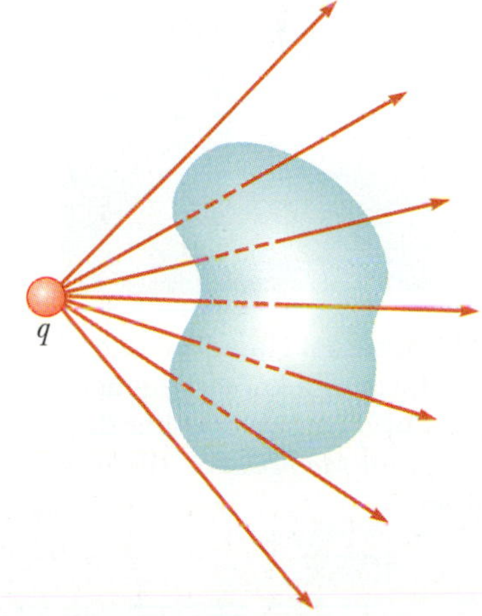

Figure 16.25
A point charge located *outside* a closed surface. The number of lines entering the surface equals the number leaving the surface.

Let us extend these arguments to the generalized case of many point charges, which we consider to be a continuous distribution of charge. We shall again make use of the superposition principle. That is, we can express the flux through any closed surface as

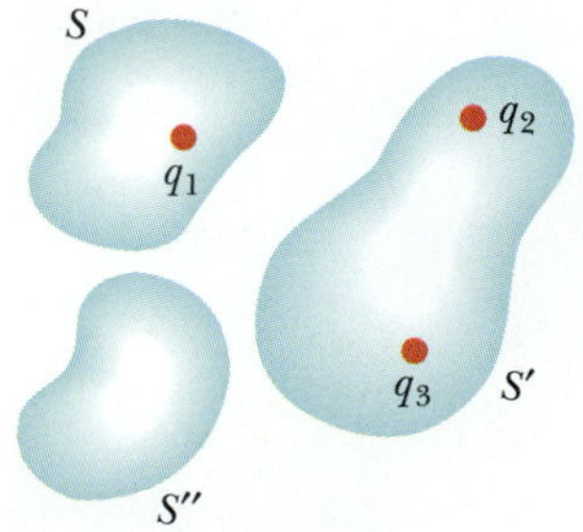

Figure 16.26
The net electric flux through any closed surface depends only on the charge *inside* that surface. The net flux through surface S is q_1/ϵ_0, the net flux through surface S' is $(q_2 + q_3)/\epsilon_0$, and the net flux through surface S'' is zero.

$$\oint \mathbf{E} \cdot d\mathbf{A} = \oint (\mathbf{E}_1 + \mathbf{E}_2 + \mathbf{E}_3) \cdot d\mathbf{A}$$

where **E** is the total electric field at any point on the surface and $\mathbf{E}_1$, $\mathbf{E}_2$, and $\mathbf{E}_3$ are the fields produced by the individual charges at that point. Consider the system of charges shown in Figure 16.26. The surface S surrounds only one charge, q_1; hence, the net flux through S is q_1/ϵ_0. The flux through S due to the charges outside it is zero since each electric field line that enters S at one point leaves it at another. The surface S′ surrounds charges q_2 and q_3; hence, the net flux through S′ is $(q_2 + q_3)/\epsilon_0$. Finally, the net flux through surface S″ is zero since there is no charge inside this surface. That is, *all* lines that enter S″ at one point leave S″ at another.

Gauss's law, which is a generalization of the foregoing discussion, states that the net flux through *any* closed surface is

$$\Phi_c = \oint \mathbf{E} \cdot d\mathbf{A} = \frac{q_{\text{in}}}{\epsilon_0} \qquad [16.15]$$

where q_{in} represents the *net charge inside* the gaussian surface and **E** represents the electric field at any point on the gaussian surface. In words,

Gauss's law

Gauss's law states that the net electric flux through any closed surface is equal to the net charge inside the surface divided by ϵ_0.

When using Equation 16.15, note that, although the charge q_{in} is the net charge inside the gaussian surface, the **E** that appears in Gauss's law represents the *total electric field,* which includes contributions from charges both inside and outside the gaussian surface. This point is often neglected or misunderstood.

In principle, Gauss's law can always be used to calculate the electric field of a system of charges or a continuous distribution of charge. In practice, however, *the technique is useful only in a limited number of situations where there is a high degree of symmetry.* As we shall see in the next section, *Gauss's law can be used to evaluate the electric field for charge distributions that have spherical, cylindrical, or plane symmetry.* If the gaussian surface surrounding the charge distribution is chosen carefully, the integral in Equation 16.15 will be easy to evaluate. Also note that a gaussian surface is a mathematical surface and need not coincide with any real physical surface.

This unusual closed surface in the form of a knot was generated by a computer using the program *Mathematica. (Wolfram Research)*

16.9 Application of Gauss's Law to Charged Insulators

In this section we give some examples of how to use Gauss's law to calculate **E** for a given charge distribution. It is important to recognize that *Gauss's law is useful when there is a high degree of symmetry in the charge distribution, as in the cases of uniformly charged spheres, long cylinders, and plane sheets.* In such cases, it is possible to find a simple gaussian surface over which the surface integral given by Equation 16.15 is easily evaluated.

Gauss's law is useful for evaluating E when the charge distribution has symmetry

A surface should always be chosen to take advantage of the symmetry of the charge distribution.

Example 16.7 The Electric Field Due to a Point Charge

Starting with Gauss's law, calculate the magnitude of the electric field due to an isolated point charge, q, and show that Coulomb's law follows from this result.

Solution For this situation we choose a spherical gaussian surface of radius r and centered on the point charge, as in Figure 16.27. The electric field of a positive point charge is radial outward by symmetry, and is therefore normal to the surface at every point. That is, **E** is parallel to $d\mathbf{A}$ at each point, and so $\mathbf{E}\cdot d\mathbf{A} = E\,dA$ and Gauss's law gives

$$\Phi_c = \oint \mathbf{E}\cdot d\mathbf{A} = \oint E\,dA = \frac{q}{\epsilon_0}$$

By symmetry, E is constant everywhere on the surface, and so it can be removed from the integral. Therefore,

$$\oint E\,dA = E\oint dA = E(4\pi r^2) = \frac{q}{\epsilon_0}$$

where we have used the fact that the surface area of a sphere is $4\pi r^2$. Hence, the magnitude of the field at a distance of r from the charge q is

$$E = \frac{q}{4\pi\epsilon_0 r^2} = k_e\frac{q}{r^2}$$

If a second point charge, q_0, is placed at a point where the field is E, the electrostatic force on this charge has a magnitude given by

$$F = q_0 E = k_e\frac{qq_0}{r^2}$$

This, of course, is Coulomb's law. Note that this example is logically circular. It does, however, demonstrate the equivalence of Coulomb's law and Gauss's law.

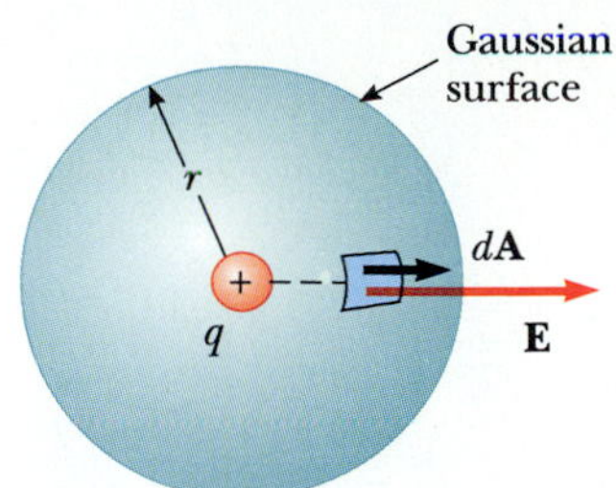

Figure 16.27 (Example 16.7) The point charge q is at the center of the spherical gaussian surface, and E is parallel to dA at every point on the surface.

Example 16.8 A Spherically Symmetric Charge Distribution

An insulating sphere of radius a has a uniform charge density of ρ and a total positive charge of Q (Fig. 16.28).

(a) Calculate the electric field strength at a point *outside* the sphere, that is, for $r > a$.

Solution Since the charge distribution is spherically symmetric, we again select a spherical gaussian surface of radius r, concentric with the sphere, as in Figure 16.28a. Following the line of reasoning given in Example 16.7, we find that

$$E = k_e\frac{Q}{r^2} \qquad (\text{for } r > a)$$

Note that this result is identical to that obtained for a point charge. Therefore, we conclude that, for a uniformly charged sphere, the field in the region external to the sphere is *equivalent* to that of a point charge located at the center of the sphere.

(b) Find the electric field strength at a point *inside* the sphere, that is, for $r < a$.

Solution In this case we select a spherical gaussian surface with radius $r < a$, concentric with the charge distribution (Fig. 16.28b). To apply Gauss's law in this situation, it is important to recognize that the charge q_{in} *within* the gaussian surface of volume V' is *less* than the total charge Q. To calculate the charge q_{in}, we use the

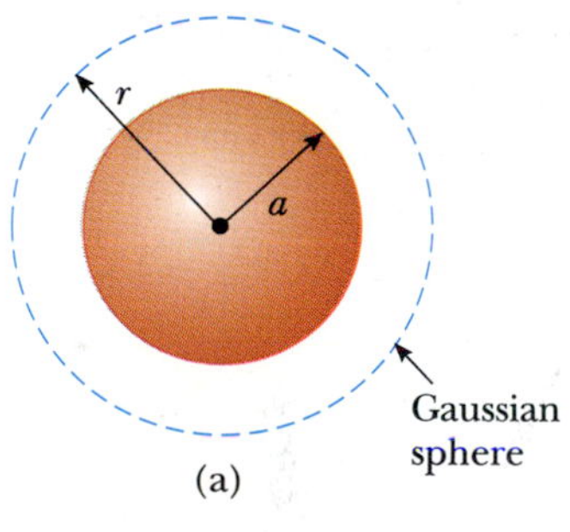

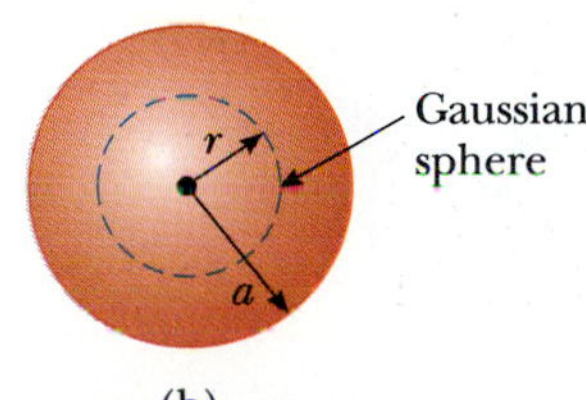

Figure 16.28 (Example 16.8) A uniformly charged insulating sphere of radius a and total charge Q. (a) The field at a point exterior to the sphere is k_eQ/r^2. (b) The field inside the sphere is due only to the charge *within* the gaussian surface and is given by $(k_eQ/a^3)r$.

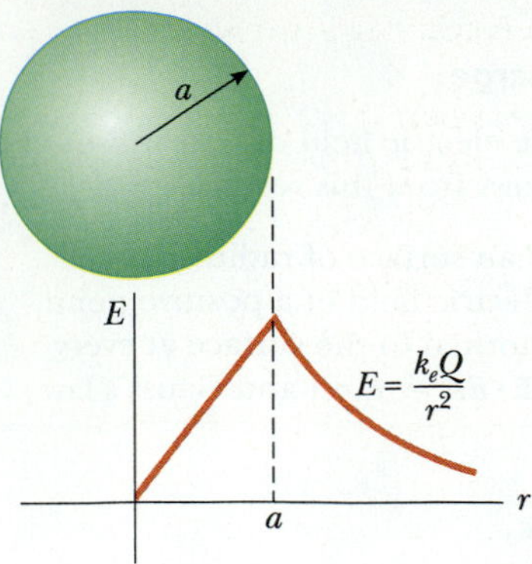

Figure 16.29 (Example 16.8) A plot of E versus r for a uniformly charged insulating sphere. The field inside the sphere ($r < a$) varies linearly with r. The field outside the sphere ($r > a$) is the same as that of a point charge, Q, located at the origin.

fact that $q_{\text{in}} = \rho V'$, where ρ is the charge per unit volume and V′ is the volume enclosed by the gaussian surface, given by $V' = \frac{4}{3}\pi r^3$ for a sphere. Therefore,

$$q_{\text{in}} = \rho V' = \rho\left(\tfrac{4}{3}\pi r^3\right)$$

As in Example 16.7, the electric field is constant in magnitude everywhere on the spherical gaussian surface and is normal to the surface at each point. Therefore, Gauss's law in the region $r < a$ gives

$$\oint E\,dA = E\oint dA = E\,(4\pi r^2) = \frac{q_{\text{in}}}{\epsilon_0}$$

Solving for E gives

$$E = \frac{q_{\text{in}}}{4\pi\epsilon_0 r^2} = \frac{\rho\frac{4}{3}\pi r^3}{4\pi\epsilon_0 r^2} = \frac{\rho}{3\epsilon_0}\,r$$

Since by definition $\rho = Q/(\frac{4}{3}\pi a^3)$, this can be written

$$E = \frac{Qr}{4\pi\epsilon_0 a^3} = \frac{k_eQ}{a^3}\,r \qquad (\text{for } r < a)$$

Note that this result for E differs from that obtained in (a). It shows that $E \rightarrow 0$ as $r \rightarrow 0$, as you might have guessed based on the spherical symmetry of the charge distribution. Therefore, the result fortunately eliminates the singularity that would exist at $r = 0$ if E varied as $1/r^2$ inside the sphere. That is, if $E \propto 1/r^2$, the field would be infinite at $r = 0$, which is clearly a physical impossibility. A plot of E versus r is shown in Figure 16.29.

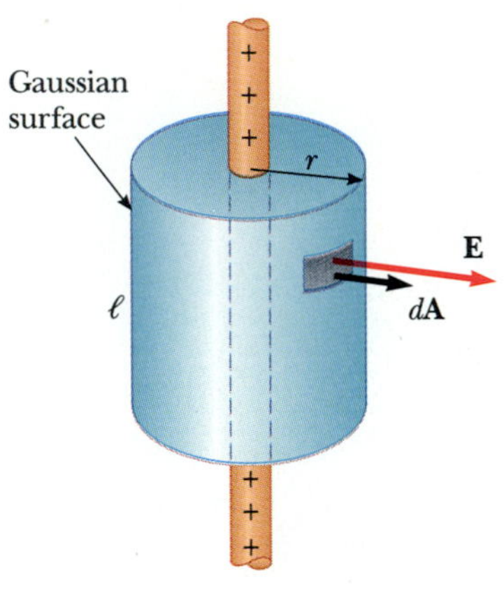

(a)

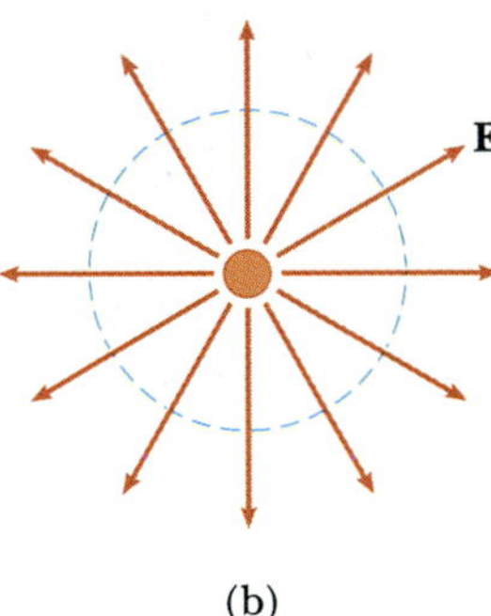

(b)

Figure 16.30 (Example 16.9) (a) An infinite line of charge surrounded by a cylindrical gaussian surface that is concentric with the line charge. (b) The field on the cylindrical surface is constant in magnitude and perpendicular to the surface.

Example 16.9 A Cylindrically Symmetric Charge Distribution

Find the magnitude and direction of the electric field at a distance of r from a uniform positive line charge of infinite length whose charge per unit length is λ = constant (Fig. 16.30).

Solution The symmetry of the charge distribution shows that **E** must be perpendicular to the line charge and directed outward, as in Figure 16.30a. The end view of the line charge shown in Figure 16.30b should help you visualize the directions of the electric field lines. In this situation, we select a cylindrical gaussian surface of radius r and length ℓ that is coaxial with the line charge. For the curved part of this surface, **E** is constant in magnitude and perpendicular to the surface at each point.

Furthermore, the flux through the *ends* of the gaussian cylinder is *zero* since **E** is *parallel* to these surfaces.

The total charge inside our gaussian surface is $\lambda\ell$, where λ is the charge per unit length and ℓ is the length of the cylinder. Applying Gauss's law and noting that **E** is parallel to $d\mathbf{A}$ everywhere on the cylindrical surface, we find that

$$\Phi_c = \oint \mathbf{E}\cdot d\mathbf{A} = E\oint dA = \frac{q_{\text{in}}}{\epsilon_0} = \frac{\lambda\ell}{\epsilon_0}$$

But the area of the curved surface is $A = 2\pi r\ell$; therefore,

$$E(2\pi r\ell) = \frac{\lambda\ell}{\epsilon_0}$$

$$E = \frac{\lambda}{2\pi\epsilon_0 r} = 2k_e\frac{\lambda}{r} \qquad [16.16]$$

Thus, we see that the field of a cylindrically symmetric charge distribution varies as $1/r$, whereas the field external to a spherically symmetric charge distribution varies as $1/r^2$. Equation 16.16 can also be obtained using Coulomb's law and integration; however, the mathematical techniques necessary for that calculation are more cumbersome than those used here.

If the line charge has a finite length, the result for E is *not* given by Equation 16.16. For points close to the line charge and far from the ends, Equation 16.16 gives a good approximation of the value of the field. It turns out that Gauss's law is *not useful*, for calculating E for a finite line charge. This is because the electric field is no longer constant in magnitude over the surface of the gaussian cylinder. Furthermore, **E** is not perpendicular to the cylindrical surface at all points. When there is little symmetry in the charge distribution, as in this situation, it is necessary to calculate **E** using Coulomb's law.

It is left to Problem 45 at the end of the chapter to show that the **E** field *inside* a uniformly charged rod of finite thickness is proportional to r.

Example 16.10 A Nonconducting Plane Sheet of Charge

Find the electric field due to a nonconducting, infinite plane with uniform charge per unit area σ.

Solution The symmetry of the situation shows that **E** must be perpendicular to the plane and that the direction of **E** on one side of the plane must be opposite its direction on the other side, as in Figure 16.31. It is convenient to choose for our gaussian surface a small cylinder whose axis is perpendicular to the plane and whose ends each have an area of A and are equidistant from the plane. Here we see that, since **E** is parallel to the cylindrical surface, there is no flux through this surface. The flux out of *each* end of the cylinder is EA (since **E** is perpendicular to the ends); hence, the *total* flux through our gaussian surface is $2EA$. Noting that the total charge *inside* the surface is σA, we use Gauss's law to get

$$\Phi_c = 2EA = \frac{q_{\text{in}}}{\epsilon_0} = \frac{\sigma A}{\epsilon_0}$$

$$E = \frac{\sigma}{2\epsilon_0} \qquad [16.17]$$

Since the distance between surfaces and plane does not appear in Equation 16.17, we conclude that $E = \sigma/2\epsilon_0$ at *any* distance from the plane. That is, the field is *uniform* everywhere.

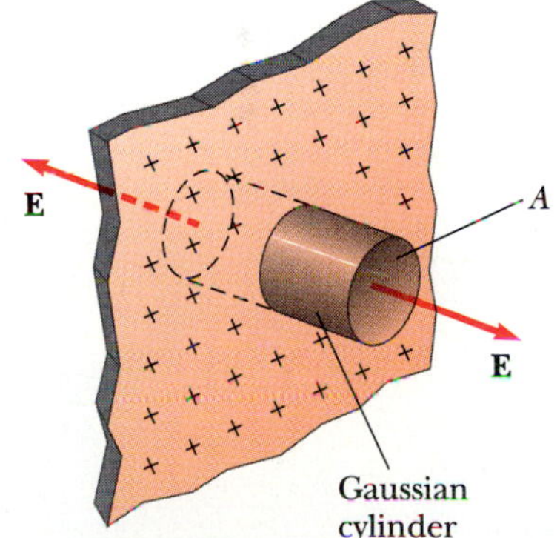

Figure 16.31 (Example 16.10) A cylindrical gaussian surface penetrating an infinite sheet of charge. The flux through each end of the gaussian surface is EA. There is no flux through the cylinder's curved surface.

An important configuration related to this example is the case of two parallel sheets of charge, with charge densities σ and $-\sigma$, respectively (Problem 65). In this situation, the electric field is σ/ϵ_0 *between* the sheets and approximately zero elsewhere.

Problem-Solving Strategy and Hints

Gauss's law may seem mysterious to you, and it is usually one of the most challenging concepts in introductory physics. However, as we have seen, Gauss's law is very powerful in the solution of problems with a high degree of symmetry. In this section, we have used Gauss's law to solve problems with three kinds of symmetry: plane, cylindrical, and spherical. It is important to review Examples 16.7 through 16.10 and to use the following procedure:

1. **Select a gaussian surface that *has the same symmetry as the charge distribution.* For point charges or spherically symmetric charge distributions, the gaussian surface should be a sphere centered on the charge, as in Examples 16.7 and 16.8. For uniform line charges or uniformly charged cylinders, the gaussian surface should be a cylindrical surface that is coaxial with the line charge or cylinder, as in Example 16.9. For sheets of charge having plane symmetry, the gaussian surface should be a small cylinder that straddles the sheet, as in Example 16.10. Note that in all cases, the gaussian surface is selected so that the electric field has the same magnitude everywhere on the surface and is directed perpendicular to the surface or is parallel to the surface. This enables you to easily evaluate the surface integral that appears on the left side of Gauss's law, and this integral represents the total electric flux through that surface.**
2. **Evaluate the right side of Gauss's law, which amounts to calculating the total electric charge, q_{in}, *inside* the gaussian surface. If the charge density is uniform, as is usually the case (that is, if λ, σ, or ρ is constant), simply multiply that charge density by the length, area, or volume enclosed by the gaussian surface. However, if the charge distribution is *nonuniform,* you must integrate the charge density over the region enclosed by the gaussian surface. For example, if the charge were distributed along a line, you would integrate the expression $dq = \lambda\, dx$, where dq is the charge on an infinitesimal element dx and λ is the charge per unit length. For a plane of charge, you would integrate $dq = \sigma\, dA$, where σ is the charge per unit area and dA is an infinitesimal element of area. For a volume of charge, you would integrate $dq = \rho\, dV$, where ρ is the charge per unit volume and dV is an infinitesimal element of volume.**
3. **Once the left and right sides of Gauss's law have been evaluated, you can calculate the electric field on the gaussian surface, assuming the charge distribution is given in the problem. Conversely, if the electric field is known, you can calculate the charge distribution that produces the field.**

16.10 Conductors in Electrostatic Equilibrium

A good electrical conductor, such as copper, contains charges (electrons) that are not bound to any atom and are free to move about within the material. When no *net* motion of charge occurs within the conductor, the conductor is in **electrostatic equilibrium.** As we shall see, *an isolated conductor (one that is insulated from ground) in electrostatic equilibrium* has the following properties:

1. The electric field is zero everywhere inside the conductor.
2. Any excess charge on the conductor must reside entirely on its surface.
3. The electric field just outside the charged conductor is perpendicular to the conductor's surface and has the magnitude σ/ϵ_0, where σ is the charge per unit area at that point.
4. On an irregularly shaped conductor, charge tends to accumulate at locations where the radius of curvature of the surface is the smallest, that is, at sharp points.

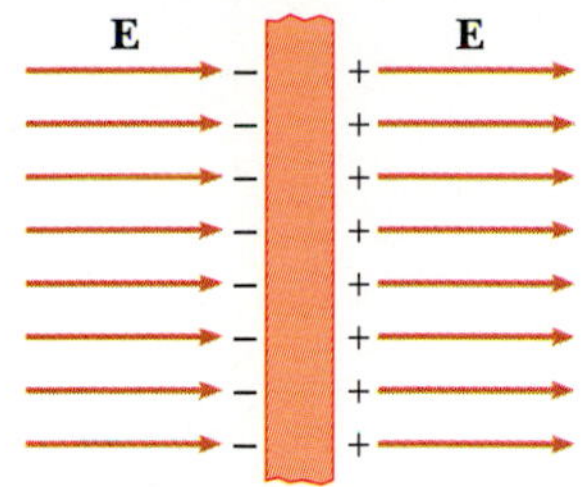

Figure 16.32
A conducting slab in an external electric field, **E**. The charges induced on the surfaces of the slab produce an electric field that opposes the external field, giving a resultant field of zero in the conductor.

The first property can be understood by considering a conducting slab placed in an external field, **E** (Fig. 16.32). In electrostatic equilibrium, the electric field *inside* the conductor must be zero. If this were not the case, the free charges would accelerate under the action of an electric field. Before the external field is applied, the electrons are uniformly distributed throughout the conductor. When the external field is applied, the free electrons accelerate to the left, causing a buildup of negative charge on the left surface (excess electrons) and of positive charge on the right (where electrons have been removed). These charges create their own electric field, which *opposes* the external field. The surface charge density increases until the magnitude of the electric field set up by these charges equals that of the external field, giving a net field of zero *inside* the conductor. In a good conductor, the time it takes the conductor to reach equilibrium is on the order of 10^{-16} s, which for most purposes can be considered instantaneous.

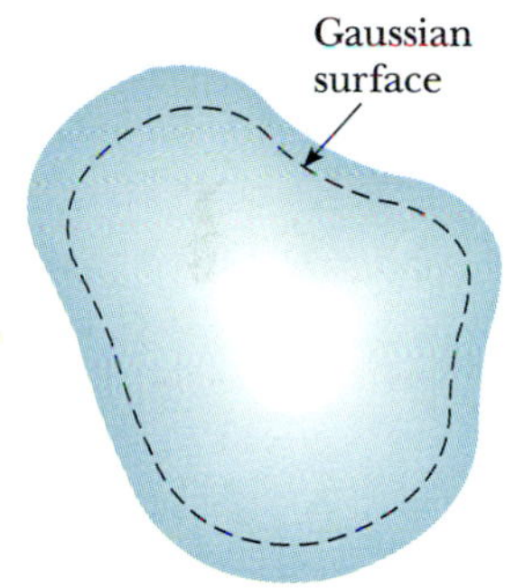

Figure 16.33
An insulated conductor of arbitrary shape. The broken line represents a gaussian surface just inside the conductor.

We can use Gauss's law to verify the second and third properties. Figure 16.33 shows an arbitrarily shaped insulated conductor. A gaussian surface is drawn inside the conductor as close to the surface as we wish. As we have just shown, the electric field everywhere inside the conductor is zero when it is in electrostatic equilibrium. Since the electric field is therefore zero at *every* point on the gaussian surface, we see that the net flux through this surface is zero. From this result and Gauss's law, we conclude that the net charge inside the gaussian surface is zero. Since there can be no net charge inside the gaussian surface (which is arbitrarily close to the conductor's surface), *any net charge on the conductor must reside on its surface.* Gauss's law does *not* tell us how this excess charge is distributed on the surface.

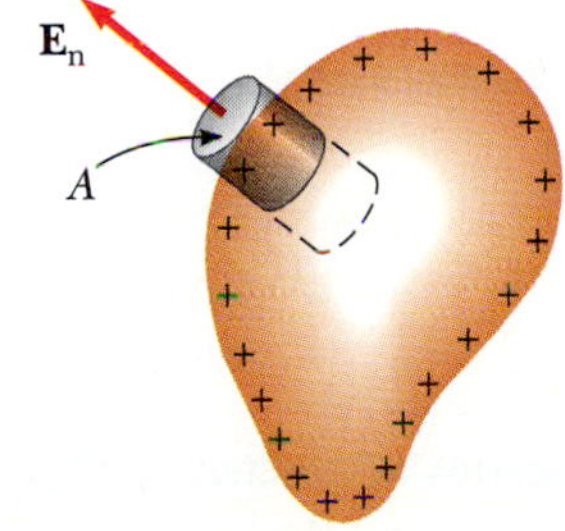

Figure 16.34
A gaussian surface in the shape of a small cylinder is used to calculate the electric field just outside a charged conductor. The flux through the gaussian surface is E_nA. Note that **E** is zero inside the conductor.

Examining the third property, we can use Gauss's law to relate the electric field just outside the surface of a charged conductor in equilibrium to the charge distribution on the conductor. To do this, it is convenient to draw a gaussian surface in the shape of a small cylinder with end faces parallel to the surface (Fig. 16.34). Part of the cylinder is just outside the conductor, and part is inside. There is no flux through the face on the inside of the cylinder since $\mathbf{E} = 0$ inside the conductor. Furthermore, the field outside the conductor is normal to the surface. If **E** had a tangential component, the free charges would move along the surface, creating surface currents, and the conductor would not be in equilibrium. There is no flux

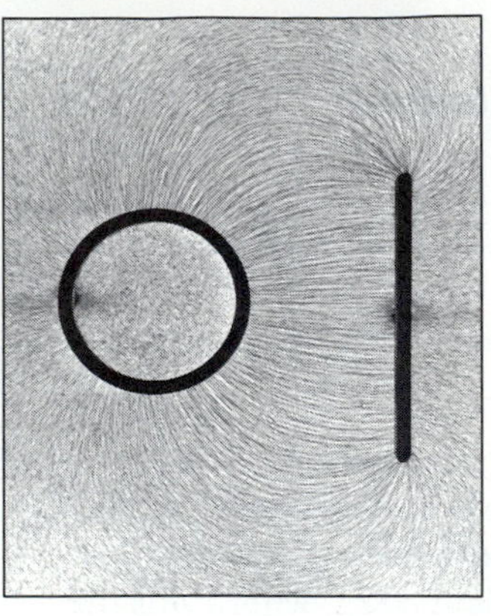

The electric field pattern of a charged conducting plate near an oppositely charged conducting cylinder. Small pieces of thread suspended in oil align with the electric field lines. Note that (1) the electric field lines are perpendicular to the conductors and (2) there are no lines inside the cylinder ($E = 0$). *(Courtesy of Harold M. Waage, Princeton University)*

through the cylindrical face of the gaussian surface since **E** is tangent to this surface. Hence, the net flux through the gaussian surface is E_nA, where E_n is the electric field just outside the conductor. Applying Gauss's law to this surface gives

$$\Phi_c = \oint E_n \, dA = E_nA = \frac{q_{\text{in}}}{\epsilon_0} = \frac{\sigma A}{\epsilon_0}$$

We have used the fact that the charge inside the gaussian surface is $q_{\text{in}} = \sigma A$, where A is the area of the cylinder's face and σ is the (local) charge per unit area. Solving for E_n gives

Electric field just outside a charged conductor

$$E_n = \frac{\sigma}{\epsilon_0} \qquad [16.18]$$

Summary

Electric charges have the following important properties:

Properties of electric charges

1. Unlike charges attract one another, and like charges repel one another.
2. Electric charge is always conserved.
3. Charge is quantized; that is, the charge on any object is an integral multiple of the electronic charge.
4. The force between charged particles varies as the inverse square of their separation.

Conductors are materials in which charges move freely. **Insulators** are materials that do not readily transport charge.

Coulomb's law states that the electrostatic force between two stationary, charged particles separated by a distance of r has the magnitude

$$F = k_e \frac{|q_1||q_2|}{r^2} \qquad [16.1]$$

Coulomb's law

where the Coulomb constant $k_e = 8.99 \times 10^9 \text{ N} \cdot \text{m}^2/\text{C}^2$.

The electric field due to the point charge q at a distance of r from the charge is

$$\mathbf{E} = k_e \frac{q}{r^2} \hat{\mathbf{r}} \qquad [16.4]$$

Electric field of the point charge q

where $\hat{\mathbf{r}}$ is a unit vector directed from the charge to the point in question. The electric field is directed radially outward from a positive charge and is directed *toward* a negative charge.

The *electric field* due to a group of charges can be obtained using the superposition principle. That is, the total electric field equals the *vector sum* of the electric fields of all the charges at some point:

$$\mathbf{E} = k_e \sum_i \frac{q_i}{r_i^2} \hat{\mathbf{r}}_i \qquad [16.6]$$

Electric field of a group of charges

Similarly, the electric field of a continuous charge distribution at some point is

$$\mathbf{E} = k_e \int \frac{dq}{r^2} \hat{\mathbf{r}} \qquad [16.6]$$

Electric field of a continuous charge distribution

where dq is the charge on one element of the charge distribution and r is the distance from the element to the point in question.

Electric field lines are useful for describing the electric field in any region of space. The electric field vector, $\mathbf{E}$, is always tangent to the electric field lines at every point. Furthermore, the number of lines per unit area through a surface perpendicular to the lines is proportional to the magnitude of $\mathbf{E}$ in that region.

Electric flux is a measure of the number of electric field lines that penetrate a surface. If the electric field is uniform and makes an angle of θ with the normal to the surface, the electric flux through the surface is

$$\Phi = EA \cos \theta \qquad [16.11]$$

Flux through a surface in a uniform electric field

In general, the electric flux through a surface is defined by the expression

$$\Phi = \int_{\text{surface}} \mathbf{E} \cdot d\mathbf{A} \qquad [16.12]$$

Definition of electric flux

Gauss's law says that the net electric flux, Φ_c, through any closed gaussian surface is equal to the *net* charge *inside* the surface divided by ϵ_0:

$$\Phi_c = \oint \mathbf{E} \cdot d\mathbf{A} = \frac{q_{\text{in}}}{\epsilon_0} \qquad [16.15]$$

Gauss's law

Using Gauss's law, one can calculate the electric field due to various symmetric charge distributions.

A **conductor in electrostatic equilibrium** has the following properties:

Properties of a conductor in electrostatic equilibrium

1. The electric field is zero everywhere inside the conductor.
2. Any excess charge on an isolated conductor must reside entirely on its surface.

3. The electric field just outside the conductor is perpendicular to its surface and has the magnitude σ/ϵ_0, where σ is the charge per unit area at that point.
4. On an irregularly shaped conductor, charge tends to accumulate where the radius of curvature of the surface is the smallest, that is, at sharp points.

Questions and Conceptual Exercises

1. Explain from an atomic viewpoint why charge is usually transferred by electrons.
2. A balloon is negatively charged by rubbing, and then clings to a wall. Does this mean that the wall is positively charged? Why does the balloon eventually fall?
3. A charged comb often attracts small bits of dry paper that then fly away when they touch the comb. Explain.
4. If a suspended object, A, is attracted to object B, which is charged, can we conclude that object A is charged? Explain.
5. Operating-room personnel must wear special conducting shoes while working around oxygen. Why? Contrast this procedure with what might happen if personnel wore rubber shoes.
6. Would life be different if the electron were positively charged and the proton were negatively charged? Could we identify a difference? Would any physical laws be different?
7. When defining the electric field, why is it necessary to specify that the magnitude of the test charge be very small (i.e., $q \to 0$ taken as the limit)?
8. Consider two equal point charges separated by some distance, d. At what point (other than ∞) would a third test charge experience no net force?
9. An uncharged, metallic-coated Ping-Pong ball is placed in the region between two vertical parallel metal plates. If the two plates are charged, one positive and one negative, describe the motion the Ping-Pong ball undergoes.
10. A negative point charge, $-q$, is placed at the point P near the positively charged ring shown in Figure 16.12 (Example 16.5). If $x \ll a$, describe the motion of the point charge if it is released from rest.
11. Explain the differences among linear, surface, and volume charge densities, and give examples of when each would be used.
12. If the net flux through a gaussian surface is zero, which of the following statements are true? (a) There are no charges inside the surface. (b) The net charge inside the surface is zero. (c) The electric field is zero everywhere on the surface. (d) The number of electric field lines entering the surface equals the number leaving the surface.
13. A very large, thin, flat plate of aluminum of area A has a total charge of Q uniformly distributed over its surfaces. The same charge is spread uniformly over only the *upper* surface of an otherwise identical glass plate. Compare the electric fields just above the centers of the plates' upper surfaces.
14. If more electric field lines leave a gaussian surface than enter the surface, what can you conclude about the *net* charge enclosed by that surface?
15. A uniform electric field exists in a region of space in which there are no charges. What can you conclude about the *net* electric flux through a gaussian surface placed in this region of space?
16. Explain why Gauss's law cannot be used to calculate the electric field of each of the following: (a) an electric dipole, (b) a charged disk, (c) a charged ring, (d) three point charges at the corners of a triangle.
17. If the total charge inside a closed surface is known but the distribution of the charge is unspecified, can you use Gauss's law to find the electric field? Explain.
18. Use Gauss's law to explain why electric field lines must begin and end on electric charges. (*Hint:* Change the size of the gaussian surface.)
19. A spherical gaussian surface surrounds a point charge, q. Describe what happens to the flux through the surface if (a) the charge is tripled; (b) the volume of the sphere is doubled; (c) the shape of the surface is changed to that of a cube; (d) the charge is moved to another position inside the surface.
20. Using the repulsive nature of the force between like charges and the freedom of motion of charge within the conductor, explain why excess charge on an isolated conductor must reside on its surface.
21. A person is placed in a large, hollow metallic sphere that is insulated from ground. If a large charge is placed on the sphere, will the person be harmed upon touching the inside of the sphere? Explain what will happen if the person also has an initial charge whose sign is opposite that of the charge on the sphere.
22. Richard Feynman once said that if two persons stood at arm's length from each other and each person had 1% more electrons than protons, the force of repulsion between the two people would be enough to lift a "weight" equal to that of the entire Earth. Carry out an order-of-magnitude calculation to substantiate this assertion.

Problems

Section 16.4 Coulomb's Law Revisited

1. Two point charges of magnitude 3 nC and 6 nC are separated by a distance of 0.3 m. Find the electric force of repulsion between them.
2. Two identical conducting small spheres are placed with their centers 0.3 m apart. One is given a charge of 12 nC and the other a charge of −18 nC. (a) Find the electrostatic force exerted on one sphere by the other. (b) The spheres are connected by a conducting wire. After equilibrium has occurred, find the electrostatic force between the two.
3. Determine what the mass of a pair of protons would be if the gravitational and electrical forces between them were equal in magnitude.
4. An electron is released above the surface of the Earth. A second electron directly below it exerts an electrostatic force on the first that is just great enough to cancel the gravitational force on it. How far below the first electron is the second?
5. Calculate the magnitude and direction of the Coulomb force on each of the three charges in Figure 16.35.

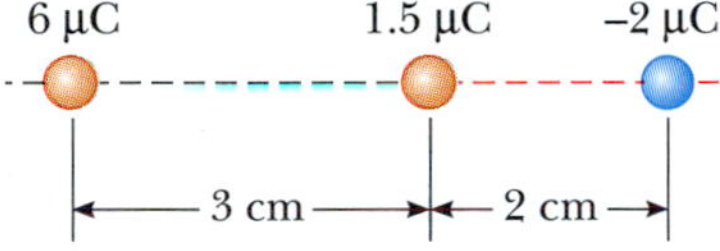

Figure 16.35 (Problem 5)

6. A molecule of DNA (deoxyribonucleic acid) is 2.17 μm long. The ends of the molecule become singly ionized—negative on one end, positive on the other. The helical molecule acts like a spring and compresses 1.0% upon becoming charged. Determine the effective spring constant of the molecule. Assume the helix is linear.
7. Three charges are arranged as shown in Figure 16.36. Find the magnitude and direction of the electrostatic force on the charge at the origin.
8. In the Bohr theory of the hydrogen atom, an electron moves in a circular orbit about a proton, where the radius of the orbit is 0.51×10^{-10} m. (a) Find the electrostatic force between the two. (b) If this force serves as the centripetal force on the electron, what is the speed of the electron?
9. Three charges are arranged as shown in Figure 16.37. Find the magnitude and direction of the electrostatic force on the 6-nC charge.

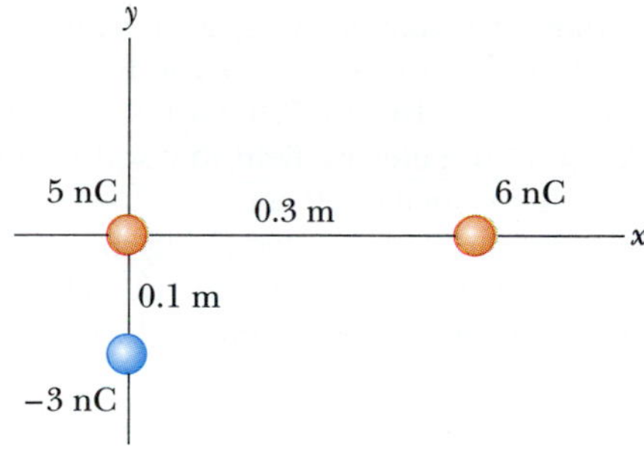

Figure 16.36 (Problem 7)

10. Two small silver spheres, each with a mass of 10 g, are separated by 1 m. Calculate the fraction of the electrons in one sphere that must be transferred to the other in order to produce an attractive force of 10^4 N (about a ton) between the spheres. (The number of electrons per atom of silver is 47, and the number of atoms per gram is Avogadro's number divided by the atomic weight of silver, 107.87.)
11. A disk of radius R has a uniform charge per unit area σ. Calculate the electric field along the axis of the disk, a distance x from the center of its surface. (*Hint:* Treat the disk as a set of concentric rings.)
12. A piece of aluminum foil of mass 5×10^{-2} kg is suspended by a string in an electric field directed vertically upward. If the charge on the foil is 3 μC, find the strength of the field that will reduce the tension in the string to zero.
13. A continuous line of charge lies along the x axis, extending from $x = +x_0$ to $+\infty$. This line carries a uniform linear charge density λ_0. What are the magnitude and direction of the electric field at the origin?

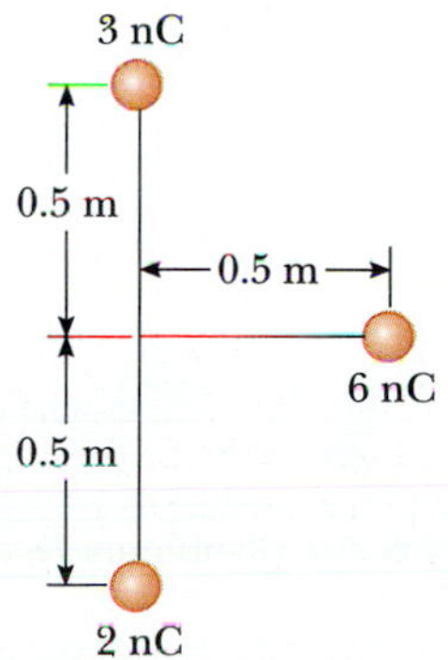

Figure 16.37 (Problem 9)

14. A proton accelerates from rest in a uniform electric field of 640 N/C. At some later time, its speed is 1.20×10^6 m/s. (a) Find the acceleration of the proton. (b) How long does it take the proton to reach this speed? (c) How far has it moved in this time? (d) What is its kinetic energy at this time?
15. Each of the electrons in a particle beam has a kinetic energy of 1.6×10^{-17} J. What are the magnitude and direction of the electric field that will stop these electrons in a distance of 10 cm?
16. An electric dipole is located along the x axis as in Figure 16.38. Show that the electric field at a distant point along the x axis is given by the expression $E_x = 4k_e qa/x^3$.

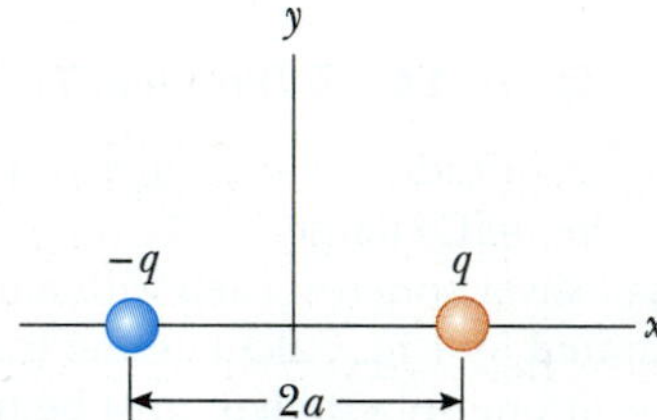

Figure 16.38 (Problem 16)

17. In Figure 16.39, determine the point (other than infinity) at which the electric field is zero.

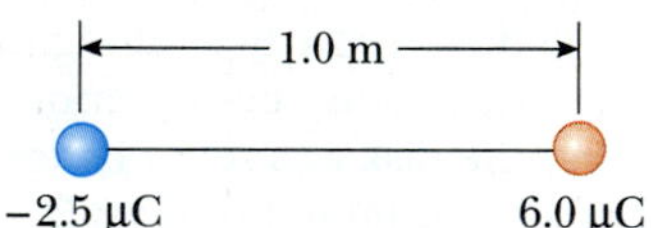

Figure 16.39 (Problem 17)

18. Three identical point charges ($q = +2.7\ \mu$C) are placed on the corners of an equilateral triangle with 35-cm sides (see Fig. 16.40). What is the magnitude of the resultant electric field at the center of the triangle?

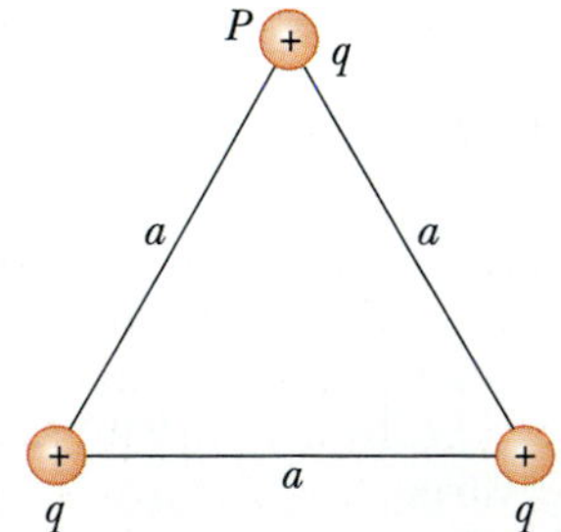

Figure 16.40 (Problems 18 and 19)

19. Three equal positive charges, q, are at the corners of an equilateral triangle with sides of length a, as in Figure 16.40. (a) At what points in the plane of the charges (other than ∞) is the electric field zero? (b) What are the magnitude and direction of the electric field at the point P due to the two charges at the base of the triangle?
20. A rod 14 cm long is uniformly charged and has a total charge of $-22\ \mu$C. Determine the magnitude and direction of the electric field along the axis of the rod, at a point 36 cm from the center of the rod.
21. A uniformly charged ring of radius 10 cm has a total charge of 75 μC. Find the electric field on the *axis* of the ring at (a) 1 cm, (b) 5 cm, (c) 30 cm, and (d) 100 cm from the center of the ring.
22. Show that the maximum field strength, E_m, along the axis of a uniformly charged ring occurs at $x = a/\sqrt{2}$ (see Fig. 16.12) and has the value $Q/(6\sqrt{3}\,\pi\epsilon_0 a^2)$.
23. A sphere of radius 4 cm has a net charge of $+39\ \mu$C. (a) If this charge is uniformly distributed throughout the volume of the sphere, what is the volume charge density? (b) If this charge is uniformly distributed on the sphere's surface, what is the surface charge density?
24. A 10-g piece of Styrofoam carries a net charge of $-0.7\ \mu$C and "floats" above the center of a very large horizontal sheet of plastic which has a uniform charge density on its surface. What is the charge per unit area on the plastic sheet?
25. A uniformly charged insulating rod of length 14 cm is bent into the shape of a semicircle, as in Figure 16.41. If the rod has a total charge of $-7.5\ \mu$C, find the magnitude and direction of the electric field at O, the center of the semicircle.

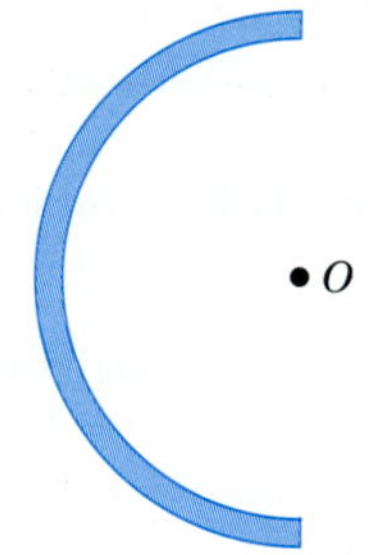

Figure 16.41 (Problem 25)

Section 16.7 Electric Flux

26. A flat surface having an area of 3.2 m² is rotated in a uniform electric field of intensity $E = 6.2 \times 10^5$ N/C. Calculate the electric flux through this area when the electric field (a) is perpendicular to the surface; (b) is parallel to the surface; (c) makes an angle of 75° with the plane of the surface.
27. An electric field of intensity 3.5×10^3 N/C is applied

along the x axis. Calculate the electric flux through a rectangular plane 0.35 m wide and 0.70 m long if (a) the plane is parallel to the yz plane; (b) the plane is parallel to the xy plane; (c) the plane contains the y axis, and its normal makes an angle of 40° with the x axis.

28. Consider a closed triangular box resting within a horizontal electric field, $E = 7.8 \times 10^4$ N/C, as shown in Figure 16.42. Calculate the electric flux through (a) the left-hand vertical surface (A'), (b) the slanted surface (A), and (c) the entire surface of the box.

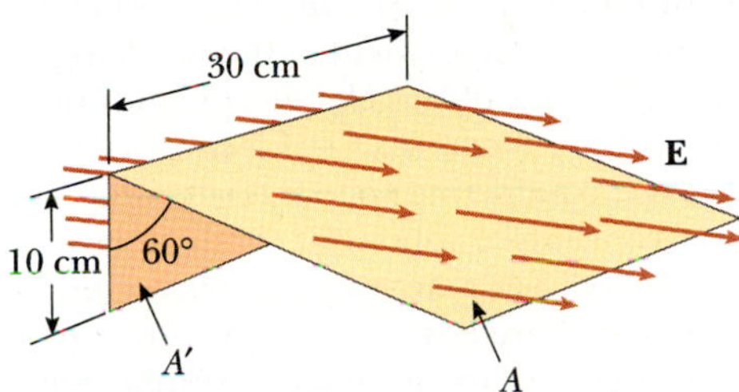

Figure 16.42 (Problem 28)

29. A 40-cm-diameter loop is rotated in a uniform electric field until the position of maximum electric flux is found. The flux in this position is measured to be 5.2×10^5 Nm²/C. What is the electric field strength?

30. A pyramid with a 6-m square base and height of 4 m is placed in a vertical electric field of 52 N/C. Calculate the total electric flux through the pyramid's four slanted surfaces.

31. A cone with a circular base of radius R stands upright so that its axis is vertical. A uniform electric field, **E**, is applied in the vertical direction. Show that the flux through the cone's surface (not counting its base) is given by $\pi R^2 E$.

Section 16.8 Gauss's Law

32. A point charge of $+5\ \mu$C is located at the center of a sphere with a radius of 12 cm. What is the electric flux through the surface of this sphere?

33. The electric field in the Earth's atmosphere is $E =$ 100 N/C, pointing downward. Determine the electric charge on the Earth.

34. A charge of 12 μC is at the geometric center of a cube. What is the electric flux through one of the cube's faces?

35. Four closed surfaces, S_1 through S_4, together with the charges $-2Q$, $+Q$, and $-Q$, are sketched in Figure 16.43. Find the electric flux through each surface.

36. A point charge of 0.0462 μC is inside a pyramid.

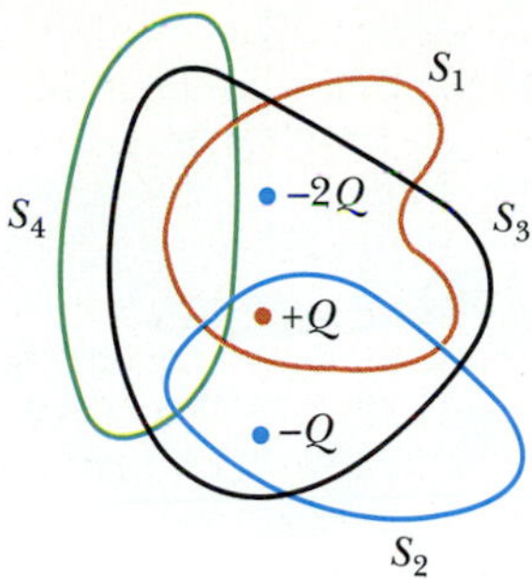

Figure 16.43 (Problem 35)

Determine the total electric flux through the surface of the pyramid.

37. The following charges are located inside a submarine: $+5\ \mu$C, $-9\ \mu$C, $+27\ \mu$C, and $-84\ \mu$C. Calculate the net electric flux through the submarine hull. Compare the number of electric field lines leaving the submarine with the number entering it.

38. The electric field everywhere on the surface of a thin spherical shell of radius 0.75 m is measured to be equal to 8.90×10^2 N/C and points radially toward the center of the sphere. (a) What is the net charge within the sphere's surface? (b) What can you conclude about the nature and distribution of the charge inside the spherical shell?

39. A thin, spherical shell of radius a has a total charge of Q distributed uniformly over its surface (Fig. 16.44). Find the electric field at points inside and outside the shell.

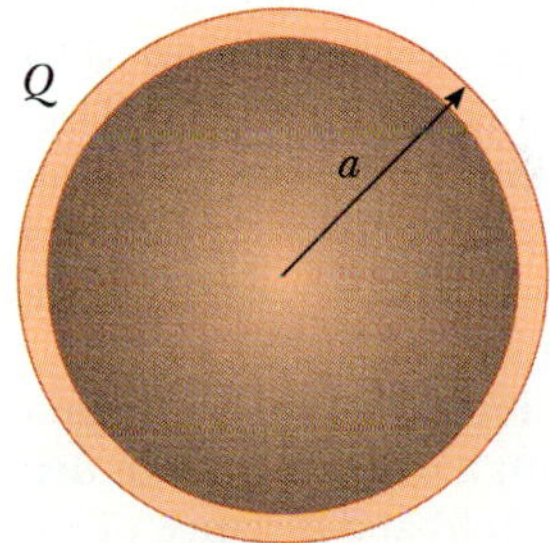

Figure 16.44 (Problem 39)

40. The total electric flux through a closed surface in the shape of a cylinder is 8.60×10^4 N·m²/C. (a) What is the net charge within the cylinder? (b) From the information given, what can you say about the charge within the cylinder? (c) How would your answers to (a) and (b) change if the net flux were -8.60×10^4 N·m²/C?

41. A point charge, Q, is located just above the center of the flat face of a hemisphere of radius R, as shown in Figure 16.45. (a) What is the electric flux through the

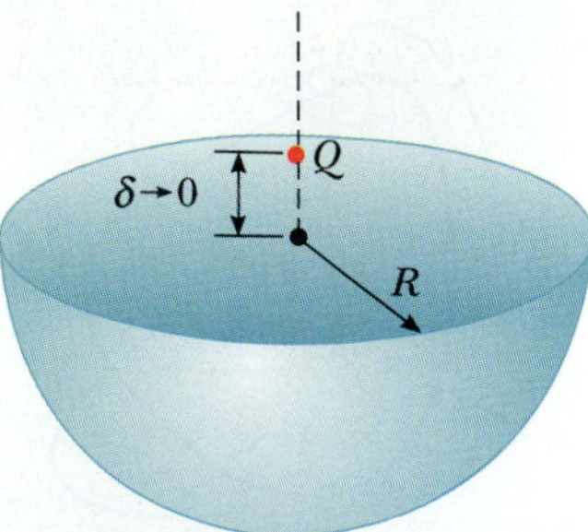

Figure 16.45 (Problem 41)

curved surface of this hemisphere? (b) What is the electric flux through the flat face of this hemisphere?

Section 16.9 Application of Gauss's Law to Charged Insulators

42. A solid sphere of radius 40 cm has a total positive charge of 26 μC uniformly distributed throughout its volume. Calculate the electric field intensity at the following distances from the center of the sphere: (a) 0 cm, (b) 10 cm, (c) 40 cm, (d) 60 cm.
43. An insulating sphere of radius 10 mm has a uniform charge density of 6×10^{-3} C/m^3. Calculate the electric flux through a concentric spherical surface with the following radii: (a) $r = 5$ mm, (b) $r = 10$ mm, (c) $r = 25$ mm.
44. A cylindrical shell of radius 7 cm and length 240 cm has its charge uniformly distributed on its surface. The electric field intensity at a point 19 cm radially outward from its axis (measured from the midpoint of the shell) is 3.6×10^4 N/C. Use approximate relations to find (a) the net charge on the shell and (b) the electric field at a point 4 cm from the axis, measured radially outward from the midpoint of the shell.
45. Consider a long cylindrical charge distribution of radius R with a uniform charge density of ρ. Find the electric field at a distance of r from the axis, where $r < R$.
46. A nonconducting wall carries a uniform charge density of 8.6 μC/cm^2. What is the electric field at a distance of 7 cm from the wall? Does the result change as the distance from the wall is varied?

Section 16.10 Conductors in Electrostatic Equilibrium

47. A conducting spherical shell of radius 15 cm carries a net charge of -6.4 μC uniformly distributed on its surface. Find the electric field at points (a) just outside the shell and (b) inside the shell.
48. A long, straight metal rod has a radius of 5 cm and a charge per unit length of 30 nC/m. Find the electric field at the following distances from the axis of the rod: (a) 3 cm, (b) 10 cm, (c) 100 cm.
49. A square plate of copper with 50-cm sides is placed in an extended electric field of 8×10^4 N/C directed *perpendicularly* to the plate. Find (a) the charge density of each face of the plate and (b) the total charge on each face.
50. A solid, conducting sphere of radius 2 cm has a positive charge of $+8$ μC. A conducting spherical shell of inner radius 4 cm and outer radius 5 cm is concentric with the solid sphere and has a net charge of -4 μC. Find the electric field at the following distances from the center of this charge configuration: (a) $r =$ 1 cm, (b) $r =$ 3 cm, (c) $r =$ 4.5 cm, (d) $r =$ 7 cm.
51. A hollow, conducting sphere is surrounded by a larger concentric, spherical, conducting shell. The inner sphere has a net negative charge of $-Q$, and the outer sphere has a net positive charge of $+3Q$. The charges are in electrostatic equilibrium. Using Gauss's law, find the charges and the electric fields everywhere.
52. A *long*, straight wire is surrounded by a hollow metallic cylinder whose axis coincides with that of the wire. The solid wire has a charge per unit length of $+\lambda$, and the hollow cylinder has a *net* charge per unit length of $+2\lambda$. From this information, use Gauss's law to find (a) the charge per unit length on the inner and outer surfaces of the hollow cylinder and (b) the electric field outside the hollow cylinder, a distance of r from the axis.
53. The electric field on the surface of an irregularly shaped conductor varies from 5.6×10^4 N/C to 2.8×10^4 N/C. Calculate the local surface charge density at the point on the surface where the radius of curvature of the surface is (a) greatest and (b) smallest.
54. A Geiger counter is like an electroscope that discharges whenever ions formed by a radioactive particle produce a conducting path. A typical Geiger counter consists of a thin, conducting wire of radius 0.002 cm stretched along the axis of a conducting cylinder of radius 2.0 cm. The wire and the cylinder carry equal and opposite charges of 8×10^{-10} C uniformly distributed along their length of 10.0 cm. What, therefore, is the magnitude of the electric field at the surface of the wire?

Additonal Problems

55. Three point charges are aligned along the x axis as shown in Figure 16.46. Find the electric field at (a) the position (2, 0) and (b) the position (0, 2).
56. Three point charges lie along the y axis. A charge of

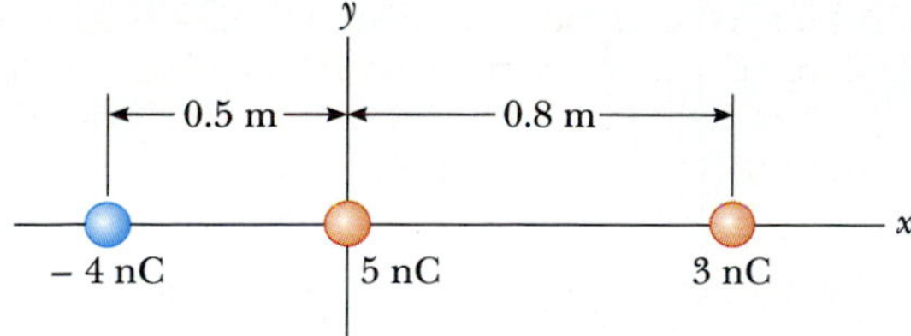

Figure 16.46 (Problem 55)

$q_1 = -9\ \mu C$ is at $y = 6.0$ m, and a charge of $q_2 = -8\ \mu C$ is at $y = -4.0$ m. Where must a third positive charge, q_3, be placed so that the resultant force on it is zero?

57. A small, 2-g plastic ball is suspended by a 20-cm-long string in a uniform electric field, as shown in Figure 16.47. If the ball is in equilibrium when the string makes a 15° angle with the vertical as indicated, what is the net charge on the ball?

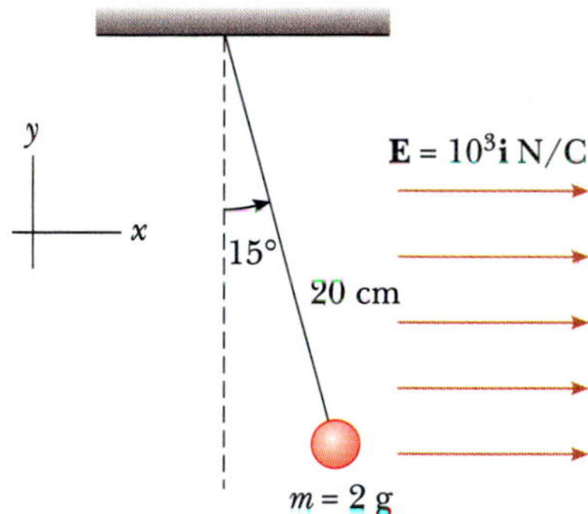

Figure 16.47 (Problem 57)

58. Two small spheres, each of mass 2 g, are suspended by light strings 10 cm in length (Fig. 16.48). A uniform electric field is applied in the x direction. If the spheres have charges of -5×10^{-8} C and $+5 \times 10^{-8}$ C, determine the electric field intensity that enables the spheres to be in equilibrium at $\theta = 10°$.

59. A charged cork ball of mass 1 g is suspended on a light string in the presence of a uniform electric field, as shown in Figure 16.49. The electric field has components $E_x = 3 \times 10^5$ N/C and $E_y = 5 \times 10^5$ N/C. The ball is in equilibrium at $\theta = 37°$. Find (a) the charge on the ball and (b) the tension in the string.

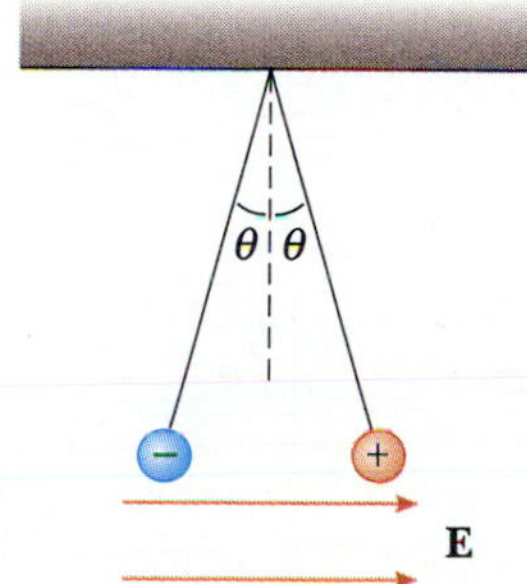

Figure 16.48 (Problem 58)

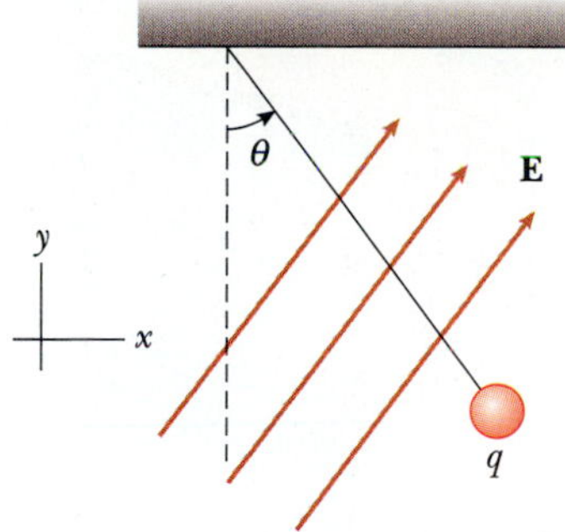

Figure 16.49 (Problem 59)

60. Four identical point charges ($q = +10\ \mu C$) are located on the corners of a rectangle, as shown in Figure 16.50. The dimensions of the rectangle are $L = 60$ cm and $W = 15$ cm. Calculate the magnitude and direction of the net electrostatic force exerted on the charge at the lower left corner of the rectangle by the other three charges.

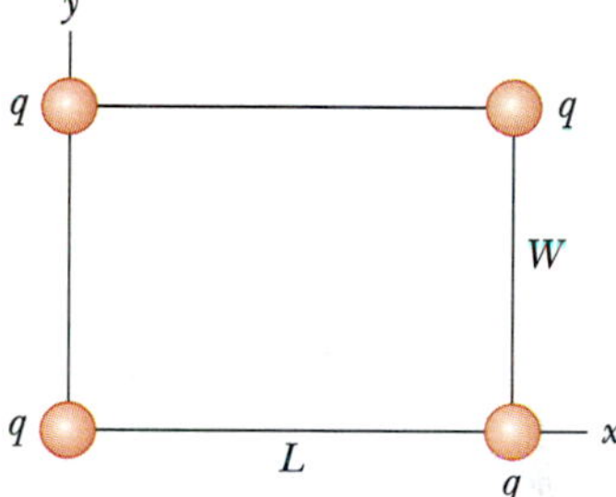

Figure 16.50 (Problem 60)

61. A solid, *insulating* sphere of radius a has a uniform charge density of ρ and a total charge of Q. Concentric with this sphere is an *uncharged, conducting* hollow sphere whose inner and outer radii are b and c, as in Figure 16.51. (a) Find the electric field intensity in

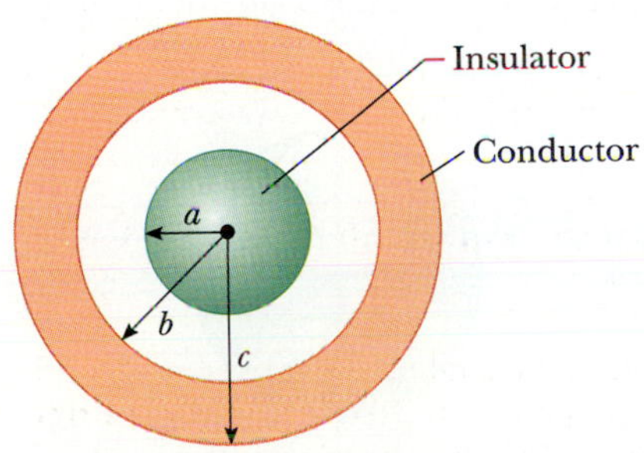

Figure 16.51 (Problem 61)

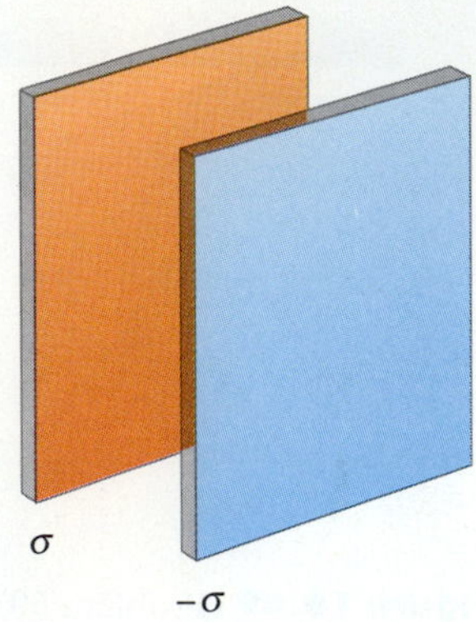

Figure 16.52 (Problems 65 and 66)

the regions $r < a$, $a < r < b$, $b < r < c$, and $r > c$. (b) Determine the induced charge per unit area on the inner and outer surfaces of the hollow sphere.

62. The nucleus of lead-208, $^{208}_{82}\text{Pb}$, has 82 protons within a sphere of radius 6.34×10^{-15} m. Each proton has a charge of 1.6×10^{-19} C. Calculate the electric field at the surface of the nucleus.
63. At the point of fission, visualize a nucleus of uranium-238 (with 92 protons) as two smaller spheres in contact, each with 46 protons and a radius of 5.9×10^{-15} m. What is the repulsive force pushing the two spheres apart? (The mass of the ^{238}U nucleus is 3.98×10^{-25} kg.)
64. Consider a solid insulating sphere of radius b with nonuniform charge density $\rho = Cr$. Find the charge contained within the radius (a) when $r < b$ and (b) when $r > b$. (*Note:* The volume element dV for a spherical shell of radius r and thickness dr is equal to $4\pi r^2\, dr$.)
65. Two infinite, nonconducting sheets of charge are parallel to each other as in Figure 16.52. The sheet on the left has a uniform surface charge density of σ, and the one on the right has a uniform charge density of $-\sigma$. Calculate the value of the electric field at points (a) to the left of, (b) in between, and (c) to the right of the two sheets. (*Hint:* See Example 16.10.)
66. Repeat the calculations for Problem 65 when both sheets have *positive* uniform charge densities of value σ.
67. A sphere of radius $2a$ is made of an insulator that has a uniform volume charge density of ρ. (Assume that the material does not affect the electric field.) A spherical cavity of radius a is now removed from the sphere, as shown in Figure 16.53. Show that the electric field within the cavity is uniform and is given by $E_x = 0$ and $E_y = \rho a/3\epsilon_0$. (*Hint:* The field within the cavity is the superposition of the field due to the original uncut sphere, plus the field due to a sphere the size of the cavity with the uniform negative charge density $-\rho$.)

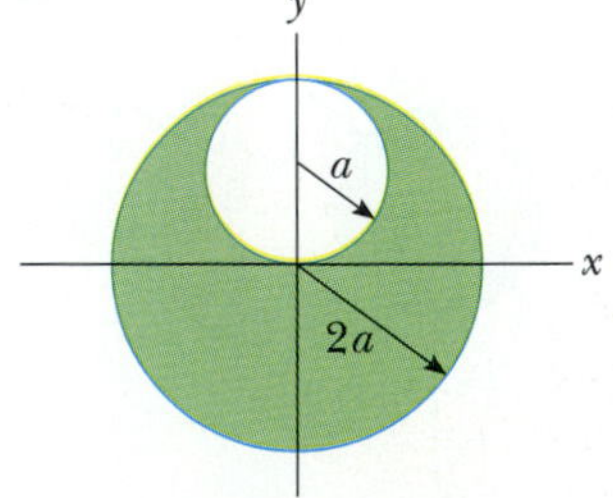

Figure 16.53 (Problem 67)

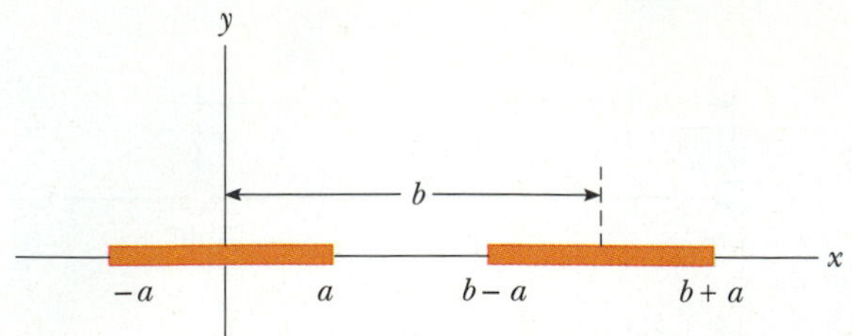

Figure 16.54 (Problem 69)

68. Three identical small Styrofoam balls ($m = 2$ g) are suspended from a fixed point by three nonconducting threads, each 50 cm long and with negligible mass. At equilibrium the three balls form an equilateral triangle with 30-cm sides. What is the common charge, q, that is carried by each ball?
69. Identical thin rods of length $2a$ carry equal charges, $+Q$, uniformly distributed along their lengths. The rods lie along the x axis with their centers separated by a distance of $b > 2a$ (Fig. 16.54). Show that the force exerted on the right rod is given by

$$F = \left(\frac{k_e Q^2}{4a^2}\right) \ln\left(\frac{b^2}{b^2 - 4a^2}\right)$$

70. A line of positive charge is formed into a semicircle of radius $R = 60$ cm, as shown in Figure 16.55. The charge per unit length along the semicircle is described by the expression

$$\lambda = \lambda_0 \cos\theta$$

The total charge on the semicircle is 12 μC. Calculate the total force on a charge of 3 μC placed at the center of curvature of the semicircle.

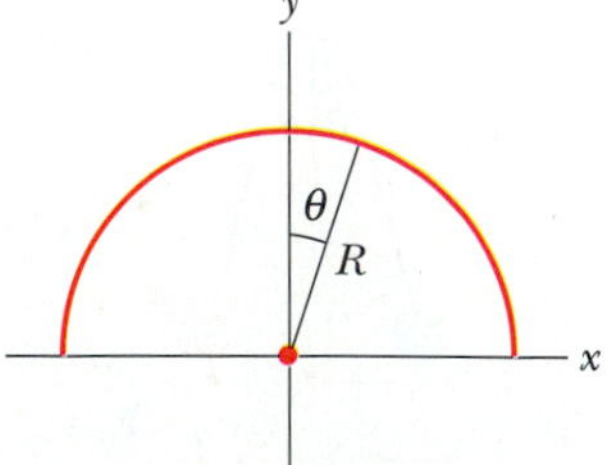

Figure 16.55 (Problem 70)

Jennifer holds onto a charged sphere that reaches a potential of about 100 000 volts. The device that generates this high potential is called a Van de Graaff generator. Why do you suppose Jennifer's hair stands on end like the needles of a porcupine? Why is it important that she stand on a pedestal, insulated from ground? *(Courtesy of Henry Leap and Jim Lehman)*

Electric Potential and Capacitance

CHAPTER 17

The concept of potential energy was introduced in Chapter 8 in connection with such conservative forces as the force of gravity, the elastic force of a spring, and the electrostatic force. By using the principle of conservation of energy, we were often able to avoid working directly with forces when solving mechanical problems. In this chapter we shall use the energy concept in our study of electricity. Because the electrostatic force (given by Coulomb's law) is conservative, electrostatic phenomena can conveniently be described in terms of an *electric* potential energy function. This concept enables us to define a quantity called *electric potential*, which is a scalar function of position and thus leads to a simpler method of describing some electrostatic phenomena than the electric field method. Although the potential is clearly an easier path for many problems, one often requires **E**. In some situations, a knowledge of **E** provides the simplest path to calculation of the potential. As we shall see in subsequent chapters, the concept of electric potential is of great practical value. For example, the measured voltage between any two points in an electrical circuit is simply the difference in electric potential between the points.

This chapter is also concerned with the properties of capacitors, devices that store charge. Capacitors are commonly used in a variety of electrical circuits. For instance, they are used to tune the frequency of radio receivers, as filters in power supplies, to eliminate unwanted sparking in automobile ignition systems, and as energy-storing devices in electronic flash units.

A capacitor basically consists of two conductors separated by an insulator. We shall see that the capacitance of a given device depends on its geometry and on the material called a *dielectric,* separating the charged conductors.

17.1 Potential Difference and Electric Potential

When a test charge, q_0, is placed in an electrostatic field, $\mathbf{E}$, the electric force on the charge is $q_0\mathbf{E}$. This force is the vector sum of the individual forces exerted on q_0 by the various charges producing the field $\mathbf{E}$. It follows that the force $q_0\mathbf{E}$ is conservative, since the individual forces governed by Coulomb's law are conservative. When a charge is moved within an electric field, the work done on q_0 by the electric field is equal to the negative of the work done by the external agent causing the displacement. For an infinitesmal displacement, $d\mathbf{s}$, the work done by the electric field is $\mathbf{F}\cdot d\mathbf{s} = q_0\mathbf{E}\cdot d\mathbf{s}$. This decreases the potential energy of the electric field by an amount $dU = -q_0\mathbf{E}\cdot d\mathbf{s}$. For a finite displacement of the test charge between points A and B, the **change in potential energy** is

Change in potential energy

$$\Delta U = U_B - U_A = -q_0 \int_A^B \mathbf{E}\cdot d\mathbf{s} \qquad [17.1]$$

The integral in Equation 17.1 is performed along the path by which q_0 moves from A to B and is called either a *path integral* or a *line integral.* Since the force $q_0\mathbf{E}$ is conservative, *this integral does not depend on the path taken between A and B.*

The potential energy per unit charge, U/q_0, is independent of the value of q_0 and has a unique value at every point in an electric field. The quantity U/q_0 is called the **electric potential** (or simply the **potential**), V. Thus, the electric potential at any point in an electric field is

$$V = \frac{U}{q_0} \qquad [17.2]$$

Because potential energy is a scalar, electric potential is also a scalar quantity.

The **potential difference,** $\Delta V = V_B - V_A$, between the points A and B is defined as the change in potential energy divided by the test charge q_0:

Potential difference

$$\Delta V = \frac{\Delta U}{q_0} = -\int_A^B \mathbf{E}\cdot d\mathbf{s} \qquad [17.3]$$

Potential difference should not be confused with potential energy. The potential difference is *proportional* to the potential energy, and we see from Equation 17.3 that the two are related by $\Delta U = q_0\, \Delta V$. The potential is characteristic of the field, independent of the charges that may be placed in the field. The potential energy resides in the field, also. However, because we are interested in the potential at the location of a charge, and in the potential energy caused by the interaction of the charge with the field, we follow the common convention of speaking of the potential energy *as if* it belonged to the charge, except when we specifically require information about transfer of energy between the field and the charge.

Because, as already noted, the change in the potential energy of the charge is the negative of the work done by the electric force, the potential difference, ΔV, equals the work per unit charge that an external agent must perform to move a test charge from A to B without a change in kinetic energy.

Equation 17.3 defines potential difference only. That is, only *differences* in V are meaningful. The electric potential function is often taken to be zero at some convenient point. We shall usually set the potential at zero for a point at infinity (that is, a point infinitely remote from the charges producing the electric field). With this choice, we can say that the *electric potential at an arbitrary point equals the work required per unit charge to bring a positive test charge from infinity to that point.* Thus, if we take $V_A = 0$ at infinity in Equation 17.3, then the potential at any point P is

$$V_P = -\int_{\infty}^{P} \mathbf{E} \cdot d\mathbf{s} \qquad [17.4]$$

In reality, V_P represents the potential difference between the point P and a point at infinity. (Equation 17.4 is a special case of Equation 17.3.)

Since potential difference is a measure of energy per unit charge, the SI units of potential are joules per coulomb, called the **volt** (V):

Definition of a volt

$$1 \text{ V} \equiv 1 \text{ J/C}$$

That is, 1 J of work must be done to take a 1-C charge through a potential difference of 1 V. Equation 17.3 shows that the potential difference also has the same units as electric field times distance. From this, it follows that the SI units of electric field, newtons per coulomb, can also be expressed as volts per meter.

$$1 \text{ N/C} = 1 \text{ V/m}$$

As we learned in Chapter 10, Section 10.7, a unit of energy commonly used in atomic and nuclear physics is the **electron volt** (eV):

The electron volt

$$1 \text{ eV} = 1.60 \times 10^{-19} \text{ C}\cdot\text{V} = 1.60 \times 10^{-19} \text{ J} \qquad [17.5]$$

For instance, an electron in the beam of a typical TV picture tube has a speed of 5×10^7 m/s. This corresponds to a kinetic energy of 1.1×10^{-15} J, which is equivalent to 7.1×10^3 eV. Such an electron has to be accelerated from rest through a potential difference of 7.1 kV to reach this speed.

17.2 Potential Differences in a Uniform Electric Field

In this section we shall describe the potential difference between any two points in a *uniform* electric field. The potential difference is independent of the path between the two points; that is, the work done in taking a test charge from point A to point B is the same along all paths. This confirms that a static, uniform electric field is conservative.

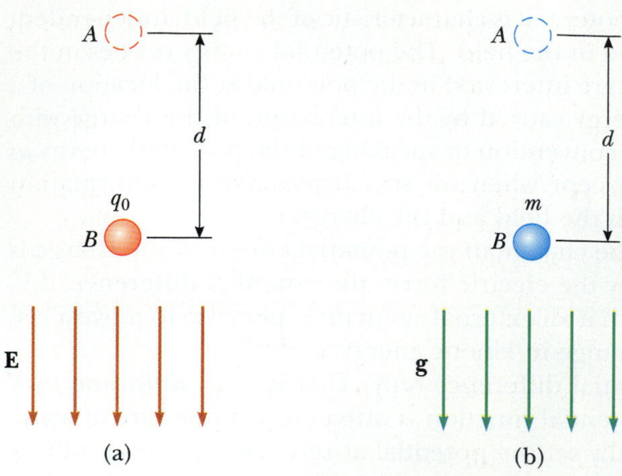

Figure 17.1
(a) When the electric field, **E**, is directed downward, point B is at a lower electric potential than point A. A positive test charge that moves from A to B loses electric potential energy. (b) A mass, m, moving downward in the direction of the gravitational field, **g**, loses gravitational potential energy.

First, consider a uniform electric field directed along the negative y axis, as in Figure 17.1a. Let us calculate the potential difference between two points, A and B, separated by the distance d, where d is measured parallel to the field lines. If we apply Equation 17.3 to this situation, we get

$$V_B - V_A = \Delta V = -\int_A^B \mathbf{E} \cdot d\mathbf{s} = -\int_A^B E \cos 0^\circ \, ds = -\int_A^B E \, ds$$

Since E is constant, it can be removed from the integral sign, giving

Potential difference in a uniform E field

$$\Delta V = -E \int_A^B ds = -Ed \qquad [17.6]$$

The minus sign results from the fact that point B is at a lower potential than point A; that is, $V_B < V_A$. In general, *electric field lines always point in the direction of decreasing electric potential,* as shown in Figure 17.1a.

Now suppose that a test charge, q_0, moves from A to B. The change in its electric potential energy can be found from Equations 17.3 and 17.6:

$$\Delta U = q_0 \, \Delta V = -q_0 E d \qquad [17.7]$$

From this result, we see that if q_0 is positive, ΔU is negative. This means that *an electric field does work on a positive charge when the positive charge moves in the direction of the electric field.* (This is analogous to the work done by the gravitational field on a falling mass, as shown in Fig. 17.1b.) If a positive test charge is released from rest in the electric field, it experiences an electric force, $q_0\mathbf{E}$, in the direction of **E** (downward in Fig. 17.1a). Therefore, it accelerates downward, gaining kinetic energy. *As the charged particle gains kinetic energy, it loses an equal amount of potential energy.*

If q_0 is negative, then ΔU is positive and the situation is reversed. A *negative charge gains electric potential energy when it moves in the direction of the electric field.* If a negative charge is released from rest in the field **E**, it accelerates in a direction opposite the electric field.

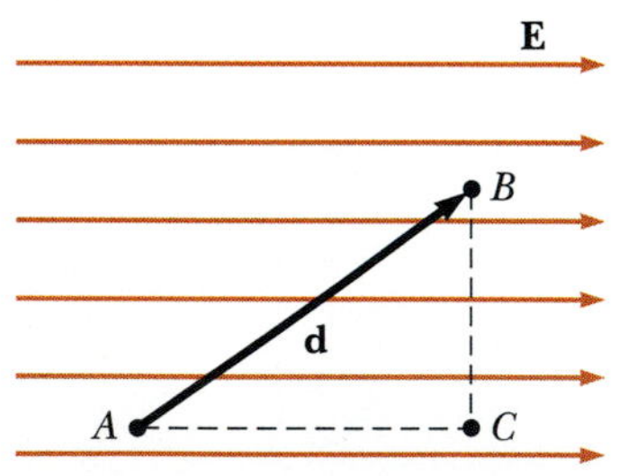

Figure 17.2
A uniform electric field directed along the positive x axis. Point B is at a lower electric potential than point A. Points B and C are at the *same* electric potential.

Now consider the more general case of a charged particle moving between any two points in a uniform electric field directed along the x axis, as in Figure 17.2. If **d** represents the displacement vector between points A and B, Equation 17.3 gives

$$\Delta V = -\int_A^B \mathbf{E} \cdot d\mathbf{s} = -\mathbf{E} \cdot \int_A^B d\mathbf{s} = -\mathbf{E} \cdot \mathbf{d} \qquad [17.8]$$

where again we can remove E from the integral, since it is constant. Furthermore, the change in electric potential energy of the charge is

$$\Delta U = q_0\,\Delta V = -q_0\,\mathbf{E}\cdot\mathbf{d} \qquad [17.9]$$

Finally, our results show that all points in a plane *perpendicular* to a uniform electric field are at the same potential. This can be seen in Figure 17.2, where the potential difference $V_B - V_A$ is *equal* to $V_C - V_A$. Therefore, $V_B = V_C$.

An equipotential surface

The name **equipotential surface** is given to any surface consisting of a continuous distribution of points having the same electric potential.

Note that since $\Delta U = q_0\,\Delta V$, *no* work is done in moving a test charge between any two points on an equipotential surface. The equipotential surfaces of a uniform electric field consist of a family of planes, all *perpendicular* to the field. Equipotential surfaces for fields with other symmetries will be described in later sections.

Example 17.1 The Electric Field Between Two Parallel Plates of Opposite Charge

A 12-V battery is connected between two parallel plates as in Figure 17.3. The separation between the plates is 0.3 cm, and the electric field is assumed to be uniform. (This assumption is reasonable if the plate separation is small relative to the plate size and if we do not consider points near the edges of the plates.) Find the magnitude of the electric field between the plates.

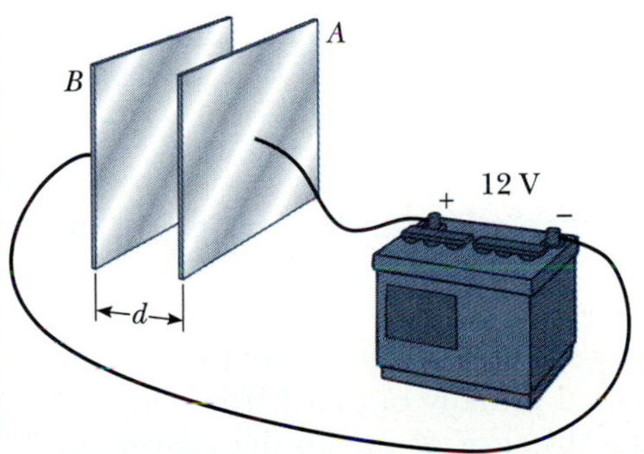

Figure 17.3 (Example 17.1) A 12-V battery connected to two parallel plates. The electric field between the plates has a magnitude given by the potential difference divided by the plate separation, *d*.

Solution The electric field is directed from the positive plate toward the negative plate. Since field lines always point toward decreasing potential, we know that the positive plate is at a *higher* potential than the negative plate. Note that the potential difference between the plates must equal the potential difference between the battery terminals. This can be understood by noting that all points on a conductor in equilibrium are at the same potential, and hence there is no potential difference between a terminal of the battery and any portion of the plate to which it is connected. Therefore, the magnitude of the electric field between the plates is

$$E = \frac{|V_B - V_A|}{d} = \frac{12\text{ V}}{0.3 \times 10^{-2}\text{ m}} = 4.0 \times 10^3\text{ V/m}$$

This configuration, called a *parallel-plate capacitor,* will be examined in more detail later in this chapter.

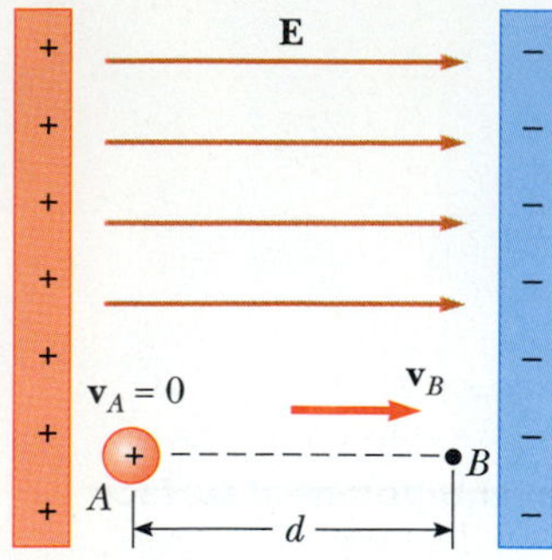

Figure 17.4 (Example 17.2) A proton accelerates from A to B in the direction of the uniform electric field.

Example 17.2 Motion of a Proton in a Uniform Electric Field

A proton is released from rest in a uniform electric field of magnitude 8×10^4 V/m, directed along the positive x axis (Fig. 17.4). The proton undergoes a displacement of 0.5 m in the direction of the field.

(a) Find the *change* in the electric potential of the proton as a result of this displacement.

Solution From Equation 17.6, we have

$$\Delta V = V_B - V_A = -Ed = -(8 \times 10^4 \text{ V/m})(0.5 \text{ m}) = -4 \times 10^4 \text{ V}$$

Thus, the electric potential of the proton *decreases* as it moves from A to B.

(b) Find the change in electric potential energy of the proton for this displacement.

Solution

$$\Delta U = q_0 \, \Delta V = e \, \Delta V = (1.6 \times 10^{-19} \text{ C})(-4 \times 10^4 \text{ V}) = -6.4 \times 10^{-15} \text{ J}$$

The negative sign here means that the electric potential energy of the proton decreases as it moves in the direction of the electric field. This makes sense since, as the proton accelerates in the direction of the field, it gains kinetic energy and at the same time the field loses potential energy (energy is conserved).

Exercise Apply the principle of energy conservation to find the speed of the proton after it has moved 0.5 m, starting from rest.

Answer $v_f = 2.77 \times 10^6$ m/s.

17.3 Electric Potential and Electric Potential Energy Due to Point Charges

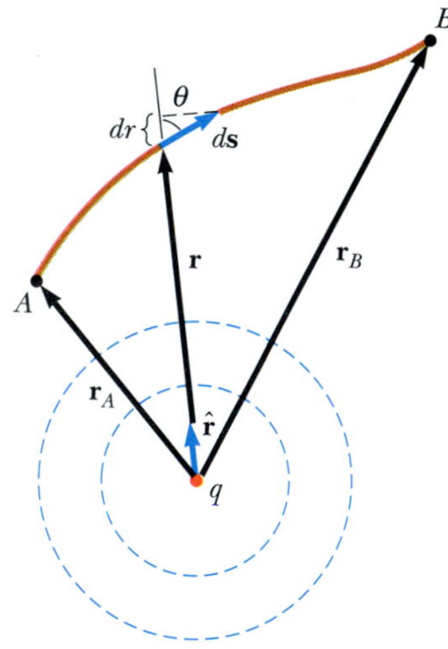

Figure 17.5 The potential difference between points A and B due to the point charge q depends *only* on the initial and final radial coordinates, $\mathbf{r}_A$ and $\mathbf{r}_B$. The two dashed circles represent equipotential surfaces.

Consider an isolated positive point charge, q (Fig. 17.5). Recall that such a charge produces an electric field that is directed radially outward from the charge. In order to find the electric potential at a distance of r from the charge, we begin with the general expression for potential difference:

$$V_B - V_A = -\int_A^B \mathbf{E} \cdot d\mathbf{s}$$

Since the electric field due to the point charge is given by $\mathbf{E} = kq\hat{\mathbf{r}}/r^2$, where $\hat{\mathbf{r}}$ is a unit vector directed from the charge to the field point, the quantity $\mathbf{E} \cdot d\mathbf{s}$ can be expressed as

$$\mathbf{E} \cdot d\mathbf{s} = k_e \frac{q}{r^2} \hat{\mathbf{r}} \cdot d\mathbf{s}$$

The dot product $\hat{\mathbf{r}} \cdot d\mathbf{s}$ equals $ds \cos \theta$, where θ is the angle between $\hat{\mathbf{r}}$ and $d\mathbf{s}$ as in Figure 17.5. Furthermore, note that $ds \cos \theta$ is the projection of $d\mathbf{s}$ onto $\mathbf{r}$, and so $ds \cos \theta = dr$. That is, any displacement $d\mathbf{s}$ produces the change dr in the magnitude of **r**. With these substitutions, we find that $\mathbf{E} \cdot d\mathbf{s} = (k_e q/r^2)\, dr$, and so the expression for the potential difference becomes

$$V_B - V_A = -\int E_r\, dr = -k_e q \int_{r_A}^{r_B} \frac{dr}{r^2} = \frac{k_e q}{r}\bigg]_{r_A}^{r_B}$$

$$V_B - V_A = k_e q\left[\frac{1}{r_B} - \frac{1}{r_A}\right] \qquad [17.10]$$

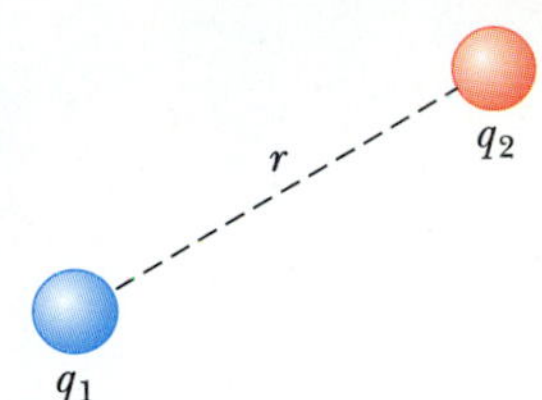

Figure 17.6
If two point charges are separated by the distance r, the potential energy of the pair is $k_e q_1 q_2/r$.

The line integral of $\mathbf{E} \cdot d\mathbf{s}$ is *independent* of the path between A and B—as it must be, because the electric field of a point charge is conservative. Furthermore, Equation 17.10 expresses the important result that the potential difference between any two points A and B depends *only* on the *radial* coordinates r_A and r_B. As we learned in Section 17.1, it is customary to choose the reference of potential to be zero at $r_A = \infty$. With this choice, the electric potential due to a point charge at any distance r from the charge is

Potential of a point charge

$$V = k_e \frac{q}{r} \qquad [17.11]$$

From this we see that V is constant on a spherical surface of radius r. Hence, we conclude that *the equipotential surfaces for an isolated point charge consist of a family of spheres concentric with the charge,* as shown in Figure 17.5. Note that the equipotential surfaces are perpendicular to the lines of electric force, as was the case for a uniform electric field.

The electric potential of two or more point charges is obtained by applying the superposition principle. That is, the total potential at some point P due to several point charges is the sum of the potentials due to the individual charges. For a group of charges, we can write the total potential at P in the form

The potential of several point charges

$$V = k_e \sum_i \frac{q_i}{r_i} \qquad [17.12]$$

where the potential is again taken to be zero at infinity and r_i is the distance from the point P to the charge q_i. Note that the sum in Equation 17.12 is an *algebraic sum* of scalars rather than a vector sum (which is used to calculate the electric field of a group of charges). Thus, it is much easier to evaluate V than to evaluate **E**.

We now consider the electric potential energy of interaction of a system of charged particles. If V_1 is the electric potential due to charge q_1 at point P, then the work required to bring a second charge, q_2, from infinity to point P without acceleration is $q_2 V_1$. By definition, this work equals the potential energy, U, of the two-particle system when the particles are separated by a distance of r (Fig. 17.6). We can therefore express the potential energy as

Electric potential energy of two charges

$$U = q_2 V_1 = k_e \frac{q_1 q_2}{r} \qquad [17.13]$$

which is the same as Equation 8.23 in Chapter 8.

Note that if the charges are of the same sign, U is positive. This is consistent with the fact that like charges repel, and so positive work must be done *on* the system to bring the two charges near one another. If the charges are of opposite sign, the force is attractive and U is negative. This means that negative work must be done to bring the unlike charges near one another.

If the system consists of more than two charged particles, the total electric potential energy can be obtained by calculating U for every pair of charges and summing the terms algebraically. The total electric potential energy of a system of point charges is equal to the work required to bring the charges, one at a time, from an infinite separation to their final positions.

Example 17.3 An Example of Two Charges

Two charges of 2 μC and −6 μC are located at the positions (0, 0) m and (0, 3) m, respectively, as in Figure 17.7.

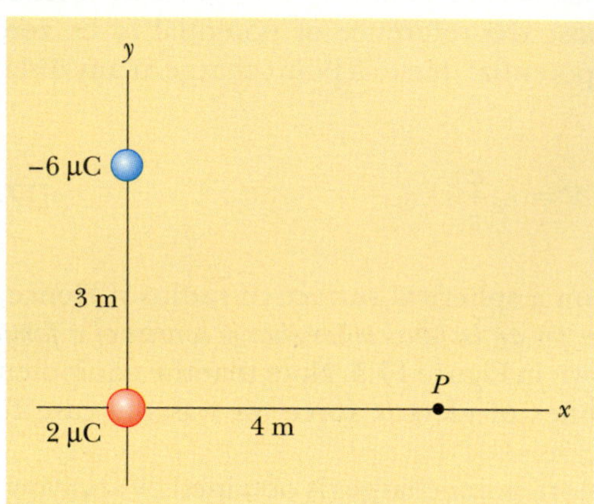

Figure 17.7 (Example 17.3)

(a) Find the total electric potential due to these charges at the point P, located at (4, 0) m.

Solution For two charges, the sum in Equation 17.12 gives

$$V = k_e\left[\frac{q_1}{r_1} + \frac{q_2}{r_2}\right]$$

In this example, $q_1 = 2 \times 10^{-6}$ C, $r_1 = 4$ m, $q_2 = -6 \times 10^{-6}$ C, and $r_2 = 5$ m. Therefore, V at P reduces to

$$V_p = 8.99 \times 10^9\ \text{N}\cdot\frac{\text{m}^2}{\text{C}^2}\left[\frac{2 \times 10^{-6}\ \text{C}}{4\ \text{m}} - \frac{6 \times 10^{-6}}{5\ \text{m}}\right] = -6.29 \times 10^3\ \text{V}$$

(b) How much work is required to bring a 3-μC charge from ∞ to P?

Solution Using Equation 17.13, we find that, since the work required equals the potential energy of the system when the particles are assembled,

$$W = q_3V_p = (3 \times 10^{-6}\ \text{C})(-6.29 \times 10^3\ \text{V}) = -18.9 \times 10^{-3}\ \text{J}$$

The negative sign means that work is done by the field on the charge as it is displaced from ∞ to P. Therefore, positive work would have to be done by an external force to remove the charge from P back to ∞.

Exercise What is the electric potential energy of the system of three charges?

Answer -5.48×10^{-2} J.

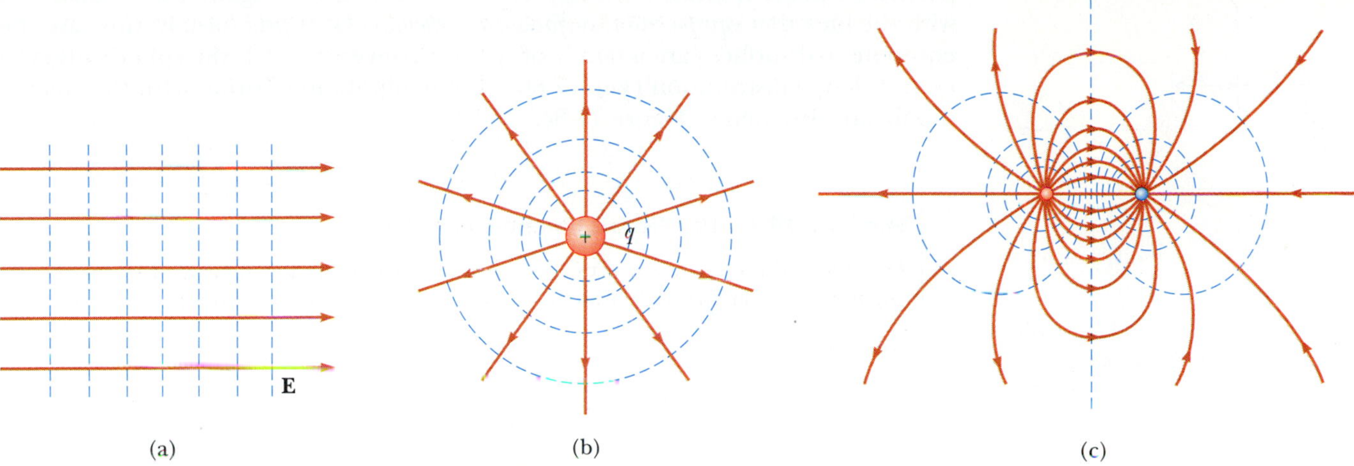

Figure 17.8
Equipotential surfaces (dashed blue lines) and electric field lines (red lines) for (a) a uniform electric field produced by an infinite sheet of charge, (b) a point charge, and (c) an electric dipole. In all cases, the equipotential surfaces are *perpendicular* to the electric field lines at every point.

17.4 Obtaining E from the Electric Potential

The electric field, **E**, and the electric potential, V, are related by Equation 17.3. Both quantities are determined by a specific charge distribution. We now show how to calculate the electric field if the electric potential is known in a certain region.

From Equation 17.3 we can express the potential difference, dV, between two points a distance of ds apart as

$$dV = -\mathbf{E}\cdot d\mathbf{s} \qquad [17.14]$$

If the electric field has only *one* component, E_x, then $\mathbf{E}\cdot d\mathbf{s} = E_x\,dx$. Therefore, Equation 17.14 becomes $dV = -E_x\,dx$, or

$$E_x = -\frac{dV}{dx} \qquad [17.15]$$

That is, the electric field is equal to the negative of the derivative of the electric potential with respect to some coordinate. Note that the potential change is zero for any displacement perpendicular to the electric field. This is consistent with the notion of equipotential surfaces being perpendicular to the field, as in Figure 17.8a.

If the charge distribution has *spherical symmetry, where the charge density depends only on the radial distance, r,* then the electric field is radial. In this case, $\mathbf{E}\cdot d\mathbf{s} = E_r\,dr$, and so we can express dV in the form $dV = -E_r\,dr$. Therefore,

Equipotential surfaces are always perpendicular to the electric field lines

$$E_r = -\frac{dV}{dr} \qquad [17.16]$$

For example, the potential of a point charge is $V = k_eq/r$. Since V is a function of r only, the potential function has spherical symmetry. Applying Equation 17.16, we find that the electric field due to the point charge is $E_r = k_eq/r^2$, a familiar result. Note that the potential changes only in the radial direction, not in a direction

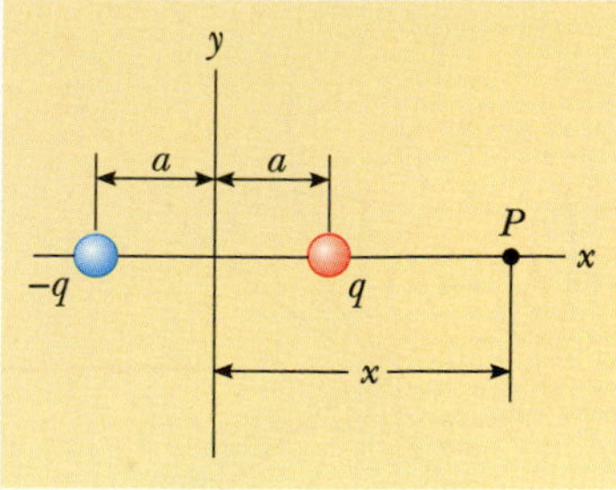

Figure 17.9
(Example 17.4) An electric dipole located on the x axis.

perpendicular to r. Thus, V (like E_r) is a function only of r. Again, this is consistent with the idea that *equipotential surfaces are perpendicular to field lines.* In this case the equipotential surfaces are a family of spheres concentric with the spherically symmetric charge distribution (Fig. 17.8b). The equipotential surfaces for the electric dipole are sketched in Figure 17.8c.

Example 17.4 The Electric Potential of a Dipole

An electric dipole consists of two equal and opposite charges separated by the distance $2a$, as in Figure 17.9. Calculate the electric potential and the electric field at P.

Solution

$$V = k_e \sum \frac{q_i}{r_i} = k_e\left(\frac{q}{x-a} - \frac{q}{x+a}\right) = \frac{2k_e qa}{x^2 - a^2}$$

If P is far from the dipole, so that $x \gg a$, then a^2 can be neglected in the term $x^2 - a^2$, and V becomes

$$V \approx \frac{2k_e qa}{x^2} \qquad (x \gg a)$$

Using Equation 17.15 and this result, we calculate the electric field at P:

$$E_x = -\frac{dV}{dx} = \frac{4k_e qa}{x^3} \qquad \text{for } x \gg a$$

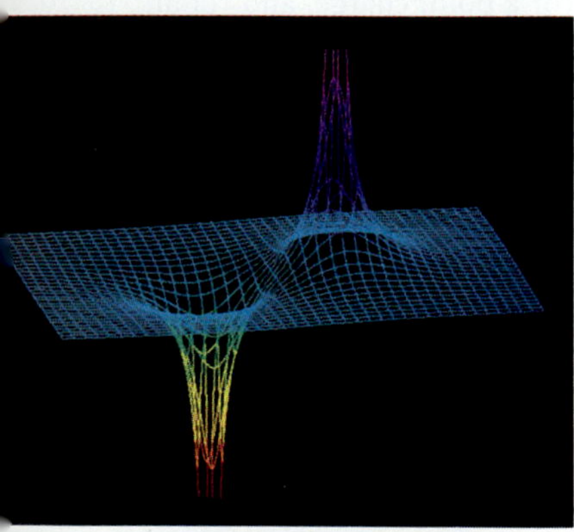

Computer-generated plot of the electric potential associated with an electric dipole. The changes lie in the horizontal plane, at the centers of the potential spikes. The contour lines help visualize the size of the potential, whose values are plotted vertically. *(Richard Megna 1990, Fundamental Photographs)*

17.5 Electric Potential Due to Continuous Charge Distributions

The electric potential due to a continuous charge distribution can be calculated in two ways. If the charge distribution is known, we can start with Equation 17.11 for the potential of a point charge. We then consider the potential due to a small charge element, dq, treating this element as a point charge (Fig. 17.10). The potential dV at some point P due to the charge element dq is

$$dV = k_e \frac{dq}{r} \qquad [17.17]$$

where r is the distance from the charge element to P. To get the total potential at P, we integrate Equation 17.17 to include contributions from all elements of the charge distribution. Since each element is, in general, at a different distance from P and since k_e is a constant, we can express V as

$$V = k_e \int \frac{dq}{r} \qquad [17.18]$$

In effect, we have replaced the sum in Equation 17.12 with an integral. Note that this expression for V uses a particular reference: the potential is taken to be zero when P is infinitely far from the charge distribution.

The second method for calculating the potential of a continuous charge distribution makes use of Equation 17.3. This procedure is useful when the electric field is already known from other considerations, such as Gauss's law. If the charge distribution is highly symmetric, we first evaluate **E** at any point, using Gauss's law, and then substitute the obtained value into Equation 17.3 to determine the potential difference between any two points. We then choose V to be zero at some convenient point. Let us illustrate both methods with several examples.

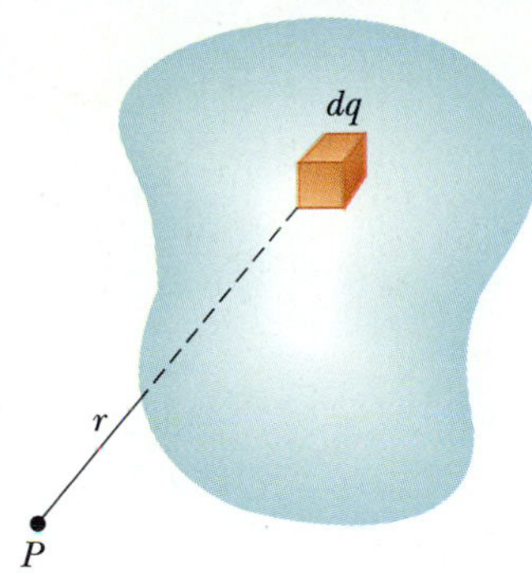

Figure 17.10 The electric potential at the point P due to a continuous charge distribution can be calculated by dividing the charged body into segments of charge dq and summing the potential contributions over all segments.

Example 17.5 Potential Due to a Uniformly Charged Ring

Find the electric potential at a point, P, located on the axis of a uniformly charged ring of radius a and total charge Q. The plane of the ring is perpendicular to the x axis (Fig. 17.11).

Solution Let us take P to be at a distance of x from the center of the ring. The charge element, dq, is at a distance of $\sqrt{x^2 + a^2}$ from P. Hence, we can express V as

$$V = k_e \int \frac{dq}{r} = k_e \int \frac{dq}{\sqrt{x^2 + a^2}}$$

In this case, each element dq is at the same distance from P. Therefore, the term $\sqrt{x^2 + a^2}$ can be removed from the integral, and V reduces to

$$V = \frac{k_e}{\sqrt{x^2 + a^2}} \int dq = \frac{k_e Q}{\sqrt{x^2 + a^2}}$$

The only variable in this expression for V is x. This is not surprising, since our calculation is valid only for points along the x axis. From symmetry, we see that along the x axis **E** can have only an x component. Therefore, we can use Equation 17.15 to find the electric field at P:

$$E_x = -\frac{dV}{dx} = -k_e Q \frac{d}{dx}(x^2 + a^2)^{-1/2}$$

$$= -k_e Q(-\tfrac{1}{2})(x^2 + a^2)^{-3/2}(2x)$$

$$= \frac{k_e Q x}{(x^2 + a^2)^{3/2}}$$

This result agrees with that obtained by direct integration (see Chapter 16, Example 16.5). Note that $E_x = 0$ at $x = 0$ (the center of the ring). Could you have guessed this from Coulomb's law?

Exercise What is the electric potential at the center of the uniformly charged ring? What does the field at the center imply about this result?

Answer $V = k_e Q/a$ at $x = 0$. Because $E = 0$, V must have a maximum or minimum value; it is, in fact, a maximum.

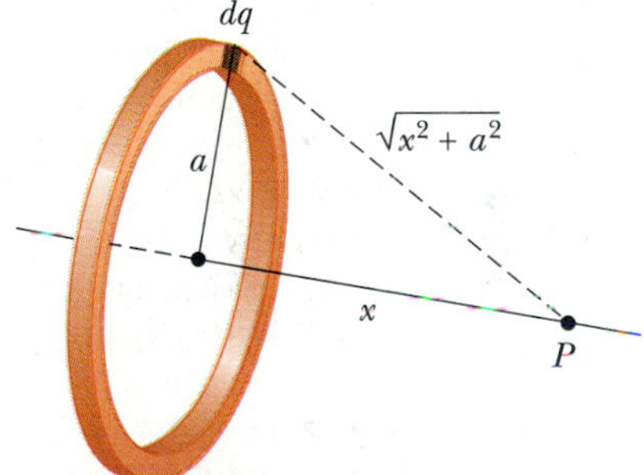

Figure 17.11 (Example 17.5) A uniformly charged ring of radius a, whose plane is perpendicular to the x axis. All segments of the ring are at the same distance from any axial point P.

Example 17.6 Potential of a Uniformly Charged Sphere

An insulating solid sphere of radius R has a uniform positive charge density with a total charge of Q (Fig. 17.12a).

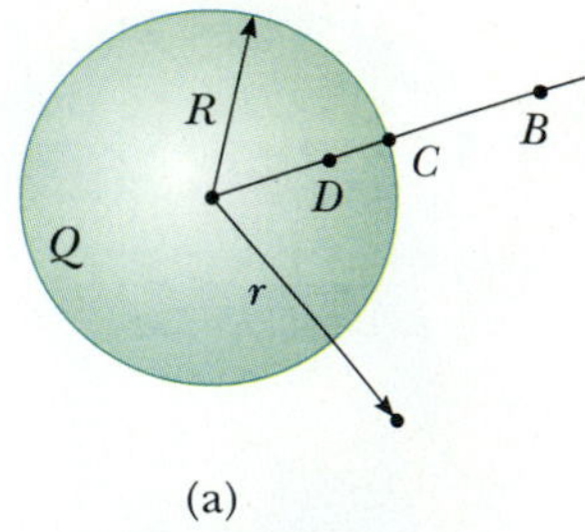

(a)

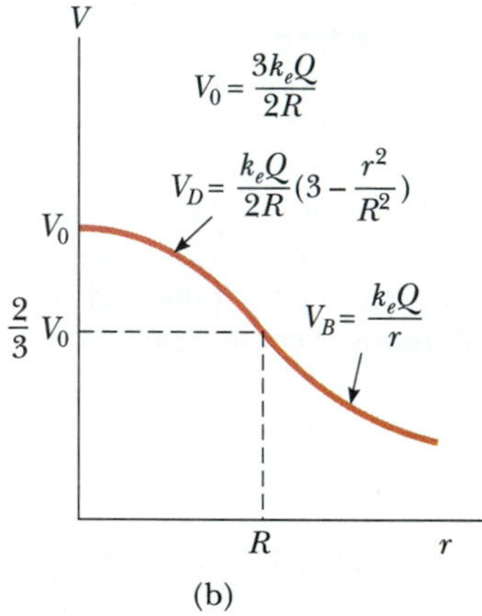

(b)

Figure 17.12 (Example 17.6) (a) A uniformly charged insulating sphere of radius R and total charge Q. The electric potential at points B and C is equivalent to that of a point charge, Q, located at the center of the sphere. (b) A plot of the electric potential, V, versus the distance r from the center of a uniformly charged, insulating sphere of radius R. The curve for V_D inside the sphere is parabolic and joins smoothly with the curve for V_B outside the sphere, which is a hyperbola. The potential has the maximum value V_0 at the center of the sphere.

(a) Find the electric potential at a point *outside* the sphere, that is, for $r > R$. Take the potential to be zero at $r = \infty$.

Solution Recall from Chapter 16, Example 16.8, that the magnitude of the electric field outside a uniformly charged sphere is

$$E_r = k_e \frac{Q}{r^2} \qquad \text{(for } r > R\text{)}$$

where the field is directed radially outward when Q is positive. To obtain the potential at an exterior point, such as B in Figure 17.12a, we substitute this expression for E into Equation 17.4. Since $\mathbf{E}\cdot d\mathbf{s} = E_r\,dr$ in this case, we get

$$V_B = -\int_\infty^r E_r\,dr = -k_eQ\int_\infty^r \frac{dr}{r^2}$$

$$V_B = k_e\frac{Q}{r} \qquad \text{(for } r > R\text{)}$$

Note that the result is identical to that for the electric potential due to a point charge. Since the potential must be continuous at $r = R$, we can use this expression to obtain the potential at the surface of the sphere. That is, the potential at a point such as C in Figure 17.12a is

$$V_C = k_e\frac{Q}{R} \qquad \text{(for } r = R\text{)}$$

(b) Find the potential at a point inside the charged sphere, that is, for $r < R$.

Solution In Example 16.8 we found that the electric field inside a uniformly charged sphere is

$$E_r = \frac{k_eQ}{R^3}r \qquad \text{(for } r < R\text{)}$$

We can use this result and Equation 17.3 to evaluate the potential difference $V_D - V_C$, where D is an interior point:

$$V_D - V_C = -\int_R^r E_r\,dr = -\frac{k_eQ}{R^3}\int_R^r r\,dr = \frac{k_eQ}{2R^3}(R^2 - r^2)$$

Substituting $V_C = k_eQ/R$ into this expression and solving for V_D, we get

$$V_D = \frac{k_eQ}{2R}\left(3 - \frac{r^2}{R^2}\right) \qquad \text{(for } r < R\text{)}$$

At $r = R$, this expression gives a result that agrees with that for the potential at the surface, that is, V_C. A plot of V versus r for this charge distribution is given in Figure 17.12b.

Exercise What are the electric field and electric potential at the center of a uniformly charged sphere?

Answer At $r = 0$, $\mathbf{E} = 0$ and $V_0 = 3k_eQ/2R$.

Problem-Solving Strategy and Hints

1. When working problems involving electric potential, remember that it is a *scalar quantity*, and so there are no components to worry about. Therefore, when using the superposition principle to evaluate the electric potential at a point due to a system of point charges, simply take the algebraic sum of the potentials due to each charge. You must keep track of signs, however. The potential for each positive charge ($V = k_e q/r$) is positive, whereas the potential for each negative charge is negative.
2. Just as in mechanics, only *changes* in electric potential are significant; hence, the point where the potential is set at zero is arbitrary. When dealing with point charges or a finite-sized charge distribution, we usually define $V = 0$ to be at a point infinitely far from the charges. However, if the charge distribution itself extends to infinity, some other nearby point must be selected as the reference point.
3. The electric potential at some point P due to a continuous distribution of charge can be evaluated by dividing the charge distribution into infinitesimal elements of charge dq located at a distance of r from the point P. This element is then treated as a point charge, and so the potential at P due to the element is $dV = k_e\, dq/r$. The total potential at P is obtained by integrating dV over the entire charge distribution. For most problems, it is necessary in performing the integration to express dq and r in terms of a single variable. In order to simplify the integration, it is important to give careful consideration to the geometry involved in the problem. Review Example 17.5 as a guide for using this method.
4. Another method that can be used to obtain the potential due to a finite continuous charge distribution is to start with the definition of the potential difference given by Equation 17.3. If $\mathbf{E}$ is known or can be obtained easily (say, from Gauss's law), then the line integral of $\mathbf{E} \cdot d\mathbf{s}$ can be evaluated. Example 17.6 uses this method.
5. Once you know the electric potential at a point, it is possible to obtain the electric field at that point by remembering that *the electric field is equal to the negative of the derivative of the potential with respect to some coordinate.* Example 17.5 illustrates how to use this procedure.

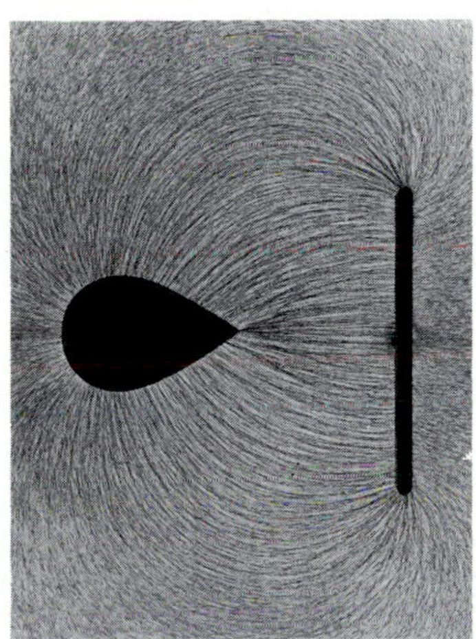

The electric field pattern of a charged conducting plate near an oppositely charged pointed conductor. Small pieces of thread suspended in oil align with the electric field lines. Note that the field is most intense near the pointed part of the conductor and at other points where the radius of curvature is small. *(Courtesy of Harold M. Waage, Princeton University)*

17.6 Electric Potential of a Charged Conductor

In Chapter 16 we found that when a solid conductor in electrostatic equilibrium carries a net charge, the charge resides on the outer surface of the conductor. Furthermore, we showed that the electric field just outside the surface of a conductor in equilibrium is perpendicular to the surface, whereas the field *inside* the conductor is zero. If the electric field had a component parallel to the surface, this would cause surface charges to move, creating a current and nonequilibrium.

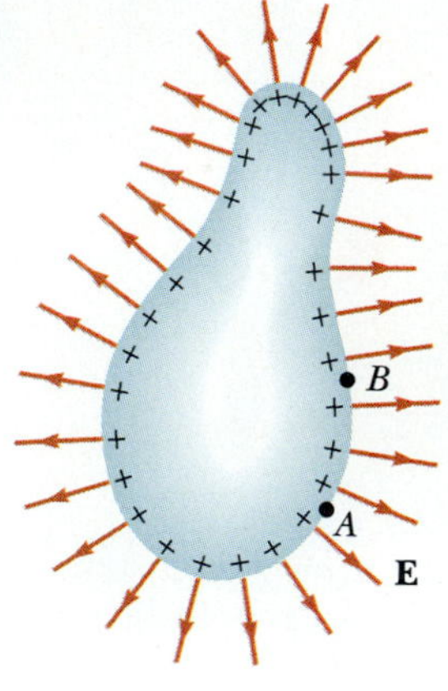

Figure 17.13
An arbitrarily shaped conductor with an excess positive charge. When the conductor is in electrostatic equilibrium, all of the charge resides at the surface, $\mathbf{E} = 0$ inside, and the electric field just outside is perpendicular to the surface. The electric potential is constant inside and is equal to the potential at the surface. The surface charge density is nonuniform.

We shall now show that *every point on the surface of a charged conductor in electrostatic equilibrium is at the same electric potential.* Consider two points, A and B, on the surface of a charged conductor, as in Figure 17.13. Along a surface path connecting these points, $\mathbf{E}$ is always perpendicular to the displacement, $d\mathbf{s}$; therefore, $\mathbf{E} \cdot d\mathbf{s} = 0$. Using this result and Equation 17.3, we conclude that the potential difference between A and B is necessarily zero. That is,

$$V_B - V_A = -\int_A^B \mathbf{E} \cdot d\mathbf{s} = 0$$

This result applies to *any* two points on the surface. Therefore, V is constant everywhere on the surface of a charged conductor in equilibrium, and so such a surface is an equipotential surface. Furthermore, since the electric field is zero inside the conductor, we conclude that the potential is constant everywhere inside the conductor and equal to its value at the surface. It follows that no work is required to move a test charge from the interior of a charged conductor to its surface. (Note that the potential is not zero inside the conductor, even though the electric field is zero.)

For example, consider a solid metal sphere of radius R and total positive charge Q, as shown in Figure 17.14a. The electric field outside the sphere is k_eQ/r^2 and points radially outward. Following Example 17.6, we see that the potential at the interior and surface of the sphere must be k_eQ/R relative to infinity. The potential outside the sphere is k_eQ/r. Figure 17.14b is a plot of the potential as a function of r, and Figure 17.14c shows the variations of the electric field with r.

When a net charge is placed on a spherical conductor, the surface charge density is uniform, as indicated in Figure 17.14a. However, if the conductor is nonspherical, as in Figure 17.13, the surface charge density is high where the radius of curvature is small and low where the radius of curvature is large. Since the electric field just outside a charged conductor is proportional to the surface charge density, σ, we see that *the electric field is large near points having a small convex radius of curvature and reaches very high values at sharp points.*

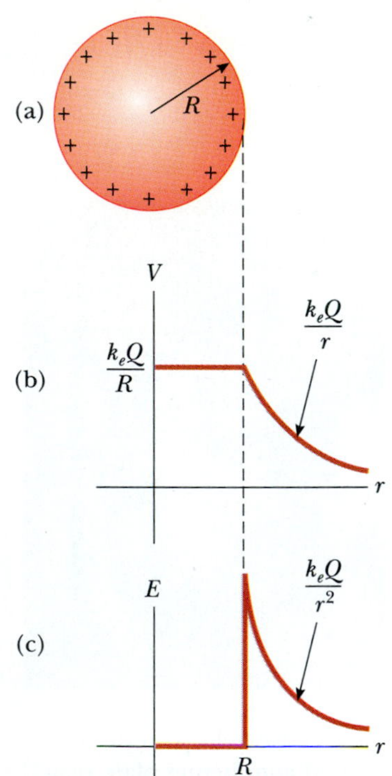

Figure 17.14
(a) The excess charge on a conducting sphere of radius R is uniformly distributed on its surface. (b) The electric potential versus the distance, r, from the center of the charged conducting sphere. (c) The electric field intensity versus the distance r.

A Cavity Within a Conductor

Now consider a conductor of arbitrary shape containing a cavity, such as that in Figure 17.15. Let us assume there are no charges inside the cavity. We shall show that *the electric field inside the cavity must be zero,* regardless of the charge distribution on the outside surface of the conductor. Furthermore, the field in the cavity is zero even if an electric field exists outside the conductor.

In order to prove this point, we shall use the fact that every point on the conductor is at the same potential, and therefore any two points A and B on the surface of the cavity must be at the same potential. Now imagine that a field, **E**, exists in the cavity, and evaluate the potential difference $V_B - V_A$, defined by the expression

$$V_B - V_A = -\int_A^B \mathbf{E} \cdot d\mathbf{s}$$

If **E** is nonzero, we can invariably find a path between A and B for which $\mathbf{E} \cdot d\mathbf{s}$ is always a positive number (a path along the direction of **E**), and so the integral must be positive. However, since $V_B - V_A = 0$, the integral must also be zero. This contradiction can be reconciled only if $\mathbf{E} = 0$ inside the cavity. Thus, we conclude that a cavity surrounded by conducting walls is a field-free region as long as there are no charges inside the cavity.

This result has some interesting applications. For example, it is possible to shield an electronic circuit or even an entire laboratory from external fields by surrounding it with conducting walls. Shielding is often necessary during highly sensitive electrical measurements.

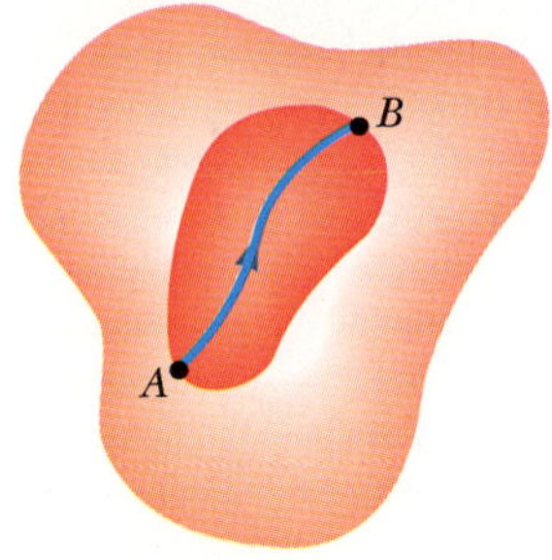

Figure 17.15
A conductor in electrostatic equilibrium containing an empty cavity. The electric field in the cavity is *zero*, regardless of the charge on the conductor.

Corona Discharge

A phenomenon known as **corona discharge** is often observed near sharp points of a conductor raised to a high potential. It is often visible to the naked eye as a greenish glow.

In this process, air becomes a conductor as a result of the ionization of air molecules in regions of high electric fields. At standard temperature and pressure, the discharge occurs at electric field strengths equal to or greater than about 3×10^6 V/m. Since air contains a small number of ions (produced, for example, by cosmic rays), a charged conductor attracts ions of the opposite sign from the air. Near sharp points, where the field is very high, the ions in the air are accelerated to high speeds and collide with other air molecules, producing more ions and an increase in conductivity of the air.

Problem-Solving Strategy and Hints

Until now, we have been using the symbols V to represent the electric potential at some point and ΔV to represent the potential difference between two points. In descriptions of electrical devices, however, it is common practice to use the symbol V to represent the potential difference across the device. Hence, in this book both symbols will be used to denote potential differences, depending on the circumstances.

In practice, a variety of phrases are used to describe the potential difference between two points, the most common being "voltage." A voltage *applied to* a device or *across* a device has the same meaning as the potential difference across the device. For example, if we say that the voltage across a certain capacitor is 12 volts, we mean that the potential difference between the capacitor's plates is 12 volts.

17.7 Capacitance

Consider two conductors having a potential difference of V between them. Let us assume that the conductors have equal and opposite charges, as in Figure 17.16. This can be accomplished by connecting two uncharged conductors to the terminals of a battery. As we learned in Example 17.1, such a combination of two conductors is called a *capacitor*. The potential difference, V, is found to be proportional to the magnitude of the charge, Q, on the capacitor. The **capacitance,** C, of a capacitor is defined as the ratio of the magnitude of the charge on either conductor to the magnitude of the potential difference between them:

Definition of capacitance

$$C \equiv \frac{Q}{V} \qquad [17.19]$$

Note that by definition *capacitance is always a positive quantity*. Furthermore, since the potential difference increases as the stored charge increases (see Eq. 17.7), the ratio Q/V is constant for a given capacitor. Therefore, the capacitance of a device is a measure of its ability to store charge and electrical potential energy.

From Equation 17.19, we see that capacitance has the SI units coulombs per volt, called the **farad** (F) in honor of Michael Faraday. That is,

$$[\text{Capacitance}] = 1\ \text{F} = 1\ \text{C/V}$$

The farad is a very large unit of capacitance. In practice, typical devices have capacitances ranging from microfarads to picofarads. Capacitors are often labeled "mF" for microfarads and "mmF" for micromicrofarads (picofarads).

As you will soon see, the capacitance of a device depends on the geometric arrangement of the conductors. To illustrate this point, let us calculate the capacitance of an isolated spherical conductor of radius R and charge Q. (The second conductor can be taken as a concentric hollow, conducting sphere of infinite radius.) Since the sphere's potential is simply k_eQ/R (where $V = 0$ at infinity), its capacitance is

$$C = \frac{Q}{V} = \frac{Q}{k_eQ/R} = \frac{R}{k_e} = 4\pi\epsilon_0 R \qquad [17.20]$$

(Remember from Chapter 16, Section 16.4, that the Coulomb constant $k_e = 1/4\pi\epsilon_0$.) Equation 17.20 shows that the capacitance of an isolated charged sphere is proportional to the sphere's radius and is independent of both the charge and the potential difference.

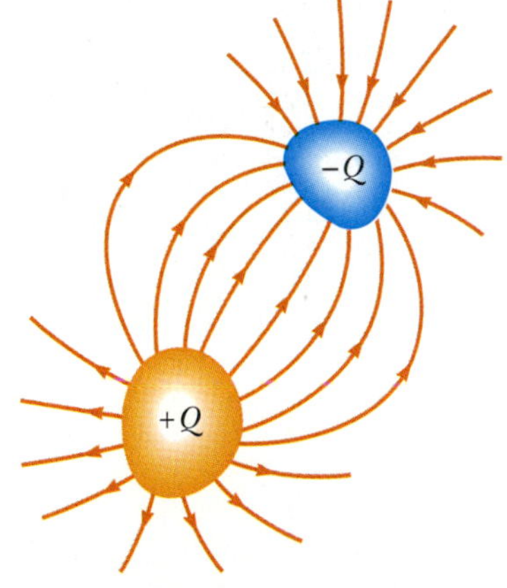

Figure 17.16
A capacitor consists of two conductors isolated from each other and from their surroundings. Once the capacitor is charged, the two conductors carry equal and opposite charges.

Calculation of Capacitance

The capacitance of a pair of oppositely charged conductors can be calculated in the following manner. A convenient charge of magnitude Q is assumed, and the potential difference is calculated using the techniques described in Section 17.5. One then simply uses $C = Q/V$ to evaluate the capacitance. As you might expect, the calculation is relatively easy if the geometry of the capacitor is simple.

Let us illustrate this with two geometries with which we are all familiar: parallel plates and concentric cylinders. In these examples, we shall assume that the charged conductors are separated by a vacuum. (The effect of a dielectric material between the conductors will be treated in Section 17.10.)

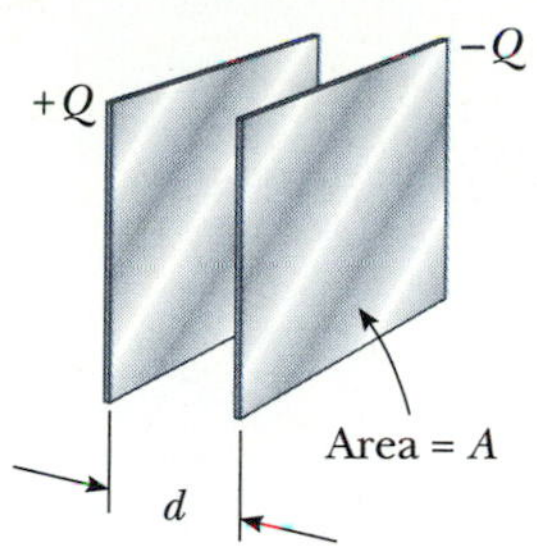

Figure 17.17
A parallel-plate capacitor consists of two parallel plates, each of area A, separated by the distance d. The plates carry equal and opposite charges.

The Parallel-Plate Capacitor

Two parallel plates of equal area A are separated by a distance of d, as in Figure 17.17. One plate has charge $+Q$, the other, charge $-Q$. The charge per unit area on either plate is $\sigma = Q/A$. If the plates are very close together (compared with their length and width), we assume that the electric field is uniform between the plates and zero elsewhere. According to Example 16.10, the electric field between the plates is

$$E = \frac{\sigma}{\epsilon_0} = \frac{Q}{\epsilon_0 A}$$

The potential difference between the plates equals Ed; therefore,

$$V = Ed = \frac{Qd}{\epsilon_0 A}$$

Substituting this result into Equation 17.19, we find that the capacitance is

$$C = \frac{Q}{V} = \frac{Q}{Qd/\epsilon_0 A}$$

$$C = \frac{\epsilon_0 A}{d} \qquad [17.21]$$

That is *the capacitance of a parallel-plate capacitor is proportional to the area of its plates and inversely proportional to the plate separation.*

As you can see from the definition of capacitance, $C = Q/V$, the amount of charge a given capacitor can store for a given potential difference across its plates increases as the capacitance increases. Therefore, it seems reasonable that a capacitor constructed from plates having large areas should be able to store a large charge. The amount of charge needed to produce a given potential difference increases with decreasing plate separation.

A careful inspection of the electric field lines for a parallel-plate capacitor reveals that the field is uniform in the central region between the plates. However, the field is nonuniform at the edges of the plates. Figure 17.18 is a photograph of the electric field pattern of a parallel-plate capacitor, showing the nonuniform field lines at the plates' edges.

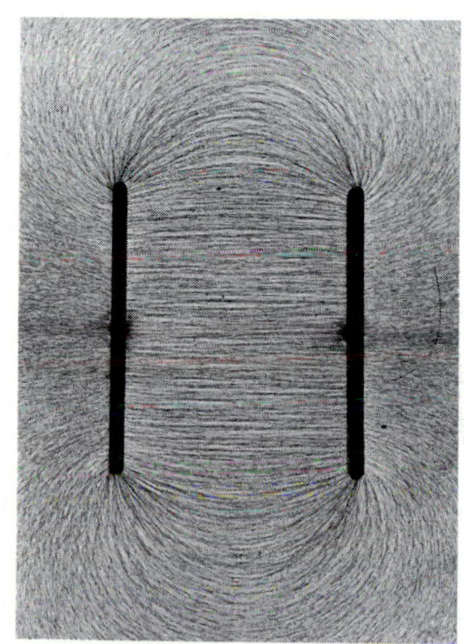

Figure 17.18
Small pieces of thread on an oil surface align with the electric field that exists between the plates of a parallel-plate capacitor. Note the nonuniform nature of the electric field at the edges of the plates. Such edge effects can be neglected if the plate separation is small relative to the lengths of the plates.
(Courtesy of Harold M. Waage, Princeton University)

Example 17.7 A Parallel-Plate Capacitor

A parallel-plate capacitor has the area $A = 2\ \text{cm}^2$ and a plate separation $d = 1$ mm. Find its capacitance.

Solution From Equation 17.21, we find

$$C = \epsilon_0 \frac{A}{d} = \left(8.85 \times 10^{-12} \frac{\text{C}^2}{\text{N}\cdot\text{m}^2}\right)\left(\frac{2 \times 10^{-4}\ \text{m}^2}{1 \times 10^{-3}\ \text{m}}\right)$$

$$= 1.77 \times 10^{-12}\ \text{F} = 1.77\ \text{pF}$$

Exercise If the plate separation of this capacitor is increased to 3 mm, find its capacitance.

Answer 0.59 pF.

The Cylindrical Capacitor

A cylindrical conductor of radius a and charge $+Q$ is concentric with a larger cylindrical shell of radius b and charge $-Q$ (Fig. 17.19a). Let us find the capacitance of this cylindrical capacitor if its length is ℓ.

If we assume that ℓ is long compared with a and b, we can neglect edge effects. In this case, the field is perpendicular to the axis of the cylinders and is confined to the region between them (Fig. 17.19b). We must first calculate the potential difference between the two cylinders, which in general is

$$V_b - V_a = -\int_a^b \mathbf{E} \cdot d\mathbf{s}$$

where $\mathbf{E}$ is the electric field in the region $a < r < b$. In Chapter 16 we showed, using Gauss's law, that the electric field of a cylinder of charge per unit length λ is $2k_e\lambda/r$. The same result applies here, since the outer cylinder does not contribute to the electric field inside it. Using this result and noting that $\mathbf{E}$ is along r in Figure 17.19b, we find that

$$V_b - V_a = -\int_a^b E_r\, dr = -2k_e\lambda \int_a^b \frac{dr}{r} = -2k_e\lambda \ln\left(\frac{b}{a}\right)$$

Substituting this into Equation 17.19 and using the fact that $\lambda = Q/\ell$, we get

$$C = \frac{Q}{V} = \frac{Q}{\dfrac{2k_eQ}{\ell}\ln\left(\dfrac{b}{a}\right)}$$

$$C = \frac{\ell}{2k_e \ln\left(\dfrac{b}{a}\right)} \qquad \textbf{[17.22]}$$

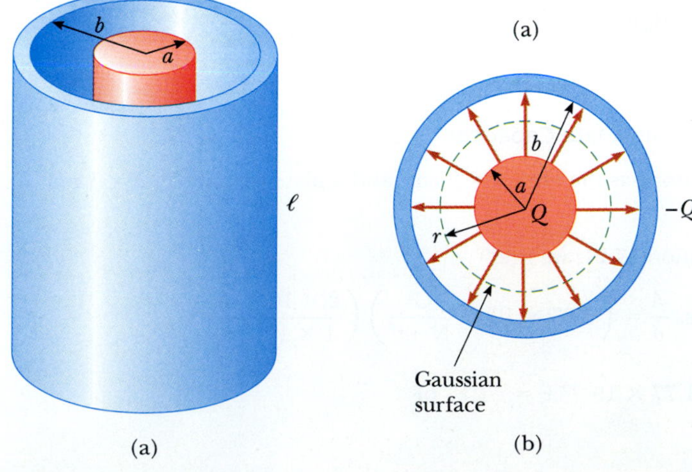

Figure 17.19
(a) A cylindrical capacitor consists of a cylindrical conductor of radius a and length ℓ, surrounded by a coaxial cylindrical shell of radius b. (b) The end view of a cylindrical capacitor. The blue dashed line represents the end of the cylindrical gaussian surface of radius r and length ℓ.

Note that V is the magnitude of the potential difference given by $2k_e\lambda \ln(b/a)$, a *positive* quantity. That is, $V = V_a - V_b$ is *positive* since the inner cylinder is at the higher potential.

Our result for C makes sense because it shows that the capacitance is proportional to the length of the cylinders. As you might expect, the capacitance also depends on the radii of the two cylinders. As an example, a coaxial cable consists of two concentric cylindrical conductors of radii a and b separated by an insulator. The cable carries currents in opposite directions in the inner and outer conductors. Such a geometry is especially useful for shielding an electrical signal from external influences.

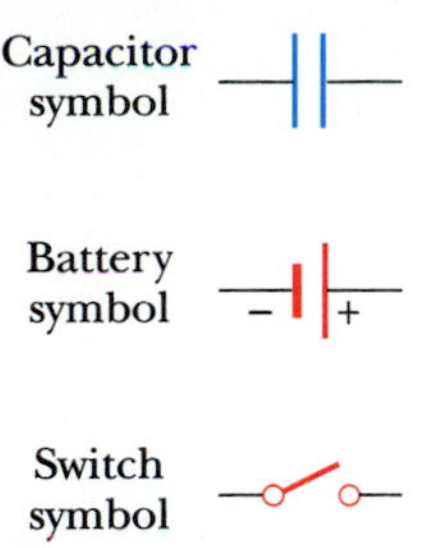

Figure 17.20
Circuit symbols for capacitors, batteries, and switches. Note that capacitors are in blue, and batteries and switches are in red.

17.8 Combinations of Capacitors

Two or more capacitors are often combined in circuits; this can be done in several ways. The equivalent capacitance of certain combinations can be calculated using methods described in this section. Figure 17.20 shows the circuit symbols for capacitors and batteries, together with their color codes. The positive terminal of the battery is at the higher potential and is represented by the longer vertical line in the battery symbol.

Parallel Combination

Two capacitors connected as shown in Figure 17.21a are known as a *parallel combination* of capacitors. The left plates of the capacitors are connected by a conducting wire to the positive terminal of the battery, and therefore both are at the same

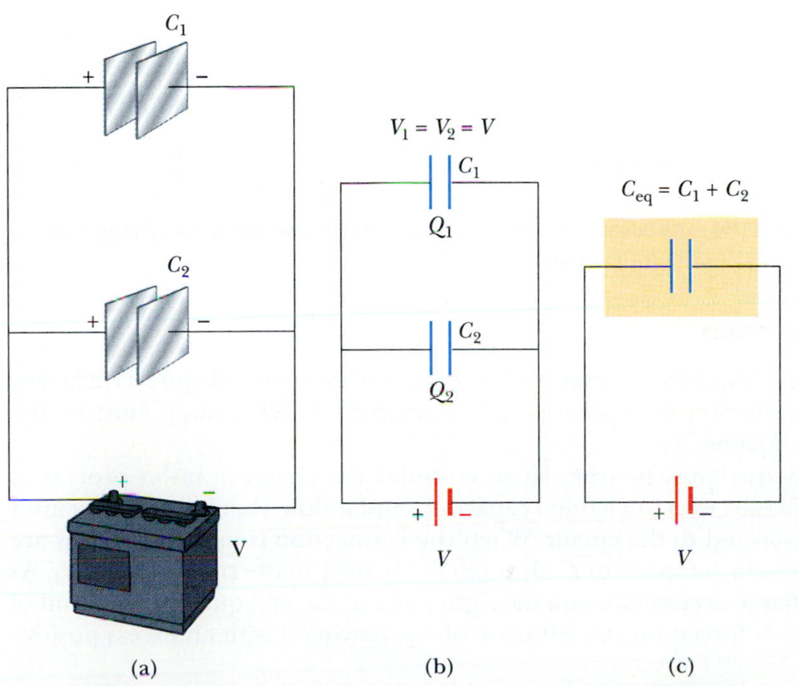

Figure 17.21
(a) A parallel connection of two capacitors. (b) The circuit diagram for the parallel combination. The potential differences across the capacitors are the same and are equal to those across the battery terminals. (c) The potential difference across the equivalent capacitance is also V.

potential. Likewise, the right plates are connected to the negative terminal of the battery. When the capacitors are first connected in the circuit, electrons are transferred through the battery from the left plates to the right plates, leaving the left plates positively charged and the right plates negatively charged. The energy source for this charge transfer is the internal chemical energy stored in the battery, which is converted to electrical energy. The flow of charge ceases when the voltage across the capacitors is equal to that of the battery. The capacitors reach their maximum charge when the flow of charge ceases. Let us call the maximum charges on the two capacitors Q_1 and Q_2. Then the *total charge,* Q, stored by the two capacitors is

$$Q = Q_1 + Q_2 \qquad [17.23]$$

Suppose we wish to replace these two capacitors with one equivalent capacitor having the capacitance C_{eq}. This equivalent capacitor must have exactly the same external effect on the circuit as the original two. That is, it must store Q units of charge. We see from Figure 17.21b that in the parallel circuit the potential differences across the two capacitors are the same, and each is equal to the voltage of the battery, V. From Figure 17.21c, we see that the voltage across the equivalent capacitor is also V. Thus, we have

$$Q_1 = C_1V \qquad Q_2 = C_2V$$

and, for the equivalent capacitor,

$$Q = C_{eq}V$$

Substitution of these relations into Equation 17.23 gives

$$C_{eq}V = C_1V + C_2V$$

or

$$C_{eq} = C_1 + C_2 \qquad \left(\begin{matrix}\text{parallel} \\ \text{combination}\end{matrix}\right) \qquad [17.24]$$

If we extend this treatment to three or more capacitors connected in parallel, the equivalent capacitance is

$$C_{eq} = C_1 + C_2 + C_3 + \ldots \qquad \left(\begin{matrix}\text{parallel} \\ \text{combination}\end{matrix}\right) \qquad [17.25]$$

Thus, we see that *the equivalent capacitance of a parallel combination of capacitors is larger than any of the individual capacitances.*

Series Combination

Now consider two capacitors connected in *series,* as illustrated in Figure 17.22a. For this series combination of capacitors, the magnitude of the charge must be the same on all the plates.

To see why this must be true, let us consider the charge transfer process in some detail. We start with uncharged capacitors and follow what happens just after a battery is connected to the circuit. When the connection is made, electrons are transferred from the left plate of C_1 through the battery to the right plate of C_2. As this negative charge accumulates on the right plate of C_2, an equivalent amount of negative charge is forced off the left plate of C_2, leaving it with an excess positive

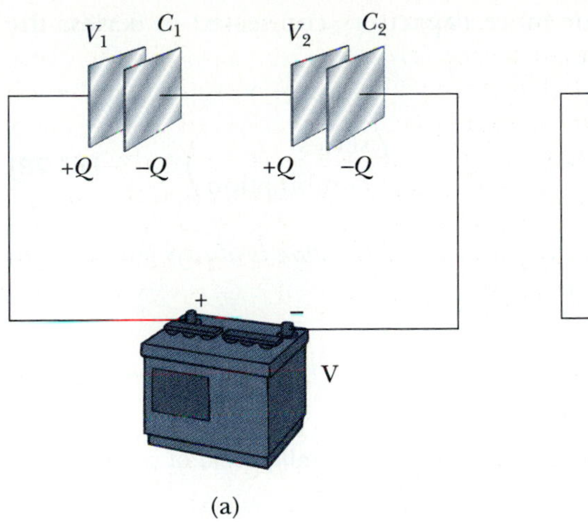

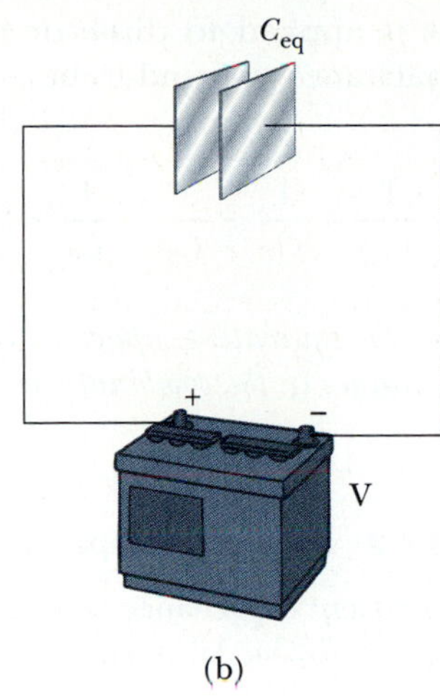

Figure 17.22
A series connection of two capacitors. The charges on the capacitors are the same, and the equivalent capacitance can be calculated from the relation $1/C_{eq} = 1/C_1 + 1/C_2$.

charge. The negative charge leaving the left plate of C_2 accumulates on the right plate of C_1, where an equivalent amount of negative charge has left the left plate. The result is that *each of the right plates gains a charge of* $-Q$ *while each of the left plates has a charge of* $+Q$.

Suppose an equivalent capacitor performs the same function as the series combination. After it is fully charged, *the equivalent capacitor must end up with a charge of* $-Q$ *on its right plate and* $+Q$ *on its left plate.* By applying the definition of capacitance to the circuit shown in Figure 17.22b, we have

$$V = \frac{Q}{C_{eq}}$$

where V is the potential difference between the terminals of the battery and C_{eq} is the equivalent capacitance. From Figure 17.22a, we see that

$$V = V_1 + V_2 \qquad \textbf{[17.26]}$$

where V_1 and V_2 are the potential differences across capacitors C_1 and C_2. In general the potential difference across any number of capacitors in series is equal to the sum of the potential differences across the individual capacitors. Since $Q = CV$ can be applied to each capacitor, the potential difference across each is

$$V_1 = \frac{Q}{C_1} \qquad V_2 = \frac{Q}{C_2}$$

Substituting these expressions into Equation 17.26, and noting that $V = Q/C_{eq}$, we have

$$\frac{Q}{C_{eq}} = \frac{Q}{C_1} + \frac{Q}{C_2}$$

Cancelling Q, we arrive at the relationship

$$\frac{1}{C_{eq}} = \frac{1}{C_1} + \frac{1}{C_2} \quad \begin{pmatrix}\text{series}\\ \text{combination}\end{pmatrix} \qquad \textbf{[17.27]}$$

If this analysis is applied to three or more capacitors connected in series, the equivalent capacitance is found to be

$$\frac{1}{C_{eq}} = \frac{1}{C_1} + \frac{1}{C_2} + \frac{1}{C_3} + \cdots \quad \left(\begin{array}{l}\text{series}\\ \text{combination}\end{array}\right) \qquad [17.28]$$

This shows that *the equivalent capacitance of a series combination is always less than any individual capacitance in the combination.*

Example 17.8 Equivalent Capacitance

Find the equivalent capacitance between a and b for the combination of capacitors shown in Figure 17.23a. All capacitances are in microfarads.

Solution Using Equations 17.25 and 17.28, we reduce the combination step by step, as indicated in the figure. The 1-μF and 3-μF capacitors are in *parallel* and combine according to $C_{eq} = C_1 + C_2$. Their equivalent capacitance is 4 μF. Likewise, the 2-μF and 6-μF capacitors are also in *parallel* and have an equivalent capacitance of 8 μF. The upper branch in Figure 17.23b now consists of two 4-μF capacitors in *series,* which combine according to

$$\frac{1}{C_{eq}} = \frac{1}{C_1} + \frac{1}{C_2} = \frac{1}{4\ \mu\text{F}} + \frac{1}{4\ \mu\text{F}} = \frac{1}{2\ \mu\text{F}}$$

$$C_{eq} = 2\ \mu\text{F}$$

Likewise, the lower branch in Figure 17.23b consists of two 8-μF capacitors in *series,* which give an equivalent capacitance of 4 μF. Finally, the 2-μF and 4-μF capacitors in Figure 17.23c are in *parallel* and have an equivalent capacitance of 6 μF. Hence, the equivalent capacitance of the circuit is 6 μF.

Exercise Consider three capacitors with capacitances of 3 μF, 6 μF, and 12 μF. Find their equivalent capacitance if they are connected (a) in parallel; (b) in series.

Answer (a) 21 μF, (b) 1.71 μF.

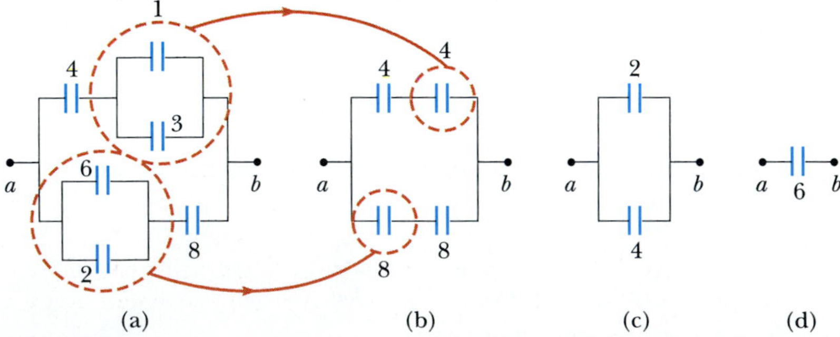

Figure 17.23
(Example 17.8) To find the equivalent capacitance of the capacitors in (a), the various combinations are reduced in steps in (b), (c), and (d), using the series and parallel rules described in the text.

17.9 Energy Stored in a Charged Capacitor

Almost everyone who works with electronic equipment has at some time verified that a capacitor can store energy. If the plates of a charged capacitor are connected by a conductor, such as a wire, charge transfers from one plate to the other until the two are uncharged. The discharge can often be observed as a visible spark. If you accidentally touched the opposite plates of a charged capacitor, your fingers would act as pathways by which the capacitor could discharge, resulting in an electric shock. The degree of shock would depend on the capacitance and voltage applied to the capacitor. Where high voltages are present, such as in the power supply of a television set, the shock could be fatal.

Consider a parallel-plate capacitor that is initially uncharged, so that the initial potential difference across the plates is zero. Now imagine that the capacitor is connected to a battery and develops a maximum charge of Q. We shall assume that the capacitor is charged *slowly* so that the problem can be treated as an electrostatic system. The final potential difference across the capacitor is $V = Q/C$. Since the initial potential difference is zero, the *average* potential difference during the charging process is $V/2 = Q/2C$. From this we might conclude that the work needed to charge the capacitor is $W = QV/2 = Q^2/2C$. Although this result is correct, a more detailed proof is desirable.

Suppose that q is the charge on the capacitor at some instant during the charging process. At the same instant, the potential difference across the capacitor is $V = q/C$. The work that is necessary[1] to transfer an increment of charge dq from the plate of charge $-q$ to the plate of charge q (which is at the higher potential) is

$$dW = V\,dq = \frac{q}{C}\,dq$$

Thus, the total work required to charge the capacitor from $q = 0$ to some final charge, $q = Q$, is

$$W = \int_0^Q \frac{q}{C}\,dq = \frac{Q^2}{2C}$$

But the work done in charging the capacitor can be considered as potential energy, U, stored in the capacitor. Using $Q = CV$, we can express the electrostatic energy stored in a charged capacitor in the following alternative forms:

$$U = \frac{Q^2}{2C} = \tfrac{1}{2}QV = \tfrac{1}{2}CV^2 \qquad [17.29]$$

Energy stored in a charged capacitor

This result applies to *any* capacitor, regardless of its geometry. We see that the stored energy increases as C increases and as the potential difference increases. In practice, there is a limit to the maximum energy (or charge) that can be stored. This is because electrical discharge ultimately occurs between the plates of the capacitor at a sufficiently large value of V. For this reason, capacitors are usually labeled with a maximum operating voltage.

[1] One mechanical analog of this process is the work required to stretch or compress a spring through some small distance.

The energy stored in a capacitor can be thought of as stored in the electric field created between the plates as the capacitor is charged. This is reasonable in view of the fact that the magnitude of the electric field is proportional to the charge on the capacitor. For a parallel-plate capacitor, the potential difference is related to the electric field through the relationship $V = Ed$. Furthermore, the capacitance is $C = \epsilon_0 A/d$. Substituting these expressions into Equation 17.29 gives

Energy stored in a parallel-plate capacitor

$$U = \frac{1}{2}\frac{\epsilon_0 A}{d}(E^2 d^2) = \frac{1}{2}(\epsilon_0 A d)E^2 \qquad [17.30]$$

Since the volume of a parallel-plate capacitor that is occupied by the electric field is Ad, the *energy per unit volume,* $u = U/Ad$—called the *energy density*—is

Energy density in an electric field

$$u = \frac{1}{2}\epsilon_0 E^2 \qquad [17.31]$$

Although Equation 17.31 was derived for a parallel-plate capacitor, the expression is generally valid. That is, the *energy density in any electrostatic field is proportional to the square of the electric field intensity at a given point.*

Example 17.9 Rewiring Two Charged Capacitors

Two capacitors, C_1 and C_2 (where $C_1 > C_2$), are charged to the same potential difference, V_0, but with opposite polarity. The charged capacitors are removed from the battery, and their plates are connected as shown in Figure 17.24a. The switches S_1 and S_2 are then closed as in Figure 17.24b.

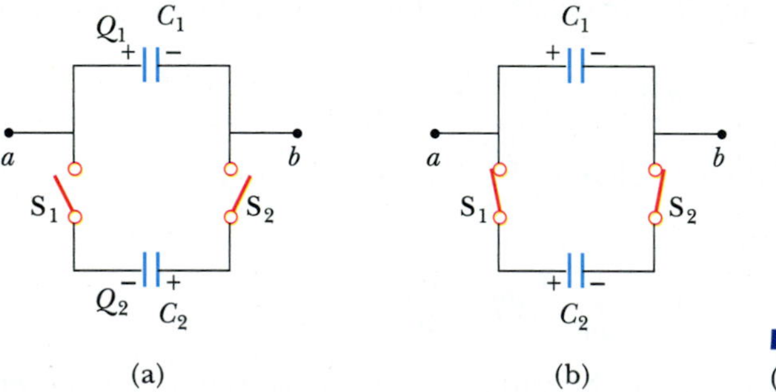

Figure 17.24 (Example 17.9)

(a) Find the final potential difference between a and b after the switches are closed.

Solution The charges on the left plates of the capacitors *before* the switches are closed are

$$Q_1 = C_1 V_0 \quad \text{and} \quad Q_2 = -C_2 V_0$$

The negative sign for Q_2 is necessary since this capacitor's polarity is *opposite* that of capacitor C_1. After the switches are closed, the charges on the plates redistribute until the total charge, Q, shared by the capacitors is

$$Q = Q_1 + Q_2 = (C_1 - C_2)V_0$$

The two capacitors are now in *parallel,* and so the final potential differences across them are the *same*:

$$V = \frac{Q}{C_1 + C_2} = \left(\frac{C_1 - C_2}{C_1 + C_2}\right)V_0$$

(b) Find the total energy stored in the capacitors before and after the switches are closed.

Solution Before the switches are closed, the total energy stored is

$$U_i = \tfrac{1}{2}C_1V_0^2 + \tfrac{1}{2}C_2V_0^2 = \tfrac{1}{2}(C_1 + C_2)V_0^2$$

After the switches are closed and the capacitors have reached an equilibrium charge, the total energy is

$$U_f = \tfrac{1}{2}C_1V^2 + \tfrac{1}{2}C_2V^2 = \tfrac{1}{2}(C_1 + C_2)V^2$$

$$= \tfrac{1}{2}(C_1 + C_2)\left(\frac{C_1 - C_2}{C_1 + C_2}\right)^2 V_0^2 = \left(\frac{C_1 - C_2}{C_1 + C_2}\right)^2 U_i$$

Therefore, the ratio of the final to the initial energy stored is

$$\frac{U_f}{U_i} = \left(\frac{C_1 - C_2}{C_1 + C_2}\right)^2$$

This shows that the final energy is *less* than the initial energy. At first you might think energy conservation has been violated, but that is not the case since the calculated values are only for initial and final states, neglecting the intermediate processes. Part of the missing energy appears as heat energy in the connecting wires, and part of the energy is radiated away in the form of electromagnetic waves.

17.10 Capacitors with Dielectrics

A *dielectric* is a nonconducting material such as rubber, glass, or waxed paper. When a dielectric material is inserted between the plates of a capacitor, the capacitance increases. If the dielectric completely fills the space between the plates, the capacitance increases by the dimensionless factor κ, called the **dielectric constant.**

The following experiment can be performed to illustrate the effect of a dielectric in a capacitor. Consider a parallel-plate capacitor of charge Q_0 and capacitance C_0 in the absence of a dielectric. The potential difference across the capacitor, as measured by a voltmeter, is $V_0 = Q_0/C_0$ (Fig. 17.25a). Notice that the

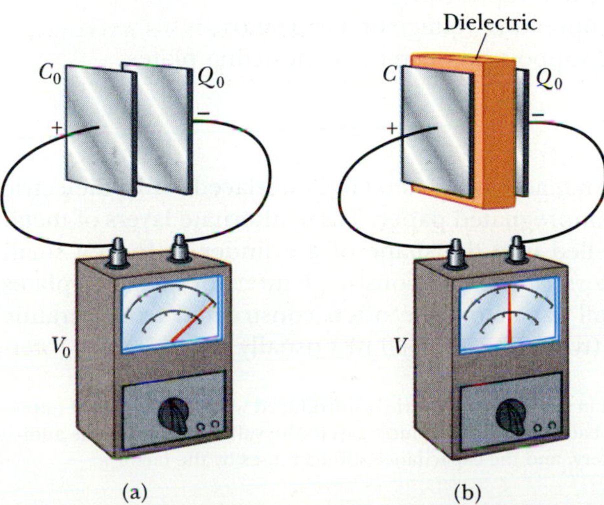

Figure 17.25 When a dielectric is inserted between the plates of a charged capacitor, the charge on the plates remains unchanged, but the potential difference as recorded by an electrostatic voltmeter is reduced from V_0 to $V = V_0/\kappa$. Thus, the capacitance *increases* in the process by the factor κ.

capacitor circuit is *open;* that is, the plates of the capacitor are *not* connected to a battery and charge cannot flow through an ideal voltmeter. Hence, there is *no* path by which charge can flow and alter the charge on the capacitor. If a dielectric is now inserted between the plates as in Figure 17.25b, it is found that the voltmeter reading *decreases* by a factor of κ to the value V, where

$$V = \frac{V_0}{\kappa}$$

Since $V < V_0$, we see that $\kappa > 1$.

Since the charge Q_0 on the capacitor *does not change,* we conclude that the capacitance must change to the value

$$C = \frac{Q_0}{V} = \frac{Q_0}{V_0/\kappa} = \kappa \frac{Q_0}{V_0}$$

$$C = \kappa C_0 \qquad [17.32]$$

where C_0 is the capacitance in the absence of the dielectric. That is, the capacitance *increases* by the factor κ when the dielectric completely fills the region between the plates.[2] For a parallel-plate capacitor, where $C_0 = \epsilon_0 A/d$, we can express the capacitance when the capacitor is filled with a dielectric as

The capacitance of a filled capacitor is greater than that of an empty one by the factor κ

$$C = \kappa \frac{\epsilon_0 A}{d} \qquad [17.33]$$

From this result, it would appear that the capacitance could be made very large by decreasing d, the distance between the plates. In practice, the lowest value of d is limited by the electrical discharge that could occur through the dielectric medium separating the plates. For any given separation d, the maximum voltage that can be applied to a capacitor without causing a discharge depends on the *dielectric strength* (maximum electric field intensity) of the dielectric, which for air is equal to 3×10^6 V/m. If the field strength in the medium exceeds the dielectric strength, the insulating properties break down and the medium begins to conduct. Most insulating materials have dielectric strengths and dielectric constants greater than those of air, as Table 17.1 indicates. Thus, we see that a dielectric provides the following advantages:

1. It increases the capacitance of a capacitor.
2. It increases the maximum operating voltage of a capacitor.
3. It may provide mechanical support between the conducting plates.

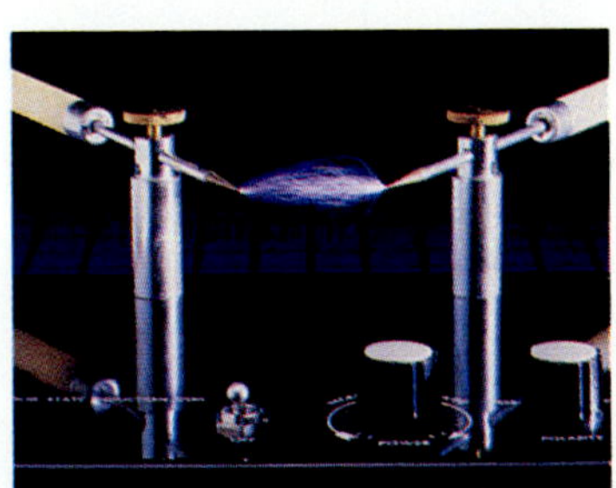

This photograph illustrates dielectric breakdown in air. Sparks are produced when a large alternating voltage is applied across the electrodes by way of a high-voltage induction coil power supply. *(Courtesy of CENCO)*

Types of Capacitors

Commercial capacitors are often made using metal foil interlaced with a dielectric such as thin sheets of paraffin-impregnated paper. These alternate layers of metal foil and dielectric are then rolled into the shape of a cylinder to form a small package. High-voltage capacitors commonly consist of interwoven metal plates immersed in silicone oil. Small capacitors are often constructed from ceramic materials. Variable capacitors (typically 10 to 500 pF) usually consist of two inter-

[2] If another experiment is performed in which the dielectric is introduced while the potential difference is held constant by means of a battery, the charge increases to the value $Q = \kappa Q_0$. The additional charge is supplied by the battery, and the capacitance still increases by the factor κ.

Table 17.1
Dielectric Constants and Dielectric Strengths of Various Materials at Room Temperature

Material	Dielectric Constant κ	Dielectric Strength[a] (V/m)
Vacuum	1.00000	—
Air	1.00059	3×10^6
Bakelite	4.9	24×10^6
Fused quartz	3.78	8×10^6
Pyrex glass	5.6	14×10^6
Polystyrene	2.56	24×10^6
Teflon	2.1	60×10^6
Neoprene rubber	6.7	12×10^6
Nylon	3.4	14×10^6
Paper	3.7	16×10^6
Strontium titanate	233	8×10^6
Water	80	—
Silicone oil	2.5	15×10^6

[a] The dielectric strength equals the maximum electric field that can exist in a dielectric without electrical breakdown.

woven sets of metal plates, one fixed and the other movable, with air as the dielectric.

An electrolytic capacitor is often used to store large amounts of charge at relatively low voltages. This device consists of a metal foil in contact with an electrolyte—a solution that conducts electricity by virtue of the motion of ions contained in the solution. When a voltage is applied between the foil and the electrolyte, a thin layer of metal oxide (an insulator) is formed on the foil, and this layer serves as the dielectric. Very large capacitance values can be attained because the dielectric layer is very thin.

When electrolytic capacitors are used in circuits, the polarity must be installed properly. If the polarity of the applied voltage is opposite what is intended, the oxide layer will be removed and the capacitor will conduct electricity rather than store charge.

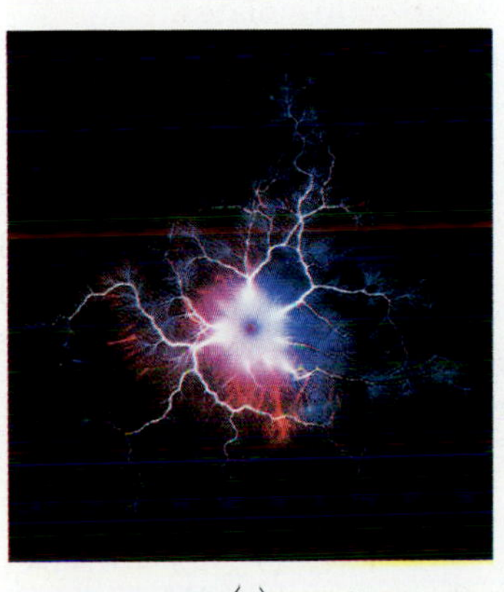
(a)

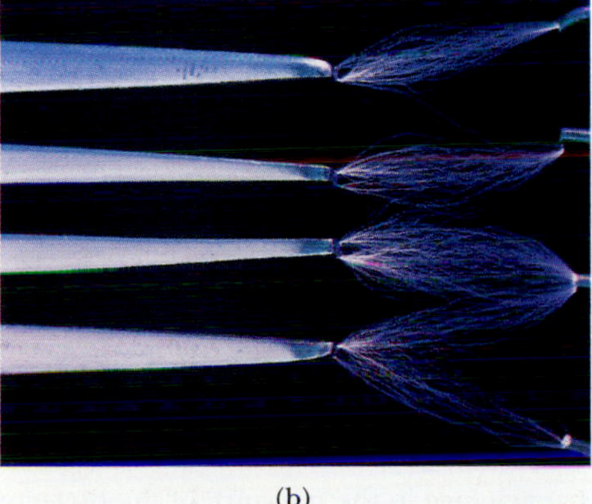
(b)

(a) Kirlian photograph created by dropping a steel ball into a high-energy electric field. This technique is also known as electrophotography. *(Henry Dakin/Science Photo Library)* (b) Sparks from static electricity discharge between a fork and four electrodes. Many sparks were used to make this image, because only one spark will form for a given discharge. Each spark follows the line of least resistance through the air at the time. Note that the bottom prong of the fork forms discharges to both electrodes at bottom right. The light of each spark is created by the ionization of gas atoms along its path. *(Adam Hart-Davis/Science Photo Library)*

Example 17.10 A Paper-Filled Capacitor

A parallel-plate capacitor has plates of dimensions 2 cm × 3 cm. The plates are separated by a 1-mm thickness of paper.

(a) Find the capacitance of this device.

Solution Since $\kappa = 3.7$ for paper (Table 17.1), we get

$$C = \kappa \frac{\epsilon_0 A}{d} = 3.7 \left(8.85 \times 10^{-12} \frac{\text{C}^2}{\text{N}\cdot\text{m}^2}\right)\left(\frac{6 \times 10^{-4}\ \text{m}^2}{1 \times 10^{-3}\ \text{m}}\right)$$

$$= 19.6 \times 10^{-12}\ \text{F} = \boxed{19.6\ \text{pF}}$$

(b) What is the maximum charge that can be placed on the capacitor?

Solution From Table 17.1 we see that the dielectric strength of paper is 16×10^6 V/m. Since the thickness of the paper is 1 mm, the maximum voltage that can be applied before breakdown occurs is

$$V_{\text{max}} = E_{\text{max}} d = (16 \times 10^6\ \text{V/m})\ (1 \times 10^{-3}\ \text{m})$$

$$= 16 \times 10^3\ \text{V}$$

Hence, the maximum charge is given by

$$Q_{\text{max}} = CV_{\text{max}} = (19.6 \times 10^{-12}\ \text{F})(16 \times 10^3\ \text{V})$$

$$= \boxed{0.31\ \mu\text{C}}$$

Exercise What is the maximum energy that can be stored in the capacitor?

Answer 2.5×10^{-3} J.

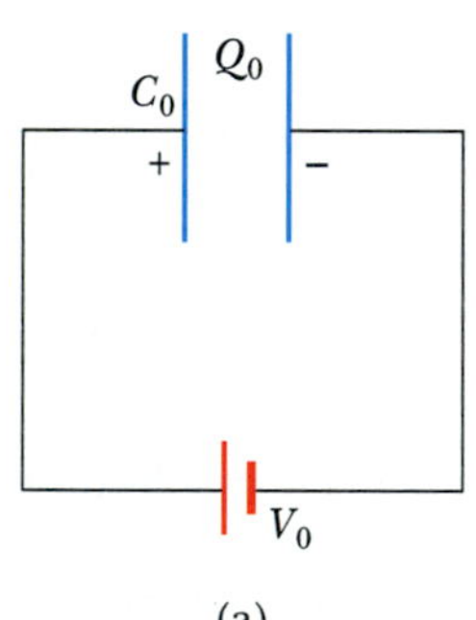

(a)

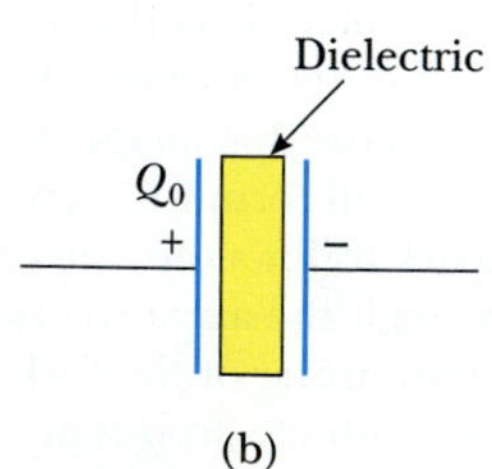

(b)

Figure 17.26 (Example 17.11)

Example 17.11 Energy Stored Before and After

A parallel-plate capacitor is charged with a battery to the charge Q_0, as in Figure 17.26a. The battery is then removed, and a slab of dielectric constant κ is inserted between the plates, as in Figure 17.26b. Find the energy stored in the capacitor before and after the dielectric is inserted.

Solution The energy stored in the capacitor in the absence of the dielectric is

$$U_0 = \tfrac{1}{2} C_0 V_0^2$$

Since $V_0 = Q_0/C_0$, this can be expressed as

$$U_0 = \frac{Q_0^2}{2C_0}$$

After the battery is removed and the dielectric is inserted between the plates, *the charge on the capacitor remains the same.* Hence, the energy stored in the presence of the dielectric is

$$U = \frac{Q_0^2}{2C}$$

But the capacitance in the presence of the dielectric is $C = \kappa C_0$, and so U becomes

$$U = \frac{Q_0^2}{2\kappa C_0} = \frac{U_0}{\kappa}$$

Since $\kappa > 1$, we see that the final energy is *less* than the initial energy by the factor $1/\kappa$. This missing energy can be accounted for by noting that when the dielectric is inserted into the capacitor, it gets pulled into the device. The external agent must do negative work to keep the slab from accelerating. This work is simply the difference $U - U_0$. (Alternatively, the positive work done by the system on the external agent is $U_0 - U$.)

Exercise Suppose that the capacitance in the absence of a dielectric is 8.50 pF, and the capacitor is charged to a potential difference of 12.0 V. The battery is disconnected, and a slab of polystyrene ($\kappa = 2.56$) is inserted between the plates. Calculate the energy difference $U - U_0$.

Answer 373 pJ.

Problem-Solving Strategy and Hints

1. **Be careful with your choice of units. To calculate capacitance in farads, make sure that distances are in meters and use the SI value of ϵ_0. When checking consistency of units, remember that the units for electric fields can be either newtons per coulomb or volts per meter.**
2. **When two or more unequal capacitors are connected in series, they carry the same charge, but their potential differences are not the same. The capacitances add as reciprocals, and the equivalent capacitance of the combination is always less than the smallest individual capacitor.**
3. **When two or more capacitors are connected in parallel, the potential differences across them are the same. The charge on each capacitor is proportional to its capacitance; hence, the capacitances add directly to give the equivalent capacitance of the parallel combination.**
4. **A dielectric increases capacitance by the factor κ (the dielectric constant) because induced surface charges on the dielectric reduce the electric field inside the material from E to E/κ.**
5. **Be careful about problems in which you may be connecting or disconnecting a battery to a capacitor. It is important to note whether modifications to the capacitor are being made while the capacitor is connected to the battery or after it is disconnected. If the capacitor remains connected to the battery, the voltage across the capacitor necessarily remains the same (equal to the battery voltage), and the charge is proportional to the capacitance, *however it may be modified* (say, by insertion of a dielectric). On the other hand, if you disconnect the capacitor from the battery before making any modifications to the capacitor, then its charge remains the same. In this case, as you vary the capacitance, the voltage across the plates changes in inverse proportion to capacitance, according to $V = Q/C$.**

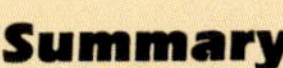

Summary

When a positive test charge, q_0, is moved between points A and B in an electrostatic field, **E**, the **change in potential energy** is

Change in potential energy

$$\Delta U = -q_0 \int_A^B \mathbf{E} \cdot d\mathbf{s} \qquad [17.1]$$

The **potential difference,** ΔV, between points A and B in an electrostatic field, **E**, is defined as the change in potential energy divided by the test charge, q_0.

Potential difference

$$\Delta V = \frac{\Delta U}{q_0} = -\int_A^B \mathbf{E} \cdot d\mathbf{s} \qquad [17.3]$$

where the electric potential V is a scalar and has the units joules per coulomb, defined as 1 volt (V).

The potential difference between two points, A and B, in a uniform electric field, **E**, is

$$\Delta V = -\mathbf{E} \cdot \mathbf{d} \qquad [17.8]$$

where **d** is the displacement vector between A and B.

Equipotential surfaces are surfaces on which the electric potential remains constant. Equipotential surfaces are *perpendicular* to the electric field lines.

The electric potential due to a point charge, q, at any distance r from the charge is

Potential of a point charge

$$V = k_e \frac{q}{r} \qquad [17.11]$$

The electric potential due to a group of point charges is obtained by summing the potentials due to the individual charges. Since V is a scalar, the sum is a simple algebraic operation.

The **electric potential energy of a pair of point charges** separated by a distance of r is

Electric potential energy of two charges

$$U = k_e \frac{q_1 q_2}{r} \qquad [17.13]$$

This represents the work required to bring the charges from an infinite separation to the separation r. The potential energy of a distribution of point charges is obtained by summing terms like Equation 17.13 over all *pairs* of particles.

If the electric potential is known as a function of coordinates x, y, z, the components of the electric field can be obtained by taking the negative derivative of the potential with respect to the coordinates. For example, the x component of the electric field is

$$E_x = -\frac{dV}{dx} \qquad [17.15]$$

The **electric potential due to a continuous charge distribution** is

Electric potential due to a continuous charge distribution

$$V = k_e \int \frac{dq}{r} \qquad [17.18]$$

Every point on the surface of a charged conductor in electrostatic equilibrium is at the same potential. Furthermore, the potential is constant everywhere inside the conductor and equal to its value at the surface.

A *capacitor* consists of two equal and oppositely charged conductors spaced very close together compared to their size, with a potential difference of V between them. The **capacitance,** C, of any capacitor is defined to be the ratio of the magnitude of the charge Q on either conductor to the magnitude of the potential difference, V:

$$C \equiv \frac{Q}{V} \qquad [17.19]$$

Definition of capacitance

The SI units of capacitance are coulombs per volt, or the farad (F), and $1\ \text{F} = 1\ \text{C/V}$.

If two or more capacitors are connected in parallel, the potential differences across them must be the same. The equivalent capacitance of a parallel combination of capacitors is

$$C_{eq} = C_1 + C_2 + C_3 + \ . \ . \ . \qquad [17.25]$$

Parallel combination

If two or more capacitors are connected in series, the charges on them are the same, and the equivalent capacitance of the series combination is

$$\frac{1}{C_{eq}} = \frac{1}{C_1} + \frac{1}{C_2} + \frac{1}{C_3} + \ . \ . \ . \qquad [17.28]$$

Series combination

Work is required to charge a capacitor, since the charging process consists of transferring charges from one conductor at a lower potential to another conductor at a higher potential. The work done in charging the capacitor to the charge Q equals the electrostatic potential energy, U, stored in the capacitor, where

$$U = \frac{Q^2}{2C} = \tfrac{1}{2}QV = \tfrac{1}{2}CV^2 \qquad [17.29]$$

Energy stored in a charged capacitor

When a dielectric material is inserted between the plates of a capacitor, the capacitance generally increases by the dimensionless factor κ, called the **dielectric constant.** That is,

$$C = \kappa C_0 \qquad [17.32]$$

where C_0 is the capacitance in the absence of the dielectric.

Questions and Conceptual Exercises

1. If a proton is released from rest in a uniform electric field, does its electric potential energy increase or decrease?
2. If the potential is constant in a certain region, what is the nature of the electric field in that region?
3. In your own words, distinguish between electric potential and electrical potential energy.
4. If the electric potential at some point is zero, can you conclude that there are no charges in the vicinity of that point? Explain.
5. Give a physical explanation of the fact that the potential energy of a pair of like charges is positive, whereas the potential energy of a pair of unlike charges is negative.
6. Explain why, under static conditions, all points in a conductor must be at the same electric potential.

7. Why is it important to avoid sharp edges, or points, on conductors used in high-voltage equipment?
8. In what type of weather would a car battery be more likely to discharge, and why?
9. How would you shield an electronic circuit or laboratory from stray electric fields? Why does this work?
10. Why is it relatively safe to stay in an automobile with a metal body during a severe thunderstorm?
11. A farad is a very large unit of capacitance. Calculate the length of one side of a square air-filled capacitor with a plate separation of 1 cm. Assume it has a capacitance of 1 farad.
12. If you are given three different capacitors, C_1, C_2, and C_3, how many different combinations of capacitance can you produce?
13. The plates of a capacitor are connected to a battery. What happens to the charge on the plates if the connecting wires are removed from the battery? What happens to the charge if the wires are removed from the battery and connected to each other?
14. What is the difference between dielectric strength and the dielectric constant?
15. If you want to increase the maximum operating voltage of a parallel-plate capacitor, describe how you can do this for a fixed plate separation.
16. What happens to the charge on a capacitor if the potential difference between the conductors is doubled?
17. Why is it dangerous to touch the terminals of a high-voltage capacitor even after the applied voltage has been turned off? What could be done to make the capacitor safe to handle after the voltage source has been removed?
18. Since the charges on the plates of a parallel-plate capacitor are equal and opposite, they attract each other. Hence, it takes positive work to increase the plate separation. What happens to the external work done in this process? (Assume the capacitor plates are removed from the charging battery.)
19. If the potential difference across a capacitor is doubled, by what factor does the stored energy change?
20. If you were asked to design a capacitor in which small size and large capacitance were required, what factors would be important in your design?
21. Explain why a dielectric increases the maximum operating voltage of a capacitor although the physical size of the capacitor does not change.
22. A pair of capacitors is connected in parallel; an identical pair is connected in series. Which pair would be more dangerous to handle after both were connected to the same voltage source?
23. The energy stored in a particular capacitor is increased fourfold. What is the accompanying change in (a) the charge? (b) the potential difference across the capacitor?

Problems

Section 17.1 Potential Difference and Electric Potential
Section 17.2 Potential Differences in a Uniform Electric Field

1. (a) How much work is done on a proton by a uniform electric field of 200 N/C as the charge moves a distance of 2 cm in the field? (b) What is the difference in potential energy between these two points?
2. A uniform electric field of magnitude 250 V/m is directed in the positive x direction. A $+12\text{-}\mu C$ charge moves from the origin to the point $(x, y) = (20\text{ cm}, 50\text{ cm})$. (a) What was the change in the potential energy of this charge? (b) Through what potential difference did the charge move?
3. The difference in potential between the accelerating plates of a TV set is about 25 000 V. If the distance between these plates is 1.5 cm, find the magnitude of the uniform electric field in this region.
4. A capacitor consists of two parallel plates separated by a distance of 0.3 mm. If a 20-V potential difference is maintained between those plates, calculate the electric field strength in the region between the plates.
5. How much work is done (by a battery, generator, or some other source of electrical energy) in moving Avogadro's number of electrons from a point where the electric potential is 6 V to a point where the electric potential is -10 V?
6. A spherical cell is 3.6 μm in diameter, and its outer membrane is 0.11 μm thick. The potential difference between the inner and outer surfaces of the cell is 90 mV, and the inner surface is negative. How much work is required to eject a positive sodium ion (Na^+) from the interior of the cell?
7. An ion, after being accelerated through a potential difference of 60 V, experiences an increase in potential energy of 1.92×10^{-17} J. Calculate the charge on the ion.
8. A proton is between two plates, separated by a dis-

tance of 2 cm, across which there is a potential difference of 5000 V. Find (a) the electrical force exerted on the proton and (b) its acceleration.

9. (a) Calculate the speed of a proton that is accelerated from rest through a potential difference of 120 V. (b) Calculate the speed of an electron that is accelerated through the same potential difference.
10. What potential difference is needed to stop an electron with an initial speed of 4.2×10^5 m/s?
11. (1) Through what potential difference would an electron need to accelerate in order to achieve a speed of 60% of the speed of light, starting from rest? The speed of light is 3×10^8 m/s. (b) Repeat your calculation for a proton.
12. An electron moves from one plate to another. The potential difference between the plates is 2000 V. (a) Find the speed with which the electron strikes the positive plate. (b) Repeat part (a) for a proton moving from the positive toward the negative plate.

Section 17.2 Potential Differences in a Uniform Electric Field

13. In a tandem Van de Graaff accelerator a proton is accelerated through a potential difference of 14 MV. Assuming that the proton starts from rest, calculate its (a) final kinetic energy in joules, (b) final kinetic energy in MeV, and (c) final speed.
14. Suppose an electron is released from rest in a uniform electric field whose strength is 5.9×10^3 V/m. (a) Through what potential difference will it have passed after moving 1 cm? (b) How fast will the electron be moving after it has traveled 1 cm?
15. An electron moving parallel to the x axis has an initial speed of 3.7×10^6 m/s at the origin. Its speed is reduced to 1.4×10^5 m/s at the point $x = 2$ cm. Calculate the potential difference between the origin and the point $x = 2$ cm. Which point is at the higher potential?
16. A positron has the same charge as a proton, but the same mass as an electron. Suppose a positron moves 5.2 cm in the direction of a uniform 480-V/m electric field. (a) How much potential energy does it gain or lose? (b) How much kinetic energy does it gain or lose?
17. A proton moves in a region of a uniform electric field. The proton experiences an increase in kinetic energy of 5×10^{-18} J after being displaced 2 cm in a direction parallel to the field. What is the magnitude of the electric field?
18. A uniform electric field of magnitude 325 V/m is directed in the *negative* y direction in Figure 17.27. The coordinates of point A are $(-0.2, -0.3)$ m, and those of point B are (0.4, 0.5) m. Calculate the potential $V_B - V_A$, using the blue path.

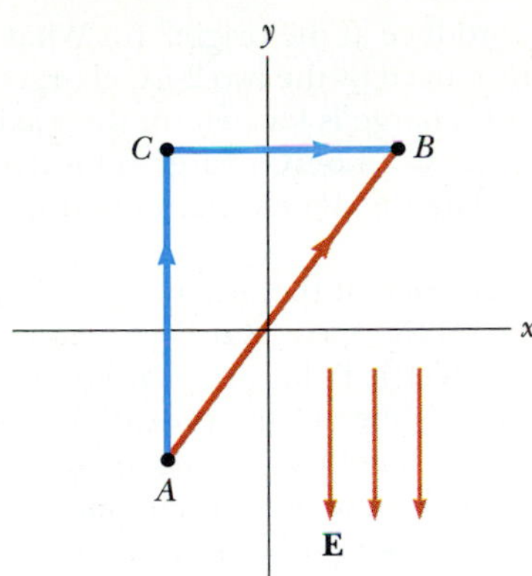

Figure 17.27 (Problem 18)

Section 17.3 Electric Potential and Electric Potential Energy Due to Point Charges

Note: Assume a reference level for potential as $V = 0$ *at* $r = \infty$ *unless the statement of the problem requires otherwise.*

19. (a) Find the potential at a distance of 1 cm from a proton. (b) What is the potential difference between two points that are 1 cm and 2 cm from a proton? (c) Repeat parts (a) and (b) for an electron.
20. Two point charges are located along the y axis. One charge of magnitude 3 nC is at the origin, and a second of magnitude 6 nC is at the point $y = 30$ cm. Calculate the potential (a) at $y = 60$ cm and (b) at $y = -60$ cm. (c) Repeat this problem assuming the 6-nC charge is replaced by a -6-nC charge.
21. A 6-nC charge is located at the point (0, 0.15), where the coordinates are in meters, and a second charge, -4 nC, is at $(-0.45, 0)$. Find the potential (a) at $(0, -0.45)$ and (b) at (0.15, 0).
22. A charge of -3 nC is located at the origin of a coordinate system, and a charge of 8 nC is on the x axis at $x = 2$ m. At what two locations on the x axis is the electric potential equal to zero?
23. At a distance of r from a point charge, q, the electrical potential is $V = 400$ V and the magnitude of the electric field is $E = 150$ N/C. Determine the value of q and r.
24. Given two 2-μC charges, as shown in Figure 17.28, and a positive test charge, $q = 1.28 \times 10^{-18}$ C, at the origin, (a) what is the net force exerted on q by the two 2-μC charges? (b) What field, **E**, do the two 2-μC

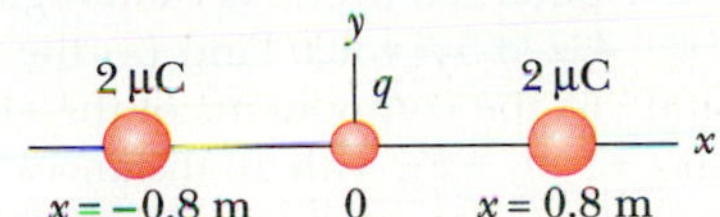

Figure 17.28 (Problem 24)

charges produce at the origin? (c) What is the potential, V, produced by the two 2-μC charges at the origin?

25. A +2.8-μC charge is located on the y axis at $y = +1.6$ m, and a -4.6-μC charge is located at the origin. Calculate the net electric potential at the point (0.4 m, 0).
26. The charge $+q$ is at the origin. The charge $-2q$ is at $x = 2.0$ m on the x axis. (a) For what finite value(s) of x is the electric field zero? (b) For what finite value(s) of x is the electric potential zero?
27. The Bohr model of the hydrogen atom states that the electron can exist only in certain allowed orbits. The radius of each Bohr orbit is given by the expression $r = n^2(0.0529 \text{ nm})$ where $n = 1, 2, 3, \ldots$. Calculate the electric potential energy of a hydrogen atom when the electron (a) is in the first allowed orbit, $n = 1$; (b) is in the second allowed orbit, $n = 2$; (c) has escaped from the atom, $r = \infty$. Express your answers in electron volts.
28. Show that the amount of work required to assemble four identical point charges of magnitude Q at the corners of a square with sides of length s is given by $5.41 k_eQ^2/s$.
29. Two point charges, $Q_1 = +5$ nC and $Q_2 = -3$ nC, are separated by 35 cm. (a) What is the potential energy of the pair? What is the significance of the algebraic sign of your answer? (b) What is the electric potential at a point midway between the charges?
30. Four charges are located at the corners of a rectangle as in Figure 17.29. How much energy would be expended in removing the two 4-μC charges to infinity?

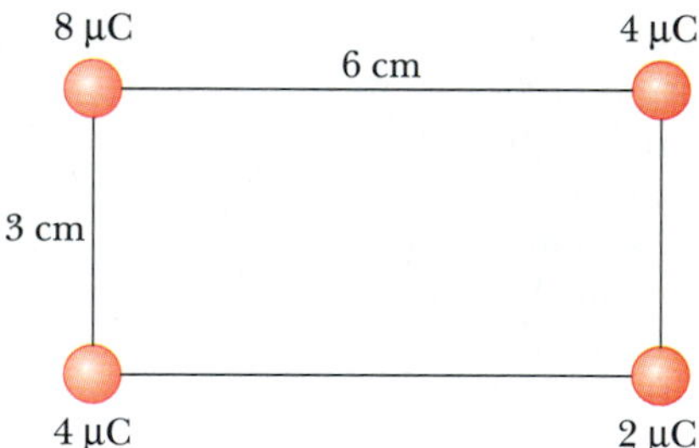

Figure 17.29 (Problem 30)

*Section 17.4 Obtaining E from the Electric Potential

31. The electric potential over a certain region of space is $V = 3x^2y - 4xz - 5y^2$ volts. Find (a) the electric potential and (b) the components of the electric field at the point (+1, 0, +2), with all distances in meters.
32. Over a certain region of space, the electric potential is $V = 5x - 3x^2y + 2yz^2$. Find the expressions for the x, y, and z components of the electric field over this region. What is the magnitude of the field at the point P, which has coordinates (in meters) (1, 0, −2)?
33. The potential in a region between $x = 0$ and $x = 6$ m is $V = a + bx$, where $a = 10$ V and $b = -7$ V/m. Determine (a) the potential at $x = 0$, 3 m, and 6 m, and (b) the magnitude and direction of the electric field at $x = 0$, 3 m, and 6 m.
34. The electric potential in a certain region is given by

$$V = ax^2 + bx + c$$

$$a = 12 \text{ V/m}^2 \qquad b = -10 \text{ V/m} \qquad c = 62 \text{ V}$$

Determine (a) the magnitude and direction of the electric field at $x = +2$ m and (b) the position where the electric field is zero.

35. The electric potential inside a charged spherical conductor of radius R is $V = k_eQ/R$; outside it is $V = k_eQ/r$. Using $E_r = -\dfrac{dV}{dr}$, derive the electric field both (a) inside ($r < R$) and (b) outside ($r > R$) this charge distribution.
36. The electric potential inside a uniformly charged spherical insulator of radius R is

$$V = \frac{k_eQ}{2R}\left(3 - \frac{r^2}{R^2}\right)$$

Outside it is

$$V = \frac{k_eQ}{r}$$

Use $E_r = -\dfrac{dV}{dr}$ to derive the electric field both (a) inside ($r < R$) and (b) outside ($r > R$) this charge distribution.

Section 17.5 Electric Potential Due to Continuous Charge Distributions

37. Consider a ring of radius R with the total charge Q spread uniformly over its perimeter. What is the potential difference between the point at the center of the ring and a point on the axis of the ring at a distance of $2R$ from the center of the ring?
38. Consider a Helmholtz pair consisting of two coaxial rings of 30-cm radius, separated by a distance of 30 cm. (a) Calculate the electric potential at a point on their common axis midway between the two rings, assuming that each ring carries a uniformly distributed charge of +5 μC. (b) What is the potential at this point if the two rings carry equal and opposite charges?
39. A rod of length L (Fig. 17.30) lies along the x axis with its left end at the origin and has a *nonuniform*

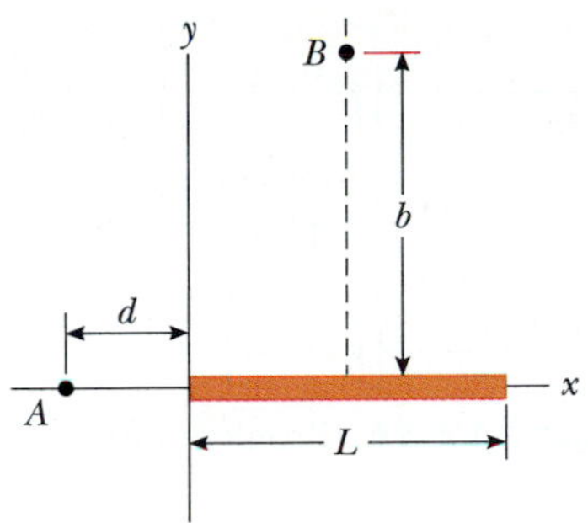

Figure 17.30 (Problems 39 and 40)

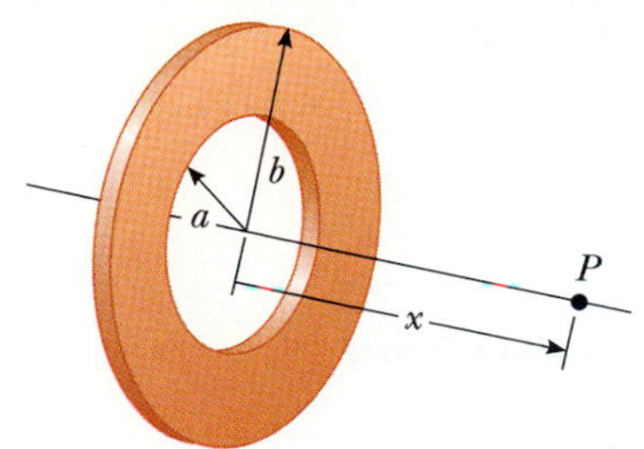

Figure 17.31 (Problem 41)

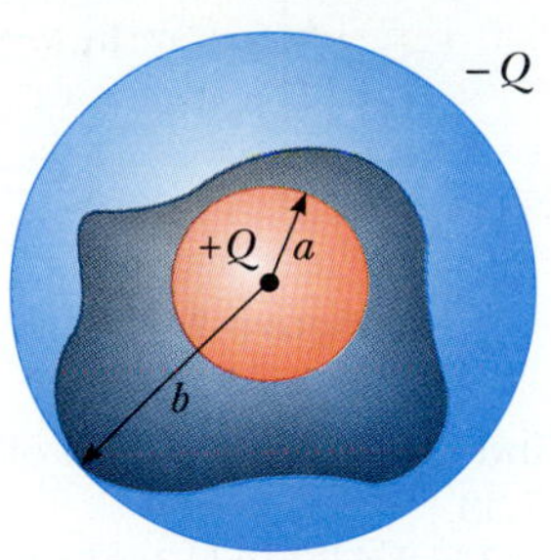

Figure 17.32 (Problem 46)

charge density of $\lambda = \alpha x$ (where α is a positive constant). (a) What are the units of the constant α? (b) Calculate the electric potential at point A, a distance of d from the left end of the rod.

40. For the arrangement described in Problem 39, calculate the electric potential at point B on the perpendicular bisector of the rod, a distance of b above the x axis. Note that the rod has the *nonuniform* charge density $\lambda = \alpha x$.

41. Calculate the electric potential at point P on the axis of the annulus shown in Figure 17.31, which has the uniform charge density α and inner and outer radii a and b, respectively.

Section 17.7 Capacitance

42. (a) How much charge is on each plate of a 4-μF capacitor when it is connected to a 12-V battery? (b) If this same capacitor is connected to a 1.5-V battery, what charge is stored?

43. To what difference in potential would a 2-μF capacitor have to be connected in order to store 98 μC of charge?

44. The plates of a parallel-plate capacitor are separated by 0.1 mm. If the material between the plates is air, what plate area is required to provide a capacitance of 2 pF?

45. Is it feasible to construct a parallel-plate capacitor with its two plates separated by 0.10 mm and a capacitance of 1.0 F?

46. A spherical capacitor consists of a spherical conducting shell of radius b and charge $-Q$ that is concentric with a smaller conducting sphere of radius a and charge $+Q$ (Fig. 17.32). (a) Show that its capacitance is

$$C = \frac{ab}{k_e(b - a)}$$

(b) Show that as b approaches infinity, the capacitance approaches the value $a/k_e = 4\pi\epsilon_0 a$.

47. Regarding the Earth and a cloud layer 800 m above the Earth as the "plates" of a capacitor, calculate the capacitance if the cloud layer has an area of 1.0 km^2. If an electric field of 3.0×10^6 N/C makes the air break down and conduct electricity (that is, causes lightning), what is the maximum charge the cloud can hold?

48. A 1-megabit computer memory chip contains many 6×10^{-14} F capacitors. Each capacitor has a plate area of 21×10^{-12} m^2. Determine the plate separation of such a capacitor (assume an air-filled parallel-plate configuration). The characteristic atomic diameter is 0.1 nm. Express the plate separation in nanometers.

49. An air-filled capacitor consists of two parallel plates, each with an area of 7.6 cm^2, separated by a distance of 1.8 mm. If a 20-V potential difference is applied to these plates, calculate (a) the electric field between the plates, (b) the surface charge density, (c) the capacitance, and (d) the charge on each plate.

50. A 50-m length of coaxial cable has an inner conductor with a diameter of 2.58 mm and a charge of +8.1 μC. The surrounding conductor has an inner diameter of 7.27 mm and a charge of −8.1 μC. (a) What is the capacitance of this cable? (b) What is the potential difference between the two conductors?

Section 17.8 Combinations of Capacitors

51. Two capacitors, $C_1 = 5$ μF and $C_2 = 12$ μF, are connected in parallel, and the resulting combination is connected to a 9-V battery. (a) What is the value of the equivalent capacitance of the combination? What is (b) the potential difference across each capacitor and (c) the charge stored on each capacitor?

52. The two capacitors of Problem 51 are now connected in series and to a 9-V battery. Find (a) the value of the equivalent capacitance of the combination, (b) the voltage across each capacitor, and (c) the charge on each capacitor.

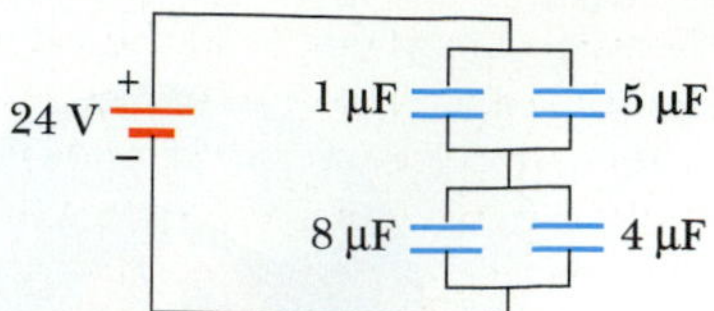

Figure 17.33 (Problem 53)

53. Find the charge on each of the capacitors in Figure 17.33.
54. (a) Find the equivalent capacitance of the group of capacitors shown in Figure 17.34. (b) Find the charge on and the potential difference across each.

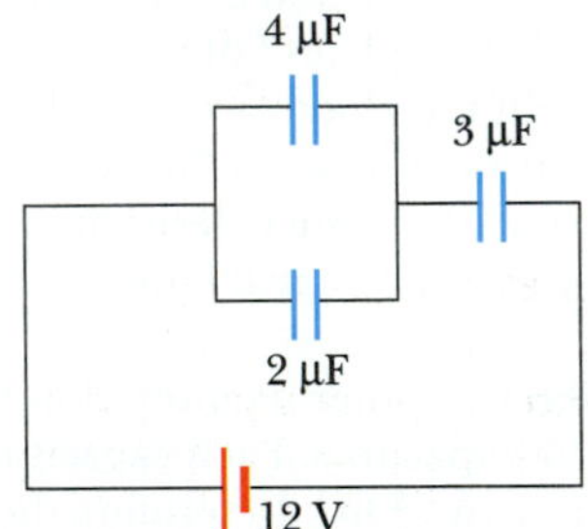

Figure 17.34 (Problem 54)

55. (a) Determine the equivalent capacitance for the capacitor network shown in Figure 17.35. (b) If the network is connected to a 12-V battery, calculate the potential difference across each capacitor and the charge on each capacitor.
56. Evaluate the effective capacitance of the configuration shown in Figure 17.36. Each of the capacitors is identical and has capacitance C.
57. Consider the circuit shown in Figure 17.37, where $C_1 = 6\ \mu\text{F}$, $C_2 = 3\ \mu\text{F}$, and $V = 20$ V. C_1 is first charged by the closing of switch S_1. Switch S_1 is then opened, and the charged capacitor is connected to the uncharged capacitor by the closing of S_2. Calculate the initial charge acquired by C_1 and the final charge on each of the two capacitors.
58. Consider the group of capacitors shown in Figure 17.38. (a) Find the equivalent capacitance between points a and b. (b) Determine the charge on each capacitor when the potential difference between a and b is 12 V.

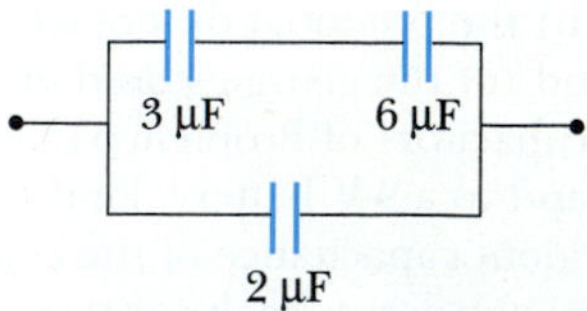

Figure 17.35 (Problem 55)

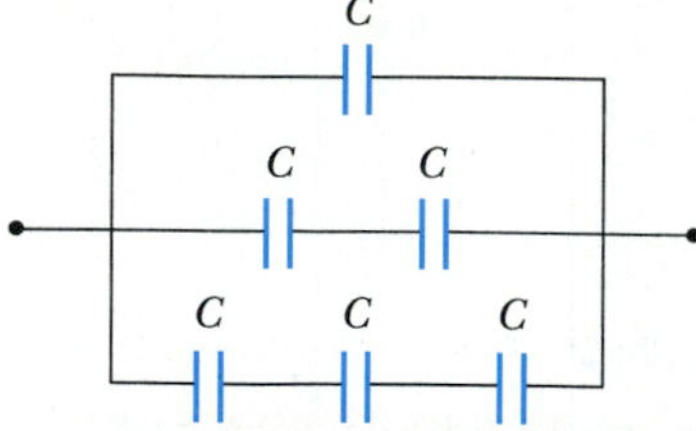

Figure 17.36 (Problem 56)

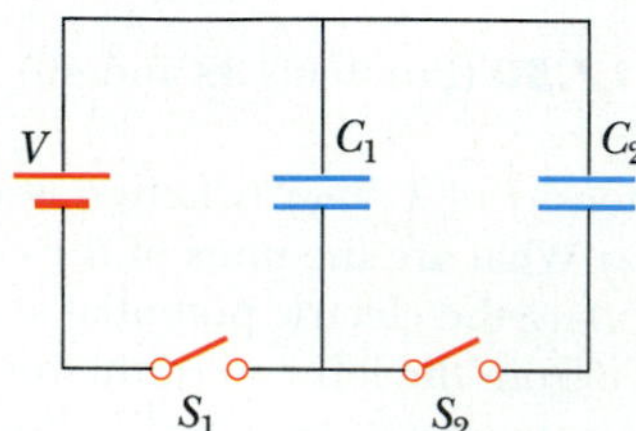

Figure 17.37 (Problem 57)

59. Consider the combination of capacitors shown in Figure 17.39. (a) What is the equivalent capacitance between points a and b? (b) Determine the charge on each capacitor if $V_{ab} = 4.8$ V.
60. How should four 2-μF capacitors be connected to have a total capacitance of (a) 8 μF, (b) 2 μF, (c) 1.5 μF, and (d) 0.5 μF?
61. A conducting slab with thickness d and area A is inserted into the space between the plates of a parallel-plate capacitor with spacing s and area A, as shown in Figure 17.40. What is the value of the capacitance of the system?

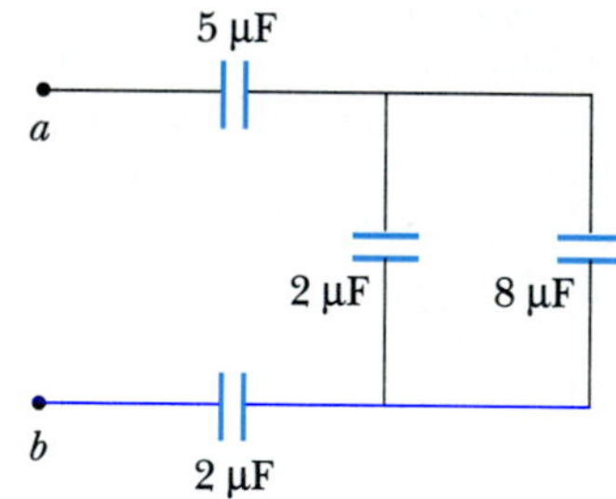

Figure 17.38 (Problem 58)

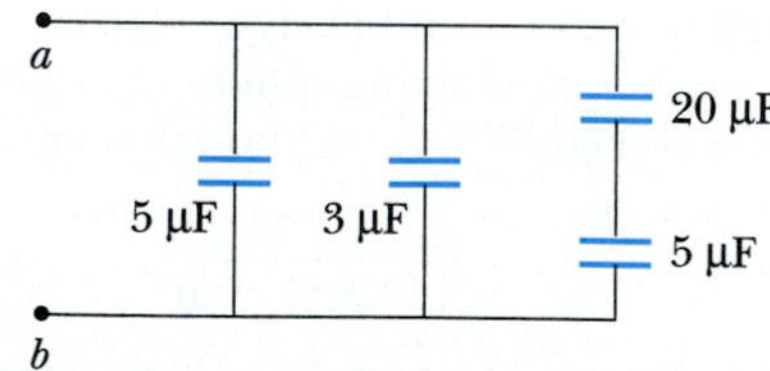

Figure 17.39 (Problem 59)

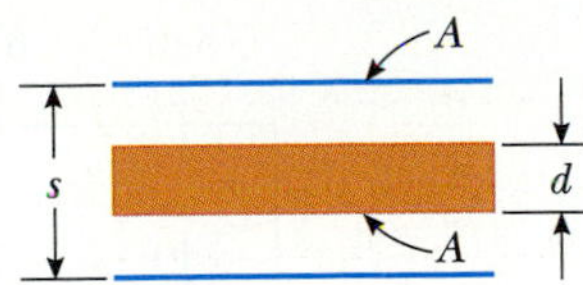

Figure 17.40 (Problem 61)

Section 17.9 Energy Stored in a Charged Capacitor

62. (a) A 3-μF capacitor is connected to a 12-V battery. How much energy is stored in the capacitor? (b) If the capacitor had been connected to a 6-V battery, how much energy would have been stored?
63. A capacitor is connected to a 120-V source and holds a charge of 36 μC. (a) What is the capacitance of the capacitor? (b) Find the energy stored by the capacitor by the use of at least two different equations.
64. A parallel-plate capacitor is charged and then disconnected from a battery. By what fraction does the stored energy change (increase or decrease) when the plate separation is doubled?
65. A parallel-plate capacitor has plates of area 2 cm^2, a separation of 5 mm, and air between the plates. If a 12-V battery is connected to this capacitor, how much energy does it store?
66. Two capacitors, $C_1 = 25\ \mu$F and $C_2 = 5\ \mu$F, are connected in parallel and charged with a 100-V power supply. (a) Calculate the total energy stored in the two capacitors. (b) What potential difference would be required across the same two capacitors connected in series in order that the combination store the same energy as in (a)?
67. A parallel-plate capacitor has the charge Q and plates of area A. Show that the force exerted on each plate by the other is $F = Q^2/2\epsilon_0 A$. (*Hint:* Let $C = \epsilon_0 A/x$ for an arbitrary plate separation, x; then require that the work done in separating the two charged plates be $W = \int F\,dx$.)
68. A uniform electric field, $E = 3000$ V/m, exists within a certain region. What volume of space would contain an energy equal to 10^{-7} J? Express your answer in cubic meters and in liters.

Section 17.10 Capacitors with Dielectrics

69. Find the capacitance of a parallel-plate capacitor that uses Bakelite as a dielectric, if each of the plates has an area of 5 cm^2 and the plate separation is 2 mm.
70. A capacitor that has air between its plates is connected across a potential difference of 12 V and stores 48 μC of charge. It is then disconnected from the source while still charged. (a) Find the capacitance of the capacitor. (b) A piece of Teflon is inserted between the plates. Find the voltage and charge on the capacitor. (c) Find its new capacitance.
71. A capacitor with air between the plates is charged to 100 V and then disconnected from the battery. When a piece of glass is placed between the plates, the voltage across the capacitor drops to 25 V. What is the dielectric constant of the glass? Assume that the glass completely fills the space between the plates.
72. Determine (a) the capacitance and (b) the maximum voltage that can be applied to a Teflon-filled parallel-plate capacitor having a plate area of 1.75 cm^2 and insulation thickness of 0.04 mm.
73. (a) How much charge can be placed on a capacitor with air between the plates before it breaks down, if the area of each of the plates is 5 cm^2? (b) Find the maximum charge if polystyrene is used between the plates instead of air.

Additional Problems

74. At a distance of r away from the point charge q, the electric potential is $V = 600$ V and the magnitude of the electric field is $E = 200$ N/C. Determine the values of q and r.
75. The gap between electrodes in a spark plug is 0.06 cm. In order to produce an electric spark in a gasoline-air mixture, an electric field of 3×10^6 V/m must be achieved. On starting a car, what minimum voltage must be supplied by the ignition circuit?
76. To recharge a 12-V battery, a battery charger must move 3.6×10^5 C of charge from the negative terminal to the positive terminal. What amount of work is done by the battery charger? How many kilowatt-hours is this?
77. A reasonable model of the proton charge distribution in an atomic nucleus is a uniformly charged sphere. A nucleus of lead-208 has a radius of 6.34×10^{-15} m and contains 82 protons, each with a charge of 1.6×10^{-19} C. Calculate the electric potential at the surface of this nucleus.
78. How much electrical charge is needed to raise an isolated metal sphere of radius 1.0 m to a potential of 1.0×10^6 V?
79. The charge distribution shown in Figure 17.41 is referred to as a linear quadrupole. (a) Find the exact expression for the potential at a point on the x axis where $x > d$. (b) Show that the expression obtained in (a) reduces to

$$V = \frac{2k_e Q d^2}{x^3}$$

when $x \gg d$.
80. The x axis is the symmetry axis of a uniformly charged ring of radius R and charge Q (Fig. 17.42). A point charge, Q, of mass M is located at the center of

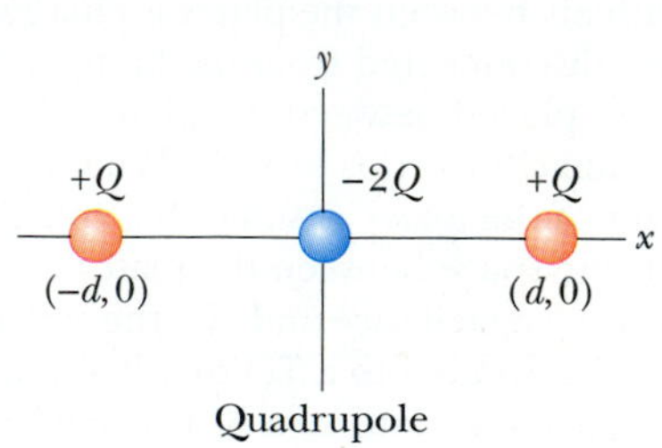

Figure 17.41 (Problem 79)

Figure 17.42 (Problem 80)

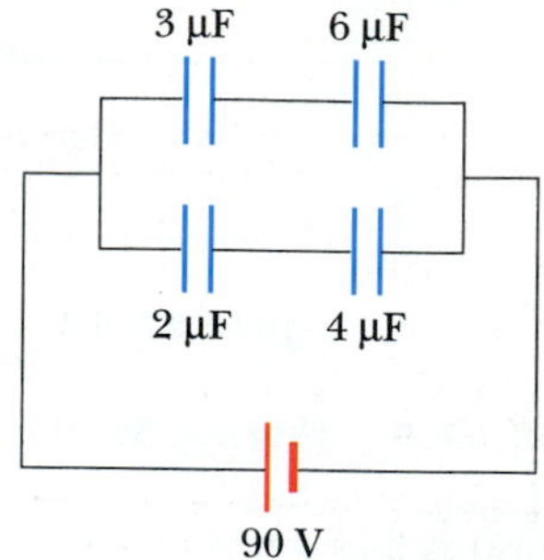

Figure 17.43 (Problem 82)

the ring. When it is displaced slightly, the point charge accelerates along the x axis to infinity. Show that the ultimate speed of the point charge is

$$v = \left(\frac{2k_eQ^2}{MR}\right)^{1/2}$$

81. When two capacitors are connected in parallel, the equivalent capacitance is 4 μF. If the same capacitors are reconnected in series, the equivalent capacitance is one-fourth the capacitance of one of the two capacitors. Determine the two capacitances.
82. For the system of capacitors shown in Figure 17.43, find (a) the equivalent capacitance of the system, (b) the potential across each capacitor, (c) the charge on each capacitor, and (d) the total energy stored by the group.
83. The energy stored in a 52-μF capacitor is used to melt a 6-mg sample of lead. To what voltage must the capacitor be initially charged, assuming the initial temperature of the lead is 20°C?
84. A 2-nF parallel-plate capacitor is charged to the initial potential difference $V_i = 100$ V and then isolated. The dielectric material between the plates is mica ($\kappa = 5$). (a) How much work is required to withdraw the mica sheet? (b) What is the potential difference of the capacitor after the mica is withdrawn?
85. A parallel-plate capacitor is constructed using a dielectric material whose dielectric constant is 3 and whose dielectric strength is 2×10^8 V/m. The desired capacitance is 0.25 μF, and the capacitor must withstand a maximum potential difference of 4000 V. Find the minimum area of the capacitor plates.
86. A parallel-plate capacitor is constructed using three different dielectric materials, as shown in Figure 17.44. (a) Find an expression for the capacitance of the device in terms of the plate area, A, and d, κ_1, κ_2, and κ_3. (b) Calculate the capacitance using the values $A = 1$ cm^2, $d = 2$ mm, $\kappa_1 = 4.9$, $\kappa_2 = 5.6$, and $\kappa_3 = 2.1$.
87. The liquid-drop model of the nucleus suggests that high-energy oscillations of certain nuclei can split the nucleus into two unequal fragments plus a few neutrons. The fragments acquire kinetic energy from their mutual Coulombic repulsion. Calculate the Coulomb potential energy (in MeV) of two spherical fragments from a uranium nucleus having the following charges and radii: $+38e$ and radius 5.5×10^{-15} m; $+54e$ and radius 6.2×10^{-15} m. Assume that the charge is distributed uniformly throughout the volume of each spherical fragment and that their surfaces are initially in contact at rest. (The electrons surrounding the nucleus can be neglected.)
88. Figure 17.45 shows two capacitors in series. The rigid center section of length b is movable horizontally, and the area of each plate is A. Show that the capacitance of the series combination is independent of the position of the center section and is

$$C = \frac{\epsilon_0 A}{a - b}$$

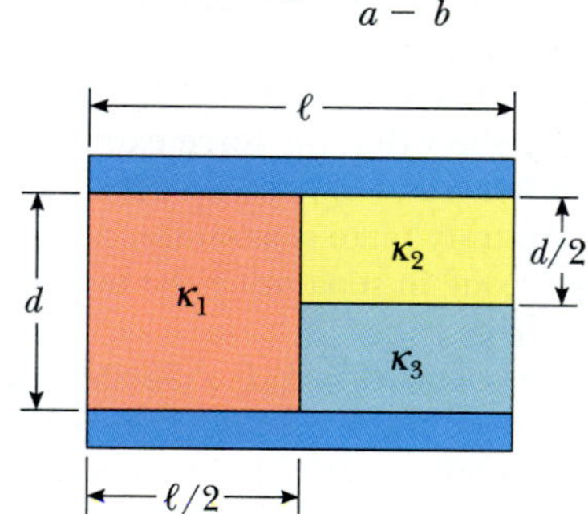

Figure 17.44 (Problem 86)

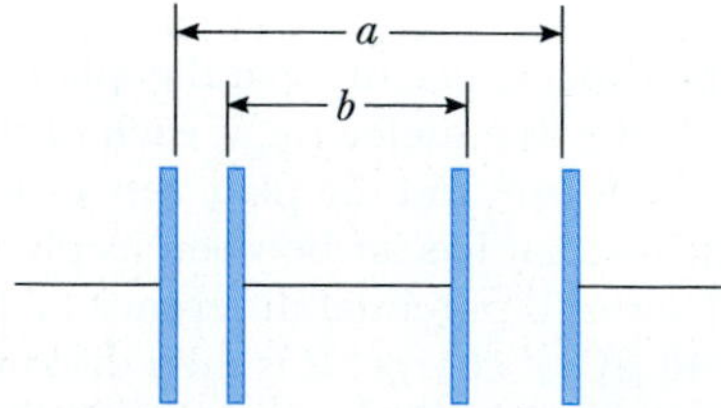

Figure 17.45 (Problem 88)

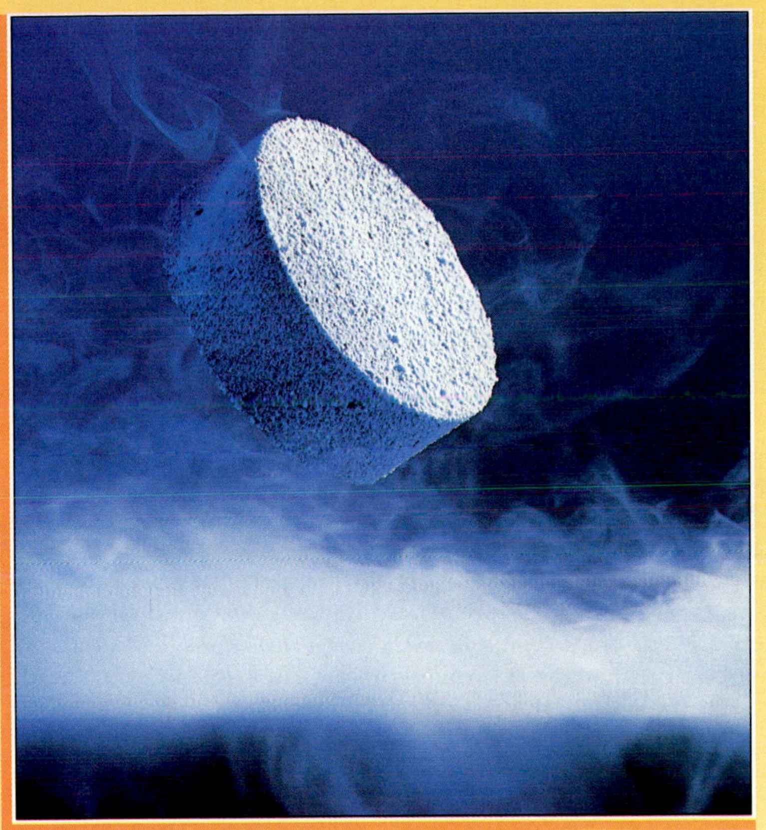

Photograph of a small permanent magnet levitated above a disk of the superconductor $YBa_2Cu_3O_7$, which is at 77 K. This superconductor has zero electric resistance at temperatures below 92 K and expels any applied magnetic field. (The superconductor is not visible in the photograph.) *(© Tony Stone/ Worldwide)*

Current and Direct Current Circuits

CHAPTER 18

Thus far, our discussion of electrical phenomena has been confined to charges at rest, or electrostatics. We shall now consider situations involving electric charges in motion. The term *electric current,* or simply *current,* is used to describe the rate of flow of charge through some region of space. Most practical applications of electricity involve electric currents. For example, the battery of a flashlight supplies current to the filament of the bulb when the switch is turned on. A variety of home appliances operate on alternating current. In these common situations, the flow of charge takes place in a conductor, such as a copper wire. It is also possible for currents to exist outside a conductor. For instance, a beam of electrons in a TV picture tube constitutes a current.

In this chapter we shall first define current and current density. A microscopic description of current will be given, and some of the factors that contribute to

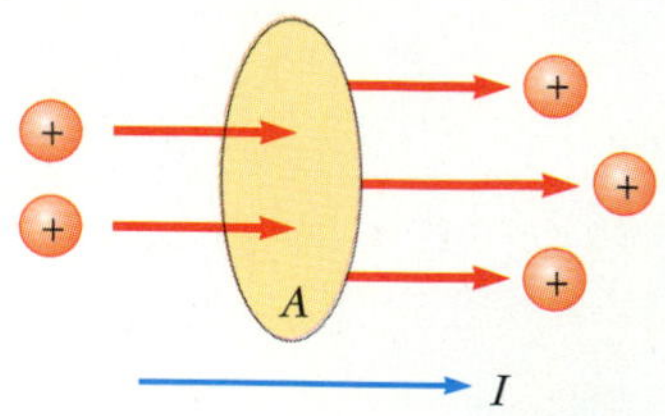

Figure 18.1
Charges in motion through an area, A. The time rate of flow of charge through the area is defined as the current I. The direction of the current is the direction in which positive charge would flow if it were free to do so.

resistance to the flow of charge in conductors will be discussed. Mechanisms responsible for the electrical resistances of various materials depend on the materials' compositions and on temperature. A classical model is used to describe electrical conduction in metals; we shall point out some of the limitations of this model.

This chapter is also concerned with the analysis of some simple circuits whose elements include batteries, resistors, and capacitors in varied combinations. The analysis of these circuits is simplified by the use of two rules known as *Kirchhoff's rules,* which follow from the laws of conservation of energy and conservation of charge. Most of the circuits analyzed are assumed to be in *steady state,* which means that the currents are constant in magnitude and direction. We shall close with a discussion of circuits containing resistors and capacitors, in which the current varies with time.

18.1 Electric Current

Whenever electric charges of like sign move, a *current* is said to exist. To define current more precisely, suppose the charges are moving perpendicularly to a surface of area A, as in Figure 18.1. (This area could be the cross-sectional area of a wire, for example.) The **current** is *the rate at which charge flows through this surface.* If ΔQ is the amount of charge that passes through this area in a time interval, Δt, the **average current,** I_{av}, is the ratio of the charge to the time interval:

$$I_{av} = \frac{\Delta Q}{\Delta t} \qquad [18.1]$$

If the rate at which charge flows varies in time, the current also varies in time. We define the **instantaneous current,** I, as the differential limit of the preceding expression:

Electric current

$$I \equiv \frac{dQ}{dt} \qquad [18.2]$$

The SI unit of current is the **ampere** (A):

$$1\text{ A} = 1\text{ C/s} \qquad [18.3]$$

That is, 1 A of current is equivalent to 1 C of charge passing through a surface in 1 s.

The direction of the current

When charges flow through a surface as in Figure 18.1, they can be positive, negative, or both. *It is conventional to give the current the same direction as the flow of positive charge.* In a common conductor, such as copper, the current is due to the motion of the negatively charged electrons. Therefore, when we speak of current in such a conductor, *the direction of the current is opposite the direction of flow of electrons.* On the other hand, if one considers a beam of positively charged protons in an accelerator, the current is in the direction of motion of the protons. In some cases—gases and electrolytes, for example—the current is the result of the flow of both positive and negative charges. It is common to refer to a moving charge (whether it is positive or negative) as a mobile *charge carrier*. For example, the charge carriers in a metal are electrons.

It is instructive to relate current to the motions of the charged particles. To illustrate this point, consider the current in a conductor of cross-sectional area A (Fig. 18.2). The volume of an element of the conductor of length Δx is $A\,\Delta x$. If n represents the number of mobile charge carriers per unit volume, then the number of carriers in the volume element is $nA\,\Delta x$. Therefore, the charge ΔQ in this element is

$$\Delta Q = \text{number of carriers} \times \text{charge per carrier} = (nA\,\Delta x)q$$

where q is the charge on each carrier. If the carriers move with a speed of v_d, the distance they move in the time Δt is $\Delta x = v_d\,\Delta t$. Therefore, we can write ΔQ in the form

$$\Delta Q = (nAv_d\,\Delta t)q$$

If we divide both sides of this equation by Δt, we see that the current in the conductor is

$$I = \frac{\Delta Q}{\Delta t} = nqv_dA \qquad [18.4]$$

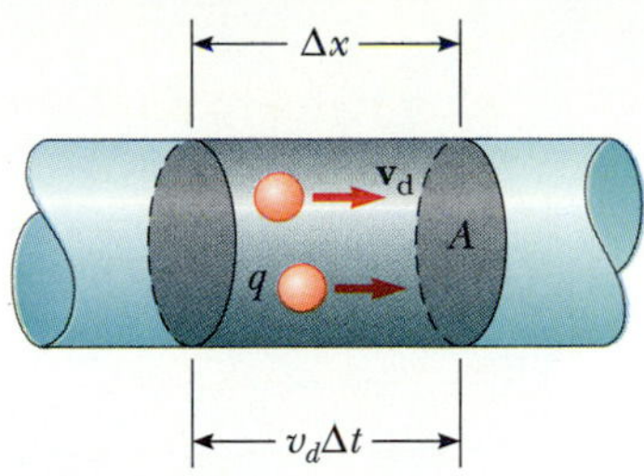

Figure 18.2 A section of a uniform conductor of cross-sectional area A. The charge carriers move with a speed of v_d, and the distance they travel in the time Δt is given by $\Delta x = v_d\,\Delta t$. The number of mobile charge carriers in the section of length Δx is given by $nAv_d\,\Delta t$, where n is the number of mobile carriers per unit volume.

The speed of the charge carriers, v_d, is an average speed called the **drift speed.** To understand its meaning, consider a conductor in which the charge carriers are free electrons. If the conductor is isolated, these electrons undergo random motion similar to that of gas molecules. When a potential difference is applied across the conductor (say, by means of a battery), an electric field is set up in the conductor, which creates an electric force on the electrons and hence a current. In reality, the electrons do not simply move in straight lines along the conductor. Instead, they undergo repeated collisions with the metal atoms, and the result is a complicated zigzag motion (Fig. 18.3). The energy transferred from the electrons to the metal atoms during collision causes an increase in the vibrational energy of the atoms and a corresponding increase in the temperature of the conductor. However, despite the collisions, the electrons move slowly along the conductor (in a direction opposite **E**) with the drift velocity, $\mathbf{v}_d$. The work done by the field on the electrons exceeds the average loss in energy due to collisions, and this work provides a steady current. One can think of the collisions within a conductor as being an effective internal friction (or drag force) similar to that experienced by the molecules of a liquid flowing through a pipe stuffed with steel wool.

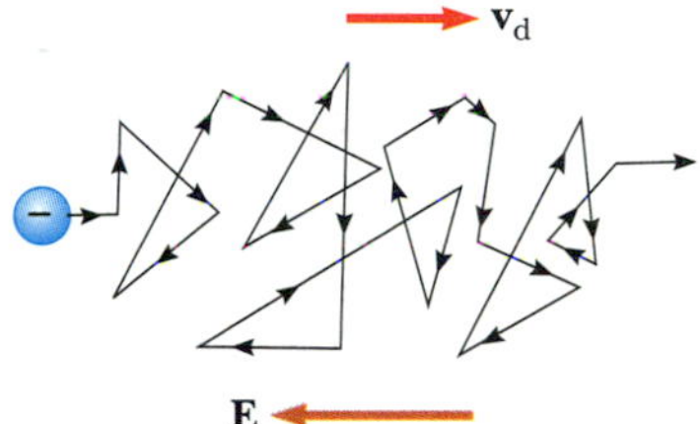

Figure 18.3 A schematic representation of the zigzag motion of a charge carrier in a conductor. The changes in direction are due to collisions with atoms in the conductor. Note that the net motion of electrons is opposite the direction of the electric field.

The **current density,** J, in the conductor is defined to be the current per unit area. Since $I = nqv_dA$, the current density is

Current density

$$J \equiv \frac{I}{A} = nqv_d \qquad [18.5]$$

where J has the SI units amperes per square meter. In general, the current density is a *vector quantity*. That is,

$$\mathbf{J} = nq\mathbf{v}_d \qquad [18.6]$$

From this definition, we see that the current density is in the direction of motion of positive charge carriers and opposite the direction of motion of negative charge carriers. Since the drift velocity is proportional to the electric field, **E**, in the conductor, we conclude that the current density is also proportional to **E**.

Example 18.1 The Drift Velocity in a Copper Wire

A copper wire of cross-sectional area $3 \times 10^{-6}\ \text{m}^2$ carries a current of 10 A. Find the drift speed of the electrons in this wire. The density of copper is $8.95\ \text{g/cm}^3$.

Solution From the periodic table of the elements, we find that the atomic mass of copper is 63.5 g/mol. Knowing the density of copper enables us to calculate the volume occupied by 63.5 g of copper:

$$V = \frac{m}{\rho} = \frac{63.5\ \text{g}}{8.95\ \text{g/cm}^3} = 7.09\ \text{cm}^3$$

Recall that one atomic mass of any substance contains Avogadro's number of atoms, 6.02×10^{23} atoms.

If we now assume that each copper atom contributes one free electron to this volume of copper, we find that the number of electrons per unit volume, n, is

$$n = \frac{6.02 \times 10^{23}\ \text{electrons}}{7.09\ \text{cm}^3} = 8.48 \times 10^{22}\ \text{electrons/cm}^3$$

$$= \left(8.48 \times 10^{22}\ \frac{\text{electrons}}{\text{cm}^3}\right)\left(10^6\ \frac{\text{cm}^3}{\text{m}^3}\right) = 8.48 \times 10^{28}\ \text{electrons/m}^3$$

From Equation 18.4, we have

$$v_d = \frac{I}{nqA} = \frac{10\ \text{C/s}}{(8.48 \times 10^{28}\ \text{m}^{-3}/\text{m}^3)(1.6 \times 10^{-19}\ \text{C})(3 \times 10^{-6}\ \text{m}^2)}$$

$$= 2.46 \times 10^{-4}\ \text{m/s}$$

Example 18.1 shows that typical drift speeds are very small. In fact, the drift speed is much smaller than the average speed between collisions. For instance, electrons traveling with this speed would take about 68 min to travel 1 m! In view of this low speed, you might wonder why a light turns on almost instantaneously when a switch is thrown. This can be explained by considering the flow of water through a pipe. If a drop of water is forced in one end of a pipe that is already filled with water, a drop must be pushed out the other end. While it may take individual drops of water a long time to make it through the pipe, a flow initiated at one end produces a similar flow at the other end very quickly. In a conductor, the electric field that drives the free electrons travels through the conductor with a speed close to that of light. Thus, when you flip a light switch, the message for the electrons to start moving through the wire (the electric field) reaches them at a speed on the order of 10^7 m/s.

18.2 Resistance and Ohm's Law

When a voltage (potential difference), V, is applied across the ends of a metallic conductor as in Figure 18.4, the current in the conductor is found to be proportional to the applied voltage; that is, $I \propto V$. If the proportionality is exact, we can

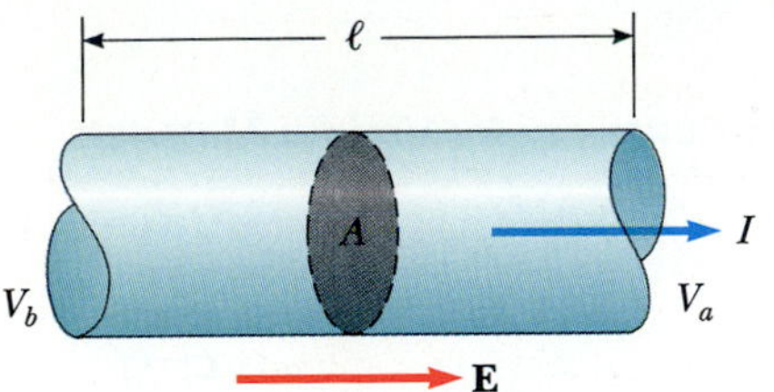

Figure 18.4
A uniform conductor of length ℓ and cross-sectional area A. The current I in the conductor is proportional to the applied voltage, $V = V_b - V_a$. The electric field, **E**, set up in the conductor is also proportional to the current.

write $V = IR$, where the proportionality constant R is called the resistance of the conductor. In fact, we define this **resistance** as the ratio of the voltage across the conductor to the current it carries:

$$R \equiv \frac{V}{I} \quad [18.7]$$

Resistance

Resistance has the SI units volts per ampere, called **ohms** (Ω). Thus, if a potential difference of 1 V across a conductor produces a current of 1 A, the resistance of the conductor is 1 Ω. For example, if an electrical appliance connected to a 120-V source carries a current of 6 A, its resistance is 20 Ω.

It is useful to compare the concepts of electric current, voltage, and resistance with the flow of water in a river. As water flows downhill in a river of constant width and depth, the flow rate (water current) depends on the angle of flow and the effects of rocks, the river bank, and other obstructions. Likewise, electric current in a uniform conductor depends on the applied voltage and the resistance of the conductor caused by collisions of the electrons with atoms in the conductor.

For many materials, including most metals, experiments show that *the resistance is constant over a wide range of applied voltages.* This statement is known as *Ohm's law* after Georg Simon Ohm (1789–1854), who was the first to conduct a systematic study of electrical resistance.

Ohm's law is *not* a fundamental law of nature, but an empirical relationship that is valid only for certain materials. Materials that obey Ohm's law, and hence have a constant resistance over a wide range of voltages, are said to be *ohmic.* Materials that do not obey Ohm's law are *nonohmic.* Ohmic materials have a linear current-voltage relationship over a large range of applied voltages (Fig. 18.5a). Nonohmic materials have a nonlinear current-voltage relationship (Fig. 18.5b). One common semiconducting device that is nonohmic is the diode. Its resistance is small for currents in one direction (positive V) and large for currents in the reverse direction (negative V). Most modern electronic devices, such as transis-

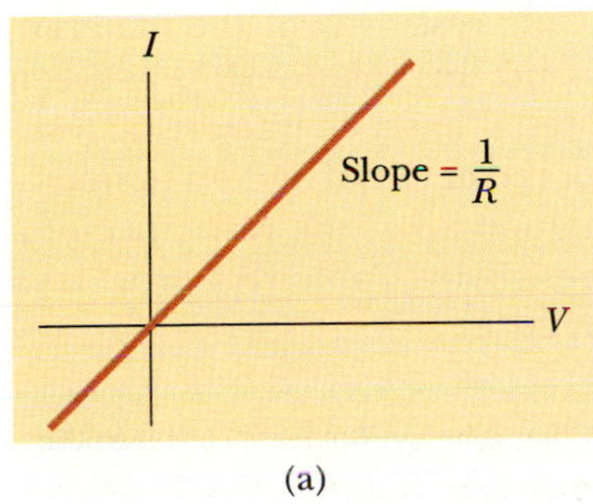

(a)

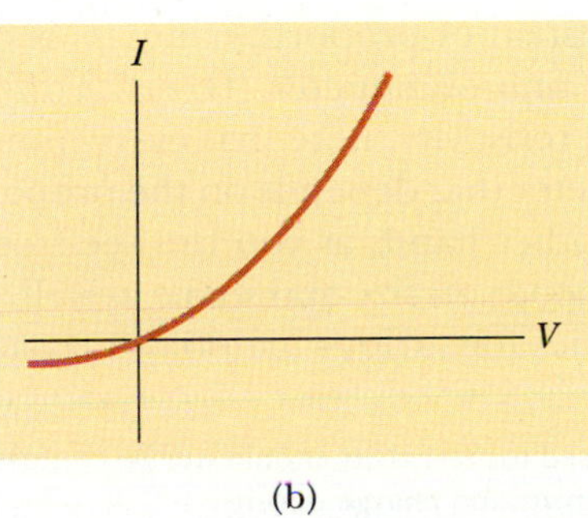

(b)

Figure 18.5
(a) The current-voltage curve for an ohmic material. The curve is linear, and the slope gives the resistance of the conductor. (b) A nonlinear current-voltage curve for a semiconducting diode. This device does not obey Ohm's law.

Table 18.1
Resistivities and Temperature Coefficients of Resistivity for Various Materials

Material	Resistivity[a] ($\Omega \cdot m$)	Temperature Coefficient of Resistivity α [$(°C)^{-1}$]
Silver	1.59×10^{-8}	3.8×10^{-3}
Copper	1.7×10^{-8}	3.9×10^{-3}
Gold	2.44×10^{-8}	3.4×10^{-3}
Aluminum	2.82×10^{-8}	3.9×10^{-3}
Tungsten	5.6×10^{-8}	4.5×10^{-3}
Iron	10×10^{-8}	5.0×10^{-3}
Platinum	11×10^{-8}	3.92×10^{-3}
Lead	22×10^{-8}	3.9×10^{-3}
Nichrome[b]	150×10^{-8}	0.4×10^{-3}
Carbon	3.5×10^{-5}	-0.5×10^{-3}
Germanium	0.46	-48×10^{-3}
Silicon	640	-75×10^{-3}
Glass	$10^{10} - 10^{14}$	
Hard rubber	$\approx 10^{13}$	
Sulfur	10^{15}	
Quartz (fused)	75×10^{16}	

[a] All values are at 20°C.
[b] A nickel-chromium alloy commonly used in heating elements.

tors, have nonlinear current-voltage relationships; their operation depends on the particular ways in which they violate Ohm's law.

It is common practice to express Ohm's law as

$$V = IR \qquad [18.8]$$

where R is understood to be independent of V. We shall continue to use this traditional form of Ohm's law when discussing electrical circuits. A **resistor** is a simple circuit element that provides a specified resistance in an electrical circuit. The symbol for a resistor in circuit diagrams is a zigzag line, —ʌʌʌ—.

The resistance of an ohmic conducting wire is found to be proportional to its length and inversely proportional to its cross-sectional area. That is,

$$R = \rho \frac{\ell}{A} \qquad [18.9]$$

where the constant of proportionality, ρ, is called[1] the **resistivity** of the material, which has the unit ohm-meter ($\Omega \cdot m$). To understand this relationship between resistance and resistivity, note that every ohmic material has a characteristic resistivity, a parameter that depends on the properties of the material and on temperature. On the other hand, as you can see from Equation 18.7, the resistance of a conductor depends on size and shape as well as on resistivity. Table 18.1 provides a list of resistivities for various materials measured at 20°C.

[1] The symbol ρ used for resistivity should not be confused with the same symbol used earlier in the text for mass density and charge density.

The inverse of the resistivity is defined[2] as the **conductivity**, σ. Hence, the resistance of an ohmic conductor can also be expressed in terms of its conductivity, as

$$R = \frac{\ell}{\sigma A} \qquad [18.10]$$

where $\sigma(= 1/\rho)$ has the unit $(\Omega \cdot \text{m})^{-1}$.

Equation 18.10 shows that the resistance of a cylindrical conductor is proportional to its length and inversely proportional to its cross-sectional area. This is analogous to the flow of liquid through a pipe. As the length of the pipe is increased, the resistance to liquid flow increases because of a gain in friction between the fluid and the walls of the pipe. As its cross-sectional area is increased, the pipe can transport more fluid in a given time interval, so its resistance drops.

George Simon Ohm (1789–1854), German physicist. *(Courtesy of AIP Niels Bohr Library, E. Scott Barr Collection)*

Example 18.2 The Resistance of Nichrome Wire

(a) Calculate the resistance per unit length of a 22-gauge Nichrome wire of radius 0.321 mm.

Solution The cross-sectional area of this wire is

$$A = \pi r^2 = \pi(0.321 \times 10^{-3}\text{m})^2 = 3.24 \times 10^{-7} \text{ m}^2$$

The resistivity of Nichrome is $1.5 \times 10^{-6}\ \Omega \cdot \text{m}$ (Table 18.1). Thus, we can use Equation 18.9 to find the resistance per unit length:

$$\frac{R}{\ell} = \frac{\rho}{A} = \frac{1.5 \times 10^{-6}\ \Omega\cdot\text{m}}{3.24 \times 10^{-7}\text{ m}^2} = \boxed{4.6\ \Omega/\text{m}}$$

(b) If a potential difference of 10 V is maintained across a 1-m length of the Nichrome wire, what is the current in the wire?

Solution Since a 1-m length of this wire has a resistance of 4.6 Ω, Ohm's law gives

$$I = \frac{V}{R} = \frac{10\text{ V}}{4.6\ \Omega} = \boxed{2.2\text{ A}}$$

Note from Table 18.1 that the resistivity of Nichrome is about 100 times that of copper, a typical "good" conductor. Therefore, a copper wire of the same radius would have a resistance per unit length of only 0.052 Ω/m. A 1-m length of copper wire of the same radius would carry the same current (2.2 A) with an applied voltage of only 0.11 V.

Because of its high resistivity and its resistance to oxidation, Nichrome is often used for heating elements in toasters, irons, and electric heaters.

Exercise What is the resistance of a 6-m length of 22-gauge Nichrome wire? How much current does it carry when connected to a 120-V source?

Answer 28 Ω; 4.3 A.

Exercise Calculate the current density and electric field in the wire, assuming that it carries a current of 2.2 A.

Answer $6.7 \times 10^6\text{ A/m}^2$; 10 N/C.

[2] Again, do not confuse the symbol σ for conductivity with the same symbol used for surface charge density.

Change in Resistivity with Temperature

Resistivity depends on a number of factors, one of which is temperature. For most metals, resistivity increases approximately linearly with increasing temperature over a limited temperature range, according to the expression

Variation of ρ with temperature

$$\rho = \rho_0[1 + \alpha(T - T_0)] \qquad [18.11]$$

where ρ is the resistivity at some temperature, T (in degrees Celsius), ρ_0 is the resistivity at some reference temperature, T_0 (usually 20°C), and α is called the **temperature coefficient of resistivity.** From Equation 18.11, we see that α can also be expressed as

Temperature coefficient of resistivity

$$\alpha = \frac{1}{\rho_0}\frac{\Delta\rho}{\Delta T} \qquad [18.12]$$

where $\Delta\rho = \rho - \rho_0$ is the change in resistivity in the temperature interval $\Delta T = T - T_0$.

The resistivities and temperature coefficients of certain materials are listed in Table 18.1. Note the enormous range in resistivities, from very low values for good conductors, such as copper and silver, to very high values for good insulators, such as glass and rubber. An ideal, or "perfect," conductor would have zero resistivity, and an ideal insulator would have infinite resistivity.

Since resistance is proportional to resistivity according to Equation 18.9, the temperature variation of the resistance can be written

$$R = R_0[1 + \alpha(T - T_0)] \qquad [18.13]$$

Precise temperature measurements are often made using this property, as shown in the following example.

Example 18.3 A Platinum Resistance Thermometer

A resistance thermometer, which measures temperature by measuring the change in resistance of a conductor, is made from platinum and has a resistance of 50.0 Ω at 20°C. When it is immersed in a vessel containing melting indium, its resistance increases to 76.8 Ω. From this information, find the melting point of indium.

Solution Solving Equation 18.13 for ΔT and getting α from Table 18.1, we have

$$\Delta T = \frac{R - R_0}{\alpha R_0} = \frac{76.8\ \Omega - 50.0\ \Omega}{[3.92 \times 10^{-3}\ (°\text{C})^{-1}](50.0\ \Omega)} = 137\ \text{C°}$$

Since $\Delta T = T - T_0$ and $T_0 = 20°\text{C}$, we find that $T =$ 157°C.

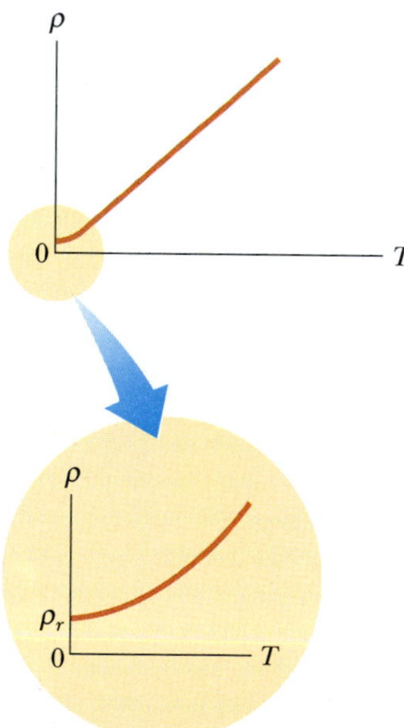

Figure 18.6
Resistivity versus temperature for a normal metal, such as copper. The curve is linear over a wide range of temperatures, and ρ increases with increasing temperature. As T approaches absolute zero (inset), the resistivity approaches a finite value, ρ_r.

For several metals, resistivity is nearly proportional to temperature, as shown in Figure 18.6. In reality, however, there is always a nonlinear region at very low temperatures, and the resistivity usually approaches some finite value near absolute zero (see the magnified inset in Fig. 18.6). This residual resistivity near absolute zero is due primarily to collisions of electrons with impurities and to imperfections in the metal. In contrast, the high-temperature resistivity (the linear region) is dominated by collisions of electrons with the metal atoms. We shall describe this process in more detail in Section 18.4.

Semiconductors, such as silicon and germanium, have intermediate resistivity values. Their resistivity generally decreases with increasing temperature, corresponding to a negative temperature coefficient of resistivity (Fig. 18.7). This is due to the increase in the density of charge carriers at the higher temperatures. Since the charge carriers in a semiconductor are often associated with impurity atoms, the resistivity is very sensitive to the type and concentration of such impurities. A **thermistor** is a semiconducting thermometer that makes use of the large changes in its resistivity with temperature.

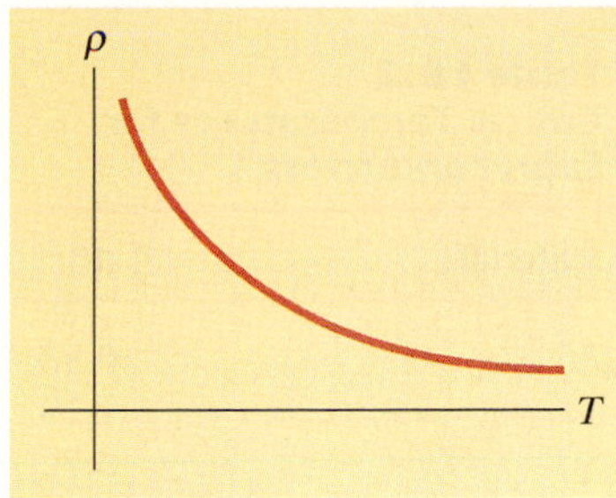

Figure 18.7
Resistivity versus temperature for a pure semiconductor, such as silicon or germanium.

18.3 Superconductors

There is a class of metals and compounds whose resistances go to zero below certain *critical temperatures,* T_c. These materials are known as **superconductors.** The resistance-temperature graph for a superconductor follows that of a normal metal at temperatures above T_c (Fig. 18.8). When the temperature is at or below T_c, the resistivity drops suddenly to zero. This phenomenon was discovered by the Dutch physicist Heike Kamerlingh Onnes in 1911 as he worked with mercury, which is a superconductor below 4.2 K. Recent measurements have shown that the resistivities of superconductors below T_c are less than $4 \times 10^{-25}\ \Omega\cdot\text{m}$, which is around 10^{17} times smaller than the resistivity of copper and considered to be zero in practice.

Today thousands of superconductors are known. Such common metals as aluminum, tin, lead, zinc, and indium are superconductors. Table 18.2 lists the critical temperatures of several superconductors. The value of T_c is sensitive to chemical composition, pressure, and crystalline structure. It is interesting to note that copper, silver, and gold, which are excellent conductors, do not exhibit superconductivity.

One of the truly remarkable features of superconductors is the fact that once a current is set up in them, it persists *without any applied voltage* (since $R = 0$). In fact, steady currents have been observed to persist in superconducting loops for several years with no apparent decay!

An important recent development in physics that has created much excitement in the scientific community has been the discovery of high-temperature copper-oxide-based superconductors. The excitement began with a 1986 publication by Georg Bednorz and K. Alex Müller, scientists at the IBM Zurich Research Laboratory in Switzerland, in which they reported evidence for superconductivity at a temperature near 30 K in an oxide of barium, lanthanum, and copper. Bednorz and Müller were awarded the Nobel Prize in 1987 for their remarkable discovery. Shortly thereafter, a new family of compounds was open for investigation, and research activity in the field of superconductivity proceeded vigorously. In early 1987, groups at the University of Alabama at Huntsville and the University of Houston announced the discovery of superconductivity at about 92 K in an oxide of yttrium, barium, and copper ($YBa_2Cu_3O_7$). Late in 1987, teams of scientists from Japan and the United States reported superconductivity at 105 K in an oxide of bismuth, strontium, calcium, and copper. More recently, scientists have reported superconductivity at temperatures as high as 125 K in an oxide containing thallium. At this point one cannot rule out the possibility of room-temperature superconductivity, and the search for novel superconducting materials continues.

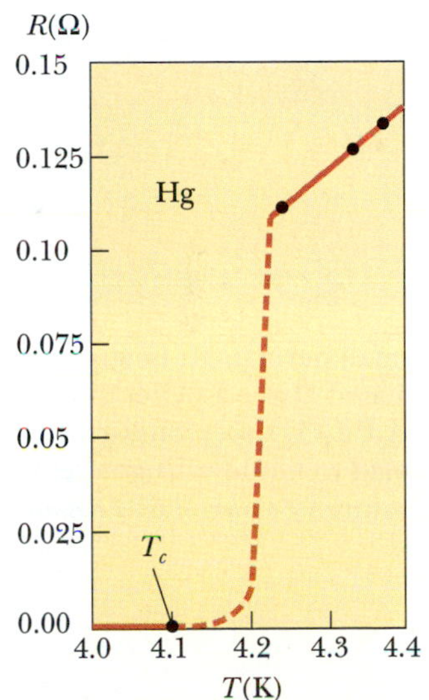

Figure 18.8
Resistance versus temperature for mercury. The graph follows that of a normal metal above the critical temperature, T_c. The resistance drops to zero at the critical temperature, which is 4.15 K for mercury.

Table 18.2
Critical Temperatures for Superconductors

Material	T_c (K)
Zn	0.88
Al	1.19
Sn	3.72
Hg	4.15
Pb	7.18
Nb	9.46
Nb_3Sn	18.05
Nb_3Ge	23.2
$YBa_2Cu_3O_7$	92
Bi-Sr-Ca-Cu-O	105
Tl-Ba-Ca-Cu-O	125

A small permanent magnet levitated above a pellet of the $YBa_2Cu_3O_7$ superconductor cooled to liquid nitrogen temperature. *(Courtesy of IBM Research)*

It is an important search both for scientific reasons and because practical applications become more probable and widespread as the critical temperature is raised.

An important and useful application is superconducting magnets in which the magnetic field strengths are about ten times greater than those of the best normal electromagnets. Such superconducting magnets are being considered as a means of storing energy. The idea of using superconducting power lines for transmitting power efficiently is also receiving some consideration. Modern superconducting electronic devices consisting of two thin-film superconductors separated by a thin insulator have been constructed. They include magnetometers (a magnetic-field measuring device) and various microwave devices.

18.4 A Model for Electrical Conduction

The classical model of electrical conduction in metals leads to Ohm's law and shows that resistivity can be related to the motion of electrons in metals.

Consider a conductor as a regular array of atoms containing free electrons (sometimes called *conduction* electrons). Such electrons are free to move through the conductor (as we learned in our discussion of drift speed in Section 18.1) and are approximately equal in number to the atoms. In the absence of an electric field, the free electrons move in random directions with average speeds on the order of 10^6 m/s. The situation is similar to the motion of gas molecules confined in a vessel. In fact, some scientists refer to conduction electrons in a metal as an *electron gas.*

The conduction electrons are not totally free, because they are confined to the interior of the conductor and undergo frequent collisions with the array of atoms. The collisions are the predominant mechanism for the resistivity of a metal at normal temperatures. Note that there is no current through a conductor in the absence of an electric field, since the average velocity of the free electrons is zero. In other words, just as many electrons move in one direction as in the opposite direction, on the average, and so there is no net flow of charge.

The situation is modified when an electric field is applied to the metal. In addition to random thermal motion, the free electrons drift slowly in a direction opposite that of the electric field, with an average drift speed of v_d, which is much less (typically 10^{-4} m/s; see Example 18.1) than the average speed between collisions (typically 10^6 m/s). Figure 18.9 provides a crude depiction of the motion of free electrons in a conductor. In the absence of an electric field, there is no net displacement after many collisions (Fig. 18.9a). An electric field, **E**, modifies the random motion and causes the electrons to drift in a direction opposite that of **E** (Fig. 18.9b). The slight curvature in the paths in Figure 18.9b results from the acceleration of the electrons between collisions, caused by the applied field. One mechanical system somewhat analogous to this situation is a ball rolling down a slightly inclined plane through an array of closely spaced, fixed pegs (Fig. 18.10). The ball represents a conduction electron, the pegs represent defects in the crystal lattice, and the component of the gravitational force along the incline represents the electric force, $e\mathbf{E}$.

In our model, we shall assume that the excess energy acquired by the electrons in the electric field is lost to the conductor in the collision process. The energy given up to the atoms in the collisions increases the vibrational energy of the

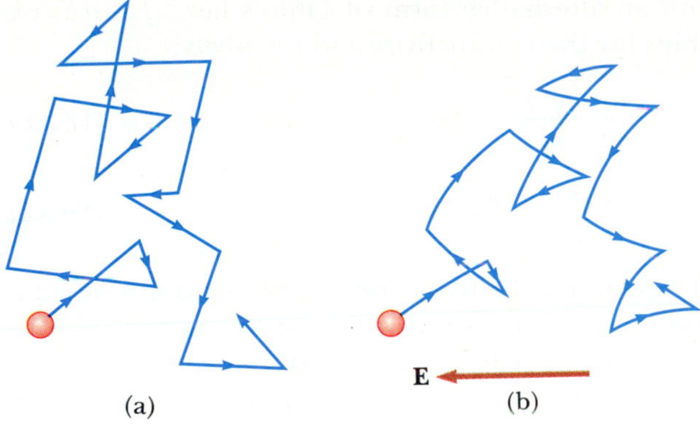

Figure 18.9
(a) A schematic diagram of the random motion of a charge carrier in a conductor in the absence of an electric field. The drift velocity is zero. (b) The motion of a charge carrier in a conductor in the presence of an electric field. Note that the random motion is modified by the field, and the charge carrier has a drift velocity.

atoms, causing the conductor to heat up. The model also assumes that an electron's motion after a collision is independent of its motion before the collision.

We are now in a position to obtain an expression for the drift velocity. When a mobile, charged particle of mass m and charge q is subjected to an electric field, $\mathbf{E}$, it experiences a force of $q\mathbf{E}$. Since $\mathbf{F} = m\mathbf{a}$, we conclude that the acceleration of the particle is

$$\mathbf{a} = \frac{q\mathbf{E}}{m} \quad [18.14]$$

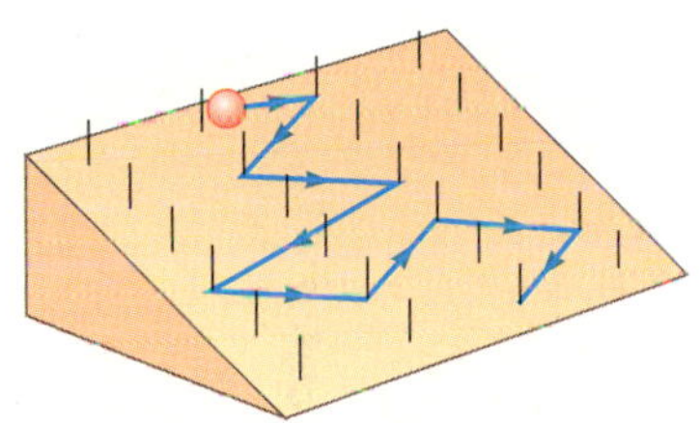

Figure 18.10
A mechanical system somewhat analogous to the motion of charge carriers in the presence of an electric field. The collisions of the ball with the pegs represent the resistance to the ball's motion down the incline.

This acceleration, which occurs for only a short time between collisions, enables the electron to acquire a small drift velocity. If t is the time interval since the last collision and $\mathbf{v}_0$ is the initial velocity, then the velocity of the electron after the time t is

$$\mathbf{v} = \mathbf{v}_0 + \mathbf{a}t = \mathbf{v}_0 + \frac{q\mathbf{E}}{m}t \quad [18.15]$$

We now take the average value of $\mathbf{v}$ over all possible times t and all possible values of $\mathbf{v}_0$. If the initial velocities are assumed to be randomly distributed in space, we see that the average value of $\mathbf{v}_0$ is zero. The term $(qE/m)t$ is the velocity added by the field at the end of one trip between atoms. If the electron starts with zero velocity, the average value of the second term of Equation 18.15 is $(q\mathbf{E}/m)\tau$, where τ is the *average time between collisions.* Because the average of $\mathbf{v}$ is equal to the drift velocity,[3] we have

Drift velocity

$$\mathbf{v}_d = \frac{q\mathbf{E}}{m}\tau \quad [18.16]$$

Substituting this result into Equation 18.6, we find that the magnitude of the current density is

Current density

$$J = nqv_d = \frac{nq^2E}{m}\tau \quad [18.17]$$

[3] Since the collision process is random, each collision event is *independent* of what happened earlier. This is analogous to the random process of throwing a die. The probability of rolling a particular number on one throw is independent of the result of the previous throw. On the average, it would take six throws to come up with that number, starting at any arbitrary time.

Comparing this expression with an alternative form of Ohm's law,[4] $J = \sigma E$, we obtain the following relationships for the conductivity and resistivity:

Conductivity

$$\sigma = \frac{nq^2\tau}{m} \qquad [18.18]$$

Resistivity

$$\rho = \frac{1}{\sigma} = \frac{m}{nq^2\tau} \qquad [18.19]$$

According to this classical model, conductivity and resistivity do not depend on the electric field. This feature is characteristic of a conductor obeying Ohm's law. The model shows that the conductivity can be calculated from a knowledge of the density of the charge carriers, their charge and mass, and the average time between collisions, which is related to the average distance between collisions, ℓ (the mean free path), and the average thermal speed, $\bar{v}$, through the expression[5]

$$\tau = \frac{\ell}{\bar{v}} \qquad [18.20]$$

Example 18.4 Electron Collisions in Copper

(a) Using data and results from Example 18.1 and the classical model of electron conduction, estimate the average time between collisions for electrons in copper at 20°C.

Solution From Equation 18.19 we see that

$$\tau = \frac{m}{nq^2\rho}$$

where $\rho = 1.7 \times 10^{-8}\ \Omega\cdot\text{m}$ for copper, and the carrier density $n = 8.48 \times 10^{28}$ electrons/m^3 for the wire described in Example 18.1. Substitution of these values into the preceding expression gives

$$\tau = \frac{(9.11 \times 10^{-31}\ \text{kg})}{(8.48 \times 10^{-28}\ \text{m}^{-3})(1.6 \times 10^{-19}\ \text{C})^2(1.7 \times 10^{-8}\ \Omega\cdot\text{m})}$$

$$= 2.5 \times 10^{-14}\ \text{s}$$

(b) Assuming the mean thermal speed for free electrons in copper to be 1.6×10^6 m/s, and using the result from (a), calculate the mean free path for electrons in copper.

Solution

$$\ell = \bar{v}\tau = (1.6 \times 10^6\ \text{m/s})(2.5 \times 10^{-14}\ \text{s})$$

$$= 4.0 \times 10^{-8}\ \text{m}$$

[4] The relation $J = \sigma E$ can be derived as follows: The potential difference across a conductor of length ℓ is $V = E\ell$, and from Ohm's law, $V = IR$. Using these relations, together with Equations 18.5 and 18.10, we find that the magnitude of the current density is $J = I/A = V/RA = E\ell/RA = \sigma E$.

[5] Recall that the thermal speed is the speed a particle has as a consequence of the temperature of its surroundings (Chapter 13).

which is equivalent to 40 nm (compared with atomic spacings of about 0.2 nm). Thus, although the interval between collisions is very short, the electrons travel about 200 atomic distances before colliding with an atom.

Although this classical model of conduction is consistent with Ohm's law, it is not satisfactory for explaining some important phenomena. For example, classical calculations for $\overline{v}$ using the ideal-gas model are about a factor of 10 smaller than the true values. Furthermore, according to Equations 18.19 and 18.20, the temperature variation of the resistivity is predicted to vary as $\overline{v}$, which, according to an ideal-gas model (Chapter 13, Eq. 13.15), is proportional to $\sqrt{T}$. This is in disagreement with the linear dependence of resistivity with temperature for pure metals (Fig. 18.5a). It is possible to account for such observations only by using a quantum mechanical model, which we shall describe briefly.

According to quantum mechanics, electrons have wavelike properties. If the array of atoms in a conductor is regularly spaced (that is, periodic), the wavelike character of the electrons makes it possible for them to move freely through the conductor, and a collision with an atom is unlikely. For an idealized conductor there would be no collisions, the mean free path would be infinite, and the resistivity would be zero. Electron waves are scattered only if the atomic arrangement is irregular (not periodic)—for example, as a result of structural defects or impurities. At low temperatures, the resistivity of metals is dominated by scattering caused by collisions between the electrons and impurities. At high temperatures, the resistivity is dominated by scattering caused by collisions between the electrons and the atoms of the conductor, which are continuously displaced as a result of thermal agitation. The thermal motion of the atoms makes the structure irregular (compared with an atomic array at rest), thereby reducing the electron's mean free path.

18.5 Electrical Energy and Power

If a battery is used to establish an electric current in a conductor, there occurs a continuous transformation of chemical energy stored in the battery to kinetic energy of the charge carriers. This kinetic energy is quickly lost as a result of collisions between the charge carriers and the lattice ions, resulting in an increase in the temperature of the conductor. Thus, the chemical energy stored in the battery is continuously transformed into thermal energy.

In order to understand the process of energy transfer in a simple circuit, consider a battery whose terminals are connected to a resistor, R, as shown in Figure 18.11. (Remember that the positive terminal of the battery is always at the higher potential.) Now imagine following a positive quantity of charge ΔQ around the circuit from point a through the battery and resistor and back to a. Point a is a reference point that is grounded (the ground symbol is ⏚), and its potential is taken to be zero. As the charge moves from a to b through the battery, its electrical potential energy increases by the amount $V\,\Delta Q$ (where V is the potential at b) while the chemical potential energy in the battery decreases by the same amount. (Re-

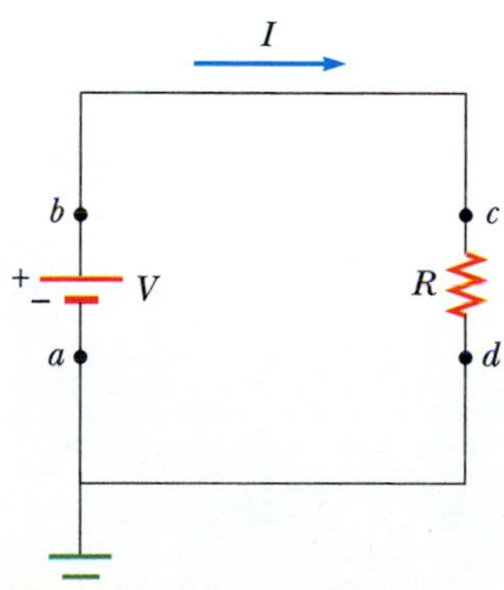

Figure 18.11
A circuit consisting of a battery and a resistor. Positive charge flows clockwise, from the positive to the negative terminal of the battery. Points a and d are grounded.

call from Chapter 17 that $\Delta U = q\,\Delta V$.) However, as the charge moves from c to d through the resistor, it loses this electrical potential energy during collisions with atoms in the resistor, thereby producing thermal energy. Note that, if we neglect the resistance of the interconnecting wires, no loss in energy occurs for paths bc and da. When the charge returns to point a, it must have the same potential energy (zero) as it had at the start.[6]

The rate at which the charge ΔQ loses potential energy in going through the resistor is

$$\frac{\Delta U}{\Delta t} = \frac{\Delta Q}{\Delta t} V = IV$$

where I is the current in the circuit. Of course, the charge regains this energy when it passes through the battery. Since the rate at which the charge loses energy equals the power, P, dissipated in the resistor, we have

Power

$$P = IV \qquad [18.21]$$

In this case, the power is supplied to a resistor by a battery. However, Equation 18.21 can be used to determine the power transferred from a battery to *any* device carrying a current, I, and having a potential difference, V, between its terminals.

Using Equation 18.21 and the fact that $V = IR$ for a resistor, we can express the power dissipated by the resistor in the alternative forms

Power loss in a resistor

$$P = I^2R = \frac{V^2}{R} \qquad [18.22]$$

When I is in amperes, V in volts, and R in ohms, the SI unit of power is the watt, introduced in Chapter 7. The power dissipated as heat in a conductor of resistance R is called *joule heating*;[7] it is also often referred to as an *I^2R loss.*

As we learned in Chapter 7, Section 7.5, the unit of energy the electric company uses to calculate energy consumption, the kilowatt-hour, is the energy consumed in 1 h at the constant rate of 1 kW. Since 1 W = 1 J/s, we have

$$1 \text{ kWh} = (10^3 \text{ W})(3600 \text{ s}) = 3.6 \times 10^6 \text{ J} \qquad [18.23]$$

Photograph of a hydroelectric generating plant at the Grand Coulee Dam. Water released to lower levels from the dam is used to drive the turbines that generate electricity. *(The Stock Market)*

Example 18.5 Electrical Rating of a Lightbulb

A lightbulb is rated 75 W at 120 V. That is, its operating voltage is 120 V and it has a power rating of 75 W. The bulb is powered by a 120-V power supply. Find the current in the bulb and its resistance.

Solution Since the power rating of the bulb is 75 W and the operating voltage is 120 V, we can use $P = IV$ to find the current:

$$I = \frac{P}{V} = \frac{75 \text{ W}}{120 \text{ V}} = 0.625 \text{ A}$$

[6] Note that when the current reaches its steady-state value, there is *no* change with time in the kinetic energy associated with the current.

[7] It is called *joule heating* even though its dimensions are *energy per unit time,* which are dimensions of power.

Using Ohm's law, $V = IR$, the resistance is calculated to be

$$R = \frac{V}{I} = \frac{120\text{ V}}{0.625\text{ A}} = 192\ \Omega$$

Exercise What would the resistance be in a lamp rated at 120 V and 100 W?

Answer 144 Ω.

Example 18.6 The Cost of Operating a Lightbulb

How much does it cost to burn a 100-W lightbulb for 24 h if electricity costs 8 cents per kilowatt-hour?

Solution A 100-W lightbulb is equivalent to a 0.1-kW bulb. Since the energy consumed equals power × time, the amount of energy you must pay for, expressed in kilowatt-hours, is

$$\text{Energy} = (0.10\text{ kW})\,(24\text{ h}) = 2.4\text{ kWh}$$

If energy is purchased at 8 cents per kWh, the 24-h cost is

$$\text{Cost} = (2.4\text{ kWh})\,(\$0.08/\text{kWh}) = \$0.19$$

This is \$6 a month. Survey your energy use. What appliance has the highest number of watts stamped on it, and how long do you run it?

Exercise If electricity costs 8 cents per kilowatt-hour, what does it cost to operate an electric oven, which operates at 20 A and 220 V, for 5 h?

Answer \$1.76.

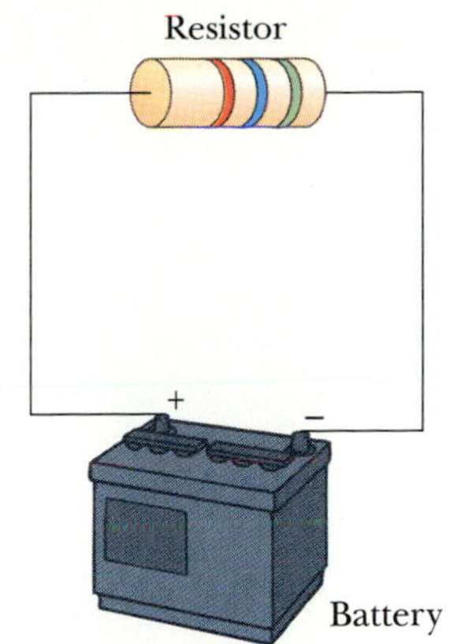

Figure 18.12 A circuit consisting of a resistor connected to the terminals of a battery.

18.6 Sources of emf

The source that maintains the constant current in Figure 18.13 is called an "emf."[8] Sources of emf are any devices (such as batteries and generators) that increase the potential energy of charges circulating in circuits. One can think of a source of emf as a "charge pump" that forces electrons to move in a direction opposite the electrostatic field inside the source. The emf, $\mathcal{E}$, of a source describes the work done per unit charge, and hence the SI unit of emf is the volt.

Consider the circuit shown in Figure 18.12, consisting of a battery connected to a resistor. We shall assume that the connecting wires have no resistance. If we neglect the internal resistance of the battery, the potential difference across the battery (the terminal voltage) equals the emf of the battery. However, because a real battery always has some internal resistance, r, the terminal voltage is not equal to the emf. The circuit shown in Figure 18.12 can be described by the circuit diagram in Figure 18.13a. The battery, represented by the dashed rectangle, consists of a source of emf, $\mathcal{E}$, in series with internal resistance, r. Now imagine a

[8] The term was originally an abbreviation for *electromotive force,* but it is not a force, so the long form is discouraged.

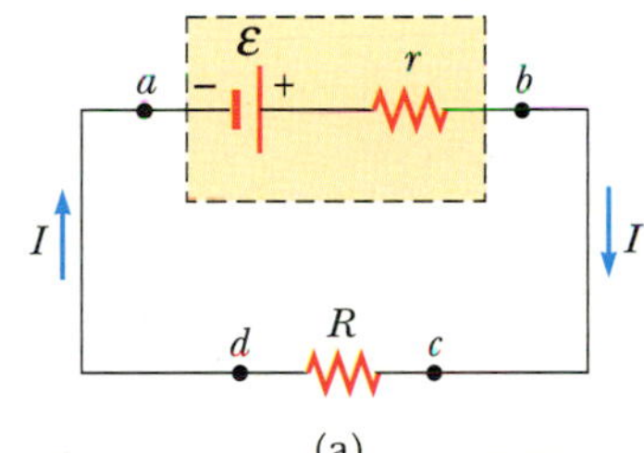

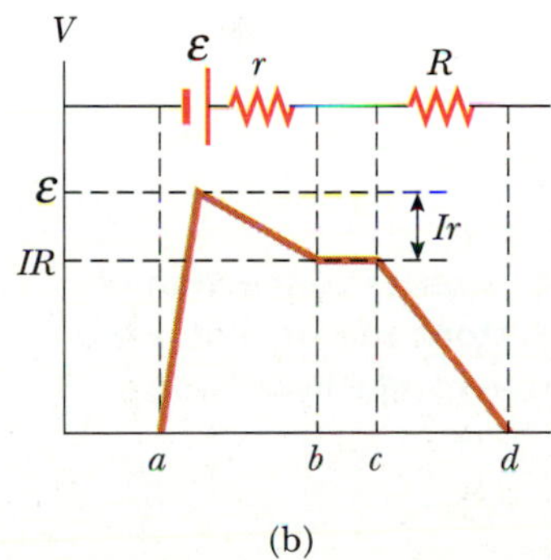

Figure 18.13 (a) Circuit diagram of a source of emf, $\mathcal{E}$, of internal resistance r connected to an external resistor, R. (b) A graphical representation showing how the potential changes as the series circuit in (a) is traversed clockwise.

positive charge moving from a to b in Figure 18.13a. As the charge passes from the negative to the positive terminal within the battery, the potential of the charge increases by $\mathcal{E}$. However, as it moves through the resistance, r, its potential decreases by an amount Ir, where I is the current in the circuit. Thus, the terminal voltage of the battery, $V = V_b - V_a$, is[9]

$$V = \mathcal{E} - Ir \tag{18.24}$$

Note from this expression that $\mathcal{E}$ is equivalent to the **open-circuit voltage,** that is, the *terminal voltage when the current is zero.* Figure 18.13b, is a graphical representation of the changes in potential as the circuit is traversed clockwise. By inspecting Figure 18.13a, we see that the terminal voltage, V, must also equal the potential difference across the external resistance, R—often called the **load resistance;** that is, $V = IR$. Combining this with Equation 18.24, we see that

$$\mathcal{E} = IR + Ir \tag{18.25}$$

Solving for the current gives

$$I = \frac{\mathcal{E}}{R + r}$$

This shows that the current in this simple circuit depends on both the resistance external to the battery and the internal resistance. If R is much greater than r, we can neglect r in our analysis. In many circuits we shall ignore this internal resistance.

If we multiply Equation 18.25 by the current, I, we get

$$I\mathcal{E} = I^2R + I^2r$$

This equation tells us that the total power output of the source of emf, $I\mathcal{E}$, is converted into power that is dissipated as joule heat in the load resistance, I^2R, *plus* power that is dissipated in the internal resistance, I^2r. Again, if $r \ll R$, most of the power delivered by the battery is transferred to the load resistance.

18.7 Resistors in Series and in Parallel

For a series connection of resistors, the current is the same in all the resistors

When two or more resistors are connected together so that they have only one common point per pair, they are said to be in *series.* Figure 18.14 shows two resistors connected in series. Note that the current is the same through the two resistors, since any charge that flows through R_1 must also flow through R_2. Because the potential drop from a to b in Figure 18.14b equals IR_1 and the potential drop from b to c equals IR_2, the potential drop from a to c is

$$V = IR_1 + IR_2 = I(R_1 + R_2)$$

Therefore, we can replace the two resistors in series with a single *equivalent resistance,* R_{eq}, whose value is the *sum* of the individual resistances:

$$R_{eq} = R_1 + R_2 \tag{18.26}$$

[9] The terminal voltage in this case is less than the emf by the amount Ir. In some situations, the terminal voltage may *exceed* the emf by the amount Ir. This happens when the direction of the current is *opposite* that of the emf, as when a battery is charged with another source of emf.

(a)

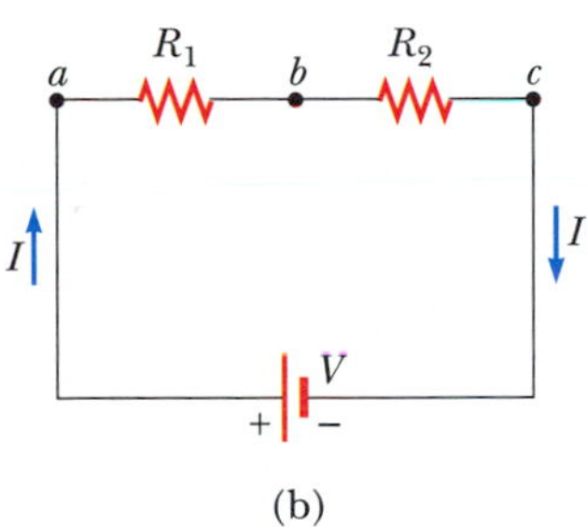

(b)

Figure 18.14
A series connection of two resistors, R_1 and R_2. The currents in the resistors are the same.

The resistance R_{eq} is equivalent to the series combination $R_1 + R_2$ in the sense that the circuit current is unchanged when R_{eq} replaces $R_1 + R_2$. The equivalent resistance of three or more resistors connected in series is simply

$$R_{eq} = R_1 + R_2 + R_3 + \cdots \qquad [18.27]$$

Therefore, *the equivalent resistance of a series connection of resistors is always greater than any individual resistance.*

Note that if the filament of one lightbulb in Figure 18.14 were to break, or "burn out," the circuit would no longer be complete (an open-circuit condition would exist) and the second bulb would also go out. Some Christmas-tree light sets (especially older ones) are connected in this way, and the tedious task of determining which bulb is burned out is a familiar one.

In many circuits, fuses are used in series with other circuit elements for safety purposes. The conductor in the fuse is designed to melt and open the circuit at some maximum current, the value of which depends on the nature of the circuit. If a fuse were not used, excessive currents could damage circuit elements, overheat wires, and perhaps cause a fire. In modern home construction, circuit breakers are used in place of fuses. When the current in a circuit exceeds some value (typically 15 A), the circuit breaker acts as a switch and opens the circuit.

Now consider two resistors connected in *parallel,* as shown in Figure 18.15. In this case, the potential differences across the resistors are equal. However, the currents are generally not the same. When the current I reaches point a (called a *junction*) in Figure 18.15b, it splits into two parts, I_1 going through R_1 and I_2 going through R_2. If R_1 is greater than R_2, then I_1 is less than I_2. That is, the charge tends to take the path of least resistance. Clearly, since charge must be conserved, the current, I, that enters point a must equal the total current leaving point b:

$$I = I_1 + I_2$$

The potential drops across the resistors must be the *same,* and so Ohm's law gives

$$I = I_1 + I_2 = \frac{V}{R_1} + \frac{V}{R_2} = V\left(\frac{1}{R_1} + \frac{1}{R_2}\right) = \frac{V}{R_{eq}}$$

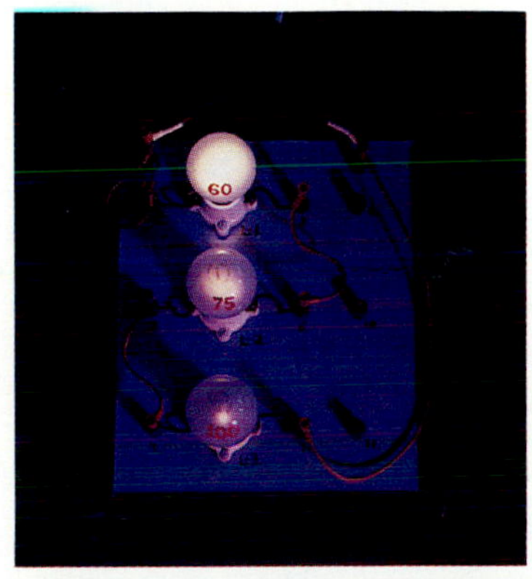

A series connection of three lamps, all rated at 120 V, with power ratings of 60 W, 75 W, and 200 W. Why do the intensities of the lamps differ? Which lamp has the greatest resistance? How would their relative intensities differ if they were connected in parallel? *(Courtesy of Henry Leap and Jim Lehman)*

This versatile circuit enables the experimentor to examine the properties of circuit elements such as capacitors and resistors and their effects on circuit behavior. *(Courtesy of CENCO)*

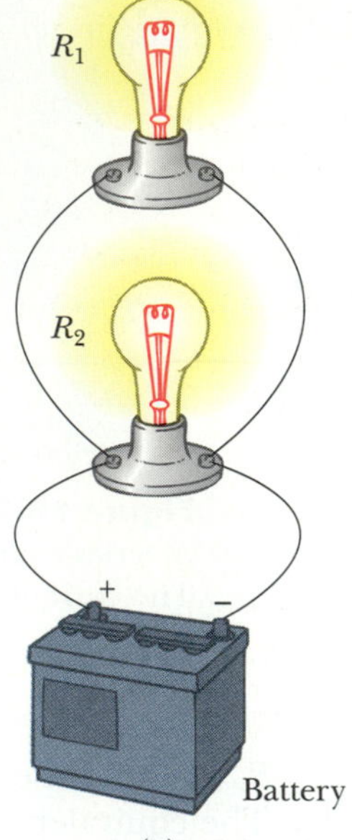

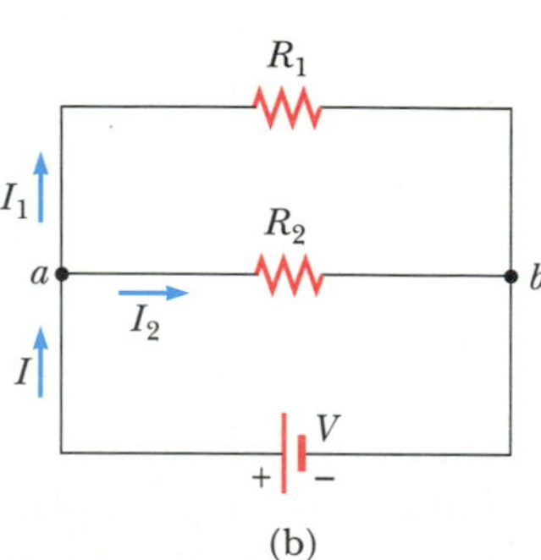

Figure 18.15 A parallel connection of two resistors, R_1 and R_2. The potential differences across the resistors are the same, and the equivalent resistance of the combination is $R_{eq} = R_1R_2/(R_1 + R_2)$.

From this result, we see that the equivalent resistance of two resistors in parallel is

$$\frac{1}{R_{eq}} = \frac{1}{R_1} + \frac{1}{R_2} \qquad [18.28]$$

This can be rearranged to become

$$R_{eq} = \frac{R_1R_2}{R_1 + R_2}$$

An extension of this analysis to three or more resistors in parallel yields the following general expression:

The equivalent resistance of several resistors in parallel

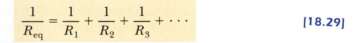

$$\frac{1}{R_{eq}} = \frac{1}{R_1} + \frac{1}{R_2} + \frac{1}{R_3} + \cdots \qquad [18.29]$$

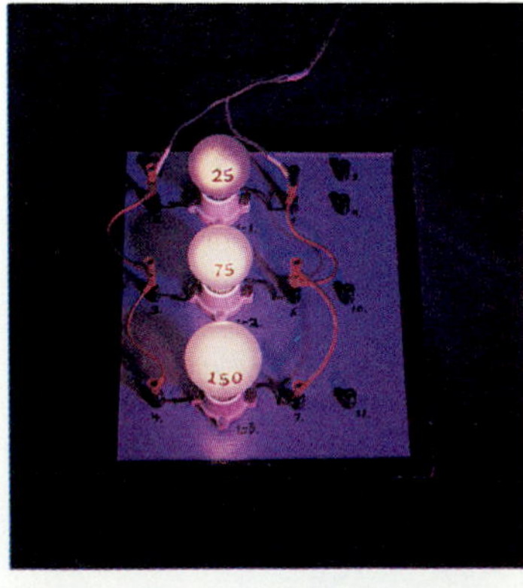

Three incandescent lamps with power ratings of 25 W, 75 W, and 150 W, connected in parallel to a voltage source of about 100 V. All lamps are rated at the same voltage. Why do the intensities of the lamps differ? Which lamp draws the most current? Which has the least resistance? *(Courtesy of Henry Leap and Jim Lehman)*

From this expression it can be seen that the equivalent resistance of two or more resistors connected in parallel is always *less* than the smallest resistance in the group.

Household circuits are always wired so that the lightbulbs (or appliances, or whatever) are connected in parallel, as in Figure 18.15a. In this manner, each device operates independently of the others, so that if one is switched off, the others remain on. Equally important, each device operates on the same voltage.

Finally, it is interesting to note that parallel resistors combine in the same way that series capacitors combine, and vice versa.

Problem-Solving Strategy: Resistors

1. **When two or more unequal resistors are connected in *series*, they carry the same current, but the potential differences across them are not the same. The resistors add directly to give the equivalent resistance of the series combination.**
2. **When two or more unequal resistors are connected in *parallel*, the potential differences across them are the same. Since the current is inversely proportional to the resistance, the currents through them are not the same. The equivalent resistance of a parallel combination of resistors is found through reciprocal addition, and the equivalent resistor is always *less* than the smallest individual resistor.**
3. **A complicated circuit consisting of resistors can often be reduced to a simple circuit containing only one resistor. To do so, examine the initial circuit and replace any resistors in series or any in parallel using the procedures outlined in Steps 1 and 2. Draw a sketch of the new circuit after these changes have been made. Examine the new circuit and replace any series or parallel combinations. Continue this process until a single equivalent resistance is found.**
4. **If the current through or the potential difference across a resistor in the complicated circuit is to be found, start with the final circuit found in Step 3 and gradually work your way back through the circuits, using $V = IR$ and the rules of Steps 1 and 2.**

Example 18.7 Equivalent Resistance

Four resistors are connected as shown in Figure 18.16a.

(a) Find the equivalent resistance between a and c.

Solution The circuit can be reduced in steps. The 8-Ω and 4-Ω resistors are in series, and so the equivalent resistance between a and b is 12 Ω (Eq. 18.26). The 6-Ω and 3-Ω resistors are in parallel, and so from Equation 18.28 we find that the equivalent resistance from b to c is 2 Ω. Hence, the equivalent resistance from a to c is 14 Ω.

(b) What is the current in each resistor if a potential difference of 42 V is maintained between a and c?

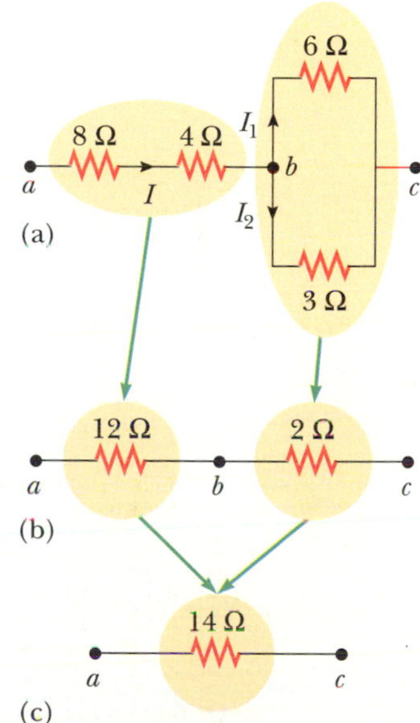

Figure 18.16 (Example 18.7) The equivalent resistance of the four resistors shown in (a) can be reduced in steps to an equivalent 14-Ω resistor.

Solution The current I in the 8-Ω and 4-Ω resistors is the same since they are in series. Using Ohm's law and the results from (a), we get

$$I = \frac{V_{ac}}{R_{eq}} = \frac{42\text{ V}}{14\ \Omega} = \boxed{3\text{ A}}$$

When this current enters the junction at b, it splits. Part of it passes through the 6-Ω resistor (I_1) and part goes through the 3-Ω resistor (I_2). Since the potential differences across these resistors, V_{bc}, are the same (they are in parallel), $6I_1 = 3I_2$, or $I_2 = 2I_1$. Using this result and the fact that $I_1 + I_2 = 3$ A, we find that $I_1 = 1$ A and $I_2 = 2$ A. We could have guessed this from the start by noting that the current through the 3-Ω resistor has to be twice the current through the 6-Ω resistor in view of their relative resistances and the fact that the same voltage is applied to each of them.

As a final check, note that $V_{bc} = 6I_1 = 3I_2 = 6$ V and $V_{ab} = 12I = 36$ V; therefore, $V_{ac} = V_{ab} + V_{bc} = 42$ V, as it must.

Example 18.8 Three Resistors in Parallel

Three resistors are connected in parallel as in Figure 18.17. A potential difference of 18 V is maintained between points a and b.

(a) Find the current in each resistor.

Solution The resistors are in parallel, and the potential difference across each is 18 V. Applying $V = IR$ to each resistor gives

$$I_1 = \frac{V}{R_1} = \frac{18\text{ V}}{3\ \Omega} = \boxed{6\text{ A}}$$

$$I_2 = \frac{V}{R_2} = \frac{18\text{ V}}{6\ \Omega} = \boxed{3\text{ A}}$$

$$I_3 = \frac{V}{R_3} = \frac{18\text{ V}}{9\ \Omega} = \boxed{2\text{ A}}$$

(b) Calculate the power dissipated by each resistor and the total power dissipated by the three resistors.

Solution Applying $P = I^2R$ to each resistor gives

3-Ω: $P_1 = I_1^2R_1 = (6\text{ A})^2(3\ \Omega) = \boxed{108\text{ W}}$

6-Ω: $P_2 = I_2^2R_2 = (3\text{ A})^2(6\ \Omega) = \boxed{54\text{ W}}$

9-Ω: $P_3 = I_3^2R_3 = (2\text{ A})^2(9\ \Omega) = \boxed{36\text{ W}}$

This shows that the smallest resistor dissipates the most power since it carries the most current. (Note that you can also use $P = V^2/R$ to find the power dissipated by each resistor.) Summing the three quantities gives a total power of 198 W.

Exercise Calculate the equivalent resistance of the three resistors, and from this result find the total power dissipated.

Answer $\frac{18}{11}$ Ω; 198 W.

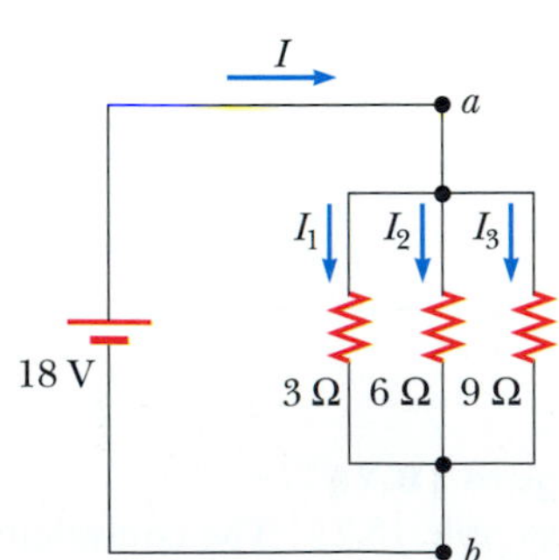

Figure 18.17 (Example 18.8) Three resistors connected in parallel. The voltage across each resistor is 18 V.

18.8 Kirchhoff's Rules and Simple DC Circuits

As indicated in the preceding section, we can analyze simple circuits using Ohm's law and the rules for series and parallel combinations of resistors. However, there are many ways in which resistors can be connected so that the circuits formed cannot be reduced to a single equivalent resistor. The procedure for analyzing such complex circuits is greatly simplified by the use of two simple rules called **Kirchhoff's rules:**

1. The sum of the currents entering any junction must equal the sum of the currents leaving that junction. (This rule is often referred to as the **junction rule.**)
2. The sum of the potential differences across each element around any closed circuit loop must be zero. (This rule is usually called the **loop rule.**)

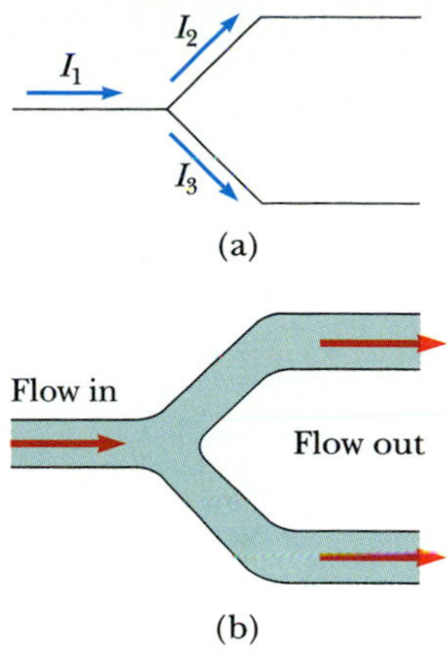

Figure 18.18
(a) A schematic diagram illustrating Kirchhoff's junction rule. Conservation of charge requires that whatever current enters a junction must leave that junction. Therefore, in this case, $I_1 = I_2 + I_3$. (b) A hydraulic analog of the junction rule: the flow out must equal the flow in.

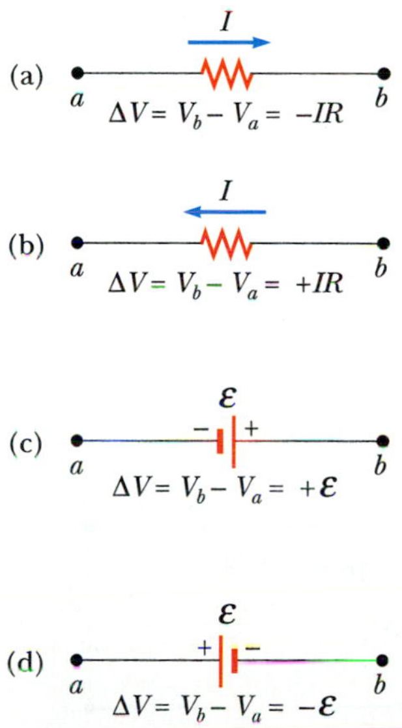

The junction rule is a statement of *conservation of charge.* Whatever current enters a given point in a circuit must leave that point, because charge cannot build up or disappear at a point. If we apply this rule to the junction in Figure 18.18a, we get

$$I_1 = I_2 + I_3$$

Figure 18.18b represents a hydraulic analog to this situation in which water flows through a branched pipe with no leaks. The flow rate into the pipe equals the total flow rate out of the two branches.

The second rule is equivalent to the law of *conservation of energy.* Any charge that moves round any closed loop in a circuit (starting and ending at the same point) must gain as much energy as it loses. It gains energy from a battery. Its energy may decrease in the form of a potential drop, $-IR$, across a resistor or as a result of flowing backward through a source of emf, that is, from the positive to the negative terminal inside the battery. In the latter case, electrical energy is converted to chemical energy as the battery is charged.

As an aid in applying the loop rule, the following points should be noted. They are summarized in Figure 18.19, where it is assumed that movement is from point a toward point b:

1. If a resistor is traversed in the direction of the current, the change in potential across the resistor is $-IR$ (Fig. 18.19a).
2. If a resistor is traversed in the direction *opposite* the current, the change in potential across the resistor is $+IR$ (Figure 18.19b).
3. If a source of emf is traversed in the direction of the emf (from $-$ to $+$ on the terminals), the change in potential is $+\mathcal{E}$ (Fig. 18.19c).
4. If a source of emf is traversed in the direction opposite the emf (from $+$ to $-$ on the terminals), the change in potential is $-\mathcal{E}$ (Fig. 18.19d).

(a) a $\Delta V = V_b - V_a = -IR$ b

(b) a $\Delta V = V_b - V_a = +IR$ b

(c) a $\Delta V = V_b - V_a = +\mathcal{E}$ b

(d) a $\Delta V = V_b - V_a = -\mathcal{E}$ b

Figure 18.19
Rules for determining the potential changes across a resistor and a battery, assuming the battery has no internal resistance.

There are limitations on the use of the junction rule and the loop rule. You may use the junction rule as often as needed so long as each time you write an equation, you include in it a current that has not been used in a previous junction rule equation. In general, the number of times the junction rule must be used is one fewer than the number of junction points in the circuit. The loop rule can be used as often as needed so long as a new circuit element (a resistor or battery) or a new current appears in each new equation. In general, *the number of independent equations you need must equal the number of unknowns in order to solve a particular circuit problem.*

The following examples illustrate the use of Kirchhoff's rules in analyzing circuits. In all cases, it is assumed that the circuits have reached steady-state conditions—that is, the currents in the various branches are constant. If a capacitor is included as an element in one of the branches, *it acts as an open circuit*: the current in the branch containing the capacitor is zero under steady-state conditions.

Problem-Solving Strategy and Hints: Kirchhoff's Rules

1. **First, draw the circuit diagram and assign labels to all the known quantities, and symbols to all the unknown quantities. You must assign *directions* to the currents in each part of the circuit. Do not be alarmed if you guess the direction of a current incorrectly; the result will have a negative value, but *its magnitude will be correct*. Although the assignment of current directions is arbitrary, you must adhere *rigorously* to the directions you assigned when you apply Kirchhoff's rules.**
2. **Apply the junction rule (Kirchhoff's first rule) to any junction in the circuit; doing so provides a relation between the currents. (This step is easy!)**
3. **Now apply the loop rule (Kirchhoff's second rule) to as many loops in the circuit as are needed to solve for the unknowns. In order to apply this rule, you must correctly identify the change in potential as you cross each element in traversing the closed loop (either clockwise or counterclockwise). Watch out for signs!**
4. **Solve the equations simultaneously for the unknown quantities. Be careful in your algebraic steps, and check your numerical answers for consistency.**

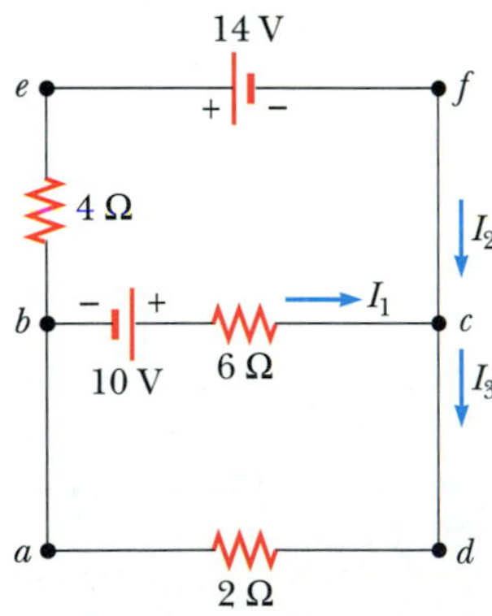

Figure 18.20 (Example 18.9) A circuit containing three loops.

Example 18.9 Applying Kirchhoff's Rules

Find I_1, I_2, and I_3 in Figure 18.20.

Solution We shall choose the directions of the currents as shown in the figure. Applying Kirchhoff's first rule to junction *c* gives

$$(1) \qquad I_1 + I_2 = I_3$$

The circuit has three loops: *abcda, befcb,* and *aefda*. We need only two loop equations to determine the unknown currents. The third loop equation would give no new information. Applying Kirchhoff's second rule to loops *abcda* and *befcb* and traversing these loops clockwise, we obtain the following expressions:

$$(2) \quad \text{Loop } abcda\text{: } 10\text{ V} - (6\ \Omega)I_1 - (2\ \Omega)I_3 = 0$$

$$(3) \quad \text{Loop } befcb\text{: } -14\text{ V} + (6\ \Omega)I_1 - 10\text{ V} - (4\ \Omega)I_2 = 0$$

Note that in loop *befcb*, a positive sign is obtained when the 6-Ω resistor is traversed, since the direction of the path is opposite the direction of the current I_1. A third loop equation for *aefda* gives $-14 - 2I_3 - 4I_2 = 0$, which is just the sum of (2) and

(3). Expressions (1), (2), and (3) represent three linear, independent equations with three unknowns. We can solve the problem as follows: Substitution of (1) into (2) gives, with units ignored for the moment,

$$10 - 6I_1 - 2(I_1 + I_2) = 0$$

$$(4) \qquad 10 = 8I_1 + 2I_2$$

Dividing each term in (3) by 2 and rearranging the equation gives

$$(5) \qquad -12 = -3I_1 + 2I_2$$

Subtracting (5) from (4) eliminates I_2, giving

$$22 = 11I_1$$

$$I_1 = 2\text{ A}$$

Using this value of I_1 in (5) yields a value for I_2:

$$2I_2 = 3I_1 - 12 = 3(2) - 12 = -6$$

$$I_2 = -3\text{ A}$$

Finally, $I_3 = I_1 + I_2 = -1$ A. Hence, the currents have the values

$$I_1 = 2\text{ A} \qquad I_2 = -3\text{ A} \qquad I_3 = -1\text{ A}$$

The fact that I_2 and I_3 are both negative indicates only that we chose the wrong direction for these currents. The numerical values are correct.

Exercise Find the potential difference between points b and c.

Answer $V_b - V_c = 2$ V.

Example 18.10 **A Multiloop Circuit**

(a) Find the values of I_1, I_2, and I_3 in Figure 18.21 under steady-state conditions.

Solution First note that *the capacitor represents an open circuit, and hence there is no current along path ghab under steady-state conditions.* Therefore, $I_{fg} = I_1$. Applying Kirchhoff's first rule to junction c, we get

$$(1) \qquad I_1 + I_2 = I_3$$

Kirchhoff's second rule applied to loops *defcd* and *cfgbc* gives, when the direction is clockwise,

$$(2)\ \text{Loop } defcd: \qquad 4\text{ V} - (3\ \Omega)I_2 - (5\ \Omega)I_3 = 0$$

$$(3)\ \text{Loop } cfgbc: \qquad 8\text{ V} + (3\ \Omega)I_2 - (5\ \Omega)I_1 = 0$$

From (1) we see that $I_1 = I_3 - I_2$, which, substituted into (3), gives

$$(4) \qquad 8\text{ V} - (5\ \Omega)I_3 + (8\ \Omega)I_2 = 0$$

Subtracting (4) from (2), we eliminate I_3 and find

$$I_2 = -\tfrac{4}{11}\text{ A} = -0.364\text{ A}$$

Because the value we calculate for I_2 is negative, we conclude that the I_2 direction shown in Figure 18.21 is wrong and should be from c to f through the 3-Ω resistor. Using this value of I_2 in (3) and (1) gives the following values for I_1 and I_3:

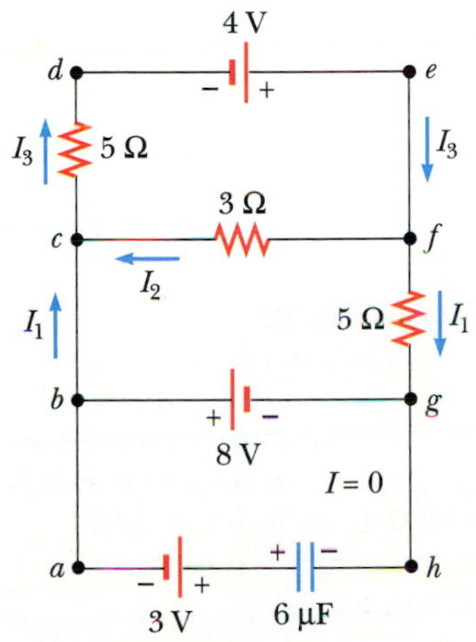

Figure 18.21 (Example 18.10) A multiloop circuit. Note that Kirchhoff's loop equation can be applied to *any* closed loop, including one containing the capacitor.

$I_1 = \boxed{1.38\text{ A}} \qquad I_3 = \boxed{1.02\text{ A}}$

(b) What is the charge on the capacitor?

Solution We can apply Kirchhoff's second rule to loop *abgha* (or any other loop that contains the capacitor) to find the potential difference across the capacitor, V_c:

$$-8\text{ V} + V_c - 3\text{ V} = 0$$

$$V_c = 11.0\text{ V}$$

Since $Q = CV_c$, the charge on the capacitor is

$$Q = (6\ \mu\text{F})(11.0\text{ V}) = \boxed{66.0\ \mu\text{C}}$$

Why is the left side of the capacitor positively charged?

Exercise Find the voltage across the capacitor by traversing any other loop, such as *adeha.*

Answer 11.0 V.

18.9 *RC* Circuits

So far we have been concerned with circuits with constant currents, or so-called *steady-state circuits.* We shall now consider circuits containing capacitors, in which the currents may vary in time.

Charging a Capacitor

Consider the series circuit shown in Figure 18.22. Let us assume that the capacitor is initially uncharged. There is no current when switch S is open (Fig. 18.22b). If the switch is closed at $t = 0$, charges begin to flow, setting up a current in the circuit, and the capacitor begins to charge (Fig. 18.22c). Note that during the

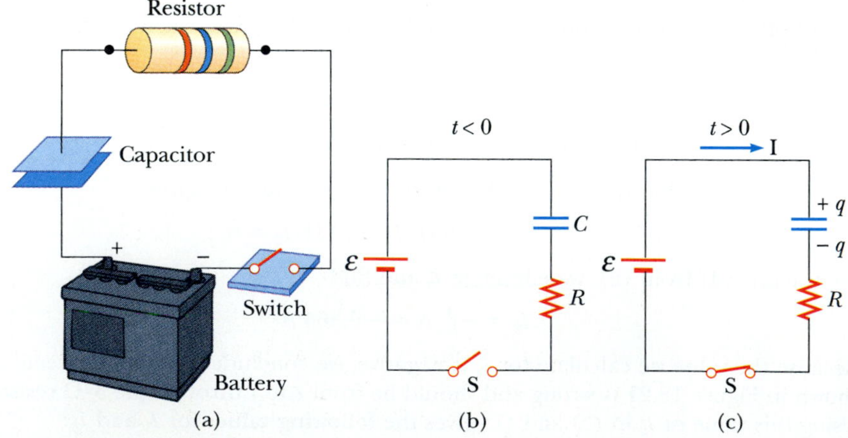

Figure 18.22 (a) A capacitor in series with a resistor, battery, and switch. (b) A circuit diagram representing this system before the switch is closed, $t < 0$. (c) A circuit diagram after the switch is closed, $t > 0$.

charging, charges do not jump across the plates of the capacitor, since the gap between the plates represents an open circuit. Instead, positive charge is transferred from the bottom plate to the top plate only by moving through the resistor, switch, and battery until the capacitor is fully charged. The value of the maximum charge depends on the emf of the battery. Once the maximum charge is reached, the current in the circuit is zero.

To put this discussion on a quantitative basis, let us apply Kirchhoff's second rule to the circuit after the switch is closed. Choosing clockwise as our direction around the circuit, we get

$$\mathcal{E} - \frac{q}{C} - IR = 0 \qquad [18.30]$$

where q/C is the potential drop across the capacitor and IR is the potential drop across the resistor. Note that q and I are *instantaneous* values of the charge and current, respectively, as the capacitor is charged.

We can use Equation 18.30 to find the initial current in the circuit and the maximum charge on the capacitor. At $t = 0$, when the switch is closed, the charge on the capacitor is zero, and from Equation 18.30 we find that the initial current in the circuit, I_0, is a maximum and equal to

$$I_0 = \frac{\mathcal{E}}{R} \qquad \text{(current at } t = 0\text{)} \qquad [18.31]$$

Maximum current

At this time, *the potential drop is entirely across the resistor.* Later, when the capacitor is charged to its maximum value, Q, charges cease to flow, the current in the circuit is zero, and *the potential drop is entirely across the capacitor.* Substituting $I = 0$ into Equation 18.30 yields the following expression for Q:

$$Q = C\mathcal{E} \qquad \text{(maximum charge)} \qquad [18.32]$$

Maximum charge on the capacitor

To determine analytical expressions for the time dependence of the charge and current, we must solve Equation 18.30, a single equation containing two variables, q and I. In order to do this, let us substitute $I = dq/dt$ and rearrange the equation:

$$\frac{dq}{dt} = \frac{\mathcal{E}}{R} - \frac{q}{RC}$$

An expression for q may be found in the following way. Rearrange the equation by placing terms involving q on the left side and those involving t on the right side. Then integrate both sides:

$$\frac{dq}{(q - C\mathcal{E})} = -\frac{1}{RC}\,dt$$

$$\int_0^q \frac{dq}{(q - C\mathcal{E})} = -\frac{1}{RC}\int_0^t dt$$

$$\ln\left(\frac{q - C\mathcal{E}}{C\mathcal{E}}\right) = -\frac{t}{RC}$$

From the definition of the natural logarithm, we can write this expression as

$$q(t) = C\mathcal{E}[1 - e^{-t/RC}] = Q[1 - e^{-t/RC}] \qquad [18.33]$$

Charging versus time for a capacitor being charged through a resistor

where e is the base of the natural logarithm and $Q = C\mathcal{E}$ is the *maximum* charge on the capacitor.

An expression for the charging current may be found by differentiating Equation 18.33 with respect to time. Using $I = dq/dt$, we obtain

Current versus time

$$I(t) = \frac{\mathcal{E}}{R} e^{-t/RC} \qquad [18.34]$$

where $\mathcal{E}/R$ is the initial current in the circuit.

Plots of charge and current versus time are shown in Figure 18.23. Note that the charge is zero at $t = 0$ and approaches the maximum value of $C\mathcal{E}$ as $t \rightarrow \infty$ (Fig. 18.23a). Furthermore, the current has its maximum value of $I_0 = \mathcal{E}/R$ at $t = 0$ and decays exponentially to zero as $t \rightarrow \infty$ (Fig. 18.23b). The quantity RC, which appears in the exponential of Equations 18.33 and 18.34, is called the **time constant,** τ, of the circuit. It represents the time it takes the current to decrease to $1/e$ of its initial value; that is, in the time τ, $I = e^{-1} I_0 = 0.37 I_0$. In a time of 2τ, $I = e^{-2} I_0 = 0.135 I_0$, and so forth. Likewise, in the time τ the charge increases from zero to $C\mathcal{E}[1 - e^{-1}] = 0.63 C\mathcal{E}$.

The following dimensional analysis shows that τ has units of time:

$$[\tau] = [RC] = \left[\frac{V}{I} \times \frac{Q}{V}\right] = \left[\frac{Q}{Q/T}\right] = [T]$$

The work done by the battery during the charging process is $Q\mathcal{E} = C\mathcal{E}^2$. After the capacitor is fully charged, the energy stored in it is $\frac{1}{2}Q\mathcal{E} = \frac{1}{2}C\mathcal{E}^2$, which is just half the work done by the battery. It is left to an end-of-chapter problem to show that the remaining half of the energy supplied by the battery goes into joule heat in the resistor (Problem 80).

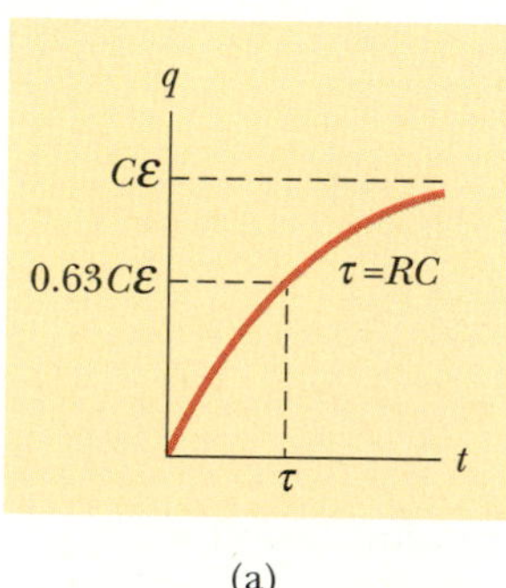

(a)

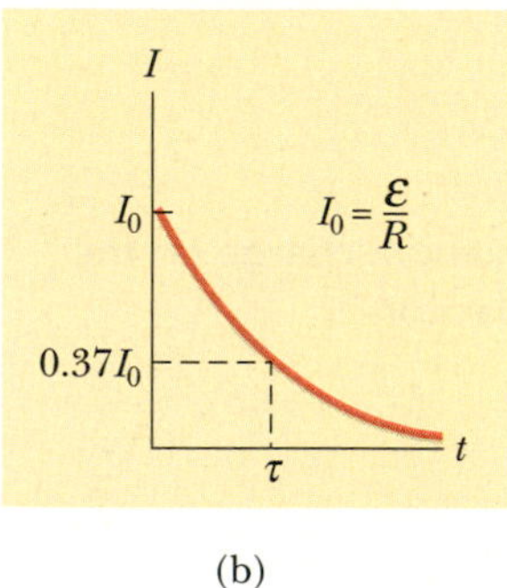

(b)

Figure 18.23
(a) A plot of capacitor charge versus time for the circuit shown in Figure 18.22. After one time constant, τ, the charge is 63% of the maximum value, $C\mathcal{E}$. The charge approaches its maximum value as t approaches infinity.
(b) A plot of current versus time for the RC circuit shown in Figure 18.22. The current has its maximum value, $I_0 = \mathcal{E}/R$, at $t = 0$ and decays to zero exponentially as t approaches infinity. After one time constant, τ, the current decreases to 37% of its initial value.

Discharging a Capacitor

Now consider the circuit in Figure 18.24, consisting of a capacitor with an initial charge of Q, a resistor, and a switch. When the switch is open (Fig. 18.24a), there is a potential difference of Q/C across the capacitor and zero potential difference across the resistor, since $I = 0$. If the switch is closed at $t = 0$, the capacitor begins to discharge through the resistor. At some time during the discharge, the current in the circuit is I and the charge on the capacitor is q (Fig. 18.24b). From Kirchhoff's second rule, we see that the potential drop across the resistor, IR, must equal the potential difference across the capacitor, q/C:

$$IR = \frac{q}{C} \qquad [18.35]$$

However, the current in the circuit must equal the rate of *decrease* of charge on the capacitor. That is, $I = -dq/dt$, and so Equation 18.35 becomes

$$-R\frac{dq}{dt} = \frac{q}{C}$$

$$\frac{dq}{q} = -\frac{1}{RC}\,dt$$

Integrating this expression, using the fact that $q = Q$ at $t = 0$, gives

$$\int_{Q}^{q} \frac{dq}{q} = -\frac{1}{RC}\int_{0}^{t} dt$$

$$\ln\left(\frac{q}{Q}\right) = -\frac{t}{RC}$$

$$q(t) = Q\,e^{-t/RC} \qquad [18.36]$$

Charge versus time for a discharging capacitor

Differentiating Equation 18.36 with respect to time gives the current as a function of time:

$$I(t) = -\frac{dq}{dt} = I_0\,e^{-t/RC} \qquad [18.37]$$

Current versus time for a discharging capacitor

where the initial current is $I_0 = Q/RC$. Thus we see that both the charge on the capacitor and the current decay exponentially at a rate characterized by the time constant $\tau = RC$.

Example 18.11 Charging a Capacitor in an *RC* Circuit

An uncharged capacitor and a resistor are connected in series to a battery as in Figure 18.22. If $\mathcal{E} = 12$ V, $C = 5\ \mu$F, and $R = 8 \times 10^5\ \Omega$, find the time constant of the circuit, the maximum charge on the capacitor, the maximum current in the circuit, and the charge and current as a function of time.

Solution The time constant of the circuit is $\tau = RC = (8 \times 10^5\ \Omega)(5 \times 10^{-6}\ \text{F}) = 4$ s. The maximum charge on the capacitor is $Q = C\mathcal{E} = (5 \times 10^{-6}\ \text{F})(12\ \text{V}) = 60\ \mu$C. The maximum current in the circuit is $I_0 = \mathcal{E}/R = (12\ \text{V})/(8 \times 10^5\ \Omega) = 15\ \mu$A. Using these values and Equations 18.33 and 18.34, we find that

$$q(t) = 60[1 - e^{-t/4}]\ \mu\text{C}$$

$$I(t) = 15\,e^{-t/4}\ \mu\text{A}$$

Exercise Calculate the charge on the capacitor and the current in the circuit after one time constant has elapsed.

Answer 37.9 μC; 5.52 μA.

(a)

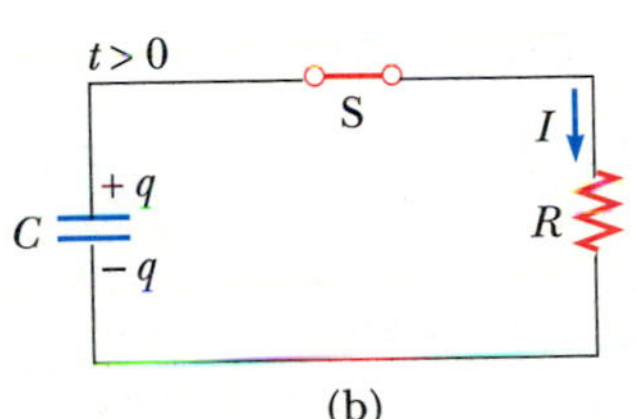

Figure 18.24 (a) A charged capacitor connected to a resistor and a swtich, which is open at $t < 0$. (b) After the switch is closed, a nonsteady current is set up in the direction shown, and the charge on the capacitor decreases exponentially with time.

Example 18.12 Discharging a Capacitor in an *RC* Circuit

Consider a capacitor, *C*, being discharged through a resistor, *R*, as in Figure 18.24.

(a) After how many time constants will the charge on the capacitor drop to one fourth of its initial value?

Solution The charge on the capacitor varies with time according to Equation 18.36,

$$q(t) = Qe^{-t/RC}$$

where Q is the initial charge on the capacitor. To find the time it takes the charge q to drop to one fourth of its initial value, we substitute $q(t) = Q/4$ into this expression and solve for t:

$$\tfrac{1}{4}Q = Qe^{-t/RC}$$

$$\tfrac{1}{4} = e^{-t/RC}$$

Taking logarithms of both sides, we find

$$-\ln 4 = -\frac{t}{RC}$$

$$t = RC \ln 4 = 1.39RC$$

(b) The energy stored in the capacitor decreases with time as the capacitor discharges. After how many time constants will this stored energy drop to one fourth of its initial value?

Solution Using Equations 18.27 and 18.36, we can express the energy stored in the capacitor at any time t as

$$U = \frac{q^2}{2C} = \frac{Q^2}{2C}e^{-2t/RC} = U_0e^{-2t/RC}$$

where U_0 is the initial energy stored in the capacitor. As in part (a), we now set $U = U_0/4$ and solve for t:

$$\tfrac{1}{4}U_0 = U_0e^{-2t/RC}$$

$$\tfrac{1}{4} = e^{-2t/RC}$$

Again, taking logarithms of both sides and solving for t gives

$$t = \tfrac{1}{2}RC \ln 4 = 0.693RC$$

Exercise After how many time constants will the current in the RC circuit drop to one half of its initial value?

Answer $0.693RC$.

Summary

The **electric current,** I, in a conductor is defined as

Electric current

$$I \equiv \frac{dQ}{dt} \quad [18.2]$$

where dQ is the charge that passes through a cross-section of the conductor in the time dt. The SI unit of current is the ampere (A); 1 A = 1 C/s.

The current in a conductor is related to the motion of the charge carriers through the relationship

Current in a conductor

$$I = nqv_dA \quad [18.4]$$

where n is the density of charge carriers, q is their charge, v_d is the drift speed, and A is the cross-sectional area of the conductor.

The **current density,** J, in a conductor is defined as the current per unit area:

$$J \equiv \frac{I}{A} = nqv_d \qquad [18.5]$$

Current density

The **resistance,** R, of a conductor is defined as the ratio of the potential difference across the conductor to the current:

$$R \equiv \frac{V}{I} \qquad [18.7]$$

Resistance of a conductor

The SI units of resistance are volts per ampere, defined as ohms (Ω). That is, $1\ \Omega = 1\ \text{V/A}$.

If the resistance is independent of the applied voltage, the conductor obeys Ohm's law, and conductors that have a constant resistance over a wide range of voltages are said to be ohmic. It is common practice to write Ohm's law as

$$V = IR \qquad [18.8]$$

Ohm's law

If a conductor has a uniform cross-sectional area of A and a length of ℓ, its resistance is

$$R = \rho \frac{\ell}{A} \qquad [18.9]$$

Resistance of a uniform conductor

where ρ is called the **resistivity** of the conductor. The inverse of the resistivity is defined as the **conductivity,** σ. That is, $\sigma = 1/\rho$.

The resistivity of a conductor varies with temperature in an approximately linear fashion; that is,

$$\rho = \rho_0[1 + \alpha(T - T_0)] \qquad [18.11]$$

Variation of ρ with temperature

where α is the temperature coefficient of resistivity and ρ_0 is the resistivity at some reference temperature T_0.

In a classical model of electronic conduction in a metal, the electrons are treated as molecules of a gas. In the absence of an electric field, the average velocity of the electrons is zero. When an electric field is applied, the electrons move (on the average) with a **drift velocity,** $\mathbf{v}_d$, which is opposite the electric field:

$$\mathbf{v}_d = \frac{q\mathbf{E}}{m}\tau \qquad [18.16]$$

Drift velocity

where τ is the average time between collisions with the atoms of the metal. The resistivity of the material according to this model is

$$\rho = \frac{m}{nq^2\tau} \qquad [18.19]$$

Resistivity

where n is the number of free electrons per unit volume.

If a potential difference, V, is maintained across a resistor, the **power,** or rate at which energy is supplied to the resistor, is

$$P = IV \qquad [18.21]$$

Power

Since the potential difference across a resistor is $V = IR$, we can express the power dissipated in a resistor in the form

$$P = I^2R = \frac{V^2}{R} \qquad [18.22]$$

Power loss in a resistor

The electrical energy supplied to a resistor appears in the form of thermal energy in the resistor.

The **emf** of a battery is the voltage across its terminals when the current is zero. That is, the emf is equivalent to the open-circuit voltage of the battery.

The **equivalent resistance** of a set of resistors connected in **series** is

Equivalent resistance of resistors in series

$$R_{eq} = R_1 + R_2 + R_3 + \cdots \quad [18.27]$$

The **equivalent resistance** of a set of resistors connected in *parallel* is

Equivalent resistance of resistors in parallel

$$\frac{1}{R_{eq}} = \frac{1}{R_1} + \frac{1}{R_2} + \frac{1}{R_3} + \cdots \quad [18.29]$$

Kirchhoff's rules

Complex circuits involving more than one loop are conveniently analyzed using two simple rules called **Kirchhoff's rules:**

1. The sum of the currents entering any junction must equal the sum of the currents leaving that junction.
2. The sum of the potential differences across the elements around any closed-circuit loop must be *zero.*

The first rule is a statement of **conservation of charge;** the second rule is equivalent to a statement of **conservation of energy.**

When a resistor is traversed in the direction of the current, the change in potential, ΔV, across the resistor is $-IR$. If a resistor is traversed in the direction opposite the current, $\Delta V = +IR$.

If a source of emf is traversed in the direction of the emf (negative to positive) the change in potential is $+\mathcal{E}$. If it is traversed opposite the emf (positive to negative), the change in potential is $-\mathcal{E}$.

If a capacitor is charged with a battery of emf $\mathcal{E}$ through a resistance, R, the charge on the capacitor and the current in the circuit vary in time according to the expressions

Charge versus time

$$q(t) = Q[1 - e^{-t/RC}] \quad [18.33]$$

Current versus time

$$I(t) = \frac{\mathcal{E}}{R} e^{-t/RC} \quad [18.34]$$

where $Q = C\mathcal{E}$ is the *maximum* charge on the capacitor. The product RC is called the **time constant** of the circuit.

If a charged capacitor is discharged through a resistance, R, the charge and current decrease exponentially in time according to the expressions

$$q(t) = Q\,e^{-t/RC} \quad [18.36]$$

$$I(t) = I_0\,e^{-t/RC} \quad [18.37]$$

where $I_0 = Q/RC$ is the initial current in the circuit and Q is the initial charge on the capacitor.

Questions and Conceptual Exercises

1. In an analogy between automobile traffic flow and electrical current, what would correspond to the charge Q? What would correspond to the current I?
2. What factors affect the resistance of a conductor?
3. We have seen that an electric field must exist inside a conductor that carries a current. How is this possible in

view of the fact that in *electrostatics,* we concluded that E must be zero inside a conductor?

4. Two wires of circular cross-section, A and B, are made of the same metal and have equal lengths, but the resistance of wire A is three times greater than that of wire B. What is the ratio of their cross-sectional areas? How do their radii compare?
5. Use the atomic theory of matter to explain why the resistance of a material should increase as its temperature increases.
6. Explain how a current can persist in a superconductor without any applied voltage.
7. What single experimental requirement makes superconducting devices expensive to operate? In principle, can this limitation be overcome?
8. What would happen to the drift speed of the electrons in a wire and to the current in the wire if the electrons could move freely, without resistance, through the wire?
9. If charges flow very slowly through a metal, why does it not take several hours for a light to come on after you throw the switch?
10. If you were to design an electric heater using Nichrome wire as the heating element, what parameters of the wire could you vary to meet a specific power output, such as 1000 W?
11. Car batteries are often rated in ampere-hours. Does this designate the amount of current, power, energy, or charge that can be drawn from the battery?
12. How would you connect resistors in order for the equivalent resistance to be greater than the individual resistances? Give an example.
13. How would you connect resistors in order for the equivalent resistance to be less than the individual resistances? Give an example.
14. Explain why a bird can roost on a high-voltage power line with no ill effects. Is there any connection between the bird-on-the-wire effect and the admonition given to experimenters to "keep one hand in the pocket" when working around high voltage?

(Question 14) Bird on a high-voltage wire. *(Superstock)*

15. A "short circuit" is a circuit containing a path of very low resistance in parallel with some other part of the circuit. Discuss the effect of a short circuit on the portion of the circuit it parallels. Use a lamp with a frayed line cord as an example.
16. A series circuit consists of three identical lamps connected to a battery, as shown in Figure 18.25. When switch S is closed, (a) what happens to the intensities of lamps A and B? (b) What happens to the intensity of lamp C? (c) What happens to the current in the circuit? (d) Does the power dissipated in the circuit increase, decrease, or remain the same?

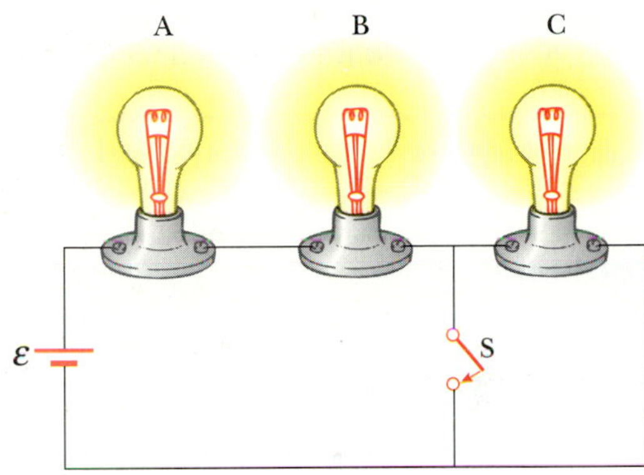

Figure 18.25 (Question 16)

17. If materials could be produced that would be superconducting at room temperature, list some ways in which such materials could benefit mankind.
18. Two lightbulbs operate from 120 V, but one has a power rating of 25 W and the other a power rating of 100 W. Which bulb has the higher resistance? Which carries the greater current?
19. If electrical power is transmitted over long distances, the resistance of the wires becomes significant. Why? Which mode of transmission would result in less energy loss—high current and low voltage or low current and high voltage? Discuss.
20. Two different sets of Christmas-tree lights are available. In set A, when one bulb is removed (or burns out), the remaining bulbs remain illuminated. In set B, when one bulb is removed, the remaining bulbs do not operate. Explain the difference in wiring of the two sets of lights. (See also *The Physics Teacher,* December 1992.)
21. Are the two headlights on a car wired in series or in parallel? How can you tell?
22. Embodied in Kirchhoff's rules are two conservation laws. What are they?
23. With reference to Figure 18.26, describe what happens to the lightbulb after the switch is closed. Assume the capacitor has a large capacitance and is initially un-

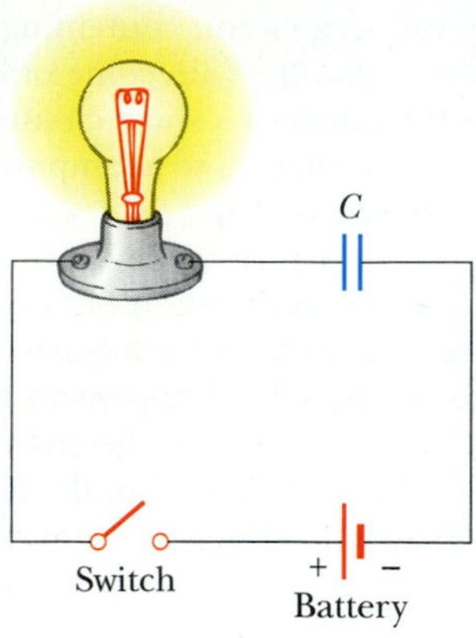

Figure 18.26 (Question 23)

charged, and assume that the light will illuminate when connected directly across the battery terminals.

24. Would a fuse work successfully if it were placed in parallel with the device it is supposed to protect?
25. What procedure would you use to try to save a person "frozen" to a live high-voltage wire without endangering your own life?
26. If it is the current flowing through a body that determines how serious the shock will be, why do we see warnings of high voltage rather than high current near electric equipment?
27. How many 100-W bulbs can you connect in parallel in a 120-V household circuit without tripping the 20-A circuit breaker?

Problems

Section 18.1 Electric Current

1. If Avogadro's number of electrons pass by a given cross-sectional area in one hour, find the current in the conductor.
2. In a particular television picture tube, the measured beam current is 60 μA. How many electrons strike the screen every second?
3. If a current of 80.0 mA exists in a metal wire, how many electrons flow past a given cross-section of the wire in 10.0 min? Sketch the direction of current and the direction of the electrons' net motion.
4. If 3.25×10^{-3} kg of gold is deposited on the negative electrode of an electrolytic cell in a period of 2.78 h, what is the current through the cell in this period? Assume the gold ions carry one elementary unit of positive charge.
5. A teapot with a surface area of 700 cm^2 is to be silver plated. It is attached to the negative electrode of an electrolytic cell containing silver nitrate ($Ag^+NO_3^-$). If the cell is powered by a 12.0-V battery and has a resistance of 1.8 Ω, how long does it take to build up a 0.133-mm layer of silver on the teapot? (Density of silver = 10.5×10^3 kg/m^3.)
6. Calculate the number of free electrons per cubic meter for gold, assuming one free electron per atom. (Density of gold = 19.3×10^3 kg/m^3.)
7. An aluminum wire having a cross-sectional area of 4×10^{-6} m^2 carries a current of 5 A. Find the drift speed of the electrons in the wire. The density of aluminum is 2.7 g/cm^3. (Assume one electron is supplied by each atom.)
8. A metal wire 1.0 mm in diameter contains 2.50×10^{22} free electrons per cubic centimeter. If the electrons travel through the wire with an average drift speed of 0.5 mm/s, what is the current in the wire?
9. The current I (in amperes) in a conductor depends on time as $I = 2t^2 - 3t + 7$, where t is in seconds. What quantity of charge moves across a section through the conductor during the interval $t = 2$ s to $t = 4$ s?
10. Suppose that the current through a conductor decreases exponentially with time according to

$$I(t) = I_0 e^{-t/\tau}$$

where I_0 is the initial current (at $t = 0$), and τ is a constant having dimensions of time. Consider a fixed observation point within the conductor. (a) How much charge passes this point between $t = 0$ and $t = \tau$? (b) How much charge passes this point between $t = 0$ and $t = 10\tau$? (c) How much charge passes this point between $t = 0$ and $t = \infty$?

Section 18.2 Resistance and Ohm's Law

11. A wire with a resistance of R is lengthened to 1.25 times its original length by pulling it through a small hole. Find the resistance of the wire after it is stretched.
12. Suppose that you wish to fabricate a uniform wire out of 1 g of copper. If the wire is to have a resistance of $R = 0.5\ \Omega$, and all of the copper is to be used, what will be (a) the length and (b) the diameter of this wire?

13. Calculate the diameter of a 2-cm length of tungsten filament in a small lightbulb if its resistance is 0.05 Ω.
14. A lightbulb has a resistance of 240 Ω when operating at a voltage of 120 V. What is the current through the lightbulb?
15. A typical color television draws about 2.5 A when connected to a 120-V source. What is the effective resistance of the TV set?
16. A 0.9-V potential difference is maintained across a 1.5-m length of tungsten wire that has a cross-sectional area of 0.6 mm^2. What is the current in the wire?
17. The electron beam emerging from a certain high-energy electron accelerator has a circular cross-section of radius 1 mm. (a) If the beam current is 8 μA, find the current density in the beam, assuming that it is uniform throughout. (b) The speed of the electrons is so close to the speed of light that it can be taken as $c = 3 \times 10^8$ m/s with negligible error. Find the electron density in the beam. (c) How long does it take for Avogadro's number of electrons to emerge from the accelerator?
18. The resistance of a platinum wire is to be calibrated for low-temperature measurements. A platinum wire with resistance 1 Ω at 20°C is immersed in liquid nitrogen at 77 K (−196°C). If the temperature response of the platinum wire is linear, what is the expected resistance of the platinum wire at −196°C? ($\alpha_{\text{platinum}} = 3.92 \times 10^{-3}$/°C.)
19. An aluminum rod has a resistance of 1.234 Ω at 20°C. Calculate the resistance of the rod at 120°C by accounting for the changes in both the resistivity and the dimensions of the rod.
20. At what temperature does tungsten have a resistivity four times that of copper? (Assume that the copper is at 20°C.)
21. If a silver wire has a resistance of 10 Ω at 20°C, what resistance does it have at 40°C? Neglect any change in length or cross-sectional area due to the change in temperature.
22. At 20°C the carbon resistor in an electric circuit has a resistance of 200 Ω and is connected to a 5-V battery. What is the current in the circuit when the temperature of the carbon rises to 80°C?
23. A wire 3 m in length and 0.45 mm^2 in cross-sectional area has a resistance of 41 Ω at 20°C. If the resistance of the wire increases to 41.4 Ω at 29°C, what is the temperature coefficient of resistivity?
24. Calculate the percentage change in the resistance of a carbon filament when it is heated from 20°C to 160°C.
25. A certain lightbulb has a tungsten filament with a resistance of 19 Ω when cold and 140 Ω when hot. Assume that Equation 18.11 can be used over the large temperature range involved here, and find the temperature of the filament when hot. Assume an initial temperature of 20°C.

Section 18.4 A Model for Electrical Conduction

26. Calculate the current density in a gold wire in which an electric field of 0.74 V/m exists.
27. If the drift speed of free electrons in a copper wire is 7.84×10^{-4} m/s, calculate the electric field in the conductor.
28. If the current through a given conductor is doubled, what happens to the (a) charge carrier density? (b) current density? (c) electron drift speed? (d) average time between collisions?
29. Use data from Example 18.4 to calculate the collision mean free path of electrons in copper if the average thermal speed of conduction electrons is 8.6×10^5 m/s.

Section 18.5 Electrical Energy and Power

30. A 10-V battery is connected to a 120-Ω resistor. Neglecting the internal resistance of the battery, calculate the power dissipated in the resistor.
31. Suppose that a voltage surge produces 140 V for a moment. By what percentage will the output of a 120-V, 100-W lightbulb increase, assuming the bulb's resistance does not change?
32. A toaster is rated at 600 W when connected to a 120-V source. What current does the toaster carry, and what is its resistance?
33. What is the required resistance of an immersion heater that will increase the temperature of 1.5 kg of water from 10°C to 50°C in 10 min while operating at 120 V?
34. Determine the loss of electrical power per meter of a copper wire 2.0 mm in diameter if a current of 40.0 A exists in the wire.
35. In a hydroelectric installation, a turbine delivers 1500 hp to a generator, which in turn converts 80% of the mechanical energy into electrical energy. Under these conditions, what current will the generator deliver at a terminal potential difference of 2000 V?
36. Suppose that you want to install a heating coil that will convert electric energy to heat at a rate of 300 W for a current of 1.5 A. (a) Determine the resistance of the coil. (b) The coil wire's resistivity is 10^{-6} Ω·m, and its diameter is 0.3 mm. Determine its length.
37. An electric heater with a resistance of 20 Ω requires 100 V across its terminals. A built-in switching circuit repetitively turns the heater on for 1 s and off for 4 s. (a) How much energy is produced by the heater in 1 h? (b) What is the average power delivered by the heater over a period of one cycle?

Section 18.6 Sources of emf

38. A battery with an emf of 12 V and internal resistance of 0.9 Ω is connected across a load resistor, R. If the current in the circuit is 1.4 A, what is the value of R?
39. (a) What is the current in a 5.6-Ω resistor connected to a battery with a 0.2-Ω internal resistance if the terminal voltage of the battery is 10 V? (b) What is the emf of the battery?
40. If the emf of a battery is 15 V and a current of 60 A is measured when the battery is shorted, what is the internal resistance of the battery?
41. A typical fresh AA dry cell has an emf of 1.50 V and an internal resistance of 0.311 Ω. (a) Find the terminal voltage of the battery when it supplies 58 mA to a circuit. (b) What is the resistance, R, of the external circuit?
42. A battery has an emf of 15 V. The terminal voltage of the battery is 11.6 V when it is delivering 20 W of power to an external load resistor, R. (a) What is the value of R? (b) What is the internal resistance of the battery?
43. Two 1.50-V batteries—with their positive terminals in the same direction—are inserted in series into the barrel of a flashlight. One battery has an internal resistance of 0.255 Ω, the other an internal resistance of 0.153 Ω. When the switch is closed, a current of 0.6 A occurs in the lamp. (a) What is the lamp's resistance? (b) What fraction of the power dissipated is dissipated in the batteries?

Section 18.7 Resistors in Series and in Parallel

44. A television repairman needs a 100-Ω resistor to repair a malfunctioning set. He is temporarily out of resistors of this value. All he has in his tool box are a 500-Ω resistor and two 250-Ω resistors. How can the desired resistance be obtained from the resistors on hand?
45. The current in a circuit is tripled by connecting a 500-Ω resistor in parallel with the resistance of the circuit. Determine the resistance of the circuit in the absence of the 500-Ω resistor.
46. (a) Find the equivalent resistance between points a and b in Figure 18.27. (b) If a potential difference of 34 V is applied between points a and b, calculate the current in each resistor.

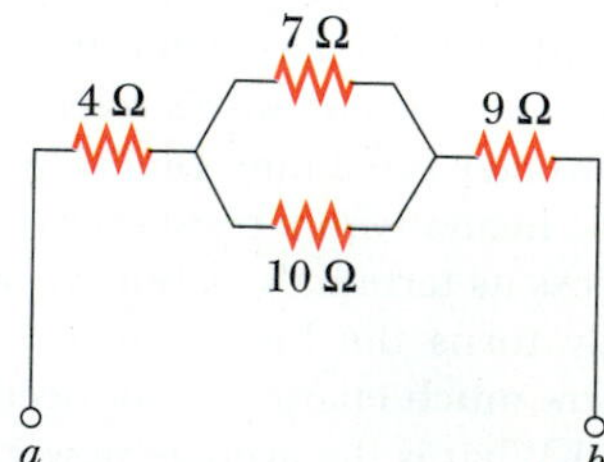

Figure 18.27 (Problem 46)

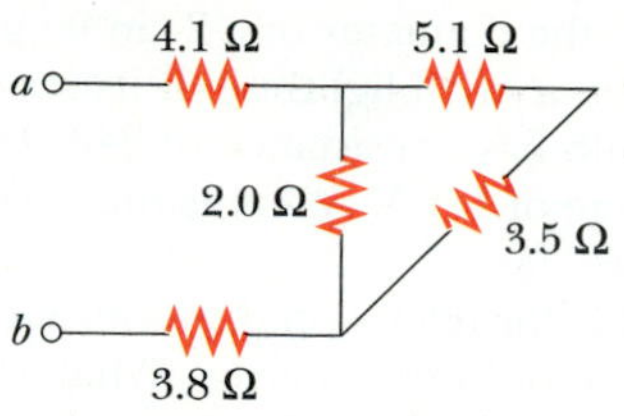

Figure 18.28 (Problem 47)

47. (a) Find the equivalent resistance between points a and b in Figure 18.28. (b) If a potential difference of 24 V is applied between points a and b, calculate the current in the 4.1-Ω resistor.
48. Consider the circuit shown in Figure 18.29. Find (a) the current in the 20-Ω resistor and (b) the potential difference between points a and b.
49. Three 100-Ω resistors are connected as shown in Figure 18.30. The maximum power that can be dissipated in any one of the resistors is 25 W. (a) What is the maximum voltage that can be applied to the terminals a and b? For the voltage determined in (a), what is the power dissipation in each resistor? What is the total power dissipation?

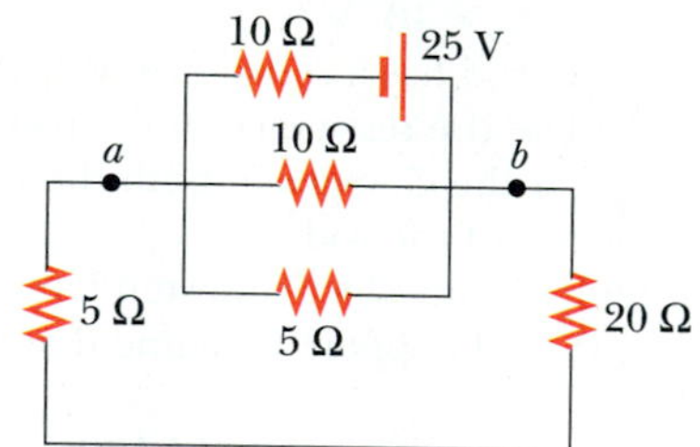

Figure 18.29

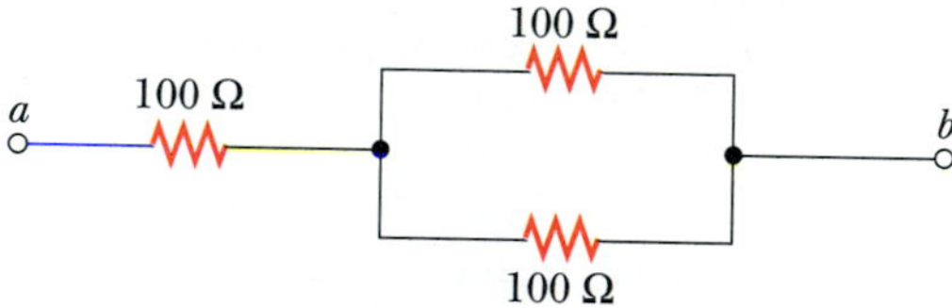

Figure 18.30 (Problem 49)

Section 18.8 Kirchhoff's Rules

The currents are not necessarily in the directions shown for some circuits.

50. Consider the circuit shown in Figure 18.31. Find (a) the potential difference between points a and b and (b) the currents I_1, I_2, and I_3.

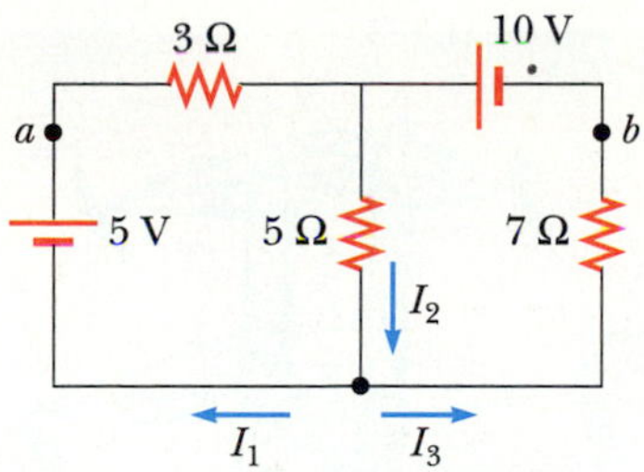

Figure 18.31 (Problem 50)

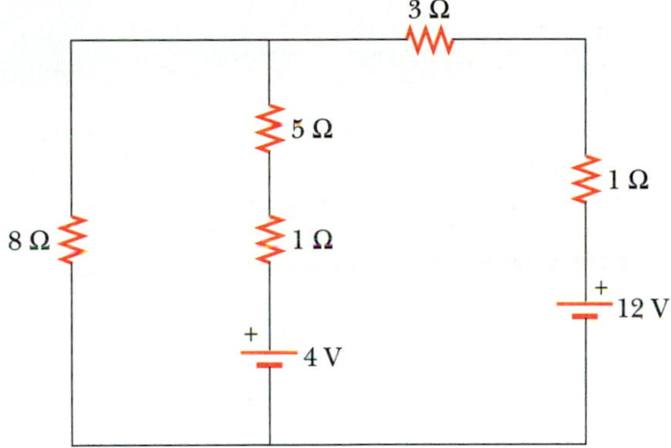

Figure 18.32 (Problem 51)

51. Determine the current in each of the branches of the circuit shown in Figure 18.32.
52. A dead battery is "charged" by being connected to the live battery of another car (Fig. 18.33). Determine the current in the starter and in the dead battery.
53. The ammeter in the circuit shown in Figure 18.34 reads 2 A. Find the currents I_1 and I_2 and the value of $\mathcal{E}$.
54. Using Kirchhoff's rules, (a) find the current in each of the resistors in the circuit shown in Figure 18.35, and (b) find the potential difference between points *c* and *f*. Which is at the higher potential?
55. Consider the circuit shown in Figure 18.36. Find the values of I_1, I_2, and I_3.
56. (a) Find the values of I_1 and I_3 in the circuit of Figure 18.36 if the 4-V battery is replaced by a 5-μF capacitor. (b) Determine the charge on the 5-μF capacitor.
57. Calculate the power dissipated in each resistor in the circuit of Figure 18.37.

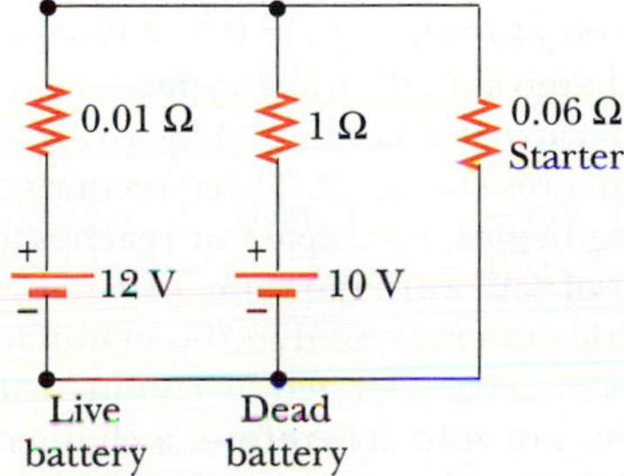

Figure 18.33 (Problem 52)

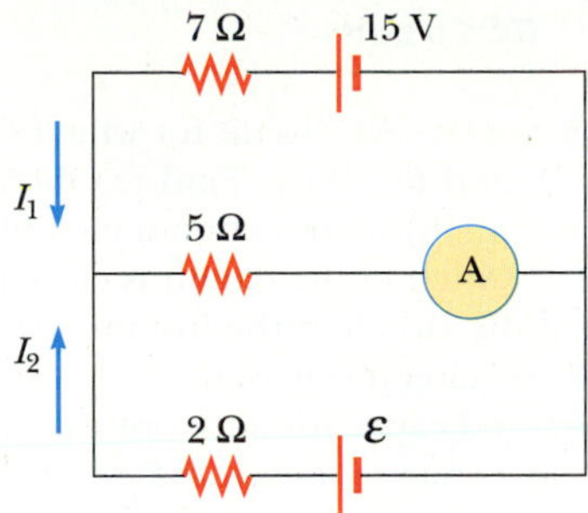

Figure 18.34 (Problem 53)

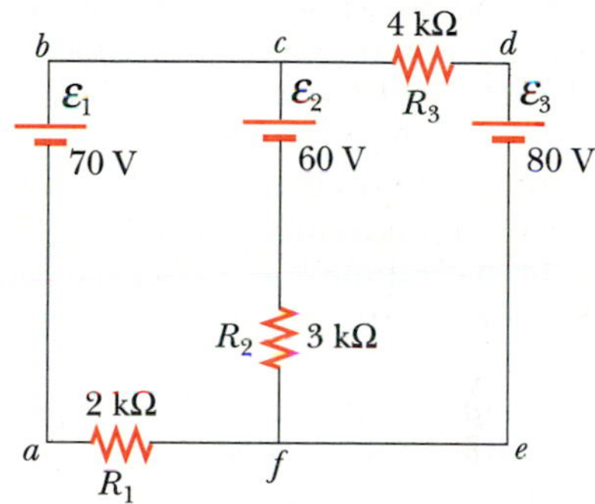

Figure 18.35 (Problem 54)

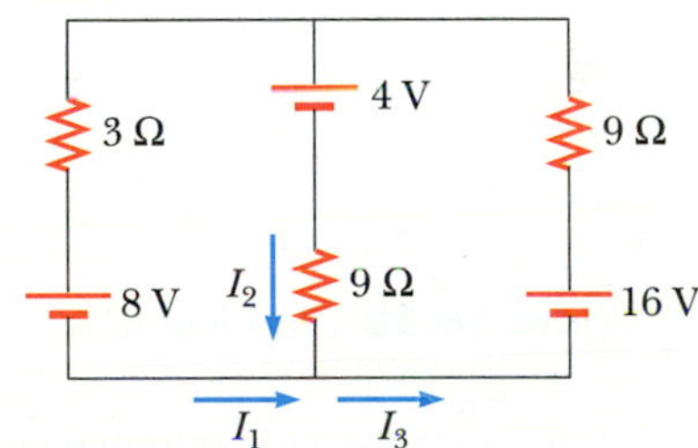

Figure 18.36 (Problems 55 and 56)

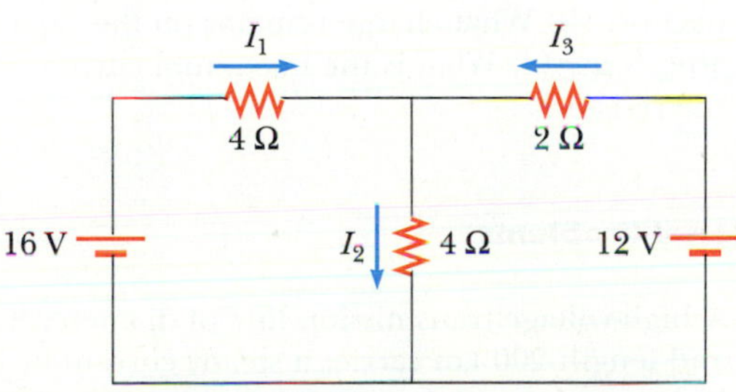

Figure 18.37 (Problem 57)

Section 18.9 *RC* Circuits

58. Consider a series *RC* circuit for which $C = 6\ \mu F$, $R = 2 \times 10^6\ \Omega$, and $\mathcal{E} = 20$ V. Find (a) the time constant of the circuit, (b) the maximum charge on the capacitor after a switch in the circuit is closed, and (c) the current in the circuit at the instant just after the switch in the circuit is closed.
59. An uncharged capacitor and a resistor are connected in series to a source of emf. If $\mathcal{E} = 9$ V, C = 20 μF, and $R = 100\ \Omega$, find (a) the time constant of the circuit, (b) the maximum charge on the capacitor, (c) the maximum current in the circuit, (d) the charge on the capacitor after one time constant, and (e) the current in the circuit after one time constant.
60. Consider a series *RC* circuit (Fig. 18.22) for which $R = 1\ M\Omega$, $C = 5\ \mu F$, and $\mathcal{E} = 30$ V. Find (a) the time constant of the circuit, (b) the *maximum* charge on the capacitor after the switch is closed, and (c) the current in the resistor *R* 10 s after the switch is closed.
61. A 4-MΩ resistor and a 3-μF capacitor are connected in series with a 12-V power supply. (a) What is the time constant for the circuit? (b) Express the current in the circuit and the charge on the capacitor as functions of time.
62. A circuit has been connected as shown in Figure 18.38 a "long" time. (a) What is the voltage across the capacitor? (b) If the battery is disconnected, how long does it take for the capacitor to discharge to 1/10 of its initial voltage?

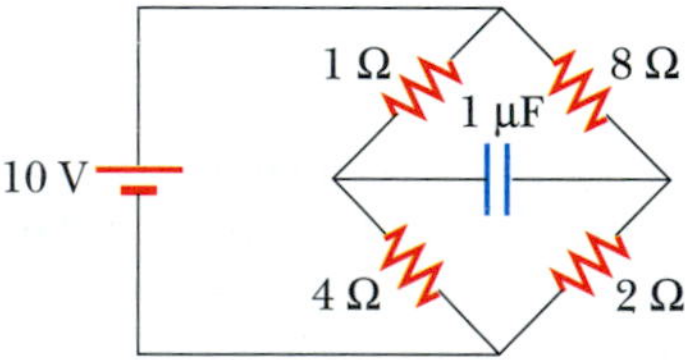

Figure 18.38 (Problem 62)

63. A 2×10^{-3}-μF capacitor with an initial charge of 5.1 μC is discharged through a 1300-Ω resistor. (a) Calculate the current through the resistor 9 μs after the resistor is connected across the terminals of the capacitor. (b) What charge remains on the capacitor after 8 μs? (c) What is the maximum current through the resistor?

Additional Problems

64. A high-voltage transmission line of diameter 2 cm and length 200 km carries a steady current of 1000 A. If the conductor is copper wire with a free charge density of 8×10^{28} electrons/m^3, how long does it take one electron to travel the full length of the cable?

(Problems 64 and 65) High-voltage transmission lines which transport electrical energy often operate at 765 kV. *(Fred R. Palmer, Stock Boston)*

65. A high-voltage transmission line carries 1000 A starting at 700 kV for a distance of 100 miles. If the resistance in the wire is 0.5 Ω/mi, what is the power loss due to resistive losses?
66. The heating coil of a water heater has a resistance of 20 Ω and operates at 210 V. If electrical energy costs 5.5 cents/kWh, what does it cost to raise the 200 kg in the tank from 15°C to 80°C? (The specific heat of water is 4186 J/kg·°C.)
67. A copper cable is to be designed to carry a current of 300 A with a power loss of only 2 W/m. What is the required radius of the copper cable?
68. Four 1.5-V AA batteries in series are used to power a transistor radio. If the batteries can move a charge of 240 C, how long will they last if the radio has a resistance of 200 Ω?
69. Two resistors, R_1 and R_2, have an equivalent resistance of 690 Ω when they are connected in series and an equivalent resistance of 150 Ω when they are connected in parallel. What are R_1 and R_2?
70. A 5000-Ω resistor and a 50-μF capacitor are connected in series at $t = 0$ with a 6-V battery. The capacitor is initially uncharged. What is the current in the circuit at $t = 0$? At $t = 0.5$ s? What is the maximum charge stored on the capacitor?
71. A 10-μF capacitor is charged by a 10-V battery through a resistance, *R*. Three seconds after the charging begins, the capacitor reaches a potential difference of 4 V. Find the value of *R*.
72. Dielectric materials used in the manufacture of capacitors are characterized by conductivities that are small but not zero. Therefore, a charged capacitor slowly loses its charge by "leaking" across the dielec-

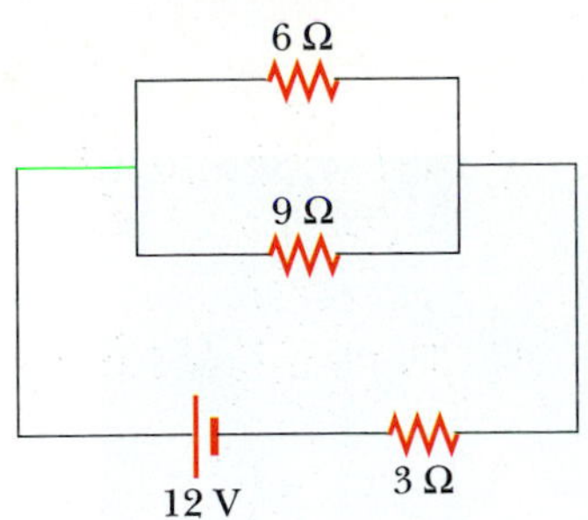

Figure 18.39 (Problem 76)

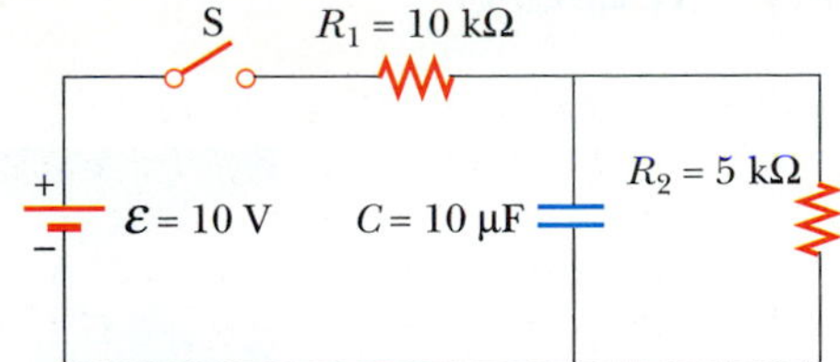

Figure 18.40 (Problem 77)

tric. If a certain 3.6-μF capacitor leaks charge so that the potential difference decreases to half its initial value in 4 s, what is the equivalent resistance of the dielectric?

73. The headlights on a car are rated at 80 W. If these are connected to a fully charged 90 amp·h, 12-V battery, how long will it take the battery to completely discharge?

74. An electric car is designed to run off a bank of 12-V batteries with total energy storage of 2×10^7 J. (a) If the electric motor draws 8 kW, what is the current delivered to the motor? (b) If the electric motor draws 8 kW as the car moves at a steady speed of 20 m/s, how far will the car travel before it is "out of juice"?

75. An electric heater is rated at 1500 W, a toaster is rated at 750 W, and an electric grill is rated at 1000 W. The three appliances are connected to a common 120-V circuit. (a) How much current does each appliance draw? (b) Is a 25-A circuit sufficient in this situation? Explain.

76. Find the current in each resistor of Figure 18.39 by (a) the rules for resistors in series and parallel and (b) by the use of Kirchhoff's rules.

77. Before the switch is closed in the circuit in Figure 18.40, no charge is stored by the capacitor. Determine the currents in R_1, R_2, and C (a) at the instant the switch is closed (that is, $t = 0$), and (b) after the switch has been closed for a long period of time (that is, as $t \rightarrow \infty$).

△78. An experiment is conducted to measure the electrical resistivity of Nichrome in the form of wires with different lengths and cross-sectional areas. For one set of measurements, a student uses 30-gauge wire, which has a cross-sectional area of 7.3×10^{-8} m^2. The voltage across the wire and the current in the wire are measured with a voltmeter and an ammeter, respectively. For each of the measurements in the following table, which were taken on wires of three different lengths, calculate the resistances of the wires and the corresponding values of resistivity. What is the average value of the resistivity, and how does it compare with the value given in Table 18.1?

L (m)	V (V)	I (A)	R (Ω)	ρ (Ω·m)
0.54	5.22	0.500		
1.028	5.82	0.276		
1.543	5.94	0.187		

79. A general definition of the temperature coefficient of resistivity is

$$\alpha = \frac{1}{\rho}\frac{d\rho}{dT}$$

where ρ is the resistivity at temperature T. (a) Assuming that α is constant, show that

$$\rho = \rho_0 e^{\alpha(T - T_0)}$$

where ρ_0 is the resistivity at temperature T_0. (b) Using the series expansion ($e^x \approx 1 + x$; $x \ll 1$), show that the resistivity is given approximately by the expression $\rho = \rho_0[1 + \alpha(T - T_0)]$ for $\alpha(T - T_0) \ll 1$.

80. A battery is used to charge a capacitor through a resistor as in Figure 18.22. Show that in the process of charging the capacitor, half of the energy supplied by the battery is dissipated as heat in the resistor and half is stored in the capacitor.

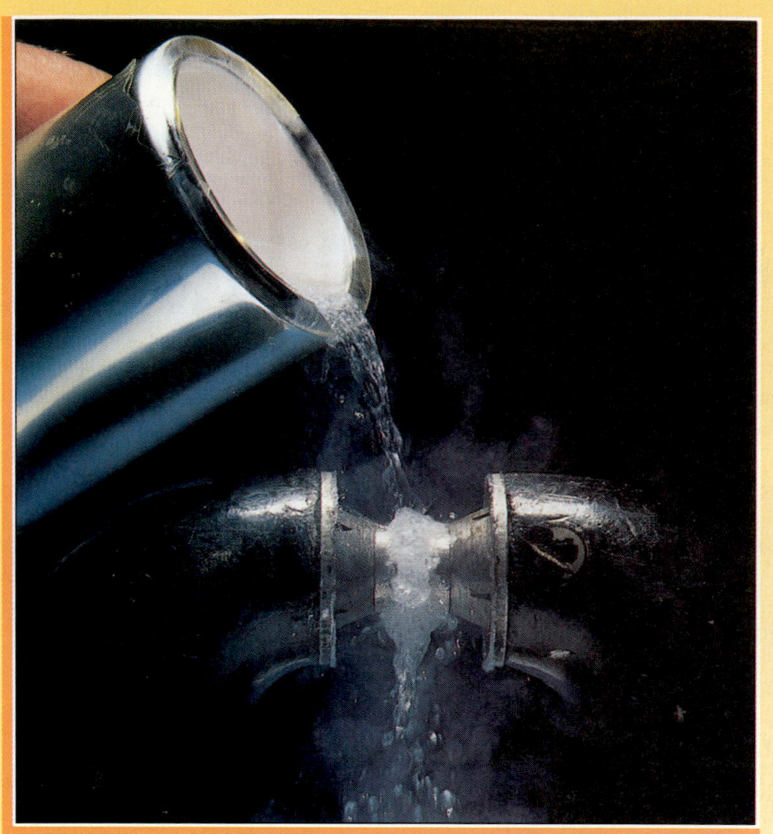

Oxygen, a paramagnetic substance, is attracted to a magnetic field. The liquid oxygen in this photograph is suspended between the poles of a permanent magnet. Paramagnetic substances contain atoms (or ions) that have permanent magnetic dipole moments. These dipoles interact weakly with each other and are randomly oriented in the absence of an external magnetic field. When the substance is placed in an external magnetic field, its atomic dipoles tend to line up with the field. *(Courtesy of Leon Lewandowski)*

Magnetism

CHAPTER 19

The list of important technological applications of magnetism is very long. For instance, large electromagnets are used to pick up heavy loads. Magnets are also used in such devices as meters, motors, and loudspeakers. Magnetic tapes are routinely used in sound and video recording equipment, and for computer memory, and magnetic recording material is used on computer disks. Intense magnetic fields generated by superconducting magnets are currently being used as a means of containing the plasmas (heated to temperatures of the order of 10^8 K) used in controlled nuclear fusion research.

As we investigate magnetism in this chapter, you will find that the subject cannot be divorced from electricity. For example, magnetic fields affect moving charges and moving charges produce magnetic fields. Ultimately we shall find that the source of all magnetic fields is electric current, whether it be the current in a wire or the current produced by the motion of charges within atoms or molecules.

Hans Christian Oersted (1777–1851), a Danish physicist. *(The Bettmann Archive)*

19.1 Historical Overview; Magnets

Many historians of science believe that the compass, which uses a magnetic needle, was used in China as early as the 13th century B.C., its invention being of Arab or Indian origin. The phenomenon of magnetism was known to the Greeks as early as about 800 B.C. They discovered that certain stones, now called magnetite (Fe_3O_4), attracted pieces of iron. Legend ascribes the name *magnetite* to the shepherd Magnes, "the nails of whose shoes and the tip of whose staff stuck fast in a magnetic field while he pastured his flocks." In 1269 Pierre de Maricourt mapped out the directions taken by a needle when it was placed at a variety of points on the surface of a spherical natural magnet. He found that the directions formed lines that encircled the sphere and passed through two points diametrically opposite each other, which he called the *poles* of the magnet. Subsequent experiments have shown that every magnet, regardless of its shape, has two poles, called *north* and *south poles,* which exhibit forces on each other in a manner analogous to electrical charges. That is, like poles repel each other and unlike poles attract each other. The poles received their names because of the behavior of a magnet in the presence of the Earth's magnetic field. If a bar magnet is suspended from its midpoint by a piece of string so that it can swing freely in a horizontal plane, it will rotate until its "north" pole points to the north of the Earth and its "south" pole points to the south. (The same idea is used to construct a simple compass.)

In 1600 William Gilbert extended these experiments to a variety of materials. Using the fact that a compass needle orients in preferred directions, he suggested that magnets are attracted to land masses. In 1750 John Michell (1724–1793) used a torsion balance to show that magnetic poles exert attractive or repulsive forces on each other and that these forces vary as the inverse square of their separation. Although the force between two magnetic poles is similar to the force between two electric charges, an important difference exists. Electric charges can be isolated (witness the electron and proton), whereas *magnetic poles cannot be isolated.* That is, *magnetic poles are always found in pairs.* No matter how many times a permanent magnet is cut, each piece always has a north pole and a south pole. (There is some theoretical basis for speculating that magnetic monopoles—isolated north or

An assortment of commercially available magnets. Some of the magnets are made of metallic alloys, and others are ceramic compounds.

south poles—may exist in nature, and attempts to detect them currently make up an active experimental field of investigation. However, none of these attempts has proven successful.)

The relationship between magnetism and electricity was discovered in 1819 when, while preparing for a lecture demonstration, the Danish scientist Hans Oersted found that an electric current in a wire deflected a nearby compass needle. Shortly thereafter, André Ampère (1775–1836) deduced quantitative laws of magnetic force between current-carrying conductors. He also suggested that electric current loops of molecular size are responsible for *all* magnetic phenomena.

In the 1820s, further connections between electricity and magnetism were identified by Faraday and, independently, Joseph Henry (1797–1878). They showed that an electric current could be produced in a circuit either by moving a magnet near the circuit or by changing the current in another, nearby circuit. Their observations demonstrated that a changing magnetic field produces an electric field. Years later, theoretical work by Maxwell showed that the reverse is also true: a changing electric field gives rise to a magnetic field.

There is a similarity between electric and magnetic effects that has given rise to methods of making permanent magnets. In Chapter 16 we learned that when rubber and wool are rubbed together, both become charged, one positively and the other negatively. In a somewhat analogous fashion, an unmagnetized piece of iron can be magnetized by stroking it with a magnet. Magnetism can also be induced in iron (and other materials) by other means. For example, if a piece of unmagnetized iron is placed near a strong permanent magnet, the piece of iron eventually becomes magnetized. The process of magnetizing the piece of iron in the presence of a strong external field can be accelerated either by heating and cooling the iron or by hammering.

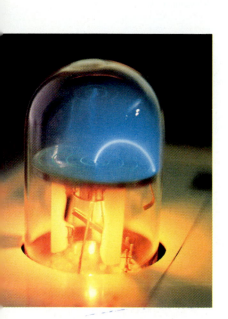

The white arc in this photograph indicates the circular path followed by an electron beam moving in a magnetic field. The vessel contains gas at very low pressure, and the beam is made visible as the electrons collide with the gas atoms, which in turn emit visible light. The magnetic field is produced by two coils (not shown). The apparatus can be used to measure the charge/mass ratio for the electron. *(Courtesy of CENCO).*

19.2 The Magnetic Field

In earlier chapters we found it convenient to describe the interaction between charged objects in terms of electric fields. Recall that an electric field surrounds any electric charge. The region of space surrounding a *moving* charge includes a magnetic field in addition to the electric field. A magnetic field also surrounds any magnetic material.

In order to describe any type of field, we must define its magnitude, or strength, and its direction. The direction of the magnetic field, **B**, at any location is the direction in which the north pole of a compass needle points at that location. Figure 19.1a shows how the magnetic field of a bar magnet can be traced with the aid of a compass. Several magnetic field lines of a bar magnet traced out in this manner are shown in Figure 19.1b. Magnetic field patterns can be displayed by small iron filings, as shown in Figure 19.2.

As you will recall from Chapter 6, we can define a magnetic field vector, **B** (sometimes called the *magnetic induction*), at some point in space in terms of the magnetic force exerted on an appropriate test object. Our test object is taken to be a charged particle moving with a velocity of **v**. For the time being, let us assume

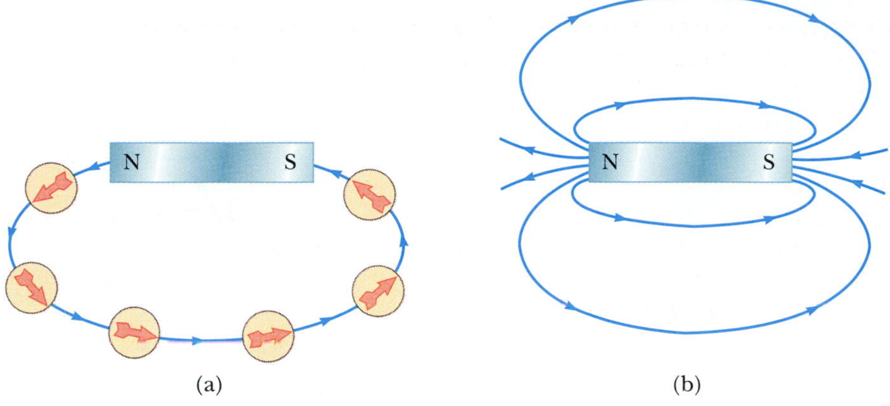

Figure 19.1
(a) Tracing the magnetic field of a bar magnet. (b) Several magnetic field lines of a bar magnet.

that no electric or gravitational fields are present in the region of the charge. Experiments on the motions of various charged particles in a magnetic field give the following results:

Properties of the magnetic force on a charge moving in a B field

1. The magnetic force is proportional to the charge, q, and speed, v, of the particle.
2. The magnitude and direction of the magnetic force depend on the velocity of the particle and on the magnitude and direction of the magnetic field.
3. When a charged particle moves parallel to the magnetic field vector, the magnetic force, **F**, on the charge is zero.

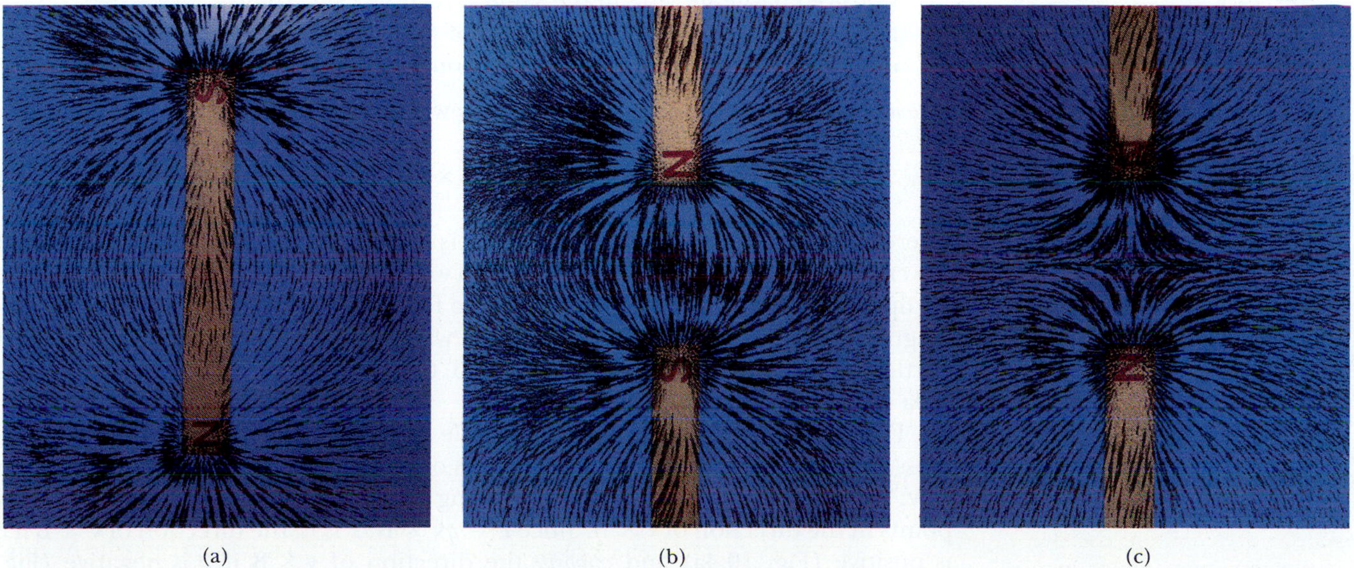

Figure 19.2
(a) Magnetic field pattern of a bar magnet as displayed by iron filings on a sheet of paper. (b) Magnetic field patterns surrounding two bar magnets as displayed with iron filings. This demonstrates the magnetic field pattern between *unlike* poles. (c) This demonstrates the magnetic field pattern between two *like* poles. *(Courtesy of Henry Leap and Jim Lehman)*

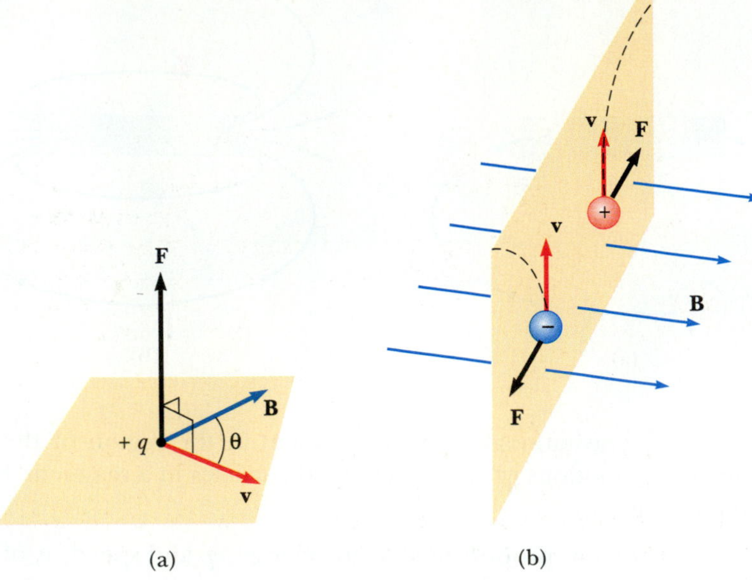

Figure 19.3
The direction of the magnetic force on a charged particle moving with a velocity **v** in the presence of a magnetic field. (a) When **v** is at an angle θ to **B**, the magnetic force is perpendicular to both **v** and **B**. (b) In the presence of a magnetic field, the moving charged particles are deflected as indicated by the dotted lines.

4. When the velocity vector makes an angle of θ with the magnetic field, the magnetic force acts in a direction perpendicular to both **v** and **B**; that is, **F** is perpendicular to the plane formed by **v** and **B** (Fig. 19.3a).
5. The magnetic force on a positive charge is directed opposite the force on a negative charge moving in the same direction (Fig. 19.3b).
6. If the velocity vector makes an angle of θ with the magnetic field, the magnitude of the magnetic force is proportional to sin θ.

These observations can be summarized by writing the magnetic force in the form

Magnetic force on a charged particle in a magnetic field

$$\mathbf{F} = q\mathbf{v} \times \mathbf{B} \qquad [19.1]$$

where the direction of the magnetic force is that of $\mathbf{v} \times \mathbf{B}$, which, by definition of the cross product, is perpendicular to both **v** and **B**. We can regard this equation as an operational definition of the magnetic field at a point in space. That is, the magnetic field is defined in terms of a sideways force acting on a moving charged particle. Recall from Chapter 6, Section 6.5, that the SI unit of magnetic field is the **tesla** (T); $1\ \text{T} = 1\ \text{N}\cdot\text{s/C}\cdot\text{m}$.

Figure 19.4 reviews the right-hand rule for determining the direction of the cross product $\mathbf{v} \times \mathbf{B}$. Point the four fingers of your right hand along the direction of **v**, then turn them until they point along the direction of **B**. The thumb then points in the direction of $\mathbf{v} \times \mathbf{B}$. Since $\mathbf{F} = q\mathbf{v} \times \mathbf{B}$, **F** is in the direction of $\mathbf{v} \times \mathbf{B}$ if q is positive (Fig. 19.4a) and *opposite* the direction of $\mathbf{v} \times \mathbf{B}$ if q is negative (Fig. 19.4b). The magnitude of the magnetic force is

$$F = qvB \sin \theta \qquad [19.2]$$

where θ is the angle between **v** and **B**. From this expression, we see that F is zero when **v** is either parallel or antiparallel to **B** ($\theta = 0$ or $180°$). Furthermore, the force has its maximum value, $F = qvB$, when **v** is perpendicular to **B** ($\theta = 90°$).

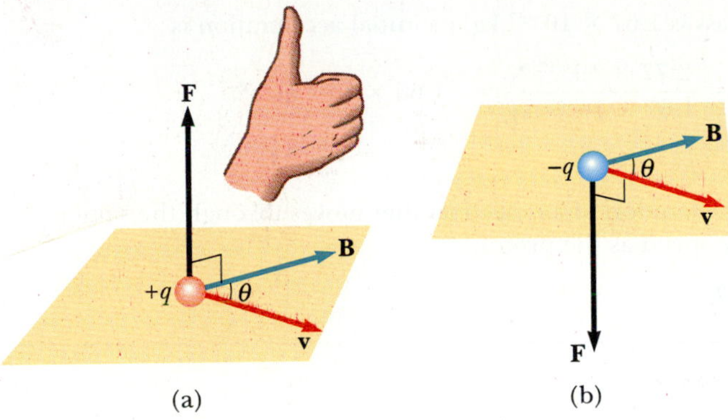

Figure 19.4 The right-hand rule for determining the direction of the magnetic force **F** acting on a charge q moving with a velocity **v** in a magnetic field **B**. If q is positive, **F** is upward in the direction of the thumb. If q is negative, **F** is downward.

There are several important differences between electric and magnetic forces on charged particles:

1. The electric force is always in the direction of the electric field, whereas the magnetic force is perpendicular to the magnetic field.
2. The electric force acts on a charged particle independent of the particle's velocity, whereas the magnetic force acts on a charged particle only when the particle is in motion.
3. The electric force does work in displacing a charged particle, whereas the magnetic force associated with a steady magnetic field does *no* work when a charged particle is displaced.

Differences between electric and magnetic fields

This last statement is a consequence of the fact that when a charge moves in a steady magnetic field, the magnetic force is always *perpendicular* to the displacement. That is, $\mathbf{F} \cdot d\mathbf{s} = (\mathbf{F} \cdot \mathbf{v})\ dt = 0$, since the magnetic force is a vector perpendicular to **v**. From this property and the work-energy theorem, we conclude that the kinetic energy of a charged particle *cannot* be altered by a magnetic field alone. In other words, when a charge moves with a velocity of **v**, an applied magnetic field can alter the direction of the velocity vector, but it cannot change the speed of the particle.

A magnetic field cannot change the speed of a particle

Example 19.1 A Proton Moving in a Magnetic Field

A proton moves with a speed of 8×10^6 m/s along the x axis. It enters a region where there is a magnetic field of magnitude 2.5 T, directed at an angle of 60° to the x axis and lying in the xy plane (Fig. 19.5). Calculate the initial magnetic force and the acceleration of the proton.

Solution From Equation 19.2, we get

$$F = qvB \sin \theta$$
$$= (1.6 \times 10^{-19}\ \text{C})(8 \times 10^6\ \text{m/s})(2.5\ \text{T})(\sin 60°)$$
$$= 2.77 \times 10^{-12}\ \text{N}$$

Since $\mathbf{v} \times \mathbf{B}$ is in the positive z direction (the right-hand rule), and since the charge is positive, the force **F** is in the positive z direction.

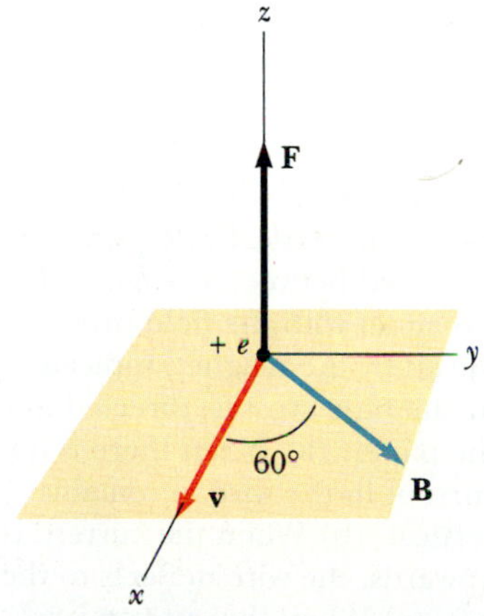

Figure 19.5 (Example 19.1) The magnetic force **F** on a proton is in the positive z direction when **v** and **B** lie in the xy plane.

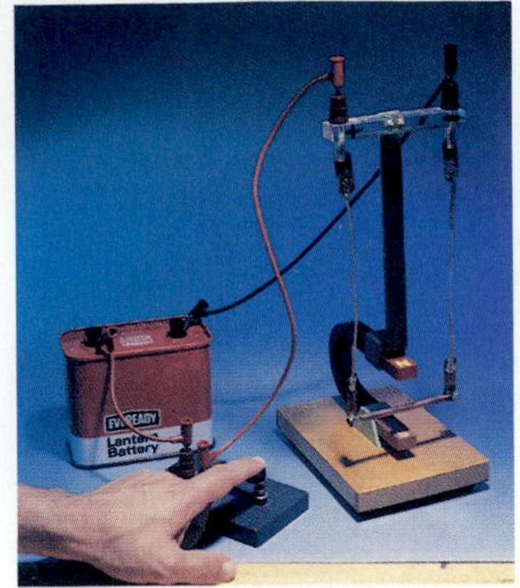

This apparatus demonstrates the force on a current-carrying conductor in an external magnetic field. Why does the bar swing *into* the magnet after the switch is closed? *(Courtesy of Henry Leap and Jim Lehman)*

Since the proton's mass is 1.67×10^{-27} kg, its initial acceleration is

$$a = \frac{F}{m} = \frac{2.77 \times 10^{-12}\ \text{N}}{1.67 \times 10^{-27}\ \text{kg}} = 1.66 \times 10^{15}\ \text{m/s}^2$$

is the positive z direction.

Exercise Calculate the acceleration of an electron that moves through the same magnetic field at the same speed as the proton.

Answer $3.04 \times 10^{18}\ \text{m/s}^2$.

19.3 Magnetic Force on a Current-Carrying Conductor

If a force is exerted on a single charged particle when it moves through an external magnetic field, it should not surprise you to find that a current-carrying wire also experiences a force when placed in an external magnetic field. This follows from the fact that the current represents a collection of many charged particles in motion; hence, the resultant force on the wire is due to the sum of the individual forces on the charged particles. The force on the particles is transmitted to the "bulk" of the wire through collisions with the atoms making up the wire.

Before we continue our discussion, some explanation is in order concerning notation in many of our figures. To indicate the direction of **B**, we use the following convention. If **B** is directed into the page, as in Figure 19.6, we use a series of blue crosses, which represent the tails of arrows. If **B** is directed out of the page, we use a series of blue dots, which represent the tips of arrows. If **B** lies in the plane of the page, we use a series of blue field lines with arrowheads.

The force on a current-carrying conductor can be demonstrated by hanging a wire between the faces of a magnet as in Figure 19.6, where the magnetic field is

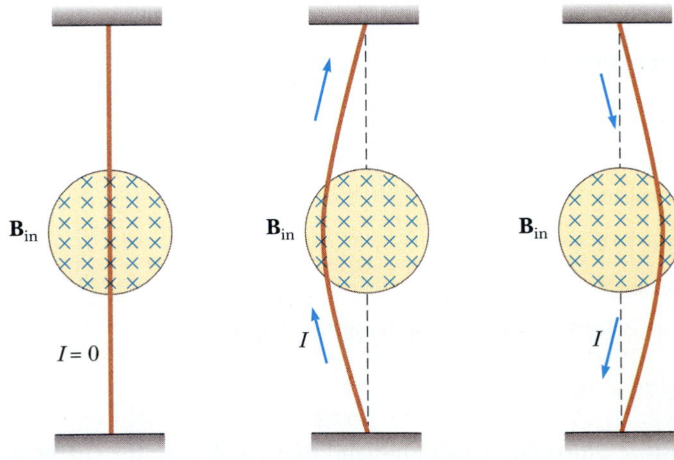

Figure 19.6
A flexible vertical wire partially stretched between the faces of a magnet with the field produced by the magnet, indicated by the blue crosses, directed into the paper. (a) When there is no current in the wire, it remains vertical. (b) When the current is upwards, the wire deflects to the left. (c) When the current is downwards, the wire deflects to the right.

directed into the page. The wire deflects to the right or left when a current is passed through it.

Let us quantify this discussion by considering a straight segment of wire of length ℓ and cross-sectional area A, carrying a current, I, in a uniform external magnetic field, $\mathbf{B}$, as in Figure 19.7. The magnetic force on the charge q moving with the drift velocity $\mathbf{v}_d$ is $q\mathbf{v}_d \times \mathbf{B}$. To find the total force on the wire segment, we multiply the force on one charge by the number of charges in the segment. Since the volume of the segment is $A\ell$, the number of charges in the segment is $nA\ell$, where n is the number of charges per unit volume. Hence, the total magnetic force on the wire of length ℓ is

$$\mathbf{F} = (q\mathbf{v}_d \times \mathbf{B})nA\ell$$

This can be written in a more convenient form by noting that, from Equation 18.4, the current in the wire is $I = nqv_dA$. Therefore, $\mathbf{F}$ can be expressed as

$$\mathbf{F} = I\boldsymbol{\ell} \times \mathbf{B} \qquad [19.3]$$

where $\boldsymbol{\ell}$ is a vector in the direction of the current I; the magnitude of $\boldsymbol{\ell}$ equals the length of the segment. Note that this expression applies only to a straight segment of wire in a uniform external magnetic field. Furthermore, we have neglected the internal magnetic field produced by the current.

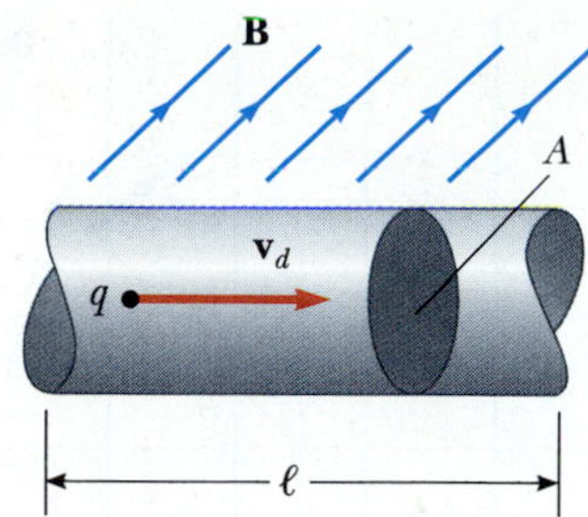

Figure 19.7
A section of a wire containing moving charges in an external magnetic field $\mathbf{B}$. The magnetic force on each charge is $q\mathbf{v} \times \mathbf{B}$, and the net force on a straight wire segment of length ℓ is $I\boldsymbol{\ell} \times \mathbf{B}$.

Now consider an arbitrarily shaped wire of uniform cross-section in an external magnetic field, as in Figure 19.8. It follows from Equation 19.3 that the magnetic force on a very small segment, $d\mathbf{s}$, in the presence of an external field, $\mathbf{B}$, is

$$d\mathbf{F} = I\, d\mathbf{s} \times \mathbf{B} \qquad [19.4]$$

where $d\mathbf{F}$ is directed out of the page for the directions assumed in Figure 19.8. We can consider Equation 19.4 as an alternative definition of $\mathbf{B}$. That is, the field $\mathbf{B}$ can be defined in terms of a measurable force on a current element, where the force is a maximum when $\mathbf{B}$ is perpendicular to the element and zero when $\mathbf{B}$ is parallel to the element.

To get the total force $\mathbf{F}$ on the wire, we integrate Equation 19.4 over the length of the wire:

$$\mathbf{F} = I\int_a^b d\mathbf{s} \times \mathbf{B} \qquad [19.5]$$

In this expression, a and b represent the end points of the wire. When this integration is carried out, the magnitude of the magnetic field and the direction the field makes with the vector $d\mathbf{s}$ (that is, the element orientation) may not be the same at each point.

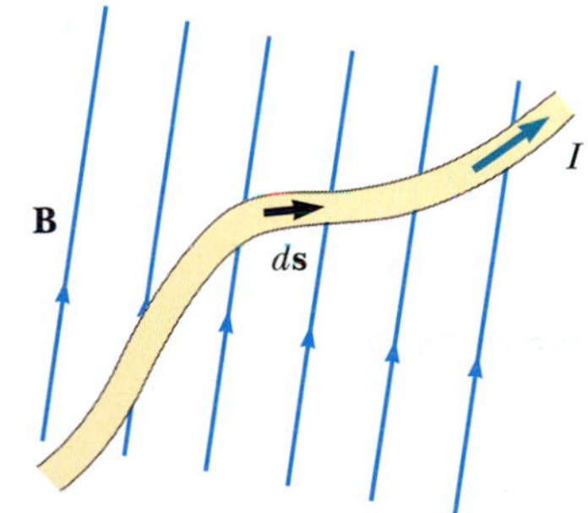

Figure 19.8
A wire of arbitrary shape carrying a current I in an external magnetic field $\mathbf{B}$ experiences a magnetic force. The force on any segment $d\mathbf{s}$ is given by $I\, d\mathbf{s} \times \mathbf{B}$ and is directed *out* of the page. You should use the right-hand rule to confirm this direction.

Example 19.2 **Force on a Semicircular Conductor**

A wire bent into the shape of a semicircle of radius R forms a closed circuit and carries the current I. The circuit lies in the xy plane, and a uniform external magnetic field is present along the positive y axis, as in Figure 19.9. Find the magnetic forces on the straight and curved portions of the wire.

Solution The force on the straight portion of the wire has the magnitude $F_1 = I\ell B = 2IRB$, since $\ell = 2R$ and the wire is perpendicular to $\mathbf{B}$. The direction of $\mathbf{F}_1$ is

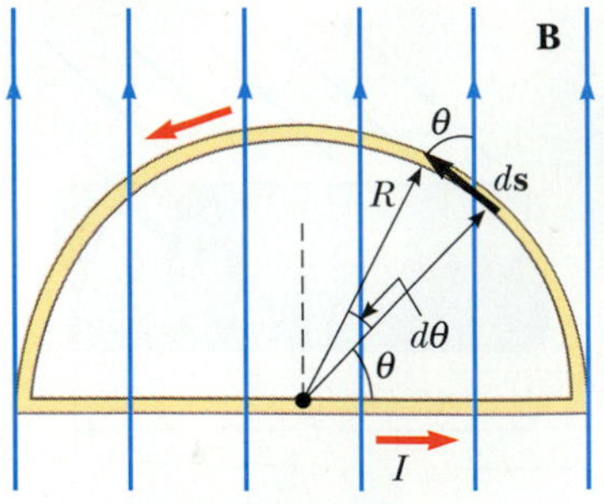

Figure 19.9
(Example 19.2) The net force on a closed current loop in a uniform magnetic field is zero. In this case, the force on the straight portion is $2IRB$ out of the plane of the page, while the force on the curved portion is also $2IRB$ but is directed into the plane of the page.

out of the paper since $\boldsymbol{\ell} \times \mathbf{B}$ is outward. (That is, $\boldsymbol{\ell}$ is to the right in the direction of the current, and so, by the rule of cross products, $\boldsymbol{\ell} \times \mathbf{B}$ is outward.)

To find the force on the curved part, we must first write an expression for the force dF_2 on the element ds. If θ is the angle between $\mathbf{B}$ and ds in Figure 19.9, then the magnitude of dF_2 is

$$dF_2 = I|\,ds \times \mathbf{B}\,| = IB \sin\theta \, ds$$

where ds is the length of the small element, measured along the circular arc. In order to integrate this expression, we must express ds in terms of θ. Since $s = R\theta$, $ds = R\,d\theta$, and the expression for dF_2 can be written

$$dF_2 = IRB \sin\theta \, d\theta$$

To get the total force $\mathbf{F}_2$ on the curved portion, we can integrate this expression to account for contributions from all elements. Note that the direction of the force on every element is the same: into the paper (since $ds \times \mathbf{B}$ is inward). Therefore, the resultant force $\mathbf{F}_2$ on the curved wire must also be into the paper. Integration of dF_2 over the limits $\theta = 0$ to $\theta = \pi$ (that is, the entire semicircle) gives

$$F_2 = IRB \int_0^{\pi} \sin\theta \, d\theta = IRB \Big[-\cos\theta \Big]_0^{\pi}$$

$$= -IRB(\cos\pi - \cos 0) = -IRB(-1 - 1) = 2IRB$$

Since $F_2 = 2IRB$ and the force $\mathbf{F}_2$ is directed *into* the paper while the force on the straight wire, $F_1 = 2IRB$, is *out* of the paper, we see that the net force on the closed loop is zero. In fact, one can show that *the total magnetic force on any closed current loop in a uniform external magnetic field is zero* (see Problem 18).

19.4 Torque on a Current Loop in a Uniform Magnetic Field

In the preceding section we showed how a force is exerted on a current-carrying conductor when the conductor is placed in an external magnetic field. With this as a starting point, we shall show that a torque is exerted on a current loop placed in a magnetic field. The results of this analysis will be of great practical value when we discuss generators in Chapter 20.

Consider a rectangular loop carrying a current, I, in the presence of a uniform external magnetic field *in the plane of the loop*, as in Figure 19.10a. The forces on the sides of length a are zero since these wires are parallel to the field; hence, $ds \times \mathbf{B} = 0$ for these sides. The magnitude of the forces on the sides of length b, however, is

$$F_1 = F_2 = IbB$$

The direction of $\mathbf{F}_1$, the force on the left side of the loop, is out of the paper, and that of $\mathbf{F}_2$, the force on the right side of the loop, is into the paper. If we view the loop from an end, as in Figure 19.10b, we see the forces directed as shown. If we assume that the loop is pivoted so that it can rotate about point O, we see that these two forces produce a torque about O that rotates the loop clockwise. The magnitude of this torque, $\tau_{\max}$, is

$$\tau_{max} = F_1 \frac{a}{2} + F_2 \frac{a}{2} = (IbB)\frac{a}{2} + (IbB)\frac{a}{2} = IabB$$

where the moment arm about O is $a/2$ for each force. Since the area of the loop is $A = ab$, the torque can be expressed as

$$\tau_{max} = IAB \qquad [19.6]$$

Remember that this result is valid only when the field **B** is parallel to the plane of the loop. The sense of the rotation is clockwise when the loop is viewed from the bottom end, as indicated in Figure 19.10b. If the current were reversed, the forces would reverse their directions and the rotational tendency would be counterclockwise.

Now suppose the uniform magnetic field makes an angle of θ with a line perpendicular to the plane of the loop, as in Figure 19.10c. For convenience, we shall assume that the field **B** is perpendicular to the sides of length b. (The end view of these sides is shown in Fig. 19.9c.) In this case, the forces on the sides of length a cancel each other and produce no torque, since they pass through a common origin. However, the forces acting on the sides of length b, $\mathbf{F}_1$ and $\mathbf{F}_2$, form a couple and hence produce a torque about *any point.* Referring to the end view in Figure 19.10b, we note that the moment arm of $\mathbf{F}_1$ about O is $(a/2)\sin\theta$. Likewise, the moment arm of $\mathbf{F}_2$ about O is also $(a/2)\sin\theta$. Since $F_1 = F_2 = IbB$, the net torque about O has the magnitude

$$\begin{aligned}\tau &= F_1 \frac{a}{2}\sin\theta + F_2 \frac{a}{2}\sin\theta \\ &= IbB\left(\frac{a}{2}\sin\theta\right) + IbB\left(\frac{a}{2}\sin\theta\right) = IabB\sin\theta \\ &= IAB\sin\theta\end{aligned}$$

where $A = ab$ is the area of the loop. This result shows that the torque has its maximum value, IAB, when the field is parallel to the plane of the loop ($\theta = 90°$) and is zero when the field is perpendicular to the plane of the loop ($\theta = 0$). As we see in Figure 19.10c, the loop tends to rotate in the direction of decreasing values of θ (that is, so that the normal to the plane of the loop rotates toward the direction of the magnetic field).

A convenient vector expression for the torque is

$$\boldsymbol{\tau} = I\mathbf{A} \times \mathbf{B} \qquad [19.7]$$

where **A**, a vector perpendicular to the plane of the loop, has a magnitude equal to the area of the loop. The sense of **A** is determined by the right-hand rule illus-

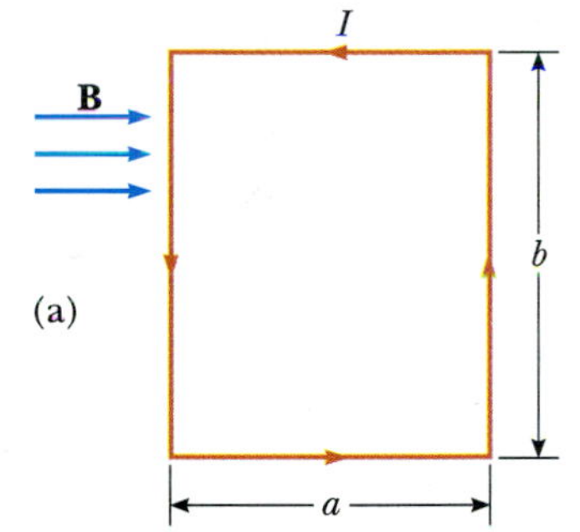

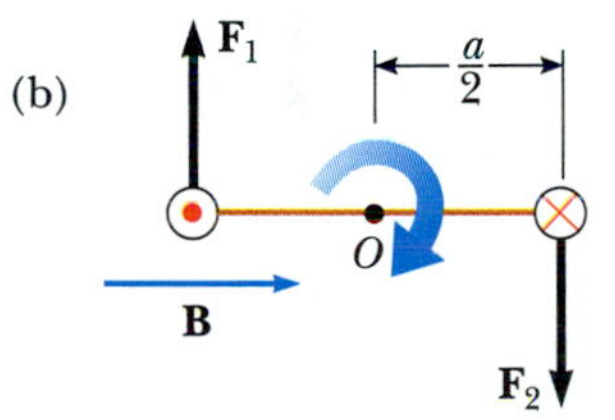

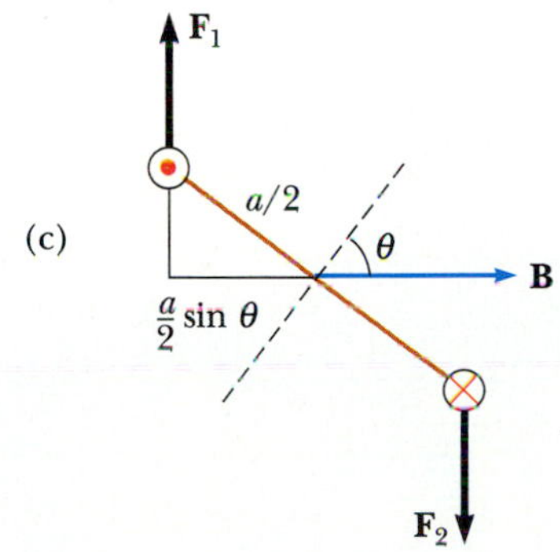

Figure 19.10
(a) Top view of a rectangular loop in a uniform external magnetic field **B**. There are no magnetic forces on the sides of length a parallel to **B**, but there are forces acting on the sides of length b. (b) Front view of the rectangular loop shows that the forces $\mathbf{F}_1$ and $\mathbf{F}_2$ on the sides of length b create a torque that tends to twist the loop clockwise as shown. (c) If **B** is at an angle θ with respect to a line perpendicular to the plane of the loop, the torque is $IAB\sin\theta$.

trated in Figure 19.11. When the four fingers of the right hand are rotated in the direction of the current in the loop, the thumb points in the direction of **A**. The product $I\mathbf{A}$ is defined to be the **magnetic moment, $\boldsymbol{\mu}$,** of the loop:

$$\boldsymbol{\mu} = I\mathbf{A} \qquad [19.8]$$

The SI unit of magnetic moment is the ampere-meter² ($\mathrm{A \cdot m^2}$). Using this definition, the torque can be expressed as

Torque on a current loop

$$\boldsymbol{\tau} = \boldsymbol{\mu} \times \mathbf{B} \qquad [19.9]$$

Although the torque was obtained for a particular orientation of **B** with respect to the loop, Equation 19.9 is valid for any orientation. Furthermore, although the torque expression was derived for a rectangular loop, the result is valid for a loop of any shape.

If a coil consists of N turns of wire, each carrying the same current and having the same area, the total magnetic moment of the coil is the product of the number of turns and the magnetic moment for one turn. The torque on an N-turn coil is N times greater than that on a one-turn coil.

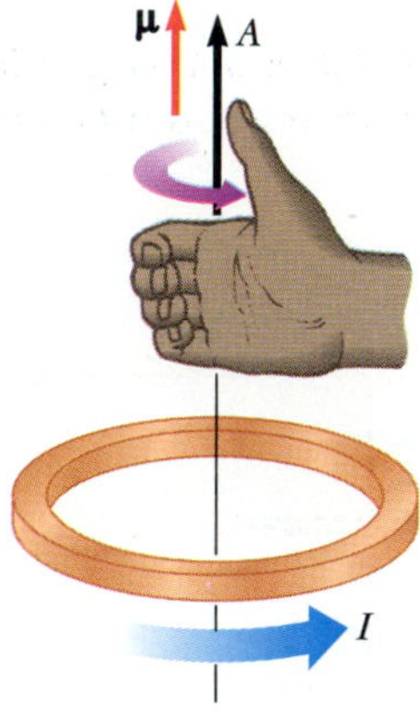

Figure 19.11 A right-hand rule for determining the direction of the vector **A**. The magnetic moment $\boldsymbol{\mu}$ is also in the direction of **A**.

Example 19.3 The Magnetic Moment of a Coil

A rectangular coil of dimensions 5.40 cm × 8.50 cm consists of 25 turns of wire. The coil carries a current of 15 mA. (a) Calculate the magnitude of the magnetic moment of the coil.

Solution The magnitude of the magnetic moment of one current loop is $\mu = IA$ (Eq. 19.8), where A is the area of the loop. In this case, $A = (0.0540\ \mathrm{m})(0.0850\ \mathrm{m}) = 4.59 \times 10^{-3}\ \mathrm{m^2}$. Since the coil has 25 turns, and assuming that each turn has the same area, A, we have

$$\mu_{\text{coil}} = NIA = (25)(15 \times 10^{-3}\ \mathrm{A})(4.59 \times 10^{-3}\ \mathrm{m^2}) = 1.72 \times 10^{-3}\ \mathrm{A \cdot m^2}$$

(b) Suppose a magnetic field of magnitude 0.350 T is applied parallel to the plane of the coil. What is the magnitude of the torque acting on the coil?

Solution In general, the torque is $\boldsymbol{\tau} = \boldsymbol{\mu} \times \mathbf{B}$, where $\boldsymbol{\mu}$ is directed perpendicular to the plane of the coil. In this case, **B** is perpendicular to $\boldsymbol{\mu}_{\text{coil}}$, so that

$$\tau = \mu_{\text{coil}} B = (1.72 \times 10^{-3}\ \mathrm{A \cdot m^2})(0.350\ \mathrm{T})$$

$$= 6.02 \times 10^{-4}\ \mathrm{N \cdot m}$$

Exercise Show that the units $\mathrm{A \cdot m^2 \cdot T}$ reduce to $\mathrm{N \cdot m}$.

Exercise Calculate the magnitude of the torque on the coil when the 0.350-T magnetic field makes angles of (a) 60° and (b) 0° with $\boldsymbol{\mu}$.

Answer (a) $5.21 \times 10^{-4}\ \mathrm{N \cdot m}$; (b) zero.

19.5 The Biot-Savart Law

From their investigations on the force between a current-carrying conductor and a magnet, Jean-Baptiste Biot and Félix Savart arrived at an expression for the magnetic field at some point in space in terms of the current that produces the field. The *Biot-Savart law* says that if a wire carries a steady current, *I*, then at point *P* the magnetic field *d*B associated with a wire element, *ds* (Fig. 19.12), has the following properties:

Properties of the magnetic field due to a current element

1. The vector $d\mathbf{B}$ is perpendicular both to $d\mathbf{s}$ (which is in the direction of the current) and to the unit vector, $\hat{\mathbf{r}}$, directed from the element to *P*.
2. The magnitude of $d\mathbf{B}$ is inversely proportional to r^2, where r is the distance from the element to *P*.
3. The magnitude of $d\mathbf{B}$ is proportional to the current and to the length, *ds*, of the element.
4. The magnitude of $d\mathbf{B}$ is proportional to $\sin\theta$, where θ is the angle between *ds* and $\hat{\mathbf{r}}$.

The **Biot-Savart law** can be summarized in the following convenient form:

Biot-Savart law

$$d\mathbf{B} = k_m \frac{I\, d\mathbf{s} \times \hat{\mathbf{r}}}{r^2} \qquad [19.10]$$

where k_m is a constant that in SI units is exactly $10^{-7}\ \mathrm{T \cdot m/A}$. The constant k_m is usually written $\mu_0/4\pi$, where μ_0 is another constant, called the **permeability of free space:**

Permeability of free space

$$\frac{\mu_0}{4\pi} = k_m = 10^{-7}\ \mathrm{T \cdot m/A} \qquad [19.11]$$

$$\mu_0 = 4\pi k_m = 4\pi \times 10^{-7}\ \mathrm{T \cdot m/A} \qquad [19.12]$$

Hence, the Biot-Savart Law, Equation 19.10, can also be written

$$d\mathbf{B} = \frac{\mu_0}{4\pi} \frac{I\, d\mathbf{s} \times \hat{\mathbf{r}}}{r^2} \qquad [19.13]$$

It is important to note that the Biot-Savart law gives the magnetic field at a point only for a small element of the conductor. To find the total magnetic field **B** at some point due to a conductor of finite size, we must sum up contributions from all current elements making up the conductor. That is, we must evaluate **B** by integrating Equation 19.13.

There are two similarities between the Biot-Savart law of magnetism and Coulomb's law of electrostatics, and one important difference. The current element *I ds* produces a magnetic field, whereas the point charge *q* produces an electric field. Furthermore, the magnitude of the magnetic field varies as the inverse square of the distance from the current element, as does the electric field due to a point charge. However, the directions of the two fields are quite different. The electric field due to a point charge is radial; in the case of a positive point charge, **E** is directed from the charge to the field point. The magnetic field due to a current element is perpendicular to both the current element and the radius vector.

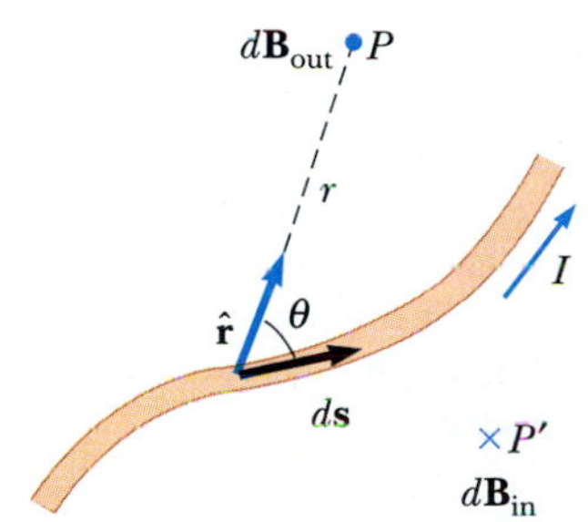

Figure 19.12
The magnetic field *d*B at a point *P* due to a wire element *ds* is given by the Biot-Savart law, Equation 19.10. The field points out of the page at *P* and into the page at *P*′.

Hence, if the conductor lies in the plane of the page, as in Figure 19.12, $d\mathbf{B}$ points out of the page at the point P and into the page at P'.

The examples that follow illustrate the use of the Biot-Savart law for calculating the magnetic field vectors of several important geometric arrangements. It is important that you recognize that the magnetic field described in these calculations is *the field due to a given current-carrying conductor.* This is not to be confused with any *external* field that may be applied to the conductor.

Example 19.4 Magnetic Field of a Thin, Straight Conductor

Consider a thin, straight wire carrying a constant current, I, and placed along the x axis as in Figure 19.13. Let us find an expression for the total magnetic field at the point P, located a distance of a from the wire.

Solution The element ds is a distance of r from P. The direction of the field at P due to this element is out of the page, since $ds \times \hat{\mathbf{r}}$ is out of the page. In fact, *all* elements give contributions directly out of the page at P. Therefore, we have only to determine the magnitude of the field at P.

Since $ds = \mathbf{i}\ dx$ in this case, we see that $|ds \times \hat{\mathbf{r}}| = dx \sin\theta$. Using this in Equation 19.13, we get

$$(1) \qquad dB = \frac{\mu_0 I}{4\pi}\frac{dx \sin\theta}{r^2}$$

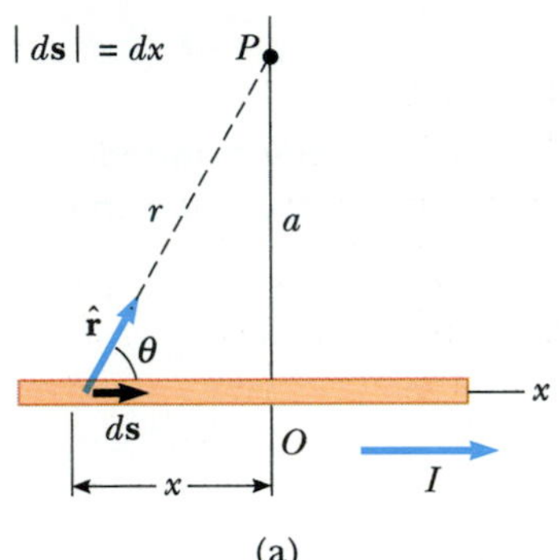

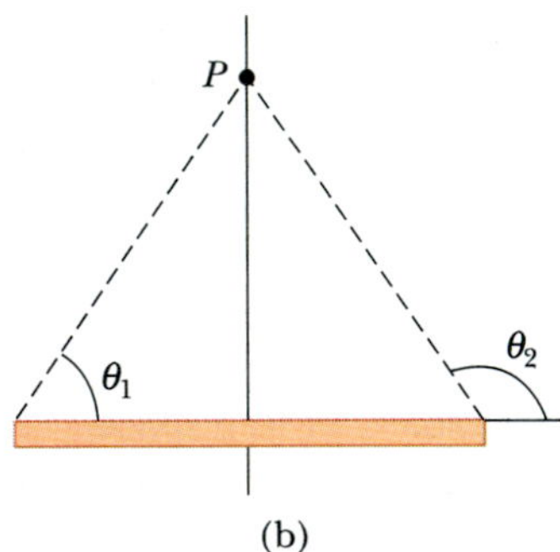

Figure 19.13 (Example 19.4) (a) A straight wire segment carrying a current I. The magnetic field at P due to each element ds is out of the paper, and so the net field is also out of the paper. (b) The limiting angles θ_1 and θ_2 for this geometry.

In order to integrate this expression, we must relate the variables θ, x, and r. One approach is to express x and r in terms of θ. From the geometry in Figure 19.13a and some simple differentiation, we obtain the following relationship:

$$(2) \qquad r = \frac{a}{\sin\theta} = a \csc\theta$$

Since $\tan\theta = -a/x$ from the right triangle in Figure 19.13a, we have $x = -a \cot\theta$, and so

$$(3) \qquad dx = a \csc^2\theta\ d\theta$$

Substitution of (2) and (3) into (1) gives

$$(4) \qquad dB = \frac{\mu_0 I}{4\pi a}\sin\theta\ d\theta$$

Thus, we have reduced the expression to one involving only the variable θ. We can now obtain the total field at P by integrating (4) over all elements that subtend angles ranging from θ_1 to θ_2, as defined in Figure 19.13b. This gives

$$B = \frac{\mu_0 I}{4\pi a}\int_{\theta_1}^{\theta_2} \sin\theta\ d\theta = \frac{\mu_0 I}{4\pi a}(\cos\theta_1 - \cos\theta_2) \qquad [19.14]$$

We can apply this result to find the magnetic field of any straight wire—if we know the geometry and hence the angles θ_1 and θ_2.

Consider the special case of an infinitely long, straight wire. In this case, $\theta_1 = 0$ and $\theta_2 = \pi$ for segments ranging from $x = -\infty$ to $x = +\infty$. Since $(\cos\theta_1 - \cos\theta_2) = (\cos 0 - \cos\pi) = 2$, Equation 19.14 becomes

$$B = \frac{\mu_0 I}{2\pi a} \qquad [19.15]$$

Figure 19.14 is a three-dimensional view of the direction of **B** for a long, straight wire. *The field lines are circles concentric with the wire and are in a plane perpendicular to the wire.* The magnitude of **B** is constant on any circle of radius a and is given by Equation 19.15. A convenient method of determining the direction of **B** is to grasp the wire with the right hand, with the thumb pointed in the direction of the current. The four fingers wrap in the direction of the magnetic field.

Equation 19.15 shows that the magnitude of the magnetic field is proportional to the current and decreases as the distance from the wire increases, as one might intuitively expect.

Exercise Calculate the magnetic field of a long, straight wire carrying a current of 5 A, at a distance of 4 cm from the wire.

Answer 2.5×10^{-5} T.

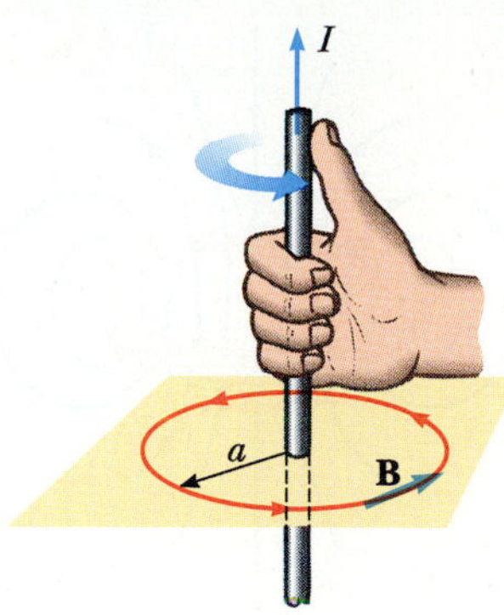

Figure 19.14
The right-hand rule for determining the direction of the magnetic field due to a long, straight wire. Note that the magnetic field lines form circles around the wire.

Example 19.5 Magnetic Field on the Axis of a Circular Current Loop

Consider a circular loop of wire, of radius R, located in the yz plane and carrying a steady current, I, as in Figure 19.15. Let us calculate the magnetic field at an axial point P, a distance of x from the center of the loop.

Solution In this situation, note that any loop element $d\mathbf{s}$ is perpendicular to $\hat{\mathbf{r}}$. Furthermore, all elements around the loop are at the same distance, r, from P, where $r^2 = x^2 + R^2$. Hence, the *magnitude* of $d\mathbf{B}$ due to the element $d\mathbf{s}$ is

$$dB = \frac{\mu_0 I}{4\pi} \frac{|d\mathbf{s} \times \hat{\mathbf{r}}|}{r^2} = \frac{\mu_0 I}{4\pi} \frac{ds}{(x^2 + R^2)}$$

The direction of $d\mathbf{B}$ due to $d\mathbf{s}$ is perpendicular to the plane formed by $\hat{\mathbf{r}}$ and $d\mathbf{s}$, as shown in Figure 19.15. The vector $d\mathbf{B}$ can be resolved into components dB_x and dB_y. When the y components are summed over the whole loop, the result is zero. That is, by symmetry, any element on one side of the loop sets up a perpendicular component that cancels the component set up by a diametrically opposite element. Therefore, we see that *the resultant field at P must be along the x axis* and can be found

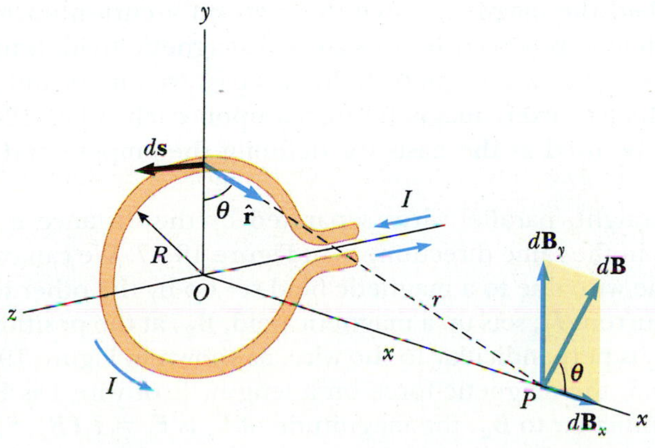

Figure 19.15
(Example 19.5) The geometry for calculating the magnetic field at an axial point P for a current loop. Note that by symmetry the total field **B** is along the x axis.

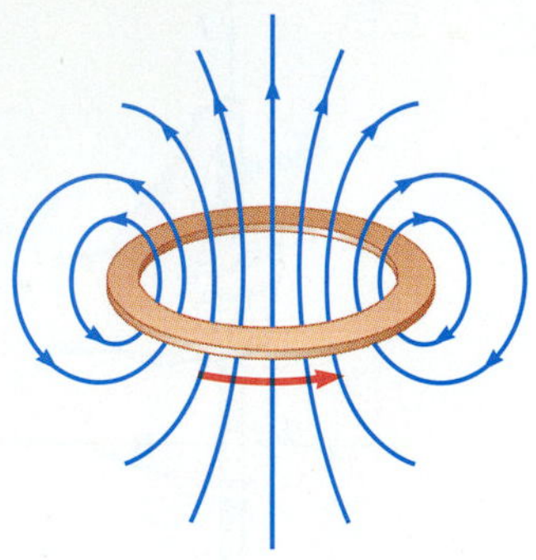
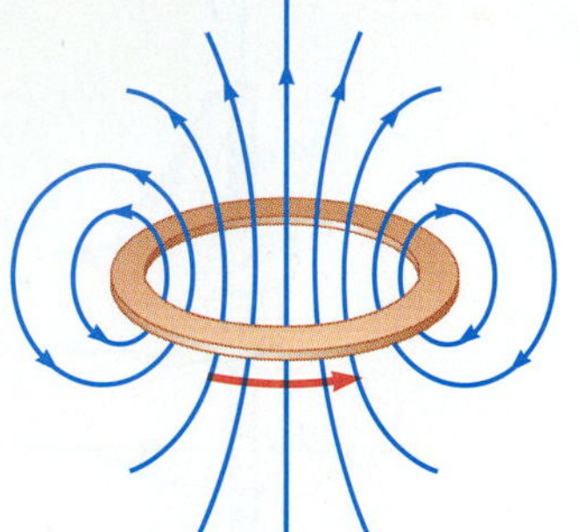

Figure 19.16
Magnetic field lines for a current loop. Far from the loop, the field lines are identical in form to those of an electric dipole.

by integrating the components $dB_x = dB\ \cos\theta$; this expression is obtained by resolving the vector $d\mathbf{B}$ into its components (Fig. 19.15). That is, $\mathbf{B} = \mathbf{i}B_x$, where B_x is

$$B_x = \oint dB \cos\theta = \frac{\mu_0 I}{4\pi}\oint \frac{ds \cos\theta}{x^2 + R^2}$$

and the integral must be taken over the entire loop. Since θ, x, and R are constants for all elements of the loop and since $\cos\theta = R/(x^2 + R^2)^{1/2}$, we get

$$B_x = \frac{\mu_0 IR}{4\pi(x^2 + R^2)^{3/2}}\oint ds = \frac{\mu_0 IR^2}{2(x^2 + R^2)^{3/2}} \qquad [19.16]$$

Here we have used the fact that $\oint ds = 2\pi R$ (the circumference of the loop).

To find the magnetic field at the center of the loop, we set $x = 0$ in Equation 19.16. At this special point, we get

$$B = \frac{\mu_0 I}{2R} \qquad (\text{at } x = 0) \qquad [19.17]$$

It is also interesting to determine the behavior of the magnetic field at large distances from the loop—that is, when x is large compared with R. In this case, we can neglect the term R^2 in the denominator of Equation 19.16 and obtain

$$B \approx \frac{\mu_0 IR^2}{2x^3} \qquad (\text{for } x \gg R) \qquad [19.18]$$

Since the magnitude of the magnetic moment, $\boldsymbol{\mu}$, of the loop is defined as the product of the current and the area (Eq. 19.8), $\mu = I(\pi R^2)$ and we can express Equation 19.18 in the form

$$B = \frac{\mu_0}{2\pi}\frac{\mu}{x^3} \qquad [19.19]$$

This result is similar in form to the expression for the electric field due to an electric dipole, $E = k_e 2qa/y^3$ (see Example 19.3).

The pattern of the magnetic field lines for a circular loop is shown in Figure 19.16. For clarity, the lines are drawn only for one plane, which contains the axis of the loop. The field pattern is axially symmetric.

19.6 The Magnetic Force Between Two Parallel Conductors

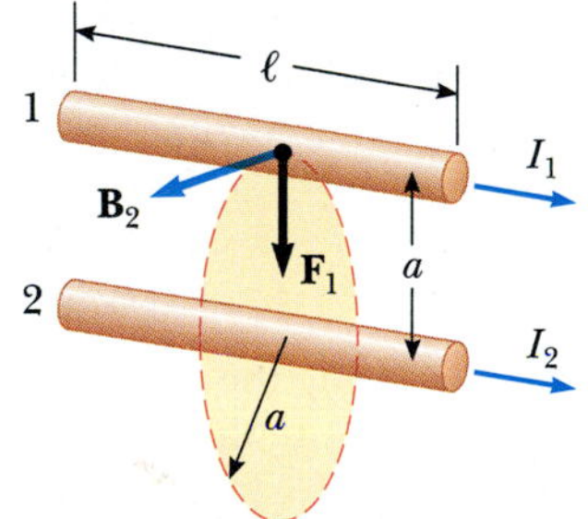

Figure 19.17
Two parallel wires, each carrying a steady current, exert a force on each other. The field $\mathbf{B}_2$ at wire 1 due to wire 2 produces a force on wire 1 given by $\mathbf{F}_1 = I_1 \ell \mathbf{B}_2$. The force is attractive if the currents are parallel as shown and repulsive if the currents are antiparallel.

In Section 19.3 we described the magnetic force that acts on a current-carrying conductor when the conductor is placed in an external magnetic field. Since a current in a conductor sets up its own magnetic field, it is easy to understand that two current-carrying conductors exert magnetic forces upon each other. As we shall see, such forces can be used as the basis for defining the ampere and the coulomb.

Consider two long, straight, parallel wires separated by the distance a and carrying currents I_1 and I_2 in the same direction, as in Figure 19.17. We can easily determine the force on one wire due to a magnetic field set up by the other wire. Wire 2, which carries the current I_2, sets up a magnetic field, $\mathbf{B}_2$, at the position of wire 1. The direction of $\mathbf{B}_2$ is perpendicular to the wire, as shown in Figure 19.17. According to Equation 19.3, the magnetic force on a length, ℓ, of wire 1 is $\mathbf{F}_1 = I_1 \boldsymbol{\ell} \times \mathbf{B}_2$. Since $\boldsymbol{\ell}$ is perpendicular to $\mathbf{B}_2$, the magnitude of $\mathbf{F}_1$ is $F_1 = I_1 \ell B_2$. Since the field due to wire 2 is given by Equation 19.15,

$$B_2 = \frac{\mu_0 I_2}{2\pi a}$$

we see that

$$F_1 = I_1 \ell B_2 = I_1 \ell \left(\frac{\mu_0 I_2}{2\pi a}\right) = \frac{\ell \mu_0 I_1 I_2}{2\pi a}$$

We can rewrite this in terms of the force per unit length, as

$$\frac{F_1}{\ell} = \frac{\mu_0 I_1 I_2}{2\pi a} \qquad [19.20]$$

The direction of $\mathbf{F}_1$ is downward, toward wire 2, because $\boldsymbol{\ell} \times \mathbf{B}_2$ is downward. If one considers the field set up at wire 2 due to wire 1, the force $\mathbf{F}_2$ on wire 2 is found to be equal to and opposite $\mathbf{F}_1$. That is what one would expect, because Newton's third law of action-reaction must be obeyed.

When the currents are in opposite directions, the forces are reversed and the wires repel each other. Hence, we find that parallel conductors carrying currents in the same direction attract each other, whereas parallel conductors carrying currents in opposite directions repel each other.

The force between two parallel wires, each carrying a current, is used to define the **ampere:** if two long, parallel wires 1 m apart carry the same current and the force per unit length on each wire is 2×10^{-7} N/m, then the current is defined to be 1 A. The numerical value of 2×10^{-7} N/m is obtained from Equation 19.20, with $I_1 = I_2 = 1$ A and $a = 1$ m.

The SI unit of charge, the **coulomb,** can now be defined in terms of the ampere: If a conductor carries a steady current of 1 A, then the quantity of charge that flows through a cross-section of the conductor in 1 s is 1 C.

19.7 Ampère's Law

A simple experiment first carried out by Oersted in 1820 clearly demonstrates that a current-carrying conductor produces a magnetic field. In this experiment, several compass needles are placed in a horizontal plane near a long vertical wire, as in Figure 19.18a. When there is no current in the wire, all needles point in the

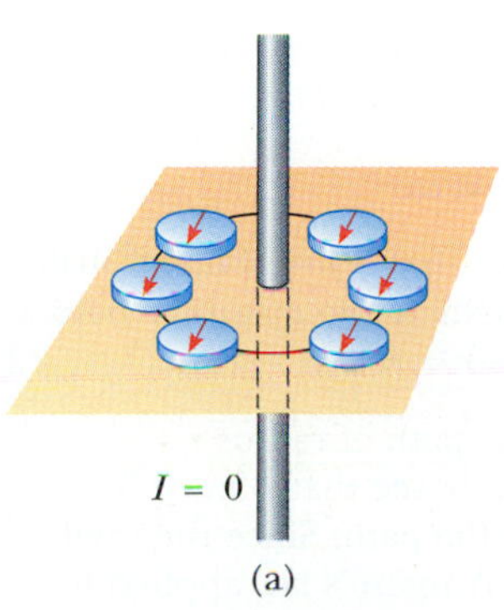

(a)

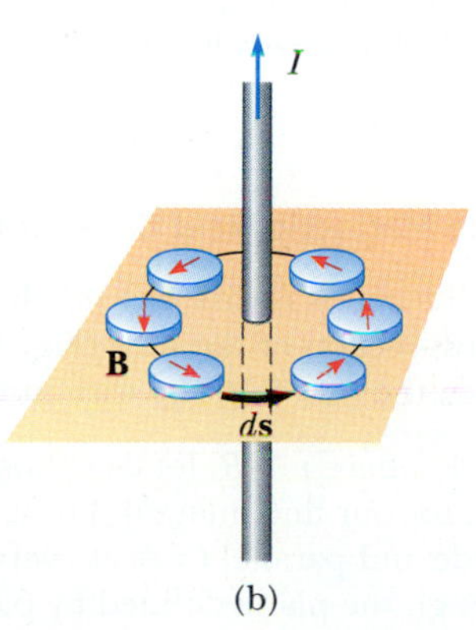

(b)

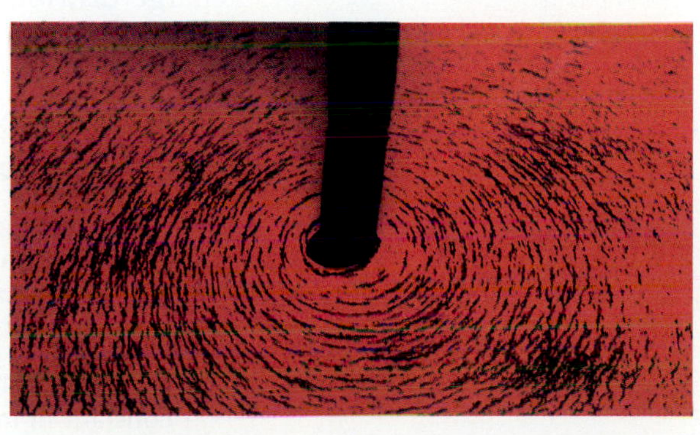
(c)

Figure 19.18 (a) When there is no current in the vertical wire, all compass needles point in the same direction. (b) When the wire carries a strong current, the needles deflect in a direction tangent to the circle, which is the direction of **B** due to the current. (c) Circular magnetic field line surrounding a current-carrying conductor as displayed with iron filings. The photograph was taken using 30 parallel wires each carrying a current of $\frac{1}{2}$ A. *(Photo courtesy of Henry Leap and Jim Lehman)*

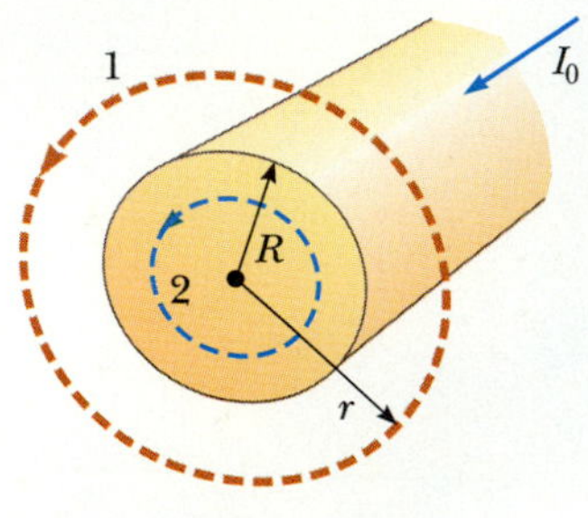

Figure 19.19 (Example 19.6) A long, straight wire of radius R carrying a steady current I_0 uniformly distributed across the wire. The magnetic field at any point can be calculated from Ampère's law using a circular path for our line integral that has radius r, and is concentric with the wire.

same direction (that of the Earth's field), as one would expect. However, when the wire carries a strong, steady current, the needles all deflect in a direction tangent to the circle, as in Figure 19.18b. These observations show that the direction of **B** is consistent with the right-hand rule described in Section 19.6.

If the wire is grasped in the right hand with the thumb in the direction of the current, the fingers will curl in the direction of B.

When the current is reversed, the needles in Figure 19.18b also reverse.

Since the needles point in the direction of **B**, we conclude that the lines of **B** form circles about the wire, as discussed in Section 19.5. By symmetry, the magnitude of **B** is the same everywhere on a circular path that is centered on the wire and lies in a plane perpendicular to the wire. By varying the current and distance from the wire, one finds that **B** is proportional to the current and inversely proportional to the distance from the wire.

Now let us evaluate the product $\mathbf{B} \cdot d\mathbf{s}$ and sum the products $B\,ds$ over the closed circular path centered on the wire. Along this path, the vectors $d\mathbf{s}$ and **B** are parallel at each point (Fig. 19.18b), so that $\mathbf{B} \cdot d\mathbf{s} = B\,ds$. Furthermore, **B** is constant in magnitude on this circle and is given by Equation 19.15. Therefore, the sum of the products $B\,ds$ over the closed path, which is equivalent to the line integral of $\mathbf{B} \cdot d\mathbf{s}$, is

$$\oint \mathbf{B} \cdot d\mathbf{s} = B \oint ds = \frac{\mu_0 I}{2\pi r}(2\pi r) = \mu_0 I \qquad [19.21]$$

where $\oint ds = 2\pi r$ is the circumference of the circle.

This result, known as **Ampère's law,** was calculated for the special case of a circular path surrounding a wire. However, it can also be applied in the general case in which an arbitrary closed path is threaded by a steady current. That is, Ampère's law says that the line integral of $\mathbf{B} \cdot d\mathbf{s}$ around any closed path equals $\mu_0 I$, where I is the total steady current passing through any surface bounded by the closed path.

Ampère's law

$$\oint \mathbf{B} \cdot d\mathbf{s} = \mu_0 I \qquad [19.22]$$

Ampère's law is valid only for steady currents. Furthermore, *it is useful only for calculating the magnetic fields of current configurations with high degrees of symmetry,* just as Gauss's law is useful only for calculating the electric fields of highly symmetric charge distributions. The following examples illustrate some symmetric current configurations for which Ampère's law is useful.

Example 19.6 The Magnetic Field Created by a Long Wire

A long, straight wire of radius R carries a steady current, I_0, that is uniformly distributed through the cross-section of the wire (Fig. 19.19). Calculate the magnetic field at a distance of r from the center of the wire in the regions $r \geq R$ and $r < R$.

Solution In region 1, where $r \geq R$, let us choose a circular path of radius r, centered at the wire, for our line integral. From symmetry, we see that **B** must be constant in magnitude and parallel to $d\mathbf{s}$ at every point on the path. Since the total current passing through the plane defined by path 1 is I_0, Ampère's law applied to the path gives

$$\oint \mathbf{B} \cdot d\mathbf{s} = B \oint ds = B(2\pi r) = \mu_0 I_0$$

$$B = \frac{\mu_0 I_0}{2\pi r} \qquad \text{(for } r \geq R\text{)}$$

which is identical to Equation 19.15.

Now consider the interior of the wire—that is, region 2, where $r < R$. In this case, note that the current I enclosed by the path is *less* than the total current, I_0. Since the current is assumed to be uniform over the cross-section of the wire, we see that the fraction of the current enclosed by the path of radius $r < a$ must equal the ratio of the area πr^2 enclosed by path 2 to the cross-sectional area, πR^2, of the wire. That is,

$$\frac{I}{I_0} = \frac{\pi r^2}{\pi R^2} \qquad \text{or} \qquad I = \frac{r^2}{R^2} I_0$$

Following the same procedure as for path 1, we can now apply Ampère's law to path 2:

$$\oint \mathbf{B} \cdot d\mathbf{s} = B(2\pi r) = \mu_0 I = \mu_0 \left(\frac{r^2}{R^2} I_0 \right)$$

$$B = \left(\frac{\mu_0 I_0}{2\pi R^2} \right) r \qquad \text{(for } r < R\text{)} \qquad [19.23]$$

The magnetic field versus r for this configuration is sketched in Figure 19.20. Note that inside the wire, $B \rightarrow 0$ as $r \rightarrow 0$. This result is similar in form to that of the electric field inside a uniformly charged rod.

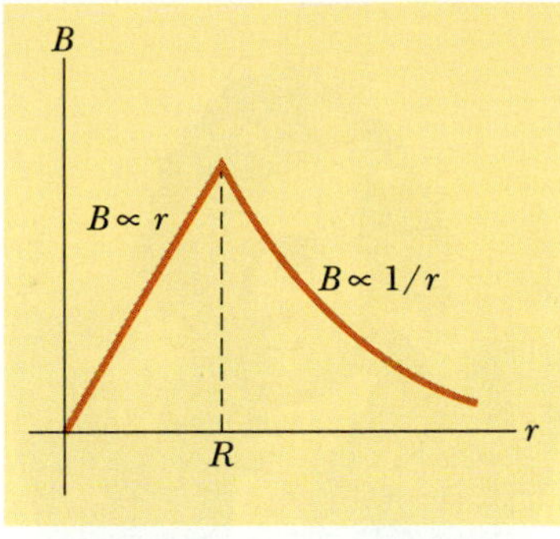

Figure 19.20 A sketch of the magnetic field versus r for the wire described in Example 19.6. The field is proportional to r inside the wire and varies as $1/r$ outside the wire.

Example 19.7 The Magnetic Field Created by a Toroidal Coil

The *toroidal coil* consists of N turns of wire wrapped around a torus, as in Figure 19.21. Assuming that the turns are closely spaced, calculate the magnetic field inside the coil, a distance of r from the center of the torus.

Solution To calculate the magnetic field inside the coil, we evaluate the line integral of $\mathbf{B} \cdot d\mathbf{s}$ over a circle of radius r. By symmetry, we see that the field is constant in magnitude on this path and tangent to it, so that $\mathbf{B} \cdot d\mathbf{s} = B\,ds$. Furthermore, note that the closed path threads N loops of wire, each of which carries current I. Therefore, the right side of Ampère's law, Equation 19.22, is $\mu_0 NI$ in this case. Ampère's law applied to this path then gives

$$\oint \mathbf{B} \cdot d\mathbf{s} = B \oint ds = B(2\pi r) = \mu_0 NI$$

$$B = \frac{\mu_0 NI}{2\pi r} \qquad [19.24]$$

This result shows that B varies as $1/r$ and hence is nonuniform within the coil. However, if r is large compared with a, where a is the cross-sectional radius of the torus, then the field is approximately uniform inside the coil. Furthermore, for the ideal toroidal coil, where the turns are closely spaced, the field anywhere outside the torus is zero. This can be seen by noting that the net current threaded by any circular path

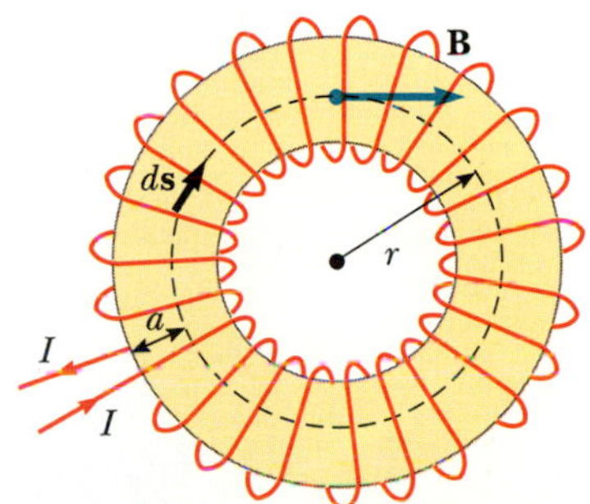

Figure 19.21 (Example 19.7) A toroidal coil consists of many turns of wire wrapped around a doughnut-shaped structure called a torus. If the coils are closely spaced, the field inside the toroidal coil is tangent to the dashed circular path and varies as $1/r$, and the exterior field is zero.

André-Marie Ampère (1775–1836), a French mathematician, chemist, and philosopher. *(Photo courtesy of AIP Niels Bohr Library)*

lying outside the toroidal coil is zero (including the region of the "hole in the doughnut"). Therefore, from Ampère's law, one finds that $\mathbf{B} = 0$ in the regions exterior to the toroidal coil. In reality, the turns of a toroidal coil form a helix rather than circular loops (the ideal case). As a result, there is always a small field outside the coil.

Example 19.8 The Magnetic Force on a Current Segment

A long, straight wire, wire 1, oriented along the y axis carries a steady current, I_1, as in Figure 19.22. A rectangular circuit to the right of the wire carries current I_2. Find the magnetic force on wire 2, which is the upper horizontal segment of the circuit and runs from $x = a$ to $x = a + b$.

Solution You may be tempted to use Equation 19.20 to obtain the force. However, that result applies *only* to two *parallel* wires and so cannot be used here. The correct approach is to start with the force on a small segment of wire 2, given by $d\mathbf{F} = I\,d\mathbf{s} \times \mathbf{B}$ (Eq. 19.4), where, in this case, $I = I_2$ and $\mathbf{B}$ is the magnetic field due to wire 1 at the position of the segment of wire 2 of length dx. From Ampère's law, the field at the distance x from wire 1 is

$$\mathbf{B} = \frac{\mu_0 I_1}{2\pi x}\,(-\mathbf{k})$$

where the field points into the page, as indicated by the unit vector notation $(-\mathbf{k})$, and crosses to the right of wire 1 in the drawing. (The dots to the left of wire 1 indicate that the field is out of the page on this side, which is consistent with the fact that the magnetic field lines around a long, straight wire form circles.) Taking the length of our segment as $d\mathbf{s} = dx\,\mathbf{i}$, we find

$$d\mathbf{F} = \frac{\mu_0 I_1 I_2}{2\pi x}\,[\mathbf{i} \times (-\mathbf{k})]\,dx = \frac{\mu_0 I_1 I_2}{2\pi}\,\frac{dx}{x}\,\mathbf{j}$$

Integrating this equation over the limits $x = a$ to $x = a + b$ gives

$$\mathbf{F} = \frac{\mu_0 I_1 I_2}{2\pi}\left[\ \ln x\ \right]_a^{a+b}\mathbf{j} = \frac{\mu_0 I_1 I_2}{2\pi}\ln\left(1 + \frac{b}{a}\right)\mathbf{j}$$

The force points upward, as indicated by the notation $\mathbf{j}$ and as shown in Figure 19.22.

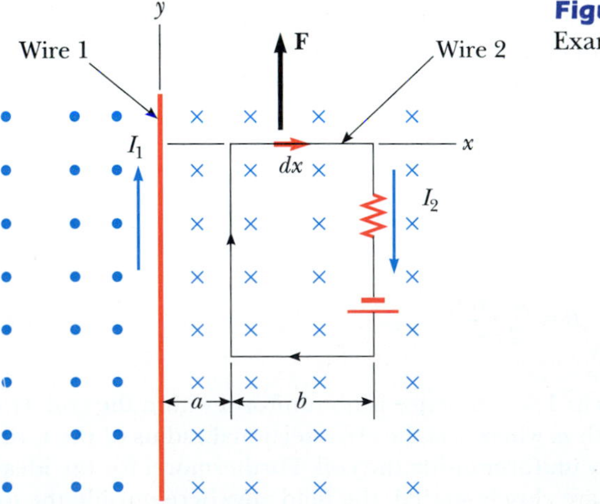

Figure 19.22 Example 19.8

Exercise What is the force on the lower horizontal segment of the circuit?

Answer The force has the same magnitude as that on the upper segment but is directed downward.

19.8 The Magnetic Field of a Solenoid

A solenoid is a long wire wound in the form of a helix. If the turns are closely spaced, this configuration can produce a reasonably uniform magnetic field within a small volume of the solenoid's interior region. Each of the turns can be regarded as a circular loop, and the net magnetic field is the vector sum of the fields due to all the turns.

Figure 19.23a shows the magnetic field lines of a loosely wound solenoid. Note that the interior field lines are nearly parallel, uniformly distributed, and close together. This indicates that the field in the solenoid's interior is uniform. The field lines between turns tend to cancel each other. The field outside the solenoid is both nonuniform and weak. To see why, consider the point P in Figure 19.23a. The field is nonuniform because of the finite length of the solenoid and a resultant field distribution that resembles that of a bar magnet. The field is weak because the field due to current elements on the upper portions tends to cancel the field due to current elements on the lower portions.

If the turns are closely spaced and the solenoid is of finite length, the field lines are as shown in Figure 19.23b. In this case, the field lines diverge from one end and converge at the opposite end. An inspection of this field distribution exterior to the solenoid shows a similarity to the field of a bar magnet (see Fig. 19.1b). Hence, one end of the solenoid behaves like the north pole of a magnet while the opposite end behaves like the south pole. As the length of the solenoid

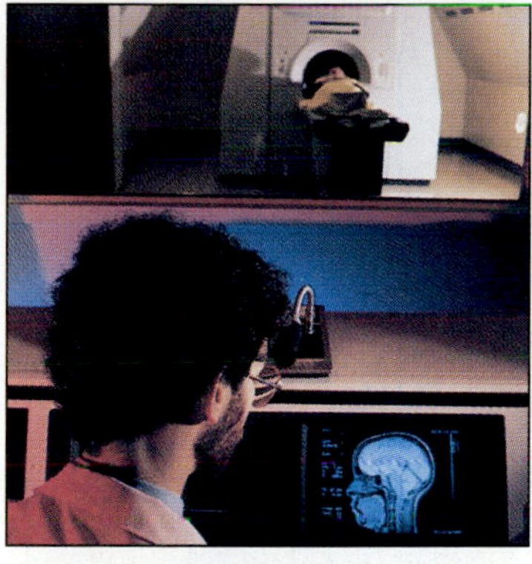

A technician studies the scan of a head. The scan was obtained using a medical diagnostic technique known as magnetic resonance imaging (MRI). This instrument makes use of strong magnetic fields produced by superconducting solenoids. *(Hank Morgan, Science Source)*

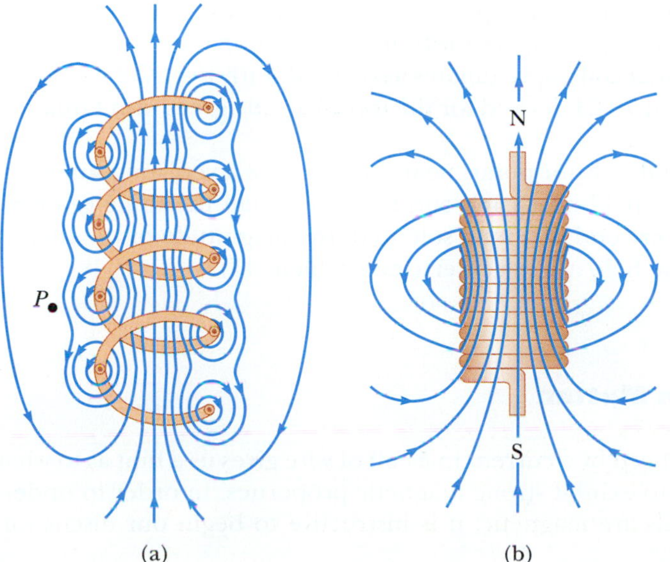

Figure 19.23
(a) The magnetic field lines for a loosely wound solenoid. (b) Magnetic field lines for a tightly wound solenoid of finite length carrying a steady current. The field inside the solenoid is nearly uniform and strong. The field lines resemble those of a bar magnet, so that the solenoid effectively has north and south poles.

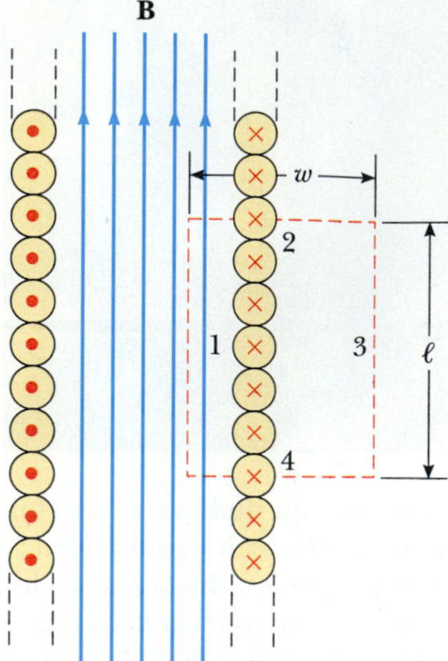

Figure 19.24
A cross-sectional view of a tightly wound solenoid. If the solenoid is long relative to its radius, we can assume that the magnetic field is uniform inside and zero outside. Ampère's law applied to the red dashed rectangular path can then be used to calculate the field inside the solenoid.

increases, the field within it becomes more and more uniform. When the solenoid's turns are closely spaced and its length is great compared with its radius, it approaches the case of an *ideal solenoid.* The field outside the solenoid is weak compared with the field inside, and the field inside is uniform over a large volume.

We can use Ampère's law to obtain an expression for the magnetic field inside an ideal solenoid. A longitudinal cross-section of part of our ideal solenoid (Fig. 19.24) carries current I. **B** inside the ideal solenoid is uniform and parallel to the axis, and **B** outside is zero. Consider a rectangular path of length ℓ and width w, as shown in Figure 19.24. We can apply Ampère's law to this path by evaluating the integral of $\mathbf{B} \cdot d\mathbf{s}$ over each of the four sides of the rectangle. The contribution along side 3 is clearly zero, since $\mathbf{B} = 0$ in this region. The contributions from sides 2 and 4 are both zero since **B** is perpendicular to $d\mathbf{s}$ along these paths. Side 1, whose length is ℓ, gives a contribution of $B\ell$ to the integral since **B** along this path is uniform and parallel to $d\mathbf{s}$. Therefore, the integral over the closed rectangular path has the value

$$\oint \mathbf{B} \cdot d\mathbf{s} = \int_{\text{path 1}} \mathbf{B} \cdot d\mathbf{s} = B \int_{\text{path 1}} ds = B\ell$$

The right side of Ampère's law involves the *total* current that passes through the area bound by the path of integration. In our case, the total current through the rectangular path equals the current through each turn of the solenoid multiplied by the number of turns. If N is the number of turns in the length ℓ, then the total current through the rectangle equals NI. Therefore, Ampère's law applied to this path gives

$$\oint \mathbf{B} \cdot d\mathbf{s} = B\ell = \mu_0 NI$$

$$B = \mu_0 \frac{N}{\ell} I = \mu_0 nI \qquad [19.25]$$

where $n = N/\ell$ is the number of turns *per unit length* (not to be confused with N).

We also could obtain this result in a simpler manner by reconsidering the magnetic field of a toroidal coil (Example 19.10). If the radius, r, of the toroidal coil containing N turns is large compared with its cross-sectional radius, a, then a short section of the toroidal coil approximates a solenoid, with $n = N/2\pi r$. In this limit, we see that Equation 19.24 derived for the toroidal coil agrees with Equation 19.25.

Equation 19.25 is valid only for points near the center of a very long solenoid. As you might expect, the field near each end is smaller than the value given by Equation 19.25. At the very end of a long solenoid, the magnitude of the field is about one-half that of the field at the center (see Problem 47).

19.9 Magnetism in Matter

The magnetic field produced by a current in a coil of wire gives us a hint as to what causes certain materials to exhibit strong magnetic properties. In order to understand why some materials are magnetic, it is instructive to begin our discussion with the classical model of the atom, in which electrons are assumed to move in circular orbits about the much more massive nucleus. In this model, each electron, with its charge of 1.6×10^{-19} C, circles the atom once in about 10^{-16} s. If we divide the electronic charge by this time interval, we find that the orbiting electron is equivalent to a current of 1.6×10^{-3} A. Therefore, each orbiting electron is viewed as a tiny current loop with a corresponding magnetic moment. Such a current loop produces a magnetic field of about 20 T at the center of the circular path (see Problem 26).

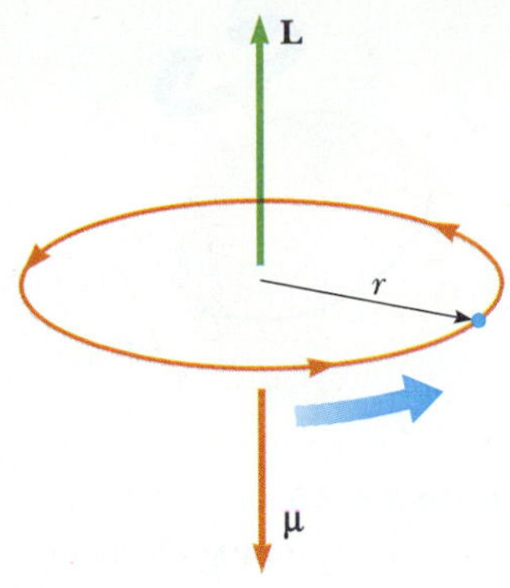

Figure 19.25
An electron moving in a circular orbit of radius r has an angular momentum **L** and a magnetic moment $\boldsymbol{\mu}$ which are in *opposite* directions.

Consider an electron moving with a constant speed of v in a circular orbit of radius r about the nucleus, as in Figure 19.25. Since the electron travels a distance of $2\pi r$ (the circumference of the circle) in the time T, which is the time required for one revolution, the electron's orbital speed is $v = 2\pi r/T$. The current associated with this orbiting electron equals its charge divided by the time for one revolution:

$$I = \frac{e}{T} = \frac{e}{2\pi r/v} = \frac{ev}{2\pi r}$$

The magnitude of the magnetic moment associated with this current loop is given by $\mu = IA$, where $A = \pi r^2$ is the area of the orbit. Therefore,

$$\mu = IA = \left(\frac{ev}{2\pi r}\right)\pi r^2 = \tfrac{1}{2}evr$$

Since the magnitude of the orbital angular momentum of the electron is $L = mvr$ (Eq. 11.24), the magnitude of the magnetic moment can be written as

Orbital magnetic moment

$$\mu = \left(\frac{e}{2m}\right)L \qquad [19.26]$$

This result says that *the magnetic moment of an orbiting electron is proportional to its orbital angular momentum.* Note that, since the electron is negatively charged, the vectors $\boldsymbol{\mu}$ and **L** point in opposite directions. Both vectors are perpendicular to the plane of the orbit, as indicated in Figure 19.25.

As we learned in Chapter 11, Section 11.9, a fundamental outcome of quantum physics is that the orbital angular momentum must be quantized and is always some integer multiple of $\hbar = h/2\pi = 1.06 \times 10^{-34}$ J·s, where h is Planck's constant. That is,

Angular momentum is quantized

$$L = 0, \hbar, 2\hbar, 3\hbar, \ldots$$

Hence, the smallest nonzero value of the magnetic moment of an orbiting electron is

$$\mu = \frac{e}{2m}\hbar \qquad [19.27]$$

In most substances, the magnetic moment of one electron in an atom is canceled by that of another electron in the atom, orbiting in the opposite direction. The net result is that *the magnetic effect produced by the orbital motion of the electrons is either zero or very small for most materials.*

As mentioned earlier, an electron has another intrinsic property called *spin,* which also contributes to its magnetic moment. In this regard, one can view the electron as a sphere of charge spinning about its axis as it orbits the nucleus, as in

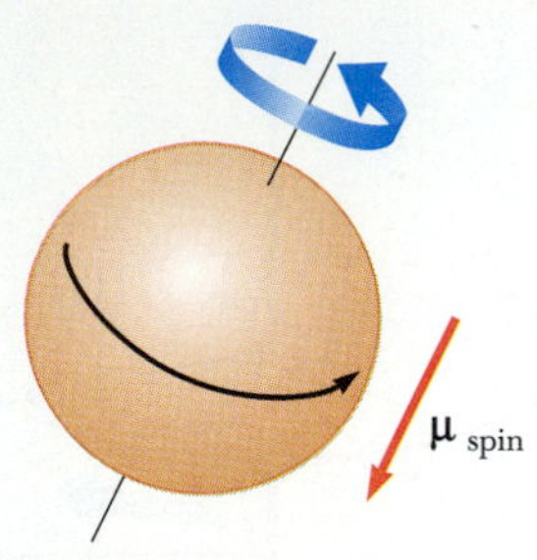

Figure 19.26
Model of a spinning electron.

Figure 19.26. (This classical description of a spinning electron should not be taken literally. The property of spin can be understood only through a relativistic model.) This spinning motion produces an effective current loop and hence a magnetic moment of the same order of magnitude as that due to the orbital motion. The magnitude of the spin angular momentum predicted by quantum theory is

$$S = \frac{\hbar}{2} = 5.2729 \times 10^{-35}\ \text{J}\cdot\text{s}$$

Every electron has an intrinsic magnetic moment associated with its spin with the value

$$\mu_B = \frac{e}{2m}\hbar = 9.27 \times 10^{-24}\ \text{J/T} \qquad [19.28]$$

Bohr magneton

which is called the **Bohr magneton.**

In atoms or ions containing multiple electrons, the electrons usually pair up with their spins opposite each other, an arrangement that results in a cancellation of the spin magnetic moments. However, an atom with an odd number of electrons must have at least one "unpaired" electron and a corresponding spin magnetic moment. The magnetic moments of several atoms and ions are listed in Table 19.1.

Table 19.1 Magnetic Moments of Some Atoms and Ions, in Units of Bohr Magnetons

Atom (or Ion)	Magnet Moment per Atom or Ion (μ/μ_B)
H	1
He	0
Ne	0
Fe	2.22
Co	1.72
Ni	0.606
Gd	7.1
Dy	10.0
Co^{2+}	4.8
Ni^{2+}	3.2
Fe^{2+}	5.4
Ce^{3+}	2.14
Yb^{3+}	4.00

Ferromagnetic Materials

Iron, cobalt, nickel, gadolinium, and dysprosium are strongly magnetic materials and are said to be **ferromagnetic.** Ferromagnetic substances, used to fabricate permanent magnets, contain spin magnetic moments that tend to align parallel to each other even in a weak external magnetic field. Once the moments are aligned, the substance remains magnetized after the external field is removed. This permanent alignment is due to strong coupling between neighboring atoms, which can only be understood using quantum physics.

All ferromagnetic materials contain microscopic regions called **domains,** within which all magnetic moments are aligned. The domains range from about 10^{-12} to 10^{-8} m^3 in volume and contain 10^{17} to 10^{21} atoms. The boundaries between domains having different orientations are called **domain walls.** In an unmagnetized sample, the domains are randomly oriented so that the net magnetic moment is zero, as in Figure 19.27a. When the sample is placed in an external magnetic field, the domains tend to align with the field, which results in a magnetized sample, as in Figure 19.27b. Observations show that domains initially oriented along the external field grow in size at the expense of the less favorably oriented domains. When the external field is removed, the sample may retain most of its magnetism.

The extent to which a ferromagnetic substance retains its magnetism depends on whether it is classified as being magnetically **hard** or **soft.** Soft magnetic materials, such as iron, are easily magnetized but also tend to lose their magnetism easily. When a soft magnetic material is magnetized and the external magnetic field is removed, thermal agitation produces domain motion and the material quickly returns to an unmagnetized state. In contrast, hard magnetic materials, such as cobalt and nickel, are difficult to magnetize but tend to retain their magnetism, and domain alignment persists in them after the external magnetic field is re-

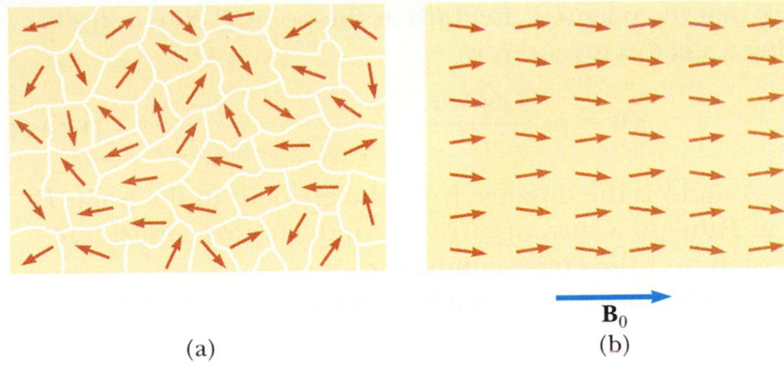

Figure 19.27
(a) Random orientation of domains in an unmagnetized substance. (b) When an external magnetic field, $\mathbf{B}_0$, is applied, the domains tend to align with the magnetic field.

moved. Such hard magnetic materials are referred to as **permanent magnets.** Rare-earth permanent magnets, such as samarium-cobalt, are now regularly used in industry.

Summary

The **magnetic force** that acts on a charge, q, moving with a velocity of v in an external magnetic field, B, is

Magnetic force on a charged particle moving in a magnetic field

$$\mathbf{F} = q\mathbf{v} \times \mathbf{B} \qquad [19.1]$$

This force is in a direction perpendicular both to the velocity of the particle and to the field. The magnitude of the magnetic force is

$$F = qvB \sin \theta \qquad [19.2]$$

where θ is the angle between **v** and **B**. **F** = 0 when **v** is either parallel or antiparallel to **B**, and $F = qvB$ when **v** is perpendicular to **B**.

If a straight conductor of length ℓ carries a current, I, the force on that conductor when placed in a uniform *external* magnetic field, **B**, is

Force on a straight wire carrying a current

$$\mathbf{F} = I\boldsymbol{\ell} \times \mathbf{B} \qquad [19.3]$$

where $\boldsymbol{\ell}$ is in the direction of the current and $|\boldsymbol{\ell}| = \ell$.

If an arbitrarily shaped wire carrying current I is placed in an *external* magnetic field, the force on a very small segment, $d\mathbf{s}$, is

Force on a current element

$$d\mathbf{F} = I\, d\mathbf{s} \times \mathbf{B} \qquad [19.4]$$

To determine the total force on the wire, one has to integrate Equation 19.4.

The **magnetic moment,** μ, of a current loop carrying current I is

Magnetic moment of a current loop

$$\boldsymbol{\mu} = I\mathbf{A} \qquad [19.8]$$

where **A** is perpendicular to the plane of the loop and $|\mathbf{A}|$ is equal to the area of the loop. The SI unit of $\boldsymbol{\mu}$ is the ampere-meter².

The torque, $\boldsymbol{\tau}$, on a current loop when the loop is placed in a uniform *external* magnetic field, **B**, is

Torque on a current loop

$$\boldsymbol{\tau} = \boldsymbol{\mu} \times \mathbf{B} \qquad [19.9]$$

The **Biot-Savart law** says that the magnetic field $d\mathbf{B}$ at the point P, due to the wire element $d\mathbf{s}$ carrying a steady current, I, is

Biot-Savart law

$$d\mathbf{B} = k_m \frac{I\, d\mathbf{s} \times \hat{\mathbf{r}}}{r^2} \qquad [19.10]$$

where $k_m = 10^{-7}\ \text{T}\cdot\text{m/A}$ and r is the distance from the element to the point P. To find the total field at P due to a current-carrying conductor, one must integrate this vector expression over the entire conductor.

The **magnetic field** at a distance of a from a long, straight wire carrying current I is

Magnetic field at a distance a of an infinitely long wire

$$B = \frac{\mu_0 I}{2\pi a} \qquad [19.15]$$

where $\mu_0 = 4\pi \times 10^{-7}\ \text{T}\cdot\text{m/A}$ is the **permeability of free space.** The field lines are circles concentric with the wire.

The force per unit length between two parallel wires separated by a distance of a and carrying currents I_1 and I_2 has the magnitude

Force per unit length between two wires separated by a distance a

$$\frac{F_1}{\ell} = \frac{\mu_0 I_1 I_2}{2\pi a} \qquad [19.20]$$

The force is attractive if the currents are in the same direction and repulsive if they are in opposite directions.

Ampère's law says that the line integral of $\mathbf{B}\cdot d\mathbf{s}$ around any closed path equals $\mu_0 I$, where I is the total steady current passing through any surface bounded by the closed path:

Ampère's law

$$\oint \mathbf{B}\cdot d\mathbf{s} = \mu_0 I \qquad [19.22]$$

Using Ampère's law, one finds that the fields inside a toroidal coil and solenoid are

Magnetic field inside a toroid

$$B = \frac{\mu_0 N I}{2\pi r} \quad \text{(toroid)} \qquad [19.24]$$

Magnetic field inside a solenoid

$$B = \mu_0 \frac{N}{\ell} I = \mu_0 n I \quad \text{(solenoid)} \qquad [19.25]$$

where N is the total number of turns and n is the number of turns per unit length.

The fundamental sources of all magnetic fields are the magnetic dipole moments associated with atoms. The atomic dipole moments can arise both from the orbital motions of the electrons and from an intrinsic property of electrons known as *spin.*

Questions and Conceptual Exercises

1. Two charged particles are projected into a region where there is a magnetic field perpendicular to their velocities. If the charges are deflected in opposite directions, what can you say about them?
2. If a charged particle moves in a straight line through some region of space, can you say that the magnetic field in that region is zero?
3. A current-carrying conductor experiences no magnetic force when placed in a certain manner in a uniform magnetic field. Explain.

4. Is it possible to orient a current loop in a uniform magnetic field so that the loop will not tend to rotate? Explain.
5. The north-seeking pole of a magnet is attracted toward the geographic north pole of the Earth, yet like poles repel. What is the way out of this dilemma?
6. Which way would a compass point if you were at the north magnetic pole?
7. A magnet attracts a piece of iron. The iron can then attract another piece of iron. Explain, on the basis of alignment of the domains, what happens in each piece of iron.
8. Will a nail be attracted to either pole of a magnet? Explain what is happening inside the nail.
9. Why does hitting a magnet with a hammer cause its magnetism to be reduced?
10. A charged particle moves in a circular path because of an applied magnetic field. Does the particle gain energy from the magnetic field? Explain.
11. Explain why it is not possible to determine the charge and mass of a charged particle separately via electric and magnetic forces.
12. How can a current loop be used to determine the presence of a magnetic field in a given region of space?
13. It is found that charged particles from outer space, called cosmic rays, strike the Earth more frequently at the poles than at the equator. Why?
14. Explain why two parallel wires carrying currents in opposite directions repel each other.
15. A hollow copper tube carries a current. Why is $\mathbf{B} = 0$ inside the tube? Is $\mathbf{B}$ nonzero outside the tube?
16. Describe the change in the magnetic field inside a solenoid that carries a steady current, I, if (a) the length of the solenoid is doubled, but the number of turns remains the same; (b) the number of turns is doubled, but the length remains the same.
17. Consider an electron near the magnetic equator. In which direction will it tend to be deflected if its velocity is directed (a) downward? (b) northward? (c) westward? (d) southeastward?
18. You are an astronaut stranded on a planet with no test equipment or minerals around. The planet does not even have a magnetic field. You have two bars of iron in your possession; one is magnetized, one is not. How could you determine which is magnetized?
19. A Hindu ruler once suggested that he be entombed in a magnetic coffin with the polarity arranged so that he would be forever suspended between heaven and Earth. Is such magnetic levitation possible? Discuss.
20. What is the difference between hard and soft ferromagnetic materials?
21. Should the surface of a computer disk be made from a "hard" or a "soft" ferromagnetic substance?
22. Figure 19.28 shows two permanent magnets with holes through their centers. Note that the upper magnet is levitated above the lower magnet. (a) How does this occur? (b) What purpose does the pencil serve? (c) What can you say about the poles of the magnets from this observation? (d) If the upper magnet were inverted, what do you suppose would happen?

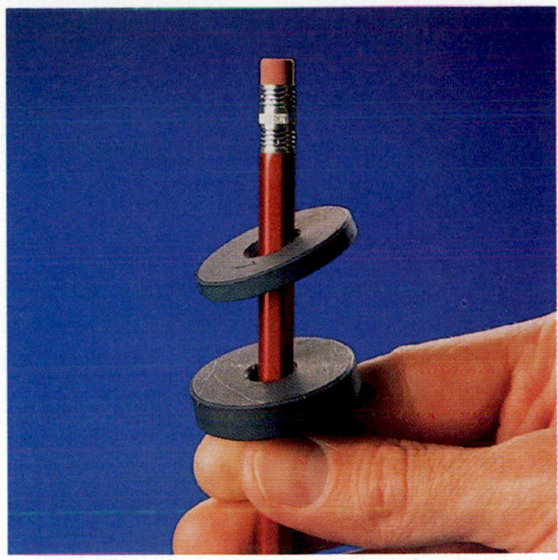

Figure 19.28
(Question 22) Magnetic levitation using two ceramic magnets. *(Courtesy of CENCO)*

23. The *bubble chamber* is a device for observing the tracks of particles that pass through it. The chamber is immersed in a magnetic field. If some of the tracks are spirals and others are straight lines, what can you say about the particles?
24. The electron beam in Figure 19.29 is projected to the right. The beam deflects downward in a 0.001-T magnetic field that is produced by a pair of current-carrying coils. (a) What is the direction of the magnetic field? (b) If the diameter of the sphere is 0.1 m, estimate the speed of the electrons in the beam.

Figure 19.29
(Question 24) Electrons in a magnetic field. *(Courtesy of CENCO)*

Problems

Section 19.2 The Magnetic Field

1. A proton moves with a speed of 2.5×10^6 m/s horizontally, at right angles to a magnetic field. (a) How strong a field is required to just balance the weight of the proton and keep it moving horizontally? (b) Should the direction of the magnetic field be in a horizontal or a vertical plane?
2. A proton travels with a speed of 3×10^6 m/s at an angle of 37° with the direction of a magnetic field of 0.3 T in the $+y$ direction. What are (a) the magnitude of the magnetic force on the proton and (b) its acceleration?
3. A proton moving with a speed of 5×10^7 m/s through a magnetic field of 2 T experiences a magnetic force of 3×10^{-12} N.What is the angle between the proton's velocity and the field?
4. An electron is accelerated through 2400 V from rest and then enters a region where there is a uniform 1.7-T magnetic field. What are (a) the maximum and (b) the minimum values of the magnetic force this charge can experience?
5. A proton moves perpendicularly to a uniform magnetic field, **B**, with a speed of 10^7 m/s and experiences an acceleration of 2×10^{13} m/s^2 in the $+x$ direction when its velocity is in the $+z$ direction. Determine the magnitude and direction of the field.
6. Show that the work done by the magnetic force on a charged particle moving in a uniform magnetic field is zero for any displacement of the particle.
7. A cosmic-ray proton in interstellar space has an energy of 10 MeV and executes a circular orbit with a radius equal to that of Mercury's orbit around the Sun (5.8×10^{10} m). What is the galactic magnetic field in that region of space?
8. At the Fermilab accelerator in Weston, Illinois, protons with momentum of 4.8×10^{-16} kg·m/s are held in a horizontal circular orbit of radius 1 km by an upward magnetic field. What upward magnetic field must be used to maintain the protons in this orbit?
9. Singly charged uranium ions are accelerated through a potential difference of 2 kV and enter a uniform magnetic field of 1.2 T, directed perpendicular to their velocities. Determine the radius of the circular path followed by these ions, assuming that they are (a) U^{238} ions; (b) U^{235} ions. How does the ratio of these path radii depend on the accelerating voltage and the magnetic field strength?
10. The picture tube in a television uses magnetic deflection coils rather than electric deflection plates. Suppose an electron beam is accelerated through a 50-kV potential difference and then passes, for 1 cm, through a uniform magnetic field produced by these coils. The screen is 10 cm from the center of the coils and is 50 cm wide. When the field is turned off, the electron beam hits the center of the screen. What field strength is necessary to deflect the beam to the side of the screen?

Section 19.3 Magnetic Force on a Current-Carrying Conductor

11. Calculate the magnitude of the force per unit length exerted on a conductor carrying a current of 22 A in a region where a uniform magnetic field has a magnitude of 0.77 T and is directed perpendicular to the conductor.
12. A wire carries a steady current of 2.4 A. A straight section of the wire, with a length of 0.75 m along the x axis, lies in a uniform magnetic field, $\mathbf{B} = (1.6\ T)\mathbf{k}$. If the current flows in the $+x$ direction, what is the magnetic force on the section of wire?
13. A conductor suspended by two flexible wires as in Figure 19.30 has a mass per unit length of 0.04 kg/m. What current must exist in the conductor in order for the tension in the supporting wires to be zero if the magnetic field over the region is 3.6 T into the page? What is the required direction for the current?

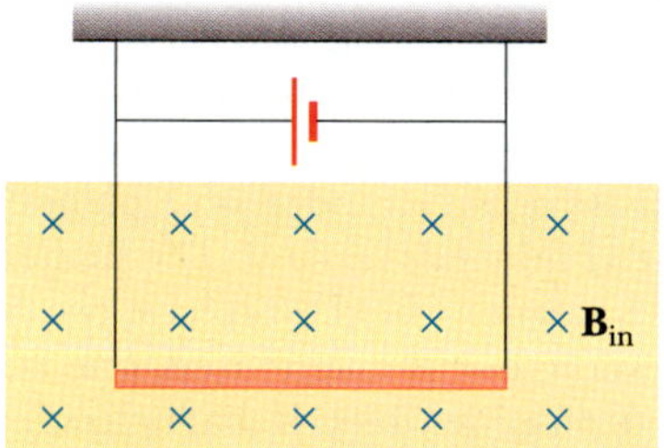

Figure 19.30 (Problem 13)

14. A wire with a linear mass density of 0.5 g/cm carries a 2-A current horizontally to the south. What are the direction and magnitude of the minimum magnetic field needed to lift this wire vertically upward?
15. A rectangular loop with dimensions 10 cm × 20 cm is suspended by a string, and the lower horizontal section of the loop is immersed in a magnetic field con-

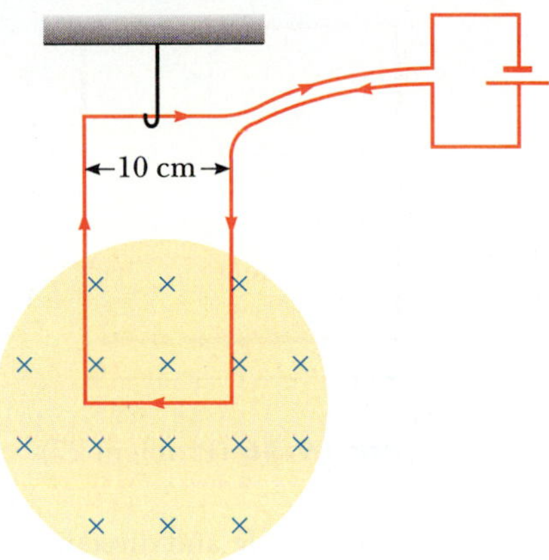

Figure 19.31 (Problem 15)

fined to a circular region (Fig. 19.31). If a current of 3 A is maintained in the loop in the direction shown, what are the direction and magnitude of the magnetic field required to produce a tension of 4×10^{-2} N in the supporting string? (Neglect the mass of the loop.)

16. In Figure 19.32, the cube is 40 cm on each edge. Four straight segments of wire—*ab, bc, cd,* and *da*—form a closed loop that carries a current of $I = 5$ A, as shown. A uniform magnetic field, $B = 0.02$ T, is in the positive y direction. Make a table showing the magnitude and direction of the force on each segment, listing them in the preceding order.

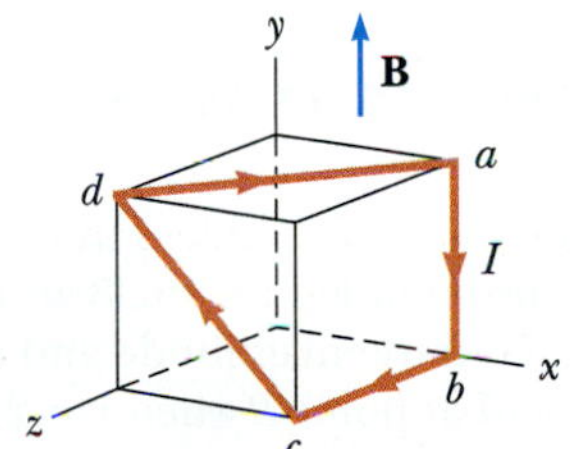

Figure 19.32 (Problem 16)

17. A current of $I = 15$ A is directed along the positive x axis in a wire perpendicular to a magnetic field. The wire experiences a magnetic force per unit length of 0.63 N/m in the negative y direction. Calculate the magnitude and direction of the magnetic field in the region through which the current passes.

18. An arbitrarily shaped, closed loop carrying current I is placed in a uniform external magnetic field, **B**, as in Figure 19.33. Show that the net magnetic force on the loop is zero.

19. A strong magnet is placed under a horizontal conducting ring of radius r that carries current I, as shown in Figure 19.34. If the magnetic lines of force make an angle of θ with the vertical at the ring's location, what are the magnitude and direction of the resultant force on the ring?

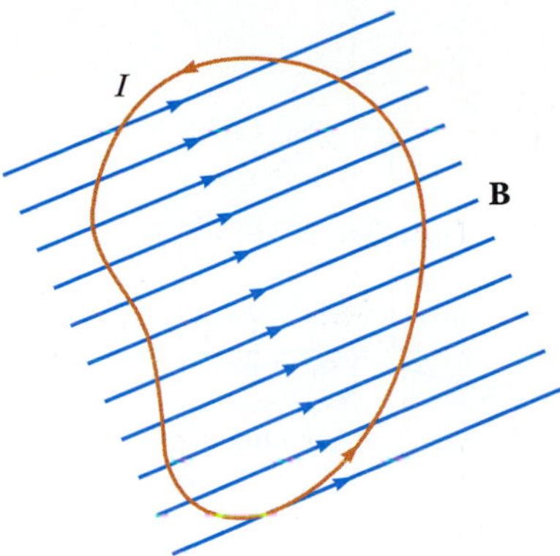

Figure 19.33 (Problem 18)

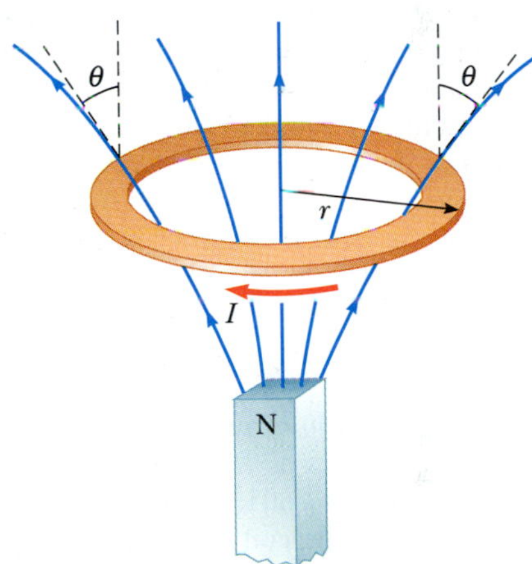

Figure 19.34 (Problem 19)

Section 19.4 Torque on a Current Loop in a Uniform Magnetic Field

20. A current of 17 mA is maintained in a single circular loop with a 2-m circumference. An external magnetic field of 0.8 T is directed parallel to the plane of the loop. (a) Calculate the magnetic moment of the current loop. (b) What is the magnitude of the torque exerted on the loop by the magnetic field?

21. A rectangular loop consists of 100 closely wrapped turns and has the dimensions 0.4 m × 0.3 m. The loop is hinged along the y axis, and the plane of the coil makes an angle of 30° with the x axis (Fig.

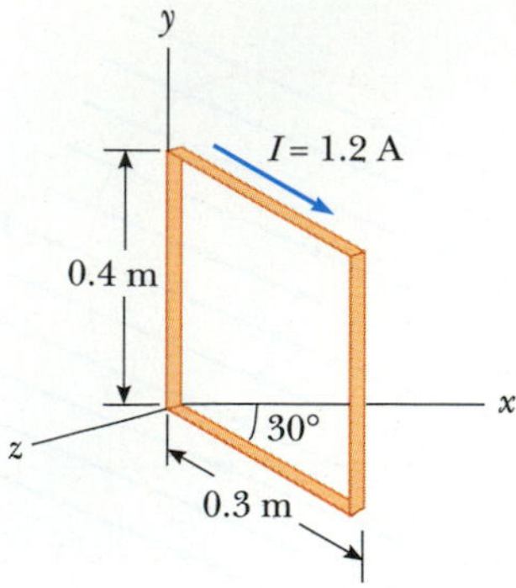

Figure 19.35 (Problem 21)

19.35). What is the magnitude of the torque exerted on the loop by a uniform magnetic field of 0.8 T, directed along the x axis, when the current in the windings has a value of 1.2 A in the direction shown? What is the expected direction of rotation of the loop?

22. A small bar magnet is suspended in a uniform 0.25-T magnetic field. The maximum torque experienced by the bar magnet is 4.6×10^{-3} N·m. Calculate the magnetic moment of the bar magnet.
23. A rectangular coil with 225 turns and an area of 0.45 m^2 is in a uniform magnetic field of 0.21 T. Measurements indicate that the maximum torque exerted on the loop by the field is 8×10^{-3} N·m. (a) Calculate the current in the coil. (b) Would the value found for the required current be different if the 225 turns of wire were used to form a single-turn coil with the same shape but a greater area? Explain.
24. A circular coil of 100 turns has a radius of 0.025 m and carries a current of 0.1 A while in a uniform external magnetic field of 1.5 T. How much work must be done to rotate the coil from a position where the magnetic moment is parallel to the field to a position where the magnetic moment is opposite the field?

Section 19.5 The Biot-Savart Law

25. A wire in which there is a current of 5 A is to be formed into a circular loop of one turn. If the required value of the magnetic field at the *center* of the loop is 10 μT, what is the required radius of the loop?
26. In Neils Bohr's 1913 model of the hydrogen atom, an electron circles the proton at a distance of 5.3×10^{-11} m with a speed of 2.2×10^6 m/s. Compute the magnetic field strength produced by the electron's motion at the location of the proton.
27. A conductor in the shape of a square of edge length $\ell = 0.4$ m carries a current of $I = 10$ A (Fig. 19.36). Calculate the magnitude and direction of the magnetic field produced at the *center* of the square.

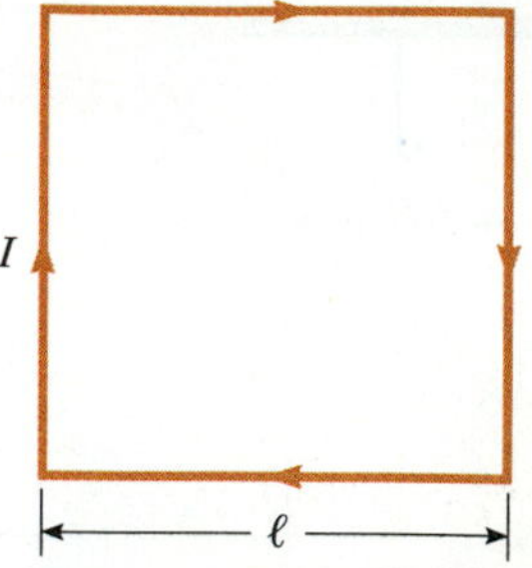

Figure 19.36 (Problem 27)

28. A 12 cm × 16 cm rectangular loop of superconducting wire carries a current of 30 A. What is the magnetic field at the center of the loop?
29. Determine the magnetic field at a point, P, that is a distance of x from the corner of an infinitely long wire that is bent at a right angle, as shown in Figure 19.37. The wire carries a steady current, I.

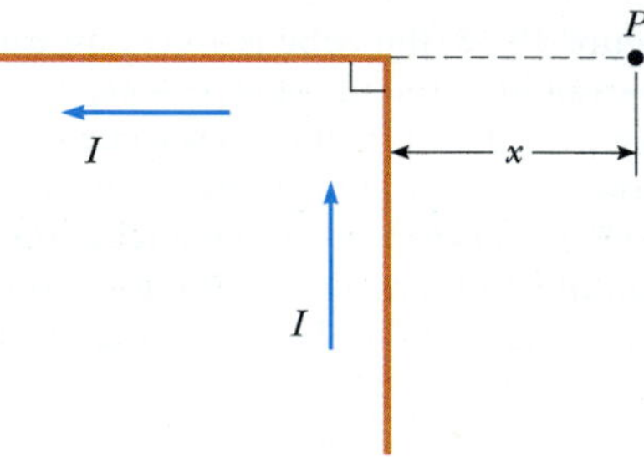

Figure 19.37 (Problem 29)

30. A segment of wire of total length $4r$ is formed into the shape shown in Figure 19.38 and carries a current of $I = 6$ A. Find the magnitude and direction of the magnetic field at point P when $r = 2\pi$ cm.

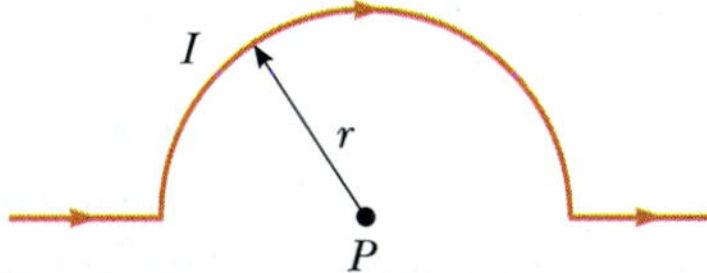

Figure 19.38 (Problem 30)

31. A current path shaped as shown in Figure 19.39 produces a magnetic field at P, the center of the arc. If the arc subtends an angle of 30° and the radius of the arc is 0.6 m, what are the magnitude and direction of the field produced at P if the current is 3 A?
32. Consider the current-carrying loop shown in Figure

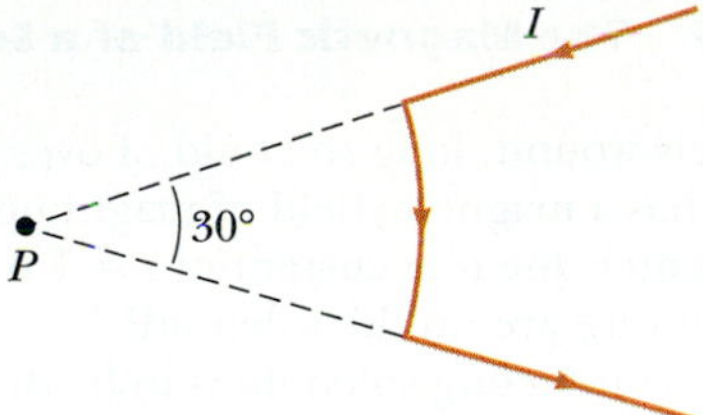

Figure 19.39 (Problem 31)

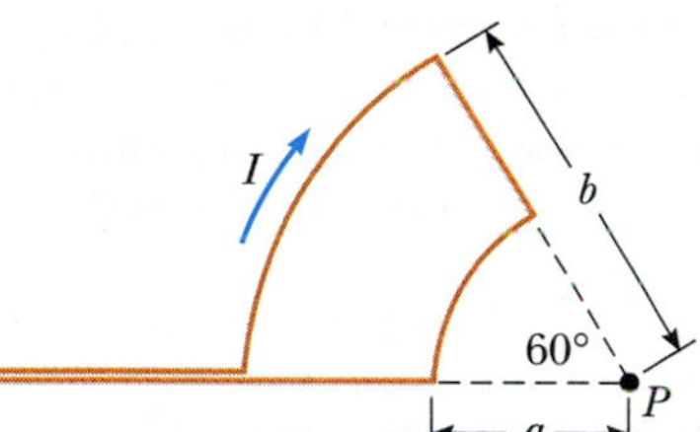

Figure 19.40 (Problem 32)

19.40, formed of radial lines and segments of circles whose centers are at point *P*. Find the magnitude and direction of the magnetic field **B** at *P*.

Section 19.6 The Magnetic Force Between Two Parallel Conductors

33. Two long parallel conductors, separated by a distance of $a = 10$ cm, carry currents in the same direction. If $I_1 = 5$ A and $I_2 = 8$ A, what is the force per unit length exerted on each conductor by the other?
34. Two long parallel wires, each having a mass per unit length of 40 g/m, are supported in a horizontal plane by strings 6 cm long, as shown in Figure 19.41. Each wire carries the same current, *I*, causing the wires to repel each other so that the angle, θ, between the supporting strings is 16°. (a) Are the currents in the same or opposite directions? (b) Find the magnitude of each current.
35. At what distance from a long, straight wire carrying a current of 5 A is the magnetic field due to the wire equal to the strength of the Earth's field—approximately 5×10^{-5} T?
36. For the arrangement shown in Figure 19.42, the current in the long, straight conductor has the value $I_1 = 5$ A and lies in the plane of the rectangular loop, which carries a current of $I_2 = 10$ A. The dimensions are $c = 0.1$ m, $a = 0.15$ m, and $\ell = 0.45$ m. Find the magnitude and direction of the *net force* exerted on the rectangle by the magnetic field of the straight current-carrying conductor.

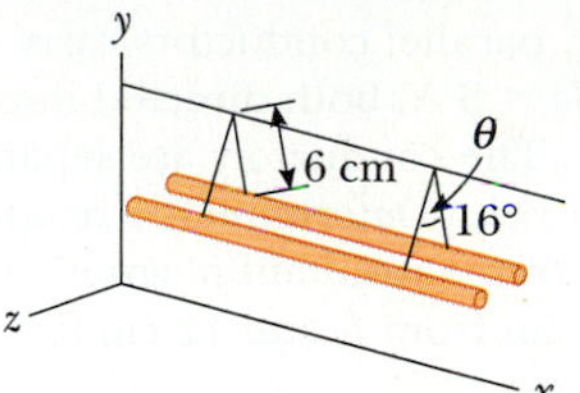

Figure 19.41 (Problem 34)

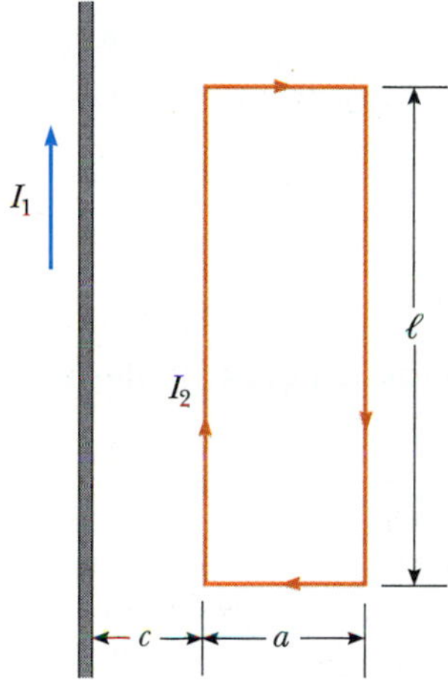

Figure 19.42 (Problem 36)

Section 19.7 Ampère's Law

37. Four long, parallel conductors carry equal currents, $I = 5$ A. Figure 19.43 is an end view of the conductors. The current direction is into the page at points *A* and *B* (indicated by the crosses) and out of the page at points *C* and *D* (indicated by the dots). Calculate the magnitude and direction of the magnetic field at point *P*, located at the center of the square of edge length 0.2 m.

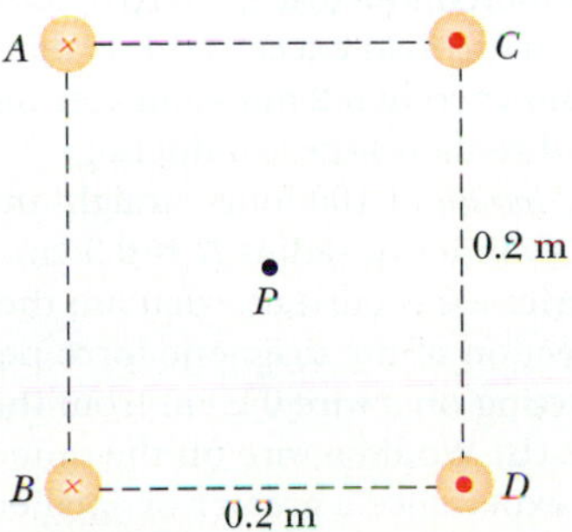

Figure 19.43 (Problem 37)

38. Two long, parallel conductors carry currents of $I_1 = 3$ A and $I_2 = 3$ A, both directed into the page in Figure 19.44. The conductors are separated by a distance of 13 cm. Determine the magnitude and direction of the resultant magnetic field at point P, located 5 cm from I_1 and 12 cm from I_2.

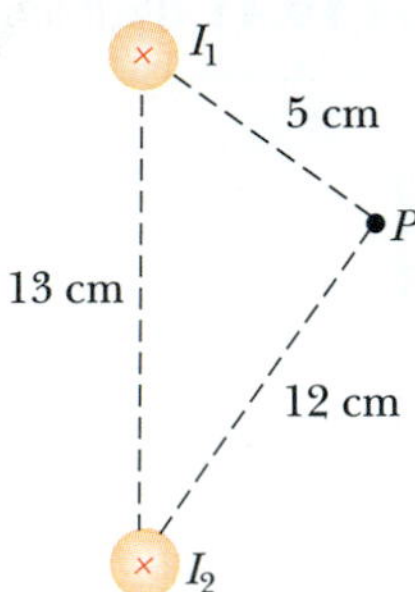

Figure 19.44 (Problem 38)

39. Each of the magnetic coils of a Tokamak fusion reactor is in the shape of a toroid, having an inner radius of 0.7 m and an outer radius of 1.3 m. Inside the toroid is the plasma. If the toroid has 900 turns of large-diameter wire, each of which carries a current of 14 000 A, find the magnetic field strength along (a) the inner radius of the toroid and (b) the outer radius of the toroid.

40. A cylindrical conductor of radius $R = 2.5$ cm carries a current of $I = 2.5$ A along its length; this current is uniformly distributed throughout the cross-section of the conductor. (a) Calculate the magnetic field midway along the radius of the wire (that is, at $r = R/2$). (b) Find the distance beyond the surface of the conductor at which the magnitude of the magnetic field has the same value as the magnitude of the field at $r = R/2$.

41. Niobium metal becomes a superconductor (with electrical resistance equal to zero) when cooled below 9 K. If superconductivity is destroyed when the surface magnetic field exceeds 0.1 T, determine the maximum current a 2-mm-diameter niobium wire can carry and remain superconducting.

42. A *packed bundle* of 100 long, straight insulated wires forms a cylinder of radius $R = 0.5$ cm. (a) If each wire carries a 2-A current, what are the magnitude and direction of the magnetic force per unit length that is acting on a wire 0.2 cm from the center of the bundle? (b) Would a wire on the outer edge of the bundle experience a greater or smaller force than the wire 0.2 cm from the center?

Section 19.8 The Magnetic Field of a Solenoid

43. A closely wound, long solenoid of overall length 30 cm has a magnetic field of magnitude 5×10^{-4} T at its center due to a current of $I = 1$ A. How many turns of wire are on the solenoid?

44. A superconducting solenoid is to be designed to generate a magnetic field of 10 T. (a) If the solenoid winding has 2000 turns per meter, what is the required current? (b) What force per unit length is exerted on the solenoid windings by this magnetic field?

45. What current is required in the windings of a long solenoid that has 1000 turns uniformly distributed over a length of 0.4 m, in order to produce a magnetic field of magnitude 1.0×10^{-4} T at the center of the solenoid?

46. Some superconducting alloys at very low temperatures can carry very high currents. For example, Nb_3Sn wire at 10 K can carry 10^3 A and maintain its superconductivity. Determine the maximum B field which can be achieved in a solenoid of length 25 cm if 1000 turns of Nb_3Sn wire are wrapped on the outside surface.

47. Consider a solenoid of length ℓ and radius R, containing N closely spaced turns and carrying a steady current, I. (a) Using these parameters, find the magnetic field along the axis as a function of distance a from the end of the solenoid. (b) Show that as ℓ becomes very long, B approaches $\mu_0 NI/2\ell$ at each end of the solenoid.

*Section 19.9 Magnetism in Matter

48. A magnetized cylinder of iron has a magnetic field, $\mathbf{B} = 0.04$ T, in its interior. The magnet is 3 cm in diameter and 20 cm long. If the same magnetic field is to be produced by a 5-A current carried by an air-core solenoid having the same dimensions as the cylindrical magnet, how many turns of wire must be on the solenoid?

49. In Bohr's 1913 model of the hydrogen atom, the electron is in a circular orbit of radius 5.3×10^{-11} m, and its speed is 2.2×10^6 m/s. (a) What is the magnitude of the magnetic moment due to the electron's motion? (b) If the electron orbits counterclockwise in a horizontal circle, what is the direction of this magnetic moment vector?

50. At saturation, the alignment of spins in iron can contribute as much as 2 T to the total magnetic field $\mathbf{B}$. This is equivalent to a total magnetization of magnitude 1.6×10^6 A·m^2, where magnetization is defined as the magnetic moment per cubic meter. If each electron contributes a magnetic moment of $9.27 \times$

10^{-24} A·m² (one Bohr magneton), how many electrons per atom contribute to the saturated field of iron? (*Hint:* There are 8.5 × 10^{28} iron atoms in each cubic meter.)

51. The magnetic moment of the Earth is approximately 8.7 = 10^{22} A·m². (a) If this were caused by the complete magnetization of a huge iron deposit, how many unpaired electrons would this correspond to? (b) At two unpaired electrons per iron atom, how many kilograms of iron would this correspond to? (The density of iron is 7900 kg/m³, and there are approximately 8.5 × 10^{28} iron atoms in each cubic meter.)

Additional Problems

52. A lightning bolt may carry a current of 10^4 A for a short period of time. What is the resulting magnetic field at a point 100 m from the bolt?
53. Measurements of the magnetic field of a large tornado were made at the Geophysical Observatory in Tulsa, Oklahoma, in 1962. If the tornado's field was **B** = 1.5 × 10^{-8} T, pointing north, when the tornado was 9 km east of the observatory, what current was carried up or down the funnel of the tornado?
54. Two long, parallel conductors carry currents in the same direction, as in Figure 19.45. Conductor A carries a current of 150 A and is held firmly in position. Conductor B carries a current of I_B and is allowed to slide freely up and down (parallel to A) between a set of nonconducting guides. If the linear mass density of conductor B is 0.10 g/cm, what value of current I_B will result in equilibrium when the distance between the two conductors is 2.5 cm?
55. The magnitude of the Earth's magnetic field at either pole is about magnitude 7 × 10^{-5} T. Using a model in which you assume that this field is produced by a current loop around the equator, determine the current that would generate such a field. (R_e = 6.37 × 10^6 m.)

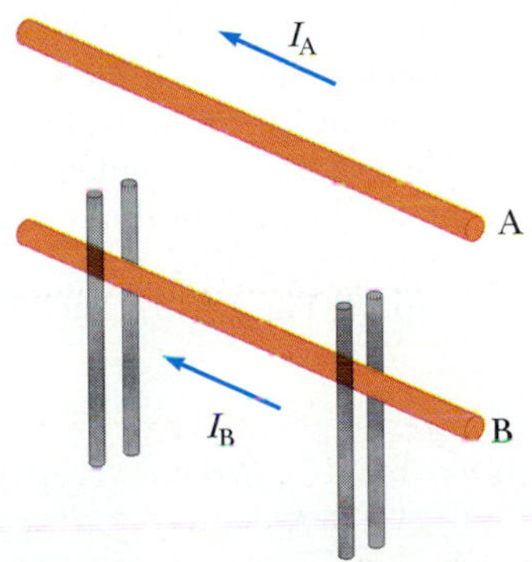

Figure 19.45 (Problem 54)

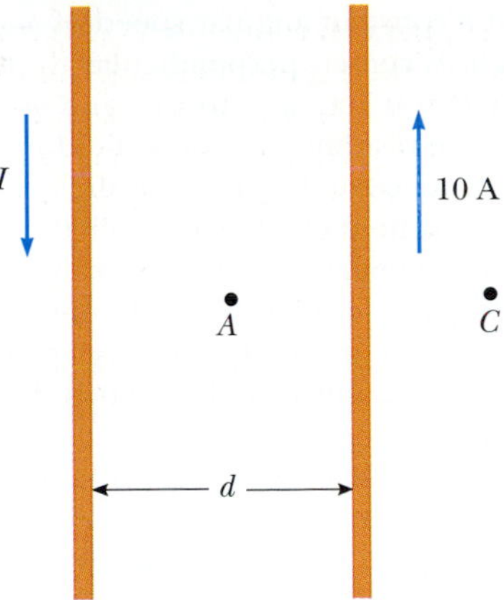

Figure 19.46 (Problem 56)

56. Two parallel conductors carry current in opposite directions, as shown in Figure 19.46. One conductor carries a current of 10 A. Point *A* is at the *midpoint* between the wires and point *C* is a distance of $d/2$ to the right of the 10-A current. If d = 18 cm and I is adjusted so that the magnetic field at *C* is zero, find (a) the value of current I and (b) the value of the magnetic field at *A*.
57. A very long, thin strip of metal of width w carries current I along its length as in Figure 19.47. Find an expression for the magnetic field in the *plane* of the strip (at an external point, *P*) a distance of b from one edge.
58. A nonconducting ring of radius R is uniformly charged with a total positive charge of q. The ring ro-

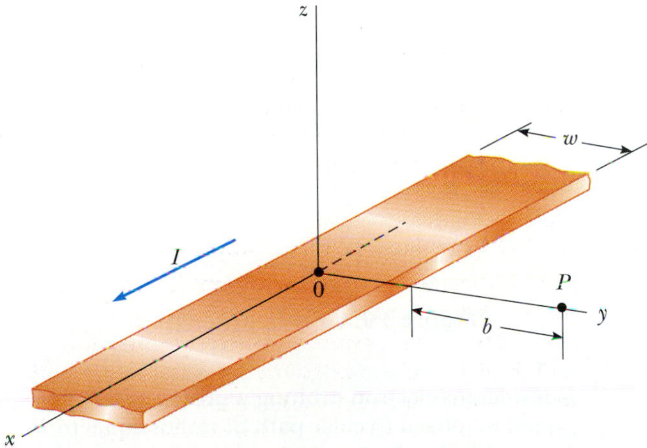

Figure 19.47 (Problem 57)

tates at a constant angular speed of ω about an axis through its center, perpendicular to the plane of the ring. If $R = 0.1$ m, $q = 10\ \mu$C, and $\omega = 20$ rad/s, what is the resulting magnetic field on the axis of the ring a distance of 0.05 m from the center?

59. A magnetic field of 2.0 T is applied to a bubble chamber to make the tracks of protons and other charged particles identifiable by the radii of the circles they move in. If a high-energy proton makes an arc of a 3.3-m radius circle, what is the momentum of the proton?

60. A stream of electrons passes through a velocity filter where the crossed magnetic and electric fields are 0.02 T and 5×10^4 V/m, respectively. Find the kinetic energy (in electron volts) of the electrons that emerge undeviated in passing through the filter.

61. A singly charged heavy ion is observed to complete five revolutions in a uniform magnetic field of magnitude 5×10^{-2} T in 1.50 ms. Calculate the (approximate) mass of the ion in kilograms.

62. Sodium melts at 210°F. Liquid sodium, an excellent thermal conductor, is used in some nuclear reactors to remove thermal energy from the reactor core. The liquid sodium can be moved through pipes by pumps that exploit the force on a moving charge in a magnetic field. The principle is as follows: imagine the liquid metal to be in a pipe having a rectangular cross-section of width w and height h. A uniform magnetic field perpendicular to the pipe affects a section of length L (Fig. 19.48). An electric current directed perpendicular to the pipe and to the magnetic field produces a current density of **J**. (a) Explain why this arrangement produces a force on the liquid that is directed along the length of the pipe. (b) Show that the section of liquid in the magnetic field experiences a pressure increase equal to JLB.

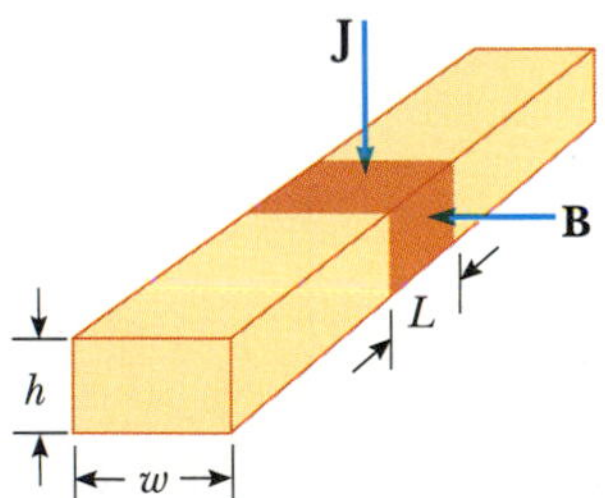

Figure 19.48 (Problem 62)

63. Consider an electron orbiting a proton and maintained in a fixed circular path of radius equal to $R = 5.29 \times 10^{-11}$ m by the Coulomb force of mutual attraction. Treating the orbiting charge as a current loop, calculate the resulting torque when the system is in an external magnetic field of 0.4 T, directed perpendicular to the magnetic moment of the orbiting electron.

64. Two circular coils of radius R are each perpendicular to a common axis. The coil centers are a distance of R apart, and a steady current, I, flows in the same direction around each coil, as shown in Figure 19.49. (a) Show that the magnetic field on the axis at the distance x from the center of one coil is

$$B = \frac{\mu_0 I R^2}{2}\left[\frac{1}{(R^2 + x^2)^{3/2}} + \frac{1}{(2R^2 + x^2 - 2Rx)^{3/2}}\right]$$

(b) Show that $\frac{dB}{dx}$ and $\frac{d^2B}{dx^2}$ are both zero at a point *midway* between the coils. This means the magnetic field in the region midway between the coils is *uniform*. Coils in this configuration are called **Helmholtz coils.**

65. Each of two identical, flat, circular coils of wire has 100 turns and a radius of 0.50 m. If these coils are arranged as a set of Helmholtz coils, and each coil carries a current of 10 A, determine the magnitude of the magnetic field at a point halfway between the coils on their axis. (See Fig. 19.49.)

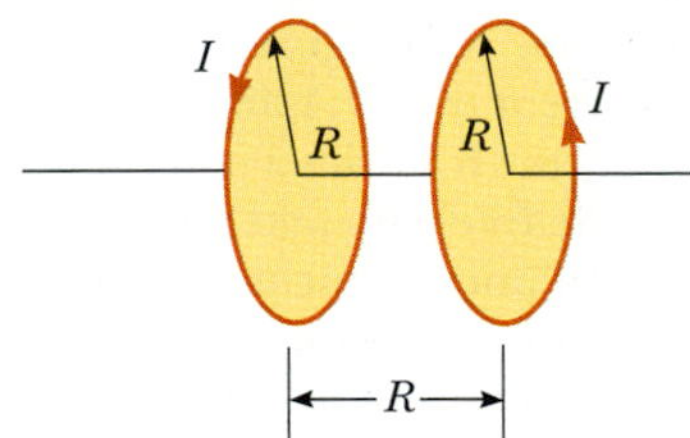

Figure 19.49 (Problems 64 and 65)

66. For a research project, a student needs a solenoid that produces an interior magnetic field of 0.03 T. She decides to use a current of 1.0 A and a wire 0.50 mm in diameter. She winds the solenoid as layers on an insulating form 1.0 cm in diameter and 10.0 cm long. Determine the number of layers of wire needed and the total length of the wire.

67. Two circular loops are parallel, coaxial, and almost in contact, 1 mm apart (Fig. 19.50). Each loop is 10 cm

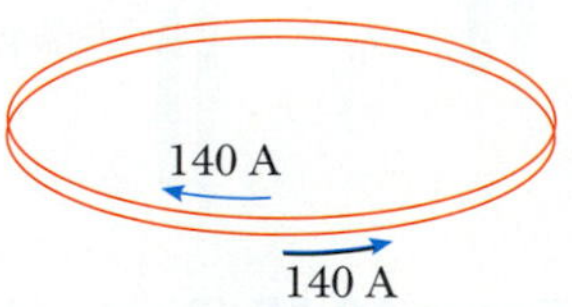

Figure 19.50 (Problem 67)

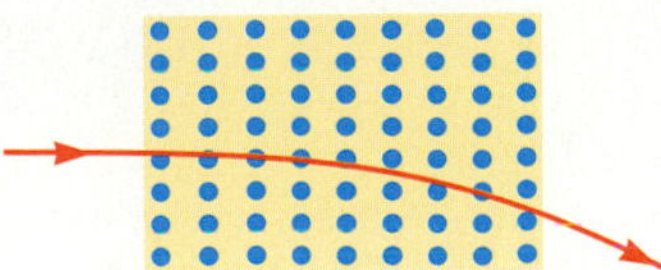

Figure 19.51 (Problem 68)

in radius. The top loop carries a current of 140 A clockwise. The bottom loop carries 140 A counterclockwise. (a) Calculate the magnetic force that the bottom loop exerts on the top loop. (b) The upper loop has a mass of 0.021 kg. Calculate its acceleration, assuming that the only forces acting on it are the force in part (a) and its weight.

68. Protons with kinetic energy of 5 MeV are moving in the positive x direction and enter a magnetic field $\mathbf{B} = (0.05 \text{ T})\mathbf{k}$ directed out of the plane of the page and extending from $x = 0$ to $x = 1$ m as shown in Figure 19.51. (a) Calculate the y component of the protons' momentum as they leave the magnetic field at $x = 1$ m. (b) Find the angle, α, between the initial velocity vector of the proton beam and the velocity vector after the beam emerges from the field. (*Hint:* Neglect relativistic effects.)

69. An infinite sheet of current lying in the yz plane carries a surface current of density $\mathbf{J}_s$. The current is in the y direction, and J_s represents the current per unit length measured along the z axis. Figure 19.52 is an edge view of the sheet. Find the magnetic field near the sheet. (*Hint:* Use Ampère's law, and evaluate the line integral that takes a rectangular path around the sheet, represented by the dashed line in Fig. 19.52.)

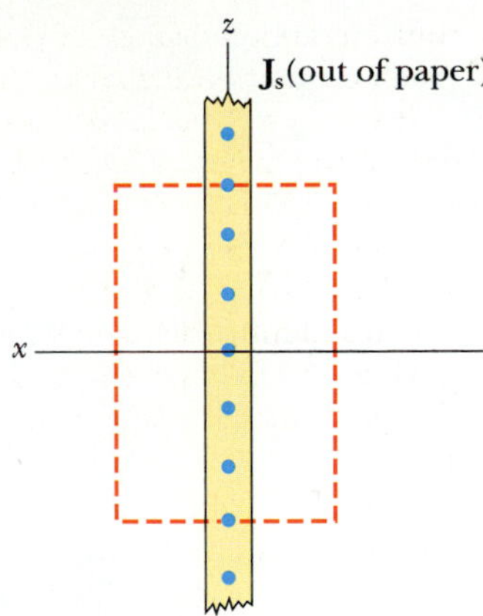

Figure 19.52 (Problem 69) A top view of an infinite current sheet lying in the yz plane, where the current is in the y direction (out of the paper).

Michael Faraday in his laboratory at the Royal Institute, London, in 1870. *(Courtesy of The Bettmann Archive)*

Faraday's Law and Inductance

CHAPTER 20

Our studies so far have been concerned with the electric fields due to stationary charges and the magnetic fields produced by moving charges. This chapter deals with electric fields that originate in changing magnetic fields.

As we learned in Chapter 19, Section 19.1, experiments conducted by Michael Faraday in England in the early 1800s and, independently, by Joseph Henry in the United States showed that an electric current could be induced in a circuit by a changing magnetic field. The results of those experiments led to a very basic and important law of electromagnetism known as *Faraday's law of induction.* It says that the magnitude of the emf induced in a circuit is proportional to the time rate of change of the magnetic field through the circuit.

With the exploration of Faraday's law, we complete our introduction to the

With the exploration of Faraday's law, we complete our introduction to the fundamental laws of electromagnetism. These laws can be summarized in a set of four equations called *Maxwell's equations.* Together with the Lorentz force law, which we shall discuss briefly, they represent a complete theory for describing the interaction of charged objects. Maxwell's equations relate electric and magnetic fields to each other and to their ultimate source: electric charges.

The phenomenon of electromagnetic induction has some practical consequences. In particular, we shall describe an effect known as *self-induction,* in which a time-varying current in a conductor induces an emf that opposes the external emf that set up the current. This phenomenon is the basis of a circuit element known as an *inductor,* which plays an important role in circuits with time-varying currents. Finally, we shall discuss the energy stored in the magnetic field of an inductor and the energy density associated with a magnetic field.

20.1 Faraday's Law of Induction

We begin by describing two simple experiments that demonstrate that an electric current can be produced by a changing magnetic field. First, consider a loop of wire connected to a galvanometer, a device that measures current, as in Figure 20.1. If a magnet is moved toward the loop, the galvanometer needle deflects in one direction, as in Figure 20.1a. If the magnet is moved away from the loop, the galvanometer needle deflects in the opposite direction, as in Figure 20.1b. If the magnet is held stationary relative to the loop, no deflection is observed. Finally, if the magnet is held stationary and the coil is moved either toward or away from it, the needle deflects. From these observations comes the conclusion that *an electric*

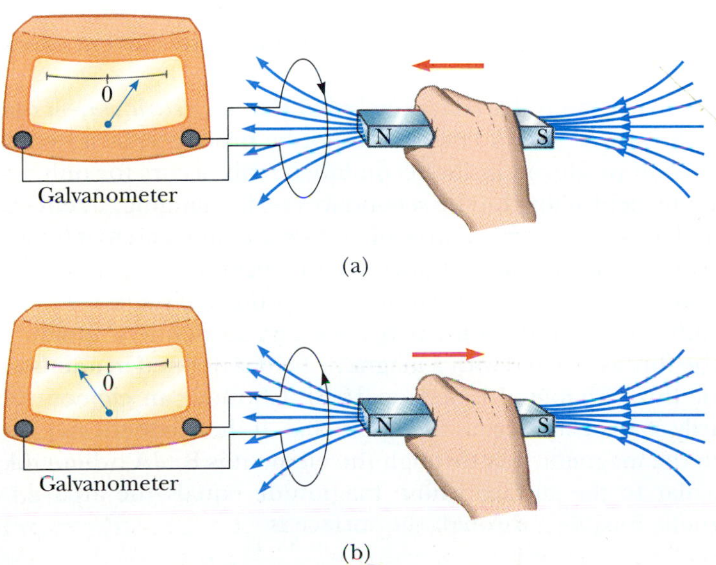

Figure 20.1
(a) When a magnet is moved toward a loop of wire connected to a galvanometer, the galvanometer deflects as shown. This shows that a current is induced in the loop. (b) When the magnet is moved away from the loop, the galvanometer deflects in the opposite direction, indicating that the induced current is opposite that shown in (a).

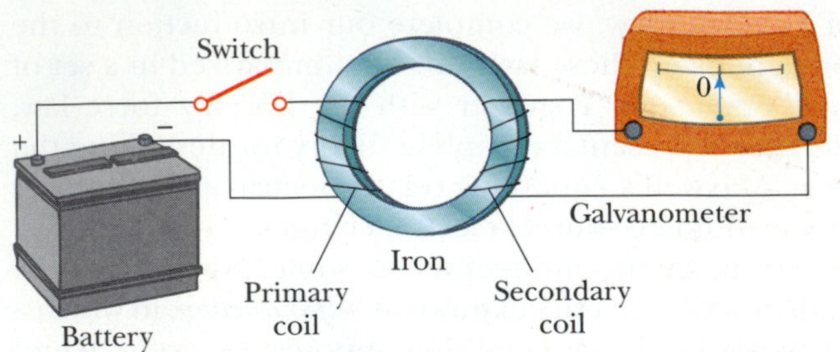

Figure 20.2 Faraday's experiment. When the switch in the primary circuit is closed, the galvanometer in the secondary circuit deflects momentarily. The emf induced in the secondary circuit is caused by the changing magnetic field through the secondary coil in this circuit.

current is set up in the galvanometer circuit as long as there is relative motion between the magnet and the coil.[1]

These results are quite remarkable in view of the fact that *a current is set up in the circuit even though there are no batteries in the circuit!* We call such a current an *induced current,* and it is produced by an *induced emf.*

Now let us describe an experiment, first conducted by Faraday, that is illustrated in Figure 20.2. Part of the apparatus consists of a coil connected to a switch and a battery. We shall refer to this coil as the *primary coil* of wire and to the corresponding circuit as the primary circuit. The coil is wrapped around an iron ring to intensify the magnetic field produced by the current through the coil. A second coil, at the right, is also wrapped around the iron ring and is connected to a galvanometer. We shall refer to this as the *secondary coil* and to the corresponding circuit as the secondary circuit. There is no battery in the secondary circuit, and the secondary coil is not connected to the primary coil. The only purpose of this circuit is to detect any current that might be generated by a change in the magnetic field produced by the primary circuit.

At first sight, you might guess that no current would ever be detected in the secondary circuit. However, something quite amazing happens when the switch in the primary circuit is suddenly closed or opened. At the instant the switch is closed, the galvanometer deflects in one direction and then returns to zero. When the switch is opened, the galvanometer deflects in the opposite direction and again returns to zero. Finally, the galvanometer reads zero when a steady current exists in the primary circuit.

As a result of these observations, Faraday concluded that *an electric current can be produced by a changing magnetic field.* A current cannot be produced by a steady magnetic field. The current produced in the secondary circuit occurs for only an instant while the magnetic field acting on the secondary coil is changing. In effect, the secondary circuit behaves as though a source of emf were connected to it for an instant. It is customary to say that an emf is induced in the secondary circuit by the changing magnetic field produced by the current in the primary circuit.

In order to quantify such observations, it is necessary to define a quantity called *magnetic flux.* The flux associated with a magnetic field is defined in a similar manner to the electric flux (Chapter 16, Section 16.8). Consider an element of area dA on an arbitrarily shaped surface, as in Figure 20.3. If the magnetic field at this element is $\mathbf{B}$, then the magnetic flux through the element is $\mathbf{B} \cdot d\mathbf{A}$, where $d\mathbf{A}$ is a vector perpendicular to the surface whose magnitude equals the area dA. Hence, the total magnetic flux, Φ_m, through the surface is

Michael Faraday (1791–1867) Michael Faraday was a British physicist and chemist who is often regarded as the greatest experimental scientist of the 1800s. His many contributions to the study of electricity include the invention of the electric motor, electric generator, and transformer, as well as the discovery of electromagnetic induction and the laws of electrolysis. *(By kind permission of the President and Council of the Royal Society)*

[1] The magnitude of the current depends on the particular resistance of the circuit, but the existence of the current (or its algebraic sign) does *not.*

$$\Phi_m = \int \mathbf{B} \cdot d\mathbf{A} \qquad [20.1]$$

The SI unit of magnetic flux is the tesla-meter squared, which is named the *weber* (Wb): $1\ \text{Wb} = 1\ \text{T}\cdot\text{m}^2$.

The two experiments illustrated in Figures 20.1 and 20.2 have one thing in common. In both cases, an emf is induced in a circuit when the *magnetic flux* through the circuit *changes with time.* In fact, a general statement summarizes such experiments involving induced emfs:

> The emf induced in a circuit is directly proportional to the time rate of change of magnetic flux through the circuit.

This statement, known as **Faraday's law of induction,** can be written

Faraday's law

$$\mathcal{E} = -\frac{d\Phi_m}{dt} \qquad [20.2]$$

where Φ_m is the magnetic flux threading the circuit, given by Equation 20.1. The negative sign in Equation 20.2 is a consequence of Lenz's law and will be discussed in Section 20.3. If the circuit is a coil consisting of N loops all of the same area and if the flux threads all loops, the induced emf is

$$\mathcal{E} = -N\frac{d\Phi_m}{dt} \qquad [20.3]$$

Suppose the magnetic field is uniform over a loop of area A lying in a plane as in Figure 20.4. In this case, the flux through the loop is equal to $BA\cos\theta$; hence, the induced emf is

$$\mathcal{E} = -\frac{d}{dt}(BA\cos\theta) \qquad [20.4]$$

From this expression, we see that an emf can be induced in a circuit in several ways: (1) the magnitude of **B** can vary with time; (2) the area of the circuit can change with time; (3) the angle θ between **B** and the normal to the plane can change with time; and (4) any combination of these can occur.

The following examples illustrate cases where an emf is induced in a circuit as a result of a time variation of the magnetic field.

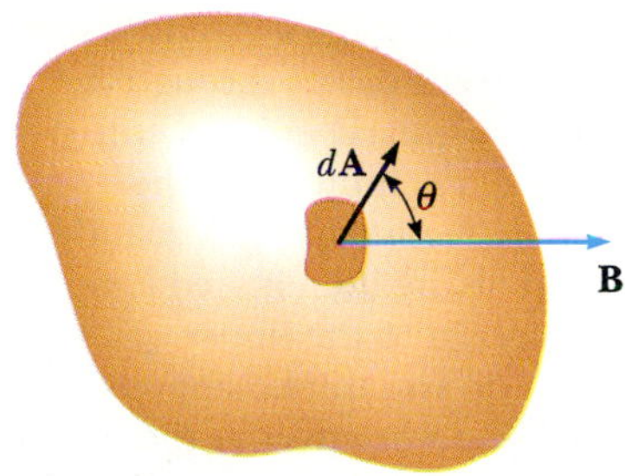

Figure 20.3 The magnetic flux through an area element $d\mathbf{A}$ is given by $\mathbf{B}\cdot d\mathbf{A} = BdA\cos\theta$. Note that $d\mathbf{A}$ is perpendicular to the surface.

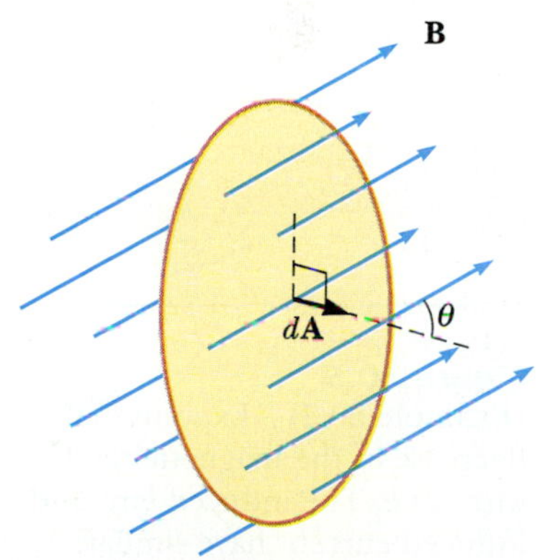

Figure 20.4 A conducting loop of area A in the presence of a uniform magnetic field **B**, which is at an angle θ with the normal to the loop.

Example 20.1 Application of Faraday's Law

A coil is 200 turns of wire around the perimeter of a square frame with sides of length 18 cm. Each turn has the same area, equal to that of the frame, and the total resistance of the coil is 2 Ω. A uniform magnetic field is applied perpendicular to the plane of the coil. If the field changes uniformly from 0 to 0.5 T in 0.8 s, find the magnitude of the induced emf in the coil while the field is changing.

Solution The area of the coil is $(0.18\ \text{m})^2 = 0.0324\ \text{m}^2$. The magnetic flux Φ_{mi} through the coil at $t = 0$ is zero because $B = 0$. At $t = 0.8$ s, the magnetic flux Φ_{mf} through the coil is

$$\Phi_{mf} = BA = (0.5\ \text{T})(0.0324\ \text{m}^2) = 0.0162\ \text{T}\cdot\text{m}^2$$

Therefore, the *change* in flux through the coil during the 0.8-s interval is

$$\Delta\Phi_m = \Phi_{mf} - \Phi_{mi} = 0.0162\ \text{T}\cdot\text{m}^2$$

Faraday's law of induction enables us to find the magnitude of the induced emf:

$$|\mathcal{E}| = N\frac{\Delta\Phi_m}{\Delta t} = (200\ \text{turns})\left(\frac{0.0162\ \text{T}\cdot\text{m}^2}{0.8\ \text{s}}\right) = 4.05\ \text{V}$$

(Note that $1\ \text{T}\cdot\text{m}^2 = 1\ \text{V}\cdot\text{s}$.)

Exercise Find the magnitude of the induced current in the coil while the field is changing.

Answer 2.03 A.

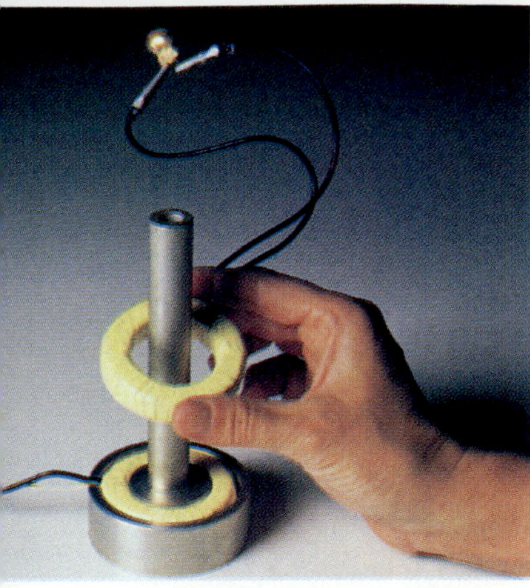

A demonstration of electromagnetic induction. As a time-varying current passes through the lower (primary) coil, an emf (voltage) is induced in the upper (secondary) coil as indicated by the illuminated lamp connected to the upper coil. What do you think happens to the lamp's intensity as the upper coil is moved up or down over the vertical tube? *(Courtesy of CENCO)*

Example 20.2 An Exponentially Decaying *B* Field

A flat loop of wire of area A is placed in a region where the magnetic field is perpendicular to the plane of the loop. The magnitude of **B** varies in time according to the expression $B = B_0e^{-at}$, where a is some constant. That is, at $t = 0$ the field is B_0, and for $t > 0$, the field decreases exponentially in time (Fig. 20.5). Find the induced emf in the loop as a function of time.

Solution Since **B** is perpendicular to the plane of the loop, the magnetic flux through the loop at time $t > 0$ is given by

$$\Phi_m = BA = AB_0e^{-at}$$

Also, since the coefficient AB_0 and the parameter a are constants, the induced emf can be calculated from Equation 20.2:

$$\mathcal{E} = -\frac{d\Phi_m}{dt} = -AB_0\frac{d}{dt}e^{-at} = aAB_0e^{-at}$$

That is, the induced emf decays exponentially in time. Note that the maximum emf occurs at $t = 0$, where $\mathcal{E}_{max} = aAB_0$. Why is this true? The plot of $\mathcal{E}$ versus t is similar to Figure 20.5.

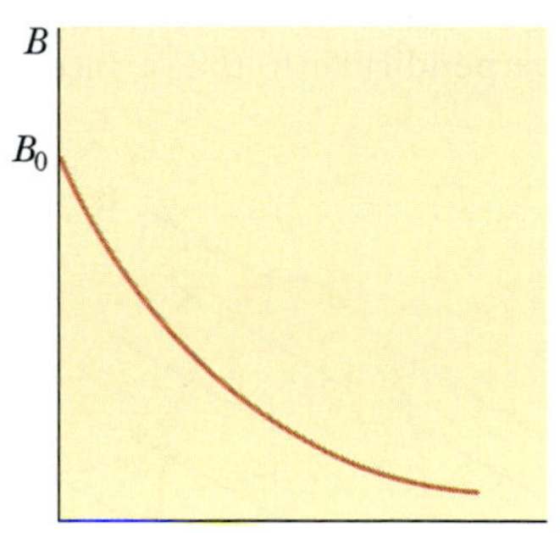

Figure 20.5
(Example 20.2) Exponential decrease of the magnetic field with time. The induced emf and induced current have similar time variations.

20.2 Motional emf

In Examples 20.1 and 20.2, we considered cases in which an emf is produced in a circuit when the magnetic field changes with time. In this section we describe motional emf, which is the emf induced in a conductor moving through a magnetic field.

First, consider a straight conductor of length ℓ moving with constant velocity through a uniform magnetic field directed into the page, as in Figure 20.6. For simplicity, we shall assume that the conductor is moving perpendicularly to the field. The electrons in the conductor experience a force along the conductor of $\mathbf{F} = q\mathbf{v} \times \mathbf{B}$. Under the influence of this force, the electrons move to the lower end and accumulate there, leaving a net positive charge at the upper end. As a result of this charge separation, an electric field is produced within the conductor. The charge at the ends builds up until the magnetic force qvB is balanced by the electric force qE. At this point, charge stops flowing, and the condition for equilibrium requires that the electric force must balance the magnetic force:

$$qE = qvB \quad \text{or} \quad E = vB$$

Since the electric field produced in the conductor is constant, it is related to the potential difference across the ends according to the relation $V = E\ell$. Thus,

$$V = E\ell = B\ell v$$

where the upper end is at a higher potential than the lower end. Thus, *a potential difference is maintained as long as the conductor is moving through the magnetic field. If the motion is reversed, the polarity of V is also reversed.*

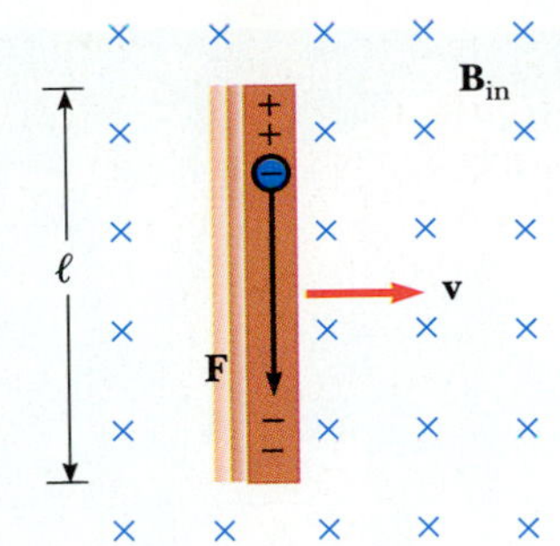

Figure 20.6
A straight conducting bar of length ℓ moving with a velocity **v** through a uniform magnetic field **B** directed perpendicular to **v**. An emf equal to $B\ell v$ is induced between the ends of the bar.

A more interesting situation occurs if we now consider what happens when the moving conductor is part of a closed circuit. This circumstance is particularly useful for illustrating how a changing magnetic flux can cause an induced current in a closed circuit. Consider a circuit consisting of a conducting bar of length ℓ sliding along two fixed parallel conducting rails as in Figure 20.7a. For simplicity, we assume that the moving bar has zero electrical resistance and that the stationary part of the circuit has a resistance of R. A uniform and constant magnetic field, **B**, is applied perpendicular to the plane of the circuit.

As the bar is pulled to the right with a velocity of **v**, under the influence of an applied force, $\mathbf{F}_{app}$, free charges in the bar experience a magnetic force along the length of the bar. This magnetic force sets up an induced current, since the charges on which the force is acting are free to move in the circuit. In this case, the rate of change of magnetic flux through the loop and the accompanying induced emf across the moving bar are proportional to the change in loop area as the bar moves through the magnetic field. As we shall see, if the bar is pulled to the right with a constant velocity, the work done by the applied force is dissipated in the form of joule heating in the circuit's resistive element.

Since the area of the circuit at any instant is ℓx, the external magnetic flux through the circuit is

$$\Phi_m = B\ell x$$

where x is the width of the circuit, which changes with time. Using Faraday's law, we find that the induced emf is

$$\mathcal{E} = -\frac{d\Phi_m}{dt} = -\frac{d}{dt}(B\ell x) = -B\ell\frac{dx}{dt}$$

$$\mathcal{E} = -B\ell v \qquad [20.5]$$

Since the resistance of the circuit is R, the magnitude of the induced current is

$$I = \frac{|\mathcal{E}|}{R} = \frac{B\ell v}{R} \qquad [20.6]$$

The equivalent circuit diagram for this example is shown in Figure 20.7b.

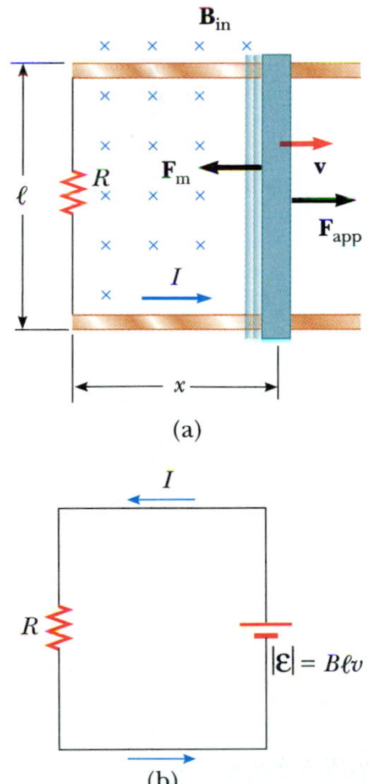

Figure 20.7
(a) A conducting bar sliding with a velocity **v** along two conducting rails under the action of an applied force $\mathbf{F}_{app}$. The magnetic force $\mathbf{F}_m$ opposes the motion, and a counterclockwise current is induced in the loop. (b) The equivalent circuit of (a).

Let us examine the system using energy considerations. Because there is no battery in the circuit, one might wonder about the origin of the induced current and the electrical energy in the system. We can understand this by noting that the external force does work on the conductor, thereby moving charges through a magnetic field. This causes the charges to move along the conductor with some average drift velocity, and hence a current is established. From the viewpoint of energy conservation, the total work done by the applied force during some time

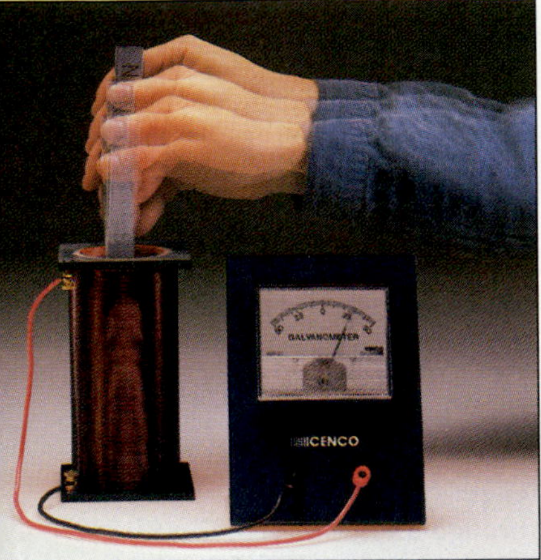

When a strong magnet is moved toward or away from the coil attached to a galvanometer, an electric current is induced, indicated by the momentary deflection of the galvanometer during the movement of the magnet. *(Richard Megna, Fundamental Photographs)*

interval should equal the electrical energy that the induced emf supplies in that same period. Furthermore, if the bar moves with constant speed, the work done must equal the energy dissipated as heat in the resistor during this time interval.

As the conductor of length ℓ moves through the uniform magnetic field **B**, it experiences a magnetic force, $\mathbf{F}_m$, of magnitude $I\ell B$ (Eq. 19.3 in Chapter 19). The direction of this force is opposite the motion of the bar, or to the left in Figure 20.7a.

If the bar is to move with a *constant* velocity, the applied force must be equal to and opposite the magnetic force, or to the right in Figure 20.7a. (If the magnetic force acted in the direction of motion, it would cause the bar to accelerate once it was in motion, thereby increasing its velocity. This state of affairs would represent a violation of the principle of energy conservation.) Using Equation 20.6 and the fact that $F_{app} = I\ell B$, we find that the power delivered by the applied force is

$$P = F_{app} v = (I\ell B) v = \frac{B^2\ell^2v^2}{R} \qquad [20.7]$$

This power is equal to the rate at which energy is dissipated in the resistor, I^2R, as we would expect. It is also equal to the power $I\mathcal{E}$ supplied by the induced emf. This example is a clear demonstration of the conversion of mechanical energy to electrical energy and finally to thermal energy (joule heating).

Example 20.3 Emf Induced in a Rotating Bar

A conducting bar of length ℓ rotates with a constant angular velocity, ω, about a pivot at one end. A uniform magnetic field, **B**, is directed perpendicular to the plane of rotation, as in Figure 20.8. Find the magnitude of the emf induced between the ends of the bar.

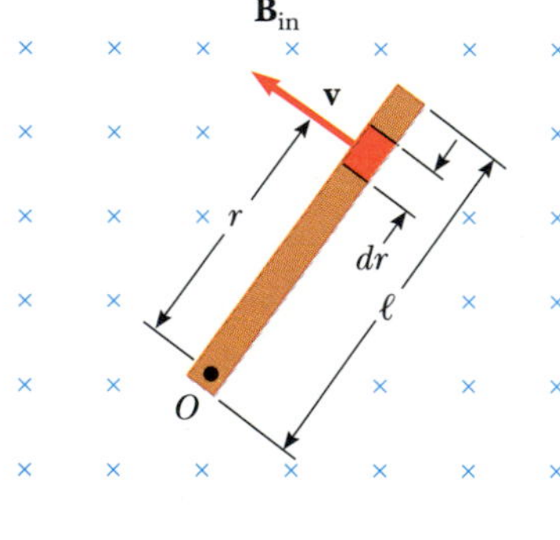

Figure 20.8 (Example 20.3) A conducting bar rotating about a pivot at one end in a uniform magnetic field that is perpendicular to the plane of rotation. An emf is induced across the ends of the bar.

Solution Consider a segment of the bar of length dr, whose velocity is **v**. According to Equation 20.5, the magnitude of the emf induced in a conductor of this length moving perpendicularly to a field, **B**, is

$$(1) \qquad |d\mathcal{E}| = Bv\,dr$$

Each segment of the bar is moving perpendicularly to **B**, and so an emf is generated across each segment; its value is given by (1). Summing up the emfs induced across all elements, which are in series, gives the total emf between the ends of the bar:

$$|\mathcal{E}| = \int Bv\,dr$$

In order to integrate this expression, note that the linear speed of an element is related to the angular speed, ω, through the relationship $v = r\omega$. Therefore, since B and ω are constants, we find that

$$|\mathcal{E}| = B\int v\,dr = B\omega\int_0^\ell r\,dr = \tfrac{1}{2}B\omega\ell^2$$

Example 20.4 Magnetic Force on a Sliding Bar □

A bar of mass m and length ℓ moves on two frictionless parallel rails in the presence of a uniform magnetic field directed into the page (Fig. 20.9). The bar is given an initial velocity, $\mathbf{v}_0$, to the right and released. Find the velocity of the bar as a function of time.

Solution First note that the induced current is counterclockwise and the magnetic force is $\mathbf{F}_m = -I\ell\mathbf{B}$, where the negative sign denotes that the force is to the left and so retards the motion. This is the only horizontal force acting on the bar, and hence Newton's second law applied to motion in the horizontal direction gives

$$F_x = ma = m\frac{dv}{dt} = -I\ell B$$

Since the induced current is given by Equation 20.6, $I = B\ell v/R$, we can write this expression as

$$m\frac{dv}{dt} = -\frac{B^2\ell^2}{R}v$$

$$\frac{dv}{v} = -\left(\frac{B^2\ell^2}{mR}\right)dt$$

Integrating this last equation using the initial condition that $v = v_0$ at $t = 0$, we find that

$$\int_{v_0}^{v}\frac{dv}{v} = \frac{-B^2\ell^2}{mR}\int_0^t dt$$

$$\ln\left(\frac{v}{v_0}\right) = -\left(\frac{B^2\ell^2}{mR}\right)t = -\frac{t}{\tau}$$

where the constant $\tau = mR/B^2\ell^2$. From this, we see that the speed can be expressed in the exponential form

$$v = v_0 e^{-t/\tau}$$

Therefore, the speed of the bar decreases exponentially with time under the action of the magnetic retarding force. Furthermore, if we substitute this result into Equations 20.6 and 20.7, we find that the induced emf and induced current also decrease exponentially with time:

$$\mathcal{E} = IR = B\ell v_0 e^{-t/\tau}$$

$$I = \frac{B\ell v}{R} = \frac{B\ell v_0}{R}e^{-t/\tau}$$

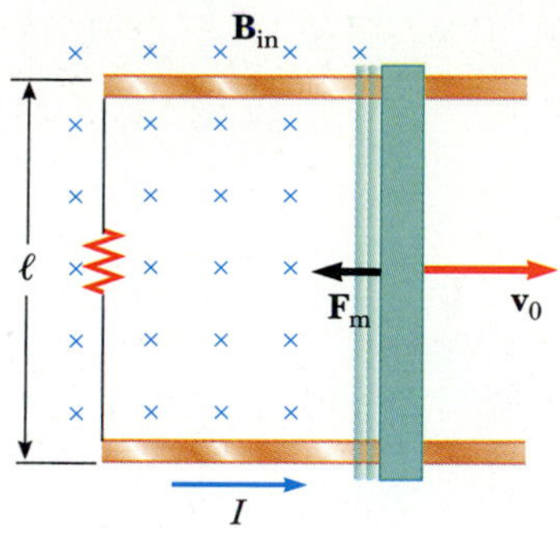

Figure 20.9 (Example 20.4) A conducting bar of length ℓ sliding on two fixed conducting rails is given an initial velocity $\mathbf{v}_0$ to the right.

The Alternating-Current Generator

The alternating current generator (or ac generator), a device that converts mechanical energy to electrical energy, consists of a coil of wire rotated by some external means in an external magnetic field. In commercial power plants, the energy required to rotate the loop can be derived from a variety of sources. For example, in a hydroelectric plant, falling water directed against the blades of a turbine produces the rotary motion; in a coil-fired plant, the heat produced by burning coal is used to convert water to steam and this steam is directed against the turbine blades. As the loop rotates, the magnetic flux through it changes with time, inducing an emf and a current in an external circuit.

Suppose that the coil has N turns, all of the same area A, and suppose that the coil rotates with a constant angular speed ω about an axis perpendicular to the magnetic field. If θ is the angle between the magnetic field and the direction perpendicular to the plane of the coil, then the magnetic flux through the loop at any time t is given by

$$\Phi_m = BA\cos\theta = BA\cos\omega t$$

where we have used the relationship between angular displacement and angular speed, $\theta = \omega t$. (We have set the clock so that $t = 0$ when $\theta = 0$.) Hence, the induced emf in the coil is given by

$$\mathcal{E} = -N\frac{d\Phi_m}{dt} = -NAB\frac{d}{dt}(\cos \omega t) = NAB\omega \sin \omega t \qquad [20.8]$$

This result shows that the emf varies sinusoidally with time. From Equation 20.8 we see that the maximum emf has the value $\mathcal{E}_{max} = NAB\omega$ which occurs when $\omega t = 90°$ or $270°$. In other words, $\mathcal{E} = \mathcal{E}_{max}$ when the magnetic field is in the plane of the coil, and the time rate of change of flux is a maximum. Furthermore, the emf is *zero* when $\omega t = 0$ or $180°$, that is, when **B** is perpendicular to the plane of the coil, and the time rate of change of flux is zero.

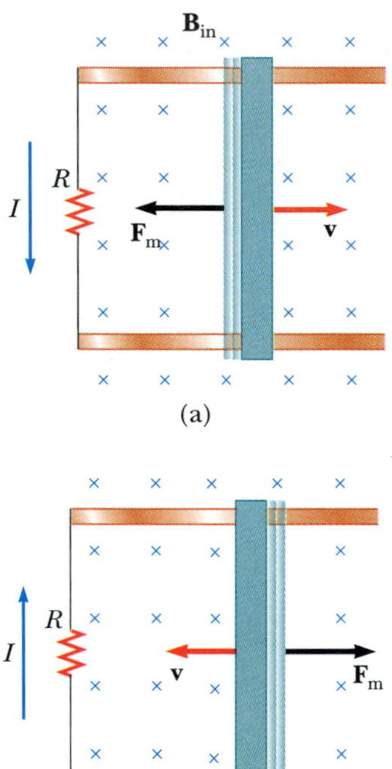

Figure 20.10
(a) As the conducting bar slides on the two fixed conducting rails, the magnetic flux through the loop increases in time. By Lenz's law, the induced current must be *counterclockwise* so as to produce a counteracting flux *out* of the paper. (b) When the bar moves to the left, the induced current must be *clockwise.* Why?

20.3 Lenz's Law

The direction of the induced emf and induced current can be found from **Lenz's law:**

> The polarity of the induced emf is such that it produces a current whose magnetic field opposes the change in magnetic flux through the loop. That is, the induced current tends to maintain the original flux through the circuit.

In other words, the induced current tends to keep the original flux through the circuit from changing. Interpretation of this statement depends on the circumstances. As we shall see, this law is a consequence of the law of conservation of energy.

In order to attain a better understanding of Lenz's law, let us return to the example of a bar moving to the right on two parallel rails in the presence of a uniform magnetic field directed into the page (Fig. 20.10a). As the bar moves to the right, the magnetic flux through the circuit increases with time, since the area of the loop increases. Lenz's law says that the induced current must be in such a direction that the flux *it* produces opposes the change in the magnetic flux caused by the external magnetic field. Since the flux due to the external field is increasing into the page, the induced current, if it is to oppose the change, must produce a flux out of the page. Hence, the induced current must be counterclockwise when the bar moves to the right to give a counteracting flux out of the page in the region inside the loop. (Use the right-hand rule to verify this direction.) If the bar is moving to the left, as in Figure 20.10b, the magnetic flux through the loop decreases with time. Since the flux is into the page, the induced current has to be clockwise to produce a flux into the page inside the loop. In either case, the induced current tends to maintain the original flux through the circuit.

Let us examine this situation from the viewpoint of energy considerations. Suppose that the bar is given a slight push to the right. In the preceding analysis, we found that this motion leads to a counterclockwise current in the loop. Let us

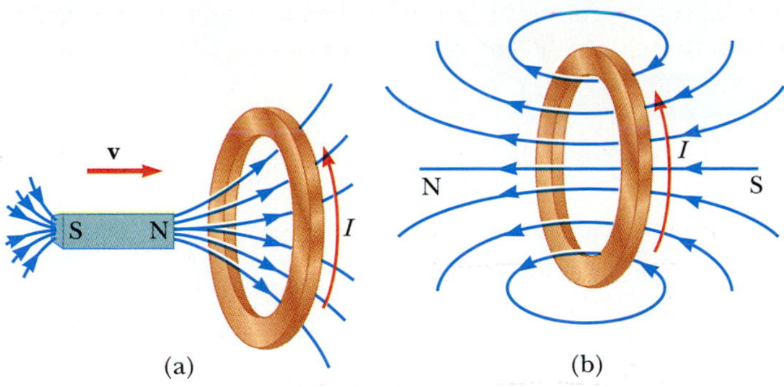

Figure 20.11
(a) When the magnet is moved toward the stationary conducting loop, a current is induced in the direction shown. (b) This induced current produces its own magnetic flux to the left to counteract the increasing external magnetic flux directed to the right.

see what happens if we assume that the current is clockwise. For a clockwise current, I, the direction of the magnetic force, $BI\ell$, on the sliding bar would be to the right. This force would accelerate the rod and increase its speed, which would in turn cause the area of the loop to increase more rapidly. This would increase the induced current, which would increase the force, which would increase the current, which would. . . . In effect, the system would acquire energy with no additional energy input. This result is clearly inconsistent with all experience and with the law of conservation of energy. Thus, we are forced to conclude that the current must be counterclockwise.

Consider another situation, one in which a bar magnet is moved to the right toward a stationary loop of wire, as in Figure 20.11a. As the magnet moves toward the loop, the magnetic flux through the loop increases with time. To counteract this increase in flux to the right, the induced current produces a flux to the left, as in Figure 20.11b; hence, the induced current is in the direction shown. Note that the magnetic field lines associated with the induced current oppose the motion of the magnet. Therefore, the left face of the current loop is a north pole and the right face is a south pole.

If the magnet moved to the left, its flux through the loop, which is toward the right, would decrease in time. Under these circumstances, the induced current in the loop would be in a direction to set up a field through the loop from left to right, in an effort to maintain a constant number of flux lines. Hence, the direction of the induced current in the loop would be opposite that shown in Figure 20.11b. In this case, the left face of the loop would be a south pole and the right face would be a north pole.

Example 20.5 Application of Lenz's Law

A metal ring is placed near an electromagnet, as shown in Figure 20.12a. Find the direction of the induced current in the ring (a) at the instant the switch is closed, (b) after the switch has been closed for several seconds, and (c) when the switch is opened again.

Solution (a) At the moment the switch is closed, the situation changes from a condition in which no lines of magnetic flux pass through the ring to one in which lines of flux pass through in the direction shown in Figure 20.12b. To counteract this change in the number of lines, the ring must set up a field from left to right in the figure. This requires a clockwise current, as shown in Figure 20.12b.

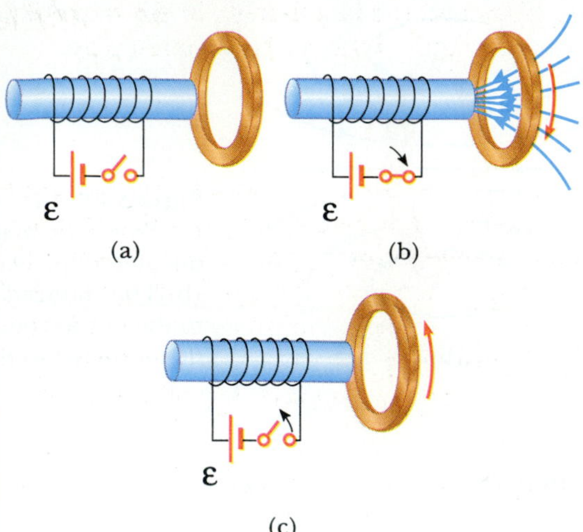

Figure 20.12 (Example 20.5)

(b) After the switch has been closed for several seconds, there is no change in the number of lines through the loop; hence, the induced current is zero.

(c) Opening the switch causes the magnetic field to change from a condition in which flux lines thread through the coil from right to left to a condition of zero flux. The induced current must then be as shown in Figure 20.12c, so as to set up its own magnetic field (and hence magnetic flux lines) from right to left.

20.4 Induced emfs and Electric Fields

We have seen that a changing magnetic flux induces an emf and a current in a conducting loop. We therefore must conclude that *an electric field is created in the conductor as a result of changing magnetic flux.* In fact, the law of electromagnetic induction demonstrates that *an electric field is always generated by a changing magnetic flux,* even in free space where no charges are present. However, this induced electric field has properties that are quite different from those of the electrostatic field produced by stationary charges.

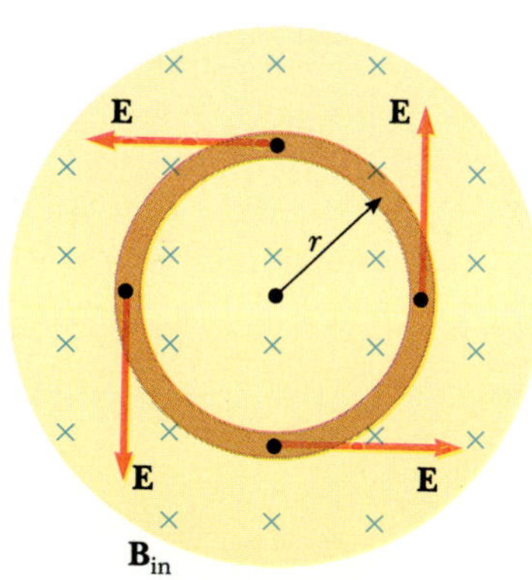

Figure 20.13 A loop of radius r in a uniform magnetic field perpendicular to the plane of the loop. If **B** changes in time, an electric field is induced in a direction tangent to the loop.

We can illustrate this point by considering a conducting loop of radius r, situated in a uniform magnetic field that is perpendicular to the plane of the loop, as in Figure 20.13. If the magnetic field changes with time, then Faraday's law tells us that an emf of $\mathcal{E} = -d\Phi_m/dt$ is induced in the loop. The induced current that is produced implies the presence of an induced electric field, E, which must be tangent to the loop since all points on the loop are equivalent. The work done in moving a test charge, q, once around the loop is equal to $q\mathcal{E}$. Since the electric force on the charge is qE, the work done by that force in moving the charge once around the loop is $qE(2\pi r)$, where $2\pi r$ is the circumference of the loop. These two expressions for the work must be equal; therefore, we see that

$$q\mathcal{E} = qE(2\pi r)$$

$$E = \frac{\mathcal{E}}{2\pi r}$$

Using this result along with Faraday's law and the fact that $\Phi_m = BA = \pi r^2 B$ for a circular loop, we find that the induced electric field can be expressed as

$$E = -\frac{1}{2\pi r}\frac{d\Phi_m}{dt} = -\frac{r}{2}\frac{dB}{dt}$$

This expression can be used to calculate the induced electric field if the time variation of the magnetic field is specified. The negative sign indicates that the induced electric field, **E**, opposes the change in the magnetic field. It is important to understand that *this result is also valid in the absence of a conductor.* That is, a free charge placed in a changing magnetic field experiences the same electric field.

In general, the emf for any closed path can be expressed as the line integral of $\mathbf{E} \cdot d\mathbf{s}$ over that path. Hence, the general form of Faraday's law of induction is

$$\mathcal{E} = \oint \mathbf{E} \cdot d\mathbf{s} = -\frac{d\Phi_m}{dt} \qquad [20.9]$$

Faraday's law in general form

It is important to recognize that *the induced electric field,* **E**, *that appears in Equation 20.9 is a nonconservative, time-varying field that is generated by a changing magnetic field.* The field **E** that satisfies Equation 20.9 could not possibly be an electrostatic field, for the following reason. If the field were electrostatic, and hence conservative, the line integral of $\mathbf{E} \cdot d\mathbf{s}$ over a closed loop would be zero, contrary to Equation 20.9.

20.5 Maxwell's Wonderful Equations

In this section we present four equations that can be regarded as the bases of all electrical and magnetic phenomena. These relationships, known as Maxwell's equations after James Clerk Maxwell, are as fundamental to electromagnetic phenomena as Newton's laws are to mechanical phenomena. In fact, the theory developed by Maxwell was more far-reaching than even he imagined, because it was shown by Einstein in 1905 to be in agreement with the special theory of relativity. As we shall see, Maxwell's equations represent laws of electricity and magnetism that have already been discussed. However, the equations have additional important consequences, in that they predict the existence of electromagnetic waves (traveling patterns of electric and magnetic fields), which travel with a speed of $c = 1/\sqrt{\mu_0 \epsilon_0} \approx 3 \times 10^8$ m/s, the speed of light. Furthermore, Maxwell's equations show that such waves are radiated by accelerating charges.

For simplicity, we present **Maxwell's equations** as applied to free space—that is, in the absence of any dielectric or magnetic material. The four equations are:

$$\oint \mathbf{E} \cdot d\mathbf{A} = \frac{Q}{\epsilon_0} \qquad [20.10]$$

Gauss's law

$$\oint \mathbf{B} \cdot d\mathbf{A} = 0 \qquad [20.11]$$

Gauss's law in magnetism

$$\oint \mathbf{E} \cdot d\mathbf{s} = -\frac{d\Phi_m}{dt} \qquad [20.12]$$

Faraday's law

$$\oint \mathbf{B} \cdot d\mathbf{s} = \mu_0 I + \epsilon_0 \mu_0 \frac{d\Phi_e}{dt} \qquad [20.13]$$

Ampère-Maxwell law

Equation 20.10 is *Gauss's law,* which states that *the total electric flux through any closed surface equals the net charge inside that surface divided by* ϵ_0. This law relates the electric field to the charge distribution, where electric field lines originate on positive charges and terminate on negative charges.

Equation 20.11, which can be considered *Gauss's law in magnetism,* says that *the net magnetic flux through a closed surface is zero.* That is, the number of magnetic field lines that enter a closed volume must equal the number that leave that volume. This implies that magnetic field lines cannot begin or end at any point. If they did, this would mean that isolated magnetic monopoles existed at those points. The fact that isolated magnetic monopoles have not been observed in nature can be taken as a basis of Equation 20.11.

Equation 20.12 is *Faraday's law of induction* (Eq. 20.2), which describes the relationship between an electric field and a changing magnetic flux. This law states that *the line integral of the electric field around any closed path (which equals the emf) equals the rate of change of magnetic flux through any surface area bounded by that path.* One consequence of Faraday's law is the current induced in a conducting loop placed in a time-varying magnetic field.

Equation 20.13 is the generalized form of Ampère's law (Eq. 19.23) and describes a relationship between magnetic and electric fields and electric currents. That is, *the line integral of the magnetic field around any closed path is determined by the sum of the net current through that path and the rate of change of electric flux through any surface bounded by that path.*

Once the electric and magnetic fields are known at some point in space, the force those fields exert on a particle of charge q can be calculated from the expression

The Lorentz force

$$\mathbf{F} = q\mathbf{E} + q\mathbf{v} \times \mathbf{B} \qquad [20.14]$$

This is called the **Lorentz force.** Maxwell's equations, together with this force law, give a complete description of all classical electromagnetic interactions.

It is interesting to note the symmetry of Maxwell's equations. Equations 20.10 and 20.11 are symmetric, apart from the absence of a magnetic monopole term in Equation 20.11. Furthermore, Equations 20.12 and 20.13 are symmetric in that the line integrals of **E** and **B** around a closed path are related to the rate of change of magnetic flux and electric flux, respectively.

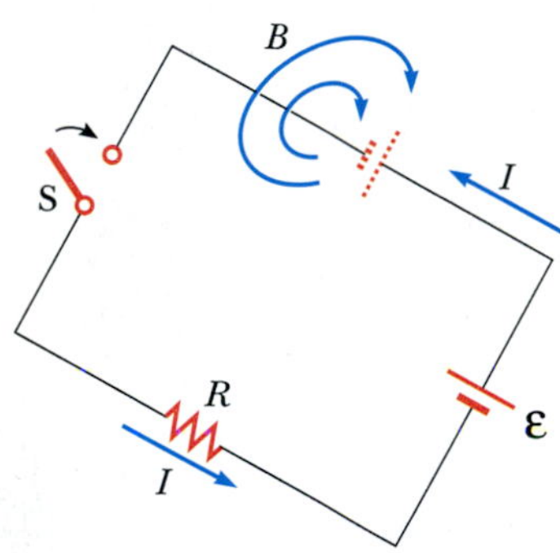

Figure 20.14
After the switch in the circuit is closed, the current produces a magnetic flux through the loop. As the current increases toward its equilibrium value, the flux changes in time and induces an emf in the loop as indicated by the dashed battery symbol.

20.6 Self-Inductance

Consider an isolated circuit consisting of a switch, a resistor, and a source of emf, as in Figure 20.14. When the switch is closed, the current doesn't immediately jump from zero to its maximum value, $\mathcal{E}/R$; the law of electromagnetic induction (Faraday's law) prevents this from occurring. What happens is that, as the current increases with time, the magnetic flux through the loop due to the current also increases with time. This increasing flux induces an emf in the circuit that opposes the change in the net magnetic flux through the loop. By Lenz's law, the induced electric field in the wires must therefore be opposite the direction of the current, and this opposing emf results in a *gradual* increase in the current. This effect is called *self-induction* since the changing flux through the circuit arises from the circuit itself. The emf that is set up in this case is called a **self-induced emf.**

To obtain a quantitative description of self-induction, we recall from Faraday's law that the induced emf is the negative time rate of change of the magnetic flux. The magnetic flux is proportional to the magnetic field, which in turn is proportional to the current in the circuit. Therefore, *the self-induced emf is always proportional to the time rate of change of the current.* For a closely spaced coil of N turns of fixed geometry (a toroidal coil or the ideal solenoid), we find that

$$\mathcal{E} = -N\frac{d\Phi_m}{dt} = -L\frac{dI}{dt} \qquad [20.15]$$

Self-induced emf

where L is a proportionality constant, called the **inductance** of the device, that depends on the geometric features of the circuit and other physical characteristics. From this expression, we see that the inductance of a coil containing N turns is

$$L = \frac{N\Phi_m}{I} \qquad [20.16]$$

Inductance of an *N*-turn coil

where it is assumed that the same flux passes through each turn. Later we shall use this equation to calculate the inductance of some special current geometries.

From Equation 20.15 we can also write the inductance as the ratio

$$L = -\frac{\mathcal{E}}{dI/dt} \qquad [20.17]$$

Inductance

This is usually taken to be the defining equation for the inductance of any coil, regardless of its shape, size, or material characteristics. Just as resistance is a measure of opposition to current, inductance is a measure of opposition to the *change* in current.

The SI unit of inductance is the **henry** (H), which, from Equation 20.17, is seen to be equal to 1 volt-second per ampere:

$$1\ \text{H} = 1\ \frac{\text{V}\cdot\text{s}}{\text{A}}$$

As we shall see, *the inductance of a device depends on its geometry.* Because inductance calculations can be quite difficult for complicated geometries, the following examples involve simple situations for which inductances are easily evaluated.

Joseph Henry (1797–1878)
An American physicist, Henry was the first director of the Smithsonian Institution and first president of the Academy of Natural Science. He improved the design of the electromagnet and constructed one of the first motors. He also discovered the phenomenon of self-induction but failed to publish his findings. The unit of inductance, the henry, is named in his honor.

Example 20.6 Inductance of a Solenoid

Find the inductance of a uniformly wound solenoid with N turns and length ℓ. Assume that ℓ is long compared with the radius and that the core of the solenoid is air.

Solution In this case, we can take the interior magnetic field to be uniform and given by Equation 19.26 in Chapter 19,

$$B = \mu_0 nI = \mu_0 \frac{N}{\ell} I$$

where n is the number of turns per unit length, N/ℓ. The flux through each turn is

$$\Phi_m = BA = \mu_0 \frac{NA}{\ell} I$$

where A is the cross-sectional area of the solenoid. Using this expression and Equation 20.16, we find that

$$L = \frac{N\Phi_m}{I} = \frac{\mu_0 N^2 A}{\ell}$$

This shows that L depends on geometric factors and is proportional to the square of the number of turns. Since $N = n\ell$, we can also express the result in the form

$$L = \mu_0 \frac{(n\ell)^2}{\ell} A = \mu_0 n^2 A\ell$$

where $A\ell$ is the volume of the solenoid.

Example 20.7 Inductance and Self-Induced emf of a Solenoid

Calculate the inductance of a solenoid containing 300 turns if the length of the solenoid is 25 cm and its cross-sectional area is $4 \text{ cm}^2 = 4 \times 10^{-4} \text{ m}^2$.

Solution Using the result of Example 20.6, we get

$$L = \frac{\mu_0 N^2 A}{\ell} = (4\pi \times 10^{-7} \text{ T}\cdot\text{m/A}) \frac{(300)^2(4 \times 10^{-4} \text{ m}^2)}{25 \times 10^{-2} \text{ m}}$$

$$= 1.81 \times 10^{-4} \text{ T}\cdot\text{m}^2/A = 0.181 \text{ mH}$$

Exercise Calculate the self-induced emf in the solenoid if the current through it is decreasing at the rate of 50 A/s.

Answer 9.05 mV.

20.7 *RL* Circuits

A circuit that contains a coil, such as a solenoid, has a self-inductance that prevents the current from increasing or decreasing instantaneously. A circuit element that has a large inductance is called an **inductor.** The circuit symbol for an inductor is ⅇⅇⅇⅇ. We shall always assume that the self-inductance of the remainder of the circuit is negligible compared with that of the inductor.

Consider the circuit, consisting of a resistor, inductor, and battery, shown in Figure 20.15. The internal resistance of the battery will be neglected. Suppose the switch S is closed at $t = 0$. The current will begin to increase, and, due to the increasing current, the inductor will produce an emf (sometimes referred to as a *back emf*) that opposes the increasing current. In other words, the inductor acts like a battery whose polarity is opposite that of the real battery in the circuit. The back emf produced by the inductor is

$$\mathcal{E}_L = -L\frac{dI}{dt}$$

Since the current is increasing, dI/dt is positive; therefore $\mathcal{E}_L$ is negative. This corresponds to the fact that there is a potential drop from a to b across the inductor. For this reason, point a is at a higher potential than point b, as illustrated in Figure 20.15.

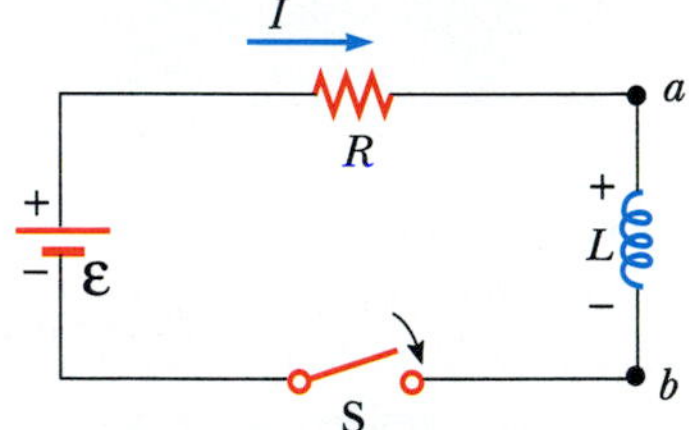

Figure 20.15
A series *RL* circuit. As the current increases toward its maximum value once the switch is closed, the inductor produces an emf that opposes the increasing current.

With this in mind, we can apply Kirchhoff's loop rule to this circuit. If we begin at the battery and travel clockwise, we have

$$\mathcal{E} - IR - L\frac{dI}{dt} = 0 \qquad \textbf{[20.18]}$$

where IR is the voltage drop across the resistor. We must now look for a solution to this differential equation, which is similar to Equation 18.30 (Chapter 18) for the *RC* circuit.

To obtain a mathematical solution of Equation 20.18, it is convenient to change variables by letting $x = (\mathcal{E}/R) - I$, so that $dx = -dI$. With these substitutions, Equation 20.18 can be written

$$x + \frac{L}{R}\frac{dx}{dt} = 0$$

$$\frac{dx}{x} = -\frac{R}{L}\,dt$$

Integrating this last expression gives

$$\ln\frac{x}{x_0} = -\frac{R}{L}t$$

where the integrating constant is taken to be $-\ln x_0$. Taking the antilog of this result gives

$$x = x_0 e^{-Rt/L}$$

where the current is zero at $t = 0$ and $x_0 = \mathcal{E}/R$. Hence, the last expression is equivalent to

$$\frac{\mathcal{E}}{R} - I = \frac{\mathcal{E}}{R}e^{-Rt/L}$$

$$I = \frac{\mathcal{E}}{R}(1 - e^{-Rt/L})$$

which represents the solution of Equation 20.18.

This mathematical solution of Equation 20.18, which represents the current as a function of time, can also be written

$$I(t) = \frac{\mathcal{E}}{R}(1 - e^{-Rt/L}) = \frac{\mathcal{E}}{R}(1 - e^{-t/\tau}) \qquad \textbf{[20.19]}$$

where the constant τ is the **time constant** of the *RL* circuit:

$$\tau = L/R \qquad \textbf{[20.20]}$$

Time constant of the *RL* circuit

It is left to an exercise to show that the dimension of τ is time. Physically, τ is the time it takes the current to reach $(1 - e^{-1}) = 0.63$ of its final value, $\mathcal{E}/R$.

Figure 20.16 plots the current versus time, where $I = 0$ at $t = 0$. Note that the final equilibrium value of the current, which occurs at $t = \infty$, is $\mathcal{E}/R$. This can be seen by setting dI/dt equal to zero in Equation 20.18 (at equilibrium, the change in the current is zero) and solving for the current. Thus, we see that the current rises very fast initially and then gradually approaches the equilibrium value $\mathcal{E}/R$ as $t \to \infty$.

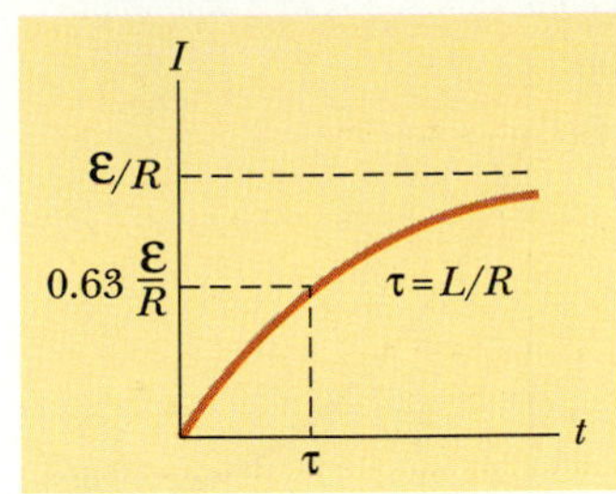

Figure 20.16
Plot of the current versus time for the *RL* circuit shown in Figure 20.15. The switch is closed at $t = 0$, and the current increases toward its maximum value, $\mathcal{E}/R$. The time constant τ is the time it takes I to reach 63% of its maximum value.

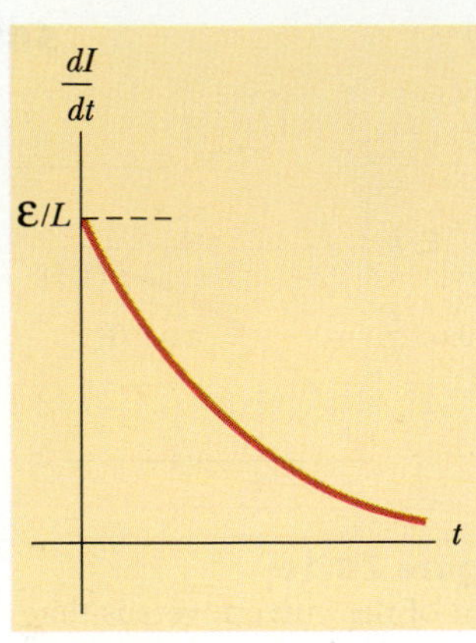

Figure 20.17 Plot of dI/dt versus time for the RL circuit shown in Figure 20.15. The rate of change of current is a maximum at $t = 0$ at the instant the switch is closed and decreases exponentially with time as I increases toward its maximum value.

Taking the first time derivative of Equation 20.19, we get

$$\frac{dI}{dt} = \frac{\mathcal{E}}{L} e^{-t/\tau} \qquad [20.21]$$

From Equation 20.21 we see that the rate of increase of current, dI/dt, is a *maximum* (equal to $\mathcal{E}/L$) at $t = 0$ and falls off exponentially to zero as $t \rightarrow \infty$ (Fig. 20.17).

Now consider the RL circuit arranged as shown in Figure 20.18. The circuit contains two switches that operate so that when one is closed, the other is opened. Now suppose that S_1 is closed for a long enough time to allow the current to reach its equilibrium value, $\mathcal{E}/R$. If S_1 is now opened and S_2 is closed at $t = 0$, we have a circuit with no battery ($\mathcal{E} = 0$). If we apply Kirchhoff's loop rule to this circuit, we obtain

$$IR + L\frac{dI}{dt} = 0 \qquad [20.22]$$

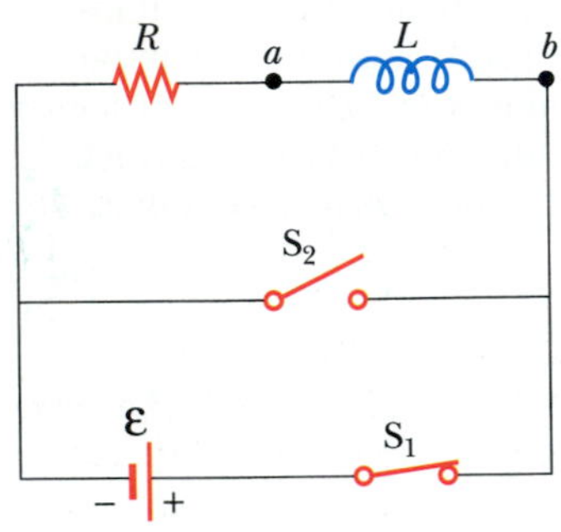

Figure 20.18 An RL circuit containing two switches. When S_1 is closed and S_2 is open as shown, the battery is in the circuit. At the instant S_2 is closed, S_1 is opened and the battery is thereby removed from the circuit.

It is left to Problem 40 to show that the solution of this differential equation is

$$I(t) = \frac{\mathcal{E}}{R} e^{-t/\tau} = I_0 e^{-t/\tau} \qquad [20.23]$$

where the current at $t = 0$ is $I_0 = \mathcal{E}/R$ and $\tau = L/R$.

The graph of current versus time for the circuit of Figure 20.18 (Fig. 20.19) shows that the current is continuously decreasing with time, as one would expect. Furthermore, note that the slope, dI/dt, is always negative and has its maximum value at $t = 0$. The negative slope signifies that $\mathcal{E}_L = -L(dI/dt)$ is now *positive*; that is, point a in Figure 20.18 is at a lower potential than point b.

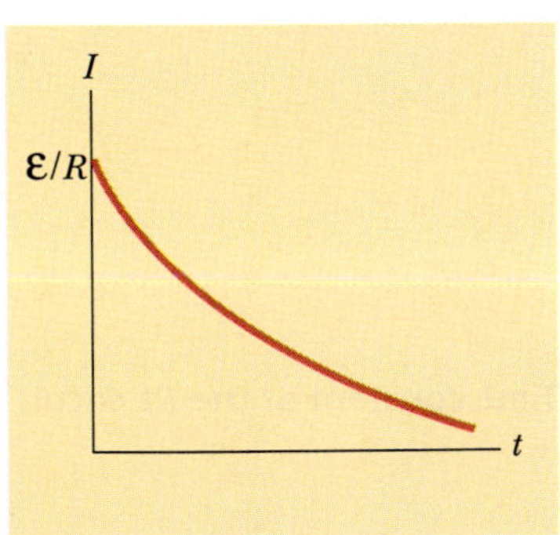

Figure 20.19 Current versus time for the circuit shown in Figure 20.18. At $t < 0$, S_1 is closed and S_2 is open. At $t = 0$, S_2 is closed, S_1 is open, and the current has its maximum value $\mathcal{E}/R$.

Example 20.8 Time Constant of an *RL* Circuit

The switch in the circuit shown in Figure 20.20a is closed at $t = 0$.

(a) Find the time constant of the circuit.

Solution The time constant is

$$\tau = \frac{L}{R} = \frac{30 \times 10^{-3}\ \text{H}}{6\ \Omega} = 5.00\ \text{ms}$$

(b) Calculate the current in the circuit at $t = 2$ ms.

Solution Using Equation 20.19 for the current as a function of time (with t and τ in milliseconds), we find that at $t = 2$ ms,

$$I = \frac{\mathcal{E}}{R}(1 - e^{-t/\tau}) = \frac{12\ \text{V}}{6\ \Omega}(1 - e^{-0.4}) = 0.659\ \text{A}$$

A plot of Equation 20.19 for this circuit is given in Figure 20.20b.

Exercise Calculate the current in the circuit and the voltage across the resistor after one time constant has elapsed.

Answer 1.26 A, 7.56 V.

Exercise Show that the inductive time constant τ has SI units of seconds.

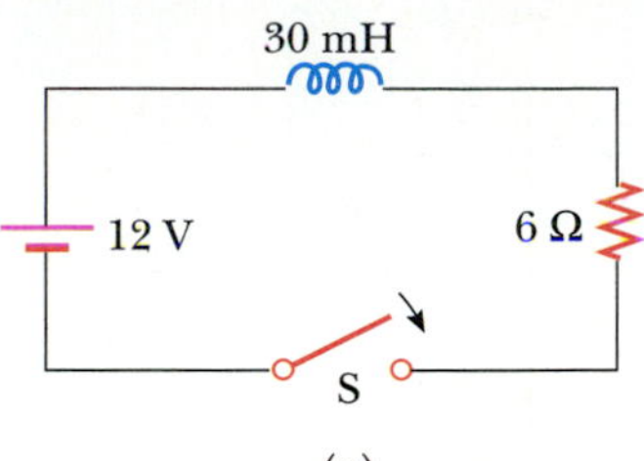

(a)

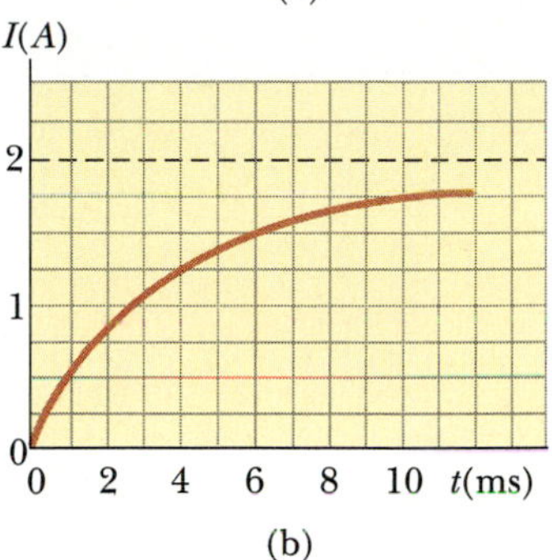

(b)

Figure 20.20 (Example 20.8) (a) The switch in this *RL* circuit is closed at $t = 0$. (b) A graph of the current versus time for the circuit in (a).

20.8 Energy Stored in a Magnetic Field

In the preceding section we found that the induced emf set up by an inductor prevents a battery from establishing an instantaneous current. Hence, a battery has to do work against an inductor to create a current. Part of the energy supplied by the battery goes into joule heat dissipated in the resistor, while the remaining energy is stored in the inductor. If we multiply each term in Equation 20.18 by the current I and rearrange the expression, we get

$$I\mathcal{E} = I^2R + LI\frac{dI}{dt} \qquad [20.24]$$

This expression tells us that the rate at which energy is supplied by the battery, $I\mathcal{E}$, equals the sum of the rate at which joule heat is dissipated in the resistor, I^2R, and the rate at which energy is stored in the inductor, $LI(dI/dt)$. Thus, Equation 20.24 is simply an expression of energy conservation. If we let U_m denote the energy stored in the inductor at any time, then the rate dU_m/dt at which energy is stored in the inductor can be written

$$\frac{dU_m}{dt} = LI\frac{dI}{dt}$$

To find the total energy stored in the inductor, we can rewrite this expression as $dU_m = LI\,dI$ and integrate:

$$U_m = \int_0^{U_m} dU_m = \int_0^I LI\,dI$$

$$U_m = \tfrac{1}{2}LI^2 \qquad [20.25]$$

Energy stored in an inductor

where L is constant and so has been removed from the integral. Equation 20.25 represents the energy stored as magnetic energy in the field of the inductor when the current is I. Note that it is similar to the equation for the energy stored in the electric field of a capacitor, $Ue = Q^2/2C$ (Eq. 17.29 in Chapter 17). In either case, we see that it takes work to establish a field.

We can also determine the energy per unit volume, or energy density, stored in a magnetic field. For simplicity, consider a solenoid whose inductance is $L = \mu_0 n^2 A\ell$ (see Example 20.6). The magnetic field of the solenoid is $B = \mu_0 nI$. Substituting the expression for L and $I = B/\mu_0 n$ into Equation 20.25 gives

$$U_m = \tfrac{1}{2}LI^2 = \tfrac{1}{2}\mu_0 n^2 A\ell\left(\frac{B}{\mu_0 n}\right)^2 = \frac{B^2}{2\mu_0}(A\ell) \qquad [20.26]$$

Because $A\ell$ is the volume of the solenoid, the energy stored per unit volume in a magnetic field—in other words, the *energy density*—is

$$u_m = \frac{U_m}{A\ell} = \frac{B^2}{2\mu_0} \qquad [20.27]$$

Magnetic energy density

Although Equation 20.27 was derived for the special case of a solenoid, *it is valid for any region of space in which a magnetic field exists.* Note that it is similar to the equation for the energy per unit volume stored in an electric field, given by $\frac{1}{2}\epsilon_0 E^2$ (Eq. 17.29). In both cases, the energy density is proportional to the square of the field strength.

Example 20.9 What Happens to the Energy in the Inductor?

Consider once again the RL circuit in Figure 20.18, in which switch S_2 is closed at the instant S_1 is opened (at $t = 0$). Recall that the current in the upper loop decays exponentially with time according to the expression $I = I_0 e^{-t/\tau}$, where $I_0 = \mathcal{E}/R$ is the initial current in the circuit and $\tau = L/R$ is the time constant. The energy stored in the magnetic field of the inductor gradually dissipates as thermal energy in the resistor. Let us show explicitly that all the energy stored in the inductor gets dissipated as heat in the resistor.

Solution The rate at which energy is dissipated in the resistor, dU/dt (in other words, the power), is equal to I^2R, where I is the instantaneous current:

$$\frac{dU}{dt} = I^2R = (I_0 e^{-Rt/L})^2 R = I_0^2 R e^{-2Rt/L} \quad (1)$$

To find the total energy dissipated in the resistor, we integrate this expression over the limits $t = 0$ to $t = \infty$. (The upper limit of ∞ is used because it takes an infinite time for the current to reach zero.) Hence,

$$U = \int_0^\infty I_0^2 R e^{-2Rt/L}\, dt = I_0^2 R \int_0^\infty e^{-2Rt/L}\, dt \quad (2)$$

The value of the definite integral is $L/2R$, and so U becomes

$$U = I_0^2 R\left(\frac{L}{2R}\right) = \tfrac{1}{2} L I_0^2$$

Note that this is equal to the initial energy stored in the magnetic field of the inductor (Eq. 20.25), as we set out to prove.

Exercise Evaluate the integral on the right side of Equation (2) and show that it has the value $L/2R$.

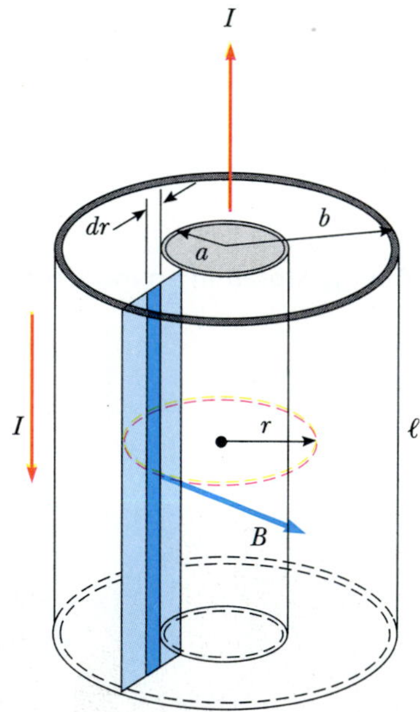

Figure 20.21 (Example 20.10) Section of a long coaxial cable. The inner and outer conductors carry equal and opposite currents.

Example 20.10 The Coaxial Cable

A long coaxial cable consists of two concentric cylindrical conductors of radii a and b and length ℓ, as in Figure 20.21. The inner conductor is a thin cylindrical shell. Each conductor carries a current, I (the outer one being a return path). (a) Calculate the self-inductance, L, of this cable.

Solution To obtain L, we must know the magnetic flux through any cross-section between the two conductors. From Ampère's law (Chapter 19, Section 19.8), it is easy to see that the magnetic field between the conductors is $B = \mu_0 I/2\pi r$. Furthermore, the field is zero outside the conductors and zero inside the inner hollow conductor. The field is zero outside because the net current through a circular path surrounding both wires is zero, and hence, from Ampère's law, $\oint \mathbf{B}\cdot d\mathbf{s} = 0$. The field is zero inside the inner conductor, since it is hollow and there is no current within a radius $r < a$.

The magnetic field is perpendicular to the shaded rectangular strip of length ℓ and width $(b - a)$. This is the cross-section of interest. Dividing this rectangle into strips of width dr, we see that the area of each strip is $\ell\, dr$ and the flux through each strip is $B\, dA = B\ell\, dr$. Hence, the *total* flux through any cross-section is

$$\Phi_m = \int B\, dA = \int_a^b \frac{\mu_0 I}{2\pi r}\ell\, dr = \frac{\mu_0 I\ell}{2\pi}\int_a^b \frac{dr}{r} = \frac{\mu_0 I\ell}{2\pi}\ln\left(\frac{b}{a}\right)$$

Using the result, we find that the self-inductance of the cable is

$$L = \frac{\Phi_m}{I} = \frac{\mu_0 \ell}{2\pi}\ln\left(\frac{b}{a}\right)$$

Furthermore, the self-inductance per unit length is $(\mu_0/2\pi)\ \ln(b/a)$.

(b) Calculate the total energy stored in the magnetic field of the cable.

Solution Using Equation 20.25 and the results to (a), we get

$$U_m = \tfrac{1}{2}LI^2 = \frac{\mu_0 \ell I^2}{4\pi}\ln\left(\frac{b}{a}\right)$$

Summary

Faraday's law of induction states that the emf induced in a circuit is directly proportional to the time rate of change of magnetic flux through the circuit:

$$\mathcal{E} = -N\frac{d\Phi_m}{dt} \qquad [20.3]$$

Faraday's law

where N is the number of turns and Φ_m is the magnetic flux through each, given by

$$\Phi_m = \int \mathbf{B}\cdot d\mathbf{A} \qquad [20.1]$$

Magnetic flux

When a conducting bar of length ℓ moves through a magnetic field, **B**, with a velocity of **v** so that **v** is perpendicular to **B**, the emf induced in the bar (called the **motional emf**) is

$$\mathcal{E} = -B\ell v \qquad [20.5]$$

Motional emf

Lenz's law states that the induced current and induced emf in a conductor are in such a direction as to oppose the change that produced them.

A general form of **Faraday's law of induction** is

$$\mathcal{E} = \oint \mathbf{E}\cdot d\mathbf{s} = -\frac{d\Phi_m}{dt} \qquad [20.9]$$

Faraday's law in general form

where **E** is a nonconservative, time-varying electric field that is produced by the changing magnetic flux.

When used with the Lorentz force law—$\mathbf{F} = q\mathbf{E} + q\mathbf{v}\times\mathbf{B}$—**Maxwell's equations** describe *all* electromagnetic phenomena:

$$\oint \mathbf{E}\cdot d\mathbf{A} = \frac{Q}{\epsilon_0} \qquad [20.10]$$

Gauss's law (electricity)

Gauss's law (magnetism)

$$\oint \mathbf{B} \cdot d\mathbf{A} = 0 \qquad [20.11]$$

Faraday's law

$$\oint \mathbf{E} \cdot d\mathbf{s} = -\frac{d\Phi_m}{dt} \qquad [20.12]$$

Ampère-Maxwell law

$$\oint \mathbf{B} \cdot d\mathbf{s} = \mu_0 I + \epsilon_0 \mu_0 \frac{d\Phi_e}{dt} \qquad [20.13]$$

The last two equations are of particular importance in the context of this chapter. Faraday's law describes how an electric field can be induced by a changing magnetic flux. The Ampère-Maxwell law describes how a magnetic field can be produced by either a conduction current or a changing electric flux.

When the current in a coil changes with time, an emf is induced in the coil according to Faraday's law. The **self-induced emf** is defined by the expression

Self-induced emf

$$\mathcal{E} = -L\frac{dI}{dt} \qquad [20.15]$$

where L is the *inductance* of the coil. Inductance is a measure of the opposition of a device to a change in current. Inductance has the SI unit the **henry** (H), where $1 \text{ H} = 1 \text{ V}\cdot\text{s/A}$.

The **inductance** of a coil is

Inductance of an *N*-turn coil

$$L = \frac{N\Phi_m}{I} \qquad [20.16]$$

where Φ_m is the magnetic flux through the coil and N is the total number of turns.

If a resistor and inductor are connected in series to a battery of emf $\mathcal{E}$, as shown in Figure 20.15, and a switch in the circuit is closed at $t = 0$, the current in the circuit varies in time according to the expression

Current in an *RL* circuit

$$I(t) = \frac{\mathcal{E}}{R}(1 - e^{-t/\tau}) \qquad [20.19]$$

where $\tau = L/R$ is the *time constant* of the RL circuit. That is, the current rises to an equilibrium value of $\mathcal{E}/R$ after a time interval that is long compared with τ.

If the battery is removed from an RL circuit, as in Figure 20.18 with S_1 open and S_2 closed, the current decays exponentially with time according to the expression

$$I(t) = \frac{\mathcal{E}}{R}e^{-t/\tau} \qquad [20.23]$$

where $\mathcal{E}/R$ is the initial current in the circuit.

The **energy** *stored in the magnetic field of an inductor* carrying a current I is

Energy stored in an inductor

$$U_m = \tfrac{1}{2}LI^2 \qquad [20.25]$$

The **energy per unit volume** (or energy density) at a point where the magnetic field is B is

Magnetic energy density

$$u_m = \frac{B^2}{2\mu_0} \qquad [20.27]$$

Questions and Conceptual Exercises

1. A loop of wire is placed in a uniform magnetic field. For what orientation of the loop is the magnetic flux a maximum? For what orientation is the flux zero?
2. A circular loop is located in a uniform and constant magnetic field. Describe how an emf can be induced in the loop in this situation.
3. A spacecraft circling the Earth has a coil of wire in it. An astronaut notes that there is a current in the coil even though no battery is connected to it and there are no magnets on the spacecraft. What is causing the current?
4. As the conducting bar in Figure 20.22 moves to the right, an electric field directed downward is set up in the conductor. Explain why the electric field would be upward if the bar were moving to the left.

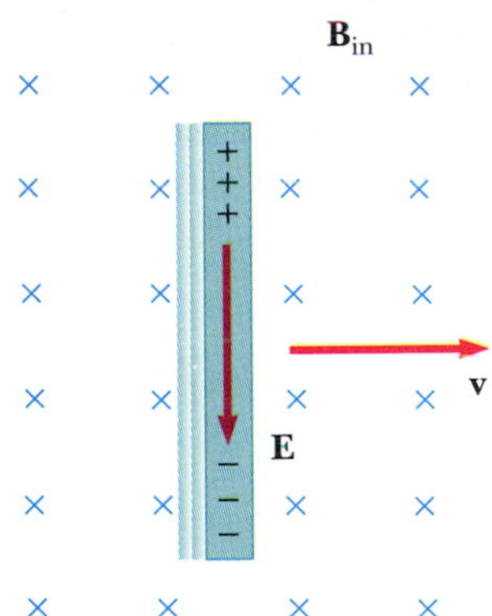

Figure 20.22 (Questions 4 and 5)

5. As the bar in Figure 20.22 moves perpendicular to the magnetic field, is an external force required to keep it moving with constant velocity? Explain.
6. Magnetic storms on the Sun can cause difficulties with communications here on the Earth. Why do these sunspots affect us in this way?
7. Wearing a metal bracelet in a region of strong magnetic field could be hazardous. Discuss.
8. How is electrical energy produced in dams (that is, how is the energy of motion of the water converted to ac electricity)?
9. A piece of aluminum is dropped vertically downward between the poles of an electromagnet. Does the magnetic field affect the velocity of the aluminum?
10. The bar in Figure 20.23 moves on rails to the right with the velocity v, and the uniform, constant magnetic field is *outward*. Why is the induced current clockwise? If the bar were moving to the left, what would be the direction of the induced current? Explain why an external force is necessary to keep the bar moving with a constant velocity.

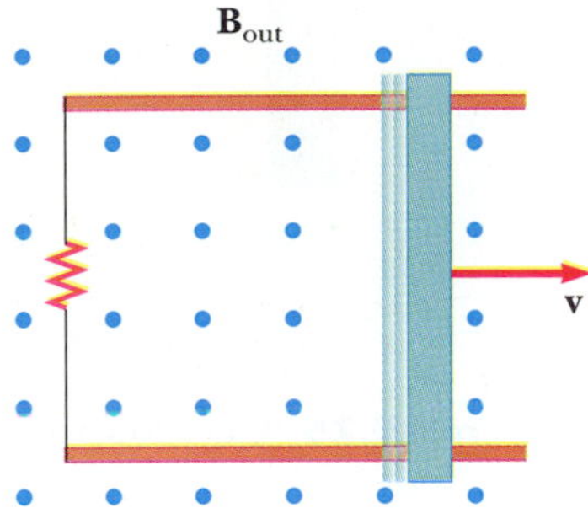

Figure 20.23 (Question 10)

11. In a beam balance scale, an aluminum plate is sometimes used to slow the oscillations of the beam near equilibrium. The plate is mounted at the end of the beam, and moves between the poles of a small horseshoe magnet attached to the frame. Why are the oscillations of the beam strongly damped?
12. When the switch in the circuit shown in Figure 20.24a is closed, a current is set up in the coil and the metal ring springs upward (see Fig. 20.24b). Explain this behavior.
13. Assume that the battery in Figure 20.24a is replaced by an alternating current source and the switch S is held closed. If the metal ring on top of the solenoid is held down, it will become *hot*. Why?
14. Identify the individuals generally associated with each of Maxwell's four equations.
15. Do Maxwell's equations (Section 20.5) allow for the existence of magnetic "charges"—that is, isolated N or S poles?

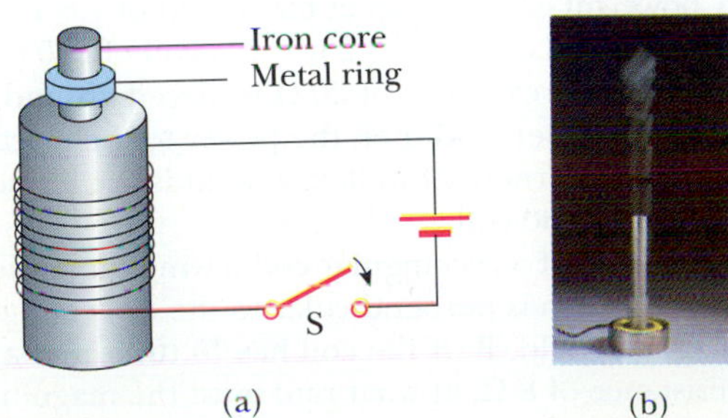

Figure 20.24 (Questions 12 and 13) *(Courtesy of CENCO)*

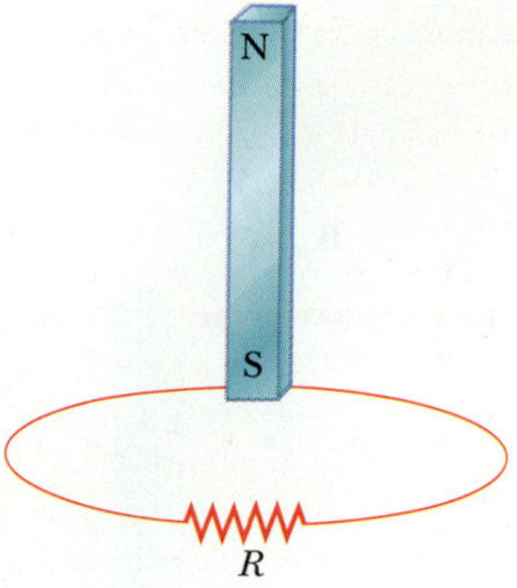

Figure 20.25 (Question 16)

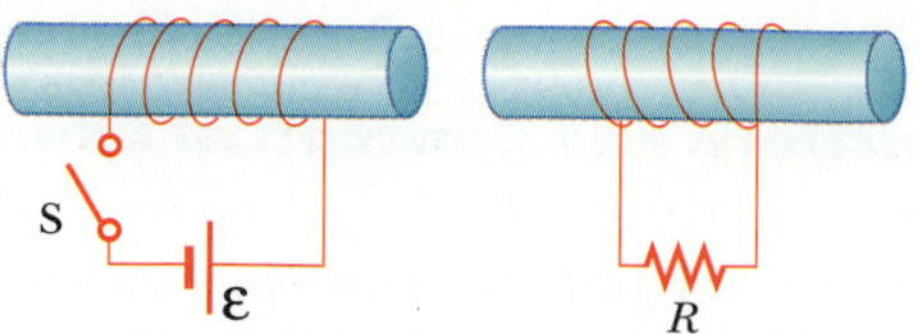

Figure 20.26 (Question 17)

16. A bar magnet is held above a loop of wire in a horizontal plane, as shown in Figure 20.25. The south end of the magnet is toward the loop of wire. The magnet is dropped toward the loop. Find the direction of the current through the resistor (a) while the magnet is falling toward the loop and (b) after the magnet has passed through the loop and moves away from it.
17. Find the direction of the current through the resistor in Figure 20.26 (a) at the instant the switch is closed, (b) after the switch has been closed for several minutes, and (c) at the instant the switch is opened.
18. Discuss the similarities between the energy stored in the electric field of a charged capacitor and the energy stored in the magnetic field of a current-carrying coil.
19. What is the effective inductance of two isolated inductors, connected in series?
20. If the current in an inductor is doubled, by what factor does the stored energy change?
21. Suppose the switch in the *RL* circuit in Figure 20.15 has been closed for a long time and is suddenly opened. Does the current instantaneously drop to zero? Why does a spark tend to appear at the switch contacts when the switch is opened?
22. A "Slinky toy" spring has a radius of 4 cm and an inductance of 125 μH when extended to a length of 2 m. What is the total number of turns in the spring? Assume the coils are closely spaced.

Problems

Section 20.1 Faraday's Law of Induction

1. A plane loop of wire consisting of a single turn of cross-sectional area 8.0 cm^2 is perpendicular to a magnetic field that increases uniformly in magnitude from 0.5 T to 2.5 T in a time of 1.0 s. What is the resulting induced current if the coil has a total resistance of 2 Ω?
2. A powerful electromagnet has a field of 1.6 T and a cross-sectional area of 0.2 m^2. If a coil of 200 turns with a total resistance of 20 Ω is placed around the electromagnet, and then the power to the electromagnet is turned off in 0.02 s, what is the induced current in the coil?
3. The plane of a rectangular coil having dimensions of 5 cm × 8 cm is perpendicular to the direction of a magnetic field, **B**. If the coil has 75 turns and a total resistance of 8 Ω, at what rate must the magnitude of **B** change in order to induce a current of 0.1 A in the windings of the coil?
4. A plane loop of wire of area 14 cm^2 with two turns is perpendicular to a magnetic field whose magnitude decays in time according to $B = (0.5\text{ T})\,e^{-t/7}$. What is the induced emf as a function of time?
5. A rectangular loop of area A is placed in a region where the magnetic field is perpendicular to the plane of the loop. The magnitude of the field is allowed to vary in time according to $B = B_0 e^{-t/\tau}$, where B_0 and τ are constants. The field has a value of B_0 at $t \leq 0$. (a) Use Faraday's law to show that the emf induced in the loop is given by

$$\mathcal{E} = \frac{AB_0}{\tau}\, e^{-t/\tau}$$

(b) Obtain a numerical value for $\mathcal{E}$ at $t = 4$ s when $A = 0.16$ m^2, $B_0 = 0.35$ T, and $\tau = 2$ s. (c) For the values of A, B_0, and τ given in (b), what is the *maximum* value of $\mathcal{E}$?
6. A magnetic field of 0.2 T exists within a solenoid of

500 turns and a diameter of 10 cm. How rapidly (that is, within what period of time) must the field be reduced to zero magnitude if the average magnitude of the induced emf within the coil during this time interval is to be 10 kV?

7. A long, straight wire carries a current that varies with time as $I = I_0 \sin(\omega t + \delta)$ and lies in the plane of a rectangular loop of N turns of wire, as shown in Figure 20.27. The quantities I_0, ω, and δ are all constants. Determine the emf induced in the loop by the magnetic field due to the current in the straight wire. Assume $I_0 = 50$ A, $\omega = 200\pi$ s^{-1}, $N = 100$, $a = b = 5$ cm, and $\ell = 20$ cm.

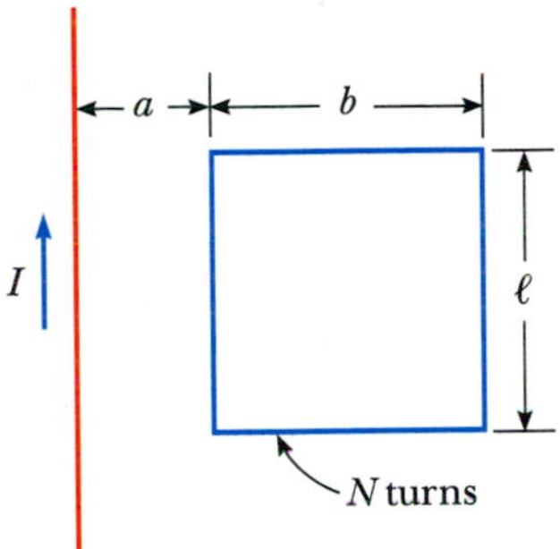

Figure 20.27 (Problem 7)

8. A circular loop of radius R consists of N tight turns of wire and is penetrated by an external magnetic field directed perpendicular to the plane of the loop. The magnitude of the field in the plane of the loop is $B = B_0(1 - r/2R) \cos \omega t$, where R is the radius of the loop and r is measured from the center of the loop, as shown in Figure 20.28. Determine the induced emf in the loop.

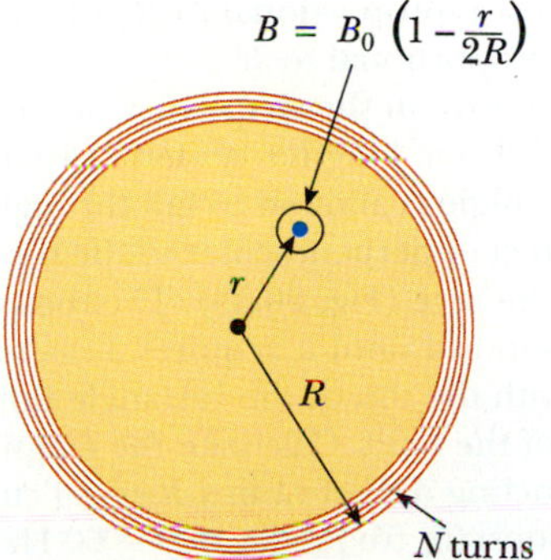

Figure 20.28 (Problem 8)

9. A 500-turn circular-loop coil 15.0 cm in diameter is initially aligned so that its axis is parallel to the Earth's magnetic field. In 2.77 ms the coil is flipped so that its axis is perpendicular to the Earth's magnetic field. If a voltage of 0.166 V is induced in the coil, what is the value of the Earth's magnetic field?

10. A square coil of wire having a single turn 1 cm on a side is placed inside a solenoid that has a circular cross-section of radius 3 cm, as shown in Figure 20.29. The solenoid is 20 cm long and wound with 100 turns of wire. (a) Find the flux through the coil when the current in the solenoid is 3 A. (b) If the current in the solenoid is reduced to zero in 3 s, find the magnitude of the average induced emf in the coil.

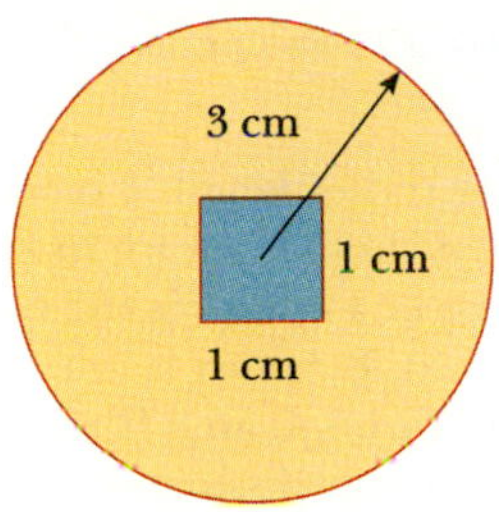

Figure 20.29 (Problem 10)

11. A bolt of lightning strikes the ground 200 m from a 100-turn coil that is oriented vertically with its plane pointing toward the lightning strike (Fig. 20.30). The radius of the coil is 0.80 m, and the current in the lightning bolt falls from 6.02×10^6 A to zero in a time interval of 10.5 μs. What is the voltage induced in the coil over this time period?

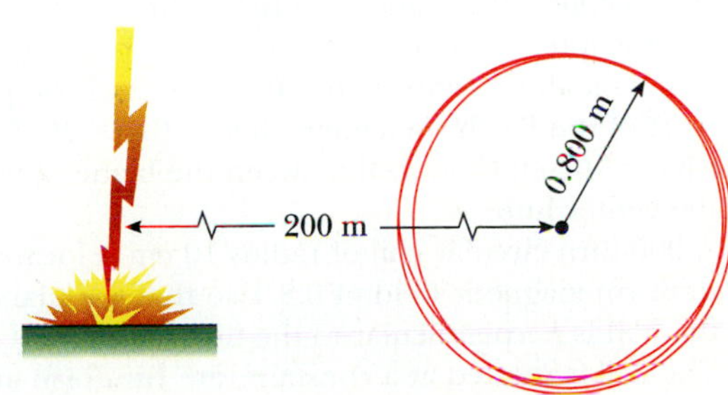

Figure 20.30 (Problem 11)

Section 20.2 Motional emf
Section 20.3 Lenz's Law

12. Consider the arrangement shown in Figure 20.31. Assume that $R = 6\ \Omega$, $\ell = 1.2$ m, and a uniform 2.5-T magnetic field is directed *into* the page. At what speed should the bar be moved to produce a current of 0.5 A in the resistor?

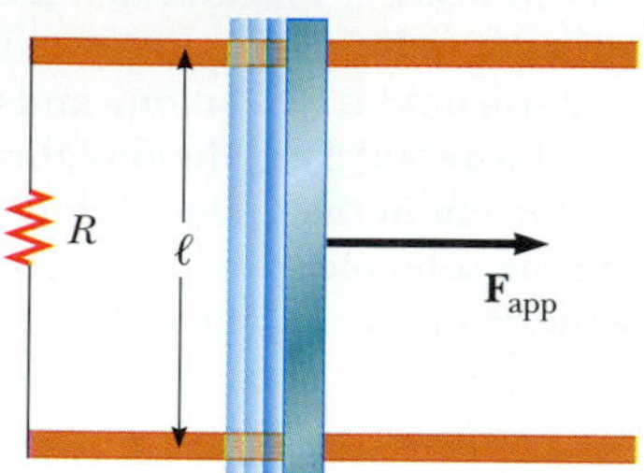

Figure 20.31 (Problems and 12 and 13)

13. A conducting rod of length ℓ moves on two horizontal, frictionless rails, as shown in Figure 20.31. If a constant force of 1 N moves the bar at 2 m/s through a magnetic field, **B**, which is into the page, (a) what is the current through an 8-Ω resistor R? (b) What is the rate of energy dissipation in the resistor? (c) What is the mechanical power delivered by the force **F**?
14. Over a region where the *vertical* component of the Earth's magnetic field is 40 μT directed downward, a 5-m length of wire is held along an east-west direction and moved horizontally to the north with a speed of 10 m/s. Calculate the potential difference between the ends of the wire and determine which end is positive.
15. A small airplane with a wing span of 14 m flies due north at a speed of 70 m/s over a region where the vertical component of the Earth's magnetic field is 1.2 μT. (a) What potential difference is developed between the wing tips? (b) How would the answer to (a) change if the plane were flying due east?
16. A helicopter has blades of length 3 m, rotating at 2 rev/s about a central hub. If the vertical component of the Earth's magnetic field is 0.5×10^{-4} T, what is the emf induced between the blade tip and the center hub?
17. A 200-turn circular coil of radius 10 cm is located in a uniform magnetic field of 0.8 T so that the plane of the coil is perpendicular to the direction of the field. The coil is rotated at a constant rate (uniform angular velocity) through 90° in a time of 1.5 s, so that the plane of the coil is finally parallel to the direction of the field. (a) Calculate the *average* emf induced in the coil as a result of the rotation. (b) What is the instantaneous value of the emf in the coil at the moment the plane of the coil makes an angle of 45° with the magnetic field?
18. Use Lenz's law to answer the following questions concerning the direction of induced currents. What is the direction of the induced current in resistor R (a) when the bar magnet in Figure 20.32a is moved to the left? (b) when the magnet in Figure 20.32a is moved to the right? (c) when the current I in Figure 20.32b decreases rapidly to zero?

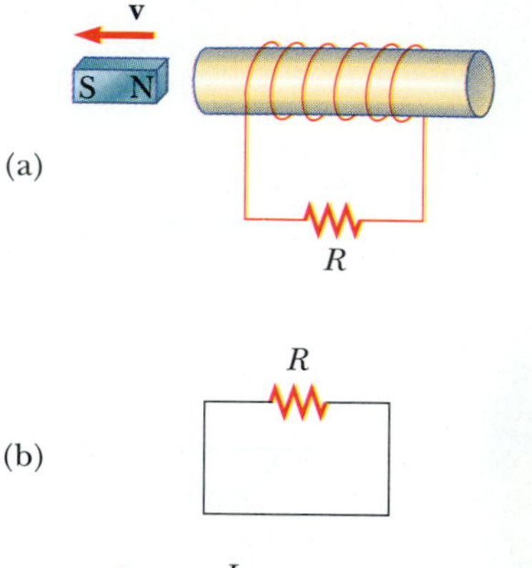

Figure 20.32 (Problem 18)

19. A conducting rectangular loop of mass M, resistance R, and dimensions w wide by ℓ long falls from rest into a magnetic field, **B**, as shown in Figure 20.33. The loop accelerates until it reaches terminal speed, v_t. (a) Show that

$$v_t = \frac{MgR}{B^2w^2}$$

(b) Why is v_t proportional to R? (c) Why is it inversely proportional to B^2?
20. A 0.15-kg wire in the shape of a closed rectangle 1 m wide and 1.5 m long has a total resistance of 0.75 Ω. The rectangle is allowed to fall through a magnetic field directed perpendicular to the direction of motion of the wire (Fig. 20.33). The rectangle accelerates downward until it acquires a *constant* speed of 2 m/s with the top of the rectangle not yet in that region of the field. Calculate the magnitude of **B**.
21. A conducting disc of radius $R = 8.0$ cm is rotated with a constant frequency of $f = 60$ Hz about an axis through its center in a uniform magnetic field, **B**, that is parallel to the axis of rotation. The leads of a voltmeter are contacted by brushes to the center and

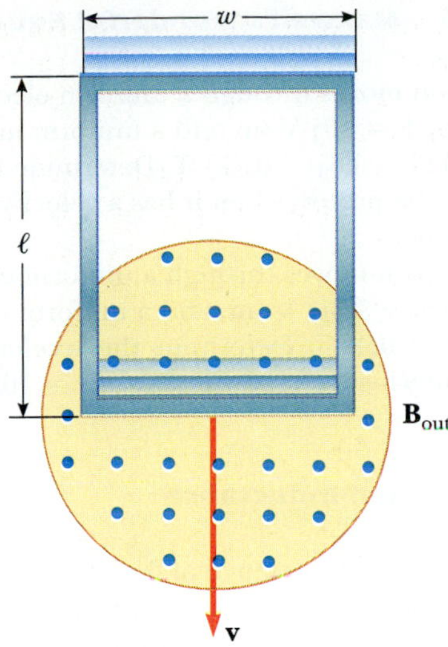

Figure 20.33 (Problems 19 and 20)

the outer edge of the disc. If $B = 8.5 \times 10^{-3}$ T, what is the reading of the voltmeter?

22. Coils rotating in a magnetic field are often used to measure unknown magnetic fields. As an example, consider a coil of radius 1 cm with 50 turns that is rotated about an axis perpendicular to the field at a rate of 20 Hz. If the maximum induced emf in the coil is 3 V, find the strength of the magnetic field.

23. A square coil (20 cm × 20 cm) that consists of 100 turns of wire rotates about a vertical axis at 1500 rpm, as indicated in Figure 20.34. The horizontal component of the Earth's magnetic field at the location of the loop is 2×10^{-5} T. Calculate the maximum emf induced in the coil by the Earth's field.

Section 20.4 Induced emfs and Electric Fields

24. The current in a solenoid is increasing at a rate of 10 A/s. The cross-sectional area of the solenoid is π cm^2, and there are 300 turns on its 15-cm length. What is the induced emf that acts to oppose the increasing current?

25. A coil of 15 turns and radius 10 cm surrounds a long solenoid of radius 2 cm and 10^3 turns/m (Fig. 20.35). If the current in the inner solenoid changes as $I = (5\text{ A})\sin(120t)$, what is the induced emf in the 15-turn coil?

26. A solenoid has a radius of 2 cm and 1000 turns/m. The current varies with time according to the expression $I = 3e^{0.2t}$, where I is in amperes and t is in seconds. Calculate the electric field 5 cm from the axis of the solenoid at $t = 10$ s.

27. A magnetic field directed into the page changes with time according to $B = (0.03t^2 + 1.4)$ T, where t is in seconds. The field has a circular cross-section of radius $R = 2.5$ cm (Fig. 20.36). What are the magnitude and direction of the electric field at point P_1 when $t = 3$ s and $r_1 = 0.02$ m?

28. For the situation depicted in Figure 20.36, the magnetic field varies as $B = (2t^3 - 4t^2 + 0.8)$ T, and $r_2 = 2R = 5$ cm. (a) Calculate the magnitude and direction of the force exerted on an electron located at point P_2 when $t = 2$ s. (b) At what time is the force equal to zero?

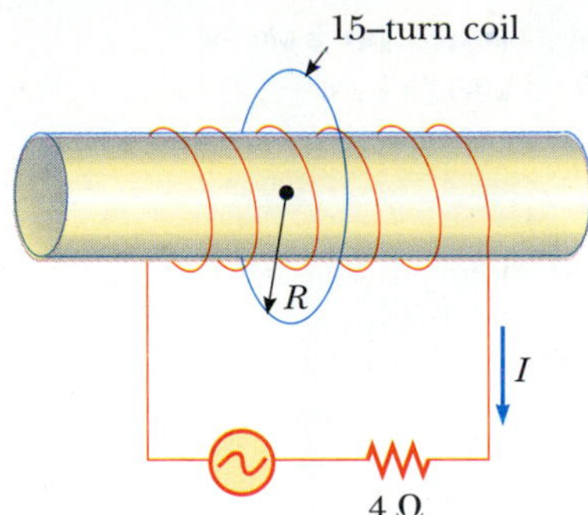

Figure 20.35 (Problem 25)

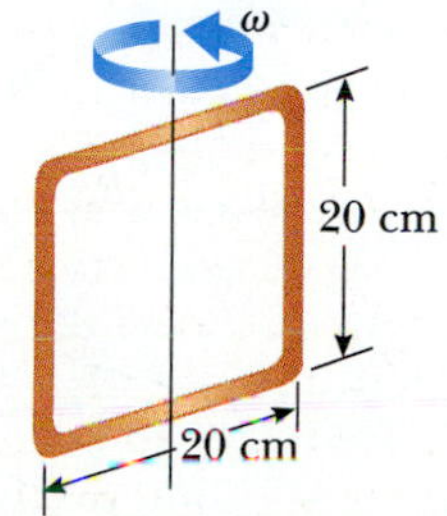

Figure 20.34 (Problem 23)

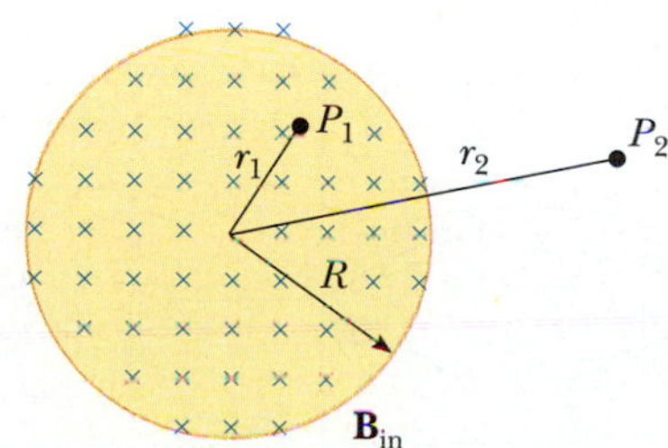

Figure 20.36 (Problems 27 and 28)

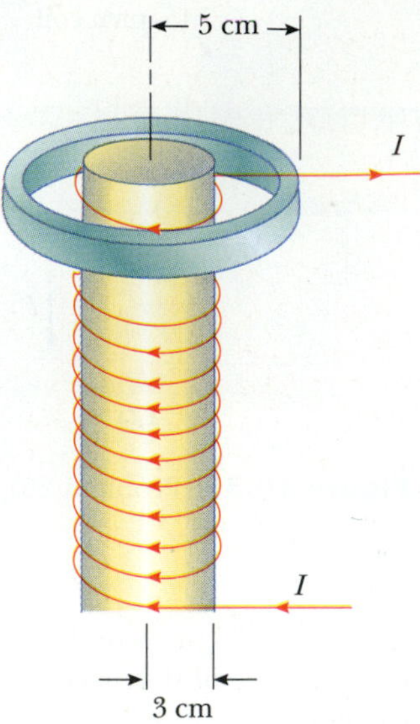

Figure 20.37 (Problem 29)

29. An aluminum ring having a radius of 5 cm and resistance $3 \times 10^{-4}\ \Omega$ is placed on top of a long air-core solenoid with 1000 turns per meter and radius 3 cm, as shown in Figure 20.37. At the location of the ring, the magnetic field due to the current in the solenoid is one-half that at the center of the solenoid. If the current in the solenoid is *increasing* at a rate of 270 A/s, (a) what is the induced current in the ring? (b) At the center of the ring, what is the magnetic field produced by the induced current in the ring? (c) What is the direction of the field in (b)?
30. A circular coil enclosing an area of 100 cm² is made of 200 turns of copper wire, as shown in Figure 20.38. Initially, a 1.1-T uniform magnetic field points perpendicularly *upward* through the plane of the coil. The direction of the field then reverses so that the final magnetic field has a magnitude of 1.1 T pointing *downward* through the coil. During the time the field is changing its direction, how much charge flows through the coil if the coil is connected to a 5-Ω resistor as shown?

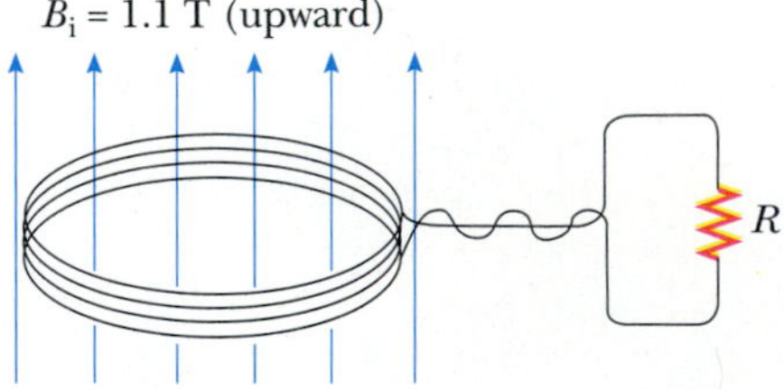

Figure 20.38 (Problem 30)

Section 20.5 Maxwell's Wonderful Equations

31. A proton moves through a uniform electric field given by $\mathbf{E} = 50\mathbf{j}$ V/m and a uniform magnetic field, $\mathbf{B} = (0.2\mathbf{i} + 0.3\mathbf{j} + 0.4\mathbf{k})$ T. Determine the acceleration of the proton when it has a velocity given by $\mathbf{v} = 200\mathbf{i}$ m/s.
32. An electron moves through a uniform electric field, $\mathbf{E} = (2.5\mathbf{i} + 5.0\mathbf{j})$ V/m, and a uniform magnetic field, $\mathbf{B} = 0.4\mathbf{k}$ T. Determine the acceleration of the electron when it has a velocity of $\mathbf{v} = 10\mathbf{i}$ m/s.

Section 20.6 Self-Inductance

33. A coil has an inductance of 3 mH, and a current through it changes from 0.2 A to 1.5 A in a time of 0.2 s. Find the magnitude of the average induced emf in the coil during this time.
34. An emf of 24 mV is induced in a 500-turn coil at an instant when the current is 4 A and is changing at the rate of 10 A/s. What is the magnetic flux through each turn of the coil?
35. A 0.388-mH inductor has a length that is four times its diameter. If it is wound with 22 turns per centimeter, what is its length?
36. The current in a 90-mH inductor changes with time as $I = t^2 - 6t$ (in SI units). Find the magnitude of the induced emf at (a) $t = 1$ s and (b) $t = 4$ s. (c) At what time is the emf zero?
37. A current, $I = I_0 \sin \omega t$, with $I_0 = 5$ A and $\omega/2\pi = 60$ Hz, flows through an inductor whose inductance is 10 mH. What is the back emf as a function of time?
38. A toroid has a major radius of R and a minor radius of r, and is tightly wound with N turns of wire, as shown in Figure 20.39. If $R \gg r$, the magnetic field inside the toroid is essentially that of a long solenoid that has been bent into a large circle of radius R. Using the uniform field of a long solenoid, show that

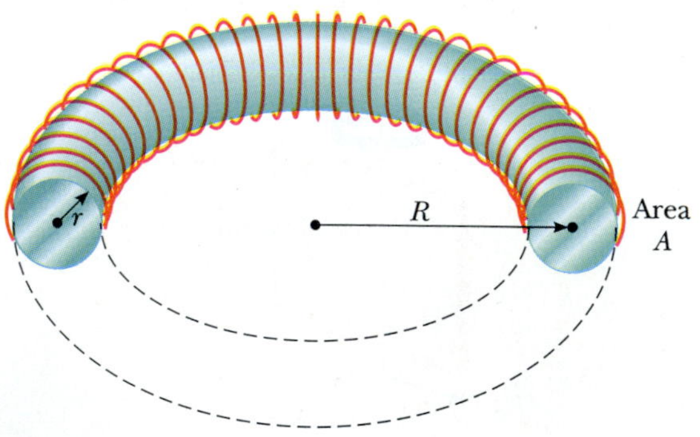

Figure 20.39 (Problem 38)

the self-inductance of such a toroid is given (approximately) by

$$L \cong \frac{\mu_0 N^2 A}{2\pi R}$$

39. A solenoidal inductor contains 420 turns, is 16 cm in length, and has a cross-sectional area of 3 cm^2. What uniform rate of decrease of current through the inductor will produce an induced emf of 175 μV?

Section 20.7 *RL* Circuits

40. Show that $I = I_0 e^{-t/\tau}$ is a solution of the differential equation

$$IR + L\frac{dI}{dt} = 0$$

where $\tau = L/R$ and $I_0 = \mathcal{E}/R$ is the value of the current at $t = 0$.

41. Calculate the inductance in an *RL* circuit in which $R = 0.5\ \Omega$ and the current increases to one-fourth its final value in 1.5 s.

42. An inductor with an inductance of 15 H and resistance of 30 Ω is connected across a 100-V battery. (a) What is the *initial* rate of increase of current in the circuit? (b) At what rate is the current changing at $t = 1.5$ s?

43. A 12-V battery is about to be connected to a series circuit containing a 10-Ω resistor and a 2-H inductor. (a) How long will it take the current to reach 50% of its final value? (b) How long will it take to reach 90% of its final value?

44. Consider the circuit shown in Figure 20.40, taking $\mathcal{E} = 6$ V, $L = 8$ mH, and $R = 4\ \Omega$. (a) What is the inductive time constant of the circuit? (b) Calculate the current in the circuit 250 μs after the switch is closed. (c) What is the value of the final steady-state current? (d) How long does it take the current to reach 80% of its maximum value?

45. When the switch in Figure 20.40 is closed, the current takes 3.0 ms to reach 98% of its final value. If the resistance $R = 10\ \Omega$, what is the inductance, L?

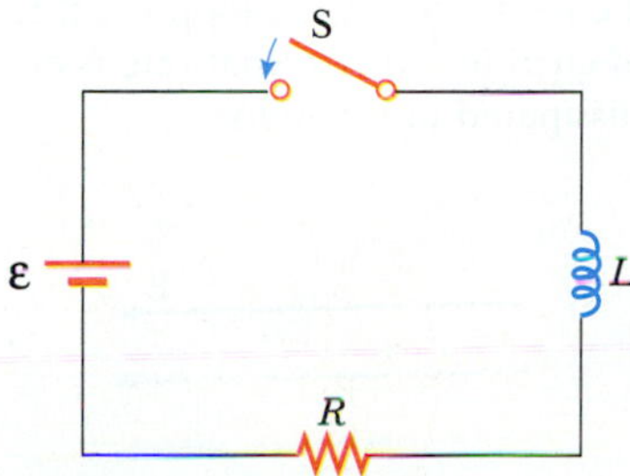

Figure 20.40 (Problems 44, 45, 46, 47)

46. For the *RL* circuit shown in Figure 20.40, let $L = 3$ H, $R = 8\ \Omega$, and $\mathcal{E} = 36$ V. (a) Calculate the ratio of the potential difference across the resistor to that across the inductor when $I = 2$ A. (b) Calculate the voltage across the inductor when $I = 4.5$ A.

47. In the circuit shown in Figure 20.40, let $L = 7$ H, $R = 9\ \Omega$, and $\mathcal{E} = 120$ V. What is the self-induced emf 0.2 s after the switch is closed?

48. One application of an *RL* circuit is the generation of high-voltage transients from a low-voltage dc source, as shown in Figure 20.41. (a) What is the current in the circuit a long time after the switch has been in position A? (b) Now the switch is thrown quickly from A to B. Compute the initial voltage across each resistor and the inductor. (c) How much time elapses before the voltage across the inductor drops to 12 V?

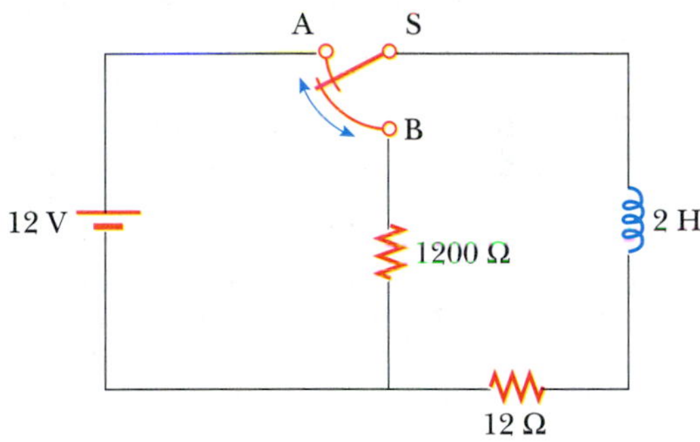

Figure 20.41 (Problem 48)

Section 20.8 Energy Stored in a Magnetic Field

49. Calculate the energy associated with the magnetic field of a 200-turn solenoid in which a current of 1.75 A produces a flux of 3.7×10^{-4} Wb in each turn.

50. Consider the circuit shown in Figure 20.42. What energy is stored in the inductor when the current reaches its final equilibrium value after the switch is closed?

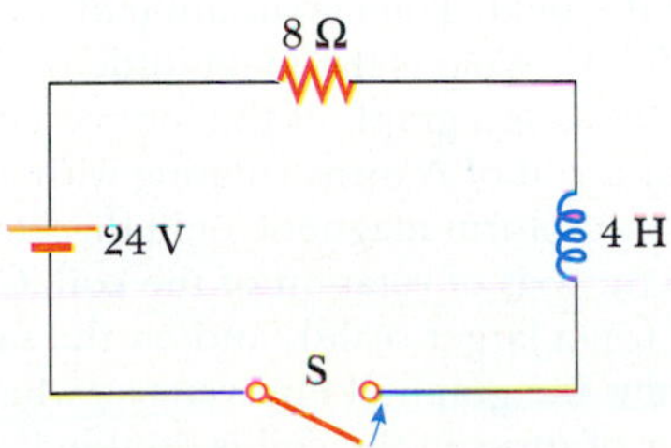

Figure 20.42 (Problems 50 and 55)

51. A 10-V battery, a 5-Ω resistor, and a 10-H inductor are connected in series. After the current in the circuit has reached its maximum value, calculate (a) the power supplied to the circuit by the battery, (b) the power dissipated in the resistor, (c) the power dissipated in the inductor, and (d) the energy stored in the magnetic field of the inductor.

52. At $t = 0$, a source of emf, $\mathcal{E} = 500$ V, is applied to a coil that has an inductance of 0.80 H and a resistance of 30 Ω. (a) Find the energy stored in the magnetic field when the current reaches half its maximum value. (b) How long after the emf is connected does it take for the current to reach this value?

53. The magnetic field inside a superconducting solenoid is 4.5 T. The solenoid has an inner diameter of 6.2 cm and a length of 26 cm. (a) Determine the magnetic energy density in the field. (b) Determine the magnetic energy stored in the magnetic field within the solenoid.

54. On a clear day, there is a vertical electric field near the Earth's surface with a magnitude of 100 V/m. At the same time, the Earth's magnetic field has a magnitude of approximately 0.5×10^{-4} T. Compute the energy density of the two fields.

55. The switch in the circuit of Figure 20.42 is closed at $t = 0$. (a) Calculate the *rate* at which energy is being stored in the inductor after an elapsed time equal to the time constant of the circuit. (b) At what rate is energy being dissipated as joule heat in the resistor at this time? (c) What is the total energy stored in the inductor at this time?

56. The magnitude of the magnetic field outside a sphere of radius R is given by $B = B_0(R/r)^2$, where B_0 is a constant. Determine the total energy stored in the magnetic field outside the sphere and evaluate your result for $B_0 = 5 \times 10^{-5}$ T and $R = 6 \times 10^6$ m, values appropriate for the Earth's magnetic field.

Additional Problems

57. A 50-turn rectangular coil of dimensions 0.2 m × 0.3 m is rotated at 90 rad/s in a magnetic field so that the axis of the coil is perpendicular to the direction of the field. The maximum emf induced in the coil is 0.5 V. What is the magnitude of the field?

58. Figure 20.43 is a graph of the induced emf versus time for a coil of N turns rotating with angular velocity ω in a uniform magnetic field directed perpendicular to the axis of rotation of the coil. Copy this sketch (on a larger scale), and on the same set of axes show the graph of emf versus t when (a) the number of turns in the coil is doubled; (b) the angular velocity is doubled; (c) the angular velocity is doubled while the number of turns in the coil is halved.

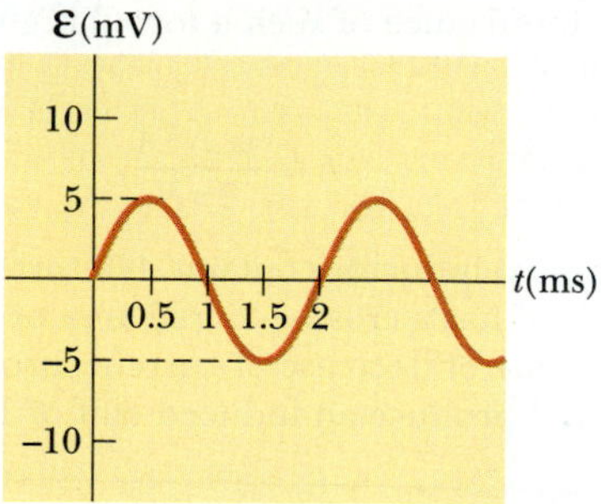

Figure 20.43 (Problem 58)

59. A solenoid wound with 2000 turns/m is supplied with current that varies in time according to $I = 4 \sin(120\pi t)$, where I is in A and t is in s. A small coaxial circular coil of 40 turns and radius $r = 5$ cm is located inside the solenoid near its center. (a) Derive an expression that describes the manner in which the emf in the small coil varies in time. (b) At what average rate is energy dissipated in the small coil if the windings have a total resistance of 8 Ω?

60. A loop of area 0.1 m² is rotating at 60 rev/s with the axis of rotation perpendicular to a 0.2-T magnetic field. (a) If there are 1000 turns on the loop, what is the maximum voltage induced in the loop? (b) When the maximum induced voltage occurs, what is the orientation of the loop with respect to the magnetic field?

61. A long solenoid with 1000 turns/m and radius 2 cm carries an oscillating current given by the expression $I = (5 \text{ A}) \sin(100\pi t)$. What is the electric field induced at a radius $r = 1$ cm from the axis of the solenoid? What is the direction of this electric field when the current is *increasing* counterclockwise in the coil?

62. A circular loop of wire 5 cm in radius is in a spatially uniform magnetic field, with the plane of the circular loop perpendicular to the direction of the field (Fig. 20.44). The magnetic field varies with time:

$$B(t) = a + bt \qquad a = 0.20 \text{ T} \qquad b = 0.32 \text{ T/s}$$

(a) Calculate the magnetic flux through the loop at $t = 0$. (b) Calculate the emf induced in the loop. (c) If the resistance of the loop is 1.2 Ω, what is the induced current? (d) At what rate is electric energy being dissipated in the loop?

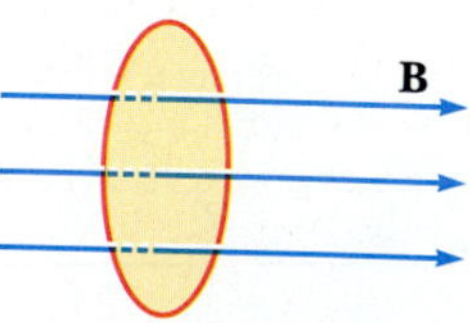

Figure 20.44 (Problem 62)

63. A steel beam 12 m in length is accidentally dropped by a construction crane from a height of 9 m. The horizontal component of the Earth's magnetic field over the region is 18 μT. What is the induced emf in the beam just before impact with the Earth, assuming its long dimension remains in a horizontal plane, oriented perpendicular to the horizontal component of the Earth's magnetic field?

64. A conducting rod moves with a constant velocity, **v**, perpendicular to a long, straight wire carrying a current of I, as in Figure 20.45. Show that the emf generated between the ends of the rod is given by

$$|\mathcal{E}| = \frac{\mu_0 v I}{2\pi r}\ell$$

In this case, note that the emf decreases with increasing r, as you might expect.

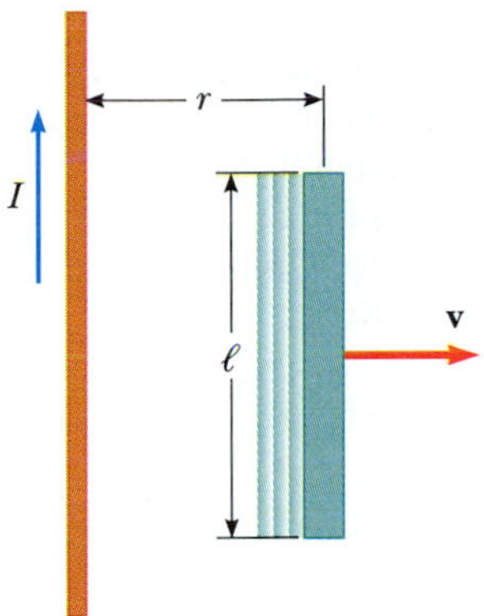

Figure 20.45 (Problem 64)

65. A long solenoid has n turns per meter and carries a current of $I = I_0(1 - e^{-\alpha t})$, with $I_0 = 30$ A and $\alpha = 1.6\ \text{s}^{-1}$. Inside the solenoid and coaxial with it is a loop that has a radius of $R = 6$ cm and consists of a total of N turns of fine wire. What emf is induced in the loop by the changing current? Take $n = 400$ turns/m and $N = 250$ turns. (Assume that the loop is at the center of the solenoid, where the magnetic field is uniform and perpendicular to the plane of the loop.)

66. A wire of mass m, length d, and resistance R slides without friction on parallel rails, as shown in Figure 20.46. A battery that maintains a constant emf, $\mathcal{E}$, is connected between the rails, and a constant magnetic field, **B**, is directed perpendicular to the plane of the page. If the wire starts from rest, show that at time t it moves with a speed of

$$v = \frac{\mathcal{E}}{Bd}(1 - e^{-B^2d^2t/mR})$$

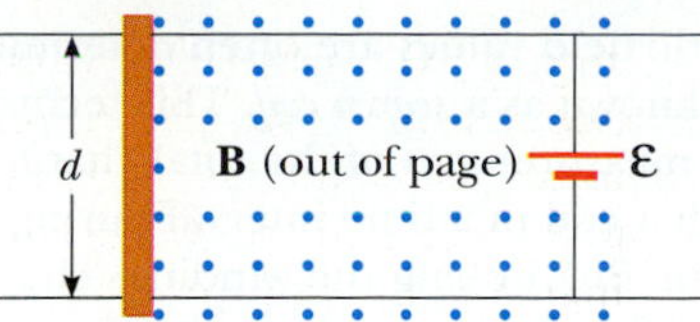

Figure 20.46 (Problem 66)

67. An automobile has a vertical radio antenna 1.2 m long. The automobile travels at 65 km/h on a horizontal road where the Earth's magnetic field is 50 μT directed downward (toward the north) at an angle of 65° below the horizontal. (a) Specify the direction in which the automobile should move in order to generate the maximum motional emf in the antenna, with the top of the antenna positive relative to the bottom. (b) Calculate the magnitude of this induced emf.

68. A long, straight wire is parallel to one edge and is in the plane of a single turn rectangular loop, as in Figure 20.47. (a) If the current in the long wire varies in time as $I = I_0 e^{-t/\tau}$, show that the induced emf in the loop is given by

$$\mathcal{E} = \frac{\mu_0 b}{2\pi}\frac{I}{\tau}\ln\left(1 + \frac{a}{d}\right)$$

(b) Calculate the value for the induced emf at $t = 5$ s, taking $I_0 = 10$ A, $d = 3$ cm, $a = 6$ cm, $b = 15$ cm, and $\tau = 5$ s.

69. To monitor the breathing of a hospital patient, a thin belt is placed about the patient's chest. The belt is a 200-turn coil. During inhalation, the area within the coil increases by 39 cm². The Earth's magnetic field is 50 μT and makes an angle of 28° with the plane of the coil. If a patient takes 1.80 s to inhale, find the average induced emf in the coil while the patient is inhaling.

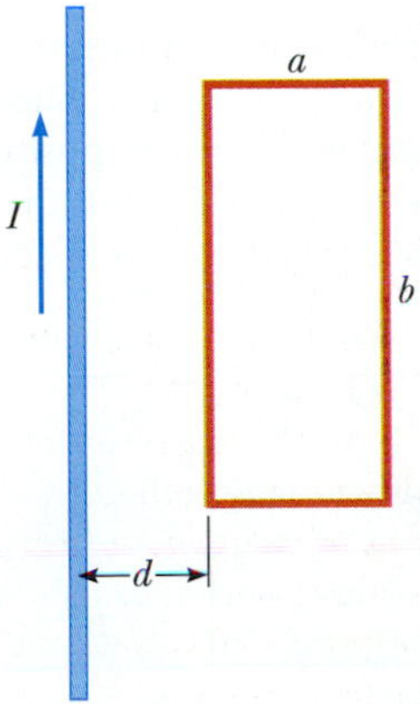

Figure 20.47 (Problem 68)

70. Magnetic field values are often determined by using a device known as a *search coil.* This technique depends on the measurement of the total charge passing through a coil in a time interval during which the magnetic flux linking the windings changes either because of the motion of the coil or because of a change in the value of B. (a) Show that if the flux through the coil changes from Φ_1 to Φ_2, the charge transferred through the coil between t_1 and t_2 will be given by $Q = N(\Phi_2 - \Phi_1)/R$, where R is the resistance of the coil and associated circuitry (galvanometer) and N is the number of turns. (b) As a specific example, calculate B when a 100-turn coil of resistance 200 Ω and cross-sectional area 40 cm² produces the following results. A total charge of 5×10^{-4} C passes through the coil when it is rotated in a uniform field from a position where the plane of the coil is perpendicular to the field to a position where the coil's plane is parallel to the field.

71. The magnetic flux threading a metal ring varies with time t according to

$$\Phi_m = 3(at^3 - bt^2)\ \text{T}\cdot\text{m}^2 \qquad a = 2\ \text{s}^{-3} \qquad b = 6\ \text{s}^{-2}$$

The resistance of the ring is 3 Ω. Determine the *maximum current* induced in the ring during the interval from $t = 0$ to $t = 2$ s.

72. The inductor in the circuit in Figure 20.48 has negligible resistance. When the switch is opened after having been closed for a long time, the current in the inductor drops to 0.25 A in 0.15 s. What is the inductance of the inductor?

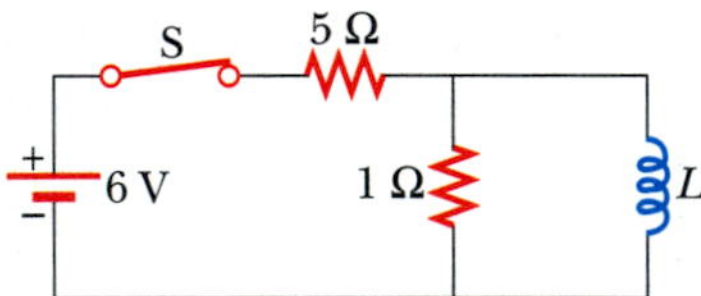

Figure 20.48 (Problem 72)

73. Assume that the switch in the circuit shown in Figure 20.49 is initially in position 1. Show that if the switch is thrown from position 1 to position 2, all the energy stored in the magnetic field of the inductor will be dissipated as thermal energy in the resistor.

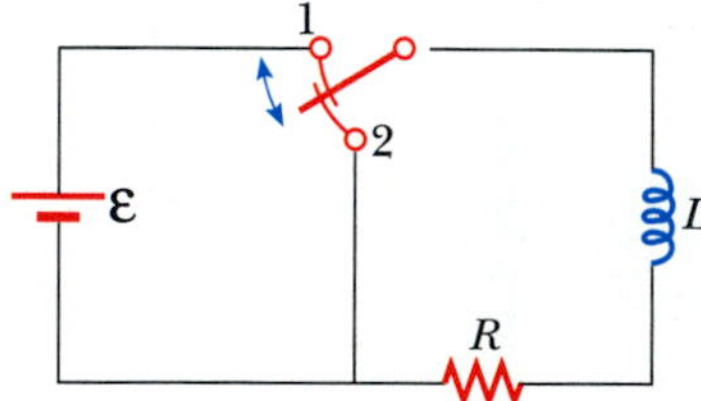

Figure 20.49 (Problem 73)

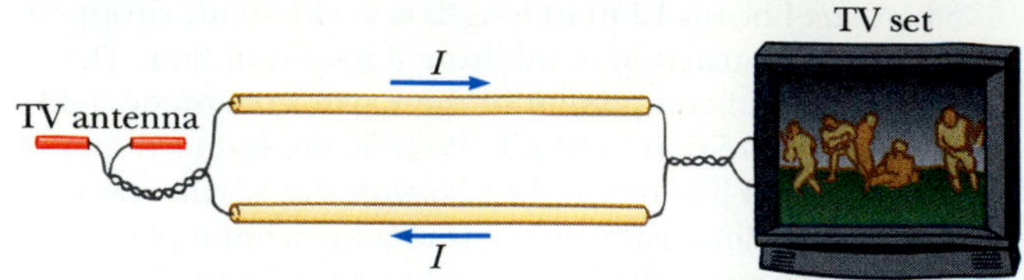

Figure 20.50 (Problem 74)

74. The lead-in wires from a TV antenna are often constructed in the form of two parallel wires (Fig. 20.50). (a) Why does this configuration of conductors have an inductance? (b) What constitutes the flux loop for this configuration? (c) Neglecting any magnetic flux inside the wires, show that the inductance of a length x of this type of lead-in is

$$L = \frac{\mu_0 x}{\pi} \ln\left(\frac{w - a}{a}\right)$$

where a is the radius of the wires and w is the center-to-center separation of the wires.

75. At $t = 0$, the switch in Figure 20.51 is closed. By using Kirchhoff's laws for the instantaneous currents and voltages in this two-loop circuit, show that the current through the inductor is

$$I(t) = \frac{\mathcal{E}}{R_1}[1 - e^{-(R'/L)t}]$$

where $R' = R_1R_2/(R_1 + R_2)$.

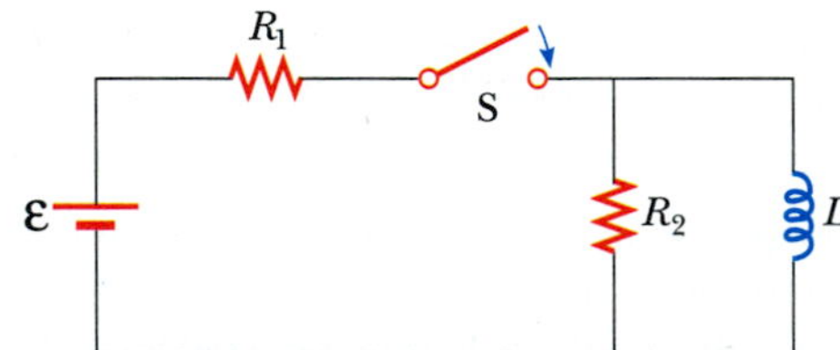

Figure 20.51 (Problem 75)

76. The toroidal coil shown in Figure 20.52 consists of N turns and has a rectangular cross-section. Its inner and outer radii are a and b, respectively. (a) Show that the self-inductance of the coil is

$$L = \frac{\mu_0 N^2 h}{2\pi} \ln(b/a)$$

(b) Using this result, compute the self-inductance of a 500-turn toroid with $a = 10$ cm, $b = 12$ cm, and $h = 1$ cm. (c) In Problem 38, an approximate formula for the inductance of a toroid with $R \gg r$ was derived. To get a feel for the accuracy of this result, use the expression in Problem 38 to compute the

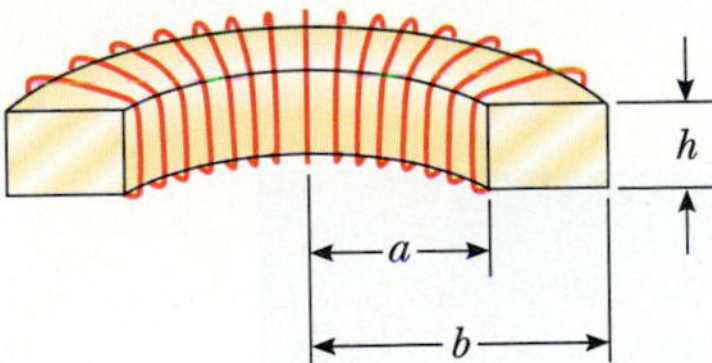

Figure 20.52 (Problem 76)

(approximate) inductance of the toroid described in part (b).

77. A novel method of storing electrical energy has been proposed. A huge underground superconducting coil, 1.0 km in diameter, would be fabricated. It would carry a maximum current of 50 kA through each winding of a 150-turn Nb_3Sn solenoid. (a) If the inductance of this huge coil were 50 H, what would be the total energy stored? (b) What would be the compressive force per meter length acting between two adjacent windings 0.25 m apart?

78. In Figure 20.53, the rolling axle, 1.5 m long, is pushed along horizontal rails at a constant speed of $v = 3$ m/s. A resistor, $R = 0.4\ \Omega$, is connected to the rails at points a and b, directly opposite each other. (The wheels make good electrical contact with the rails, so the axle, rails, and R form a complete, closed-loop circuit. The only significant resistance in the circuit is R.) There is a uniform magnetic field, $B = 0.08$ T, vertically downward. (a) Find the induced current I in the resistor. (b) What horizontal force **F** is required to keep the axle rolling at constant speed? (c) Which end of the resistor, a or b, is at the higher electric potential? (d) After the axle rolls past the resistor, does the current in R reverse direction?

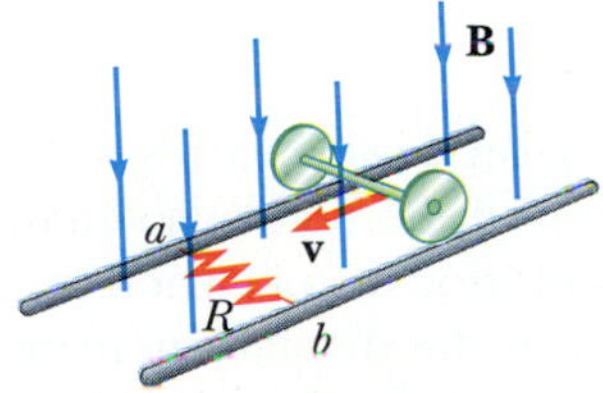

Figure 20.53 (Problem 78)

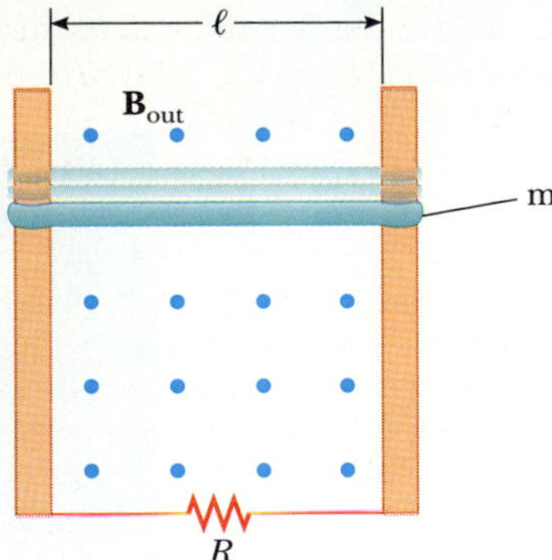

Figure 20.54 (Problem 79)

79. A horizontal wire is free to slide on the vertical rails of a conducting frame as in Figure 20.54. The wire has mass m and length ℓ, and the resistance of the circuit is R. If a uniform magnetic field is directed perpendicular to the frame, what is the terminal velocity of the wire as it falls under the force of gravity? (Neglect friction.)

80. A thin metal strip is allowed to slide down parallel frictionless rails of negligible resistance, connected at the bottom end and elevated at an angle of 30° above the horizontal, as in Figure 20.55. A uniform magnetic field of 2.60 T is directed vertically upward throughout the region. The strip has a mass of $m =$ 40 g, resistance of $R = 30\ \Omega$, and a length between the rails of $\ell = 0.2$ m. Calculate the terminal speed achieved by the strip sliding along the incline.

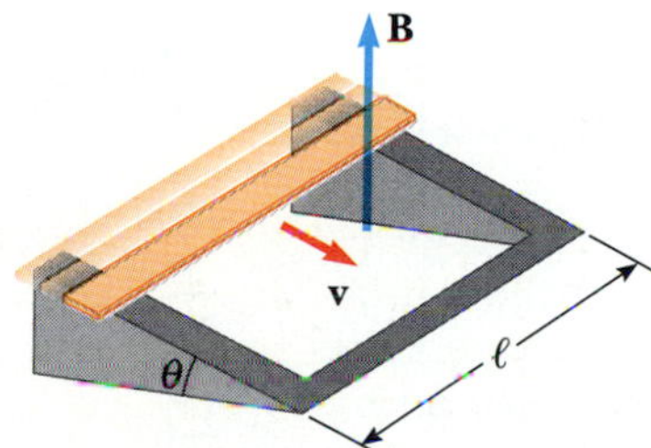

Figure 20.55 (Problem 80)

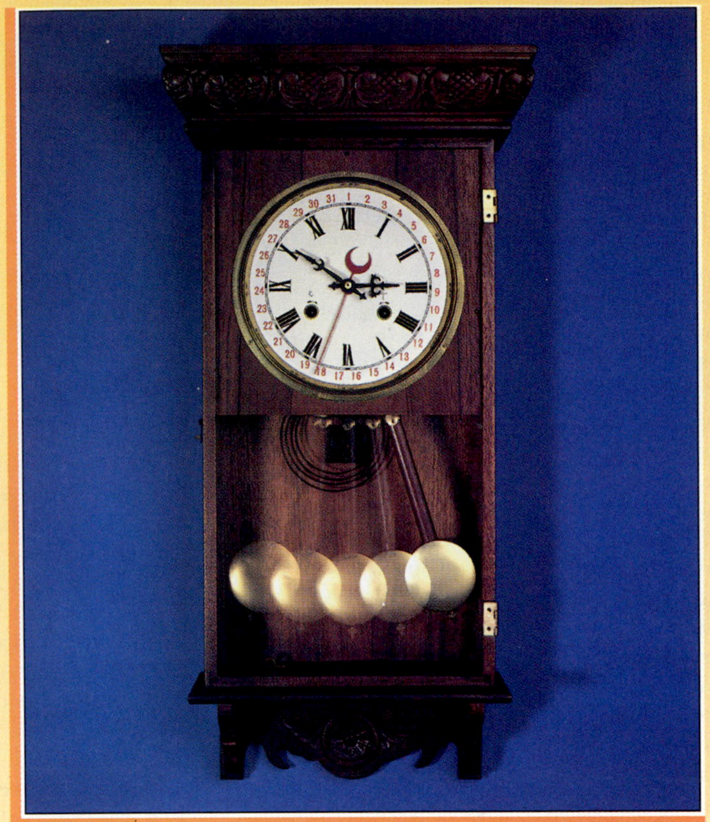

Multiflash photograph of a pendulum clock. The period of motion of the pendulum (the time for one complete oscillation) depends on the length of the pendulum and on the acceleration due to gravity. *(© Robert Mathena 1990, Fundamental Photographs)*

Oscillatory Motion

CHAPTER 21

If a force acting on a body varies in time, the velocity and acceleration of the body also change with time. A special kind of motion occurs when the force on a body is proportional to the displacement of the body from equilibrium. If this force always acts toward the equilibrium position of the body, a repetitive back-and-forth motion about that position results. The motion is an example of what is called *periodic* or *oscillatory* motion.

You are most likely familiar with several examples of periodic motion, such as the oscillations of a mass on a spring, the motion of a pendulum, and the vibrations of a stringed musical instrument. Numerous systems exhibit oscillatory motion. For example, the molecules in a solid oscillate about their equilibrium positions; electromagnetic waves, such as light waves, radar, and radio waves, are characterized by oscillating electric and magnetic field vectors; in alternating-current circuits, voltage, current, and electrical charge vary periodically with time.

Much of the material in this chapter deals with *simple harmonic motion.* In this type of motion, an object oscillates between two spatial positions for an indefinite period of time with no loss in mechanical energy. In real mechanical systems, retarding (frictional) forces are always present. Such forces reduce the mechanical energy of the system with time, and the oscillations are said to be *damped.* If an external driving force is applied so that the energy loss is balanced by the energy input, we call the motion a *forced oscillation.* We conclude this chapter with a discussion of damped and forced oscillations in series RLC circuits.

21.1 Simple Harmonic Motion

A particle moving along the x axis is said to exhibit **simple harmonic motion** when x, its displacement from equilibrium, varies in time according to the relationship

Displacement versus time for simple harmonic motion

$$x = A\cos(\omega t + \phi) \qquad [21.1]$$

where A, ω, and ϕ are constants of the motion. In order to give physical significance to these constants, it is convenient to plot x as a function of t, as in Figure 21.1. First, we note that A, called the **amplitude** of the motion, is simply the *maximum displacement* of the particle in either the positive or negative x direction. The constant ω is called the *angular frequency* (defined in Eq. 21.4). The constant angle ϕ is called the **phase constant** (or phase angle) and, along with the amplitude A, is determined uniquely by the initial displacement and velocity of the particle. The constants ϕ and A tell us what the displacement was at time $t = 0$. The quantity $(\omega t + \phi)$ is called the **phase** of the motion and is useful for comparing the motions of two systems of particles. Note that the function x is periodic and repeats itself each time ωt increases by 2π radians.

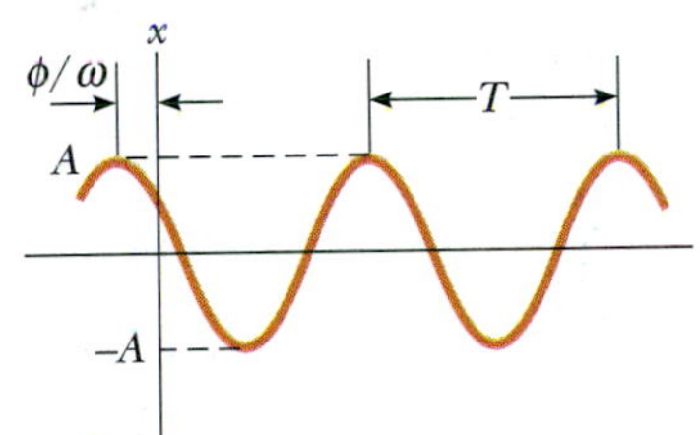

Figure 21.1
Displacement versus time for a particle undergoing simple harmonic motion. The amplitude of the motion is A, and the period is T.

The **period,** T, of the motion is the time required for the particle to go through one full cycle of its motion. That is, the value of x at time t equals the value of x at time $t + T$. We can show that the period is given by $T = 2\pi/\omega$ by using the fact that the phase increases by 2π radians in a time of T:

$$\omega t + \phi + 2\pi = \omega(t + T) + \phi$$

Simplifying this expression, we see that $\omega T = 2\pi$, or

Period

$$T = \frac{2\pi}{\omega} \qquad [21.2]$$

The inverse of the period is called the **frequency** of the motion, f. The frequency represents the *number of oscillations the particle makes per unit time:*

Frequency

$$f = \frac{1}{T} = \frac{\omega}{2\pi} \qquad [21.3]$$

The units of f are cycles per second, or hertz (Hz).

Rearranging Equation 21.3 gives

Angular frequency

$$\omega = 2\pi f = \frac{2\pi}{T} \qquad [21.4]$$

The constant ω is called the **angular frequency**, as noted above, and has the units radians per second. We shall discuss the geometric significance of ω in Section 21.4.

We can obtain the velocity of a particle undergoing simple harmonic motion by differentiating Equation 21.1 with respect to time:

Velocity in simple harmonic motion

$$v = \frac{dx}{dt} = -\omega A \sin(\omega t + \phi) \qquad [21.5]$$

The acceleration of the particle is dv/dt:

Acceleration in simple harmonic motion

$$a = \frac{dv}{dt} = -\omega^2 A \cos(\omega t + \phi) \qquad [21.6]$$

Since $x = A\cos(\omega t + \phi)$, we can express Equation 21.6 in the form

$$a = -\omega^2 x \qquad [21.7]$$

From Equation 21.5 we see that, since the sine and cosine functions oscillate between ± 1, the extreme values of v are $\pm\omega A$. Equation 21.6 tells us that the extreme values of the acceleration are $\pm\omega^2 A$. Therefore, the *maximum* values of the velocity and acceleration are

Maximum values of velocity and acceleration in simple harmonic motion

$$v_{\text{max}} = \omega A \qquad [21.8]$$

$$a_{\text{max}} = \omega^2 A \qquad [21.9]$$

Figure 21.2a plots displacement versus time for an arbitrary value of the phase constant. The velocity and acceleration-versus-time curves are illustrated in Figures 21.2b and 21.2c. They show that the phase of the velocity differs from the phase of the displacement by $\pi/2$ rad, or 90°. That is, when x is a maximum or a minimum, the velocity is zero. Likewise, when x is zero, the speed is a maximum. Furthermore, note that the phase of the acceleration differs from the phase of the displacement by π radians, or 180°. That is, when x is a maximum, a is a maximum in the opposite direction.

As we stated earlier, $x = A\cos(\omega t + \phi)$ is a general expression for the displacement of the particle from equilibrium, where the phase constant ϕ and the amplitude A must be chosen to meet the initial conditions of the motion. Suppose that the initial position, x_0, and initial velocity, v_0, of a single oscillator are given; that is, at $t = 0$, $x = x_0$ and $v = v_0$. Under these conditions, Equations 21.1 and 21.5 give

$$x_0 = A\cos\phi \quad \text{and} \quad v_0 = -\omega A \sin\phi \qquad [21.10]$$

Dividing these two equations eliminates A, giving $\frac{v_0}{x_0} = -\omega \tan\phi$, or

$$\tan\phi = -\frac{v_0}{\omega x_0} \qquad [21.11]$$

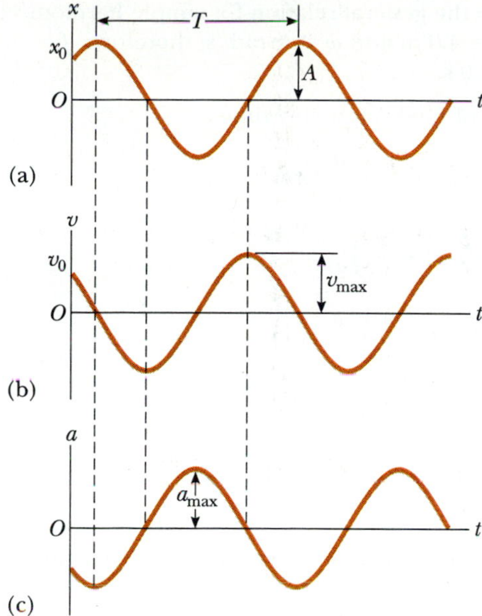

Figure 21.2
Graphical representation of simple harmonic motion: (a) the displacement versus time, (b) the velocity versus time, and (c) the acceleration versus time. Note that the velocity is 90° out of phase with the displacement and the acceleration is 180° out of phase with the displacement.

Furthermore, if we take the sum $x_0{}^2 + (v_0/\omega)^2 = A^2 \cos^2 \phi + A^2 \sin^2 \phi$ (where we have used Eq. 21.10) and solve for A, we find that

The phase angle ϕ and amplitude A can be obtained from the initial conditions

$$A = \sqrt{x_0{}^2 + \left(\frac{v_0}{\omega}\right)^2} \qquad [21.12]$$

Thus, we see that ϕ and A are known if x_0, ω, and v_0 are specified. We shall treat a few specific cases in the next section.

We conclude this section by pointing out the following important properties of a particle moving in simple harmonic motion:

Properties of simple harmonic motion

1. The displacement, velocity, and acceleration all vary sinusoidally with time but are not in phase, as shown in Figure 21.2.
2. The acceleration is proportional to the displacement, but in the opposite direction.
3. The frequency and the period of motion are independent of the amplitude.

▼▼▼

Example 21.1 An Oscillating Body

A body oscillates with simple harmonic motion along the x axis. Its displacement varies with time according to the equation

$$x = (4.0 \text{ m}) \cos\left(\pi t + \frac{\pi}{4}\right)$$

where t is in seconds and the units of the two terms in the parentheses are in radians.

(a) Determine the amplitude, frequency, and period of the motion.

Solution By comparing this equation with the general relation for simple harmonic motion, $x = A\cos(\omega t + \phi)$, we see that $A = 4.0$ m and $\omega = \pi$ rad/s; therefore, $f = \omega/2\pi = \pi/2\pi = 0.50\ \text{s}^{-1}$ and $T = 1/f = 2.0$ s.

(b) Calculate the velocity and acceleration of the body at any time t.

$$v = \frac{dx}{dt} = -4.0\sin\left(\pi t + \frac{\pi}{4}\right)\frac{d}{dt}(\pi t)$$

$$= -(4\pi\ \text{m/s})\sin\left(\pi t + \frac{\pi}{4}\right)$$

$$a = \frac{dv}{dt} = -4\pi\cos\left(\pi t + \frac{\pi}{4}\right)\frac{d}{dt}(\pi t)$$

$$= -(4\pi^2\ \text{m/s}^2)\cos\left(\pi t + \frac{\pi}{4}\right)$$

(c) Using the results to (b), determine the position, velocity, and acceleration of the body at $t = 1$ s.

Solution Noting that the angles in the trigonometric functions are in radians, we get, at $t = 1$ s,

$$x = (4.0\ \text{m})\cos\left(\pi + \frac{\pi}{4}\right) = (4.0\ \text{m})\cos\left(\frac{5\pi}{4}\right)$$

$$= (4.0\ \text{m})(-0.707) = \boxed{-2.83\ \text{m}}$$

$$v = -(4\pi\ \text{m/s})\sin\left(\frac{5\pi}{4}\right) = -(4\pi\ \text{m/s})(-0.707) = \boxed{8.89\ \text{m/s}}$$

$$a = -(4\pi^2\ \text{m/s}^2)\cos\left(\frac{5\pi}{4}\right) = -(4\pi^2\ \text{m/s}^2)(-0.707)$$

$$= \boxed{27.9\ \text{m/s}^2}$$

(d) Determine the maximum speed and maximum acceleration of the body.

Solution From the general relations for v and a found in (b), we see that the maximum values of the sine and cosine functions are unity. Therefore, v varies between $\pm 4\pi$ m/s, and a varies between $\pm 4\pi^2\ \text{m/s}^2$. Thus, $v_{\text{max}} = 4\pi$ m/s and $a_{\text{max}} = 4\pi^2\ \text{m/s}^2$. The same results are obtained using $v_{\text{max}} = \omega A$ and $a_{\text{max}} = \omega^2 A$, where $A = 4.0$ m and $\omega = \pi$ rad/s.

(e) Find the displacement of the body between $t = 0$ and $t = 1$ s.

Solution The x coordinate at $t = 0$ is

$$x_0 = (4.0\ \text{m})\cos\left(0 + \frac{\pi}{4}\right) = (4.0\ \text{m})(0.707) = 2.83\ \text{m}$$

In (c), we found that the coordinate at $t = 1$ s was -2.83 m; therefore, the displacement between $t = 0$ and $t = 1$ s is

$$\Delta x = x - x_0 = -2.83\ \text{m} - 2.83\ \text{m} = \boxed{-5.66\ \text{m}}$$

Because the particle's velocity changes sign during the first second, the magnitude of Δx is *not* the same as the distance traveled in the first second.

21.2 Motion of a Mass Attached to a Spring

Consider a physical system consisting of a mass, m, attached to the end of a spring, where it is free to move on a horizontal, frictionless track (Fig. 21.3). We know from experience that such a system oscillates back and forth if disturbed from the equilibrium position $x = 0$, where the spring is unstretched. Since the surface is frictionless, the mass exhibits simple harmonic motion. One experimental arrangement that clearly demonstrates that such a system exhibits simple harmonic motion is illustrated in Figure 21.4. A mass oscillating vertically on a spring has a marking pen attached to it. While the mass is in motion, a sheet of paper is moved horizontally as shown, and the marking pen traces out a sinusoidal pattern. We can understand this qualitatively by first recalling that when the mass is displaced a small distance, x, from equilibrium, the spring exerts a force on it, given by **Hooke's law,**

$$F = -kx \quad [21.13]$$

where k is the force constant of the spring. We call this a **linear restoring force** because it is linearly proportional to the displacement and always directed toward the equilibrium position, *opposite* the displacement. That is, when the mass is displaced to the right in Figure 21.3, x is positive and the restoring force is to the left. When the mass is displaced to the left of $x = 0$, then x is negative and F is to the right.

If we apply Newton's second law to the motion of the mass in the x direction, we get

$$F = -kx = ma$$

$$a = -\frac{k}{m}x \quad [21.14]$$

That is, just as we learned in Section 21.1, *the acceleration is proportional to the displacement of the mass from equilibrium and is in the opposite direction.* If the mass is displaced a maximum distance, $x = A$, at some initial time and released from rest, its *initial* acceleration is $-kA/m$ (that is, the acceleration has its extreme negative value). When the mass passes through the equilibrium position, $x = 0$ and its acceleration is zero. At this instant, its speed is a maximum. It then continues to travel to the left of equilibrium and finally reaches $x = -A$, at which time its acceleration is kA/m (maximum positive) and its speed is again zero. Thus, we see that the mass oscillates between the turning points $x = \pm A$. In one full cycle of its motion it travels a distance of $4A$.

Let us now describe this motion in a quantitative fashion. Recall that $a = dv/dt = d^2x/dt^2$, and so we can express Equation 21.14 as

$$\frac{d^2x}{dt^2} = -\frac{k}{m}x \quad [21.15]$$

If we denote the ratio k/m with the symbol ω^2,

$$\omega^2 = \frac{k}{m} \quad [21.16]$$

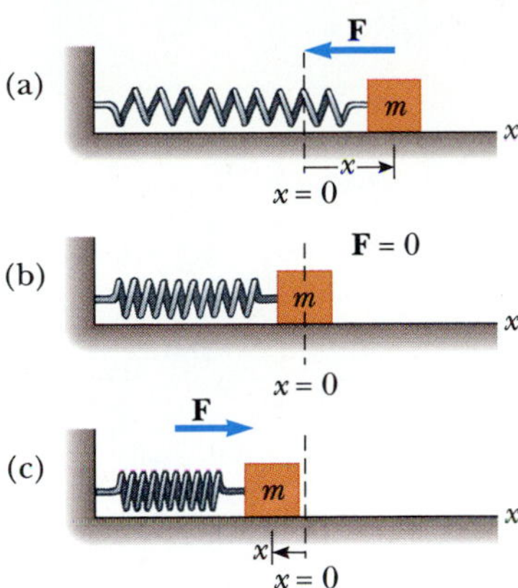

Figure 21.3
A mass attached to a spring on a frictionless track exhibits simple harmonic motion. (a) When the mass is displaced to the right of equilibrium, the displacement is positive and the acceleration is negative. (b) At the equilibrium position, $x = 0$, the acceleration is zero but the speed is a maximum. (c) When the displacement is negative, the acceleration is positive.

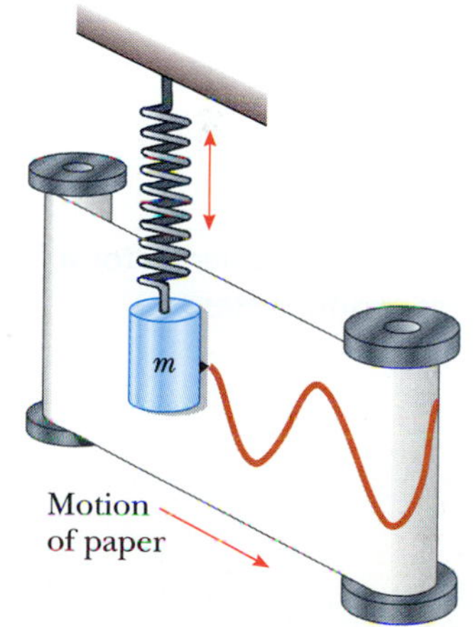

Figure 21.4
An experimental apparatus for demonstrating simple harmonic motion. A pen attached to the oscillating mass traces out a sine wave on the moving chart paper.

then Equation 21.15 can be written in the form

$$\frac{d^2x}{dt^2} = -\omega^2 x \qquad [21.17]$$

What we now require is a solution to Equation 21.17—that is, a function $x(t)$ that satisfies this second-order differential equation. Since Equations 21.17 and 21.7 are equivalent, we see that the solution must be that of simple harmonic motion:

$$x(t) = A\cos(\omega t + \phi)$$

To see this explicitly, note that if

$$x = A\cos(\omega t + \phi)$$

then

$$\frac{dx}{dt} = A\frac{d}{dt}\cos(\omega t + \phi) = -\omega A\sin(\omega t + \phi)$$

$$\frac{d^2x}{dt^2} = -\,\omega A\frac{d}{dt}\sin(\omega t + \phi) = -\,\omega^2 A\cos(\omega t + \phi)$$

Comparing the expressions for x and d^2x/dt^2, we see that $d^2x/dt^2 = -\omega^2 x$ and Equation 21.17 is satisfied.

The following general statement can be made based on the foregoing discussion:

Whenever the force acting on a particle is linearly proportional to the displacement and in the opposite direction, the particle exhibits simple harmonic motion.

We shall give additional physical examples in subsequent sections.

Since the period of simple harmonic motion is $T = 2\pi/\omega$ and the frequency is the inverse of the period, we can express the period and frequency of the motion for this mass-spring system as

Period and frequency for a mass-spring system

$$T = \frac{2\pi}{\omega} = 2\pi\sqrt{\frac{m}{k}} \qquad [21.18]$$

$$f = \frac{1}{T} = \frac{1}{2\pi}\sqrt{\frac{k}{m}}\frac{\omega}{2\pi} \qquad [21.19]$$

That is, the period and frequency depend *only* on the mass and on the force constant of the spring. As we might expect, the frequency is larger for a stiffer spring (larger value of k) and decreases with increasing mass.

Special Case I In order to better understand the physical significance of our solution of the equation of motion, let us consider the following special case. Suppose we pull the mass from equilibrium by a distance of A and release it from rest in this stretched position, as in Figure 21.5. We must then require that our solution for $x(t)$ obey the initial conditions that at $t = 0$, $x_0 = A$ and $v_0 = 0$. These conditions are met if we choose $\phi = 0$, giving $x = A\cos\omega t$ as our solution. To check this solution, we note that it satisfies the condition that $x_0 = A$ at $t = 0$,

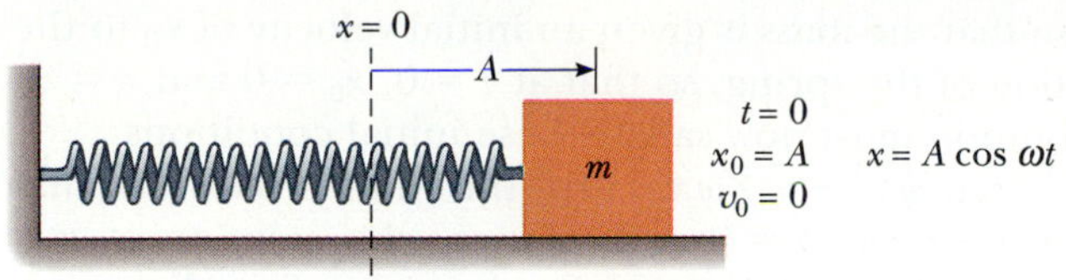

Figure 21.5 A mass-spring system that starts from rest at $x_0 = A$. In this case, $\phi = 0$, and so $x = A \cos \omega t$.

since cos 0 = 1. Thus, we see that A and ϕ contain the information on initial conditions.

Now let us investigate the behavior of the velocity and acceleration in this special case. Since $x = A \cos \omega t$, we have

$$v = \frac{dx}{dt} = -\omega A \sin \omega t$$

and

$$a = \frac{dv}{dt} = -\omega^2 A \cos \omega t$$

From the preceding velocity expression we see that at $t = 0$, $v_0 = 0$, as we require. The expression for the acceleration tells us that at $t = 0$, $a = -\omega^2 A$. Physically this makes sense, since the force on the mass is to the left when the displacement is positive. In fact, at this position $F = -kA$ (to the left), and the initial acceleration is $-kA/m$.

We could also use a more formal approach to show that $x = A \cos \omega t$ is the correct solution by using the relation $\tan \phi = -v_0/\omega x_0$ (Eq. 21.10). Since $v_0 = 0$ at $t = 0$, $\tan \phi = 0$ and so $\phi = 0$.

The displacement, velocity, and acceleration versus time are plotted in Figure 21.6 for this special case. Note that the acceleration reaches extreme values of $\mp \omega^2 A$ when the displacement has extreme values of $\pm A$. Furthermore, the velocity has extreme values of $\pm \omega A$, which both occur at $x = 0$. Hence, the quantitative solution agrees with our qualitative description of this system.

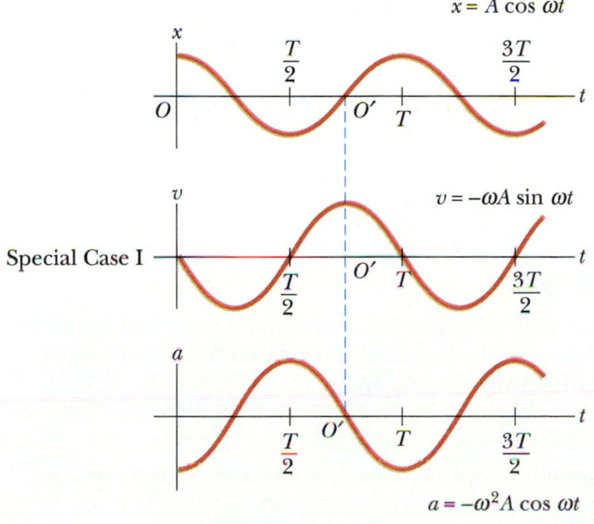

Figure 21.6 Displacement, velocity, and acceleration versus time for a mass undergoing simple harmonic motion under the initial conditions that at $t = 0$, $x_0 = A$ and $v_0 = 0$.

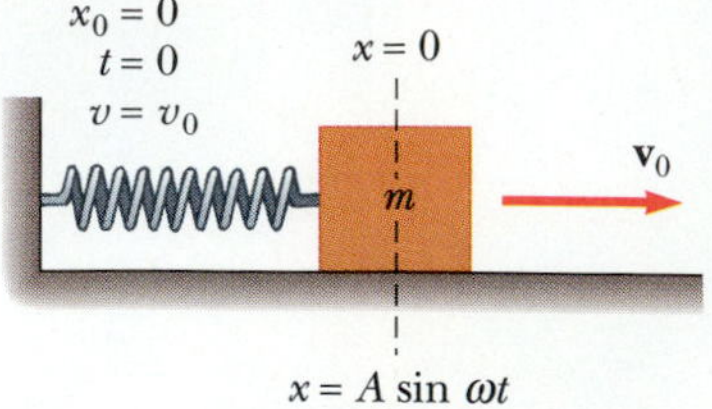

Figure 21.7 The mass-spring system starts its motion at the equilibrium position $x_0 = 0$ at $t = 0$. If its initial velocity is v_0 to the right, its x coordinate varies as $x = A \sin \omega t$.

Special Case II Now suppose that the mass is given an initial velocity of $\mathbf{v}_0$ to the *right* at the unstretched position of the spring, so that at $t = 0$, $x_0 = 0$ and $v = v_0$ (Fig. 21.7). Our particular solution must now satisfy these initial conditions.

Applying Equation 21.11, $\tan \phi = -v_0/\omega x_0$, and the initial condition that $x_0 = 0$ at $t = 0$ gives $\tan \phi = -\infty$ or $\phi = -\pi/2$. Hence, the solution is $x = A \cos(\omega t - \pi/2)$, which can be written $x = A \sin \omega t$. Furthermore, from Equation 21.11 we see that $A = v_0/\omega$; therefore, we can express our solution as

$$x = \frac{v_0}{\omega} \sin \omega t$$

The velocity and acceleration in this case are

$$v = \frac{dx}{dt} = v_0 \cos \omega t$$

$$a = \frac{dv}{dt} = -\omega v_0 \sin \omega t$$

This is consistent with the fact that the mass always has its maximum speed at $x = 0$, while the force and acceleration are zero at that position. The graphs of these functions versus time in Figure 21.6 correspond to the origin at O'. What would be the solution for x if the mass were initially moving to the left in Figure 21.7?

Example 21.2 That Car Needs a New Set of Shocks

A car of mass 1300 kg is constructed using a frame supported by four springs. Each spring has a force constant of 20 000 N/m. If two people riding in the car have a combined mass of 160 kg, find the frequency of vibration of the car when it is driven over a pothole in the road.

Solution We shall assume that the mass is evenly distributed. Thus, each spring supports one fourth of the load. The total mass supported by the springs is 1460 kg, and therefore each spring supports 365 kg. Hence, the frequency of vibration is, from Equation 21.19,

$$f = \frac{1}{2\pi}\sqrt{\frac{k}{m}} = \frac{1}{2\pi}\sqrt{\frac{20\ 000\ \text{N/m}}{365\ \text{kg}}} = \boxed{1.18\ \text{Hz}}$$

Exercise How long does it take the car to execute two complete vibrations?

Answer 1.70 s.

Example 21.3 A Mass-Spring System

A mass of 200 g is connected to a light spring of force constant 5 N/m and is free to oscillate on a horizontal, frictionless track. The mass is displaced 5 cm to the right from equilibrium and released from rest, as in Figure 21.5.

(a) Find the period of its motion.

Solution This situation corresponds to Special Case I, where $x = A \cos \omega t$ and $A = 5 \times 10^{-2}$ m. Thus,

$$\omega = \sqrt{\frac{k}{m}} = \sqrt{\frac{5 \text{ N/m}}{200 \times 10^{-3} \text{ kg}}} = 5 \text{ rad/s}$$

Therefore,

$$T = \frac{2\pi}{\omega} = \frac{2\pi}{5} = \boxed{1.26 \text{ s}}$$

(b) Determine the maximum speed of the mass.

$$v_{\max} = \omega A = (5 \text{ rad/s})(5 \times 10^{-2} \text{ m}) = \boxed{0.250 \text{ m/s}}$$

(c) What is the maximum acceleration of the mass?

$$a_{\max} = \omega^2 A = (5 \text{ rad/s})^2(5 \times 10^{-2} \text{ m}) = \boxed{1.25 \text{ m/s}^2}$$

(d) Express the displacement, speed, and acceleration as functions of time.

Solution The expression $x = A \cos \omega t$ is our solution for Special Case I, and so we can use the results from (a), (b), and (c) to get

$$x = A \cos \omega t = \boxed{(0.05 \text{ m}) \cos 5t}$$

$$v = -\omega A \sin \omega t = \boxed{-(0.25 \text{ m/s}) \sin 5t}$$

$$a = -\omega^2 A \cos \omega t = \boxed{-(1.25 \text{ m/s}^2) \cos 5t}$$

21.3 Energy of the Simple Harmonic Oscillator

Let us examine the mechanical energy of the mass-spring described in Figure 21.5. Since the surface is frictionless, we expect that the total mechanical energy is constant. We can use Equation 21.5 to express the kinetic energy as

$$K = \tfrac{1}{2}mv^2 = \tfrac{1}{2}m\omega^2 A^2 \sin^2(\omega t + \phi) \qquad [21.20]$$

Kinetic energy of a simple harmonic oscillator

The elastic potential energy stored in the spring for any elongation x is $\frac{1}{2}kx^2$. Using Equation 21.1, we get

$$U = \tfrac{1}{2}kx^2 = \tfrac{1}{2}kA^2 \cos^2(\omega t + \phi) \qquad [21.21]$$

Potential energy of a simple harmonic oscillator

We see that K and U are always positive quantities. Since $\omega^2 = k/m$, we can express the *total energy* of the simple harmonic oscillator as

$$E = K + U = \tfrac{1}{2}kA^2[\sin^2(\omega t + \phi) + \cos^2(\omega t + \phi)]$$

But $\sin^2 \theta + \cos^2 \theta = 1$; therefore, this equation reduces to

$$\boxed{E = \tfrac{1}{2}kA^2} \qquad [21.22]$$

Total energy of a simple harmonic oscillator

That is, the energy of a simple harmonic oscillator is a constant of the motion and proportional to the square of the amplitude. In fact, the total mechanical

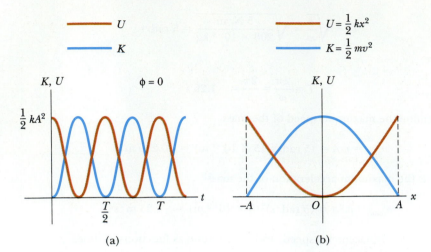

Figure 21.8
(a) Kinetic energy and potential energy versus time for a simple harmonic oscillator with $\phi = 0$. (b) Kinetic energy and potential energy versus displacement for a simple harmonic oscillator. Note that $K + U =$ constant in both plots.

energy is just equal to the maximum potential energy stored in the spring when $x = \pm A$. At these points, $v = 0$ and there is no kinetic energy. At the equilibrium position, $x = 0$ and $U = 0$, so that the total energy is all in the form of kinetic energy. That is, at $x = 0$, $E = \frac{1}{2}mv_{\text{max}}^2 = \frac{1}{2}m\omega^2A^2$.

Plots of the kinetic and potential energies versus time are shown in Figure 21.8a, where $\phi = 0$. In this situation, both K and U are always positive, and their sum at all times is a constant equal to $\frac{1}{2}kA^2$, the total energy of the system. The variations of K and U with displacement are plotted in Figure 21.8b. Energy is continuously being transferred between potential energy stored in the spring and the kinetic energy of the mass. Figure 21.9 illustrates the position, velocity, acceleration, kinetic energy, and potential energy of the mass-spring system for one full period of the motion. Most of the ideas discussed so far are incorporated in this important figure. We suggest that you study it carefully.

Finally, we can use energy to obtain the velocity for an arbitrary displacement, x, expressing the total energy at some arbitrary position as

$$E = K + U = \tfrac{1}{2}mv^2 + \tfrac{1}{2}kx^2 = \tfrac{1}{2}kA^2$$

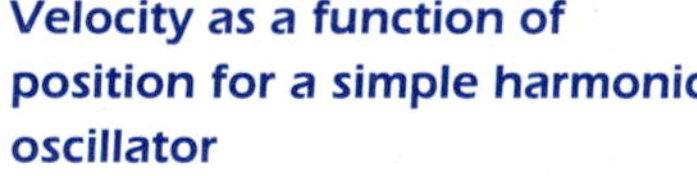

Velocity as a function of position for a simple harmonic oscillator

$$v = \pm\sqrt{\frac{k}{m}(A^2 - x^2)} = \pm\,\omega\sqrt{A^2 - x^2} \qquad [21.23]$$

Again, this expression substantiates the fact that the speed is a maximum at $x = 0$ and zero at the turning points, $x = \pm A$.

Example 21.4 Oscillations on a Horizontal Surface

A 0.5-kg mass connected to a light spring of force constant 20 N/m oscillates on a horizontal, frictionless surface. (a) Calculate the total energy of the system and the maximum speed of the mass if the amplitude of the motion is 3 cm.

Solution Using Equation 21.22, we get

$$E = \tfrac{1}{2}kA^2 = \tfrac{1}{2}(20\text{ N/m})\,(3 \times 10^{-2}\text{ m})^2$$
$$= 9.00 \times 10^{-3}\text{ J}$$

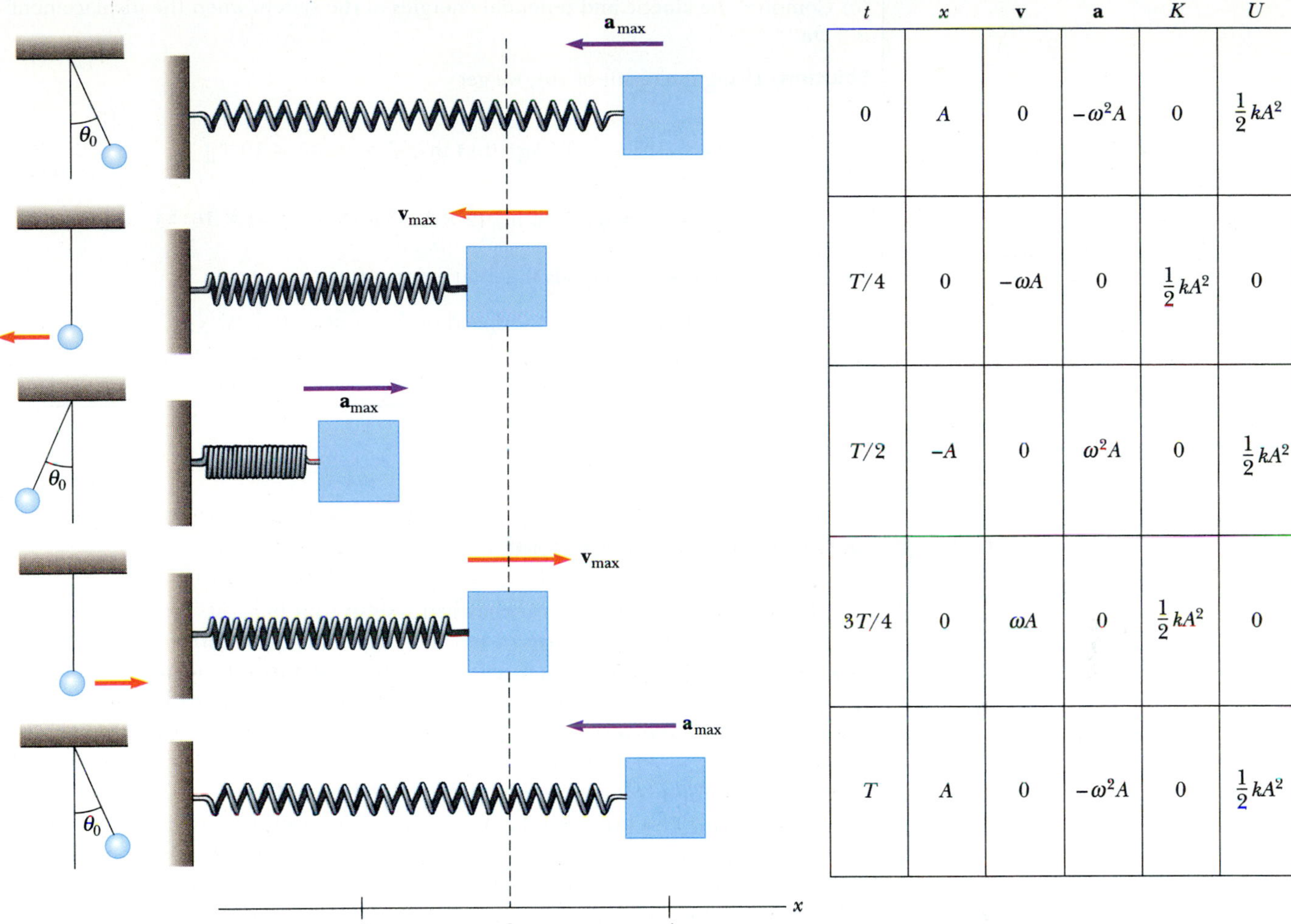

t	x	$\mathbf{v}$	$\mathbf{a}$	K	U
0	A	0	$-\omega^2 A$	0	$\frac{1}{2}kA^2$
$T/4$	0	$-\omega A$	0	$\frac{1}{2}kA^2$	0
$T/2$	$-A$	0	$\omega^2 A$	0	$\frac{1}{2}kA^2$
$3T/4$	0	ωA	0	$\frac{1}{2}kA^2$	0
T	A	0	$-\omega^2 A$	0	$\frac{1}{2}kA^2$

Figure 21.9 Simple harmonic motion for a mass-spring system and its analogy to the motion of a simple pendulum. The parameters in the table refer to the mass-spring system, assuming that at $t = 0$, $x = A$ so that $x = A\cos\omega t$ (Special Case I).

When the mass is at $x = 0$, $U = 0$ and $E = \frac{1}{2}mv_{\text{max}}^2$; therefore,

$$\tfrac{1}{2}mv_{\text{max}}^2 = \boxed{9.00 \times 10^{-3}\ \text{J}}$$

$$v_{\text{max}} = \sqrt{\frac{18 \times 10^{-3}\ \text{J}}{0.5\ \text{kg}}} = \boxed{0.190\ \text{m/s}}$$

(b) What is the speed of the mass when the displacement is 2 cm?

Solution We can apply Equation 21.23 directly:

$$v = \pm\sqrt{\frac{k}{m}(A^2 - x^2)} = \pm\sqrt{\frac{20}{0.5}(3^2 - 2^2) \times 10^{-4}}$$

$$= \boxed{\pm 0.141\ \text{m/s}}$$

The positive and negative signs indicate that the mass could be moving to the right or left at this instant.

(c) Compute the kinetic and potential energies of the system when the displacement is 2 cm.

Solution Using the result of (b), we get

$$K = \tfrac{1}{2}mv^2 = \tfrac{1}{2}(0.5\text{ kg})(0.14\text{ m/s})^2 = 5.00 \times 10^{-3}\text{ J}$$

$$U = \tfrac{1}{2}kx^2 = \tfrac{1}{2}(20\text{ N/m})\ (2 \times 10^{-2}\text{ m})^2 = 4.00 \times 10^{-3}\text{ J}$$

Note that the sum $K + U$ equals the total energy, E.

Exercise For what values of x is the speed of the mass 0.10 m/s?

Answer ± 2.55 cm.

21.4 Motion of a Pendulum

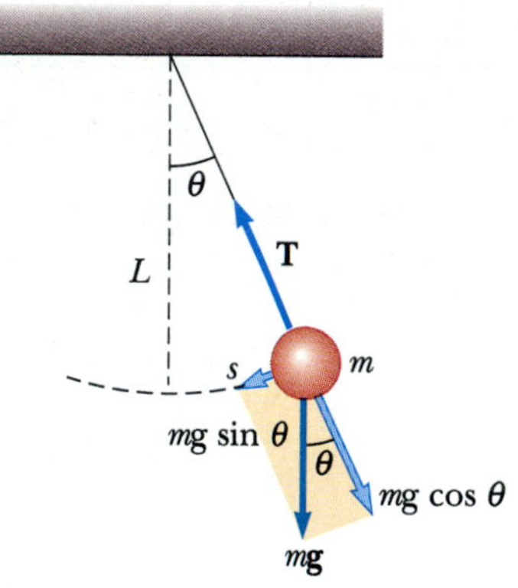

Figure 21.10 When θ is small, the simple pendulum oscillates with simple harmonic motion about the equilibrium position ($\theta = 0$). The restoring force is $mg \sin \theta$, the component of the weight tangent to the circular arc.

The **simple pendulum** is another mechanical system that exhibits periodic motion. It consists of a point mass, m, suspended by a light string of length L, where the upper end of the string is fixed as in Figure 21.10. The motion occurs in a vertical plane and is driven by the force of gravity. We shall show that the motion is that of a simple harmonic oscillator, provided the angle, θ, that the pendulum makes with the vertical is small.

The forces acting on the mass are the tension force, **T**, acting along the string, and the weight, $m\mathbf{g}$. The tangential component of the weight, $mg \sin \theta$, always acts toward $\theta = 0$, opposite the displacement. Therefore, the tangential force is a restoring force, and we can use Newton's second law to write the equation of motion in the tangential direction as

$$F_t = -mg \sin \theta = m\frac{d^2s}{dt^2}$$

where s is the displacement measured along the arc in Figure 21.10, and the minus sign indicates that F_t acts toward the equilibrium position. Since $s = L\theta$ (from Eq. 11.1 in Chapter 11) and L is constant, this equation reduces to

$$\frac{d^2\theta}{dt^2} = -\frac{g}{L}\sin \theta$$

The right side is proportional to $\sin \theta$ rather than to θ; hence, we conclude that the motion is not simple harmonic motion since it is not of the form of Equation 21.17. However, if we assume that θ is *small*, we can use the approximation $\sin \theta \approx \theta$, where θ is measured in radians, and the equation of motion becomes

Equation of motion for the simple pendulum (small θ)

$$\frac{d^2\theta}{dt^2} = -\frac{g}{L}\theta \qquad [21.24]$$

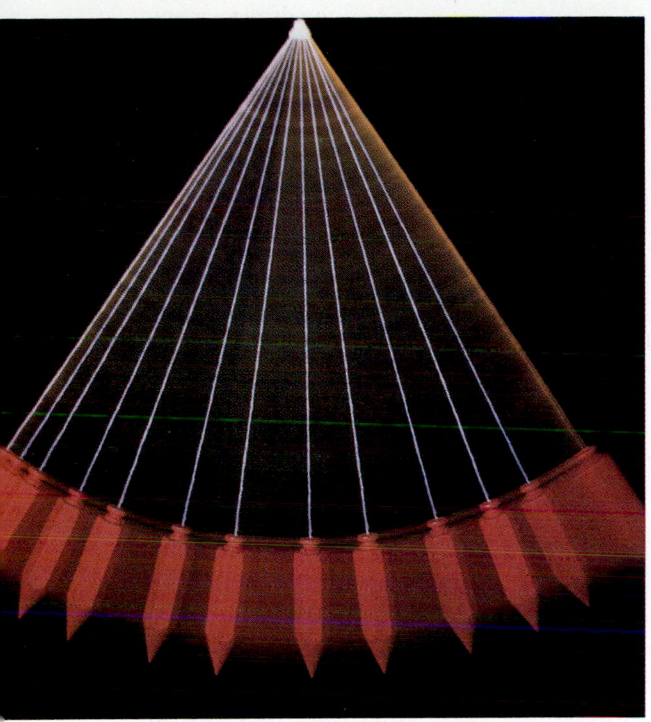

A multiflash photograph of a swinging pendulum. Is the oscillating motion simple harmonic in this case? *(Paul Silverman, Fundamental Photographs)*

The Foucault pendulum at the Smithsonian Institution in Washington, D.C. This type of pendulum was first used by the French physicist Jean Foucault to verify the Earth's rotation experimentally. The pendulum's plane of oscillation appears to rotate, as the bob successively knocks over the red indicators arranged in a horizontal circle. In reality, the pendulum's plane of motion is fixed in space, while the Earth rotates beneath it. *(Courtesy of the Smithsonian Institution)*

Now we have an expression with exactly the same form as Equation 21.16, and so we conclude that the motion is simple harmonic motion. Therefore, θ can be written as $\theta = \theta_0 \cos(\omega t + \phi)$, where θ_0 is the *maximum angular displacement* and the angular frequency ω is

$$\omega = \sqrt{\frac{g}{L}} \qquad [21.25]$$

Angular frequency of motion for a simple pendulum

The period of the motion is

$$T = \frac{2\pi}{\omega} = 2\pi\sqrt{\frac{L}{g}} \qquad [21.26]$$

Period of motion for a simple pendulum

In other words, *the period and angular frequency of a simple pendulum depend only on the length of the string and the free-fall acceleration.* Since the period is *independent* of the mass, we conclude that *all* simple pendulums of equal length at the same location oscillate with equal periods. The analogy between the motion of a simple pendulum and the mass-spring system is illustrated in Figure 21.9.

The simple pendulum can be used as a timekeeper. It is also a convenient device for making precise measurements of the free-fall acceleration. Such measurements are important since variations in local values of g can provide information on the locations of oil and other valuable underground resources.

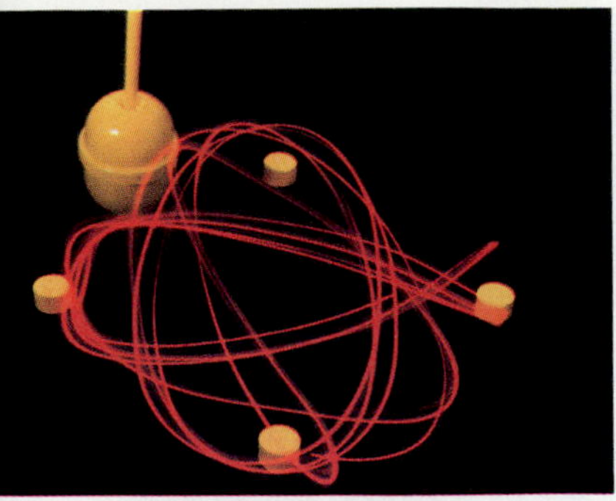

This is a computer-generated simulation of the chaotic pattern of a driven pendulum. The nonlinearity in the motion is provided by the gravitational torque, which is proportional to the sine of the angle. An important research problem is to predict the conditions under which different physical systems exhibit chaotic behavior. *(© Yoav Levy/Phototake)*

Example 21.5 What Is the Height of That Tower?

A man enters a tall tower, needing to know its height. He notes that a long pendulum extends from the ceiling almost to the floor and that its period is 12 s. How tall is the tower?

Solution If we solve $T = 2\pi\sqrt{L/g}$ for L, we get

$$L = \frac{gT^2}{4\pi^2} = \frac{(9.80 \text{ m/s}^2)(12 \text{ s})^2}{4\pi^2} = \boxed{35.7 \text{ m}}$$

Exercise If this pendulum were taken to the Moon, where the free-fall acceleration is 1.67 m/s^2, what would its period be?

Answer 29.1 s.

*The Physical Pendulum

If a hanging object oscillates about a fixed axis that does not pass through its center of mass, and the object cannot be accurately approximated as a point mass, then it must be treated as a physical, or compound, pendulum. Consider a rigid body pivoted at a point O that is a distance of d from the center of mass (Fig. 21.11). The torque about O is provided by the force of gravity, and its magnitude is $mgd \sin \theta$. Using the fact that $\tau = I\alpha$, where I is the moment of inertia about the axis through O, we get

$$-mgd \sin \theta = I\frac{d^2\theta}{dt^2}$$

The minus sign on the left indicates that the torque about O tends to decrease θ. That is, the force of gravity produces a restoring torque.

If we again assume that θ is small, then the approximation $\sin \theta \approx \theta$ is valid and the equation of motion reduces to

$$\frac{d^2\theta}{dt^2} = -\left(\frac{mgd}{I}\right)\theta = -\omega^2\theta \qquad [21.27]$$

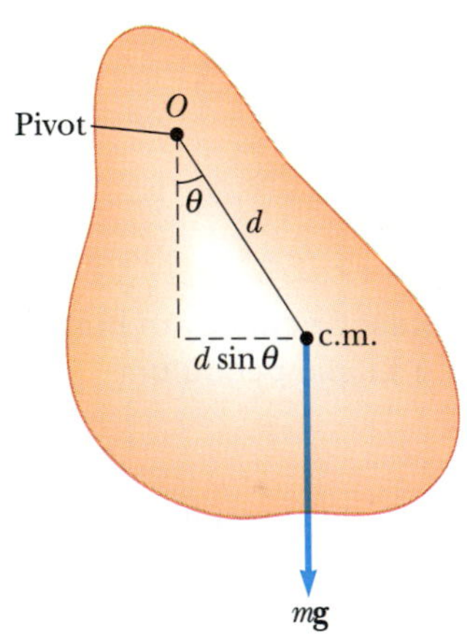

Figure 21.11 The physical pendulum consists of a rigid body pivoted at the point O, and not through the center of mass. At equilibrium, the weight vector passes through O, corresponding to $\theta = 0$. The restoring torque about O when the system is displaced through an angle θ is $mgd \sin \theta$.

Note that the equation has the same form as Equation 21.17, and so the motion is simple harmonic motion. That is, the solution of Equation 21.27 is $\theta = \theta_0 \cos(\omega t + \phi)$, where θ_0 is the maximum angular displacement and

$$\omega = \sqrt{\frac{mgd}{I}}$$

The period is

$$T = \frac{2\pi}{\omega} = 2\pi\sqrt{\frac{I}{mgd}} \qquad [21.28]$$

One can use this result to measure the moment of inertia of a planar rigid body. If the location of the center of mass and, hence, the distance d are known, the moment of inertia can be obtained through a measurement of the period. Finally, note that Equation 21.28 reduces to the period of a simple pendulum (Eq. 21.26) when $I = md^2$, that is, when all the mass is concentrated at the center of mass.

Example 21.6 A Swinging Rod

A uniform rod of mass M and length L is pivoted about one end and oscillates in a vertical plane (Fig. 21.12). Find the period of oscillation if the amplitude of the motion is small.

Solution The moment of inertia of a uniform rod about an axis through one end is $\frac{1}{3}ML^2$. The distance d from the pivot to the center of mass is $L/2$. Substituting these quantities into Equation 21.28 gives

$$T = 2\pi\sqrt{\frac{\frac{1}{3}ML^2}{Mg\frac{L}{2}}} = 2\pi\sqrt{\frac{2L}{3g}}$$

Comment In one of the early Moon landings, an astronaut walking on the Moon's surface had a belt hanging from his spacesuit, and the belt oscillated as a physical pendulum. A scientist on Earth observed this motion on TV and from it was able to estimate the free-fall acceleration on the Moon. How do you suppose he did it?

Exercise Calculate the period of a meter stick pivoted about one end and oscillating in a vertical plane as in Figure 21.12.

Answer 1.64 s.

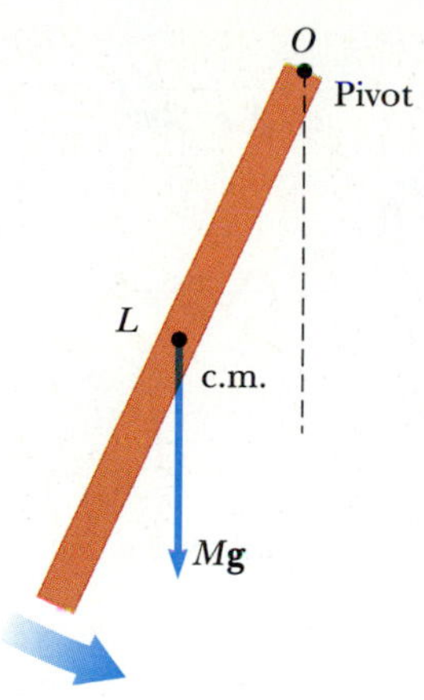

Figure 21.12 (Example 21.6) A rigid rod oscillating about a pivot through one end is a physical pendulum with $d = L/2$ and $I_0 = \frac{1}{3}ML^2$.

*21.5 Damped Oscillations

The oscillatory motions we have considered so far have occurred in the context of an ideal system, that is, one that oscillates indefinitely under the action of a linear restoring force. In realistic systems, dissipative forces, such as friction, are present and retard the motion of the system. Consequently, the mechanical energy of the system diminishes in time, and the motion is said to be *damped.*

One common type of drag force, which we discussed in Chapter 5, is proportional to the velocity and acts in the direction opposite the motion. This type of drag is often observed when an object is oscillating in air, for instance. Because the drag force can be expressed as $\mathbf{R} = -b\mathbf{v}$, where b is a constant, and the restoring force of the system is $-kx$, we can write Newton's second law as

$$\sum F_x = -kx - bv = ma_x$$

$$-kx - b\frac{dx}{dt} = m\frac{d^2x}{dt^2} \qquad [21.29]$$

The solution of this equation requires mathematics that may not yet be familiar to you, and so it will simply be stated without proof. When the drag force is small compared with kx—that is, when b is small—the solution to Equation 21.29 is

$$x = A\,e^{-\frac{b}{2m}t}\cos(\omega t + \phi) \qquad [21.30]$$

where the angular frequency of motion is

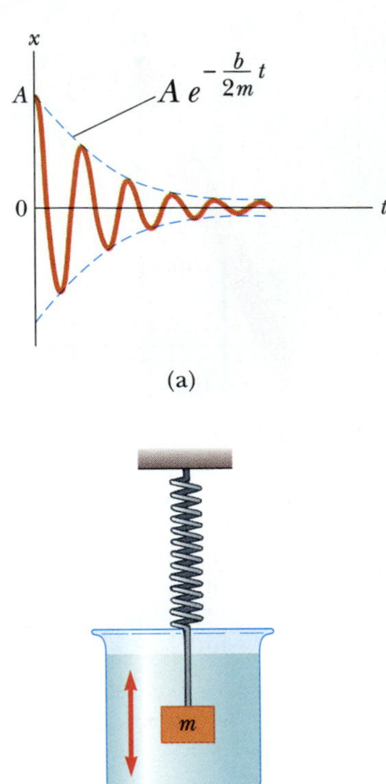

Figure 21.13
(a) Graph of the displacement versus time for an underdamped oscillator. Note the decrease in amplitude with time. (b) One example of a damped oscillator is a mass on a spring submersed in a liquid.

$$\omega = \sqrt{\frac{k}{m} - \left(\frac{b}{2m}\right)^2} \quad [21.31]$$

This result can be verified by substituting Equation 21.30 into Equation 21.29. Figure 21.13a shows the displacement as a function of time in this case. We see that, *when the drag force is small compared with the restoring force, the oscillatory character of the motion is preserved but the amplitude of vibration decreases in time* and the motion ultimately ceases. This is known as an **underdamped oscillator.** The dashed blue lines in Figure 21.13a, which outline the *envelope* of the oscillatory curve, represent the exponential factor that appears in Equation 21.30. The exponential factor shows that *the amplitude decays exponentially with time.* For motion with a given spring constant and particle mass, the oscillations dampen more rapidly as the maximum value of the drag force approaches the maximum value of the restoring force. One example of a damped harmonic oscillator is a mass immersed in a fluid as in Figure 21.13b.

It is convenient to express the angular frequency of vibration of a damped system (Eq. 21.31) in the form

$$\omega = \sqrt{\omega_0^2 - \left(\frac{b}{2m}\right)^2}$$

where $\omega_0 = \sqrt{k/m}$ represents the angular frequency of oscillation in the absence of a drag force (the undamped oscillator). In other words, when $b = 0$, the drag force is zero and the system oscillates with its natural frequency, ω_0. As the magnitude of the drag force approaches the value of the restoring force in the spring, the oscillations dampen more rapidly. When b reaches a critical value, b_c, so that $b_c/2m = \omega_0$, the system does not oscillate and is said to be **critically damped.** In this case it returns to equilibrium in an exponential manner with time, as in Figure 21.14.

If the medium is so viscous that the drag force is greater than the restoring force—that is, if $b/2m > \omega_0$—the system is **overdamped.** Again, the displaced system does not oscillate but simply returns to its equilibrium position. As the damping increases, the time it takes the displacement to reach equilibrium also increases, as indicated in Figure 21.14. In any case, when a drag force is present, the energy of the oscillator eventually falls to zero. The lost mechanical energy dissipates into thermal energy in the resistive medium.

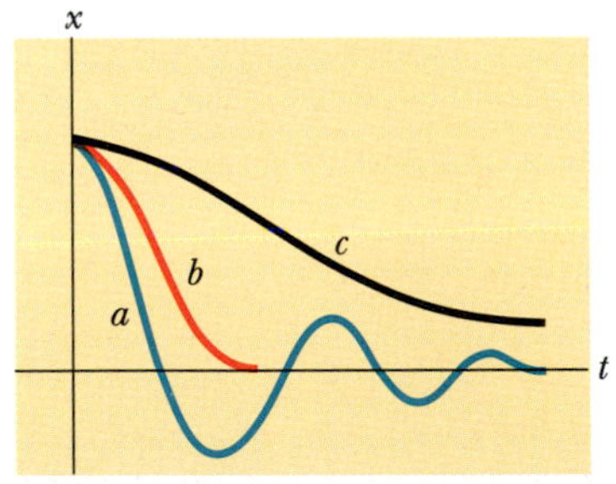

Figure 21.14
Plots of displacement versus time for (*a*) an underdamped oscillator, (*b*) a critically damped oscillator and (*c*) an overdamped oscillator.

*21.6 Forced Oscillations

We have seen that the energy of a damped oscillator decreases in time as a result of the drag force. It is possible to compensate for this energy loss by applying an external force that does positive work on the system. At any instant, energy can be put into the system by an applied force that acts in the direction of motion of the oscillator. For example, a child on a swing can be kept in motion by appropriately timed "pushes." The amplitude of motion remains constant if the energy input per cycle of motion exactly equals the energy lost as a result of air drag.

A common example of a forced oscillator is a damped oscillator driven by an external force that varies periodically, such as $F = F_0 \cos \omega t$, where ω is the angular frequency of the force and F_0 is a constant. Adding this driving force to the left side of Equation 21.29 gives

$$F_0 \cos \omega t - b\frac{dx}{dt} - kx = m\frac{d^2x}{dt^2} \quad \text{[21.32]}$$

Again, the solution of this equation is rather lengthy and will not be presented. However, after a sufficiently long period of time, when the energy input per cycle equals the energy lost per cycle, a steady-state condition is reached in which the oscillations proceed with constant amplitude. At this time, Equation 21.32 has the solution

$$x = A\cos(\omega t + \phi) \quad \text{[21.33]}$$

where

$$A = \frac{F_0/m}{\sqrt{(\omega^2 - \omega_0{}^2)^2 + \left(\frac{b\omega}{m}\right)^2}} \quad \text{[21.34]}$$

and where $\omega_0 = \sqrt{k/m}$ is the angular frequency of the undamped oscillator ($b = 0$). From a physical point of view, one can argue that in a steady state the oscillator must have the same angular frequency as the driving force, and so the solution given by Equation 21.33 is expected.

Equation 21.34 shows that the motion of the forced oscillator is not damped since it is being driven by an external force. That is, the external agent provides the energy necessary to overcome the losses due to the drag force. Note that the mass oscillates at the angular frequency of the driving force, ω. For small damping, the amplitude becomes large when the frequency of the driving force is near the natural frequency of oscillation, or when $\omega \approx \omega_0$. The dramatic increase in amplitude near the natural frequency is called **resonance,** and the angular frequency ω_0 is called the **resonance frequency** of the system.

Physically, the reason for large-amplitude oscillations at the resonance frequency is that energy is being transferred to the system under the most favorable conditions. This can be better understood by taking the first time derivative of x, which gives an expression for the velocity of the oscillator. In doing so, one finds that v is proportional to $\sin(\omega t + \phi)$. When the applied force is in phase with **v**, the rate at which work is done on the oscillator by the force **F** (in other words, the power) equals Fv. Since the quantity Fv is always positive when **F** and **v** are in phase, we conclude that *at resonance the applied force is in phase with the velocity, and the power transferred to the oscillator is a maximum.*

Figure 21.15 is a graph of amplitude as a function of frequency for the forced oscillator, with and without a resistive force. Note that the amplitude increases with decreasing damping ($b \rightarrow 0$) and that the resonance curve broadens as the damping increases. Under steady-state conditions, and at any driving frequency, the energy transferred into the system equals the energy lost because of the damping force; hence, the average total energy of the oscillator remains constant. In the absence of a damping force ($b = 0$), we see from Equation 21.34 that the steady-state amplitude approaches infinity as $\omega \rightarrow \omega_0$. In other words, if there are no losses in the system, and we continue to drive an initially motionless oscillator with a sinusoidal force that is in phase with the velocity, the amplitude of motion will build up without limit (Fig. 22.15). This does not occur in practice because some damping is always present. That is, at resonance the amplitude is large but finite for small damping.

One experiment that demonstrates a resonance phenomenon is illustrated in Figure 21.16. Several pendula of different lengths are suspended from a stretched

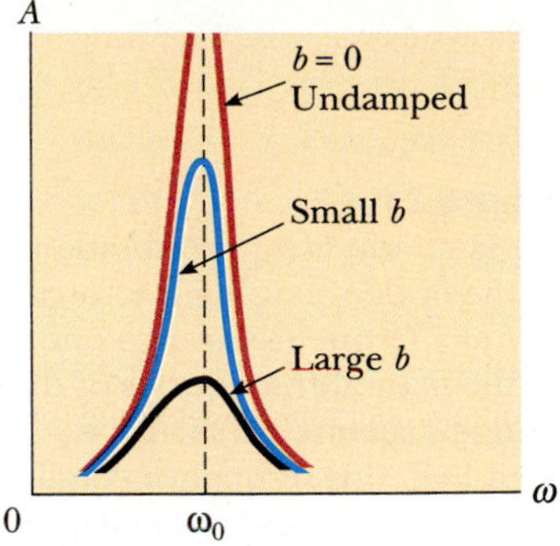

Figure 21.15 Graph of the amplitude versus frequency for a damped oscillator when a periodic driving force is present. When the frequency of the driving force equals the natural frequency, ω_0, resonance occurs. Note that the shape of the resonance curve depends on the size of the damping coefficient, b.

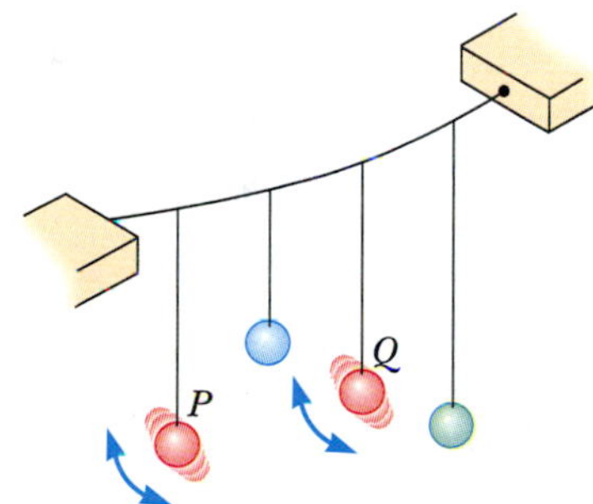

Figure 21.16 If pendulum P is set into oscillation, pendulum Q will eventually oscillate with the greatest amplitude because of the coupling between them and the fact that they have the same natural frequency of vibration. The pendula are oscillating in a direction perpendicular to the plane formed by the stationary strings.

(a)

(b)

Figure 21.17
(a) High winds set up vibrations in the bridge, causing it to oscillate at a frequency near to one of the natural frequencies of the bridge structure. (b) Once established, this resonance condition led to the bridge's collapse. *(UPI/Bettmann Newsphotos)*

string. If one of them, such as P, is set swinging in the plane perpendicular to the plane of the string, the others will begin to oscillate, since they are coupled by the stretched string. Of those that are forced into oscillation by this coupling, pendulum Q, which is the same length as P (hence, the two pendula have the same natural frequency), will oscillate with the greatest amplitude.

Resonance appears in other areas of physics. For example, certain electrical circuits have natural (resonant) frequencies. A structure such as a bridge has natural frequencies and can be set into resonance by an appropriate driving force. A striking example of such a structural resonance occurred in 1940, when the Tacoma Narrows Bridge in Washington was destroyed by resonant vibrations (Fig. 21.17). The winds were not particularly strong on that occasion, but the bridge still collapsed because vortices (turbulences) generated by the wind blowing through the bridge occurred at a frequency that matched the natural frequency of the bridge.

Many other examples of resonant vibrations can be cited. A resonant vibration you may have experienced is the "singing" of telephone wires in the wind. Machines often break if one vibrating part is at resonance with some other moving part. Finally, soldiers marching in cadence across a bridge have been known to set up resonant vibrations in the structure, causing it to collapse. In one famous accident, which occurred in France in 1850, a collapsed suspension bridge resulted in the death of 226 soldiers.

21.7 Oscillations in Circuits

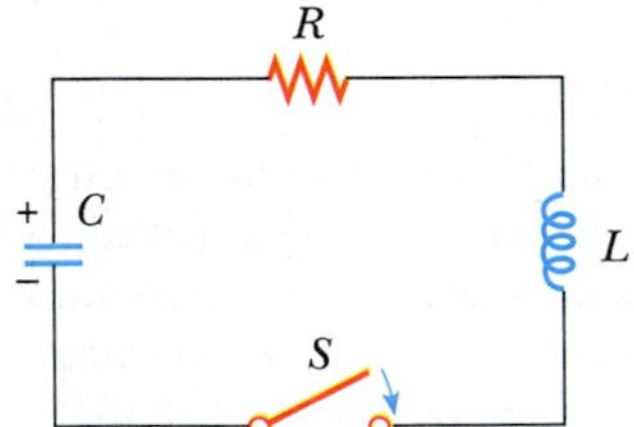

Figure 21.18
A series RLC circuit. The capacitor is charged at $t = 0$ when the switch is being closed.

Consider a circuit consisting of an inductor, a capacitor, and a resistor connected in series, as in Figure 21.18. In typical situations, the current is supplied by a battery. In this case, however, the current is started by the charged capacitor, and we shall assume that a current I is present at some time. Applying Kirchhoff's loop rule to this circuit, we find

$$-L\frac{dI}{dt} - IR - \frac{q}{C} = 0 \qquad [21.35]$$

where the first term is the voltage drop across the inductor, the second term is the voltage drop across the resistor, and the last term is the voltage drop across the capacitor. Taking the derivative of Equation 21.35 with respect to time, and using the fact that $I = dq/dt$, gives

$$L\frac{d^2I}{dt^2} + R\frac{dI}{dt} + \frac{I}{C} = 0 \qquad [21.36]$$

The solution of this differential equation, together with the appropriate initial conditions, gives the current as a function of time. Let us consider two special cases:

The *LC* Circuit ($R = 0$)

If the resistance in the circuit is negligible, we can set $R = 0$ in Equation 21.36, and divide each term by L to give

$$\frac{d^2I}{dt^2} + \frac{1}{LC}I = 0 \qquad [21.37]$$

This equation is identical in form to the equation of motion for the simple harmonic oscillator (Eq. 21.17), $d^2x/dt^2 - {\omega_0}^2x = 0$, where x is replaced by I and

$$\omega_0 = \frac{1}{\sqrt{LC}} \qquad [21.38]$$

Thus, we conclude that the current in the LC circuit oscillates according to the expression $I = I_m \sin(\omega t + \phi)$, with an angular frequency given by Equation 21.38, and a phase angle that depends on the initial conditions. Note that the angular frequency depends solely on the inductance and capacitance of the circuit.

The sustained oscillations in the LC circuit can be described from an energy transfer viewpoint as follows. When the capacitor is fully charged, the total energy in the circuit is stored in the electric field of the capacitor. At this time, the current is zero and there is no energy stored in the inductor. As the capacitor begins to discharge, the energy stored in the electric field decreases. At the same time, the current increases and part of the energy is now stored in the magnetic field of the inductor. Thus, energy is transferred from the electric field of the capacitor to the magnetic field of the inductor. When the capacitor is fully discharged, it stores no energy, the current reaches its maximum value, and all of the energy is now stored in the inductor. The energy continues to transfer between the inductor and capacitor indefinitely, corresponding to oscillations in the current and charge.

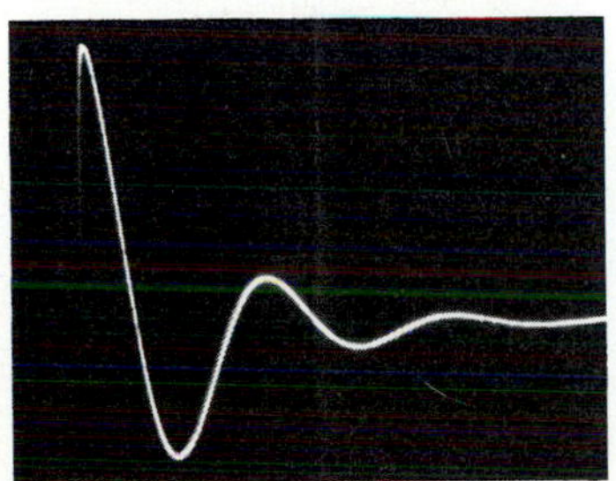

Oscilloscope pattern showing the decay in the oscillations of an RLC circuit. The parameters used were $R = 75\ \Omega$, $L =$ 10 mH, $C = 0.19\ \mu$F, and $f =$ 300 Hz. *(Courtesy of J. Rudmin)*

The *RLC* Circuit

Now consider the more general case where the resistance in the circuit is not zero. In this case, note that Equation 21.36 is identical in form to the equation of motion for the damped oscillator, given by Equation 21.29. It is useful to note the following correspondences between parameters in both systems:

$$R \leftrightarrow b \qquad L \leftrightarrow m \qquad C \leftrightarrow \frac{1}{k}$$

If R is sufficiently small, the solution of Equation 21.36 is

$$I = I_m e^{-(R/2L)t}\sin(\omega t + \phi) \qquad [21.39]$$

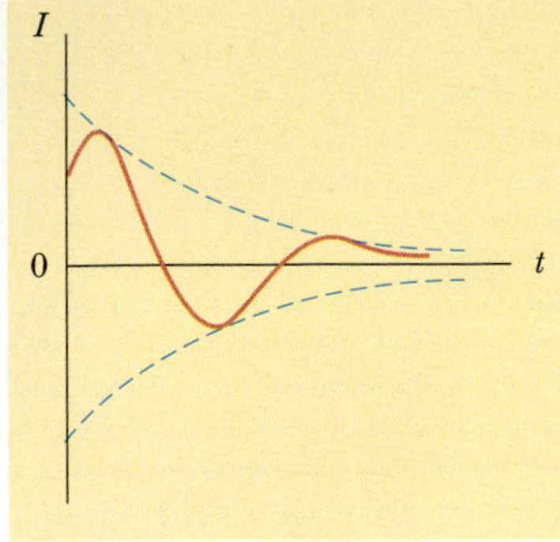

Figure 21.19
Current versus time for a damped *RLC* circuit. This occurs for $R \ll \sqrt{4L/C}$. The I versus t curve represents a plot of Equation 21.39.

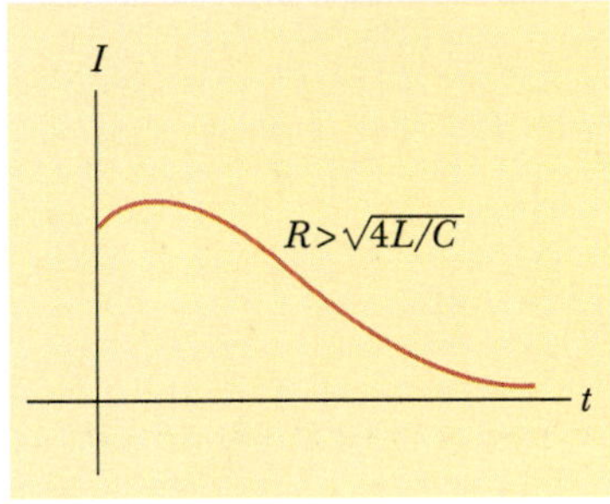

Figure 21.20
Plot of I versus t for an overdamped *RLC* circuit, which occurs for values of $R > \sqrt{4L/C}$.

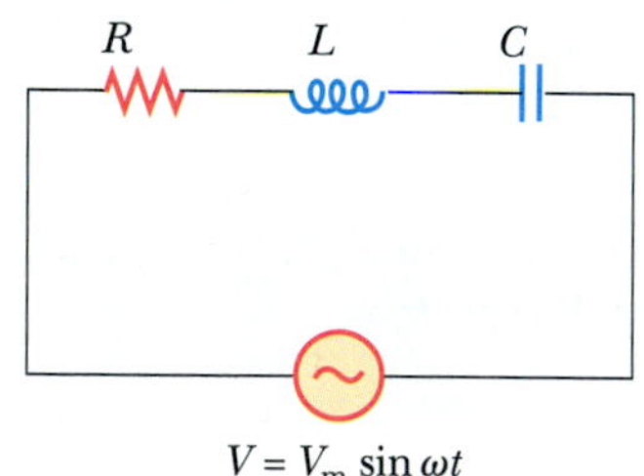

Figure 21.21
A series circuit consisting of a resistor, inductor, and capacitor connected to an alternating-current generator, denoted by the circuit symbol ⊖.

where

$$\omega = \left[\frac{1}{LC} - \left(\frac{R}{2L}\right)^2\right]^{1/2} \qquad \textbf{[21.40]}$$

That is, the current oscillates with *damped harmonic motion* in analogy with a mass-spring system moving in a viscous medium. From Equation 21.40, we see that when $R \ll \sqrt{4L/C}$ the frequency ω of the damped oscillator will be close to that of the undamped oscillator, $1/\sqrt{LC}$. A plot of the current versus time is shown in Figure 21.19. Note that the maximum value of I decreases after each oscillation, just as the amplitude of a damped harmonic oscillator decreases in time.

When we consider larger values of R, we find that the oscillations damp out more rapidly; in fact, there exists a critical resistance value R_c above which *no* oscillations occur. The critical value is given by $R_c = \sqrt{4L/C}$. A system with $R = R_c$ is said to be *critically damped*. When R exceeds R_c, the system is said to be *overdamped* and the current decreases gradually without oscillation (Fig. 21.20).

Alternating Current Circuits: Forced Oscillations

Now consider a circuit consisting of a resistor, an inductor, and a capacitor connected in series across an alternating voltage source as in Figure 21.21, where the applied voltage varies sinusoidally with time. That is, $V = V_m \sin \omega t$, where V_m is the voltage amplitude. Applying Kirchoff's loop rule to this circuit gives a differential equation involving the current (similar to Eq. 21.36), whose solution is

$$I = I_m \sin(\omega t - \phi) \qquad \textbf{[21.41]}$$

where the current amplitude I_m and phase angle ϕ between the current and applied voltage are

$$I_m = \frac{V_m}{\sqrt{R^2 + \left(\omega L - \dfrac{1}{\omega C}\right)^2}} \qquad \textbf{[21.42]}$$

$$\tan \phi = \frac{\omega L - \dfrac{1}{\omega C}}{R} \qquad \textbf{[21.43]}$$

It is common to define the **impedance** of the circuit to be

$$Z = \sqrt{R^2 + \left(\omega L - \frac{1}{\omega C}\right)^2} \qquad \textbf{[21.42]}$$

where Z has units of ohms. Therefore, Equation 21.42 can be written in the form $I_m = V_m/Z$, which can be regarded as a generalized Ohm's law applied to an alternating-current circuit.

From Equation 21.43, we see that when $\omega L > 1/\omega C$, the phase angle is positive, signifying that the current lags behind the voltage. When $\omega L < 1/\omega C$, the phase angle is negative, signifying that the current leads the applied voltage. Finally, when $\omega L = 1/\omega C$, the phase angle is zero. In this case, the impedance equals the resistance and the current has its maximum value, given by V_m/R. The frequency at which this occurs is called the **resonance frequency** of the circuit, given by $\omega_0 = (LC)^{-1/2}$. Again, you should note the interesting analogy between the alternating-current circuit and the forced mechanical oscillator discussed in Section 21.6.

When the power losses in an alternating-current circuit are analyzed, one finds that there is no power loss associated with the inductor or capacitor. The average power P_{av} delivered by the generator is dissipated as heat in the resistor, where

$$P_{av} = \frac{1}{2} I_m V_m \cos \phi \qquad [21.45]$$

Furthermore, the average power has its maximum value at the resonance frequency, corresponding to the phase angle $\phi = 0$.

Summary

The position of a simple harmonic oscillator varies periodically in time according to the relation

Displacement versus time for simple harmonic motion

$$x = A \cos(\omega t + \phi) \qquad [21.1]$$

where A is the amplitude of the motion, ω is the angular frequency, and ϕ is the phase constant. The value of ϕ depends on the initial position and velocity of the oscillator.

The time for one complete oscillation is called the **period** of the motion, defined by

Period

$$T = \frac{2\pi}{\omega} \qquad [21.2]$$

The inverse of the period is the **frequency** of the motion, which equals the number of oscillations per second.

The **velocity** and **acceleration** of a simple harmonic oscillator are

Velocity in simple harmonic motion

$$v = \frac{dx}{dt} = -\omega A \sin(\omega t + \phi) \qquad [21.5]$$

Acceleration in simple harmonic motion

$$a = \frac{dv}{dt} = -\omega^2 A \cos(\omega t + \phi) \qquad [21.6]$$

Thus, the maximum velocity is ωA, and the maximum acceleration is $\omega^2 A$. The velocity is zero when the oscillator is at its turning points, $x = \pm A$, and the speed is a maximum at the equilibrium position, $x = 0$. The magnitude of the acceleration is a maximum at the turning points and is zero at the equilibrium position.

A mass-spring system moving on a frictionless track exhibits simple harmonic motion with the period

Period of motion for a mass-spring system

$$T = \frac{2\pi}{\omega} = 2\pi\sqrt{\frac{m}{k}} \qquad [21.18]$$

where k is the force constant of the spring and m is the mass attached to the spring.

The kinetic energy and potential energy of a simple harmonic oscillator vary with time and are

Kinetic and potential energy of a simple harmonic oscillator

$$K = \tfrac{1}{2} mv^2 = \tfrac{1}{2} m\omega^2 A^2 \sin^2(\omega t + \phi) \qquad [21.20]$$

$$U = \tfrac{1}{2} kx^2 = \tfrac{1}{2} kA^2 \cos^2(\omega t + \phi) \qquad [21.21]$$

The **total energy** of a simple harmonic oscillator is a constant of the motion and is

Total energy of a simple harmonic oscillator

$$E = \tfrac{1}{2}kA^2 \quad [21.22]$$

The potential energy of a simple harmonic oscillator is a maximum when the particle is at its turning points (maximum displacement from equilibrium) and is zero at the equilibrium position. The kinetic energy is zero at the turning points and is a maximum at the equilibrium position.

A **simple pendulum** of length L exhibits simple harmonic motion for small angular displacements from the vertical, with a period of

Period of motion for a simple pendulum

$$T = 2\pi\sqrt{\frac{L}{g}} \quad [21.26]$$

That is, the period is independent of the suspended mass.

A **physical pendulum** exhibits simple harmonic motion about a pivot that does not go through the center of mass. The period of this motion is

Period of motion for a physical pendulum

$$T = 2\pi\sqrt{\frac{I}{mgd}} \quad [21.28]$$

where I is the moment of inertia about an axis through the pivot and d is the distance from the pivot to the center of mass.

Damped oscillations occur in a system in which a drag force opposes the linear restoring force. If such a system is set in motion and then left to itself, its mechanical energy decreases in time because of the presence of the nonconservative drag force. It is possible to compensate for this loss in energy by driving the system with an external periodic force that is in phase with the motion of the system. When the frequency of the driving force matches the natural frequency of the *undamped* oscillator that starts its motion from rest, energy is continuously transferred to the oscillator and its amplitude increases without limit.

When a charged capacitor is discharged through an inductor, and the resistance of the circuit is assumed to be zero, the current oscillates in a sinusoidal manner, with a frequency that depends on the inductance and capacitance. In a series RLC circuit, for small values of R, the current undergoes damped oscillations. When a series RLC circuit is connected across an alternating voltage which varies as $V_m \sin \omega t$, the current in the circuit also varies sinusoidally, but is in general out of phase with the applied voltage.

Questions and Conceptual Exercises

1. Does the acceleration of a simple harmonic oscillator remain constant during its motion? Is the acceleration ever zero? Explain.
2. Explain why the kinetic and potential energies of a mass-spring system can never be negative.
3. What is the total distance traveled by an object moving with simple harmonic motion in a time equal to its period, if its amplitude is A?
4. Given a spring, a clock, and a single known mass, devise an experiment to measure the force constant of the spring.
5. A mass-spring system undergoes simple harmonic motion with an amplitude of A. Does the total energy change if the mass is doubled but the amplitude is not changed? Do the kinetic and potential energies depend on the mass? Explain.
6. If a mass-spring system is hung vertically and set into oscillation, why does the motion eventually stop?
7. If a pendulum clock keeps perfect time at the base of a mountain, will it also keep perfect time when moved to the top of the mountain? Explain.
8. If a grandfather clock were running slowly, how could

one adjust the "length" of the pendulum to correct the time?

9. A pendulum bob is made from a ball filled with water. What happens to the frequency of vibration of this pendulum if the ball has a hole that allows water to slowly leak out?
10. A simple pendulum is suspended from the ceiling of a stationary elevator, and the period is determined. Describe the changes, if any, in the period if the elevator (a) accelerates upward; (b) accelerates downward; (c) moves with constant velocity.
11. Give a few examples of damped oscillations that are commonly observed.
12. Is it possible to have damped oscillations when a system is at resonance? Explain.
13. A platoon of soldiers marches in step along a road. Why are they ordered to break step when crossing a bridge?
14. Give as many examples as you can of workings of an automobile whose motions are simple harmonic or damped.
15. A pendulum is to be used in a clock. What length should the pendulum be so that its period of vibration is 1 s?
16. (a) Find the ratio of the period of a pendulum on Earth to the period of an identical pendulum on the Moon, where free-fall acceleration is one-sixth that on Earth. (b) If the period of the pendulum is 2.5 s on Earth, what will be its period on the Moon?
17. A "seconds" pendulum is one that moves through its equilibrium position once each second. (The period of the pendulum is 2 s.) The length of a seconds pendulum is 0.9927 m at Tokyo and 0.9942 m at Cambridge, England. What is the ratio of the free-fall acceleration at these two locations?
18. If the length of a simple pendulum is quadrupled, what happens to (a) its frequency and (b) its period?
19. The amplitude of a system moving with simple harmonic motion is doubled. Determine the changes in (a) the total energy, (b) the maximum velocity, (c) the maximum acceleration, and (d) the period.
20. A simple harmonic oscillator has a total energy of E. (a) Determine the kinetic and potential energies when the displacement equals one-half the amplitude. (b) For what value of the displacement does the kinetic energy equal the potential energy?
21. A spring-mass oscillator has $k = 4.72$ N/m, $m = 0.93$ kg. If it is driven by a periodic driving force, what must be the period of the force if it is to excite resonance?

Problems

Section 21.1 Simple Harmonic Motion

1. The displacement of a particle is given by the expression $x = (4 \text{ m}) \cos(3\pi t + \pi)$, where x is in meters and t is in seconds. Determine (a) the frequency and period of the motion, (b) the amplitude of the motion, (c) the phase constant, and (d) the displacement of the particle at $t = 0.25$ s.
2. A particle oscillates with simple harmonic motion so that its displacement varies according to the expression $x = (5 \text{ cm}) \cos(2t + \pi/6)$, where x is in centimeters and t is in seconds. At $t = 0$, find (a) the displacement of the particle, (b) its velocity, and (c) its acceleration. (d) Find the period and amplitude of the motion.
3. A particle moving with simple harmonic motion travels a total distance of 20 cm in each cycle of its motion, and its maximum acceleration is 50 m/s^2. Find (a) the angular frequency of the motion and (b) the maximum speed of the particle.
4. A piston in an automobile engine is in simple harmonic motion. If its amplitude of oscillation from centerline is ± 5 cm, and the mass of the piston is 2 kg, find the maximum velocity and the maximum acceleration of the piston when the auto engine is running at the rate of 3600 rev/min.
5. The displacement of a body is given by the expression $x = (8.0 \text{ cm}) \cos(2t + \pi/3)$, where x is in centimeters and t is in seconds. Calculate (a) the velocity and acceleration at $t = \pi/2$ s, (b) the maximum speed and the earliest time ($t > 0$) at which the particle has this speed, and (c) the maximum acceleration and the earliest time ($t > 0$) at which the particle has this acceleration.
6. A 20-g particle moves in simple harmonic motion with a frequency of 3 oscillations/s and an amplitude of 5 cm. (a) Through what total distance does the particle move during one cycle of its motion? (b) What is its maximum speed? Where does this occur? (c) Find the maximum acceleration of the particle. Where in the motion does the maximum acceleration occur?
7. A particle moving along the x axis with simple harmonic motion starts from the origin at $t = 0$ and moves toward the right. The amplitude of its motion is 2 cm, and the frequency is 1.5 Hz. (a) Show that the displacement of the particle is given by $x = (2 \text{ cm}) \sin(3\pi t)$. Determine (b) the maximum speed

and the earliest time ($t > 0$) at which the particle has this speed, (c) the maximum acceleration and the earliest time ($t > 0$) at which the particle has this acceleration, and (d) the total *distance* traveled between $t = 0$ and $t = 1$ s.

Section 21.2 Motion of a Mass Attached to a Spring

8. (a) A 400-g mass is suspended from a spring hanging vertically, and the spring is found to stretch 8 cm. Find the spring constant. (b) How much will the spring stretch if the suspended mass is 575 g?
9. A 3-kg mass is attached to a spring and pulled out horizontally to a maximum displacement from equilibrium of 0.5 m. What spring constant must the spring have if the mass is to achieve an acceleration equal to the free-fall acceleration g?
10. The mat of a trampoline is held by 32 springs, each having a spring constant of 5000 N/m. A person with a mass of 40.0 kg jumps onto the trampoline from a platform 1.93 m high. Determine the stretch of each of the springs. (Assume the springs were initially unstretched and that they stretch equally.)
11. An archer pulls his bow string back 0.4 m by exerting a force that increases uniformly from zero to 230 N. (a) What is the equivalent spring constant of the bow? (b) How much work is done in pulling the bow?
12. A 1-kg mass attached to a spring of force constant 25 N/m oscillates on a horizontal, frictionless track. At $t = 0$, the mass is released from rest at $x = -3$ cm. (That is, the spring is compressed by 3 cm.) Find (a) the period of its motion, (b) the maximum values of its speed and acceleration, and (c) the displacement, velocity, and acceleration as functions of time.
13. A 0.5-kg mass attached to a spring of force constant 8 N/m vibrates with simple harmonic motion with an amplitude of 10 cm. Calculate (a) the maximum value of its speed and acceleration, (b) the speed and acceleration when the mass is at $x = 6$ cm from the equilibrium position, and (c) the time it takes the mass to move from $x = 0$ to $x = 8$ cm.
14. A particle that hangs from an ideal spring has an angular frequency for oscillations of $\omega_0 = 2.0$ rad/s. The spring is suspended from the ceiling of an elevator car and hangs motionless (relative to the elevator car) as the car descends at a constant velocity of 1.5 m/s. The car then stops suddenly. (a) With what amplitude does the particle oscillate? (b) What is the equation of motion for the particle? (Choose the upward direction to be positive.)

Section 21.3 Energy of the Simple Harmonic Oscillator

Neglect spring masses.

15. A 200-g mass is attached to a spring and executes simple harmonic motion with a period of 0.25 s. If the total energy of the system is 2 J, find (a) the force constant of the spring and (b) the amplitude of the motion.
16. A mass-spring system oscillates with an amplitude of 3.5 cm. If the spring constant is 250 N/m and the mass is 0.5 kg, determine (a) the mechanical energy of the system, (b) the maximum speed of the mass, and (c) the maximum acceleration.
17. A bullet of mass 10 g is fired into and embeds in a 2-kg block attached to a spring with $k = 19.6$ N/m. How far will the spring be compressed if the speed of the bullet just before striking the block is 300 m/s and the block slides on a frictionless track? (*Hint:* You must use conservation of momentum in this problem. Why?)
18. The spring constant of the spring in Figure 21.22 is 19.6 N/m, and the mass of the object is 1.5 kg. The spring is unstretched and the track is frictionless. A constant 20-N force is applied to the object horizontally, as shown. Find the speed of the object after it has moved a distance of 0.3 m from equilibrium.

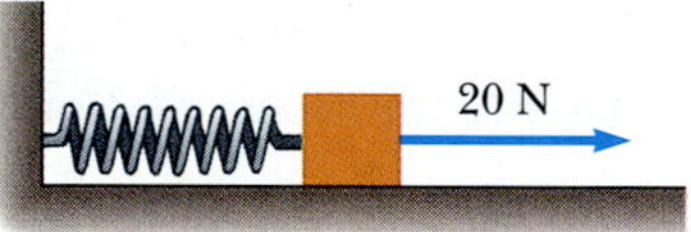

Figure 21.22 (Problem 18)

19. A 50-g mass, connected to a light spring of force constant 35 N/m, oscillates on a horizontal track with an amplitude of 4 cm. Friction is negligible. Find (a) the total energy of the oscillating system and (b) the speed of the mass when the displacement is 1 cm. When the displacement is 3 cm, find (c) the kinetic energy and (d) the potential energy.
20. A 2-kg mass is attached to a spring and placed on a horizontal, frictionless track. A horizontal force of 20 N is required to hold the mass at rest when it is pulled 0.2 m from its equilibrium position (the origin of the x axis). The mass is then released from rest with an initial displacement of $x_0 = 0.2$ m, and it subsequently undergoes simple harmonic oscillations. Find (a) the force constant, k, of the spring; (b) the frequency, f, of the oscillations; and (c) the maximum speed, v_{max}, of the mass. Where does this maxi-

mum speed occur? (d) Find the maximum acceleration, a_{max}, of the mass. Where does it occur? (e) Find the total energy, E, of the oscillating system. When the displacement, x, is one-third the maximum value, find (f) the velocity and (g) the acceleration.

21. An automobile ($m = 10^3$ kg) is driven into a brick wall in a safety test. The bumper behaves like a spring ($k = 5 \times 10^6$ N/m) and is observed to compress a maximum distance of 3.16 cm as the car is brought to rest. What was the initial speed of the automobile?

22. An ore car of mass 4000 kg starts from rest and rolls downhill from the mine on tracks. At the end of the tracks, 10 m lower in elevation, is a spring with $k = 400\,000$ N/m. How much is the spring compressed in stopping the ore car? Ignore friction.

Section 21.4 Motion of a Pendulum

23. A simple pendulum has a period of 2.50 s. (a) What is its length? (b) What would its period be on the Moon, where $g_m = 1.67$ m/s²?

24. A simple pendulum has a mass of 0.25 kg and a length of 1 m. It is displaced through an angle of 15° and then released. What is (a) the maximum velocity? (b) the maximum angular acceleration? (c) the maximum restoring force?

25. A visitor to a lighthouse wishes to determine the height of the tower. She has a spool of thread that she unwinds and uses to support a rock, making a simple pendulum to hang down the center of the spiral staircase of the tower. The period of oscillation is 9.4 s. What is the height (in meters) of the tower?

26. The angular displacement of a pendulum is represented by the equation $\theta = 0.32 \cos \omega t$, where θ is in radians and $\omega = 4.43$ rad/s. Determine the period and the length of the pendulum.

27. A simple pendulum has a length of 3.00 m. Determine the *change* in its period if it is taken from a point where $g = 9.80$ m/s² to a higher elevation, where the acceleration due to gravity decreases to $g = 9.79$ m/s².

△28. A mass is attached to the end of a light string to form a simple pendulum. The period of its harmonic motion is measured for small angular displacements, using three different lengths and timing the motion with a stopwatch for 50 complete oscillations. For lengths of 1.00 m, 0.75 m, and 0.50 m, total times of 99.8 s, 86.6 s, and 71.1 s are measured. (a) Determine the period of motion for each of these lengths. (b) Determine the mean value of g obtained from these three independent measurements, and compare it with the accepted value of g. (c) Make a plot of T^2 versus L, and obtain a value for g from the slope of your best-fit straight-line graph. Compare this value with that obtained in part (b).

29. A "seconds" pendulum is one that moves through its equilibrium position once each second. (The period of the pendulum is 2 s.) The length of a seconds pendulum is 0.9927 m at Tokyo and 0.9942 m at Cambridge, England. What is the ratio of the free-fall acceleration at these two locations?

30. A particle of mass m slides inside a hemispherical bowl of radius R. Show that for small displacements from equilibrium, the particle exhibits simple harmonic motion with an angular frequency equal to that of a simple pendulum of length R. That is, $\omega = \sqrt{g/R}$.

31. When the simple pendulum illustrated in Figure 21.23 makes an angle of θ with the vertical, its speed is v. (a) Calculate the total mechanical energy of the pendulum as a function of v and θ. (b) Show that when θ is small, the potential energy can be expressed as $\frac{1}{2}mgL\theta^2 = \frac{1}{2}m\omega^2 s^2$. (*Hint:* In part (b), use the small angle approximation $\cos \theta \approx 1 - \theta^2/2$.)

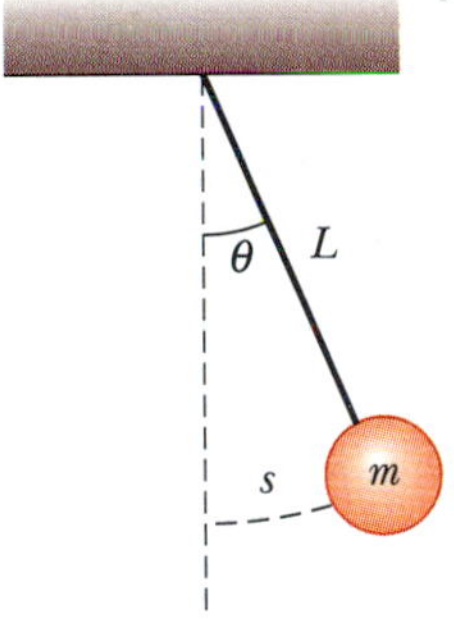

Figure 21.23 (Problem 31)

Section 21.5 Damped Oscillations

32. Show that the damping constant, b, has the units kilograms per second.

33. Show that Equation 21.30 is a solution of Equation 21.29 provided that $b^2 < 4mk$.

34. Show that the time rate of change of mechanical energy for a damped, undriven oscillator is given by $dE/dt = -bv^2$ and hence is *always negative.* (*Hint:* Differentiate the expression for the mechanical energy of an oscillator, $E = \frac{1}{2}mv^2 + \frac{1}{2}kx^2$, and make use of Eq. 21.29.)

35. A pendulum of length 1 m is released from an initial angle of 15°. After 1000 s, its amplitude is reduced by friction to 5.5°. What is the value of $b/2m$?

Section 21.6 Forced Oscillations

36. A 2-kg mass attached to a spring is driven by an external force, $F = (3\text{ N})\cos(2\pi t)$. If the force constant of the spring is 20 N/m, determine (a) the period and (b) the amplitude of the motion. (*Hint:* Assume that there is *no* damping—that is, $b = 0$—and make use of Eq. 21.34.)
37. Calculate the resonant frequencies of the following systems: (a) a 3-kg mass attached to a spring of force constant 240 N/m; (b) a simple pendulum 1.5 m in length.
38. Consider an *undamped*, forced oscillator ($b = 0$), and show that Equation 21.33 is a solution of Equation 21.32, with an amplitude given by Equation 21.34.
39. A weight of 40 N is suspended from a spring with force constant 200 N/m. The system is undamped and is subjected to a harmonic force of frequency 10 Hz, resulting in a forced-motion amplitude of 2 cm. Determine the maximum value of the impressed force.

Section 21.7 Oscillations in Circuits

40. A 1.0-μF capacitor is charged by a 40-V dc power supply. The fully-charged capacitor is then discharged through a 10-mH inductor. Find the *maximum* current that occurs in the resulting oscillations.
41. An LC circuit consists of a 20-mH inductor and a 0.5-μF capacitor. If the maximum instantaneous current in this circuit is 0.1 A, what is the greatest potential difference that appears across the capacitor?
42. An LC circuit has an inductance of 0.57 mH and a capacitance of 15 pF. The capacitor is charged to its maximum value by a 32-V battery. The battery is then removed from the circuit and the capacitor discharged through the inductor. (a) If all resistance in the circuit is neglected, determine the maximum value of the current in the oscillating circuit. (b) At what frequency does the circuit oscillate? (c) What is the maximum energy stored in the magnetic field of the inductor?
43. Consider the circuit shown in Figure 21.18. Let $R = 7.6\ \Omega$, $L = 2.2$ mH, and $C = 1.8\ \mu$F. (a) Calculate the frequency of the damped oscillation of the circuit. (b) What is the value of the critical resistance in the circuit?
44. Consider an RLC series circuit consisting of a charged 500-μF capacitor connected to a 32-mH inductor and a resistor R. Calculate the frequency of the oscillations (in Hz) that result for the following values of R: (a) $R = 0$ (no damping); (b) $R = 16\ \Omega$ (critical damping: $R = \sqrt{4L/C}$); (c) $R = 4\ \Omega$ (underdamped: $R < \sqrt{4L/C}$); (d) $R = 64\ \Omega$ (overdamped: $R > \sqrt{4L/C}$).
45. Consider a series LC circuit ($L = 2.18$ H, $C = 6$ nF). What is the maximum value of a resistor that, if inserted in series with L and C, will allow the circuit to continue to oscillate?
46. A sinusoidal voltage $\mathcal{E} = (40\text{ V})\sin(100t)$ is applied to a series RLC circuit with $L = 160$ mH, $C = 99\ \mu$F, and $R = 68\ \Omega$. (a) What is the impedance of the circuit? (b) What is the current amplitude? (c) Determine the numerical values for I_m, ω, and ϕ in the equation $I(t) = I_m \sin(\omega t - \phi)$.
47. An RLC circuit consists of a 150-Ω resistor, a 21-μF capacitor, and a 460-mH inductor, connected in series with a 120-V, 60-Hz power supply. (a) What is the phase angle between the current and the applied voltage? (b) Does the current or voltage reach its peak earlier?
48. A resistor ($R = 900\ \Omega$), a capacitor ($C = 0.25\ \mu$F), and an inductor ($L = 2.5$ H) are connected in series across a 240-Hz ac source for which $V_m = 140$ V. Calculate the (a) impedance of the circuit, (b) peak current delivered by the source, and (c) phase angle between the current and voltage. (d) Is the current leading or lagging behind the voltage?
49. A coil with an inductance of 18.1 mH and a resistance of 7 Ω is connected to a *variable*-frequency ac generator. At what frequency will the voltage across the coil lead the current by 45°?

Additional Problems

50. A car with bad shock absorbers bounces up and down with a period of 1.5 s after hitting a bump. The car has a mass of 1500 kg and is supported by four springs of equal force constant, k. Determine a value for k.
51. A large passenger of mass 150 kg sits in the car (Problem 50) with bad shocks. The mass of the car is now 1650 kg. What is the new period of oscillation?
52. A block rests on a flat plate that executes vertical simple harmonic motion with a period of 1.2 s. What is the maximum amplitude of the motion for which the block will not separate from the plate?
53. A horizontal platform vibrates with horizontal simple harmonic motion, with a period of 2 s. A body on the platform starts to slide when the amplitude of vibration reaches 0.3 m. Find the coefficient of static friction between the body and the platform.
54. A horizontal plank of mass m and length L is pivoted at one end, and the opposite end is attached to a

spring of force constant k (Fig 21.24). The moment of inertia of the plank about the pivot is $\frac{1}{3}mL^2$. If the plank is displaced a *small* angle θ from its equilibrium position and released, show that it will move with simple harmonic motion with an angular frequency given by $\omega = \sqrt{3k/m}$.

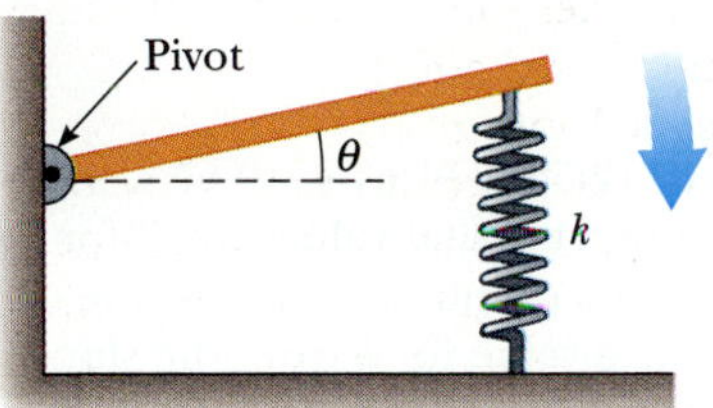

Figure 21.24 (Problem 54)

55. A pendulum of length L and mass M has a spring of force constant k connected to it at a distance of h below its point of suspension (Fig. 21.25). Find the frequency of vibration of the system for small values of the amplitude (small θ). (Assume the vertical suspension of length L is rigid, but neglect its mass.)

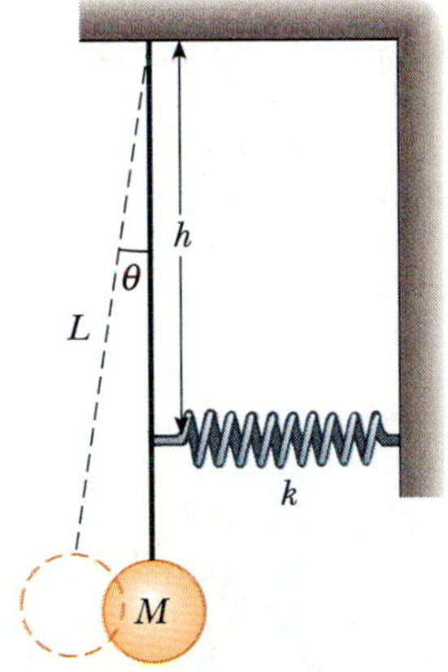

Figure 21.25 (Problem 55)

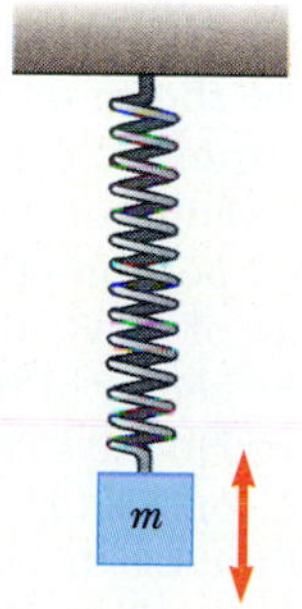

Figure 21.26 (Problems 56 and 64)

56. A mass, m, oscillates freely on a linear spring (Fig. 21.26). When $m = 0.81$ kg, the period is 0.91 s. An unknown mass on the same spring is observed to have a period of 1.16 s. Determine (a) the spring constant, k, and (b) the unknown mass.

57. A flat plate, P, executes horizontal simple harmonic motion by sliding across a frictionless surface with a frequency of $f = 1.5$ Hz. A block, B, rests on the plate, as shown in Figure 21.27, and the coefficient of static friction between the block and the plate is $\mu_s = 0.60$. What maximum amplitude of oscillation can the plate-block system have if the block is not to slip on the plate?

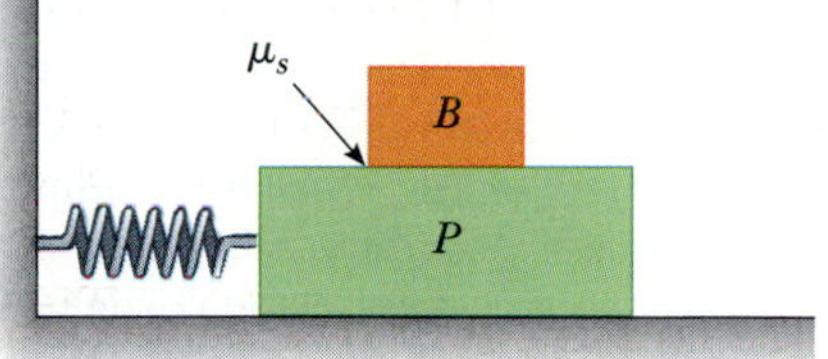

Figure 21.27 (Problem 57)

58. The mass of the deuterium molecule (D_2) is twice that of the hydrogen molecule (H_2). If the vibrational frequency of H_2 is 1.3×10^{14} Hz, what is the vibrational frequency of D_2, assuming that the "spring constant" of attracting forces is the same?

59. A simple pendulum with a length of 2.23 m and a mass of 6.74 kg is given an initial speed of 2.06 m/s at its equilibrium position. Assume that it undergoes simple harmonic motion, and determine its (a) period, (b) total energy, and (c) maximum angular displacement.

60. A mass, M, is connected to a spring of mass m and oscillates in simple harmonic motion on a horizontal, frictionless track (Fig. 21.28). The force constant of the spring is k and the equilibrium length is ℓ. Find (a) the kinetic energy of the system when the mass has a speed of v, and (b) the period of oscillation. (*Hint:* Assume that all portions of the spring oscillate in phase and that the velocity of a segment dx is proportional to the distance from the fixed end; that is,

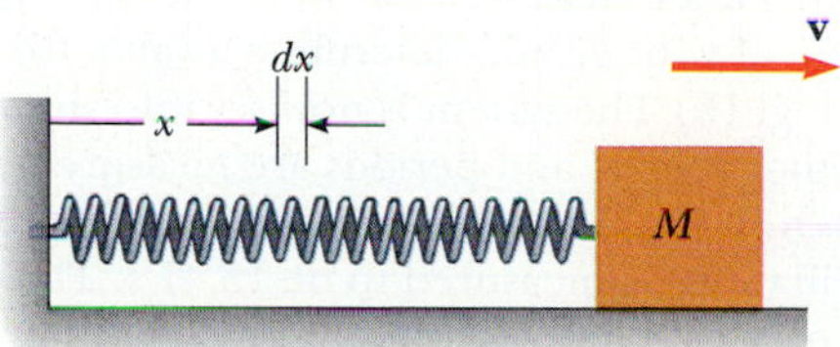

Figure 21.28 (Problem 60)

$v_x = (x/\ell)v$. Also, note that the mass of a segment of the spring is $dm = (m/\ell)\ dx$.

61. One end of a hacksaw blade is clamped in a vise, and the other end is set into vibration. The free end makes 5 complete vibrations each second with an amplitude of 0.4 cm. Find the speed of the end of the hacksaw blade when the displacement from the equilibrium position is 0.1 cm.

62. A spring stretches 3.0 cm from its relaxed length when a force of 7.5 N is applied. A particle with a mass of 0.5 kg rests on a frictionless, horizontal track and is attached to the free end of the spring. The particle is pulled horizontally so that it stretches the spring 5.0 cm and is then released from rest at $t = 0$. (a) What is the force constant of the spring? (b) What are the period, frequency (f), and angular frequency (ω) of the motion? (c) What is the total energy of the system? (d) What is the amplitude of the motion? (e) What are the maximum velocity and the maximum acceleration of the particle? (f) Determine the displacement, x, of the particle from the equilibrium position at $t = 0.5$ s.

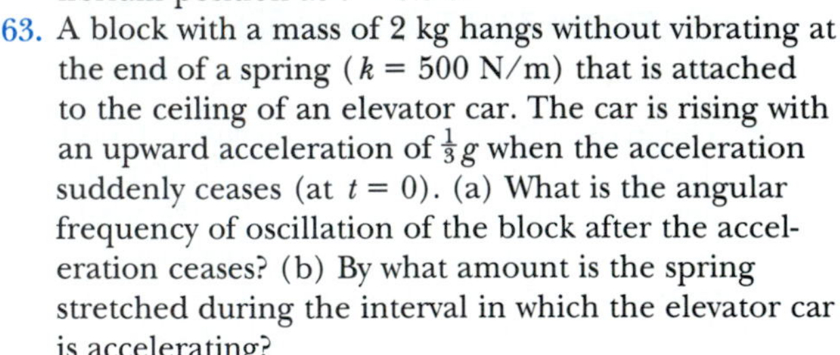

63. A block with a mass of 2 kg hangs without vibrating at the end of a spring ($k = 500$ N/m) that is attached to the ceiling of an elevator car. The car is rising with an upward acceleration of $\frac{1}{3}g$ when the acceleration suddenly ceases (at $t = 0$). (a) What is the angular frequency of oscillation of the block after the acceleration ceases? (b) By what amount is the spring stretched during the interval in which the elevator car is accelerating?

△64. When a mass, M, connected to the end of a spring of mass m_s and force constant k is set into simple harmonic motion, the period of its motion (Problem 60) is

$$T = 2\pi\sqrt{\frac{M + (m_s/3)}{k}}$$

A two-part experiment is conducted with a spring whose mass is measured to be 7.4 g. Various masses are suspended vertically from the spring, as in Figure 21.26. (a) Displacements of 17 cm, 29.3 cm, 35.3 cm, 41.3 cm, 47.1 cm, and 49.3 cm are measured for M values of 20 g, 40 g, 50 g, 60 g, 70 g, and 80 g, respectively. Construct a graph of Mg versus x, and perform a linear least-squares fit to the data. From the slope of your graph, determine a value for k for this spring. (b) The system is now set into simple harmonic motion, and periods are measured with a stopwatch. With $M = 80$ g, the total time for 10 complete oscillations is measured to be 13.41 s. The experiment is repeated with M values of 70 g, 60 g, 50 g, 40 g, and 20 g, with corresponding times for 10 oscillations of 12.52 s, 11.67 s, 10.67 s, 9.62 s, and 7.03 s. Obtain experimental values for T for each of these M values. Plot a graph of T^2 versus the quantity M, and determine a value for k from the slope of the linear least-squares fit through the data points. Compare this value of k with that obtained in part (a). (c) Obtain a value for m_s from your graph and compare it with the measured value.

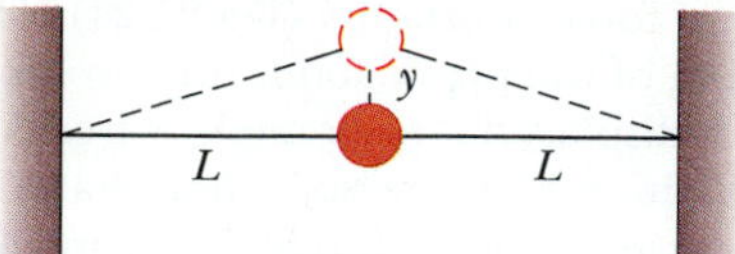

Figure 21.29 (Problem 65)

65. A mass, m, is connected to two rubber bands of length L, each under tension F, as in Figure 21.29. The mass is displaced vertically by a *small* distance, y. Assuming that the tension does not change appeciably, show that (a) the restoring force is $-(2F/L)y$ and (b) the system exhibits simple harmonic motion with an angular frequency of $\omega = \sqrt{2F/mL}$.

66. Electrical oscillations are initiated in a series circuit with a capacitance C, inductance L, and resistance R. (a) If $R \ll \sqrt{4L/C}$ (weak damping), how much time elapses before the current amplitude in the circuit falls off to 50% of its initial value? (b) How long does it take for the energy in the circuit to decrease to 50% of its initial value?

67. Consider a series RLC circuit with the following circuit parameters: $R = 200\ \Omega$, $L = 663$ mH, and $C = 26.5\ \mu$F. The applied voltage has an amplitude of 50 V and a frequency of 60 Hz. Find the following amplitudes: (a) The current, including its phase constant ϕ relative to the applied voltage; (b) the voltage V_R across the resistor and its phase relative to the current; (c) the voltage V_C across the capacitor and its phase relative to the current; and (d) the voltage V_L across the inductor and its phase relative to the current.

68. Imagine that a hole is drilled through the center of the Earth. It can be shown that an object of mass m at a distance of r from the center of the Earth is pulled toward the center of the Earth only by the mass in the shaded portion of Figure 21.30. Write down Newton's law of gravitation for an object at the distance r from the center of the Earth and show that the force on it is of Hooke's law form, $F = kr$, where the effective force constant is $k = (4/3)\pi\rho Gm$, where ρ is the density of the Earth, assumed uniform, and G is the gravitational constant.

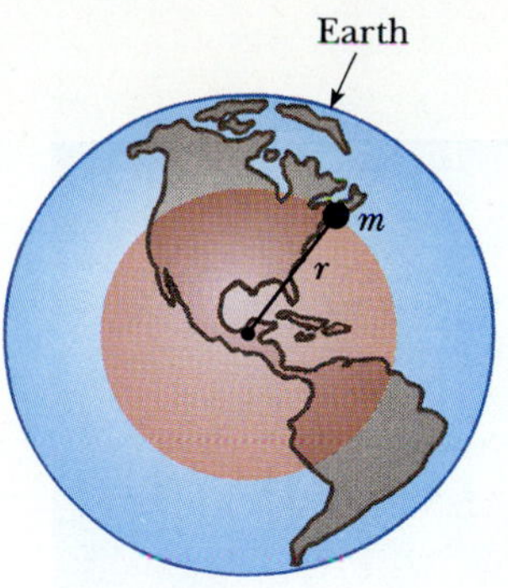

Figure 21.30 (Problem 68)

Calculator/Computer Problems

69. An object attached to the end of a spring vibrates with an amplitude of 20 cm. Find the position of the object at these times: 0, $T/8$, $T/4$, $3T/8$, $T/2$, $5T/8$, $3T/4$, $7T/8$, and T, where T is the period of vibration. Plot your results (position along the vertical axis and time along the horizontal axis).

70. A body oscillates with simple harmonic motion according to the equation $x = (-7 \text{ cm}) \cos(2\pi t)$. (a) Determine the velocity and acceleration as functions of time. (b) Make a table of x, v, and a versus t for the interval $t = 0$ to $t = 1$ s, in steps of 0.1 s. (c) Plot x, v, and a versus time for this interval.

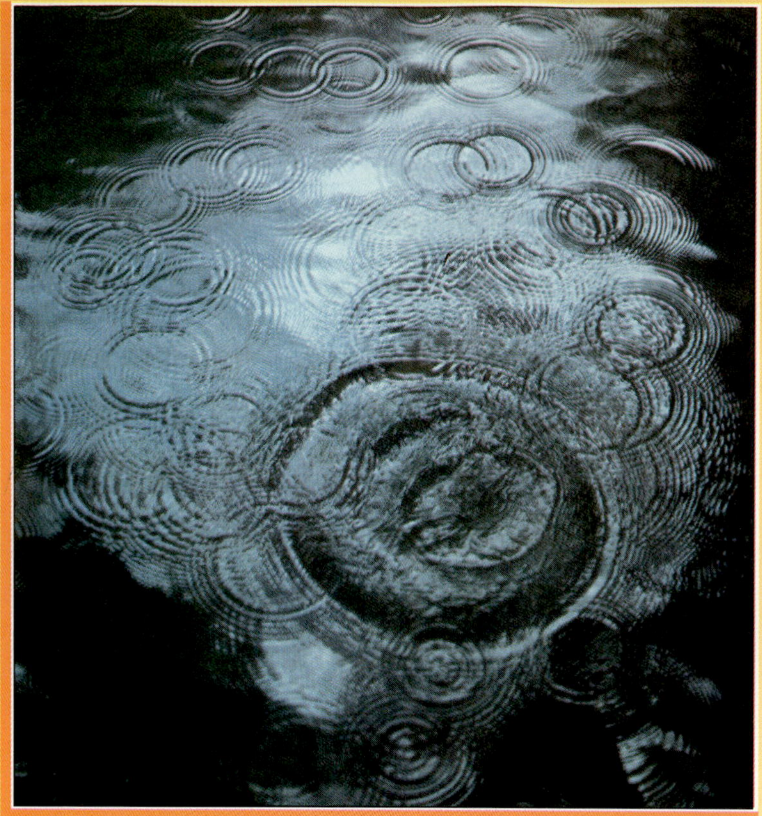

An interesting pattern, called an interference pattern, is observed as raindrops fall on a water surface. The phenomenon of wave interference will be discussed in this chapter. *(Stock)*

Wave Motion

CHAPTER 22

Most of us experienced waves as children, when we dropped pebbles into a pond. The disturbance created by a pebble manifests itself as ripple waves that move outward, finally reaching the shore. If you were to carefully examine the motion of a leaf floating near the point where the pebble entered the water, you would see that the leaf moves up and down and sideways about its original position but does not undergo any net displacement away from or toward the source of the disturbance. The water wave (the disturbance) moves from one place to another, *yet the water is not carried with it.*

The world is full of other kinds of waves, including sound waves, waves on strings, earthquake waves, radio waves, and x-rays. Most waves can be placed in one of two categories. **Mechanical waves** are waves that disturb and propagate through a medium; sound waves, for which air is the medium, and earthquake waves, for which the Earth's crust is the medium, are two examples of mechanical waves. **Electromagnetic waves** are a special class of waves that do not require a medium in order to propagate; light waves and radio waves are two familiar examples. In this chapter we shall confine our attention to mechanical waves, deferring our study of electromagnetic waves to Chapter 24.

The wave concept is abstract. When we observe a water wave, what we see is a rearrangement of the water's surface. Without the water, there would be no wave. In the case of this or any other mechanical wave, what we interpret as a wave corresponds to the disturbance of a medium. Therefore, we can consider a mechanical wave to be *the motion of a disturbance.* This motion is not to be confused with the motion of the particles that make up the medium. In general, we describe mechanical wave motion by specifying the positions of all particles of the disturbed medium as a function of time.

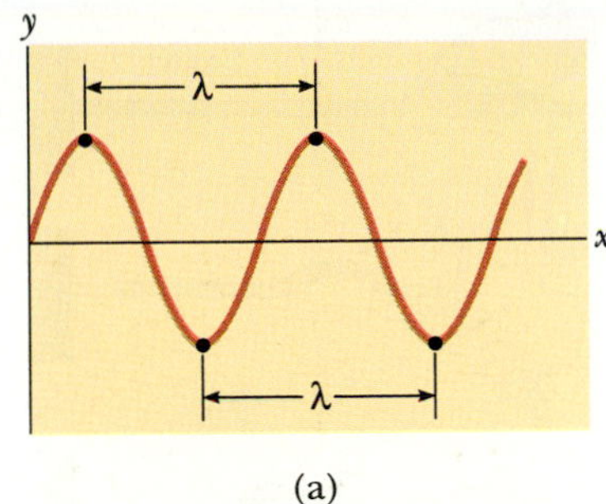

(a)

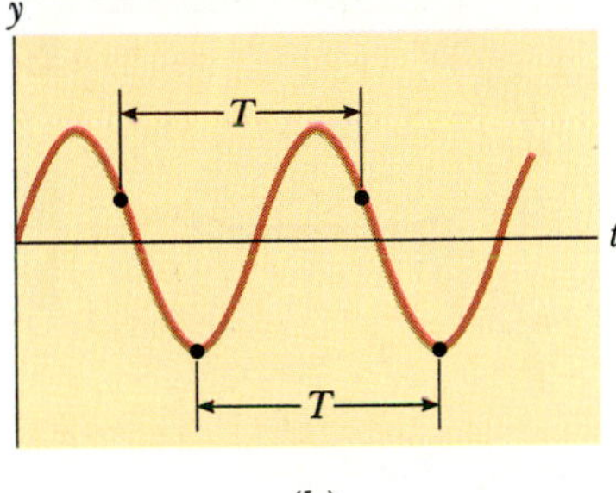

(b)

Figure 22.1
(a) The wavelength λ of a wave is the distance between adjacent crests or adjacent troughs.
(b) The period T of a wave is the time it takes the wave to travel one wavelength.

22.1 Three Wave Characteristics

All mechanical waves require (1) some source of disturbance, (2) a medium that can be disturbed, and (3) some physical mechanism through which particles of the medium can influence each other. All waves carry energy, but the amount of energy transmitted through a medium and the mechanism responsible for the energy transport differ from case to case. For instance, the power of ocean waves during a storm is much greater than the power of sound waves generated by a musical instrument.

Three physical characteristics are important in describing waves: wavelength, frequency, and wave speed. One **wavelength** is *the minimum distance between any two points on a wave that behave identically*—for example, adjacent crests or adjacent troughs. Figure 22.1a shows displacement versus position for a sinusoidal wave at a specific time. The symbol λ (Greek lambda) is used to denote wavelength.

Most waves are periodic. The **frequency** of such waves is *the rate at which the disturbance repeats itself.* As we learned in Chapter 21, the period of the wave is the minimum time it takes the disturbance to repeat itself, and is equal to the inverse of the frequency. Figure 22.1b shows displacement versus time for a sinusoidal wave at a fixed position, where the period is the time between identical displacements of a string element.

Waves travel, or *propagate,* with a specific speed, which depends on the properties of the medium being disturbed. For instance, sound waves travel through air at 20°C with a speed of about 343 m/s, whereas the speed of sound in most solids is higher than 343 m/s.

22.2 Types of Waves

One way to demonstrate wave motion is to flip the free end of a long rope that is under tension and has its opposite end fixed, as in Figure 22.2. In this manner, a single wave bump (or pulse) is formed and travels (to the right in Fig. 22.2) with a

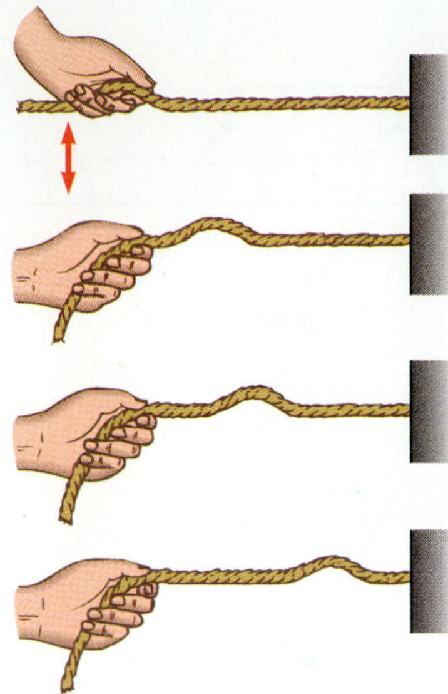

Figure 22.2
A wave pulse traveling down a stretched rope. The shape of the pulse is approximately unchanged as it travels along the rope.

definite speed. This type of disturbance is called a **traveling wave,** and the rope is the medium through which it travels. Figure 22.2 represents four consecutive "snapshots" of the traveling wave. The shape of the wave pulse changes very little as it travels along the rope.[1]

Note that, as the wave pulse travels, *each rope segment that is disturbed moves in a direction perpendicular to the wave motion.* Figure 22.3 illustrates this point for a particular segment, labeled *P*. Note that there is no motion of any part of the rope in the direction of the wave. A traveling wave such as this, in which the particles of the disturbed medium move perpendicularly to the wave velocity, is called a **transverse wave.** Figure 22.4a illustrates the formation of transverse waves on a long spring.

In another class of waves, called **longitudinal waves,** the particles of the medium undergo displacements *parallel* to the direction of wave motion. Sound waves in air, for instance, are longitudinal. Their disturbance corresponds to a series of high- and low-pressure regions that may travel through air or through any material medium with a certain speed. A longitudinal pulse can be easily produced in a stretched spring, as in Figure 22.4b. The free end is pumped back and forth along the length of the spring. This action produces compressed and stretched regions of the coil that travel along the spring, parallel to the wave motion.[2]

Some waves are neither transverse nor longitudinal, but a combination of the two. Surface water waves are a good example. Figure 22.5 shows the motion of water particles at the surface as a wave moves to the right. Each particle moves in a

[1] Strictly speaking, the pulse changes its shape and gradually spreads out during the motion. This effect is called *dispersion* and is common to many mechanical waves.

[2] In the case of longitudinal waves in a gas, each compressed area is a region of higher-than-average pressure and density, and each stretched region is a region of lower-than-average pressure and density.

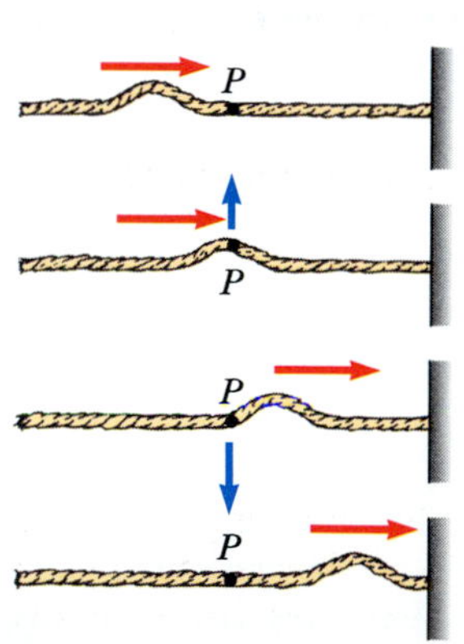

Figure 22.3
A pulse traveling on a stretched rope is a transverse wave. That is, any element *P* on the rope moves in a direction perpendicular to the wave motion.

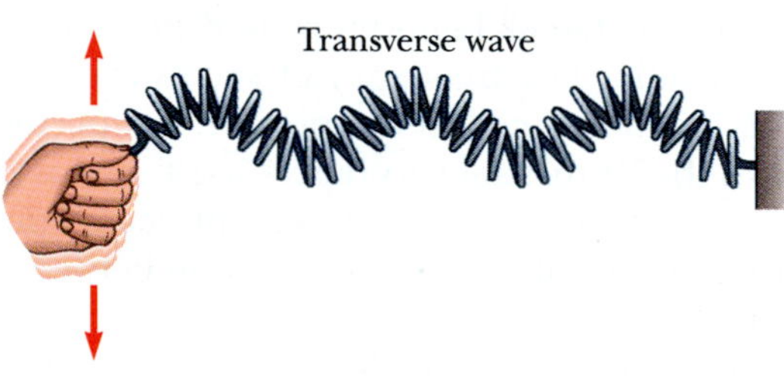

(a) Transverse wave

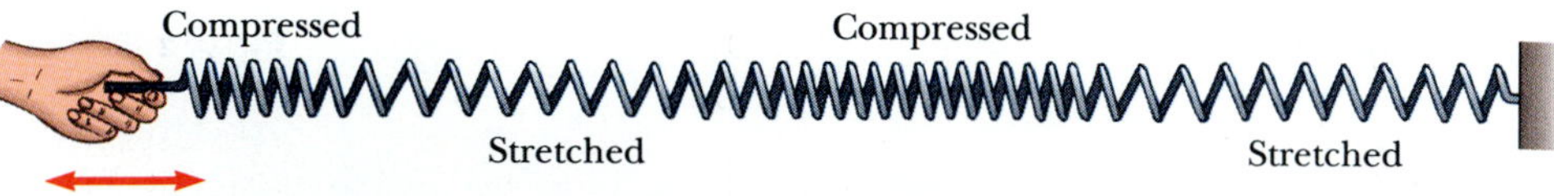

(b) Longitudinal wave

Figure 22.4
(a) A transverse wave is set up in a spring by moving one end of the spring perpendicular to its length. (b) A longitudinal pulse along a stretched spring. The displacement of the coils is in the direction of the wave motion. For the starting motion described in the text, the compressed region is followed by a stretched region *R*.

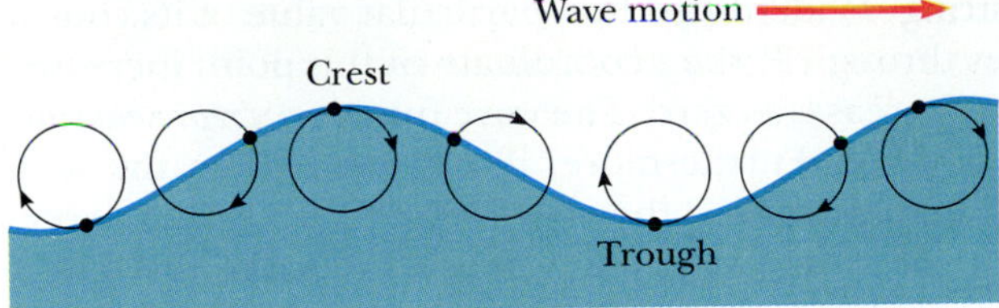

Figure 22.5
Wave motion on the surface of water. The particles at the water's surface move in nearly circular paths. Each particle is displaced horizontally and vertically from its equilibrium position, and the displacement is represented by circles.

circular path, and hence the disturbance has both transverse and longitudinal components. As the wave passes, water particles at the crests are moving in the direction of the wave while particles at the troughs move in the opposite direction. Hence, no *net* displacement of water particles takes place. A cork bobbing on a pond surface as a wave passes exhibits this circular motion.

22.3 One-Dimensional Traveling Waves

So far we have provided only verbal and graphical descriptions of a traveling wave. Let us now develop a mathematical description of a one-dimensional traveling wave. Consider a wave pulse traveling to the right with constant speed, v, along a long, stretched string, as in Figure 22.6. The pulse moves along the x axis (the axis of the string), and the transverse (up-and-down) displacement of the string is measured with the coordinate y.

Figure 22.6a represents the shape and position of the pulse at time $t = 0$. At this time, the shape of the pulse, whatever it may be, can be represented as $y = f(x)$. That is, y is some definite function of x. The *maximum displacement* of the string, y_m, is called the **amplitude** of the wave. Since the speed of the wave pulse is v, the pulse travels to the right a distance of vt in the time t (Fig. 22.6b). If the shape of the wave pulse doesn't change with time, we can represent the displacement y for all later times, measured in a stationary frame with the origin at O, as

$$y = f(x - vt) \qquad [22.1]$$

Wave traveling to the right

Similarly, if the wave pulse travels to the left, the displacement of the string is

$$y = f(x + vt) \qquad [22.2]$$

Wave traveling to the left

The displacement y, sometimes called the *wave function,* depends on the two variables x and t. For this reason, it is often written $y(x, t)$, which is read "y as a function of x and t." It is important to understand the meaning of y.

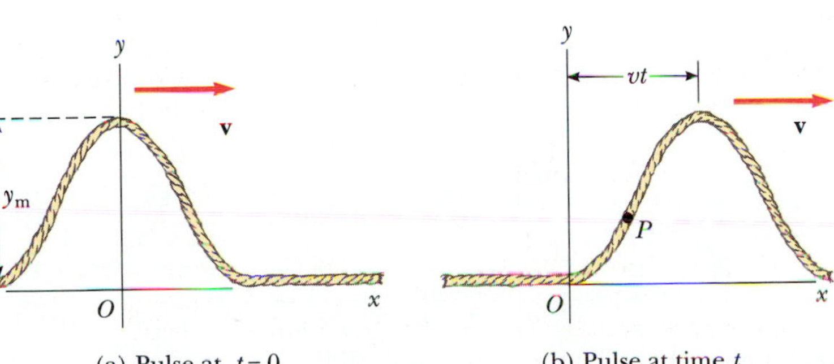

Figure 22.6
A one-dimensional wave pulse traveling to the right with a speed v. (a) At $t = 0$, the shape of the pulse is given by $y = f(x)$. (b) At some later time t, the shape remains unchanged and the vertical displacement is given by $y = f(x - vt)$.

Consider a point on the string, P, identified by a particular value of its coordinates. As the wave pulse passes through P, the y coordinate of this point increases, reaches a maximum, and then decreases to zero. The wave function *y represents the y coordinate of any point P at any time t.* Furthermore, if t is fixed, then the wave function y as a function of x, sometimes called the *waveform, defines a curve representing the actual shape of the pulse at that time.* This is equivalent to a "snapshot" of the pulse at time t.

For a pulse that moves without changing its shape, the velocity of the pulse is the same as the velocity of any point along the pulse, such as the crest. To find the velocity of the pulse, we can calculate how far the crest moves horizontally in a short time and then divide that distance by the time interval. In order to follow the motion of the crest, some particular value, say x_0, must be substituted for $x - vt$ in Equation 22.1. (The value x_0 is called the *argument* of the function y.) Regardless of how x and t change individually, we must require that $x - vt = x_0$ in order to stay with the crest. This equation therefore represents the motion of the crest. At $t = 0$, the crest is at $x = x_0$; after an interval of dt, the crest is at $x = x_0 + v\,dt$. Therefore, in the time dt the crest has moved a distance of $dx = (x_0 + v\,dt) - x_0 = v\,dt$. Clearly, the wave speed, often called the **phase speed,** is

Phase speed

$$v = \frac{dx}{dt} \qquad [22.3]$$

The wave velocity, or phase velocity must not be confused with the transverse velocity of a particle in the medium (which is always perpendicular to the wave velocity).

The following example illustrates how a specific wave function is used to describe the motion of a traveling wave pulse.

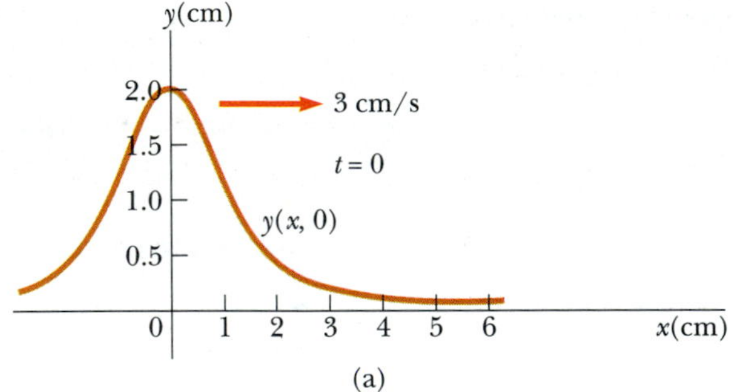

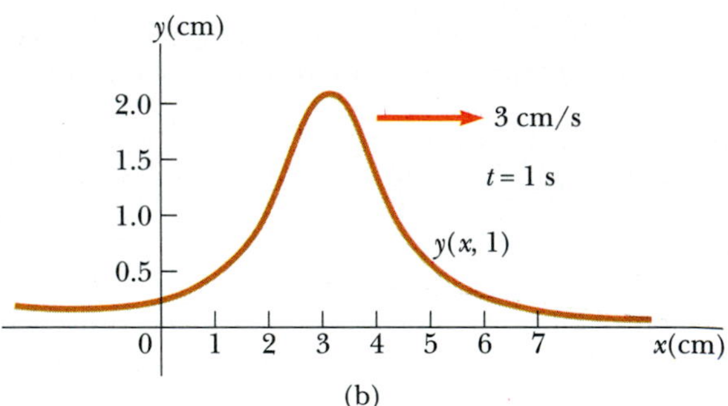

Figure 22.7
(Example 22.1) Graphs of the function $y(x, t) = 2/[(x - 3t)^2 + 1]$. (a) $t = 0$, (b) $t = 1$ s, and (c) $t = 2$ s.

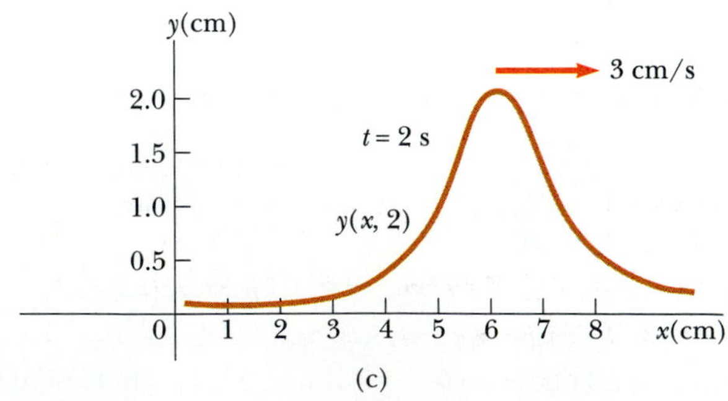

Example 22.1 A Pulse Moving to the Right

A wave pulse traveling to the right along the x axis is represented by the wave function

$$y(x, t) = \frac{2}{(x - 3t)^2 + 1}$$

where x and y are measured in centimeters and t in seconds. Let us plot the waveform at $t = 0$, $t = 1$ s, and $t = 2$ s.

Solution First, note that this function is of the form $y = f(x - vt)$ because of the term $(x - 3t)^2$ in the denominator. Comparing the term $(x - 3t)^2$ with the general function $f(x - vt)$, we see that the pulse speed is $v = 3$ cm/s. Furthermore, the wave amplitude (the maximum value of y) is $y_m = 2$ cm. At times $t = 0$, $t = 1$ s, and $t = 2$ s, the wave function expressions are

$$y(x, 0) = \frac{2}{x^2 + 1} \qquad \text{at } t = 0$$

$$y(x, 1) = \frac{2}{(x - 3)^2 + 1} \qquad \text{at } t = 1 \text{ s}$$

$$y(x, 2) = \frac{2}{(x - 6)^2 + 1} \qquad \text{at } t = 2 \text{ s}$$

We can now use these expressions to plot the wave function versus x at these times. For example, let us evaluate $y(x, 0)$ at $x = 0.5$ cm:

$$y(0.5, 0) = \frac{2}{(0.5)^2 + 1} = 1.60 \text{ cm}$$

Likewise, $y(1, 0) = 1.0$ cm, $y(2, 0) = 0.40$ cm, and so on. A continuation of this procedure for other values of x yields the waveform shown in Figure 22.7a. In a similar manner, one obtains the graphs of $y(x, 1)$ and $y(x, 2)$ shown in Figures 22.7b and 22.7c, respectively. These snapshots show that the wave pulse moves to the right without changing its shape and has a constant speed of 3 cm/s.

22.4 Superposition and Interference of Waves

A vibrating guitar string. *(E. R. Degginger)*

Many interesting wave phenomena in nature cannot be described by a single moving pulse. Instead, one must analyze complex waveforms in terms of a combination of many traveling waves. To analyze such wave combinations, one can make use of the **superposition principle,** which was introduced in Chapter 16 in connection with calculating the electric field of many charged particles. This principle, as it applies to wave motion, states that

Linear waves obey the superposition principle

> if two or more traveling waves are moving through a medium, the resultant wave function at any point is the algebraic sum of the wave functions of the individual waves.

This rather striking property is exhibited by many waves in nature. Waves that obey this principle are called *linear waves,* and they are generally characterized by small amplitudes. Waves that violate the superposition principle are called *nonlinear*

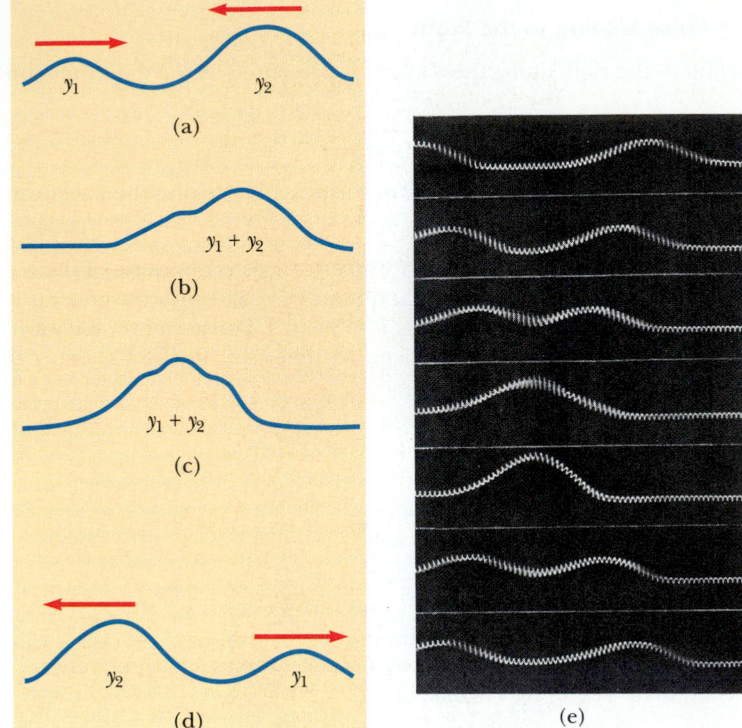

Figure 22.8 (Left) Two wave pulses traveling on a stretched string in opposite directions pass through each other. When the pulses overlap, as in (b) and (c), the net displacement of the string equals the sum of the displacements of each pulse. (Right) Photograph of superposition of two equal and symmetric pulses traveling in opposite directions on a stretched spring. *(Photo, Education Development Center, Newton, Mass.)*

waves and are often characterized by large amplitudes. In this book, we shall deal only with linear waves.

One consequence of the superposition principle is that *two traveling waves can pass through each other without being destroyed or even altered.* For instance, when two pebbles are thrown into a pond, the expanding circular surface waves do not destroy each other. In fact, the ripples pass through each other. The complex pattern that is observed can be viewed as two independent sets of expanding circles. Likewise, when sound waves from two sources move through air, they also can pass through each other. The sound one hears at a given point is the resultant of both disturbances.

A simple pictorial representation of the superposition principle is obtained by considering two pulses traveling in opposite directions on a stretched string, as in Figure 22.8. The wave function for the pulse moving to the right is y_1, and the wave function for the pulse moving to the left is y_2. The pulses have the same speed but different shapes. Each pulse is assumed to be symmetric (although this is not a necessary condition), and in both cases the vertical displacements of the string are taken to be positive. When the waves begin to overlap (Fig. 22.8b), the resulting complex waveform is given by $y_1 + y_2$. When the crests of the pulses exactly coincide (Fig. 22.8c), the resulting waveform, $y_1 + y_2$, is symmetric. The two pulses finally separate and continue moving in their original directions (Fig. 22.8d). Note that the final waveforms remain unchanged, as if the two pulses had never met! The combination of separate waves in the same region of space to produce a resultant wave is called **interference**.

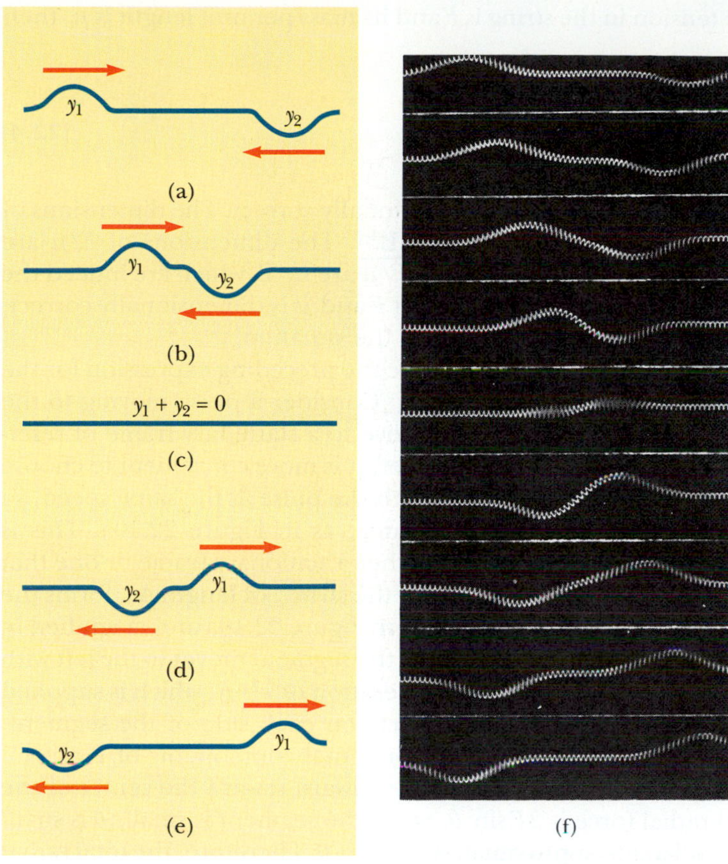

Figure 22.9
(Left) Two wave pulses traveling in opposite directions with equal but opposite displacements. When the two overlap, their displacements subtract from each other. In frame (c), the displacement is zero for all values of x. (Right) Photograph of superposition of two symmetric pulses traveling in opposite directions, where one is inverted relative to the other. *(Photo, Education Development Center, Newton, Mass.)*

For the two pulses shown in Figure 22.8, the vertical displacements caused by the pulses are in the same direction, and so the resultant waveform (when the pulses overlap) exhibits a displacement greater than those of the individual pulses. Now consider two identical pulses, again traveling in opposite directions on a stretched string, but this time one pulse is inverted relative to the other, as in Figure 22.9. In this case, when the pulses begin to overlap, the resultant waveform is the *difference* between the two separate displacements. Again, the two pulses pass through each other. When they exactly overlap, they *cancel* each other. At this time (Fig. 22.9c), the string is horizontal and the energy associated with the disturbance is contained in the string's kinetic energy, where the string segments move *vertically*.

22.5 The Speed of Waves on Strings

For linear mechanical waves, *the speed depends only on the properties of the medium through which the disturbance travels.* In this section we shall focus our attention on determining the speed of a transverse pulse traveling on a stretched string. We

Speed of a wave on a stretched string

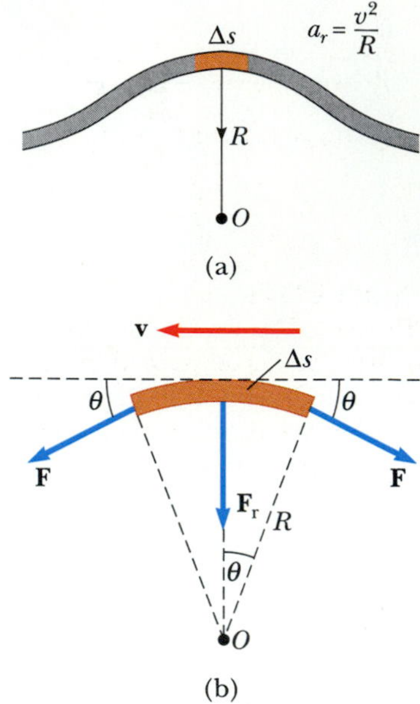

Figure 22.10
(a) To obtain the speed, v, of a wave on a stretched string, it is convenient to describe the motion of a small segment of the string in a moving frame of reference. (b) The net force on a small segment of length Δs is in the radial direction. The horizontal components of the tension force cancel.

shall show that if the tension in the string is F and its mass per unit length is μ, then the wave speed, v, is

$$v = \sqrt{\frac{F}{\mu}} \qquad [22.4]$$

First, we verify that this expression is dimensionally correct. The dimensions of F are MLT^{-2}, and the dimensions of μ are ML^{-1}. The dimensions of F/μ are therefore L^2/T^2, and hence the dimensions of $\sqrt{F/\mu}$ are L/T, which are indeed the dimensions of velocity. No other combination of F and μ is dimensionally correct, assuming they are the only variables relevant to the situation.

Now let us use a mechanical analysis to derive the preceding expression for the speed of a pulse traveling on a stretched string. Consider a pulse moving to the right with a uniform speed of v, measured relative to a stationary frame of reference. Instead of using a stationary frame, however, it is more convenient to choose as our reference frame one that moves along with the pulse at the same speed, so that the pulse appears to be at rest in the frame, as in Figure 22.10a. This is permitted because Newton's laws are valid in either a stationary frame or one that moves with constant velocity. A small segment of the string, of length Δs, forms the approximate arc of a circle of radius R, as shown in Figure 22.10a and magnified in Figure 22.10b. In the pulse's frame of reference, the segment moves to the left with the speed v. This segment has a centripetal acceleration of v^2/R, which is supplied by the tension, F, in the string. The force **F** acts on each side of the segment, tangent to the arc, as in Figure 22.10b. The horizontal components of **F** cancel, and each vertical component $F \sin\theta$ acts radially inward toward the center of the arc. Hence, the total radial force is $2F \sin\theta$. Since the segment is small, θ is small and we can use the small-angle approximation $\sin\theta \approx \theta$. Therefore, the total radial force can be expressed as

$$F_r = 2F \sin\theta \approx 2F\theta$$

The segment has the mass $m = \mu\Delta s$, where μ is the mass per unit length of the string. Since the segment forms part of a circle and subtends an angle of 2θ at the center, $\Delta s = R(2\theta)$, and hence

$$m = \mu\,\Delta s = 2\mu R\theta$$

If we apply Newton's second law to this segment, the radial component of motion gives

$$F_r = \frac{mv^2}{R} \qquad \text{or} \qquad 2F\theta = \frac{2\mu R\theta v^2}{R}$$

where F_r is the force that supplies the centripetal acceleration of the segment and maintains the curvature of this segment.

Solving for v gives

$$v = \sqrt{\frac{F}{\mu}}$$

Notice that this derivation is based on the assumption that the pulse height is small relative to the length of the string. Using this assumption, we were able to use the approximation $\sin\theta \approx \theta$. Furthermore, the model assumes that the tension, F, is not affected by the presence of the pulse, so that F is the same at all points on the string. Finally, this proof does *not* assume any particular shape for the pulse. Therefore, we conclude that a pulse of *any shape* will travel on the string with speed $v = \sqrt{F/\mu}$, without changing its shape.

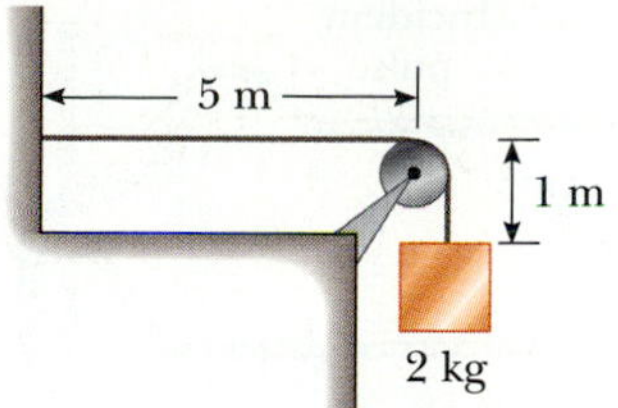

Figure 22.11 (Example 22.2) The tension F in the cord is maintained by the suspended mass. The wave speed is calculated using the expression $v = \sqrt{F/\mu}$.

Example 22.2 The Speed of a Pulse on a Cord

A uniform cord has a mass of 0.3 kg and a length of 6 m. Tension is maintained in the cord by suspending a 2-kg mass from one end (Fig. 22.11). Find the speed of a pulse on this cord.

Solution The tension, F, in the cord is equal to the weight of the suspended 2-kg mass:

$$F = mg = (2\text{ kg})(9.80\text{ m/s}^2) = 19.6\text{ N}$$

(This calculation of the tension neglects the small mass of the cord. Strictly speaking, the cord can never be exactly horizontal, and therefore the tension is not uniform.)

The mass per unit length, μ, is

$$\mu = \frac{m}{\ell} = \frac{0.3\text{ kg}}{6\text{ m}} = 0.05\text{ kg/m}$$

Therefore, the wave speed is

$$v = \sqrt{\frac{F}{\mu}} = \sqrt{\frac{19.6\text{ N}}{0.05\text{ kg/m}}} = 19.8\text{ m/s}$$

Exercise Find the time it takes the pulse to travel from the wall to the pulley.

Answer 0.253 s.

22.6 Reflection and Transmission of Waves

Whenever a traveling wave pulse reaches a boundary, part or all of the pulse is *reflected.* Any part not reflected is said to be *transmitted* through the boundary. Consider a pulse traveling on a string that is fixed at one end (Fig. 22.12). When the pulse reaches the fixed boundary, it is reflected. Since the support attaching the string to the wall is rigid, none of the pulse is transmitted through the wall.

Note that the reflected pulse has exactly the same amplitude as the incoming pulse but is inverted. The inversion can be explained as follows. When the pulse meets the end of the string that is fixed at the support, the string produces an upward force on the support. By Newton's third law, the support must exert an equal and opposite reaction force on the string. This downward force causes the pulse to invert upon reflection.

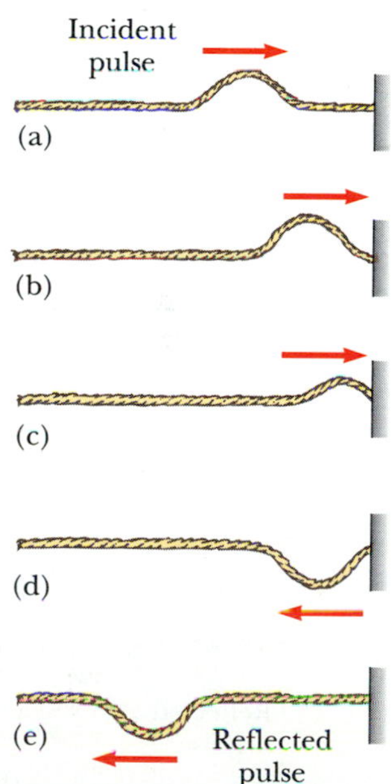

Figure 22.12 The reflection of a traveling wave pulse at the fixed end of a stretched string. Note that the reflected pulse is inverted, but its shape remains the same.

Now consider another situation in which there is total reflection and zero transmission. In this case the pulse arrives at the end of a string that is free to move vertically, as in Figure 22.13. The tension at the free end is maintained by tying the string to a ring of negligible mass that is free to slide vertically on a frictionless post. Again, the pulse is reflected, but this time it is not inverted. As the pulse reaches the post, it exerts a force on the free end, causing the ring to accelerate upward. In the process, the ring has upward momentum as it reaches the top of its motion,

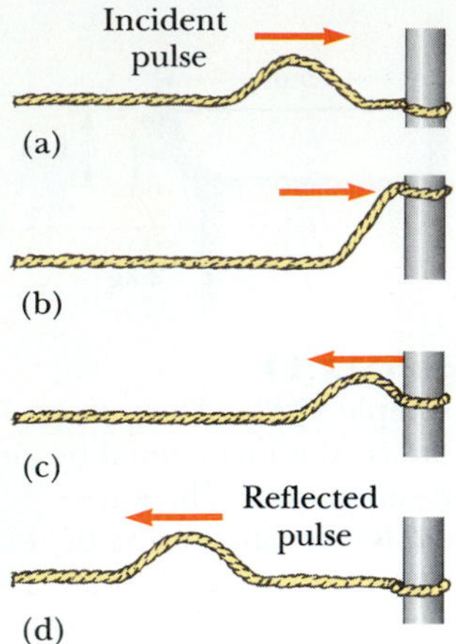

Figure 22.13
The reflection of a traveling wave pulse at the free end of a stretched string. In this case, the reflected pulse is not inverted.

and is then returned to its original position by the downward component of the tension force. This produces a reflected pulse that is not inverted, whose amplitude is the same as that of the incoming pulse.

Finally, we may have a situation in which the boundary is intermediate between these two extreme cases; that is, it is neither completely rigid nor completely free. In this case, part of the wave is transmitted and part is reflected. For instance, suppose a light string is attached to a heavier string as in Figure 22.14. When a pulse traveling on the light string reaches the boundary between the two strings, part of the pulse is reflected and inverted and part is transmitted to the heavier string. As one would expect, both the reflected pulse and the transmitted one have a smaller amplitude than the incident pulse. The inversion in the reflected wave is similar to the behavior of a pulse meeting a rigid boundary.

When a pulse traveling on a heavy string strikes the boundary of a lighter string, as in Figure 22.15, again part is reflected and part transmitted. This time, however, the reflected pulse is not inverted.

In both the lighter-to-heavier case and the heavier-to-lighter case, the relative heights of the reflected and transmitted pulses depend on the relative densities of the two strings.

In the preceding section, we found that the speed of a wave on a string increases as the mass per unit length of the string decreases. In other words, a pulse travels more slowly on a heavy string than on a light string if both are under the same tension. The following general rules apply to reflected waves: *When a wave pulse travels from medium A to medium B and $v_A > v_B$ (that is, when B is more massive than A), the reflected part of the pulse is inverted upon reflection. When a wave pulse travels from medium A to medium B and $v_A < v_B$ (A is more massive than B), the reflected pulse is not inverted.*

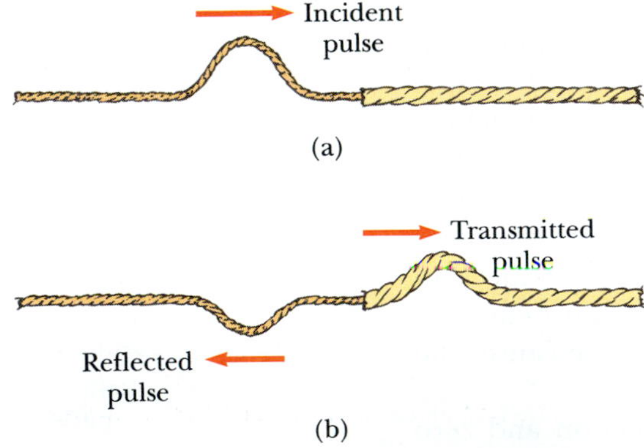

Figure 22.14
(a) A pulse traveling to the right on a light string tied to a heavier string. (b) Part of the incident pulse is reflected (and inverted), and part is transmitted to the heavier string. (Note that the change in pulse width is not shown.)

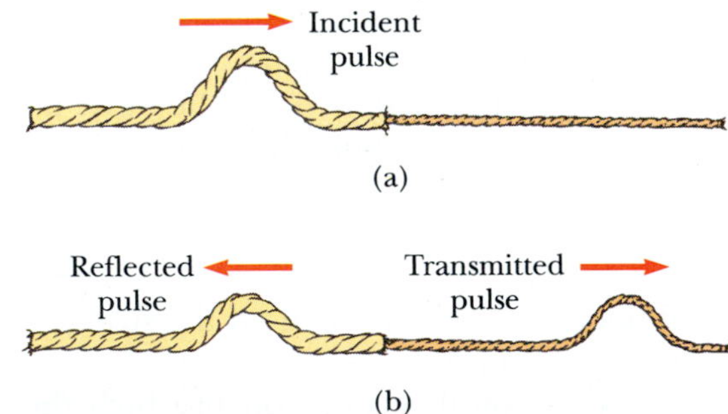

Figure 22.15
(a) A pulse traveling to the right on a heavy string tied to a lighter string. (b) The incident pulse is partially reflected and partially transmitted. In this case, the reflected pulse is not inverted. (Note that the change in pulse width is not shown.)

22.7 Sinusoidal Waves

In this section we introduce an important periodic waveform known as a **sinusoidal wave** (Fig. 22.16). The red curve represents a snapshot of a traveling sinusoidal wave at $t = 0$, and the blue curve represents a snapshot of the wave at some later t. At $t = 0$, the displacement of the curve can be written

$$y = A \sin\left(\frac{2\pi}{\lambda} x\right) \quad [22.5]$$

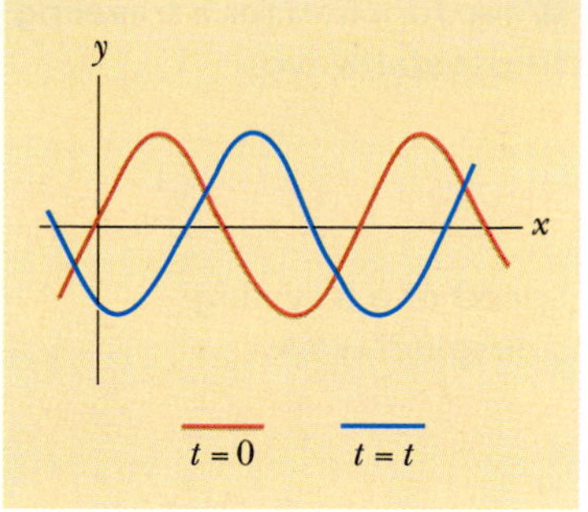

Figure 22.16
A one-dimensional harmonic wave traveling to the right with a speed v. The red curve represents a snapshot of the wave at $t = 0$, and the blue curve at some later time t.

where the amplitude, A, as usual, represents the maximum value of the displacement, and λ is the wavelength as defined in Figure 22.1a. Thus, we see that the displacement repeats itself when x is increased by an integral multiple of λ. If the wave moves to the right with a phase speed of v, the wave function at some later t is

$$y = A \sin\left[\frac{2\pi}{\lambda}(x - vt)\right] \quad [22.6]$$

That is, the sinusoidal wave moves to the right a distance of vt in the time t, as in Figure 22.16. Note that the wave function has the form $f(x - vt)$ and represents a wave traveling to the right. If the wave were traveling to the left, the quantity $x - vt$ would be replaced by $x + vt$, just as in the case of the traveling pulse described by Equations 22.1 and 22.2.

Since the period T is the time it takes the wave to travel a distance of one wavelength, the phase speed, wavelength, and period are related by

$$v = \frac{\lambda}{T} \quad [22.7]$$

or

$$\lambda = vT \quad [22.8]$$

Substituting Equation 22.7 into Equation 22.6, we find that

$$y = A \sin\left[2\pi\left(\frac{x}{\lambda} - \frac{t}{T}\right)\right] \quad [22.9]$$

This form of the wave function clearly shows the periodic nature of y. That is, at any given time t (a snapshot of the wave), y has the same value at the positions x, $x + \lambda$, $x + 2\lambda$, and so on. Furthermore, at any given position x, the value of y at times t, $t + T$, $t + 2T$, and so on, is the same.

We can express the sinusoidal wave function in a convenient form by defining two other quantities: **angular wave number**, k, and **angular frequency**, ω:

$$k \equiv \frac{2\pi}{\lambda} \quad [22.10]$$

Angular wave number

$$\omega \equiv \frac{2\pi}{T} = 2\pi f \quad [22.11]$$

Angular frequency

Note that in Equation 22.11, we used the definition of frequency, $f = 1/T$. Using these definitions, we see that Equation 22.9 can be written in the more compact form

Wave function for a traveling sinusoidal wave

$$y = A\sin(kx - \omega t) \qquad [22.12]$$

We shall use this form most frequently.

Using Equations 22.10 and 22.11, we can express the phase speed, v, in the alternative forms

Speed of a traveling sinusoidal wave

$$v = \frac{\omega}{k} \qquad [22.13]$$

$$v = f\lambda \qquad [22.14]$$

The wave function given by Equation 22.12 assumes that the displacement, y, is zero at $x = 0$ and $t = 0$. This need not be the case. If the transverse displacement is not zero at $x = 0$ and $t = 0$, we generally express the wave function in the form

General relation for a traveling sinusoidal wave

$$y = A\sin(kx - \omega t - \phi) \qquad [22.15]$$

where ϕ is called the **phase constant** and can be determined from the initial conditions.

Example 22.3 A Traveling Sinusoidal Wave

A sinusoidal wave traveling in the positive x direction has an amplitude of 15 cm, a wavelength of 40 cm, and a frequency of 8 Hz. The displacement of the wave at $t = 0$ and $x = 0$ is also 15 cm, as shown in Figure 22.17.

(a) Find the angular wave number, period, angular frequency, and phase speed of the wave.

Solution Using Equations 22.10, 22.11, and 22.14 and given the information that $\lambda = 40$ cm and $f = 8$ Hz, we find the following:

$$k = \frac{2\pi}{\lambda} = \frac{2\pi}{40\text{ cm}} = 0.157\text{ rad/cm}$$

$$T = \frac{1}{f} = \frac{1}{8\text{ s}^{-1}} = 0.125\text{ s}$$

$$\omega = 2\pi f = 2\pi(8\text{ s}^{-1}) = 50.3\text{ rad/s}$$

$$v = f\lambda = (8\text{ s}^{-1})(40\text{ cm}) = 320\text{ cm/s}$$

(b) Determine the phase constant, ϕ, and write a general expression for the wave function.

Solution Since the amplitude $A = 15$ cm and since it is given that $y = 15$ cm at $x = 0$ and $t = 0$, substitution into Equation 22.15 gives

$$15 = 15\sin(-\phi) \quad \text{or} \quad \sin(-\phi) = 1$$

Since $\sin(-\phi) = -\sin\phi$, we see that $\phi = -\pi/2$ rad (or $-90°$). Hence, the wave function is of the form

$$y = A\sin\left(kx - \omega t + \frac{\pi}{2}\right) = A\cos(kx - \omega t)$$

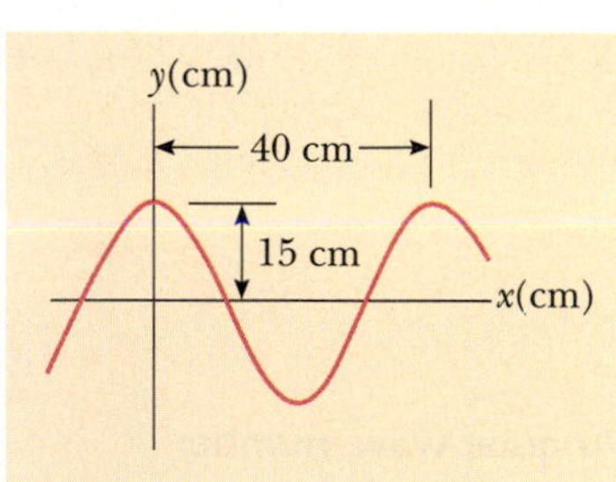

Figure 22.17 (Example 22.3) A harmonic wave of wavelength $\lambda = 40$ cm and amplitude $A = 15$ cm. The wave function can be written in the form $y = A\cos(kx - \omega t)$.

This can be seen by inspection, noting that the cosine argument is displaced by 90° from the sine function. Substituting the values for A, k, and ω into this expression gives

$$y = (15 \text{ cm}) \cos(0.157x - 50.3t)$$

Sinusoidal Waves on Strings

One method of producing a traveling sinusoidal wave on a very long string is shown in Figure 22.18. One end of the string is connected to a blade that is set in vibration. As the blade oscillates vertically with simple harmonic motion, a traveling wave moving to the right is set up on the string. Figure 22.18 represents snapshots of the wave at intervals of one quarter of a period. Note that *each particle of the string, such as P, oscillates vertically in the y direction with simple harmonic motion.* This must be the case because each particle follows the simple harmonic motion of the blade. Therefore, every segment of the string can be treated as a simple harmonic oscillator vibrating with a frequency equal to the frequency of vibration of the blade that drives the string.[3] Note that although each segment oscillates in the y direction, the wave (or disturbance) travels in the x direction with a speed of v. Of course, this is the definition of a transverse wave. In this case, the energy carried by the traveling wave is supplied by the vibrating blade.

If the waveform at $t = 0$ is as described in Figure 22.18b, then the wave function can be written

$$y = A \sin(kx - \omega t)$$

[3] In this arrangement, we are assuming that the segment always oscillates in a vertical line. The tension in the string would vary if the segment were allowed to move sideways. Such a motion would make the analysis very complex.

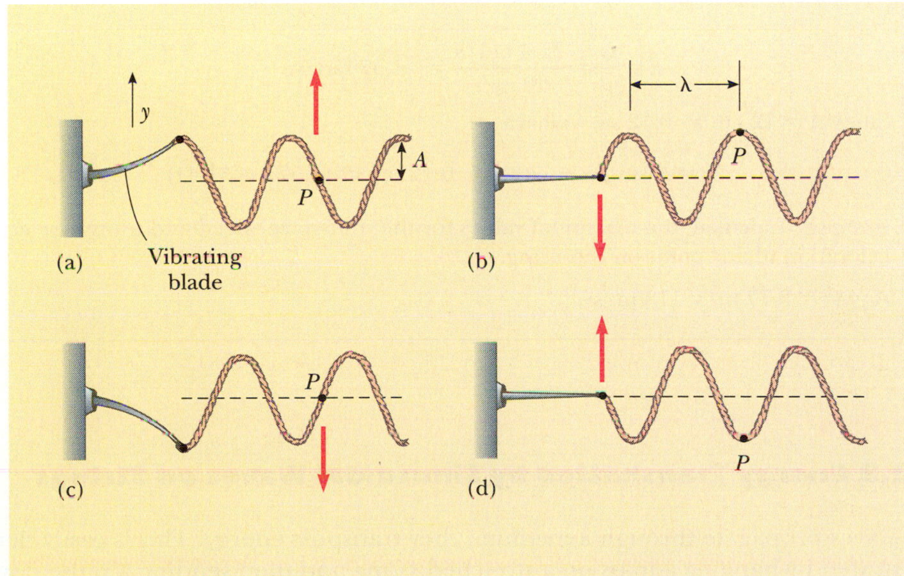

Figure 22.18 One method for producing harmonic waves on a string. The left end of the string is connected to a blade that is set into vibration. Note that every segment, such as P, oscillates with simple harmonic motion in the vertical direction.

We can use this expression to describe the motion of any point on the string. The point P (or any other point on the string) moves vertically, and so its *x coordinate remains constant.* Therefore, the *transverse speed,* v_y, of the point P (not to be confused with the wave speed, v) and its *transverse acceleration,* a_y, are

$$v_y = \left.\frac{dy}{dt}\right]_{x\,=\,\text{constant}} = \frac{\partial y}{\partial t} = -\omega A \cos(kx - \omega t) \qquad [22.16]$$

$$a_y = \left.\frac{dv_y}{dt}\right]_{x\,=\,\text{constant}} = \frac{\partial v_y}{\partial t} = -\omega^2 A \sin(kx - \omega t) \qquad [22.17]$$

The maximum values of these quantities are simply the absolute values of the coefficients of the cosine and sine functions:

$$(v_y)_{\max} = \omega A \qquad [22.18]$$

$$(a_y)_{\max} = \omega^2 A \qquad [22.19]$$

You should recognize that the transverse speed and transverse acceleration of any point on the string do not reach their maximum values simultaneously. In fact, the transverse speed reaches its maximum value (ωA) when the displacement $y = 0$, whereas the transverse acceleration reaches its maximum value ($\omega^2 A$) when $y = -A$. Finally, Equations 22.18 and 22.19 are identical to the corresponding equations for simple harmonic motion.

Example 22.4 A Harmonically Driven String

The string shown in Figure 22.18 is driven at one end at a frequency of 5 Hz. The amplitude of the motion is 12 cm, and the wave speed is 20 m/s. Determine the angular frequency and wave number for this wave, and write an expression for the wave function.

Solution Use of Equations 22.11 and 22.13 gives

$$\omega = \frac{2\pi}{T} = 2\pi f = 2\pi(5\text{ Hz}) = 31.4\text{ rad/s}$$

$$k = \frac{\omega}{v} = \frac{31.4\text{ rad/s}}{20\text{ m/s}} = 1.57\text{ rad/m}$$

Since $A = 12\text{ cm} = 0.12\text{ m}$, we have

$$y = A\sin(kx - \omega t) = 0.12\text{ m})\sin(1.57x - 31.4t)$$

Exercise Calculate the maximum values for the transverse speed and transverse acceleration of any point on the string.

Answer 3.77 m/s; 118 m/s^2.

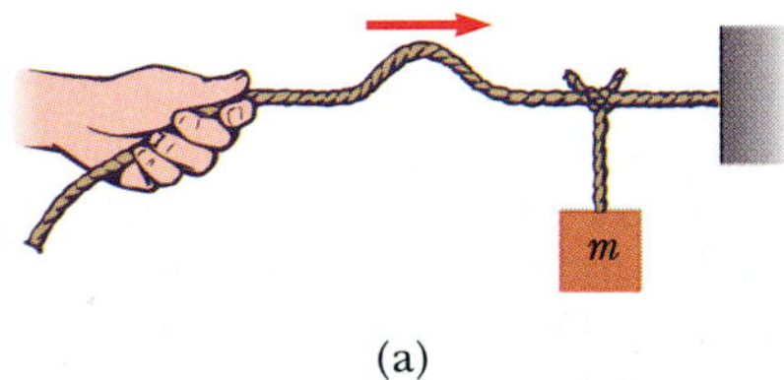

(a)

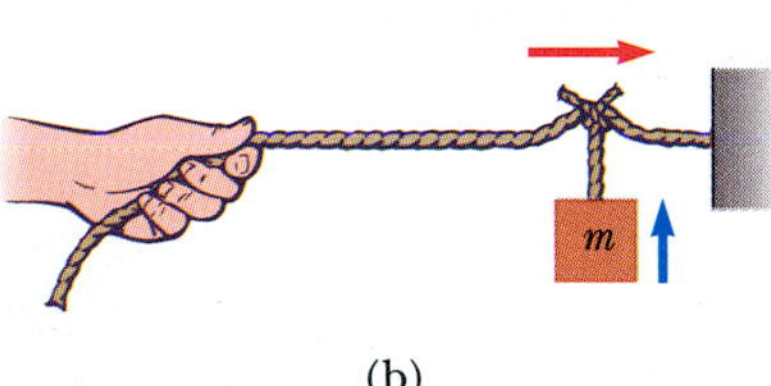

(b)

Figure 22.19
(a) A pulse traveling to the right on a stretched string on which a mass has been suspended.
(b) Energy is transmitted to the suspended mass when the pulse arrives.

22.8 Energy Transmitted by Sinusoidal Waves on Strings

As waves propagate through a medium, they transport energy. This is easily demonstrated by hanging a mass on a stretched string and then sending a pulse down the string, as in Figure 22.19. When the pulse meets the suspended mass, the mass

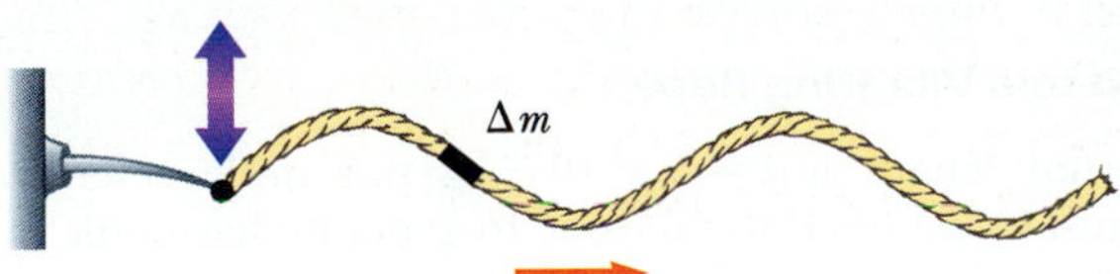

Figure 22.20
A harmonic wave traveling along the x axis on a stretched string. Every segment moves vertically, and each has the same total energy. The power transmitted by the wave equals the energy contained in one wavelength divided by the period of the wave.

is momentarily displaced, as in Figure 22.19b. In the process, energy is transferred to the mass, since work must be done in moving it upward.

This section examines the rate at which energy is transported along a string. We shall assume that this one-dimensional wave is sinusoidal when we calculate the power transferred. Later we shall extend the same ideas to three-dimensional waves.

Consider a sinusoidal wave traveling on a string (Fig. 22.20). The source of the energy is some external agent at the left end of the string, which does work in producing the oscillations. Let us focus our attention on an element of the string of length Δx and mass Δm. Each such segment moves vertically with simple harmonic motion. Furthermore, each segment has the same angular frequency, ω, and the same amplitude, A. As we found in Chapter 21, the total energy, E, associated with a particle moving with simple harmonic motion is $E = \frac{1}{2}kA^2 = \frac{1}{2}m\,\omega^2A^2$ (Eq. 21.22), where k is the effective force constant of the restoring force. If we apply this equation to the element of length Δx, we see that the total energy of this element is

$$\Delta E = \tfrac{1}{2}(\Delta m)\,\omega^2A^2$$

If μ is the mass per unit length of the string, then the element of length Δx has a mass, Δm, that is equal to $\mu\,\Delta x$. Hence, we can express the energy as

$$\Delta E = \tfrac{1}{2}(\mu\,\Delta x)\,\omega^2A^2 \qquad [22.20]$$

If the wave travels from left to right, as in Figure 22.20, the energy ΔE arises from the work done on the element Δm by the string element to the left of Δm. Similarly, the element Δm does work on the element to its right, so we see that energy is transmitted to the right. The rate at which energy is transmitted along the string, or the power (P), is dE/dt. If we let Δx approach 0, Equation 22.20 gives

$$P = \frac{dE}{dt} = \frac{1}{2}\left(\mu\,\frac{dx}{dt}\right)\omega^2A^2$$

Since dx/dt is equal to the wave speed, v, we have

$$P = \tfrac{1}{2}\mu\omega^2A^2\,v \qquad [22.21]$$

Power

This shows that the power transmitted by a sinusoidal wave on a string is proportional to (a) the wave speed, (b) the square of the frequency, and (c) the square of the amplitude. In fact, *all* sinusoidal waves have the following general property: *the power transmitted by any sinusoidal wave is proportional to the square of the angular frequency and to the square of the amplitude.*

Thus, we see that a wave traveling through a medium corresponds to energy transport through the medium, with no net transfer of matter. An oscillating source provides the energy and produces a harmonic disturbance of the medium. The disturbance is able to propagate through the medium as the result of the interaction between adjacent particles. In order to verify Equation 22.20 by direct experiment, one would have to design some device at the far end of the string to extract the energy of the wave without producing any reflections.

Example 22.5 **Power Supplied to a Vibrating Rope**

A stretched rope having mass per unit length of $\mu = 5 \times 10^{-2}$ kg/m is under a tension of 80 N. How much power must be supplied to the rope to generate harmonic waves at a frequency of 60 Hz and an amplitude of 6 cm?

Solution The wave speed on the stretched rope (Eq. 22.4) is

$$v = \sqrt{\frac{F}{\mu}} = \left(\frac{80 \text{ N}}{5 \times 10^{-2} \text{ kg/m}}\right)^{1/2} = 40 \text{ m/s}$$

Since $f = 60$ Hz, the angular frequency, ω, of the harmonic waves on the string is

$$\omega = 2\pi f = 2\pi(60 \text{ Hz}) = 377 \text{ rad/s}$$

Using these values in Equation 22.21, with $A = 6 \times 10^{-2}$ m, gives

$$\begin{aligned} P &= \tfrac{1}{2}\mu\omega^2 A^2 v \\ &= \tfrac{1}{2}(5 \times 10^{-2} \text{ kg/m})(377 \text{ s}^{-1})^2(6 \times 10^{-2} \text{ m})^2 (40 \text{ m/s}) \\ &= 512 \text{ W} \end{aligned}$$

22.9 Sound Waves

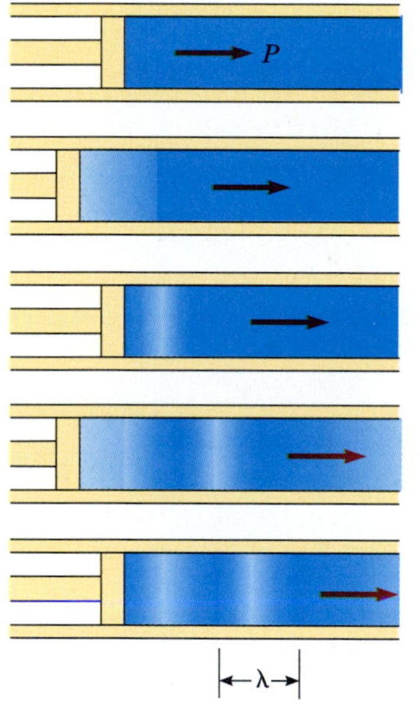

Figure 22.21 A harmonic longitudinal wave propagating down a tube filled with a compressible gas. The source of the wave is a vibrating piston at the left. The high- and low-pressure regions are dark and light, respectively.

Let us turn our attention from transverse waves to longitudinal ones. As stated in Section 22.2, longitudinal waves are waves in which the particles of the medium undergo displacements parallel to the direction of wave motion. Sound waves are the most important example of longitudinal waves. They can travel through any material medium, their speed depending on the properties of that medium.

The displacements accompanying a sound wave in air are longitudinal displacements of individual molecules from their equilibrium positions. Such displacements result if the source of the waves, such as the diaphragm of a loudspeaker, oscillates with simple harmonic motion. For instance, one can produce a one-dimensional harmonic sound wave in a long, narrow tube containing a gas by means of a vibrating piston at one end, as in Figure 22.21. The darker regions in the figure represent regions where the gas is compressed and, consequently, the density and pressure are *above* their equilibrium values. Such a compressed layer of gas, called a **condensation,** is formed when the piston is being pushed into the tube. The condensation moves down the tube as a pulse, continuously compressing the layers in front of it. When the piston is withdrawn from the tube, the gas in front of it expands and consequently the pressure and density in this region fall below their equilibrium values. These low-pressure regions, called **rarefactions,** are represented by the lighter areas in Figure 22.21. The rarefactions also propagate along the tube, following the condensations. Both regions move with a speed equal to the speed of sound in that medium. The speed of sound waves in air at 20°C is about 343 m/s.

As the piston oscillates back and forth in a sinusoidal fashion, regions of condensation and rarefaction are continuously set up. The distance between two successive condensations (or two successive rarefactions) equals the wavelength, λ.

As these regions travel down the tube, any small volume of the medium moves with simple harmonic motion parallel to the direction of the wave (in other words, longitudinally). If $s(x, t)$ is the displacement of a small volume element measured from its equilibrium position, we can express this harmonic displacement function as

$$s(x, t) = s_m \cos(kx - \omega t) \qquad [22.22]$$

where s_m is the *maximum displacement from equilibrium* (the displacement amplitude), k is the angular wave number, and ω is the angular frequency of the piston. Note that the displacement of the volume element is along x, the direction of motion of the sound wave, which of course means we are describing a longitudinal wave. The variation in the pressure of the gas, ΔP, measured from its equilibrium value is also harmonic; it is given by

$$\Delta P = \Delta P_m \sin(kx - \omega t) \qquad [22.23]$$

The **pressure amplitude,** ΔP_m, is the *maximum change in pressure from the equilibrium value.* It is proportional to the displacement amplitude, s_m:

$$\Delta P_m = \rho v \omega s_m \qquad [22.24]$$

Pressure amplitude

where ρ is the density of the medium, v is the wave speed, and ωs_m is the maximum longitudinal speed of the medium in front of the piston.

Thus, we see that a sound wave may be considered as either a displacement wave or a pressure wave. A comparison of Equations 22.22 and 22.23 shows that *the pressure wave is 90° out of phase with the displacement wave.* Graphs of these functions are shown in Figure 22.22. Note that the pressure variation is a maximum when the displacement is zero, whereas the displacement is a maximum when the pressure variation is zero. Since the pressure is proportional to the density of the medium, the expression that describes how that density varies from its equilibrium value is similar to Equation 22.23.

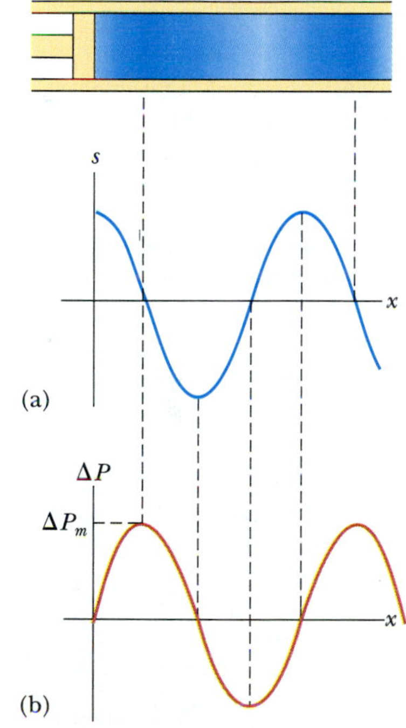

Figure 22.22 (a) Displacement amplitude versus position and (b) pressure amplitude versus position for a harmonic longitudinal wave. The displacement wave is 90° out of phase with the pressure wave.

22.10 The Doppler Effect

When a vehicle sounds its horn as it travels along a highway, the frequency of the sound you hear is higher as the vehicle approaches you than it is as the vehicle moves away from you. This is one example of the **Doppler effect,** named after the Austrian physicist Christian Johann Doppler (1803–1853). **In general, a Doppler effect is experienced whenever there is relative motion between the source of sound and the observer. When the source and observer are moving toward each other, the observer hears a frequency that is higher than the true frequency of the source. When the source and observer are moving away from each other, the observer hears a frequency that is lower than the true frequency of the source.**[4] The Doppler effect is used in police radar systems to measure the speeds of motor vehicles. Likewise,

[4] Although the Doppler effect is most commonly experienced with sound waves, it is a phenomenon common to all harmonic waves. For example, a shift in frequencies of light waves (electromagnetic waves) is produced by the relative motion of source and observer. In fact, in 1842 Doppler first reported the frequency shift in connection with light emitted by stars revolving about each other in double-star systems.

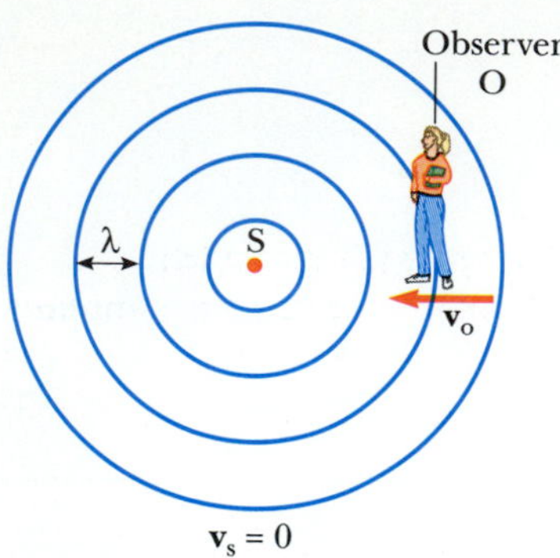

Figure 22.23
An observer O moving with a speed v_O toward a stationary point source S hears a frequency f' that is greater than the true source frequency.

astronomers use the effect to determine the relative motions of stars, galaxies, and other celestial objects.

First, let us consider the case where the observer, O, is moving with a speed of v_O and the sound source, S, is stationary, as in Figure 22.23. For simplicity, we shall assume that the air is stationary and that the observer moves directly toward the source (considered as a point source). We shall take the frequency of the source to be f, the wavelength to be λ, and the speed of sound to be v. A stationary observer would detect f vibrations each second, where $f = v/\lambda$. (That is, when the source and observer are both at rest, the observed frequency must equal the true frequency of the source.) However, the observer moving toward the source with the speed v_O receives an additional number of vibrations per second, equal to v_O/λ. Hence, the frequency the observer hears, f', is *increased:* $f' = v/\lambda + v_O/\lambda$. Since $\lambda = v/f$, we can express f' as

$$f' = f\left(\frac{v + v_O}{v}\right) \quad \text{(observer moving toward source)} \qquad [22.25]$$

Now consider the situation in which the source moves with a speed of v_S relative to the medium, and the observer is at rest. If the source moves directly toward observer A in Figure 22.24a, the wavefronts seen by the observer are closer to each other than they would be if the source were at rest. As a result, the wavelength λ' measured by observer A is shorter than the true wavelength, λ, of the source. During each vibration, with a duration of T (the period), the source moves a distance of $v_S T = v_S/\lambda$. Therefore, the wavelength is shortened by this amount, and the observed wavelength has the shorter value $\lambda' = \lambda - v_S/\lambda$. Since $\lambda = v/f$, the frequency heard by observer A is

$$f' = \frac{v}{\lambda'} = f\left(\frac{v}{v - v_S}\right) \quad \text{(source moving toward observer)} \qquad [22.26]$$

That is, the frequency is *increased* when the source moves toward the observer.

Figure 22.24
(a) A source S moving with a speed v_S toward a stationary observer A and away from a stationary observe B. Observer A hears an increased frequency, and observer B hears a decreased frequency. (b) The Doppler effect in water observed in a ripple tank. *(Courtesy Educational Development Center, Newton, Mass.)*

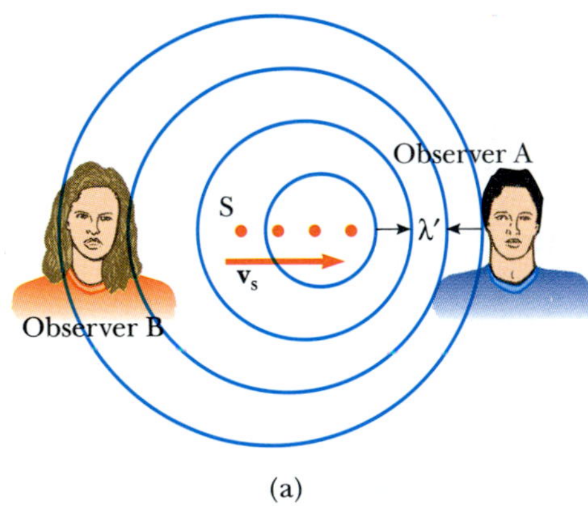

(a)

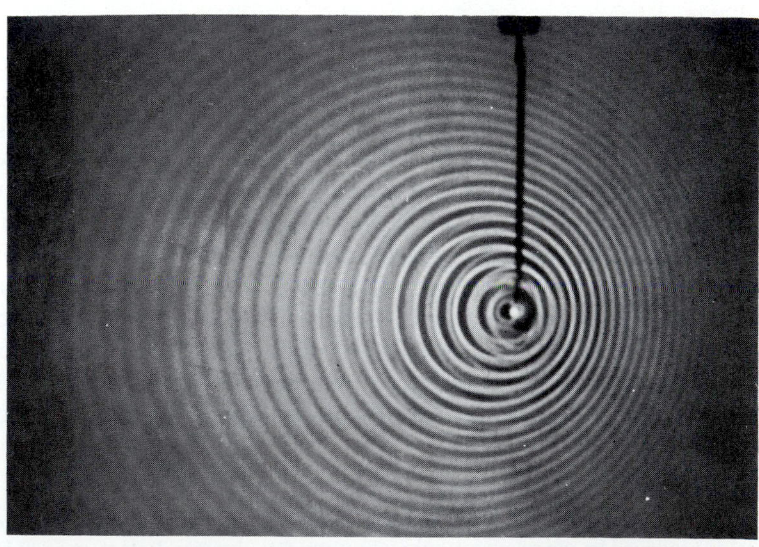

(b)

In a similar manner, if the motion of the source is toward observer B in Figure 22.24a, the signs of v_0 and v_S are reversed.

Finally, if both the source and the observer are in motion, one finds the following general formula for the observed frequency:

$$f' = f\left(\frac{v \pm v_0}{v \mp v_S}\right) \quad [22.27]$$

Frequency heard with observer and source in motion

In this expression, the upper signs ($+v_0$ and $-v_S$) refer to motion of the source or observer toward the other, and the lower signs ($-v_0$ and $+v_S$) refer to motion of one away from the other.

When working with any Doppler effect problem, a convenient rule to remember concerning signs is the following: The word *toward* is associated with an *increase* in the observed frequency. The words *away from* are associated with a *decrease* in the observed frequency.

"I love hearing that lonesome wail of the train whistle as the magnitude of the frequency of the wave changes due to the Doppler effect."

*22.11 Shock Waves

Now let us consider what happens when the source speed, v_S, exceeds the wave speed, v, in which case Equation 22.27 is not valid. This situation is depicted graphically in Figure 22.25. The circles represent spherical wavefronts emitted by the source at various times during its motion. At $t = 0$, the source is at S_0, and at some later time t, the source is at S_n. In the time interval t, the wavefront centered at S_0 reaches the radius vt. In this same interval, the source travels a distance of v_St to S_n. At the instant the source is at S_n, a new wave is just beginning to be generated and so the wavefront has zero radius at this point. The line drawn from S_n to the wavefront centered on S_0 is tangent to all other wavefronts generated at interme-

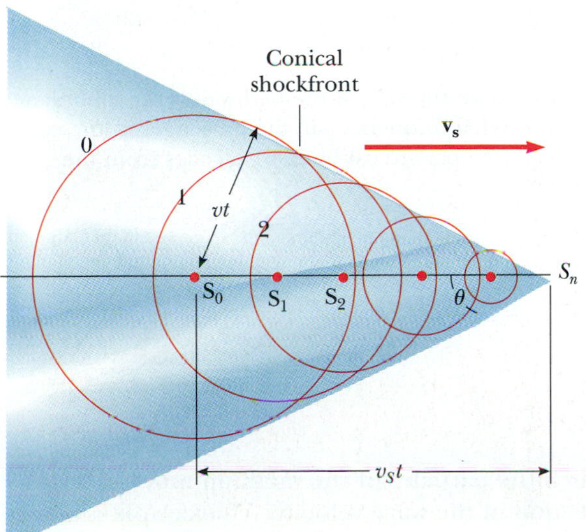

Figure 22.25 Representation of a shock wave produced when a source moves from S_0 to S_n with a speed v_S, that is greater than the wave speed v in that medium. The envelope of the wavefronts forms a cone whose apex angle is given by $\sin\theta = v/v_S$.

diate times. Thus, we see that the envelope of these waves is a cone whose apex angle, θ, is

$$\sin \theta = \frac{v}{v_S} \qquad [22.28]$$

The V-shaped wavefront that follows the duck is analogous to shock waves produced by airplanes traveling at supersonic speeds. *(© Harry Engels)*

The ratio v_S/v is referred to as the *Mach number.* The conical wavefront produced when $v_S > v$ (supersonic speeds) is known as a *shock wave.* An interesting analogy to shock waves is the V-shaped wavefronts produced by a duck (the bow wave) when the duck's speed exceeds the speed of the surface water waves as shown in the photograph.

Example 22.6 The Noisy Siren

An ambulance travels down a highway at 33.5 m/s. Its siren emits sound at a frequency of 400 Hz. What is the frequency heard by a passenger in a car traveling at 24.6 m/s in the opposite direction as the car approaches the ambulance and after the two vehicles have passed each other?

Solution Let us take the speed of sound in air to be $v = 343$ m/s. We can use Equation 22.27 in both cases. As the ambulance and car approach each other, the observed apparent frequency is

$$f' = f\left(\frac{v + v_0}{v - v_S}\right) = (400 \text{ Hz})\left(\frac{343 \text{ m/s} + 24.6 \text{ m/s}}{343 \text{ m/s} - 33.5 \text{ m/s}}\right)$$

$$= 475 \text{ Hz}$$

As they recede from each other, a passenger in the car hears the frequency

$$f' = f\left(\frac{v - v_0}{v + v_S}\right) = (400 \text{ Hz})\left(\frac{343 \text{ m/s} - 24.6 \text{ m/s}}{343 \text{ m/s} + 33.5 \text{ m/s}}\right)$$

$$= 338 \text{ Hz}$$

The change in frequency detected by the passenger in the car is 475 − 338 = 137 Hz, which is more than 30% of the true frequency emitted.

Exercise Suppose that the car is parked on the side of the highway as the ambulance travels down the highway at 33.5 m/s. What frequency will the passenger in the car hear as the ambulance (a) approaches the parked car and (b) recedes from the parked car?

Answer (a) 443 Hz; (b) 364 Hz.

Summary

Transverse wave

A **transverse wave** is a wave in which the particles of the medium move in a direction perpendicular to the direction of the wave velocity. An example is a wave on a stretched string.

Longitudinal wave

Longitudinal waves are waves in which the particles of the medium move parallel to the direction of the wave speed. Sound waves in air are longitudinal.

Any one-dimensional wave traveling with a speed of v in the positive x direction can be represented by a wave function of the form $y = f(x - vt)$. Likewise, the wave function for a wave traveling in the negative x direction has the form $y = f(x + vt)$. The shape of the wave at any instant (a snapshot of the wave) is obtained by holding t constant.

The **superposition principle** says that when two or more waves move through a medium, the resultant wave function equals the algebraic sum of the individual wave functions. Waves that obey this principle are said to be *linear*. When two waves combine in space, they interfere to produce a resultant wave. — Superposition principle

The speed of a wave traveling on a stretched string of mass per unit length μ and tension F is

Speed of a wave on a stretched string

$$v = \sqrt{\frac{F}{\mu}} \qquad [22.4]$$

When a pulse traveling on a string meets a fixed end, the pulse is reflected and inverted. If the pulse reaches a free end, it is reflected but not inverted.

The wave function for a one-dimensional harmonic wave traveling to the right can be expressed as

Wave function for a harmonic wave

$$y = A \sin\left[\frac{2\pi}{\lambda}(x - vt)\right] = A \sin(kx - \omega t) \qquad [22.6, 22.12]$$

where A is the amplitude, λ is the wavelength, k is the angular wave number, and ω is the angular frequency. If T is the period and f is the frequency, then v, k, and ω can be written

$$v = \frac{\lambda}{T} = f\lambda \qquad [22.7, 22.14]$$

Angular wave number

$$k \equiv \frac{2\pi}{\lambda} \qquad [22.10]$$

Angular frequency

$$\omega \equiv \frac{2\pi}{T} = 2\pi f \qquad [22.11]$$

The **power** transmitted by a harmonic wave on a stretched string is

Power

$$P = \tfrac{1}{2}\mu\omega^2 A^2 v \qquad [22.21]$$

The change in frequency heard by an observer whenever there is relative motion between a wave source and the observer is called the **Doppler effect.** When the source and observer are moving toward each other, the observer hears a frequency that is higher than the true frequency of the source. When the source and observer are moving away from each other, the observer hears a frequency that is lower than the true frequency of the source.

Questions and Conceptual Exercises

1. Why is a wave pulse traveling on a string considered a transverse wave?
2. How would you set up a longitudinal wave in a stretched spring? Would it be possible to set up a transverse wave in a spring?
3. By what factor would you have to increase the tension in a stretched string in order to double the wave speed?
4. Can two pulses traveling in opposite directions on the same string reflect from each other? Explain.
5. Harmonic waves are generated on a string under con-

stant tension by a vibrating source. If the power delivered to the string is doubled, by what factor does the amplitude change? Does the wave speed change under these circumstances?

6. If a long rope is hung from a ceiling and waves are sent up the rope from its lower end, the waves do not ascend with constant speed. Explain.
7. What happens to the wavelength of a wave on a string when the frequency is doubled? Assume the tension in the string remains the same.
8. What happens to the speed of a wave on a string when the frequency is doubled? Assume the tension in the string remains the same.
9. How do transverse waves differ from longitudinal waves?
10. When all the strings on a guitar are stretched to the same tension, will the speed of a wave along the more massive bass strings be faster or slower than the speed of a wave on the lighter strings?
11. If you stretch a rubber hose and pluck it, you can observe a pulse traveling up and down the hose. What happens to the speed if you stretch the hose tighter? if you fill the hose with water?
12. In a longitudinal wave in a spring, the coils move back and forth in the direction the wave travels. Does the speed of the wave depend on the maximum speed of each of the coils?
13. When two waves interfere, can the resultant wave be larger than either of the two original waves? Under what conditions?
14. In an earthquake, both S (shear) and P (pressure) waves are sent out. The S (transverse) waves travel through the Earth more slowly than the P (longitudinal) waves (5 km/s versus 9 km/s). By detecting the times of arrival of the waves, how may one determine how far away the epicenter of the quake was? How many detection centers are necessary to pinpoint the location of the epicenter?
15. How is it that water waves originating in a South Pacific storm can carry energy and momentum 10 000 km to the California shore, and yet none of the water in the South Pacific actually travels to California?
16. As a result of a distant explosion, an observer senses a ground tremor and then hears the explosion. Explain.
17. Explain how the Doppler effect is used with microwaves to determine the speed of an automobile.
18. If you are in a moving vehicle, explain what happens to the frequency of your echo as you move *toward* a canyon wall. What happens to the frequency as you move *away* from the wall?
19. Suppose an observer and a source of sound are both at rest, and a strong wind blows from a source toward the observer. Describe the effects (if any) of the wind on (a) the observed wavelength, (b) the observed frequency, and (c) the wave speed.
20. A binary star system consists of two stars revolving about each other. If we observe the light reaching us from one of these stars as it makes one complete revolution about the other, what does the Doppler effect predict will happen to this light?
21. How could an object move with respect to an observer so that the sound from it is not shifted in frequency?
22. If the wavelength of a sound source is reduced by a factor of 2, what happens to its frequency? its speed?
23. A sound wave travels in air at a frequency of 500 Hz. If part of the wave travels from the air into water, does its frequency change? Does its wavelength change? Justify your answers. Note that the speed of sound in air is about 340 m/s, whereas the speed of sound in water is about 1500 m/s.
24. Explain how the distance to a lightning bolt may be determined by counting the seconds between the flash and the sound of the thunder. Does the speed of the light signal have to be taken into account?

Problems

Section 22.3 One-Dimensional Traveling Waves

1. Ocean waves with a crest-to-crest distance of 10 m can be described by

$$\psi(x, t) = (0.8 \text{ m}) \sin[0.63(x - vt)] \qquad v = 1.2 \text{ m/s}$$

(a) Sketch $\psi(x, t)$ at $t = 0$. This corresponds to a snapshot of the wave at $t = 0$. (b) Sketch $\psi(x, t)$ at $t = 2$ s. Note how the entire wave form has shifted 2.4 m in the positive x direction in this time interval.

2. Two wave pulses, A and B, are moving in *opposite* di-

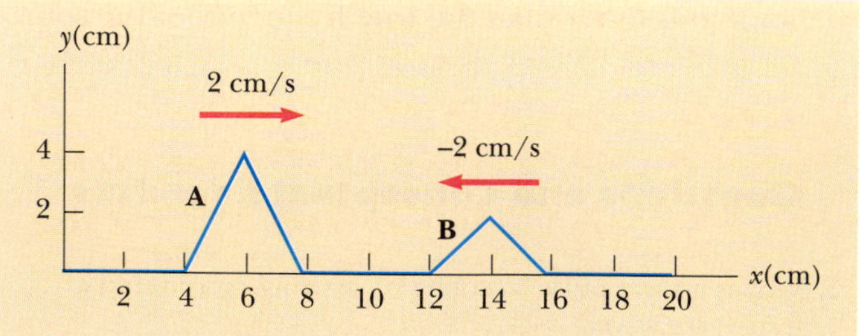

Figure 22.26 (Problem 2)

rections along a stretched string with a speed of 2 cm/s. The amplitude of A is twice the amplitude of B. The pulses are shown in Figure 22.26 at $t = 0$. Sketch the shape of the string at $t = 1$, 1.5, 2, 2.5, and 3 s.

3. At $t = 0$, a transverse wave pulse in a wire is described by the function

$$y = \frac{6}{x^2 + 3}$$

where x and y are in meters. Write the function $y(x, t)$ that describes this wave if it is traveling in the positive x direction with a speed of 4.5 m/s.

4. A wave pulse traveling to the right along the x axis is represented by the following wave function:

$$y(x, t) = \frac{4}{2 + (x - 4t)^2}$$

where x and y are measured in centimeters and t is in seconds. Plot the shape of the waveform at $t = 0$, 1, and 2 s.

5. A wave pulse traveling to the left is described by the following function:

$$y = \frac{10}{5 + (x + 2t)^4}$$

where x and y are in centimeters and t is in seconds. (a) Compute the speed of this wave. (b) Plot this wave at two different times and show that it is traveling to the left.

Section 22.4 Superposition and Interference of Waves

6. Two waves in one string are described by the following relationships:

$$y_1 = 3\cos(4x - 5t) \qquad y_2 = 4\sin(5x - 2t)$$

Find the superposition of the waves $y_1 + y_2$ at the points (a) $x = 1$, $t = 1$; (b) $x = 1$, $t = 0.5$; (c) $x = 0.5$, $t = 0$. (Remember that the arguments of the trigonometric functions are in radians.)

7. Two harmonic waves in a string are defined by the following functions:

$$y_1 = 2 \text{ cm} \sin(20x - 30t)$$

$$y_2 = 2 \text{ cm} \sin(25x - 40t)$$

where the y's and x are in centimeters and t is in seconds. (a) What is the phase difference between these two waves at the point $x = 5$ cm at $t = 2$ s? (b) What is the positive x value closest to the origin for which the two phases will differ by $\pm\pi$ at $t = 2$ s? (This is where the two waves will add to zero.)

8. Two waves are described by $y_1 = 1/(ax - bt)$ and $y_2 = 1/(ax + bt)$, where a and b are constants. Find a relationship for the superposition of the two waves $y_1 + y_2$. What points along this wave would be physically unrealistic, and why?

9. One wave pulse in a string is described by the equation

$$y_1 = \frac{5}{(3x - 4t)^2 + 2}$$

A second wave pulse in the same string is described by

$$y_2 = \frac{-5}{(3x + 4t - 6)^2 + 2}$$

(a) In which direction does each pulse travel? (b) At what time will the two waves exactly cancel everywhere? (c) At what point do the two waves always cancel?

Section 22.5 The Speed of Waves on Strings

10. A piano string with a mass per unit length of 0.005 kg/m is under a tension of 1350 N. Find the speed with which a wave travels on this string.

11. A circus performer stretches a tightrope between two towers. He strikes one end of the rope and sends a wave along it toward the other tower. He notes that it takes the wave 0.8 s to reach the opposite tower, which is located 20 m from the performer. If one meter of the rope has a mass of 0.35 kg, find the tension in the tightrope.

12. A phone cord is 4.0 m long. The cord has a mass of 0.2 kg. If a transverse wave pulse travels from the receiver to the phone box in 0.1 s, what is the tension in the cord?

13. Transverse waves travel with a speed of 20 m/s in a string that is under a tension of 6 N. What tension is required for a wave speed of 30 m/s in the same string?

14. Transverse pulses travel with a speed of 200 m/s along a taut copper wire whose diameter is 1.5 mm. What is the tension in the wire? (The density of copper is 8.92 g/cm^3.)

15. A 30-m steel wire and a 20-m copper wire, both with 1-mm diameters, are connected end to end and stretched to a tension of 150 N. How long will it take a transverse wave to travel the entire length of the two wires?

16. A light string with a mass per unit length of 8 g/m has its ends tied to two walls that are separated by a distance equal to 3/4 the length of the string (see Fig. 22.27). A mass, m, is suspended from the center of the string, putting a tension in the string. (a) Find an expression for the transverse wave speed in the string as a function of the hanging mass. (b) How

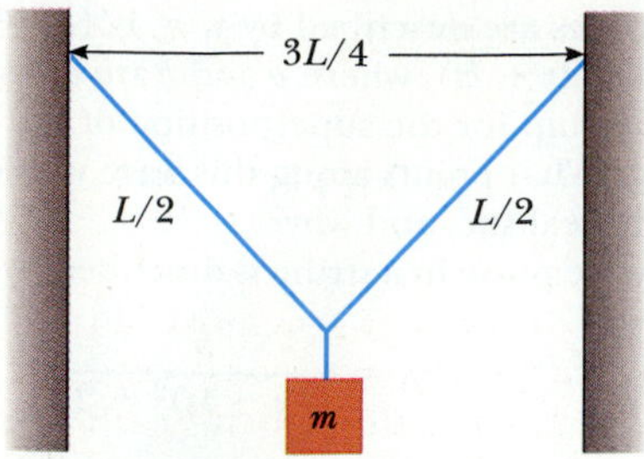

Figure 22.27 (Problem 16)

much mass should be suspended from the string to achieve a wave speed of 60 m/s?

Section 22.7 Harmonic Waves

17. A harmonic wave is traveling along a rope. It is observed that the oscillator that generates the wave completes 40 vibrations in 30 s. Also, a given maximum travels 425 cm along the rope in 10 s. What is the wavelength?
18. For a certain transverse wave, it is observed that the distance between two successive maxima is 1.2 m. It is also noted that eight crests, or maxima, pass a given point along the direction of travel every 12 s. Calculate the wave speed.
19. A transverse wave moving along a string in the positive x direction with a speed of 200 m/s has an amplitude of 0.7 mm and a wavelength of 20 cm. Determine (in SI units) the values of A, k, and ω in the equation describing the wave: $y = A\sin(kx - \omega t)$.
20. In Example 22.3, the harmonic wave was found to be described by $y = (15 \text{ cm})\cos(0.157x - 50.3t)$, where x and y are in centimeters and t is in seconds. (a) Plot y versus x at $t = 0$ and $t = 0.050$ s. (b) Determine the wave speed from this plot and compare your result with the value found in Example 22.3.
21. A harmonic wave train is described by

$$y = (0.25 \text{ m})\sin(0.3x - 40t)$$

where x and y are in meters and t is in seconds. Determine for this wave the (a) amplitude, (b) angular frequency, (c) wave number, (d) wavelength, (e) wave speed, and (f) direction of motion.

22. (a) Write the expression for y as a function of x and t for a sinusoidal wave traveling along a rope in the *negative* x direction with the following characteristics: $y_{max} = 8$ cm, $\lambda = 80$ cm, $f = 3$ Hz, $y(0, t) = 0$ at $t = 0$. (b) Write the expression for y as a function of x for the wave in (a), assuming that $y(x, 0) = 0$ at the point $x = 10$ cm.
23. A transverse wave on a string is described by

$$y = (0.12 \text{ m})\sin \pi\left(\frac{x}{8} + 4t\right)$$

(a) Determine the transverse speed and acceleration at $t = 0.20$ s for the point on the string located at $x = 1.6$ m. (b) What are the wavelength, period, and speed of propagation of this wave?

24. A transverse harmonic wave has a period $T = 25$ ms and travels in the negative x direction with a speed of 30 m/s. At $t = 0$, a particle on the string located at $x = 0$ has a displacement of 2.0 cm and a velocity of $v = -2.0$ m/s. (a) What is the amplitude of the wave? (b) What is the initial phase angle? (c) What is the magnitude of the maximum transverse speed? (d) Write the wave function for the wave.
25. A wave form is described by

$$\psi = (2.0 \text{ cm})\sin(kx - \omega t)$$
$$k = 2.11 \text{ rad/m}$$
$$\omega = 3.62 \text{ rad/s}$$

where x is the position along the wave form (in meters) and t is the time (in seconds). Determine the amplitude, wave number, wavelength, angular frequency, and speed of the wave.

26. A sinusoidal wave on a string is described by

$$\psi = (0.51 \text{ cm})\sin(kx - \omega t)$$
$$k = 3.1 \text{ rad/cm}$$
$$\omega = 9.3 \text{ rad/s}$$

How far does a wave crest move in 10 s? Does it move in the positive x direction or in the negative x direction?

27. A transverse traveling wave on a stretched wire has an amplitude of 0.2 mm and a frequency of 500 Hz, and travels with a speed of 196 m/s. (a) Write an equation, in SI units, of the form $y = A\sin(kx - \omega t)$ for this wave. (b) The mass per unit length of the wire is 4.10 g/m. Find the tension in the wire.

Section 22.8 Energy Transmitted by Harmonic Waves on Strings

28. A stretched rope has a mass of 0.18 kg and a length of 3.6 m. What power must be supplied in order to generate harmonic waves having an amplitude of 0.1 m and a wavelength of 0.5 m and traveling with a speed of 30 m/s?
29. Harmonic waves 5 cm in amplitude are to be transmitted along a string that has a linear density of 4×10^{-2} kg/m. If the maximum power delivered by the source is 300 W and the string is under a tension of

100 N, what is the highest vibrational frequency at which the source can operate?

30. Transverse waves are being generated on a rope under *constant tension.* By what factor will the required power be increased or decreased if (a) the length of the rope is doubled and the angular frequency remains constant? (b) the amplitude is doubled and the angular frequency is halved? (c) both the wavelength and the amplitude are doubled? (d) both the length of the rope and the wavelength are halved?

31. A harmonic wave in a string is described by the equation

$$y = (0.15 \text{ m}) \sin(0.8x - 50t)$$

where x and y are in meters and t is in seconds. If the mass per unit length of this string is 12 g/m, determine (a) the speed of the wave, (b) the wavelength, (c) the frequency, and (d) the power transmitted to the wave.

32. It is found that a 6-m segment of a long string contains 4 complete waves and has a mass of 180 g. The string is vibrating sinusoidally with a frequency of 50 Hz and a peak-to-peak displacement of 15 cm. ("Peak to peak" means the vertical distance from the farthest positive displacement to the farthest negative displacement.) (a) Write down the function that describes this wave traveling in the positive x direction. (b) Determine the power being supplied to the string.

Section 22.9 Sound Waves

(In this section, use the following values as needed unless otherwise specified: the equilibrium density of air, $\rho = 1.2 \text{ kg/m}^3$; the speed of sound in air, $v = 343 \text{ m/s}$. Also, pressure variations, ΔP, are measured relative to atmospheric pressure.)

33. Calculate the pressure amplitude of a 2000-Hz sound wave in air if the displacement amplitude is equal to 2×10^{-8} m.

34. A sound wave in air has a pressure amplitude equal to 4×10^{-3} Pa. Calculate the displacement amplitude of the wave at a frequency of 10 kHz.

35. An experimenter wishes to generate in air a sound wave that has a displacement amplitude equal to 5.5×10^{-6} m. The pressure amplitude is to be limited to 8.4×10^{-1} Pa. What is the minimum wavelength the sound wave can have?

36. A sound wave in air has a pressure amplitude of 4 Pa and a frequency of 5000 Hz. $\Delta P = 0$ at the point $x = 0$ when $t = 0$. (a) What is ΔP at $x = 0$ when $t = 2 \times 10^{-4}$ s? (b) What is ΔP at $x = 0.02$ m when $t = 0$?

37. Write an expression that describes the pressure variation as a function of position and time for a harmonic sound wave in air if its wavelength is 0.1 m and $\Delta P_m = 0.2$ Pa.

*Section 22.10 The Doppler Effect

38. Standing at a crosswalk, you hear a frequency of 560 Hz from the siren of an approaching police car. After the police car passes, the observed frequency of the siren is 480 Hz. Determine the car's speed from these observations.

39. A commuter train passes a passenger platform at a constant speed of 40 m/s. The train horn is sounded at its characteristic frequency of 320 Hz. (a) What change in frequency is detected by a person on the platform as the train passes? (b) What wavelength is detected by a person on the platform as the train approaches?

40. A band is playing on a moving truck. The band strikes the note middle C (262 Hz), but it is heard by spectators ahead of the truck as C sharp (C#; 277 Hz). How fast is the truck moving? Use 340 m/s for the speed of sound in air.

41. A tuning fork vibrating at 512 Hz falls from rest and accelerates at 9.80 m/s^2. How far below the point of release is the tuning fork when waves of frequency 485 Hz reach the release point? Take the speed of sound in air to be 340 m/s.

42. A bat, flying at 5 m/s, emits a chirp at 30 kHz. If this sound pulse is reflected by a wall in front of the bat, what is the frequency of the echo received by the bat? ($v_{\text{sound}} = 340 \text{ m/s}$.)

43. A driver travels north on a highway at a speed of 25 m/s. A police car, driving south at a speed of 40 m/s, approaches with its siren sounding at a base frequency of 2500 Hz. (a) What frequency is heard by the driver as the police car approaches? (b) What frequency is heard by the driver after the police car passes him? (c) Repeat (a) and (b) for the case when the police car is traveling northbound.

44. An airplane traveling at half the speed of sound emits sound waves of frequency 5000 Hz. At what frequencies does a stationary listener hear the sound as the plane approaches and after it passes by?

*Section 22.11 Shock Waves

45. A bullet fired from a rifle travels at Mach 1.38 (that is, 1.38 times the speed of sound in air). What angle does the shock front make with the path of the bullet?

46. At what speed should a supersonic aircraft fly so that the conical wavefront has an apex half-angle of 40°?

47. A jet fighter plane travels in horizontal flight at Mach 1.2 (that is, 1.2 times the speed of sound in air). At the instant an observer on the ground hears the

shock wave, what is the angle her line of sight makes with the horizontal as she looks at the plane?

48. When high-energy charged particles move through a transparent medium with a speed greater than the speed of light in that medium, a shock wave, or bow wave, of light is produced. This phenomenon is called the *Cerenkov effect* and can be observed in the vicinity of the core of a swimming pool reactor due to high-speed electrons moving through the water. In a particular case, the Cerenkov radiation produces a wavefront with an apex half-angle of 53°. Calculate the speed of the electrons in the water. (Use 2.25×10^8 m/s as the speed of light in water.)

49. A supersonic jet traveling at Mach 3 at an altitude of 20 000 m is directly overhead at time $t = 0$, as in Figure 22.28. (a) How long will it be before the observer encounters the shock wave? (b) Where will the plane be when it is finally heard? (Assume the speed of sound in air is 335 m/s.)

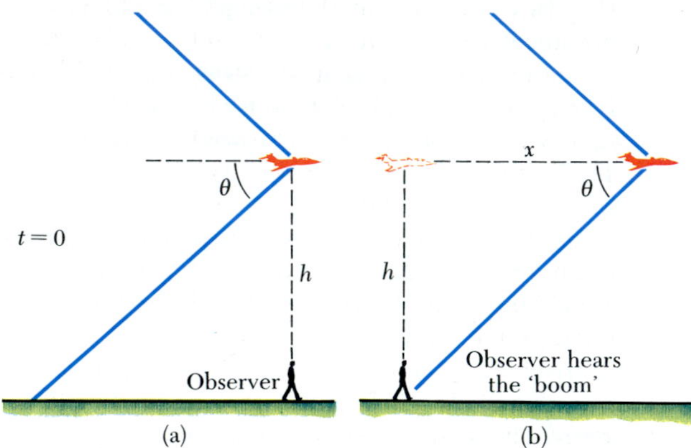

Figure 22.28 (Problem 49)

Additional Problems

50. A traveling wave propagates according to the expression $y = (4.0 \text{ cm}) \sin(2.0x - 3.0t)$, where x is in centimeters and t is in seconds. Determine (a) the amplitude, (b) the wavelength, (c) the frequency, (d) the period, and (e) the direction of travel of the wave.

51. The wave function for a transverse wave on a taut string is (in SI units)

$$y(x, t) = (0.35 \text{ m}) \sin\left(10\pi t - 3\pi x + \frac{\pi}{4}\right)$$

(a) What is the velocity of the wave? (Give the speed and the direction.) (b) What is the displacement at $t = 0$, $x = 0.1$ m? (c) What are the wavelength and frequency of the wave? (d) What is the maximum magnitude of the transverse velocity of the string?

52. (a) Determine the speed of transverse waves on a stretched string that is under a tension of 80 N if the string has a length of 2 m and a mass of 5 g. (b) Calculate the power required to generate these waves if they have a wavelength of 16 cm and are 4 cm in amplitude.

53. A stone is dropped into a deep canyon and is heard to strike the bottom 10.2 s after release. The speed of sound waves in air is 343 m/s. How deep is the canyon? What would be the percentage error in the depth if the time required for the sound to reach the canyon rim were ignored?

54. The ocean floor is underlain by a layer of basalt that constitutes the crust, or uppermost layer of the Earth, in this region. Below this crust is found denser periodotite rock, which forms the Earth's mantle. The boundary between these two layers is called the Mohorovičić discontinuity ("Moho" for short). If an explosive charge is set off at the surface of the basalt, it generates a seismic wave that is reflected back at the Moho. If the speed of this wave in basalt is 6.5 km/s, and the two-way travel time is 1.85 s, what is the thickness of this oceanic crust?

55. An earthquake on the ocean floor in the Gulf of Alaska induces a *tsunami* (sometimes called a "tidal wave") that reaches Hilo, Hawaii, 4450 km distant, in a time of 9 h 30 min. Tsunamis have enormous wavelengths (100 to 200 km), and for such waves the propagation speed is $v \approx \sqrt{g\bar{d}}$, where $\bar{d}$ is the average depth of the water. From the information given, find the average wave speed and the average ocean depth between Alaska and Hawaii. (This method was used in 1856 to estimate the average depth of the Pacific Ocean long before soundings were made to give a direct determination.)

56. An earthquake emits both P waves and S waves, which travel at different speeds through the Earth. A P wave travels at a speed of 9000 m/s, and an S wave travels at 5000 m/s. If P waves are received at a seismic station 1 minute before an S wave arrives, how far away is the earthquake center?

57. A worker strikes a steel pipeline with a hammer, generating both longitudinal and transverse waves. Reflected waves return 2.4 s apart. How far away is the reflection point? (For steel, $v_{\text{long.}} = 6.2$ km/s and $v_{\text{trans.}} = 3.2$ km/s.)

58. A rope of total mass m and length L is suspended vertically. Show that a transverse wave pulse will travel the length of the rope in the time $t = 2\sqrt{L/g}$. (*Hint:* First find an expression for the speed at any point a distance of x from the lower end of the rope, by considering the tension in the rope as resulting from the weight of the segment below that point.)

59. As a sound wave travels through the air, it produces pressure variations (above and below atmospheric

pressure) given by $P = 1.27 \sin \pi(x - 240t)$ in SI units. Find (a) the amplitude of the pressure variations, (b) the frequency, (c) the wavelength in air, and (d) the speed of the sound wave.

60. A train whistle ($f = 400$ Hz) sounds higher or lower in pitch, depending on whether it approaches or recedes. (a) Prove that the difference in frequency between the approaching and receding train whistle is

$$\Delta f = \frac{2f\left(\frac{u}{v}\right)}{1 - \frac{u^2}{v^2}} \qquad \begin{array}{l} u = \text{speed of train} \\ v = \text{speed of sound} \end{array}$$

(b) Calculate this difference for a train moving at a speed of 130 km/h. Take the speed of sound in air to be 340 m/s.

61. In order to be able to determine her speed, a skydiver carries a tone generator. A friend on the ground at the landing site has equipment for receiving and analyzing sound waves. While the skydiver is falling at terminal speed, her tone generator emits a steady tone of 1800 Hz. (Assume that the air is calm and that the sound speed is 343 m/s, independent of altitude.) (a) If her friend on the ground (directly beneath the skydiver) receives waves of frequency 2150 Hz, what is the skydiver's speed of descent? (b) If the skydiver were also carrying sound-receiving equipment sensitive enough to detect waves reflected from the ground, what frequency would she receive?

62. A wave pulse traveling along a string of linear mass density μ is described by the relationship

$$y = [A_0 e^{-bx}] \sin(kx - \omega t)$$

where the factors in brackets before the sine are said to be the amplitude. (a) What is the power, $P(x)$, carried by this wave at a point x? (b) What is the power carried by this wave at the origin? (c) Compute the ratio of $\dfrac{P(x)}{P(0)}$.

63. A bat, moving at 5 m/s, is chasing a flying insect. If the bat emits a 40-kHz chirp and receives back an echo at 40.4 kHz, at what velocity is the insect moving toward or away from the bat? (Take the speed of sound in air to be $v = 340$ m/s.)

64. A supersonic aircraft is flying parallel to the ground. When the aircraft is directly overhead, an observer sees a rocket fired from the aircraft. Ten seconds later the observer hears the sonic boom, followed 2.8 s later by the sound of the rocket engine. What is the Mach number of the aircraft?

(Problem 63) (© *Merlin Tuttle, Science Source/Photo Researchers*)

65. (a) Show that the speed of longitudinal waves along a spring of force constant k is $v = \sqrt{kL/\mu}$, where L is the unstretched length of the spring and μ is the mass per unit length. (b) A spring of mass 0.4 kg has an unstretched length of 2 m and a force constant of 100 N/m. Using the results of (a), determine the speed of longitudinal waves along this spring.

Calculator/Computer Problem

66. Two transverse wave pulses traveling in opposite directions along the x axis are represented by the following wave functions:

$$y_1(x, t) = \frac{6}{(x - 3t)^2} \qquad y_2(x, t) = -\frac{3}{(x + 3t)^2}$$

where x and y are measured in centimeters and t is in seconds. Write a program that will enable you to obtain the shape of the composite waveform $y_1 + y_2$ as a function of time. Use your program and make plots of the waveform at $t = 0, 0.5, 1, 1.5, 2, 2.5,$ and 3 s.

Even when silent, this organ at the Mormon Tabernacle conveys a sense of the power of sound waves. *(Courtesy of Henry Leap)*

Superposition and Standing Waves

CHAPTER 23

An important aspect of waves is their combined effect when two or more travel in the same medium. For instance, what happens to a string when a wave traveling toward a fixed end is reflected back on itself?

In a linear medium—that is, one in which the restoring force of the medium is proportional to the displacement of the medium—one can apply the *principle of superposition* to obtain the resultant disturbance. This principle can be applied to many types of waves, including waves on strings, sound waves, surface water waves, and electromagnetic waves. The superposition principle states that the displacement of any part of the disturbed medium equals the vector sum of the displacements caused by the individual waves. We discussed this principle in Chapter 16, Section 16.4, as it applies to electrostatic forces, and in Chapter 22, Section 22.4, as it applies to waves. In the last discussion, the term *interference* was used to describe the effect produced by combining two waves that are moving simultaneously through a medium.

This chapter is concerned with the superposition principle as it applies to harmonic waves. If the harmonic waves that combine in a given medium have the same frequency and wavelength, one finds that a stationary pattern, called a *standing wave,* can be produced at certain frequencies under certain circumstances. For example, a stretched string fixed at both ends has a discrete set of oscillation patterns, called *modes of vibration,* that depend upon the tension and the mass per unit length of the string. These modes of vibration are found in the strings of stringed muscial instruments. Other musical instruments, such as the organ and flute, make use of the natural frequencies of sound waves in hollow pipes. Such frequencies depend upon the length and shape of the pipe and upon whether one end is open or closed.

In this chapter we also consider the superposition and interference of waves with different frequencies and wavelengths. When two sound waves with nearly the same frequency interfere, one hears variations in loudness called *beats.* The beat frequency corresponds to the rate of alternation between constructive and destructive interference. Finally, we describe how any complex periodic waveform can, in general, be described by a sum of sine and cosine functions.

23.1 Superposition and Interference of Harmonic Waves

Let us apply the superposition principle to two harmonic waves traveling in the same direction in a medium. If the two waves are traveling to the right and have the same frequency, wavelength, and amplitude but differ in phase, we can express their individual wave functions as

$$y_1 = A\sin(kx - \omega t) \qquad \text{and} \qquad y_2 = A\sin(kx - \omega t - \phi)$$

Hence, the resultant wave function, y, is

$$y = y_1 + y_2 = A[\sin(kx - \omega t) + \sin(kx - \omega t - \phi)]$$

In order to simplify this expression, it is convenient to use the trigonometric identity

$$\sin a + \sin b = 2\cos\left(\frac{a-b}{2}\right)\sin\left(\frac{a+b}{2}\right)$$

If we let $a = kx - \omega t$ and $b = kx - \omega t - \phi$, the resultant wave, y, reduces to

$$y = \left(2A\cos\frac{\phi}{2}\right)\sin\left(kx - \omega t - \frac{\phi}{2}\right) \qquad [23.1]$$

Resultant of two traveling harmonic waves

This result has several important features. The resultant wave function, y, is also harmonic and has the *same* frequency and wavelength as the individual waves. The amplitude of the resultant wave is $2A\cos(\phi/2)$, and the phase is $\phi/2$. If the phase constant, ϕ, equals 0, then $\cos(\phi/2) = \cos 0 = 1$ and the amplitude of the resultant wave is $2A$. In other words, the amplitude of the resultant wave is twice the amplitude of either individual wave. In this case, the waves are said to be every-

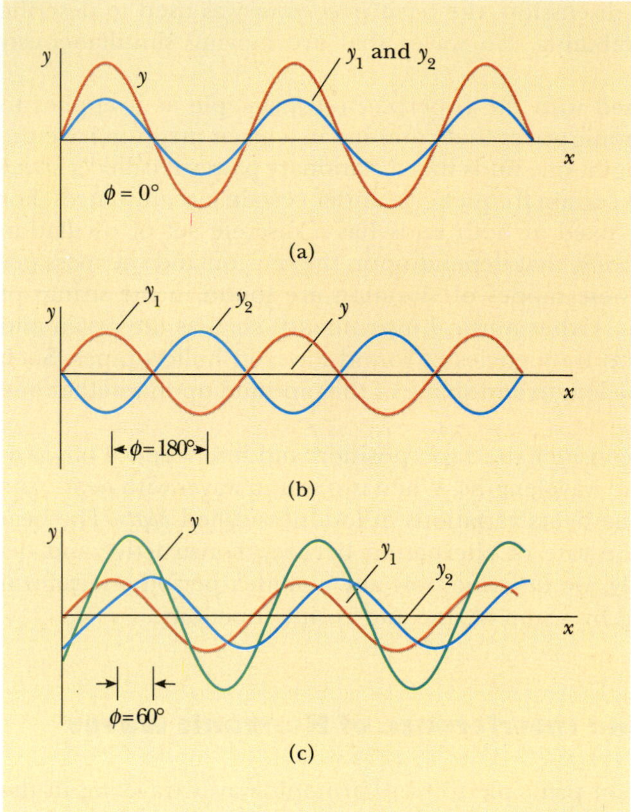

Figure 23.1 The superposition of two waves with amplitudes y_1 and y_2, where $y_1 = y_2$. (a) When the two waves are in phase, the result is constructive interference. (b) When the two waves are 180° out of phase, the result is destructive interference. (c) When the phase angle lies in the range $0 < \phi < 180°$, the resultant y falls somewhere between that shown in (a) and that shown in (b).

Constructive interference

where *in phase* and to thus *interfere constructively*. That is, the crests and troughs of the individual waves occur at the same positions, as is shown by the blue lines in Figure 23.1a. In general, constructive interference occurs when $\cos(\phi/2) = \pm 1$, or when $\phi = 0, 2\pi, 4\pi, \ldots$. On the other hand, if ϕ is equal to π radians, or to any *odd* multiple of π, then $\cos(\phi/2) = \cos(\pi/2) = 0$ and the resultant wave has *zero* amplitude everywhere. In this case, the two waves *interfere destructively*. That is, the crest of one wave coincides with the trough of the second (Fig. 23.1b) and their displacements cancel at every point. Finally, when the phase constant has an arbitrary value between 0 and π, as in Figure 23.1c, the resultant wave has an amplitude whose value is somewhere between 0 and $2A$.

Interference of Sound Waves

One simple device for demonstrating interference of sound waves is illustrated in Figure 23.2. Sound from speaker S is sent into a tube at P, where there is a T-shaped junction. Half the sound power travels in one direction, and half in the opposite direction. Thus, the sound waves that reach receiver R at the other side can travel along either of two paths. The total distance from speaker to receiver is called the *path length, r*. The length of the lower path is fixed at r_1. The upper path length, r_2, can be varied by sliding the U-shaped tube (similar to that on a slide trombone). When the difference in the path lengths, $\Delta r = |r_2 - r_1|$, is either zero or some integral multiple of the wavelength λ, the two waves reaching the receiver

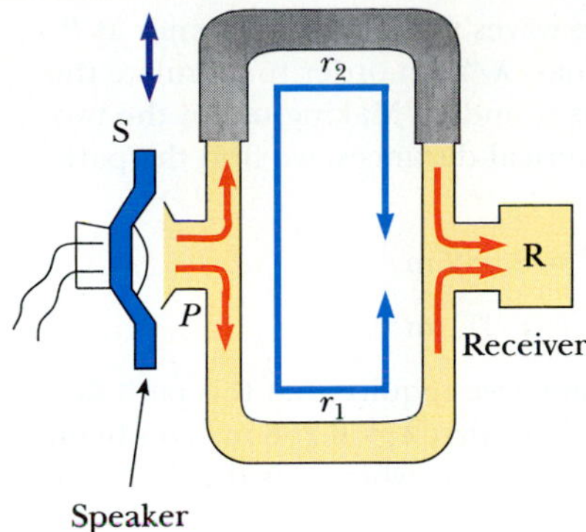

Figure 23.2
An acoustical system for demonstrating interference of sound waves. Sound from the speaker propagates into a tube and splits into two parts at P. The two waves, which superimpose at the opposite side, are detected at R. Note that the upper path length, r_2, can be varied by the sliding section.

are in phase and interfere constructively, as in Figure 23.1a. In this case, a maximum in the sound intensity is detected at the receiver. If path length r_2 is adjusted so that Δr is $\lambda/2$, $3\lambda/2$, ... $n\lambda/2$ (for n odd), the two waves are exactly 180° out of phase at the receiver and hence cancel each other. In this case of completely destructive interference, no sound is detected at the receiver. This simple experiment is a striking illustration of interference. In addition, it demonstrates the fact that a phase difference may arise between two waves generated by the same source when they travel along paths of unequal lengths.

It is often useful to express the path difference in terms of the phase difference, ϕ, between the two waves. Since a path difference of one wavelength corresponds to a phase difference of 2π radians, we obtain the ratio $\lambda/2\pi = \Delta r/\phi$, or

Relationship between path difference and phase angle

$$\Delta r = \frac{\lambda}{2\pi}\phi \qquad [23.2]$$

Nature provides many other examples of interference phenomena. Later in the text we shall describe several interesting interference effects involving light waves.

Example 23.1 Two Speakers Driven by the Same Source

A pair of speakers separated by 3 m are driven by the same oscillator, as in Figure 23.3. The listener is originally at point O, 8 m from the midpoint of the speakers. The listener walks a distance of 0.35 m *perpendicularly* to the center line before reaching the first minimum in sound intensity (the point P in Fig. 23.3). What is the frequency of the oscillator?

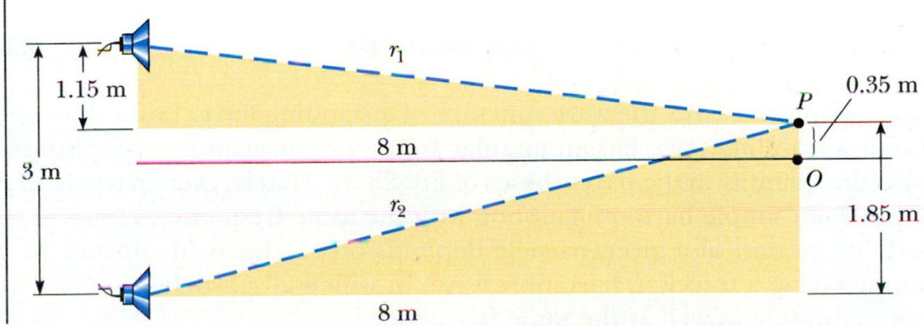

Figure 23.3 (Example 23.1).

Solution The first minimum occurs when the two waves reaching the listener at P are 180° out of phase, or their path difference equals $\lambda/2$. In order to calculate the path difference, we must first find the path lengths r_1 and r_2. Making use of the two shaded triangles in Figure 23.3 and the given numerical distances, we find the path lengths to be

$$r_1 = \sqrt{(8\text{ m})^2 + (1.15\text{ m})^2} = 8.082\text{ m}$$

$$r_1 = \sqrt{(8\text{ m})^2 + (1.85\text{ m})^2} = 8.211\text{ m}$$

Hence, the path difference is $r_2 - r_1 = 0.129$ m. Since we require that this path difference be equal to $\lambda/2$ for the first minimum, we find that $\lambda = 0.258$ m. To obtain the oscillator frequency, we can use Equation 22.14, $v = f\lambda$, where v is the speed of sound in air, 343 m/s:

$$f = \frac{v}{\lambda} = \frac{343\text{ m/s}}{0.258\text{ m}} = \boxed{1330\text{ Hz}}$$

Exercise If the frequency of the oscillator is adjusted so that the listener hears the first minimum at a distance of 0.75 m from O, what is the new frequency of the oscillator?

Answer 625 Hz.

23.2 Standing Waves

If a stretched string is clamped at both ends, waves traveling in both directions are reflected from the ends. The incident and reflected waves combine according to the superposition principle.

Consider two sinusoidal waves, in the same medium, that have the same amplitude, frequency, and wavelength but are traveling in *opposite* directions. Their wave functions can be written

$$y_1 = A\sin(kx - \omega t) \quad \text{and} \quad y_2 = A\sin(kx + \omega t)$$

where y_1 represents a wave traveling to the right and y_2 represents a wave traveling to the left. Adding these two functions gives the resultant wave function, y:

$$y = y_1 + y_2 = A\sin(kx - \omega t) + A\sin(kx + \omega t)$$

where $k = 2\pi/\lambda$ and $\omega = 2\pi f$, as usual. Using the trigonometric identity $\sin(a \pm b) = \sin a \cos b \pm \cos a \sin b$, this reduces to

Wave function for a standing wave

$$y = (2A\sin kx)\cos\omega t \qquad [23.3]$$

This expression represents the wave function of a **standing wave.** From this result, we see that a standing wave has an angular frequency of ω and an amplitude of $2A\sin kx$ (the quantity in the parentheses of Eq. 23.3). That is, every particle of the string vibrates in simple harmonic motion with the same frequency. However, the amplitude of motion of a given particle depends on x. This is in contrast to the situation involving a traveling harmonic wave, in which all particles oscillate with the same amplitude as well as the same frequency.

Because the amplitude of the standing wave at any value of x is equal to $2A \sin kx$, we see that the *maximum* amplitude has the value $2A$. This occurs when the coordinate x satisfies the condition $\sin kx = 1$, or when

$$kx = \frac{\pi}{2}, \frac{3\pi}{2}, \frac{5\pi}{2}, \ldots$$

Since $k = 2\pi/\lambda$, the positions of maximum amplitude, called **antinodes,** are

$$x = \frac{\lambda}{4}, \frac{3\lambda}{4}, \frac{5\lambda}{4}, \ldots = \frac{n\lambda}{4} \quad [23.4]$$

Position of antinodes

where $n = 1, 3, 5, \ldots$. Note that *adjacent antinodes are separated by a distance of* $\lambda/2$.

Similarly, the standing wave has a *minimum* amplitude of zero when x satisfies the condition $\sin kx = 0$, or when

$$kx = \pi, 2\pi, 3\pi, \ldots$$

giving

$$x = \frac{\lambda}{2}, \lambda, \frac{3\lambda}{2}, \ldots = \frac{n\lambda}{2} \quad [23.5]$$

Position of nodes

where $n = 1, 2, 3, \ldots$. These points of zero amplitude, called **nodes,** *are also spaced by* $\lambda/2$. The distance between a node and an adjacent antinode is $\lambda/4$.

The standing wave patterns produced at various times by two waves traveling in opposite directions are depicted graphically in Figure 23.4. The upper part of each figure represents the individual traveling waves, and the lower part represents the standing wave patterns. The nodes of the standing wave are labeled N, and the antinodes are labeled A. At $t = 0$ (Fig. 23.4a), the two waves are spatially identical, giving a standing wave of maximum amplitude, $2A$. One-quarter of a period later, at $t = T/4$ (Fig. 23.4b), the individual waves have moved one-quarter of a wavelength (one to the right and the other to the left). At this time, the individual displacements are equal and opposite for all values of x, and hence the resultant wave has zero displacement everywhere. At $t = T/2$ (Fig. 23.4c), the individual waves are again identical spatially, producing a standing wave pattern that is inverted relative to the $t = 0$ pattern.

It is instructive to describe the energy associated with the motion of a standing wave. To illustrate this point, consider a standing wave formed on a stretched

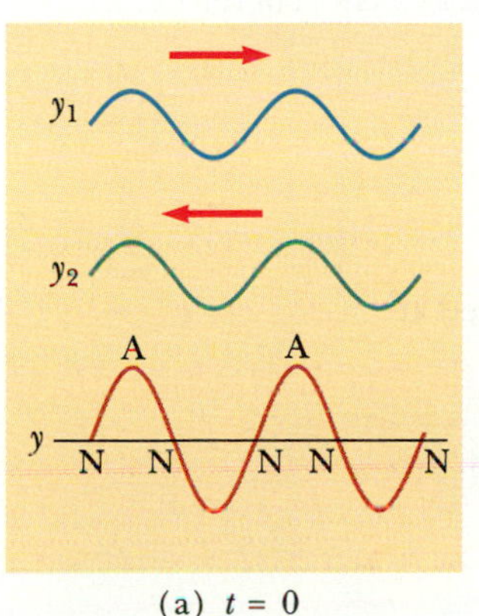

(a) $t = 0$

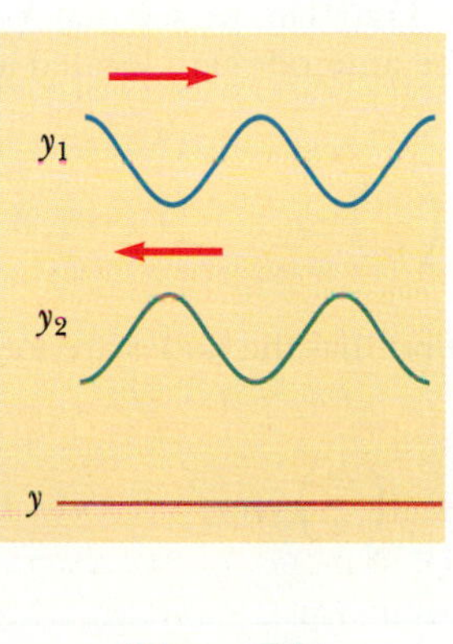

(b) $t = T/4$

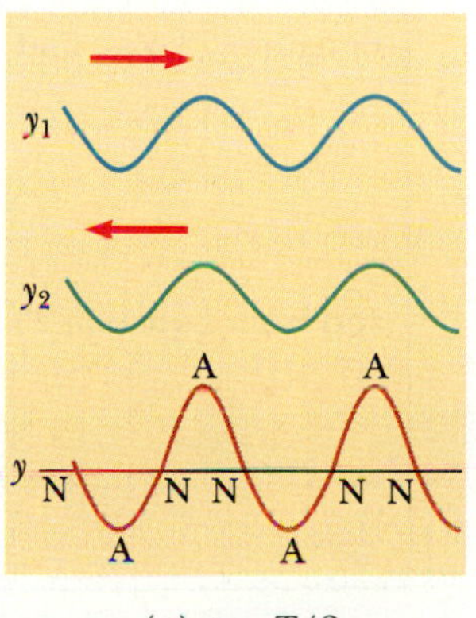

(c) $t = T/2$

Figure 23.4 Standing wave patterns at various times produced by two waves of equal amplitude traveling in opposite directions. For the resultant wave y, the nodes (N) are points of zero displacement, and the antinodes (A) are points of maximum displacement.

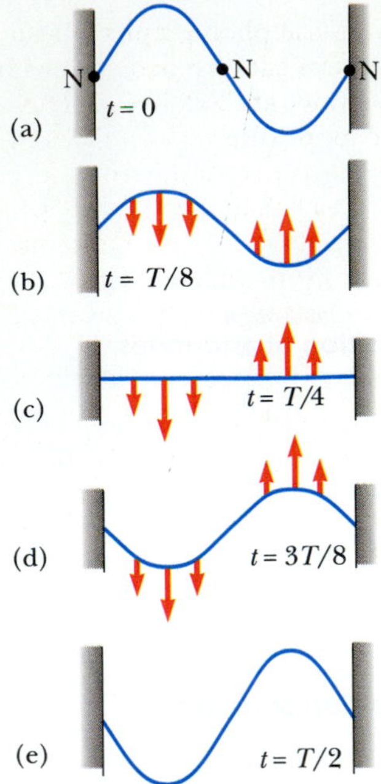

Figure 23.5 A standing wave pattern in a stretched string, showing snapshots during one half cycle. (a) At $t = 0$, the string is momentarily at rest, and so $K = 0$ and all of the energy is potential energy U associated with the vertical displacements of the string segments. (b) At $t = T/8$, the string is in motion, and the energy is half kinetic and half potential. (c) At $t = T/4$, the string is horizontal (undeformed) and therefore $U = 0$; all of the energy is kinetic. The motion continues as indicated, and ultimately the initial configuration (a) is repeated.

string that is fixed at each end, as in Figure 23.5. Except for the nodes, which are stationary, all points on the string oscillate vertically with the same frequency. Furthermore, different points have different amplitudes of motion. Figure 23.5 represents snapshots of the standing wave at various times over one-half cycle. The nodes represent points on the string that are always at rest, and the locations of the maxima and minima never change. Each point on the string executes simple harmonic motion in the vertical direction. That is, one can view the standing wave as a large number of oscillators vibrating parallel to each other. The energy of the vibrating string continually alternates between elastic potential energy, at which time the string is momentarily stationary (Fig. 23.5a), and kinetic energy, at which time the string is horizontal and the particles have their maximum speed (Fig. 23.5c). The string particles have both potential energy and kinetic energy at intermediate times (Figs. 23.5b and 23.5d).

Example 23.2 Formation of a Standing Wave

Two waves traveling in opposite directions produce a standing wave. The individual wave functions are

$$y_1 = (4\text{ cm})\sin(3x - 2t)$$

$$y_2 = (4\text{ cm})\sin(3x + 2t)$$

where x and y are in centimeters.

(a) Find the maximum displacement of the motion at $x = 2.3$ cm.

Solution When the two waves are summed, the result is a standing wave whose function is given by Equation 23.3, with $A = 4$ cm and $k = 3$ rad/cm:

$$y = (2A\sin kx)\cos\omega t = [(8\text{ cm})\sin 3x]\cos\omega t$$

Thus, the *maximum* displacement of the motion at the position $x = 2.3$ cm is

$$y_{\text{max}} = (8\text{ cm})\sin 3x|_{x=2.3}$$

$$= (8\text{ cm})\sin(6.9\text{ rad}) = \boxed{4.63\text{ cm}}$$

(b) Find the positions of the nodes and antinodes.

Solution Since $k = 2\pi/\lambda = 3$ rad/cm, we see that $\lambda = 2\pi/3$ cm. Therefore, from Equation 23.4 we find that the antinodes are located at

$$x = \boxed{n\left(\frac{\pi}{6}\right)\text{ cm}} \qquad (n = 1, 3, 5, \ldots)$$

and from Equation 23.5 we find that the nodes are located at

$$x = n\frac{\lambda}{2} = \boxed{n\left(\frac{\pi}{3}\right)\text{ cm}} \qquad (n = 1, 2, 3, \ldots)$$

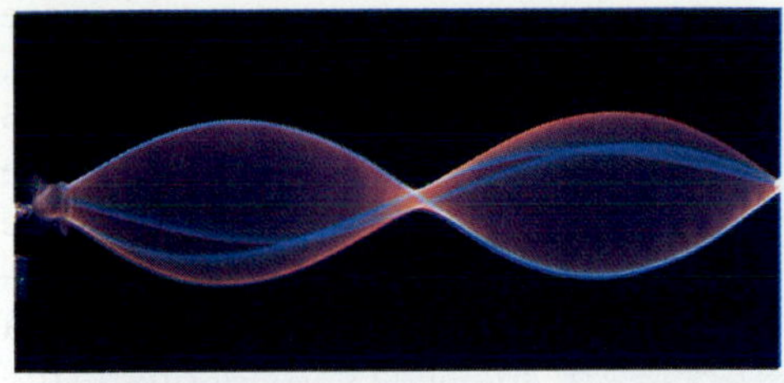

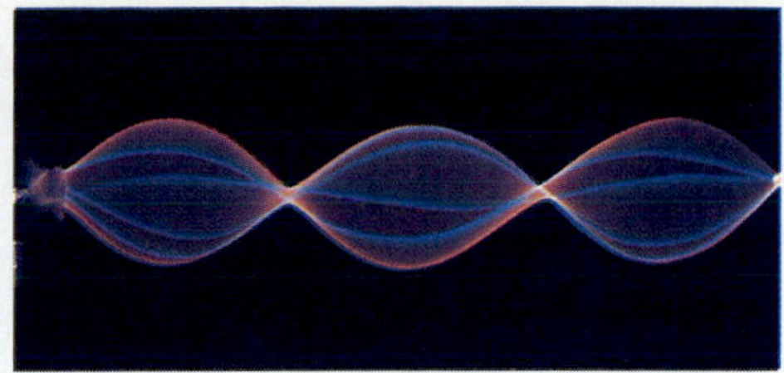

Multiflash photographs of standing wave patterns in a cord with a vibrator at the left end. The two-loop pattern at the left represents the second harmonic ($n = 2$), while the three-loop pattern at the right represents the third harmonic ($n = 3$). *(© Richard Megna 1991, Fundamental Photographs)*

23.3 Natural Frequencies in a Stretched String

Consider a string that has a length of L and is fixed at both ends, as in Figure 23.6. Standing waves are set up in the string by a continuous superposition of waves incident on and reflected from the ends. The string has a number of natural patterns of vibration, called **normal modes.** Each of these modes has a characteristic frequency; the frequencies are easily calculated.

First, note that the ends of the string must be nodes, since these points are fixed. If the string is displaced at its midpoint and released, a vibration is produced in which the center of the string is an antinode (Fig. 23.6b). For this normal mode, the length of the string equals $\lambda/2$ (the distance between nodes):

$$L = \lambda_1/2 \qquad \text{or} \qquad \lambda_1 = 2L$$

The next normal mode, of wavelength λ_2 (Fig. 23.6c), occurs when the length of the string equals one wavelength, that is, when $\lambda_2 = L$. The third normal mode (Fig. 23.6d) corresponds to the case where the length equals $3\lambda/2$; therefore, $\lambda_3 = 2L/3$. In general, the wavelengths of the various normal modes can be conveniently expressed as

$$\lambda_n = \frac{2L}{n} \qquad (n = 1, 2, 3, \ldots) \qquad [23.6]$$

Wavelengths of normal modes

Figure 23.6
(a) A string of length L fixed at both ends. The normal modes of vibration, shown as multiple exposures, form a harmonic series; (b) the fundamental frequency, or first harmonic; (c) the second harmonic; and (d) the third harmonic.

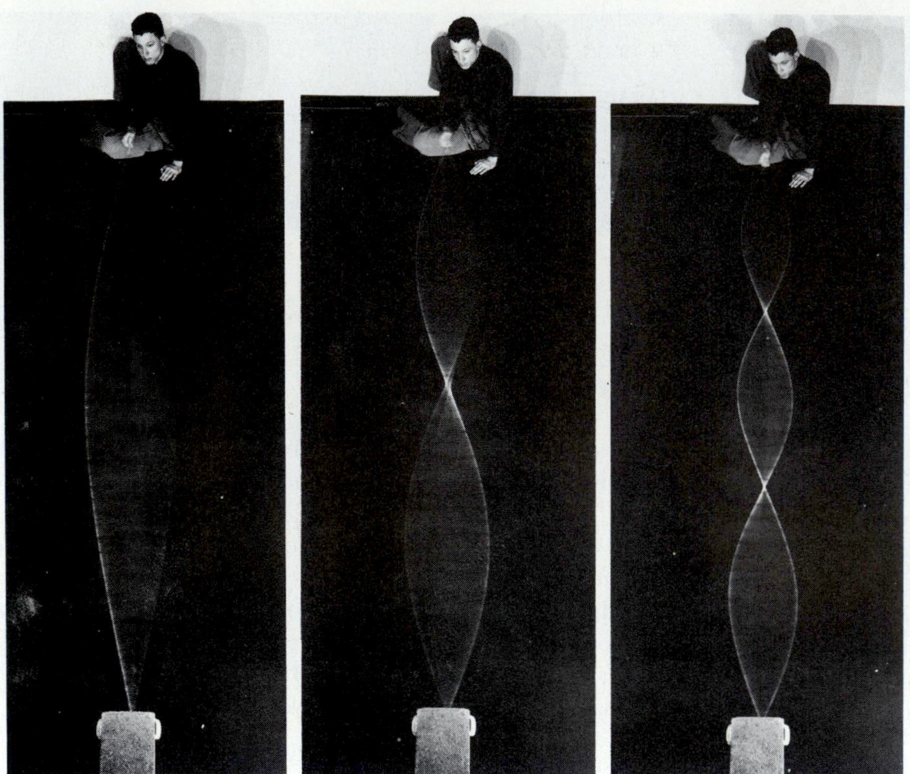

Photographs of standing waves. As one end of the rope is moved from side to side with increasing frequency, patterns with more and more loops are formed; only certain definite frequencies of the man's hand produce standing waves. *(Education Development Center, Newton, Mass.)*

where the index n refers to the nth mode of vibration. The natural frequencies associated with these modes are obtained from the relationship $f = v/\lambda$, where the *wave speed, v, is the same for all frequencies.* Using Equation 23.6, we find that the frequencies of the normal modes are

Frequencies of normal modes

$$f_n = \frac{v}{\lambda_n} = \frac{n}{2L}v \qquad (n = 1, 2, 3, \ldots) \qquad \text{[23.7]}$$

Because $v = \sqrt{F/\mu}$ (Eq. 23.4), where F is the tension in the string and μ is its mass per unit length, we can express the natural frequencies of a stretched string (sometimes called *harmonics*) as

Normal modes of a stretched string fixed at both ends

$$f_n = \frac{n}{2L}\sqrt{\frac{F}{\mu}} \qquad (n = 1, 2, 3, \ldots) \qquad \text{[23.8]}$$

The lowest frequency, corresponding to $n = 1$, is called the *fundamental* or the **fundamental frequency,** f_1, and is

Fundamental frequency of a stretched string

$$f_1 = \frac{1}{2L}\sqrt{\frac{F}{\mu}} \qquad \text{[23.9]}$$

Clearly, the frequencies of the remaining modes are integral multiples of the fundamental frequency—that is, $2f_1$, $3f_1$, $4f_1$, and so on. These higher natural frequencies, together with the fundamental frequency, form a **harmonic series.** The fundamental, f_1, is the first harmonic; the frequency $f_2 = 2f_1$ is the second harmonic; the frequency f_n is the nth harmonic.

We can obtain the foregoing results in an alternative manner. Since we require that the string be fixed at $x = 0$ and $x = L$, the wave function $y(x, t)$ given by Equation 23.3 must be *zero* at these points for *all* times. That is, the boundary conditions require that $y(0, t) = 0$ and $y(L, t) = 0$ for all values of t. Since $y = (2A \sin kx) \cos \omega t$, the first condition, $y(0, t) = 0$, is automatically satisfied because $\sin kx = 0$ at $x = 0$. To meet the second condition, $y(L, t) = 0$, we require that $\sin kL = 0$. This condition is satisfied when the angle kL equals an integral multiple of π (180°). Therefore, the allowed values of k are[1]

$$k_n L = n\pi \qquad (n = 1, 2, 3, \ldots) \qquad [23.10]$$

Since $k_n = 2\pi/\lambda_n$, we find that

$$\left(\frac{2\pi}{\lambda_n}\right) L = n\pi \qquad \text{or} \qquad \lambda_n = \frac{2L}{n}$$

which is identical to Equation 23.6.

When a stretched string is distorted to a shape that corresponds to any one of its harmonics, after being released it will vibrate at the frequency of that harmonic. However, if the string is plucked or bowed, the resulting vibration will include frequencies of various harmonics, including the fundamental. In effect, the string "selects" the normal-mode frequencies when disturbed by a nonharmonic disturbance (for example, a finger plucking it).

Figure 23.7 Multiple exposures of a stretched string vibrating simultaneously in its first and second harmonics.

Figure 23.7 shows a stretched string vibrating with its first and second harmonics simultaneously. In this figure, the combined vibration is the superposition of the two vibrations shown in Figures 23.6b and 23.6c. The larger loop corresponds to the fundamental frequency of vibration, f_1, and the smaller loops correspond to the second harmonic, f_2. In general, the resulting motion, or displacement, of the string can be described by a superposition of the various harmonic wave functions, with different frequencies and amplitudes. Hence, the sound that one hears corresponds to a complex waveform associated with these modes of vibration. (We shall return to this point in Section 23.8.)

The frequency of a stringed instrument can be changed either by varying the strings' tension, F, or by changing their length, L. For example, the tension in the strings of guitars and violins is adjusted by a screw mechanism or by turning pegs on the neck of the instrument. As the tension increases, the frequency of the normal modes increases according to Equation 23.8. Once the instrument is "tuned," the player varies the frequency by moving his or her fingers along the neck, thereby changing the length of the vibrating portion of the string. As this length is reduced, the frequency increases, since the normal-mode frequencies are inversely proportional to (vibrating) string length.

Example 23.3 Give Me a C Note

A middle C string of the C-major scale on a piano has a fundamental frequency of 264 Hz, and the A note has a fundamental frequency of 440 Hz.

(a) Calculate the frequencies of the next two harmonics of the C string.

Solution Since $f_1 = 264$ Hz, we can use Equations 23.8 and 23.9 to find the frequencies f_2 and f_3:

[1] We exclude $n = 0$ since this corresponds to the trivial case where no wave exists ($k = 0$).

$$f_2 = 2f_1 = \boxed{528 \text{ Hz}}$$

$$f_3 = 3f_1 = \boxed{792 \text{ Hz}}$$

(b) If the two piano strings for the A and C notes are assumed to have the same mass per unit length and the same length, determine the ratio of tensions in the two strings.

Solution Use of Equation 23.8 for the two strings vibrating at their fundamental frequencies gives

$$f_{1A} = \frac{1}{2L}\sqrt{F_A/\mu} \quad \text{and} \quad f_{1C} = \frac{1}{2L}\sqrt{F_C/\mu}$$

$$f_{1A}/f_{1C} = \sqrt{F_A/F_C}$$

$$F_A/F_C = (f_{1A}/f_{1C})^2 = (440/264)^2 = \boxed{2.78}$$

(c) While the string densities are, in fact, equal, the A string is only 64% as long as the C string. What is the ratio of their tensions?

$$\frac{f_{1A}}{f_{1C}} = \left(\frac{L_C}{L_A}\right)\sqrt{\frac{F_A}{F_C}} = \left(\frac{100}{64}\right)\sqrt{\frac{F_A}{F_C}}$$

$$\frac{F_A}{F_C} = (0.64)^2\left(\frac{440}{264}\right)^2 = \boxed{1.14}$$

23.4 Resonance

We have seen that a system such as a stretched string is capable of oscillating in one or more natural modes of vibration. *If a periodic force is applied to such a system, the resulting amplitude of motion of the system will be large when the frequency of the applied force is equal or nearly equal to one of the natural frequencies of the system.* We have already discussed this phenomenon, known as *resonance,* for the mechanical systems of Chapter 22, Section 22.6. The natural frequencies of oscillation of the system are often referred to as **resonant frequencies.** Figure 23.8 shows the response of a system to various frequencies, where the peak of the curve represents the resonant frequency, f_0. Note that the amplitude is greatest when the frequency of the driving force equals the resonant frequency. When the frequency of the driving force exactly matches one of the resonant frequencies, the amplitude of the motion is limited by friction in the system. Once maximum amplitude is reached, the work done by the periodic force is used only to overcome friction. A system is said to be *weakly damped* when the amount of friction is small. Such a system undergoes a large amplitude of motion when driven at one of its resonant frequencies. The oscillations in such a system persist for a long time after the driving force is removed. On the other hand, a system with considerable friction—that is, one that is *strongly damped*—undergoes small-amplitude oscillations that decrease rapidly once the driving force is removed.

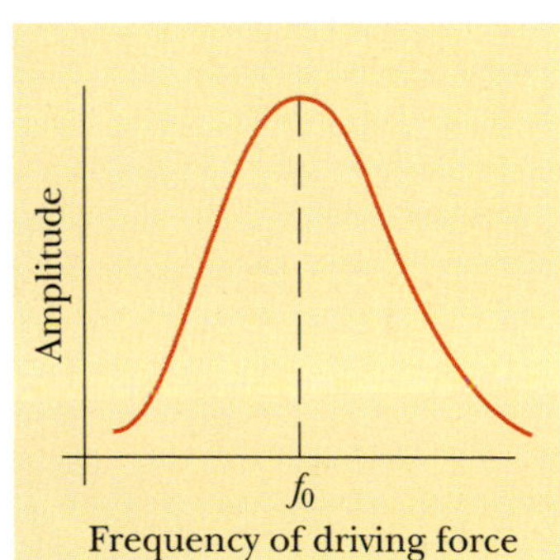

Figure 23.8 The amplitude (response) versus driving frequency for an oscillating system. The amplitude is a maximum at the resonance frequency, f_0.

Resonant systems abound in the natural world. To consider only one example, think of a playground swing. It is a pendulum with a natural frequency that depends on its length. Whenever we push a child in a swing with a series of regular impulses, the swing goes higher if the frequency of the periodic force equals the natural frequency of the swing.

23.5 Standing Waves in Air Columns

Standing longitudinal waves can be set up in a tube of air, such as an organ pipe, as the result of interference between longitudinal waves traveling in opposite directions. The phase relationship between the incident wave and the wave reflected from one end depends on whether that end is open or closed. This is analogous to the phase relationships between incident and reflected transverse waves at the ends of a string. *The closed end of an air column is a displacement node,* just as the fixed end of a vibrating string is a displacement node. As a result, the reflected wave at a closed end of a tube of air is 180° out of phase with the incident wave. Furthermore, since the pressure wave is 90° out of phase with the displacement wave (Section 23.9), *the closed end of an air column corresponds to a pressure antinode* (that is, a point of maximum pressure variation).

If the end of an air column is open to the atmosphere, the air molecules have complete freedom of motion. The wave reflected from an open end is nearly in phase with the incident wave when the tube's diameter is small relative to the wavelength of the sound. Consequently, *the open end of an air column is approximately a displacement antinode and a pressure node.*

Strictly speaking, the open end of an air column is not exactly an antinode. A condensation does not reach full expansion until it passes somewhat beyond an open end. For a thin-walled tube of circular cross-section, this end correction is about $0.6R$, where R is the tube's radius. Hence, the effective length of the tube is somewhat greater than the true length, L.

The first three modes of vibration of a pipe that is open at both ends are shown in Figure 23.9a. When air is directed into the pipe from the left, longitudinal standing waves are formed and the pipe resonates at its natural frequencies. All modes of vibration are excited simultaneously (although not with the same amplitude). Note that the ends are displacement antinodes (approximately). In the fundamental mode, the wavelength is twice the length of the pipe, and hence the frequency of the fundamental, f_1, is $v/2L$. Similarly, the frequencies of the higher harmonics are $2f_1, 3f_1, \ldots$. Thus, in a pipe that is open at both ends, the natural frequencies of vibration form a harmonic series; that is, the higher harmonics are integral multiples of the fundamental frequency. Since all harmonics are present, we can express the natural frequencies of vibration as

Natural frequencies of a pipe open at both ends

$$f_n = n\frac{v}{2L} \qquad (n = 1, 2, 3, \ldots) \qquad [23.11]$$

where v is the speed of sound in air.

If a pipe is closed at one end and open at the other, the closed end is a displacement node (Fig. 23.9b). In this case, the wavelength for the fundamental mode is four times the length of the tube. Hence, the fundamental, f_1, is equal to

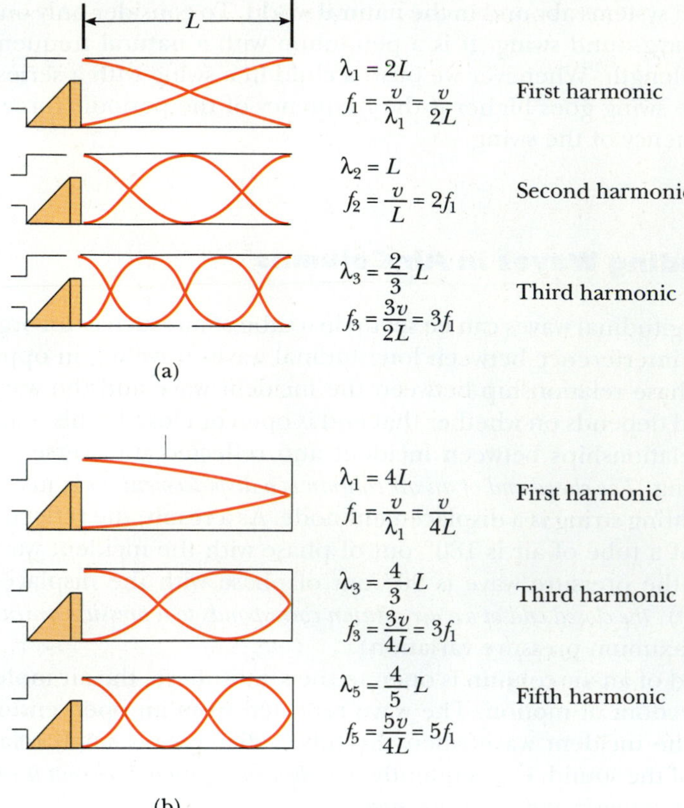

Figure 23.9 (a) Standing longitudinal waves in an organ pipe open at both ends. The natural frequencies that form a harmonic series are $f_1, 2f_1, 3f_1, \ldots$ (b) Standing longitudinal waves in an organ pipe closed at one end. Only the *odd* harmonics are present, and so the natural frequencies are f_1, $3f_1, 5f_1, \ldots$

$v/4L$, and the frequencies of the higher harmonics are equal to $3f_1, 5f_1, \ldots$ That is, **in a pipe that is closed at one end, only odd harmonics are present,** and these are

Natural frequencies of a pipe closed at one end

$$f_n = n\frac{v}{4L} \qquad (n = 1, 3, 5, \ldots) \qquad [23.12]$$

Standing waves in air columns are the primary sources of the sounds produced by wind instruments.

Example 23.4 Resonance in a Pipe

A pipe has a length of 1.23 m.

(a) Determine the frequencies of the first three harmonics if the pipe is open at both ends. Take $v = 344$ m/s as the speed of sound in air.

Solution The first harmonic of an open pipe is

$$f_1 = \frac{v}{2L} = \frac{344 \text{ m/s}}{2(1.23 \text{ m})} = 140 \text{ Hz}$$

Since all harmonics are present, the second and third harmonics are $f_2 = 2f_1 = 280$ Hz and $f_3 = 3f_1 = 420$ Hz, respectively.

(b) What are the three frequencies determined in (a) if the pipe is closed at one end?

Solution The fundamental frequency of a pipe that is closed at one end is

$$f_1 = \frac{v}{4L} = \frac{344 \text{ m/s}}{4(1.23 \text{ m})} = 70 \text{ Hz}$$

In this case, only odd harmonics are present, and so the next two resonances have frequencies $f_3 = 3f_1 = 210$ Hz and $f_5 = 5f_1 = 350$ Hz, respectively.

(c) For the open pipe, how many harmonics are present in the normal human hearing range (20 to 20 000 Hz)?

Solution Since all harmonics are present, $f_n = nf_1$. For $f_n = 20\ 000$ Hz, we have $n = 20\ 000/140 = 142$, and so 142 harmonics are present in the audible range. Actually, only the first few harmonics have sufficient amplitude to be heard.

Example 23.5 Measuring the Frequency of a Tuning Fork

A simple apparatus for demonstrating resonance in a tube is described in Figure 23.10a. A long, vertical tube that is open at both ends is partially submerged in water, and a vibrating tuning fork of unknown frequency is placed near the top. The length of the air column, L, is adjusted by moving the tube vertically. The sound waves generated by the fork are reinforced when the length of the column corresponds to one of the resonant frequencies of the tube. The smallest value of L for which a peak occurs in the sound intensity is 9 cm. From this measurement, determine the frequency of the turning fork and the value of L for the next two resonant modes.

Solution Once the tube is placed in water, this setup represents a pipe that is closed at one end. The fundamental has a frequency of $v/4L$ (Fig. 23.10b). Taking $v = 344$ m/s for the speed of sound in air, and $L = 0.09$ m, we get

$$f_1 = \frac{v}{4L} = \frac{344 \text{ m/s}}{4(0.09 \text{ m})} = 956 \text{ Hz}$$

From this information about the fundamental mode, we see that the wavelength is $\lambda = 4L = 0.36$ m. Since the frequency of the source is constant, we see that the next two resonance modes (Fig. 23.10b) correspond to lengths of $3\lambda/4 = 0.27$ m and $5\lambda/4 = 0.45$ m, respectively.

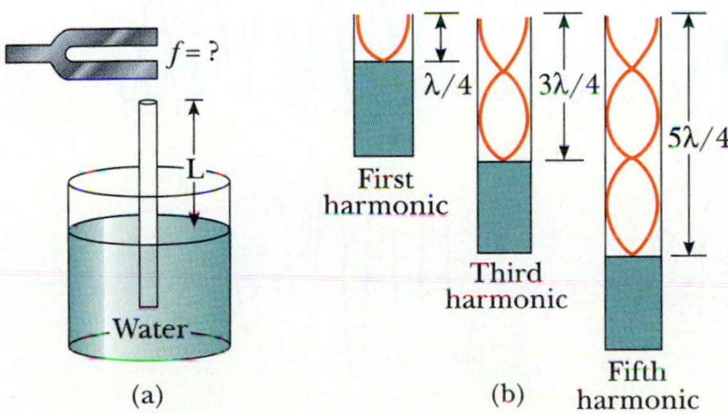

Figure 23.10 (Example 23.5) (a) Apparatus for demonstrating the resonance of sound waves in a tube closed at one end. The length, L, of the air column is varied by moving the tube vertically while it is partially submerged in water. (b) The first three normal modes of the system shown in (a).

*23.6 Beats: Interference in Time

The interference phenomena with which we have been dealing so far involve the superposition of two or more waves with the same frequency, traveling in opposite directions. Since the resultant waveform in this case depends on the coordinates of the disturbed medium, we can refer to the phenomenon as *spatial interference.* Standing waves in strings and pipes are common examples of spatial interference.

We now consider another type of interference effect, one that results from the superposition of two waves with slightly *different frequencies,* traveling in the *same direction.* In this case, when the two waves are observed at a given point, they are periodically in and out of phase. That is, there occurs a temporal alternation between constructive and destructive interference. We refer to this phenomenon as *interference in time* or *temporal interference.*

For example, if two tuning forks of slightly different frequencies are struck, one hears a sound of pulsating intensity, called a **beat.**

Definition of beats

A beat is the periodic variation in intensity at a given point due to the superposition of two waves with slightly different frequencies.

The number of beats one hears per second, or *beat frequency,* equals the difference in frequency between the two sources. The maximum beat frequency that the human ear can detect is about 20 beats/s. When the beat frequency exceeds this value, it blends indistinguishably with the compound sounds producing the beats.

One can use beats to tune a stringed instrument, such as a piano, by beating a note against a reference tone of known frequency. The string can then be adjusted to equal the frequency of the reference by tightening or loosening it until the beats become too infrequent to notice.

Consider two waves with equal amplitudes, traveling through a medium in the same direction but with slightly different frequencies, f_1 and f_2. We can represent the displacement that each wave would produce at a point as

$$y_1 = A \cos 2\pi f_1 t \qquad \text{and} \qquad y_2 = A \cos 2\pi f_2 t$$

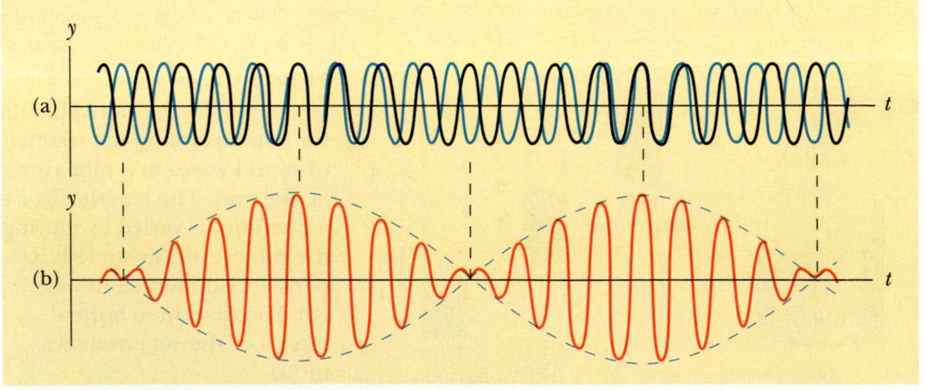

Figure 23.11 Beats are formed by the combination of two waves of slightly different frequencies traveling in the same direction. (a) The individual waves. (b) The combined wave has an amplitude (broken line) that oscillates in time.

Using the superposition principle, we find that the resultant displacement at that point is given by

$$y = y_1 + y_2 = A(\cos 2\pi f_1 t + \cos 2\pi f_2 t)$$

It is convenient to write this in a form that uses the trigonometric identity

$$\cos a + \cos b = 2\cos\left(\frac{a-b}{2}\right)\cos\left(\frac{a+b}{2}\right)$$

Letting $a = 2\pi f_1 t$ and $b = 2\pi f_2 t$, we find that

$$y = 2A\cos 2\pi\left(\frac{f_1 - f_2}{2}\right)t \cos 2\pi\left(\frac{f_1 + f_2}{2}\right)t \qquad [23.13]$$

Resultant of two waves of different frequencies but equal amplitude

Graphs demonstrating the individual waveforms as well as the resultant wave are shown in Figure 23.11. From the factors in Equation 23.13, we see that the resultant vibration at a point has an effective frequency equal to the average frequency, $(f_1 + f_2)/2$, and an amplitude of

$$A = 2A\cos 2\pi\left(\frac{f_1 - f_2}{2}\right)t \qquad [23.14]$$

That is, the *amplitude varies in time* with a frequency of $(f_1 - f_2)/2$. When f_1 is close to f_2, this amplitude variation is slow, as illustrated by the envelope (broken line) of the resultant waveform in Figure 23.11b.

Note that a beat, or a maximum in amplitude, will be detected whenever

$$\cos 2\pi\left(\frac{f_1 - f_2}{2}\right)t = \pm 1$$

That is, there are *two* maxima in each cycle. Since the amplitude varies with frequency as $(f_1 - f_2)/2$, the number of beats per second, or the beat frequency, f_b, is twice this value:

$$f_b = f_1 - f_2 \qquad [23.15]$$

For instance, if two tuning forks vibrate individually at frequencies of 438 Hz and 442 Hz, respectively, the resultant sound wave of the combination has a frequency of 440 Hz (the musical note A) and a beat frequency of 4 Hz. That is, the listener hears the 440-Hz sound wave go through an intensity maximum four times every second.

*23.7 Complex Waves

The sound wave patterns produced by most instruments are very complex. Some characteristic waveforms produced by a tuning fork, a harmonic flute, and a clarinet—each playing the same pitch—are shown in Figure 23.12. Although each instrument has its own characteristic pattern, Figure 23.12 shows that all three waveforms are periodic. Furthermore, note that a struck tuning fork produces only one harmonic (the fundamental), whereas the flute and clarinet pro-

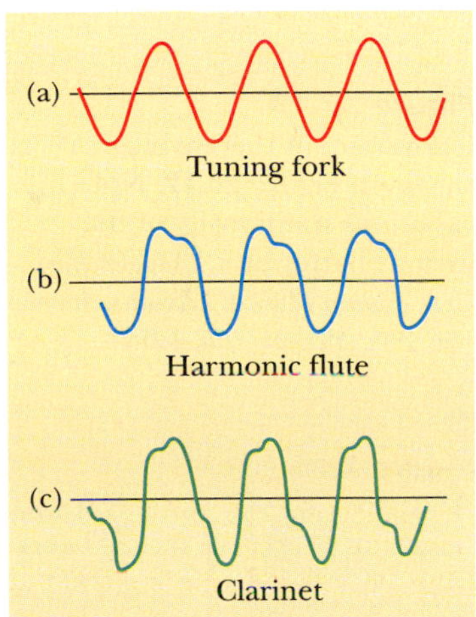

Figure 23.12 Waveforms produced by (a) a tuning fork, (b) a harmonic flute, and (c) a clarinet, each at approximately the same frequency. *(Adapted from C. A. Culver, Musical Acoustics, 4th ed., New York, McGraw-Hill, 1956, p. 128.)*

duce many frequencies, which include the fundamental and various harmonics. Thus, the complex waveforms produced by a violin or clarinet, and the corresponding richness of musical tones, are the result of the superposition of various harmonics. This is in contrast to the drum, in which the overtones do not form a harmonic series.

It is interesting to investigate what happens to the frequencies of a flute and a violin during a concert as the temperature rises. The flute goes sharp (increases in frequency) as it warms up, since the speed of sound increases. In contrast, the violin goes flat (decreases in frequency) as the strings expand thermally, with a consequent decrease in tension.

Analysis of complex waveforms appears at first sight to be a formidable task. However, if the waveform is periodic, it can be represented with arbitrary precision by the combination of a sufficiently large number of sinusoidal waves that form a harmonic series. In fact, one can represent any periodic function or any function over a finite interval as a series of sine and cosine terms by using a mathematical technique based on *Fourier's theorem.* The corresponding sum of terms that represents the periodic waveform is called a **Fourier series.**

Let $y(t)$ be any function that is periodic in time, with a period of T, so that $y(t+T) = y(t)$. **Fourier's theorem** states that this function can be written

Fourier's theorem

$$y(t) = \sum_n (A_n \sin 2\pi f_n t + B_n \cos 2\pi f_n t) \qquad [23.16]$$

where the lowest frequency is $f_1 = 1/T$.

The higher frequencies are integral multiples of the fundamental, and so $f_n = nf_1$. The coefficients A_n and B_n represent the amplitudes of the various waves. The amplitude of the nth harmonic is proportional to $\sqrt{A_n^2 + B_n^2}$, and its intensity is proportional to $A_n^2 + B_n^2$.

Figure 23.13 represents a harmonic analysis of the waveforms shown in Figure 23.12. Note the variation of relative intensity with harmonic content for the flute and clarinet. In general, any musical sound (of definite pitch) contains components that are members of a harmonic set with varying relative intensities.

As an example of *Fourier synthesis,* consider the periodic square wave shown in Figure 23.14. The square wave is synthesized by a series of *odd* harmonics of the

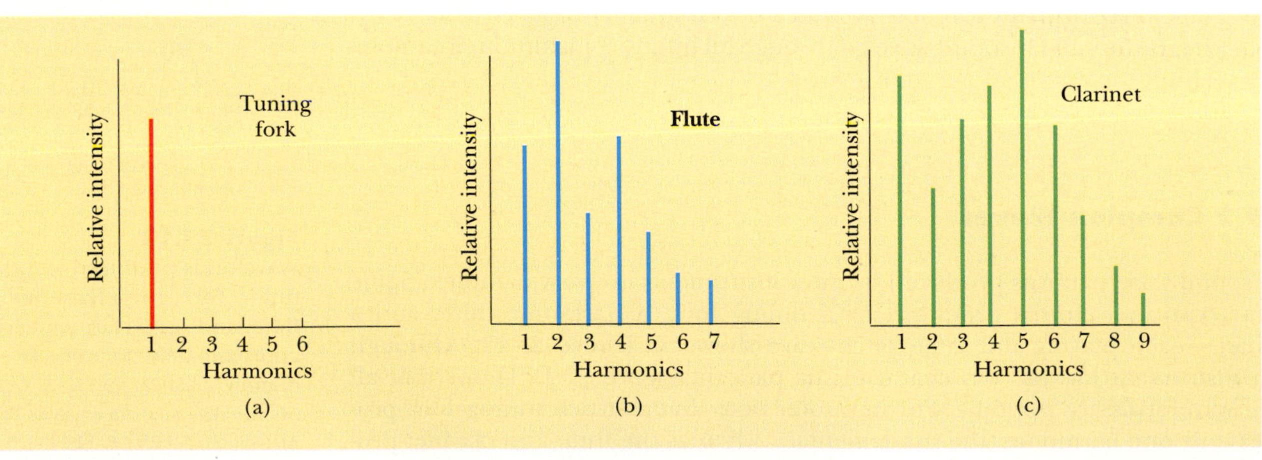

Figure 23.13
Harmonics of the waveforms shown in Figure 23.12. Note the variations in intensity of the various harmonics. *(Adapted from C. A. Culver, Musical Acoustics, 4th ed., New York, McGraw-Hill, 1956.)*

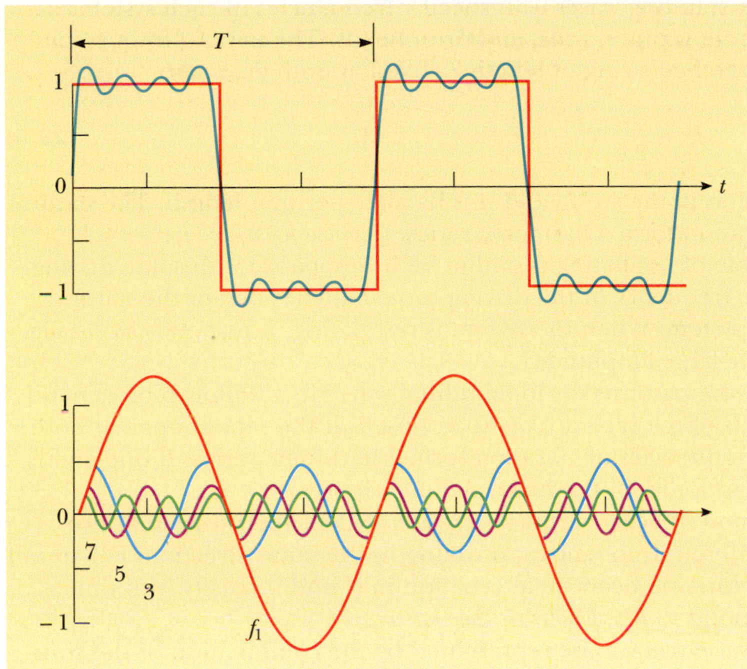

Figure 23.14
Harmonic synthesis of a square wave, which can be represented by the sum of odd harmonics of the fundamental. *(From M. L. Warren, Introductory Physics, San Francisco, W. H. Freeman, 1979, p. 178; by permission of the publisher.)*

fundamental. The series contains only sine functions (that is, $B_n = 0$ for all n). Only the first four odd harmonics and their respective amplitudes are shown. One obtains a better fit to the true waveform by adding more harmonics.

Using modern technology, one can produce a variety of musical tones by mixing harmonics with varying amplitudes. Electronic music synthesizers that do this are now widely used.

Summary

When two waves with equal amplitudes and frequencies superimpose, the resultant wave has an amplitude that depends on the phase angle, ϕ, between the two waves. **Constructive interference** occurs when the two waves are *in phase* everywhere, corresponding to $\phi = 0, 2\pi, 4\pi, \ldots$. **Destructive interference** occurs when the two waves are 180° out of phase everywhere, corresponding to $\phi = \pi, 3\pi, 5\pi, \ldots$.

Standing waves are formed from the superposition of two harmonic waves that have the same frequency, amplitude, and wavelength but are traveling in *opposite* directions. The resultant standing wave is described by the wave function

Wave function for a standing wave

$$y = (2A \sin kx) \cos \omega t \qquad [23.3]$$

Hence, its amplitude varies as sin kx. The maximum amplitude points (called **antinodes**) are separated by a distance $\lambda/2$. Halfway between antinodes are points of zero amplitude (called nodes). The distance between adjacent nodes is $\lambda/2$.

One can set up standing waves with specific frequencies in such systems as stretched strings, hollow pipes, rods, and drumheads. The natural frequencies of vibration of a stretched string of length L, fixed at both ends, are

Normal modes of a stretched string

$$f_n = \frac{n}{2L}\sqrt{\frac{F}{\mu}} \qquad (n = 1, 2, 3, \ldots) \qquad [23.8]$$

where F is the tension in the string and μ is its mass per unit length. The natural frequencies of vibration form a **harmonic series,** that is, $f_1, 2f_1, 3f_1, \ldots$.

A system capable of oscillating is said to be in **resonance** with some driving force whenever the frequency of the driving force matches one of the natural frequencies of the system. When the system is resonating, it responds by oscillating with a relatively large amplitude.

The standing wave patterns for longitudinal waves in a hollow pipe depend on whether the ends of the pipe are open or closed. If the pipe is open at both ends, the natural frequencies of vibration form a harmonic series. If one end is closed, only odd harmonics of the fundamental are present.

The phenomenon of **beats** occurs as a result of the superposition of two waves of slightly different frequencies, traveling in the same direction. For sound waves at a given point, one hears an alternation in sound intensity with time. Thus, beats correspond to *interference as time passes.*

Any periodic waveform can be represented by the combination of the sinusoidal waves that form a harmonic series. The process is called *Fourier synthesis* and is based upon *Fourier's theorem.*

Questions and Conceptual Exercises

1. When two waves interfere constructively or destructively, is there any gain or loss in energy? Explain.
2. Does the phenomenon of wave interference apply only to harmonic waves?
3. What are the conditions necessary for the production of standing waves with well-defined nodes?
4. Some singers claim to be able to shatter a wineglass by maintaining a certain vocal pitch over a period of several seconds (Fig. 23.15). What mechanism causes the glass to break? (The glass must be very clean in order to break.)
5. What limits the amplitude of motion of a real vibrating system that is driven at one of its resonant frequencies?
6. If the temperature of the air increases, the speed of sound increases. What happens to the resonance frequencies of an organ pipe when the air temperature increases?
7. Explain why your singing voice sounds better than usual in the shower.
8. What is the purpose of the slide on a trombone or the valves on a trumpet?
9. Explain why all harmonics are present in an organ pipe that is open at both ends, but only the odd harmonics are present in a pipe that is closed at one end.

Figure 23.15 (Question 4) A wine glass shattered by the amplified sound of a human voice. *(© Ben Rose 1992/The IMAGE Bank)*

10. Explain how a musical instrument such as a piano can be tuned using the phenomenon of beats.
11. An airplane mechanic notices that the sound from a twin-engine aircraft rapidly varies in loudness when both engines are running. What could be causing this variation?
12. Why does a vibrating guitar string sound louder when

placed on the instrument than it would if allowed to vibrate in the air while off the instrument?

13. When the base of a vibrating tuning fork is placed against a chalkboard, the sound becomes louder due to resonance. How does this affect the length of time the fork vibrates? Does this agree with conservation of energy?
14. Stereo speakers are supposed to be "phased" when set up. That is, the waves emitted from them should be in phase with each other. What would the sound be like along the centerline of the speakers if one speaker were wired backward, that is, out of phase?
15. To keep deer away from their cars, some people mount short, thin pipes on the fenders. The pipes give out a high-pitched wail when the cars are moving. How do they create the sound?
16. If you wet your fingers and lightly ran them around the rim of a fine wineglass, you would hear a high-pitched sound. Why? How could you produce varied musical notes with a set of wine glasses?
17. When a bell is rung, standing waves are set up around the bell's circumference. What boundary conditions must be satisfied by the resonant wavelengths? How does a crack, such as that in the Liberty Bell, affect the bell's ability to satisfy the boundary conditions, and the sound emanating from the bell?
18. Despite a reasonably steady hand, a certain person often spills his coffee when carrying it to his seat. Discuss resonance as a possible cause of this difficulty, and devise a means for solving the problem.
19. A soft-drink bottle resonates as air is blown across the top. What happens to the resonant frequency as the level of fluid in the bottle decreases?

Problems

Section 23.1 Superposition and Interference of Harmonic Waves

1. Two harmonic waves are described by

$$y_1 = (0.5 \text{ m}) \sin[\pi(4x - 1200t)]$$

$$y_2 = (0.5 \text{ m}) \sin[\pi(4x - 1200t - 0.25)]$$

where x, y_1, and y_2 are in meters and t is in seconds. (a) What is the amplitude of the resultant wave? (b) What is the frequency of the resultant wave?

2. Two harmonic waves are described by

$$y_1 = (0.06 \text{ m}) \sin\left(\frac{\pi}{15}x - \frac{\pi}{0.005}t\right)$$

$$y_2 = (0.06 \text{ m}) \sin\left(\frac{\pi}{15}x - \frac{\pi}{0.005}t - \phi\right)$$

where x, y_1, and y_2 are in meters and t is in seconds. (a) What is the amplitude of the resultant wave when $\phi = (\pi/6)$ rad? (b) For what values of ϕ will the amplitude of the resultant wave have its maximum value?

3. A harmonic wave is described by

$$y_1 = (0.08 \text{ m}) \sin[2\pi(0.1x - 80t)]$$

where y_1 and x are in meters and t is in seconds. Write an expression for a wave that has the same frequency, amplitude, and wavelength as y_1, but when added to y_1 gives a resultant with an amplitude of $8\sqrt{3}$ cm.

4. Two speakers are arranged similarly to those shown in Figure 23.3. The distance between the speakers is 2 m, and they are driven at a frequency of 1500 Hz. An observer is initially at a point 6 m along the perpendicular bisector of the line joining the two speakers. (a) What distance must the observer move along a line parallel to the line joining the two speakers before reaching the first minimum in intensity? (Use $v = 343$ m/s.) (b) At what distance from the perpendicular bisector will the observer find the first relative maximum in intensity?

5. Two identical sound sources are located along the y axis. Source S_1 is located at (0, 0.1) m, and source S_2 is located at (0, −0.1) m. The two sources radiate isotropically at a frequency of 1715 Hz, and the amplitude of each wave separately is assumed to be A. A listener is located along the y axis a distance of 5 m from source S_1. (a) What is the phase difference between the sound waves at the position of the listener? (b) What is the amplitude of the resultant wave at the location of the listener? (Use $v = 343$ m/s.)

6. A tuning fork generates sound waves with a frequency of 246 Hz. The waves travel in opposite directions along a hallway, are reflected by walls, and return. What is the phase difference between the reflected waves when they meet? The corridor is 47 m long and the tuning fork is 14 m from one end. The speed of sound in air is 343 m/s.

7. Two speakers are driven by a common oscillator at 800 Hz and face each other at a distance of 1.25 m. Locate the points where relative minima would be expected along a line joining the two speakers. (Use $v = 343$ m/s.)

Section 23.2 Standing Waves

8. Use the trigonometric identity

$$\sin(a \pm b) = \sin a \cos b \pm \cos a \sin b$$

to show that the resultant of two wave functions—each of amplitude A, angular frequency ω, and angular wave number k, traveling in opposite directions—can be written

$$y = (2A \sin kx) \cos \omega t$$

9. Two waves in a long string are

$$y_1 = (0.015 \text{ m}) \cos\left(\frac{x}{2} - 40t\right)$$

$$y_2 = (0.015 \text{ m}) \cos\left(\frac{x}{2} + 40t\right)$$

where the y's and x are in meters and t is in seconds. (a) Determine the positions of the nodes of the resulting standing wave. (b) What is the maximum displacement at the position $x = 0.4$ m?

10. The wave function for a standing wave in a string is

$$y = (0.03 \text{ m}) \sin(0.25x) \cos(120\pi t)$$

where x is in meters and t is in seconds. Determine the wavelength and frequency of the interfering traveling waves.

11. Two harmonic waves traveling in opposite directions interfere to produce a standing wave described by

$$y = (0.15 \text{ m}) \sin(0.4x) \cos(200t)$$

where x is in meters and t is in seconds. Determine the wavelength, frequency, and speed of the interfering waves.

12. Two waves that set up a standing wave in a long string are given by

$$y_1 = A \sin(kx - \omega t + \phi)$$

$$y_2 = A \sin(kx + \omega t)$$

Show (a) that the addition of the arbitrary phase angle will change only the position of the nodes, and (b) that the distance between nodes remains constant.

13. A student wants to establish a standing wave on a wire that is 1.8 m long and clamped at both ends. The wave speed is 540 m/s. What is the minimum frequency the student should apply to set up standing waves?

14. A standing wave is formed by the interference of two traveling waves, each of which has an amplitude $A = \pi$ cm, angular wave number $k = (\pi/2)$rad/cm, and angular frequency $\omega = 10\pi$ rad/s. (a) Calculate the distance between the first two antinodes. (b) What is the amplitude of the standing wave at $x = 0.25$ cm?

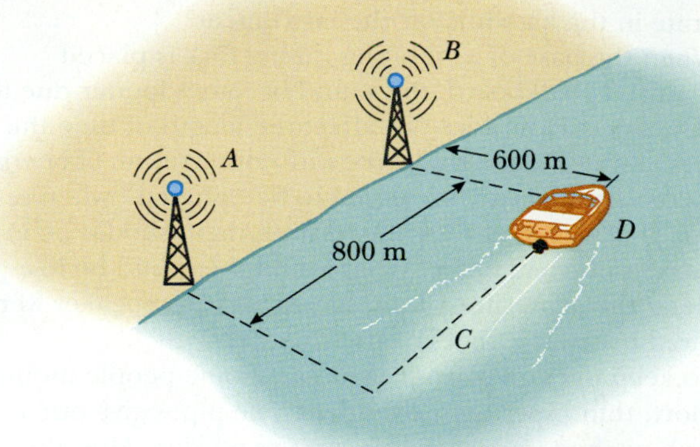

Figure 23.16 (Problem 15)

15. The ship in Figure 23.16 travels along a straight line that is parallel to the shore and 600 m from the shore. The ship's radio receives simultaneous signals of the same frequency from antennas A and B. The signals interfere constructively at point C, which is equidistant from A and B. The signal goes through the first minimum at point D. Determine the wavelength of the radio waves.

Section 23.3 Natural Frequencies in a Stretched String

16. A string 50 cm long has a mass per unit length of 20×10^{-5} kg/m. To what tension should this string be stretched if its fundamental frequency is to be (a) 20 Hz and (b) 4500 Hz?

17. A steel wire in a harp has a length of 0.70 m and a mass of 4.3×10^{-3} kg (Fig. 23.17). To what tension must this wire be stretched in order that the fundamental vibration correspond to middle C ($f_C =$ 261.6 Hz)?

18. A standing wave is established in a 120-cm-long string that is fixed at both ends. When driven at 120 Hz, the string vibrates in four segments. (a) Determine the wavelength. (b) What is the fundamental frequency?

19. A cello A-string vibrates in its fundamental mode with a frequency of 220 vibrations/s. The vibrating segment is 70 cm long and has a mass of 1.2 g. (a) Find the tension in the string. (b) Determine the frequency of the harmonic that causes the string to vibrate in three segments.

20. A string of length L, mass per unit length μ, and tension F is vibrating at its fundamental frequency. What effect will the following have on the fundamental frequency? (a) The length of the string is doubled, with

Figure 23.17 (Problem 17)

all other factors held constant. (b) The mass per unit length is doubled, with all other factors held constant. (c) The tension is doubled, with all other factors held constant.

21. A 60-cm guitar string under a tension of 50 N has a mass per unit length of 0.1 g/cm. What is the highest resonant frequency that can be heard by a person who is capable of hearing frequencies up to 20 000 Hz?
22. A stretched wire vibrates in its fundamental mode at a frequency of 400 vibrations/s. What would be the fundamental frequency if the wire were half as long, with twice the diameter and four times the tension?
23. A violin string has a length of 0.35 m and is tuned to concert G, $f_G = 392$ Hz. Where must the violinist place her finger to play concert A, $f_A = 440$ Hz? If this position is to remain correct to one-half the width of a finger (i.e., to within 0.6 cm), by what fraction may the string tension be allowed to slip?

Section 23.4 Resonance

24. A child's swing has a length of 2 m. Find the frequency at which this swing should be pushed to achieve maximum amplitude.
25. Standing-wave vibrations are set up in a crystal goblet with two nodes and two antinodes equally spaced around the 20-cm circumference of its rim. If transverse waves move around the glass at 900 m/s, an opera singer would have to produce a high harmonic with what frequency in order to shatter the glass with a resonant vibration?

Section 23.5 Standing Waves in Air Columns

(In this section, unless otherwise indicated, assume that the speed of sound in air is 344 m/s.)

26. The fundamental frequency of an open organ pipe corresponds to middle C (261.6 Hz on the chromatic musical scale). The third resonance of a closed organ pipe has the same frequency. What are the lengths of the two pipes?
27. Middle C has a frequency of 261.6 Hz. (a) What is the length of the shortest pipe, open at both ends, that would produce a sound of this frequency? (b) What would be the required length if the pipe were closed at one end?
28. A tuning fork whose frequency is f is used to set up a resonance condition in a pipe. Write an expression for the length that will cause the pipe to resonate in its nth mode if it is (a) open at both ends; (b) closed at one end. (Assume that the speed of sound is v.)
29. If an organ pipe is to resonate at 20 Hz, what is its required length if it is (a) open at both ends? (b) closed at one end?
30. A tuning fork of frequency 512 Hz is placed near the top of the tube shown in Figure 23.10a. The water level is lowered so that the length, L, slowly increases from an initial value of 20 cm. Determine the next two values of L that correspond to resonant modes.
31. A glass tube (open at both ends) of length L is positioned near an audio speaker of frequency $f = 0.68$ kHz. For what values of L will the tube resonate with the speaker?
32. Calculate the minimum length for a pipe that has a fundamental frequency of 240 Hz if the pipe is (a) closed at one end; (b) open at both ends.
33. A tunnel beneath a river is approximately 2 km long. At what frequencies can this tunnel resonate?
34. The overall length of a piccolo is 32 cm. The resonating air column vibrates as a pipe that is open at both ends. (a) Find the frequency of the lowest note a piccolo can play, assuming the speed of sound in air is 340 m/s. (b) Opening holes in the side effectively shortens the length of the resonant column. If the highest note a piccolo can sound is 4000 Hz, find the distance between adjacent nodes for this mode of vibration.
35. The human ear canal is about 2.8 cm long. If the canal is regarded as a tube that is open at one end and closed at the eardrum, at what fundamental frequency would you expect hearing to be best?

36. A shower stall measures 86 cm × 86 cm × 210 cm. When you sing in the shower, which frequencies sound the richest (resonate the most), assuming the shower acts as a pipe that is closed at both ends (nodes at both sides)? Assume that the human voice (not necessarily one individual's voice) ranges from 130 Hz to 2000 Hz. Let the speed of sound in the hot shower stall be 355 m/s.

*Section 23.6 Beats: Interference in Time

37. In certain ranges of a piano keyboard, two or more strings are tuned to the same note to provide extra loudness. For example, two strings are tuned to 110 Hz. If one of the two slips from its normal tension, 600 N, to 540 N, what beat frequency will be heard when the two strings are struck simultaneously?

38. Two waves with equal amplitude but slightly different frequencies are traveling in the same direction through a medium. At a given point the separate displacements are

$$y_1 = A_0 \cos \omega_1 t \qquad \text{and} \qquad y_2 = A_0 \cos \omega_2 t$$

Use the trigonometric identity

$$\cos a + \cos b = 2 \cos\left(\frac{a-b}{2}\right) \cos\left(\frac{a+b}{2}\right)$$

to show that the resultant displacement due to the two waves is

$$y = 2A_0\left[\cos\left(\frac{\omega_1 - \omega_2}{2}\right)t\right]\left[\cos\left(\frac{\omega_1 + \omega_2}{2}\right)t\right]$$

39. A student holds a tuning fork that is oscillating at 256 Hz. He walks toward a wall at a constant speed of 1.33 m/s. (a) What beat frequency does he observe between the tuning fork and its echo? (b) How fast must he walk away from the wall to observe a beat frequency of 5 Hz?

40. A flute is designed so that it plays a frequency of 261.6 Hz, middle C, when all the holes are covered and the temperature is 20°C. (a) Consider the flute as a pipe that is open at both ends; find the length of the flute, assuming that middle C above is the fundamental. (b) A second player, nearby in a colder room, also attempts to play middle C on an identical flute. A beat frequency of 3 Hz is heard. What is the temperature of the room? The speed of sound in air is described by

$$v = (331 \text{ m/s})\sqrt{1 + \frac{T}{273°}}$$

where T is the Celsius temperature.

41. While attempting to tune a C note at 523 Hz, a piano tuner hears 3 beats per second between the oscillator and the string. (a) What are the possible frequencies of the string? (b) By what percentage should the tension in the string be changed to bring the string into tune?

Additional Problems

42. (a) What is the fundamental frequency of a steel piano wire with a mass of 0.005 kg and a length of 1 m, under a tension of 1350 N? (b) What is the fundamental frequency of an organ pipe, 1 m in length, that is closed at the bottom and open at the top?

43. Two loudspeakers are placed on a wall, 2 m apart. A listener stands directly in front of one of the speakers, 3 m from the wall. The speakers are being driven by a single oscillator at a frequency of 300 Hz. (a) What is the phase difference between the two waves when they reach the observer? (b) What is the frequency closest to 300 Hz to which the oscillator may be adjusted so that the observer will hear minimal sound?

44. On a marimba, the wooden bar that sounds a tone when struck vibrates as a transverse standing wave with three antinodes and two nodes. The lowest-frequency note is 87 Hz, produced by a bar 40 cm long. (a) Find the speed of transverse waves on the bar. (b) The loudness and duration of the emitted sound are enhanced by a resonant pipe suspended vertically below the center of the bar. If the pipe is open at the top end only and the speed of sound in air is 340 m/s, what is the pipe length required for the pipe to resonate with the bar in part (a)?

45. If two adjacent natural frequencies of an organ pipe

(Problem 44) Marimba players in Mexico City.
(Murray Greenberg)

are determined to be 0.55 kHz and 0.65 kHz, calculate the fundamental frequency and length of this pipe. (Use $v = 340$ m/s.)

46. A speaker at the front of a room and an identical speaker at the rear of the room are being driven by the same oscillator at 456 Hz. A student walks at a uniform rate of 1.5 m/s along the length of the room. How many beats does the student hear per second?

47. While waiting for Stan Speedy to arrive on a late passenger train, Kathy Kool notices beats occurring as a result of two trains blowing their whistles simultaneously. One train is at rest and the other is approaching her at a speed of 20 km/h. Assume that the two whistles have the same frequency and that the speed of sound is 344 m/s. If Kathy hears 4 beats per second, what is the frequency of the whistles?

48. A string (mass = 4.8 g, length = 2.0 m, and tension = 48 N), fixed at both ends, vibrates in its second ($n = 2$) natural mode. What is the wavelength in air of the sound emitted by this vibrating string?

49. Two train whistles have identical frequencies of 180 Hz. When one train is at rest in the station, sounding its whistle, a beat frequency of 2 Hz is heard from a moving train. What two possible speeds and directions can the moving train have?

50. The speed of sound in air is described by

$$v = (331 \text{ m/s})\sqrt{1 + \frac{T}{273°}}$$

where T is the Celsius temperature. What is the lowest frequency of the standing wave of sound that can be set up between two walls that are 8 m apart, if the temperature is 22°C?

51. A pipe that is open at both ends has a fundamental frequency of 300 Hz when the speed of sound in air is 333 m/s. (a) What is the length of the pipe? (b) What is the frequency of the second harmonic when the temperature of the air is increased so that the speed of sound in the pipe is 344 m/s?

52. A standing wave is established in a string that is 240 cm long and fixed at both ends. The string vibrates in four segments when driven at 120 Hz. (a) Determine the wavelength. (b) What is the fundamental frequency?

53. A string with a mass of 8 g and a length of 5 m has one end attached to a wall; the other end is draped over a pulley and attached to a hanging mass of 4 kg. If the string is plucked, what is the fundamental frequency of vibration?

54. In a major chord on the physical pitch muscial scale, the frequencies are in the ratios 4:5:6:8. A set of pipes, closed at one end, is to be cut so that, when sounded in their fundamental mode, the pipes will sound out a major chord. (a) What are the ratios of the lengths of the pipes? (b) What pipe lengths are needed if the lowest frequency of the chord is 256 Hz? (c) What are the frequencies of this chord?

55. Two wires are welded together. They are of the same material, but one is twice the diameter of the other. The wires are subjected to a tension of 4.6 N. The thin wire has a length of 40 cm and a linear mass density of 2 g/m. The combination is fixed at both ends and vibrated in such a way that two antinodes are present, with the central node being right at the weld. (a) What is the frequency of vibration? (b) How long is the thick wire?

56. Two identical strings, each fixed at both ends, are near each other. It is observed that, if string A starts oscillating in its fundamental mode, string B begins vibrating in its third ($n = 3$) natural mode. Determine the ratio of the tension of string B to the tension of string A.

57. A standing wave is set up in a string of variable length and tension by a vibrator of variable frequency. When the vibrator has a frequency of f in a string of length L and tension F, n antinodes are set up in the string. (a) If the length of the string is doubled, by what factor should the frequency be changed to get the same number of antinodes? (b) If the frequency and length are held constant, what tension will produce $n + 1$ antinodes? (c) If the frequency is tripled and the length halved, by what factor should the tension be changed to get twice as many antinodes?

58. Radar detects the speed of a car, using the Doppler shift of microwaves that are reflected off the moving car, by beating the received wave with the transmitted wave and measuring the difference. The Doppler shift for light is given by

$$f = f_0\sqrt{\frac{c + v}{c - v}}$$

where f_0 is the transmitted frequency, c is the speed of light (3×10^8 m/s) and v is the relative speed of the two objects. (a) Show that the wave that reflects back to the source has a frequency of

$$f = f_0\frac{(c + v)}{(c - v)}$$

(b) Show that the expression for the beat frequency of the microwaves may be written $f_b = 2v/\lambda$. (Since the beat frequency is much smaller than the transmitted frequency, use the approximation $f + f_0 = 2f_0$.) (c) What beat frequency is measured for a speed of 30 m/s (67 mph) if the microwaves have a frequency of 10 GHz? (1 GHz = 10^9 Hz.) (d) If the beat frequency measurement is accurate to ± 5 Hz, how accurate is the velocity measurement?

Calculator/Computer Problems

59. Sketch the resultant waveform due to the interference of the two waves, y_1 and y_2, in Problem 2 at $t = 0$ s for (a) $\phi = 0$, (b) $\phi = 90°$, and (c) $\phi = 270°$. Let x range over the interval 0 to 30 m.
60. A standing wave is described by the function

$$y = (0.06 \text{ m}) \sin(\pi x/2) \cos(100\pi t)$$

where x and y are in meters and t is in seconds. (a) Plot $y(x)$ versus x for $t = 0$, 0.0005 s, 0.001 s, 0.0015 s, and 0.002 s. (b) What is the frequency of the wave? (c) What is the wavelength, λ?

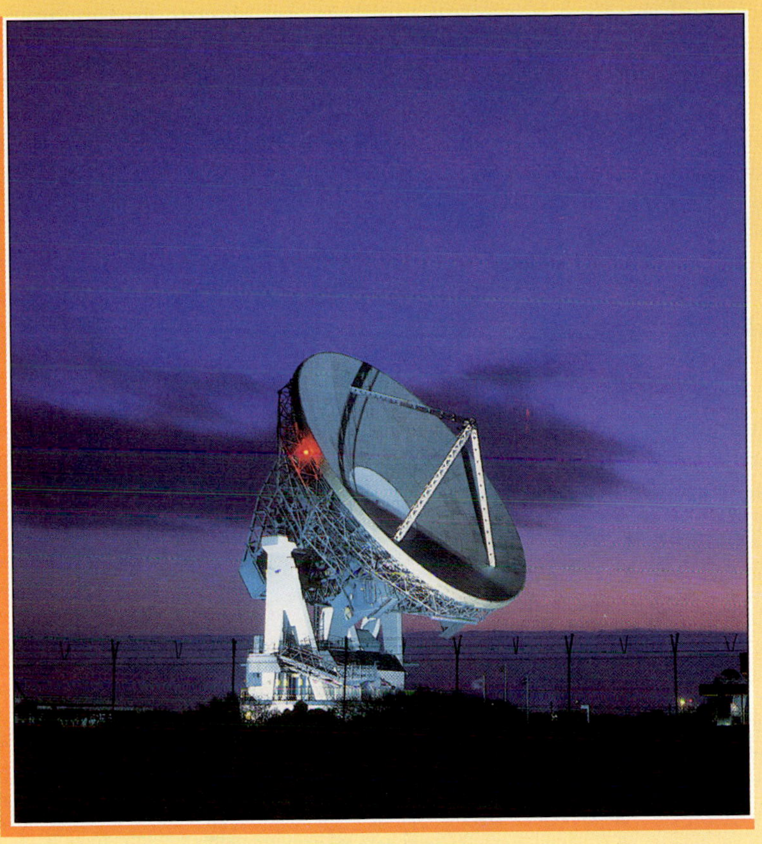

Satellite receiver-transmitter dish at night. The photograph was taken at the Land Earth Station at Goonhilly in Cornwall, United Kingdom. This Earth Station serves the INMARSAT (International Maritime Satellite) organization. It provides telephone, telex, data and facsimile operations to the shipping, aviation, offshore and land mobile industries. *(Photo Researchers, Inc.)*

Electromagnetic Waves

CHAPTER 24

Although we are not always aware of their presence, electromagnetic waves permeate our environment. In the form of visible light, they enable us to view the world around us; infrared waves warm our environment; radio-frequency waves carry our favorite television and radio programs; microwaves cook our food and are used in radar communication systems. The list goes on and on. The waves we have described until now have all been mechanical waves, which, by definition, can exist only if a medium is present. Electromagnetic waves, in contrast, can propagate through a vacuum.

The purpose of this chapter is to explore the properties of electromagnetic waves. The fundamental laws of electricity and magnetism—Maxwell's equations—form the basis of all electromagnetic phenomena. One of the equations predicts that a time-varying electric field produces a magnetic field just as a time-varying magnetic field produces an electric field. From this generalization, Maxwell provided the final important link between electric and magnetic fields. The most dramatic prediction of his equations is the existence of electromagnetic waves that propagate through empty space with the speed of light. This discovery led to many practical applications, such as radio and television, and to the realization that light is one form of electromagnetic radiation. We shall close the chapter by discussing how electromagnetic waves carry energy and momentum.

24.1 Maxwell's Equations and Electromagnetic Waves

The fundamental laws governing the behavior of electric and magnetic fields are Maxwell's equations, discussed in Chapter 20, Section 20.5. In his unified theory of electromagnetism, Maxwell showed that electromagnetic waves are a natural consequence of these fundamental laws. Recall that *Maxwell's equations* are

Maxwell's equations

$$\oint \mathbf{E} \cdot d\mathbf{A} = \frac{Q}{\epsilon_0} \qquad [24.1]$$

$$\oint \mathbf{B} \cdot d\mathbf{A} = 0 \qquad [24.2]$$

$$\oint \mathbf{E} \cdot d\mathbf{s} = -\frac{d\Phi_{\mathrm{m}}}{dt} \qquad [24.3]$$

$$\oint \mathbf{B} \cdot d\mathbf{s} = \mu_0 I + \mu_0 \epsilon_0 \frac{d\Phi_{\mathrm{e}}}{dt} \qquad [24.4]$$

In empty space ($Q = 0$, $I = 0$), these equations permit a wavelike solution, where the *wave speed equals the measured speed of light.* This result led Maxwell to the prediction that light waves are, in fact, a form of electromagnetic radiation.

The properties of electromagnetic waves can be deduced from Maxwell's equations. One approach is to solve the second-order differential equation obtained from Maxwell's third and fourth equations. Such a rigorous mathematical treatment is beyond the scope of this text. To circumvent the problem, we shall assume that the electric and magnetic field vectors have specific space-time behavior that is consistent with Maxwell's equations.

First, we shall assume that the electromagnetic wave is a *plane wave*; that is, it travels in one direction only. The plane wave we are describing has the following properties. It travels in the x direction (the direction of propagation), the electric field **E** is in the y direction, and the magnetic field **B** is in the z direction, as in Figure 24.1. Waves in which the electric and magnetic fields are restricted to being parallel to certain lines in the yz plane are said to be **linearly polarized waves.**[1] Furthermore, we assume that at any point P, E and B depend upon x and t only, and not upon the y or z coordinate.

[1] Waves with other particular patterns of vibrations of E and B include *circularly polarized waves.* The most general polarization pattern is *elliptical.*

We can relate E and B to each other by using Maxwell's third and fourth equations (Eqs. 24.3 and 24.4). In empty space, where $Q = 0$ and $I = 0$, these equations are

$$\oint \mathbf{E} \cdot d\mathbf{s} = -\frac{d\Phi_m}{dt} \qquad [24.5]$$

$$\oint \mathbf{B} \cdot d\mathbf{s} = \epsilon_0 \mu_0 \frac{d\Phi_e}{dt} \qquad [24.6]$$

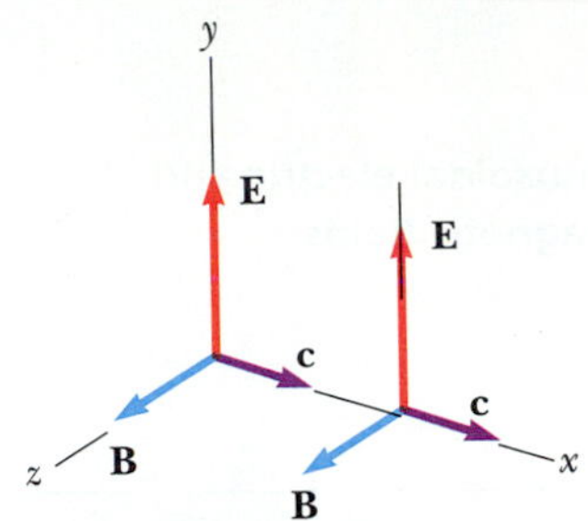

Figure 24.1
A plane-polarized electromagnetic wave traveling in the positive x direction. The electric field is along the y direction, and the magnetic field is along the z direction. These fields depend only on x and t.

Using these expressions and the plane wave assumption, one obtains the following differential equations relating E and B. For simplicity of notation, we have dropped the subscripts on the components E_y and B_z:

$$\frac{\partial E}{\partial x} = -\frac{\partial B}{\partial t} \qquad [24.7]$$

$$\frac{\partial B}{\partial x} = -\mu_0 \epsilon_0 \frac{\partial E}{\partial t} \qquad [24.8]$$

Note that the derivatives here are partial derivatives. For example, when $\partial E/\partial x$ is being evaluated, we assume that t is constant. Likewise, when $\partial B/\partial t$ is being evaluated, x is held constant. Taking the derivative of Equation 24.7 and using this with Equation 24.8, we get

$$\frac{\partial^2 E}{\partial x^2} = -\frac{\partial}{\partial x}\left(\frac{\partial B}{\partial t}\right) = -\frac{\partial}{\partial t}\left(\frac{\partial B}{\partial x}\right) = -\frac{\partial}{\partial t}\left(-\mu_0 \epsilon_0 \frac{\partial E}{\partial t}\right) \qquad [24.9]$$

Wave equations for electromagnetic waves in free space

$$\frac{\partial^2 E}{\partial x^2} = \mu_0 \epsilon_0 \frac{\partial^2 E}{\partial t^2} \qquad [24.10]$$

In the same manner, taking the derivative of Equation 24.8 and using Equation 24.7, we get

$$\frac{\partial^2 B}{\partial x^2} = \mu_0 \epsilon_0 \frac{\partial^2 B}{\partial t^2} \qquad [24.11]$$

Equations 24.10 and 24.11 both have the form of the general wave equation,[2] with a speed, c, of

$$c = \frac{1}{\sqrt{\mu_0 \epsilon_0}} \qquad [24.12]$$

Taking $\mu_0 = 4\pi \times 10^{-7}$ Wb/A·m and $\epsilon_0 = 8.85418 \times 10^{-12}$ C²/N·m² in Equation 24.12, we find that

The speed of electromagnetic waves

$$c = 2.99792 \times 10^8 \text{ m/s} \qquad [24.13]$$

Since this speed is precisely the same as the speed of light in empty space,[3] one is led to believe (correctly) that light is an electromagnetic wave.

[2] The general wave equation is of the form $(\partial^2 g/\partial x^2) = (1/v^2)(\partial^2 g/\partial t^2)$, where v is the speed of the wave and g is the wave amplitude.

[3] Because of the redefinition of the meter in 1983, the speed of light is now a *defined* quantity with an *exact* value of $c = 2.99792458 \times 10^8$ m/s.

The simplest plane wave solution is a sinusoidal wave, for which the field amplitudes E and B vary with x and t according to the expressions

Sinusoidal electric and magnetic fields

$$E = E_{\text{max}} \cos(kx - \omega t) \qquad [24.14]$$

$$B = B_{\text{max}} \cos(kx - \omega t) \qquad [24.15]$$

where E_{max} and B_{max} are the maximum values of the fields. The angular wave number $k = 2\pi/\lambda$, where λ is the wavelength, and the angular frequency $\omega = 2\pi f$, where f is the number of cycles per second. Since $v = f\lambda$ from Equation 22.14 in Chapter 22, the ratio ω/k equals the speed, c:

$$\frac{\omega}{k} = \frac{2\pi f}{2\pi/\lambda} = f\lambda = c$$

James Clerk Maxwell (1831–1879) James Clerk Maxwell is generally regarded as the greatest theoretical physicist of the 19th century. Born in Edinburgh to a well-known Scottish family, he entered the University of Edinburgh at age 15. He was appointed to his first professorship, at Aberdeen, in 1856. This was the beginning of a career during which Maxwell would develop the electromagnetic theory of light, the kinetic theory of gases, and explanations of the nature of Saturn's rings and of color vision. *(Photo courtesy of Emilio Segré Visual Archives)*

Figure 24.2 represents one instant of a sinusoidal, linearly polarized plane wave moving in the positive x direction.

Taking partial derivatives of Equations 24.14 and 24.15, we find that

$$\frac{\partial E}{\partial x} = -kE_{\text{max}} \sin(kx - \omega t)$$

$$-\frac{\partial B}{\partial t} = -\omega B_{\text{max}} \sin(kx - \omega t)$$

Since these must be equal, according to Equation 24.7, we find that at any instant

$$kE_{\text{max}} = \omega B_{\text{max}}$$

$$\frac{E_{\text{max}}}{B_{\text{max}}} = \frac{\omega}{k} = c$$

(The minus sign is ignored here because we are interested only in comparing the amplitudes.) Using this last identity together with Equations 24.14 and 24.15, we see that

$$\frac{E_{\text{max}}}{B_{\text{max}}} = \frac{E}{B} = c \qquad [24.16]$$

That is, *at every instant the ratio of the electric field to the magnetic field of an electromagnetic wave equals the speed of light.*

Finally, one should note that electromagnetic waves obey the *superposition principle,* since the differential equations involving E and B are *linear* equations. For

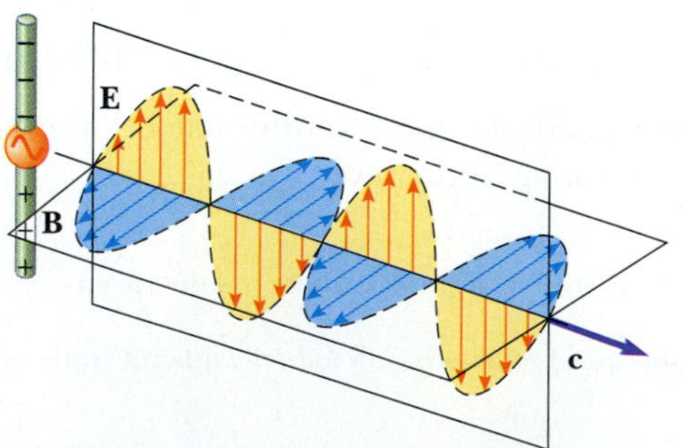

Figure 24.2 An electromagnetic wave sent out by oscillating charges in an antenna. This represents the wave at one instant of time. Note that the electric field is perpendicular to the magnetic field, and both are perpendicular to the direction of wave propagation.

example, two waves traveling in opposite directions with the same frequency could be added by simply adding the wave fields algebraically.

Let us summarize the properties of electromagnetic waves as we have described them:

Properties of electromagnetic waves

1. The solutions of Maxwell's third and fourth equations are wavelike, where both E and B satisfy the same wave equation.
2. Electromagnetic waves travel through empty space with the speed of light, $c = 1/\sqrt{\epsilon_0\mu_0}$.
3. The electric and magnetic field components of plane electromagnetic waves are perpendicular to each other and also perpendicular to the direction of wave propagation. The latter property can be summarized by saying that electromagnetic waves are transverse waves.
4. The relative magnitudes of **E** and **B** in empty space are related by $E/B = c$.
5. Electromagnetic waves obey the superposition principle.

Example 24.1 An Electromagnetic Wave

A plane electromagnetic sinusoidal wave of frequency 40 MHz travels in free space in the x direction, as in Figure 24.3. At some point and at some instant, the electric field **E** has its *maximum* value of 750 N/C and is along the y axis.

(a) Determine the wavelength and period of the wave.

Solution Since $c = \lambda f$ and $f = 40 \text{ MHz} = 4 \times 10^7 \text{ s}^{-1}$, we get

$$\lambda = \frac{c}{f} = \frac{3 \times 10^8 \text{ m/s}}{4 \times 10^7 \text{ s}^{-1}} = 7.50 \text{ m}$$

The period of the wave T equals the inverse of the frequency:

$$T = \frac{1}{f} = \frac{1}{4 \times 10^7 \text{ s}^{-1}} = 2.5 \times 10^{-8} \text{ s}$$

(b) Calculate the magnitude and direction of **B** when **E** = 750**j** N/C.

Solution From Equation 24.16 we see that

$$B_{\text{max}} = \frac{E_{\text{max}}}{c} = \frac{750 \text{ N/C}}{3 \times 10^8 \text{ m/s}} = 2.50 \times 10^{-6} \text{ T}$$

Since **E** and **B** must be perpendicular to each other and both must be perpendicular to the direction of wave propagation (x in this case), we conclude that **B** is in the z direction.

(c) Write expressions for the space-time variation of the electric and magnetic field components for this wave.

Solution We can apply Equations 24.14 and 24.15 directly:

$$E = E_{\text{max}} \cos(kx - \omega t) = (750 \text{ N/C}) \cos(kw - \omega t)$$

$$B = B_{\text{max}} \cos(kx - \omega t) = (2.50 \times 10^{-6} \text{ T}) \cos(kx - \omega t)$$

where

$$\omega = 2\pi f = 2\pi(4 \times 10^7 \text{ s}^{-1}) = 8\pi \times 10^7 \text{ rad/s}$$

$$k = \frac{2\pi}{\lambda} = \frac{2\pi}{7.5 \text{ m}} = 0.838 \text{ rad/m}$$

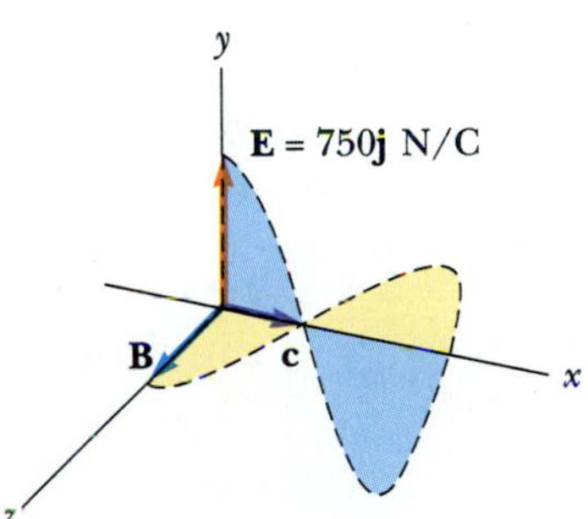

Figure 24.3 (Example 24.1) At some instant, a plane electromagnetic wave moving in the x direction has a maximum electric field of 750 N/C in the positive y direction. The corresponding magnetic field at that point has a magnitude E/c and is in the z direction.

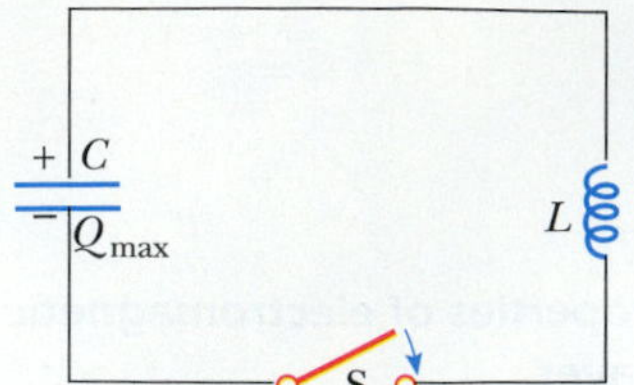

Figure 24.4
A simple LC circuit. The capacitor has an initial charge Q_{max} and the switch is closed at $t = 0$.

24.2 Hertz's Discoveries

In 1888, Heinrich Hertz (1857–1894) was the first to generate and detect electromagnetic waves in a laboratory setting. In order to appreciate the details of his experiment, let us first examine the properties of an LC circuit. In such a circuit, a charged capacitor is connected to an inductor, as in Figure 24.4. When the switch is closed, both the current in the circuit and the charge on the capacitor oscillate. If resistance is neglected, no energy is lost to heat and the oscillations continue indefinitely.

Let us assume that the capacitor has an initial charge of Q_{max} and that the switch is closed at $t = 0$. It is convenient to describe what happens from an energy viewpoint. When the capacitor is fully charged, the total energy in the circuit is stored in the electric field of the capacitor and is equal to $Q_{max}^2/2C$. At this time, the current is zero and so no energy is stored in the inductor. As the capacitor begins to discharge, the energy stored in its electric field decreases. At the same time, the current increases and an amount of energy equal to $LI^2/2$ is now stored in the magnetic field of the inductor. Thus, we see that energy is transferred from the electric field of the capacitor to the magnetic field of the inductor. When the capacitor is fully discharged, it stores no energy. At this time, the current reaches its maximum value and all of the energy is stored in the inductor. The process then repeats in the reverse direction. The energy continues to transfer between the inductor and the capacitor, corresponding to oscillations of both current and charge.

A representation of this energy transfer is shown in Figure 24.5. The behavior of the circuit is analogous to that of the oscillating mass-spring system studied in Chapter 21. The potential energy stored in a stretched spring, $kx^2/2$, is analogous to the potential energy stored in the capacitor, $Q_{max}^2/2C$. The kinetic energy of the moving mass, $mv^2/2$, is analogous to the energy stored in the inductor, $LI^2/2$, which requires the presence of moving charges. In Figure 24.5a, all of the energy is stored as potential energy in the capacitor at $t = 0$ (because $I = 0$). In Figure 24.5b, all of the energy is stored as "kinetic" energy in the inductor, $LI_{max}^2/2$, where I_{max} is the maximum current. At intermediate points, part of the energy is potential and part is kinetic.

The frequency of oscillation of an LC circuit, called the *resonance frequency,* is[4]

$$f_0 = \frac{1}{2\pi\sqrt{LC}} \qquad \textbf{[24.17]}$$

The circuit Hertz used in his investigations of electromagnetic waves is similar to that already discussed and is shown schematically in Figure 24.6. An induction coil (a large coil of wire) is connected to two metal spheres with a narrow gap between them to form a capacitor. Oscillations are initiated in the circuit by sending short voltage pulses via the coil to the spheres, charging one positive, the other negative. Because L and C are quite small in this circuit, the frequency of oscillation is quite high, $f \approx 100$ MHz. This circuit is called a transmitter because it produces electromagnetic waves.

[4] For a more detailed discussion of resonance, see Chapter 21, Section 21.7 (on RLC circuits), and set $R = 0$.

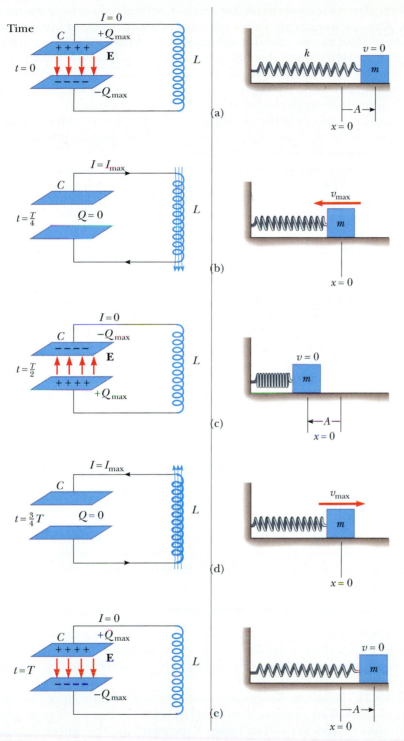

Figure 24.5
Energy transfer in a resistanceless LC circuit. The capacitor has a charge Q_{max} at $t = 0$ when the switch is closed. The mechanical analog of this circuit, the mass-spring system, is shown at the right.

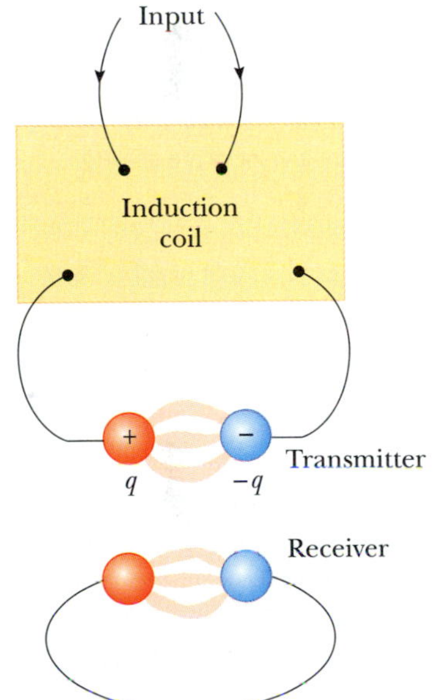

Figure 24.6
Schematic diagram of Hertz's apparatus for generating and detecting electromagnetic waves. The transmitter consists of two spherical electrodes connected to an induction coil, which provides short voltage surges to the spheres, setting up oscillations in the discharge. The receiver is a nearby single-wire loop.

Heinrich Hertz was born in 1857 in Hamburg, Germany. He studied physics under Helmholtz and Kirchhoff at the University of Berlin. In 1885, Hertz accepted the position of Professor of Physics at Karlsruhe; it was here that he discovered radio waves in 1888, his most important accomplishment.

Discovering radio waves, demonstrating their generation, and determining their velocity are among Hertz's many achievements. After finding that the velocity of a radio wave was the same as that of light, Hertz showed that radio waves, like light waves, could be reflected, refracted, and diffracted.

Hertz died of blood poisoning at the age of 36. During his short life, he made many contributions to science. The hertz, equal to one complete vibration or cycle per second, is named after him.

Hertz placed a second circuit, the receiver, several meters from the transmitter circuit. This receiver circuit, which consisted of a single loop of wire connected to two spheres, had its own effective inductance, capacitance, and natural frequency of oscillation. Hertz found that energy was being sent from the transmitter to the receiver when the resonance frequency of the receiver was adjusted to match that of the transmitter.[5] The energy transfer was detected when the voltage across the spheres in the receiver circuit became high enough to ionize air molecules, which caused sparks to appear in the air gap separating the spheres. Hertz's experiment is analogous to the mechanical phenomenon in which one tuning fork picks up the vibrations from an identical vibrating fork.

Hertz assumed that the energy transferred from the transmitter to the receiver was carried in the form of waves, which are now known to have been electromagnetic waves. In a series of experiments, he also showed that the radiation generated by the transmitter exhibited the wave properties of interference, diffraction, reflection, refraction, and polarization. As we shall see shortly, all of these properties are exhibited by light. Thus, it became evident that the waves observed by Hertz had properties similar to those of light waves and differed only in frequency and wavelength.

Perhaps Hertz's most convincing experiment was his measurement of the speed of the waves from the transmitter. Waves of known frequency from the transmitter were reflected from a metal sheet so that an interference pattern was set up, much like the standing wave pattern on a stretched string. As we saw in our discussion of standing waves, the distance between nodes is $\lambda/2$, and so Hertz was able to determine the wavelength, λ. Using the relationship $v = f\lambda$, Hertz found that v was close to 3×10^8 m/s, the known speed of visible light. Thus, Hertz's experiments provided the first evidence in support of Maxwell's theory.

24.3 Production of Electromagnetic Waves by an Antenna

The energy transfer that takes place in an LC circuit continues for prolonged periods only when the changes occur slowly. If the current alternates rapidly, the circuit loses some of its energy in the form of electromagnetic waves. In fact, electromagnetic waves are radiated by *any* circuit carrying an alternating current. The fundamental mechanism responsible for this radiation is the acceleration of a charged particle.

Whenever a charged particle undergoes an acceleration, it must radiate energy.

An alternating voltage applied to the wires of an antenna forces an electric charge in the antenna to oscillate. This is a common technique for accelerating charged particles and is the source of the radio waves emitted by the antenna of a radio station.

Figure 24.7 illustrates how an electromagnetic wave is produced by oscillating electric charges in an antenna. Two metal rods are connected to an ac generator, which causes charges to oscillate between the two rods. The output voltage of the

[5] Following Hertz's discoveries, Marconi succeeded in developing a practical, long-range radio communication system.

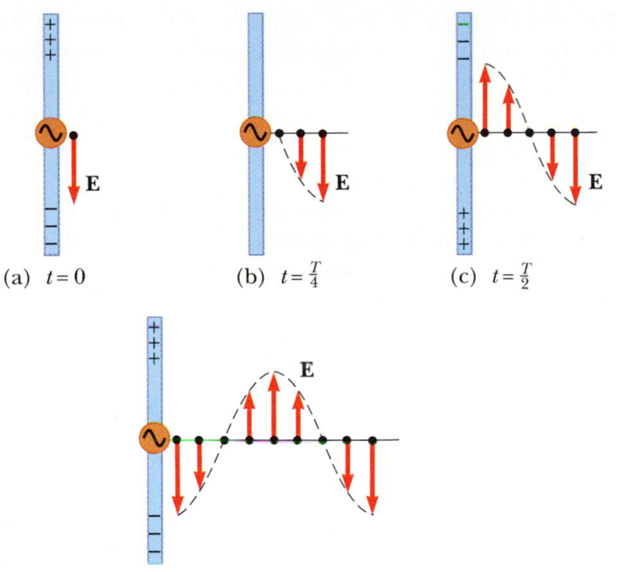

Figure 24.7
The electric field set up by oscillating charges in an antenna. The field moves away from the antenna at the speed of light.

generator is sinusoidal. At $t = 0$, the upper rod is given a maximum positive charge and the bottom rod an equal negative charge, as in Figure 24.7a. The electric field near the antenna at this instant is also shown in Figure 24.7a. As the charges oscillate, the rods become less charged, the field near the rods decreases in strength, and the downward-directed maximum electric field produced at $t = 0$ moves away from the rod. When the charges are neutralized, as in Figure 24.7b, the electric field has dropped to zero. This occurs at a time equal to one quarter of the period of oscillation. Continuing in this fashion, the upper rod soon obtains a maximum negative charge and the lower rod becomes positive, as in Figure 24.7c, resulting in an electric field directed upward. This occurs after a time equal to one-half the period of oscillation. The oscillations continue as indicated in Figure 24.7d.

The electric field near the antenna oscillates in phase with the charge distribution. That is, the field points down when the upper rod is positive and up when the upper rod is negative. Furthermore, the magnitude of the field at any instant depends on the amount of charge on the rods at that instant. As the charges continue to oscillate (and accelerate) between the rods, the electric field set up by the charges moves away from the antenna at the speed of light. One cycle of charge oscillation produces one full wavelength in the electric field pattern.

Because the oscillating charges create a current in the rods, a magnetic field is also generated when the current in the rods is upward, as shown in Figure 24.8. The magnetic field lines circle the antenna and are *perpendicular to the electric field at all points.* As the current changes with time, the magnetic field lines spread out from the antenna. At great distances from the antenna, the electric and magnetic fields become very weak. However, at these great distances, it is necessary to take into account the facts that (1) a changing magnetic field produces a changing electric field and (2) a changing electric field produces a changing magnetic field, as predicted by Maxwell. These induced electric and magnetic fields are in phase:

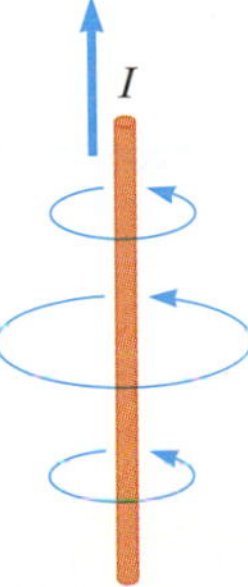

Figure 24.8
Magnetic field lines around an antenna carrying a changing current. Why do the circles have different radii?

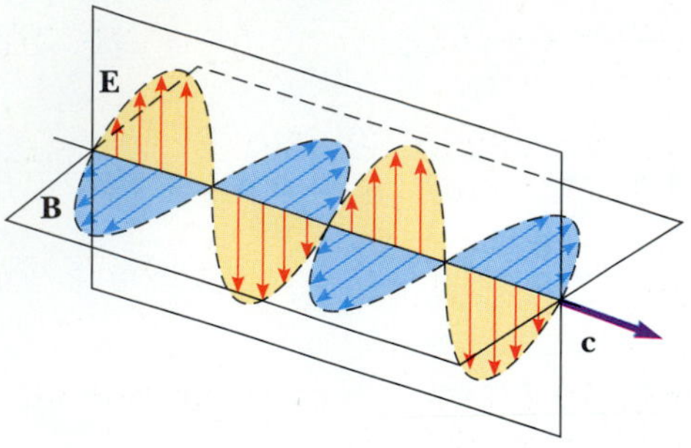

Figure 24.9
An electromagnetic wave sent out by oscillating charges in an antenna. This drawing represents the wave at one instant of time.

at any point, the two fields reach their maximum values at the same instant. This is illustrated for one instant of time in Figure 24.9.

24.4 Energy Carried by Electromagnetic Waves

Electromagnetic waves carry energy, and as they propagate through space they can transfer energy to objects placed in their path. The rate of flow of energy in an electromagnetic wave is described by a vector, **S**, called the **Poynting vector.** It is defined by the expression

Poynting vector

$$\mathbf{S} \equiv \frac{1}{\mu_0} \mathbf{E} \times \mathbf{B} \qquad [24.18]$$

The magnitude of the Poynting vector represents the rate at which energy flows through a unit surface area perpendicular to the flow. The direction of **S** is the direction of wave propagation (Fig. 24.10). The SI units of the Poynting vector are $J/s \cdot m^2 = W/m^2$ (The Poynting vector must have these units since it represents power per unit area, where the unit area is oriented at right angles to the direction of wave propagation.)

As an example, let us evaluate the magnitude of **S** for a plane electromagnetic wave, where $|\mathbf{E} \times \mathbf{B}| = EB$. In this case,

Poynting vector for a plane wave

$$S = \frac{EB}{\mu_0} \qquad [24.19]$$

Since $B = E/c$, we can also express this as

$$S = \frac{E^2}{\mu_0 c} = \frac{c}{\mu_0} B^2 \qquad [24.20]$$

These equations for S apply at any instant of time.

What is of more interest for a sinusoidal plane electromagnetic wave is the time average of S taken over one or more cycles. When such an average is taken, an expression is obtained that involves the time average of $\cos^2(kx - \omega t)$, which equals $\frac{1}{2}$. Hence, the average value of S, which equals the average power per unit area (wave intensity), is

$$S_{av} = \frac{E_{max} B_{max}}{2\mu_0} = \frac{E_{max}^2}{2\mu_0 c} = \frac{c}{2\mu_0} B_{max}^2 \quad [24.21]$$

Average power per unit area

Note that E_{max} and B_{max} represent *maximum* values of the fields. The constant $\mu_0 c$, called the **impedance of free space,** has the SI units ohms and has the value

Impedance of free space

$$\mu_0 c = \sqrt{\frac{\mu_0}{\epsilon_0}} = 120\pi\ \Omega \approx 377\ \Omega$$

Recall that the energy per unit volume, u_e, the instantaneous energy density associated with an electric field (Chapter 17, Section 17.9), is

$$u_e = \tfrac{1}{2}\epsilon_0 E^2 \quad [24.22]$$

and the instantaneous energy density, u_m, associated with a magnetic field (Chapter 20, Section 20.8) is

$$u_m = \frac{B^2}{2\mu_0} \quad [24.23]$$

Because E and B vary with time for an electromagnetic wave, the energy densities also vary with time. With use of the relationships $B = E/c$ and $c = 1/\sqrt{\epsilon_0 \mu_0}$, Equation 24.23 becomes

$$u_m = \frac{(E/c)^2}{2\mu_0} = \frac{\epsilon_0 \mu_0}{2\mu_0} E^2 = \tfrac{1}{2}\epsilon_0 E^2$$

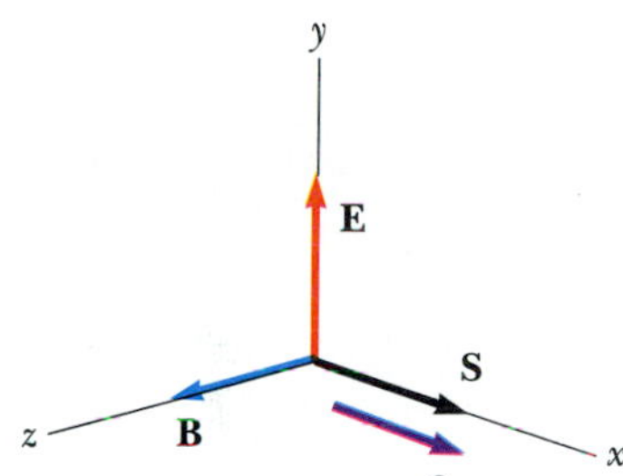

Figure 24.10 The Poynting vector **S** for a plane electromagnetic wave is along the direction of propagation.

Comparing this result with Equation 24.22, we see that

$$u_m = u_e = \tfrac{1}{2}\epsilon_0 E^2 = \frac{B^2}{2\mu_0} \quad [24.24]$$

That is, *for an electromagnetic wave the instantaneous energy density associated with the magnetic field equals the instantaneous energy density associated with the electric field.* Hence, in a given volume the energy is equally shared by the two fields.

The **total instantaneous energy density,** u, is equal to the sum of the energy densities associated with the electric and magnetic fields:

Total energy density

$$u = u_e + u_m = \epsilon_0 E^2 = \frac{B^2}{\mu_0} \quad [24.25]$$

When this is averaged over one or more cycles of an electromagnetic wave, we again get a factor of $\frac{1}{2}$. Hence, the total *average* energy per unit volume of an electromagnetic wave is

Average energy density of an electromagnetic wave

$$u_{av} = \epsilon_0 (E^2)_{av} = \tfrac{1}{2}\epsilon_0 E_{max}^2 = \frac{B_{max}^2}{2\mu_0} \quad [24.26]$$

Comparing this result with Equation 24.21 for the average value of S, we see that

$$S_{av} = c u_{av} \quad [24.27]$$

In other words, *the average power per unit area, or intensity, of an electromagnetic wave equals the average energy density multiplied by the speed of light.*

Example 25.2 Fields Due to a Point Source

A point source of electromagnetic radiation has an average power output of 800 W. Calculate the *maximum* values of the electric and magnetic fields at a point 3.50 m from the source.

Solution If we assume that the point source emits spherical waves, the average power per unit area at a distance of r from the source is

$$S_{av} = \frac{P_{av}}{4\pi r^2}$$

where P_{av} is the average power output of the source and $4\pi r^2$ is the area of a sphere of radius r centered on the source. Since the average power per unit area of an electromagnetic wave is also given by Equation 24.21, we have

$$\frac{P_{av}}{4\pi r^2} = \frac{E_{max}^2}{2\mu_0 c}$$

Solving for the maximum electric field, E_{max}, gives

$$\begin{aligned} E_{max} &= \sqrt{\frac{\mu_0 c P_{av}}{2\pi r^2}} \\ &= \sqrt{\frac{(4\pi \times 10^{-7}\ \text{N/A}^2)(3.00 \times 10^8\ \text{m/s})(800\ \text{W})}{2\pi(3.50\ \text{m})^2}} \\ &= 62.6\ \text{V/m} \end{aligned}$$

We can easily calculate the maximum value of the magnetic field using the preceding result and the relation $B_{max} = E_{max}/c$ (Eq. 24.16):

$$B_{max} = \frac{E_{max}}{c} = \frac{62.6\ \text{V/m}}{3.00 \times 10^8\ \text{m/s}} = 2.09 \times 10^{-7}\ \text{T}$$

Exercise Calculate the value of the energy density at the point 3.50 m from the point source.

Answer $1.73 \times 10^{-8}\ \text{J/m}^3$

24.5 Momentum and Radiation Pressure

Electromagnetic waves transport linear momentum as well as energy. Hence, it follows that pressure (radiation pressure) is exerted on a surface when an electromagnetic wave impinges on it. In what follows, we shall assume that the electromagnetic wave transports a total energy of U to a surface in the time t. If the surface *absorbs all* the incident energy, U, in this time, Maxwell showed that the total momentum, p, delivered to this surface has the magnitude

Momentum delivered to an absorbing surface

$$p = \frac{U}{c} \quad \text{(complete absorption)} \qquad [24.28]$$

Furthermore, if the Poynting vector of the wave is **S**, the *radiation pressure, P* (force per unit area), exerted on the perfect absorbing surface is

$$P = \frac{S}{c} \qquad [24.29]$$

Radiation pressure exerted on a perfect absorbing surface

We can apply these results to a perfect black body, defined as one for which *all* of the incident energy is absorbed (none is reflected).

If the surface is a perfect reflector (for example, a mirror with a 100% reflecting surface), then the momentum delivered in a time of t for normal incidence is *twice* that given by Equation 24.28, or $2U/c$. That is, a momentum equal to U/c is delivered by the incident wave and U/c is delivered by the reflected wave, analogous to a ball colliding elastically with a wall. Therefore,

$$p = \frac{2U}{c} \quad \text{(complete reflection)} \qquad [24.30]$$

Momentum delivered to a perfectly reflecting surface

The momentum delivered to an arbitrary surface has a value between U/c and $2U/c$, depending on the properties of the surface. Finally, the radiation pressure exerted on a perfect reflecting surface for normal incidence of the wave is[6]

$$P = \frac{2S}{c} \qquad [24.31]$$

Pressure exerted on a perfectly reflecting surface

Although radiation pressures are very small (about 5×10^{-6} N/m^2 for direct sunlight), they have been measured using torsion balances such as the one shown in Figure 24.11. Light is allowed to strike either a mirror or a black disk, both of which are suspended from a fine fiber. Light striking the black disk is completely absorbed, and so all of its momentum is transferred to the disk. Light striking the mirror (normal incidence) is totally reflected, hence the momentum transfer is twice as great as that transferred to the disk. The radiation pressure is determined by measuring the angle through which the horizontal portion rotates. The apparatus must be placed in a high vacuum to eliminate the effects of air currents.

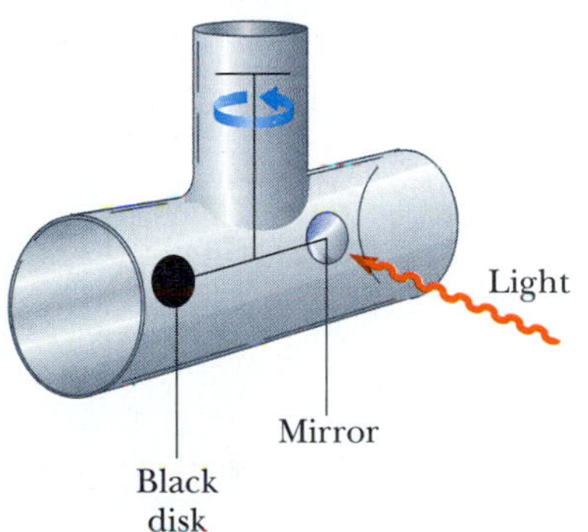

Figure 24.11 An apparatus for measuring the radiation pressure of light. In practice, the system is contained in a high vacuum.

Example 24.3 Solar Energy

The Sun delivers about 1000 W/m^2 of electromagnetic flux to the Earth's surface.

(a) Calculate the total power incident on a roof of dimensions 8 m $\times$ 20 m.

Solution The Poynting vector has a magnitude of $S = 1000$ W/m^2, which represents the average power per unit area, or the light intensity. Assuming the radiation is incident *normal* to the roof (Sun directly overhead), we get

$$\text{Power} = SA = (1000 \text{ W/m}^2)(8 \times 20 \text{ m}^2)$$
$$= 1.60 \times 10^5 \text{ W}$$

If this power could *all* be converted into electrical energy, it would provide more than enough power for the average home. However, solar energy is not easily harnessed, and the prospects for large-scale conversion are not as bright as they may appear from this simple calculation. For example, the conversion efficiency from solar to electrical energy is typically 10% for photovoltaic cells. Roof systems for converting solar energy to *thermal* energy have been built with efficiencies of around 50%; however, with solar energy other practical problems must be considered, such as overcast days, geographic location, and energy storage.

[6] For *oblique* incidence, the momentum transferred is $2U\cos\theta/c$ and the pressure is given by $P = 2S\cos\theta/c$, where θ is the angle between the normal to the surface and the direction of propagation.

(b) Determine the radiation pressure and radiation force on the roof, assuming the roof covering is a perfect absorber.

Solution Using Equation 24.29 with $S = 1000\ \text{W/m}^2$, we find that the radiation pressure is

$$P = \frac{S}{c} = \frac{1000\ \text{W/m}^2}{3 \times 10^8\ \text{m/s}} = \boxed{3.33 \times 10^{-6}\ \text{N/m}^2}$$

Because pressure equals force per unit area, this corresponds to a radiation force of

$$F = PA = (3.33 \times 10^{-6}\ \text{N/m}^2)(160\ \text{m}^2)$$

$$= \boxed{5.33 \times 10^{-4}\ \text{N}}$$

Of course, this "load" is *far* less than the other loads one must contend with on roofs, such as the roof's own weight or a layer of snow.

Exercise How much solar energy (in joules) is incident on the roof in 1 h?

Answer 5.76×10^8 J.

Example 24.4 Poynting Vector for a Wire

A long, straight wire of resistance R, radius a, and length ℓ carries a constant current, I, as in Figure 24.12. Calculate the Poynting vector for this wire.

Solution First, let us find the electric field, **E**, along the wire. If V is the potential difference across the ends of the wire, then $V = IR$ and

$$E = \frac{V}{\ell} = \frac{IR}{\ell}$$

Recall that the magnetic field at the surface of the wire (Example 19.4) is

$$B = \frac{\mu_0 I}{2\pi a}$$

The vectors **E** and **B** are mutually perpendicular, as shown in Figure 24.12, and therefore $|\mathbf{E} \times \mathbf{B}| = EB$. Hence, the Poynting vector, **S**, is directed radially inward and has the magnitude

$$S = \frac{EB}{\mu_0} = \frac{1}{\mu_0}\frac{IR}{\ell}\frac{\mu_0 I}{2\pi a} = \frac{I^2R}{2\pi a\ell} = \frac{I^2R}{A}$$

where $A = 2\pi a\ell$ is the *surface* area of the wire and the total area through which S passes. From this result, we see that

$$SA = I^2R$$

where SA has units of power (J/s = W). That is, *the rate at which electromagnetic energy flows into the wire, SA, equals the rate of energy (or power) dissipated as joule heat,* I^2R.

Exercise A heater wire of radius 0.3 mm and resistance 5 Ω carries a current of 2 A. Determine the magnitude and direction of the Poynting vector for this wire.

Answer $1.06 \times 10^4\ \text{W/m}^2$, directed radially inward.

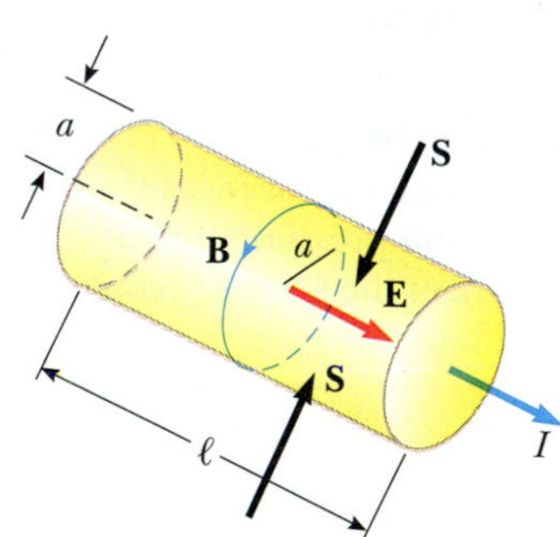

Figure 24.12 (Example 24.4) A wire of length l, resistance R, and radius a carrying a current I. The Poynting vector **S** is directed radially *inward*.

24.6 The Spectrum of Electromagnetic Waves

We have seen that all electromagnetic (EM) waves travel in a vacuum with the speed of light, c. These waves transport energy and momentum from some source to a receiver. In 1888, Hertz successfully generated and detected the radio-frequency electromagnetic waves predicted by Maxwell. Maxwell himself had recognized as EM waves both visible light and the near infrared radiation discovered in 1800 by William Herschel. It is now known that other forms of electromagnetic waves exist; they are distinguished by their frequencies and wavelengths.

Since all electromagnetic waves travel through vacuum with the speed c, their frequency, f, and wavelength, λ, are related by the important expression

$$c = f\lambda \qquad [24.32]$$

The various types of electromagnetic waves, all produced by accelerating charges, are listed in Figure 24.13. Note the wide range of frequencies and wavelengths. For instance, a radio wave of frequency 5 MHz (a typical value) has the wavelength

$$\lambda = \frac{c}{f} = \frac{3 \times 10^{8} \text{ m/s}}{5 \times 10^{6} \text{ s}^{-1}} = 60 \text{ m}$$

Let us briefly describe the wave types shown in Figure 24.13.

Radio waves

Radio waves, which were discussed in the preceding section, are the result of charges accelerating through conducting wires. They are generated by such electronic devices as LC oscillators and are used in radio and television communication systems.

Microwaves

Microwaves (short-wavelength radio waves) have wavelengths ranging between about 1 mm and 30 cm and are also generated by electronic devices. Because of their short wavelengths, they are well suited for radar systems used in aircraft navigation and for studying the atomic and molecular properties of matter. Microwave ovens are an interesting domestic application of these waves.

Infrared waves

Infrared waves (sometimes called *heat waves*) have wavelengths ranging from about 1 mm to the longest wavelength of visible light, 7×10^{-7} m. These waves, produced by hot bodies and molecules, are readily absorbed by most materials. The infrared energy absorbed by a substance appears as thermal energy because the energy agitates the atoms of the body, increasing their vibrational or translational motion, which results in a temperature rise. Infrared radiation has many practical and scientific applications, including physical therapy, infrared photography, and vibrational spectroscopy.

Visible light

Visible light, the most familiar form of electromagnetic waves, is that part of the spectrum the human eye can detect. Light is produced by the rearrangement of electrons in atoms and molecules. The wavelengths of visible light are classified by color, ranging from violet ($\lambda \approx 4 \times 10^{-7}$ m) to red ($\lambda \approx 7 \times 10^{-7}$ m). The eye's sensitivity is a function of wavelength and is a maximum at a wavelength of about 5.6×10^{-7} m (yellow-green). Light is the basis of the science of optics and optical instruments, to be discussed later.

Ultraviolet waves

Ultraviolet light covers wavelengths ranging from about 3.8×10^{-7} m (380 nm) down to 6×10^{-10} m (0.6 nm). The Sun is an important source of ultraviolet waves, which is the main cause of suntans. Most of the ultraviolet (uv) waves from

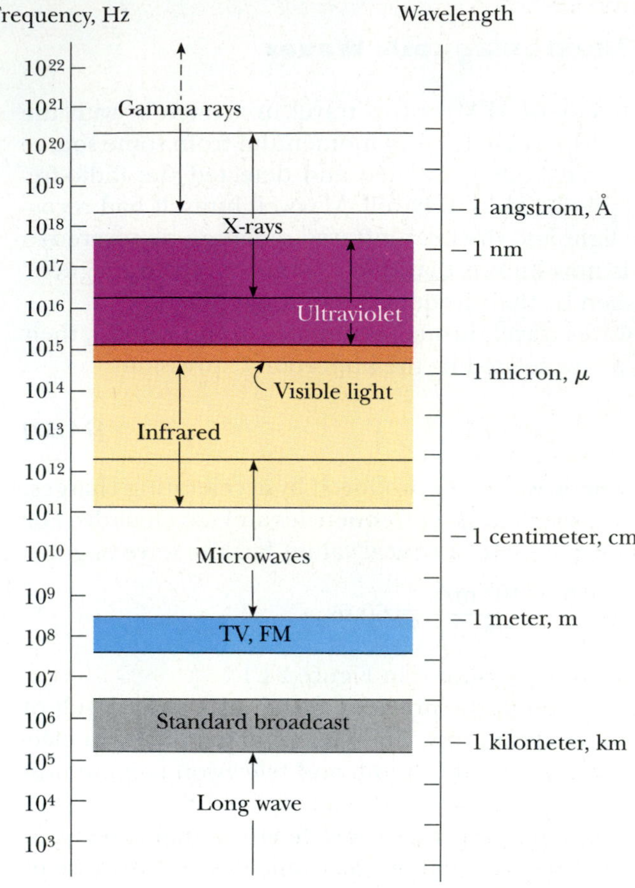

Figure 24.13 The electromagnetic spectrum. Note the overlap between one type of wave and the next. There is no sharp division between the two types.

the Sun is absorbed by atoms in the stratosphere. (This is fortunate, since uv waves in large quantities have harmful effects on humans.) One important constituent of the stratosphere is ozone (O_3), which results from reactions of oxygen with ultraviolet radiation. This ozone shield converts lethal high-energy ultraviolet radiation to thermal energy, which raises the temperature of the stratosphere. Recently, a great deal of controversy has arisen concerning the depletion of the protective ozone layer by the use of Freon in aerosol spray cans and as refrigerants.

X-rays

X-rays are electromagnetic waves with wavelengths in the range of about 10^{-8} m (10 nm) down to 10^{-13} m (10^{-4} nm). The most common source of x-rays is the acceleration of high-energy electrons bombarding a metal target. X-rays are used as a diagnostic tool in medicine and as a treatment for certain forms of cancer. Since x-rays damage or destroy living tissues and organisms, care must be taken to avoid unnecessary exposure and overexposure. X-rays are also used in the study of crystal structure; x-ray wavelengths are comparable to the atomic separation distances ($\approx$ 0.1 nm) in solids.

Gamma rays

Gamma rays are electromagnetic waves emitted by radioactive nuclei (such as ^{60}Co and ^{137}Cs) and during certain nuclear reactions. They have wavelengths ranging from about 10^{-10} m to less than 10^{-14} m. Gamma rays are highly penetrating and produce serious damage when absorbed by living tissues. Consequently, those working near such dangerous radiation must be protected with heavily absorbing materials, such as layers of lead.

Summary

Electromagnetic waves, which are predicted by Maxwell's equations, have the following properties:

1. The electric and magnetic fields satisfy the following wave equations, which can be obtained from Maxwell's third and fourth equations:

$$\frac{\partial^2 E}{\partial x^2} = \mu_0 \epsilon_0 \frac{\partial^2 E}{\partial t^2} \quad [24.10]$$

$$\frac{\partial^2 B}{\partial x^2} = \mu_0 \epsilon_0 \frac{\partial^2 B}{\partial t^2} \quad [24.11]$$

Wave equations

2. Electromagnetic waves travel through a vacuum with the speed of light, c, where

$$c = \frac{1}{\sqrt{\mu_0 \epsilon_0}} = 3.00 \times 10^8 \text{ m/s} \quad [24.12]$$

The speed of electromagnetic waves

3. The electric and magnetic fields of an electromagnetic wave are perpendicular to each other and perpendicular to the direction of wave propagation. (Hence electromagnetic waves are transverse waves.)
4. The instantaneous magnitudes of $|\mathbf{E}|$ and $|\mathbf{B}|$ in an electromagnetic wave are related by the expression

$$\frac{E}{B} = c \quad [24.16]$$

5. Electromagnetic waves carry energy. The rate of flow of energy crossing a unit area is described by the Poynting vector **S**, where

$$\mathbf{S} \equiv \frac{1}{\mu_0} \mathbf{E} \times \mathbf{B} \quad [24.18]$$

Poynting vector

6. Electromagnetic waves carry momentum and hence can exert pressure on surfaces. If an electromagnetic wave whose Poynting vector is **S** is completely absorbed by a surface upon which it is normally incident, the radiation pressure on that surface is

$$P = \frac{S}{c} \quad \text{(complete absorption)} \quad [24.29]$$

If the surface totally reflects a normally incident wave, the pressure is doubled.

The electric and magnetic fields of a sinusoidal plane electromagnetic wave propagating in the positive x direction can be written

$$E = E_{\max} \cos(kx - \omega t) \quad [24.14]$$

$$B = B_{\max} \cos(kx - \omega t) \quad [24.15]$$

Sinusoidal electric and magnetic fields

where ω is the angular frequency of the wave and k is the angular wave number. These equations represent special solutions to the wave equations for **E** and **B**. Since $\omega = 2\pi f$ and $k = 2\pi/\lambda$, where f and λ are the frequency and wavelength, respectively, one finds that

$$\omega/k = f\lambda = c$$

The average value of the Poynting vector for a plane electromagnetic wave has the magnitude

Average power per unit area

$$S_{av} = \frac{E_{max}B_{max}}{2\mu_0} = \frac{E_{max}^2}{2\mu_0 c} = \frac{c}{2\mu_0}B_{max}^2 \quad [24.21]$$

The average power per unit area (intensity) of a sinusoidal plane electromagnetic wave equals the average value of the Poynting vector taken over one or more cycles.

The fundamental sources of electromagnetic waves are *accelerating electric charges.*

The electromagnetic spectrum includes waves covering a broad range of frequencies and wavelengths. The frequency, f, and wavelength, λ, of a given wave are related by

$$c = f\lambda \quad [24.32]$$

Questions and Conceptual Exercises

1. What is the fundamental source of electromagnetic radiation?
2. Electrical engineers often speak of the *radiation resistance* of an antenna. What do you suppose they mean by this phrase?
3. In your own words, describe the physical significance of the Poynting vector.
4. If a high-frequency current is passed through a solenoid containing a metallic core, the core heats up by induction. Explain why the materials heat up in these situations.
5. Certain orientations of the receiving antenna on a TV give better reception than others. Furthermore, the best orientation varies from station to station. Explain these observations.
6. If you charge a comb by running it through your hair and then hold the comb next to a bar magnet, do the electric and magnetic fields produced constitute an electromagnetic wave?
7. An empty plastic or glass dish removed from a microwave oven is cool to the touch. How is this possible?
8. What does a radio wave do to the charges in the receiving antenna to provide a signal for your car radio?
9. When light (or other electromagnetic radiation) travels across a given region, what is it that moves?
10. Suppose a creature from another planet had eyes that were sensitive to infrared radiation. Describe what he would see if he looked around the room you are now in. That is, what would be bright and what would be dim?
11. Why should an infrared photograph of a person look different from a photograph taken with visible light?
12. A welder must wear protective glasses and clothing to prevent eye damage and sunburn. What does this imply about the light produced by the welding?
13. Radio stations often advertise "instant news." If what they mean is that you hear the news at the instant they speak it, is their claim true? About how long would it take for a message to travel across this country by radio waves, assuming that these waves could travel this great distance and still be detected?
14. Light from the Sun takes approximately $8\frac{1}{3}$ minutes to reach the Earth. During this time the Earth has continued to move in its orbit around the Sun. How far is the actual location of the Sun from its image in the sky?

Problems

Section 24.1 Maxwell's Equations and Electromagnetic Waves

1. An electromagnetic wave in vacuum has an electric field amplitude of 220 V/m. Calculate the amplitude of the corresponding magnetic field.
2. The speed of an electromagnetic wave traveling in a nonmagnetic transparent substance is given by $v = 1/\sqrt{\kappa\mu_0\epsilon_0}$, where κ is the dielectric constant of the substance. Determine the speed of light in water, which has a dielectric constant of 1.78 at optical frequencies.
3. The magnetic field amplitude of an electromagnetic wave is 5.4×10^{-7} T. Calculate the electric field amplitude if the wave is traveling (a) in free space; (b) in a nonmagnetic medium in which the speed of the wave is $0.8c$.
4. Verify that the following pair of equations for E and B are solutions of Equations 25.7 and 25.8.

$$E = \frac{A}{\sqrt{\epsilon_0\mu_0}} e^{a(x - ct)} \quad \text{and} \quad B = Ae^{a(x - ct)}$$

5. Write down expressions for the electric and magnetic fields of a sinusoidal plane electromagnetic wave with a frequency of 3 GHz, traveling in the positive x direction in a vacuum. The amplitude of the electric field is 300 V/m.
6. Figure 24.3 shows a plane electromagnetic sinusoidal wave propagating in the x direction. The wavelength is 50 m, and the electric field vibrates in the xy plane with an amplitude of 22 V/m. Calculate (a) the sinusoidal frequency and (b) the magnitude and direction of **B** when the electric field has its maximum value in the negative y direction. (c) Write an expression for B in the form $B = B_{max} \cos(kx - \omega t)$, with numerical values for B_{max}, k, and ω.
7. The electric field in an electromagnetic wave, in SI units, is described by $E_y = 100 \sin(10^7 x - \omega t)$. Find (a) the amplitude of the corresponding magnetic wave, (b) the wavelength, λ, and (c) the frequency, f.
8. Verify that the following equations are solutions to Equations 24.10 and 24.11, respectively:

$$E = E_{max} \cos(kx - \omega t)$$

$$B = B_{max} \cos(kx - \omega t)$$

Section 24.2 Hertz's Discoveries

9. If the coil in the resonant circuit of a radio has an inductance of 2 μH, what range of values must the tuning capacitor have in order to cover the complete range of FM frequencies? (The FM range of frequencies is 88 MHz to 108 MHz.)
10. A standing wave interference pattern is set up by radio waves between two metal sheets 2 m apart. This is the shortest distance between the plates that will produce a standing wave pattern. What is the fundamental frequency?

Section 24.3 Production of Electromagnetic Waves by an Antenna

11. What is the length of a half-wave antenna designed to broadcast 20-MHz radio waves?
12. An AM radio station broadcasts isotropically with an average power of 4 kW. A dipole receiving antenna, 65 cm long, is located 4 miles from the transmitter. Compute the emf induced by this signal between the ends of the receiving antenna.
13. A television set uses a dipole receiving antenna for VHF channels and a loop antenna for UHF channels. The UHF antenna produces a voltage from the changing *magnetic* flux through the loop. (a) Using Faraday's law, derive an expression for the amplitude of the voltage that appears in a single-turn circular loop antenna with a radius of r. The TV station broadcasts a signal with a frequency of f, and the signal has an electric field amplitude of E_{max} and a magnetic field amplitude of B_{max} at the receiving antenna's location. (b) If the electric field in the signal points vertically, what should be the orientation of the loop for best reception?
14. Two hand-held radio transceivers with dipole antennas are separated by a large fixed distance. Assuming a vertical transmitting antenna, what fraction of the maximum received power will occur in the receiving antenna when it is inclined from the vertical by (a) 15°? (b) 45°? (c) 90°?
15. Two radio-transmitting antennas are separated by half the broadcast wavelength and are driven in phase with each other. (a) In which direction is the strongest signal radiated? (b) In which direction is the weakest signal radiated?

Section 24.4 Energy Carried by Electromagnetic Waves

16. How much electromagnetic energy per cubic meter is contained near the Earth's surface if the intensity of sunlight under clear skies is 1000 W/m^2?

17. The light from a 5-mW laser is spread out in a cylindrical beam whose diameter is about 0.5 cm. What are the peak values of **E** and **B** in this beam?
18. At a distance of 10 km from a radio transmitter, the amplitude of the **E** field is 0.20 V/m. What is the total power emitted by the radio transmitter, assuming it emits spherical waves?
19. What is the average magnitude of the Poynting vector at a distance of 5 miles from a radio transmitter emitting spherical waves, broadcasting with an average power of 250 kW?
20. The Sun radiates energy at a rate of 3.79×10^{26} W. Its radius is 7.0×10^{8} m. If the distance from the Earth to the Sun is 1.5×10^{11} m, what is the intensity of solar radiation at the top of the Earth's atmosphere?
21. A community plans to build a facility to convert solar radiation to electrical power. They require 1 MW of power, and the system to be installed has an efficiency of 30% (that is, 30% of the solar energy incident on the surface is converted to electrical energy). What must be the effective area of a perfectly absorbing surface used in such an installation, assuming a constant power per unit area of 1000 W/m² in the incident solar radiation?
22. The Earth's magnetic field near its surface is about 0.5×10^{-4} T. How much energy is stored in 1 m³ of the atmosphere because of this field?
23. The filament of an incandescent lamp has a 150-Ω resistance, and carries a dc current of 1 A. The filament is 8 cm long and 0.9 mm in radius. (a) Calculate the Poynting vector at the surface of the filament. (b) Find the magnitude of the electric and magnetic fields at the surface of the filament.
24. A monochromatic light source emits 100 W of electromagnetic power uniformly in all directions. (a) Calculate the average electric-field energy density 1 m from the source. (b) Calculate the average magnetic-field energy density at the same distance from the source. (c) Find the wave intensity at this location.
25. A helium-neon laser intended for instructional use operates at a power of 5.0 mW. (a) Determine the maximum value of the electric field at a point where the cross-section of the beam is 4 mm². (b) Calculate the electromagnetic energy in a 1-m length of the beam.
26. At one location on the Earth, the rms value of the magnetic field due to solar radiation is 1.8 μT. From this value calculate (a) the average electric field due to solar radiation, (b) the average energy density of the solar component of electromagnetic radiation at this location, and (c) the magnitude of the Poynting vector for the Sun's radiation.

Section 24.5 Momentum and Radiation Pressure

27. A radio wave transmits 25 W/m² of power per unit area. A plane surface of area A is perpendicular to the direction of propagation of the wave. Calculate the radiation pressure on the surface if the surface is a perfect absorber.
28. A 100-mW laser beam is reflected back upon itself by a mirror. Calculate the force on the mirror.
29. A 15-mW helium-neon laser (λ = 632.8 nm) emits a beam of circular cross-section whose diameter is 2 mm. Assume the beam has a uniform intensity over its cross section. (a) Find the maximum electric field in the beam. (b) What total energy is contained in a 1-m length of the beam? (c) Find the momentum carried by a 1-m length of the beam.
30. A plane electromagnetic wave of intensity 6 W/m² strikes a small pocket mirror, of area 40 cm², held perpendicular to the approaching wave. (a) What momentum does the wave transfer to the mirror each second? (b) Find the force that the wave exerts on the mirror.
31. A possible means of space flight is to place a perfectly reflecting aluminized sheet into Earth orbit and use the light from the Sun to push this solar sail. If such a sail of area 6×10^{5} m² and mass 6000 kg were placed into orbit and turned toward the Sun, what would be the force exerted on it? (Assume a solar intensity of 1340 W/m².)
32. The intensity of the sunlight incident on the Earth is 1340 W/m². (a) Assuming that the Earth absorbs all the sunlight incident upon it, find the total force that the Sun exerts on the Earth due to radiation pressure. (b) Compare this value with the Sun's gravitational attraction.

Section 24.6 The Spectrum of Electromagnetic Waves

33. What is the frequency of an electromagnetic wave that has a wavelength of (a) 2 m? (b) 20 m? (c) 200 m?
34. What are the wavelength ranges in (a) the AM radio band (540–1600 kHz) and (b) the FM radio band (88–108 MHz)?
35. The eye is most sensitive to light having a wavelength of 5.5×10^{-7} m, which is in the green-yellow region of the electromagnetic spectrum. What is the frequency of this light?
36. A "rabbit ears" television antenna consists of a pair of adjustable rods. If we set each rod to a quarter-wavelength for channel 3, which has a central frequency of 63.0 MHz, how long are the rods?
37. A radio receiver is located 200 km from a transmitter.

A signal travels to the receiver via a direct line-of-sight path and also via a wave from an ionospheric layer at an altitude of 100 km. Make the approximation of a flat Earth and calculate the time difference between the arrivals of the two signals.

38. A singer's voice is transmitted by a radio wave to a person 100 km away. (a) How much time passes before the distant listener hears the sound? (b) By the time the radio message reaches a listener, how far from the singer has the sound wave moved in the auditorium? Assume that the speed of sound is 345 m/s.

39. A radar pulse returns to the receiver after a total travel time of 4×10^{-4} s. How far away is the object that reflected the wave?

Additional Problems

40. Assume that the solar radiation incident on the Earth is 1340 W/m^2. (This is the value of the solar flux above the Earth's atmosphere.) (a) Calculate the total power radiated by the Sun, taking the average Earth-Sun separation to be 1.49×10^{11} m. (b) Determine the maximum values of the electric and magnetic fields at the Earth's surface due to solar radiation.

41. A microwave source produces pulses of 20-GHz radiation, with each pulse lasting 1.0 ns. A parabolic reflector (R = 6 cm) is used to focus these into a parallel beam of radiation, as shown in Figure 24.14. The average power during each pulse is 25 kW. (a) What is the wavelength of these microwaves? (b) What is the total energy contained in each pulse? (c) Compute the average energy density inside each pulse. (d) Determine the amplitude of the electric and magnetic fields in these microwaves. (e) If this pulsed beam strikes an absorbing surface, compute the force exerted on the surface during the 1-ns duration of each pulse.

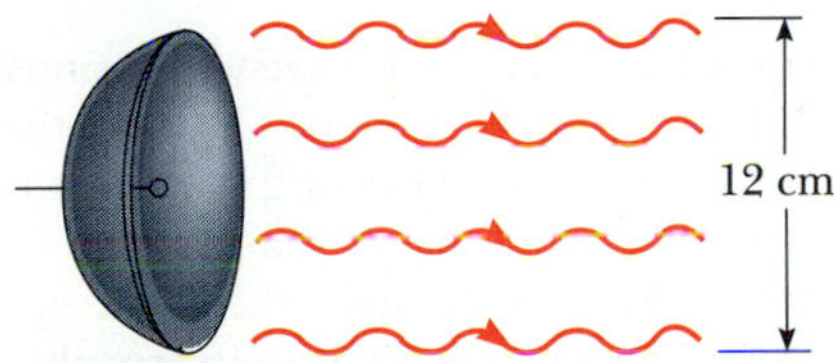

Figure 24.14 (Problem 41).

42. A radio station high on a hill sends out spherical waves with an output of 100 000 W. What are the maximum values of E and B at 10 km from the station?

43. A dish antenna with a diameter of 20 m receives (at normal incidence) a radio signal from a distant source, as shown in Figure 24.15. The radio signal is a continuous sinusoidal wave with amplitude $E_0 = 0.2$ μV/m. Assume the antenna absorbs all the radiation that falls on the dish. (a) What is the amplitude of the magnetic field in this wave? (b) What is the intensity of the radiation received by this antenna? (c) What is the power received by the antenna? (d) What force is exerted on the antenna by the radio waves?

Figure 24.15 (Problem 43).

44. Show that the instantaneous value of the Poynting vector has the magnitude

$$S = \frac{c}{2}\left(\epsilon_0 E^2 + \frac{B^2}{\mu_0}\right)$$

45. In 1965, Penzias and Wilson discovered the cosmic microwave radiation left over from the Big Bang expansion of the Universe. The energy density of this radiation is 4×10^{-14} J/m^3. Determine the corresponding electric field amplitude.

46. The intensity of solar radiation at the top of the Earth's atmosphere is 1340 W/m^3. Assuming that 60% of the arriving solar energy reaches the Earth's surface and assuming that you absorb 50% of the incident energy, *estimate* the amount of solar energy you absorb in a 60-min sunbath.

47. An astronaut in a spacecraft moving with constant velocity wishes to increase the speed of the craft by using a laser beam attached to the spaceship. The laser beam emits 100 J of electromagnetic energy per pulse, and the laser is pulsed at the rate of 0.2 pulse/s. If the mass of the spaceship plus its contents is 5000 kg, for how long a time must the beam be on in order to increase the speed of the vehicle by 1 m/s in the direction of its initial motion? In what direction should the beam be pointed to achieve this?

48. Consider a small, spherical particle of radius r located in space a distance of R from the Sun. (a) Show that the ratio $F_{\text{rad}}/F_{\text{grav}} \propto 1/r$, where F_{rad} = the force

due to solar radiation and F_{grav} = the force of gravitational attraction. (b) The result of (a) means that for a sufficiently small value of r the force exerted on the particle due to solar radiation will exceed the force of gravitational attraction. Calculate the value of r for which the particle will be in equilibrium under the two forces. (Assume that the particle has a perfectly absorbing surface and a mass density of 1.5 g/cm^3. Let the particle be located 3.75×10^{11} m from the Sun, and use 214 W/m^2 as the value of the solar flux at that point.)

49. As the spaceship *Voyager* nears Jupiter, a radio command to turn on the ship's TV cameras is sent to *Voyager* from Earth. If Jupiter is 630×10^6 km away, what is the minimum time required for a TV signal to be received on Earth?

50. A microwave transmitter emits electromagnetic waves of a single wavelength. The maximum electric field 1 km from the transmitter is 6.0 V/m. Assuming the transmitter is a point source, and neglecting waves reflected from the Earth, calculate (a) the maximum magnetic field at this distance and (b) the total power emitted by the transmitter.

51. A linearly polarized microwave of wavelength 1.5 cm is directed along the positive x axis. The electric field vector has a maximum value of 175 V/m and vibrates in the xy plane. (a) Assume that the magnetic field component of the wave can be written in the form $B = B_{max} \sin(kx - \omega t)$ and give values for B_{max}, k, and ω. Also, determine in which plane the magnetic field vector vibrates. (b) Calculate the Poynting vector for this wave. (c) What radiation pressure would this wave exert if directed at normal incidence onto a perfectly reflecting sheet? (d) What acceleration would be imparted to a 500-g sheet (perfectly reflecting and at normal incidence) with dimensions 1 m $\times$ 0.75 m?

52. A police radar unit transmits at a frequency of 10.525 GHz, better known as the "X-band." (a) Show that when this radar wave is reflected from the front of a moving car, the reflected wave is shifted in frequency by the amount $\Delta f \cong 2fv/c$, where v is the speed of the car. (*Hint:* Treat the reflection of a wave as two separate processes. First, the motion of the car produces a "moving-observer" Doppler shift in the frequency of the waves striking the car, and then these Doppler-shifted waves are re-emitted by a moving source. Also, automobile speeds are low enough that you can use the binomial expansion $(1 + x)^n \cong 1 + nx$ in the acoustic Doppler formulas derived in Chapter 23.) (b) The unit is usually calibrated before and after an arrest for speeds of 35 mph and 80 mph. Calculate the frequency shift produced by these two speeds.

53. An astronaut, stranded in space "at rest" 10 m from his spacecraft, has a mass (including equipment) of 110 kg. He has a 100-W light source that forms a directed beam, so he decides to use the beam of light as a photon rocket to propel himself continuously toward the spacecraft. (a) Calculate how long it will take him to reach the spacecraft by this method. (b) Suppose, instead, that he decides to throw the light source away in a direction opposite to the spacecraft. If the mass of the light source is 3 kg and, after being thrown, moves with a speed of 12 m/s *relative to the recoiling astronaut,* how long will the astronaut take to reach the spacecraft?

54. Assuming that the antenna of a 10-kW radio station radiates spherical electromagnetic waves, compute the maximum value of the magnetic field 5 km from the antenna, and compare this value with the magnetic field of the earth at the surface of the Earth.

55. Consider the situation shown in Figure 24.16. An electric field of 300 V/m is confined to a circular area 10 cm in diameter and directed outward from the plane of the figure. If the field is increasing at a rate of 20 V/m·s, what are the direction and magnitude of the magnetic field at the point P, 15 cm from the center of the circle?

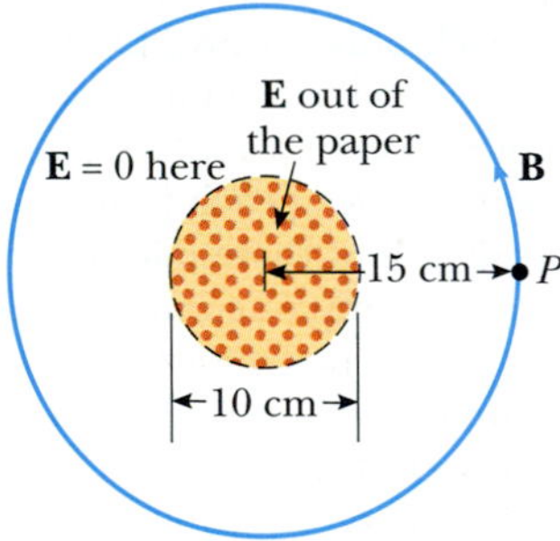

Figure 24.16 (Problem 55).

56. A plane electromagnetic wave varies sinusoidally at 90 MHz as it travels along the $+x$ direction. The electric field has a peak value of 2 mV/m and is directed along the $\pm y$ direction. (a) Find the wavelength, the period, and the peak value, B_0, of the magnetic field. (b) Write expressions, in SI units, for the space and time variations of the electric field and of the magnetic field. Include numerical values as well as subscripts to indicate coordinate directions. (c) Find the average power per unit area that this wave propagates through space. (d) Find the average energy density in the radiation (in joules per cubic meter). (e) What radiation pressure would this wave exert upon a perfectly reflecting surface at normal incidence?

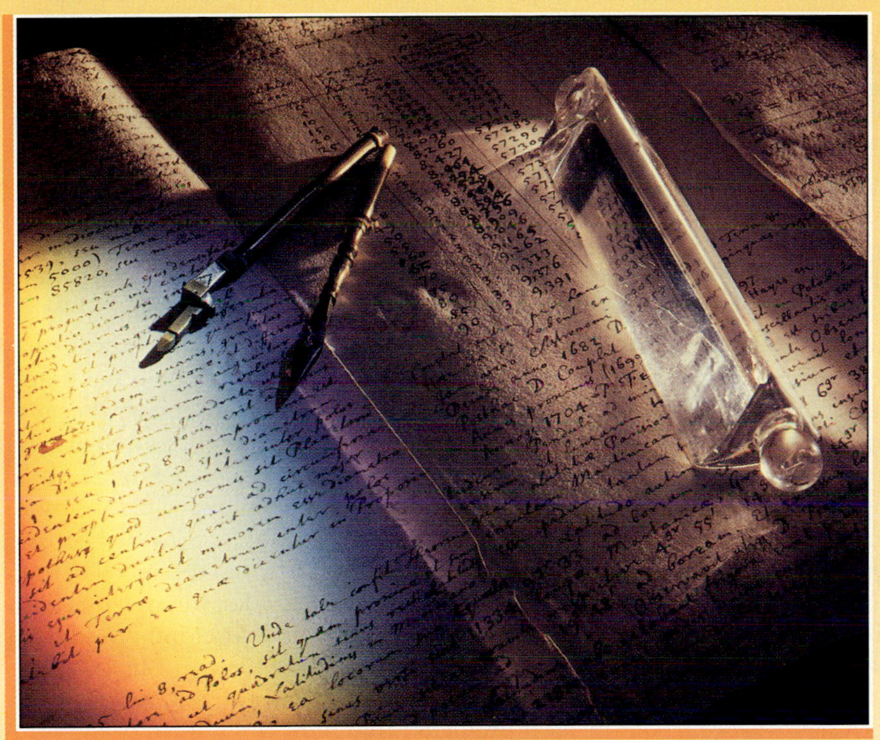

White light directed through a triangular glass prism creates "Pheaenomena of Colours" on an old manuscript. *(© Erich Lessing, Magnum Photos)*

Reflection and Refraction of Light

CHAPTER 25

25.1 The Nature of Light

Until the beginning of the 19th century, light was thought to be a stream of particles that were emitted by a light source and stimulated the sense of sight upon entering the eye. The chief architect of this particle theory of light was Isaac Newton. The theory provided a simple explanation of some known experimental facts concerning the nature of light—namely the laws of reflection and refraction.

Most scientists accepted the particle theory of light. However, during Newton's lifetime, another theory—that light might be some sort of wave motion—was proposed. In 1678 a Dutch physicist and astronomer, Christian Huygens (1629–1695), showed that a wave theory of light could also explain the laws of reflection and refraction.

The wave theory did not receive immediate acceptance, for several reasons. All the waves known at the time (sound, water, and so on) traveled through some sort of medium, but light from the Sun could travel to us through empty space. Furthermore, it was argued that if light were some form of wave motion, the waves would be able to bend around obstacles; hence, we should be able to see around corners. It is now known that light does indeed bend around the edges of objects. This phenomenon, known as *diffraction,* is not easy to observe because light waves have such short wavelengths. Even though experimental evidence for the diffraction of light was discovered by Francesco Grimaldi (1618–1663) around 1660, most scientists for more than a century rejected the wave theory and adhered to Newton's particle theory. This was, for the most part, due to Newton's great reputation as a scientist.

The first clear demonstration of the wave nature of light was provided in 1801 by Thomas Young (1773–1829), who showed that, under appropriate conditions, light exhibits interference behavior. That is, light waves coming from a single source arriving at a point by two different paths can combine and cancel each other by destructive interference, with the energy appearing at some other point. Such behavior could not be explained at that time by a particle theory, because scientists thought that there was no conceivable way in which two or more particles could come together and cancel one another. Several years later, a French physicist, Augustin Fresnel (1788–1829), performed a number of experiments on interference and diffraction. In 1850 Jean Foucault (1791–1868) provided further evidence of the inadequacy of the particle theory by showing that the speed of light in liquids is less than that in air; according to the particle model, the speed of light would be *higher* in glasses and liquids than in air. Additional developments during the 19th century led to the general acceptance of the wave theory of light.

The most important development concerning the theory of light was the work of James Clerk Maxwell, who in 1865 predicted that light was a form of high-frequency electromagnetic wave. His theory predicted that light waves should have a speed of about 3×10^8 m/s, in agreement with measured values. As discussed in Chapter 24, Hertz in 1887 provided experimental confirmation of Maxwell's theory by producing and detecting electromagnetic waves. Furthermore, Hertz and other investigators showed that *these waves exhibited reflection, refraction, and all the other characteristic properties of waves.*

Although the classical theory of electricity and magnetism could explain most known properties of light, some experiments could not be accounted for by the assumption that light was a wave. The most striking of these was the *photoelectric effect,* discovered by Hertz, in which electrons are ejected from a metal whose surface is exposed to light. As one example of the difficulties that arose, experiments showed that the kinetic energy of an ejected electron is *independent* of the light intensity. This finding contradicted the wave theory, which held that a more intense beam of light should add more energy to the electron. An explanation proposed by Einstein in 1905 used the concept of quantization developed in 1900 by Max Planck (1858–1947). Einstein's prediction that the kinetic energy of the ejected electrons would depend on the frequency of the radiation, but not on the intensity, was initially rejected. However, experiments performed between 1905 and 1912 eventually led to the acceptance of Einstein's theory. The quantization model assumes that the energy of a light wave is present in bundles of energy called *photons;* hence, the energy is said to be *quantized.* More specifically, Einstein showed that the energy of a photon is proportional to the frequency of the electromagnetic wave.

$$E = hf \qquad [25.1]$$

Energy of a photon

where $h = 6.63 \times 10^{-34}$ J·s is *Planck's constant*. This theory retains some features of both the wave theory and the particle theory of light. As we shall discuss later, the photoelectric effect is the result of energy transfer from a single photon to an electron in the metal. That is, the electron interacts with one photon of light as if it, the electron, had been struck by a particle. Yet this photon has wave-like characteristics because its energy is determined by the frequency (wave-like quantity).

In view of these developments, light must be regarded as having a *dual nature: in some cases it acts like a wave and in others it acts like a particle.* Classical electromagnetic wave theory provides an adequate explanation of light propagation and interference, whereas the photoelectric effect and other experiments involving the interaction of light with matter are best explained by assuming that light is a particle. Light is light, to be sure. The question "Is light a wave or a particle?" is inappropriate; sometimes it acts like one, sometimes like the other. Fortunately, it never acts like both in the same experiment.

25.2 The Ray Approximation in Geometric Optics

In studying geometric optics here and in Chapter 26, we shall use what is called the *ray approximation.* To understand it, first note that the direction of energy flow of a wave, corresponding to the direction of wave propagation, is called a ray. The rays of a given wave are straight lines that are perpendicular to the wavefronts, as illustrated in Figure 25.1 for a plane wave. In the ray approximation, a ray is a line drawn in the direction in which the light is traveling. For example, a beam of sunlight passing through a darkened room traces out the path of a ray.

If the wave meets a barrier containing a circular opening whose diameter is large relative to the wavelength, as in Figure 25.2a, the wave emerging from the opening continues to move in a straight line (apart from some small edge effects); hence, the ray approximation continues to be valid. If the diameter of the opening is on the order of the wavelength, as in Figure 25.2b, the waves (and, consequently, the rays we draw) spread out from the opening in all directions. We say that the outgoing wave is noticeably *diffracted.* If the opening is small relative to the wavelength, the opening can be approximated as a point source of waves (Fig. 25.2c).

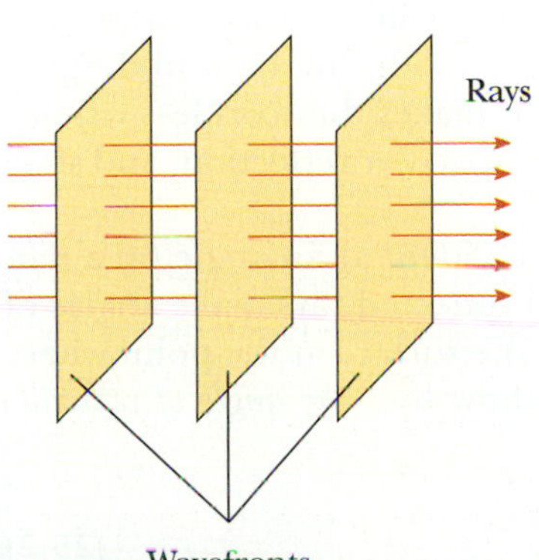

Figure 25.1
A plane wave propagating to the right. Note that the rays, corresponding to the direction of wave motion, are straight lines perpendicular to the wavefronts.

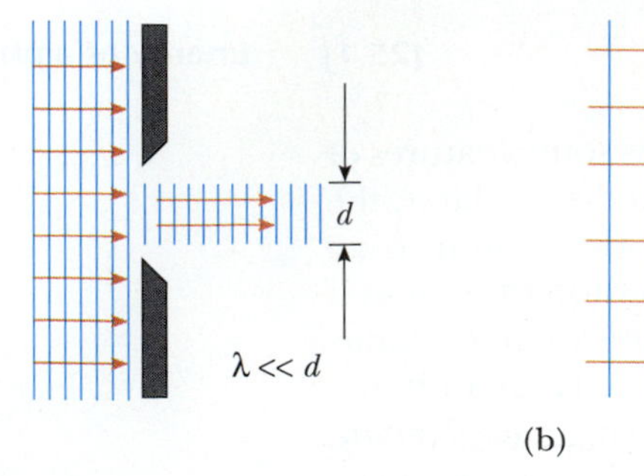

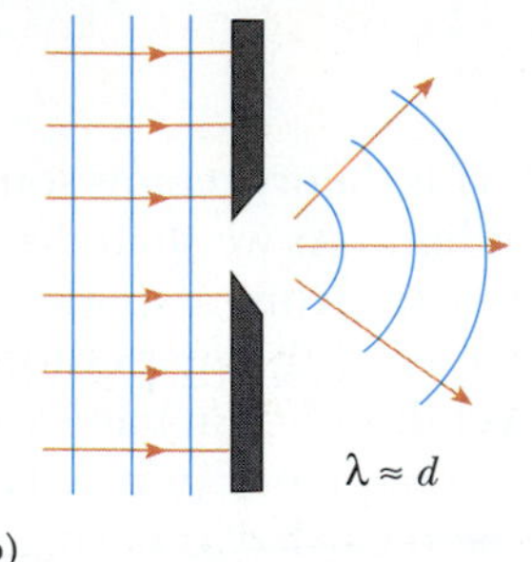

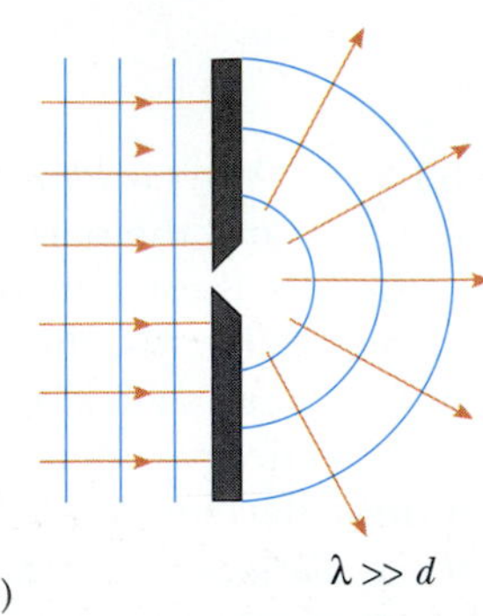

Figure 25.2
A plane wave of wavelength λ is incident on a barrier containing a hole of diameter d. (a) When $\lambda \ll d$, there is almost no observable diffraction and the ray approximation remains valid. (b) When $\lambda \approx d$, diffraction becomes significant. (c) When $\lambda \gg d$, the opening behaves as a point source emitting spherical waves.

Thus, diffraction is more pronounced as the ratio d/λ approaches zero. Similar effects are seen when waves encounter an opaque circular object. In this case, when $\lambda \ll d$, the object casts a sharp shadow.

We will use the ray approximation and the assumption that $\lambda \ll d$ here and in Chapter 26 in discussions of geometric optics. The approximation is very good for the study of mirrors, lenses, prisms, and associated optical instruments, such as telescopes, cameras, and eyeglasses. We shall return to the subject of diffraction (where $\lambda \geq d$) in Chapter 27.

25.3 Reflection and Refraction

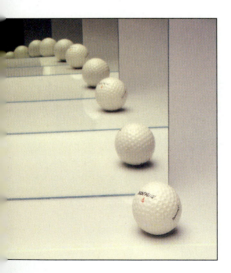

This interesting photograph shows multiple reflections of objects placed between two parallel plane mirrors. *(© Richard Megna 1986, Fundamental Photographs)*

Reflection of Light

Figure 25.3a shows several rays of a beam of light incident on a smooth, reflecting surface. The reflected rays are parallel to each other, as indicated in the figure. Reflection of light from such a smooth surface is called **specular reflection.** If the reflecting surface is rough, as in Figure 25.3b, it will reflect the rays in various directions. Reflection from any rough surface is known as **diffuse reflection.** A surface behaves as a smooth surface as long as the surface variations are small compared with the wavelength of the incident light. Figure 25.3c and d are photographs of specular reflection and diffuse reflection, using laser light.

For instance, consider the two types of reflection from a road surface that you observe when you drive at night. On a dry night, light from oncoming vehicles is scattered off the road in different directions (diffuse reflection) and the road is quite visible. On a rainy night, all the little irregularities in the road surface are filled with water. Because the water surface is quite smooth, the light undergoes specular reflection and the glare from reflected light makes the road less visible.

In this book we shall concern ourselves only with specular reflection, and shall use the term *reflection* to mean specular reflection.

Consider a light ray that is traveling in air and incident at an angle on a flat, smooth surface, as in Figure 25.4. The incident and reflected rays make angles of θ_1 and θ_1', respectively, with a line drawn normal to the surface at the point where the incident ray strikes the surface. Experiments show that *the angle of reflection equals the angle of incidence*; that is,

Law of reflection

$$\theta_1' = \theta_1 \qquad [25.2]$$

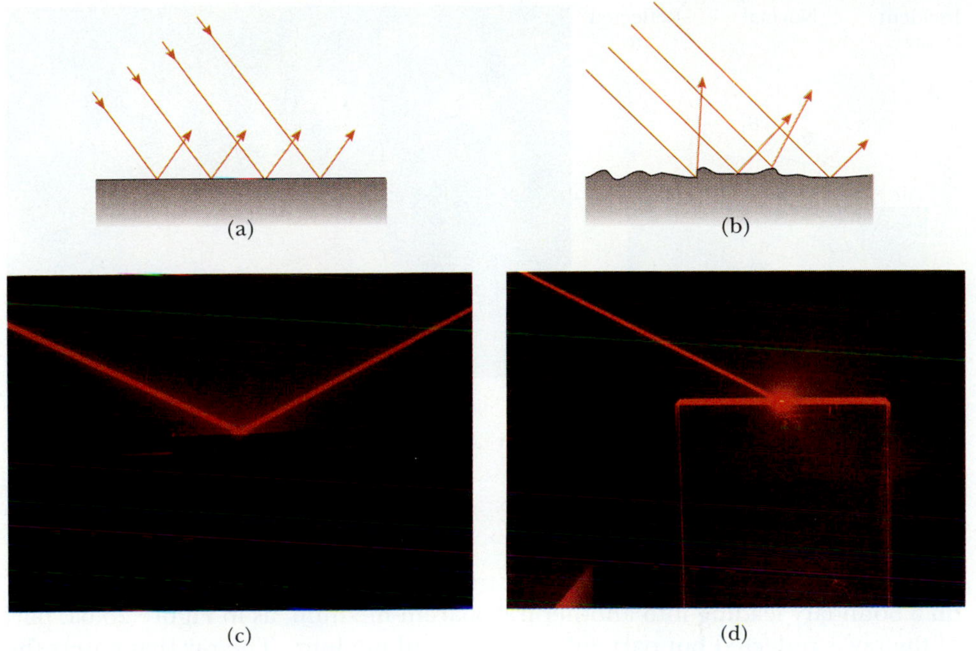

Figure 25.3
A schematic representation of (a) specular reflection, where the reflected rays are all parallel to each other, and (b) diffuse reflection, where the reflected rays travel in random directions. (c, d) Photographs of specular and diffuse reflection using laser light. *(Photographs courtesy of Henry Leap and Jim Lehman)*

Example 25.1 The Double-Reflected Light Ray

Two mirrors make an angle of 120° with each other, as in Figure 25.5. A ray is incident on mirror M_1 at an angle of 65° to the normal. Find the direction of the ray after it is reflected from mirror M_2.

Solution From the law of reflection, we see that the first reflected ray also makes an angle of 65° with the normal. Thus, it follows that this same ray makes an angle of 90 − 65°, or 25°, with the horizontal. From the triangle made by the first reflected ray and the two mirrors, we see that the first reflected ray makes an angle of 35° with M_2 (since the sum of the interior angles of any triangle is 180°). This means that this ray makes an angle of 55° with the normal to M_2. Hence, from the law of reflection, it follows that the second reflected ray makes an angle of 55° with the normal to M_2. Finally, by comparing the direction of the ray incident on M_1 with its direction after reflecting from M_2, we see that the angle between them is 120°, which corresponds to the angle between the mirrors.

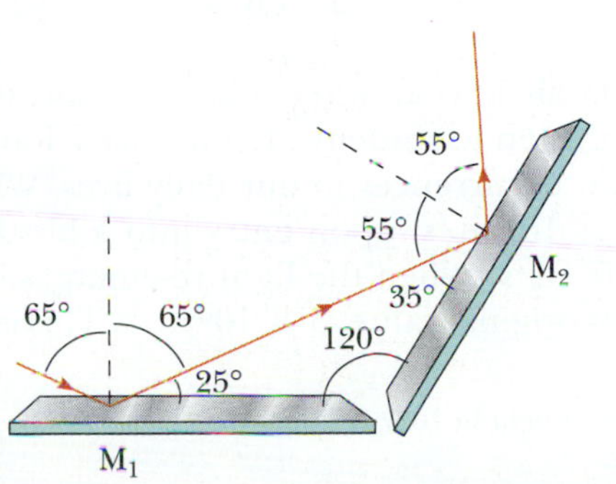

Figure 25.5
(Example 25.1) Mirrors M_1 and M_2 make an angle of 120° with each other.

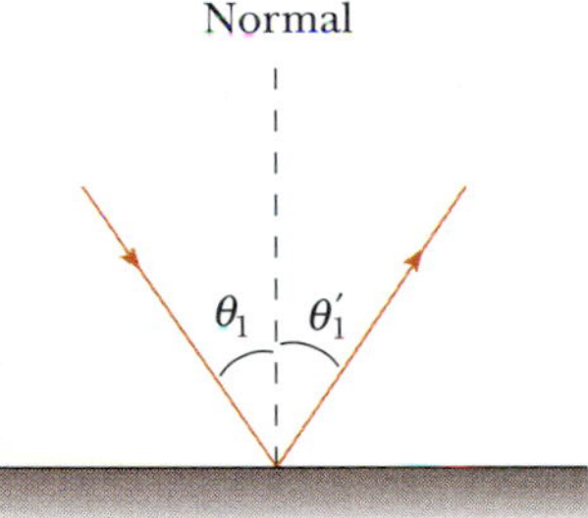

Figure 25.4
According to the law of reflection, $\theta_1 = \theta_1'$. The incident ray, the reflected ray, and the normal all lie in the same plane.

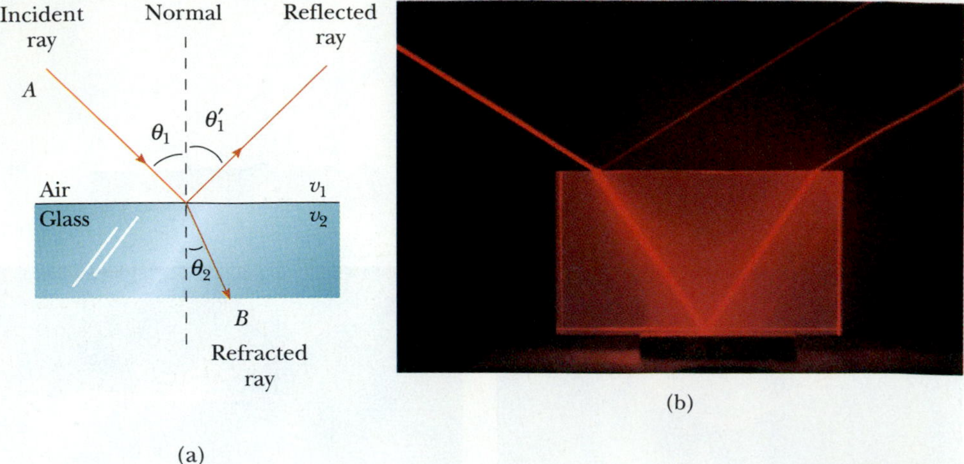

Figure 25.6 (a) A ray obliquely incident on an air-glass interface. The refracted ray is bent toward the normal since $n_2 > n_1$ and $v_2 < v_1$. All rays and the normal lie in the same plane. (b) Light incident on a Lucite block bends when it enters the block and also when it leaves the block. *(Photograph courtesy of Henry Leap and Jim Lehman)*

Refraction of Light

When a ray of light traveling through a transparent medium is obliquely incident on a boundary leading into another transparent medium, as in Figure 25.6a, part of the ray is reflected but part enters the second medium. The ray that enters the second medium is bent at the boundary and is said to be **refracted.** *The incident ray, the reflected ray, and the refracted ray all lie in the same plane.* The **angle of refraction,** θ_2 in Figure 25.6, depends on the properties of the two media and on the angle of incidence, through the relationship

$$\frac{\sin \theta_2}{\sin \theta_1} = \frac{v_2}{v_1} = \text{constant} \quad [25.3]$$

where v_1 is the speed of light in medium 1 and v_2 is the speed of light in medium 2. The experimental discovery of this relationship is usually credited to Willebrord Snell (1591–1627) and is therefore known as **Snell's law.**[1]

The path of a light ray through a refracting surface is reversible. For example, the ray in Figure 25.6 travels from point *A* to point *B*. If the ray originated at *B*, it would follow the same path to reach point *A*. In the latter case, however, the reflected ray would be in the glass.

When light moves from a material in which its speed is high to a material in which its speed is lower, the angle of refraction, θ_2, is less than the angle of incidence, and so the refracted ray is bent toward the normal, as shown in Figure 25.7a. If the ray moves from a material in which it travels slowly to a material in which it travels more rapidly, θ_2 is greater than θ_1, and so the ray is bent away from the normal, as shown in Figure 25.7b.

The behavior of light as it passes from air into another substance and then re-emerges into air is often a source of confusion to students. Let us take a look at why this behavior is so different from other occurrences in our daily lives. When light travels in air, its speed is equal to 3×10^8 m/s; upon entry into a block of glass, its speed is reduced to about 2×10^8 m/s. When the light re-emerges into air, its speed instantaneously increases to its original value, 3×10^8 m/s. This is far

[1] The same law was deduced from the particle theory of light in 1637 by René Descartes (1596–1650) and hence is known as *Descartes' law* in France.

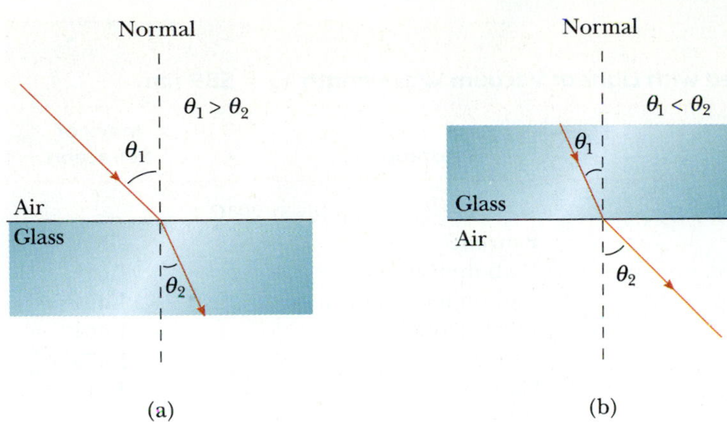

Figure 25.7
(a) When the light beam moves from air into glass, its path is bent toward the normal. (b) When the beam moves from glass into air, its path is bent away from the normal.

different from what happens when a bullet is fired through a block of wood, for example. In this case, the speed of the bullet is reduced as it moves through the wood because some of its original energy is used to tear apart the fibers of the wood. When the bullet enters the air once again, it emerges at the speed it had just before leaving the block of wood.

In order to see why light behaves as it does, consider Figure 25.8, which represents a beam of light entering a piece of glass from the left. Once inside the glass, the light may encounter an electron bound to an atom, represented by point *A* in the figure. Let us assume that light is absorbed by the atom, which causes the electron to oscillate. The oscillating electron then acts as an antenna and radiates (emits) the beam of light toward an atom at point *B*, where the light is again absorbed by an atom. The details of these absorptions and emissions are best explained in terms of quantum mechanics, a subject we shall study in Chapter 29. For now, it is sufficient to think of the process as one in which the light passes from one atom to another through the glass. (The situation is somewhat analogous to a relay race in which a baton is passed between runners on the same team.) Although light travels from one atom to another with a speed of 3×10^8 m/s, the absorptions and emissions of light by the atoms cause the overall speed of light in the glass to be lowered. Once the light emerges into the air, the absorptions and emissions cease and the light's speed returns to its original value.

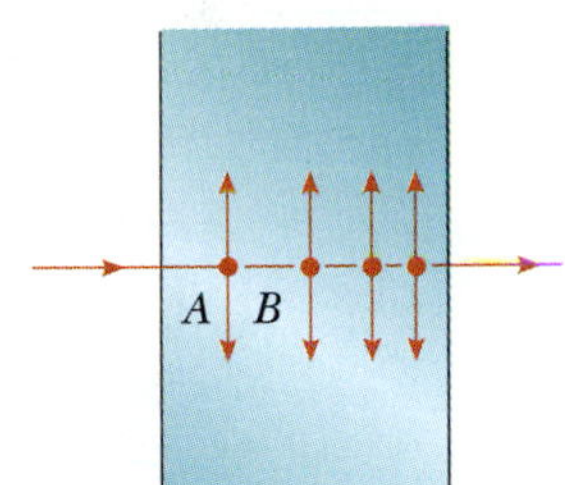

Figure 25.8
Light passing from one atom to another in a material medium.

The Law of Refraction

Light passing from one medium to another is refracted because the speed of light is different in the two media. In fact, *light travels at its maximum speed in vacuum.* It is convenient to define the **index of refraction,** n, of a medium to be the ratio

$$n \equiv \frac{\text{speed of light in vacuum}}{\text{speed of light in a medium}} = \frac{c}{v} \qquad [25.4]$$

Index of refraction

From this definition we see that the index of refraction is a dimensionless number that is greater than unity, because v is usually less than c. Furthermore, n is equal to unity for vacuum. The indices of refraction for various substances measured with respect to vacuum are listed in Table 25.1.

Table 25.1
Index of Refraction for Various Substances Measured with Light of Vacuum Wavelength λ_0 = 589 nm

Substance	Index of Refraction	Substance	Index of Refraction
Solids at 20°C		Liquids at 20°C	
Cubic zirconia	2.21	Benzene	1.501
Diamond (C)	2.419	Carbon disulfide	1.628
Fluorite (CaF_2)	1.434	Carbon tetrachloride	1.461
Fused quartz (SiO_2)	1.458	Ethyl alcohol	1.361
Glass, crown	1.52	Glycerine	1.473
Glass, flint	1.66	Corn syrup	2.21
Ice (H_2O)	1.309	Water	1.333
Polystyrene	1.49		
Sodium chloride (NaCl)	1.544	Gases at 0°C, 1 atm	
Zircon	1.923	Air	1.000293
		Carbon dioxide	1.00045

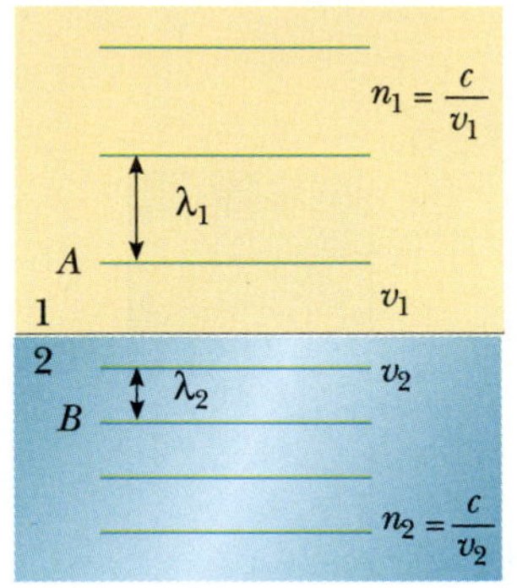

Figure 25.9
As the wave moves from medium 1 to medium 2, its wavelength changes but its frequency remains constant.

As light travels from one medium to another, *its frequency does not change.* To see why this is so, consider Figure 25.9. Wavefronts pass an observer at point A in medium 1 with a certain frequency and are incident on the boundary between medium 1 and medium 2. The frequency with which the wavefronts pass an observer at point B in medium 2 must equal the frequency at which they arrive at point A. If this were not the case, either wavefronts would pile up at the boundary or they would be destroyed or created at the boundary. Since there is no mechanism to facilitate such behaviors, the frequency must be a constant as a light ray passes from one medium into another.

Therefore, because the relation $v = f\lambda$ must be valid in both media and because $f_1 = f_2 = f$, we see that

$$v_1 = f\lambda_1 \qquad \text{and} \qquad v_2 = f\lambda_2$$

where the subscripts refer to the two media. A relationship between index of refraction and wavelength can be obtained by dividing these two equations and making use of the definition of the index of refraction given by Equation 25.4:

$$\frac{\lambda_1}{\lambda_2} = \frac{v_1}{v_2} = \frac{c/n_1}{c/n_2} = \frac{n_2}{n_1} \qquad \textbf{[25.5]}$$

which gives

$$\lambda_1 n_1 = \lambda_2 n_2 \qquad \textbf{[25.6]}$$

If medium 1 is vacuum or, for all practical purposes, air, then $n_1 = 1$. Hence, it follows from Equation 25.5 that the index of refraction of any medium can be expressed as the ratio

$$n = \frac{\lambda_0}{\lambda_n} \qquad \textbf{[25.7]}$$

where λ_0 is the wavelength of light in vacuum and λ_n is the wavelength in the medium whose index of refraction is n. A schematic representation of this reduction in wavelength is shown in Figure 25.10.

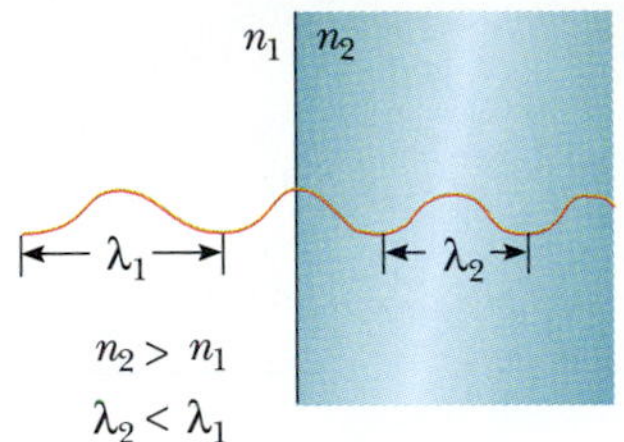

Figure 25.10
Schematic diagram of the *reduction* in wavelength when light travels from a medium of low index of refraction to one of higher index of refraction.

We are now in a position to express Snell's law (Eq. 25.3) in an alternative form. If we substitute Equation 25.5 into Equation 25.3, we get

$$n_1 \sin \theta_1 = n_2 \sin \theta_2 \qquad [25.8]$$

Snell's law

This form of Snell's law is practical because it is expressed in terms of n values.

Example 25.2 A Measurement of Index of Refraction

A beam of light of wavelength 550 nm, traveling in air, is incident on a slab of transparent material. The incident beam makes an angle of 40° with the normal, and the refracted beam makes an angle of 26° with the normal. Find the index of refraction of the material.

Solution Snell's law of refraction (Eq. 25.8) with the given data—$\theta_1 = 40°$, $n_1 = 1.00$ for air, and $\theta_2 = 26°$—gives

$$n_2 = \frac{n_1 \sin \theta_1}{\sin \theta_2} = (1.00)\frac{\sin 40°}{\sin 26°} = \frac{0.643}{0.438} = 1.47$$

If we compare this value with the data in Table 25.1, we see that the material may be fused quartz.

Exercise What is the wavelength of light in the material?

Answer 374 nm.

Example 25.3 The Speed of Light in Fused Quartz

Light of wavelength 589 nm in vacuum passes through a piece of fused quartz ($n = 1.458$).

(a) Find the speed of light in fused quartz.

Solution The speed of light in quartz can be obtained easily from Equation 25.4:

$$v = \frac{c}{n} = \frac{3 \times 10^8 \text{ m/s}}{1.458} = 2.058 \times 10^8 \text{ m/s}$$

(b) What is the wavelength of this light in fused quartz?

Solution We can use $\lambda_n = \lambda_0/n$ (Eq. 25.7) to calculate the wavelength in fused quartz, noting that we are given the wavelength in vacuum at $\lambda_0 = 589$ nm:

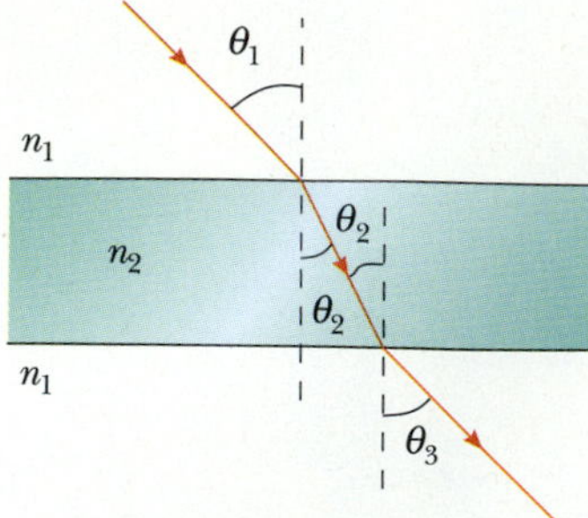

Figure 25.11 (Example 25.4) When light passes through a flat slab of material, the emerging beam is parallel to the incident beam, and therefore $\theta_1 = \theta_3$.

$$\lambda_n = \frac{\lambda_0}{n} = \frac{589 \text{ nm}}{1.458} = \boxed{404 \text{ nm}}$$

Exercise Find the frequency of the light passing through the fused quartz.

Answer 5.09×10^{14} Hz.

Example 25.4 Light Passing Through a Slab

A light beam passes from medium 1, with an index of refraction of n_1, through a thick slab of material whose index of refraction is n_2, and emerges from the slab into medium 1 (Fig. 25.11). Show that the emerging beam is parallel to the incident beam.

Solution First, let us apply Snell's law to the upper boundary:

$$(1) \qquad \sin \theta_2 = \frac{n_1}{n_2} \sin \theta_1$$

Applying Snell's law to the lower boundary gives

$$(2) \qquad \sin \theta_3 = \frac{n_2}{n_1} \sin \theta_2$$

Substituting (1) into (2) gives

$$\sin \theta_3 = \frac{n_2}{n_1}\left(\frac{n_1}{n_2} \sin \theta_1\right) = \sin \theta_1$$

That is, $\theta_3 = \theta_1$, and so the slab does not alter the direction of the beam. It does, however, produce a displacement of the beam. The same result is obtained when light passes through multiple layers of materials.

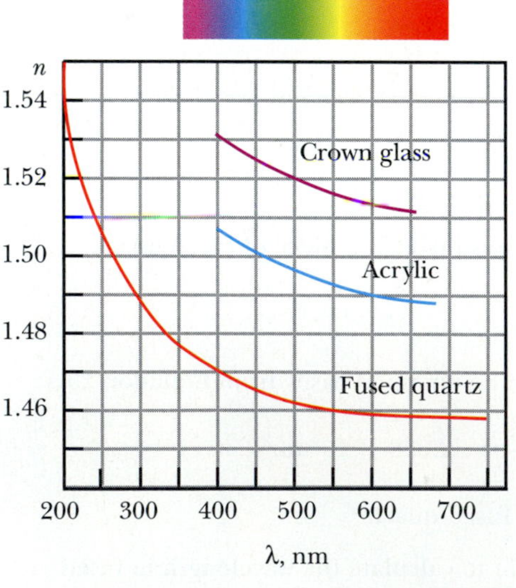

Figure 25.12 Variations of index of refraction with vacuum wavelength for three materials.

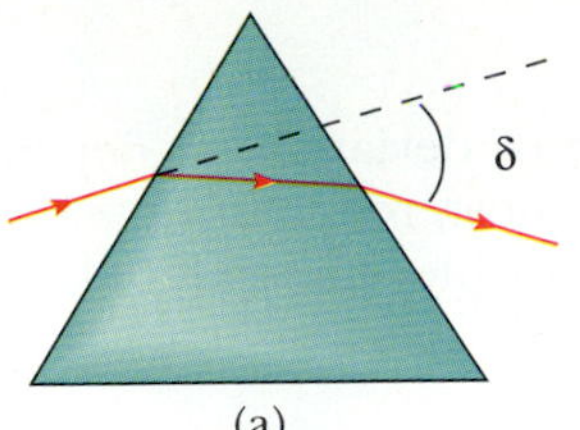

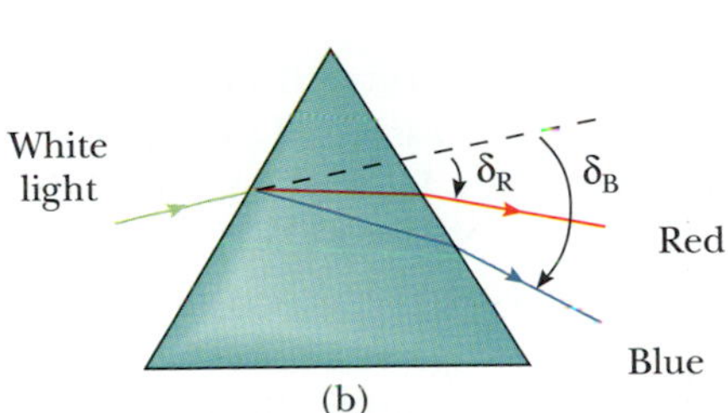

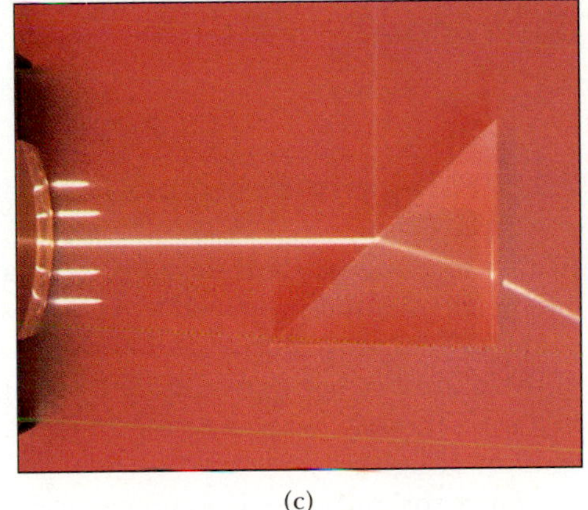

Figure 25.13
(a) A prism refracts a light ray and deviates the light through an angle of δ. (b) When light is incident on a prism, the blue light is bent more than the red. (c) Photograph of light being refracted by a prism.

25.4 Dispersion and Prisms

An important property of the index of refraction is that its value in anything but vacuum depends on the wavelength of light. This phenomenon is called **dispersion** (Fig. 25.12). Since n is a function of wavelength, Snell's law indicates that light of *different wavelengths* is bent at *different angles* when incident on a refracting material. As we see from Figure 25.12, the index of refraction decreases with increasing wavelength. This means that blue light ($\lambda \cong 470$ nm) bends more than red light ($\lambda \cong 650$ nm) when passing into a refracting material.

To understand how dispersion can affect light, let us consider what happens when light strikes a prism, as in Figure 25.13a. A single ray of light that is incident on the prism from the left emerges bent away from its original direction of travel by an angle, δ, called the **angle of deviation.** Now suppose a beam of white light (a combination of all visible wavelengths) is incident on a prism, as in Figure 25.13b. Because of dispersion, the blue component of the incident beam is bent more than the red component, and the rays that emerge from the second face of the prism spread out in a series of colors known as a **visible spectrum,** as shown in Figure 25.14. These colors, in order of decreasing wavelength, are red, orange, yellow,

Figure 25.14
(a) Dispersion of white light by a prism. Since n varies with wavelength, the prism disperses the white light into its various spectral components. (b) Different colors are refracted at different angles because the index of refraction of the glass depends on wavelength. The blue light deviates the most; red light deviates the least. *(Photograph courtesy of Bausch and Lomb)*

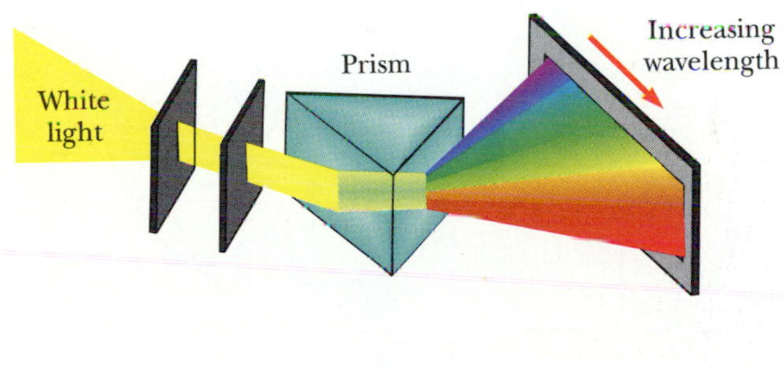

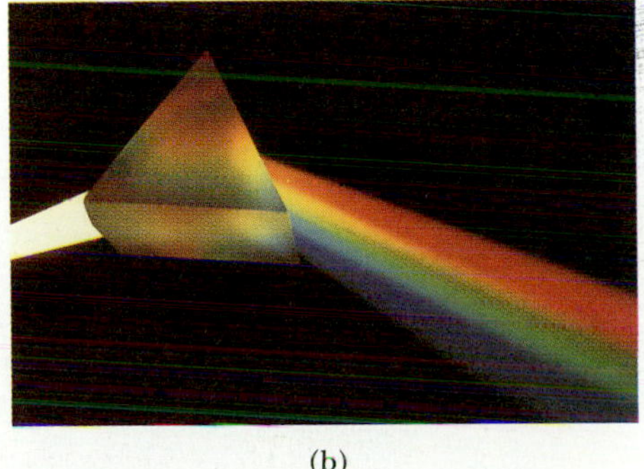

Christian Huygens (1629–1695). *(Courtesy of Rijksmuseum voor de Geschiedenis der Natuurwetenschappen. Courtesy AIP Niels Bohr Library)*

green, blue, indigo, and violet. Clearly, the angle of deviation, δ, depends on wavelength. Violet light ($\lambda \cong 400$ nm) deviates the most, red light ($\lambda \cong 650$ nm) deviates the least, and the remaining colors in the visible spectrum fall between these extremes.

25.5 Huygens' Principle

In this section, we shall develop the laws of reflection and refraction by using a geometric method proposed by Huygens in 1678. Huygens assumed that light was some form of wave motion rather than a stream of particles. He had no knowledge of the nature of light or of its electromagnetic character. Nevertheless, his simplified wave model is adequate for understanding many practical aspects of the propagation of light.

Huygens' principle is a geometric construction for determining the position of a wavefront from a knowledge of an earlier wavefront. In Huygens' construction,

all points on a given wavefront are taken as point sources for the production of spherical secondary waves, called wavelets, that propagate outward with speeds characteristic of waves in that medium. After some time has elapsed, the new position of the wavefront is the surface tangent to the wavelets.

Figure 25.15 illustrates two simple examples of Huygens' construction. First, consider a plane wave moving through free space, as in Figure 25.15a. At $t = 0$, the wavefront is indicated by the plane labeled AA'. Each point on this wavefront is a point source for a wavelet. Showing three of these points, we draw circles, each of radius $c\,\Delta t$, where c is the speed of light in free space and Δt is the time of propagation from one wavefront to the next. The surface drawn tangent to the wavelets is the plane BB', which is parallel to AA'. In a similar manner, Figure 25.15b shows Huygens' construction for an outgoing spherical wave.

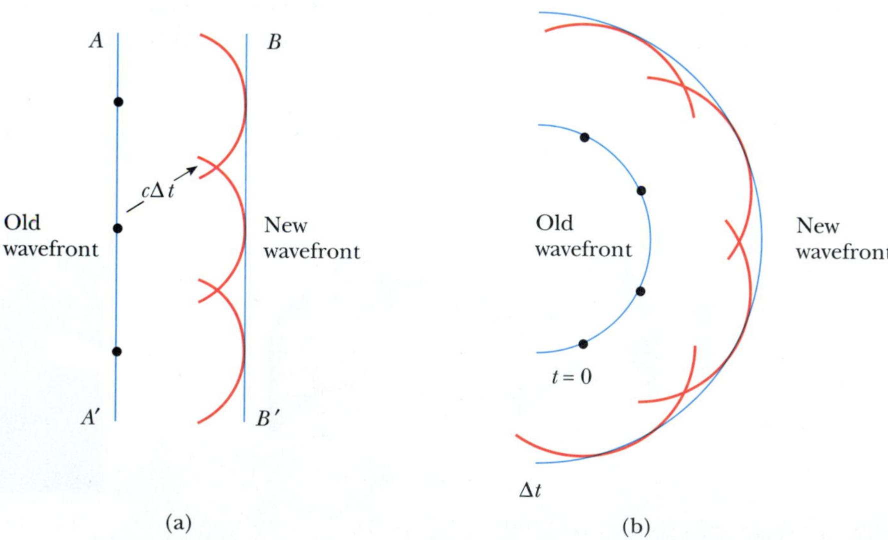

Figure 25.15 Huygens' constructions for (a) a plane wave propagating to the right and (b) a spherical wave.

A convincing demonstration of the existence of Huygens' wavelets is obtained with water waves in a shallow tank (called a ripple tank), as in Figure 25.16. Plane waves produced incident on a barrier with a small opening emerge from the opening as two-dimensional circular waves propagating outward.

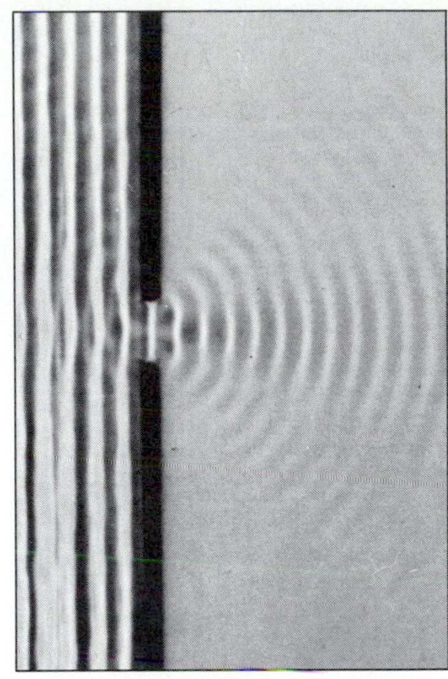

Figure 25.16 Water waves in a ripple tank, which demonstrates Huygens' wavelets. A plane wave is incident on a barrier with a small opening. The opening acts as a source of circular wavelets. *(Photograph courtesy of Education Development Center, Newton, Mass.)*

Huygens' Principle Applied to Reflection and Refraction

The laws of reflection and refraction were stated earlier in this chapter without proof. We shall now derive them using Huygens' principle. Figure 25.17a will be used in our consideration of the law of reflection. The line AA' represents a wavefront of the incident light. As ray 3 travels from A' to C, ray 1 reflects from A and produces a spherical wavelet of radius AD. (Recall that the radius of a Huygens wavelet is equal to vt.) Since the two wavelets having radii $A'C$ and AD are in the same medium, they have the same speed, v, and thus $AD = A'C$. Meanwhile, the spherical wavelet centered at B has spread only half as far as the one centered at A, since ray 2 strikes the surface later than ray 1.

From Huygens' principle, we find that the reflected wavefront is CD, a line tangent to all the outgoing spherical wavelets. The remainder of the analysis depends upon geometry, as summarized in Figure 25.17b. Note that the right triangles ADC and $AA'C$ are identical because they have the same hypotenuse, AC, and because $AD = A'C$. From Figure 25.17b we have

$$\sin\theta_1 = \frac{A'C}{AC} \quad \text{and} \quad \sin\theta_1' = \frac{AD}{AC}$$

Thus,

$$\sin\theta_1 = \sin\theta_1'$$

$$\theta_1 = \theta_1'$$

which is the law of reflection.

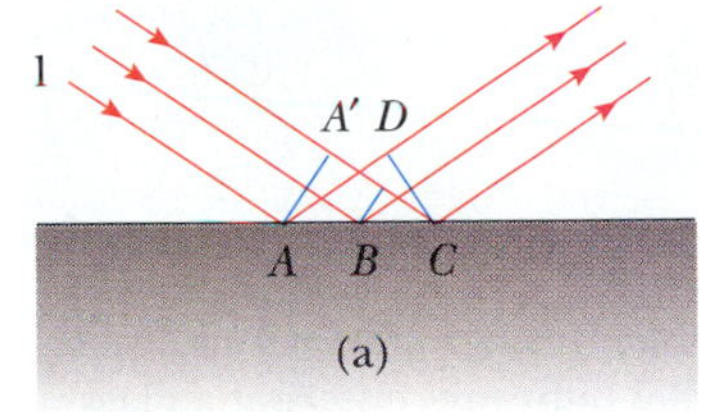

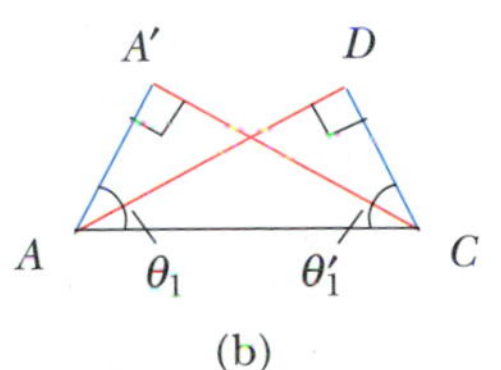

Figure 25.17 (a) Huygens' construction for proving the law of reflection. (b) Triangle ADC is identical to triangle $AA'C$.

Now let us use Huygens' principle and Figure 25.18a to derive Snell's law of refraction. Note that in the time interval Δt, ray 1 moves from A to B and ray 2 moves from A' to C. The radius of the outgoing spherical wavelet centered at A is equal to $v_2\,\Delta t$. The distance $A'C$ is equal to $v_1\,\Delta t$. Geometric considerations show that angle $A'AC$ equals θ_1 and angle ACB equals θ_2. From triangles $AA'C$ and ACB, we find that

$$\sin\theta_1 = \frac{v_1\,\Delta t}{AC} \quad \text{and} \quad \sin\theta_2 = \frac{v_2\,\Delta t}{AC}$$

If we divide these two equations, we get

$$\frac{\sin\theta_1}{\sin\theta_2} = \frac{v_1}{v_2}$$

But from Equation 25.4 we know that $v_1 = c/n_1$ and $v_2 = c/n_2$. Therefore,

$$\frac{\sin\theta_1}{\sin\theta_2} = \frac{c/n_1}{c/n_2} = \frac{n_2}{n_1}$$

$$n_1 \sin\theta_1 = n_2 \sin\theta_2$$

which is the law of refraction.

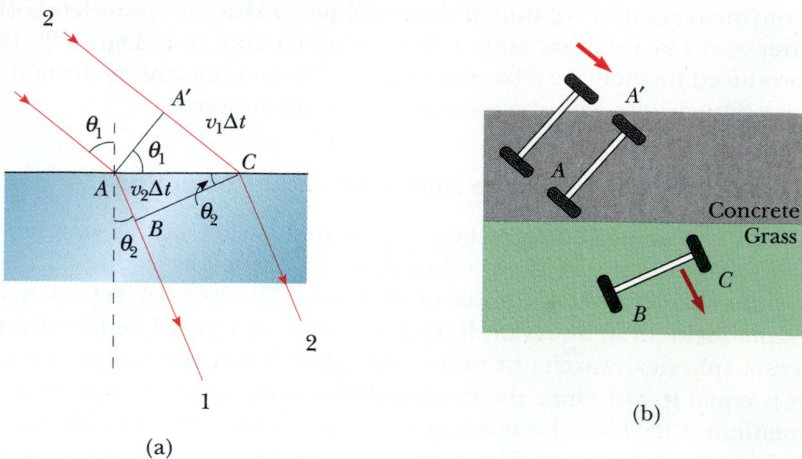

Figure 25.18
(a) Huygens' construction for proving the law of refraction. (b) A mechanical analog of refraction.

A mechanical analog of refraction is shown in Figure 25.18b. The wheels on a device such as a wagon change their direction as they move from a concrete surface to a grass surface.

This photograph shows nonparallel light rays entering a glass prism. The bottom two rays undergo total internal reflection at the longest side of the prism. The top three rays are refracted at the longest side as they leave the prism. *(Courtesy of Henry Leap and Jim Lehman)*

25.6 Total Internal Reflection

An interesting effect called *total internal reflection* can occur when light travels from a medium with a high index of refraction to one with a *lower* index of refraction. Consider a light beam traveling in medium 1 and meeting the boundary between mediums 1 and 2, where $n_1 > n_2$ (Fig. 25.19). Various possible directions of the beam are indicated by rays 1 through 5. The refracted rays are bent away from the normal because $n_1 > n_2$. Furthermore, you should remember that when light refracts at the interface between the two media, it is also partially reflected. For example, the rays labeled 2, 3, and 4 in Figure 25.19 are partially reflected back into medium 1, but the reflected components are not shown. At some particular angle of incidence, θ_c, called the **critical angle,** the refracted light ray will move parallel to the boundary so that $\theta_2 = 90°$ (Fig. 25.19b). *For angles of incidence greater than* θ_c, the beam is entirely reflected at the boundary, as is ray 5 in Figure 25.19a. This ray is reflected at the boundary as though it had struck a perfectly reflecting surface. It and all rays like it obey the law of reflection; that is, the angle of incidence equals the angle of reflection.

We can use Snell's law to find the critical angle. When $\theta_1 = \theta_c$, $\theta_2 = 90°$ and Snell's law (Eq. 26.8) gives

$$n_1 \sin \theta_c = n_2 \sin 90° = n_2$$

Critical angle

$$\sin \theta_c = \frac{n_2}{n_1} \qquad (\text{for } n_1 > n_2) \qquad [25.9]$$

This equation can be used only when n_1 is greater than n_2. That is,

total internal reflection occurs only when light travels from a medium of high index of refraction to a medium of lower index of refraction.

If n_1 were less than n_2, Equation 25.9 would give $\sin\theta_c > 1$, which is an absurd result because the sine of an angle can never be greater than unity.

When medium 1 is air, the critical angle for a substance in air is small for substances with a large index of refraction, such as diamond, where $n = 2.42$ and $\theta_c = 24°$. For crown glass, $n = 1.52$ and $\theta_c = 41°$. In fact, this property combined with proper faceting causes diamonds and crystal glass to sparkle.

One can use a prism and the phenomenon of total internal reflection to alter the direction of travel of a light beam. Two such possibilities are illustrated in Figure 25.20. In one case the light beam is deflected by 90° (Fig. 25.20a), and in the second case the path of the beam is reversed (Fig. 25.20b). A common application of total internal reflection is in a submarine periscope. In this device, two prisms are arranged as in Figure 25.20c so that an incident beam of light follows the path shown and the periscope user is able to "see around corners."

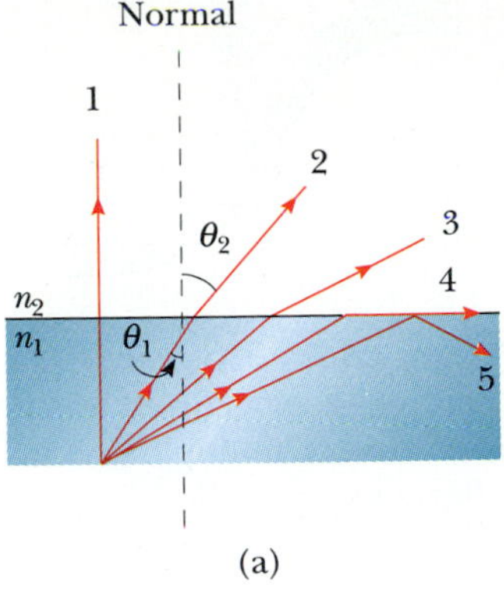

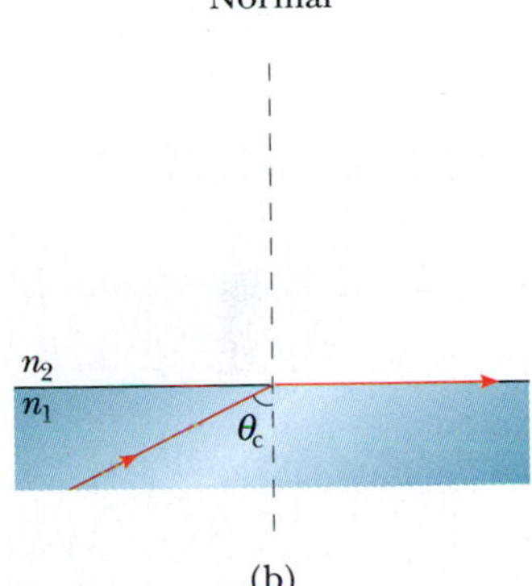

Figure 25.19 (a) A ray from a medium of index of refraction n_1 to a medium of index of refraction n_2, where $n_1 > n_2$. As the angle of incidence increases, the angle of refraction increases until θ_2 is 90° (ray 4). For even larger angles of incidence, total internal reflection occurs (ray 5). (b) The angle of incidence producing an angle of refraction equal to 90° is often called the *critical angle*, θ_c.

Example 25.5 A View from the Fish's Eye

(a) Find the critical angle for a water-air boundary if the index of refraction of water is 1.33.

Solution Applying Equation 25.9, we find that the critical angle is

$$\sin\theta_c = \frac{n_2}{n_1} = \frac{1}{1.33} = 0.752$$

$$\theta_c = 48.8°$$

(b) Use the results of (a) to predict what a fish will see if it looks upward toward the water surface at angles of 40°, 49°, and 60°.

Solution Because the path of a light ray is reversible, the fish can see out of the water if it looks toward the surface at an angle less than the critical angle. Thus, at 40°, the fish can see into the air above the water. At an angle of 49°, the critical angle for water, the light that reaches the fish has to skim along the water surface before being refracted to the fish's eye. At angles greater than the critical angle, the light reaching the fish comes via internal reflection at the surface. Thus, at 60°, the fish sees a reflection of some object on the bottom of the pool.

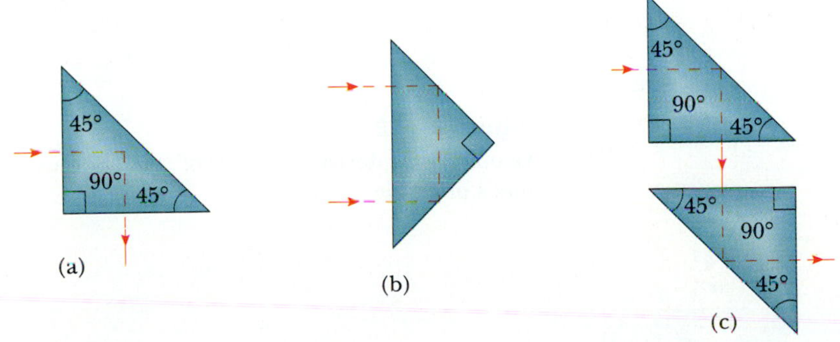

Figure 25.20 Internal reflection in a prism. (a) The ray is deviated by 90°. (b) The direction of the ray is reversed. (c) Two prisms used as a periscope.

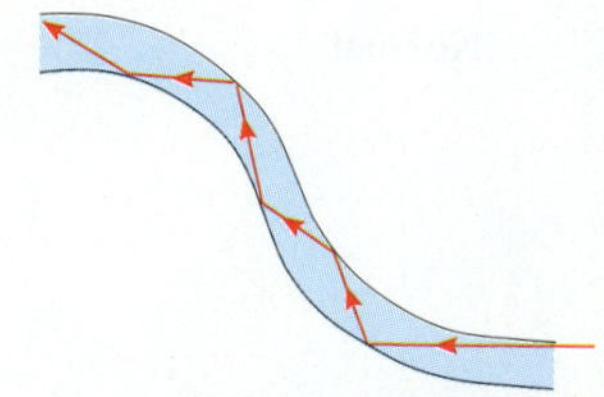

Figure 25.21
Light travels in a curved transparent rod by multiple internal reflections.

Fiber Optics

Another interesting application of total internal reflection is the use of glass or transparent plastic rods to "pipe" light from one place to another. As indicated in Figure 25.21, light is limited to traveling within the rods, even around gentle curves, as the result of successive internal reflections. Such a "light pipe" can be flexible if thin fibers are used rather than thick rods. If a bundle of parallel fibers is used to construct an optical transmission line, images can be transferred from one point to another.

This technique is used in the industry known as *fiber optics.* Very little light intensity is lost in the fibers as a result of reflections on the sides. Any loss in intensity is due essentially to reflections from the two ends and absorption by the fiber material. Fiber optic devices are particularly useful for viewing an image produced at an inaccessible location. Physicians often use fiber optic cables to aid in the diagnosis and repair of certain medical conditions, without the intrusive effects of major surgery. For example, a fiber optic cable can be threaded through the esophagus and into the stomach to enable the physician to look for ulcers. In this application, the cable actually consists of two fiber optic lines, one to transmit a beam of light into the stomach for illumination and the other to allow this light to be transmitted out of the stomach. The resulting image can be viewed directly by the physician in some cases but most often is displayed on a television monitor or captured on film. The field of fiber optics is also finding increasing use in telecommunications because the fibers can carry a much higher volume of telephone calls, or other forms of communication, than electrical wires.

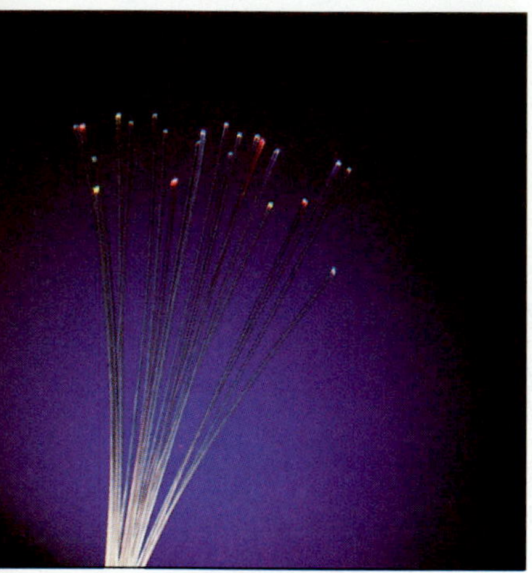

Strands of glass optical fibers are used to carry voice, video, and data signals in telecommunication networks. Typical fibers have a diameter of 60 μm.
(© Richard Megna 1983, Fundamental Photographs)

*25.7 Fermat's Principle

A general principle for determining paths of light rays was developed by Pierre de Fermat (1601–1665). **Fermat's principle** states that when a light ray travels between any two points, P and Q, its path will be the one that requires the least time. This is sometimes called the *principle of least time.* An obvious consequence of Fermat's principle is that when the rays travel in a single, homogeneous medium, the paths are straight lines because a straight line is the shortest distance between two points. Let us illustrate how to use Fermat's principle to derive the law of refraction.

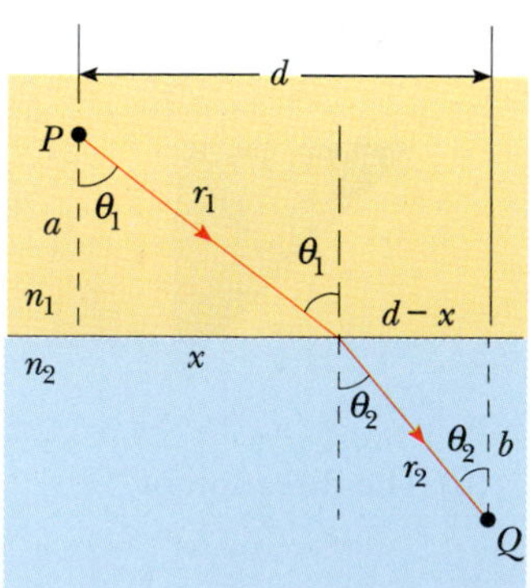

Figure 25.22
Geometry for deriving the law of refraction using Fermat's principle.

Suppose a light ray is to travel from P to Q, where P is in medium 1 and Q is in medium 2 (Fig. 25.22). The points P and Q are at perpendicular distances a and b, respectively, from the interface. The speed of light is c/n_1 in medium 1 and c/n_2 in medium 2. Using the geometry of Figure 25.22, we see that the time it takes the ray to travel from P to Q is

$$t = \frac{r_1}{v_1} + \frac{r_2}{v_2} = \frac{\sqrt{a^2 + x^2}}{c/n_1} + \frac{\sqrt{b^2 + (d - x)^2}}{c/n_2}$$

We obtain the least time, or the minimum value of t, by taking the derivative of t with respect to x (the variable) and setting the derivative equal to zero:

$$\frac{dt}{dx} = \frac{n_1}{c}\frac{d}{dx}(a^2 + x^2)^{1/2} + \frac{n^2}{c}\frac{d}{dx}[b^2 + (d - x)^2]^{1/2}$$

$$= \frac{n_1}{c}\left(\frac{1}{2}\right)\frac{2x}{(a^2 + x^2)^{1/2}} + \frac{n_2}{c}\left(\frac{1}{2}\right)\frac{2(d - x)(-1)}{[b^2 + (d - x)^2]^{1/2}}$$

$$\frac{dt}{dx} = \frac{n_1 x}{c(a^2 + x^2)^{1/2}} - \frac{n_2(d - x)}{c[b^2 + (d - x)^2]^{1/2}} = 0$$

Using this result and Figure 26.22, which gives relations for $\sin\theta_1$ and $\sin\theta_2$, we find that

$$n_1 \sin\theta_1 = n_2 \sin\theta_2$$

which is Snell's law of refraction.

It is a simple matter to use a similar procedure to derive the law of reflection. The calculation is left for you to carry out (Problem 45).

Summary

In geometric optics, we use the **ray approximation,** in which we assume that a wave travels through a medium in straight lines in the direction of the rays of that wave. We neglect diffraction effects, which is a good approximation as long as the wavelength is short compared with any aperture dimensions.

The **index of refraction** of a material, n, is defined as

$$n \equiv \frac{c}{v} \qquad [25.4]$$

Index of refraction

where c is the speed of light in a vacuum and v is the speed of light in the material. In general, n varies with wavelength as

$$n = \frac{\lambda_0}{\lambda_n} \qquad [25.7]$$

Index of refraction and wavelength

where λ_0 is the wavelength of the light in vacuum and λ_n is its wavelength in the material.

The **law of reflection** states that a wave reflects from a surface so that the *angle of reflection,* θ_1', equals the *angle of incidence,* θ_1.

The **law of refraction,** or **Snell's law,** states that

Snell's law of refraction

$$n_1 \sin \theta_1 = n_2 \sin \theta_2 \qquad [25.8]$$

Huygens' principle

Huygens' principle states that all points on a wavefront can be taken as point sources for the production of secondary wavelets. At some later time, the new position of the wavefront is the surface tangent to these secondary wavelets.

Total internal reflection can occur when light travels from a medium of high index of refraction to one of lower index of refraction. The minimum angle of incidence, θ_c, for which total reflection occurs at an interface is

Critical angle for total internal reflection

$$\sin \theta_c = \frac{n_2}{n_1} \quad (\text{where } n_1 > n_2) \qquad [25.9]$$

Fermat's principle

Fermat's principle states that when a light ray travels between two points, its path will be the one that requires the least time.

Questions and Conceptual Exercises

1. You can make a corner reflector by placing three plane mirrors in the corner of a room where the ceiling meets the walls. Show that no matter where you are in the room, you can see yourself reflected in the mirrors—upside down.
2. In 1969, several corner reflectors were left on the Moon's Sea of Tranquility by the astronauts of Apollo 11. How can scientists even today utilize a laser beam sent from Earth, along with the reflectors, to determine the precise distance from the Earth to the Moon?
3. The rectangular aquarium sketched in Figure 25.23 contains only one goldfish. When the fish is near a corner of the aquarium and is viewed along a direction which makes an equal angle with two adjacent faces, the observer sees two fish mirroring each other, as shown. Explain this observation.
4. As light travels from one medium to another, does its wavelength change? Does its frequency change? Does its speed change? Explain.
5. Explain why an oar in water appears to be bent.
6. Explain why a diamond loses most of its sparkle when submerged in carbon disulfide, and why an imitation diamond of cubic zirconia loses all of its sparkle in corn syrup.
7. A laser beam passing through a nonhomogeneous sugar solution is observed to follow a curved path. Explain what happens to the beam if the concentration of sugar increases with depth.
8. Light of wavelength λ is incident on a slit of width d. Under what conditions is the ray approximation valid? Under what circumstances will the slit produce significant diffraction?
9. A laser beam (λ = 632.8 nm) is incident on a piece of Lucite, as in Figure 25.24. Part of the beam is reflected and part is refracted. What information can you get from this photograph?
10. The level of water in a clear, colorless glass is easily observed with the naked eye. The level of liquid helium in a clear glass vessel is extremely difficult to see with the naked eye. Explain.
11. Describe an experiment in which internal reflection is used to determine the index of refraction of a medium.

Figure 25.23 (Question 3)

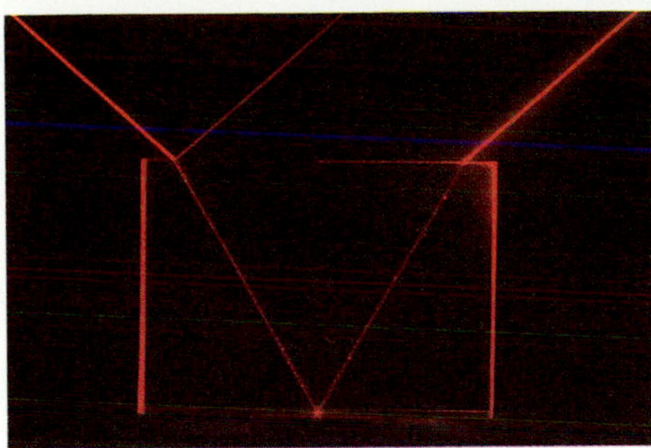

Figure 25.24 (Question 9) Light from a helium-neon laser beam ($\lambda = 632.8$ nm) is incident on a block of Lucite. The photograph shows both reflected and refracted rays. Can you identify the incident, reflected, and refracted rays? From this photograph, estimate the index of refraction of Lucite at this wavelength. *(Courtesy of Henry Leap and Jim Lehman)*

12. Why does a diamond show flashes of color when observed under ordinary white light?
13. Explain why a diamond shows more "sparkle" than a glass crystal of the same shape and size.
14. Why do astronomers looking at distant galaxies talk about looking backward in time?
15. A solar eclipse occurs when the Moon gets between the Earth and the Sun. Use a diagram to show why some areas of the Earth see a total eclipse, other areas see a partial eclipse, and most areas see no eclipse.
16. Figure 25.25 represents sunlight striking a drop of water in the atmosphere. Use the laws of refraction and reflection and the fact that sunlight consists of a wide range of wavelengths to discuss the formation of rainbows.
17. Why does the arc of a rainbow appear to an observer to have red colors on top and violet hues on the bottom?
18. How is it possible that a complete *circular* rainbow can sometimes be seen from an airplane?
19. What are the conditions for the production of a mirage? On a hot day, what is it that we are seeing when we observe "water on the road"?

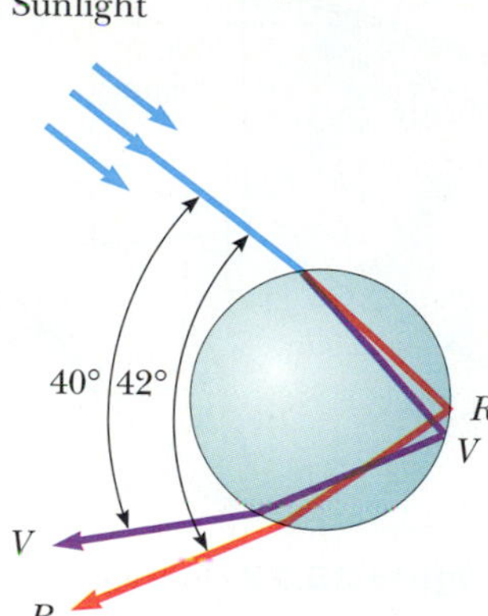

Figure 25.25 (Questions 16 and 17) Refraction of sunlight by a spherical raindrop.

20. The "professor in the box" shown in Figure 25.26 appears to be balancing himself on a few fingers, with his feet elevated from the floor. How do you suppose this illusion was created?

Figure 25.26 (Question 20) The "professor in the box." *(Photograph courtesy of Henry Leap and Jim Lehman)*

Problems

Section 25.3 Reflection and Refraction

(Note: In this section if an index of refraction value is not given, refer to Table 25.1.)

1. Yellow light of vacuum wavelength 589 nm is incident on a smooth surface of water at an angle of 35° to the vertical. Determine its angle of refraction and its wavelength in water.
2. The wavelength of red helium-neon laser light in air is 632.8 nm. (a) What is its frequency? (b) What is its wavelength in glass with an index of refraction of 1.5? (c) What is its speed in glass?
3. A soda straw is stuck into water at an angle of 36° to the vertical. To have the illusion of sighting down along the submerged part of the straw, what angle with the vertical must you look at?

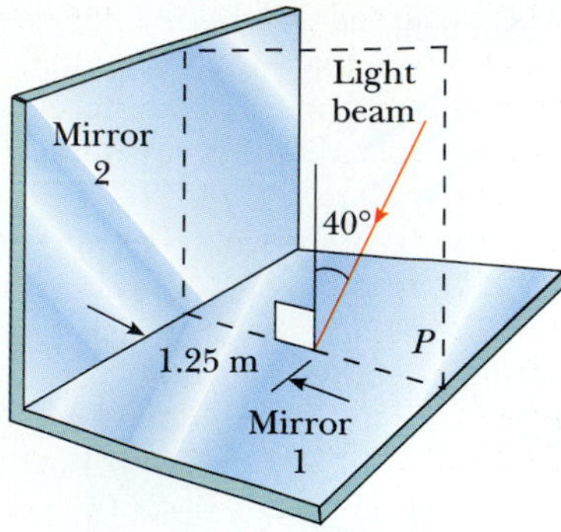

Figure 25.27 (Problem 10)

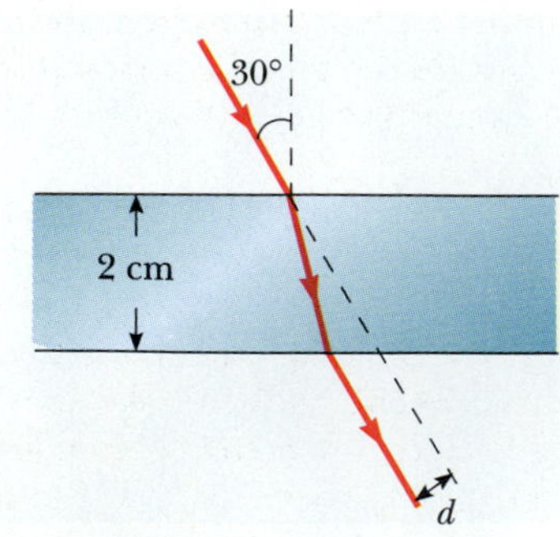

Figure 25.29 (Problems 14 and 15)

4. A laser beam is incident at an angle of 30° to the vertical onto a solution of corn syrup in water. If the beam is refracted to 19.24° to the vertical, what is the index of refraction of the syrup solution?
5. An underwater scuba diver sees the Sun at an apparent angle of 45° from the vertical. Where is the Sun?
6. A ray of light strikes a flat block of glass ($n = 1.50$), of thickness 2 cm, at an angle of 30° with the normal. Trace the light beam through the glass, and find the angles of incidence and refraction at each surface.
7. Find the speed of light in (a) flint glass, (b) water, and (c) zircon.
8. Light of wavelength 436 nm in air enters a fishbowl filled with water, then exits through the crown glass wall of the container. What is the wavelength of the light (a) while in the water and (b) while in the glass?
9. A cylindrical tank with an open top has a diameter of 3 m and is completely filled with water. When the setting Sun reaches an angle of 28° above the horizon, sunlight ceases to illuminate the bottom of the tank. How deep is the tank?
10. The angle between the two mirrors in Figure 25.27 is a right angle. The beam of light in the vertical plane, *P*, strikes mirror 1 as shown, (a) Determine the distance the reflected light beam travels before striking mirror 2. (b) In what direction does the light beam travel after being reflected from mirror 2?
11. How many times will the incident beam shown in Figure 25.28 be reflected by each of the parallel mirrors?
12. The wavelength of sodium light in air is 589 nm. Find its wavelength in ethyl alcohol.
13. A beam of light strikes the surface of mineral oil at an angle of 23.1° with the normal to the surface. If the light travels with a speed of 2.17×10^8 m/s through the oil, what is the angle of refraction?
14. When the light in Figure 25.29 passes through the glass block, it is shifted laterally by the distance *d*. If $n = 1.50$, find the value of *d*.
15. Find the time required for the light to pass through the glass block described in Problem 14.
16. The light beam shown in Figure 25.30 makes an angle of 20.0° with the normal line, *NN*′, in the linseed oil. Determine the angles θ and θ'. (The index of refraction for linseed oil is 1.48.)
17. A submarine is 300 m horizontally out from the shore and 100 m beneath the surface of the water. A laser beam is sent from the sub so that it strikes the surface

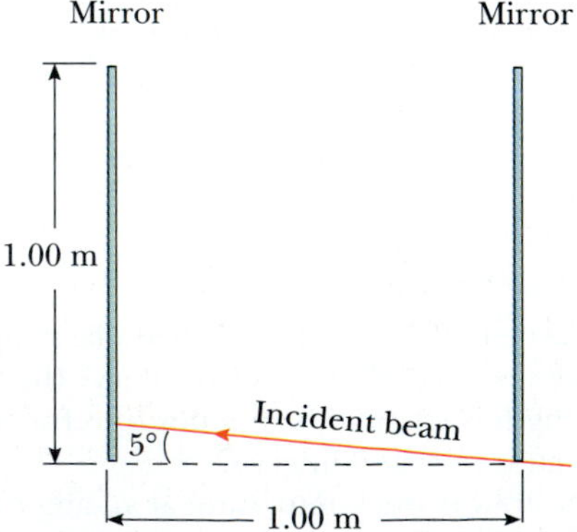

Figure 25.28 (Problem 11)

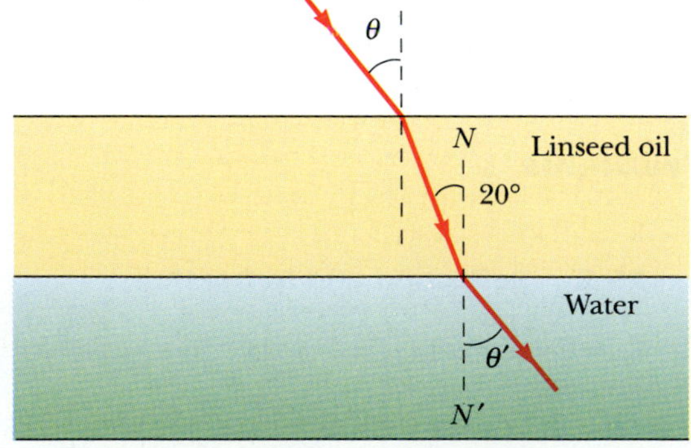

Figure 25.30 (Problem 16)

of the water at a point 210 m from the shore. If the beam just strikes the top of a building standing exactly at the water's edge, find the height of the building.

18. Two light pulses are emitted simultaneously from a source. Both pulses travel to a detector, but one first passes through 6.2 m of ice. Determine the difference in the pulses' times of arrival at the detector.

19. How far does a beam of light travel in water in the same time it takes it to travel 10 m in glass of index of refraction 1.5?

20. A drinking glass is 4 cm wide at the bottom, as shown in Figure 25.31. When an observer is positioned as shown, he or she sees the edge of the bottom of the glass. When this glass is filled with water, the observer sees the center of the bottom of the glass. Find the height of the glass.

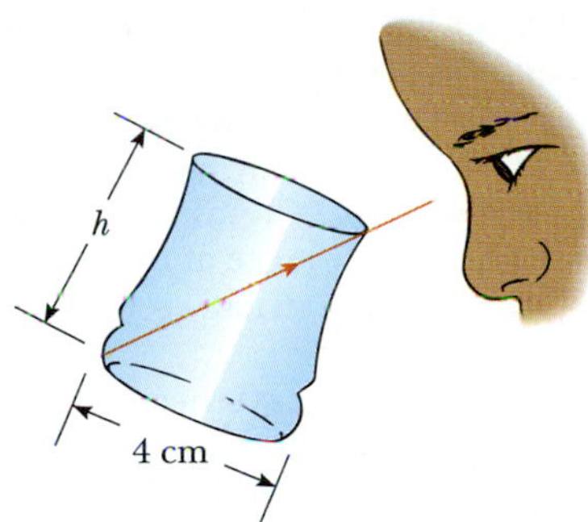

Figure 25.31 (Problem 20)

*Section 25.4 Dispersion and Prisms

21. A certain kind of glass has an index of refraction of 1.6500 for blue light of wavelength 430 nm and an index of 1.6150 for red light of wavelength 680 nm. If a beam containing these two colors is incident at an angle of 30° on a piece of this glass, what is the angle between the two beams inside the glass?

22. A ray of light strikes the midpoint of one face of an equiangular glass prism ($n = 1.50$) at an angle of incidence of 30°. Trace the path of the light ray through the glass and find the angles of incidence and refraction at each surface.

23. The index of refraction for red light in water is 1.331; that for blue light is 1.340. If a ray of white light enters the water at an angle of incidence of 83°, what are the underwater angles of refraction for the blue and red components of the light?

24. Light of wavelength 700 nm is incident on the face of a fused quartz prism at an angle of 75° (with respect to the normal to the surface). The apex angle of the prism is 60°. Use the value $n = 1.455$ and calculate the angles (a) of refraction at this (first) surface, (b) of incidence at the second surface, (c) of refraction at the second surface, and (d) between the incident and emerging rays.

25. A triangular glass prism with apex angle 60° has an index of refraction of $n = 1.5$. (a) What is the smallest angle of incidence, θ_1, for which a light ray can emerge from the other side? (See Fig. 25.13.) (b) For what angle of incidence, θ_1, does the light ray leave at the same angle, θ_1?

26. The index of refraction for violet light in silica flint glass is 1.66, and that for red light is 1.62. What is the angular dispersion of visible light passing through a prism of apex angle 60° if the angle of incidence is 50°? (See Fig. 25.32.)

27. A ray of light passes from air into water. In order for its deviation angle $\delta = |\theta_1 - \theta_2|$ to be 10°, what must be its angle of incidence? You may need to use a calculator.

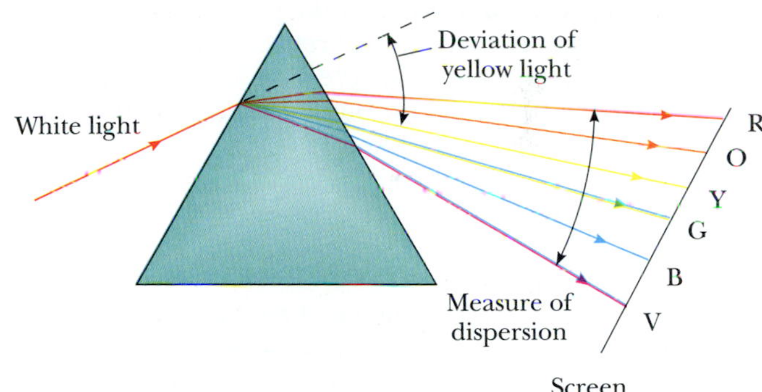

Figure 25.32 (Problem 26)

*Section 25.6 Total Internal Reflection

28. A large Lucite cube ($n = 1.59$) has a small air bubble (a defect in the casting process) below one surface. When a penny (diameter 1.9 cm) is placed directly over the bubble, the bubble cannot be seen by looking down into the cube at an angle. However, when a dime (diameter 1.75 cm) is placed directly over it, the bubble can be seen by looking down into the cube. What is the range of the possible depths of the air bubble beneath the surface?

29. A fiber optic cable ($n = 1.50$) is submerged in water ($n = 1.33$). What is the critical angle for light to stay inside the cable?

30. Calculate the critical angles for the following materials when surrounded by air: (a) diamond, (b) flint glass, and (c) ice. (Assume that $\lambda = 589$ nm.)

31. Consider a common mirage formed by heated air just above the roadway. If an observer viewing from 2 m

above the road (where $n = 1.0003$) sees water up the road at $\theta_1 = 88.8°$, find the index of refraction of the air above the road surface. (*Hint:* Treat this as a problem in total internal reflection.)

32. An optical fiber is made of a clear plastic with index of refraction $n = 1.50$. For what angles with the surface will light remain contained within the plastic "guide"?

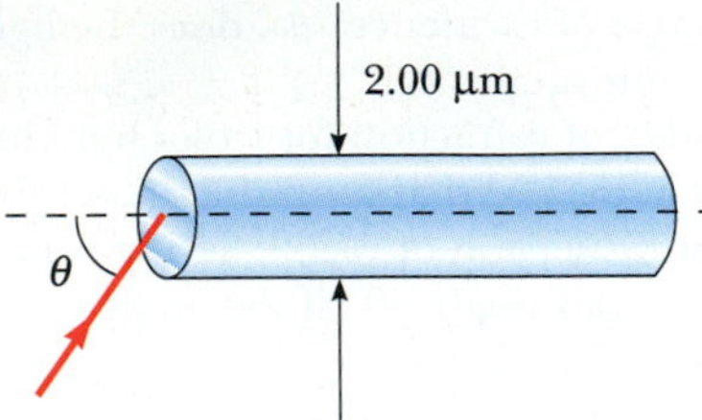

Figure 25.33 (Problem 33)

33. Determine the maximum angle, θ, for which the light rays incident on the end of the pipe in Figure 25.33 are subject to total internal reflection along the walls of the pipe. Assume that the pipe has an index of refraction of 1.36 and the outside medium is air.

Additional Problems

34. A layer of ice floats on water. If light is incident on the upper surface of the ice at an angle of 30°, what is the angle of refraction in the water?
35. The Sun is 10° above the horizon. If you are swimming beneath the surface of a pool of water, at what angle above the horizon does the Sun appear to be?
36. A light beam is incident on a water surface from air. What is the maximum possible value for the angle of refraction?
37. A small underwater pool light is 1 m below the surface. What is the diameter of the circle of light on the surface where light emerges from the water?
38. A narrow beam of light is incident from air onto a glass surface of index of refraction 1.56. Find the angle of incidence for which the corresponding angle of refraction is one-half the angle of incidence. (*Hint:* You might want to use the trigonometric identity $\sin 2\theta = 2 \sin \theta \cos \theta$.)
39. Light is incident normally on a 1-cm layer of water that lies on top of a flat Lucite plate with a thickness of 0.5 cm. How much longer does light take to pass through this double layer than to traverse the same distance in air? ($n_{\text{Lucite}} = 1.59$.)
40. One technique to measure the angle of a prism is shown in Figure 25.34. A parallel beam of light is directed on the angle so that the beam reflects from opposite sides. Show that the angular separation of the two beams is given by $B = 2A$.

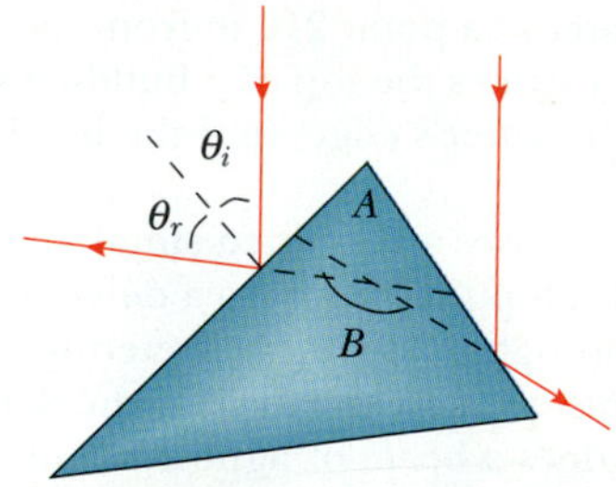

Figure 25.34 (Problem 40)

41. The laws of refraction and reflection are the same for sound as for light. The speed of sound in air is 340 m/s, and that in water is 1510 m/s. If a sound wave approaches a plane water surface at an angle of incidence of 12°, what is the angle of refraction?
42. A layer of kerosene ($n = 1.45$) is floating on water. For what angles of incidence at the kerosene-water interface is light totally internally reflected within the kerosene?
43. When the Sun is directly overhead, a narrow shaft of light enters a cathedral through a small hole in the ceiling and forms a spot on the floor 10.0 m below. (a) At what speed (in centimeters per minute) does the spot move across the (flat) floor? (b) If a mirror is placed on the floor to intercept the light, at what speed does the reflected spot move across the ceiling?
44. Show that the time required for light to travel from point source A in air, a distance of h_1 above a water surface, to point B at h_2 below the water surface is

$$t = \frac{h_1 \sec \theta_1}{c} + \frac{h_2 n \sec \theta_2}{c}$$

where n is the index of refraction of water, θ_1 is the angle of incidence, and θ_2 is the angle of refraction.

45. Derive the law of reflection (Eq. 25.2) from Fermat's principle of least time. (See the procedure outlined in Section 25.7 for the derivation of the law of refraction from Fermat's principle.)
46. A light ray of wavelength 589 nm is incident at an angle of θ on the top surface of a block of polystyrene, as shown in Figure 25.35. (a) Find the maximum value of θ for which the refracted ray will undergo *total* internal reflection at the left vertical face of the block. (b) Repeat the calculation for the case in which the polystyrene block is immersed in water. (c) What happens if the block is immersed in carbon disulfide?

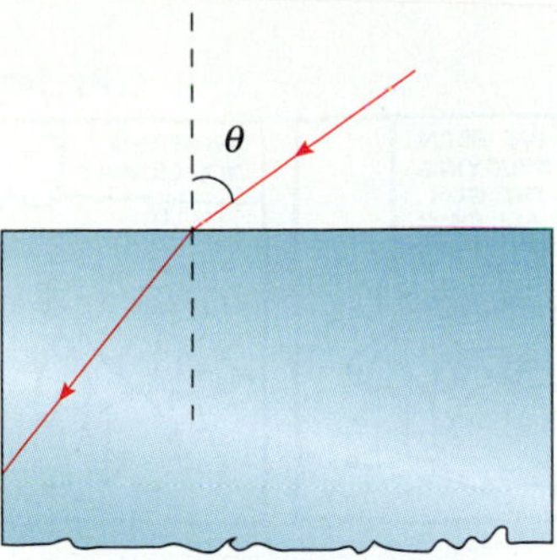

Figure 25.35 (Problem 46)

47. A hiker stands on a mountain peak near sunset and observes a (primary) rainbow caused by water droplets in the air about 8 km away. The valley is 2 km below the mountain peak and is entirely flat. What fraction of the complete circular arc of the rainbow is visible to the hiker? (See Question 16.)

48. A fish is at a depth of d underwater. Show that when viewed from an angle of incidence, θ_1, the *apparent depth*, z, of the fish is

$$z = \frac{3d\cos\theta_1}{\sqrt{7 + 9\cos^2\theta_1}}$$

49. The prism shown in Figure 25.36 has an index of refraction of 1.55. Light is incident at an angle of 20°. Determine the angle, θ, at which the light emerges.

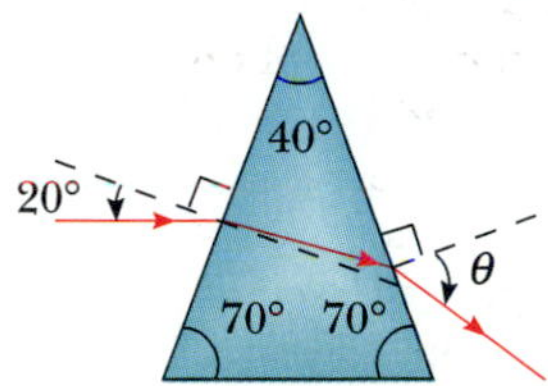

Figure 25.36 (Problem 49)

50. A light ray is incident on a prism and refracted at the first surface, as shown in Figure 25.37. Let Φ represent the apex angle of the prism and n its index of refraction. Find, in terms of n and Φ, the smallest allowed value of the angle of incidence at the first surface for which the refracted ray will *not* undergo internal reflection at the second surface.

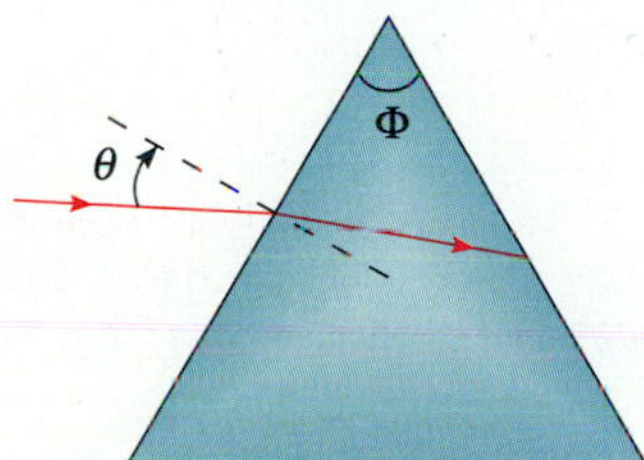

Figure 25.37 (Problem 50)

51. A laser beam strikes one end of a slab of material, as shown in Figure 25.38. The index of refraction of the slab is 1.48. Determine the number of internal reflections of the beam before it emerges from the opposite end of the slab.

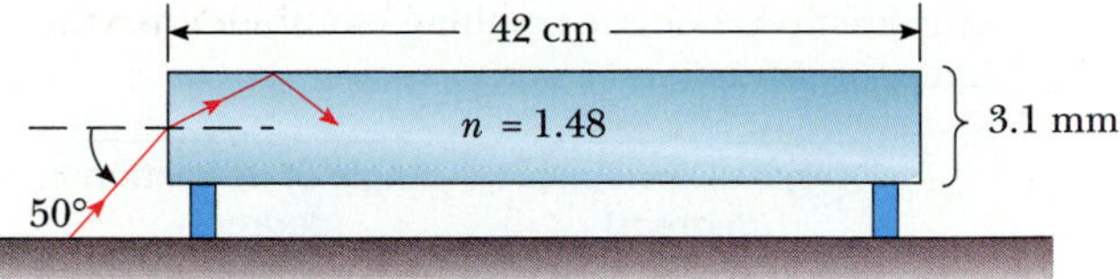

Figure 25.38 (Problem 51)

52. Figure 25.39 shows a top view of a square enclosure. The inner surfaces are plane mirrors. A ray of light enters a small hole in the center of one mirror. (a) At what angle, θ, must the ray enter in order to exit through the hole after being reflected once by each of the other three mirrors? (b) Are there other values of θ for which the ray can exit after multiple reflections? If so, make a sketch of one of the ray's paths.

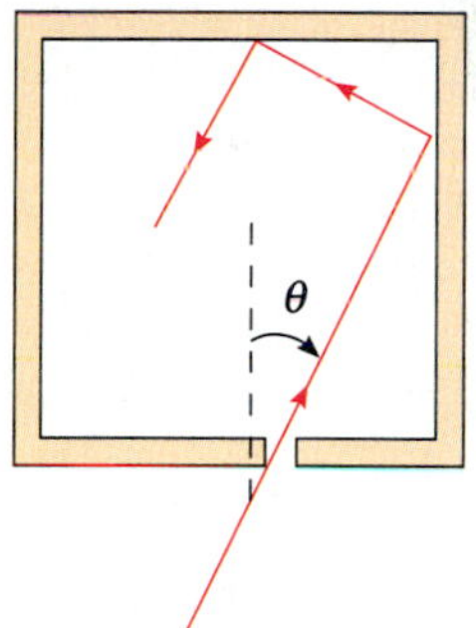

Figure 25.39 (Problem 52)

53. The light beam in Figure 25.40 strikes surface 2 at the critical angle. Determine the angle of incidence, θ_1.

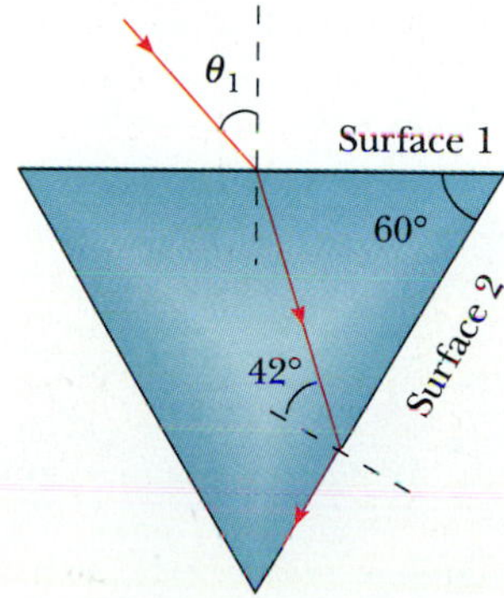

Figure 25.40 (Problem 53)

△54. Students in a laboratory allow a narrow beam of laser light to strike a water surface. They arrange to measure the angle of refraction for selected angles of incidence, and record the data in the following table. Use the data to verify Snell's law by plotting the sine of the angle of incidence versus the sine of the angle of refraction. Use the resulting plot to deduce the index of refraction of water.

Angle of Incidence (degrees)	Angle of Refraction (degrees)
10.0	7.5
20.0	15.1
30.0	22.3
40.0	28.7
50.0	35.2
60.0	40.3
70.0	45.3
80.0	47.7

By permission of John Hart and Field Enterprises, Inc.

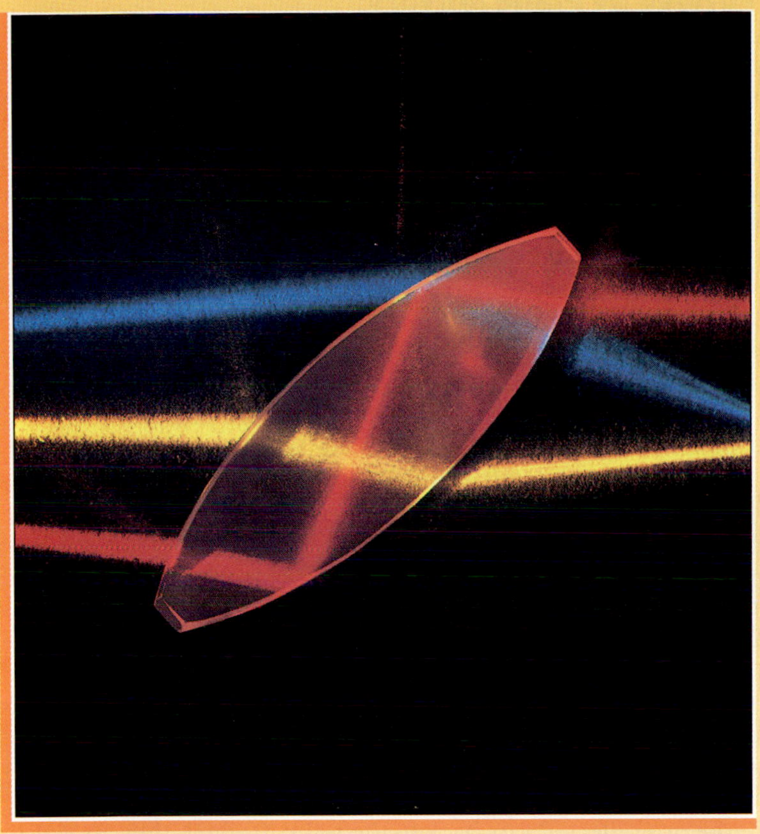

Three beams of light are incident on a biconvex lens. Part of the beams are reflected, and part refracted, by the glass lens at the left air-glass interface. The refracted beams entering the lens are then partially reflected and partially refracted at the opposite glass-air interface. How do you explain the fact that most of the red beam is reflected at the second glass-air interfaces, while the yellow and blue beams are mostly refracted at this interface? *(Richard Megna, Fundamental Photographs)*

Mirrors and Lenses

CHAPTER 26

This chapter is concerned with the formation of images when plane or spherical waves fall on plane and spherical surfaces. We shall find that images can be formed either by reflection or by refraction. Mirrors and lenses form images in both ways. In our study of mirrors and lenses, we shall continue to use the ray approximation and to assume that light travels in straight lines (in other words, we shall ignore diffraction). This chapter completes our study of geometric optics. In Chapters 27 and 28 we shall concern ourselves with interference and diffraction effects—the field of wave optics.

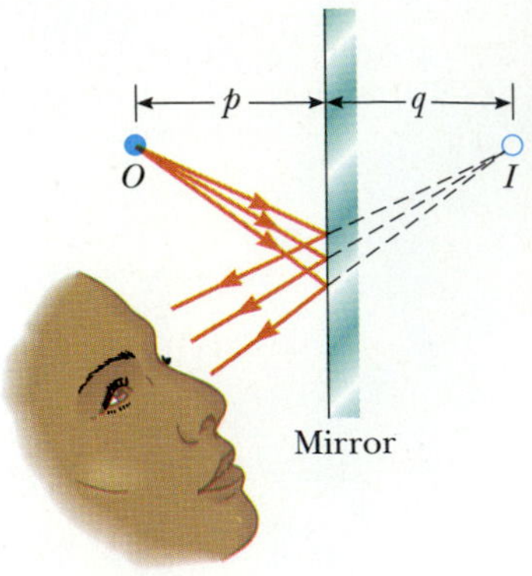

Figure 26.1
An image formed by reflection from a plane mirror. The image point, I, is located behind the mirror at a distance of q (called the image distance) which is equal to the object distance, p.

26.1 Images Formed by Plane Mirrors

The main objective of this chapter is to discuss the manner in which optical elements such as lenses and mirrors form images. We shall begin our investigation by considering the simplest possible mirror, the plane mirror.

Consider a point source of light placed at O in Figure 26.1, a distance of p in front of a plane mirror. The distance p is called the **object distance.** Light rays leave the source and are reflected from the mirror. After reflection, the rays diverge (spread apart), but they appear to the viewer to come from a point, I, behind the mirror. Point I is called the **image** of the object at O. Regardless of the system under study, images are always formed in the same way. *Images are formed at the point where rays of light intersect or at the point at which they appear to originate.* Since the rays in Figure 26.1 appear to originate at I, which is a distance of q behind the mirror, this is the location of the image. The distance q is called the **image distance.**

Images are classified as real or virtual. *A* ***real image*** *is one in which light actually intersects, or passes through, the image point; a* ***virtual image*** *is one in which the light does not really pass through the image point but appears to diverge from that point.* The image formed by the mirror in Figure 26.1 is a virtual image. The images seen in plane mirrors are *always virtual* for real objects. Real images can usually be displayed on a screen (as at a movie), but virtual images cannot be displayed on a screen.

We shall examine some of the properties of the images formed by plane mirrors by using the simple geometric techniques shown in Figure 26.2. In order to find out where an image is formed, it is always necessary to follow at least two rays of light as they reflect from the mirror. One of those rays starts at P, follows the horizontal path PQ to the mirror, and reflects back on itself. The second ray follows the oblique path PR and reflects as shown. An observer to the left of the mirror would trace the two reflected rays back to the point at which they appear to have originated, point P'. A continuation of this process for points other than P on the object would result in a virtual image (drawn as a yellow arrow) to the right of the mirror. Since triangles PQR and $P'QR$ are identical, $PQ = P'Q$. Hence, we conclude that *the image formed by an object placed in front of a plane mirror is as far behind the mirror as the object is in front of the mirror.* Geometry also shows that the object height, h, equals the image height, h'. Let us define **magnification,** M, as follows:

$$M \equiv \frac{\text{image height}}{\text{object height}} = \frac{h'}{h} \qquad [26.1]$$

This is a general definition of the magnification of any type of mirror. $M = 1$ for a plane mirror because $h' = h$ in this case.

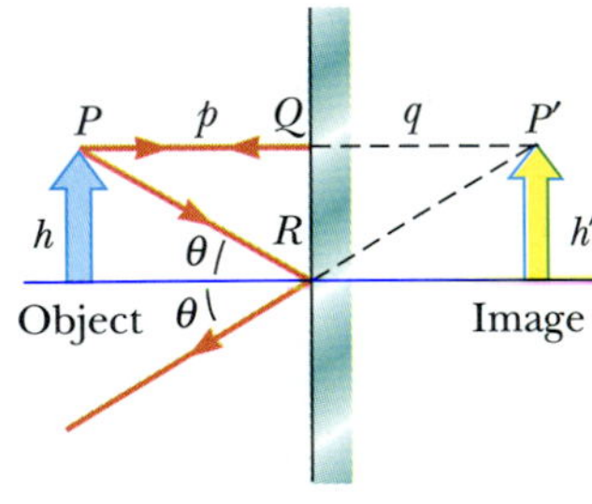

Figure 26.2
A geometric construction that is used to locate the image of an object placed in front of a plane mirror. Because the triangles PQR and $P'QR$ are identical, $p = q$ and $h = h'$.

It is also interesting to note that the observer assumes that the image formed by a plane mirror has right-left reversal. This reversal can be seen by standing in front of a mirror and raising your right hand. The image you see raises its left hand. Likewise, your hair appears to be parted on the opposite side and a mole on your right cheek appears to be on your left cheek.

In summary, the image formed by a plane mirror has the following properties:

1. **The image is as far behind the mirror as the object is in front.**
2. **The image is unmagnified, virtual, and erect. (By *erect* we mean that, if the object arrow points upward as in Figure 26.2, so does the image arrow. The opposite of an erect image is an inverted image.)**

Example 26.1 Multiple Images Formed by Two Mirrors

Two plane mirrors are at right angles to each other, as in Figure 26.3, and an object is placed at point O. In this situation, multiple images are formed. Locate the positions of those images.

Solution The image of the object is at I_1 in mirror 1 and at I_2 in mirror 2. In addition, a third image is formed at I_3, which will be considered the image of I_1 in mirror 2 or, equivalently, the image of I_2 in mirror 1. That is, the image at I_1 (or I_2) serves as the object for I_3. When viewing I_3, note that the rays reflect twice after leaving the object at O.

Exercise Sketch the rays that correspond to viewing the images at I_1 and I_2, and show that the light is reflected only once in these cases.

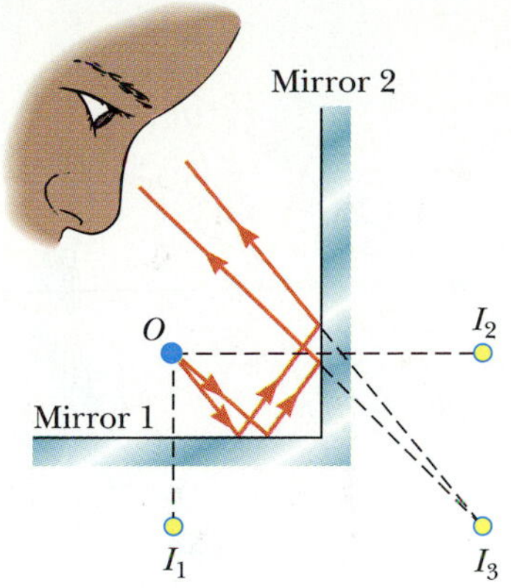

Figure 26.3 (Example 26.1) When an object is placed in front of two mutually perpendicular mirrors, three images are formed.

26.2 Images Formed by Spherical Mirrors

Concave Mirrors

A **spherical mirror,** as its name implies, has the shape of a segment of a sphere. Figure 26.4 shows the cross-section of a spherical mirror with light reflecting from its surface, represented by the solid curved line. Such a mirror, in which light is reflected from the inner, concave surface, is called a **concave mirror.** The mirror's radius of curvature is R, and its center of curvature is at point C. Point V is the center of the spherical segment, and a line drawn from C to V is called the **principal axis** of the mirror.

Now consider a point source of light placed at point O in Figure 26.4, on the principal axis and outside point C. Several diverging rays that originate at O are shown. After reflecting from the mirror, these rays converge and meet at I, called the **image point.** They then continue to diverge from I as if there were an object there. As a result, a real image is formed. *Whenever reflected light actually passes through a point, a real image is formed there.*

In what follows, we shall assume that all rays that diverge from an object make a small angle with the principal axis. Such rays, called **paraxial rays,** always reflect through the image point, as in Figure 26.4. Rays that are far from the principal

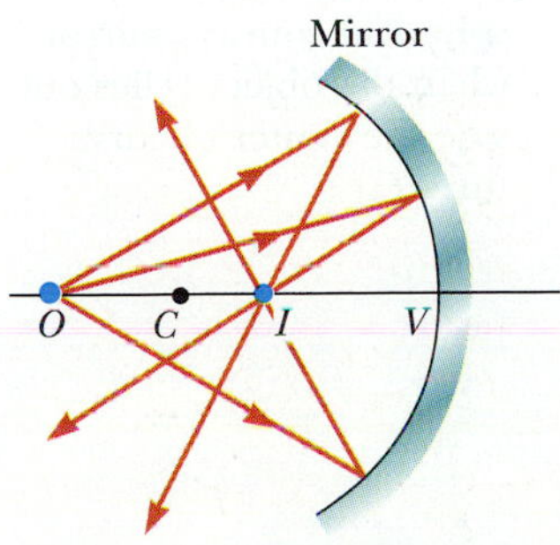

Figure 26.4 A point object at O, outside the center of curvature of a concave spherical mirror, forms a real image at I. If the rays diverage from O at small angles, they all reflect through the same image point.

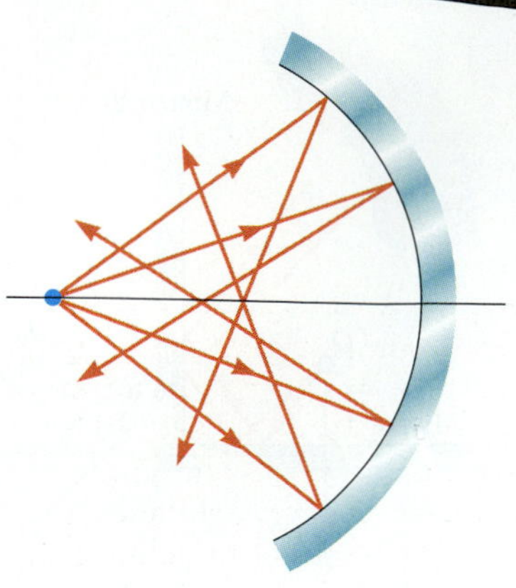

Figure 26.5
Rays at large angles from the principal axis reflect from a spherical concave mirror to intersect the principal axis at different points, resulting in a blurred image. This is called *spherical aberration.*

axis, as in Figure 26.5, converge at other points on the principal axis, producing a blurred image. This effect, called **spherical aberration,** is produced to some extent by any spherical mirror and will be discussed in Section 26.5.

We can use the geometry shown in Figure 26.6 to calculate the image distance, q, from the object distance, p, and radius of curvature, R. By convention, these distances are measured from point V. Figure 26.6 shows two rays of light leaving the tip of the object. One passes through the center of curvature, C, of the mirror, hitting the mirror head on (perpendicular to the mirror surface) and reflecting back on itself. The second ray strikes the mirror at the center, point V, and reflects as shown, obeying the law of reflection. The image of the tip of the arrow is at the point where these two rays intersect. From the largest right triangle in Figure 26.6 (colored gold), we see that $\tan\theta = h/p$, whereas the light blue triangle gives $\tan\theta = -h'/q$. The negative sign signifies that the image is inverted, so h' is negative. Thus, from Equation 26.1 and these results, we find that the magnification of the mirror is

$$M = \frac{h'}{h} = -\frac{q}{p} \tag{26.2}$$

We also note, from the two other right triangles in the figure, that

$$\tan\alpha = \frac{h}{p-R} \qquad \text{and} \qquad \tan\alpha = -\frac{h'}{R-q}$$

from which we find that

$$\frac{h'}{h} = -\frac{R-q}{p-R} \tag{26.3}$$

If we compare Equations 26.2 and 26.3, we see that

$$\frac{R-q}{p-R} = \frac{q}{p}$$

Simple algebra reduces this to

Mirror equation

$$\frac{1}{p} + \frac{1}{q} = \frac{2}{R} \tag{26.4}$$

This expression is called the **mirror equation.** It is applicable only to paraxial rays.

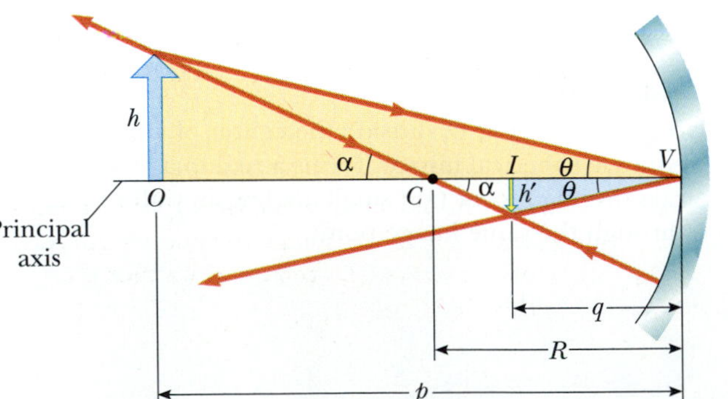

Figure 26.6
The image formed by a spherical concave mirror when the object O lies outside the center of curvature, C.

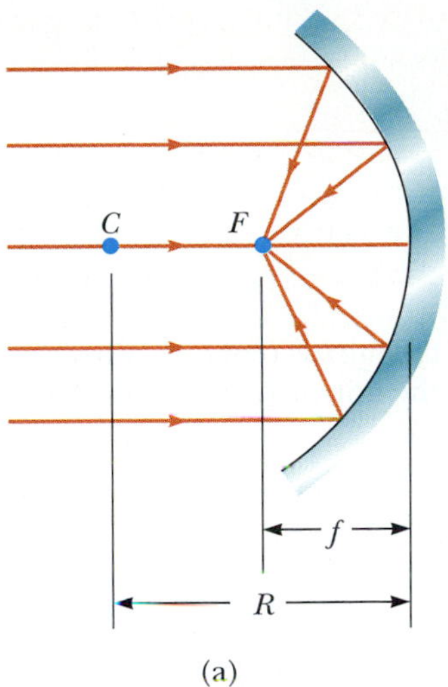

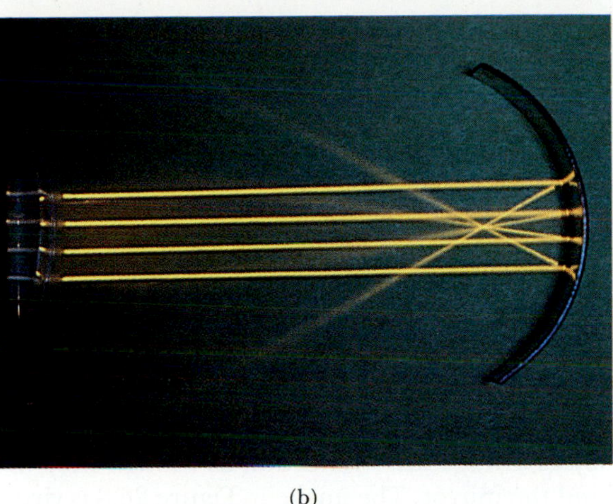

Figure 26.7
(a) Light rays from a distant object ($p = \infty$) reflect from a concave mirror and intersect each other at the focal point, F. In this case, the image distance is $q = R/2 = f$, where f is the focal length of the mirror. (b) Photograph of the reflection of parallel rays from a concave mirror. *(Courtesy of Henry Leap and Jim Lehman)*

If the object is very far from the mirror—that is, if the object distance, p, is great enough compared with R that p can be said to approach infinity, then $1/p \approx 0$, and we see from Equation 26.4 that $q \approx R/2$. In other words, when the object is very far from the mirror, *the image point is halfway between the center of curvature and the center of the mirror,* as in Figure 26.7a. The rays are essentially parallel in this figure because their source, the object, is assumed to be very far from the mirror. In this special case we call the image point the **focal point,** F, and the image distance the **focal length,** f, where

$$f = \frac{R}{2} \qquad [26.5]$$

Focal length

The mirror equation can therefore be expressed in terms of the focal length:

$$\frac{1}{p} + \frac{1}{q} = \frac{1}{f} \qquad [26.6]$$

Mirror equation

Note that objects at infinity are always focused at the focal point.

Convex Mirrors

Figure 26.8 shows the formation of an image by a **convex mirror**—a mirror that is silvered so that light is reflected from the outer, convex surface. Convex mirrors are sometimes called **diverging mirrors** because the rays from any point on a real

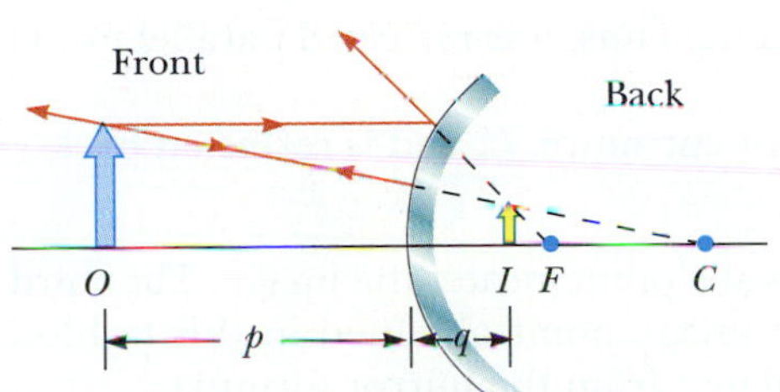

Figure 26.8
Formation of an image by a spherical convex mirror. The image formed by the real object is virtual, erect, and smaller than the object.

Table 26.1
Sign Conventions for Mirrors

p is + if the object is in front of the mirror (real object).
p is − if the object is in back of the mirror (virtual object).
q is + if the image is in front of the mirror (real image).
q is − if the image is in back of the mirror (virtual image).

Both f and R are + if the center of curvature is in front of the mirror (concave mirror).
Both f and R are − if the center of curvature is in back of the mirror (convex mirror).

If M is positive, the image is erect.
If M is negative, the image is inverted.

object diverge after reflection as though they were coming from some point behind the mirror. The image in Figure 26.8 is virtual rather than real because it lies behind the mirror at the point where the reflected rays appear to originate. In general, as shown in the figure, the image formed by a convex mirror is always erect, virtual, and smaller than the object.

We shall not derive any equations for convex spherical mirrors. If we did, we would find that the equations developed for concave mirrors can be used with convex mirrors if we adhere to a particular sign convention. Let us refer to the region in which light rays move as the *front side* of the mirror, and the other side, where virtual images are formed, as the *back side.* For example, in Figures 26.6 and 26.8, the side to the left of the mirror is the front side, and that to the right of the mirror is the back side. Table 26.1 summarizes the sign conventions for all the necessary quantities.

Ray Diagrams for Mirrors

The positions and sizes of images formed by mirrors can be conveniently determined by constructing *ray diagrams* similar to the ones we have been using. These graphical constructions tell us the overall nature of the image and can be used to check parameters calculated from the mirror and magnification equations. To make a ray diagram, one needs to know the position of the object and the location of the center of curvature. In order to locate the image, three rays are constructed (rather than just the two we have been constructing so far), as shown by the examples in Figure 26.9. All three rays start from the same object point; in these examples, the top of the arrow was chosen. For the concave mirrors in Figure 26.9a and b, the rays are drawn as follows:

1. **Ray 1 is drawn parallel to the principal axis and is reflected back through the focal point, F.**
2. **Ray 2 is drawn through the focal point. Thus, it is reflected parallel to the principal axis.**
3. **Ray 3 is drawn through the center of curvature, C, and is reflected back on itself.**

The intersection of any *two* of these rays at a point locates the image. The third ray serves as a check of construction. The image point obtained in this fashion must always agree with the value of q calculated from the mirror formula.

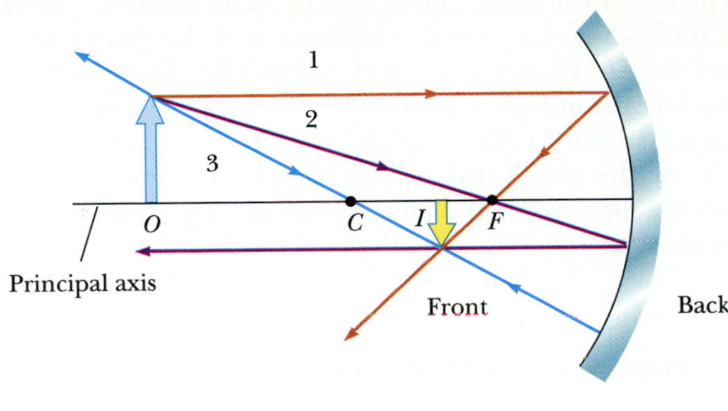

(a)

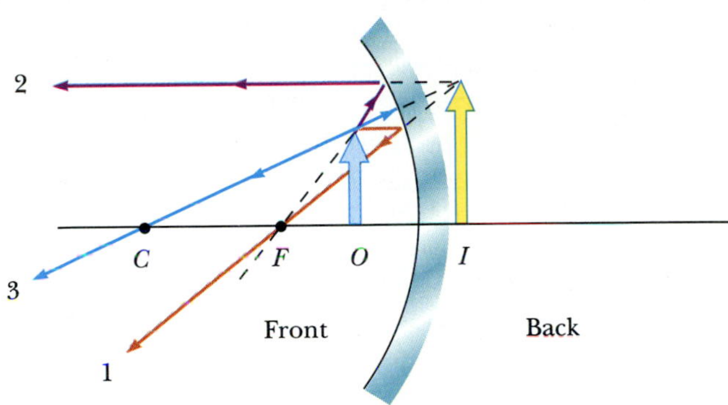

(b)

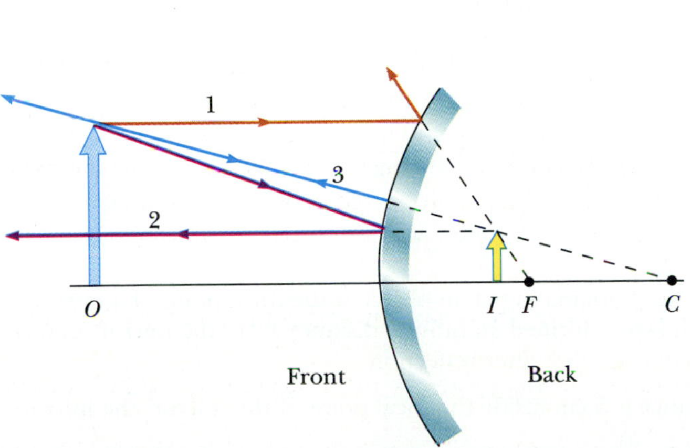

(c)

Figure 26.9
Ray diagrams for spherical mirrors. Different colors are used for the three rays so they can easily be followed. (a) An object outside the center of curvature of a spherical concave mirror. (b) An object between the spherical concave mirror and the focal point, *F*, gives a virtual, erect image. (c) An object located anywhere in front of a spherical convex mirror gives a virtual, erect image.

In the case of a concave mirror, note what happens as the object is moved closer to the mirror. The real, inverted image in Figure 26.9a moves to the left as the object approaches the focal point. When the object is at the focal point, the image is infinitely far to the left. However, when the object lies between the focal point and the mirror surface, as in Figure 26.9b, the image is virtual and erect.

For the convex mirror shown in Figure 26.9c, the image of a real object is always virtual and erect. As the object distance increases, the virtual image gets smaller and approaches the focal point as p approaches infinity. You should construct a ray diagram to verify this.

Example 26.2 The Image for a Concave Mirror

Assume that a certain concave spherical mirror has a focal length of 10 cm. Find the locations of the image for object distances of (a) 25 cm, (b) 10 cm, and (c) 5 cm. Describe the image in each case.

Solution (a) For an object distance of 25 cm, we find the image distance using the mirror equation:

$$\frac{1}{p} + \frac{1}{q} = \frac{1}{f}$$

$$\frac{1}{25\text{ cm}} + \frac{1}{q} = \frac{1}{10\text{ cm}}$$

$$q = \boxed{16.7\text{ cm}}$$

The magnification is given by Equation 26.2:

$$M = -\frac{q}{p} = -\frac{16.7\text{ cm}}{25\text{ cm}} = \boxed{-0.668}$$

Thus, the image is smaller than the object. Furthermore, the image is inverted because M is negative. Finally, because q is positive, the image is on the front side of the mirror and is therefore real. This situation is pictured in Figure 26.9a.

(b) When the object distance is 10 cm, the object is located at the focal point. Substituting in the mirror equation, we find

$$\frac{1}{10\text{ cm}} + \frac{1}{q} = \frac{1}{10\text{ cm}}$$

$$q = \boxed{\infty}$$

Thus, we see that rays of light originating at an object at the focal point of a mirror are reflected so that the image is formed an infinite distance from the mirror; that is, the rays travel parallel to one another after reflection.

(c) When the object distance is 5 cm inside the focal point of the mirror, the mirror equation gives

$$\frac{1}{5\text{ cm}} + \frac{1}{q} = \frac{1}{10\text{ cm}}$$

$$q = \boxed{-10\text{ cm}}$$

That is, the image is virtual since it is behind the mirror. The magnification is

$$M = -\frac{q}{p} = -\left(\frac{-10 \text{ cm}}{5 \text{ cm}}\right) = 2$$

From this, we see that the image is magnified by a factor of 2, and the positive sign indicates that the image is erect (Fig. 26.9b).

Note the characteristics of the images formed by a concave spherical mirror. When the object is outside the focal point, the image is inverted and real; at the focal point, the image is formed at infinity; inside the focal point, the image is erect and virtual.

Exercise If the object distance is 20 cm, find the image distance and the magnification of the mirror.

Answer $q = 20$ cm, $M = -1$.

Convex cylindrical mirror: reflection of parallel lines. The image of any object in front of the mirror is virtual, erect, and diminished in size. *(© Richard Megna 1990, Fundamental Photographs)*

Example 26.3 **The Image for a Convex Mirror**

An object 3 cm high is placed 20 cm from a convex mirror having a focal length of 8 cm. Find (a) the position of the final image and (b) the magnification of the mirror.

Solution (a) Since the mirror is convex, its focal length is negative. To find the image position, we use the mirror equation:

$$\frac{1}{p} + \frac{1}{q} = \frac{1}{f} = -\frac{1}{8 \text{ cm}}$$

$$\frac{1}{q} = -\frac{1}{8 \text{ cm}} - \frac{1}{20 \text{ cm}}$$

$$q = -5.71 \text{ cm}$$

The negative value of q indicates that the image is virtual, or behind the mirror, as in Figure 26.9c.

(b) The magnification of the mirror is

$$M = -\frac{q}{p} = -\left(\frac{-5.71 \text{ cm}}{20 \text{ cm}}\right) = 0.286$$

The image is erect because M is positive.

Exercise Find the height of the image.

Answer 0.857 cm.

26.3 Images Formed by Refraction

In this section we shall describe how images are formed by the refraction of rays at a spherical surface of a transparent material. Consider two transparent media with indices of refraction n_1 and n_2, where the boundary between the two media is a

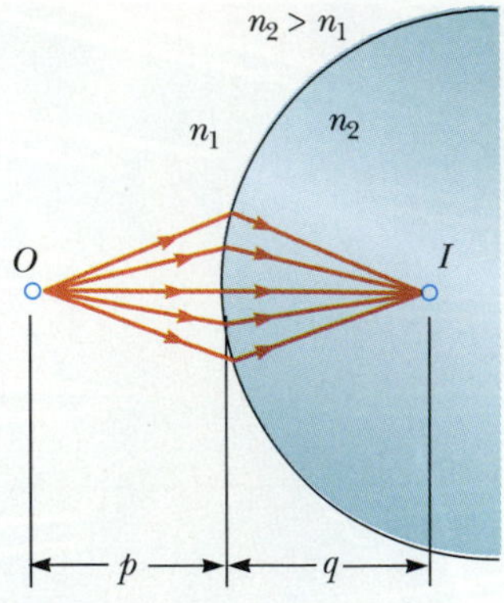

Figure 26.10
An image formed by refraction at a spherical surface. Paraxial rays making small angles with the axis diverge from a point object at O and pass through the image point I.

spherical surface whose radius of curvature is R (Fig. 26.10). We shall assume that the object at point O is in the medium whose index of refraction is n_1. Furthermore, let us consider only paraxial rays leaving O. As we shall see, all such rays are refracted at the spherical surface and focus at a single point, I, the image point.

Let us proceed by considering the geometric construction in Figure 26.11, which shows a single ray leaving point O and focusing at point I. Snell's law applied to this refracted ray gives

$$n_1 \sin \theta_1 = n_2 \sin \theta_2$$

Because the angles θ_1 and θ_2 are assumed to be small, we can use the approximation $\sin \theta \approx \theta$ (angles in radians). Therefore, Snell's law becomes

$$n_1 \theta_1 = n_2 \theta_2$$

Now we make use of the fact that an exterior angle of any triangle equals the sum of the two opposite interior angles. Applying this to the triangles OPC and PIC in Figure 26.11 gives

$$\theta_1 = \alpha + \beta$$

$$\beta = \theta_2 + \gamma$$

If we combine the last three equations and eliminate θ_1 and θ_2, we find

$$n_1 \alpha + n_2 \gamma = (n_2 - n_1)\beta \qquad [26.7]$$

In the small angle approximation, $\tan \theta \approx \theta$, and so we can write the approximate relations

$$\alpha = \frac{d}{p} \qquad \beta = \frac{d}{R} \qquad \gamma = \frac{d}{q}$$

where d is the distance shown in Figure 26.11. We substitute these into Equation 26.7 and divide through by d to give

$$\frac{n_1}{p} + \frac{n_2}{q} = \frac{n_2 - n_1}{R} \qquad [26.8]$$

For a fixed object distance of s, the image distance, q, is independent of the angle that the ray makes with the axis. This tells us that all paraxial rays focus at the same point, I. The magnification of a refracting surface is

$$M = \frac{h'}{h} = -\frac{n_1 q}{n_2 p} \qquad [26.9]$$

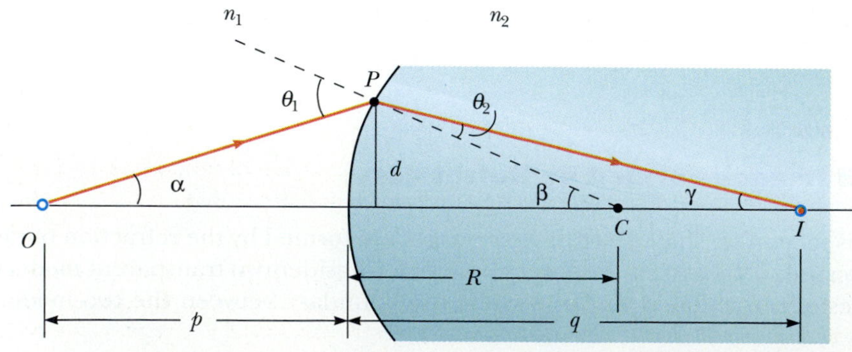

Figure 26.11
Geometry used to derive Equation 26.8.

Table 26.2
Sign Conventions for Refracting Surfaces

p is + if the object is in front of the surface (real object).
p is − if the object is in back of the surface (virtual object).

q is + if the image is in back of the surface (real image).
q is − if the image is in front of the surface (virtual image).

R is + if the center of curvature is in back of the surface.
R is − if the center of curvature is in front of the surface.

As we did with mirrors, we must use a sign convention if we are to apply Equation 26.8 to a variety of circumstances. First note that real images are formed on the side of the surface that is *opposite* the side from which the light comes. This is in contrast to mirrors, where real images are formed on the side where the light is. Therefore, *the sign conventions for spherical refracting surfaces are similar to the conventions for mirrors, recognizing the change in sides of the surface for real and virtual images.* For example, in Figure 26.11, p, q, and R are all positive.

The sign conventions for spherical refracting surfaces are summarized in Table 26.2. The same conventions will be used for thin lenses, discussed in the next section. As with mirrors, we assume that the front of the refracting surface is the side from which the light approaches the surface.

Plane Refracting Surfaces

If the refracting surface is a plane, then R approaches infinity and Equation 26.8 reduces to

$$\frac{n_1}{p} = -\frac{n_2}{q}$$

or

$$q = -\frac{n_2}{n_1}p \qquad [26.10]$$

From Equation 26.10 we see that the sign of q is opposite that of p. Thus, *the image formed by a plane refracting surface is on the same side of the surface as the object.* This is illustrated in Figure 26.12 for the situation in which n_1 is greater than n_2, where a virtual image is formed between the object and the surface. Note that the refracted ray bends *away* from the normal in this case, because $n_1 > n_2$.

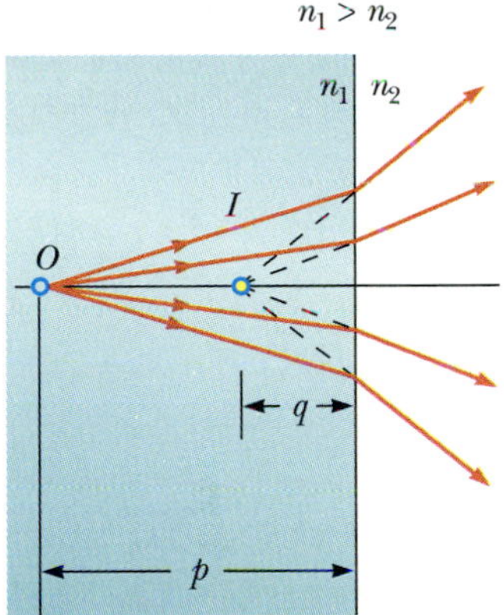

Figure 26.12
The image formed by a plane refracting surface is virtual; that is, it forms to the left of the refracting surface. All rays are assumed to be paraxial.

Example 26.4 Gaze into the Crystal Ball

A coin 2 cm in diameter is embedded in a solid glass ball of radius 30 cm (Fig. 26.13). The index of refraction of the ball is 1.5, and the coin is 20 cm from the surface. Find the position of the image.

Solution Because they are moving from a medium of high index of refraction to a medium of lower index of refraction, the rays originating at the object are refracted away from the normal at the surface and diverge outward. The image is formed in

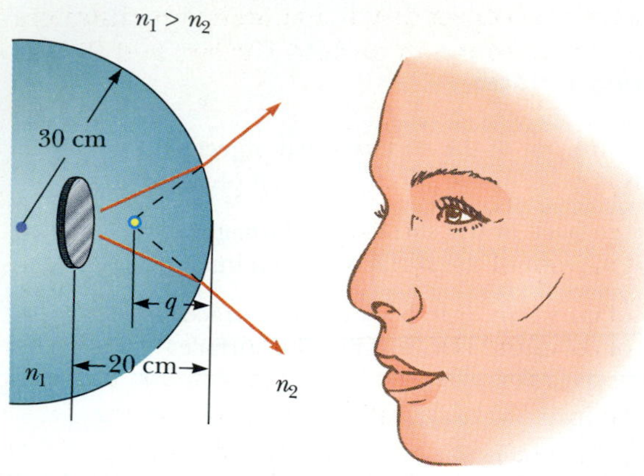

Figure 26.13 (Example 26.4) A coin embedded in a glass ball forms a virtual image between the coin and the glass surface.

the glass and is virtual. Applying Equation 26.8 and taking $n_1 = 1.5$, $n_2 = 1$, $p = 20$ cm, and $R = -30$ cm, we get

$$\frac{1.5}{20 \text{ cm}} + \frac{1}{q} = \frac{1 - 1.5}{-30 \text{ cm}}$$

$$q = -17.1 \text{ cm}$$

The negative sign indicates that the image is in the same medium as the object (the side of incident light), in agreement with our ray diagram. Because the image is in the same medium as the object, it must be virtual.

Exercise Find the diameter of the coin's image.

Answer 2.58 cm.

Example 26.5 The One That Got Away

A small fish is swimming at a depth of d below the surface of a pond (Fig. 26.14). What is the apparent depth of the fish as viewed from directly overhead?

Solution In this example, the refracting surface is a plane, and so R is infinite. Hence, we can use Equation 26.10 to determine the location of the image. Using the facts that $n_1 = 1.33$ for water and $p = d$, we find

$$q = -\frac{n_2}{n_1} p = -\frac{1}{1.33} d = -0.750d$$

Again, since q is negative, the image is virtual, as indicated in Figure 26.14. The apparent depth is three-fourths the actual depth. For instance, if $d = 4$ m, then $q = -3$ m.

Exercise If the fish is 10 cm long, how long is its image?

Answer 10 cm.

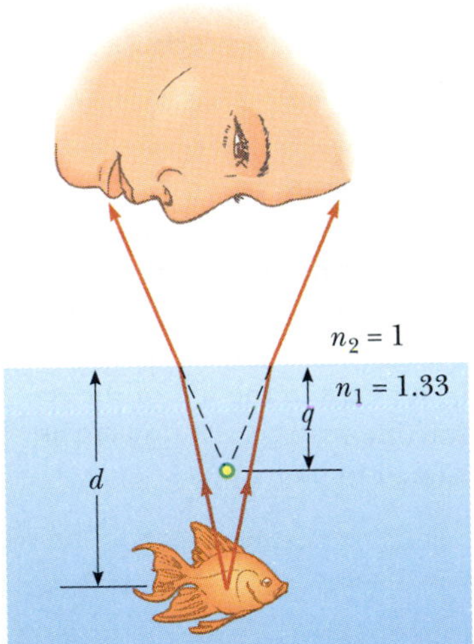

Figure 26.14 (Example 26.5) The apparent depth, q, of the fish is less than the true depth, d. All rays are assumed to be paraxial.

26.4 Thin Lenses

A typical **thin lens** consists of a piece of glass or plastic, ground so that its two surfaces are either segments of spheres or planes. Lenses are commonly used in optical instruments, such as cameras, telescopes, and microscopes, to form images by refraction. The equation that relates object distances and image distances for a lens is virtually identical to the mirror equation derived earlier, and the method used to derive it is also similar.

Figure 26.15 shows some representative shapes of lenses. Note that we have placed these lenses in two groups. Those in Figure 26.15a are thicker at the center than at the rim, and those in Figures 26.15b are thinner at the center than at the rim. The lenses in the first group are examples of **converging lenses,** and those in the second group are called **diverging lenses.** The reason for these names will become apparent shortly.

As we did with mirrors, it is convenient to define a point called the **focal point** for a lens. For example, in Figure 26.16a a group of rays parallel to the principal axis passes through the focal point, F, after being converged by the lens. The distance from the focal point to the lens is again called the **focal length,** f. *The focal length is the image distance that corresponds to an infinite object distance.* Recall that we are considering the lens to be very thin. As a result, it makes no difference whether we take the focal length to be the distance from the focal point to the surface of the lens or the distance from the focal point to the center of the lens, because the difference in these two lengths is negligible. A thin lens has *two* focal points, as illustrated in Figure 26.16b, corresponding to parallel light rays traveling from the left or right.

Rays parallel to the axis diverge after passing through a lens of the shape shown in Figure 26.16b. In this case, the focal point is defined as the point at which

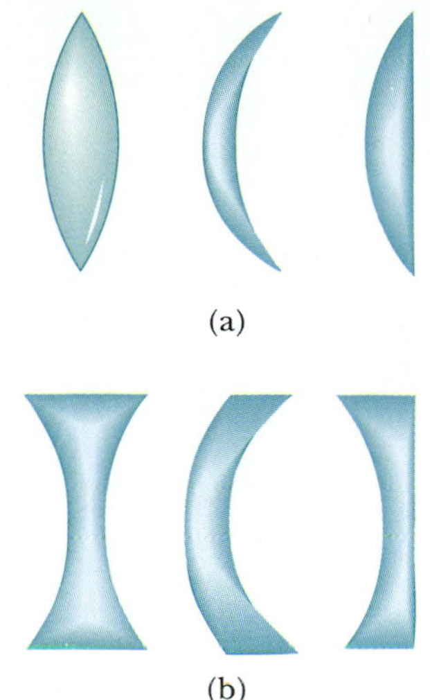

Figure 26.15
Lens shapes. (a) Converging lenses have positive focal lengths and are thickest at the middle. From left to right, these are biconvex, convex-concave, and plano-convex. (b) Diverging lenses have negative focal lengths and are thickest at the edges. From left to right are biconcave, convex-concave, and plano-concave lenses.

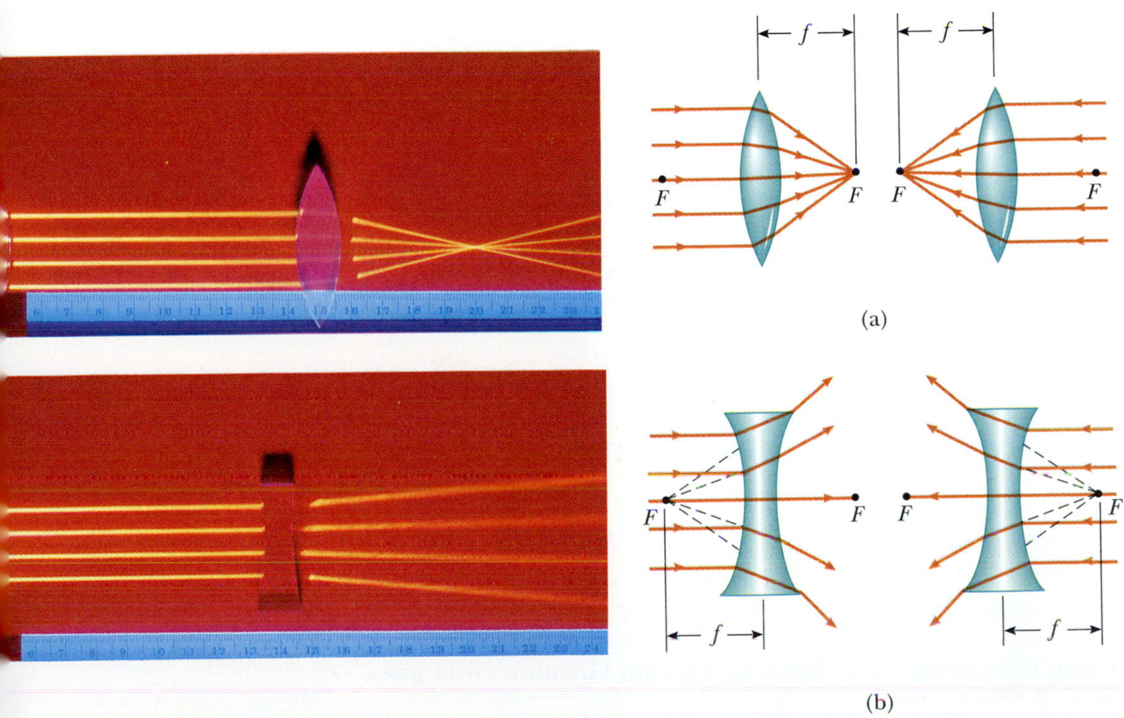

Figure 26.16
(Left) Photographs of the effect of converging and diverging lenses on parallel rays. *(Courtesy of Jim Lehman/James Madison University)* *(Right)* The focal points of (a) the biconvex lens and (b) the biconcave lens.

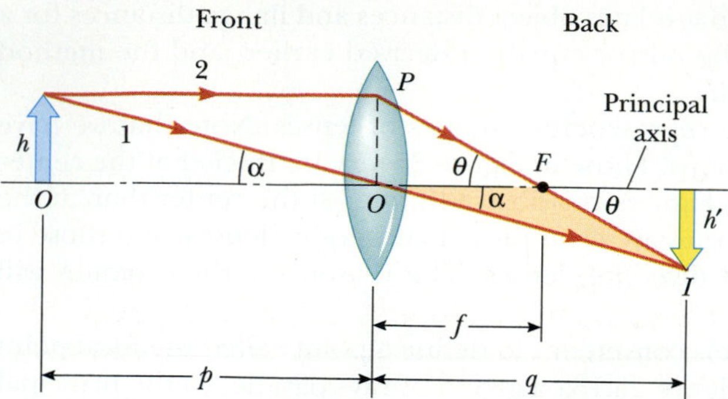

Figure 26.17
A geometric construction for developing the thin-lens equation.

the diverged rays appear to originate, labeled F in the figure. Figures 26.16a and 26.16b indicate why the names *converging* and *diverging* are applied to these lenses.

Consider a ray of light passing through the center of a lens, shown as ray 1 in Figure 26.17. If we apply Snell's law at both surfaces, we find that this ray is deflected from its original direction of travel by a distance of δ, shown in Figure 26.18. Here, in order to avoid the complications arising from this deflection, we shall make what is called the *thin-lens approximation: the thickness of the lens is assumed to be negligible.* As a result, δ becomes vanishingly small and we see that the ray passes through the lens undeflected. Ray 2 in Figure 26.17 is parallel to the principal axis of the lens (the horizontal axis passing through O), and as a result it passes through the focal point, F, after refraction. The point at which these two rays intersect is the image point.

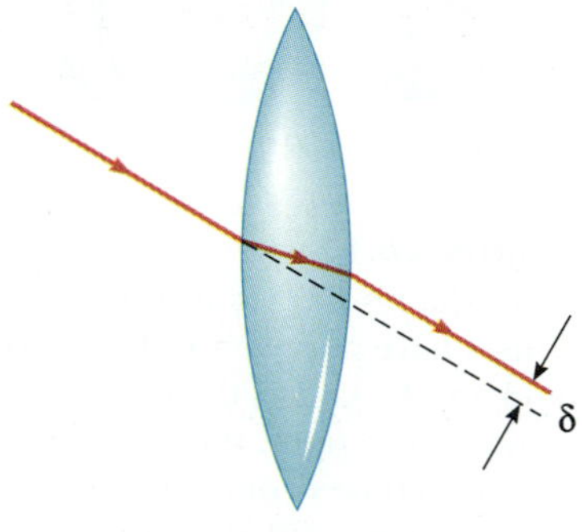

Figure 26.18
A ray passing through the center of the lens is deviated from its path by a distance of δ. In the thin-lens approximation, this deviation is ignored.

We first note that the tangent of the angle α can be found by using the shaded triangles in Figure 26.17:

$$\tan \alpha = \frac{h}{p} \quad \text{or} \quad \tan \alpha = -\frac{h'}{q}$$

from which

$$M = \frac{h'}{h} = -\frac{q}{p} \qquad \textbf{[26.11]}$$

Thus, the equation for magnification by a lens is the same as the equation for magnification by a mirror. (Eq. 26.2). We also note from Figure 26.17 that the tangent of θ is

$$\tan \theta = \frac{PO}{f} \quad \text{or} \quad \tan \theta = -\frac{h'}{q-f}$$

However, the height PO is the same as h. Therefore,

$$\frac{h}{f} = -\frac{h'}{q-f}$$

$$\frac{h'}{h} = -\frac{q-f}{f}$$

Using this expression in combination with Equation 26.11 gives

$$\frac{q}{p} = \frac{q-f}{f}$$

Table 26.3
Sign Conventions for Thin Lenses

p is + if the object is in front of the lens.
p is − if the object is in back of the lens.

q is + if the image is in back of the lens.
q is − if the image is in front of the lens.

R_1 and R_2 are + if the center of curvature for each surface is in back of the lens.
R_1 and R_2 are − if the center of curvature for each surface is in front of the lens.

f is + for a converging lens.
f is − for a diverging lens.

p = object distance; q = image distance; R_1 = radius of curvature of front surface; R_2 = radius of curvature of back surface; f = focal length.

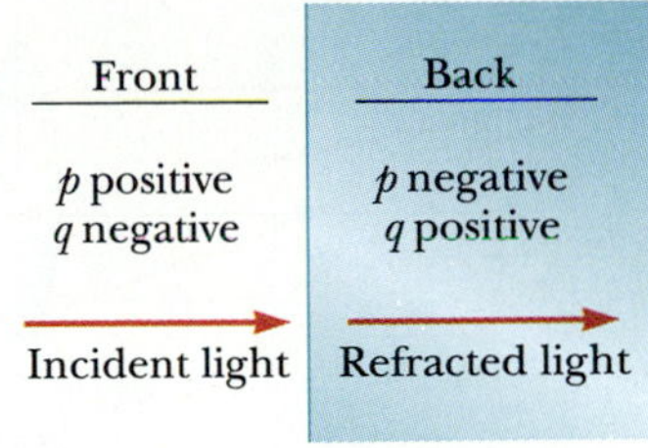

Figure 26.19
Diagram for obtaining the signs of p and q for a thin lens or a refracting surface. Although the interface between the two media is shown flat here, the diagram is also valid for convex and concave surfaces.

which reduces to

$$\frac{1}{p} + \frac{1}{q} = \frac{1}{f} \quad [26.12]$$

Thin-lens equation

This equation, called the **thin-lens equation,** can be used with either converging or diverging lenses if we adhere to a set of sign conventions. Figure 26.19 is useful for obtaining the signs of p and q, and the complete sign conventions for lenses are provided in Table 26.3. Note that *a converging lens has a positive focal length* under this convention and *a diverging lens has a negative focal length.* Hence, the names *positive* and *negative* are often given to these lenses.

The focal length for a lens in air is related to the curvatures of its surfaces and to the index of refraction, n, of the lens material by

$$\frac{1}{f} = (n - 1)\left(\frac{1}{R_1} - \frac{1}{R_2}\right) \quad [26.13]$$

Lens makers' equation

where R_1 is the radius of curvature of the front surface and R_2 is the radius of curvature of the back surface. (As with mirrors, we arbitrarily call the side from which the light approaches the *front* of the lens.) Equation 26.13 enables us to calculate the focal length from the known properties of the lens. It is called the **lens makers' equation.**

Ray Diagrams for Thin Lenses

Ray diagrams are very convenient for locating the image of a thin lens or system of lenses. They should also help clarify the sign conventions we have already discussed. Figure 26.20 illustrates this method for three single-lens situations. To locate the image of a converging lens (Fig. 26.20a and b), the following three rays are drawn from the top of the object:

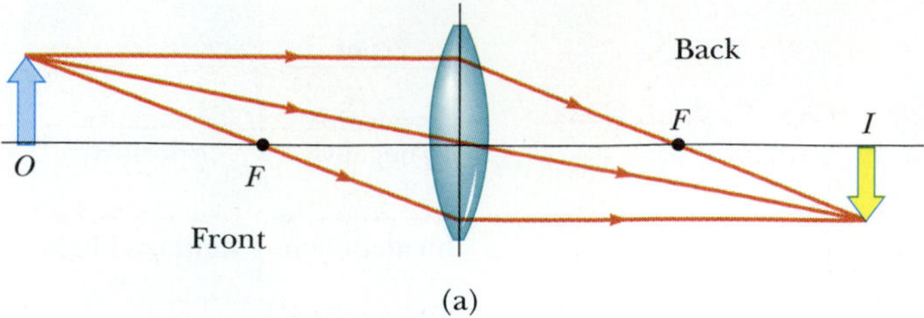

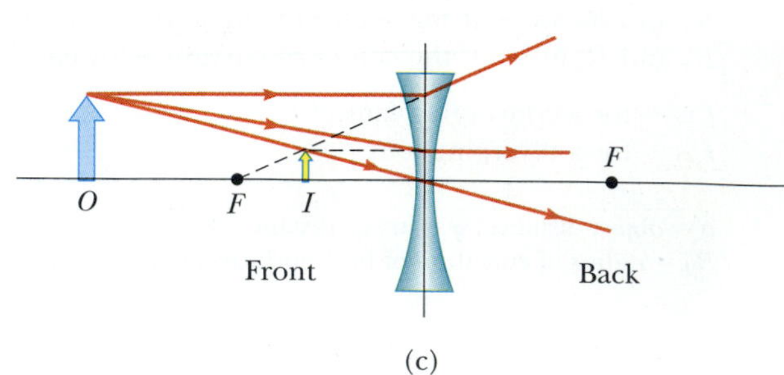

Figure 26.20
Ray diagrams for locating the image formed by thin lenses. (a) The object is outside the front focal point of a converging lens. (b) The object is inside the focal point of a converging lens. (c) The object is outside the front focal point of a diverging lens.

1. **The first ray is drawn parallel to the principal axis. After being refracted by the lens, this ray passes through (or appears to come from) one of the focal points.**
2. **The second ray is drawn through the center of the lens. This ray continues in a straight line.**
3. **The third ray is drawn through the focal point, *F*, and emerges from the lens parallel to the principal axis.**

A similar construction is used to locate the image of a diverging lens, as shown in Figure 26.20c. The point of intersection of *any two* of the rays in these diagrams can be used to locate the image. The third ray serves as a check of construction.

For the converging lens in Figure 26.20a, where the object is *outside* the front focal point ($p > f$), the image is real and inverted. When the real object is *inside* the front focal point ($p < f$), as in Figure 26.20b, the image is virtual and erect. For the diverging lens of Figure 26.20c, the image is virtual and erect.

Problem-Solving Strategy: Lenses and Mirrors

Success or failure in working lens and mirror problems is largely determined by whether or not you make sign errors when substituting into the equations, and the only way to ensure that you don't make such errors is to become adept at using the sign conventions. The best way to do this is to work a multitude of problems on your own. Watching your instructor or reading the example problems is no substitute for practice.

Example 26.6 The Lens Makers' Equation

The biconvex lens of Figure 26.21 has an index of refraction of 1.50. The radius of curvature of the front surface is $R_1 = 10$ cm, and that of the back surface is $R_2 = -15$ cm. Find the focal length of the lens.

Solution From the sign conventions in Table 26.3 we find that $R_1 = +10$ cm and $R_2 = -15$ cm. Thus, using the lens makers' equation, we have

$$\frac{1}{f} = (n-1)\left(\frac{1}{R_1} - \frac{1}{R_2}\right) = (1.5-1)\left(\frac{1}{10\text{ cm}} - \frac{1}{-15\text{ cm}}\right)$$

$$f = \boxed{12\text{ cm}}$$

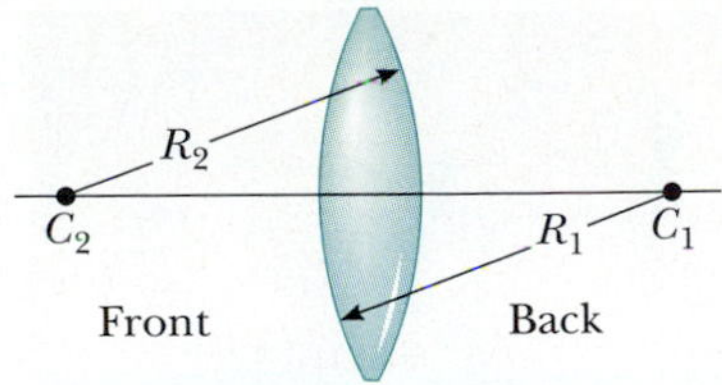

Figure 26.21 (Example 26.6) This lens has two curved surfaces with radii of curvature R_1 and R_2.

Example 26.7 An Image Formed by a Converging Lens

A converging lens of focal length 10 cm forms images of objects placed (a) 30 cm, (b) 10 cm, and (c) 5 cm in front of the lens. In each case, find the image distance and describe the image.

Solution (a) The thin-lens equation, Equation 26.12, can be used to find the image distance:

$$\frac{1}{p} + \frac{1}{q} = \frac{1}{f}$$

$$\frac{1}{30\text{ cm}} + \frac{1}{q} = \frac{1}{10\text{ cm}}$$

$$q = \boxed{15\text{ cm}}$$

The positive sign for the image distance tells us that the image is on the back side of the lens (Table 26.3). The magnification is

$$M = -\frac{q}{p} = -\frac{15\text{ cm}}{30\text{ cm}} = \boxed{-0.50}$$

The image is reduced in size by one half, and the negative sign for M tells us that the image is inverted. The situation is like that depicted in Figure 26.20a.

(b) No calculation is necessary for this case because we know that, when the object is at the focal point, the image is formed at infinity. This is readily verified by substituting $p = 10$ cm into the lens equation.

(c) We now move inside the focal point, to an object distance of 5 cm. In this case the lens equation gives

$$\frac{1}{5\text{ cm}} + \frac{1}{q} = \frac{1}{10\text{ cm}}$$

$$q = \boxed{-10\text{ cm}}$$

$$M = -\frac{q}{p} = -\left(\frac{-10\text{ cm}}{5\text{ cm}}\right) = \boxed{2}$$

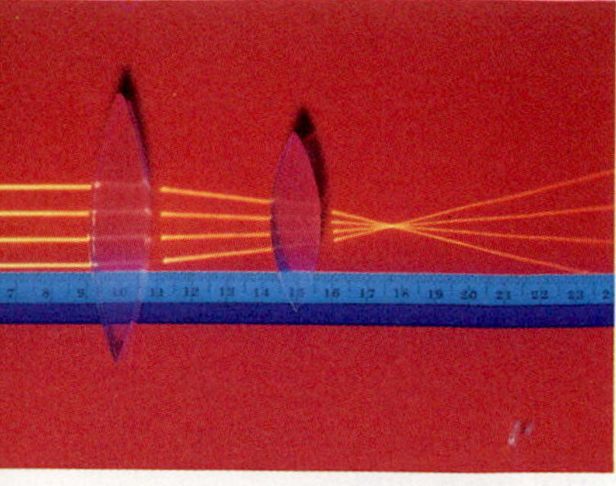

Light from a distant object brought into focus by two converging lenses. *(Courtesy of Henry Leap and Jim Lehman)*

The negative image distance tells us that the image is formed on the side of the lens from which the light is incident, the front side (Table 26.3). The image is enlarged, and the positive sign for M tells us that the image is erect, as shown in Figure 26.20b.

Exercise Consider a diverging lens of focal length -10 cm. What are the image distance and magnification for an object placed (a) 10 cm and (b) 5 cm from the lens?

Answer (a) -5 cm, 0.50; (b) -3.33 cm, 0.66.

Combination of Thin Lenses

If two thin lenses are used to form an image, the system can be treated in the following manner. First, the image of the first lens is calculated as though the second lens were not present. The light then approaches the second lens *as if* it had come from the image formed by the first lens. Hence, the image of the first lens is treated as the object of the second lens. The image of the second lens is the final image of the system. If the image of the first lens lies on the back side of the second lens, then the image is treated as a virtual object for the second lens (that is, p is negative). The same procedure can be extended to a system of three or more lenses. The overall magnification of a system of thin lenses equals the *product* of the magnifications of the separate lenses.

Example 26.8 Two Lenses in a Row

Two converging lenses are placed 20 cm apart, as shown in Figure 26.22. If the first lens has a focal length of 10 cm, and the second has a focal length of 20 cm, locate the final image of an object 30 cm in front of the first lens. Find the magnification of the system.

Solution The location of the image formed by the first lens is found by the thin-lens equation:

$$\frac{1}{30\text{ cm}} + \frac{1}{q} = \frac{1}{10\text{ cm}}$$

$$q = 15\text{ cm}$$

and the magnification of this lens is

$$M_1 = -\frac{q}{p} = -\frac{15\text{ cm}}{30\text{ cm}} = -\frac{1}{2}$$

The image formed by this lens becomes the object for the second lens. Thus, the object distance for the second lens is 5 cm (in front of the lens). We now apply the thin-lens equation again to find the location of the final image.

$$\frac{1}{5\text{ cm}} + \frac{1}{q} = \frac{1}{20\text{ cm}}$$

$$q = -6.67\text{ cm}$$

The magnification of the second lens is

$$M_2 = -\frac{q}{p} = -\frac{(-6.67\text{ cm})}{5\text{ cm}} = 1.33$$

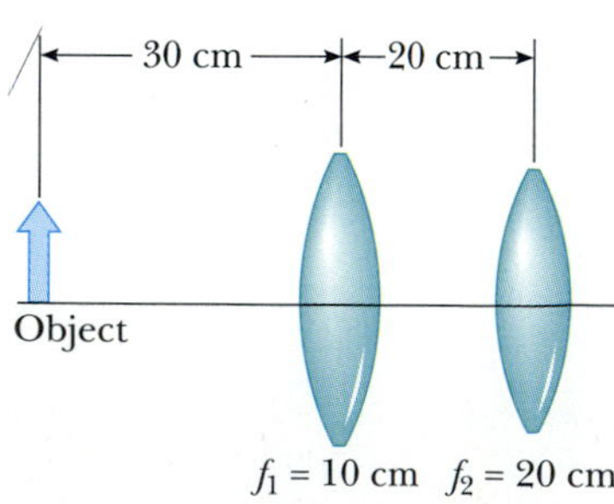

Figure 26.22 (Example 26.8)

Thus, the final image is 6.67 cm in front of the second lens, and the overall magnification of the system is

$$M = M_1 M_2 = \left(-\frac{1}{2}\right)(1.33) = \boxed{-0.667}$$

The negative sign indicates that the final image is inverted with respect to the initial object.

Exercise If the two lenses in Figure 26.22 are separated by 10 cm, find the location of the final image and the magnification of the system.

Answer 4 cm behind the second lens; $M = -0.40$.

*26.5 Lens Aberrations

One of the basic problems of lenses and lens systems is the imperfect quality of the images, which is largely the result of defects in shape and form. The simple theory of mirrors and lenses assumes that rays make small angles with the principal axis and that all rays reaching the lens or mirror from a point source are focused at a single point, producing a sharp image. Clearly, this is not always true in the real world. Where the approximations used in this theory do not hold, imperfect images are formed.

If one wishes to precisely analyze image formation, it is necessary to trace each ray, using Snell's law, at each refracting surface. This procedure shows that there is no single point image; instead, the image is *blurred.* The departures of real (imperfect) images from the ideal predicted by the simple theory are called **aberrations.** Two types of aberrations will now be described.

Spherical Aberrations

Spherical aberrations result from the fact that the focal points of light rays far from the principal axis of a spherical lens (or mirror) are different from the focal points of rays of the same wavelength passing near the axis. Figure 26.23 illustrates spherical aberration for parallel rays passing through a converging lens. Rays near the middle of the lens are imaged farther from the lens than rays at the edges. Hence, there is no single focal length for a lens.

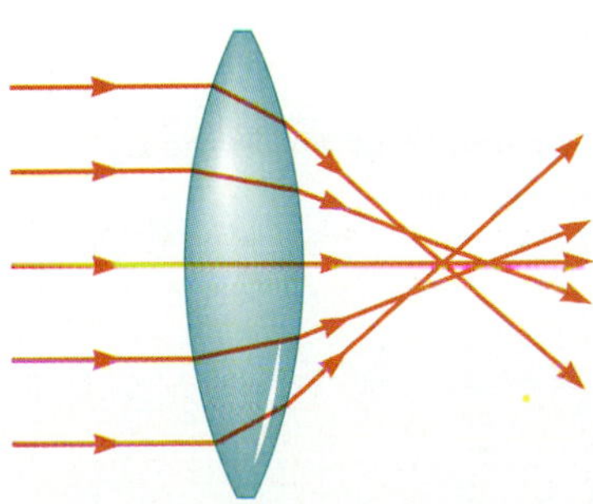

Figure 26.23 Spherical aberration caused by a converging lens. Does a diverging lens cause spherical aberration?

Most cameras are equipped with an adjustable aperture to control the light intensity and reduce spherical aberration when possible. (An aperture is an opening that controls the amount of light transmitted through the lens.) Sharper images are produced as the aperture size is reduced, since only the central portion of the lens is exposed to the incident light when the aperture is very small. At the same time, however, less light is imaged. To compensate for this loss, a longer exposure time is used. An example of the good results obtained with a small aperture is the sharp image produced by a "pinhole" camera, with an aperture size of approximately 1 mm.

In the case of mirrors used for very distant objects, one can eliminate, or at least minimize, spherical aberration by employing a parabolic surface rather than a spherical surface. Parabolic surfaces are not used in many applications, however,

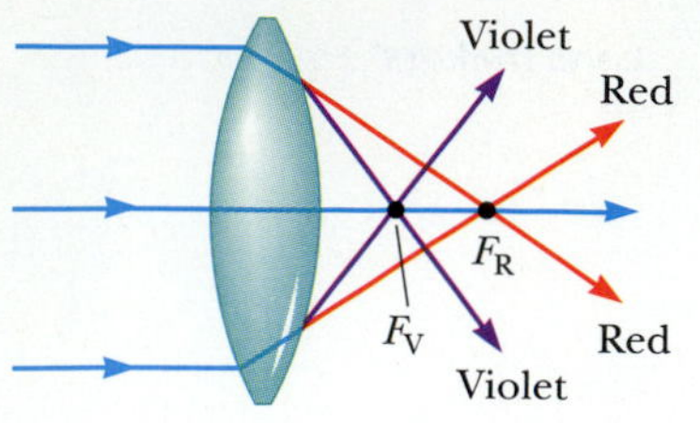

Figure 26.24
Chromatic aberration caused by a converging lens. Rays of different wavelengths focus at different points.

because they are very expensive to make with high-quality optics. Parallel light rays incident on such a surface focus at a common point. Parabolic reflecting surfaces are used in many astronomical telescopes in order to enhance the image quality. They are also used in searchlights, where a nearly parallel light beam is produced from a small lamp placed at the focus of the reflecting surface.

Chromatic Aberrations

The fact that different wavelengths of light refracted by a lens focus at different points gives rise to *chromatic aberrations.* In Chapter 25 we described how the index of refraction of a material varies with wavelength. When white light passes through a lens, one finds, for example, that violet light rays are refracted more than red light rays (Fig. 26.24). From this we see that the focal length is larger for red light than for violet light. Other wavelengths (not shown in Fig. 26.24) would have intermediate focal points. The chromatic aberration for a diverging lens is opposite that for a converging lens. Chromatic aberration can be greatly reduced by using a combination of a converging and diverging lens made from two different types of glass.

Summary

The **magnification**, M, of a mirror or lens is defined as the ratio of the image height, h', to the object height, h:

Magnification of a mirror

$$M = \frac{h'}{h} = -\frac{q}{p} \qquad [26.2, 26.11]$$

In the paraxial ray approximation, the object distance, p, and image distance, q, for a spherical mirror of radius R are related by the **mirror equation,**

Mirror equation

$$\frac{1}{p} + \frac{1}{q} = \frac{2}{R} = \frac{1}{f} \qquad [26.4, 26.6]$$

where $f = R/2$ is the **focal length** of the mirror.

An image can be formed by refraction from a spherical surface of radius R. The object and image distances for refraction from such a surface are related by

Formation of an image by refraction

$$\frac{n_1}{p} + \frac{n_2}{q} = \frac{n_2 - n_1}{R} \qquad [26.8]$$

where the light is incident in the medium of index of refraction n_1 and is refracted in the medium whose index of refraction is n_2.

For a thin lens, and in the paraxial ray approximation, the object and image distances are related by the **thin lens equation:**

Thin-lens equation

$$\frac{1}{p} + \frac{1}{q} = \frac{1}{f} \qquad [26.12]$$

The **focal length,** f, of a thin lens in air is related to the curvature of its surfaces and to the index of refraction, n, of the lens material by

false. (a) A virtual image can be a virtual object. (b) A virtual image can be a real object. (c) A real image can be a virtual object. (d) A real image can be a real object.

16. Discuss the type of aberration involved in each of the following situations. (a) The edges of the image appear reddish. (b) A clear focus of the image's central portion cannot be obtained. (c) A clear focus of the image's outer portion cannot be obtained. (d) The central portion of the image is enlarged relative to the outer portions.

17. Consider a spherical, concave mirror with the object positioned to the left of the mirror beyond the focal point. Using ray diagrams, show that the image of the object moves to the left as the object approaches the focal point.

18. In a Jules Verne novel, a piece of ice is shaped to form a magnifying lens that will focus sunlight to start a fire. Is this possible?

19. A solar furnace can be constructed by using a concave mirror to reflect and focus sunlight into a furnace enclosure. What factors in the design of the reflecting mirror will guarantee that very high temperatures can be achieved?

Problems

Section 26.1 Images Formed by Plane Mirrors

1. Two plane mirrors, A and B, are in contact along one edge, and their planes make an angle of 45° (Fig. 26.26). A point object is placed at P along the bisector of the angle between the two mirrors. Make a sketch similar to Figure 26.26 to a suitable scale. (a) Graphically locate the image of P in mirror A and the image of P in mirror B. (b) Label the images found in (a) P'_A and P'_B, respectively, and locate the image of P'_A in mirror B and the image of P'_B in mirror A. (c) Determine the total number of images for the arrangement described.

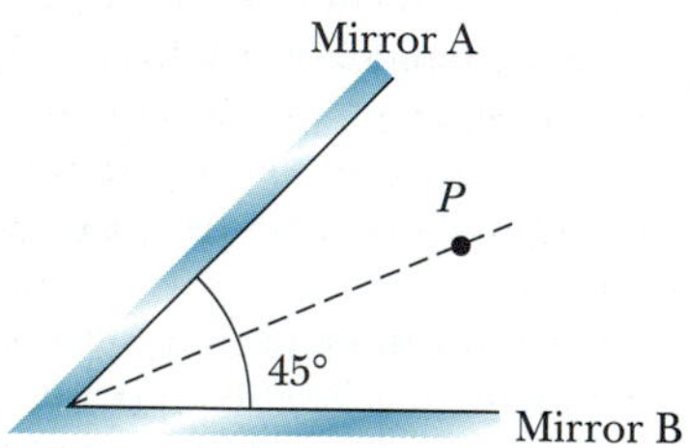

Figure 26.26 (Problem 1)

2. In a physics laboratory experiment, a torque is applied to a small-diameter wire that is suspended vertically under tensile stress. It is necessary to accurately measure the small angle through which the wire turns as a consequence of the net torque. This is accomplished by attaching a small mirror to the wire and reflecting a beam of light off the mirror and onto a circular scale. Such an arrangement is known as an *optical lever* and is shown from the top in Figure 26.27. Show that when the mirror turns through an angle of θ, the reflected beam is rotated by an angle of 2θ.

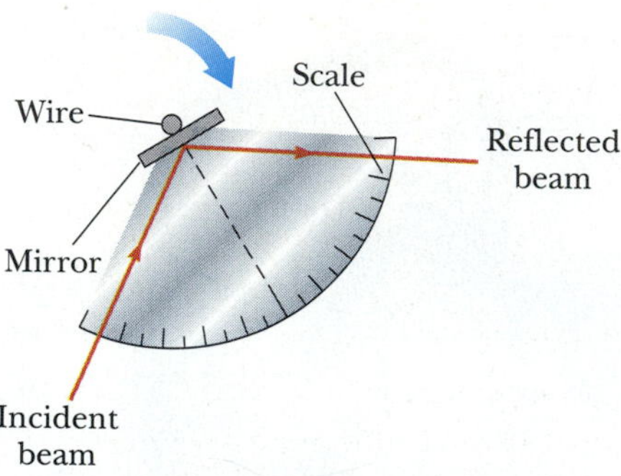

Figure 26.27 (Problem 2)

3. Determine the minimum height of a vertical plane mirror in which a person 5′10″ in height can see his or her full image. (A ray diagram would be helpful.)

4. Two plane mirrors have their reflecting surfaces facing one another, with an edge of one mirror in contact with an edge of the other so that the angle between the mirrors is α. When an object is placed between the mirrors, a number of images are formed. In general, if the angle α between the two mirrors is such that $n\alpha = 360°$, where n is an integer, the number of images formed is $n - 1$. Graphically, find all of the image positions for the case $n = 6$ when a point object is between the mirrors (but not on the angle bisector).

5. A person walks into a room with two plane mirrors on opposite walls, which produce multiple images. When the person is 5 ft from the mirror on the left wall and 10 ft from the mirror on the right wall, find the distance from the person to the first three images seen in the mirror on the left.

Section 26.2 Images Formed by Spherical Mirrors

6. A concave spherical mirror has a radius of curvature of 20 cm. Find the location of the image for object distances of (a) 40 cm, (b) 20 cm, and (c) 10 cm. For each case, state whether the image is real or virtual and erect or inverted. Find the magnification in each case.
7. A convex mirror monitors the aisles in a store. The mirror has a radius of curvature of 0.55 m. Locate and describe the image of a customer 10.0 m from the mirror. Determine the magnification.

(Problem 7) Convex mirrors, often used for security in department stores, provide wide-angle viewing. *(© Paul Silverman 1990, Fundamental Photographs)*

8. A child holds a candy bar 10 cm in front of a convex mirror and notices that the image is reduced by one half. What is the radius of curvature of the mirror?
9. A concave mirror has a focal length of 40 cm. Determine the object position for which the resulting image will be erect and four times the size of the object.
10. An object 2 cm in height is placed 3 cm in front of a concave mirror. If the image is 5 cm in height and virtual, what is the focal length of the mirror?
11. A concave mirror with a radius of curvature of 1 m is illuminated by a candle that is on the symmetry axis 3 m from the mirror. Where is the image of the candle?
12. A convex mirror has a focal length of $-$ 20 cm. Determine an object location for which the image will be one-half the size of the object.
13. A spherical mirror is to be used to form an image five times the size of an object on a screen positioned 5 m from the object. (a) Describe the type of mirror required. (b) Where should the mirror be positioned relative to the object?
14. A spherical convex mirror has a radius of 40 cm. Determine the position of the virtual image and magnification of the mirror for object distances of (a) 30 cm and (b) 60 cm. (c) Are the images erect or inverted?
15. An object is 15 cm from the surface of a reflective spherical Christmas-tree ornament 6 cm in diameter. What are the magnification and position of the image?

Section 26.3 Images Formed by Refraction

16. A swimming pool is 2 m deep. How deep does it appear to be (a) when completely filled with water? (b) when filled halfway with water?
17. A cubical block of ice 50 cm on a side is placed on a level floor over a speck of dust. Find the location of the image of the speck if the index of refraction of ice is 1.309.
18. A colored marble is dropped into a large tank filled with benzene ($n = 1.50$). What is the depth of the tank if the apparent depth of the marble, when viewed from directly above the tank, is 35 cm?

A paperweight is made of a solid glass hemisphere with an index of refraction of 1.50. The radius of the circular cross-section is 4 m. The center of the hemisphere is placed directly over a line drawn to a length of 2.5 mm on a sheet of paper. What is the length of the line as seen from vertically above the hemisphere?

20. A simple model of the human eye ignores its lens entirely. Most of what the eye does to light happens at the transparent cornea. Assume that this outer surface has a radius of curvature 6 mm, and assume that the eyeball contains just one fluid of index of refraction 1.4. Prove that a very distant object will be imaged on the retina, 21 mm behind the cornea. Describe the image.
21. A glass sphere ($n = 1.50$) of radius 15 cm has a tiny air bubble 5 cm from the center. The sphere is viewed along a direction parallel to the radius containing the bubble. What is the apparent depth of the bubble below the surface of the sphere?
22. A flint glass plate ($n = 1.66$) rests on the bottom of an aquarium tank. The plate is 8 cm thick (vertical dimension) and is covered with water ($n = 1.33$) to a depth of 12 cm. Calculate the apparent thickness of the plate as viewed from above the water. (Assume nearly normal incidence.)
23. A glass hemisphere is used as a paperweight, with its flat face resting on a stack of papers. The radius of the circular cross-section is 4 cm, and the index of refraction of the glass is 1.55. The center of the hemisphere is directly over a letter "O" that is 2.5 mm in diameter. What is the diameter of the image of the letter as seen from above along a vertical radius?
24. A goldfish is swimming in water inside a spherical plastic bowl with an index of refraction of 1.33. If the goldfish is 10 cm from the wall of the 15-cm-radius bowl, where does the goldfish appear to an observer outside the bowl?

25. A transparent sphere of unknown composition is observed to form an image of the Sun on the surface of the sphere opposite the Sun. What is the refractive index of the sphere material?

Section 26.4 Thin Lenses

26. An object 32 cm in front of a lens forms an image on a screen 8 cm behind the lens. (a) Find the focal length of the lens. (b) Determine the magnification. (c) Is the lens converging or diverging?
27. The left face of a biconvex lens has a radius of curvature of 12 cm, and the right face has a radius of curvature of 18 cm. The index of refraction of the glass is 1.44. (a) Calculate the focal length of the lens. (b) Calculate the focal length if the radii of curvature of the two faces are interchanged.
28. A magnifying glass is a converging lens of focal length 15 cm. At what distance from a postage stamp should you hold this lens to get a magnification of +2?
29. A contact lens is made of plastic with an index of refraction of 1.50. The lens has an outer radius of curvature of +2.0 cm and an inner radius of curvature of +2.5 cm. What is the focal length of the lens?
30. When the full Moon is viewed from the Earth, its diameter subtends an angle of about 0.5°. A photograph of the full Moon is obtained with a camera lens having a focal length of 120 mm. Find the diameter of the Moon's image on the film.
31. Suppose an object is placed 6 cm in front of a lens that has a focal length of 4 cm. Where is the image located? What are the magnification and the nature of the image?
32. The nickel's image in Figure 26.28 has twice the diameter of the nickel and is 2.84 cm from the lens. Determine the focal length of the lens.

Figure 26.28 (Problem 32)

33. A microscope slide is placed in front of a converging lens with a focal length of 2.44 cm. The lens forms an image of the slide 12.9 cm from the slide. How far is the lens from the slide if the image is (a) real? (b) virtual?
34. A thin lens has a focal length of 25.0 cm. Locate the image when the object is placed (a) 26.0 cm and (b) 24.0 cm in front of the lens. Describe the image in each case.
35. A lens with radii of curvature of 52.5 cm and −61.9 cm has a focal length of +60 cm. Find its index of refraction.
36. An object is placed 30 cm from a lens, and a virtual image is formed 15 cm from the lens. (a) What is the focal length of the lens? (b) Is the lens converging or diverging?
37. A diverging lens is used to form a virtual image of an object. The object is 80 cm to the left of the lens, and the image is 40 cm to the left of the lens. Determine the focal length of the lens.
38. A converging lens has a focal length of 20 cm. Locate the image for object distances of (a) 40 cm, (b) 20 cm, and (c) 10 cm. For each case, state whether the image is real or virtual and erect or inverted. Find the magnification in each case.
39. A person looks at a gem with a jeweler's microscope—a converging lens with a focal length of 12.5 cm. The microscope forms a virtual image 30.0 cm from the lens. Determine the magnification of the lens. Is the image upright or inverted?
40. Suppose an object has thickness dp, so that it extends from object distance p to $p + dp$. Prove that the thickness dq of its image is given by $(-q^2/p^2)\,dp$, so that the longitudinal magnification $dq/dp = -M^2$, where M is the lateral magnification.
41. An object is placed 50 cm from a screen. Where should a converging lens with a 10-cm focal length be placed in order to form an image on the screen? Find the magnification(s).

Additional Problems

42. The lens and mirror in Figure 26.29 have focal lengths of +80 cm and −50 cm, respectively. An object is placed 1.0 m to the left of the lens as drawn. Locate the final image. State whether the image is erect or inverted, and determine the overall magnification.
43. The object in Figure 26.30 is midway between the lens and the mirror. The mirror's radius of curvature is 20.0 cm, and the lens has a focal length of

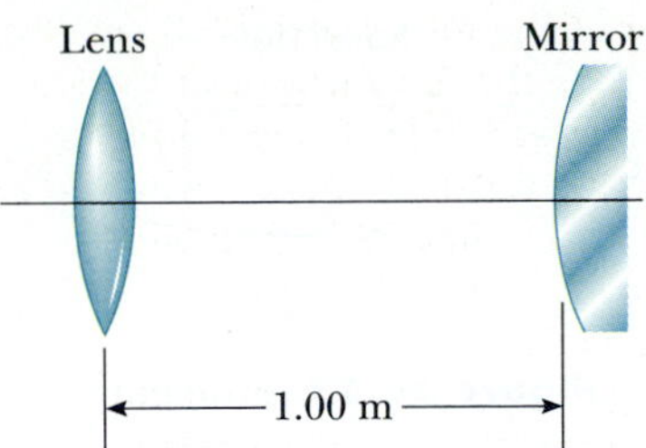

Figure 26.29 (Problem 42)

−16.7 cm. Considering only the light that leaves the object and travels first toward the mirror, locate the final image formed by this system. Is this image real or virtual? Is it erect or inverted? What is the overall magnification?

44. An object placed 10 cm from a concave spherical mirror produces a real image 8 cm from the mirror. If the object is moved to a new position 20 cm from the mirror, what is the position of the image? Is the final image real or virtual?

45. A real object is located at the zero end of a meter stick. A large concave mirror at the 100-cm end of the meter stick forms an image of the object at the 70-cm position. A small convex mirror placed at the 20-cm position forms a final image at the 10-cm point. What is the radius of curvature of the convex mirror?

46. A 1.7-m-tall woman stands 5 m in front of a camera equipped with a 50-mm focal length lens. What is the size of the image formed on the film?

47. A philatelist examines the printing detail on a stamp using a convex lens of focal length 10 cm as a simple magnifier. The lens is held close to the eye, and the lens-to-object distance is adjusted so that the virtual image is formed at the normal near point, 25 cm from the eye. Calculate the expected magnification.

48. In a small room used as a photography studio, people as tall as 2.0 m stand 3.20 m from the camera, and their head-to-foot images must fit onto film frames 35 mm long. What focal length should the camera lens have?

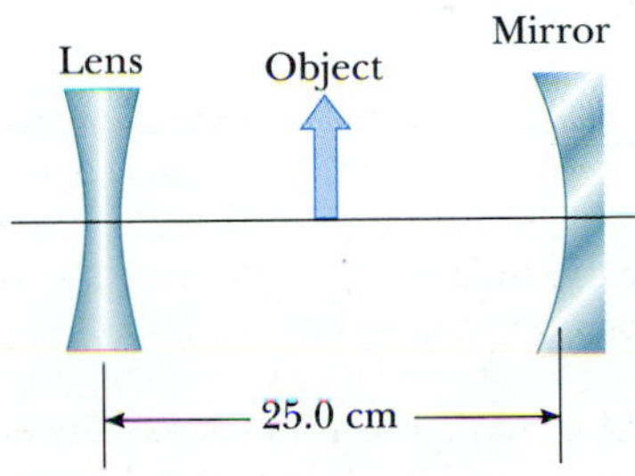

Figure 26.30 (Problem 43)

49. An object is in a fixed position in front of a screen. A thin lens, placed between the object and the screen, produces a sharp image on the screen when it is in either of two positions that are 10 cm apart. The sizes of images in the two situations are in the ratio 3 : 2. (a) What is the focal length of the lens? (b) What is the distance from the screen to the object?

50. A thin lens of focal length 30 cm lies on a horizontal front-surfaced plane mirror. How far above the lens should an object be held if its image is to coincide with the object?

51. A parallel beam of light enters a glass hemisphere perpendicular to the flat face, as shown in Figure 26.31. The radius is $R = 6$ cm, and the index of refraction is $n = 1.560$. Determine the point at which the beam is focused. (Assume paraxial rays.)

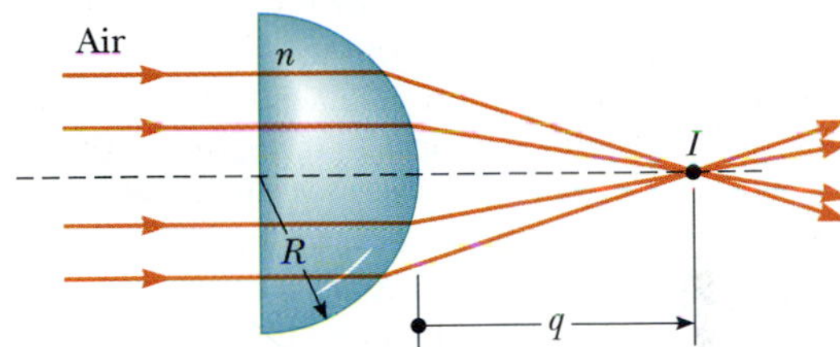

Figure 26.31 (Problem 51)

52. Derive the lens-makers' equation as follows: consider an object in vacuum at $p_1 = \infty$ from a first refracting surface of radius of curvature R_1. Locate its image. Use this image as the object for the second refracting surface, which has nearly the same location as the first, since the lens is thin. Locate the final image, proving it is at the image distance q_2 given by

$$\frac{1}{q_2} = (n-1)\left(\frac{1}{R_1} - \frac{1}{R_2}\right)$$

53. An object is placed 12 cm to the left of a diverging lens of focal length −6 cm. A converging lens of focal length 12 cm is placed a distance d to the right of the diverging lens. Find the distance d such that the final image is at infinity. Draw a ray diagram for this case.

54. A converging lens has a focal length of 20 cm. Find the position of the image for a real object at distances of (a) 50 cm, (b) 30 cm, and (c) 10 cm. (d) Determine the magnification of the lens for these object distances, and whether the image is erect or inverted. (e) Draw ray diagrams to locate the images for these object distances.

55. The disk of the Sun subtends an angle of 0.5 degrees at the Earth. What are the position and diameter of the solar image formed by a concave spherical mirror of radius 3 m?

56. An object 1 cm in height is placed 4 cm to the left of a converging lens of focal length 8 cm. A diverging lens of focal length −16 cm is positioned 6 cm to the right of the converging lens. Find the position and size of the final image. Is the image inverted or erect? Real or virtual?

57. The cornea of the eye has a radius of curvature of 0.80 cm. (a) What is the focal length of the reflecting surface of the eye? (b) If a $20 gold piece 3.4 cm in diameter is held 25 cm from the cornea, what are the size and location of the reflected image?

58. Two converging lenses having focal lengths of 10.0 cm and 20.0 cm are placed 50.0 cm apart, as shown in Figure 26.32. The final image is to be located between the lenses at the position indicated. (a) How far to the left of the first lens should the object be positioned? (b) What is the overall magnification? (c) Is the final image erect or inverted?

59. In a darkened room, a burning candle is placed 1.5 m from a white wall. A lens is placed between the candle and wall at a location that causes a larger, inverted image of the candle to form on the wall. When the lens is moved 90 cm toward the wall, another image of the candle is formed. Find (a) the two object distances that produce the images just described and (b) the focal length of the lens. (c) Characterize the second image.

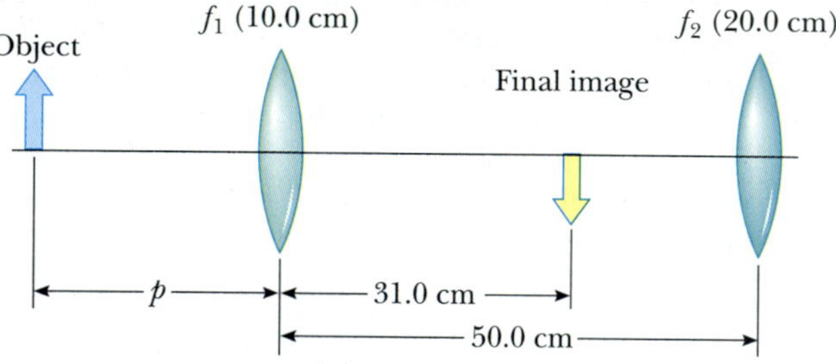

Figure 26.32 (Problem 58)

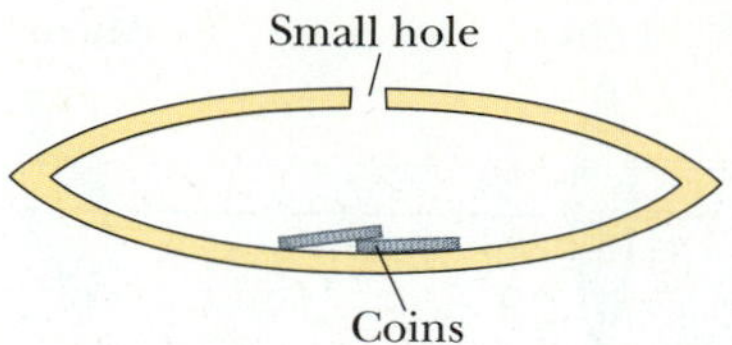

Figure 26.33 (Problem 60)

60. A "floating coin" illusion consists of two parabolic mirrors, each with a focal length of 7.5 cm, facing each other so that their centers are 7.5 cm apart (Fig. 26.33). If a few coins are placed on the lower mirror, an image of the coins is formed at the small opening at the center of the top mirror. Show that the final image is formed at that location, and describe its characteristics. (*Note:* A very startling effect is to shine a flashlight beam on these *images*. Even at a glancing angle, the incoming light beam is seemingly reflected off the *images* of the coins! Why?)

61. Lens L_1 in Figure 26.34 has a focal length of 15.0 cm, whereas lens L_2 has a focal length of 13.3 cm. The distance, d, of lens L_2 from the film plane can be varied from 5.0 cm to 10.0 cm. Determine the range of distances for which objects can be focused on the film.

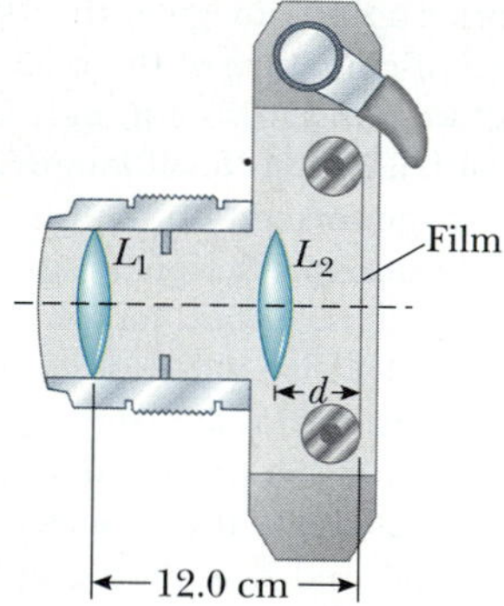

Figure 26.34 (Problem 61)

A layer of bubbles on water, produced by soap film. The colors, produced just before the bubbles burst, are due to interference between light rays reflected from the front and back of the thin film of water making the bubble. The colors depend on the thickness of the film, ranging from black where the film is at its thinnest to red as the film gets thicker. *(Dr. Jeremy Burgess/Science Photo Library)*

Interference of Light Waves

CHAPTER 27

In the preceding chapter on geometric optics, we used the concept of light rays to examine what happens when light passes through a lens or reflects from a mirror. This chapter explains certain interference effects associated with light. The following two chapters are concerned with the subject of *wave optics,* which addresses the phenomena of interference, diffraction, and polarization of light. These phenomena cannot be adequately explained with ray (geometric) optics, but we shall discuss how the wave theory of light leads to satisfying descriptions of such events.

27.1 Conditions for Interference

In our discussion of wave interference of mechanical waves in Chapter 23, we found that two waves could add together either constructively or destructively. In constructive interference, the amplitude of the resultant wave is greater than that of either individual wave, whereas in destructive interference, the resultant amplitude is less than that of either individual wave. Electromagnetic waves also undergo interference. Fundamentally, all interference associated with electromagnetic waves arises as a result of combining the electromagnetic fields that constitute the individual waves.

Interference effects in visible electromagnetic waves are not easy to observe because of the short wavelengths involved (from about 4×10^{-7} m to 7×10^{-7} m). In order to observe sustained interference in light waves, the following conditions must be met:

Conditions for interference

1. The sources must be **coherent;** that is, they must maintain a constant phase with respect to each other.
2. The sources must be of identical wavelength.
3. The superposition principle must apply.

Let us examine the characteristics of coherent sources. Two sources (producing two traveling waves) are needed to create interference. However, in order to produce a stable interference pattern, *the individual waves must maintain a constant phase with one another.* When this situation prevails, the sources are said to be *coherent.* As an example, the sound waves emitted by two side-by-side loudspeakers driven by a single amplifier can produce interference because the two speakers respond to the amplifier in the same way at the same time.

Now, if two light sources are placed side by side, no interference effects are observed because the light waves from one source are emitted independently of the other source; hence, the emissions from the two sources do not maintain a constant phase relationship with each other over the time of observation. An ordinary light source undergoes random changes about once every 10^{-8} s. Therefore, the conditions for constructive interference, destructive interference, or some intermediate state last for times on the order of 10^{-8} s. The result is that no interference effects are observed, since the eye cannot follow such short-term changes. Such light sources are said to be **noncoherent.**

A common method for producing two coherent light sources is to use one single wavelength source to illuminate a screen containing two small slits. The light emerging from the two slits is coherent because a single source produces the original light beam and the two slits serve only to separate the original beam into two parts (which is exactly what was done to the sound signal just discussed). A random change in the light emitted by the source will occur in the two separate beams at the same time, and interference effects can still be observed.

27.2 Young's Double-Slit Experiment

Interference in light waves from two slits was first demonstrated by Thomas Young in 1801. A schematic diagram of the apparatus used in this experiment is shown in Figure 27.1a (Young used pinholes in his original experiments, rather than slits).

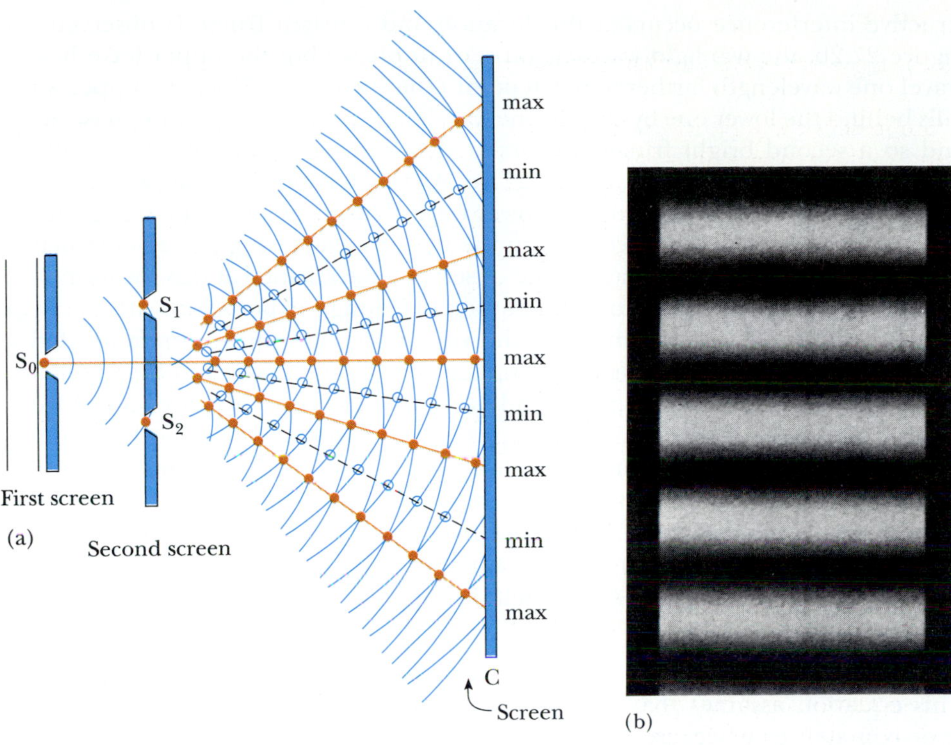

Figure 27.1 (a) Schematic diagram of Young's double-slit experiment. The narrow slits S_1 and S_2 act as coherent sources of light waves and produce an interference pattern on screen C. (Note that this drawing is not to scale.) (b) The fringe pattern formed on screen C could look like this.

Light is incident on a screen in which there is a narrow slit, S_0. The light waves emerging from this slit arrive at a second screen that contains two narrow, parallel slits, S_1 and S_2. These two slits serve as a pair of coherent light sources because waves emerging from them originate at the same wavefront and therefore maintain a constant phase relationship. The light from the two slits produces a visible pattern on screen C; the pattern consists of a series of bright and dark parallel bands called **fringes** (Fig. 27.1b). When the light from S_1 and S_2 arrives at a point on screen C so that constructive interference occurs at that location, a bright line appears. When the light from the two slits combines destructively at any location on the screen, a dark line results.

Figure 27.2 is a schematic diagram of some of the ways the two waves in Young's experiment can combine at screen C. In Figure 27.2a, the two waves, which leave the two slits in phase, strike the screen at the central point, *P*. Since these waves travel equal distances, they arrive in phase at *P*, and as a result con-

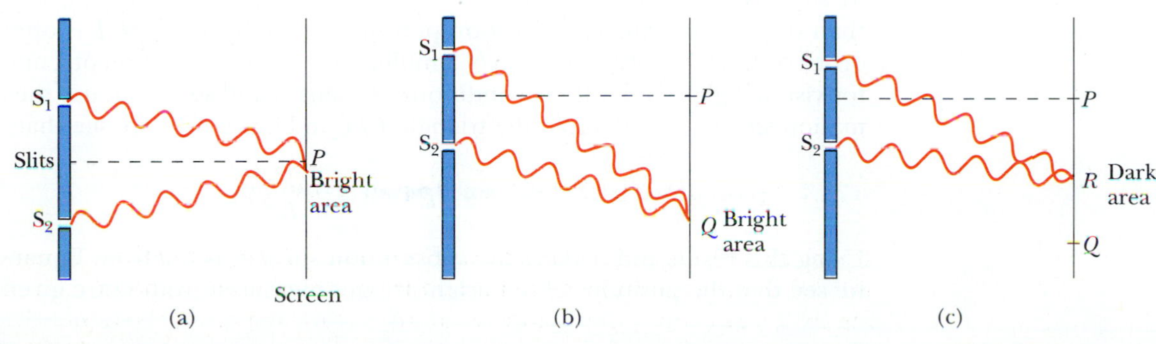

Figure 27.2 (a) When the two waves combine, constructive interference occurs at *P*, the point on the screen that is at the same distance from S_1 and S_2. (b) Constructive interference also occurs at *Q*. (c) Destructive interference occurs at *R* because the wave from the upper slit falls half a wavelength behind the wave from the lower slit. (Note that these figures are not drawn to scale.)

structive interference occurs at this location and a bright fringe is observed. In Figure 27.2b, the two light waves again start in phase, but the upper wave has to travel one wavelength farther to reach point Q on the screen. Since the upper wave falls behind the lower one by exactly one wavelength, they still arrive in phase at Q, and so a second bright fringe appears at this location. Now consider point R, midway between P and Q in Figure 27.2c. At this location, the upper wave has fallen half a wavelength behind the lower wave. This means that the trough from the lower wave overlaps the crest from the upper wave, giving rise to destructive interference at R. For this reason, one observes a dark fringe at this location.

We can obtain a quantitative description of Young's experiment with the help of Figure 27.3. Consider point P on the viewing screen; the screen is located a perpendicular distance of L from the screen containing slits S_1 and S_2, which are separated by a distance of d, and r_1 and r_2 are the distances the waves travel from slit to screen. Let us assume that the source is monochromatic. Under these conditions, the waves emerging from S_1 and S_2 have the same frequency and amplitude and are in phase. The light intensity on the screen at P is the resultant of the light coming from both slits. Note that a wave from the lower slit travels farther than a wave from the upper slit by an amount equal to $d \sin \theta$. This distance is called the **path difference,** δ (lowercase Greek delta), where

Path difference

$$\delta = r_2 - r_1 = d \sin \theta \qquad [27.1]$$

This equation assumes that the two waves are parallel to each other, which is approximately true because L is much greater than d. As noted earlier, the value of this path difference determines whether or not the two waves are in phase when they arrive at P. If the path difference is either zero or some integral multiple of the wavelength, the two waves are in phase at P and constructive interference results. Therefore, the condition for bright fringes, or **constructive interference,** at P is

Constructive interference

$$\delta = d \sin \theta = m\lambda \qquad (m = 0, \pm 1, \pm 2, \ldots) \qquad [27.2]$$

The number m is called the **order number.** The central bright fringe at $\theta = 0$ ($m = 0$) is called the *zeroth-order maximum.* The first maximum on either side, when $m = \pm 1$, is called the *first-order maximum,* and so forth.

Similarly, when the path difference is an odd multiple of $\lambda/2$, the two waves arriving at P are 180° out of phase and give rise to destructive interference. Therefore, the condition for dark fringes, or **destructive interference,** at P is

Destructive interference

$$\delta = d \sin \theta = (m + \tfrac{1}{2})\lambda \qquad (m = 0, \pm 1, \pm 2, \ldots) \qquad [27.3]$$

It is useful to obtain expressions for the positions of the bright and dark fringes measured vertically from O to P. In addition to our assumption that $L \gg d$, we shall assume that $d \gg \lambda$; that is, the distance between the two slits is much larger than the wavelength. This situation prevails in practice because L is often on the order of 1 m while d is a fraction of a millimeter and λ is a fraction of a micrometer for visible light. Under these conditions, θ is small, and so we can use the approximation $\sin \theta \approx \tan \theta$. From the triangle OPQ in Figure 27.3, we see that

$$\sin \theta \approx \tan \theta = \frac{y}{L} \qquad [27.4]$$

Using this result and making the substitution $\sin \theta = m\lambda/d$ from Equation 27.2, we see that the positions of the bright fringes measured from O are given by

$$y_{\text{bright}} = \frac{\lambda L}{d} m \qquad [27.5]$$

Similarly, using Equations 27.3 and 27.4, we find that the dark fringes are located at

$$y_{\text{dark}} = \frac{\lambda L}{d} \left(m + \tfrac{1}{2}\right) \qquad [27.6]$$

As we shall demonstrate in Example 27.1, Young's double-slit experiment provides a method for measuring the wavelength of light. In fact, Young used this technique to make the first measurement of the wavelength of light. Additionally, the experiment gave the wave model of light a great deal of credibility. Today we still use the phenomenon of interference to describe many observations of wave-like behavior.

An interference pattern involving water waves is produced by two vibrating sources at the water's surface. The pattern is analogous to that observed in Young's double-slit experiment. Note the regions of constructive and destructive interference. *(Richard Megna, Fundamental Photographs)*

Example 27.1 Measuring the Wavelength of a Light Source

A screen is separated from a double-slit source by 1.2 m. The distance between the two slits is 0.03 mm. The second-order bright fringe ($m = 2$) is 4.5 cm from the centerline.

(a) Determine the wavelength of the light.

Solution We can use Equation 27.5, with $m = 2$, $y_2 = 4.5 \times 10^{-2}$ m, $L = 1.2$ m, and $d = 3 \times 10^{-5}$ m:

$$\lambda = \frac{dy_2}{mL} = \frac{(3 \times 10^{-5}\text{ m})(4.5 \times 10^{-2}\text{ m})}{2 \times 1.2\text{ m}}$$

$$= 5.62 \times 10^{-7}\text{ m} = \boxed{560\text{ nm}}$$

(b) Calculate the distance between adjacent bright fringes.

Solution From Equation 27.5 and the results to (a), we get

$$y_{m+1} - y_m = \frac{\lambda L(m+1)}{d} - \frac{\lambda L m}{d}$$

$$= \frac{\lambda L}{d} = \frac{(5.62 \times 10^{-7}\text{ m})(1.2\text{ m})}{3 \times 10^{-5}\text{ m}}$$

$$= 2.25 \times 10^{-2}\text{ m} = \boxed{2.25\text{ cm}}$$

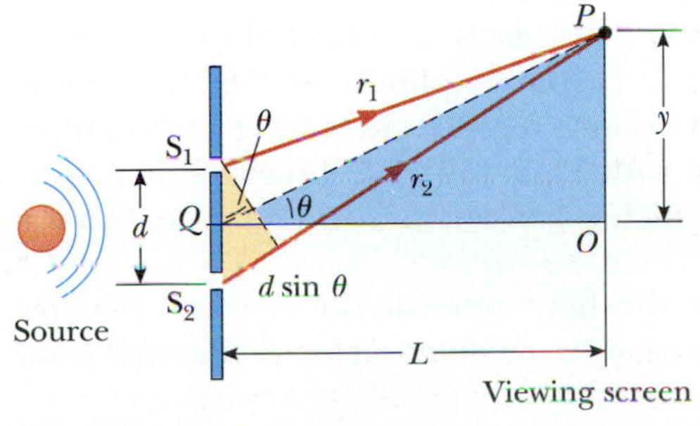

Figure 27.3 A geometric construction for describing Young's double-slit experiment. The path difference between the two rays is $r_2 - r_1 = d \sin \theta$. (Note that this figure is not drawn to scale.)

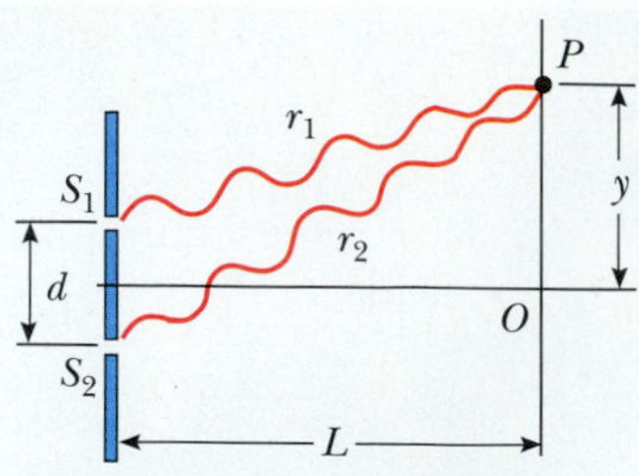

Figure 27.4
Construction for analyzing the double-slit interference pattern. A bright region, or intensity maximum, is observed at O.

27.3 Intensity Distribution of the Double-Slit Interference Pattern

We shall now calculate the distribution of light intensity (the energy delivered by the wave per unit area per unit time) associated with the double-slit interference pattern. Again, suppose that the two slits represent coherent sources of sinusoidal waves. Hence, the two waves have the same angular frequency, ω, and a constant phase difference, ϕ. The total electric field at the point P on the screen in Figure 27.4 is the *vector superposition* of the two waves. Assuming the two waves have the same amplitude, E_0, we can write the electric field at P due to each wave separately as

$$E_1 = E_0 \sin \omega t \qquad \text{and} \qquad E_2 = E_0 \sin(\omega t + \phi) \qquad [27.7]$$

Although the waves have equal phase at the slits, *their phase difference, ϕ, at P depends on the path difference, $\delta = r_2 - r_1 = d \sin \theta$.* Since a path difference of λ corresponds to a phase difference of 2π rad (constructive interference), whereas a path difference of $\lambda/2$ corresponds to a phase difference of π rad (destructive interference), we obtain the ratio

$$\frac{\delta}{\phi} = \frac{\lambda}{2\pi}$$

Phase difference

$$\phi = \frac{2\pi}{\lambda}\delta = \frac{2\pi}{\lambda} d \sin \theta \qquad [27.8]$$

This equation tells us precisely how the phase difference ϕ depends on the angle θ.

Using the superposition principle and Equation 27.7, we can obtain the resultant electric field at the point P:

$$E_P = E_1 + E_2 = E_0[\sin \omega t + \sin(\omega t + \phi)] \qquad [27.9]$$

To simplify this expression, we use the trigonometric identity

$$\sin A + \sin B = 2 \sin\left(\frac{A + B}{2}\right) \cos\left(\frac{A - B}{2}\right)$$

Taking $A = \omega t + \phi$ and $B = \omega t$, we can write Equation 27.9 in the form

$$E_P = 2E_0 \cos\left(\frac{\phi}{2}\right) \sin\left(\omega t + \frac{\phi}{2}\right) \qquad [27.10]$$

Hence, the electric field at P has the same frequency ω as the original two waves, but its amplitude is multiplied by the factor $2 \cos(\phi/2)$. To check the consistency of this result, note that if $\phi = 0, 2\pi, 4\pi, \ldots$, the amplitude at P is $2E_0$, corresponding to the condition for constructive interference. Referring to Equation 27.8, we find that our result is consistent with Equation 27.2. Likewise, if $\phi = \pi, 3\pi, 5\pi, \ldots$, the amplitude at P is zero, which is consistent with Equation 27.3 for destructive interference.

Finally, to obtain an expression for the light intensity at P, recall that *the intensity of a wave is proportional to the square of the resultant electric field at that point*

(Chapter 24, Section 24.4). Using Equation 27.10, we can therefore express the intensity at P as

$$I \propto E_P^2 = 4E_0^2 \cos^2(\phi/2) \sin^2\left(\omega t + \frac{\phi}{2}\right)$$

Since most light-detecting instruments measure the time-averaged light intensity, and the time-averaged value of $\sin^2(\omega t + \phi/2)$ over one cycle is 1/2, we can write the average intensity at P as

$$I_{av} = I_0 \cos^2(\phi/2) \qquad [27.11]$$

where I_0 is the *maximum* possible time average light intensity. [Note that $I_0 = (E_0 + E_0)^2 = (2E_0)^2 = 4E_0^2$.] Substituting Equation 27.8 into Equation 27.11, we find that

Average light intensity for the double-slit interference pattern

$$I_{av} = I_0 \cos^2\left(\frac{\pi d \sin\theta}{\lambda}\right) \qquad [27.12]$$

Alternatively, since $\sin\theta \approx y/L$ for small values of θ, we can write Equation 27.12 in the form

$$I_{av} = I_0 \cos^2\left(\frac{\pi d}{\lambda L} y\right) \qquad [27.13]$$

Constructive interference, which produces intensity maxima, occurs when the quantity $\pi y d/\lambda L$ is an integral multiple of π, corresponding to $y = (\lambda L/d)m$. This is consistent with Equation 27.5. Intensity distribution versus θ is plotted in Figure 27.5. Note that the interference pattern consists of equally spaced fringes of equal

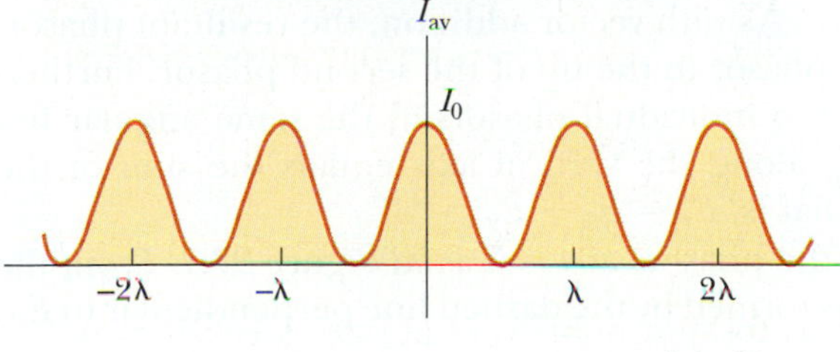

Figure 27.5
Intensity distribution versus $d \sin\theta$ for the double-slit pattern when the screen is far from two slits ($L \gg d$). *(Photo from M. Cagnet, M. Francon, and J. C. Thierr, Atlas of Optical Phenomena, Berlin, Springer-Verlag, 1962)*

intensity. However, the result is valid only if the slit-to-screen distance, L, is large relative to the slit separation, and only for small values of θ.

We have seen that the interference phenomena arising from two coherent sources depend on the relative phase of the waves at a given point. Furthermore, the phase difference at a given point depends on the *path difference* between the two waves. The *resultant intensity at a point is proportional to the square of the resultant amplitude.* That is, the intensity is proportional to $(E_1 + E_2)^2$. It would be *incorrect* to calculate the resultant intensity by adding the intensities of the individual waves. This procedure would give a different quantity, namely $E_1^2 + E_2^2$. Finally, $(E_1 + E_2)^2$ has the same *average* value as $E_1^2 + E_2^2$, when the time average is taken over all values of the phase difference between E_1 and E_2. Hence, the principle of energy conservation is not violated.

27.4 Phasor Addition of Waves

In the preceding section we combined two waves algebraically to obtain the resultant wave amplitude at some point on a screen. Unfortunately, this analytical procedure becomes cumbersome when several amplitudes have to be added. Since we shall eventually be interested in combining a large number of waves, we now describe a graphical procedure for this purpose.

Again, consider a sinusoidal wave whose electric field component is expressed by

$$E_1 = E_0 \sin \omega t$$

where E_0 is the wave amplitude and ω is the angular frequency. This wave disturbance can be represented graphically with a vector of magnitude E_0, *rotating* counterclockwise about the origin with an angular frequency of ω, as in Figure 27.6a. Such a rotating vector is called a *phasor* and is commonly used in electrical engineering. Note that the phasor makes an angle of ωt with the horizontal axis. The projection of the phasor on the vertical axis represents E_1, the magnitude of the wave disturbance at some time t. Hence, as the phasor rotates in a circle, the projection E oscillates along the vertical axis about the origin.

Now consider a second sinusoidal wave whose electric field is

$$E_2 = E_0 \sin(\omega t + \phi)$$

The phasor representing the wave E_2 is shown in Figure 27.6b. The resultant wave, which is the sum of E_1 and E_2, can be obtained graphically by redrawing the phasors end to end, as in Figure 27.6c, where the tail of the second phasor is placed at the tip of the first phasor. As with vector addition, the resultant phasor, $\mathbf{E}_R$, runs from the tail of the first phasor to the tip of the second phasor. Furthermore, $\mathbf{E}_R$ rotates along with the two individual phasors at the same angular frequency, ω. The projection of $\mathbf{E}_R$ along the vertical axis equals the sum of the projections of the two phasors. That is, $E_P = E_1 + E_2$.

It is convenient to construct the phasors at $t = 0$ as in Figure 27.7. From the geometry of the two right triangles formed by the dashed line perpendicular to E_R, we see that

$$E_R = E_0 \cos \alpha + E_0 \cos \alpha = 2E_0 \cos \alpha$$

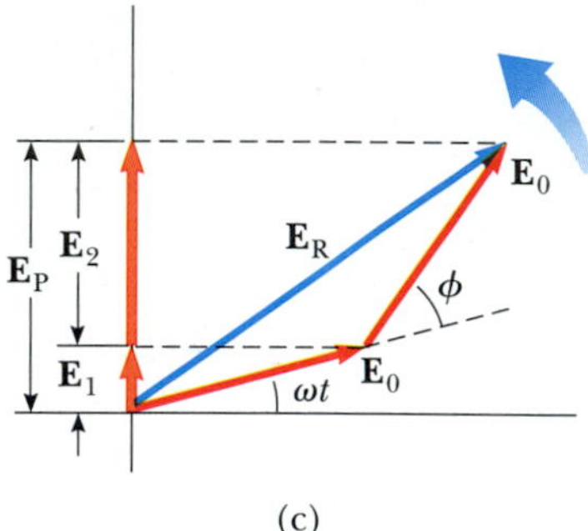

(c)

Figure 27.6 (a) Phasor diagram for the wave disturbance $E_1 = E_0 \sin \omega t$. The phasor is a vector of length E_0, rotating counterclockwise. (b) Phasor diagram for the wave $E_2 = E_0 \sin(\omega t + \phi)$. (c) E_R is the resultant phasor formed from the individual phasors shown in (a) and (b).

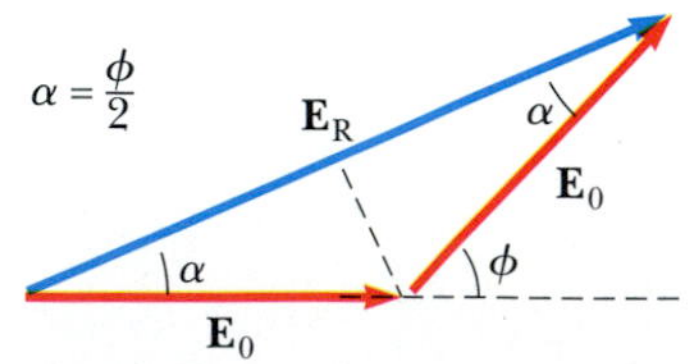

Figure 27.7 A reconstruction of the resultant phasor E_R. From the geometry, note that $\alpha = \phi/2$.

Since the sum of the two opposite interior angles equals the exterior angle, ϕ, we see that $\alpha = \phi/2$, so that

$$E_R = 2E_0 \cos\left(\frac{\phi}{2}\right)$$

Hence, the projection of the phasor $\mathbf{E_R}$ along the vertical axis at any time t is

$$E_P = E_R \sin\left(\omega t + \frac{\phi}{2}\right) = 2E_0 \cos\left(\frac{\phi}{2}\right) \sin\left(\omega t + \frac{\phi}{2}\right)$$

This is consistent with the result obtained algebraically, Equation 27.10. The resultant phasor has the amplitude $2E_0 \cos(\phi/2)$ and makes an angle of $\phi/2$ with the first phasor. Furthermore, the average intensity at P, which varies as $E_P{}^2$, is proportional to $\cos^2(\phi/2)$, as described in Equation 27.11.

We can now discuss how to obtain the resultant of several waves that have the same frequency:

1. Draw the phasors representing each wave end to end, as in Figure 27.8, remembering to maintain the proper phase relationship between waves.
2. The resultant represented by the phasor $\mathbf{E_R}$ is the vector sum of the individual phasors. At each instant, the projection of $\mathbf{E_R}$ along the vertical axis represents the time variation of the resultant wave. The phase angle, α, of the resultant wave is the angle between $\mathbf{E_R}$ and the first phasor. From the construction in Figure 27.8, drawn for four phasors, we see that the phasor of the resultant wave is given by $E_P = E_R \sin(\omega t + \alpha)$.

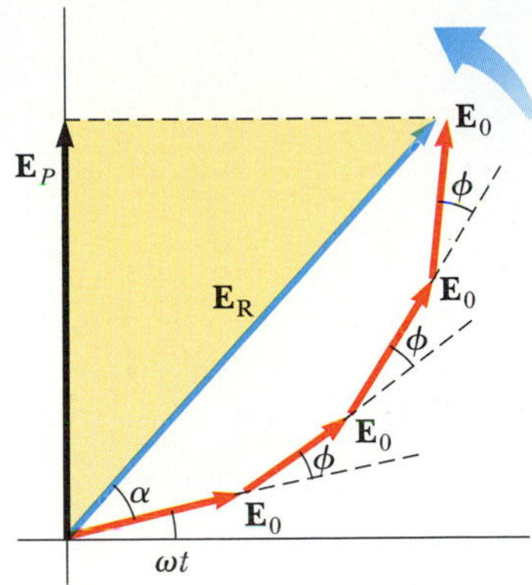

Figure 27.8
The phasor $\mathbf{E_R}$ is the resultant of four phasors of equal amplitude, $\mathbf{E}_0$. The phase of $\mathbf{E_R}$ with respect to the first phasor is α.

Phasor Diagrams for Two Coherent Sources

As an example of the phasor method, consider again the interference pattern produced by two coherent sources. Figure 27.9 represents the phasor diagrams for some values of the phase difference ϕ and the corresponding values of the path difference δ, which are obtained using Equation 27.8.

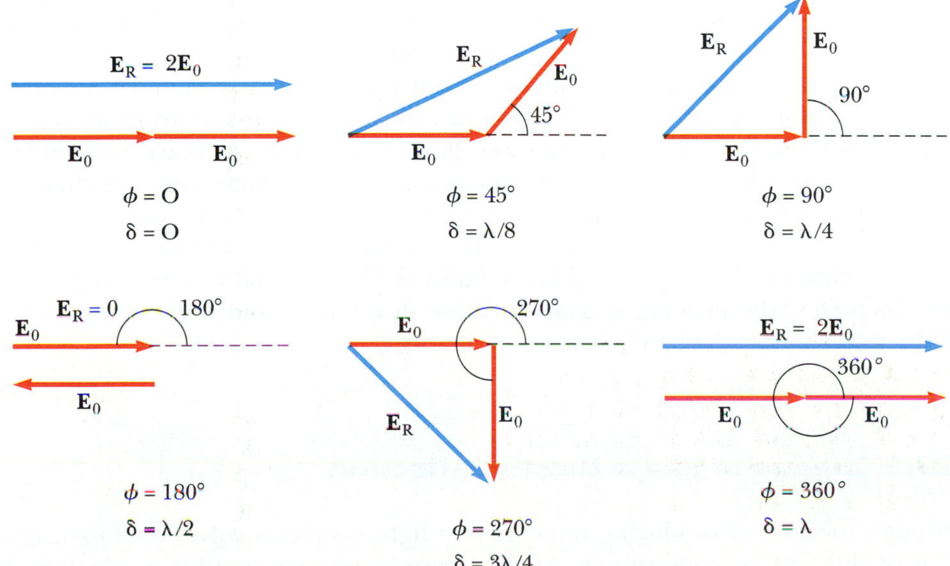

Figure 27.9
Phasor diagrams for the double-slit interference pattern. The resultant phasor $\mathbf{E_R}$ is a maximum when $\phi = 0, 2\pi, 4\pi, \ldots$ and is zero when $\phi = \pi, 3\pi, 5\pi, \ldots$.

$$\frac{1}{f} = (n - 1)\left(\frac{1}{R_1} - \frac{1}{R_2}\right) \qquad [26.13]$$

Lens makers' equation

Converging lenses have positive focal lengths, and **diverging lenses** have negative focal lengths.

Questions and Conceptual Exercises

1. Why do some emergency vehicles have the symbol ƎƆИA⅃UᙠMA written on the front?
2. The side-view mirror on late-model cars warns the user that objects may be closer than they appear. What kind of mirror is being used, and why was that type selected?

(Question 2). *(© Junebug Clark 1988, Photo Researchers, Inc.)*

3. Consider a concave spherical mirror with a real object. Is the image always inverted? Is the image always real? Give conditions for your answers.
4. Why does a clear stream always appear to be shallower than it actually is?
5. A person spearfishing from a boat sees a fish that is 3 m from the boat at a depth of 1 m. In order to spear the fish, should the person aim at, above, or below the image of the fish?
6. Explain why a fish in a spherical goldfish bowl appears larger than it really is.
7. If a cylinder of solid glass or clear plastic is placed above the words LEAD OXIDE and viewed from the side, as shown in Figure 26.25, the word LEAD appears inverted but the word OXIDE does not. Explain.

Figure 26.25 (Question 7) *(Courtesy of Henry Leap and Jim Lehman)*

8. A mirage forms when the air gets gradually cooler as the height above the ground increases. What might happen if the air became gradually warmer as the height was increased? This often happens over bodies of water or snow-covered ground: the effect is called *looming*.
9. It is well known that distant objects viewed underwater with the naked eye appear blurred and out of focus. On the other hand, the use of goggles provides the swimmer with a clear view of objects. Explain this, using the fact that the indices of refraction of the cornea, water, and air are 1.376, 1.333, and 1.00029, respectively.
10. Lenses used in eyeglasses, whether converging or diverging, are always designed so that the middle of the lens curves away from the eye, like the center lenses in Figure 26.15a and b. Why?
11. Describe lenses that can be used to start a fire.
12. Explain this statement: "The focal point of a lens is the location of the image of a point object at infinity." Discuss the notion of infinity in real terms as it applies to object distances. Based on this statement, can you think of a "quick and dirty" method for determining the focal length of a positive lens?
13. Consider the image formed by a thin converging lens. Under what conditions will the image be (a) inverted? (b) erect? (c) real? (d) virtual? (e) larger than the object? (f) smaller than the object?
14. Discuss the proper position of a slide relative to the lens in a slide projector. What type of lens must the slide projector have?
15. Discuss why each of the following statements is true or

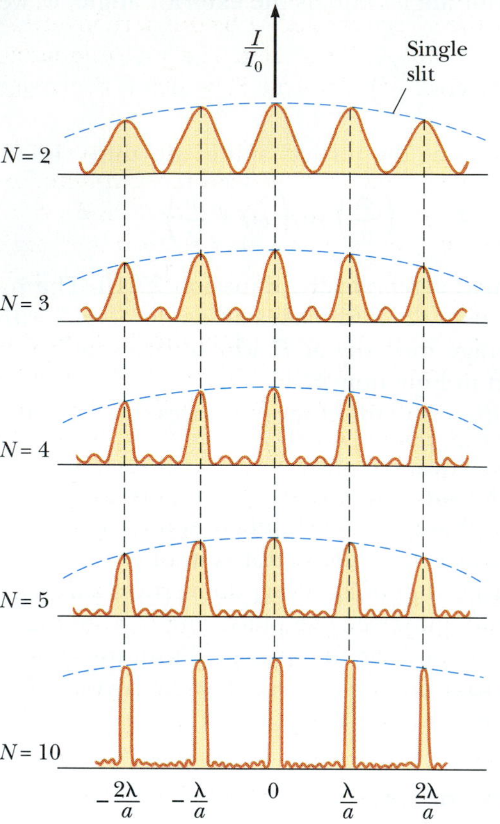

Figure 27.10 Multiple-slit interference patterns. As the number of slits is increased, the primary maxima (the most intense bands) become narrower but remain fixed in position, and the number of secondary maxima increases. The decrease in intensity of the maxima, as indicated by the blue dashed lines, is due to the phenomenon of diffraction, which is discussed in Chapter 29.

From Figure 27.9 we see that the intensity at a point is a maximum when $\mathbf{E}_R$ is a maximum. This occurs at $\phi = 0$, 2π, 4π, and so on. Likewise, we see that the intensity at a point is zero when $\mathbf{E}_R$ is zero. The first zero-intensity point occurs at $\phi = 180°$, corresponding to $\delta = \lambda/2$, and the other zero points (not shown) occur at $\delta = 3\lambda/2$, $5\lambda/2$, and so on. These results are in complete agreement with the analytical procedure described in the preceding section.

Figure 27.10 shows multiple-slit interference patterns for a number of configurations. These patterns represent plots of intensity for the various primary and secondary maxima. For the case of three slits, note that the primary maxima are nine times more intense than the secondary maxima. This is because the intensity varies as E_R^2. Figure 27.10 also shows that as the number of slits increases, the number of secondary maxima increases. In fact, the number of secondary maxima is always equal to $N - 2$, where N is the number of slits. Finally, as the number of slits increases, the primary maxima increase in intensity and become narrower, while the secondary maxima decrease in intensity.

27.5 Change of Phase Due to Reflection

Young's method of producing two coherent light sources involves illuminating a pair of slits with a single source. Another simple arrangement for producing an interference pattern with a single light source is known as *Lloyd's mirror.* A light source is placed at point S close to a mirror, as illustrated in Figure 27.11. Waves can reach the viewing point, P, either by the direct path SP or by the path involving reflection from the mirror. The reflected ray can be treated as a ray originating from a source at S', located behind the mirror. This source S', which is the image of S, can be considered a virtual source.

At points far from the source, one would expect an interference pattern due to waves from S and S', just as is observed for two real coherent sources. An interference pattern is indeed observed. However, the positions of the dark and bright fringes are *reversed* relative to the pattern of two real coherent sources (Young's experiment). This is because the coherent sources at S and S' differ in phase by 180°. This 180° phase change is produced upon reflection.

To illustrate this further, consider the point P', where the mirror meets the screen. This point is equidistant from S and S'. If path difference alone were responsible for the phase difference, one would expect to see a bright fringe at P' (since the path difference is zero for this point), corresponding to the central fringe of the two-slit interference pattern. Instead, one observes a *dark* fringe at P' because of the 180° phase change produced by reflection. In general, an electromagnetic wave undergoes a phase change of 180° upon reflection from a medium of higher index of refraction than the one in which it was traveling.

It is useful to draw an analogy between reflected light waves and the reflections of a transverse wave on a stretched string when the wave meets a boundary (Chapter 22, Section 22.6), as in Figure 27.12. The reflected pulse on a string undergoes a phase change of 180° when it is reflected from the boundary of a denser medium, such as a heavier string, and no phase change when it is reflected from the boundary of a less dense medium. Similarly, an electromagnetic wave undergoes a 180° phase change when reflected from the boundary of a medium of higher

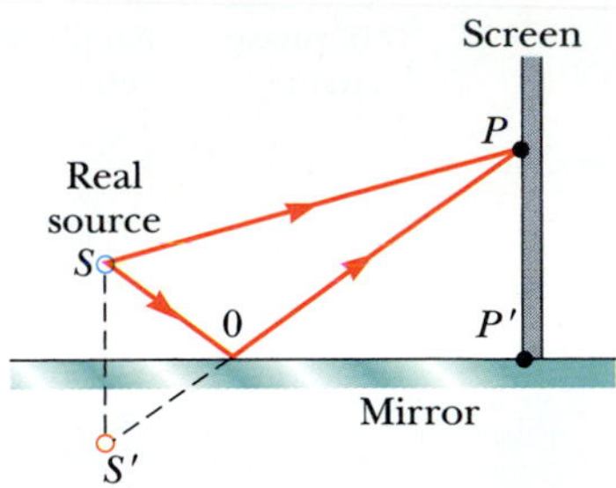

Figure 27.11 Lloyd's mirror. An interference pattern is produced on a screen at P as a result of the combination of the direct ray and the reflected ray. The reflected ray undergoes a phase change of 180°.

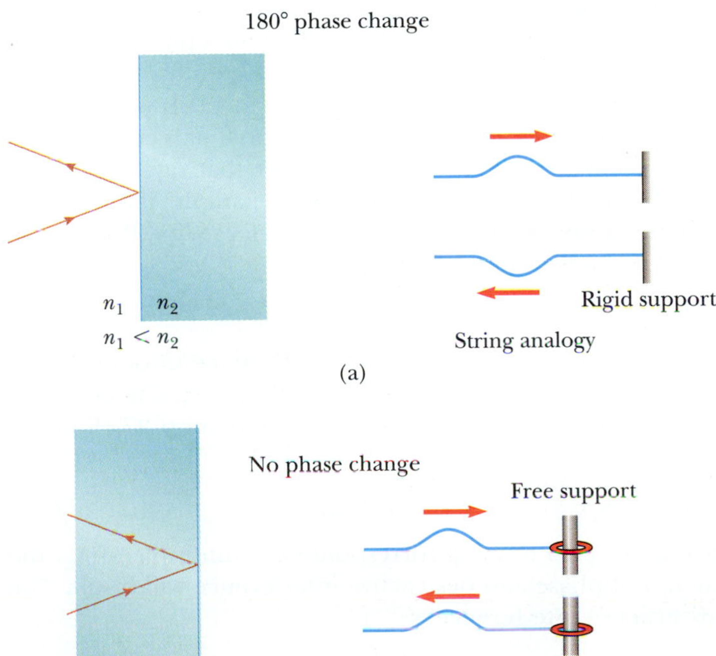

Figure 27.12 (a) A ray reflecting from a medium of higher refractive index undergoes a 180° phase change. The right side shows the analogy with a reflected pulse on a string. (b) A ray reflecting from a medium of lower refractive index undergoes *no* phase change.

index of refraction than the one in which it was traveling. There is no phase change when the wave is reflected from a boundary leading to a medium of lower index of refraction. The part of the wave that crosses the boundary undergoes no phase change.

27.6 Interference in Thin Films

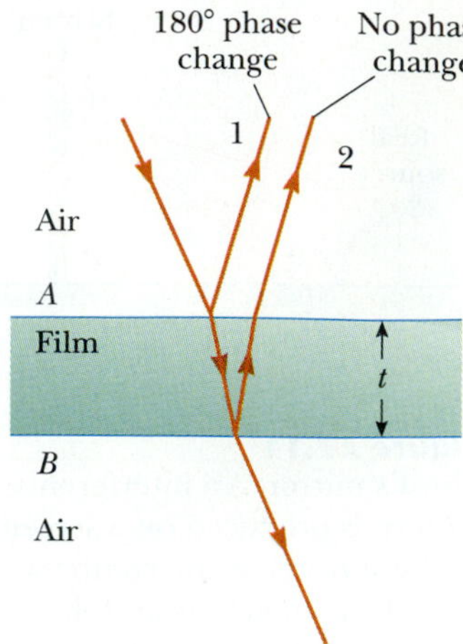

Figure 27.13 Interference in light reflected from a thin film is due to a combination of rays reflected from the upper and lower surfaces of the film.

Interference effects are commonly observed in thin films, such as soap bubbles and thin layers of oil on water. The varied colors observed with ordinary white light result from the interference of waves reflected from the opposite surfaces of the film.

Consider a film of uniform thickness t and index of refraction n, as in Figure 27.13. Let us assume that the light rays traveling in air are nearly normal to the two surfaces of the film. To determine whether the reflected rays interfere constructively or destructively, we must first note the following facts:

1. An electromagnetic wave traveling from a medium of index of refraction n_1 toward a medium of index of refraction n_2 undergoes a 180° phase change upon reflection when $n_2 > n_1$. There is no phase change in the reflected wave if $n_2 < n_1$.
2. The wavelength of light, λ_n, in a medium whose index of refraction n is

$$\lambda_n = \frac{\lambda}{n} \qquad [27.14]$$

where λ is the wavelength of light in free space.

Let us apply these rules to the film of Figure 27.15. According to the first rule, ray 1, which is reflected from the upper surface (A), undergoes a phase change of 180° with respect to the incident wave. Ray 2, which is reflected from the lower surface (B), undergoes no phase change with respect to the incident wave. Therefore, ray 1 is 180° out of phase with respect to ray 2, a situation that is equivalent to a path difference of $\lambda_n/2$. However, we must also consider that ray 2 travels an extra distance equal to $2t$ before the waves recombine. For example, if $2t = \lambda_n/2$, rays 1 and 2 will recombine in phase and constructive interference will result. In general, the condition for constructive interference is

$$2t = (m + \tfrac{1}{2})\lambda_n \qquad (m = 0, 1, 2, \ldots) \qquad [27.15]$$

This condition takes into account two factors: (a) the difference in optical path length for the two rays (the term $m\lambda_n$) and (b) the 180° phase change upon reflection (the term $\lambda_n/2$). Since $\lambda_n = \lambda/n$, we can write Equation 27.15 in the form

Constructive interference

$$2nt = (m + \tfrac{1}{2})\lambda \qquad (m = 0, 1, 2, \ldots) \qquad [27.16]$$

If the extra distance $2t$ traveled by ray 2 corresponds to a multiple of λ_n, the two waves will combine out of phase and destructive interference will result. The general equation for destructive interference is

Destructive interference

$$2nt = m\lambda \qquad (m = 0, 1, 2, \ldots) \qquad [27.17]$$

It is important that you realize that two factors influence interference: (1) phase changes and (2) differences in travel distance. The preceding conditions for constructive and destructive interference are valid only when the medium above the top surface of the film is the same as the medium below the bottom surface. The surrounding medium may have a refractive index less than or greater than that of the film. In either case, the rays reflected from the two surfaces will be out of phase by 180°. If the film is placed between two *different* media, one of lower refractive index and one of higher refractive index, the conditions for constructive and destructive interference are *reversed.* In this case, either there is a phase change of 180° for both ray 1 reflecting from surface A and ray 2 reflecting from surface B, or there is no phase change for either ray; hence, the net change in relative phase due to the reflections is *zero.*

Newton's Rings

Another method for observing interference of light waves is to place a plano-convex lens on top of a flat glass surface, as in Figure 27.14a. With this arrangement, the air film between the glass surfaces varies in thickness from zero at the point of contact to some value t at P. If the radius of curvature of the lens R is very large compared with the distance r, and if the system is viewed from above using light of wavelength λ, a pattern of light and dark rings is observed (Fig. 27.14b). These circular fringes are called **Newton's rings** after their discoverer. Although Newton could have calculated the wavelength of light from his observations, he explained them in terms of his particle model.

The interference is due to the combination of ray 1, reflected from the plate, with ray 2, reflected from the lower surface of the lens. Ray 1 undergoes a phase change of 180° upon reflection, because it is reflected from a boundary leading into a medium of higher refractive index, whereas ray 2 undergoes no phase change. Hence, the conditions for constructive and destructive interference are given by Equations 27.16 and 27.17, respectively, with $n = 1$ since the "film" is air. Here again, one might guess that the contact point O would be bright, corresponding to constructive interference. Instead, it is dark, as seen in Figure 27.14b,

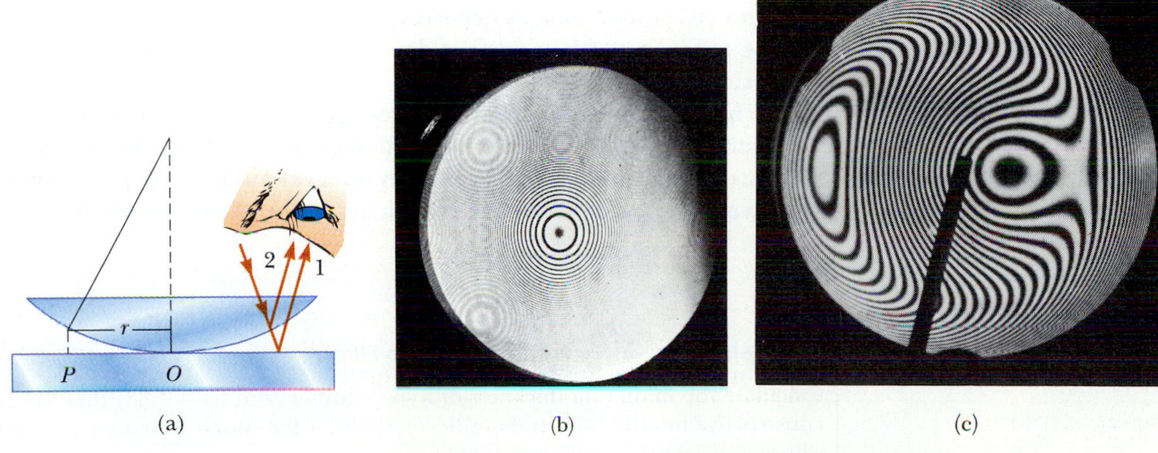

Figure 27.14 (a) The combination of rays reflected from the glass plate and the curved surface of the lens gives rise to an interference pattern known as Newton's rings. (b) Photograph of Newton's rings. *(Courtesy of Bausch and Lomb Optical Co.)* (c) This asymmetrical interference pattern indicates imperfections in the lens, producing the skewed set of Newton's rings. *(From Physical Science Study Committee, College Physics, Lexington, Mass., Heath, 1968)*

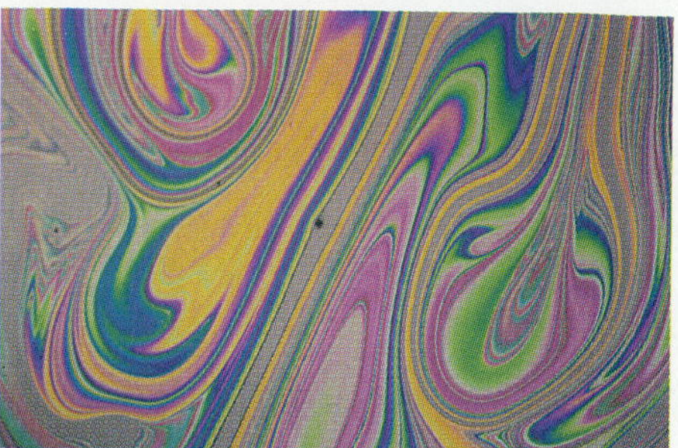

(Left) Interference in soap bubbles. White light incident on soap bubbles (thin films of soap) forms a beautiful pattern of colors as a result of interference in the films, as described in Section 27.6. (Right) Thin film interference. A thin film of oil on water displays interference, as shown by the pattern of colors when white light is incident on the film. The film thickness varies, thereby producing the interesting color pattern. (Left, *Peter Aprahamian/Science Photo Library;* right © *Tom Branch 1984, Photo Researchers*)

because ray 1, reflected from the plate, undergoes a 180° phase change with respect to ray 2. Using the geometry shown in Figure 27.14a, one can obtain expressions for the radii of the bright and dark bands in terms of the radius of curvature R and vacuum wavelength λ. For example, the dark rings have radii given by $r \approx \sqrt{m\lambda R/n}$. The details are left as a problem for the reader (Problem 36).

One of the important uses of Newton's rings is in the testing of optical lenses. A circular pattern like that in Figure 27.14b is obtained only when the lens is ground to a perfectly symmetric curvature. Variations from such symmetry might produce a pattern like that in Figure 27.14c. These variations give an indication of how the lens must be ground and polished in order to remove the imperfections.

Problem-Solving Strategy: Thin Film Interference

The following features should be kept in mind while working thin film interference problems:

1. **Identify the thin film causing the interference.**
2. **Which type of interference occurs is determined by the phase relationship between the portion of the wave reflected at the upper surface of the film and the portion reflected at the lower surface.**
3. **Phase differences between the two portions of the wave have two causes: (a) differences in the distances traveled by the two portions and (b) phase changes occurring upon reflection. *Both* causes must be considered when you are determining which type of interference occurs.**
4. **When distance and phase changes upon reflection are both taken into account, the interference will be constructive if the path difference between the two waves is an integral multiple of λ, and destructive if the path difference is $\lambda/2$, $3\lambda/2$, $5\lambda/2$, and so forth.**

Example 27.2 Interference in a Soap Film

Calculate the minimum thickness of a soap bubble film ($n = 1.33$) that will result in constructive interference in the reflected light, if the film is illuminated with light whose wavelength in free space is 600 nm.

Solution The minimum film thickness for constructive interference corresponds to $m = 0$ in Equation 27.16. This gives $2nt = \lambda/2$, or

$$t = \frac{\lambda}{4n} = \frac{600 \text{ nm}}{4(1.33)} = 113 \text{ nm}$$

Exercise What other film thicknesses will produce constructive interference?

Answer 338 nm, 564 nm, 789 nm, and so on.

Example 27.3 Nonreflecting Coatings for Solar Cells

Semiconductors such as silicon are used to fabricate solar cells, devices that generate electricity when exposed to sunlight. Solar cells are often coated with a transparent thin film, such as silicon monoxide (SiO; $n = 1.45$), in order to minimize reflective losses (Fig. 27.15). A silicon solar cell ($n = 3.5$) is coated with a thin film of silicon monoxide for this purpose. Determine the minimum thickness of the film that will produce the least reflection at a wavelength of 550 nm.

Solution Reflection is least when rays 1 and 2 in Figure 27.15 meet the condition of destructive interference. Note that *both* rays undergo a 180° phase change upon reflection. Hence, the net change in phase is zero due to reflection, and the condition for a reflection *minimum* requires a path difference of $\lambda_n/2$; therefore, $2t = \lambda/2n$, or

$$t = \frac{\lambda}{4n} = \frac{550 \text{ nm}}{4(1.45)} = 94.8 \text{ nm}$$

Typically, such coatings reduce the reflective loss from 30% (with no coating) to 10% (with coating), thereby increasing the cell's efficiency since more light is available to create charge carriers in the cell. In reality, the coating is never perfectly nonreflecting, because the required thickness is wavelength-dependent and the incident light covers a wide range of wavelengths.

Glass lenses used in cameras and other optical instruments are usually coated with a transparent thin film, such as magnesium fluoride (MgF_2), to reduce or eliminate unwanted reflection. More important, such coatings enhance the transmission of light through the lenses.

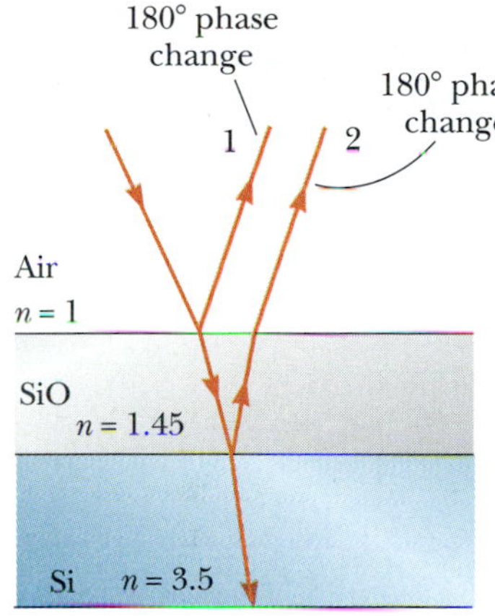

Figure 27.15 (Example 27.3) Reflective losses from a silicon solar cell are minimized by coating the cell with a thin film of silicon monoxide.

Example 27.4 Interference in a Wedge-Shaped Film

A thin, wedge-shaped film of refractive index n is illuminated with monochromatic light of wavelength λ, as illustrated in Figure 27.16. Describe the interference pattern observed for this case.

Solution The interference pattern is that of a thin film of variable thickness surrounded by air. Hence, the pattern will be a series of alternating bright and dark parallel bands. A dark band corresponding to destructive interference appears at point O, the apex, since the upper reflected ray undergoes a 180° phase change while the lower one does not. According to Equation 27.17, other dark bands appear when $2nt = m\lambda$, so that $t_1 = \lambda/2n$, $t_2 = \lambda/n$, $t_3 = 3\lambda/2n$, and so on. Similarly, bright bands are observed when the thickness satisfies the condition $2nt = (m + \frac{1}{2})\lambda$, corresponding to thicknesses of $\lambda/4n$, $3\lambda/4n$, $5\lambda/4n$, and so on. If white light is used, bands of different colors will be observed at different points, corresponding to the different wavelengths of light.

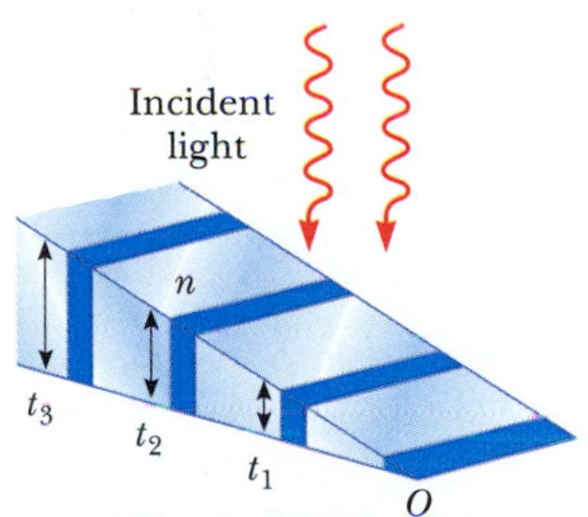

Figure 27.16 (Example 27.4) Interference bands in reflected light can be observed by illuminating a wedge-shaped film with monochromatic light. The dark blue areas correspond to positions of destructive interference.

Summary

Interference of light waves is the result of the linear superposition of two or more waves at a given point. A sustained interference pattern is observed if (1) the sources are coherent, (2) the sources have identical wavelengths, and (3) the superposition principle is applicable.

In Young's double-slit experiment, two slits separated by a distance of d are illuminated by a monochromatic light source. An interference pattern consisting of bright and dark fringes is observed on a screen that is a distance of L from the slits. The condition for **constructive interference** is

Conditions for constructive interference

$$d \sin \theta = m\lambda \qquad (m = 0, \pm 1, \pm 2, \ldots) \qquad [27.2]$$

The condition for **destructive interference** is

Conditions for destructive interference

$$d \sin \theta = (m + \tfrac{1}{2})\lambda \qquad (m = 0, \pm 1, \pm 2, \ldots) \qquad [27.3]$$

The number m is called the **order number** of the fringe.

The **average intensity** of the double-slit interference pattern is

$$I_{av} = I_0 \cos^2\left(\frac{\pi d \sin \theta}{\lambda}\right) \qquad [27.12]$$

where I_0 is the maximum intensity on the screen.

When a series of N slits is illuminated, a diffraction pattern is produced that can be viewed as interference arising from the superposition of a large number of waves. It is convenient to use phasor diagrams to simplify the analysis of interference from three or more equally spaced slits.

An electromagnetic wave traveling from a medium of index of refraction n_1 toward a medium of index of refraction n_2 undergoes a 180° phase change upon reflection when $n_2 > n_1$. There is no phase change in the reflected wave if $n_2 < n_1$.

The wavelength of light, λ_n, in a medium whose refractive index is n is

$$\lambda_n = \frac{\lambda}{n} \qquad [27.14]$$

where λ is the wavelength of light in free space.

The condition for constructive interference in a film of thickness t and refractive index n with the same medium on both sides of the film is given by

Constructive interference in thin films

$$2nt = (m + \tfrac{1}{2})\lambda \qquad (m = 0, 1, 2, \ldots) \qquad [27.16]$$

Similarly, the condition for destructive interference is

Destructive interference in thin films

$$2nt = m\lambda \qquad (m = 0, 1, 2, \ldots) \qquad [27.17]$$

Questions and Conceptual Exercises

1. What is the necessary condition on path length difference between two waves that interfere (a) constructively? (b) destructively?
2. Explain why two flashlights held close together do not produce an interference pattern on a distant screen.
3. A simple way of observing an interference pattern is to

look at a distant light source through a stretched handkerchief or an opened umbrella. Explain how this works.

4. Why is it so much easier to perform interference experiments with a laser than with an ordinary light source?
5. In Young's double-slit experiment, why do we use monochromatic light? If white light were used, how would the pattern change?
6. If Young's double-slit experiment were performed underwater, how would the observed interference pattern be affected?
7. If a soap film on a wire loop is held in air, it appears black in the thinnest regions when observed by reflected light, and shows a variety of colors in thicker regions, as in Figure 27.17. Explain.

Figure 27.17 (Question 7)

8. In the process of evaporation, a soap bubble appears black just before it breaks. Explain this phenomenon in terms of the phase changes that occur upon reflection from the two surfaces.
9. If an oil film is observed on water, the film appears brightest at the outer regions, where it is thinnest. From this information, what can you say about the index of refraction of the oil relative to that of water?
10. In order to observe interference in a thin film, why must the film not be very thick (on the order of a few wavelengths)?
11. A lens with outer radius of curvature R and index of refraction n rests on a flat glass plate. It is illuminated with white light from above. Is there a dark spot or a light spot at the center of the lens? What does it mean if the observed rings are noncircular?
12. Why is the lens on a good-quality camera coated with a thin film?
13. Would it be possible to place a nonreflective coating on an airplane to cancel radar waves of wavelength 3 cm?
14. Lenses with a "nonreflective" coating usually appear to have a purplish cast. Explain the color.
15. Washed dishes that are not rinsed well often have colored bands or rings across them. Discuss the interference effect that causes this.
16. Suppose we use reflected white light to observe a thin, transparent coating on glass as the coating material is gradually deposited by evaporation in a vacuum. Describe possible color changes that might occur during the process of building up the thickness of the coating.
17. What change, if any, would occur in the pattern of Newton's rings if the space between the lens and the plate were filled with water?

Problems

Section 27.2 Young's Double-Slit Experiment

1. A laser beam ($\lambda = 632.8$ nm) is incident upon two slits 0.2 mm apart. Approximately how far apart will the bright interference lines be on a screen 5 m from the double slits?
2. A Young's interference experiment is performed with blue-green argon laser light. The separation between the slits is 0.50 mm, and the interference pattern on a screen 3.3 m away shows the first maximum at a distance of 3.4 mm from the center of the pattern. What is the wavelength of argon laser light?
3. Light from a helium-cadmium laser ($\lambda = 442$ nm) passes through a double-slit system with a slit separation of $d = 0.40$ mm. Determine how far away a screen must be placed in order that a dark fringe appear directly opposite each slit.
4. Coherent light of wavelength $\lambda = 587.5$ nm is allowed to fall on a plane containing parallel slits that are 0.2 mm apart. A screen is placed so that the second bright band in the interference pattern is at a distance equal to 10 slit spacings from the central maximum. What is the distance between the source plane and the screen?
5. Two radio antennas separated by 300 m, as shown in Figure 27.18, simultaneously transmit identical signals (assume waves) on the same wavelength. A radio in a car traveling due north receives the signals. (a) If the car is at the position of the second maximum, what is the wavelength of the signals? (b) How much farther must the car travel to encounter the next minimum in reception? (*Caution:* Avoid small-angle approximations in this problem.)
6. Light of wavelength 546 nm (the intense green line

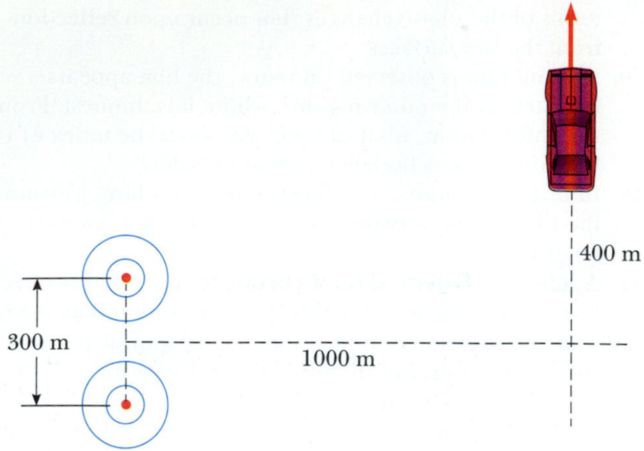

Figure 27.18 (Problem 5)

from a mercury discharge tube) produces a Young's interference pattern in which the second-order minimum is along a direction that makes an angle of 18 minutes of arc relative to the direction of the central maximum. What is the distance between the parallel slits?

7. In a double-slit arrangement like that illustrated in Figure 27.4, $d = 0.15$ mm, $L = 140$ cm, $\lambda = 643$ nm, and $y = 1.8$ cm. (a) What is the path difference, δ, for the two slits at the point P? (b) Express this path difference in terms of the wavelength. (c) Will point P correspond to a maximum, a minimum, or an intermediate condition?

8. Waves broadcast by a 1500-kHz radio station arrive at a home receiver by two paths. One is a direct path, and the second is from reflection off an airplane directly above the home receiver. The airplane is approximately 100 m above the home receiver, and the direct distance from the station to the home is 20 km. What is the exact height of the airplane if destructive interference is occurring? (Assume no phase change occurs on reflection from the plane.)

9. An oscillator drives two loudspeakers 35 cm apart, which vibrate in phase at a frequency of 2 kHz. At what angles, measured from the perpendicular bisector of the line joining the speakers, would a distant observer hear maximum sound intensity? Minimum? (Take the speed of sound as 340 m/s.)

10. One of the bright bands in Young's interference pattern is located 12 mm from the central maximum. The screen is 119 cm from the pair of slits that serve as sources. The slits are 0.241 mm apart and are illuminated by the blue light from a hydrogen discharge tube ($\lambda = 486$ nm). How many bright lines are observed between the central maximum and the 12-mm position?

Section 27.3 Intensity Distribution of the Double-Slit Interference Pattern

11. In the arrangement of Figure 27.3, let $L = 120$ cm and $d = 0.25$ cm. The slits are illuminated with light of wavelength 600 nm. Calculate the distance y above the central maximum for which the average intensity on the screen will be 75% of the maximum.

12. In an arrangement similar to that illustrated in Figure 27.3, let $L = 140$ cm and $y = 8$ mm. Find the value of the ratio d/λ for which the average intensity at point P will be 60% of the maximum intensity.

13. In Figure 27.3, let $L = 1.2$ m and $d = 0.12$ mm, and assume that the slit system is illuminted with monochromatic light of wavelength 500 nm. Calculate the phase difference between the two wavefronts arriving at point P from S_1 and S_2 when (a) $\theta = 0.5°$ and (b) $y = 5$ mm.

14. For the situation described in Problem 13, what is the value of θ for which (a) the phase difference will be equal to 0.333 rad? (b) the path difference will be $\lambda/4$?

15. The intensity on the screen at a certain point in a double-slit interference pattern is 64% of the maximum value. (a) What is the minimum phase difference (in radians) between sources that will produce this result? (b) Express the phase difference calculated in (a) as a path difference if the wavelength of the incident light is 486.1 nm (H_β line).

16. At a particular location in a Young's interference pattern, the intensity on the screen is 6.4% of maximum. (a) Calculate the minimum phase difference in this case. (b) If the wavelength of light is 587.5 nm (from a helium discharge tube), determine the path difference.

17. Coherent light from a helium-neon laser ($\lambda = 632.8$ nm) is incident on two parallel slits 0.2 mm apart. What is the distance to the first maximum and its intensity (relative to the central maximum) on a screen 2 m beyond the slits?

18. Two narrow parallel slits are separated by 0.85 mm and are illuminated by coherent light with $\lambda = 6000$ Å. (a) What is the phase difference between the two interfering waves on a screen (2.8 m away) at a point 2.50 mm from the central bright fringe? (b) What is the ratio of the intensity at this point to the intensity at the center of a bright fringe?

Section 27.4 Phasor Addition of Waves

19. The electric fields from three coherent sources are described by $E_1 = E_0 \sin \omega t$, $E_2 = E_0 \sin(\omega t + \phi)$, and $E_3 = E_0 \sin(\omega t + 2\phi)$. Let the resultant field be represented by $E_P = E_R \sin(\omega t + \alpha)$. Use the phasor

method to find E_R and α when (a) $\phi = 20°$, (b) $\phi = 60°$, (c) $\phi = 120°$.

20. Use the method of phasors to find the resultant (magnitude and phase angle) of two fields represented by $E_1 = 12 \sin \omega t$ and $E_2 = 18 \sin(\omega t + 60°)$. (Note that in this case the amplitudes of the two fields are unequal.)
21. Determine the resultant of the two waves $E_1 = 6 \sin(100\ \pi t)$ and $E_2 = 8 \sin(100\ \pi t + \pi/2)$.
22. Two coherent waves are described by

$$E_1 = E_0 \sin\left(\frac{2\pi x_1}{\lambda} - 2\pi f t + \frac{\pi}{6}\right)$$

$$E_2 = E_0 \sin\left(\frac{2\pi x_2}{\lambda} - 2\pi f t + \frac{\pi}{8}\right)$$

Determine the relationship between x_1 and x_2 that produces constructive interference when the two waves are superposed.
23. Sketch a phasor diagram to illustrate the resultant of $E_1 = E_{01} \sin \omega t$ and $E_2 = E_{02} \sin(\omega t + \phi)$, where $E_{02} = 1.5E_{01}$ and $\pi/6 \leq \phi \leq \pi/3$. Use the sketch and the law of cosines to show that, for two coherent waves, the resultant *intensity* can be written in the form $I_R = I_1 + I_2 + 2\sqrt{I_1 I_2} \cos \phi$.

Section 27.6 Interference in Thin Films

24. A material having an index of refraction of 1.30 is used to coat a piece of glass ($n = 1.50$). What should be the minimum thickness of this film in order to minimize reflected light at a wavelength of 500 nm?
25. A soap bubble ($n = 1.33$) is floating in air. If the thickness of the bubble wall is 115 nm, what is the wavelength of the light that is most strongly reflected?
26. A uniform film of oil ($n = 1.31$) is floating on water. When sunlight in air is incident normally on the film, an observer finds that the reflected light has a maximum at $\lambda = 450$ nm and a minimum at $\lambda = 600$ nm. What is the thickness of the oil film?
27. A thin film of MgF_2, 10^{-5} cm thick ($n = 1.38$), is used to coat a camera lens. Will any wavelengths in the visible spectrum be intensified in the reflected light?
28. A soap bubble of index of refraction 1.33 strongly reflects both red and green colors in white light. What thickness of soap bubble allows this to happen? (In air, λ red = 700 nm, λ green = 500 nm.)
29. A thin layer of liquid methylene iodide ($n = 1.756$) is sandwiched between two flat parallel plates of glass. What must be the thickness of the liquid layer if normally incident light with $\lambda = 600$ nm is to be strongly reflected?
30. A beam of light of wavelength 580 nm passes through two closely spaced glass plates, as shown in Figure 27.19. For what minimum nonzero value of the plate separation, d, will the transmitted light be bright?

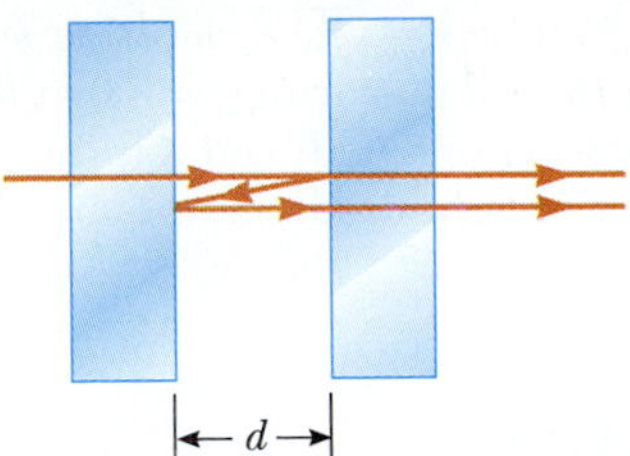

Figure 27.19 (Problem 30)

31. An oil film ($n = 1.45$) floating on water is illuminated by white light at normal incidence. The film is 280 nm thick. Find (a) the dominant observed color in the reflected light and (b) the dominant color in the transmitted light. Explain your reasoning.
32. A possible means for making an airplane radar-invisible is to coat it with an antireflective polymer. If radar waves have a wavelength of 3 cm and the index of refraction of the polymer is $n = 1.5$, how thick would you make the coating?
33. Two rectangular, optically flat glass plates ($n = 1.52$) are in contact along one end and are separated along the other end by a sheet of paper that is 4×10^{-3} cm thick (Fig. 27.20). The top plate is illuminated by monochromatic light ($\lambda = 546.1$ nm). Calculate the number of dark parallel bands crossing the top plate (include the dark band at zero thickness along the edge of contact between the two plates).

Figure 27.20 (Problems 33 and 34)

34. An air wedge is formed between two glass plates separated at one edge by a very fine wire as in Figure 27.20. When the wedge is illuminated from above by light with a wavelength of 600 nm, 30 dark fringes are observed. Calculate the radius of the wire.
35. When a liquid is introduced into the air space between the lens and the plate in a Newton's-rings apparatus, the diameter of the tenth ring changes from 1.50 to 1.31 cm. Find the index of refraction of the liquid.
36. If the Newton's-ring arrangement is such that the radius of curvature of the lens R is much greater than the thickness of the air gap, the radius of the dark

rings is given by $r = \sqrt{m\lambda R/n}$ where m is an integer counting the rings, starting with 0 at the central spot and increasing outward, and λ is the vacuum wavelength of the light used. Show that this relationship for r is valid.

Additional Problems

37. A certain crude oil has an index of refraction n of 1.25. A ship dumps 1.0 m^3 of this oil into the ocean, and the oil spreads into a thin slick. If the film produces a first-order maximum of light of wavelength 500 nm normally incident on it, how much surface area of the ocean does the oil slick cover? Assume the index of refraction of the ocean is 1.34.

38. Interference effects are produced at point P on a screen as a result of direct rays from a source of wavelength 500 nm and of reflected rays off the mirror, as shown in Figure 27.21. If the source is 100 m to the left of the screen, and 1 cm above the mirror, find the distance y (in millimeters) to the first dark band above the mirror.

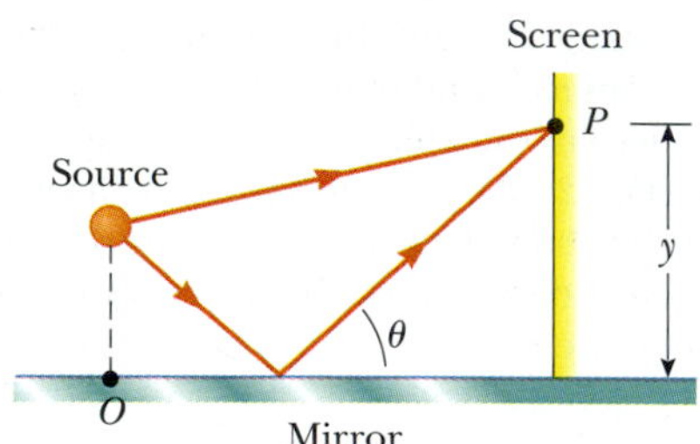

Figure 27.21 (Problem 38)

39. In an application of interference effects to radioastronomy, Australian astronomers observed a 60-MHz radio source both directly from the source and from its reflection from the sea. If the receiving dish is 20 m above sea level, what is the angle of the radio source above the horizon at first maximum?

40. The waves from a radio station can reach a home receiver by two different paths. One is a straight-line path from the transmitter to the home, a distance of 30 km. The second path is by reflection from the ionosphere (a layer of ionized air molecules near the top of the atmosphere). Assume this reflection takes place at a point midway between receiver and transmitter. If the wavelength broadcast by the radio station is 350 m, find the minimum height of the ionospheric layer that would produce destructive interference between the direct and reflected beams. (Assume no phase changes on reflection.)

41. Measurements are made of the intensity distribution in a Young's interference pattern (as illustrated in Figure 27.5). At a particular value of y (distance from the center of the screen), it is found that $I/I_0 = 0.81$ when light of wavelength 600 nm is used. What wavelength of light should be used to reduce the relative intensity at the same location to 64%?

42. In a Young's interference experiment, the two slits are separated by 0.15 mm, and the incident light includes light of wavelengths $\lambda_1 = 540$ nm and $\lambda_2 = 450$ nm. The overlapping interference patterns are formed on a screen 1.4 m from the slits. Calculate the minimum distance from the center of the screen to the point where a bright line of the λ_1 light coincides with a bright line of the λ_2 light.

43. Young's double-slit experiment is performed with sodium yellow light ($\lambda = 5890$ Å) and with a slits-to-screen distance of 2.0 m. The tenth interference minimum (dark fringe) is observed to be 7.26 mm from the central maximum. Determine the spacing of the slits.

44. A thin film of oil ($n = 1.38$) having a thickness of d floats on water. (a) Show that the equation

$$\lambda = \frac{2nd}{(m + \frac{1}{2})} \qquad (m = 0, 1, 2, \ldots)$$

where d is the film thickness, gives the wavelengths that exhibit constructive interference for reflected light. (b) What color is the film at points where $d = 300$ nm? (The film is viewed from above in white light.)

45. A hair is placed at one edge between two flat glass plates 8 cm long. When this arrangement is illuminated with yellow light of wavelength 600 nm, a total of 121 dark bands are counted, starting at the point of contact of the two plates. How thick is the hair?

46. A glass plate ($n = 1.61$) is covered with a thin, uniform layer of oil ($n = 1.2$). A nonmonochromatic light beam in air is incident normally on the oil surface. Observation of the reflected beam shows destructive interference at 500 nm and constructive interference at 750 nm, with no intervening maxima or minima. Calculate the thickness of the oil film.

47. A piece of transparent material with index of refraction n is cut into a wedge shape, as shown in Figure 27.22. The angle of the wedge is small, and monochromatic light of wavelength λ is normally incident from above. If the height of the wedge is h and the width is ℓ, show that bright fringes occur at the positions $x = \lambda\ell(m + \frac{1}{2})/2hn$ and dark fringes occur at the positions $x = \lambda\ell m/2hn$, where $m = 0, 1, 2, \ldots$ and x is measured as shown.

48. An air wedge is formed between two glass plates that are in contact along one edge and slightly separated at the opposite edge. When illuminated with monochromatic light from above, the reflected light reveals a total of 85 dark fringes. Calculate the number of

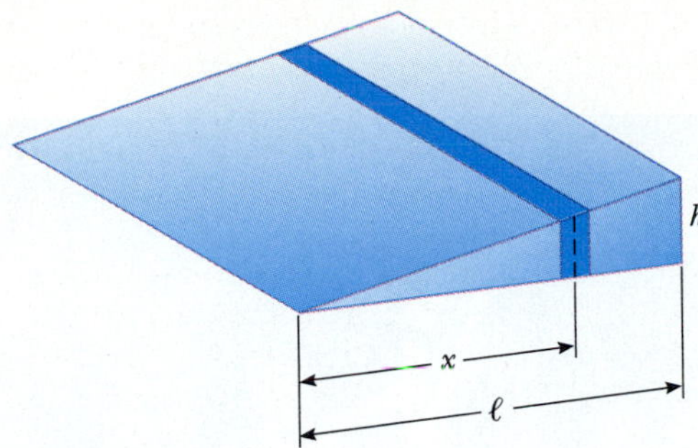

Figure 27.22 (Problem 47)

dark fringes that would appear if water ($n = 1.33$) were to replace the air between the plates.

49. Our discussion of the techniques for determining constructive and destructive interference by reflection from a thin film in air has been confined to rays striking the film at nearly normal incidence. Assume that a ray is incident at an angle of 30° (relative to the normal) on a film with index of refraction 1.38. Calculate the minimum thickness for constructive interference if the light is sodium light with a wavelength of 590 nm.

50. Use the method of phasor addition to find the resultant amplitude and phase constant when the following three harmonic functions are combined: $E_1 = \sin(\omega t + \pi/6)$, $E_2 = 3\sin(\omega t + 7\pi/2)$, $E_3 = 6\sin(\omega t + 4\pi/3)$.

51. A soap film ($n = 1.33$) is contained within a rectangular wire frame. The frame is held vertically so that the film drains downward due to gravity and becomes thicker at the bottom than at the top, where the thickness is essentially zero. The film is viewed in white light with near-normal incidence, and the first violet ($\lambda = 420$ nm) interference band is observed 3 cm from the top edge of the film. (a) Locate the first red ($\lambda = 680$ nm) interference band. (b) Determine the film thickness at the positions of the violet and red bands. (c) What is the wedge angle of the film?

52. A soap film such as that in Problem 51 is viewed in yellow light ($\lambda = 589$ nm) with near-normal incidence. Interference fringes with a uniform spacing of 4.5 mm are observed. What is the variation in thickness of the film per centimeter of vertical distance?

53. We can produce interference fringes using a *Lloyd's mirror* arrangement with a single monochromatic source, with $\lambda = 606$ nm, as in Figure 27.23. The image S' of the source formed by the mirror acts as a second coherent source that interferes with S_0. If fringes spaced 1.2 mm apart are formed on a screen 2 m from the source S_0, find the vertical distance h of the source above the plane of the reflecting surface.

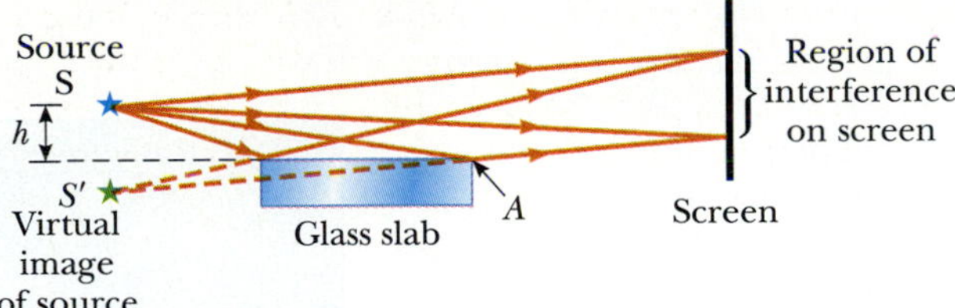

Figure 27.23 (Problem 53)

54. (a) Both sides of a uniform film of index of refraction n and thickness d are in contact with air. For normal incidence of light, an intensity minimum is observed in the reflected light at λ_2, and an intensity maximum is observed at λ_1, where $\lambda_1 > \lambda_2$. If no intensity minima are observed between λ_1 and λ_2, show that the integer m that appears in Equations 27.16 and 27.17 is given by $m = \lambda_1/2(\lambda_1 - \lambda_2)$. (b) Determine the thickness of the film if $n = 1.40$, $\lambda_1 =$ 500 nm, and $\lambda_2 = 370$ nm.

55. Consider the double-slit arrangement shown in Figure 27.24, where the separation of the slits, d, is 0.30 mm and the distance L to the screen is 1 m. A thin sheet of transparent plastic, of thickness 0.050 mm (about the thickness of this page) and refractive index $n = 1.50$, is placed over only the upper slit. As a result, the central maximum of the interference pattern moves upward a distance of y'. Find this distance.

56. A light source emits light of two wavelengths in the visible region, given by $\lambda = 430$ nm and $\lambda' = 510$ nm. The source is used in a double-slit interference experiment in which $L = 1.5$ m and $d = 0.025$ mm. Find the separation between the third-order bright fringes corresponding to these wavelengths.

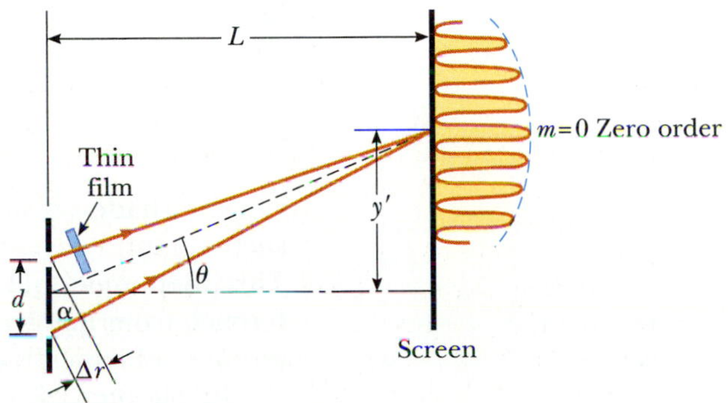

Figure 27.24 (Problem 55)

A plastic model of an arch structure under load conditions observed between two crossed polarizers. The stress pattern is produced in those regions where the stresses are greatest. Such patterns and models are useful in the optimum design of architectural components. *(Peter Aprahamian/Sharples Stress Engineers Ltd./Science Photo Library)*

Diffraction and Polarization

CHAPTER 28

In this chapter, we continue our treatment of wave optics with a discussion of diffraction and polarization phenomena. When coherent light waves pass through a small aperture, an interference pattern rather than a sharp spot of light is observed. This shows that, once it passes through the aperture, the light spreads out into regions where a shadow would be expected if light traveled in straight lines. Other waves, such as sound waves and water waves, also have this ability to bend around corners. This phenomenon, known as *diffraction,* can be regarded as a consequence of interference from many coherent wave sources. In other words, diffraction and interference are basically equivalent.

In Chapter 24, we discussed the properties of electromagnetic waves and the fact that they are transverse. That is, the electric and magnetic field vectors associated with the wave are perpendicular to the direction of propagation. Under certain conditions, light waves can be plane-polarized. Although ordinary light is usually not polarized, we shall discuss various methods for producing polarized light, such as the use of polarizing sheets.

28.1 Introduction to Diffraction

Suppose a light beam is incident on two slits, as in Young's double-slit experiment. If the light truly traveled in straight-line paths after passing through the slits, as in Figure 28.1a, the waves would not overlap and no interference pattern would be seen. Instead, Huygens' principle requires that the waves spread out from the slits, as shown in Figure 28.1b. In other words, the light deviates from a straight-line path and enters the region that would otherwise be shadowed. This divergence of light from its initial line of travel is called **diffraction.**

In general, diffraction occurs when waves pass through small openings, around obstacles, or by sharp edges. For example, when a narrow slit is placed between a distant light source (or a laser beam) and a screen, the light produces a diffraction pattern like that in Figure 28.2. The pattern consists of a broad, intense central band, the **central maximum,** flanked by a series of narrower, less intense secondary bands (called **secondary maxima**) and a series of dark bands, or **minima.** This cannot be explained within the framework of geometric optics, which says that light rays traveling in straight lines should cast a sharp rendition of the slit on the screen.

Figure 28.3 shows the diffraction pattern and shadow of a penny. The pattern consists of the shadow, a bright spot at its center, and a series of bright and dark bands of light near the edge of the shadow. The bright spot at the center (called the *Arago bright spot* after its discoverer, Dominique Arago) can be explained through the wave theory of light, which predicts constructive interference at this

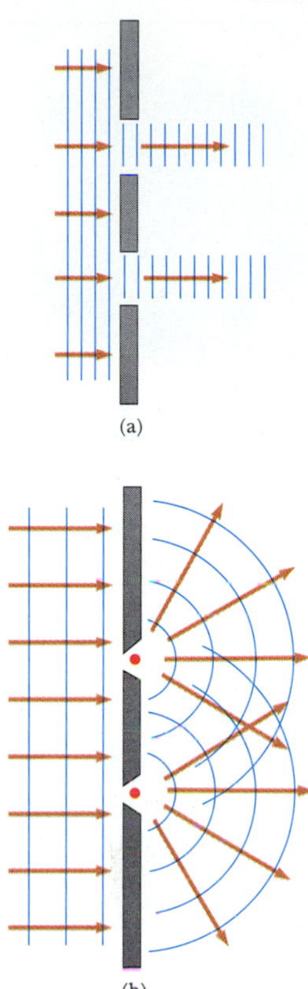

Figure 28.1
(a) If light waves did not spread out after passing through the slits, no interference would occur. (b) The light waves from the two slits overlap as they spread out, producing interference fringes.

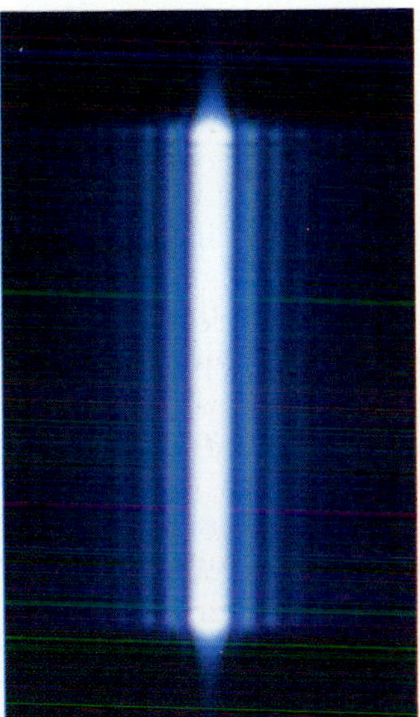

Figure 28.2
The diffraction pattern that appears on a screen when light passes through a narrow vertical slit. The pattern consists of a broad central band and a series of less intense and narrower side bands.

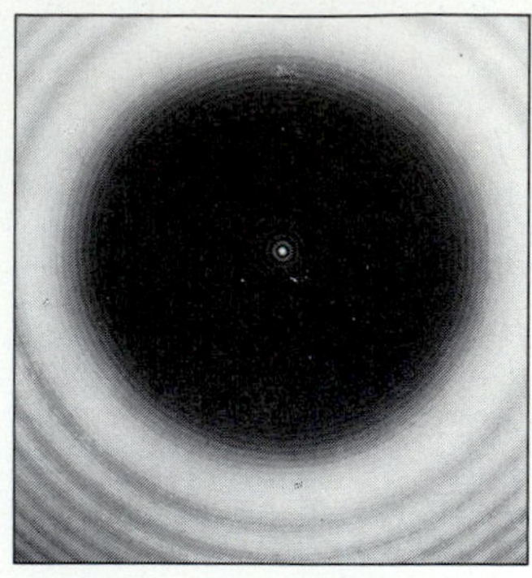

Figure 28.3
The diffraction pattern of a penny. *(Courtesy of P. M. Rinard, from Am. J. Phys. 44:70, 1976)*

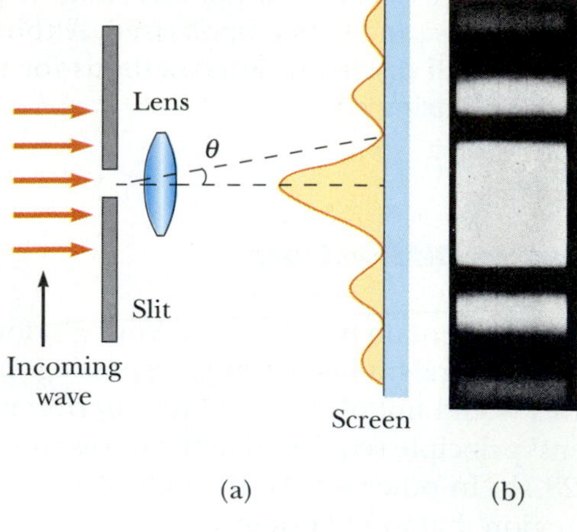

Figure 28.4
(a) A Fraunhofer diffraction pattern of a single slit. The parallel rays are brought into focus on the screen with a converging lens. The pattern consists of a central bright region flanked by much weaker alternating bright and dark bands. (Note that this drawing is not to scale.) (b) Photograph of a single-slit Fraunhofer diffraction pattern. *(From M. Cagnet, M. Francon, and J. C. Thierr, Atlas of Optical Phenomena, Berlin, Springer-Verlag, 1962, plate 18)*

point. In contrast, from the viewpoint of geometric optics, the center of the pattern would be completely screened by the penny, and so one would never observe a central bright spot.

Diffraction phenomena are usually classified into two types, named after the men who first explained them. The first type, called **Fraunhofer diffraction,** occurs when the rays reaching the observing screen are approximtely parallel. This can be achieved experimentally either by placing the observing screen far from the slit or by using a converging lens to focus the parallel rays on the screen, as in Figure 28.4a. A bright fringe is observed along the axis at $\theta = 0$, with alternating dark and bright fringes on each side of the central bright fringe. Figure 28.4b is a photograph of a single-slit Fraunhofer diffraction pattern.

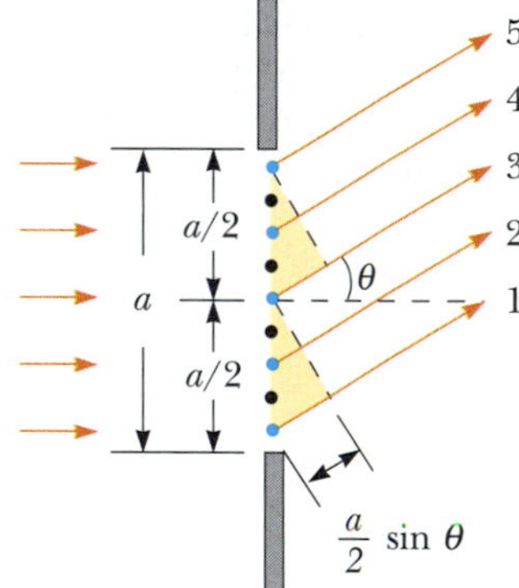

Figure 28.5
Diffraction of light by a narrow slit of width a. Each portion of the slit acts as a point source of waves. The path difference between rays 1 and 3 or between rays 3 and 5 is equal to $(a/2)\sin\theta$. (Note that this drawing is not to scale.)

28.2 Single-Slit Diffraction

Until now we have assumed that slits are point sources of light. In this section we shall determine how their finite widths are the basis for understanding the nature of the Fraunhofer diffraction pattern produced by a single slit.

We can deduce some important features of this problem by examining waves coming from various portions of the slit, as shown in Figure 28.5. According to Huygens' principle, *each portion of the slit acts as a source of waves.* Hence, *light from one portion of the slit can interfere with light from another portion,* and the resultant intensity on the screen depends on the direction θ.

To analyze the diffraction pattern, it is convenient to divide the slit into two halves, as in Figure 28.5. All the waves that originate at the slit are in phase. Consider waves 1 and 3, which originate at the bottom and center of the slit, respectively. To reach the same point on the viewing screen, wave 1 travels farther than wave 3 by an amount equal to the path difference $(a/2)\sin\theta$, where a is the width of the slit. Similarly, the path difference between waves 3 and 5 is also $(a/2)\sin\theta$. If the path difference is exactly one half of a wavelength (corresponding to a phase difference of 180°), the two waves cancel each other and destructive interference results. This is true, in fact, for any two waves that originate at points separated by half the slit width, because the phase difference between two such points is 180°. Therefore, waves from the upper half of the slit interfere *destructively* with waves from the lower half of the slit when

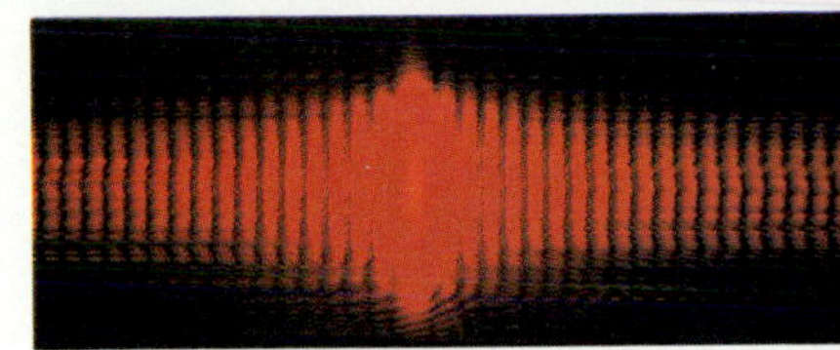

Diffraction pattern produced by a single slit illuminated with red light. *(© Yoav Levy/Phototake, Inc.)*

$$\frac{a}{2}\sin\theta = \frac{\lambda}{2}$$

or when

$$\sin\theta = \frac{\lambda}{a}$$

If we divide the slit into four parts rather than two and use similar reasoning, we find that the screen is also dark when

$$\sin\theta = \frac{2\lambda}{a}$$

Likewise, we can divide the slit into six parts and show that darkness occurs on the screen when

$$\sin\theta = \frac{3\lambda}{a}$$

Therefore, the general condition for **destructive interference** is

Condition for destructive interference

$$\sin\theta = m\frac{\lambda}{a} \qquad (m = \pm 1, \pm 2, \pm 3, \ldots) \qquad [28.1]$$

Equation 28.1 gives the values of θ for which the diffraction pattern has zero intensity—that is, a dark fringe is formed. However, Equation 28.1 tells us nothing about the variation in intensity along the screen. The general features of the intensity distribution are shown in Figure 28.6: a broad central bright fringe flanked by much weaker, alternating bright fringes. The various dark fringes (points of zero

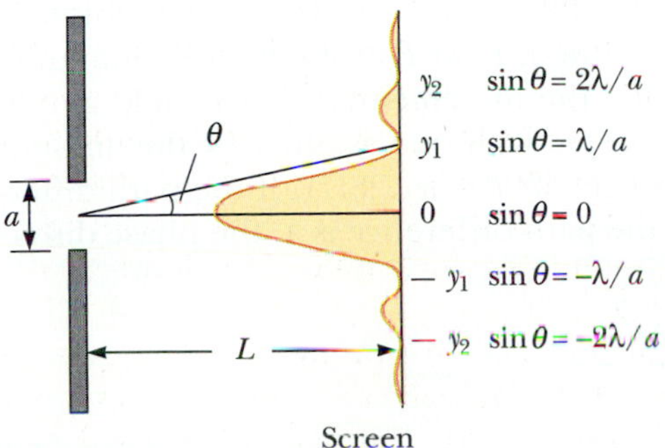

Figure 28.6
Positions of the minima for the Fraunhofer diffraction pattern of a single slit of width a. (Note that this drawing is not to scale.)

intensity) occur at the values of θ that satisfy Equation 28.1. The position of the points of constructive interference lie approximately halfway between the dark fringes. Note that the central bright fringe is twice as wide as the weaker maxima.

Example 28.1 Where Are the Dark Fringes?

Coherent light of wavelength 580 nm is incident on a slit of width 0.30 mm. The observing screen is placed 2 m from the slit. Find the positions of the first dark fringes and the width of the central bright fringe.

Solution The first dark fringes that flank the central bright fringe correspond to $m = \pm 1$ in Equation 28.1:

$$\sin\,\theta = \pm\frac{\lambda}{a} = \pm\frac{5.8 \times 10^{-7}\text{ m}}{0.3 \times 10^{-3}\text{ m}} = \pm\,1.93 \times 10^{-3}$$

From the triangle in Figure 28.6, note that $\tan\,\theta = y_1/L$. Since θ is very small, we can use the approximation $\sin\,\theta \approx \tan\,\theta$, so that $\sin\,\theta \approx y_1/L$. Therefore, the positions of the first minima, measured from the central axis, are

$$y_1 \approx L\sin\,\theta = \pm\,L\frac{\lambda}{a} = \pm\,3.87 \times 10^{-3}\text{ m}$$

The positive and negative signs correspond to the dark fringes on either side of the central bright fringe. Hence, the width of the central bright fringe is equal to $2|y_1| = 7.73 \times 10^{-3}\text{ m} = 7.73\text{ mm}$. Note that this value is *much greater* than the width of the slit. However, as the width of the slit is *increased,* the diffraction pattern will *narrow,* corresponding to smaller values of θ. In fact, for large values of a, the maxima and minima are so closely spaced that only a large central bright area is observed, which resembles the geometric image of the slit. This matter is of great importance in the design of lenses used in telescopes, microscopes, and other optical instruments.

Exercise Determine the width of the first-order bright fringe.

Answer 3.87 mm.

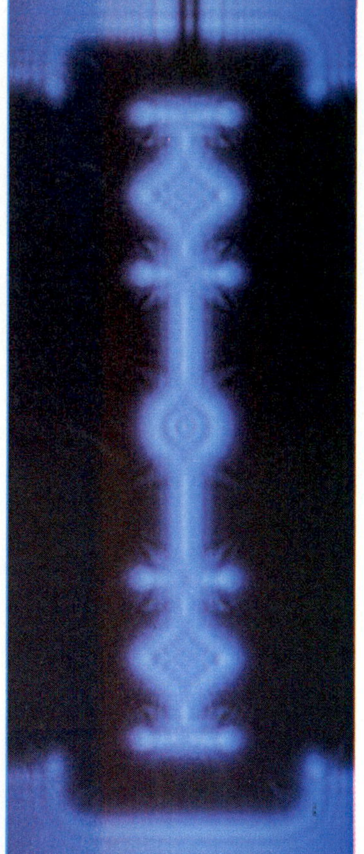

The diffraction pattern of a razor blade observed under monochromatic light. *(© Ken Kay, 1987, Fundamental Photographs)*

Intensity of the Single-Slit Diffraction Pattern

We can use phasors to determine the intensity distribution for the single-slit diffraction pattern. Imagine a slit divided into a large number of small zones, each of width Δy, as in Figure 28.7. Each zone acts as a source of coherent radiation, and each contributes an incremental electric field amplitude of ΔE at some point, P, on the screen. The total electric field amplitude, E, at P is obtained by summing the contributions from all zones. Note that the incremental electric field amplitudes between adjacent zones are out of phase with one another by the amount $\Delta\beta$, corresponding to a path difference of $\Delta y\sin\,\theta$ (Fig. 28.7). If the path difference is $\lambda/2$, the phase difference is π; if the path difference is λ, the phase difference is 2π; and so on. In general,

$$\frac{\Delta\beta}{2\pi} = \frac{\Delta y\sin\,\theta}{\lambda}$$

or

$$\Delta\beta = \frac{2\pi}{\lambda}\Delta y \sin\theta \qquad \textbf{[28.2]}$$

To find the total electric field amplitude on the screen at any angle θ, we sum the incremental amplitudes, ΔE, due to each zone. For small values of θ, we can assume that all the amplitudes are the same. It is convenient to use the phasor diagrams for various angles, as in Figure 28.8. When $\theta = 0$, all phasors are aligned as in Figure 28.8a, since the waves from each zone are in phase. In this case, the total amplitude at the center of the screen is $E_0 = N\Delta E$, where N is the number of zones. The amplitude E_θ at some small angle θ is shown in Figure 28.8b, where each phasor differs in phase from an adjacent one by the amount $\Delta\beta$. In this case, note that E_θ is the *vector sum* of the incremental amplitudes, and hence is given by the length of the chord. Therefore, $E_\theta < E_0$. The total phase difference between waves from the top and bottom portions of the slit is

$$\beta = N\Delta\beta = \frac{2\pi}{\lambda}N\Delta y \sin\theta = \frac{2\pi}{\lambda}a\sin\theta \qquad \textbf{[28.3]}$$

where $a = N\Delta y$ is the width of the slit.

As θ increases, the chain of phasors eventually forms a closed path as in Figure 28.8c. At this point, the vector sum is zero, so $E_\theta = 0$, corresponding to the first minimum on the screen. Noting that $\beta = N\Delta\beta = 2\pi$ in this situation, we see from Equation 28.3 that

$$2\pi = \frac{2\pi}{\lambda}a\sin\theta$$

$$\sin\theta = \frac{\lambda}{a}$$

That is, the first minimum in the diffraction pattern occurs when $\sin\theta = \lambda/a$, which agrees with Equation 28.1.

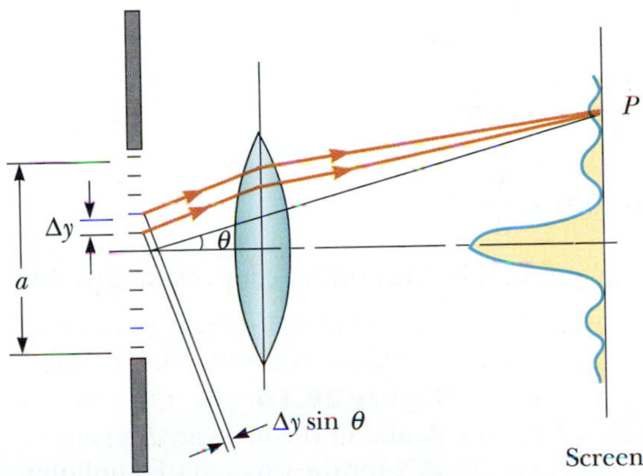

Figure 28.7
Fraunhofer diffraction by a single slit. The intensity at the point P on the screen is the resultant of all the incremental electric fields from zones of width Δy leaving the slit.

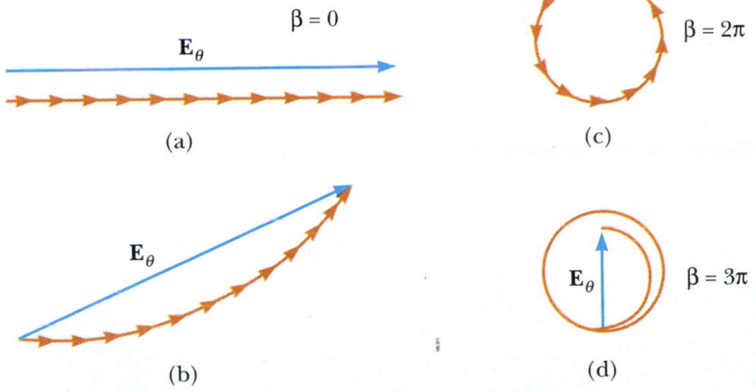

Figure 28.8
Phasor diagrams for obtaining the various maxima and minima of the single-slit diffraction pattern.

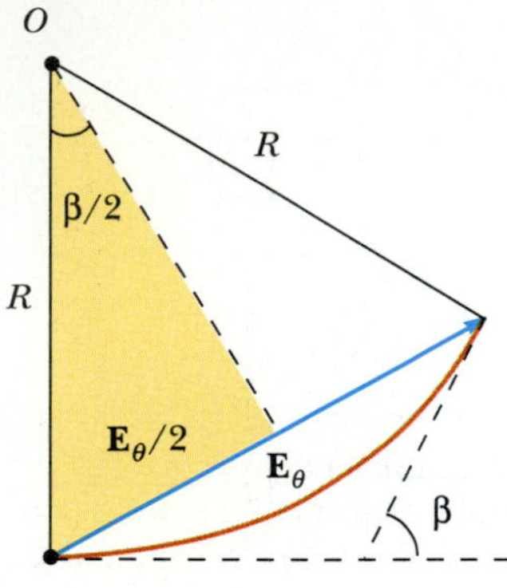

Figure 28.9
Phasor diagram for a large number of coherent sources. Note that the ends of the phasors lie on a circular arc of radius R. The resultant amplitude, E_θ, equals the length of the chord.

At even larger values of θ, the spiral chain of phasors continues. For example, Figure 28.8d represents the situation corresponding to the second maximum, which occurs when $\beta = 360° + 180° = 540°$ (3π rad). The second minimum (two complete spirals not shown) corresponds to $\beta = 720°$ (4π rad), which satisfies the condition $\sin\theta = 2\lambda/a$.

The total amplitude and intensity at any point on the screen can now be obtained by considering the limiting case where Δy becomes infinitesimal (dy) and $N \rightarrow \infty$. In this limit, the phasor diagrams in Figure 28.8 become smooth curves, as in Figure 28.9. From this figure we see that at some angle θ, the wave amplitude on the screen, E_θ, is equal to the chord length, and E_0 is the arc length. From the triangle whose angle is $\beta/2$, we see that

$$\sin\frac{\beta}{2} = \frac{E_\theta/2}{R}$$

where R is the radius of curvature. But the arc length, E_0, is equal to the product $R\beta$, where β is in radians. Combining this with the preceding expression gives

$$E_\theta = 2R\sin\frac{\beta}{2} = 2\left(\frac{E_0}{\beta}\right)\sin\frac{\beta}{2}$$

or

$$E_\theta = E_0\left[\frac{\sin(\beta/2)}{\beta/2}\right]$$

Since the resultant intensity I_θ at P is proportional to the square of the amplitude, E_θ, we find that

Intensity of a single-slit Fraunhofer diffraction pattern

$$I_\theta = I_0\left[\frac{\sin(\beta/2)}{\beta/2}\right]^2 \quad [28.4]$$

where I_0 is the intensity at $\theta = 0$ (the central maximum), and $\beta = 2\pi a\sin\theta/\lambda$. Substitution of this expression for β into Equation 28.4 gives

$$I_\theta = I_0\left[\frac{\sin(\pi a\sin\theta/\lambda)}{\pi a\sin\theta/\lambda}\right]^2 \quad [28.5]$$

From this result, we see that minima occur when

$$\frac{\pi a\sin\theta}{\lambda} = m\pi$$

Condition for intensity minima

$$\sin\theta = m\frac{\lambda}{a}$$

where $m = \pm1, \pm2, \pm3, \ldots$. This is in agreement with our earlier result, given by Equation 28.1.

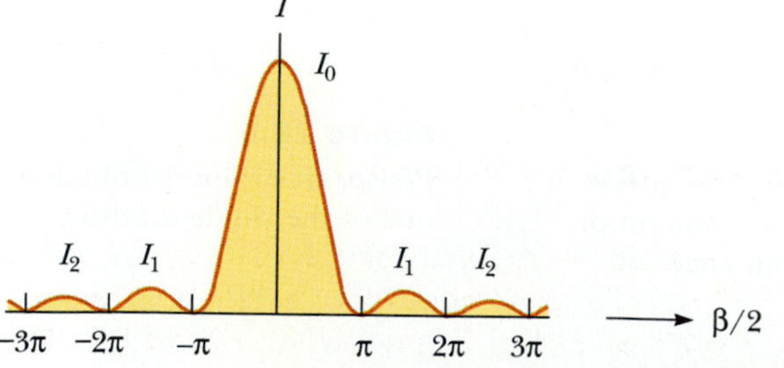

Figure 28.10
A plot of the intensity I versus $\beta/2$ for the single-slit Fraunhofer diffraction pattern.

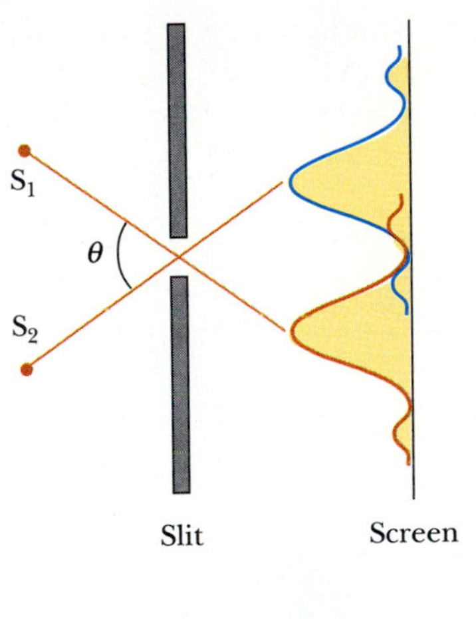

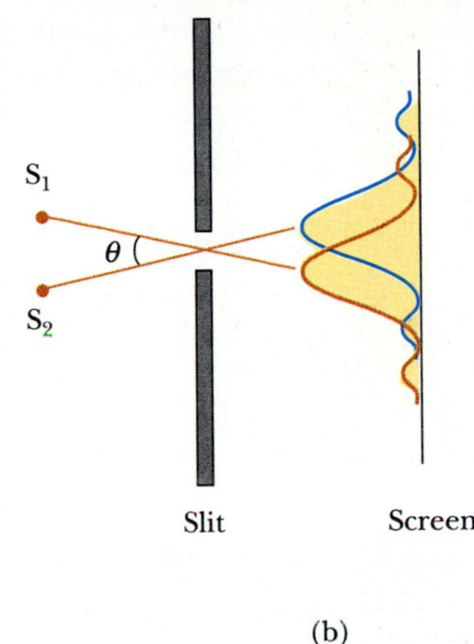

Figure 28.11 Two point sources far from a small aperture each produce a diffraction pattern. (a) The angle subtended by the sources at the aperture is large enough that the diffraction patterns are distinguishable. (b) The angle subtended by the sources is so small that the diffraction patterns overlap and the images are not well resolved. (Note that the angles are greatly exaggerated.)

Figure 28.10 is a plot of Equation 28.5. Most of the light intensity is concentrated in the central bright fringe.

Example 28.2 Relative Intensities of the Maxima

Find the ratio of the intensities of the secondary maxima to the intensity of the central maximum for the single-slit Fraunhofer diffraction pattern.

Solution To a good approximation, we can assume that the secondary maxima lie midway between the centers of the dark bands, which have zero intensities. From Figure 28.11a, we see that this corresponds to $\beta/2$ values of $3\pi/2$, $5\pi/2$, $7\pi/2$, ... Substituting these into Equation 28.4 gives, for the first two ratios,

$$\frac{I_1}{I_0} = \left[\frac{\sin(3\pi/2)}{(3\pi/2)}\right]^2 = \frac{1}{9\pi^2/4} = 0.045$$

$$\frac{I_2}{I_0} = \left[\frac{\sin(5\pi/2)}{(5\pi/2)}\right]^2 = \frac{1}{25\pi^2/4} = 0.016$$

That is, the secondary maximum adjacent to the central maximum has a peak intensity that is 4.5% of that of the central maximum, and the intensity of the next secondary maximum is 1.6% of that of the central maximum.

Exercise Determine an intensity of the secondary maximum that corresponds to $m = 3$ relative to the central maximum.

Answer 0.0083.

28.3 Resolution of Single-Slit and Circular Apertures

The ability of optical systems to distinguish between closely spaced objects is limited because of the wave nature of light. To understand this difficulty, consider Figure 28.11, which shows two light sources far from a narrow slit of width a. The

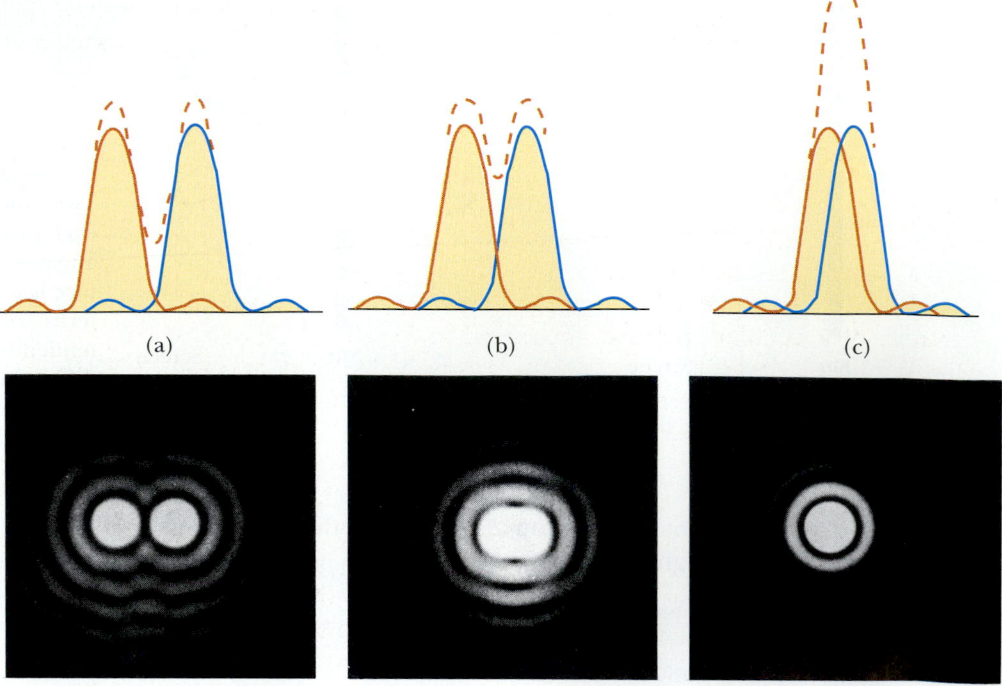

Figure 28.12 The diffraction patterns of two point sources (solid curves) and the resultant pattern (dashed curves), for various angular separations of the sources. In each case, the dashed curve is the sum of the two solid curves. (a) The sources are far apart, and the patterns are well resolved. (b) The sources are closer together, and the patterns are just resolved. (c) The sources are so close together that the patterns are not resolved. *(From M. Cagnet, M. Francon, and J. C. Thierr, Atlas of Optical Phenomena, Berlin, Springer-Verlag, 1962, plate 16)*

sources can be considered as two point sources, S_1 and S_2, that are noncoherent. For example, they could be two distant stars. If no diffraction occurred, one would observe two distinct bright spots (or images) on the screen at the right in the figure. However, because of diffraction, each source is imaged as a bright central region flanked by weaker bright and dark bands. What is observed on the screen is the sum of two diffraction patterns, one from S_1 and the other from S_2.

If the two sources are far enough apart to ensure that their central maxima do not overlap, as in Figure 28.11a, their images can be distinguished and are said to be *resolved.* If the sources are close together, however, as in Figure 28.11b, the two central maxima may overlap and the images are *not resolved.* To decide when two images are resolved, the following condition is often used:

> When the central maximum of one image falls on the first minimum of another image, the images are said to be just resolved. This limiting condition of resolution is known as **Rayleigh's criterion.**

Figure 28.12 shows the diffraction patterns for three situations. When the objects are far apart, their images are well resolved (Fig. 28.12a). The images are just resolved when their angular separation satisfies Rayleigh's criterion (Fig. 28.12b). Finally, the images are not resolved in Figure 28.12c.

From Rayleigh's criterion, we can determine the minimum angular separation, θ_m, subtended by the sources at the slit such that their images are just resolved. In Section 28.2, we found that the first minimum in a single-slit diffraction pattern occurs at the angle that satisfies the relationship

$$\sin \theta = \frac{\lambda}{a}$$

where a is the width of the slit. According to Rayleigh's criterion, this expression gives the smallest angular separation for which the two images are resolved. Because $\lambda \ll a$ in most situations, $\sin\theta$ is small and we can use the approximation $\sin\theta \approx \theta$. There, the limiting angle of resolution for a slit of width a is

Limiting angle of resolution for a slit

$$\theta_m = \frac{\lambda}{a} \qquad [28.6]$$

where θ_m is expressed in radians. Hence, the angle subtended by the two sources at the slit must be *greater* than λ/a if the images are to be resolved.

Many optical systems use circular apertures rather than slits. The diffraction pattern of a circular aperture, illustrated in Figure 28.13, consists of a central circular bright disk surrounded by progressively fainter rings. Analysis shows that the limiting angle of resolution of the circular aperture is

$$\theta_m = 1.22\frac{\lambda}{D} \qquad [28.7]$$

where D is the diameter of the aperture. Note that Equation 28.7 is similar to Equation 28.6 except for the factor of 1.22, which arises from a complex mathematical analysis of diffraction from the circular aperture.

Example 28.3 Resolution of a Telescope

The Hale telescope at Mount Palomar has a diameter of 200 in. What is its limiting angle of resolution at a wavelength of 600 nm?

Solution Because $D = 200$ in. $= 5.08$ m and the wavelength $\lambda = 6 \times 10^{-7}$ m, Equation 28.7 gives

$$\theta_m = 1.22\frac{\lambda}{D} = 1.22\left(\frac{6 \times 10^{-7}\text{ m}}{5.08\text{ m}}\right)$$

$$= 1.44 \times 10^{-7}\text{ rad} = 0.03\text{ s of arc}$$

Therefore, any two stars that subtend an angle greater than or equal to this value will be resolved (assuming ideal atmospheric conditions).

Exercise The large radio telescope at Arecibo, Puerto Rico, has a diameter of 305 m and is designed to detect radio waves at a wavelength of 0.75 m. Calculate the minimum angle of resolution for this telescope, and compare it with that for the Hale telescope.

Answer 3×10^{-3} rad (10 min 19 s of arc), which is more than 10 000 times larger than the calculated minimum angle for the Hale telescope.

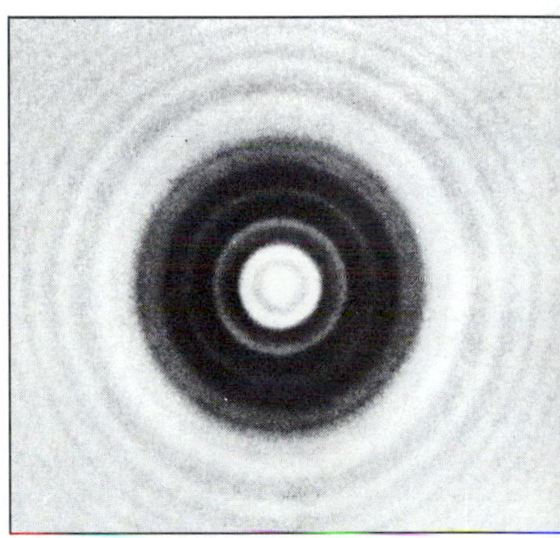

Figure 28.13 The diffraction pattern of a circular aperture consists of a central bright disk surrounded by concentric bright and dark rings. *(From M. Cagnet, M. Francon, and J. C. Thierr, Atlas of Optical Phenomena, Berlin, Springer-Verlag, 1962, plate 34)*

Example 28.4 Resolution of the Eye

Calculate the limiting angle of resolution for the eye, assuming a pupil diameter of 2 mm, a wavelength of 500 nm in air, and an index of refraction for the eye equal to 1.33.

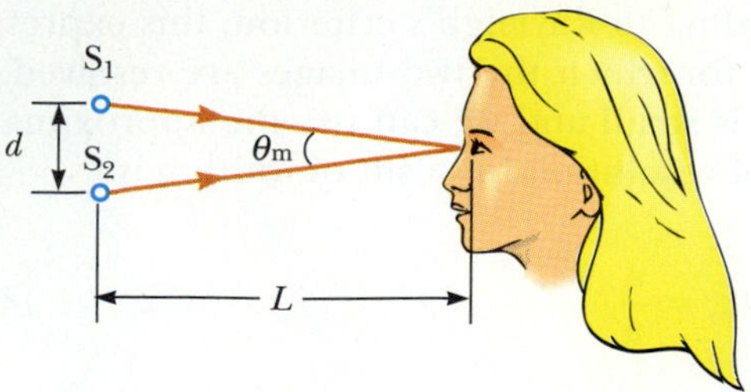

Figure 28.14 (Example 28.4) Two point sources separated by the distance d, as observed by the eye.

Solution We can use Equation 28.7, noting that λ is the wavelength in the medium containing the aperture. Once the light enters the eye, its wavelength is reduced by the index of refraction of the eye medium, and we find that $\lambda = (500 \text{ nm})/1.33 = 376$ nm. Therefore, Equation 28.7 gives

$$\theta_m = 1.22\frac{\lambda}{D} = 1.22\left(\frac{3.76 \times 10^{-7} \text{ m}}{2 \times 10^{-3} \text{ m}}\right)$$

$$= 2.29 \times 10^{-4} \text{ rad} = 0.0131°$$

We can use this result to calculate the minimum separation, d, between two point sources that the eye can distinguish if they are at the distance L from the observer (Fig. 28.14). Since θ_m is small, we see that

$$\sin \theta_m \approx \theta_m \approx d/L$$

$$d = L\theta_m$$

For example, if the objects are positioned at a distance of 25 cm from the eye (the near point), then

$$d = (25 \text{ cm})(2.29 \times 10^{-4} \text{ rad}) = 5.73 \times 10^{-3} \text{ cm}$$

This is approximately equal to the thickness of a human hair.

Exercise If the eye is dilated to a diameter of 5 mm, what is the minimum separation distance it can distinguish at a distance of 40 cm from the eye?

Answer 3.67×10^{-3} cm.

28.4 The Diffraction Grating

The diffraction grating, a useful device for analyzing light sources, consists of a large number of equally spaced parallel slits. A grating can be made by cutting parallel, equally spaced grooves on a glass or metal plate with a precision ruling machine. In a transmission grating, the spaces between lines are transparent to the light and hence act as separate slits. Gratings with many lines very close to each other can have very small slit spacings. For example, a grating ruled with 5000 lines/cm has a slit spacing of $d = (1/5000)$ cm $= 2 \times 10^{-4}$ cm.

Figure 28.15 is a schematic diagram of a section of a flat diffraction grating. A plane wave is incident from the left, normal to the plane of the grating. A converging lens can be used to bring the rays together at the point P. The pattern observed on the screen is the result of the combined effects of interference and diffraction.

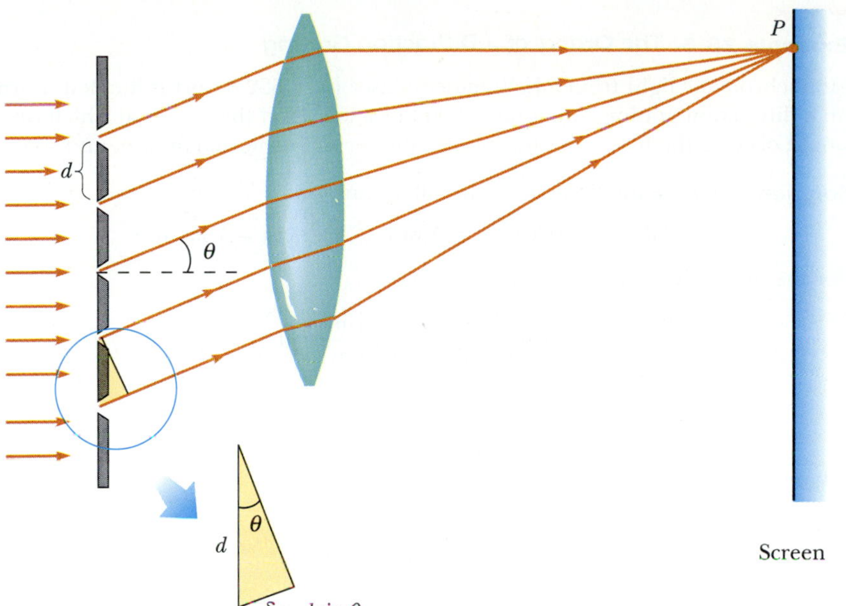

Figure 28.15
Side view of a diffraction grating. The slit separation is d, and the path difference between adjacent slits is $d \sin \theta$.

Each slit produces diffraction, and the diffracted beams interfere with each other to produce the final pattern. Moreover, each slit acts as a source of waves, with all waves starting at the slits in phase. However, for some arbitrary direction θ measured from the horizontal, the waves must travel different path lengths before reaching a particular point P on the screen. From Figure 28.15, note that the path difference between waves from any two adjacent slits is equal to $d \sin \theta$. If this path difference equals one wavelength or some integral multiple of a wavelength, waves from all slits will be in phase at P and a bright line will be observed. Therefore, when the light is incident normal to the plane of the grating, the condition for *maxima* in the interference pattern at the angle θ is

$$d \sin \theta = m\lambda \qquad (m = 0, 1, 2, 3, \ldots) \qquad [28.8]$$

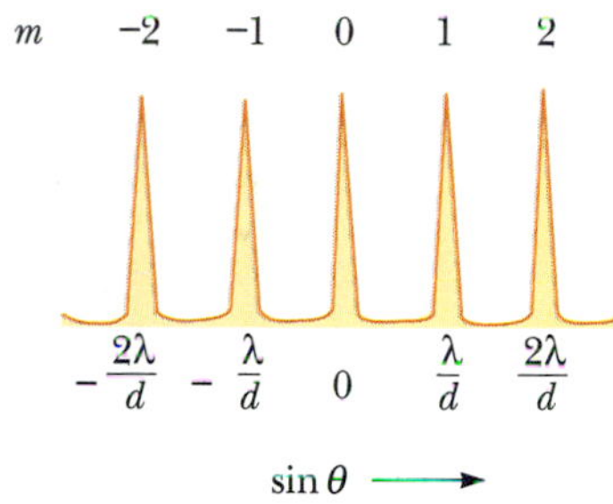

Figure 28.16
Intensity versus $\sin \theta$ for a diffraction grating. The zeroth-, first-, and second-order maxima are shown.

This expression can be used to calculate the wavelength from a knowledge of the grating spacing and the angle of deviation, θ. If the incident radiation contains several wavelengths, the mth-order maximum for each wavelength occurs at a specific angle. All wavelengths are seen at $\theta = 0$, corresponding to $m = 0$.

The intensity distribution for a diffraction grating is shown in Figure 28.16. If the source contains various wavelengths, a spectrum of lines at different positions for different order numbers will be observed. Note the sharpness of the principal maxima and the broad range of dark areas. This is in contrast to the broad, bright fringes characteristic of the two-slit interference pattern (Fig. 28.1).

A simple arrangement for measuring orders of the diffraction pattern is shown in Figure 28.17. This is a form of a diffraction grating spectrometer. The light to be analyzed passes through a slit, and a parallel beam of light exits from the collimator perpendicular to the grating. The diffracted light leaves the grating at angles that satisfy Equation 28.8. A telescope is used to view the image of the slit. The wavelength can be determined by measuring the precise angles at which the images of the slit appear for the various orders.

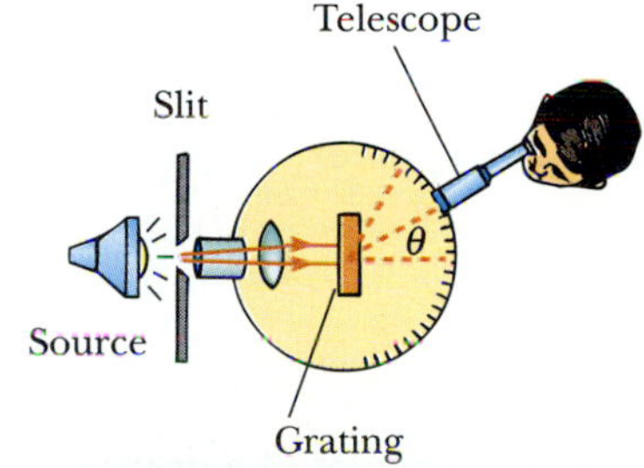

Figure 28.17
Diagram of a diffraction grating spectrometer. The collimated beam incident on the grating is diffracted into the various orders at the angles θ that satisfy the equation $d \sin \theta = m\lambda$, where $m = 0, 1, 2, \ldots$.

Example 28.5 The Orders of a Diffraction Grating

Monochromatic light from a helium-neon laser ($\lambda = 632.8$ nm) is incident normally on a diffraction grating containing 6000 lines/cm. Find the angles at which one would observe the first-order maximum, the second-order maximum, and so forth.

Solution First, we must calculate the slit separation:

$$d = (1/6000)\text{ cm} = 1.667 \times 10^{-4}\text{ cm} = 1667\text{ nm}$$

For the first-order maximum ($m = 1$), we get

$$\sin\theta_1 = \frac{\lambda}{d} = \frac{632.8\text{ nm}}{1667\text{ nm}} = 0.3797$$

$$\theta_1 = 22.31°$$

Likewise, for $m = 2$, we find that

$$\sin\theta_2 = \frac{2\lambda}{d} = \frac{2(632.8\text{ nm})}{1667\text{ nm}} = 0.7592$$

$$\theta_2 = 49.41°$$

However, for $m = 3$, we find that $\sin\theta_3 = 1.139$. Since $\sin\theta$ cannot exceed unity, this does not represent a realistic solution. Hence, only zeroth-, first-, and second-order maxima will be observed in this situation.

Resolving Power of the Diffraction Grating

The diffraction grating is most useful for taking accurate wavelength measurements. Like the prism, the diffraction grating can be used to disperse a spectrum into its components. Of the two devices, the grating is more precise if one wants to distinguish between two closely spaced wavelengths.

If λ_1 and λ_2 are the two nearly equal wavelengths between which the spectrometer can just barely distinguish, the **resolving power,** R, is

Resolving power

$$R \equiv \frac{\lambda}{\lambda_2 - \lambda_1} = \frac{\lambda}{\Delta\lambda} \qquad [28.9]$$

where $\lambda = (\lambda_1 + \lambda_2)/2$ and $\Delta\lambda = \lambda_2 - \lambda_1$. It is left to you, in an end-of-chapter problem, to show that if N lines of the grating are illuminated, the resolving power in the mth-order diffraction equals the product Nm:

Resolving power of a grating

$$R = Nm \qquad [28.10]$$

Thus, resolving power increases with increasing order number.

Furthermore, R is large for a grating with a large number of illuminated slits. Note that for $m = 0$, $R = 0$, which signifies that *all wavelengths are indistinguishable* for the zeroth-order maximum. However, consider the second-order diffraction pattern ($m = 2$) of a grating that has 5000 rulings illuminated by the light source. The resolving power of such a grating in second order is $R = 5000 \times 2 = 10\ 000$. Therefore, the *minimum* wavelength separation between two spectral lines that can be just resolved, assuming a mean wavelength of 600 nm, is $\Delta\lambda = \lambda/R = 6 \times 10^{-2}$ nm. For the third-order principal maximum, we find that $R = 15\ 000$ and $\Delta\lambda = 4 \times 10^{-2}$ nm, and so on.

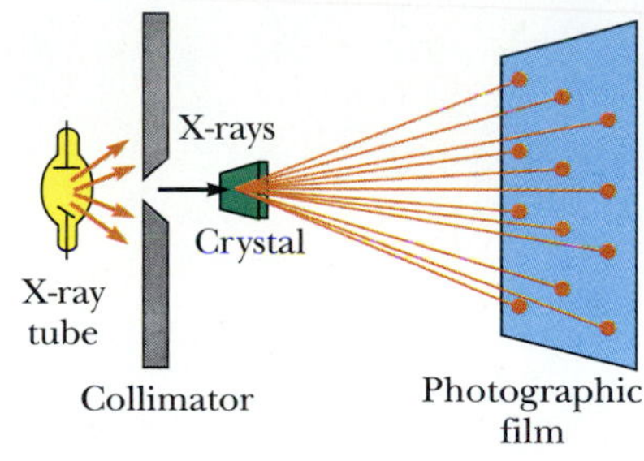

Figure 28.18
Schematic diagram of x-ray diffraction by a single crystal. The array of spots formed on the film is called a Laue pattern.

Example 28.6 Resolving the Sodium Spectral Lines

Two strong lines in the spectrum of sodium have wavelengths of 589.00 nm and 589.59 nm.

(a) What must the resolving power of the grating be in order to distinguish these wavelengths?

$$R = \frac{\lambda}{\Delta\lambda} = \frac{589.30\text{ nm}}{589.59\text{ nm} - 589.00\text{ nm}} = \frac{589.30}{0.59} = 999$$

(b) In order to resolve these lines in the second-order spectrum ($m = 2$), how many lines of the grating must be illuminated?

From Equation 28.10 and the results of (a), we find that

$$N = \frac{R}{m} = \frac{999}{2} = 500\text{ lines}$$

*28.5 Diffraction of X-Rays by Crystals

In principle, the wavelength of any electromagnetic wave can be determined if a grating of the proper spacing (on the order of λ) is available. **X-rays,** discovered in 1895 by W. Roentgen (1845–1923), are electromagnetic waves with very short wavelengths (on the order of 1 Å $= 10^{-10}$ m $= 0.1$ nm). Obviously, it would be impossible to construct a grating with such a small spacing. However, the atomic spacing in a solid is known to be about 10^{-10} m. In 1913, Max von Laue (1879–1960) suggested that the regular array of atoms in a crystal could act as a three-dimensional diffraction grating for x-rays. Subsequent experiments confirmed his prediction. The diffraction patterns that are observed are complicated because of the three-dimensional nature of the crystal. Nevertheless, x-ray diffraction is an invaluable technique for elucidating crystalline structures and for understanding the structure of matter.

Figure 28.18 is one experimental arrangement for observing x-ray diffraction from a crystal. A collimated beam of x-rays with a continuous range of wavelengths is incident on a crystal. The diffracted beams are very intense in certain directions, corresponding to constructive interference from waves reflected from layers of atoms in the crystal. The diffracted beams can be detected by a photographic film, and they form an array of spots known as a Laue pattern. The crystalline structure is deduced by analyzing the positions and intensities of the various spots in the pattern.

The arrangement of atoms in a crystal of NaCl is shown in Figure 28.19. The red spheres represent Na^+ ions, and the blue spheres represent Cl^- ions. Each unit cell (the set of atoms that repeats through the crystal) contains four Na^+ and four Cl^- ions. The unit cell is a cube whose edge length is a.

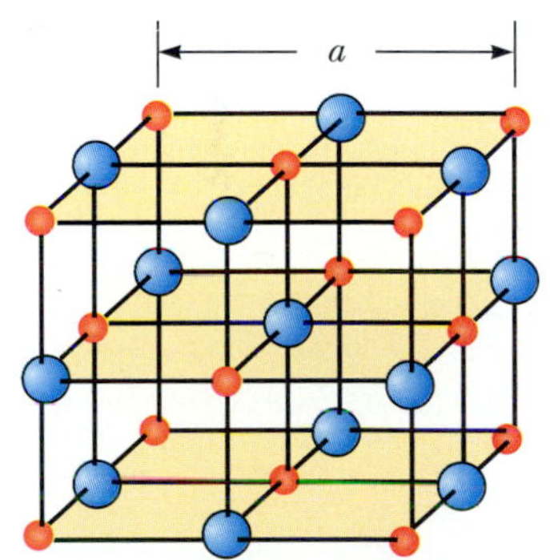

Figure 28.19
A model of the cubic crystalline structure of sodium chloride. The blue spheres represent the Cl^- ions, and the red spheres represent the Na^+ ions. The length of the cube edge is $a = 0.562737$ nm.

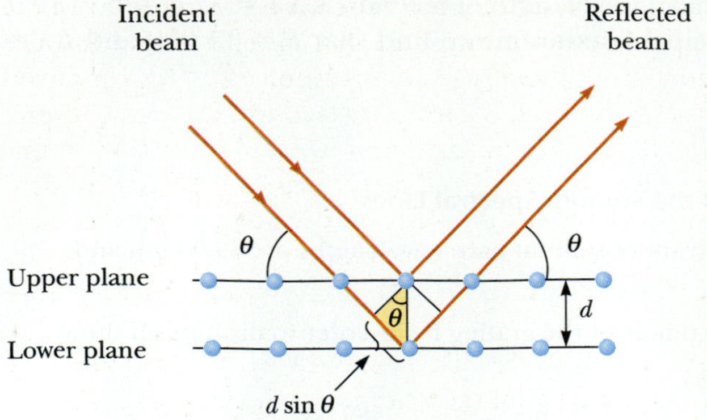

Figure 28.20
A two-dimensional representation of the reflection of an x-ray beam from two parallel crystalline planes separated by the distance d. The beam reflected from the lower plane travels farther than the one reflected from the upper plane by a distance of $2d \sin \theta$.

The ions in a crystal lie in various planes, as shown by the shaded areas in Figure 28.19. Now suppose an incident x-ray beam makes an angle of θ with one of the planes, as in Figure 28.20. The beam can be reflected from both the upper plane and the lower one. However, the geometric construction in Figure 28.20 shows that the beam reflected from the lower surface travels farther than the beam reflected from the upper surface. The path difference between the two beams is $2d \sin \theta$. The two beams will reinforce each other (constructive interference) when this path difference equals some integral multiple of the wavelength λ. The same is true of reflection from the entire family of parallel planes. Hence, the condition for constructive interference (maxima in the reflected wave) is given by

Bragg's law

$$2d \sin \theta = m\lambda \qquad (m = 1, 2, 3, \ldots) \qquad [28.11]$$

This condition is known as **Bragg's law** after W. L. Bragg (1890–1971), who first derived the relationship. If the wavelength and diffraction angle are measured, Equation 28.11 can be used to calculate the spacing between atomic planes.

28.6 Polarization of Light Waves

In Chapter 24 we described the transverse nature of electromagnetic waves. Figure 28.21 shows that the electric and magnetic vectors associated with an electromagnetic wave are at right angles to each other and also to the direction of wave propagation. The phenomenon of polarization, described in this section, is firm evidence of the transverse nature of electromagnetic waves.

An ordinary beam of light consists of a large number of waves emitted by the atoms or molecules of the light source. Each atom produces a wave with its own orientation, **E**, as in Figure 28.21, corresponding to the direction of atomic vibra-

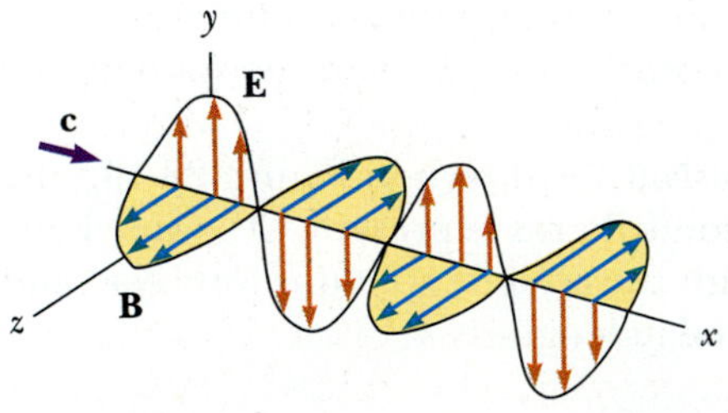

Figure 28.21
Schematic diagram of an electromagnetic wave propagating in the x direction. The electric field vector, **E**, vibrates in the xy plane, and the magnetic field vector, **B**, vibrates in the xz plane. This is an example of a linearly polarized wave.

tion. The direction of polarization of the electromagnetic wave is defined to be the direction in which **E** is vibrating. However, since all directions of vibration are possible, the resultant electromagnetic wave is a superposition of waves produced by the individual atomic sources. The result is an **unpolarized** light wave, represented schematically in Figure 28.22a. The direction of wave propagation in this figure is perpendicular to the page. Note that *all* directions of the electric field vector, lying in a plane perpendicular to the direction of propagation, are equally probable. At any given point and at some instant of time, there is only one resultant electric field; hence, you should not be misled by the meaning of Figure 28.22a.

A wave is said to be **linearly polarized** if **E** vibrates in the same direction *at all times* at a particular point, as in Figure 28.22b. (Sometimes such a wave is described as *plane-polarized* or simply *polarized.*) The wave described in Figure 28.21 is an example of a wave linearly polarized in the y direction. As the field propagates in the x direction, **E** is always in the y direction. The plane formed by **E** and the direction of propagation is called the *plane of polarization* of the wave. In Figure 28.21, the plane of polarization is the xy plane. It is possible to obtain a linearly polarized wave from an unpolarized wave by removing from the unpolarized wave all components of electric field vectors except those that lie in a single plane. We shall now discuss three processes for doing this: (1) selective absorption, (2) reflection, and (3) scattering.

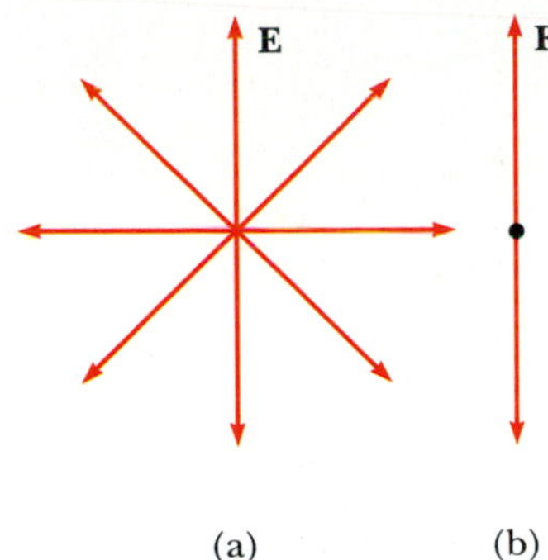

Figure 28.22
(a) An unpolarized light beam viewed along the direction of propagation (perpendicular to the page). The electric field vector can vibrate in any direction with equal probability. (b) A linearly polarized light beam with the electric field vector vibrating in the vertical direction.

Polarization by Selective Absorption

The most common technique for polarizing light is to send it through a material that passes only components of electric field vectors that are parallel to a characteristic direction of the material called the *polarizing* direction.

In 1938 E. H. Land discovered a material, which he called **polaroid,** that polarizes light through selective absorption by oriented molecules. This material is fabricated in thin sheets of long-chain hydrocarbons, which are stretched during manufacture so that the molecules align. After a sheet is dipped into a solution containing iodine, the molecules become good electrical conductors. However, the conduction takes place primarily along the hydrocarbon chains, since the valence electrons of the molecules can move easily only along the chains (recall that valence electrons are "free" electrons that can readily move through the conductor). As a result, the molecules readily *absorb* light whose electric field vector is parallel to their length and *transmit* light whose electric field vector is perpendicular to their length. It is common to refer to the direction perpendicular to the molecular chains as the **transmission axis.** An ideal polarizer passes the components of electric vectors that are parallel to the transmission axis. Components that are perpendicular to the transmission axis are absorbed. If light passes through several polarizers, whatever is transmitted has the plane of polarization parallel to the polarizing direction of the last polarizer it passed.

Let us now obtain an expression for the intensity of light that passes through a polarizing material. In Figure 28.23, an unpolarized light beam is incident on the first polarizing sheet, called the **polarizer,** where the transmission axis is as indicated. The light that is passing through this sheet is polarized vertically, and the transmitted electric field vector is $\mathbf{E}_0$. A second polarizing sheet, called the **analyzer,** intercepts this beam with its transmission axis at an angle of θ to the axis of the polarizer. The component of $\mathbf{E}_0$ that is perpendicular to the axis of the analyzer is completely absorbed, and the component parallel to that axis is $E_0 \cos\,\theta$.

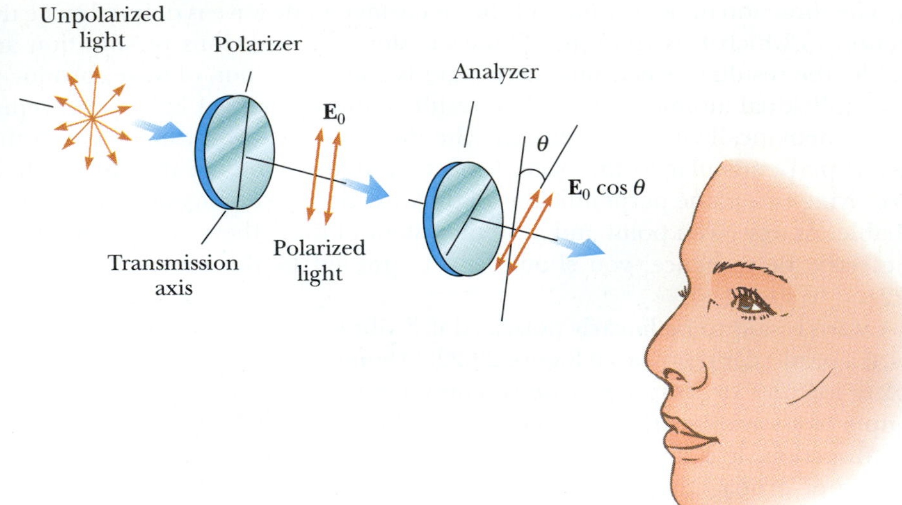

Figure 28.23 Two polarizing sheets whose transmission axes make an angle of θ with each other. Only a fraction of the polarized light incident on the analyzer is transmitted.

We know from Equation 28.11 that the transmitted intensity varies as the *square* of the transmitted amplitude, and so we conclude that the intensity of the transmitted (polarized) light varies as

$$I = I_0 \cos^2 \theta \qquad [28.12]$$

where I_0 is the intensity of the polarized wave incident on the analyzer. This expression, known as **Malus's law,** applies to any two polarizing materials whose transmission axes are at an angle of θ to each other. From this expression, note that the transmitted intensity is a maximum when the transmission axes are parallel ($\theta = 0$ or $180°$), and zero (complete absorption by the analyzer) when the transmission axes are perpendicular to each other. This variation in transmitted intensity through a pair of polarizing sheets is illustrated in Figure 28.24.

Polarization by Reflection

Another method for polarizing light is reflection. When an unpolarized light beam is reflected from a surface, the reflected light is completely polarized, partially polarized, or unpolarized, depending on the angle of incidence. If the angle of

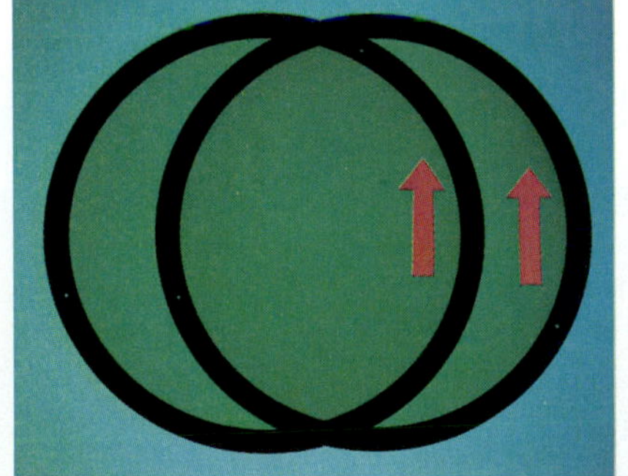

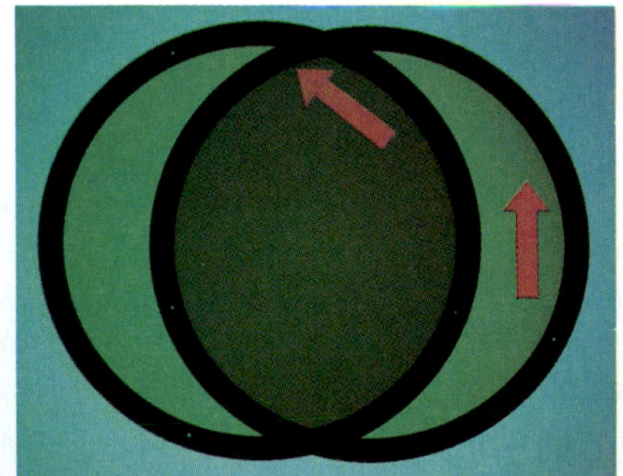

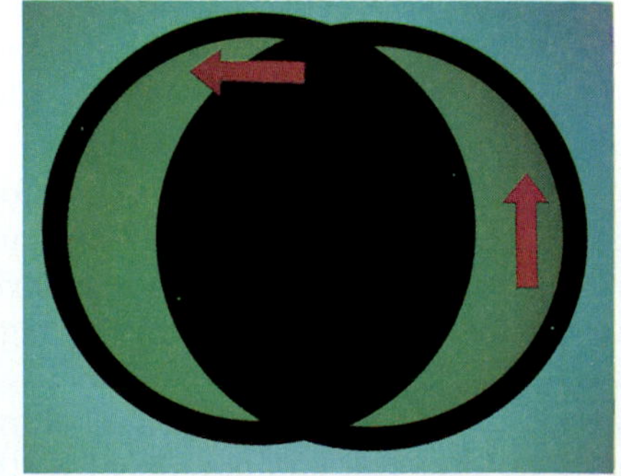

Figure 28.24 The intensity of light transmitted through two polarizers depends on the relative orientation of their transmission axes. (a) The transmitted light has *maximum* intensity when the transmission axes are *aligned* with each other. (b) The transmitted light intensity diminishes when the transmission axes are at an angle of 45° with each other. (c) The transmitted light intensity is a *minimum* when the transmission axes are at *right angles* to each other. *(Photographs courtesy of Henry Leap)*

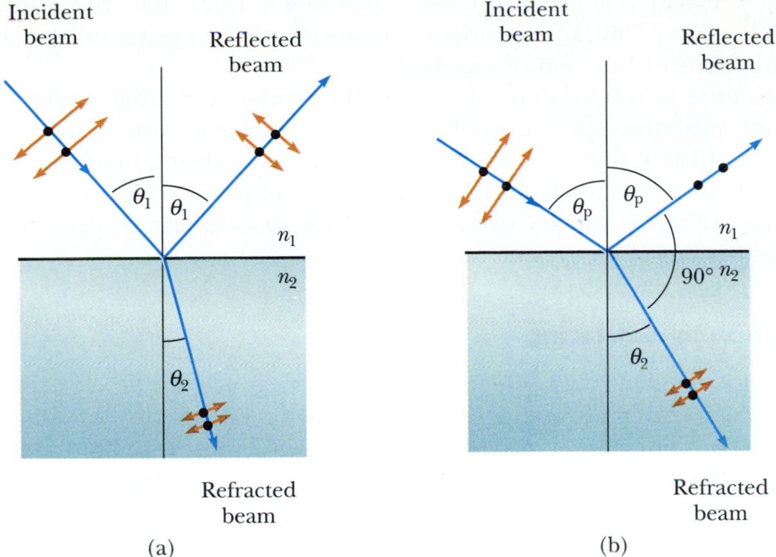

Figure 28.25
(a) When unpolarized light is incident on a reflecting surface, the reflected and refracted beams are partially polarized. (b) The reflected beam is completely polarized when the angle of incidence equals the polarizing angle, θ_p, which satisfies the equation $n = \tan\theta_p$.

incidence is either 0 or 90°, the reflected beam is unpolarized. However, for angles of incidence between 0 and 90°, the reflected light is polarized to some extent. For one particular angle of incidence, the reflected light is completely polarized. Let us now investigate that special angle.

Suppose an unpolarized light beam is incident on a surface as in Figure 28.25a. The beam can be described by two electric field components, one parallel to the surface (represented by the dots) and the other perpendicular to the first component (represented by the red arrows) and to the direction of propagation. It is found that the parallel component reflects more strongly than the other component, and this results in a partially polarized beam. Furthermore, the refracted beam is also partially polarized.

Now suppose the angle of incidence, θ_1, is varied until the angle between the reflected and refracted beams is 90° (Fig. 28.25b). At this particular angle of incidence, the reflected beam is completely polarized with its electric field vector parallel to the surface, while the refracted beam is only partially polarized. The angle of incidence at which this occurs is called the **polarizing angle,** θ_p.

The polarizing angle

An expression relating the polarizing angle to the index of refraction of the reflecting substance can be obtained by using Figure 28.25b. From this figure, we see that at the polarizing angle, $\theta_p + 90° + \theta_2 = 180°$, so that $\theta_2 = 90° - \theta_p$. Using Snell's law and taking $n_1 = n_{air} = 1.00$ and $n_2 = n$, we have

$$n = \frac{\sin\theta_1}{\sin\theta_2} = \frac{\sin\theta_p}{\sin\theta_2}$$

Because $\sin\theta_2 = \sin(90° - \theta_p) = \cos\theta_p$, the expression for n can be written

$$n = \frac{\sin\theta_p}{\cos\theta_p} = \tan\theta_p \qquad [28.13]$$

Brewster's law

This expression is called **Brewster's law,** and the polarizing angle θ_p is sometimes called **Brewster's angle,** after its discoverer, Sir David Brewster (1781–1868). For

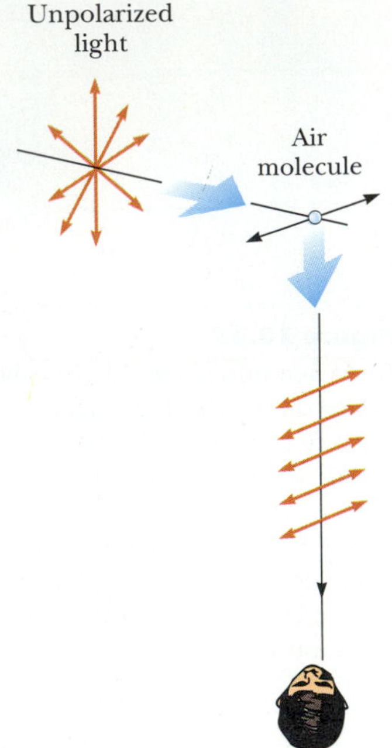

Figure 28.26 The scattering of unpolarized sunlight by air molecules. The light observed at right angles is plane-polarized because the vibrating molecule has a horizontal component of vibration.

example, Brewster's angle for crown glass ($n = 1.52$) has the value $\theta_p = \tan^{-1}(1.52) = 56.7°$. Because n varies with wavelength for a given substance, Brewster's angle is also a function of wavelength.

Polarization by reflection is a common phenomenon. Sunlight reflected from water, glass, and snow is partially polarized. If the surface is horizontal, the electric field vector of the reflected light will have a strong horizontal component. Sunglasses made of polarizing material reduce the glare of reflected light. The transmission axes of the lenses are oriented vertically so as to absorb the strong horizontal component of the reflected light.

Polarization by Scattering

When light is incident on a system of particles, such as gas, the electrons in the medium can absorb and reradiate part of the light. The absorption and reradiation of light by the medium, called **scattering,** is what causes sunlight reaching an observer on the Earth to be partially polarized. You can observe this effect by looking directly up through a pair of sunglasses whose lenses are made of polarizing material. Less light passes through at certain orientations of the lenses than at others.

Figure 28.26 illustrates how the sunlight becomes partially polarized. The left side of the figure shows an incident unpolarized beam of sunlight, traveling horizontally, on the verge of striking an air molecule. When this beam strikes the air molecule, it sets the electrons of the molecule into vibration. These vibrating charges act like the vibrating charges in an antenna, except that they vibrate in a complicated pattern. The horizontal part of the electric field vector in the incident wave causes the charges to vibrate horizontally, and the vertical part of the vector simultaneously causes them to vibrate vertically. A horizontally polarized wave is emitted by the electrons as a result of their horizontal motion, and a vertically polarized wave is emitted parallel to the Earth as a result of their motion.

Some phenomena involving the scattering of light in the atmosphere can be understood as follows. When light of various wavelengths λ is incident on air molecules of size d, where $d \ll \lambda$, the relative intensity of the scattered light varies as $1/\lambda^4$. The condition $d \ll \lambda$ is satisfied for scattering from O_2 and N_2 molecules in the atmosphere, whose diameters are about 0.2 nm. Hence, shorter wavelengths (blue light) are scattered more efficiently than longer wavelengths (red light). Because sunlight contains wavelengths from the entire visible spectrum, the sky appears to be blue. (This also explains why the sky is black in outer space, where there is no atmosphere to scatter the sunlight!)

Optical Activity

Many important practical applications of polarized light involve the use of materials that display the property of **optical activity.** A substance is said to be optically active if it rotates the plane of polarization of transmitted light. The angle through which the light is rotated by the material depends on the length of the sample and on the concentration if the substance is in solution. One optically active material is a solution of sucrose (common sugar). A standard method for determining the concentration of sugar solutions is to measure the rotation produced by a fixed length of the solution.

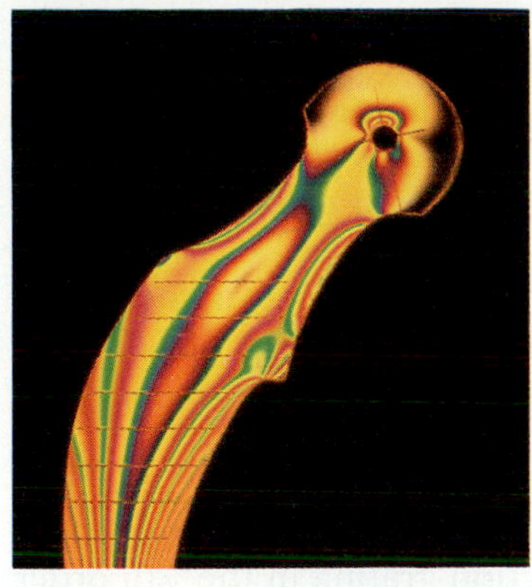

Figure 28.27
Photograph showing strain distribution in a plastic model of a hip replacement used in a medical research laboratory. The pattern is produced when the plastic model is placed between two crossed polarizers. *(© Sepp Seitz 1981. All rights reserved)*

Optical activity occurs in a material because of an asymmetry in the shape of its constituent molecules. For example, some proteins are optically active because of their spiral shape. Other materials, such as glass and plastic, become optically active when placed under stress. If polarized light is passed through an unstressed piece of plastic and then through an analyzer whose axis is perpendicular to that of the polarizer, none of the polarized light is transmitted. However, if the plastic is placed under stress, the regions of greatest stress produce the largest angles of rotation of polarized light. Hence, one observes a series of bright and dark bands in the transmitted light, with the bright bands corresponding to regions of greatest stress. Engineers often use this technique, called *optical stress analysis,* to assist them in the design of structures ranging from bridges to small tools and machine parts. A plastic model is built and analyzed under different load conditions to determine regions of potential weakness and failure under stress. An example of a plastic model under stress is shown in Figure 28.27.

Summary

Diffraction, defined as the deviation of light from a straight-line path when the light passes through an aperture or around obstacles, arises from the interference of a large number or continuous distribution of coherent sources.

The **Fraunhofer diffraction pattern** produced by a *single slit* of width a on a distant screen consists of a central, bright maximum and alternating bright and dark regions of much lower intensities. The angles θ at which the diffraction pattern has *zero* intensity are given by

Condition for intensity minima in the single-slit diffraction pattern

$$\sin\theta = m\frac{\lambda}{a} \qquad (m = \pm 1, \pm 2, \pm 3, \ldots) \qquad [28.1]$$

The variation of intensity, I, with angle θ is given by

Intensity of a single-slit Fraunhofer diffraction pattern

$$I_\theta = I_0\left[\frac{\sin(\beta/2)}{\beta/2}\right]^2 \qquad [28.4]$$

where $\beta = 2\pi a \sin\theta/\lambda$ and I_0 is the intensity at $\theta = 0$.

Rayleigh's criterion, which is a limiting condition of resolution, says that two images formed by an aperture are just distinguishable if the central maximum of the diffraction pattern for one image falls on the first minimum of the other image. The limiting angle of resolution for a slit of width a is given by $\theta_m = \lambda/a$, and the limiting angle of resolution for a circular aperture of diameter D is given by $\theta_m = 1.22\lambda/D$.

A **diffraction grating** consists of a large number of equally spaced, identical slits. The condition for intensity maxima in the interference pattern of a diffraction grating for normal incidence is

Condition for intensity maxima for a grating

$$d \sin \theta = m\lambda \qquad (m = 0, 1, 2, 3, \ldots) \qquad [28.8]$$

where d is the spacing between adjacent slits and m is the order number of the diffraction pattern. The resolving power of a diffraction grating in the mth order of the diffraction pattern is $R = Nm$, where N is the number of rulings in the grating.

When polarized light of intensity I_0 is incident on a polarizing film, the light transmitted through the film has an intensity equal to $I_0 \cos^2 \theta$, where θ is the angle between the transmission axis of the polarizer and the electric field vector of the incident light.

In general, reflected light is partially polarized. However, reflected light is completely polarized when the angle of incidence is such that the angle between the reflected and refracted beams is 90°. This angle of incidence, called the **polarizing angle,** θ_p, satisfies **Brewster's law:**

Brewster's law

$$n = \tan \theta_p \qquad [28.13]$$

where n is the index of refraction of the reflecting medium.

Questions and Conceptual Exercises

1. Observe the shadow of your book or some other straight edge when it is held a few inches above a table with a lamp several feet above the book. Why is the shadow of the book somewhat fuzzy at the edges?
2. If you place your thumb and index finger very close together and view light passing between them when they are a few centimeters in front of your eye, dark lines parallel to your thumb and finger will appear. Explain.
3. Although we can hear around corners, we cannot see around corners. How can you explain this in view of the fact that sound and light are both waves?
4. Describe the change in width of the central maximum of the single-slit diffraction pattern as the width of the slit is made smaller.
5. Assuming that the headlights of a car are point sources, and assuming no glare (i.e., no reflection inside the eyeball), estimate the maximum distance, from an observer to the car, at which the headlights are distinguishable from each other.
6. A laser beam is incident at a shallow angle on a machinist's ruler that has a finely calibrated scale. The rulings on the scale give rise to a diffraction pattern on a screen. Discuss how you can use this technique to obtain a measurement of the wavelength of the laser light.
7. Certain sunglasses use a polarizing material to reduce the intensity of light reflected from shiny surfaces, such as water or the hood of a car. What orientation of polarization should the material have in order to be most effective?
8. Why is the sky black when viewed from the Moon?
9. The diffraction grating effect is easily observed with everyday equipment. For example, a compact disk can be held so that light is reflected from it at a glancing angle (Fig. 28.28), and various colors in the reflected light can be seen. Furthermore, the observation depends on the orientation of the disk relative to the eye and light source. Explain how this works.

Figure 28.28
(Question 9) A compact disk acts as a diffraction grating when observed under white light. *(© Kristen Brochmann 1991, Fundamental Photographs)*

Figure 28.29
Iridescence in peacock feathers. *(© Diane Schiumo 1988, Fundamental Photographs)*

10. The path of a light beam from a helium-neon laser can be made visible by placing chalk dust in the air (perhaps by shaking a blackboard eraser in the path of the light beam). Explain why the beam can be seen under these circumstances.
11. Is light from the sky polarized? Why is it that clouds seen through Polaroid glasses stand out in bold contrast to the sky?
12. If a coin is glued to a glass sheet and this arrangement is held in front of a helium-neon laser, the projected shadow shows diffraction rings around the coin and a bright spot in the center of the shadow. How is this possible?
13. A pinhole camera can be constructed by punching a small hole in one side of a cardboard box. If the opposite side is cut out and replaced with white paper, the image formed by the box can be examined. When the box is pointed toward an open window with the pinhole facing the incoming light from a distant object, you should be able to see an image on the white paper. No lens is involved here. Can you explain why the image is formed? Why is the image sharp?
14. If a fine wire is stretched across the path of a laser beam, is it possible to produce a diffraction pattern?
15. How would one determine the index of refraction of a flat piece of dark obsidian?
16. The brilliant colors of peacock feathers (Fig. 28.29) are due to a phenomenon known as *iridescence.* The melanin fibers in the feathers act as a natural transmission grating. How do you explain the different colors? Why do the colors often change as the bird moves?

Problems

Section 28.2 Single-Slit Diffraction

1. Helium-neon laser light ($\lambda = 632.8$ nm) is sent through a 0.3-mm-wide single slit. What is the width of the central maximum on a screen 1 m in back of the slit?
2. Light of wavelength 587.5 nm illuminates a single slit 0.75 mm in width. (a) At what distance from the slit should a screen be placed if the first minimum in the diffraction pattern is to be 0.85 mm from the center of the screen? (b) What is the width of the central maximum?
3. A screen is placed 50 cm from a single slit, which is illuminated with light of wavelength 690 nm. If the distance between the first and third minima in the diffraction pattern is 3.0 mm, what is the width of the slit?
4. In daylight, the pupil of a cat's eye narrows to a slit of width 0.5 mm. What is the angular resolution? (Let the wavelength of light in the cat's eye be 500 nm.)
5. In Equation 28.4, let $\beta/2 \equiv \phi$ and show that $I = 0.5I_0$ when $\sin\phi = \phi/\sqrt{2}$.
6. The equation $\sin\phi = \phi/\sqrt{2}$ found in Problem 5 is known as a *transcendental equation.* One method of

solving such an equation is the graphical method. To illustrate this, let $\phi = \beta/2$, $y_1 = \sin \phi$, and $y_2 = \phi/\sqrt{2}$. Plot y_1 and y_2 on the same set of axes over a range from $\phi = 1$ rad to $\phi = \pi/2$ rad. Determine ϕ from the point of intersection of the two curves.

7. A beam of green light from a helium-cadmium laser is diffracted by a slit of width 0.55 mm. The diffraction pattern forms on a wall 2.06 m beyond the slit. The distance between the positions of zero intensity ($m = \pm 1$) is 4.1 mm. Estimate the wavelength of the laser light. The helium-cadmium laser can produce light at wavelengths of 441.2 nm or 537.8 nm.
8. A diffraction pattern is formed on a screen 120 cm away from a 0.4-mm-wide slit. Monochromatic light of 546.1 nm is used. Calculate the fractional intensity, I/I_0, at a point on the screen 4.1 mm from the center of the principal maximum.
9. A slit of width 0.5 mm is illuminated with coherent light of wavelength 500 nm, and a screen is placed 120 cm in front of the slit. Find the widths of the first and second maxima on each side of the central maximum.
10. Coherent microwaves of wavelength 5.0 cm enter a long, narrow window in a building otherwise essentially opaque to the microwaves. If the window is 36.0 cm wide, what is the distance from the central maximum to the first-order minimum along a wall 6.5 m from the window?

Section 28.3 Resolution of Single-Slit and Circular Apertures

11. A helium-neon laser emits light with a wavelength of 632.8 nm. The circular aperture through which the beam emerges has a diameter of 0.50 cm. Estimate the diameter of the beam at a distance of 10 km from the laser.
12. The Moon is approximately 400 000 km from the Earth. Can two lunar craters 50 km apart be resolved by a telescope on the Earth whose mirror has a diameter of 15 cm? Can craters 1 km apart be resolved? Take the wavelength to be 700 nm, and justify your answers with approximate calculations.
13. If we were to send a ruby laser beam ($\lambda = 694.3$ nm) outward from the barrel of a 2.7-m-diameter telescope, what would be the diameter of the big red spot when the beam hit the Moon 384 000 km away? (Neglect atmospheric dispersion.)
14. Find the radius of a star image formed on the retina of the eye if the aperture diameter (the pupil) at night is 0.7 cm, and the length of the eye is 2 cm. Assume that the wavelength of starlight in the eye is 500 nm.
15. At what distance could one theoretically distinguish two automobile headlights separated by 1.4 m? Assume a pupil diameter of 6 mm and yellow headlights ($\lambda = 580$ nm). The index of refraction in the eye is approximately 1.33.
16. A binary star system in the constellation Orion has an angular separation between the two stars of 10^{-5} radians. If $\lambda = 500$ nm, what is the smallest-diameter telescope that can just resolve the two stars?
17. The Impressionist painter Georges Seurat created paintings with enormous numbers of dots of pure pigment, about 2 mm in diameter. The idea was to put colors such as red and green next to each other to form a scintillating canvas. Outside what distance would one be *un*able to discern individual dots on the canvas? (Assume $\lambda = 500$ nm within the eye and a pupil diameter of 4 mm.)
18. The angular resolution of a radio telescope is to be 0.1° when it operates at a wavelength of 3 mm. Approximately what minimum diameter is required for the telescope's receiving dish?
19. A circular radar antenna on a navy ship has a diameter of 2.1 m and radiates at a frequency of 15 GHz. Two small boats are positioned 9 km away from the ship. How close together could the boats be and still be detected as *two* objects?
20. Suppose a 5-m-diameter telescope were constructed on the Moon. The seeing there would be excellent. As an example, what would be the separation between two objects that could just be resolved on the planet Mars in 500-nm light? (The distance to Mars at closest approach is 50 million miles.)
21. Three discrete spectral lines occur at angles of 10.09°, 13.71°, and 14.77° in the first-order spectrum of a grating spectroscope. (a) If the grating has 3660 slits/cm, what are the wavelengths of the light? (b) At what angles are these lines found in the second-order spectra?
22. Coherent light from a hydrogen source is incident on a diffraction grating. The incident light contains four wavelengths: $\lambda_1 = 410.1$ nm, $\lambda_2 = 434.0$ nm, $\lambda_3 = 486.1$ nm, and $\lambda_4 = 656.3$ nm. The diffraction grating has 410 lines/nm. Calculate the angles between (a) λ_1 and λ_4 in the first-order spectrum and (b) λ_1 and λ_3 in the third-order spectrum.
23. When Mars is nearest the Earth, the distance separating the two planets is 88.6×10^6 km. Mars is viewed through a telescope whose mirror has a diameter of 30 cm. (a) If the wavelength of the light is 590 nm, what is the angular resolution of the telescope? (b) What is the smallest distance that can be resolved between two points on Mars?

Section 28.4 The Diffraction Grating

In the following problems, assume that the light is incident normally on the gratings.

24. Light from an argon laser strikes a diffraction grating with 5310 lines per centimeter. The central and first-

order principal maxima are separated by a distance of 0.488 m on a wall that is 1.72 m from the grating. Determine the wavelength of the laser light.

25. A source emits light with wavelengths of 531.62 nm and 531.81 nm. (a) What is the minimum number of lines required for a grating that resolves the two wavelengths in the first-order spectrum? (b) Determine the slit spacing for a grating 1.32 cm wide that has the required minimum number of lines.

26. A helium-neon laser (λ = 632.8 nm) is used to calibrate a diffraction grating. If the first-order maximum occurs at 20.5°, what is the line spacing, d?

27. White light is spread out into spectral hues by a diffraction grating. If the grating has 2000 lines per cm, at what angle will red light (λ = 640 nm) appear in first order?

28. Two spectral lines in a mixture of hydrogen (H_2) and deuterium (D_2) gas have wavelengths of 656.30 nm and 656.48 nm. What is the minimum number of lines a diffraction grating must have to resolve these two wavelengths in first order?

29. Monochromatic light from a helium-neon laser (λ = 632.8 nm) is incident on a diffraction grating containing 400 lines/cm. Determine the angle of the first-order maximum.

30. A diffraction grating with 4000 lines/cm is illuminated by light from the Sun. The solar spectrum is spread out on a white wall across the room. At what angle from centerline is blue light (λ = 400 nm)? At what angle from centerline does red light (λ = 650 nm) appear?

Section 28.5 Diffraction of X-Rays by Crystals

31. Potassium iodide (KI) has the same crystalline structure as NaCl, with d = 0.353 nm. A monochromatic x-ray beam shows a diffraction maximum when the angle between the incident beam and the surface is 7.6°. Calculate the x-ray wavelength. (Assume first order.)

32. Monochromatic x-rays of the K_α line of potassium from a nickel target (λ = 0.166 nm) are incident on a KCl crystal surface. The interplanar distance in KCl is 0.314 nm. At what angle (relative to the surface) should the beam be directed in order that a second-order maximum be observed?

33. A monochromatic x-ray beam is incident on an NaCl crystal surface that has an interplanar spacing of 0.281 mm. The second-order maximum in the reflected beam is found when the angle between the incident beam and the surface is 20.5°. Determine the wavelength of the x-rays.

34. A wavelength of 0.129 nm characterizes K_β x-rays from zinc. When a beam of these x-rays is incident on the surface of a crystal whose structure is similar to that of NaCl, a first-order maximum is observed at an angle of 8.15°. Based on this information, calculate the interplanar spacing.

35. If an interplanar spacing of NaCl is 0.281 nm, what is the predicted angle at which x-rays of wavelength 0.14 nm will be diffracted in a first-order maximum?

36. In an x-ray diffraction experiment using x-rays of λ = 0.5×10^{-10} m, a first-order maximum occurred at 5°. Find the crystal plane spacing.

Section 28.6 Polarization of Light Waves

37. Light is reflected from a smooth ice surface, and the reflected ray is completely polarized. Determine the angle of incidence. (n = 1.309 for ice.)

38. A light beam is incident on heavy flint glass (n = 1.65) at the polarizing angle. Calculate the angle of refraction for the transmitted ray.

39. How far above the horizon is the Moon when its image reflected in calm water is completely polarized? (n_{water} = 1.33.)

40. The transmitting antenna on a submarine is 5 m above the ocean surface when the ship is surfaced. The captain wishes to transmit a message to a receiver on a 90-m-tall cliff at the ocean shore. If his signal is to be completely polarized by reflection off the ocean surface, how far must the ship be from the shore?

41. Unpolarized light passes through two polaroid sheets. The axis of the first is vertical, and that of the second is at 30° to the vertical. What fraction of the initial light is transmitted?

42. Three polarizing disks whose planes are parallel are centered on a common axis. The direction of the transmission axis in each case is shown, in Figure 28.30, relative to the common vertical direction. A plane-polarized beam of light with E_0 parallel to the vertical reference direction is incident from the left on the first disk with intensity I_i = 10 units (arbitrary). Calculate the transmitted intensity, I_f, when (a) θ_1 = 20°, θ_2 = 40°, and θ_3 = 60°; (b) θ_1 = 0°, θ_2 = 30°, and θ_3 = 60°.

43. A stack of polarizing sheets will rotate the direction of polarization of incident, linearly polarized light if

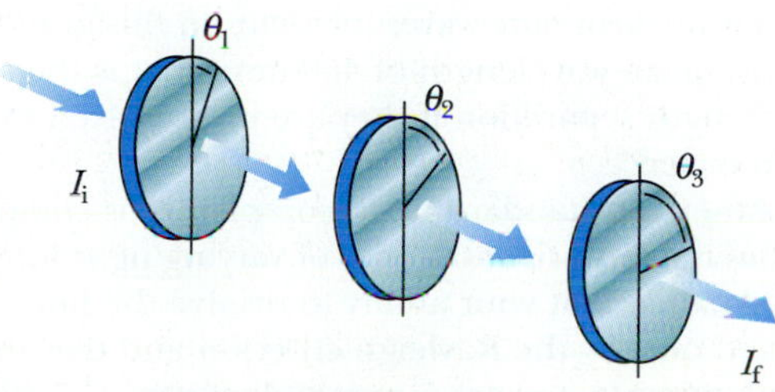

Figure 28.30 (Problems 42 and 56)

each successive sheet is oriented at an angle of θ (in the desired direction) with respect to the previous sheet. Using 10 ideal sheets to produce a 90° rotation, determine the maximum percentage of the incident, polarized light intensity, I_0, that will be transmitted through the tenth sheet.

44. The critical angle for sapphire surrounded by air is 34.4°. Calculate the polarizing angle for sapphire.
45. For a particular transparent medium surrounded by air, show that the critical angle for internal reflection and the polarizing angle are related by $\cot \theta_p = \sin \theta_c$.
46. Light strikes a flat water surface ($n = 1.333$) at Brewster's angle. Calculate the angle between the reflected ray and the refracted ray.
47. If the polarizing angle for cubic zirconia (ZrO_2) is 65.6°, what is the index of refraction of this material?

Additional Problems

48. A solar eclipse is projected through a pinhole of diameter 0.5 mm and strikes a screen 2 meters away. (a) What is the diameter of the projected image? (b) What is the radius of the first diffraction minimum? Both the Sun and Moon have angular diameters of very nearly 0.5 degrees. (Assume $\lambda = 550$ nm.)
49. The hydrogen spectrum has a red line at 656 nm and a blue line at 434 nm. What is the angular separation between two spectral lines obtained if the light is normally incident on a diffraction grating with 4500 lines/cm?
50. What are the approximate dimensions of the smallest object on Earth that the astronauts can resolve by eye at a 250-km height from the space shuttle? Assume $\lambda = 500$ nm light in the eye and a pupil diameter of 0.005 m.
51. Grote Reber, a pioneer in radio astronomy, constructed a radio telescope with a 10-m-diameter receiving dish. What was the telescope's angular resolution for radio waves with a wavelength of 2 m?
52. An unpolarized beam of light is reflected from a glass surface. It is found that the reflected beam is linearly polarized when the light is incident from air at an angle of 58.6°. What is the refractive index of the glass?
53. A diffraction grating of length 4 cm contains 6000 rulings over a width of 2 cm. (a) What is the resolving power of this grating in the first three orders? (b) If two monochromatic waves incident on this grating have a mean wavelength of 400 nm, what is their wavelength separation if they are just resolved in the third order?
54. An American standard television picture is composed of about 485 horizontal lines of varying light intensity. Assume that your ability to resolve the lines is limited only by the Rayleigh criterion and that the pupils of your eyes are 5 mm in diameter. Calculate the ratio of minimum viewing distance to the vertical dimension of the picture such that you will not be able to resolve the lines. Assume that the average wavelength of the light coming from the screen is 550 nm.
55. Derive Equation 28.10 for the resolving power of a grating, $R = Nm$, where N is the number of lines illuminated and m is the order in the diffraction pattern. Remember that Rayleigh's criterion (Section 28.3) states that two wavelengths will be resolved when the principal maximum for one falls on the first minimum for the other.
56. In Figure 28.30, suppose that the transmission axes of the left and right polarizing disks are perpendicular to each other. Also, let the center disk be rotated on the common axis with an angular velocity of ω. Show that if *unpolarized* light is incident on the left disk with an intensity of I_0, the intensity of the beam emerging from the right disk will be

$$I = \frac{1}{16} I_0 (1 - \cos 4\omega t)$$

This means that the intensity of the emerging beam will be modulated at a rate that is four times the rate of rotation of the center disk. (*Hint:* Use the trigonometric identities $\cos^2 \theta = (1 + \cos 2\theta)/2$ and $\sin^2 \theta = (1 - \cos 2\theta)/2$, and recall that $\theta = \omega t$.)
57. Suppose that the single-slit opening in Figure 28.7 is 6 cm wide and is placed in front of a microwave source operating at a frequency of 7.5 GHz. (a) Calculate the angle subtended by the first minimum in the diffraction pattern. (b) What is the relative intensity I/I_0 at $\theta = 15°$? (c) Consider the case when there are *two* such sources, separated laterally by 20 cm, behind the single slit. What is the maximum distance between the plane of the sources and the slit if the diffraction patterns are to be resolved? (In this case, the approximation that $\sin \theta \approx \tan \theta$ may not be valid because of the relatively small value of the ratio a/λ.)

△58. Light from a helium-neon laser (wavelength = 632.8 nm) illuminates a single slit, and a diffraction pattern is formed on a screen 1.00 m from the slit. The following data record measurements of the relative intensity as a function of distance from the line formed by the undiffracted laser beam. Make a plot of relative intensity versus distance. Choose an appropriate value for the slit width a, and plot the theoretical expression for the relative intensity

$$\frac{I_\theta}{I_0} = \frac{\sin^2(\beta/2)}{(\beta/2)^2}$$

on the same graph used for the experimental data. What value of a gives the best fit of theory and experiment?

Relative Intensity	Distance (mm)
0.95	0.8
0.80	1.6
0.60	2.4
0.39	3.2
0.21	4.0
0.079	4.8
0.014	5.6
0.003	6.5
0.015	7.3
0.036	8.1
0.047	8.9
0.043	9.7
0.029	10.5
0.013	11.3
0.002	12.1
0.0003	12.9
0.005	13.7
0.012	14.5
0.016	15.3
0.015	16.1
0.010	16.9
0.0044	17.7
0.0006	18.5
0.0003	19.3
0.003	20.2

Calculator/Computer Problems

59. Figure 28.10 shows the relative intensity of a single-slit Fraunhofer diffraction pattern as a function of the parameter $\beta/2 = (\pi a \sin\theta)/\lambda$. Make a plot of the relative intensity I/I_0 as a function of θ, the angle subtended by a point on the screen at the slit, when (a) $\lambda = a$, (b) $\lambda = 0.5a$, (c) $\lambda = 0.1a$, (d) $\lambda = 0.05a$. Let θ range over the interval from 0 to 20° and choose a number of steps appropriate for each case.
60. From Equation 28.4, show that, in the Fraunhofer diffraction pattern of a single slit, the angular width of the central maximum at the point where $I = 0.5I_0$ is $\Delta\theta = 0.886\lambda/a$. (*Hint:* In Equation 28.4, let $\beta/2 = \phi$ and solve the resulting transcendental equation graphically; see Problem 6.)
61. Another method of solving the equation $\phi = \sqrt{2}\sin\phi$ in Problem 6 is to use a scientific calculator, guess a first value of ϕ, see if it fits, and continue to update your estimate until the equation balances. How many steps (iterations) did this take? [Another approach is to apply the Newton-Raphson method to find the roots of $f(\phi) = \phi - \sqrt{2}\sin\phi$. In this approach, $\phi_2 = \phi_1 - f(\phi_1)/f'(\phi_1)$, where f' is the first derivative of f.)

False-color scanning tunnelling microscope (STM) image of a fatty acid bilayer deposited on a graphite substrate. The STM provides a detailed image of the surfaces of certain samples, allowing superficial atoms to be identified. Basically, the image is formed by moving a fine point just above the sample surface and electronically recording the height of the point as it scans. This is possible due to *tunnelling,* whereby an exchange of electrons occurs when the electron clouds surrounding superficial nuclei in the STM's point and sample surface overlap as they approach each other. STM studies to date include DNA, bacteriophages, and many inorganic molecules. *(Phillippe Plailly/Science Photo Library)*

Quantum Physics

CHAPTER 29

Although many problems were resolved by Einstein's special theory of relativity in the early part of the 20th century, many other problems remained unanswered. Attempts to apply the laws of classical physics to the behavior of matter on the atomic scale were consistently unsuccessful. Various phenomena, such as black-body radiation, the photoelectric effect, and the emission of sharp spectral lines by atoms in a gas discharge, could not be understood within the framework of classical physics. Between 1900 and 1930, however, a modern version of mechanics called *quantum mechanics* was highly successful in explaining the behavior of atoms, molecules, and nuclei. Moreover, the quantum theory reduces to classical physics when applied to macroscopic systems. Like relativity, the quantum theory requires a modification of our ideas concerning the physical world.

An extensive study of quantum theory is certainly beyond the scope of this book, and therefore this chapter is simply an introduction to its underlying ideas. We shall also discuss two simple applications of quantum theory: the photoelectric effect and the Compton effect.

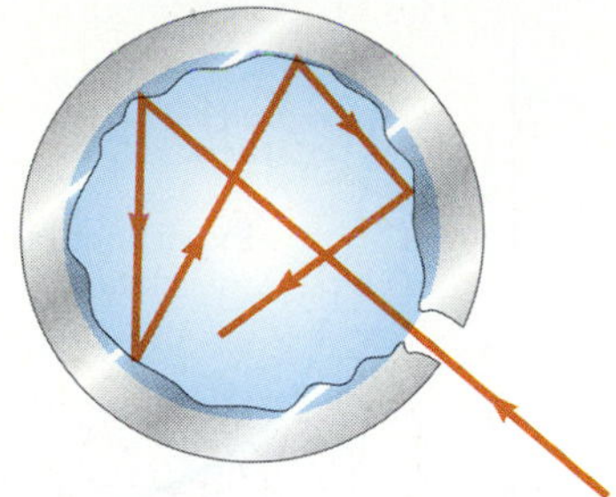

Figure 29.1
The opening in the cavity inside a body is a good approximation of a black body. Light entering the small opening strikes the far wall, where some of it is absorbed but some is reflected at some random angle. The light continues to be reflected, and each time it is, part of it is absorbed by the cavity walls. After many reflections, essentially all of the incident energy is absorbed.

29.1 Black-Body Radiation and Planck's Theory

An object at any temperature emits radiation that is sometimes referred to as **thermal radiation.** The characteristics of this radiation depend on the temperature and properties of the object. At low temperatures, the wavelengths of the thermal radiation are mainly in the infrared region and hence are not observed by the eye. As the temperature of the object is increased, the object eventually begins to glow red. At sufficiently high temperatures, it appears to be white, like the glow of the hot tungsten filament of a lightbulb. A careful study of thermal radiation shows that it consists of a continuous distribution of wavelengths from the infrared, visible, and ultraviolet portions of the spectrum.

From a classical viewpoint, thermal radiation originates in accelerated charged particles near the surface of the object; those charges emit radiation much as small antennas do. The thermally agitated charges can have a distribution of accelerations, which accounts for the continuous spectrum of radiation emitted by the object. By the end of the 19th century, it had become apparent that this classical explanation of thermal radiation was inadequate. The basic problem was in understanding the observed distribution of wavelengths in the radiation emitted by a black body, which, as we learned in Chapter 25, Section 25.5, is an ideal system that absorbs all radiation incident on it. A good approximation of a black body is the inside of a hollow object, as shown in Figure 29.1. The nature of the radiation emitted through a small hole leading to the cavity depends only on the temperature of the cavity walls.

Experimental data for the distribution of energy in black-body radiation at three temperatures are shown in Figure 29.2. The radiated energy varies with wavelength and temperature. As the temperature of the black body increases, the total amount of energy it emits increases. Also, with increasing temperatures, the peak of the distribution shifts to shorter wavelengths. This shift was found to obey the following relationship, called **Wien's displacement law:**

$$\lambda_{\text{max}} T = 0.2898 \times 10^{-2} \text{ m} \cdot \text{K} \qquad [29.1]$$

where λ_{max} is the wavelength at which the curve peaks and T is the absolute temperature of the object emitting the radiation.

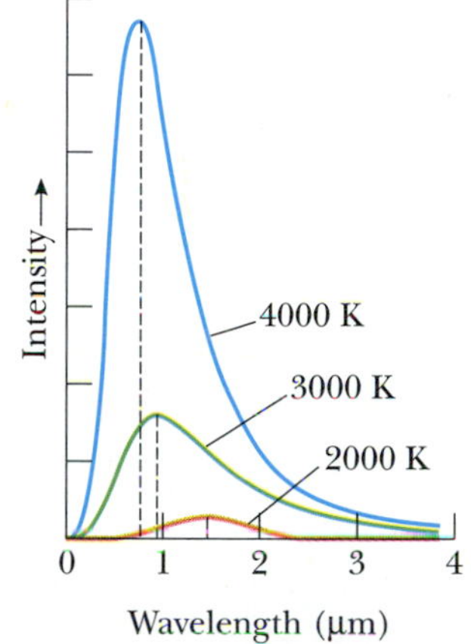

Figure 29.2
Intensity of black-body radiation versus wavelength at three temperatures. Note that the amount of radiation emitted (the area under a curve) increases with increasing temperature.

Early attempts to use classical ideas to explain the shapes of the curves in Figure 29.2 failed. Figure 29.3 shows an experimental plot of the black-body radiation spectrum (red) together with the curve predicted by classical theory (blue). At long wavelengths, classical theory is in good agreement with the experimental data. At short wavelengths, however, major disagreement exists between theory and experiment. This contradiction is often called the **ultraviolet catastrophe.**

In 1900 Planck developed a formula for black-body radiation that was in complete agreement with experiments at all wavelengths. Planck's analysis led to the curve shown in red in Figure 29.3. In his theory, Planck made two bold and con-

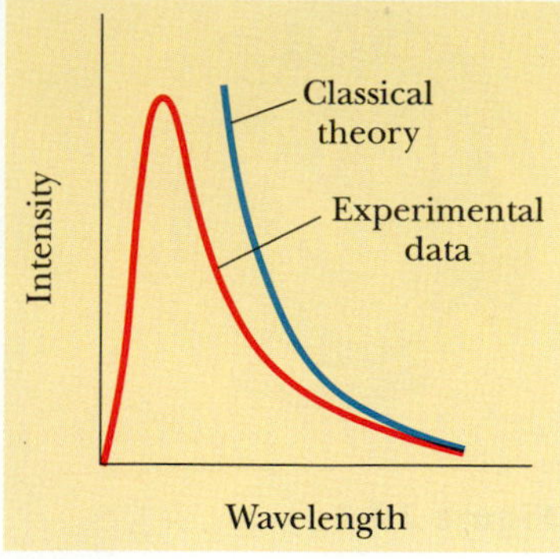

Figure 29.3
Comparison of the experimental results with the curve predicted by classical theory for the distribution of black-body radiation.

troversial assumptions concerning the nature of the oscillating charges at the surface of the black body:

1. The oscillating molecules that emit the radiation can have only certain *discrete* amounts of energy,

$$E_n = nhf \qquad [29.2]$$

where, as we learned in Chapter 10, Section 10.9, n is a positive integer called a quantum number, f is the frequency of oscillation of the molecules, and h is Planck's constant. Because the energy contained by each molecule can have only discrete values given by Equation 29.2, we say the energy is *quantized.* Each discrete value represents a different *quantum state,* with each value of n representing a specific quantum state. When the molecule is in the $n = 1$ quantum state, its energy is hf; when it is in the $n = 2$ quantum state, its energy is $2hf$; and so on.
2. The molecules emit energy in discrete units called **quanta** (today called **photons**). A suggestive image of several photons, not to be taken too literally, is shown in Figure 29.4. The molecules emit these quanta by "jumping" from one quantum state to another. If the jump is from one state to an adjacent state—say, from the $n = 3$ state to the $n = 2$ state—Equation 29.2 shows that the amount of energy radiated by the molecule equals hf. Hence, the energy of one light quantum corresponds to the energy difference between two adjacent levels and is given by

$$E = hf \qquad [29.3]$$

A molecule radiates or absorbs energy only when it changes quantum states. If it remains in one quantum state, no energy is absorbed or emitted. Figure 29.5 shows the quantized energy levels and allowed transitions proposed by Planck.

The key point in Planck's theory is the radical assumption of quantized energy states. This development marked the birth of the quantum theory. When Planck presented his theory, most scientists (including Planck!) did not consider the quantum concept to be realistic. Hence, Planck and others continued to search for a more rational explanation of black-body radiation. However, subsequent developments showed that a theory based on the quantum concept (rather than on classical concepts) had to be used to explain a number of other phenomena at the atomic level.

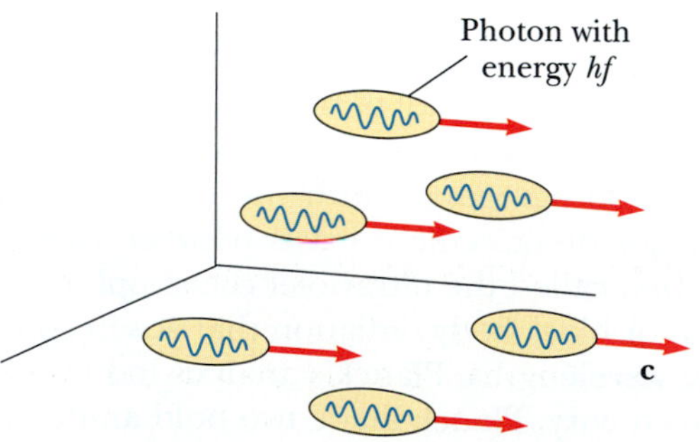

Figure 29.4
A representation of Einstein's light quanta. Each photon has a discrete energy, hf.

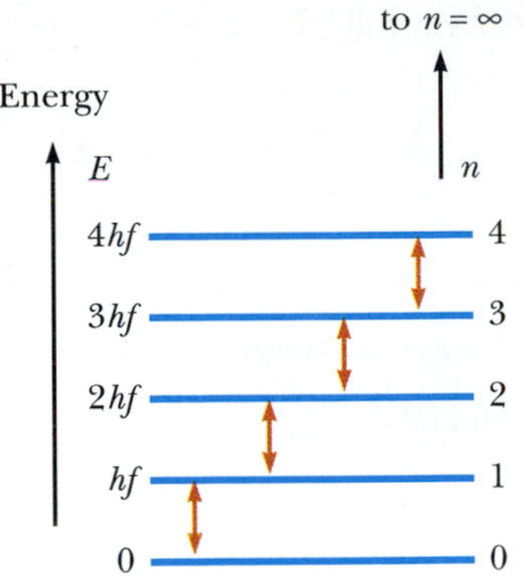

Figure 29.5 Allowed energy levels for a molecule that oscillates at a natural frequency, f. Allowed transitions are indicated by vertical arrows.

Example 29.1 Thermal Radiation from the Human Body

The temperature of your skin is approximately 35°C. What is the peak wavelength of the radiation it emits?

Solution From Wien's displacement law (Eq. 29.1), we have

$$\lambda_{\text{max}} T = 0.2898 \times 10^{-2} \text{ m}\cdot\text{K}$$

Solving for λ_{max}, noting that 35°C corresponds to an absolute temperature of 308 K, we have

$$\lambda_{\text{max}} = \frac{0.2898 \times 10^{-3} \text{ m}\cdot\text{K}}{308 \text{ K}} = \boxed{9.40\ \mu\text{m}}$$

This radiation is in the infrared region of the spectrum.

Max Planck (1858–1947)
Planck introduced the concept of a "quantum of action" (Planck's constant, h) in an attempt to explain the spectral distribution of black-body radiation, which laid the foundations for quantum theory. In 1918 he was awarded the Nobel Prize for this discovery of the quantized nature of energy. The work leading to the "lucky" black-body radiation formula was described by Planck in his Nobel Prize acceptance speech (1920): "But even if the radiation formula proved to be perfectly correct, it would after all have been only an interpolation formula found by lucky guesswork and, thus, would have left us rather unsatisfied." *(Photo courtesy of AIP Niels Bohr Library, W. F. Meggers Collection)*

Example 29.2 The Quantized Oscillator

A 2-kg mass is attached to a massless spring of force constant $k = 25$ N/m. The spring is stretched 0.4 m from its equilibrium position and released.

(a) Find the total energy and frequency of oscillation according to classical calculations.

Solution From Chapter 21, Equation 21.22, the total energy of a simple harmonic oscillator having an amplitude of A is $\frac{1}{2}kA^2$. Therefore

$$E = \tfrac{1}{2}kA^2 = \tfrac{1}{2}(25 \text{ N/m})(0.40 \text{ m})^2 = \boxed{2.0 \text{ J}}$$

The frequency of oscillation is, from Equation 21.18,

$$f = \frac{1}{2\pi}\sqrt{\frac{k}{m}} = \frac{1}{2\pi}\sqrt{\frac{25 \text{ N/m}}{2.0 \text{ kg}}} = \boxed{0.56 \text{ Hz}}$$

(b) Assume that the energy is quantized and find the quantum number, n, for the system.

Solution If the energy is quantized, we have $E_n = nhf$, and from the result of (a) we have

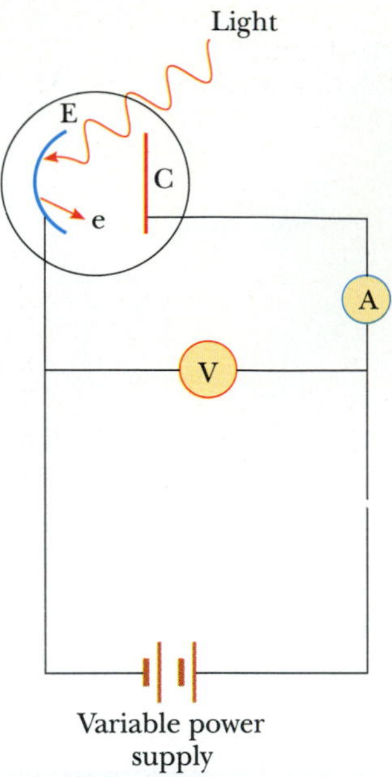

Figure 29.6
A circuit diagram for observing the photoelectric effect. When light strikes the plate, E (the emitter), photoelectrons are ejected from the plate. Electrons collected at C (the collector) constitute a current in the circuit.

$$E_n = nhf = n(6.626 \times 10^{-34}\,\text{J}\cdot\text{s})(0.56\text{ Hz}) = 2.0\text{ J}$$

Therefore,

$$n = \boxed{5.4 \times 10^{33}}$$

(c) How much energy is carried away in a one-quantum change?

Solution The energy carried away in a one-quantum change of energy is

$$E = hf = (6.63 \times 10^{-34}\,\text{J}\cdot\text{s})(0.56\text{ Hz})$$

$$= \boxed{3.7 \times 10^{-34}\text{ J}}$$

The energy carried away by a one-quantum change is such a small fraction of the total energy of the oscillator that we could never expect to see such a small change in the system. Thus, even though the decrease in energy of a spring-mass system is quantized and proceeds by small quantum jumps, our senses perceive the decrease as continuous. Quantum effects become important and measurable on the submicroscopic level of atoms and molecules.

29.2 The Photoelectric Effect

In the latter part of the 19th century, experiments showed that light incident on certain metallic surfaces caused electrons to be emitted from the surfaces. As mentioned in Chapter 25, Section 25.1, this phenomenon is known as the **photoelectric effect,** and the emitted electrons are called **photoelectrons.** The first discovery of the photoelectric effect was made by Hertz.

Figure 29.6 is a schematic diagram of a photoelectric effect apparatus. An evacuated glass tube contains a metal plate, E, connected to the negative terminal of a battery. Another metal plate, C, is maintained at a positive potential by the battery. When the tube is kept in the dark, the ammeter reads zero, indicating that there is no current in the circuit. However, when monochromatic light of the appropriate wavelength shines on plate E, a current is detected by the ammeter, indicating a flow of charges across the gap between E and C. This current arises from electrons emitted from the negative plate (the emitter) and collected at the positive plate (the collector).

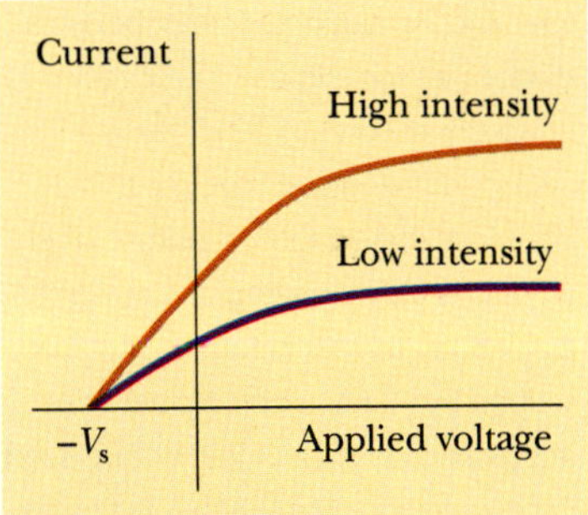

Figure 29.7
Photoelectric current versus applied voltage for two light intensities. The current increases with intensity but reaches a saturation level for large values of V. At voltages equal to or less than $-V_s$, the current is zero.

Figure 29.7 is a plot of the photoelectric current versus the potential difference, V, between E and C for two light intensities. Note that for large values of V, the current reaches a maximum value. In addition, the current increases as the incident light intensity increases, as you might expect. Finally, when V is negative—that is, when the battery polarity is reversed to make E positive and C negative—the current drops to a very low value because most of the emitted photoelectrons are repelled by the negative collecting plate, C. Only those electrons having a kinetic energy greater than eV will reach C, where e is the charge on the electron. When V is less than or equal to V_s, the **stopping potential,** no electrons reach C and the current is zero. The stopping potential is *independent* of the radiation intensity. The maximum kinetic energy of the photoelectrons is related to the stopping potential through the relation

$$K_{\max} = eV_s \quad [29.4]$$

Several features of the photoelectric effect cannot be explained with classical physics or with the wave theory of light:

1. No electrons are emitted if the incident light frequency falls below some **cutoff frequency,** f_c, which is characteristic of the material being illuminated. This is inconsistent with the wave theory, which predicts that the photoelectric effect should occur at any frequency, provided the light intensity is high enough.
2. If the light frequency exceeds the cutoff frequency, a photoelectric effect is observed and the number of photoelectrons emitted is proportional to the light intensity. However, the maximum kinetic energy of the photoelectrons is independent of light intensity, a fact that cannot be explained by the concepts of classical physics.
3. The maximum kinetic energy of the photoelectrons increases with increasing light frequency.
4. Electrons are emitted from the surface almost instantaneously (less than 10^{-9} s after the surface is illuminated), even at low light intensities. Classically, one would expect that the electrons would require some time to absorb the incident radiation before they acquired enough kinetic energy to escape from the metal.

Table 29.1
Work Functions of Selected Metals

Metal	ϕ (eV)
Na	2.28
Al	4.08
Cu	4.70
Zn	4.31
Ag	4.73
Pt	6.35
Pb	4.14
Fe	4.50

Einstein successfully explained the photoelectric effect in 1905, the same year he published his special theory of relativity. As part of a general paper on electromagnetic radiation, for which he received the Nobel Prize in 1921, Einstein extended Planck's concept of quantization to electromagnetic waves. He assumed that light (or any other electromagnetic wave) of frequency f can be considered a stream of photons (remember, *photon* is just another word for "quantum"). Each photon has an energy E given by Equation 29.3. Einstein's simple view of the photoelectric effect was that a photon of the incident light gives *all* its energy, hf, to a single electron in the metal. Electrons emitted from the surface of the metal possess the maximum kinetic energy, $K_{\max}$. According to Einstein, the maximum kinetic energy for these liberated electrons is

$$K_{\max} = hf - \phi \quad [29.5]$$

Photoelectric effect equation

where ϕ is called the **work function** of the metal. The work function represents the minimum energy with which an electron is bound in the metal, and is on the order of a few electron volts. Table 29.1 lists selected values.

With the photon theory of light, one can explain the features of the photoelectric effect that cannot be understood using classical concepts:

1. The fact that the effect is not observed below a certain cutoff frequency follows from the fact that the photon energy must be $\geq \phi$. If the energy of the incoming photon is not $\geq \phi$, the electrons will never be ejected from the surface, regardless of light intensity.
2. The fact that $K_{\max}$ is independent of the light intensity can be understood with the following argument. If the light intensity is doubled, the number of photons is doubled, which doubles the number of photoelectrons emitted. However, their kinetic energy, which equals $hf - \phi$, depends only on the light frequency and the work function, not on the light intensity.

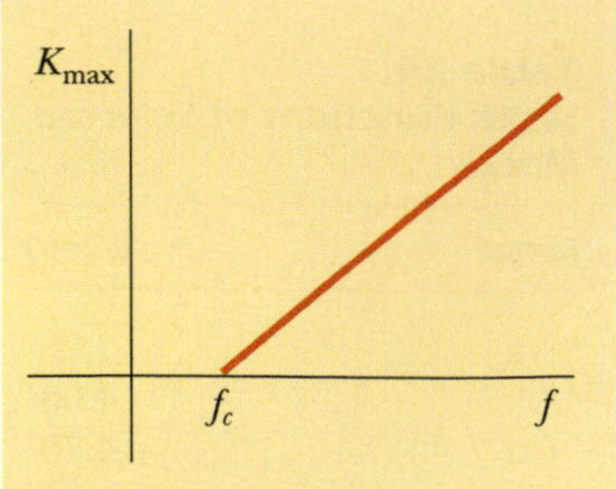

Figure 29.8
A sketch of K_{max} versus frequency of incident light for photoelectrons in a typical photoelectric effect experiment. Photons with frequency less than f_c do not have sufficient energy to eject an electron from the metal.

3. The fact that K_{max} increases with increasing frequency is easily understood with Equation 29.5.
4. The fact that the electrons are emitted almost instantaneously is consistent with the particle theory of light, in which the incident energy appears in small packets and there is a one-to-one interaction between photons and electrons. This is in contrast to the energy of the photons being distributed uniformly over a large area.

Experimental observation of a linear relationship between f and K_{max} would be a final confirmation of Einstein's theory. Indeed, such a linear relationship is observed, as sketched in Figure 29.8. The slope of this curve is h. The intercept on the horizontal axis is the cutoff frequency, which is related to the work function through the relation $f_c = \phi/h$. This corresponds to a **cutoff wavelength** of

$$\lambda_c = \frac{c}{f_c} = \frac{c}{\phi/h} = \frac{hc}{\phi} \quad [29.6]$$

where c is the speed of light $(3.00 \times 10^8 \text{ m/s})$. Wavelengths *greater* than λ_c incident on a material with a work function of ϕ do not result in the emission of photoelectrons.

Example 29.3 The Photoelectric Effect for Sodium

A sodium surface is illuminated with light of wavelength 300 nm. The work function for sodium metal is 2.46 eV. Find (a) the maximum kinetic energy of the ejected photoelectrons and (b) the cutoff wavelength for sodium.

Solution

(a) The energy of the illuminating light beam is

$$E = hf = \frac{hc}{\lambda} = \frac{(6.626 \times 10^{-34}\text{ J}\cdot\text{s})(3.00 \times 10^8\text{ m/s})}{300 \times 10^{-9}\text{ m}}$$

$$= 6.63 \times 10^{-19}\text{ J} = 4.14\text{ eV}$$

Use of Equation 29.5 gives

$$K_{max} = hf - \phi = 4.14\text{ eV} - 2.46\text{ eV} = \boxed{1.68\text{ eV}}$$

(b) The cutoff wavelength can be calculated from Equation 29.6 after we convert ϕ from electron volts to joules:

$$\phi = 2.46\text{ eV} = 3.94 \times 10^{-19}\text{ J}$$

$$\lambda_c = \frac{hc}{\phi} = \frac{(6.626 \times 10^{-34}\text{ J}\cdot\text{s})(3.00 \times 10^8\text{ m/s})}{3.94 \times 10^{-19}\text{ J}}$$

$$= 5.05 \times 10^{-7}\text{ m} = \boxed{505\text{ nm}}$$

This wavelength is in the green region of the visible spectrum.

Exercise Calculate the maximum speed of the photoelectrons under the conditions described in this example.

Answer 7.68×10^5 m/s.

29.3 The Compton Effect

In 1919 Einstein concluded that a photon of energy E travels in a single direction (unlike a spherical wave) and carries a momentum equal to $E/c = hf/c$. In his own words, "If a bundle of radiation causes a molecule to emit or absorb an energy packet hf, then momentum of quantity hf/c is transferred to the molecule, directed along the line of the bundle for absorption and opposite the bundle for emission." In 1923 Arthur Holly Compton (1892–1962) and Peter Debye (1884–1966) independently carried Einstein's idea of photon momentum farther. They realized that the scattering of x-ray photons from electrons could be explained by treating photons as point-like particles with energy hf and momentum hf/c, and by conserving the energy and momentum of the photon-electron pair in a collision.

Prior to 1922, Compton and his coworkers had accumulated evidence that showed that the classical wave theory of light failed to explain the scattering of x-rays from electrons. According to classical theory, incident electromagnetic waves of frequency f_0 should accelerate electrons, forcing them to oscillate and reradiate at a frequency of $f < f_0$ as in Figure 29.9a. Furthermore, according to

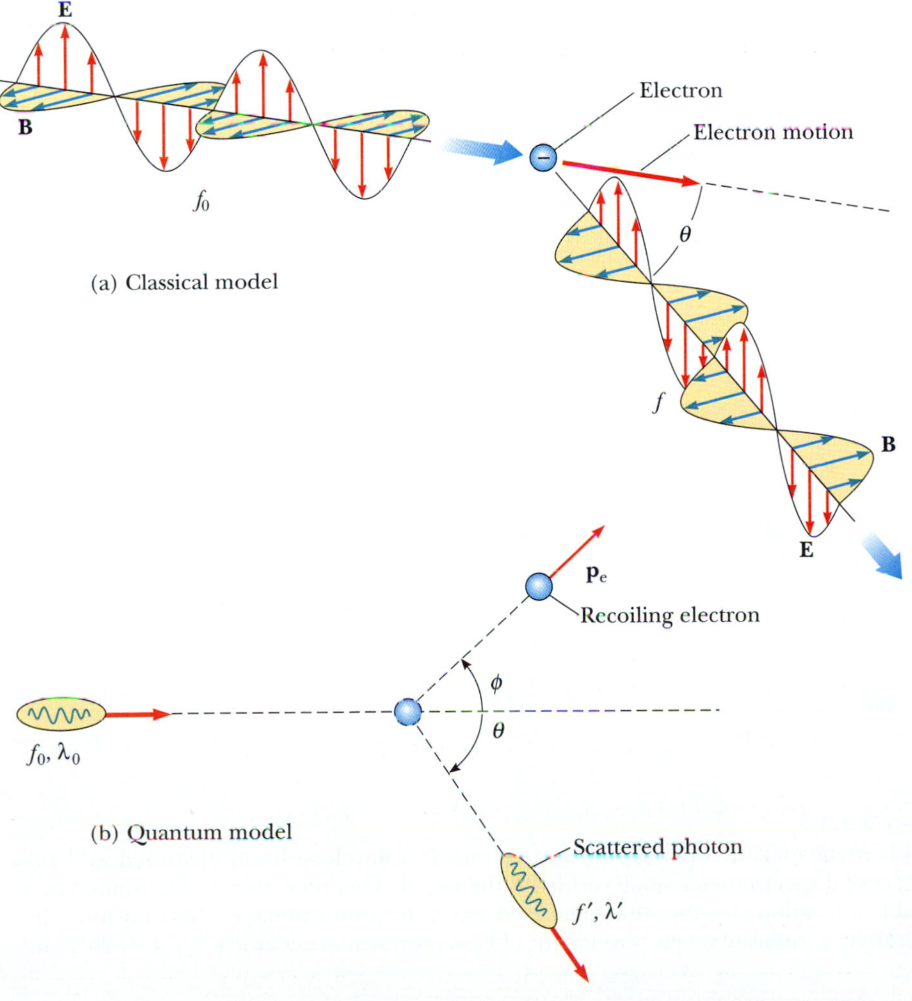

Figure 29.9 X-ray scattering from an electron: (a) the classical model and (b) the quantum model.

classical theory, the frequency of the reradiated radiation should depend on how long the sample is exposed to the incident radiation and on the intensity of the incident radiation. Contrary to these predictions, Compton's experiment showed that the frequency shift of x-rays scattered at a given angle depends *only on the scattering angle.* Figure 29.9b shows the quantum picture of the exchange of momentum and energy between an individual x-ray photon and an electron.

Figure 29.10a is a schematic diagram of the apparatus used by Compton. In his original experiment Compton measured how scattered x-ray intensity depended on wavelength at three scattering angles. The incident beam consisted of monochromatic x-rays of wavelength $\lambda_0 = 0.071$ nm. The experimental plots of intensity versus wavelength obtained by Compton for four scattering angles are shown in Figure 29.10b. They show two peaks, one at λ_0 and a shifted peak at λ'. The peak at λ_0 is caused by x-rays scattered from electrons that are tightly bound to the target atoms, and the shifted peak at λ' is caused by x-rays scattered from free electrons in the target. In his analysis, Compton predicted that the shifted peak should depend on scattering angle θ as follows:

Compton shift equation

$$\lambda' - \lambda_0 = \frac{h}{mc}(1 - \cos\theta) \qquad [29.7]$$

In this expression, known as the **Compton shift equation,** m is the mass of the electron; h/mc is called the **Compton wavelength,** λ_c, of the electron and has a currently accepted value of

Compton wavelength

$$\lambda_c = \frac{h}{mc} = 0.00243 \text{ nm}$$

Compton's measurements were in excellent agreement with the predictions of Equation 29.7. It is fair to say that these were the first experimental results to convince most physicists of the fundamental validity of the quantum theory!

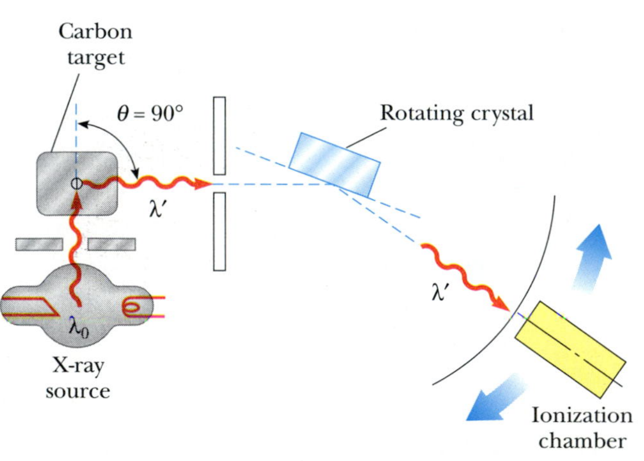

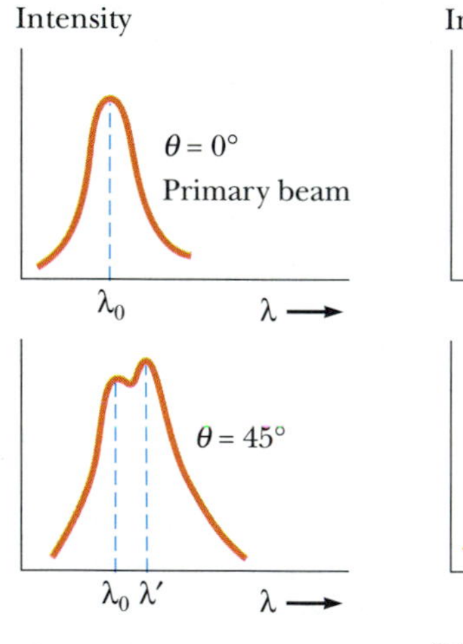

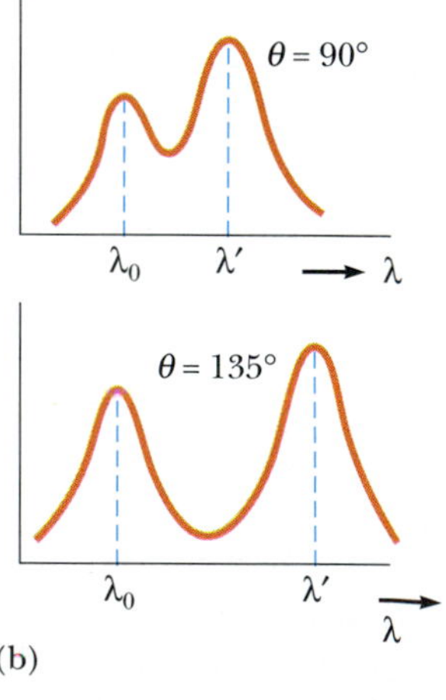

Figure 29.10
(a) Schematic diagram of Compton's apparatus. The wavelength was measured with a rotating crystal spectrometer using carbon as the target. The intensity was determined by a movable ionization chamber that generated a current proportional to the x-ray intensity. (b) Scattered intensity versus wavelength of Compton scattering at $\theta = 0°$, $45°$, $90°$, and $135°$.

Arthur Holly Compton (1892–1962), American physicist. *(Courtesy of AIP Niels Bohr Library)*

Example 29.4 Compton Scattering at 45°

X-rays of wavelength $\lambda_0 = 0.20$ nm are scattered from a block of material. The scattered x-rays are observed at an angle of 45° to the incident beam. Calculate the wavelength of the x-rays scattered at this angle.

Solution The shift in wavelength of the scattered x-rays is given by Equation 29.7:

$$\Delta\lambda = \frac{h}{mc}(1 - \cos\theta) = \frac{6.626 \times 10^{-34}\ \text{J}\cdot\text{s}}{(9.11 \times 10^{-31}\ \text{kg})(3.00 \times 10^{8}\ \text{m/s})}(1 - \cos 45°)$$

$$= 7.10 \times 10^{-13}\ \text{m} = 0.000710\ \text{nm}$$

Hence, the wavelength of the scattered x-rays at this angle is

$$\lambda' = \Delta\lambda + \lambda_0 = \boxed{0.200710\ \text{nm}}$$

Exercise Find the fraction of energy lost by the photon in this collision.

Answer Fraction $= \Delta\lambda/\lambda_0 = 0.00354$.

29.4 Photons and Electromagnetic Waves

An explanation of phenomena such as the photoelectric effect and the Compton effect presents very convincing evidence in support of the photon (or particle) concept of light. These phenomena offer ironclad evidence that when it interacts with matter, light behaves as if it were composed of particles with energy hf and momentum h/λ. An obvious question that arises at this point is "How can light be considered a photon when it exhibits wave-like properties?" On the one hand, we describe light in terms of photons having energy and momentum. On the other hand, we must also recognize that light and other electromagnetic waves exhibit interference and diffraction effects, which are consistent only with a wave interpretation.

Which model is correct? Is light a wave or a particle? The answer depends on the phenomenon being observed. Some experiments can be explained better, or solely, with the photon concept, whereas others are best described, or can be described only, with a wave model. The end result is that *we must accept both models and admit that the true nature of light is not describable in terms of a single classical picture.* However, you should recognize that the same light beam that can eject photoelectrons from a metal can also be diffracted by a grating. In other words, *the photon theory and the wave theory of light complement each other.*

All forms of electromagnetic radiation can be described from two points of view. At one extreme, electromagnetic waves describe the overall interference pattern or probability distribution, formed by photons. At the other extreme, the photon description is natural in the context of a highly energetic photon of very short wavelength. Hence, light has a dual nature: it exhibits both wave and photon characteristics.

The dual nature of light

The success of the particle model of light in explaining the photoelectric effect and the Compton effect raises many other questions. Since the photon is a particle, what is the meaning of its "frequency" and "wavelength," and which determines its energy and momentum? Is light in some sense simultaneously a wave and a particle? Although photons have no rest mass (a nonobservable quantity), is there a simple expression for the effective mass of a "moving" photon? If a "moving" photon has mass, do photons experience gravitational attraction? What is the spatial extent of a photon, and how does an electron absorb or scatter one photon? Some of these questions can be answered, but others demand a view of atomic processes that is too pictorial and literal. Furthermore, many of these questions stem from classical analogies such as colliding billiard balls and water waves breaking on a shore. Quantum mechanics gives light a more fluid and flexible nature by treating the particle model and wave model of light as both necessary and as complementary. Neither model can be used exclusively to describe all properties of light. A complete understanding of the observed behavior of light can be attained only if the two models are combined in a complementary manner.

We can perhaps understand why photons are compatible with electromagnetic waves in the following manner. We may suspect that long-wavelength waves do not exhibit particle characteristics. Consider, for instance, radio waves at a frequency of 2.5 MHz. The energy of a photon having this frequency is only about 10^{-8} eV. From a practical viewpoint, this energy is too small to be detected as a single photon. A sensitive radio receiver might require as many as 10^{10} of these photons to produce a detectable signal. Such a large number of photons would appear, on the average, as a continuous wave. With so many photons reaching the detector every second, it is unlikely that any graininess would appear in the detected signal. That is, with 2.5-MHz waves one would not be able to detect the individual photons striking the antenna.

Now consider what happens with shorter wavelengths. In the visible region, it is possible to observe both the photon and the wave characteristics of light. As we mentioned earlier, a beam of visible light shows interference phenomena (thus, it is a wave) and at the same time can produce photoelectrons (thus, it is a particle). At even shorter wavelengths, the momentum and energy of the photon increase. Consequently, the photon nature of light becomes more evident than its wave nature. For example, absorption of an x-ray photon is easily detected as a single event. However, as the wavelength decreases, wave effects become more difficult to observe.

29.5 The Wave Properties of Particles

Students who are being introduced to the dual nature of light often find the concept very difficult to accept. Even more disconcerting, however, is that *matter* has a dual nature as well!

In 1923, in his doctoral dissertation, Louis Victor de Broglie postulated that, *because photons have wave and particle characteristics, and electrons have particle characteristics and properties typical of waves, perhaps all forms of matter have wave as well as particle properties.* This was a highly revolutionary idea with no experimental confirmation at that time. According to de Broglie, electrons have a dual particle–wave nature. Accompanying every electron was a wave (not an electromagnetic wave!) that guided the electrons through space. He explained the source of this assertion in his 1929 Nobel Prize acceptance speech:

> On the one hand the quantum theory of light cannot be considered satisfactory since it defines the energy of a light corpuscle by the equation $E = hf$

Louis de Broglie (1892–1987), a French physicist, was awarded the Nobel Prize in 1929 for his discovery of the wave nature of electrons. "It would seem that the basic idea of quantum theory is the impossibility of imaging an isolated quantity of energy without associating with it a certain frequency." *(AIP Niels Bohr Library)*

containing the frequency f. Now a purely corpuscular theory contains nothing that enables us to define a frequency; for this reason alone, therefore, we are compelled, in the case of light, to introduce the idea of a corpuscle and that of periodicity simultaneously. On the other hand, determination of the stable motion of electrons in the atom introduces integers, and up to this point the only phenomena involving integers in physics were those of interference and of normal modes of vibration. This fact suggested to me the idea that electrons too could not be considered simply as corpuscles, but that periodicity must be assigned to them also.

In Chapter 10 we found that the relationship between energy and momentum for a photon, which has a rest mass of zero, is $p = E/c$. We also know from Equation 29.3 that the energy of a photon is $E = hf = hc/\lambda$. Thus, the momentum of a photon can be expressed as

$$p = \frac{E}{c} = \frac{hc}{c\lambda} = \frac{h}{\lambda} \qquad [29.8]$$

Momentum of a photon

From this equation we see that the photon wavelength can be specified by its momentum, or $\lambda = h/p$. De Broglie suggested that material particles of momentum p should also have wave properties and a corresponding wavelength. Because the momentum of a particle of mass m and speed[1] v is $p = mv$, the **de Broglie wavelength** of a particle is

$$\lambda = \frac{h}{p} = \frac{h}{mv} \qquad [29.9]$$

De Broglie wavelength

Furthermore, in analogy with photons, de Broglie postulated that the frequencies of matter waves (that is, waves associated with particles of nonzero rest mass) obey the Einstein relation $E = hf$, so that

$$f = \frac{E}{h} \qquad [29.10]$$

Frequency of matter waves

The dual nature of matter is apparent in these two equations. That is, each equation contains both particle concepts (mv and E) and wave concepts (λ and f). The fact that these relationships have been established experimentally for photons makes the de Broglie hypothesis that much easier to accept.

Quantization of Angular Momentum in the Bohr Model

Bohr's model of the atom, discussed in Chapter 12, has many shortcomings and problems. For example, as the electrons revolve around the nucleus, how does one visualize the fact that only certain electron energies are allowed? Why do all atoms of a given element have precisely the same physical properties regardless of the infinite variety of starting velocities and positions of the electrons in each atom?

De Broglie's great insight was to recognize that wave theories of matter use interference phenomena to handle these problems neatly. Recall from Chapter 23

[1] This speed, v, the speed of the particle, is the group speed of the wave (the speed at which a signal may travel). It is not the same as the phase speed of the wave, related to the index of refraction and wavelength. See L. de Broglie, *Matter and Light*, Dover, pp. 168–171.

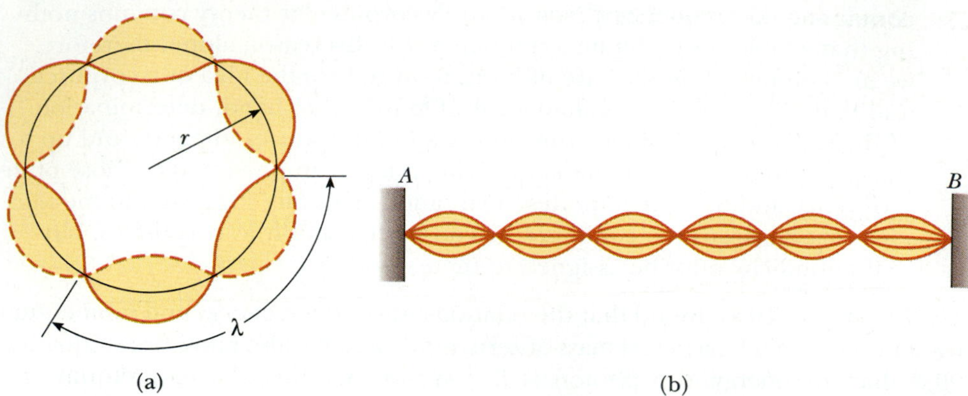

Figure 29.11
(a) The standing wave pattern for an electron wave in a stable orbit of hdyrogen. There are three full wavelengths in this orbit. (b) The standing wave pattern for a vibrating, stretched string fixed at its ends. This pattern has three full wavelengths.

that a plucked guitar string, while initially subjected to a wide range of wavelengths, will support only those standing wave patterns that have nodes at both of its ends. Any free vibration of the string consists of a superposition of varied amounts of many standing waves. This same reasoning can be applied to electron matter waves bent into a circle around the nucleus. All of the possible states of the electron are standing wave states, each with its own wavelength, speed, and energy. Figure 29.11a illustrates this point when three complete wavelengths are contained in one circumference of the orbit. Similar patterns can be drawn for orbits containing two wavelengths, four wavelengths, and so forth. This situation is analogous to that of standing waves on a string that has preferred (resonant) frequencies of vibration. Figure 29.11b shows a standing wave pattern containing three wavelengths for a string that is fixed at each end. Now imagine that the vibrating string is removed from its supports at A and B and bent into a circular shape such that points A and B are connected. The end result is a pattern like that shown in Figure 29.11a.

Another aspect of the Bohr theory that is easier to visualize by using de Broglie's hypothesis is the quantization of angular momentum. One simply needs to assume that the allowed Bohr orbits arise because the electron matter waves form standing waves when an integral number of wavelengths exactly fits into the circumference of a circular orbit. Thus,

$$n\lambda = 2\pi r$$

where r is the radius of the orbit. Since the de Broglie wavelength is $\lambda = h/mv$, we can write the preceding condition as $n(h/mv) = 2\pi r$, or

$$mvr = n\hbar$$

Note that this is precisely the Bohr condition of quantization of angular momentum (see Chapter 11, Section 11.5).[2] Thus, electron waves that fit the orbits are standing waves because of the boundary conditions imposed. These standing waves have discrete frequencies, corresponding to the allowed wavelengths. If $n\lambda \neq 2\pi r$, a standing-wave pattern can never form a closed circular orbit as in Figure 29.11a.

[2] Note that de Broglie's analysis still failed to explain the fact that states having an orbital angular momentum quantum number of zero do exist. Furthermore, the actual states of the hydrogen atom are not planar, but have spherical symmetry.

The Davisson–Germer Experiment

De Broglie's proposal that any kind of particle exhibits both wave and particle properties was first regarded as pure speculation. If particles such as electrons had wave-like properties, then under the correct conditions they should exhibit diffraction effects. In 1927, three years after de Broglie published his work, C. J. Davisson and L. H. Germer of the United States succeeded in measuring the wavelength of electrons. Their important discovery provided the first experimental confirmation of the matter waves proposed by de Broglie.

Interestingly, the intent of the initial Davisson–Germer experiment was not to confirm the de Broglie hypothesis. In fact, their discovery was made by accident (as is often the case). The experiment involved the scattering of low-energy electrons (about 54 eV) shot at a nickel target in a vacuum. During one experiment, the nickel surface was badly oxidized because of an accidental break in the vacuum system. After the nickel target was heated in a flowing stream of hydrogen to remove the oxide coating, electrons scattered by it exhibited intensity maxima and minima at specific angles. The experimenters finally realized that the nickel had formed large crystal regions upon heating and that the regularly spaced planes of atoms in the crystalline regions served as a diffraction grating for electron matter waves.

Shortly thereafter, Davisson and Germer performed more extensive diffraction measurements on electrons scattered from single-crystal targets. Their results showed conclusively the wave nature of electrons and confirmed the de Broglie relation $p = h/\lambda$. In the same year, G. P. Thomson of Scotland also observed electron diffraction patterns by passing electrons through very thin gold foils. Diffraction patterns have since been observed for helium atoms, hydrogen atoms, and neutrons. Hence, the universal nature of matter waves has been established in a variety of ways.

Example 29.5 The Wavelength of an Electron

Calculate the de Broglie wavelength for an electron ($m = 9.11 \times 10^{-31}$ kg) moving with a speed of 10^7 m/s.

Solution Equation 29.9 gives

$$\lambda = \frac{h}{mv} = \frac{6.63 \times 10^{-34}\ \text{J}\cdot\text{s}}{(9.11 \times 10^{-31}\ \text{kg})(10^7\ \text{m/s})}$$

$$= 7.28 \times 10^{-11}\ \text{m}$$

This wavelength corresponds to that of typical x-rays.

Exercise Find the de Broglie wavelength of a proton moving with a speed of 10^7 m/s.

Answer 3.97×10^{-14} m.

Example 29.6 The Wavelength of a Rock

A rock of mass 50 g is thrown with a speed of 40 m/s. What is its de Broglie wavelength?

Solution From Equation 29.12 we have

$$\lambda = \frac{h}{mv} = \frac{6.63 \times 10^{-34}\ \text{J}\cdot\text{s}}{(50 \times 10^{-3}\ \text{kg})(40\ \text{m/s})} = 3.32 \times 10^{-34}\ \text{m}$$

Notice that this wavelength is much smaller than any aperture through which the rock could possibly pass. This means that we could not observe diffraction effects, and as a result the wave properties of large-scale objects cannot be observed.

Example 29.7 **An Accelerated Charge**

A particle of charge q and mass m is accelerated from rest through a 50-V potential difference. Find its de Broglie wavelength.

Solution When a charge is accelerated from rest through a potential difference of V, its gain in kinetic energy, $\frac{1}{2}mv^2$, must equal its loss in potential energy, qV:

$$\tfrac{1}{2}mv^2 = qV$$

Since $p = mv$, we can express this in the form

$$\frac{p^2}{2m} = qV \qquad \text{or} \qquad p = \sqrt{2mqV}$$

Substituting this expression for p in the de Broglie relation $\lambda = h/p$ gives

$$\lambda = \frac{h}{\sqrt{2mqV}}$$

$$= \frac{6.63 \times 10^{-34}\ \text{J}\cdot\text{s}}{\sqrt{2(9.11 \times 10^{-31}\ \text{kg})(1.6 \times 10^{-19}\ \text{C})(50\ \text{V})}}$$

$$= 1.74 \times 10^{-10}\ \text{m} = 0.174\ \text{nm}$$

This wavelength is of the order of atomic dimensions and the spacing between atoms in a solid. Such low-energy electrons are normally used in electron diffraction experiments.

29.6 The Double-Slit Experiment Revisited

One way to crystallize our ideas about the electron's wave–particle duality is to consider a double-slit electron diffraction experiment. This experiment shows the impossibility of *simultaneously* measuring both wave and particle properties, and illustrates the use of the wave function in the determination of interference effects.

Consider a parallel beam of monoenergetic electrons that is incident on a double slit, like that in Figure 29.12. Because the slit openings are much smaller than the slit separation D, no structure is expected from single-slit diffraction. An electron detector is positioned far from the slits at a distance much greater than D. *If the detector collects electrons for a long enough time, one finds a typical wave interference pattern for the counts per minute, or probability of arrival of electrons.* This experiment is analogous to the interference of monochromatic light waves in Young's double-slit experiment, described in Chapter 27. Such an interference pattern would not be expected if the electrons behaved as classical particles.

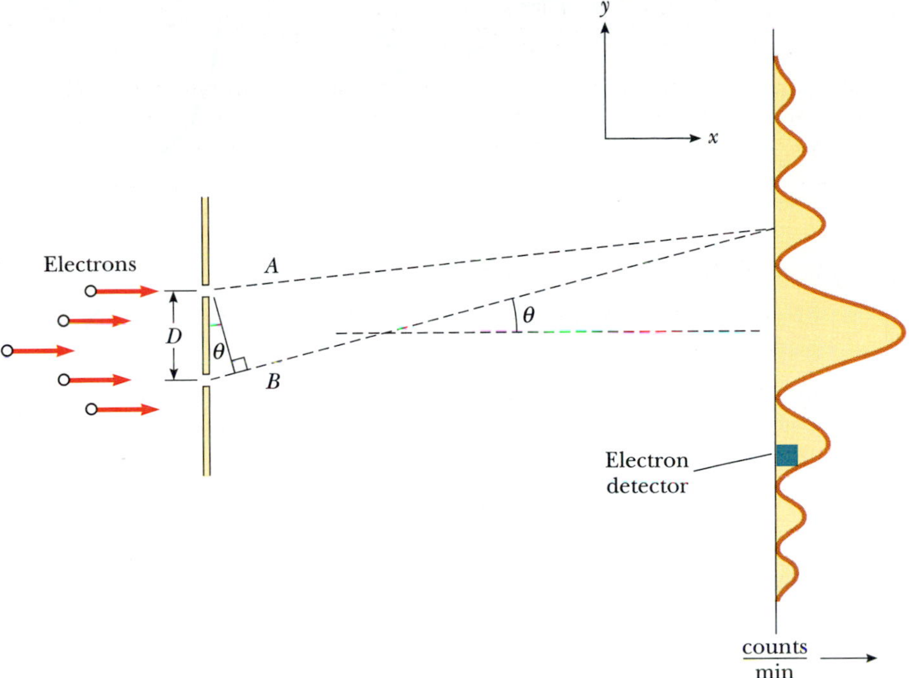

Figure 29.12
Electron diffraction. The slit separation D is much greater than the individual slit widths and much less than the distance between the slit and the detector. The electron detector is movable along the y-direction in the drawing and so can detect electrons diffracted at different values of θ. The detector acts like the "viewing screen" of Young's double-slit experiment with light discussed in Chapter 27.

(a) After 28 electrons

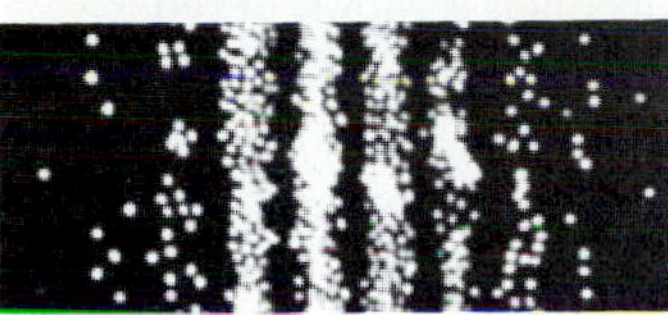

(b) After 1000 electrons

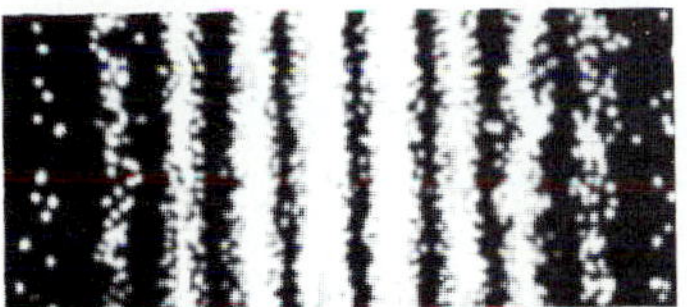

(c) After 10,000 electrons

(d) Two-slit electron pattern

Figure 29.13
(a, b, c) Computer-simulated interference patterns for a beam of electrons incident on a double slit. *(From E. R. Huggins, Physics I, New York, W. A. Benjamin, 1968)* (d) Photograph of a double-slit intereference pattern produced by electrons. *(From C. Jönsson, Zeitschrift fur Physik 161:454, 1961; used with permission)*

If the experiment is carried out at lower beam intensities, the interference pattern is still observed if the time of the measurement is sufficiently long. This is illustrated by the computer-simulated patterns in Figure 29.13. Note that the interference pattern becomes clearer as the number of electrons reaching the screen increases.

If one imagines a single electron producing in-phase "wavelets" as it reaches one of the slits, standard wave theory can be used to find the angular separation, θ, between the central probability maximum and its neighboring minimum. The minimum occurs when the path length difference between A and B is half a wavelength, or

$$D \sin \theta = \frac{\lambda}{2}$$

As the electron's wavelength is given by $\lambda = h/p_x$, we see that

$$\sin \theta \approx \theta = \frac{h}{2p_xD}$$

for small θ. Thus, the dual nature of the electron is clearly shown in this experiment: *whereas the electrons are detected as particles at a localized spot at some instant of time, the probability of arrival at that spot is determined by finding the intensity of two interfering matter waves.*

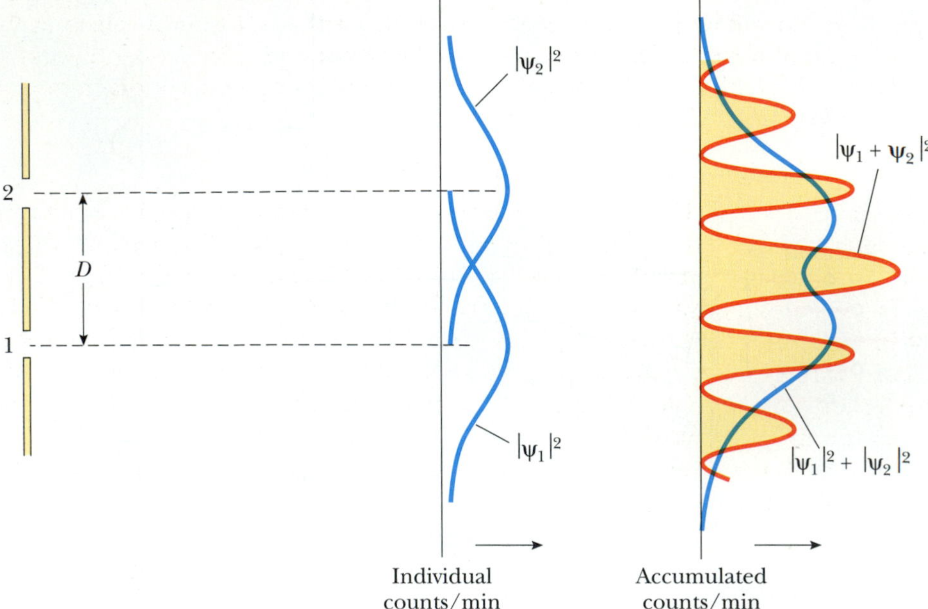

Figure 29.14
Accumulated results of the two-slit electron diffraction experiment with each slit closed half the time. The result with both slits open is shown in red.

But there is more. What happens if one slit is covered during the experiment? In this case one obtains a symmetric curve peaked around the center of the open slit. Plots of the counts per minute (probability of arrival of electrons) with the lower or upper slit closed are shown in the central part of Figure 29.14. These are expressed as the appropriate square of the absolute value of some wave function, $|\psi_1|^2 = \psi_1^*\psi_1$ or $|\psi_2|^2 = \psi_2^*\psi_2$, where ψ_1 and ψ_2 represent the cases of the electron passing through slit 1 and slit 2, respectively. If an experiment is now performed with slit 2 blocked half of the time and then slit 1 blocked during the remaining time, the accumulated pattern of counts/min shown by the blue curve on the right side of Figure 29.14 is completely different from the case with both slits open. There is no longer a maximum probability of arrival of an electron at $\theta = 0$. In fact, *the interference pattern has been lost and the accumulated result is simply the sum of the individual results.* When only one slit is open at a time, we know the electron has the same localizability and indivisibility at the slits as we measure at the detector, since the electron clearly goes through slit 1 or slit 2. Thus, the total must be analyzed as the sum of those electrons that come through slit 1, $|\psi_1|^2$, and those that come through slit 2, $|\psi_2|^2$.

When both slits are open, it is tempting to assume that the electron goes through either slit 1 or slit 2, and that the counts/min are again given by $|\psi_1|^2 + |\psi_2|^2$. We know, however, that the experimental results indicated by the red interference pattern in Figure 29.14 contradict this. Hence, our assumption that the electron is localized and goes through only one slit when both slits are open must be wrong (a painful conclusion!). Because it exhibits interference, we must conclude that—somehow—*the electron must be simultaneously present at both slits.* In order to find the probability of detecting the electron at a particular point on the screen with both slits open, we may say that the electron is in a *superposition state* given by

$$\psi = \psi_1 + \psi_2$$

Thus, the probability of detecting the electron at the screen is equal to the quantity $|\psi_1 + \psi_2|^2$ and not $|\psi_1|^2 + |\psi_2|^2$. Since matter waves that start out in phase at the slits in general travel different distances to the screen (see Fig. 29.15), ψ_1 and ψ_2 possess a relative phase difference of ϕ at the screen. Using a phasor diagram (Fig. 29.15) to find $|\psi_1 + \psi_2|^2$ immediately yields

$$|\psi|^2 = |\psi_1 + \psi_2|^2 = |\psi_1|^2 + |\psi_2|^2 + 2|\psi_1||\psi_2| \cos \phi$$

where $|\psi_1|^2$ is the probability of detection if slit 1 is open and slit 2 is closed, and $|\psi_2|^2$ is the probability of detection if slit 2 is open and slit 1 is closed. The term $2|\psi_1||\psi_2| \cos \phi$ is the interference term, which arises from the relative phase, ϕ, of the waves in analogy with the phasor addition used in wave optics (Chapter 28).

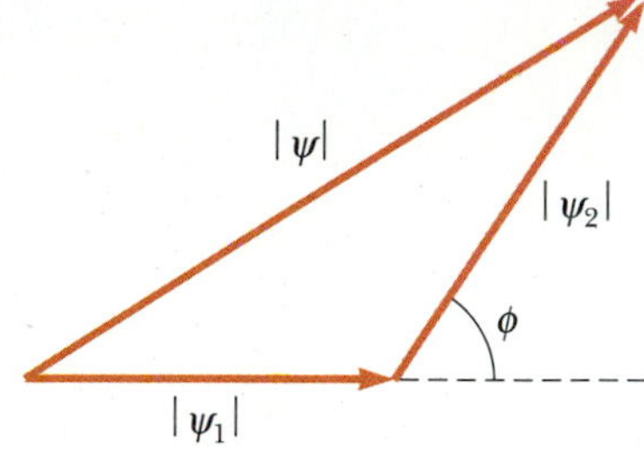

Figure 29.15
A phasor diagram representing the addition of two complex quantities, ψ_1 and ψ_2.

In order to interpret these results, we are forced to conclude that *an electron interacts with both slits simultaneously*. If we attempt to determine experimentally which slit the electron goes through, the act of measuring will destroy the interference pattern. It is impossible to determine which slit the electron will go through. In effect, *we can say only that the electron passes through both slits!* The same arguments apply to photons.

29.7 The Uncertainty Principle

Whenever one measures the position and velocity of a particle at any instant, experimental uncertainties are built into the measurements. According to classical mechanics, there is no fundamental barrier to an ultimate refinement of the apparatus or experimental procedures. In other words, it is possible, in principle, to make such measurements with arbitrarily small uncertainty. Quantum theory predicts, however, that *it is fundamentally impossible to make simultaneous measurements of a particle's position and momentum with infinite accuracy*.

In 1927, Werner Heisenberg (1901–1976) introduced this notion, which is now known as the **Heisenberg uncertainty principle:**

Uncertainty principle

If a measurement of position is made with precision Δx and a simultaneous measurement of momentum is made with precision Δp_x, then the product of the two uncertainties can never be smaller than $\hbar/2$, where $\hbar = h/2\pi$:

$$\Delta x \, \Delta p_x \geq \frac{\hbar}{2} \qquad [29.11]$$

Heisenberg was careful to point out that the inescapable uncertainties Δx and Δp_x do not arise from imperfections in practical measuring instruments. Rather, they arise from the quantum structure of matter, from effects such as the *unpredictable* recoil of an electron when struck by an *indivisible* photon, or the diffraction of light or electrons passing through a small opening.

In order to understand the uncertainty principle, consider the following thought experiment introduced by Heisenberg. Suppose you wish to measure the position and momentum of an electron as accurately as possible. You might be able to do this by viewing the electron with a powerful light microscope. In order for you to see the electron, and thus determine its position, at least one photon of light must bounce off the electron and pass through the microscope into your eye,

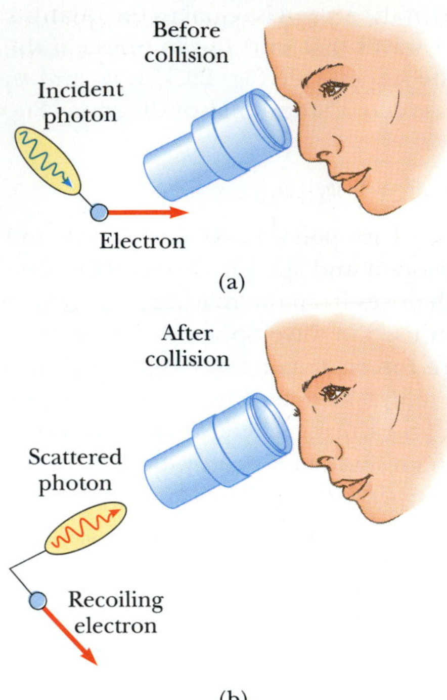

Figure 29.16
A thought experiment for viewing an electron with a powerful microscope. (a) The electron is viewed before it collides with the photon. (b) The electron recoils (is disturbed) as a result of the collision with the photon.

as shown in Figure 29.16a. When it strikes the electron, however, the photon transfers some unknown amount of its own momentum to the electron. Thus, in the process of locating the electron's position very accurately (that is, making Δx very small), the very light that allows you to succeed changes the electron's momentum to some undeterminable extent (making Δp_x very large).

Let us analyze the collision between the photon and the electron by first noting that the incoming photon has momentum h/λ. As a result of the collision, the photon transfers all of its momentum along the x axis to the electron. Thus, the uncertainty in the electron's momentum after the collision is as large as the momentum of the incoming photon. That is, $\Delta p_x = h/\lambda$. Furthermore, since light also has wave properties, we would expect to be able to determine the position of the electron to within one wavelength of the light being used to view it, so that $\Delta x = \lambda$. Multiplying these two uncertainties gives

$$\Delta x\, \Delta p_x = \lambda\left(\frac{h}{\lambda}\right) = h$$

This represents the minimum in the products of the uncertainties. Since the uncertainty can always be greater than this minimum, we have

$$\Delta x\, \Delta p_x \gtrsim h$$

This agrees with Equation 29.11 (apart from a small numerical factor introduced by Heisenberg's more precise analysis).

Heisenberg's uncertainty principle enables us to better understand the dual wave–particle nature of light and matter. We have seen that the wave description is quite different from the particle description. Therefore, if an experiment (such as the photoelectric effect) is designed to reveal the particle character of an electron, its wave character will become less apparent. Likewise, if the experiment (such as diffraction from a crystal) is designed to accurately measure the electron's wave properties, its particle character will become less apparent.

Werner Heisenberg (1901–1976) A German theoretical physicist, Heisenberg obtained his PhD in 1923 at the University of Munich, where he studied under Arnold Sommerfeld and became an enthusiastic mountain climber and skier. While physicists such as de Broglie and Schrödinger tried to develop physical models of the atom, Heisenberg developed an abstract mathematical model called matrix mechanics to explain the wavelengths of spectral lines. The more successful wave mechanics by Schrödinger, announced a few months later, was shown to be equivalent to Heisenberg's approach. Heisenberg made many other significant contributions to physics, including his famous uncertainty principle, for which he received the Nobel Prize in 1932; the prediction of two forms of molecular hydrogen; and theoretical models of the nucleus.

Example 29.8 Locating an Electron

The speed of an electron is measured to be 5.00×10^3 m/s ± 0.003%. Within what limits could one determine the position of this electron?

Solution The momentum is

$$p = mv = (9.11 \times 10^{-31}\text{ kg})(5.00 \times 10^3\text{ m/s})$$
$$= 4.56 \times 10^{-27}\text{ kg}\cdot\text{m/s}$$

Because the uncertainty is 0.003% of this value, we get

$$\Delta p = 0.00003p = 1.37 \times 10^{-31}\text{ kg}\cdot\text{m/s}$$

The uncertainty in position can now be calculated by using this value of Δp and Equation 29.11:

$$\Delta x \geqslant \frac{\hbar}{2\Delta p} = \frac{1.05 \times 10^{-34}\text{ J}\cdot\text{s}}{2.74 \times 10^{-31}\text{ kg}\cdot\text{m/s}}$$
$$= 0.384 \times 10^{-3}\text{ m} = \boxed{0.384\text{ mm}}$$

This is the *lower limit* on the uncertainty with which we could have determined the electron's position.

29.8 An Interpretation of Quantum Mechanics

Matter waves are described by the complex-valued wave function ψ, introduced in Section 29.6. The absolute square $|\psi|^2 = \psi^*\psi$ gives the probability of finding a particle at a given point at some instant, where ψ^* is the complex conjugate of ψ. The wave function contains within it all the information that can be known about the particle.

This interpretation of de Broglie's matter waves was first suggested by Max Born (1882–1970) in 1928. In 1926, Erwin Schrödinger (1887–1961) proposed a wave equation that described the manner in which matter waves change in space and time. The *Schrödinger wave equation*, which we will examine in Section 29.10, represents a key element in the theory of quantum mechanics.

The concepts of quantum mechanics, strange as they sometimes may seem, developed from classical ideas. In fact, when the techniques of quantum mechanics are applied to macroscopic systems, the results are essentially identical to those of classical physics. This blending of the two theories occurs when the de Broglie wavelength is small compared with the dimensions of the system. The situation is similar to the agreement between relativistic mechanics and classical mechanics when $v \ll c$.

A number of experiments show that matter has both a wave nature and a particle nature. A question that arises quite naturally in this regard is the follow-

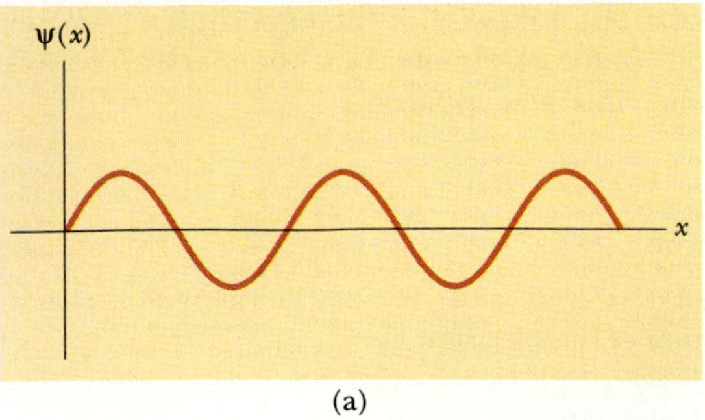

(a)

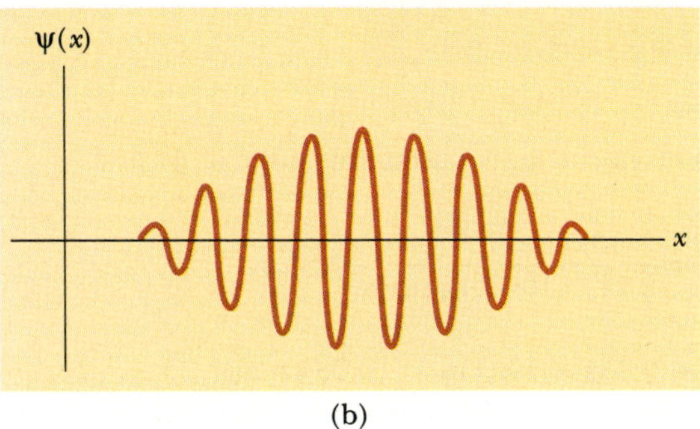

(b)

Figure 29.17
(a) A portion of the wave function for a particle whose wavelength is precisely known. (b) The wave function for a particle whose wavelength is not precisely known and hence whose momentum is known only over some range of values. This pattern is formed by a sum of functions as in (a) having a continuous distribution of wavelengths.

ing: If we are describing a particle, how do we view what is waving? In the cases of waves on strings, water waves, and sound waves, the wave is represented by some quantity that varies with time and position. In a similar manner, matter waves (de Broglie waves) are represented by the **wave function,** Ψ. In general, Ψ depends on the positions of all the particles in a system and on time, and therefore is often written $\Psi(x, y, z, t)$. If Ψ is known for a particle, then the particular properties of that particle can be described. In fact, a fundamental problem of quantum mechanics is this: given the wave function at some instant, find the wave function at some later time, t.

In Section 29.5, we found that the de Broglie equation relates the momentum of a particle to its wavelength through the relation $p = h/\lambda$. If a free particle has a precisely known momentum, p_x, its wave function is an uninterrupted sinusoidal wave of wavelength $\lambda = h/p_x$, and the particle has equal probability of being at any point along the x axis. The wave function for such a free particle moving along the x axis can be written in the form

$$\psi(x) = A \sin\left(\frac{2\pi x}{\lambda}\right) = A \sin(kx) \qquad [29.12]$$

where $k = 2\pi/\lambda$ is the wave number and A is a constant.[3] As we mentioned earlier, the wave function is generally a function of both position and time. Equation 29.12 represents the part of the wave function that depends on position only. For this reason, one can view $\psi(x)$ as a "snapshot" of the wave at a given instant, as shown in Figure 29.17a. The wave function for a particle whose wavelength is not precisely

[3] In general, ψ has a real part and an imaginary part, and we would write the wave function for the particle in the form Ae^{ikx}, where the imaginary part of the function describes the phase of the wave.

defined is shown in Figure 29.17b. Since the wavelength is not precisely defined, it follows that the momentum is only approximately known. That is, if one were to measure the momentum of the particle, the result would have any value over some range, determined by the spread in wavelength.

Although we cannot measure ψ, we can measure the quantity $|\psi|^2$, which can be interpreted as follows. If ψ represents a single particle, then $|\psi|^2$ is the relative probability per unit volume that the particle will be found at any given point in the volume. This interpretation, first suggested by Born in 1928, can also be stated in the following manner. If dV is a small volume element surrounding some point, then the probability of finding the particle in that volume element is $|\psi|^2\,dV$. In this section we deal only with one-dimensional systems, where the particle must be located along the x axis, and thus we replace dV with dx. In this case, the probability, $P(x)\,dx$, that the particle will be found in the infinitesimal interval dx about the point x is

$$P(x)\,dx = |\psi|^2\,dx \qquad [29.13]$$

Since the particle must be somewhere along the x axis, the sum of the probabilities over all values of x must be 1:

$$\int_{-\infty}^{\infty} |\psi|^2\,dx = 1 \qquad [29.14]$$

Normalization condition on ψ

Any wave function satisfying Equation 29.14 is said to be *normalized.* The quantity $|\psi|^2$ is often called the **probability density.** Note that the normalization condition on ψ is simply a statement that the particle exists at some point at all times. If the probability were zero, the particle would not exist. Therefore, although it is not possible to specify the position of a particle with complete certainty, it is possible, through $|\psi|^2$, to specify the probability of observing it. Furthermore, *the probability of finding the particle in the internal* $a \leq x \leq b$ *is*

$|\psi|^2$ equals the probability density

$$P_{ab} = \int_a^b |\psi|^2\,dx \qquad [29.15]$$

The probability P_{ab} is the area under the curve of $|\psi|^2$ versus x between the points $x = a$ and $x = b$, as in Figure 29.18.

Experimentally, there is a finite probability of finding a particle in an interval near some point at some instant. The value of that probability must lie between the limits 0 and 1. For example, if the probability is 0.3, there is a 30% chance of finding the particle.

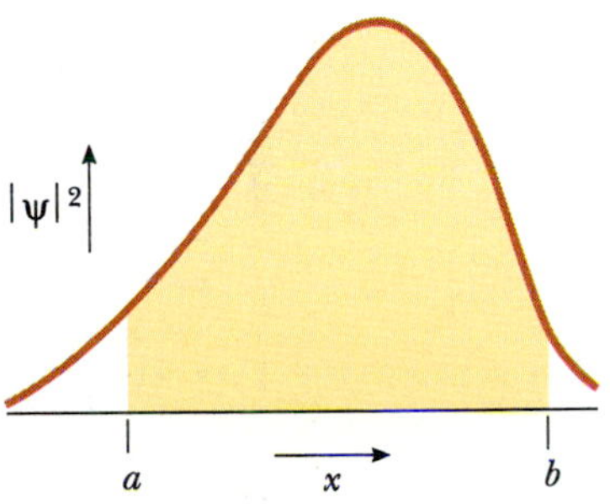

Figure 29.18
The probability that a particle is in the interval $a \leq x \leq b$ is represented by the area under the curve from a to b of the probability density function $|\psi(x, t)|^2$.

The wave function ψ satisfies a wave equation, just as the electric field associated with an electromagnetic wave satisfies a wave equation that follows from Maxwell's equations. The wave equation satisfied by ψ, however, which is the Schrödinger equation, cannot be derived from any more fundamental laws. However, ψ can be computed from it. Although ψ is not a measurable quantity, all the measurable quantities of a particle, such as its energy and momentum, can be derived from a knowledge of $|\psi|^2$. For example, once the wave function for a particle is known, it is possible to calculate the average position, x, of the particle after many experimental trials. This average position is called the **expectation value** of x and is defined by the equation

$$\langle x \rangle \equiv \int_{-\infty}^{\infty} x|\psi|^2\,dx \qquad [29.16]$$

Expectation value of x

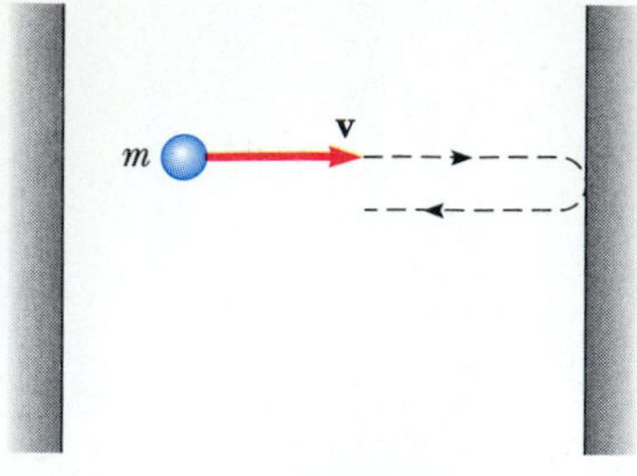

Figure 29.19
A particle of mass m and velocity v, confined to bouncing between two impenetrable walls.

This expression implies that the particle is in a definite state, so that the probability density is time-independent. Note that the expectation value is equivalent to the average value of x that one would obtain when dealing with a large number of particles in the same state. Furthermore, one can find the expectation value of *any* function $f(x)$ by using Equation 29.16 with the integrand replaced by $\psi^* f(x) \psi$.

29.9 A Particle in a Box

From a classical viewpoint, if a particle is confined to moving along the x axis and to bouncing back and forth between two impenetrable walls (Fig. 29.19), its motion is easy to describe. If the speed of the particle is v, then the magnitude of its momentum (mv) remains constant, as does its kinetic energy. Furthermore, classical physics places no restrictions on the values of a particle's momentum and energy. The quantum mechanics approach to this problem is quite different and requires that we find the appropriate wave function consistent with the conditions of the situation.[4]

Because the walls are impenetrable, the wave function $\psi(x)$ is zero at the walls and outside the walls. That is, $\psi(x) = 0$ for $x \leqslant 0$ and for $x \geqslant L$, where L is the distance between the two walls. Only the wave functions that satisfy this condition are allowed. In analogy with standing waves on a string (Eq. 23.3), the allowed wave functions are sinusoidal and are given by

Allowed wave functions for a particle in a box

$$\psi(x) = A \sin\left(\frac{n\pi x}{L}\right) \qquad n = 1, 2, 3, \ldots \qquad [29.17]$$

where A is the maximum value of the wave function. This expression shows that, for a particle confined to a box and having a well-defined de Broglie wavelength, ψ is represented by a sinusoidal wave. The allowed wavelengths are those for which the length L is equal to an integral number of half-wavelengths, that is, $L = n\lambda/2$. These allowed states of the system are called **stationary states** because they are constant with time.

Figure 29.20 plots ψ versus x and $|\psi|^2$ versus x for $n = 1$, 2, and 3. As we shall soon see, these states correspond to the three lowest allowed energies for the particle. Note that although ψ can be positive or negative, $|\psi|^2$ is always positive. From any viewpoint, a negative value for $|\psi|^2$ is meaningless.

Further inspection of Figure 29.20b shows that $|\psi|^2$ is always zero at the boundaries, indicating that it is impossible to find the particle at these points. In addition, $|\psi|^2$ is zero at other points, depending on the value of n. For $n = 2$, $|\psi|^2 = 0$ at the midpoint and at $x = L/2$; for $n = 3$, $|\psi|^2 = 0$ at $x = L/3$ and at $x = 2L/3$. For $n = 1$, however, the probability of finding the particle is a maximum at $x = L/2$. For $n = 4$, $|\psi|^2$ has maxima also at $x = L/4$ and at $x = 3L/4$, and so on.

Since the wavelengths of the particle are restricted by the condition $\lambda = 2L/n$, the magnitude of the momentum is also restricted to specific values:

$$p = \frac{h}{\lambda} = \frac{h}{2L/n} = \frac{nh}{2L}$$

[4] Before continuing, you should review Chapter 23, Sections 23.2 and 23.3 on standing mechanical waves.

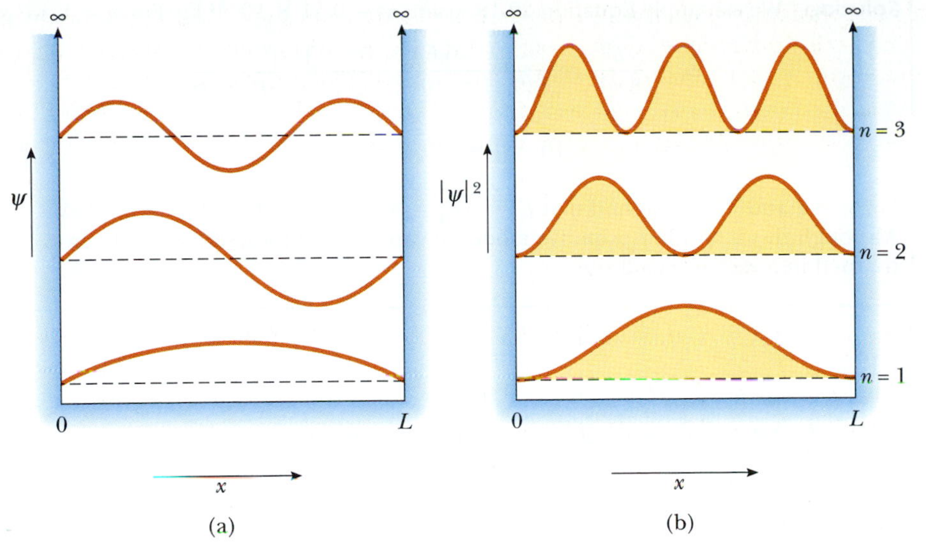

Figure 29.20
The first three allowed stationary states for a particle confined to a one-dimensional box. (a) The wave functions for $n = 1$, 2, and 3. (b) The probability distributions for $n = 1$, 2, and 3.

Using $p = mv$, we find that the allowed values of the energy are

$$E_n = \tfrac{1}{2}mv^2 = \frac{p^2}{2m} = \frac{(nh/2L)^2}{2m}$$

$$E_n = \left(\frac{h^2}{8mL^2}\right) n^2 \qquad n = 1, 2, 3, \ldots \qquad \textbf{[29.18]}$$

Allowed energies for a particle in a box

As we see from this expression, *the energy of the particle is quantized,* as we would expect. The lowest allowed energy corresponds to $n = 1$, for which $E_1 = h^2/8mL^2$. Since $E_n = n^2E_1$, the excited states corresponding to $n = 2, 3, 4, \ldots$ have energies given by $4E_1, 9E_1, 16E_1, \ldots$. Figure 29.21 is an energy level diagram describing the positions of the allowed states. Note that the state $n = 0$ is not allowed. This means that according to quantum mechanics, the particle can never be at rest. The least energy it can have, corresponding to $n = 1$, is called the *zero-point energy*. This result is clearly contradictory to the classical viewpoint, in which $E = 0$ is an acceptable state, as are all positive values of E. In this situation, only positive values of E are considered, because the total energy E equals the kinetic energy, and the potential energy is zero.

The energy levels are of special importance for the following reason. If the particle is electrically charged, it can emit a photon when it drops from an excited state, such as E_3, to one of the lower-lying states, such as E_2. It can also absorb a photon whose energy matches the difference in energy between two allowed states. For example, if the photon frequency is f, the particle will jump from the state E_1 to the state E_2 if $hf = E_2 - E_1$. The processes of photon emission and absorption can be observed by spectroscopy, in which spectral wavelengths are a direct measure of such energy differences.

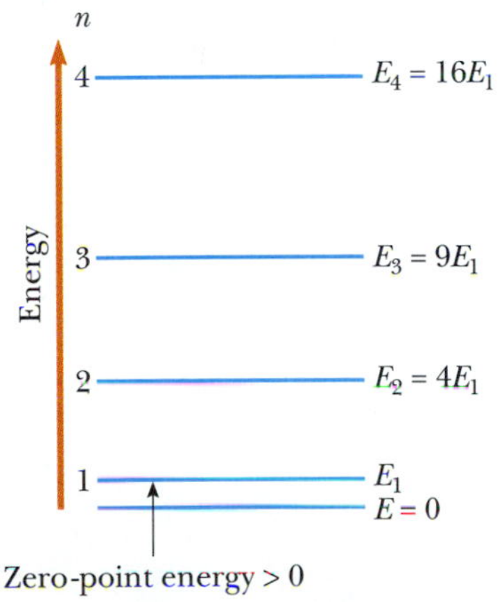

n
4 — $E_4 = 16E_1$
3 — $E_3 = 9E_1$
2 — $E_2 = 4E_1$
1 — E_1
$E = 0$
Energy
Zero-point energy > 0

Figure 29.21
An energy level diagram for a particle confined to a one-dimensional box of width L. The lowest allowed energy is E_1 and has the value $E_1 = h^2/8mL^2$.

Example 29.9 A Bound Electron

An electron is confined between two impenetrable walls 0.2 nm apart. Determine the energy levels for $n = 1$, 2, and 3.

Solution We can apply Equation 29.18, using $m = 9.11 \times 10^{-31}$ kg. For $n = 1$, we get

$$E_1 = \frac{h^2}{8\,mL^2} = \frac{(6.63 \times 10^{-34}\ \text{J}\cdot\text{s})^2}{8(9.11 \times 10^{-31}\ \text{kg})(2 \times 10^{-10}\ \text{m})^2}$$

$$= 1.51 \times 10^{-18}\ \text{J} = \boxed{9.42\ \text{eV}}$$

For $n = 2$ and $n = 3$, we find that $E_2 = 4E_1 = 37.7$ eV and $E_3 = 9E_1 = 84.8$ eV. Although this is a rather primitive model, it can be used to describe an electron trapped in a vacant crystal site.

Example 29.10 Energy Quantization for a Macroscopic Object

A 1-mg object is confined to moving between two rigid walls separated by 1 cm.

(a) Calculate its minimum speed.

Solution The minimum speed corresponds to the state for which $n = 1$. Using Equation 29.18 with $n = 1$ gives the zero-point energy:

$$E_1 = \frac{h^2}{8mL^2} = \frac{(6.63 \times 10^{-34}\ \text{J}\cdot\text{s})^2}{8(1 \times 10^{-6}\ \text{kg})(1 \times 10^{-2}\ \text{m})^2} = 5.49 \times 10^{-58}\ \text{J}$$

Since $E = \frac{1}{2}mv^2$, we can find v as follows:

$$\tfrac{1}{2}mv^2 = 5.49 \times 10^{-58}\ \text{J}$$

$$v = \left[\frac{2(5.49 \times 10^{-58}\ \text{J})}{1 \times 10^{-6}\ \text{kg}}\right]^{1/2} = \boxed{3.31 \times 10^{-26}\ \text{m/s}}$$

This result is so small that the object can be considered to be at rest, which is what one would expect for a macroscopic object.

(b) If the speed of the object is 3×10^{-2} m/s, find the corresponding value of n.

Solution The kinetic energy of the object is

$$E = \tfrac{1}{2}mv^2 = \tfrac{1}{2}(1 \times 10^{-6}\ \text{kg})(3 \times 10^{-2}\ \text{m/s})^2$$

$$= 4.5 \times 10^{-10}\ \text{J}$$

Since $E_n = n^2E_1$ and $E_1 = 5.49 \times 10^{-58}$ J, we find that

$$n^2E_1 = 4.5 \times 10^{-10}\ \text{J}$$

$$n = \left(\frac{4.5 \times 10^{-10}\ \text{J}}{E_1}\right)^{1/2} = \left(\frac{4.5 \times 10^{-10}\ \text{J}}{5.49 \times 10^{-58}\ \text{J}}\right)^{1/2}$$

$$\approx \boxed{9.05 \times 10^{23}}$$

This value of n is so large that we would never be able to distinguish the quantized nature of the energy levels. That is, the difference in energy between the two states $n_1 = 9.05 \times 10^{23}$ and $n_2 = (9.05 \times 10^{23}) + 1$ is too small to be detected experimentally. This is another example that illustrates the working of the correspondence principle; that is, as m and/or L become large, the quantum description must agree with the classical result.

29.10 The Schrödinger Equation

As we mentioned earlier, the wave function for de Broglie waves must satisfy an equation developed by Schrödinger in 1926. One of the methods of quantum mechanics is to determine a solution to this equation, which in turn yields the allowed wave functions and energy levels of the system under consideration. Proper manipulation of the wave functions enables one to calculate *all* measurable features of the system.

The Schrödinger equation as it applies to a particle confined to moving along the x axis is

Time-independent Schrödinger equation

$$\frac{\partial^2 \psi}{\partial x^2} = -\frac{2m}{\hbar^2}(E - U)\psi \qquad [29.19]$$

Since this equation is independent of time, it is commonly referred to as the *time-independent Schrödinger equation.* (We shall not discuss the time-dependent Schrödinger equation in this text.)

In principle, if the potential energy $U(x)$ for the system is known, one can solve Equation 29.19 and obtain the wave functions and energies for the allowed states. Since U may vary with position, it may be necessary to solve the equation in pieces. In the process, the wave functions for the different regions must join smoothly at the boundaries. In the language of mathematics, we require that $\psi(x)$ be *continuous.* Furthermore, in order that $\psi(x)$ obey the normalization condition, we require that $\psi(x)$ approach zero as x approaches $\pm\infty$. Finally, $\psi(x)$ must be *single-valued* and $d\psi/dx$ must also be continuous for finite values of $U(x)$.

The task of solving the Schrödinger equation may be very difficult, depending on the form of the potential energy function. As it turns out, the Schrödinger equation has been extremely successful in explaining the behavior of atomic and nuclear systems, whereas classical physics has failed to do so. Furthermore, when wave mechanics is applied to macroscopic objects, the results agree with classical physics, as required by the correspondence principle.

The Particle in a Box

Let us solve the Schrödinger equation for our particle in a one-dimensional box of width L (Fig. 29.22). The walls are infinitely high, corresponding to $U(x) = \infty$ for $x = 0$ and $x = L$. The potential energy is constant within the box, and it is convenient to choose $U = 0$ as its value. Hence, in the region $0 < x < L$, we can express the Schrödinger equation in the form

$$\frac{d^2\psi}{dx^2} = -\frac{2mE}{\hbar^2}\psi = -k^2\psi \qquad [29.20]$$

$$k = \frac{\sqrt{2mE}}{\hbar}$$

Since the walls are infinitely high, the particle cannot exist outside the box. Consequently, $\psi(x)$ must be zero outside the box and at the walls. The solution of Equation 29.20 that meets the boundary conditions $\psi(x) = 0$ at $x = 0$ and $x = L$ is

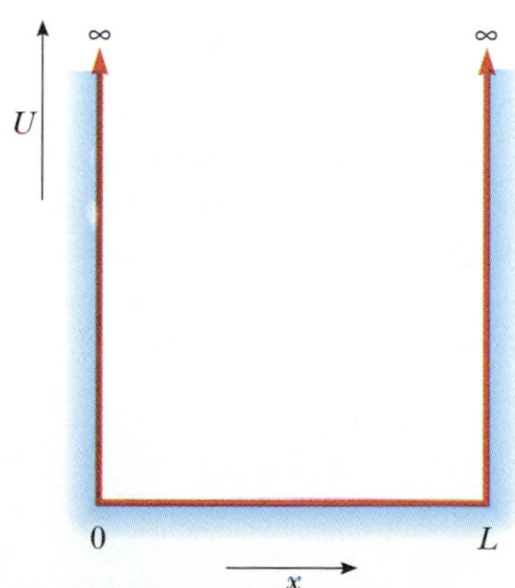

Figure 29.22
Diagram of a one-dimensional box of width L and infinitely high walls.

Erwin Schrödinger (1887–1961)
An Austrian theoretical physicist, Schrödinger is best known as the creator of wave mechanics. He also produced important papers in the fields of statistical mechanics, color vision, and general relativity. Schrödinger did much to hasten the universal acceptance of quantum theory by demonstrating the mathematical equivalence between his wave mechanics and the more abstract matrix mechanics developed by Heisenberg. In 1927 Schrödinger accepted the chair of theoretical physics at the University of Berlin, where he formed a close friendship with Max Planck. In 1933 he left Germany and eventually settled at the Dublin Institute of Advanced Study, where he spent 17 happy, creative years working on problems in general relativity, cosmology, and the application of quantum physics to biology. In 1956 he returned home to Austria and to his beloved Tirolean mountains, where he died in 1961.

$$\psi(x) = A\sin(kx) \qquad \textbf{[29.21]}$$

This can easily be verified by substitution into Equation 29.20. Note that the first boundary condition, $\psi(0) = 0$, is satisfied by Equation 29.21 since $\sin 0° = 0$. The second boundary condition, $\psi(L) = 0$, is satisfied only if kL is an integral multiple of π, that is, if $kL = n\pi$, where n is an integer. Since $k = \sqrt{2mE}/\hbar$, we get

$$kL = \frac{\sqrt{2mE}}{\hbar}L = n\pi$$

Solving for the allowed energies E gives

$$E_n = \left(\frac{h^2}{8mL^2}\right)n^2 \qquad \textbf{[29.22]}$$

Likewise, the allowed wave functions are given by

$$\psi_n(x) = A\sin\left(\frac{n\pi x}{L}\right) \qquad \textbf{[29.23]}$$

These results agree with those obtained in the preceding section. It is left to an end-of-chapter problem (Problem 56) to show that A for this solution is equal to $(2/L)^{1/2}$.

*29.11 Tunneling Through a Barrier

A very interesting and peculiar phenomenon occurs when a particle strikes a barrier of finite height and width. Consider a particle of energy E that is incident on a rectangular barrier of height U and width L, where $E < U$ (Fig. 29.23). Classically, the particle is reflected by the barrier since it does not have sufficient energy to cross or even penetrate it. Thus, regions II and III are classically *forbidden* to the particle. According to quantum mechanics, however, *all regions are accessible to the particle, regardless of its energy,* since the amplitude of the matter wave associated with the particle is nonzero everywhere (except at certain points). A typical waveform for this case, illustrated in Figure 29.23, shows the penetration of the wave into the barrier and beyond. The wave functions are sinusoidal to the left (region I) and right (region II) of the barrier and join smoothly with an exponentially decaying function within the barrier (region II). Since the probability of locating the particle is proportional to $|\psi|^2$, we conclude that the chance of finding the particle beyond the barrier in region III is nonzero. This barrier penetration is in complete disagreement with classical physics. The possibility of finding the particle on the far side of the barrier is called **tunneling** or **barrier penetration.** Any attempt to observe the particle inside the barrier *and* confirm the value of its energy is frustrated by the uncertainty principle. If tunneling is to take place, the barrier must be sufficiently narrow and low, to allow a measurement of the particle's position and momentum consistent with expectations. This surprise result arises from our classical view of the barrier as being continuous. In practice, barriers are typically formed by particles of uncertain position, much like wandering sentinels outside camp. Occasionally, intruders are able to penetrate the barriers of sentinels and enter the camp.

The probability of tunneling can be described with a *transmission coefficient, T,* and a *reflection coefficient, R.* The transmission coefficient measures the probability that the particle penetrates to the other side of the barrier, and the reflection coefficient is the probability that the particle is reflected by the barrier. Since the incident particle is either reflected or transmitted, we must require that $T + R = 1$. An approximate expression for the transmission coefficient that is obtained when $T \ll 1$ (a high or wide barrier) is given by

$$T \cong e^{-2KL} \qquad [29.24]$$

where

$$K = \frac{\sqrt{2m(U - E)}}{\hbar} \qquad [29.25]$$

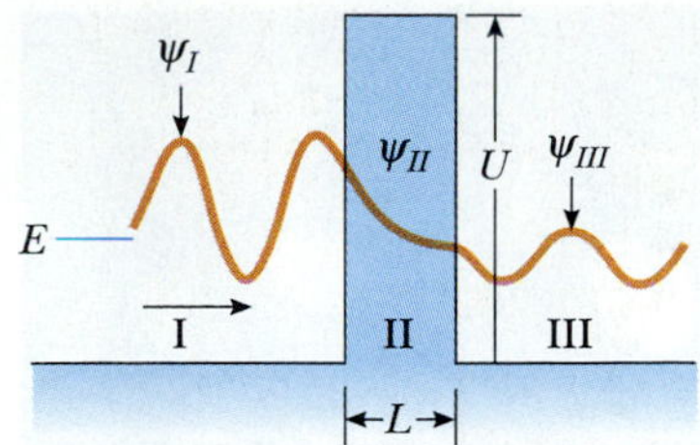

Figure 29.23
Wave function for a particle incident from the left on a barrier of height U. Note that the wave function is sinusoidal in regions I and III but is exponentially decaying in region II. Both amplitude of ψ and energy are plotted along the vertical axis.

Example 29.11 Transmission Coefficient for an Electron

A 30-eV electron is incident on a square barrier of height 40 eV. What is the probability that the electron will tunnel through the barrier if its width is (a) 1 nm? (b) 0.1 nm?

Solution (a) In this situation, the quantity $U - E$ has the value

$$U - E = (40 \text{ eV} - 30 \text{ eV}) = 10 \text{ eV} = 1.6 \times 10^{-18} \text{ J}$$

Using Equation 29.25, and given that $L = 1$ nm, the quantity $2KL$ is

$$2KL = 2\,\frac{\sqrt{2(9.11 \times 10^{-31} \text{ kg})(1.6 \times 10^{-18} \text{ J})}}{1.054 \times 10^{-34} \text{ J}\cdot\text{s}}\,(1 \times 10^{-9} \text{ m}) = 32.4$$

Thus, the probability of tunneling through the barrier is

$$T \cong e^{-2KL} = e^{-32.4} = 8.49 \times 10^{-15}$$

That is, the electron has only about 1 chance in 10^{14} to tunnel through the 1-nm-wide barrier.

(b) For $L = 0.1$ nm, we find $2KL = 3.24$, and

$$T \cong e^{-2KL} = e^{-3.24} = 0.0392$$

This result shows that the electron has a high probability (4% chance) of penetrating the 0.1-nm barrier. Thus, reducing the width of the barrier by only one order of magnitude has increased the probability of tunneling by about 12 orders of magnitude!

Applications of Tunneling

As we have seen, tunneling is a quantum phenomenon and is a manifestation of the wave nature of matter. There are many examples in nature, on the atomic and nuclear scales, for which tunneling is very important. We shall briefly describe four such examples.

1. **Tunnel diode.** The tunnel diode is a semiconductor device consisting of two oppositely charged regions separated by a very narrow neutral region. The current in this device is largely due to tunneling of electrons through the neutral region. The current, or rate of tunneling, can be controlled over a wide range by varying the bias voltage, which changes the height of the barrier.

The surface of $TaSe_2$, "viewed" with a scanning tunneling microscope. The photograph is actually a charge density wave contour of the surface, where different colors indicate regions of different charge densities. *(Courtesy of Prof. R. V. Coleman, University of Virginia)*

2. **Josephson junction.** The Josephson junction consists of two superconductors separated by a thin insulating oxide layer, 1 to 2 nm thick. Under appropriate conditions, electrons in the superconductors travel as pairs and tunnel from one superconductor to the other through the oxide layer. Several effects have been observed in this type of junction. For example, a dc current is observed across the junction *in the absence of electric and magnetic fields.* The current is proportional to sin ϕ, where ϕ is the phase difference between the wave functions in the two superconductors. When a bias voltage, V, is applied across the junction, one observes oscillations in the current, with a frequency of $f = 2eV/h$, where e is the charge on the electron.
3. **Alpha decay.** One form of radioactive decay is the emission of alpha particles (the nuclei of helium atoms) by unstable, heavy nuclei. In order for the alpha particle to escape from the nucleus, it must penetrate a barrier that arises from a combination of the attractive nuclear force and the Coulomb repulsion between the alpha particle and the remaining part of the nucleus. Occasionally an alpha particle tunnels through the barrier, which explains the basic mechanism for this type of decay and the large variations in the mean lifetimes of various radioactive nuclei.
4. **Scanning tunneling microscope.** The scanning tunneling microscope, or STM, is a remarkable device that uses tunneling to create images of surfaces with resolution comparable to the size of a single atom. A small probe with a very fine tip is made to scan very close to the surface of a specimen. A tunneling current is maintained between the probe and specimen; the current is very sensitive to the separation between the tip and specimen. With the maintenance of a constant tunneling current, a feedback signal is obtained that is used to raise and lower the probe as the surface is scanned. Since the vertical motion of the probe follows the contour of the specimen's surface, an image of the surface is obtained.

Summary

The characteristics of *black-body radiation* cannot be explained by classical concepts. Planck introduced the *quantum concept* when he assumed that the atomic oscillators responsible for this radiation existed only in discrete states.

In the **photoelectric effect,** electrons are ejected from a metallic surface when light is incident on that surface. Einstein provided a successful explanation of this effect by extending Planck's quantum theory to the electromagnetic field. In this model, light is viewed as a stream of particles called *photons,* each with energy $E = hf$, where f is the frequency and h is Planck's constant. The maximum kinetic energy of the ejected photoelectron is given by

Photoelectric effect equation

$$K_{max} = hf - \phi \qquad [29.5]$$

where ϕ is the work function of the metal.

X-rays striking a target are scattered at various angles by electrons in the target. A shift in wavelength is observed for the scattered x-rays, and the phenomenon is known as the **Compton effect.** Classical physics does not explain this effect. If the x-ray is treated as a photon, conservation of energy and momentum applied to the photon-electron collisions yields for the Compton shift the expression

$$\lambda^1 - \lambda_0 = \frac{h}{mc}(1 - \cos\theta) \quad [29.7]$$

Compton shift equation

where m is the mass of the electron, c is the speed of light, and θ is the scattering angle.

Every object of mass m and momentum p has wave-like properties, with a wavelength given by the de Broglie relation

$$\lambda = \frac{h}{p} \quad [29.9]$$

The de Broglie wavelength of an object with momentum p.

By applying this wave theory of matter to electrons in atoms, de Broglie was able to explain the appearance of quantization in the Bohr model of hydrogen as a standing wave phenomenon.

The **uncertainty principle** states that if a measurement of position is made with precision Δx and a *simultaneous* measurement of momentum is made with precision Δp_x, then the product of the two uncertainties can never be less than a number on the order of $\hbar$.

$$\Delta x\, \Delta p_x \geq \frac{\hbar}{2} \quad [29.11]$$

The uncertainty principle

Matter waves are represented by the wave function $\Psi(x, y, z, t)$. The probability per unit volume that a particle will be found at a point is $|\psi|^2$. If the particle is confined to moving along the x axis, then the probability that it will be located in an interval dx is given by $|\psi|^2\, dx$. Furthermore,

$$\int_{-\infty}^{\infty} |\psi|^2\, dx = 1 \quad [29.14]$$

Normalization condition on ψ

The measured position x of the particle, averaged over many trials, is called the **expectation value** of x and is defined by

$$\langle x \rangle \equiv \int_{-\infty}^{\infty} x|\psi|^2\, dx \quad [29.16]$$

Expectation value of x

If a particle of mass m is confined to moving in a one-dimensional box of width L whose walls are perfectly rigid, the **allowed wave functions** for the particle are

$$\psi(x) = A\sin\left(\frac{n\pi x}{L}\right) \qquad n = 1, 2, 3, \ldots \quad [29.17]$$

Allowed wave functions for a particle in a box

where A is the maximum value of ψ. The particle has a well-defined wavelength λ whose values are such that the width of the box L is equal to an integral number of half wavelengths, that is, $L = n\lambda/2$. These allowed states are called **stationary states** of the system. The energies of a particle in a box are quantized and are given by

$$E_n = \left(\frac{h^2}{8mL^2}\right) n^2 \qquad n = 1, 2, 3, \ldots \quad [29.18]$$

Allowed energies for a particle in a box

The wave function must satisfy the **Schrödinger equation.** The time-independent Schrödinger equation for a particle confined to moving along the x axis is

Time-independent Schrödinger equation

$$\frac{\partial^2 \psi}{\partial x^2} = -\frac{2m}{\hbar^2}(E - U)\psi \qquad [29.19]$$

where E is the total energy of the system and U is the potential energy.

When a particle of energy E meets a barrier of height U, where $E < U$, the particle has a finite probability of penetrating the barrier. Part of the incident wave is transmitted through the barrier, and part is reflected. This process, called **tunneling,** is the basic mechanism that explains the operation of the Josephson junction and the phenomenon of alpha decay in some radioactive nuclei.

Questions and Conceptual Exercises

1. Why is it impossible to simultaneously measure the position and velocity of a particle with infinite accuracy?
2. If matter has a wave-like nature, why is this character not observable in our daily experiences?
3. An electron and a proton are accelerated from rest through the same potential difference. Which particle has the longer wavelength?
4. In describing the passage of electrons through a slit and their arrival at a screen, Feynman said that "electrons arrive in lumps, like particles, but the probability of arrival of these lumps is determined as the intensity of the waves would be. It is in this sense that the electron behaves sometimes like a particle and sometimes like a wave." Elaborate on this point in your own words. (For a further discussion of this point, see R. Feynman, *The Character of Physical Law,* Cambridge, Mass., MIT Press, 1980, Chapter 6.)
5. In the photoelectric effect, explain why the photocurrent depends on the intensity of the light source but not on the frequency.
6. In the photoelectric effect, explain why the stopping potential depends on the frequency of light but not on the intensity.
7. Why does the existence of a cutoff frequency in the photoelectric effect favor a particle theory of light rather than a wave theory?
8. Is light a wave or a particle? Support your answer by citing specific experimental evidence.
9. Suppose a photograph were made of a person's face using only a few photons. Would the result be simply a very faint image of the face? Discuss.
10. Which has more energy, a photon of ultraviolet radiation or a photon of yellow light?
11. Some stars are observed to be reddish, and some are blue. Which group has the higher surface temperatures? Explain.
12. An x-ray photon is scattered by an electron. What happens to the frequency of the scattered photon relative to that of the incident photon?
13. Using Wien's law, calculate the wavelength of highest intensity given off by a human body. Using this information, explain why an infrared detector would be a useful alarm for security work.
14. Why is an electron microscope more suitable than an optical microscope for "seeing" objects of an atomic size?
15. Why was the Davisson–Germer diffraction of electrons an important experiment?
16. What is the significance of the wave function ψ?
17. What is the Schrödinger equation? How is it useful in describing atomic phenomena?
18. If the surface temperature of the Sun is 5800 K, estimate the wavelength that corresponds to the maximum rate of energy emission from the Sun.
19. Figure 29.24 shows the spectrum of light emitted by a firefly. Determine the temperature of a black body that would emit radiation peaked at the same frequency. Based on your result, would you say firefly radiation is black-body radiation?
20. The power output of the Sun is 3.8×10^{26} W. If the Sun were displaced to 50 lightyears, it would be barely visible. Estimate the number of yellow photons that you would see per second (with a pupil diameter of 7 mm) if the Sun were at this distance.

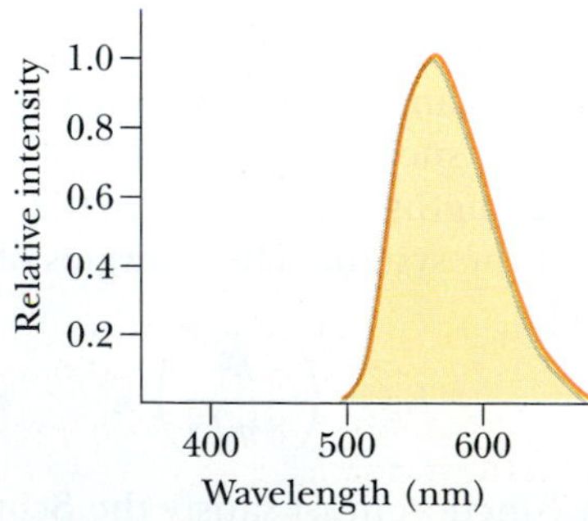

Figure 29.24 (Question 19)

Problems

Section 29.1 Black-body Radiation and Planck's Theory

1. Calculate the energy of a photon whose frequency is (a) 6.2×10^{14} Hz; (b) 3.1 GHz; (c) 46 MHz. Express your answers in electron volts.
2. An FM radio transmitter has a power output of 150 kW and operates at a frequency of 99.7 MHz. How many photons per second does the transmitter emit?
3. The average power generated by the Sun is equal to 3.74×10^{26} W. Assuming the average wavelength of the Sun's radiation to be 500 nm, find the number of photons emitted by the sun in 1 s.
4. Using Wien's displacement law, calculate the surface temperature of a red giant star that radiates with a peak wavelength of $\lambda_{max} = 650$ nm.
5. What is the peak wavelength emitted by the human body? Assume a body temperature of 98.6°F and use the Wien displacement law. In what part of the electromagnetic spectrum does this wavelength lie?
6. A tungsten filament is heated to a temperature of 800°C. What is the wavelength of the most intense radiation?
7. The human eye is most sensitive to light with a wavelength of $\lambda = 560$ nm. A black body of what temperature would radiate most intensely at this wavelength?
8. The average threshold of dark-adapted (scotopic) vision is 4.0×10^{-11} W/m^2 at a central wavelength of 500 nm. If light having this intensity and wavelength enters the eye and the pupil is open to its maximum diameter of 8.5 mm, how many photons per second enter the eye?

Section 29.2 The Photoelectric Effect

9. The photocurrent of a photocell is stopped by a retarding potential of 0.54 V for radiation of wavelength 750 nm. Find the work function for the material.
10. The work function for potassium is 2.24 eV. If potassium metal is illuminated with light of wavelength 480 nm, find (a) the maximum kinetic energy of the photoelectrons and (b) the cutoff wavelength.
11. Molybdenum has a work function of 4.2 eV. (a) Find the cutoff wavelength and threshold frequency for the photoelectric effect. (b) Calculate the stopping potential if the incident light has a wavelength of 180 nm.
12. When cesium metal is illuminated with light of wavelength 500 nm, the photoelectrons emitted have a maximum kinetic energy of 0.57 eV. Find (a) the work function of cesium and (b) the stopping potential if the incident light has a wavelength of 600 nm.
13. From the scattering of sunlight, Thomson calculated the classical radius of the electron as having a value of 2.82×10^{-15} m. If sunlight with an intensity of 500 W/m^2 falls on a disk with this radius, estimate the time required to accumulate 1.0 eV of energy. Assume that light is a classical wave and that the light striking the disk is completely absorbed. How does your estimate compare with the observation that photoelectrons are promptly (within 10^{-9} s) emitted?
14. Ultraviolet light is incident normally on the surface of a certain substance. The binding energy of the electrons in this substance is 3.44 eV. The incident light has an intensity of 0.055 W/m^2. The electrons are photoelectrically emitted with a maximum speed of 4.2×10^5 m/s. How many electrons are emitted from a square centimeter of the surface? Assume 100% of the photons are absorbed.
15. A metal has a work function of 2×10^{-19} J. If yellow light of wavelength 600 nm falls on the surface of the metal, find (a) the maximum kinetic energy of the ejected electrons and (b) the cutoff wavelength for the metal.
16. Electrons are ejected from a metallic surface with speeds ranging up to 4.6×10^5 m/s when light with a wavelength of $\lambda = 625$ nm is used. (a) What is the work function of the surface? (b) What is the cutoff frequency for this surface?
17. Consider the metals lithium, beryllium, and mercury, which have work functions of 2.3 eV, 3.9 eV, and 4.5 eV, respectively. If light of wavelength 400 nm is incident on each of these metals, determine (a) which metals exhibit the photoelectric effect and (b) the maximum kinetic energy for the photoelectron in each case.
18. Light of wavelength 300 nm is incident on a metallic surface. If the stopping potential for the photoelectric effect is 1.2 V, find (a) the maximum energy of the emitted electrons, (b) the work function, and (c) the cutoff wavelength.

Section 29.3 The Compton Effect

19. Calculate the energy and momentum of a photon of wavelength 700 nm.
20. X-rays of wavelength 0.200 nm are scattered from a

block of carbon. If the scattered radiation is detected at 60° to the incident beam, find (a) the Compton shift, $\Delta\lambda$, and (b) the kinetic energy imparted to the recoiling electron.

21. A gamma-ray photon with an energy equal to the rest energy of an electron (511 keV) collides with an electron that is initially at rest. Calculate the kinetic energy acquired by the electron if the photon is scattered 30° from its original line of approach.

22. X-rays with an energy of 300 keV undergo Compton scattering from a target. If the scattered rays are detected at 37° relative to the incident rays, find (a) the Compton shift at this angle, (b) the energy of the scattered x-ray, and (c) the energy of the recoiling electron.

23. A metal target is placed in a beam of 662-keV gamma rays emitted by a radioactive isotope of cesium (^{137}Cs). Find the energy of those photons that are scattered through an angle of 90°. The electrons in the target may be considered as free electrons.

24. A 0.5-nm x-ray photon is deflected through a 134° angle in a Compton scattering event from a stationary free electron. At what angle (with respect to the incident beam) is the recoiling electron found?

25. A 0.0016-nm photon scatters from a free electron at rest. For what (photon) scattering angle will the recoiling electron and scattered photon have the same kinetic energy?

Section 29.4 Photons and Electromagnetic Waves
Section 29.5 The Wave Properties of Particles

26. Calculate the de Broglie wavelength for a proton moving with a speed of 10^6 m/s.

27. Calculate the de Broglie wavelength for an electron with kinetic energy of (a) 50 eV; (b) 50 keV.

28. The "seeing" ability, or resolution, of radiation is determined by its wavelength. If the size of an atom is on the order of 0.1 nm, how fast must an electron travel to have a wavelength small enough to "see" an atom?

29. Calculate the de Broglie wavelength of a proton that is accelerated through a potential difference of 10 MV.

30. Show that the de Broglie wavelength of an electron accelerated from rest through a potential difference of V is given by $\lambda = 1.226/\sqrt{V}$ nm, where V is in volts.

31. The distance between adjacent atoms in crystals is on the order of 1.0 Å. The use of electrons in diffraction studies of crystals requires that the de Broglie wavelength of the electrons be on the order of the distance between atoms of the crystals. What must be the minimum energy (in electron volts) of electrons to be used for this purpose?

32. An electron has a de Broglie wavelength equal to the diameter of the hydrogen atom. What is the kinetic energy of the electron? How does this energy compare with the ground-state energy of the hydrogen atom?

33. In order for an electron to be confined to a nucleus, its de Broglie wavelength has to be less than 10^{-14} m. (a) What would be the kinetic energy of an electron confined to this region? (b) On the basis of this result, would you expect to find an electron in a nucleus? Explain.

34. In the Davisson–Germer experiment, 54-eV electrons were diffracted from a nickel lattice. If the first maximum in the diffraction pattern was observed at $\phi = 50°$ (Fig. 29.25), what was the lattice spacing, d, of nickel atoms?

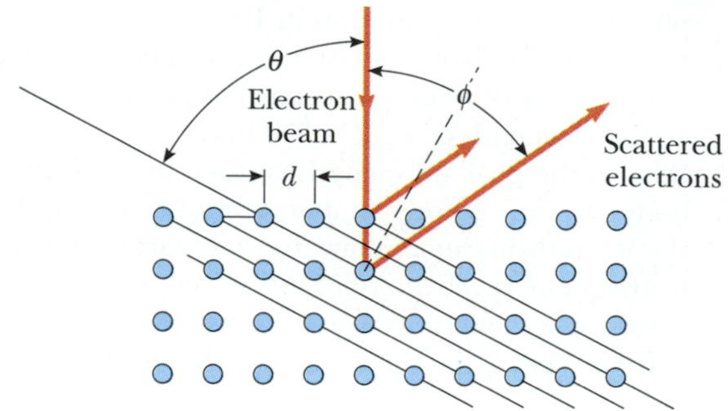

Figure 29.25 (Problem 34)

35. Robert Hofstadter won the 1961 Nobel Prize in physics for his pioneering work in scattering 20-GeV electrons from nuclei. (a) What is the γ-factor for a 20-GeV electron, where $\gamma = (1 - v^2/c^2)^{-1/2}$? What is the momentum of the electron in kg·m/s? (b) What is the wavelength of a 20-GeV electron, and how does it compare with the size of a nucleus?

36. Electrons are accelerated through 40 000 V in an electron microscope. What, theoretically, is the smallest distance between objects that can be observed?

Section 29.6 The Double-Slit Experiment Revisited

37. Through what potential difference would an electron have to be accelerated to have a de Broglie wavelength of 10^{-10} m?

38. A monoenergetic beam of electrons is incident on a single slit of width 0.5 nm. A diffraction pattern is formed on a screen 20 cm from the slit. If the distance between successive minima of the diffraction pattern is 2.1 cm, what is the energy of the incident electrons?

39. A neutron beam with a selected speed of 0.4 m/s is directed through a double slit with a 1-mm separation. An array of detectors is placed 10 m from the slit. (a) What is the de Broglie wavelength of the neutrons? (b) How far off axis is the first zero-intensity point on the detector array? (c) Can we say which slit the neutron passed through? Explain.

40. The resolution of a microscope is proportional to the wavelength used. If one wished to use a microscope to "see" an atom, a resolution of approximately 10^{-11} m (0.1 Å) would have to be obtained. (a) If electrons are used (electron microscope), what minimum kinetic energy is required for the electrons? (b) If photons are used, what minimum photon energy is needed to obtain 10^{-11} m resolution?

Section 29.7 The Uncertainty Principle

41. A light source is used to determine the location of an electron in an atom to a precision of 0.05 nm. What is the uncertainty in the component of the velocity of the electron parallel to the slit width?

42. Suppose Fuzzy, the quantum-mechanical duck, lives in a world in which $h = 2\pi$ J·s. Fuzzy has a mass of 2.0 kg and is initially known to be within a region 1.0 m wide. (a) What is his minimum uncertainty in speed if Fuzzy is moving parallel to the 1.0-m region? (b) Assuming this uncertainty in speed to prevail for 5.0 s, determine the uncertainty in position after this time.

43. An electron ($m = 9.11 \times 10^{-31}$ kg) and a bullet ($m = 0.02$ kg) each have a speed of 500 m/s, accurate to within 0.01%. Within what limits could we determine the position of the paths of the objects?

44. A 0.5-kg block rests on the icy surface of a frozen pond, which we can assume to be frictionless. If the location of the block is measured to a precision of 0.5 cm, what is the uncertainty in the speed of the block that arises in this measurement?

45. In the ground state of hydrogen the uncertainty in the position of the electron is roughly 0.1 nm. If the speed of the electron is on the order of the uncertainty in the speed, how fast is the electron moving?

46. A small boy on a ladder drops small pellets toward a spot on the floor. (a) Show that, according to the uncertainty principle, the spread of hit locations must be at least

$$\Delta x = \left(\frac{\hbar}{2m}\right)^{1/2}\left(\frac{H}{2g}\right)^{1/4}$$

where H is the initial vertical distance of each pellet above the floor and m is the mass of each pellet. (b) If $H = 2$ meters and $m = \frac{1}{2}$ gram, what is Δx?

Section 29.8 An Interpretation of Quantum Mechanics

47. A free electron has a wave function

$$\psi(x) = A \sin(5 \times 10^{10}\, x)$$

where x is measured in meters. Find (a) the electron's de Broglie wavelength, (b) the electron's momentum, and (c) the electron's energy in electron volts.

48. An electron has the wave function

$$\psi(x) = \sqrt{\frac{2}{L}}\sin\left(\frac{2\pi x}{L}\right)$$

Determine the probability of finding the electron between $x = 0$ and $x = L/4$.

Section 29.9 A Particle in a Box

49. An electron is confined to a one-dimensional region in which its ground-state ($n = 1$) energy is 2 eV. (a) What is the width of the region? (b) How much energy is required to "promote" the electron to its first excited state?

50. Use the particle-in-a-box model to calculate the first three energy levels of a neutron trapped in a nucleus 2×10^{-5} nm in diameter. Are the energy level differences realistic?

51. An electron with an energy of approximately 6 eV moves between rigid walls exactly 1 nm apart. (a) Find the quantum number n for the energy state occupied by the electron. (b) Find the exact value for the electron's energy.

52. An alpha particle in a nucleus can be modeled as a particle moving in a box of width 10^{-14} m (the approximate diameter of a nucleus). Using this model, estimate the energy and momentum of an alpha particle in its lowest energy state. (The mass of an alpha particle is $4 \times 1.66 \times 10^{-27}$ kg.)

53. An electron is contained in a one-dimensional box of width 0.1 nm. (a) Draw an energy level diagram for the electron for levels up to $n = 4$. (b) Find the wavelengths of *all* photons that can be emitted by the electron in making transitions that would eventually get it from the $n = 4$ state to the $n = 1$ state.

54. Consider a particle moving in a one-dimensional box with walls at $x = -L/2$ and $x = L/2$. (a) Write the wave functions and probability densities for the states $n = 1$, $n = 2$, and $n = 3$. (b) Sketch the wave functions and probability densities. (*Hint:* Make an analogy with the case of a particle in a box with walls at $x = 0$ and $x = L$.)

55. A ruby laser emits light of wavelength 694.3 nm. If

this light is due to transitions from the $n = 2$ state to the $n = 1$ state of an electron in a box, find the width of the box.

Section 29.10 The Schrödinger Equation

56. The wave function for a particle confined to moving in a one-dimensional box is given by

$$\psi(x) = A \sin\left(\frac{n\pi x}{L}\right)$$

Use the normalization condition on ψ to show that the constant A is given by

$$A = \sqrt{\frac{2}{L}}$$

Hint: Since the particle is confined to the box of width L, the normalization condition (Eq. 29.14) in this case becomes

$$\int_0^L |\psi|^2\, dx = 1$$

57. A particle in the space $-a \leq x \leq a$ may be represented by either of the following wave functions:

$$\psi_1 = A \cos\left(\frac{\pi x}{2a}\right) \quad \text{or} \quad \psi_2 = B \sin\left(\frac{\pi x}{a}\right)$$

Using the normalization condition on ψ, find A and B. (See Problem 56 for a helpful hint.)

58. The wave function of a particle is given by

$$\psi(x) = A\cos(kx) + B\sin(kx)$$

where A, B, and k are constants. Show that ψ is a solution of the Schrödinger equation (Eq. 29.19), assuming the particle is free ($U = 0$), and find the corresponding energy E of the particle.

59. Show that the wave function $\psi = Ae^{ikx}$ is a solution to the Schrödinger equation (Eq. 29.19) where $k = 2\pi\lambda$.

60. A particle with 7 eV of kinetic energy moves from a region where the potential is zero into one in which $U = 5$ eV (Fig. 29.26). Classically, one would expect the particle to continue on, although with less kinetic energy. According to quantum mechanics, the particle has a probability of being transmitted and a probability of being reflected. What are these probabilities?

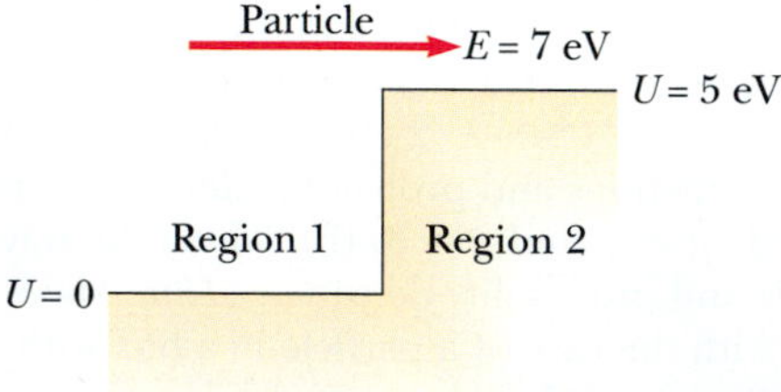

Figure 29.26 (Problem 60)

*Section 29.11 Tunneling Through a Barrier

61. A 5-eV electron is incident on a barrier 0.2 nm thick and 10 eV high (Fig. 29.27). (a) What is the probability that the electron will tunnel through the barrier? (b) What is the probability that the electron will be reflected?

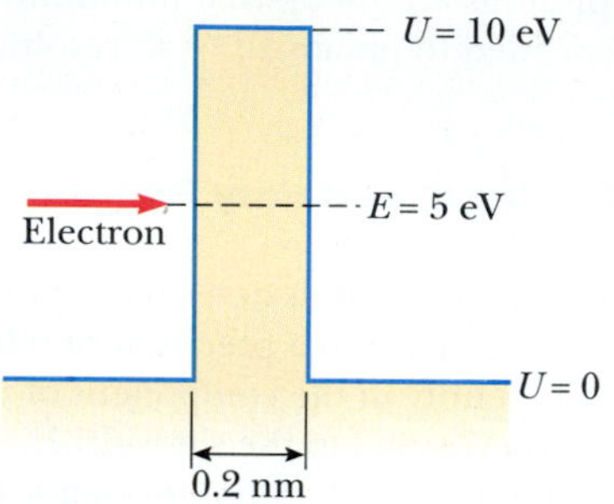

Figure 29.27 (Problem 61)

62. The *transmission coefficient* T gives the probability that a particle of mass m approaching the rectangular potential barrier of Figure 29.28 may "tunnel" through the barrier:

$$T = e^{-2KL} \qquad \text{where } K = \frac{\sqrt{2m(U - E)}}{\hbar}$$

Consider a barrier with $U = 5$ eV and having a width of $L = 950$ pm. Suppose that an electron with energy $E = 4.5$ eV approaches the barrier. Classically, the electron could not pass through the barrier because $E < U$. However, quantum-mechanically there is a finite probability of tunneling. Calculate this probability.

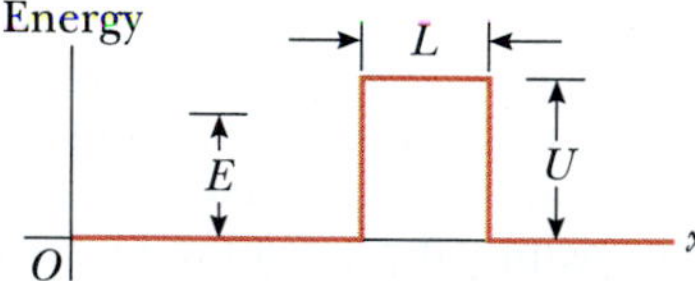

Figure 29.28 (Problems 62 and 63)

63. In Problem 62, by how much would the width L of the potential barrier have to be increased for the chance of an incident 4.5-eV electron tunneling through the barrier to be one in a million?

Additional Problems

64. The nuclear potential that binds protons and neutrons in the nucleus of an atom is often approximated by a square well. Imagine a proton confined in an infinite square well of width 10^{-5} nm, a typical nuclear diameter. Calculate the wavelength and energy associated with the photon that is emitted when the proton undergoes a transition from the first excited state ($n = 2$) to the ground state ($n = 1$). In what region of the electromagnetic spectrum does this wavelength belong?

65. Figure 29.29 shows the stopping potential versus incident photon frequency for the photoelectric effect for sodium. Use these data points to find (a) the work function, (b) the ratio h/e, and (c) the cutoff wavelength. (Data taken from R. A. Millikan, *Phys. Rev.* 7:362, 1916.)

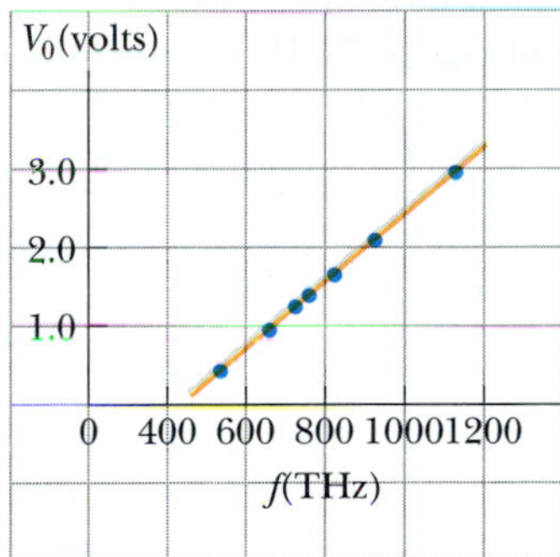

Figure 29.29 (Problem 65)

66. Photons of wavelength 450 nm are incident on a metal. The most energetic electrons ejected from the metal are bent into a circular arc of radius 20 cm by a magnetic field whose strength is equal to 2×10^{-5} T. What is the work function of the metal?

67. A two-slit electron diffraction experiment is done with slits of *unequal* widths. When only slit 1 is open, the number of electrons reaching the screen per second is 25 times the number of electrons reaching the screen per second when only slit 2 is open. When both slits are open, an interference pattern results in which the destructive interference is not complete. Find the ratio of the probability of an electron arriving at an interference maximum to the probability of an electron arriving at an adjacent interference minimum. (*Hint:* Use the superposition principle.)

68. The neutron has a mass of 1.67×10^{-27} kg. Neutrons emitted in nuclear reactions can be slowed down via collisions with matter. They are referred to as thermal neutrons once they come into thermal equilibrium with their surroundings. The average kinetic energy ($3kT/2$) of a thermal neutron is approximately 0.04 eV. Calculate the de Broglie wavelength of a neutron with a kinetic energy of 0.04 eV. How does it compare with the characteristic atomic spacing in a crystal? Would you expect thermal neutrons to exhibit diffraction effects when scattered by a crystal?

69. A particle of mass 2×10^{-28} kg is confined to a one-dimensional box of width 10^{-10} m (1 Å). For $n = 1$, (a) What is the particle wavelength? (b) Its momentum? (c) Its ground-state energy?

70. A particle is described by the wave function

$$\psi(x) = \begin{cases} A\cos\left(\dfrac{2\pi x}{L}\right) & \text{for } -\dfrac{L}{4} \leq x \leq \dfrac{L}{4} \\ 0 & \text{for other values of } x \end{cases}$$

(a) Determine the normalization constant, A.
(b) What is the probability that the particle will be found between $x = 0$ and $x = L/8$ if a measurement of its position is made? (*Hint:* Use Eq. 29.13.)

△ 71. The following table shows data obtained in a photoelectric experiment. (a) Using these data, make a graph similar to Figure 29.8. From the graph, determine (b) an experimental value for Planck's constant (in joule seconds) and (c) the work function (in electron volts) for the surface. (Two significant figures for each answer are sufficient.)

Wavelength (nm)	Maximum Kinetic Energy of Photoelectrons (eV)
588	0.67
505	0.98
445	1.35
399	1.63

72. Particles incident from the left are confronted with a step potential as shown in Figure 29.30. The step has a height of U_0, and the particles have energy $E > U_0$. Classically, all the particles would pass into the region of higher potential at the right. However, according

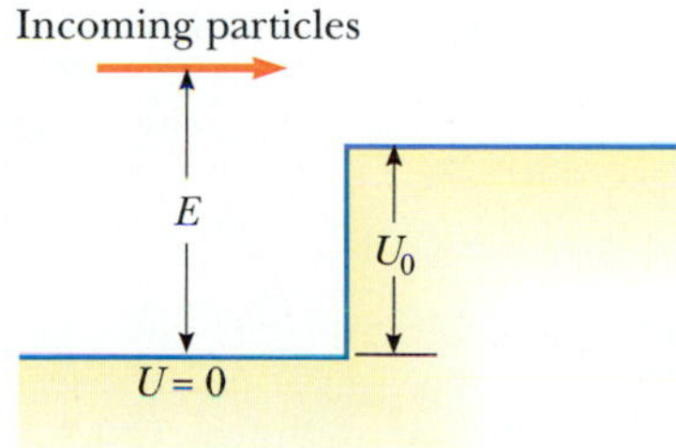

Figure 29.30 (Problem 72)

to quantum mechanics, one finds that a fraction of the particles are reflected at the barrier. The probability that a particle will be reflected, called the *reflection coefficient, R*, is given by

$$R = \frac{(k_1 - k_2)^2}{(k_1 + k_2)^2}$$

where $k_1 = 2\pi/\lambda_1$ and $k_2 = 2\pi/\lambda_2$ are the wave numbers for the incident and transmitted particles, respectively. If $E = 2U_0$, what fraction of the incident particles are reflected? (This situation is analogous to the partial reflection and transmission of light striking an interface between two different media.)

73. A particle has a wave function given by

$$\psi(x) = \begin{cases} \sqrt{\frac{2}{a}}\, e^{-x/a} & \text{for } x > 0 \\ 0 & \text{for } x < 0 \end{cases}$$

(a) Find and sketch the probability density. (b) What is the probability that the particle will be found anywhere with $x < 0$? (c) Show that ψ is normalized, and then find the probability that the particle will be found between $x = 0$ and $x = a$.

74. An electron is confined to one-dimensional motion between two rigid walls separated by a distance of L. (a) What is the probability of finding the electron within the interval $x = 0$ to $x = L/3$ from one wall if the electron is in its $n = 1$ state? (b) Compare this value with the classical probability.

75. An electron is represented by the time-independent wave function

$$\psi(x) = \begin{cases} Ae^{-\alpha x} & \text{for } x > 0 \\ Ae^{+\alpha x} & \text{for } x < 0 \end{cases}$$

(a) Sketch the wave function as a function of x. (b) Sketch the probability that the electron is found between x and $x + dx$. (c) Why do you suppose this is a physically reasonable wave function? (d) Normalize the wave function. (e) Determine the probability of finding the electron somewhere in the range.

$$x_1 = -\frac{1}{2\alpha} \text{ to } x_2 = \frac{1}{2\alpha}$$

76. An electron is trapped at a defect in a crystal. The defect may be modeled as a one-dimensional, rigid-walled box of width 1 nm. (a) Sketch the wave functions and probability densities for the $n = 1$ and $n = 2$ states. (b) For the $n = 1$ state, determine the probability of finding the electron between $x_1 = 0.15$ nm and $x_2 = 0.35$ nm, where $x = 0$ is the left side of the box. (c) Repeat (b) for the $n = 2$ state. (d) Calculate the energies, in electron volts, of the $n = 1$ and $n = 2$ states. *Hint:* For (b) and (c), use Equation 29.13 and note that

$$\int \sin^2 ax\, dx = \tfrac{1}{2}x - \frac{1}{4a} \sin 2ax$$

77. ***Model of an atom.*** An atom can be viewed as several electrons moving around a positively charged nucleus, where the electrons are subject mainly to the Coulomb attraction of the nucleus (which is actually partially "screened" by the inner-core electrons.) The potential well "seen" by each electron is sketched in Figure 29.31. Assume that the atom has a radius of 0.1 nm. (a) Use the model of a particle in a box to *estimate* the energy (in electron volts) required to raise an electron from the state $n = 1$ to the state $n = 2$. (b) Calculate the wavelength of the photon that would cause this transition.

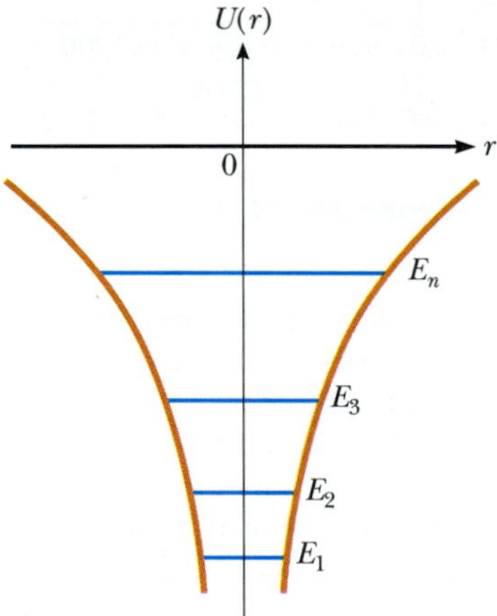

Figure 29.31 (Problem 77) A model of potential energy versus r for the one-electron atom.

This image combines the real lower portion of a broken champagne glass with the hologram of the whole glass before it was broken. A hologram is produced by splitting a laser into two beams. One beam is reflected off the object to be photographed, and the other is used as a reference beam. The two beams are then combined to form an interference pattern that is recorded on a photographic film. A three-dimensional replica of the object is observed when the film is illiminated with a laser. *(Philippe Plailly/Science Photo Library)*

Atomic Physics

CHAPTER 30

In Chapter 29 we introduced some of the basic concepts and techniques used in quantum mechanics, along with their applications to various simple systems. This chapter describes the application of quantum mechanics to the real world of atomic structure.

A large portion of this chapter is concerned with the study of the hydrogen atom from the viewpoint of quantum mechanics. Although the hydrogen atom is the simplest atomic system, it is an especially important system to understand, for several reasons:

1. Much of what we learn about the hydrogen atom, with its single electron, can be extended to such single-electron ions as He^+ and Li^{2+}.
2. The hydrogen atom is an ideal system for performing precise tests of theory against experiment and for improving our overall understanding of atomic structure.

3. The quantum numbers used to characterize the allowed states of hydrogen can be used to describe approximately the allowed states of more complex atoms. This characterization enables us to understand the periodic table of the elements, which is one of the greatest triumphs of quantum mechanics.
4. The basic ideas about atomic structure must be well understood before we attempt to deal with the complexities of molecular structures and the electronic structures of solids.

The full mathematical solution of the Schrödinger equation as applied to the hydrogen atom gives a complete and beautiful description of the atom's properties. However, the mathematical procedures that make up the solution are beyond the scope of this text, and so the details will be omitted. The solutions for some states of hydrogen will be discussed, together with the quantum numbers used to characterize allowed stationary states. We shall also discuss the physical significance of the quantum numbers and the effect of a magnetic field on certain quantum states.

The *exclusion principle,* also presented in this chapter, is extremely important to an understanding of the properties of multielectron atoms and the arrangement of elements in the periodic table. The implications of the exclusion principle are almost as far-reaching as those of the Schrödinger equation. Finally, we shall apply our knowledge of atomic structure to describe the mechanisms involved in the production of x-rays and in the operation of a laser.

30.1 Early Models of the Atom

The model of the atom in Newton's day was a tiny, hard, indestructible sphere. This model was a good basis for the kinetic theory of gases. However, new models had to be devised when later experiments revealed the electrical nature of atoms. J. J. Thomson suggested as a model of the atom a volume of positive charge with electrons embedded throughout it, much like the seeds in a watermelon (Fig. 30.1).

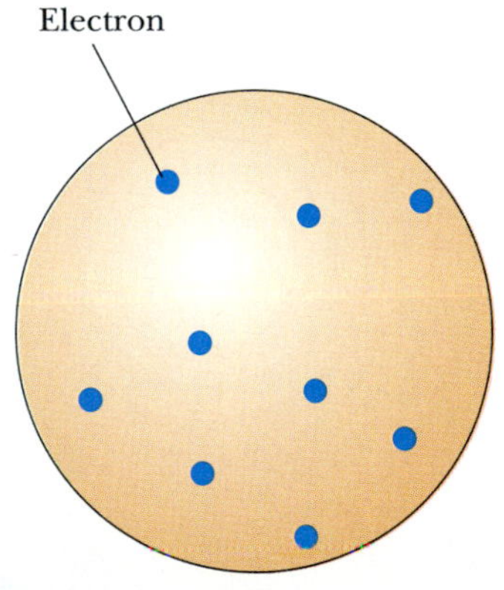

Figure 30.1
Thomson's model of the atom with the electrons embedded inside the positive charge like seeds in a watermelon.

In 1911 Ernest Rutherford and his students Hans Geiger and Ernst Marsden performed a critical experiment that showed that Thomson's model could not be correct. In this experiment, a beam of positively charged alpha particles was projected into a thin metal foil, as in Figure 30.2a. The results of the experiment were astounding. Most of the particles passed through the foil as if it were empty space. Furthermore, a small fraction of those alpha particles that were deflected from their original direction of travel were scattered through very large angles. Some particles were even deflected back toward the source. As Rutherford wrote, "It was quite the most incredible event that has ever happened to me in my life. It was almost as incredible as if you fired a 15-inch shell at a piece of tissue paper and it came back and hit you."

Such large deflections were not expected on the basis of Thomson's model. According to this model, a positively charged alpha particle would never come close enough to a large enough positive charge to cause any large-angle deflections. Rutherford explained his astounding results by assuming that the positive charge was concentrated in a region that was small relative to the size of the atom. He called this concentration of positive charge the **nucleus** of the atom. Any electrons belonging to the atom were assumed to be in the relatively large volume

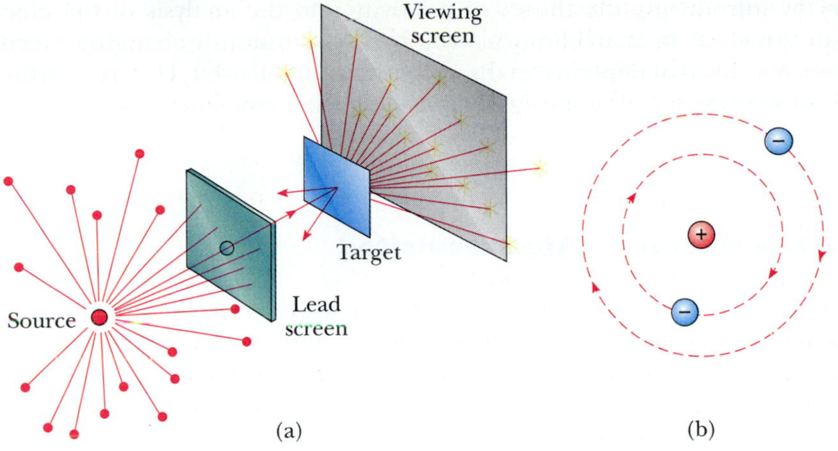

Figure 30.2
(a) Rutherford's technique for observing the scattering of alpha particles from a thin foil target. The source is a naturally occurring radioactive substance, such as radium. (b) Rutherford's planetary model of the atom.

outside the nucleus. In order to explain why these electrons were not pulled into the nucleus, the electrons were viewed as moving in orbits about the positively charged nucleus in the same manner as the planets orbit the Sun, as in Figure 30.2b.

There are two basic difficulties with Rutherford's planetary model. As we saw in Chapter 12, an atom emits certain characteristic frequencies of electromagnetic radiation and no others; the Rutherford model is unable to explain this phenomenon. A second difficulty is that Rutherford's electrons are undergoing a centripetal acceleration. According to Maxwell's theory of electromagnetism, centripetally accelerated charges revolving with frequency f should radiate electromagnetic waves of frequency f. Unfortunately, this classical model leads to disaster when applied to the atom. As the electron radiates energy, the radius of its orbit steadily decreases and its frequency of revolution increases. This leads to an ever increasing frequency of emitted radiation and a rapid collapse of the atom as the electron plunges into the nucleus (Fig. 30.3).

Now the stage was set for Bohr! In order to circumvent the erroneous deductions of electrons falling into the nucleus and a continuous emission spectrum from elements, Bohr postulated that classical radiation theory does not hold for atomic-sized systems. He overcame the problem of a classical electron that continuously loses energy by applying Planck's ideas of quantized energy levels to orbiting atomic electrons. Thus, as described in Section 12.5, Bohr postulated that electrons in atoms are generally confined to stable, nonradiating energy levels and orbits called stationary states. Furthermore, he applied Einstein's concept of the photon to arrive at an expression for the frequency of light emitted when the electron jumps from one stationary state to another.

One of the first indications that there was a need to modify the Bohr theory arose when improved spectroscopic techniques were used to examine the spectral lines of hydrogen. It was found that many of the lines in the Balmer and other series were not single lines at all. Instead, each was a group of lines spaced very close together. An additional difficulty arose when it was observed that, in some situations, certain single spectral lines were split into three closely spaced lines when the atoms were placed in a strong magnetic field.

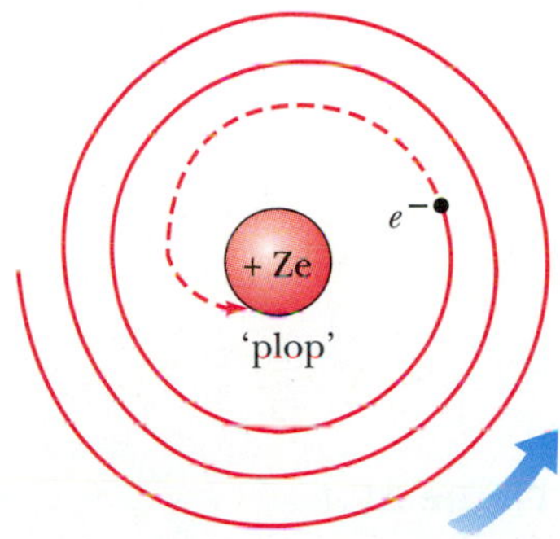

Figure 30.3
The classical model of the nuclear atom.

Efforts to explain these deviations from the Bohr model led to improvements in the theory. One of the changes introduced was the classical concept that the electron could spin on its axis. Also, Arnold Sommerfeld improved the Bohr

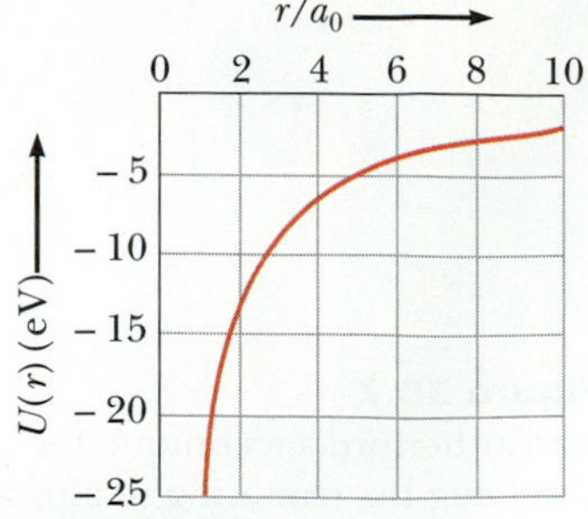

Figure 30.4
The potential energy $U(r)$ versus the ratio r/a_0 for the hydrogen atom. The constant a_0 is the Bohr radius, and r is the electron-proton separation.

theory by introducing the theory of relativity into the analysis of the electron's motion. An electron in an elliptical orbit has a continuously changing speed, with an average value that depends on the eccentricity of the orbit. Different orbits have different average speeds and slightly different relativistic energies.

30.2 The Hydrogen Atom Revisited

The de Broglie model gave electrons a wave nature by viewing them as standing waves in allowed orbits. This standing wave description removed the objections to Bohr's postulates, which were quite arbitrary. However, the de Broglie model created other problems. The exact nature of the de Broglie wave was unspecified, and the model implied unlikely electron densities at great distances from the nucleus. Fortunately, these difficulties are removed when the methods of quantum mechanics are used to describe atoms.

The potential energy function for the hydrogen atom is

$$U(r) = -k_e\frac{e^2}{r} \quad [30.1]$$

where k_e is the Coulomb constant and r is the radial distance from the proton (situated at $r = 0$) to the electron. Figure 30.4 is a plot of this function versus r/a_0 where a_0 is the Bohr radius, 0.0529 nm.

The formal procedure for solving the problem of the hydrogen atom would be to substitute $U(r)$ into the Schrödinger equation and find appropriate solutions to the equation, as we did in Chapter 29. The current problem is more complicated, however, because it is three-dimensional and because U depends on the radial coordinate r. We shall not attempt to carry out these solutions. Rather, we shall simply describe their properties and some of their implications with regard to atomic structure.

According to quantum mechanics, the energies of the allowed states for the hydrogen atom are

Allowed energies for the hydrogen atom

$$E_n = -\left(\frac{k_e e^2}{2a_0}\right)\frac{1}{n^2} = -\frac{13.6}{n^2}\text{ eV} \qquad n = 1, 2, 3, \ldots \quad [30.2]$$

This result is in exact agreement with the Bohr theory. Note that the allowed energies depend only on the quantum number n.

In one-dimensional problems, only one quantum number is needed to characterize a stationary state. In the three-dimensional hydrogen atom, three quantum numbers are needed for each stationary state, corresponding to three independent degrees of freedom for the electron. The three quantum numbers that emerge from the theory are represented by the symbols n, ℓ, and m_ℓ. The quantum number n is called the **principal quantum number,** ℓ is called the **orbital quantum number,** and m_ℓ is called the **orbital magnetic quantum number,** where

n can range from 1 to ∞
ℓ can range from 0 to $n - 1$
m_ℓ can range from $-\ell$ to ℓ

Table 30.1
Three Quantum Numbers for the Hydrogen Atom

Quantum Number	Name	Allowed Values	Number of Allowed States
n	Principal quantum number	$1, 2, 3, \ldots$	Any number
ℓ	Orbital quantum number	$0, 1, 2, \ldots, n-1$	n
m_ℓ	Orbital magnetic quantum number	$-\ell, -\ell+1, \ldots, 0, \ldots, \ell-1, \ell$	$2\ell+1$

Table 30.1 summarizes the rules for determining the allowed values of ℓ and m_ℓ for a given value of n.

For historical reasons, *all states with the same principal quantum number are said to form a* **shell.** Shells are identified by the letters K, L, M, . . . , which designate the states for which $n = 1, 2, 3, \ldots$. Likewise, *all states having the same values of n and ℓ are said to form a* **subshell.** The letters *s, p, d, f, g, h,* . . . are used to designate the subshells for which $\ell = 0, 1, 2, 3, \ldots$. For example, the state designated by $3p$ has the quantum numbers $n = 3$ and $\ell = 1$; the $2s$ state has the quantum numbers $n = 2$ and $\ell = 0$. These notations are summarized in Table 30.2.

States that violate the rules given in Table 30.1 cannot exist. For instance, the $2d$ state, which would have $n = 2$ and $\ell = 2$, cannot exist because the highest allowed value of ℓ is $n - 1$, or 1 in this case. Thus, for $n = 2$, $2s$ and $2p$ are allowed states but $2d$, $2f$, . . . are not. For $n = 3$, the allowed states are $3s$, $3p$, and $3d$.

Table 30.2
Atomic Shell and Subshell Notations

n	Shell Symbol	ℓ	Subshell Symbol
1	K	0	*s*
2	L	1	*p*
3	M	2	*d*
4	N	3	*f*
5	O	4	*g*
6	P	5	*h*
. . .		. . .	

Example 30.1 The $n = 2$ Level of Hydrogen

For a hydrogen atom, determine the number of orbital states corresponding to the principal quantum number $n = 2$, and calculate the energies of these states.

Solution When $n = 2$, ℓ can be 0 or 1. For $\ell = 0$, m_ℓ can only be 0; for $\ell = 1$, m_ℓ can be -1, 0, or 1. Hence, we have a state designated as the $2s$ state associated with the quantum numbers $n = 2$, $\ell = 0$, and $m_\ell = 0$, and three orbital states designated as $2p$ states for which the quantum numbers are $n = 2$, $\ell = 1$, $m_\ell = -1$; $n = 2$, $\ell = 1$, $m_\ell = 0$; and $n = 2$, $\ell = 1$, $m_\ell = 1$.

Because all these states have the same principal quantum number, they also have the same energy, which can be calculated with Equation 30.2, taking $n = 2$:

$$E_2 = -\frac{13.6}{2^2}\text{ eV} = -3.40\text{ eV}$$

Exercise How many possible states are there for the $n = 3$ level of hydrogen? For the $n = 4$ level?

Answer 9 and 16.

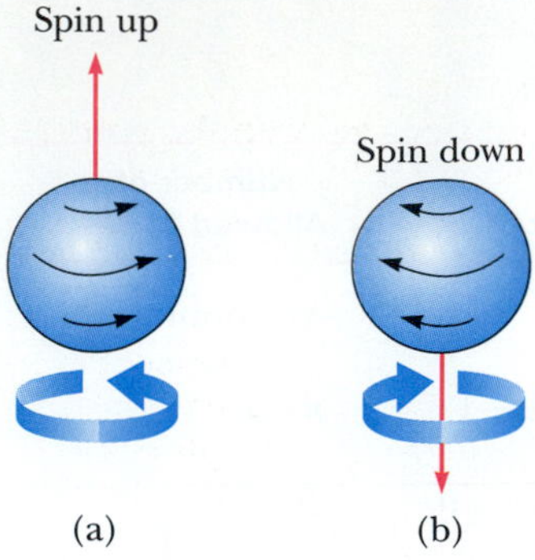

Figure 30.5
The spin of an electron can be either (a) up or (b) down.

30.3 The Spin Magnetic Quantum Number

Example 30.1 was presented to give you practice in manipulating quantum numbers, but, as we shall see in this section, there are *eight* electron states for $n = 2$ rather than the four we found. These extra states can be explained by requiring a fourth quantum number for each state; the **spin magnetic quantum number,** m_s.

The need for this new quantum number came about because of an unusual feature in the spectra of certain gases such as sodium vapor. Close examination of one of the prominent lines of sodium shows that it is, in fact, two very closely spaced lines called a doublet. The wavelengths of these lines occur in the yellow region at 589.0 nm and 589.6 nm. In 1925, when this doublet was first noticed, atomic theory could not explain it. To resolve this dilemma, Samuel Goudsmidt and George Uhlenbeck, following a suggestion by the Austrian physicist Wolfgang Pauli, proposed that a fourth quantum number, called the spin quantum number, be required to describe a quantum state.

In order to describe the spin quantum number, it is convenient (but incorrect) to think of the electron as spinning on its axis as it orbits the nucleus, just as the Earth spins on its axis as it orbits the Sun. There are only two ways in which the electron can spin as it orbits the nucleus, as shown in Figure 30.5. If the direction of spin is as shown in Figure 30.5a, the electron is said to have "spin up." If the direction of spin is reversed, as in Figure 30.5b, the electron is said to have "spin down." The energy of the electron is slightly different for the two spin directions, and this energy difference accounts for the sodium doublet. The quantum numbers associated with electron spin are $m_s = \frac{1}{2}$ for the spin-up state and $m_s = -\frac{1}{2}$ for the spin-down state. As we shall see in Example 30.2, this added quantum number doubles the number of allowed states specified by the quantum numbers n, ℓ, and m_ℓ.

This classical description of electron spin is incorrect, because quantum mechanics tells us that a rotational degree of freedom would require too many quantum numbers, and more recent theory indicates that the electron is a point particle, without spatial extent. The electron cannot be considered to be spinning as pictured in Figure 30.5. In spite of this conceptual difficulty, all experimental evidence supports the fact that the electron does have some intrinsic property that can be described by the spin quantum number. The origin of this fourth quantum number was shown by Sommerfeld and Dirac to lie in the relativistic properties of the electron, which requires four quantum numbers to describe its location in four-dimensional space-time.

Example 30.2 **Adding Some Spin on Hydrogen**

For a hydrogen atom, determine the quantum numbers associated with the possible states that correspond to the principal quantum number $n = 2$.

Solution With the addition of the spin quantum number, we have the possibilities given in the following table.

n	ℓ	m_ℓ	m_s	Subshell	Shell	Number of Electrons in Subshell
2	0	0	$\frac{1}{2}$	$2s$	L	2
2	0	0	$-\frac{1}{2}$			
2	1	1	$\frac{1}{2}$	$2p$	L	6
2	1	1	$-\frac{1}{2}$			
2	1	0	$\frac{1}{2}$			
2	1	0	$-\frac{1}{2}$			
2	1	-1	$\frac{1}{2}$			
2	1	-1	$-\frac{1}{2}$			

Exercise Show that for $n = 3$, there are 18 possible states. (This follows from the restrictions that the maximum number of electrons in the $3s$ state is 2, the maximum number in the $3p$ state is 6, and the maximum number in the $3d$ state is 10.)

30.4 The Wave Functions for Hydrogen

Neglecting electron spin for the present, the potential energy of the hydrogen atom depends only on the radial distance r. We therefore expect that some of the allowed states for this atom can be represented by wave functions that depend only on r. This indeed is the case. The simplest wave function for hydrogen is the one that describes the $1s$ state and is designated $\psi_{1s}(r)$:

$$\psi_{1s}(r) = \frac{1}{\sqrt{\pi a_0^3}} e^{-r/a_0} \quad [30.3]$$

Wave function for hydrogen in its ground state

where a_0 is the Bohr radius. This wave function satisfies the condition that it approach zero as r approaches ∞ and is normalized as presented. Furthermore, since ψ_{1s} depends only on r, it is *spherically symmetric*. This, in fact, is true for all s states.

Recall that the probability density of finding the electron in any region is equal to an integral of $|\psi|^2$ over the region, if ψ is normalized. The probability density for the $1s$ state is

$$|\psi_{1s}|^2 = \left(\frac{1}{\pi a_0^3}\right) e^{-2r/a_0} \quad [30.4]$$

Furthermore, the probability of finding the electron in a volume element, dV, is $|\psi|^2\, dV$. It is convenient to define the radial probability density function, $P(r)$, as the probability of finding the electron in a spherical shell of radius r and thickness

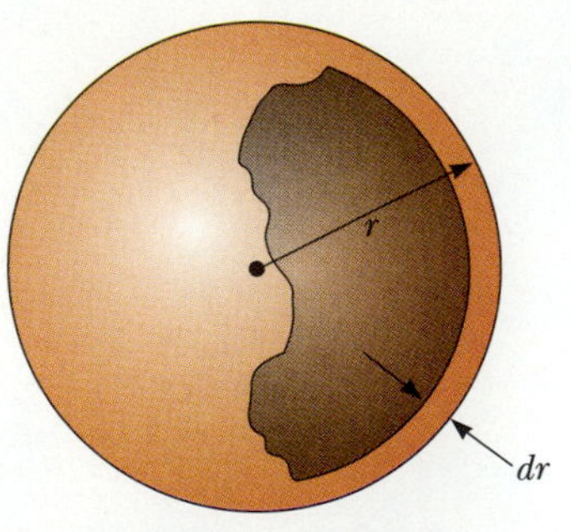

Figure 30.6
A spherical shell of radius r and thickness dr has a volume equal to $4\pi r^2\ dr$.

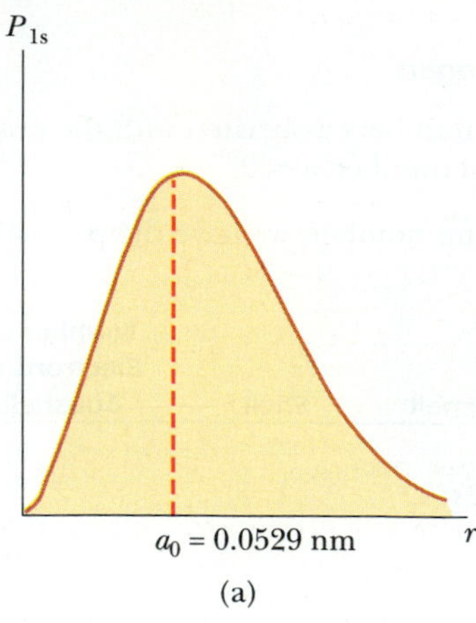

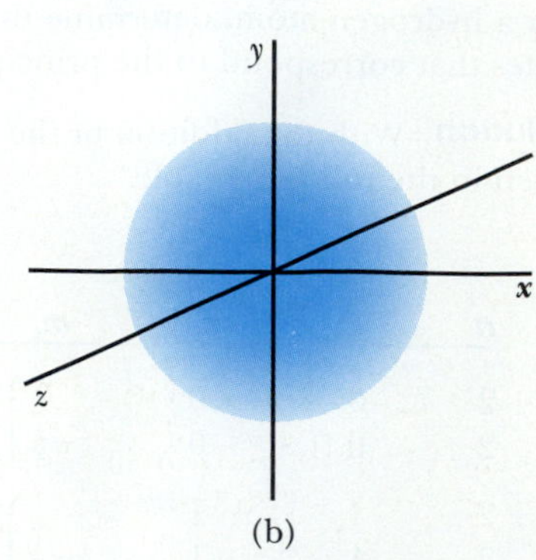

Figure 30.7
(a) The probability of finding the electron as a function of distance from the nucleus for the hydrogen atom in the 1s (ground) state. Note that the probability has its maximum value when r equals the first Bohr radius, a_0. (b) The spherical electron cloud for the hydrogen atom in its 1s state.

dr. The volume of such a shell equals its surface area, $4\pi r^2$, multiplied by the shell thickness, dr (Fig. 30.6), so that we get

$$P(r)\ dr = |\psi|^2\ dV = |\psi|^2 4\pi r^2\ dr \qquad [30.5]$$

$$P(r) = 4\pi r^2 |\psi|^2 \qquad [30.6]$$

Substituting Equation 30.4 into Equation 30.6 gives the radial probability density function for the hydrogen atom in its ground state:

Radial probability density for the 1s state of hydrogen

$$P_{1s}(r) = \left(\frac{4r^2}{{a_0}^3}\right) e^{-2r/a_0} \qquad [30.7]$$

A plot of the function $P_{1s}(r)$ versus r is presented in Figure 30.7a. The peak of the curve corresponds to the most probable value of r for this particular state. The spherical symmetry of the distribution function is shown in Figure 30.7b.

Example 30.3 The Ground State of Hydrogen

Calculate the most probable value of r for an electron in the ground state of the hydrogen atom.

Solution The most probable value of r corresponds to the peak of the plot of $P(r)$ versus r. The slope of the curve at this point is zero, and so we can evaluate the most probable value of r by setting $dP/dr = 0$ and solving for r. Using Equation 30.7, we get

$$\frac{dP}{dr} = \frac{d}{dr}\left[\left(\frac{4r^2}{{a_0}^3}\right) e^{-2r/a_0}\right] = 0$$

Carrying out the derivative operation and simplifying the expression, we get

$$e^{-2r/a_0}\frac{d}{dr}(r^2) + r^2\frac{d}{dr}(e^{-2r/a_0}) = 0$$

$$2re^{-2r/a_0} + r^2(-2/a_0)e^{-2r/a_0} = 0$$

$$2r[1 - (r/a_0)]e^{-2r/a_0} = 0$$

This expression is satisfied if

$$1 - \frac{r}{a_0} = 0$$

$$r = a_0$$

Example 30.4 Probabilities for the Electron in Hydrogen

Calculate the probability that the electron in the ground state of hydrogen will be found outside the Bohr radius.

Solution The probability is found by integrating the radial probability density for this state, $P_{1s}(r)$, from the Bohr radius, a_0, to ∞. Using Equation 30.7,

$$P = \int_{a_0}^{\infty} P_{1s}(r)\ dr = \frac{4}{{a_0}^3}\int_{a_0}^{\infty} r^2 e^{-2r/a_0}\ dr$$

We can put the integral in dimensionless form by changing variables from r to $z = 2r/a_0$. Noting that $z = 2$ when $r = a_0$, and that $dr = (a_0/2)\ dz$, we get

$$P = \tfrac{1}{2}\int_{2}^{\infty} z^2 e^{-z}\ dz = -\tfrac{1}{2}(z^2 + 2z + 2)e^{-z}\Big|_{2}^{\infty}$$

$$P = 5e^{-2} = 0.677 \qquad \text{or} \qquad 67.7\%$$

Example 30.3 shows that, for the ground state of hydrogen, the most probable value of r equals the Bohr radius, a_0. It turns out that the average value of r for the ground state of hydrogen is $\frac{3}{2}a_0$, which is 50% larger than the most probable value of r (see Problem 38). The reason for this is the large asymmetry in the radial distribution function shown in Figure 30.7a. According to quantum mechanics, there is no sharply defined boundary to the atom. One could view the probability distribution for the electron as an effective "electron cloud."

The next-simplest wave function for the hydrogen atom is the one corresponding to the $2s$ state ($n = 2$, $\ell = 0$). The normalized wave function for this state is given by

$$\psi_{2s}(r) = \frac{1}{4\sqrt{2\pi}}\left(\frac{1}{a_0}\right)^{3/2}\left[2 - \frac{r}{a_0}\right]e^{-r/2a_0} \qquad [30.8]$$

Wave function for hydrogen in the $2s$ state

Like the ψ_{1s} function, ψ_{2s} depends only on r and is spherically symmetric. The energy corresponding to this state is $E_2 = -(13.6/4)$ eV $= -3.4$ eV. This represents the first excited state of hydrogen. Plots of the radial distribution function for this state and several other states of hydrogen are shown in Figure 30.8. The plot for the $2s$ state, which applies to a single electron, has two peaks. In this case, the most probable value corresponds to that value of r with the higher value of $P(\approx 5a_0)$. An electron in the $2s$ state would be much farther from the nucleus (on the average) than an electron in the $1s$ state. The average value of r is even greater for the $3d$, $3p$, and $4d$ states.

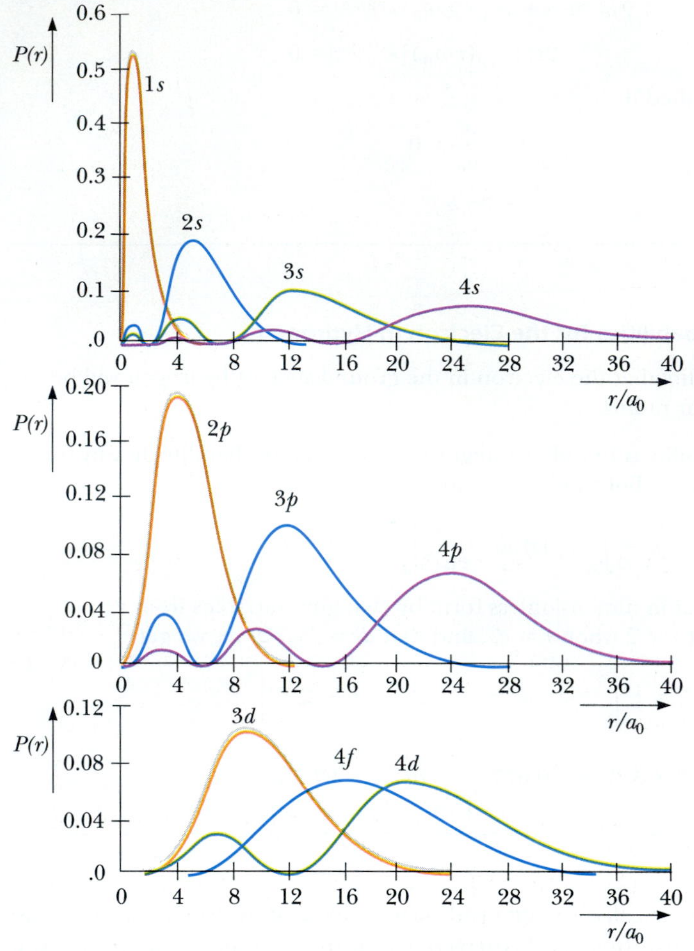

Figure 30.8 The radial probability density function versus r/a_0 for several states of the hydrogen atom. *(From E. U. Condon and G. H. Shortley, The Theory of Atomic Spectra, Cambridge, Cambridge University Press, 1953; used with permission)*

As we have mentioned, all *s* states have spherically symmetric wave functions. The other states are not spherically symmetric. For example, the three wave functions corresponding to the states for which $n = 2$, $\ell = 1$ ($m_\ell = 1, 0,$ or -1) can be expressed as appropriate linear combinations of the three *p* states. Although quantum mechanics limits our knowledge of angular momentum to the projection along any one axis at a time, these *p* states may be described mathematically as linear combinations of mutually perpendicular functions p_x, p_y, and p_z, as represented in Figure 30.9, where *only* the angular dependence of these functions is shown. Note that the three clouds have identical structure but differ in orientation with respect to the *x*, *y*, and *z* axes. The nonspherical wave functions for these states are

Wave functions for the 2*p* state

$$\psi_{2p_x} = xF(r)$$
$$\psi_{2p_y} = yF(r) \qquad [30.9]$$
$$\psi_{2p_z} = zF(r)$$

where $F(r)$ is some exponential function of *r*. Wave functions with a highly directional character, such as these, are convenient for bookkeeping descriptions of chemical bonding, the formation of molecules, and chemical properties.

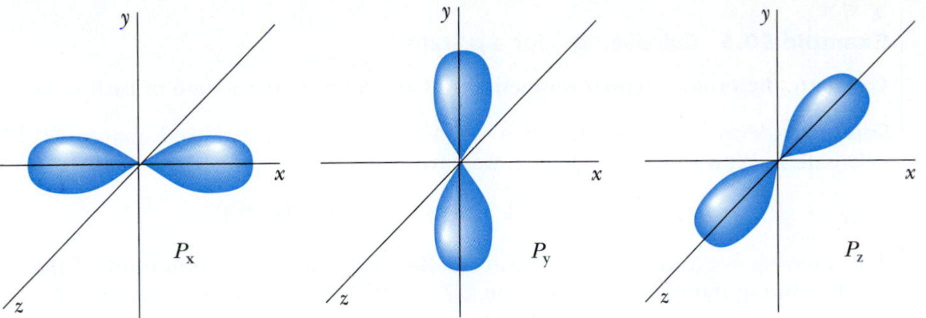

Figure 30.9 Angular dependence of the electron charge distribution for an electron in a p state. The three charge distributions p_x, p_y, and p_z have the same structure and differ only in their orientation in space.

30.5 The "Other" Quantum Numbers

The energy of a particular state depends primarily on the principal quantum number. Now let us see what the other three quantum numbers contribute to our atomic model.

The Orbital Quantum Number

If a particle moves in a circle of radius r, the magnitude of its angular momentum relative to the center of the circle is $L = mvr$. The direction of **L** is perpendicular to the plane of the circle, and the sense of **L** is given by a right-hand rule.[1] According to classical physics, L can have any value. However, the Bohr model of hydrogen postulates that the angular momentum is restricted to multiples of $\hbar$; that is, $mvr = n\hbar$. This model must be modified because it predicts (incorrectly) that the ground state of hydrogen ($n = 1$) has one unit of angular momentum. Furthermore, if L is taken to be zero in the Bohr model, one is forced to accept the electron as a *particle* oscillating along a straight line through the nucleus. This is a physically unacceptable situation.

These difficulties are resolved with the quantum mechanical model of the atom. According to quantum mechanics, an atom in a state whose principal quantum number is n can take on the following *discrete* values of orbital angular momentum:

$$L = \sqrt{\ell(\ell + 1)}\,\hbar \qquad \ell = 0, 1, 2, \ldots, n - 1 \qquad [30.10]$$

Allowed values of L

Because ℓ is restricted to these values, $L = 0$ (corresponding to $\ell = 0$) is an acceptable value of the angular momentum. The fact that L can be zero in this model points out the difficulties inherent in any attempt to describe results based on quantum mechanics in terms of a purely particle-like model. In the quantum mechanical interpretation, the electron cloud for the $L = 0$ state is spherically symmetric and has no fundamental axis of revolution.

[1] See Sections 11.7 and 11.8 for details on angular momentum if you have forgotten this material.

Example 30.5 Calculating *L* for a *p* State

Calculate the orbital angular momentum of an electron in a *p* state of hydrogen.

Solution Because we know that $\hbar = 1.054 \times 10^{-34}$ J·s, we can use Equation 30.10 to calculate L. With $\ell = 1$ for a p state, we have

$$L = \sqrt{1(1+1)}\,\hbar = \sqrt{2}\,\hbar = 1.49 \times 10^{-34}\ \text{J}\cdot\text{s}$$

This number is extremely small relative to the orbital angular momentum of the Earth orbiting the Sun, which is about 2.7×10^{40} J·s. The quantum number that describes L for macroscopic objects, such as the Earth, is so large that the separation between adjacent states cannot be measured. Once again, the correspondence principle is upheld.

The Magnetic Orbital Quantum Number

Since angular momentum is a vector, its direction must also be specified. Recall from Chapter 19 that an orbiting electron can be considered an effective current loop with a corresponding magnetic moment. Such a moment placed in a magnetic field, **B**, will interact with the field. Suppose a weak magnetic field is applied along the z axis so that it defines a direction in space. According to quantum mechanics, L^2 and L_z, the projection of **L** along the z axis, can have discrete values. The magnetic orbital quantum number m_ℓ specifies the allowed values of L_z according to the expression

Allowed values of L_z

$$L_z = m_\ell \hbar \qquad [30.11]$$

The fact that the direction of **L** is quantized with respect to an external magnetic field is often referred to as **space quantization.**

Let us look at the possible orientations of **L** for a given value of ℓ. Recall that m_ℓ can have values ranging from $-\ell$ to ℓ. If $\ell = 0$, then $m_\ell = 0$ and $L_z = 0$. In order for L_z to be zero, **L** must be perpendicular to **B**. If $\ell = 1$, then the possible values of m_ℓ are -1, 0, and 1, so that L_z may be $-\hbar$, 0, or $\hbar$. If $\ell = 2$, m_ℓ can be -2, -1, 0, 1, or 2, corresponding to L_z values of $-2\hbar$, $-\hbar$, 0, $\hbar$, or $2\hbar$, and so on.

Space quantization

A vector model describing space quantization for $\ell = 2$ is shown in Figure 30.10a. Note that **L** can never be aligned parallel or antiparallel to **B** since L_z must

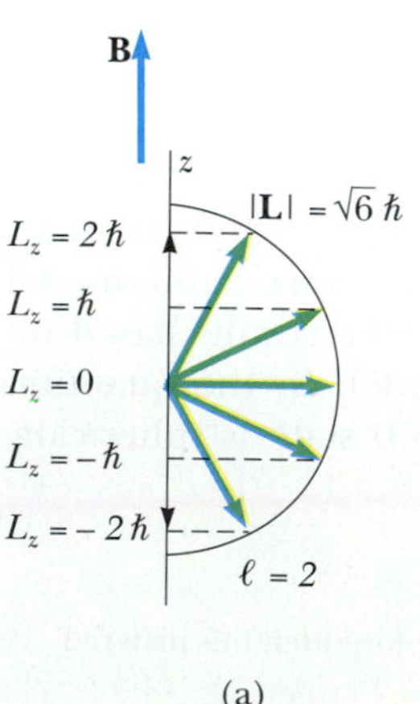

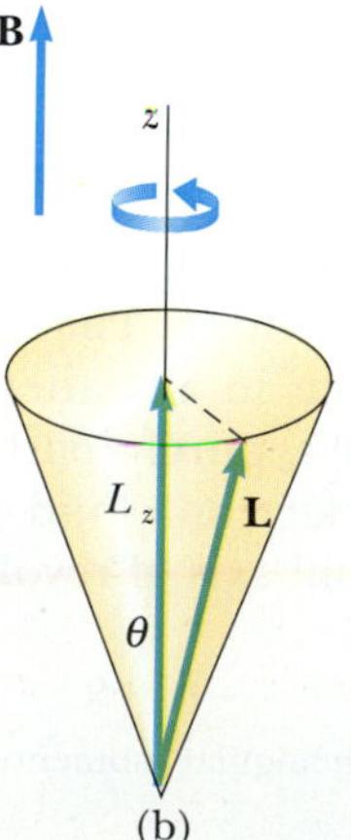

Figure 30.10 (a) The allowed projections of the orbital angular momentum for the case $\ell = 2$. (b) The orbital angular momentum vector lies on the surface of a cone and precesses about the z axis when a magnetic field **B** is applied in this direction.

be smaller than the total angular momentum, L. From a three-dimensional viewpoint, $\mathbf{L}$ must lie on the surface of a cone that makes an angle of θ with the z axis, as shown in Figure 30.10b. From the figure, we see that θ is also quantized and that its values are specified through the relation

$$\cos\theta = \frac{L_z}{|\mathbf{L}|} = \frac{m_\ell}{\sqrt{\ell(\ell+1)}} \qquad [30.12]$$

θ is quantized

Note that m_ℓ is never greater than ℓ, and therefore θ can never be zero. (Classically, θ can have any value.)

Because of the uncertainty principle, $\mathbf{L}$ does not point in a specific direction but rather traces out a cone in space. If $\mathbf{L}$ had a definite value, then all three components L_x, L_y, and L_z would be exactly specified. For the moment, let us assume that this is the case, and let us suppose that the electron moves in the xy plane, so that $\mathbf{L}$ is in the z direction and $p_z = 0$. This means that p_z is precisely known, which is in violation of the uncertainty principle, $\Delta p_z \Delta z \gtrsim \hbar/2$. In reality, only the magnitude of $\mathbf{L}$ and one component (say, L_z) can have definite values. In other words, quantum mechanics allows us to specify L and L_z but not L_x and L_y. Since the direction of $\mathbf{L}$ is constantly changing as it precesses about the z axis, the average values of L_x and L_y are zero and L_z maintains a fixed value of $m_\ell\hbar$.

Example 30.6 Space Quantization for Hydrogen

For the hydrogen atom in the $\ell = 3$ state, calculate the magnitude of $\mathbf{L}$ and the allowed values of L_z and θ.

Solution We use Equation 30.10 with $\ell = 3$:

$$L = \sqrt{\ell(\ell+1)}\hbar = \sqrt{3(3+1)}\hbar = 2\sqrt{3}\hbar$$

The allowed values of L_z are $L_z = m_\ell\hbar$ with $m_\ell = -3, -2, -1, 0, 1, 2$, and 3:

$$L_z = -3\hbar, -2\hbar, -\hbar, 0, \hbar, 2\hbar, 3\hbar$$

Finally, we use Equation 30.12 to calculate the allowed values of θ. Since $\ell = 3$, $\sqrt{\ell(\ell+1)} = 2\sqrt{3}$, and we have

$$\cos\theta = \frac{m_\ell}{2\sqrt{3}}$$

Substitution of the allowed values of m_ℓ gives

$$\cos\theta = \pm 0.866, \pm 0.577, \pm 0.289, 0$$

$$\theta = 30.0°, 54.8°, 73.2°, 90.0°, 107°, 125°, 150°$$

Electron Spin

In 1921 Stern and Gerlach performed an experiment that demonstrated space quantization. However, their results were not in quantitative agreement with the theory that existed at that time. In their experiment, a beam of neutral silver atoms sent through a nonuniform magnetic field was split into two components (Fig. 30.11). The experiment was repeated using other atoms, and in each case the beam split into two or more components. The classical argument is as follows: If the z direction is chosen to be the direction of the maximum inhomogeneity of $\mathbf{B}$,

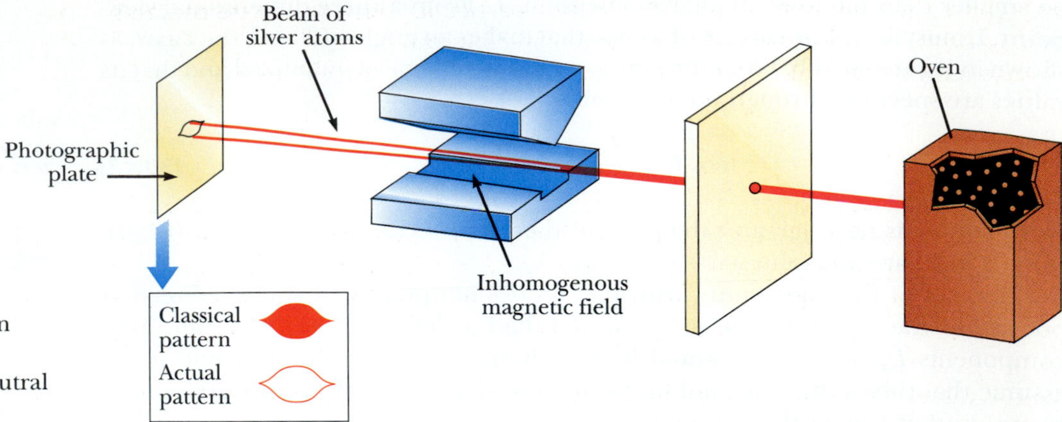

Figure 30.11 The apparatus used by Stern and Gerlach to verify space quantization. A beam of neutral silver atoms is split into two components by a nonuniform magnetic field, as shown by the actual pattern in the box.

the net magnetic force on the atom is along the z axis and is proportional to the magnetic moment in the direction of $\boldsymbol{\mu}_z$. Classically, $\boldsymbol{\mu}_z$ can have any orientation, and so the deflected beam should be spread out continuously. According to quantum mechanics, however, the deflected beam has several components, and the number of components determines the possible values of $\boldsymbol{\mu}_z$. Hence, because the Stern–Gerlach experiment showed split beams, space quantization was at least qualitatively verified.

For the moment, let us assume that $\boldsymbol{\mu}_z$ is due to the orbital angular momentum. Since μ_z is proportional to m_ℓ, the number of possible values of μ_z is $2\ell + 1$. Furthermore, since ℓ is an integer, the number of values of μ_z is always odd. This prediction is clearly not consistent with the observations of Stern and Gerlach, who observed only two components in the deflected beam of silver atoms. Hence, one is forced to conclude that either quantum mechanics is incorrect or the model is in need of refinement.

In 1927 Phipps and Taylor repeated the Stern–Gerlach experiment using a beam of hydrogen atoms. This experiment is important because it deals with an atom with a single electron in its ground state, for which the theory makes reliable predictions. Recall that $\ell = 0$ for hydrogen in its ground state, and so $m_\ell = 0$. Hence, one would not expect the beam to be deflected by the field, since μ_z would be zero. However, the beam in the Phipps–Taylor experiment is again split into two components. On the basis of this result, one can conclude only one thing: there is some contribution to the magnetic moment other than the orbital motion.

In 1925 Goudsmit and Uhlenbeck proposed that the electron has an intrinsic angular momentum apart from its orbital angular momentum. From a classical viewpoint, this intrinsic angular momentum is attributed to the charged electron spinning about its own axis, and hence is called *electron spin*.[2] In other words, the

[2] Physicists often use the word *spin* when referring to *spin angular momentum*. For example, it is common to use the statement "The electron has a spin of one half." The spin angular momentum of the electron *never changes*. This notion contradicts classical laws, which would hold that a rotating charge slows down in the presence of an applied magnetic field because of the Faraday emf that accompanies the changing field. Furthermore, if the electron is viewed as a spinning ball of charge subject to classical laws, parts of it near its surface would be rotating with velocities exceeding the speed of light. Thus, the classical picture must not be pressed too far; ultimately, the spinning electron is a quantum entity defying any simple classical description. It is more accurate to describe spin as a relativistic effect (which was treated classically by Sommerfeld and quantum mechanically by Dirac).

total angular momentum of the electron in a particular electronic state contains both an orbital contribution, **L**, and a spin contribution, **S**.

The magnitude of the **spin angular momentum, S**, for the electron is

Spin angular momentum of an electron

$$S = \sqrt{s(s+1)}\hbar = \frac{\sqrt{3}}{2}\hbar \qquad [30.13]$$

Like orbital angular momentum, spin angular momentum is quantized in space, as described in Figure 30.12. It can have two orientations, specified by the spin magnetic quantum number m_s, where m_s has two possible values, $\pm\frac{1}{2}$. The z component of spin angular momentum is

$$S_z = m_s\hbar = \pm\tfrac{1}{2}\hbar \qquad [30.14]$$

The two values $\pm\hbar/2$ for S_z correspond to the two possible orientations for **S** shown in Figure 30.12.

The spin magnetic moment of the electron, $\boldsymbol{\mu}_s$, is related to its spin angular momentum, **S**, by the expression

$$\boldsymbol{\mu}_s = -\frac{e}{m}\mathbf{S} \qquad [30.15]$$

Since $S_z = \pm\frac{1}{2}\hbar$, the z component of the spin magnetic moment can have the values

$$\mu_{sz} = \pm\frac{e\hbar}{2m} \qquad [30.16]$$

The quantity $e\hbar/2m$ is the **Bohr magneton**, μ_B, and has the numerical value 9.27×10^{-24} J/T (see Eq. 19.29). Note that the spin contribution to the angular momentum is *twice* the contribution of the orbital motion.

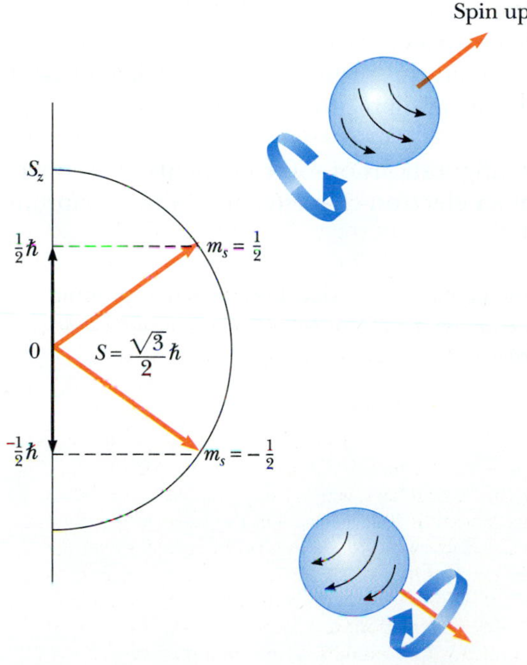

Figure 30.12
The spin angular momentum also exhibits space quantization. This figure shows the two allowed orientations of the spin vector S for a spin ½ particle, such as the electron.

Today physicists explain the outcome of the Stern–Gerlach experiment as follows. The observed moments for both silver and hydrogen are due to spin angular momentum and not to orbital angular momentum. A single-electron atom such as hydrogen has its electron quantized in the magnetic field in such a way that its z component of spin angular momentum is either $\frac{1}{2}\hbar$ or $-\frac{1}{2}\hbar$, corresponding to $m_s = \pm\frac{1}{2}$.

Electrons with spin $+\frac{1}{2}$ are deflected in one direction, and those with spin $-\frac{1}{2}$ are deflected in the opposite direction. The Stern–Gerlach experiment provided two important results. First, it verified the concept of space quantization. Second, it showed that spin angular momentum existed even though this property was not recognized until long after the experiments were performed.

30.6 The Exclusion Principle and the Periodic Table

Because any electron in any atom is specified by four quantum numbers, n, ℓ, m_ℓ, and m_s, an obvious and important question is "How many electrons can have a particular set of quantum numbers?" Pauli provided an answer in 1925 in a powerful statement known as the **exclusion principle:**

Exclusion principle

No two electrons in an atom can ever be in the same quantum state; that is, no two electrons in the same atom can have the same set of quantum numbers.

It is interesting to note that if this principle were not valid, every electron would end up in the lowest energy state of the atom and the chemical behavior of the elements would be grossly modified. Nature as we know it would not exist! In reality, we can view the electronic structure of complex atoms as a succession of filled levels increasing in energy, where the outermost electrons are primarily responsible for the chemical properties of the element.

As a general rule, the order of filling of an atom's subshells with electrons is as follows. Once one subshell is filled, the next electron goes into the vacant subshell that is lowest in energy. One can understand this principle by recognizing that if the atom were not in the lowest energy state available to it, it would radiate energy until it reached this state.

Before we discuss the electronic configurations of some elements, it is convenient to define an *orbital* as the state of an electron characterized by the quantum

Wolfgang Pauli and Niels Bohr watch a spinning top. *(Courtesy of AIP Niels Bohr Library, Margarethe Bohr Collection)*

Table 30.3
Allowed Quantum Numbers for an Atom up to n = 3

n	1	2				3								
ℓ	0	0	1			0	1			2				
m_ℓ	0	0	1	0	−1	0	1	0	−1	2	1	0	−1	−2
m_s	↑↓	↑↓	↑↓	↑↓	↑↓	↑↓	↑↓	↑↓	↑↓	↑↓	↑↓	↑↓	↑↓	↑↓

Wolfgang Pauli (1900–1958)
An extremely talented Austrian theoretical physicist who made important contributions in many areas of modern physics, Pauli gained public recognition at the age of 21 with a masterful review article on relativity, which is still considered one of the finest and most comprehensive introductions to the subject. Other major contributions were the discovery of the exclusion principle, the explanation of the connection between particle spin and statistics, and theories of relativistic quantum electrodynamics, the neutrino hypothesis, and the hypothesis of nuclear spin. *(Photo taken by S.A. Goudsmit, AIP Niels Bohr Library)*

numbers n, ℓ, and m_ℓ. From the exclusion principle, we see that *there can be only two electrons in any orbital.* One of these electrons has $m_s = +\frac{1}{2}$, and the other has $m_s = -\frac{1}{2}$. Because each orbital is limited to two electrons, the numbers of electrons that can occupy the levels are also limited.

Table 30.3 shows the numbers of allowed quantum states for an atom up to $n = 3$. Each square in the bottom row of the table represents one orbital, with the ↑ arrows representing $m_s = \frac{1}{2}$ and the ↓ arrows representing $m_s = -\frac{1}{2}$. The $n = 1$ shell can accommodate only two electrons, since only one orbital is allowed with $m_\ell = 0$. The $n = 2$ shell has two subshells, with $\ell = 0$ and $\ell = 1$. The $\ell = 0$ subshell is limited to only two electrons since $m_\ell = 0$. The $\ell = 1$ subshell has three allowed orbitals, corresponding to $m_\ell = 1$, 0, and −1. Since each orbital can accommodate two electrons, the $\ell = 1$ subshell can hold six electrons (and the $n = 2$ shell can hold eight). The $n = 3$ shell has three subshells and nine orbitals and can accommodate up to 18 electrons. Each shell can accommodate up to $2n^2$ electrons.

The exclusion principle can be illustrated by an examination of the electronic arrangement in a few of the lighter atoms.

Hydrogen has only one electron, which, in its ground state, can be described by either of two sets of quantum numbers: 1, 0, 0, $\frac{1}{2}$ or 1, 0, 0, $-\frac{1}{2}$. The electronic configuration of this atom is often designated as $1s^1$. The notation $1s$ refers to a state for which $n = 1$ and $\ell = 0$, and the superscript indicates that one electron is present in the s subshell.

Neutral *helium* has two electrons. In the ground state, the quantum numbers for these two electrons are 1, 0, 0, $\frac{1}{2}$ and 1, 0, 0, $-\frac{1}{2}$. There are no other possible combinations of quantum numbers for this level, and we say that the K shell is filled. Helium is designated as $1s^2$.

The electronic configurations of some successive elements are given in Figure 30.13. Neutral *lithium* has three electrons. In the ground state, two of these are in the $1s$ subshell and the third is in the $2s$ subshell, because this subshell is lower in energy than the $2p$ subshell. Hence, the electronic configuration for lithium is $1s^2\,2s^1$.

Note that the electronic configuration of *beryllium,* with its four electrons, is $1s^22s^2$, and *boron* has a configuration of $1s^22s^22p^1$. The $2p$ electron in boron may be described by one of six sets of quantum numbers, corresponding to six states of equal energy.

Carbon has six electrons, and a question arises concerning how to assign the two $2p$ electrons. Do they go into the same orbital with paired spins (↑ ↓), or do they occupy different orbitals with unpaired spins (↑ ↑)? Experimental data show that the most stable configuration (that is, the one that is energetically preferred)

Table 30.4
Electronic Configuration of the Elements

Z	Symbol	Ground Configuration	Ionization Energy (eV)	Z	Symbol	Ground Configuration	Ionization Energy (eV)
1	H	$1s^1$	13.595	25	Mn	$3d^54s^2$	7.432
2	He	$1s^2$	24.581	26	Fe	$3d^64s^2$	7.87
				27	Co	$3d^74s^2$	7.86
3	Li	[He] $2s^1$	5.39	28	Ni	$3d^84s^2$	7.633
4	Be	$2s^2$	9.320	29	Cu	$3d^{10}4s^1$	7.724
5	B	$2s^22p^1$	8.296	30	Zn	$3d^{10}4s^2$	9.391
6	C	$2s^22p^2$	11.256	31	Ga	$3d^{10}4s^24p^1$	6.00
7	N	$2s^22p^3$	14.545	32	Ge	$3d^{10}4s^24p^2$	7.88
8	O	$2s^22p^4$	13.614	33	As	$3d^{10}4s^24p^3$	9.81
9	F	$2s^22p^5$	17.418	34	Se	$3d^{10}4s^24p^4$	9.75
10	Ne	$2s^22p^6$	21.559	35	Br	$3d^{10}4s^24p^5$	11.84
				36	Kr	$3d^{10}4s^24p^6$	13.996
11	Na	[Ne] $3s^1$	5.138				
12	Mg	$3s^2$	7.644	37	Rb	[Kr] $5s^1$	4.176
13	Al	$3s^23p^1$	5.984	38	Sr	$5s^2$	5.692
14	Si	$3s^23p^2$	8.149	39	Y	$4d^15s^2$	6.377
15	P	$3s^23p^3$	10.484	40	Zr	$4d^25s^2$	
16	S	$3s^23p^4$	10.357	41	Nb	$4d^45s^1$	6.881
17	Cl	$3s^23p^5$	13.01	42	Mo	$4d^55s^1$	7.10
18	Ar	$3s^23p^6$	15.755	43	Tc	$4d^55s^2$	7.228
				44	Ru	$4d^75s^1$	7.365
19	K	[Ar] $4s^1$	4.339	45	Rh	$4d^85s^1$	7.461
20	Ca	$4s^2$	6.111	46	Pd	$4d^{10}$	8.33
21	Sc	$3d^14s^2$	6.54	47	Ag	$4d^{10}5s^1$	7.574
22	Ti	$3d^24s^2$	6.83	48	Cd	$4d^{10}5s^2$	8.991
23	V	$3d^34s^2$	6.74	49	In	$5p^1$	
24	Cr	$3d^54s^1$	6.76	50	Sn	$4d^{10}5s^25p^2$	7.342

Note: The bracket notation is used as a shorthand method to avoid repetition in indicating inner-shell electrons. Thus, [He] represents $1s^2$, [Ne] represents $1s^22s^22p^6$, [Ar] represents $1s^22s^22p^63s^23p^6$, and so on.

is the latter, where the spins are unpaired. Hence, the two $2p$ electrons in carbon and the three $2p$ electrons in nitrogen have unpaired spins (Fig. 30.13). The general rule that governs such situations, called **Hund's rule,** states that when an atom has orbitals of equal energy, the order in which they are filled by electrons is such that a maximum number of electrons will have unpaired spins.

Hund's rule

A complete list of electronic configurations is provided in Table 30.4. An early attempt to find some order among the elements was made by a Russian chemist, Dmitri Mendeleev, in 1871. He arranged the atoms in a table (similar to that in Appendix C) according to their atomic weights and chemical similarities. The first table Mendeleev proposed contained many blank spaces, and he boldly stated that the gaps were there only because the elements had not yet been discovered. By noting the columns in which these missing elements should be located, he was able to make rough predictions about their chemical properties. Within 20 years of Mendeleev's announcement, the missing elements were indeed discovered.

Table 30.4
Electronic Configuration of the Elements *(Continued)*

Z	Symbol	Ground Configuration	Ionization Energy (eV)	Z	Symbol	Ground Configuration	Ionization Energy (eV)
51	Sb	$4d^{10}5s^25p^3$	8.639	78	Pt	$4f^{14}5d^86s^2$	8.88
52	Te	$4d^{10}5s^25p^4$	9.01	79	Au	[Xe] $4f^{14}5d^{10}6s^1$	9.22
53	I	$4d^{10}5s^25p^5$	10.454	80	Hg	$6s^2$	10.434
54	Xe	$4d^{10}5s^25p^6$	12.127	81	Tl	$6s^26p^1$	6.106
				82	Pb	$6s^26p^2$	7.415
55	Cs	[Xe] $6s^1$	3.893	83	Bi	$6s^26p^3$	7.287
56	Ba	$6s^2$	5.210	84	Po	$6s^26p^4$	8.43
57	La	$5d^16s^2$	5.61	85	At	$6s^26p^5$	
58	Ce	$4f^15d^16s^2$	6.54	86	Rn	$6s^26p^6$	10.745
59	Pr	$4f^36s^2$	5.48				
60	Nd	$4f^46s^2$	5.51	87	Fr	[Rn] $7s^1$	
61	Pm	$4f^56s^2$		88	Ra	$7s^2$	5.277
62	Fm	$4f^66s^2$	5.6	89	Ac	$6d^17s^2$	6.9
63	Eu	$4f^76s^2$	5.67	90	Th	$6d^27s^2$	
64	Gd	$4f^75d^16s^2$	6.16	91	Pa	$5f^26d^17s^2$	
65	Tb	$4f^96s^2$	6.74	92	U	$5f^36d^17s^2$	4.0
66	Dy	$4f^{10}6s^2$		93	Np	$5f^46d^17s^2$	
67	Ho	$4f^{11}6s^2$		94	Pu	$5f^67s^2$	
68	Er	$4f^{12}6s^2$		95	Am	$5f^77s^2$	
69	Tm	$4f^{13}6s^2$		96	Cm	$5f^76d^17s^2$	
70	Yb	$4f^{14}6s^2$	6.22	97	Bk	$5f^86d^17s^2$	
71	Lu	$4f^{14}5d^16s^2$	6.15	98	Cf	$5f^{10}7s^2$	
72	Hf	$4f^{14}5d^26s^2$	7.0	99	Es	$5f^{11}7s^2$	
73	Ta	$4f^{14}5d^36s^2$	7.88	100	Fm	$5f^{12}7s^2$	
74	W	$4f^{14}5d^46s^2$	7.98	101	Mv	$5f^{13}7s^2$	
75	Re	$4f^{14}5d^56s^2$	7.87	102	No	$5f^{14}7s^2$	
76	Os	$4f^{14}5d^66s^2$	8.7	103	Lw	$5f^{14}6d^17s^2$	
77	Ir	$4f^{14}5d^76s^2$	9.2	104	Ku	$5f^{14}6d^27s^2$	

The elements in the periodic table are arranged so that all those in a vertical column have similar chemical properties. For example, consider the elements in the last column: He (helium), Ne (neon), Ar (argon), Kr (krypton), Xe (xenon), and Rn (radon). The outstanding characteristic of these elements is that they do not normally take part in chemical reactions—that is, they do not join with other atoms to form molecules—and they are therefore classified as being inert. Because of this aloofness, they are referred to as the noble gases. We can partially understand this behavior by looking at the electronic configurations in Table 30.4. This table also lists the ionization energies for certain elements. The element helium is one in which the electronic configuration is $1s^2$—in other words, one shell is filled. Additionally, it is found that the electrons in this filled shell are considerably separated in energy from the next available level, the $2s$ level.

The electronic configuration for neon is $1s^22s^22p^6$. Again, the outermost shell is filled and there is a gap in energy between the $2p$ level and the $3s$ level. Argon has

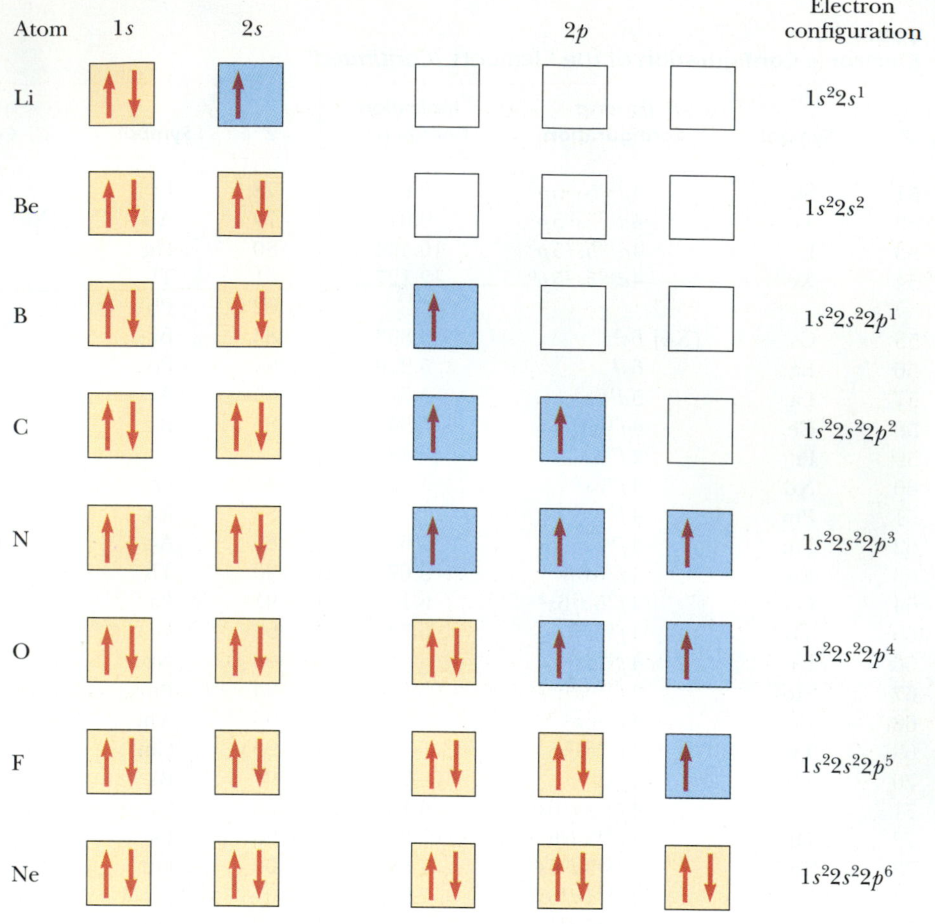

Atom
1s
2s
2p
Electron configuration
Li
$1s^22s^1$
Be
$1s^22s^2$
B
$1s^22s^22p^1$
C
$1s^22s^22p^2$
N
$1s^22s^22p^3$
O
$1s^22s^22p^4$
F
$1s^22s^22p^5$
Ne
$1s^22s^22p^6$

Figure 30.13
The filling of electronic states must obey the Pauli exclusion principle and Hund's rule.

the configuration $1s^22s^22p^63s^23p^6$. Here, the $3p$ subshell is filled and there is a gap in energy between the $3p$ subshell and the $3d$ subshell. We could continue this procedure through all the noble gases; the pattern remains the same. A noble gas is formed when either a shell or a subshell is filled and there is a gap in energy before the next possible level is encountered.

30.7 Atomic Spectra: Visible and X-Ray

In Chapter 12 we briefly discussed the origin of the spectral lines for hydrogen and hydrogen-like ions. Recall that an atom will emit electromagnetic radiation if an electron in an excited state makes a transition to a lower energy state.

The energy level diagram for hydrogen is shown in Figure 30.14. The diagonal lines represent allowed transitions between stationary states. Whenever an electron makes a transition from a higher energy state to a lower one, a photon of light is emitted. The frequency of this photon is $f = \Delta E/h$, where ΔE is the energy difference between the two levels and h is Planck's constant. The allowed transitions are those for which ℓ changes by 1. That is, the **selection rule** *for the allowed*

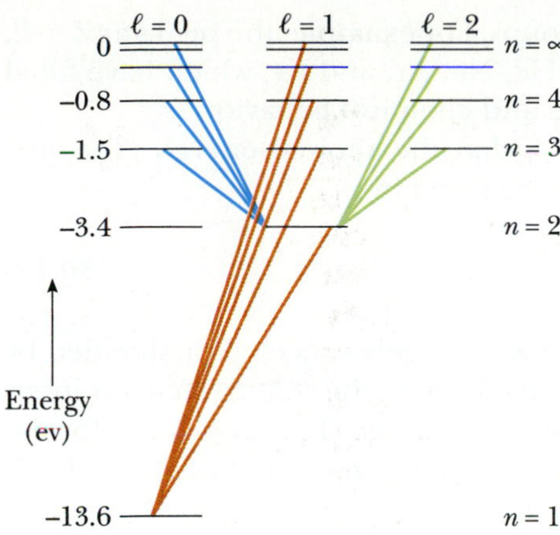

Figure 30.14
Some allowed electronic transitions for hydrogen, represented by the colored lines. These transitions must obey the selection rule $\Delta\ell = \pm 1$.

transitions is $\Delta\ell = \pm 1$. Transitions for which $\Delta\ell \neq \pm 1$ are forbidden (they can occur, but their probability is negligible relative to the probability of the allowed transitions).

Selection rule for allowed atomic transitions

Since the orbital angular momentum of an atom changes when a photon is emitted or absorbed (that is, as a result of a transition) and since angular momentum must be conserved, we conclude that *the photon involved in the process must carry angular momentum.* In fact, the photon has an angular momentum equivalent to that of a particle with a spin of 1. Also, the angular momentum of the photon is consistent with the classical description of electromagnetic radiation. Hence, a photon has energy, linear momentum, and angular momentum.

The photon carries angular momentum

It is interesting to plot ionization energy versus the atomic number Z, as in Figure 30.15. Note the pattern 2, 8, 8, 18, 18, 32 for the ionization energies. This pattern follows from the Pauli exclusion principle and helps explain why the ele-

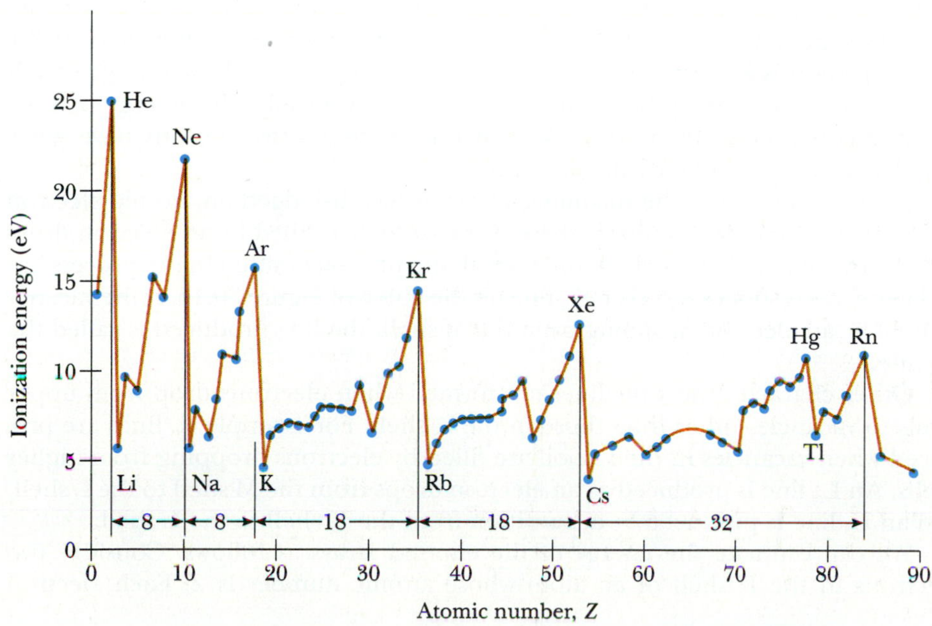

Figure 30.15
Ionization energy of the elements versus atomic number Z. *(Adapted from J. Orear, Physics, New York, Macmillan, 1979)*

ments repeat their chemical properties in groups. For example, the peaks at $Z = 2$, 10, 18, and 36 correspond to the elements He, Ne, Ar, and Kr, which have filled shells. These elements have similar energies and chemical behavior.

Recall from Chapter 12, Equation 12.23, that the allowed energies for one-electron atoms, such as hydrogen and He^+, are given by

Allowed energies for one-electron atoms

$$E_n = -\frac{13.6Z^2}{n^2}\text{ eV} \qquad [30.17]$$

For multielectron atoms, the nuclear charge *Ze* is largely canceled or shielded by the negative charge of the inner-core electrons. Hence, the outer electrons interact with a net charge on the order of the electronic charge. The expression for the allowed energies for multielectron atoms has the same form as Equation 30.17, with Z replaced by an effective atomic number, Z_{eff},

Allowed energies for multielectron atoms

$$E_n = -\frac{13.6Z_{\text{eff}}^2}{n^2}\text{ eV} \qquad [30.18]$$

where Z_{eff} depends on n and ℓ. For the higher energy states, this reduction in charge increases and $Z_{\text{eff}} \to 1$.

X-Ray Spectra

X-rays are emitted from a metal target when it is bombarded by high-energy electrons. The x-ray spectrum typically consists of a broad continuous band and a series of sharp lines that depend on the type of material used for the target, as shown in Figure 30.16. These lines, called **characteristic x-rays,** were discovered in 1908, but their origin remained unexplained until the details of atomic structure were brought to light.

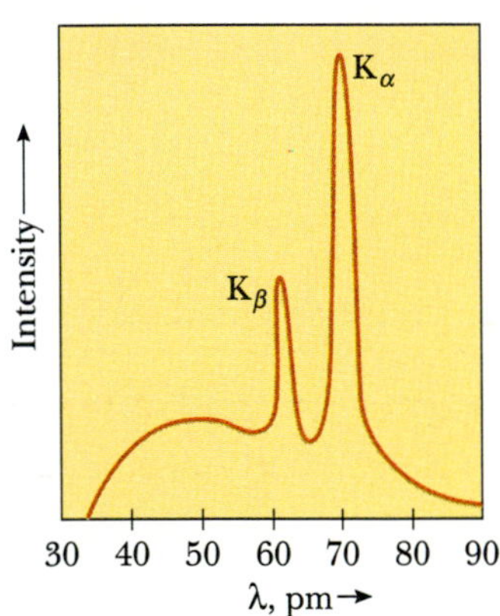

Figure 30.16
The x-ray spectrum of a metal target consists of a broad continuous spectrum plus a number of sharp lines, which are due to *characteristic x-rays.* The data shown were obtained when 35-keV electrons bombarded a molybdenum target. Note that 1 pm = 10^{-12} m = 10^{-3} nm.

The first step in the production of characteristic x-rays occurs when a bombarding electron collides with an electron in an inner shell of a target atom with sufficient energy to remove the electron from the atom. The vacancy created in the shell is filled when an electron in a higher level drops down into the level containing the vacancy. The time it takes for this to happen is very short, less than 10^{-9} s. This transition is accompanied by the emission of a photon whose energy equals the difference in energy between the two levels. Typically, the energy of such transitions is greater than 1000 eV, and the emitted x-ray photons have wavelengths in the range of 0.01 nm to 1 nm.

Let us assume that the incoming electron has dislodged an atomic electron from the innermost shell, the K shell. If the vacancy is filled by an electron dropping from the next higher shell, the L shell, the photon emitted in the process has an energy corresponding to the K_α line on the curve of Figure 30.16. If the vacancy is filled by an electron dropping from the M shell, the line produced is called the K_β line.

Other characteristic x-ray lines are formed when electrons drop from upper levels to vacancies other than those in the K shell. For example, L lines are produced when vacancies in the L shell are filled by electrons dropping from higher shells. An L_α line is produced as an electron drops from the M shell to the L shell, and an L_β line is produced by a transition from the N shell to the L shell.

We can estimate the energy of the emitted x-rays as follows. Consider two electrons in the K shell of an atom whose atomic number is Z. Each electron partially shields the other from the charge of the nucleus, Ze, and so each electron is subject to an effective nuclear charge of $Z_{\text{eff}} = (Z - 1)e$. We can now use Equation 30.18 to estimate the energy of either electron:

$$E_K = -(Z - 1)^2(13.6 \text{ eV}) \qquad [30.19]$$

As we shall show in the following example, one can estimate the energy of an electron in an L or M shell in a similar fashion. Taking the energy difference between these two levels, one can then calculate the energy and wavelength of the emitted photon.

In 1914 Henry G. J. Moseley found that $1/\lambda$ gave a nearly straight line when the elements were arranged approximately according to their atomic weights. From this, he defined the *atomic number, Z,* which proved to be a better measure for placing the atoms in the periodic table than the atomic weight used by Mendeleev and others. Figure 30.17 is a plot of the Z values for a number of elements versus $1/\lambda$, where λ is the wavelength of the K_α line for each element.

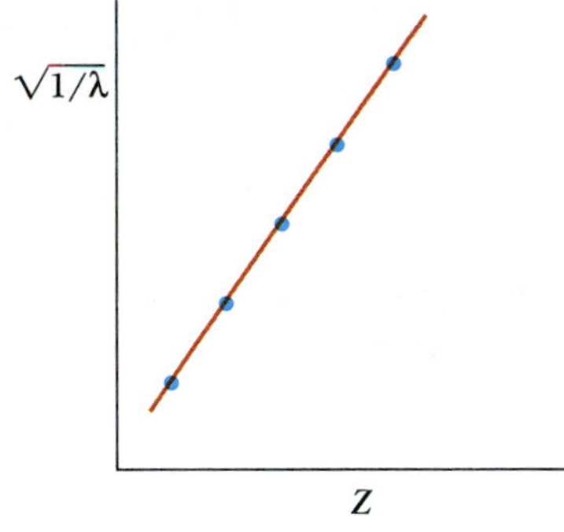

Figure 30.17
A Moseley plot for the K_α x-ray lines of a number of elements.

Example 30.7 Estimating the Energy of an X-Ray

Estimate the energy of the characteristic x-ray emitted from a tungsten target when an electron drops from an M shell ($n = 3$ state) to a vacancy in the K shell ($n = 1$ state).

Solution The atomic number for tungsten is $Z = 74$. Using Equation 30.19, we get

$$E_K = -(74 - 1)^2(13.6 \text{ eV}) = -72{,}500 \text{ eV}$$

The electron in the M shell is subject to an effective nuclear charge that depends on the number of electrons in the $n = 1$ and $n = 2$ states, which shield the nucleus. Because there are eight electrons in the $n = 2$ state and one electron in the $n = 1$ state, roughly nine electrons shield the nucleus, and so $Z_{\text{eff}} = Z - 9$. Hence, the energy of an electron in the M shell, following Equation 30.18, is

$$E_M = -Z_{\text{eff}}^2 E_3 = -(Z - 9)^2 \frac{E_0}{3^2}$$

$$= -(74 - 9)^2 \frac{(13.6 \text{ eV})}{9} = -6380 \text{ eV}$$

where E_3 is the energy of an electron in the M shell of the hydrogen atom, and E_0 is the ground-state energy. Therefore, the emitted x-ray has an energy equal to $E_M - E_K = -6380 \text{ eV} - (-72\,500 \text{ eV}) = 66\,100 \text{ eV}$. Note that this energy difference is also equal to $hf = hc/\lambda$, where λ is the wavelength of the emitted x-ray.

Exercise Calculate the wavelength of the emitted x-ray for this transition.

Answer 0.0188 nm.

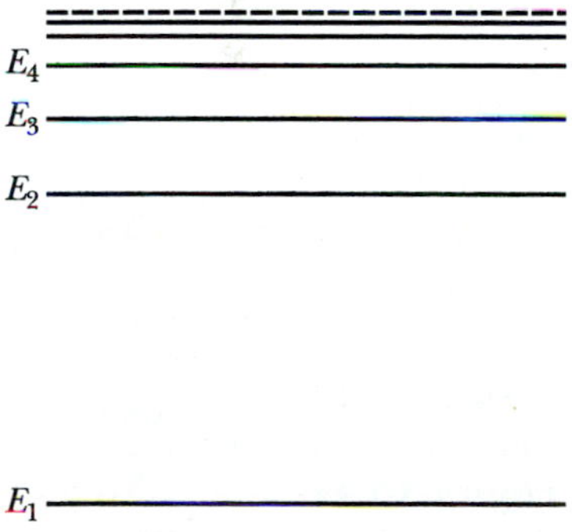

30.8 Atomic Transitions

We have seen that an atom emits radiation only at certain frequencies that correspond to the energy separations between the allowed states. Consider an atom with many allowed energy states, labeled $E_1, E_2, E_3, \ldots$ in Figure 30.18. When light

Figure 30.18
Energy level diagram of an atom with various allowed states. The lowest energy state, E_1, is the ground state. All others are excited states.

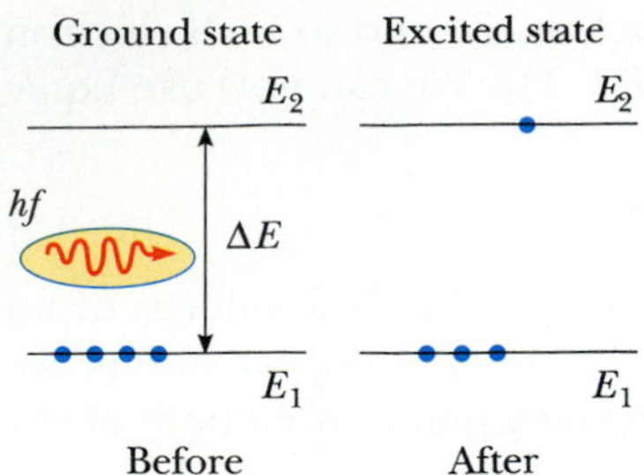

Figure 30.19
Diagram representing the *stimulated absorption* of a photon by an atom. The dots represent electrons. One electron is transferred from the ground state to the excited state when the atom absorbs a photon whose energy $hf = E_2 - E_1$.

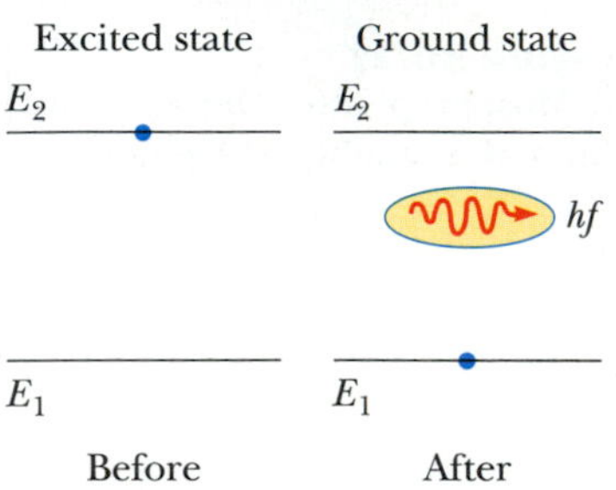

Figure 30.20
Diagram representing the *spontaneous emission* of a photon by an atom that is initially in the excited state E_2. When the electron falls to the ground state, the atom emits a photon whose energy $hf = E_2 - E_1$.

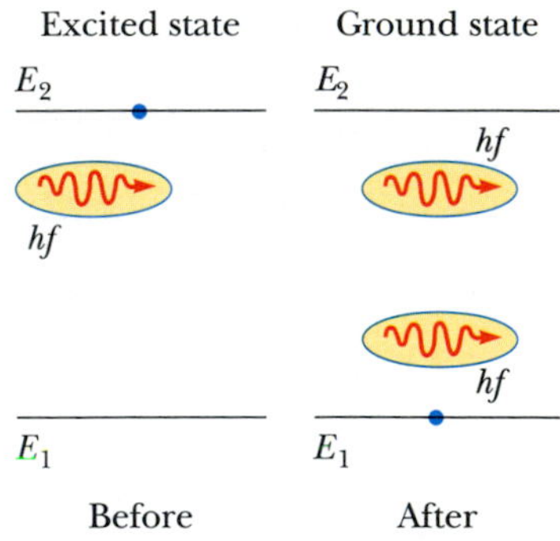

Figure 30.21
Diagram representing the *stimulated emission* of a photon by an incoming photon of energy *hf*. Initially, the atom is in the excited state. The incoming photon stimulates the atom to emit a second photon of energy $hf = E_2 - E_1$.

is incident on the atom, only those photons whose energy *hf* matches the energy separation, ΔE, between two levels can be absorbed by the atom. Figure 30.19 is a schematic diagram representing this **stimulated absorption.** At ordinary temperatures, most of the atoms are in the ground state. If a vessel containing many atoms of a gaseous element is illuminated with a light beam containing all possible photon frequencies (that is, a continuous spectrum), only those photons of energies $E_2 - E_1$, $E_3 - E_1$, $E_4 - E_1$, and so on, can be absorbed. As a result of this absorption, some atoms are raised to allowed higher energy levels called **excited states.**

Once an atom is in an excited state, there is a certain probability that the excited electron will jump back to a lower level by emitting a photon, as shown in Figure 30.20. This process is known as **spontaneous emission.** Typically, an atom remains in an excited state for only about 10^{-8} s.

Finally, there is a third process, **stimulated emission,** that is of importance in lasers. Suppose an atom is in the excited state E_2, as in Figure 30.21, and a photon with energy $hf = E_2 - E_1$ is incident on it. The incoming photon increases the probability that the excited electron will return to the ground state and thereby emit a second photon having the same energy *hf*. Note that the incident photon is not absorbed, so after the stimulated emission, there are two nearly identical photons—the incident photon and the emitted photon. The emitted photon is exactly in phase with the incident photon. These photons can stimulate other atoms to emit photons in a chain of similar processes. The many photons produced in this fashion are the source of the intense, coherent light in a laser.

*30.9 Lasers and Holography

We have described how an incident photon can cause atomic transitions either upward (stimulated absorption) or downward (stimulated emission). The two processes are equally probable. When light is incident on a system of atoms, there is usually a net absorption of energy because, when the system is in thermal equilibrium, there are many more atoms in the ground state than in excited states. However, if one can invert the situation so that there are more atoms in an excited state than in the ground state, a net emission of photons can result. Such a condition is called **population inversion.**

This is the fundamental principle involved in the operation of a **laser,** an acronym for *l*ight *a*mplification by *s*timulated *e*mission of *r*adiation. The amplification corresponds to a buildup of photons in the system as the result of a chain

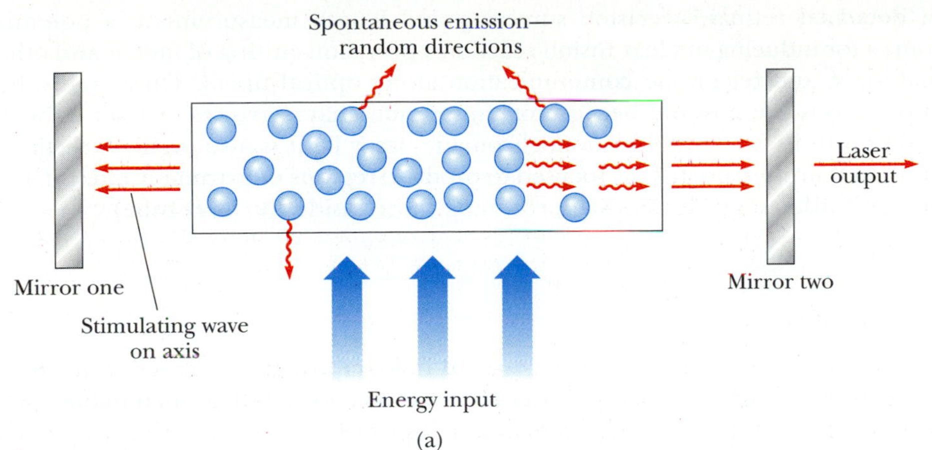

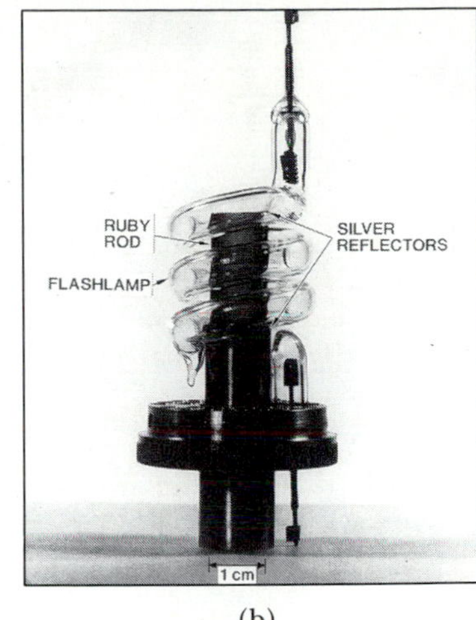

Figure 30.22
(a) A schematic of a laser design. The tube contains atoms, which represent the active medium. An external source of energy (optical, electrical, etc.) is needed to "pump" the atoms to excited energy states. The parallel end mirrors provide the feedback of the stimulating wave. (b) Photograph of the first ruby laser showing the flash lamp surrounding the ruby rod. *(Courtesy of Hughes Aircraft Company)*

reaction of events. The following three conditions must be satisfied in order to achieve laser action:

1. The system must be in a state of population inversion (that is, more atoms in an excited state than in the ground state).
2. The excited state of the system must be a *metastable state,* which means its lifetime must be long compared with the usually short lifetimes of excited states. When such is the case, stimulated emission will occur before spontaneous emission.
3. The emitted photons must be confined in the system long enough to enable them to stimulate further emission from other excited atoms. This is achieved by the use of reflecting mirrors at the ends of the system. One end is made totally reflecting, and the other is slightly transparent to allow the laser beam to escape (Fig. 30.22a).

One device that exhibits stimulated emission of radiation is the helium-neon gas laser. Figure 30.23 is an energy level diagram for the neon atom in this system. The mixture of helium and neon is confined to a glass tube that is sealed at the ends by mirrors. An oscillator connected to the tube causes electrons to sweep through the tube, colliding with the atoms of the gases and raising them into excited states. Neon atoms are excited to state E_3 through this process and also as a result of collisions with excited helium atoms. Stimulated emission occurs as the neon atoms make a transition to state E_2 and neighboring excited atoms are stimulated. This results in the production of coherent light at a wavelength of 632.8 nm.

Since the development of the first laser in 1960, laser technology has experienced tremendous growth. Lasers that cover wavelengths in the infrared, visible, and ultraviolet regions are now available. Applications include surgical "welding"

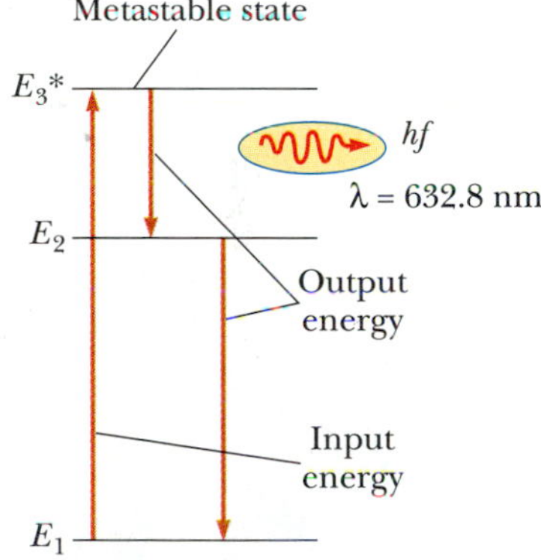

Figure 30.23
Energy level diagram for the neon atom, which emits photons at a wavelength of 632.8 nm through stimulated emission. The photon at this wavelength arises from the transition $E_3^* \rightarrow E_2$. This is the source of coherent light in the helium-neon gas laser.

of detached retinas, precision surveying and length measurement, a potential source for inducing nuclear fusion reactions, precision cutting of metals and other materials, and telephone communication along optical fibers. These and other applications are possible because of the unique characteristics of laser light. In addition to its being highly monochromatic, laser light is also highly directional and can therefore be sharply focused to produce regions of extremely intense light energy (with energy densities approaching those inside the laser tube).

Holography

One interesting application of the laser is **holography,** the production of three-dimensional images of objects. Figure 30.24a shows how a hologram is made. Light from the laser is split into two parts by a half-silvered mirror at *B*. One part of the beam reflects off the object to be photographed and strikes an ordinary photographic film. The other half of the beam is diverged by lens L_2, reflects from mirrors M_1 and M_2, and finally strikes the film. The two beams overlap to form an extremely complicated interference pattern on the film. Such an interference pattern can be produced only if the phase relationship of the two waves is constant throughout the exposure of the film. This condition is met by illuminating the scene with light coming through a pinhole or with coherent laser radiation. The hologram records not only the intensity of the light scattered from the object (as in a conventional photograph), but also the phase difference between the reference beam and the beam scattered from the object. Because of this phase difference, an interference pattern is formed that produces an image with full three-dimensional perspective.

A hologram is best viewed by allowing coherent light to pass tnrough the developed film as one looks back along the direction from which the beam comes. Figure 30.24b is a photograph of a hologram made using a cylindrical film.

Figure 30.24
(a) Experimental arrangement for producing a hologram. (b) Photograph of a hologram which uses a cylindrical film. Note the detail of the Volkswagen image. *(Courtesy of CENCO)*

Summary

The methods of quantum mechanics can be applied to the hydrogen atom using the appropriate potential energy function, $U(r) = -k_e e^2/r$, in the Schrödinger equation. The solution to this equation yields the wave functions for the allowed states and the allowed energies, given by

$$E_n = -\left(\frac{k_e e^2}{2a_0}\right)\frac{1}{n^2} = -\frac{13.6}{n^2}\text{ eV} \qquad n = 1, 2, 3, \ldots \qquad [30.2]$$

Allowed energies for the hydrogen atom

This is precisely the result obtained in the Bohr theory. The allowed energy depends only on the **principal quantum number,** n. The allowed wave functions depend on three quantum numbers, n, ℓ, and m_ℓ, where ℓ is the **orbital quantum number** and m_ℓ is the **orbital magnetic quantum number.** The restrictions on the quantum numbers are as follows:

$$n = 1, 2, 3, \ldots$$
$$\ell = 0, 1, 2, \ldots, (n-1)$$
$$m_\ell = -\ell, -\ell+1, \ldots, \ell-1, \ell$$

Allowed values for the quantum numbers

All states with the same principal quantum number n form a **shell,** identified by the letters K, L, M, . . . (corresponding to $n = 1, 2, 3, \ldots$). All states with the same values of both n and ℓ form a **subshell,** designated by the letters, s, p, d, f, . . . (corresponding to $\ell = 0, 1, 2, 3, \ldots$).

In order to completely describe a quantum state of the hydrogen atom, it is necessary to include a fourth quantum number, m_s, called the **spin magnetic quantum number.** This quantum number can have only two values, $\pm\frac{1}{2}$. In effect, this doubles the number of allowed states specified by the quantum numbers n, ℓ, and m_ℓ.

Spin magnetic quantum number

An atom in a state characterized by a specific n can have the following values of **orbital angular momentum,** L:

$$L = \sqrt{\ell(\ell+1)}\hbar \qquad \ell = 0, 1, 2, \ldots, n-1 \qquad [30.10]$$

Allowed values of L

The allowed values of the projection of **L** along the z axis are given by

$$L_z = m_\ell \hbar \qquad [30.11]$$

Allowed values of L_z

where m_ℓ is restricted to integer values lying between $-\ell$ and ℓ. Only discrete values of L_z are allowed, and these are determined by the restrictions on m_ℓ. This quantization of L_z is referred to as **space quantization.**

The electron has an intrinsic angular momentum called **spin angular momentum.** That is, the total angular momentum of an electron in an atom can have two contributions, one arising from the spin of the electron (**S**) and one arising from the orbital motion of the electron (**L**).

Electronic spin can be described by a single quantum number, $s = \frac{1}{2}$. The **magnitude of the spin angular momentum** is

$$S = \frac{\sqrt{3}}{2}\hbar \qquad [30.13]$$

Spin angular momentum of an electron

and the z component of **S** is given by

The z component of spin angular momentum

$$S_z = m_s\hbar = \pm \tfrac{1}{2}\hbar \qquad [30.14]$$

That is, the spin angular momentum is also quantized in space, as specified by the **spin magnetic quantum number,** $m_s = \pm \frac{1}{2}$.

The magnetic moment $\boldsymbol{\mu}_s$ associated with the spin angular momentum of an electron is

Relation between spin magnetic moment and spin angular momentum

$$\boldsymbol{\mu}_s = -\frac{e}{m}\mathbf{S} \qquad [30.15]$$

which is *twice* as large as the orbital magnetic moment. The z component of $\boldsymbol{\mu}_s$ can have the values

The z component of spin magnetic moment

$$\mu_{sz} = \pm \frac{e\hbar}{2m} \qquad [30.16]$$

The **exclusion principle** states that *no two electrons in an atom can ever have the same set of quantum numbers n, ℓ, m_ℓ,* and m_s. Using this principle and the principle of minimum energy, one can determine the electronic configuration of the elements. This serves as a basis for understanding atomic structure and the chemical properties of the elements.

Selection rule for allowed atomic transitions

The allowed electronic transitions between any two levels in an atom are governed by the selection rule $\Delta\ell = \pm 1$.

X-rays are emitted by atoms when an electron undergoes a transition from an outer shell into an electron vacancy in one of the inner shells. Transitions into a vacant state in the K shell give rise to the K series of spectral lines; transitions into a vacant state in the L shell create the L series of lines; and so on. The x-ray spectrum of a metal target consists of a set of sharp characteristic lines superimposed on a broad, continuous spectrum.

Questions and Conceptual Exercises

1. Does the light emitted by a neon sign constitute a continuous spectrum or only a few colors? Defend your answer.
2. Must an atom first be ionized before it can emit light? Discuss.
3. Discuss why the term *electron clouds* is used to describe the electronic arrangement in the quantum mechanical view of the atom.
4. It was stated in the text that, if the exclusion principle were not valid, every electron would end up in the lowest energy state of the atom and the chemical behavior of the elements would be grossly modified. Explain this statement.
5. When a hologram is produced, the system (including light source, object, beam splitter, and so on) must be held motionless within a quarter of a wavelength. Why?
6. Why is an inhomogeneous magnetic field used in the Stern–Gerlach experiment?
7. Could the Stern–Gerlach experiment be performed with ions rather than neutral atoms? Explain.
8. Describe some experiments that would support the conclusion that the spin quantum number for electrons can only have the values $\pm \frac{1}{2}$.
9. Discuss some of the consequences of the exclusion principle.
10. Why do lithium, potassium, and sodium exhibit similar chemical properties?
11. From Table 30.4, we find that the ionization energies for Li, Na, K, Rb, and Cs are 5.390, 5.138, 4.339, 4.176, and 3.893 eV, respectively. Explain why these values are to be expected in terms of the atomic structures.
12. Does the intensity of light from a laser fall off as $1/r^2$?

13. How is it possible that electrons, with a probability distribution around a nucleus, can exist in states of definite energy (e.g., 1*s*, 2*p*, 3*d*, . . .)?
14. It is easy to understand how two electrons (one spin up, one spin down) can fill the 1*s* shell for a helium atom. How is it possible that eight more electrons can fit into the 2*s*, 2*p* level to complete the $1s^2 2s^2 2p^6$ shell for a neon atom?
15. In 1914 Henry Moseley was able to determine the atomic number of an element from its characteristic x-ray spectrum. How was this possible? (*Hint:* See Figs. 30.16 and 30.17.)
16. Why is *stimulated emission* so important in the operation of a laser? (Interestingly, the concept of stimulated emission was first discussed by Albert Einstein 43 years before the earliest successful laser.)
17. The efficiencies of most solid-state lasers are on the order of 1 to 2%. Although the laser output is monochromatic and highly directional, can you use Figures 30.22 and 30.23 to determine why the energy input must exceed laser energy output by a factor of 50 to 100?

Problems

Section 30.2 The Hydrogen Atom Revisited

1. (a) Determine the quantum numbers ℓ and m_ℓ for the He^+ ion in the state corresponding to $n = 3$. (b) What is the energy of this state?
2. The Balmer series for the hydrogen atom corresponds to electronic transitions that terminate in the state of quantum number $n = 2$, as shown in Figure 30.25. (a) Find the longest-wavelength photon emitted and determine its energy. (b) Find the shortest-wavelength photon emitted and determine its energy.

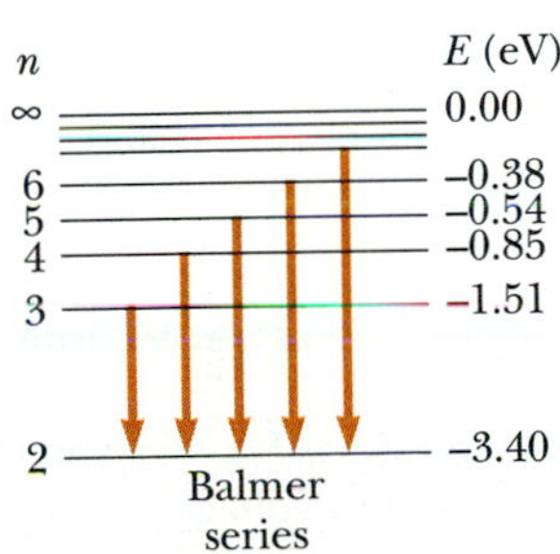

Figure 30.25 (Problem 2) An energy-level diagram for hydrogen showing the Balmer series.

3. A general expression for the energy levels of one-electron atoms is

$$E_n = -\left(\frac{\mu k_e^2 q_1^2 q_2^2}{2\hbar^2}\right)\frac{1}{n^2}$$

where k_e is the Coulomb constant, q_1 and q_2 are the charges of the two particles, and μ is the reduced mass given by $\mu = m_1 m_2/(m_1 + m_2)$. In Problem 2 we found that the wavelength for the $n = 3$ to $n = 2$ transition of the hydrogen atom is 656.3 nm (visible red light). What are the wavelengths for this same transition in (a) positronium, which consists of an electron and a positron, and (b) singly ionized helium? (*Note:* A positron is a positively charged electron.)
4. The energy of an electron in a hydrogen atom is

$$E = \frac{p^2}{2m_e} - \frac{k_e e^2}{r}$$

According to the uncertainty principle, if the electron is localized within r, the magnitude p may be as large as its momentum $\hbar/2r$. Use this principle to find the *minimum* values of E and r. Compare your results to Bohr's.

Section 30.3 The Spin Magnetic Quantum Number
Section 30.4 The Wave Functions for Hydrogen

5. List the possible sets of quantum numbers for electrons in the 3*d* subshell.
6. List the possible sets of quantum numbers for electrons in the 3*p* subshell.
7. When the principal quantum number is $n = 4$, how many different values of (a) ℓ and (b) m_ℓ are possible?
8. Make plots of the wave function $\psi_{1s}(r)$ (Eq. 30.3) and the radial probability density function $P_{1s}(r)$ (Eq. 30.7) for hydrogen. Let r range from 0 to $1.5a_0$, where a_0 is the Bohr radius.
9. The wave function for an electron in a 2*p* state in hydrogen is

$$\psi_{2p} = \frac{1}{\sqrt{3}(2a_0)^{3/2}}\frac{r}{a_0}e^{-r/2a_0}$$

What is the most likely distance from the H nucleus to an electron in the $2p$ state? (See Fig. 30.8.)

10. Show that the $1s$ wave function for an electron in hydrogen,

$$\psi(r) = \frac{1}{\sqrt{\pi a_0{}^3}} e^{-r/a_0}$$

satisfies the radially symmetric Schrödinger equation:

$$-\frac{\hbar^2}{2m}\left(\frac{d^2\psi}{dr^2} + \frac{2}{r}\frac{d\psi}{dr}\right) - \frac{k_e e^2}{r}\psi = E\psi$$

11. If a muon (a negatively charged particle with mass 206 times the electron's mass) is captured by a lead nucleus, $Z = 82$, the resulting system behaves like a one-electron atom. (a) What is the "Bohr radius" for a muon captured by a lead nucleus? (*Hint:* Use Eq. 30.4.) (b) Using Equation 30.2 with e replaced by Ze, calculate the ground-state energy of a muon captured by a lead nucleus. (c) What is the transition energy of a muon descending from the $n = 2$ to the $n = 1$ level in a muonic lead atom?

Section 30.5 The "Other" Quantum Numbers

12. Calculate the orbital angular momentum for an electron (a) in the $4d$ state; (b) in the $6f$ state.
13. If an electron has an orbital angular momentum of 4.714×10^{-34} J · s, what is the orbital quantum number for this state of the electron?
14. Two electrons in the same atom both have $\ell = 1$ and $n = 3$. (a) List the possible states. (b) How many states would be possible if the exclusion principle were inoperative?
15. How many different sets of quantum numbers are possible for an electron for which (a) $n = 1$? (b) $n = 2$? (c) $n = 3$? (d) $n = 4$? (e) $n = 5$? Check your results to show that they agree with the general rule that the number of different sets of quantum numbers is equal to $2n^2$.
16. (a) Write out the electronic configuration for oxygen ($Z = 8$). (b) Write out the values for the set of quantum numbers n, ℓ, m_ℓ, and m_s for each of the electrons in oxygen.
17. Determine the number of electrons that can occupy the $n = 3$ shell.
18. Find the possible values of L, L_z, and θ for an electron in a $3d$ state of hydrogen.
19. All objects, large and small, behave quantum-mechanically. (a) Estimate the quantum number ℓ for the Earth in its orbit about the Sun. (b) What energy change (in joules) would occur if the Earth made a transition to an adjacent allowed state?
20. Like the electron, the nucleus of an atom has spin angular momentum and a corresponding magnetic moment. The z-component of the spin magnetic moment for a nucleus is characterized by the *nuclear magneton* $\mu_n = e\hbar/2m_p$, where m_p is the proton mass. (a) Calculate the value of μ_n in J/T and in eV/T. (b) Determine the ratio μ_n/μ_B, and comment on your result.
21. The z-component of the electron's spin magnetic moment is given by the Bohr magneton, $\mu_B = e\hbar/2m$. Show that the Bohr magneton has the numerical value 9.27×10^{-24} J/T, or 5.79×10^{-5} eV/T.

Section 30.6 The Exclusion Principle and the Periodic Table

22. Which electronic configuration has the lower energy, $[\text{Ar}]3d^44s^2$ or $[\text{Ar}]3d^54s^1$? Identify this element and discuss Hund's rule in this case.
23. Which electronic configuration has the lesser energy and the greater number of unpaired spins, $[\text{Kr}]4d^95s^1$ or $[\text{Kr}]4d^{10}$? Identify this element and discuss Hund's rule in this case. (*Note:* The notation [Kr] represents the filled configuration for Kr.)
24. Devise a table similar to that shown in Figure 30.13 for atoms with 11 through 19 electrons. Use Hund's rule and educated guesswork.
25. (a) Scanning through Table 30.4 in order of increasing atomic number, note that the electrons fill the subshells in such a way that those subshells with the lowest values of $n + \ell$ are filled first. If two subshells have the same value of $n + \ell$, the one with the lower value of n is filled first. Using these two rules, write the order in which the subshells are filled through $n = 7$. (b) Predict the chemical valence for elements with atomic numbers 15, 47, and 86, and compare them with the actual valences.

Section 30.7 Atomic Spectra: Visible and X-Ray

26. If you wish to produce 1-Å x-rays in the laboratory, what is the minimum voltage you must use in accelerating the electrons?
27. A tungsten target is struck by electrons that have been accelerated from rest through a 40-kV potential difference. Find the shortest wavelength of the radiation emitted.
28. Find the wavelength of the K_α x-ray line that is emitted when electrons strike an iron target. Note that since the innermost shells are involved, Z_{eff} is approximately $Z - 1$.
29. Use the method illustrated in Example 30.7 to calculate the wavelength of the x-ray emitted from a molybdenum target ($Z = 42$) when an electron undergoes a transition from the L shell ($n = 2$) to the K shell ($n = 1$).

30. The K series of the discrete spectrum of tungsten contains wavelengths of 0.0185 nm, 0.0209 nm, and 0.0215 nm. The K-shell ionization energy is 69.5 keV. Determine the ionization energies of the L, M, and N shells. Sketch the transitions.

Section 30.8 Atomic Transitions

31. The familiar yellow light from a sodium vapor street lamp results from the $3p \rightarrow 3s$ transition in ^{11}Na. Evaluate the wavelength of the light, given the energy difference $E_{3p} - E_{3s} = 2.1$ eV.
32. The wavelength of coherent ruby laser light is 694.3 nm. What is the energy difference (in electron volts) between the upper, excited state and the lower, unexcited energy state?
33. A ruby laser delivers a 10-ns pulse of 1 MW average power. If all the photons are of wavelength 694.3 nm, how many photons are contained in the pulse?
34. A Nd:YAG laser used in eye surgery emits a 3-mJ pulse in 1 ns, focused to a spot 30 μm in diameter on the retina. (a) Find (in SI units) the power per unit area at the retina. (This quantity is called the *irradiance*.) (b) What energy is delivered to an area of molecular size—say, a circular area 0.6 nm in diameter?

Additional Problems

35. (a) How much energy is required to cause an electron in hydrogen to move from the $n = 1$ state to the $n = 2$ state? (b) If the electrons gain this energy by interacting with hydrogen atoms at a high temperature, find the minimum temperature of the heated hydrogen gas. The thermal energy of the heated atoms is given by $3k_BT/2$, where k_B is the Boltzmann constant.
36. Zirconium has two unpaired electrons in the d subshell. (a) What are all possible values of ℓ and s for each electron? (b) What are all possible values of n, m_ℓ, and m_s? (c) What is the electron configuration in zirconium?
37. In the technique known as electron spin resonance (ESR), a sample containing unpaired electrons is placed in a magnetic field. Consider the simplest situation, that in which there is only one electron and therefore only two possible energy states exist, corresponding to $m_s = \pm \frac{1}{2}$. In ESR, the electron's spin magnetic moment is "flipped" from a lower energy state to a higher energy state by the absorption of a photon. (The lower energy state corresponds to the case where the magnetic moment μ_s is aligned against the magnetic field, and the higher energy state corresponds to the case where μ_s is aligned with the field.) What is the photon frequency required to excite an ESR transition in a magnetic field of 0.35 T?
38. Show that the average value of r for the $1s$ state of hydrogen is $3a_0/2$. (*Hint:* Use Eq. 30.7.)
39. The carbon dioxide (CO_2) laser is one of the most powerful lasers ever developed. The energy difference between the two laser levels is 0.117 eV. Determine the frequency and wavelength of the radiation emitted by this laser. In what portion of the electromagnetic spectrum is this radiation found?
40. Show that the wave function for an electron in the $2s$ state in hydrogen,

$$\psi(r) = \frac{1}{4\sqrt{2\pi}} \left(\frac{1}{a_0}\right)^{3/2} \left(2 - \frac{r}{a_0}\right) e^{-r/2a_0}$$

satisfies the radially symmetric Schrödinger equation:

$$-\frac{\hbar^2}{2m}\left(\frac{d^2\psi}{dr^2} + \frac{2}{r}\frac{d\psi}{dr}\right) - \frac{k_e e^2}{r}\psi = E\psi$$

41. For the ground state of hydrogen, what is the probability of finding the electron closer to the nucleus than the Bohr radius corresponding to $n = 1$?
42. A pulsed ruby laser emits light at 694.4 nm. For a 14-ps pulse containing 3 J of energy, find (a) the physical length of the pulse as it travels through space and (b) the number of photons in the pulse. (c) If the beam has a circular cross-section of 0.6 cm diameter, find the number of photons per cubic millimeter in the beam.
43. The number N of atoms in a particular state is called the *population* of that state. This number depends on the energy of that state and on the temperature. The equilibrium population of atoms in a state of energy E_n is given by a Boltzmann distribution expression:

$$N = N_0 e^{-E_n/k_BT}$$

where N_0 is the population of the state as $T \rightarrow \infty$. (a) Find the ratio of populations of the states E_3^* to E_2 for the laser in Figure 30.23, assuming $T = 27°$C. (b) Find the ratio of the populations of the two states in a ruby laser that produces a light beam of wavelength 694.3 nm at 4 K.
44. The force on a magnetic moment, μ_z, in a nonuniform magnetic field, B_z, is given by

$$F_z = \mu_z \frac{dB_z}{dz}$$

If a beam of silver atoms travels a horizontal distance of 1 m through such a field and each atom has a speed of 100 m/s, how strong must the field gradient dB_z/dz be in order to deflect the beam 1 mm?
45. *Positronium* is a hydrogen-like atom consisting of a positron (a positively charged electron) and an electron revolving around each other. Using the Bohr model,

find the allowed radii (relative to the center of mass of the two particles) and the allowed energies of the system.

46. *The Auger process.* An electron in chromium makes a transition from the $n = 2$ state to the $n = 1$ state without emitting a photon. Instead, the excess energy is transferred to an outer electron (in the $n = 4$ state), which is ejected by the atom. [This is called an *Auger* (pronounced 'Oh-zhay') *process,* and the ejected electron is referred to as an *Auger electron.*] Use the Bohr theory to find the kinetic energy of the Auger electron.

47. A *muon* is a particle with a charge of $-e$ and a mass equal to 207 times the mass of an electron (see Problem 11). Muonic lead is formed when a ^{208}Pb nucleus captures a muon. According to the Bohr theory, what are the radius and energy of the ground state of muonic lead?

48. A muon (Problem 47) is captured by a deuteron to form a muonic atom. (a) Find the energy of the ground state and the first excited state. (b) What is the wavelength of the photon that is emitted when the atom makes a transition from the first excited state to the ground state?

49. Use Bohr's model of the hydrogen atom to show that when the atom makes a transition from the state n to the state $n - 1$, the frequency of the emitted light is given by

$$f = \frac{2\pi^2 m k_e^2 e^4}{h^3}\left(\frac{2n-1}{(n-1)^2 n^2}\right)$$

50. Suppose the ionization energy of an atom is 4.1 eV. In this same atom, we observe emission lines with wavelengths 310 nm, 400 nm, and 1377.8 nm. Use this information to construct the energy-level diagram with the least number of levels. Assume the higher energy levels are closer together.

51. Calculate the classical frequency for the light emitted by an atom. To do so, note that the frequency of revolution is $v/2\pi r$, where r is the Bohr radius. Show that as n approaches infinity in the equation of Problem 49, the expression given there varies as $1/n^3$ and reduces to the classical frequency. (This is an example of the correspondence principle, which requires that the classical and quantum models agree for large values of n.)

52. A dimensionless number that often appears in atomic physics is the *fine-structure constant,* α, given by

$$\alpha = \frac{k_e e^2}{\hbar c}$$

where k_e is the Coulomb constant. (a) Obtain a numerical value for $1/\alpha$. (b) In terms of α, what is the ratio of the Bohr radius, a_0, to the Compton wavelength, $\lambda = h/mc$?

53. (a) Calculate the most probable radius for an electron in the $2s$ state of hydrogen. (*Hint:* Let $x = r/a_0$, find an equation for x, and show that $x = 5.236$ is a solution to this equation.) (b) Show that the wave function given by Equation 30.8 is normalized.

54. All atoms are roughly the same size. (a) To show this, estimate the diameters for aluminum, with molar atomic mass of 27 g/mol and density of 2.70 g/cm^3, and uranium, with molar atomic mass of 238 g/mol and density of 18.9 g/cm^3. (b) What do the results imply about the wave functions for inner-shell electrons as we progress to higher and higher atomic weight atoms? (*Hint:* The molar volume is roughly proportional to $D^3 N_A$, where D is the atomic diameter and N_A is Avogadro's number.)

55. For hydrogen in the $1s$ state, what is the probability of finding the electron farther than $2.50a_0$ from the nucleus?

56. According to classical physics, an accelerated charge, e, radiates at the rate

$$\frac{dE}{dt} = -\frac{1}{6\pi\epsilon_0}\frac{e^2a^2}{c^3}$$

where a is the acceleration of the charge. (a) Show that an electron in a classical hydrogen atom (see Fig. 30.3) will spiral into the nucleus at the rate

$$\frac{dr}{dt} = -\frac{e^4}{12\pi^2\epsilon_0{}^2 r^2 m^2 c^3}$$

(b) Find the time required for the electron to reach $r = 0$, starting from $r_0 = 2 \times 10^{-10}$ m.

57. Light from a certain He-Ne laser has a power output of 1.0 mW and a cross-sectional area of 10 mm². The entire beam is incident on a metal target that requires 1.5 eV to remove an electron from its surface. (a) Perform a classical calculation to determine how long it would take one atom in the metal to absorb 1.5 eV from the incident beam. (*Hint:* Assume that the area of an atom is 1 Å² = 10^{-20} m², and first calculate the energy incident on each atom per second.) (b) Compare the (wrong) answer obtained in (a) to the actual response time for photoelectric emission ($\approx 10^{-9}$ s), and discuss the reasons for the large discrepancy.

58. In interstellar space, atomic hydrogen produces the sharp spectral line called the *21-cm radiation,* which astronomers find most helpful in detecting clouds of hydrogen between stars. This radiation is useful because interstellar dust that obscures visible wavelengths is transparent to these radio wavelengths. The radiation is not generated by an electron transition between energy states characterized by n. Instead, in the ground state ($n = 1$), the electron and proton spins may be *parallel* or *antiparallel,* with a resultant slight difference in these energy states. (a) Which condition has the higher energy? (b) The line is actually at 21.11 cm. What is the energy difference between the states?

Particle tracks generated from the collision of two protons with two antiprotons. A subatomic particle known as omega meson was "born" in the collision. The tracks form in liquid hydrogen inside the bubble chamber. *(©Lawrence Radiation Laboratory/Science Source, Photo-researchers, Inc.)*

Nuclear Physics

CHAPTER 31

In 1896, the year that marked the birth of nuclear physics, Antoine Henri Becquerel (1852–1908) discovered radioactivity in uranium compounds. A great deal of research followed as scientists attempted to understand the radiation emitted by radioactive nuclei. Pioneering work by Rutherford showed that the radiation was of three types, which he called alpha, beta, and gamma rays. These types are classified according to the natures of their electric charges and according to their ability to penetrate matter and ionize air. Later experiments showed that alpha rays are helium nuclei, beta rays are electrons, and gamma rays are high-energy photons.

As we saw in Chapter 30, Section 30.1, the 1911 experiments of Rutherford and his students Geiger and Marsden established that the nucleus of an atom could be regarded as essentially a point mass and point charge and that most of the atomic mass is contained in the nucleus. Furthermore, such studies demonstrated a new type of force, the *nuclear force,* which is predominant at distances of less than about 10^{-14} m and zero at great distances (Chapter 6, Section 6.1).

Other milestones in the development of nuclear physics include:

1. The observation of nuclear reactions by Cockroft and Walton in 1930
2. The discovery of the neutron by Chadwick in 1932
3. The discovery of artificial radioactivity by Joliot and Irene Curie in 1933
4. The discovery of nuclear fission by Hahn and Strassman in 1938
5. The development of the first controlled fission reactor by Fermi and his collaborators in 1942

In this chapter we shall examine the properties and structure of the atomic nucleus. We start by describing the basic properties of nuclei, then discuss nuclear forces and binding energy, nuclear models, and the phenomenon of radioactivity. We shall also address nuclear reactions and the processes by which nuclei decay.

31.1 Some Properties of Nuclei

All nuclei are composed of two types of particles: protons and neutrons. The only exception is the ordinary hydrogen nucleus, which is a single proton. In describing some of the properties of nuclei, we shall use atomic number Z and mass number A, as described in Chapter 10, Section 10.8, plus **neutron number,** N, which is the number of neutrons in the nucleus. Recall that $A = N + Z$.

The symbol we use to represent nuclei is ${}^{A}_{Z}\text{X}$, where X represents the chemical symbol for the element. For example, ${}^{27}_{13}\text{Al}$ has the mass number 27 and the atomic number 13; therefore, it contains 13 protons and 14 neutrons. When no confusion is likely to arise, we omit the subscript Z because the chemical symbol can always be used to determine Z.

The nuclei of all atoms of a particular element must contain the same number of protons but can contain different numbers of neutrons. Nuclei that are related in this way are called **isotopes** (Appendix A). The isotopes of an element have the same Z value but different N and A values. The natural abundances of isotopes can differ substantially. For example, ${}^{11}_{6}\text{C}$, ${}^{12}_{6}\text{C}$, ${}^{13}_{6}\text{C}$, and ${}^{14}_{6}\text{C}$ are four isotopes of carbon. The natural abundance of ${}^{12}_{6}\text{C}$ is about 98.9%, whereas that of ${}^{13}_{6}\text{C}$ is only about 1.1%. Some isotopes do not occur naturally but can be produced in the laboratory through nuclear reactions. Even the simplest element, hydrogen, has isotopes: ${}^{1}_{1}\text{H}$, hydrogen; ${}^{2}_{1}\text{H}$, deuterium; and ${}^{3}_{1}\text{H}$, tritium.

Charge and Mass

The proton carries a single positive charge, $+e$, the electron carries a single negative charge, $-e$, where $e = 1.6 \times 10^{-19}$ C, and the neutron is electrically neutral. Because the neutron has no charge, it is difficult to detect. The proton is about 1836 times as massive as the electron, and the masses of the proton and the neutron are almost equal (Table 31.1).

Because the rest energy of a particle is $E_0 = mc^2$ (Chapter 10, Section 10.7), it is often convenient to express the unified mass unit in terms of its energy equivalence. For a proton, we have

Ernest Rutherford (1871–1937) Rutherford, a New Zealander, was awarded the Nobel Prize in 1908 for his studies of radioactivity. When he discovered that atoms can be broken apart by alpha rays, he remarked: "On consideration, I realized that this scattering backward must be the result of a single collision, and when I made calculations I saw that it was impossible to get anything of that order of magnitude unless you took a system in which the greater part of the mass of the atom was concentrated in a minute nucleus. It was then that I had the idea of an atom with a minute massive center carrying a charge." *(Photo courtesy of AIP Niels Bohr Library)*

Table 31.1
Masses of the Proton, Neutron, and Electron in Various Units

Particle	Mass kg	Mass u	Mass MeV/c^2
Proton	1.6726×10^{-27}	1.007276	938.28
Neutron	1.6750×10^{-27}	1.008665	939.57
Electron	9.109×10^{-31}	5.486×10^{-4}	0.511

$$E_0 = mc^2 = (1.67 \times 10^{-27}\text{ kg})(3 \times 10^8\text{ m/s})^2$$
$$= 1.50 \times 10^{-10}\text{ J} = 9.38 \times 10^8\text{ eV} = 938\text{ MeV}$$

The rest energy of an electron is 0.511 MeV, and that of a neutron is 940 MeV.

Nuclear physicists often express the unified mass unit in terms of the unit MeV/c^2, where

$$1\text{ u} \equiv 1.660559 \times 10^{-27}\text{ kg} = 931.50\text{ MeV}/c^2$$

The Size of Nuclei

The size and structure of nuclei were first investigated in the scattering experiments of Rutherford, discussed in Section 30.1. Using the principle of conservation of energy, Rutherford found an expression for how close an alpha particle moving directly toward the nucleus can come to the nucleus before being turned around by Coulomb repulsion.

In such a head-on collision, the kinetic energy of the incoming alpha particle must be converted completely to electrical potential energy when the particle stops at the point of closest approach and turns around (Fig. 31.1). If we equate the initial kinetic energy of the alpha particle to the maximum electrical potential energy of the system (alpha particle plus target nucleus), we have

$$\tfrac{1}{2}mv^2 = k_e\frac{q_1q_2}{r} = k_e\frac{(2e)(Ze)}{d}$$

where d is the distance of closest approach. Solving for d, we get

$$d = \frac{4k_eZe^2}{mv^2}$$

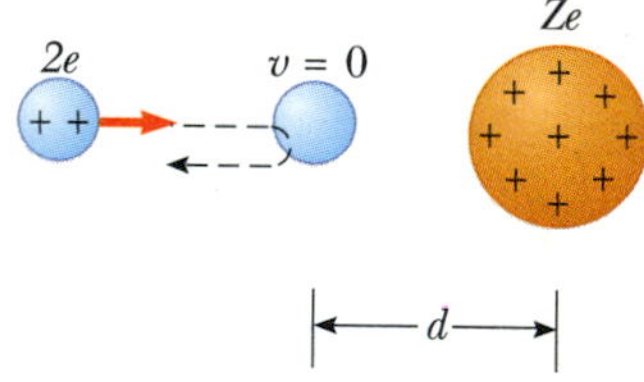

Figure 31.1
An alpha particle on a head-on collision course with a nucleus of charge *Ze*. Because of the Coulomb repulsion between the like charges, the alpha particle stops instantaneously at a distance *d* from the target nucleus, called the distance of closest approach.

From this expression, Rutherford found that alpha particles approached to within 3.2×10^{-14} m of a nucleus when the foil was made of gold. Thus, the radius of the gold nucleus must be less than this value. For silver atoms, the distance of closest approach was 2×10^{-14} m. From these results, Rutherford concluded that the positive charge in an atom is concentrated in a small sphere, which he called the nucleus, whose radius is no greater than about 10^{-14} m. Because such small lengths are common in nuclear physics, a convenient unit of length is the *femtometer* (fm), sometimes called the **fermi,** defined as

$$1\text{ fm} \equiv 10^{-15}\text{ m}$$

Since the time of Rutherford's scattering experiments, a multitude of other experiments have shown that most nuclei are approximately spherical and have an average radius of

$$r = r_0 A^{1/3} \quad [31.1]$$

Radius of a nucleus

where A is the mass number and r_0 is a constant equal to 1.2×10^{-15} m. Because the volume of a sphere is proportional to the cube of the radius, it follows from Equation 31.1 that the volume of a nucleus (assumed to be spherical) is directly proportional to A, the total number of nucleons. This suggests that *all nuclei have nearly the same density*. Nucleons combine to form a nucleus *as though* they were tightly packed spheres (Fig. 31.2).

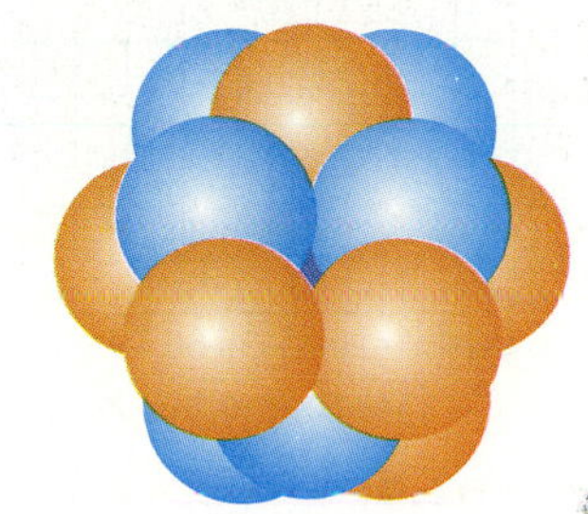

Figure 31.2
A nucleus can be visualized as a cluster of tightly packed spheres, where each sphere is a nucleon.

Example 31.1 Nuclear Volume and Density

Find (a) an approximate expression for the mass of a nucleus of mass number A, (b) an expression for the volume of this nucleus in terms of the mass number, and (c) a numerical value for its density.

Solution (a) The mass of the proton is approximately equal to that of the neutron. Thus, if the mass of one of these particles is m, the mass of the nucleus is approximately Am.

(b) Assuming the nucleus is spherical and using Equation 31.1, we find that the volume is

$$V = \tfrac{4}{3}\pi r^3 = \tfrac{4}{3}\pi r_0^{\,3} A$$

(c) The nuclear density is

$$\rho_n = \frac{\text{mass}}{\text{volume}} = \frac{Am}{\tfrac{4}{3}\pi r_0^{\,3} A} = \frac{3m}{4\pi r_0^{\,3}}$$

$$= \frac{3(1.67 \times 10^{-27}\ \text{kg})}{4\pi(1.2 \times 10^{-15}\ \text{m})^3} = 2.3 \times 10^{17}\ \text{kg/m}^3$$

Note that the nuclear density is about 2.3×10^{14} times as great as the density of water (10^3 kg/m^3)!

Nuclear Stability

Given that the nucleus consists of a closely packed collection of protons and neutrons, you might be surprised that it can exist. The very large repulsive electrostatic forces between protons should cause the nucleus to fly apart. However, nuclei are stable because of the presence of another, short-range (about 2 fm) force, the **nuclear force.** This is an attractive force that acts between all nuclear particles. The protons attract each other via the nuclear force, and at the same time they repel each other through the Coulomb force. The nuclear force also acts between pairs of neutrons and between neutrons and protons.

The nuclear force dominates the Coulomb repulsive force within the nucleus (at short ranges). If this were not the case, stable nuclei would not exist. Moreover, the strong nuclear force is nearly independent of charge. In other words, the

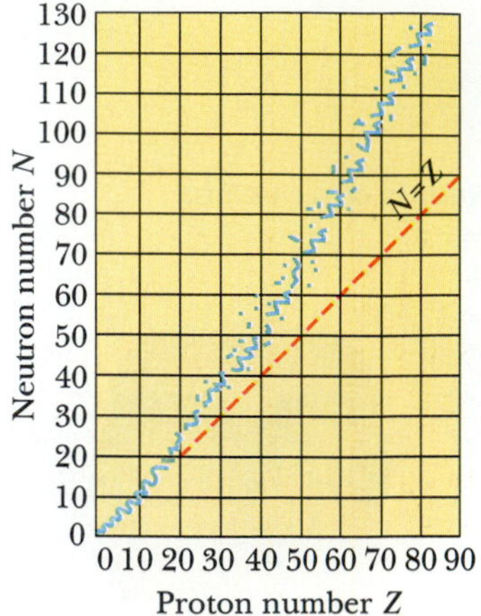

Figure 31.3
A plot of neutron number, N, versus atomic number, Z, for the stable nuclei. The dashed line corresponds to the condition $N = Z$.

nuclear forces associated with the proton-proton, proton-neutron, and neutron-neutron interactions are approximately the same, apart from the additional repulsive Coulomb force for the proton-proton interaction.

There are about 400 stable nuclei; hundreds of others have been observed but are unstable. A plot of N versus Z for a number of stable nuclei is given in Figure 31.3. Note that light nuclei are most stable if they contain equal numbers of protons and neutrons—that is, if $N = Z$—but heavy nuclei are more stable if $N > Z$. This can be partially understood by recognizing that, as the number of protons increases, the strength of the Coulomb force increases, which tends to break the nucleus apart. As a result, more neutrons are needed to keep the nucleus stable, since neutrons experience only the attractive nuclear forces. Eventually, when $Z = 83$, the repulsive forces between protons cannot be compensated by the addition of more neutrons. In effect, the additional neutrons "dilute" the nuclear charge density. Elements that contain more than 83 protons do not have stable nuclei.

It is interesting that most stable nuclei have even values of A. Furthermore, only eight have Z and N both odd. In fact, certain values of Z and N correspond to nuclei with unusually high stability. These values of N and Z, called **magic numbers,** are

$$Z \text{ or } N = 2, 8, 20, 28, 50, 82, 126 \qquad [31.2]$$

Nuclear Spin and Magnetic Moment

In Chapter 30 we discussed the fact that an electron has an intrinsic angular momentum called *spin.* A nucleus also has an intrinsic spin angular momentum. The magnitude of the *nuclear angular momentum* is $\sqrt{I(I+1)}\hbar$, where I is a quantum number called the *nuclear spin* and may be an integer or a half-integer. The maximum component of the nuclear angular momentum projected along any direction is $I\hbar$. Figure 31.4 illustrates the possible orientations of the nuclear spin and its projections along the z axis for the case where $I = \frac{3}{2}$.

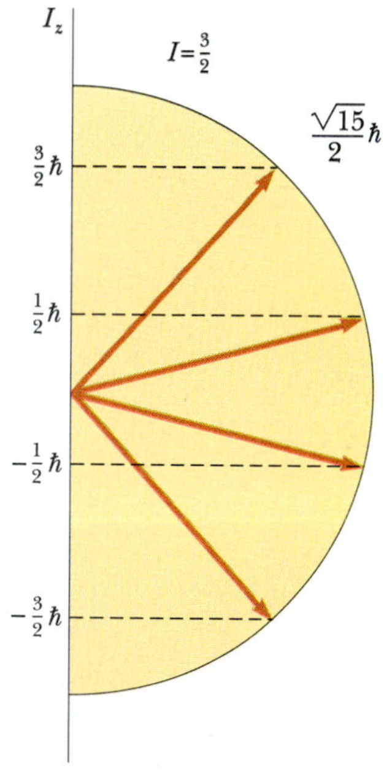

Figure 31.4
The possible orientations of the nuclear spin and its projections along the z axis for the case $I = \frac{3}{2}$.

The nuclear angular momentum has a nuclear magnetic moment associated with it. The *magnetic moment* of a nucleus is measured in terms of the **nuclear magneton,** μ_n, a unit of moment defined as

$$\mu_n \equiv \frac{e\hbar}{2m_p} = 5.05 \times 10^{-27} \text{ J/T} \qquad [31.3]$$

This definition is analogous to Equation 19.29 for the Bohr magneton, μ_B. Note that μ_n is smaller than μ_B by a factor of about 2000, due to the large difference in masses of the proton and electron.

The magnetic moment of a free proton is not μ_n but $2.7928\mu_n$. Unfortunately, there is no general theory of nuclear magnetism that explains this value. Another surprising point is the fact that a neutron also has a magnetic moment, which has a value of $-1.9135\mu_n$. The minus sign indicates that the moment is opposite the neutron's spin angular momentum.

It is interesting that nuclear magnetic moments (as well as electronic magnetic moments) precess in an external magnetic field. The frequency at which they precess, called the **Larmor precessional frequency,** ω_p, is directly proportional to the magnetic field. This is described schematically in Figure 31.5a, where the magnetic field is along the z axis. For example, the Larmor frequency of a proton in a magnetic field of 1 T is equal to 42.577 MHz. Recall that the potential energy

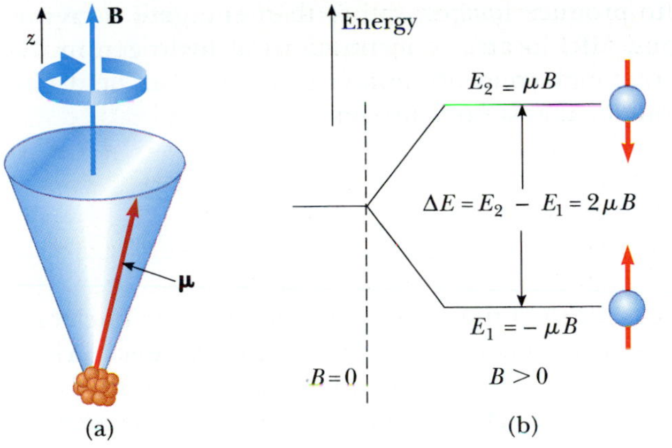

Figure 31.5
(a) When a nucleus is placed in an external magnetic field, the nuclear magnetic moment precesses about the magnetic field with a frequency that is proportional to the field. (b) A proton, whose spin is $\frac{1}{2}$, can occupy one of two energy states when placed in an external magnetic field. The lower energy state, E_1, corresponds to the case in which the spin is aligned with the field, and the higher energy state, E_2, corresponds to the case in which the spin is opposite the field. The reverse is true for electrons.

of a magnetic dipole moment in an external magnetic field is $-\boldsymbol{\mu}\cdot\mathbf{B}$. When the projection of $\boldsymbol{\mu}$ is along the field, the potential energy of the dipole moment is $-\mu B$; that is, it has its minimum value. When the projection of $\boldsymbol{\mu}$ is against the field, the potential energy is μB and it has its maximum value. These two energy states for a nucleus with a spin of $\frac{1}{2}$ are shown in Figure 31.5b.

It is possible to observe transitions between these two spin states by using a technique known as **nuclear magnetic resonance.** A dc magnetic field is introduced to align the magnetic moments (Fig. 31.5a), along with a second, weak, oscillating magnetic field oriented perpendicular to **B**. When the frequency of the oscillating field is adjusted to match the Larmor precessional frequency, a torque acting on the precessing moments causes them to "flip" between the two spin states. These transitions result in a net absorption of energy by the spin system. A diagram of the apparatus used to detect a nuclear magnetic resonance signal is illustrated in Figure 31.6. The absorbed energy is supplied by the generator producing the oscillating magnetic field.

Nuclear magnetic resonance

Nuclear magnetic resonance and electron spin resonance are extremely important methods for studying nuclear and atomic systems and the interactions of these systems with their surroundings. An important modern application of nuclear magnetic resonance is a procedure called *magnetic resonance imaging* (MRI), used in medical diagnostics. The main advantage of MRI over other imaging techniques is that it causes minimal damage to cellular structures, since it uses low-en-

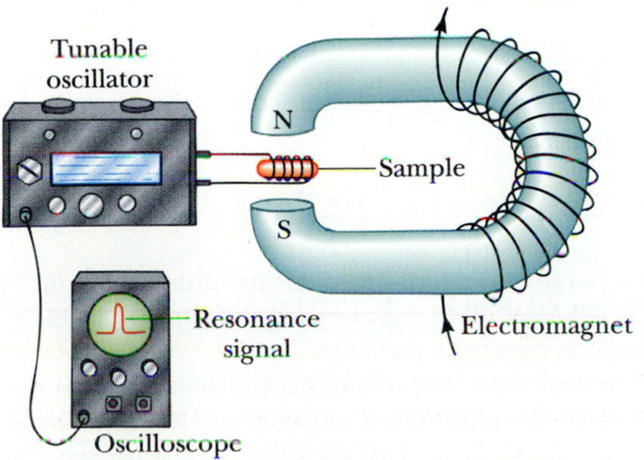

Figure 31.6
An experimental arrangement for nuclear magnetic resonance. The radio-frequency magnetic field of the coil, provided by the variable-frequency oscillator, must be perpendicular to the dc magnetic field. When the nuclei in the sample meet the resonance condition, the spins absorb energy from the field of the coil, which changes the response of the circuit in which the coil is included.

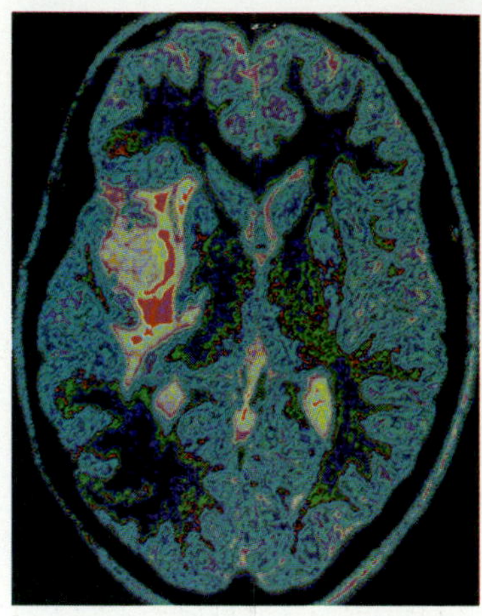

A computer-digitized color-enhanced MRI of a brain with a glioma tumor. *(© Scott Camazine/ Science Source)*

ergy radio-frequency signals to produce images, rather than energetic x-rays or gamma rays. In most situations, MRI locates concentrations of hydrogen in the body. Thus tumors, which have a high concentration of hydrogen, can be distinguished from bones, which contain almost no hydrogen.

31.2 Binding Energy

The mass of a nucleus is always less than the sum of the masses of its nucleons. According to the Einstein mass-energy relationship, if the mass difference, Δm, is multiplied by c^2, we obtain the binding energy of the nucleus. In other words, *the energy of the bound system (the nucleus) is less than the sum of the energies of the separated nucleons.* Therefore, in order to separate a nucleus into protons and neutrons, energy must be delivered to the system.

In addition to radioactive decay, two other important processes result in energy release from the nucleus. In **nuclear fission,** a nucleus splits into two or more fragments. In **nuclear fusion,** two or more nuclei combine to form a heavier nucleus. In any fission or fusion reaction, the total rest mass of the products is less than that of the reactants. The decrease in rest mass is accompanied by a decrease of energy of the system. The following example illustrates this point for the deuteron.

Example 31.2 The Binding Energy of the Deuteron

The nucleus of the deuterium atom, called the **deuteron,** consists of a proton and a neutron. Calculate the deuteron's binding energy, given that its mass is 2.014102 u.

Solution We know from Table 31.1 that

$$m_p = 1.007825 \text{ u} \qquad m_n = 1.008665 \text{ u}$$

Note that the masses used for the proton and deuteron in this example are actually atomic masses. We can use atomic masses to calculate mass differences, since the electron masses cancel. Therefore,

$$m_p + m_n = 2.016490 \text{ u}$$

To calculate the mass difference, we subtract the deuteron mass from this value:

$$\Delta m = (m_p + m_n) - m_d$$
$$= 2.016490 \text{ u} - 2.014102 \text{ u} = 0.002388 \text{ u}$$

Since 1 u corresponds to 931.50 MeV (that is, $1 \text{ u} \cdot c^2 = 931.50$ MeV), the mass difference corresponds to the binding energy

$$E_b = (0.002388 \text{ u})(931.5 \text{ MeV/u}) = 2.224 \text{ MeV}$$

This result tells us that, in order to separate a deuteron into its constituent proton and neutron, it is necessary to add 2.224 MeV of energy. One way of supplying this energy is by bombarding the nucleus with energetic particles.

If the binding energy of a nucleus were zero, the nucleus could separate into its constituent protons and neutrons without the addition of any energy; that is, it could spontaneously break apart.

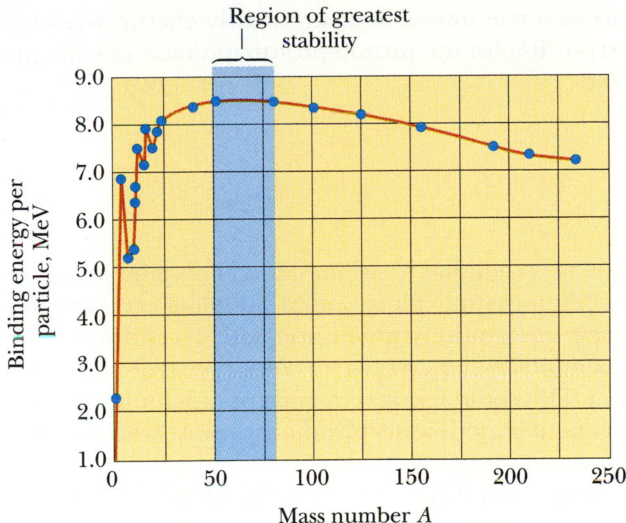

Figure 31.7
A plot of binding energy per nucleon versus mass number for nuclei that lie along the line of stability in Figure 31.3.

The binding energy of any nucleus can be calculated from

Binding energy of a nucleus

$$E_b\ (\text{MeV}) = [Zm_H + Nm_n - M(^A_Z\text{X})] \times 931.50\ \text{MeV/u} \qquad [31.4]$$

where m_H is the atomic mass of hydrogen, $M(^A_Z\text{X})$ represents the atomic mass of the nucleus, and the masses are all in unified mass units. Note that *atomic* masses rather than *nuclear* masses are used in these calculations, because tables usually give atomic masses.[1]

It is interesting to examine a plot of binding energy per nucleon, E_b/A, as a function of mass number for various stable nuclei (Fig. 31.7). Except for the lighter nuclei, the average binding energy per nucleon is about 8 MeV. Note that the curve peaks in the vicinity of $A = 60$. That is, nuclei with mass numbers greater or less than 60 are not as strongly bound as those near the middle of the periodic table. As we shall see later, this fact allows energy to be released in fission and fusion reactions. The curve is slowly varying for $A > 40$, which suggests that the nuclear force saturates. In other words, a particular nucleon can interact with only a limited number of other nucleons, which can be viewed as the "nearest neighbors" in the close-packed structure illustrated in Figure 31.2.

Scattering experiments establish that the proton-proton interaction can be represented by the potential energy curve in Figure 31.8a. For large values of the proton-proton separation, r, the Coulomb repulsive force is clearly dominant, corresponding to a positive potential energy. When r is reduced to about 3 fm, a break to negative values of potential energy occurs in the curve. That is, as the proton-proton separation approaches the nuclear radius, the attractive nuclear force overcomes the repulsive Coulomb force. In contrast, the proton-neutron interaction can be represented by the diagram in Figure 31.8b. In this case, because there is no Coulomb force, the potential energy is zero at large values of r. However, when the proton-neutron separation is about 2 fm, the nucleons are attracted by the nuclear force. The depths of the potential energy minima (corresponding to the nuclear binding energies) are about the same for proton-proton

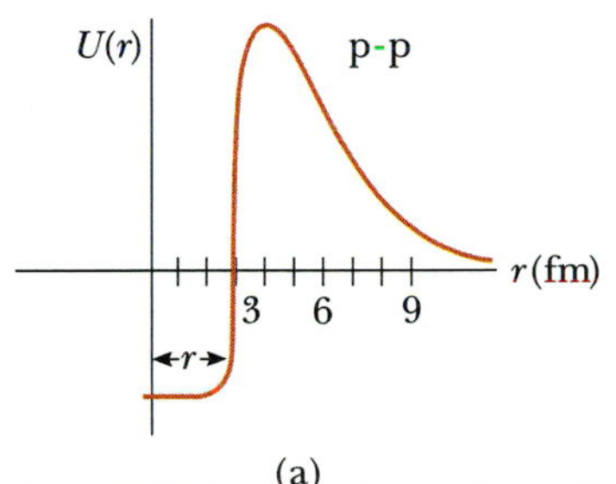

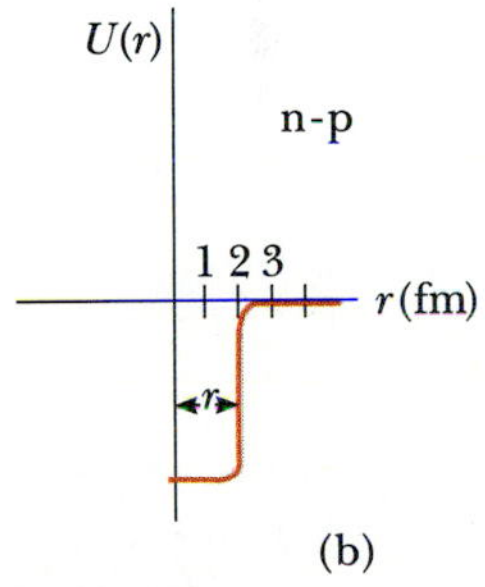

Figure 31.8
(a) Potential energy versus separation for the proton-proton system. (b) Potential energy versus separation for the neutron-proton system. The difference in the two curves is due mainly to the large Coulomb repulsion in the case of the proton-proton interaction.

[1] It is possible to do this because electron masses cancel in such calculations. One exception is the β^+ decay process.

Marie Curie (1867–1934)
A Polish scientist, Marie Curie shared the Nobel prize in 1903 with her husband, Pierre, and with Becquerel for their work on spontaneous radioactivity and the radiation emitted by radioactive substances. "I persist in believing that the ideas that then guided us are the only ones which can lead to the true social progress. We cannot hope to build a better world without improving the individual. Toward this end, each of us must work toward his own highest development, accepting at the same time his share of responsibility in the general life of humanity."

and proton-neutron curves, because the interactions are nearly charge-independent. However, the Coulomb repulsion for the proton-proton interaction explains the larger value for r in this case.

31.3 Radioactivity

In 1896 Becquerel accidentally discovered that uranium salt crystals emit an invisible radiation that can darken a photographic plate even if the plate is covered to exclude light. After several such observations under controlled conditions, he concluded that the radiation emitted by the crystals was of a new type, one that required no external stimulation. This spontaneous emission of radiation was soon to be called **radioactivity.** Subsequent experiments by other scientists showed that other substances were also radioactive.

The most significant investigations of this type were conducted by Marie and Pierre Curie. After several years of careful and laborious chemical separation processes on tons of pitchblende, a radioactive ore, the Curies reported the discovery of two previously unknown elements, both of which were radioactive. These were named polonium and radium. Subsequent experiments, including Rutherford's famous work on alpha-particle scattering, suggested that radioactivity was the result of the decay, or disintegration, of unstable nuclei.

Three types of radiation can be emitted by a radioactive substance: alpha (α) rays, where the emitted particles are ^{4}He nuclei; beta (β) rays, in which the emitted particles are either electrons or positrons; and gamma (γ) rays, in which the emitted rays are high-energy photons. A positron is a particle similar to the electron in all respects except that it has a charge of $+e$ (the positron is said to be the **antiparticle** of the electron). In discussions of beta decay, the symbol β^- is used to designate an electron and β^+ to designate a positron.

It is possible to distinguish these three forms of radiation using the scheme illustrated in Figure 31.9. The radiation from a radioactive sample is directed into a region in which there is a magnetic field. The beam splits into three components, two bending in opposite directions and the third experiencing no change in direc-

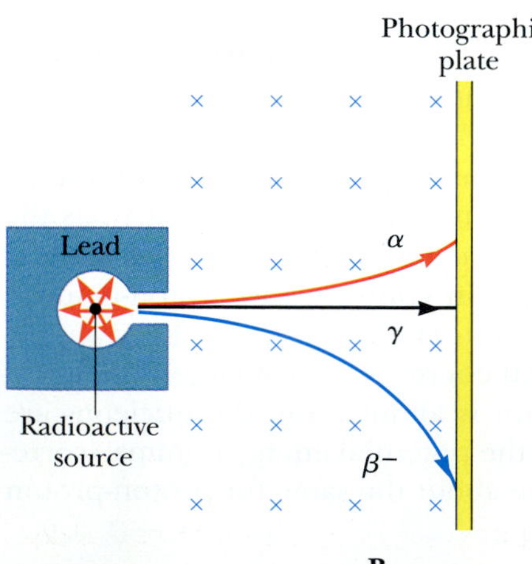

Figure 31.9
The radiation from a radioactive source can be separated into three components by a magnetic field. The photographic plate at the right records the events.

tion. From this simple observation, one can conclude that the radiation of the undeflected beam carries no charge (the gamma ray), the component deflected upward corresponds to positively charged particles (alpha particles), and the component deflected downward corresponds to negatively charged particles (β^-). If the beam includes a positron (β^+), it is deflected upward.

The three types of radiation have quite different penetrating powers. Alpha particles barely penetrate a sheet of paper, beta particles can penetrate a few millimeters of aluminum, and gamma rays can penetrate several centimeters of lead.

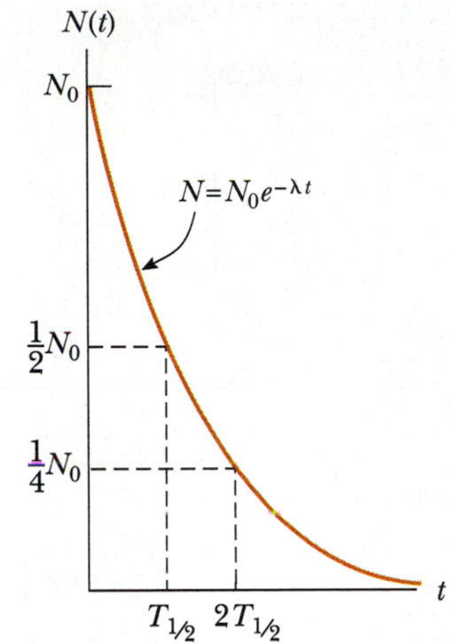

Figure 31.10
A plot of the exponential decay law for radioactive nuclei. The vertical axis represents the number of radioactive nuclei present at any time t, and the horizontal axis is time. The time $T_{1/2}$ is the half-life of the sample.

The rate at which a particular decay process occurs in a radioactive sample is proportional to the number of radioactive nuclei present (that is, those nuclei that have not yet decayed). If N is the number of radioactive nuclei present at some instant, the rate of change of N is

$$\frac{dN}{dt} = -\lambda N \qquad [31.5]$$

where λ is called either the **decay constant** or the **disintegration constant.** The minus sign indicates that dN/dt is negative; that is, N is decreasing in time.

If we write Equation 31.5 in the form

$$\frac{dN}{N} = -\lambda\, dt$$

we can integrate:

$$\int_{N_0}^{N} \frac{dN}{N} = -\lambda \int_0^t dt$$

$$\ln\left(\frac{N}{N_0}\right) = -\lambda t$$

$$N = N_0 e^{-\lambda t} \qquad [31.6]$$

Exponential decay

The constant N_0 represents the number of radioactive nuclei at $t = 0$.

The **decay rate,** R $(= |dN/dt|)$, can be obtained by differentiating Equation 31.6 with respect to time:

$$R = \left|\frac{dN}{dt}\right| = N_0\lambda e^{-\lambda t} = R_0 e^{-\lambda t} \qquad [31.7]$$

Decay rate

where $R = N\lambda$ and $R_0 = N_0\lambda$ is the decay rate at $t = 0$. The decay rate of a sample is often referred to as its **activity.** Note that both N and R decrease exponentially with time. The plot of N versus t in Figure 31.10 illustrates the exponential decay law.

Another parameter that is useful for characterizing radioactive decay is the half-life, $T_{1/2}$. The **half-life** of a radioactive substance is the time it takes half of a given number of radioactive nuclei to decay. Setting $N = N_0/2$ and $t = T_{1/2}$ in Equation 31.6 gives

$$\frac{N_0}{2} = N_0 e^{-\lambda T_{1/2}}$$

Writing this in the form $e^{\lambda T_{1/2}} = 2$ and taking the natural logarithm of both sides, we get

Half-life equation

$$T_{1/2} = \frac{\ln 2}{\lambda} = \frac{0.693}{\lambda} \quad [31.8]$$

This is a convenient expression relating the half-life to the decay constant. Note that after an elapsed time of one half-life, $N_0/2$ radioactive nuclei remain (by definition); after two half-lives, half of these will have decayed and $N_0/4$ radioactive nuclei will be left; after three half-lives, $N_0/8$ will be left; and so on. In general, after n half-lives, the number of radioactive nuclei remaining is $N_0/2^n$. Thus, we see that nuclear decay is independent of the past history of the sample.

The unit of activity is the **curie** (Ci), defined as

The curie

$$1 \text{ Ci} \equiv 3.7 \times 10^{10} \text{ decays/s}$$

This unit was selected as the original activity unit because it is the approximate activity of 1 g of radium. The SI unit of activity is called the **becquerel** (Bq):

The becquerel

$$1 \text{ Bq} \equiv 1 \text{ decay/s}$$

Therefore, 1 Ci = 3.7×10^{10} Bq. The most commonly used units of activity are millicuries and microcuries.

Example 31.3 The Activity of Radium

The half-life of the radioactive nucleus $^{226}_{86}\text{Ra}$ is 1.6×10^3 years. If a sample contains 3×10^{16} such nuclei, determine the activity at this time.

Solution First, let us convert the half-life to seconds:

$$T_{1/2} = (1.6 \times 10^3 \text{ years})(3.16 \times 10^7 \text{ s/year}) = 5.0 \times 10^{10} \text{ s}$$

Now we can use this value in Equation 31.8 to get the decay constant:

$$\lambda = \frac{0.693}{T_{1/2}} = \frac{0.693}{5.0 \times 10^{10} \text{ s}} = 1.4 \times 10^{-11} \text{ s}^{-1}$$

We can calculate the activity of the sample at $t = 0$ using $R_0 = \lambda N_0$, where R_0 is the decay rate at $t = 0$ and N_0 is the number of radioactive nuclei present at $t = 0$:

$$R_0 = \lambda N_0 = (1.4 \times 10^{-11} \text{ s}^{-1})(3 \times 10^{16}) = 4.1 \times 10^5 \text{ decays/s}$$

Since 1 Ci = 3.7×10^{10} decays/s, the activity, or decay rate, at $t = 0$ is

$$R_0 = 11.1 \ \mu\text{Ci}$$

Example 31.4 The Activity of Carbon

A radioactive sample contains 3.50 μg of pure $^{11}_{6}\text{C}$, which has a half-life of 20.4 min.

(a) Determine the number of nuclei present initially.

Solution The atomic mass of $^{11}_{6}\text{C}$, is approximately 11, and therefore 11 g contains Avogadro's number of nuclei. Hence, 3.50 μg contains N nuclei, where

$$\frac{N}{6.02 \times 10^{23} \text{ nuclei/mol}} = \frac{3.50 \times 10^{-6} \text{ g}}{11 \text{ g/mol}}$$

$$N = \boxed{1.92 \times 10^{17} \text{ nuclei}}$$

(b) What is the activity of the sample initially and after 8 h?

Solution Since $T_{1/2} = 20.4 \text{ min} = 1224 \text{ s}$, the decay constant is

$$\lambda = \frac{0.693}{T_{1/2}} = \frac{0.693}{1224 \text{ s}} = 5.66 \times 10^{-4} \text{ s}^{-1}$$

Therefore, the initial activity of the sample is

$$R_0 = \lambda N_0 = (5.66 \times 10^{-4} \text{ s}^{-1})(1.92 \times 10^{17})$$
$$= 1.08 \times 10^{14} \text{ decays/s}$$

We can use Equation 31.7 to find the activity at any time t. For $t = 8 \text{ h} = 2.88 \times 10^4 \text{ s}$, we see that $\lambda t = 16.3$, and so

$$R = R_0 e^{-\lambda t} = (1.09 \times 10^{14} \text{ decays/s}) e^{-16.3}$$

$$= \boxed{8.96 \times 10^6 \text{ decays/s}}$$

Exercise Calculate the number of radioactive nuclei remaining after 8 h.

Answer $N = 1.58 \times 10^{10}$ nuclei.

The hands and numbers of this luminous watch contain minute amounts of radium salt. The radioactive decay of radium causes the watch to glow in the dark. *(© Richard Megna 1990, Fundamental Photographs)*

31.4 The Decay Processes

As we stated in the preceding section, a radioactive nucleus spontaneously decays via alpha, beta, or gamma decay. Let us discuss these three processes in more detail.

Alpha Decay

If a nucleus emits an alpha particle (^4_2He), it loses two protons and two neutrons. Therefore, N decreases by 2, Z decreases by 2, and A decreases by 4. The decay can be written symbolically as

Alpha decay

$$^A_Z\text{X} \longrightarrow {}^{A-4}_{Z-2}\text{Y} + {}^4_2\text{He} \quad [31.9]$$

where X is called the **parent nucleus** and Y the **daughter nucleus.** As examples, ^{238}U and ^{226}Ra are both alpha emitters and decay according to the schemes

$$^{238}_{92}\text{U} \longrightarrow {}^{234}_{90}\text{Th} + {}^4_2\text{He} \quad [31.10]$$

$$^{226}_{88}\text{Ra} \longrightarrow {}^{222}_{86}\text{Rn} + {}^4_2\text{He} \quad [31.11]$$

The half-life for ^{238}U decay is 4.47×10^9 years, and the half-life for ^{226}Ra decay is 1.60×10^3 years. In both cases, note that the A of the daughter nucleus is 4 less

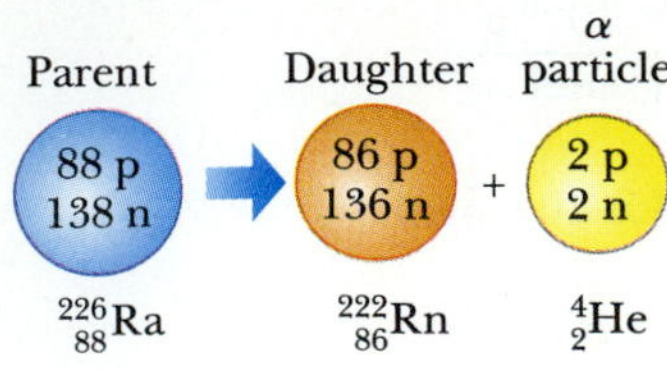

Figure 31.11
Alpha decay of the $^{226}_{88}Ra$ nucleus.

than that of the parent nucleus. Likewise, Z is reduced by 2. The differences are accounted for in the emitted alpha particle (the ^{4}He nucleus).

The decay of ^{226}Ra is shown in Figure 31.11. When one element changes into another, as happens in alpha decay, the process is called *spontaneous decay*. As a general rule, (1) the sum of the mass numbers A must be the same on both sides of the equation, and (2) the sum of the atomic numbers Z must be the same on both sides of the equation. In addition, the total energy must be conserved. If we call M_X the mass of the parent nucleus, M_Y the mass of the daughter nucleus, and M_α the mass of the alpha particle, we can define the **disintegration energy, Q**:

$$Q \equiv (M_X - M_Y - M_\alpha)c^2 \qquad [31.12]$$

Note that Q will be in joules if the masses are in kilograms, and c is the usual 3×10^8 m/s. However, when the nuclear masses are expressed in the more convenient atomic mass unit u, the value of Q can be calculated in MeV units using the expression

$$Q = (M_X - M_Y - M_\alpha) \times 931.50 \text{ MeV/u} \qquad [31.13]$$

It is left to end-of-chapter Problem 31 to show that 931.50 MeV/u is the correct conversion factor.

The disintegration energy Q appears in the form of kinetic energy of the daughter nucleus and the alpha particle. The quantity given by Equation 31.12 is sometimes referred to as the Q value of the nuclear reaction. In the case of the ^{226}Ra decay described in Figure 31.11, if the parent nucleus decays at rest, the residual kinetic energy of the products is 4.87 MeV. Most of the kinetic energy is associated with the alpha particle because this particle is much less massive than the recoiling daughter nucleus. That is, since momentum must be conserved, the lighter alpha particle recoils with a much higher speed than the daughter nucleus. Generally, light particles carry off most of the energy in nuclear decays.

Finally, it is interesting to note that if one assumed that ^{238}U (or other alpha emitters) decayed by emitting protons and neutrons, the mass of the decay products would exceed that of the parent nucleus, corresponding to negative Q values. Therefore, such spontaneous decays do not occur.

Enrico Fermi (1901–1954)
An Italian physicist, Fermi was awarded the Nobel prize in 1938 for producing the transuranic elements by neutron irradiation and for his discovery of nuclear reactions brought about by slow neutrons. He made many other outstanding contributions to physics including his theory of beta decay, the free electron theory of metals, and the development of the world's first fission reactor in 1942. Fermi was truly a gifted theoretical and experimental physicist. He was also well known for his ability to present physics in a clear and exciting manner. "Whatever Nature has in store for mankind, unpleasant as it may be, men must accept, for ignorance is never better than knowledge."

Example 31.5 The Energy Liberated When Radium Decays

The ^{226}Ra nucleus undergoes alpha decay according to Equation 31.11. Calculate the Q value for this process. Take the mass of ^{226}Ra to be 226.025406 u, the mass of ^{222}Rn to be 222.017574 u, and the mass of $^4_2$He to be 4.002603 u, as found in Table A.3.

Solution

$$Q = (M_X - M_Y - M_\alpha) \times 931.50\text{ MeV/u}$$

$$= (226.025406\text{ u} - 222.017574\text{ u} - 4.002603\text{ u}) \times 931.50\text{ MeV/u}$$

$$= (0.005229\text{ u}) \times \left(931.50\,\frac{\text{MeV}}{\text{u}}\right) = 4.87\text{ MeV}$$

It is left to end-of-chapter Problem 67 to show that the kinetic energy of the alpha particle is about 4.8 MeV, whereas that of the recoiling daughter nucleus is only about 0.1 MeV.

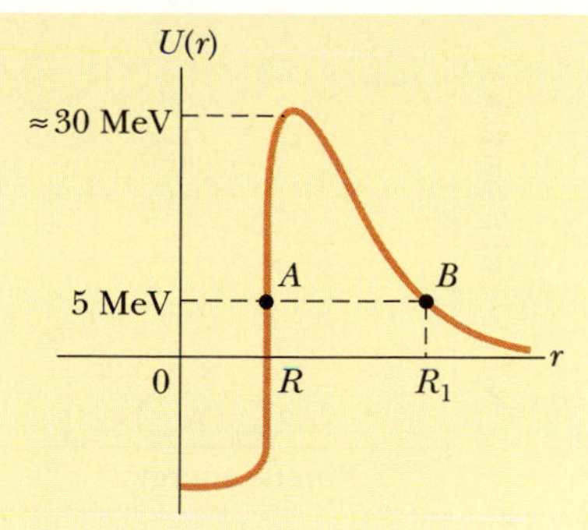

Figure 31.12 Potential energy versus separation for the alpha particle–nucleus system. Classically, the energy of the alpha particle is not sufficient to overcome the barrier, and so the particle should not be able to escape the nucleus.

We now turn to the mechanism of alpha decay. Figure 31.12 is a plot of the potential energy versus distance r from the nucleus for the alpha particle–nucleus system, where R is the range of the nuclear force. The curve represents the combined effects of (1) the Coulomb repulsive energy, which gives the positive peak for $r > R$, and (2) the nuclear attractive force, which causes the curve to be negative for $r < R$. As we saw in Example 31.5, the disintegration energy is about 5 MeV, which is the approximate kinetic energy of the alpha particle, represented by the lower dotted line in Figure 31.12. According to classical physics, the alpha particle is trapped in the potential well. How, then, does it ever escape from the nucleus?

The answer to this question was provided by Gamow and, independently, Gurney and Condon in 1928, using quantum mechanics. Briefly, the view of quantum mechanics is that there is always some probability that the particle can penetrate (or tunnel) through the barrier (Section 29.11). Recall that the probability of locating the particle depends on its wave function, ψ, and that the probability of tunneling is measured by $|\psi|^2$. Figure 31.13 is a sketch of the wave function for a particle of energy E, meeting a square barrier of finite height, which approximates the nuclear barrier. Note that the wave function is oscillating both inside and outside the barrier but is greatly reduced in amplitude because of the barrier. As the energy E increases, the probability of escaping also increases. Furthermore, the probability increases as the width of the barrier is decreased.

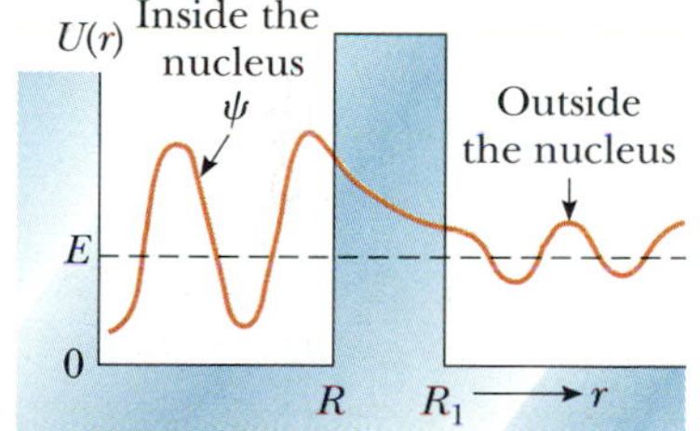

Figure 31.13 The nuclear potential energy is modeled as a square barrier. The energy of the alpha particle is E, which is less than the height of the barrier. According to quantum mechanics, the alpha particle has some chance of tunneling through the barrier, as indicated by the finite size of the wave function for $r < R_1$.

Beta Decay

When a radioactive nucleus undergoes beta decay, the daughter nucleus has the same number of nucleons as the parent nucleus, but the atomic number is changed by 1:

$$^{A}_{Z}\text{X} \longrightarrow {}_{Z+1}^{\;\;A}\text{Y} + \beta^- \qquad [31.14]$$

$$^{A}_{Z}\text{X} \longrightarrow {}_{Z-1}^{\;\;A}\text{Y} + \beta^+ \qquad [31.15]$$

Again, note that the nucleon number and total charge are both conserved in these decays. However, as we shall see later, *these processes are not described completely by these expressions.* We shall give reasons for this shortly.

It is important to note that the electron or positron involved in these decays is created within the nucleus as an initial step in the decay process. This is equivalent

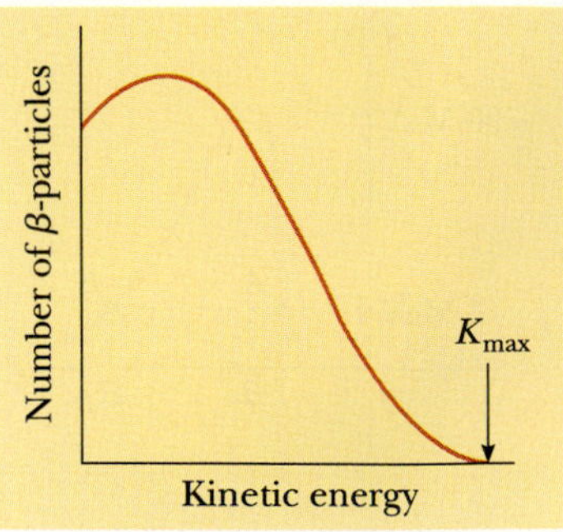

Figure 31.14
A typical beta decay curve. The maximum kinetic energy observed for the beta particles corresponds to the value of Q for the reaction.

to saying that, during beta decay, a neutron in the nucleus is transformed into a proton. Indeed, the process $n \rightarrow p + \beta^-$ does occur, although it is extremely improbable outside the nucleus.

Now consider the energy of the system before and after the decay. As with alpha decay, energy must be conserved. Experimentally, one finds that the beta particles are emitted over a continuous range of energies (Fig. 31.14). The kinetic energy of the electrons must be balanced by the decrease in mass of the system, that is, the Q value. However, since all decaying nuclei have the same initial mass, *the Q value must be the same for each decay.* In view of this, why do the emitted electrons have different kinetic energies? The principle of conservation of energy seems to be violated! Further analysis shows that, according to the decay processes given by Equations 31.14 and 31.15, the principles of conservation of both angular momentum (spin) and linear momentum are also violated!

After a great deal of experimental and theoretical study, Pauli in 1930 proposed that a third particle must be present to carry away the "missing" energy and momentum. Fermi later named this particle the **neutrino** (little neutral one) since it had to be electrically neutral and have little or no rest mass. Although it eluded detection for many years, the neutrino (symbol ν) was finally detected experimentally in 1956. It has the following properties:

Properties of the neutrino

1. It has zero electric charge.
2. It has a rest mass smaller than that of the electron, and in fact its mass may be zero (although modern experiments suggest that this may not be true).
3. It has a spin of $\frac{1}{2}$, which satisfies the law of conservation of angular momentum.
4. It interacts very weakly with matter and is therefore quite difficult to detect.

We can now write the beta decay processes in their correct form:

$$^{14}_{6}\text{C} \longrightarrow {}^{14}_{7}\text{N} + \beta^- + \bar{\nu} \qquad [31.16]$$

$$^{12}_{7}\text{N} \longrightarrow {}^{12}_{6}\text{C} + \beta^+ + \nu \qquad [31.17]$$

$$\text{n} \longrightarrow \text{p} + \beta^- + \bar{\nu} \qquad [31.18]$$

where $\bar{\nu}$ represents the **antineutrino,** the antiparticle to the neutrino. We shall discuss antiparticles further in Chapter 32. For now, it suffices to say that *a neutrino is emitted in positron decay, and an antineutrino is emitted in electron decay.*

Carbon Dating

The beta decay of ^{14}C to $^{14}_{7}$N plus an electron, β^-, is used to date organic samples. Cosmic rays (high-energy particles from outer space) in the upper atmosphere create ^{14}C. The ratio of ^{14}C to ^{12}C in the carbon dioxide molecules of our atmosphere has a constant value of about 1.3×10^{-12}. All living organisms have the same ratio of ^{14}C to ^{12}C because they continuously exchange carbon dioxide with their surroundings. When an organism dies, however, it no longer absorbs ^{14}C from the atmosphere, and so the ratio of ^{14}C to ^{12}C decreases as the result of the beta decay of ^{14}C. It is therefore possible to determine the age of a material by measuring its activity per unit mass due to the decay of ^{14}C. Using carbon dating, samples of wood, charcoal, bone, and shell have been identified as having lived 1000 to 25 000 years ago.

Example 31.6 Radioactive Dating

A piece of charcoal of mass 25 g is found in the ruins of an ancient city. The sample shows a ^{14}C ($T_{1/2}$ = 5730 years) activity of 250 decays/min. How long has the tree from which this charcoal came been dead?

Solution First, let us calculate the decay constant for ^{14}C:

$$\lambda = \frac{0.693}{T_{1/2}} = \frac{0.693}{(5730 \text{ y})(3.16 \times 10^{7} \text{ s/y})} = 3.83 \times 10^{-12} \text{ s}^{-1}$$

The number of ^{14}C nuclei can be calculated in two steps. First, the number of ^{12}C nuclei in 25 g of carbon is

$$N(^{12}\text{C}) = \frac{6.02 \times 10^{23} \text{ nuclei/mol}}{12 \text{ g/mol}}(25 \text{ g}) = 1.25 \times 10^{24} \text{ nuclei}$$

Assuming that the ratio of ^{14}C to ^{12}C is 1.3×10^{-12}, we see that the number of ^{14}C nuclei in 25 g *before* decay is

$$N_0(^{14}\text{C}) = (1.3 \times 10^{-12})(1.26 \times 10^{24}) = 1.63 \times 10^{12} \text{ nuclei}$$

Hence, the initial activity of the sample is

$$R_0 = N_0\lambda = (1.63 \times 10^{12} \text{ nuclei})(3.83 \times 10^{-12} \text{ s}^{-1})$$

$$= 6.25 \text{ decays/s} = 375 \text{ decays/min}$$

We can now calculate the age of the charcoal using Equation 31.7, which relates the activity R at any time t to the initial activity, R_0:

$$R = R_0 e^{-\lambda t} \qquad \text{or} \qquad e^{-\lambda t} = \frac{R}{R_0}$$

Since it is given that R = 250 decays/min, and since we found that R_0 = 375 decays/min, we can calculate t by taking the natural logarithm of both sides of the last equation:

$$-\lambda t = \ln\left(\frac{R}{R_0}\right) = \ln\left(\frac{250}{375}\right) = -0.405$$

$$t = \frac{0.405}{\lambda} = \frac{0.405}{3.84 \times 10^{-12} \text{ s}^{-1}}$$

$$= 1.06 \times 10^{11} \text{ s} = \boxed{3350 \text{ years}}$$

This technique of radioactive-carbon dating has been successfully used to measure the ages of many organic relics up to 25 000 years old.

A process that competes with beta decay is called **electron capture.** This occurs when a parent nucleus captures one of its own orbiting electrons and emits a neutrino. The final product after decay is a nucleus whose charge is $Z - 1$:

$$^{A}_{Z}\text{X} + {}^{0}_{-1}\text{e} \longrightarrow {}_{Z-1}^{\;\;\;A}\text{X} + \nu \qquad [31.19]$$

Electron capture process

In most cases it is a K-shell electron that is captured, and the event is termed K capture.

Table 31.2
Decay Pathways

Alpha decay	${}^{A}_{Z}\text{X} \longrightarrow {}^{A-4}_{Z-2}\text{X} + {}^{4}_{2}\text{He}$
Beta decay (β^-)	${}^{A}_{Z}\text{X} \longrightarrow {}_{Z+1}^{A}\text{X} + \beta^- + \bar{\nu}$
Beta decay (β^+)	${}^{A}_{Z}\text{X} \longrightarrow {}_{Z-1}^{A}\text{X} + \beta^+ + \nu$
Electron capture	${}^{A}_{Z}\text{X} + {}^{0}_{-1}\text{e} \longrightarrow {}_{Z-1}^{A}\text{X} + \nu$
Gamma decay	${}^{A}_{Z}\text{X}^* \longrightarrow {}^{A}_{Z}\text{X} + \gamma$

Gamma Decay

Very often, a nucleus that undergoes radioactive decay is left in an excited energy state. It can then undergo a second decay to a lower energy state, perhaps to the ground state, by emitting a photon. Photons emitted in such a de-excitation process are called gamma rays. Such photons have very high energy (in the range of 1 MeV to 1 GeV) relative to the energy of visible light (about 1 eV). Recall from Chapter 30 that the energy of photons emitted (or absorbed) by an atom equals the difference in energy between the two electronic states involved in the transition. Similarly, a gamma ray photon has an energy, hf, that equals the energy difference, ΔE, between two nuclear energy levels. When a nucleus decays by emitting a gamma ray, the nucleus does not change, apart from the fact that it ends up in a lower energy state. We can represent gamma decay as

Gamma decay

$$ {}^{A}_{Z}\text{X}^* \longrightarrow {}^{A}_{Z}\text{X} + \gamma \quad [31.20] $$

where X* indicates a nucleus in an excited state.

A nucleus may reach an excited state as the result of a violent collision with another particle. However, it is more common for a nucleus to be in an excited state after it has undergone an alpha or beta decay. The following sequence of events represents a typical situation in which gamma decay occurs:

$$ {}^{12}_{5}\text{B} \longrightarrow {}^{12}_{6}\text{C}^* + {}^{0}_{-1}\text{e} + \bar{\nu} \quad [31.21] $$

$$ {}^{12}_{6}\text{C}^* \longrightarrow {}^{12}_{6}\text{C} + \gamma \quad [31.22] $$

Figure 31.15 shows the decay scheme for ^{12}B, which undergoes beta decay to either of two levels of ^{12}C. It can either (1) decay directly to the ground state of ^{12}C by emitting a 13.4-MeV electron or (2) undergo β^- decay to an excited state of ^{12}C*, followed by gamma decay to the ground state. The latter process results in the emission of a 9.0-MeV electron and a 4.4-MeV photon.

The pathways by which a radioactive nucleus can undergo decay are summarized in Table 31.2.

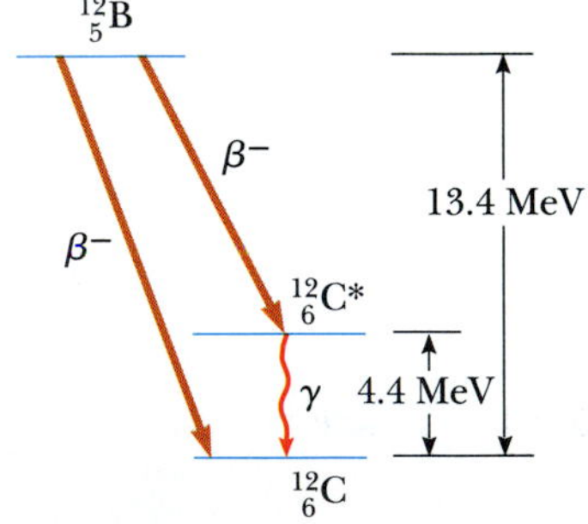

Figure 31.15
The ^{12}B nucleus undergoes β^- decay to two levels of ^{12}C. The decay to the excited level, ^{12}C*, is followed by gamma decay to the ground state.

31.5 Natural Radioactivity

Radioactive nuclei are generally classified into two groups: (1) unstable nuclei that are found in nature, which give rise to what is called **natural radioactivity,** and (2) nuclei produced in the laboratory through nuclear reactions, which exhibit **artificial radioactivity.**

Table 31.3
The Four Radioactive Series

Series		Starting Isotope	Half-Life (years)	Stable End Product
Uranium	Natural	$^{238}_{92}U$	4.47×10^9	$^{206}_{82}Pb$
Actinium	Natural	$^{235}_{92}U$	7.04×10^8	$^{207}_{82}Pb$
Thorium	Natural	$^{232}_{90}Th$	1.41×10^{10}	$^{208}_{82}Pb$
Neptunium		$^{237}_{93}Np$	2.14×10^6	$^{209}_{83}Bi$

Three series of radioactive nuclei occur naturally (Table 31.3). Each series starts with a specific long-lived radioactive isotope whose half-life exceeds that of any of its descendents. The fourth series in Table 31.3 begins with ^{237}Np, a transuranic element (one having an atomic number greater than that of uranium) not found in nature. This element has a half-life of "only" 2.14×10^6 years.

The two uranium series are somewhat more complex than the ^{232}Th series (Fig. 31.16). Also, there are several naturally occurring radioactive isotopes, such as ^{14}C and ^{40}K, that are not part of either decay series.

Natural radioactivity constantly resupplies our environment with radioactive elements that would otherwise have disappeared long ago. For example, because the Solar System is about 5×10^9 years old, the supply of ^{226}Ra (whose half-life is only 1600 years) would have been depleted by radioactive decay long ago if it were not for the decay series that starts with ^{238}U, with a half-life of 4.47×10^9 years.

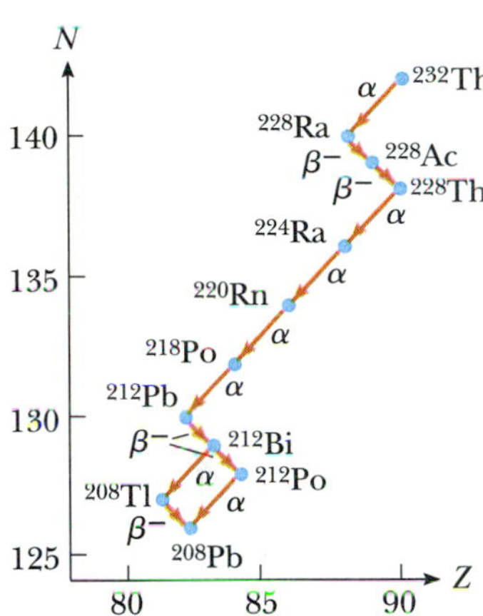

Figure 31.16
Successive decays for the ^{232}Th series.

31.6 Nuclear Reactions

It is possible to change the structures and properties of nuclei by bombarding them with energetic particles. Such changes are called **nuclear reactions.** In 1919 Rutherford was the first to observe nuclear reactions, using naturally occurring radioactive sources for the bombarding particles. Since then, thousands of nuclear reactions have been observed following the development of charged-particle accelerators in the 1930s. With today's advanced technology in particle accelerators and particle detectors, it is possible to achieve particle energies of at least 1000 GeV = 1 TeV. These high-energy particles are used to create new particles whose properties are helping to solve the mysteries of the nucleus.

Consider a reaction in which a target nucleus, X, is bombarded by a particle, a, resulting in a nucleus, Y, and a particle, b:

Nuclear reaction

$$a + X \longrightarrow Y + b \quad [31.23]$$

Sometimes this reaction is written in the more compact form

$$X(a, b)Y$$

Reaction energy *Q*

In the preceding section, the Q value, or disintegration energy, associated with radioactive decay was defined as the energy released during the decay process. Likewise, we define the **reaction energy,** Q, associated with a nuclear reaction as *the total energy released as the result of the reaction:*

$$Q = (M_a + M_X - M_Y - M_b)c^2 \quad [31.24]$$

Exothermic reaction

A reaction such as this, for which Q is positive, is called **exothermic.** After the reaction, this energy appears as an increase in kinetic energy of (Y, b). The loss of energy (and mass) of the system is balanced by an increase in the kinetic energy (and mass) of the escaping particles. A reaction for which Q is negative is called **endothermic.** An endothermic reaction will not occur unless the bombarding particle has a kinetic energy greater than Q. The minimum energy necessary for such a reaction to occur is called the **threshold energy.**

Endothermic reaction

Threshold energy

Nuclear reactions must obey the principle of conservation of linear momentum. That is, the total linear momentum of the system of interacting particles must be the same before and after the reaction. This assumes that the only force acting on the interacting particles is their mutual force of interaction; that is, no external accelerating electric fields are present near the colliding particles.

If a nuclear reaction occurs in which particles a and b are identical, so that X and Y are also necessarily identical, the reaction is called a *scattering event.* If kinetic energy is constant as a result of the reaction (that is, if $Q = 0$), the event is classified as *elastic scattering.* On the other hand, if $Q \neq 0$, kinetic energy is not constant and the reaction is called *inelastic scattering.* This terminology is identical to that used in dealing with collisions between macroscopic objects (Chapter 9, Section 9.4).

In addition to energy and momentum, the total charge and total number of nucleons must be conserved in any nuclear reaction. For example, consider the reaction $^{19}\text{F}(\text{p}, \alpha)^{16}\text{O}$, which has a Q value of 8.124 MeV. We can show this reaction more completely as

$$^{1}_{1}\text{H} + {}^{19}_{9}\text{F} \longrightarrow {}^{16}_{8}\text{O} + {}^{4}_{2}\text{He}$$

We see that the total number of nucleons before the reaction $(1 + 19 = 20)$ is equal to the total number after the reaction $(16 + 4 = 20)$. Furthermore, the total charge $(Z = 10)$ is the same before and after the reaction.

Summary

A nuclear species can be represented by $^{A}_{Z}\text{X}$, where A is the **mass number,** the total number of nucleons, and Z is the **atomic number,** the total number of protons. The total number of neutrons in a nucleus is the **neutron number,** N, where $A = N + Z$. Elements with the same Z but different A and N values are called **isotopes.**

Assuming that a nucleus is spherical, its radius is

Radii of most nuclei

$$r = r_0 A^{1/3} \quad [31.1]$$

where $r_0 = 1.2$ fm.

Nuclei are stable because of the **nuclear force** between nucleons. This short-range force dominates the Coulomb repulsive force at distances of less than about 2 fm and is nearly independent of charge.

Light nuclei are most stable when the number of protons equals the number of neutrons. Heavy nuclei are most stable when the number of neutrons exceeds the number of protons. In addition, many stable nuclei have Z and N values that are both even. Nuclei with unusually high stability have Z or N values of 2, 8, 20, 28, 50, 82, and 126, called **magic numbers.**

Properties of stable nuclei

Nuclei have an intrinsic spin angular momentum of magnitude $\sqrt{I(I+1)}\hbar$, where I is the **nuclear spin.** The magnetic moment of a nucleus is measured in terms of the **nuclear magneton,** μ_n, where

$$\mu_n = 5.05 \times 10^{-27}\ \text{J/T} \qquad [31.3]$$

Nuclear magneton

The difference in mass between the separate nucleons and the nucleus containing these nucleons, when multiplied by c^2, gives the **binding energy,** E_b, of the nucleus: $E_b = \Delta mc^2$. We can calculate the binding energy of any nucleus, ${}^{A}_{Z}\text{X}$, using the expression

$$E_b\ (\text{MeV}) = [\text{Z}m_H + Nm_n - M({}^{A}_{Z}\text{X})] \times 931.50\ \text{MeV/u} \qquad [31.4]$$

Binding energy of a nucleus

A radioactive substance can undergo alpha decay, beta decay, or gamma decay. An alpha particle is the ^{4}He nucleus; a beta particle is either an electron (β^-) or a positron (β^+); a gamma particle is a high-energy photon.

Radioactive decay

If a radioactive material contains N_0 radioactive nuclei at $t = 0$, the number, N, of nuclei remaining after a time t has elapsed is

$$N = N_0 e^{-\lambda t} \qquad [31.6]$$

Exponential decay

where λ is the **decay constant,** or **disintegration constant.** The *decay rate,* or *activity,* of a radioactive substance is given by

$$R = \left|\frac{dN}{dt}\right| = R_0 e^{-\lambda t} \qquad [31.7]$$

Decay rate

where $R_0 = N_0\lambda$ is the activity at $t = 0$. The **half-life,** $T_{1/2}$, is defined as the time it takes half of a given number of radioactive nuclei to decay, where

$$T_{1/2} = \frac{0.693}{\lambda} \qquad [31.8]$$

Half-life

Alpha decay can occur because, according to quantum mechanics, some nuclei have barriers that can be penetrated by the alpha particles (the tunneling process). This process is energetically more favorable for those nuclei having large excesses of neutrons. A nucleus can undergo beta decay in two ways. It can emit either an electron (β^-) and an antineutrino ($\overline{\nu}$) or a positron (β^+) and a neutrino (ν). In the electron capture process, the nucleus of an atom absorbs one of its own electrons (usually from the K shell) and emits a neutrino. In gamma decay, a nucleus in an excited state decays to its ground state and emits a gamma ray.

Nuclear reactions can occur when a target nucleus, X, is bombarded by a particle, a, resulting in a nucleus, Y, and a particle, b:

$$\text{a} + \text{X} \longrightarrow \text{Y} + \text{b} \qquad \text{or} \qquad \text{X(a, b)Y} \qquad [31.23]$$

Nuclear reaction

The energy released in such a reaction, called the **reaction energy,** Q, is

$$Q = (M_a + M_X - M_Y - M_b)c^2 \qquad [31.24]$$

Reaction energy *Q*

Questions and Conceptual Exercises

1. In Rutherford's experiment, assume that an alpha particle is headed directly toward the nucleus of an atom. Why doesn't the alpha particle make physical contact with the nucleus?
2. Estimate the mass of a pinhead composed entirely of densely packed nuclear matter.
3. Explain the main differences among alpha, beta, and gamma rays.
4. Why do heavier elements require more neutrons in order to maintain stability?
5. A proton precesses with a frequency of ω_p in the presence of a magnetic field. If the magnetic field intensity is doubled, what happens to the precessional frequency?
6. Why do nearly all the naturally occurring isotopes lie *above* the $N = Z$ line in Figure 31.3?
7. If a nucleus has a half-life of one year, does this mean that it will be completely decayed after two years? Explain.
8. What fraction of a radioactive sample has decayed after two half-lives have elapsed?
9. Two samples of the same radioactive nuclide are prepared. Sample A has twice the initial activity of sample B. How does the half-life of A compare to the half-life of B? After each has passed through five half-lives, what is the ratio of their activities?
10. If no more people were to be born, the law of population growth would strongly resemble the radioactive decay law. Discuss this statement.
11. If a nucleus captures a slow-moving neutron, the product is left in a highly excited state with an energy approximately 8.0 MeV above the ground state. Explain the source of the excitation energy.
12. Use the analogy of a bullet and a rifle to explain why the recoiling nucleus carries off only a very small fraction of the disintegration energy in the α decay of a nucleus.
13. If a nucleus such as ^{226}Ra that is initially at rest undergoes alpha decay, which has more kinetic energy after the decay, the alpha particle or the daughter nucleus?
14. Explain why many heavy nuclei undergo alpha decay but do not spontaneously emit neutrons or protons.
15. If an alpha particle and an electron have the same kinetic energy, which undergoes the greater deflection when passed through a magnetic field?
16. If film is kept in a box, alpha particles from a radioactive source outside the box cannot expose the film, but beta particles can. Explain.
17. Explain in detail how you can determine the age of a sample by carbon dating.
18. Suppose it could be shown that the cosmic ray intensity was much greater 10 000 years ago. How would this affect the current values for the ages of ancient samples of once-living matter?
19. The radioactive nucleus $^{222}_{88}$Ra has a half-life of about 1.6×10^3 years. Given that the Solar System is about 5 billion years old, how can you explain why we still can find this nucleus in nature?
20. The compressed core of a star formed in the wake of a supernova explosion can consist of pure nuclear material and is called a *pulsar* or *neutron star.* Estimate the mass of 10 cm^3 of a pulsar.
21. From Table A.3, identify the four stable nuclei that have magic numbers in both Z and N.
22. Consider the hydrogen atom to be a sphere with a radius equal to the Bohr radius, a_0, and calculate the approximate value of the ratio of the nuclear density to the atomic density.

Problems

Table 31.4 will be useful for many of these problems. A more complete list of atomic masses is given in Appendix A.3.

Section 31.1 Some Properties of Nuclei

1. Find the radius of a nucleus of (a) $^4_2$He; (b) $^{238}_{92}$U.
2. Find the nucleus that has a radius approximately equal to one-half the radius of uranium, $^{238}_{92}$U.
3. Find the diameter of a sphere of nuclear matter that would have a mass equal to that of the Earth. Base your calculation on an Earth radius of 6.37×10^6 m and an average Earth density of 5.52×10^3 kg/m^3.
4. What would be the gravitational force between two golf balls (each with a 4.3-cm diameter), 1 meter apart, if they were made of nuclear matter?
5. Use energy methods to calculate the distance of closest approach for a head-on collision between an

Table 31.4
Some Atomic Masses

Element	Atomic Mass (u)	Element	Atomic Mass (u)
$^{4}_{2}He$	4.002603	$^{27}_{13}Al$	26.981541
$^{7}_{3}Li$	7.016004	$^{30}_{15}P$	29.978310
$^{9}_{4}Be$	9.012182	$^{40}_{20}Ca$	39.962591
$^{10}_{5}B$	10.012938	$^{42}_{20}Ca$	41.95863
$^{12}_{6}C$	12.000000	$^{43}_{20}Ca$	42.958770
$^{13}_{6}C$	13.003355	$^{56}_{26}Fe$	55.934939
$^{14}_{7}N$	14.003074	$^{64}_{30}Zn$	63.929145
$^{15}_{7}N$	15.000109	$^{64}_{29}Cu$	63.929599
$^{15}_{8}O$	15.003065	$^{93}_{41}Nb$	92.906378
$^{17}_{8}O$	16.999131	$^{197}_{79}Au$	196.966560
$^{18}_{8}O$	17.999159	$^{202}_{80}Hg$	201.970632
$^{18}_{9}F$	18.000937	$^{216}_{84}Po$	216.001790
$^{20}_{10}Ne$	19.992439	$^{220}_{86}Rn$	220.011401
$^{23}_{11}Na$	22.989770	$^{234}_{90}Th$	234.043583
$^{23}_{12}Mg$	22.994127	$^{238}_{92}U$	238.050786

alpha particle with an initial energy of 0.5 MeV and a gold nucleus (^{197}Au) at rest. Assume the gold nucleus remains at rest during the collision.

6. (a) Find the speed an alpha particle must have in order to come within 3.2×10^{-14} m of a gold nucleus. (b) Find the energy of the alpha particle in MeV units.
7. The nucleus of an iron atom has a radius equal to 4.60×10^{-15} m. What must be the minimum speed of an α particle if it is to reach the nucleus? Disregard the effect of the outer electrons.
8. An α particle ($Z = 2$, mass 6.64×10^{-27} kg) approaches to within 1.0×10^{-14} m of a carbon nucleus ($Z = 6$). What are (a) the maximum Coulomb force on the α particle, (b) the acceleration of the α particle at this point, and (c) the potential energy of the α particle at this point?
9. Certain stars are thought to collapse at the ends of their lives, combining their protons and electrons to form a neutron star. Such a star could be thought of as a giant atomic nucleus. If a star with a mass equal to that of the Sun ($M = 1.99 \times 10^{30}$ kg) collapsed into neutrons ($m_n = 1.67 \times 10^{-27}$ kg), what would be the radius of such a star? (Assume that $r = r_0 A^{1/3}$.)
10. The Larmor precessional frequency is

$$f = \frac{\Delta E}{h} = \frac{2\mu B}{h}$$

Calculate the radio-wave frequencies at which resonance absorption will occur for (a) free neutrons in a magnetic field of 1 T, (b) free protons in a magnetic field of 1 T, and (c) free protons in the Earth's magnetic field at a location where the field strength is equal to 50 μT.

11. How much energy (in MeV units) must an α particle have to reach the surface of a gold nucleus ($Z = 79$, $A = 197$)?

Section 31.2 Binding Energy

12. Calculate the binding energy per nucleon for (a) ^{2}H, (b) ^{4}He, (c) ^{56}Fe, and (d) ^{238}U.
13. In Example 31.2, the binding energy of the deuteron was calculated to be 2.224 MeV. This corresponds to a value of 1.112 MeV/nucleon. What is the binding energy per nucleon for the heaviest isotope of hydrogen, ^{3}H (called *tritium*)?
14. Using the value of the atomic mass of $^{56}_{26}Fe$ given in Table 31.4, find its binding energy. Then compute the binding energy per nucleon and compare your result with Figure 31.7.
15. Calculate the minimum energy required to remove a neutron from the $^{43}_{20}Ca$ nucleus.
16. Two nuclei having the same mass number are known as *isobars*. Calculate the difference in binding energy per nucleon for the isobars $^{23}_{11}Na$ and $^{23}_{12}Mg$. How do you account for the difference?
17. The $^{139}_{57}La$ isotope of lanthanum is stable. A radioactive isobar (see Problem 16) of this lanthanum isotope, $^{139}_{59}Pr$, is located above the line of stable nuclei in Figure 31.3 and decays by β^{+} emission. Another

radioactive isobar of ^{139}La, $^{139}_{55}Cs$, decays by β^- emission and is located below the line of stable nuclei in Figure 31.3. (a) Which of these three isobars has the highest neutron-to-proton ratio? (b) Which has the greatest binding energy per nucleon? (c) Which of the two radioactive nuclei (^{139}Pr or ^{139}Cs) do you expect to be heavier?

Section 31.3 Radioactivity

18. How much time elapses before 90.0% of the radioactivity of a sample of $^{72}_{33}As$ disappears, as measured by the activity? The half-life of $^{72}_{33}As$ is 26 h.
19. A sample of radioactive material contains 10^{15} atoms and has an activity of 6.00×10^{11} Bq. What is the half-life for this material?
20. The half-life of radioactive ^{131}I is eight days. Find the number of ^{131}I nuclei necessary to produce a sample of activity 1.0 μCi.
21. A pure sample of ^{226}Ra contains 2.0×10^{14} atoms of the isotope. If the half life of ^{226}Ra is 1.6×10^3 years, what is the decay rate of this sample?
22. The half-life of ^{131}I is eight days. On a certain day, the activity of an I-131 sample is 6.4 mCi. What is its activity 40 days later?
23. A freshly prepared sample of a certain radioactive isotope has an activity of 10 mCi. After an elapsed time of 4 h, its activity is 8 mCi. (a) Find the decay constant and half-life of the isotope. (b) How many atoms of the isotope were contained in the freshly prepared sample? (c) What is the sample's activity 30 h after it is prepared?
24. Determine the disintegration rate (activity) of 1 gram of ^{60}Co. The half-life of ^{60}Co is 5.24 years.
25. Tritium has a half-life of 12.33 years. What percentage of the nuclei in a tritium sample will decay in five years?
26. A radioactive sample contains 30.0×10^{-15} kg of $^{108}_{47}Ag$ (half-life = 2.42 min) at some instant. What is the activity of this sample, in millicuries?
27. A building has accidentally become contaminated with radioactivity. The longest-lived material in the building is strontium-90 ($^{90}_{38}Sr$, atomic mass 89.9077). If the building initially contained 5.0 kg of this substance and the safe level is less than 10.0 counts/min, how long will the building be unsafe?

Section 31.4 The Decay Processes

28. Identify the missing nucleus (X) in each of the following reactions.

(a) $X \longrightarrow {}^{65}_{28}Ni + \gamma$

(b) ${}^{215}_{84}Po \longrightarrow X + \alpha$

(c) $X \longrightarrow {}^{55}_{26}Fe + \beta^+ + \nu$

(d) ${}^{109}_{48}Cd + X \longrightarrow {}^{109}_{47}Ag + \nu$

(e) ${}^{14}N(\alpha, X){}^{17}O$

29. Find the energy released in the alpha decay of $^{238}_{92}U$:

$$^{238}_{92}U \longrightarrow {}^{234}_{90}Th + {}^{4}_{2}He$$

You will find the following mass values useful:

$$M(^{238}_{92}U) = 238.050786 \text{ u}$$
$$M(^{234}_{90}Th) = 234.043583 \text{ u}$$
$$M(^{4}_{2}He) = 4.002603 \text{ u}$$

30. A living specimen of organic material in equilibrium with the atmosphere contains one atom of ^{14}C (half-life = 5730 years) for every 7.7×10^{11} stable carbon atoms. An archeological sample of wood (cellulose, $C_{12}H_{22}O_{11}$) contains 21.0 mg of carbon. When the sample is placed inside a beta counter with a counting efficiency of 88%, 837 counts are accumulated in one week. Assuming that the cosmic-ray flux and the Earth's atmosphere have not changed appreciably since the sample was formed, find the age of the sample.
31. Perform a calculation to verify the accuracy of the conversion factor used in Equation 31.13. In addition to the definition value for u, you will need the following constants: $c = 2.99793 \times 10^8$ m/s and $e = 1.60219 \times 10^{-19}$ C.
32. Calculate the energy released in the alpha decay of ^{210}Po.
33. A ^{239}Pu nucleus at rest undergoes alpha decay, leaving a ^{235}U nucleus in its ground state. Determine the kinetic energy of the alpha particle.
34. Is it energetically possible for a $^{8}_{4}Be$ nucleus to spontaneously decay into two alpha particles? Explain. (See Appendix A.3 for information on $^{8}_{4}Be$.)
35. Starting with $^{235}_{92}U$, the following sequence of decays is observed, ending with the stable isotope $^{207}_{82}Pb$ (Fig.

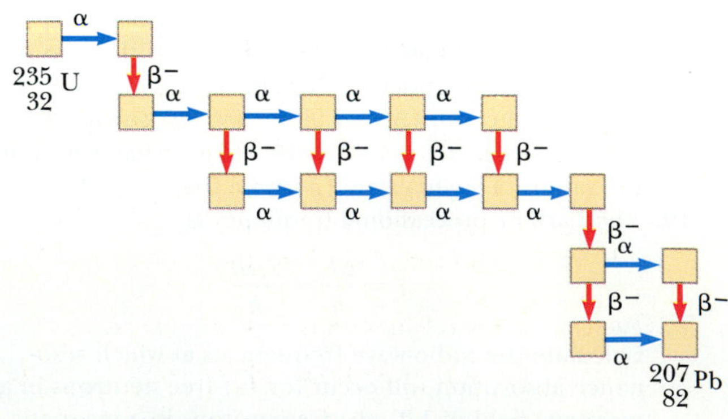

Figure 31.17 (Problem 35)

31.17). Enter the correct isotope symbol in each open square.

36. The radiocarbon content of ^{14}C decreases after the death of a living system with a half-life of 5730 years. If an archaeologist working a dig finds an ancient fire pit containing some partially consumed firewood and the wood contains only 0.125 as much ^{14}C as is found in an equal carbon sample from a present-day tree, what is the age of the ancient site?

37. A piece of charcoal used for cooking is found at the remains of an ancient campsite. A 1000-g sample of carbon from the wood is found to have an activity of 2000 decays per minute. Find the age of the charcoal.

38. Show that β^- decay of ^{255}Md cannot occur.

39. The nucleus $^{15}_{8}O$ decays by electron capture. Write (a) the basic nuclear process and (b) the decay process referring to neutral atoms. (c) Determine the energy of the neutrino. Disregard the daughter's recoil.

Section 31.6 Nuclear Reactions

40. The following reaction, first observed in 1930, led to the discovery of the neutron by Chadwick:

$$^{9}_{4}Be(\alpha, n)^{12}_{6}C$$

Calculate the Q value of this reaction.

41. The following is the first known reaction (achieved in 1934) in which the product nucleus is radioactive:

$$^{27}_{13}Al(\alpha, n)^{30}_{15}P$$

Calculate the Q value of this reaction.

42. Determine the Q associated with the spontaneous fission of ^{236}U into the fragments ^{90}Rb and ^{143}Cs with masses 89.914811 u and 142.927220 u, respectively. The masses of the other reacting particles are given in Table A.3.

43. Natural gold has only one isotope, $^{197}_{79}Au$. If natural gold is irradiated by a flux of slow neutrons, β^- particles are emitted. (a) Write the appropriate reaction equations. (b) Calculate the maximum energy of the emitted beta particles. The mass of $^{198}_{80}Hg$ is 197.96675 u.

44. Complete the following nuclear reactions:

$$? + ^{14}_{7}N \longrightarrow ^{1}_{1}H + ^{17}_{8}O$$

$$^{7}_{3}Li + ^{1}_{1}H \longrightarrow ^{4}_{2}He + ?$$

45. When a $^{6}_{3}Li$ nucleus is struck by a proton, an alpha particle and a product nucleus are released. (a) What is the product nucleus? (b) Find the Q value of the reaction.

46. Find the energy released in the fission reaction

$$n + ^{235}_{92}U \longrightarrow ^{98}_{40}Zr + ^{135}_{52}Te + 3n$$

The atomic masses of the fission products are: $^{98}_{40}Zr$, 97.9120 u; $^{135}_{52}Te$, 134.9087 u.

47. A beam of 6.61 MeV protons is incident on a target of $^{27}_{13}Al$. Those that collide produce the reaction

$$p + ^{27}_{13}Al \longrightarrow ^{27}_{14}Si + n$$

($^{27}_{14}Si$ has a mass of 26.986721 u.) Neglecting any recoil of the product nucleus, determine the kinetic energy of the emerging neutrons.

48. (a) Suppose $^{10}_{5}B$ is struck by an alpha particle, releasing a proton and a product nucleus in the reaction. (a) What is the product nucleus? (b) An alpha particle and a product nucleus are produced when $^{13}_{6}C$ is struck by a proton. What is the product nucleus?

49. Find the threshold energy that the incident neutron must have to produce the reaction

$$^{1}_{0}n + ^{4}_{2}He \longrightarrow ^{2}_{1}H + ^{3}_{1}H$$

Additional Problems

50. A sample of organic material (bones) is found to contain 18 g of carbon. Based on samples of pottery taken from the site, the investigators believe the bones to be 20 000 years old. If they are right, what is the expected activity of the sample?

51. Is it energetically possible for a $^{12}_{6}C$ nucleus to spontaneously decay into three alpha particles? Explain.

52. A sample of 200 mCi of a radioactive isotope is purchased by a medical supply house. If the sample has a half-life of 14 days, how long can they keep the sample before its activity is reduced to 20 mCi?

53. One method of producing neutrons for experimental use is to bombard $^{7}_{3}Li$ with protons. The neutrons are emitted according to the following reaction:

$$^{1}_{1}H + ^{7}_{3}Li \longrightarrow ^{7}_{4}Be + ^{1}_{0}n$$

What is the minimum kinetic energy the incident proton must have if this reaction is to occur?

54. A medical laboratory stock solution is prepared with an initial activity due to ^{24}Na of 2.5 mCi/ml, and 10 ml of the stock solution is diluted at $t_0 = 0$ to a working solution with a total volume of 250 ml. After 48 h, a 5-ml sample of the working solution is monitored with a counter. What is the measured activity? (Note that 1 ml = 1 milliliter.)

55. A by-product of some fission reactors is the isotope $^{239}_{94}Pu$, which is an alpha emitter with a half-life of 24 000 years:

$$^{239}_{94}Pu \longrightarrow ^{235}_{92}U + \alpha$$

Consider a sample of 1 kg of pure $^{239}_{94}Pu$ at $t = 0$. Calculate (a) the number of $^{239}_{94}Pu$ nuclei present at $t = 0$ and (b) the initial activity of the sample. (c) How long does the sample have to be stored if a "safe" activity level is 0.1 Bq?

56. A fission reactor is hit by a nuclear weapon and evaporates 5×10^6 Ci of ^{90}Sr ($T_{1/2} = 28.7$ years) into the

air. The ^{90}Sr falls out over an area of 10^4 km^2. How long will it take the activity of the ^{90}Sr to reach the agriculturally "safe" level of 2 μCi/m^2?

57. During the manufacture of a steel engine component, radioactive iron (^{59}Fe) is included in the total mass of 0.2 kg. The component is placed in a test engine when the activity due to the isotope is 20 μCi. After a 1000-hour test period, oil is removed from the engine and found to contain enough ^{59}Fe to produce 800 disintegrations/min per liter of oil. The total volume of oil in the engine is 6.5 liters. Calculate the total mass worn from the engine component per hour of operation. (The half-life of ^{59}Fe is 45.1 days.)

58. △ The activity of a sample of radioactive material was measured over 12 h, and the following *net* count rates were obtained at the times indicated:

Time (h)	Counting Rate (counts/min)
1	3100
2	2450
4	1480
6	910
8	545
10	330
12	200

(a) Plot the activity curve on semilog paper. (b) Determine the disintegration constant and half-life of the radioactive nuclei in the sample. (c) What counting rate would you expect for the sample at $t = 0$? (d) Assuming the efficiency of the counting instrument to be 10%, calculate the number of radioactive atoms in the sample at $t = 0$.

59. A large nuclear power reactor produces about 3000 MW of thermal power in its core. Three months after a reactor is shut down, the thermal power in the core is 10 MW, due to radioactive by-products. Assuming that each emission delivers 1 MeV of energy to the thermal power, estimate the activity, in becquerels, three months after the reactor is shut down.

60. In a piece of rock from the Moon the ^{87}Rb content is assessed to be 1.82×10^{10} atoms per gram of material. In a piece of the same rock, the ^{87}Sr content is found to be 1.07×10^9 atoms per gram. (a) Determine the age of the rock. (b) Could the material in the rock actually be much older? What assumption is implicit in using the radioactive dating method? (The relevant decay is $^{87}\text{Rb} \rightarrow {}^{87}\text{Sr} + e^-$. The half-life of the decay is 4.8×10^{10} years.)

61. (a) Find the radius of the $^{12}_{6}$C nucleus. (b) Find the force of repulsion between a proton at the surface of a $^{12}_{6}$C nucleus and the remaining five protons. (c) How much work (in MeV units) has to be done to overcome this electrostatic repulsion to put the last proton into the nucleus? (d) Repeat (a), (b), and (c) for $^{238}_{92}$U.

62. (a) Why is the following inverse beta decay forbidden for a free proton?

$$\text{p} \longrightarrow \text{n} + \beta^+ + \nu$$

(b) Why is the same reaction possible if the proton is bound in a nucleus? For example, the following reaction occurs:

$$^{13}_{7}\text{N} \longrightarrow {}^{13}_{6}\text{C} + \beta^+ + \nu$$

(c) How much energy is released in the reaction given in (b)? [Take the masses to be $m(\beta^+) = 0.000549$ u, $M(^{13}\text{C}) = 13.003355$ u, and $M(^{13}\text{N}) = 13.005739$ u.]

63. Carbon detonations are powerful nuclear reactions that temporarily tear apart the cores of massive stars late in their lives. These blasts are produced by carbon fusion, which requires a temperature of about 6×10^8 K to overcome the strong Coulomb repulsion between carbon nuclei. (a) Estimate the repulsive energy barrier to fusion, using the required ignition temperature for carbon fusion. (In other words, what is the kinetic energy for a carbon nucleus at a temperature of 6×10^8 K?) (b) Calculate the energy (in MeV units) released in each of these "carbon-burning" reactions:

$$^{12}\text{C} + {}^{12}\text{C} \longrightarrow {}^{20}\text{Ne} + {}^{4}\text{He}$$

$$^{12}\text{C} + {}^{12}\text{C} \longrightarrow {}^{24}\text{Mg} + \gamma$$

(c) Calculate the energy (in kilowatt-hours) given off when 2 kg of carbon completely fuses according to the first reaction.

64. When a material of interest is irradiated by neutrons, radioactive atoms are produced continually, and some decay according to their given half-lives. (a) If radioactive atoms are produced at a constant rate, R, and their decay is governed by the conventional radioactive decay law, show that the number of radioactive atoms accumulated after an irradiation time of t is

$$N = \frac{R}{\lambda}(1 - e^{-\lambda t})$$

(b) What is the maximum number of radioactive atoms that can be produced?

65. Consider a model of the nucleus in which the positive charge (Ze) is uniformly distributed throughout a sphere of radius R. By integrating the energy density of the electrostatic field, $\frac{1}{2}\epsilon_0 E^2$, over all space, show that the electrostatic energy may be written

$$U = \frac{3Z^2e^2}{20\pi\epsilon_0 R}$$

66. When, after a reaction or disturbance of any kind, a nucleus is left in an excited state, it can return to its normal (ground) state by emission of a gamma-ray photon (or several photons). This process is illustrated by Equation 31.23. The emitting nucleus must recoil in order to conserve both energy and momentum. (a) Show that the recoil energy of the nucleus is given by

$$E_r = \frac{(\Delta E)^2}{2Mc^2}$$

where ΔE is the difference in energy between the excited and ground states of a nucleus of mass M. (b) Calculate the recoil energy of the ^{57}Fe nucleus when it decays by gamma emission from the 14.4-keV excited state. For this calculation, take the mass to be 57 u. (*Hint:* When writing the equation for conservation of energy, use $(Mv)^2/2M$ for the kinetic energy of the recoiling nucleus. Also, assume that $hf \ll Mc^2$ and use the binomial expansion.)

67. The decay of an unstable nucleus by alpha emission is represented by Equation 31.9. The disintegration energy, Q, given by Equation 31.12 must be shared by the alpha particle and the daughter nucleus in order to conserve both energy and momentum in the decay process. (a) Show that Q and K_α, the kinetic energy of the alpha particle, are related by the expression

$$Q = K_\alpha\left(1 + \frac{M_\alpha}{M}\right)$$

where M is the mass of the daughter nucleus. (b) Use the result of (a) to find the energy of the alpha particle emitted in the decay of ^{226}Ra. (See Example 31.5 for the calculation of Q.)

68. The theory of nuclear astrophysics assumes that all the heavy elements, such as uranium, are formed in the interiors of massive stars. These stars eventually explode, releasing the elements into space. If we assume that at the time of the explosion there were equal amounts of ^{235}U and ^{238}U, how long ago did the star(s) explode that released the elements that formed our Earth? The present ^{235}U/^{238}U ratio is 0.007. (The half-lives of ^{235}U and ^{238}U are 0.7×10^9 years and 4.47×10^9 years.)

△69. The rate of decay of a radioactive sample is measured at 10-s intervals, beginning at $t = 0$. The following data, in counts per second, are obtained: 1137, 861, 653, 495, 375, 284, 215, 163. (a) Plot these data on semilog graph paper and determine the best-fit straight line. (b) From the graph, determine the half-life of the sample.

△70. *Student Determination of the Half-Life of* ^{137}Ba. The radioactive barium isotope (^{137}Ba) has a relatively short half-life and can easily be extracted from a solution containing radioactive cesium (^{137}Cs). This barium isotope is commonly used in an undergraduate laboratory exercise for demonstrating the radioactive decay law. The data presented in Figure 31.18 were taken by undergraduate students using modest experimental equipment. Determine the half-life for the decay of ^{137}Ba, using their data.

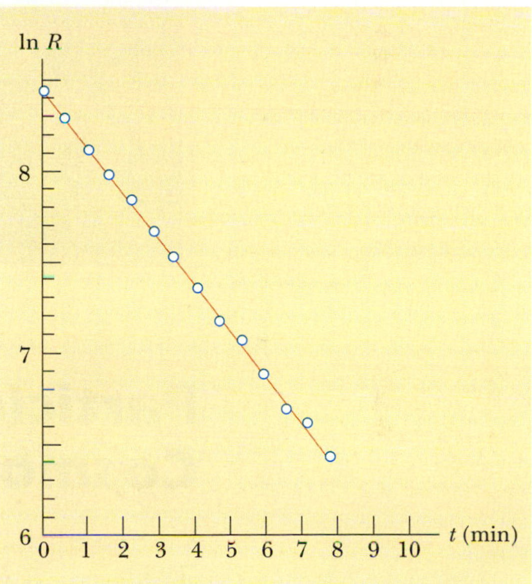

Figure 31.18 (Problem 70)

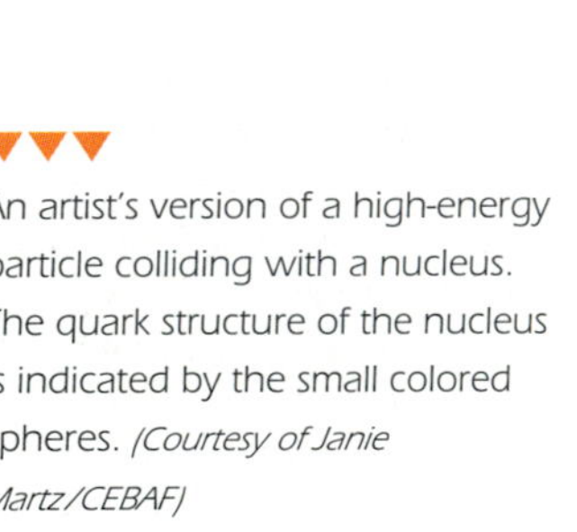

An artist's version of a high-energy particle colliding with a nucleus. The quark structure of the nucleus is indicated by the small colored spheres. *(Courtesy of Janie Martz/CEBAF)*

Particle Physics and Cosmology

CHAPTER 32

In this concluding chapter, we shall examine the known subatomic particles and the fundamental interactions that govern their behavior. We shall also discuss the current theory of elementary particles, in which all matter in nature is believed to be constructed from only two families of particles, quarks and leptons. Finally, we shall discuss how clarifications of such models may help scientists understand the evolution of the Universe.

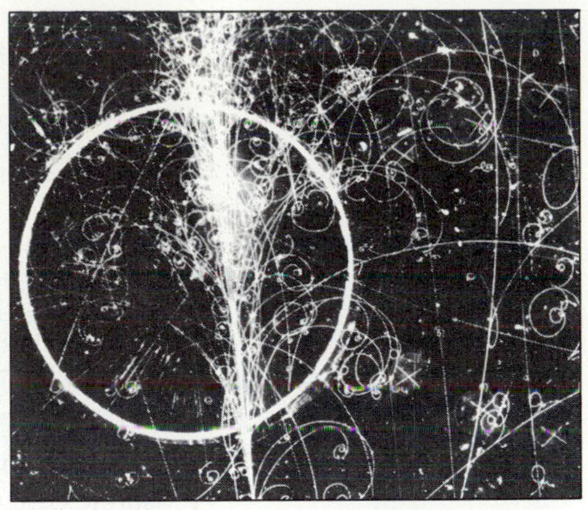

(Left) The main tunnel at Fermilab, now housing both the older ring of conventional magnets (red and blue, above) and the new ring of superconducting magnets (yellow, below). In its present mode of operation as the Tevatron, the superconducting ring receives injections of protons and antiprotons from the conventional ring. The two beams are accelerated in opposite directions around the ring to their final energy of 1 TeV and then collide with each other at interaction points. The Tevatron is the most powerful accelerator in the world today. (Right) Photograph of a particle interaction in the 15-ft bubble chamber at Fermilab. The photographs showing particle "tracks" are studied by scanners and experimenters. *(Photographs courtesy of Fermi National Accelerator Laboratory)*

32.1 Introduction

The word *atom* is from the Greek word *atomos,* which means indivisible. At one time atoms were thought to be the indivisible constituents of matter; that is, they were regarded as **elementary particles.** Discoveries in the early part of the 20th century revealed that the atom has as its constituents protons, neutrons, and electrons. In 1932, physicists viewed these three constituent particles as elementary because, with the exception of the free neutron, they are very stable. The theory soon fell apart, however, and beginning in 1945, many new particles were discovered in experiments involving high-energy collisions between known particles. These new particles are characteristically unstable and have very short half-lives, ranging between 10^{-6} and 10^{-23} s. So far, more than 300 of them have been catalogued!

Until the 1960s, physicists were bewildered by the large number and variety of subatomic particles being discovered. They wondered whether the particles were like animals in a zoo with no systematic relationship connecting them, or whether a pattern was emerging that would provide a better understanding of the elaborate structure in the subnuclear world. In the last 25 years, physicists have tremendously advanced our knowledge of the structure of matter by recognizing that all of the 300 or so particles discovered over the past few decades (with the exception of electrons, photons, and a few others) are made of smaller particles called *quarks.* Thus, protons and neutrons are not elementary particles but are systems of tightly bound quarks. The quark model has reduced the bewildering array of particles to a manageable number and has successfully predicted new quark combinations that were later found in many experiments.

32.2 The Fundamental Force in Nature Revisited

As you will recall from Chapter 6, all natural phenomena can be described by four fundamental forces between elemental particles. In order of decreasing strength, these are the **strong** force, the **electromagnetic** force, the **weak** force, and the **gravitational** force.

Table 32.1
Particle Interactions

Interaction (Force)	Relative Strength	Range of Force	Mediating Field Particle
Strong	1	Short ($\approx$1 fm)	Gluon
Electromagnetic	10^{-2}	Long ($\propto 1/r^2$)	Photon
Weak	10^{-13}	Short ($<10^{-3}$ fm)	$W^{\pm}$ and Z bosons
Gravitational	10^{-38}	Long ($\propto 1/r^2$)	Graviton

The strong force represents the glue that holds nucleons together to form stable nuclei. It is very short-range and is negligible for separations greater than about 10^{-15} m (the approximate diameter of the nucleus). The electromagnetic force, which binds atoms and molecules together to form ordinary matter has about 10^{-2} times the strength of the strong force. The electromagnetic force is a long-range force that decreases in strength as the inverse square of the separation between interacting particles. The weak force is a short-range force that tends to produce instability in certain nuclei. It is responsible for most radioactive decay processes, and its strength is only about 10^{-13} times that of the strong force. Finally, the gravitational force is a long-range force whose strength is only about 10^{-38} times that of the strong force. Although this familiar interaction is the force that holds the planets, stars, and galaxies together, its effect on elementary particles is negligible.

In modern physics, one often describes the interactions between particles in terms of the actions of "field particles" or "quanta." In the case of the familiar electromagnetic interactions, the field particles are photons. In the language of modern physics, one can say that the electromagnetic force is *mediated* by photons, which are the quanta of the electromagnetic field. Likewise, the strong force is mediated by field particles called *gluons,* the weak force is mediated by field particles called the W and Z *bosons,* and the gravitational force is mediated by quanta of the gravitational field called *gravitons.* These interactions, their ranges, and their relative strengths are summarized in Table 32.1.

32.3 Positrons and Other Antiparticles

Paul Adrien Maurice Dirac (1902–1984), winner of the Nobel Prize for Physics in 1933, is shown speaking at Yeshiva University. *(UPI/Bettmann)*

In the 1920s, the theoretical physicist Paul Adrien Maurice Dirac (1902–1984) developed a version of quantum mechanics that incorporated special relativity. Dirac's theory successfully explained the origin of the electron's spin and its magnetic moment. But it had one major problem: its relativistic wave equation required solutions corresponding to negative energy states, and if negative energy states existed, one would expect an electron in a state of positive energy to make a rapid transition to one of these states, emitting a photon in the process.

Dirac circumvented this difficulty by postulating that all negative energy states are filled. The electrons that occupy the negative energy states are said to be in the "Dirac sea" and are not directly observable because the Pauli exclusion principle does not allow them to react to external forces. However, if one of these negative

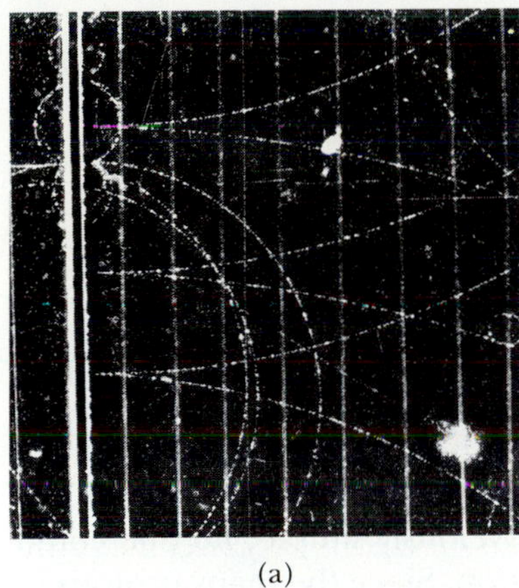

(a)

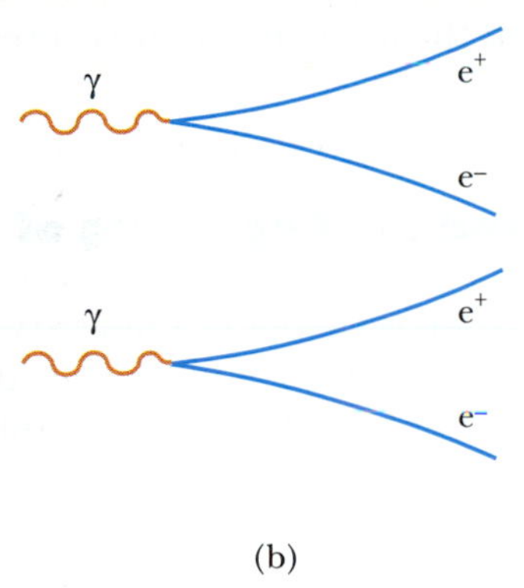

(b)

Figure 32.1
(a) Bubble-chamber tracks of electron-positron pairs produced by 300-MeV γ-rays striking a lead sheet. *(Courtesy of Lawrence Berkeley Laboratory, University of California)* (b) Sketch of the pertinent pair-production events. Note that the positrons deflect upward while the electrons deflect downward in an applied magnetic field that is directed into the diagram.

energy states is vacant, leaving a hole in the sea of filled states, the hole can react to external forces and therefore is observable. The profound implication of this theory is that, *for nearly every particle, there is an antiparticle.* The antiparticle has the same mass as the particle, but opposite charge. For example, the electron's antiparticle (now called a *positron*) has a mass of 0.511 MeV and a positive charge of 1.6×10^{-19} C. As noted in Chapter 31, we designate an antiparticle with a bar over the symbol for the particle. Thus, $\bar{p}$ denotes the antiproton and $\bar{\nu}$ the antineutrino.

The positron was discovered by Carl Anderson in 1932, and in 1936 he was awarded the Nobel Prize for his achievement. Anderson discovered the positron while examining tracks created by electron-like particles of positive charge in a cloud chamber. (These early experiments used cosmic rays—mostly energetic protons passing through interstellar space—to initiate high-energy reactions on the order of several GeV.) In order to discriminate between positive and negative charges, the cloud chamber was placed in a magnetic field, causing moving charges to follow curved paths. Anderson noted that some of the electron-like tracks deflected in a direction corresponding to a positively charged particle.

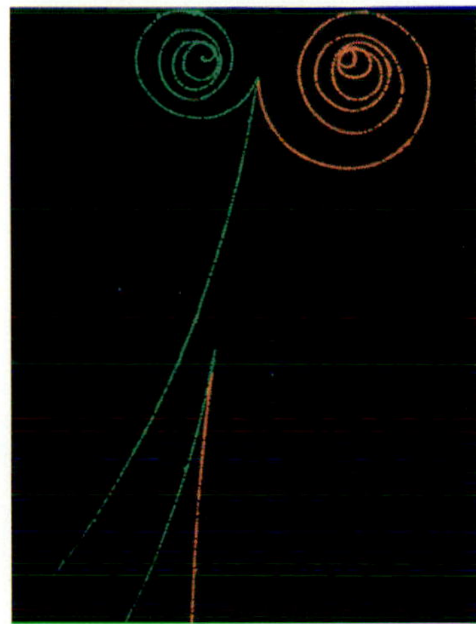

Bubble chamber photograph of electron (green) and positron (red) tracks produced by energetic gamma rays. The highly curved tracks at the top are due to an electron-positron pair that bend in opposite directions in the magnetic field. The lower tracks are produced by more energetic electrons and a positron. *(Lawrence Berkeley Laboratory/Science Photo Library, Photo Researchers, Inc.)*

Since Anderson's initial discovery, the positron has been observed in other experiments. Perhaps the most common process for producing positrons is **pair production.** In this process, a gamma-ray photon with sufficiently high energy collides with a nucleus, and an electron-positron pair is created. Since the rest energy of the electron-positron pair is $2mc^2 = 1.02$ MeV (where m is the mass of the electron), the photon must have at least this much energy to create an electron-positron pair. Figure 32.1 shows tracks of electron-positron pairs.

The reverse process can also occur. Under the proper conditions, an electron and positron can annihilate each other and produce two gamma-ray photons that have a combined energy of at least 1.02 MeV:

$$e^- + e^+ \longrightarrow 2\gamma$$

Very rarely, a proton and antiproton also annihilate each other to produce two gamma-ray photons.

Practically every known elementary particle has an antiparticle. Among the exceptions are the photon and the neutral pion (π^0). Following the construction of high-energy accelerators in the 1950s, many of these antiparticles were discovered. These included the antiproton discovered by Emilio Segre and Owen Chamberlain in 1955, and the antineutron discovered shortly thereafter.

Hideki Yukawa (1907–1981), a Japanese physicist, was awarded the Nobel Prize in 1949 for predicting the existence of mesons. This photograph of Yukawa at work was taken in 1950 in his office at Columbia University. *(UPI/Bettmann)*

32.4 Mesons and the Beginning of Particle Physics

The structure of matter as viewed by physicists in the mid-1930s was fairly simple. The building blocks were the proton, the electron, and the neutron. Three other particles were known at the time: the photon, the neutrino, and the positron. These six particles were considered the fundamental constituents of matter. Although the accepted picture of the world was marvelously simple, no one could provide an answer to the following important question. Since the many protons in proximity in any nucleus should strongly repel each other, what is the nature of the force that holds the nucleus together? Scientists recognized that this mysterious force must be much stronger than anything encountered in nature up to that time.

The first theory to explain the nature of the strong force was proposed in 1935 by the Japanese physicist Hideki Yukawa (1907–1981), an effort that later earned him the Nobel Prize. In order to understand Yukawa's theory, it is useful to recall that *two atoms can form a covalent chemical bond by sharing electrons.* Similarly, in the modern views of electromagnetic interactions, *charged particles interact by sharing a photon.* Yukawa used this idea to explain the strong force by proposing a new particle that is shared by nucleons in the nucleus to produce the strong force. Furthermore, he established that the range of the force is inversely proportional to the mass of this particle, and predicted that the mass would be about 200 times the mass of the electron. Since the new particle would have a mass between that of the electron and proton, it was called a **meson** (from the Greek *meso,* meaning "middle").

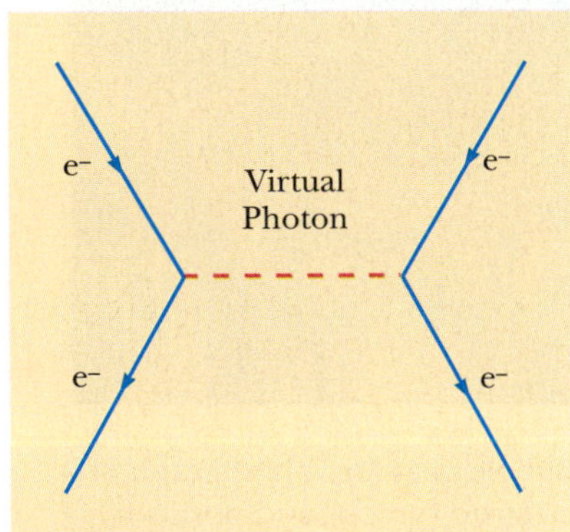

Figure 32.2
Feynman diagram showing how a photon mediates the electromagnetic forces between two interacting electrons. The top half represents the two electrons before the interaction, and the bottom half represents the same electrons after the interaction.

In an effort to substantiate Yukawa's predictions, physicists began looking for the meson by studying cosmic rays that enter the Earth's atmosphere. In 1937 Carl Anderson and his collaborators discovered a particle whose mass was 106 MeV/c^2, about 207 times the mass of the electron. However, subsequent experiments showed that the particle interacted very weakly with matter, and hence could not be the carrier of the strong force. The puzzling situation inspired several theoreticians to propose that there are two mesons with slightly different masses. This idea was confirmed in 1947 with the discovery of the pi meson (π), or simply *pion,* in cosmic rays by Cecil Frank Powell (1903–1969) and Guiseppe P. S. Occhialini (1907–). The lighter meson discovered earlier by Anderson, now called a *muon* (μ), has only weak and electromagnetic interaction and plays no role in the strong interaction.

The pion comes in three varieties, corresponding to three charge states: π^+, π^-, and π^0. The π^+ and π^- particles have masses of 139.6 MeV/c^2, and the π^0 has a mass of 135.0 MeV/c^2. Pions and muons are very unstable particles. For example, the π^-, which has a mean lifetime of 2.6×10^{-8} s, decays into a muon and an antineutrino. The muon, which has a mean lifetime of 2.2 μs, then decays into an electron, a neutrino, and an antineutrino:

$$\pi^- \longrightarrow \mu^- + \bar{\nu} \qquad [32.1]$$

$$\mu^- \longrightarrow e^- + \nu + \bar{\nu}$$

The interaction between two particles can be represented in a *Feynman diagram,* developed by Richard P. Feynman (1918–1988). Figure 32.2 is such a diagram for the electromagnetic interaction between two electrons. In this simple case, a photon is the field particle that mediates the electromagnetic force between the electrons. The photon transfers energy and momentum from one electron to the other in this interaction. Such a photon, called a *virtual photon,* can never be detected directly because it is absorbed by the second electron very shortly after being emitted by the first electron. The existence of a virtual photon would violate the law of conservation of energy, but because of the uncertainty principle and its very short lifetime, Δt, the photon's excess energy is less than the uncertainty in its energy, given by $\Delta E \approx \hbar/\Delta t$.

Now consider a pion exchange between a proton and a neutron via the strong force (Fig. 32.3). One can reason that the energy, ΔE, needed to create a pion of mass m_π is $\Delta E = m_\pi c^2$. Again, the existence of the pion avoids violation of conservation of energy only if this energy is surrendered in a time of Δt, the time it takes the pion to transfer from one nucleon to the other. From the uncertainty principle, $\Delta E\, \Delta t \approx \hbar$, we get

$$\Delta t \approx \frac{\hbar}{\Delta E} = \frac{\hbar}{m_\pi c^2} \qquad [32.2]$$

Since the pion cannot travel faster than the speed of light, the maximum distance, d, it can travel in a time of Δt is $c\,\Delta t$. Using Equation 32.2 and $d = c\,\Delta t$, we find this maximum distance to be

$$d \approx \frac{\hbar}{m_\pi c} \qquad [32.3]$$

From Chapter 31, we know that the range of the strong force is about 1.5×10^{-15} m. Using this value for d in Equation 32.3, the rest energy of the pion is calculated to be

$$m_\pi c^2 \approx \frac{\hbar c}{d} = \frac{(1.05 \times 10^{-34}\ \text{J}\cdot\text{s})(3 \times 10^8\ \text{m/s})}{1.5 \times 10^{-15}\ \text{m}}$$

$$= 2.1 \times 10^{-11}\ \text{J} \cong 130\ \text{MeV}$$

This corresponds to a mass of 130 MeV/c^2 (about 250 times the mass of the electron), which is in good agreement with the observed mass of the pion.

The concept we have just described is quite revolutionary. In effect, it says that a proton can change into a proton plus a pion, as long as it returns to its original state in a very short time. High-energy physicists often say that a nucleon undergoes "fluctuations " as it emits and absorbs pions. As we have seen, these fluctuations are a consequence of a combination of quantum mechanics (through the uncertainty principle) and special relativity (through Einstein's energy-mass relation $E = mc^2$).

This section has dealt with the particles that mediate the strong force, namely the pions, and the mediators of the electromagnetic force, photons. The graviton, which is the mediator of the gravitational force, has yet to be observed. The W and Z particles that mediate the weak force were discovered in 1983 by Carlo Rubbia (1934–) and his associates, using a proton-antiproton collider. Rubbia and Simon van der Meer, both at CERN, shared the 1984 Nobel Prize in physics for the discovery of the $W^\pm$ and Z^0 particles and the development of the proton-antiproton collider.

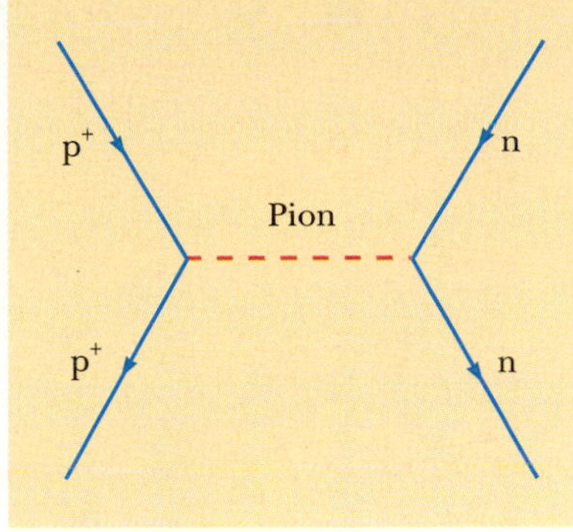

Figure 32.3
Feynman diagram representing a proton interacting with a neutron via the strong force. In this case, the pion mediates the strong force.

Richard Feynman (1918–1988) with his son, Carl, after winning the Nobel Prize for Physics in 1965. The prize was shared by Feynman, Julian Schwinger, and Sin Itiro Tomonaga. *(UPI Telephotos)*

32.5 Classification of Particles

All particles other than photons can be placed in two broad categories, hadrons and leptons, according to their interactions.

Hadrons

Particles that interact through the strong force are called **hadrons.** There are two classes of hadrons, mesons and baryons, distinguished by their masses and spins. *Mesons* all have zero or integral spins (0 or 1), with masses that lie between the mass of the electron and the mass of the proton. All mesons are known to decay finally into electrons, positrons, neutrinos, and photons. The pion is the lightest known meson, with a mass of about 140 MeV/c^2 and a spin of 0. Another is the K meson, with a mass of about 500 MeV/c^2 and spin 0.

Table 32.2
A Table of Some Particles and Their Properties

Category	Particle Name	Symbol	Anti-particle	Rest Mass (MeV/c^2)	Spin ($\hbar$)	B	L_e	L_μ	L_τ	S	Lifetime (s)	Typical Decay Modes[a]
Photon	Photon	γ	Self	0	1	0	0	0	0	0	Stable	
Leptons	Electron	e^-	e^+	0.511	1/2	0	+1	0	0	0	Stable	
	Neutrino (e)	ν_e	$\bar{\nu}_e$	0(?)	1/2	0	+1	0	0	0	Stable	
	Muon	μ^-	μ^+	105.7	1/2	0	0	+1	0	0	2.20×10^{-6}	$e^-\bar{\nu}_e\nu_\mu$
	Neutrino (μ)	ν_μ	$\bar{\nu}_\mu$	0(?)	1/2	0	0	+1	0	0	Stable	
	Tau	τ^-	τ^+	1784	1/2	0	0	0	−1	0	$<4 \times 10^{-13}$	$\mu^-\bar{\nu}_\mu\nu_\tau$, $e^-\bar{\nu}_e\nu_\tau$
	Neutrino (τ)	ν_τ	$\bar{\nu}_\tau$	0(?)	1/2	0	0	0	−1	0	Stable	
Hadrons												
Mesons	Pion	π^+	π^-	139.6	0	0	0	0	0	0	2.60×10^{-8}	$\mu^+\nu_\mu$
		π^0	Self	135.0	0	0	0	0	0	0	0.83×10^{-16}	2γ
	Kaon	K^+	K^-	493.7	0	0	0	0	0	+1	1.24×10^{-8}	$\mu^+\nu_\mu$, $\pi^+\pi^0$
		K_S^0	$\bar{K}_S^0$	497.7	0	0	0	0	0	+1	0.89×10^{-10}	$\pi^+\pi^-$, $2\pi^0$
		K_L^0	$\bar{K}_L^0$	497.7	0	0	0	0	0	+1	5.2×10^{-8}	$\pi^\pm e^\mp \bar{\nu}_e$
	Eta	η^0	Self	548.8	0	0	0	0	0	0	$<10^{-18}$	2γ, 3μ
Baryons	Proton	p^+	p^-	938.3	1/2	+1	0	0	0	0	Stable	
	Neutron	n	n	939.6	1/2	+1	0	0	0	0	920	$p^+e^-\bar{\nu}_e$
	Lambda	Λ^0	$\bar{\Lambda}^0$	1115.6	1/2	+1	0	0	0	−1	2.6×10^{-10}	$p^-\pi^-$, $n\pi^0$
	Sigma	Σ^+	$\bar{\Sigma}^-$	1189.4	1/2	+1	0	0	0	−1	0.80×10^{-10}	$p^-\pi^0$, $n\pi^+$
		Σ^0	$\bar{\Sigma}^0$	1192.5	1/2	+1	0	0	0	−1	6×10^{-20}	$\Lambda^0\gamma$
		Σ^-	$\bar{\Sigma}^+$	1197.3	1/2	+1	0	0	0	−1	1.5×10^{-10}	$n\pi^-$
	Xi	Ξ^0	$\bar{\Xi}^0$	1315	1/2	+1	0	0	0	−2	2.9×10^{-10}	$\Lambda^0\pi^0$
		Ξ^-	$\bar{\Xi}^+$	1321	1/2	+1	0	0	0	−2	1.64×10^{-10}	$\Lambda^0\pi^-$
	Omega	Ω^-	Ω^+	1672	3/2	+1	0	0	0	−3	0.82×10^{-10}	$\Xi^0\pi^-$, Λ^0K^-

[a] A notation in this column such as $p^+\pi^-$, $n\pi^0$ means two possible decay modes. For example, $p^+\pi^-$, $n\pi^0$ means that the Λ^0 particle can have two possible decays, $\Lambda^0 \rightarrow p^+ + \pi^-$ and $\Lambda^0 \rightarrow n + \pi^0$.

Baryons have masses equal to or greater than the proton mass (the word *baryon* means "heavy" in Greek), and their spins are always noninteger values (1/2, 3/2). Protons and neutrons are baryons, as are many other particles. With the exception of the proton, all baryons decay in such a way that the end products include a proton. For example, the baryon called the Ξ hyperon first decays to the baryon called the Λ^0 in about 10^{-10} s. The Λ^0 then decays to a proton and a π^- in about 3×10^{-10} s.

Today it is believed that hadrons are composed of quarks. (Later we shall have more to say about the quark model.) Some of the important properties of hadrons are listed in Table 32.2.

Leptons

Leptons (from the Greek *leptos,* meaning "small" or "light") are a group of particles that participate in the weak interaction. All leptons have spins of 1/2. Included in this group are electrons, muons, and neutrinos, which are less massive than the lightest hadron. Although hadrons have size and structure, leptons appear to be truly elementary, with no structure (that is, point-like).

Quite unlike hadrons, the number of known leptons is small. Currently, scientists believe there only are six leptons (each having an antiparticle)—the electron, the muon, the tau, and a neutrino associated with each:

$$\begin{pmatrix} e^- \\ \nu_e \end{pmatrix} \quad \begin{pmatrix} \mu^- \\ \nu_\mu \end{pmatrix} \quad \begin{pmatrix} \tau^- \\ \nu_\tau \end{pmatrix}$$

The neutrino associated with the tau has not yet been observed in the laboratory. The tau lepton, discovered in 1975, has a mass about twice that of the proton.

Although neutrinos are thought to be massless, there is a possibility that they have some small nonzero mass. As we shall see later, a firm knowledge of the neutrino's mass could have great significance in cosmological models and the future of the Universe.

32.6 Conservation Laws

A number of conservation laws are important in the study of elementary particles. Although the two described here have no theoretical foundation, they are supported by abundant empirical evidence.

Baryon Number

The **law of conservation of baryon number** tells us that whenever a baryon is created in a reaction or decay, an antibaryon is also created. This can be quantified by assigning a baryon number: $B = +1$ for all baryons, $B = -1$ for all antibaryons, and $B = 0$ for all other particles. Thus, whenever a nuclear reaction or decay occurs, the sum of the baryon numbers before the process must equal the sum of the baryon numbers after the process.

Note that if the baryon number is absolutely conserved, the proton must be absolutely stable. If it were not for the law of conservation of baryon number, the proton could decay to a positron and a neutral pion. However, such a decay has never been observed. At the present, we can say only that the proton has a half-life of at least 10^{31} years (the estimated age of the Universe is about 10^{10} years). In one version of a *grand unified theory,* physicists have predicted that the proton is actually unstable. According to this theory, the baryon number (sometimes called the baryonic charge) cannot be absolutely conserved, whereas electric charge is always conserved.

Example 32.1 Checking Baryon Numbers

Determine whether or not the following reactions can occur based on the law of conservation of baryon number.

$$p^+ + n \longrightarrow p^+ + p^+ + n + p^- \quad [1]$$

$$p^+ + n \longrightarrow p^+ + p^+ + p^- \quad [2]$$

Solution Recall that $B = +1$ for baryons and $B = -1$ for antibaryons. Hence, the left side of (1) gives a baryon number of $1 + 1 = 2$. The right side gives a baryon number of $1 + 1 + 1 + (-1) = 2$. Thus, the reaction can occur provided the incoming proton has sufficient energy.

The left side of (2) gives a baryon number of $1 + 1 = 2$. However, the right side gives $1 + 1 + (-1) = 1$. Since the baryon number is not conserved, the reaction cannot occur.

Lepton Number

There are three conservation laws involving lepton numbers, one for each variety of lepton. The **law of conservation of electron-lepton number** states that the sum of the electron-lepton numbers before a reaction or decay must equal the sum of the electron-lepton numbers after the reaction or decay. The electron and the electron neutrino are assigned a positive lepton number, $L_e = +1$; the antileptons e^+ and $\bar{\nu}_e$ are assigned the lepton number $L_e = -1$; and all other particles have $L_e = 0$. For example, consider the decay of the neutron,

Neutron decay

$$n \longrightarrow p^+ + e^- + \bar{\nu}_e$$

Before the decay, the electron-lepton number is $L_e = 0$; after the decay it is $0 + 1 + (-1) = 0$. Thus, the electron-lepton number is conserved. It is important to recognize that the baryon number must also be conserved. This can easily be seen by noting that before the decay $B = +1$, whereas after the decay $B = +1 + 0 + 0 = +1$.

Similarly, when a decay involves muons, the muon-lepton number, L_μ, is conserved. The μ^- and the ν_μ are assigned $L_\mu = +1$, the antimuons μ^+ and $\bar{\nu}_\mu$ are assigned $L_\mu = -1$, and all other particles have $L_\mu = 0$. Finally, the tau-lepton number, L_τ, is conserved, and similar assignments can be made for the τ lepton and its neutrino.

Example 32.2 Checking Lepton Numbers

Determine which of the following decay schemes can occur on the basis of conservation of lepton number.

$$\mu^- \longrightarrow e^- + \bar{\nu}_e + \nu_\mu \quad [1]$$

$$\pi^+ \longrightarrow \mu^+ + \nu_\mu + \nu_e \quad [2]$$

Solution Since decay 1 involves a muon and an electron, L_μ and L_e must both be conserved. Before the decay, $L_\mu = +1$ and $L_e = 0$. After the decay, $L_\mu = 0 + 0 + 1 = +1$, and $L_e = +1 - 1 + 0 = 0$. Thus, both numbers are conserved, and on this basis the decay mode is possible.

Before decay 2, $L_\mu = 0$ and $L_e = 0$. After the decay, $L_\mu = -1 + 1 + 0 = 0$, but $L_e = +1$. Thus, the decay is not possible because the electron-lepton number is not conserved.

Exercise Determine whether the decay $\mu^- \rightarrow e^- + \bar{\nu}_e$ can occur.

Answer No. The muon-lepton number is +1 before the decay and is 0 after.

32.7 Strange Particles and Strangeness

Many particles discovered in the 1950s were produced by the nuclear interaction of pions with protons and neutrons in the atmosphere. A group of these particles, namely the K, Λ, and Σ, were found to exhibit unusual properties in their production and decay, and hence were called *strange particles.*

One unusual property is that these particles are always produced in pairs. For example, when a pion collides with a proton, two neutral strange particles are produced with high probability:

$$\pi^- + p^+ \longrightarrow K^0 + \Lambda^0$$

On the other hand, the reaction $\pi^- + p^+ \rightarrow K^0 + n$ never occurred, even though no known conservation laws were violated and the energy of the pion was sufficient to initiate the reaction.

The second peculiar feature of strange particles is that, although they are produced by the strong interaction at a high rate, they do not decay into particles that interact via the strong force at a very high rate. Instead, they decay very slowly, which is characteristic of the weak interaction. Their half-lives are in the range 10^{-10} s to 10^{-8} s; most other particles that interact via the strong force have lifetimes on the order of 10^{-23} s.

To explain these unusual properties of strange particles, a law called **conservation of strangeness** was introduced, together with a new quantum number, S, called **strangeness.** The strangeness numbers for some particles are given in Table 32.2. The production of strange particles in pairs is explained by assigning $S = +1$ to one of the particles and $S = -1$ to the other. All nonstrange particles are assigned strangeness $S = 0$. Thus, whenever a nuclear reaction or decay occurs, the sum of the strangeness numbers before the process must equal the sum of the strangeness numbers after the process.

One can explain the slow decay of strange particles by assuming that the strong and electromagnetic interactions obey the law of conservation of strangeness, whereas the weak interaction does not. Since the decay reaction involves the loss of one strange particle, it violates strangeness conservation and hence proceeds slowly via the weak interaction.

Example 32.3 Is Strangeness Conserved?

(a) Determine whether the following reaction occurs on the basis of conservation of strangeness.

$$\pi^0 + n \longrightarrow K^+ + \Sigma^-$$

Solution The initial state has strangeness $S = 0 + 0 = 0$, and the final state is $+1 - 1 = 0$. Thus, strangeness is conserved and the reaction is allowed.

(b) Show that the following reaction does not conserve strangeness.

$$\pi^- + p^+ \longrightarrow \pi^- + \Sigma^+$$

Solution The initial state has strangeness $S = 0 + 0 = 0$, and the final state has strangeness $S = 0 + (-1) = -1$. Thus, strangeness is not conserved.

Exercise Show that the observed reaction

$$p^+ + \pi^- \longrightarrow K^0 + \Lambda^0$$

obeys the law of conservation of strangeness.

32.8 The Eightfold Way

As we have seen, conserved quantities such as spin, baryon number, lepton number, and strangeness are labels we associate with particles. Many classification schemes have been proposed that group particles into families based on such labels. First, consider the first eight baryons listed in Table 32.2, all having a spin of 1/2. If we plot their strangenesses versus their charges using a sloping coordinate system, as in Figure 32.4, a fascinating pattern is observed. Six of the baryons form a hexagon, and the remaining two are at the hexagon's center.

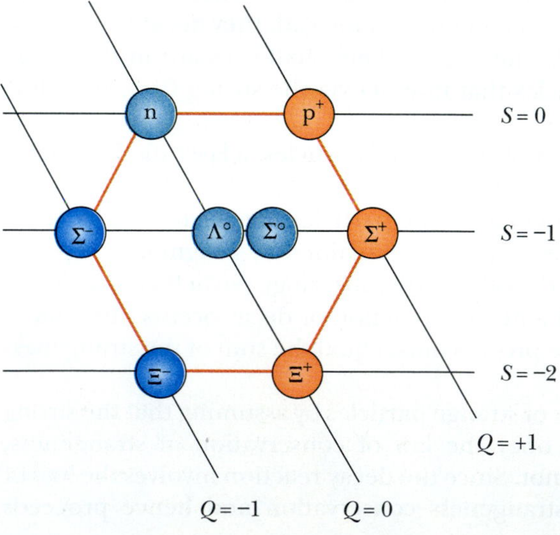

Figure 32.4 The hexagonal eightfold way pattern for the eight spin-1/2 baryons. This strangeness-versus-charge plot uses a sloping axis for the charge number Q.

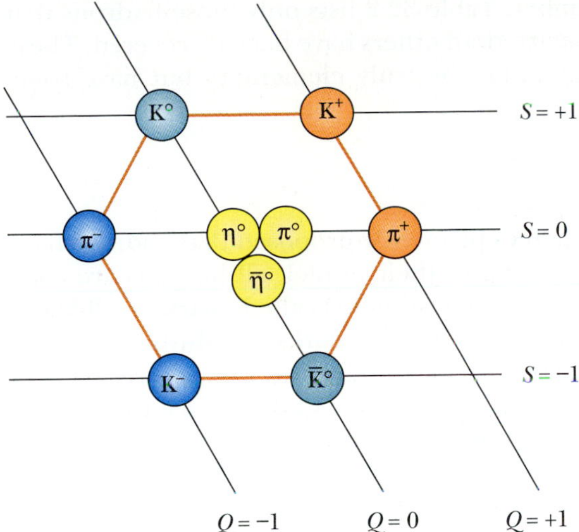

Figure 32.5
The eightfold way pattern for the nine spin-zero mesons.

American physicist Murray Gell-Mann won the Nobel Prize in 1969 for his theoretical studies dealing with subatomic particles. *(Photo courtesy of Michael R. Dressler)*

Now consider the family of mesons listed in Table 32.2, with spins of zero. If we count both particles and antiparticles, there are nine such mesons. Figure 32.5 is a plot of strangeness versus charge for this family. Again a fascinating hexagonal pattern emerges. In this case, the particles on the perimeter lie opposite their antiparticles, and the remaining three (which form their own antiparticles) are at its center. These and related symmetric patterns, called the **eightfold way,** were proposed independently in 1961 by Murray Gell-Mann and Yuval Ne'eman.

The groups of baryons and mesons can be displayed in many other symmetric patterns within the framework of the eightfold way. For example, the family of spin- 3/2 baryons contains ten particles arranged in a pattern like the tenpins in a bowling alley. After the pattern was proposed, one of the particles was missing—it had yet to be discovered. Gell-Mann predicted that the missing particle, which he called the omega minus (Ω^-), should have a spin of 3/2, a charge of -1, a strangeness of -3, and a rest energy of about 1680 MeV. Shortly thereafter, in 1964, scientists at the Brookhaven National Laboratory found the missing particle through careful analyses of bubble chamber photographs, and confirmed all its predicted properties.

The patterns of the eightfold way have much in common with the pattern of the periodic table. Whenever a vacancy (a missing particle or element) occurs in the organized patterns, experimentalists have a guide for their investigations. Furthermore, the existence of the eightfold way patterns suggest that baryons and mesons have a more elemental substructure, to which we now turn.

32.9 Quarks—Finally

As we have noted, leptons appear to be truly elementary particles because they have no measurable size or internal structure, are limited in number, and do not seem to break down into smaller units. Hadrons, on the other hand, are complex particles having size and structure. Furthermore, we know that hadrons decay into other hadrons and are many in number. Table 32.2 lists only those hadrons that are stable against hadronic decay; hundreds of others have been discovered. These facts strongly suggest that hadrons cannot be truly elementary but have some substructure.

The Original Quark Model

In 1963 Gell-Mann and George Zweig independently proposed that hadrons have a more elemental substructure. According to their model, all hadrons are composite systems of two or three fundamental constituents called **quarks.** Gell-Mann borrowed the word *quark* from the passage "Three quarks for Muster Mark" in James Joyce's book *Finnegans Wake.* In the original model, there were three types of quarks designated by the symbols *u, d,* and *s.* These were given the arbitrary names *up, down,* and *sideways* (or, now more commonly, *strange*).

An unusual property of quarks is that they have fractional electronic charges, as shown—along with other properties—in Table 32.3. Associated with each quark is an antiquark of opposite charge, a baryon number, and a strangeness. The compositions of all hadrons known when Gell-Mann and Zweig presented their models could be completely specified by three simple rules:

1. Mesons consist of one quark and one antiquark, giving them a baryon number of 0, as required.
2. Baryons consist of three quarks.
3. Antibaryons consist of three antiquarks.

Table 32.3
Properties of Quarks and Antiquarks

				Quarks				
Name	*Symbol*	*Spin*	*Charge*	*Baryon Number*	*Strangeness*	*Charm*	*Bottomness*	*Topness*
Up	u	1/2	$+\frac{2}{3}e$	$\frac{1}{3}$	0	0	0	0
Down	d	1/2	$-\frac{1}{3}e$	$\frac{1}{3}$	0	0	0	0
Strange	s	1/2	$-\frac{1}{3}e$	$\frac{1}{3}$	−1	0	0	0
Charmed	c	1/2	$+\frac{2}{3}e$	$\frac{1}{3}$	0	+1	0	0
Bottom	b	1/2	$-\frac{1}{3}e$	$\frac{1}{3}$	0	0	+1	0
Top	t	1/2	$+\frac{2}{3}e$	$\frac{1}{3}$	0	0	0	+1
				Antiquarks				
Name	*Symbol*	*Spin*	*Charge*	*Baryon Number*	*Strangeness*	*Charm*	*Bottomness*	*Topness*
Up	$\bar{u}$	1/2	$-\frac{2}{3}e$	$-\frac{1}{3}$	0	0	0	0
Down	$\bar{d}$	1/2	$+\frac{1}{3}e$	$-\frac{1}{3}$	0	0	0	0
Strange	$\bar{s}$	1/2	$+\frac{1}{3}e$	$-\frac{1}{3}$	+1	0	0	0
Charmed	$\bar{c}$	1/2	$-\frac{2}{3}e$	$-\frac{1}{3}$	0	−1	0	0
Bottom	$\bar{b}$	1/2	$+\frac{1}{3}e$	$-\frac{1}{3}$	0	0	−1	0
Top	$\bar{t}$	1/2	$-\frac{2}{3}e$	$-\frac{1}{3}$	0	0	0	−1

Table 32.4 lists the quark compositions of several mesons and baryons. Note that just two of the quarks, u and d, are contained in all hadrons encountered in ordinary matter (protons and neutrons). The third quark, s, is needed only to construct strange particles with a strangeness of either +1 or −1. Figure 32.6 is a pictorial representation of the quark compositions of several particles.

Table 32.4 Quark Composition of Several Hadrons

Particle	Quark Composition
Mesons	
π^+	$u\bar{d}$
π^-	$\bar{u}d$
K^+	$u\bar{s}$
K^-	$\bar{u}s$
K^0	$d\bar{s}$
Baryons	
p^+	uud
n	udd
Λ^0	uds
Σ^+	uus
Σ^0	uds
Σ^-	dds
Ξ^0	uss
Ξ^-	dss
Ω^-	sss

Charm and Other Recent Developments

Although the original quark model was highly successful in classifying particles into families, there were some discrepancies between predictions of the model and certain experimental decay rates. Consequently, a fourth quark was proposed by several physicists in 1967. They argued that if there were four leptons (as was thought at the time), then there should also be four quarks because of an underlying symmetry in nature. The fourth quark, denoted by c, was given a property called **charm.** A *charmed* quark would have the charge $+2e/3$, but its charm would distinguish it from the other three quarks. The new quark would have a charm of $C = +1$, its antiquark would have a charm of $C = -1$, and all the other quarks would have $C = 0$, as indicated in Table 32.3. Charm, like strangeness, would be conserved in strong and electromagnetic interactions but not in weak interactions.

In 1974 a new heavy meson called the J/Ψ particle (or simply Ψ) was discovered independently by a group led by Burton Richter at the Stanford Linear Accelerator and another group led by Samuel Ting at the Brookhaven National Laboratory. Richter and Ting were awarded the Nobel Prize in 1976 for this work. The J/Ψ particle did not fit into the three-quark model but had the properties of a combination of a charmed quark and its antiquark ($c\bar{c}$). It was much more massive than the other known mesons (~3100 MeV/c^2), and its lifetime was much longer than those of other particles that decay via the strong force. Soon, related charmed mesons were discovered, corresponding to such quark combinations as $\bar{c}d$ and $c\bar{d}$, all of which have large masses and long lifetimes. In 1975 researchers at Stanford University reported strong evidence for the tau (τ) lepton, with a mass of 1784 MeV/c^2. Such discoveries led to more elaborate quark models and the proposal of

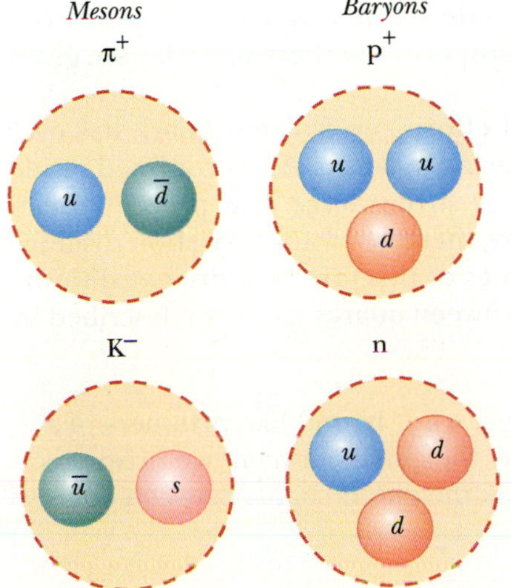

Figure 32.6
The quark compositions of several particles. Note that the mesons on the left contain two quarks each, whereas the baryons on the right contain three quarks each.

Table 32.5
The Fundamental Particles and Some of Their Properties

Particle	Rest Energy	Charge
Quarks		
u	360 MeV	$+\frac{2}{3}e$
d	360 MeV	$-\frac{1}{3}e$
c	1500 MeV	$+\frac{2}{3}e$
s	540 MeV	$-\frac{1}{3}e$
t (?)	~100 GeV	$+\frac{2}{3}e$
b	5 GeV	$-\frac{1}{3}e$
Leptons		
e^-	511 keV	$-e$
μ^-	107 MeV	$-e$
τ^-	1784 MeV	$-e$
ν_e	<30 eV	0
ν_μ	<0.5 MeV	0
ν_τ	<250 MeV	0

two new quarks, named *top* (t) and *bottom* (b). (Some physicists prefer the whimsical names *truth* and *beauty*.) To distinguish these quarks from the old ones, quantum numbers called *topness* and *bottomness* were assigned to these new particles and are included in Table 32.3. In 1977 researchers at the Fermi National Laboratory, under the direction of Leon Lederman, reported the discovery of a very massive new meson, Υ, whose composition is considered to be $b\bar{b}$.

You are probably wondering whether such discoveries will ever end. How many "building blocks" of matter really exist? At the present, physicists believe that the fundamental particles in nature include six quarks and six leptons (together with their antiparticles). Some of the properties of these particles are given in Table 32.5.

In spite of many extensive experimental efforts, no isolated quark has ever been observed. Physicists now believe that quarks are permanently confined inside ordinary particles because of an exceptionally strong force that prevents them from escaping. This force, called the *color force,* increases with separation distance (similar to the force of a spring); the properties of this force are discussed in the next section. The great strength of the force between quarks has been described by one author as follows:[1]

> Quarks are slaves of their own color charge, . . . bound like prisoners of a chain gang. . . . Any locksmith can break the chain between two prisoners, but no locksmith is expert enough to break the gluon chains between quarks. Quarks remain slaves forever.

[1] Harald Fritzsch, *Quarks, the Stuff of Matter,* London, Allen Lane, 1983.

32.10 The Standard Model

Shortly after the quark model was proposed, scientists recognized that certain particles have quark compositions that violate the Pauli exclusion principle. Because all quarks have spin 1/2, they are expected to follow the exclusion principle. One example of a particle that violates the exclusion principle is the Ω^- (*sss*) baryon predicted by Gell-Mann, which contains three *s* quarks having parallel spins, giving it a total spin of 3/2. Other examples are baryons that consist of identical quarks with parallel spins, such as the Δ^{++} (*uuu*) and the Δ^- (*ddd*). To resolve this problem, it was suggested that quarks possess a property called **color.** This property is similar in many respects to electric charge except that it occurs in three varieties called red, green, and blue. Of course, the antiquarks have the colors antired, antigreen, and antiblue. In order to satisfy the exclusion principle, all three quarks in a baryon must have different colors. A meson consists of a quark of one color and an antiquark of the corresponding anticolor. The result is that baryons and mesons are always colorless (or white). Furthermore, the property of color increases the number of quarks by a factor of three.

Although the concept of color in the quark model was originally conceived to satisfy the exclusion principle, it also explains certain experimental results. For example, the modified theory correctly predicts the lifetime of the π^0 meson. The theory of how quarks interact with each other is called **quantum chromodynamics,** or QCD, to parallel the name *quantum electrodynamics* (the theory of interaction between electric charges). In QCD, the quark is said to carry a *color charge,* in analogy with electric charge. As mentioned, the force between quarks is often called the **color force.**

As stated earlier, the strong interaction between hadrons is mediated by massless particles called **gluons.** According to the theory, there are eight gluons, six of which have color charge. Because of their color charges, quarks can attract each other and form composite particles. When a quark emits or absorbs a gluon, its color changes. For example, a blue quark that emits a gluon may become a red quark, and the red quark that absorbs this gluon becomes a blue quark. The color force between quarks is analogous to the electric force between charges; like colors repel and opposite colors attract. Therefore, two red quarks repel each other, but a red quark will be attracted to an antired quark. The attraction between quarks of opposite color to form a meson ($q\bar{q}$) is indicated in Figure 32.7a. Differently colored quarks also attract each other, but with less intensity than opposite colors of quark and antiquark. For example, a cluster of red, blue, and green quarks all attract each other to form baryons as indicated in Figure 32.7b. Thus, all baryons contain three quarks, each of which has a different color.

Recall that the weak force is believed to be mediated by the W^+, W^-, and Z^0 bosons (spin 1 particles). These particles are said to have *weak charge* just as a quark has color charge. Thus, each elementary particle can have mass, electric charge, color charge, and weak charge. Of course, one or more of these could be zero. Scientists now believe that the truly elementary particles are leptons and quarks, and the force mediators are the gluon, the photon, $W^{\pm}$, Z^0, and the graviton.

In 1979 Sheldon Glashow, Abdus Salam, and Steven Weinberg won a Nobel Prize for developing a theory that unified the electromagnetic and weak interactions. This **electroweak theory** postulates that the weak and electromagnetic inter-

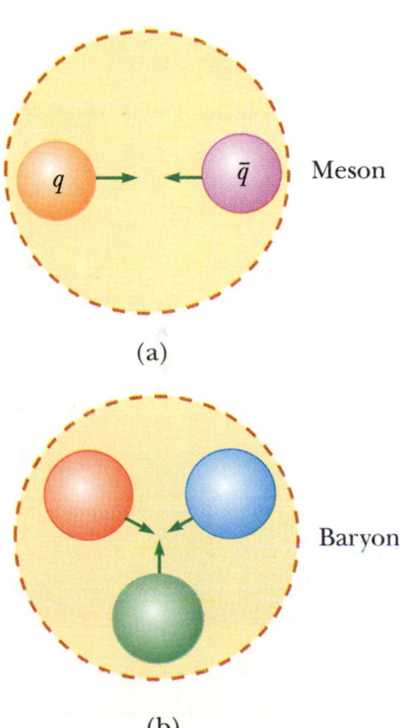

Figure 32.7
(a) A red quark is attracted to an antired quark. They form a meson whose quark structure is ($q\bar{q}$). (b) Three different colored quarks attract each other to form a baryon.

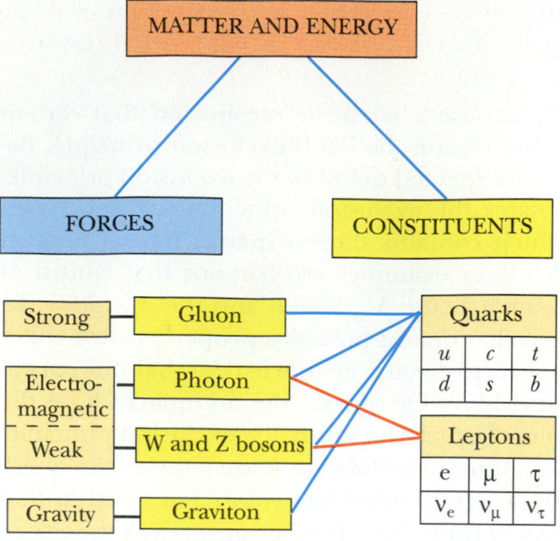

Figure 32.8
The standard model of particle physics.

actions have the same strength at very high particle energies. Thus, the two interactions are viewed as two manifestations of a single unifying electroweak interaction. The combination of the electroweak theory and QCD for the nuclear interaction is referred to as the "standard model." Although the details of the standard model are complex, its essential ingredients can be summarized with the help of Figure 32.8. The strong force, mediated by gluons, holds quarks together to form composite particles such as protons, neutrons, and mesons. Leptons participate only in the electromagnetic and weak interactions.

A technician works on one of the particle detectors at CERN, the European center for particle physics near Geneva, Switzerland. Electrons and positrons accelerated to an energy of 50 GeV collide in a circular tunnel 2 km in circumference, located 100 m underground. *(David Parker/Science Photo Library, Photo Researchers, Inc.)*

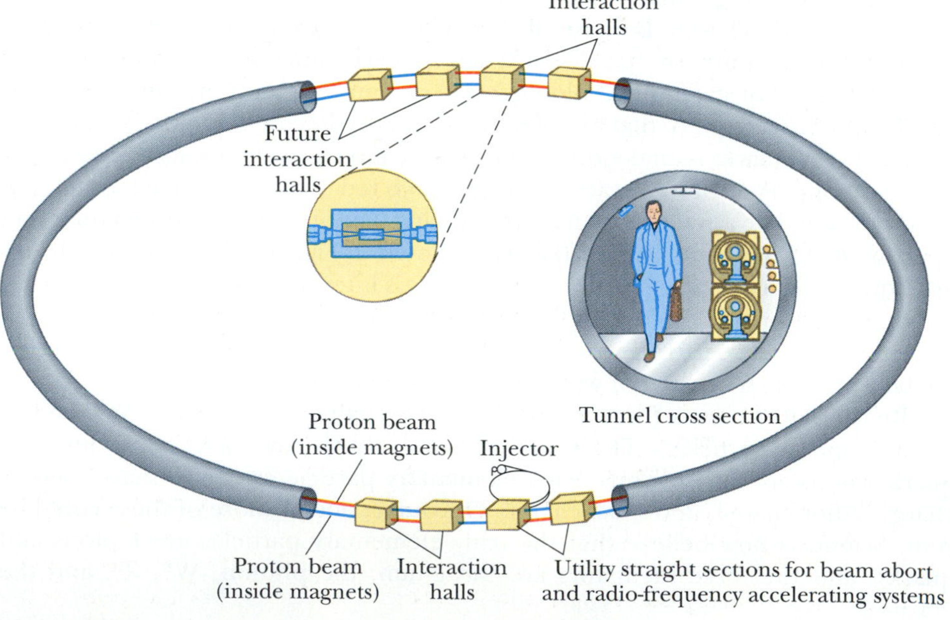

Figure 32.9
The proposed layout of the SSC is 52 mi in circumference. The experimental halls and other components on the ring are not to scale. Access and service points are placed at regular intervals around the ring.

However, the standard model does not answer all questions. A major problem is why the photon has no mass but the W and Z bosons do. Because of this mass difference, the electromagnetic and weak forces are quite distinct at low energies but become similar in nature at very high energies. This behavior as one goes from low to high energies, called *symmetry breaking,* leaves open the question of the origin of particle masses. In order to resolve this problem, a hypothetical particle called the *Higgs boson* has been proposed, which provides a mechanism for breaking the electroweak symmetry. The standard model, including the Higgs mechanism, provides a logically consistent explanation of the massive nature of the W and Z bosons. Unfortunately, the Higgs boson has not yet been found, but physicists know that its mass should be less than 1 TeV (10^{12} eV).

In order to determine whether the Higgs boson exists, two quarks of at least 1 TeV of energy must collide, but calculations show that this requires injecting 40 TeV of energy within the volume of a proton.

Scientists are convinced that because of the limited energy available in conventional accelerators using fixed targets, it is necessary to build colliding-beam accelerators called **colliders.** The concept of colliders is straightforward. Particles with equal masses and kinetic energies, traveling in opposite directions in an accelerator ring, collide head on to produce the required reaction and the formation of new particles. Because the total momentum of the interacting particles is zero, all of their kinetic energy is available for the reaction. The Large Electron-Positron Storage ring at CERN near Geneva, Switzerland and the Stanford Linear Collider in California collide both electrons and protons. The Super Proton Synchrotron at CERN accelerates protons and antiprotons to energies of 270 GeV, while the world's highest-energy proton accelerator, the Tevatron, at the Fermi National Lab in Illinois (see photo on p. 905) produces protons at almost 1000 GeV (or 1 TeV). The Superconducting Super Collider (SSC) which was being built in Texas, is an accelerator designed to produce 20-TeV protons in a ring 52 mi in circumference (Fig. 32.9). After much debate in Congress, and an investment of almost 2 billion dollars, the SSC project was canceled by the U.S. Department of Energy in October, 1993. Figure 32.10 shows the evolution of the stages of matter that scientists have been able to investigate with various types of microscopes.

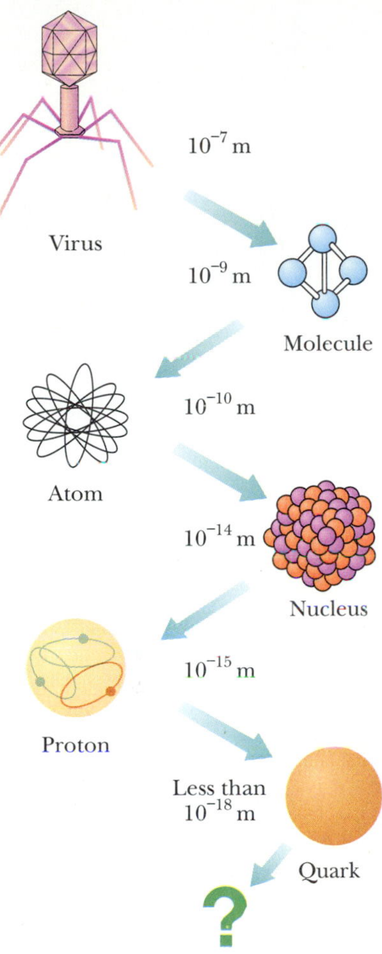

Figure 32.10 Observation of matter with microscopes reveals structures ranging in size from the smallest living thing, a virus, down to a quark, which has not yet been seen as an isolated particle.

32.11 The Cosmic Connection

As we have seen, the world around us is dominated by protons, electrons, neutrons, and neutrinos. Some of the other more exotic particles can be seen in cosmic rays. However, most of the new particles are produced using large, expensive machines that accelerate protons and electrons to energies in the GeV and TeV range. These energies are enormous when compared to the thermal energy in today's Universe. For example, the thermal energy kT at the center of the Sun is only about 1 keV, but the temperature of the early Universe was high enough to reach energies higher than 1 TeV.

In this section we shall describe one of the most fascinating theories in all of science—the Big Bang theory of the creation of the Universe—and the experimental evidence that supports it. This theory of cosmology states that the Universe had a beginning and that this beginning was so cataclysmic that it is impossible to look back beyond it. According to the theory, the Universe erupted from a point-like singularity about 15 to 20 billion years ago. The first few minutes after the Big

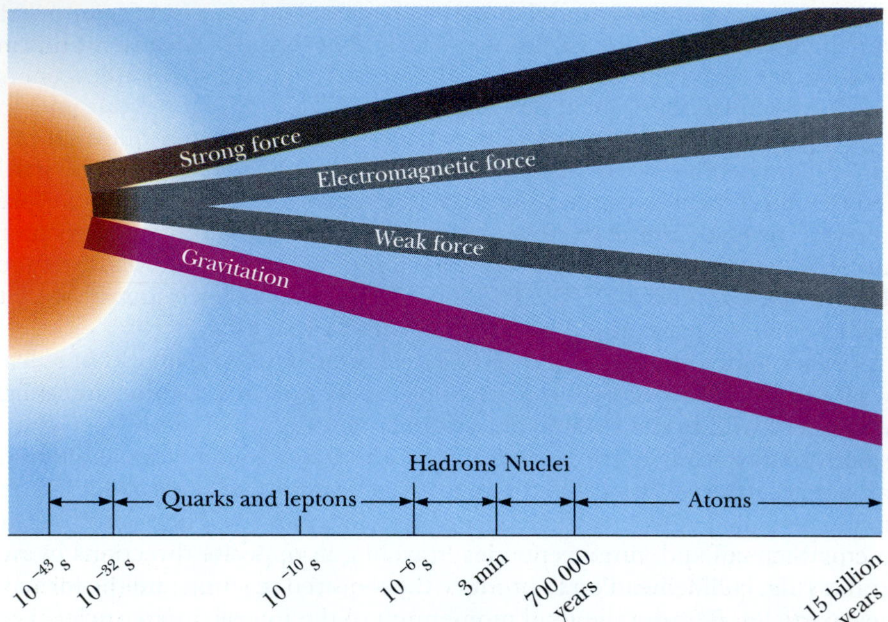

Figure 32.11
A brief history of the Universe from the Big Bang to the present. The fundamental forces became distinguishable during the first microsecond. Following this, all the quarks combined to form the strongly interacting particles. The leptons, however, remained separate and exist as individually observable particles to this day.

Bang saw such extremes of energy that it is believed that the fundamental interactions of physics were unified and that all matter melted down into an undifferentiated "quark soup."

The evolution of the fundamental forces from the Big Bang to the present is shown in Figure 32.11. During the first 10^{-43} s following the Big Bang (the ultrahot epoch during which $T \approx 10^{32}$ K), it is presumed that the strong, electroweak, and gravitational forces were joined to form a completely unified force. Between 10^{-43} s and 10^{-32} s following the Big Bang (the hot epoch, $T \approx 10^{29}$ K), gravity broke free of this unification while the strong and electroweak forces remained as one (they are described by a grand unification theory). This was a period when particle energies were so great ($>10^{16}$ GeV) that very massive particles as well as quarks, leptons, and their antiparticles existed. Then the Universe rapidly expanded and cooled during the warm epoch, when the temperatures ranged from 10^{29} to 10^{15} K, the strong and electroweak forces parted company, and the grand unification scheme was broken. As the Universe continued to cool, the electroweak force split into the weak force and the electromagnetic force about 10^{-10} s after the Big Bang.

Until about 700 000 years after the Big Bang, the Universe was dominated by radiation: ions absorbed and reemitted photons, thereby ensuring thermal equilibrium of radiation and matter. Energetic radiation also prevented matter from forming clumps or even single hydrogen atoms. By the time the Universe was about 700 000 years old, it had expanded and cooled to about 3000 K, and protons could bind to electrons to form hydrogen atoms. Since neutral atoms do not appreciably scatter photons, the Universe suddenly became transparent to photons. Radiation no longer dominated the Universe, and clumps of neutral matter steadily grew—first atoms, followed by molecules, gas clouds, stars, and finally galaxies.

Observation of Radiation from the Primordial Fireball

In 1965 Arno A. Penzias and Robert W. Wilson of Bell Labs were testing a sensitive microwave receiver and made an amazing discovery. A pesky signal producing a faint background hiss was interfering with their satellite communications experiments. In spite of their valiant efforts, the signal remained. Ultimately, it became clear that they were perceiving a microwave background radiation (at a wavelength of 7.35 cm) representing the leftover glow from the Big Bang.

The microwave horn that served as their receiving antenna is shown in Figure 32.12. The intensity of the detected signal remained unchanged as the antenna was pointed in different directions. The fact that the radiation had equal strengths in all directions suggested that the entire Universe was the source of this radiation. Eviction of a flock of pigeons from the 20-foot horn and cooling of the microwave detector both failed to remove the "spurious" signal. Through a casual conversation, Penzias and Wilson discovered that a group at Princeton had predicted the residual radiation from the Big Bang and were planning an experiment seeking to confirm the theory. The excitement in the scientific community was high when Penzias and Wilson announced that they had already observed an excess microwave background compatible with a 3-K blackbody source.

Figure 32.12
Robert W. Wilson *(left)* and Arno A. Penzias with a Bell Telephone Laboratories horn-reflector antenna. *(AT&T Bell Laboratories)*

Because the measurements of Penzias and Wilson were taken at a single wavelength, they did not completely confirm the radiation as 3-K blackbody radiation. Subsequent experiments by other groups added intensity data at different wavelengths, as shown in Figure 32.13. The results confirm that the radiation is that of a black body at 2.9 K. This figure is, perhaps, the most clearcut evidence for the Big Bang theory. The 1978 Nobel Prize in physics was awarded to Penzias and Wilson for their important discovery.

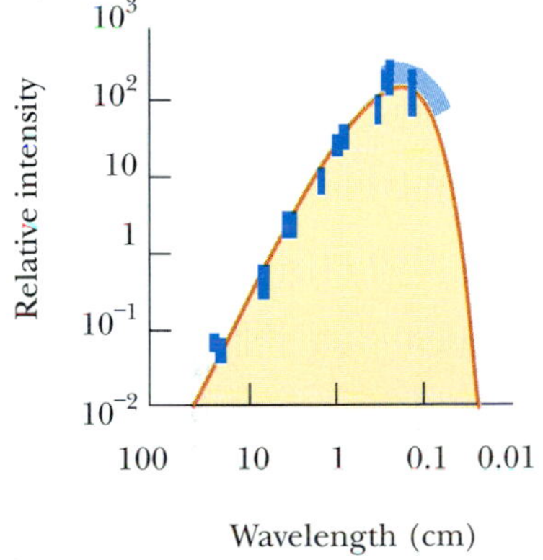

Figure 32.13
The radiation spectrum of the Big Bang. The shaded areas are experimental results. The solid line is the spectrum calculated for a black body at 2.9 K.

Other Evidence for the Expanding Universe

Most of the key discoveries supporting the theory of an expanding Universe, and indirectly the Big Bang theory of cosmology, were made in the 20th century. Vesto Melvin Slipher, an American astronomer, reported that most nebulae are receding from the Earth at speeds up to several million miles per hour. Slipher was one of the first to use Doppler shifts in the spectral lines due to radiation from these distant objects to measure velocities.

In the late 1920s, Edwin P. Hubble made the bold assertion that the whole Universe is expanding. From 1928 to 1936, he and Milton Humason toiled at Mount Wilson to prove this assertion until they reached the limits of the 100-inch telescope. The results of this work and its continuation on a 200-inch telescope in the 1940s showed that the speeds of galaxies increase in direct proportion to their distance from us, R (Fig. 32.14). This linear relation, known as **Hubble's law,** may be written

Hubble's law

$$v = HR$$

where H, called the **Hubble parameter,** has the approximate value

$$H = 17 \times 10^{-3} \text{ m/(s·lightyear)}$$

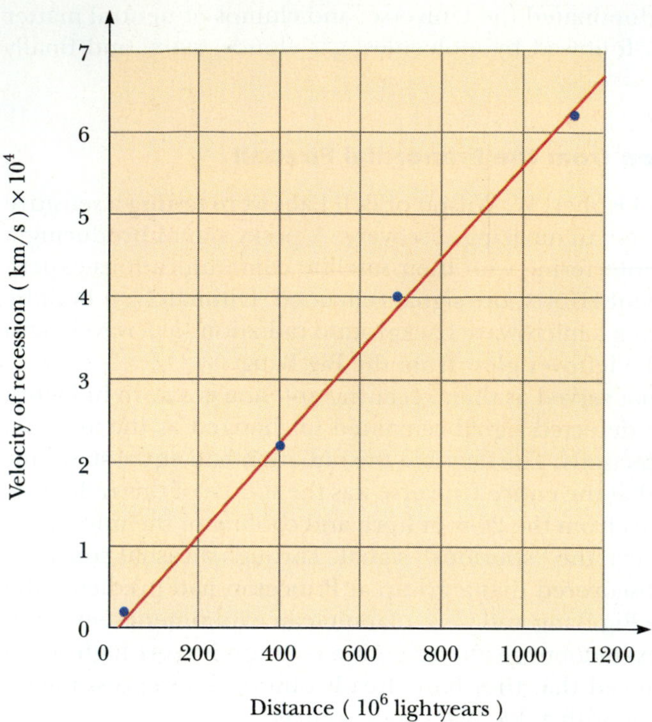

Figure 32.14
Hubble's law: A plot of velocity of recession versus distance for four galaxies.

Example 32.4 Recession of a Quasar

A quasar is a star-like object that is very distant from the Earth. Its speed can be determined from Doppler shift measurements in the light it emits. A certain quasar recedes from the Earth at a speed of $0.55c$. How far away is it?

Solution We can find the distance from Hubble's law:

$$R = \frac{v}{H} = \frac{(0.55)(3 \times 10^8 \text{ m/s})}{17 \times 10^{-3} \text{ m/(s} \cdot \text{lightyear)}}$$

$$= 9.7 \times 10^9 \text{ lightyears}$$

Exercise Assuming that the quasar has moved with the speed $0.55c$ ever since the Big Bang, estimate the age of the Universe.

Answer $t = R/v = 1/H \approx 18$ billion years, which is in good agreement with other calculations.

Will the Universe Expand Forever?

In the 1950s and 1960s, Allan R. Sandage used the 200-inch telescope at Mount Palomar to measure the speeds of galaxies at distances of up to 6 billion lightyears. These measurements showed that these very distant galaxies were moving about 10 000 km/s faster than the Hubble law predicted. According to this result, the

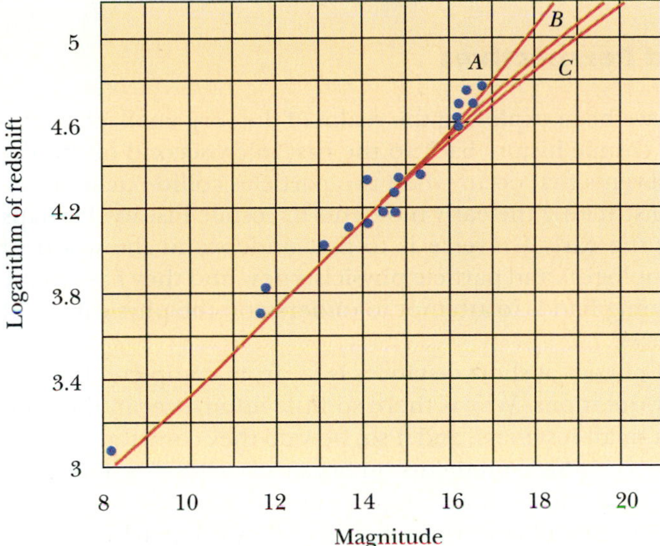

Figure 32.15
Red shift or speed of recession versus apparent magnitude or distance of 18 faint clusters. Curve A is the trend suggested by the six faintest clusters of galaxies. Line C corresponds to a Universe with a constant rate of expansion. If the data fall between B and C, the expansion slows but never stops. If the data fall to the left of B, expansion stops and contraction occurs.

Universe must have been expanding more rapidly 1 billion years ago and, consequently, the expansion is slowing (see Fig. 32.15); today astronomers and physicists are trying to determine the rate of slowing. If the average mass density of the Universe is less than some critical density ρ_c (current research gives a value of $\rho_c = 3 \text{ atoms/m}^3$), the galaxies will slow in their outward rush but still escape to infinity. If the average density exceeds the critical value, the expansion will eventually stop and contraction will begin, possibly leading to a superdense state and another expansion or an oscillating universe.

George Gamow (1904–1968), one of the first physicists to take the first half hour of the Universe seriously. In a mostly overlooked paper published in 1948, he and his students Ralph Alpher and Robert Herman made truly remarkable cosmological predictions. They correctly calculated the abundances of hydrogen and helium after the first half hour (75% H and 25% He), and predicted that radiation from the Big Bang should still be present with an apparent temperature of about 5 K.

Missing Mass in the Universe(?)

The visible matter in galaxies averages out to 5×10^{-33} g/cm^3. The radiation in the Universe has a mass equivalent of about 2% of the visible matter. Nonluminous matter (such as interstellar gas or black holes) may be estimated from the speeds of galaxies orbiting each other in a cluster. The higher the galaxy speeds, the more mass in the cluster. Results from measurements on the Coma cluster of galaxies, surprisingly, indicate that the amount of invisible matter is 20 to 30 times the amount present in stars and luminous gas clouds. Yet even this large invisible component, if applied to the Universe as a whole, leaves the observed mass density a factor of 10 less than ρ_c. This so-called *missing mass* (or *dark matter*) has been the subject of intense theoretical and experimental work, with exotic particles such as axions, photinos, and superstring particles suggested as candidates for the missing mass. More mundane proposals have been that the missing mass is present in certain galaxies as neutrinos. In fact, neutrinos are so abundant that a tiny neutrino rest mass on the order of 20 eV would furnish the missing mass and "close" the Universe.

Although we are a bit more sure about the beginning of the Universe, we are uncertain about its end. Will the Universe expand forever? Will it collapse and repeat its expansion in an endless series of oscillations? Results and answers to these questions remain inconclusive, and the exciting controversy continues.

32.12 Problems and Perspectives

While particle physicists have been exploring the realm of the very small, cosmologists have been exploring cosmic history back to the first microsecond of the Big Bang. Observation of the events that occur when two particles collide in an accelerator is essential for reconstructing the early moments in cosmic history. Perhaps the key to understanding the early Universe is to first understand the world of elementary particles. Cosmologists and particle physicists now find they have many common goals and are joining hands to attempt to understand the physical world at its most fundamental level.

Our understanding of physics at short distances is far from complete. Particle physics is faced with many questions. Why is there so little antimatter in the Universe? Do neutrinos have a small rest mass, and if so, how do they contribute to the "dark matter" of the Universe? (Measurements on Supernova 1987A established an upper limit of 16 eV for the neutrino mass.) Is it possible to unify the strong and electroweak theories in a logical and consistent manner? Why do quarks and leptons form three similar but distinct families? Are muons the same as electrons (apart from their different mass), or do they have other subtle differences that have not yet been detected? Why are some particles charged and others neutral? Why do quarks carry a fractional charge? What determines the masses of the fundamental constituents? Can isolated quarks exist? The questions go on and on. Because of the rapid advances and new discoveries in the field of particle physics, some of these questions will likely have been resolved by the time you read this book, while others may have emerged.

An important and obvious question that remains is whether leptons and quarks have a substructure. If they do, one could envision an infinite number of deeper structure levels. However, if leptons and quarks are indeed the ultimate constituents of matter, as physicists today tend to believe, we should be able to construct a final theory of the structure of matter, as Einstein dreamed of doing. In the view of many physicists, the end of the road is in sight, but how long it will take to reach that goal is anyone's guess.

Summary

Every fundamental interaction is said to be mediated by the exchange of field particles. The electromagnetic interaction is mediated by the photon; the weak interaction is mediated by the $W^{\pm}$ and Z^0 bosons; the gravitational interaction is mediated by gravitons; the strong interaction is mediated by gluons.

An antiparticle and a particle have the same mass but opposite charge, and other properties may also have opposite values, such as lepton number and baryon number. It is possible to produce particle-antiparticle pairs in nuclear reactions if the available energy is greater than $2mc^2$, where m is the rest mass of the particle (or antiparticle).

Particles other than photons are classified as hadrons or leptons. **Hadrons** interact primarily through the strong force. They have size and structure and hence are not elementary particles. There are two types of hadrons, *baryons* and *mesons.* Mesons have a baryon number of zero and have either zero or integral spin. Baryons, which generally are the most massive particles, have nonzero baryon numbers and spins of 1/2 or 3/2. The neutron and proton are examples of baryons.

Leptons have no structure or size and are considered truly elementary particles. Leptons interact only through the weak and electromagnetic forces. There are six leptons: the electron, e^-, the muon, μ^-; the tau, τ^-; and their neutrinos, ν_e, ν_μ, and ν_τ.

In all reactions and decays, quantities such as energy, linear momentum, angular momentum, electric charge, baryon number, and lepton number are strictly conserved. Certain particles have properties called **strangeness** and **charm.** These unusual properties are conserved only in those reactions and decays that occur via the strong force.

Recent theories postulate that all hadrons are composed of smaller units known as **quarks,** which have fractional electric charges and baryon numbers of 1/3, and come in six "flavors": up, down, strange, charmed, top, and bottom. Each baryon contains three quarks, and each meson contains one quark and one antiquark.

According to the theory of **quantum chromodynamics,** quarks have a property called **color,** and the strong force between quarks is referred to as the **color force.**

Observation of background microwave radiation by Penzias and Wilson strongly suggested that the Universe started with a Big Bang about 15 billion years ago. The background radiation is equivalent to that of a black body at a temperature of about 3 K.

A variety of astronomical measurements strongly suggest that the Universe is expanding. According to **Hubble's law,** distant galaxies are receding from the Earth at a speed $v = HR$, where R is the distance to the galaxy and H is **Hubble's parameter,** which has the approximate value 17×10^{-3} m/(s · lightyear).

Questions and Conceptual Exercises

1. Name the four fundamental interactions and the particles that mediate each interaction.
2. Identify the particle decays in Table 32.2 that occur by the weak interaction. Justify your answers.
3. Identify the particle decays in Table 32.2 that occur by the electromagnetic interaction. Justify your answers.
4. The family of K mesons all decay into final states that contain no protons or neutrons. What is the baryon number of the K mesons?
5. Particles known as resonances have very short lifetimes, on the order of 10^{-23} s. From this information, would you guess that they are hadrons or leptons? Explain.
6. An antibaryon interacts with a meson. Can a baryon be produced in such an interaction? Explain.
7. The Ξ^0 particle decays by the weak interaction according to the decay mode $\Xi^0 \rightarrow \Lambda^0 + \pi^0$. Would you expect this decay to be fast or slow? Explain.
8. Two protons in a nucleus interact via the strong interaction. Are they also subject to the weak interaction?
9. When an electron and a positron meet at low speeds in empty space, why is it that *two* gamma rays with energy of 0.511 MeV are produced, rather than *one* gamma ray with energy of 1.02 MeV?
10. Why is it that the neutron (which decays in free space in 900 s) is stable inside the nucleus?
11. Discuss the quark model of hadrons, and describe the properties of quarks.
12. What is the quark composition of the Ξ^- particle? (See Table 32.4.)
13. In the theory of quantum chromodynamics, quarks come in three colors. How would you justify the statement that "all baryons and mesons are colorless"?
14. Discuss the essential features of the standard model of particle physics.

15. Which baryon was predicted to exist by Murray Gell-Mann in 1961? What is the supposed quark composition of this particle?
16. How many quarks are there in (a) a baryon, (b) an antibaryon, (c) a meson, and (d) an antimeson? How do you account for the fact that baryons have half-integral spins and mesons have spins of 0 or 1? (*Hint:* Quarks have spins of 1/2.)
17. The W and Z bosons were first produced (by having a beam of protons and a beam of antiprotons meet at high energy) in 1983 at CERN. Why was this an important discovery?
18. How did Edwin Hubble in 1928 determine that the Universe is expanding?
19. How will the Hubble Space Telescope (when properly repaired in 1995) help determine the large-scale nature of the Universe?
20. The sky is illuminated with a uniform background radiation corresponding to a temperature of 2.9 K. Discuss the spectrum of this radiation, thought to be a remnant of the "Big Bang" that took place as the Universe began.

Problems

Section 32.3 Positrons and Other Antiparticles

1. A photon produces a proton-antiproton pair according to the reaction $\gamma \rightarrow p^+ + p^-$. What is the frequency of the photon? What is its wavelength?
2. Two photons are produced when a proton and an antiproton annihilate each other. What are the minimum frequency and corresponding wavelength of each proton?
3. A photon with an energy of $E_\gamma = 2.09$ GeV creates a proton-antiproton pair in which the proton has a kinetic energy of 95 MeV. What is the kinetic energy of the antiproton? ($m_pc^2 = 938.3$ MeV.)

Section 32.4 Mesons and the Beginning of Particle Physics

4. Occasionally, high-energy muons collide with electrons and produce two neutrinos according to the reaction $\mu^+ + e^- \rightarrow 2\nu$. What kind of neutrinos are these?
5. One of the mediators of the weak interaction is the Z^0 boson, whose mass is 96 GeV/c^2. Use this information to find an approximate value for the range of the weak interaction.
6. When a high-energy proton or pion traveling near the speed of light collides with a nucleus, it travels an average distance of 3×10^{-15} m before interacting. From this information, estimate the time of the strong interaction.
7. A neutral pi meson at rest decays into two gamma-ray photons according to

$$\pi^0 \longrightarrow \gamma + \gamma$$

Find the energy, momentum, and frequency of each gamma-ray photon.

Section 32.5 Classification of Particles

8. Calculate the range of the force that might be produced by the virtual exchange of a proton.
9. Name one possible decay mode (see Table 32.2) of each of the following particles: Ω^+, $\overline{K}^0$, $\overline{\Lambda}^0$, $\overline{n}$.
10. Identify the unknown particle on the left side of the following reaction:

$$? + p^+ \longrightarrow n + \mu^+$$

Section 32.6 Conservation Laws

11. Each of the following reactions is forbidden. Determine a conservation law that is violated by each reaction.
(a) $p^+ + p^- \rightarrow \mu^+ + e^-$
(b) $\pi^- + p^+ \rightarrow p^+ + \pi^+$
(c) $p^+ + p^+ \rightarrow p^+ + \pi^+$
(d) $p^+ + p^+ \rightarrow p^+ + p^+ + n$
(e) $\gamma + p^+ \rightarrow n + \pi^0$
12. (a) Show that baryon number and charge are conserved in the following reactions of a pion with a proton.

$$\pi^+ + p^+ \longrightarrow K^+ + \Sigma^+ \qquad [1]$$

$$\pi^+ + p^+ \longrightarrow \pi^+ + \Sigma^+ \qquad [2]$$

(b) The first reaction is observed, but the second never occurs. Explain.
13. The following reactions or decays involve one or more neutrinos. Supply the missing neutrinos (ν_e, ν_μ, or ν_τ).
(a) $\pi^- \rightarrow \mu^- + ?$
(b) $K^+ \rightarrow \mu^+ + ?$
(c) $? + p^+ \rightarrow n + e^+$
(d) $? + n \rightarrow p^+ + e^-$
(e) $? + n \rightarrow p^+ + \mu^-$
(f) $\mu^- \rightarrow e^- + ? + ?$

14. For the following two reactions, the first may occur but the second cannot. Explain.

$$K^0 \longrightarrow \pi^+ + \pi^- \quad \text{(can occur)}$$

$$\Lambda^0 \longrightarrow \pi^+ + \pi^- \quad \text{(cannot occur)}$$

15. Determine which of the following reactions can occur. For those that cannot occur, determine the conservation law (or laws) violated by each.
(a) $p^+ \rightarrow \pi^+ + \pi^0$ (d) $\pi^+ \rightarrow \mu^+ + \nu_\mu$
(b) $p^+ + p^+ \rightarrow p^+ + p^+ + \pi^0$ (e) $n \rightarrow p^+ + e^- + \overline{\nu}_e$
(c) $p^+ + p^+ \rightarrow p^+ + \pi^+$ (f) $\pi^+ \rightarrow \mu^+ + n$

16. A K^0 particle at rest decays into a π^+ and a π^-. What will be the speed of each of the pions? The mass of the K^0 is 497.7 MeV/c^2, and the mass of each π is 139.6 MeV/c^2.

Section 32.7 Strange Particles and Strangeness

17. Determine whether or not strangeness is conserved in the following decays and reactions.
(a) $\Lambda^0 \rightarrow p^+ + \pi^-$ (d) $\pi^- + p^+ \rightarrow \pi^- + \Sigma^+$
(b) $\pi^- + p^+ \rightarrow \Lambda^0 + K^0$ (e) $\Xi^- \rightarrow \Lambda^0 + \pi^-$
(c) $p^- + p^+ \rightarrow \overline{\Lambda}^0 + \Lambda^0$ (f) $\Xi^0 \rightarrow p^+ + \pi^-$

18. The neutral ρ meson decays by the strong interaction into two pions according to $\rho^0 \rightarrow \pi^+ + \pi^-$, with a half-life of about 10^{-23} s. The neutral K meson also decays into two pions according to $K^0 \rightarrow \pi^+ + \pi^-$, but with a much longer half-life—about 10^{-10} s. How do you explain these observations?

19. Each of the following decays is forbidden. For each process, determine a conservation law that is violated.
(a) $\mu^- \rightarrow e^- + \gamma$ (d) $p^+ \rightarrow e^+ + \pi^0$
(b) $n \rightarrow p^+ + e^- + \nu_e$ (e) $\Xi^0 \rightarrow n + \pi^0$
(c) $\Lambda^0 \rightarrow p^+ + \pi^0$

Section 32.9 Quarks—Finally

20. The quark composition of the proton is *uud*, and that of the neutron is *udd*. Show that in each case the charge, baryon number, and strangeness of the particle equal the sum of these numbers for the quark constituents.

21. The quark compositions of the K^0 and Λ^0 particles are $d\overline{s}$ and *uds*, respectively. Show that the charge, baryon number, and strangeness of each of these particles equals the sum of these numbers for the quark constituents.

22. Neglecting binding energies, estimate the mass of the *u* and *d* quarks from the masses of the proton and neutron.

23. Analyze each of the reactions in terms of its constituent quarks:
(a) $\pi^- + p^+ \rightarrow K^0 + \Lambda^0$
(b) $\pi^+ + p^+ \rightarrow K^+ + \Sigma^+$
(c) $K^- + p^+ \rightarrow K^+ + K^0 + \Omega^-$
(d) $p^+ + p^+ \rightarrow K^0 + p^+ + \pi^+ + ?$
In the last reaction, identify the mystery particle.

Section 32.11 The Cosmic Connection

24. Using Hubble's law (Eq. 32.4), estimate the wavelength of the 590-nm sodium line emitted by galaxies (a) 2×10^6 lighyears away, (b) 2×10^8 lightyears away, and (c) 2×10^9 lightyears away. (*Hint:* Use the relativistic Doppler formula for wavelength λ' of light emitted from a moving source, as follows.)

$$\lambda' = \lambda\sqrt{\frac{1 + v/c}{1 - v/c}}$$

25. A distant quasar is moving away from the Earth at such high speed that the blue 434-nm hydrogen line is observed at 650 nm, in the red portion of the spectrum. (a) How fast is the quasar receding? (See the hint in the preceding problem.) (b) Using Hubble's law, determine the distance from Earth to this quasar.

Additional Problems

26. The strong interaction has a range of approximately 1.4×10^{-15} m. It is thought that an elementary particle is exchanged between the protons and neutrons in the nucleus, leading to an attractive force. (a) Utilize the uncertainty principle, $\Delta E\, \Delta t \gtrsim \hbar/2$, to estimate the mass of the elementary particle if it moves at nearly the speed of light. (b) Using Table 32.2, identify the particle.

27. Name at least one conservation law that prevents each of the following reactions:
(a) $\pi^- + p^+ \longrightarrow \Sigma^+ + \pi^0$
(b) $\mu^- \longrightarrow \pi^- + \nu_e$
(c) $p^+ \longrightarrow \pi^+ + \pi^+ + \pi^-$

28. Supernova 1987A, located about 170 000 lightyears from the Earth, is estimated to have emitted a burst of $\sim 10^{46}$ J of neutrinos. Assuming an average neutrino energy of 6 MeV and a 5000-cm^2 cross-sectional area for your body, how many of these neutrinos passed through you?

29. A gamma-ray photon strikes a stationary electron. Determine the minimum gamma-ray energy to make this reaction go:

$$\gamma + e^- \longrightarrow e^- + e^- + e^+$$

30. The energy flux of neutrinos from the Sun is estimated to be on the order of 0.4 W/m^2 at the Earth's surface. Estimate the fractional mass loss of the Sun

over 10^9 years due to the radiation of neutrinos. (The mass of the Sun is about 2×10^{30} kg. The distance from the Earth to the Sun is 1.5×10^{11} m.)

31. What are the kinetic energies of the proton and pi meson resulting from the decay of a Λ^0 at rest?

$$\Lambda^0 \longrightarrow p^+ + \pi^-$$

32. A Σ^0 particle at rest decays according to

$$\Sigma^0 \longrightarrow \Lambda^0 + \gamma$$

Find the energy of the gamma-ray photon.

33. If a K^0 meson at rest decays in 0.9×10^{-10} s, how far will a K^0 meson travel if it is moving at $0.96c$ through a bubble chamber?

34. Two protons approach each other with equal and opposite velocities. What is the minimum kinetic energy of each of the protons if they are to produce a π^+ meson at rest in the following reaction?

$$p^+ + p^+ \longrightarrow p^+ + n + \pi^+$$

35. A π-meson at rest decays according to $\pi^- \rightarrow \mu^- + \overline{\nu}_\mu$. What is the energy carried off by the neutrino? (Assume the neutrino moves off with the speed of light.) $m_\pi c^2 = 139.5$ MeV, $m_\mu c^2 = 105.7$ MeV, $m_\nu = 0$.

36. What processes are described by the Feynman diagrams in Figure 32.16? What is the exchanged particle in each process?

"Particles, particles, particles."

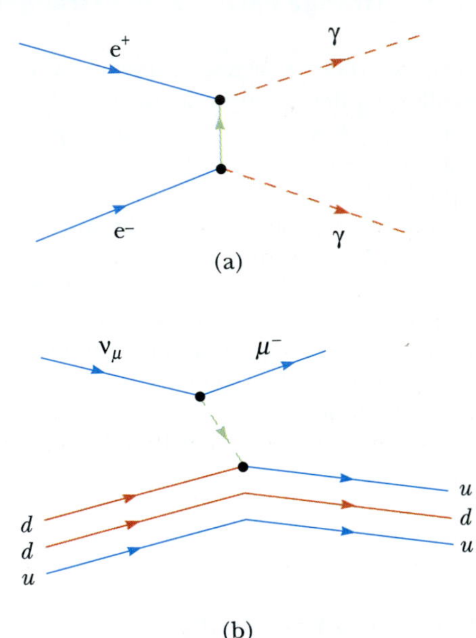

Figure 32.16 (Problem 36)

Appendix A

Table A.1
Conversion Factors

Length						
	m	cm	km	in.	ft	mi
1 meter	1	10^{2}	10^{-3}	39.37	3.281	6.214×10^{-4}
1 centimeter	10^{-2}	1	10^{-5}	0.3937	3.281×10^{-2}	6.214×10^{-6}
1 kilometer	10^{3}	10^{5}	1	3.937×10^{4}	3.281×10^{3}	0.6214
1 inch	2.540×10^{-2}	2.540	2.540×10^{-5}	1	8.333×10^{-2}	1.578×10^{-5}
1 foot	0.3048	30.48	3.048×10^{-4}	12	1	1.894×10^{-4}
1 mile	1609	1.609×10^{5}	1.609	6.336×10^{4}	5280	1

Mass				
	kg	g	slug	u
1 kilogram	1	10^{3}	6.852×10^{-2}	6.024×10^{26}
1 gram	10^{-3}	1	6.852×10^{-5}	6.024×10^{23}
1 slug	14.59	1.459×10^{4}	1	8.789×10^{27}
1 atomic mass unit	1.660×10^{-27}	1.660×10^{-24}	1.137×10^{-28}	1

Time					
	s	min	h	day	year
1 second	1	1.667×10^{-2}	2.778×10^{-4}	1.157×10^{-5}	3.169×10^{-8}
1 minute	60	1	1.667×10^{-2}	6.994×10^{-4}	1.901×10^{-6}
1 hour	3600	60	1	4.167×10^{-2}	1.141×10^{-4}
1 day	8.640×10^{4}	1440	24	1	2.738×10^{-3}
1 year	3.156×10^{7}	5.259×10^{5}	8.766×10^{3}	365.2	1

Speed				
	m/s	cm/s	ft/s	mi/h
1 meter/second	1	10^{2}	3.281	2.237
1 centimeter/second	10^{-2}	1	3.281×10^{-2}	2.237×10^{-2}
1 foot/second	0.3048	30.48	1	0.6818
1 mile/hour	0.4470	44.70	1.467	1

Note: 1 mi/min = 60 mi/h = 88 ft/s.

Table A.1 (Continued)

Force			
	N	dyn	lb
1 newton	1	10^5	0.2248
1 dyne	10^{-5}	1	2.248×10^{-6}
1 pound	4.448	4.448×10^5	1

Work, Energy, Heat			
	J	erg	ft·lb
1 joule	1	10^7	0.7376
1 erg	10^{-7}	1	7.376×10^{-8}
1 ft·lb	1.356	1.356×10^7	1
1 eV	1.602×10^{-19}	1.602×10^{-12}	1.182×10^{-19}
1 cal	4.186	4.186×10^7	3.087
1 Btu	1.055×10^3	1.055×10^{10}	7.779×10^2
1 kWh	3.600×10^6	3.600×10^{13}	2.655×10^6

	eV	cal	Btu	kWh
1 joule	6.242×10^{18}	0.2389	9.481×10^{-4}	2.778×10^{-7}
1 erg	6.242×10^{11}	2.389×10^{-8}	9.481×10^{-11}	2.778×10^{-14}
1 ft·lb	8.464×10^{18}	0.3239	1.285×10^{-3}	3.766×10^{-7}
1 eV	1	3.827×10^{-20}	1.519×10^{-22}	4.450×10^{-26}
1 cal	2.613×10^{19}	1	3.968×10^{-3}	1.163×10^{-6}
1 Btu	6.585×10^{21}	2.520×10^2	1	2.930×10^{-4}
1 kWh	2.247×10^{25}	8.601×10^5	3.413×10^2	1

Pressure			
	Pa	dyn/cm^2	atm
1 pascal	1	10	9.869×10^{-6}
1 dyne/centimeter^2	10^{-1}	1	9.869×10^{-7}
1 atmosphere	1.013×10^5	1.013×10^6	1
1 centimeter mercury*	1.333×10^3	1.333×10^4	1.316×10^{-2}
1 pound/inch^2	6.895×10^3	6.895×10^4	6.805×10^{-2}
1 pound/foot^2	47.88	4.788×10^2	4.725×10^{-4}

	cm Hg	lb/in.^{2}	lb/ft^2
1 newton/meter^2	7.501×10^{-4}	1.450×10^{-4}	2.089×10^{-2}
1 dyne/centimeter^2	7.501×10^{-5}	1.450×10^{-5}	2.089×10^{-3}
1 atmosphere	76	14.70	2.116×10^3
1 centimeter mercury*	1	0.1943	27.85
1 pound/inch^2	5.171	1	144
1 pound/foot^2	3.591×10^{-2}	6.944×10^{-3}	1

* At 0°C and at a location where the free-fall acceleration has its "standard" value, 9.80665 m/s^2.

Table A.2
Symbols, Dimensions, and Units of Physical Quantities

Quantity	Common Symbol	Unit*	Dimensions†	Unit in Terms of Base SI Units
Acceleration	a	m/s^2	L/T^2	m/s^2
Amount of substance	n	mole		mol
Angle	θ, ϕ	radian (rad)	1	
Angular acceleration	α	rad/s^2	T^{-2}	s^{-2}
Angular frequency	ω	rad/s	T^{-1}	s^{-1}
Angular momentum	L	kg · m^2/s	ML^2/T	kg · m^2/s
Angular velocity	ω	rad/s	T^{-1}	s^{-1}
Area	A	m^2	L^2	m^2
Atomic number	Z			
Capacitance	C	farad (F) (= Q/V)	Q^{2}T^2/ML^2	A^2 · s^4/kg · m^2
Charge	q, Q, e	coulomb (C)	Q	A · s
Charge density				
Line	λ	C/m	Q/L	A · s/m
Surface	σ	C/m^2	Q/L^2	A · s/m^2
Volume	ρ	C/m^3	Q/L^3	A · s/m^3
Conductivity	σ	1/Ω · m	Q^{2}T/ML^3	A^2 · s^3/kg · m^3
Current	I	**AMPERE**	Q/T	A
Current density	$\mathbf{J}$	A/m^2	Q/T^2	A/m^2
Density	ρ	kg/m^3	M/L^3	kg/m^3
Dielectric constant	κ			
Displacement	$\mathbf{s}$	**METER**	L	m
Distance	d, h			
Length	ℓ, L			
Electric dipole moment	$\mathbf{p}$	C · m	QL	A · s · m
Electric field	$\mathbf{E}$	V/m	ML/QT^2	kg · m/A · s^3
Electric flux	Φ	V · m	ML^3/QT^2	kg · m^3/A · s^3
Electromotive force	$\mathcal{E}$	volt (V)	ML^2/QT^2	kg · m^2/A · s^3
Energy	E, U, K	joule (J)	ML^2/T^2	kg · m^2/s^2
Entropy	S	J/K	ML^2/T^2 · K	kg · m^2/s^2 · K
Force	$\mathbf{F}$	newton (N)	ML/T^2	kg · m/s^2
Frequency	f, ν	hertz (Hz)	T^{-1}	s^{-1}
Heat	Q	joule (J)	ML^2/T^2	kg · m^2/s^2
Inductance	L	henry (H)	ML^2/Q^2	kg · m^2/A^2 · s^2
Magnetic dipole moment	μ	N · m/T	QL^2/T	A · m^2
Magnetic field	$\mathbf{B}$	tesla (T) (= Wb/m^2)	M/QT	kg/A · s^2
Magnetic flux	Φ_m	weber (Wb)	ML^2/QT	kg · m^2/A · s^2
Mass	m, M	**KILOGRAM**	M	kg
Molar specific heat	C	J/mol · K		kg · m^2/s^2 · mol · K
Moment of inertia	I	kg · m^2	ML^2	kg · m^2
Momentum	$\mathbf{p}$	kg · m/s	ML/T	kg · m/s
Period	T	s	T	s
Permeability of space	μ_0	N/A^2 (= H/m)	ML/Q^{2}T	kg · m/A^2 · s^2
Permittivity of space	ϵ_0	C^2/N · m^2 (= F/m)	Q^{2}T^2/ML^3	A^2 · s^4/kg · m^3
Potential (voltage)	V	volt (V) (= J/C)	ML^2/QT^2	kg · m^2/A · s^3
Power	P	watt (W) (= J/s)	ML^2/T^3	kg · m^2/s^3
Pressure	P, p	pascal (Pa) = (N/m^2)	M/LT^2	kg/m · s^2

(Table continues)

* The base SI units are given in uppercase letters.

† The symbols M, L, T, and Q denote mass, length, time, and charge, respectively.

Table A.2 (Continued)

Quantity	Common Symbol	Unit*	Dimensions†	Unit in Terms of Base SI Units
Resistance	R	ohm (Ω)(= V/A)	ML^2/Q^2T	$kg \cdot m^2/A^2 \cdot s^3$
Specific heat	c	J/kg·K	$L^2/T^2 \cdot K$	$m^2/s^2 \cdot K$
Temperature	T	**KELVIN**	K	K
Time	t	**SECOND**	T	s
Torque	τ	N·m	ML^2/T^2	$kg \cdot m^2/s^2$
Speed	v	m/s	L/T	m/s
Volume	V	m^3	L^3	m^3
Wavelength	λ	m	L	m
Work	W	joule (J)(= N·m)	ML^2/T^2	$kg \cdot m^2/s^2$

* The base SI units are given in uppercase letters.

† The symbols M, L, T, and Q denote mass, length, time, and charge, respectively.

Table A.3
Table of Selected Atomic Masses*

Atomic Number, Z	Element	Symbol	Mass Number, A	Atomic Mass†	Percent Abundance, or Decay Mode (if radioactive)‡	Half-Life (if radioactive)
0	(Neutron)	n	1	1.008665	β^-	10.6 min
1	Hydrogen	H	1	1.007825	99.985	
	Deuterium	D	2	2.014102	0.015	
	Tritium	T	3	3.016049	β^-	12.33 y
2	Helium	He	3	3.016029	0.00014	
			4	4.002603	≈100	
3	Lithium	Li	6	6.015123	7.5	
			7	7.016005	92.5	
4	Beryllium	Be	7	7.016930	EC, γ	53.3 days
			8	8.005305	2α	6.7×10^{-17} s
			9	9.012183	100	
5	Boron	B	10	10.012938	19.8	
			11	11.009305	80.2	
6	Carbon	C	11	11.011433	β^+, EC	20.4 min
			12	12.000000	98.89	
			13	13.003355	1.11	
			14	14.003242	β^-	5730 y
7	Nitrogen	N	13	13.005739	β^+	9.96 min
			14	14.003074	99.63	
			15	15.000109	0.37	
8	Oxygen	O	15	15.003065	β^+, EC	122 s
			16	15.994915	99.759	
			18	17.999159	0.204	
9	Fluorine	F	19	18.998403	100	
10	Neon	Ne	20	19.992439	90.51	
			22	21.991384	9.22	

Table A.3 (Continued)

Atomic Number, *Z*	Element	Symbol	Mass Number, *A*	Atomic Mass†	Percent Abundance, or Decay Mode (if radioactive)‡	Half-Life (if radioactive)
11	Sodium	Na	22	21.994435	β^+, EC, γ	2.602 y
			23	22.989770	100	
			24	23.990964	β^-, γ	15.0 h
12	Magnesium	Mg	24	23.985045	78.99	
13	Aluminum	Al	27	26.981541	100	
14	Silicon	Si	28	27.976928	92.23	
			31	30.975364	β^-, γ	2.62 h
15	Phosphorus	P	31	30.973763	100	
			32	31.973908	β^-	14.28 days
16	Sulfur	S	32	31.972072	95.0	
			35	34.969033	β^-	87.4 days
17	Chlorine	Cl	35	34.968853	75.77	
			37	36.965903	24.23	
18	Argon	Ar	40	39.962383	99.60	
19	Potassium	K	39	38.963708	93.26	
			40	39.964000	β^-, EC, γ, β^+	1.28×10^9 y
20	Calcium	Ca	40	39.962591	96.94	
21	Scandium	Sc	45	44.955914	100	
22	Titanium	Ti	48	47.947947	73.7	
23	Vanadium	V	51	50.943963	99.75	
24	Chromium	Cr	52	51.940510	83.79	
25	Manganese	Mn	55	54.938046	100	
26	Iron	Fe	56	55.934939	91.8	
27	Cobalt	Co	59	58.933198	100	
			60	59.933820	β^-, γ	5.271 y
28	Nickel	Ni	58	57.935347	68.3	
			60	59.930789	26.1	
			64	63.927968	0.91	
29	Copper	Cu	63	62.929599	69.2	
			64	63.929766	β^-, β^+	12.7 h
			65	64.927792	30.8	
30	Zinc	Zn	64	63.929145	48.6	
			66	65.926035	27.9	
31	Gallium	Ga	69	68.925581	60.1	
32	Germanium	Ge	72	71.922080	27.4	
			74	73.921179	36.5	
33	Arsenic	As	75	74.921596	100	
34	Selenium	Se	80	79.916521	49.8	
35	Bromine	Br	79	78.918336	50.69	
36	Krypton	Kr	84	83.911506	57.0	
			89	88.917563	β^-	3.2 min
37	Rubidium	Rb	85	84.911800	72.17	

(Table continues)

* Data are taken from *Chart of the Nuclides,* 12th ed., General Electric, 1977, and from C. M. Lederer and V. S. Shirley, eds., *Table of Isotopes,* 7th ed., New York, John Wiley & Sons, 1978.

† The masses given are those for the neutral atom, including the *Z* electrons.

‡ EC = electron capture.

Table A.3 (Continued)

Atomic Number, Z	Element	Symbol	Mass Number, A	Atomic Mass†	Percent Abundance, or Decay Mode (if radioactive)‡	Half-Life (if radioactive)
38	Strontium	Sr	86	85.909273	9.8	
			88	87.905625	82.6	
			90	89.907746	β^-	28.8 y
39	Yttrium	Y	89	88.905856	100	
40	Zirconium	Zr	90	89.904708	51.5	
41	Niobium	Nb	93	92.906378	100	
42	Molybdenum	Mo	98	97.905405	24.1	
43	Technetium	Tc	98	97.907210	β^-, γ	4.2×10^6 y
44	Ruthenium	Ru	102	101.904348	31.6	
45	Rhodium	Rh	103	102.90550	100	
46	Palladium	Pd	106	105.90348	27.3	
47	Silver	Ag	107	106.905095	51.83	
			109	108.904754	48.17	
48	Cadmium	Cd	114	113.903361	28.7	
49	Indium	In	115	114.90388	95.7, β^-	5.1×10^{14} y
50	Tin	Sn	120	119.902199	32.4	
51	Antimony	Sb	121	120.903824	57.3	2×10^{21} y
52	Tellurium	Te	130	129.90623	34.5, β^-	
53	Iodine	I	127	126.904477	100	
			131	130.906118	β^-, γ	8.04 days
54	Xenon	Xe	132	131.90415	26.9	
			136	135.90722	8.9	
55	Cesium	Cs	133	132.90543	100	
56	Barium	Ba	137	136.90582	11.2	
			138	137.90524	71.7	
			144	143.922673	β^-	11.9 s
57	Lanthanum	La	139	138.90636	99.911	
58	Cerium	Ce	140	139.90544	88.5	
59	Praseodymium	Pr	141	140.90766	100	
60	Neodymium	Nd	142	141.90773	27.2	
61	Promethium	Pm	145	144.91275	EC, α, γ	17.7 y
62	Samarium	Sm	152	151.91974	26.6	
63	Europium	Eu	153	152.92124	52.1	
64	Gadolinium	Gd	158	157.92411	24.8	
65	Terbium	Tb	159	158.92535	100	
66	Dysprosium	Dy	164	163.92918	28.1	
67	Holmium	Ho	165	164.93033	100	
68	Erbium	Er	166	165.93031	33.4	
69	Thulium	Tm	169	168.93423	100	
70	Ytterbium	Yb	174	173.93887	31.6	
71	Lutecium	Lu	175	174.94079	97.39	
72	Hafnium	Hf	180	179.94656	35.2	
73	Tantalum	Ta	181	180.94801	99.988	
74	Tungsten (wolfram)	W	184	183.95095	30.7	
75	Rhenium	Re	187	186.95577	62.60, β^-	4×10^{10} y

Table A.3 (Continued)

Atomic Number, *Z*	Element	Symbol	Mass Number, *A*	Atomic Mass†	Percent Abundance, or Decay Mode (if radioactive)‡	Half-Life (if radioactive)
76	Osmium	Os	191	190.96094	β^-, γ	15.4 days
			192	191.96149	41.0	
77	Iridium	Ir	191	190.96060	37.3	
			193	192.96294	62.7	
78	Platinum	Pt	195	194.96479	33.8	
79	Gold	Au	197	196.96656	100	
80	Mercury	Hg	202	201.97063	29.8	
81	Thallium	Tl	205	204.97441	70.5	
			208	207.981988	β^-, γ	3.053 min
82	Lead	Pb	204	203.973044	β^-, 1.48	1.4×10^{17} y
			206	205.97446	24.1	
			207	206.97589	22.1	
			208	207.97664	52.3	
			210	209.98418	α, β^-, γ	22.3 y
			211	210.98874	β^-, γ	36.1 min
			212	211.99188	β^-, γ	10.64 h
			214	213.99980	β^-, γ	26.8 min
83	Bismuth	Bi	209	208.98039	100	
			211	210.98726	α, β^-, γ	2.15 min
84	Polonium	Po	210	209.98286	α, γ	138.38 days
			214	213.99519	α, γ	164 μs
85	Astatine	At	218	218.00870	α, β^-	$\approx$2 s
86	Radon	Rn	222	222.017574	α, γ	3.8235 days
87	Francium	Fr	223	223.019734	α, β^-, γ	21.8 min
88	Radium	Ra	226	226.025406	α, γ	1.60×10^3 y
			228	228.031069	β^-	5.76 y
89	Actinium	Ac	227	227.027751	α, β^-, γ	21.773 y
90	Thorium	Th	228	228.02873	α, γ	1.9131 y
			232	232.038054	100, α, γ	1.41×10^{10} y
91	Protactinium	Pa	231	231.035881	α, γ	3.28×10^4 y
92	Uranium	U	232	232.03714	α, γ	72 y
			233	233.039629	α, γ	1.592×10^5 y
			235	235.043925	0.72, α, γ	7.038×10^8 y
			236	236.045563	α, γ	2.342×10^7 y
			238	238.050786	99.275, α, γ	4.468×10^9 y
			239	239.054291	β^-, γ	23.5 min
93	Neptunium	Np	239	239.052932	β^-, γ	2.35 days
94	Plutonium	Pu	239	239.052158	α, γ	2.41×10^4 y
95	Americium	Am	243	243.061374	α, γ	7.37×10^3 y
96	Curium	Cm	245	245.065487	α, γ	8.5×10^3 y
97	Berkelium	Bk	247	247.07003	α, γ	1.4×10^3 y

(Table continues)

* Data are taken from *Chart of the Nuclides,* 12th ed., General Electric, 1977, and from C. M. Lederer and V. S. Shirley, eds., *Table of Isotopes,* 7th ed., New York, John Wiley & Sons, 1978.

† The masses given are those for the neutral atom, including the *Z* electrons.

‡ EC = electron capture.

Table A.3 (Continued)

Atomic Number, Z	Element	Symbol	Mass Number, A	Atomic Mass†	Percent Abundance, or Decay Mode (if radioactive)‡	Half-Life (if radioactive)
98	Californium	Cf	249	249.074849	α, γ	351 y
99	Einsteinium	Es	254	254.08802	α, γ, β^-	276 days
100	Fermium	Fm	253	253.08518	EC, α, γ	3.0 days
101	Mendelevium	Md	255	255.0911	EC, α	27 min
102	Nobelium	No	255	255.0933	EC, α	3.1 min
103	Lawrencium	Lr	257	257.0998	α	$\approx$35 s
104	Unnilquadium	Rf	261	261.1087	α	1.1 min
105	Unnilpentium	Ha	262	262.1138	α	0.7 min
106	Unnilhexium		263	263.1184	α	0.9 s
107	Unnilseptium		261	261	α	1–2 ms

* Data are taken from *Chart of the Nuclides,* 12th ed., General Electric, 1977, and from C. M. Lederer and V. S. Shirley, eds., *Table of Isotopes,* 7th ed., New York, John Wiley & Sons, 1978.

† The masses given are those for the neutral atom, including the Z electrons.

‡ EC = electron capture.

Appendix B

Mathematics Review

These mathematics appendices are intended as a brief review of operations and methods. Early in your course, you should be facile with basic algebraic techniques, analytic geometry, and trigonometry. The appendices on differential and integral calculus are more detailed and are intended for those students who have difficulty applying calculus concepts to physical situations. Mathematical symbols used in the text are given in the following table.

Mathematical Symbols Used in the Text and Their Meanings

Symbol	Meaning
$=$	is equal to
$\equiv$	is defined as
$\neq$	is not equal to
$\propto$	is proportional to
$>$	is greater than
$<$	is less then
$\gg (\ll)$	is much greater (less) than
$\approx$	is approximately equal to
Δx	the change in x
$\sum_{i=1}^{N} x_i$	the sum of all quantities x_i from $i = 1$ to $i = N$
$\|x\|$	the magnitude of x (always a positive quantity)
$\Delta x \rightarrow 0$	Δx approaches zero
$\frac{dx}{dt}$	the derivative of x with respect to t
$\frac{\partial x}{\partial t}$	the partial derivative of x with respect to t
$\int$	integral

B.1 Scientific Notation

Many quantities that scientists deal with often have very large or very small values. For example, the speed of light is about 300 000 000 m/s, and the ink required to make the dot over an i in this textbook has a mass of about 0.000 000 001 kg. Obviously, it is very cumbersome to read, write, and keep track of numbers such as these. We avoid this problem by using the powers of the number 10:

$$10^0 = 1$$

$$10^1 = 10$$

$$10^2 = 10 \times 10 = 100$$

$$10^3 = 10 \times 10 \times 10 = 1000$$

$$10^4 = 10 \times 10 \times 10 \times 10 = 10\ 000$$

$$10^5 = 10 \times 10 \times 10 \times 10 \times 10 = 100\ 000$$

and so on. The number of zeros corresponds to the power to which 10 is raised, called the **exponent** of 10. For example, the speed of light, 300 000 000 m/s, can be expressed as 3×10^8 m/s.

For numbers less than one, we note the following:

$$10^{-1} = \frac{1}{10} = 0.1$$

$$10^{-2} = \frac{1}{10 \times 10} = 0.01$$

$$10^{-3} = \frac{1}{10 \times 10 \times 10} = 0.001$$

$$10^{-4} = \frac{1}{10 \times 10 \times 10 \times 10} = 0.0001$$

$$10^{-5} = \frac{1}{10 \times 10 \times 10 \times 10 \times 10} = 0.00001$$

In these cases, the number of places that the decimal point lies to the left of the digit 1 equals the value of the (negative) exponent. Numbers that are expressed as some power of 10 multiplied by another number between 1 and 10 are said to be in **scientific notation.** For example, the scientific notation for 5 943 000 000 is 5.943×10^9, and that for 0.0000832 is 8.32×10^{-5}.

When numbers expressed in scientific notation are being multiplied, the following general rule is very useful:

$$10^n \times 10^m = 10^{n+m} \quad \text{[B.1]}$$

where n and m can be *any* numbers (not necessarily integers). For example, $10^2 \times 10^5 = 10^7$. The rule also applies if one of the exponents is negative. For example, $10^3 \times 10^{-8} = 10^{-5}$.

When dividing numbers expressed in scientific notation, note that

$$\frac{10^n}{10^m} = 10^n \times 10^{-m} = 10^{n-m} \quad \text{[B.2]}$$

Exercises

With help from the above rules, verify the answers to the following:

1. $86\ 400 = 8.64 \times 10^4$
2. $9\ 816\ 762.5 = 9.8167625 \times 10^6$
3. $0.0000000398 = 3.98 \times 10^{-8}$

4. $(4 \times 10^8)(9 \times 10^9) = 3.6 \times 10^{18}$
5. $(3 \times 10^7)(6 \times 10^{-12}) = 1.8 \times 10^{-4}$
6. $\dfrac{75 \times 10^{-11}}{5 \times 10^{-3}} = 1.5 \times 10^{-7}$
7. $\dfrac{(3 \times 10^6)(8 \times 10^{-2})}{(2 \times 10^{17})(6 \times 10^5)} = 2 \times 10^{-18}$

B.2 Algebra

Some Basic Rules

When algebraic operations are performed, the laws of arithmetic apply. Symbols such as x, y, and z are usually used to represent quantities that are not specified—the **unknowns.**

First, consider the equation

$$8x = 32$$

If we wish to solve for x, we can divide (or multiply) each side of the equation by the same factor without destroying the equality. In this case, if we divide both sides by 8, we have

$$\frac{8x}{8} = \frac{32}{8}$$

$$x = 4$$

Next consider the equation

$$x + 2 = 8$$

In this type of expression, we can add or subtract the same quantity from each side. If we subtract 2 from each side, we get

$$x + 2 - 2 = 8 - 2$$

$$x = 6$$

In general, if $x + a = b$, then $x = b - a$.

Now consider the equation

$$\frac{x}{5} = 9$$

If we multiply each side by 5, we are left with x on the left by itself and 45 on the right:

$$\left(\frac{x}{5}\right)(5) = 9 \times 5$$

$$x = 45$$

In all cases, *whatever operation is performed on the left side of the equality must also be performed on the right side.*

The following rules for multiplying, dividing, adding, and subtracting fractions should be recalled, where a, b, and c are three numbers:

	Rule	Example
Multiplying	$\left(\frac{a}{b}\right)\left(\frac{c}{d}\right) = \frac{ac}{bd}$	$\left(\frac{2}{3}\right)\left(\frac{4}{5}\right) = \frac{8}{15}$
Dividing	$\frac{(a/b)}{(c/d)} = \frac{ad}{cb}$	$\frac{2/3}{4/5} = \frac{(2)(5)}{(4)(3)} = \frac{10}{12}$
Adding	$\frac{a}{b} \pm \frac{c}{d} = \frac{ad \pm bc}{bd}$	$\frac{2}{3} - \frac{4}{5} = \frac{(2)(5) - (4)(3)}{(3)(5)} = -\frac{2}{15}$

Exercises

In the following exercises, solve for x:

	Answers
1. $a = \frac{1}{1 + x}$	$x = \frac{1 - a}{a}$
2. $3x - 5 = 13$	$x = 6$
3. $ax - 5 = bx + 2$	$x = \frac{7}{a - b}$
4. $\frac{5}{2x + 6} = \frac{3}{4x + 8}$	$x = -\frac{11}{7}$

Powers

When powers of a given quantity x are multiplied, the following rule applies:

$$x^n x^m = x^{n+m} \quad \text{[B.3]}$$

For example, $x^2 x^4 = x^{2+4} = x^6$.

When dividing the powers of a given quantity, the rule is

$$\frac{x^n}{x^m} = x^{n-m} \quad \text{[B.4]}$$

For example, $x^8/x^2 = x^{8-2} = x^6$.

A power that is a fraction, such as $\frac{1}{3}$, corresponds to a root as follows:

$$x^{1/n} = \sqrt[n]{x} \quad \text{[B.5]}$$

For example, $4^{1/3} = \sqrt[3]{4} = 1.5874$. (A scientific calculator is useful for such calculations.)

Finally, any quantity x^n that is raised to the mth power is

$$(x^n)^m = x^{nm} \quad \text{[B.6]}$$

Table B.1 summarizes the rules of exponents.

Table B.1
Rules of Exponents

$x^0 = 1$
$x^1 = x$
$x^n x^m = x^{n+m}$
$x^n/x^m = x^{n-m}$
$x^{1/n} = \sqrt[n]{x}$
$(x^n)^m = x^{nm}$

Exercises

Verify the following:

1. $3^2 \times 3^3 = 243$
2. $x^5 x^{-8} = x^{-3}$
3. $x^{10}/x^{-5} = x^{15}$
4. $5^{1/3} = 1.709975$ (Use your calculator.)
5. $60^{1/4} = 2.783158$ (Use your calculator.)
6. $(x^4)^3 = x^{12}$

Factoring

Some useful formulas for factoring an equation are

$$ax + ay + az = a(x + y + z) \qquad \text{Common factor}$$

$$a^2 + 2ab + b^2 = (a + b)^2 \qquad \text{Perfect square}$$

$$a^2 - b^2 = (a + b)(a - b) \qquad \text{Differences of squares}$$

Quadratic Equations

The general form of a quadratic equation is

$$ax^2 + bx + c = 0 \qquad \textbf{[B.7]}$$

where x is the unknown quantity and a, b, and c are numerical factors referred to as **coefficients** of the equation. This equation has two roots, given by

$$x = \frac{-b \pm \sqrt{b^2 - 4ac}}{2a} \qquad \textbf{[B.8]}$$

If $b^2 \geqslant 4ac$, the roots are real.

Example 1

The equation $x^2 + 5x + 4 = 0$ has the following roots, corresponding to the two signs of the square-root term:

$$x = \frac{-5 \pm \sqrt{5^2 - (4)(1)(4)}}{2(1)} = \frac{-5 \pm \sqrt{9}}{2} = \frac{-5 \pm 3}{2}$$

That is,

$$x_+ = \frac{-5 + 3}{2} = -1 \qquad x_- = \frac{-5 - 3}{2} = -4$$

where x_+ denotes the root corresponding to the positive sign and x_- denotes the root corresponding to the negative sign.

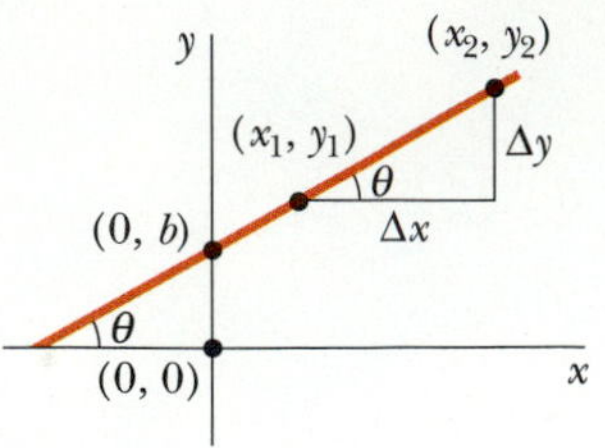

Figure B.1

Exercises

Solve the following quadratic equations:

	Answers	
1. $x^2 + 2x - 3 = 0$	$x_+ = 1$	$x_- = -3$
2. $2x^2 - 5x + 2 = 0$	$x_+ = 2$	$x_- = \frac{1}{2}$
3. $2x^2 - 4x - 9 = 0$	$x_+ = 1 + \sqrt{22}/2$	$x_- = 1 - \sqrt{22}/2$

Linear Equations

A linear equation has the general form

$$y = ax + b \qquad \text{[B.9]}$$

where a and b are constants. This equation is referred to as linear because the graph of y versus x is a straight line, as shown in Figure B.1. The constant b, called the **intercept**, represents the value of y at which the straight line intersects the y axis. The constant a is equal to the slope of the straight line and is also equal to the tangent of the angle that the line makes with the x axis. If any two points on the straight line are specified by the coordinates (x_1, y_1) and (x_2, y_2), as in Figure B.1, then the **slope** of the straight line can be expressed as

$$\text{Slope} = \frac{y_2 - y_1}{x_2 - x_1} = \frac{\Delta y}{\Delta x} = \tan\theta \qquad \text{[B.10]}$$

Note that a and b can have either positive or negative values. If $a > 0$, the straight line has a *positive* slope, as in Figure B.1. If $a < 0$, the straight line has a *negative* slope. In Figure B.1, both a and b are positive. Three other possible situations are shown in Figure B.2: $a > 0$, $b < 0$; $a < 0$, $b > 0$; and $a < 0$, $b < 0$.

Exercises

1. Draw graphs of the following straight lines:
   (a) $y = 5x + 3$ (b) $y = -2x + 4$ (c) $y = -3x - 6$.
2. Find the slopes of the straight lines described in Exercise 1.

Answers (a) 5 (b) −2 (c) −3

3. Find the slopes of the straight lines that pass through the following sets of points:
   (a) (0, −4) and (4, 2); (b) (0, 0) and (2, −5); (c) (−5, 2) and (4, −2).

Answers (a) 3/2 (b) −5/2 (c) −4/9

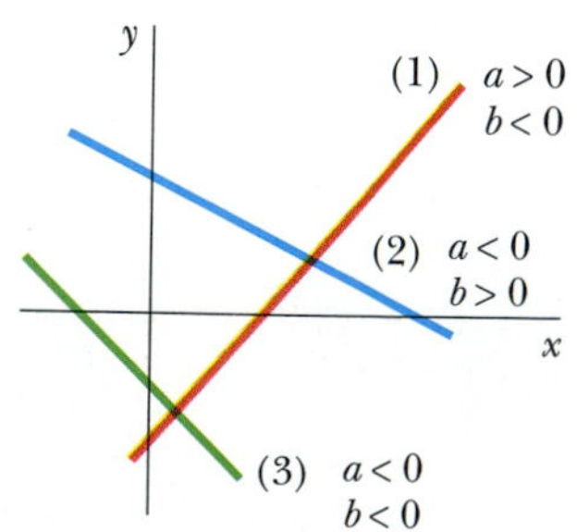

Figure B.2

Solving Simultaneous Linear Equations

Consider an equation such as $3x + 5y = 15$, which has two unknowns, x and y. Such an equation does not have a unique solution. That is, $(x = 0, y = 3)$, $(x = 5, y = 0)$, and $(x = 2, y = 9/5)$ are all solutions to this equation.

If a problem has two unknowns, a unique solution is possible only if there are *two* equations. In general, if a problem has n unknowns, its solution requires n equations. In order to solve two simultaneous equations involving two unknowns, x and y, we solve one of the equations for x in terms of y and substitute this expression into the other equation.

Example 2

Solve the following two simultaneous equations:

$$(1)\ 5x + y = -8$$

$$(2)\ 2x - 2y = 4$$

Solution From (2), $x = y + 2$. Substitution of this into (1) gives

$$5(y + 2) + y = -8$$

$$6y = -18$$

$$y = -3$$

$$x = y + 2 = \boxed{-1}$$

Alternative Solution Multiply each term in (1) by the factor 2 and add the result to (2):

$$10x + 2y = -16$$

$$\underline{2x - 2y = 4}$$

$$12x = -12$$

$$x = -1$$

$$y = x - 2 = \boxed{-3}$$

Two linear equations with two unknowns can also be solved by a graphical method. If the straight lines corresponding to the two equations are plotted in a conventional coordinate system, the intersection of the two lines represents the solution. For example, consider the two equations

$$x - y = 2$$

$$x - 2y = -1$$

These are plotted in Figure B.3. The intersection of the two lines has the coordinates $x = 5$, $y = 3$. This represents the solution to the equations. You should check this solution by the analytical technique discussed earlier.

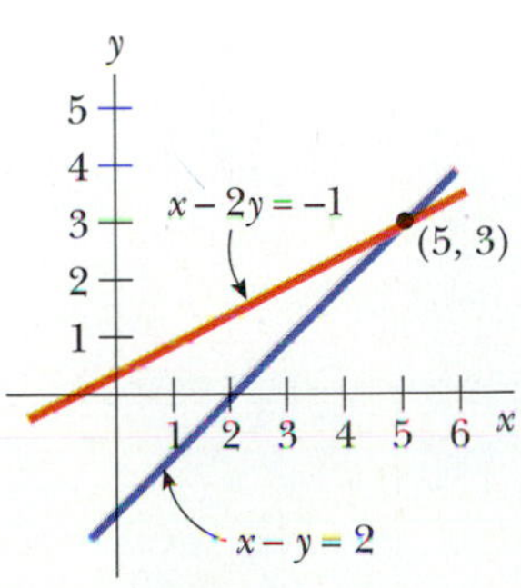

Figure B.3

Exercises

Solve the following pairs of simultaneous equations involving two unknowns:

		Answers
1.	$x + y = 8$ $x - y = 2$	$x = 5,\ y = 3$
2.	$98 - T = 10a$ $T - 49 = 5a$	$T = 65,\ a = 3.27$
3.	$6x + 2y = 6$ $8x - 4y = 28$	$x = 2,\ y = -3$

Logarithms

Suppose that a quantity, x, is expressed as a power of some quantity a:

$$x = a^y \tag{B.11}$$

The number a is called the **base** number. The **logarithm** of x with respect to the base a is equal to the exponent to which the base must be raised in order to satisfy the expression $x = a^y$:

$$y = \log_a x \tag{B.12}$$

Conversely, the **antilogarithm** of y is the number x:

$$x = \text{antilog}_a\, y \tag{B.13}$$

In practice, the two bases most often used are base 10, called the *common* logarithm base, and base $e = 2.718\ldots$, called the *natural* logarithm base. When common logarithms are used,

$$y = \log_{10} x \qquad (\text{or } x = 10^y) \tag{B.14}$$

When natural logarithms are used,

$$y = \ln_e x \qquad (\text{or } x = e^y) \tag{B.15}$$

For example, $\log_{10} 52 = 1.716$, and so $\text{antilog}_{10}\, 1.716 = 10^{1.716} = 52$. Likewise, $\ln_e 52 = 3.951$, so $\text{antiln}_e\, 3.951 = e^{3.951} = 52$.

In general, note that you can convert between base 10 and base e with the equality

$$\ln_e x = (2.302585) \log_{10} x \tag{B.16}$$

Finally, some useful properties of logarithms are as follows:

$$\log(ab) = \log a + \log b$$

$$\log(a/b) = \log a - \log b$$

$$\log(a^n) = n \log a$$

$$\ln e = 1$$

$$\ln e^a = a$$

$$\ln\left(\frac{1}{a}\right) = -\ln a$$

B.3 Geometry

The **distance**, d, between two points whose coordinates are (x_1, y_1) and (x_2, y_2) is

$$d = \sqrt{(x_2 - x_1)^2 + (y_2 - y_1)^2} \tag{B.17}$$

The arc length, s, of a circular arc (Fig. B.4) is proportional to the radius, r, for a fixed value of θ (in radians)—the **radian measure:**

$$s = r\theta$$

$$\theta = \frac{s}{r} \qquad \textbf{[B.18]}$$

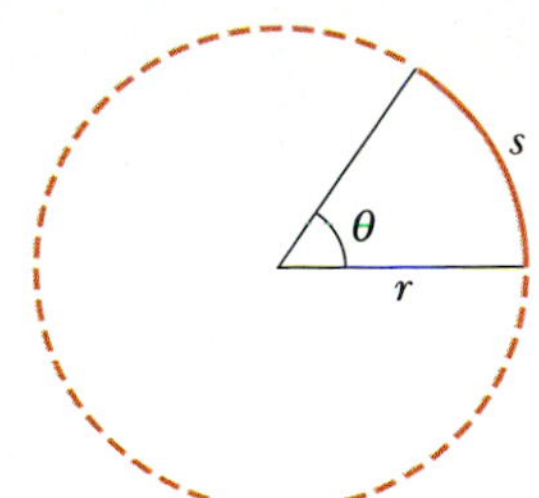

Figure B.4

Table B.2 gives the areas and volumes for several geometric shapes used throughout this book:

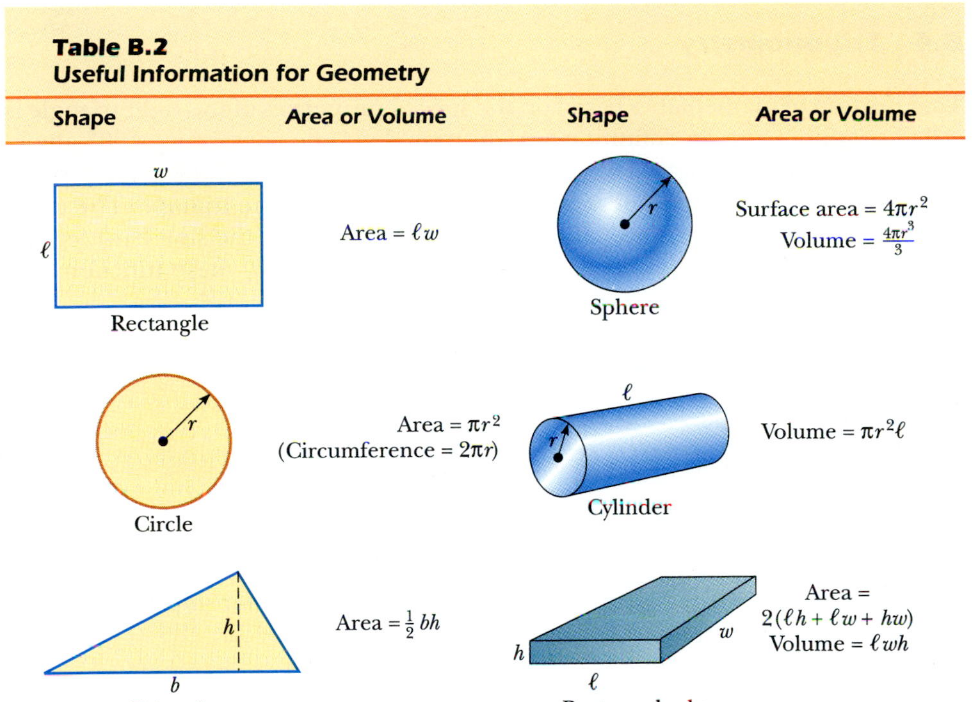

Table B.2
Useful Information for Geometry

Shape	Area or Volume	Shape	Area or Volume
Rectangle (w, ℓ)	Area = ℓw	Sphere (r)	Surface area = $4\pi r^2$ Volume = $\frac{4\pi r^3}{3}$
Circle (r)	Area = πr^2 (Circumference = $2\pi r$)	Cylinder (ℓ, r)	Volume = $\pi r^2 \ell$
Triangle (h, b)	Area = $\frac{1}{2}bh$	Rectangular box (h, w, ℓ)	Area = $2(\ell h + \ell w + hw)$ Volume = ℓwh

The equation of a **straight line** (Fig. B.5) is

$$y = mx + b \qquad \textbf{[B.19]}$$

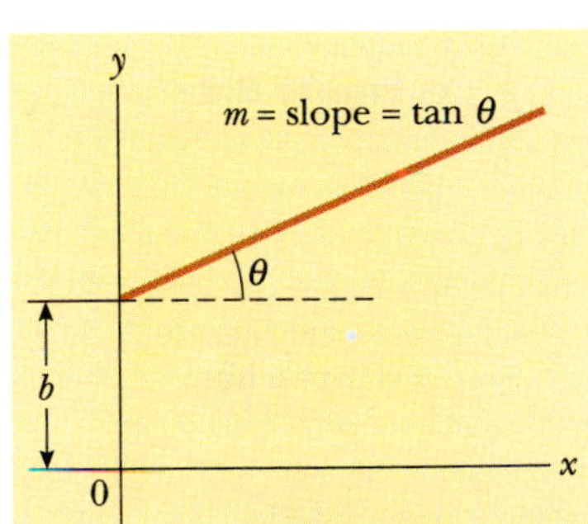

Figure B.5

where b is the y intercept and m is the slope of the line.

The equation of a **circle** of radius R centered at the origin is

$$x^2 + y^2 = R^2 \qquad \textbf{[B.20]}$$

The equation of an **ellipse** with the origin at its center (Fig. B.6) is

$$\frac{x^2}{a^2} + \frac{y^2}{b^2} = 1 \qquad \textbf{[B.21]}$$

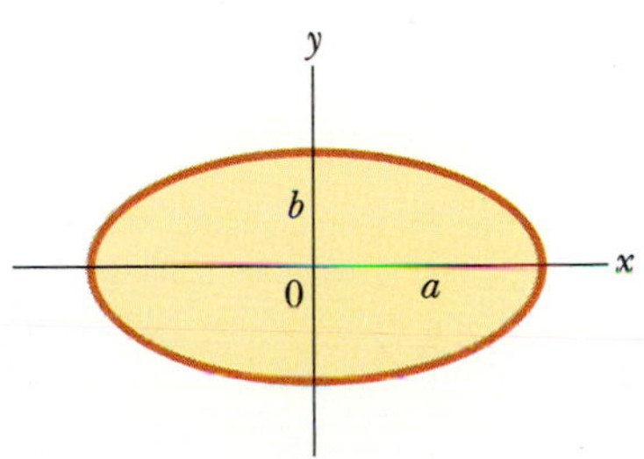

Figure B.6

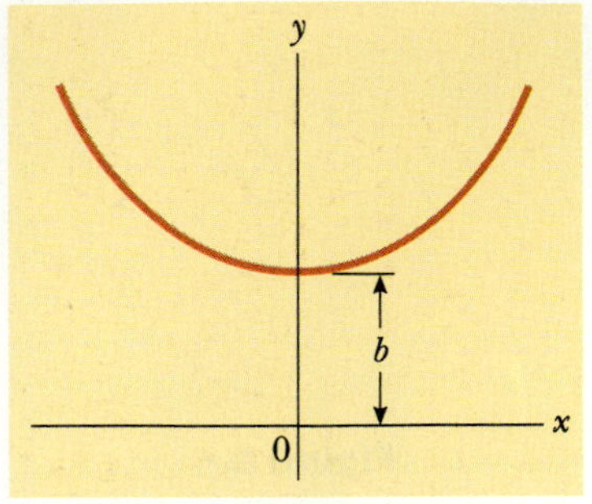

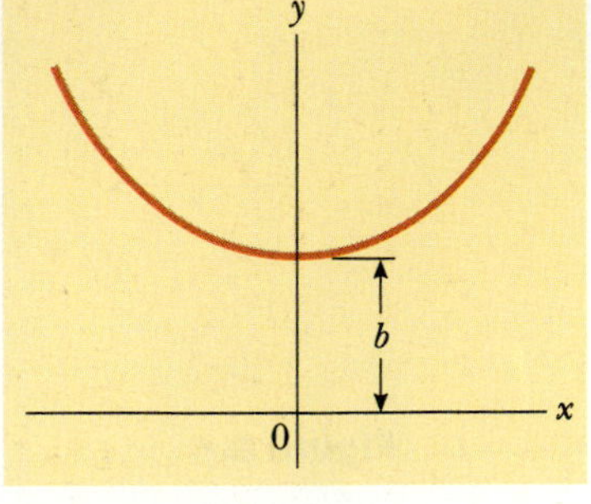

Figure B.7

where a is the length of the semi-major axis and b is the length of the semi-minor axis.

The equation of a **parabola** whose vertex is at $y = b$ (Fig. B.7) is

$$y = ax^2 + b \quad \text{[B.22]}$$

The equation of a **rectangular hyperbola** (Fig. B.8) is

$$xy = \text{constant} \quad \text{[B.23]}$$

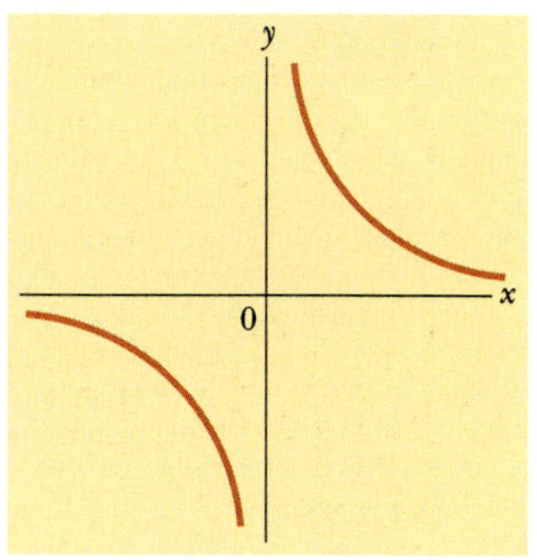

Figure B.8

B.4 Trigonometry

That portion of mathematics based on the special properties of the right triangle is called trigonometry. By definition, a right triangle is one containing a 90° angle. Consider the right triangle in Figure B.9, where side a is opposite the angle θ, side b is adjacent to the angle θ, and side c is the hypotenuse of the triangle. The three basic trigonometric functions defined by such a triangle are the sine (sin), cosine (cos), and tangent (tan) functions. In terms of the angle θ, these functions are defined by

$$\sin\theta \equiv \frac{\text{side opposite } \theta}{\text{hypotenuse}} = \frac{a}{c} \quad \text{[B.24]}$$

$$\cos\theta \equiv \frac{\text{side adjacent to } \theta}{\text{hypotenuse}} = \frac{b}{c} \quad \text{[B.25]}$$

$$\tan\theta \equiv \frac{\text{side opposite } \theta}{\text{side adjacent to } \theta} = \frac{a}{b} \quad \text{[B.26]}$$

a = opposite side
b = adjacent side
c = hypotenuse

c, 90°−θ, a, 90°, θ, b

Figure B.9

The Pythagorean theorem provides the following relationship between the sides of a right triangle:

$$c^2 = a^2 + b^2 \quad \text{[B.27]}$$

From the preceding definitions and the Pythagorean theorem, it follows that

$$\sin^2\theta + \cos^2\theta = 1$$

$$\tan\theta = \frac{\sin\theta}{\cos\theta}$$

The cosecant, secant, and cotangent functions are defined by

$$\csc\theta \equiv \frac{1}{\sin\theta} \qquad \sec\theta \equiv \frac{1}{\cos\theta} \qquad \cot\theta \equiv \frac{1}{\tan\theta}$$

The following relations come directly from the right triangle shown in Figure B.9:

$$\begin{cases} \sin\theta = \cos(90° - \theta) \\ \cos\theta = \sin(90° - \theta) \\ \cot\theta = \tan(90° - \theta) \end{cases}$$

Table B.3
Some Trigonometric Identities

$\sin^2 \theta + \cos^2 \theta = 1$	$\csc^2 \theta = 1 + \cot^2 \theta$
$\sec^2 \theta = 1 + \tan^2 \theta$	$\sin^2 \frac{\theta}{2} = \frac{1}{2}(1 - \cos \theta)$
$\sin 2\theta = 2 \sin \theta \cos \theta$	$\cos^2 \frac{\theta}{2} = \frac{1}{2}(1 + \cos \theta)$
$\cos 2\theta = \cos^2 \theta - \sin^2 \theta$	$1 - \cos \theta = 2 \sin^2 \frac{\theta}{2}$
$\tan 2\theta = \frac{2 \tan \theta}{1 - \tan^2 \theta}$	$\tan \frac{\theta}{2} = \sqrt{\frac{1 - \cos \theta}{1 + \cos \theta}}$

$$\sin(A \pm B) = \sin A \cos B \pm \cos A \sin B$$

$$\cos(A \pm B) = \cos A \cos B \mp \sin A \sin B$$

$$\sin A \pm \sin B = 2 \sin[\tfrac{1}{2}(A \pm B)] \cos[\tfrac{1}{2}(A \mp B)]$$

$$\cos A + \cos B = 2 \cos[\tfrac{1}{2}(A + B)] \cos[\tfrac{1}{2}(A - B)]$$

$$\cos A - \cos B = 2 \sin[\tfrac{1}{2}(A + B)] \sin[\tfrac{1}{2}(B - A)]$$

Some properties of trigonometric functions are as follows:

$$\begin{cases} \sin(-\theta) = -\sin \theta \\ \cos(-\theta) = \cos \theta \\ \tan(-\theta) = -\tan \theta \end{cases}$$

The following relations apply to *any* triangle, as shown in Figure B.10:

$$\alpha + \beta + \gamma = 180^\circ$$

$$\text{Law of cosines} \begin{cases} a^2 = b^2 + c^2 - 2bc \cos \alpha \\ b^2 = a^2 + c^2 - 2ac \cos \beta \\ c^2 = a^2 + b^2 - 2ab \cos \gamma \end{cases}$$

$$\text{Law of sines} \left\{ \frac{a}{\sin \alpha} = \frac{b}{\sin \beta} = \frac{c}{\sin \gamma} \right.$$

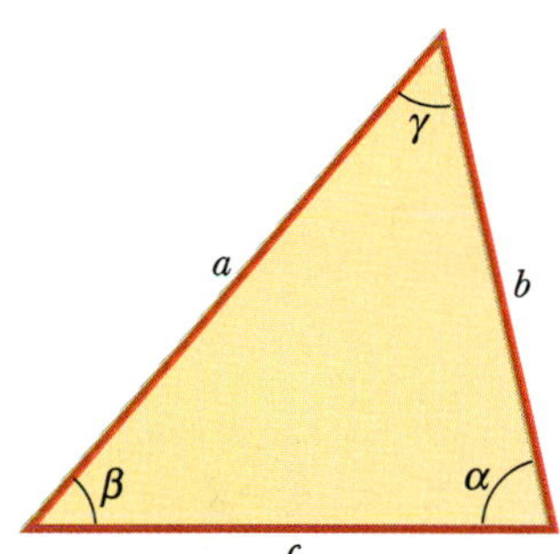

Figure B.10

Table B.3 lists a number of useful trigonometric identities.

Example 3

Consider the right triangle in Figure B.11, in which $a = 2$, $b = 5$, and c is unknown. From the Pythagorean theorem, we have

$$c^2 = a^2 + b^2 = 2^2 + 5^2 = 4 + 25 = 29$$

$$c = \sqrt{29} = 5.39$$

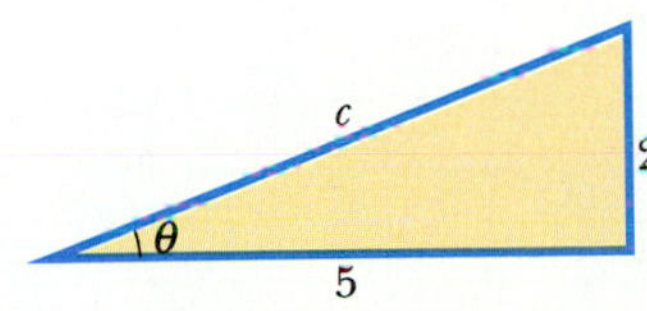

Figure B.11

To find the angle θ, note that

$$\tan\theta = \frac{a}{b} = \frac{2}{5} = 0.400$$

From a table of functions or from a calculator, we have

$$\theta = \tan^{-1}(0.400) = \boxed{21.8°}$$

where $\tan^{-1}(0.400)$ is the notation for "angle whose tangent is 0.400," sometimes written as arctan(0.400).

Exercises

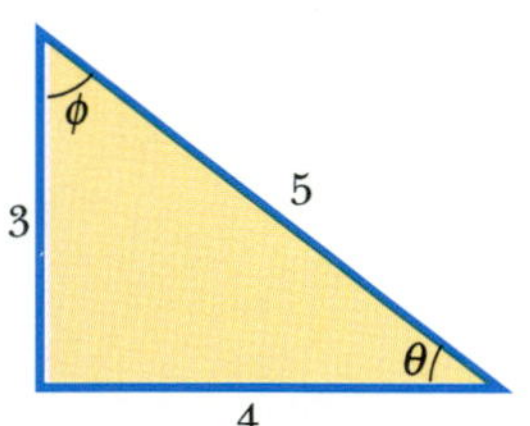

Figure B.12

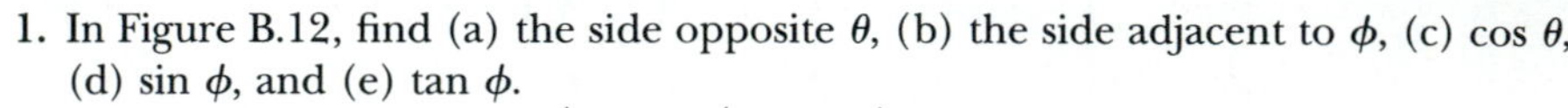

1. In Figure B.12, find (a) the side opposite θ, (b) the side adjacent to ϕ, (c) $\cos\theta$, (d) $\sin\phi$, and (e) $\tan\phi$.

Answers (a) 3, (b) 3, (c) $\frac{4}{5}$, (d) $\frac{4}{5}$, (e) $\frac{4}{3}$

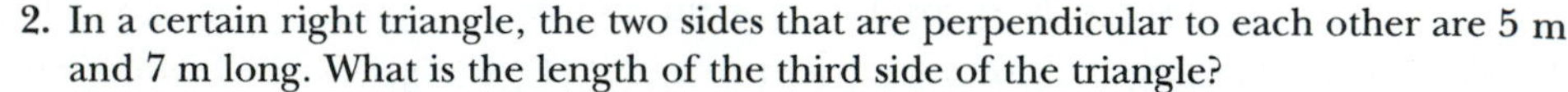

2. In a certain right triangle, the two sides that are perpendicular to each other are 5 m and 7 m long. What is the length of the third side of the triangle?

Answer 8.60 m

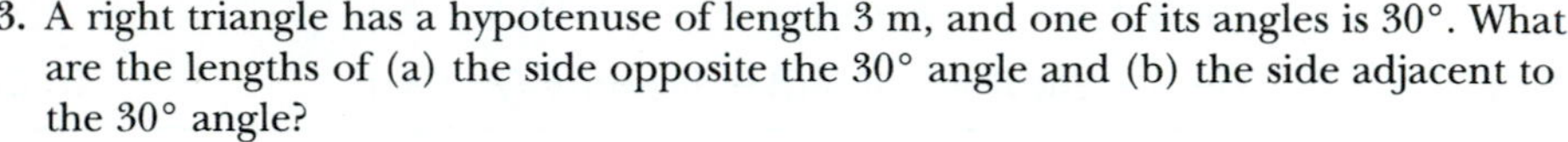

3. A right triangle has a hypotenuse of length 3 m, and one of its angles is 30°. What are the lengths of (a) the side opposite the 30° angle and (b) the side adjacent to the 30° angle?

Answers (a) 1.5 m, (b) 2.60 m

B.5 Series Expansions

$$(a+b)^n = a^n + \frac{n}{1!}a^{n-1}b + \frac{n(n-1)}{2!}a^{n-2}b^2 + \cdots$$

$$(1+x)^n = 1 + nx + \frac{n(n-1)}{2!}x^2 + \cdots$$

$$e^x = 1 + x + \frac{x^2}{2!} + \frac{x^3}{3!} + \cdots$$

$$\ln(1 \pm x) = \pm x - \tfrac{1}{2}x^2 \pm \tfrac{1}{3}x^3 - \cdots$$

$$\left.\begin{aligned} \sin x &= x - \frac{x^3}{3!} + \frac{x^5}{5!} - \cdots \\ \cos x &= 1 - \frac{x^2}{2!} + \frac{x^4}{4!} - \cdots \\ \tan x &= x + \frac{x^3}{3} + \frac{2x^5}{15} + \cdots \qquad |x| < \pi/2 \end{aligned}\right\} x \text{ in radians}$$

For $x \ll 1$, the following approximations can be used:

$$(1+x)^n \approx 1 + nx \qquad \sin x \approx x$$

$$e^x \approx 1 + x \qquad \cos x \approx 1$$

$$\ln(1 \pm x) \approx \pm x \qquad \tan x \approx x$$

B.6 Differential Calculus

In various branches of science, it is sometimes necessary to use the basic tools of calculus, first invented by Newton, to describe physical phenomena. The use of calculus is fundamental in the treatment of a variety of problems in newtonian mechanics, electricity, and magnetism. In this section, we simply state some basic properties and "rules of thumb," which should be a useful review.

First, a **function** must be specified that relates one variable to another (such as a coordinate as a function of time). Suppose one of the variables is called y (the dependent variable), and the other, x (the independent variable). We might have a function relation such as

$$y(x) = ax^3 + bx^2 + cx + d$$

If a, b, c, and d are specified constants, then y can be calculated for any value of x. We usually deal with continuous functions, that is, those for which y varies "smoothly" with x.

The **derivative** of y with respect to x is defined as the limit of the slopes of chords drawn between two points on the y versus x curve as Δx approaches zero. Mathematically, we write this definition as

$$\frac{dy}{dx} = \lim_{\Delta x \to 0} \frac{\Delta y}{\Delta x} = \lim_{\Delta x \to 0} \frac{y(x + \Delta x) - y(x)}{\Delta x} \qquad \text{[B.28]}$$

where Δy and Δx are defined as $\Delta x = x_2 - x_1$ and $\Delta y = y_2 - y_1$ (see Fig. B.13).

A useful expression to remember when $y(x) = ax^n$, where a is a *constant* and n is *any* positive or negative number (integer or fraction), is

$$\frac{dy}{dx} = nax^{n-1} \qquad \text{[B.29]}$$

If $y(x)$ is a polynomial or algebraic function of x, we apply Equation B.29 to *each* term in the polynomial and take $da/dx = 0$. It is important to note that dy/dx *does not* mean dy divided by dx, but is simply a notation of the limiting process of the derivative as defined by Equation B.28. In Examples 4 through 7, we evaluate the derivatives of several well-behaved functions.

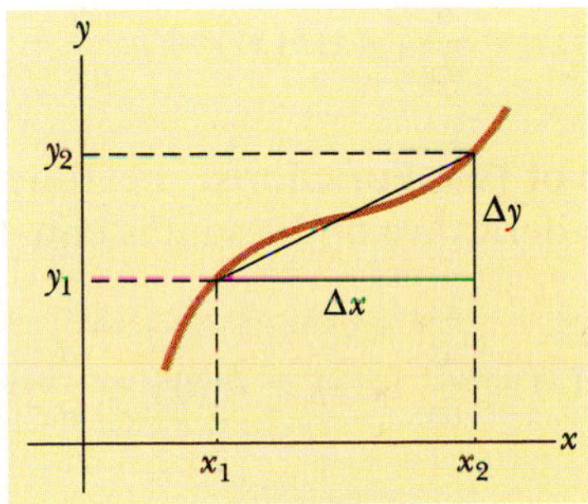

Figure B.13

Example 4

Suppose $y(x)$ (that is, y as a function of x) is given by

$$y(x) = ax^3 + bx + c$$

where a and b are constants. Then it follows that

$$y(x + \Delta x) = a(x + \Delta x)^3 + b(x + \Delta x) + c$$
$$y(x + \Delta x) = a(x^3 + 3x^2\Delta x + 3x\Delta x^2 + \Delta x^3) + b(x + \Delta x) + c$$

so

$$\Delta y = y(x + \Delta x) - y(x) = a(3x^2\Delta x + 3x\Delta x^2 + \Delta x^3) + b\Delta x$$

Substituting this into Equation B.28 gives

$$\frac{dy}{dx} = \lim_{\Delta x \to 0} \frac{\Delta y}{\Delta x} = \lim_{\Delta x \to 0} [3ax^2 + 3x\Delta x + \Delta x^2] + b$$

$$\frac{dy}{dx} = 3ax^2 + b$$

Example 5

$$y(x) = 8x^5 + 4x^3 + 2x + 7$$

Solution Applying Equation B.29 to each term independently, and remembering that d/dx (constant) $= 0$, we have

$$\frac{dy}{dx} = 8(5)x^4 + 4(3)x^2 + 2(1)x^0 + 0$$

$$\frac{dy}{dx} = 40x^4 + 12x^2 + 2$$

Special Properties of the Derivative

A. Derivative of the Product of Two Functions. If a function, y, is given by the product of two functions—say, $g(x)$ and $h(x)$—then the derivative of y is defined as

$$\frac{d}{dx}f(x) = \frac{d}{dx}[g(x)h(x)] = g\frac{dh}{dx} + h\frac{dg}{dx} \qquad \text{[B.30]}$$

B. Derivative of the Sum of Two Functions. If a function, y, is equal to the sum of two functions, then the derivative of the sum is equal to the sum of the derivatives:

$$\frac{d}{dx}f(x) = \frac{d}{dx}[g(x) + h(x)] = \frac{dg}{dx} + \frac{dh}{dx} \qquad \text{[B.31]}$$

C. Chain Rule of Differential Calculus. If $y = f(x)$ and x is a function of some other variable, z, then dy/dx can be written as the product of two derivatives:

$$\frac{dy}{dx} = \frac{dy}{dz}\frac{dz}{dx} \quad \text{[B.32]}$$

D. The Second Derivative. The second derivative of y with respect to x is defined as the derivative of the function dy/dx (that is, the derivative of the derivative). It is usually written

$$\frac{d^2y}{dx^2} = \frac{d}{dx}\left(\frac{dy}{dx}\right) \quad \text{[B.33]}$$

Example 6

Find the first derivative of $y(x) = x^3/(x+1)^2$ with respect to x.

Solution We can rewrite this function as $y(x) = x^3(x+1)^{-2}$ and apply Equation B.30 directly:

$$\frac{dy}{dx} = (x+1)^{-2}\frac{d}{dx}(x^3) + x^3\frac{d}{dx}(x+1)^{-2}$$

$$= (x+1)^{-2}\,3x^2 + x^3(-2)(x+1)^{-3}$$

$$\frac{dy}{dx} = \frac{3x^2}{(x+1)^2} - \frac{2x^3}{(x+1)^3}$$

Example 7

A useful formula that follows from Equation B.30 is the derivative of the quotient of two functions. Show that the expression is given by

$$\frac{d}{dx}\left[\frac{g(x)}{h(x)}\right] = \frac{h\dfrac{dg}{dx} - g\dfrac{dh}{dx}}{h^2}$$

Solution We can write the quotient as gh^{-1} and then apply Equations B.29 and B.30:

$$\frac{d}{dx}\left(\frac{g}{h}\right) = \frac{d}{dx}(gh^{-1}) = g\frac{d}{dx}(h^{-1}) + h^{-1}\frac{d}{dx}(g)$$

$$= -gh^{-2}\frac{dh}{dx} + h^{-1}\frac{dg}{dx}$$

$$= \frac{h\dfrac{dg}{dx} - g\dfrac{dh}{dx}}{h^2}$$

Some of the most commonly used derivatives of functions are listed in Table B.4.

Table B.4
Derivatives for Several Functions

$$\frac{d}{dx}(a) = 0$$

$$\frac{d}{dx}(ax^n) = nax^{n-1}$$

$$\frac{d}{dx}(e^{ax}) = ae^{ax}$$

$$\frac{d}{dx}(\sin ax) = a\cos ax$$

$$\frac{d}{dx}(\cos ax) = -a\sin ax$$

$$\frac{d}{dx}(\tan ax) = a\sec^2 ax$$

$$\frac{d}{dx}(\cot ax) = -a\csc^2 ax$$

$$\frac{d}{dx}(\sec x) = \tan x\sec x$$

$$\frac{d}{dx}(\csc x) = -\cot x\csc x$$

$$\frac{d}{dx}(\ln ax) = \frac{1}{x}$$

Note: The letters a and n are constants.

B.7 Integral Calculus

We think of integration as the inverse of differentiation. As an example, consider the expression

$$f(x) = \frac{dy}{dx} = 3ax^2 + b$$

which was the result of differentiating the function

$$y(x) = ax^3 + bx + c$$

in Example 4. We can write the first expression, $dy = f(x)\,dx = (3ax^2 + b)\,dx$, and obtain $y(x)$ by "summing" over all values of x. Mathematically, we write this inverse operation

$$y(x) = \int f(x)\ dx$$

For the function $f(x)$ already given,

$$y(x) = \int (3ax^2 + b)\,dx = ax^3 + bx + c$$

where c is a constant of the integration. This type of integral is called an *indefinite integral* because its value depends on the choice of the constant c.

A general **indefinite integral**, $I(x)$, is defined as

$$I(x) = \int f(x)\ dx \qquad \text{[B.34]}$$

where $f(x)$ is called the *integrand* and $f(x) = \dfrac{dI(x)}{dx}$.

For a *general continuous* function $f(x)$, the integral can be described as the area under the curve bounded by $f(x)$ and the x axis, between two specified values of x—say, x_1 and x_2—as in Figure B.14.

The area of the shaded element is approximately $f_i\,\Delta x_i$. If we sum all these area elements from x_1 to x_2 and take the limit of this sum as $\Delta x_i \rightarrow 0$, we obtain the *true* area under the curve bounded by $f(x)$ and x, between the limits x_1 and x_2:

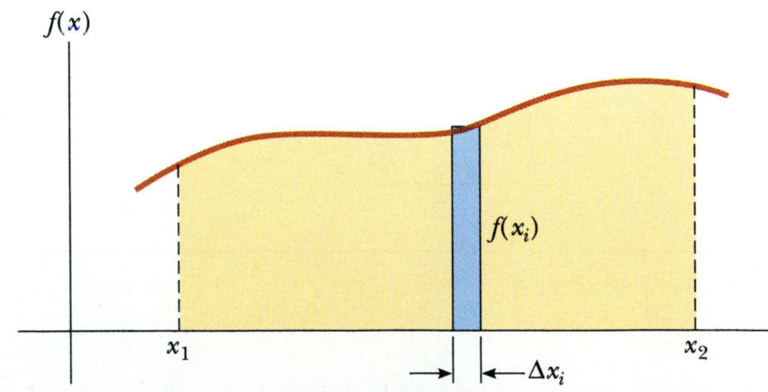

Figure B.14

$$\text{Area} = \lim_{\Delta x_i \to 0} \sum_i f(x_i)\, \Delta x_i = \int_{x_1}^{x_2} f(x)\, dx \qquad \textbf{[B.35]}$$

Integrals of the type defined by Equation B.35 are called **definite integrals.**

One of the common types of integrals that arise in practical situations has the form

$$\int x^n\, dx = \frac{x^{n+1}}{n+1} + c \qquad (n \neq -1) \qquad \textbf{[B.36]}$$

This result is obvious since differentiation of the right-hand side with respect to x gives $f(x) = x^n$ directly. If the limits of the integration are known, this integral becomes a *definite integral* and is written

$$\int_{x_1}^{x_2} x^n\, dx = \frac{x_2{}^{n+1} - x_1{}^{n+1}}{n+1} \qquad (n \neq -1) \qquad \textbf{[B.37]}$$

Examples

1. $\displaystyle\int_0^a x^2\, dx = \frac{x^3}{3}\bigg]_0^a = \frac{a^3}{3}$

2. $\displaystyle\int_0^b x^{3/2} dx = \frac{x^{5/2}}{5/2}\bigg]_0^b = \frac{2}{5}\, b^{5/2}$

3. $\displaystyle\int_3^5 x\, dx = \frac{x^2}{2}\bigg]_3^5 = \frac{5^2 - 3^2}{2} = 8$

Partial Integration

Sometimes it is useful to apply the method of *partial integration* to evaluate certain integrals. The method uses the property that

$$\int u\, dv = uv - \int v\, du \qquad \textbf{[B.38]}$$

where u and v are *carefully* chosen so as to reduce a complex integral to a simpler one. In many cases, several reductions have to be made. Consider the example

$$I(x) = \int x^2 e^x\, dx$$

This can be evaluated by integrating by parts twice. First, if we choose $u = x^2$, $v = e^x$, we get

$$\int x^2 e^x\, dx = \int x^2\, d(e^x) = x^2 e^x - 2\int e^x x\, dx + c_1$$

Now, in the second term, choose $u = x$, $v = e^x$, which gives

$$\int x^2 e^x\, dx = x^2 e^x - 2\, x e^x + 2\int e^x\, dx + c_1$$

or

$$\int x^2 e^x\, dx = x^2 e^x - 2\, x e^x + 2\, e^x + c_2$$

Table B.5
Some Indefinite Integrals (an arbitrary constant should be added to each of these integrals)

$$\int x^n\,dx = \frac{x^{n+1}}{n+1} \quad (\text{provided } n \neq -1)$$

$$\int \frac{dx}{x} = \int x^{-1}\,dx = \ln x$$

$$\int \frac{dx}{a+bx} = \frac{1}{b}\ln(a+bx)$$

$$\int \frac{dx}{(a+bx)^2} = -\frac{1}{b(a+bx)}$$

$$\int \frac{dx}{a^2+x^2} = \frac{1}{a}\tan^{-1}\frac{x}{a}$$

$$\int \frac{dx}{a^2-x^2} = \frac{1}{2a}\ln\frac{a+x}{a-x} \quad (a^2-x^2>0)$$

$$\int \frac{dx}{x^2-a^2} = \frac{1}{2a}\ln\frac{x-a}{x+a} \quad (x^2-a^2>0)$$

$$\int \frac{x\,dx}{a^2\pm x^2} = \pm\tfrac{1}{2}\ln(a^2\pm x^2)$$

$$\int \frac{dx}{\sqrt{a^2-x^2}} = \sin^{-1}\frac{x}{a} = -\cos^{-1}\frac{x}{a} \quad (a^2-x^2>0)$$

$$\int \frac{dx}{\sqrt{x^2\pm a^2}} = \ln(x+\sqrt{x^2\pm a^2})$$

$$\int \frac{x\,dx}{\sqrt{a^2-x^2}} = -\sqrt{a^2-x^2}$$

$$\int \frac{x\,dx}{\sqrt{x^2\pm a^2}} = \sqrt{x^2\pm a^2}$$

$$\int \sqrt{a^2-x^2}\,dx = \tfrac{1}{2}\left(x\sqrt{a^2-x^2}+a^2\sin^{-1}\frac{x}{a}\right)$$

$$\int x\sqrt{a^2-x^2}\,dx = -\tfrac{1}{3}(a^2-x^2)^{3/2}$$

$$\int \sqrt{x^2\pm a^2}\,dx = \tfrac{1}{2}[x\sqrt{x^2\pm a^2}\pm a^2\ln(x+\sqrt{x^2\pm a^2})]$$

$$\int x(\sqrt{x^2\pm a^2})\,dx = \tfrac{1}{3}(x^2\pm a^2)^{3/2}$$

$$\int e^{ax}\,dx = \frac{1}{a}e^{ax}$$

$$\int \ln ax\,dx = (x\ln ax) - x$$

$$\int xe^{ax}\,dx = \frac{e^{ax}}{a^2}(ax-1)$$

$$\int \frac{dx}{a+be^{cx}} = \frac{x}{a} - \frac{1}{ac}\ln(a+be^{cx})$$

$$\int \sin ax\,dx = -\frac{1}{a}\cos ax$$

$$\int \cos ax\,dx = \frac{1}{a}\sin ax$$

$$\int \tan ax\,dx = -\frac{1}{a}\ln(\cos ax) = \frac{1}{a}\ln(\sec ax)$$

$$\int \cot ax\,dx = \frac{1}{a}\ln(\sin ax)$$

$$\int \sec ax\,dx = \frac{1}{a}\ln(\sec ax + \tan ax) = \frac{1}{a}\ln\left[\tan\left(\frac{ax}{2}+\frac{\pi}{4}\right)\right]$$

$$\int \csc ax\,dx = \frac{1}{a}\ln(\csc ax - \cot ax) = \frac{1}{a}\ln\left(\tan\frac{ax}{2}\right)$$

$$\int \sin^2 ax\,dx = \frac{x}{2} - \frac{\sin 2ax}{4a}$$

$$\int \cos^2 ax\,dx = \frac{x}{2} + \frac{\sin 2ax}{4a}$$

$$\int \frac{dx}{\sin^2 ax} = -\frac{1}{a}\cot ax$$

$$\int \frac{dx}{\cos^2 ax} = \frac{1}{a}\tan ax$$

$$\int \tan^2 ax\,dx = \frac{1}{a}(\tan ax) - x$$

$$\int \cot^2 ax\,dx = -\frac{1}{a}(\cot ax) - x$$

$$\int \sin^{-1} ax\,dx = x(\sin^{-1} ax) + \frac{\sqrt{1-a^2x^2}}{a}$$

$$\int \cos^{-1} ax\,dx = x(\cos^{-1} ax) - \frac{\sqrt{1-a^2x^2}}{a}$$

$$\int \frac{dx}{(x^2+a^2)^{3/2}} = \frac{x}{a^2\sqrt{x^2+a^2}}$$

$$\int \frac{x\,dx}{(x^2+a^2)^{3/2}} = -\frac{1}{\sqrt{x^2+a^2}}$$

The Perfect Differential

Another useful method to remember is the use of the *perfect differential.* That is, we should sometimes look for a change of variable such that the differential of the function is the differential of the independent variable appearing in the integrand. For example, consider the integral

$$I(x) = \int \cos^2 x \sin x \, dx$$

This becomes easy to evaluate if we rewrite the differential as $d(\cos x) = -\sin x\, dx$. The integral then becomes

$$\int \cos^2 x \sin x \, dx = -\int \cos^2 x \, d(\cos x)$$

If we now change variables, letting $y = \cos x$, we get

$$\int \cos^2 x \sin x \, dx = -\int y^2 dy = -\frac{y^3}{3} + c = -\frac{\cos^3 x}{3} + c$$

Table B.5 lists some useful indefinite integrals. Table B.6 gives Gauss's probability integral and other definite integrals. A more complete list can be found in various handbooks, such as *The Handbook of Chemistry and Physics,* CRC Press.

Table B.6
Gauss's Probability Integral and Related Integrals

$$I_0 = \int_0^\infty e^{-\alpha x^2}\, dx = \tfrac{1}{2}\sqrt{\frac{\pi}{\alpha}} \qquad \text{(Gauss's probability integral)}$$

$$I_1 = \int_0^\infty x e^{-\alpha x^2}\, dx = \frac{1}{2\alpha}$$

$$I_2 = \int_0^\infty x^2 e^{-\alpha x^2}\, dx = -\frac{dI_0}{d\alpha} = \frac{1}{4}\sqrt{\frac{\pi}{\alpha^3}}$$

$$I_3 = \int_0^\infty x^3 e^{-\alpha x^2}\, dx = -\frac{dI_1}{d\alpha} = \frac{1}{2\alpha^2}$$

$$I_4 = \int_0^\infty x^4 e^{-\alpha x^2}\, dx = \frac{d^2 I_0}{d\alpha^2} = \frac{3}{8}\sqrt{\frac{\pi}{\alpha^5}}$$

$$I_5 = \int_0^\infty x^5 e^{-\alpha x^2}\, dx = \frac{d^2 I_1}{d\alpha^2} = \frac{1}{\alpha^3}$$

$$\vdots$$

$$I_{2n} = (-1)^n \frac{d^n}{d\alpha^n} I_0$$

$$I_{2n+1} = (-1)^n \frac{d^n}{d\alpha^n} I_1$$

Appendix C

Periodic Table of the Elements*

Symbol	Ca 20	Atomic number
Atomic mass †	40.08	
	$4s^2$	Electron configuration

Group I	Group II	Transition elements						
H 1 1.0080 $1s^1$								
Li 3 6.94 $2s^1$	**Be** 4 9.012 $2s^2$							
Na 11 22.99 $3s^1$	**Mg** 12 24.31 $3s^2$							
K 19 39.102 $4s^1$	**Ca** 20 40.08 $4s^2$	**Sc** 21 44.96 $3d^14s^2$	**Ti** 22 47.90 $3d^24s^2$	**V** 23 50.94 $3d^34s^2$	**Cr** 24 51.996 $3d^54s^1$	**Mn** 25 54.94 $3d^54s^2$	**Fe** 26 55.85 $3d^64s^2$	**Co** 27 58.93 $3d^74s^2$
Rb 37 85.47 $5s^1$	**Sr** 38 87.62 $5s^2$	**Y** 39 88.906 $4d^15s^2$	**Zr** 40 91.22 $4d^25s^2$	**Nb** 41 92.91 $4d^45s^1$	**Mo** 42 95.94 $4d^55s^1$	**Tc** 43 (99) $4d^55s^2$	**Ru** 44 101.1 $4d^75s^1$	**Rh** 45 102.91 $4d^85s^1$
Cs 55 132.91 $6s^1$	**Ba** 56 137.34 $6s^2$	57-71*	**Hf** 72 178.49 $5d^26s^2$	**Ta** 73 180.95 $5d^36s^2$	**W** 74 183.85 $5d^46s^2$	**Re** 75 186.2 $5d^56s^2$	**Os** 76 190.2 $5d^66s^2$	**Ir** 77 192.2 $5d^76s^2$
Fr 87 (223) $7s^1$	**Ra** 88 (226) $7s^2$	89-103**	**Unq** 104 (261) $6d^27s^2$	**Unp** 105 (262) $6d^37s^2$	**Unh** 106 (263)	**Uns** 107 (262)	**Uno** 108 (265)	**Une** 109 (266)

*Lanthanide series	**La** 57 138.91 $5d^16s^2$	**Ce** 58 140.12 $5d^14f^16s^2$	**Pr** 59 140.91 $4f^36s^2$	**Nd** 60 144.24 $4f^46s^2$	**Pm** 61 (147) $4f^56s^2$	**Sm** 62 150.4 $4f^66s^2$
**Actinide series	**Ac** 89 (227) $6d^17s^2$	**Th** 90 (232) $6d^27s^2$	**Pa** 91 (231) $5f^26d^17s^2$	**U** 92 (238) $5f^36d^17s^2$	**Np** 93 (239) $5f^46d^17s^2$	**Pu** 94 (239) $5f^66d^07s^2$

◻ Atomic mass values given are averaged over isotopes in the percentages in which they exist in nature.
† For an unstable element, mass number of the most stable known isotope is given in parentheses.

			Group III	Group IV	Group V	Group VI	Group VII	Group 0
							H 1 1.0080 $1s^1$	**He** 2 4.0026 $1s^2$
			B 5 10.81 $2p^1$	**C** 6 12.011 $2p^2$	**N** 7 14.007 $2p^3$	**O** 8 15.999 $2p^4$	**F** 9 18.998 $2p^5$	**Ne** 10 20.18 $2p^6$
			Al 13 26.98 $3p^1$	**Si** 14 28.09 $3p^2$	**P** 15 30.97 $3p^3$	**S** 16 32.06 $3p^4$	**Cl** 17 35.453 $3p^5$	**Ar** 18 39.948 $3p^6$
Ni 28 58.71 $3d^84s^2$	**Cu** 29 63.54 $3d^{10}4s^2$	**Zn** 30 65.37 $3d^{10}4s^2$	**Ga** 31 69.72 $4p^1$	**Ge** 32 72.59 $4p^2$	**As** 33 74.92 $4p^3$	**Se** 34 78.96 $4p^4$	**Br** 35 79.91 $4p^5$	**Kr** 36 83.80 $4p^6$
Pd 46 106.4 $4d^{10}$	**Ag** 47 107.87 $4d^{10}5s^1$	**Cd** 48 112.40 $4d^{10}5s^2$	**In** 49 114.82 $5p^1$	**Sn** 50 118.69 $5p^2$	**Sb** 51 121.75 $5p^3$	**Te** 52 127.60 $5p^4$	**I** 53 126.90 $5p^5$	**Xe** 54 131.30 $5p^6$
Pt 78 195.09 $5d^96s^1$	**Au** 79 196.97 $5d^{10}6s^1$	**Hg** 80 200.59 $5d^{10}6s^2$	**Tl** 81 204.37 $6p^1$	**Pb** 82 207.2 $6p^2$	**Bi** 83 208.98 $6p^3$	**Po** 84 (210) $6p^4$	**At** 85 (218) $6p^5$	**Rn** 86 (222) $6p^6$

Eu 63 152.0 $4f^76s^2$	**Gd** 64 157.25 $5d^14f^76s^2$	**Tb** 65 158.92 $5d^14f^86s^2$	**Dy** 66 162.50 $4f^{10}6s^2$	**Ho** 67 164.93 $4f^{11}6s^2$	**Er** 68 167.26 $4f^{12}6s^2$	**Tm** 69 168.93 $4f^{13}6s^2$	**Yb** 70 173.04 $4f^{14}6s^2$	**Lu** 71 174.97 $5d^14f^{14}6s^2$
Am 95 (243) $5f^76d^07s^2$	**Cm** 96 (245) $5f^76d^17s^2$	**Bk** 97 (247) $5f^86d^17s^2$	**Cf** 98 (249) $5f^{10}6d^07s^2$	**Es** 99 (254) $5f^{11}6d^07s^2$	**Fm** 100 (253) $5f^{12}6d^07s^2$	**Md** 101 (255) $5f^{13}6d^07s^2$	**No** 102 (255) $6d^07s^2$	**Lr** 103 (257) $6d^17s^2$

Appendix D

SI Units

Table D.1
SI Base Units

Base Quantity	SI Base Unit: *Name*	SI Base Unit: *Symbol*
Length	meter	m
Mass	kilogram	kg
Time	second	s
Electric current	ampere	A
Temperature	kelvin	K
Amount of substance	mole	mol
Luminous intensity	candela	cd

Table D.2
Some Derived SI Units

Quantity	Name	Symbol	Expression in Terms of Base Units	Expression in Terms of Other SI Units
Plane angle	radian	rad	$\mathrm{m/m}$	
Frequency	hertz	Hz	$\mathrm{s^{-1}}$	
Force	newton	N	$\mathrm{kg \cdot m/s^2}$	$\mathrm{J/m}$
Pressure	pascal	Pa	$\mathrm{kg/m \cdot s^2}$	$\mathrm{N/m^2}$
Energy: work	joule	J	$\mathrm{kg \cdot m^2/s^2}$	$\mathrm{N \cdot m}$
Power	watt	W	$\mathrm{kg \cdot m^2/s^3}$	$\mathrm{J/s}$
Electric charge	coulomb	C	$\mathrm{A \cdot s}$	
Electric potential (emf)	volt	V	$\mathrm{kg \cdot m^2/A \cdot s^3}$	$\mathrm{W/A}$
Capacitance	farad	F	$\mathrm{A^2 \cdot s^4/kg \cdot m^2}$	$\mathrm{C/V}$
Electric resistance	ohm	Ω	$\mathrm{kg \cdot m^2/A^2 \cdot s^3}$	$\mathrm{V/A}$
Magnetic flux	weber	Wb	$\mathrm{kg \cdot m^2/A \cdot s^2}$	$\mathrm{V \cdot s}$
Magnetic field intensity	tesla	T	$\mathrm{kg/A \cdot s^2}$	$\mathrm{Wb/m^2}$
Inductance	henry	H	$\mathrm{kg \cdot m^2/A^2 \cdot s^2}$	$\mathrm{Wb/A}$

Answers to Odd-Numbered Problems

Chapter 1

1. 2800 kg/m^3
3. 0.242
5. (a) 7.14×10^{-2} gal/s (b) 2.7×10^{-4} m^3/s (c) 1.03 h
7. The expression is dimensionally correct.
11. The units of G are (m^3/kg·s^2)
13. 4045 m^2
15. 1.2×10^{57}
17. (a) 1 mph = 1.609 km/h (b) 88.5 km/h (c) 16.1 km/h
19. 1.51×10^{-4} m
21. 1.00×10^{10} pounds
23. 1.79×10^{-9} m
25. 2.86 cm
27. Approximately 10^9 quarts and Approximately \$1 billion
29. 3.84×10^8 m
31. 34.1 m
33. (a) 22 cm (b) 67.9 cm^2
35. (a) 3 (b) 4 (c) 3 (d) 2
37. (a) 8.6 m (b) $r = 4.47$ m, $\theta = -63.4°$ (c) $r = 4.24$ m, $\theta = 135°$
39. $x = -2.75$ m, $y = -4.76$ m
41. 70 m
43. 9.54 N at 57° above the x axis
45. $x = 420$ ft, $y = -19$ ft, $d = 421$ ft at $-2.63°$
47. $x = 86.6$ m, $y = -50.0$ m
49. $d = 82.5$ m at 33° N of W
51. 47.2 units at an angle of 122° from the x axis
53. (a) $6\mathbf{i} + 7\mathbf{j}$ (b) 9.22 at $\theta = 49.4°$
55. $v = 390$ mph at 7.37° N of E
57. $B = 196$ cm at $\theta = -14.3°$
59. (a) $R_x = 49.5$, $R_y = 27.1$ (b) $R = 56.4$ at $\theta = 28.7°$
61. 0.449%
63. 4.50 m^2
65. 2.29 km
67. $R = 240$ m at $\theta = 237°$
69. (a) $\mathbf{R}_1 = a\mathbf{i} + b\mathbf{j}$ $|\mathbf{R}_1| = \sqrt{a^2 + b^2}$ (b) $\mathbf{R}_2 = a\mathbf{i} + b\mathbf{j} + c$ $|\mathbf{R}_2| = \sqrt{a^2 + b^2 + c^2}$

Chapter 2

1. (a) 2.3 m/s (b) 16.1 m/s (c) 11.5 m/s
3. (a) 5 m/s (b) 1.25 m/s (c) −2.5 m/s (d) −3.3 m/s (e) 0
5. 2.5 m/s (b) −2.27 m/s (c) 0
7. (a) a straight-line graph (b) v = slope = 1.6 m/s
9. (a) 5 m/s (b) −2.5 m/s (c) 0 (d) 5 m/s
11. −4 m/s^2
13. (a) 1.6 m/s^2 (b) 0.8 m/s^2
15. (a) 2 m (b) −3 m/s (c) −2 m/s^2
17. (a) 1.33 m/s^2 (b) at $t = 3$ s, $a = 2$ m/s^2 (c) $a = 0$ at $t = 6$ s and $t > 10$ s (d) max negative acceleration at $t = 8$ s, and is −1.5 m/s^2
19. −16 cm/s^2
21. (a) 12.7 m/s (b) −2.3 m/s
23. (a) 20 s (b) No, it would overshoot the runway by 200 m.
25. (a) 8.94 s (b) 89.4 m/s
27. $a = 4.89$ m/s^2 ($0.499g$)
29. 23.8 s
31. (a) 3×10^{-10} s (b) 1.26×10^{-4} m
33. (a) −662 ft/s^2 (b) 649 ft
35. $t = 1.79$ s
37. (a) 2.64 s (b) −20.9 m/s (c) 1.62 s; −20.9 m/s
39. $v_0 = 98$ m/s (b) at $t = 10$ s, $h = 490$ m
41. on the Moon, $t = 1.11$ s; on the Earth, $t = 0.45$ s
43. (a) 7.82 m (b) 0.782 s
45. (a) 7 m/s (b) −5.35 m/s (c) −9.8 m/s^2
47. (a) −6.26 m/s (b) 6.02 m/s (c) 1.25 s
49. 4.63 m
51. (a) 6.46 s (b) 334 ft (c) 103 ft/s and 89.5 ft/s
53. 697 m
55. (a) 26.4 m (b) Ignoring sound travel time, d = 28.2 m, and the error in calculation is 6.88%.
57. (a) 5.43 m/s^2, 3.83 m/s^2 (b) 10.86 m/s, 11.49 m/s (c) Maggie is ahead by 2.62 m.
59. $t = 155$ s, $t_1 = 129$ s
61. $\dfrac{dy}{dt} = \dfrac{v}{\sqrt{3}}$

63. (a) $a \cong 2\ \text{m/s}^2$ at 0.5 s, 4 m/s² at 1.5 s (b) $x \cong$ 34.5 m at 4.5 s, 42.6 m at 5 s
65. when $t = 5.7$ s, $v = 19.8$ m/s and $x = 44.3$ m

Chapter 3

1. $(8a + 2b)\mathbf{i} + 2c\mathbf{j}$
3. (a) 4.87 km (b) 23.3 m/s (c) 13.5 m/s
5. $(2\mathbf{i} + 3\mathbf{j})\ \text{m/s}^2$ (b) $x = (3t + t^2)$ m; $y = (1.5t^2 - 2t)$ m
7. (a) $a_x = 0.8\ \text{m/s}^2$, $a_y = -0.3\ \text{m/s}^2$ (b) 339.4° from $+x$ axis (c) at $t = 25$ s, $x = 360$ m, $y = -72.7$ m, $\theta = -15°$
9. (a) $\mathbf{v} = (5t\mathbf{i} + 1.5t^2\mathbf{j})$ m (b) $x = 10$ m, $y = 6$ m, $v = 7.81$ m/s
11. (a) 3.34 m/s (b) 50.9°
13. $t = 7$ s, $v = 143$ m/s
15. (a) 2.67 s (b) 29.9 m/s (c) $v_x = 29.9$ m/s, $v_y = -26.2$ m/s
17. (a) The ball clears by 0.89 m; (b) the ball is descending.
19. (a) 27.1 m/s (b) 3.91 s
21. $x = 7230$ m, $y = 1680$ m
23. 2.7 m/s²
25. 32 m/s²
27. 377 m/s²
29. 19.8 m/s
31. (a) 13.0 m/s² (b) 5.70 m/s (c) 7.50 m/s²
33. 1.48 m/s²
35. 30.8 m/s² down; 70.4 m/s² up
37. 0.85 m/s
39. 153 km/h at 11.3° N of W
41. (a) 36.9° (b) 41.6° (c) 3 minutes
43. (a) $\mathbf{v} = 4\mathbf{i}$ m/s (b) $x = 3.14$ m, $y = 6$ m
45. When $x = 45$ m, $y = 3.14$ m.
47. $v = 8.94$ m/s at −63° (below horizontal)
49. $v = 7580$ m/s, $T = 96.7$ min
51. 87.3 m upstream; 567.3 m downstream
53. 10.8 m
55. (a) 6804 m (b) The plane will be 2000 m above the impact point. (c) 66.2°
57. 700 m/s
59. (a) 55.5 s (b) 33.4 s
61. (a) 2 km/h due East (b) $\mathbf{v} = (4\mathbf{i} + 3\mathbf{j})$ km/h
63. (a) 43.2 m (b) $v_x = 9.66$ m/s; $v_y = -25.5$ m/s
65. (a) 46.5 m/s (b) −77.6° (c) 6.34 s
67. By using the long jump technique, he will make it to safety.
71. (a) 0.511 m (b) minimum separation at $t = 0.74$ s

Chapter 4

1. (a) $\frac{1}{3}$ (b) 0.750 m/s²
3. (a) 5 m/s² (b) 19.6 N (c) 10 m/s
5. $(49.0\ \text{mN})\mathbf{j} + (49.0\ \text{mN})(-\mathbf{j}) + (312\ \text{N})\mathbf{i}$
7. $(6\mathbf{i} + 15\mathbf{j})$N, 16.2 N
9. (a) 533.76 N (b) 54.4 kg
11. 70.5 kg
13. $f = 9.6$ N
15. (a) $(0.200\ \text{m/s}^2)^2$ (b) 10.0 m (c) $(2.00\ \text{m/s})\mathbf{i}$
17. 444 s
19. (a) 2 m/s² (b) 160 N (c) 3.2 m/s²
21. 600 N
23. (a) 3.6×10^{-18} N (b) 8.9×10^{-30} N
25. 2156 N
27. 8.66 N (East)
29. 613 N
31. (a) $(-375\ \text{N})\mathbf{i} + (1.47\ \text{N})\mathbf{j}$ (b) $\mathbf{F} = (375\mathbf{i} - 1.47\mathbf{j})$ N
33. −551 N
35. $F_1 = 100.2$ N; $F_2 = 203.8$ N
37. 2.21 m/s² at 73.7° N of W
39. (a) 36.8 N (b) 2.45 m/s² (c) 1.23 m
41. 3.73 m
43. 112 N
45. (a) 8930 N (b) 1.05 m/s² (c) 369 kg
47. (a) 84.9 N upward (b) 84.9 N
49. (a) 3.35 m/s² (b) 2.44 s
51. 2260 N
53. 1.66×10^6 N
55. 1180 N
57. (a) $(-45\mathbf{i} + 15\mathbf{j})$ m/s (b) 162° from $+x$ axis (c) $(-225\mathbf{i} + 75\mathbf{j})$ m (d) $(-227\mathbf{i} + 79\mathbf{j})$ m
59. 12.8 s

Chapter 5

1. $\mu_s = 0.306$ $\mu_k = 0.245$
3. 0.456
5. (a) 16.3 N (b) 8.07 N up the incline
7. $t = 7.0$ s
9. (a) 1.78 m/s² (b) 0.368 (c) 9.37 N (d) 2.67 m/s
11. 0.732
13. (a) 55.15° (b) 167.3 N
15. $f = -6400$ N
17. 60.0 N
19. $0 < v < 8.08$ m/s
21. $v < 14.3$ m/s
23. (a) $T = 9.80$ N (b) $F = 9.80$ N (c) $v = 6.26$ m/s
25. (a) static friction (b) 0.085
27. 3.13 m/s
29. (a) $v = 4.81$ m/s (b) $n = 700$ N
31. 2.87 N down
33. He doesn't make it across, because the rope breaks.
35. (a) 6.27 m/s² downward (b) 784 N up (c) 283 N
37. (a) $0.0347\ \text{s}^{-1}$ (b) 2.50 m/s (c) $-cv$
39. (a) 1.47 N·s/m (b) 2.04×10^{-3} s (c) 2.94×10^{-2} N

41. (a) 13.7 m/s (b) the hailstone reaches 99.95% of v_t at $t = 5.0$ s
43. (a) $v_t = -49.5$ m/s with 'chute closed and $v_t = -4.95$ m/s with 'chute open. (b) The parachute opens at 676 m when $v = 47.8$ m/s. Jumper reaches 0 m at 14.5 s, with $v = 4.95$ m/s.
45. Angle for greatest distance = 40.5° with time-of-flight 5.38 s. (c) On the order of 40 m
47. 0.685 m/s^2
49. (a) 0.822 m/s^2 (b) 37.0 N (c) 0.084
51. $a_r = 0.493$ m/s^2, $a_t = -0.524$ m/s^2, $a = 0.720$ m/s^2
53. 37.5 N
55. (a) 0.61 rev/s (b) $v = 0.766$ m/s $a = 2.93$ m/s^2
57. 20.1°
59. 12.8 N
61. (b) 23.6 rpm
63. (a) $F_{\text{min}} = \dfrac{\mu_s W \sec\phi}{1 - \mu_s \tan\phi}$
65. (a) 5.19 m/s (b) 555 N

Chapter 6

1. 2.97×10^{-9} N
3. (a) 5590 m/s (b) 239 min (c) 735 N down
5. 35 N
7. 1.26×10^{32} kg (63 solar masses for each star)
9. 5.10×10^5 N
11. 2.03×10^{-3} N
13. 0.613 m/s^2 down
15. 2.14 rev/min
17. A 600-N person's weight is reduced by 0.365 N.
19. -3.6×10^6 *N* (down on the top and up on the bottom)
21. 1.49×10^{-3} kg
23. 5.54×10^{11} N/C (directed away from proton)
25. 1.28×10^5 N/C opposite to the motion
27. (a) 6.14×10^{10} m/s^2 (b) 19.5 μs (c) 11.7 m
29. $t = 0.783$ μs
31. (a) 0.111 μs (b) 5.67 mm (c) (450 km/s)i + (102 km/s)j
33. 1.88×10^8 m/s (63% the speed of light)
35. **B** is in the negative *z*-direction
37. 48.8°
39. (a) up (b) out (c) no deflection (d) in
41. (a) 4.97×10^{-17} N (b) 1.29 km
43. $r = 3.02$ m $t = 0.715$ μs
45. (a) 3.45×10^7 m/s (b) 2.05×10^{-8} s
47. $r = 31.3$ km. The proton will not hit the Earth.
49. $F = 1.99 \times 10^{20}$ N, toward the Earth.
51. 1.90×10^{27} kg
53. 86.5 ns
55. $F_g = 8.93 \times 10^{-30}$ N down, $F_e = 1.60 \times 10^{-17}$ N up, $F_m = 4.80 \times 10^{-17}$ N down

Chapter 7

1. 30.6 m
3. 15.0 MJ
5. 1590 J
7. (a) 5.00 (b) 71.6°
9. 28.9
11. 16.0
13. 6.00, 59.0°
15. 83.4°
17. 3.00 J
19. (a) 24.0 J (b) −3.0 J (c) 21.0 J
21. (a) (b) −12.0 J

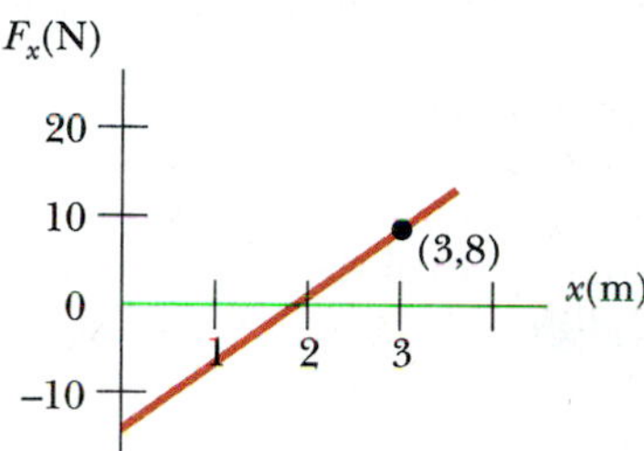

23. (a)

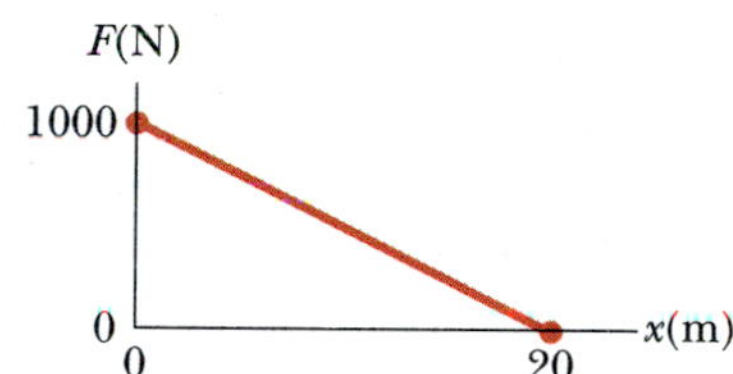

(b) (1000 − 25*x*) N (c) 10.0 kJ
25. (a) 417 N/m (b) 3.0 J
27. (a) 1.2 J (b) 5.0 m/s (c) 6.3 J
29. 2.45×10^{10} J
31. (a) 60.0 J (b) 60.0 J
33. (a) 2.94×10^5 J (b) All this energy is lost to friction.
35. (a) 329 J (b) 0 (c) 0 (d) −185 J
37. 2.25×10^5 N (b) 1.33×10^{-4} s
39. 2.04 m
41. 875 W
43. 3.27 kW
45. 830 N
47. (a) 131.0 kJ (b) 1.64 kW
49. 90.0 J
51. (a) 3.28×10^{-2} J (b) -3.28×10^{-2} J
53. 3.70 m/s
57. 8.78×10^5 N
59. (a) 4.12 m (b) 3.35 m
61. 1.94 kJ
63. 1.68 m/s
65. $W = 509$ J

Chapter 8

1. 147 J
3. (a) $U_A = 2.59 \times 10^5$ J (b) $U_B = -2.59 \times 10^5$ J
5. (a) −147.0 J (b) −147.0 J (c) −147.0 J
7. (a) 40.0 J (b) −40.0 J (c) 62.5 J
9. 2.94 m/s
11. (a) 19.8 m/s (b) 294 J (c) $\mathbf{v} = (30.0 \text{ m/s})\mathbf{i} - (39.6 \text{ m/s})\mathbf{j}$
13. (a) $v = \sqrt{gh + v_0^2}$ (b) $\mathbf{v} = 0.60v_0\mathbf{i} - \sqrt{0.64v_0^2 + gh}\mathbf{j}$
15. 1.84 m
17. (a) 4.43 m/s (b) 5.00 m
19. (a) −160 J (b) 73.5 J (c) 28.8 N (d) 0.679
21. 489 kJ
23. 3.68 m/s
25. 2060 N
27. 3.74 m/s
29. 289 m
31. (a) -1.67×10^{-14} J (b) at the center of the equilateral triangle
33. (a) 1.84×10^9 kg/m³ (b) 3.27×10^6 m/s² (c) -2.08×10^{13} J
35. 0.288 J
37. 0.116 J
39. −3.96 J
41. 3.53 J
43. F_x is zero at *A*, *C*, and *E* positive at *B*, and negative at *D* (b) *C* is stable; *A* and *E* are unstable.

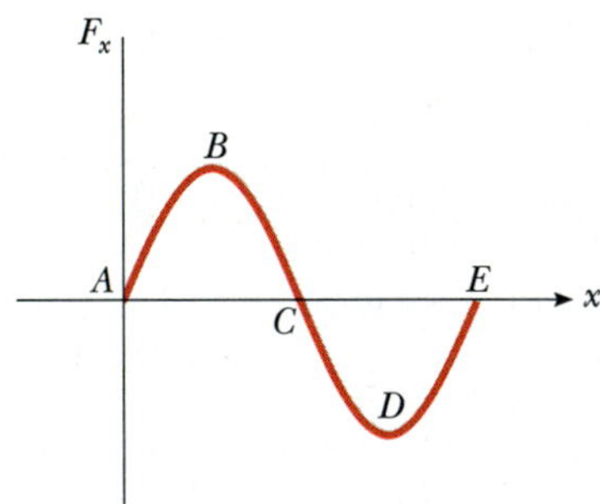

45. (a) 5.94 m/s and 7.67 m/s (b) 147.0 J
47. A/r^2
49. (a) 0.588 J (b) 0.588 J (c) 2.42 m/s (d) 0.196 J and 0.392 J
51. 4.8 cm
53. (a) 349 J, 676 J, 741 J (b) 175 N, 338 N, 370 N (c) yes
55. 0.327
57. 0.115
59. 1.25 m/s
61. (a) 0.400 m (b) 4.10 m/s (c) the block stays on the track.
63. 31.1 N/m
65. 914 N/m

Chapter 9

1. 40.0 m/s
3. (a) 8.35×10^{-21} kg·m/s (b) 4.50 kg·m/s (c) 750 kg·m/s (d) 1.78×10^{29} kg·m/s
5. -7.5×10^4 N
7. -6.25×10^{-2} m/s
9. 31.0 m/s
11. −2.67 m/s, west
13. (a) 12.0 N·s (b) 6.00 m/s (c) 4.00 m/s
15. (a) 13.5 N·s (b) 9000 N (c) 18.0 kN
17. (a) 13.5 kg·m/s (b) 6.75 kN toward the pitcher
19. −260**i** N
21. (a) 3.33 m/s (b) −83.4 J
23. 301 m/s
25. (a) 20.93 m/s (b) 8.68 kJ
27. (a) 0.284 (b) $K_C = 4.54 \times 10^{-14}$ J; $K_n = 1.15 \times 10^{-13}$ J
29. $v_1 = -0.500$ m/s; $v_2 = 2.00$ m/s
31. $\dfrac{4M}{m}\sqrt{gl}$
33. 0.556 m
35. 2.66 m/s
37. $v = 0.816$ m/s
39. (3.00**i** − 1.20**j**) m/s
41. 2.50 m/s at −60.0°
43. 3.01 m/s; 3.99 m/s
45. (a) $v_x = -9.33 \times 10^6$ m/s, $v_y = -8.33 \times 10^6$ m/s (b) 4.39×10^{-13} J
47. $x = 0$ $y = 1$ m
49. $x = 3.85$ cm; $y = 6.85$ cm
51. 454 km
53. $x = 11.7$ cm; $y = 13.3$ cm
55. (a) (1.40**i** + 2.40**j**) m/s (b) (7.00**i** + 12.0**j**) kg·m/s
57. 0.70 m
59. (a) 3.9×10^7 N (b) 3.2 m/s²
61. 0.595 m³/s
63. 291 N at 45.0°
65. 1390 m/s
67. 450,000 N; 1.84*g*; 2750 m/s
69. 210 s
71. $H = 0.98$ m
73. (a) The center of mass of the system doesn't move. (b) 5.55 m (c) no
75. (a) 100 m/s (b) −374 J

Chapter 10

1. (a) 60 m/s (b) 20 m/s (c) 44.7 m/s
3. Since the acceleration is the same, the force is the same.
5. In the rest frame, $p = 0.9$ kg·m/s; In the moving frame, $p' = 1.9$ kg·m/s
7. 0.866*c*
9. 1.54 ns

11. $0.436\ L_0$
13. (a) 2.18×10^{-6} s (b) 649 m
15. $0.1404c$
17. $0.960c$
19. (a) at rest, 20 m (b) in motion, 19 m (c) $v = 0.312c$
21. 0.088
23. (a) $v = 0.85c$ (b) The result is the same.
25. $v_2 = 0.285c$
27. (a) $0.141c$ (b) $0.436c$
29. (a) 939.4 MeV (b) 3008 MeV (c) 2069 MeV
31. $0.864c$
33. (a) 2.25×10^{22} J (b) 2.5×10^5 kg
35. 3.18×10^{-12} kg
37. (a) 9.53×10^{22} fissions/s (b) 1.17×10^6 kg/y
39. 5800 km
41. (a) 64.9 pulses/min (b) 10.6 pulses/min
43. 0.7%
45. (a) 8.33×10^{-8} s (b) 23.7 m
47. The train is Lorentz-contracted to 31.2 m; it fits.
49. $0.37c$
51. (a) 7.05 μs (b) 11 250 muons
53. 1.47 km

Chapter 11

1. (a) 4.00 rad/s (b) 18.0 rads
3. (a) 1.99×10^{-7} rad/s (b) 2.65×10^{-6} rad/s
5. (a) 5.24 s (b) 27.4 rad
7. (a) 54.3 rev (b) 12.1 rev/s
9. 50 rev
11. -226 rad/s^2
13. (a) 0.180 rad/s (b) 8.10 m/s^2, toward the center of the track
15. (a) 25 rad/s (b) 39.8 rad/s^2 (c) 0.628 s
17. 40.0 rad/s
19. (a) 143 kg·m^2 (b) 2570 J
21. -3.55 N·m
23. 168 N·m (clockwise)
25. (a) -17**k** (b) 70.5°
27. (a) (-7**k**) N/m (b) (11**k**) N/m
29. 2**k** N·m
31. (a) 24 N·m (b) 0.0356 rad/s^2 (c) 1.07 m/s^2
33. (60**k**) kg·m^2/s
35. (17.5 kg·m^2/s)k
37. 0.294
39. (a) 3.58 rad/s (b) 539 J
41. (a) -0.36 rad/s (b) 99.9 J
43. 1/160
45. 276 J
47. 149 rad/s
49. 2/7
51. $v = \left[\dfrac{2(m_2 - m_1)gh}{\left(m_1 + m_2 + \dfrac{I}{R^2}\right)}\right]^{1/2}$; $\omega = \dfrac{v}{R}$
53. (a) 1.03 s (b) 10.3 rev
55. 227 rad/s
57. 1.41×10^{-46} kg·m^2
59. (a) $h = 2.7(R - r)$
61. 139 m/s
63. (a) $T_1 = 118$ N; $T_2 = 156$ N (b) $I = 1.19$ kg·m^2
65. (a) 2160 N·m (b) 439 W

Chapter 12

1. (a) 1.50×10^{11} m (b) 3.55×10^{22} N toward Earth
3. 2/3
5. 2.67×10^{-7} m/s^2
7. 1.27
9. 5.90×10^4 m/s
11. 89,800 km above the planet
13. (a) 1.63 m/s^2 (b) 1680 m/s (c) 1.81 h
15. (a) -4.77×10^9 J (b) 568 N
17. 1.66×10^4 m/s
19. (a) 1.88×10^{11} J (b) 103 kW
23. 7.35 km/s
25. (a) 10.0 m/s^2 (b) 21.8 km/s
27. (a) 2.19×10^6 m/s (b) 2.18×10^{-18} J (c) -4.36×10^{-18} J
29. (a) 13.6 eV (b) 1.51 eV
31. (a) 0.212 nm (b) 9.95×10^{-25} kg·m/s (c) 2.11×10^{-34} kg·m^2/s (d) 3.40 eV (e) -6.80 eV (f) -3.40 eV
35. 0.0572 rad/s or 1 rev/110 s
37. 7.41×10^{-10} N
39. (a) $k = 1$ y^2/(A.U.)^3 (b) 11.9 y
41. (b) $K_1 = 1.07 \times 10^{32}$ J; $K_2 = 2.67 \times 10^{31}$ J
43. (a) 1.33×10^{12} m/s^2 (b) 9.29×10^{13} N (c) 2.22×10^{-11} J
45. 119 km
47. (a) transition B (b) transition A (c) transitions B and C
51. 1.5×10^{16} m (approximately 1.6 lightyears)
53. $v = \sqrt{2\left(Rg + \dfrac{gR^2}{r}\right)}$

Chapter 13

1. 5.27×10^{18} kg
3. (a) -251°C (b) 1.358 atm
5. (a) 37°C; 310 K (b) -20.6°C; 252.5 K
9. (a) -321°F (b) 139 R (c) 77.3 K
11. 40.6°C to -31.6°C
13. 0.313 m
15. $+3.27$ cm
17. (a) -179°C, attainable with liquid nitrogen (b) -396°C, unattainable
19. $+0.109$ cm^2
21. 1.08 liters
23. (a) 0.176 mm (b) 8.78×10^{-4} cm (c) 0.093 cm^3

25. 6.85 atm
27. 1.50×10^{29} molecules
29. 472 K
31. (a) 4.00×10^5 Pa (absolute) (b) 4.48×10^5 Pa
33. (a) 900 K (b) 1200 K
35. 16.0 cm^3
37. (a) 468 kg for a 4-m by 10-m by 8.5 m house; (b) 84 kg must leave
39. 3.65×10^4 N
41. (a) 8.76×10^{-21} J; (b) for He, for Ar, 514 m/s
43. (a) 2.02×10^4 K (b) 904 K
45. (a) 4.14×10^{-16} J (b) 704 km/s
47. 1.10 cm
49. 0.523 kg
53. (b) 5×10^{-5} °C^{-1}
55. 208°C
57. (a) $h = \dfrac{nRT}{(mg + P_aA)}$ (b) 0.662 m
59. (a) 0.169 m (b) 1.35×10^5 Pa
61. 804°C
63. The copper bullet wins.
65. 28.7°C
67. 5.31 h
69. 466 J

Chapter 14

1. 3740 m
3. 0.105°C
5. 234 J/kg·°C
7. 29.6°C
9. 361 pellets
11. (a) 9.88×10^{-3} °C (b) The remaining energy is absorbed by the surface.
13. 59.4°C
15. (a) 25.8°C (b) no
17. 1.22×10^5 J
19. (a) 40.4°C (b) 42 g of ice is left
21. 0.414 kg
23. 223 grams
25. (a) 0°C (b) 114 grams
27. 2.99×10^{-3} grams
29. 0.26 grams if the bullet is at 0°C
31. (a) 810 J (b) 507 J (c) 203 J
33. 1.18×10^6 J
35. (a) 0.00765 m^3 (b) 32°C
37. −720 J, transferred from the system
39. Q, W and ΔU are positive for process AB
41. 37.6 kJ
43. (a) 71.2 kJ (b) 0.702 m^3
45. (a) 0.041 m^3 (b) −5.48 kJ (c) −5.48 kJ
47. (a) 48.6 mJ (b) 16.2 kJ (c) 16.2 kJ
49. none; $T_f = 1.96$°C
51. 20.8 W
53. 7.75 h
55. 781 kg
57. (a) $-P_0V_0/2$ (b) $-1.39P_0V_0$ (c) 0
59. The bullet will partially melt.
61. 285 J

Chapter 15

1. (a) 0.067 (b) 350 J
3. 467 J
5. (a) 10,667 J (b) 0.533 s
7. (a) 8.69×10^8 J (b) 3.3×10^8 J
9. 63.3%
11. (a) 0.268 (b) 0.423
13. (a) 741 J (b) 459 J
15. (a) 9.10 kW (b) 1.19×10^4 J
17. 1.17 J of heat for each 1 J of work done
19. (a) 2.46×10^9 J (b) +0.49°C
21. COP = 9
23. 72.25 J for each J of heat removed
25. 195 J/K
27. 4.88 kJ/kg·K
29. 1020 J/K
31. (a) 1.38 cal/K (b) There is no change in temperature.
33. +0.085 J/K
35. (a) 53.0°C (b) 7.34 J/k
37. (a) lowest entropy: all red or all green (b) highest entropy: 3 red and 2 green or 2 red and 3 green
39. (a) 1/52 (b) 1/13 (c) 1/4
41. COP = 11.7; $W = 85.3$ J
43. 8.36×10^6 J/K
45. (a) 5000 W (b) 763 W
47. (a) 13.4 J/K (b) 98.19°F; 10.9 J/k
49. 29.3 J
51. (a) $2nRT_0 \ln 2$ (b) 0.273
53. (a) $10.5RT_0$ (b) $8.5RT_0$ (c) 0.190 (d) 0.833

Chapter 16

1. 1.8×10^{-6} N
3. 1.89×10^{-9} kg per "proton"
5. (−46.8 N)i on 6 μC; (157.5 N)i on 1.5 μC; (−110.7 N)i on −2 μC
7. 1.38×10^{-5} N at 77.5°
9. 3.90×10^{-7} N at 11.3°
11. $\mathbf{E} = 2\pi k_e\sigma\left(\dfrac{x}{|x|} - \dfrac{x}{(x^2 + R^2)^{1/2}}\right)\mathbf{i}$
13. $\dfrac{k_e\lambda_0}{x_0}(-\mathbf{i})$
15. 1000 N/C in the direction of electron's motion
17. $d = 1.82$ m to the left of the −2.5 μC charge
19. ($\bar{a}$) at the center of the triangle (b) $\mathbf{E} = 1.73\dfrac{k_eq}{a^2}\mathbf{j}$
21. (a) $(6.65 \times 10^6$ N/C)i (b) $(2.41 \times 10^7$ N/C)i (c) $(6.40 \times 10^6$ N/C)i (d) $(6.65 \times 10^5$ N/C)i

23. (a) 0.145 C/m^3 (b) 1.94×10^{-3} C/m^2
25. $(-2.16 \times 10^7 \text{ N/C})\mathbf{i}$
27. (a) 858 $N \cdot m^2/C$ (b) zero (c) 657 $N \cdot m^2/C$
29. 4.14×10^6 N/C
33. -4.6×10^5 C
35. The flux through S_1 is $-Q/\epsilon_0$. The flux through S_3 is $-2Q/\epsilon_0$. The flux through S_1 and S_4 is zero.
37. -6.89×10^6 $N \cdot m^2/C$
39. The shell produces no field inside. For points outside, behaves as if all its charge were at the center.
41. (a) $+Q/2\epsilon_0$
43. (a) 355 $N \cdot m^2/C$ (b) 2.84 $kN \cdot m^2/C$ (c) 2.84 $kN \cdot m^2/C$
45. $E = \dfrac{\rho}{2\epsilon_0} r$
47. (a) 2.56×10^6 N/C (b) 0
49. (a) 1.42×10^{-6} C/m^2 (b) 0.355 μC
51. $\mathbf{E} = 0$ inside the sphere; $\mathbf{E} = \dfrac{1}{4\pi\epsilon_0}\dfrac{Q}{r^2}(-\mathbf{r})$ between the sphere and shell; $\mathbf{E} = \dfrac{1}{4\pi\epsilon_0}\dfrac{2Q}{r^2}(\mathbf{r})$ outside the shell.
53. (a) 0.496 $\mu C/m^2$ (b) 0.248 $\mu C/m^2$
55. (a) 24.2 N/C $\mathbf{i}$ (b) 9.43 N/C at 63.5° above $-x$ axis
57. 5.25 μC
59. $T = 5.43 \times 10^{-3}$ N; $g = 1.09 \times 10^{-8}$ C
61. For $r < a$; $\mathbf{E} = \dfrac{\rho r}{3\epsilon_0}\mathbf{r}$ For $a < r < b$ $\mathbf{E} = k_e \dfrac{Q}{r^2}\mathbf{r}$

    For $b \leq r \leq c$: $\mathbf{E} = 0$ For $r > c$ $\mathbf{E} = \dfrac{k_e Q}{r^2}\mathbf{r}$
63. 3500 N
65. (a) zero (b) σ/ϵ_0 to the right (c) zero

Chapter 17

1. (a) 6.40×10^{-19} J (b) -6.40×10^{-19} J
3. 1.67×10^6 N/C
5. 1.54×10^6 J
7. 3.2×10^{-19} C
9. (a) 1.52×10^5 m/s (b) 6.49×10^6 m/s
11. (a) 9.22×10^4 V (b) 1.69×10^8 V
13. (a) 2.24×10^{-12} J (b) 14 MeV (c) 5.18×10^7 m/s
15. +38.9 V; the origin
17. 1.56×10^3 N/C
19. (a) 1.44×10^{-7} V (b) -7.2×10^{-8} V (c) -1.44×10^{-7} V and 7.2×10^{-8} V
21. (a) 33.4 V (b) 195 V
23. $r = 2.67$ m; $q = 1.19 \times 10^{-7}$ C
25. -8.82×10^4 V
27. (a) -27.3 eV (b) -6.81 eV (c) 0
29. (a) -3.86×10^{-7} J (b) 103 V
31. (a) -8.0 V (b) $+8.0$ V/m (c) -3.0 V/m (d) $+4.0$ V/m
33. (a) 10 V, -11 V, -32 V (b) 7 N/C along $+x$
35. (a) 0 (b) $(k_e Q/r^2)\mathbf{r}$
37. $-0.533 k_e Q/R$
39. $V = k_e \alpha \left[L - d \ln\left(1 + \dfrac{L}{d}\right)\right]$
41. $V = 2\pi k_e \sigma[\sqrt{x^2 + b^2} - \sqrt{x^2 + a^2}]$
43. 49 V
45. 1.13×10^7 m^2
47. $C = 11.1$ nF; $Q = 17.8$ C
49. (a) 1.11×10^4 V/m (b) 9.83×10^{-8} C/m^2 (c) 3.74 pF (d) 74.7×10^{-12} C
51. (a) 17 μF (b) 9 V (c) 108 μC
53. $Q_1 = 16$ μC; $Q_5 = 80$ μC; $Q_8 = 64$ μC; $Q_4 = 32$ μC
55. (a) $C_{eq} = 4$ μF (b) $Q_2 = Q_3 = Q_6 = 24$ μC; $V_3 = 8$ V; $V_6 = 4$ V and $V_2 = 12$ V
57. $Q = 120$ μC; $Q_1 = 80$ μC and $Q_2 = 40$ μC
59. (a) 12.0 μF (b) $Q_5 = 24.0$ μC; $Q_3 = 14.4$ μC; $Q_{20} = 19.2$ μC
61. $C = \dfrac{\epsilon_0 A}{s - d}$
63. (a) 0.3 μF (b) 2.16×10^{-3} J
65. 2.55×10^{-11} J
69. 10.8 pF
71. 4.0
73. (a) 13.3 nC (b) 272 nC
75. 1800 V
77. 18.6 MV
79. $V = \dfrac{2k_e Q d^2}{x^3 - x d^2}$
81. $C_1 = 1.00$ μF; $C_2 = 3.00$ μF
83. 121 V
85. 0.188 m^2
87. 253 MeV

Chapter 18

1. 26.8 A
3. 3.0×10^{20} electrons
5. 3.64 hours
7. 0.13 mm/s
9. 33.3 C
11. $1.56 R_0$
13. 1.69×10^{-4} m
15. 48 Ω
17. (a) 2.55 A/m^2 (b) 5.31×10^{10} m^{-3} (c) 1.2×10^{10} s
19. 1.711 Ω
21. 10.8 Ω
23. 1.084×10^{-3}/C°
25. 1435 C°
27. 0.180 V/m
29. 2.12×10^{-8} m
31. 36.1%
33. 34.4 Ω
35. 448 A
37. (a) 360 kJ (b) 100 W

39. (a) 1.79 A (b) 10.4 V
41. (a) 1.48 V (b) 25.6 Ω
43. (a) 4.59 Ω (b) 8.16%
45. 1000 Ω
47. (a) 9.52 Ω (b) 2.52 A
49. (a) 75 V (b) $P_1 = 25$ W, $P_2 = P_3 = 6.25$ W, 37.5 W
51. $I_1 = 11/13$ A, $I_2 = 6/13$ A, $I_3 = 17/13$ A
53. $I_1 = 0.714$ A, $I_2 = 1.29$ A, $\mathcal{E} = 12.6$ V
55. $I_1 = 0.266$ A, $I_2 = 0.533$ A, $I_3 = 0.799$ A
57. $P(I_1) = 9.0$ W, $P(I_2) = 25.0$ W, $P(I_3) = 2.0$ W
59. (a) 2×10^{-3} s (b) 180 μC (c) 0.09 A (d) $Q_1 = 0.63Q_{max} = 113$ μC (e) $I_1 = 0.37I_{max} = 33.3$ mA
61. (a) 12.0 s (b) $I = 3\,e^{-t/12}$ μA, $Q = 36(1 - e^{-t/12})$ μC
63. (a) 61.5 mA (b) 0.235 μC (c) 1.96 A
65. 50.0 MV
67. 1.56 cm
69. 470 Ω and 220 Ω
71. 587 kΩ
73. 13.5 h
75. (a) 12.5 A (heater), 6.25 A (toaster), 8.33 A (grill) (b) 27.1 A total; the 25-A fuse is insufficient
77. (a) 1.0 mA, 0, 1.0 mA (b) 667 μA, 667 μA, 0

Chapter 19

1. (a) 4.09×10^{-14} T (b) in the horizontal plane
3. 10.8°
5. 2.09×10^{-2} T in the $-y$ direction
7. 7.88×10^{-12} T
9. (a) 8.28 cm (b) 8.23 cm
11. 16.9 N/m
13. $I = 0.109$ A to the right
15. 0.133 T
17. $\mathbf{B} = (0.042\mathbf{k})$ T
19. $F = 2\pi rIB \sin\theta$, upward
21. $\tau = 9.98$ N·m, clockwise
23. (a) $I = 3.76 \times 10^{-4}$ A (b) $I = 1.67 \times 10^{-6}$ A
25. 31.4 cm
27. 28.3 μT
29. $\dfrac{\mu_0 I}{4\pi x}$ into the plane of the paper
31. 261 nT, into the paper
33. 8.0×10^{-5} N/m
35. 2 cm
37. 20.0 μT, upwards
39. (a) $B_{inner} = 3.60$ T (b) $B_{outer} = 1.94$ T
41. 500 A
43. 120 turns
45. 32.0 mA
47. $B = \dfrac{\mu_0 NI}{2l}\left[\dfrac{a}{\sqrt{a^2 r R^2}} - \dfrac{(a-l)}{\sqrt{(a-l)^2 + R^2}}\right]$
49. (a) 9.3×10^{-24} A·m² (b) μ is downward
51. (a) 9.39×10^{45} electrons (b) 4.36×10^{20} kg of iron
53. 675 A, down
55. $\mathrm{I} = 2.0 \times 10^{9}$ A
57. $\mathbf{B} = \dfrac{\mu_0 I}{2\pi w}\ln\left(\dfrac{b+w}{b}\right)\mathbf{k}$
59. 1.06×10^{-18} kg·m/s
61. 3.82×10^{-25} kg
63. 3.70×10^{-24} N·m
65. 1.80×10^{-3} T
67. (a) 2.46 N (b) 107.3 m/s²
69. $B = \dfrac{\mu_0 J_s}{2}$

Chapter 20

1. −800 μA
3. 2.67 T/s
5. (b) 3.79 mV (c) 28.0 mV
7. (87.1 mV) $\cos[200\pi t + \delta]$
9. 5.2×10^{-5} T
11. +115 kV
13. (a) 0.5 A (b) 2.0 W (c) 2.0 W
15. (a) 1.18 mV (b) same emf with north end positive
17. (a) 3.35 V (b) 3.72 V
21. 10.3 mV
23. 1.52 T
25. $\mathcal{E} = 14.2\cos(120t)$ mV
27. 1.8×10^{-3} N/C, ⊥ to r_1 and clockwise
29. (a) 1.60 A (b) 2.01×10^{-5} T (c) upward
31. $(-2.87 \times 10^{9}\,\mathbf{j} + 5.75 \times 10^{9}\,\mathbf{k})$ m/s²
33. 19.5 mV
35. 0.109 m
37. $(18.8 \cos 377t)$ V
39. −0.421 A/s
41. 2.61 H
43. (a) 0.139 s (b) 0.461 s
45. 7.67 mH
47. 92.8 V
49. 64.8 mJ
51. (a) 20 W (b) 20 W (c) 0 (d) 20 J
53. (a) 8.06×10^{6} J/m² (b) 6.32 kJ
55. (a) 16.7 W (b) 28.8 W (c) 7.19 J
57. 1.85×10^{-3} T
59. (a) $1.19\cos(120\pi t)$ V (b) 88.5 mW
61. $(9.87 \times 10^{-3} \cos 100\pi t)$ V/m, clockwise
63. 2.87 mV
65. $68.2\,e^{-1.62t}$ mV
67. (a) The auto must move east. (b) 0.458 mV
69. −10.2 μV
71. $I_{max} = 6.0$ A
77. (a) 6.25×10^{10} J (b) 2000 N/m
79. mgR/L^2B^2

Chapter 21

1. (a) $f = 1.50$ Hz; $T = 0.667$ s (b) A = 4.0 m (c) π rad (d) 2.83 m

3. (a) 31.6 rad/s (b) 1.58 m/s
5. (a) $v = 13.9$ cm/s, $a = 16.0$ cm/s^2
   (b) $v_{max} = 16.0$ cm/s at $t = 0.262$ s
   $a_{max} = 32.0$ cm/s^2 at $t = 1.05$ s
7. (b) $v_{max} = 6\pi$ cm/s at $t = 0$ and $t = 0.33$ s
   (c) $a_{max} = 18\pi^2$ cm/s^2 at $t = 0.50$ s
   (d) $x = 12.0$ cm
9. $k = 58.8$ N/m
11. (a) 575 N/m (b) 46 J
13. (a) $v_{max} = 40.0$ cm/s, $a_{max} = 160$ cm/s^2
    (b) $v(6) = 32.0$ cm/s, $a(6) = -96.0$ cm/s^2
    (c) $\Delta t = 0.232$ s
15. (a) 126 N/m (b) 17.8 cm
17. 0.478 m
19. (a) 0.028 J (b) 1.02 m/s (c) 12.2 mJ (d) 15.8 mJ
21. 2.23 m/s
23. (a) 1.55 m (b) 6.06 s
25. 21.9 m
27. 1.78×10^{-3} s
29. 1.0015
31. (a) $E = \frac{1}{2}mv^2 + mgL(1 - \cos\theta)$
35. 1.00×10^{-3} s^{-1}
37. (a) 1.42 Hz (b) 0.407 Hz
39. 322 N
41. 20.0 V
43. (a) 2.51 Hz (b) 69.9 Ω
45. 38.1 kΩ
47. (a) 17.4° (b) V leads I
49. 61.6 Hz
51. 1.57 s
53. 0.302
55. $f = \dfrac{1}{2\pi}\sqrt{\dfrac{MgL + kh^2}{ML^2}}$
57. 6.62 cm
59. (a) 3.00 s (b) 14.3 J (c) 25.5°
61. 12.2 cm/s
63. (a) 15.8 rad/s (b) 5.23 cm
67. (a) $I = 0.2$ A, $\phi = 36.8°$ (b) V = 40 V; $\phi = 0°$
    (c) 20 V; $\phi = -90°$ (d) 40 V, $\phi = +90°$

Chapter 22

1.

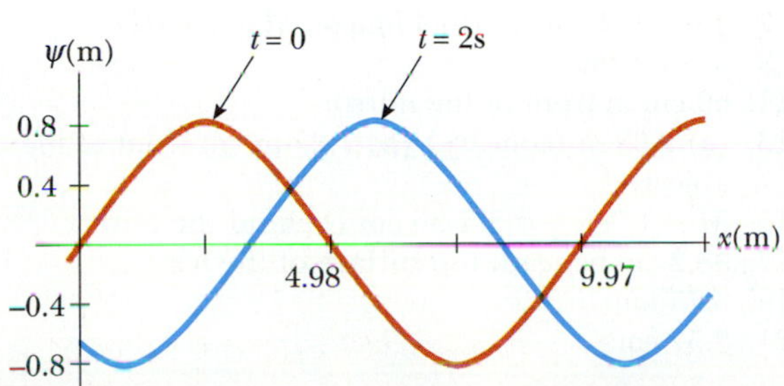

3. $y = \dfrac{6}{(x - 4.5t)^2 + 3}$
5. (a) $v = -2$ cm/s
7. (a) 5 radians (b) 0.858 cm
9. (a) Wave 1 travels in the $+x$ direction; wave 2 travels in the $-x$ direction. (b) 0.75 s (c) $x = 1.00$ m
11. 219 N
13. 13.5 N
15. 0.329 s
17. 0.319 m
19. $A = 0.7$ mm; $k = 31.4$ rad/m; $\omega = 6280$ rad/s
21. (a) 0.25 m (b) 40.0 rad/s (c) 0.3 m^{-1} (d) 20.9 m (e) 133 m/s (f) along $+x$
23. (a) $v = -1.51$ m/s; $a = 0$ (b) $\lambda = 16.0$ m; T = 0.50 s; $v = 32.0$ m/s
25. $A = 2.0$ cm; $k = 2.11$ rad/m; $\lambda = 2.98$ m; $\omega = 3.62$ rad/s; $v = 1.72$ m/s
27. (a) $y = (0.2 \text{ mm})\sin[16x - 3140t]$ (b) 158 N
29. 55.1 Hz
31. (a) 62.5 m/s (b) 7.85 cm (c) 7.96 Hz (d) 21.1 W
33. 0.103 Pa
35. 5.81 m
37. $(0.2)\sin[62.8x - 2.16 \times 10^4 t]$ Pa
39. (a) 75 Hz (b) 0.948 m
41. 19.3 m
43. (a) 3.04 kHz (b) 2.08 kHz (c) 2.62 kHz and 2.40 kHz
45. 46.4°
47. 56.4°
49. (a) 56.3 s (b) (56.6 km)i + (20.0 km)j from the observer
51. (a) 3.33 m/s, along $+x$ (b) −5.48 cm (c) $\lambda = 0.667$ m; $f = 5.00$ Hz (d) 11.0 m/s
53. 397 m; 28.0%
55. 130 m/s; 1.72 km
57. 7.9 km
59. (a) 1.27 Pa (b) 170 Hz (c) 2.00 m (d) 340 m/s
61. (a) 55.8 m/s (b) 2500 Hz
63. The bat is gaining at 1.7 m/s.
65. (b) 31.6 m/s

Chapter 23

1. $A = 0.924$ m; $f = 600$ Hz
3. $y_2 = (0.08 \text{ m})\sin[2\pi(0.1x - 80t - 0.167)]$
5. (a) 6.04 rad (b) 1.985 A
7. 0.089 m, 0.303 m, 0.518 m, 0.732 m, 0.947 m, 1.161 m
9. (a) nodes at $x = \pi, 3\pi, 5\pi, \ldots$ (b) $y_{max} = 2.94$ cm
11. $\lambda = 15.7$ m; $f = 31.8$ Hz; $v = 500$ m/s
13. 150 Hz
15. 800 m
17. 824 N
19. (a) 163 N (b) 660 Hz
21. 19.976 kHz
23. 3.9%
25. 4.50 kHz

27. (a) 0.66 m (b) 0.33 m
29. (a) 8.55 m (b) 4.28 m
31. 0.253 m; 0.506 m; 0.759 m; . . . n(0.253 m)
33. 0.0858 Hz
35. 3.07 kHz
37. 5.6 beats/s
39. (a) 1.99 Hz (b) 3.38 m/s
41. (a) 520 Hz or 526 Hz (b) 1.15%
43. (a) 3.33 rad (b) 283 Hz
45. $f = 50.0$ Hz; $L = 1.7$ m
47. 243 Hz
49. 3.87 m/s away from the station or 3.77 m/s toward station
51. (a) 0.555 m (b) 619.8 Hz
53. 15.7 Hz
55. (a) 34.4 Hz (b) 20 cm
57. (a) 0.5 (b) $\dfrac{n^2F}{(n+1)^2}$ (c) $\dfrac{F'}{F} = \dfrac{9}{16}$

Chapter 24

1. 7.33×10^{-7} T
3. (a) 162 V/m (b) 130 V/m
5. $E = (300\text{ V/m})\cos[62.8x - 1.88 \times 10^{10}\,t]$
   $B = (1.0\ \mu\text{T})\cos[62.8x - 1.88 \times 10^{10}\,T]$
7. (a) 3.33×10^{-7} T (b) 6.28×10^{-7} m (c) 4.78×10^{14} Hz
9. 1.09 pF $< C <$ 1.64 pF
11. 7.50 m
13. (a) $\mathcal{E}_m = 2\pi^2 r^2 f B_m \cos\theta$, where θ is the angle between the magnetic field and the normal to the loop. (b) Vertical, with its plane pointing toward the broadcast antenna
15. (a) strong signal at 90° and 270° (b) weak signal at 0° and 180°
17. $E_{max} = 438$ N/C; $B_{max} = 1.46\ \mu$T
19. 3.07×10^{-4} W/m^2
21. 3330 m^2
23. (a) 3.32×10^5 W/m^2 (b) 1880 V/m and 222 μT
25. (a) 971 V/m (b) 1.67×10^{-11} J
27. 8.33×10^{-8} N/m^2
29. (a) 1.90×10^3 N/C (b) 5.0×10^{-11} J (c) 1.67×10^{-19} kg·m/s
31. 5.36 N
33. (a) 150 MHz (b) 15 Mhz (c) 1.5 Mhz
35. 5.45×10^{14} Hz
37. 0.276 ms
39. 60 km
41. (a) 1.5 cm (b) 25.0 μJ (c) 7.37×10^{-3} J/m^3 (d) $E_{max} = 4.08 \times 10^4$ V/m; $B_{max} = 1.36 \times 10^{-4}$ T (e) 8.33×10^{-5} N
43. (a) 6.67×10^{-16} T (b) 5.31×10^{-17} W/m^2 (c) 1.67×10^{-14} W (d) 5.56×10^{-23} N
45. 95.1 mV/m
47. 7.5×10^{10} s = 2370 y; out the back
49. It is 35 light-minutes from Jupiter to Earth.
51. (a) $B_0 = 583$ nT, $K = 419$ rad/m, $\omega = 1.26 \times 10^{11}$ rad/s; xz (b) 40.6 W/m^2 in average value (c) 271 nPa (d) 406 nm/s^2
53. (a) 22.6 h (b) 30.6 s
55. 1.85×10^{-18} T

Chapter 25

1. 25.5°; 442 nm
3. 51.4°
5. 19.5° above the horizon
7. (a) 1.81×10^8 m/s (b) 2.25×10^8 m/s (c) 1.56×10^8 m/s
9. 3.4 m
11. six times from mirror 1, five times from mirror 2
13. 16.5°
15. 1.06×10^{-10} s
17. 106 m
19. 11.25 m
21. 0.395°
23. $\theta_{red} = 48.2°$; $\theta_{blue} = 47.8°$
25. (a) 27.9° (b) 48.6°
27. 36.7°
29. 62.45°
31. 1.00008
33. 67.2°
35. 42.4° above the horizontal
37. 2.27 m
39. 2.09×10^{-11} s
41. 67.4°
43. (a) 4.36 cm/min (b) 8.72 cm/min
47. 62.2%
49. 45.2°
51. 82 reflections
53. 28.7°

Chapter 26

1. Seven images are observed.
3. 2′11″
5. first image: 5 ft behind left mirror;
   second image: 25 ft behind left mirror;
   third image: 35 ft behind left mirror.
7. $q = -0.267$ m, virtual image, $M = +0.027$
9. $p = 30$ cm
11. 60 cm in front of the mirror
13. (a) 2.08 m (concave) (b) 1.25 m (in front of the object)
15. $M = 1/11$, $q = -1.36$ cm (behind the mirror)
17. 38.2 cm beneath top surface of the ice
19. 3.75 mm
21. 8.57 cm

23. 3.88 mm
25. 2
27. (a) 16.4 cm (b) 16.4 cm
29. +20 cm
31. 12 cm beyond the 4 cm lens, $M = -2$, the image is real and inverted
33. (a) 9.63 cm or 3.27 cm (b) 2.1 cm
35. 1.473
37. $f = -80$ cm
39. $M = +3.40$, upright image
41. $p = 36.2$ cm, $M = -0.382$
    $p = 13.8$ cm, $M = -2.62$
43. 25.3 cm to right of mirror, virtual, erect, $M = 8.05$
45. −25 cm
47. $M = 3.5$
49. (a) $f = 24.5$ cm (b) $p + q = 99.0$ cm
51. 10.7 cm beyond the curved surface
53. $d = 8$ cm
55. $q = 1.5$ m; $h' = -1.31$ cm
57. (a) −4.0 mm (b) $h' = 0.535$ mm, $q = -3.94$ mm
59. (a) 30.0 cm and 120 cm (b) 24.0 cm (c) real, inverted, diminished
61. $0.23 \text{ m} < p < 187 \text{ m}$

Chapter 27

1. 1.58 cm
3. 36.0 cm
5. (a) $\lambda = 55.25$ m (b) 123.2 m
7. (a) 1.93×10^{-6} m (b) $\delta = 3\lambda$ (c) a maximum at P
9. Maxima at 0°, 29.1°, 76.2° and minima at 14.1°, 46.8°
11. 4.80×10^{-5} m
13. (a) 13.2 rad (b) 6.28 rad
15. (a) 1.29 rad (b) 9.98×10^{-8} m
17. ±6.33 mm, equal intensity at center of each maximum
19. (a) $2.88E_0 \sin(\omega t + 0.349)$ (b) $2.06E_0 \sin\left(\omega t + \frac{\pi}{3}\right)$ (c) 0
21. $10 \sin(100\pi t + 0.93)$
23.

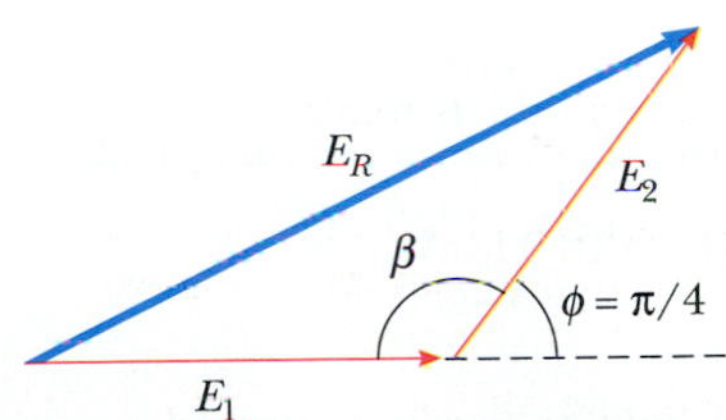

25. 612 nm (red)
27. no reflection maxima in the visual spectrum
29. 85.4 nm
31. (a) yellow (b) violet
33. 147 dark fringes
35. 1.31
37. 5 km² = 5×10^6 m²
39. 3.58°
41. 421 nm
43. 1.54 mm
45. 3.6×10^{-5} m
49. 115 nm
51. (a) 4.86 cm (b) 78.9 nm and 127.8 nm (c) 2.63×10^{-6} rad
53. 0.505 mm
55. 8.33 cm

Chapter 28

1. 4.22 mm
3. 2.3×10^{-4} m
7. 547 nm
9. 1.2 mm; 1.2 mm
11. 3.09 m
13. 240 m
15. 15.7 km
17. 13 m
19. 105 m
21. (a) 478.7 nm; 647.6 nm; 696.6 nm (b) 20.51°, 28.30°, 30.66°
23. (a) 2.4×10^{-6} rad (b) 213 km
25. (a) 2800 lines (b) 4.72 μm
27. 7.35°
29. 14.7°
31. 0.934 Å
33. 0.984 Å
35. 14.4°
37. 52.6°
39. 36.87° above the horizon
41. $0.375I_0$
43. 78.1%
47. 2.2
49. 5.9°
51. 0.244 rad (or 14°)
53. (a) 12000; 24000; 36000 (b) 0.0111 nm
57. (a) 41.8° (b) 0.593 (c) 0.262 m
61. $\phi = 1.392$ rad; from 5 to 7 steps depending on the starting value of ϕ

Chapter 29

1. (a) 2.57 eV (b) 12.8 μeV (c) 1.91×10^{-7} eV
3. 9.40×10^{44} photons/s
5. 9.35 μm (infrared)
7. 5180 K
9. 1.12 eV
11. (a) $\lambda_c = 296$ nm; $f = 1.01 \times 10^{15}$ Hz (b) $V_0 = 2.71$ V
13. 1.28×10^7 s

15. (a) 0.0822 eV (b) $\lambda_c = 995$ nm
17. (a) only lithium (b) 0.808 eV
19. $E = 1.78$ eV; $p = 9.47 \times 10^{-28}$ kg·m/s
21. 60.4 keV
23. 288 keV
25. $\theta = 70.1°$
27. (a) 0.174 nm (b) 5.5×10^{-12} m
29. 9.06×10^{-15} m
31. 151 eV
33. (a) 124 MeV (b) No, the kinetic energy is too large.
35. (a) $\gamma = 39{,}139$ $p = 1.07 \times 10^{-17}$ kg·m/s
(b) $\lambda = 6.2 \times 10^{-17}$ m, much less than the diameter of a nucleus $\cong 10^{-14}$ m
37. 151 V
39. (a) 9.93×10^{-7} m (b) 4.97 mm
(c) only if we interact with the neutron (and thus destroy the interference pattern)
41. 1.6×10^6 m/s
43. For the electron, 1.16 mm; For the bullet, 5.28×10^{-32} m
45. Approximately 10^5 m/s
47. (a) 1.26×10^{-10} m (b) 5.26×10^{-24} kg·m/s
(c) 95 eV
49. (a) 4.34×10^{-10} m (b) 6 eV
51. (a) $n = 4$ (b) 6.03 eV
53. (a) 37.3 eV, 151 eV, 339 eV, 603 eV
(b) 2.20 nm, 2.75 nm, 4.71 nm, 4.12 nm, 6.59 nm, 11.0 nm
55. 7.93 Å
57. $A = a^{-1/2}$ $B = a^{-1/2}$
61. (a) $T = e^{-4.6} = 0.01$ (b) $R = 0.99$
63. $\Delta L = 0.96$ nm
65. (a) $\phi = 1.7$ eV (b) $h/e = 4.14 \times 10^{-15}$ V·s
(c) $\lambda_c = 731$ nm
67. 2.25
69. $\lambda = 2 \times 10^{-10}$ m, $p = 3.31 \times 10^{-24}$ kg·m/s,
$E = 0.172$ eV
71. (a)

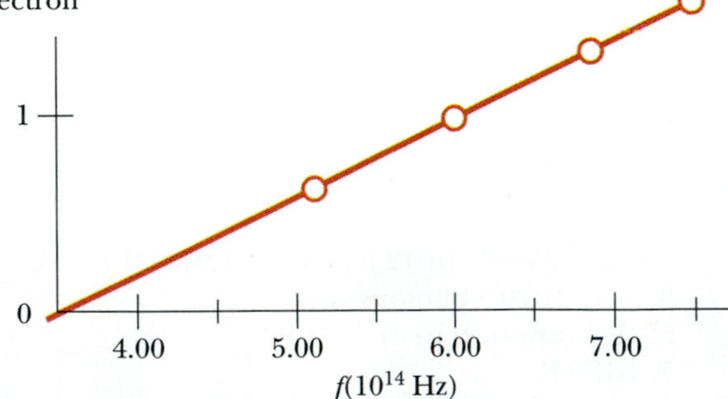

(b) 6.4×10^{-34} J·s (c) 1.4 eV

73. (a)

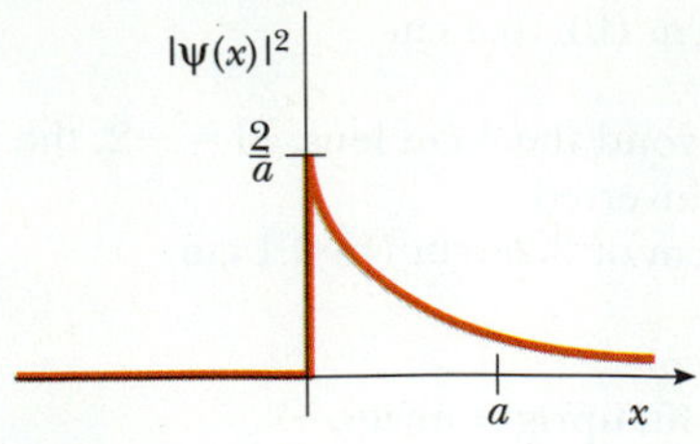

(b) zero (c) 0.865

Chapter 30

1. (a) $\ell = 0$, $m_\ell = 0$
$\ell = 1$, $m_\ell = -1, 0, 1$
$\ell = 2$, $m_\ell = -2, -1, 0, 1, 2$
(b) All have energy -6.05 eV
3. (a) 1312 nm (b) 164 nm
5. $n = 3$, $\ell = 2$, $m_\ell = 2, 1, 0, -1, -2$; $m_s = +1/2$ or $-1/2$
7. (a) 4 (b) 7
9. $r = 4a_0$
11. (a) 0.000257 nm (b) -18.8 MeV (c) 14.1 MeV
13. 4
15. (a) 2 (b) 8 (c) 18 (d) 32
17. 18, for $n = 3$: $3s^2\,3p^6\,3d^{10}$
19. (a) 2.53×10^{74} (b) 2.10×10^{-41} J
21. 9.27×10^{-24} J/T
23. (a) [Kr] $4d^{10}$ has the lower energy
(b) All spins are paired for [Kr] $4d^{10}$ and two are unpaired for [Kr]$4d^9 5s^1$.
(c) The element is Pd.
25. P +3 or −5; Ag −1 (prediction fails); Rn 0
27. 0.031 nm
29. $\lambda = 0.725$ Å
31. 590 nm
33. 3.5×10^{16} photons
35. (a) 1.63×10^{-18} J = 10.2 eV (b) 7.88×10^4 K
37. 9.79 GHz
39. $f = 2.83 \times 10^{13}$ Hz; $\lambda = 10.6$ μm (infra-red)
41. 0.323
43. (a) 1.22×10^{-33} (b) zero
45. $r_n = (1.06\ \text{Å})n^2$, $E_n = -\dfrac{6.80\ \text{eV}}{n^2}$, $n = 1, 2, 3 \ldots$
47. (a) 3.12 fm (b) -18.9 MeV
53. $5.24a_0$
55. 0.125
57. (a) 0.24 s (b) light energy is quantized

Chapter 31

1. (a) 1.9×10^{-15} m (b) 7.4×10^{-15} m
3. 368 m

5. 4.55×10^{-13} m
7. 2.80×10^{7} m/s
9. 12.7 km
11. 30.7 MeV
13. 2.83 MeV/nucleon
15. 7.93 MeV
17. (a) ^{139}Cs (b) ^{139}La (c) ^{139}Cs
19. 1160 s
21. 2750 Bq
23. (a) 1.55×10^{-5} s^{-1}; 12.4 h (b) 2.39×10^{13} atoms (c) 1.87 mCi
25. 24.5%
27. 1645 y
29. 4.28 MeV
33. 5.16 MeV
35.

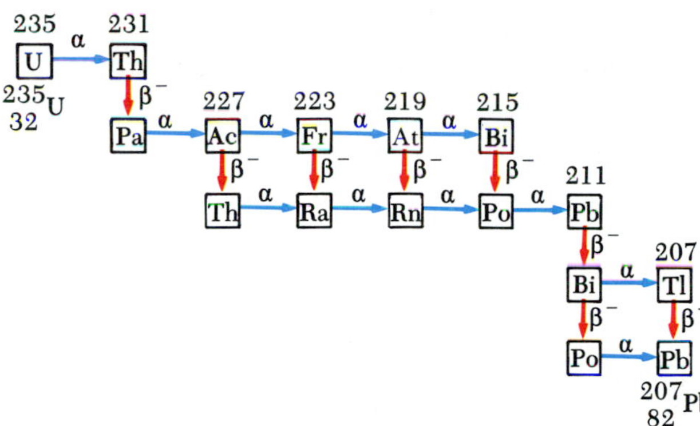

37. 16,600 y
39. (b) $e^- + {}^{15}O \rightarrow {}^{15}N + \nu$
41. −2.64 MeV
43. (b) 7.89 MeV
45. (a) ^{3}He (b) 4.02 MeV
47. 0.989 MeV
49. 22.0 MeV
51. $Q = -7.27$ MeV, no
53. 1.88 MeV
55. (a) 2.52×10^{24} (b) 2.306×10^{12} decays/s (c) 1.07 million years
57. 4.45×10^{-8} kg/h
59. 6.25×10^{19} Bq
61. (a) 2.75 fm (b) 152 N (c) 2.62 MeV (d) $R = 7.44$ fm; $F = 379$ N; $U = 17.6$ MeV
63. (a) 77.0 keV (b) 4.62 MeV; 13.9 MeV (c) $10^{3} \times 10^{6}$ kWh
67. (b) 4.79 MeV
69. (b) 24.994 s ± 0.1%

Chapter 32

1. $f = 4.53 \times 10^{23}$ Hz; $\lambda = 6.62 \times 10^{-16}$ m
3. 118.4 MeV
5. 2.06×10^{-18} m
7. $E = 67.5$ MeV/c; 1.63×10^{22} Hz
9. $\Omega^+ \rightarrow \overline{\Lambda}^\circ + K^+$
   $\overline{K}^\circ \rightarrow \pi^+ + \pi^-$
   $\overline{\Lambda}^\circ \rightarrow \overline{p} + \pi^+$
   $\overline{n} \rightarrow \overline{p} + e^+ + \nu_c$
11. (a) lepton number (b) charge (c) baryon number (d) baryon number (e) charge
13. (a) $\overline{\nu}_\mu$ (b) ν_μ (c) $\overline{\nu}_e$ (d) ν_e (e) ν_μ (f) $\overline{\nu}_e$; ν_μ
15. (a), (c), (f) baryon number violation. (f) lepton number violation (d) and (e) can occur
17. Strangeness is *not* conserved in (a), (d), (e), (f).
19. (a) lepton number (b) lepton number (c) strangeness and charge (d) baryon number (e) strangeness
23. (a) $d\overline{u} + u\overline{u}d \rightarrow d\overline{s} + uds$
   (b) $\overline{d}u + uud \rightarrow u\overline{s} + uus$
   (c) $\overline{u}s + uud \rightarrow \overline{u}s + d\overline{s} + sss$
   (d) $uud + vud \rightarrow d\overline{s} + uud + u\overline{d} + uds$
25. (a) 0.384c (b) 6.7×10^{9} lightyears
27. (a) charge (b) energy (c) baryon number
29. 2 MeV
31. $K_p = 5.4$ MeV; $K_\pi = 32.3$ MeV
33. 9.26 cm
35. $E_\nu = 29.7$ MeV

Index

Page numbers in *italics* indicate illustrations; page numbers followed by "n" indicate footnotes; page numbers followed by "t" indicate tables.

B

C

Q

T

U

V

W

Solar System Data

Body	Mass (kg)	Mean Radius (m)	Period (s)	Distance from Sun (m)
Mercury	3.18×10^{23}	2.43×10^{6}	7.60×10^{6}	5.79×10^{10}
Venus	4.88×10^{24}	6.06×10^{6}	1.94×10^{7}	1.08×10^{11}
Earth	5.98×10^{24}	6.37×10^{6}	3.156×10^{7}	1.496×10^{11}
Mars	6.42×10^{23}	3.37×10^{6}	5.94×10^{7}	2.28×10^{11}
Jupiter	1.90×10^{27}	6.99×10^{7}	3.74×10^{8}	7.78×10^{11}
Saturn	5.68×10^{26}	5.85×10^{7}	9.35×10^{8}	1.43×10^{12}
Uranus	8.68×10^{25}	2.33×10^{7}	2.64×10^{9}	2.87×10^{12}
Neptune	1.03×10^{26}	2.21×10^{7}	5.22×10^{9}	4.50×10^{12}
Pluto	$\approx 1.4 \times 10^{22}$	$\approx 1.5 \times 10^{6}$	7.82×10^{9}	5.91×10^{12}
Moon	7.36×10^{22}	1.74×10^{6}	—	—
Sun	1.991×10^{30}	6.96×10^{8}	—	—

Physical Data Often Used[a]

Average earth-moon distance	3.84×10^{8} m
Average earth-sun distance	1.496×10^{11} m
Average radius of the earth	6.37×10^{6} m
Density of air (20°C and 1 atm)	1.20 kg/m^3
Density of water (20°C and 1 atm)	1.00×10^{3} kg/m^3
Free-fall acceleration	9.80 m/s^2
Mass of the earth	5.98×10^{24} kg
Mass of the moon	7.36×10^{22} kg
Mass of the sun	1.99×10^{30} kg
Standard atmospheric pressure	1.013×10^{5} Pa

[a] These are the values of the constants as used in the text.

Some Prefixes for Powers of Ten

Power	Prefix	Abbreviation	Power	Prefix	Abbreviation
10^{-18}	atto	a	10^{1}	deka	da
10^{-15}	femto	f	10^{2}	hecto	h
10^{-12}	pico	p	10^{3}	kilo	k
10^{-9}	nano	n	10^{6}	mega	M
10^{-6}	micro	μ	10^{9}	giga	G
10^{-3}	milli	m	10^{12}	tera	T
10^{-2}	centi	m	10^{15}	peta	P
10^{-1}	deci	d	10^{18}	exa	E